GEOLOGY AND RESOURCE POTENTIAL OF THE CONTINENTAL MARGIN OF WESTERN NORTH AMERICA AND ADJACENT OCEAN BASINS— BEAUFORT SEA TO BAJA CALIFORNIA

Circum-Pacific Council
for Energy and Mineral Resources
Earth Science Series, Volume 6

GEOLOGY AND RESOURCE POTENTIAL OF THE CONTINENTAL MARGIN OF WESTERN NORTH AMERICA AND ADJACENT OCEAN BASINS—BEAUFORT SEA TO BAJA CALIFORNIA

Compiled and Edited by

David W. Scholl
Arthur Grantz
and
John G. Vedder

United States Geological Survey,
Menlo Park, California, U.S.A. 94025

Publication made possible through a Grant
from the Department of Energy, Office of Basic
Energy Sciences, Geosciences Program, Washington, D.C. U.S.A.;

Editorial support provided by the
American Association of Petroleum Geologists, Tulsa, Oklahoma, U.S.A.

Published by the
Circum-Pacific Council for Energy and Mineral Resources, Houston, Texas, U.S.A.

Published December, 1987
Printed in the United States of America
Circum-Pacific Publications are available from:
AAPG Bookstore
P.O. Box 979
Tulsa, Oklahoma
74101-0979, U.S.A.

Library of Congress Cataloging-in-Publication Data

Geology and resource potential of the continental
margin of western North America and adjacent ocean
basins--Beaufort Sea to Baja California.

(Earth science series ; v. #6)
Bibliography: p.
Includes index.
1. Geology -- North Pacific Ocean. 2. Marine mineral
resources--North Pacific Ocean. 3. Continental margin--
North America. I. Scholl, David W. II. Grantz, Arthur,
1927- III. Vedder, J. G. (John Graham), 1926-
IV. Series: Earth science series (Houston, Tex.) ; no. 6.
QE350.42.N65G47 1987 551.4′1 88-3338
ISBN 0-933687-06-0

CPCEMR *Earth Science Series Project Editors:* F.L. Wong and H.G. Greene

Circum-Pacific Council for Energy and Mineral Resources Earth Science Series

1. **Tectonostratigraphic Terranes of the Circum-Pacific Region**
   Edited by D.G. Howell
2. **Geology and Offshore Resources of Pacific Island Arcs—Tonga Region**
   Compiled and edited by D.W. Scholl and T.L. Vallier
3. **Investigations of the Northern Melanesian Borderland**
   Edited by T.M. Brocher
4. **Geology and Offshore Resdources of Pacific Island Arcs—Central and Western Solomon Islands**
   Edited by J.G. Vedder, K.S. Pound, and S.Q. Boundy

5A. **The Antarctic Continental Margin: Geology and Geophysics of Offshore Wilkes Land**
   Edited by S.L. Eittreim and M.A. Hampton

5B. **The Antarctic Continental Margin: Geology and Geophysics of the Western Ross Sea**
   Edited by A.K. Cooper and F.J. Daviey

6. **Geology and Resource Potential of the Continental Margin of Western North America and Adjacent Ocean Basins—Beaufort Sea to Baja California**
   Edited by D.W. Scholl, A. Grantz, and J.G. Vedder
7. **Marine Geology, Geophysics, and Geochemistry of the Woodlark Basin—Solomon Islands**
   Edited by B. Taylor and N.F. Exon

Cover Design: Gary M. Mann, U.S. Geological Survey.

DEDICATION

This volume is dedicated to the memory and scientific foresight of our friend and colleague, Vincent E. McKelvey, Director of the United States Geological Survey, 1971-1978. Vince's entire career reflected his deep philosophical commitment to providing government and the public with timely and accurate information about the Nation's resources. This commitment led him, in the early 1970s, to initiate an orderly program of scientific investigations of the Nation's continental margins and adjacent ocean basins. The results of the fact-finding mission that he set in motion, and the insights it provided concerning the energy and mineral resource potentials of these extensive regions, are largely chronicled in this volume. As the volume was being prepared for publication, Vince passed away at his home in St. Cloud, Florida, on January 23, 1987. Each chapter of this volume serves as an acknowledgement of his leadership and vision.

CONTENTS

PREFACE

This collection of papers summarizes the geologic framework and resource potential of offshore western North America from the Arctic Ocean to Baja California, including ocean regions of the United States Exclusive Economic Zone (EEZ). The idea that these papers should be prepared and brought together in a single volume arose late in 1982. At that time the U.S. Geological Survey's program of marine studies, which began in the early 1960s, completed its regional reconnaissance of the stratigraphy and structure of almost the entire margin of western North America and adjacent sea floor. It was therefore deemed timely and appropriate to summarize the important results of this effort and provide a systematic, region-by-region description of the framework geology and resource potential of this extensive, submerged region. It also appeared—as economic conditions ultimately become more favorable for the commercial development of marine georesources—that such a summary would be useful to the petroleum and mineral industries, and to public agencies and officials concerned with management of the offshore domains of the United States.

During the past two decades, USGS scientists have regularly published the results of marine geologic investigations in widely available technical journals. But much additional information and thinking have remained relatively inaccessible in the so-called gray literature of internal administrative reports of the Geological Survey and the Department of the Interior. A major function of this volume is therefore to provide a medium for the blending of published—but widely scattered—data with that mostly veiled in government reports. The intended product is therefore a comprehensive as well as orderly summary of framework investigations for the entire offshore of Alaska and the western conterminous United States. For the purpose of regional completeness, areal coverage was extended beyond the borders of the United States. To achieve this goal, and also to provide topical balance, contributions were solicited from colleagues in other organizations. Examples are the chapters on the continental margin of western Canada and on submerged geothermal resources.

Historically, the USGS engaged in marine studies primarily to provide public scientific knowledge of offshore lands of interest to the Nation, especially information on the stratigraphy and structure of submerged rock units and their corresponding resource potentials, and on sea-floor processes and related geohazards. Early efforts to carry out this mission in the Pacific-Arctic region gathered widely spaced and irregularly distributed geologic information over a vast territory that was then, and remains, the largest frontier region under United States jurisdiction. The exploration and discovery programs until about 1973 depended largely on vessels of opportunity, short-term charter vessels, or those supplied by cooperating institutions, agencies, and colleagues. These efforts provided adequate coverage of some areas, but in others the coverage was sporadic and many key areas were not investigated at all. Nonetheless, during the course of these studies, virtually all of the major geological basins of offshore Alaska were discovered and reported.

In the early 1970s, the combined effects of the oil embargo and the pricing policies of OPEC caused the U.S. Government to energetically promote the leasing of the outer continental shelf for the purpose of encouraging domestic petroleum exploration and recovery. This policy created an immediate need for a greatly expanded data base of public information about the resource potentials of the Nation's marine frontiers. One response to this demand was the establishment of a coherent program of offshore framework, resources, and geohazard studies within the U.S. Geological Survey by Director Vincent E. McKelvey. Vital to this effort was programmatic guidance and interagency coordination provided by U.S.G.S. Associate Director William A. Radlinski. With the aid of two dedicated research vessels—*R/V Lee* and *R/V Sea Sounder*—

orderly public investigations off Alaska, Washington, Oregon, and California were instituted. By 1982, a large quantity of new marine stratigraphic and structural data had been gathered, processed, interpreted, and reported. The bulk of the information presented in this volume was therefore gathered between 1975 and 1982. Since this period of orderly scientific exploration, the program has been greatly contracted in step with the oversupply—assuredly temporary—of petroleum in world markets and a commensurately lessened demand for marine framework studies.

This volume is organized into three sections. The first section presents the framework geology and resource potential of tectonically and geographically defined sectors of continental and island-arc margins and their bordering shelves. The second section groups chapters describing the more deeply submerged regions of the Bering Sea and nearshore north Pacific Basin. Specific resources are treated topically in the third section. The regionally oriented chapters are arranged in geographic order, from north to south, beginning with the Beaufort Sea and progressing to the Pacific offshore of Baja California. The topical chapters summarize or provide broad exploration perspectives for a variety of potential offshore resource commodities. Some topical chapters are a response to that vital part of the Survey's public mission that looks ahead and reports on the availability of resources that for prevailing technological or economic circumstances are not considered of contemporary importance.

As editors, we provided direction for the thematic focus of each chapter, and endeavored to insist on general rules concerning style and format. Authors were otherwise given free rein to determine what was important to say and how this information should be expressed and presented. We emphasize that the quality and completeness of the data base that different authors could call upon were unequal. For this reason, the comprehensiveness of subject treatment is also uneven. Thus some chapters present strongly supportable interpretations, whereas others are necessarily more speculative and theoretical. Vexingly, owing to a serious interruption in financing, publication of submitted galley proofs was delayed for two years. As a consequence, all chapters suffer from less-than-current reference citations.

It is hardly a surprise that contributors to this volume repeatedly observe that the amount of important knowledge yet to be gained far exceeds that in hand. Many important research directions with significant resource implications remain. In fact, vast tracts of the Pacific-Arctic region remain virtually unexplored, including much of the continental slope and rise and most of the deep oceanic areas of the EEZ. For the purpose of resolving remaining geologic questions concerning structures and formations that might contain overlooked or unconventional petroleum and hard-mineral resources, a balanced program of revitalized studies calls for the reexamination of the better known areas with new tools, methods, and concepts. The approach to reexamine is especially important, because new ideas about the effects and consequences of geological processes evolve rapidly in this particularly dynamic era in the growth of earth sciences.

We trust that this volume will serve as much to point out the promise of new areas and new topics of resource-oriented marine framework investigations as it does to summarize some of the knowledge gleaned thus far.

Dave Scholl
Art Grantz
Jack Vedder

— Menlo Park, California
April, 1986

FOREWORD

Beginning with the *Challenger* expedition in the latter part of the 19th century, scientific exploration of the ocean basins proceeded for the next several decades at the steady but relatively slow pace of academic investigations. During the last 20 years, however, marine geologic studies have intensified around the world. The main stimulus for this accelerated activity has been the growing interest in developing the petroleum and other mineral resources of the continental margins and deep ocean floor. Another contributing factor was the discovery of sea-floor spreading in the early 1960s and the realization that sea-floor geology holds the key to many important geologic processes.

Although scientific publications on marine geology have followed in welcome proliferation, they have been largely topical or have focused on the geology of relatively small areas. Breaking new ground, therefore, is this volume describing the subsea geology and resource potential offshore the whole of western North America, and also part of the Pacific overseas United States Exclusive Economic Zone.

One would expect that a volume of this scope might consist largely of papers reviewing and synthesizing previously published work. Its papers do stand on the shoulders of earlier workers, of course, but they present much previously unpublished information acquired by the writers in the course of Geological Survey studies and those of cooperating organizations and other governmental agencies. Much of the Survey's investigations were undertaken in conjunction with subsea geologic exploration and with various actions involved in the U.S. Department of the Interior's oil and gas leasing program.

The resource potential is examined in the context of the geology of each subregion and also in summary form by commodities or groups of commodities. Only a few commodities—such as oil and gas, sand and gravel, and high-quality placer deposits—are likely to be producible economically in the near future, but the potential for the others is large when scarcities from other sources or advances in technology justify their exploitation.

Complementing the papers in this volume, and vice versa, are the geologic, tectonic, geodynamic, and energy and mineral resource maps of the northeast quadrant of the Pacific Basin recently published at a scale of 1:10,000,000 by the Circum-Pacific Map Project. The two combine to give government and industry a splendid basis for identifying and analyzing resource potentials and prospects and for planning the initial stages of prospecting and leasing programs.

The seabottom offshore western North America is still largely a frontier area, but this volume adds light and understanding that will guide and facilitate its future exploration.

V.E. McKelvey
St. Cloud, Florida
November, 1985

CHAPTER ONE

TECTONICS AND BASIN EVOLUTION OF WESTERN NORTH AMERICA—AN OVERVIEW

David G. Howell
George W. Moore
Thomas J. Wiley

U.S. Geological Survey
Menlo Park, California

ABSTRACT

The nucleation of the craton of western North America was completed by about the middle of the Proterozoic. Near the end of the Proterozoic, a major rifting event occurred, and this was followed during most or all of the Paleozoic by a long episode of passive-margin deposition toward the west. With the breakup of Pangea beginning in the early Mesozoic, and a consequent realignment of plate motion, terrane accretion became the dominant tectonic mode. Terrane accretion, accompanied by dispersion along transform faults and by rifting, has created a variety of hydrocarbon-bearing basin types. In this chapter, they are classified into four categories—broad pliant transform, narrow transform, convergent, and rifted or passive-margin basins.

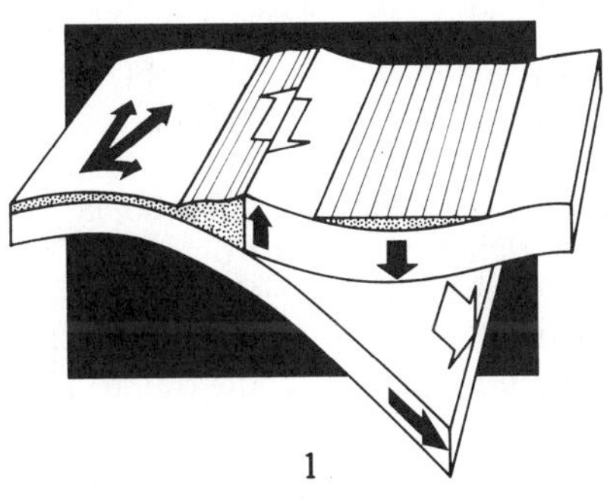

INTRODUCTION

The North American continent represents more than 2.5 b.y. (billion years) of earth history. The Cordilleran region, extending from the Beaufort Sea southward to the peninsula of Baja California and west of the Rocky Mountain front, contains crustal fragments that contributed to continental growth since the early Mesozoic. The geologic development of the Cordillera can be described in terms of continental rifting, subduction, magmatism, terrane accretion, and crustal dispersion. Viewing the geology in these terms helps explain the formation of sedimentary basins as well as the distribution of hydrocarbons.

The tectonic evolution of North America has progressed through three major stages—nucleation of the craton, accretion of Cordilleran terranes, and development of marginal basins and uplifts. The cratonal core of the North American continent consists of rocks 2.5+ b.y. old. In the remainder of the craton, episodic accretionary events as young as 1.0 b.y. occurred along the Ouachita-Appalachian trend, but such events are no younger than about 1.5 b.y. adjacent to the Cordillera (Condie, 1982). The Archean and Proterozoic rocks that constitute the craton are everywhere truncated by numerous rifts. In the Late Proterozoic, 600 to 800 Ma (million years ago), the craton was completely surrounded by trailing margins, much as Antarctica has been since the late Mesozoic. This configuration indicates that North America was once in the interior of a supercontinent, just as Antarctica lay at the core of Gondwana.

In the Cordillera, the passive-margin mode persisted for most of the Paleozoic. In the Arctic region, from the Mackenzie delta to northern Greenland, a mid-Paleozoic orogenic event indicates that there was a collision of microcontinents such as Baltica or Peryea with North America. The so-called Franklinian sequence of the Brooks Range (Bird, 1981; Grantz, May, and Dinter, this volume) may consist of terranes derived from this or some other pre-Mississippian orogenic event.

Terrane accretion reached its climax during the late Mesozoic. The Cordilleran orogen began to form after the Permian and Triassic. The Golconda allochthon and the then-offshore volcanic island areas such as Stikinia indicate subduction and collision. Accretions of Jurassic terranes have been recognized in the Sierra Nevada and along the Rocky Mountains of eastern British Columbia (Nokleberg, 1983; Monger, Price, and Tempelman-Kluit, 1982). Stratal relationships in the Arctic, however, suggest that the Jurassic was a time of local continental rifting. A popular model envisions the Brooks Range area as scissoring away from the Ellesmere Island area of northern Canada; alternatively, the Brooks Range strata may have been displaced from eastern Asia or from a part of the Canadian Cordillera (Grantz and May, 1983; Nilsen, 1981; Grantz, May, and Dinter, this volume). In the southwestern part of the continent, left slip of as much as 800 km, associated with the Jurassic opening of the Gulf of Mexico, reshuffled part of the continent along the Mojave-Sonora megashear (Silver and Anderson, 1974).

The wholesale assemblage of tectonostratigraphic terranes in Alaska began in the Early Cretaceous and was largely completed by the end of the Cretaceous (Fisher, Detterman, and Magoon, this volume). In southern Alaska, however, the Prince William terrane accreted in the Tertiary, and the Yakutat terrane is still accreting (von Huene, Fisher, and Bruns, this volume; Plafker, this volume). Terrane accretion in British Columbia was largely completed by the end of the Cretaceous. In the western United States, particularly in southwestern California, emplacement of terranes affected continental growth until the Miocene.

Igneous bodies of Mesozoic and Cenozoic age are widely distributed throughout the Cordillera. All pre-Tertiary batholiths have moved, either owing to simple rotation or to rotation and transcurrent slip (Beck, 1980), with the possible exception of the Cretaceous batholiths in the central Sierra Nevada (Frei, Magill, and Cox, 1984). Plutonism generally results from subduction; the tectonic phenomena associated with postplutonic crustal movements, however, are of special concern in studies of basin formation.

Basin formation is intimately related to petroleum accumulation. According to the plate-tectonic concept, azimuthal crustal motions of one plate relative to another form a continuum from head-on collision, through transcurrent slip, to orthogonal rifting. Compression and transpressional slip (Harland, 1971) lead to accretion and orogenic processes. Deep basins may form in front of a region of crustal loading and thickening—for example, the Colville Trough in front of the Brooks Range (Grantz, May, and Dinter, this volume). In another kinematic mode, a crustal sag may form because of transtensional slip (Harland, 1971) and rifting.

Basins that form in transcurrent regimes are relatively short lived, because changes in plate motion can lead to transpression, squeezing, and basin eversion. The Neogene basins of southern California are now being squeezed. The resulting anticlines form good petroleum traps, but vertical crustal movements are so fast that in 1 or 2 m.y. all structures might be breached. Basin eversion has already occurred in similar Upper Cretaceous and lower Tertiary strata in Washington and British Columbia—for example, in the Methow-Tyaughton Trough.

Regions of crustal rifting generally survive for long periods. In the Cordillera, Late Proterozoic and Paleozoic "miogeosynclinal" sequences are passive-margin accumulations. Since the mid-Cretaceous, the Arctic has been a passive margin, resulting in a thick and generally tectonically undisrupted sequence (Grantz, May, and Dinter, this volume; Grantz and May, this volume).

Sedimentary strata of the Cordillera can be grouped into three categories, each with a different petroleum potential—(1) sediment that is deposited before or during subduction or accretionary episodes has a low potential; (2) overlap strata that fill depressions which remain after subduction or accretionary episodes have a low to moderate potential; and (3) basins associated with active rifting and transcurrent dispersive processes have the highest potential.

Sedimentary strata that predate accretionary episodes are generally disrupted and consequently have a low potential for conventional petroleum prospects. Examples include the Jurassic and Cretaceous flysch of Alaska (Manley and Kahiltna terranes), the graywacke of the Franciscan Complex of California, and the continental-rise sediments of the Golconda allochthon of Nevada. Alternatively, most areas of known petroleum deposits or areas with high petroleum potential contain sediment deposited in basins that resulted from crustal sagging or dispersion; for example, Jurassic and Cretaceous strata of the North Slope of Alaska (Grantz, May, and Dinter, this volume), Cretaceous and Tertiary strata of the Navarin Basin on the Bering Sea Shelf (Cooper, Marlow, and Scholl, this volume; Marlow, Cooper, and Fisher, this volume), and Miocene and Pliocene strata of southern California (Vedder, this volume).

Strata of the Jurassic Bowser Basin of British Columbia, the Cretaceous of the northern Great Valley, California, and the Tertiary of Cook Inlet, Alaska (Fisher, Detterman, and Magoon, this volume) are depositional infillings of basins that remained after accretionary episodes—so-called successor basins or overlap sequences. Although some oil and gas is extracted from these strata, the amount represents a small proportion of that obtained from strata associated with rifting and dispersion (2 vs. at least 30 billion barrels of recoverable oil).

PLATE-TECTONIC CONSTRUCTION OF NORTHWESTERN NORTH AMERICA

In spite of, and in part stimulated by, the complex tectonic relationships in northwestern North America, this area and the adjacent sea floor have been used to develop several of geology's most far-reaching and useful tectonic concepts. As a forerunner to the plate-tectonic conceptual revolution, Hill and Dibblee (1953) mapped an offset of 560 km along the San Andreas fault and validated large lithospheric displacements as a factor in global tectonics. Benioff (1954) interpreted seismic data from the Aleutian volcanic arc and from other arc-trench pairs as indicating slip along deep thrust faults that extend so far down into the earth that they suggest movement of a much thicker layer than was implied by the earlier continental-drift theory. On the basis of further work in the Aleutians, Coats (1962) developed a largely overlooked but essentially modern view of subduction zones.

The first map of magnetic lineations covered a sea-floor area off the northwest coast of the United States (Mason and Raff, 1961). This area was one used by Vine and Matthews (1963) to formulate their hypothesis that the lineations were formed by magnetization of lava which had welled up along sea-floor-spreading axes. Also using data from off the northwest coast of the United States, Wilson (1965) developed the concept of transform faults as connectors between spreading axes and subduction zones, thereby completing the basic framework of the theory of plate tectonics.

Working with data from the western United States, Hamilton (1969) showed that the Sierra Nevada batholith constitutes the roots of volcanoes related to a former subduction zone, and that the Franciscan Complex consists of underthrust oceanic material from such a zone. On the basis of displaced fossil faunas, Monger and Ross (1971) suggested that a terrane in British Columbia is exotic, having been brought in and attached at a subduction zone. Subsequently, Jones, Silberling, and Hillhouse (1977) broadened that concept to show that an enormous area of western North America consists of exotic terranes, and that the accretion of such tectonostratigraphic terranes is a fundamental process of continental development.

Nuclear Western North America up to Triassic Time

The concept of tectonostratigraphic terranes modifies our understanding of the pre-Triassic development of western North America (Figure 1). A regional rifting event in the late Precambrian created an Atlantic-type margin. This passive or trailing margin received shelf and slope (miogeosynclinal) sediment throughout the Paleozoic (Stewart, 1972). A poorly understood tectonic disturbance—locally called the Antler orogeny—is recorded by massive piles of Devonian and Mississippian clastic strata in western North America and the Arctic Islands area, and a second event—the Sonoma orogeny—occurred in the Triassic. Most investigators infer that the rocks which record these tectonic events are para-autochthonous in the sense that they were deposited on or near North America. But this conclusion requires the juxtaposition of seemingly incompatible tectonic processes, such as simultaneous or rapidly alternating thrusting and rifting (Nilsen and Stewart, 1980). Moreover, the evidence that links the strata representing the Antler orogeny to North America is equivocal (Ketner and Smith, 1982). Therefore, we suggest that some and maybe all of the mid-Paleozoic tectonic events occurred in oceanic island-arc settings unrelated to North America, located in the paleo-Pacific (Panthalassa) Basin. With this model, it is not until the Triassic that exotic volcanic-arc material was underthrust by, and accreted to, the western edge of North America.

While Paleozoic arc-related rocks in the Panthalassa Basin were being deposited, imbricated, uplifted, and overlapped by younger beds, shelf-edge deposition occurred at the western margin of nuclear North America. Reef limestone of the same age as the flysch was deposited as far west as the Utah-Nevada border (Poole and Sandberg, 1977). Meanwhile, at the Atlantic side of North America during this period, the proto-Atlantic (Iapaetus) Basin was closing. The late Paleozoic Alleghenian orogeny in the southern Appalachians and the Ouachitan orogeny in the Gulf of Mexico area indicate the collision of Africa and South America with North America.

During the Triassic, as the Pacific Coast of North America collided with the volcanic-arc rocks, the Atlantic region began to spread apart again. We are uncertain as to whether a cause-and-effect relationship existed between these events, whether their synchrony was coincidental, or whether yet another process was the cause of both. The absolute motion of North America toward the west, as indicated by paleomagnetic Euler poles (Gordon, Cox, and O'Hare, 1984), accelerated during the Early Triassic after Africa collided with North America, declined during the Late Triassic coincident with volcanic-arc accretion in the Cordillera, and accelerated slightly during the Jurassic when rapid initial spreading occurred in the Atlantic. After the Pacific Coast collision, the subduction zone there changed from west dipping to east dipping. Clear magmatic evidence for the east-dipping zone at the Pacific Coast appeared in the Jurassic, and the westerly movement of North America has continued since.

Relationship of the Pacific Basin to the Breakup of Pangea

After the collision between Africa and North America during the Pennsylvanian and Permian, the supercontinents of Laurasia and Gondwana briefly joined to form Pangea. The subsequent breakup of Pangea was presaged in the area of the North Atlantic Ocean and the Gulf of Mexico during the Late Triassic by a stretching of the continental crust to form rift valleys along the Atlantic Coast, such as those containing the Newark Supergroup. During the Jurassic, about 40 m.y. after this continental-stretching phase began, the continen-

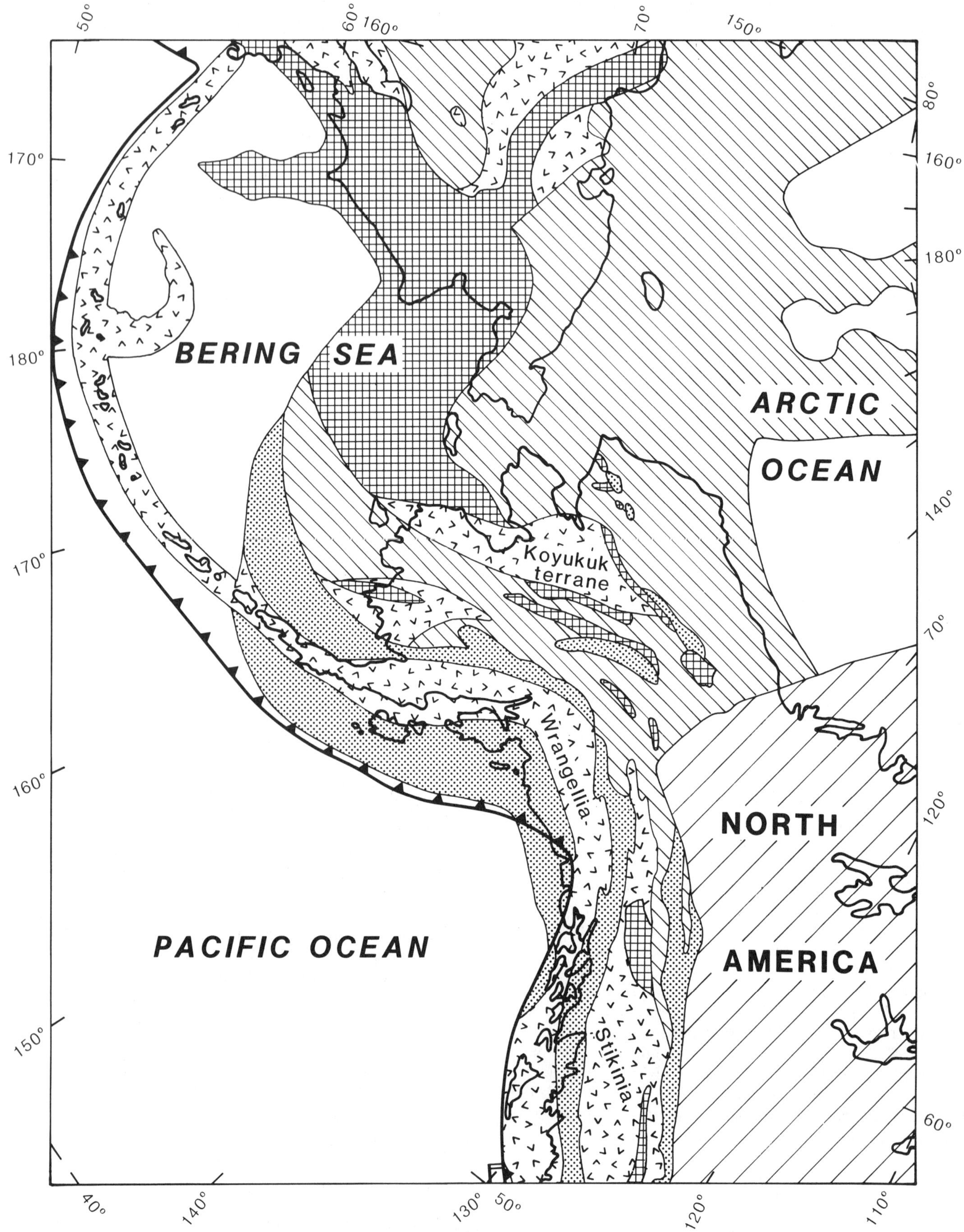

Figure 1. Predominant types of terranes that make up the Cordillera of North America. Active plate boundaries: Double line, spreading axis; barbed line, subduction zone; heavy line, transform fault.

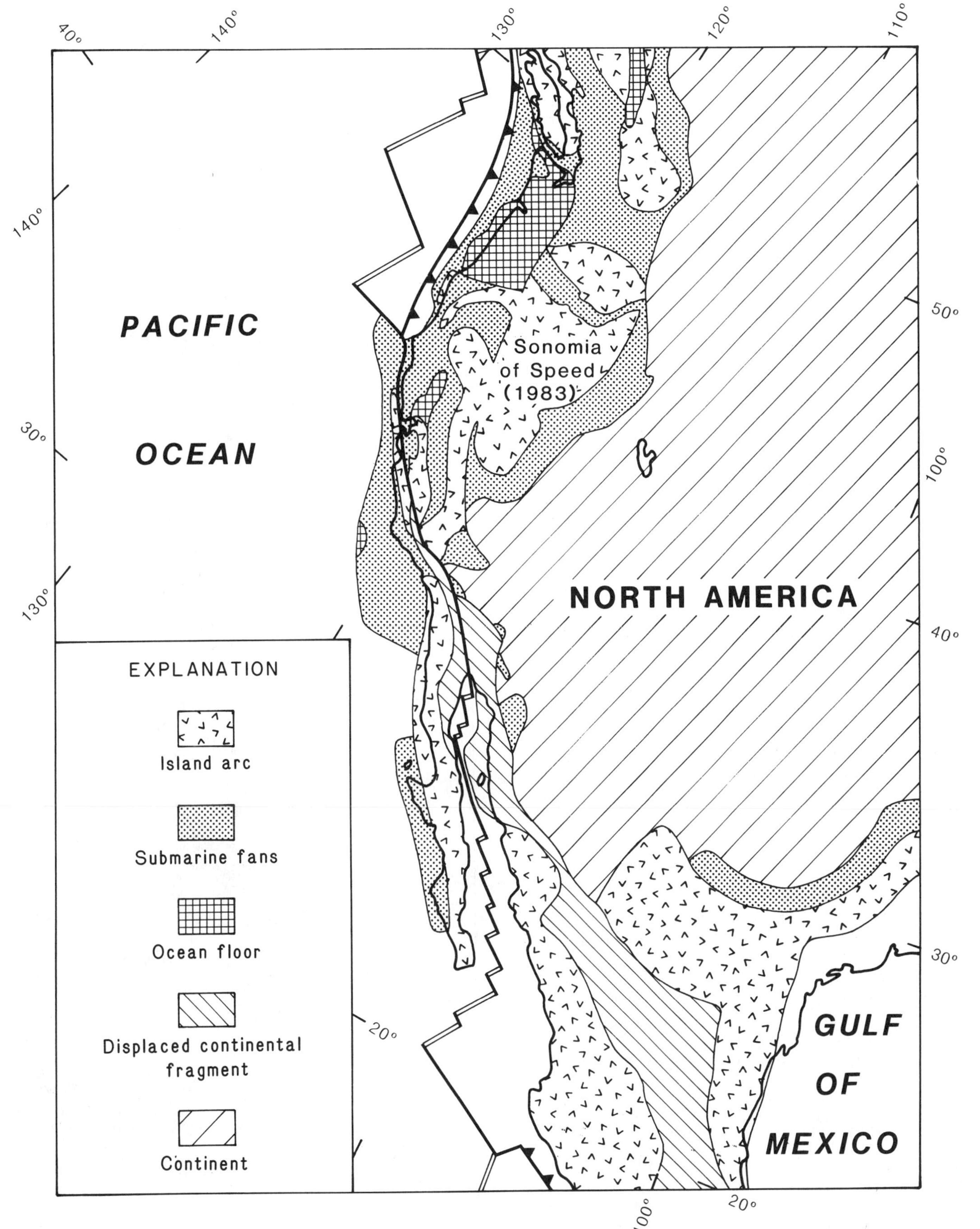

Figure 1. (continued).

tal crust ruptured completely, new seas began to form, and deformation extended from the North Atlantic through the Gulf of Mexico and Caribbean into what is now the Pacific Basin.

The initial extension of sea-floor spreading across the Americas from the Atlantic toward the Pacific, starting about 170 Ma (Middle Jurassic), stepped toward the northwest along two major transform faults—one on the west side of Florida (Moore and del Castillo, 1975), and one from the northwest corner of the Gulf of Mexico toward the present position of southeastern California (Silver and Anderson, 1974). This initial transform-fault connection across Mexico to California occurred at the same time that the east-dipping subduction zones were established along the Pacific Coast.

Sea-floor spreading across the Americas continued until about 102 Ma (Early Cretaceous), when the spreading in the Atlantic jumped to a new position between Africa and South America. Before that, but after the Gulf of Mexico had opened, the locus of spreading jumped to the south side of Yucatan, where the main part of the separation between North and South America occurred (Pindell and Dewey, 1982). At that time, beginning about 155 Ma (Middle Jurassic), Atlantic rifting probably extended most deeply toward the southwest into the south Pacific Basin.

During this episode of spreading through what is now the Caribbean Sea, the north side of the south Pacific Basin was displaced toward the north. At that time the most exotic of the younger terranes that are now accreted to western North America—such as Wrangellia—probably began their northward transit. Subsequently, the bulk of the Pacific Ocean floor was created; it is now represented in part by a north-trending belt of Cretaceous crust that is about 4500 km wide. But this crust is only the western half of the Cretaceous crust that was formed during this period. A matching eastern belt of Cretaceous crust is believed to have been subducted beneath North America. During this long period of sea-floor spreading, the southern terranes moved ahead of the now-subducted eastern belt of Cretaceous oceanic crust, and at about 80 Ma (Late Cretaceous) Wrangellia became attached to North America.

Mesozoic Subduction, Accretion, and Magmatism

The Triassic Sonoma orogeny marks the beginning of a protracted series of accretionary events that affected nuclear western North America. This event marks the time when an east-dipping subduction zone was established and North America became the overriding plate, acquiring the possibility of being uplifted and deformed during the arrival and accretion of extracontinental terranes.

Several terranes, such as Stikinia, arrived during the Middle Jurassic, after the east-dipping subduction zone was established and before formation of the Cretaceous Sierra Nevada batholith (Monger, Price, and Tempelman-Kluit, 1982).

Once a succession of terranes began to be accreted to North America, a crude horizontal stratigraphy became established among the terranes, which could only come in from the general direction of the Pacific Basin (Figure 2). Although this horizontal stratigraphy is locally contused by strike-slip dispersion, overall it is a great aid in establishing the chronology of the accretion. Terranes that were formerly island arcs are particularly useful, because their rocks are distinctive, they are typically fossiliferous, and many are known from their paleomagnetic latitudes to have formed at

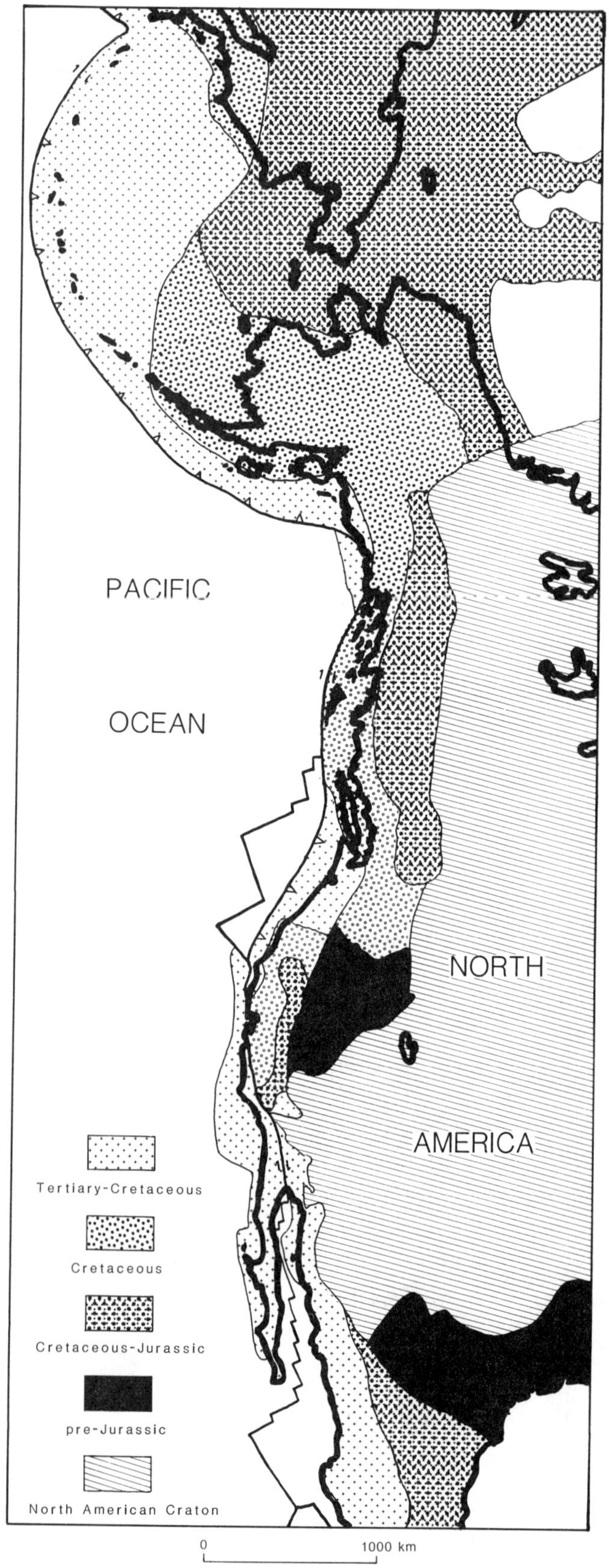

Figure 2. Age of terrane accretion to western North America.

distant places. Besides Stikinia, island-arc terranes accreted during the Jurassic include what are now the host rocks of batholiths in the eastern Klamath Mountains, the Sierra Nevada, and southern California. During the Cretaceous, a broad sweep of island-arc terranes was accreted including, for example, the Koyukuk terrane in north-central Alaska, the Alexander terrane in British Columbia, and the Guerrero terrane in Baja California (Howell et al., 1985). The Koyukuk terrane, directly south of the Brooks Range, is dated by the presence of Lower Cretaceous limestone (Patton, 1973), which shows that this terrane and all of central Alaska south of it was accreted by collision and lateral strike-slip transport during or after the Cretaceous.

Immediately after the collision of the Koyukuk terrane during the Early Cretaceous, major tectonic events occurred in northwestern North America during a brief mid-Cretaceous time interval. The most notable event was the opening of the Alaskan sector of the deep Arctic Ocean. Two other events, possibly related, were the closing off of the Koyukuk terrane to the south by the Ruby terrane—a probable continental fragment—and the rhombochasmic opening of eastern Oregon and Washington. The latter event permitted the Late Cretaceous accretion of the southern part of Wrangellia directly adjacent to the Precambrian craton of North America.

The arrival of Wrangellia signaled the start of an extremely complex phase of tectonic activity in western North America. Although North America itself swung fairly steadily toward the southwest, there were profound changes in the Pacific Basin. These changes occurred because the direction of movement of individual plates changed (as indicated by changes in the alignment of hotspot traces), and because some plates passed beyond various parts of North America or became inactive, so that other plates with different motions interacted with the continent.

Tertiary Strike–Slip Dispersion

The plate-tectonic map of the Pacific margin of North America represents an instant in time of a geologic kaleidoscope (Figure 1). The four major plates that now affect the area, and also other plates, interacted during the Tertiary at rates of as much as 2000 km in 10 m.y. (Moore, 1984). Strike-slip dispersion affected native North American materials as well as terranes accreted from distant sources. Consequently, multiple deformations have caused a great complexity that has not yet been fully decoded, but considerable progress has been made by means of structural, stratigraphic, paleontologic, and paleomagnetic analysis. This research has enhanced our understanding of how movements of the dispersed terranes have affected the older accreted terranes, which in turn has helped us to define the processes of basin evolution in western North America.

DISTRIBUTION OF TECTONOSTRATIGRAPHIC TERRANES

The western margin of North America comprises numerous terranes consisting of fragments of island arcs, oceanic basins, seamounts, pieces of continents, and piles of continentally derived sedimentary strata. The heterogeneous mixture of terranes, as shown by Figure 1, produces new continental crust. Although the so-called Franklinian sequence of the Arctic and possibly other terranes may reflect Paleozoic accretion, most Cordilleran terranes have accreted since the Triassic.

Paleontologic and paleomagnetic data indicate that some terranes have moved great distances. The wholly oceanic character of other terranes also suggests a truly exotic origin; nonetheless, many terranes are probably portions of the American continents or nearby parts of the Asian continent that have been reshuffled by dispersion and accretion.

OVERLAP SEQUENCES

Postaccretionary sequences that cross terrane boundaries provide important horizons for dating the amalgamation of two or more terranes or the accretion of terranes to the continent. Defining the stratigraphy of overlap sequences is challenging, because successive sequences may record several amalgamation events before accretion to a continent. Brief accounts of some of the better known overlap sequences in western North America are given below.

In west-central Baja California, the oldest unit that laps across the submarine-fan terranes of the Cedros-Vizcaino-Magdelaena area is the late Albian and younger Valle Formation. The provenance of debris composing this sequence links it to the Allisitos volcanic arc of the core region of the Baja California Peninsula, suggesting accretion of the oceanic terrane to Mexico by at least the mid-Cretaceous.

The southern California borderland and the adjoining Peninsular Ranges and Santa Moncia Mountains comprise several terranes with Jurassic basement. Upper Cretaceous lithofacies can be correlated among the terranes, but direct linkage is equivocal. Middle Eocene strata, with their unique conglomeratic assemblage (Poway Group), establish contiguity of the terranes, but post-Eocene dispersion within the assembled terranes involves several hundred kilometers of slip. Furthermore, paleomagnetic data indicate that the entire amalgam lay far to the south in the Eocene (Champion, Howell, and Marshall, 1981). Miocene strata are the oldest units that lap across all the terranes, as well as part of the North American craton.

In central California, the Salinia, Stanley Mountain, and Tujunga terranes are all linked by similar upper Campanian strata, but not until the early Eocene do strata lap from these terranes across the San Simeon terrane and the North American craton. In the Klamath-Sierra Nevada region, Cretaceous strata of the Great Valley sequence link the terranes of these two mountain ranges.

Strata older than Late Cretaceous in northwestern Washington are tectonically disrupted. Accretion was active until the Late Cretaceous or early Tertiary, when strata of the Nanaimo and Kuganut Formations lapped across the assembled terranes from Vancouver Island eastward to the high Cascades.

In British Columbia, the Late Jurassic Bowser Basin is referred to as a successor basin, although it lies wholly on Stikinia, but the provenance of debris from the adjoining Cache Creek terrane indicates contiguity of these two terranes. Throughout central Alaska, the oldest beds that lap across broad regions of assembled terranes are the Late Cretaceous Kuskokwim Group and the Paleocene Cantwell Formation; locally, these units are highly folded, indicating postaccretionary consolidation and possibly dispersion along the Denali and related fault systems.

Although we do not know the exact age range of the flat-lying strata on the Bering Sea Shelf (Marlow, Cooper, and Fisher, this volume), data from COST (Continental Offshore Stratigraphic Test) wells suggest a not-later-than Eocene age for accretion. Along the Arctic margin, the Cretaceous Brookian sequence heralds the onset of deposition from a southern source[1] (Grantz, May, and Dinter, this volume). This sequence follows (1) large-scale, north-directed[1] forelandlike thrusting that stacks a variety of terranes composing the Brooks Range, and (2) the dispersion or wholesale movement of this Brooks Range region from some previous location, either the Arctic Islands and northeast Asia, or British Columbia and Yukon Territory.

BASINAL STRATA ALONG THE MARGIN OF NORTHWESTERN NORTH AMERICA

Sedimentary rocks of Mesozoic and Cenozoic age crop out in a nearly continuous belt stretching from the Beaufort Sea to Baja California (Figure 3). Within this belt, basins with thick sedimentary sequences (> 1 km) alternate with basement highs on which sedimentary cover is thin or absent. Both the thick basinal sequences and the intervening basement highs have many characteristics that can be related to their plate-tectonic setting.

It is important to note that the term "sedimentary basin" has been applied to most thick sedimentary sections regardless of their size or depositional history. The term is used to describe shelf-progradational sequences, open and closed basinal sequences, and oceanic sequences. Distinctions between sedimentary basins and related topographic basins are not always made. The lack of a specific definition for "sedimentary basin" causes problems of scale that complicate comparisons between different basins; hence one author's basin is another's subbasin. For these reasons, quantitative comparisons of basin characteristics are difficult to present. Nevertheless, sedimentary deposits from similar tectonic settings share many characteristics.

The plate boundaries that form the continental margin of western North America have changed markedly since the late Mesozoic. Not only have the rates of tectonic processes changed, but also the types of processes have changed. Regions of plate convergence have become regions of transform motion, oceanic ridges have been subducted, and triple junctions of a variety of configurations have migrated along North America's western margin. The strata that fill the sedimentary basins record many of these changes, and indeed, many of the basins seem to have formed as a response to major plate reorganizations. Unfortunately, each tectonic phase in the history of a basin makes interpretation of the composite basin-fill and basin-bounding structures more complicated, and makes development of reasonable basin-classification schemes more difficult.

In the sections that follow, sedimentary basins of the western margin of North America are grouped according to their present tectonic setting. Groups of basins are classified as broad pliant transform, narrow transform, convergent, and rifted or passive-margin basins. Where the tectonic regime has changed with time, an attempt is made to isolate the effects of each phase in the history of the basins. Hydrocarbon potential of the basins is summarized by McLean and Wiley (this volume).

Broad Pliant Transform Margins

In southern California and in the northeastern Gulf of Alaska, the continental margin consists of a complex system of northwest-trending right-slip faults. Within both of these systems, zones of transpression and transtension form ridges and basins (Table 1) bounded by major strike-slip faults and associated secondary structures. Each of these broad transform-margin segments lies south (toward a spreading ridge) of a major westward bend in the continental margin. In southern California, the San Andreas fault curves to the west at the Transverse Ranges north of the broad pliant margin that forms the California Continental Borderland. In the northern Gulf of Alaska, the broad transform margin intersects the northeastern end of the Aleutian convergent margin.

Basement consists of displaced and dispersed tectonostratigraphic terranes that are commonly juxtaposed along the same major strike-slip faults that border the basins. Mesozoic and Paleogene sedimentary rocks, originally deposited in laterally extensive convergent-margin basins, underlie many of the Neogene basin sequences.

Local sedimentation and erosion rates may vary greatly in both time and space within the wide zone of faulting that forms these broad pliant transform margins. Restricted areas of uplift or subsidence regulate the position of source areas, amount of relief, and location of sediment-dispersal paths. At any one time, a single fault may act as a master fault and be responsible for a large percentage of the total transform motion. Near Los Angeles, the present master fault is the San Andreas. About 3 cm/yr of the 6 cm/yr of transform motion taking place on this part of the margin occurs as offset along the San Andreas fault itself. Reconstructions of the history of the southern California margin suggest that, in general, the location of the master fault has shifted eastward with time.

The position of the shoreline influences the type and amount of sediment to a greater degree on broad pliant transform margins than on other types of margins. Because the basins are generally small and isolated, continentally derived sediment follows well-defined transport paths and forms major depocenters in only a few basins at a given time. Accordingly, terrigenous sedimentation rates vary greatly in basins with comparable subsidence rates (Howell et al., 1980).

Basins of southern California and the adjacent California Continental Borderland can be grouped into three categories—onshore, nearshore, and offshore. Onshore basins, such as Ridge Basin (a small basin northeast of Ventura Basin; Table 1), fill with continental and shallow-marine sediment. Sedimentation rates in onshore basins are largely dependent on the size of the source area. Lower rates are commonly found in basins where the source area is closely restricted to the basin flanks, and higher sedimentation rates occur in basins located along regional drainage paths. Along the fault-bounded margins of many of the basins, thick conglomerate sequences form wedges that thin toward the basin center; for example, the Violin Breccia of Ridge Basin (Crowell and Link, 1982).

The largest accumulations of sediment in the southern California area are in basins along or near the shoreline (Figure

[1]Present coordinates.

3). These nearshore basins have histories of rapid filling, with overall shallowing-upward sequences of deep-marine to continental sediments (Vedder, this volume). Where subsidence continues for long periods, these basins may contain extremely thick sedimentary sections, such as the Neogene sections of the Los Angeles and Santa Barbara-Ventura basins. In most cases, sediment that fills nearshore basins has a predominantly terrigenous source. Longshore transport, along with other shelf and nearshore processes, somewhat reduces the influence of local tectonics upon the location and type of sediment-dispersal paths and depocenters within nearshore basins.

Broad pliant transform margin basins that lie farthest offshore receive only small amounts of terrigenous sediment, and the little they do receive is mostly fine grained. Most continentally derived clastic debris is trapped in onshore and nearshore basins and is transported farther offshore only when those basins are filled and bypassed. Accordingly, the sediment that fills offshore basins is primarily biogenic, sedimentation rates are low, and the depth to the sea floor is likely to be great.

Nearshore and offshore basins on broad transform margins may be silled. These sills restrict circulation of bottom water and can result in anoxic bottom conditions favorable to the formation of petroleum source beds.

Narrow Transform Margins

In central California and in southeastern Alaska to northern coastal British Columbia, the edge of North America is characterized by narrow transform margins. Although called "narrow," these margin segments actually consist of systems of near-parallel right-slip faults. Elongate basins and intervening ridges are parallel or en echelon to these margins (Table 1). The seaward side of such basins may or may not be silled by shelf-edge ridges. In central California, many of the basin-bounding faults are reverse faults (Blake et al., 1978). Basins may form on either side of a master fault. Narrow transform-margin basins are generally larger and more elongate than broad transform-margin basins, although a continuum exists between the two. Transform margins that lie along the boundary between continental and oceanic crust—for example, the margin from southeastern Alaska to northern coastal British Columbia—are more sharply defined than intracontinental transform margins (Howell and von Huene, 1981).

Sediment-dispersal paths follow troughs and ridges that are parallel or en echelon to the margin. Large drainages may form continentward of the margin—the Sacramento and San Joaquin River system of California is a good example. These supply disproportionate amounts of sediment to basins near the point where they intersect the shoreline—as in the Sacramento Delta and San Francisco Bay system. Seaward of the shoreline, narrow continental shelves are associated with narrow transform margins. These shelves are cut by numerous submarine canyons, some of which head in water less than 20 m deep, that shunt large amounts of sediment away from the coastal zone. This sediment bypasses the shelf to fill inactive trenches and form deep-sea fans such as the Arguello, Monterey, and Delgada fans off central California and the Baranof fan off southeastern Alaska (Stevenson and Embley, this volume).

The size, shape, and sediment provenance of submarine fans formed along transform margins—especially narrow transform margins—are greatly influenced by the relative position of submarine canyons and adjacent strike-slip faults. The simple model depicted in Figure 4 shows how strike-slip movement landward of a submarine canyon allows a stable canyon-fan system to develop; a relatively large lobate fan is produced, with vertical changes in provenance indicating the passage of different source terranes landward of the fault. In contrast, where a major strike-slip fault lies at the base of a submarine canyon, lateral displacement of the can-

Table 1. Typical transform-margin basins of western North America.

Basin Name	Type*	Location (lat., long.)	Approximate Area (km^2)	Maximum Thickness (km)	Age of Sedimentary Section
Amak	TP	55°N, 165°W	2,400	4	Cenozoic
Anadyr	TP	64°N, 178°W	40,000	9	Cretaceous and younger
Bodega	NT	38°N, 123°W	7,700	>3.2	Eocene(?) and younger
California Continental Borderland	BT	31°N, 117°W	100,000	6	Cretaceous and younger
Kresta	TP	65°N, 179°W	4,500	5	Cretaceous(?) and Cenozoic
Los Angeles	BT	33°55′N, 118°W	2,900	9	Late Cretaceous and younger
Navarin	TP	60°N, 177°W	62,000	>12	Cretaceous(?) and younger
Oomyousik	TP	62°20′N, 178°W	500	11	Cretaceous(?) and younger
Point Arena	NT	39°N, 124°W	7,700	>3.6	Eocene and younger
Queen Charlotte	NT	52°N, 131°W	35,000	5	Neogene and younger
				5	Mesozoic and Paleogene
Ridge	BT	34°40′N, 118°50′W	200	3	Neogene
Saint George	TP	56°N, 166°W	12,000	>10	Cenozoic
Salinas	NT	36°N, 120°W	14,000	4	Cretaceous and younger
Santa Barbara	BT	34°15′N, 120°W	7,700	20	Late Jurassic and younger
Santa Cruz	NT	37°N, 122°W	2,500	3	Cretaceous and younger
Santa Maria	NT	35°N, 121°W	13,600	3.5	Cretaceous and younger
Southeastern Alaska	NT	56°N, 135°W	25,000	6	Neogene
Ventura	TP	34°15′N, 119°W	4,150	20	Cretaceous and younger
Yakataga	BT	59°30′N, 143°W	15,000	7	Eocene and younger
Yakutat	BT	59°N, 141°W	35,000	>9	Cenozoic

*Basin types: BT—Broad transform margin basin, NT—Narrow transform margin basin, TP—Transpressive margin basin.

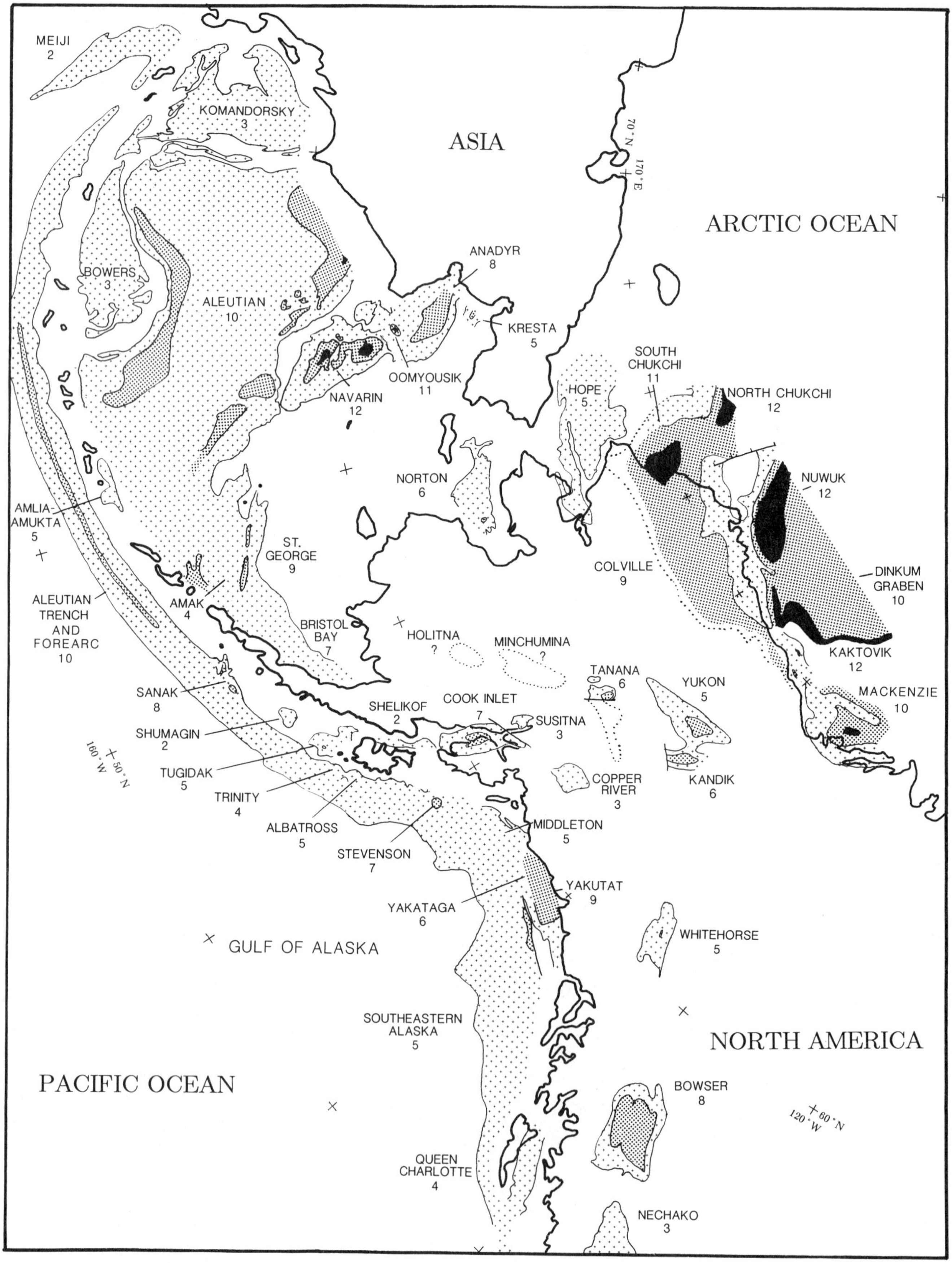

Figure 3. Sedimentary basins of western North America. Numerals below basin names indicate maximum thickness of sedimentary section, in kilometers. Basin outlines dotted where inferred. Light screen pattern depicts sections above economic basement 1 to 5 km thick; dark screen pattern, 5 to 10 km; black, >10 km.

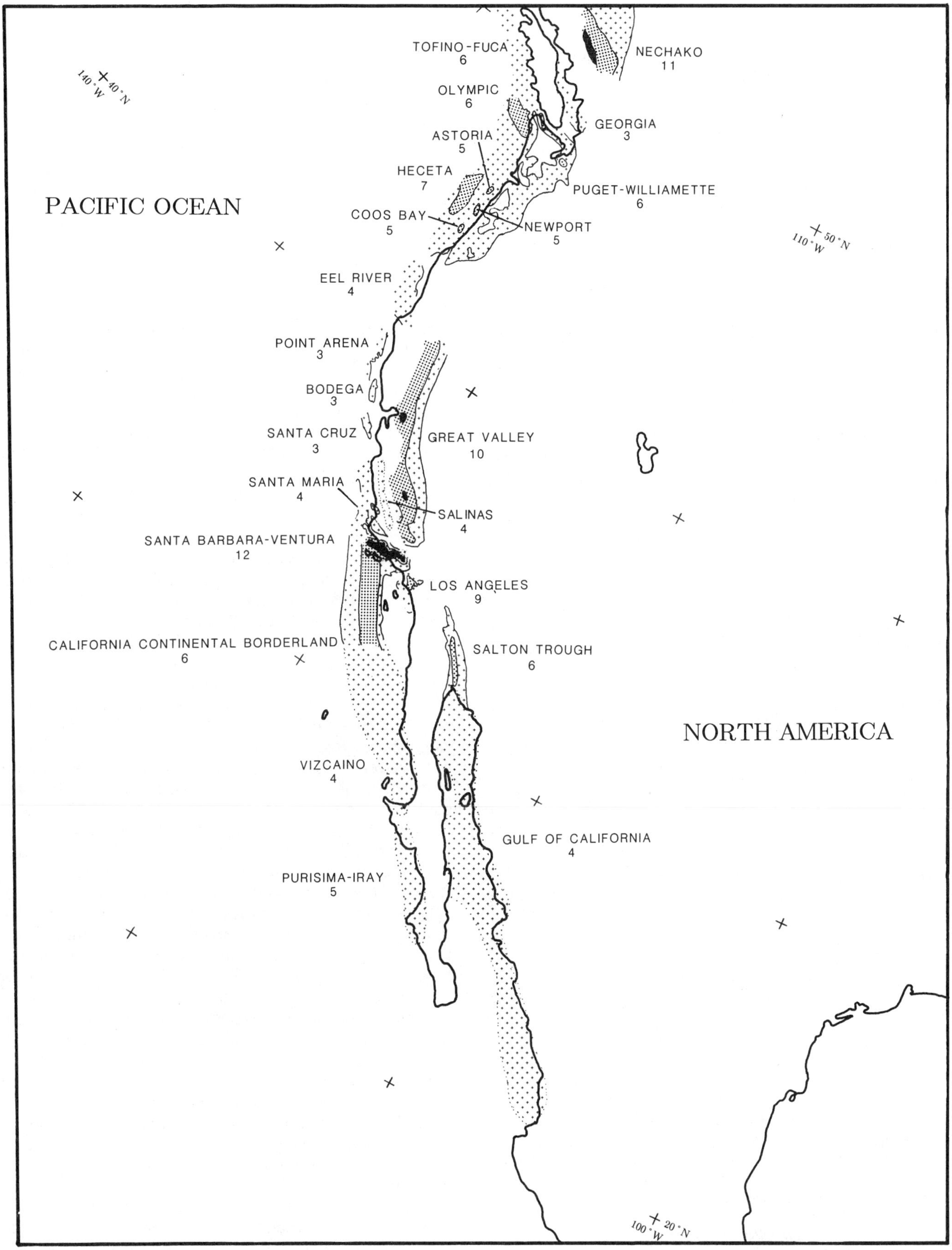

Figure 3. (continued).

yon and adjacent source terrane results in the formation of a comparatively thin and narrow submarine fan, elongate parallel to the margin. This fan shows lateral changes in provenance that indicate an unroofing of the adjacent source area (Howell and von Huene, 1981).

Convergent–Margin Basins

The northwestern North American margin includes two convergent segments. One lies along the northwestern Gulf of Alaska and the eastern Aleutian Islands, where the Pacific plate is being subducted beneath the North America plate at the Aleutian Trench. The other lies off the coast of southern British Columbia, Washington, Oregon, and northern California, where the North America plate and the Juan de Fuca plate converge.

At these convergent margins, sedimentary basins form on the continental shelf, along the slope, in the trench, and beyond the trench (where it has been filled). These basins (Table 2), together with oceanic sediment and accreted fragments of oceanic crust, are folded, cut by thrust faults, and imbricated to form an accretionary complex. Imbrication and tectonic underplating result in vertical rotation and uplift that occurs rapidly near the trench and lower slope, and more gradually on the upper slope (Howell and von Huene, 1981). As a result, bathymetry shallows with time. This process is affected by changes in the rate of subduction and by changes in the age of the subducted crust. Since these factors affect entire margin segments, regional unconformities may develop.

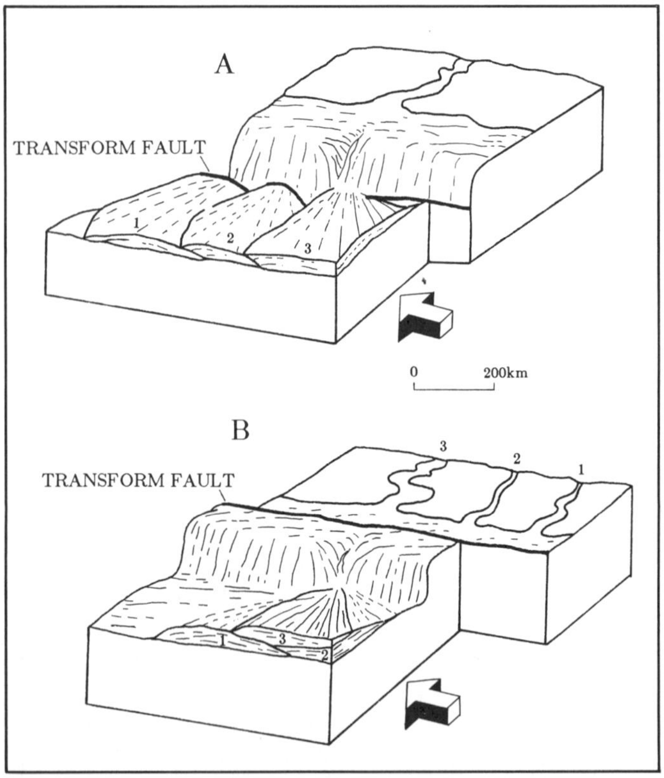

Figure 4. Models showing two configurations of deep-sea fans that can evolve along transform margins. In model A (example: Baranof fan west of southern Alaska; Stevenson and Embley, this volume), a submarine canyon is landward of a transform fault; thus with slip on the fault successive submarine fans are transported away from a single source (stages 1 to 3). In model B (examples: Delgada and Monterey fans of California), the submarine canyon is seaward of a transform fault; slip on the fault results in overlapping development of submarine fans with distinct source terranes. To recognize these conditions each river system must have a characteristic provenance. In A, lithology is uniform in fans 1, 2, and 3; in B, detritus is distinctly different in fan lobes 1, 2, and 3 (modified from Howell, Alpha, and Joyce, 1980).

Basins on the shelf typically overlie deformed rocks of the accretionary complex or of older framework rocks underthrust by the complex. They are generally not as large or elongate as trench basins. Shelf basins are commonly separated by poorly understood ridges that trend perpendicular to the margin, such as the Portlock anticline on the Kodiak Shelf and the Augustine-Seldovia arch in Cook Inlet (Fisher, 1980; Fisher and Magoon, 1978; Fisher, Detterman, and Magoon, this volume; von Huene, Fisher, and Bruns, this volume).

Slope basins can form in depressions on or structurally behind the accretionary complex and may eventually be incorporated into it themselves. Submarine canyons and channels, slump blocks, and diapiric ridges may cut basin-filling strata and complicate stratigraphic relationships.

Typical trench basins are arcuate, narrow, and elongate. Trench-filling sediment may be deposited on oceanic crust, on oceanic sedimentary rocks overlying the crust, or on material scraped off the subducting plate. In contrast, "subduction erosion"—the downward removal of material under trench slopes—may be a significant process in some trench basins (Howell and von Huene, 1981). Trench-filling strata are typically turbiditic and may have distant source terranes, particularly where sediment transport is along the trench axis, as it is in the Aleutian Trench.

Where sediment fills the trench, deposition continues onto the adjacent ocean floor and forms deep-sea fans. This process has occurred along the southern British Columbia to northern California segment of the margin, where Nitinat and Astoria fans build across a filled trench and into the Cascadia Basin. If a seamount chain, sea-floor rise, or other positive-relief feature approaches a convergent margin, large bodies of sediment may pond in front of it. In the Cascadia Basin, the seaward limits of the Nitinat and Astoria fans lie along the Juan de Fuca Ridge to the east and the Blanco fracture zone to the south (Carlson and Nelson, this volume). The distribution of sediment is similarly restricted in Gorda Basin by the Gorda Ridge and the Mendocino Escarpment. Where sedimentary terranes approach the margin, allochthonous and autochthonous sedimentary sequences may be juxtaposed; eventually an overlap sequence or successor basin forms, for example, Cook-Shelikof Basin (Fisher, Detterman, and Magoon, this volume).

Compression associated with subduction and accretion may also be noted at a considerable distance from the actual plate boundary. Although the mechanism is not completely understood, the formation of some fold-thrust belts and related foredeeps is tied to accretionary and collision events along the continental margin. In western North America, systems of back-arc basins and fold-thrust belts have formed parallel to the margin many hundreds of kilometers inland from the active plate boundary. These basins commonly form deep in the continental interior and fill with predominantly nonmarine sedimentary sequences. Where a continent narrows, however, as North America does at Alaska and Mexico, back-arc basins and fold-thrust belts can be superimposed on the unrelated continental margin on the opposite coast. An example is the crustal sag of the Colville foredeep formed in response to tectonic thickening in the Brooks Range, yet any genetic linkage between hinterland tectonics along the Pacific Basin and rifting on the adjacent Arctic margin remains equivocal (Grantz and May, 1983).

The petroleum potential of convergent-margin basins is generally limited by the lack of good reservoirs and by low heat flow. Productive areas such as the Cook Inlet oil and gas province in Alaska have developed locally where reservoirs in the basin fill overlie mature source rocks in the basement.

Rifted and Passive Margins

A continuum exists between incipient rifts and passive margins. At least in the earliest stages of their development, rifted continental margins can be considered tectonically active. As crustal extension begins, the development of normal faults results in horst-and-graben topography. If extension continues, heat flow and volcanism increase, followed first by the formation of intermediate crust and last by the formation of oceanic crust. Only after an ocean has formed and the loci of spreading has moved away from the margin is the term "passive" strictly appropriate. Once formed, this type of margin is relatively stable and may be extremely long lived.

At present, western North America has two rifted margins. One, the North Slope of Alaska, formed during the late Mesozoic and has subsequently been buried by a thick shelf-progradational sequence of Cretaceous and younger sediment (Table 3). The second, located along the Salton Trough and the Gulf of California, is a young, actively rifting margin.

During early phases of continental rifting, the typical rift basin forms as a graben or half-graben. At this stage, nonmarine sediment—largely composed of lacustrine deposits, alluvial deposits, or evaporites—covers the basin floor. As rifting progresses and the basin deepens, these continental and paralic deposits commonly are overlain by a marine sequence, indicating deposition at increasing water depths. If the basin opens to form a small sea, marine progradational sequences form at the basin margins and cover older shallow-marine and nonmarine sequences.

Margins that are tectonically passive may also form in other ways. When motion on a convergent or transform margin stops, a passive margin forms. In contrast to passive margins that slowly evolve from actively rifting margins, formation of this type of passive margin occurs rapidly and requires a major reorganization of plate boundaries.

North American examples of this type of passive margin include the western margin of the Baja California Peninsula and the outer Bering Sea Shelf. Along the western side of Baja California, Mesozoic and early Tertiary convergence stopped as the Rivera triple junction migrated southward along the coast. Subsequently, the active plate boundary jumped inland to form the rift in the Gulf of California. In the

Table 2. Typical convergent-margin basins of western North America.

Basin Name	Type*	Location (lat., long.)	Approximate Area (km^2)	Maximum Thickness (km)	Age of Sedimentary Section
Albatross	FA	56°30′N, 153°W	6,000	5	Late Miocene and younger
Aleutian Basin	O&T	57°N, 180°	640,000	10	Cenozoic (and Cretaceous?)
Aleutian forearc	FA	53°N, 179°E	60,000	8	Eocene and younger
Aleutian Trench	T	51°N, 179°E	75,000	4	Cenozoic
Amlia-Amukta	IA	52°30′N, 172°W	5,700	5	Late Miocene (?) and younger
Astoria	FA	45°30°N, 124°10′W	3,250	5	Eocene and younger
Bowers	O	53°30°N, 176°E	80,000	3	Cenozoic (and Cretaceous?)
Bristol Bay	BA	56°N, 162°W	18,900 >1 km thick	>7	Jurassic and younger
Colville foredeep	FD	70°N, 160°W	165,000	9	Cretaceous and younger
Cook-Shelikof	FA	60°N, 152°W	40,000	7	Tertiary
				12	Mesozoic
Coos Bay	FA	44°N, 124°30′W	4,000	>6	Eocene and younger
Eel River	FA	41°N, 124°30′W	14,000	>4	Tertiary
Great Valley	FA†	37°N, 120°W	62,000	>10	Jurassic and younger
Heceta	FA	45°N, 125°W	13,700	>7	Late Miocene and younger
Humboldt (onshore Eel River)	FA	40°30′N, 124°W	2,300	>4	Tertiary
Komandorsky	O	57°N, 168°E	122,000	3	Tertiary (and Cretaceous?)
Middleton	FA	60°N, 146°W	16,000	>5	Middle Eocene and younger
Newport	FA	44°40′N, 124°10′W	1,550	>6	Eocene and younger
Olympic	FA	48°N, 125°20′W	8,550	>6	Late Eocene and younger
Puget-Williamette Trough	FA	46°N, 122°W	37,500	>6	Eocene and younger
Sanak	FA	54°N, 161°W	4,000	>8	Cenozoic
Shumagin	FA	55°30′N, 158°30′W	3,800	>2.5	Late Miocene and younger
Stevenson	FA	58°N, 149°W	7,500	7	Late Miocene(?) and younger
Tofino-Fuca	FA	49°N, 126°W	8,000	6	Eocene and younger
Trinity	FA	56°N, 154°W	2,200	4	Late Miocene and younger
Tugidak	FA	56°15′N, 155°30′W	3,000	5	Late Miocene and younger
Willapa	FA	47°N, 125°W	8,100	>4	Eocene and younger

*Basin types: BA—Back-arc basin, FA—Fore-arc basin, FD—Foredeep, IA—Intra-arc basin, O—Oceanic basin (deep basins with sedimentary sections overlying relatively undeformed segments of oceanic crust that have accreted to North America), T—Trench basin.

†The southern part of the Great Valley has been affected by Neogene transform tectonics.

Table 3. Typical rift and passive-margin basins of western North America.

Basin Name	Type*	Location (lat., long.)	Approximate Area (km^2)	Maximum Thickness (km)	Age of Sedimentary Section
Dinkum graben	R	70°30′N, 148°W	4,250	>10	Jurassic and younger
Gulf of California	R	26°N, 110°W	140,000	4	Neogene
Hope	E	67°N, 168°W	63,000	5	Cenozoic (and Cretaceous?)
Kaktovik	SP	70°N, 143°W	14,000	>12	Jurassic and younger
North Chukchi	SP	73°N, 167°W	10,000	>12	Cretaceous and younger
Norton	E	64°N, 166°W	55,000	6	Cenozoic
Nuwuk	SP	71°40′N, 153°W	15,000	>12	Cretaceous and younger
Purisima-Iray	YP	25°N, 113°W	20,000	5	Cretaceous and younger
Salton Trough	RT	33°N, 116°W	26,000	6	Neogene
Vizcaino	YP	28°N, 115°W	35,000	>6	Cretaceous and younger

*Basin types: E—Extensional basin not on the continental margin, R—Rift basin, RT—Rift and transform basin, SP—Shelf progradational sequence, YP—Young passive-margin basins of western Baja California (most of the basin fill was deposited in a convergent fore-arc setting).

Late Cretaceous or early Tertiary, this type of passive margin formed along the outer Bering Sea Shelf when motion on the extant transform or oblique-convergent margin stopped. This transformation apparently accompanied the plate-boundary jump to the south that formed the Aleutian convergent margin (Cooper, Marlow, and Scholl, this volume; Scholl, Vallier, and Stevenson, this volume).

Sedimentary sequences deposited on passive margins of any origin are typified by progradational sequences. Progradational sequences are commonly elongate parallel to the margin and consist of sediment prisms that prograde toward the ocean. These shallowing-upward (regressive) sequences typically include deep-marine strata overlain by shallow-marine and deltaic units capped by fluvial or other nonmarine deposits.

SUMMARY

The North American continent began forming more than 2.5 b.y. ago. The Cordillera, however, seems to represent tectonic processes that took place less than 200 m.y. ago. Terranes within the Cordillera may consist of much older strata and, indeed, many of the terranes first formed in exotic regions far distant from their current locations. Nonetheless, the late Mesozoic and Cenozoic events of amalgamation, accretion, and dispersion have sculptured the Cordillera into its present configuration.

The crustal mobility of the Cordillera resulted in a host of transient basin configurations with different degrees of preservation. Stratigraphic and emerging paleomagnetic data indicate the degrees of allochthoneity, and sedimentologic analyses indicate the complex interplay of ongoing tectonic and eustatic processes. The basin classification used in this chapter conveys the complexity and variations in basin styles; other chapters give more detail on the evolution of selected basins throughout the expanse of the Cordillera, from the Beaufort Sea to the peninsula of Baja California.

REFERENCES CITED

Beck, M. E., Jr., 1980, Paleomagnetic record of plate-margin tectonic processes along the western edge of North America: Journal of Geophysical Research, v. 85, p. 7115–7131.

Benioff, H., 1954, Orogenesis and deep crustal structure; additional evidence from seismicity: GSA Bulletin, v. 65, p. 385–400.

Bird, K. J., 1981, Petroleum exploration of the North Slope, Alaska, *in* J. F. Mason, ed., Petroleum geology in China: New York, United Nations, p. 233–248.

Blake, M. C., Jr., R. H. Campbell, T. W. Dibblee, Jr., D. G. Howell, T. H. Nilsen, W. R. Normark, J. G. Vedder, and E. A. Silver, 1978, Neogene basin formation in relation to plate-tectonic evolution of San Andreas fault system, California: AAPG Bulletin, v. 62, p. 344–372.

Carlson, P. R., and C. H. Nelson, 1987, Marine geology and resource potential of the Cascadia Basin, this volume.

Champion, D. E., D. G. Howell, and M. C. Marshall, 1981, Paleomagnetic evidence for 3800 km of northwestward translation of San Miguel Island, southern California borderland: Eos (American Geophysical Union Transactions), v. 62, p. 855.

Coats, R. R., 1962, Magma type and crustal structure in the Aleutian arc: American Geophysical Union Geophysical Monograph 6, p. 92–109.

Condie, K. C., 1982, Plate-tectonics model for Proterozoic continental accretion in the southwestern United States: Geology, v. 10, p. 37–42.

Cooper, A. K., M. S. Marlow, and D. W. Scholl, 1987, Geologic framework of the Bering Sea crust, this volume.

Crowell, J. C., and M. H. Link (eds.), 1982, Geologic history of Ridge Basin, southern California: SEPM, Pacific Section, Los Angeles, California, 304 p.

Fisher, M. A., 1980, Petroleum geology of Kodiak Shelf, Alaska: AAPG Bulletin, v. 64, p. 1140–1157.

Fisher, M. A., and L. B. Magoon, 1978, Geologic framework of lower Cook Inlet, Alaska: AAPG Bulletin, v. 62, p. 373–402.

Fisher, M. A., R. L. Detterman, and L. B. Magoon, 1987, Tectonics and petroleum geology of the Cook-Shelikof Basin, this volume.

Frei, L. S., J. R. Magill, and A. Cox, 1984, Paleomagnetic results from the central Sierra Nevada: Constraints on reconstructions of the western United States: Tectonics, v. 3, p. 157–177.

Gordon, R. G., A. Cox, and S. O'Hare, 1984, Paleomagnetic Euler poles and the apparent polar wander and absolute motion of North America since the Carboniferous: Tectonics, v. 3, p. 499–537.

Grantz, A., and S. D. May, 1983, Rifting history and structural development of the continental margin north of Alaska: AAPG Memoir 34, p. 77–100.

Hamilton, W., 1969, Mesozoic California and the underflow of Pacific mantle: GSA Bulletin, v. 80, p. 2409–2429.

Harland, W. B., 1971, Tectonic transpression in Caledonian Spitzbergen: Geological Magazine, v. 108, p. 27–42.

Hill, M. L., and T. W. Dibblee, Jr., 1953, San Andreas, Garlock, and Big Pine faults, California—A study of the character, history, and tectonic significance of their displacements: GSA Bulletin, v. 64, p. 443–458.

Howell, D. G., and R. von Huene, 1981, Tectonics and sediment, along active continental margins, *in* R. G. Douglass, I. P. Colburn, and D. S. Gorsline, eds., Depositional systems of active continental margin basins: SEPM, Pacific Section, p. 1–13.
Howell, D. G., T. R. Alpha, and J. M. Joyce, 1980, Plate tectonics, physiography, and sedimentation of the northeast Pacific region, *in* M. E. Field, R. G. Douglas, A. H. Bouma, J. C. Ingle, and I. P. Colburn, eds., Quaternary depositional environments of the Pacific coast: SEPM, Pacific Section, Pacific Coast Paleogeography Symposium 4, p. 43–53.
Howell, D. G., J. K. Crouch, H. G. Greene, D. S. McCulloch, and J. G. Vedder, 1980, Basin development along the late Mesozoic and Cainozoic California margin—A plate tectonic margin of subduction, oblique subduction and transform tectonics: International Association of Sedimentologists Special Publication, v. 4, p. 43–62.
Howell, D. G., E. R. Schermer, D. L. Jones, Z. Ben-Avraham, and E. Scheibner, 1985, Preliminary tectonostratigraphic terrane map of the Circum-Pacific region: AAPG, scale 1:17,000,000.
Jones, D. L., N. J. Silberling, and J. Hillhouse, 1977, Wrangellia—A displaced terrane in northwestern North America: Canadian Journal of Earth Sciences, v. 14, p. 2565–2577.
Ketner, K. B., and J. F. Smith, Jr., 1982, Mid-Paleozoic age of the Roberts thrust unsettled by new data from northern Nevada: Geology, v. 10, p. 298–303.
Marlow, M. S., A. K. Cooper, and M. A. Fisher, 1987, Petroleum geology of the Beringian continental shelf, this volume.
Mason, R. G., and A. D. Raff, 1961, A magnetic survey off the west coast of North America: GSA Bulletin, v. 72, p. 1259–1265.
McLean, H., and T. J. Wiley, 1987, Geologic potential for hydrocarbons in unexplored offshore basins of western North America, this volume.
Monger, J. W. H., and C. A. Ross, 1971, Distribution of fusulinaceans in the Canadian Cordillera: Canadian Journal of Earth Sciences, v. 8, p. 259–278.
Monger, J. W. H., R. A. Price, and D. J. Tempelman-Kluit, 1982, Tectonic accretion and origin of the two major metamorphic and plutonic welts in the Canadian Cordillera: Geology, v. 10, p. 70–75.
Moore, G. W., 1984, Tertiary dismemberment of western North America: Third Circum-Pacific Energy and Mineral Resources Conference Transactions, p. 607–612.
Moore, G. W., and L. del Castillo, 1974, Tectonic evolution of the southern Gulf of Mexico: GSA Bulletin, v. 85, p. 607–618.
Nilsen, T. H., 1981, Upper Devonian and Lower Mississippian redbeds, Brooks Range, Alaska, *in* A. D. Miall, ed., Sedimentation and tectonics in alluvial basins: Geological Association of Canada Special Paper 23, p. 187–219.
Nilsen, T. H., and J. H. Stewart, 1980, The Antler orogeny—Mid-Paleozoic tectonism in western North America: Geology, v. 8, p. 298–302.
Nokleberg, W. J., 1983, Wallrocks of the central Sierra Nevada batholith, California—A collage of accreted tectono-stratigraphic terranes: U.S. Geological Survey Professional Paper 1255, 28 p.
Patton, W. W., Jr., 1973, Reconnaissance geology of the northern Yukon-Koyukuk province, Alaska: U.S. Geological Survey Professional Paper 774–A, 17 p.
Pindell, J., and J. F. Dewey, 1982, Permo-Triassic reconstruction of western Pangea and the evolution of the Gulf of Mexico/Caribbean region: Tectonics, v. 1, p. 179–211.
Plafker, G., 1987, Regional geology and petroleum potential of the northern Gulf of Alaska continental margin, this volume.
Poole, F. G., and C. A. Sandberg, 1977, Mississippian paleogeography and tectonics of the western United States, *in* J. H. Stewart, C. H. Stevens, and A. E. Fritsche, eds., Paleozoic paleogeography of the western United States: SEPM, Pacific Section, Pacific Coast Paleogeography Symposium 1, p. 67–85.
Scholl, D. W., T. L. Vallier, and A. J. Stevenson, 1987, Geologic evolution and petroleum geology of the Aleutian Ridge, this volume.
Silver, L. T., and T. H. Anderson, 1974, Possible left-lateral early to middle Mesozoic disruption of the southwestern North American craton margin (abs.): GSA Abstracts with Programs, v. 6, p. 955–956.
Speed, R. C., 1983, Evolution of the sialic margin in the central and western United States: AAPG Memoir 34, p. 457–468.
Stevenson, A. J., and R. Embley, 1987, Deep-sea fan bodies, terrigenous turbidite sedimentation, and petroleum geology, Gulf of Alaska, this volume.
Stewart, J. H., 1972, Initial deposits in the Cordilleran geosyncline—Evidence of a late Precambrian (850 m.y.) continental separation: GSA Bulletin, v. 83, p. 1345–1360
Vedder, J. G., 1987, Regional geology and petroleum potential of the southern California borderland, this volume.
Vine, F. J., and D. H. Matthews, 1963, Magnetic anomalies over oceanic ridges: Nature, v. 199, p. 947–949.
von Huene, R., M. A. Fisher, and T. R. Bruns, 1987, Geology and evolution of the Kodiak margin, Gulf of Alaska, this volume.
Wilson, J. T., 1965, A new class of faults and their bearing on continental drift: Nature, v. 207, p. 343–347.

CHAPTER TWO

REGIONAL GEOLOGY AND PETROLEUM POTENTIAL OF THE UNITED STATES BEAUFORT AND NORTHEASTERNMOST CHUKCHI SEAS

Arthur Grantz
Steven D. May
David A. Dinter

U.S. Geological Survey
Menlo Park, California

ABSTRACT

The United States Beaufort and northeast Chukchi Seas overlie a passive margin of late Neocomian age and sedimentary prisms that range from less than one-half to more than 10 km in thickness. Prerift strata commonly consist of strongly deformed and mildly metamorphosed sedimentary rocks of Cambrian to Devonian age that constitute economic basement and of northerly sourced, stable-shelf clastic and carbonate rocks of Mississippian to Neocomian age of the sequence that contains the large petroleum deposits at Prudhoe Bay. Postrift strata consist of southerly sourced prodelta and intradelta clastic rocks of Aptian or Albian to Tertiary age that were deposited in the Colville foredeep in front of the Brooks Range orogen. These rocks overstep the Mississippian to Neocomian stable-shelf sequence and rest on the pre-Mississippian sequence (economic basement) beneath the present middle or outer Beaufort Shelf. Seismic character and structural and stratigraphic continuity with northern Alaska suggest that most of the Beaufort and adjacent northeast Chukchi Seas are prospective for petroleum.

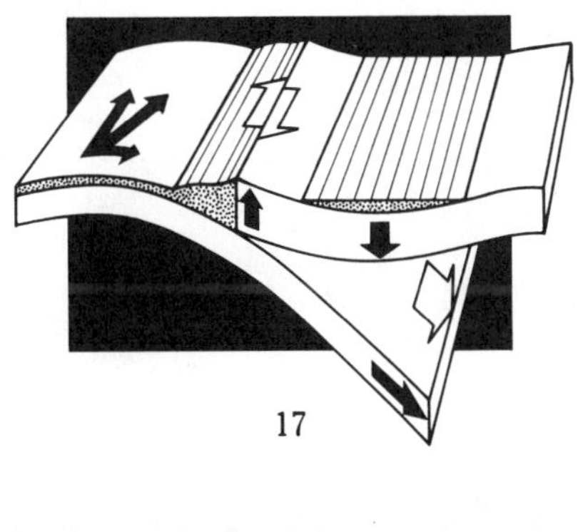

INTRODUCTION

The Beaufort and northeasternmost Chukchi Seas (Figure 1) offer petroleum explorationists a terrane that is promising for medium to giant oil and gas accumulations, but is the most environmentally difficult region in the nation for petroleum exploration and development. Despite the difficulties, however, the National Petroleum Council (1981, p. 5) estimates that "Proven technology and sufficient information and technical expertise for advanced design work is available for the industry to proceed confidently with operations in water as deep as 650 feet in the southern Bering Sea and to about 200 feet in the more severely ice-covered areas of the northern Bering, Chukchi, and Beaufort Seas." The region lies adjacent to the North Slope of Alaska, which is served by the Trans-Alaska Pipeline System (TAPS). This common carrier will provide an outlet for oil brought to its northern terminus near Prudhoe Bay. If the proposed Alaska Natural Gas Transportation System, a pipeline with a terminus near Prudhoe Bay, is also built, it would provide the means for carrying offshore natural gas to market. The feasibility of economically transporting large volumes of oil or gas by tanker year round from the Beaufort Sea to markets in the conterminous United States has yet to be demonstrated.

This report presents an overview of the geologic framework and petroleum potential of the United States Beaufort and northeasternmost Chukchi Seas (Figure 1). The principal data base consists of the multichannel seismic-reflection profiles and accompanying high-resolution profiles shown in Figure 1 and some additional single-channel seismic-reflection profiles, sonobuoy refraction measurements, and bathymetric data. Areas shallower than 200 ft (60 m), where the National Petroleum Council considers that the petroleum industry can now confidently proceed with operations, underlie most of the outer continental shelf (i.e., the continental shelf out to the 200-ft or 60-m isobath) north of Alaska. Sea-ice conditions in this area during late September, the usual period of maximum retreat, are shown in Figure 2.

GEOLOGIC FRAMEWORK

The continental shelf in the study area is divided into two sectors of contrasting geologic structure. The western (Bar-

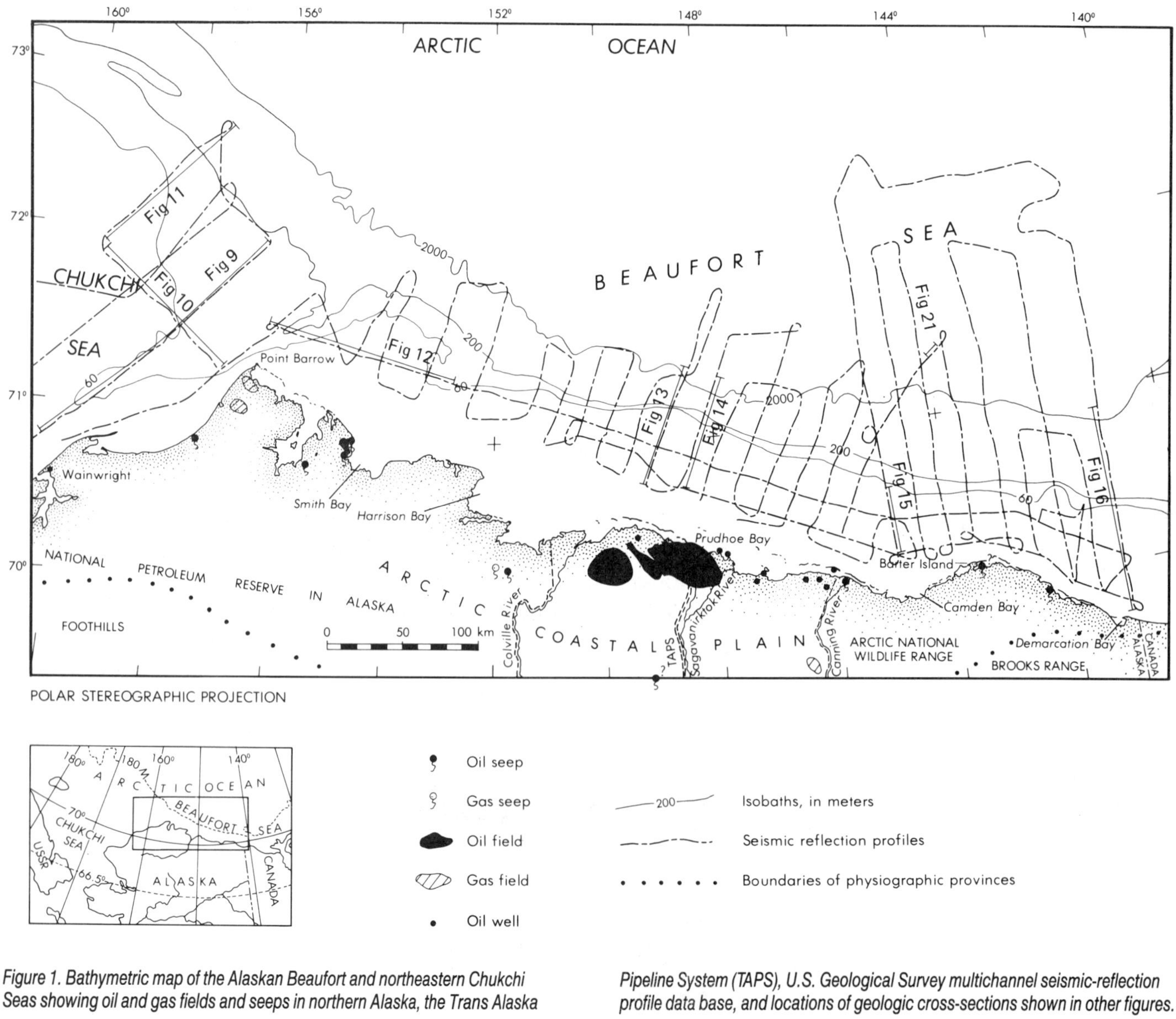

Figure 1. Bathymetric map of the Alaskan Beaufort and northeastern Chukchi Seas showing oil and gas fields and seeps in northern Alaska, the Trans Alaska Pipeline System (TAPS), U.S. Geological Survey multichannel seismic-reflection profile data base, and locations of geologic cross-sections shown in other figures.

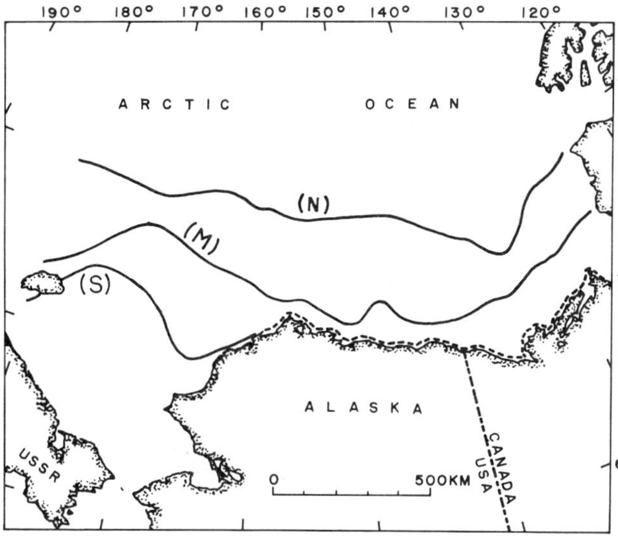

Figure 2. Map showing most northerly (N), most southerly (S), and median (M) positions of the southern edge of the Arctic ice pack north of Alaska during the usual period of maximum retreat, September 16 to 30. Based on data from 1954 through 1970 (after Brower et al., 1977).

row) sector, dominated by the Barrow arch, extends eastward from the northeast Chukchi Sea to near long. 147°W. The eastern (Barter Island) sector extends eastward from near long. 147°W to the Canadian border. For convenience in describing its petroleum potential, the region has been divided into six provinces and five subprovinces. The areal extent and relationship of these provinces to the major physiographic features of the region are shown in Figure 3. Figures 4 and 5 show the general geology, stratigraphy, and depositional history of northern Alaska and of the Beaufort and northeast Chukchi Shelves. Figure 6 gives the approximate relationship between seismic-reflection times and depth to the geologic horizons shown in Figure 5 and other illustrations. The thickness of sedimentary rocks above the Franklinian sequence, which is considered to be basement for oil and gas deposits on the Beaufort Shelf, is given in Figure 7, and typical cross-sections are presented in Figures 8–16. Data for this report are mainly from Grantz and May (1983); Grantz, Eittreim, and Dinter (1979); Grantz, Eittreim, and Whitney (1981); and Eittreim and Grantz (1979). Data for the onshore are from Alaska Geological Society (1971, 1972, 1977, and 1981); Brosgé and Tailleur (1971); Jones and

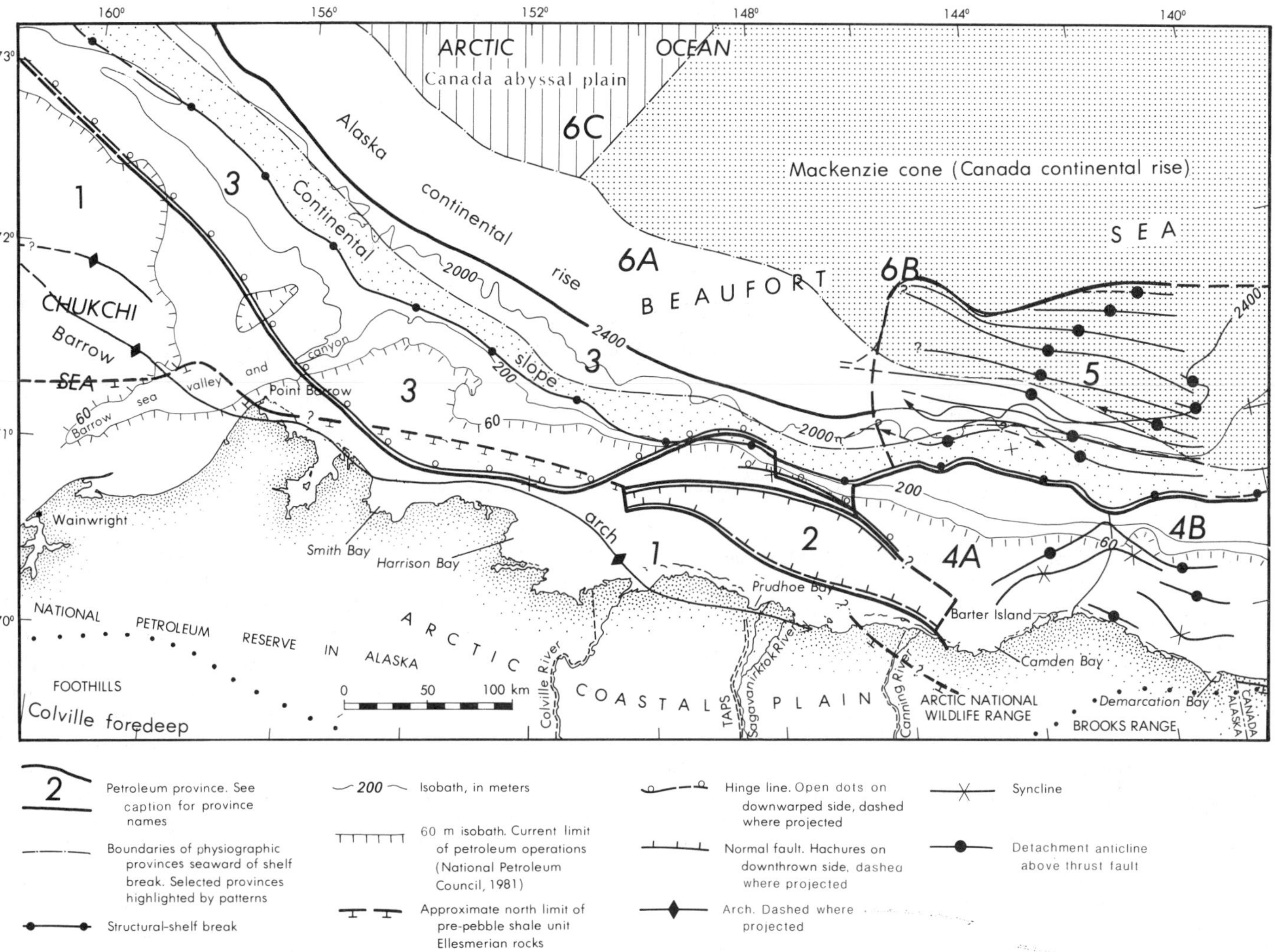

Figure 3.. Geologic provinces of the United States Beaufort and northeast Chukchi Seas: 1. Arctic platform; 2. Dinkum graben west of Kaktovik Basin; 3. Nuwuk Basin; 4A. West Kaktovik Basin; 4B. East Kaktovik Basin; 5. Detachment fold province; 6. Canada Basin; 6A. Alaska continental rise seaward of 2400-m isobath; 6B. Mackenzie cone of Canada continental rise exclusive of detachment fold province; 6C. Canada Abyssal Plain.

AGE			SEQUENCE	STRATIGRAPHY (South — North)	THICKNESS(m)	GENERALIZED LITHOLOGY	DEPOSITIONAL CHARACTERISTICS
CENOZOIC	QUATERNARY		BROOKIAN	GUBIK FORMATION	10-200	Marine sand, gravel, silt, and clay.	Sediment derived from the Brooks Range, the Arctic foothills, wave erosion of sea cliffs, and melting icebergs.
CENOZOIC	TERTIARY	NEOGENE; PALEOGENE	BROOKIAN	SAGAVANIRKTOK FM. (eastern North Slope only)	0-2,500 W - E	Poorly consolidated nonmarine and marine shale, sandstone, and conglomerate, with some carbonaceous shale, lignite, and bentonite.	Sediment mostly prograded northeastward from the Brooks Range into the southward-deepening Colville foredeep, an east-west-elongate trough created when the Arctic platform tilted southward, probably as a result of loading of Brooks Range thrust sheets and clastic sediment on the south part of the platform. When the Colville foredeep was filled, Cretaceous and Tertiary sediments overtopped the Barrow arch and prograded northward onto the western Beaufort shelf, where they thicken northward.
MESOZOIC	CRETACEOUS	UPPER	BROOKIAN	COLVILLE GROUP (central and eastern North Slope only)	0-3,600 W - E	Predominantly nonmarine, with coal in the west. Shallow- to deep-marine clastic rocks in the east.	
MESOZOIC	CRETACEOUS	LOWER	BROOKIAN	NANUSHUK GROUP (W. North Slope)	0-3,300+; 200-1,200+	Marine and nonmarine shale, siltstone, sandstone, coal, conglomerate and bentonite.	
MESOZOIC	CRETACEOUS	LOWER	BROOKIAN	FORTRESS MOUNTAIN FM. (W. North Slope); TOROK FM.	400-3,000; 1000-3,000	Marine shale, sandstone, and conglomerate. Marine shale, sandstone, and siltstone turbidites.	
MESOZOIC	CRETACEOUS	LOWER	ELLESMERIAN	PEBBLE SHALE UNIT, KONGAKUT FM., and KEMIK SANDSTONE MBR.	0-700	Shelf and basinal marine shale and siltstone containing rounded quartz grains and chert pebbles. Coquinoid to south; quartzose sandstone at base in east.	The Ellesmerian sequence on the Alaskan North Slope was derived from a northerly source terrane called Barrovia by Tailleur (1973). The constituent formations generally thin and coarsen northward, and onlap the uplifted northern Arctic platform in the crestal region of the Barrow arch.
MESOZOIC	JURASSIC		ELLESMERIAN	KINGAK SHALE (locally includes KUPARUK RIVER SANDS[1/] at top)	0-1,200+	Marine shale, siltstone, and chert, locally containing glauconitic sandstone (in the west). Shallower water facies are apparently the northerly ones.	
MESOZOIC	TRIASSIC		ELLESMERIAN	SHUBLIK FORMATION	0-225	To the north, marine shale, carbonate, and sandstone. As shown, includes the Sag River Sandstone. To the south, shale, chert, limestone, and oil shale.	
PALEOZOIC	PERMIAN		ELLESMERIAN	SADLEROCHIT GROUP (and SIKSIKPUK FM. on western North Slope)	0-700+	Eastern North Slope: marine and nonmarine sandstone, shale, and conglomerate; marine sandstone, siltstone, and shale. Western North Slope: sandstone, conglomerate, and shale to the north; argillite, chert, and shale to the south.	
PALEOZOIC	PENNSYLVANIAN		ELLESMERIAN	?; LISBURNE GROUP	0-2,000+	Fossiliferous marine limestone and dolomite with some chert and local volcanic rock. Basal beds commonly sandstone, shale, and conglomerate.	
PALEOZOIC	MISSISSIPPIAN		EO-ELLESMERIAN	ENDICOTT GROUP	0-5,000	Dominantly nonmarine sandstone, shale, and coal. Some limestone in upper beds.	Probably derived from both a northerly source terrane (Barrovia) and from local sources. Deposited in local deep basins that in places are fault-bounded.
PALEOZOIC	DEVONIAN		EO-ELLESMERIAN	? ?; ENDICOTT(?)GP.		Nonmarine sandstone, carbonaceous shale with coal partings, and conglomerate.	
PALEOZOIC	SILURIAN AND OLDER		FRANKLINIAN	Marine sedimentary rocks (includes IVIAGIK GROUP of Martin (1970) on Lisburne Peninsula)	Thousands of meters	Eastern North Slope: argillite, graywacke, limestone, dolomite, chert, quartzose sandstone, shale, and metamorphic equivalents. Western North Slope: Ordovician and Silurian argillite and graywacke with graptolites.	Deposited during Middle Cambrian to Silurian time in the Franklinian geosyncline, which trended generally parallel to the Arctic margin of North America. North and northwestern facies are mostly eugeoclinal, south and southeastern facies mostly miogeoclinal. Probably extends northward beneath the Beaufort and Chukchi shelves.

1/ Local usage

Figure 4. Generalized stratigraphy of subsurface Alaskan North Slope and adjacent continental shelf.

Speers (1976); Tailleur, Pessel, and Engwicht (1978); Miller et al. (1979); and Guldenzopf et al. (1980).

Sedimentary Sequences

The sedimentary strata of northern Alaska and the Beaufort Shelf are conveniently grouped into four regionally extensive sequences of contrasting lithology, tectonic characteristics, and hydrocarbon potential (Figures 4, 8). The Cambrian to Devonian Franklinian sequence, consisting of mildly to strongly metamorphosed sedimentary and some volcanic rocks that are in places intruded by granitic plutons, is inferred to constitute economic basement. Following mild metamorphism and regional deformation, an extensive platform was cut across the Franklinian rocks in the Early Devonian and the Late Devonian to Early Mississippian. This, the Arctic platform, was a surface of low relief that remained relatively stable from the Devonian to the Neocomian (Early Cretaceous), an interval of more than 200 m.y. During this time, lithologically diverse suites of clastic and carbonate sedimentary rocks—the Eo-Ellesmerian and Ellesmerian sequences—were deposited on the platform. In the central and western North Slope, the Eo-Ellesmerian sequence (Grantz and May, 1984) consists of two units of dominantly nonmarine clastic sedimentary rocks. Both lie unconformably on the more strongly deformed marine sedimentary rocks of the Franklinian sequence. The older unit, known from two test wells (Topogaruk-1 and South Meade-1) about 90 km south of Barrow, is most likely Middle Devonian (Collins and Robinson, 1967). It is more strongly folded than a younger and more extensive unit which is Lower and lower Upper Mississippian. Both units are overlain with low angular unconformity by the Ellesmerian sequence. The unconformities represent late phases of the Ellesmerian orogeny of northern Alaska and the Canadian Arctic Archipelago. The

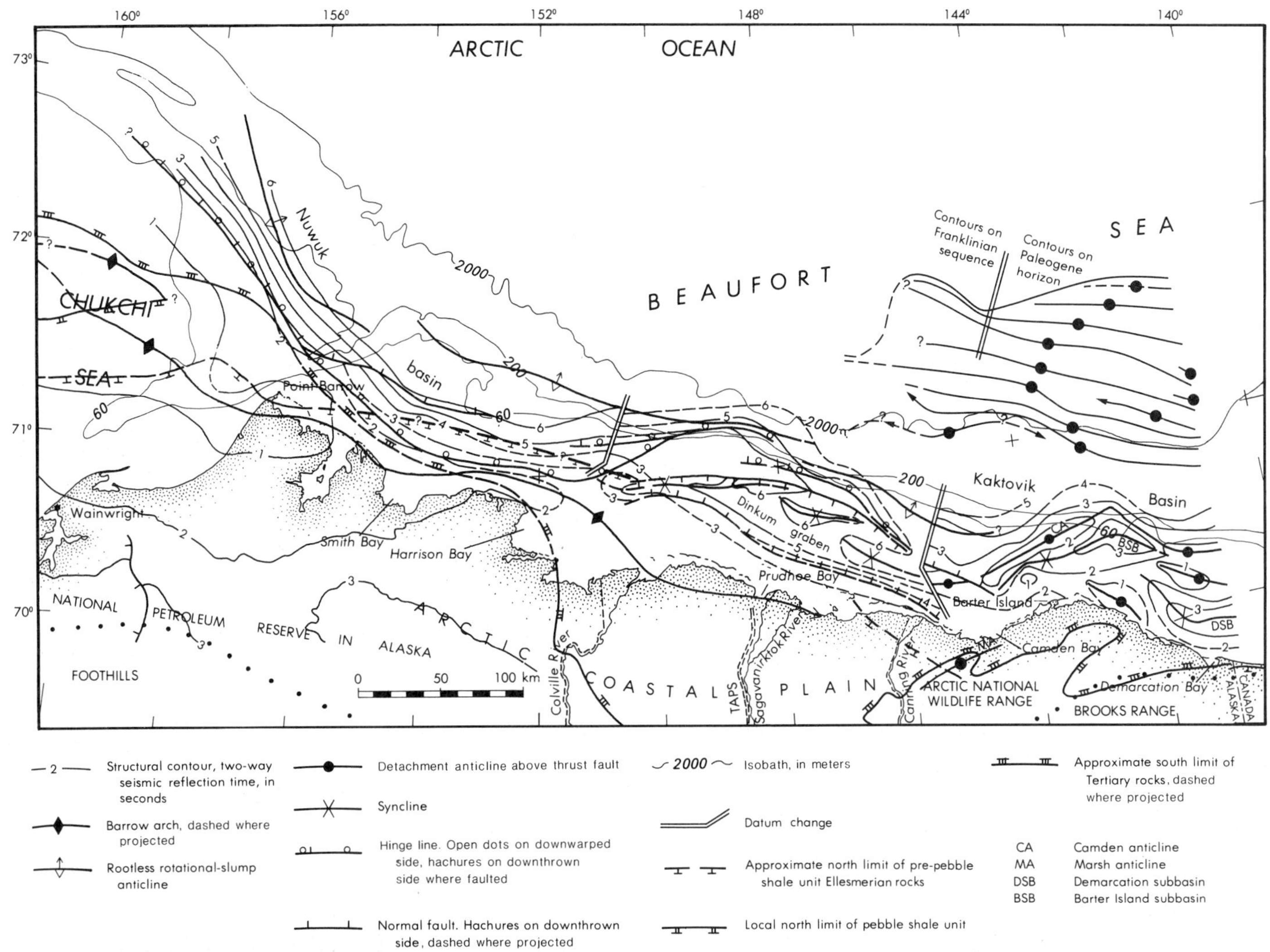

Figure 5. Structural geologic map of areas of the United States Beaufort and northeast Chukchi Seas covered by U.S. Geological Survey multichannel seismic-reflection data (see Figure 1). Onshore data west of the Colville River from Miller et al. (1979).

overlying Ellesmerian sequence contains both marine and nonmarine beds, including stratigraphically condensed organic-rich shale, texturally mature sandstone and conglomerate, and some dolomitized limestone. Clastic components were derived from a source terrane that lay beneath and north of the present outer Beaufort Shelf. The Eo-Ellesmerian sequence is differentiated from the dominantly marine Ellesmerian sequence because it is thicker, at least mainly nonmarine, occurs only in localized basins, and is separated from the Ellesmerian beds by a low angular unconformity.

In the Early Jurassic the northern part of the Arctic platform, the source terrane for the Ellesmerian Shelf sediments, was separated from northern Alaska by an incipient rifting event that disrupted earlier sedimentation patterns in the eastern Beaufort Sea (Grantz and May, 1983). This event was followed in the late Neocomian (mid-Early Cretaceous) by the main rifting event that created the present continental margin of northern Alaska. Beginning in the latest Jurassic or earliest Cretaceous, the Arctic platform was tilted down to the south and overridden by nappes from newly formed tec-

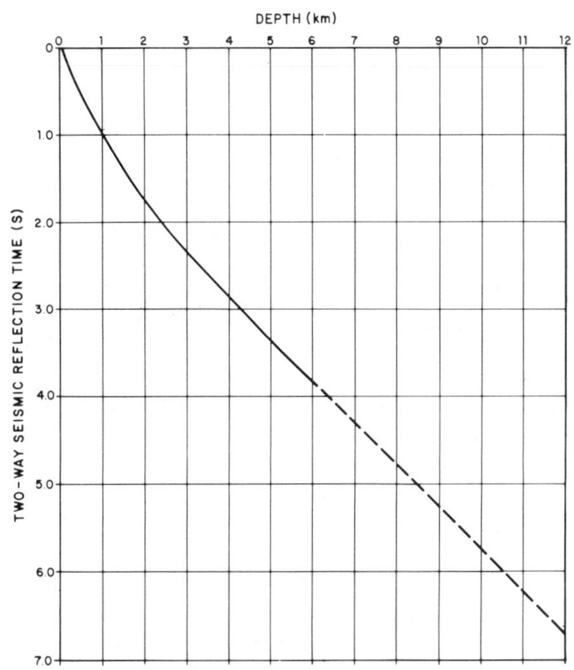

Figure 6. Generalized average seismic-reflection time as a function of depth for the Beaufort Shelf, derived from seismic-stacking-velocity measurements. Dashed where inferred.

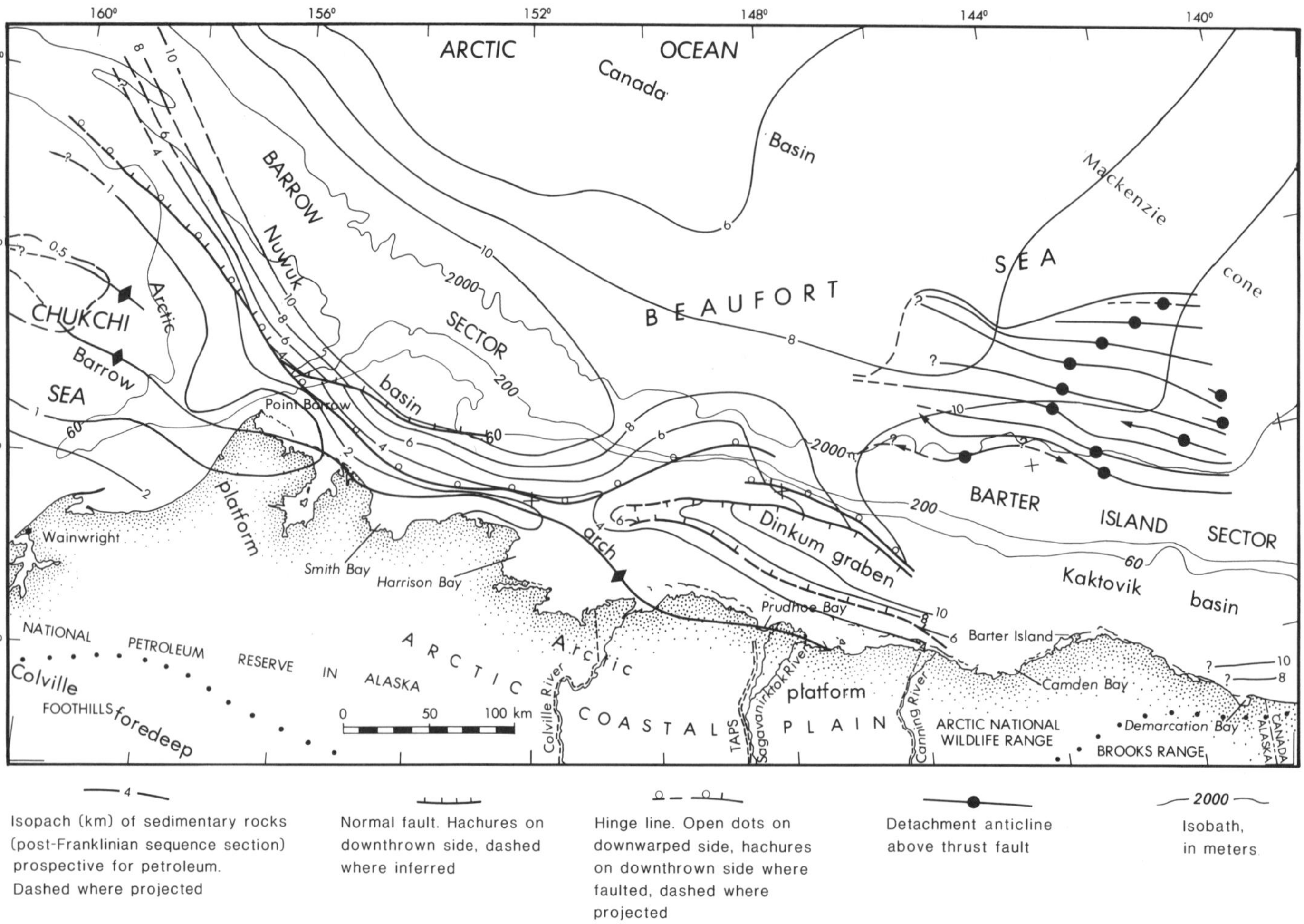

Figure 7. Interpretive map showing thickness of sedimentary rocks having petroleum potential beneath the United States Beaufort and northeast Chukchi Seas.

tonic highlands in the area of the present Brooks Range. An asymmetric foreland basin, the Colville foredeep, formed on the southward-tilted platform and received large volumes of sediment from the nascent Brooks Range to the south. Because of their provenance, these deposits have been named the Brookian sequence by Lerand (1973). The section in the Colville foredeep consists of Upper Jurassic(?) and Cretaceous and, on the eastern North Slope, of Tertiary clastic sedimentary rocks. In the southern part of the asymmetric foredeep, where it consists mainly of Cretaceous flysch and molasse, the Brookian section is >9000 m thick. The sequence thins northward to between 500 and 2000 m because of downlapping and erosion, and its basal beds are Albian where it crosses the Barrow arch. The younger beds of the Brookian sequence overtop the Barrow arch and thicken northward beneath the outer continental shelf.

Barrow Sector

Barrow arch, a broad basement high which trends parallel to the coast, is the dominant structure in the Barrow sector of the Beaufort Shelf. Its south flank is defined by the south-sloping Arctic platform, and its north flank is defined by the Arctic platform where it dips northward beneath the Beaufort Shelf. The north dip resulted from early uplift and erosion adjacent to the Early Cretaceous rift, and from subsequent rapid subsidence caused by postrift crustal cooling and sediment loading. Synrift and early postrift faulting, erosion, and tilting on the arch created the oil-bearing structure at Prudhoe Bay.

The position of the tectonic hinge line produced by postrift subsidence with respect to the structural shelf break off northern Alaska is the major determinant of the distribution of geologic provinces—the Arctic platform, the Dinkum graben, and the Nuwuk Basin—beneath the continental shelf and slope in the Barrow sector (Figure 3). Some of the geologic features of the Barrow sector are known from test wells on the North Slope of Alaska. Others, however, are absent from the North Slope and are known only from offshore seismic-reflection and sonobuoy-refraction data.

Arctic Platform

The Arctic platform (Figures 7, 9–13) extends offshore from northern Alaska with typically gentle dips to a tectonic hinge line (Figures 5, 7, 11–14), along which it is bent down to the north and is faulted in places. The crest of the broad Barrow arch (Figures 5, 7, 9), which formed initially during mid-Early Cretaceous rifting, underlies the continental shelf west of Point Barrow and between Smith Bay and the Colville River Delta. Erosional thinning of Albian beds over the arch indicates that its initial structural relief was augmented by Late Cretaceous and/or early Tertiary uplift. The northern wedge edge of the Ellesmerian sequence also extends offshore along much of the Beaufort Shelf. A rough estimate

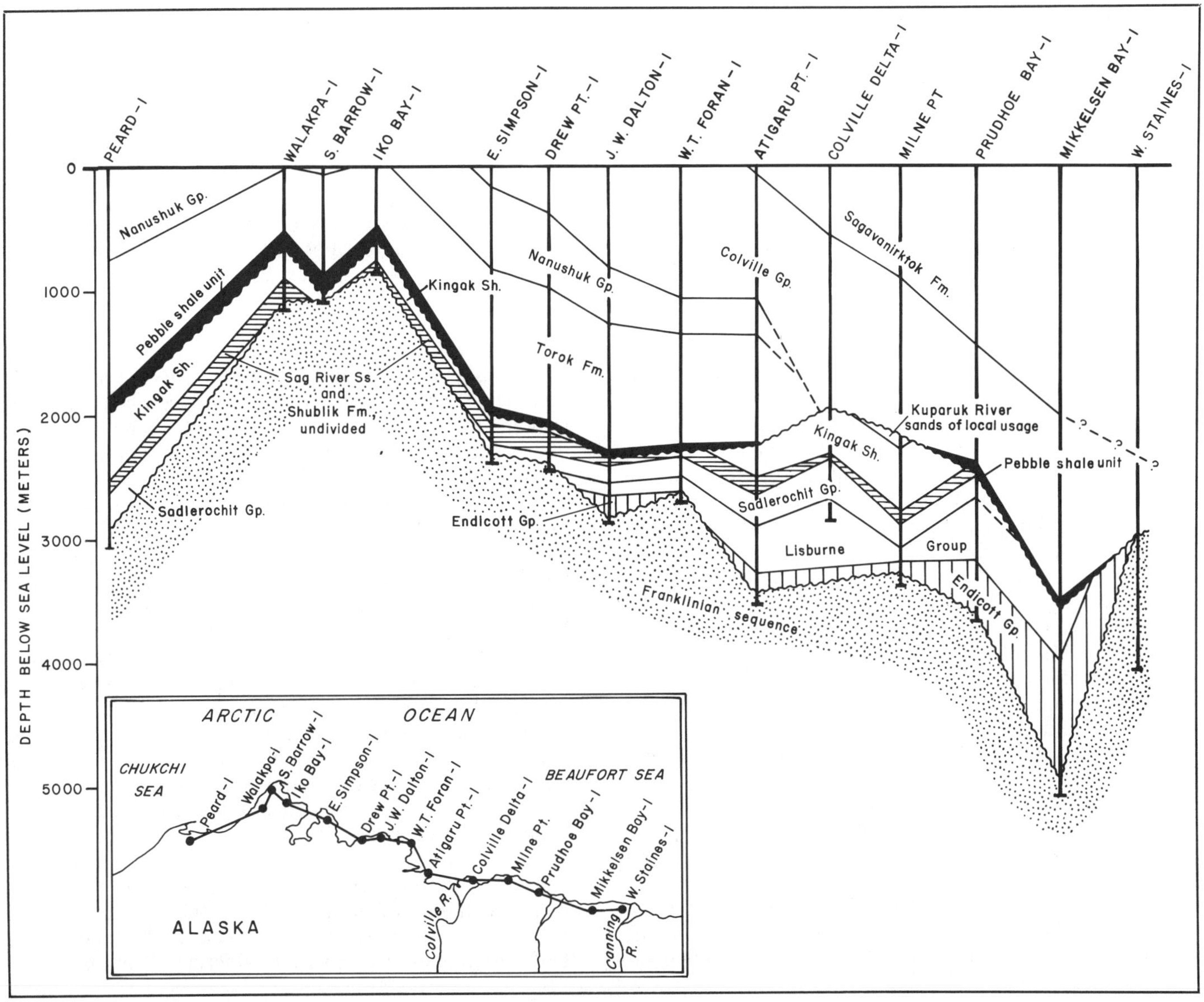

Figure 8. Correlated well section in northern Alaska adjacent to the Beaufort and northeast Chukchi Seas.

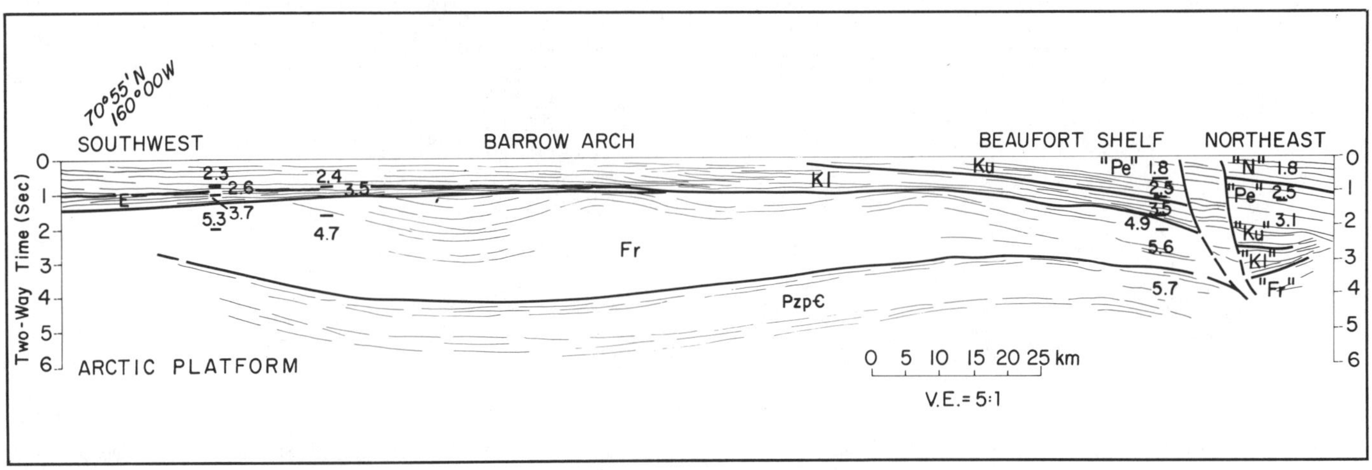

Figure 9. Northeast–southwest regional geologic section, based on seismic-reflection profiles across Barrow arch in the northeastern Chukchi Sea. Pzp€, lower Paleozoic or Precambrian bedded rocks; Fr, Franklinian sequence; E, Ellesmerian sequence; "Kl," "Ku," "Pe" and "N" are Lower Cretaceous, Upper Cretaceous, Paleogene(?) and Neogene(?) sedimentary rocks, respectively. Numbers are sonobuoy velocities in km/sec. See Figure 1 for location of section.

of the offshore position of this edge, based mainly on projection of onshore trends, is shown in Figures 3 and 5. Seaward of the wedge edge of the Ellesmerian beds, the Arctic platform and Franklinian sequence are directly overlain by upper Lower Cretaceous to Tertiary rocks of the Brookian sequence. The Ellesmerian strata on the Arctic platform range upward in thickness from the wedge edge to 1000 m or more near the coast (Alaska Geological Society, 1977). The overlying Brookian rocks range in thickness from 0.3 km or less in the crestal region of the Barrow arch northwest of Point Barrow to as much as 6 km on the outer continental shelf east of Prudhoe Bay.

Structurally, the Arctic platform province is dominated by the Barrow arch (Figures 5, 7, 9), and dips in the province are generally low (Figures 10–12). Belts of normal faults, however, offset the platform and the underlying Franklinian basement at and near the tectonic hinge line (Figures 11, 12, 14), at the margins of the Dinkum graben (Figures 13, 14), and on the north side of the faulted anticline at Prudhoe Bay (Jamison, Brockett, and McIntosh, 1980). The faults north of the Prudhoe Bay structure extend beneath the inner Beaufort Shelf, probably as far as the Dinkum graben. In addition to the normal faults that offset the Arctic platform, an extensive system of large suprabasement growth faults offsets the Brookian sedimentary rocks over and seaward of the tectonic hinge line. Some of these faults extend updip into the Brookian prism over the outer (seaward) part of the Arctic platform province (Figures 3, 5, 12–14).

The normal faults associated with the tectonic hinge line displace basement, the Arctic platform, and only the lowest part of the overlying Cretaceous section above the synrift breakup unconformity (Figures 11, 12), and are thus prerift or synrift features. A large normal fault with many splays that begins about 10 km northeast of Point Barrow and trends east–southeast along the coast to within 35 km of Harrison Bay (Miller et al., 1979) may also be related to the hinge line. This fault, like some of the hinge-line faults, appears to cut basement and only the lower part of the overlying Lower Cretaceous section.

Jamison, Brockett, and McIntosh (1980) indicate that the system of basement-cutting, northwest- and west-striking down-to-the-north normal faults on the north side of the Prudhoe Bay faults are truncated by the mid-Neocomian unconformity underlying the pebble-shale unit (Figures 4, 8), which is also interpreted to be the breakup unconformity. Small normal faults with similar stratigraphic relationships also occur nearby on the south side of the Dinkum graben (Figure 13). The similarities between the Prudhoe Bay and Dinkum graben faults suggest that they may belong to the same structural set.

Dinkum Graben

The eastern part of the Arctic platform province is disrupted by the strongly asymmetric east–southeast-trending Dinkum half-graben (Figures 5, 13, 14). Franklinian basement beneath the deepest parts of the graben lies >6 sec (10.5 km) below sea level. The oldest seismic-stratigraphic unit within the graben consists of fairly strong reflectors postulated to represent Jurassic or older Ellesmerian sandstone and/or carbonate rocks (unit MzPz in Figures 13, 14). This unit is overlain by a thick section of weaker reflectors that may be the Jurassic and early Neocomian (Early Cretaceous) Kingak Shale (unit KJk, Figures 13, 14). These units are tilted to the north and downdropped against a large fault system (Figures 13, 14) that appears to be largely of pre-pebble-shale age (unit Kps in Figures 10–14). Post-pebble-shale displacement has augmented the structural relief on this fault system, which is the main graben-forming structure. This late displacement, in combination with residual structural relief and perhaps locally renewed displacement on older (pre-pebble-shale) strands of the fault system, produced a late-stage basin in the graben. A post-pebble-shale Cretaceous and Paleogene fill locally as much as 1.5 km thick was deposited in this basin. A shallow syncline in this fill near the north-graben fault system suggests renewed displacement of the fault system, and perhaps some differential compaction, in

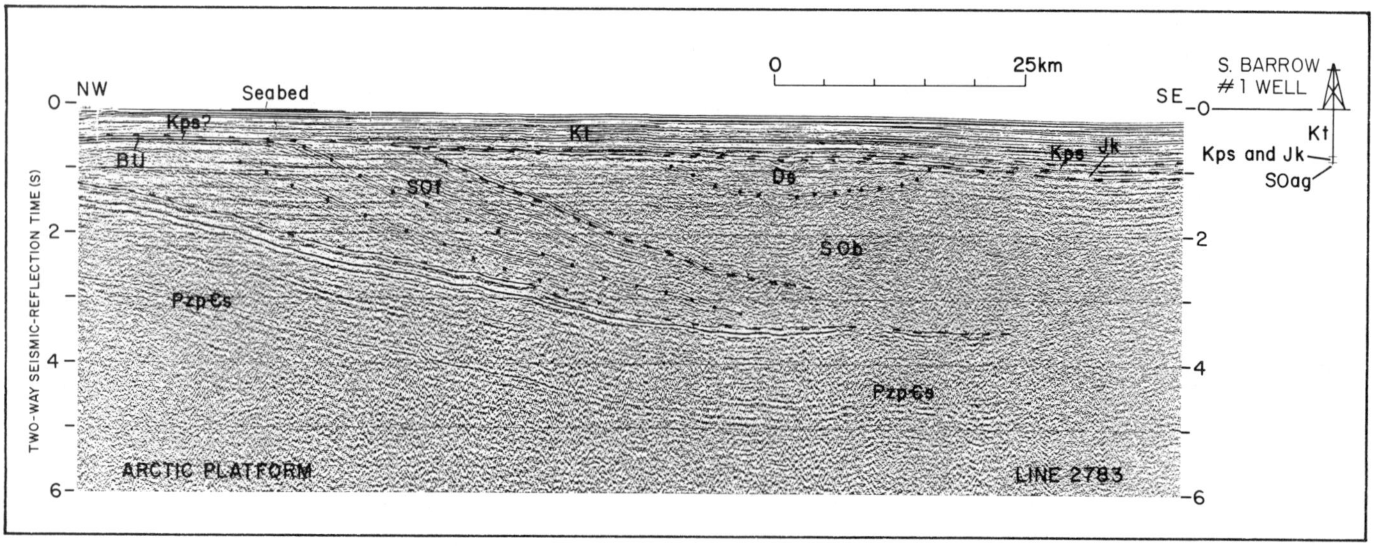

Figure 10. CDP seismic-reflection profile 2783, showing pre-Jurassic clastic deposits (SOf) beneath the Arctic platform west of Point Barrow. BU, breakup unconformity. Inferred acoustic units: Kt, Torok Formation (Albian foreset beds); Kps, pebble-shale unit (Lower Cretaceous); Jk, shale, siltstone, and sandstone in the lower part (Jurassic) of the Kingak Shale of Jurassic and early Neocomian age; SOf, inferred submarine-fan(?) deposits, and SOb, marine basinal deposits correlated with Ordovician and Silurian marine argillite and graywacke (SOag) of subsurface northern Alaska; Ds, Devonian(?) bedded, presumed sedimentary rocks; PzpЄs, upper Precambrian(?) and lower Paleozoic bedded, presumed sedimentary rocks. See Figure 1 for location of profile.

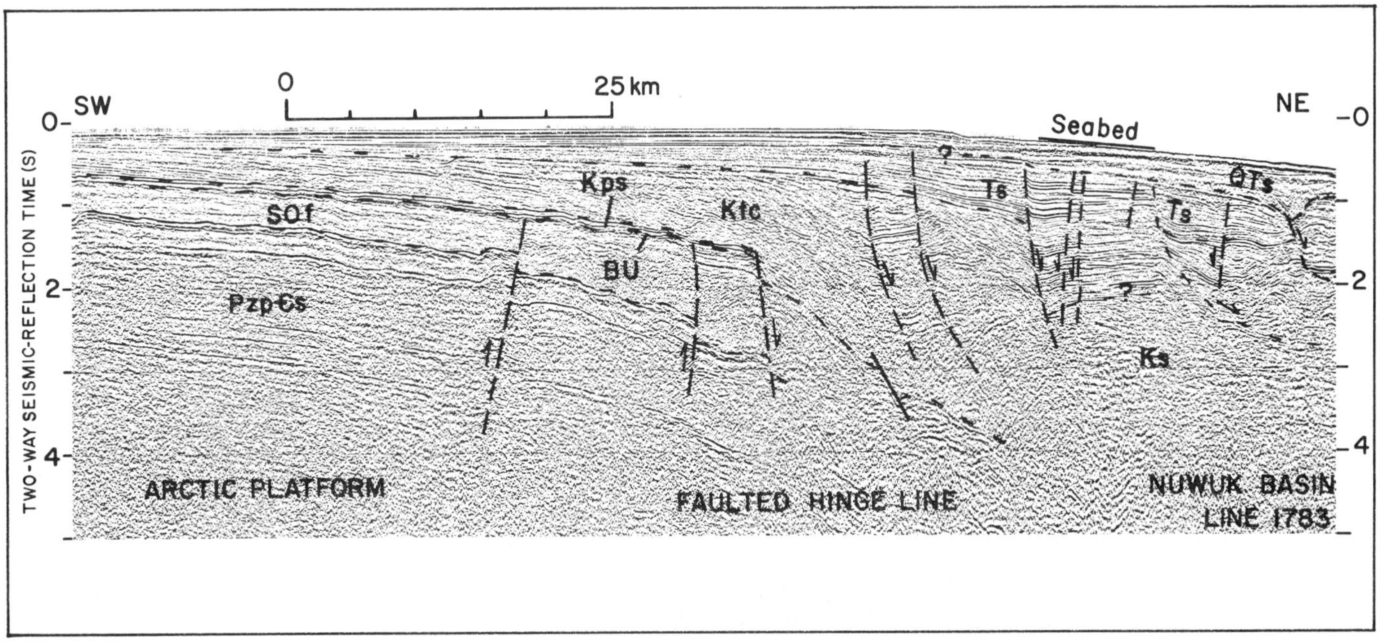

Figure 11. CDP seismic-reflection profile 1783, a dip line across rifted continental margin northwest of Point Barrow. Note rifted edge of Arctic platform, breakup unconformity (BU), faulted subsidence hinge line, deep Nuwuk Basin seaward of hinge line, and growth faults beneath outer shelf. Inferred acoustic units: QTs, Quaternary and upper Tertiary(?) clastic sedimentary rocks; Ts, Tertiary and possibly Upper Cretaceous clastic sedimentary rocks; Ktc, Torok Formation (Albian foreset beds), Colville Group (Upper Cretaceous) and possibly Tertiary clastic sedimentary rocks, undivided; Kps, probably the pebble-shale unit (Hauterivian to Barremian condensed shale), and possibly bottomset beds of basal part of the Torok Formation (Albian); Ks, Lower and Upper Cretaceous clastic sedimentary rocks of the Nuwuk Basin; SOf, inferred submarine-fan deposits tentatively correlated with Ordovician and Silurian marine argillite and graywacke of subsurface northern Alaska; PzpЄs, upper Precambrian(?) and lower Paleozoic bedded, presumed sedimentary rocks. See Figure 1 for location of profile.

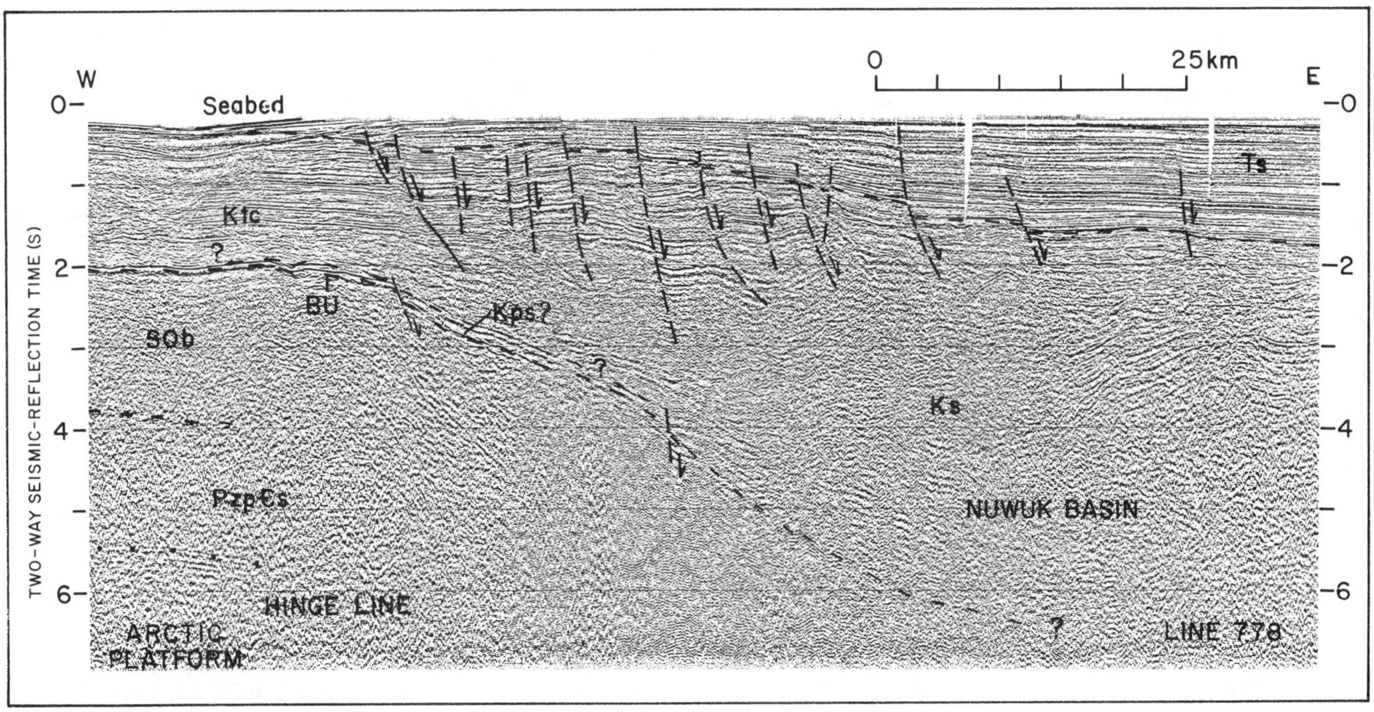

Figure 12. Oblique crossing of rifted continental margin along CDP seismic-reflection profile 778 northeast of Point Barrow, showing Arctic platform, breakup unconformity (BU), subsidence hinge line, and Nuwuk Basin. Inferred acoustic units: Ts, Tertiary and possibly Upper Cretaceous clastic sedimentary rocks; Ktc, Torok Formation (Albian) and Colville Group (Upper Cretaceous) marine and possibly some paralic sedimentary rocks, undivided; Kps, probably pebble-shale unit (Hauterivian to Barremian condensed shale); Ks, Lower and Upper Cretaceous clastic sedimentary rocks of the Nuwuk Basin; SOb, inferred basinal deposits, tentatively correlated with Ordovician and Silurian marine argillite and graywacke of subsurface northern Alaska; PzpЄs, upper Precambrian(?) and lower Paleozoic bedded, presumed sedimentary rocks. Note concentration of growth faults over steep slope north of subsidence hinge line, and extensive slumping and channeling in Cretaceous beds of the Nuwuk Basin. See Figure 1 for location of profile.

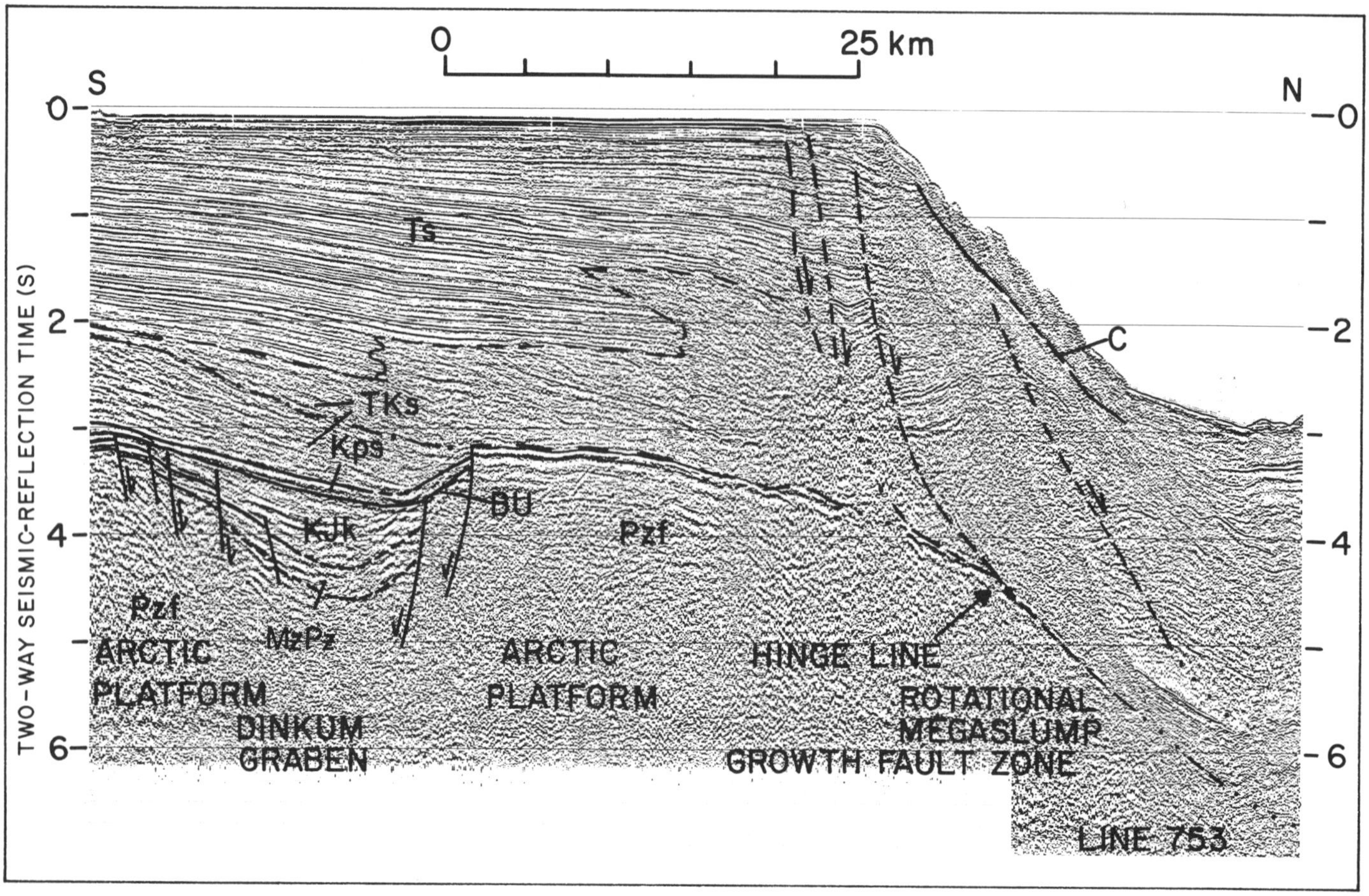

Figure 13. CDP seismic-reflection profile 753, a dip line across the continental margin northwest of Prudhoe Bay showing rifted Arctic platform, Dinkum graben, subsidence hinge line, breakup unconformity (BU), and growth faults and associated rotational megaslump near shelf break. Interpreted acoustic units: Ts, topset beds in Tertiary regressive sequence in upper part of postbreakup progradational sedimentary prism; TKs, Cretaceous and lower Tertiary foreset and some bottomset beds in mainly transgressive lower part of postbreakup progradational sedimentary prism. Dot-dashed line within unit TKs is a rough estimate for base of Tertiary, projected from onshore; Kps, pebble-shale unit and possibly condensed bottomset beds at base of Torok Formation; KJk, weakly reflective beds (Kingak Shale?) in Dinkum graben; MzPz, moderately reflective beds (sandstone? and possibly carbonate rocks) of Ellesmerian sequence in Dinkum graben (Jurassic or older); Pzf, Franklinian sequence (Cambrian to Devonian); C, basal reflector of zone of gas hydrate. See Figure 1 for location of profile.

the Paleogene. The graben appears to die out, and its sedimentary fill to thin, where it strikes west–southwest out of our area of seismic coverage. The east end is structurally well developed where, near Camden Bay, it becomes acoustically hidden beneath the thick sedimentary section of the Kaktovik Basin.

Nuwuk Basin

From the Prudhoe Bay area west, the outer continental shelf and slope are underlain by a deep sedimentary basin that contains > 12 km of post-pebble-shale (mid-Neocomian) Cretaceous and Tertiary (Brookian) clastic sediment. The sediment was prograded across the subsided Arctic platform and underlying basement seaward of the tectonic hinge line (Figures 5, 9, 11, 12), and it constitutes a progradational continental-terrace sedimentary prism rather than a true sedimentary basin. Beneath the continental slope these sediments probably step from the continental basement of the Arctic platform onto transitional or oceanic crust of the Arctic Basin.

The Nuwuk Basin contains foreset and topset beds, and large slump excavations and slump fills. Structurally, it is dominated by large multistrand growth faults that preliminary studies suggest have been active since the Late Cretaceous or early Tertiary. The youngest faults of this system displace Holocene sediment, and in places the seabed. The faults appear to be restricted to the Brookian sequence and curve into, but do not displace, the upper surface of pre-Brookian bedrock. The growth faults bound two rootless anticlines (rotational megaslumps) of regional extent that were produced by movement on the growth faults (Figures 5, 11–14). The profiles, although widely spaced (Figure 1), suggest that the rootless anticlines are continuous or semicontinuous features as much as 200 km long that have variable structural relief that commonly exceeds $1/2$ km.

Barter Island Sector

The deep Kaktovik Basin (Figures 3, 5), which contains large detachment thrust faults and folds, characterizes the Barter Island sector. The field of thrust fault-related detachment folds of the continental slope and rise of the eastern Alaskan Beaufort Sea (Figures 3, 5, 7) is limited to this sector.

Kaktovik Basin

Between Prudhoe Bay and the Canning River (Figures 3, 5, 7), the tectonic hinge line swings southeast from a position near the shelf break toward the coast of eastern Camden

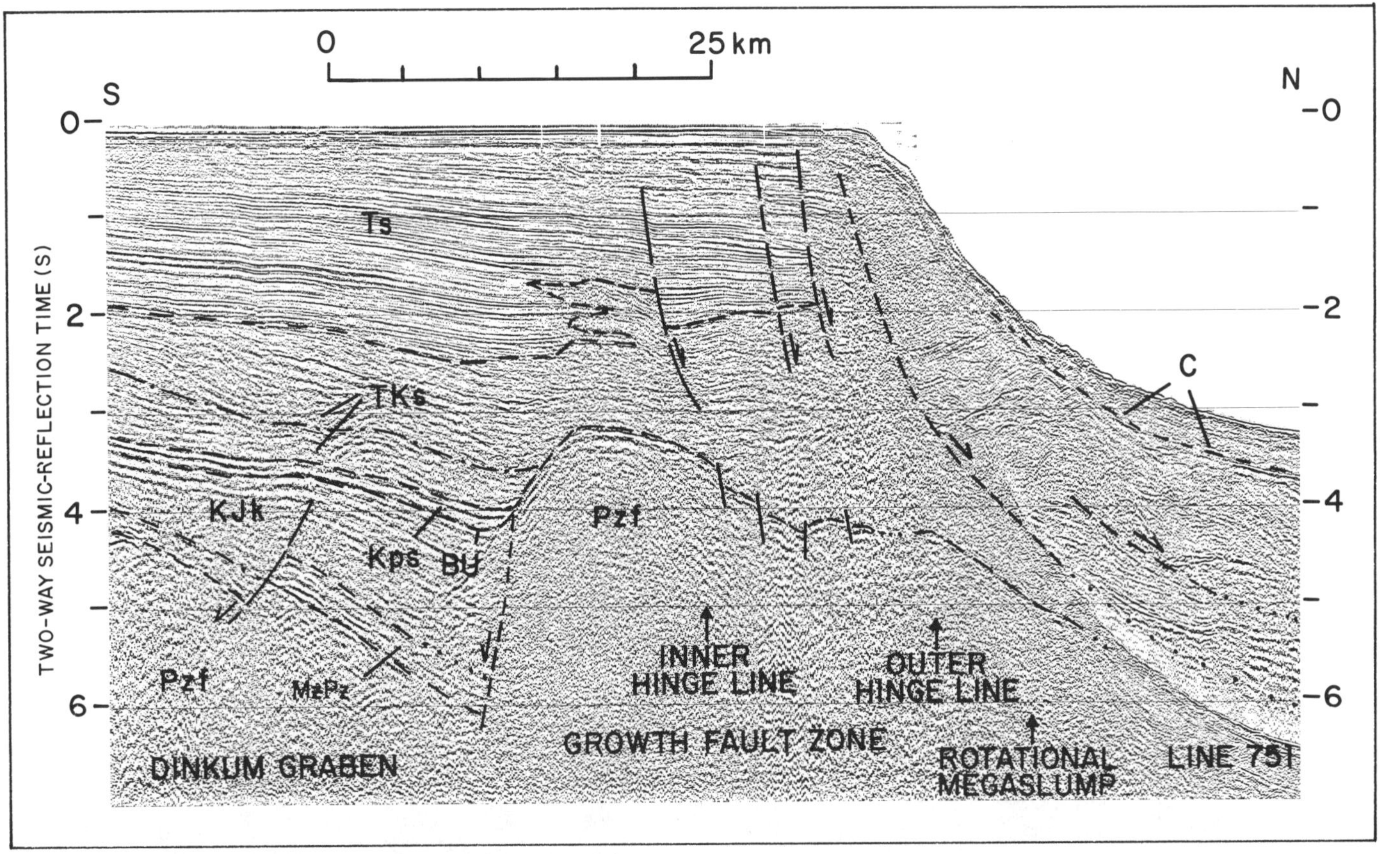

Figure 14. CDP seismic-reflection profile 751, a dip line across the continental margin north of Prudhoe Bay, showing Dinkum graben, compound subsidence hinge line, breakup unconformity (BU), and zone of growth faults and associated rotational megaslump overlying the hinge zone. Interpreted acoustic units same as in Figure 13. See Figure 1 for location of profile.

Bay. In the same area the structural shelf break changes trend from east–southeast to east–northeast, and as a consequence, the continental terrace for some distance to the east is wider than it is to the west (Figure 3). The widened continental terrace, east of the swing in trend of the hinge line and the structural shelf break, is underlain by the Kaktovik Basin (Figures 3, 15, 16). Much of the basin probably rests on transitional and oceanic crust. The thick pre-pebble-shale sedimentary prism of the Dinkum graben trends beneath the Kaktovik Basin, where it is too deep to map on our seismic records. The pebble-shale and younger (Brookian) graben fill and the Brookian sedimentary prism of the Arctic platform province are stratigraphically continuous with the sedimentary fill in the Kaktovik Basin. The total thickness of suprabasement sedimentary rocks beneath the basin exceeds 12.5 km, and local subbasins (BSB and DSB in Figure 5) are filled with Tertiary strata as much as 7.2 km thick. The seismic profiles record Franklinian basement beneath only a small part of the Kaktovik Basin; therefore, the horizon that is contoured in the area of the basin in Figure 5 is mid-Paleogene (middle or upper Eocene), for reasons discussed by Grantz and May (1983). Onshore outcrops in Alaska and subsurface data on the Canadian Beaufort Shelf (Hea et al., 1980) suggest that, if Jurassic to mid-Upper Cretaceous beds underlie the Kaktovik Basin, they are marine. Uppermost Cretaceous beds and Paleocene beds are likely to be nonmarine near the coast, and paralic and probably marine beneath the outer shelf. Higher Tertiary beds are probably both marine and nonmarine.

Contrasting structural styles distinguish the eastern and western parts of the Kaktovik Basin. West of the general vicinity of Barter Island the lower part of the basin, below the horizon contoured in Figure 5, appears to deepen uniformly northward and eastward; the upper part of the section is broken by the large, northeast-striking Camden detachment anticline (CA on Figure 5) and its basal thrust fault, and by a few smaller parallel folds. The Camden anticline (Figure 15) appears to be detached at about 4 sec subsea, probably near the base of the Tertiary. The involvement of beds as young as Holocene in the fold and the distribution of shallow earthquakes in the Camden Bay area (Figure 16) suggest that the thrust fault and fold are active (Grantz, Dinter, and Biswas, 1983). In places a number of large, northward-dipping growth faults, at least in part related to the hinge line, offset the basin sediments. A large fold, inferred to be a rootless rotational-slump anticline related to growth faulting in the Brookian strata north of the hinge line, occurs beneath the western part of the Kaktovik Basin seaward of the Camden anticline (Figure 15). The growth faults in this area appear to merge at depth with the basal thrust fault beneath the Camden anticline.

East of the vicinity of Barter Island the structure of the Kaktovik Basin is dominated by large, in part diapiric, detachment anticlines (Figures 5, 16), and Tertiary sedimentary subbasins (BSB and DSB on Figure 5) that formed contemporaneously with the detachment folds. Multiple unconformities observed over one of the structures (Figure 16) record at least three episodes of movement. At the oldest of these unconformities (U-1 on Figure 16), the core sediments of the fold were raised above wave base, and perhaps

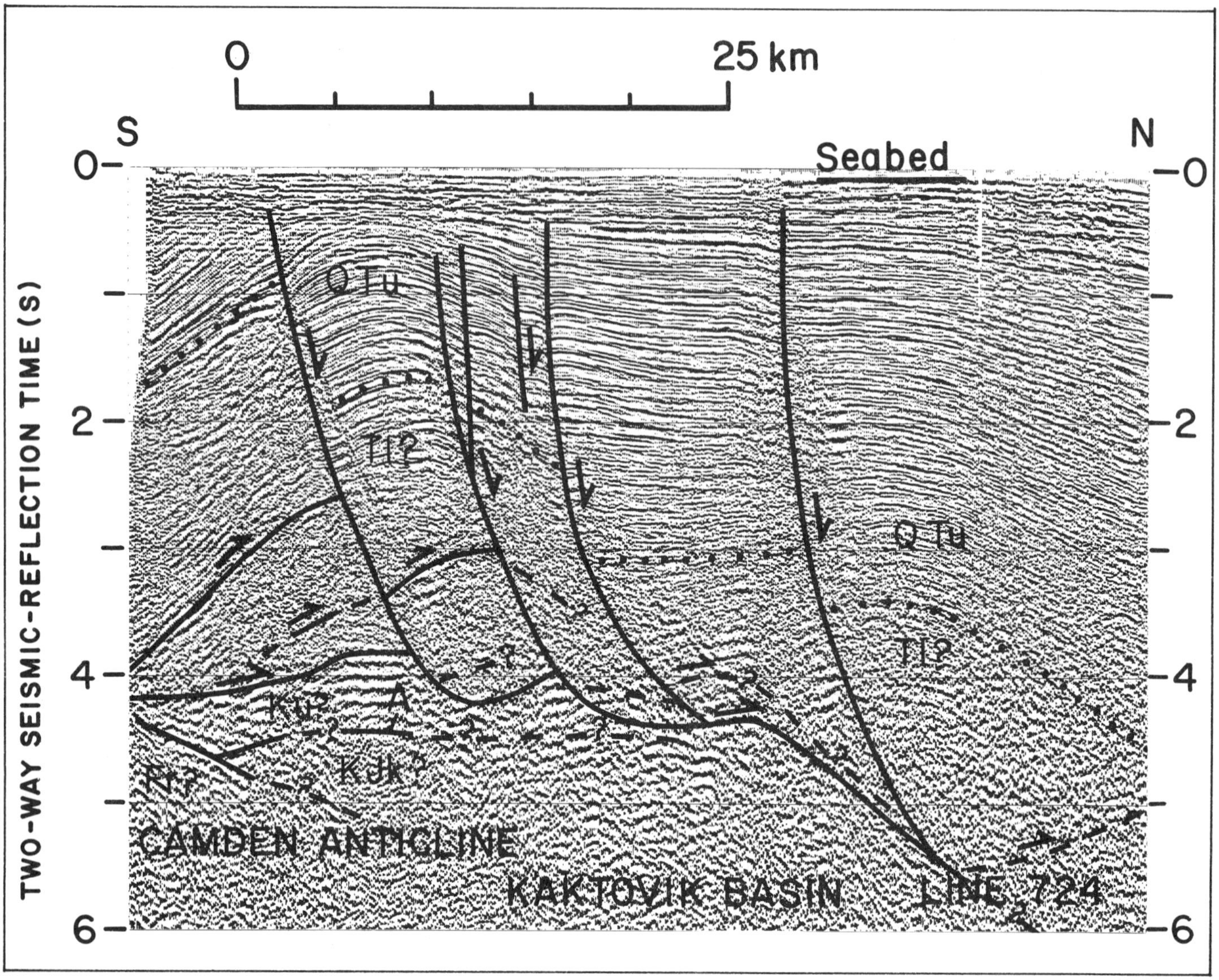

Figure 15. CDP seismic-reflection profile 724 across Camden anticline, a large east–northeast-striking detachment fold off Camden Bay. Fold is inferred, from discordance between horizon A and higher beds, to be detached at a thrust fault near 4-sec isochron. Horizon A is estimated by rough projection from onshore to be near the base of Tertiary sedimentary rocks (unit Ts). Horizon B may be at the top of a thin sequence of Lower Cretaceous bottomset shale beds that possibly overlies the Jurassic and Lower Cretaceous Kingak Shale (KJk). Unit Ks may be Cretaceous marine sedimentary rocks. See Figure 1 for location of profile.

above sea level. In places, delta-like topset-foreset-bottomset sedimentary packets were built out from structurally high areas in the subunconformity section. Regional geologic relationships suggest that the cores of these structures contain Jurassic, Cretaceous, and lower Paleogene clastic sedimentary rocks, and that the overlying, less strongly deformed beds may be upper-Eocene and younger.

Detachment Fold Province

Numerous anticlines separated by generally broader, less sharply folded synclines underlie the continental slope and rise in the Barter Island sector east of long. 146°W (Figures 5, 16). These anticlines are much smaller, and have more regular cross-sectional geometries, than do the large detachment folds of the eastern Kaktovik Basin. They are symmetrical or mildly asymmetrical anticlines rather than piercement structures or irregular shale ridges, and they strike east–west and are moderately elongate (Figures 3, 5, 7). The anticlines are thought to be detachment folds related to underlying thrust faults, and some of them appear to be cored by splay faults that rise from the underlying thrusts.

The detachment folds are typically 2 to 10 km wide on seismic crossings, and the largest have at least 1.5 sec (1.25 km) of structural relief in two-way reflection time. The most northerly structures, about 100 km north of the structural shelf break (Figure 3), die out upward, and are overlain by undisturbed beds about 2 sec (2 km) below the seabed. The tops of the folds extend progressively higher in the section landward; beginning about 50 or 60 km from the shelf break, they buckle the seabed and act as dams for clastic sediment and landslide debris moving down the continental slope.

Seismic-reflection interval velocities of about 1.8 to 3.5 km/sec in the folded beds (Eittreim and Grantz, 1979), and lateral tracing of reflectors, suggest that these detachment structures penetrate Tertiary and Quaternary beds and that the core material is sedimentary. Lateral tracing of reflectors also indicates that those folds which are closest to the shelf, and which are strongly developed in the shallow part of the section and disrupt the seabed, deform lower Tertiary and possibly deeper beds. The folds farthest from the continental shelf, which die out upward at about 2-sec two-way

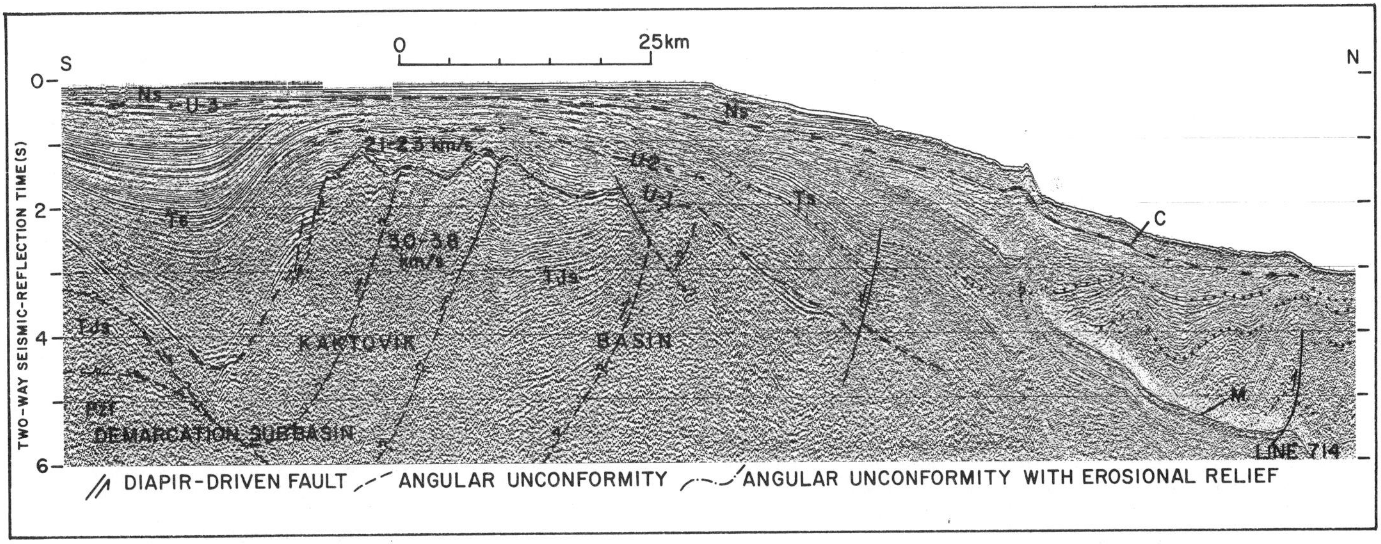

Figure 16. CDP seismic-reflection profile 714 across Kaktovik Basin north of Alaska-Yukon Territory boundary, showing Demarcation subbasin, diapiric detachment fold that dominates shelf structure in this area, and smaller detachment folds of continental slope and rise. Inferred acoustic units: Ns, Neogene nonmarine and marine sedimentary rocks; Ts, upper Paleogene marine and nonmarine(?) sedimentary rocks; TJs, Jurassic and Cretaceous marine and lower Paleogene marine or nonmarine sedimentary rocks, undivided; Pzf, Franklinian(?) sequence; D, detachment fold; C, base of gas hydrate; M, seabed multiple; km/sec, sonobuoy seismic velocity (in kilometers per second). Note that three angular unconformities (dashed lines U-1, U-2, and U-3) cross top of diapiric detachment anticline. The lowest (U-1) shows rough erosional relief and, on south flank of the detachment anticline, is disrupted by normal faults related to growth of the structure. Unconformities and axial thickening of reflector packets in the Demarcation subbasin demonstrate that growth of the basin and anticline, and sedimentation were penecontemporaneous. See Figure 1 for location of profile.

reflection-time subbottom, are no more than broad warps at 3.5-sec subbottom. The folds extend below the strong deep-water multiple on our seismic records, and possibly disrupt beds as old as the early postrift Lower Cretaceous sedimentary deposits of the Canada Basin.

Canada Basin

Water depths exceeding 3500 m and seismic-refraction measurements (Mair and Lyons, 1981; Grantz, Eittreim, and Whitney, 1981) indicate that the Canada Basin is underlain by oceanic crust and was therefore formed by sea-floor spreading. The inferred thickness of sediment in the southern part of the basin is shown in Figure 7, and the three major physiographic provinces into which it can be subdivided are shown in Figure 3. All three provinces—the Alaska continental rise, the Mackenzie cone, and the Canada Abyssal Plain—are the result of rapid sedimentation from the continental landmasses that surround the basin. Their relative areal extent, and the relative thicknesses of the sedimentary prisms which underlie them, are the result of the positions of the major rivers and glaciers which entered the Canada Basin since it was formed in the Cretaceous.

Alaska Continental Rise

The Alaska continental rise, at water depths of about 2000 to 3800 m, is underlain by clastic sedimentary rocks of the Brookian sequence, which were derived from sources to the south, in continental Alaska, and probably in Eurasia. The strata are inferred to consist dominantly of foreset and bottomset lutites, but probably also include some turbidite sandstones and submarine slump deposits. As deposited, these sediments were in stratigraphic continuity with the beds of the progradational continental terrace sedimentary prisms, herein called the Nuwuk and Kaktovik Basins. A few of the seismic profiles shown in Figure 1 extend over the rise and show it to be underlain by well-bedded, uniformly basinward-dipping strata. These strata are internally deformed beneath their upslope margin by submarine slumps and the distal ends of listric growth faults that originate on the continental slope and outer shelf seaward of the tectonic hinge line. The age of the main sedimentary prism is inferred to range from late Neocomian (mid-Early Cretaceous) to Quaternary. The distribution of Brookian rocks and facies in northern Alaska suggests that the sedimentary prism may consist mainly of Albian strata near Point Barrow and mainly Tertiary strata in the Barter Island sector. In addition, Jurassic and early Early Cretaceous beds of the rift valley stage of rifting on the Alaskan continental margin (Grantz and May, 1983) underlie the Neocomian beds beneath the eastern Beaufort Sea and probably part of the Canada Basin. The thickness of the sedimentary prism beneath the Alaska continental rise is estimated to range from 6 km to >10 km (Figure 7).

Mackenzie Cone

The Mackenzie cone is that portion of the Canada continental rise lying off the mouth of the Mackenzie River. Water depths on the cone in the Beaufort Sea (Figure 3) range from about 1800 m to 3700 m. The cone occupies about three-fifths of the southern Canada Basin north of the Alaska continental rise. The areal extent of the cone and the thickness of its underlying sedimentary prism (Figure 7) are the result of the proximity of the cone to the delta of the large Mackenzie River system and to the recurringly heavily glaciated Canadian Arctic Islands and Canadian shield. These sources supplied voluminous sediment to the cone, and in the Tertiary the cone grew much more rapidly than the Alaska continental rise sedimentary prism, which it overwhelmed. The boundary between the Alaska continental rise and the Mackenzie cone is marked by a change in trend of the depth contours (Figure 3). The oldest sediments in the cone are inferred to be possibly Early Jurassic and certainly late Early

Cretaceous; the youngest are Quaternary. Beyond the detachment folds, seismic-reflection profiles show the sediments to consist of flat basinward-dipping strata that produce some well-defined, but mainly weak, seismic reflections. The sequence is interpreted to consist dominantly of lutite with some turbidite beds, and to range in thickness from < 6 km to > 10 km (Figure 7).

Canada Abyssal Plain

The abyssal plain of the Canada Basin (Figure 3) occupies about two-fifths of the southern Canada Basin north of the Alaska continental rise. Water depths over the plain range from about 3700 to about 4000 m. We have no seismic-reflection data over the abyssal plain and can only speculate on the characteristics of the underlying sedimentary prism. Presumably, it is composed of fine-grained sediment, the distal elements of turbidity current deposits that originated at the Mackenzie delta and were ponded in the deepest part of the Canada Basin between the foot of the Mackenzie cone and the Chukchi continental borderland. In addition, there may have been a contribution from the smaller Alaska continental rise. Extrapolation of seismic-refraction data indicates that the sedimentary section beneath the portion of the abyssal plain adjacent to northern Alaska is a little less than 6 km thick.

NEOTECTONICS

Young Faults

High-angle normal faults, some of which deform Holocene deposits and the seabed, and monoclines overlying such faults are abundant off Camden Bay (Figure 17). These features are described in some detail by Grantz, Dinter and Biswas (1983). The faults lie within the western end of a terrane of large-amplitude, late Cenozoic detachment anticlines that characterize the western part of the Barter Island sector of the Beaufort Shelf. These folds overlie a low-angle thrust fault and appear to represent the active front of the thrust system that produced them. The youthfulness and local abundance of these features, and their areal coincidence with Holocene folds and earthquakes, suggest that some of them may be active.

North-dipping, down-to-basin, listric normal faults also occur beneath the middle and outer shelf outside the zone of detachment folds. These faults offset Quaternary deposits and the seabed and include many growth faults with a long history of activity. The outer-shelf faults of this set displace Holocene deposits and the seabed as much as 15 to 20 m, and in one area possibly as much as 70 m. Those on the mid-shelf displace sediments no younger than late Pleistocene. These gravity faults are active in the sense that they formed in the present tectonic environment and displace Pleistocene or Holocene sediments. However, they have not generated earthquakes of sufficient magnitude to be detected by the regional and local seismograph networks in place since 1968.

Earthquakes

The Beaufort and northeast Chukchi continental margin has been historically aseismic, except for a zone of concentrated seismic activity in the vicinity of Barter Island, and an anomalous earthquake located by the worldwide network about 200 km north of the Colville River delta. Epicenters of representative earthquakes of magnitude 3.0 or greater on the Beaufort continental margin, recorded by the worldwide seismograph network through 1980 and by the Canadian network and a local seismographic network in northeast Alaska from 1968 to 1977, are shown in Figure 17 (adapted from Grantz, Dinter, and Biswas, 1983). The earthquakes in the vicinity of Barter Island appear to record a northeastern extension of the central Alaska seismic zone. Seismic-reflection data on the eastern Alaskan Beaufort Shelf suggest that the earthquakes are associated with Quaternary movement on Camden and Marsh detachment anticlines and their underlying detachment thrust fault(s) (Figure 15). The largest of the earthquakes recorded (in 1968) had a magnitude of 5.3.

PETROLEUM POTENTIAL

Commercial deposits of oil and gas probably will be found beneath the United States Beaufort and northeast Chukchi Seas. Large economic and potentially economic oil and gas pools of several ages have been developed or discovered along the Beaufort Sea coast in the Prudhoe Bay area (Jamison, Brockett, and McIntosh, 1980) and between the Sagavanirktok and Canning Rivers (Figure 1). Some of these are known, and others can be surmised, to extend offshore. It is remarkable that every major stratigraphic unit of formation or group rank beneath the Arctic coastal plain of northern Alaska has been found to contain commercial or potentially commercial pools or strong shows of oil and gas (Figure 18). All of these major stratigraphic units can be interpreted to extend offshore beneath the Beaufort Shelf, suggesting that oil and gas have been generated there. The chances for potentially commercial discoveries are considered good because, in addition to favorable sedimentary rocks, the area contains many structures and stratigraphic configurations of types that have been found to trap economic petroleum deposits in other regions. The chief concern is whether sufficiently large deposits can be found to support the anticipated high cost of offshore exploration and development in the Beaufort and Chukchi Seas and adjacent Arctic Ocean, and to support the development of new technologies to make such operations possible.

Thermal Maturity of Postrift Sediments

The presence of suitable organic material and a thermal history conducive to the chemical transformation of such organic material into petroleum hydrocarbons are basic to the generation of petroleum in sedimentary basins. Figure 18 suggests that suitable organic matter is indeed abundant in the strata that underlie the Arctic coastal plain of northern Alaska. In the absence of thermal data from the Beaufort Shelf, we estimate its thermal maturity by using a method developed by Royden, Sclater, and von Herzen (1980) for estimating the thermal history of passive continental margins. The method is based on a relationship between the amount of subsidence a passive margin undergoes following breakup of a continent, and the heat flow through time in that area. If the depth to the breakup unconformity can be measured, a thermal history of the area can be reconstructed.

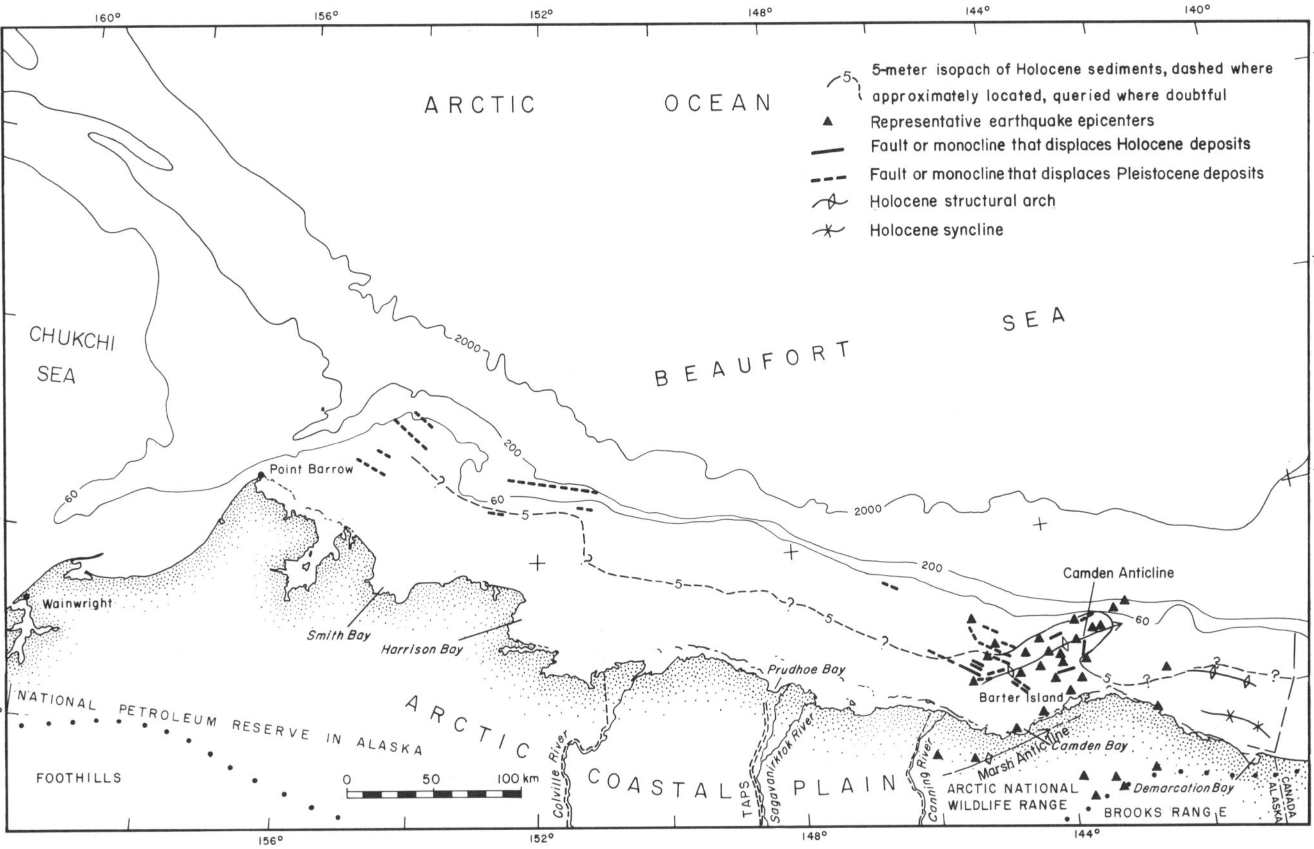

Figure 17. Map showing isopach of Holocene sediments, Holocene and Pleistocene faults, earthquake epicenters, and structural axes on the continental shelf north of Alaska.

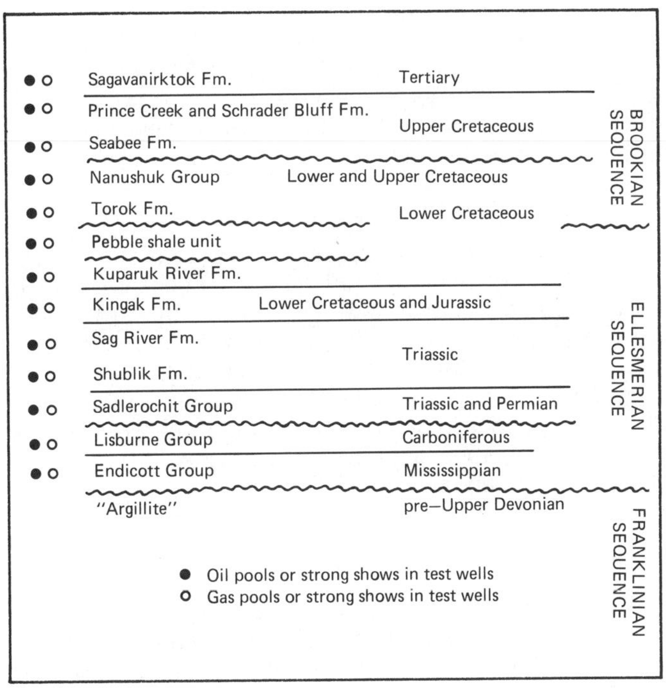

Figure 18. Generalized stratigraphic column beneath the Arctic coastal plain of northern Alaska, showing position of oil and gas pools and of strong shows of oil and gas encountered in test wells. (Data from Alaska Geological Society, 1971 and 1972; Brosgé and Tailleur, 1971; Jamison, Brockett, and McIntosh, 1980; Jones and Speers, 1976; and industry announcements).

This method was used to estimate the depth range at which petroleum hydrocarbons may have been generated in the postbreakup (post-Neocomian) sedimentary rocks of the Beaufort and northeast Chukchi Seas, as shown in Figure 19.

Arctic Platform

Most of his province (area 1 in Figure 3) is prospective for petroleum. Thermal history calculations indicate that the Cretaceous rocks are within the oil window (Figure 19), and the Ellesmerian rocks are known to be productive onshore. The areas of greatest potential lie along the broad crest of the Barrow arch from Prudhoe Bay to Harrison Bay, where significant sections of Ellesmerian rocks can be extrapolated offshore. The possibility that the pre-pebble-shale faults on the north side of the Prudhoe Bay structure extend to the south side of the Dinkum graben further enhances the prospects of that area.

Additional prospects in the Arctic platform province are fault-block plays in Cretaceous rocks and possibly, in places, Ellesmerian rocks on the north side of hinge-line-related faults that lie off the coast between Point Barrow and Harrison Bay. Hinge-line-associated growth faults and rollovers in Brookian rocks in the northern part of the province may also have potential. Interdigitated topset and foreset seismic-stratigraphic units of inferred Cretaceous and Paleogene age in appropriate structural configurations may offer prospects in parts of the Arctic platform province, but it is difficult to infer from our data how large such stratigraphically controlled accumulations might be.

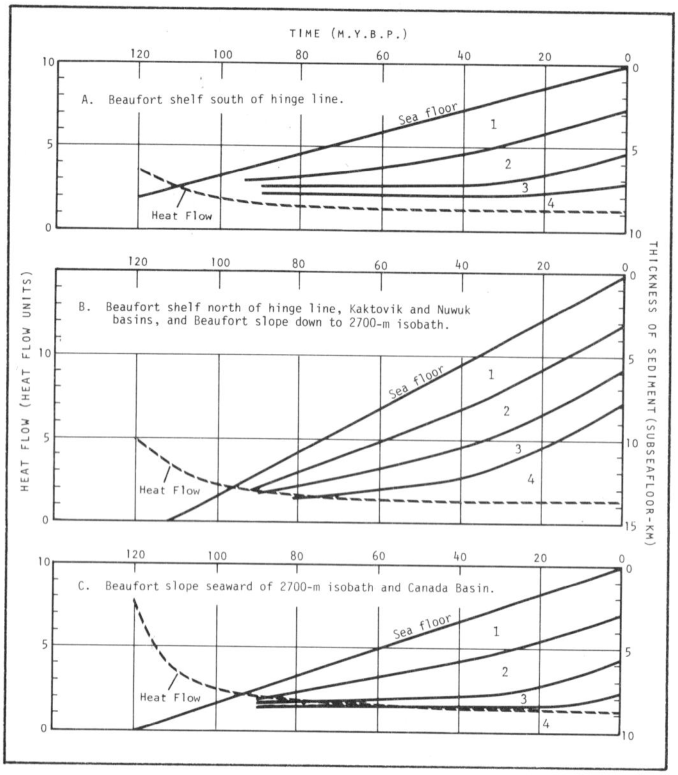

Figure 19. Graphs showing thermal maturity of postrift sediments in the study area as a function of time and present depth of burial. Numbered zones represent: (1) Immature zone—generation of biogenic gas only; (2) Oil zone—generation of oil and associated gas; (3) Thermal gas zone—generation of thermal gas and breakdown of liquid hydrocarbons into gas; and (4) Overly mature zone—temperatures above that at which hydrocarbons are stable. Heat-flow curves, used to determine the temperature history of the sediments, calculated from heat-flow measurements in the Canada Basin by Lachenbruch and Marshall (1966) and from the Geothermal Gradient Map of North America (American Association of Petroleum Geologists and U.S. Geological Survey, 1976). Heat-flow curves are extrapolated back in time, using the method of Royden, Sclater, and von Herzen (1980).

Dinkum Graben

From Figure 19 we estimate that the Cretaceous and lowest Tertiary strata in the thick sedimentary section of the Dinkum graben (area 2 in Figure 3) are now within the oil window. The underlying pre-Jurassic(?) or Jurassic(?) Ellesmerian beds within the deep central parts of the graben range from just below the oil window into the overly mature zone, and would be of interest mainly as a source of oil or gas that migrated out of these beds during the Cretaceous or Paleogene. Good prospects may exist where the Cretaceous and lower Tertiary beds of the graben fill are upturned against pregraben rocks at buttress unconformities or fault contacts, provided these beds are no deeper than about 5 km. However, our data do not allow a reasonable speculation on the size of any such accumulations.

Nuwuk Basin

The section between roughly 3 km and 6 km subseabed in the thick (>12 km) Brookian sedimentary prism in the Nuwuk Basin (area 3 in Figure 3) is in the oil window. Thus, the many early-formed growth faults and associated large rootless anticlines (rotational megaslumps) and smaller rollover anticlines were potentially available to trap migrating hydrocarbons. The presence of topset, as well as foreset and bottomset, seismic-stratigraphic units in the Cretaceous and lower Tertiary beds of the basin holds out the possibility that reservoir sands may be present in these structures. Although the seismic lines are too widely spaced to demonstrate closure on the possible structures and stratigraphic traps recognized on seismic profiles, they suggest that large closed structures could be present in most parts of the Nuwuk Basin.

Kaktovik Basin

Heat-flow considerations (Figure 19) suggest that the section between roughly 3 and 6 km below the seabed in the Kaktovik Basin is in the oil window. The window includes the upper (but not the highest) parts of the diapiric detachment anticlines and the deeper parts of the Tertiary subbasins in the eastern part of the province (area 4B in Figure 3). The subbasins formed contemporaneously with the diapiric detachment anticlines, and the buttress unconformities at their upturned and locally faulted margins may offer prospects for large accumulations of oil or gas. In addition, updip migration of hydrocarbon fluids might have created local accumulations in the tops of the detachment anticlines.

The discovery of oil deposits in Cretaceous and lower Tertiary rocks at or near the coast between the Sagavanirktok and Canning Rivers, and the presence offshore of the large Camden detachment anticline, indicates that the western part of the Kaktovik Basin (area 4A in Figure 3) also has good prospects. The Camden anticline (Figures 3, 5, 15) may not be as attractive, however, as its substantial length, width, and structural relief might suggest. The anticline is apparently still growing, and it is probably an entirely Neogene and perhaps even Quaternary structure. If the fold is indeed this young, it may substantially postdate the migration of oil from the lower Tertiary and older beds of the basin, and there may not have been sufficient time to accumulate large quantities of hydrocarbons generated in the Tertiary strata. Because of its youthfulness, the Camden anticline may offer more promise for gas than for oil.

Detachment Fold Province

Most of the numerous anticlinal structures of the detachment fold province (area 5 in Figure 3) lie beneath 1500 to 3000 m of water in areas of the Arctic Basin that are seldom ice free (Figure 2). They offer, for possible future exploration, many structural traps within the oil window, and updip migration may have carried hydrocarbon fluids into higher structural positions on the folds. A large unknown in this province is whether suitable hydrocarbon source beds and reservoir rocks are present.

Canada Basin

Water depths exceeding 2400 m, perennial ice cover (Figure 2), and uncertainties concerning the presence of source and reservoir rocks appear to assure that petroleum exploration in the Canada Basin (areas 6A, 6B, and 6C in Figure 3) will not be undertaken in this century. However, the sedimentary rocks of the basin, which are interpreted to be of Brookian (and locally of Jurassic) age, are potentially prospective because they are from 5 km to >10 km thick and a substantial part of the section is estimated to be in the oil win-

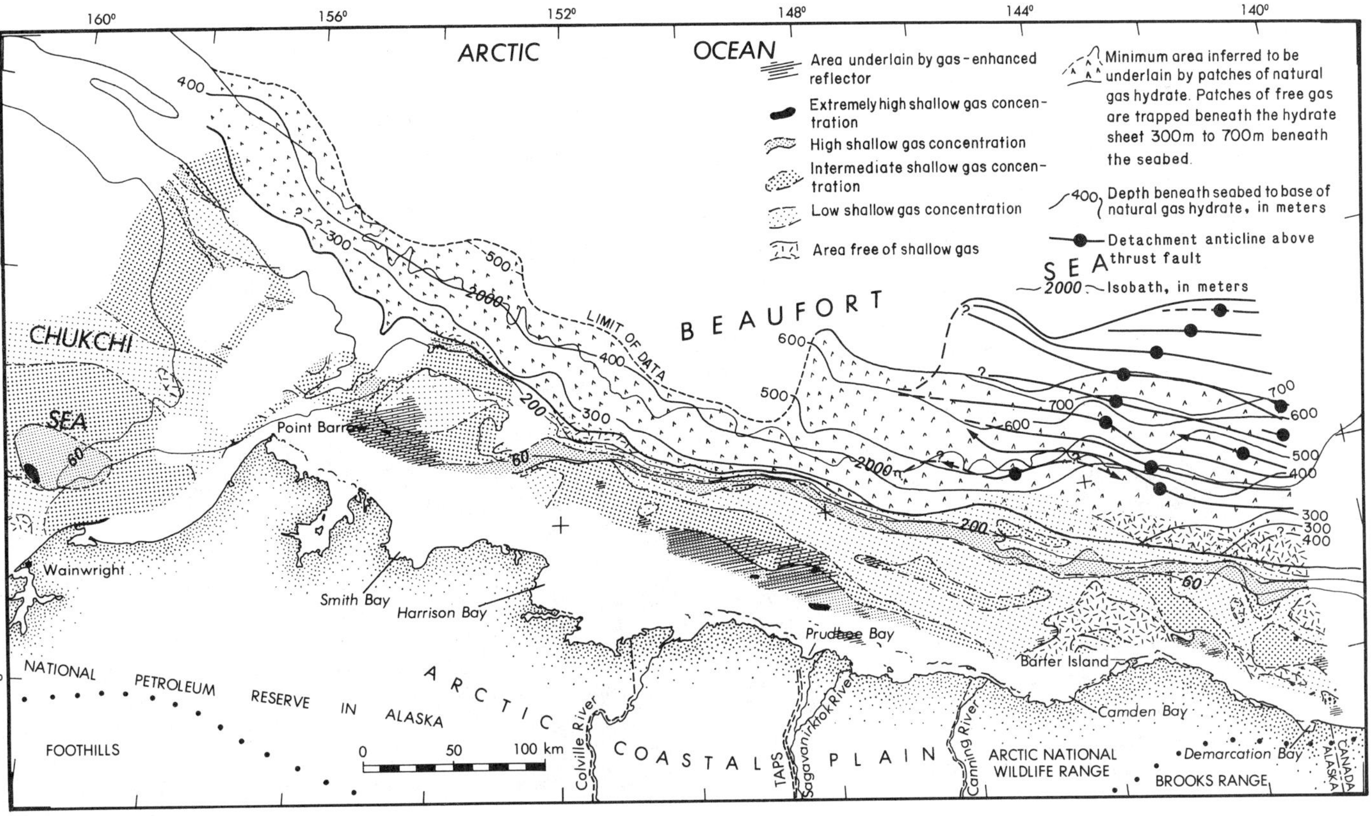

Figure 20. Map showing detachment folds, shallow free-gas concentrations, depths to base of natural gas hydrate, and area possibly underlain by overpressured shale offshore of northern Alaska.

dow (Figure 19). The adequacy of hydrocarbon source and reservoir rocks is not known, but suitable source beds and porous turbidite sands could be present. Seismic-reflection profiles across part of the Alaskan continental rise subprovince (area 6A in Figure 3) demonstrate the presence of low-angle faults and buckled beds and fault planes at the toes of growth faults and deep slumps. These low-angle faults originated in the Nuwuk Basin and beneath the continental slope. Such complex structures might have formed suitable traps for the accumulation of hydrocarbons, but the seismic lines are too widely spaced to demonstrate structural closures.

Gas Hydrate

Where water depths exceed about 400 m, the top 300 m to 700 m of sediment beneath the sea floor in the Beaufort and northeast Chukchi Seas is in the temperature-pressure range for stability of natural gas hydrate (Figure 20). Evidence for the occurrence of gas hydrate was found on about 75% of our seismic-reflection coverage in the area between the 400-m isobath and the north end of our seismic lines or the 2800-m isobath. The zone of gas hydrate may also extend beyond the 2800-m isobath, but the evidence is difficult to recognize in the nearly flat sedimentary strata that underlie the Canada Basin at those depths.

It is not clear from our seismic data what portion of the potential zone of occurrence is actually cemented with gas hydrate. Hydrate formation requires either the in-situ generation of gas in the appropriate pressure-temperature zone, or the introduction of gas by migration from outside the zone. The completeness of hydrate cementation within the hydrate pressure-temperature zone depends on the availability of gas or of suitable organic matter for the generation of gas, the permeability of the sediment, and the existence of appropriate pathways for gas migration. It is assumed that the hydrate contains mostly methane, because this is the most abundant gas in naturally occurring gas hydrates, but this has yet to be demonstrated. Until the hydrate zone is sampled, its resource potential can only be conjectured, but theoretical calculations suggest that the potential resource could be very large. To date, however, attempts onshore in the Soviet Union and Canada to decompose subpermafrost hydrate in situ have apparently not led to a method for producing hydrate-bound methane gas commercially. Subsurface gas hydrate is of greater current interest as a potential geohazard than as a resource. Under the environmental conditions of the Arctic Ocean, development of the gas hydrate resource will probably be undertaken first, if at all, in conjunction with the development of deeper, conventional deposits of oil and gas.

A more likely resource in the region of gas hydrate is the free gas that, in many places, may immediately underlie the hydrate zone. Many of the strongest concentrations occur under pseudoanticlines formed where the base of the gas hydrate, which tends to parallel the seabed, underlies bathymetric highs (Figure 21). Large volumes of gas at low to moderate pressure could be present in such structures and other subhydrate accumulations. However, sampling to determine the properties of the potential reservoirs, the composition of the gas, the thickness of the gas zone, and the gas content of the associated connate water, as well as additional seismic-reflection profiles, will be required to estimate the potential resource.

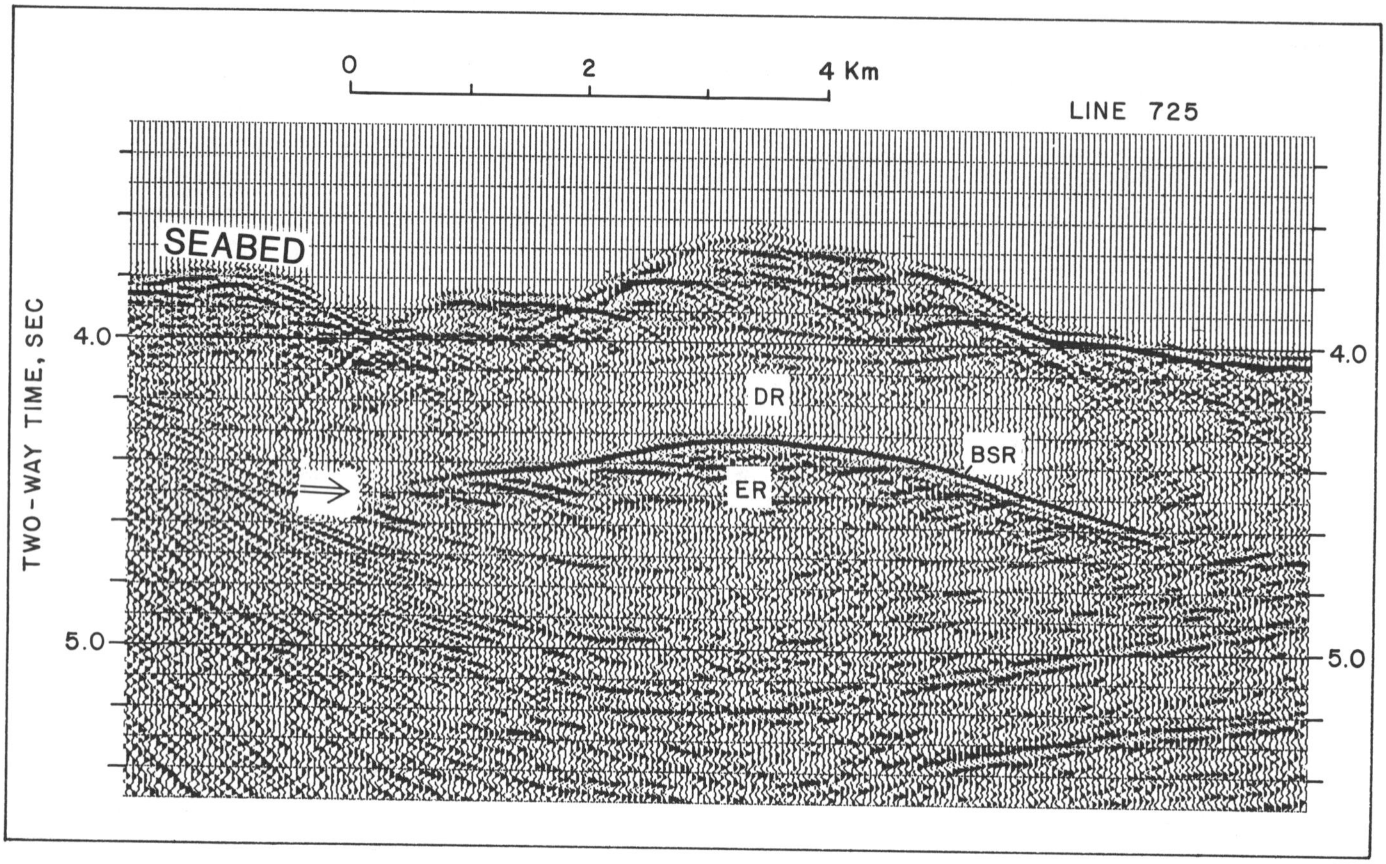

Figure 21. CDP seismic-reflection profile 725, on the continental slope at 2670– to 2935–depths off Camden Bay, showing zone of diminished reflections (DR) thought to be caused by cementation with gas hydrate; and the underlying zone of enhanced reflections (ER) interpreted to represent gas-charged sediment beneath the gas hydrate. The strong bottom-simulating reflector (BSR) is interpreted to represent the interface between the hydrated zone and the underlying, hydrate-free gas-charged sediment. Note that the hydrate-related reflections are best developed under a bathymetric high, where they form a pseudoanticline that is superimposed on the synclinal structure of the bedded country rock.

Relative Petroleum Potential of the Provinces

Prospects for the occurrence of large petroleum deposits, without regard to economic realities, appear to be highest in the Arctic platform province east from Smith Bay, and in the Kaktovik and Nuwuk Basins. Also included as prospective, although somewhat less favorable for large deposits, are the Dinkum graben, the Arctic platform near the hinge line (including the area with hinge-line-related faults between Point Barrow and Harrison Bay), the Arctic platform south of the wedge-out of Ellesmerian strata west of Point Barrow, and probably the detachment fold province. The structurally shallow area of the Arctic platform in the broad crestal region of the Barrow arch northwest of the Ellesmerian wedge-out in the Chukchi Sea has poor prospects. Data are insufficient and suitable analogs are lacking from which to rank the subprovinces of the Canada Basin or the gas hydrate zone. The best prospects in the Canada Basin, exclusive of the detachment folds, may lie in the structurally disturbed and stratigraphically complex zone that lies at the foot of the listric growth faults within the upslope (landward) part of the Alaska continental rise sedimentary prism.

REFERENCES CITED

Alaska Geological Society, 1971, West to east stratigraphic correlation section, Point Barrow to Ignek Valley, Arctic North Slope, Alaska: Anchorage, Alaska Geological Society.

Alaska Geological Society, 1972, Northwest to southeast stratigraphic correlation section, Prudhoe Bay to Ignek Valley, Arctic North Slope, Alaska: Anchorage, Alaska Geological Society.

Alaska Geological Society, 1977, North to south stratigraphic correlation section, Beaufort Sea-Prudhoe Bay-Nora No. 1, Arctic Slope, Alaska: Anchorage, Alaska Geological Society.

Alaska Geological Society, 1981, West to east stratigraphic correlation section, N.P.R.A.-Prudhoe Bay, Arctic Slope, Alaska: Anchorage, Alaska Geological Society.

American Association of Petroleum Geologists and United States Geological Survey, 1976, Geothermal gradient map of North America: 2 sheets, scale 1:5,000,000.

Brosgé, W. P., and I. L. Tailleur, 1971, Northern Alaska petroleum province, *in* I. H. Cram, ed., Future petroleum provinces of the United States—Their geology and potential: AAPG Memoir 15, p. 68–99.

Brower, W. A., Jr., H. F. Diaz, A. S. Prechtel, H. W. Searby, and J. L. Wise, 1977, Climatic atlas of the outer continental shelf waters and coastal regions of Alaska; Vol. III, Chukchi-Beaufort Sea: Arctic Environmental Information and Data Center Publication B-77, 409 p.

Collins, F. R., and F. M. Robinson, 1967, Subsurface stratigraphy, structural and economic geology, northern Alaska: U.S. Geological Survey Open-File Report 287, 259 p.

Eittreim, S., and A. Grantz, 1979, CDP seismic sections of the western Beaufort continental margin: Tectonophysics, v. 59, p. 251–262.

Grantz, A., and S. D. May, 1983, Rifting history and structural development of the continental margin north of Alaska, *in* J. S. Watkins and C. L. Drake, eds., Studies in continental margin geology: AAPG Memoir 34, p. 77–100.

Grantz, A., and S. D. May, 1984, Sedimentary basins and geologic structure of the continental margin north of Alaska, in Arctic Geology (Colloquium 04): Reports of the 17th International Geologic Congress, Moscow, August 4–14, 1984: Moscow, Publishing Office "Nauka," v. 4, p. 125–142.

Grantz, A., D. A. Dinter, and N. N. Biswas, 1983, Map, cross sections, and chart showing late Quaternary faults, folds, and earthquake epicenters on the Alaskan Beaufort Shelf: U.S. Geological Survey Miscellaneous Investigations Series Map I-11B2-C, scale 1:500,000.

Grantz, A., S. Eittreim, and D. A. Dinter, 1979, Geology and tectonic development of the continental margin north of Alaska: Tectonophysics, v. 59, p. 263–291.

Grantz, A., S. Eittreim, and O. T. Whitney, 1981, Geology and physiography of the continental margin north of Alaska and implications for origin of the Canada Basin, *in* A. E. M. Nairn, M. Churkin, Jr., and F. G. Stehli, eds., The ocean basins and margins: New York, Plenum Press, v. 5, p. 439–492.

Guldenzopf, E. C., M. B. Orlovsky, D. A. Higgs, C. G. Freytag, and E. R. Ovalle, 1980, National Petroleum Reserve, Alaska; Summary geological report fiscal year 1980: Tetra Tech., Inc., prepared for Husky Oil NPRA Operations, Inc., under contract to U.S. Geological Survey, 168 p.

Hea, J. P., J. Arcuri, G. R. Cambell, I. Fraser, M. O. Fuglem, J. J. O'Bertos, D. R. Smith, and M. Zayat, 1980, Post-Ellesmerian basins of Arctic Canada; their depocentres, rates of sedimentation, and petroleum potential, *in* A. D. Miall, ed., Facts and principles of world petroleum occurrence: Canadian Society of Petroleum Geologists Memoir 6, p. 447–488.

Jamison, H. C., L. B. Brockett, and R. A. McIntosh, 1980, Prudhoe Bay—A 10-year perspective, *in* M. T. Halbouty, ed., Giant oil and gas fields of the decade 1968–1978: AAPG Memoir 30, p. 289–314.

Jones, H. P., and R. G. Speers, 1976, Permo-Triassic reservoirs of Prudhoe Bay field, North Slope, Alaska: AAPG Memoir 24, p. 23–50.

Lachenbruch, A. H., and B. V. Marshall, 1966, Heat flow through the Ocean floor—The Canada Basin-Alpha Rise boundary: Journal of Geophysical Research, v. 71, p. 1223–1248.

Lerand, M., 1973, Beaufort Sea, *in* R. G. McCrossan, ed., The future petroleum provinces of Canada—Their geology and potential: Canadian Society of Petroleum Geologists Memoir 1, p. 315–386.

Mair, J. A., and J. A. Lyons, 1981, Crustal structure and velocity anistropy beneath the Beaufort Sea: Canadian Journal of Earth Sciences, v. 18, p. 724–741.

Martin, A. J., 1970, Structure and tectonic history of the western Brooks Range, De Long Mountains and Lisburne Hills, northern Alaska: GSA Bulletin, v. 81, p. 3605–3622.

Miller, C. C., J. E. Pickard, E. C. Taber, and R. B. Rogers, 1979, National Petroleum Reserve, Alaska; Summary geophysical report fiscal year 1979: Tetra Tech., Inc., prepared for Husky Oil NPR Operations, Inc., under contract to U.S. Geological Survey, 47 p.

National Petroleum Council, 1981, U.S. Arctic oil and gas: Washington, D. C., National Petroleum Council, 286 p.

Royden, L., J. G. Sclater, and R. P. von Herzen, 1980, Continental margin subsidence and heat flow; important parameters in formation of petroleum hydrocarbons: AAPG Bulletin, v. 64, p. 173–187.

Tailleur, I. L., 1973, Probable rift origin of Canada Basin: AAPG Memoir 19, p. 526–535.

Tailleur, I. L., (ed), G. H. Pessel and S. G. Engwicht, 1978, Maps showing land status and well locations and tables of well data, eastern North Slope petroleum province, Alaska: U.S. Geological Survey Miscellaneous Field Studies Map MF–928–A, 5 sheets, scale 1:250,000.

CHAPTER THREE

REGIONAL GEOLOGY AND PETROLEUM POTENTIAL OF THE UNITED STATES CHUKCHI SHELF NORTH OF POINT HOPE

Arthur Grantz
Steven D. May

U.S. Geological Survey
Menlo Park, California

ABSTRACT

The United States Chukchi Shelf north of Point Hope is divided into seven structural provinces. In the north, the Arctic platform culminates in the broad Barrow arch, which in places is overlain by less than 0.5 km of strata prospective for petroleum. South of the arch a southwest-thickening wedge of clastic and carbonate strata locally 13 km thick overlies the platform. A large north-trending downwarp in the platform in the central Chukchi Shelf, Hanna Trough, contains 10 km or more of sedimentary rocks that thin westward to less than 1 km on the extensively normal-faulted Chukchi platform. Near Cape Lisburne the Arctic and Chukchi platforms are overridden by the northeastward-directed Herald arch overthrust zone, which placed Paleozoic over Cretaceous beds of the Colville foredeep. Post-Albian foreland folding related to the overthrust zone produced large detachment folds in the sedimentary rocks of the foredeep. On the north, the Arctic and Chukchi platforms are disrupted by rifts related to late Neocomian opening of the Canada Basin of the Arctic Ocean and late Neocomian and Laramide opening of the North Chukchi Basin. Both the North Chukchi Basin and Nuwuk Basin, which prograded across the rifted margin that faces the Canada Basin, contain more than 12 km of clastic sedimentary rock. Seismic data and extrapolation of geologic data from northern Alaska suggest that parts of the Arctic and Chukchi platforms, Hanna Trough, and the Nuwuk and North Chukchi basins are more prospective for petroleum than the overthrust and foreland fold belts.

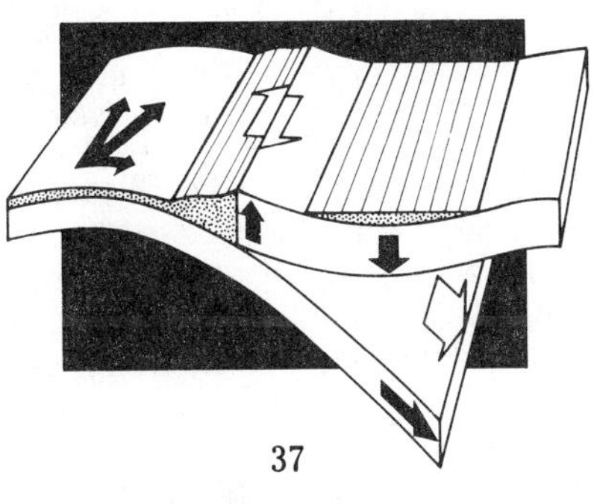

INTRODUCTION

An extensive frontier terrain that is prospective for petroleum but as yet is incompletely explored and entirely untested underlies the United States Chukchi Shelf north of Point Hope (Figure 1). The area is in most places underlain by a thick section of sedimentary rocks prospective for oil and gas and contains diverse geologic structures and stratigraphic features that may have trapped hydrocarbon fluids. The prospective sedimentary section includes every geologic system from the Mississippian to the Tertiary, and includes a number of formations that contain petroleum deposits or strong shows of oil or gas on the North Slope of Alaska. These formations have proved disappointing, however, where tested in a few exploratory wells in the western part of the National Petroleum Reserve in Alaska (NPRA)—an area adjacent to the Chukchi Shelf.

Petroleum exploration and development in the Chukchi Sea will be hindered by the Arctic climate, polar ice pack, winter darkness, absence of harbors in Alaska north of the Aleutian Islands, and remoteness from exploration support facilities and supplies. Pack ice entirely covers the area from the months of October or November to June, and part of the area is ice free for a short time in late summer only in favorable ice years. The bathymetry, however, is favorable for exploration because most of the area is < 60 m (200 ft) below sea level. The National Petroleum Council (1981, p. 5) estimates that "Proven technology and sufficient information and technical expertise for advanced design work is available for the industry to proceed confidently with operations in water as deep as 650 ft in the southern Bering Sea and to about 200 ft in the more severely ice-covered areas of the northern Bering, Chukchi, and Beaufort Seas." Thus, the National Petroleum Council indicates that essentially all of the United States Chukchi Shelf is thought to be technologically accessible to petroleum exploration and development at the present time.

Our data base consists mainly of the U.S. Geological Survey multichannel seismic-reflection profiles and accompanying high-resolution seismic-reflection profiles shown in Figure 1, and some additional Geological Survey single-channel seismic-reflection profiles, sonobuoy refraction measurements, and seabed samples. The multichannel seismic profiles, on which the interpretations presented here are mainly based, are mostly 30 to 90 km apart, with typical line spacings of 40 to 50 km. Ice conditions during data acquisition were such, however, that profile coverage in the north and northwest parts of the Chukchi Sea is sparse (Figure 1). Because of the wide spacing and irregular distribution of the profiles, our interpretations are reconnaissance in character. Strong artifacts on some of the profiles, particularly in the southwestern part of the study area, further limit the usefulness of the affected profiles for geologic interpretation and resource assessment.

GEOLOGIC FRAMEWORK

The northern United States Chukchi Shelf encompasses seven major structural provinces (Figure 2; also see Figure 12), each with more or less distinct geologic characteristics and petroleum potential. The general stratigraphy and lithology of the region is summarized in Figure 4 of Grantz, May, and Dinter (this volume). A cross section based on correlated test wells in the adjacent onshore is presented in Figure 3. The offshore geologic data are mostly new, but some material was taken from Grantz, Eittreim, and Whitney (1981). Onshore data are mainly from Alaska Geological Society (1981), Lathram (1965), Miller et al. (1979), Guldenzopf et al. (1980), and Tetra Tech, Inc. (1982).

Depth to seismic horizons and the thickness of seismic-stratigraphic units are given in seconds of two-way seismic-reflection traveltime. An approximate conversion of time to depth can be made by referring to the graph in Figure 4, which represents the average regional velocity structure of the Chukchi Shelf and the North Chukchi Basin. The graph for the North Chukchi Basin also approximates conditions in the northern Hope Basin and the young sedimentary sequence in the Tertiary canyon fill (Figure 2). The graph for the Chukchi Shelf is biased by the Tertiary canyon fill and gives depths that are somewhat shallower than the actual depths where the canyon fill is not present. Thickness of sediment in the Nuwuk Basin is based on a time-depth function derived for the Beaufort Sea (Figure 6 of Grantz, May, and Dinter, this volume).

Stratigraphy

The sedimentary strata of northern Alaska and the United States Chukchi Shelf are conveniently grouped into three regionally extensive sequences—the Franklinian, Ellesmerian, and Brookian—each having contrasting lithology, tectonic characteristics, and hydrocarbon potential (Figure 4 of Grantz, May, and Dinter, this volume). The Cambrian to Devonian Franklinian sequence, which regionally consists of slightly to strongly metamorphosed sedimentary and some volcanic rocks, is inferred to constitute the economic basement for the Chukchi Shelf. Following mild metamorphism and regional deformation, an extensive platform was cut across the Franklinian rocks in the Early Devonian and the Late Devonian and Early Mississippian. This, the Arctic platform, is a low-gradient surface that has remained remarkably stable beneath northern Alaska and the Chukchi Shelf, and is the foundation for the potentially petroliferous strata of the region. A diverse suite of clastic and carbonate sedimentary rocks—the Ellesmerian sequence—was deposited on the Arctic platform. The sequence contains both marine and nonmarine beds, including stratigraphically condensed organic-rich shale, texturally mature sandstone and conglomerate, and some dolomitized limestone. Clastic components were derived from a northern source land that lay beneath and north of the outer part of the present continental shelf. In a following section on the Hanna Trough, it is suggested that the Chukchi platform of the west-central Chukchi Sea (Figure 2) may also have been a western source land for the Ellesmerian clastic strata of the United States Chukchi Shelf. In the Chukchi Shelf, the main Ellesmerian sequence appears to be underlain, beneath a mild angular unconformity, by a thick sedimentary unit that filled in structurally low areas of the Arctic platform. This unit is called the Eo-Ellesmerian sequence in this report.

Eo-Ellesmerian Sequence

Structural sags and faulted depressions in the angular unconformity (the Arctic platform) at the top of the Franklinian sequence beneath the Chukchi Shelf are filled with sedimentary prisms that coalesce at their edges to form a

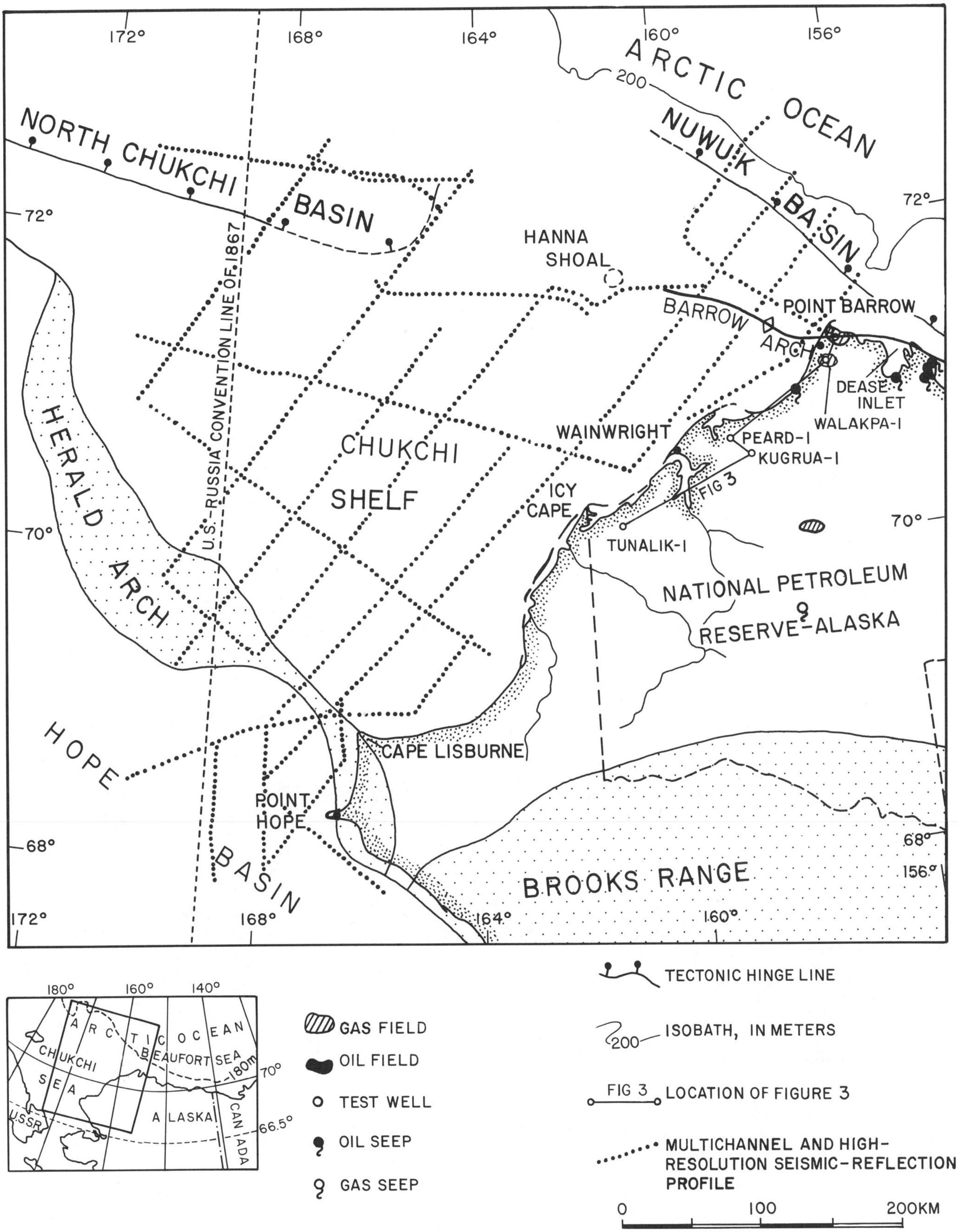

Figure 1. Map of northern United States Chukchi Sea and vicinity showing major geological features, petroleum development, oil and gas seeps, and locations of seismic-reflection profiles.

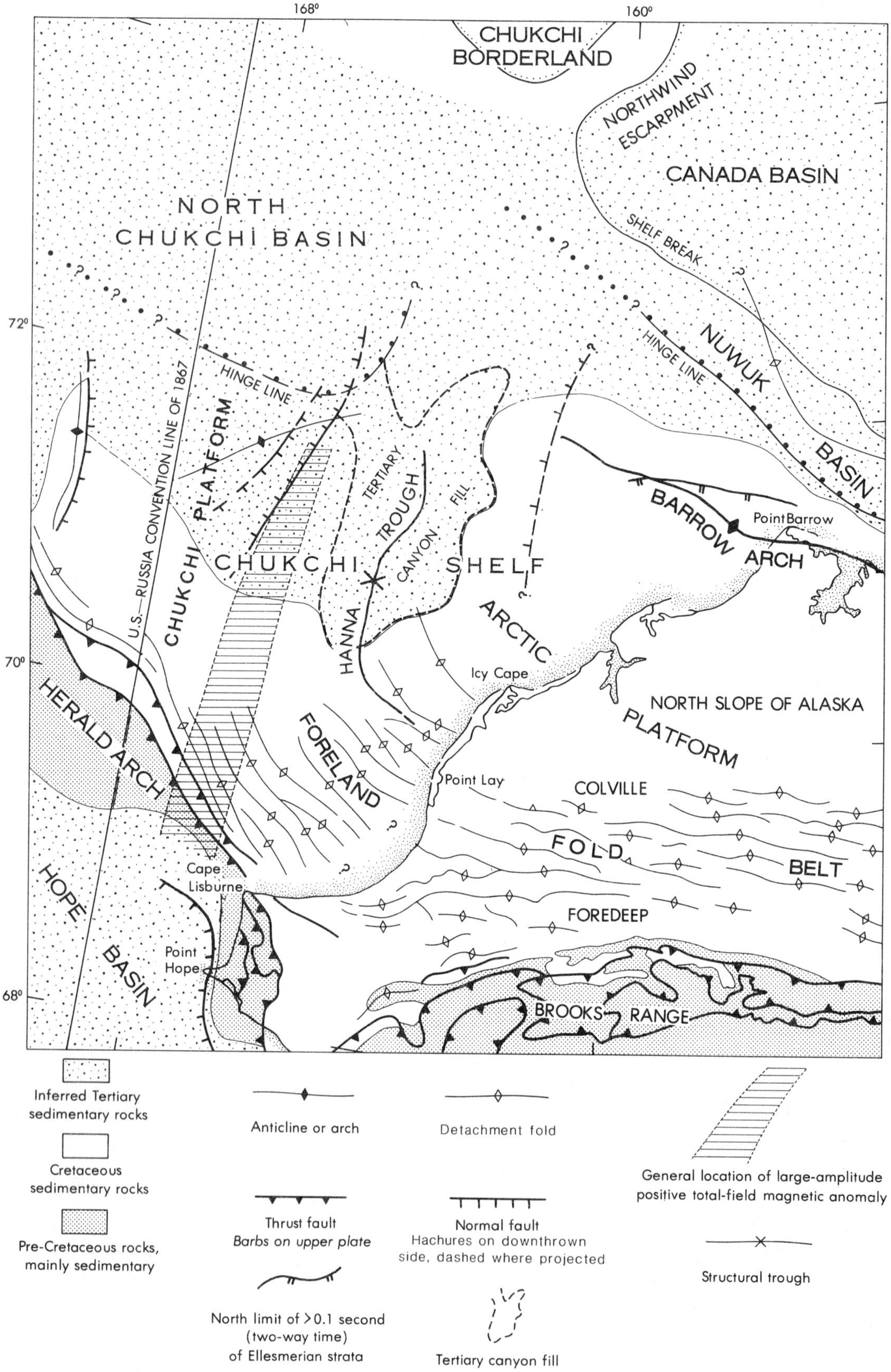

Figure 2. Tectonic map of the United States Chukchi Sea and vicinity.

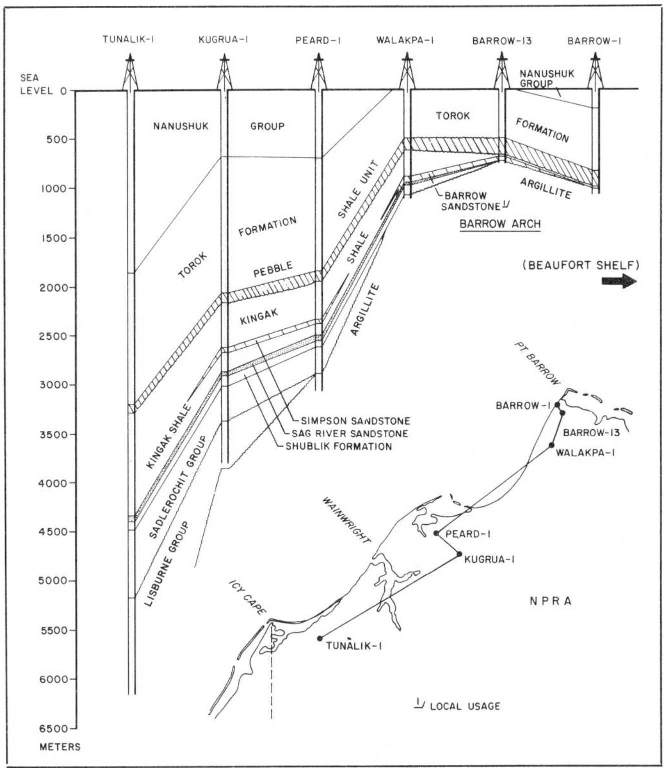

Figure 3. Geologic cross-section based on correlated test wells in the National Petroleum Reserve in Alaska that lie near the northern Chukchi Sea. Data from Bird (1982).

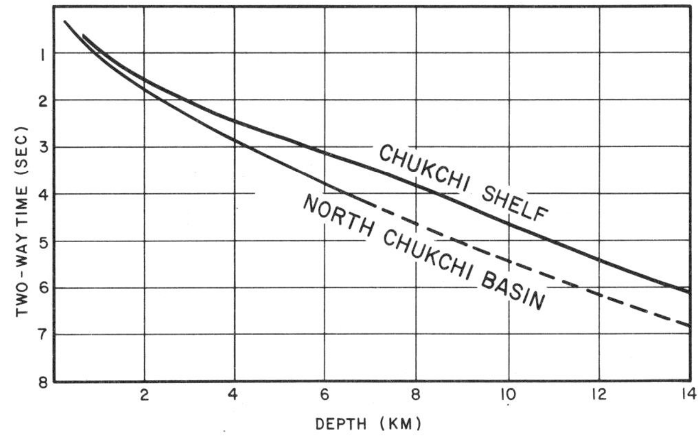

Figure 4. Generalized average seismic-reflection time as a function of depth for the Chukchi Shelf and North Chukchi Basin, derived from seismic-stacking-velocity measurements.

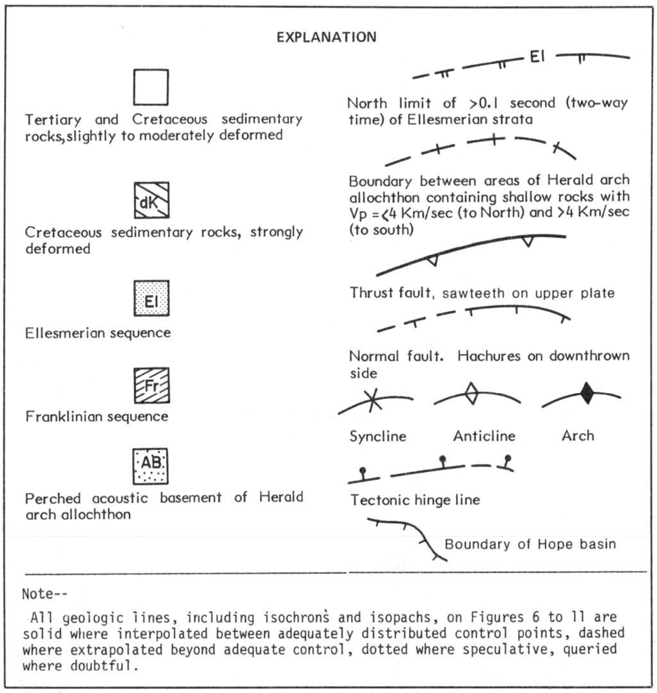

Figure 5. Explanation of map symbols used on Figures 6–12.

regional stratigraphic sequence of irregular outline, having marked local differences in thickness (Figures 5, 6). The deepest of these depressions, located about 50 km west of Icy Cape, contains as much as 5 km of section on one seismic profile. These sedimentary prisms are closely related to the Ellesmerian sequence, because they produce coherent seismic reflections over a large area and are only slightly more deformed than that sequence. However, there is commonly a low-angle, but pronounced, angular unconformity between these rocks and the Ellesmerian beds, and in places fault-bounded prisms of these sediments are overstepped by Ellesmerian strata. These sediments are here referred to as the Eo-Ellesmerian sequence because of their structurally close, but unconformable, relationship with the overlying Ellesmerian sequence.

The characteristics of the Eo-Ellesmerian beds can only be inferred, because they cannot be definitively correlated with rocks observed in outcrops or test wells. Comparison with onshore seismic-stratigraphic and subsurface units recognized in the NPRA (Guldenzopf et al., 1980; Alaska Geological Society, 1981; Tetra Tech, Inc., 1982) suggests, however, that the Eo-Ellesmerian beds belong to the Endicott Group, which has a somewhat sporadic distribution in the NPRA. Where encountered in the subsurface or interpreted from seismic-reflection profiles in the NPRA, the Endicott consists of a thick sequence of pre-Lisburne Group Mississippian and perhaps Upper Devonian clastic rocks, mainly nonmarine and in part coal bearing. Its principal development is in the northwest-elongated, compound Ikpikpuk-Umiat Basin of the eastern NPRA. However, seismic-reflection profiles (C. E. Kirschner, unpublished data, 1984) suggest that a wedge of Endicott clastic rocks may also occur in the Meade Basin, which underlies the western NPRA between about long. 158° and 159°W, and lat. 69°30′ and 70°30′N. The extent of the inferred Endicott beds in the Meade Basin is much smaller than in the Ikpikpuk-Umiat Basin to the east.

Several characteristics suggest correlation of the Eo-Ellesmerian seismic-stratigraphic sequence with the subsurface Endicott Group of the NPRA, mainly the stratigraphic position of the Eo-Ellesmerian beds beneath a mild angular unconformity and a seismic-stratigraphic unit that correlates with the Lisburne Group of the Ellesmerian sequence in the western NPRA. The shape of the Eo-Ellesmerian prism (Figures 5, 6) also closely resembles that of the Endicott Group. Both are coalesced, locally thick subbasins that in the aggregate are less extensive and less uniform in thickness than the

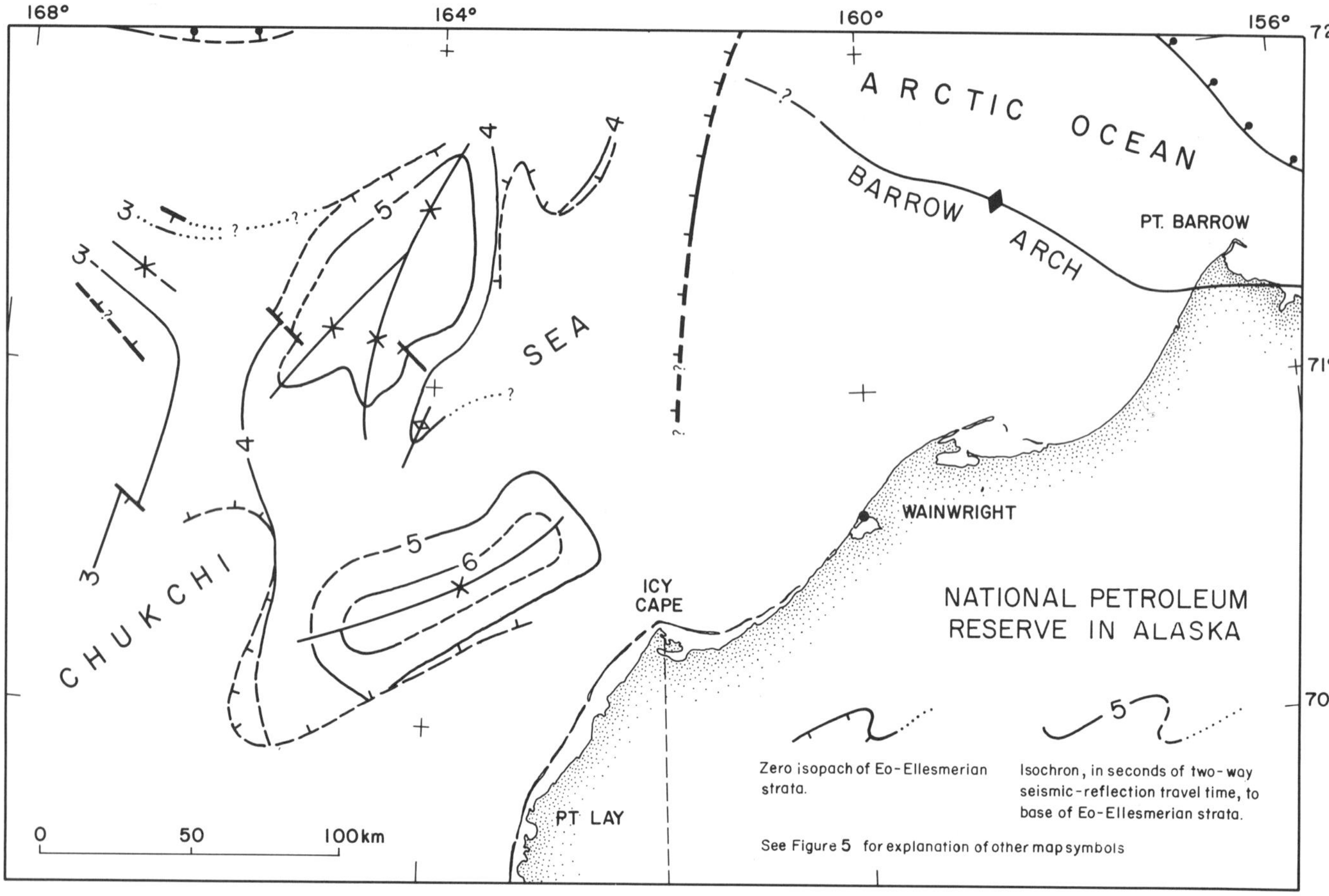

Figure 6. Isochrons on base of Eo-Ellesmerian strata, top of the Franklinian sequence, beneath the northern part of the United States Chukchi Shelf. The northeast limit of the Eo-Ellesmerian sedimentary prism was not recognized on the seismic-reflection profiles.

overlying Ellesmerian sequence (Figures 7, 8). Both also have significant structural relief at the base and considerable thickness locally. Maximum observed thickness of the Eo-Ellesmerian prism is 5 km, and of the Endicott prism in the subsurface of the NPRA more than 3 km (C. E. Kirschner, unpublished data, 1984). Although the seismic-stratigraphic correlation of the Eo-Ellesmerian sequence with the Endicott Group of the NPRA appears well supported, these prisms are not contiguous. The principal development of the Endicott Group is in the Ikpikpuk-Umiat Basin, more than 200 km east of the Eo-Ellesmerian prism of the Chukchi Shelf. The smaller Endicott Group prism in the Meade Basin is at least 100 km east of the Eo-Ellesmerian prism, and no pre-Lisburne Endicott rocks have been reported to date from the westernmost North Slope.

Ellesmerian Sequence

The dominantly clastic Ellesmerian sequence of the western North Slope, which was derived from a northern source land, underlies most, but not all, of the United States Chukchi Shelf north of Point Hope (Figures 7, 8). Beneath the eastern part of the shelf the sequence rests unconformably on the south-sloping Arctic platform and has the same general west strike, south dip, and southward increase in thickness that characterize it on the North Slope. In this area the sequence thickens from 0.1 sec (about 100–150 m) at a depth of about 1 sec (1 km) on the Barrow arch, to about 2 sec (5 km) at a depth of 3 sec, to slightly more than 5 sec (6 to 11 km) in a southeast-striking structural low off Point Lay. This structural low merges with the Hanna Trough—a north-striking structural low and sedimentary thick near long. 164°W. The base of the Ellesmerian sequence is more than 4.85 sec (10 km) deep, and the sequence is more than 3.0 sec (7.7 km) thick near lat. 71°15′N. A northward-deepening, probably genetically related structural low north of lat. 71°30′N near long. 162°W contains a similar or greater thickness of Ellesmerian strata. West of the Hanna Trough the Ellesmerian sequence becomes progressively shallower and thinner. Beyond long. 167°W it is generally less than 0.5 sec (1.3 km) thick, and in places beyond about long. 168° W it appears to be entirely absent. The decrease in thickness is due to both stratal thinning and to erosional truncation and overlap, suggesting the possibility that a platform or source land lay to the west. On the other hand, sections of tentatively identified Ellesmerian rocks as thick as 2 sec (5.5 km) are preserved in fault blocks near Cape Lisburne, and these sections appear to extend west of the 0.5-sec isopach of Ellesmerian rocks in the region north of the fault blocks. A strong seismic reflector of regional extent is interpreted to mark the top of the Ellesmerian sequence.

In the eastern part of the Chukchi Shelf the reflectors correlated with the Ellesmerian sequence consist of an upper unit of relatively weak, less continuous reflectors and a lower unit that contains a number of strong reflectors of considerable lateral extent. A strong reflector, which typically lies stratigraphically 30 to 50% below the top of the Ellesmerian reflector packet on the seismic profiles, marks the

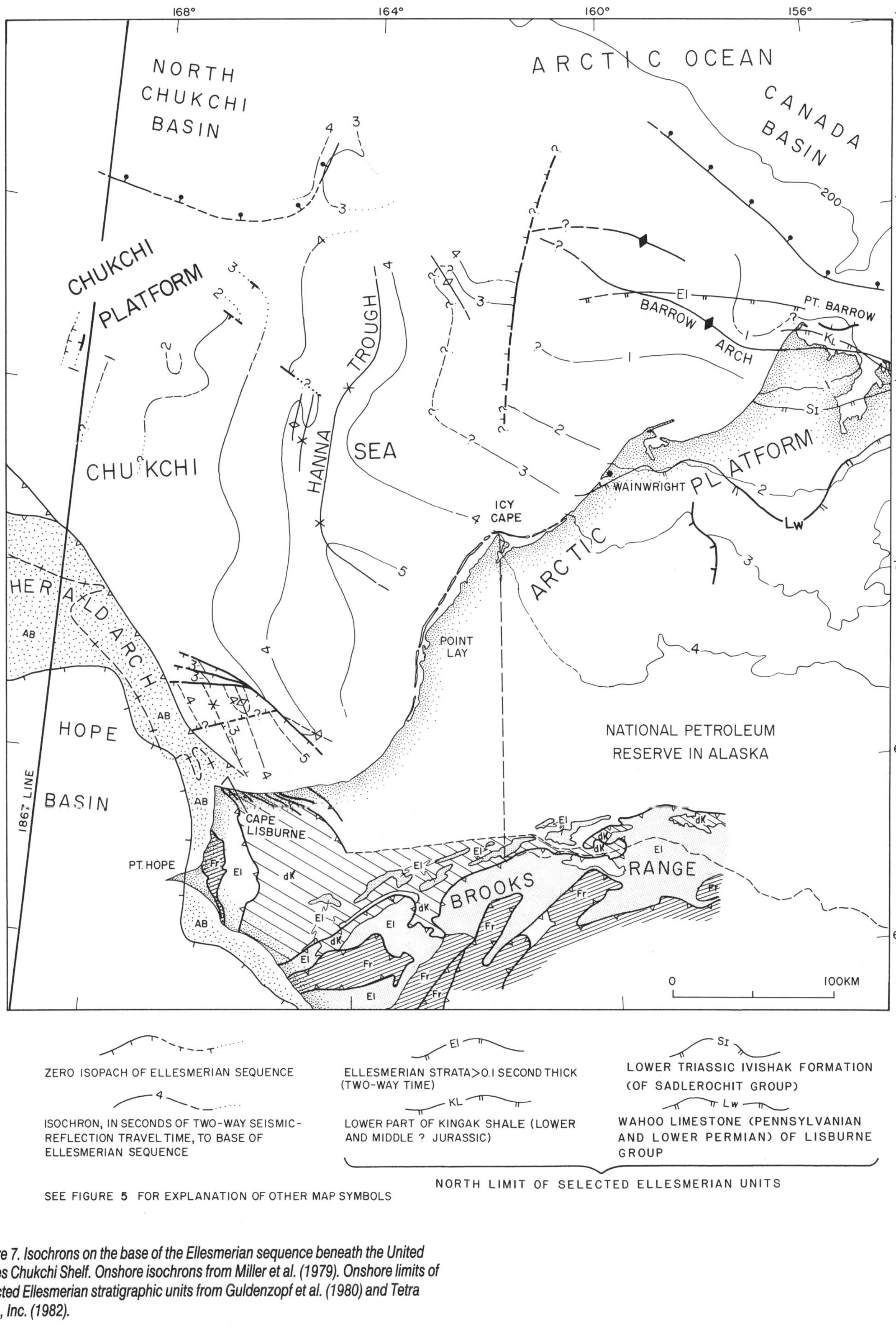

Figure 7. Isochrons on the base of the Ellesmerian sequence beneath the United States Chukchi Shelf. Onshore isochrons from Miller et al. (1979). Onshore limits of selected Ellesmerian stratigraphic units from Guldenzopf et al. (1980) and Tetra Tech, Inc. (1982).

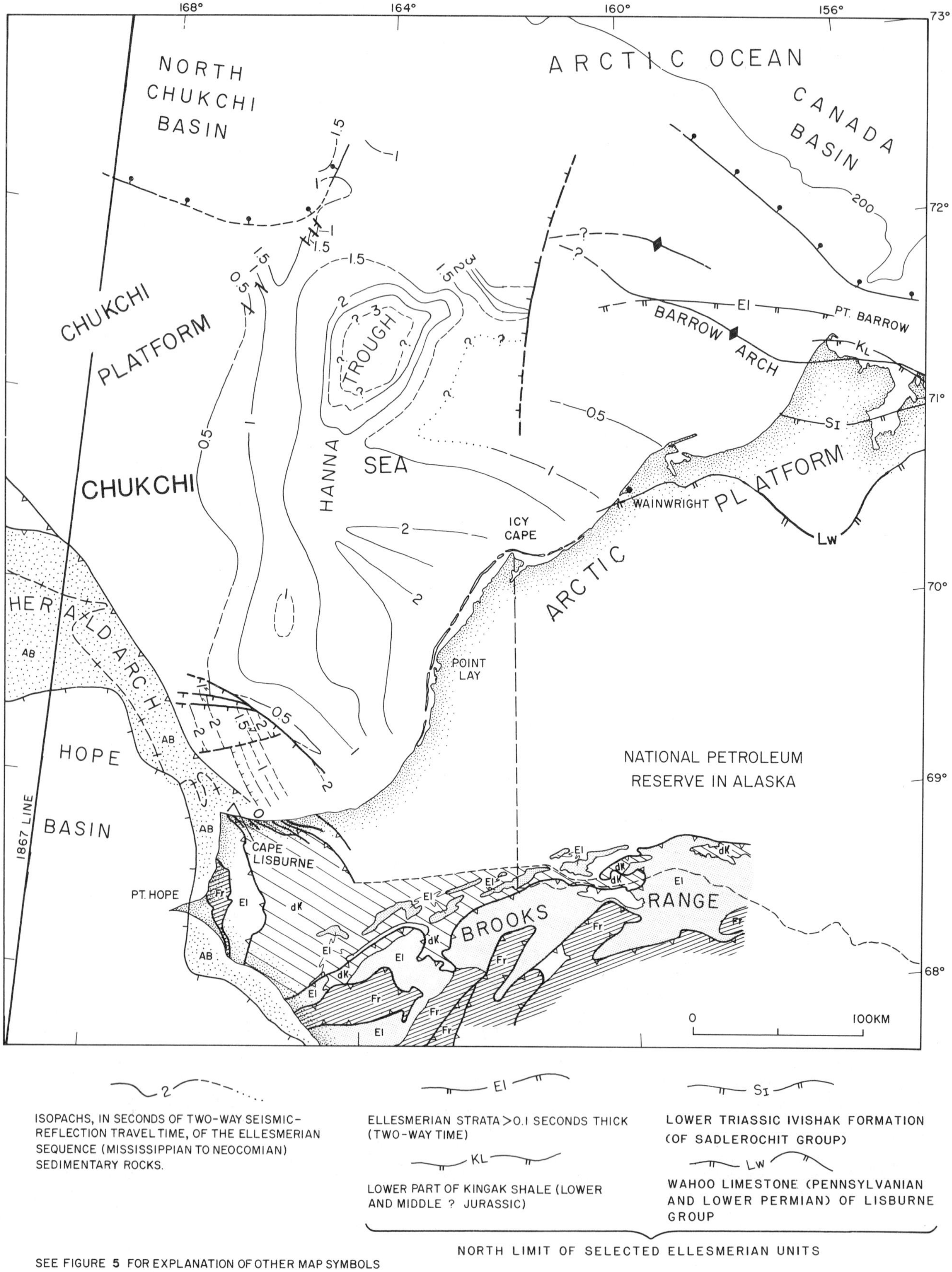

Figure 8. Isopachs, in seconds of two-way seismic-reflection time, of the Ellesmerian sequence beneath the United States Chukchi Shelf. Onshore limits of selected Ellesmerian stratigraphic units from Guldenzopf et al. (1980) and Tetra Tech, Inc. (1982).

top of the lower unit. The relative thickness of the units changes markedly to the west, however, and in places one or both are absent. Comparison of our seismic-stratigraphic units with those of the western NPRA, as shown by Tetra Tech, Inc. (1982) in the area between Icy Cape and the Barrow arch, suggests that the upper unit consists of the Lower Cretaceous (Neocomian) pebble-shale unit and the Jurassic and Lower Cretaceous Kingak Shale. North of Icy Cape, however, the interval interpreted to be the Kingak contains a few stronger reflectors in its lower part that may represent sandstone beds. The lower Ellesmerian unit is thought to represent the Lisburne Group (Pennsylvanian and Permian in the western NPRA; K. J. Bird, 1985, oral communication), the Permian and Triassic Sadlerochit Group, and the Triassic Shublik Formation and Sag River Sandstone. The upper unit (Kingak Shale and pebble-shale unit) appears to be relatively thinner, and the lower unit relatively thicker offshore than they are beneath the western North Slope.

The northern limits of key stratigraphic components of the Ellesmerian sequence in the western NPRA (taken from Guldenzopf et al., 1980, and Tetra Tech, Inc., 1982), where there is a progressive south to north overlap of older by younger stratigraphic units, are shown in Figures 7 and 8. The seismic data indicate that the same progressive overlap occurs beneath the eastern Chukchi Sea.

Lower Part of Brookian Sequence

The lower part of the Brookian sequence of the United States Chukchi Shelf consists of a thick succession of seismic reflectors of weak to moderate strength that are interpreted to represent Cretaceous marine and nonmarine clastic sedimentary rocks of the Colville foredeep. Structural contours drawn at the base of the Brookian sequence on the eastern Chukchi Shelf are shown in Figures 9 and 10. The sequence rests, typically by downlap, on a strong regional seismic reflector that is thought to mark the top of the pebble-shale unit (Neocomian), at the top of the Ellesmerian sequence. The seismic reflectors and comparison with onshore data (Molenaar, 1981; Tetra Tech, Inc., 1982) suggest that the lower Brookian sequence of the Chukchi Shelf consists of marine prodelta clastic deposits of the intertonguing Fortress Mountain and Torok Formations and nonmarine and shallow-marine intradelta deposits of the Nanushuk Group (Figure 3; and Figure 4 of Grantz, May, and Dinter, this volume). The overall age of these units is Aptian(?), Albian, and possibly earliest Late Cretaceous (Cenomanian). Foreset beds, as interpreted from our seismic records, and paleocurrent data (Molenaar, 1981) suggest that these beds were derived from a source land that lay to the southwest. Near lat. 70°N on the North Slope, and possibly in places in the western and northern parts of the Chukchi Shelf, the Albian beds are overlain by nonmarine and shallow-marine intradelta deposits of the Colville Group (Upper Cretaceous). The Colville Group is interpreted to intertongue north and northeastward with marine prodelta deposits beneath the northern Chukchi and Beaufort shelves.

The lower Brookian sequence thickens north and south from a minimum of 0.25 to 0.67 sec (0.25 to 0.7 km) on the Barrow arch. North of the arch, the base of the sequence deepens gradually to about 1.5 to 2 sec (2 to 3 km) at the faulted tectonic hinge line (Figures 1, 2, 9), which is thought to have formed during the opening of the Canada Basin by sea-floor spreading in the late Neocomian (Grantz and May, 1983; Grantz, May, and Dinter, this volume). At the hinge line the slope increases markedly, and the base of the Brookian sequence lies more than 12 km below sea level beneath the outer continental shelf. The overlying sedimentary prism, the Nuwuk Basin (Figures 1, 2, 9), is progradational, and near the shelf break it probably steps from the continental basement of the Arctic platform onto transitional or oceanic crust of the Arctic Basin. The Nuwuk Basin contains both foreset and topset beds, and large erosional channels and channel fills. Structurally it is dominated by large, multistrand, down-to-the-north growth faults that preliminary studies suggest were active since the Late Cretaceous or early Tertiary. The faults appear to be restricted to the Brookian sequence and to curve into, but not displace, the upper surface of pre-Brookian bedrock.

Southward from the Barrow arch, beneath the eastern Chukchi Shelf, the lower Brookian sequence thickens rather uniformly to a maximum of 3.7 sec (about 7.5 km) beneath the axis of the Colville foredeep between Icy Cape and Point Lay. From this structural low and sedimentary thick the axis of the foredeep plunges southeastward beneath the foothills of the Brooks Range. Up-plunge, the axis merges with the north-trending Hanna Trough in which the base of the Brookian sequence lies 2.4 to 3.0 sec (about 4 to 5.5 km) below sea level. The thickness of the beds in the lower part of the Brookian sequence in the trough is reduced, however, by a paleocanyon filled with a thick, presumed Tertiary sedimentary prism belonging to the upper part of the Brookian sequence (Figures 2, 11). West of the Hanna Trough and south of lat. 71°30′N, the lower Brookian sedimentary sequence thins to less than 1.0 sec (about 1 km); however, noise artifacts in the data caused by high-velocity rocks near the surface in the area northwest of Cape Lisburne preclude determining whether or not the sequence wedges out entirely. Northwest of the Hanna Trough the lower Brookian sequence thickens sharply as it drops into the eastern part of the North Chukchi Basin, where the sequence is extensively deformed by listric normal faults of large displacement. The base of the unit lies more than 6 sec (about 12 km) below sea level in this basin, and is 4.5 km to perhaps 8.5 km or more in thickness.

The lower Brookian sequence of the North Slope and eastern Chukchi Sea was derived from a source land that lay to the southwest and filled the Colville foredeep by progressive downlap on the top of the Ellesmerian sequence and by onlap against an early phase of the Barrow arch to the northeast. The lower Brookian rocks also thicken into the Hanna Trough, which has somewhat less structural relief at the base of the Brookian than at the base of the underlying Ellesmerian rocks. The suggested, though incompletely documented, westward reduction in the thickness of the lower Brookian sequence west of the Hanna Trough by stratal thinning and erosional truncation implies that a source land, or at least a structurally positive area, lay to the west of the trough in early Brookian time.

Upper Part of Brookian Sequence

The Beaufort and northeastern Chukchi shelves north of the Barrow arch, the Hanna Trough, the shelf west of the Hanna Trough north of the latitude of Icy Cape, and the North Chukchi Basin are underlain by an irregularly shaped body of young sedimentary rocks. Structural contours at the base of these rocks, which are called the upper Brookian

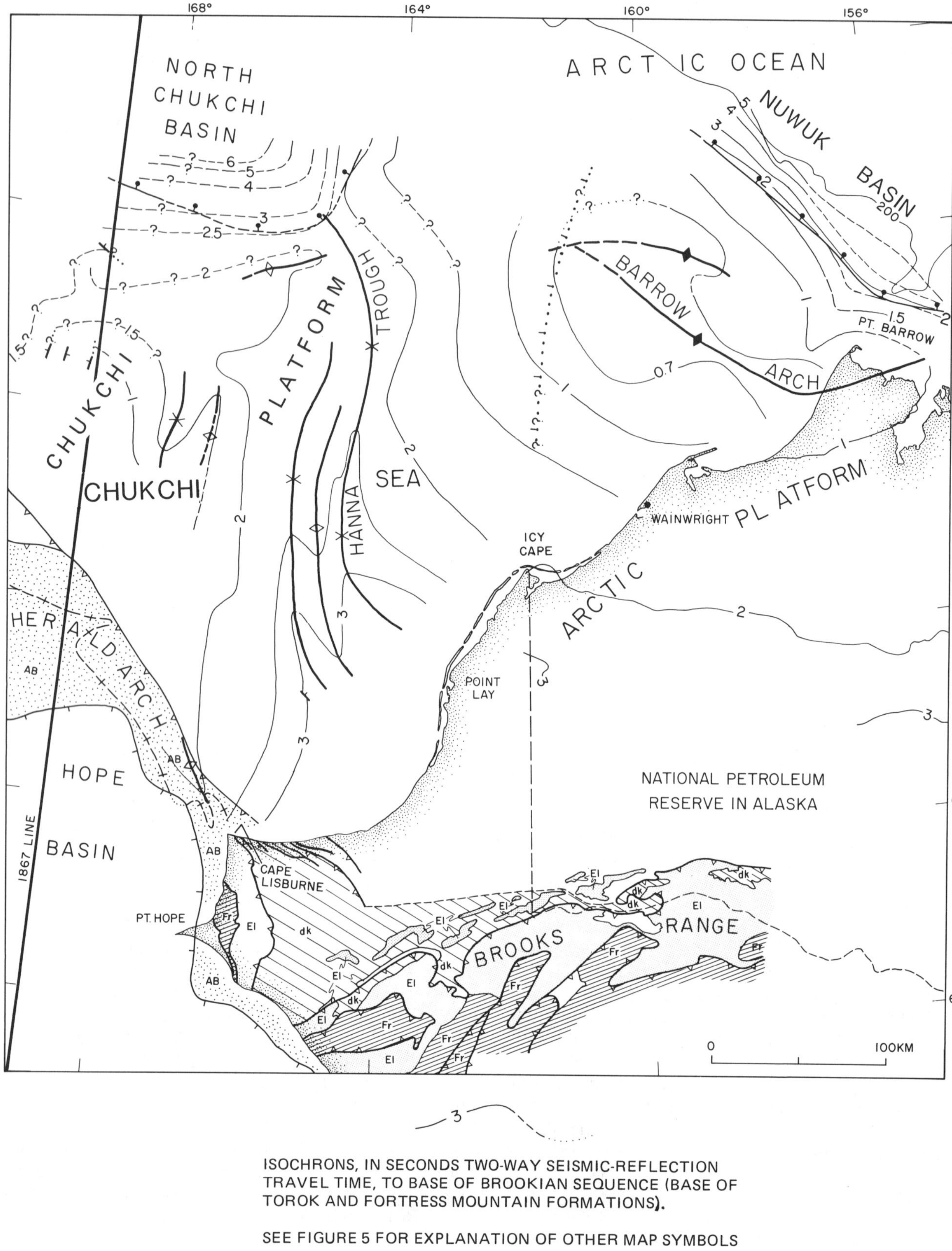

Figure 9. Isochrons on base of the lower Brookian sequence (Torok and Fortress Mountain Formations) beneath the United States Chukchi Shelf. Onshore data from Miller et al. (1979).

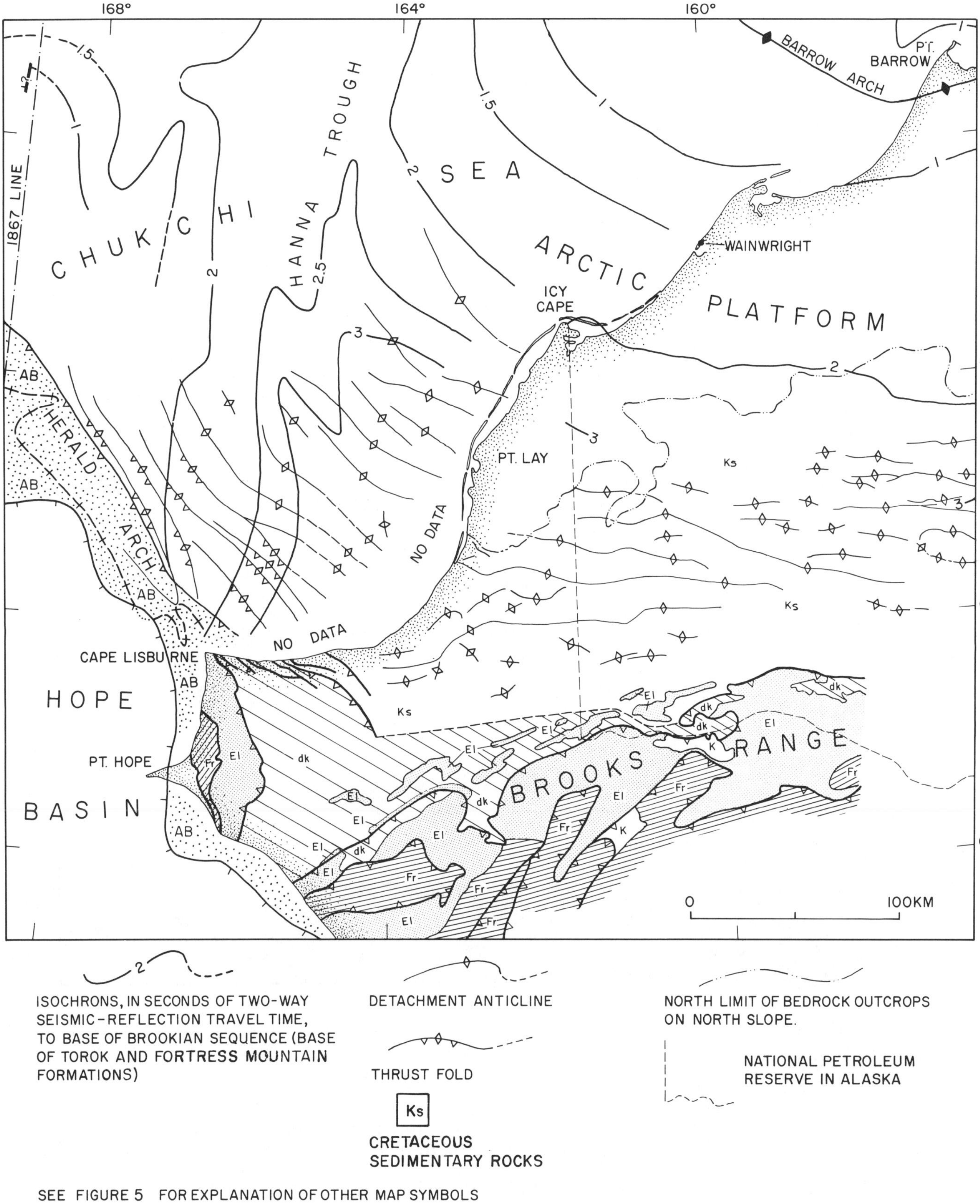

Figure 10. Detachment folds and thrust folds in Cretaceous sedimentary rocks of the Colville foredeep on the United States Chukchi Shelf and western North Slope. Beneath the Chukchi Shelf the folds are detached at, or close to, the base of the Torok and Fortress Mountain Formations, which is contoured in isochrons. Onshore isochrons from Miller et al. (1979); onshore folds from Lathram (1965).

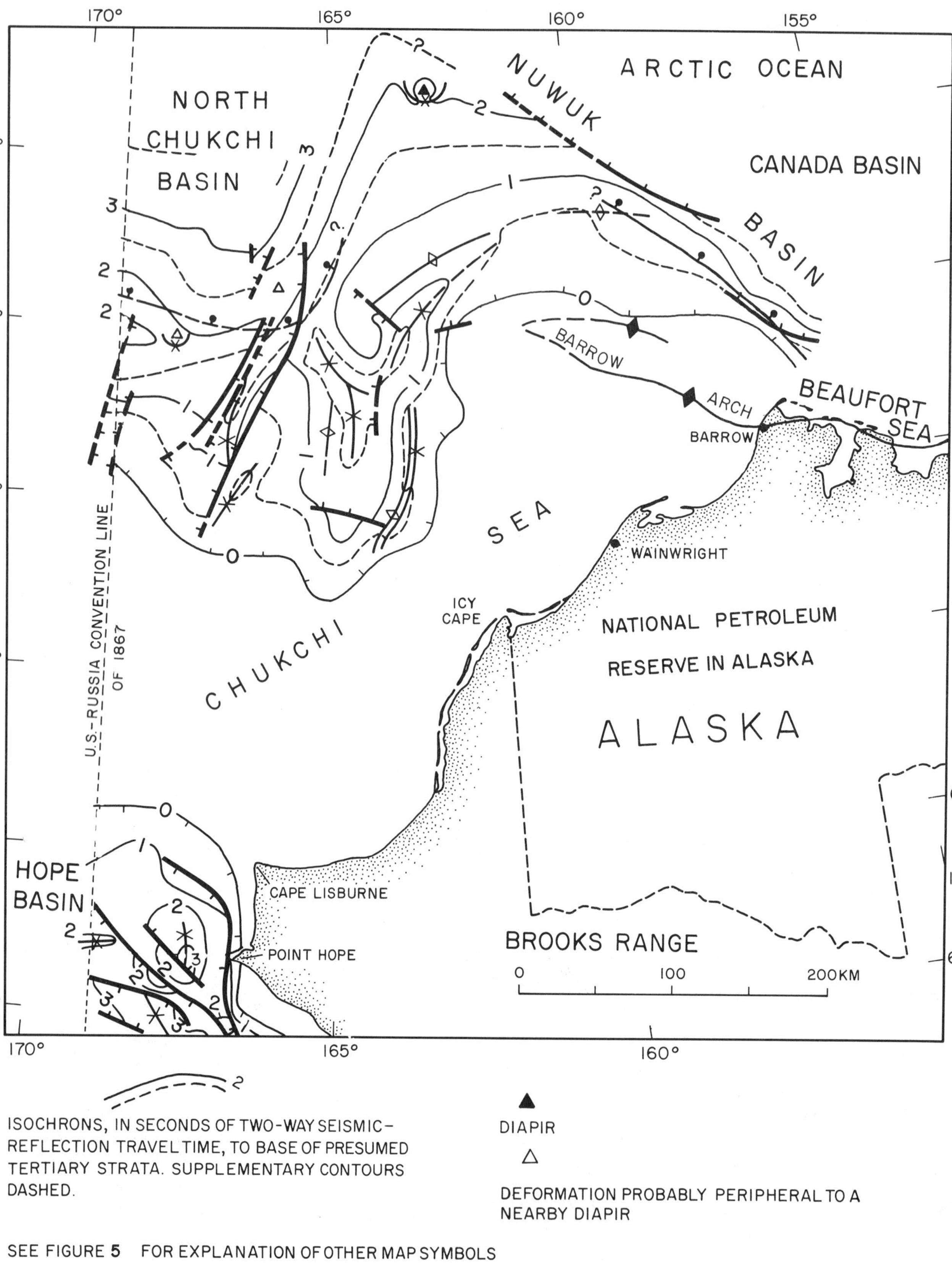

Figure 11. Isochrons on base of the upper Brookian (presumed Tertiary) sequence beneath the northern United States Chukchi Sea. Isochrons north of the multichannel profiles (Figure 1) are based only on shallow-penetration single-channel seismic-reflection profiles, and are speculative.

sequence in this report, are shown in Figure 11. They overlie the lower Brookian rocks with angular unconformity, have relatively low seismic-reflection velocities, and constitute the youngest bedrock stratigraphic unit on the Chukchi Shelf, from which we infer that they are largely or entirely Tertiary. Although much of the section can be traced into Tertiary beds of the Beaufort Shelf and North Slope east of Harrison Bay, the basal contact or unconformity has not been tied to the Cretaceous-Tertiary boundary. The sequence may therefore lack the lowest part of the Tertiary or, alternatively, the lowest beds may be uppermost Cretaceous. Because the upper Brookian rocks are the highest stratigraphic sequence in the area, and their upper beds lie at or near the floor of the shallow Chukchi Sea, the structural contours on the base of these rocks (Figure 11) are essentially isopachs.

The upper Brookian rocks are as much as 2.0 sec (2.2 km) thick in the subbasin that overlies the Hanna Trough, 1.5 to 1.9 sec (1.6 to 2.1 km) thick in a filled channel that connects the subbasin with the North Chukchi Basin, more than 1.5 sec (1.6 km) thick in the Nuwuk Basin, and more than 3.6 sec (5.6 km) thick in the North Chukchi Basin itself. The morphology of the subbasin that overlies the Hanna Trough, and the fact that the subbasin is linked to the deep North Chukchi Basin by a buried channel (Figure 11), suggests that it is a submarine canyon that was tributary to the North Chukchi Basin before both were filled by Tertiary sedimentary rocks. In morphology and size the postulated submarine canyon resembles the modern Navarinsky and Zhemchug canyons of the outer Bering Shelf (Carlson and Karl, 1984).

Low seismic velocities, the characteristics of Tertiary strata on the North Slope, and the geometry of the seismic reflectors suggest that the upper Brookian sequence of the Chukchi Shelf is composed of clastic sedimentary rocks. On the east side of the subbasin overlying the Hanna Trough the seismic reflectors are conformable with the basal unconformity. On the more steeply sloping west side of the basin, however, reflectors within the subbasin buttress against the basal unconformity and the underlying basin wall. Because the height of the west wall of the subbasin, and the thickness of the buttressing beds, is at least 1.25 sec (about 1.75 km), much of the section in the subbasin is probably marine. Presumably, the even thicker section in the North Chukchi Basin is also dominantly marine. The Tertiary beds of the North Slope, the Beaufort Shelf, and probably those of the Chukchi Shelf grade shoreward into nonmarine beds.

Structural Provinces

Most of the United States Chukchi Shelf north of Point Hope is underlain by the Arctic platform (see previous section on Stratigraphy), which has been remarkably stable since its development in the Devonian and Early Mississippian. The platform has been tilted and warped, but high-gradient folds have not been recognized and faults of significant displacement are common only in the Chukchi platform of the western part of the United States Chukchi Shelf. Although structural gradients in the tilted and warped areas of the platform are very low, the affected areas are extensive and consequently the amplitude of the resultant deformation is large. Most of the major structural provinces of the shelf (Figure 12) are recorded in the morphology of the Arctic and Chukchi platforms.

Barrow Arch and Arctic Platform

Structurally, the northern part of the Arctic platform east of Hanna Shoal is dominated by the Barrow arch (Figures 2, 7–11). Franklinian basement lies within 0.25 sec (0.3 km) of the sea surface in one locality along the broad crest of the arch, and flank dips are very low. The south-sloping Arctic platform forms the south flank of the arch. The crest and north flank are also underlain by the Arctic platform, but the platform in these areas was modified by uplift and erosion adjacent to the late Neocomian rift that created the present continental margin and Canada Basin north of Alaska (Grantz and May, 1983) and by subsequent sedimentary loading of the margin north of the tectonic hinge line (Figures 5–9, 11, 12). The tectonic hinge line, which lies north of the arch and is thought to have resulted from rifting, is the north boundary of the Barrow arch and of the Arctic platform province in offshore northern Alaska.

The Arctic platform slopes uniformly south–southwest, with a gradient of about 2.5°, from the crest of Barrow arch to the present structural axis of the Colville foredeep near Point Lay (Figures 2, 7, 9). The sedimentary section on the platform near Point Lay consists of about 5.5 km of Ellesmerian clastic and carbonate strata and about 7.5 km of lower Brookian marine and nonmarine strata. Off Icy Cape a local basin that strikes northeastward contains about 1.7 sec (5 km) of Eo-Ellesmerian strata (Figure 6). The northeast limit of these beds is uncertain. The Ellesmerian sequence thins northward by stratal thinning and progressive northward overlap of older by younger formations (Figures 3, 6, 7), and it is only 0 to 200 m thick on the crest of the Barrow arch. The lower Brookian sequence thins northward by the progressive northward downlap of foreset beds on its substrate and by erosion at the present land surface following Tertiary uplift along the Barrow arch (Figure 3). The foreset beds formed in a prograding prodelta that filled the Colville foredeep from the southwest. The lower Brookian sequence on the Chukchi Shelf consists mainly of the marine Torok Formation and the marine and nonmarine Nanushuk Group. In places, <0.25 sec (about 0.3 km) of these very thick units remain on the Barrow arch.

The Barrow arch, as expressed at the base of the Brookian sequence (Figure 9), dies out at Hanna Trough near long. 163°W. A large fault near long. 162°W is an important component of the boundary between the inferred thick Ellesmerian section in the Hanna Trough and its northern extension, and the thin Ellesmerian section on the Barrow arch. The fault was observed on only one multichannel profile and we are therefore uncertain of its strike, but it appears to die out to the south near lat. 71°N. Other than the Barrow arch and the south tilt of strata south of the arch, the only significant deformational structures on the Arctic platform are the detachment folds of the foreland fold belt structural province (Figures 2, 10).

Hanna Trough

Downwarping and downfaulting of the Arctic platform from depths as shallow as 0.25 sec (0.3 km) on the Barrow arch to depths as great as 4.85 sec (about 10 km) in the vicinity of long. 164°W, created the east flank of the Hanna Trough (Figure 7). The change in structural level is largely accomplished by downwarping, but the major north-striking fault near long. 162°W forms the boundary of a northeast structural extension or subbasin of the trough near lat.

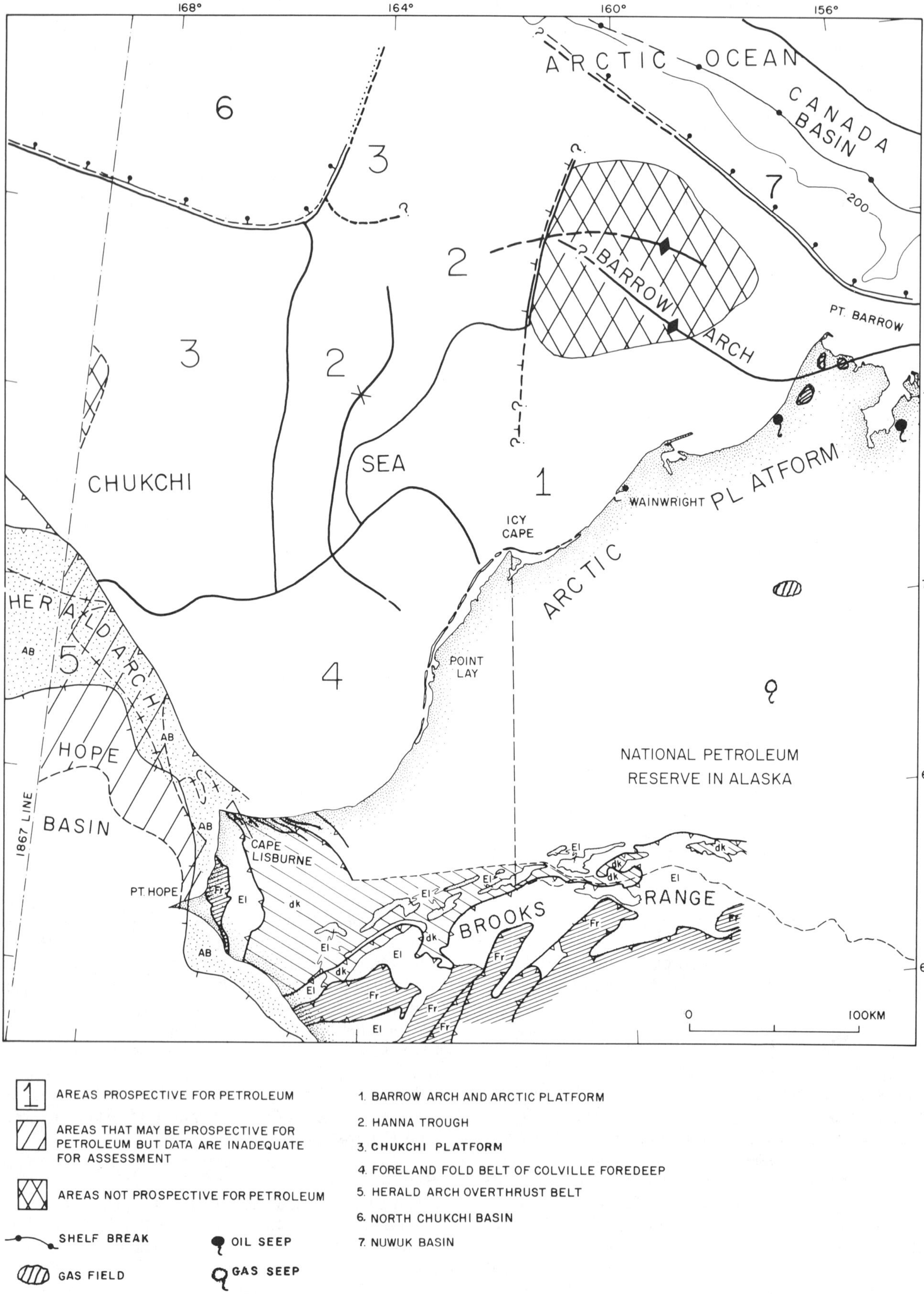

Figure 12. Structural provinces and areas prospective for petroleum on the United States Chukchi Shelf north of Point Hope.

71°40′N. Except for the large north-striking fault, and a broad arch that separates the trough from its northeast extension, the Arctic platform beneath the east flank of the Hanna Trough is almost lacking in secondary structures (Figures 2, 7). In contrast to the east flank, Franklinian basement beneath the axial region of the trough and the Chukchi platform to the west of it are in many places warped or folded and broken by faults. To the south, the trough loses its identity where its trend changes from north–south to northwest-southeast off Point Lay and it merges with the Ellesmerian Basin of the North Slope. The north end lies beyond our multichannel profiles, but gravity data suggest that the trough or a related subbasin extends to the north–northeast.

At least 0.4 sec (1 km) of Eo-Ellesmerian strata and >3 sec (7 to 8 km) of Ellesmerian strata were deposited in the deepest parts of the Hanna Trough and its northeast extension, or subbasin (Figure 8). The Eo-Ellesmerian beds were deposited in structural downwarps, in places faulted, in the Arctic platform. The overlying Ellesmerian strata in the trough are areally more extensive, and less variable in thickness. As previously noted, Ellesmerian strata in the west flank of the Hanna Trough thin westward by stratal thinning and by erosional thinning and overlap; this suggests that the Chukchi platform was a structural high, and perhaps a source land for Ellesmerian clastic materials. As the Ellesmerian strata on the Arctic platform also thin toward a source land, which lay to the north, the the Hanna Trough is a depositional basin as well as a structural trough (Figure 8).

The Hanna Trough is also expressed in the thickness of the Brookian sequence and the structure of its basal contact (Figure 9). However, the flanks of the trough in these rocks have very low gradients, and the structural relief at their basal contact is less than that in the underlying Ellesmerian rocks. The maximum thickness of Brookian strata in the trough on our profiles is about 2.5 sec (4 km). Our present data do not indicate whether an extension or subbasin of the Hanna Trough, as expressed in Brookian rocks, also continues to the north–northeast, as is inferred for the Ellesmerian rocks. The lower Brookian thick in the Hanna Trough formed in part by infilling an existing broad sag, and in part by postdepositional faulting and downwarping. Crustal loading by the underlying Ellesmerian sedimentary rocks, Cretaceous uplift of the Barrow arch and the Chukchi platform, and perhaps some differential compaction in Ellesmerian rocks may have all contributed to the structural relief on lower Brookian rocks in the Hanna Trough.

Upper Brookian strata that fill a postulated submarine canyon form the highest sedimentary sequence in the Hanna Trough. The axis of this canyon fill and the channel connecting it to the North Chukchi Basin coincide generally with the axis of the trough as defined by the base of the lower Brookian sequence (Figure 9), even though these prisms are thought to have quite different origins. The canyon fill onlaps against the steep west wall of the canyon, which is clearly an erosional feature. To the east, however, the base of the presently preserved canyon fill is nearly conformable with the underlying lower Brookian strata. A steep canyon wall has not been identified in this area. Thus the canyon was broad and the east side had a gentle slope, or a formerly existing steep east wall was removed by subsequent uplift and erosion.

The maximum height of the west wall of the filled canyon is at least 1.25 sec (about 1.75 km), and the total thickness of the upper Brookian section is 2 sec (about 2.5 km). Thus most of the relief at the base of these rocks is due to erosion and canyon filling, but some downwarping of strata into the basin added a small structural component to the present depth of the infilled paleocanyon.

Structural deformation may also have created the sill in the buried channel that appears to connect the filled paleocanyon with the North Chukchi Basin. The sill is 0.5 sec (about 0.75 km) higher than the deepest part of the filled paleocanyon on our profiles and creates an obvious difficulty for the canyon-fill hypothesis. The sill may be partly explained, however, by the large north–northeast-trending graben that obliquely crosses the channel near its junction with the North Chukchi Basin. Isochrons on the base of the upper Brookian sequence (Figure 11) indicate that the basal surface has been tilted east in the footwall of the fault bounding the east side of the graben. The tilt is interpreted to represent actual, as well as relative, uplift of the footwall, and this uplift is thought to have contributed to the elevation of the adjacent sill. Additional relief on the sill was created by the synwarping and post-upper Brookian downwarping of the paleocanyon fill that is recorded in post-canyon-fill strata.

Chukchi Platform

Westward shallowing of the acoustic basement, westward overlap of older by younger Ellesmerian beds, westward stratal thinning within the Ellesmerian and lower Brookian sequences, and extensive high-angle normal faulting characterize the Chukchi platform, which underlies the United States Chukchi Shelf west of the Hanna Trough (Figures 7–9). The platform (top of Franklinian basement) shallows from depths of 4.5 to 5.5 sec (about 9 to 12 km) at the axis of the Hanna Trough and the Colville foredeep to <1 sec (about 1.0 to 1.25 km) in places near the U.S.-Russia Convention line of 1867. As noted in a previous section, the westward overlap of older by younger Ellesmerian units against the westward-shallowing Chukchi platform indicates that an Ellesmerian structural high and perhaps a source land lay to the west. In places, Ellesmerian rocks wedge out entirely in the western part of the United States Chukchi Shelf (Figure 8), and lower Brookian rocks thin to <1 km (Figure 9). The numerous normal faults west of the Hanna Trough are young features because they offset Brookian rocks (Figures 9, 11).

Upper Brookian rocks on the Chukchi platform constitute a gently northward-dipping and northward-thickening sedimentary wedge that is broken by some large high-angle normal faults (Figure 11). Southward, these rocks wedge out near lat. 71°30′N. Northward, they thicken to about 2 sec (2.3 km) at the hinge line that separates the platform from the North Chukchi Basin. The Brookian rocks are sufficiently shallow in many places to be mapped from single-channel profiles. Accordingly, many structural features such as a north–northeast-striking graben near long. 166°W are shown in these rocks (Figure 11) that could not be delineated in older sequences. The graben has as much as 1.2 sec (about 1.4 km) of structural relief and a total sedimentary section that is 2 sec (about 2.3 km) thick. It is aligned with the east end of the North Chukchi Basin and is possibly related to the formation of the basin.

Foreland Fold Belt of Colville Foredeep

Large, high-amplitude detachment folds and thrust folds

characterize the Chukchi Shelf between Icy Cape on the northeast and Cape Lisburne and Herald arch on the southwest (Figure 10). These folds have flank dips of 1° to 15°, amplitudes as much as 1.2 km, wavelengths generally in the range of 10 to 30 km, and strike lengths of 10 to >120 km. They are the continuation of the foreland fold province on the north side of the Brooks Range, where they involve mainly lower Brookian beds, but apparently also involve Ellesmerian beds. The strike of these folds swings from east–west on the western North Slope to west–northwest and northwest offshore. The folds die out about 70 km off Icy Cape and about 140 km off Cape Lisburne. Offshore, the folds appear to be restricted to the lower Brookian sequence and to die out abruptly at the top of the Ellesmerian sequence, probably in the pebble-shale unit of Neocomian age.

Their geometry indicates that these are detachment folds in an allochthonous plate of a regional low-angle detachment fault system. Immediately above the detachment zone the cores of the folds contain thickened wedges of poorly reflective sediment—presumably tectonically mobilized shale. The folds become progressively more complex toward the Herald arch; near the arch they are thrust folds with cores that are appressed and contain upward-steepening listric thrust faults which rise northeastward from the detachment fault to the seabed. Northeast vergence and southwest increase in structural complexity indicate a relative northeast movement of the folds and the upper plate of the detachment fault. As observed on incompletely processed multichannel profiles, the folds appear to die out to the northwest by a decrease of displacement and shortening in the upper plate, rather than by extension of the basal detachment fault to the surface or its transformation into tear faults.

The faulted, folded, and therefore structurally thickened lower Brookian prism is from 2 sec to >3.6 sec (about 2.8 to 7.3 km) thick in the fold belt. Beneath this prism and its basal detachment fault lies 0.5 to >2 sec (about 1 to >5 km) of Ellesmerian section (Figures 7, 8). Near Cape Lisburne, tentatively identified Ellesmerian and older rocks are broken by a system of large normal faults whose trends and structural effects are presented conjecturally in Figures 7 and 8. These faults, and the faulted blocks of Ellesmerian(?) and older rocks, terminate upward at the low-dipping detachment surface at the base of the folded Brookian sequence. The Ellesmerian(?) rocks in the fault blocks appear to range in thickness from a feather edge between Franklinian and Brookian rocks near Cape Lisburne to slightly >2 sec (about 5.5 km) in the next fault block to the north.

Herald Arch Overthrust

A belt of perched acoustic basement which trends northwest across the Chukchi Shelf from Cape Lisburne has been named the Herald arch (Grantz, Holmes, and Kososki, 1975). This belt lies between the folded lower Brookian rocks of the Colville foredeep to the northeast, which it overlies structurally, and the young sedimentary rocks of the Hope Basin, which overlie it depositionally. The fault plane at the base of the perched acoustic basement was not recognized on the seismic profiles. However, a reflector inferred to represent the detachment fault at the base of the folded lower Brookian rocks can be traced at least 60 km southwest of Cape Lisburne, generally down the inferred dip of the fault plane.

The perched acoustic basement is tentatively interpreted to be the upper, allochthonous plate of a large overthrust fault system that dips southwestward at a low angle from the north side of the Herald arch (Figure 2). The folded lower Brookian rocks below the overthrust fault are thought to form a lower allochthonous plate that in turn rests on the detachment fault discussed in the preceding section. The lower Brookian rocks extend at least 15 km southwestward beneath the acoustic basement of the upper allochthonous plate, and a significant area and thickness of Ellesmerian rocks appears to underlie the lower allochthonous plate and its basal detachment fault. The sole faults underlying the upper and lower allochthons appear to merge west of the Lisburne Peninsula. The Tertiary Hope Basin, which overlies the upper allochthonous plate west and south of Cape Lisburne, is thought to postdate the thrust faulting because it unconformably overlies an interpreted extension of the Herald arch and fault zone south of Point Hope (Figure 10).

Shallow seismic-refraction measurements obtained with a multichannel streamer divide the perched acoustic basement of the Herald arch into two domains of shallow bedrock (0 to 0.5 km deep) that have contrasting ranges of compressional-wave seismic velocities (Vp) (Figures 5, 9, 10). In the northern domain, Vp ranges from 3.1 to 4.0 km/sec; in the southern domain from 3.7 to 7.3 km/sec. These velocities, a few dredge and dart core samples, and projection of onshore geologic boundaries from the Lisburne Peninsula (Martin, 1970) suggest that the domain of shallow bedrock with lower Vp consists mainly of strongly deformed lower Brookian and perhaps upper Ellesmerian (post-Lisburne) sedimentary rocks. They also suggest that the domain of shallow bedrock with higher Vp consists, in places, of the Lisburne Group, Endicott Group, and Franklinian sequence and that these rocks are more common in this domain than Brookian rocks.

North Chukchi Basin

A tectonic hinge line, at which the dip of Ellesmerian and Brookian beds, and the thickness of the lower Brookian sequence increase to the north and west, marks the south and east boundaries of the eastern part of the North Chukchi Basin (Figures 2, 7, 9, 11). The position of the hinge line is imperfectly defined by three multichannel profiles (Figure 1), which do not adequately relate the hinge line to the lower Brookian isochrons.

The base of the Brookian sequence deepens, and the sequence thickens, from 2 to 3.5 sec (about 2.5 to 5.5 km) on the Chukchi plattorm near the hinge line to >6 sec (12 km) on our northernmost multichannel lines in the basin. Most of the increase in thickness is in the lower Brookian sequence, but some, especially in the east end of the basin, is in upper Brookian rocks. The former attains a thickness of at least 7 km, and the latter at least 5 km in the basin. On one profile near the eastern margin of the basin, about 4 km of Ellesmerian rocks appear to underlie the lower Brookian beds.

Contrasting structural deformation in the tentatively identified lower and upper Brookian sequences of the North Chukchi Basin suggests that the basin formed in two stages of rifting. The first stage is recorded in the tectonic hinge line and the basinward thickening of the lower Brookian beds. It presumably correlates with the late Neocomian rifting event

(Grantz and May, 1983; Grantz, May, and Dinter, this volume) that produced the Canada Basin and the continental margin of the Alaskan Beaufort Sea. The second stage is recorded by numerous listric normal faults in the lower Brookian rocks of the eastern part of the North Chukchi Basin. These faults are commonly antithetic, bound-rotated fault blocks, and appear to extend into pre-Brookian rocks at moderate angles of dip. The faulting and rotation produced a structurally and paleomorphologically complex seabed surface that was smoothed soon after deformation by sedimentary infilling of the structurally produced low spots. The smoothed seabed was then covered by the only slightly faulted basin-fill sequence contoured in Figure 11. Both the infilling sedimentary rocks and the subsequent basin fill are thought to belong to the upper Brookian sequence. If our correlation of the youngest sedimentary sequences of North Chukchi Basin as lower and upper Brookian is correct, the second stage of rifting originated in the latest Cretaceous or earliest Tertiary.

Crustal stretching related to the initial stages of a sea-floor spreading event according to the model of Montadert et al. (1979), as observed in the Bay of Biscay, is thought to have induced the second stage of rifting observed in the eastern part of the North Chukchi Basin. Stretching, expressed in the upper crust as listric normal faulting and rotated fault blocks, resulted in crustal thinning, subsidence of the sea floor over the thinned crust, and deepening of the basin. Initiation of the second stage of basin subsidence coincided with development of the listric faults in post-early Brookian, pre-late Brookian time. Faulting was in part older than, and in part coeval with, deposition of the basal upper Brookian lenticular sedimentary sequence that smoothed the block-faulted sea floor.

Single-channel seismic-reflection profiles show diapirs intruding the sedimentary rocks within or adjacent to the North Chukchi Basin. Two diapirs were actually traversed, and two others are inferred from isolated upwarps in otherwise flat-lying upper Brookian beds (Grantz, Holmes, and Kososki, 1975). The upwarps resemble drag peripheral to the diapirs. The diapirs are piercement structures that rise to within 100 m of the seabed. The best-studied diapir, located near lat. 73°N, long. 163°W (Figure 11), is about 2 km in diameter and extends at least 3 km beneath the seabed. Low sonobuoy velocities, apparent lack of strong gravity or magnetic anomalies, and regional stratigraphy suggest that the diapirs are probably shale rather than salt, gypsum, or igneous rock. On the basis of single-channel data, the diapirs are tentatively interpreted to have originated in the lower Brookian or possibly the upper Ellesmerian sequence.

Nuwuk Basin

A thick clastic sedimentary prism, the Nuwuk Basin, was prograded northward across the faulted tectonic hinge line and continent-ocean boundary on the outermost northeastern Chukchi Shelf (Figures 1, 2, 9). The basin contains at least 12 km of lower and upper Brookian clastic sedimentary rocks derived from source lands to the south and southwest. Lower Brookian rocks appear to comprise two-thirds or more of the section north of the Barrow arch. Numerous large listric growth faults and related rotational megaslumps, or roll-over anticlines, dominate the structure of the Nuwuk Basin. Additional information on the basin is given in Grantz, May, and Dinter (this volume). The extent of the Nuwuk Basin to the northwest and its relationship to the North Chukchi Basin are not shown on our seismic lines.

PETROLEUM POTENTIAL

An assessment of the petroleum potential of the northern United States Chukchi Shelf on the basis of reconnaissance seismic-reflection data and a few dredge samples, but no subsurface information, is necessarily speculative. Nevertheless, the seismic data, augmented by extrapolation of the results of petroleum exploration in the NPRA, permit a preliminary evaluation of the relative petroleum potential of various parts of the Chukchi Shelf. The assessment is relatively more objective in the eastern part of the shelf, adjacent to the NPRA, than elsewhere. A synopsis of general lithologic conditions relevant to the petroleum potential of the western North Slope, taken mainly from K. J. Bird and C. E. Kirschner (unpublished data, 1984), Bird (1981), Bird and Andrews (1979), Carter et al. (1977), Guldenzopf et al. (1980), and Tetra Tech, Inc. (1982), is presented below as background for evaluating the potential of the Chukchi Shelf.

A preliminary map of petroleum prospectiveness of the Chukchi Shelf is presented in Figure 12. Most of the shelf is considered prospective because seismic data show that most of it is underlain by > 1 km of Ellesmerian and Brookian sedimentary rocks. Small areas near the U.S.-Russia Convention line and on the crest of the Barrow arch, incompletely outlined by our reconnaissance data, are considered not prospective because the sedimentary section in these areas appears to be < 1 km thick. One kilometer is probably less than the minimum thickness of sedimentary rock that is prospective for petroleum because, barring truly exceptional discoveries, offshore oil and gas pools at shallower depths would require too many costly production platforms for development to be economically feasible. Prospectiveness in Figure 12 is based solely on sedimentary thickness, because subsurface data are lacking on the Chukchi Shelf.

Western North Slope

Strata

The general subsurface stratigraphy and structure, and the location and stratigraphic position of oil and gas pools and shows in the western NPRA, are shown in Figures 1, 3, 13. The Permian Echooka Formation (basal unit of the Sadlerochit Group) and all formations from the Shublik (Middle and Upper Triassic) to the Nanushuk Group (Albian and Cenomanian?) have yielded shows of oil or gas in test wells. However, neither the Lisburne Group (Pennsylvanian and Permian in the western NPRA) nor the Lower Triassic Ivishak Formation (upper unit of the Sadlerochit Group) have as yet yielded oil or gas in test wells west of Dease Inlet (Figure 1). These stratigraphic units do, however, produce oil or gas shows in test wells east of Dease Inlet, and they contain the major oil and gas accumulations at Prudhoe Bay. Hydrocarbon occurrences west of Dease Inlet include structurally trapped, subeconomic gas pools in Lower Jurassic sandstones in the South Barrow and East Barrow fields, which are 15 to 20 km south of Point Barrow, and additional small accumulations of similar type probably occur in the vicinity (Lantz, 1981). Sands of Early Cretaceous (pebble-shale unit) age may also form pools in combination stratigraphic-structural traps at the Iko Bay-1 and Walakpa-1

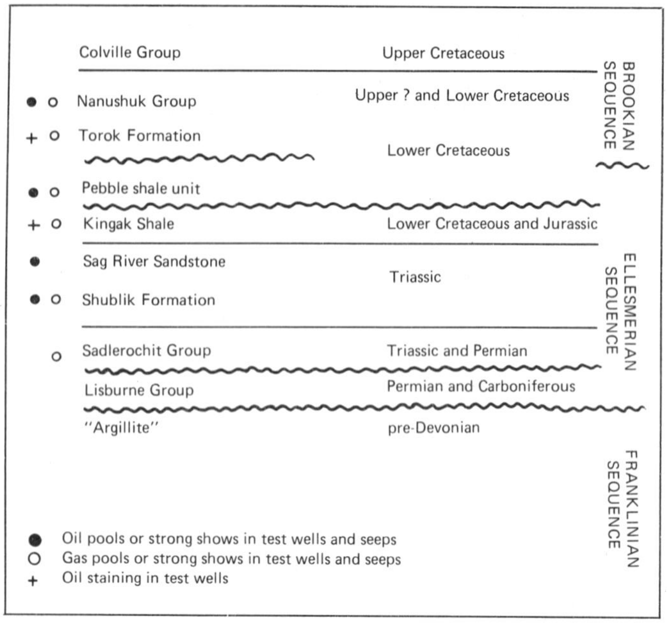

Figure 13. Generalized stratigraphic column beneath the Arctic coastal plain of northwestern Alaska west of Dease Inlet showing stratigraphic position of oil and gas pools or strong shows of oil and gas, and oil staining encountered in test wells and seeps. (Data from Alaska Geological Society, 1981; Bird, 1982; and K. J. Bird, unpublished data, 1984).

and -2 wells, 30 to 40 km southeast and southwest of Point Barrow, but these occurrences have not been adequately evaluated (Bird, 1981; C. E. Kirschner, unpublished data, 1984). Pre-Lisburne (Endicott Group) and post-Nanushuk (Colville Group and Sagavanirktok Formation) sedimentary strata have not been encountered in test wells west of Dease Inlet, even though many of the wells reached Ordovician and Silurian argillite.

In general, reservoir quality is better in the Ellesmerian Shelf and shelf basin deposits than in the Brookian sequence. The Ellesmerian sandstones are thicker, coarser, and cleaner to the north, toward their source land, and most of them shale out southward in an east-trending zone across the central NPRA. Reservoir quality in the Brookian rocks is harder to characterize. The Brookian sequence consists of thick deltaic and prodeltaic deposits that filled a foredeep and were deposited much more rapidly than the Ellesmerian rocks. The best reservoirs may be expected in the paralic and shallow-marine facies of the intradelta, and in submarine-fan turbidites at the foot of the prodelta slope.

Geochemical studies at the Prudhoe Bay field (Seifert, Moldowan, and Jones, 1979) indicate that the oil there was derived from the Shublik Formation, Kingak Shale, and pebble-shale unit, which correlates well with the occurrence of oil and gas shows in the western NPRA (Figure 13). The lack of shows in the Lisburne in the test wells of the western NPRA may be because permeability pathways have not connected them with petroleum source beds, or because they consist of the nondolomitic, light-colored Wahoo Limestone rather than the locally dolomitic and vuggy, dark-colored and shaly Alapah Limestone that contains the major Lisburne reservoirs at Prudhoe Bay. It is not certain, however, that the prospects drilled in western NPRA are traps at the Lisburne level. Thermal-alteration studies on the Inigok-1 well, in the eastern NPRA, indicate that the Endicott and Lisburne Groups are apt to be postmature and gas prone below 3600 m on the Arctic platform (Magoon and Claypool, 1982). These rocks lie at similar or greater depths in the western NPRA where the thermal maturity of the surface formations, as estimated from vitrinite reflectance studies (L. B. Magoon and G. E. Claypool, unpublished data, 1984), tends to be even higher than at Inigok-1.

The Lower Triassic Ivishak Formation of the Sadlerochit Group, the main reservoir at Prudhoe Bay, is represented by lutite in the western NPRA (Guldenzopf et al., 1980). The only significant sandstone in the Sadlerochit of western NPRA occurs in a 30-km-wide band in the Upper Permian Ikiakpaurak Member of the Echooka Formation (Guldenzopf et al., 1980). This sandstone yielded minor gas shows in the Peard-1 well. It is about 50 to 60 m thick in the Peard-1 and Kugrua-1 wells and extends from a seismically mapped pinch-out about 10 km north of Wainwright to a shale-out several kilometers north of the Tunalik-1 well. It is separated from the source rocks of the Shublik Formation by the Kavik Member (of the Ivishak Formation) and the lutite facies that constitutes the Ivishak in the western NPRA. The gas shows may therefore have originated in the adjacent shales of the Sadlerochit Group.

The Middle and Upper Triassic Shublik Formation and Upper Triassic Sag River Sandstone, reservoirs for oil and gas at Prudhoe Bay, are possibly more prospective than the Sadlerochit in the western NPRA and the eastern Chukchi Shelf. These units yield oil and gas shows in test wells west of Dease Inlet, and they contain sandstones that are in depositional contact with the Shublik Formation and Kingak Shale, which contain petroleum source rocks. Sandstone in the Shublik occurs in a west–southwest-striking belt 40 km or more wide that onlaps the Barrow arch on the north and shales out a few kilometers north of the Kugrua-1 and Tunalik-1 wells. Sand-shale ratios in the Shublik are highest in the western NPRA, reaching values of 5.0 and 5.1 in the Walakpa-1 and Peard-1 wells, where the formation is about 90 m thick. The overlying Sag River Sandstone, mainly a lutite in the western NPRA, is 30 to 40 m thick and contains sandy, barlike complexes of unknown extent at the Peard-1 and Kugrua-1 wells (Guldenzopf et al., 1980).

Good prospects may also be found in the Kingak Shale, a westerly thickening Jurassic and Neocomian lutite that ranges in thickness from 500 m to >1000 m in the western NPRA. The Kingak, a source rock at Prudhoe Bay, includes several sandstone units west of Dease Inlet, and some of these contain subeconomic or incompletely evaluated gas pools at the South Barrow and East Barrow fields and the Tunalik-1 test well. The sandstones, which may include transgressive nearshore and offshore barlike deposits and deep-water turbidites (Guldenzopf et al., 1980), appear to have been derived from northerly and northwesterly sources, and in general become more abundant toward the Chukchi Shelf. In the western NPRA, the Kingak is truncated by the unconformity at the base of the upper Neocomian pebble-shale unit, which could serve as both a source rock and reservoir seal for hydrocarbon deposits in Kingak sandstones in contact with the unconformity. Several thin sand bodies in the pebble-shale unit may also have reservoir potential. One such sand body in the Peard-1 well, where the pebble-shale unit is 116 m thick, is inferred to extend into the Chukchi Shelf north of Icy Cape (Guldenzopf et al., 1980).

The Brookian rocks of the western NPRA were derived

from source lands to the south and southwest and consist of the interfingering Fortress Mountain and Torok Formations and the partly interfingering, partly overlying Nanushuk Group of Albian and Cenomanian? age. The mainly intradeltaic Nanushuk and the prodeltaic Torok both yield shows of oil and gas in the western NPRA. The shows in the Torok west of Dease Inlet are from foreset and bottomset turbidite and submarine-fan deposits (C. E. Kirschner, unpublished data, 1984). The Nanushuk Group contains several small oil and gas fields in the eastern NPRA and contains a small gas field (Meade) west of Dease Inlet. It is dominantly nonmarine and contains thick coal beds south of Point Lay; it thins and becomes paralic and then shallow-marine to the northeast (Chapman and Sable, 1960; Ahlbrandt, 1979). These conditions should improve reservoir quality in that direction. Accordingly, the Nanushuk is expected to be gas prone and contain less capable reservoirs to the southwest, and to be more likely to contain oil and have more capable reservoirs to the north or northeast. An oil seep from the Nanushuk at Skull Cliff, 50 km southwest of Barrow (McKinnery, Garton, and Schwartz, 1959), may support this contention.

Thermal Maturation

Vitrinite reflectance data and the rank of coal in test wells and outcrops provide some insight into the thermal history and possible stage of hydrocarbon generation and preservation in the sedimentary rocks of the western NPRA, but they do not reveal whether the affected rocks actually generated oil or gas. The vitrinite reflectance data are reported by Magoon and Claypool (1979, 1982), and we are grateful to them for access to their additional unpublished data. Data on coal rank were taken from Barnes (1967) and Conwell and Triplehorn (1976). Relationship of coal rank to stages of hydrocarbon generation is from Heroux, Chagnon, and Bertrand (1979), but we follow Magoon and Claypool (1982) in using vitrinite reflectance values of 0.6 and 2.0 for the top and base of the liquid window—the zone of oil and gas condensate generation.

Vitrinite reflectance data indicate that equivalent stages of thermal maturation become shallower from the eastern and central parts to the western part of the NPRA, suggesting more uplift on the west (Magoon and Claypool, 1979). The data also suggest that the top of the liquid window (the highest rocks that have experienced a thermal history capable of generating oil) lies about 0.3 to 1.0 km below the surface in the western NPRA, between the latitude of Point Lay and the Peard-1 well (Figure 1). The rank of outcropping Albian coal in this area is subbituminous, which corresponds to the zone of wet-gas generation above the liquid window. North of Peard-1, the deepest prospective strata on basement are at the top of, or above, the liquid window in the overlying zone of wet-gas generation. The base of the liquid window between the latitudes of Point Lay and Peard-1 in the western NPRA, as interpreted from vitrinite reflectance measurements, lies 2.3 to 3.4 km below the surface in the three wells that reached the underlying zone of overmature rocks. One test well and the rank of outcropping Albian coal suggest that surface rocks in the foothills of the Brooks Range south of the latitude of Point Lay are in the liquid window, and that the base of the liquid window lies about 2.0 km below the surface. The rank of coal in the Endicott Group on the Lisburne Peninsula suggests that the Paleozoic rocks of the western Brooks Range and the upper allochthon of Herald arch are overmature for oil and gas condensate, and that former hydrocarbon liquids in these beds would now be dry gas. The vitrinite reflectance values also suggest that, in the eastern half of the United States Chukchi Shelf, upper Paleozoic and Triassic strata are overmature from Icy Cape or Wainwright south, and that the lower part of the lower Brookian sequence and older strata are overmature from the general latitude of Point Lay southward.

United States Chukchi Shelf North of Point Hope

Barrow Arch and Arctic Platform

The petroleum potential of the Arctic platform (area 1 in Figure 12) is much like that of the western NPRA, which it adjoins. Offshore seismic data and projection of onshore stratigraphy and thermal maturation data suggest that in area 1 sandstones from the Sadlerochit Group (Permian and Triassic) to the lower part of the Nanushuk Group (Albian and Cenomanian?) may be in contact with adequate source rocks and could contain significant pools of oil and, especially, gas. To date, only some subeconomic or inadequately evaluated gas deposits have been found in the adjacent onshore. Subsurface data (Guldenzopf et al., 1980; and C. E. Kirschner, unpublished data, 1984) suggest, however, that sandstone beds in the Middle Triassic to Lower Cretaceous section, consisting of the Shublik, Sag River, Kingak, and pebble-shale units, which locally attain significant thickness in the western NPRA, may be thicker, more numerous, and perhaps better reservoirs in area 1. If structural traps exist in this area, they have very low amplitude and have not been recognized on our reconnaissance seismic network. Stratigraphic traps caused by permeability gradients or shale oversteps at one of several unconformities in the section possibly occur. If present, such traps could be very large because of the uniform but low regional dip. Eo-Ellesmerian and Lisburne strata in area 1 are probably overmature and at best would contain dry gas. A large but incompletely defined area on the crest of the Barrow arch is underlain by <1 km of post-Franklinian sedimentary rocks, and is therefore considered nonprospective.

The petroleum potential of the pre-Ellesmerian strata that underlie the Barrow arch in the northeastern Chukchi Sea is equivocal. These beds, interpreted by Grantz and May (1983) to be of early Paleozoic and perhaps Precambrian age, are well bedded, slightly deformed, and >10 km thick. Seismic velocities (Vp) measured in the upper part of this section with sonobuoys range from 4.15 to at least 5.35 km/sec, and in the lower part from 5.70 to 7.30 km/sec (Houtz, Eittreim, and Grantz, 1981). These velocities equal or exceed those found in the strongly deformed Ordovician and Silurian argillite and graywacke that underlie the Ellesmerian rocks of the North Slope. The reflections which define these strata beneath the northern Chukchi Sea lose their coherence near the Alaskan coast; they may represent an undeformed equivalent of the Ordovician, Silurian, and underlying older rocks of the North Slope. In any case, the pre-Ellesmerian (and therefore pre-Ellesmerian-orogeny) age of these strata and their high seismic velocities suggest that their petroleum potential is slight.

Hanna Trough

Economically significant oil and gas deposits could be

present in the Hanna Trough (area 2 in Figure 12), because it contains a very thick section of sedimentary rocks. Some large basement-involved folds, locally numerous faults, and rotated fault blocks provide possible oil traps. In addition to the area shown in Figure 12, the trough may extend north of lat. 72°N, but we have no multichannel seismic-reflection data for that area.

If the thermal history of the sedimentary rocks in the Hanna Trough is similar to that found in the Tunalik-1 well, the Eo-Ellesmerian and most of the Ellesmerian section is overmature and could only contain dry gas. Ellesmerian and Brookian strata thicken into the Hanna Trough (Figures 7–11), but the nature of any accompanying facies changes is conjectural. The marked thickening of the Ellesmerian beds into the trough (Figure 8) suggests that it may there be in a basinal facies, perhaps dominated by lutite, and possibly containing turbidite beds. Although these thickened deposits are probably themselves now overmature, they could have generated oil or gas at an earlier time—oil and gas that migrated out of the trough into possible reservoirs on the flanks, where subsequent thermal alteration was less intense.

The presence of westward-increasing amounts of sandstone of north and northwest provenance in the interval between the Shublik and the pebble-shale unit in the western NPRA suggests that the broad crestal region of the present Barrow arch east of the the Hanna Trough was formerly part of a more extensive source land for detrital sediment through the late Ellesmerian. This source land might also have furnished reservoir sands to the upper Ellesmerian section of the Hanna Trough. Most of the upper Ellesmerian rocks are probably in the liquid window in most parts of the Hanna Trough.

The thickening of the lower and upper Brookian sequences into the Hanna Trough (Figures 9–11) may have also created conditions favorable for the generation and entrapment of oil and gas. Extrapolation of thermal conditions and facies patterns from the western NPRA suggests that at least part of the lower Brookian sequence in the Hanna Trough is in the liquid window, and that in places the section probably contains paralic and shallow-marine beds that could include winnowed sands. In addition, the deeper parts of the fill in the upper Brookian Canyon (Figure 11) might also be in the upper part of the liquid window, but we have no thermal data from which to judge. The upper Brookian Canyon fill and the adjacent wall rocks could contain hydrocarbon traps where reservoir sands or sealing shale beds of the fill are in contact with lower Brookian source beds or sands in the canyon walls. Faults and rotated fault blocks, which are locally numerous in the canyon fill and adjacent wall rocks, may have created structural traps in both the lower and upper Brookian beds.

Chukchi Platform

The westward thinning and shallowing of Ellesmerian and Brookian strata in area 3 on the Chukchi platform (Figure 12) suggest that a structural high or former sediment source land lay to the west. If it did, these rocks could contain good reservoir beds analogous to the north-sourced sandstone and conglomerate beds that are productive at Prudhoe Bay and elsewhere on the North Slope. Hydrocarbon fluids could have entered these beds from interlayered source beds or migrated into them from shales downdip in the Hanna Trough. Much of the Ellesmerian and lower Brookian section in the eastern part of area 3 may lie in the liquid window, but extrapolating thermal conditions this far offshore from the western NPRA is speculative at best. If source and reservoir beds are present in area 3, the presence of numerous normal faults, overlaps at unconformities, some folds, and possibly updip pinch-outs may provide traps, but our reconnaissance seismic grid does not demonstrate whether closed structures are present. The eastern and central parts of area 3 are more prospective than the western part, where the total post-Franklinian sedimentary section thins to < 1 km.

Foreland Fold Belt

Extrapolation of thermal maturation conditions from the western NPRA indicates that the top of the liquid window in the foreland fold belt (area 4 in Figure 12) is at the surface, or has been removed by erosion, from near Point Lay southward. North of Point Lay, the top of the window deepens to about 1 km at the latitude of Icy Cape. The vertical height of the liquid window in the Tunalik-1 and Awuna-1 wells of the western NPRA (Magoon and Claypool, unpublished data, 1984) is about 2.0 to 2.5 km. This suggests that most of the Ellesmerian sequence, and in places the lower part of the lower Brookian sequence in the broad structural low off Point Lay, is overmature and now contains only dry gas. All or most of the lower Brookian sequence, on the other hand, appears to lie within the liquid window and could have generated and preserved oil and gas condensate.

Broad folds and fault blocks in Ellesmerian strata below the detachment fault that underlies area 4 offer possible structural traps for gas, but the extrapolation of stratigraphic trends from the western NPRA suggests that adequate reservoirs may be absent. The numerous large folds and thrust folds in the lower Brookian rocks above the detachment fault are probably the most prospective structures in area 4. Thrust faults, however, disrupt the cores of many of the folds in the central and southern parts of area 4 and may reduce their capacity to hold hydrocarbon fluids. The folds may also be largely within the Nanushuk intradelta, where the prospects for finding oil are relatively poorer and where the occurrence of winnowed sandstones of paralic and neritic facies is less likely than in the eastern and north-central NPRA, where the Nanushuk contains several small oil and gas fields.

Herald Arch Overthrust Belt

Brookian rocks in the Herald arch overthrust belt (area 5 in Figure 12) are probably too strongly deformed by listric thrusts and thrust folds, related to the underlying detachment fault zone, to have significant petroleum potential. Extrapolation of onshore data indicates that both the perched Ellesmerian rocks in the overthrust, and those that underlie the detachment fault, are overmature for hydrocarbon liquids in the southeastern part of area 5; the overlying Ellesmerian rocks are also probably too strongly faulted to be prospective. If Ellesmerian rocks are present beneath the detachment fault zone in the northwest part of area 5, where the Brookian sequence is thinner, they may have escaped deep burial and overmaturation and may be comparable in potential to the Ellesmerian rocks of the Chukchi platform (area 3).

North Chukchi Basin

The Brookian sequences of the eastern North Chukchi Basin (area 6 in Figure 12) are believed to be represented by a substantial or dominant component of marine sedimentary rock, and therefore likely contain source beds for oil and gas. Strong reflectors in both sequences further suggest that many sandstone and perhaps conglomerate beds that could serve as reservoirs may also be present. This interpretation is supported by the observed westward thinning of Brookian sedimentary units on the Chukchi platform. A source land in that direction would have been much closer to the North Chukchi Basin than the Brooks Range source land was to the Brookian clastic deposits of the North Slope, and could have resulted in the delivery of coarser detritus to the basin. The numerous listric normal faults and rotated fault blocks that characterize the lower Brookian sequence in the eastern North Chukchi Basin may provide large structural traps. The potential of these traps may be enhanced by the locally thick, early postfaulting infilling sedimentary unit that unconformably overlies the fault blocks. The infilling unit has weaker seismic reflectors than the sequences above and below, and probably contains more shale. It could have served as a seal for hydrocarbon fluids in the underlying faulted rocks, and also could represent a hydrocarbon source. Where the infilling unit is absent, the base of the upper Brookian sequence unconformably overlies the faulted beds, and in some places sealing or reservoir beds may overlie truncated sandstones and shales of the lower sequence. The upper Brookian rocks offer possible traps (1) against small normal faults, (2) in a few very low amplitude folds (closure not demonstrated), (3) against possible updip permeability barriers or pinch-outs, and (4) around shale(?) diapirs.

Assessment of the petroleum potential of the North Chukchi Basin is particularly sensitive to its thermal history. In the absence of data, we have applied the position of the liquid window in the partially analogous Norton Basin (Marlow, Cooper, and Fisher, this volume) to the North Chukchi Basin. Fisher (1982) estimates that in the assumed moderately high heat-flow extensional, epicontinental Norton Basin the liquid window might lie about 2.8 to 5.0 km subsea. In the North Chukchi Basin, a liquid window in this range would intersect the structurally important boundary between the extensively faulted lower Brookian and the little-faulted upper Brookian sequences. If the Norton Basin model or one that assumes a less intense thermal history is realistic, then a substantial part of the faulted lower Brookian section and perhaps the deeper part of the little-faulted upper Brookian section are in the liquid window. However, if the basin had two episodes of crustal thinning and subsidence, as proposed in an earlier section, then most of the lower Brookian sequence and the deeper part of the upper Brookian section may be overmature, and the liquid window might encompass the lower part of the upper sequence and only the structurally highest parts of the lower sequence. The Norton Basin or "cooler" models would thus place substantially more of the very prospective block-faulted lower Brookian sequence in the liquid window than the two-stage model. However, even the latter model places a large volume of lower Brookian rocks at the margins of the basin in the liquid window, and the overmature rocks could contain substantial deposits of dry gas. In addition, some faults cut both the lower and upper Brookian sequences, and these faults might have provided pathways for the migration of hydrocarbon fluids out of the lower sequence into higher reservoirs before the lower sequence became overmature. Updip migration of hydrocarbon fluids in the little-faulted upper sequence might have achieved a similar result.

Nuwuk Basin

A thick clastic sedimentary section, numerous growth faults and roll-over anticlines, and some strong seismic reflectors that may represent sandstone beds suggest that the Nuwuk Basin (area 7 in Figure 12) has good prospects for petroleum. A more complete discussion of the petroleum potential of the basin, which is better known on the Beaufort Shelf, is given in Grantz, May, and Dinter (this volume).

Relative Potential of the Structural Provinces

Ranking the relative petroleum potential of the seven structural provinces of the northern United States Chukchi Shelf from available data is highly speculative. Subsurface information in areas 2, 3, and 6 are especially needed. Nevertheless, despite the logistic and engineering problems posed by the Arctic climate and the polar ice pack, the Arctic platform east of the Hanna tough (area 1), the Hanna Trough (area 2), the Chukchi platform (area 3), the eastern North Chukchi Basin (area 6), and the Nuwuk Basin (area 7) appear to have potential for significant deposits of oil or gas. However, in large parts of these areas, difficult-to-define stratigraphic traps will have to be sought. Exploration in the western NPRA suggests that in at least the eastern part of area 1 the potential may be greater for gas than for oil. The foreland fold belt (area 4) is estimated to have modest potential for significant deposits of oil and gas, and may be gas prone. Structural complexity and estimated thermal conditions suggest that most of the Herald arch overthrust belt (area 5) has a low potential for significant deposits of oil or gas. The northwest part of area 5 may have some potential in Ellesmerian rocks beneath the overthrust faults, but our data are inadequate for assessment.

REFERENCES CITED

Ahlbrandt, T. S. (ed.), 1979, Preliminary geologic, petrologic, and paleontologic results of the study of Nanushuk Group rocks, North Slope, Alaska: U.S. Geological Survey Circular 794, 163 p.

Alaska Geological Society, 1981, West to east stratigraphic correlation section, N.P.R.A.-Prudhoe Bay, Arctic Slope, Alaska: Anchorage, Alaska Geological Society.

Barnes, F. F., 1967, Coal resources of the Cape Lisburne-Colville River region, Alaska: U.S. Geological Survey Bulletin 1242-E, 37 p.

Bird, K. J., 1981, Petroleum exploration of the North Slope in Alaska, U.S.A.: U.S. Geological Survey Open-File Report 81–227, 43 p.

Bird, K. J., 1982, Rock-unit reports of 228 wells drilled on the North Slope, Alaska: U.S. Geological Survey Open-File Report 82–278, 105 p.

Bird, K. J., and J. Andrews, 1979, Subsurface studies of the Nanushuk Group, North Slope, Alaska, *in* T. S. Ahlbrandt, ed., Preliminary geologic, petrologic, and paleontologic results of the study of Nanushuk Group rocks, North Slope, Alaska: U.S. Geological Survey Circular 794, p. 32–41.

Carlson, P. R., and H. A. Karl, 1984, Discovery of two new large submarine canyons in the Bering Sea: Marine Geology, v. 56, p. 159–179.

Carter, R. D., C. G. Mull, K. J. Bird, and R. B. Powers, 1977, The petroleum geology and hydrocarbon potential of Naval Petroleum Reserve No. 4, North Slope, Alaska: U.S. Geological Survey Open-File Report 77-475, 61 p.
Chapman, R. M., and E. C. Sable, 1960, Geology of the Utukok-Corwin region, northwestern Alaska: U.S. Geological Survey Professional Paper 303–C, p. 47–167.
Conwell, C. N., and D. M. Triplehorn, 1976, High-quality coal near Point Hope, northwestern Alaska, *in* Short notes on Alaskan geology—1976: Alaska Division of Geological and Geophysical Surveys Geologic Report 51, p. 31–35.
Fisher, M. A., 1982, Petroleum geology of Norton Basin, Alaska: AAPG Bulletin, v. 66, p. 286–301.
Grantz, A., and S. D. May, 1983, Rifting history and structural development of the continental margin north of Alaska, *in* J. S. Watkins and C. L. Drake, eds., Studies in continental margin geology: AAPG Memoir 34, p. 77–100.
Grantz, A., S. Eittreim, and O. T. Whitney, 1981, Geology and physiography of the continental margin north of Alaska and implications for origin of the Canada Basin, *in* A. E. M. Nairn, M. Churkin, Jr., and F. G. Stehli, eds., The ocean basins and margins: New York, Plenum Press, v. 5, p. 439–492.
Grantz, A., M. L. Holmes, and B. A. Kososki, 1975, Geologic framework of the Alaskan continental terrace in the Chukchi and Beaufort Seas, *in* C. J. Yorath, E. R. Parker, and D. J. Glass, eds., Canada's continental margins and offshore petroleum exploration: Canadian Society of Petroleum Geologists Memoir 4, p. 669–700.
Grantz, A., S. D. May, and D. A. Dinter, 1987, Regional geology and petroleum potential of the United Sates Beaufort and northeasternmost Chuckchi Seas, this volume.
Guldenzopf, E. C., M. B. Orlovsky, D. A. Higgs, C. G. Freytag, and E. R. Ovalle, 1980, National Petroleum Reserve, Alaska; Summary geological report fiscal year 1980: Tetra Tech, Inc., prepared for Husky Oil NPR Operations, Inc., under contract to U.S. Geological Survey, 168 p.
Heroux, Y., A. Chagnon, and R. Bertrand, 1979, Compilation and correlation of major thermal maturation indicators: AAPG Bulletin, v. 63, p. 2128–2144.
Houtz, R. E., S. Eittreim, and A. Grantz, 1981, Acoustic properties of northern Alaskan shelves in relation to the regional geology: Journal of Geophysical Research, v. 86, no. B5, p. 3932–3943.
Lantz, R. J., 1981, Barrow gas fields—North Slope, Alaska: Oil and Gas Journal, v. 79, n. 13, p. 197–200.
Lathram, E. H., 1965, Preliminary geological map of northern Alaska: U. S. Geological Survey Open-File Report 65-96, 2 sheets, scale 1:1,000,000.
Magoon, L. B., and G. E. Claypool, 1979, Hydrocarbon source potential of the Nanushuk Group and the Torok Formation, a preliminary report, *in* T. S. Ahlbrandt, ed., Preliminary geologic, petrologic, and paleontologic results of the study of Nanushuk Group rocks, North Slope, Alaska: U.S. Geological Survey Circular 794, p. 54–60.
Magoon, L. B., and G. E. Claypool, 1982, NPRA Inigok No. 1—Use of Lopatin's method to reconstruct thermal maturity: Pacific Petroleum Geologist Newsletter, AAPG, Pacific Section, p. 4–6.
Marlow, M. S., A. K. Cooper, and M. A. Fisher, 1987, Petroleum geology of the Beringian continental shelf, this volume.
Martin, A. J., 1970, Structure and tectonic history of the western Brooks Range, De Long Mountains and Lisburne Hills, northern Alaska: GSA Bulletin, v. 81, p. 3605–3622.
McKinnery, C. M., E. L. Garton, and F. G. Schwartz, 1959, Analyses of some crude oils from Alaska: U.S. Bureau of Mines Report of Investigations 5447, 29 p.
Miller, C. C., J. E. Pickard, E. C. Taber, and R. B. Rogers, 1979, National Petroleum Reserve, Alaska; Summary geophysical report fiscal year 1979: Tetra Tech., Inc., prepared for Husky Oil NPR Operations, Inc., under contract to U.S. Geological Survey, 47 p., 13 illus., 12 maps.
Molenaar, C. M., 1981, Depositional history and seismic stratigraphy of Lower Cretaceous rocks, National Petroleum Reserve in Alaska and adjacent areas: U.S. Geological Survey Open-File Report 81-1084, 42 p.
Montadert, L., D. G. Roberts, O. De Charpal, and P. Guennoe, 1979, Rifting and subsidence of the northern continental margin of the Bay of Biscay, *in* Initial reports of the Deep Sea Drilling Project, Vol. 48: Washington D. C., U.S. Government Printing Office, p. 1025–1060.
National Petroleum Council, 1981, U.S. Arctic oil and gas: Washington, D.C., National Petroleum Council, 286 p.
Seifert, E. K., J. M. Moldowan, and R. W. Jones, 1979, Application of biological marker chemistry to petroleum exploration, *in* World Petroleum Congress, 10th, Bucharest, 1979, Special Paper: Philadelphia, Heyden and Sons, Limited, p. 425–440.
Tetra Tech, Inc., 1982, Petroleum exploration of NPRA, 1974–1981: Houston, Texas, Tetra Tech, Inc., Energy Management Division, Report No. 8200; 3 vols., 26 tables, 132 figures, 71 seismic maps. (Prepared for Husky Oil NPRA Operations, Inc., Anchorage, Alaska, under U.S. Geological Survey-Husky Contract No. 14-08-0001-16474.)

STRUCTURE AND STRATIGRAPHY OF THE HOPE BASIN, SOUTHERN CHUKCHI SEA, ALASKA

Ralph B. Tolson

Stanford University
Stanford, California

ABSTRACT

Multichannel seismic-reflection data from the Hope Basin of the southern Chukchi Sea, Alaska, indicate as much as 5.8 km of probable Eocene and younger sedimentary rock. The basin evolved in three main stages. First, isolated half-grabens formed during the early Paleogene in response to transtensional movement of southeast–northwest-trending faults. These fractures were related to possible right-lateral slip on the Kobuk fault zone. The half-grabens filled with nonmarine and possibly marine or lacustrine sedimentary rocks. Second, during the early Miocene the basin thermally subsided and nonmarine sediment was deposited over the entire basin. Third, an extensional phase characterized by reactivation of the Eocene faults and initiation of numerous normal faults commenced during the middle to late Miocene. Extension was coeval with deposition of nonmarine and marine sediment in basin lows not necessarily coincident with the Eocene half-grabens.

Hydrocarbon potential of the Hope Basin remains low, primarily because of the abundance in the Kotzebue Sound area of thermally immature deposits that are nonmarine and gas prone. However, thermally mature marine or lacustrine sedimentary rocks may be present in some of the deeper, untested parts of the basin.

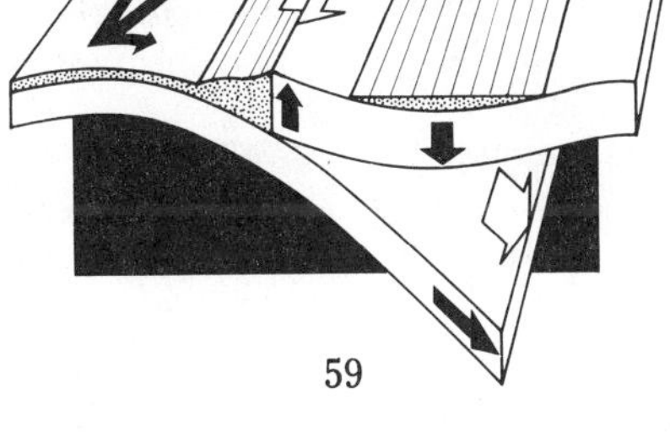

INTRODUCTION

Hope Basin is an elongate, roughly southeast–northwest-trending prism of sediment that underlies more than 75,000 km^2 of the southern and central Chukchi Sea off northwest Alaska. It is bordered on the south by the Seward Peninsula and the Bering Strait, and to the west by northeastern Siberia; to the north and east is the northwest coast of Alaska and Herald arch, a basement high trending northwest from Cape Lisburne (Figure 1). The sea floor of the Hope Basin is quite flat, averaging 45–55 m below sea level across the entire shelf (Figure 2). The basin is one of the few loci of late Tertiary subsidence and deposition in northern Alaska, and the structural contour maps and seismic data presented here provide the first description of the basin based on regional multichannel seismic-reflection data.

REGIONAL GEOLOGY

The Hope Basin is an extensional depression in pre-Upper Cretaceous rocks. Where exposed in the Lisburne and Seward Peninsulas these rocks include strongly deformed Lower Cretaceous flyschoid clastic rocks, Jurassic to Ordovician carbonate, chert, shale, and both flyschoid and paralic clastic rocks, and lower Paleozoic and older metamorphic rocks (Campbell, 1967; Sainsbury, 1975; Hudson, 1977; American Stratigraphic Co., 1977a, 1977b). Paleozoic and lower Mesozoic rocks of the DeLong Mountains, Lisburne Hills, and Herald arch lie to the north; Lower and lower Upper Cretaceous rocks of the northern Yukon-Koyukuk province crop out to the east; and Precambrian and Paleozoic rocks underlie the Seward and Chukotsk Peninsulas to the south and southwest (Figure 2).

Prominent northeast–southwest-trending folds and faults in the DeLong Mountains of the western Brooks Range are not reflected in the basement surface beneath the Hope Basin (Figure 2). To the southeast, however, a positive gravity anomaly centered along the Igichuk Hills in the southwestern Baird Mountains runs east–west from long. 160°W to Cape Krusenstern at almost long. 164°W (Figure 2; Barnes, 1977). Trending west across the Hope Basin from Cape Krusenstern almost to the Siberian coast is a positive gravity anomaly (May, 1985) that appears to be the offshore extension of the Igichuk Hills gravity high (Eittreim, Grantz, and Whitney, 1979). This gravity anomaly west of Cape Krusenstern coincides with the Kotzebue arch, the most prominent basement high in the basin (Figure 2; Grantz, Holmes, and Kososki, 1975). To the southeast of the Igichuk Hills, and extending along the south side of the Brooks Range lies the Kobuk fault zone (Figure 1). In the south-central Brooks Range, the Malamute and South Fork fault systems project westward into the Kobuk fault zone and show evidence for > 90 km of Albian–Cenomanian or younger right-lateral displacement (Dillon, 1985). East of Kotzebue Sound and south of the western Brooks Range, the Kobuk fault zone exhibits fault-related geomorphic features along the Kobuk River Valley floor (Hamilton, 1984). However, because of extensive Quaternary cover and absence of identified Tertiary sediment along the fault zone, no evidence of large-scale Tertiary movement can be seen. Perhaps faults mapped in Upper Cretaceous rocks just east of the Kobuk River delta (Patton and Miller, 1968) represent a western extension of the fault zone. To the south of Kotzebue Sound, a north–south-trending vertical fault zone extends along the eastern Seward Peninsula (Figure 1; Till, 1984). A Cretaceous pluton parallel to the zone and probable Tertiary sediment in the zone suggest that it may be a long-lived, significant structure, but it does not trace a simple suture between the Seward Peninsula and Yukon-Koyukuk geologic provinces (Till, 1984).

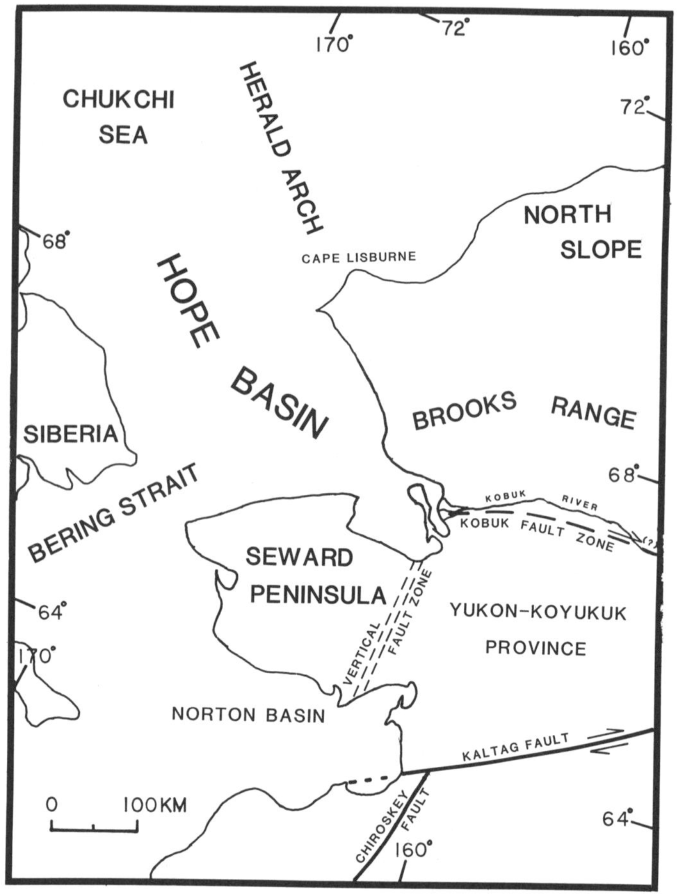

Figure 1. Geographic setting of northwestern Alaska.

The ages of sedimentary rocks that fill the Hope Basin can be inferred from isolated Tertiary and Upper Cretaceous sedimentary outcrops around Kotzebue Sound (Ellersieck et al., 1978; Grantz, Holmes, and Kososki, 1975; Hopkins and MacNeil, 1960; Hopkins, 1963; Hopkins et al., 1971; Patton, 1973; Patton and Miller, 1968; Sainsbury, 1975). Additional evidence comes from the age of strata penetrated in two Standard Oil of California (SOCAL) wells on the perimeter of the basin—the Cape Espenberg #1, located in the north-central part of the Seward Peninsula, and the Nimiuk Pt. #1, located on the eastern side of the Baldwin Peninsula (Figure 2; Alaska Oil and Gas Conservation Commission, 1981a, 1981b). Most of the Tertiary sedimentary rocks exposed around Kotzebue Sound are sandstone, siltstone, and coal, which are mostly nonmarine. Extensive tholeiitic and alkalic basalt flows of Tertiary age crop out on the Seward Peninsula (Hopkins, 1963). These volcanic rocks were erupted in two episodes, one 18–26 Ma (million years ago) and the other

Figure 2. Generalized regional geologic structure map around Hope Basin (modified from Grantz, Eittreim, and Whitney, 1981).

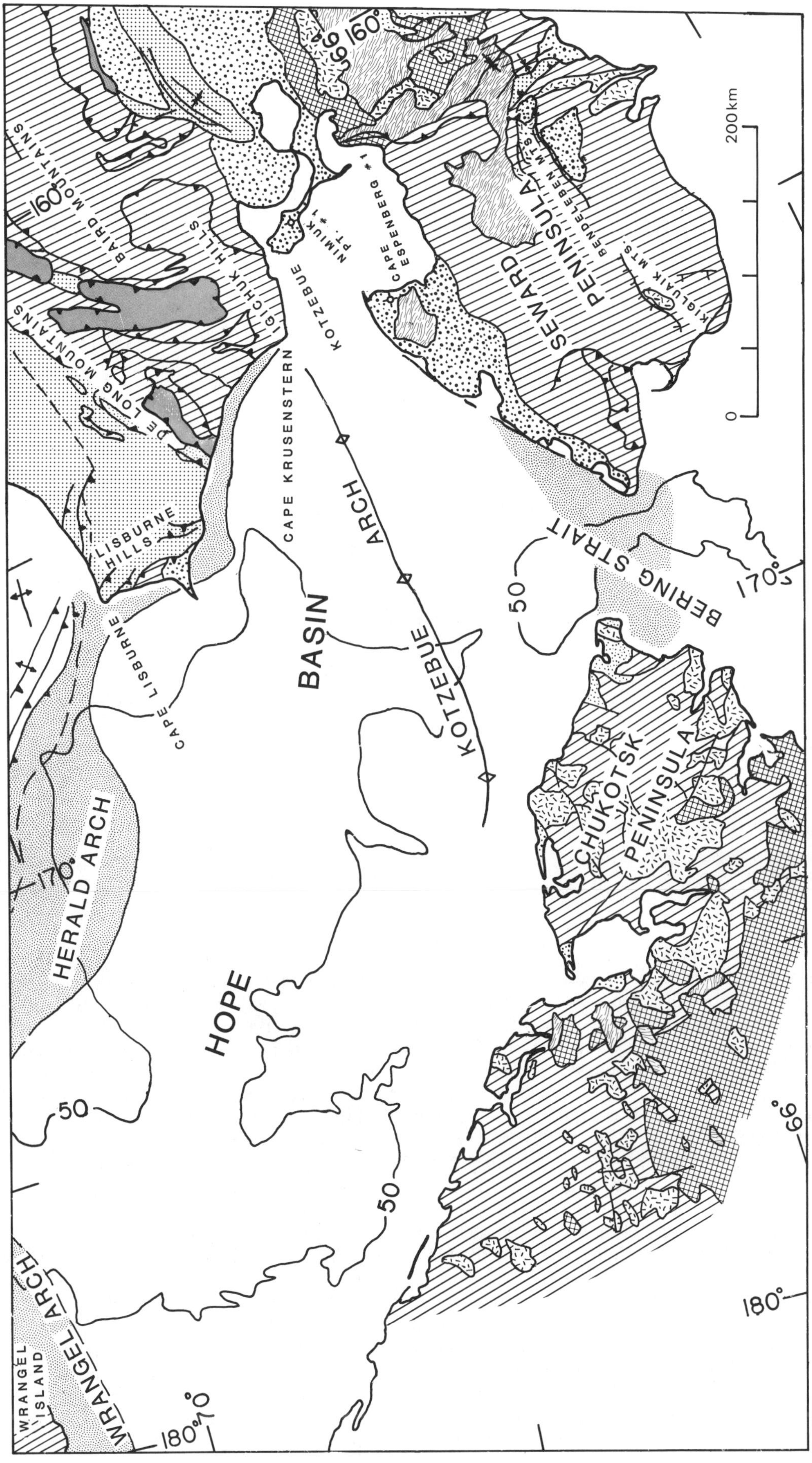
WRANGEL ISLAND
WRANGEL ARCH
HERALD ARCH
HOPE BASIN
KOTZEBUE ARCH
CAPE LISBURNE
LISBURNE HILLS
DE LONG MOUNTAINS
BAIRD MOUNTAINS
IGICHUK HILLS
CAPE KRUSENSTERN
KOTZEBUE
PT. NIMIUK #1
CAPE ESPENBERG #1
SEWARD PENINSULA
BENDELEBEN MTS.
KIGLUAIK MTS.
BERING STRAIT
CHUKOTSK PENINSULA
50
160°
170°
180°
66°
70°
0
200 km
Anticline
Arch
Thrust fault, sawteeth in upper plate
Syncline
Isobath, meters
Normal fault, hachures on down-thrown side
PRE-JURASSIC sedimentary and volcanic rocks, generally strongly deformed or metamorphosed
CRETACEOUS to UPPER PALEOZOIC plutonic rocks
JURASSIC to PERMIAN ophiolites and ultramafics
Acoustic basement-strongly deformed or metamorphosed CRETACEOUS to PRECAMBRIAN rocks
QUATERNARY and TERTIARY clastic sedimentary rocks, marine and nonmarine
QUATERNARY and TERTIARY volcanic rocks
CRETACEOUS clastic sedimentary rocks, marine and nonmarine
CRETACEOUS volcanic rocks-andesitic in Alaska, acidic in Chukotka, U.S.S.R.

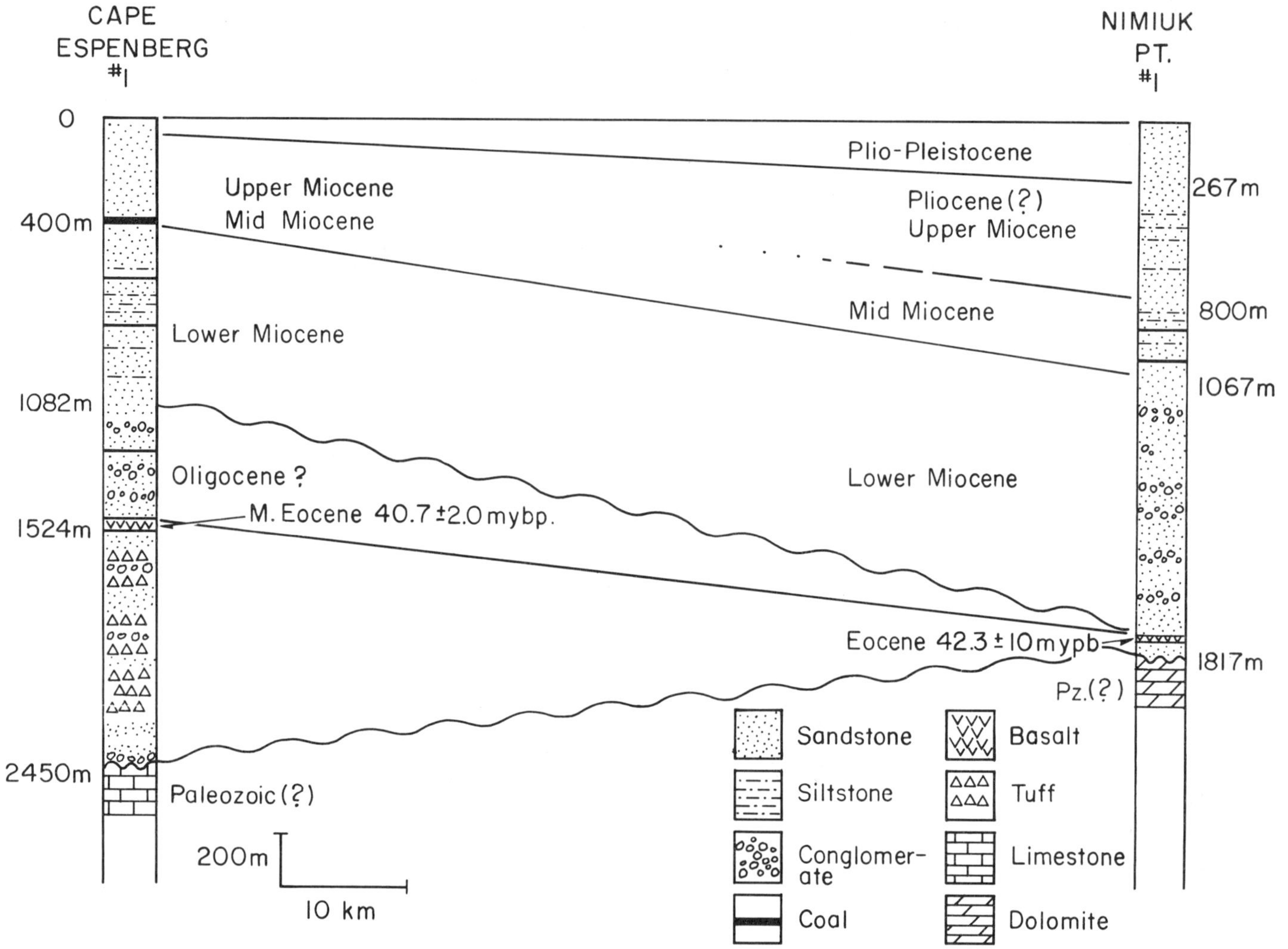

Figure 3. Lithologic characteristics and stratigraphic correlation in the SOCAL wells in the Hope Basin (after American Stratigraphic Company, 1977a, 1977b; Alaska Oil and Gas Conservation Commission, 1981a, 1981b).

from 5.8 Ma through the Holocene (Swanson et al., 1981). Four hundred kilometers southeast of the Hope Basin, in the western Yukon-Koyukuk Basin, the presence of Paleocene to middle Eocene subaerial basalt and andesite flows, rhyolite domes, and tuff characterizes the widespread early Tertiary calc-alkalic volcanic province of western and central Alaska (Moll and Patton, 1984).

Lithologic descriptions and ages in the SOCAL wells are summarized in Figure 3. The Eocene ages reported for basalt in both wells were isotopically determined by consultant Roger Denison of Dallas, Texas, and released to the State of Alaska in 1982. In the Cape Espenberg #1 well, the basalt overlies approximately 850 m of volcanic tuff and 50 m of volcanic conglomerate, which in turn rests on crystalline limestone. In the Nimiuk Pt. #1 well the basalt lies on dolomite. Thus, the oldest unmetamorphosed bedded rocks in the Cape Espenberg #1 well are Eocene or older volcaniclastic beds (Figure 3). Sedimentary rocks in both wells are predominantly nonmarine(?) Miocene sandstone, siltstone and coal (American Stratigraphic Company, 1977a, 1977b; Alaska Oil and Gas Conservation Commission, 1981a, 1981b).

Tectonism was widespread throughout western Alaska during the Late Cretaceous–Early Tertiary, and some of the regional tectonic events provide a context for interpreting the geology of the Hope Basin. The Late Cretaceous–Tertiary evolution of the Bering Sea Shelf to the south of the Hope Basin involved the cessation of arc volcanism and the development of pull-apart basins along the Beringian continental margin during the Eocene (Marlow and Cooper, 1984; Marlow, Cooper, and Fisher, this volume; Cooper, Marlow, and Scholl, this volume). This tectonic transition coincided with a change from strong convergence to transform motion between the North American and Eurasian plates (Harbert et al., 1985), a decrease in the northward velocity of the Pacific plate, and reduction of magmatism in the Aleutian Arc (Scholl, Vallier, and Stevenson, this volume). Norton Basin, to the south of the Seward Peninsula, developed in two stages along a right-stepping offset of the right-lateral, northeast-trending Kaltag fault. It consists of mostly local, fault-controlled subbasins formed in the Paleocene and Eocene, succeeded by broad thermal downwarping of the entire basin from the Miocene to present (Marlow, Cooper, and Fisher, this volume).

SEISMIC STUDIES

Data Base

Published studies of the Hope Basin are limited to interpretations based primarily on single-channel and high-resolution seismic-reflection data (Grantz et al., 1970; Grantz, Holmes, and Kososki, 1975; Holmes, 1975; Eittreim, Grantz, and Whitney, 1978), and data from the two wells drilled by SOCAL to the west and east sides of the Kotzebue Sound (Eittreim, Grantz, and Whitney, 1979). These studies identified the major structural features and principal depositional sites in the basin. The principal data set used in this study consists of approximately 3000 km of multichannel seismic data collected aboard the U.S. Geological Survey R/V S. P. Lee during the summers of 1977, 1978, and 1980. The 1977 multichannel data have been released (Grantz, Mann, and May, 1982). The reflection data were collected using a 24-channel streamer and an air-gun source.

Seismic Sequences

In the text and figures, seismic section depths are expressed in seconds of two-way seismic reflection traveltime and/or kilometers below sea level. Conversions from time to depth are made by using the time-depth curve constructed from stacking-velocity-derived seismic reflection interval-velocities (Figure 4). Because of the uniformly shallow water depths across the area, time measurements are made from sea level.

Figure 5 is a structural contour map of the top of acoustic basement in the Hope Basin, expressed in traveltime. Acoustic basement consists of the pre-Upper Cretaceous rocks that crop out in the surrounding geologic provinces. The thickest accumulation of sedimentary rock in the Hope Basin is located approximately 25 km west of Point Hope, where 3.8 sec (5.8 km) of sedimentary rock is located in a southward-tilted half-graben. Other areas of thick sedimentary fill lie farther south. One is in a north-tilted half-graben approximately 50 km southwest of Point Hope, and the other is in a west–northwest-trending trough originating 75 km south of Point Thompson. Each of these depocenters contains as much as 3.4 sec (5.1 km) of sedimentary rock. Also, sedimentary rock thickness increases to >2.6 sec (3.8 km) near the middle of Kotzebue Sound, indicating the presence of another depocenter here. This subbasin is referred to as the Kotzebue Basin. Gravity data indicate that the eastward limit of this east–west-trending basin lies under the eastern Kobuk River delta (Figure 5; Barnes, 1977).

In most of the Hope Basin, the sedimentary prism is divided into three seismic sequences or units by distinctive regional unconformities (and their correlative conformities) and highly contrasting reflection characteristics (see, for example, Mitchum, Vail, and Thompson, 1977). Other minor unconformities can be identified as well, but they cannot be regionally correlated. The lowermost sequence (Unit I) rests on acoustic basement, which has an irregular surface that can be difficult to discern at deeper structural levels (Figure 6). Unit I fills basement lows and laps onto basement highs (Figure 7). It is difficult to identify where it is thin and drapes the top of basement. Sequence thickness ranges from zero to >0.8 sec (0.7 km). Reflectors are commonly discontinuous and chaotic, but are coherent and parallel where they lap

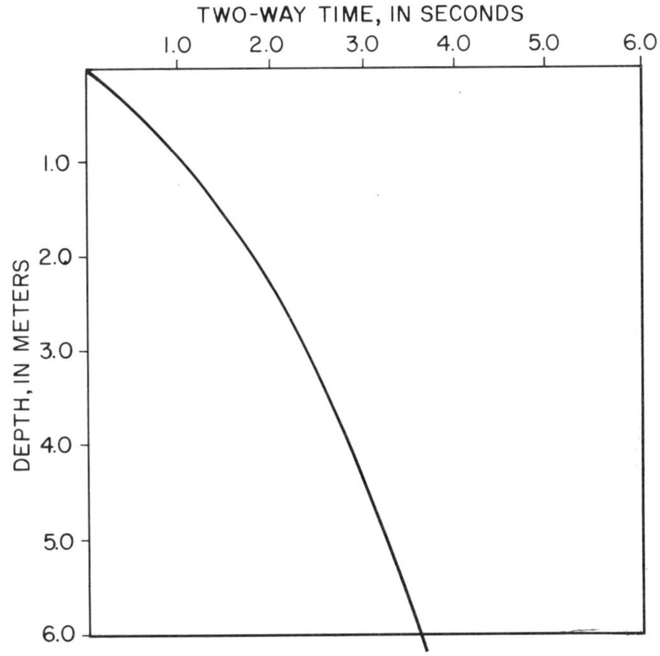

Figure 4. Generalized seismic-reflection time as a function of subsurface depth for the Hope Basin, derived from stacking-velocity measurements.

onto basement highs. Locally, reflectors of Unit I are in angular discordance with the overlying sequence. In most places the discordance is structurally induced, but in some areas it appears to be depositional and exhibits a sigmoidal prograding geometry (Figures 7, 8).

Unit II overlaps all of Unit I and the intrabasinal basement highs (Figure 6). Except at the basin margins, where it laps onto basement, the unit is 0.5 sec to 1.0 sec thick (0.4 to 0.9 km, respectively). Most reflectors in Unit II are concordant, except where they are structurally disturbed. In the lower part of this unit reflections are rather discontinuous and reflection amplitudes differ markedly. However, reflections in the upper half of the unit are fairly coherent and are typically concordant. Where the sequence is quite thick, the lower reflectors are more coherent and continuous.

The uppermost sequence is termed Unit III (Figure 7). Seismic reflections in this unit are generally discontinuous and weak just above Unit II. Reflections become stronger and more continuous upsection, but are masked by sea-floor multiples near the top of the unit. Reflections are concordant throughout most of the basin but onlap Unit II at some localities. The thickness of Unit III ranges from zero at the basin margin to 2.0 sec (2.1 km) in an east–west-trending trough whose center is approximately 75 km south of Point Hope (Figure 9).

Age and Depositional Environments of Seismic Units

Ages and depositional environments of strata represented by seismic-stratigraphic Units I, II, and III can be grossly defined using bore-hole information, geometric characteristics of the seismic reflections, and overall basin geometry. However, precise determinations remain impossible, because the Cape Espenberg #1 well is about 30 km south of the nearest seismic line used in this study (line 2019; Figures

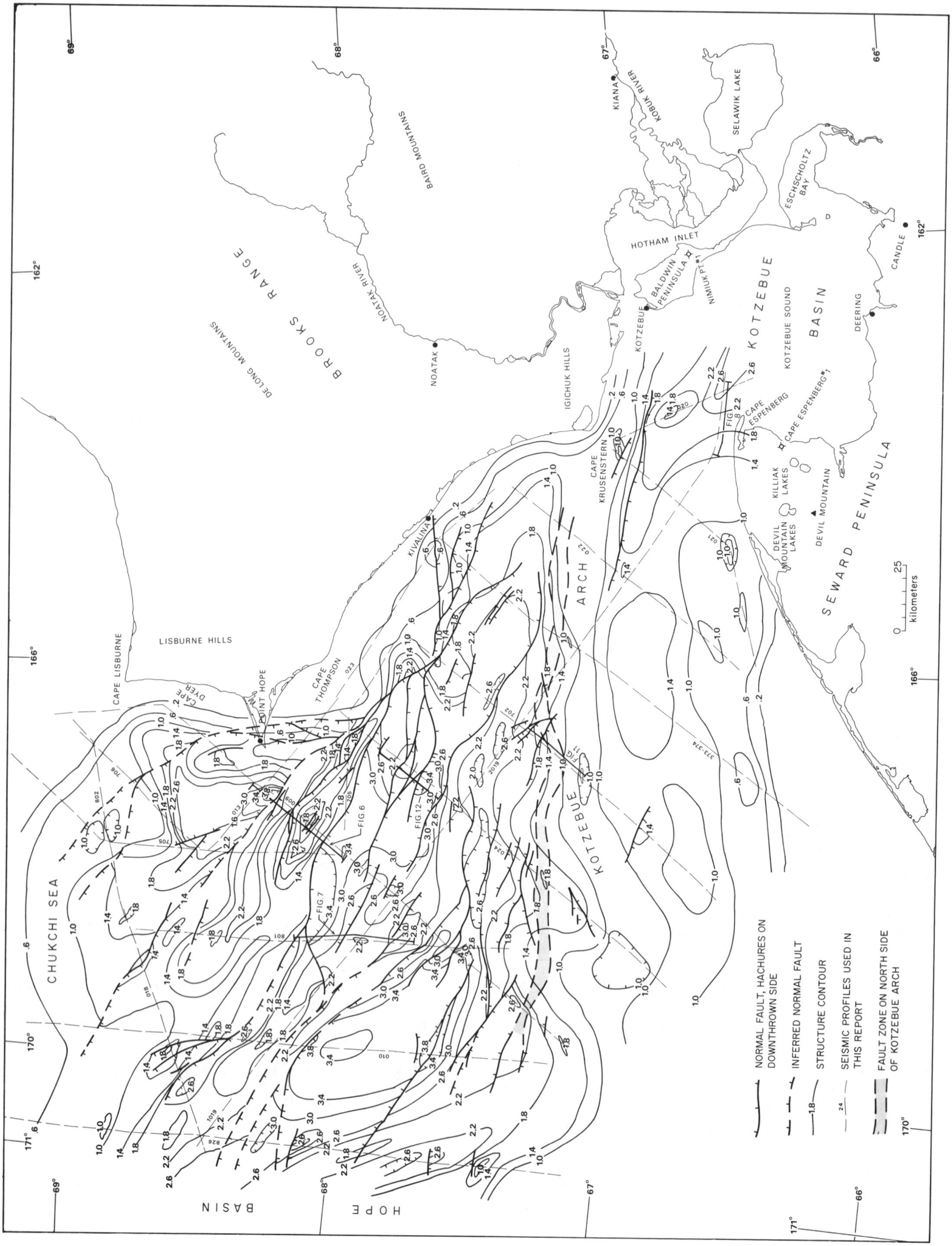
NORMAL FAULT, HACHURES ON DOWNTHROWN SIDE
INFERRED NORMAL FAULT
STRUCTURE CONTOUR
SEISMIC PROFILES USED IN THIS REPORT
FAULT ZONE ON NORTH SIDE OF KOTZEBUE ARCH
kilometers
CHUKCHI SEA
HOPE BASIN
KOTZEBUE ARCH
KOTZEBUE BASIN
SEWARD PENINSULA
BROOKS RANGE
DE LONG MOUNTAINS
BAIRD MOUNTAINS
LISBURNE HILLS
CAPE LISBURNE
CAPE DYER
POINT HOPE
CAPE THOMPSON
KIVALINA
NOATAK
NOATAK RIVER
KOBUK RIVER
KIANA
SELAWIK LAKE
ESCHSCHOLTZ BAY
HOTHAM INLET
BALDWIN PENINSULA
KOTZEBUE
NIMIUK PT. #1
KOTZEBUE SOUND
CANDLE
DEERING
IGICHUK HILLS
CAPE KRUSENSTERN
CAPE ESPENBERG
CAPE ESPENBERG #1
KILLIAK LAKES
DEVIL MOUNTAIN
DEVIL MOUNTAIN LAKES

Figure 5. Structure contour map of the top of acoustic basement in the Hope Basin.

5, 9). Despite the lack of a direct tie between seismic and well data, ages can be inferred for seismic Units I, II, and III by making some simple assumptions. Biostratigraphic data from the Cape Espenberg #1 well can be used in chronostratigraphic zonation of the section penetrated by the bore hole (Figure 3; Alaska Oil and Gas Conservation Commission, 1981a). The time-depth curve (Figure 4) and a knowledge of drill depths to the biostratigraphically controlled chronostratigraphic boundaries in the Cape Espenberg #1 well can be used to determine seismic traveltimes to those chronostratigraphic boundaries. Then, by making a projection along structural strike from the well to line 2019 (Figures 5, 9), we can tentatively assign geologic ages to the seismic sequences defined on line 2019 (Figure 8). The calculated traveltimes to the chronostratigraphic divisions of the Cape Espenberg #1 well agree reasonably well with the traveltimes to the unconformities separating the seismic sequences on line 2019. For example, the depth to the chronostratigraphic division between the middle Miocene–late Miocene and early Miocene boundary is around 400 m below sea level. This corresponds to a traveltime of about 0.5 sec (Figure 4), which agrees fairly well with a contoured traveltime of 0.5 to 0.6 sec on the structure map of the top of Unit II (Figure 9). Therefore, Unit III is probably middle to late Miocene in age, and Unit II early Miocene in age. The depositional position of Unit I between Unit II and basement indicates a Paleogene, possibly Oligocene and Eocene, age when compared with the bore-hole chronostratigraphic units. Figure 10 summarizes the interpreted correlation between chronostratigraphic intervals in the Cape Espenberg #1 well and the three seismic sequences.

Although correlation between chronostratigraphic divisions in the Cape Espenberg #1 well and seismic sequences of line 2019 can be made, as outlined above, by following the structural strike of the Kotzebue Basin, it is uncertain if the Cape Espenberg #1 well can be correlated with line 2019 along depositional strike. However, a reconnaissance analysis of the seismic sequences yields valuable information such as reflection configuration, amplitude, and continuity that are useful in determining the possible depositional environments of the seismic sequences. The most recognizable seismic facies in Unit I near the Cape Espenberg #1 well is the group of southeastward-dipping, sigmoid, clinoform reflections seen on line 2019 (Figure 8). This progradational configuration is typical of a prograding shelf or platform (Brown

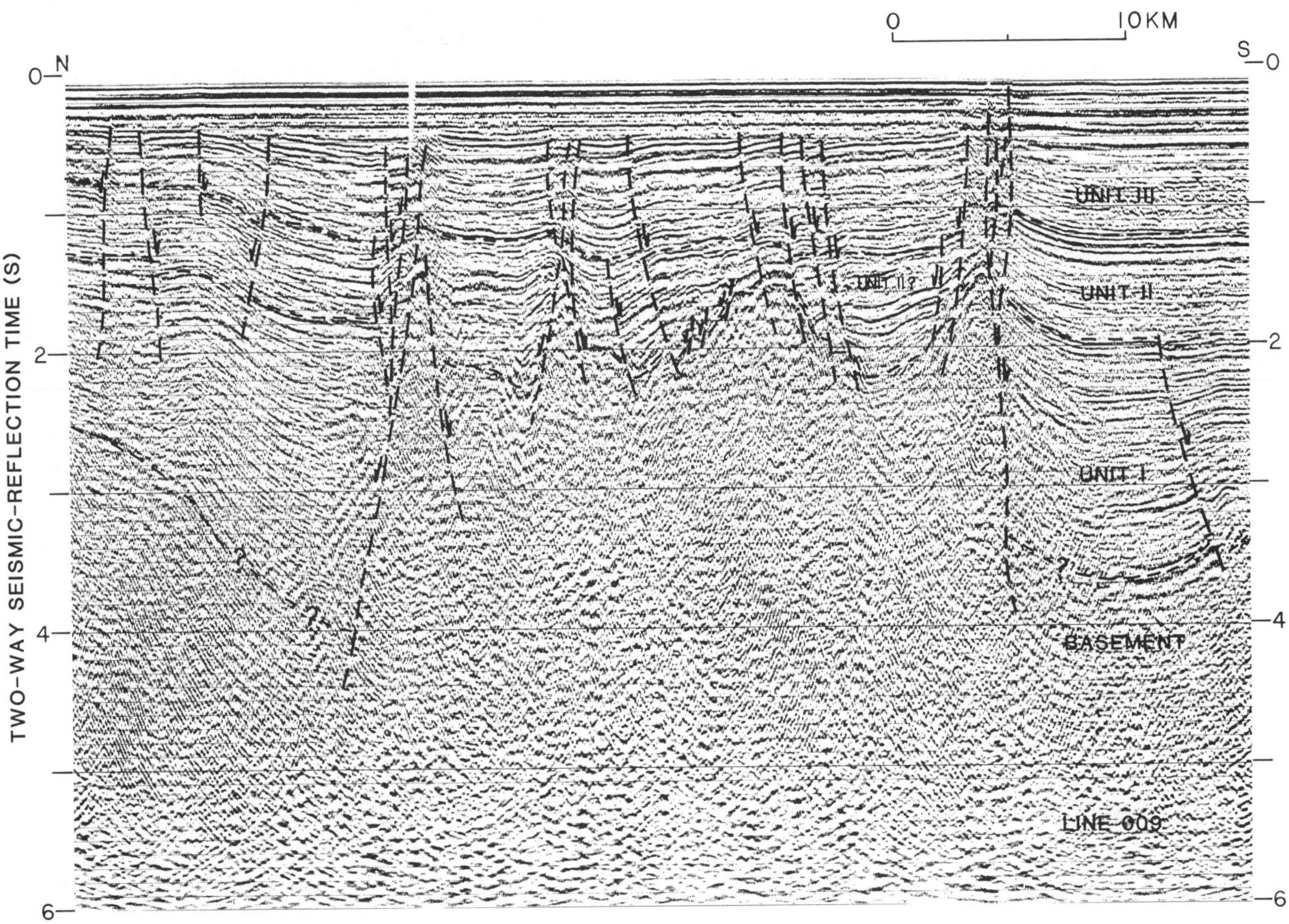

Figure 6. Migrated CDP seismic-reflection profile 009, a line showing large offsets at transtensional faults bordering half-grabens. See Figures 5 and 9 for location of profile.

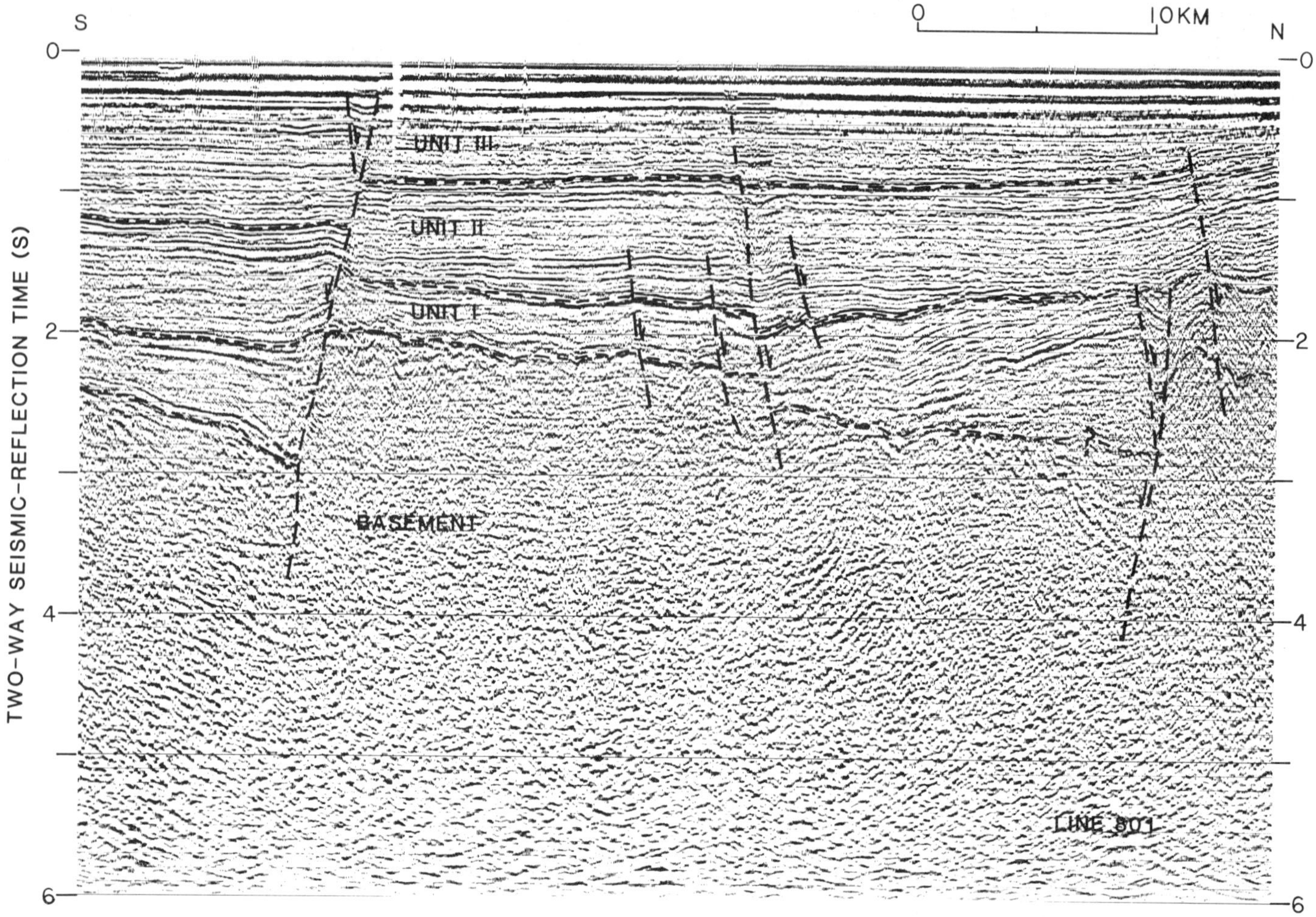

Figure 7. Migrated CDP seismic-reflection profile 801, a line showing southward prograding clinoforms of Unit I. See Figures 5 and 9 for location of profile.

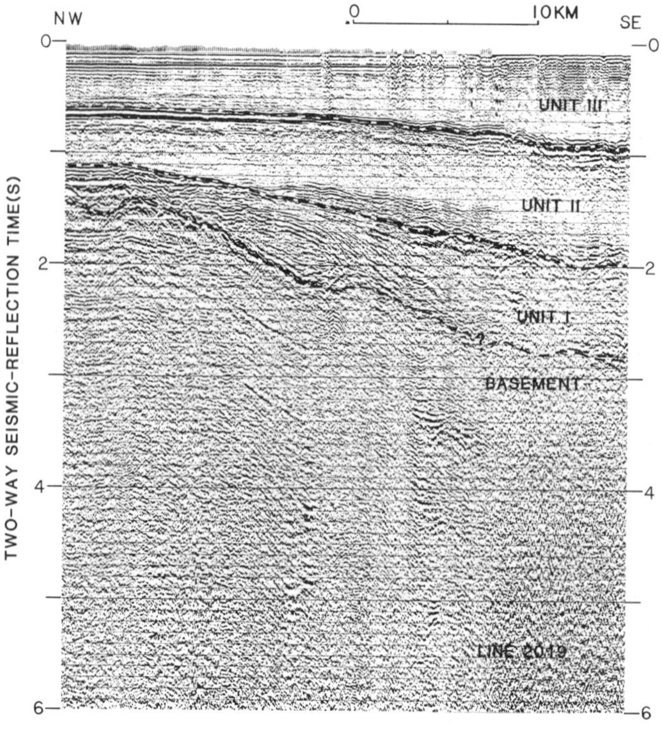

Figure 8. Unmigrated CDP seismic-reflection profile 2019, a dip line showing southeastward prograding clinoforms of Unit I. See Figures 5 and 9 for location of profile.

and Fisher, 1979). Farther east on line 2019 (Figure 8), the sequence changes downdip into low-amplitude parallel reflectors typical of a basin plain (Brown and Fisher, 1979). In the Cape Espenberg #1 well, Unit I consists mainly of volcanic tuff with some interbedded sandstone and conglomerate (American Stratigraphic Co., 1977a) and contains no reported marine microfossils (Alaska Oil and Gas Conservation Commission, 1981a). Thus, these rocks are interpreted as lithologies present in a marginal marine to nonmarine seismic facies seen updip to the west from the progradational seismic facies on line 2019 (Figure 8). Other progradational configurations can be seen in Unit I in the northwestern part of the Hope Basin (line 801; Figure 7).

Unit II in the Cape Espenberg #1 well is predominantly sandstone and siltstone with some interbedded woody coal (American Stratigraphic Co., 1977a). Few palynomorphs, and no diagnostic marine palynomorphs, were found in the Cape Espenberg well samples (Alaska Oil and Gas Conservation Commission, 1981a). The lack of any other reported marine microfossils in the Cape Espenberg #1 well, and the variable seismic characteristics of the seismic sequence in most of the data, suggest that Unit II represents an alluvial plain or paralic deposit. However, reflectors of this unit are more continuous to the northwest, indicating that the unit is possibly marine in that area.

Figure 9. Structure contour map of the top of seismic Unit II in the Hope Basin.

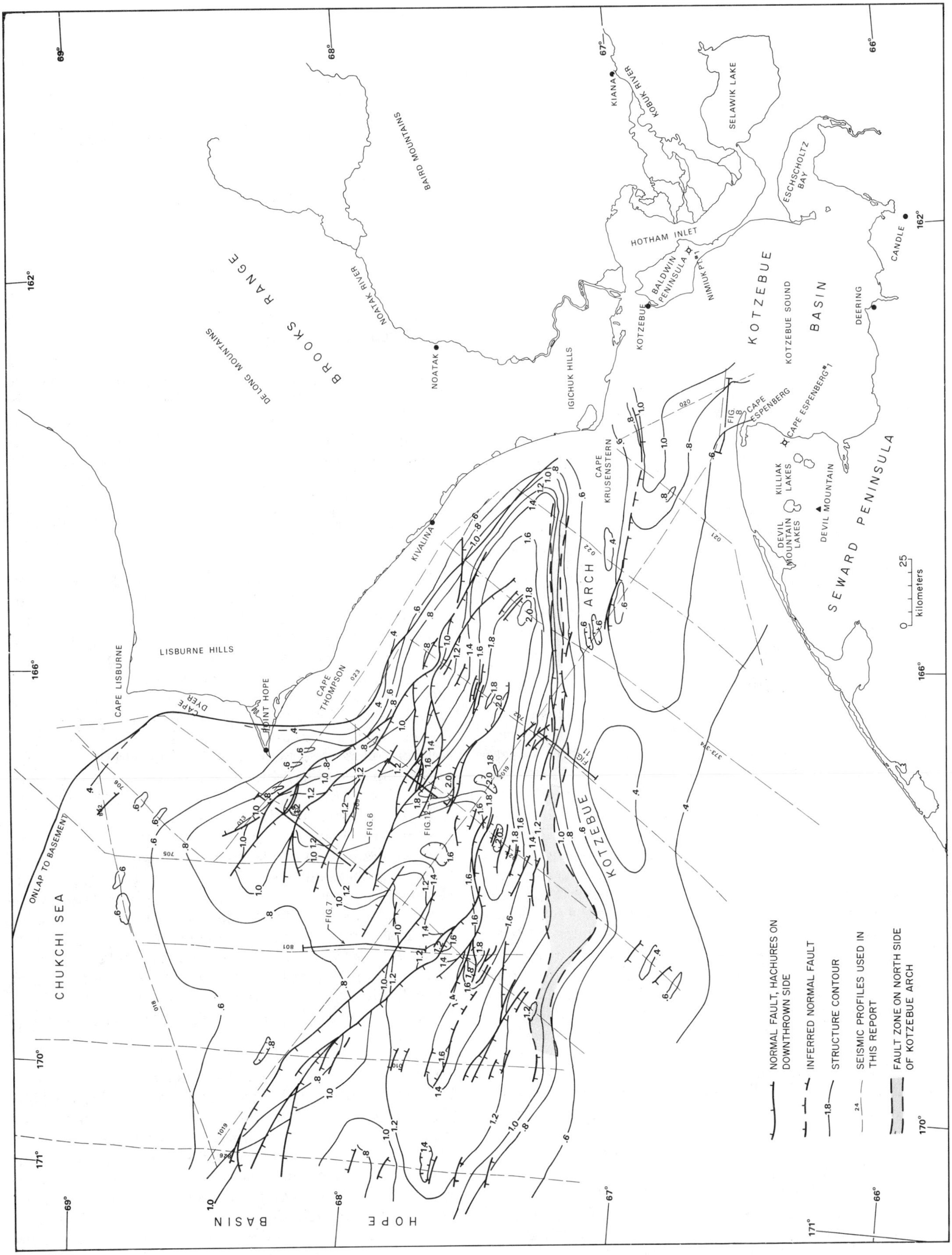
BROOKS RANGE
DE LONG MOUNTAINS
BAIRD MOUNTAINS
NOATAK RIVER
NOATAK
KIANA
KOBUK RIVER
SELAWIK LAKE
ESCHSCHOLTZ BAY
HOTHAM INLET
BALDWIN PENINSULA
NIMIUK PT. #1
KOTZEBUE
CANDLE
KOTZEBUE BASIN
KOTZEBUE SOUND
DEERING
IGICHUK HILLS
CAPE ESPENBERG
CAPE ESPENBERG #1
CAPE KRUSENSTERN
KILLIAK LAKES
DEVIL MOUNTAIN
DEVIL MOUNTAIN LAKES
SEWARD PENINSULA
KIVALINA
LISBURNE HILLS
CAPE LISBURNE
CAPE THOMPSON
CAPE DYER
POINT HOPE
KOTZEBUE ARCH
CHUKCHI SEA
ONLAP TO BASEMENT
HOPE BASIN
FIG. 6
FIG. 7
FIG. 8
FIG. 11
FIG. 12
0 25 kilometers
NORMAL FAULT, HACHURES ON DOWNTHROWN SIDE
INFERRED NORMAL FAULT
STRUCTURE CONTOUR
SEISMIC PROFILES USED IN THIS REPORT
FAULT ZONE ON NORTH SIDE OF KOTZEBUE ARCH
69°
68°
67°
66°
162°
166°
170°
171°

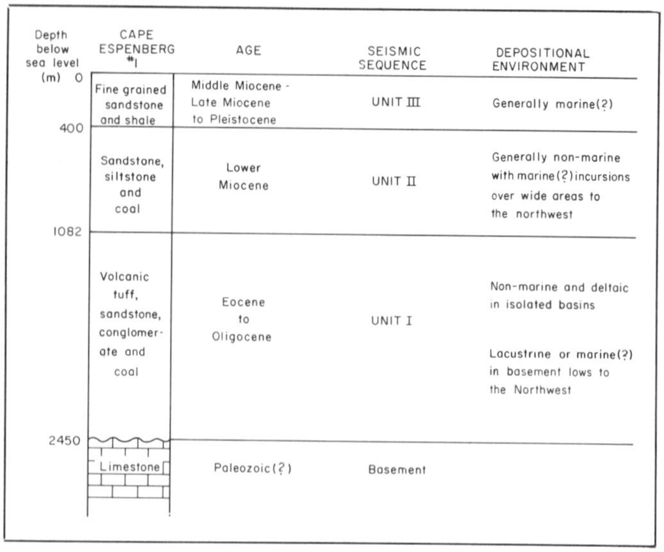

Figure 10. Correlation of chronostratigraphic intervals in the Cape Espenberg #1 well with seismic sequences and their interpreted depositional environments.

Unit III in the Cape Espenberg #1 well consists mainly of fine-grained silt and sand (American Stratigraphic Co., 1977a). Although no marine palynomorphs were reported for this sequence in the Cape Espenberg #1 well (Alaska Oil and Gas Conservation Commission, 1981a), its seismic characteristics suggest a marine or lacustrine depositional setting, especially in the thicker sections south of Point Hope where the seismic reflections are parallel and low in amplitude (Figure 9). In the Kotzebue Basin this sequence may represent a nonmarine delta-plain deposit.

Basin Structure

The overall structural style of the Hope Basin appears to be extensional. Numerous normal faults with displacements of 0.1 to >1.0 sec are oriented southeast–northwest throughout the basin. The two dominant structural features of the study area are Kotzebue arch, a broad east–west structural high extending across the entire basin from Cape Krusenstern and the Igichuk Hills to the northeast Siberian coast, and a system of southeast–northwest-trending faults that involve basement and border half-grabens just southwest of Point Hope (Grantz, Holmes, and Kososki, 1975; Figure 5). Kotzebue arch extends >200 km along the southern edge of the Hope Basin and structurally separates it from Kotzebue Sound. Unit II and the bottom of Unit III can be correlated across this high, but correlation of any part of Unit I across the arch is problematic. Structurally, the arch is characterized by a 10-km-wide disrupted zone of small antithetic faults that extend the length of the north side of the arch (Figures 5, 11). In contrast, few faults occur along the southern flank of the arch (Figure 5).

Examples of the large basement faults southwest of Point Hope can be seen in Figure 6. These faults are steeply dipping, and at deeper structural levels commonly exhibit large basement offsets (>0.5 sec). At shallower levels, a series of upward-bifurcating faults penetrates the overlying sedimentary package and generally shows small offsets (0.1 sec–0.2 sec). This structural style is similar to the "flower structures" observed on seismic lines across wrench faults (Harding and Lowell, 1979; Harding, 1985). In map view, these faults and others south of Point Hope extend over distances as great as 150 km, exhibiting changes in throw and, in some areas, sense of throw along their courses (Figure 5). This evidence suggests that these faults have a component of strike-slip, as well as dip-slip movement. Their strike and tectonic style indicate they may be offshore offsets of the Kobuk fault zone. However, data are unavailable from the center of Kotzebue Sound, where the Kobuk fault zone and the north–south-trending Seward Peninsula-Yukon Koyukuk vertical fault zone project offshore.

BASIN TECTONIC EVOLUTION

As a result of plate realignment in the Late Cretaceous and early Tertiary, compressional tectonism associated with strong convergence between the North American and Eurasian plate was superseded by an extensional/transtensional regime. Many sedimentary basins along the Bering Sea margin and western Alaska originated during the early Paleogene and correlate with this change in tectonic style (Harbert et al., 1985). Right-lateral movement of the Kaltag fault during the early Tertiary (Patton and Hoare, 1968) coincided with this plate reorganization and is inferred to have led to the opening of the Norton Basin (Marlow, Cooper, and Fisher, this volume). Similarly, right-lateral movement along the Kobuk fault zone may have formed the Hope Basin.

Seismic Unit I, its distribution limited by actively growing structures, represents the initial stage of basin filling. Figure 8 shows Unit I underlying the lower Miocene Unit II. Because Eocene volcaniclastic strata underlie correlative lower Miocene Unit II in the Cape Espenberg #1 well, the seismic unit overlying basement on line 2019 is likely Eocene or older. As mentioned earlier, correlating this unit across Kotzebue arch is difficult, but Unit I on both sides of the arch is probably similar in age and origin. In Kotzebue Basin, Unit I shows southeast-dipping clinoforms and thickens to the southeast, away from the arch. A similar geometry characterizes Unit I in seismic data from the northwest part of the Hope Basin.

Structurally controlled basement topography to the northwest of Kotzebue Sound also evolved during the Eocene. Large-scale faulting of basement commenced at this time, with a predominantly southwest–northeast sense of transtensional separation. Basement topography consists of grabens and half-grabens produced by the basin-forming structural event. During the Eocene and Oligocene, these basement lows were filled by sediments of Unit I. The Middle Eocene basalt associated with Unit I in the Cape Espenberg #1 well, and the upper Oligocene volcanic exposures on Seward Peninsula (Swanson et al., 1981), characterize and date the transtensional tectonism that formed the Hope Basin.

In the early Miocene the entire basin area subsided, faulting waned, and Unit II blanketed basement highs, as well as the depocenters of Unit I. Renewed faulting commenced during the middle to late Miocene as subsidence continued and Unit III was deposited. Faults formed earlier in the basin's history were reactivated, and new extensional faults with a more westerly orientation formed. This reorientation apparently resulted in some component of strike-slip motion along the major Eocene–Oligocene basement faults. Growth on the Miocene faults is easily recognized along many lines

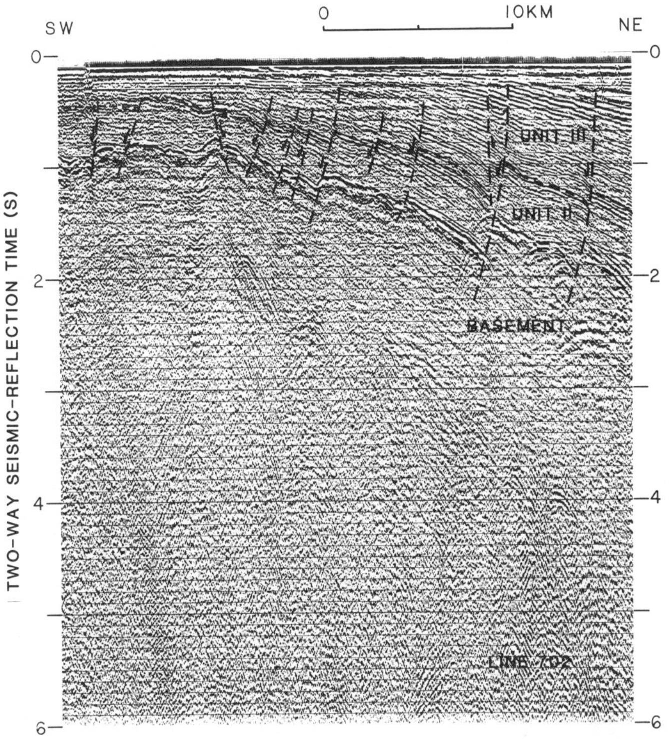

Figure 11. Unmigrated CDP seismic-reflection profile 702, a dip line illustrating the antithetic faults along the north side of the Kotzebue arch. See Figures 5 and 9 for location of profile.

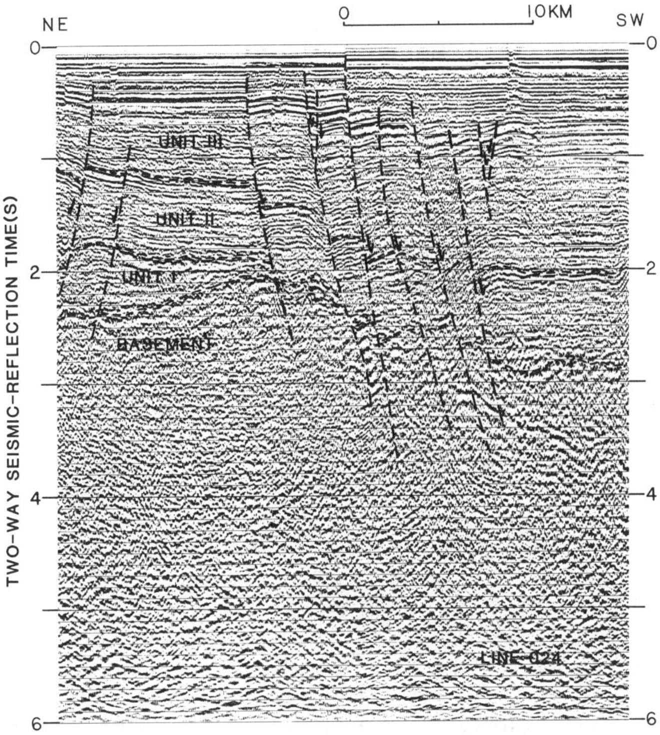

Figure 12. Migrated CDP seismic-reflection profile 024, a dip line showing interpreted southwest-dipping Miocene growth faults. See Figures 5 and 9 for location of profile.

(Figure 12). During this time, the Kotzebue arch assumed its present form. This uplift gave rise to the antithetic faults along the north side of the arch.

Quaternary time has been marked by continued faulting and seismicity in the region. The formation of large subaerial basalt fields in the Seward Peninsula and the Yukon-Koyukuk Basin (Hopkins, 1963; Swanson et al., 1981; Moll and Patton, 1984), vertical fault movements along the Kobuk River Valley to the east (Hamilton, 1984), and extensional faulting in the Bendeleben and Kilawik Mountains of the Seward Peninsula to the south (Hudson and Plafker, 1981; Figure 2) are additional evidence of the ongoing tectonism in this area.

HYDROCARBON POTENTIAL

Although the Hope Basin remains one of the few untested, frontier basins of offshore Alaska, its hydrocarbon potential can be generally characterized. Vitrinite reflectance studies (Alaska Oil and Gas Conservation Commission, 1981a, 1981b) of both SOCAL wells indicate that sediment in both is thermally immature with respect to oil generation. The total organic carbon (TOC) is generally 1% or less in both the Miocene chronostratigraphic units of each well but is slightly higher (2%) in some samples from the Oligocene and older chronostratigraphic units. Pyrolysis data indicate that there is little hydrocarbon potential in any of the samples, and that gas is the hydrocarbon most likely to have been generated. However, these analyses are from two wells on the perimeter of the Kotzebue Basin; the hydrocarbon potential for basin interiors, both in the Kotzebue Basin and the larger Hope Basin, may be significantly different. The seismic characteristics and thicknesses of sedimentary rocks in the Hope Basin suggest that they may contain rocks of marine or lacustrine depositional environments. An estimated geothermal gradient of 50°C/km derived from both the SOCAL wells suggests that peak oil generation should occur between 2.5 or 3.0 km and 5.0 km below the sea floor (Fisher, 1982). Sediment >5.0 km thick is present in some of the deeper areas of the Hope Basin (Figure 5). If organic-rich, oil-prone source rocks and adequate reservoir rocks are present in these areas, structural and stratigraphic traps observed on the seismic data could provide habitats for significant hydrocarbon accumulation. However, if volcaniclastic rock cored in the Cape Espenberg #1 well constitutes much of the sedimentary fill in the deeper depocenters of the basin, the porosity and permeability of any reservoir rock may be reduced.

CONCLUSIONS

The Hope Basin is a predominantly extensional Tertiary Basin cut by numerous west- to northwest-trending faults. The basin locally contains as much as 5.8 km of sedimentary rock in graben and half-graben structures. Seismic-reflection profiles define three seismic sequences, each representing an episode of the basin evolution. Northwest–southeast-trending transtensional faults were likely initiated during the Paleogene by right-lateral movement along the Kobuk fault zone. Localized topographic lows on the basement surface were filled by nonmarine and possibly marine or lacustrine sediment of seismic Unit I during the Eocene and Oligocene. Regional subsidence and deposition of predominantly nonmarine deltaic platform sediment in seismic Unit II occurred during the early Miocene. Reactivation of the Eocene faults and initiation of new east–west-trending normal faults led to increased subsidence and formation of a predominantly marine seismic Unit III in the middle Miocene.

Timing of the formation of the Hope Basin is coincident with the formation of many other basins on the Bering Sea margin and along western Alaska and is probably related to plate reorganizations between North America and Eurasia (Harbert et al., 1985). These plate movements caused a relaxation of compressive forces in the Bering Strait region and were followed by right-lateral movement along the Kobuk fault zone as Arctic Alaska moved to the northeast. This movement is inferred to have initiated transtensional faulting and graben formation in the Hope Basin.

ACKNOWLEDGMENTS

Financial support for this study came from U.S. Geological Survey Grant 14-08-0001-G-955 and Petroleum Research Fund Grant 17061-AC2. The author would like to thank Stephen Eittreim, Steven D. May, Denise Davies, and David A. Dinter of the U.S. Geological Survey for their help on earlier versions of this study and Arthur Grantz and David W. Scholl of the U.S. Geological Survey and Stephan A. Graham of Stanford University for their many reviews of this manuscript.

REFERENCES CITED

Alaska Oil and Gas Conservation Commission, 1981a, Visual kerogen, thermal maturation, TOC and rock-eval analyses, SOCAL Cape Espenberg #1, Kotzebue, Alaska: Geochemical report prepared by Mobil Exploration for the Alaska Oil and Gas Conservation Commission, Department of Natural Resources, Division of Geological and Geophysical Surveys, Eagle River, Alaska, 22 p.

Alaska Oil and Gas Conservation Commission, 1981b, Visual kerogen, thermal maturation, TOC and rock-eval analyses, SOCAL Nimiuk Pt. #1, Kotzebue, Alaska: Geochemical report prepared by Mobil Exploration for the Alaska Oil and Gas Conservation Commission, Department of Natural Resources, Division of Geological and Geophysical Surveys, Eagle River, Alaska, 22 p.

American Stratigraphic Company, 1977a, Log number A–299, Standard Oil of California No. 1 Cape Espenberg: Denver, American Stratigraphic Company, 1 sheet.

American Stratigraphic Company, 1977b, Log number A–296, Standard Oil of California No. 1 Nimiuk Point: Denver, American Stratigraphic Company, 1 sheet.

Barnes, D. F., 1977, Bouguer gravity map of Alaska: U.S. Geological Survey Geophysical Investigations Map GP–913, scale 1:2,500,000.

Brown, L. F., Jr., and W. L. Fisher, 1980, Seismic stratigraphic interpretation and petroleum exploration: AAPG Continuing Education Course Note Series 16, 192 p.

Campbell, R. H., 1967, Areal geology in the vicinity of the Chariot site, Lisburne Peninsula, northwestern Alaska: U.S. Geological Survey Professional Paper 395, 71 p.

Cooper, A. K., M. S. Marlow, and D. W. Scholl, 1987, Geologic framework of the Bering Sea crust, this volume.

Dillon, J. T., 1985, Structure and stratigraphy of southern Brooks Range and northern Koyukuk Basin near the Dalton Highway, *in* C. G. Mull, ed., Guidebook to bedrock geology along the Dalton Highway, Yukon River to Prudhoe Bay, Alaska: AAPG Pacific Section, Guidebook 7, p. 41–46.

Eittreim, S., A. Grantz, and O. T. Whitney, 1978, Isopach maps of sediment thickness in Hope Basin, southeast Chukchi Sea, Alaska: U.S. Geological Survey Miscellaneous Field Studies Map MF–906, scale 1:1,000,000.

Eittreim, S., A. Grantz, and O. T. Whitney, 1979, Cenozoic sedimentation and tectonics of Hope Basin, southern Chukchi Sea, *in* A. Sisson, ed., The relationship of plate tectonics to Alaskan geology and resources: Anchorage, Alaska Geological Society Symposium Proceedings, April 1977, p. B1–B11.

Ellersieck, I., I. L. Tailleur, C. F. Mayfield, and S. M. Curtis, 1978, A new find of Upper Cretaceous or Tertiary sedimentary rocks in the Noatak Valley: U.S. Geological Survey Circular 804-B, p. B-13.

Fisher, M. A., 1982, Petroleum geology of Norton Basin, Alaska: AAPG Bulletin, v. 66, p. 286–301.

Grantz, A., S. Eittreim, and O. T. Whitney, 1981, Geology and physiography of the continental margin north of Alaska and implications for the origin of the Canada Basin, *in* A. E. M. Nairn, M. Churkin, Jr., and F. G. Stehli, eds., The ocean basins and margins: New York, Plenum Publishing Corporation, v. 5, p. 439–492.

Grantz, A., M. L. Holmes, and B. A. Kososki, 1975, Geologic framework of the Alaskan continental terrace in the Chukchi and Beaufort Seas, *in* C. J. Yorath, E. R. Parker, and D. J. Glass, eds., Canada's continental margins and offshore petroleum exploration: Canadian Society of Petroleum Geologists Memoir 4, p. 669–700.

Grantz, A., D. M. Mann, and S. D. May, 1982, Tracklines of multichannel seismic reflection data collected by the U.S. Geological Survey in the Beaufort and Chukchi Seas in 1977 for which profiles and stack tapes are available: U.S. Geological Survey Open-File Report 82-0735, 1 oversize sheet.

Grantz, A., S. C. Wolf, L. Breslau, T. C. Johnson, and W. F. Hanna, 1970, Reconnaissance geology of the Chukchi Sea as determined by acoustic and magnetic profiling, *in* W. L. Adkison and M. M. Brosgé, eds., Proceeding of the geological seminar on the North Slope of Alaska: AAPG, Pacific Section meeting, Los Angeles, California, p. F1–F28.

Hamilton, T. D., 1984, Late Quaternary offsets along the Kobuk and related fault zones, northwestern Alaska (abs.): GSA Abstracts with Programs, Cordilleran Section meeting, Anchorage, Alaska, p. 288.

Harbert, W. P., L. S. Frei, A. Cox, and D. C. Engebretson, 1985, Relative motions between Eurasia and North America in Bering Sea region (abs.): AAPG Program and Abstracts, Pacific Section meeting, Anchorage, Alaska, p. 44.

Harding, T. P., 1985, Seismic characteristics and identification of negative flower structures, positive flower structures, and positive structural inversion: AAPG Bulletin, v. 69, p. 582–600.

Harding, T. P., and J. D. Lowell, 1979, Structural styles, their plate-tectonic habitats, and hydrocarbon traps in petroleum provinces: AAPG Bulletin, v. 63, p. 1016–1058.

Holmes, M. L., 1975, Tectonic framework and geologic evolution of the southern Chukchi Sea continental shelf: Seattle, University of Washington unpublished Ph.D. thesis, 143 p.

Hopkins, D. M., 1963, Geology of the Imuruk Lake area, Seward Peninsula, Alaska: U.S. Geological Survey Bulletin 1141-C, 101 p.

Hopkins, D. M., and F. S. MacNeil, 1960, A marine fauna probably of Late Pliocene age near Kivalina, Alaska: U.S. Geological Survey Professional Paper 400-B, p. B339–B342.

Hopkins, D. M., J. V. Matthews, J. A. Wolfe, and M. L. Silberman, 1971, A Pliocene flora and insect fauna from the Bering Strait region: Paleogeography, Paleoclimatology, Paleoecology, v. 9, p. 211–231.

Hudson, T., 1977, Preliminary geologic map of Seward Peninsula, Alaska: U.S. Geological Survey Open-File Report OF-77-796A, scale 1:1,000,000.

Hudson, T., and G. Plafker, 1981, Kigluaik and Bendeleben faults, Seward Peninsula: U.S. Geological Survey Circular 772-B, p. B47–B50.

Marlow, M. S., and A. K. Cooper, 1984, New discoveries along the

Bering Sea continental margin (abs.): GSA Abstracts with Programs, Cordilleran Section meeting, Anchorage, Alaska, p. 320.

Marlow, M. S., A. K. Cooper, and M. A. Fisher, 1987, Petroleum geology of the Beringian continental shelf, this volume.

May, S. D., 1985, Free-air gravity anomaly map of the Chukchi and Alaskan Beaufort Seas, Arctic Ocean: U.S. Geological Survey Miscellaneous Investigations Series Map I-1182-E, scale 1:1,000,000.

Mitchum, R. M., Jr., P. R. Vail, and S. Thompson, III, 1977, Seismic stratigraphy and global changes of sea level, part 2: The depositional sequence as a basic unit for stratigraphic analysis, *in* C. E. Payton, ed., Seismic stratigraphy—Applications to hydrocarbon exploration: AAPG Memoir 26, p. 53–62.

Moll, E. J., and W. W. Patton, Jr., 1984, Volcanic history of the southern Yukon-Koyukuk Depression: Evidence from the Unalakleet quadrangle, Alaska (abs.): GSA Abstracts with Programs, Cordilleran Section meeting, Anchorage, Alaska, p. 322.

Patton, W. W., Jr., 1973, Reconnaissance geology of the northern Yukon-Koyukuk province, Alaska: U.S. Geological Survey Professional Paper 774-A, p. A1–A17.

Patton, W. W., Jr., and J. M. Hoare, 1968, The Kaltag fault, west-central Alaska: U.S. Geological Survey Professional Paper 600-D, p. D147–D153.

Patton, W. W. Jr., and T. P. Miller, 1968, Regional geologic map of the Selawik and southeastern Baird Mountains quadrangles, Alaska: U.S. Geological Survey Miscellaneous Geologic Investigations Map I-530, scale 1:250,000.

Sainsbury, C. L., 1975, Geology, ore deposits, and mineral potential of the Seward Peninsula, Alaska: U.S. Bureau of Mines Open-File Report 73-75, 108 p.

Scholl, D. W., T. L. Vallier, and A. J. Stevenson, 1987, Geologic evolution and petroleum geology of the Aleutian Ridge, this volume.

Swanson, S. E., D. L. Turner, R. B. Forbes, and D. M. Hopkins, 1981, Petrology and geochronology of Tertiary and Quaternary basalts from the Seward Peninsula, western Alaska (abs.): GSA Abstracts with Programs, Cincinnati, Ohio, v. 13, p. 563.

Till, A. B., 1984, Low-grade metamorphic rocks of Seward Peninsula, Alaska (abs.): GSA Abstracts with Programs, Cordilleran Section meeting, Anchorage, Alaska, p. 337.

CHAPTER FIVE

GEOLOGIC FRAMEWORK OF THE BERING SEA CRUST

A. K. Cooper
M. S. Marlow
D. W. Scholl

U.S. Geological Survey
Menlo Park, California

ABSTRACT

The Bering Sea includes the broad (500-km-wide) continental shelf, abyssal basins, and large bathymetric ridges that lie north of the Aleutian Ridge between western Alaska and eastern Siberia. The sea evolved when it was separated from the Pacific Basin by the formation of the Aleutian Ridge in the early Tertiary.

The crust of the broad Beringian Shelf is in part composed of the diverse geologic and structural belts and terranes of Paleozoic to Cenozoic age that trend toward the Bering Sea from surrounding onshore regions. This structural framework formed by the amalgamation of allochthonous terranes in the middle and late Mesozoic. Since the late Mesozoic, the structural fabric of the shelf has been modified by the formation of major sedimentary basins.

The abyssal basins are underlain by oceanic crust. In the Aleutian Basin, the crust is believed to be Mesozoic in age, and in the Bowers and Komandorsky Basins the crust is younger, probably Cenozoic. Large volcanic ridges separate the basins and are the site of postulated subduction along the north side of Bowers Ridge and possibly the west side of Shirshov Ridge. The abyssal basins may have been part of the Pacific Basin before the construction of the Aleutian Ridge; since then, back-arc spreading in the Komandorsky and possibly Bowers Basins has altered or displaced pre-existing oceanic crust. The Aleutian Basin has been a relatively stable depocenter since the Paleogene. Thick sedimentary sections are buried beneath the continental rise along Bowers Ridge and the Beringian-Koryak margins. These possibly ancient trench deposits are only slightly deformed.

New interpretations of geophysical data indicate that at least two major structural boundaries lie beneath the Beringian Shelf. These boundaries may corres-

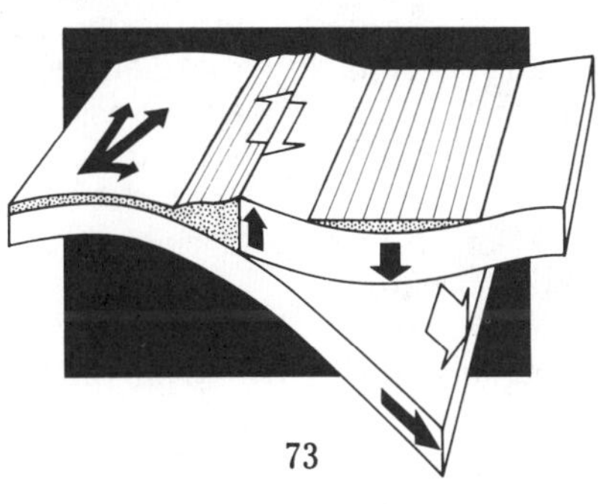

pond to (a) the northern landward edge of the Peninsular terrane of Alaska, and (b) the southwestern seaward edge of a Late Cretaceous to Paleogene volcanic arc that extends from Siberia to the Alaska Peninsula. These subparallel boundaries may have formed in the late Mesozoic to early Tertiary, when the Peninsular terrane was emplaced against southern Alaska and arc magmatism commenced at a more southern, seaward location. Older terranes of the inner shelf probably were truncated during emplacement of the Peninsular terrane. Later regional compression and oroclinal bending, as suggested by recent paleomagnetic data, further deformed and bent inner-shelf structures as well as the Peninsular terrane.

Since the Mesozoic, major overlapping sedimentary bodies have been deposited across the shelf, slope, and abyssal basins. Many of these depocenters may be the sites of significant hydrocarbon deposits.

INTRODUCTION

The Bering Sea includes an area of nearly 2.6 million km^2, nearly the size of the western third of the conterminous United States. Over the past two decades, government, industry, and academic groups have conducted extensive geologic and geophysical studies over the continental shelves and enclosed abyssal basins in the Bering Sea. These studies have documented a variety of geologic structures and overlying sedimentary bodies that are believed to have significant economic potential. Only within the past 10 years has the offshore petroleum resource of the region been assessed and tested.

In this chapter, we describe the geologic framework, tectonic history, and briefly, the hydrocarbon potential of the principal structural provinces of the Bering Sea region. We also give regional compilations of geophysical data and present new interpretations on the deep structure of the Beringian continental shelf and on the origins and continuation of foldbelts across the shelf from Alaska to Siberia. Companion chapters describe and assess, in greater detail, the resource potential for other areas within and adjacent to the Bering Sea—the Aleutian Ridge (Scholl, Vallier, and Stevenson, this volume), the Aleutian and Bowers Basins (Cooper, Scholl, and Marlow, this volume), and the Beringian Shelf (Marlow, Cooper, and Fisher, this volume; Stauffer, this volume).

PHYSIOGRAPHIC SETTING

Nearly half of the Bering Sea region (Figure 1) is underlain by continental shelf where water depths are less than 200 m. The largest shelf area borders western Alaska and is referred to as the Beringian Shelf. Most of the remaining area lies in abyssal depths of 2800 to 3600 m (the Aleutian, Bowers, and Komandorsky Basins). The Beringian continental slope, which is adjacent to the Beringian Shelf, is incised by immense submarine canyons.

The arcuate Aleutian Ridge extends nearly 2500 km, forming the southern boundary of the Bering Sea, and separating the Bering Sea physiographically from the Pacific Ocean. At its eastern end, the Aleutian Ridge joins the Beringian continental shelf near the flat-topped Umnak Plateau. Two large submerged mountain chains, Bowers and Shirshov Ridges, divide the abyssal Bering Sea Basin into the separate Bowers, Komandorsky, and Aleutian Basins. Bowers Ridge curves northward from the central Aleutian Ridge and nearly connects with the north-trending Shirshov Ridge, which terminates at Cape Olyutorsky on the southern Siberian margin.

GEOLOGIC FRAMEWORK

Many scientific papers have been published on the geology and geophysics of the Bering Sea and surrounding regions. A partial list includes publications on the Alaska Peninsula and Aleutian Ridge by Burk (1965), Scholl, Vallier, and Stevenson (1983); on western Alaska by Patton et al. (1976), Patton and Tailleur (1977), Fisher, Patton, and Holmes, (1982), Gemuts, Steefel, and Puchner (1983); on eastern Siberia by Fujita (1978), Zinkevich (1978), McLean (1979a), Fujita and Newberry (1982); on the offshore Bering Sea by Scholl, Buffington, and Marlow (1975), Marlow et al. (1976a), Cooper, Scholl, and Marlow (1976), Marlow and Cooper (1980), Marlow et al. (1982), Steffy et al. (1985); and the entire region by Beikman (1980), Stone (1983, and in press). The following summary is based on this information.

Onshore Geology

The Bering Sea and its flanking land areas are composed of numerous tectonostratigraphic and lithotectonic terranes (Figure 2A). These terranes, as well as superimposed volcanic and plutonic rocks, can be grouped into structurally aligned rock belts (Figure 2B) commonly marked by characteristic geophysical signatures and structural styles. The rock belts generally trend southwest across western Alaska and across the central Koryak Range of Siberia. At the northern and southern ends of the Beringian Shelf, these trends curve to the southeast-northwest and parallel the outer Beringian margin. Most terranes of onshore Alaska and Siberia (Figure 2A) are allocthonous and form the basement framework on which successor basins and younger volcanic and plutonic rocks are emplaced.

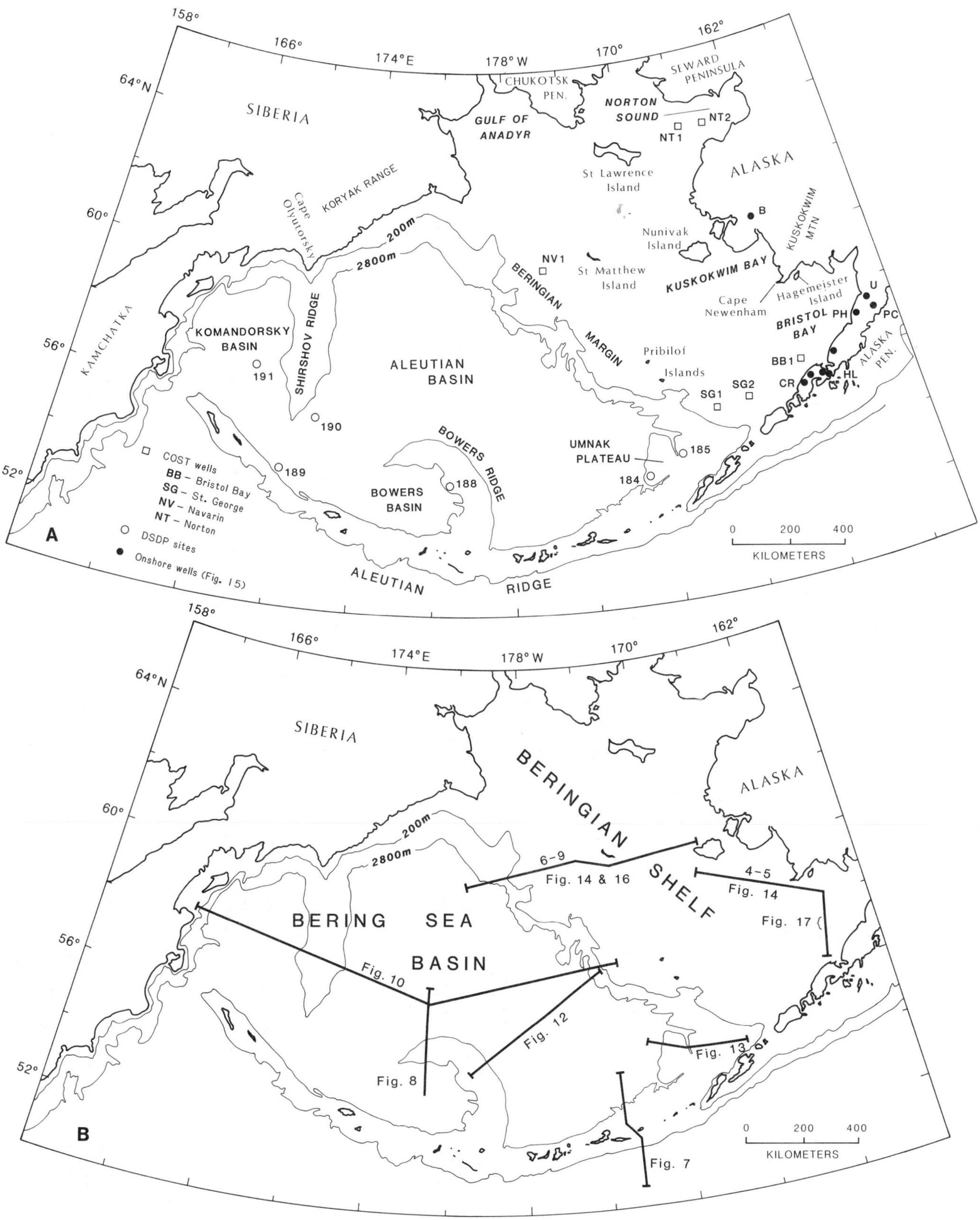

Figure 1. Index maps of the Bering Sea showing: A, place names and locations of offshore and onshore wells; and B, locations of seismic sections and drawings on other figures. Albers equal-area projection used for this and all other figures.

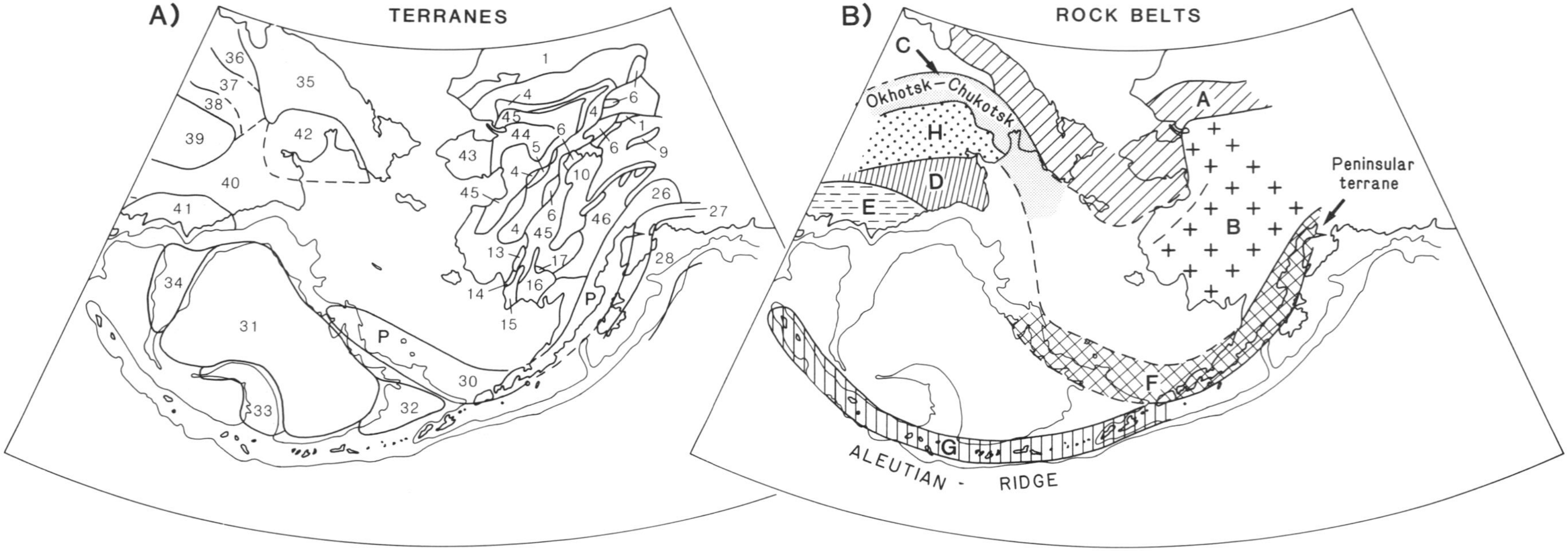

1	North America	26	Peninsular (P)	38	Alazeya
4	Ruby	27	Chugach	39	Omolon
5	Angayucham	28	Prince William	40	Koryak
6	Innoko	30	Peninsular (offshore)	41	Olyutorsk
9	70 Mile	31	Aleutia	42	Anadyr
10	Nixon Fork	32	Umnak	43	Seward
13	Nyak	33	Bowers	44	Koyukuk
14	Kilbuck	34	Shirshov	45	Late Cretaceous deposits
15	Goodnews	35	Chukotka	46	Kahiltna
16	Togiak	36	North Anyui		
17	Tikchik	37	South Anyui		

BELT	CENOZOIC	MESOZOIC	PALEOZOIC	P€	PRINCIPAL ROCK-TYPE
A					Sedimentary and volcanic
B					Sedimentary
B					Volcanic and plutonic
C					Volcanic and plutonic
D					Melange
E		?			Flysch
F					Sed./volc./plut.
G					Volcanic
H					Melange and volc./plut.

REFERENCES
(List below also used on Figure 3)

1) Zinkevich (1978); 2) Beikman (1980); 3) Turner et al. (1983a, 1983b, 1984a, 1984b, and 1984c); 4) Marlow, Cooper, and Fisher (this volume); 5) Gemuts, Steefel, and Puchner (1983); 6) Fisher, Patton, and Holmes (1982); 7) Bogdanov, Vishnevskaia, and Sukhov (1983); 8) Hoare and Coonrad (1980); 9) Cooper and Marlow (1984); 10) Hudson (1979a, 1979b); 11) Scholl, Vallier, and Stevenson (1983); 12) Moll and Patton (1982); 13) Patton et al. (1976); 14) Marlow and Cooper (in press); 15) Bogdanov and Neprochnov (1984); 16) Jacob (1972); 17) Box (1985); 18) Fisher and von Huene (1984); 19) Minster et al. (1974); 20) Cooper et al. (1979b); 21) Meyerhoff (1982); 22) Marlow et al. (1979a, 1979b); 23) Wallace and Engebretson (1984).

Figure 2. Distribution of terranes and rocks in belts in the Bering Sea region. Numbers in parentheses under B (below) are keyed to the references listed below the figure; these references are also used for Figure 3. A, Tectonostratigraphic and lithotectonic terranes. Terranes 1–29 are from Jones and Silberling (1979); 30–31 from Marlow and Cooper (1983); 32–34 from Ben-Avraham and Cooper (1981); 35–42 from Fujita and Newberry (1982); and 43–45 from Jones et al. (1984). B, Selected rock belts A through H discussed in text. Maps C–F, sedimentary, metamorphic, ultramafic, and undifferentiated sedimentary and volcanic rocks; C, Tertiary (1,2,3,4); D, Early and Middle Cretaceous (1,3,4,5,6,7); E, Paleozoic through Jurassic (1,2,3,4,5); F, Precambrian and Paleozoic (1,2,5,6). Maps G–J, volcanic and plutonic rocks; G, Late Cenozoic (2,6,8,9); H, Middle Tertiary (1,10,11); I, Late Cretaceous to Paleogene (1,10,12,13,14,23); J, Early Jurassic and Early Cretaceous (1,6,10).

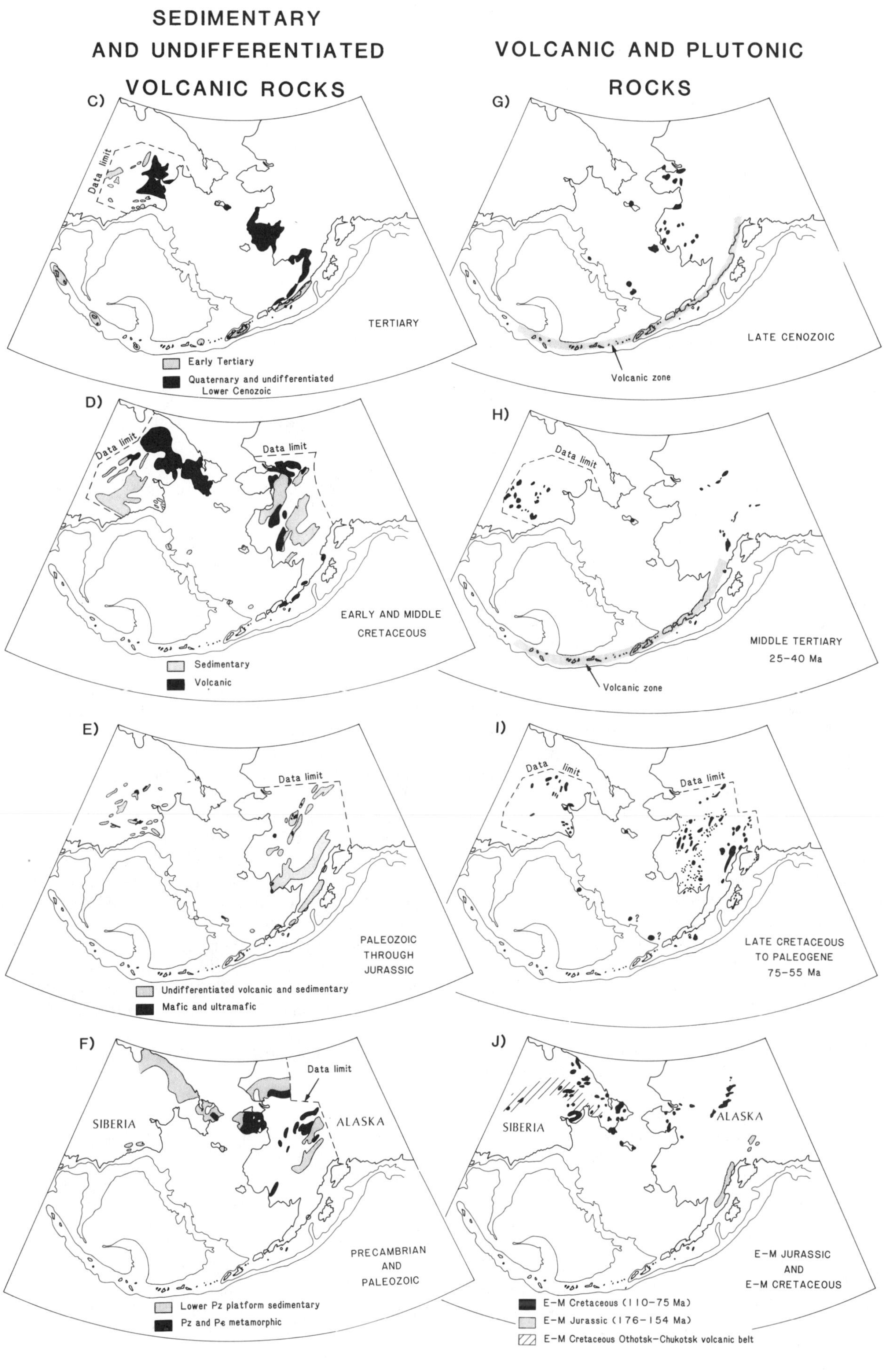
SEDIMENTARY AND UNDIFFERENTIATED VOLCANIC ROCKS
VOLCANIC AND PLUTONIC ROCKS
C)
Data limit
TERTIARY
Early Tertiary
Quaternary and undifferentiated Lower Cenozoic
G)
LATE CENOZOIC
Volcanic zone
D)
Data limit
Data limit
EARLY AND MIDDLE CRETACEOUS
Sedimentary
Volcanic
H)
Data limit
MIDDLE TERTIARY 25–40 Ma
Volcanic zone
E)
Data limit
PALEOZOIC THROUGH JURASSIC
Undifferentiated volcanic and sedimentary
Mafic and ultramafic
I)
Data limit
Data limit
LATE CRETACEOUS TO PALEOGENE 75–55 Ma
F)
Data limit
SIBERIA
ALASKA
PRECAMBRIAN AND PALEOZOIC
Lower Pz platform sedimentary
Pz and Pє metamorphic
J)
SIBERIA
ALASKA
E–M JURASSIC AND E–M CRETACEOUS
E–M Cretaceous (110–75 Ma)
E–M Jurassic (176–154 Ma)
E–M Cretaceous Othotsk–Chukotsk volcanic belt

The rocks surrounding and underlying the Bering Sea are principally Mesozoic and Cenozoic (Figure 2). Limited exposures of Paleozoic sedimentary and metamorphic rocks are, however, mapped on the Alaska Peninsula. Precambrian rocks are exposed onshore on the north side of Bristol Bay, in the Koryak melange belt, and in a belt of miogeoclinal nonvolcanic rocks that extends across the northern part of the Bering Sea (Seward Peninsula, St. Lawrence Island, and the Chukotsk Peninsula; belt A, Figure 2B; Figure 2F). Other Paleozoic and lower Mesozoic rocks include sedimentary, mafic volcanic, and ultramafic rocks along the flanks of the Ruby geanticline (western Alaska, Figure 3A) which extend southwestward to Cape Newenham (Figure 3A; belt B, Figure 2B). These rocks are considered to be an obducted and dismembered ophiolite. Similar Paleozoic and early Mesozoic rocks have not been mapped in northeastern Siberia, although ultramafic and ophiolotic rocks are present in the Mesozoic melange complexes of the Koryak Mountains.

Extensive deposits of Jurassic and Cretaceous, marine and nonmarine, sedimentary and volcanic rocks occur throughout western Alaska, the Alaska Peninsula, and northeastern Siberia (belts B, F, Figure 2B; Figures 2D, 2E). These sedimentary units are dominantly graywacke and flysch deposits with interbedded volcaniclastic rocks. Most Jurassic to mid-Cretaceous deposits are part of oceanic and arc terranes which were accreted to western Alaska and northern Siberia by the mid-Cretaceous and to southern Alaska and Siberia in the Late Cretaceous and early Tertiary. Mid-Cretaceous to Late Cretaceous and younger strata were deposited in successor basins (for example, the Yukon-Koyukuk and Kuskokwim Basins) where sediment thicknesses are in excess of 12 km (Kuskokwim Basin, Decker, 1984, Moore and Wallace, 1985; Figures 2D, 3C).

Belts of Mesozoic rocks cross northeastern Siberia and form the basement of the northern Bering Sea (Figures 2B, 2D). From north to south they include middle and Upper Cretaceous volcanic rocks (Okhotsk-Chukotsk belt C, Figure 2B; Figure 2D) that are juxtaposed on the south by a deformed convergent-margin complex. This margin complex includes melange, mafic, and ultramafic masses of Early Cretaceous to early Tertiary age, as well as younger nonmarine sedimentary rocks (belts D and H, Figure 2B). Cretaceous flysch deposits (belt E, Figure 2B) form the southernmost rocks.

Undifferentiated pre-Quaternary Cenozoic sedimentary and volcanic rocks are found in restricted outcrops in several areas, including the Alaska Peninsula and western Alaska, the Aleutian Ridge, the Beringian Shelf islands (St. Lawrence, St. Matthew, Pribilof), and the Koryak Mountains (Figures 2C, 2G).

Magmatic belts of plutonic and volcanic rocks are superimposed on, or largely constitute, the rock belts framing the Bering Sea region. These magmatic belts include (a) Lower to Middle Jurassic (176–154 Ma) plutonic rocks in southern Alaska (belt F, Figure 2B; Figure 2J); (b) Early to middle Cretaceous (110–75 Ma) granitic plutons of northern Alaska and Siberia (belts A and B, Figure 2B, Figure 2J); (c) Upper Cretaceous to lower Tertiary (75–55 Ma) calc-alkaline volcanic and plutonic rocks of western Alaska and Siberia (belts B and C, Figure 2B; Figure 2I); (d) middle Tertiary (50–25 Ma) volcanic rocks of southern Alaska, the Aleutian Ridge (belt G), and Siberia (Figures 2B, 2H); and (e) late Cenozoic alkalic and calc-alkalic rocks of western and southern Alaska and the Aleutian Ridge (belt G, Figure 2B; Figure 2G).

Onshore Structure

Faults and other structures in the Bering Sea region can be grouped into regionally extensive structural zones on the basis of type and age of deformation (Figure 3A, B). These zones commonly overlap, reflecting the different episodes of deformation that have affected the region. In western Alaska, faults occur in at least three zones (zones A, B, and C, Figure 3B) that trend west-southwest toward the inner Beringian Shelf. These fault zones include the middle Mesozoic thrust faults along the flanks of the Ruby geanticline (Figure 3A; zone B, Figure 3B), the Late Cretaceous to Paleogene (Neogene?) strike-slip faults of western Alaska (zone C, Figure 3B; Grantz, 1966) and the Alaska Peninsula, and the faults within the Brooks Range rocks of northern Alaska (zone A, Figure 3B). Rocks of the Brooks Range, which extend into eastern Siberia, were subjected to north-northwest-directed thrusting in the Late Jurassic to Early Cretaceous and were oroclinally bent in the Late Cretaceous to early Tertiary (Patton and Tailleau, 1977).

The Koryak region of northeast Siberia contains principally west- to southwest-striking thrust faults that dip to the north (zones I, J, and K, Figure 3B). Compressional deformation in these belts is related to the middle Mesozoic to early Cenozoic convergence and subduction of oceanic lithosphere along the northern margin of the Aleutian and Komandorsky Basins.

Offshore Geology

Little is known of the geology of the offshore areas of the Bering Sea, because few rocks have been dredged and drilled from beneath the region. Geologic inferences have been based principally on seismic-reflection data. In the following discussion, the terms bedrock and basement are used. Bedrock, or acoustic basement in seismic-reflection profiles, may include sedimentary as well as volcanic rocks. These rocks lack internal reflective coherence, probably because they are deformed or have a massive structure. Basement refers to rock units that are probably composed of great thicknesses of volcanic, mafic igneous, plutonic, or metamorphic rocks.

The geology of the Beringian Shelf, like that of onshore Alaska, is dominated by the suspected Paleozoic and Mesozoic basement terranes that are incised by Late Cretaceous and younger successor basins and intruded by coeval and younger igneous rocks (Marlow, Cooper, and Fisher, this volume). In offshore localities, only Jurassic and younger rocks have been recovered. Paleozoic and possibly older rocks have been found only in two Continental Offshore Stratigraphic Test (COST) wells drilled in Norton Basin (Figure 1).

Bedrock sampled along the outer edge of the Beringian Shelf is composed of Jurassic, Cretaceous, and Paleogene(?) sedimentary rocks as well as Eocene and possibly older volcanic rocks (Figure 2D, 2E; belt F, Figure 2B). Rocks that probably form the basement beneath the inner shelf are exposed on St. Lawrence, St. Matthew, and the Pribilof Islands, as well as on the coast of southwestern Alaska. In these areas, early Tertiary and older volcanic, mafic, ultramafic, and few metamorphosed continental rocks are found. Large successor basins, filled mostly with Cenozoic sedimen-

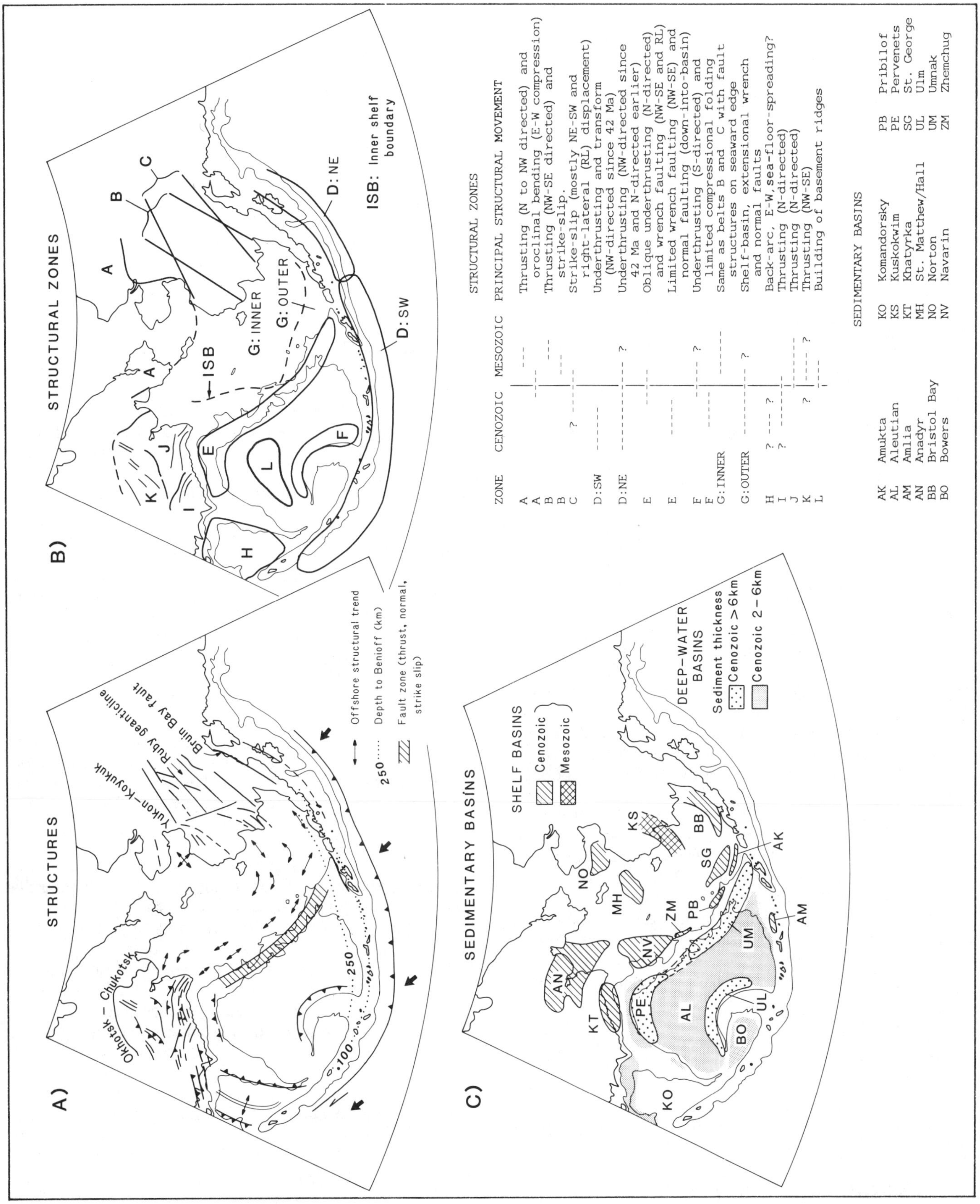

STRUCTURAL ZONES

ZONE	CENOZOIC	MESOZOIC	PRINCIPAL STRUCTURAL MOVEMENT
A		---	Thrusting (N to NW directed) and
A	--	--	oroclinal bending (E-W compression)
B		---	Thrusting (NW-SE directed) and
B		---	strike-slip.
C	? -----	-	Strike-slip (mostly NE-SW and right-lateral (RL) displacement)
D:SW	------		Underthrusting and transform (NW-directed since 42 Ma)
D:NE	----------	-- ?	Underthrusting (NW-directed since 42 Ma and N-directed earlier)
E	---	--	Oblique underthrusting (N-directed) and wrench faulting (NW-SE and RL)
E	-----		Limited wrench faulting (NW-SE) and normal faulting (down-into-basin)
F	--	--- ?	Underthrusting (S-directed) and
F	-----		limited compressional folding
G:INNER		------	Same as belts B and C with fault structures on seaward edge
G:OUTER	--------	-- ?	Shelf-basin, extensional wrench and normal faults
H	? --- ?		Back-arc, E-W, sea-floor-spreading?
I	? ------	-	Thrusting (N-directed)
J		-------	Thrusting (N-directed)
K	?	---- ?	Thrusting (NW-SE)
L	-	---	Building of basement ridges

SEDIMENTARY BASINS

AK	Amukta	KO	Komandorsky	PB	Pribilof
AL	Aleutian	KS	Kuskokwim	PE	Pervenets
AM	Amlia	KT	Khatyrka	SG	St. George
AN	Anadyr	MH	St. Matthew/Hall	UL	Ulm
BB	Bristol Bay	NO	Norton	UM	Umnak
BO	Bowers	NV	Navarin	ZM	Zhemchug

Figure 3. Structural features of the Bering Sea region. Numbers in parentheses are keyed to the references listed on Figure 2. A, Major faults, spreading ridges, present Pacific plate motion, and depth to Benioff zone (1,2,5,15,16,17,18,19). B, Major structural zones, their approximate ages, and types of structural movement. Fault locations and trends are highly generalized. See text for description and references. C, Cenozoic and Mesozoic structural basins of the offshore areas. Outlined basins and dotted areas have >2 km of sedimentary rock above acoustic basement. Names applied to basins beneath the continental slope (PE, UM) and Bowers Ridge (UL) are new (4,9,20,21,22).

tary rocks that have been sampled in COST wells (Turner et al., 1983a, 1983b; Turner et al., 1984a, 1984b, 1984c) and dredge sites (Marlow et al., 1979, 1983; Jones et al., 1981; Figure 1), lie along the outer parts of the shelf (Marlow, Cooper, and Fisher, this volume).

The deep-water regions of the Bring Sea appear to be underlain by oceanic basement rocks. Late Cenozoic basement rocks have been sampled at DSDP (Deep Sea Drilling Project) site 191 in the Komandorsky Basin (Creager et al., 1973; a more precise K/Ar date on basalt recovered at site 191 is 9.3 $\pm$0.8 Ma from our files) and Upper Cretaceous rocks have been sampled from basement beneath Shirshov Ridge (Bogdanov and Neprochnov, 1984). Mid-Cenozoic and undated volcanic rocks have been dredged from the southern Shirshov Ridge (Scholl, Buffington, and Marlow, 1975) and Bowers Ridge. Basement beneath the Aleutian and Bowers Basins has not been sampled, but marine magnetic data indicate that Mesozoic oceanic crust probably underlies the Aleutian Basin. Thick sedimentary sections fill the abyssal basins, but only the upper parts have been sampled and include Miocene and younger rocks (Scholl and Creager, 1973).

Offshore Structure

The offshore areas, like those onshore, can be divided into structural zones (Figure 3B). We combine the faults and other structures beneath the Beringian continental shelf into one structural zone (zone G: inner and outer, Figure 3B). Zone G encompasses the Late Cretaceous to early Tertiary structures associated with the successor sedimentary basins as well as those associated with underlying Mesozoic and possibly older rocks (Figure 3C). The outer edge of this zone, by our definition, is contiguous with Mesozoic to Cenozoic thrust (?) and transform (wrench) faults as well as Paleogene to Holocene extensional faults of the outer Beringian margin (zone E, Figure 3B; Marlow and Cooper, 1980; Cooper et al., 1984).

The structural zone that lies at the frontal part of the Aleutian Ridge and Alaska Peninsula (zones D:SW and D:NE, Figure 3B; McCarthy and Scholl, 1985; Scholl, Vallier, and Stevenson, this volume; Bruns et al., 1985; Bruns et al., this volume) includes the northwest-directed thrust faults of the active subduction zone. Thrust faults related to Mesozoic and early Cenozoic periods of subduction may exist at depth along the northern side of Bowers Ridge (zone F, Figure 3B) and along the base of the Siberian and Beringian continental margins (zone E, Figure 3B).

In the abyssal areas of the Bering Sea Basin, the structural zones include the postulated Cenozoic sea-floor-spreading ridge and associated structures in the Komandorsky Basin (zone H, Figure 3A, B; Bogdanov and Neprochnov, 1984) and Mesozoic and Cenozoic basement ridges in the central Aleutian Basin (zone L, Figure 3B; Cooper et al., 1979a, 1979b).

REGIONAL GEOPHYSICAL DATA

In this section, we present compilations of geophysical data for the Bering Sea that include published, unpublished, and newly acquired data. In a later section, we discuss these data in terms of geologic structures within the Bering Sea.

Magnetic Data

Several catagories of magnetic anomalies occur in the Bering Sea region (Figure 4). These include irregular, short-wavelength, and high-amplitude anomalies associated with zones that extend across western Alaska (Godson et al., 1984), northeastern Siberia (Fujita and Newberry, 1982, Figure 3; Cooper et al., 1976), and the Beringian Shelf (Pratt et al., 1972; Childs, Cooper, and Wright, 1981; Bailey et al., 1976; Marlow et al., 1976a, 1983; unpublished map, authors' files). These zones commonly have sharp boundaries, especially the southwestern side of the high-amplitude zone that crosses the Beringian Shelf (Figure 4). Similar anomaly patterns also occur locally over the axial regions of the volcanic masses of the Aleutian, Bowers, and Shirshov Ridges (Cooper et al., 1976; Cooper, Marlow, and Scholl, 1976).

A pattern of linear anomalies of intermediate wavelength (25 to 100 km) exists over the oceanic crustal section of the Aleutian, Bowers, and Komandorsky Basins (Cooper et al., 1976; Cooper, Marlow, and Scholl, 1976). This pattern is interpreted to be sea-floor-spreading anomalies from a Mesozoic fragment of the Kula plate underlying the Aleutian Basin (Cooper, Scholl, and Marlow, 1976) and from a younger Cenozoic crust beneath the Komandorsky Basin (Bogdanov and Neprochnov, 1984).

Long-wavelength anomalies (200 to 400 km) are observed over the Bering Shelf and the Aleutian Basin (unpublished map, authors' files; Cooper, Marlow, and Scholl, 1977). The anomaly measured over the Aleutian Basin is also recorded in satellite magnetic data and is thought to be caused by deep crustal processes (Cooper, Marlow, and Scholl, 1977).

Gravity Data

Free-air gravity data for the offshore Bering Sea region have been compiled by Childs, Magistrale, and Cooper (1985; Figure 5). In areas of large bathymetric relief, such as over the continental slope and rise (Figure 5B), these anomalies partly mimic the sea floor (for example, bathymetric highs correspond to positive gravity anomalies). Free-air gravity anomalies, measured over the central Aleutian Basin where the sea floor is flat, are close to zero, indicating that the crust is nearly isostatically compensated. Gravity values become negative near the continental margin and north of Bowers Ridge, in part because of bathymetric relief and the thick sedimentary sections that lie beneath these areas.

On the shelf (Figure 5A), where there is little bathymetric relief, gravity anomalies indicate changes in rock types and structures. Gravity lows on the shelf are commonly associated with the thick basinal sections of Mesozoic and Cenozoic sedimentary rocks (compare Figures 3C, 5A). Other parts of the Beringian Shelf have moderate gravity anomalies but have no apparent subsurface geologic structures. Regionally, gravity values are about 40 mgal higher over the southern half of the Beringian Shelf than over the northern half. A similar regional increase in gravity anomalies does not occur over the abyssal regions of the Bering Sea Basin (Figure 5B); hence, the increase is likely to be caused by geologic factors, such as regional changes in mantle depth. Significantly, the largest gravity anomalies on the Beringian Shelf are offshore from Cape Newenham, along a southwest trend similar to that of the onshore foldbelts and structures associated with the Ruby geanticline (Figures 3A, 5A). Furthermore, the pattern of gravity anomalies extending offshore from the Kuskokwim Bay region continues along a southwest trend to a point about midway between the Alaskan coast and the edge of the Beringian continental shelf. Here, the pattern turns west and then northwest, a direction that generally parallels that of the zone of high-amplitude magnetic anomalies of the southern Beringian Shelf (Figure 4A). These anomalies may be related to major structural boundaries beneath the shelf, as discussed below.

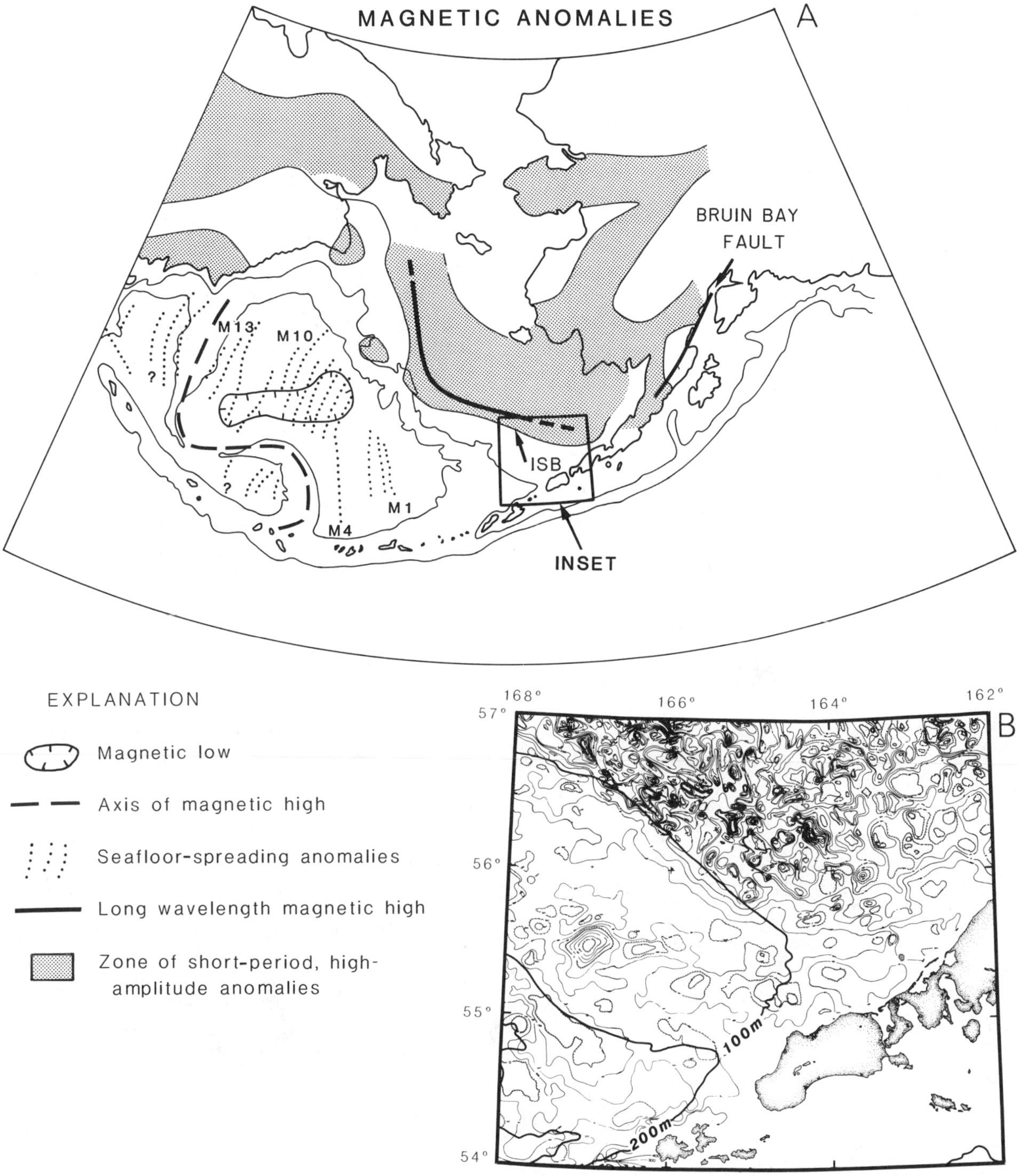

Figure 4. Total field magnetic anomalies in the Bering Sea region. A, Distribution of three catagories of magnetic anomalies (see text for references). (1) Linear anomalies in abyssal basins are oceanic sea-floor-spreading anomalies. (2) Zones of short-period, high-amplitude anomalies located over the inner Beringian Shelf and adjacent onshore areas. Onshore, these zones generally coincide with volcanic and plutonic belts. (3) Long-wavelength anomalies that include a linear positive anomaly at the seaward edge of the inner Beringian Shelf magnetic zone and a broad magnetic low centered over the Aleutian Basin. B, Magnetic anomaly map from Childs, Cooper, and Wright (1981). ISB, the inner shelf boundary, delineates the offshore location where a major change occurs in the characteristics of short-period magnetic anomalies. ISB may be a major structural boundary at the edge of a late Mesozoic to early Tertiary volcanic arc.

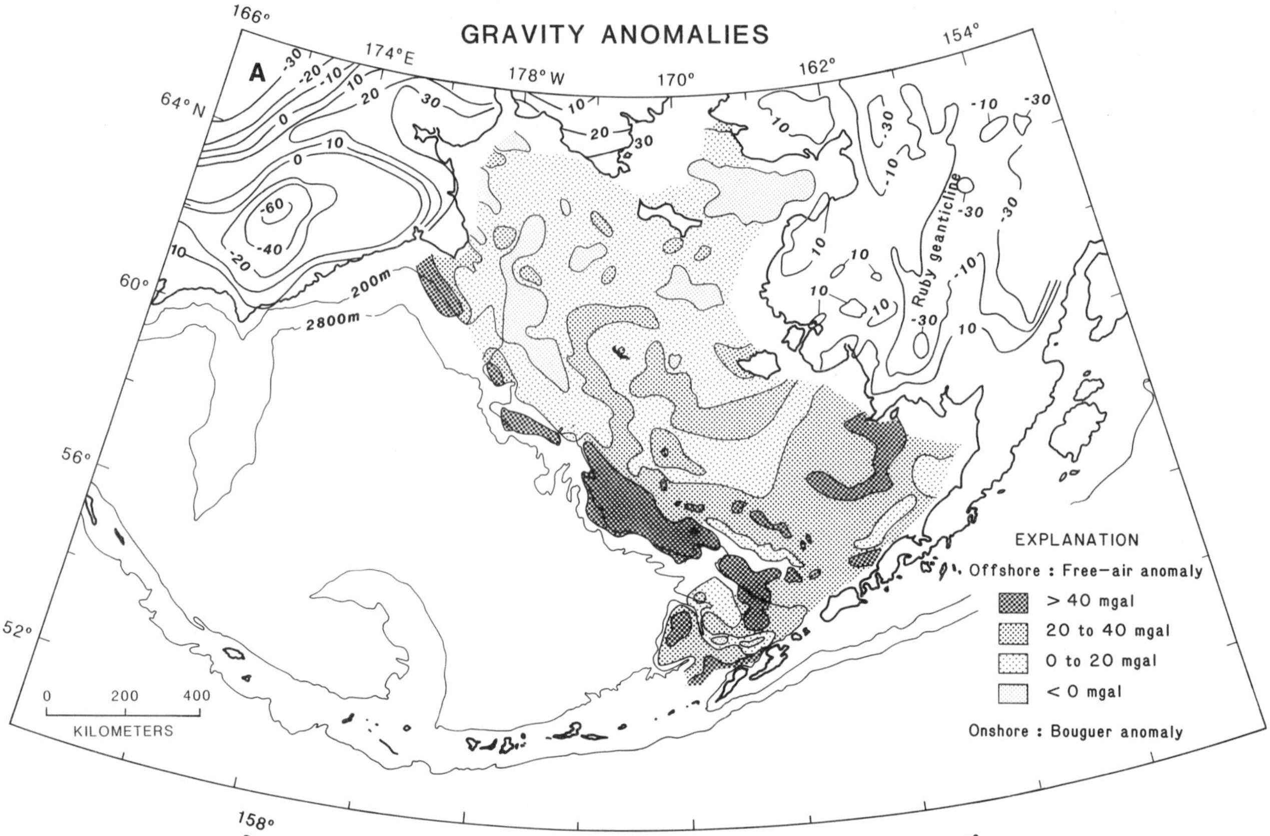

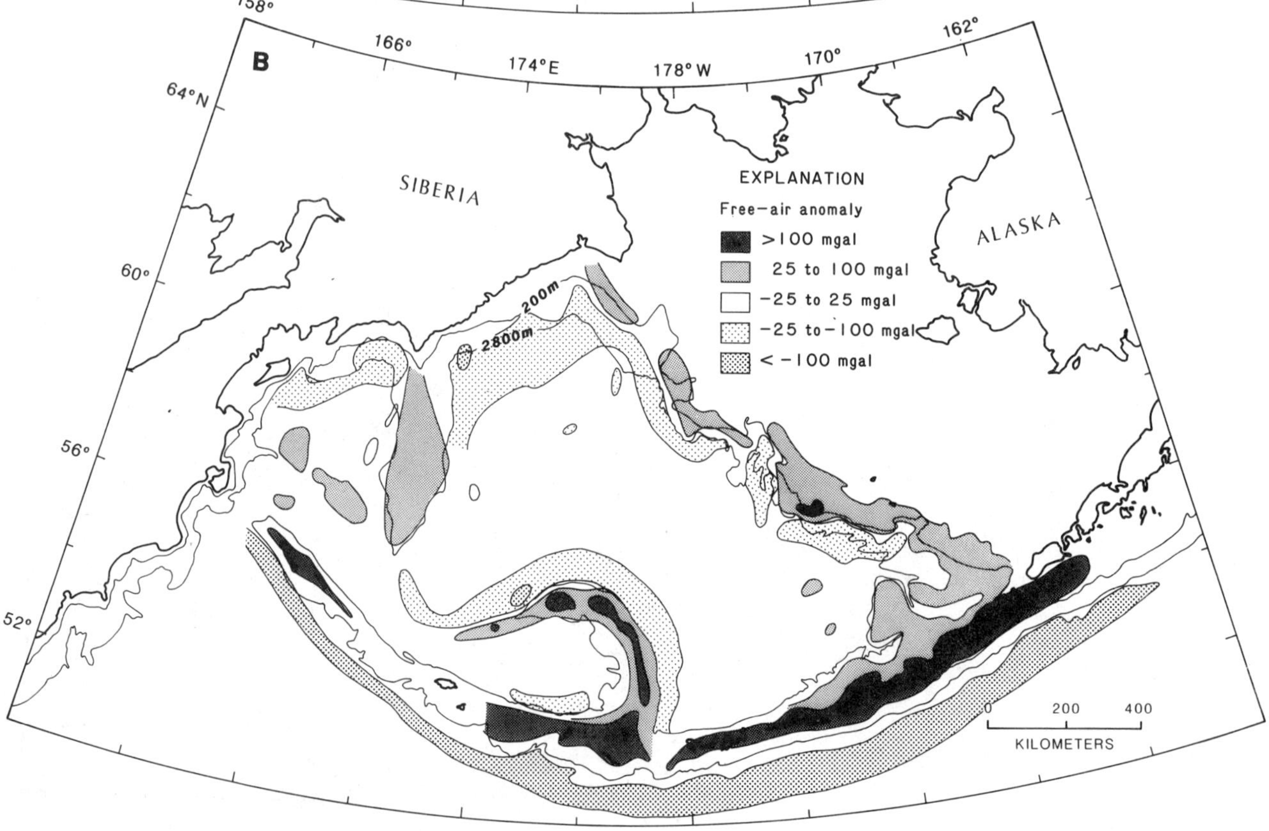

Figure 5. Free-air gravity anomalies for the Bering Sea (from Childs, Magistrale, and Cooper, 1985) and Complete-Bouguer gravity anomalies for southwestern Alaska (from Barnes, 1977) and Siberia (from Stone, 1983). Note that different contour intervals are used for the two maps. A, Gravity anomalies over the Beringian Shelf, southwestern Alaska, and Siberia. Contour interval 10 mgal. Anomaly values are regionally about 40 mgal larger on the southern part of the shelf, where onshore Mesozoic foldbelts trend southwest toward the shelf; in this same area, anomalies change trend midway across the shelf, probably in response to a truncation or bending of the foldbelts. B, Gravity anomalies in the Bering Sea Basin. Contour interval 25 mgal. Here, most large-amplitude anomalies reflect bathymetric relief. Gravity lows around the perimeter of the Aleutian Basin are due in part to 8- to 10-km-thick sedimentary wedges.

Seismic Data

Seismic-reflection, seismic-refraction, earthquake-seismic, and gravity-modelling studies summarize what is known about the deep structure of the Bering Sea crust (Figure 6). Map 6A shows the depth to bedrock or acoustic basement based on seismic-reflection data (Cooper et al., 1979a). In the Bering Sea Basin, these depths are to the top of igneous oceanic crust; for the Beringian Shelf, the depths are to an acoustic horizon that overlies deformed late Mesozoic or early Tertiary rocks.

The total depth to subshelf horizons having refraction velocities > 6.0 km/sec—a velocity that is likely to indicate mafic volcanic, metamorphic, or plutonic rocks—is shown on map 6B (Figure 6). These depths are based on unreversed sonobuoy refraction measurements (Marlow, Cooper, and Childs, 1983; Cooper and Marlow, 1984; Cooper et al., 1980, unpublished data, authors' files) and on a depth to granitic basement measured in a well on the Alaska Peninsula (McLean, 1977).

Map 6C (Figure 6) shows depths to mantle derived from two-ship refraction studies (Shor, 1964; Shor and Fornari, 1976; Ludwig et al., 1971), unreversed sonobuoy measurements (Cooper et al., 1980; Cooper et al., 1979b), ocean-bottom seismometer surveys (Bogdanov and Neprochnov, 1984), earthquake shear-wave traveltimes (Jin and Herren, 1980), and gravity model estimates that use refraction velocity data for upper crustal rocks to control density assignments (Kienle 1971; Grow 1973; unpublished estimates based on Figures 14 and 16 of this chapter).

MAJOR STRUCTURES OF THE BERING SEA CRUST

The major structural provinces of the Bering Sea include (1) the large bathymetric ridges—Aleutian, Bowers, and Shirshov Ridges—of the Bering Sea Basin; (2) the abyssal basins—Aleutian, Bowers, and Komandorsky Basins—and their flanking continental margins and marginal plateaus (Umnak Plateau); and (3) the extensive Beringian continental shelf.

Bathymetric Ridges

Aleutian Ridge

The Aleutian Arc includes the Alaska Peninsula and the Aleutian Ridge; the latter is a classic example of an ensimatic magmatic arch (Scholl, Vallier, and Stevenson, this volume). A northward traverse of the Aleutian Ridge near Amlia Island (Figure 7) shows that structurally the ridge forms a massive subduction complex beneath the landward trench slope—a broad fore-arc basin (Aleutian Terrace)—and the massive volcanic cordillera of the Aleutian Arc. The upper crustal rocks beneath the ridge are divided into three rock series—(1) the lower series of mostly Eocene volcanic and intrusive rocks that form the bedrock core of the ridge; (2) the middle series of Oligocene through Miocene igneous and sedimentary rocks; and (3) the upper series of Pliocene and Quaternary sedimentary and eruptive sequences (Scholl, Vallier, and Stevenson, 1983; this volume).

From gravity and seismicity studies, Grow (1973) shows the Aleutian Ridge as a 25-km-thick crustal welt that is underlain on the south by a northward-dipping Benioff zone and on the north by flat-lying, sediment-thickened oceanic crust. The thick crustal welt is primarily constructed of tholeiitic and calc-alkaline igneous and volcanic rocks of the lower series.

Initial construction of the Aleutian Ridge is believed to have occurred in the early Tertiary, with major volcanic outpourings along the ridge ending in the late Eocene (Scholl, Vallier, and Stevenson, this volume). The Oligocene and Miocene were primarily marked by epizonal plutonism, regional erosion, and deposition of the middle series of sedimentary rocks on the flanks of the arc. Subsidence and collapse of the ridge crest and flanks and the infilling of extensional summit basins with as much as 3 km of upper series strata began in the Pliocene. The massive subduction complex of the landward trench slope, and the basinal structure of the adjacent Aleutian Terrace, also formed initially in the Pliocene.

Bowers and Shirshov Ridges

Bowers Ridge—a massive arcuate bathymetric feature—extends northward from its structural connection with the Aleutian Ridge and curves westward, forming a series of narrow subsurface basement ridges that merge with Shirshov Ridge. Although only undated albitized arc-type volcanic rocks have been dredged from Bowers Ridge, the geomorphology, magnetic signature, and structural setting of the ridge suggest that it is a volcanic arc similar in composition and age (?) to the Aleutian Ridge (Scholl, Buffington, and Marlow, 1985; Cooper, Marlow, and Ben-Avraham, 1981).

Multichannel seismic-reflection data across Bowers Ridge show that it is a flat-topped (wave-cut) structure with a buried trench on its north side, which is filled with a 9- to 10-km-thick sedimentary sequence. A similar buried trench does not occur on the south side of the ridge (Figure 8; Cooper, Scholl, and Marlow, this volume). Early Tertiary and perhaps older rocks are locally deformed at the ridgeward side of the trench and probably beneath the ridge flank (Kienle, 1971; Cooper, Marlow, and Ben-Avraham, 1981). Refraction data (Ludwig et al., 1971) and gravity models (Kienle, 1971) suggest that the 21- to 29-km-deep root of Bowers Ridge is similar to the root of the connecting Aleutian Ridge (Figure 6). Bowers Ridge is presently aseismic, yet the filled trench flanking the ridge and associated deformational structures in the lower trench fill indicate an early Tertiary or older period of southward-directed subduction of Aleutian Basin crust beneath Bowers Ridge (Cooper, Marlow, and Ben-Avraham, 1981).

Shirshov Ridge is a curvilinear submerged mountain range that extends from its southern bathymetric, and probably structural, connection with Bowers Ridge to its northern junction with onshore thrust and flysch belts of eastern Siberia (Figure 9). The ridge lies along a major crustal boundary separating the high-standing (younger and warmer) crust of the Komandorsky Basin from the structurally lower (older and colder) crust of the Aleutian Basin. Rocks dredged from acoustic basement beneath the southern part of the ridge have island-arc affinities and include albitized palagonitic tuff and unaltered fragmental andesite of late Oligocene age. [The age of the dated specimen 70-B70-187 was reported as 16.8 Ma in Scholl, Buffington, and Marlow (1975), but plagio-

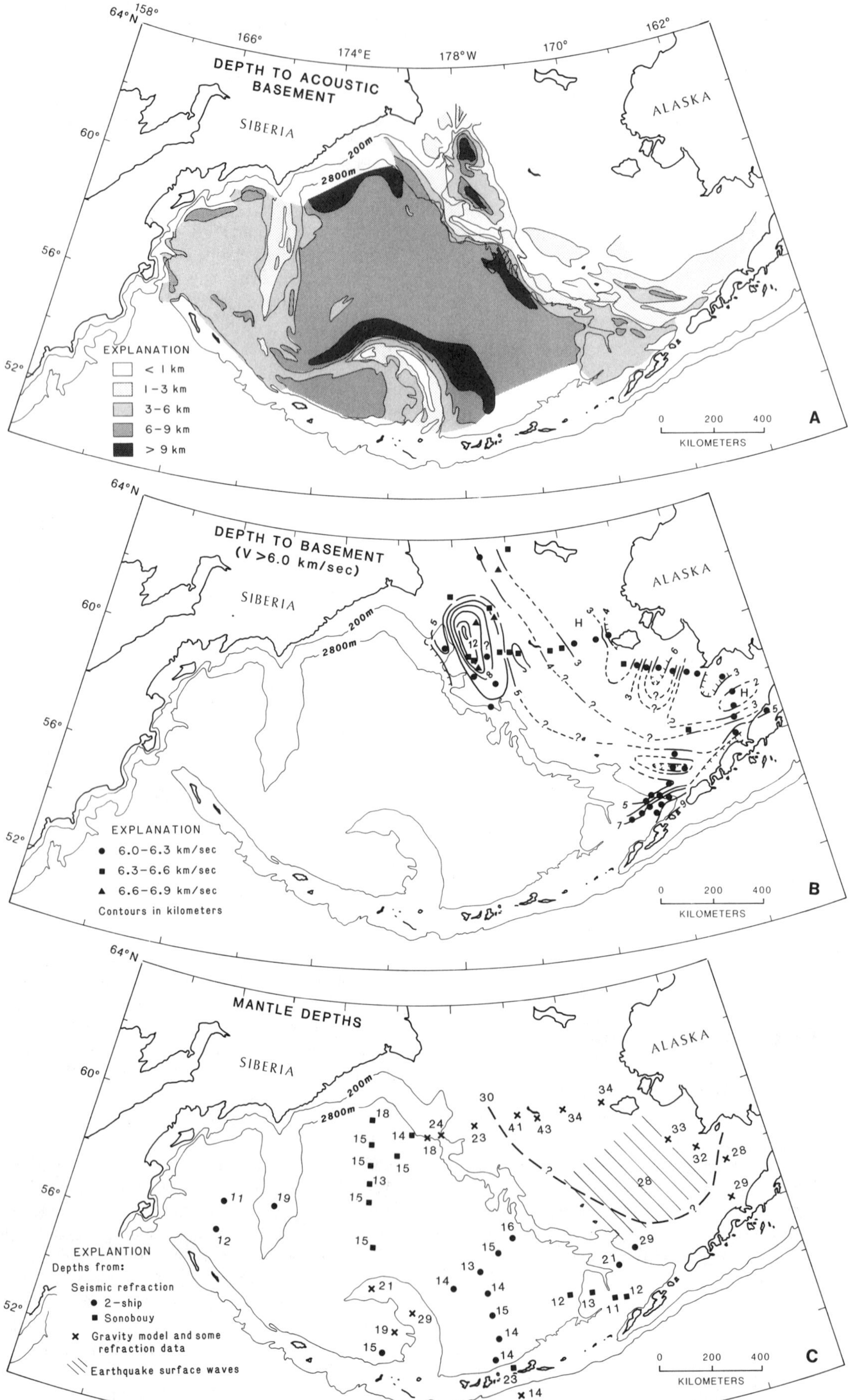

DEPTH TO ACOUSTIC BASEMENT
SIBERIA
ALASKA
64°N
158°
166°
174°E
178°W
170°
162°
60°
56°
52°
200m
2800m
EXPLANATION
< 1 km
1-3 km
3-6 km
6-9 km
> 9 km
0 200 400
KILOMETERS
A
DEPTH TO BASEMENT (V >6.0 km/sec)
SIBERIA
ALASKA
EXPLANATION
6.0-6.3 km/sec
6.3-6.6 km/sec
6.6-6.9 km/sec
Contours in kilometers
B
MANTLE DEPTHS
SIBERIA
ALASKA
EXPLANTION
Depths from:
Seismic refraction
2-ship
Sonobouy
Gravity model and some refraction data
Earthquake surface waves
C

Figure 6. Structural maps showing total depths (in km) to different horizons. A, Depths to bedrock or acoustic basement (from Cooper et al., 1979a). B, Depths to shallowest crustal horizon with a refraction velocity >6.0 km/sec, which is thought to be a velocity for igneous or metamorphic basement of undifferentiated age in the Bering Sea region. Depths are from unreversed sonobuoy-seismic measurements. C, Depths to mantle based on seismic-refraction and earthquake surface-wave measurements, and on gravity-model estimates. See text for data sources.

clase separates from this specimen dated in U.S. Geological Survey laboratories are 27.8 ±1.1 Ma. The latter age is the accepted value. The rock specimen was thickly coated with a manganese oxide encrustation and thus was judged to represent outcrop rather than ice-rafted debris. Severely altered andesitic rocks, probably of ice-rafted origin and collected in the same dredge haul, provide minimum whole-rock ages of Late Cretaceous. Ice-rafted debris, especially from the nearby Koryak drainage, is a major contaminant that hinders the proper identification of basement rock in the Bering Sea.]

Samples of bedrock from the central part of Shirshov Ridge have oceanic-crustal affinities and include basalt, diabase, and highly lithified cherts (bearing radiolarians of Late Cretaceous to Paleocene age; Bogdanov, Vishnevskaia, and Sukhov, 1983). Seismic-reflection profiles (Figure 9) reveal that the ridge's acoustic basement is extensionally faulted and that grabens parallel to the axis of the ridge are filled with thick deposits of Cenozoic clastic and pelagic sediment (Rabinowitz and Cooper, 1977).

The origin of Bowers and Shirshov Ridges is not understood; it has been suggested that they represent ancient spreading centers, island-arcs, or allochthonous terranes (Table 1).

Table 1. Postulated origins for Bowers and Shirshov Ridges and Umnak Plateau (modified from Ben-Avraham and Cooper, 1981)

Feature and Postulated Origin	References
BOWERS RIDGE	
Ancient island arc	Kienle (1971), Scholl, Buffington, and Marlow (1975)
Remnant arc	Karig (1972)
Outgrowth of Aleutian Ridge	Scholl, Buffington, and Marlow (1975)
Microcontinent	Nur and Ben-Avraham (1978)
Allochthonous protostructure	Ben-Avraham and Cooper (1981)
SHIRSHOV RIDGE	
Ancient spreading center	Kienle (1971)
Ancient island arc	Scholl, Buffington, and Marlow (1975), Dickinson (1978), Bogdanov and Neprochnov (1984)
Remnant arc	Karig (1972)
Continuation of onshore Koryak Mountains	Scholl, Buffington, and Marlow (1975)
Microcontinent	Nur and Ben-Avraham (1978)
Allochthonous protostructure	Ben-Avraham and Cooper (1981)
UMNAK PLATEAU	
Subsided continental or arc margin	Scholl, Buffington, and Hopkins (1968); Scholl, Vallier, and Stevenson (1983)
Uplifted oceanic crust	Cooper et al. (1980)
Microcontinent	Emery and Skinner (1977)
Outgrowth of Aleutian Ridge	Scholl, Vallier, and Stevenson (1983)
Allochthonous protostructure	Ben-Avraham and Cooper (1981)

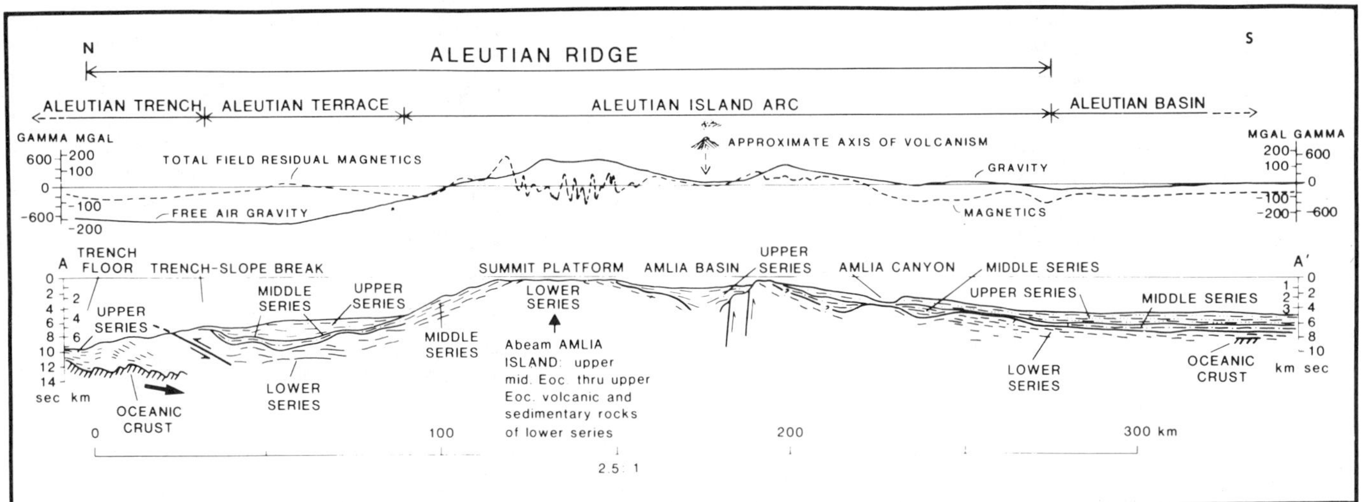

Figure 7. Interpretive cross-section of the Aleutian Ridge near Amlia Island (Scholl, Vallier, and Stevenson, 1983) showing geology and upper crustal structure. See Figure 1B for location.

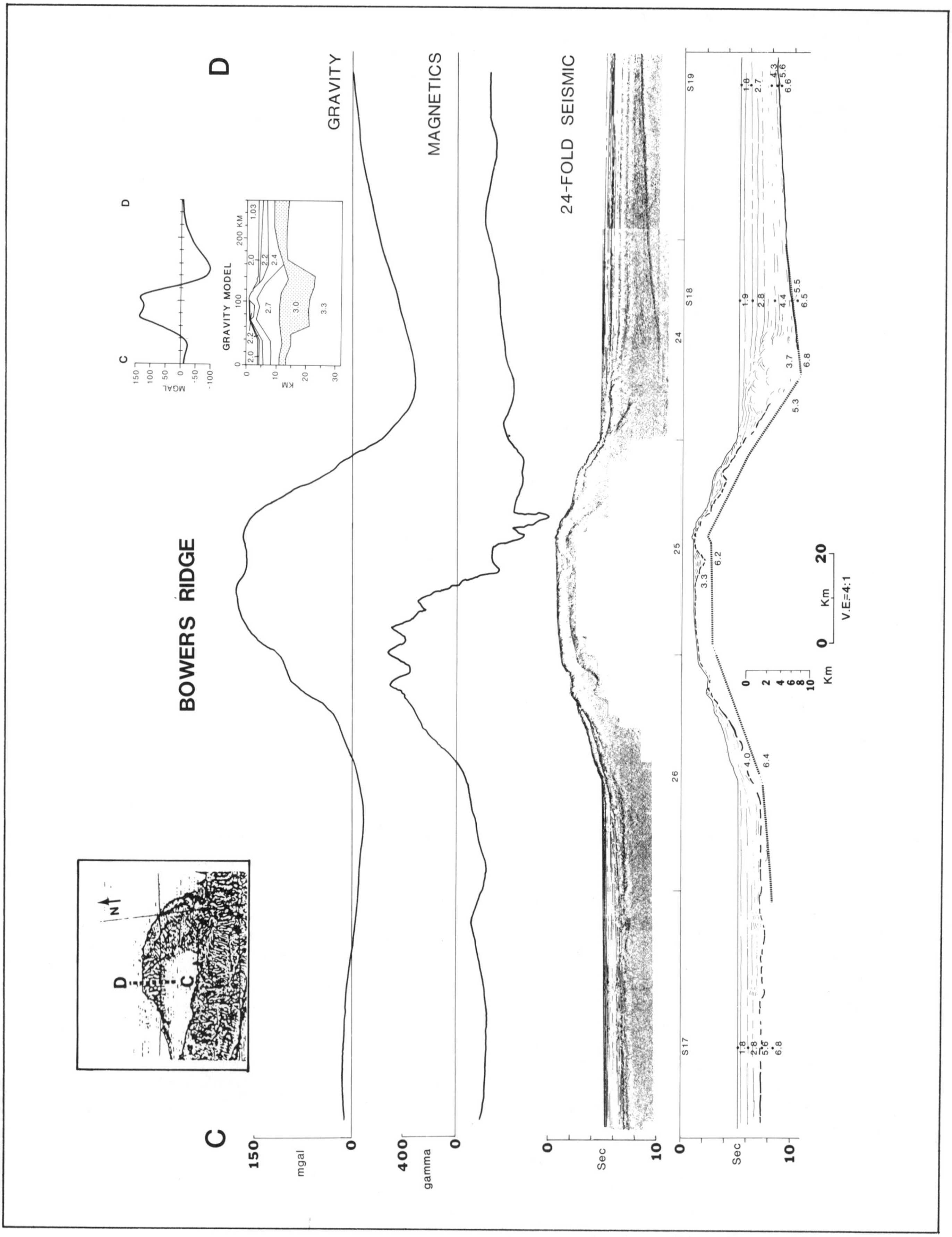

Figure 8. Transect across Bowers Ridge showing the possibly late Mesozoic to early Tertiary buried trench on the north, but not south, side of the ridge (from Cooper, Marlow, and Ben-Avraham, 1981). Gravity model in inset (from Kienle, 1971) is along the same transect. See Figure 1B for location.

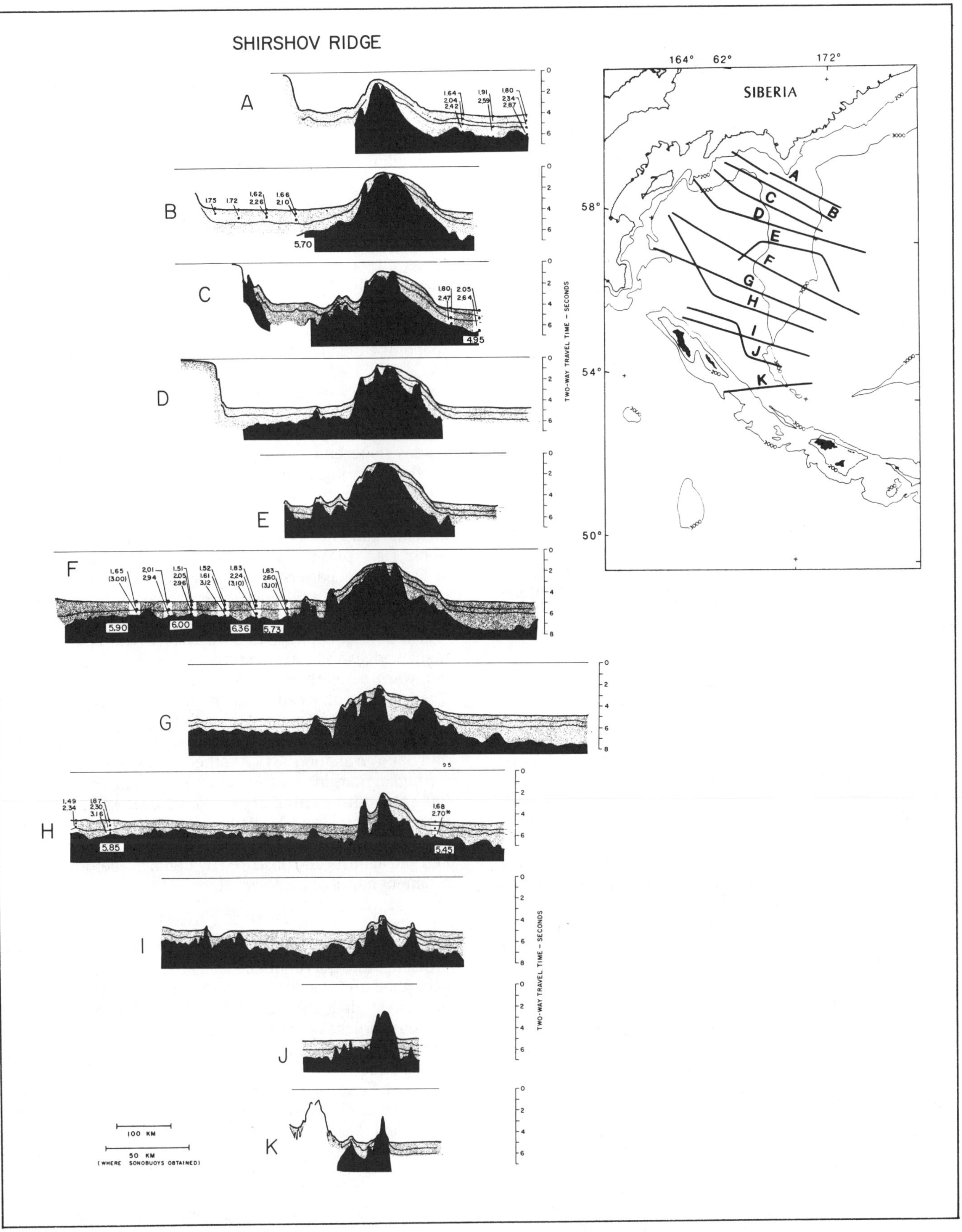

Figure 9. Line drawings of seismic-reflection profiles and sonobuoy velocity measurements across Shirshov Ridge (from Rabinowitz and Cooper, 1977). Black is acoustic basement and velocities are in km/sec.

Abyssal Basins and Basin Margins

Aleutian Basin

The Aleutian Basin constitutes nearly one-third of the Bering Sea and lies in water depths of 2800 to 3600 m. The basin is underlain by oceanic crust of presumed Early Cretaceous age (Cooper, Marlow, and Scholl, 1976), which is covered by 2 to 9 km of sedimentary rock (Figure 6). At DSDP site 190 near the eastern flank of Shirshov Ridge, the age of the upper 900 m of the nearly flat-lying sedimentary section that overlies a buried basement high is middle Miocene(?) and younger (Creager et al., 1973). Deeper horizons are thought to be early Tertiary to late Mesozoic in age, based on regional geologic considerations (Scholl, Buffington, and Marlow, 1975; Cooper et al., 1980) and on dredge samples from the Beringian margin (Marlow et al., 1979b) and the Aleutian Ridge (Scholl, Vallier, and Stevenson, 1983).

Oceanic crust in the Aleutian Basin is thinnest in the central section of the basin (Figures 6, 10), coincident with higher basement relief, slightly higher heat flow, and a long-wavelength magnetic anomaly (Figure 4). Deep crust and upper-mantle convective processes have been proposed to explain these observations (Cooper, Marlow, and Scholl, 1977); however, there is no evidence that sea-floor spreading has occurred in the Aleutian Basin since deposition of the thick sedimentary section. The crust of the Aleutian Basin is, instead, believed to be a remnant of an oceanic plate that was trapped in the Bering Sea when the Aleutian Arc formed, probably in the Eocene (Scholl, Buffington, and Marlow, 1975; Cooper, Scholl, and Marlow, 1976; Scholl, Vallier, and Stevenson, this volume).

Bowers and Komandorsky Basins

Bowers and Komandorsky Basins are small back-arc basins in water depths equivalent to that of the much larger Aleutian Basin. These smaller basins are also underlain by sediment-thickened oceanic crust.

Bowers Basin lies south of the Bowers Ridge and appears, on multifold seismic-reflection data, to have a thinner sedimentary section than the Aleutian Basin (Figure 8). Only the upper few hundred meters of the upper Miocene and younger sedimentary section draping Bowers Ridge and leading into the 2- to 3-km-thick sedimentary section in the basin have been sampled at DSDP site 188 (Figure 1A; Creager et al., 1973). Although the total thickness of oceanic crust is similar in both basins, deep crustal layers in the eastern part of Bowers Basin differ in both thickness and velocity (Ludwig et al., 1971) from those in the Aleutian Basins (Figure 6).

In Bowers Basin, the relatively thin sedimentary section, the distinctive crustal properties, as well as the poorly lineated magnetic anomalies (Figure 4) support the hypothesis that Bowers Basin may have formed by back-arc sea-floor spreading (Ludwig et al., 1971; Cooper, Marlow, and Scholl, 1976; Rubenstone, 1984).

The Komandorsky Basin is the westernmost of the Bering Sea abyssal basins and is structurally distinct from the adjacent Aleutian Basin. The 1- to 2-km-thick sediment fill in the central and southern parts of the basin has been sampled along the western flank of Shirshov Ridge at DSDP site 191 (Figure 1A). Core samples indicate that the section comprises upper Miocene and younger clastic and pelagic deposits that overlie tholeiitic basalt basement. The Komandorsky Basin may have formed more recently than the Aleutian Basin, possibly by sea-floor spreading (Figure 3A; Cormier, 1975; Gnibidenko, 1980; Bogdanov and Neprochnov, 1984). Pertinent observations include high heat-flow values (Watanabe, Lanseth, and Anderson, 1977), high basement relief (Rabinowitz and Cooper, 1977), thickened oceanic crust (Shor and Fornari, 1976), and Neogene igneous and sedimentary rocks (Creager et al., 1973).

Beringian and Koryak Continental Slopes

The Beringian and Koryak continental slopes form the eastern and northern boundaries of the Aleutian Basin and extend 1600 km from the Aleutian Ridge to Cape Oluytorsky. These slopes are incised by submarine canyons (Scholl, Buffington, and Marlow, 1975; Carlson and Karl, 1984), and are underlain by thick sedimentary sequences that drape the upper slope and merge with the upper part of the 4- to 10-km-thick sediment wedges, or rise prisms, beneath the lower slope and continental rise (Figure 11; Cooper et al., 1979b).

Neogene and younger sediment prisms beneath the continental slope and rise are thickest along the Koryak and northern Beringian margins, although deeper sedimentary layers are probably present. Paleogene strata, however, appear to be thickest along the southern and central Beringian margin. The regional difference in thickness of the Neogene rise prisms is thought to reflect a change in sedimentation patterns caused by the middle Miocene to Pliocene uplift of the Koryak and coastal mountain belts (Cooper, Scholl, and Marlow, this volume).

Mesozoic igneous oceanic crust lies beneath the thick sediment wedges and can be traced on seismic-reflection profiles landward beneath Mesozoic continental crust (Figure 12). The presence of oceanic basement beneath continental bedrock may result from late Mesozoic to early Tertiary subduction beneath the central Beringian margin (Cooper, Marlow, and Ben-Avraham, 1981). At other places along the northern Beringian margin, oceanic basement has not been detected beneath continental crust on seismic-reflection data (Cooper, Scholl, and Marlow, this volume). In these areas, the basement overlap may not exist or may have been modified and acoustically obscured by Cenozoic deformation and sediment burial of the lower slope.

Umnak Plateau

The Umnak Plateau is a triangular bathymetric high that protrudes into the Aleutian Basin at the junction of the Beringian Shelf and the Aleutian Ridge (Figure 1A). The plateau lies at a depth of 1600 to 2800 m. Mesozoic sedimentary and igneous(?) rocks of the Peninsular terrane are thought to lie beneath the continental margin on the northeast side of the plateau (belt F, Figure 2B; Marlow and Cooper, 1980), whereas Cenozoic island-arc rocks of the Aleutian Ridge form the southeast boundary of the plateau (belt G, Figure 2B; Scholl, Vallier, and Stevenson, 1983). On the west, the Umnak Plateau is flanked by Mesozoic igneous crust and Cenozoic sedimentary rocks that underlie the Aleutian Basin (Cooper et al., 1979b). The plateau consists of a 9- to 14-km-thick transitional crust (Cooper et al., 1980; Figures 6 and 13).

The origin of the Umnak Plateau is unknown, and many hypotheses have been proposed (Table 1). It may have formed in the late Mesozoic to early Tertiary by (1) uplift of

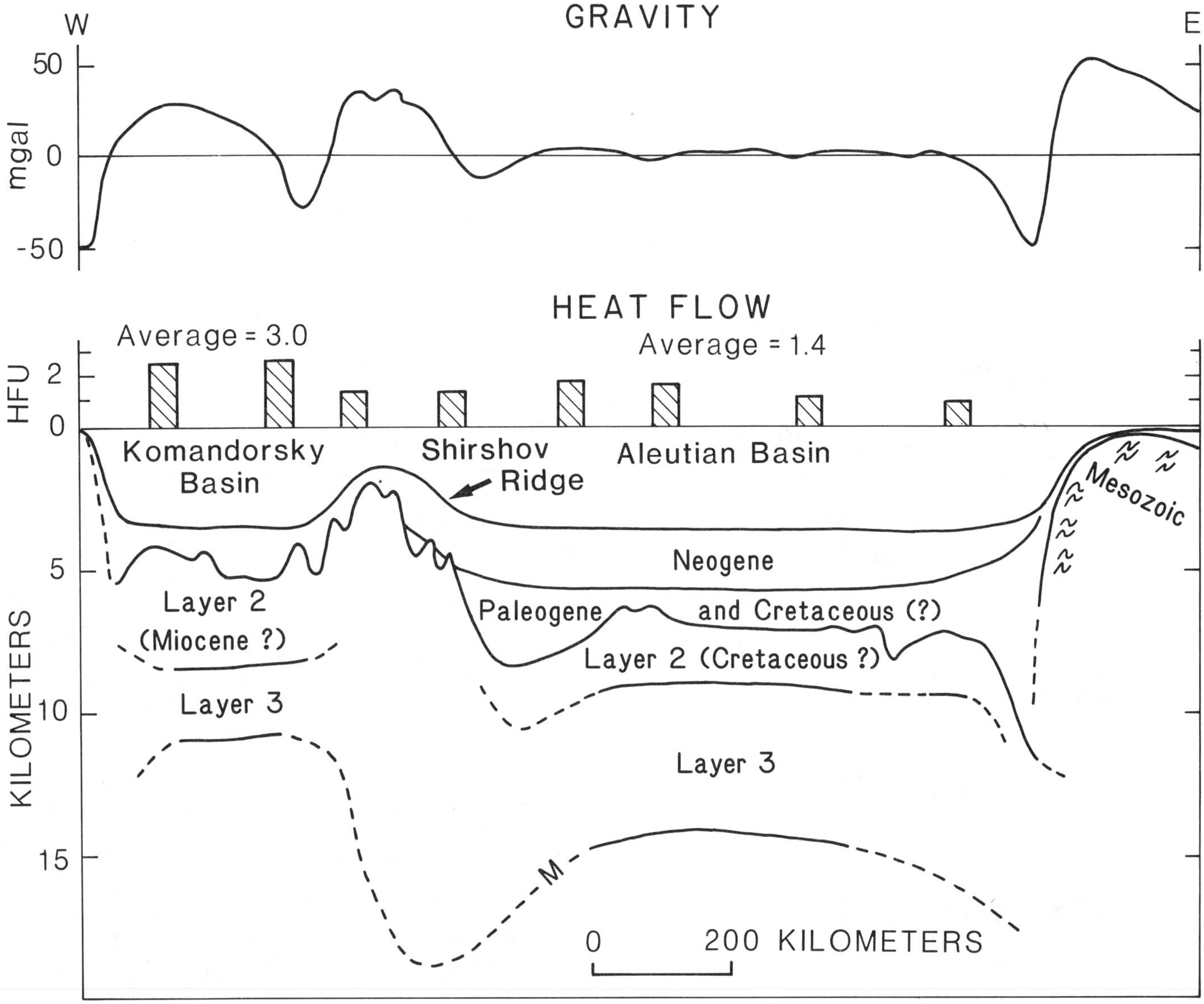

Figure 10. Interpretive cross-section of the Aleutian and Komandorsky Basins based on structural depths shown in Figure 6, sonobuoy refraction from Cooper et al, (1979b), ages from Figure 12, and heat flow from Langseth, Hobart, and Horai (1980). M is mantle. Note the difference in age and geophysical signature between the Komandorsky and Aleutian Basins. See Figure 1B for location.

oceanic crust in the Aleutian Basin (Cooper et al., 1980); (2) collision of an oceanic plateau or microcontinent (on the Kula plate) with the southern Beringian margin prior to development of the Aleutian Ridge (Ben-Avraham and Cooper, 1981); (3) subsidence of an outlying sector of the Aleutian Ridge (Scholl, Vallier, Stevenson, 1983); or (4) erosion and subsidence of the outermost edge of the Beringian Shelf (Scholl, Buffington, and Hopkins, 1968).

Large structural depressions developed along the east and northeast flanks of the plateau, and have been filled by a 1- to 9-km-thick sedimentary sequence derived, in part, from the Aleutian Ridge and Alaskan drainages. Large submarine canyons that incise the plateau were cut most recently in the late Cenozoic, and may have existed at various times throughout the Tertiary (Scholl et al., 1970; Cooper et al., 1980).

Beringian Shelf

The Beringian continental shelf, which is the large shallow-water area that lies adjacent to western Alaska and south of the Bering Strait (Figure 1B), can be divided into the inner-shelf and outer-shelf provinces. The inner-shelf province extends west and southwest from mainland Alaska to the northeastern edge of the large structural basins, Navarin and St. George, that border the edge of the Beringian Shelf (Figures 1A, 3C). The outer shelf includes the remaining outboard areas of the Beringian Shelf. A major structural boundary separates the two provinces near the Navarin and St. George basins.

Inner Shelf

The inner shelf is a nearly flat shallow-water (10- to 80-m-deep) platform underlain by a veneer of as much as 500 m of Cenozoic deposits. These deposits unconformably overlie a wave-planed bedrock of Mesozoic to early Tertiary (?) sedimentary and volcanic units. In the Kuskokwim Bay and Norton Sound areas, the bedrock has subsided, forming structural troughs that have filled with as much as 6 km of

Cretaceous and younger sedimentary rocks (Marlow, Cooper, and Fisher, this volume; Cooper and Marlow, 1984).

Multichannel seismic-reflection and sonobuoy seismic-refraction transects across the inner shelf show that bedrock is shallow (horizon AB, Figure 14) and includes rocks with velocities ranging from 3.6 to 4.9 km/sec. The rocks may be deformed, because they are often acoustically opaque. Multichannel seismic-reflection profiles rarely show deep reflections. High-velocity rocks (6.0 to 7.2 km/sec) lie at relatively shallow depths of 1 to 4 km beneath the bedrock surface over large parts of the inner shelf (Figures 6, 14).

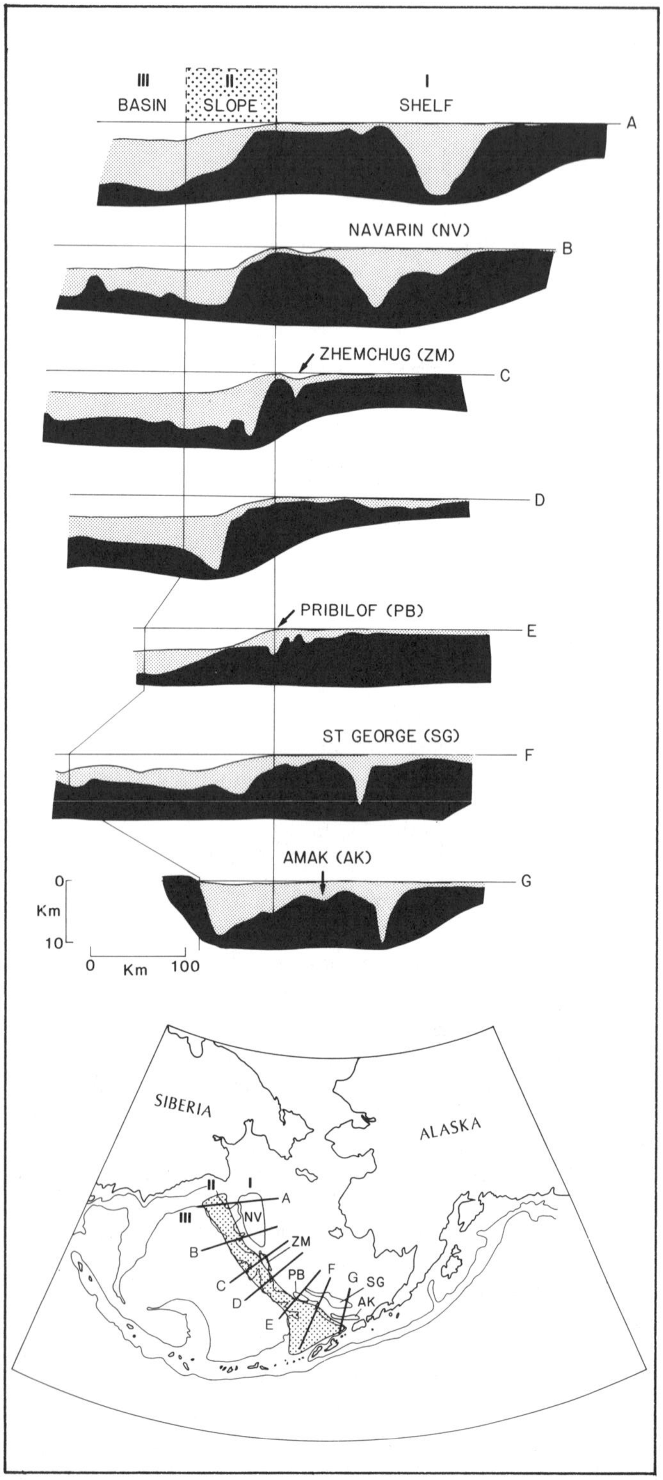

Figure 11. Simplified cross-sections of the Beringian margin based on structural map in Figure 6A. Black is bedrock or acoustic basement; the stippled layer is sedimentary.

Compilations of sonic velocities in wells drilled in western Alaska (Figure 15; Fisher, Patton, and Holmes, 1982), indicate that unmetamorphosed sedimentary and volcanic rocks have velocities lower than 6.0 km/sec. The only rocks sampled that have velocities higher than 6.0 km/sec are plutonic and metamorphic units. Rocks with similarly high velocities, but from areas other than Alaska, are metasedimentary and mafic igneous. Hence, we believe that rocks beneath the shelf with refraction velocities >6.0 km/sec are plutonic, mafic igneous, or metamorphic basement units.

Bedrock units overlying basement therefore have an average thickness of <4 km beneath large areas of the inner shelf (Figure 6, 14). The thickest bedrock sequence lies beneath Kuskokwim Bay, where nearly 6 km of probably sedimentary rocks of late Mesozoic age occur within a structural depression that is probably the offshore continuation of the onshore Kuskokwim Basin. This successor basin is filled, onshore, with more than 12 km of Upper Cretaceous marine turbidite and shallow-marine strata (Decker, 1984).

High-amplitude and short-period magnetic anomalies, characteristic of shallow magnetic sources, are common over the inner shelf (Pratt et al., 1972; Marlow et al., 1976a; McGeary and Ben-Avraham, 1981; Cooper and Marlow, 1984; Figures 4, 14). Along transect 4-5 (Figure 14), these anomalies can tentatively be correlated with onshore exposures that include near-surface, early Tertiary volcanic rocks on Hagemeister (A) and St. Matthew Island (E, transect 6-9), mafic and ultramafic rocks on Cape Newenham (B), buried middle Mesozoic (?) volcanic units off Kuskokwim Bay (C), and Quaternary volcanic bodies offshore from Nunivak Island (D).

The seaward edge of the inner-shelf province corresponds to a major structural (possibly a fault) boundary associated with abrupt changes in the character of the gravity and magnetic anomalies. Herein, we name this structure the inner-shelf boundary (ISB). The ISB can be traced magnetically across most of the Beringian Shelf, from Bristol Bay southwest, and then northwest to the Gulf of Anadyr (Figure 4). Along the landward edge of the Navarin and St. George basins, the boundary follows a structural hinge line. At the hinge line, bedrock and basement horizons drop steeply southwestward into the outer-shelf basins, and, at the Navarin Basin, mantle dips eastward beneath the inner shelf (Figures 6, 14, 16).

In Bristol Bay, the ISB is not clearly defined by distinct changes in the magnetic data. If the ISB is projected northeastward beneath Bristol Bay Basin along the regional structural grain (Figures 3A, 5A, 6A, 6B), then the ISB may connect with the Bruin Bay fault, which is mapped on the Alaska Peninsula (Figures 3A, 4). This projection of the ISB places it about 50 to 100 km south of the hinge line for the Bristol Bay Basin. Hence, the boundary between the inner- and outer- shelf provinces in Bristol Bay, if it exists, has not yet been well defined.

The inner-shelf boundary may also mark a fundamental change in the type of bedrock or basement rock, from principally igneous or metamorphic beneath the inner shelf, to mostly sedimentary beneath the outer shelf. In the St. George Basin, the ISB lies within the suspected limits of the

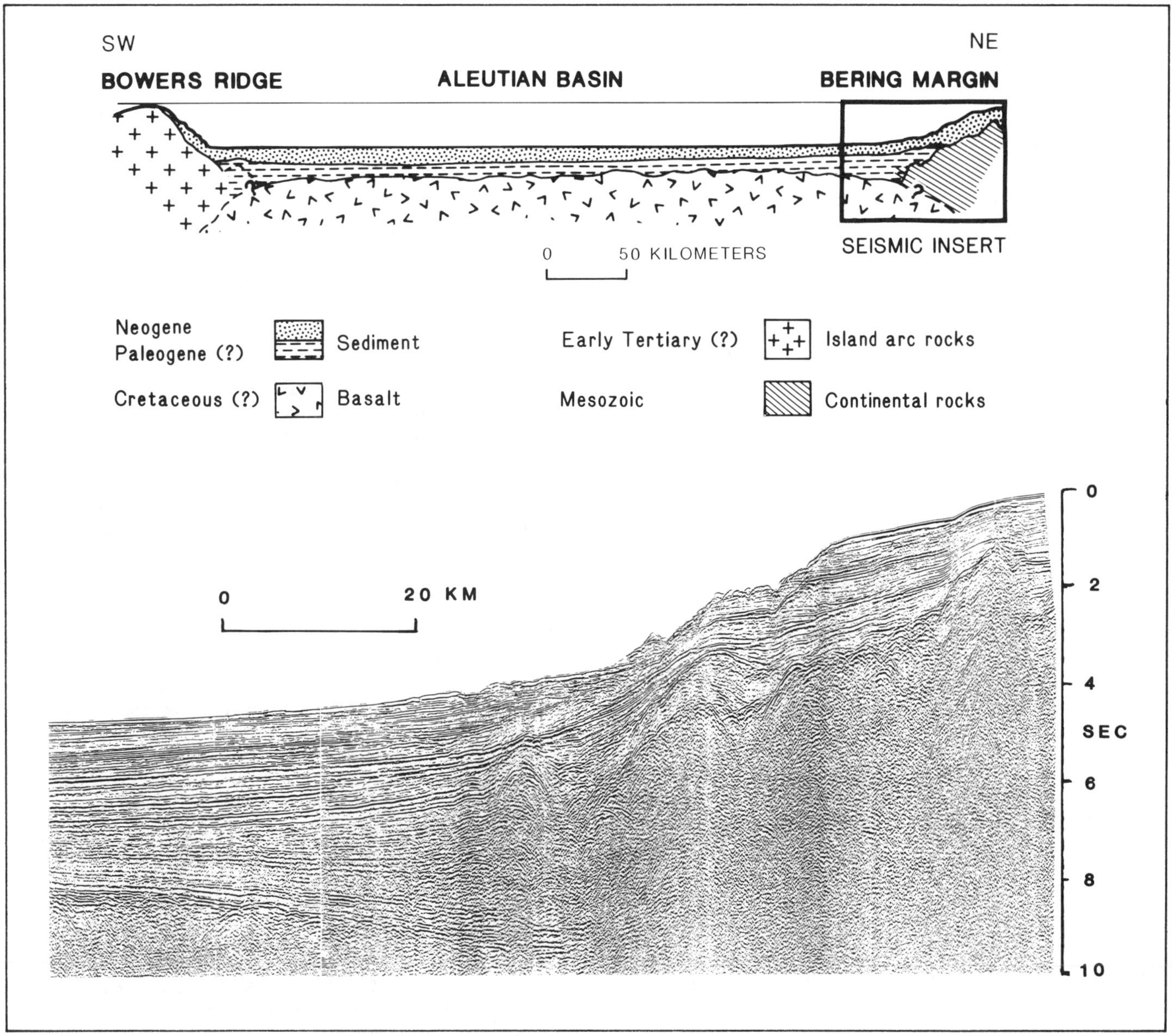

Figure 12. Interpretive section and multichannel seismic-reflection data across the Aleutian Basin (from Cooper, Marlow, and Ben-Avraham, 1981). Note the geometric similarity between the thick sediment wedges at the base of Bowers Ridge and the Bering margin, as well as the absence of a spreading ridge in the Aleutian Basin. See Figure 1B for location.

Peninsular terrane and may separate Mesozoic and Tertiary volcanic and plutonic rocks on the north side from Mesozoic sedimentary rocks on the south side. This structural configuration may be equivalent to that of the Bruin Bay fault, where volcanic-plutonic rocks (north side) are juxtaposed against sedimentary rocks (south side; Beikman, 1980; Detterman and Reed, 1980).

Numerous investigators have suggested that coeval foldbelts in Alaska and Siberia may be related and connect beneath the inner Beringian Shelf (Burk, 1965; Scholl, Buffington, and Marlow, 1975; Marlow et al., 1976a; Patton and Tailleur, 1977); but, as discussed by McLean (1979c), there is great uncertainty about the connection between the volcanic and plutonic belts, principally because of such limited geologic and geophysical data. More recent data (Figures 2, 4, 5, 6) provide stronger evidence that many of the Mesozoic and older rocks thought to underlie the Beringian Shelf are part of the same rock belts exposed in western Alaska.

Northern Edge of the Peninsular Terrane

In southwest Alaska, the Peninsular terrane consists of a southern Mesozoic sedimentary province in fault contact with a northern volcanic and plutonic source province of late Paleozoic and Mesozoic age. The northwestern province is buried beneath the north side of the Alaska Peninsula (Figures 2A, 2B), and consequently we do not know the northwestern limit of the Peninsular terrane either onshore or beneath Bristol Bay. For reasons discussed below, we suspect that it lies beneath the middle of Bristol Bay and trends southwestward toward St. George Basin.

Midway between the Alaska Peninsula and Cape Newenham (Figure 1) offshore gravity anomalies associated with the rocks of the Ruby geanticline and Kuskokwim Basin (Figure 5) are either truncated or turn northwestward

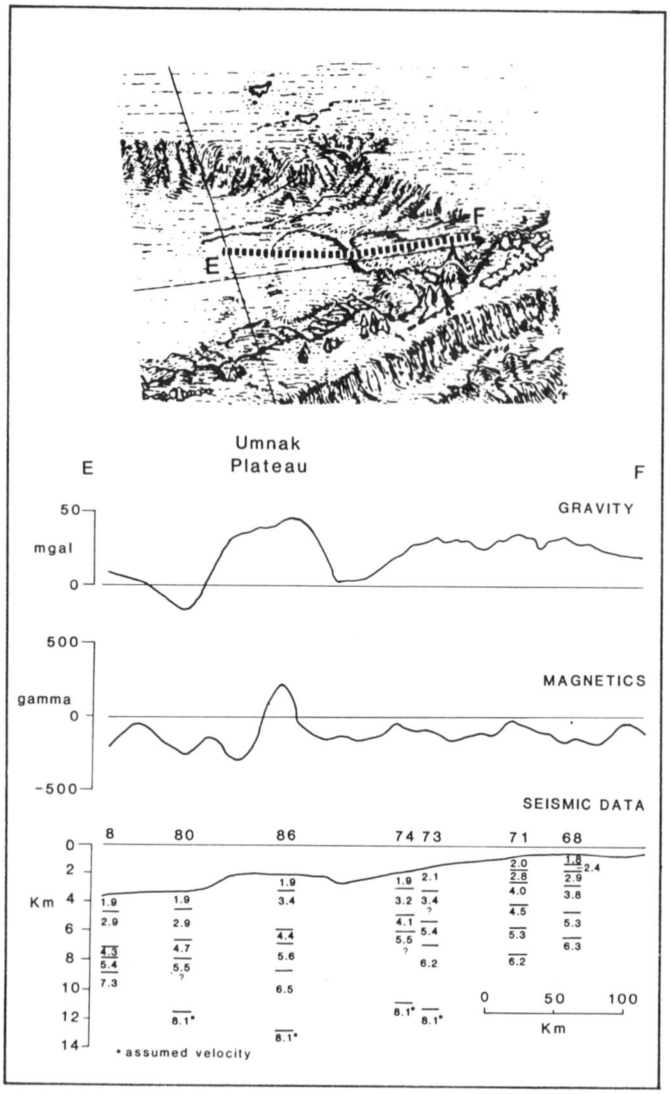

Figure 13. Transect across the Umnak Plateau (from Ben-Avraham and Cooper, 1981). Velocities are in km/sec. See Figure 1B for location and comparison with physiographic inset.

toward Siberia. At this location, bedrock and basement units begin to dip southeast into Bristol Bay Basin at a hinge line that trends toward St. George Basin (Figures 6, 14). Multichannel seismic-reflection and sonobuoy seismic data indicate deformation of the bedrock and basement units at the hinge line (point A, Figure 17), where small local magnetic and gravity anomalies also occur (Figure 14).

Because seismic data are sparse, we cannot map the extent of the deformed zone. However, magnetic and gravity maps suggest that the northern edge of the Peninsular terrane lies beneath the inner-shelf, within the zone of high-amplitude magnetic anomalies. This edge of the Peninsular terrane trends southwestward across Bristol Bay, and parallel to the ISB. Near St. George Basin, the trend of the postulated terrane edge changes to northwest and continues to at least Zhemchug Canyon, and possibly beyond. In the vicinity of the Pribilof Islands, the Peninsular terrane may therefore lie offshore between the Alaska Peninsula and the region of the inner shelf where gravity anomalies abruptly change trend from southwest to northwest (Figure 5A).

Arc Complex Beneath the Inner Shelf.

We suspect that the abrupt southwestern edge of the zone of high-amplitude magnetic anomalies, which covers most of the inner Beringian Shelf (Figure 4), marks the structural edge of a magmatic arc complex. Onshore, volcanic and plutonic belts composed principally of calc-alkaline rocks of Mesozoic and early Tertiary age overlap and strike toward the magnetic zone (Figures 2D, 2E, 2I, 4). The offshore parts of the onshore belts have not been sampled. It is possible that the magnetic zone is caused by igneous and volcanic bodies in the overlapping magmatic belts.

Areas directly northeast of the Navarin and St. George basins may be underlain by magmatic bodies associated with a Late Cretaceous to Paleogene volcanic arc. The east side of the Navarin Basin appears, from gravity model studies (Figure 16B), to have thicker crust than other parts of the inner shelf. Also, a long-wavelength magnetic anomaly (Figure 4) lies above the thicker crust and extends along most of the Beringian margin. Calc-alkaline rocks of early Tertiary age are found near the Navarin Basin on St. Matthew Island (Figures 1, 2I); these rocks lie within 100 to 200 km of the presumed late Mesozoic and early Tertiary oblique-subduction or transform zone at the base of the Beringian margin.

The high-amplitude magnetic zone of the inner Beringian Shelf is significantly wider near Alaska than near Siberia (Figure 4). Likewise, the onshore late Mesozoic and early Tertiary volcanic-plutonic belt of southwest Alaska (belt, B, Figure 2B) is broader at the coast than corresponding belts lying close to the coast in easternmost Siberia (belts C and H, Figure 2B). The greater width may be related to the accretion of the Peninsular terrane against southwestern Alaska. Arc magmatism, on the north side of a small intervening ocean basin, may have shifted southward when the ocean basin closed with the "docking" of the Peninsular terrane in the late Cretaceous (Figures 18A, 19; Wallace and Engebretson, 1984).

Outer Shelf

The outer shelf includes those parts of the Beringian Shelf underlain by the large, sediment-filled St. George and Navarin basins. These basins, as described by Marlow, Cooper, and Fisher (this volume), developed initially during the late Mesozoic or early Tertiary as extensional structures. Subsequent Cenozoic subsidence along extensional growth faults, augmented by sediment infilling, depressed the underlying Mesozoic bedrock surface to depths of 2 to 13 km.

Generalized cross-sections of the basins along the Beringian margin (Figure 11) show that basins of the northern shelf are generally larger than those of the southern shelf. In the north, Navarin Basin is a broad feature composed of at least three subbasins. Late Cenozoic compressional structures are common in the wide northern end of the Navarin Basin (profile A, Figure 11), but Neogene and older extensional wrench-fault structures are prevalent at the narrower southern end of the basin (profiles B and C, Figure 11; Turner et al., 1985; Steffy et al., 1985; Marlow, Cooper, and Fisher, this volume). The central Beringian margin has a few small extensional basins with thin sedimentary sections (profiles C, D, E, Figure 11). Two of these basins, Zhemchug and Pribilof, were breached by faults and eroded on their seaward sides, possibly in the Paleogene, and now form the T-shaped submarine canyons that incise the margin. The southern

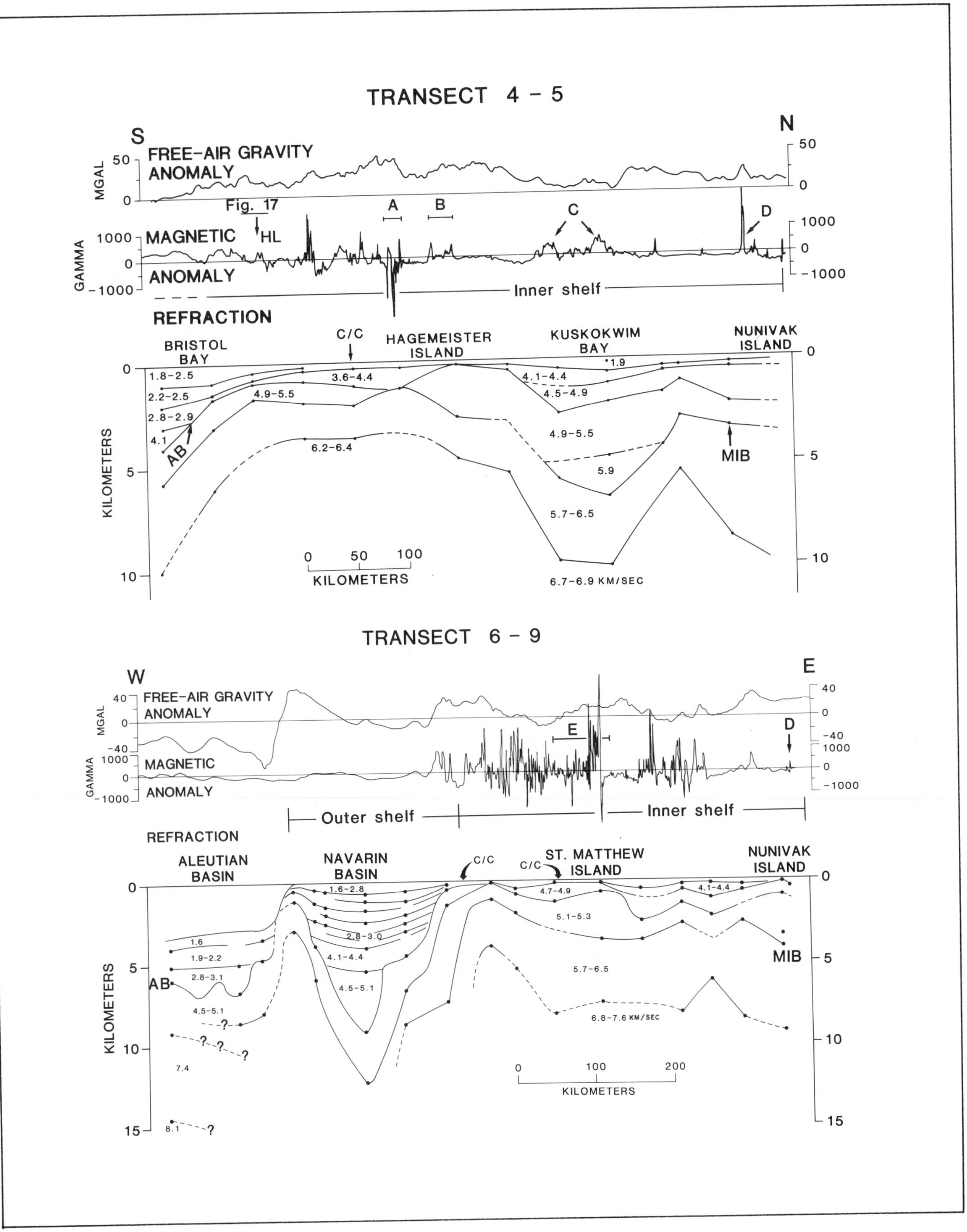

Figure 14. Crustal transects across the Beringian Shelf (from Cooper and Marlow, 1984). Numbers are refraction velocities measured from unreversed sonobuoys and large air-gun (1200 to 2500 cu in.) sound source. AB is acoustic basement and bedrock; MIB is metamorphic/igneous(?) basement. Transect 4–5 is across the southern and inner Beringian Shelf. Transect 6–9 is across the central Beringian Shelf from western Alaska to the Aleutian Basin. See Figure 1B for location.

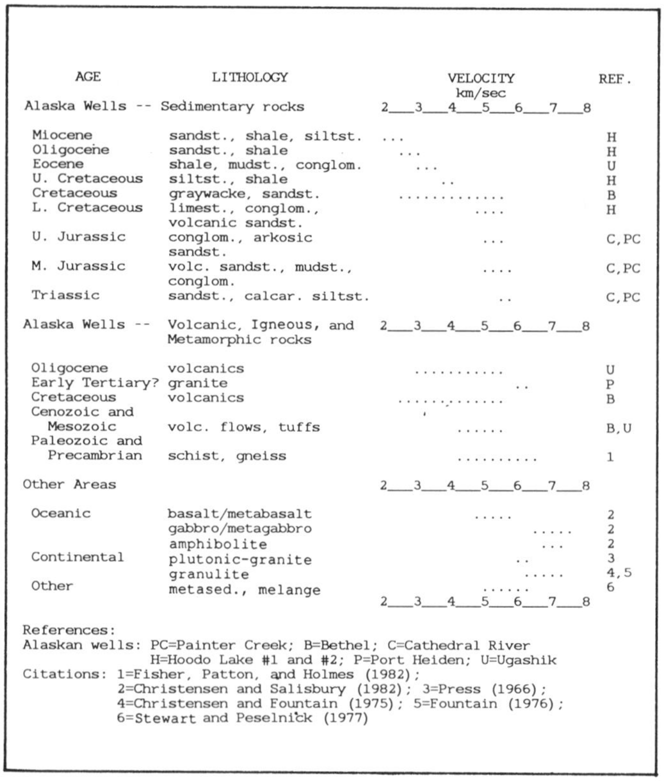

Figure 15. Rock velocities from in-situ measurements in several Alaskan wells and from laboratory measurements on rocks from other areas. Alaskan sonic logs are from the 6 wells marked by letters in Figure 1A. In-situ velocities for sedimentary and volcanic rocks, buried at depths to 4 km, are < 5.8 km/sec; rocks with velocities > 6.0 km/sec are igneous, metamorphic, or undifferentiated melange rocks.

margin is incised along Neogene and older growth faults into the deep subbasins of St. George Basin (profiles F, G, Figure 11; Marlow and Cooper, 1980).

Navarin and St. George basins are flanked on their seaward sides by bedrock ridges of Paleogene and Mesozoic sedimentary rocks that separate the basins from the 5- to 10-km-thick sedimentary prisms beneath the continental slope and rise. Bedrock beneath St. George Basin is believed to be Mesozoic strata of the Peninsular terrane (Marlow and Cooper, 1983). Coeval rocks, possibly of the Peninsular terrane, probably underlie the Navarin Basin.

The deeper structures of the northern and southern parts of the outer Beringian Shelf differ (Table 2). In the north, the crust beneath the Navarin Basin is about 23 km thick (Figures 6C, 16). The crust thickens rapidly to about 42 km on the landward side of the basin. This zone of rapid thickening occurs beneath the structural boundary that delineates the hinge line of the basin. In the south, however, the crust beneath St. George Basin is about 29 km thick (Figure 6C; Shor, 1964), and the zone of rapid crustal thickening occurs on the seaward side of the basin. The structural hinge line along the landward side of the St. George Basin does not appear to be associated with additional thickening of the crust.

Change in tectonic regime may be the main factor that controls the differences in basin sizes, in thicknesses of the crust, in the position of the zone of rapid crustal thickening, and in the orientation of basin hinge lines between the northern and southern outer-shelf areas. These differences may indicate variations in tectonic stress caused by northwest-directed transform motion along the outer Beringian margin, which was initiated in the late Mesozoic to early Tertiary. The narrow extensional basins, such as St. George Basin, occur in shelf areas that may have been subjected to convergent wrench faulting (toward the inner shelf). The wider extensional basins, such as the Navarin Basin, are in areas possibly affected by divergent wrench faulting (away from the inner shelf).

The Anadyr and Bristol Bay Basins appear to lie outside the extensional regime of the outer Beringian margin and appear to be broad crustal sags associated with magmatic arcs. Anadyr Basin is considered a fore-arc basin of the Okhotsk-Chukotsk volcanic arc in Siberia (Figures 3A, 3C; McLean, 1979a; Marlow, Cooper, and Childs, 1983), and the Bristol Bay Basin lies behind the magmatic arcs of the Alaska Peninsula.

EVOLUTIONARY MODELS

Current evolutionary models for the Bering Sea region rely heavily on the concept that allochthonous tectonostratigraphic terranes have amalgamated to form a framework of

Table 2. Characteristics of Navarin and St. George Basins.

Characteristic	Navarin Basin	St. George Basin	Reference
Basin width	40–120 km	15–40 km	Figure 3C
Sedimentary fill	13–14 km	9–10 km	Figure 6A
Depth to mantle:			
-Beneath basin	20–22 km	28–30 km	Figures 6C, 16
-On landward side of basin	41–43 km	28–29 km	Figures 6C, 16
Location of maximum zone of crustal thickening	Landward edge of basin	Seaward edge of basin	Figures 6C, 16
Structural trend of basin hinge line	N-NW	W-NW	Figure 3C
Postulated rock type beneath basin	Peninsular terrane? (SE) Melange (NW)	Peninsular terrane	
Structural setting	Extensional at SE end of basin (divergent wrench faulting?) Compressional at NW end of basin	Extensional (convergent wrench faulting?)	

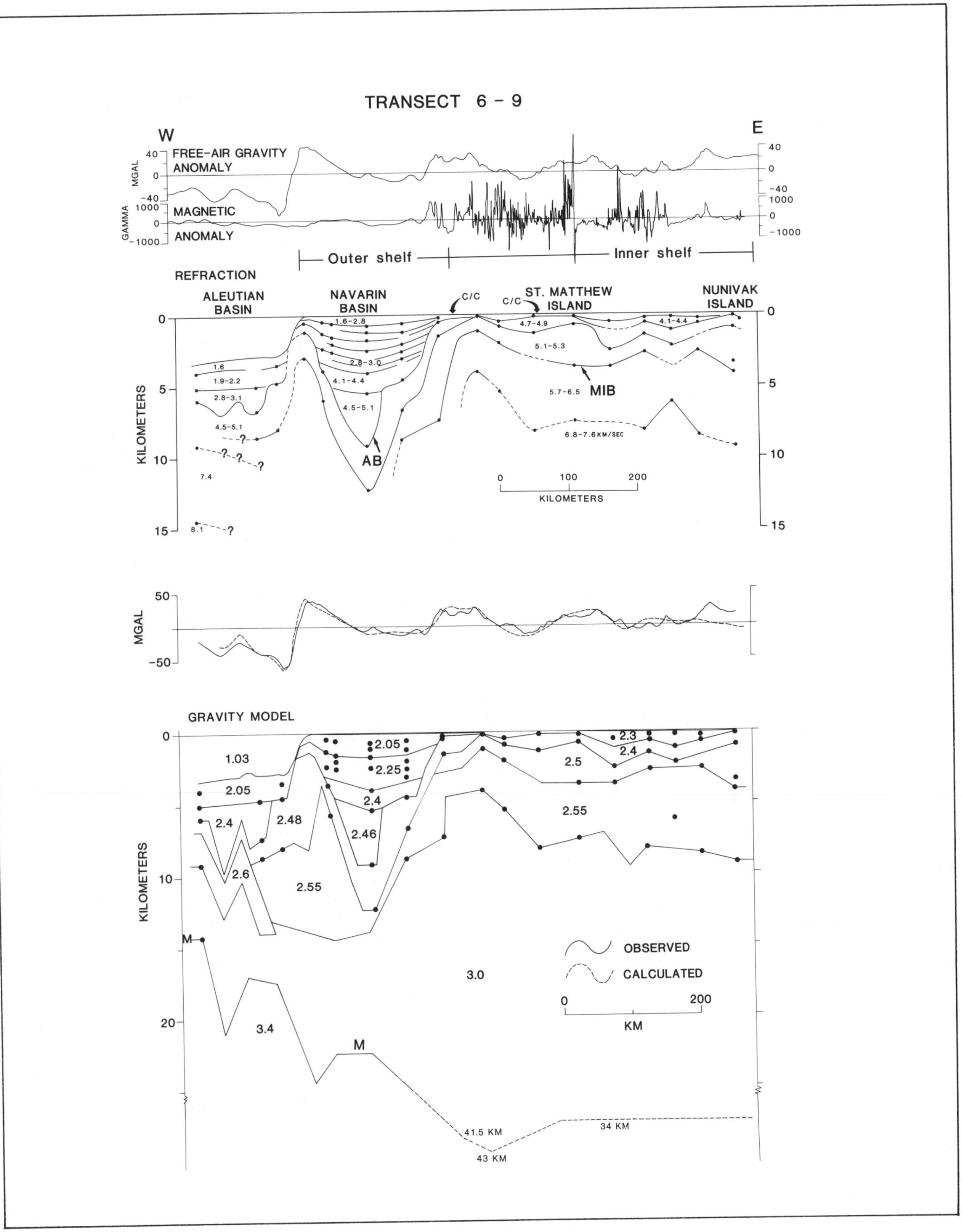

Figure 16. Crustal transect 6-9 and gravity model across the central Beringian Shelf (see Figure 1B for location and Figure 14 title for explanation). Gravity model has densities based on refraction velocities measured along crustal transect 6-9 and velocity-density curves of Ludwig, Nafe, and Drake (1970). Densities are in gm/cm^3. M is mantle.

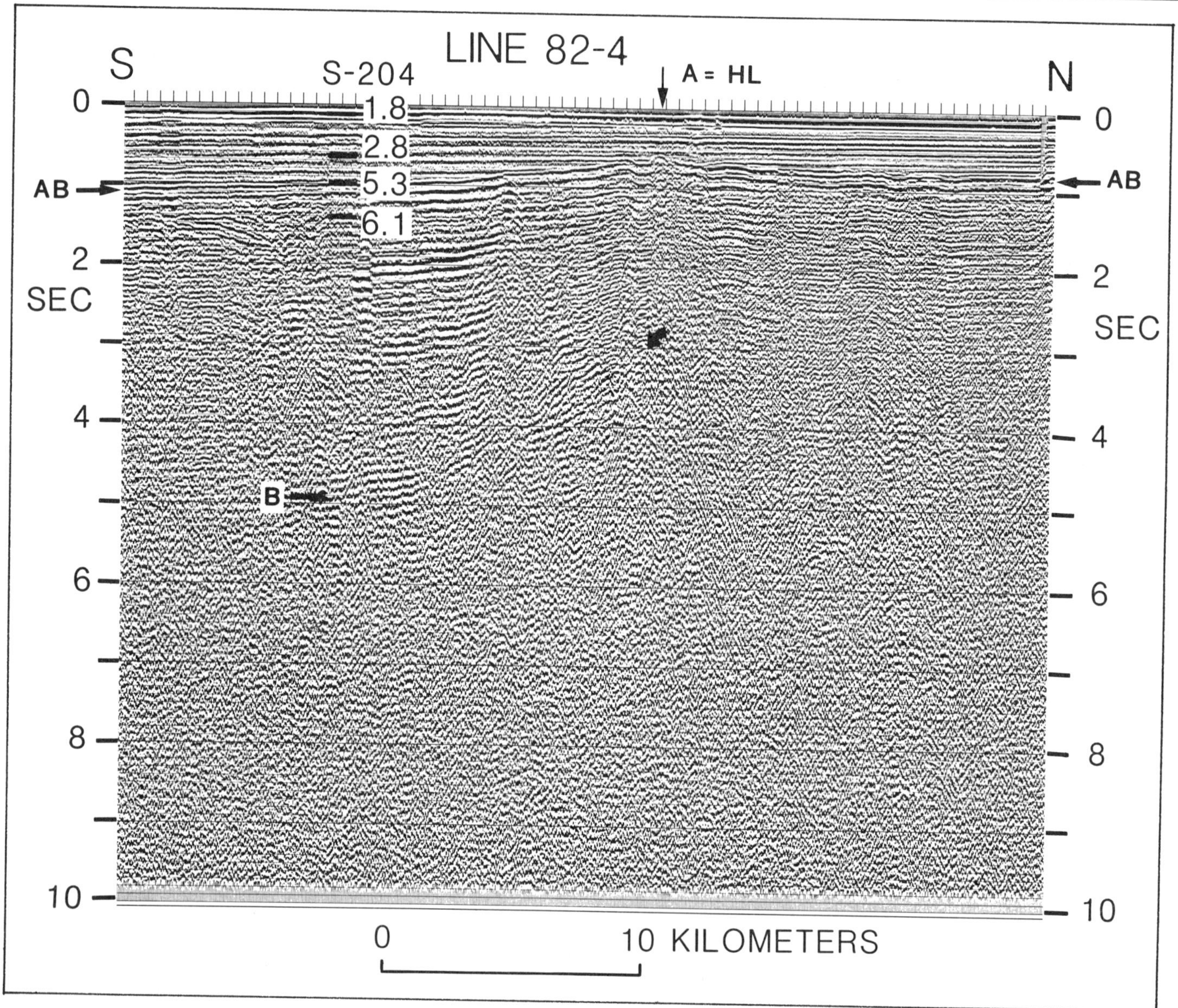

Figure 17. Multichannel seismic-reflection profile across the apparent Bristol Bay Basin hinge line (HL), which is suspected to be the structural boundary between the Peninsular terrane (on the south) and nuclear(?) Alaska (on the north). AB is acoustic basement and bedrock. Shallow and deformed bedrock 0.7 sec beneath HL is associated with a deeper fault offset or intrusives (near arrow at 3.0 sec). Sonobuoy S-204 reflection arrivals from large depths (near B) also indicate structures near HL. See Figures 1B and 14 for location.

accreted belts in the Alaska-Bering Sea-Siberia region (Figure 2A). Many of these terranes, including the Peninsular terrane, appear to have moved several thousand kilometers northward during the Jurassic through Late Cretaceous, and emplacement at present latitudes occurred before the Paleocene. The large horizontal displacements and regional orogenic deformations of the Bering Sea area are thought to be controlled by the motions and interactions of crustal plates (North American, Pacific, Eurasian, Kula, Farallon) that have taken place primarily since the early Mesozoic.

The important tectonic events affecting the circum-Bering Sea region include:

Late Triassic to Middle Jurassic: The volcanic and plutonic arc and associated sedimentary bodies of the Peninsular terrane evolved. These rocks formed at low magnetic latitude and moved northward into their present position in southern Alaska by the early Tertiary (Figures 2A, 2B, 2J; Moore and Connelly, 1977; Hudson, 1979a and 1979b); Stone, Panuska, and Packer, 1982).

Late Jurassic to Early Cretaceous: A major event involving compression and collision formed the Yukon-Koyukuk Basin and caused obduction and amalgamation of Paleozoic to Early Cretaceous oceanic and island-arc crustal rocks in southwestern Alaska and the innermost Beringian Shelf (Figure 18A; Patton and Tailleur, 1977; Box, 1985).

Middle to Late Cretaceous: Northward-moving terranes were accreted to southern and southwestern Alaska (Jones and Silberling, 1979). Subduction occurred along southern Alaska, the Beringian margin, and northeastern Siberia (Figure 18; Marlow and Cooper, in press). Thick terrigenous sediments were deposited in the Yukon-Koyukuk successor basin. Andean-type volcanism occurred along the Okhotsk-Chukotsk belt (Patton and Tailleur, 1977).

Late Cretaceous to early Tertiary: A major orogenic epi-

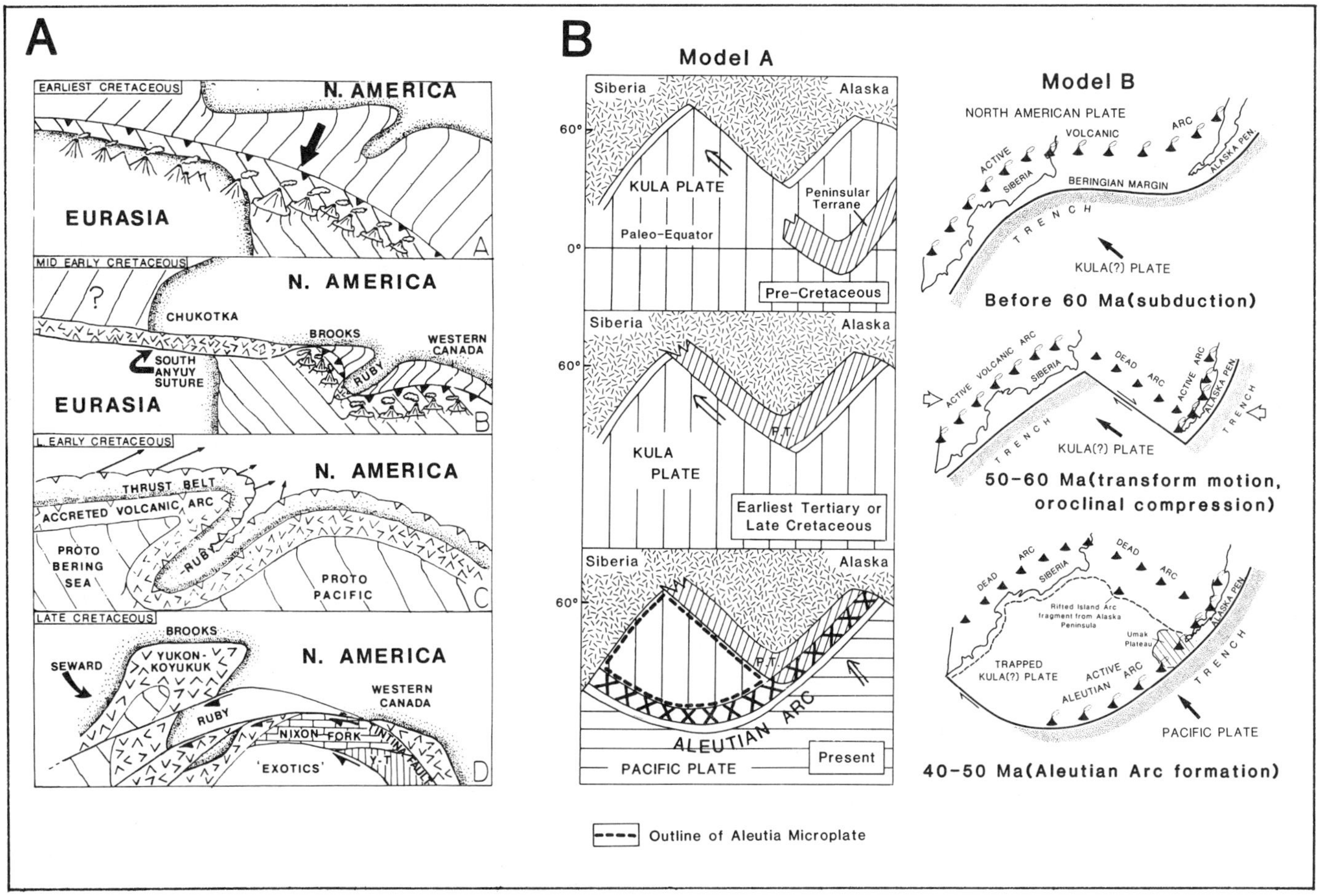

Figure 18. Models for the evolution of the Bering Sea region, A, Middle to late Mesozoic evolution of western Alaska and the Beringian shelf (Box, 1985). B, Two models for the late Mesozoic to early Tertiary evolution of the Bering Sea. Model A shows the docking of the Peninsular terrane (Marlow and Cooper, 1983), and model B shows the oroclinal compression of the Bering Sea region (Marlow and Cooper, in press).

sode was marked by (a) regional compression and oroclinal bending of foldbelts in northern Alaska and Siberia (Patton and Tailleur, 1977) and possibly foldbelts across southwestern Alaska and the Beringian Shelf (Grantz, 1966; Patton *in* Marlow and Cooper, 1980; Coe et al., 1985; Figure 18B); (b) docking of the Peninsular terrane against southern Alaska and the southern Beringian Shelf before the Eocene (Marlow and Cooper, 1983; Hillhouse, Grommé, and Csejtey, 1985; Panuska, 1985; Figure 18B); (c) formation of the Aleutian Ridge and isolation of a piece of oceanic crust in the Aleutian and possibly Bowers basins in the Eocene (Scholl, Buffington, and Marlow, 1975; Cooper, Scholl, and Marlow, 1976; Scholl, Vallier, and Stevenson, this volume); (d) initial formation of the large basins of the Beringian Shelf (Marlow et al., 1976a; Fisher, Patton, and Holmes, 1982; Marlow, Cooper, and Fisher, this volume); (e) emplacement of an extensive belt of calc-alkaline intrusive and volcanic rocks across western Alaska (Moll and Patton, 1982) and Siberia (Zinkevich, 1978); and (f) regional erosion and planation of the Beringian Shelf (Marlow et al., 1976b; Marlow, Cooper, and Fisher, this volume).

Middle to late Cenozoic: Uplift and volcanism along the southern (Alaska Peninsula) and western (Koryak and Kamchatka) Bering Sea margins (Scholl, Buffington, and Marlow, 1975), but diminished magmatic growth and subsidence of the Aleutian Ridge (Scholl, Vallier, and Stevenson, this volume).

The Bering Sea therefore did not exist, as it is known today, until the amalgamation of allochthonous terranes in the middle to late Mesozoic formed the bedrock and basement fabric of the Beringian Shelf, the contiguous margins, and the flanking mountains; and, until the formation of the Aleutian Ridge in the early Tertiary isolated a former sector of Pacific oceanic crust in the Aleutian Basin.

HYDROCARBON POTENTIAL

Geophysical exploration in the frontier offshore regions of the Bering Sea in the late 1960s and early 1970s led to the discovery of the large sedimentary basins in Norton Sound and along the outer Beringian margin (Scholl, Buffington, and Hopkins, 1968; Scholl and Hopkins, 1969; Scholl, Buffington, and Marlow, 1975). Several hundred thousand kilometers of marine seismic-reflection, gravity, and magnetic data as well as hundreds of sea-floor rock samples have since been collected by industry and the U.S. Geological Survey to assess the resource potential of this frontier area.

There are presently no producing oil and gas fields adjacent to the Bering Sea. The only shows of oil and gas are reported from wells on the Alaska Peninsula close to the Bristol Bay Basin and in the onshore Anadyr Basin of northeast Siberia (McLean, 1979b). Offshore drill holes in the Ber-

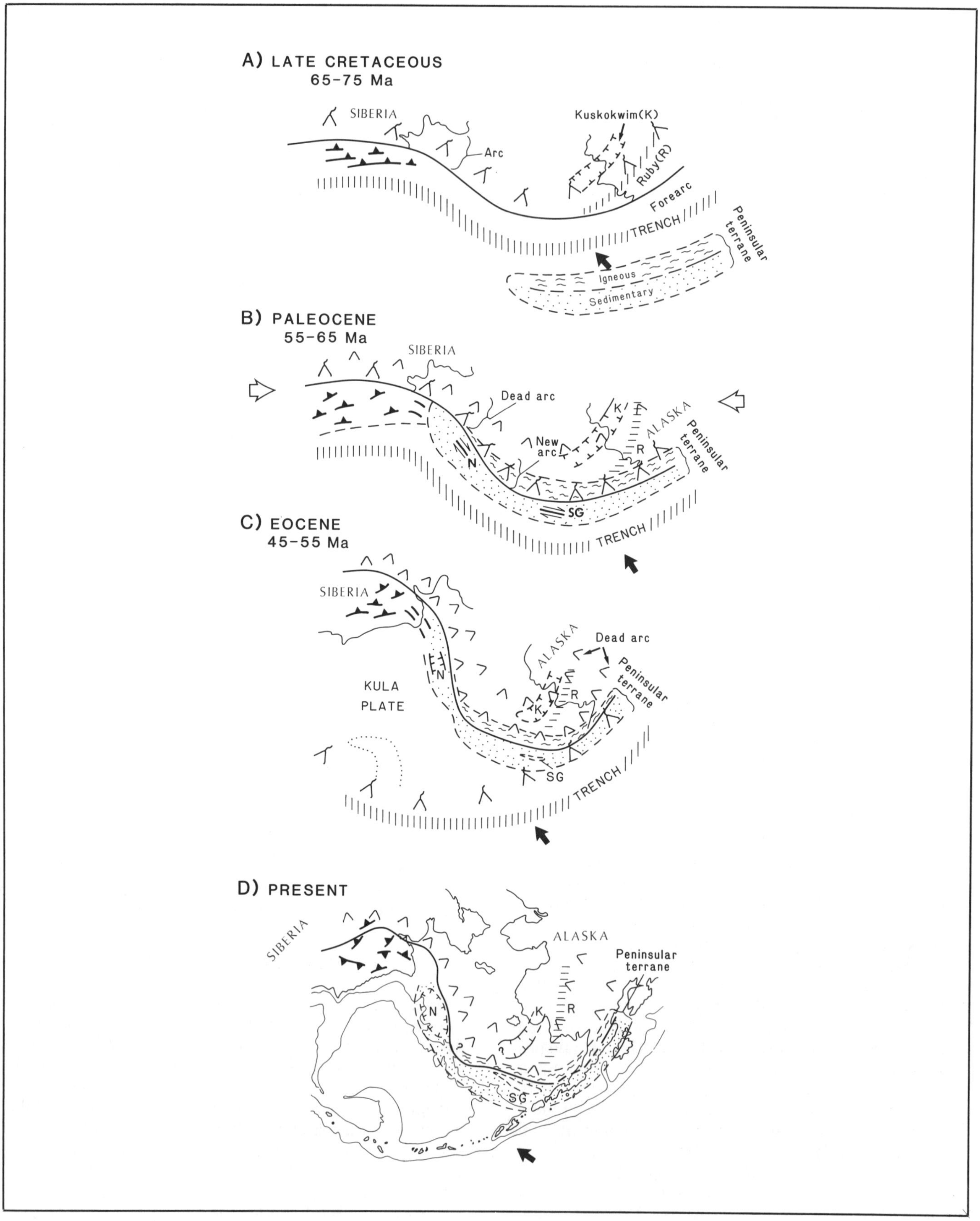

Figure 19. Speculative model for the origin of foldbelts beneath the Beringian Shelf. A, Late Cretaceous—before emplacement of Peninsular terrane (PT) at subduction zone along southern Alaska and Siberia. B, Paleocene—after PT emplacement, the locus of volcanism shifts to a more southerly location. Regional E-W compression, oroclinal bending, and development of Navarin (N) and St. George (SG) basins along a transform (wrench-fault) margin. C, Eocene—at completion of oroclinal compression of shelf foldbelts, development of Aleutian Ridge and trapping of a piece of the Kula plate. D, Present.

ing Sea (Figure 1) include the shallow DSDP holes (Creager et al., 1973), the deep COST stratigraphic wells (Turner, 1983a, 1983b; Turner et al., 1984a, 1984b, 1984c), and several deep proprietary exploratory wells in the St. George, Navarin, and Norton basins. Economic discoveries of oil or gas have not been reported from any of the offshore wells.

The potential for future hydrocarbon discoveries offshore in the Bering Sea, however, may be good. Each of the major hydrocarbon provinces (shelf basins, deep-water basins, and island-arc summit basins) cover thousands of square kilometers and have the characteristics required for generation and accumulation of hydrocarbons (Marlow, Cooper, and Fisher, this volume; Cooper, Scholl, and Marlow, this volume; Scholl, Vallier, and Stevenson, this volume). In general, sedimentary bodies >2 or 3 km thick have been mapped over large areas of the Bering Sea (Figure 3C), where moderate to high temperature gradients have been measured in COST wells (27° to 45°C/km in shelf basins) and in shallow cores (57°C/km in the Aleutian Basin). These gradients and thicknesses are adequate to reach the normal temperatures required for the onset of hydrocarbon generation (50° to 100°C), especially in the older and thicker Tertiary sedimentary sections. Antiforms and faulted structures that exhibit closure, sediment-covered (sealed?) angular unconformities, and stratigraphic onlap sequences provide traps for potential migrating hydrocarbons. However, source beds with high concentrations of amorphous organic material and reservoir beds with good porosity and permeability have not been found together in the COST wells and are not common in rocks dredged from the Beringian margin.

Diatom-rich sequences of Oligocene to Holocene age that form a major part of the sedimentary section in the Bering Sea are possible source rocks. These rocks have concentrations of organic carbon (as high as 0.8%) that are adequate for the generation of hydrocarbons. Marine sands and fractured siliceous rocks that have been sampled in the region may be potential reservoir rocks.

Regions of particular interest for petroleum exploration are the Cenozoic basins of the outer Beringian Shelf and the basins beneath its inner bays and sounds, principally because of their large size and the shallow water depths. We believe the older Mesozoic basins beneath the inner eastern Bering Sea Shelf may be unproductive because the highly heated, unproductive rocks found onshore are likely to extend offshore. Areas for future hydrocarbon exploration, as deep-water production technology improves, are as large or larger than those presently being explored. The sediment wedges or rise prisms along the Beringian margin, the expansive basin-plain deposits of the Aleutian Basin, and the structurally complex basins on the crest and flanks of the Aleutian Ridge are prospective exploration targets for the future.

SUMMARY

The geologic framework of the Bering Sea consists of a diverse suite of crustal rocks that include sediment-thickened oceanic crust, such as the Bering Sea Basin; variable-thickness continental crust typified by the inner Beringian Shelf; transitional crust beneath the Beringian margin and the Umnak Plateau; island-arc crust of the Aleutian Ridge, and possibly the remnant island-arc crust of Bowers and Shirshov Ridges. Most of the northward transport and accretion of allochthonous basement terranes occurred during the middle to late Mesozoic. Such terranes form the rock framework of much of the Beringian Shelf. Also, coeval and subsequent deformation and magmatic activity created the regional fabric of the structural and igneous rock belts that strike toward the Bering Sea and are believed to continue offshore. For example, structures south of the Yukon-Koyukuk Basin extend southwestward offshore and are truncated near the middle of the Beringian Shelf by the northwest-trending segment of the Peninsular terrane. Structural trends of the southern Koyukuk Basin extend offshore and may bend northward to merge with rock belts in Siberia; but rocks with Brooks Range affinities north of the Koyukuk Basin continue across the northern shelf from Alaska to Siberia via St. Lawrence Island (Patton and Tailleur, 1977). Belts of Upper Cretaceous to lower Paleogene magmatic rocks have overprinted older rock belts and trend in a broad zone generally northwestward across the Beringian Shelf.

An Upper Cretaceous to Paleogene volcanic-plutonic belt is probably the cause of the high-amplitude magnetic zone of the central Beringian Shelf. The southwestern seaward edge of this zone probably delineates the magmatic front of the volcanic arc that formed after the docking of the Peninsular terrane in the Late Cretaceous. Before the docking, a possibly older northern arc may have existed across the southern shelf. Volcanic and plutonic rocks emplaced by the two arcs may explain the greater width of the magnetic zone on the southern shelf. Near the St. George Basin, the younger volcanic front may lie within the Peninsular terrane and parallel to an older and possibly reactivated fault, similar to the onshore Bruin Bay fault.

Oroclinal bending of Alaska has been widely discussed (Carey, 1955; Grantz, 1966; Freeland and Dietz, 1973; Stone, 1980; Coe et al., 1985; Hillhouse, Gromme, and Csejtey, 1985). Coe et al. (1985) believe, from their analysis of Alaskan paleomagnetic data, that western Alaska has rotated about 40° counterclockwise since the latest Cretaceous as a result of the convergence between Eurasia and North America. Similar evidence is given by Hillhouse, Grommé, and Csejtey (1985), indicating about 40° of counterclockwise rotation of southwestern Alaska (near Bristol Bay) since the Paleocene.

Oroclinal bending and earlier emplacement of the Peninsular terrane along an obliquely convergent or transform margin could explain (a) the offshore truncation or bending of inner shelf terranes, (b) the structural bending of the Peninsular terrane near the tip of the Alaska Peninsula, (c) the initiation of rift (wrench) basins of the outer shelf, and (d) the apparent continuity of magmatic belts from northern Alaska to Siberia. Development of the Aleutian Ridge and trapping of oceanic crust in the Aleutian Basin would postdate this oroclinal bending (Figure 19).

Since the early Tertiary, the abyssal regions of the Bering Sea have been a relatively stable site of continuous and rapid sediment deposition. Also, paleomagnetic data indicate that little or no northward movement of basement rocks on the Aleutian Islands, the Alaska Peninsula, the Beringian Shelf, or western Alaska has occurred since the early Tertiary. Vertical movements since the early Tertiary, however, have developed the Cenozoic basins beneath the Beringian Shelf, the continental slope and rise, and the summit and upper flanks of the Aleutian Ridge.

In general, during the Mesozoic Era horizontal offsets, emplacement, and deformation of crustal blocks occurred in the Bering Sea. With the possible exception of the Koryak region, the Cenozoic Era was marked principally by vertical and probable strike-slip movements. The tectonic history of the Bering Sea is closely related to the complex interactions of oceanic and continental plates around the rim of the North Pacific (Wallace and Engebretson, 1984).

ACKNOWLEDGMENTS

We thank Hugh McLean, Mike Fisher, Bill Patton, and Terry Bruns for their thoughtful reviews of the manuscript. We also thank Phyllis Swenson and Lee Clark for their drafting and technical support.

REFERENCES CITED

Bailey, D. K., A. K. Cooper, M. S. Marlow, and D. W. Scholl, 1976, Preliminary residual magnetic map of the Bering Shelf and parts of Alaska: U.S. Geological Survey Miscellaneous Field Studies Map MF-716, scale 1:2,500,000.

Barnes, D. F., 1977, Bouguer gravity map of Alaska: U.S. Geological Survey Geophysical Investigations Map GP-913, scale 1:2,500,000.

Beikman, H. M., 1980, Geologic map of Alaska: U.S. Geological Survey map, scale 1:250,000.

Ben-Avraham Z., and A. K. Cooper, 1981, Early evolution of the Bering Sea by collision of oceanic rises and North Pacific subduction zones: GSA Bulletin, v. 92, p. 485–495.

Bogdanov, N. A., and Yu. P. Neprochnov, 1984, Geology of the Bering Sea deep basins, *in* Origin and history of marginal and inland seas: Proceedings of the 27th International Geological Congress, Moscow, VNU Science Press, v. 23, p. 1–17.

Bogdanov, N. A., V. S. Vishnevskaia, and A. N. Sukhov, 1983, Upper Cretaceous formations of the Shirshov Ridge (Bering Sea): Doklady Akademii Nauk SSSR, v. 273, p. 1183–1187.

Box, S. E., 1985, Early Cretaceous orogenic belt in northwestern Alaska: Internal organization, lateral extent, and tectonic interpretation: AAPG Earth Science Series, v. 1, p. 137–145.

Bruns, T. R., R. von Huene, R. D. Culotta, and S. D. Lewis, 1985, Summary geologic report for the Shumagin Outer Continental Shelf Planning Area, Alaska: U.S. Geological Survey Open-File Report 85-32.

Bruns, T. R., R. von Huene, R. D. Culotta, S. D. Lewis, and J. W. Ladd, 1987, Geology and petroleum potential of the Shumagin margin, Alaska, this volume.

Burk, C. A., 1965, Geology of the Alaska Peninsula island arc and continental margin: GSA Memoir 99, 250 p.

Carey, S. W., 1955, The orocline hypothesis in geotectonics: Proceedings of the Royal Society of Tasmania, v. 89, p. 255–288.

Carlson, P. R., and H. A. Karl, 1984, Discovery of two new large submarine canyons in the Bering Sea: Marine Geology, v. 56, p. 159–179.

Childs, J. R., A. K. Cooper, and A. W. Wright, 1981, Residual magnetic map of Umnak Plateau region, southeastern Bering Sea: U.S. Geological Survey Geophysical Investigations Map GP-939, scale 1:1,000,000.

Childs, J. R., H. W. Magistrale, and A. K. Cooper, 1985, Free-air gravity anomaly map of the Bering Sea: U.S. Geological Survey Miscellaneous Field Studies Map MF–1728, scale 1:2,500,000.

Christensen, N. I., and D. M. Fountain, 1975, Constitution of the lower continental crust based on experimental studies of seismic velocities in granulite: GSA Bulletin, v. 86, p. 227–236.

Christensen, N. I., and M. H. Salisbury, 1982, Lateral heterogeneity in the seismic structure of the oceanic crust inferred from velocity studies in the Bay of Islands ophiolite, Newfoundland: Royal Astronomical Society Geophysical Journal, v. 68, p. 675–688.

Coe, R. S., B. R. Globerman, P. W. Plumley, and G. A. Thrupp, 1985, Paleomagnetic results from Alaska and their tectonic implications: AAPG Earth Science Series, v. 1, p. 85–108.

Cooper, A. K., and M. S. Marlow, 1984, Geologic and geophysical studies of the inner Bering Sea Shelf during 1982: U.S. Geological Survey in Alaska: Accomplishments in 1982: U.S. Geological Survey Circular 939, p. 123–126.

Cooper, A. K., M. S. Marlow, and Z. Ben-Avrahma, 1981, Multichannel seismic evidence bearing on the origin of Bowers Ridge, Bering Sea: GSA Bulletin, v. 92, p. 474–484.

Cooper, A. K., M. S. Marlow, and D. W. Scholl, 1976, Mesozoic magnetic lineations in the Bering Sea marginal basin: Journal of Geophysical Research, v. 81, p. 1916–1934.

Cooper, A. K., M. S. Marlow, and D. W. Scholl, 1977, Bering Sea: A multifarious marginal basin, *in* Island arcs, deep sea trenches and back-arc basins: American Geophysical Union, v. 1, p. 437–450.

Cooper, A. K., M. S. Marlow, and T. L. Vallier, 1984, Summary geological report for the St. George Outer Continental Shelf Planning Area, Bering Sea, Alaska: U.S. Geological Survey Open-File Report 84-418, 81 p.

Cooper, A. K., H. McLean, and M. S. Marlow, 1985, Crustal structure of the Bristol Bay region, Alaska (abs.): AAPG, Pacific Section meeting, Anchorage, Alaska, Abstracts with Programs, p. 44.

Cooper, A. K., D. W. Scholl, and M. S. Marlow, 1976, Plate tectonic model for the evolution of the eastern Bering Sea Basin: GSA Bulletin, v. 87, p. 1119–1126.

Cooper, A. K., D. W. Scholl, and M. S. Marlow, 1987, Structural framework, sedimentary sequences, and hydrocarbon potential of the Aleutian and Bowers Basins, Bering Sea, this volume.

Cooper, A. K., M. S. Marlow, A. W. Parker, and J. R. Childs, 1979a, Structure contour map on acoustic basement in the Bering Sea: U.S. Geological Survey Miscellaneous Field Investigations Map MF-1165, scale 1:2,500,000.

Cooper, A. K., D. W. Scholl, T. L. Vallier, and E. W. Scott, 1980, Resource report for the deep-water areas of proposed Outer Continental Shelf lease sale #70, St. George Basin, Alaska: U.S. Geological Survey Open-File Report 80-246, p. 90.

Cooper, A. K., D. W. Scholl, M. S. Marlow, J. R. Childs, G. D. Redden, K. A. Kvenvolden, and A. J. Stevenson, 1979b, Hydrocarbon potential of Aleutian Basin, Bering Sea: AAPG Bulletin, v. 63, p. 2070–2087.

Cooper, A. K., K. A. Bailey, J. I. Howell, M. S. Marlow, and D. W. Scholl, 1976, Preliminary residual magnetic map of the Bering Sea Basin and Kamchatka Peninsula: U.S. Geological Survey Miscellaneous Field Studies Map MF-715, scale 1:2,500,000.

Cormier, V. F., 1975, Tectonics near the junction of the Aleutian and Kuril-Kamchatka arcs and a mechanism for middle Tertiary magmatism in the Kamchatka Basin: GSA Bulletin, v. 86, p. 443–453.

Creager, J. S., D. W. Scholl, et al., 1973, Initial reports of the Deep Sea Drilling Project, Vol. 19: Washington, D.C., U.S. Government Printing Office, 913 p.

Decker, J., 1984, The Kuskokwim Group: A post-accretionary successor basin in southwest Alaska (abs.): GSA Abstracts with Programs, v. 16, p. 277.

Detterman, R. L., and B. L. Reed, 1980, Stratigraphy, structure, and economic geology of the Iliamn quadrangle, Alaska: U.S. Geological Survey Bulletin 1368-B, p. B1-B86, 1 sheet, scale 1:250,000.

Dickinson, W. R., 1978, Plate tectonic evolution of north Pacific rim: Journal of the Physics of the Earth, v. 26, p. S1–S19.

Emery, K. O., and B. J. Skinner, 1977, Mineral deposits of the deep ocean floor: Marine Mining, v. 1, p. 1–71.

Fisher, M. A., and R. von Huene, 1984, Geophysical investigation of a suture zone: The Border Ranges fault of southern Alaska: Journal of Geophysical Research, v. 89, p. 11,333–11,351.

Fisher, M. A., W. W. Patton Jr., and M. L. Holmes, 1982, Geology of

Norton Basin and continental shelf beneath northwestern Bering Sea, Alaska: AAPG Bulletin, v. 66, p. 225–285.

Fountain, D. M., 1976, The Ivrea-Verbano and Strona-Ceneri zones, Northern Italy: A cross-section of the continental crust—New evidence from seismic velocities of rock samples: Tectonophysics, v. 33, p. 145–165.

Freeland, G. L., and R. S. Dietz, 1973, Rotation history of Alaskan tectonic blocks: Tectonophysics, v. 18, p. 379–389.

Fujita, K., 1978, Pre-Cenozoic tectonic evolution of northeast Siberia: Journal of Geology, v. 86, p. 159–172.

Fujita, K., and J. T. Newberry, 1982, Tectonic evolution of northeastern Siberia and adjacent regions: Tectonophysics, v. 89, p. 337–357.

Gemuts, I., C. I. Steefel, and C. C. Puchner, 1983, Regional geology and tectonic history of western Alaska: Journal of the Alaska Geological Society, v. 3, p. 67–86.

Gnibidenko, G. S., 1980, Tectonics of the Komandorsky Basin, *in* V. B. Kurnosov and I. O. Murdmaa, eds. Geology of the Komandorsky Basin: Vladivostok, Akad. Nauk SSSR, Dal'nevostochnyi nauchnyi Tsentr, 73 p. (In Russian.)

Godson, R. H., A. K. Cooper, J. D. Corbett, M. D. Fuller, W. F. Hanna, J. R. Heirtzler, W. J. Hinze, J. A. Schwartz, and I. Zietz, 1984, Composite magnetic anomaly map of the United States—Part B: Alaska and Hawaii: U.S. Geological Survey Map GP-954-B, 2 sheets, 4 p., scale 1:2,500,000.

Grantz, A., 1966, Strike-slip faults in Alaska: U.S. Geological Survey Open-File Report 267, 82 p.

Grow, J. A., 1973, Crustal and upper mantle structure of the central Aleutian Arc: GSA Bulletin, v. 84, p. 2169–2192.

Hillhouse, J. W., C. S. Grommé, and B. Csejtey, 1985, Tectonic implications of paleomagnetic poles from lower Tertiary volcanic rocks, south central Alaska: Journal of Geophysical Research, v. 90, p. 12, 523–12,535.

Hoare, J. M., and W. L. Coonrad, 1980, The Togiak Basalt, a new formation in southwestern Alaska: U.S. Geological Survey Circular 1482-C, p. C1–C11.

Hudson, T., 1979a, Mesozoic plutonic belts of southern Alaska: Geology, v. 7, p. 230–234.

Hudson, T., 1979b, Calc-alkaline plutonism along the Pacific rim of southern Alaska: U.S. Geological Survey Open-File Report 79-953, 31 p.

Jacob, K. H., 1972, Global tectonic implications of anomalous seismic p-travel times from the nuclear explosion Longshot: Journal of Geophysical Research, v. 77, p. 2556–2573.

Jin, D. J., and E. Herrin, 1980, Surface wave studies of the Bering Sea and Alaska area: Seismological Society of America Bulletin, v. 70, p. 2117–2144.

Jones, D. L., and N. J. Silberling, 1979, Mesozoic stratigraphy—The key to tectonic analysis of southern and central Alaska: U.S. Geological Survey Open-File Report 79-1200, 37 p.

Jones, D. L., N. J. Silberling, P. J. Coney, and G. Plafker, 1984, Part A: Lithotectonic terrane map of Alaska west of the 141st meridian, *in* N. J. Silberling and D. L. Jones, eds., Lithotectonic terrane maps of the North American cordillera: U.S. Geological Survey Open-File Report 84-523, 12 p.

Jones, D. M., M. J. Kingston, M. S. Marlow, A. K. Cooper, J. A. Barron, F. H. Wingate, and R. E. Arnal, 1981, Age, mineralogy, physical properties, and geochemistry of dredge samples from the Bering Sea continental margin: U.S. Geological Survey Open-File Report 81-1297, 68 p.

Karig, D., 1972, Remnant arcs: GSA Bulletin, v. 83, p. 1057–1968.

Kienle, J., 1971, Gravity and magnetic measurements over Bowers Ridge and Shirshov Ridge, Bering Sea: Journal of Geophysical Research, v. 76, p. 7138–7153.

Langseth, M. G., M. A. Hobart, and K. Horai, 1980, Heat flow in the Bering Sea: Journal of Geophysical Research, v. 85, p. 3740–3750.

Langseth, M. G., J. E. Nafe, and C. L. Drake, 1970, Seismic refraction, *in* A. E. Maxwell, ed., The sea, v. 4, pt. 1,: New York, Wiley Interscience, p. 53–84.

Ludwig, W. J., S. Murauchi, N. Den, M. Ewing, H. Hotta, R. W. Houtz, T. Yoshii, T. Asanuma, K. Hagiwara, T. Saito, and S. Ando, 1971, Structure of Bowers Ridge, Bering Sea: Journal of Geophysical Research, v. 76, p. 6350–6366.

Marlow, M. S., and A. K. Cooper, 1980, Mesozoic and Cenozoic structural trends under southern Bering Sea Shelf: AAPG Bulletin, v. 64, p. 2139–2155.

Marlow, M. S., and A. K. Cooper, 1983, Wandering terranes in southern Alaska: The Aleutia microplate and implications for the Bering Sea: Journal of Geophysical Research, v. 88, p. 3439–3446.

Marlow, M. S., and A. K. Cooper, in press, Regional geology of the Beringian margin: Proceedings of the OJI international seminar on the formation of ocean margins, University of Tokyo, Tokyo, Japan.

Marlow, M. S., A. K. Cooper, and J. R. Childs, 1983, Tectonic evolution of Gulf of Anadyr and formation of Anadyr and Navarin Basins: AAPG Bulletin, v. 67, p. 646–665.

Marlow, M. S., A. K. Cooper, and M. A. Fisher, 1987, Petroleum geology of the Beringian continental shelf, this volume.

Marlow, M. S., A. K. Cooper, A. W. Parker, and J. R. Childs, 1979a, Isopach map of strata above acoustic basement in the Bering Sea, U.S. Geological Survey Miscellaneous Field Studies Map 1164, scale 1:2,500,000.

Marlow, M. S., A. K. Cooper, D. W. Scholl, and H. McLean, 1982, Ancient plate boundaries in the Bering Sea region, *in* J. L. Leggett, ed., Trench-forearc geology: Sedimentation and tectonics on modern and ancient active plate margins: Geological Society of London Special Publication 10, p. 201–211.

Marlow, M. S., D. W. Scholl, A. K. Cooper, and E. C. Buffington, 1976a, Structure and evolution of Bering Sea Shelf south of St. Lawrence Island: AAPG Bulletin, v. 60, p. 161–183.

Marlow, M. S., A. K. Cooper, D. W. Scholl, T. L. Vallier, and H. McLean, 1979b, Description of dredge samples from the Bering Sea continental margin: U.S. Geological Survey Open-File Report 79-1139, 5 p.

Marlow, M. S., T. L. Vallier, A. K. Cooper, J. A. Barron, and F. H. Wingate, 1983, A description of dredge samples collected in 1982 from the Bering Sea continental margin west of Navarin Basin: U.S. Geological Survey Open-File Report 83-325, 4 p.

Marlow, M. S., H. McLean, T. L. Vallier, D. W. Scholl, J. V. Gardner, and R. B. Powers, 1976b, Prenomination report on the regional geology, oil and gas potential and environmental hazards of the Bering Sea Shelf south of St. Lawrence Island, Alaska: U.S. Geological Survey Open-File Report 76-785, 60 p.

McCarthy, J., and D. W. Scholl, 1985, Mechanism of subduction accretion along the central Aleutian Trench: GSA Bulletin, v. 96, p. 691–701.

McGeary, S. E., and Z. Ben-Avraham, 1981, Allochthonous terranes in Alaska: Implications for the structure and evolution of the Bering Sea Shelf: Geology, v. 9, p. 608–613.

McLean, H., 1977, Organic geochemistry, lithology, and paleontology of Tertiary and Mesozoic rocks from wells on the Alaska Peninsula: U.S. Geological Survey Open-File Report 77-813, 63 p.

McLean, H., 1979a, Review of petroleum geology of Anadyr and Khatyrka basins, USSR: AAPG Bulletin, v. 63, p. 1467–1477.

McLean, H., 1979b, Pribilof segment of the Bering Sea continental margin: A reinterpretation of Upper Cretaceous dredge samples: Geology, v. 7, p. 307–310.

McLean, H., 1979c, Tertiary stratigraphy and petroleum potential of Cold Bay-False Pass area, Alaska Peninsula: AAPG Bulletin, v. 63, p. 1522–1526.

Meyerhoff, A. A., 1982, Hydrocarbon resources in arctic and subarctic regions, *in* A. F. Embry and H. R. Bakwill, eds. Arctic geology and geophysics: Canadian Society of Petroleum Geologists Memoir 8, p. 451–552.

Minster, J. B., T. H. Jordan, P. Molnar, and E. Haines, 1974, Numerical modeling on instantaneous plate tectonics: Royal Astronomi-

cal Society Geophysical Journal, v. 36, p. 541–576.
Moll, E. J., and W. W. Patton, Jr., 1982, Preliminary report on the Late Cretaceous and early Tertiary volcanic and related plutonic rocks in western Alaska, *in* W. L. Coonrad, ed., U.S. Geological Survey in Alaska—Accomplishments during 1980: U.S. Geological Survey Circular 844, p. 73–76.
Moore, J. C., and W. Connelly, 1977, Tectonic history of the continental margin of southwestern Alaska: Late Triassic to earliest Tertiary, *in* A. Sisson, ed., The relationship of plate tectonics to Alaskan geology and resources: Anchorage, Alaskan Geological Society, p. H-1.
Moore, T. E., and W. K. Wallace, 1985, Stratigraphy and sedimentology of Kuskokwim Group in vicinity of Cairn Mountain, southwestern Alaska (abs.): AAPG, Pacific Section meeting, Anchorage, Alaska, Program and Abstracts, p. 57.
Nur, A., and Z. Ben-Avraham, 1978, Speculations on mountain building and the lost Pacifica continent: Journal of the Physics of the Earth, v. 26, p. S21–S37.
Panuska, B., 1985, Paleomagnetic evidence for a post-Cretaceous accretion of Wrangellia: Geology, v. 13, p. 880–883.
Patton, W. W., Jr., and I. L. Tailleur, 1977, Evidence in the Bering Strait region for differential movement between North America and Eurasia: GSA Bulletin, v. 88, p. 1298–1304.
Patton, W. W., Jr., M. A. Lanphere, T. P. Miller, and R. A. Scott, 1976, Age and tectonic significance of volcanic rocks on St. Matthew Island, Bering Sea, Alaska: U.S. Geological Survey Journal of Research, v. 4, p. 67–73.
Pratt, R. M., M. S. Rutstein, W. F. Walton, and J. A. Buschur, 1972, Extension of Alaska structural trends beneath Bristol Bay, Bering Shelf, Alaska: Journal of Geophysical Research, v. 77, p. 4994–4999.
Press, F., 1966, Seismic velocities, *in* S. P. Clark, Jr., ed., Handbook of physical constants: GSA Memoir 97, p. 195–218.
Rabinowitz, P. D., and A. K. Cooper, 1977, Structure and sediment distribution in the western Bering Sea Basin: Marine Geology, v. 24, p. 309–320.
Rubenstone, J. L., 1984, Early Tertiary volcanism in the Aleutian Islands: Geochemical variations and tectonic implications (abs.): GSA Abstracts with Programs, v. 16, p. 331.
Scholl, D. W., and J. S. Creager, 1973, Geologic synthesis of leg 19 (DSDP) results: Far North Pacific, Aleutian Ridge and Bering Sea, *in* Deep Sea Drilling Project Initial Reports: Washington D.C., U.S. Government Printing Office, v. 19, p. 897–913.
Scholl, D. W., and D. M. Hopkins, 1969, Newly discovered Cenozoic basins, Bering Sea Shelf, Alaska: AAPG Bulletin, v. 53, p. 2067–2078.
Scholl, D. W., E. C. Buffington, and D. M. Hopkins, 1968, Geologic history of the continental margin of North America in Bering Sea: Marine Geology, v. 6, p. 297–330.
Scholl, D. W., E. C. Buffington, and M. S. Marlow, 1975, Plate tectonics and the structural evolution of the Aleutian-Bering Sea region, *in* R. B. Forbes, ed., Contributions to the geology of the Bering Sea Basin and adjacent regions: GSA Special Paper 151, p. 1–32.
Scholl, D. W., T. L. Vallier, and A. J. Stevenson, 1983, Geologic evolution of the Aleutian Ridge—Implications for petroleum resources: Journal of the Alaska Geological Society, v. 3, p. 33–46.
Scholl, D. W., T. L. Vallier, and A. J. Stevenson, 1987, Geologic evolution and petroleum geology of the Aleutian Ridge, this volume.
Scholl, D. W., E. C. Buffington, D. M. Hopkins, and T. R. Alpha, 1970, The structure and origin of the large submarine canyons of the Bering Sea: Marine Geology, v. 8, p. 187–210.
Shor, G. G., 1964, Structure of the Bering Sea and the Aleutian Ridge: Marine Geology, v. 1, p. 213–219.
Shor, G. C., and D. J. Fornari, 1976, Seismic refraction measurements in the Kamchatka Basin, western Bering Sea: Journal of Geophysical Research, v. 81, p. 5260–5266.
Stauffer, P. H., 1987, Quaternary depositional history and potential sand and gravel resources of the Alaskan continental margins, this volume.
Steffy, D. A., R. F. Turner, G. C. Martian, and T. O. Flett, 1985, Evolution and petroleum geology of the Navarin Basin, Bering Sea, Alaska: Oil and Gas Journal, August 5, p. 116–124.
Stewart, R., and L. Peselnick, 1977, Velocity of compressional waves in dry Franciscan rocks to 8 kbar and 300°C: Journal of Geophysical Research, v. 82, p. 2027–2039.
Stone, D. B., 1980, The Alaskan orocline, the paleomagnetism and the paleogeography of Alaska: Tectonophysics, v. 63, p. 63–73.
Stone, D. B., 1983, Present day plate boundaries in Alaska and the Arctic: Journal of the Alaska Geological Society, Proceedings of the 1982 Symposium v. 3, p. 1–14.
Stone, D. B., in press, Bering Sea, Aleutian arc, Alaska, *in* A. E. M. Nairn, F. G. Stehli, and S. Uyeda, eds., Ocean basin and margins, Vol. 7B—Pacific: New York, Plenum Press.
Stone, D. B., B. C. Panuska, and D. R. Packer, 1982, Paleolatitudes versus time for southern Alaska: Journal of Geophysical Research, v. 87, p. 3697–3707.
Turner, R. F., G. C. Martin, T. O. Flett, and D. A. Steffy, 1985, Geologic report for the Navarin Basin Planning Area, Bering Sea, Alaska: Outer Continental Shelf Report MMS 85-0045, 156 p.
Turner, R. F., J. G. Bolm, C. M. McCarthy, D. A. Steffy, P. Lowry, T. O. Flett, and D. Blunt, 1983a, Geological and operational summary, Norton Sound, Alaska: U.S. Geological Survey Open-File Report 83-124, 167 p.
Turner, R. F., J. G. Bolm, C. M. McCarthy, D. A. Steffy, P. Lowry, T. O. Flett, and D. Blunt, 1983b, Geological and operational summary, Norton Sound COST No. 2 well, Norton Sound, Alaska: U.S. Geological Survey Open-File Report 83-557, 162 p.
Turner, R. F., C. M. McCarthy, C. D. Comer, J. A. Larson, J. G. Bolm, A. C. Banet, Jr., and A. J. Adams, 1984a, Geological and operational summary, St. George Basin COST No. 1 well, Bering Sea, Alaska: Outer Continental Shelf Report MMS 84-0016, 109 p.
Turner, R. F., C. M. McCarthy, C. D. Comer, J. A. Larson, J. G. Bolm, A. C. Banet, Jr., and A. J. Adams, 1984b, Geological and operational summary, St. George Basin COST No. 2 well, Bering Sea, Alaska: Outer Continental Shelf Report MMS 84-0018, 103 p.
Turner, R. F., C. M. McCarthy, D. A. Steffy, M. B. Lynch, G. C. Martin, K. W. Sherwood, T. O. Flett, and A. J. Adams, 1984c, Geological and operational summary, Navarin Basin COST No. 1 well, Bering Sea, Alaska: Outer Continental Shelf Report MMS 84-0031, 246 p.
Wallace, W. K., and D. C. Engebretson, 1984, Relationships between plate motions and Late Cretaceous to Paleogene magmatism in southwestern Alaska: Tectonics, v. 3, no. 2, 295-315.
Watanabe, T., M. G. Langseth, and R. N. Anderson, 1977, Heat flow in back-arc basins of the western Pacific, *in* Island arcs, deep sea trenches, and back-arc basins: American Geophysical Union, v. 1, p. 137–162.
Zinkevich, V. P., 1978, Orogenic structures in the northern part of the Koryak-Kamchatka region and their role in the development of the continental crust: Geotectonics, v. 12 p. 132–141.

CHAPTER SIX

PETROLEUM GEOLOGY OF THE BERINGIAN CONTINENTAL SHELF

Michael S. Marlow
Alan K. Cooper
Michael A. Fisher

U.S. Geological Survey
Menlo Park, California

ABSTRACT

The Beringian Shelf south of the Bering Strait is underlain by five large sedimentary basins that encompass 220,000 km^2 of the shelf. Three of the basins—Norton, Anadyr, and Bristol Bay Basins—are large structural sags in the earth's crust, formed by block faulting and downwarping of the bedrock or basement beneath the basins. The Navarin and St. George Basins are grabens and half-grabens, which formed by extension and collapse of the outer Beringian margin. These latter basins are filled with 10–15 km of strata, thought to be mainly Cenozoic in age. Except for the Anadyr Basin, the shelf basins are mainly Cenozoic features that developed after the amalgamation of terranes formed most of southern Alaska by the end of the Mesozoic. The oldest sedimentary units recovered from a well in the Norton Basin are Eocene in age. The oldest sedimentary bodies drilled in the Anadyr Basin are Cretaceous, and those drilled in the Navarin Basin are Late Cretaceous. The lowermost basin fill drilled in the St. George Basin is Eocene in age. The two wells were drilled along the flank of the basin and a thicker, and possibly older, section fills the central part of the basin.

Regional geologic studies onshore in western Alaska, along the Alaska Peninsula, and in eastern Siberia suggest that suitable source rocks for the generation of hydrocarbons exist offshore in the basins of the Beringian Shelf. Also, the sedimentary fill in the offshore basins is thick enough for the thermal development of hydrocarbons. Twenty onshore wells in the Anadyr Basin, nine onshore wells on the Alaska Peninsula, five offshore COST (Continental Offshore Stratigraphic Test) wells in the shelf basins, several offshore Deep Sea Drilling Project sites, and dredge samples from the Beringian margin all show that adequate reservoir beds of Cenozoic age probably exist within the shelf basins. Seismic-reflection data show large structural traps within the fills of the Beringian subshelf basins. Traps include antiforms, diapiric structures, growth faults, stratigraphic pinch-outs against the bedrock flanks of the basins, discordant overlap of the basins' lower stratigraphic section by the overlying and younger sequences, and drape structures formed by differential compaction of the basins' lower sequence over bedrock highs.

INTRODUCTION

The Beringian Shelf lies south of the Bering Strait (Figure 1) and encompasses an area of nearly 2,000,000 km^2—more than twice the combined area of California, Oregon, and Washington. The shelf area lies between western Alaska and eastern Siberia. The outer Beringian Shelf is underlain by three large basins—Bristol Bay, St. George, and Navarin—which are filled with sedimentary rocks. Bedrock ridges flanking these basins extend from the Alaska Peninsula northwestward to near Siberia (Figures 1, 2). The inner Beringian Shelf is underlain by the Norton Basin beneath Norton Sound (Figure 1; Fisher, Patton, and Holmes, 1982) and the Anadyr Basin beneath the Gulf of Anadyr adjacent to eastern Siberia (Table 1).

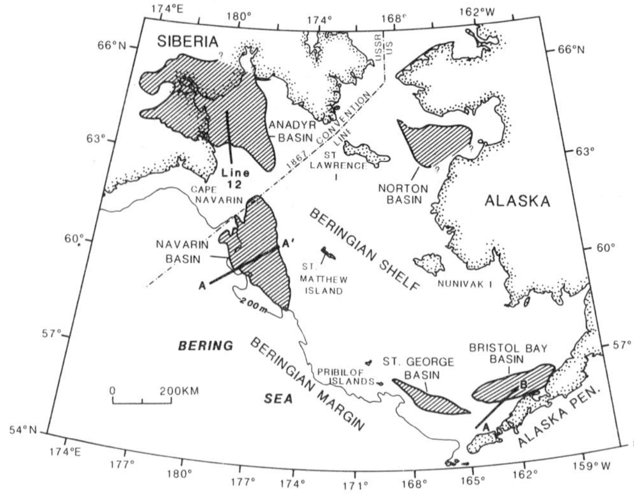

Figure 1. Outline and location of major basins in which sediment thickness exceeds 1 to 2 km beneath the Beringian Shelf. See Table 1 for the sizes of the various basins. Locations of seismic-reflection profiles illustrated in Figures 7, 11, and 13 are also shown. Albers equal-area protection.

REGIONAL FRAMEWORK OF THE BERINGIAN SHELF

The geologic framework of the Beringian Shelf can be divided into two major geologic provinces—the northeastern shelf and onshore areas surrounding Norton Sound, and the central and southern shelf and neighboring western Alaska and the Alaska Peninsula.

Northeastern Shelf

Rocks adjacent and beneath the Beringian Shelf can be divided into northern and southern belts that have distinctive rock assemblages (Patton and Dutro, 1969; Patton and Tailleur, 1977; Patton et al., 1977). Those of the northern belt are exposed in the Brooks Range, on Seward Peninsula, on eastern St. Lawrence Island, and in parts of northeast Siberia (see Figure 3 of Fisher, Patton, and Holmes, 1982). The northern assemblage consists of Precambrian, Paleozoic, and lower Mesozoic sedimentary units that were deposited on ancient continental crust; near the Norton Basin, they were intruded by Cretaceous plutons. In some areas of the Seward Peninsula, rocks of the northern belt are strongly metamorphosed to high-grade schist and gneiss. Two wells in the Norton Basin bottomed in quartzite, phyllite, and marble that are undated but appear to be lithologically similar to the sections of the northern belt. Those of the southern belt underlie the Yukon-Koyukuk province (east of the Norton Basin), western St. Lawrence Island, and extensive areas along the east coast of Siberia. The sedimentary, volcanic, and plutonic units of the belt are mainly of late Mesozoic age. Both belts lie along the coastal areas rimming the Norton Basin, but only rocks of the northern belt are known to underlie the basin.

Mesozoic Structural Trends Beneath the Southern Beringian Shelf

Large onshore geologic structures, such as major strike-slip faults and areally large tectonostratigraphic terranes, strike toward the Beringian Shelf. Two major strike-slip faults, the Kaltag and the Denali faults, trend southwest from central and western Alaska toward the Bering Sea (Figure 2). The Goodnews and Togak terranes, which similarly trend from western Alaska toward the Bering Sea (Jones and Silberling, 1979), continue offshore at least 50–100 km (Cooper and Marlow, 1983; Cooper, Marlow, and Scholl, this volume).

Gravity, magnetic, seismic-reflection, dredge, and drill data suggest that the Upper Jurassic, shallow-marine rocks exposed on the Alaska Peninsula extend west to northwest along the south flank of the St. George Basin and connect with Pribilof Ridge (Figure 2; Marlow and Cooper, 1980a). In addition, Upper Cretaceous (Campanian) rocks were dredged from the southern flank of this ridge in nearby Pribilof Canyon (Figure 2; Hopkins et al., 1969). Seismic-reflection data show that the Cretaceous dredge samples may have been recovered from an isolated, intrabasement rock sequence that is folded into the flank of Pribilof Ridge (Marlow and Cooper, 1980b). Other dredge samples along the margin indicate that Mesozoic bedrock is unconformably overlain by shallow-water diatomaceous mudstone of early Tertiary age (Jones et al., 1981).

The Upper Jurassic rocks that extend offshore beneath the Bering Shelf are coeval with the Naknek Formation exposed along the Alaska Peninsula (Burk, 1965), and both Jurassic sections are part of the Peninsular terrane of Jones and Silberling (1979). The Peninsular terrane extends about 1200 km from south-central Alaska to the western tip of the Alaska Peninsula (Figure 3). Paleomagnetic studies of Mesozoic rocks from the Peninsular terrane suggest that the terrane originated at a paleolatitude south of the equator in the Jurassic (Packer and Stone, 1972; Stone, Panuska, and Packer, 1982). Geologic data suggest that the docking of the Peninsular terrane against the ancient North American continent was complete by the end of the Mesozoic; however, Panuska (1985) has argued that accretion of his southern Alaska superterrane, which includes the Peninsular terrane, did not occur until probably the early Tertiary (see Marlow and Cooper, 1983, for a more detailed discussion of the docking time of the Peninsular terrane in the Bering Sea).

The offshore part of the Peninsular terrane can be traced by geophysical and dredge data about 1000 km along the outer Beringian Shelf and part of the adjacent margin (Figures 2, 3). Docking of the Peninsular terrane in southern

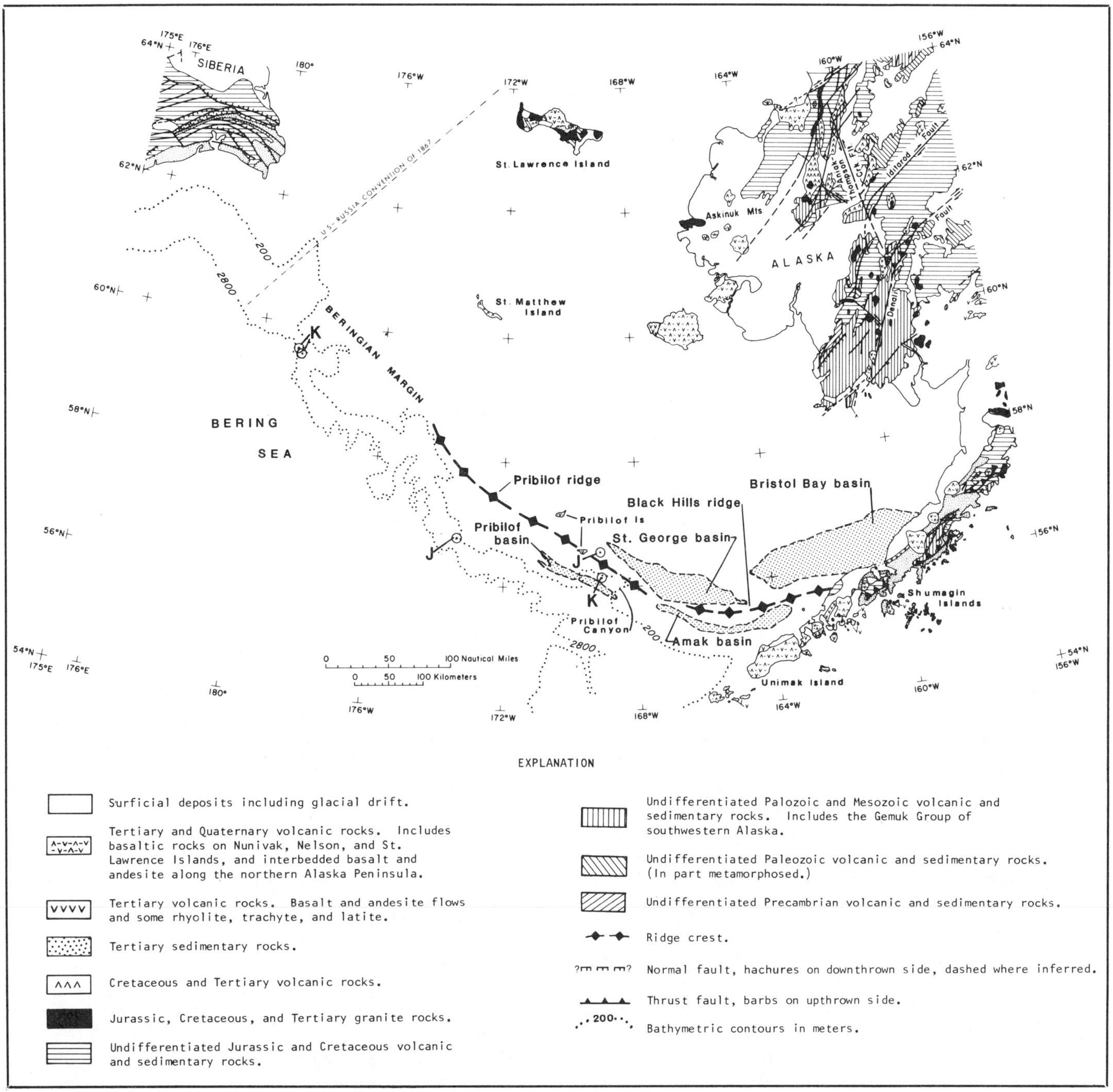

Figure 2 . Generalized geology of western Alaska, Alaska Peninsula, and eastern Siberia. Onshore geology modified from Burk (1965), Yanshin (1966), and Beikman (1974). Dotted circles show sites where Jurassic (J) and Cretaceous (K) sedimentary rocks were dredged from the continental slope. Adapted from Marlow and Cooper (1980a). Albers equal-area projection.

Alaska probably coincided in time with docking along the present-day Beringian margin. Formation of the margin at the end of the Mesozoic presumably occurred about the same time that the large outer shelf basins, such as the St. George and Navarin, began to form and fill with sediment.

We proposed in an earlier paper (Marlow et al., 1976) that a Jurassic, Cretaceous, and earliest Tertiary magmatic arc extended parallel to and inside (landward or toward Alaska) the outer shelf basins. This igneous belt is characterized by high-amplitude, high-frequency magnetic anomalies (Marlow et al., 1976). The magmatic arc is exposed as calc-alkalic volcanic and intrusive rocks of late Mesozoic and earliest Tertiary age on St. Matthew and St. Lawrence Islands (Figure 2) of the Beringian Shelf, and as similar rocks in

Table 1. Sedimentary basins of the Beringian Shelf (in part from Marlow et al., 1976).

Basin Name	Length km	Width km	Area km^2
Bristol Bay	290	75	21,750
St. George	300	50	15,000
Navarin	400	190	80,000
Anadyr	400	300	120,000
Norton	250	180	45,000

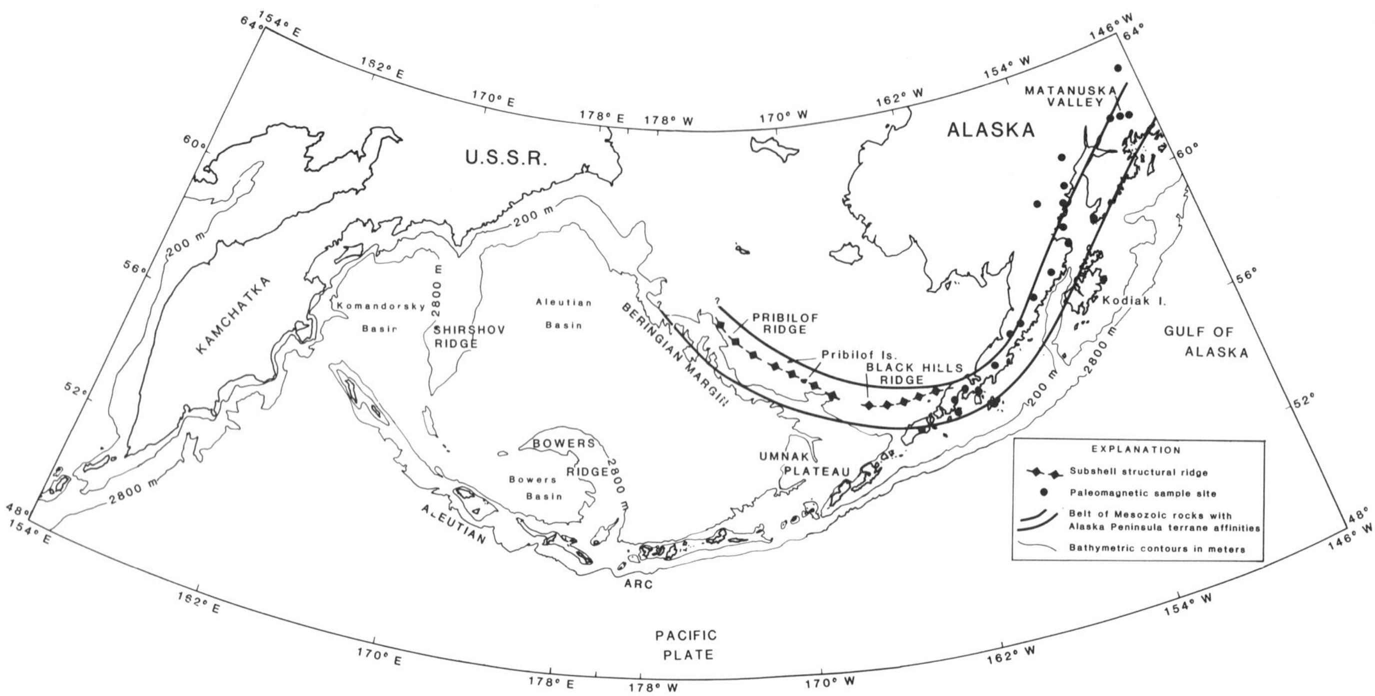

Figure 3. Locations of sample sites used in paleomagnetic studies of southern Alaska by Stone, Panuska, and Packer (1982). Offshore extension of the Peninsular terrane is shown by diamond pattern over bedrock ridges beneath the outer Bering Sea Shelf (from Marlow and Cooper, 1980a). Adapted from Marlow and Cooper (1983).

southern and western Alaska and eastern Siberia (Patton et al., 1974, 1976; Reed and Lanphere, 1973; Scholl, Buffington, and Marlow, 1975; Marlow et al., 1976). However, McGeary and Ben-Avraham (1981) and Cooper, Marlow, and Scholl (this volume) suggest that the southern Beringian Shelf is underlain in part by offshore extensions of allochthonous terranes of southern Alaska (Jones and Silberling, 1979). Whether these terranes merge with the Mesozoic igneous belt near St. Matthew Island is not known.

Cenozoic Isolation and Collapse

Plate motion along the Beringian margin apparently ceased with the formation of the Aleutian Island Arc in the earliest Tertiary. The plate boundary shifted at this time from the Bering Sea margin to a site near the present Aleutian Trench, thereby trapping a large section of oceanic plate (Kula?) within the abyssal Bering Sea (Scholl, Buffington, and Marlow, 1975; Cooper, Marlow, and Scholl, 1976; Cooper, Scholl, and Marlow, 1976; Marlow et al., 1976; Marlow and Cooper, in press). Cessation of the plate motion isolated and deactivated the Bering Sea margin. In the early Tertiary, the margin underwent extensional deformation and differential subsidence, which has continued throughout the Cenozoic. Elongate basins of great size and depth, such as the St. George and Navarin Basins, formed along the modern outer Beringian Shelf (Figure 1; Marlow et al., 1976). Parts of the inner shelf also subsided, forming Bristol Bay and Norton Basins. Extensional deformation of the folded rocks of the Mesozoic bedrock beneath the shelf, probably related to transtensional stresses, has continued to the present; this is evidenced by the normal faults flanking the outer shelf basins, which commonly offset the sea floor. These faults are growth-type structures that typically rupture the entire Cenozoic basin fill. Collapse of the outer shelf and adjacent margin may have been aided by Cenozoic sediment loading and continued subsidence of the adjacent oceanic crust (Kula? plate) that floors the Aleutian Basin of the abyssal Bering Sea Basin.

ST. GEORGE BASIN

Description

The St. George Basin is a long (300 km), narrow (30–50 km) graben whose axis strikes northwestward parallel to the trend of the Beringian margin (Figures 1, 2, 4; Table 1). The basin, located near the Priblof Islands and beneath the virtually featureless shelf, is filled with >10 km of sedimentary deposits.

A tracing of a 24-channel seismic-reflection profile, 8B, across St. George Basin is shown in Figure 5. The flat acoustic basement or bedrock surface underlying the southwest end of profile 8B (shotpoints 3200 to 2700) is Pribilof Ridge, which is overlain by an undisturbed, layered sequence 1.3 to 1.4 km thick (1.3 to 1.4 sec). Beneath the flat-lying sequence are dipping reflectors within the bedrock of Pribilof Ridge, which suggest that the ridge includes folded sedimentary beds (Jurassic Naknek? Formation). Farther north, between shotpoints 2700 and 2200, the basement or bedrock surface descends in a series of down-to-basin steps, plunging to a maximum subbottom depth of about 5.4 sec (>10 km) beneath the axis of St. George Basin. The overlying beds are broken by at least three major normal faults that dip toward the basin axis and appear to be related to offsets in the basin bedrock. Within the basin fill, the offset along the faults increases with depth, implying that these are growth structures. Strata are synclinally deformed about the structural axis of the basin. From shotpoints 2200 to about 1860, the

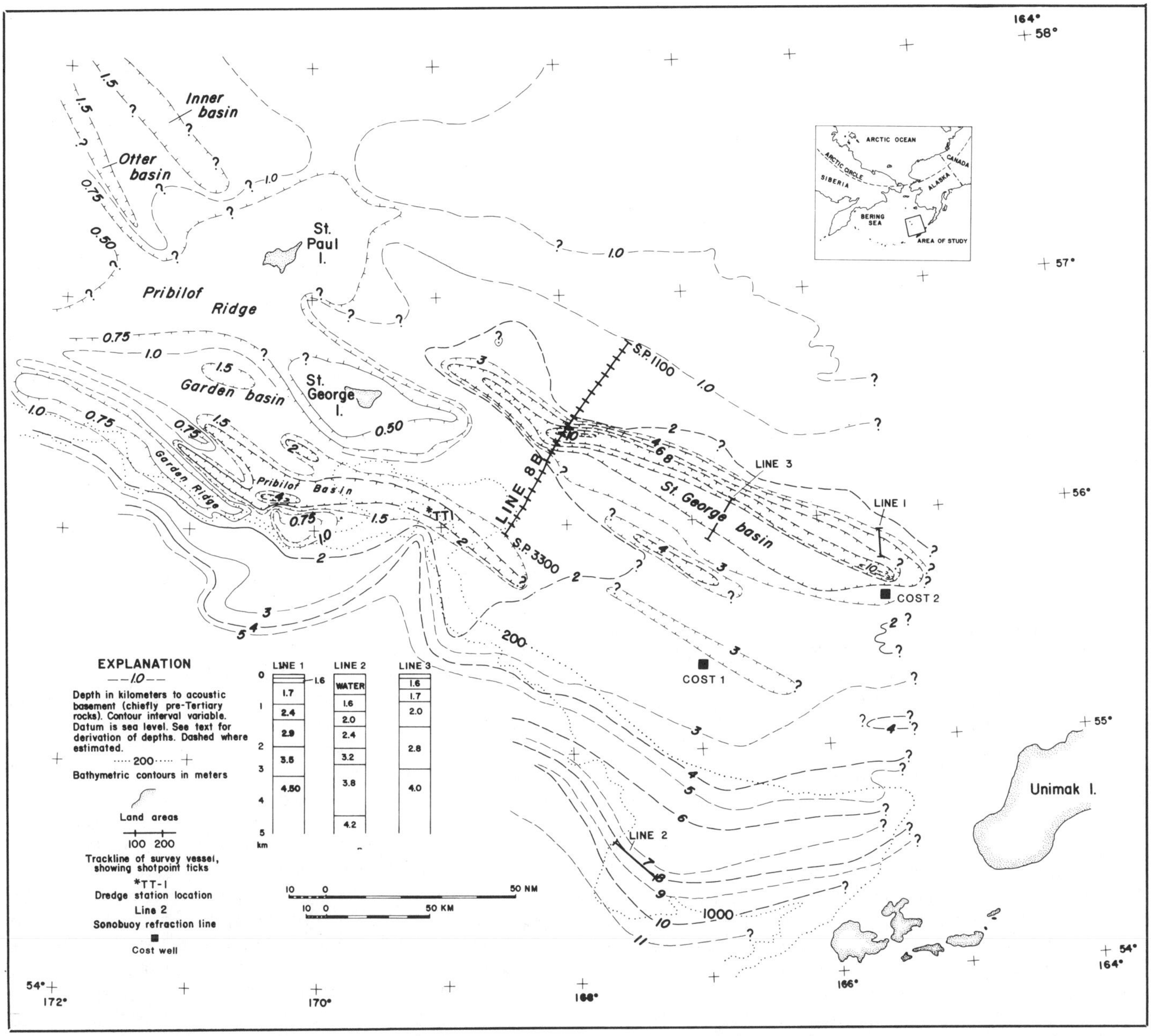

Figure 4. Structure contour map of bedrock or acoustic basement beneath the southern Bering Sea Shelf surrounding St. George Basin. Line with shotpoint ticks and numbers (S.P. 1100–3300) represents profile 8B shown on Figure 5. Lines 1, 2, and 3 correspond to sonobuoy refraction stations shown in lower left of figure (unnumbered top layer is sea water with an assumed velocity of sound of 1.5 km/sec). Cretaceous dredge site in Pribilof Canyon is marked TTI and corresponds to the site shown on Figure 2A. Adapted from Marlow, Scholl, and Cooper (1977).

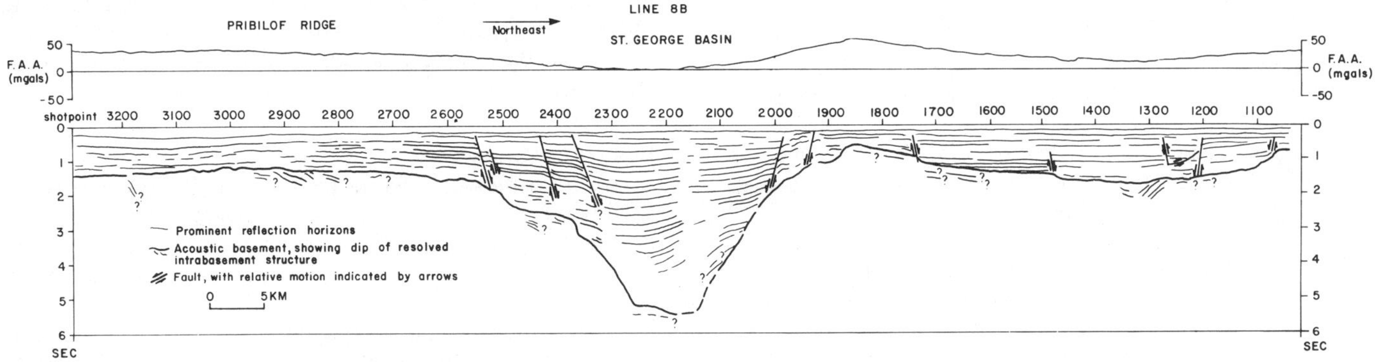

Figure 5. Interpretative time-section drawing of seismic-reflection profile 8B across St. George Basin. For location of profile, see Figure 4. Adapted from Marlow, Scholl, and Cooper (1977).

Table 2. Stratigraphic sequences penetrated at St. George Basin COST wells.

Age	Depth to Top (m)	Lithology
		COST No. 1 Well
Pliocene	487	Marine sandstone, mudstone, siltstone
Miocene	1,097	Same as above
Oligocene	1,636	Same as above, with conglomerate
Eocene	2,563	Same as above
Igneous basement: age unknown	3,163	Basalt, basalt breccia, tuff
		COST No. 2 Well
Pliocene	445	Marine diatomaceous mudstone, siltstone, muddy sandstone, minor conglomerate
Miocene	1,295	Same as above
Oligocene	1,844	Same as above
Eocene(?)	3,378	Marine to nonmarine sandstone, siltstone, conglomerate
Bedrock unconformity: Early Cretaceous or Late Jurassic	3,822	Same as above, with minor shale, coal
Late Jurassic	4,075	Marine sandstone, siltstone, and conglomerate

bedrock surface rises abruptly to a minimum depth of 500 m (0.55 sec; Figure 5). Overlying strata are broken by normal faults that dip down to the basin axis.

Age and History

Two COST wells were drilled in the vicinity of the St. George Basin province; data from these wells are described in detail by Turner et al. (1984a, 1984b). We present below, and in Table 2, a short summary of each well taken from their report. The first St. George COST well, 170 km southeast of St. George Island, Alaska, was drilled in 1976 to a total depth of 4197 m in water 135 m deep. The second St. George COST well was drilled in 1982 to a total depth of 4458 m in water 114 m deep, about 180 km northwest of the Alaska Peninsula (Figure 4).

The Cenozoic sedimentary section drilled in both wells includes interbedded sandstone, siltstone, mudstone, diatomaceous mudstone, and conglomerate. The section was derived predominantly from volcanic source terranes, and the detrital material is mechanically and chemically unstable. Porosity and permeability in the sedimentary section have been reduced by ductile deformation of individual sediment grains, and by cementation and diagenesis of the section.

The Mesozoic sedimentary section at the bottom of the COST No. 2 well consists of shallow-marine sandstone, siltstone, shale, conglomerate, and minor amounts of coal. Porosities and permeabilities in this section are also low.

Petroleum Potential

Source Beds

The two COST wells drilled in the St. George Basin area suggest that the Cenozoic section on the flanks of the basin is not prospective for either hydrocarbon source beds or reservoir beds. Possible source beds do exist within the Jurassic rocks in the eastern well (Figure 4).

Geochemical analyses of sediment samples from the first COST well show that the most common organic material present is type III humic kerogen, and that the average total organic-carbon content of the samples is low (0.5%), the maximum value being 0.74%. The calculated geothermal gradient in the well is high, 36°C/km (Turner et al., 1984a).

Geochemical analyses of sediment samples from the second COST well also show that the most common organic material present is type III humic kerogen, and that the organic-carbon content of the samples is low, generally <0.5%. Siltstone with organic-carbon content between 0.5 and 1.0%, and in a few samples >1.0 %, is present below 1550 m, and enough maturity for peak oil generation may exist below 3780 m. The section is gas prone at best, according to Turner et al. (1984b). Thermal maturation is low, and the geothermal gradient is 27°C/km.

According to Turner et al. (1984a, 1985b), strata with better source-rock and reservoir qualities than those in the two COST wells may be present in the central part of the St. George Basin, where the section is thicker.

Cretaceous mudstone dredged from the nearby continental slope in Pribilof Canyon (site TT1, Figure 4) contains as much as 1% organic carbon (Marlow et al., 1976, p. 179). Other rocks dredged from the Beringian continental slope include lithified volcanic sandstone of Late Jurassic age, mudstone of Late Cretaceous age, and less-consolidated deposits of early Tertiary age. Geochemical analyses of these rocks were described by Marlow et al. (1979). Pyrolytic analyses of these rocks show that none are good source beds for petroleum. However, submerged outcrops sampled by rock dredging are generally cemented units that may not include finer grained shales or other potential source beds possibly present within the exposed sedimentary section.

Tertiary mudstone also crops out on the continental slope; some contains >0.25% organic carbon (Marlow et al., 1976, p. 178). However, many of these rocks crop out too far down on the continental slope to be traced to coeval sections in the outer shelf basins. This is because the stratigraphically lower beds of the basin wedge out seaward against the flanks of the shelf basins.

Reservoir Beds

The porosities of Eocene to Pliocene–Pleistocene rocks dredged from 15 sites along the Beringian continental margin range from 14 to 70% and average 44% (Marlow et al., 1979; Jones et al., 1981). These Cenozoic samples are generally porous—probably because of abundant diatom frustules. However, the permeability of these rocks is variable, perhaps because of submarine weathering and cementation. The younger Cenozoic outcrops can be traced as seismic reflectors to St. George Basin, where, if the beds remain diatomaceous, good potential reservoir beds may be present.

Because sedimentation has matched basin subsidence, which averaged 100 to 200 m/m.y. during the Cenozoic (Marlow et al., 1976), Neogene reservoir beds of shallow-water origin are also likely to be present in St. George Basin. The thick Cenozoic sections of the basins beneath the Beringian Shelf accumulated near the mouths of such major Alaska and Siberian rivers as the Yukon, Kuskokwim, and Anadyr. During the Neogene, the shelf was swept by numerous marine transgressions and regressions (Hopkins, 1967; Hopkins and Scholl, 1970). All these factors suggest the possible deposition of neritic and deltaic strata and hence the possibility of good reservoir beds in the basins.

The lower section of the St. George Basin includes lower Tertiary beds (Marlow and Cooper, in press; Turner et al., 1984a, 1984b). The geologic evolution of the Beringian Shelf and margin suggests that the shelf basins, which were forming in the early Tertiary, were flanked by coastal mountains, peninsulas, and islands. These subsiding, yet high, land masses of Mesozoic rocks would have shed coarse clastic debris into the adjacent basins, and these lower Tertiary beds could also be prospective reservoirs.

Traps

Within St. George Basin, structural traps include broad anticlinal closures and tighter folds associated with faults. Closure and fault offset increase with depth; hence, growth-type structural traps formed continuously during basin filling. Stratigraphic traps may be formed by the thinning and wedging-out of the older stratigraphic sequence toward the flanks of the St. George Basin (Figure 5). These beds dip toward the axis of the basin; hence updip migrating fluids could be trapped against the much denser and presumably less permeable rocks of the Mesozoic basement. The lower stratigraphic sequence in the basin is discordantly overlapped by the younger acoustically reflective sequence. Hydrocarbons migrating updip along the lower beds might, therefore, be trapped beneath the overlapping sedimentary sequences.

BRISTOL BAY BASIN

Description

Underlying a large part of the southeastern Beringian Shelf north of the Alaska Peninsula is a sediment-filled structural depression known as the Bristol Bay Basin (Figure 1; Table 1). The sedimentary deposits along the southern flank of the basin are exposed on the northern fringe of the Alaska Peninsula. The basin trends northeastward along the peninsula, and about four-fifths of it lies offshore beneath the Beringian Shelf (Figure 2). The basin is asymmetrical in cross-section; the basement rocks of the northwest flank dip gently toward the basin axis, but the southeast flank of folded Jurassic rocks is steeply inclined, probably faulted, and crops out in the northern foothills of the Alaska Peninsula. The basin's sedimentary section, which is >6000 m thick, is composed chiefly of Cenozoic deposits that are thickest in the southeast part of the basin.

The northeast end of the Bristol Bay Basin is bounded by exposures of highly deformed, locally intruded, metamorphosed Paleozoic and Mesozoic rocks (Hatten, 1971). To the southwest, the basin is bordered by the offshore extension of the Black Hills—an anticlinal structure composed of Jurassic sedimentary rocks (Figure 2; Marlow and Cooper, 1980a). South of the basin, the economic basement consists mainly of Mesozoic and Cenozoic volcanic and plutonic units that form the core of the Alaska Peninsula. To the north, beneath the Beringian Shelf, the acoustic basement flooring the basin (as shown on geophysical records) probably consists of Mesozoic sedimentary, igneous, and metamorphic rocks.

The 24-channel seismic-reflection profile shown in Figure 6 and the interpreted section on Figure 7 illustrate the spatial relationship of such large subshelf structures as the Amak Basin, Black Hills Ridge, and Bristol Bay Basin (see Figure 8 for location). Near the center of the profile (Figure 7), at a subbottom depth of about 1.0 sec, moderately dipping strata are resolved within the core of a bedrock ridge, informally called the Black Hills Ridge after similarly deformed rocks exposed nearby in the Black Hills of the Alaska Peninsula. Strata underlying the Black Hills Ridge are cut by a major unconformity shown by a heavy line in Figure 7. The age of cutting of this unconformity may be late Cretaceous, because middle Cretaceous rocks are missing on the nearby Alaska Peninsula owing to uplift and erosion (Burk, 1965). The unconformity atop the ridge can be traced along the section to the east into the Bristol Bay Basin, where the surface of the unconformity appears to grade into a depositional contact with parallel reflectors above and below it (Figures 6, 7). If this contact is Cretaceous in age, then the layered

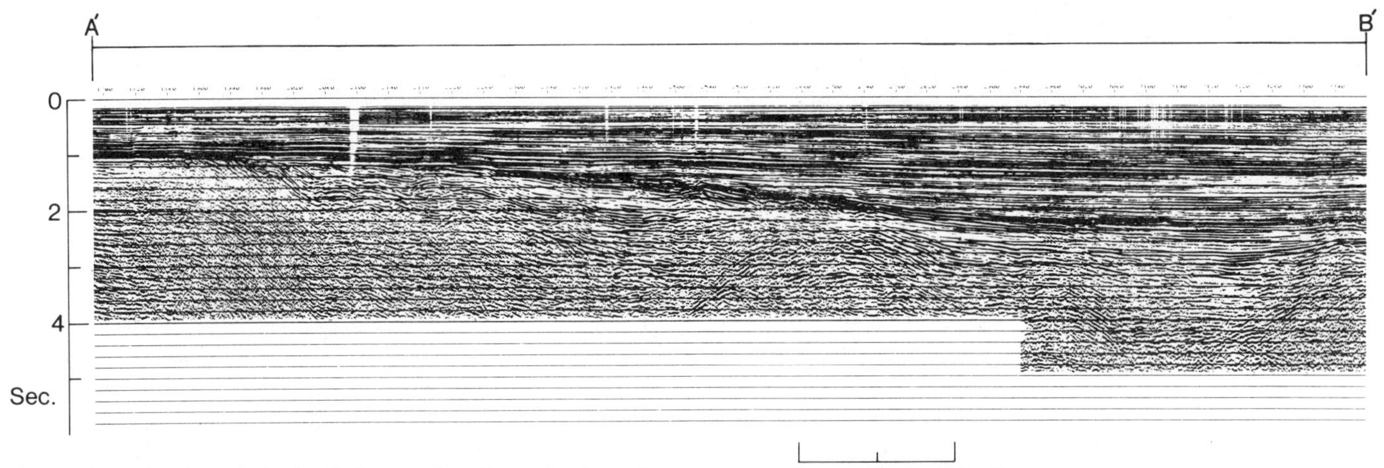

Figure 6. Twenty-four channel seismic-reflection record used in constructing part of interpretative drawing shown in Figure 7. From Marlow and Cooper (1980a).

strata beneath this horizon could be of older Cretaceous age. Hence, the initial formation and filling of the basin may have begun by the late Mesozoic. The same argument can be made for the dipping strata near the bottom of the Amak Basin along the western end of the same profile (Figure 7). We should note, however, a major unconformity along the Beringian continental margin west of the Bristol Bay Basin is early Tertiary in age (Marlow and Cooper, in press). If the unconformity along the margin is contemporaneous with the cutting of the Black Hills Ridge, then the Bristol Bay and Amak Basins may be no older than early Tertiary.

As shown by the northeastern half (A′B′) of profile AB (Figure 6), the Bristol Bay Basin is a long structural sag containing >6.6 km (4.5 sec) of moderately dipping to flat-lying

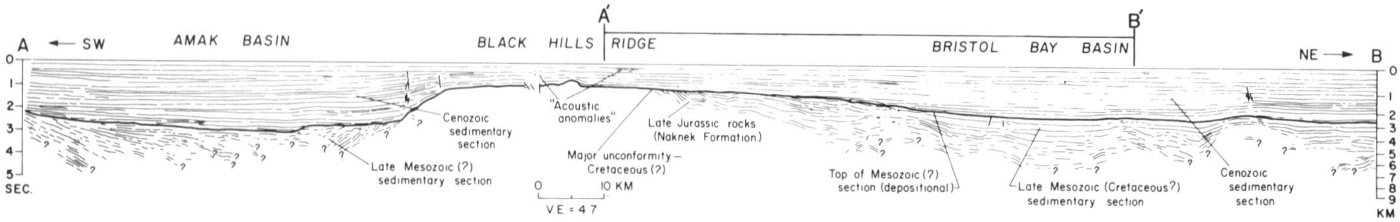

Figure 7. Interpretative drawing of 24-channel seismic-reflection profile AB across the southern Bering Sea Shelf and the western end of Bristol Bay Basin. For location of profile see Figures 1 and 8. Traveltime, in seconds, is two-way time. Seismic profile of segment A′B′ is shown in Figure 6. From Marlow and Cooper (1980a).

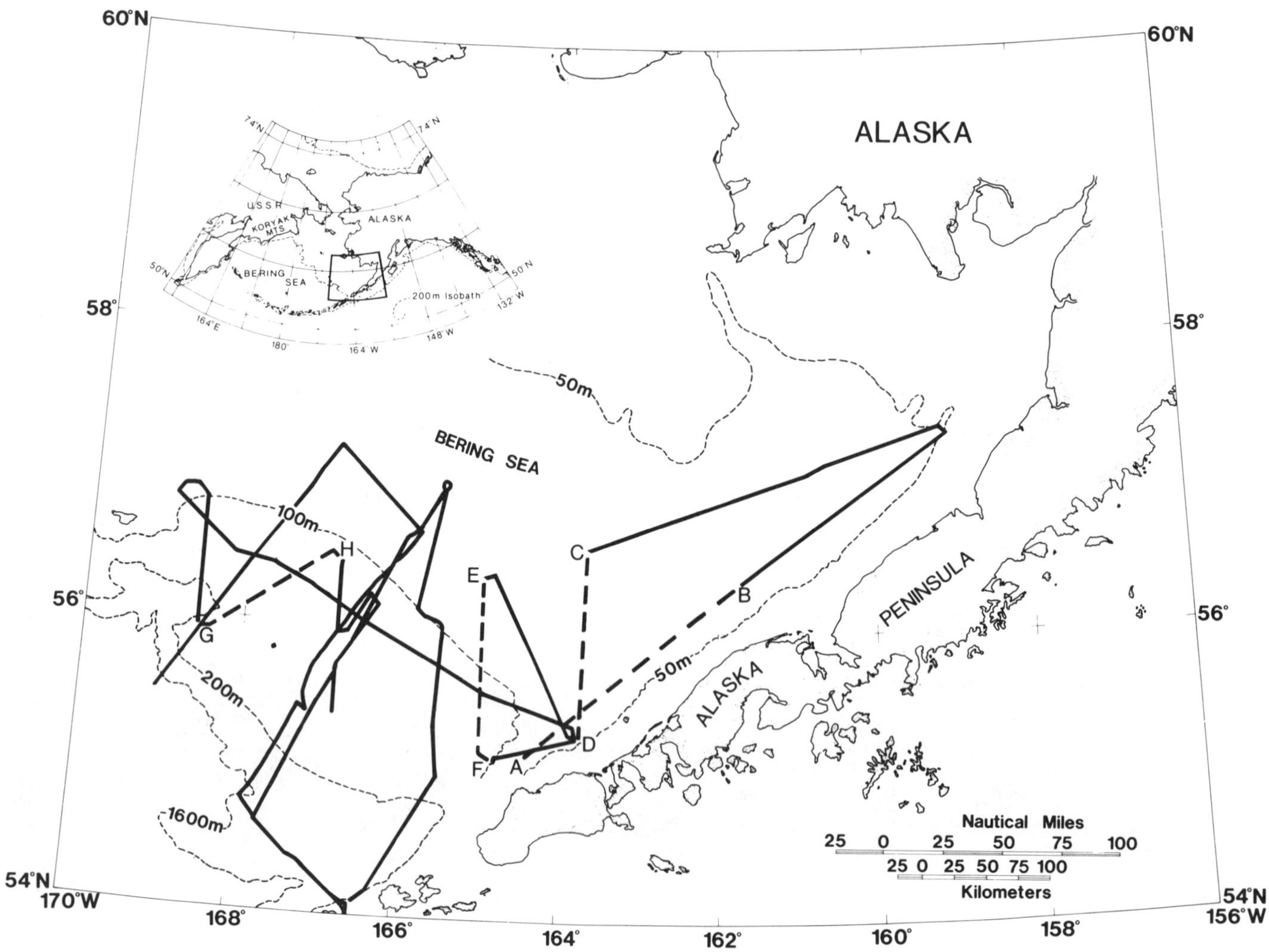

Figure 8. Track-line chart of multichannel seismic-reflection surveys across the southern Beringian Shelf. Dashed track line AB is profile shown in Figure 7. Albers equal-area projection. From Marlow and Cooper (1980a).

reflectors (Figures 6, 7). The upper half of the fill (above the heavy line) is virtually flat lying and undeformed and is interpreted to be late Cenozoic in age. The lower half of the basin fill contains discontinuous and dipping reflectors, which suggest gentle folding and draping over older basement (bedrock) highs. Our section recorded strata to a subbottom depth of 5 sec; thus the Bristol Bay Basin is filled with a section >6 to 7 km thick (Marlow and Cooper, 1980a). The fact that the nearby Gulf Sandy River No. 1 well on the Alaska Peninsula bottomed in Oligocene sedimentary rocks of the Stepovak Formation at a total depth of about 4 km without reaching a basement or bedrock complex (Hatten, 1971; Brockway et al., 1975) indicates that the adjacent Bristol Bay Basin may contain a much thicker Cenozoic sedimentary section.

Age and History

Only one COST well has been drilled in the Bristol Bay Basin, and data from this well remain proprietary. Thus, little is publicly known about the stratigraphy and age of the fill in the offshore portions of the basin. Bristol Bay Basin, as a distinct structural trough, probably began to form in the late Mesozoic or early Tertiary, and the basin has continued to subside during most of the Cenozoic. The basin probably began to form after the docking of the Peninsular terrane in the late Mesozoic and during the formation of the ancestral Aleutian Island arc in the earliest Tertiary (Marlow et al., 1976; Marlow and Cooper, 1983; Marlow and Cooper, in press).

Petroleum Potential

Source Beds

Since 1959, nine wells have been drilled along the northern coastal lowlands of the Alaska Peninsula (Figure 9). Drill cuttings from eight of these nine wells were analyzed by McLean (1977) for their organic geochemistry, lithology, and petroleum potential. The analyses show that Paleogene strata are rich in organic matter but immature. However, coeval sedimentary sections offshore in the basin may have been more deeply buried and thus may be more mature than the onshore section. In general, fine-grained Neogene sedimentary rocks have a significantly lower carbon content than Paleogene strata, suggesting that the younger strata may have been deposited in brackish to nonmarine environments (McLean, 1977).

The best source rocks in the Tertiary sequence appear to be the black marine siltstone and shale beds in the Oligocene Stepovak Formation. On the Alaska Peninsula, the Stepovak Formation is locally at least 4500 m thick (Burk, 1965). Scattered shows of oil and gas in Stepovak rocks have been reported from three Alaska Peninsula wells—Gulf Sandy River, Pan American Hoodoo Lake no. 2, and Pan American David River 1-A (Figure 9; Brockway et al.. 1975). Potential source rocks may also occur in the Mocene Bear Lake Formation because, locally, the basal part contains marine siltstone and shale that may have been buried deeply enough to generate hydrocarbons. Marine shales of Late Jurassic and Late Cretaceous age are also potential source rocks where they are in angular discordance with overlying Cenozoic sandstone reservoir beds.

Reservoir Beds and Seals

The greatest potential for oil and gas reservoirs in the offshore part of the basin is probably in the sandstone units of the Bear Lake Formation that are early(?), middle(?), and late Miocene in age. Bear Lake sandstone beds are both marine and nonmarine and contain a combination of volcanic and dioritic grains, chert, and sedimentary lithic fragments. Most of the sandstone could be classified as lithic subgraywacke and some as lithic arenite (Burk, 1965). In the Gulf Sandy River No. 1 well, Bear Lake sandstone above 1920 m has porosity values as high as 36.5% and permeability as high as 1268 md. Below 1920 m, high values for porosity and permeability are 29% and 43 md, respectively (McLean, 1977). Shows oil and gas have been reported from basal sandstone of the Bear Lake Formation in the Gulf Sandy River and Pan American David River wells.

Sandstone in two older Tertiary units—the Tolstoi and Stepovak Formations—has an abundance of volcanic detritus as well as a clay matrix. These rocks where exposed, are dense and highly indurated, and thus are not considered to be good reservoir beds. Little is known about their subsurface properties.

Traps

Most of the potential oil and gas traps in offshore Bristol Bay are probably anticlinal structures. Anticlines in the Cenozoic sequence are primarily formed by draping and differential compaction of strata over erosional and block-fault highs that developed within the acoustic-basement complex. The resultant structures are large in area but have a limited amount of closure. Structural traps in the offshore area may have formed during the early filling of the Bristol Bay Basin. Most structures decrease in amplitude upward through the Cenozoic section, and the Pliocene and Plestocene strata are flat lying and undeformed.

Stratigraphic traps formed by buttress onlap during transgression around topographic highs, by local truncation of sandstone beds, and by lenticular sandstone bodies probably occur in the Cenozoic sequence, but their size and number are unknown.

NAVARIN BASIN

Description

The Navarin Basin province underlies the northwest corner of the outer Beringian Shelf just south of Cape Navarin of eastern Siberia (Figure 1). Isopach contours of sediment thickness derived from sonobuoy and seismic-reflection data define three basins within the province that contain 10 to 15 km of strata and underlie about 80,000 km^2 of the Beringian Shelf (Figure 10, Table 1; Marlow, 1979).

A 190-km-long seismic-reflection profile, A–A′ (see location on Figure 10), transects the northwestern Beringian margin west of St. Matthew Island. An interpretative drawing of profile A–A′ (Figure 11) crosses on to the Aleutian Basin in water >3000 m (4.0 sec) deep. Underlying the basin floor are 4 to 5 km of undeformed strata that overlie a distinct acoustic basement. Magnetic and refraction-velocity data suggest that the basement is oceanic crust of Mesozoic age (Kula? plate; Cooper, Marlow, and Scholl, 1976; Cooper, Scholl, and Marlow, 1976).

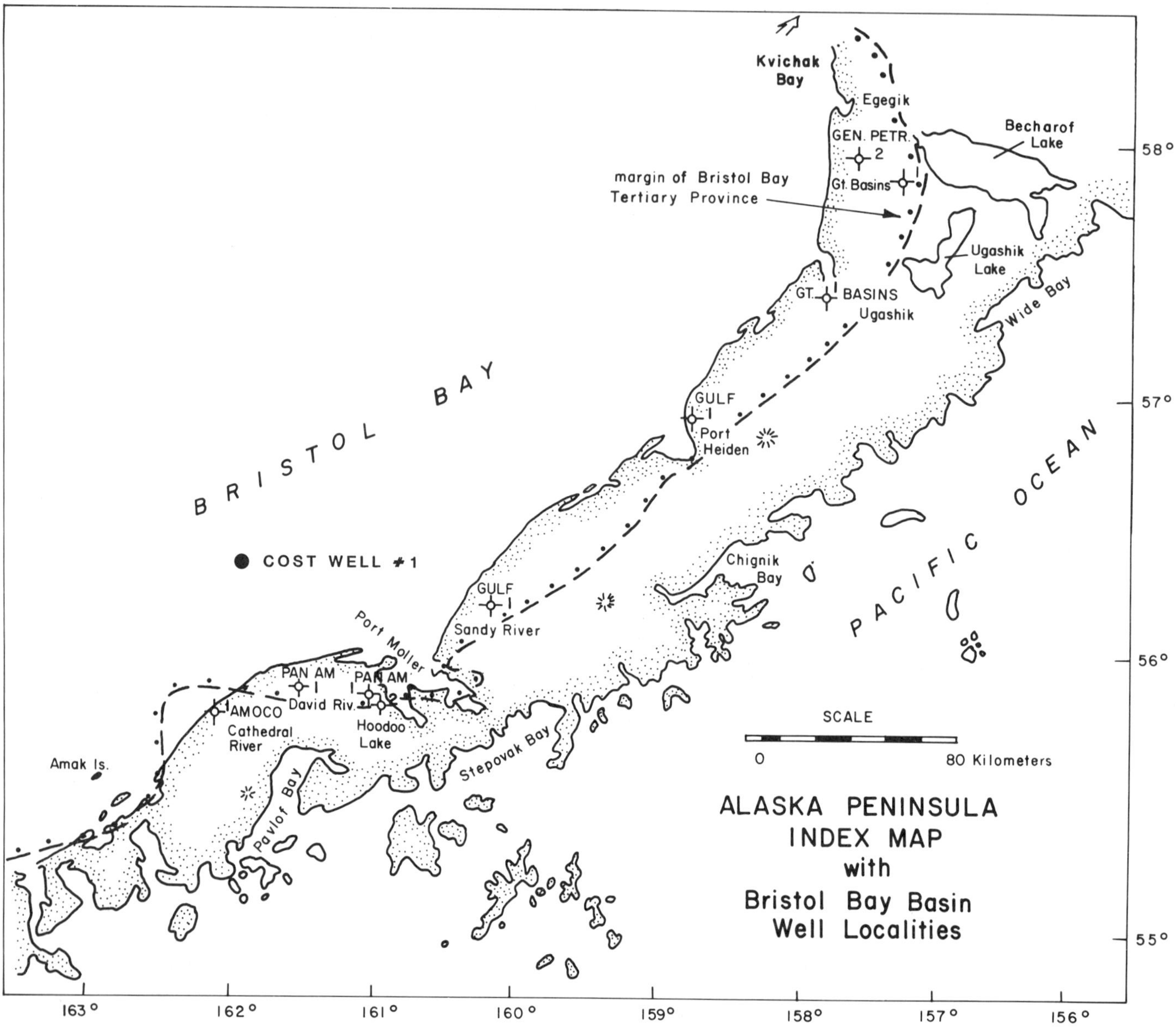

Figure 9. Index map showing sites of wells drilled on the Alaska Peninsula and in Bristol Bay. Adapted from McLean (1977).

Below the lower part of the continental slope the rock sequence is characterized by scattered and discontinuous reflectors (Figure 11). Oceanic basement was not detected. A strong subshelf basement reflector can be traced beneath the continental slope to a basement outcrop. Rocks dredged from this exposure, and the flatness of the subslope basement surface, suggest that it is a wave-base unconformity cut across broadly folded rock sequences. The dredge and seismic-reflection data attest that the subslope basement platform has subsided from a former shelf depth to at least 1500 to 1700 m since about Oligocene time (Marlow and Cooper, in press).

Basement beneath the shelf can be followed to the northeast below the thick sedimentary section filling the Navarin Basin (Figure 11). Near the northeastern end of the profile, strata in the basin are 12 km (7 sec) thick. Reflections from the upper basin fill are strong, continuous, and flat. Apparent breaks in the continuity of these reflectors are associated with columnar "wipe-out" zones beneath short, discontinuous reflectors in the upper few hundred meters of the section. The acoustic "wipe-out" zones are thought to be produced by shallow accumulations of gas that mask the lateral continuity of deeper reflectors (Marlow et al., 1982; Carlson et al., 1985). Dredge data originally indicated that the upper 3 to 4 km of beds in the Navarin Basin section included beds younger than Oligocene. Deeper strata are poorly reflective, diverge in dip from the overlying strata, and, near the base of the section, are probably early Tertiary in age (Marlow and Cooper, in press). In 1983 the basin's stratigraphic sequence was explored at a COST well described below (Turner et al., 1984c).

Age and History

The Navarin Basin province has been covered by many geophysical surveys, thus allowing the mapping of major subbasins and structures capable of trapping hydrocarbons. On the basis of our public knowledge, the geologic history of the Navarin Basin is poorly known because data from only one COST well are available for study. Many more explora-

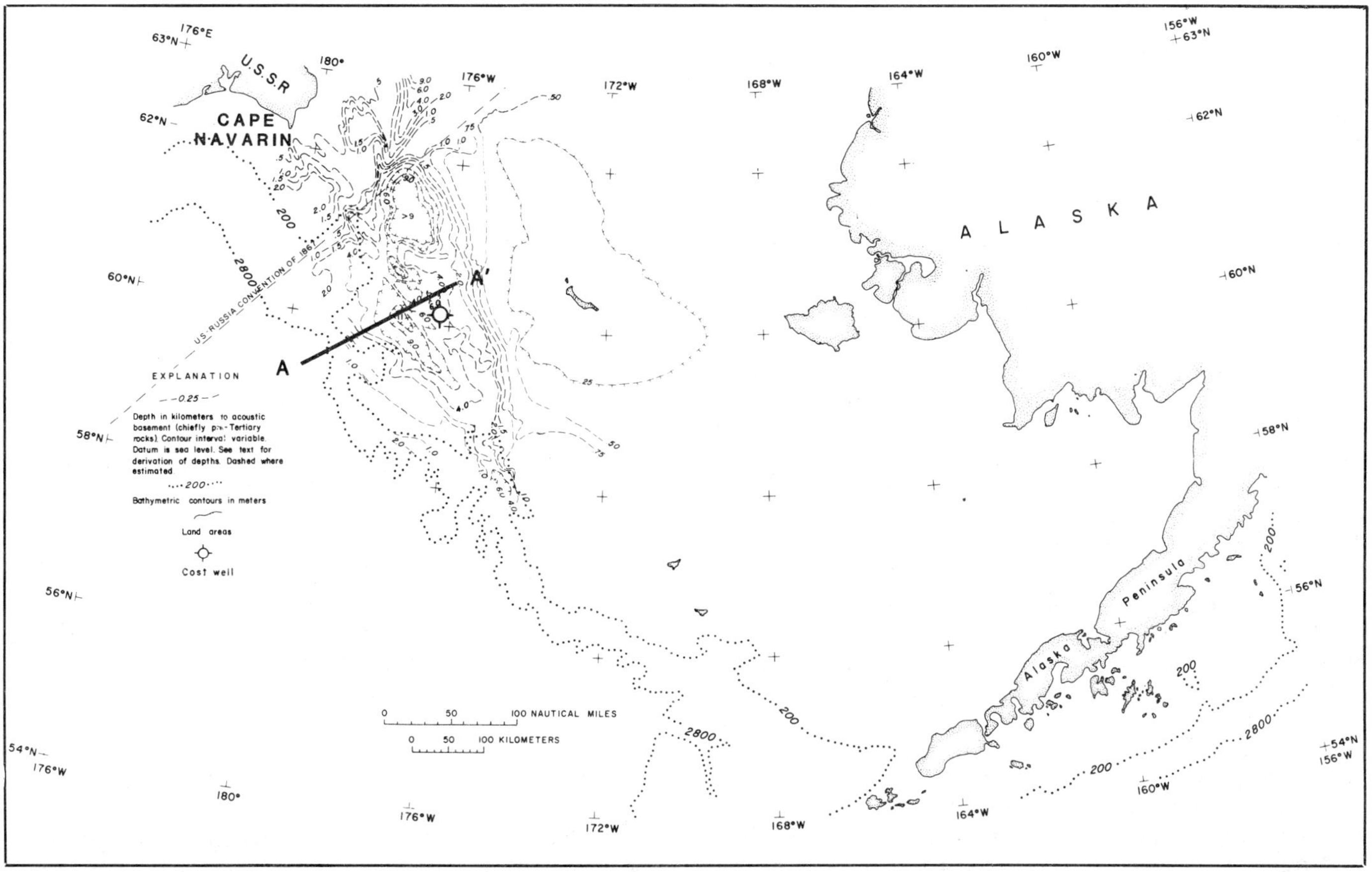

Figure 10. Structure contour map of acoustic basement or bedrock in the Navarin Basin province. From Marlow (1979). A–A′ is line of interpretative seismic section shown in Figure 11. Albers equal-area projection.

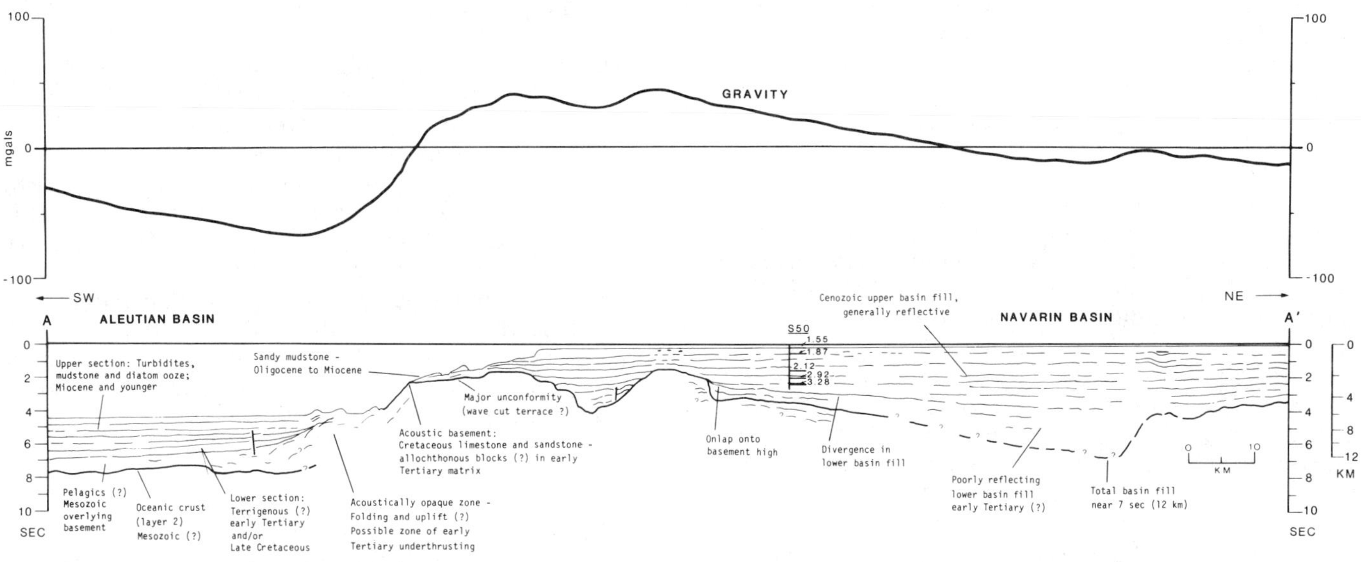

Figure 11. Interpretative drawing of seismic-reflection and gravity profiles across the northern Beringian margin and Navarin Basin. For location see A–A′ on Figures 1 and 10. The strong basal reflector beneath the Aleutian Basin can be traced beneath the base of the margin into an acoustically opaque zone. The thick sedimentary fill in the Navarin Basin extends to a subshelf depth of about 12 km. Adapted from Marlow and Cooper (1986).

tion wells have been drilled, but the resultant information remains proprietary. The COST well located on Figure 10 was drilled in 1983 to a depth of 4998 m in water 132 m deep (Table 3). Data from this well are described extensively in Turner et al. (1984c, 1985).

The Cenozoic sedimentary section sampled includes silty and sandy mudstone, diatomaceous mudstone, muddy and fine-grained sandstone interbedded with sandy mudstone, and claystone. The Upper Cretaceous section consists of siltstone, fine-grained sandstone, mudstone, claystone, tuff, and

Table 3. Stratigraphic sequence penetrated at Navarin Basin COST No. 1 well.

Age	Depth to Top (m)	Lithology
Pliocene	468	Marine diatomaceous mudstone, siltstone, claystone, and sandstone
Miocene	969	Same as above
Oligocene	1,739	Same as above
Eocene	3,743	Calcareous claystone and sandy mudstone
Bedrock unconformity: Late Cretaceous	3,895	Nonmarine and marine sandstone, mudstone, claystone, coal, tuff, with diabase and basalt sills

coal. The sedimentary section below 335 m and above 3895 m was deposited in a marine environment, whereas the section between 3895 and 4663 m was deposited in fluvial, flood-plain, and deltaic environments. The section from 3901 to 4587 m of Upper Cretaceous strata is intruded by numerous basaltic and diabasic sills. K-Ar dates on the igneous samples suggest that the emplacement of the sills was in the late Oligocene or early Miocene.

Dredge samples were recovered from the continental slope near the base of the section overlying the structural highs that flank the Navarin Basin. These samples are Paleocene to Oligocene in age (Jones et al., 1981). Although it is difficult to trace this early Tertiary section into the subbasins of the Navarin Basin province, we suspect that a thick and largely Cenozoic section fills them. The COST well results (Turner et al., 1985) establish that dipping strata observed in the lower fill of the basins are in large part early Tertiary in age. The overlying horizontal section is probably Miocene and younger. In places, the upper section is broadly folded and truncated, especially in the northwestern end of the Navarin Basin. This period of deformation may be related to late Miocene and Pliocene uplift and folding in the nearby Koryak Range of eastern Siberia (Marlow, 1979; Marlow, Cooper, and Childs, 1983). Subsidence in the Navarin Basin province has apparently been continuous during the Cenozoic, possibly extending even into the present day.

Petroleum Potential

Source Beds

Geochemical analyses of sediment samples from the Navarin COST well show that thermal maturity sufficient for the generation of crude oil begins at a depth of about 3000 m (Turner et al., 1984c). Neogene rocks to a depth of 1700 m contain predominantly type III kerogen. The Paleogene section from 1700 to 3895 m contains increasing amounts of sapropelic kerogen with depth. The section between 3876 and 3884 m is characterized by type II kerogen. Kerogen changes to type III abruptly at a depth of 3895 m, in Cretaceous strata below a major unconformity, and type III kerogen persists from there to the bottom of the hole at 4998 m. Type III kerogen generally produces gas and some condensate hydrocarbons, whereas type II kerogen can produce oil. If the organic-rich Paleogene section between 3566 and 3895 m thickens in the central part of the Navarin Basin, there may be significant amounts of oil-prone source beds in the basin (Turner et al., 1985).

The average thermal gradient in the COST well is 32°C/km between 1160 and 4998 m. Above 1160 m, the thermal gradient is an anomalously high 45°C/km, probably because of the effects of an overpressured zone of complex diagenetic alteration of diatomaceous sediment (Turner et al., 1984c). The high thermal gradients in the basin suggest that much of the lower fill in the Navarin Basin has been sufficiently heated to generate hydrocarbons.

Reservoir Beds

According to Turner et al. (1984c) the reservoir qualities of samples from the Navarin COST well are low because porosity and permeability in the section have been reduced by ductile deformation of individual grains in the sediment and by cementation and authigenesis. However, the reservoir quality of the section may be better in the thicker sedimentary sections elsewhere in the basin. The reservoir characteristics of dredge samples from the adjacent Beringian margin were discussed above under St. George Basin.

Traps

Strata in the northern subbasin of the Navarin Basin province have been folded into anticlinal structures 10 to 15 km across (Marlow, 1979) either by diapirism or by lateral compression. The crowns of the anticlines are cut by an unconformity that is overlain by a few tens of meters of flat-lying strata. High-amplitude reflection events are observed within the strata above the balded folds. Beds in the southern two subbasins are cut by normal faults along the flanks of the basins. Increase in the offset across these faults with depth shows that the faults are growth structures that formed contemporaneously with the filling of the basin.

Potential stratigraphic traps also exist within the Navarin Basin province. For example, lower beds in the basin fill thin toward the basement flanks and generally dip toward the axis of each basin. Fluids migrating updip could therefore be trapped against less permeable rocks of the basement. In addition, the lower stratigraphic sequence in the subbasins is commonly overlapped discordantly by younger beds that could trap migrating petroleum fluids. Other potential traps for hydrocarbons in the Navarin Basin province include drape structures over blocks in the basement. Offset on the basement blocks is commonly several thousands of meters, and they extend laterally >5 to 10 km.

ANADYR BASIN

Description

Anadyr Basin is a large structural depression that encompasses about 120,000 km^2 and is filled with Upper Cretaceous and Cenozoic sedimentary rocks. Onshore, the basin underlies the Anadyr lowlands and is flanked on the south and west by folded Mesozoic rocks of the Koryak Range (Figure 12; McLean, 1979). About 20 onshore exploration wells, reviewed by McLean (1979), have been drilled into the basin. To the north and east, the Anadyr Basin is flanked by the

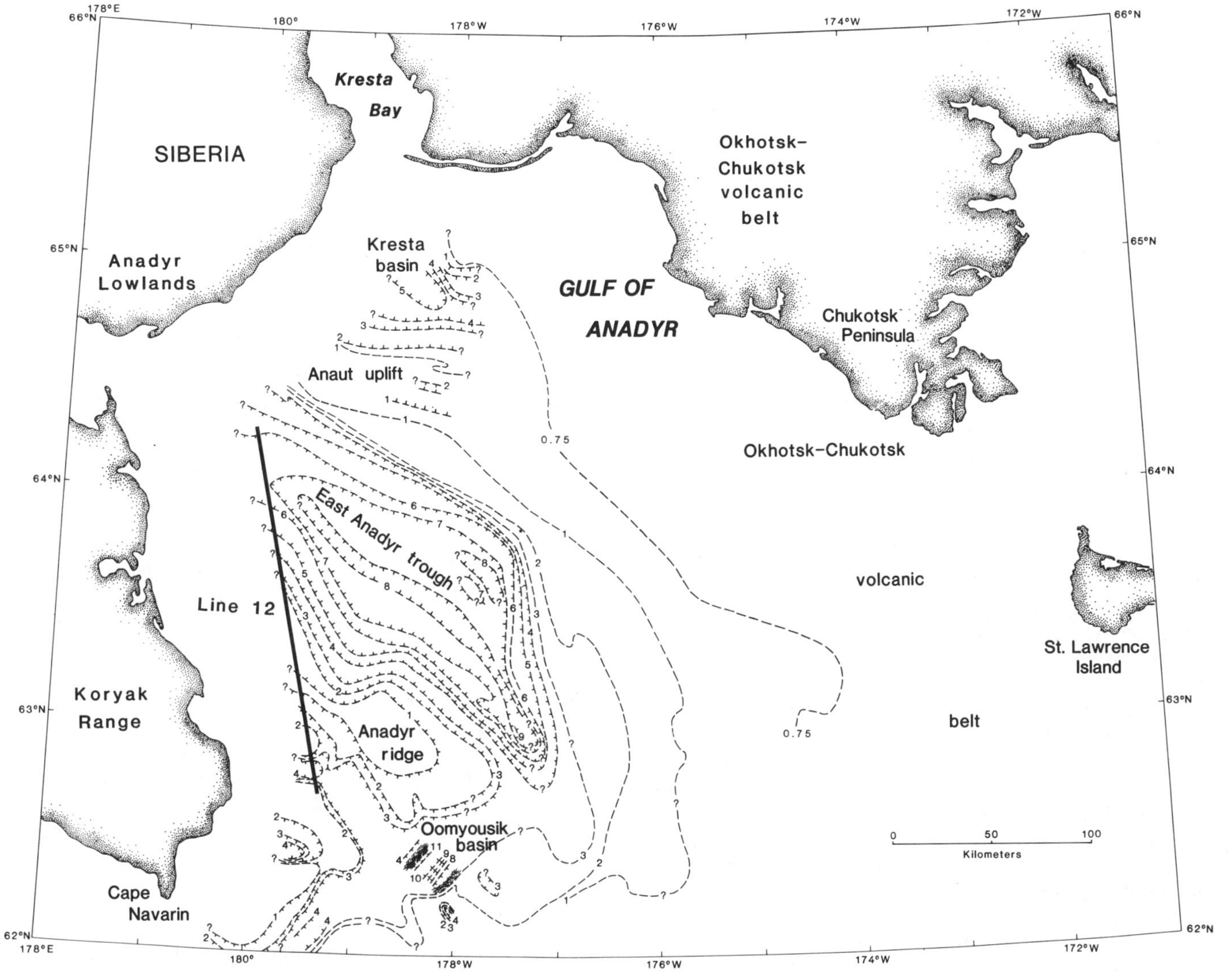

Figure 12. Structure contour map of total sediment accumulation (in km) in the offshore portions of the Anadyr Basin. Profile of line 12 shown on Figure 13. Albers equal-area projection. From Marlow, Cooper, and Childs (1983).

Okhotsk-Chukotsk volcanic belt—a broad bedrock high composed of plutonic and volcanic rocks—that extends southeastward from eastern Siberia across the Beringian Shelf at least to St. Matthew Island (Patton et al., 1974, 1976; Marlow, Cooper, and Childs, 1983). To the east and southeast, the basin extends offshore as far as 200 km (Figure 1). The thickest part of the basin fill, east Anadyr Trough, is >9 km (Figure 12; Marlow, Cooper, and Childs, 1983).

Seismic-reflection and seismic-refraction data show that the Anadyr Basin is separated from the Navarin Basin by the Anadyr Ridge, a southeast–northwest-trending bedrock high that is characterized by high-amplitude, short-wavelength magnetic anomalies (Figure 13; Marlow, Cooper, and Childs, 1983). Anadyr Ridge may be an offshore extension of the melange belt underlying the Koryak Range. Sonobuoy refraction data suggest that the velocity profile of strata in the offshore part of the Anadyr Basin is similar to that in the Navarin Basin. However, the basins are different structurally. The Navarin Basin is complex and contains both compressional and extensional elements, whereas the Anadyr Basin is a simple, broad crustal sag semicircular in outline (Figure 12).

Another smaller basement ridge underlies the south end of line 12, and exhibits an associated magnetic high (Figure 13). Reflectors above the horizon labeled beta (the presumed Paleogene–Mesozoic boundary; Marlow, Cooper, and Childs, 1983) are deformed in a diapirlike fashion toward the sea floor, suggesting Holocene(?) tectonism. The Anadyr Basin, in contrast, contains generally undeformed fill, and appears in cross-section as a saucer-shaped structural sag, undisturbed by faulting (Figure 13).

Age and History

As noted, reconstructions of plate motions for the north Pacific suggest that during the Mesozoic, before the formation of the Aleutian Arc, the Beringian continental margin was a zone of either oblique underthrusting or transform motion between the North American and Kula(?) plates. Subduction of the Kula plate beneath Siberia presumably formed the Okhotsk-Chukotsk volcanic arc (Scholl, Buffington, and Marlow, 1975; Patton et al., 1976; Marlow et al., 1976). South of the Okhotsk-Chukotsk volcanic belt, the melange and olstostrome units, now exposed as discrete tectonic terranes in the Koryak Range (Aleksandrov et al.,

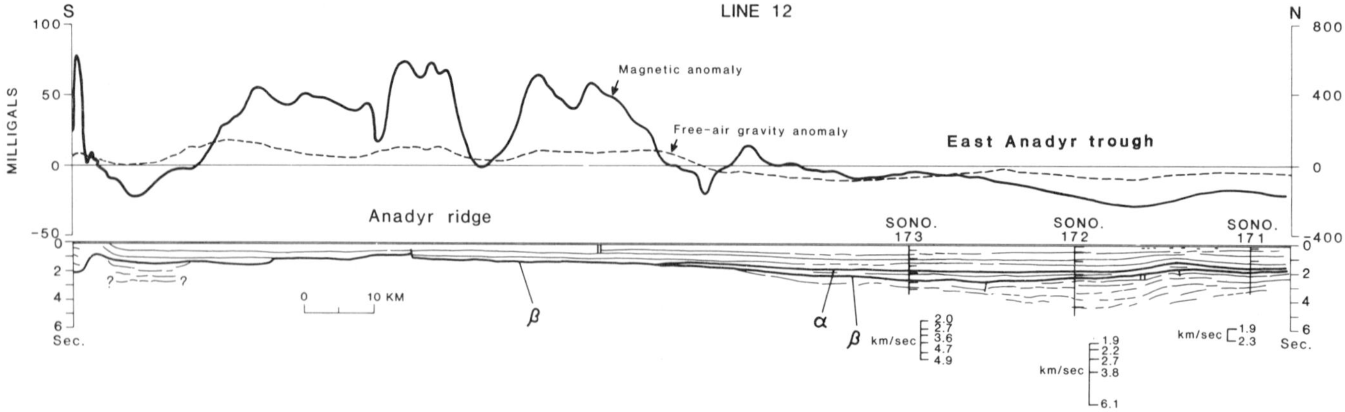

Figure 13. Gravity and magnetic profiles and interpretative drawing of 24-channel seismic-reflection line 12 across the east Anadyr Trough and Anadyr Ridge. For location of profile, see Figures 1 and 12. Sono. 173 refers to sonobuoy refraction station 173 (see Marlow, Cooper, and Childs, 1983).

1976), were formed by accretion of fragments scraped off the Kula plate (Fujita and Newberry, 1982).

Construction of the Okhotsk-Chukotsk magmatic pile, and structural thickening of the Koryak melange belt during the middle and late Cretaceous, presumably caused relative crustal downwarping of the intervening area and the initial formation of the fore-arc Anadyr Basin. The bedrock sequence in the Anadyr Basin, as deduced from seismic-reflection data and extrapolation from onshore geology, consists of strongly folded and broken tectonic blocks of Upper Jurassic and Lower Cretaceous rocks (Agapitov, Ivanov, and Krainov, 1973; Burlin, Dontsov, and Pastukhova, 1974; McLean, 1979; Marlow, Cooper, and Childs, 1983). This basement complex probably forms the lower reflecting sequence below the beta horizon in seismic-reflection profile 12 (Figure 13), and is characterized by refraction velocities of 5 km/sec or greater (Marlow, Cooper, and Childs, 1983). Offshore data further suggest that the pre-Cretaceous sequence below the Anadyr Basin is deformed into a broadly semicircular sag that trends southeast beneath the Gulf of Anadyr.

McLean (1979) compared the Anadyr Basin and the flanking Okhotsk-Chukotsk volcanic arc to the Great Valley sequence in central California and its flanking Sierra Nevadan plutonic belt. The Great Valley sequence, which fills a fore-arc basin, is flanked on its seaward side by the accretionary mass of the Franciscan Complex or assemblage of coastal California. This sequence of tectonic elements is similar to that associated with the Anadyr Basin in that it is bounded seaward by the accretionary assemblages of the Koryak Mountains of eastern Siberia.

Petroleum Potential

Source Rocks

Samples from the onshore wells show that Miocene strata contain brown humic coal (McLean, 1979). The organic content of Miocene strata is mainly humic and abundant (3 to 5%), and all the Miocene shows in the wells have been gas. Lower strata in the basin, especially Eocene and Oligocene beds, do have oil potential (McLean, 1979), and the southwestern and southern parts of the basin contain the most promising Paleogene sections (Agaptov, Ivanov, and Krainov, 1973). Upper Cretaceous beds in the basin contain abundant organic carbon (3%) but, much like coeval rocks beneath the Navarin Basin, the carbon is humic. Also, according to Link and Meyerhoff (1976), Cretaceous strata are too deeply buried to produce oil and are gas prone.

Reservoir Beds

McLean (1979) reported effective porosity values as high as 26% and permeability values of 560 md for some of the Miocene sandstone beds that are as thick as 80 m in the Anadyr Basin. Gas pressure from the sandstone beds dropped off rapidly during testing, suggesting that the sandstone beds are lenticular bodies that contain only small volumes of highly pressured gas.

To date, no economic hydrocarbon discoveries have been made by the wells drilled onshore into the Anadyr Basin. No wells have been drilled offshore. Here, the hydrocarbon potential may be greater because thicker sections should increase the possibilities for hydrocarbon maturation (McLean, 1979).

Traps

Traps within the sedimentary fill of the Anadyr Basin include anticlines that show low-amplitude closure over large areas (McLean, 1979). Along the southern edge of the basin, anticlinal structures may have formed by compressional folding. Drape structures formed by differential compaction over basement blocks occur in the central part of the onshore part of the basin. Offshore, a large anticline roughly 10 to 15 km across occurs near the mouth of the Anadyr River (Marlow, Cooper, and Childs, 1983). Within the anticline, a flat high-amplitude reflector or bright spot lies at a subbottom depth of about 1500 m. Beneath the bright spot is a velocity pulldown, and above the anticline, near the sea floor, are several seismic "wipe-out" zones. These high-amplitude anomalies are present only over the crest of the anticline and are not present to the south in other parts of the basin. These observations suggest that hydrocarbons are accumulating within the anticline and that some are leaking upward to form the shallow acoustic anomalies near the sea floor.

As with the other Beringian Shelf basins, potential stratigraphic traps are seen in the thinning and wedging of the older stratigraphic sequences deep in the basin toward its flanks (Marlow, Cooper, and Childs, 1983). Also, the lower stratigraphic sequences in the basin are discordantly overlapped by younger, overlying beds. Both of these stratigraphic situations could lead to the trapping of migrating hydrocarbons.

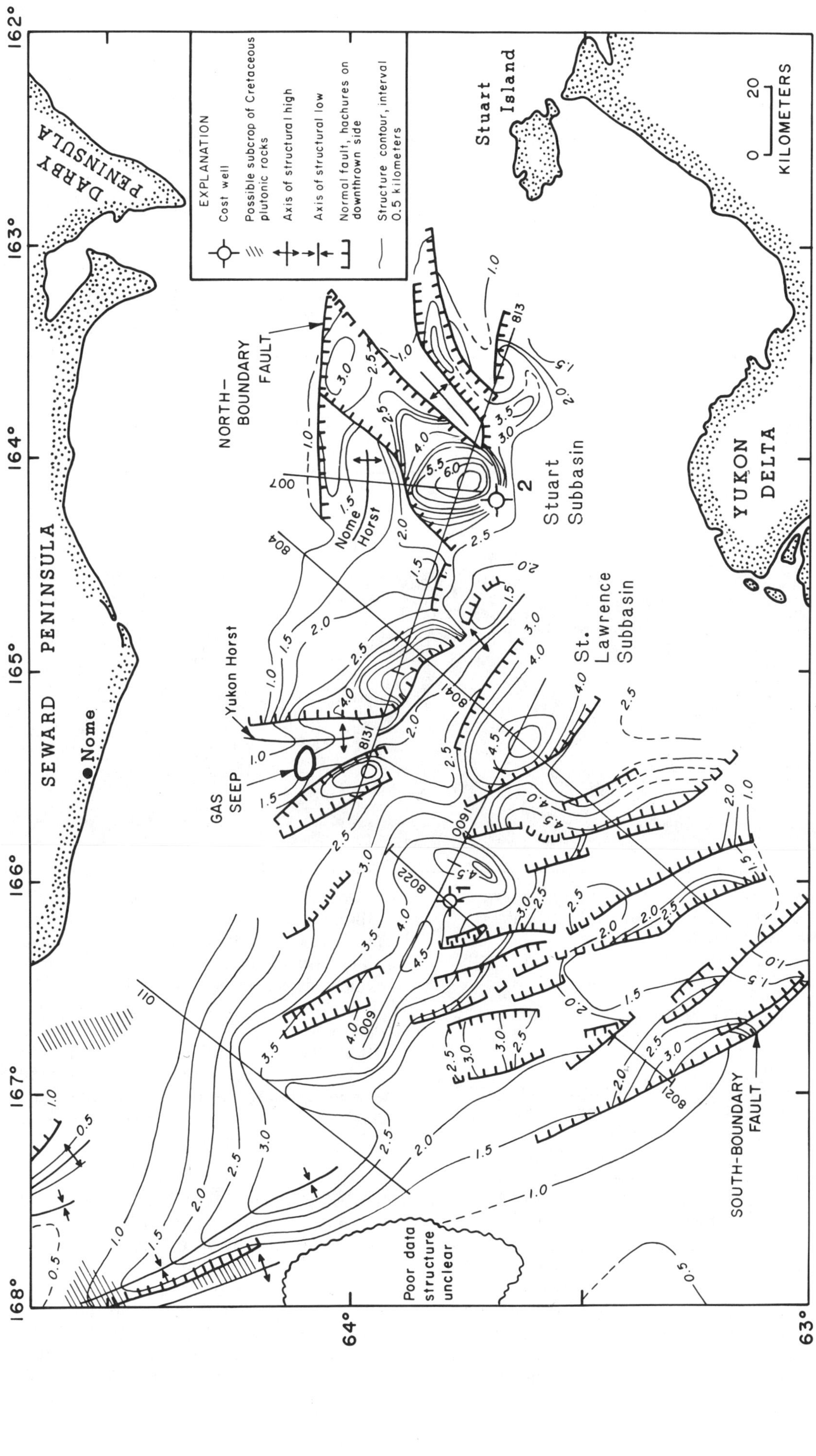

Figure 14. Structure contours on horizon A in Norton Basin. Horizon A is an unconformity between probable Precambrian and Paleozoic rocks below and uppermost Cretaceous or lower Paleogene rocks above. Lines numbered 804 and 813 are seismic sections shown in Figures 15 and 16. From Fisher, Patton, and Holmes (1982).

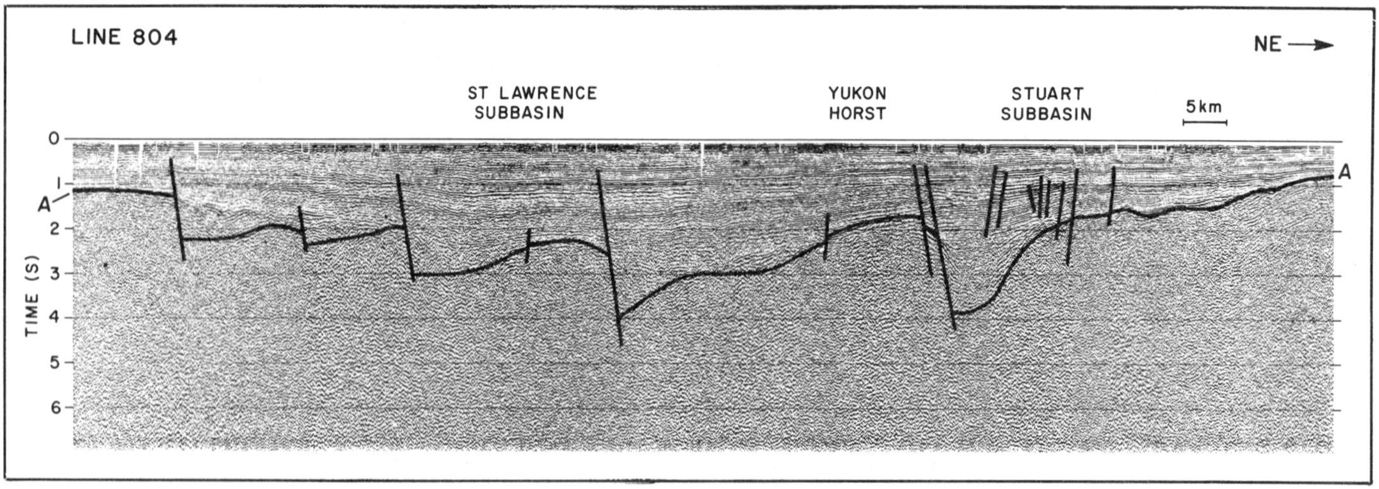

Figure 15. Line 804, a northeast–southwest seismic-reflection section through the central part of Norton Basin. Horizon A at top of basement rocks. From Fisher, Patton, and Holmes (1982).

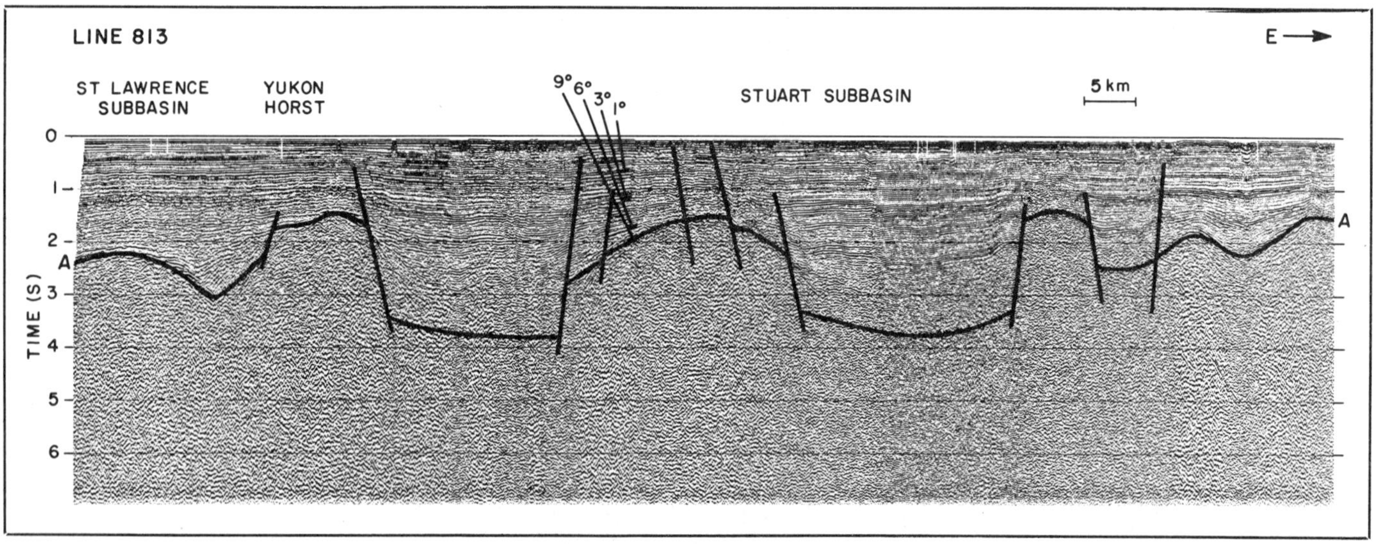

Figure 16. Line 813, an east–west seismic-reflection section through eastern and central Norton Basin. Dip of beds is shown. Horizon A is at the top of basement rocks. From Fisher, Patton, and Holmes (1982).

NORTON BASIN

Description

The Norton Basin, which underlies the northernmost part of the Beringian Shelf beneath Norton Sound, consists of the St. Lawrence and Stuart subbasins (Figures 14, 15, 16; Fisher, Patton, and Holmes, 1982). These subbasins differ in structure and thickness of fill. The St. Lawrence subbasin lies northeast of St. Lawrence Island and west of the Yukon horst—a structure that extends northwestward from the Yukon Delta toward Nome. This subbasin is surrounded by shallow basement that dips toward the basin from the Seward Peninsula and St. Lawrence Island, where Precambrian(?) and Paleozoic rocks crop out. Although major normal faults were most active adjacent to the Yukon Delta, gravity data suggest that the Norton Basin does not extend beneath the land areas covered by the delta. Large normal faults in the St. Lawrence subbasin strike northwest, and the maximum thickness of fill in this area is 5 km.

The Stuart subbasin lies east of the Yukon horst and north of Stuart Island. This subbasin is a subcircular depression that includes the deepest part of the Norton Basin, where the fill is 6.5 km thick. Fault-bounded troughs that contain basin rocks as much as 4 km thick extend northeastward and northwestward from the main basin deep. The Nome horst trends east, separating a narrow east-striking graben from the main part of the basin. Major normal faults in the Stuart subbasin strike east and northeast at right angles to faults in the St. Lawrence subbasin.

Stratigraphy

Two COST wells were drilled in the Norton Basin by a consortium of oil companies (Turner et al., 1983a, 1983b). The section penetrated by the COST No. 1 well, located in the St. Lawrence subbasin, includes about 3800 m of Cenozoic rocks; about 4400 m of Cenozoic strata were penetrated in Stuart subbasin at COST No. 2 well (Table 4).

The oldest rocks penetrated by the COST wells are quartzite, phyllite, and marble of Precambrian or Paleozoic age, an age assignment based on lithologic similarities to rocks of

Table 4. Stratigraphic sequences penetrated at Norton Basin COST wells.

Age	Depth to Top (m)	Lithology
		COST No. 1 Well
Pleistocene	?	Marine mudstone, siltstone, diatomaceous strata
Pliocene	402	Same as above
Miocene	804	Same as above
Oligocene	1,369	Marine mudstone and shale
Oligocene or older	1,369	Nonmarine sandstone, mudstone, shale, and coal
Eocene(?) or older	3,729	Same as above
Metamorphic basement: Paleozoic or older	3,824	Quartzite, phyllite, and marble
		COST No. 2 Well
Pleistocene	?	Marine mudstone, siltstone, and diatomaceous strata
Pliocene	402	Same as above
Miocene	786	Same as above, with coal
Oligocene	1,074	Nonmarine and marine siltstone, mudstone, and coal
Eocene	3,097	Nonmarine sandstone, conglomerate, and coal
Eocene(?) or older	3,871	Same as above
Metamorphic basement: Paleozoic or older	4,407	Quartzite and phyllite

these ages exposed in the York Mountains of the western Seward Peninsula. These rocks form the acoustic basement and the economic basement for oil and gas. The deepest units of the basin fill penetrated by both wells contain very few fossils, but may be Eocene or older sequences of nonmarine sandstone, with some conglomerate and coal. In the COST No. 2 well, these poorly dated rocks are separated by an unconformity from an overlying 774–m-thick section of middle and upper Eocene siltstone, mudstone, conglomerate, and coal that was deposited in nonmarine and transitional environments. Strata in the COST No. 1 well that are probably coeval with this Eocene section form a 776–m-thick unit of sandstone, mudstone, and siltstone dated as Oligocene or older that may have been deposited in a marine environment. Basalt, anomalously low in alkalies, was encountered in the lower Cenozoic section of the COST No. 1 well; the basalt yields K/Ar dates of 19.9 Ma, an early Miocene age, indicating the basalt is intrusive.

Oligocene rocks include the oldest marine strata deposited in the Norton Basin. In the COST No. 1 well, Oligocene mudstone and shale are 1585 m thick and were deposited in a middle neritic to upper bathyal environment. The Oligocene section in the COST No. 2 well is 2023 m thick. The lower half of this section is siltstone, mudstone, and some coal deposited in nonmarine or transitional environments; the upper half is mudstone, siltstone, and sandstone that accumulated in transitional and neritic environments. Miocene neritic rocks are 565 m thick in the COST No. 1 well, but they consist of mudstone, diatomaceous strata, and siltstone. In the COST No. 2 well a similar section that includes some coal is 288 m thick. Neritic diatomaceous strata, mudstone, and muddy sandstone of Pliocene and younger age are 786 m thick in the COST No. 1 well.

Age and History

Rocks exposed on the Seward Peninsula underwent a complex history of deformation and intrusion. If basement rocks beneath the Norton Basin underwent a similar history, they were last deformed in the Late Jurassic and Early Cretaceous, during the orogeny that formed the Brooks Range of northern Alaska (Patton and Tailleur, 1977; Patton et al., 1977; Roeder and Mull, 1978; Mull, 1979). In the Early and middle Cretaceous, after this mountain-building phase, the area of Norton Basin was a highlands supplying calcarenitic debris preserved in the Yukon-Koyukuk province east of the Norton Basin (Patton, 1973). During the Late Cretaceous, rocks on the Seward Peninsula and perhaps those in the basement under the Norton Basin were metamorphosed by the intrusion of granitic rocks.

On the basis of seismic stratigraphy and onshore geology, Fisher, Patton, and Holmes (1982) proposed that Norton Basin initially formed during latest Cretaceous to the end of the Paleocene (70 to 55 Ma), contemporaneously with the formation of other grabens in western Alaska that were filled with volcanic rocks. The age of the deepest strata of the basin fill is unknown. The COST wells penetrated sedimentary rocks as old as middle Eocene. But where basin fill is thickest, nearly 2 km of unsampled strata underlie the deepest horizon penetrated by either well. Extensional basins typically undergo rapid subsidence along major faults during their early histories; consequently, the thick Oligocene section (35–45% of the total fill) penetrated by both wells suggests that the basin was probably young during the Oligocene.

Seismic-reflection data show that basin subsidence occurred adjacent to large normal faults. In the St. Lawrence subbasin these faults show greatest throw near the Yukon Delta, suggesting that the Kaltag fault, a major strike-slip fault mapped east of the Norton Basin, may have contributed to basin formation.

Seismic-reflection data show that alluvial fans formed along the flanks of some of the major basement highs as the basin underwent rapid initial subsidence. During the Eocene, subsidence of the St. Lawrence subbasin was rapid enough that a marine incursion shifted the depositional environment from nonmarine to outer neritic and upper bathyal. In contrast, in the Stuart subbasin, separated from the other subbasin by the high-standing Yukon horst, a nonmarine environment of deposition continued well into the Oligocene. During the Miocene, Pliocene, and Quaternary, deposition of marine strata occurred in both subbasins. The Neogene and Quaternary rocks are considerably thinner (even without correction for compaction) than the Oligocene rocks, indicating that a general slowing of the rate of basin subsidence occurred during the past 22 m.y. This decreasing rate of infilling mirrors the decreasing offset across faults upward through the basin fill.

Petroleum Potential

Source Beds

Norton Basin contains a thick basin fill, and much of the fill in the western part of the basin is marine. This factor, and a temperature gradient as high as 37° to 45° C/km, favor the formation of hydrocarbons.

The amount of organic carbon in the COST No. 1 well generally increases linearly with depth from 0.5% in Quaternary deposits (at 384 m) to about 1.0% in lower Miocene rocks (at 1372 m). In Oligocene and Eocene(?) rocks down to the bottom of the fill (between 1373 and 3811 m), the amount of organic carbon ranges between 1.0 and 2.0%. In the pre-Mesozoic basement rocks, the amount of organic carbon decreases abruptly to about 0.3%. Organic carbon in deposits penetrated by the COST No. 2 remains nearly constant at 0.5%, except for coal-rich units in Oligocene and Miocene rocks, which contain between 1 and 60% organic carbon. The amount of organic carbon levels off at 2 to 5% in Eocene rocks but increases again to 50 to 60% in coal-rich Eocene or older rocks (below 3660 m).

In the pre-Mesozoic basement rocks, the amount of organic carbon is around 0.1%. In both COST wells, type III kerogen predominates at all stratigraphic levels. In the COST No. 1 well, where Oligocene and younger rocks contain kerogen that is mainly vitrinite, as much as 20% exinite is present locally. In Eocene or older basin fill, and in pre-Mesozoic basement rocks, the predominant kerogen type is inertinite with subsidiary amounts of vitrinite. In rocks penetrated by the COST No. 2 well, the proportion of amorphous kerogen is sufficiently high in two rock units that they could be source rocks for oil. The shallower unit (244 m thick) includes the whole Miocene section; the deeper unit (915 m thick) comprises the middle and upper Eocene and lower Oligocene section.

Both COST wells penetrated rocks that have been heated sufficiently to produce oil. In the COST No. 1 well, vitrinite reflectance values increase linearly down to the base of the Oligocene section (at 2960 m), where the kerogen is sufficiently mature to begin to generate oil. Organic matter in rocks of Eocene(?) age (below 3658 m) contain vitrinite with a reflectance of 1.0%, so that deeper rocks have been heated too much for oil generation. The reflectance of vitrinite in basement rocks is 2.44%. In the COST No. 2 well, vitrinite reflectance increases linearly downhole to 0.6%, the lower limit of oil generation, at the base of the Oligocene section. Eocene or older rocks (at and below about 4270 m) contain vitrinite with a reflectance of 1.0%, which signifies the upper limit of the oil window. Thus, source rocks for oil could be present where Paleogene rocks (between 3354 and 4268 m) contain kerogen that is rich in amorphous material and has been heated enough to produce oil. Minor amounts of gaseous and liquid hydrocarbons were encountered below the 3050-m depth.

Data from both COST wells show that the geothermal gradient through the Norton Basin increases linearly with depth down to an inflection point at 3050 m. In the COST No. 1 well, the geothermal gradient is 36.6°C/km above the inflection point and 44.4°C/km below. Temperatures were measured 36 hours after circulation stopped. In the COST No. 2 well, the gradient is 36°C/km above the inflection and 45.5°C/km below. These temperatures were measured 48.5 hours after circulation stopped.

Reservoir Beds

Data from the COST wells show that adequate reservoirs for hydrocarbons are present in the basin. Porosity decreases in both wells from 40% in shallow rocks to 10% at 3050 m. Below this depth, porosity decreases rapidly to <5% near the base of the basin fill. Permeabilities vary unsystematically with depth—between 0.05 and 0.3 md. Porosity reduction is due primarily to compaction and ductile grain deformation in matrix-rich sandstone. Authigenic cementation is a subsidiary cause for decreasing quality of reservoirs.

Traps

Hydrocarbons could be contained in four main types of traps: (1) horsts, (2) drape or compaction structures, (3) subsidiary structures developed along faults, and (4) stratigraphic traps along the flanks of horsts. Marble in basement rocks within high-standing horsts could have karst features or fractures to provide reservoir space. Drape or compaction structures over basement highs are large, have low relief, and could form the more common types of traps. Subsidiary structures along faults include small anticlines and monoclines that could be sealed at fault planes. Stratigraphic traps could be formed along the periphery of basement highs, where alluvial fans are present, or in areas where basin fill laps onto the flanks of the high.

In summary, the Norton Basin could contain oil produced from Paleogene rocks in the Stuart subbasin, but condensate and gas may be the predominant types of hydrocarbons in the basin as a whole. Basement rocks are not source rocks for oil, but they may contain hydrocarbons in fractures.

ACKNOWLEDGMENTS

We thank P. R. Carlson and M. L. Holmes for their helpful and thoughtful reviews.

REFERENCES CITED

Agapitov, D. I., V. V. Ivanov, and V. G. Krainov, 1973, New data on the geology and prospects in petroleum/gas bearing Anadyr Basin: Arademiya Nauk SSR, North-East Group Institute of Interior Center Translations, v. 49, p. 23–29. (In Russian.)

Aleksandrov, A. A., N. A. Bogdanov, S. G. Byalobzheskiy, M. S. Markov, S. M. Til'man, V. Ye. Khain, and A. D. Chekov, 1976, New data on the tectonics of the Koryak highlands: Geotectonics, v. 9, p. 292–299.

Beikman, H. M., 1974, Preliminary geologic map of the southwest quadrant of Alaska: U.S. Geological Survey Miscellaneous Field Studies Map MF–611, 2 sheets, scale 1:2,000,000.

Brockway, R., B. Alexander, P. Day, W. Lyle, R. Hiles, W. Decker, W. Polski, and B. Reed, 1975, Bristol Bay region, "Stratigraphic correlation section," southwest Alaska: Anchorage, Alaska Geological Society.

Burk, C. A., 1965, Geology of the Alaska Peninsula island arc and continental margin: GSA Memoir 99, 250 p.

Burlin, Yu. K., V. V. Dontsov, and T. N. Pastukhova, 1974, Future direction of oil and gas exploration in northeastern U.S.S.R.: Geologya Nefti Gaza, no. 5, p. 17–22 (in Russian); summary in Petroleum Geology, v. 12, no. 5, p. 222–223.

Carlson, P. R., M. Golan-Bac, H. A. Karl, and K. A. Kvenvolden, 1985, Seismic and geochemical evidence for shallow gas in sediment on Navarin continental margin, Bering Sea: AAPG Bulletin, v. 69, p. 422–436.

Cooper, A. K., and M. S. Marlow, 1983, Preliminary results of geophysical and geological studies of the Bering Sea Shelf during 1982: U.S. Geological Survey Open-File Report 83-322, 8 p.

Cooper, A. K., M. S. Marlow, and D. W. Scholl, 1976, Mesozoic magnetic lineations in the Bering Sea marginal basin: Journal of Geophysical Research, v. 81, p. 1916–1934.

Cooper, A. K., D. W. Scholl, and M. S. Marlow, 1976, Plate tectonic model for the evolution of the eastern Bering Sea Basin: GSA Bulletin, v. 87, p. 1119–1126.

Fisher, M. A., W. W. Patton, Jr., and M. L. Holmes, 1982, Geology of Norton Basin and continental shelf beneath northwestern Bering Sea, Alaska: AAPG Bulletin, v. 66, p. 225–285.

Fujita K., and J. T. Newberry, 1982, Tectonic evolution of northeastern Siberia and adjacent regions: Tectonophysics, v. 89, p. 337–357.

Hatten, C. W., 1971, Petroleum potential of Bristol Bay Basin, Alaska: AAPG Memoir 15, p. 105–108.

Hopkins, D. M., 1967, The Cenozoic history of Beringia—A synthesis, *in* D. M. Hopkins, ed., The Bering land bridge: Stanford, California, Stanford University Press, p. 451–484.

Hopkins, D. M., and D. W. Scholl, 1970, Tectonic development of Beringia, late Mesozoic to Holocene (abs.): AAPG Bulletin, v. 54, p. 2487–2488.

Hopkins, D. M., D. W. Scholl, W. O. Addicott, R. L. Pierce, P. B. Smith, J. A. Wolfe, D. Gershanovich, B. Kotenev, K. E. Lohman, J. H. Lipps, and J. Obradovich, 1969, Cretaceous, Tertiary, and early Pleistocene rocks from the continental margin in the Bering Sea: GSA Bulletin, v. 80, p. 1471–1480.

Jones, D. L., and N. J. Silberling, 1979, Mesozoic stratigraphy—The key to tectonic analysis of southern and central Alaska: U.S. Geological Survey Open-File Report 79–1200, 37 p.

Jones, D. M., M. J. Kingston, M. S. Marlow, A. K. Cooper, J. A. Barron, F. H. Wingate, and R. E. Arnal, 1981, Age, mineralogy, physical properties, and geochemistry of dredge samples from the Bering Sea continental margin: U.S. Geological Survey Open-File Report 81–1297, 68 p.

Link, P. K., and A. A. Meyerhoff, 1976, Bering Sea: Geology and hydrocarbon potential (geological-geophysical study of the Bering Sea, United States and USSR): Geneva, Switzerland, Petroconsultants S. A., v. 1, 205 p.

Marlow, M. S., 1979, Hydrocarbon prospects in Navarin Basin province, northwest Bering Sea Shelf: Oil and Gas Journal, October 29, p. 190–196.

Marlow, M. S., and A. K. Cooper, 1980a, Mesozoic and Cenozoic structural trends under southern Bering Sea Shelf: AAPG Bulletin, v. 64, p. 2139–2155.

Marlow, M. S., and A. K. Cooper, 1980b, Multichannel seismic-reflection profiles collected in 1976 in the southern Bering Sea Shelf: U.S. Geological Survey Open-File Report 80-389, 2 p.

Marlow, M. S., and A. K. Cooper, 1983, Wandering terranes in southern Alaska: The Aleutia microplate and implications for the Bering Sea: Journal of Geophysical Research, v. 88, p. 3439–3446.

Marlow, M. S., and A. K. Cooper, in press, Regional geology of the Beringian margin, *in* Proceedings of the OJI International Seminar on the Formation of Ocean Margins: University of Tokyo, Tokyo, Japan.

Marlow, M. S., A. K. Cooper, and J. R. Childs, 1983, Tectonic evolution of Gulf of Anadyr and formation of Anadyr and Navarin Basins: AAPG Bulletin, v. 67, p. 646–865.

Marlow, M. S., D. W. Scholl, and A. K. Cooper, 1977, St. George Basin, Bering Sea Shelf: A collapsed Mesozoic margin, *in* M. Talwani and W. C. Pitman, III, eds., Island arc, deep sea trenches, and back-arc basins: American Geophysical Union Maurice Ewing Series, v. 1, p. 211–220.

Marlow, M. S., D. W. Scholl, A. K. Cooper, and E. C. Buffington, 1976, Structure and evolution of Bering Sea Shelf south of St. Lawrence Island: AAPG Bulletin, v. 60, p. 161–183.

Marlow, M. S., A. K. Cooper, D. W. Scholl, T. L. Vallier, and H. McLean, 1979, Description of dredge samples from the Bering Sea continental margin: U.S. Geological Survey Open-File Report 79–1139, 5 p.

Marlow, M. S., A. K. Cooper, D. W. Scholl, T. L. Vallier, and H. McLean, 1982, Ancient plate boundaries in the Bering Sea region, *in* J. K. Leggett, ed., Trench-forearc geology—Sedimentation and tectonics on modern and ancient active plate margins: Geological Society of London Special Publication 10, p. 201–211.

McGeary, S. E., and Z. Ben-Avraham, 1981, Allochthonous terranes in Alaska: Implications for the structure and evolution of the Bering Sea Shelf: Geology, v. 9, p. 608–613.

McLean, H., 1977, Organic geochemistry, lithology, and paleontology of Tertiary and Mesozoic rocks from wells on the Alaska Peninsula: U.S. Geological Survey Open-File Report 77–813, 63 p.

McLean, H., 1979, Review of petroleum geology of Anadyr and Khatyrka Basins, USSR: AAPG Bulletin, v. 63, p. 1467–1477.

Mull, C. G., 1979, Nanushuk Group deposition and the late Mesozoic structural evolution of the central and western Brooks Range and Arctic slope: U.S. Geological Survey Circular 794, p. 5–13.

Packer, D. R., and D. B. Stone, 1972, An Alaska Jurassic paleomagnetic pole and the Alaskan orocline: Nature Physical Science, v. 237, p. 25–26.

Panuska, B. C., 1985, Paleomagnetic evidence for a post-Cretaceous accretion of Wrangella: Geology, v. 13, p. 880–883.

Patton, W. W., Jr., 1973, Reconnaissance geology of the northern Yukon-Koyukuk province, Alaska: U.S. Geological Survey Professional Paper 774–A, 17 p.

Patton, W. W., Jr., and J. T. Dutro, Jr., 1969, Preliminary report on the Paleozoic and Mesozoic sedimentary sequence on St. Lawrence Island, Alaska, *in* Geological Survey research, 1969: U.S. Geological Survey Professional Paper 650–D, p. D138–D143.

Patton, W. W., Jr., and I. L. Tailleur, 1977, Evidence in the Bering Strait region for differential movement between North America and Eurasia: GSA Bulletin, v. 88, p.1298–1304.

Patton, W. W., Jr., M. A. Lanphere, T. P. Miller, and R. A. Scott, 1974, Age and tectonic significance of volcanic rocks on St. Matthew Island, Bering Sea, Alaska (abs.): GSA Abstracts with Programs, v. 6, p. 9905–9906.

Patton, W. W., Jr., M. A. Lanphere, T. P. Miller, and R. A. Scott, 1976, Age and tectonic significance of volcanic rocks on St. Matthew Island, Bering Sea, Alaska: U.S. Geological Survey Journal of Research, v. 4, p. 67–73.

Patton, W. W., Jr., I. L. Tailleur, W. P. Brosgé, and M. A. Lanphere, 1977, Preliminary report on the ophiolites of northern and western Alaska, *in* R. G. Coleman and W. P. Irwin, eds., North American ophiolites: Oregon Department of Geology and Mineral Industry Bulletin 95, p. 51–57.

Reed, B. L., and M. A. Lanphere, 1973, Alaska-Aleutian range batholith: Geochronology, chemistry, and relation to circum-Pacific plutonism: GSA Bulletin, v. 84, p. 2583–2610.

Roeder, D., and C. G. Mull, 1978, Tectonics of Brooks Range ophiolites: AAPG Bulletin, v. 62, p. 1696–1713.

Scholl, D. W., E. C. Buffington, and M. S. Marlow, 1975, Plate tectonics and the structural evolution of the Aleutian-Bering Sea region, *in* R. B. Forbes, ed., Contributions to the geology of the Bering Sea Basin and adjacent regions: GSA Special Paper 151, p. 1–32.

Stone, D. B., B. C. Panuska, and D. R. Packer, 1982, Paleolatitudes versus time for southern Alaska: Journal of Geophysical Research, v. 87, p. 3697–3707.

Turner, R. F., G. C. Martin, T. O. Flett, and D. A. Steffy, 1985, Geologic report of the Navarin Basin planning area, Bering Sea, Alaska: Minerals Management Service OCS Report 85-0045, 153 p.

Turner, R. F., J. G. Bolm, C. M. McCarthy, O. A. Steffy, P. Lowry, and T. O. Flett, 1983a, Geological and operational summary Norton Sound COST no. 1 well, Norton Sound, Alaska: U.S. Geological Survey Open-File Report 83-124, 164 p.

Turner, R. F., J. G. Bolm, C. M. McCarthy, O. A. Steffy, P. Lowry, T. O. Flett, and D. Blunt, 1983b, Geological and operational summary, Norton Sound COST no. 2 well, Norton Sound, Alaska: U.S. Geological Survey Open-File Report 83-557, 154 p.

Turner, R. F., M. McCarthy, C. D. Comer, J. A. Larson, J. G. Bolm, A. C. Banet, and A. J. Adams, 1984a, Geological and operational summary, St. George Basin COST no. 1 well, Bering Sea, Alaska: Minerals Management Service OCS Report MMS 84-16, 105 p.

Turner, R. F., M. McCarthy, C. D. Comer, J. A. Larson, J. G. Bolm, T. O. Flett, and A. J. Adams, 1984b, Geological and operational summary, St. George Basin COST no. 2 well, Bering Sea, Alaska: Minerals Management Service OCS Report MMS 84-16, 100 p.

Turner, R. F., M. McCarthy, D. A. Steffy, M. B. Lynch, G. C. Margin, K. W. Sherwood, T. O. Flett, and A. J. Adams, 1984c, Geological and operational summary Navarin Basin COST no. 1 well, Bering Sea, Alaska: Minerals Management Service OCS Report MMS 84-31, 245 p.

Yanshin, A. L. (ed.), 1966, Tektonicbeskaia karta Euraziy (Tectonic map of Eurasia): Moscow, Akademii Nauk SSSR Geological Institute, scale 1:5,000,000.

CHAPTER SEVEN

GEOLOGIC EVOLUTION AND PETROLEUM GEOLOGY OF THE ALEUTIAN RIDGE

David W. Scholl
Tracy L. Vallier
Andrew J. Stevenson

United States Geological Survey
Menlo Park, California

ABSTRACT

Arc volcanism formed the bulk of the Aleutian Ridge antiform, probably in the early to middle Eocene (55–50 Ma). Its rapidly accumulating igneous and associated volcaniclastic masses formed the ridge's basement rocks of the lower series, which were subsequently depositionally buried or volcanically covered and intruded by Oligocene and Miocene rocks of the middle series, and, more recently, by the Pliocene and Quaternary units of the upper series. The three-tiered rock sequence records the major evolutionary phases of the Aleutian Ridge and also important changes in its tectonic setting along the northern rim of the Pacific.

Evolutionary events that affected the petroleum potential of the Aleutian Ridge included (1) a reduction in ridge-wide magmatism beginning about or before 40 Ma, (2) the subsequent accumulation of a mantle of middle and upper series sedimentary deposits over its flanks and in crestal basins, and (3) the late Cenozoic addition of a thick accretionary wedge of trench deposits to the landward slope of the Aleutian Trench. Attachment of the wedge compressively forged the broad structural trough of the forearc basin of the Aleutian Terrace, which filled with upper series terrigenous and pelagic deposits.

Collectively, the 3–5-km-thick section of upper and middle series beds underlying the terrace, and the even thicker and tectonically underlying accretionary wedge, constitute the largest (roughly 1500 × 70 km) unexplored sector of the Aleutian Ridge that has potential for petroleum accumulations.

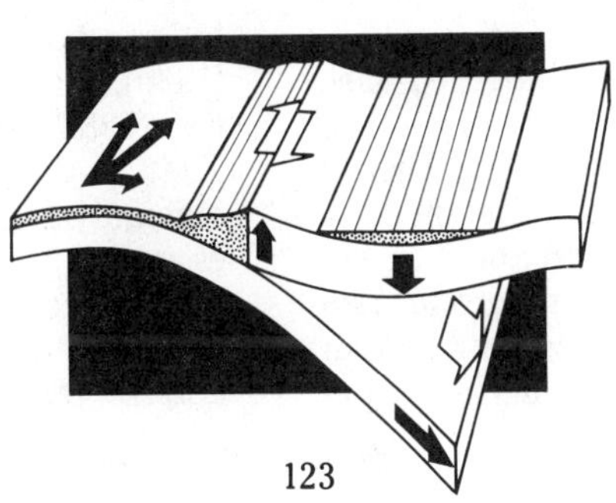

INTRODUCTION

The Aleutian Ridge forms much of the northern rim of the Pacific Basin (Figures 1, 2), separating it from the Bering Sea Basin to the north. The massive bulk and diversified physiography of the ridge record important north Pacific tectonic events and crust-forming processes.

Factual geologic knowledge about the Aleutian Ridge is sparse, partly because of its massive size and remote location. Also, field data are difficult to obtain owing to the harsh climate and short field season. The geologic evolution of the ridge—presently the best clue to regionally assessing its resource potential—can therefore be interpreted only generally.

REGIONAL PHYSIOGRAPHY AND GEOMORPHIC PROVINCES

Curving broadly southward, the Aleutian Ridge extends westward roughly 2200 km from the tip of the Alaska Peninsula to within a few hundred kilometers of Kamchatka (Figure 1). Transverse to the length of the ridge are two geomorphic provinces (Figures 2, 3). Largest and boldest is the Aleutian Island Arc, which forms the roughly 100-km-wide, island-crested cordillera of the ridge (Figures 4, 5). Companion to the arc is the adjacent deep-water province of the Aleutian fore arc, which comprises the platform of the Aleutian Terrace (30–70 km wide; Figure 4) and the bordering landward slope of the Aleutian Trench (20–40 km; Figures 6, 7). The regional geomorphic form of the terrace, which Murray (1945) originally named the Aleutian Bench, is best exhibited on the bathymetric charts of Nichols and Perry (1966).

The bulk of the ridge is submerged beneath several thousand meters of Pacific and Bering Sea waters (Figure 1). Its relief is dramatic—nearly 9300 m (measured from the +2149-m summit of Vsevidof volcano on Umnak Island to the adjacent trench floor at roughly −7150 m; Figure 2) and bold—because along much of its length the ridge drops fairly steeply (typically with a >5° slope) southward to the nearly flat surfaces of either the Aleutian Terrace (4000 m) or the trench floor (roughly 7000 m), or northward to the Bering Sea Basin (3400–4000 m; Figures 3, 4). Although thick sedimentary piles bury its lower flanks, the ridge's full structural height rises nearly 10,000 m above its Pacific base and only somewhat less—8000 m—above its northern or Bering Sea base.

The average geomorphic width of the ridge is about 160 (80–250) km; its length-to-width ratio is, therefore, large—approximately 14. Roughly and progressively, the width of the ridge decreases westward, to as narrow as 80 km in the vicinity of the Komandorsky Islands (Figures 2, 4). The distance between the trench axis and the crest of the ridge also decreases westward (Figures 2, 4).

The Aleutian Ridge rises from its deeply submerged base to a much narrower, but conspicuously flattened, summit platform (Figures 4, 5). The plane of the summit platform is interrupted by basinal depressions, scarps, and edifices of the approximately 100 islands of the Aleutian and Komandorsky groups (Figures 4, 5). These major island groups rise above the Komandorsky (long. 165°–169°E) and Aleutian (long. 172°E–164°W) segments of the ridge, which are separated by the islandless tract of the Near Pass segment (long. 169°–172°E; Figure 5). Most of these islands are composed of Eocene, Oligocene, and Miocene rocks. The younger volcanic piles of the active and dormant centers of the Aleutian volcanic chain (Figure 2) extend along the ridge only as far westward as Buldir Island (long. 176°E; Coats, 1950; see also Figure 10). The chain generally lies along the northern side of the broad crestal platform (Figures 2, 3).

BACKGROUND INFORMATION

Bathymetry

Bathymetric maps accurately depicting the submerged form of the Aleutian Ridge were first published in the 1950s by Gibson and Nichols (1953) and Gates and Gibson (1956). A decade later Nichols and Perry (1966) issued a series of charts (in fathoms) showing the submarine physiography of virtually the entire length of the Aleutian segment of the ridge (Figure 5) Good bathymetric charts, however, are still unavailable for the more western Near Pass and Komandorsky segments.

Regional charts displaying the overall form of the Aleutian Ridge have been published by Chase, Menard, and Mammerickx (1970; in fathoms) Scholl et al. (1974; in meters), and Schumacher (1976; in meters). The NOS Seamap charts (U.S. Department of Commerce, 1973; in velocity-corrected meters) of the easternmost sector (long. 180°–166°W) most accurately portray the bathymetric form of a lengthy part of the ridge (Figure 3).

Insular Geology

Although the Aleutian and Komandorsky Islands expose only a small fraction of the rock framework of the ridge (Figure 8), they reveal critical evidence about its geologic evolution during at least the past 50 m.y. (Vallier et al., in press; Scholl, Vallier, and Stevenson, 1986). Stratigraphic and structural information from the Komandorsky Islands are summarized best by Schmidt, Serova, and Dolmatova (1973), and Schmidt (1978). Borsuk and Tsvetkov (1982) describe current Soviet thinking on the origin and radiometric ages of the igneous rocks exposed on these islands, as well as on the Aleutian Islands.

Systematic regional geologic mapping and studies on the Aleutian Islands began shortly after World War II. Reports on this landmark work appeared periodically in the series of papers in Bulletin 1028 of the United States Geological Survey. The early highlights of these investigations were summarized by Gates, Fraser, and Snyder (1954), and Coats (1956a). Coats' (1962) paper on the subduction-related origin of Aleutian magmas was the most perceptive and creative statement on the cause and evolution of island-arc volcanism, although it was not realized at the time. Many investigators have continued to advance Aleutian magmatic studies, for example, Marsh (1976), Kay (1977, 1980), Perfit et al. (1980), McCulloch and Perfit (1981), Kay (1983), Conrad and Kay (1984), Kay and Kay (1982, 1985), and Marsh (in press). These authors summarize most existing information and significantly expand our knowledge about the magmatic processes that built the igneous massif of the arc.

In recent years the results of Aleutian Islands mapping and general geologic studies have continued to appear in the literature. Much of this work is either presented or reviewed

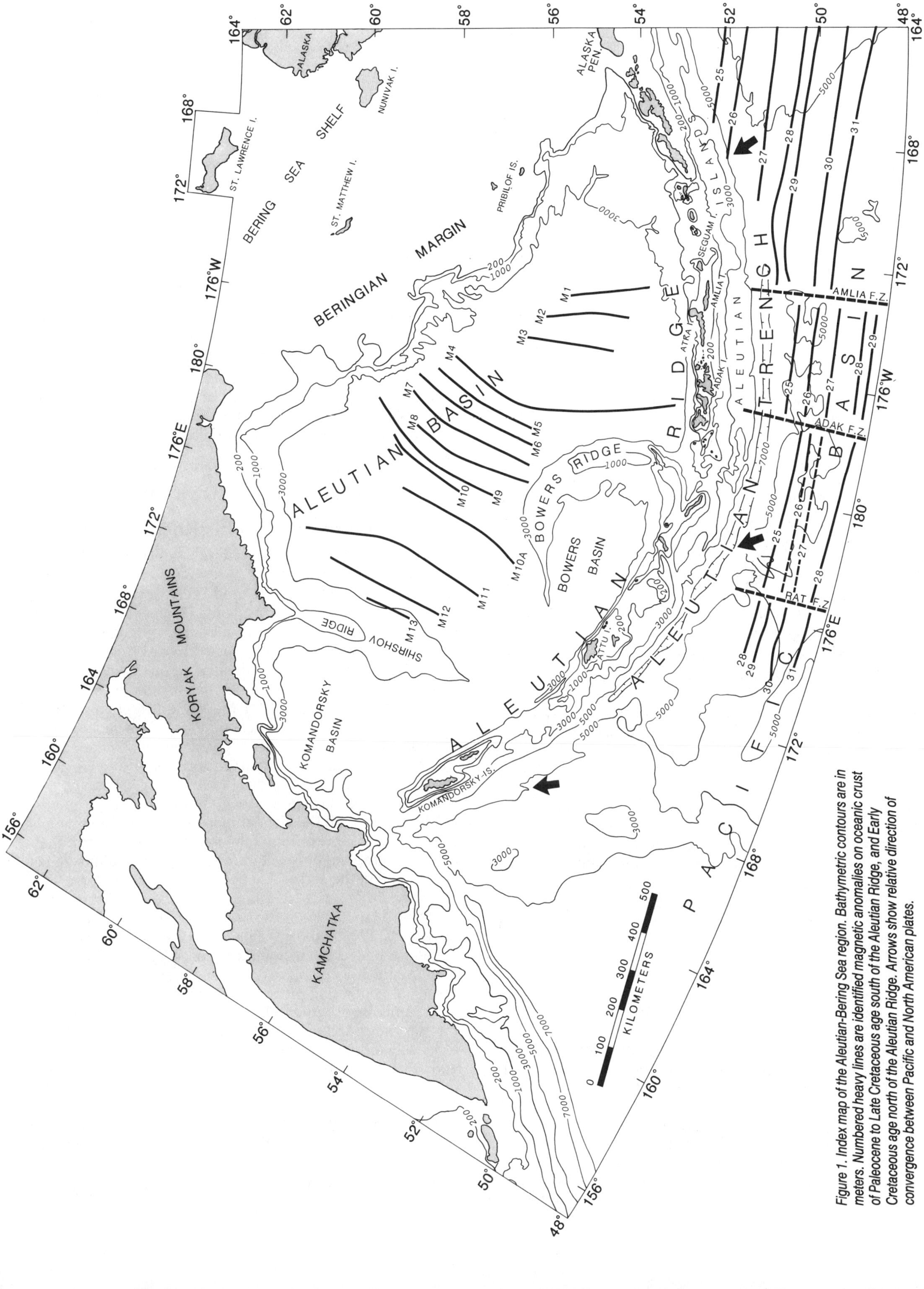

Figure 1. Index map of the Aleutian-Bering Sea region. Bathymetric contours are in meters. Numbered heavy lines are identified magnetic anomalies on oceanic crust of Paleocene to Late Cretaceous age south of the Aleutian Ridge, and Early Cretaceous age north of the Aleutian Ridge. Arrows show relative direction of convergence between Pacific and North American plates.

by Scholl, Greene, and Marlow (1970), Carr et al. (1971), DeLong, Fox, and McDowell (1978), Hein and McLean (1979), McLean and Hein (1984), McLean, Hein, and Vallier (1983), Hein, McLean, and Vallier (1984), Rubenstone (1984), Vallier et al. (in press), and Stone (in press).

Offshore Geology

Pioneering subsurface marine geological and geophysical studies of the Aleutian Ridge were effectively begun in the 1960s by Shor (1964) and Hayes and Ewing (1970). These early studies were greatly advanced by the exploratory work of Grow (1973a), who first correctly deduced the probable crustal structure of the Aleutian Ridge. Of the two crustal models he presented, model B is most consistent with multichannel reflection data presented here (Figure 6) and in the reports by Scholl, Vallier, and Stevenson (1983a, 1983b).

In 1971 DSDP (Deep Sea Drilling Project) drilling on Leg 19 ended the initial phase of offshore Aleutian investigations (Creager, Scholl, et al., 1973). The results and implications of this early work are summarized by Marlow et al. (1973) and by Scholl, Buffington, and Marlow (1975). The greatest concentration of offshore geologic data exists for the sector of the ridge between Adak and Amlia Islands (long. 173°–177°W; Figure 3).This information, as well as that collected elsewhere along the ridge during the past decade, is largely summarized or referenced by Scholl, Vallier, and Stevenson (1983a, 1983b), Vallier et al. (in press), and Stone (in press).

Deep Crustal Rocks

Knowledge about the nature of the ridge's deeper crustal rocks and underlying mantle comes from teleseismic studies, refraction seismology, implications of magma generation processes, and studies of xenoliths in young lavas. Shor (1964) initiated offshore refraction work to determine the deeper rock structure of the ridge. Grow (1973a) continued these studies and summarized much of the early information to define lower structural units. Recently, Conrad and Kay (1984) and Kay and Kay (1985) have provided new insights based on magmatic products and inclusions erupted at volcanic centers.

The geometry of the seismic Benioff zone, which signals the location and stress regime of the slab of Pacific oceanic crust underthrusting the Aleutian Ridge, has been intensively studied during the past two decades. The significant results and regional implications of much of this work are presented or summarized by Jacob (1972), Jacob, Nakamura, and Davis (1977), Spence (1977), Engdahl (1977), Davies and House (1979), Frohlich et al. (1982), House and Jacob (1983), Espinosa, (1984), and Engdahl and Billington (1986).

CENOZOIC TECTONIC SETTINGS

Regional geological and geophysical information indicates that the Aleutian Ridge lies along the boundary of the North American plate and the northwestward-moving Pacific plate to the south. The two plates converge nearly orthogonally at the eastern end of the ridge (long. 165°W), but owing to the arcuate sweep of the ridge, the angle of convergence lessens progressively to the west (Figure 1; Minster and Jordan, 1978; Engebretson, Cox and Gordon, 1985). Along the Adak-Amlia sector of the ridge; the vector of underthrusting is roughly 90 mm/yr at 320°, a direction that is approximately 30° west of the normal to the ridge axis.

The North American and Pacific plates are nearly in strike-slip contact along the Komandorsky segment (Cormier, 1975; Figure 1). However, marine geophysical data (Vallier et al., 1984, and in press), teleseismic first-motion studies (Newberry, 1983), and the relative plate motions calculated from Cox and Engebretson (1985), imply that a small angle of underthrusting occurs beneath the westernmost Aleutian Islands in the vicinity of its Near Islands or Agattu-Attu sector (long. 174°–172°E; Figures 1, 2)—and probably farther westward as well.

Because the ridge has remained generally fixed to North American crust (Harbert and Cox, 1984; Harbert, Cox, and McLean, 1984, 1985), the tectonic setting outlined above can be extended back to about 43 Ma, or to the late middle Eocene (Berggren et al., 1985[1]), when the prominent bend in the Hawaiian-Emperor Seamount chain recorded a change in motion of the Pacific plate from north–northwest to its present west–northwest direction (Clague, 1981; Clague and Dalrymple, in press). It is therefore likely that before 43 Ma the ridge was underthrust somewhat more directly to the north and, at times, at exceptionally high rates (Engebretson, Cox and Gordon, 1985).

REGIONAL GEOLOGIC ROCK FRAMEWORK

Chronostratigraphic Rock Series

Studies on Aleutian Island outcrops led Wilcox (1959) to recognize and define a two-part rocks series—a moderately deformed and regionally altered older series, and a less deformed and altered younger one. However, regional offshore structural relationships allow the recognition of a three-part rock sequence—a *lower* (LS), *middle* (MS), and *upper* (US) series, which are delineated on Figures 6 and 7. Sparse offshore seismic reflection data from the far western or Komandorsky segment of the ridge (Buffington, 1973; Gnibidenko et al., 1980), and Schmidt's (1978) geologic correlations based on the rocks exposed on the Komandorsky Islands, suggest that this tripartite rock series may be applicable to the length of the ridge (Vallier et al., in press).

The three rock groups are thought to record major phases in the evolution of the ridge (Figure 9). Hence, the series is defined as a sequence of time-stratigraphic—rather than lithostratigraphic—units. Lithologically equivalent rocks occur in all of the series. The type sections for determining the ages of the series members are based on offshore and onshore relationships observed in the Adak-Amlia sector (Figure 8). The boundary ages of rocks included within a series are subject to changes introduced by the implications of more precisely dated rock and sediment samples and the results of continuing onshore and offshore mapping in the Adak-Amlia sector (Figure 8). The acquisition of more accurate age information explains shifts in the assigned ages—especially for the beginning of late series time—given below and in Vallier et al. (in press), compared with that of Marlow et al. (1973) and Scholl, Vallier, and Stevenson (1983a, 1983b).

We recognize that existing and past lateral variations in the tectonic setting imply that growth phases are probably

[1]Numerical ages assigned to the geologic time scale are based on information provided in this reference.

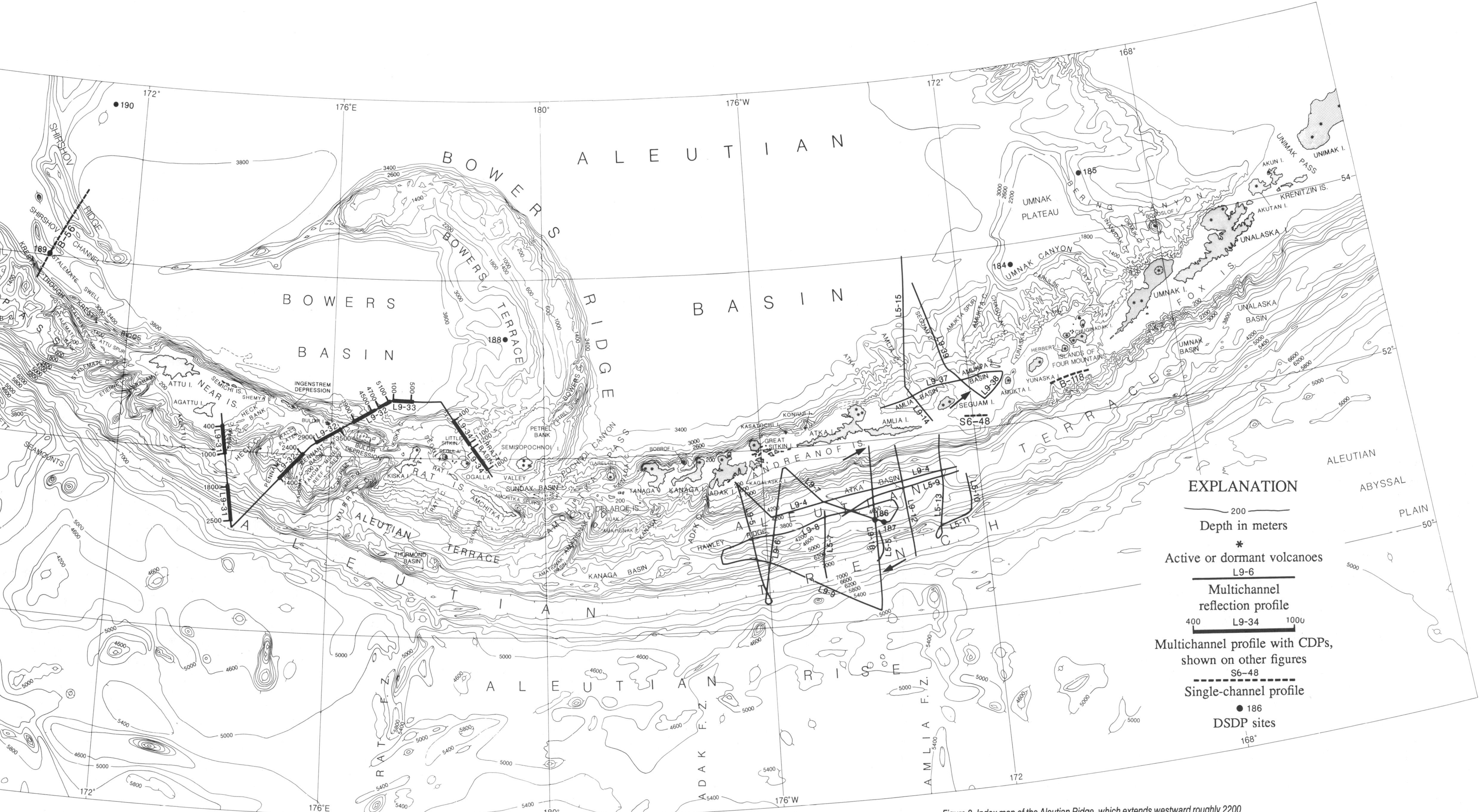

Figure 2. Index map of the Aleutian Ridge, which extends westward roughly 2200 km from Unimak Pass to Kamchatka Pass. Numbered solid lines are survey lines along which multichannel reflection profiles were collected by the R/V S.P. LEE (see also Figure 3). Insular chain of active and extinct volcanoes extends westward only as far as Buldir Island (long. 176°E). Arrows locate limits of interpreted multichannel reflection profiles L9-10 and L9-39 shown on Figure 6.

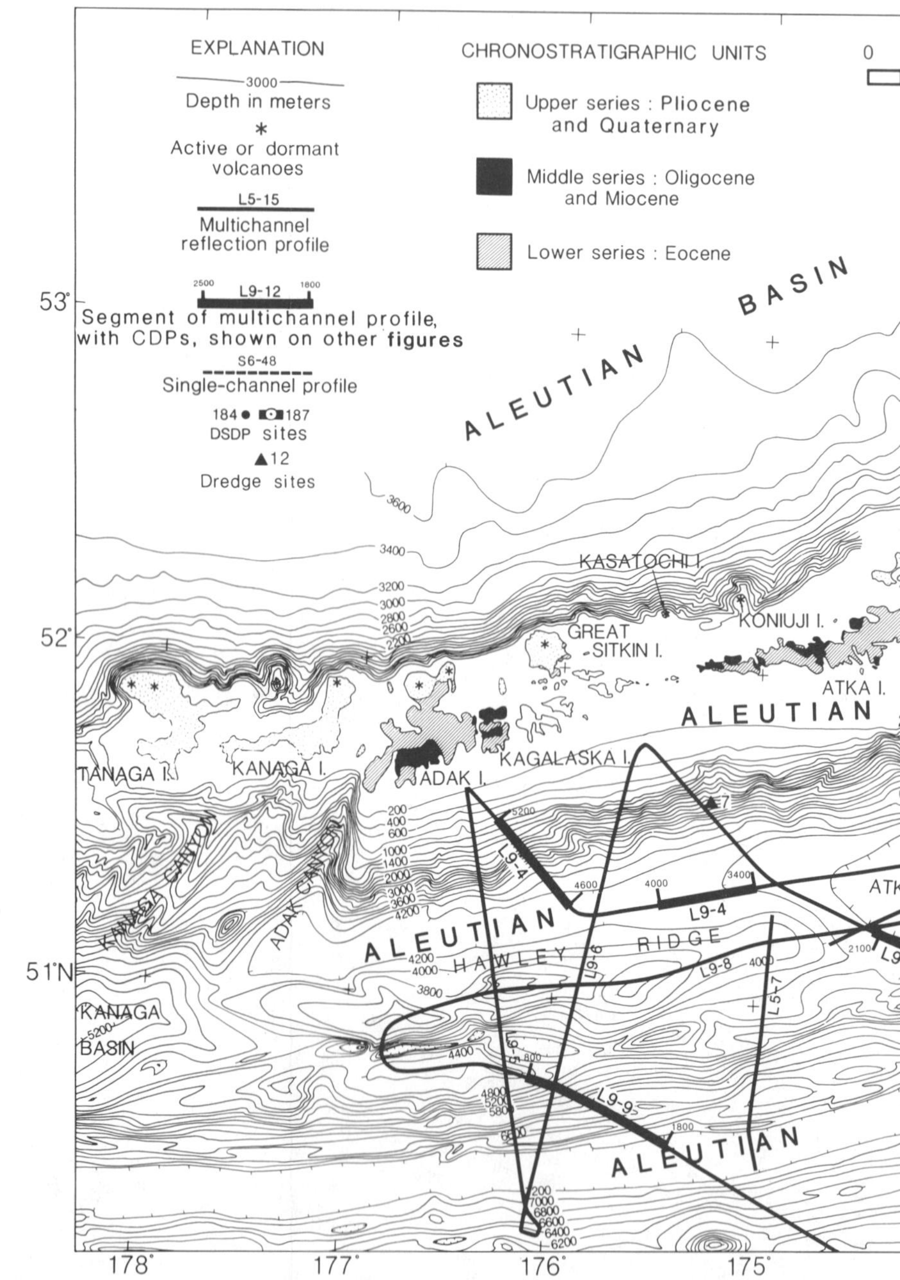

Figure 3. Index and generalized geologic map of Adak-Amlia sector of the Aleutian Ridge (see also Figure 2). Information about lithology, age, and properties of rock samples recovered at dredge stations is listed on Table 1. Bathymetric data from NOS Seamap charts (U.S. Department of Commerce, 1973).

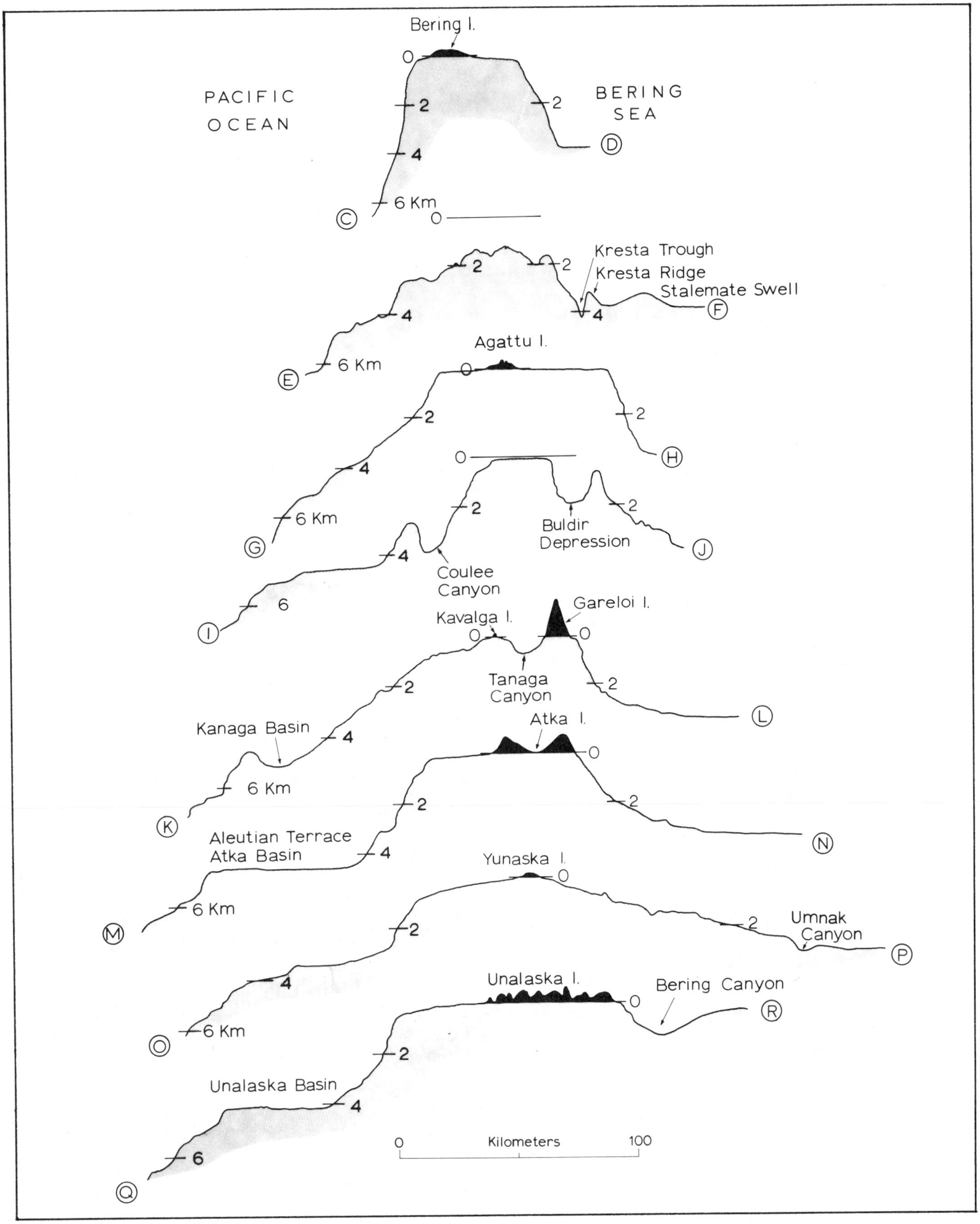

Figure 4. Transverse bathymetric profiles of the Aleutian Ridge. Profile locations are shown on Figure 5. Lower four profiles reveal the two principal geomorphic provinces of the Aleutian Ridge—the island-crested cordillera of the Aleutian Island Arc and the flanking fore-arc region, comprising the Aleutian Terrace and the adjacent landward trench slope (see also Figure 6). The flattened crest, or summit platform, of the Aleutian Island Arc is conspicuous, and lends a guyotlike appearance to the transverse profiles.

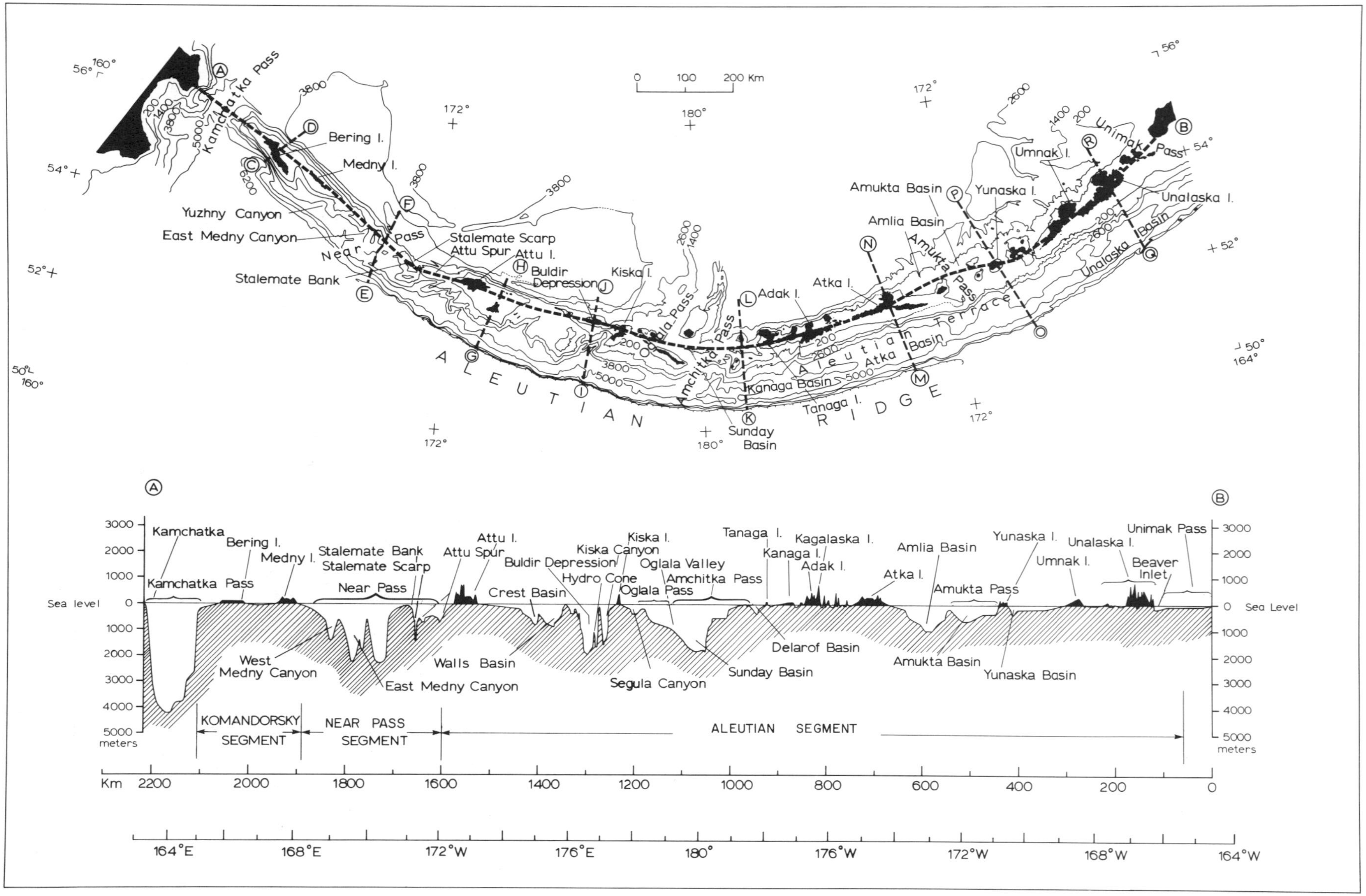

Figure 5. Longitudinal profile along the summit of the Aleutian Island Arc. Depressions shown on the profile represent grabens of summit basins, which commonly trend parallel to the length of the arc, and transverse-trending passes and canyons created by strike-slip rifting and clockwise rotation of blocklike sectors of the ridge. Transverse profiles C-D through Q-R are shown on Figure 4.

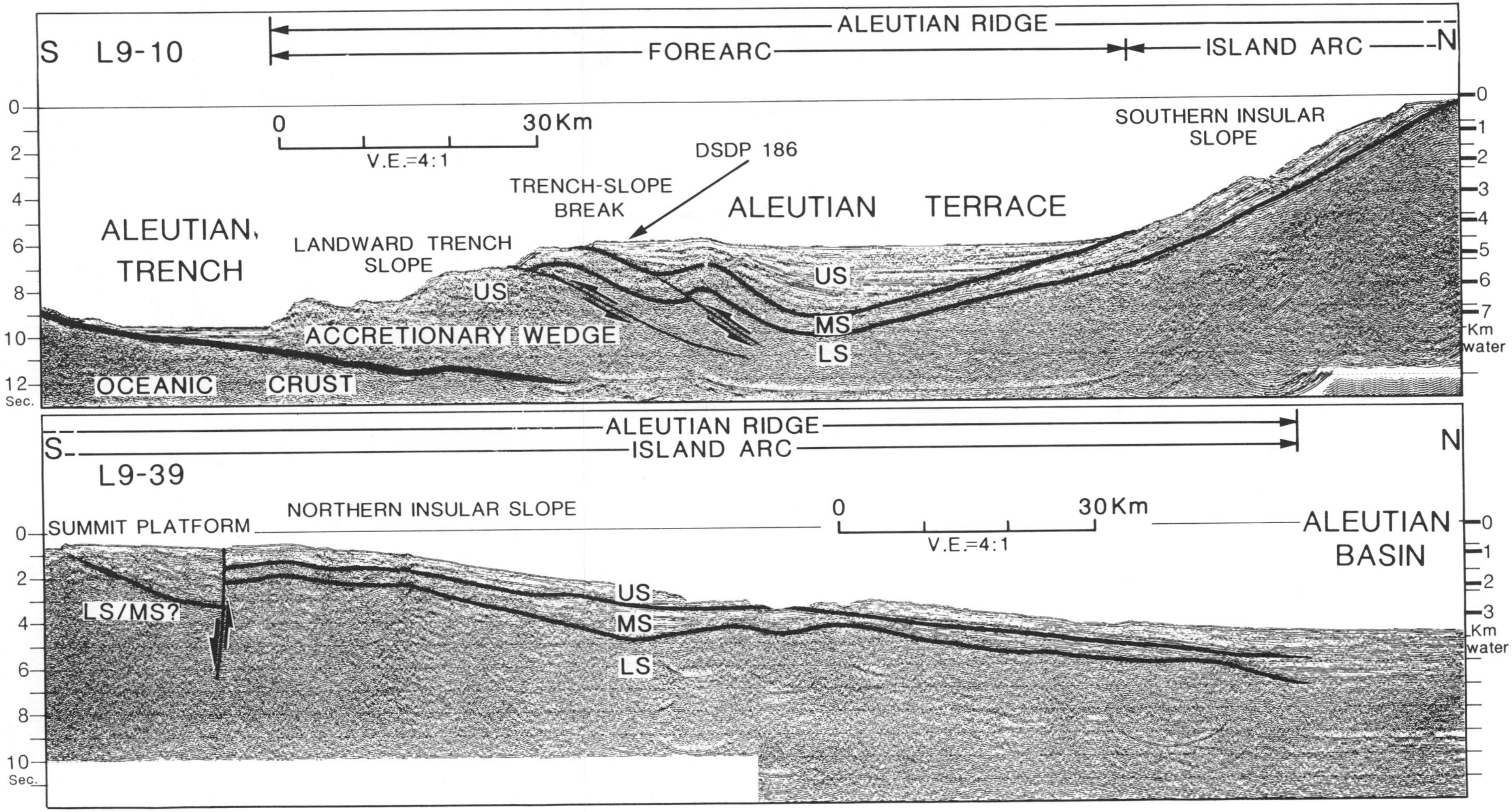

Figure 6. Interpreted multichannel sections (migrated) L9-10 and L9-39, which traverse the Pacific and Bering Sea flanks, respectively, of the Aleutian Ridge near Amlia Island (locations of sections are shown on Figures 2, 3). The schematic geologic section shown on Figure 8 roughly connects the two interpreted sections. Important physiographic features and the limits of the ridge's major geomorphic provinces—the island arc and the fore arc—are indicated. The boundaries of upper series (US: Pliocene and Quaternary), middle series (MS: Oligocene and Miocene), and lower series (LS: Eocene) rock units are traced . See Figures 7, 12, 14, 16, and 18 for enlarged segments of lines L9-10 and L9-39.

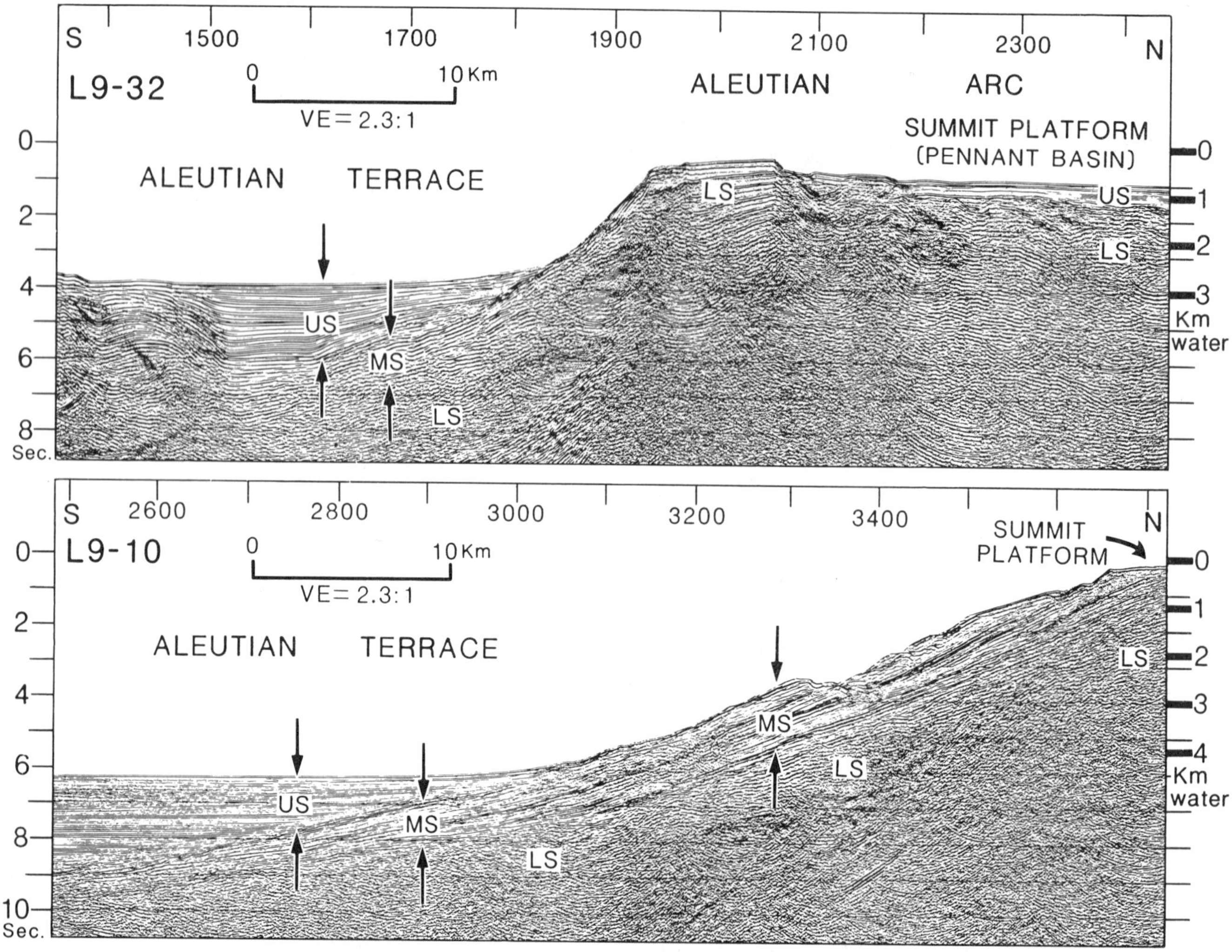

Figure 7. Migrated seismic sections L9-32 and L9-10 (see Figures 2 and 3, respectively, for locations) that exhibit the structural, depositional, and acoustical characteristics that identify the tripartite sequence of upper series (US), middle series (MS), and lower series (LS) rocks. Line L9-10 (see also Figure 6) crosses the type area for distinguishing and dating the three rock groups.

diachronous rather than coeval along the length of the ridge. The ages of growth phases recognized in the Adak-Amlia sector can therefore only be considered approximate when applied to other sectors.

Lower Series

Rocks of the lower series (LS), which are believed to be early(?) to latest Eocene in age (55?–37 Ma), effectively constitute the "basement" sequence of the Aleutian Ridge. They accumulated primarily during the natal growth phase of the ridge's magmatic arc, piling up across its 100-200-km-wide base and building the crestal region to sea level at least by the early part of the late Eocene. LS rocks also include sedimentary and igneous rocks that accrued as voluminous magmatic growth waned and before the regional burial of the ridge's submerged flanks beneath a thickening sequence of middle series (MS) sedimentary deposits.

Insular exposures of the lower series are dominantly volcanic eruptive rocks and related hypabyssal intrusive bodies and coarse volcaniclastic rubble. Interbedded sedimentary sequences, especially toward the top of the series, are common. LS rocks exhibit a wide range of metamorphic grades from zeolite to amphibolite facies. Most of the prehnite-pumpellyite and higher metamorphic grades were caused by the emplacement and crystallization of epizonal calc-alkaline plutons of Oligocene and Miocene ages (Citron et al. 1980; Kay, 1983; Vallier et al., in press). In general, LS rocks are broadly folded. They are disrupted by extensional fractures, and, toward the western end of the ridge, by suspected transcurrent offsets.

Examples of exposed LS rocks include, from west to east, the Komandorsky suite of the Komandorsky Islands (Schmidt, 1978), in the Aleutian Islands the so-called basement rocks of the Near Islands of Attu and Agattu (Gates et al., 1971), the Amchitka Formation of Amchitka Island (Carr, Quinlivan, and Gard, 1970; Carr et al., 1971), the Finger Bay Volcanics of Adak Island (Hein and McLean, 1979; Kay, 1983; Rubenstone, 1984), and the preplutonic sedimentary and volcanic rocks of Atka (Hein, McLean, and Vallier, 1984) and Amlia Islands (McLean, Hein, and Vallier, 1983).

Sediment-dominated sections within the upper part of the lower series commonly overlie volcanic-rich units. Examples of these clastic sequences include the Krugloi and Nevidiskov Formations of the Near Islands (Gates et al., 1971), the Andrew Lake Formation of Adak Island (Scholl, Greene, and Marlow, 1970; Hein and McLean, 1979), the little-deformed and little-altered but apparently pre-Oligocene section at the western end of Umnak Island (McLean and Hein, 1984), and much of the Unalaska Formation of Unalaska Island (McLean, Harbert, and Douglass, 1984). These sections are

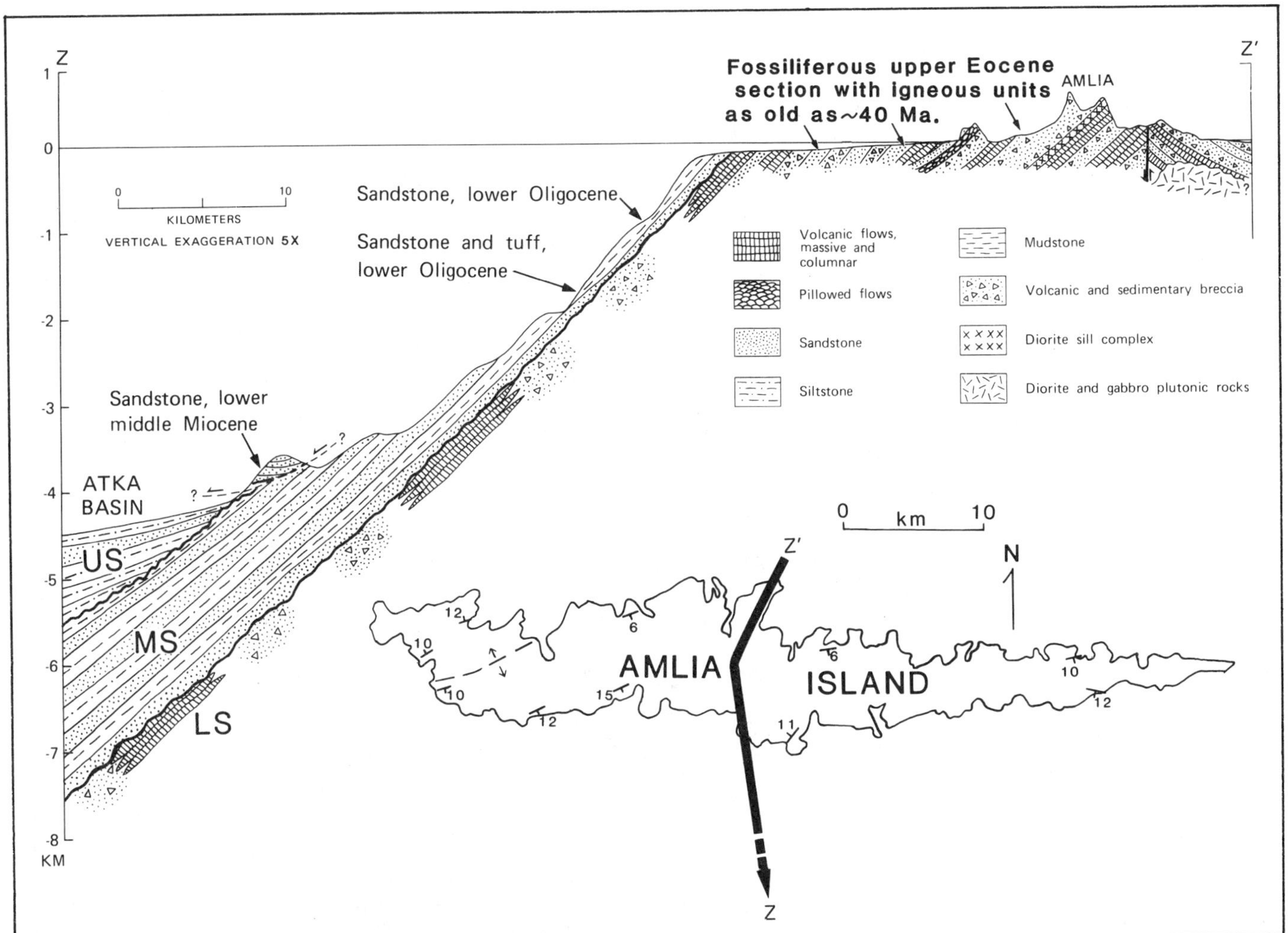

Figure 8. Idealized structural and stratigraphic section from Amlia Island southward to the northern edge of the Aleutian Terrace. The southern insular slope of the island arc is underlain by a dip-slope sequence of MS and LS rocks; the older unit (LS) crops out on Amlia Island. Lithology and age of overlying middle series rocks on the descending slope are based on dredge samples (Figure 3; Table 1). Line of section extends south of that indicated on index map.

thought to have accumulated in Paleogene summit basins adjacent to active volcanic centers during the waning phase of the arc's initial episode of rapid magmatic growth.

On seismic-reflection profiles LS rocks commonly exhibit irregular and laterally discontinuous reflection horizons (Figures 6, 7), although well-layered sequences occur locally (Figure 17, profile L9–4). Tilted and typically poorly stratified rocks of the lower series are truncated by the erosional surface of the summit platform (Figure 10; see also Figure 17). In the fashion of a large antiform, LS strata dip seaward and parallel to the slopes of both flanks of the arc (Figure 6). Beneath these slopes better stratified beds of the middle series disconformably overlie more crudely layered lower series units.

The oldest exposed rocks of the lower series may be the Finger Bay Volcanics of Adak Island, which underlie the upper Eocene strata of the Andrew Lake Formation (Scholl, Greene and Marlow, 1970; Hein and McLean, 1979). Rubenstone's (1984) ^{40}Ar-^{39}Ar preliminary age analysis of the altered and multiple-intruded rocks of the Finger Bay Volcanics indicates that this calc-alkaline, dominantly pyroclastic mass began to form before about 50 Ma, probably in the early Eocene.

Lower Tertiary mafic lava and intrusive bodies with minimum or alteration-time K-Ar ages of about 42–30 Ma are exposed widely on the Aleutian Islands (DeLong, Fox, and McDowell, 1978; Wilson, 1981; Pickthorn, Vallier, and Scholl, 1984). A recently discovered amphibolite body on Attu Island has a metamorphic K-Ar age (amphibole separate) of 41.5 Ma (Vallier et al., in press). This large rock mass consists of sills, dikes, and stocks of metamorphosed gabbro, diabase, diorite, and basalt that probably represent upper crustal plutons intruded during the early stage of arc growth and thickening. Although the oldest igneous rock dated on the Komandorsky Islands is a rhyolite of about 35 Ma (Borsuk and Tsvetkov, 1982), much older, possibly lower Eocene, fossiliferous rocks have been identified here (Schmidt, 1978).

We therefore suppose that LS rocks began to accumulate along the length of ridge at least by 55–50 Ma, and that the great bulk of the igneous massif was constructed before the end of the Eocene.

The end of early series time is defined as coincident with the shutdown of eruptive activity on the flanks of the ridge and their burial by a thickening blanket of terrigenous volcaniclastic detritus of the middle series (Figures 6, 7). Regional stratigraphic relationships, radiometric dating, and paleontological ages imply that the initial phase of expansive magmatic growth of the arc began to wane along the

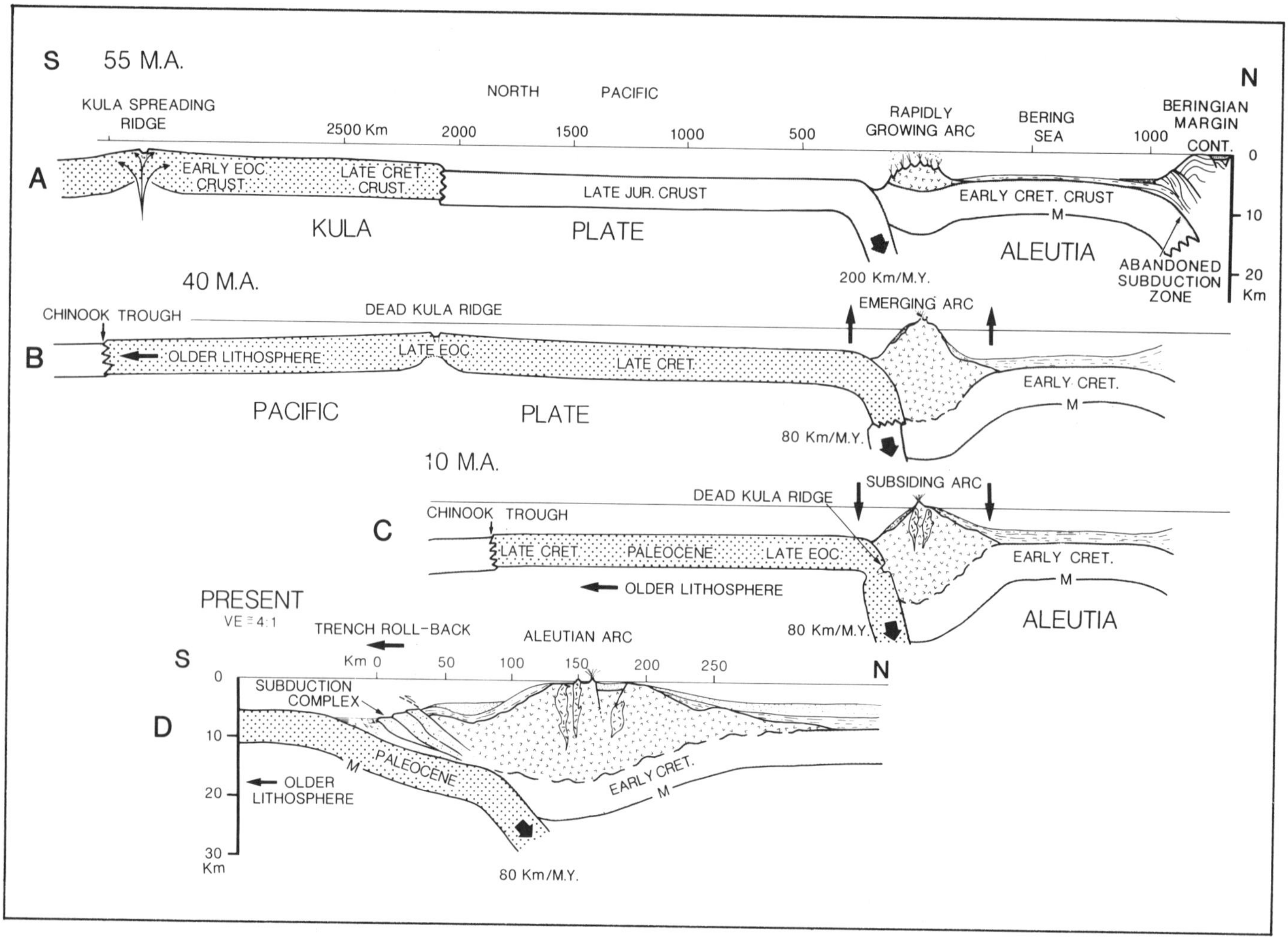

Figure 9. Schematic evolution of the Aleutian Ridge relative to changes in north Pacific plate boundary conditions. A, Initial formation of the arc in response to buckling of northward-moving Kula plate—possibly at an E–W-trending fracture zone (Scholl, Vallier, and Stevenson, 1986)—and formation of offshore subduction zone. Arc volcanism takes place over, or sweeps across, a 100–200-km-wide zone; the volcanic pile reaches sea level by the late middle Eocene. M = Moho or top of mantle.

B, Death of the Kula-Pacific spreading ridge around 43 Ma adds Kula crust to Pacific plate, which, relative to the arc, also slows and changes direction from generally northward to the present WNW convergence (Engebretson, Cox, and Gordon, 1985). The slowing, the more oblique direction of underthrusting, and the initial subduction of increasingly younger crust formed at the now-dead Kula-Pacific spreading ridge, combine to greatly reduce arc magmatic activity, which is restricted to the summit area, and to elevate large sectors of the ridge above sea level.

C, Subduction of the abandoned Kula-Pacific Ridge about 10 Ma causes increasingly older crust to underthrust the arc, which subsides and possibly collapses toward a southward-retreating trench axis.

D, During approximately the past 6–5 m.y., large volumes of terrigenous debris flooding the trench floor, and a faster and more orthogonal direction of underthrusting (Cox and Engebretson, 1985), cause a massive accretionary wedge to ram beneath the front of the arc massif. Compressional deformation and piling up of the accretionary wedge creates the trough of the prominent fore-arc basin of the Aleutian Terrace.

crest of the arc at least by 40 Ma. DeLong, Fox, and McDowell (1978) called for the beginning of the ridge's magmatic slowdown somewhat earlier at about 45 Ma, in the middle Eocene. In the Adak-Amlia sector, offshore sedimentary beds of the middle series are as old as early Oligocene, but the base of the series has not been sampled (Table 1). The unconformably underlying LS units are the downdip extension of insular exposures of volcanic rocks of middle and late Eocene age (Figure 8). We therefore provisionally place the end of early series time at the close of the Eocene Epoch, that is, at about 37 Ma. We surmise that if offshore samples of the basal beds of the middle series are collected and dated, that the age of the top of the lower series will be placed at least in the late Eocene.

Middle Series

Rocks of the middle series (MS), which are considered to be Oligocene and Miocene (37–5.3 Ma), attest to a phase of the ridge evolution that is characterized by greatly reduced and mostly restricted magmatic growth along the ridge crest, the commencement of regional dominance of destructive over constructive processes, and the regional depositional burial of the seaward-sloping flanks of the ridge. Marine sedimentary deposits that define the time span of the middle series form dip-slope sequences beneath both the arc's Pacific and Bering Sea flanks (Figures 6, 7). These deposits are locally thick (1–3 km) and only slightly deformed and diagenetically altered. They are cut by the late Cenozoic wave-planed surface of the summit platform

Table 1. Location, depth, age, lithology, and certain physical and chemical properties of rocks dredged from submerged outcrops, Aleutian Ridge.

Dredge No.[1]	Rock Series[2]	Lat. Long.	Water Depth (meters)	Age	Lithology	Sand (%)	Permeability/ Porosity[3]	Organic Carbon (%)
L5-1	US	50°59.8′N 174°06.3′W	5000	L Plio-E Quat	Ls & mudst			
L5-8	LS or MS	52°52.1′N 171°28.2′W	400–500	Unknown	Clastic breccia			
L9-3	US	51°48.9′N 172°49.5′W	2250	L Plio-E Quat	Ss & congl			
L9-4	MS & US	51°37.6′N 173°27.5′W	3680–3700	M Mioc L Plio	Ss Siltst			
L9-5/6	LS	51°53.3′N 174°00.3′W	80	Unknown (Eoc?)	Lava & ss			
L9-7	US	51°38.7′N 175°23.6′W	1800–1850	L Plio-Quat	Ss & siltst			
L9-10	MS & US	52°43.8′N 173°03.8′W	1500–1850	M Mioc L Mioc M Plio	Ss & siltst			
S6-12	MS	51°51.6′N 173°32.2′W	900–1000	E Olig	Ss	88	0.1/36	0.4
S6-13	MS	51°50.0′N 173°28.1′W	1650–1700	E Olig	Ss	95	11/52	0.8
S6-14	MS	51°48.9′N 173°56.3′W	400–500	L Olig-E Mioc	Ss	93		0.7
S6-15	US	50°53.1′N 173°12.9′W	6600–7050	E Quat	Mud	29		0.7
S6-16	MS	51°45.8′N 173°43.4′W	2450–2650	Olig	Ss & siltst	91	0.2/57	0.6
S6-17	MS	51°44.4′N 173°45.2′W	2950–3100	Olig & M Mioc	Ss	86	0.3/53	0.7
S6-19	MS	51°43.5′N 173°41.3′W	2750–2900	E-M Olig	Ss	99	38/42	0.4
S6-20	MS-US	51°48.7′N 174°00.4′W	700–930	L Mioc E Plio	Ss	97	45/44	0.1
S6-21	LS	51°52.1′N 174°12.9′W	120	Unknown (Eoc?)	Lava & ss			
S6-22	MS	52°08.9′N 171°51.9′W	300–450	M Eoc-E Olig	Ss & siltst			
S6-25	LS	52°09.6′N 171°52.2′W	260	Unknown (Eoc?)	Lava & tuff			
S6-27	US	52°39.2′N 173°07.8′W	1500–1500	E Plio-Holo	Ss & siltst			
S6-28	MS-US	52°39.7′N 173°09.9′W	1225–1350	L Mioc-E Plio	Congl & breccia	96	5/55	0.6
S6-29	US	52°39.7′N 173°09.9′W	1100–1200	L Mioc-E Plio	Ss & breccia	95	1/59	
S6-30	US	52°41.1′N 173°04.3′W	1400–1550	M Mioc	Ss, congl & breccia	99	22/44	0.2
S6-31	MS?	52°30.8′N 173°27.4′W	300–580	Unknown	Volcanic breccia			
S6-32	MS-US	52°37.5′N 172°35.2′W	900–975	L Mioc E Plio	Ss & breccia		3/14	0.3

[1]Locations of dredge stations identified by number listed after cruise designator (e.g., L5- , L9- , or S6-), are plotted on Figure 3.
[2]LS—lower series (Eocene and older ?); MS—middle series (Oligocene and Miocene); and US—upper series (Pliocene and Quaternary).
[3]Permeability in millidarcys; porosity in % (data from Core Laboratories, Inc., Dallas, Texas).

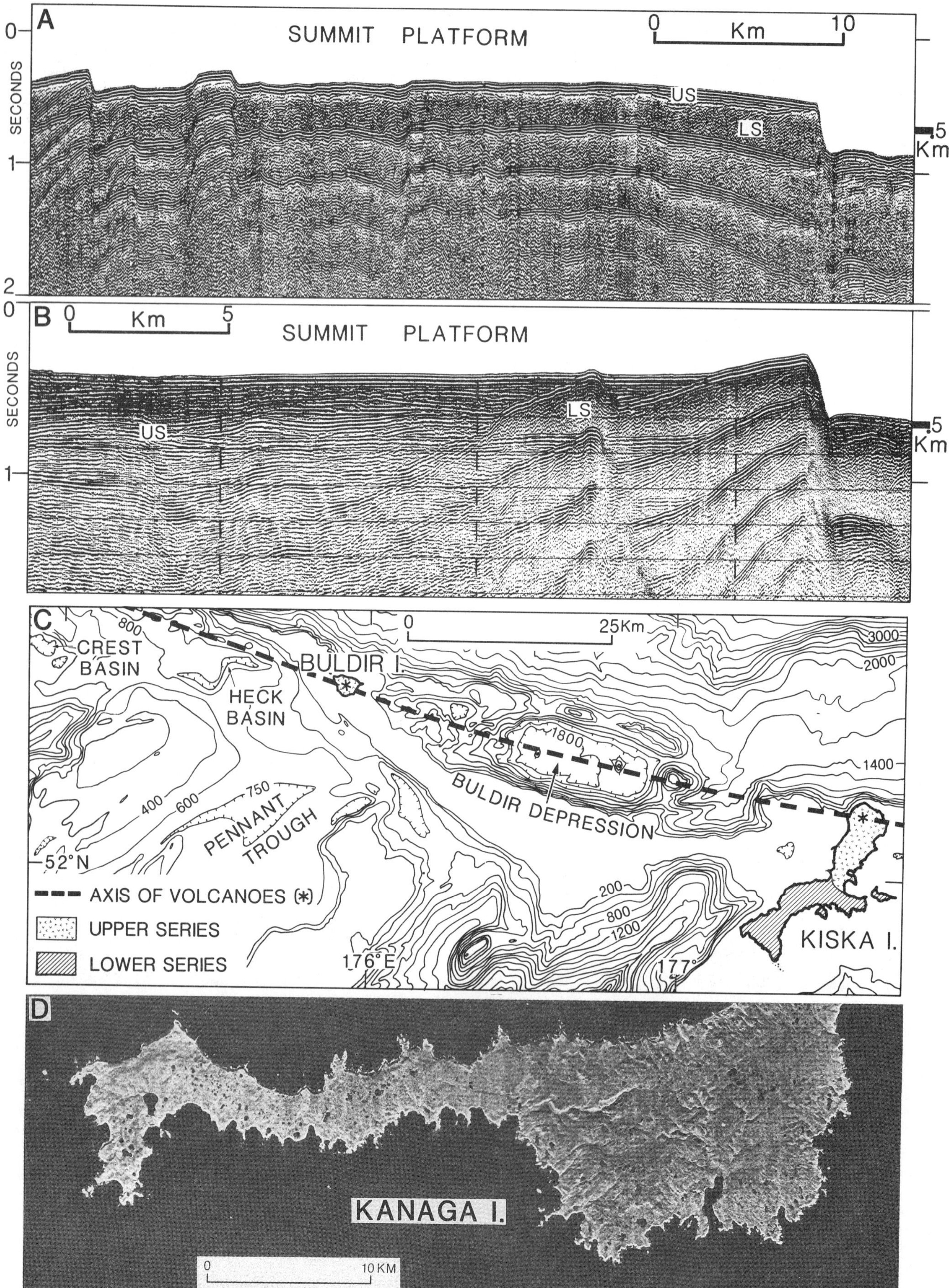
A
SUMMIT PLATFORM
0 Km 10
SECONDS
0
1
2
US
LS
.5 Km
B
0 Km 5
SUMMIT PLATFORM
SECONDS
LS
US
.5 Km
C
800
CREST BASIN
BULDIR I.
HECK BASIN
0 25Km
3000
2000
1800
1400
400
600
750
PENNANT TROUGH
BULDIR DEPRESSION
52° N
AXIS OF VOLCANOES (*)
UPPER SERIES
LOWER SERIES
200
800
1200
176° E
177°
KISKA I.
D
KANAGA I.
0 10 KM

Figure 10. Structure and geomorphology of the crestal region of the Aleutian Island Arc. A, Single-channel reflection profile B–118 southeast of Yunaska Island (Figures 2, 3), showing fault-offset surface of the summit platform. The platform, a wave-beveled surface cut across tilted lower series rocks, is here buried beneath a thin veneer of upper series sediment.

B, Reflection profile S6-48 south of Sequam Island (Figures 2, 3), showing that the summit platform is here a depositional surface underlain by a basinal sequence of US beds, which bury the platform's tilted and faulted original wave-cut surface.

C, Buldir-Kiska sector of the ridge crest and submerged summit platform, which is fragmented and depressed at numerous summit basins and also adjacent to the heads of major southwest-trending slope canyons.

D, Synthetic-aperture radar image of southern part of Kanaga Island (Figure 3), which is an elevated (100–150 m), terraced, and poorly drained sector of the summit platform (Fraser and Barnett, 1959).

and submarine canyons that scar the insular slopes of the ridge (Figure 7; see also Figure 17). Middle series beds beneath these slopes have not been traced to correlative insular exposures. Consequently, along most of the ridge, onshore members of the middle series are recognized by their middle and late Tertiary age, unconformable relationships with LS rocks, and generally unaltered and little-deformed nature.

Because dip-slope sections of MS deposits are truncated by the summit platform (Figure 7; see also Figure 17), marine sections of MS strata must have formerly been more widespread over the crestal region of the arc. Insular exposures of marine MS rocks include, from west to east, the upper Oligocene and lower Miocene Nilkol'ski suite of the Komandorsky Islands (Schmidt, 1978), the mid-Tertiary Chuniksak Formation of Attu Island (Gates et al., 1971), and, at the eastern end of the ridge, the Miocene Dutch Harbor section of Unalaska Island (Lankford and Hill, 1979; McLean, Harbert, and Douglass, 1984). Terrestrial sedimentary sections of MS deposits include the upper Miocene volcanic and wood-bearing Vodopad suite of Medny (Komandorskys) Island (Schmidt, 1978; Borsuk and Tsvetkov, 1982; Vallier et al., in press), the fossil-tree-bearing Massacre Bay and Faneto Formations of late Miocene age on Attu Island (Gates et al., 1971; DeLong and McDowell, 1975; Scholl et al., 1976), fossil-tree and wood-bearing deposits described from Tanaga Island (Fraser and Barnett, 1959[2]), and the mostly upper Miocene coal- and tree-bearing Martin Harbor section of Atka Island (McLean, Hein, and Vallier, 1983).

Virtually all of the known calc-alkaline eruptive and plutonic bodies of the Komandorsky and Aleutian Islands were emplaced from the Oligocene through the early late Miocene (35–8 Ma). These include, from west to east, the lower Oligocene through Miocene intrusive and extrusive rocks reported from the Komandorsky Islands by Borsuk and Tsvetkov (1982), the middle Miocene Chitka Point Formation and coeval plutons of Amchitka Island (Carr et al., 1971), the 33-Ma Hidden Bay pluton of Adak Island (Citron et al., 1980), Miocene intrusive bodies of Atka Island (Hein, McLean, and Vallier, 1984), the middle (30 Ma) Oligocene intrusive sills and small plutons of western Umnak Island (McLean and Hein, 1984), and the middle Miocene plutonic masses of Unalaska Island at the eastern end of the ridge (Marlow et al., 1973; Perfit and Lawrence, 1979; and Perfit et al., 1980). These igneous masses constitute a major part of the middle series rocks in the crestal region of the arc.

The oldest known sedimentary beds of the middle series—early Oligocene—have been sampled offshore of Atka and Amlia Islands (Figures 6, 8; Table 1, station S6-12, -13, and -19). In this area MS deposits are apparently as young as late Miocene (Table 1, station S6-20), but the age of the uppermost beds of the series is not well constrained. At the base of the southern insular slope, the northward-onlapping strata of the upper series that underlie the Aleutian Terrace are clearly distinguishable on reflection profiles (Figure 7; see also Figures 17, 18). Stratigraphic control extrapolated from nearby DSDP sites 186 and 187 (Creager, Scholl, et al., 1973; Figure 6; see also Figure 19) implies that lowermost Pliocene deposits—possibly uppermost Miocene beds—of the terrace's basinal fill unconformably bury the dip-slope sequence of MS beds, which therefore are somewhat older.

Adjacent to Umnak Plateau (Figures 2, 3), dredging shows that the upper part of the arc's northern insular slope is underlain by MS beds as young as middle Miocene and earliest upper Miocene. These sandy, iron-stained strata are unconformably overlain by uppermost Miocene to lowermost Pliocene deposits, which include basal conglomerate units containing clasts of igneous rocks of late middle Miocene K-Ar age (12.7 Ma; Figure 3; Table 1, stations S6-27, -28, -29, and -30; see also Figure 14).

Rocks of the middle series evidently formed during a protracted period of general crestal emergence, erosion, and episodic plutonism that lasted throughout the Oligocene and virtually all of the Miocene. In the Adak-Amlia sector, the span of middle series time is therefore defined as extending from 37 to 5.3 Ma. The ages are chosen to correspond to the base of the Oligocene and top of the Miocene, respectively (Berggren et al., 1985). This middle and late Tertiary period of general ridge-crest erosion and localized deposition was recognized initially by Gates, Fraser, and Snyder (1954).

Upper Series

Upper series (US) rocks, of Pliocene and Quaternary age (younger than 5.3 Ma) are young basinal and slope-mantling sedimentary sequences and crestal igneous masses that are mostly faulted against or unconformably overlie deposits of MS and LS rocks. During the late Cenozoic the crestal area of the ridge has been affected by regional subsidence and an accelerated rate of differential vertical and horizontal displacement of blocklike sectors of the arc. Coincident with regional subsidence, large areas of the crest of the ridge have been flattened by wave-base erosion, a process that carved large areas of its prominent summit platform (Figures 4, 5, 10). Also during the late Cenozoic, the ridge's arcuate alignment trends of dormant and active volcanic centers formed (Figure 2), and its fore-arc area was deformed through compression and in part tectonically underrun by an accretionary mass of offscraped trench deposits (Figures 6, 9).

Along the crest of the ridge, elevated marine sections of upper series beds are restricted to areally small coastal exposures. Example of these deposits are known at Amchitka Island (Allison, 1973), adjacent to the Adagdak volcanic center of Adak Island (Coats, 1956b), and as a thin mantle over

[2]Field map annotations written in 1952 by G. L. Snyder for the coastal region north of Cape Sasmik toward Cheal Harbor (long. 177°59′05″W lat. 51°39′05″N) report "zone of buried forest—carbonized and silicified wood, many logs, some flattened, many stumps, mostly nearly upright."

elevated sectors of the summit platform, such as those exposed on Kanaga, Tanaga, and Shemya Islands (Figure 10; Fraser and Barnett, 1959). Most of the insular exposures of US rocks are the eruptive masses of active and dormant volcanic edifices of the Aleutian volcanic chain (Figures 2, 3). The larger eruptive centers on Kiska Island (Panuska, 1980; Wilson, 1981), Adak Island (Marlow et al., 1973; Marsh, 1976), Atka Island (Marsh, 1980, Hein, McLean, and Vallier, 1984), and Akutan Island (Swanson and Blum, 1984) were established in the latest Miocene or earliest Pliocene (6–4 Ma).

Thick marine sequences of US beds (1–3 km) fill summit basins—many of which are localized along the volcanic chain—and thinner sequences mantle the surrounding erosional surface of the summit platform (Figures 7, 10; Scholl, Marlow, and Buffington, 1975; see also Figure 15). In general, only thin sections of upper series beds drape the arc's Pacific and Bering Sea insular slopes. However, a relatively thick (1–1.5 km) and canyon-cut section of US strata has accumulated on the northern slope of the ridge leading toward Umnak Plateau (Figure 6; see also Figures 12, 14). On the south side of the ridge, thick sequences (3–4 km) of basinal US strata underlie the floor of the Atka Basin region of the Aleutian Terrace (Figures 3, 6; see also Figure 18). Beds of presumably late series age form the accretionary complex (6–8 km) of thrust-thickened trench deposits beneath the adjacent landward slope of the Aleutian Trench (Figure 6; see also Figures 20, 21). Although technically not a depositional sequence of the Aleutian Ridge, Pleistocene deposits of late series age form the wedge-shaped body of turbidite beds underlying the floor of the Aleutian Trench (Figure 6; Scholl, Vallier, and Stevenson, 1982; McCarthy and Scholl, 1985).

RIDGE EVOLUTION AND NORTH PACIFIC TECTONICS

The three rocks series are thought to reflect important phases in the tectonic and geomorphic evolution of the Aleutian Ridge—rapid magmatic growth through initial waning (LS), diminished growth and crestal erosion (MS), and accelerated structural modification (US). These phases can be roughly linked to changes in north Pacific plate boundary conditions and relative movements. The schematic sections on Figure 9 show these evolutionary phases and their relationships to influencing north Pacific events. Grow and Atwater (1970) were the first to recognize that these events were recorded in the rock structure and fabric of the ridge.

Initial Phase (55?–37 Ma)—Growth and Early Decline

Initial accumulation of the lower series is linked to the magmatic consequences of the offshore buckling of an oceanic plate and the consequent creation of the Aleutian subduction zone and island arc (Scholl, Buffington, and Marlow, 1975). This plate buckled for unknown reasons, but it was probably the Kula plate, which formed in the north Pacific about 85 Ma (Woods and Davies, 1982; Rea and Dixon, 1983; Figure 9). In the early Tertiary the Kula plate moved rapidly northward toward a north Pacific subduction zone located along the base of the Alaskan-Siberian margin of the Bering Sea (Figure 9; Scholl, Buffington, and Marlow, 1975; Scholl, Vallier, and Stevenson, 1986).

The Aleutian Island Arc of the Aleutian Ridge evolved above the new offshore subduction zone. Early magmatic growth of the ridge may have been similar to that described by Natland and Tarney (1981) and Hawkins et al. (1984) for western Pacific oceanic arcs. The fragment of Kula plate trapped north of the growing arc was accreted to the North American crust and became the basement of both the modern Aleutian Basin and possibly the Aleutian Arc (Figure 9; Cooper, Scholl, and Marlow, 1976; Cooper, Marlow, and Scholl, 1976). The accreted terrane of Kula oceanic crust was named "Aleutia" by Marlow and Cooper (1983).

The time of capture of Aleutia can be estimated by noting when subduction-related deformation and igneous processes ceased along the Alaskan-Siberian continental margin north of the arc. Arc magmatism and processes of subduction accretion ended along this margin about 60–55 Ma (Patton et al., 1976; Wallace and Engebretson, 1984; Scholl, Vallier, and Stevenson, 1986). As noted previously, arc volcanism at the offshore Aleutian Arc began sometime before 50 Ma. The origin of the igneous massif of the Aleutian Ridge is therefore roughly synchronous with the independently estimated time of capture of Aleutia, both occurring near 55–50 Ma, in the early to middle Eocene.

Northward-moving Kula lithosphere rapidly underthrust the growing Eocene ridge (at roughly 200 km/m.y.; Engebretson, Cox, and Gordon, 1985), sustaining its areal expansion and mounting elevation until about 43 Ma (Figure 9). At this time the motion of the Pacific plate relative to the Hawaiian hotspot changed to its present west-northwest direction (Clague, 1981; Clague and Dalrymple, in press). Coincidentally, or possibly earlier (Byrne, 1979), spreading ceased at the Kula-Pacific spreading ridge (Engebretson, Cox, and Gordon, 1984), causing the Kula plate to become welded to the Pacific plate (Figure 9). Underthrusting along the length of the arc slowed markedly, and the trend of westward-increasing obliqueness of convergence was established or significantly increased (Engebretson, Cox, and Gordon, 1985).

Medial Phase (37–5.3 Ma)—Reduced Growth and Regional Elevation

The deposition of clastic debris of the middle series over the flanks of the ridge testifies to (1) a marked reduction of igneous activity and the production of related volcaniclastic rubble on the submerged flanks of the ridge, and (2) the regional elevation and concentration of arc-magmatism along the crest of the ridge. Additional evidence that large areas of the ridge crest were subaerially exposed in the middle Tertiary is the observation that mineralization adjacent to epizonal plutons intruded in the Oligocene and Miocene took place in contact with circulating meteoric waters (Perfit and Lawrence, 1979; Perfit et al., 1980; Kay, 1983; Kay and Kay, 1985).

In the Oligocene and Miocene, lower series rocks presently at or near the surface of the arc were buried generally less than 4 km beneath the summit of the arc (McLean and Hein, 1984; Kay and Kay, 1985). Thus, except locally,[3] after

[3] For example, the middle Miocene Chitka Point volcanic mass on Amchitka Island (Carr, Quinlivan, and Gard, 1970), the upper series beds filling summit basins (Figures 15 and 16), and the relatively widely spaced volcanic centers of the present Aleutian chain (Figures 2, 3; Coats, 1950; Kay and Kay, 1982; Marsh, in press).

about the close of the Eocene, thick sections of sedimentary and igneous rocks did not accumulate regionally over the crest of the arc.

Post-Eocene elevation of the summit of the arc may have been a manifestation of plutonic engorgement and a related concentration of thermal expansion along the axial crest of the arc (Anderson, 1971). Also, in comparison with the rapid rate that relatively cold lithosphere was thrust beneath the arc prior to 43 Ma, the greatly decreased rate of underthrusting after this time would have allowed the downgoing lithosphere to thermally expand beneath the arc—thus elevating it. The prominent mid-Oligocene drop in eustatic sea level (Vail, Mitchum, and Thompson, 1977; Vail and Hardenbol, 1979) must also have increased the areal exposure of the arc and consequently the supply of terrigenous debris to its sloping flanks. Adding also to the supply of sediment were the masses of ejecta and easily eroded pyroclastic debris erupted at the calc-alkaline volcanic centers overlying the crestal plutons.

Plate reconstructions for the early Tertiary imply that relatively young, and therefore bathymetrically higher and more buoyant, oceanic crust crested at the Kula-Pacific spreading ridge began to underthrust the arc in the Oligocene. This circumstance would have contributed further, and perhaps importantly, to its middle Tertiary elevation (Figure 9).

We relate the dampening of arc-wide emplacement of igneous rocks before about 40 Ma, and the transition to a phase of erosional destruction and summit-concentrated magmatism by at least 37 Ma, primarily to the effects of the 43-Ma reorganization of north Pacific plates and plate motions (Scholl, Vallier, and Stevenson, 1982; 1983a, 1983b). Secondarily, and drawing on the ideas of DeLong, Fox, and McDowell (1978), we suspect that flameout was abetted by the initial subduction of the relatively warm oceanic lithosphere created at the Kula-Pacific spreading ridge (Figure 9). By at least the beginning of the Oligocene, the circumstance of a generally stable northwest-directed underthrusting of the arc by continuously younger crust of the Pacific plate sustained the ridge's lengthy medial phase of elevation and periodic growth by crestal magmatism.

Modern Phase (Past 5.3 Ma)—Arc Subsidence and Accelerated Deformation

Crestal Collapse and Fragmentation

Wave planation induced by glacio-eustatic sea level changes cut sectors of the summit platform of the ridge in the Quaternary, especially fringing the islands. However, in many areas the depth to the platform's erosional surface is much deeper than that caused by continental glaciation (150–200 m; Figure 10) (see also Figure 15) and along its seaward edges the platform may descend to depths of 1500 m or more (see Figure 17). The depth of the platform is therefore evidence that late Cenozoic beveling of the ridge crest is chiefly a record of regional subsidence, which appears to increase toward the Pacific and Bering Sea edges of the platform. Many of the Aleutian and Komandorsky Islands are therefore effectively erosional remnants of the formerly much more elevated middle Tertiary ridge crest. Subsidence of the ridge crest has not been regionally uniform, and some areas of the platform have been elevated. Evidence for differential vertical movement includes elevated sectors of the platform on some islands, submerged fault scarps that offset the erosional surface of the platform, and deeply subsided summit basins (Figure 10).

Dramatic evidence of fragmentation of the arc west of about Adak Island (long. 176.5°W) is exhibited by the large, NE–SW- or transverse-trending Pacific-slope canyons (for example, Adak, Amatignak, Murray, and Heck Canyons; Figures 2, 3). These tectonic valleys disrupt the lateral continuity of the ridge's Pacific slope and bordering Aleutian Terrace; their tectonic origin has long been recognized (Gibson and Nichols, 1953; Gates and Gibson, 1956; Perry and Nichols, 1965; Anderson, 1971; LaForge and Engdahl, 1979; and Spence, 1977).

Evidence that fragmentation and distension of the arc, especially west of Adak Canyon, was accelerated in the late Cenozoic includes the late Cenozoic age of the many summit basins. These basins are filled with slightly deformed sequences of US deposits (Scholl, Buffington, and Marlow, 1975; Geist, Childs, and Scholl, 1985). Also the Pacific-slope canyons are the active tectonic valleys of northwestward-dipping normal faults (von Huene et al., 1971; Engdahl, 1971; Marlow et al., 1973; Kisslinger and Engdahl, 1974). The lower reaches of the canyons, and the merging structural basin of the Aleutian Terrace, are flooded with deposits of the upper series that exhibit deformational growth structures (Marlow et al., 1973).

Fore-Arc Compressional Deformation

Roughly contemporaneous with the increased rate of fragmentation and sinking of the crestal region of the ridge, the fore-arc region was deformed by compression in conjunction with the emplacement of a massive subduction complex beneath the trench's landward slope (Grow, 1973a, 1973b; Scholl, Vallier, and Stevenson, 1983a, 1983b). Although compression can be linked to underthrusting by the Pacific plate, formation of a consequent subduction complex requires the rapid deposition of turbidite deposits along the trench axis. We speculate that this process was initiated by the glaciation of Gulf of Alaska drainages in the late Miocene (McCarthy et al., 1984; McCarthy and Scholl, 1985). Presumably, therefore, the subduction complex formed only during the past 6–5 Ma.

East of about Adak Canyon (long. 177°W) the wedging of the growing subduction complex beneath the fore arc is thought to have elevated older framework rocks to form the antiformal highs that structurally underlie the trench-slope break (Figure 6; Ryan and Scholl, 1985). These highs guard the outer or seaward flanks of the Aleutian Terrace (Grow, 1973b), which is underlain by thick basinal deposits of US beds. As noted elsewhere, these deposits may be as old as latest Miocene, but samples from DSDP site 186 (Stewart, 1978) indicate that the bulk of the terrace fill accumulated only during the past 6–5 m.y.

Much less is known about the structure and age of the infilling US beds of the laterally discontinuous Aleutian Terrace west of Adak Canyon. At least as far west as the Near Islands sector (long. 173°E), the growth of an outer structural high in the late Cenozoic appears to be associated with the emplacement of a small subduction wedge (Vallier et al., 1984). However, right-lateral shearing of the arc likely is in part, or largely, responsible for the formation of the structural basins in the far western sector of the Aleutian Terrace (Vallier et al., in press).

Causes of Accelerated Deformation

Accelerated regional subsidence of the arc in the late Cenozoic probably resulted from the combined effects of (1) thermal cooling since the close of the arc's middle and late Tertiary phase of crestal plutonism; (2) subduction about 11 Ma of the long-dead (by roughly 30 m.y.) Kula-Pacific spreading ridge, which caused increasingly older and therefore less buoyant oceanic crust to underthrust the ridge (Figure 9; DeLong, Fox, and McDowell, 1978); and (3) the cooling effect on the ridge, beginning about 5 Ma (Cox and Engebretson, 1985), of more direct and faster underthrusting by the increasingly older Pacific lithosphere, a circumstance that may have been offset somewhat by the kindling of the ongoing phase of Aleutian volcanism.

Accelerated fragmentation as well as collapse of the crestal region can be related to (1) dissipation of tensional stresses along the thermally softened magmatic axis of the arc caused by trench roll-back in response to subduction of increasingly denser (older) Pacific lithosphere (Figure 9; (Geist, Childs, and Scholl, 1985)); and (2) longitudinal and transverse tearing and clockwise block rotation of arc segments forced by the obliquely converging Pacific plate. This latter process has potentially been underway since at least 43 Ma. Increased splaying and fragmentation of the arc, especially west of about Adak Canyon (long. 177°W), in the late Cenozoic may have been initiated or abetted by a combination of spreading in the Komandorsky Basin (Figure 1; Bogdanov, Vishnevskaia, and Sukhov, 1983; Bogdanov and Neprochnov, 1984), improved lower-plate lateral drag resulting from the subduction of older segments of fracture zone relief trailing southward behind the subducted Kula-Pacific spreading ridge; and, perhaps more importantly, laterally directed effects of the jamming of a rapidly thickening subduction complex beneath the landward trench slope in Pliocene and Pleistocene.

It is interesting to speculate that transform motion between the Pacific plate and the western segment of the ridge might have moved blocks and slivers of the arc northwestward relative to its eastern segment. For example, Kresta Trough (long. 170°E; Figures 2, 4, 11) in the Near Pass region, and similar longitudinal valleys and bathymetric alignments along the crest of the ridge eastward to the region of Amchitka Pass (long. 180°), may be the rift valleys of right-lateral faults (Figure 2). Clockwise rotation of ridge segments (the blocks defined by Anderson, 1971, and Spence, 1977) would be expected if caught up in a cross-ridge zone of distributed right-lateral shear (Beck, 1976; Wells and Coe, 1985). Harbert and Cox (1984) and Harbert, Cox, and McLean (1984, 1985) have compiled paleomagnetic evidence that clockwise rotation of crustal blocks has, in fact, occurred. It is likely, therefore, that the crustal fractures controlling the northeast-striking Pacific-slope canyons exhibit conjugate left-lateral offsets responding to clockwise block rotation—a supposition supported by the southwestward projections of their western walls into the physiographic province of the Aleutian Terrace (Figure 2). It is also likely that many of the ridge's summit basins are trailing-edge extensional structures created in the wake of the rotating blocks (Geist, Childs, and Scholl, 1985).

PETROLEUM GEOLOGY

Regional information bearing on the petroleum geology of the Aleutian Ridge has been reviewed by Scholl, Vallier, and Stevenson (1983b). They note that relatively thick sequences of middle and upper series beds are potential locations for petroleum deposits. Adequate reservoir beds and trapping structures exist, and samples of Tertiary rocks are sufficiently rich in organic matter to serve as potential source beds. Sedimentary sections are also sufficiently thick and warm to have generated petroleum hydrocarbons.

Notwithstanding these statements, factual information from the Aleutian Ridge is inadequate to support a reliable assessment of its petroleum potential. We highlight below observations and information along a regional north-to-south crossing of the ridge that are pertinent to evaluating the petroleum potential of its sedimentary sequences.

Bering Sea Flank

Base of Slope

In general, a relatively thick (2–3 km) sequence of sedimentary deposits lies immediately north of the structural base of the northern or Bering Sea flank of the ridge (Figures 11, 12). These deposits cannot be regarded as part of the ridge's generalized tripartite rock sequence, but instead are chiefly the infilling beds of the Komandorsky, Bowers, or Aleutian basins within the 3000–4000-m-deep Bering Sea Basin (Figures 1, 2; Cooper, Scholl, and Marlow, this volume). Reflection profiles commonly reveal that the structural mass of the island arc rises relatively abruptly from the nearly horizontal surface of the oceanic crust of these basins. The upper depositional units of the basinal section pass in transition to coeval rock sequences of the ridge flank by thinning over the rising arc massif, which is presumably composed of lower series rocks (Figures 11, 12).

Regional information extrapolated from DSDP sites 184, 185, 188, and 190 (Figure 2; Creager, Scholl, et al., 1973) and samples of submarine outcrops from the ridge (Scholl, Vallier, and Stevenson, 1983a, 1983b; Table 1) imply that north of the Aleutian Islands well-stratified base-of-slope deposits are chiefly Eocene and younger in age. Although the island arc has contributed relatively coarse debris to these sections (see drilling results, site 189, Creager, Scholl, et al., 1973; Figure 11), Alaskan and Siberian drainages also supplied fine detritus via hemipelagic and turbidity current transportation.

Drawing on the result of the DSDP sampling, Cooper et al., (1979) note that Pliocene and Pleistocene deposits are likely to be porous, permeable, and richly diatomaceous. But diagenetic conversion of opal-A to crystalline opal-CT has probably converted older Neogene beds to indurated porcellaneous mudstone. A prominent subsurface reflector on profiles L5-15 and L9-39 (Figure 12) is probably a BSR (bottom-simulating reflector) related to this diagenetic transition (Scholl and Creager, 1983; Hein et al., 1978) On reflection records, the basal 0.5–1.0 sec (roughly 1 km) of the base-of-slope section may exhibit crude bedding (profile L5-15, Figure 12); we presume this section is coarse volcaniclastic debris coeval with the arch's lower series rocks (Figure 6).

Limited information suggests that deformational structures north of the Aleutian Islands segment of the arc are not common in the base-of-slope section (Figures 11, 12). Evidently, little relative movement occurred between the arc massif and the basement rock of the Aleutian and Bowers Basins during most of the Cenozoic. However, farther west-

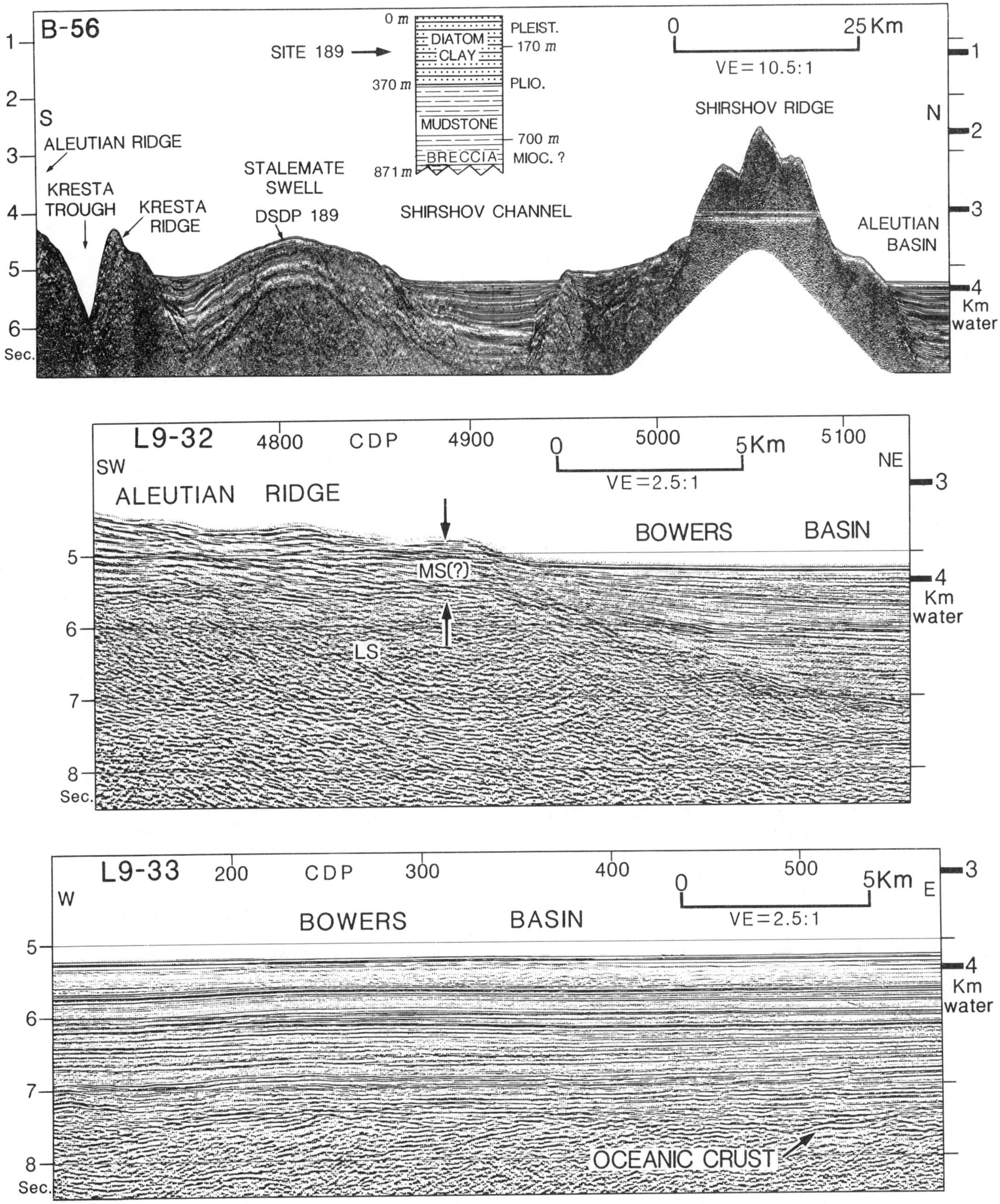

Figure 11. Seismic-reflection sections that cross and trend parallel to the base of the northern insular slope in the western Aleutian Islands (locations on Figure 2). US, MS, and LS are the upper, middle, and lower series units, respectively. Single-channel profile B–56, and the strata drilled at DSDP site 189 (Creager, Scholl, et al., 1973), establish that the Stalemate swell is an elevated section of base-of-slope beds. Deformation is related to possible right-lateral fracturing along the Kresta Trough. Migrated multichannel sections L9-32 and L9-33 exhibit, correspondingly, the ridge's base-of-slope sedimentary sequence and that of the adjacent fill of Bowers Basin.

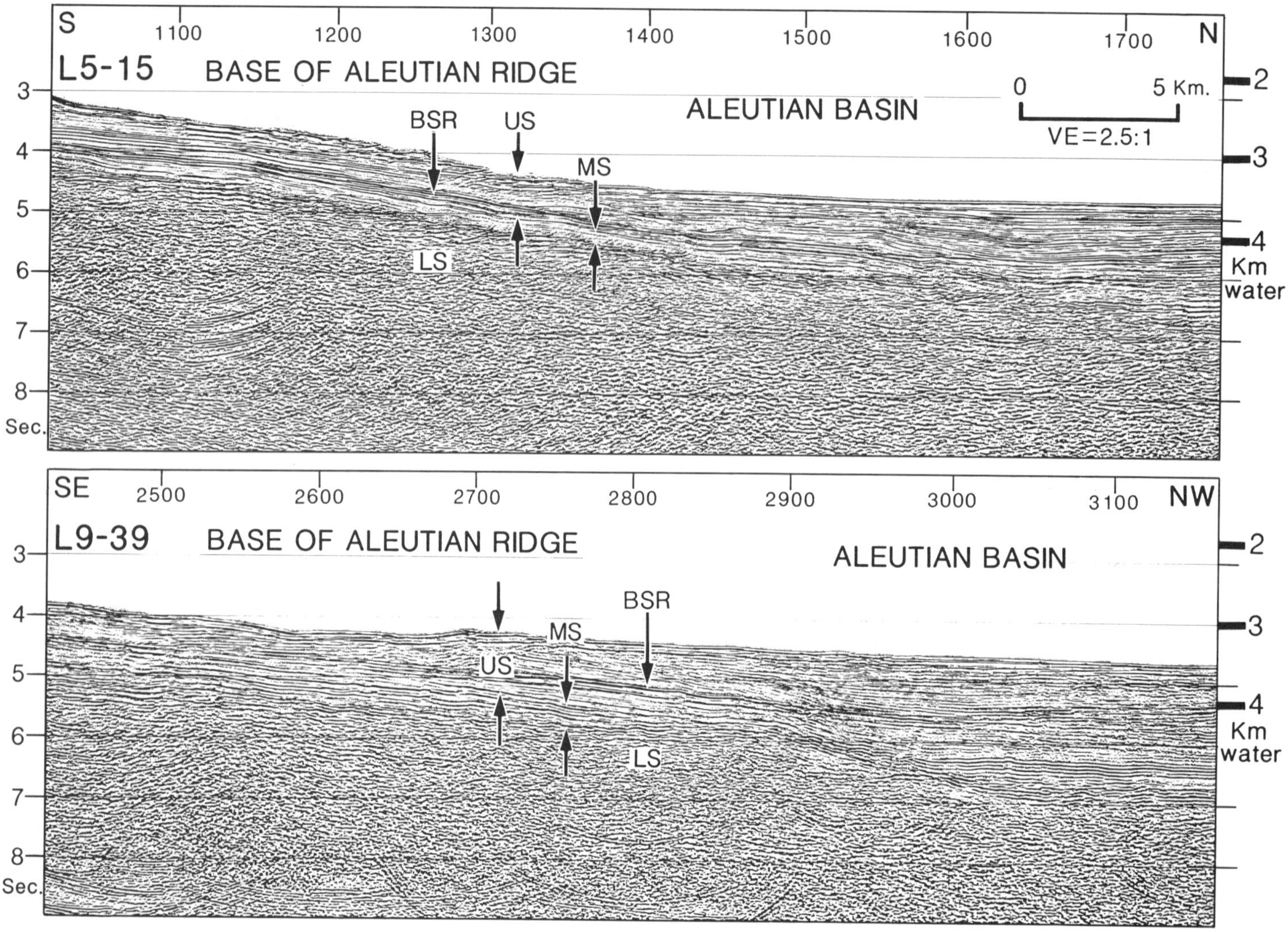

Figure 12. Migrated multichannel sections of lines L5-15 and L9-39 that cross the lower northern insular slope in the vicinity of Amlia Island (locations on Figure 3). US, MS, and LS are the upper, middle, and lower series units, respectively. The prominent BSR (bottom-simulating reflector) is related to the diagenesis of siliceous deposits of probable lower Pliocene age (Scholl and Creager, 1973; Hein et al., 1978; Cooper, Scholl, and Marlow, this volume). Separation of US and MS beds over the lower part of the slope is tenuous.

ward, opposite the Near Pass segment of the arc, base-of-slope deposits have been elevated to form the Stalemate Swell, where DSDP site 189 was drilled (Figure 11). We presume this western sector of the arc has been affected by strike-slip wrench faulting. The Kresta Trough, for example, may be the rift valley of a right-lateral wrench zone (Figure 2). The stratigraphic section penetrated at DSDP site 189 implies that uplift of the Stalemate Swell—probably in response to transpressive stresses—occurred during the Pleistocene (Creager, Scholl, et al., 1973).

Adequate stratigraphic relationships, and some structures, exist within the base-of-slope sections to trap migrating hydrocarbons expelled from the basinal deposits of the Bering Sea Basin. The regional thermal studies of Langseth, Hobart, and Horai (1980), and Cooper et al. (1979), which are reviewed by Cooper, Scholl, and Marlow (this volume), indicate that petroleum products can form below a depth of about 2 km. The existence of deep-water bright spots, termed VAMPs (velocity-amplitude anomalies) by Scholl and Cooper (1978), in Bowers Basin suggests that thermogenic gases are being generated. The base-of-slope sections likely are rich in diatomaceous debris. The characteristics of siliceous deposits that favor accumulation of petroleum (see, for example, Isaacs and Garrison, 1983; and Crain, Mero, and Patterson, 1985), including the possible formation of fracture permeability at depth, are therefore applicable (Cooper, Scholl, and Marlow, this volume). The relatively thick base-of-slope sections are therefore prospective targets for petroleum exploration (Cooper et al., 1979).

Northern Insular Slope

The northern insular slope of the Aleutian Ridge is relatively steep, descending at an average angle of 9°. However, in many areas, for example adjacent to Umnak Plateau and Bowers Ridge (Figure 2), the overall angle of descent is much less. Typically, the upper part of the slope dips northward more steeply than a basal rise-slope, which merges gradually with the nearly horizontal floor of the Bering Sea Basin. The upper part of the insular slope is also scarred by canyons and has a relatively thin mantle of US beds.

Sectors of the northern slope that are more gently inclined may be underlain by relatively thick (1–3 km) slightly deformed sequences of MS beds (Figures 13, 14). In the vicinity of the Umnak Plateau, the thickened sections appear to be confined to a mid-slope trough. MS deposits in this trough

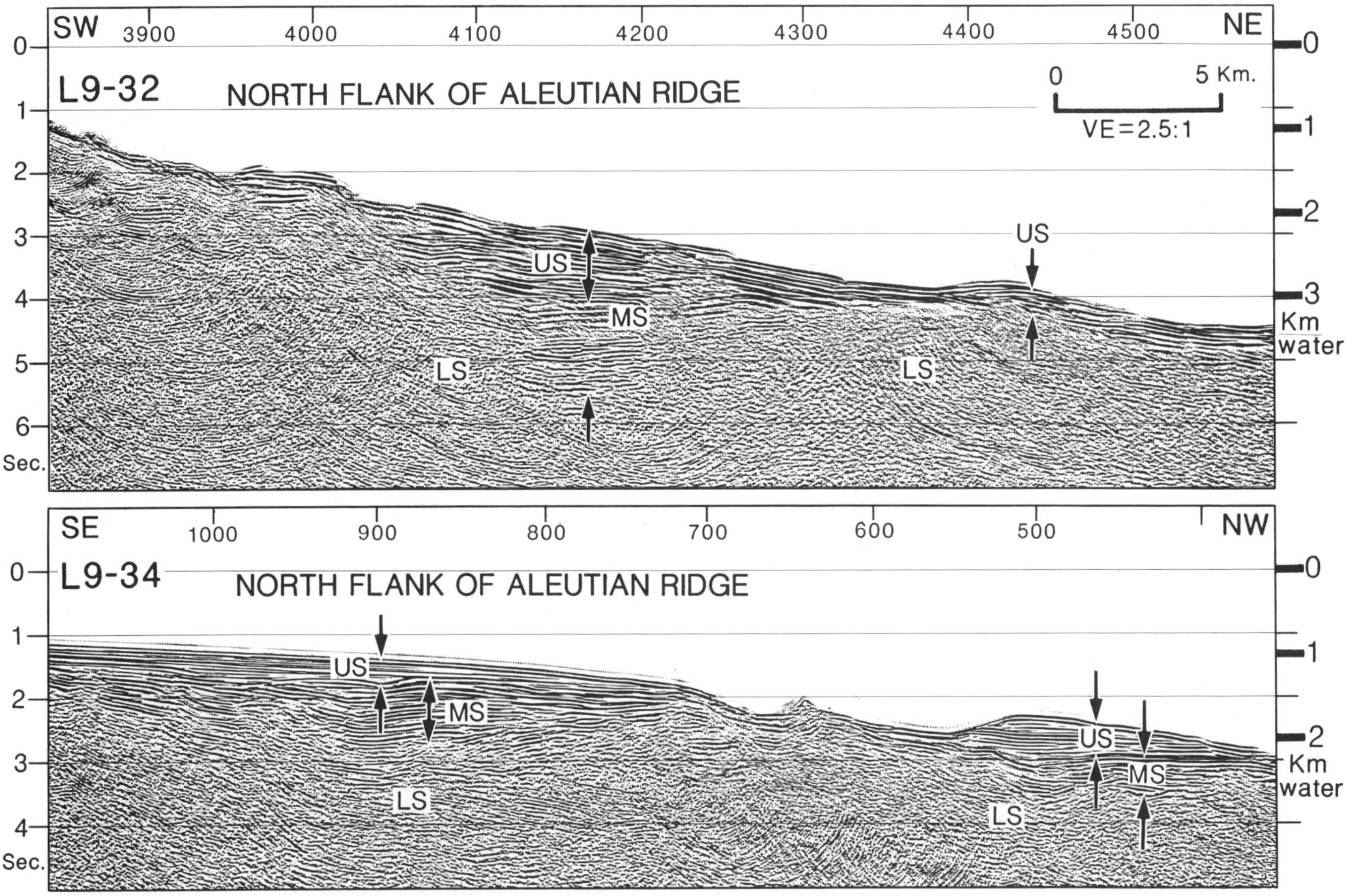

Figure 13. Migrated multichannel sections of lines L9-32 and L9-34 (locations on Figure 2) that cross the mid-slope area of the northern flank of the Aleutian Ridge, western Aleutian Islands. US, MS, and LS are the upper, middle, and lower series units, respectively. Thickened sections of presumed middle series beds unconformably underlie upper series strata. On L9-34 the unconformity may be the original surface of the wave-planed summit platform, now mantled with US sediment and deeply submerged.

thin northward over a basement high of LS rocks that builds out the lower part of the ridge flank (Figures 6, 14). The basement high shown on Figure 6 (profile L9-39) is probably an extension of the basement massif of the adjacent Umnak Plateau (Ben-Avraham and Cooper, 1981; Cooper, Marlow, and Vallier, 1984). This observation, and structures weakly revealed on single-channel seismic profiles, suggest that the mid-slope trough of MS beds may be generally continuous along the sector of the ridge bordering the Umnak Plateau. In fact, data compiled by Cooper, Marlow, and Vallier (1984) show that north of Unalaska and Akutan Islands (166°W) the combined thickness of US and MS beds may exceed 6 km (Figure 2).

Based on section thickness, the likelihood of reaching thermal maturity, and the probable occurrence of stratigraphic as well as structural traps, only the troughlike bodies of MS beds seem like prospective areas for petroleum exploration along the northern slope of the ridge. Little is known about the lithology of the MS beds, but porous, little-altered sandstone and siltstone have been recovered from the series in the vicinity of Seguam Island (long. 173°W; Figure 3, Table 1). Dredging, reflection profiles, and drilling at DSDP sites 184 and 185 establish that US beds are rich in diatomaceous debris; the underlying MS beds likely include diagenetically altered siliceous sections.

Crestal Area

Summit Basins

The arc's broad summit platform is typically a wave-planed bedrock surface that is thinly mantled by weakly consolidated US deposits (Figure 10). Locally, however, bathymetrically expressed grabens and half-grabens have formed and have been partly or almost completely filled with sequences of US beds (Figures 10, 15; Scholl, Marlow, and Buffington, 1975). Maximum water depths over the floors of these structures ranges from about 1000 to 2000 m.

Little is known about the occurrence of summit basins along the Komandorsky segment of the ridge. However, along the crest of the Aleutian segment, large summit basins include Amukta (long. 172°W), Amlia (173°W), and Sunday (180°) basins, and Pratt (179°E) and Buldir (177°E) Depressions (Figures 2, 3). These basins are typically elongate parallel to the axis of the arc and occur along or adjacent to the alignment trends of active and dormant volcanos (Figures 3, 10). However, some of the basins, for example Pennant Basin (long. 175°E), trend across the ridge parallel to the northeast strike of the large Pacific-slope canyons (Figure 2).

Reconnaissance multichannel profiles in the western Aleutian Islands reveal that US beds in Pennant Basin and Pratt Depression may be 1–2 km thick (Figure 15). Geomorphically deeper (2000 m) Buldir Depression holds only a few

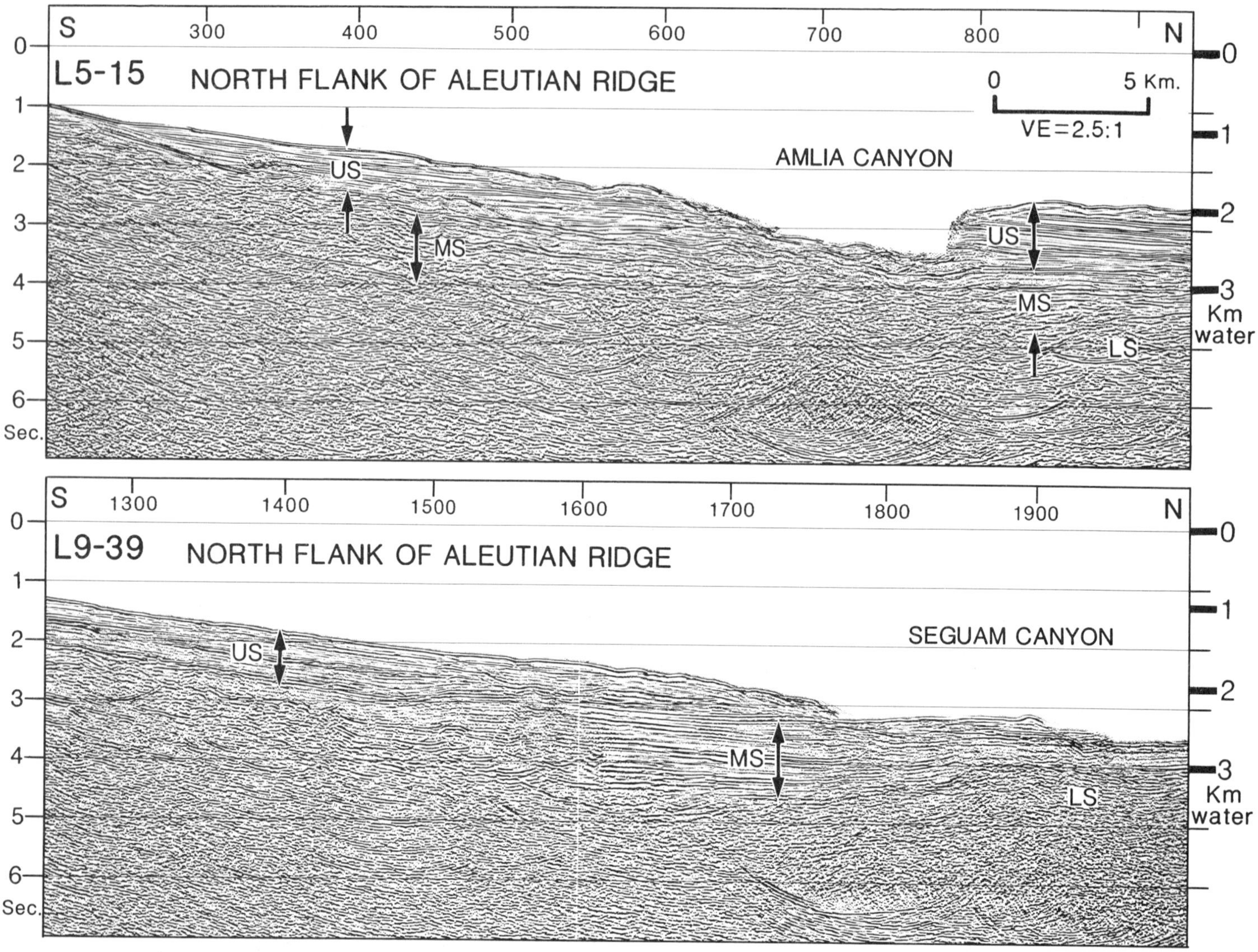

Figure 14. Migrated multichannel sections of lines L5-15 and L9-39 (locations on Figure 3) that cross the upper northern flank of the Aleutian Ridge, Amlia Island region. US, MS, and LS are the upper, middle, and lower series units, respectively. The thickened mass of MS beds may be common beneath the northern insular slope where it flanks the Umnak Plateau (Figure 2).

hundred meters of US beds (Figure 10; Marlow et al., 1970). In fact, basins with deeply submerged floors in general lack a significant fill (>500 m) of upper Cenozoic deposits (Scholl, Marlow, and Buffington, 1975). Presumably, undiscovered summit basins with exceptionally thick US deposits may exist that lack strong geomorphic expression.

Although adequate trapping structures can be identified on reflection profiles (Figure 15), the lack of thickness and relatively small volume of the sedimentary deposits in the known western summit basins argue against considering them as good prospects for petroleum accumulations. However, these western Aleutian basins, as well as the surrounding Attu-Adak sector of the summit plateau, have not been adequately explored, even at a reconnaissance level.

Amlia and Amukta Basins in the eastern sector of the ridge (Figure 3) contain a sufficient volume and thickness of US beds to be considered as possible habitats for petroleum deposits (Figures 6, 16). The combined area of these neighboring structures is approximately 5000 km^2. Reflection profiles and geopotential modeling indicate that as much as 3 km of US beds overlie basement rocks of early or possibly middle series age (Geist, Childs, and Scholl, 1985).

Both the Amlia and Amukta Basins are elongate parallel to the axis of the arc. Transverse profiles show that the Amlia Basin is a half-graben; the major eastward-striking boundary faults, located along its northern or Bering Sea side, dip southward (Scholl, Marlow, and Buffington, 1975). The Amukta Basin seems to be a similar asymmetric structure, but a major north-dipping fault appears to also border its southern side (Figure 16, profile L9–39). Seismic sections collected parallel to the major boundary faults reveal numerous northward-striking normal faults offsetting the infilling US beds as well as the sea floor (Figure 16, profile L9-37; Geist, Childs, and Scholl, 1985). Extensional faults are typically associated with growth structures; hence, sediment infilling and structural evolution of the basin occurred simultaneously.

During the past 5 m.y., the active volcanic centers of Atka, Seguam, and Amukta Islands have contributed volcaniclastic detritus to the basins. Clastic deposits of latest Miocene and younger age can be traced into the basins from the northern flank of the ridge (Figure 6). We believe that the sedimentary sections in the basins also include terrigenous deposits derived from erosion of the islands and the flanking

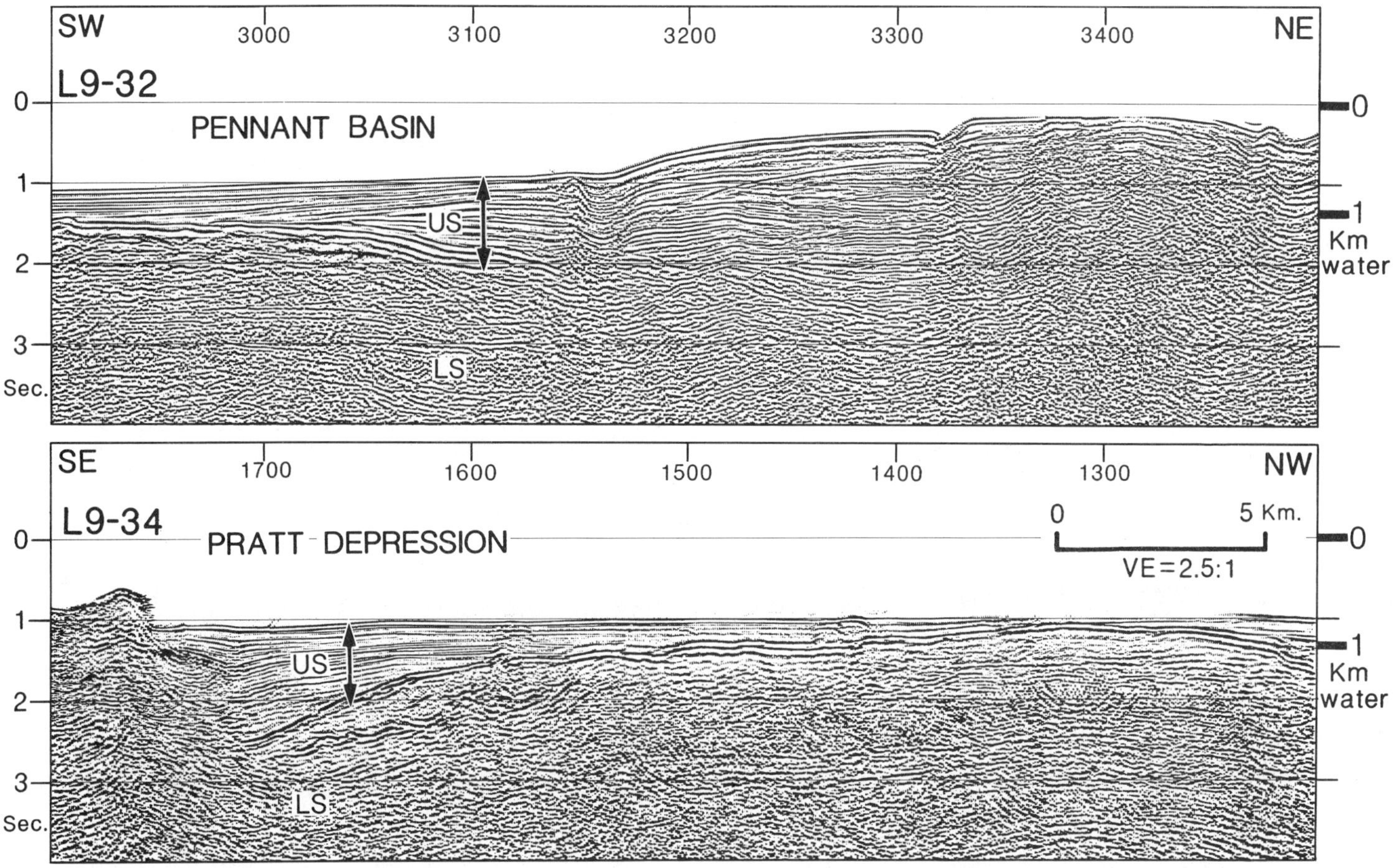

Figure 15. Migrated multichannel sections that cross summit basins in the vicinity of the western Aleutian Islands (locations on Figure 2). US, MS, and LS are the upper, middle, and lower series units, respectively. Upper series deposits are relatively thin (1–2 km) in both the Pennant Basin and the Pratt Depression. The infilling beds that bury a wave-eroded surface—the original plateau of the summit platform—cut across lower series units probably in the latest Miocene.

summit platform (Scholl, Marlow, and Buffington, 1975). The basinal sections should also have a moderate content of diatomaceous debris.

Reflection profiles reveal stratigraphic relationships and fault-related structures that are adequate for entrapment of migrating hydrocarbons. Alteration of unstable volcaniclastic debris of late Cenozoic age may be expected to reduce porosity and permeability at depth. But the occurrence at depth of relatively low interval velocities (1.8–2.5 km/sec at 1–2 km depth) implies that alteration may not have fully sealed potential reservoirs. Also, older rocks of the lower series, which supply clastic debris less susceptible to low-temperature alteration, may have contributed an important fraction to the basin fill.

Information on the possible occurrence of source beds is meager. Because surface waters along the Aleutian Ridge support an abundance of marine life, depositional conditions over the basins could have periodically favored the accumulation of organic-rich beds. Exposed LS sedimentary rocks on the nearby Amlia and Atka Islands typically contain $<0.5\%$ organic carbon. However, the coal-bearing upper Miocene beds of the Martin Harbor section of Atka Island are carbonaceous (roughly 10% C; McLean, Hein, and Vallier, 1983; Hein, McLean, and Vallier, 1984). Possibly gas-prone coaly beds underlie, or occur near the base of, the sedimentary sequences filling the Amlia and Amukta basins. Other potential sources of organic matter might be diatomaceous sections of latest Miocene and early Pliocene age.

Pacific Ridge Flank

Southern Insular Slope

The southern insular slope typically descends from an outer shelf depth near 500 m to the northern edge of the Aleutian Terrace at roughly 3400 m; or, if it is absent, to the floor of the Aleutian Trench at 6500–7000 m (Figures 2, 4). Where the insular slope leads downward to the terrace, the average slope angle is 6–7°. But adjacent to the Komandorsky Islands, the angle of descent to the trench floor is an astonishing 24°.

Regionally spaced reflection profiles generally reveal that MS beds form the principal sedimentary sequence underlying the southern insular slope (Figures 7, 17). In many areas MS and US beds have been stripped away by mass wasting and canyon-cutting processes, especially during late Cenozoic episodes of low sea level and island glaciation (Scholl, Marlow, and Buffington, 1975). In the Adak-Amlia sector, the middle series is as much as several kilometers thick and commonly forms a south-dipping dip-slope sequence that is fairly uniform in thickness as it descends toward the Aleutian Terrace. Perhaps, unlike the northern insular slope, thickened basinal accumulations of MS beds may be uncommon below the south-facing insular slope of the ridge.

Typical samples of MS strata dredged from the Adak-Amlia sector are sandstone and siltstone (Figure 3; Table 1). The rocks are little altered and only moderately consolidated. Beds of MS strata are surprisingly coarse grained to

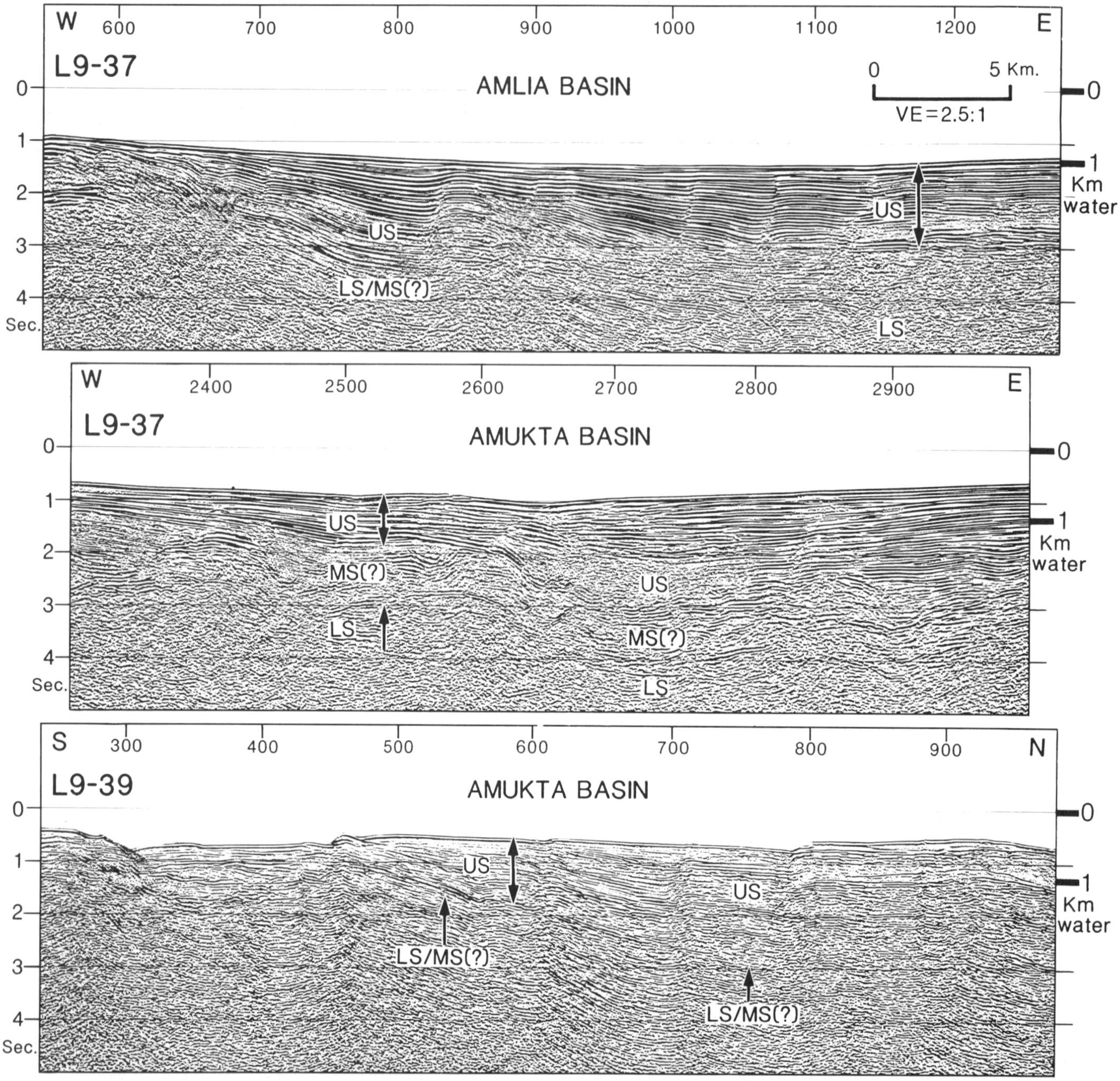

Figure 16. Migrated multichannel sections of lines L9-37 and L9-39 that cross neighboring Amlia and Amukta summit basins, eastern Aleutian Islands (locations on Figure 3). US, MS, and LS are upper, middle, and lower series units, respectively. The two L9-37 sections, which exhibit listric (?) normal growth faults, are oriented parallel to the axis of the arc and to the long-axes of the two basins. Section L9-39, a transverse crossing of the Amukta Basin, reveals a roughly orthogonal set of normal faults striking parallel to the trend of the arc.

have been deposited on a slope of 6–7°. Laterally irregular bedding and a complex depositional fabric suggest that the series is dominantly coarse terrigenous debris deposited by high-energy transport mechanisms (Figures 6, 7, 17). It seems likely that after deposition of the middle series this slope was steepened in the late Cenozoic in conjunction with the subsidence of the ridge crest and formation of the basinal structure of the adjacent Aleutian Terrace.

The apparent scarcity of structural traps, the general thinness of MS beds, and the occurrence of many fluid-venting submarine canyons cutting them, do not recommend the southern insular slope as a prospective region for petroleum accumulations. However, seaward of the base of the slope, southward-dipping MS strata are buried beneath northward-onlapping US deposits of the Aleutian Terrace (Figures 6, 7, 17, 18). The deeply buried MS beds should be thermally mature, and stratigraphic relationships updip are adequate for the containment of migrating petroleum fluids.

Aleutian Terrace

The thickest and most voluminous sedimentary body burying the arc's basement massif of lower series rocks is beneath the Aleutian Terrace (Figure 6). East of about Adak Canyon (long. 177°W), the terrace is continuous for nearly 1000 km. Its average width from the base of the southern insular slope to the antiformal highs of the trench-slope

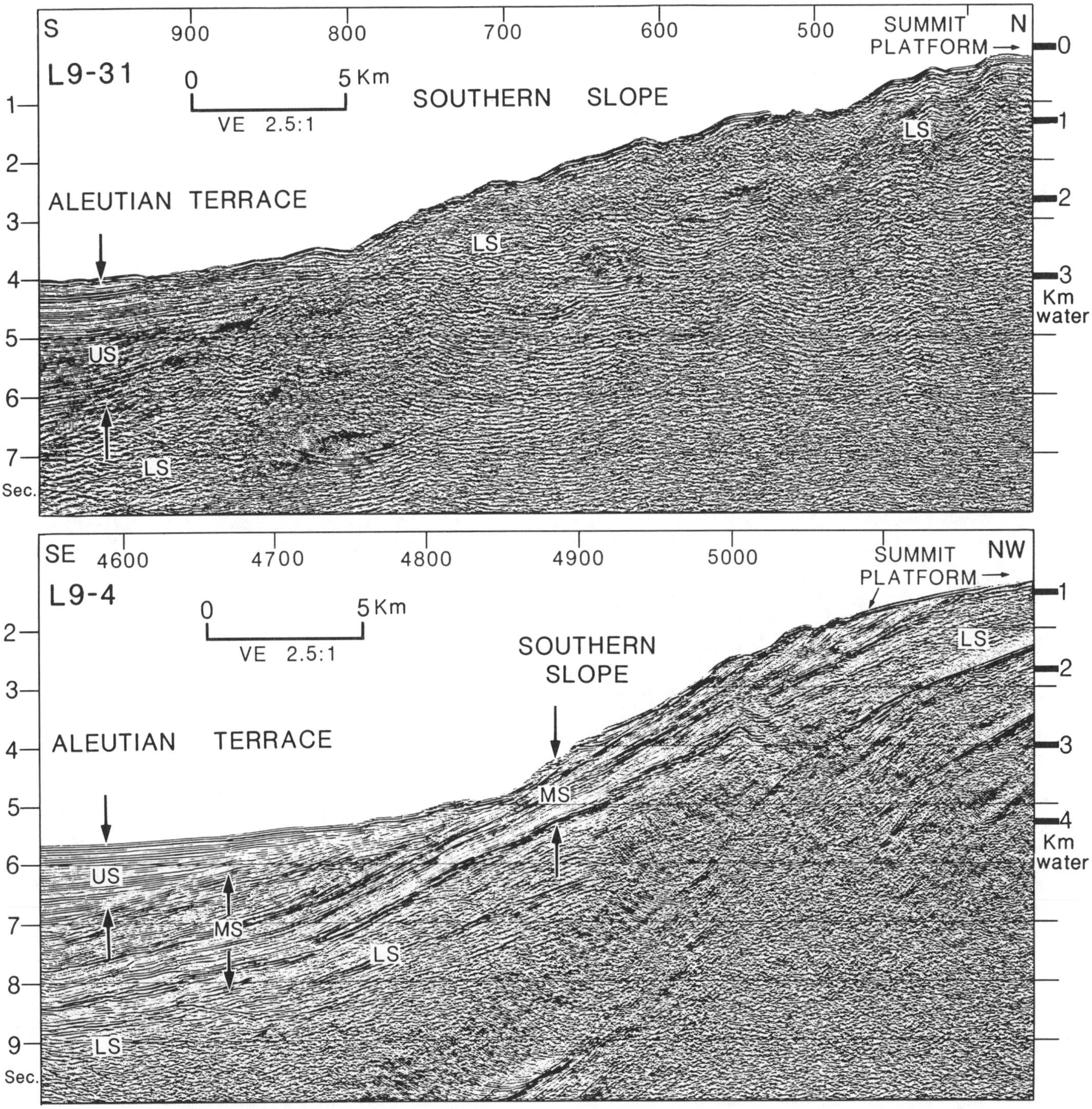

Figure 17. Migrated multichannel sections of lines L9-31 and L9-4, which cross the southern insular slope of the Aleutian Ridge. US, MS, and LS are the upper, middle, and lower series units, respectively. Middle series rocks cannot be clearly identified on section L9-31, located south of Agattu Island near the western end of the Aleutian Islands (Figure 2). However, they were detected nearby on seismic line L9-32 (Figure 7). L9-4 reveals that the insular slope south of Adak Island (Figure 3) is underlain by a relatively thick sequence of bedded middle and lower series deposits, which, southward toward the Aleutian Terrace, appear to downlap onto a surface of older, incoherently reflecting lower series rocks.

break is approximately 60 km, and its mean depth is roughly 4000 m. West of Adak Canyon, the terrace can be traced for another 1000 km; but in this area it is somewhat narrower and deeper, and its lateral continuity is disrupted by the western walls of the Pacific-slope canyons (Figures 2, 3). Adjacent to the Aleutian segment of the arc (Figures 4, 5), the volume of sedimentary deposits underlying the Aleutian Terrace is probably on the order of 150,000–200,000 km^3. Stewart (1978) and Scholl, Vallier, and Stevenson (1983b) have recently reviewed the petroleum geology of the sedimentary sequences underlying the terrace.

The structure of the terrace has been explored, by reconnaissance, with multichannel reflection data only south of the Adak-Amlia sector (Scholl, Vallier, and Stevenson, 1983a, 1983b) and along part of the western end of the Aleutian Islands (Vallier et al., in press; Figure 2). In these areas the terrace represents the surface of a sediment-filled—or nearly so—structural trough. Upper series strata as thick as 2–4 km fill this depression (Figures 6, 7, 18). The strata are presumably diatom-rich clastic deposits contributed by

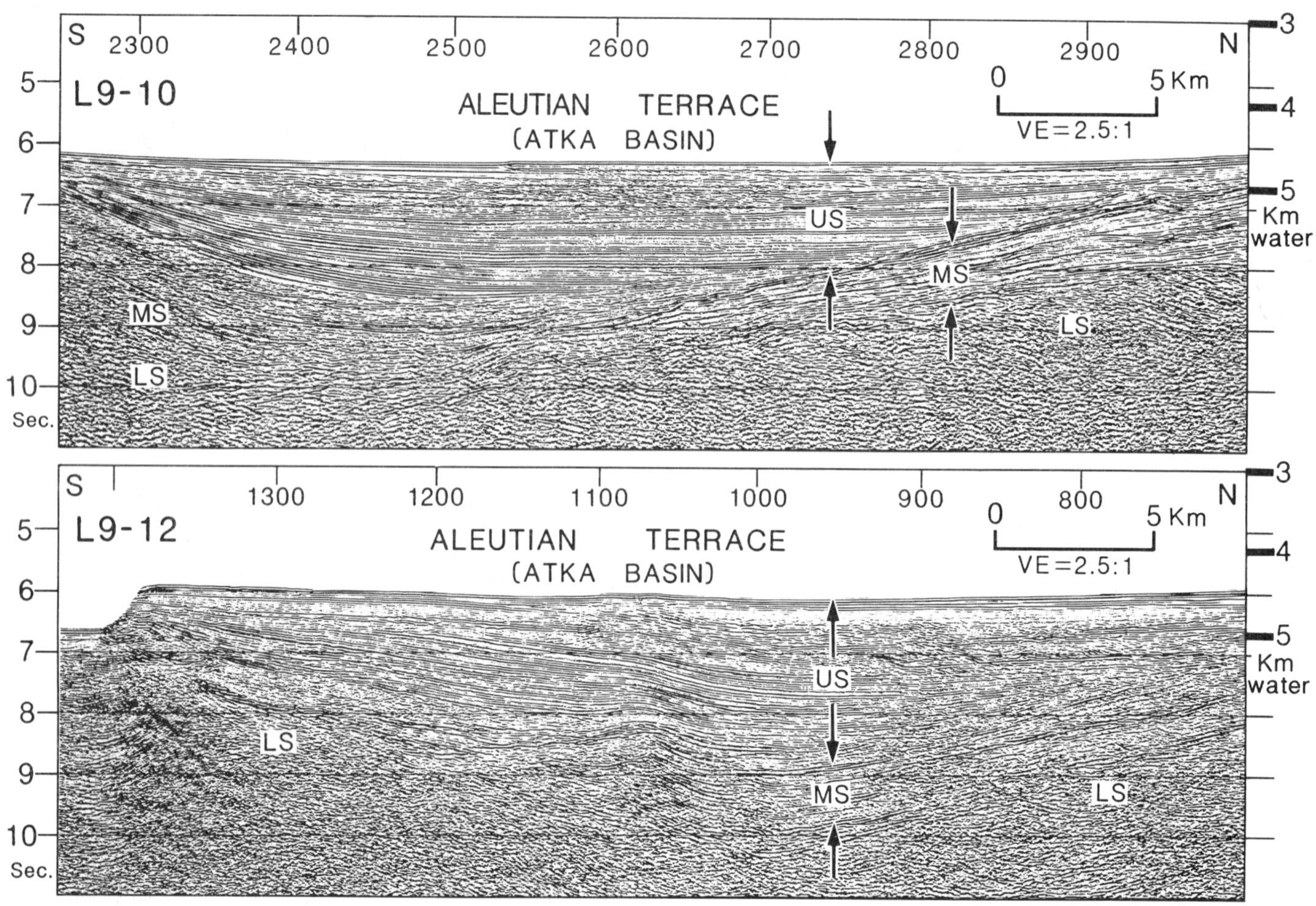

Figure 18. Migrated multichannel sections of lines L9-10 and L9-12, which traverse the Aleutian Terrace south of Amlia Island (locations on Figure 3). US, MS, and LS are the upper, middle, and lower series units, respectively. Upper series deposits, which are the basinal sequence underlying the terrace, unconformably bury middle series strata. The contact between the upper and middle series is irregular, reflecting backfilling over the canyon-scarred southern insular slope (Figures 3, 6, 17). The subsurface antiformal structure in the center of section L9-12 is presumably thrust controlled, perhaps a ramp anticline or duplex structure.

wave planation of the summit platform, stream and glacial erosion of insular areas, canyon cutting of the southern insular slope, and dispersal of ejecta from the Aleutian volcanoes (Stewart, 1978). Depositional processes were probably dominated by turbidity currents and debris flows.

The subsurface basinal form of the Aleutian Terrace is related to the relative structural rise of the outer structural highs of the trench-slope break (Figure 6; Grow, 1973b). Relative elevation of these bordering antiforms involves thrust-ramping of the arc's basement massif, presumably in response to late Cenozoic compression associated with emplacement of the adjacent accretionary complex. Elevation was probably also induced by underplating of the accretionary wedge beneath the trench-slope break (Ryan and Scholl, 1985). Arcward-directed compression may also have elastically downflexed the arc-basement rocks beneath the terrace, thus contributing to the relief of its structural basin (see Yorath, this volume).

The consequent ponding of US beds arcward of the outer structural highs has been underway since at least the earliest Pliocene. Although we define the age of the upper series as equivalent to the oldest basinal beds of the terrace fill, the age assignment is uncertain, and latest Miocene beds could be included. Growth of the antiforms and sediment infilling occurred simultaneously; slight northward migration of the zone of deformation underlying the trench-slope break caused the depositional axis of the terrace to shift arcward with time (Figures 6, 18; Harbert et al., 1982).

Beds of the middle series contribute an additional 1–2 km to the terrace's total sedimentary section. These strata are the downslope continuation of the MS sequence underlying the southern insular slope (Figures 6, 18, 19). Except locally, the MS section does not appear to thicken within the terrace's basinal structure. However, at the base of the southern slope, MS strata appear to step out or down-lap (?) southward across a surface underlain by LS rocks. A former fore-arc platform may therefore have been warped into the terrace's underlying basinal structure (Figure 6).

Middle series rocks have been acoustically traced to the antiformal highs structurally flanking the Aleutian Terrace (Figures 6, 18, 19; Ryan and Scholl, 1985). Although MS beds were not reached at DSDP sites 186 and 187, which were drilled in the vicinity of these highs (Figure 3), a displaced block of MS beds of middle Miocene age was penetrated in an upper Pliocene section at site 186 (Figure 19; Creager, Scholl, et al., 1973).

Lower series rocks can be traced from the southern insular slope to, and seaward of, the axial region of the terrace. Although locally composed of stratified units, LS rocks beneath the terrace are typically massive accumulations that exhibit an acoustic velocity higher than 4.5–5.0 km/sec. Subterrace rocks of the lower series could be a fragment of

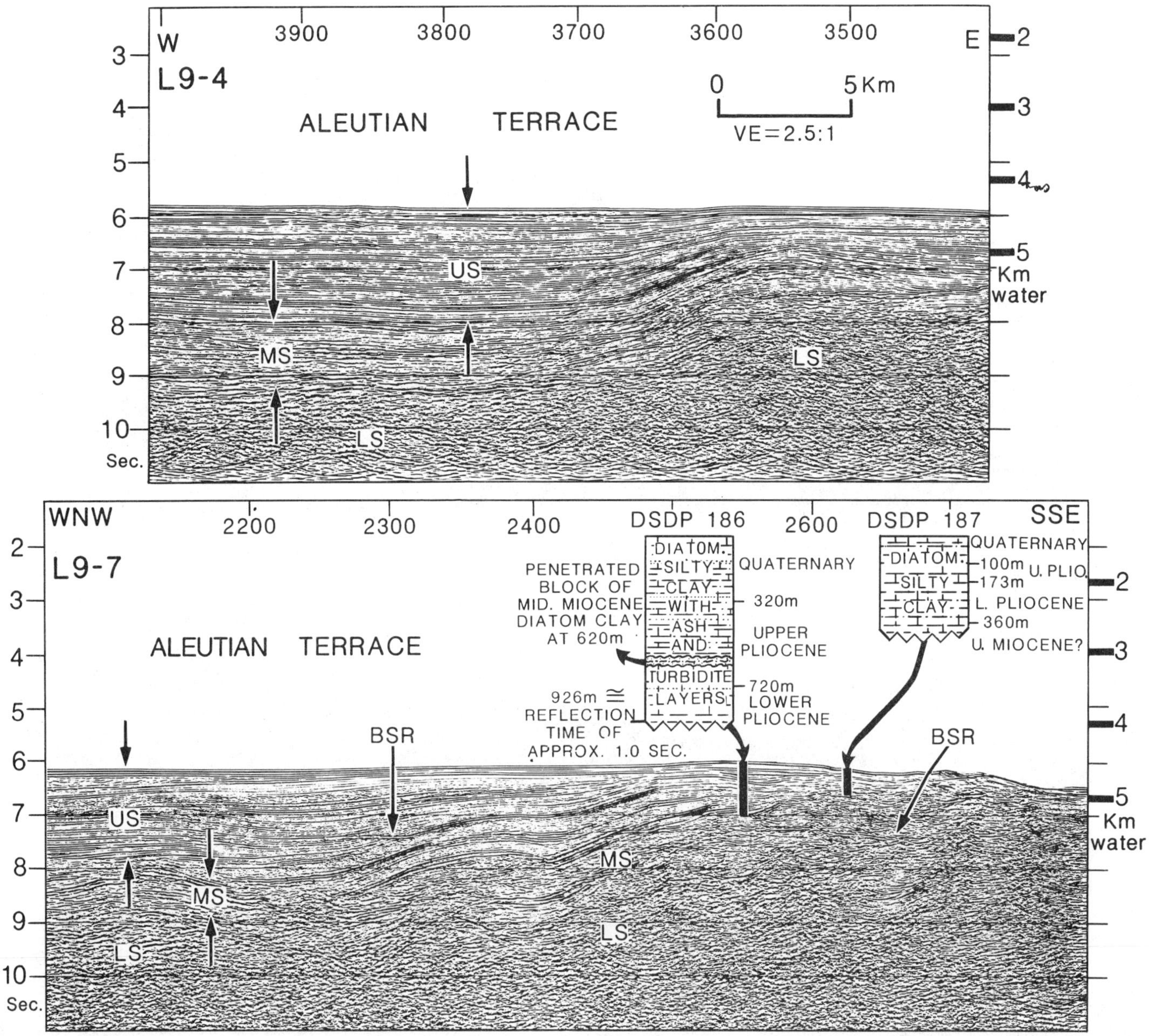

Figure 19. Migrated multichannel sections of line L9-4 and L9-7, which trend parallel and oblique, respectively, to the axis of the Aleutian Terrace south of Amlia Island (locations on Figure 3). Stratigraphic information from DSDP sites 186 and 187 determined that the age of basinal upper series beds (US) is Pliocene and Quaternary, although uppermost Miocene strata could be involved. Middle series beds (MS) were not reached at the two drill sites, but a displaced block of middle series beds of middle Miocene age was penetrated in upper Pliocene strata at site 186, and upper Miocene beds may have been recovered at site 187. The BSR (bottom-simulating reflector) is thought to be gas (methane ?) collected beneath a section containing deposits of gas hydrates (Kvenvolden, this volume). Antiformal structures on both lines are thought to be related to subduction-induced compressional deformation of the fore-arc region.

accreted oceanic crust; however, we believe they are dominantly arc-related igneous masses and associated volcaniclastic rubble (Scholl, Vallier, and Stevenson, 1983a, 1983b; Vallier et al., in press).

The arc massif of the Aleutian Ridge is therefore thought to extend seaward beneath the Aleutian Terrace, terminating structurally in the vicinity of the trench-slope break above a northward underriding accretionary wedge (Figure 6). Although the petroleum prospects of the Aleutian Terrace lie within the sedimentary section of its middle and upper series beds, petroleum fluids produced in the structurally underlying accretionary complex could arrive by leakage through the basement of LS rocks and also could be trapped within them (see von Huene, Fisher, and Bruns, this volume).

The depth to a commonly recorded bottom-simulating reflector, or BSR (Figure 19), which has been interpreted to be the base of a gas-hydrate section (Kvenvolden, this volume), along the seaward side of the Aleutian Terrace implies that the thermal gradient in its upper sedimentary section is low, roughly 22–23°C/km (McCarthy et al., 1984). If extrapolated downward, this gradient is nonetheless sufficiently high to promote the generation of petroleum hydrocarbons in the basal sequences of the terrace's US and MS sedimentary section. Rocks dredged from the southern insular slope of the arc include Oligocene and lower Neogene siltstone of the middle series with an organic carbon content between 0.5 and 1.0%. Potential source beds may therefore exist

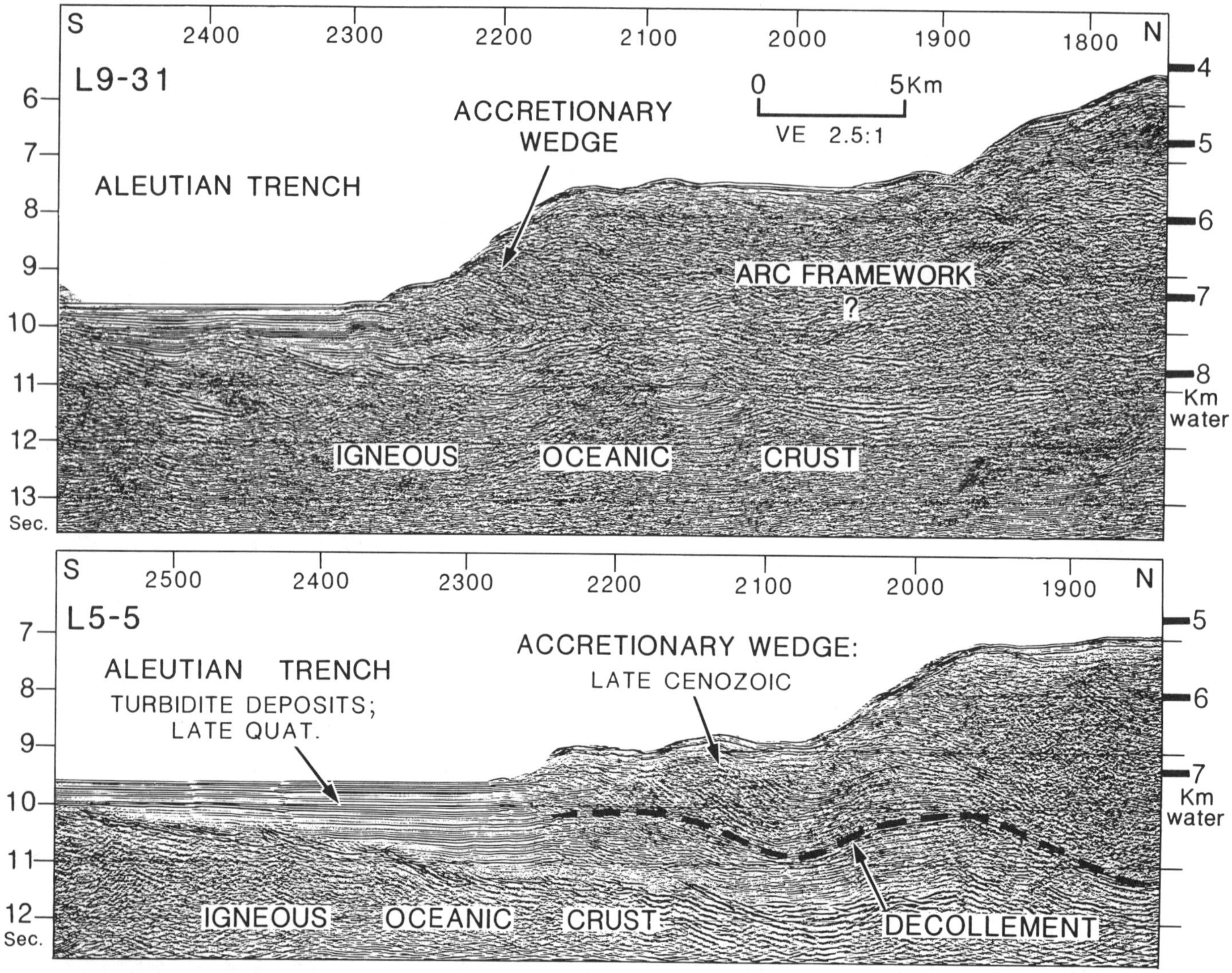

Figure 20. Migrated multichannel sections of lines that cross the accretionary wedge south of the western Aleutian Islands (L9-31; location on Figure 2) and south of Amlia Island (L5-5; location on Figure 3). The wedge is rather small off the western Aleutian Ridge, which lies nearly in strike-slip contact with the obliquely converging Pacific plate (Vallier et al., 1984, and in press). In the Amlia Island area the wedge is roughly 30 km wide (Figure 6) and characteristically exhibits a subsurface decollement. This surface separates offscraped and thrust-stacked trench-floor beds of the upper series from underlying sequences being transported farther landward and possibly subducted beneath the ridge's arc massif of lower series rocks (Figure 6). Structurally interpreted sections of the Aleutian accretionary wedge are presented by McCarthy et al.(1984) and McCarthy and Scholl (1985).

within the terrace's deeply buried sedimentary section (Table 1; Scholl, Vallier, and Stevenson, 1983b).

Downhole sampling of US deposits at DSDP site 186, and surface sampling of older MS beds on the adjacent insular slope, have demonstrated the likely occurrence of reservoir beds in the subsurface sections. However, as cautioned by Stewart (1978), reactive volcanic detritus can be expected to have reduced primary porosity and permeability in the deep subsurface. The seismic sections on Figures 17, 18, 19 outline the forms of structural and stratigraphic features capable of trapping migrating hydrocarbons. Folding of upper and middle series beds along the seaward side of the terrace is thought to be a consequence of late Cenozoic compressional deformation (Figures 6, 18, 19). Thrust faults dip shallowly to the north; and opposite-dipping high-angle reverse faults may border the northern side of Hawley Ridge, which is the largest antiformal uplift that borders the seaward side of the terrace (Figure 3).

Although reconnaissance studies of the Aleutian Terrace in the vicinity of the Near Islands (long. 174°E) and the Adak-Amlia sector of the ridge have located prospective sections for petroleum exploration (Figures 7, 17), little is known about the prospective potential of the 600-km-long terrace sector east of long. 172°W (Figure 2). This region (roughly 30,000 km^2 in areas) is the largest, virtually unexplored area of the Aleutian Ridge. The terrace sector is adjacent to some of the largest of the Aleutian Islands underlain mainly by lower series rocks—source material for terrigenous debris possibly stored in thick basinal sections of US beds and underlying MS strata.

Accretionary or Subduction Complex

The accretionary mass of offscraped trench deposits that underlies the landward slope of the trench has only been examined in reconnaissance fashion, and only adjacent to the ridge's Near Islands and Adak-Amlia sectors (Figures 2, 3). The accretionary complex is a wedge-shaped body that thickens toward the arc. Vallier et al. (in press) describe it as

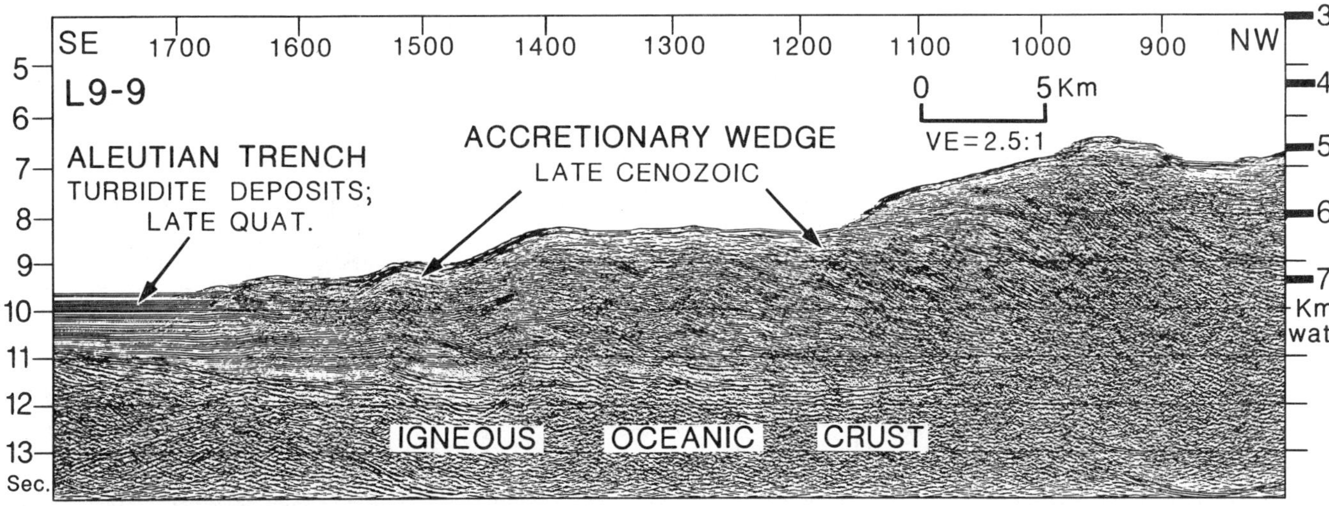

Figure 21. Migrated multichannel section of line L9-9, which obliquely traverses the accretionary wedge south of Adak Island (location on Figure 3). The oblique trend of the section reveals broad closures and the dominance of landward-dipping reflection horizons, which are the traces of bedding surfaces, thrust faults, and abandoned and elevated decollement surfaces. A decollement separating offscraped and underthrusting sedimentary sequences is not well imaged, but the surface is recorded on seismic lines L9-5 and L9-6, which cross line L9-9 and the regional trend of the wedge at high angles (Figure 3; McCarthy et al., 1984; McCarthy and Scholl, 1985). Thrust-thickened trench-floor deposits of mainly Pliocene and Quaternary age (upper series) are thought to form the wedge, which is as much as 6 km thick in the vicinity of the trench-slope break.

a relatively small accretionary complex south of the Near Islands (Figure 20). The small volume of sediment is not an inducement for petroleum exploration. However, McCarthy et al. (1984) stress that the much more massive subduction complex adjacent to the arc's Adak-Amlia section may have a petroleum potential (Figures 20, 21).

The Adak-Amlia sector of the Aleutian subduction complex is roughly 300 km long by 40 km wide, and it thickens arcward to about 6 km beneath the vicinity of the trench-slope break (McCarthy and Scholl, 1985). At this position the complex is thought to pass structurally beneath the arc's bedrock framework (Figure 6). In the Adak-Amlia sector, the volume of the complex, as acoustically resolved, is therefore about 40,000 km^3. If extrapolated along the length of the ridge, the Aleutian subduction complex consists of roughly 250,000 km^3 of thrust-thickened upper series sediment. This accretionary mass has been tectonically added to the arc's bedrock massif, largely during the Pliocene and Quaternary.

McCarthy et al. (1984) note a large variety of stratigraphic and structural traps within the complex (Figures 20, 21). Because the accretionary wedge is actively deforming, traps are subject to structural disruption and the loss of potential petroleum accumulations. Further, relief on the surface of the oceanic basement passing beneath the wedge causes continual tectonic jostling of parts of the wedge (Figures 20, 21; McCarthy and Scholl, 1985). Because the rate of deformation decreases landward in accretionary complexes (Moore, Watkins, and Shipley, 1982; and von Huene, 1984), landward structures should be more stable than those nearer the trench.

The occurrence of a structural decollement near the base of the Aleutian subduction complex testifies to an overpressured lower section, and to the isolated tectonic and compactive dewatering of the overlying package of offscraped trench deposits (Figures 20, 21). General considerations imply that the central body of the accretionary wedge is sufficiently warm to generate hydrocarbons (McCarthy et al., 1984). But the paths of fluid movement and the evolution of the wedge's thermal structure have undoubtedly been greatly complicated by tectonically forced fluid migrations, the stacking and shuffling of thrust slices, the landward thickening of the wedge by underplating or subcretion processes (see discussions by Leggett, Aoki, and Toba, 1985; von Huene, Fisher, and Bruns, this volume; Stevenson and Embley, this volume), and the tectonic "pumping" action of the passage of ocean-crust structural relief beneath the wedge.

Although little is known about the lithologic makeup of the Aleutian subduction complex, regional information implies that it includes pelagic diatomaceous sediment and fine clastic debris, in part or largely contributed by Alaskan drainages (Scholl, Vallier, and Stevenson, 1982). These sequences could include beds with excellent reservoir characteristics. Samples recovered at north Pacific DSDP sites drilled during Legs 18 and 19 imply that relatively organic-rich (about 0.6% TOC) deposits are conveyed tectonically into the subduction complex (McCarthy et al., 1984; von Huene, Fisher, and Bruns, this volume; Stevenson and Embley, this volume). If the subduction process has steadily passed different packages of sediment through the petroleum generation window, then a large amount of organic matter may have been thermally matured here during the past 6–5 Ma, which is the estimated age of the accretionary wedge.

If the technological challenge of drilling and producing in areas of truly deep water is met, the combined sedimentary masses of the tectonically juxtaposed Aleutian Terrace and the accretionary wedge represent one of the largest but least understood targets for petroleum exploration anywhere in Alaska.

ACKNOWLEDGMENTS

In preparing this paper we benefited greatly from a critical reading of the draft manuscript—which spawned fruitful discussions—by our colleagues Alan Cooper and Holly Ryan. Advanced processing of multichannel reflection records were supervised by Holly Ryan and Eric Geist. Dennis Mann managed the on-board collection of these data and tended to the production of the initial stacks. Our associate Gary Mann designed and created the illustrations; his conscientious efforts were equally essential to the completion of this project.

REFERENCES CITED

Allison, R. C., 1973, Marine paleoclimatology and paleoecology of a Pleistocene invertebrate fauna from Amchitka Island, Aleutian Islands, Alaska: Paleogeography, Paleoclimatology, Paleoecology, v. 13, p. 15–48.

Anderson, R. E., 1971, Tectonic setting of Amchitka Island, Alaska: U.S. Geological Survey Open-File Report 474-75, 46 p.

Beck, M. E., Jr., 1976, Discordant paleomagnetic pole positions as evidence of regional shear in the western cordillera of North America: American Journal of Science, v. 276, p. 694–712.

Ben-Avraham, Z., and A. K. Cooper, 1981, Early evolution of the Bering Sea by collision of oceanic rises and north Pacific subduction zones: GSA Bulletin, v. 92, p. 485–495.

Berggren, W. A., D. V. Kent, J. J. Flynn, and J. A. van Couvering, 1985, Cenozoic geochronology: GSA Bulletin, v. 96, p. 1407–1418.

Bogdanov, N. A., and Yu. P. Neprochnov, 1984, Geology of the Bering Sea deep basins, *in* Origin and history of marginal and inland seas: Proceedings of the 27th International Geological Congress, VNU Science Press, v. 23, p. 1–17.

Bogdanov, N. A., V. S. Vishnevskaia, and A. N. Sukhov, 1983, Verkhnemelovye obrazovaniia podovodnogo khrebta Shirshova (Beringnovo More) [Upper Cretaceous formations of the Shirshov Submarine Ridge (Bering Sea)]: Akademi Nauk, SSSR, Doklady, v. 273, p. 1183–1187.

Borsuk, A. M., and A. A. Tsvetkov, 1982, Magmatic associations of the western part of the Aleutian Island Arc: International Geology Review, v. 24, p. 317–327.

Buffington, E. C., 1973, The Aleutian-Kamchatka Trench convergence and investigations of lithospheric plate interaction in the light of modern geotectonic theories: Los Angeles, University of Southern California, unpublished Ph.D. dissertation, 350 p.

Byrne, T., 1979, Late Paleocene demise of the Kula plate: Geology, v. 7, p. 341-344.

Carr, W. J., W. D. Quinlivan, and L. M. Gard Jr., 1970, Age and stratigraphic relations of Amchitka, Banjo Point and Chitka Point Formations, Amchitka Island, Aleutian Islands, Alaska: U.S. Geological Survey Bulletin 1324-A, p. A16–A22.

Carr, W. J., L. M. Gard, G. D. Barth, and D. L. Healey, 1971, Earth-science studies of a nuclear test area in the western Aleutian Islands, Alaska: An interim summation of results: GSA Bulletin, v. 82, p. 669–706.

Chase, T. E., H. W. Menard, and J. Mammerickx, 1970, Bathymetry of the north Pacific: Institute of Marine Resources, Scripps Institution of Oceanography, University of California at San Diego, scale approximately 1:600,000.

Citron, G. P., R. W. Kay, S. M. Kay, L. W. Snee, and J. F. Sutter, 1980, Tectonic significance of early Oligocene plutonism on Adak Island, central Aleutian Islands, Alaska: Geology, v. 8, p. 375–379.

Clague, D. A., 1981, Linear island and seamount chains, a seismic ridges and intraplate volcanism; results from DSDP: SEPM Special Publication 32, p. 7–22.

Clague, D. A., and G. B. Dalrymple, in press, The Hawaiian-Emperor volcanic chain, Part 1: geologic evolution, *in* R. W. Decker, T. L. Wright, and P. B. Stauffer, eds., Hawaiian volcanism: U.S. Geological Survey Professional Paper 1350.

Coats, R. R., 1950, Volcanic activity in the Aleutian Arc: U.S. Geological Survey Bulletin 974-B, p. 35–47.

Coats, R. R., 1962, magma type and crustal structure in the Aleutian Arc, *in*, G. A. MacDonald and H. Kuno, eds., The crust of the Pacific Basin: American Geophysical Union Monograph 6, p. 92–109.

Coats, R. R., 1956a, Reconnaissance geology of some western Aleutian Islands, Alaska: U.S. Geological Survey Bulletin 1028-E, p. 83–100.

Coats, R. R., 1956b, Geology of northern Adak Island, Alaska: U.S. Geological Survey Bulletin 1028-C, p. 45–67.

Conrad, W. K., and R. W. Kay, 1984, Ultramafic and mafic inclusions from Adak Island: Crystallization history and implications for the nature of primary magmas and crustal evolution in the Aleutian Arc: Journal of Petrology, v. 25, p. 88–25.

Cooper, A. K., M. S. Marlow, and Z. Ben-Avraham, 1981, Multichannel seismic evidence bearing on the origin of Bowers Ridge, Bering Sea: GSA Bulletin, v. 92, p. 474–484.

Cooper, A. K., M. S. Marlow, and D. W. Scholl, 1976, Mesozoic magnetic lineations in the Bering Sea marginal basin: Journal of Geophysical Research, v. 81, p. 1916–1934.

Cooper, A. K., M. S. Marlow, and T. L. Vallier, 1984, Summary geologic report for the St. George outer continental shelf (OCS) planning area, Bering Sea, Alaska: U.S. Geological Survey Open-File Report 84-418, 81 p.

Cooper, A. K., D. W. Scholl, and M. S. Marlow, 1976, Plate tectonic model for the evolution of the Bering Sea Basin: GSA Bulletin, v. 87, p. 1119–1126.

Cooper, A. K., D. W. Scholl, and M. S. Marlow, 1987, Structural framework, sedimentary sequences and hydrocarbon potential of the Aleutian and Bowers basins, Bering Sea, this volume.

Cooper, A. K., D. W. Scholl, M. S. Marlow, J. R. Childs, G. D. Redden, K. A. Kvenvolden, and A. J. Stevenson, 1979, Hydrocarbon potential of Aleutian Basin, Bering Sea: AAPG Bulletin, v. 63, p. 2070–2087.

Cormier, V. F., 1975, Tectonics near the junction of the Aleutian and Kuril-Kamchatka Arcs and a mechanism for middle Tertiary magmatism in the Kamchatka Basin: GSA Bulletin, v. 86, p. 443–453.

Cox, A., and D. C. Engebretson, 1985, Change in motion of Pacific plate at 5 Ma: Nature, v. 313, p. 472–474.

Crain, W. E., W. E. Mero, and D. Patterson, 1985, Geology of the Point Arguello discovery: AAPG Bulletin, v. 69, p. 537–545.

Creager, J. S., D. W. Scholl, et al., 1973, Initial reports of the Deep Sea Drilling Project, Vol. 19: Washington, D.C., U.S. Government Printing Office, 913 p.

Davies, J. N., and L. House, 1979, Aleutian subduction zone seismicity, volcano-trench separation, and their relation to great thrust-type earthquakes: Journal of Geophysical Research, v. 84, p. 4583–4591.

DeLong, S. E., and F. W. McDowell, 1975, K-Ar ages from the Near Islands, western Aleutian Islands, Alaska: Indication of a mid-Oligocene thermal event: Geology, v. 3, p. 691–694.

DeLong, S. E., P. J. Fox, and F. W. McDowell, 1978, Subduction of the Kula Ridge at the Aleutian Trench: GSA Bulletin, v. 89, p.83–95.

Engdahl, E. R., 1971, Explosion effects and earthquakes in the Amchitka Island Region: Science, v. 173, p. 1232–1235.

Engdahl, E. R., 1977, Seismicity and plate subduction in the central Aleutians, *in* M. Talwani and W. C. Pitman, III, Island arcs, deep sea trenches, and back-arc basins: American Geophysical Union, Maurice Ewing Series, v. 1, p. 259–271.

Engdahl, E. R., and S. Billington, 1986, Focal depth determination of central Aleutian earthquakes: Seismological Society of America Bulletin, v. 76, p. 77-93.

Engebretson, D. C., A. Cox, and R. G. Gordon, 1984, Relative motions between ocean plates of the Pacific Basin: Journal of Geophysical Research, v. 89, p. 10,291–10,310.

Engebretson, D. C., A. Cox, and R. G. Gordon, 1985, Relative motions between oceanic and continental plates in the Pacific Basin: GSA, Special Paper 206, 59 p.

Espinosa, A. F., 1984, Seismicity of Alaska and the Aleutian Islands: U.S. Geological Survey Open-File Report OFR-84-855.

Fraser, G. D., and H. F. Barnett, 1959, Geology of the Delarof and westernmost Andreanof Islands, Alaska: U.S. Geological Survey Bulletin 1028-I, p. 211–248.

Frohlich, C., S. Billington, E. R. Engdahl, and A. Malahoff, 1982, Detection and location of earthquakes in the central Aleutian subduction zone using island and ocean bottom seismograph stations: Journal of Geophysical Research, v. 87, p. 6853–6864.

Gates, O., and W. Gibson, 1956, Interpretation of the configuration of the Aleutian Ridge: GSA Bulletin, v. 67, p. 127–146.

Gates, O., G. D. Fraser, and G. L. Snyder, 1954, Preliminary report on the geology of the Aleutian Islands: Science, v. 119, p. 446–447.

Gates, O., H. A. Powers, R. E. Wilcox, and J. P. Schafer, 1971, Geology of the Near Islands, Alaska: U.S. Geological Survey Bulletin, 1028-U, p. 709–822.

Geist, E., J. R. Childs, and D. W. Scholl, 1985, Extensional structures on the Aleutian Ridge, Alaska: EOS, (American Geophysical Union Transactions), v. 66, p. 1072.

Gibson, W., and H. Nichols, 1953, Configuration of the Aleutian Ridge Rat Islands-Semisopochnoi Island to west of Buldir Island: GSA Bulletin, v. 64, p. 1173–1186.

Gnibidenko, G. S., T. G. Bykova, O. V. Veselov, V. M. Vorobyev, K. Chung Ung, and T. Z. Tarakanov, 1980, Tektonika Kuril-Kamchatskogo glubokovodnogo zheloba (Tectonics of the Kuril-Kamchatka deep-sea trench): Nauka, Moscow, 179 p.

Grow, J. A., 1973a, Crustal and upper mantle structure of the central Aleutian Arc: GSA Bulletin, v. 84, p. 2169–2192.

Grow, J. A., 1973b, Implications of deep sea drilling, site 186 and 187, on island arc structure, *in* J. S. Creager, D. W. Scholl, et al., eds., Initial reports of the Deep Sea Drilling Project, Vol. 19: Washington, D. C., U.S. Government Printing Office, p. 799–803.

Grow, J. A., and T. Atwater, 1970, Mid-Tertiary tectonic transition in the Aleutian Arc: GSA Bulletin, v. 81, p. 3715–3722.

Harbert, W. P., and A. Cox, 1984, Preliminary paleomagnetic results from Umnak Island, Aleutian Ridge (abs.): GSA Abstracts with Programs, v. 16, p. 288.

Harbert, W. P., A. Cox, and H. McLean, 1984, Paleomagnetism of Starr Point and Driftwood Bay, Umnak Island, Alaska: EOS (American Geophysical Union Transactions), v. 65, p. 869.

Harbert, W. P., A. Cox, and H. McLean, 1985, Paleomagnetism of Eocene sediments of Amlia and Adak Islands, Aleutian Islands, Alaska: EOS (American Geophysical Union Transactions), v. 66, p. 864.

Harbert, W. P., D. W. Scholl, D. Mann, T. L. Vallier, and A. J. Stevenson, 1982, Geologic and tectonic history of Atka Basin, Aleutian Ridge: EOS (American Geophysical Union Transactions), v. 63, p. 1115.

Hawkins, J. W., S. H. Bloomer, C. A. Evans, and J. T. Melchior, 1984, Evolution of intra-oceanic arc-trench systems: Tectonophysics, v. 10, p. 175–205.

Hayes, D. E., and M. Ewing, 1970, Pacific boundary structure, *in* A. Maxwell, ed., The sea, Vol. 4; The earth beneath the sea—Concepts, part 2, Regional observations: New York, Wiley Interscience, p. 29–72.

Hein, J. R., and H. McLean, 1979, Paleogene sedimentary and volcanogenic rocks from Adak Island, central Aleutian Islands, Alaska: U.S. Geological Survey Professional Paper 1126-E, 16 p.

Hein, J. R., H. McLean, and T. Vallier, 1984, Reconnaissance geology of southern Atka Island, Aleutian Islands, Alaska: U.S. Geological Survey Bulletin 1609, 19 p.

Hein, J. R., D. W. Scholl, J. A. Barron, M. G. Jones, and J. Miller, 1978, Diagenesis of late Cenozoic diatomaceous deposits and formation of the bottom simulating reflector in the southern Bering Sea: Sedimentology, v. 25, p. 155–181.

House, L. S., and K. H. Jacob, 1983, Earthquake, plate subduction, and stress reversals in the eastern Aleutian Arc: Journal of Geophysical Research, v. 88, p. 9347–9373.

Isaacs, C. M., and R. E. Garrison, 1983, Petroleum generation and occurrence in the Miocene Monterey Formation, California: SEPM, Pacific Section, Special Publication 33, p. 228.

Jacob, K. H., 1972, Global implications of anomalous seismic P traveltimes from the nuclear explosion Longshot: Journal of Geophysical Research, v. 77, 2556–2573.

Jacob, K. H., K. Nakamura, and J. N. Davis, 1977, Trench-volcano gap along the Alaska-Aleutian Arc, *in* M. Talwani and W. C. Pitman, III, eds., Island arcs, deep sea trenches, and back-arc basins: American Geophysical Union Maurice Ewing Series, v. 1, p. 243–258.

Kay, R. W., 1977, Geochemical constraints on the origin of Aleutian magmas, *in* M. Talwani and W. C. Pitman, III, eds., Island arcs, deep sea trenches, and back-arc basins: American Geophysical Union Maurice Ewing Series, v. 1, p. 229–242.

Kay, R. W., 1980, Volcanic arc magmas; implications of a melting-mixing model for element recycling in the crust-upper mantle system: Journal of Geology, v. 88, p. 497–522.

Kay, S. M., 1983, Metamorphism in the Aleutian Arc; the Finger Bay pluton, Adak, Alaska: Canadian Mineralogist, v. 21, p. 665–681.

Kay, S. M., and R. W. Kay, 1982, Tectonic controls on tholeiitic and calc-alkaline magmatism in the Aleutian Arc: Journal of Geophysical Research, v. 87, p. 4051–4072.

Kay, S. M., and R. W. Kay, 1985, Role of crustal cumulates and the oceanic crust in the formation of the lower crust of the Aleutian Arc: Geology, v. 13, p. 461–464.

Kisslinger, C., and E. R. Engdahl, 1974, A test of the Semyenov prediction technique in the central Aleutian Islands: Tectonophysics, v. 23, p. 237–246.

Kvenvolden, K. A., 1987, Gas hydrate offshore Alaska and western continental United States, this volume.

LaForge, R., and E. R. Engdahl, 1979, Tectonic implications of seismicity in the Adak Canyon region, central Aleutians: Seismological Society of America Bulletin, v. 69, p. 1515–1532.

Langseth, M. G., M. A. Hobart, and K. Horai, 1980, Heat flow in the Bering Sea: Journal of Geophysical Research, v. 85, p. 3740–3750.

Lankford, S. M., and J. M. Hill, 1979, Stratigraphy and depositional environment of the Dutch Harbor member of the Unalaska Formation, Unalaska Island, Alaska: U.S. Geological Survey Bulletin 1457-B, 14 p.

Leggett, J., Y. Aoki, and T. Toba, 1985, Transition from frontal accretion to underplating in a part of the Nankai Trough accretionary complex off Shikoku (SW Japan) and extensional features on the lower trench slope: Marine and Petroleum Geology, v. 2, p. 131–141.

Marlow, M. S., and A. K. Cooper, 1983; Wandering terranes in southern Alaska; the Aleutia microplate and implications for the Bering Sea: Journal of Geophysical Research, v. 88, p. 3439–3446.

Marlow, M. S., D. W. Scholl, E. C. Buffington, and T. R. Alpha, 1973, Tectonic history of the central Aleutian Arc: GSA Bulletin, v. 84, p. 1555–1574.

Marlow, M. S., D. W. Scholl, E. C. Buffington, R. E. Boyce, T. R. Alpha, P. B. Smith, and C. J. Shipek, 1970, Buldir Depression—Late Tertiary graben on the Aleutian Ridge, Alaska: Marine Geology, v. 8, p. 85–108.

Marsh, B. D., 1976, Some Aleutian andesites; their nature and source: Journal of Geology, v. 84, p. 27–45.

Marsh, B. D., 1980, Geology and petrology of northern Atka, Aleutian Islands, Alaska (abs.): GSA Abstracts with Programs, v. 12, p. 476.

Marsh, B. D., in press, Age, characteristics and significance of Aleutian Arc volcanic rocks, *in* G. Plafker and D. L. Jones, eds., Cordilleran orogen, Alaska: GSA Decade on North American Geology.

McCarthy, J., and D. W. Scholl, 1985, Mechanisms of subduction accretion along the central Aleutian Trench: GSA Bulletin, v. 96, p. 691–701.

McCarthy, J., A. J. Stevenson, D. W. Scholl, and T. L. Vallier, 1984, Speculations on the petroleum geology of the accretionary body: An example from the central Aleutians: Marine and Petroleum Geology, v. 1, p. 151–167.

McCulloch, M. T., and M. R. Perfit, 1981, $^{143}Nd/^{144}Nd$, $^{87}Sr/^{86}Sr$ and trace element constraints on the petrogenesis of Aleutian Island Arc magmas: Earth and Planetary Science Letters, v. 56, p. 167–179.

McLean, H., and J. R. Hein, 1984, Paleogene geology and chronology of southwestern Umnak Island, Aleutian Islands, Alaska: Canadian Journal of Earth Sciences, v. 21, p. 171–180.

McLean, H., W. P. Harbert, and S. P. Douglass, 1984, Flysch, pillow lava, and diorite sills: A well-exposed Tertiary assemblage on Unalaska Island is widespread in the Aleutian Islands, Alaska (abs.): GSA Abstracts with Programs, v. 16, p. 590.

McLean, H., J. R. Hein, and T. L. Vallier, 1983, Reconnaissance geology of Amlia Island, Aleutian Islands, Alaska: GSA Bulletin, v. 94, p. 1020–1027.

Minster, B. J., and T. H. Jordan, 1978, Present-day plate motions: Journal of Geophysical Research, v. 83, p. 5331–5354.

Moore, J. C., J. S. Watkins, and T. H. Shipley, 1982, Summary of accretionary processes, Deep Sea Drilling Project Leg 66: Offscraping, underplating, and deformation of the slope apron, *in* J. S. Watkins, J. C. Moore, et al., Initial reports of the Deep Sea Drilling Project, Vol. 66: Washington D.C., U.S. Government Printing Office, p. 825–836.

Murray, H. W., 1945, Profiles of the Aleutian Trench: GSA Bulletin, v. 56, p. 757–782.

Natland, J. H., and J. Tarney, 1981, Petrologic evolution of the Mariana Arc and back-arc basin system—A synthesis of drilling results in the south Philippine Sea, *in* D. Hussong, S. Uyeda, et al., eds., Initial reports of the Deep Sea Drilling Project, Vol. 60: Washington D.C., U.S. Government Printing Office, p. 877–908.

Newberry, J. K., 1983, Seismicity and tectonics of the far western Aleutian Islands: East Lansing, Michigan State University, unpublished Master's thesis, 93 p.

Nichols, H., and R. B. Perry, 1966, Bathymetry of the Aleutian Arc, Alaska: U.S. Coast and Geodetic Survey Monograph 3, maps, scale 1:400,000.

Panuska, B. C., 1980, Stratigraphy and sedimentary petrology of the Kiska Harbor Formation and its relationship to the Near Islands-Amchitka lineament, Aleutian Islands: Fairbanks, University of Alaska unpublished Master's thesis, 90 p.

Patton, W. W., Jr., M. A. Lanphere, T. P. Miller, and R. A. Scott, 1976, Age and tectonic significance of volcanic rocks on St. Matthew Island, Bering Sea, Alaska: U.S. Geological Survey Journal of Research, v. 4, p. 67–73.

Perry, R. B., and H. Nichols, 1965, Bathymetry of Adak Canyon, Aleutian Arc, Alaska: GSA Bulletin, v. 76, p. 365–370.

Perfit, M. R., and J. R. Lawrence, 1979, Oxygen isotopic evidence for meteoric water interaction with the Captains Bay pluton, Aleutian Islands: Earth and Planetary Science Letters, v. 45, p. 16–22.

Perfit, M. R., B. Hrueckner, H. Lawrence, and R. W. Kay, 1980, Trace element and isotopic variations in a zoned pluton and associated rocks, Unalaska Island, Alaska: A model for fractionation in the Aleutian calc-alkaline suite: Contributions to Mineralogy and Petrology, v. 73, p. 69–87.

Pickthorn, L. B. G., T. L. Vallier, and D. W. Scholl, 1984, Geochronology of igneous rocks from the Aleutian Island Arc, Alaska (abs.): GSA Abstracts with Programs, v. 16, p. 328–329.

Rea, D. K., and J. M. Dixon, 1983, Late Cretaceous and Paleogene tectonic evolution of the north Pacific Ocean: Earth and Planetary Science Letters, v. 4, p. 67–73.

Rubenstone, J. L., 1984, Geology and geochemistry of early Tertiary submarine volcanic rocks of the Aleutian Islands, and their bearing on the development of the Aleutian Island Arc: Ithaca, New York, Cornell University unpublished Ph.D. thesis, 349 p.

Ryan, H. E., and D. W. Scholl, 1985, Formation of outer arc structural highs, central Aleutian Ridge: EOS (American Geophysical Union Transactions), v. 66, p. 1105.

Schmidt, O. A., 1978, Tektonika Komandorskikh Ostrovov i struktura Aleutskoi Griady (Tectonics of the Komandorsky Islands and structure of the Aleutian Ridge): Moscow, Akademiya Nauk Geological Institut. Trudy, v. 320, Izdatel'stvo Nauk, 100 p. (In Russian.)

Schmidt, O. A., M. Ya. Serova, and L. M. Dolmatova, 1973, Stratigraphy and paleontological features of the volcanic rock series on the Komandorsky Islands: Akademiya Nauk SSSR Izvestiya, Seriya Geologicheskaya, no. 11, p. 77–78.

Scholl, D. W., and A. K. Cooper, 1978, VAMPs: Possible hydrocarbon bearing structures in the Bering Sea Basin: AAPG Bulletin, v. 62, p. 2481–2488.

Scholl, D. W., and J. S. Creager, 1973, Geologic synthesis of Leg 19 (DSDP) results; far north Pacific, Aleutian Ridge, and Bering Sea, *in* J. S. Creager, D. W. Scholl, et al., Initial reports of the Deep Sea Drilling Project, Vol. 19: Washington D.C., U.S. Government Printing Office, p. 897–913.

Scholl, D. W., E. C. Buffington, and M. S. Marlow, 1975, Plate tectonics and the structural evolution of the Aleutian-Bering Sea region, *in* H. B. Forbes, ed., Contributions to the geology of the Bering Sea Basin and adjacent regions: GSA Special Paper 131, p. 1–31.

Scholl, D. W., H. G. Greene, and M. S. Marlow, 1970, Eocene age of the Adak Paleozoic (?) rocks, Aleutian Islands, Alaska: GSA Bulletin, v. 81, p. 3583–3592.

Scholl, D. W., M. S. Marlow, and E. C. Buffington, 1975, Summit basins of the Aleutian Ridge, north Pacific: AAPG Bulletin, v. 59, p. 799–816.

Scholl, D. W., T. L. Vallier, and A. J. Stevenson, 1982, Sedimentation and deformation in the Amlia fracture zone sector of the Aleutian Trench: Marine Geology, v. 48, p.105–134.

Scholl, D. W., T. L. Vallier, and A. J. Stevenson, 1983a, Arc, forearc, and trench sedimentation and tectonics; Amlia corridor of the Aleutian Ridge, *in* J. S. Watkins and C. L. Drake, eds., Studies in continental margin geology: AAPG Memoir 34, p. 413–439.

Scholl, D. W., T. L. Vallier, and A. J. Stevenson, 1983b, Geologic evolution of the Aleutian Ridge—Implications for petroleum resources: Journal of the Alaska Geological Society, v. 3, p. 33–46.

Scholl, D. W., T. L. Vallier, and A. J. Stevenson, 1986, Terrane accretion, production, and continental growth—A perspective based on the origin and tectonic fate of the Aleutian-Bering Sea region: Geology, v. 14, p. 68–71.

Scholl, D. W., T. R. Alpha, M. S. Marlow, and E. C. Buffington, 1974, Base map of the Aleutian-Bering Sea region: U.S. Geological Survey Miscellaneous Investigations Series Map I-879, scale 1:2,500,000.

Scholl, D. W., M. S. Marlow, N. S. MacLeod, and E. C. Buffington, 1976, Episodic Aleutian Ridge igneous activity: Implications of Miocene and younger submarine volcanism west of Buldir Island: GSA Bulletin, v. 87, p. 547–554.

Schumacher, G. M., 1976, Bathymetric map of the Aleutian Trench and Bering Sea: U.S. Geological Survey Open-File Report 76-821, 2 sheets, scale 1:2,500,000.

Shor, G. G., Jr., 1964, Structure of the Bering Sea and the Aleutian Ridge: Marine Geology, v. 1, p. 213–219.

Spence, W., 1977, The Aleutian Arc, tectonic blocks, episodic subduction, strain diffusion, and magma generation: Journal of Geophysical Research, v. 82, p. 213–230.

Stevenson, A. J., and R. Embley, 1987, Deep-sea fan bodies, terrigenous turbidite sedimentation, and petroleum geology, Gulf of Alaska, this volume.

Stewart, R. J., 1978, Neogene volcaniclastic sediments from Atka Basin, Aleutian Ridge: AAPG Bulletin, v. 62, p. 87–97.

Stone, D. B., in press, Bering Sea, Aleutian Arc, Alaska, *in* A. E. M. Nairn, F. G. Stehli, and S. Uyeda, eds., Ocean basins and margins, Vol. 7b, Pacific: New York, Plenum Press.

Swanson, S. E., and J. D. Blum, 1984, Evolution of volcanic centers in the oceanic-continental transition of the Aleutian Arc (abs.): GSA Abstracts with Programs, v. 16, p. 671.

U.S. Department of Commerce, 1973, NOS Seamap Series No. 16,648-14B, Pacific Ocean: Washington, D.C., National Oceano-

graphic and Atmospheric Administration, scale 1:1,000,000.
Vail, P. R., and J. Hardenbol, 1979, Sea level changes during the Tertiary: Oceanus, v. 22, p. 71–79.
Vail, P. R., R. M. Mitchum Jr., and S. Thompson, III, 1977, Seismic stratigraphy and global changes in sea level, part 4, Global cycles of relative changes in sea level, *in* C. E. Payton, ed., Seismic stratigraphy—Applications to hydrocarbon exploration: AAPG Memoir 26, p. 83–97.
Vallier, T. L., J. McCarthy, D. W. Scholl, A. Stevenson, and R. O'Connor, 1984, Offshore structures in the Near Islands region, western Aleutian Island Arc, Alaska (abs.): GSA Abstracts with Programs, v. 16, p. 338.
Vallier, T. L., D. W. Scholl, M. A. Fisher, R. von Huene, T. R. Bruns, and A. J. Stevenson, in press, Geological framework of the Aleutian Arc, *in* G. Plafker and D. L. Jones, eds., Cordilleran orogen, Alaska: GSA, Decade on North American Geology.
von Huene, R., 1984, Observed strain and stress gradient across some forearc areas of modern convergent margins, *in* Origin and history of marginal and inland seas: Proceedings of the 27th International Geological Congress, VNU Science Press, v. 23, p. 155–188.
von Huene, R., M. A. Fisher, and T. R. Bruns, 1987, Geology and evolution of the Kodiak margin, Gulf of Alaska, this volume.
von Huene, R., W. J. Carr, D. McManus, and M. Holmes, 1971, Marine geophysical study around Amchitka Island, Alaska: Springfield, Virginia, National Technical Information Service, U.S. Geological Survey Report 474-74, 31 p.
Wallace, W. K., and D. C. Engebretson, 1984, Relationships between plate motions and Late Cretaceous to Paleocene magmatism in southwestern Alaska: Tectonics, v. 3, p. 295–315.
Wells, R. E., and R. S. Coe, 1985, Paleomagnetism and geology of Eocene volcanic rocks of southwest Washington—Implications for mechanisms of tectonic rotation: Journal of Geophysical Research, v. 90, p. 1925–1947.
Wilcox, R. E., 1959, Igneous rocks of the Near Islands, Aleutian Islands, Alaska: Mexico City, 20th International Geological Congress Proceedings, section 11-A, Petrologia y Mineralogia, p. 365–378.
Wilson, F. H., 1981, Map and table showing radiometric ages of rocks in the Aleutian Islands and the Alaska Peninsula: U.S. Geological Survey Open-File Report 81-471, 24 p.
Woods, M. T., and G. F. Davies, 1982, Late Cretaceous genesis of the Kula plate: Earth and Planetary Science Letters, v. 58, p. 161–166.
Yorath, C. J., 1987, Petroleum geology of the Canadian Pacific continental margin, this volume.

CHAPTER EIGHT

GEOLOGY AND PETROLEUM POTENTIAL OF THE SHUMAGIN MARGIN, ALASKA

Terry R. Bruns
Roland von Huene
Raymond C. Culotta

U.S. Geological Survey
Menlo Park, California

Stephen D. Lewis
John W. Ladd

Lamont-Doherty Geological Observatory of Columbia University
Palisades, New York

ABSTRACT

Interpretations of multichannel seismic-reflection data indicate that the Shumagin continental margin seaward of the Border Ranges fault is underlain by two major seismic sequences, separated by an erosional unconformity beneath the shelf and by the time-correlative conformity seaward. Rocks above the unconformity are probably late Miocene and younger. Rocks below the unconformity can be as young as middle Miocene beneath the outer shelf and slope, seaward of a paleoshelf break, but beneath the shelf are largely Late Cretaceous turbidites of the Shumagin Formation and Paleocene granodiorite.

Late Miocene and younger structures of the Shumagin margin are created by extensional tectonics; these structures include the Shumagin, Sanak, and Unimak basins, and Unimak Ridge, a mid-slope structural high. The basins and Unimak

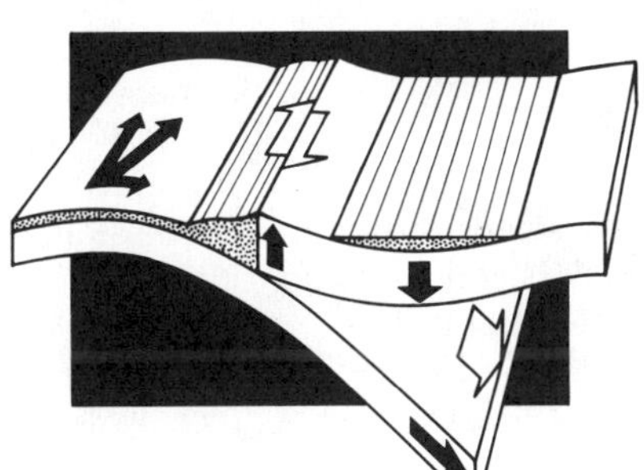

Ridge and its now-buried northeastward structural continuation have developed as a result of block faulting along landward-dipping growth faults. Strata in Sanak and Unimak basins were deposited on the subsiding outer shelf and slope, and were trapped behind the mid-slope structural highs. Sanak and Unimak basins are also in part bounded by northwest-trending extensional faults that parallel both the Late Cretaceous and early Tertiary north Pacific (Beringian) margin, and transverse tectonic and seismic-gap boundaries that segment the fore arc. These faults may have developed during strike-slip faulting and extension along the southeastward continuation of the Beringian margin.

The most promising areas of the Shumagin margin for petroleum potential are Sanak and Unimak basins, which contain strata at least 7 and 4.5 km thick, respectively, and beneath the outer shelf and slope. Paleogene source rocks like those on the adjacent Alaska Peninsula may be preserved offshore, seaward of the inferred paleoshelf break. Reservoir rocks might have formed from granitic-rich erosional products derived during Oligocene and Miocene erosion of the shelf plutons.

INTRODUCTION

This report summarizes the geology and resource potential of the Shumagin margin, Alaska, which is the 600–km-long continental shelf and slope between long. 156° and 165°W lying between the Semidi Islands and Unimak Pass (Figure 1). The geology of the Shumagin Shelf islands and of the adjacent Alaska Peninsula and Kodiak Shelf has been described in previous reports (Burk, 1965; Detterman and Hartsock, 1966; Moore, 1972, 1973a, 1973b, 1974, 1979; von Huene et al., 1976, 1979, 1983, 1985; Connelly et al., 1977; McLean, 1977, 1979; McLean, Englehardt, and Howell, 1978; von Huene, 1979; von Huene, Moore, and Moore, 1979; Moore and Connelly, 1979; Moore et al., 1983; Fisher, 1979, 1980, 1981; Moore and Allwardt, 1980; Fisher and von Huene, 1980, 1982; Fisher and Holmes, 1981; Fisher, Bruns, and von Huene, 1981; Detterman, Yount, and Case, 1981; Detterman et al., 1981, 1983; and von Huene, Fisher, and Bruns, this volume). However, geophysical and geological studies of the Shumagin shelf and slope are limited. Sedimentary basins of the Shumagin margin were partly described by Bruns and von Huene (1977), based on about 500 km of 24-fold multichannel seismic-reflection data (Bruns and Bayer, 1977). These data were augmented in 1981 and 1982 by about 4000 km of 24-fold seismic-reflection data, and by concurrently recorded gravity and magnetic data and sonobuoy-refraction profiles acquired from the U.S. Geological Survey research vessel *R/V S.P. Lee* (Figure 2). Offshore rock samples were dredged at three sites along the continental margin in 1979. Preliminary results of these geophysical and geological data were reported by Bruns et al. (1985).

GEOLOGIC FRAMEWORK

The Alaska Peninsula and the adjacent shelf islands—from east to west, the Semidi, Shumagin, and Sanak Islands—are composed of dominantly northeast–southwest-striking lithologic units ranging in age from Permian to Holocene (Figures 1, 3; Burk, 1965; Detterman, Yount, and Case, 1981; Detterman et al., 1981, 1983; Detterman, 1985). Exposed rocks of the Alaska Peninsula and the inner Shumagin Islands are mainly island-arc-related volcanic or plutonic rocks and shallow-marine and continental deposits rich in volcanic or plutonic detritus. In marked contrast, rocks of the Semidi, outer Shumagin, and Sanak Islands are deep-water Late Cretaceous volcaniclastic flysch assemblages of the Shumagin Formation, which are intruded by Paleocene and early Eocene granodiorites.

Although the contact between the Alaska Peninsula rocks and the Shumagin Formation is nowhere exposed, the regional geology leads us to believe that the contact is the southwest extension of the Border Ranges fault (Fisher, 1981; Fisher and von Huene, 1984). This fault has been traced by geophysical data southwestward from Kodiak Island to near the Semidi Islands. We infer that it continues southwestward between the inner Shumagin Islands, which are underlain by rocks related to those on the Alaska Peninsula, and the outer Shumagin Islands, which are underlain by the Shumagin Formation (Figure 1). Alternatively, Wilson, Case, and Detterman (1985) note that the fault, or splays of it, could also be inland along the south shore of the Alaska Peninsula, in the Port Moller area.

Terranes

The Border Ranges fault divides the Shumagin margin into two discrete, fault-bounded tectonostratigraphic terranes, the Peninsular and Chugach-Prince William (composite) terranes. Each terrane is characterized by rock units and a geologic history that are distinct from those of neighboring terranes (Jones et al., 1981). The geology and boundaries of the terranes are best exposed on the Alaska Peninsula and in the adjacent Kodiak Island region, but the boundaries presumably extend along strike into the Shumagin margin.

The Peninsular terrane (the Alaska Peninsula terrane of Wilson, Detterman, and Case, 1985) lies northwest of the Border Ranges fault and includes the Alaska Peninsula, the inner Shumagin Islands, and the northwest side of Kodiak Island.

The Chugach and Prince William terranes include the Cretaceous Kodiak and Shumagin Formations, which underlie

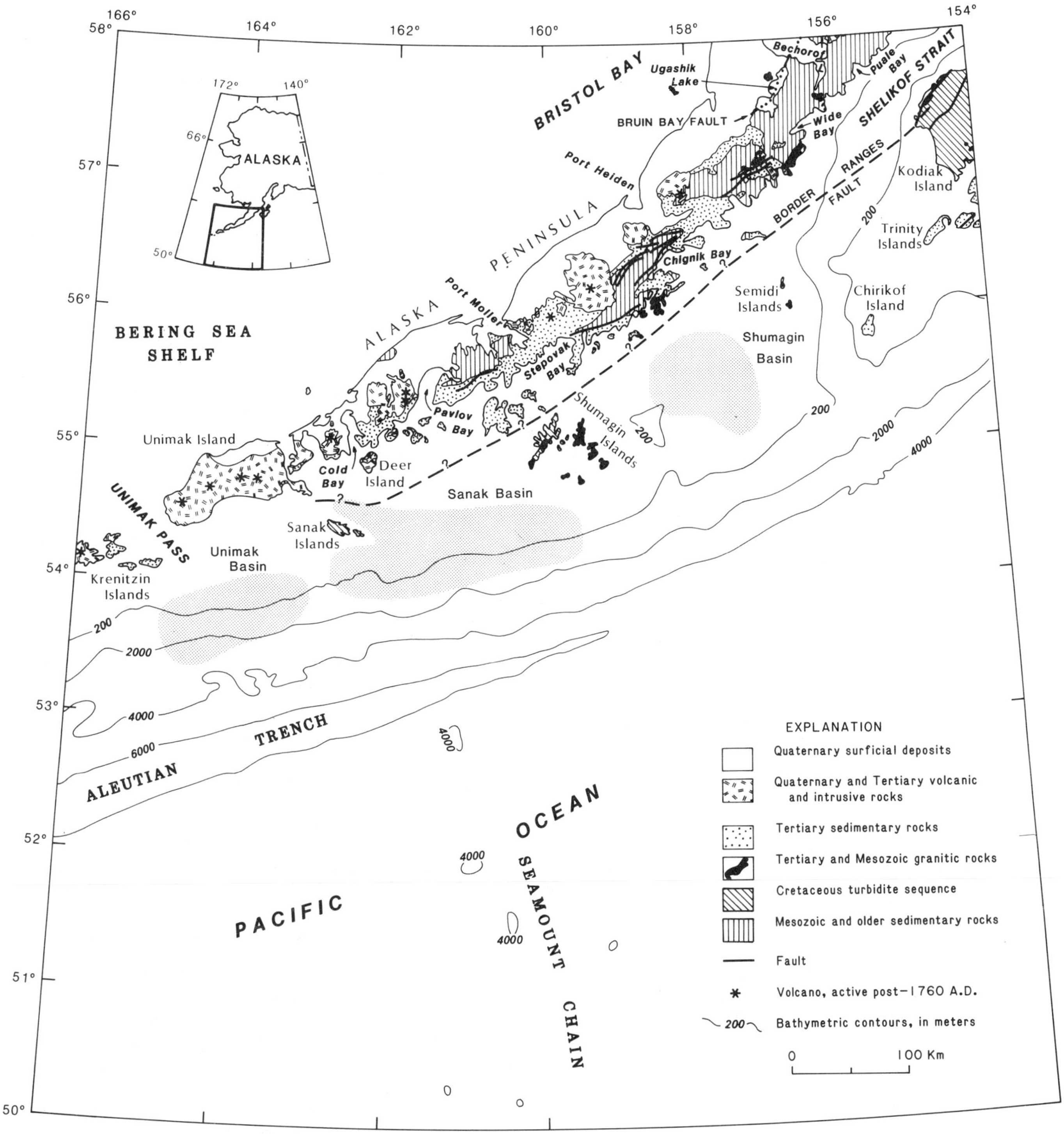

Figure 1. Index map and simplified geology of the Shumagin region. From Beikman (1980). Stippled pattern indicates areal extent of offshore basins.

much of the Kodiak, Semidi, outer Shumagin, and Sanak Islands (Chugach terrane), and the early Tertiary rocks of Kodiak Island (Prince William terrane). The early Tertiary rocks do not crop out on the Shumagin Shelf islands seaward of the inferred position of the Border Ranges fault, but almost certainly underlie the outer shelf and slope. On Kodiak Island, the Chugach and Prince William terranes are separated by the Contact fault. However, because this boundary cannot presently be traced by geological or geophysical means beneath the Shumagin margin, we combine these terranes into the Chugach-Prince William (composite) terrane. This composite terrane probably extends seaward to the accretionary wedge at the base of the continental slope.

The Pacific plate is underthrusting the Shumagin margin along the Aleutian Trench. The igneous ocean crust beneath the trench in the study area is of late Paleocene and early Eocene age, based on the dating of magnetic anomalies

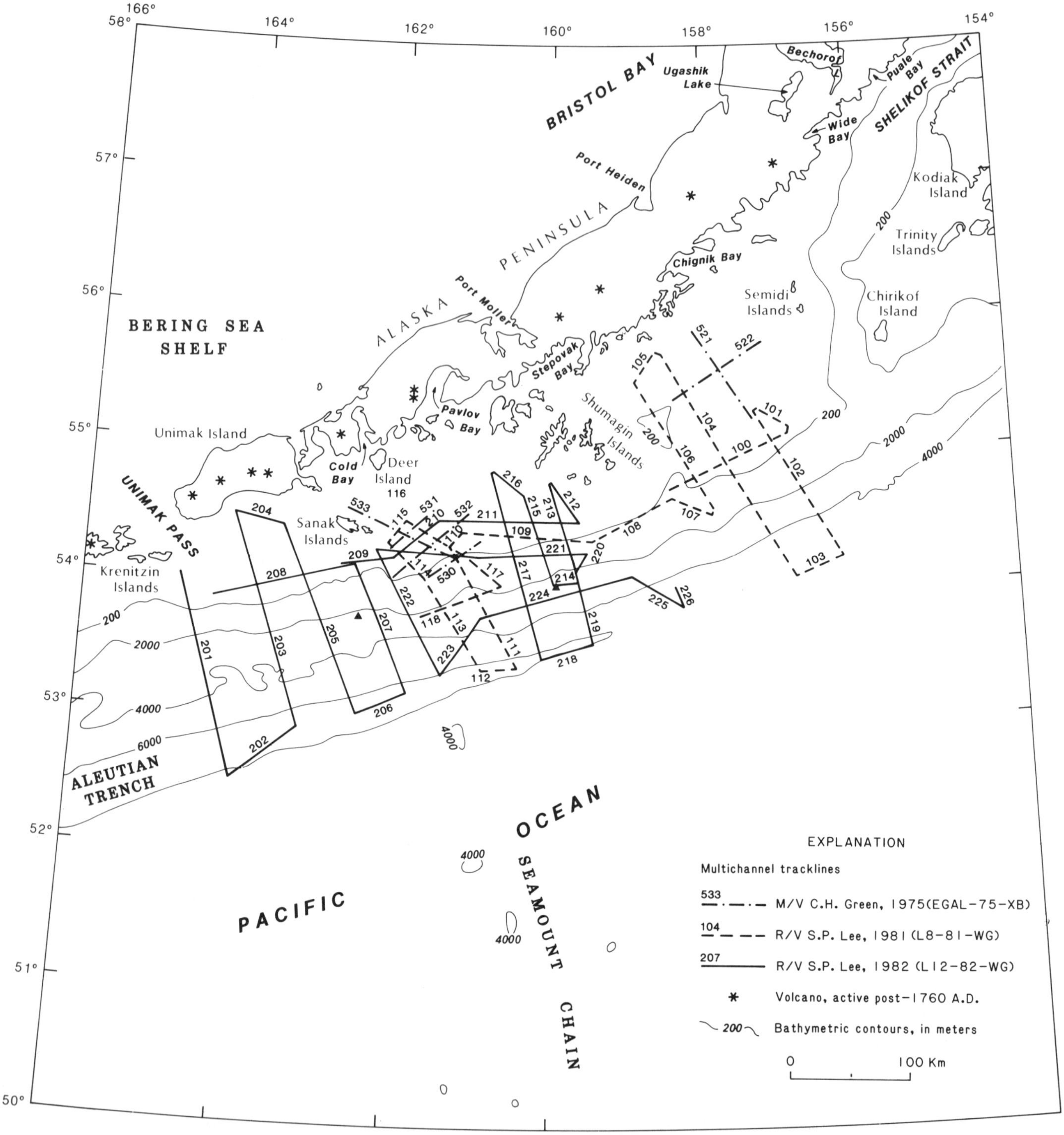

Figure 2. Tracklines of multichannel seismic data acquired by the U.S. Geological Survey in the Shumagin region, Alaska. Small triangles show dredge locations occupied in 1979 where samples were recovered; detailed locations shown on Figure 4 and given in Table 1.

(Peter, Erickson, and Grim, 1971). In the vicinity of the Shumagin margin, the Pacific and North American plates converge northwestward at a rate of about 7 cm/yr (Minster and Jordan, 1978). Results of this convergence include a high level of seismic activity along the margin, a well-defined Benioff zone, and an accretionary wedge at the base of the slope (Sykes, 1971; Davies and House, 1979; Bruns et al., 1985).

Determining when and where the Peninsular and Chugach-Prince William (composite) terranes amalgamated is difficult. On the basis of geological and geophysical constraints from the Cook Inlet and Kodiak regions, Fisher and von Huene (1984) and Pavlis (1982) argued for removal of the seaward margin of the Mesozoic Alaska Peninsula by transform faulting or by a collision and subduction mechanism along the Border Ranges fault beginning in the Early Cretaceous. The Late Cretaceous Shumagin and Kodiak Formations were accreted onto this truncated margin as autoch-

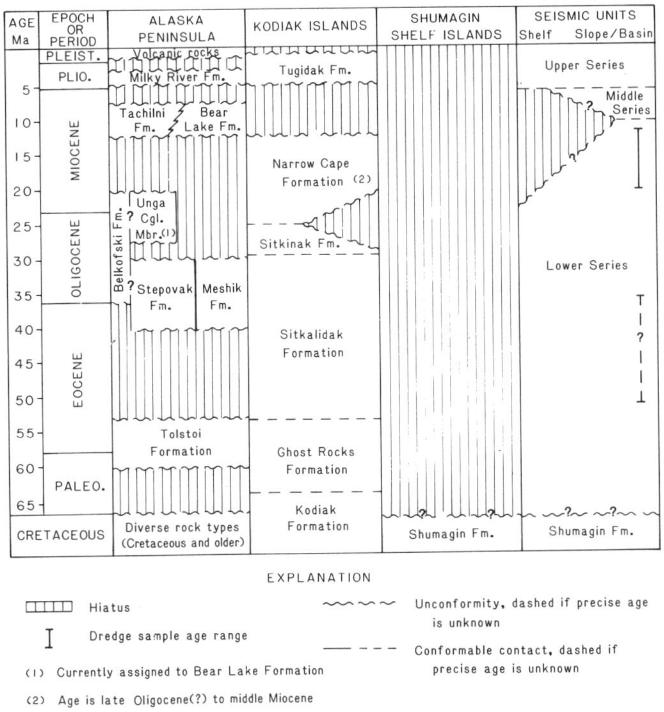

Figure 3. Sedimentary formations of the Alaska Peninsula, Shumagin, and Kodiak regions. From Burk (1965), Detterman et al. (1981a, 1983), and Vallier et al. (in press). Time scale from Palmer (1983).

thonous or parautochthonous units. However, paleomagnetic and paleontological studies within the Chugach-Prince William (composite) terrane indicate deposition and emplacement of Paleocene and Eocene rocks at latitudes 20° to 30° or more south of their present position (Plumley, Coe, and Byrne, 1983; Plumley et al., 1982; Moore et al., 1983; Keller et al., 1984). Yet, the Peninsular and Chugach-Prince William (composite) terranes are intruded by Paleocene plutons presumed to be cogenetic (Davis and Moore, 1984). Thus the two terranes were presumably close together at the time of intrusion, rather than being widely separated by several hundreds to thousands of kilometers. The Late Cretaceous and earliest Tertiary geologic history of the margin is therefore unclear. However, the history certainly includes the truncation of the Mesozoic margin of the Alaska Peninsula at the Border Ranges fault by the Late Cretaceous and early Tertiary, concurrent with or followed by accretion of the thick extensive Shumagin and Kodiak Formations (see von Huene, Fisher, and Bruns, this volume, for further discussion).

In the late Mesozoic and early Cenozoic, before the formation of the Aleutian Ridge, the north Pacific continental margin turned northward in the Sanak Island region and followed the Beringian transform or oblique convergent margin of western Alaska (Marlow and Cooper, 1980; Cooper, Marlow, and Scholl, this volume). The margin began to assume its modern form in the Eocene, with the growth of the Aleutian Ridge and subduction zone. The developing arc cut across and trapped a piece of Mesozoic oceanic crust that now lies beneath the Bering Sea, and built across the Peninsular terrane to form the Tertiary volcanic arcs of the Alaska Peninsula (Scholl, Vallier, and Stevenson, this volume).

Geology of the Alaska Peninsula

The Mesozoic and older rocks of the Alaska Peninsula are probably not present seaward of the Border Ranges fault and are therefore discussed only briefly. However, these rocks were probably part of the source terrane for the Cenozoic strata of the Shumagin Shelf and slope. Since Cenozoic rocks of the peninsula may be both a source rock and an onshore, nonmarine and shallow-marine analog for sedimentary rocks in the Shumagin margin basins, they are discussed in more detail.

Permian and Triassic volcanic rocks and limestones of the Alaska Peninsula are exposed mainly in the vicinity of Puale Bay and Becharof Lake. Lower and Middle Jurassic rocks, predominantly volcaniclastic sedimentary rocks, are exposed in areally limited outcrops on the Alaska Peninsula, but crop out over a wide area northeast of the peninsula around lower Cook Inlet (Detterman and Hartsock, 1966; Detterman and Reed, 1980; Magoon, Adkison, and Egbert, 1976). Upper Jurassic and Cretaceous rocks are as thick as 300 m, and are mainly a shallow-marine to fluvial sequence of sandstone, siltstone, and shale; also included are two deeper water sequences in the Lower Cretaceous and Upper Cretaceous, which indicate periods of transgression (Burk, 1965; Detterman, Case, and Wilson, 1979; Detterman et al., 1983).

Mesozoic rocks on the Alaska Peninsula contain large amounts of volcanic and plutonic debris, an indication that a volcanic arc was active during the middle Mesozoic. Episodes of plutonic activity occurred mainly in the Early Jurassic and Late Cretaceous. Rapid uplift along the Bruin Bay fault and erosion of the Jurassic plutons occurred as early as the Late Jurassic. Formations containing fossiliferous limestone and coal-bearing units indicate deposition in a stable shelf environment during much of the Mesozoic (Burk, 1965).

Rocks of Cenozoic age form five major unconformity-bounded stratigraphic packages composed principally of fluvial, volcaniclastic, and shallow-marine sedimentary rocks (Figure 3; Detterman et al., 1981, 1983; Detterman, 1985; Vallier et al., in press). The stratigraphic and age assignments are being modified by Detterman and his co-workers (Detterman, 1985), and some of these preliminary changes are used in this short review (Figure 3).

The oldest unit is the Tolstoi Formation of Paleocene and Eocene age, which consists predominantly of nonmarine sedimentary rocks with some coal beds. The Tolstoi Formation is as much as 1500 m thick and unconformably overlies Mesozoic rocks.

The Tolstoi Formation is overlain by a diverse package of rocks that consists of, from southwest to northeast, the Belkofski, Stepovak, and Meshik Formations. These rocks are mostly late Eocene and Oligocene in age, with the possible exception of early Miocene strata tentatively assigned to the Belkofski Formation by some workers. The Belkofski Formation is composed mainly of andesitic rocks; the Stepovak Formation consists mostly of volcaniclastic sedimentary rocks; and the Meshik Formation is composed of coarse basaltic volcanic rocks and minor amounts of volcaniclastic sedimentary rocks. The Unga Conglomerate Member may also belong to this group of formations (R. L. Detterman, 1986, personal communication; in Vallier et al., in press), although it is currently assigned to the Bear Lake Formation.

The third stratigraphic package consists of the middle and upper Miocene Tachilni and Bear Lake Formations. These formations consists of shallow-marine and nonmarine sandstone, conglomerate, siltstone, shale, and thin-bedded coal. The Bear Lake Formation also contains abundant granitic debris.

Overlying these formations is the Pliocene Milky River Formation, which is similar to the Tachilni and Bear Lake Formations, but which includes numerous volcanic flows in its upper part.

Finally, the fifth and uppermost package consists of unnamed Quaternary volcanic rocks and surficial deposits.

The abundance of volcanic debris in the Cenozoic sedimentary sequences indicates that the Aleutian volcanic arc remained episodically active throughout the Cenozoic. A middle Tertiary magmatic episode (Meshik arc; Wilson, 1985) occurred from about 40 to 20 Ma (million years ago), yielding mainly andesite and basaltic andesite. At about 22 to 25 Ma, a small amount of andesitic volcanic rock and hypabyssal andesitic intrusive rocks were emplaced seaward of the older volcanic rocks. The currently active volcanic cycle started about 10 Ma with initial basalt eruptions changing to andesite and culminating in dacite, and with the emplacement of several large granodiorite batholiths along the Pacific side of the Alaska Peninsula. The volcanic centers for this activity moved westward of the middle Tertiary centers. Many late Tertiary to Holocene volcanic flows and intrusive plugs, and numerous active or recently active volcanoes, mark the location of the present-day Aleutian volcanic arc (Figure 1; Burk, 1965; Moore et al., 1983; von Huene et al., 1985).

Geology of the Continental Shelf

The rocks exposed on the Shumagin Shelf islands seaward of the Border Ranges fault include the Shumagin Formation and Paleocene and early Eocene granodiorite plutons (Figures 1, 3; Burk, 1965; Moore, 1973a, 1974). The Shumagin Formation, exposed on the Sanak, outer Shumagin, and Semidi Islands, is part of an extensive belt of similar rocks exposed to the northeast on the Kodiak Islands (Kodiak Formation), on the Kenai Peninsula, and in the Chugach and Saint Elias Mountains of southern Alaska (Nilsen and Zuffa, 1982). The Shumagin and Kodiak Formations consist of complexly deformed deep-water turbidite sandstone, siltstone, and graywacke that probably accumulated as trench and trench-slope deposits (Burk, 1965; Moore, 1974; Nilsen and Moore, 1979; Nilsen and Zuffa, 1982); they are dated as Late Cretaceous (Maastrichtian) on the basis of rare fossils from near Anchorage, Alaska (Burk, 1965; Jones and Clark, 1973). These formations are intruded by Paleocene plutons, indicating deformation and emplacement in the latest Cretaceous and early Paleocene (Davis and Moore, 1984).

Early Tertiary rocks are not exposed on the outer islands of the Shumagin Shelf, but to the northeast are extensively exposed on strike along the southeast coast of Kodiak Island (von Huene, Fisher, and Bruns, this volume). These rocks have also been penetrated in stratigraphic test wells on the Kodiak Shelf (Herrera, 1978) and in an exploratory well near Middleton Island in the northern Gulf of Alaska (Keller et al., 1984). Early Tertiary rocks on Kodiak Island include the deep-water turbidite sequences and interbedded mid-ocean-ridge basalt of the Paleocene and early Eocene Ghost Rocks Formation, Eocene and Oligocene turbidite units of the Sitkalidak Formation, and nonmarine to shallow-marine rocks of the Oligocene Sitkinak Formation. The Ghost Rocks and Sitkalidak Formations are complexly deformed and are overlain with angular unconformity by the less deformed Sitkinak Formation.

The early Tertiary rocks are overlain by the nonmarine and shallow-marine sedimentary rocks of the late Oligocene(?) to middle Miocene Narrow Cape Formation and the late Pliocene and early Pleistocene Tugidak Formation (Figure 3; Moore, 1969; Moore et al., 1983; Nilsen and Moore, 1979; Moore and Allwardt, 1980; and von Huene, Fisher, and Bruns, this volume).

Early Tertiary rocks similar to those exposed on Kodiak Island or sampled in the offshore wells probably extend on strike southwestward and underlie the outer shelf and slope of the Shumagin margin. Also, rocks similar in age and possibly similar in lithology to those of the Narrow Cape and Tugidak Formations could constitute part or all of the basin-filling rocks of the Shumagin margin. On the basis of seismic stratigraphic studies and well data, the basin fill of the Kodiak Shelf is considered to be late Miocene and younger (Fisher and von Huene, 1980; Keller et al., 1984); this is probably also the age of the basin fill of the Shumagin Shelf.

Dredge Samples

During 1979, rocks were dredged at three sites on the Shumagin margin (Bruns et al., 1985; Figures 4, 5, 6). These few sites provide meager data with which to characterize the sedimentary rocks that underlie the margin, but they are the only dredge samples that have been acquired.

Dredge sites 2 and 3 were located on the north and south flanks, respectively, of Unimak Seamount, a prominent mid-slope bathymetric high about 60 km south of Sanak Island (Figures 4, 5). Dredge site 4 was located on the west wall of a submarine canyon south of the Shumagin Islands (Figures 4, 6).

A description of selected samples from the three dredge sites is given in Table 1. Obvious rounded or striated erratics were excluded from analysis, and the rocks described, with the one exception discussed below, are representative of several samples from the dredge hauls. Thus, we believe these samples are from outcrops and are indicative of rock units exposed on the continental slope at the dredge sites.

Four rock units were sampled from sites 2 and 3 on Unimak Seamount:

(1) Eocene. This unit is represented by a middle and/or late Eocene (late Narizian) calcareous mudstone (sample 2-10; date based on foraminifera). Deposition occurred in outer neritic to upper bathyal depths. This rock was unique in the dredge haul. It could be an erratic, because it was moderately rounded and well indurated. However, the rock was not typical of other erratics, which were hard, rounded metamorphic and igneous rocks that commonly showed glacial striations. Although we feel that this sample is from a rock unit under the continental slope, the presence of Eocene rocks will not be known for certain until confirmed by further sampling.

(2) Early and middle Miocene. This unit consists of a suite of lower and middle Miocene mudstone, siltstone (samples 2-6 and 2-8, date based on abundant diatoms), and volcaniclastic sandstone (samples 2-1, 2-2, 2-4, and 2-5). Rocks of this unit contain abundant pumice and glass shards (see samples 2-4 and 2-5), indicating proximity to an active volcanic chain

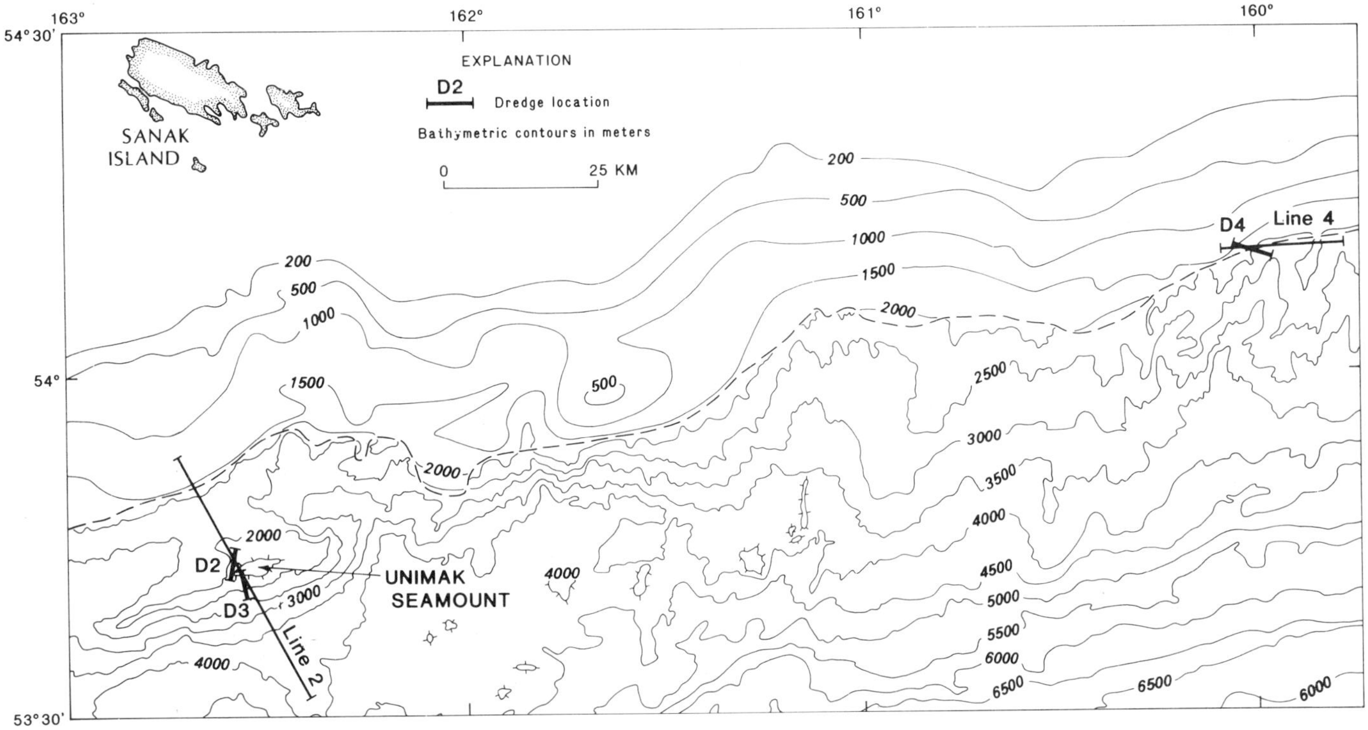

Figure 4. Locations of dredge hauls from the Shumagin margin. Bathymetry south of dashed line is from seabeam data of S. D. Lewis and J. W. Ladd, Lamont-Doherty Geological Observatory (written communication, 1985).

during deposition. None of the coarse-grained rocks included in this unit contained datable microfauna; however, the rocks are probably of early and middle Miocene age, because they occur with the abundant well-dated fine-grained rocks from site 2, all of which, except for the single Eocene sample, were late early and early middle Miocene in age. The coarse volcaniclastic sandstones contain much granitic debris, indicating a quartz diorite or granodiorite source rock. Benthic foraminifers indicate deposition in outer neritic to upper bathyal depths.

(3) Pliocene and younger. This unit, a late Pliocene to Quaternary mudstone (sample 3-2), is represented by only two cobbles.

(4) Basalt. This unit (samples 3-1 and 3-3) is of unknown age.[1] It is represented by two large (30 by 30 by 20 cm), angular pieces from site 3. Dating of these basalt samples is in progress. Geochemical analyses (Tables 2, 3; Figure 7) indicate that the samples are similar, and are arc related, as they are not mid-ocean-ridge basalt (MORB) and have characteristics unlike ocean-island basalt (D. A. Clague and T. L. Vallier, 1985, written communication). Sample 3-1 is probably arc related because of high Al_2O_3-low TiO_2 and low rare-earth elements (REE) compared with ocean-island basalt; and high Al_2O_3-low CaO, high K_2O, high Ba, and enriched light REE compared with ocean-floor basalt. The alkalics of sample 3-3 have K_2O and Na_2O values that are too high to be from an ocean-floor basalt; the sample also has high Al_2O_3 and low TiO_2 like arc rocks rather than ocean-island samples. In addition, the REE patterns are similar to calc-alkaline lava. The basalt probably underlies the sedimentary rocks described in units 1, 2, and 3, because the basalt is not typical of the abundant granitic-derived volcanic rock fragments observed in the sedimentary rocks. The basalt is probably the source rock for a pronounced magnetic anomaly coincident with Unimak Seamount and its bathymetric continuation as Unimak Ridge (Bruns and von Huene, 1977).

Site 4 yielded rocks of two kinds:

(1) Pliocene and Pleistocene. Hard nodular limestone and calcareous siltstone (samples 4-1, 4-3, and 4-6). On the basis of diatom analysis, these rocks are of late Pliocene and early Pleistocene age.

(2) Dacite. In thin section the recovered igneous rocks look like a hornblende dacite porphyry (samples 4-4 and 4-11) and dacite or andesite (samples 4-8, 4-9, and 4-12). Geochemical analyses of the rocks (Tables 2, 3; Figure 7) indicate that all samples are dacite with calc-alkaline affinities, typical of an island-arc setting. The geochemistry also indicates that the samples represent three distinct units. Samples 4-4 and 4-11 are geochemically similar, and therefore are from the same flow or intrusive unit, as are 4-8 and 4-9; whereas, 4-12 is distinct from the other four samples (Tables 2, 3; D. A. Clague and T. L. Vallier, 1985, written communication).

SEISMIC SEQUENCES

On the multichannel seismic-reflection data, we delineate three seismic sequences with different reflection characteristics. For ease of reference, these sequences are informally called the lower series (LS), middle series (MS), and upper series (US). As discussed below, these series are composed as follows: lower series = Mesozoic through middle Miocene rocks; middle series = late Miocene rocks; upper series = latest Miocene through Quaternary rocks. The sequences are divided by seismic horizons A and B (Figures 3, 8).

The major sequence division occurs at seismic horizon A;

[1]See addendum at end of chapter.

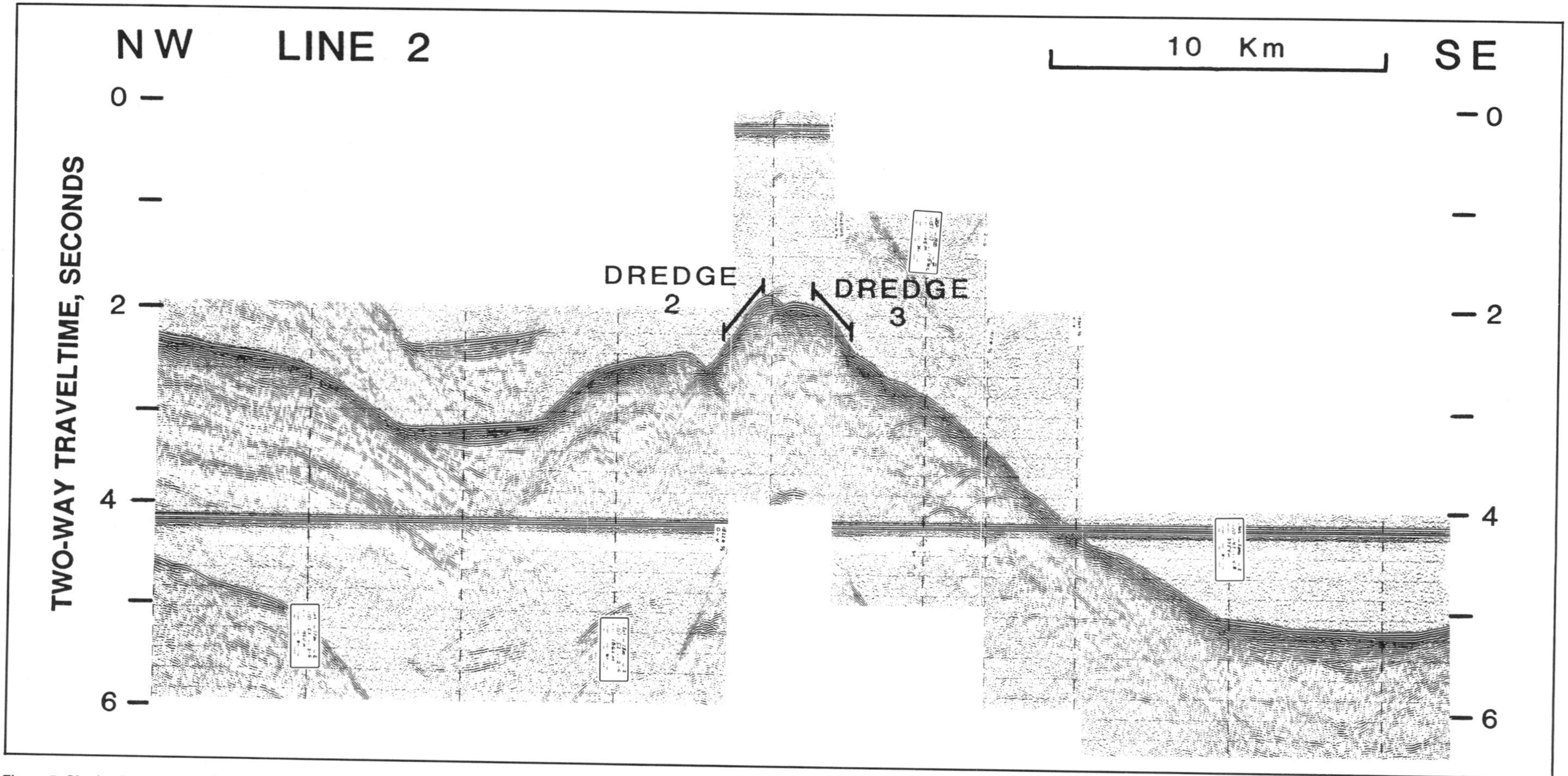

Figure 5. Single-channel seismic-reflection line showing locations of dredges 2 and 3; line location shown on Figure 4.

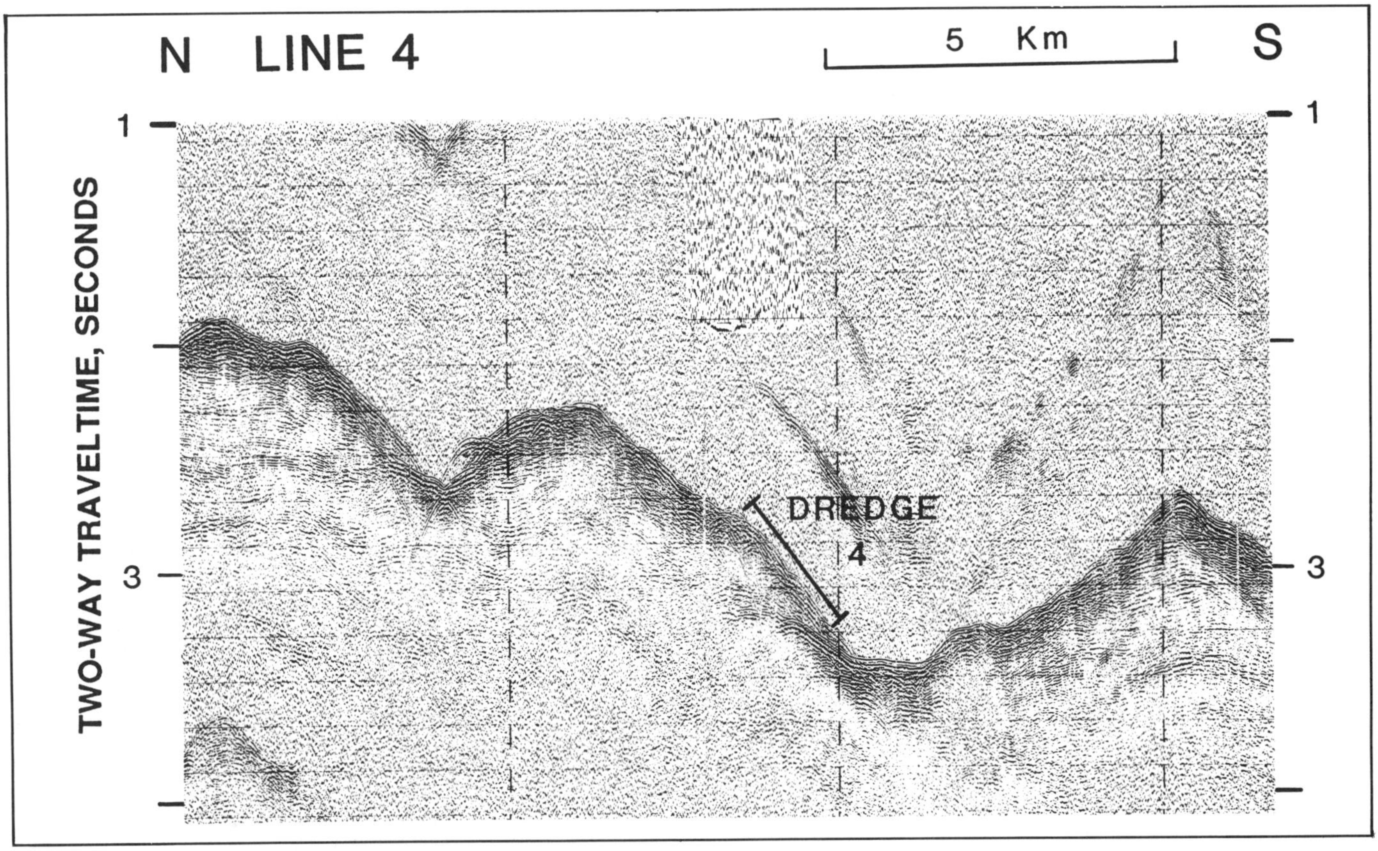

Figure 6. Single-channel seismic-reflection line showing location of dredge 4; line location shown on Figure 4.

there is also a significant variation in the character of this division from the shelf toward the middle slope. Beneath the shelf, horizon A is typically a strong regionally extensive seismic-reflection event, and is onlapped by the overlying middle and upper series rocks. Lower series rocks below horizon A are characterized by poor and discontinuous reflections, and commonly, but not always, have seismic-refraction velocities greater than about 4.6 km/sec. Horizon A is the deepest mappable horizon beneath the Shumagin Shelf.

A paleoshelf break may be present beneath the outer shelf, coincident with a marked change in the reflection characteristics of horizon A (Figure 8). In this region, horizon A merges into, and cannot everywhere be clearly distinguished from, conformable seismic reflections in the underlying and overlying strata. The transition from an unconformity to conformable reflections suggests a change from an area of shelf erosion or nondeposition to an area of continuous slope deposition. The most likely places for this change to occur are at the shelf edge and along the margins of a rapidly subsiding basin. Thus, we believe this change in character coincides with the position of the paleoshelf edge. This possible paleoshelf edge can be mapped across the outer shelf and upper slope beneath most of the Shumagin region.

Where horizon A is a strong reflection event beneath the shelf, we believe that the horizon corresponds to an erosional surface cut during regional Oligocene and early and middle Miocene erosional periods. All the characteristics of horizon A suggest that it is an angular unconformity. The horizon forms a smooth surface which truncates sparse reflections from the underlying lower series rocks. The overlying middle and upper series strata onlap horizon A as far seaward as the area where the horizon reflection character changes markedly (Figure 8). Horizon A has the same seismic expression as a similar seismic-reflection horizon beneath the Kodiak Shelf (horizon C of Fisher and von Huene, 1980), and beneath the Middleton Shelf northwest of the Kodiak Shelf (horizon M of Bruns, 1985). The Kodiak and Middleton horizons have been identified as regional unconformities cut during the Oligocene and early and middle Miocene; deposition of strata above the unconformity began in the latest Miocene (Fisher and von Huene, 1980; Keller et al., 1984). Horizon A was probably caused by the same erosional events. Thus, the entire shelf from the northern Gulf of Alaska-Prince William Sound area to Unimak Pass is probably underlain by a regional unconformity cut during the Oligocene and Miocene, with upper Miocene strata being the oldest rocks immediately overlying the unconformity.

Beneath the shelf, we correlate the lower series rocks with rocks exposed on the shelf islands. In the vicinity of the Semidi, Shumagin, and Sanak Islands, horizon A is shallow and rises toward the islands, where the deformed and metamorphosed Shumagin Formation and associated granitic intrusive rocks are exposed. These rocks are not likely to cause or return coherent reflections over large distances, and they would probably have a seismic-refraction velocity >4.6 km/sec, such as that exhibited by the lower series rocks near the islands (Figure 9). Therefore, the rocks of the lower series almost certainly include the offshore equivalent

Table 1. Location and description of selected dredge samples from the Shumagin continental margin, western Gulf of Alaska, U.S. Geological Survey cruise S7-79-WG. Dredge site locations shown on Figure 4.

Dredge Site	Location[1] Lat (°N)	Long (°W)	Recovery[2] Depth (m)	Sample Designation	Lithology[3]	Fossils	Age
2	53°43.04′ 53°41.66′	162°34.24′ 162°34.92′	1800- 1650	2-1	**Sandstone.** (Arkosic wacke). Flat 8-cm-thick slab with crude bedding on either side of a 5-cm-thick unit. Well indurated, poorly sorted, medium grained. Abundant feldspar, common quartz and hornblende, rare volcanic and plutonic rock fragments. Clay matrix. Probable quartz diorite or granodiorite source rocks.		
				2-2	**Sandstone.** (Arkosic wacke). Flat 5-cm-thick slab. Massive, well indurated, poorly sorted, medium to coarse grained. Abundant feldspar, common quartz, hornblende, and volcanic rock fragments, trace epidote, biotite, and pyroxene. Clay matrix. Probable quartz diorite or granodiorite source rocks.		
				2-4	**Sandstone.** Friable, poorly sorted, coarse grained. Common volcanic-glass shards, in part altered to clay, common flattened pumice fragments and feldspar. Clay matrix. Coarsely graded from very coarse to medium grain size in a 3-cm-thick layer.		
				2-6	**Mudstone.** Well consolidated, gray, finely laminated. Mainly silt and clay; rare feldspar and sponge spicules.	Sparse early to middle Miocene diatom assemblage, possibly late early to early middle Miocene.	Late early and early middle Miocene
				2-8	**Siltstone.** Massive, moderately indurated. Common feldspar and glauconite nodules, trace pumice and sponge spicules, rare fish teeth, volcanic rock fragments, and well-rounded pebbles.	Radiolarians, undated. Rare, very fragile foraminifers of probable middle Miocene age, outer neritic to upper bathyal. Good diatom assemblage of late early Miocene to early middle Miocene age.	Late early and early middle Miocene
				2-10	**Calcareous mudstone.** Massive, well, indurated pale brown, with rare pumice, pyrite, and volcanic-glass fragments. Less than 50% calcite.	Middle to late Eocene foraminiferal assemblage, probably late Narizian based on occurrence of *Robulus welchi*. Outer neritic to upper bathyal.	Middle and/or late Eocene
3	53°40.38′ 53°42.07′	162°31.99′ 162°33.58′	1600- 1400	3-1	**Basalt.** Altered, with plagioclase (quite fresh) and olivine (now partly clay) phenocrysts. Groundmass is plagioclase, clinopyroxene, olivine (now clay), glass (now clay), and opaques.		
				3-2	**Mudstone/siltstone.** Very friable, grayish-brown, with sand and pebbles in clay/silt matrix.	Sparse late Pliocene to Quaternary diatom assemblage.	Late Pliocene and Quaternary
				3-3	**Basalt.** Fine grained, altered, with plagioclase (now mostly clay), olivine (now iddingsite), and clinopyroxene phenocrysts in a groundmass of plagioclase, clinopyroxene, glass (now clay), and opaques.		
4	54°10.50′ 54°10.42′	159°59.70′ 160°02.99′	2200- 2000	4-1	**Limestone.** Massive, hard, gray nodule with calcite cement. Sponge spicules.	Late Pliocene to possibly earliest Pleistocene diatom and foraminiferal assemblages. Upper to middle bathyal.	Late Pliocene
				4-3	**Limestone.** Massive, hard, gray nodule with calcite cement. About 10% terrigenous input with volcanic rock fragments, feldspar, quartz, and pyrite.	Late Pliocene to early Quaternary diatom assemblage, probably early Quaternary. Pliocene to early Pleistocene foraminifers; bathyal.	Early Pleistocene
				4-6	**Carbonate sandstone.** Massive, hard gray nodule. Mainly a poorly sorted, coarse sandstone made up of sedimentary rock clasts and limestone clasts with carbonate cement. Sedimentary rock clasts include feldspar and quartz, and sandstone, volcanic and plutonic rock fragments.	Late Piocene diatom assemblage. Pliocene to early Pleistocene foraminifers; bathyal.	Late Pliocene

(continued)

Dredge Site	Location[1] Lat (°N)	Long (°W)	Recovery[2] Depth (m)	Sample Designation	Lithology[3]	Fossils	Age
				4-4 and 4-11	**Hornblende dacite porphyry**. Sample 4-4 is highly altered; 4-11 less altered.		
				4-8, 4-9 and 4-12	**Basaltic andesite.** Plagioclase, clinopyroxene, orthophyroxene, and altered orthophyroxene phenocrysts (now serpentine). Samples 4-8 and 4-9 are nearly identical; sample 4-12 has far less plagioclase.		

Notes:
1. Approximate latitude and longitude when dredge on and off bottom.
2. Approximate depth at which samples recovered, based on record from time-depth recorder, and on increased tension on dredge wire.
3. Abundant—>25%; common—10 to 25%; rare—1 to 10%; trace—<1%.

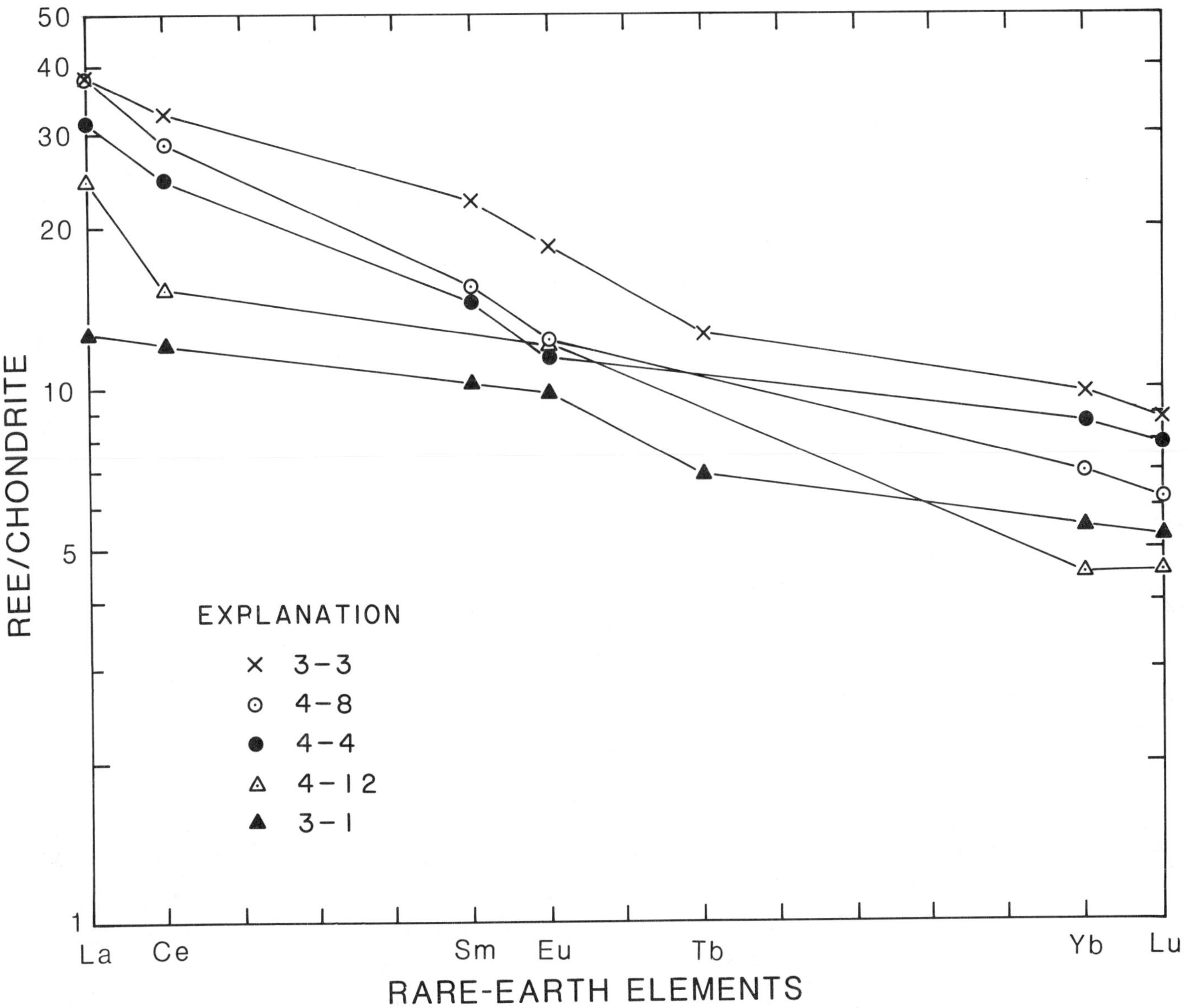

Figure 7. Plot of rare-earth elements (REE) for dredged basalts. Sample numbers correspond to those in Tables 2 and 3.

Table 2. Trace elements (ppm) of igneous rocks from dredges 3 and 4, cruise S7-79-WG. Analysts are L. Schwarz (Instrumental neutron activation analysis) and R. Johnson (X-ray fluorescence, XRF). XRF values are noted by superscript ([1]); dash (—) implies values not determined, below detection limits, or discarded because of a high coefficient of variation.

	Sample Numbers						
Element	3-1	3-3	4-4	4-8	4-9	4-11	4-12
Rb[1]	9	10	24	54	51	19	36
Sr[1]	415	349	358	400	397	354	796
Ba[1]	101	234	466	619	621	466	386
Th	0.67	2.27	3.0	5.82	5.91	3.14	4.01
U	—	.68	1.08	2.10	2.10	1.19	1.2
La	4.8	14.6	12.0	14.6	14.8	11.6	9.3
Ce	11.7	32.2	23.9	27.6	27.5	23.0	15.3
Nd	13	23	13	14	16	14	8
Sm	2.37	5.18	3.24	3.51	3.54	2.99	1.84
Eu	0.85	1.59	0.99	1.02	1.01	0.92	1.04
Tb	0.41	0.75	0.52	0.39	0.37	0.47	0.23
Yb	1.38	2.44	2.14	1.75	1.47	2.13	1.12
Lu	0.206	0.342	0.366	0.241	0.259	0.315	0.180
Y[1]	16	27	20	17	16	21	11
Zr[1]	80	138	134	110	107	134	150
Nb[1]	6	6	6	7	6	7	6
Hf	1.57	2.97	2.93	2.78	2.85	2.97	3.46
Ta	.27	.29	.35	0.22	0.27	0.36	0.31
Co	43	31	14	21	19	14	14
Cr	205	97	21	54	51	19	39
Sc	30	30	15	27	26	14	22
Zn	76	92	86	56	55	84	55

Table 3. Major and minor element oxides of igneous rocks from dredges 3 and 4, cruise S7-79-WG. Rock elemental oxides summed to 100% and H_2O^+, H_2O^- and CO_2 values included. Analyses by X-ray fluorescence and wet chemical methods; A. Bartel, D. Kobilis, and J. Pastor, analysts, U.S. Geological Survey.

	Sample Numbers						
Oxides	3-1	3-3	4-4	4-8	4-9	4-11	4-12
SiO_2	48.79	52.12	61.52	60.51	60.76	62.05	60.27
TiO_2	1.05	1.55	0.73	0.77	0.76	0.76	0.81
Al_2O_3	18.55	18.74	18.75	18.19	18.10	18.62	17.83
Fe_2O_3	3.97	5.75	3.17	3.91	3.98	3.41	3.01
FeO	4.73	3.79	3.16	1.45	1.35	2.94	2.40
MnO	0.15	0.18	0.12	0.05	0.05	0.11	0.08
MgO	8.76	5.59	2.91	2.16	2.21	2.78	3.66
CaO	10.99	7.35	5.01	7.96	7.74	4.65	7.23
Na_2O	2.51	3.67	3.45	3.11	3.07	3.53	3.27
K_2O	0.38	1.01	0.99	1.84	1.83	0.96	1.38
P_2O_5	0.12	0.25	0.19	0.15	0.15	0.19	0.06
H_2O^+	1.40	2.40	2.20	0.87	0.95	2.10	1.10
H_2O^-	0.97	2.90	1.20	1.50	1.20	1.30	1.20
CO_2	0.02	0.02	0.02	0.02	0.02	0.02	0.02
FeO*/MgO	0.02	1.61	2.07	2.25	2.23	2.16	1.40

of the Shumagin Formation and the associated Paleogene intrusive rocks.

Lower series rocks beneath the shelf may also include gently deformed and mildly metamorphosed to unmetamorphosed lower Tertiary sedimentary rocks, similar to those exposed on Kodiak Island or present beneath the Kodiak Shelf. In some areas, as for example below Shumagin Basin, layered reflections are locally present in high-velocity lower series rocks. Such reflections are more likely to come from an undeformed lower Tertiary section than from an equivalent of the deformed Shumagin Formation. Such lower Tertiary strata are not exposed on the outer Shumagin Shelf islands, but they could have been eroded away during the Miocene.

Seaward from the paleoshelf edge, deposition was presumably continuous and no well-developed unconformity is present; therefore, the lower series could include rocks as young as middle Miocene. These undeformed rocks probably have velocities of <4 km/sec, such as are found in refraction data (Figure 9), and return laterally continuous reflections on the multichannel seismic-reflection data. The lower and middle Miocene strata sampled on Unimak Seamount probably correspond to these lower series rocks.

Middle series rocks are distinguishable only in Sanak Basin

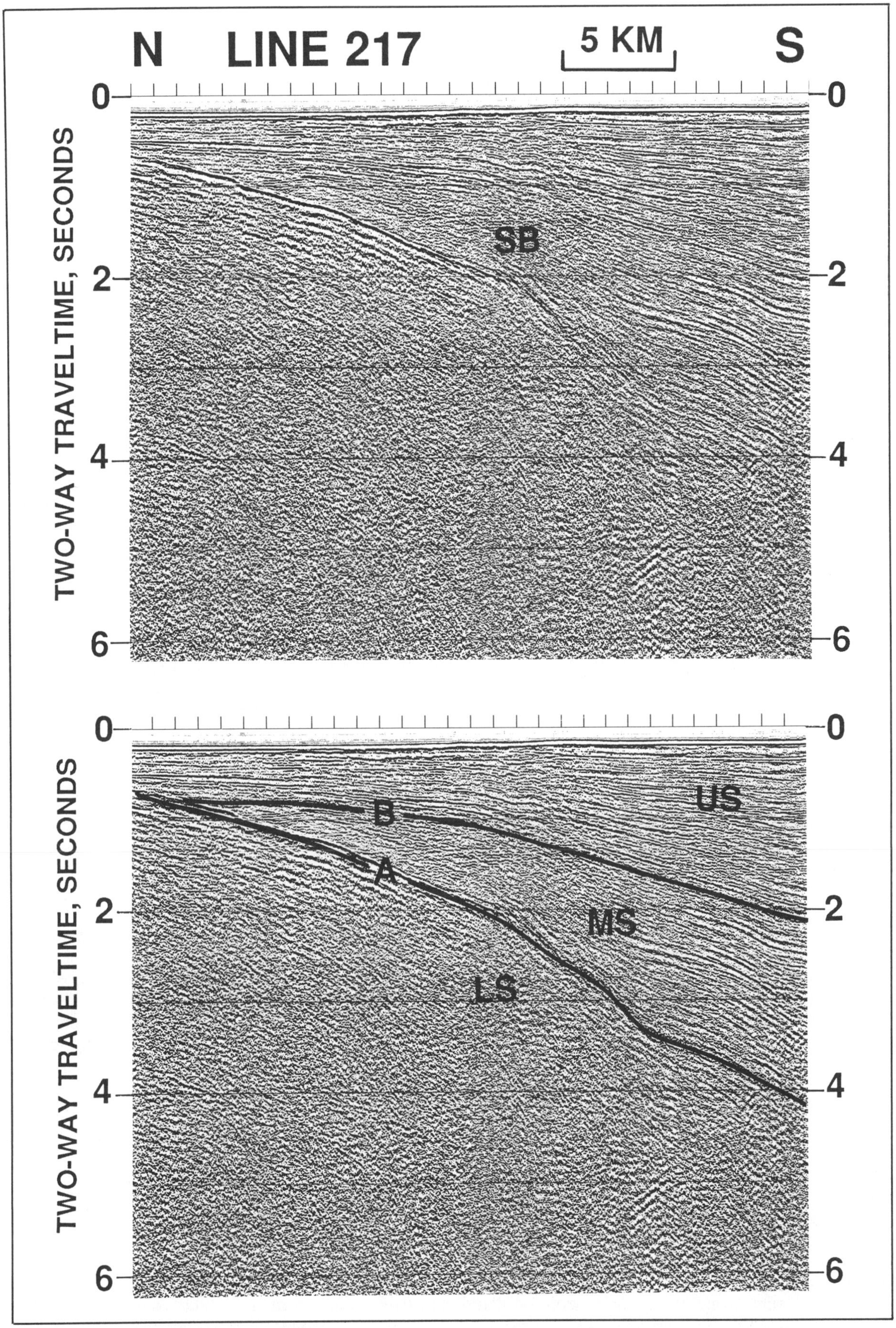

Figure 8. Part of seismic line 217 showing seismic units and character of reflection horizon A. Top—uninterpreted; bottom—interpreted. Change in reflection character of horizon A at location "SB" corresponds to possible paleoshelf edge during Miocene time. LS—lower series; MS—middle series; US—upper series.

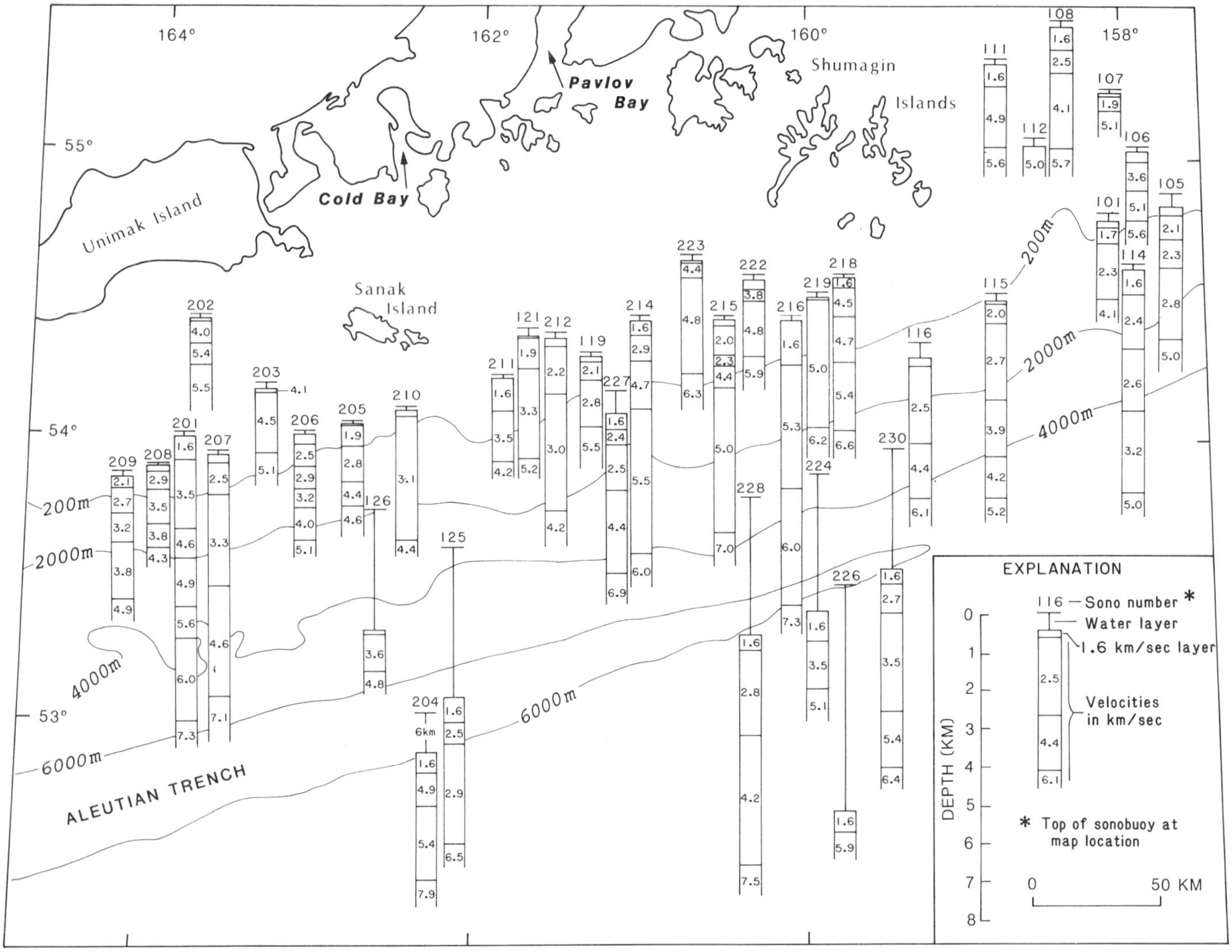

Figure 9. Layer thickness (km) and velocity (km/sec) from selected refraction sonobuoys from the Shumagin margin.

and beneath the slope. The major difference between the middle and upper series rocks is that seismic reflections from the middle series rocks are less continuous and more broken by small faults and folds than reflections from the upper series rocks (Figure 8). In Sanak Basin, the middle series also thickens markedly seaward into the growth fault bounding the seaward side of the basin at a buried basement high. Horizon B, separating the two series, is best defined in the deepest parts of Sanak Basin, and it onlaps and is truncated against horizon A near the paleoshelf edge. Middle series rocks were probably deposited seaward of the paleoshelf edge and into the rapidly subsiding basins. Inasmuch as the middle series rocks and the bounding horizon B onlap horizon A, we infer that the middle series postdates the unconformity and is largely of late Miocene age.

Upper series rocks are probably Pliocene and Quaternary in age but could include latest Miocene strata. These rocks form the upper fill in the Sanak and Unimak basins and also include most of the shelf strata overlying horizon A outside the basins, where they directly overlie lower series rocks.

In summary, our preliminary geologic correlation is that lower series rocks are composed of Shumagin Formation equivalents and, locally, Paleogene rocks beneath the shelf, but could include lower and middle Miocene rocks seaward of the inferred paleoshelf edge. Middle series rocks are about late Miocene, and probably were deposited in bathyal depths seaward of the shelf break on the slope and in the subsiding basins. Finally, we consider the upper series rocks to be of latest Miocene age and younger, and probably largely Pliocene and younger.

TIME–TO–DEPTH CONVERSION OF SEISMIC DATA

To convert seismic-reflection traveltime to depth for rocks above horizon A (middle and upper series strata), we first converted stacking velocities from the multichannel seismic-reflection data to interval velocities using the Dix equation (Dix, 1955); combined these data with seismic-refraction velocities (Figure 9); and then derived a regional time-depth (T-D) curve for the Shumagin margin (Figure 10). We also show a T-D curve from Bruns and von Huene (1977), and a

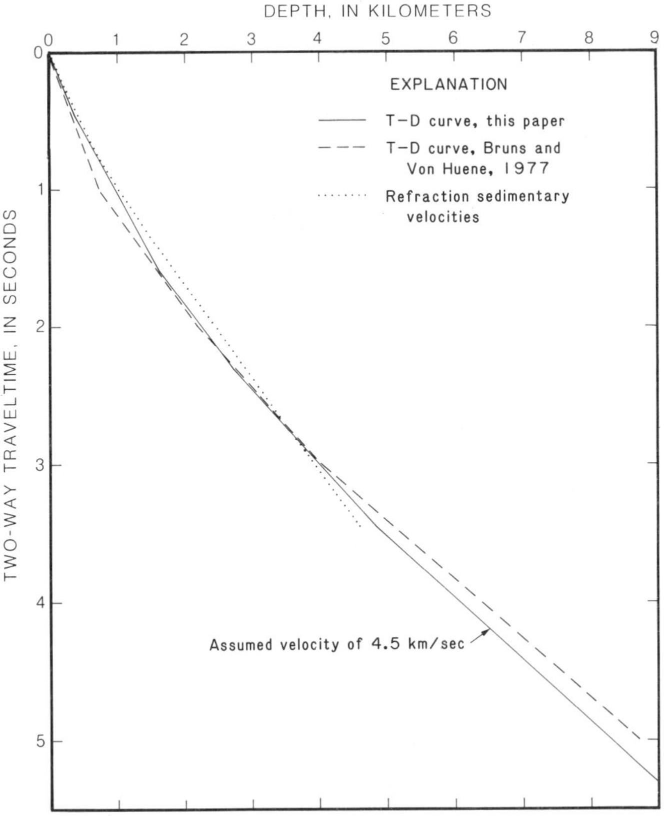

Figure 10. Curve for converting two-way subseafloor traveltime on seismic sections to depth. Water-layer thickness must be added to obtain total depth. Also shown are the curve from Bruns and von Huene (1977), and a curve based on data from refraction sonobuoys.

best-fit refraction T-D curve for comparison. The major disagreement between the refraction data and the T-D curve used in this chapter occurs below 3.5 sec; both refraction and reflection velocities are poorly constrained, and we assume a constant interval velocity of 4.5 km/sec for times >3.5 sec in the time-depth curve. The curve is valid only for middle and upper series rocks above horizon A. For conversion of time-to-depth for lower series rocks, a velocity of about 5.0 km/sec gives a reasonable estimate. Alternatively, since there is substantial variation in the lower series (basement) velocities, refraction velocities in specific areas (Figure 9) can be used.

STRUCTURE AND STRATIGRAPHY

The structure of the Shumagin margin is shown by depth contours on horizon A (Figure 11). Since we infer that this horizon separates upper Miocene and younger rocks from middle Miocene and older rocks, the structure observed in the seismic-reflection data has developed primarily as a result of late Cenozoic tectonic events. The margin is characterized by six major features: (1) Shumagin Basin, (2) Sanak Basin, (3) Unimak Basin, (4) sedimentary wedges beneath the outer shelf and upper slope, (5) Unimak Ridge and its buried structural continuation, a mid-slope structural high which includes Unimak Seamount, and (6) an accretionary complex beneath the lower slope.

An isopach map of strata above horizon A shows that, in shelf areas outside the two major basins, acoustic basement is covered by less than about 1 km, and commonly by less than 0.5 km, of post-horizon-A strata (Figure 12). These areas include much of the shelf between Unimak Pass and Sanak Island, and the shelf regions adjacent to the Alaska Peninsula, the Shumagin Islands, and the Semidi Islands.

The submerged lands adjacent to the Alaska Peninsula are probably underlain by the offshore continuation of rocks and structures observed on the peninsula. However, the seismic-reflection data do not extend into these nearshore areas, and they are not discussed further.

The shelf and slope structure of the Shumagin margin has been dominated by late Cenozoic extension, subsidence, and associated normal faulting. Major compressional deformation of the shelf, such as that observed on the Middleton Shelf to the northwest (Bruns, 1985) and near Albatross, Trinity, and Tugidak basins on the adjacent Kodiak Shelf (Fisher, 1979; Fisher and von Huene, 1980), is notably absent along the Shumagin segment landward of the accretionary complex at the base of the slope.

Shumagin Basin

Shumagin Basin (Figures 11, 12, 13) is a roughly circular subsurface depression about 75 km across and 2.5 km deep that lies between the Shumagin and Semidi Islands. Basin strata thin toward the surrounding islands and the shelf edge; the seaward margin of the basin is a structural high with low relief. Shumagin Basin has developed by subsidence of the continental shelf during the late Miocene through the Holocene.

Shumagin Basin is one of the areas of the shelf where lower Tertiary strata might be present, as inferred from the seismic data. Beneath the basin, horizon A is a strong, easily identified reflection. However, below horizon A, discontinuous reflectors and refraction velocities suggest two areas where a sedimentary section as much as 2 km thick is present; in these areas, rocks with interval velocities in the 4 to 4.4 km/sec range overlie rocks with velocities of 4.8 to 6 km/sec. The rocks with 4 to 4.4 km/sec velocities (subunit LS-T, Figure 13) may be lower Tertiary sedimentary rocks. The 4.8 to 6 km/sec rocks probably correlate with the Shumagin Formation or the Paleocene granitic intrusive rocks (subunit LS-K, Figure 13). Lower series rocks beneath Shumagin Basin are arched gently upward beneath the center of the basin, forming two separate depressions on the northwestern and southeastern flanks of the basin. Minor normal faults cut the lower series rocks but do not generally extend into the overlying section. Discontinuous seismic reflections suggest that the sedimentary section within the upper part of the lower series (subunit LS-T) is thickest beneath the depressions and may partially or entirely pinch out beneath the arch.

Rocks above horizon A are probably largely upper series rocks. These strata are flat lying except near the outer shelf, where the sequence dips gently landward, and in the lower part of the unit, where strata fill depressions in the lower series rocks and onlap the flanks of the basin with minor discordance. Only small normal faults cut the basin fill. The maximum sediment thickness above horizon A is about 2.5 km. Toward the surrounding islands and the Alaska Peninsula, the basin strata thin to <0.5 km.

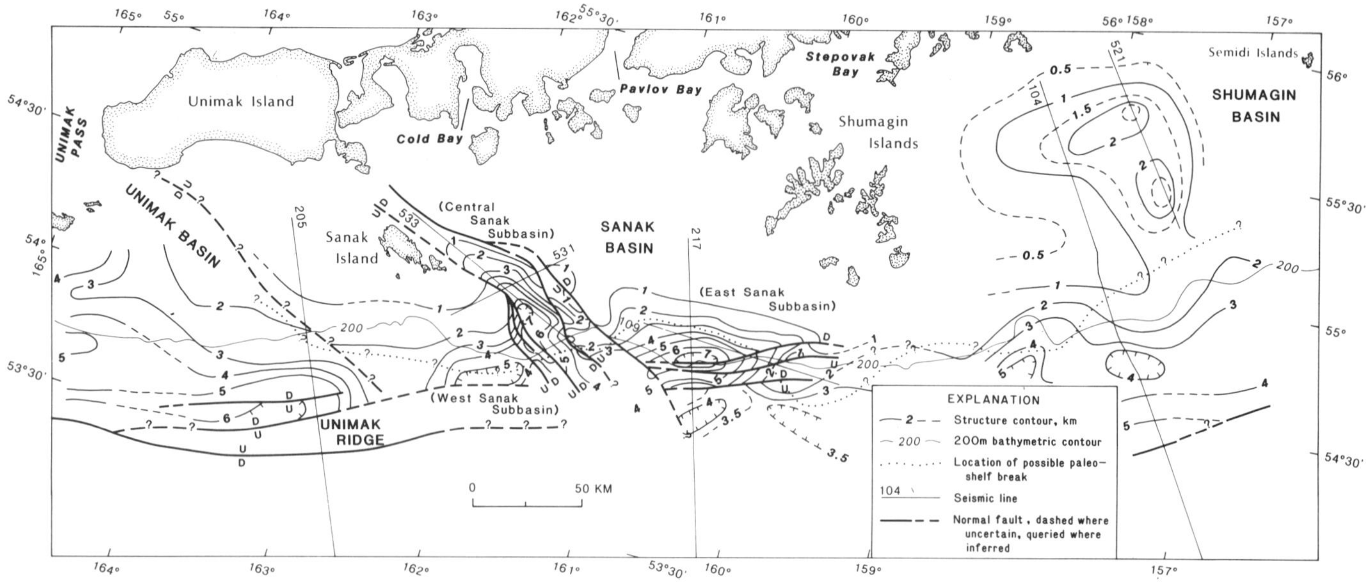

Figure 11. Structure contour map on horizon A, Shumagin margin, approximately at base of late Miocene and younger strata.

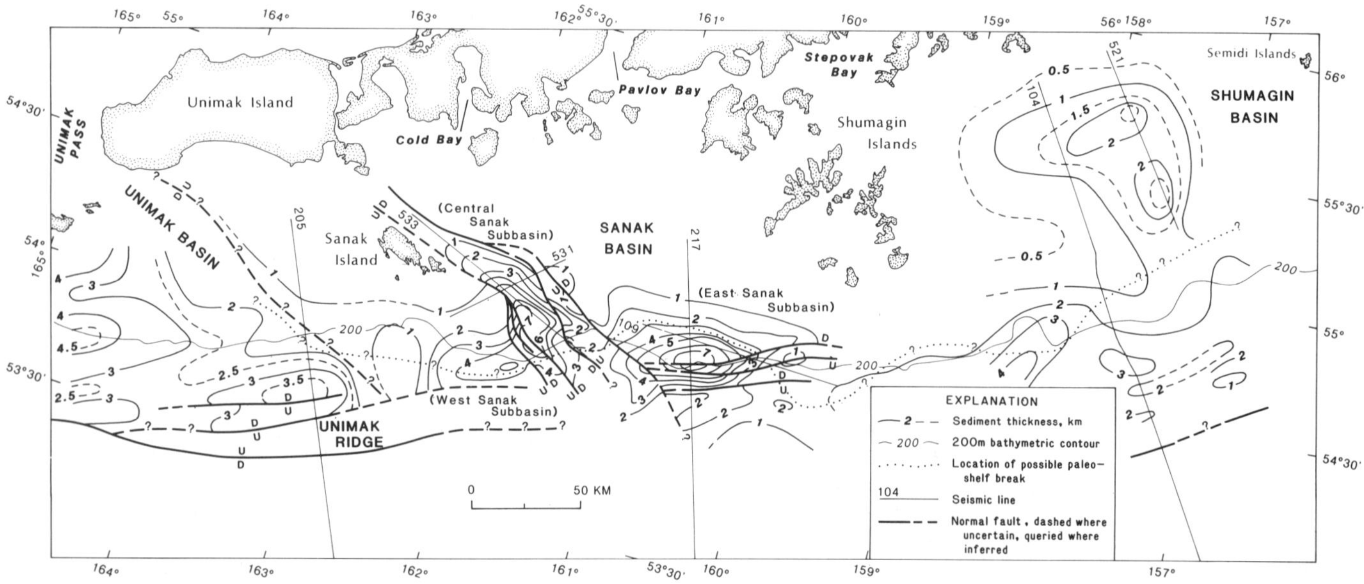

Figure 12. Isopach map of late Miocene and younger strata (middle and upper series), Shumagin margin.

Sanak Basin

Sanak Basin lies beneath the outer shelf and slope between Sanak Island and the Shumagin Islands (Figures 11, 12). The basin is segmented into west, central, and east subbasins by northwest-trending faults that form a graben beneath the central subbasin. Both the west and east subbasins trend southwest parallel to the margin, whereas the central subbasin trends northwest, oblique to the margin. The west subbasin contains 5 km, and both the central and east subbasins contain at least 7 km, of middle and upper series rocks above horizon A. The seaward side of the basin is bounded by block-faulted basement rocks, formed when the basin floor subsided along landward-dipping growth faults.

The west Sanak subbasin, the smallest of the three subbasins, lies largely beneath the upper slope south of Sanak Island (Figures 11, 12). The seaward side of the basin is formed by the structural high of Unimak Ridge (Figure 14). The basin strata form a seaward-dipping and seaward-thickening sedimentary wedge. The basin formed by subsidence of the margin landward of Unimak Ridge on a landward-dipping growth fault (Figure 14).

The central Sanak subbasin (Figures 11, 12, 15) is an elongate graben lying between bounding faults on the northeast and southwest. Sanak Island and other small islands and rock reefs encircle the northwest end of the subbasin; the southeast end extends beneath the upper continental slope. The south-

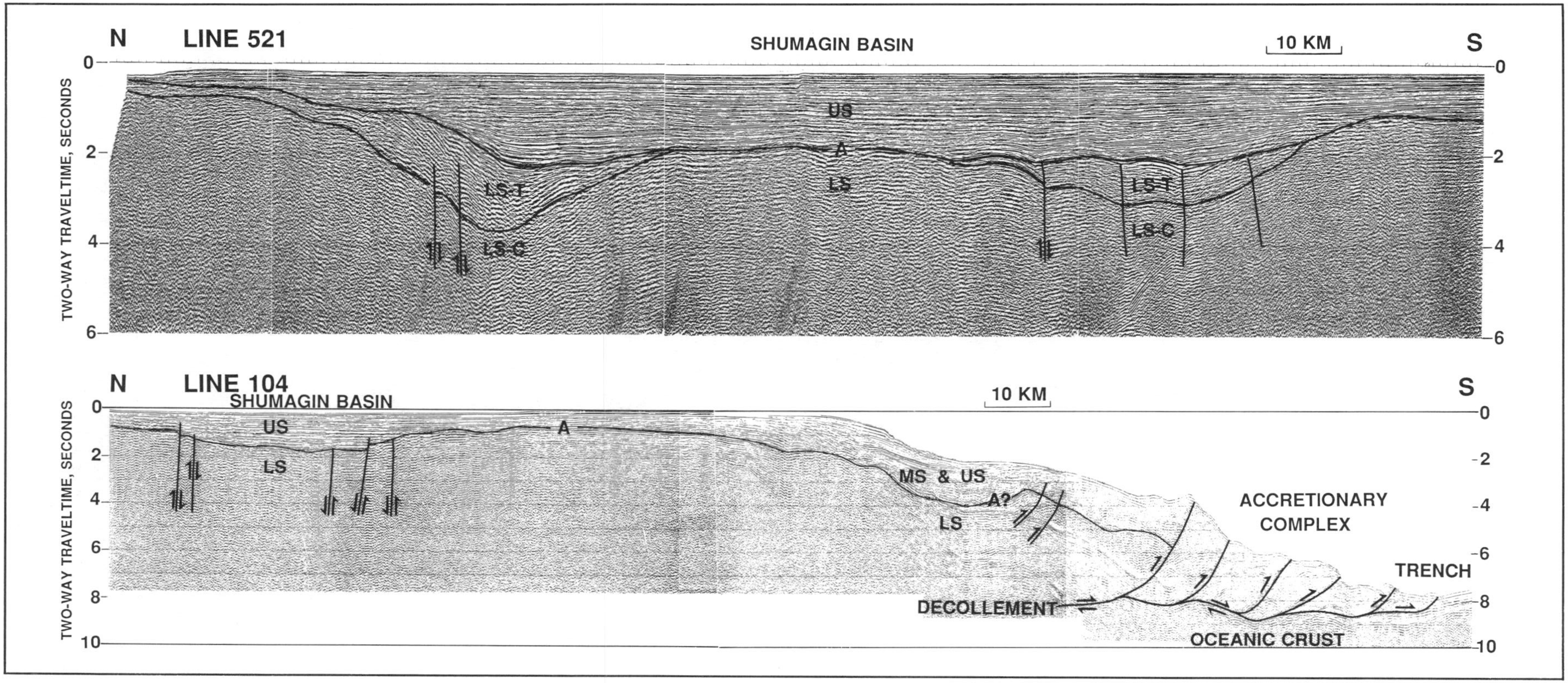

Figure 13. Seismic lines 521 and 104 across Shumagin Basin and the adjacent continental shelf and slope. Line locations shown in Figures 2, 11, and 12. LS—Lower Series; MS—Middle Series; US—Upper Series. LS-C and LS-T are Cretaceous(?) and Tertiary(?) subunits within LS. See Figure 18 for migrated section and line drawing interpretation of accretionary complex on line 104.

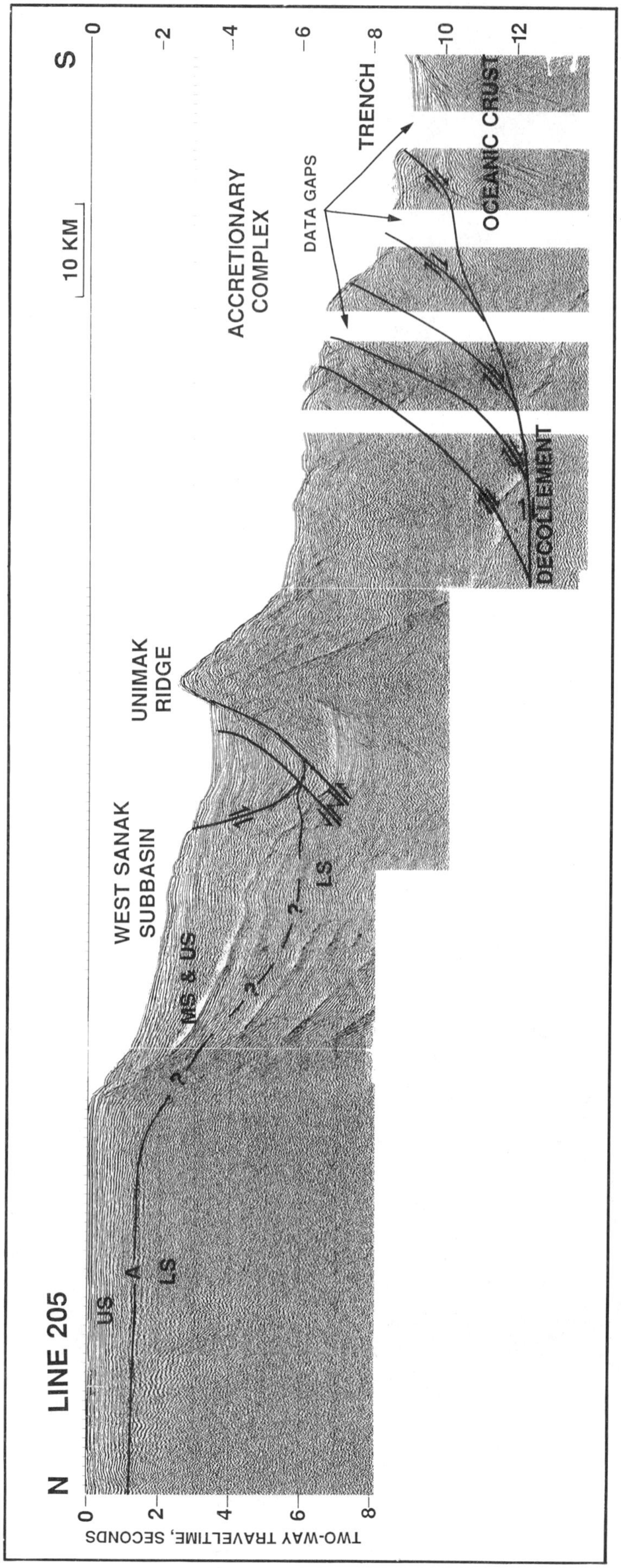

Figure 14. Seismic line 205 across west Sanak subbasin, Unimak Ridge, and continental slope. Line location shown in Figures 2, 11, and 12. Abbreviations as in Figure 13.

west side is formed by a high-angle growth fault or fault zone adjacent to Sanak Island, with as much as 5 km of vertical separation on horizon A (line 531, Figure 15). Strata within the central Sanak subbasin dip and thicken toward the fault, and are truncated along this fault. The fault must be an active feature, because the youngest basin fill is faulted and the fault has a sea-floor scarp.

A seismic line along the central subbasin axis (line 533, Figure 15) shows that the sediment fill above horizon A is at least 7 km thick, and is probably thicker, because the horizon is too deep to be well defined by the seismic data. Middle and upper series rocks crop out at or near the sea floor at the northwestern end of the basin. The middle series strata are relatively uniform in thickness from the northwestern end to about halfway into the basin, then thicken markedly seaward below the deepest part of the basin toward the basement high at the south end of the basin. Upper series strata are more uniform in thickness throughout much of the basin than middle series strata, but the upper series strata also thin and are truncated at the northwestern end. The uppermost beds show little deformation, uniform thickness, and some seaward progradation. The configuration of the middle and upper series strata at the northwestern end of the basin indicates uplift during upper series time, leading to the truncation of the basin rocks at the sea floor, and to the seaward thickening of upper series beds.

A basement high on the southeastern end of the central Sanak subbasin rises to within 1.5 km of the sea floor (line 533, Figure 15). Seismic reflectors within middle series strata thicken toward the high, and are truncated against it. These strata were deposited against the high during subsidence of the basin floor. The upper series strata cover the high and are undeformed over it. Therefore, the high developed largely during middle series time and was a barrier to seaward transport of sediment that formed the middle series rocks. Because upper series strata are mainly undeformed over the high, subsidence during upper series time has been regional, rather than localized along the growth fault. The basement high is associated with a large magnetic anomaly (Bruns and von Huene, 1977), indicating that the high probably is underlain by basalt, perhaps similar to the basalt dredged from Unimak Seamount.

The east Sanak subbasin (Figures 11, 12, 16, 17) is separated from the central Sanak subbasin by a faulted northwest-trending basement high. The maximum sediment thickness in the east Sanak subbasin is at least 7 km and could be more, because the position of horizon A is unclear on the seismic data where the section is thickest (lines 109 and 217, Figures 16, 17). At the landward side of the basin, middle and upper series strata onlap horizon A. These strata are flat lying at the north end of the line, and dip gently basinward beneath the outer shelf. There is no evidence for landward uplift, such as that observed in the central subbasin.

The east subbasin is bounded on the south by a faulted, buried basement high (line 217, Figures 16, 17). As in the central Sanak subbasin, middle series strata thicken seaward and are truncated against the high, and upper series strata are deposited seaward across the high. A migrated section of seismic line 217 (Figure 17) shows more detail of the structure of the basin and reveals rotation and thickening of middle series strata into the block-faulted basement. Subsidence along the growth fault continues to the present and affects the surface bathymetry (Figure 17). However, little deformation is observed in upper series strata against or over the high, indicating that the basement relief, and hence the basin's architecture, was largely formed during middle series time and before deposition of upper series strata.

Unimak Basin

Unimak Basin lies beneath the outer shelf and slope south of Unimak Island (Figures 11, 12). Less is known about this basin than the other basins because the region is crossed by only three, widely spaced seismic-reflection lines (lines 201, 203, and 205, Figure 2). Further, the tie-line along the shelf edge (line 208, Figure 2), which would ordinarily allow correlation of seismic units between the lines, is of very poor quality because of storm-induced acoustic noise. However, four sonobuoy refraction profiles along the line give a good estimate of sediment thickness above a layer with velocities >4.6 km/sec, a layer which we correlate with unit A (Figure 9). This information has been combined with the seismic-reflection interpretation to show the approximate basin form and sediment thickness (Figures 11, 12).

The east flank of Unimak Basin is probably controlled by a margin-oblique, down-to-the-west fault or fault zone, where shallow unit A rocks beneath the shelf west of Sanak Island change to the thick (>4.5 km) sedimentary sequence in Unimak Basin. This change is similar to that on the east flank of the central Sanak subbasin. Another feature of Unimak Basin is the ponding of sedimentary rocks in the eastern part behind Unimak Ridge, similar to that observed in the west Sanak subbasin on line 205 (Figure 14). However, to the west, the ridge diminishes and disappears, and strata below the upper slope form a seaward-prograding wedge. The location and characteristics of the west flank of Unimak Basin are unknown.

Shelf Edge and Upper Slope Sedimentary Rocks

Middle and upper series rocks (undifferentiated) extend beneath the upper continental slope, where they are 2 to 4 km thick on the seismic-reflection data (line 104, Figure 13; line 217, Figures 16, 17; Bruns, von Huene, and Childs, in press) and possibly as much as 6 km thick on seismic-refraction profiles (based on refraction velocities less than about 5 km/sec, Figure 9). Seaward of the shelf break, upper series strata are cut by a few to numerous seaward-dipping normal faults (Figures 16, 17). These are probably gravity-slide faults induced by oversteepening of the continental slope. This oversteepening in turn is caused by subsidence of the lower slope relative to the shelf, since the entire margin has been subsiding since the cutting of horizon A.

Mid-Slope Structures

Between about long. 160° and 164°W (the region seaward of Unimak Pass to the Shumagin Islands), a series of bathymetric and subsurface structural highs forms a continuous structural trend parallel to the margin at the mid-slope to upper slope region (Figure 4; line 205, Figure 14).

The bathymetric highs form the ridge herein informally called Unimak Ridge, which culminates in the formally named Unimak Seamount (Figure 4; U.S. Board on Geographic Names, 1981. Note that the name Unimak Seamount carries no connotation of process of formation or genesis, and deals only with a shape defined by bathymetric contours). The bathymetric ridge extends from seaward of Unimak Pass to

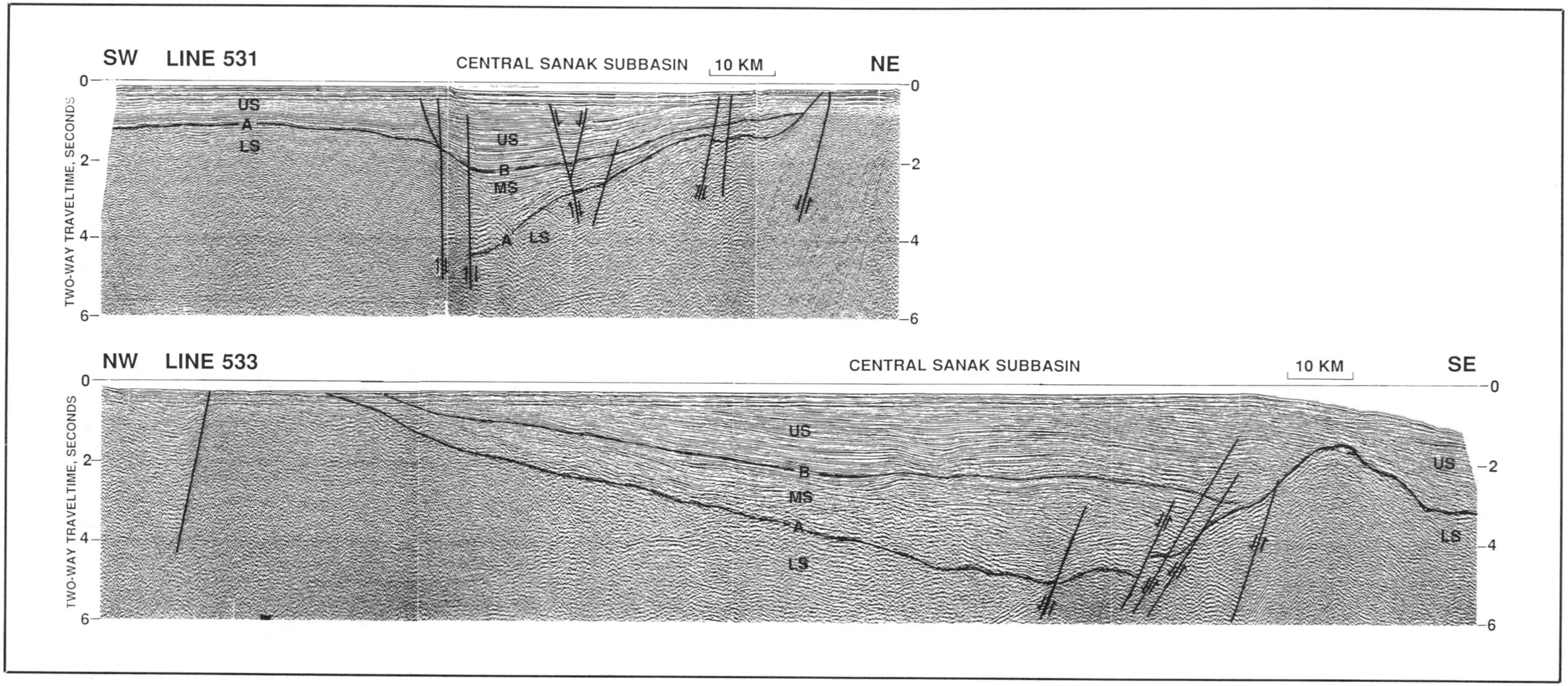

Figure 15. Seismic lines 531 and 533 across central Sanak subbasin. Line locations shown on Figures 2, 11, and 12. Abbreviations as on Figure 13.

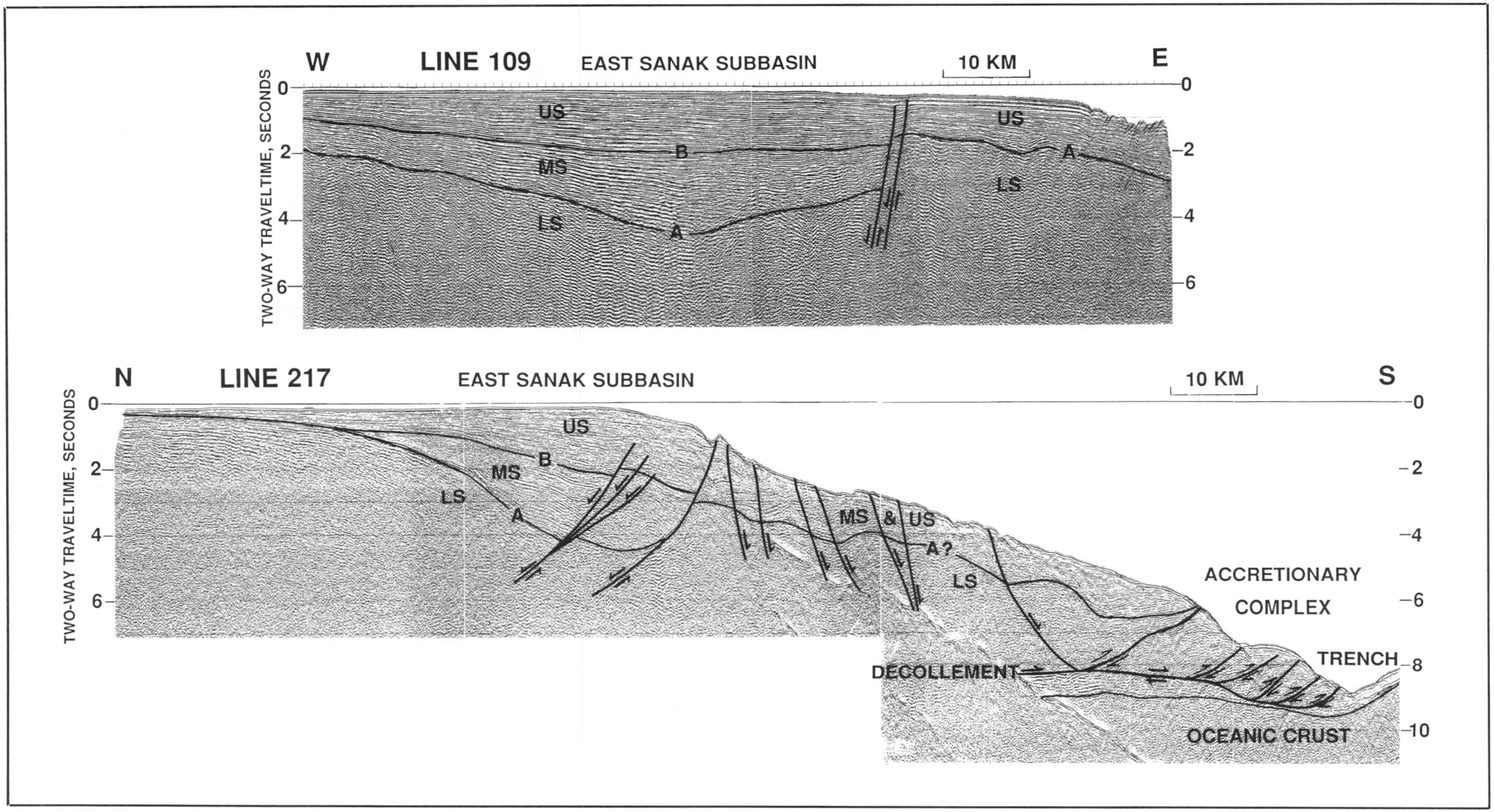

Figure 16. Seismic lines 109 and 217 across east Sanak subbasin and adjacent continental slope. Line locations shown on Figures 2, 11, and 12. Abbreviations as on Figure 13.

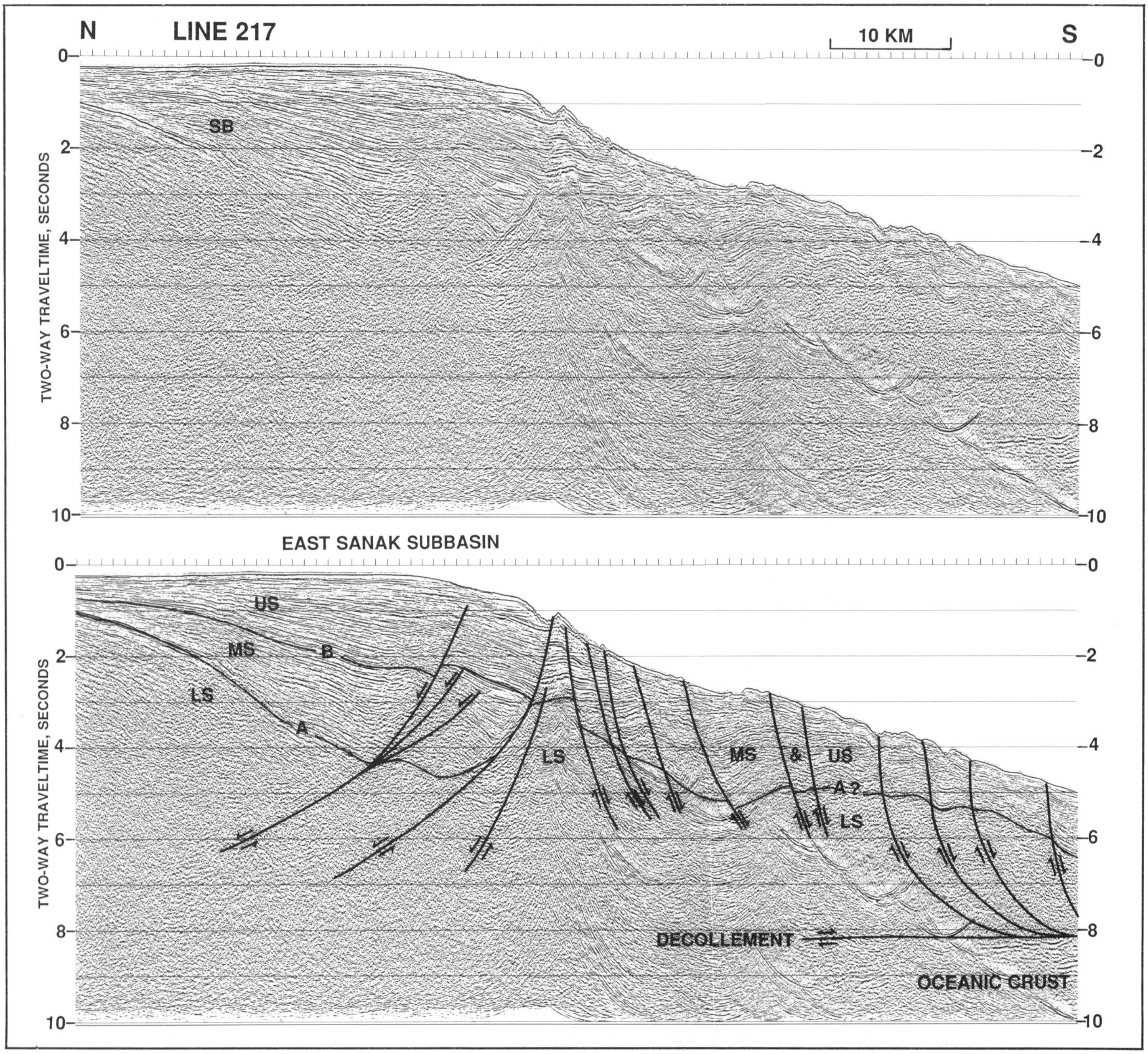

Figure 17. Migrated piece of line 217, showing details of landward- and seaward-dipping listric faults. Top—without interpretation; bottom—interpreted section. See Figure 16 for unmigrated time section. Line location shown on Figures 2, 11, and 12. Abbreviations as on Figure 13.

the seaward end of the central Sanak subbasin. Unimak Ridge is a very pronounced feature on seismic line 205 (Figure 14), which crosses the ridge southwest of Unimak Seamount. Unimak Ridge forms the outer margin of Unimak Basin and the west Sanak subbasin.

In the central and east Sanak subbasins, structural highs are not expressed bathymetrically, but are present in the subsurface where they form the outer boundary of the subbasins. As at Unimak Ridge, the subsurface structural highs apparently served as a barrier behind which middle series strata of the central and east Sanak subbasins were ponded. These subsurface structural highs are buried continuations of Unimak Ridge and, as discussed below, probably formed at about the same time.

Northeast of the Shumagin Islands, fewer bathymetric or subsurface basement highs are present. In this area, widely spaced seismic data suggest that rocks beneath the mid-slope region are disrupted primarily by small normal faults (Bruns, von Huene, and Childs, in press).

The bathymetric and subsurface highs have all developed as a result of block faulting and subsidence of lower series (basement) rocks along landward-dipping growth faults. On seismic lines that cross the highs, prominent seismic reflections mark the landward side of the highs. On a migrated seismic line (line 217, Figure 17), these reflections are resolved into probable fault-plane reflections off a growth fault. The seaward thickening of middle series strata into the fault also indicates rotation and offset along the fault during deposition. Thus, we conclude that these mid-slope highs are block-faulted basement rocks formed when the floor of the adjacent shelf basin subsided, and formed during crustal extension across the margin.

We believe that the mid-slope structures developed in post-horizon-A time, or beginning in about the late middle Miocene at the earliest. One reason for this conclusion is that the horizon A reflection appears to continue onto the tops of the basement blocks, indicating that faulting postdates development of the horizon A unconformity. Secondly, the middle series rocks, which onlap horizon A on the landward side, were deposited into the basin against the developing growth fault, indicating that the fault formed after the horizon A unconformity. Furthermore, coarse clastic rocks of late early and early middle Miocene age were dredged from the top of Unimak Seamount. Because coarse rocks are unlikely to be deposited on top of a topographic high, subsidence of the basin relative to the high must have occurred after deposition of the early middle Miocene strata, or beginning in late middle or late Miocene time.

We suggest that the mid-slope structures are underlain by autochthonous, arc-framework rocks (rocks emplaced during arc growth or deposited along the arc), rather than by allochthonous accreted or obducted oceanic rocks. We also suggest that the continental slope is underlain by lower series, arc-framework rocks as far seaward as the modern accretionary wedge. The primary evidence for these suggestions is that the geochemistry of the dredged basalts is similar to that of arc rocks, and the sedimentary rocks dredged from Unimak Seamount have clastic characteristics (no pelagic rocks). More inferential evidence is that seaward of Shumagin Basin (for example, on line 104, Figure 13), lower series rocks can be traced seaward to the accretionary complex near the trench; presumably the lower series rocks are also present southwestward beneath, and seaward of, the mid-slope structures. The mid-slope structures are therefore horst blocks developed by block faulting within the subsiding arc-framework rocks. The lower slope would be underlain by these same arc-framework rocks. Some possible causes for the structural growth are discussed below.

Accretionary Complex

The lower slope of the Shumagin margin is underlain by a thrust-faulted, highly deformed accretionary complex as much as 30 km wide (line 104, Figures 13, 18; line 205, Figure 14; line 217, Figure 16). On some seismic lines (line 104, Figures 13, 18; Bruns, von Huene, and Childs, in press; von Huene, Bruns, and Childs, in press), only the upper part of the sediment on the underthrusting Pacific plate is accreted, possibly in thrust-faulted packages that form duplex structures. The lower part of the sequence, probably composed of Paleogene sediments of the Zodiak deep-sea fan, is carried relatively undeformed well beneath the margin (Bruns, von Huene, and Childs, in press). Such a subduction style is also seen in the adjacent Kodiak and Aleutian regions (von Huene, 1979; von Huene, Moore, and Moore, 1979; von Huene et al., 1979, 1983, 1985; von Huene, Fisher, and Bruns, this volume; Scholl, Vallier, and Stevenson, 1983a, 1983b, this volume; McCarthy and Scholl, 1985; McCarthy et al., 1984). Most likely, the accretionary wedge evolved in late Miocene and younger time, as is inferred in the Aleutians to the west. A major change in the sediment-deposition rate into the Aleutian Trench occurred during the late Miocene, concurrent with pronounced uplift and glaciation of high mountains in the northern Gulf of Alaska (Plafker and Addicott, 1976; Piper, von Huene, and Duncan, 1973; von Huene and Kulm, 1973; Scholl, 1974; Scholl, Vallier, and Stevenson, 1982, 1983a; Vallier et al., in press). The emplacement of wide accretionary wedges along the Aleutian subduction zone coincides with, and is believed to be due at least in part to, increased sediment input on the subducting plate (Scholl, Vallier, and Stevenson, 1982, 1983a). The wedge could contain strata as old as Eocene from the subducting Zodiak fan (Stevenson, Scholl, and Vallier, 1982).

DISCUSSION

The Shumagin margin lies at the transition between the intraoceanic island arc of the Aleutian Ridge and the continental and accreted terranes of southern Alaska and the Alaska Peninsula. Prior to the Eocene, these two regions had markedly different geologic histories. The Aleutian Ridge developed during the Eocene, built in place on Early(?) Cretaceous oceanic crust. In marked contrast, the Alaska Peninsula-Shumagin-Kodiak region has had a complex geologic history that extends back to the middle Paleozoic and that, during the late Mesozoic and Cenozoic, was dominated by collision, accretion, and amalgamation of allochthonous terranes at a convergent margin.

Since the Eocene, the geologic history of the two regions has been similar. During the Eocene, volcanic activity led to the growth of the Aleutian Ridge beginning about 50 Ma and to the development of the volcanic Meshik arc (Wilson, 1985) across the Alaska Peninsula beginning about 40 Ma. Intense erosion of both the Aleutian Ridge and Alaska Peninsula regions during the Oligocene and Miocene resulted in deposition of thick offshore sedimentary accumulations, and in the development of a widespread unconformity at the crest of the ridge in the Aleutians, and on the shelf in the Shumagin and Kodiak regions.

This unconformity may be related to several tectonic events. First, the unconformity partly corresponds in time to marked drops in eustatic sea level during the Oligocene and late Miocene (Vail, Mitchum, and Thompson, 1977). The unconformity is also coincident with a marked change in tectonic regimes caused by a change in plate rates at about 43 Ma (Engebretson, Cox, and Gordon, in press) or possibly earlier (Byrne, 1979). At this time, spreading ceased at the Kula-Pacific spreading ridge, causing the Kula plate to move with the Pacific plate. Prior to 43 Ma, underthrusting was rapid, with rates as great as 20 cm/yr oblique to the margin. After 43 Ma, the direction changed from oblique to direct and the rate slowed to around 6 cm/yr, allowing the downgoing lithosphere to thermally expand below the margin and elevate it. Finally, subduction of the Kula-Pacific spreading ridge at around 10 Ma (Delong, Fox, and McDowell, 1978) introduced warm lithosphere beneath the margin and also elevated the margin. Each or all of these factors may have had an impact on the development of the unconformity (see extended discussions in Scholl, Vallier, and Stevenson, this volume; von Huene, Fisher, and Bruns, this volume).

During the Pliocene and Quaternary, the Aleutian Ridge was affected by differential subsidence and extreme crestal erosion as it subsided, and by the formation of a mid-slope structural uplift and deep offshore basins. The differential subsidence was accompanied by normal faulting that led to the formation of ridges, basins, and discrete tectonic blocks along the crest of the arc. During the same period, shelf and slope basins developed locally on the Kodiak and Shumagin shelves, accompanied by marked subsidence in some of the

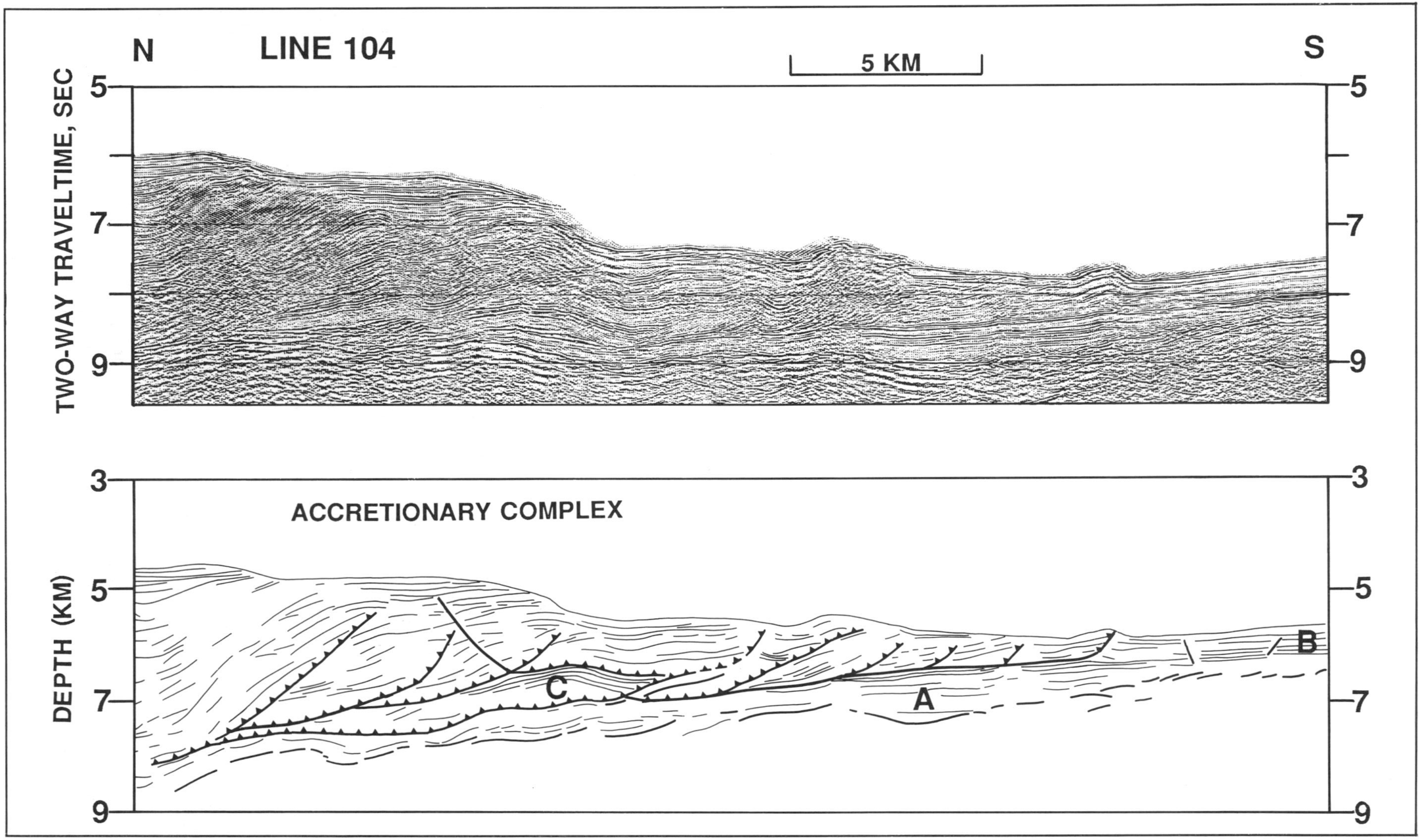

Figure 18. Migrated (top) and depth-converted, interpreted line drawing (bottom) of segment of line 104 across the lower slope accretionary complex. On bottom line drawing, strata in sediment package A are subducting and are probably composed of Eocene and Oligocene Zodiak fan material. Strata in package B are Miocene and younger, probably late Miocene and younger and largely glacially derived; these strata are being deformed by the thrust faults. Strata in package C may be uplifted package A rocks that form a duplex structure. See Figures 2, 11, and 12 for location, and Figure 13 for unmigrated time section. From von Huene, Bruns, and Childs (in press); Bruns, von Huene, and Childs, (in press).

offshore basins. In both regions, a wide accretionary wedge was emplaced during the late Miocene and later (Scholl, Vallier, and Stevenson, 1982, 1983a, this volume; Vallier et al., in press; Fisher, 1979; Fisher and von Huene, 1980; von Huene, Fisher, and Bruns, this volume).

Given these similarities and differences, what is the main feature of the Shumagin margin that seems to be confined to the arc-continent transition?

The most important characteristic of the Shumagin margin is that landward of the accretionary wedge, the late Cenozoic history of the margin is dominated by extensional tectonics, rather than by the compressional deformation commonly expected at a convergent margin. This extension is manifested in two general ways: (1) the margin-oblique subsidence of the central Sanak subbasin and possibly of Unimak Basin, which is present only at the transition from the peninsula to the Aleutian Ridge and does not have counterparts northeastward along the Aleutian convergent margin; and (2) the margin-parallel subsidence along landward-dipping growth faults that led to the development of the mid-slope structural highs of the Sanak region, and to the related massive subsidence that created the basins landward of the structural highs. The margin-oblique structures could be caused by subsidence or minor transform faulting along the underlying older Mesozoic rocks of the pre-Aleutian Ridge (Beringian) margin trend; the margin-parallel structures, however, could be due to late Cenozoic convergent-margin tectonics.

Development of Margin-Oblique Structure

Sanak and Unimak basins have developed as a result of: (1) margin-oblique extension and subsidence, forming the central Sanak Basin and possibly Unimak Basin, and (2) margin-parallel subsidence in the basins on growth faults landward of the mid-slope structural highs. The northwest structural trend of the central Sanak subbasin, and possibly Unimak Basin, is a unique feature of the Shumagin margin compared with the rest of the Alaska Peninsula convergent margin, although similar tectonic features may be present westward along the Aleutian Ridge (Scholl, Vallier, and Stevenson, 1982, 1983a, 1983b, this volume). Northeastward, structures are parallel to the margin and developed in a compressional tectonic regime (Fisher and von Huene, 1980; Bruns, 1985). Even in areas where structures are oblique to the margin, as in the Middleton Shelf region to the northeast, the shelf and slope structures are compressional (Bruns, 1985).

The extensional structure is coincident with other features that suggest segmentation of the fore arc along major transverse tectonic boundaries. Such tectonic boundaries are delineated by breaks in structural trend, by major changes in structure at the boundary, and by segmentation of earthquake aftershock zones (Fisher, Bruns, and von Huene, 1981; Davies et al., 1981). In the Shumagin region, this segmentation has led to well-defined seismic gaps—regions where major earthquakes are predicted within the next few years. The bathymetric and structural high of Unimak Ridge lies entirely between the Unimak and Shumagin seismic gaps and the corresponding bounding margin-oblique faults. Southwestward, the structure dies out. To the northeast, the mid-slope structural trend continues with diminished structural relief and forms the deeply buried outer margin of the east Sanak subbasin. The buried segment of the mid-slope structures ends seaward of the Shumagin Islands, at about the northeast margin of the Shumagin seismic gap. The position of the central Sanak subbasin is coincident with the western limit of a well-defined seismic gap, the Shumagin seismic gap, defined by Sykes (1971), Sykes et al. (1981), and Davies et al. (1981). Similarly, the eastern margin of Unimak Basin lies along the eastern limit of a suspected seismic gap, the Unimak gap (House et al., 1981; Davies et al., 1981). The faults bounding Unimak Basin and the central Sanak subbasin are thus the shallow crustal expression of the seismic-gap boundaries and of the associated transverse boundaries.

The margin-oblique faults have been active during at least the late Miocene to Holocene. Middle and upper series strata in Sanak Basin are broken by or deformed over the faults, and the faulting extends to the sea floor on a fault adjacent to Sanak Island.

The margin-oblique faults parallel, and are probably underlain by, the rock-framework of the Cretaceous and early Tertiary Beringian margin. Prior to the development of the Aleutian Ridge, the continental margin presumably turned from southwest-trending to northwest-trending in the vicinity of the present-day Sanak and Unimak Islands at the southeasternmost end of the Alaska Peninsula. Thus the faulting could be a result of structural deformation or reactivation of faults on the Beringian margin.

Alternatively, the Shumagin Formation rocks underlying Sanak Island appear to be offset northward from the regional trend of the Shumagin Formation rocks and the bounding Border Ranges fault, perhaps along right-lateral strike-slip faults. Thus, the central Sanak subbasin and Unimak Basin could have developed as pull-apart basins. Whatever the cause of the basins, they appear to have developed by a process analogous to that which created some of the Aleutian Ridge summit basins and upper-slope basins (Scholl, Vallier, and Stevenson, this volume), and possibly the Bering Shelf basins (Cooper, Scholl, and Marlow, this volume).

Development of Margin-Parallel Structure

The margin-parallel, east to northeast trends of Unimak Basin and the east and west Sanak subbasins are due to late middle or late Miocene subsidence along landward-dipping growth faults on the landward side of the mid-slope structural highs. These structures are caused by crustal extension across the margin.

These basins have the same structural form, in which an outer high forms the seaward margin of the basin. The strata in the basins were deposited onto a rapidly subsiding basin, trapped behind the structural high, and eventually overtopped it in the east Sanak subbasin. Thus, the general form of Sanak Basin developed in the middle or early late Miocene.

The latest Miocene and younger, margin-oblique northwest-trending faults that form the central Sanak subbasin are apparently superposed on and have disrupted the northeast-trending basin form, creating the present segmented basin. This transverse boundary segmentation continues to the present, because the bounding faults of the central Sanak subbasin are still active.

Although the transverse boundaries markedly segment and affect the structural relief of the margin-parallel structures, the boundaries are probably not a primary cause of the structures. Whatever the factors are that led to segmentation of the fore arc—whether related to the subducting plate or, more likely for the central Sanak subbasin and Unimak Basin, to subsidence or reactivated faulting on normal or strike-slip faults in the Mesozoic and early Tertiary north Pacific (Beringian) margin—these factors do not appear to be a direct cause of the margin-parallel basin and ridge structures.

Instead, subsidence of the margin is required to develop the observed growth faults and the depth to the buried erosional unconformity at horizon A. Subsidence is shown by the present depth to the horizon A unconformity and to the paleo-shelf edge. Originally at sea level, these features are presently buried to a depth of 2 to 2.5 km.

Microfauna also suggest margin subsidence. Foraminifera in the dredge samples are presently in water 1 to 2 km deeper than the depths where they were deposited. There are no indications in the microfauna that these samples are transported. In the dredge samples from Unimak Seamount, benthic foraminifers from both the Eocene samples and from the early and middle Miocene samples indicate an outer neritic to upper bathyal depositional depth, or less than about 600 m. These samples are now at a depth of 1650 to 1800 m. Similarly, fauna in the Pliocene and Pleistocene samples from dredge 4 were deposited in upper to middle bathyal depths (about 200 to 1200 m), and are presently at 2000 to 2200 m, suggesting at least some subsidence. Thus, the dredge samples suggest possible subsidence with respect to present sea level from 1 to 2 km since the middle Miocene. This area of subsidence stretches from at least Unimak Seamount to south of the Shumagin Islands, and probably includes the entire Unimak Ridge feature.

Subsidence could be caused by thermal effects similar to those that contributed to the subsidence of the crestal area of the Aleutian arc (Scholl, Vallier, and Stevenson, 1983a, this volume; Vallier et al., in press). An argument against a thermal model for the Shumagin region is that uplift, rather than subsidence, has occurred on the Alaska Peninsula during the late Miocene, Pliocene, and Quaternary.

Thermal subsidence could be occurring along the southeast continuation of the Beringian margin, but this would probably cause margin-oblique faulting, rather than margin-parallel structural development. Thermal subsidence could also be occurring because the passage of the Kula-Pacific spreading ridge beneath the margin at about 10 Ma (Delong, Fox, and McDowell, 1978) was followed by subduction of progressively cooler and less buoyant Pacific plate lithosphere.

We speculate that another process might also have caused the margin to subside—tectonic erosion or the removal of material from the overriding plate by the downgoing plate at a subduction zone (Scholl et al., 1980). Such an effect has been documented in the Japan arc (von Huene, Langseth, and Nasu, 1980; von Huene et al., 1982). Subduction erosion in the late Miocene and early Pliocene may have led to massive subsidence of the Shumagin margin, and to block faulting and block rotation within the subsiding margin. The initial subsidence of Sanak Basin along its bounding growth fault could be a result of this block faulting. Tectonic erosion could also have removed the outer part of the margin and thus account for the near-trench position of the island-arc-affinity basalt sampled in the dredge hauls, and for the oversteepening which has caused the development of seaward-dipping normal (listric) faults along the margin. Because subsidence of the margin continues to the present (Figure 17), this process could still be active, even while accretion occurs at the base of the slope.

In summary, the margin-oblique faults bounding the central Sanak subbasin and possibly Unimak Basin lie on margin-segmenting transverse boundaries. These basins could be pull-apart basins related to reactivated strike-slip (transform) faulting along the now-buried Late Cretaceous and early Tertiary Beringian margin. The margin-parallel growth faults bounding Unimak Ridge and its northeastward structural continuation are caused by extension across the margin. This extension may in turn be caused by subsidence over a cooling Pacific plate lithosphere, or by late Miocene to Holocene tectonic erosion-induced subsidence and block faulting of the margin.

PETROLEUM GEOLOGY

Because little is known about the Cenozoic rocks beneath the Shumagin margin, the petroleum geology must be largely inferred by analogy with the adjacent Kodiak Shelf and Alaska Peninsula regions. Exploratory drilling near the Shumagin margin has been entirely on the Alaska Peninsula. These onshore wells cannot be used to evaluate the pre-Cenozoic rocks of the Shumagin Shelf and slope, because the Mesozoic and older rocks of the Alaska Peninsula are probably truncated at the Border Ranges fault and thus would not be present beneath the Shumagin margin. However, source and reservoir rocks of Cenozoic age on the Alaska Peninsula may have coeval, lithologically similar offshore counterparts in the nearby deep basins of the Shumagin Shelf and beneath the continental slope. Thus, we will briefly discuss the wells that sampled Cenozoic rocks, but will not review the wells that sampled only Mesozoic and older rocks.

Drilling History

Since 1959, Tertiary rocks have been sampled in nine onshore wells drilled along the Bristol Bay side of the Alaska Peninsula (Figure 19; Brockway et al., 1975). Several wells have been drilled on the Gulf of Alaska side of the peninsula, but these sampled only Mesozoic and older rocks.

The northern four wells—the General Petroleum Great Basins No. 1 and 2, Great Basins Ugashik No. 1, and the Gulf-Alaskco Port Heiden No. 1 (Figure 18)—bottomed in volcanic or granitic rocks. In these wells, the thickness of the flat-lying Tertiary sequence ranges from about 1220 m to 3350 m and consists of interbedded nonmarine to shallow-marine sandstone, siltstone, claystone, and coal. Radiometric ages from the volcanic sequence underlying the flat-lying Tertiary sedimentary rocks in the Gulf Port Heiden and Great Basins Ugashik wells range from 33 to 42 Ma (Oligocene to late Eocene; Brockway et al., 1975). No significant shows of oil or gas have been reported from these wells.

The southern five wells—the Gulf Sandy River Federal No. 1, Pan American Hoodoo Lake No. 1 and 2, Pan American David River No. 1-A, and the Amoco Cathedral River No. 1—bottomed in either Tertiary or Mesozoic sedimentary rocks. The Gulf Sandy River well encountered gas associated with coal in the middle part of the Miocene Bear Lake Formation. Oil and gas shows were also noted in the Bear Lake Formation in the basal sandstone beds of the Sandy River well. However, in this well, the oil may have been derived from source rocks within the underlying late Eocene and early Oligocene Stepovak Formation (Brockway et al., 1975).

Source Beds

Core chips and drill cuttings from eight of the nine wells drilled along the north side of the Alaska Peninsula were analyzed for lithology and paleontology by McLean (1977).

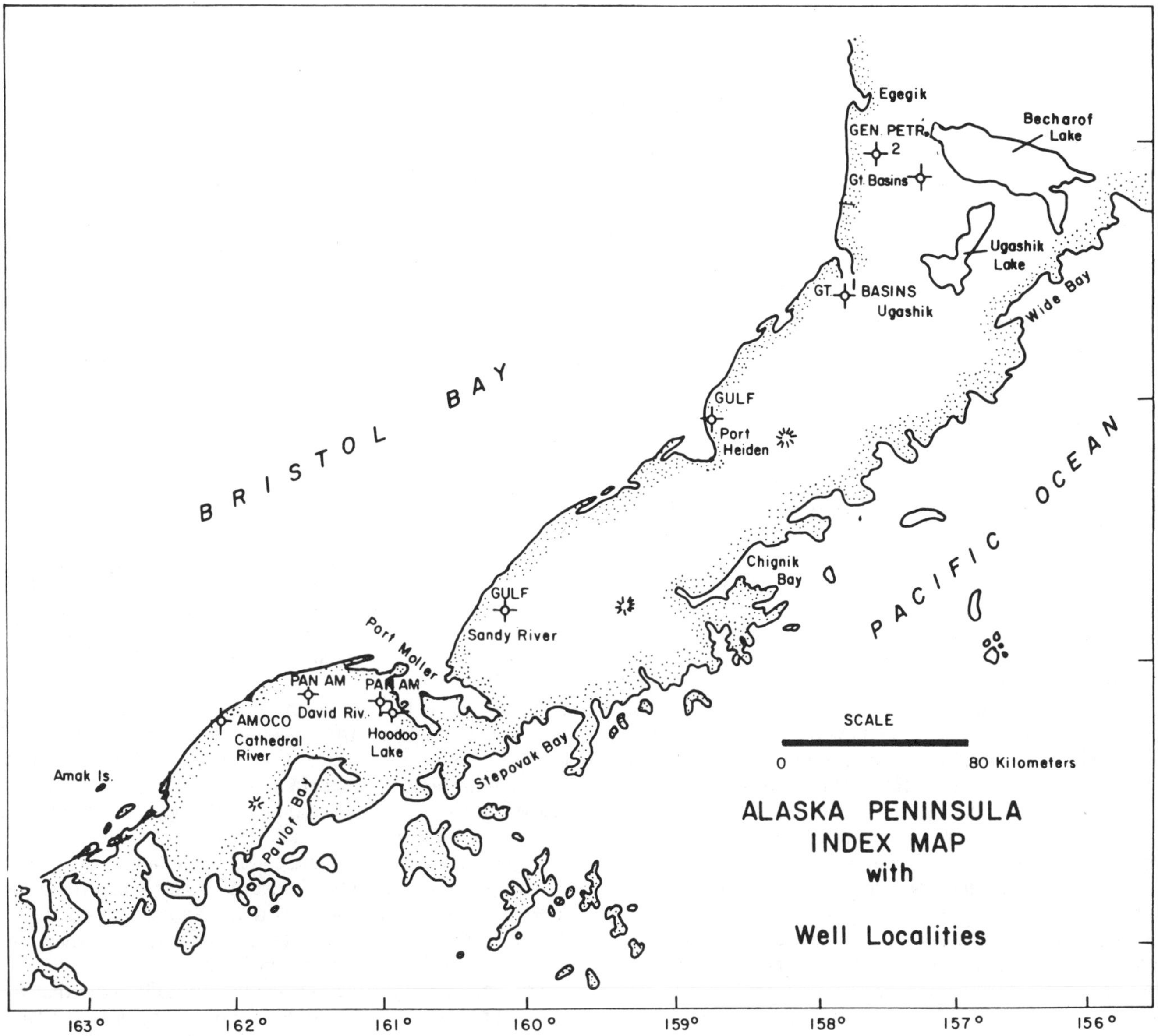

Figure 19. Index map of Alaska Peninsula showing sites of exploratory wells drilled on the peninsula which sampled Cenozoic rocks.

Results suggest that, at least locally, sedimentary rocks of Tertiary age contain potential source rocks capable of generating hydrocarbons. The Paleogene strata in these wells are rich in organic carbon but are immature. Neogene strata have significantly lower total organic-carbon content than the Paleogene rocks, possibly reflecting nonmarine or brackish-water environments of deposition. Although the Alaska Peninsula was the focus of igneous activity throughout much of the Mesozoic and Tertiary, thermal maturity indicators such as vitrinite reflectance and coal rank suggest that, on a regional scale, sedimentary rocks have not been subjected to abnormally high geothermal gradients (McLean, 1977).

Lyle et al. (1979) studied 14 stratigraphic sections totaling 5000 m along the Alaska Peninsula. In potential Tertiary source rocks, they found that the total organic carbon for their samples ranged from $<0.2\%$ to 8.0%, that the major organic constituents are herbaceous-spore debris, and that most samples are thermally immature. They concluded that dry gas is the most probable hydrocarbon to form in Tertiary source rocks on the Alaska Peninsula.

The best source rocks in Tertiary outcrops appear to be the black marine siltstone and shale beds in the late Eocene and early Oligocene Stepovak Formation (McLean, 1977). On the Alaska Peninsula, the Stepovak Formation is locally at least 4500 m thick (Burk, 1965). Scattered shows of oil and gas in the Stepovak Formation have been reported from three Alaska Peninsula wells—Gulf Sandy River, Pan American Hoodoo Lake No. 2, and Pan American David River 1-A (Brockway et al., 1975).

Organic Analysis of Dredge Samples

Because the rocks offshore have not been extensively sampled, little is known about possible source rocks from the Shumagin margin. Six dredge samples (including four not described in Table 1) were carefully picked from all available

Table 4. Geochemical analysis of selected Shumagin dredge samples.

Sample	2-9	2-10	2-12	2-13	2-16	4-3	Average for site 2
Lithology	Mudstone	Dolomite	Clastic dolomite	Mudstone	Mudstone	Nodular limestone	
Age	Early or middle Miocene	Middle or late Eocene	Early or middle Miocene	Eocene(?)	Early or middle Miocene	Early Pleistocene	
Organic Carbon (%) TEA/Rock-Eval	0.45	0.33	0.24	0.66/1.1	0.80/0.92	0.30	0.50 ± 0.23
Total Carbon (%)	0.43	9.4	9.0	0.67	0.77	6.6	
Carbonate Carbon (%)	0.0	9.1	8.8	0.01	0.0	6.3	
Total Hydrocarbon	0.03	0.05	0.03	0.05	0.06	0.03	0.04 ± 0.01
"Live Carbon" (%)	7	15	13	8	8	10	10 ± 4
Volatile Hydrocarbon (ppm)	37	46	20	61	94	76	52 ± 28
T_{Max} °C TEA/Rock-Eval	440	443	456	437/409	429/441	433	441 ± 9

samples and were analyzed for organic carbon (Table 4; Kvenvolden and von Huene, 1985; Bruns et al., 1985); five samples are from site 2, and one sample is from site 4 (locations shown on Figure 4).

All samples were analyzed by a pyrolysis technique (Thermal Evolution Analysis or TEA; Claypool and Reed, 1976), which provides information on total hydrocarbon yield (measured in percent of organic carbon), volatile hydrocarbons (measured in parts per million of the total hydrocarbon yield), and T_{max} in °C, representing the temperature at which the maximum amount of organic matter is thermally decomposed. "Live carbon" is considered to be the carbon that will, when heated, still yield additional hydrocarbon products such as methane gas.

The results of these analysis (Table 4) indicate that the rock samples are poor potential source rocks for both oil and gas. The average amount of organic carbon was $<0.6\%$, a value considered near the lower limit for potential sources of hydrocarbons (Hunt, 1979). The total hydrocarbon yield and live carbon were also low and ranged from 0.03 to 0.06% and from 7 to 15%, respectively. Live carbon values $<30\%$ indicate that the organic matter is prone to generate gas rather than oil (Magoon and Claypool, 1981). The volatile hydrocarbon for all six samples was <100 ppm. T_{max} ranged from 429° to 456°C, but the pyrograms from which these data were taken lacked distinctive peaks, indicating both that the organic matter is unstructured and probably immature, and that the T_{max} temperatures are considered unreliable for interpretive purposes.

Two samples of the six were selected for further detailed examination of organic matter type (kerogen analysis; Table 5); these samples were chosen because they appeared to be richest in organic carbon, and because there was sufficient material on which to carry out the analysis. These samples were subjected to a specialized pyrolytic technique called Rock-Eval (Tissot and Welte, 1978). In addition, the analysis included vitrinite-reflectance measurements, visual-kerogen analyses by both transmitted and incident light, and evaluation of the thermal alteration index (TAI). The Rock-Eval measurements for organic carbon and T_{max} (Table 4) are considered less reliable than TEA measurements and therefore are not considered further. Rock-Eval analysis yielded hydrogen indices (HI) and oxygen indices (OI) indicating that the organic matter in the samples is of terrestrial origin and is gas prone (Kvenvolden and von Huene, 1985; Tissot and Welte, 1978).

Vitrinite reflectance (R_o) values of primary vitrinite are 0.38 and 0.50% (Table 5). TAI evaluations range from 2.0 to 2.5 (Table 5). Both vitrinite reflectance values and TAI values indicate that the organic matter in these samples is immature with respect to oil or gas generation. Recycled vitrinite with higher R_o values was also present, and indicates reworking. Visual kerogen analysis (Table 5) indicates that, in addition to vitrinite and recycled vitrinite, the samples also contain exinite, inertinite, recycled sporinite, and other amorphous material. The distribution of these kerogen macerals suggests that most of the organic matter is terrestrial in origin.

The combined geochemical-analysis results show that the amount of organic carbon in the dredge samples is small and approaches the lower limit as a potential source of hydrocarbons. The organic matter is mainly from terrestrial sources, is gas prone, and is immature with respect to gas generation. Upon thermal evolution, the samples are expected to generate only small amounts of gas (Kvenvolden and von Huene, 1985).

These discouraging results are from a limited number of dredge samples and cannot be considered as representative of the thick strata within the basins or beneath the outer shelf and upper slope. Onshore Tertiary rocks such as the Stepovak Formation have better source-rock characteristics, and could have offshore correlatives within the lower series rocks seaward of the paleoshelf break. Also, early Tertiary rocks with better source rock potential have been found in dredge samples in the adjacent Aleutian margin (Scholl, Vallier, and Stevenson, 1983b, this volume; Cooper, Marlow, and Vallier, 1984) and the Kodiak margin (Fisher, 1980).

Geothermal Gradient

The hydrocarbon potential of a sedimentary sequence is dependent upon burial at sufficient depths and temperatures to generate hydrocarbons. Heat-flow measurements have not

Table 5. Results of kerogen analysis of selected Shumagin dredge samples.

	Sample 2-13	Sample 2-16
HI (mgHc/gOC)	160	122
OI ($mgCO_2/gOC$)	71	111
R_o (%)	0.50	0.38
TAI	2.0-2.4	2.3-2.5
Exinite (%)	5	7
Vitrinite (%)	20	15
Inertinite (%)	5	—
Recycled Vitrinite (%)	30	24
Recycled Sporinite (%)	10	16
Other Amorphous	30	40

been made along the Shumagin margin. The best indirect measurement is from MacLeod (1982), who made an estimate of about 30 °C/km based on the depth to gas hydrates beneath the Shumagin slope. Fisher (1980) estimated a geothermal gradient of 20° to 30 °C/km for the adjacent Kodiak margin, based primarily on wells in Cook Inlet and in the northern Gulf of Alaska at Middleton Island. Using techniques similar to those used by MacLeod (1982), Kvenvolden and von Huene (1985) obtained gradients in the Kodiak region ranging from about 28° to 36 °C/km, with a best estimate of about 30 °C/km; and McCarthy et al. (1984) estimated 22.7 °C/km for the Adak-Amlia sector of the central Aleutian Ridge. Other convergent margins have similar geothermal gradients, such as an average gradient of 32 °C/km along the Middle America Trench (von Huene et al., 1982) and a 24° to 32 °C/km gradient for the Japan Trench (Langseth and Burch, 1980). Thus, a gradient of 20° to 30 °C/km, and most likely around 25° to 30 °C/km, is probably a reasonable estimate for the frontal part of Shumagin margin.

Although 25° to 30 °C/km is a low gradient, with such a gradient onset of hydrocarbon generation (about 100 °C) could occur at a subsurface depth of about 3 to 4 km. Substantial thicknesses of strata, primarily in Sanak Basin, but also in Unimak Basin and below the outer shelf and upper slope, could be within the hydrocarbon generation zone.

Reservoir Beds

The rocks that have the greatest potential as reservoirs for oil and gas in the Shumagin margin are probably sandstone units equivalent to those in the onshore Miocene Bear Lake Formation. In the onshore exploratory wells, Neogene sandstone beds locally have high porosity and permeability to a depth of about 2400 m (McLean, 1977). Below this depth, the porosity and permeability values decline rapidly, because of alteration of unstable volcanic rock fragments and precipitation of chlorite and zeolite cements. Sandstone in Paleogene formations, such as the Tolstoi and Stepovak, contains abundant unstable volcanic detritus and matrix clay. These rocks are dense and highly indurated, and thus are not considered good reservoir beds (McLean, 1977).

In outcrop sections, Lyle et al. (1979) found that 63% of 14 measured stratigraphic sections contain potential Tertiary reservoir sandstone. However, in most areas the porosities and permeabilities of these sandstones are low because of pervasive pore-filling mineralization. The best-preserved and most probable reservoir rocks are sandstone beds in the Bear Lake Formation. Bear Lake sandstone beds are both marine and nonmarine and contain a combination of volcanic grains, dioritic grains, chert, and sedimentary lithic fragments. Shows of oil and gas were reported from the basal sandstone beds of the Bear Lake Formation in the Gulf Sandy River and Pan American David River wells.

No potential reservoir sandstone was found in the dredge samples from the Shumagin margin. The sandstone samples contain an abundance of unstable rock fragments and matrix clay. However, the samples are rich in dioritic or granitic material, and are therefore similar in composition to sandstone in the Bear Lake Formation. Thus, somewhat better reservoir characteristics, similar to those of the sandstone beds in the Bear Lake Formation onshore could occur landward of the dredge location, closer to the source terrane of the granitic debris.

Traps

Little is known about the location, extent, and amount of closure on potential traps beneath the Shumagin margin. Within the major basins, structural and stratigraphic traps are known or are likely to be present within the upper Miocene and younger section (middle and upper series rocks). These traps could form by draping and differential compaction of strata over block-faulted basement highs, by pinch-outs within the sedimentary sequence, or by onlap against the lower series rocks (against rocks below horizon A). The lower series rocks may also contain traps, although none have been mapped. Broad, gentle anticlinal features are present beneath and around Shumagin Basin, and may be present within Sanak Basin or beneath the slope, but have not been recognized in the widely spaced seismic data. Major anticlinal features are not present on the Shumagin margin, except possibly along the mid-slope structural trend.

The lower series rocks below horizon A may contain traps, and the middle series rocks overlying the horizon may act as a seal. Also, if Paleogene strata have generated hydrocarbons, these hydrocarbons could migrate into traps in the overlying Miocene and younger strata.

Structures within the mid-slope structural trend could be traps, either within the structures or in stratigraphic traps along the growth faults on the flanks of the structures. More detailed mapping is needed to determine the amount of faulting that may have occurred, and if closure and traps are present.

The numerous structural folds within the accretionary complex may be potential traps, as suggested by McCarthy et al. (1984) for the Aleutian accretionary body. However, structures within the complex may be geologically ephemeral—quickly deformed or removed with continued subduction. Thus, the overall potential for trapping hydrocarbons generated in place or migrating through subducted sediment is probably low.

Resource Potential

The resource potential of the Shumagin margin might be considered as relatively favorable if source and reservoir rocks equivalent to those on the Alaska Peninsula extend seaward. We speculate that lower Tertiary source rocks could be present seaward of the paleoshelf break beneath the outer shelf and slope, and in Sanak Basin. We also speculate that offshore equivalents of the onshore Bear Lake Formation, with good reservoir potential, might be present in Sanak Basin, and possibly elsewhere beneath the slope.

A possible simple model for Tertiary deposition and erosion is that Paleogene and perhaps lower and middle Miocene strata were deposited over accreted Cretaceous material and across the continental slope. Oligocene and Miocene erosional events removed part or all of this material from the shelf as far out as the inferred paleoshelf edge. However, seaward, beneath the slope, such material would still be present. In particular, the position of the paleoshelf edge (Figure 11) is such that at least part of Sanak Basin (primarily the east Sanak subbasin) and much of the present outer shelf and slope would have been seaward of the paleoshelf break. If so, then Paleogene rocks, and possibly offshore rocks equivalent to the onshore organic-rich Stepovak Formation, could be present in these areas.

During the late Miocene, subsidence occurred on the growth faults of the mid-slope structural trend (including Unimak Ridge), and sediment began to fill the developing Sanak Basin. The lower part of this late Miocene and younger sequence (middle series rocks) could provide both source and reservoir rocks.

The main reason for suggesting source-rock potential in the middle series rocks of Sanak Basin is that these strata may have been deposited in the oxygen-minimum zone of the slope or into a closed, silled, and possibly anoxic basin formed behind the basin-margin highs.

Potential reservoir rocks could be present in Sanak Basin, because at least part of the adjacent high-standing shelf rocks that were eroded include granitic batholiths of the Sanak, Shumagin, and Semidi Islands. Substantial amounts of granitic detritus could have reached Sanak Basin and the adjacent continental slope; such a source terrane is indeed required for the dredged sandstone samples from Unimak Ridge. Nilsen (1984) suggests that the Bear Lake Formation may have been derived in part from these same batholiths. Certainly Sanak Basin, immediately adjacent to the Shumagin and Sanak batholiths, would receive and trap sediment eroded from these batholiths during the Oligocene and Miocene erosional events. Such strata within Sanak Basin could contain sandstone with good reservoir potential.

Sanak and Unimak basins are the best prospects for resource potential within the Shumagin region. Positive factors include thick strata, traps, the possibility of source and reservoir rocks, and the potential for hydrocarbon generation at a subsurface depth of about 3 to 4 km. However, there is a lack of known major anticlinal traps with extensive closure, and we do not know about the presence or extent of potential source and reservoir rocks.

Outer shelf and continental slope areas with thick strata are the next most prospective areas, for the reasons noted above, and because they are more likely to contain Paleogene source rocks. Traps may be present within or along the mid-slope structural trend. A major negative feature is the greater water depth over potential target areas, thus making exploration and exploitation more difficult.

The Shumagin Basin has relatively low resource potential within the basin-fill sequence, because the fill is relatively thin (<2.5 km) and therefore is probably immature. However, lower Tertiary rocks may be present below the basin and may be deformed into potential traps; or these rocks may allow migration of hydrocarbons into traps in the overlying Miocene and younger basin fill.

Much of the shelf outside the Sanak and Shumagin basins probably has low resource potential, because only thin (<1 to 2 km) sequences of Miocene and younger strata are present; and much of the lower Tertiary sequence, if it was ever present, was probably stripped off by the Oligocene and Miocene erosional events.

The accretionary wedge has relatively low resource potential, although fluids or gas generated within subducting sediment could migrate along subduction zone faults into structures within the subduction complex, or upward into rocks overlying the subduction zone.

Much more information needs to be obtained from the Shumagin margin before we can make a valid estimate of its resource potential. Although we have some knowledge of structure and sediment thickness, we know little of ages, lithologies, source and reservoir rock potential, and thermal maturity of the sedimentary rocks. We do not know the similarities and differences between Alaska Peninsula rocks and Shumagin margin rocks. The resource potential of the Shumagin margin could be substantially greater than we currently estimate, if extensive lower Tertiary (lower series) strata have been preserved beneath the margin in areas otherwise deemed unprospective because late Miocene and younger strata are thin.

ACKNOWLEDGMENTS

We thank Michael A. Fisher and Alan K. Cooper for critical reviews. We also thank Donna Blackman for interpreting the sonobuoy refraction records; Phyllis A. Swenson for drafting support; Tracy L. Vallier and David A. Clague for identifying the dredge basalts and interpreting the geochemical data; and Tracy L. Vallier for providing descriptions of the thin sections. John Barron identified the diatoms from the dredge samples, and Gerta Keller and David Schuster provided identification of foraminifers.

REFERENCES CITED

Beikman, H. M., 1980, Geologic map of Alaska: U.S. Geological Survey, 1 sheet, scale 1:2,500,000.

Brockway, R., B. Alexander, P. Day, W. Lyle, R. Hiles, W. Decker, W. Polski, and B. Reed, 1975, Bristol Bay region, Stratigraphic correlation section, southwest Alaska: Anchorage, Alaska Geological Society.

Bruns, T. R., 1985, Tectonics of the Yakutat block, an allochthonous terrane in the northern Gulf of Alaska: U.S. Geological Survey Open-File Report 85–13, 112 p.

Bruns, T. R., and K. C. Bayer, 1977, Multichannel seismic reflection data acquired on the *M/V Cecil H. Green* in the Gulf of Alaska, June–August 1975: U.S. Geological Survey Open-File Report 77-352, 16 p.

Bruns, T. R., and R. von Huene, 1977, Sedimentary basins on the Shumagin Shelf, western Gulf of Alaska: Houston, Texas, Ninth Annual Offshore Technology Conference Proceedings, v. 1, p. 41.

Bruns, T. R., R. von Huene, and J. R. Childs, in press, The Shumagin margin—record section 104, *in* R. von Huene, ed., Seismic images of modern convergent margin tectonic structure: AAPG Studies in Geology.

Bruns, T. R., R. von Huene, R. C. Culotta, and S. D. Lewis, 1985, Summary geologic report for the Shumagin Outer Continental Shelf (OCS) planning area, Alaska: U.S. Geological Survey Open-File Report 85-32, 58 p.

Burk, C. A., 1965, Geology of Alaska Peninsula island arc and continental margin: GSA Memoir 99, 250 p.

Byrne, T., 1979, Late Paleocene demise of the Kula-Pacific spreading center: Geology, v. 7, p. 341–344.
Claypool, G. E., and P. R. Reed, 1976, Thermal analysis technique for source-rock evaluation: Quantitative estimate of organic richness and effects of lithologic variation: AAPG Bulletin, v. 60, p. 608–626.
Connelly, W., M. Hill, B. Hill, and J. C. Moore, 1977, The Uyak Complex, Kodiak Islands, Alaska: A subduction complex of early Mesozoic age, *in* M. Talwani and W. C. Pitman, eds., Island arcs, deep sea trenches and back-arc basins: American Geophysical Union Maurice Ewing Series, v. 1, p. 465.
Cooper, A. K., M. S. Marlow, and D. W. Scholl, 1986, Geological framework of the Bering Sea crust, this volume.
Cooper, A. K., M. S. Marlow, and T. L. Vallier, 1984, Summary geologic report for the St. George Outer Continental Shelf (OCS) planning area, Bering Sea, Alaska: U.S. Geological Survey Open-File Report 84–418, 81 p.
Cooper, A. K., D. W. Scholl, and M. S. Marlow, 1986, Structural framework, sedimentary sequences, and hydrocarbon potential of the Aleutian and Bowers basins, Bering Sea, this volume.
Davies, J. N., and L. House, 1979, Aleutian subduction zone seismicity, volcano-trench separation and their relation to great thrust-type earthquakes: Journal of Geophysical Research, v. 84, p. 4583–4591.
Davies, J. N., L. R. Sykes, L. House, and K. H. Jacob, 1981, Shumagin seismic gap, Alaska Peninsula: History of great earthquakes, tectonic setting, and evidence for high seismic potential: Journal of Geophysical Research, v. 86, p. 3821–3855.
Davis, D. L., and J. C. Moore, 1984, 60 m.y. intrusive rocks from the Kodiak Islands link the Peninsular, Chugach, and Prince William terranes (abs.): GSA Abstracts with Programs, v. 16, no. 5, p. 277.
Delong, S. E., P. J. Fox, and F. W. McDowell, 1978, Subduction of the Kula Ridge at the Aleutian Trench: GSA Bulletin, v. 89, p. 83–95.
Detterman, R. L., 1985, The Paleogene sequence on the Alaska Peninsula (abs.): AAPG Bulletin, v. 69, p. 661.
Detterman, R. L., and J. K. Hartsock, 1966, Geology of the Iniskin-Tuxedni region, Alaska: U.S. Geological Survey Professional Paper 512, 78 p.
Detterman, R. L., and R. L. Reed, 1980, Stratigraphy, structure, and economic geology of the Iliamna quadrangle, Alaska: U.S. Geological Survey Bulletin 1368–B, p. B1-B86, 1 sheet, scale 1:250,000.
Detterman, R. L., J. E. Case, and F. H. Wilson, 1979, Paleozoic rocks on the Alaska Peninsula: U.S. Geological Survey Circular 804–B, p. B85–B86.
Detterman, R. L., M. E. Yount, and J. E. Case, 1981, Megafossil localities, checklists, and stratigraphic sections, Chignik and Sutwik Island quadrangles, Alaska: U.S. Geological Survey Miscellaneous Field Studies Map MF–1053–N, 2 sheets, scale 1:250,000.
Detterman, R. L., T. P. Miller, M. E. Yount, and F. H. Wilson, 1981, Geologic map of the Chignik and Sutwik Island quadrangles, Alaska: U.S. Geological Survey Miscellaneous Investigations Map I–1229, scale 1:250,000.
Detterman, R. L., J. E. Case, F. H. Wilson, M. E. Yount, and W. H. Allaway, Jr., 1983, Generalized geologic map of the Ugashik, Bristol Bay, and part of Karluk quadrangles, Alaska: U.S. Geological Survey Miscellaneous Field Studies Map MF–1539–A, scale 1:250,000.
Dix, C. H., 1955, Seismic velocities from surface measurements: Geophysics, v. 20, no. 1, p. 68–86.
Engebretson, D. C., A. Cox, and R. G. Gordon, 1985, Relative motions between oceanic and continental plates in the Pacific Basin: GSA Special Paper 206, 59 p.
Fisher, M. A., 1979, Structure and tectonic setting of continental shelf southwest of Kodiak Island, Alaska: AAPG Bulletin, v. 63, 373 p.
Fisher, M. A., 1980, Petroleum geology of the Kodiak Shelf, Alaska: AAPG Bulletin, v. 64, p. 1140–1157.
Fisher, M. A., 1981, Location of the Border Ranges fault southwest of Kodiak Island, Alaska: GSA Bulletin, v. 92, p. 19–30.
Fisher, M. A., and M. Holmes, 1981, Large-scale structure of deep strata beneath Kodiak Shelf, Alaska: GSA Bulletin, v. 91, p. 218–224.
Fisher, M. A., and R. von Huene, 1980, Structure of upper Cenozoic strata beneath Kodiak Island, Alaska: AAPG Bulletin, v. 64, p. 1014–1033.
Fisher, M. A., and R. von Huene, 1982, Maps showing the geologic structure of the continental shelf southeast and southwest of Kodiak Island, Alaska from 24-fold seismic data: U.S. Geological Survey Miscellaneous Field Studies Map MF–1457, scale 1:500,000.
Fisher, M. A., and R. von Huene, 1984, Geophysical investigation of a suture zone: The Border Ranges fault of southern Alaska: Journal of Geophysical Research, v. 89, no. B12, p. 11,333–11,351.
Fisher, M. A., T. R. Bruns, and R. von Huene, 1981, Transverse tectonic boundaries near Kodiak Island, Alaska: GSA Bulletin, v. 92, p. 10–18.
Herrera, R. C., 1978, Developments in Alaska in 1977: AAPG Bulletin, v. 62, p. 1311–1321.
House, L., L. R. Sykes, J. N. Davies, and K. H. Jacob, 1981, Evidence for a possible seismic gap near Unalaska Island, *in* R. W. Simpson and P. G. Richards, eds., Earthquake prediction, an international review: American Geophysical Union Maurice Ewing Series 4, p. 81–92.
Hunt, J. M., 1979, Petroleum geochemistry and geology: San Francisco, W. H. Freeman, 617 p.
Jones, D. L., and S. H. B. Clark, 1973, Upper Cretaceous (Maestrichtian) fossils from the Kenai-Chugach Mountains, Kodiak, and Shumagin Islands, southern Alaska: U.S. Geological Survey Journal of Research, v. 1, p. 125.
Jones, D. L., N. J. Silberling, H. C. Berg, and G. Plafker, 1981, Map showing tectonostratigraphic terranes of Alaska, columnar sections, and summary description of terranes: U.S. Geological Survey Open-File Report 81–792, 20 p., 2 maps.
Keller, G., R. von Huene, K. McDougall, and T. R. Bruns, 1984, Paleoclimatic evidence for Cenozoic migration of Alaskan terranes: Tectonics, v. 3, p. 473–495.
Kvenvolden, K. A., and R. von Huene, 1985, Natural gas generation in sediments of the convergent margin of the eastern Aleutian Trench area, *in* D. G. Howell, ed., Tectonostratigraphic terranes of the Circum-Pacific region: Circum-Pacific Council for Energy and Mineral Resources, Earth Science Series no. 1, p. 31–49.
Langseth, M., and T. Burch, 1980, Geothermal observations of the Japan Trench transect, *in* Scientific Party, Initial Reports of the Deep Sea Drilling Project, Vols. 56 and 57, pt. 2: Washington, D.C., U.S. Government Printing Office, p. 1207–1210.
Lyle, W. M., J. A. Morehouse, I. F. Palmer, Jr., and J. G. Bolm, 1979, Tertiary formations and associated Mesozoic rocks in the Alaska Peninsula area, Alaska, and their petroleum-reservoir and source-rock potential: Alaska Division of Geological and Geophysical Surveys Geologic Report 62, 65 p.
MacLeod, M. K., 1982, Gas hydrates in ocean bottom sediments: AAPG Bulletin, v. 66, p. 2649–2662.
Magoon, L. B., and G. E. Claypool, 1981, Petroleum geology of Cook Inlet Basin, Alaska—An exploration model: AAPG Bulletin, v. 65, p. 355–374.
Magoon, L. B., W. L. Adkison, and R. M. Egbert, 1976, Map showing geology and petroleum operations, Cook Inlet area, Alaska: U.S. Geological Survey Miscellaneous Investigations Map I–1019, 3 sheets, scale 1:250,000.
Marlow, M. S., and A. K. Cooper, 1980, Mesozoic and Cenozoic structural trends under southern Bering Sea Shelf: AAPG Bulletin, v. 64, p. 2139–2155.
Marlow, M. S., A. J. Stevenson, D. W. Scholl, and T. L. Vallier, 1984, Speculations on the petroleum geology of the accretionary body: An example from the central Aleutians: Marine and Petroleum Geology, v. 1, p. 151–167.
McCarthy, J., and D. W. Scholl, 1985, Mechanisms of subduction—

Accretion along the central Aleutian Trench: GSA Bulletin, v. 96, p. 691-701.

McLean, H., 1977, Organic geochemistry, lithology, and paleontology of Tertiary and Mesozoic rocks from wells on the Alaska Peninsula: U.S. Geological Survey Open-File Report 77-813, 63 p.

McLean, H., 1979, Tertiary stratigraphy and petroleum potential of Cold Bay-False Pass area, Alaska Peninsula: AAPG Bulletin, v. 63, p. 1522-1526.

McLean, H., C. L. Englehardt, and D. G. Howell, 1978, Reconnaissance geologic map of the Cold Bay and False Pass quadrangles, Alaska: U.S. Geological Survey Open-File Report 78-323, scale 1:250,000.

Minster, J. B., and T. H. Jordan, 1978, Present-day plate motions: Journal of Geophysical Research, v. 83, p. 5331-5334.

Moore, G. W., 1969, New formations on Kodiak and adjacent islands, Alaska: U.S. Geological Survey Bulletin 1274-A, p. A27-A35.

Moore, J. C., 1972, Uplifted trench sediments—Southwestern Alaska—Bering Shelf edge: Science, v. 175, p. 1103-1105.

Moore, J. C., 1973a, Cretaceous continental margin sedimentation, southwestern Alaska: GSA Bulletin, v. 84, p. 596-614.

Moore, J. C., 1973b, Complex deformation of Cretaceous trench deposits southwestern Alaska: GSA Bulletin, v. 84, p. 2005-2020.

Moore, J. C., 1974, The ancient continental margin of Alaska, *in* C. A. Burk and C. L. Drake, eds., The geology of continental margins: New York, Springer-Verlag, p. 811-816.

Moore, J. C., 1978, Orientation of underthrusting during latest Cretaceous and earliest Tertiary time, Kodiak Islands, Alaska: Geology, v. 6, p. 209-213.

Moore, J. C., and A. Allwardt, 1980, Progressive deformation of a Tertiary trench slope, Kodiak Island, Alaska: Journal of Geophysical Research, v. 85, no. B9, p. 4741-4756.

Moore, J. C., and W. Connelly, 1979, Tectonic history of the continental margin of southwestern Alaska, late Triassic to earliest Tertiary, *in* A. Sisson, ed., The relationship of plate tectonics to Alaskan geology and resources: Anchorage, Alaska Geological Society, p. H1-H29.

Moore, J. C., T. Byrne, P. W. Plumley, M. Reid, H. Gibbons, and R. S. Coe, 1983, Paleogene evolution of the Kodiak Islands, Alaska: Consequences of ridge-trench interaction in a more southerly latitude: Tectonics, v. 2, no. 3, p. 265-293.

Nilsen, T. H., 1984, Miocene back-arc tidal deposits of the Bear Lake Formation, Alaska Peninsula, *in* K. M. Reed and S. Bartsch-Winkler, eds., The United States Geological Survey in Alaska: Accomplishments during 1982: U.S. Geological Survey Circular 939, p. 85-88.

Nilsen, T. H., and G. W. Moore, 1979, Reconnaissance study of Upper Cretaceous to Miocene stratigraphic units and sedimentary facies, Kodiak and adjacent islands, Alaska (with a section on sedimentary petrology by G. R. Winkler): U.S. Geological Survey Professional Paper 1093, 34 p.

Nilsen, T. H., and G. G. Zuffa, 1982, The Chugach terrane, a Cretaceous trench-fill deposit, southern Alaska, *in* J. K. Leggett, ed., Trench-forearc geology: GSA Special Publication 10, published for the Geological Society of London by Blackwell Scientific Publications, p. 213-227.

Palmer, A. R., 1983, The Decade of North American Geology 1983 geologic time scale: Geology, v. 11, no. 9, p. 503-504.

Pavlis, T. L., 1982, Origin and age of the Border Ranges fault of southern Alaska and its bearing on the Late Mesozoic tectonic evolution of Alaska: Tectonics, v. 1, p. 343-368.

Peter, G., B. H. Erickson, and P. U. Grim, 1971, Magnetic structure of the Aleutian Trench and northeast Pacific Basin, *in* A. E. Maxwell, ed., The sea: New York, Wiley-Interscience, p. 191-222.

Piper, D. J., R. von Huene, and J. R. Duncan, 1973, Late Quaternary sedimentation in the active eastern Aleutian Trench: Geology, v. 1, p. 19-22.

Plafker, G., and W. O. Addicott, 1976, Glaciomarine deposits of Miocene through Holocene age in the Yakataga Formation along the Gulf of Alaska margin, *in* T. P. Miller, ed., Recent and ancient sedimentary environments in Alaska: Anchorage, Alaska Geological Society Symposium, Proceedings, 1975, p. Q1-Q23.

Plumley, P. W., R. S. Coe, and T. Byrne, 1983, Paleomagnetism of the Paleocene Ghost Rocks Formation, Prince William terrane, Alaska: Tectonics, v. 2, p. 295-314.

Plumley, P. W., R. S. Coe, T. Byrne, M. R. Reid, and J. C. Moore, 1982, Paleomagnetism of Paleocene volcanic rocks of the Kodiak Islands indicates 23 degrees of northward latitudinal displacement: Nature, v. 300, p. 50-52.

Scholl, D. W., 1974, Sedimentary sequences in north Pacific trenches, *in* C. A. Burk and C. L. Drake, eds., The geology of continental margins: New York, Springer-Verlag, p. 493-504.

Scholl, D. W., T. L. Vallier, and A. J. Stevenson, 1982, Sedimentation and deformation in the Amlia fracture zone sector of the Aleutian Trench: Marine Geology, v. 48, p. 105-134.

Scholl, D. W., T. L. Vallier, and A. J. Stevenson, 1983a, Arc, forearc, and trench sedimentation and tectonics; Amlia Corridor of the Aleutian Ridge, *in* J. S. Watkins and C. L. Drake, eds., Studies in continental margin geology: AAPG Memoir 34, p. 413-439.

Scholl, D. W., T. L. Vallier, and A. J. Stevenson, 1983b, Geologic evolution of the Aleutian Ridge—Implications for petroleum resources: Journal of the Alaska Geological Society, v. 3, p. 33-46.

Scholl, D. W., T. L. Vallier, and A. J. Stevenson, 1986, Geologic evolution and petroleum geology of the Aleutian Ridge, this volume.

Scholl, D. W., R. von Huene, T. L. Vallier, and D. G. Howell, 1980, Sedimentary masses and concepts about tectonic processes at underthrust ocean margins: Geology, v. 8, p. 564-568.

Stevenson, A. J., D. W. Scholl, and T. L. Vallier, 1982, Tectonic and geologic implications of the Zodiak fan, Aleutian Abyssal Plain, northeast Pacific: GSA Bulletin, v. 94, p. 259-273.

Sykes, L. R., 1971, Aftershock zones of great earthquakes, seismicity gaps and earthquake prediction for Alaska and the Aleutians: Journal of Geophysical Research, v. 76, p. 8021-8041.

Sykes, L. R., J. B. Kisslinger, L. House, J. N. Davies, and K. H. Jacob, 1981, Rupture zones and repeat times of great earthquakes along the Alaska-Aleutian arc, 1784-1980, *in* D. W. Simpson and P. G. Richards, eds., Earthquake prediction, an international review: American Geophysical Union, Maurice Ewing Series 4, p. 73-80.

Tissot, B., and D. Welte, 1978, Petroleum formation and occurrence: New York, Springer-Verlag, 538 p.

U.S. Board on Geographic Names, 1981, Gazetteer of undersea features (3rd ed.): Washington D.C., Defense Mapping Agency, 125 p.

Vail, P. R., R. M. Mitchum, Jr., and S. Thompson, III, 1977, Global cycles of sea level changes, *in* C. E. Payton, ed., Seismic stratigraphy—Applications to hydrocarbon exploration: AAPG, Memoir 26, p. 83-97.

Vallier, T. L., D. W. Scholl, M. A. Fisher, T. R. Bruns, R. von Huene, and A. J. Stevenson, in press, Geologic framework of the Aleutian Arc, Alaska: GSA Decade of North American Geology.

von Huene, R., 1979, Structure of the outer continental margin off Kodiak Island, Alaska, from multichannel seismic records, *in* J. Watkins and L. Montadert, eds., Geological investigations of continental margins: AAPG Memoir 29, p. 261-272.

von Huene, R., and L. D. Kulm, 1973, Tectonic summary of Leg 18, *in* Scientific party: Initial reports of the Deep Sea Drilling Project, Vol. 18: Washington, D.C., U.S. Government Printing Office, p. 961-976.

von Huene, R., T. R. Bruns, and J. R. Childs, in press, Aleutian Trench, Shumagin segment, seismic section 104, part I, *in* R. von Huene, ed., Seismic images of modern convergent margin tectonic structure: AAPG Studies in Geology series.

von Huene, R., M. A. Fisher, and T. R. Bruns, 1986, Geology and evolution of the Kodiak margin, Gulf of Alaska, this volume.

von Huene, R., M. Langseth, and N. Nasu, 1980, Summary, Japan Trench transect, *in* Scientific party: Initial reports of the Deep Sea

Drilling Project, Vol. 56, Pt. 1: Washington, D.C., U.S. Government Printing Office, p. 473–488.
von Huene, R., G. M. Moore, and J. C. Moore, 1979, Cross-section of the Alaska Peninsula–Kodiak Island–Aleutian Trench: GSA Map and Chart series MC-28A, scale 1:250,000.
von Huene, R., M. Langseth, N. Nasu, and H. Okada, 1982, A summary of Cenozoic tectonic history along the IPOD Japan Trench transect: GSA Bulletin, v. 93, p. 829–846.
von Huene, R., J. Miller, M. A. Fisher, and G. Smith, 1983, An eastern Aleutian Trench seismic record, *in* A. W. Bally, ed., Seismic expression of structural styles—A picture and work atlas: AAPG Studies in Geology 15, v. 3, p. 3.4.2-135-139.
von Huene, R., G. M. Moore, J. C. Moore, and C. D. Stephens, 1979, Cross-section, Alaska Peninsula–Kodiak Island–Aleutian Trench: Summary, GSA Bulletin, Part 1, v. 90, p. 427–430.
von Huene, R., A. Bouma, G. W. Moore, M. A. Hampton, R. Smith, and G. L. Dolton, 1976, A summary of petroleum potential, environmental geology, and the technology, time frame, and infrastructure for exploration and development of the western Gulf of Alaska: U.S. Geological Survey Open-File Report 76–325.
von Huene, R. S. Box, R. L. Detterman, M. A. Fisher, J. C. Moore, and H. Pulpan, 1985, Continent-Ocean transect A-2: North Pacific Basin to Kuskokwim Mountains: GSA Decade of North American Geology Transect, 1 sheet.
Wilson, F. H., 1985, The Meshik arc—An Eocene to earliest Miocene magmatic arc on the Alaska Peninsula: Alaska Division of Geological and Geophysical Surveys Professional Report 88, 14 p.
Wilson, F. H., J. E. Case, and R. L. Detterman, 1985, Preliminary description of a Miocene zone of structural complexity, Port Moller and Stepovak Bay quadrangles: U.S. Geological Survey Circular 945, p. 55–56.
Wilson, F. H., R. L. Detterman, and J. E. Case, 1985, The Alaska Peninsula terrane: A definition: U.S. Geological Survey Open-File Report 85–450, 17 p.

Addendum: One of the basalt samples from site 3 was dated by the argon 40/39 total fusion technique, and yielded a minimum age of 19.8 ± 1.0 m.y. (early Miocene). The age is minimum because some argon may have been lost owing to alteration of plagioclase. The date is, however, consistent with the age of the associated lower and middle Miocene sedimentary rocks (Bruns et al., 1987).

Bruns, T. R., T. L. Vallier, L. B. Pickthorn, and R. von Huene, 1987, Volcanic-arc dacite and early Miocene basalt dredged from the Shumagin margin, Alaska, *in* Hamilton, T. D., and J. P. Galloway, eds., Geologic Studies in Alaska by the U.S. Geological Survey in 1986: USGS Circular 998, p. 143-146.

CHAPTER NINE

GEOLOGY AND EVOLUTION OF THE KODIAK MARGIN, GULF OF ALASKA

Roland von Huene
Michael A. Fisher
Terry R. Bruns

U.S. Geological Survey (USGS)
Menlo Park, California

ABSTRACT

The Kodiak margin extends landward from the eastern Aleutian Trench to the Border Ranges fault of the shelf area. The geologic record begins in the Early late Cretaceous with a record of convergent-margin tectonism that also dominated much of the Cenozoic. Intermittent arc volcanism and tectonic structures also indicate collision, tectonic erosion, and lateral slip. Early Tertiary subduction accretion was succeeded in the late Eocene by regional uplift and widespread erosion that lasted until the late Miocene, when the present tectonic phase of accretion was initiated. Geologic conditions may have been moderately favorable for the development of hydrocarbon resources, but economic accumulations have not yet been discovered. Their discovery will probably require an exploration approach to dealing with a geologic history that has no close analog to areas of present hydrocarbon production.

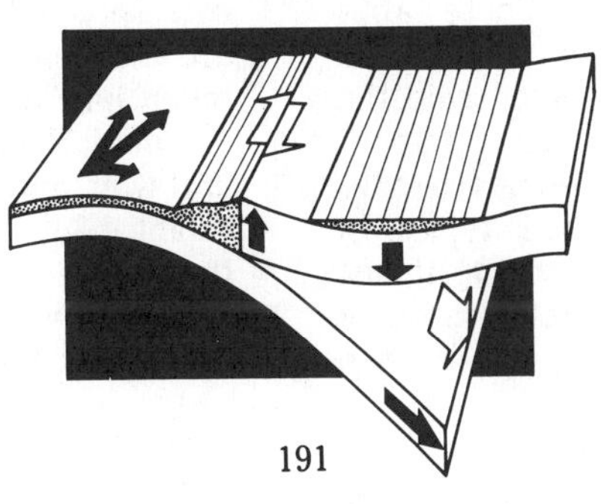

INTRODUCTION

The systematic collection of geophysical and geological data by the USGS in the Kodiak Shelf area began in 1975 and was reported in about 15 publications from 1976 to 1985. The topics of these papers ranged from estimates of hydrocarbon resources to basic research into tectonic and sedimentary processes. In this chapter, we summarize this decade of work which continues as seismic data are reprocessed with more advanced techniques than those available to us during the late 1970s, and as studies in adjacent areas or reexamination of geophysical data and samples suggest other interpretations and conclusions. No USGS offshore fieldwork has been undertaken in the area of the Kodiak margin since 1981.

Despite the lack of recent fieldwork, we have gained some new insight by reprocessing some of the USGS seismic reflection data, and by comparing the Kodiak margin tectonic history with the plate tectonic history of the north Pacific. The Kodiak margin tectonic history was recently advanced by the completion of an initial biostratigraphic study of the Kodiak and adjacent regions (Keller et al., 1984). From the temporal relationship of major north Pacific plate and Kodiak margin tectonic events, a possible cause-effect relationship, which we discuss, becomes evident. We also describe the lithostratigraphy and biostratigraphy of the shelf region of the margin, and each of its tectonic elements. It is important to understand the tectonic history of this margin in order to understand its potential for hydrocarbon resources.

REGIONAL SETTING

The Kodiak margin area includes most of the basic elements of arc-trench systems (Figure 1)—a deep linear trench, a subsiding fore-arc basin rimmed seaward by uplifted structural highs, a chain of active volcanoes, and intense earthquake seismicity. The area can also be set apart from adjacent segments of the Aleutian arc-trench system by some unique aspects in its physiography, by its upper crustal structure, by the discrete zone of aftershocks accompanying major earthquakes, and by the configuration of its Benioff zone.

The Kodiak Shelf has a bathymetry of tectonic and glacial origin that has been little modified since retreat of the ice. The physiography consists of flat areas, generally beneath 50 to 100 m of water, and erosional glacial troughs as much as 200 m deep that trend transverse to the regional NE–SW direction of the Kodiak margin and Aleutian Trench. Landward, the glacial troughs become deep fiords indenting Kodiak Island or the pass between the Kenai Peninsula and the Kodiak group of islands (Hampton, 1983).

Along the edge of the shelf are discontinuous banks with relief as high as 60 m that reflect tectonic uplift. These banks and the edge of the shelf trend northeast parallel to the trench and the volcanic arc (Figure 1). At a smaller scale, the physiographic relief near Kodiak Island includes prominent fault scarps with a northeast strike. The northeast–southwest regional trend of tectonic elements is typical of the Kodiak Shelf. The Kodiak margin ends to the northeast where the margin takes on an eastward trend south of Prince William Sound (Figure 1). Here the Aleutian Trench also ends. The shelf is traversed by the east–west trending Amatuli Trough, which projects to the break between the Kenai Peninsula and the Kodiak group of islands; its origin, whether tectonic or erosional, is not clear.

Tectonically and physiographically, the fore-arc basins of Cook Inlet and Shelikof Strait are separated from forearc basins on the continental shelf by an uplifted terrain. This terrain is the Kodiak insular area and the Kenai Peninsula, a unique structural feature in the Aleutian arc-trench system (Fisher, Detterman and Magoon, this volume). The southwest end of the Kodiak group of islands coincides with a disrupted area of the continental slope and to a jog in the gradual narrowing of the distance between the arc and the trench. The northeast and southwest boundaries of the Kodiak Shelf also correspond to offsets in the volcanic arc (Fisher, Bruns, and von Huene,, 1981).

In the offshore area of the shelf, the most extensive single stratigraphic horizon that was defined in seismic reflection data is a regional unconformity that is commonly the acoustic basement on 24-channel seismic reflection records. This horizon is the floor of at least seven basins (Figure 1) on the shelf along the eastern Aleutian arc-trench system; three are on the Kodiak Shelf, three are on the Shumagin Shelf, and one is on the Middleton Shelf (Bruns et al., this volume). Stevenson, Albatross, and Tugidak Basins, which are on the Kodiak Shelf, are filled with little-deformed sediment of presumed upper Miocene and younger age; below the unconformity, Cretaceous through earliest Miocene rocks are inferred. The basins are separated by broad basement highs trending transverse to the regional trend of the margin.

The distinctive physiography and tectonic structure of the Kodiak Shelf may be related to fundamental differences in crustal structure that set the Kodiak Shelf area apart from adjacent segments of the Aleutian arc-trench system. The difference is most distinctly indicated where the northeast end of the Kodiak Shelf area abuts the Yakutat block (Bruns, 1983). The southern edge of the Yakutat block is marked by a distinctive west-trending magnetic anomaly (Figure 1), the slope anomaly (Schwab, Bruns, and von Huene, 1980) that parallels and lies just north of the Amatuli Trough. The slope anomaly corresponds to the change in trend of the continental margin at the eastern end of the Aleutian Trench. Some thrust faults and fold axes in the Yakutat block parallel the slope anomaly, in contrast with the NE–SW strike of structures on the Kodiak Shelf. The southwest end of the Kodiak Shelf area corresponds to the boundary between the Kodiak domain of aftershock activity associated with major earthquakes and the Shumagin seismic gap (Sykes, 1971). Furthermore, the configuration of Benioff earthquake hypocenters is steepest beneath the Shumagin seismic gap area, intermediate beneath the Kodiak Shelf area, and very shallow where the Yakutat block enters the Aleutian subduction zone (Davies and House, 1979; Jacob, Nakamura and Davis, 1977). The difference in configuration of the Benioff zone corresponds to a greater separation between the volcanic arc and trench axis along the Yakutat block segment and a much narrower separation along the Shumagin segment. Although the crustal structure is known in only a very general way, and the differences between segments seem well established, the character of the crustal structural changes at the ends of the Kodiak Shelf area are unknown at depth.

Two lithostratigraphic terrane boundaries have been drawn across the Kodiak Shelf area. The boundary between

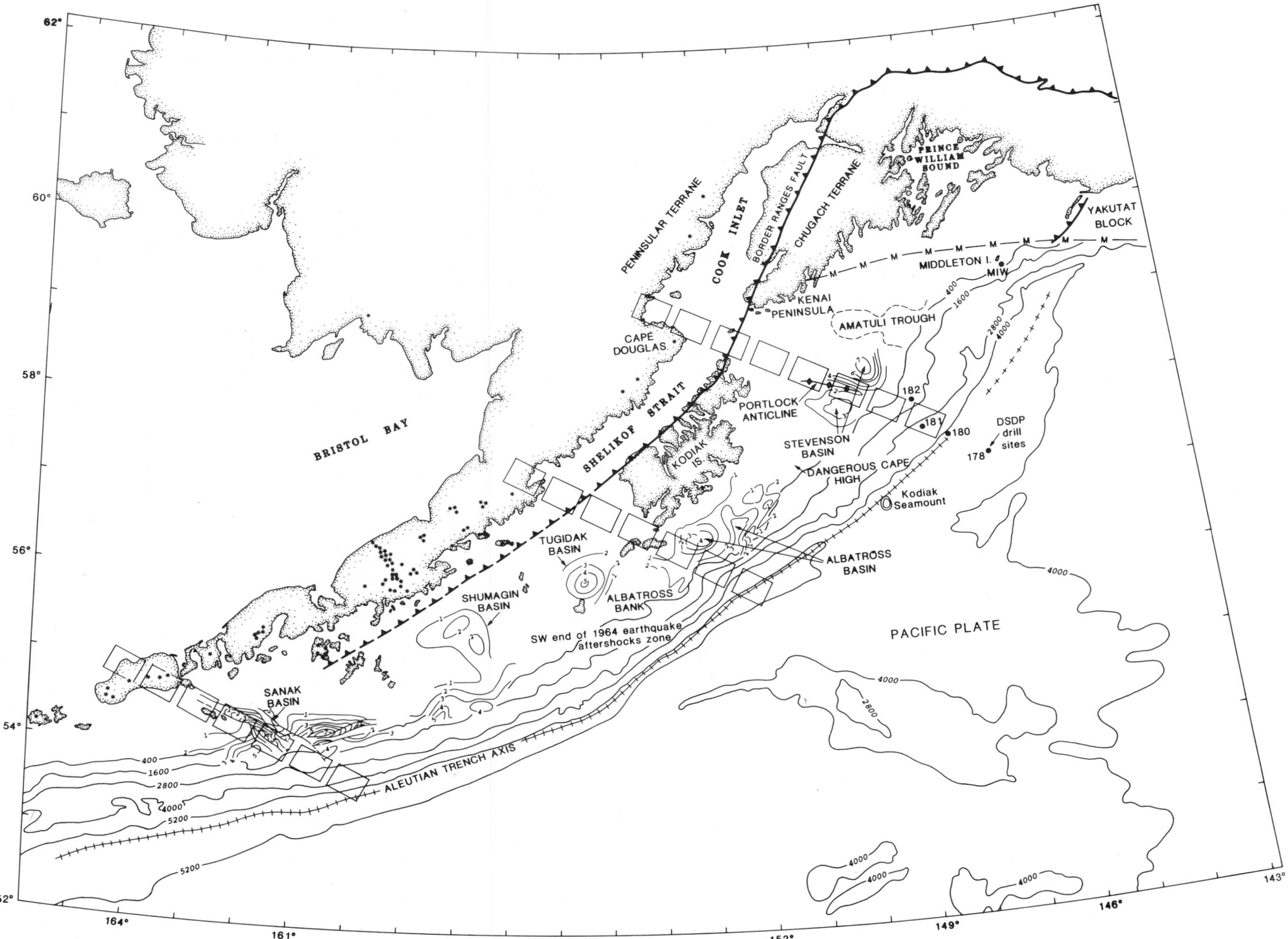

Figure 1.. Location map of the eastern Aleutian arc-trench system showing major tectonic features, lithostratigraphic terranes, end of the 1964 Earthquake aftershocks zone, Yakutat block slope anomaly (M-M-M), transverse boundaries (wide-open dashed zones), and Middleton Island well (MIW). Structure contours at 1-km intervals, for late Miocene and younger beds; bathymetric contours in meters.

the Peninsular terrane and the Chugach terrane is the Border Ranges fault (Jones et al., 1981; Fisher, Bruns and von Huene, 1984). A terrane boundary between the Chugach and southwest part of the Prince William terranes is not expressed as a significant change in geology in the Kodiak Shelf area but is rather a strike slip fault exhibiting perhaps tens of kilometers of displacement (Sample, 1984).

STRATIGRAPHY

The reconnaissance sampling in offshore areas of the Kodiak Shelf, the discontinuous onshore biostratigraphy, and the low diversity of Neogene high-latitude microfauna and flora severely constrain the scope of biostratigraphic knowledge. Our biostratigraphic record is assembled from onshore and offshore information related by regional tectonic considerations. Keller et al. (1984) reviewed scattered microfossil assemblages relative to the more complete stratigraphy of a drill hole off Middleton Island, and DSDP materials. In this way the less complete materials from various sea-floor samples and insular outcrops could be fit into a basic Gulf of Alaska biostratigraphic framework (Figure 2). Local studies of the lithostratigraphy can be projected across the islands from good shoreline exposures on either side (Nilsen and Moore, 1979, Moore and Allwardt, 1980; Sample, 1984).

We describe the regional stratigraphy beginning with the youngest rocks and sediment, which record their environment of deposition more clearly than older rocks. This environment is not necessarily that of sequences deposited in the past prior to northern-hemisphere glaciation and to the present tectonic environment. Since the stratigraphy of the Kodiak margin has been affected differently by multiple periods of tectonism, we separate our descriptions of stratigraphy and structure (Figure 3).

Quaternary

Quaternary sediment-dispersal patterns are unique to the glacial history of the Gulf of Alaska and are a poor analog for past nonglacial patterns; however, much of the Pliocene and perhaps part of the latest Miocene were times of glaciation. Upper Quaternary deposits are unconsolidated sediment of glacial and postglacial origin (Hampton, 1983). Thicknesses and textures of the unconsolidated sediment show a relationship to topography. Analysis of sediment type from 203 stations yielded five main components (Hampton, 1983). Gravel and terrigenous sand were found on banks, in channels, and in the passes draining Shelikof Strait and Cook Inlet. Mud occurs in troughs and locally on flat areas and banks. Ash from the 1912 explosive eruption of Katmai Volcano is distributed in the area nearest the eruption. Shelly sand was found on banks. An exception is Stevenson Trough, where local clean sand and gravel were cored. Thicknesses of unconsolidated sediment in troughs, as judged from seismic sections, are as much as 400 m locally, but are generally <40 m on gentle slopes.

Clay-mineral suites indicate a local source augmented by a lesser contribution from the Copper River, 400 km to the northeast. The coarse sand and gravel are probably transported by glaciers; glacial deposits may be the source for the best-sorted reworked coarse-grained sediment. Local large sand waves imply strong currents. Some uncovered banks have erosional ridges as high as 5 m, indicating a past wave-base or subareal environment. Fine-grained material is mostly winnowed debris deposited in a low-energy environment. Ash was dispersed by wind and does not reflect a topographic influence. The clean sand of Stevenson Trough probably reflects currents associated with the drainage from Cook Inlet. Seismic records show local unconformities separating the unconsolidated sediment blanket and the underlying sedimentary rock.

Figure 2. Biostratigraphy, age ranges, paleodepth, and paleoenvironment of the Prince William terrane, Yakutat block, and Pacific plate, based on sample studies for this chapter and/or earlier published reports. Age scale after Berggren et al., 1984. Plankton zones after Barron, 1980 (diatoms); Okada and Bukry, 1980 (nannofossils); and Blow, 1969 (foraminifera); and California benthic stages after Poore, 1980. Diagonal lines mark hiatuses. Codes: N, samples studied for this report; N1 (from G. Keller, unpublished data on Kodiak Shelf, 1983); N2 (from J. A. Barron, unpublished data on eastern Gulf of Alaska, 1983); N3 & N4 (from G. Keller, this report, and samples of Plafker et al., 1979, 1980); N5 (from K. McDougall, unpublished data, 1983; Kodiak Shelf and Kodiak Island-Lower Cook Inlet outcrop samples); N6 (this report, Middleton Island well); N7 (from G. Keller and K. McDougall, unpublished data, Orca Group); O, Open-File Report; 01 (80-1237, McClellan et al. [1981]); 02 (80-1089, Plafker et al. [1980]); 03 (77-747, Rau et al. [1977]); P, published reports; P1 [Tysdal and Plafker, 1978]; P2 [Lattanzi, 1979]; P3 [Byrne, 1982]; S1, State of Alaska Open File Report 114 [Lyle and Moorehouse, 1978].

Hampton (1983) suggests that little sediment presently comes from the adjacent insular or continental areas; the fiords along the coast effectively trap even much of the fine debris from present mountain-valley glaciers. This is also indicated by the rates of sedimentation in the Aleutian Trench, which were an order of magnitude less for interglacial sediment when fiords were ice free as compared with glacial marine sediment when glaciers extended onto the shelf (von Huene, Larson and Crouch, 1973).

Pliocene–Latest Miocene

Little in the lithostratigraphy distinguishes rocks of latest Miocene age from those of Pliocene age in offshore samples from the Kodiak Shelf. Almost all samples are siltstone and claystone; erratic pebbles float in the matrix, and fine sand is rare. The base of the Pliocene–upper Miocene section is a profound unconformity above which sediment is younger than about 6 Ma. The sediment sampled contains microfossils that lived in outer neritic to upper bathyal water depths.

Onshore outcrops show a greater textural variety representing the more diverse shallow and nearshore environment than offshore samples. The Narrow Cape Formation, originally assigned a middle Miocene age at Narrow Cape on the basis of its macrofauna (Moore, 1969), is now thought to contain a late Miocene microfauna. At Narrow Cape this transgressive sequence was deposited in neritic to middle neritic water depths near a surf zone.

The Pliocene and Pleistocene Tugidak Formation (Moore, 1969; Allison, 1978) is predominantly silt and clay with erratic pebbles, and local channelized gravel and sand. This sequence was deposited in neritic water depths and shows a strong glacial influence.

The formations of latest Miocene through Pliocene age are similar in texture, grain size, and structure to Quaternary deposits, and presumably the environments are closely comparable. Thus the coast has probably not shifted position much since the last marine transgression across the regional

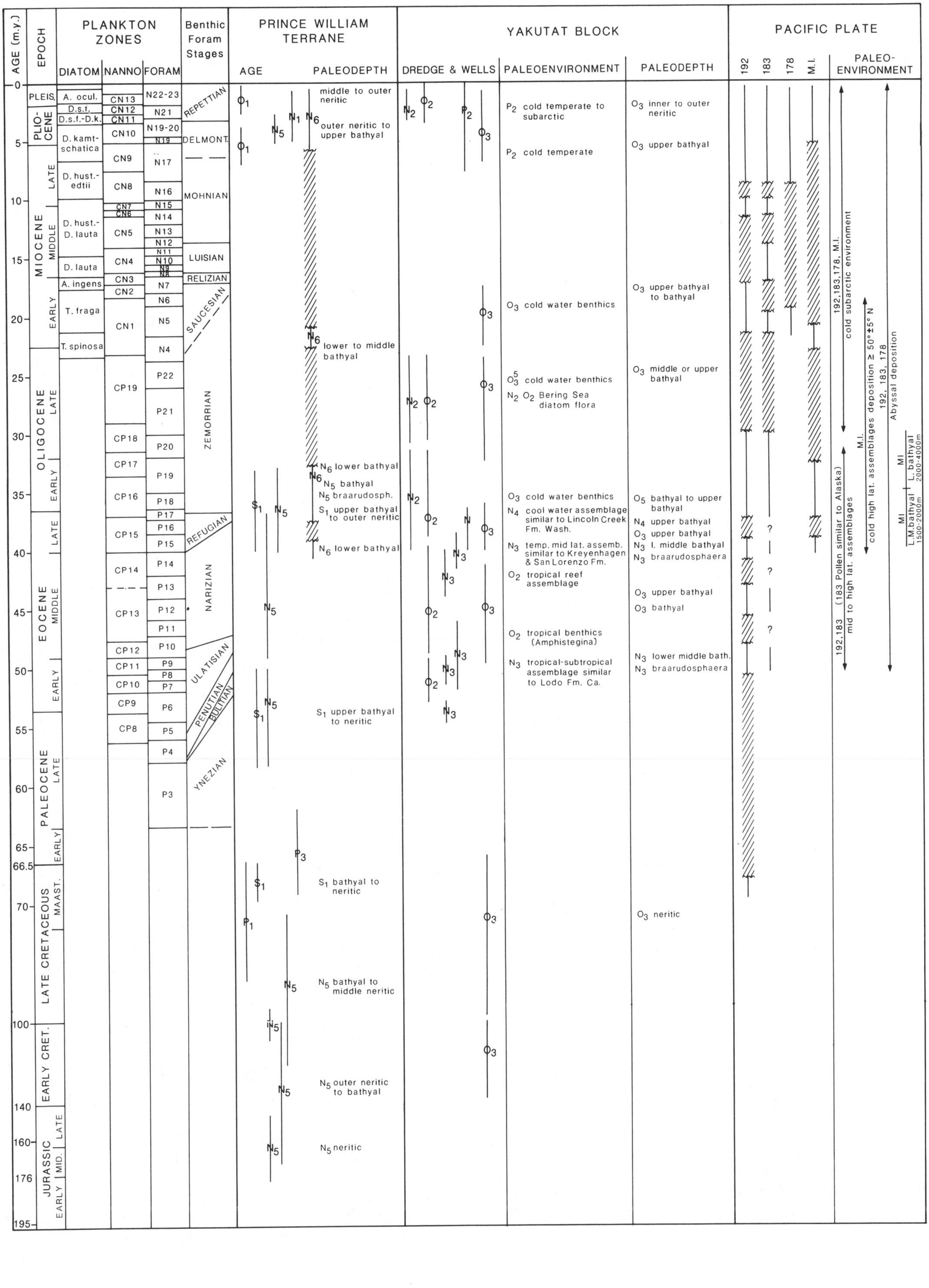
AGE (m.y.)
EPOCH
PLANKTON ZONES
DIATOM
NANNO
FORAM
Benthic Foram Stages
PRINCE WILLIAM TERRANE
AGE
PALEODEPTH
YAKUTAT BLOCK
DREDGE & WELLS
PALEOENVIRONMENT
PALEODEPTH
PACIFIC PLATE
192
183
178
M.I.
PALEO-ENVIRONMENT
PLEIS.
PLIO-CENE
MIOCENE
LATE
MIDDLE
EARLY
OLIGOCENE
EOCENE
PALEOCENE
LATE CRETACEOUS
MAAST.
EARLY CRET.
JURASSIC
MID.
A. ocul.
D.s.f.
D.s.f.-D.k.
D. kamt-schatica
D. hust.-edtii
D. hust.-D. lauta
D. lauta
A. ingens
T. fraga
T. spinosa
CN13
CN12
CN11
CN10
CN9
CN8
CN7
CN6
CN5
CN4
CN3
CN2
CN1
CP19
CP18
CP17
CP16
CP15
CP14
CP13
CP12
CP11
CP10
CP9
CP8
N22-23
N21
N19-20
N19
N17
N16
N15
N14
N13
N12
N11
N10
N9
N8
N7
N6
N5
N4
P22
P21
P20
P19
P18
P17
P16
P15
P14
P13
P12
P11
P10
P9
P8
P7
P6
P5
P4
P3
REPETTIAN
DELMONT.
MOHNIAN
LUISIAN
RELIZIAN
SAUCESIAN
ZEMORRIAN
REFUGIAN
NARIZIAN
ULATISIAN
PENUTIAN
BULITIAN
YNEZIAN
middle to outer neritic
outer neritic to upper bathyal
lower to middle bathyal
N6 lower bathyal
N5 bathyal
N5 braarudosph.
S1 upper bathyal to outer neritic
N6 lower bathyal
S1 upper bathyal to neritic
S1 bathyal to neritic
N5 bathyal to middle neritic
N5 outer neritic to bathyal
N5 neritic
P2 cold temperate to subarctic
P2 cold temperate
O3 cold water benthics
O3 cold water benthics
N2 O2 Bering Sea diatom flora
O3 cold water benthics
N4 cool water assemblage similar to Lincoln Creek Fm. Wash.
N3 temp. mid lat. assemb. similar to Kreyenhagen & San Lorenzo Fm.
O2 tropical reef assemblage
O2 tropical benthics (Amphistegina)
N3 tropical-subtropical assemblage similar to Lodo Fm. Ca.
O3 inner to outer neritic
O3 upper bathyal
O3 upper bathyal to bathyal
O3 middle or upper bathyal
O5 bathyal to upper bathyal
N4 upper bathyal
O3 upper bathyal
N3 l. middle bathyal
N3 braarudosphaera
O3 upper bathyal
O3 bathyal
N3 lower middle bath.
N3 braarudosphaera
O3 neritic
192,183,178, M.I.
cold subarctic environment
192,183 (183 Pollen similar to Alaska)
mid to high lat. assemblages
M.I.
cold high lat. assemblages deposition ≥ 50°±5° N
192, 183, 178
Abyssal deposition
MI
L.M.bathyal 1500-2000m
MI
L. bathyal 2000-4000m

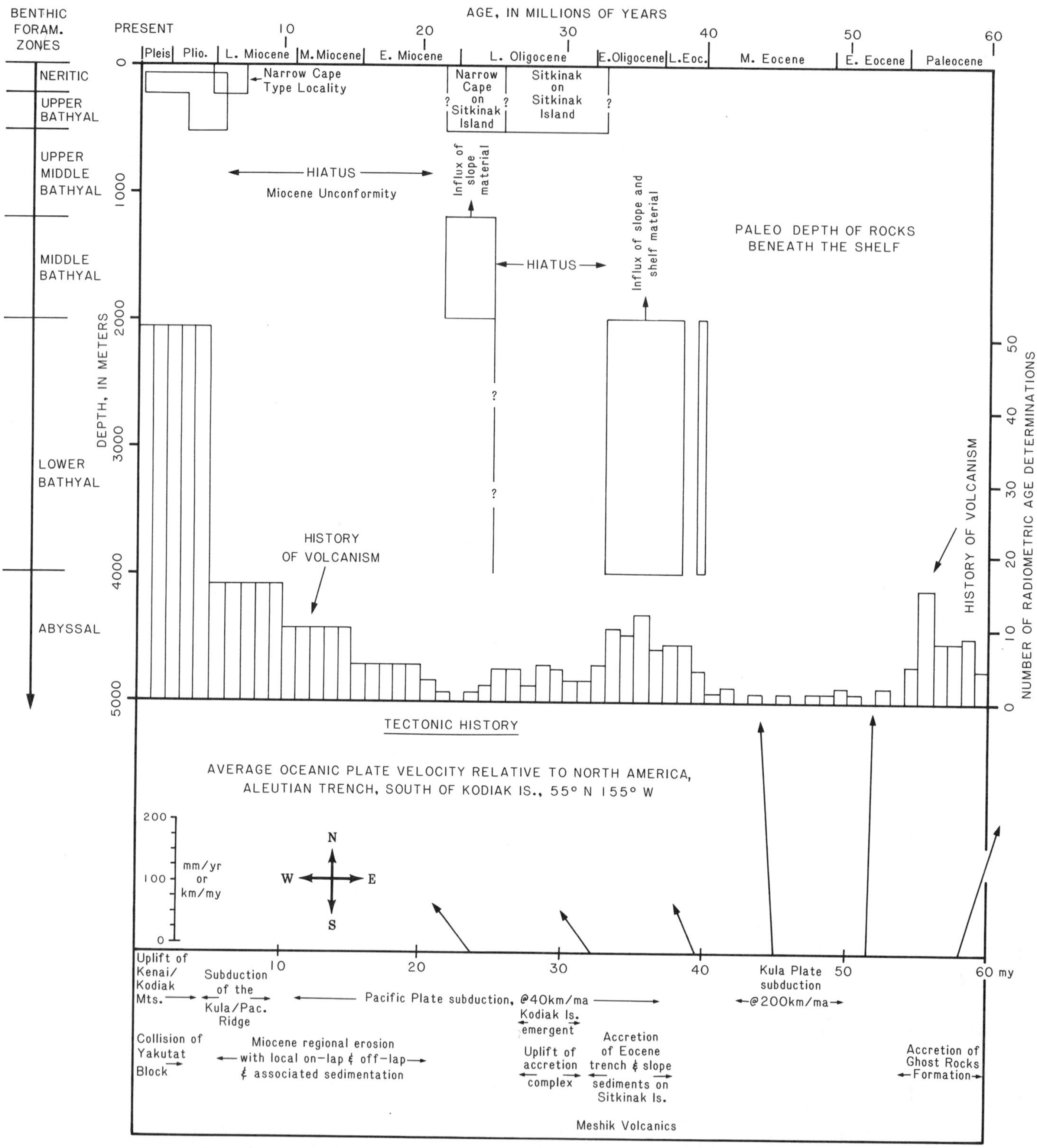

Figure 3. Lithostratigraphic summary showing a chronology of paleoenvironmental history from microfaunal studies, frequency of volcanic rock, and important tectonic events in the geologic history of the Kodiak Shelf.

Miocene unconformity about 6 Ma. If the coarse-grained sediment at Narrow Cape is representative of the late Miocene transgression, a coarse-grained transgressive sediment sequence may cover considerable areas of the major Miocene unconformity.

Early Miocene

No sediment of early and middle Miocene age has been recovered offshore from the Kodiak Shelf. A thin sequence of lower Miocene rock deposited in outer neritic to lower bathyal water depths was penetrated in the Middleton Island drill hole (Keller et al., 1984). Onshore, a sequence of transgressive upper Oligocene or lower Miocene siltstone and very fine sandstone with local coarse-grained layers occurs on Sitkinak Island (Nilsen and Moore, 1979; Allison and

Marincovich, 1980). It contains a middle or outer neritic macrofauna (Allison, 1978). This sequence is separated from the unconformably underlying upper Eocene to Oligocene nonmarine sequence by a thin pebble conglomerate (Nilsen and Moore, 1979).

Oligocene

The nonmarine upper Eocene to Oligocene Sitkinak Formation of Sitkinak Island contains conglomeratic sandstone units, carbonaceous siltstone, and coal deposited in braided delta-plain, river-channel, and mid-neritic environments (Moore, 1969; Armentrout, 1979; Wolf, 1981). The sediment sequence suggests deposition of a major fan-delta complex in a marginal marine setting (Nilsen and Moore, 1979). The section on Sitkinak Island was correlated with sediment sequences on Chirikof and Kodiak Islands by lithostratigraphy (Moore, 1969), but the Chirikof and Kodiak Island sequences are thought to be of Eocene age on the basis of sparse benthic foraminifera (Armentrout, 1979).

No offshore Oligocene samples have been reported from the Kodiak Shelf, but a thick Oligocene section deposited in outer neritic to lower bathyal water depths was penetrated in the Middleton Island well (Keller et al., 1984).

Eocene

A COST (Continental Offshore Stratigraphic Test) drill hole near the edge of the Kodiak Shelf penetrated the mid-Miocene unconformity and bottomed in material of Eocene age (Herrera, 1978). In the bottom of the Middleton Island well, also at the edge of the shelf, an Eocene sequence of chalk, mudstone, and sandstone deposited in bathyal water depths was penetrated (Keller et al., 1984). Attempts to dredge these materials from potential outcrops on the upper slope were unsuccessful.

Insular exposures of Eocene rock (Sitkalidak Formation) consist of a basin-plain to upper-slope sequence lacking volcanic clasts. This sequence accumulated as a series of submarine fans in a trench and trench slope environment (Nilsen and Moore, 1979). On Sitkinak Island, Moore and Allwardt (1980) divided the sequence into a less-deformed unit consisting of possible slope deposits over a more-deformed trench-axis fan deposit that was accreted to the margin. On Sitkinak Island the accreted sediment was uplifted above sea level by the early Oligocene.

Paleocene

Outcrops of the Paleocene Ghost Rocks Formation on Kodiak Island are composed of graywacke and shale and interbedded volcanic rocks (Moore, 1969). Locally, this formation contains thin beds of limestone and is cut by granitic rocks. The sandstone and mudstone of the Ghost Rocks Formation accumulated in a trench or trench-slope setting (Byrne, 1982). The volcanic rocks include MORB and andesite; the andesite was probably derived from MORB by addition of melted sediment (Moore, Morris, and Whelan, 1983). The Ghost Rocks Formation is cut by granodioritic plutons as old as 62 Ma that, along with plutons intruding the Kodiak Formation, are explained as mixtures of MORB and melted sediment (Hill et al., 1981).

Late Cretaceous

The Kodiak Formation of Maestrichtian age comprises a broad tract of northwest-dipping rocks exposed over most of Kodiak Island. The Kodiak Formation is represented mainly by basin-plain or slope facies, and locally by deep-sea fan facies (Nilsen and Moore, 1979). This slate and graywacke sequence probably accumulated in trench and trench-slope environments (Nilsen and Zuffa, 1982), perhaps during a relatively short period of time (Sample, 1984). The metamorphic grade suggests burial at depths of about 10 km (Moore, 1967; Connelly, 1978), and fluid inclusions suggest paleotemperatures of 250° to 300°C (J. C. Moore, written communication, 1985). Cross-cutting 60-m.y.-old plutons (Wilson and Shew, 1980) indicate deformation and emplacement prior to the latest Cretaceous or earliest Paleocene. The Kodiak Formation is considered equivalent to the Valdez Formation on the Kenai Peninsula.

Late Early Cretaceous

The Uyak Complex, considered a tectonic melange (Connelly, 1978), is dominantly blocks of chert in an argillite matrix and locally includes blocks of greenstone, graywacke, and limestone. The blocks range in age from Permian to late Early Cretaceous, and the complex is intruded by 60-m.y.-old granitic rocks. Thus, the complex was accreted between the middle Cretaceous and 60 Ma. During accretion, the complex was metamorphosed to Prehnite-Pumpellyite facies, which suggests burial 9 to 15 km deep. The Uyak Complex is considered equivalent to the McHugh Complex to the northeast on the Kenai Peninsula.

STRUCTURE

The structure of the Kodiak Shelf reflects tectonism caused by plate convergence. However, with increasing data, especially multichannel seismic reflection sections, it is difficult to explain many observations in terms of constant accretion since the Cretaceous (Dickinson and Seely, 1979). The constant accretionary model fails to account for apparent breaks in the accretionary chronology, for the effect of the accretion of far-travelled terranes, and for an episodic history of arc volcanism. Knowledge of Kodiak Shelf tectonics has progressed beyond fitting fragmentary data into a simple plate tectonic model. The brief overview here shows the general state of knowledge, which is treated more completely in the referenced publications.

Structure at the Front of the Margin

From the trench axis to the mid-slope topographic break (between 2000 m and 3500 m water depth), a complex of deep-sea-fan sequences covered by trench fill and local slope deposits have been accreted (von Huene et al., 1985; Stevenson and Embly, this volume). The structural style differs in seismic sections spaced tens of kilometers apart. A typical style, shown in a reprocessed record off southern Kodiak Island (Figure 4), consists of thrust packets above a decollement that separates the section into an accreted upper half and a subducted lower half. Strata in the subducted sequence are not significantly disrupted along the 15-km-long section in which bedding is imaged. Although a similar partitioning of the thick (3–5 km) trench fill and oceanic turbidite section is observed in record sections tens of km on either side of that illustrated on Figure 4, the structure of the accreted part is different (von Huene, 1984). In some sections the thrust

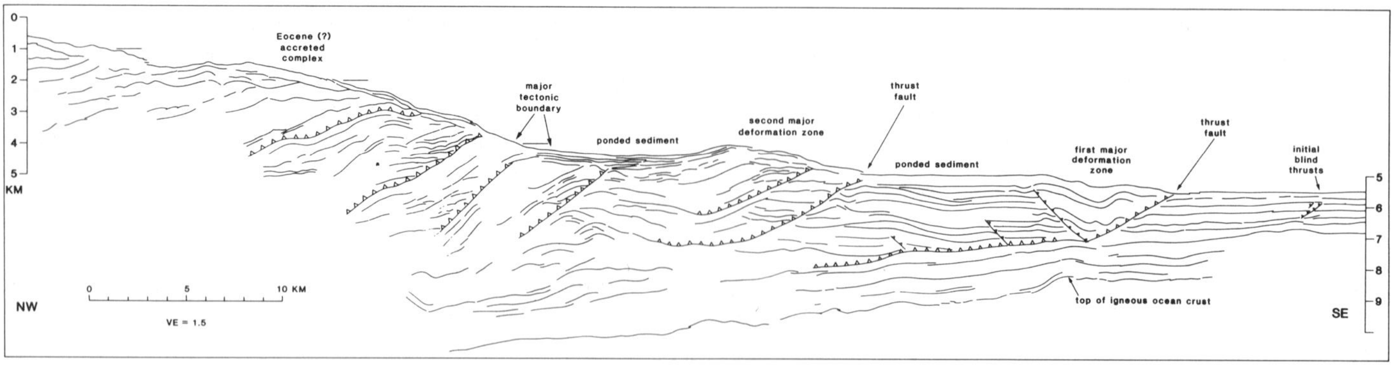

Figure 4.. Migrated seismic record 111 off southern Kodiak Island, showing the trench axis and landward trench slope. Upper part of figure is the seismic record, with time along the vertical axis; lower part is a line interpretation of the record, with depth along the vertical axis (from von Huene et al., 1986).

slices are closely spaced and thrust faults dip both landward and seaward. The great thickness of subducted sediment is of particular significance as an analog of ancient accreted deposits exposed on Kodiak Island. Subducted sediment is the source of the underplated material inferred to uplift the upper slope (von Huene et al., 1985), and probably makes up the bulk of the accretionary complex at depths >3–5 km below the sea floor.

Off central Kodiak Island, a seismic record shows a thick (1–3 km) sediment sequence under the lower slope that is generally coherent and is broken by small reverse faults (Figure 5) or local diapiric folds (von Huene, 1979). This sediment sequence appears to be subducted below the mid-slope topographic break, thereby introducing a 2–3-km-thick sediment section below the upper slope. A still different structural style characterizes the lower slope northeast of lat. 58°N, where many thrust faults dip seaward (Seely, 1977). Here, sediment from the ocean basin is locally being attached to the continent by obduction.

The abundant sediment supplied by glaciation of the high mountains surrounding the Gulf of Alaska promotes accretion. Sediment thicknesses in the trench axis are commonly 3 km and locally reach 4 to 5 km. The residence time of ponded sediment in the trench axis (about 0.4 to 0.6 Ma) exceeds the duration of most major glacial and interglacial periods. Thus sediment in the trench axis could have been as thick as perhaps 5 km since about 5 Ma, when alpine glaciation began in Gulf of Alaska drainages. The rate of Pleistocene sediment accumulation at DSDP (Deep Sea Drilling Project) site 180, on the trench floor, could readily have produced these thicknesses (von Huene, et al., 1973). Sediment supply prior to 5 Ma was considerably less, as suggested by the order-of-magnitude smaller rate of sedimentation prior to 6.5 Ma in DSDP cores from site 178 (von Huene et al., 1973).

The topographic break in the mid-slope area is structurally controlled, but these structures are difficult to image clearly in seismic sections. Off southern Kodiak Island, the break marks the change from individual thrust packets to steeper monoclinal, landward-dipping reflections, which are cut by faults subparallel to bedding planes. Reflections characteristic of the unconformity at the floor of the Neogene basins on the shelf (horizon C; Fisher and von Huene, 1980) can be followed locally 20–30 km downslope, where they merge with the slope sediment sequence. The deep landward-dipping strata are thus in the same geologic position as the Eocene horizons below the Miocene unconformity drilled in the COST hole (Herrera, 1978). This is why we have assigned an Eocene age to the landward-dipping strata beneath the upper slope (von Huene et al., 1985). We propose that a tectonic boundary at the mid-slope topographic and structural change separates the Paleogene(?) strata of the shelf and upper slope from the probable Neogene strata accreted to the lower slope. This tectonic boundary may be correlative with the unconformity observed by Moore and Allwardt (1980), which separates the deep-water facies of the Eocene Sitkalidak Formation from the overlying shallow-water Lower Oligocene Sitkinak Formation. Because Eocene rocks appear to be juxtaposed against much younger, presumably Pliocene–Pleistocene accreted trench sediment, the mid-slope structural boundary may be a truncation scar. Such a truncation scar is clearly imaged in seismic records across the adjacent Shumagin margin (Bruns et al., this volume).

Structure Beneath the Shelf

Two offshore Neogene basins, the Albatross and Stevenson basins, and the intervening transverse Dangerous Cape high are the major structures defined beneath the Kodiak Shelf (Figure 1; Fisher and von Huene, 1980). Tugidak Basin is along the transverse boundary separating the Kodiak from the Shumagin Shelf (Figure 1). Each basin is tectonically unique, but they all attest to 4 km or more of differential subsidence of the regional unconformity of late Miocene age.

Tugidak Basin

Tugidak Basin was described by Fisher (1979) as a fore-arc basin floored by lower Miocene or pre-Miocene rock and filled with gently deformed strata of late Miocene and younger age (Figures 6, 7). The basin lies south of the Kodiak group of islands (Figure 1). The sides of the basin are formed by Kodiak Island platform on the northeast; by the flank of the low structural ridge from which the Semidi Islands emerge on the southwest; and by a ridge at the edge of the shelf, Tugidak anticline, which bounds the basin on the southeast (Figures 6, 7).

The upper Miocene to Quaternary age of the strata in Tugidak Basin was estimated by comparing the sequence of unconformities in the seismic records with the sequence of tectonic events on the surrounding islands. The age estimates agree with paleontologically established sequences determined subsequently for the general tectonic history of the area (Keller et al., 1984). Vertical tectonism was rapid in the seaward part of the basin along the Tugidak anticline and considerably slower along the flank of the Kodiak Island

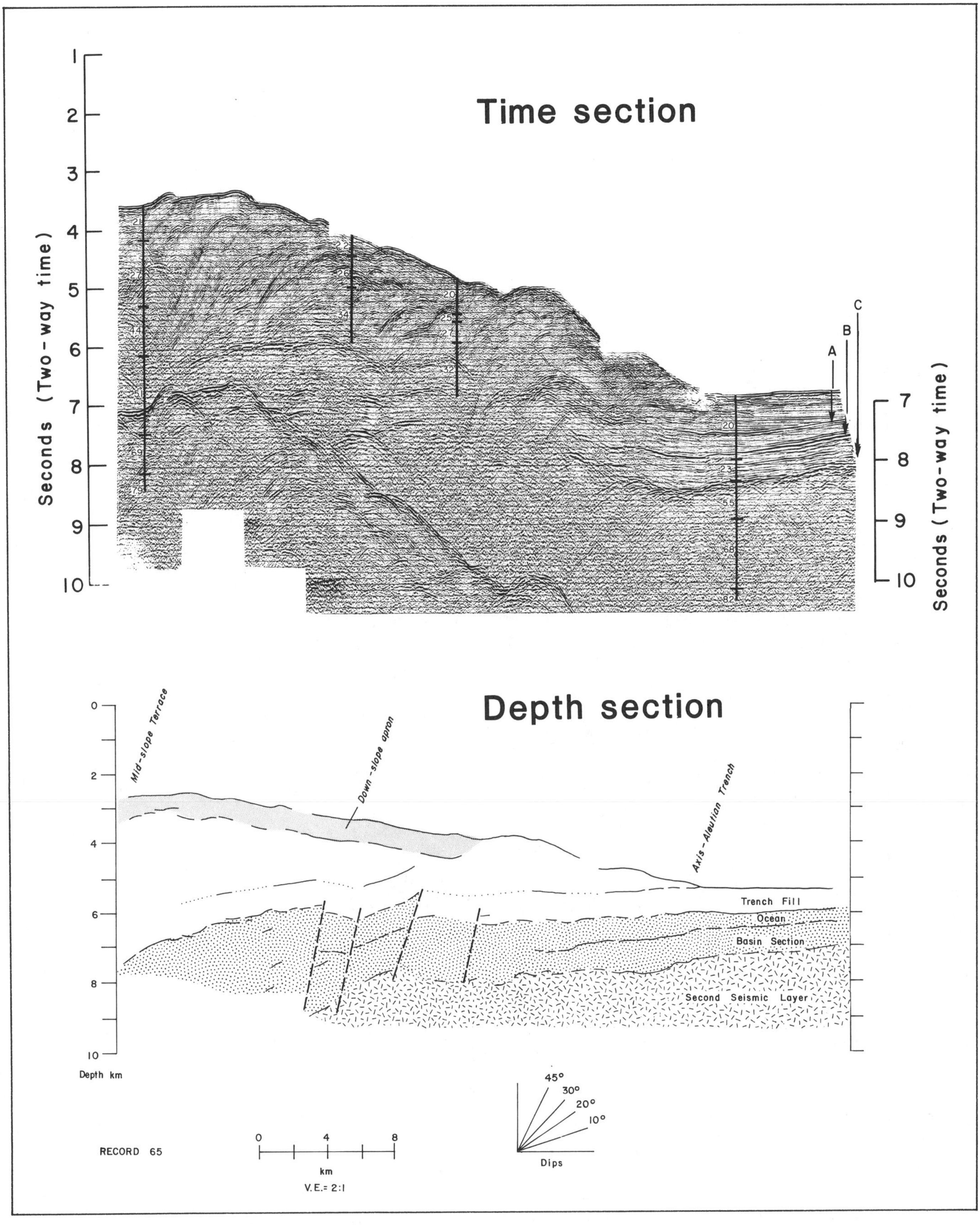

Figure 5. Seismic record 65 across the eastern Aleutian Trench off central Kodiak Island. Vertical exaggeration at sea floor approximately 5. A, at right end of seismic record, indicates base of trench fill; B indicates marker reflection in deep ocean basin section; C indicates top of second oceanic layer.

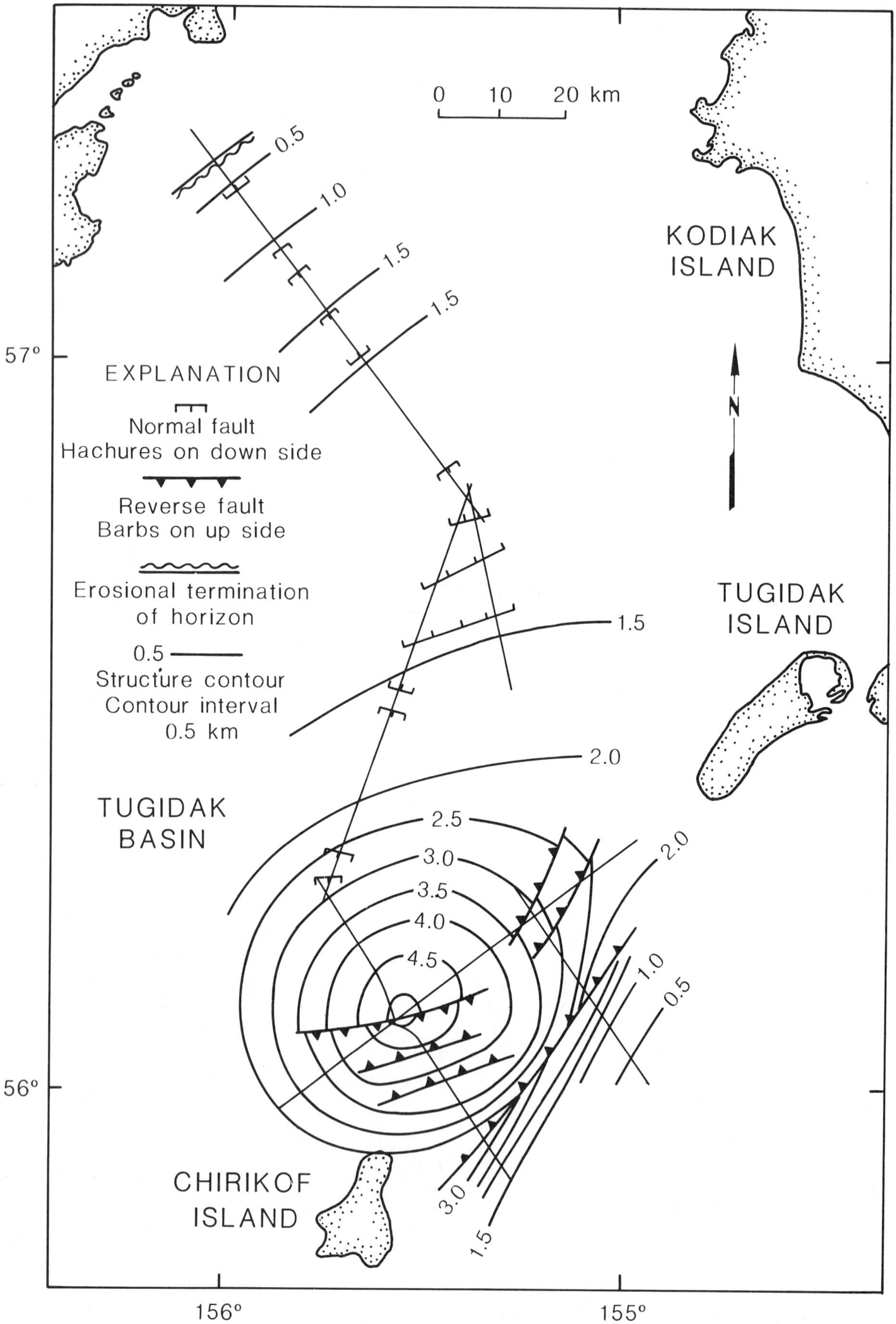

Figure 6. Structural contours on the major unconformity across acoustic basement beneath the Tugidak Basin.

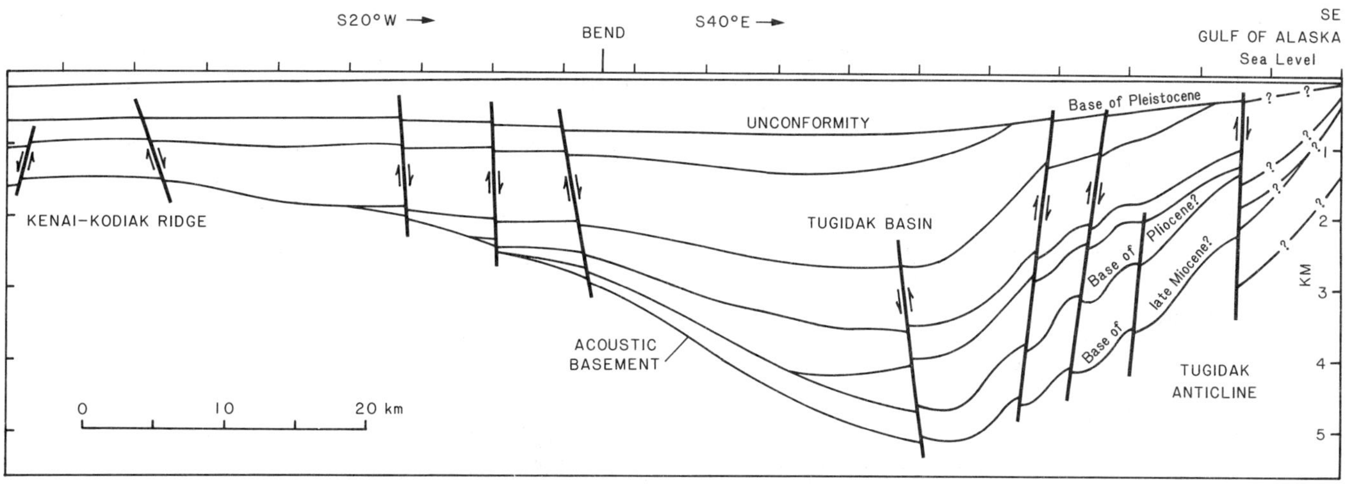

Figure 7. Section across the Tugidak Basin.

platform. Uplift of the anticline and subsidence of the basin probably were most rapid in the Pliocene (Figure 7).

The deepest part of Tugidak Basin is nearly circular and 5 km deep (Figure 6). The strike of faults is uncertain because of the wide spacing of the seismic lines used to draw the depth contours. The transition from landward-dipping to seaward-dipping faults is abrupt, and occurs along the landward flank of the basin. The zone of extensional faulting continues to the Alaska Peninsula (Figure 7).

Along the southernmost Kodiak Island platform adjacent to Tugidak Basin, some of the normal faults have increasing vertical offset with depth, suggesting faulting during basin filling. The flank of the Kodiak platform is cut by small channels, and because vigorous erosion is implied, this basement high was probably subaerial prior to 6 Ma. On the flank of the Tugidak anticline, minor folding and reverse faulting increase with depth.

The tectonic development of the Tugidak Basin (Fisher, 1979), as constructed from the analysis of seismic-reflection records, begins with an unconformity on lower Miocene(?) or older rock that was deformed into a seaward-opening trough. At least the landward part, and probably all of the unconformity surface, was subaerially eroded or eroded at wave base as suggested by channels and by contrasting seismic characteristics between rocks above and below the unconformity (Fisher, 1979). From the beginning of the Pliocene, the Tugidak anticline impeded seaward transport of sediment.

Albatross Basin

The Albatross Basin underlies the southwestern part of the Kodiak Shelf, and most of it is contained in a nearly circular depression filled by a maximum of 5 km of gently deformed sediment (Figure 8). Sitkinak Island and the southern end of Kodiak Island form the landward flank of the basin. The southwest and southeast limits of the basin are defined by the Albatross Bank. The northeastern side of the Albatross Basin is the southwest end of the Dangerous Cape high (Figure 8).

Around the Albatross Basin, the Albatross Bank consists of at least four anticlines. The most prominent anticline curves around the southeast and southwest sides of the Albatross Basin. The other anticlines strike northeast. The spacing of seismic lines is insufficient to clearly define all anticlines. Albatross Bank is the most tectonically active area of the Kodiak Shelf, as shown by deep erosion of strata turned up on the landward flank of the bank, a concentration of earthquakes, numerous slides and slumps, and a pronounced bathymetric ridge. In the core of the bank, strata dip too steeply to be resolved by the seismic-reflection technique (Figure 8). The axis of the bank is commonly broken by high-angle reverse and normal faults, as determined from scarps on the sea floor.

The basement of the Albatross Basin is difficult to pick on single seismic lines and is best located by interpretation of the grid of seismic lines. Reflections above the basement define an upper and a lower sequence of basin filling strata. The lower sequence thickens seaward and the upper sequence thickens landward. The change in the direction of thickening indicates that the basin depocenter migrated landward when the Albatross Bank began to form. Absolute uplift of the Albatross Bank is suggested by a bathyal assemblage of Miocene foraminifers in rocks dredged from the bank (von Huene, 1972), and by benthic foraminifers in Pliocene rocks that were deeply buried and subsequently uplifted to present shelf depth (McClellan and others, 1982).

Before uplift of the Albatross Bank, the floor of the basin dipped seaward, as indicated by the northwest direction of onlap of the lower sequence (Figure 8). Strata in the upper sequence, on the other hand, thin rapidly seaward and were deposited after the beginning of uplift.

Reverse faults and anticlinal folding indicate compressive deformation at the Albatross Bank, but deformation decreases landward from the tectonically active bank. In the Albatross Basin most faults are nearly vertical or normal, and near Kodiak Island faults show no compressional deformation.

The deepest part of the Albatross Basin is separated from a shallower part as the main anticline of the Albatross Bank merges with a low transverse ridge that crosses the shelf (Figure 8). The ridge separates the main part of the the Albatross Basin from a subsidiary syncline that nearly crosses the shelf and is an appendage of the Albatross Basin. The northeast flank of the syncline rises toward the Dangerous Cape high. Lower-sequence strata do not thin across the transverse ridge; hence, it formed after most of the lower sequence was deposited.

Dangerous Cape High

The Dangerous Cape high, northeast of the Albatross

Basin, is characterized by shallow burial of the Miocene regional unconformity (Figure 8). The high is shallowest at the shelf break, and it has less structural relief than the Albatross Bank. The northeast and southwest boundaries of the Dangerous Cape high are transitional and are located where the regional unconformity begins to descend into adjacent basins (Figure 8). The unconformity is commonly <1 km deep and is nearly flat in many places. Locally, intrabasement reflections are evident, especially where they diverge in dip from the unconformity, but these reflections cannot be followed throughout the seismic grid.

The shelf-edge structure consists of multiple anticlines in the southwest half of the high and a broad, low anticline in the northeast half (Figure 8). The shelf-edge structure changes from the high-relief folds along the Albatross Basin to the broad, deeply buried shelf-break structures in the northeast part of the Dangerous Cape high. High-relief structures are again present at the shelf break near Stevenson Basin.

Stevenson Basin

Stevenson Basin, northeast of the Dangerous Cape high (Figure 8), is similar to the Albatross Basin. The fill in Stevenson Basin is trapped by a shelf edge structure and is gently to moderately deformed. This shelf-edge structure, is a broad anticline. Portlock Bank is an anticline transverse to the regional structural trend that separates Stevenson Basin into southwestern and northeastern parts. The southwest part of Stevenson Basin is filled by sediment 3.5 km thick that is composed of two sequences (Figure 8). The upper sequence fills a channel cut into the lower sequence, and the sediment that now fills the channel came from the high-standing area of the Dangerous Cape high. Strata in the lower sequence in Stevenson Basin dip landward, and their updip ends terminate at an unconformity that separates the two sequences. Sediment in the lower sequence came from the Dangerous Cape high, as did strata in the upper sequence. Strata of the lower sequence thin gradually up the flank and across the crest of the Portlock anticline, showing that the anticline began to grow after deposition of part of the lower sequence and that uplift of the Portlock anticline began at about the same time as uplift of the shelf-break anticline. The northeastern subbasin may contain as much as 5 to 7 km of gently deformed strata, though velocity data are poor to establish depths >5 km.

Faults are numerous in Stevenson Basin and have vertical displacements of <200 m. The coarse spacing of the seismic grid over the basin precludes determining their strike at depth, but in a map of shallow features the faults conform to the general northeast-striking structural grain of the shelf (von Huene et al., 1980).

The Kodiak Group of Islands

The structure of pre-Miocene strata is known almost exclusively from insular exposures, because only a few samples of Eocene rock were recovered from drill holes near the edge of the shelf, where local faint landward-dipping reflections are seen in seismic records. The generally northwest-dipping onshore strata are probably repeated by folds and faults; the rocks are both pervasively sheared and in massive coherent tracts; they are separated into belts by laterally continuous faults and are cut by the Kodiak pluton and its satellite intrusions.

On Sitkinak Island upper Eocene to Oligocene sedimentary units (Sitkinak Formation of Moore, 1969; Armentrout 1979) are in fault contact with the tilted and uplifted Eocene Sitkalidak Formation of Moore (1969; Armentrout, 1979). The Eocene sequence on Sitkinak Island was divided into an intensely deformed seaward-dipping unit and a relatively less deformed sequence with stratal continuity (Moore and Allwardt, 1980). The former is interpreted as an obducted trench-filling fan sequence, the latter as a slope-basin or slope-apron sequence. These rocks were uplifted and subaerially eroded by the late Eocene or at least by the Oligocene. The shallow-water lithofacies of all Oligocene rocks, including that from offshore drilling, indicate regional uplift similar to that seen on Sitkinak Island.

The Paleocene rocks (Ghost Rocks Formation) are similarly divided into a highly disrupted trench sequence and a less disrupted slope sequence (Byrne, 1982). The intruding plutonic rocks indicate that offscraping was succeeded shortly by intrusion at about 60 Ma. The Paleocene sequence is separated from Cretaceous beds by a fault that may represent mainly strike-slip displacement over tens of km (Sample, 1984; Sample and Fisher, 1986).

The Cretaceous sequence (Kodiak Formation) crops out in a broad, generally northwest-dipping faulted and folded tract over most of central Kodiak Island (Sample and Fisher, 1986). The rocks range from regularly folded, coherently bedded units to stratally disrupted and boudinaged graywacke in a foliated argillaceous matrix. The Kodiak Formation is overthrust by a tectonic melange, the Uyak Complex, which includes blocks of argillite, ultramafic rock, pillow basalt, radiolarian chert, minor limestone, sandstone, and greenstone set in an argillaceous matrix (Connelly, 1978; Moore and Wheeler, 1978). To the north, beneath the Shelikof Strait, the Border Ranges fault separates these rocks from a truncated Mesozoic shelf sequence of the Peninsular terrane.

MIGRATION OF TECTONOSTRATIGRAPHIC TERRANES

Recent plate-tectonic models of Alaska emphasize the accretion of far-migrated blocks or terranes. The most precise data on which to base a history of travel are paleomagnetic results from multiple samples of volcanic rock; these are summarized for Alaska by Coe et al. (1985), and Hillhouse and Coe (in press). Evidence independent of the paleomagnetic results is derived from studies of microfauna, including paleoclimatic analyses (Keller et al., 1984). Latitude is derived from paleoclimatic data when the global paleocirculation and temperature is known for the period of deposition. The data developed since the first paleomagnetic studies of Packer and Stone (1974) consistently reveal that parts of Alaska were formerly much farther south than their present positions. However, the examination of all evidence leads to a bewildering array of contradictory displacement histories, many of which are at odds with other well-established geologic evidence. Consideration of only the most reliable paleomagnetic data is necessary to provide some internal consistency in the basic data, an approach taken by Coe et al. (1985) and Hillhouse and Coe (in press).

Paleomagnetic studies of the Paleocene Ghost Rocks Formation indicate deposition and volcanism at a latitude about

25° south of its present position (Plumley, Coe, and Byrne, 1983). Paleontological data indicate subsequent positions of the Chugach-Prince William terrane about 10° ±5° south of its present position at about 40–42 Ma (Keller et al., 1984; Plafker and Keller, 1985), and suggest rapid northward motion from about 60 to 40 Ma. These movements are consistent with the proposed rapid northward motion from the Kula plate of about 56 to 43 Ma, and a subsequent decrease in rate of convergence along the Alaskan margins owing to the demise of the Pacific-Kula Ridge, which caused the Pacific plate rather than the Kula plate to subduct beneath Alaska (Engebretson, Cox and Gordon, 1985). In the adjacent Alaskan interior, the Peninsular terrane had reached its present latitude by the Paleocene or Eocene (Hillhouse and Coe, in press; Thrupp and Coe, 1986). A 25° gap in latitude between the Peninsular and Chugach terranes that closed up some time in the past 40 m.y. is required by the paleomagnetic data, but a single suture along which this vast tract of lithosphere was subducted has not been found. The only major suture recognized, the Border Ranges fault, is intruded on both sides by 50–60-m.y.-old plutons which, if co-genetic, indicate cessation of major movement prior to the closing of the 25° gap. Thus, despite selection of only the most reliable paleomagnetic data, no single accommodating suture is recognized and somehow the subduction of about 2000 to 5000 km of crust has occurred. Thrupp and Coe (1986) suggest that a combination of strike-slip faulting, intraterrane shortening, and acceptance of minimum values for northward translation may provide an explanation for the best observations. Because of the remaining contradictions required to make all data compatible, we treat the migration of terranes with caution pending further information.

HISTORY OF NORTH PACIFIC PLATE MOTION

A recent treatment of relative plate motion in the north Pacific (Engebretson, Cox and Gordon, 1985) (Figure 10) confirms the general validity of Atwater and Molnar's model (1973), calling for a profound decrease in the rate of convergence after the demise of the Kula-Pacific spreading center about 43 Ma. The Kula plate converged with Alaska at rates as fast as 200 km/m.y., but the Pacific plate converged with Alaska at only 40 km/m.y. until about 5 m.y. ago. The crest of the extinct Kula-Pacific spreading ridge arrived in the Shumagin Shelf area about 10 Ma, nearly 33 m.y. after it ceased spreading and swept progressively northward beneath the shelf (Figure 10). The ridge crest arrived at the southern Kodiak Shelf about 5 Ma and is now completely subducted (Engebretson, Cox and Gordon, 1985).

Unfortunately, the plate-tectonic history of the north Pacific is insufficiently precise to know within a few m.y. the arrival of the Kula-Pacific Ridge and the Yakutat block. The extinct Kula-Pacific spreading ridge has completely disappeared beneath Alaska and is inferred from a pattern of magnetic anomalies 52 Ma and older. Its demise is associated with the major change in plate motion 43 Ma (Engebretson, Cox and Gordon, 1985). The first demonstrated arrival of the Yakutat block with the Alaskan subduction zone occurred 3 to 4 Ma, as deducted from the time required to subduct the magnetic anomaly that defines its southern boundary (Bruns, 1983). Although the collision of the Yakutat block can be observed presently, the initial subduction of the western edge of the block and the Kula-Pacific Ridge are reconstructed from circumstantial evidence. That evidence poorly constrains the time of arrival at a specific position perhaps by ±5 m.y. for the Kula-Pacific Ridge and ±3 m.y. for the Yakutat block. Thus, from magnetic anomaly and hotspot reconstruction alone, the times when major sea-floor features were subducted are known only within broad limits, and their passage down the subduction zone can be compared with the history of tectonic events at the edge of the continent.

TECTONIC HISTORY

In constructing a sequence of tectonic events that could have shaped the Kodiak Shelf since the Early Cretaceous development of the Border Ranges fault, we bridge gaps in the fragmental record with inference to build a general conceptual framework (Figure 3). Along the Alaska Peninsula, the Border Ranges fault may have begun as a strike-slip fault zone that later became the thrust zone along which the Kodiak Formation accreted in Campanian and Maestrichtian time (Fisher and von Huene, 1984). The fault truncates the mid-Mesozoic shelf section which underlies the Alaska Peninsula, and along the footwall a melange zone developed. Truncation removed the now-missing Mesozoic outer shelf and continental slope area, perhaps first by lateral translation (although the displaced continental slope has not been located) and then by subduction erosion. After removal of the Mesozoic outer shelf and slope, accretion of the Kodiak Formation began. The Border Ranges fault and its associated melange zone appear analogous to the Neogene Peruvian or Mexican margins, where the tectonically truncated continent has subsided and now serves as a buttress against which Cenozoic deep-water sediment is being accreted (Watkins et al., 1982; von Huene et al., 1985).

The accreted sedimentary rock of the Kodiak Formation is inferred to have been deposited in a trench or base-of-slope environment because of its tectonic setting, sedimentological character, petrology, and paleocurrent patterns (Nilsen, 1982; Sample, 1984, Sample and Fisher, 1986). It was derived from a nearby accretionary complex and magmatic arc which was to the present north and probably northwest, that was uplifted in the Cretaceous (Nilsen and Zuffa, 1982), and probably also from local sources on an adjacent subaeral terrain (Sample, 1984; and Sample and Fisher, 1986). Some sediment was transported longitudinally down the trench axis, and some also came down canyons from an adjacent shelf to be accreted soon after deposition (Nilsen and Zuffa, 1982). Implicit in this reconstruction is a nearby terrain of considerable size and relief and a major subduction zone of Late Cretaceous and earliest Tertiary age. Inasmuch as significant activity along the Border Ranges fault probably ceased in the Late Cretaceous, the simplest reconstruction would place the Peninsular terrane and parts of the Kodiak Formation near each other, and the Peninsular terrane would have been a source of sediment for the Kodiak Formation. This reconstruction is supported by the assumption that the 60- to 65-m.y.-old plutons and dikes that intrude the Peninsular and Chugach terranes are co-genetic and have a common source of magma (Moore et al., 1983).

During the Paleocene, accretion continued and the Ghost

Rocks Formation was emplaced. Inasmuch as the fault between the Kodiak and Ghost Rocks Formations (also the boundary between the Chugach and Prince William terranes) marks a zone of strike-slip motion on the order of perhaps tens of kilometers (Sample, 1984; and Sample and Fisher, 1986), the Chugach and Prince William terranes may have been near each other during the Paleocene, also with a sediment source from an adjacent mainland. The environments of deposition of the Ghost Rocks and Kodiak formations are essentially equivalent, and both formations are cut by 60- to 65-m.y.-old plutons; as in the Late Cretaceous, Paleocene sedimentation and accretion occurred over a relatively short period of time, near a subaerially exposed terrane.

Accretion of sediment deposited in a base-of-slope environment, and therefore inferred convergence, continued through the Eocene (Figure 3). Surprisingly, early and middle Eocene arc volcanism in the Peninsular terrane diminished to relatively low levels of activity (Wilson and Shew, 1980; Wallace and Engebretson, 1984). Accretion during the Eocene produced complex that now extends from the coast of Kodiak Island at least to the edge of the shelf, a width about equivalent to that of the Upper Cretaceous Kodiak Formation. The implied Eocene sediment volume requires a huge nearby sediment source. However, proximity of the Peninsular and Chugach-Prince William terranes is not in accord with paleomagnetic and paleontological data.

As previously noted, paleomagnetic and paleontological results from the Ghost Rocks Formation (Plumley et al., 1983; Keller et al., 1984) require about 25° of latitude separation between the Peninsular terrane and the Ghost Rocks Formation of the Chugach terrane. This separation is difficult to incorporate into an early Tertiary model where the Peninsular, Chugach, and Prince William terranes are close to their present positions. In addition, the Eocene microfossil assemblage of the Orca Formation of Prince William Sound (coeval to the Sitkinak Formation) appears to have also lived at southern latitudes (Plafker and Keller, 1985, and Keller, 1985, oral communication). If the Peninsular and Chugach-Prince William terranes were distant from each other, then the source of sediment for the Kodiak, Ghost Rocks, and Sitkinak formations was a terrain with a volcanic arc and characteristics similar to the Peninsular terrane, but some 2000 to 3000 km away and to the south. The closure of the 15° to 25° latitudinal gap, and the continual proximity to the large source of sediment requires a terrane travel path close to North America. These problems are being actively investigated (Coe et al., 1985; Thrupp and Coe, 1986).

In the early Oligocene, the accreted Eocene sequence was uplifted, as shown by microfauna in the Middleton Island well (Keller et al., 1984) and by the nonmarine Oligocene sediment covering the Eocene rocks of Sitkinak Island (Moore, 1969). The melange along the Border Ranges fault (Uyak Complex) was unroofed, because chert from it is included as clasts in the conglomerates of Oligocene age on Sitkinak Island. The early Oligocene uplift and coeval volcanism on the Alaska Peninsula (Figure 3) ended roughly 30 Ma, during which a 200-km-wide sequence of oceanic sediment was accreted to the continent. No comparable period of accretion since that time has been preserved.

North Pacific plate movement changes roughly correspond to tectonic changes on the continental margin; this circumstance encourages us to explore a cause-and-effect relationship. The two major plate reorganizations associated with spreading rates of the Kula-Pacific Ridge correspond to major vertical tectonism on the shelf at about 56 Ma when convergence increased dramatically, and between 50 and 43 Ma when the rate of convergence along Alaska decreased (from as much as 200 km/m.y. to 40 km/m.y., Engebretson, Cox and Gordon, 1985). However, rapid plate convergence between 56 and 43 Ma corresponds to apparent decreased volcanism (Figure 3), opposite to that which would be expected from a simple plate-convergence model. Wallace and Engebretson (1984) explained this by appealing to factors such as the unknown orientation of the Peninsular terrane and thus the direction of relative convergence, the configuration of the Benioff zone, and the fact that parts of the Kodiak Shelf were migrating northward from more southerly positions of origin. Perhaps general concepts regarding the transformation of subducted material to rising magmatic products and formation of island arcs do not apply where rates of convergence are much higher than any known along modern convergence zones. The configuration of the Benioff zone and its temperature field are unknown at rates of convergence above 15 mm/yr; thus, it is conceivable that magma generation for arc volcanism is suppressed. At present, this question is best addressed by numerical modeling.

The Meshik volcanic arc on the Alaska Peninsula initially developed about 40 Ma (Figure 3) and produced some of the most intense volcanism in that area. At the Middleton Island drill hole, a shallow-water (upper bathyal-outer neritic) latest Eocene–earliest Oligocene section (37–39 Ma) indicates uplift of the underlying deep-water section about 4 m.y. after the beginning of decreased rates of convergence. The greatly decreased rate of plate convergence and subduction may cause uplift. Modeling of the thermal regime in a subduction zone shows that decreased rates of subduction result in an increase in the temperature of the plates and will cause uplift (Langseth et al., 1981). To link decreased rates of subduction with uplift and development of hiatuses and erosion is appealing, because age-equivalent unconformities are seen not only in the Kodiak area but along much of the Alaska Peninsula and the Aleutian arc (Bruns et al., this volume; Bruns, 1983; Fisher and von Huene, 1980; Scholl, Vallier and Stevenson, this volume).

The absence of early and middle Miocene sedimentary rocks in the Kodiak area is consistent with early Oligocene emergence of the shelf and subsequent deep erosion. Sediment representing most of the Oligocene and Miocene is missing from the section penetrated in the Middleton Island drill hole. The missing Miocene sediment and seismic characteristics of the regional unconformity (Fisher and von Huene, 1980) suggest that the unconformity marks a 10–15-m.y.-long period of shallow-water and subareal environments across the present shelf and insular area. We speculate that multiple transgressions and regressions from changes in sea level and tectonism resulted in local buried shallow-water sediment sequences (similar to the Narrow Cape Formation at Narrow Cape), which are preserved by burial beneath the shelf but have largely been eroded (if they were present) from present insular areas.

Volcanism proceeded at a low rate after the early Oligocene, when the Meshik arc waned despite constant plate convergence (Wilson and Shew, 1980; Wallace and Engebretson, 1984); we have no explanation for the cause of this

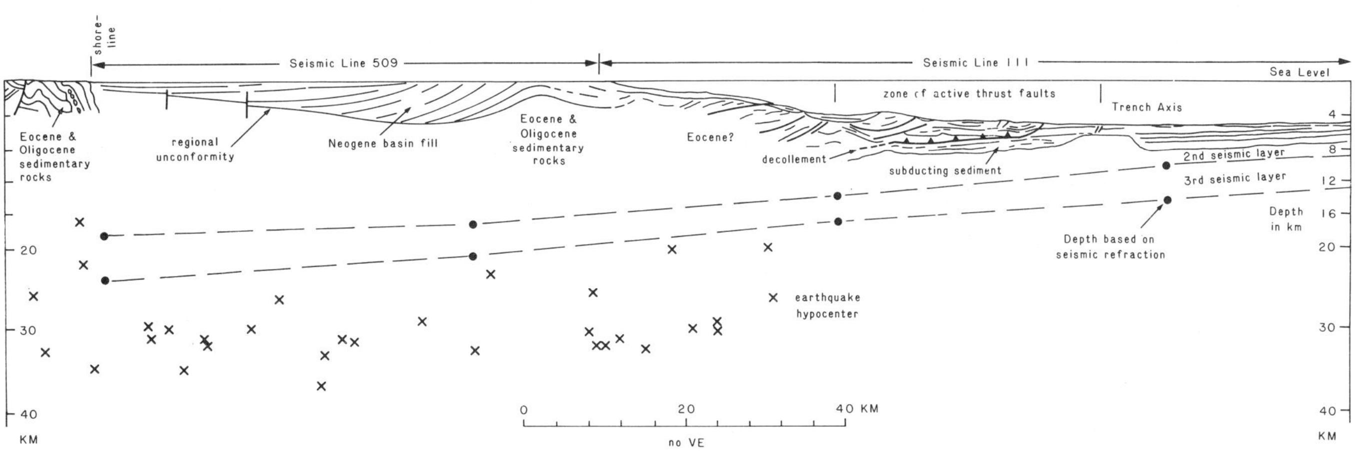

Figure 9. General section along seismic lines 509 and 111 across Kodiak Island and the adjacent shelf and slope. Earthquakes are denoted by X based on master event relocation.

decreased activity. After the extensive hiatus in the sediment record and erosion of the Miocene unconformity, upper Miocene sediment accumulation at DSDP site 178 in the Gulf of Alaska increased suddenly (Kulm et al., 1973; Keller et al., 1984); a similar increase in sediment accumulation is recorded in Cook Inlet (Kirschner and Lyon, 1973) and at Middleton Island (Keller et al., 1984). This increased sediment supply suggests not only the beginning of glacial sedimentation, but also certainly reflects an uplift of the Miocene erosion surface in insular areas where remnants of the surface form concordant summits of the present mountains. At lower elevations, the extensive areas of low relief in the northeast part of the Kodiak insular area and across the southwest part of Trinity Island could also be remnants of the Miocene erosion surface. Conversely, on the continental shelf the Miocene erosion surface was generally inundated, and areas of greater subsidence formed the Tugidak, Albatross, and Stevenson basins. General uplift of insular areas and depression of the shelf was observed during the 1964 Alaska earthquake, suggesting continuation of the general tectonic trends of the past 5 m.y. Concurrent with uplift and subsidence, increased explosive volcanism resulted in more numerous ash layers on the sea floor (Scheidegger et al., 1980) and the Alaska Peninsula began to emerge (Burk, 1965).

During the mid-Pliocene, depressions along the shelf became closed basins as ridges developed at the edge of the shelf. At this time, the continental slope of the margin may have changed from a dominantly nonaccretionary convergent margin to a dominantly accretionary convergent margin, and the Pliocene to recent accretionary complex at the front of the margin began to accumulate (von Huene et al., 1985); thus the inferred gap between Eocene and Pliocene rocks was formed (Figure 9). The tectonic mechanism by which the basins on the adjacent shelf formed remains enigmatic, because they display few compressional structures in a tectonic system dominated by the convergence of lithospheric plates and accretion. The basins are circular and are not elongate perpendicular to the direction of convergence; generally, only the seawardmost flanks contain seaward-verging folds and reverse faults. The transverse flanks of Albatross and Stevenson basins are not the result of compression from plate convergence.

Approximately coeval with these changes leading into the present tectonic regime of the Kodiak Shelf is the inferred subduction of the dormant Kula-Pacific spreading ridge, followed by initial collision of the presently subducting Yakutat block. The timing of the subduction of the Kula-Pacific Ridge and of the Yakutat block, and the beginning of various active tectonic features on the shelf, is known imprecisely, so the cause-and-effect relationship of plate tectonic events and shelf tectonism requires further constraints. However, with better resolution of shelf tectonics, the structure and morphology of the subducted Kula-Pacific Ridge and the leading edge of the Yakutat block might be hypothesized from the effects they had on the margin.

The Kula-Pacific Ridge was being subducted at Unimak Pass at the end of the Alaska Peninsula about 10 Ma (Figure 10). Perhaps about 5 Ma, the ridge was completely subducted. The Yakutat block collided with Alaska about 4 Ma ago, and has migrated northward as it subducted beneath the Kodiak and Middleton shelves (Bruns, 1983). The block is traced down the subduction zone by using the distinctive magnetic anomaly along the southern edge of the block (Figure 1). Just south of the magnetic anomaly, in the straits between the Kenai Peninsula and the Kodiak group of islands, is a series of deep subsurface reflections that could represent part of the subducted terrane (Fisher et al., 1983; von Huene et al., 1985).

Because the Yakutat block is the last terrane to collide and be subducted, and because it can be followed into the subduction zone, its history is better known than that of other terranes or the Pacific-Kula Ridge. A speculative reconstruction of the migration of the block helps explain how sediment could reach the the Zodiac fan from a source terrain on North America via a now-subducted extension of the Yakutat block (Stevenson, Scholl and Vallier, 1983) (Figure 10). Between 26 Ma and 40 Ma sediment from North America may have been channeled across the block to form Zodiac fan (von Huene et al., 1985), and perhaps a large channel system of this age can be found on the unsubducted part (Figure 10). After 26 Ma, the source of sediment was cut off or diverted as the block moved northward (Stevenson and Embly, this volume). This hypothesis for the northward migration of the Yakutat block might be a useful example for reconstructing the migration of the Chugach/Prince-William terranes. To satisfy the paleomagnetic and micropaleontological observations as well as the required proximity

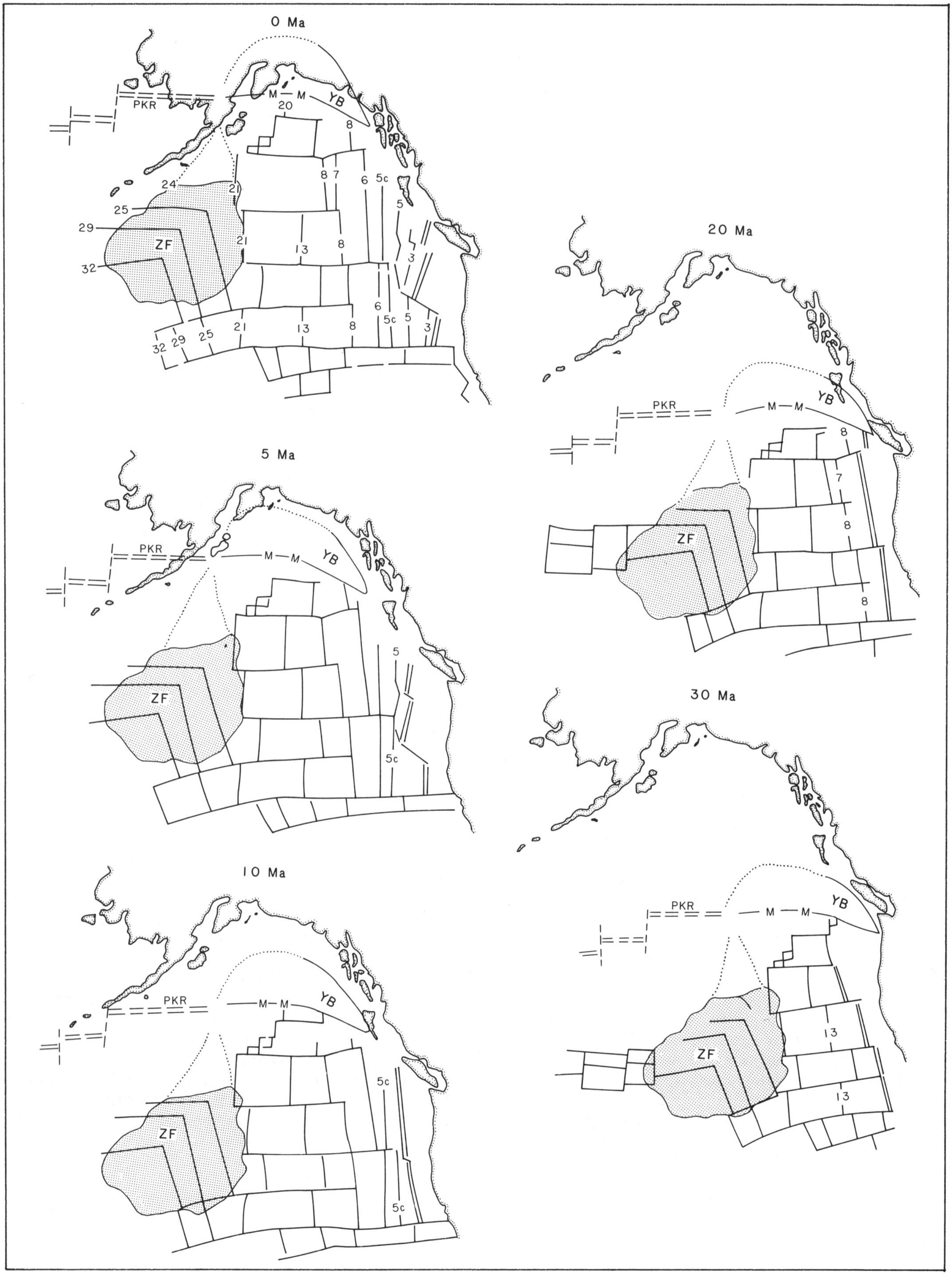

0 Ma
PKR
M — M
YB
20
8
8 7
6
5c
5
24
21
25
29
ZF
32
21
13
8
3
6
5c
5
3
2I
13
8
32 29 25
20 Ma
PKR
M — M
YB
8
7
8
ZF
8
5 Ma
PKR
M — M
YB
ZF
5
5c
30 Ma
PKR
M — M
YB
13
ZF
13
10 Ma
PKR
M — M
YB
ZF
5c
5c

Figure 10. Sequence of diagrams showing relative motion of the North American and Pacific plates during the past 30 m.y. Zodiac fan (ZF) and the Yakutat block (YB) are back-tracked by assuming a fixed position relative to the anomaly pattern, and their hypothesized subducted extensions are indicated by dotted lines. The southern edge of the Yakutat block (the slope anomaly, M-M-M) and the failed Pacific-Kula Ridge (PKR) are shown by single and tripple lines, respectively; major magnetic anomalies and transform fault patterns are also indicated. Plate motion after Engebretsen and Cox.

of these terranes to large sediment sources, the migration path had to pass close to the North American continent. A strike-slip model as proposed by Moore et al. (1983) and emphasized by Thrupp and Coe (1986) suggests a migration path close enough to North America to satisfy the sediment source requirements for part of the Cretaceous and the Paleogene sedimentary rocks of the Kodiak Shelf; however, the precision of the paleontological and magnetic data, and the absence of evidence for large lateral fault zones, make the path difficult to reconstruct. Conversely, if the data are all biased toward their limits of least possible mobility, the source of the sediment could have been from areas around the Gulf of Alaska.

The equivocal aspect of our tectonic history is apparent in the infraction of a simple one-to-one relationship between plate convergence and volcanism. Such an apparent contradiction of the simple conceptual model does not preclude the cause-and-effect relationship of plate-tectonic and margin-tectonic processes. Better constraint on timing and on the position of terranes, or more certainly in the coeval occurrence of convergence in an absence of volcanism, would make testing of simple models more compelling.

PETROLEUM GEOLOGY

An assessment of the petroleum potential of the Kodiak Shelf and upper slope depends on inference from rocks on the islands or those cropping out at the seaward edge of the shelf. Rocks from insular outcrops or from breached structures at the edge of the shelf may not have the most favorable characteristics for such an assessment, because some of the source and reservoir characteristics, altered during the uplifts that caused exposure at the surface, may be peculiar to those areas.

Analysis of organic matter in a suite of samples from nine localities onshore (Lyle et al., 1977), and three offshore (Fisher, 1980), shows that Cenozoic rocks contain generally $<0.5\%$ organic carbon. Samples from DSDP sites in the adjacent north Pacific Basin (23 samples from sites 178, 180, and 181) averaged around 0.5% also (Kvenvolden and von Huene, 1984). Sixty to 100% of the kerogen analyzed is herbaceous. Woody and coaly kerogens are present in secondary amounts (Fisher, 1980). Thus, gas and gas condensate are the most likely hydrocarbons to be generated from the Cenozoic rock of the Kodiak Shelf. However, these conclusions should first be tempered by the fact that samples from only 13 localities were analyzed, most of which were onshore. Secondly, sediment in the deep Pacific Ocean Basin, the least favorable environment for organic carbon preservation in the area, contained as much carbon as the rocks on the shelf. This suggests that many of the measurements could have been made on rocks with minimum or background hydrocarbon content, and that rocks from a more favorable environment with higher carbon content could lie buried beneath the shelf.

Strata of upper Miocene and younger age filling the basins on the shelf most likely contain kerogen that is thermally immature (Fisher, 1980). The Eocene and Oligocene strata that probably floor the basins contain mature kerogens, and locally Paleocene strata also contain mature kerogens. Some of the hydrocarbons generated may or may not have been vented, depending on the time and depth of erosion. Parts of the Paleocene and the Cretaceous rocks exposed onshore appear to represent economic basement for oil generation (Fisher, 1980), particularly those near plutons.

The present thermal gradient has been measured directly only in the Middleton Island well. Uncorrected values range from 21° to 30°C/km, and the best average is between 28°C/km and 30°C/km (Fisher, 1980). Calculations based on the depth of the base of gas hydrate reflections on the upper slope are about 30°C/km (Kvenvolden and von Huene, 1984). This thermal gradient is higher than the general gradient from analogous areas along convergent margins (Fisher, 1980), and it may reflect the regional uplift in the Miocene.

The best reservoir characteristics found in about 35 rock samples are in those of late Miocene age that constitute a transgressive sequence covering the regional unconformity (Fisher, 1980). Eocene and Oligocene rocks onshore have abundant sandstone but poor reservoir qualities. Some intervals of the Eocene and Oligocene sampled in the Middleton Island well also have high sand content (Keller et al., 1984).

The exposed rocks have marginal source and reservoir characteristics. However, the possibility that source rock characteristics improve offshore cannot be discounted. Some downdip improvement might be expected because of deposition on a former continental slope in the oxygen minimum zone where organic matter is more likely to be preserved. Also, Oligocene and Eocene strata offshore are more deeply buried than those sampled onshore, and thus may have generated more hydrocarbons.

An unconventional source of hydrocarbons may be associated with the large quantities of sediment being subducted beneath the Kodiak margin (Kvenvolden and von Huene, 1984). Although presently very speculative, the source appears worth investigating because of the potentially large volume of gas that may be generated. The basic concept is that organic matter in subducted sediment will pass through the thermal window conducive to gas generation. That gas may then migrate upward to be trapped above the subduction zone. Such a mechanism is especially interesting in the Gulf of Alaska, because the subducted sediment is that of deep-sea fans and is composed of terrigenous debris (Stevenson and Embly, this volume). Although the organic carbon content of the subducted sediment is low (about 0.5%) the volume of sediment presently being subducted is huge, and thus a correspondingly large volume of gas could potentially be generated (Kvenvolden and von Huene, 1984; Stevenson and Embly, this volume). Direct evidence for migration of gas from the subduction zone to near-surface localities was obtained off Central America on Leg 84 of the Deep Sea Drilling Project. Such direct evidence of an upward migration of gas has not yet been sought along an Alaskan margin, but it might be found because of the much greater volume of sediment involved in the Alaskan subduction zone compared with Central America. This source could thus be further investigated with sampling of the older rocks that crop out on the continental slope.

CONCLUSIONS

Assessment of the petroleum potential of the Kodiak Shelf is in a preliminary stage. The history that bears on such an assessment is marked by three major changes in tectonism: (1) development of the Border Range fault and accretion of upper Cretaceous through Eocene sediment and possibly terranes, (2) regional uplift and development of a broad erosional lowland with local erosion of structures to depths of several km; regional uplift probably resulted in development of shallow seas in the Oligocene and particularly in the middle Miocene, and (3) uplift of present insular areas and depression of the Miocene erosion surface across the continental shelf, and development of local basins in the Pliocene. These three tectonic periods appear to correspond in time to major Tertiary plate reorganization of the north Pacific and to the subduction of large features on the sea floor, including collision of the Yakutat block. However, the timing of these events is not precise enough to fit events on the continental margin to a plate-tectonic history with certainty. Thus a significant subject for future study is the precise age and duration of each tectonic period on the continent and of each North Pacific plate tectonic event to see if coincident histories support a cause-and-effect relationship.

Paleogene rocks are of great importance in assessment hydrocarbon prospects because they are the most likely source of hydrocarbons. These rocks have not been sampled *offshore* (nonexclusively) and only the Middleton Island drill hole has been reported (Keller et al., 1984). The source-rock characteristics of rocks onshore leads to a pessimistic assessment of hydrocarbon resources (Fisher, 1980). However, the Middleton Island drill hole penetrated rocks that are not as consolidated or metamorphosed as their coeval onshore counterpart. In fact, our reconstruction of the tectonic history allows for preservation of organic carbon from a Paleogene oxygen-minimum zone in sediment beneath the outer shelf and upper slope. Furthermore, it indicates periods of minimum volcanic activity during erosion of the regional unconformity, so that possible reservoir rocks could be free of altered volcanic materials that reduce porosity. Because Paleogene rocks were not recovered during reconnaissance sampling across the Kodiak Shelf, we can only speculate that the offshore Paleogene section is generally less metamorphosed, and has an organic carbon content higher than the 0.5% background, and that, locally, sorted sands associated with the subareal truncation of the shelf form migration paths and reservoirs. The Paleogene paleoshore is probably beneath the upper slope (Bruns et al., this volume), and thus the release of hydrocarbons from breaching of structure by Paleogene erosion may be minimal along parts of the shelf edge. The herbaceous characteristics of the organic carbon sampled onshore indicates predominantly gas accumulations; and unless the oxygen-minimum zone of a continental slope is incorporated into the accreted rocks beneath the upper slope, significant oil is unlikely.

The fate of organic matter in the subducting sedimentary section may reveal unconventional gas accumulations because of the potential for release of vast quantities of gas. Although much remains to be done with geophysical techniques, drilling is ultimately needed to develop and test this hypothesis to a level where it will interest industry.

The Kodiak Shelf has not been offered for lease because of discouraging results from drilling in adjacent areas. We agree with those who contend that present data provide few positive signs of commercial accumulation, particularly if only conventional targets are considered. However, present knowledge is insufficient to rule out the presence of commercial accumulation of hydrocarbons, especially if less conventional approaches are considered. We have presented some ideas that could be tested by further prelease and predrilling exploration.

ACKNOWLEDGMENTS

The reprocessing of seismic records presented here was in part supported by the Dept. of Energy Deep Source Gas Project, contract DE-A121-87MC20422.502. We were assisted in the processing by Ray Culotta and John Miller. Some of our ideas were sharpened in discussions with Casey Moore and his students.

REFERENCES CITED

Allison, R. C., 1978, Late Oligocene through Pleistocene mulluscan faunas in the Gulf of Alaska region: The Veliger, v. 21, p. 171–188.

Allison and Marincovich, 1980, USGS Professional Paper 1233.

Armentrout, J. M., 1979, Cenozoic stratigraphy, Kodiak archipelago, Alaska (abs.): GSA, Abstracts with Program, v. 11, no. 3, p. 66.

Atwater, T., and P. Molnar, 1973, Relative motion of the Pacific and North American plates deduced from seafloor spreading in the Atlantic, Indian, and South Pacific Oceans, *in* R. L. Kovach and A. Nur, eds., Proceedings of the conference on tectonic problems of the San Andreas fault system, 13, Stanford University Publications, Stanford, p. 136–148.

Berggren, W. A. et al.., in press, Paleogene geochronology and chronostratigraphy, *in* Geochronology and the geologic record: Geological Society of London Special Paper.

Bruns, T. R., 1983, Model for the origin of the Yakutat block, an accreting terrane in the northern Gulf of Alaska: Geology, v. 11, p. 718–721.

Bruns, T. R., R. von Huene, R. C. Cullota, S. D. Lewis, and J. W. Ladd, 1987, Geology and petroleum potential of the Shumagin margin, Alaska, this volume.

Burk, C. A., 1965, Geology of the Alaska Peninsula island arc and continental margin: GSA Memoir 99, 250 p.

Byrne, T., 1982, Structural evaluation of coherent terranes in the Ghost Rocks Formation, Kodiak Island, Alaska, *in* J. K. Leggett, ed., Trench-forearc geology: GSA Special Publication 10; publication for Geological Society of London by Blackwell Scientific Publishers, p. 229–242.

Coe, R. S., B. R. Globerman, P. W. Plumley, and G. A. Thrupp, 1985, Paleomagnetic results from Alaska and their tectonic implications, *in* D. G. Howell, ed., Tectonostratigraphic terranes of the Circum-Pacific region: AAPG Circum-Pacific Earth Sciences Series 1, p. 85–108.

Connelly, W., 1978, Uyak Complex, Kodiak Islands, Alaska: A Cretaceous subduction complex: GSA Bulletin, v. 89, p. 755–769.

Davies, J. N., and L. House, 1979, Aleutian subduction zone seismicity, volcano trench separation and their relation to great thrust-type earthquakes: Journal of Geophysical Research, v. 84, no. 139, p. 4583–4591.

Dickinson, W. R., and P. R. Seely, 1979, Structure and stratigraphy

of forearc regions: AAPG Bulletin, v. 63, p. 2–31.
Engebretson, D., A. Cox, and G. Gordon, 1985, Relative motions between oceanic and continental plates in the Pacific Basin: GSA, Special Publication 206.
Fisher, M. A., 1979, Structure and tectonic setting of the continental shelf southwest of Kodiak Island, Alaska: AAPG Bulletin, v. 63, p. 301–310.
Fisher, M. A., 1980, Petroleum geology of the Kodiak Shelf, Alaska: AAPG Bulletin, v. 64, p. 1140–1157.
Fisher, M. A., and R. von Huene, 1980, Structure of upper Cenozoic strata beneath Kodiak Shelf, Alaska: AAPG Bulletin, v. 64, p. 1014–1033.
Fisher, M. A., and R. von Huene, 1984, Geophysical investigation of a suture zone: The border ranges fault of southern Alaska: Journal of Geophysical Research, v. 89, p. 11,333–11,351.
Fisher, M. A., T. R. Bruns, and R. von Huene, 1981, Transverse tectonic boundaries near Kodiak Island, Alaska: GSA Bulletin, v. 92, p. 10–18.
Fisher, M. A., R. L. Detterman, and L. B. Magoon, 1987, Tectonics and petroleum geology of the Cook-Shelikof Basin, this volume.
Fisher, M. A., R. von Huene, G. L. Smith, and T. R. Bruns, 1983, Possible reflections from the down going Pacific plate, 275 kilometers arcward from the eastern Aleutian Trench: Journal of Geophysical Research, v. 88, p. 5835–5849.
Hampton, M. A., 1983, Geology of the Kodiak Shelf, Alaska: Environmental considerations for resource development: Continental Shelf Research, v. 1, no. 3, p. 253–281.
Herrera, R. C., 1978, Developments in Alaska in 1977: AAPG Bulletin, v. 62, p. 1311–1321.
Hill, M. D., J. Morris, and J. Whelan, 1981, Hybrid granodiorites intruding the accretionary prism, Kodiak, Shumagin, and Sanak Islands, southwest Alaska: Journal of Geophysical Research, v. 86, p. 10,569–10,590.
Hillhouse, J., and R. Coe, in press, Paleomagnetic data from Alaska, in Cordilleran Oregon: Alaska, GSA.
Jacob, K. H., K. Nakamura, and J. N. Davies, 1977, Trench-volcanic gap along the Alaskan-Aleutian arc: Facts and speculations on the role of terrigenous sediments, *in* M. Talwani and W. C. Pitman III, eds., Island arcs, deep-sea trenches, and back-arc basins: American Geophysical Union Maurice Ewing Series 1, p. 243–258.
Jones, D. L., Silberling, H. C. Berg. and G. Plafker, 1981, Map showing tectonostratigraphic terranes of Alaska, columnar sections, and summary description of terranes: U.S. Geological Survey Open-File Report 81-792, 20 p., 2 maps.
Keller, G., R. von Huene, K. McDougall, and T. Bruns, 1984, Paleoclimatic evidence for Cenozoic migration of Alaskan terranes: Tectonics, v. 3, no. 4, p. 473–495.
Kirschner, C. E., and C. A. Lyon, 1973, Stratigraphic and tectonic development of Cook Inlet petroleum province: AAPG Memoir 19, p. 346–407.
Kulm, L. D., R. von Huene, J. R. Duncan, J. C. Ingle, S. A. Kling, L. F. Musich, D. J. W. Piper, R. M. Pratt, H. J. Schroder, O. Weser, and S. W. Wise, 1973, Initial reports of the Deep Sea Drilling Project: Washington, D.C., U.S. Government Printing Office, v. 18, 1077 p.
Kvenvolden, K. A., and R. von Huene, 1984, Potential for generation of natural gas in sediments of the convergent margin of the Aleutian Trench area: U.S. Geological Survey Open-File Report 84-404.
Langseth, M. G., R. von Huene, N. Nasu, and H. Okada, 1981, Subsidence of the Japan Trench forearc region off northern Honshu: Oceanologica Acta, p. 173–179, Supplement to volume 4, p. 173–179.
Lyle, W., J. Morehouse, I. F. Palmer, J. G. Bolm, G. W. Moore, and T. H. Nilsen, 1977, Tertiary formations in the Kodiak Island area, Alaska, and their petroleum reservoir and source-rock potential: Alaska Geological Survey Open-File Report 114, 48 p.
McClellan, P. H., M. A. Fisher, R. von Huene, and G. M. Moore, 1982, Summary and discussion of microfossil biostratigraphy in the western Gulf of Alaska: U.S. Geological Survey Circular 844, p. 148–151.
Moore, G. W., 1967, Preliminary geologic map of Kodiak Island and vicinity, Alaska: U.S. Geological Survey Open-File Report, scale 1:250,000.
Moore, G. W., 1969, New formations on Kodiak and adjacent islands, Alaska: U.S. Geological Survey Bulletin 1274-A, p. A27–A35.
Moore, J. C., and A. Allwardt, 1980, Progressive deformation of a Tertiary trench slope, Kodiak Islands, Alaska: Journal of Geophysical Research, v. 85, no. 139, p. 4741–4756.
Moore, J. C., and R. L. Wheeler, 1978, Structural fabric of a melange, Kodiak Islands, Alaska: American Journal of Science, v. 178, p. 739–765.
Moore, J. C., T. Byrne, P. W. Plumley, M. Reid, H. Gibbons, and R. S. Coe, 1983, Paleogene evolution of the Kodiak Islands, Alaska: Consequences of ridge-trench interaction in a more southerly latitude: Tectonics, v. 2, no. 3, p. 265–293.
Nilsen, T. H., 1982, Accretion model for the Cretaceous Chugach terrane, southern Alaska: U.S. Geological Survey Circular 844, p. 93.
Nilsen, T. H., and G. W. Moore, 1979, Reconnaissance study of upper Cretaceous to Miocene stratigraphic units and sedimentary facies, Kodiak and adjacent islands, Alaska, with a section on sedimentary petrology by G. R. Winkler: U.S. Geological Survey Professional Paper 1093, 34 p.
Nilsen, T. H., and G. G. Zuffa, 1982, The Chugach terrane, a Cretaceous trench-fill deposit, southern Alaska, *in* J. K. Leggett, ed., Trench-forearc geology: GSA Special Publication 10; publication for Geological Society of London by Blackwell Scientific Publications, p. 213–227.
Packer, D. R., and D. B. Stone, 1974, Paleomagnetism of Jurassic rocks from southern Alaska, and the tectonic implications: Canadian Journal of Earth Sciences, v. 11, p. 976–997.
Plafker, G., 1969, Tectonics of the March 27, 1964, Alaska earthquake: U.S. Geological Survey Professional Paper 543-I, 74 p.
Plafker, G., and G. Keller, 1985, Paleontological data on the age of the Orca Group, Alaska: U.S. Geological Survey Open-File Report 85-429.
Plumley, P. W., R. S. Coe, and T. Byrne, 1983, Paleomagnetism of the Paleocene Ghost Rocks Formation, Prince William terrane, Alaska: Tectonics, v. 2, p. 295–314.
Pulpan, H., 1985, Seismicity, in North American continent - Continent/Ocean transect #6, A_2: Kodiak to Kuskokwim, Alaska. R. von Huene, ed., GSA, in press.
Pulpan, H., and J. Kienle, 1979, Western Gulf of Alaska seismic risk: Preprints 11th Offshore Technology Conference, p. 2209–2218.
Rau, W. W. et al., 1977, Preliminary foraminiferal biostratigraphy and correlation of selected stratigraphic sections in wells in the Gulf of Alaska Tertiary Province: U.S. Geological Survey Open-File Report 77-747, 54 p.
Sample, J., 1984, The Kodiak formation, Kodiak Islands, Alaska: A record of continuous deformation at a rapidly accreting margin: Proceedings of 97th Annual Meeting, GSA, p. 643.
Sample, J. C., and D. M. Fisher, 1986, Duplex accretion and underplating in an ancient accretionary complex, Kodiak Islands, Alaska: Geology, v. 14, no. 2.
Scheidegger, K. F., J. B. Corliss, P. A. Jezek, and D. Ninkovich, 1980, Compositions of deep-sea ash layers derived from north Pacific volcanic arcs, Variations in time and space: Journal of Volcanology and Geothermal Research, v. 7, p. 107–137.
Scholl, D. W., T. L. Vallier, and A. J. Stevenson, 1987, Geology evolution and petroleum geology of the Aleutian Ridge, this volume.
Schwab, W. C., T. R. Bruns, and R. E. von Huene, 1980, Maps showing structural interpretation of magnetic lineaments in the northern Gulf of Alaska: U.S. Geological Survey Miscellaneous Field Studies Map MF-1245, scale 1:500,000, 12 p., 3 maps.
Seely, D. R., 1977, The significance of landward vergence and

oblique-structural trends on trench inner slopes, *in* M. Talwani and W. C. Pitman, eds., Island arcs, deep-sea trenches, and back-arc basins: American Geophysical Union Maurice Ewing Series 1, p. 187–198.

Shor, G. G., Jr., and R. von Huene, 1972, Marine seismic refraction studies near Kodiak, Alaska: Geophysics, v. 37, p. 697–700.

Stauder, W., and G. A. Bollinger, 1964, The S-wave project for focal mechanisms studies: Earthquakes of 1962: Seismological Society of America Bulletin, 54, p. 2199–2208.

Stevenson, A. J., and R. Embley, 1987, Deep-sea fan bodies, terrigenous turbidite sedimentation, and petroleum geology, Gulf of Alaska, this volume.

Stevenson, A. J., D. W. Scholl, and T. L. Vallier, 1983, Tectonic and geologic implications of the Zodiac fan, Aleutian Abyssal Plain, northeast Pacific: GSA Bulletin, v. 94, p. 259–273.

Sykes, L. R., 1971, Aftershock zones of great earthquakes, seismicity gaps, and earthquake prediction for Alaska and the Aleutians: Journal of Geophysical Research, v. 70, no. 32, p. 8021–3041.

Thrupp, G. A., and R. S. Coe, 1987, Early Tertiary paleomagnetic evidence and the displacement of southern Alaska, Geology, v. 14, no. 2.

von Huene, R., 1972, Structure of the continental margin and tectonism at the eastern Aleutian Trench: GSA Bulletin, v. 83, p. 3613–3626.

von Huene, R., 1979, Structure of the outer convergent margin off Kodiak Island, Alaska, from multichannel seismic records: Geological and geophysical investigations of continental margins: AAPG Memoir 29, p. 261–272.

von Huene, R., 1984, Structural diversity along modern convergent margins and the role of overpressured pore fluids in subduction zones: Bulletin Society of Geology, France, v. 26, no. 2, p. 207–219.

von Huene, R., E. Larson, and J. Crouch, 1973, Preliminary study of ice-rafted erratics as indicators of glacial advances in the Gulf of Alaska, Initial Reports of the Deep Sea Drilling Project, v. 18, p. 835–842.

von Huene, R., G. W. Moore, and J. C. Moore, 1979, Cross section, Alaska Peninsula—Kodiak Island—Aleutian Trench: GSA Map and Chart Series MC-28A.

von Huene, R., M. A. Hampton, M. A. Fisher, D. J. Varchol, and G. R. Cochrane, 1980, Nearsurface structure map of Kodiak Shelf, Alaska: U.S. Geological Survey Miscellaneous Field Studies Map MF-1200, scale 1:500,000.

von Huene, R., S. Box, B. Detterman, M. Fisher, C. Moore, and H. Pulpan, 1985, Centennial Continent/Ocean Transect #6 A_2: A-2 Kodiak to Kuskokwim, Alaska, The Geological Society of America.

von Huene, R., L. D. Kulm, J. R. Duncan, J. C. Ingle, S. A. Kling, L. F. Musich, D. J. W. Piper, R. M. Pratt, H. Schrader, O. Weser, S. W. Wise, Jr., 1973, Initial reports of the Deep Sea Drilling Project Leg 18: Washington, D.C., U.S. Government Printing Office.

Wallace, W. K., and D. C. Engebretson, 1984, Relationships between plate motions and Late Cretaceous to Paleogene magmatism in southwestern Alaska: Tectonics, v. 3, no. 2, p. 295–315.

Watkins, J. S., J. C. Moore, S. B. Bachman, F. Beghtel, A. Butt, B. Didyk, Y. Dmitriev, J. Legget, N. Lundberg, K. McMillen, N. Niitsuma, R. Rodriguez-Torres, L. Shepard, T. Shipley, J.-F. Stephan, and H. Stradner, 1982, Initial reports of the Deep Sea Drilling Project Leg 66, U.S. Government Printing Office, Washington, D.C.

Wilson, F. H., and N. Shew, 1980, Apparent episodicity of magmatic activity deduced from radiometric age determinations, *in* The United States Geological Survey in Alaska: Accomplishments during 1980, U.S. Geological Survey Circular 844, p. 13–15.

Wolf, 1981, GSA Special Paper 184, p. 44.

Wyss, M., and J. N. Brune, 1967, The Alaska earthquake of 28 March 1964: A complex multiple rupture: Seismological Society of America Bulletin, v. 57, p. 1017–1023.

CHAPTER TEN

TECTONICS AND PETROLEUM GEOLOGY OF THE COOK-SHELIKOF BASIN, SOUTHERN ALASKA

Michael A. Fisher
Robert L. Detterman
Leslie B. Magoon

U. S. Geological Survey
Menlo Park, California

ABSTRACT

The Cook-Shelikof Basin underlies Cook Inlet, Shelikof Strait, and the northeastern Alaska Peninsula, and forms part of the Peninsular tectonostratigraphic terrane. This basin contains rocks as old as Permian. Throughout the southwest part of the basin, Cretaceous and Cenozoic sedimentary rocks are discontinuously exposed and are thin relative to Jurassic rocks; however, under the northeast part of the basin, Cenozoic rocks are more than 7 km in thickness. Mesozoic rocks form a syncline that trends northeast and spans the basin. This syncline formed during the latest Cretaceous, and the limbs were truncated by a regional unconformity at the base of Cenozoic rocks.

Twenty-five exploratory wells have been drilled southwest of upper Cook Inlet, and only oil shows have been encountered. Below upper Cook Inlet, thermal maturation of Middle Jurassic source rocks probably required burial beneath a thick Cenozoic section, but Middle Jurassic rocks southwest of this area have not been deeply buried.

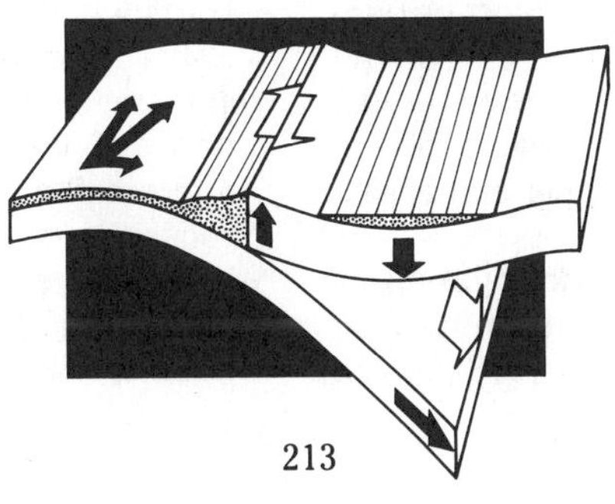

INTRODUCTION

Cook Inlet and Shelikof Strait form major marine reentrants into the coastline of south-central Alaska (Figure 1). The Cook-Shelikof Basin underlies these reentrants as well as the Kenai lowland and the northeastern Alaska Peninsula. Only the southwestern part of the Cook-Shelikof Basin is considered in this chapter—the part that extends southwestward from the Kenai lowland and underlies lower Cook Inlet, Shelikof Strait, and the part of the Alaska Peninsula northeast of Sutwik Island (Figure 2). The northwest boundary of this basin lies along the Mesozoic batholith uplifted by and exposed along the northwest side of the Bruin Bay fault. The southeast basin boundary lies along the Mesozoic crystalline rocks that abut the northeast side of the Border Ranges fault. The Cook-Shelikof Basin lies within the Peninsular tectonostratigraphic terrane (Jones et al., 1981); the Border Ranges fault separates this basin from the Chugach terrane, which includes Cretaceous melange and turbidites.

Since the Early Cretaceous, the Cook-Shelikof Basin has been within the tectonic influence of plate convergence along its southeast side, as indicated by the presence along this side of Late Cretaceous and Paleogene melange and deformed turbidite sequences of the Chugach and Prince William tectonostratigraphic terranes. Furthermore, plutonic roots of ancient magmatic arcs and active volcanoes of the modern arc occur along the northwest flank of the basin. This tectonic setting has produced the abundant hydrocarbons within the northwestern Cook-Shelikof Basin, beneath the upper Cook Inlet. The basin southwest of the Kenai lowland, however, has apparently not been endowed with significant hydrocarbon deposits, as suggested by disappointing results obtained from 25 exploratory wells.

In this chapter, we summarize the stratigraphy and structure of the Cook-Shelikof Basin. New information about this basin (since the report by Fisher and Magoon, 1978) includes the characterization of offshore rocks based on data from the Continental Offshore Stratigraphic Test (COST) No. 1 well drilled in rocks underlying lower Cook Inlet (Magoon, in press) and detailed mapping of parts of the Alaska Peninsula (Detterman et al., 1981; Detterman et al., 1983). We also review the tectonic evolution of the basin; recent proposals that large parts of southern Alaska were once affixed to mobile ocean plates have sparked reexamination of how this basin formed. Throughout this chapter we use the geologic-time scale of van Eysinga (1975).

STRATIGRAPHY

The stratigraphy of the Cook-Shelikof Basin is here divided into five major sequences—middle(?) or upper Paleozoic through middle Lower Cretaceous, upper Lower Cretaceous and Upper Cretaceous, Paleocene and lower Eocene, upper Eocene and Oligocene, and Miocene through Holocene (Figure 3). The uppermost units of all sequences, except the Paleocene and lower Eocene, were deposited during or just after periods of deformation and increased magmatic-arc activity. Thus, the stratigraphic sequences help emphasize the tectonic development of the basin.

The oldest rocks exposed near this basin are Silurian or Devonian limestone preserved as roof pendants on Jurassic intrusive rocks (Detterman, Case, and Wilson, 1979), which are part of the Peninsular tectonostratigraphic terrane (Jones et al., 1981). Outcrops of the Silurian or Devonian rocks actually lie northwest of the Bruin Bay fault (northeast of Becharof Lake; Figure 2); none occur within the basin. These rocks may underlie some parts of the Cook-Shelikof Basin, because Permian limestone crops out in a small area along the Pacific Coast of the Alaska Peninsula, near Puale Bay (Hansen, 1957), and pre-Permian limestone may lie beneath the Permian limestone. The total thickness of the Paleozoic section is unknown.

The oldest rock sequence also includes Upper Triassic through Middle Jurassic strata that are exposed around the Cook-Shelikof Basin and have been penetrated in offshore wells. These strata may locally be as thick as 5 km, and they make up much of the deep fill within this basin. Upper Triassic rocks are exposed at Puale Bay, along the northwest coasts of the Kodiak group of islands, and along the northwest (Kamishak Formation; Figure 3) and southeast shores of lower Cook Inlet (Figure 2). These rocks consist of variably metamorphosed limestone, chert, sandstone, and basalt (now mainly altered to greenstone) (Burk, 1965; Detterman and Hartsock, 1966; Magoon, Adkison, and Egbert, 1976; Connelly, 1978; Connelly and Moore, 1979; Detterman and Reed, 1980). The most highly metamorphosed rocks crop out along the margin of the Jurassic batholith that borders the basin on the northwest.

Near lower Cook Inlet, Upper Triassic rocks are unconformably overlain by Lower Jurassic rocks of the Talkeetna Formation (Figure 3), which include thick volcanic agglomerates, breccias, and lava flows. Near Puale Bay, Upper Triassic limestone is conformably overlain by limestone of Early Jurassic age. Other Lower Jurassic rocks near Puale Bay include tuffaceous sandstone and some conglomerate (Imlay and Detterman, 1977), but the abundant coarse-grained volcanic detritus evident around lower Cook Inlet is absent (Detterman et al., 1983). Along the northwest coast of lower Cook Inlet, Lower Jurassic rocks are overlain at an angular unconformity by as much as 3 km of marine Middle Jurassic graywacke, sandstone, and siltstone that are rich in volcanic debris (Tuxedni Group and Chinitna Formation). Late Middle Jurassic rocks, the uppermost part of the Tuxedni Group, are exposed along the northwest coast of lower Cook Inlet and contain more feldspar than do older rocks; this suggests that the unroofing of the Alaska-Aleutian Range batholith had already been achieved before the end of the Middle Jurassic (Egbert and Magoon, 1981). Farther southwest, on the Alaska Peninsula, Lower Jurassic rocks and Lower and Middle Jurassic rocks (Kialagvik Formation) are conformable. The Lower and Middle Jurassic rocks include nonmarine to deep-water marine sandstone, siltstone, shale, and conglomerate.

Upper Jurassic rocks unconformably overlie older rocks. The overlying rocks are feldspathic, in contrast with the underlying volcaniclastic-rich sequence (Detterman and Hartsock, 1966; Lankford and Magoon, 1977; McLean, 1979; Egbert and Magoon, 1981; Egbert, 1981). Thus the Upper Jurassic section records a continual shifting, in the provenance of basin sediment, that resulted from uplift, exposure, and erosion of the Jurassic plutonic rocks. Upper Jurassic rocks are exposed around lower Cook Inlet (Naknek Formation; Figure 3) and extensively on the Alaska Peninsula (Burk, 1965; Jones and Miller, 1976; Magoon, Adkison, and Egbert, 1976; Magoon, Egbert, and Petering, 1977; Detter-

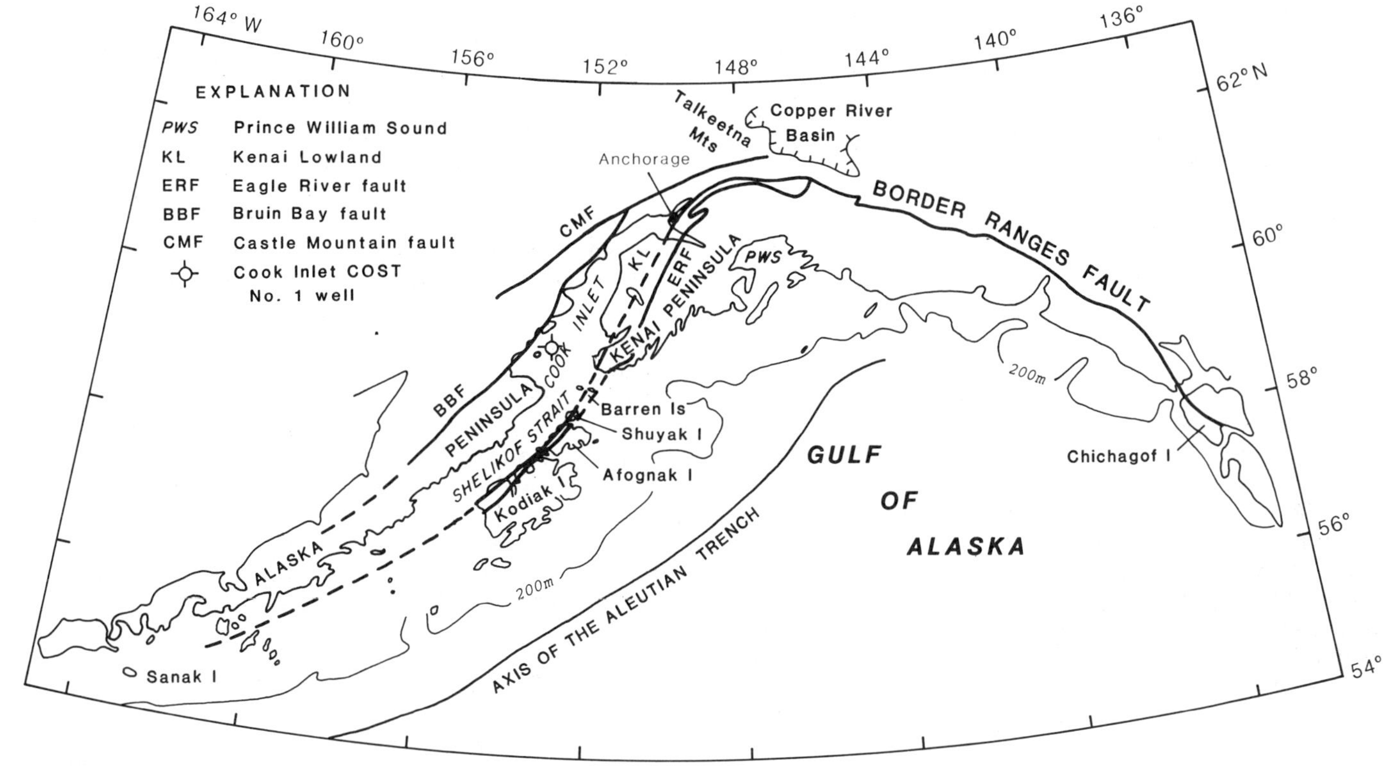

Figure 1. Regional geography and major geologic structures near Cook Inlet and Shelikof Strait.

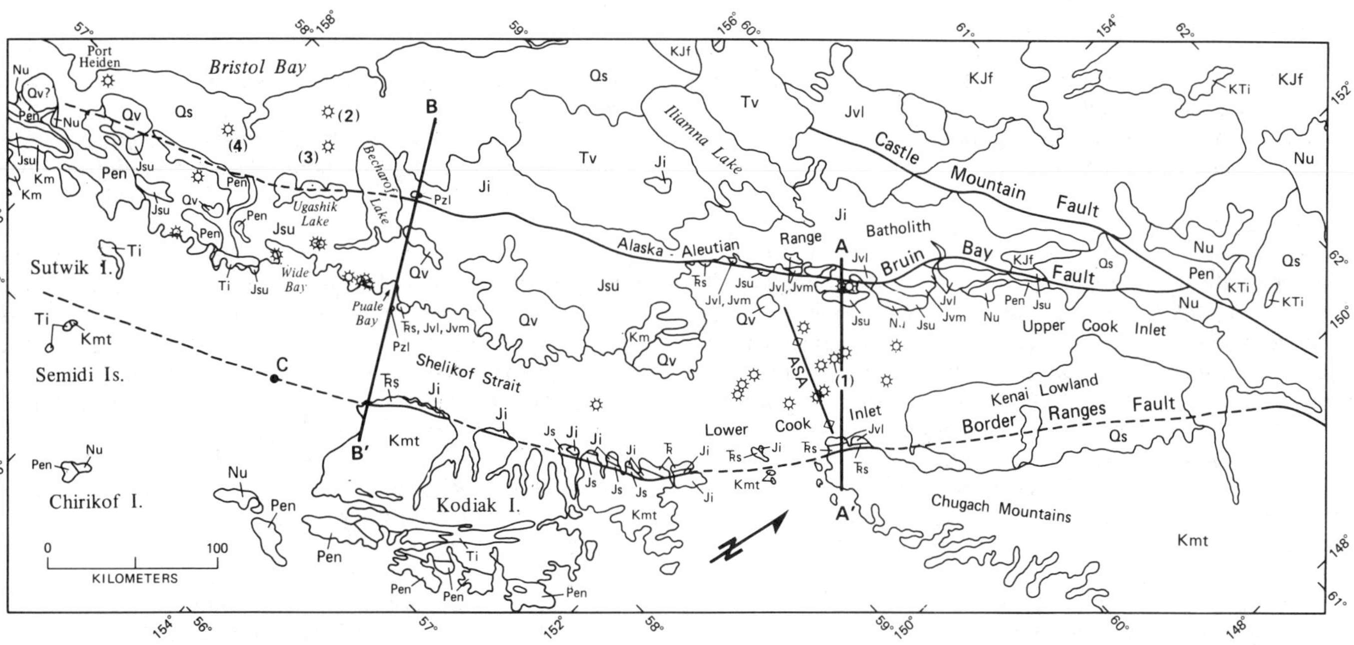

Figure 2. Generalized regional geology and locations of petroleum exploration wells. ASA = Augustine-Seldovia arch. Pzl = Silurian or Devonian and Permian limestone; Trs = Upper Triassic limestone, sandstone, and volcanic rocks; Ji = Early Jurassic and Middle and Late Jurassic intrusive rocks; Js = Early Jurassic blueschist- and greenschist-facies rocks; Jvl = Lower Jurassic volcanic and volcaniclastic rocks; Jvm = Middle Jurassic volcanigenic rocks; Jsu = Upper Jurassic basal conglomerate and feldspathic sandstone and siltstone; KJf = Jurassic and Cretaceous flysch; Km = Cretaceous marine and nonmarine rocks; Kmt = Late Cretaceous melange and deformed turbidite sequences; KTi = Cretaceous and Tertiary intrusive rocks; Czu = Cenozoic rocks undivided; Pen = Paleogene nonmarine rocks; Tv = Tertiary volcanic rocks; Ti = late Tertiary intrusive rocks; Nu = Neogene rocks, undivided; Qv = Quaternary volcanic rocks; Qs = Quaternary sedimentary rocks. A–A′ and B–B′ are lines of cross-sections shown in Figure 4. The point labeled "C" marks the abrupt decrease to the southwest of magnetic and gravity anomalies measured over the Border Ranges fault. Exploratory wells are shown by barbed circles.

man et al., 1983). They are the deepest rocks reached by the COST No. 1 well drilled in lower Cook Inlet (Magoon, in press). These rocks consist of nonmarine and marine siltstone and sandstone but also include a widespread Upper Jurassic coarse-grained, feldspathic sandstone and conglomerate, which contains chiefly clasts of metamorphic rocks and Jurassic plutonic rocks. On the Alaska Peninsula, these clasts locally exceed 2 m in diameter.

Upper Jurassic rocks are disconformably overlain by Lower Cretaceous siltstone and sandstone. These Lower Cretaceous beds are marked throughout the basin by *Inoceramus*-rich beds that form a bioclastic limestone.

In onshore areas the lowest stratigraphic sequence and the upper Lower Cretaceous and Upper Cretaceous sequence are separated by a major hiatus spanning much of the middle and Late Cretaceous (Aptian through Campanian; Figure 3; Jones and Detterman, 1966; Magoon, Adkison, and Egbert, 1976; Fisher and Magoon, 1978; Petering and Smith, 1981). Rocks representing parts of this hiatus crop out locally; for example, Albian rocks crop out near Katmai Bay on Shelikof Strait (Miller, Detterman, and Case, 1980). In contrast with this onshore hiatus, wells drilled in lower Cook Inlet show that much of the middle and Late Cretaceous hiatus is represented by Albian through Campanian shallow- to marginal-marine deposits (Hastings, Robinson, and Robinson, 1983). Seismic-reflection data obtained in lower Cook Inlet show that these Cretaceous rocks onlap to the northwest, so that a more complete section is present offshore than there is onshore near the northwest periphery of the basin. Onshore Upper Cretaceous rocks (Kaguyak Formation) consist mainly of marine sandstone and siltstone. Nonmarine rocks, however, crop out locally along the northwest side of lower Cook Inlet (Magoon, Griesbach, and Egbert, 1980), and cyclically marine and nonmarine rocks (Chignik and Hoodoo Formations) are exposed on the Alaska Peninsula (Burk, 1965; Detterman, 1977; Mancini, Deeter, and Wingate, 1978; Molenaar, 1980). Offshore Maestrichtian rocks were deposited mainly in a nonmarine environment and include coarse sandstone and conglomerate.

The upper Lower Cretaceous and Upper Cretaceous sequence is unconformably overlain by the nonmarine Cenozoic section. Cenozoic rocks below upper Cook Inlet have a maximum thickness of about 7 km (Hartman, Pessel, and McGee, 1972), whereas beneath lower Cook Inlet, the Cenozoic section is mainly <2 km thick. The lower part of this section forms the Paleocene and lower Eocene stratigraphic sequence (Tolstoi and West Foreland Formations; Figure 3). In the vicinity of Cook Inlet, Paleocene rocks include a nonmarine conglomerate that overlies the unconformity at the base of the Cenozoic section. Near lower Cook Inlet, Paleocene rocks include nonmarine tuffaceous and conglomeratic units. These rocks are dated by using the paleofloral subdivisions of Wolfe (1977) and Wolfe and Tanai (1980). Along the length of the Alaska Peninsula, Paleocene and lower Eocene rocks are nonmarine, volcaniclastic sandstone, conglomerate, siltstone, shale, and coal.

A hiatus separates the Paleocene and lower Eocene sequence from the upper Eocene and Oligocene sequence. Near Cook Inlet, the overlying rocks include nonmarine conglomerate, sandstone, and siltstone, which form the lower part of the Kenai Group (Calderwood and Fackler, 1972; Kirschner and Lyon, 1973). The Hemlock Conglomerate is the basal unit of this group and is the main reservoir for hydrocarbons below upper Cook Inlet. Conglomerate the same age as the Hemlock also crops out along the shore of Shelikof Strait (Magoon, Adkison, and Egbert, 1976; Petering and Smith, 1981). Eocene and Oligocene rocks of the Alaska Peninsula include abundant basaltic and andesitic lava and pyroclastic rocks (Meshik Formation).

An unconformity separates the upper Eocene and Oligocene sequence from the Miocene through Holocene sequence, which near lower Cook Inlet constitutes the upper part of the Kenai Group. The Miocene through Holocene sequence includes very thick nonmarine deposits derived from the Chugach Mountains and the Alaska Range, which began to rise on opposite sides of the basin during the Miocene and Pliocene (Kirschner and Lyon, 1973). Fanglomerate locally makes up much of this sequence, reflecting the rapid uplift and steep relief of areas bordering the basin. Along the Alaska Peninsula, rocks of Pliocene and Miocene age are absent from the Cook-Shelikof Basin. However, Miocene rocks (Bear Lake Formation) lie just outside the basin, abutting the northwest side of the Bruin Bay fault, the northwest basin boundary.

IGNEOUS AND METAMORPHIC ROCKS

The oldest igneous and metamorphic rocks exposed near the Cook-Shelikof Basin are Early Jurassic diorite (184–193 Ma) and schist (161–191 Ma) that abut the Border Ranges fault along the southeast boundary of this basin, from Seldovia to the southwest end of Kodiak Island (Forbes and Lanphere, 1973; Magoon, Adkison, and Egbert, 1976; Carden and Decker, 1977; Carden et al., 1977; Connelly, 1978; Cowan and Boss, 1978; Connelly and Moore, 1979; Forbes et al., 1979; Hudson, 1979). The schist includes intimately intercalated high-temperature-blueschist-facies and greenschist-facies layers; amphibolite-facies rocks lie along the margin of the diorite.

The largest exposures of igneous rocks constitute the Alaska-Aleutian Range batholith that lies along the northwest side of the Cook-Shelikof Basin, mostly northwest of the Bruin Bay fault (Reed and Lanphere, 1972; Reed and Lanphere, 1973; Reed, Miesch, and Lanphere, 1983). This batholith is mainly composed of quartz diorite that can be divided into three distinct generations—late Early but mainly Middle Jurassic (176–154 Ma), which is about 10 m.y. younger than the diorite along the southeast flank of the basin), Late Cretaceous and Paleocene (83–58 Ma), and Oligocene (38–26 Ma). As far southwest as Becharof Lake, exposed rocks of this batholith are mainly of Jurassic age. Plutonic rocks extend southwest of Becharof Lake, because such rocks were penetrated by three exploratory wells (wells numbered 2, 3, and 4 on Figure 2) drilled into rocks of the peninsula southwest and west of the Cook-Shelikof Basin (Brockway et al., 1975; McLean, 1977). K-Ar dates from plutonic rocks in one well (well number 3 on Figure 2) are Early Jurassic (177 Ma); the other sampled rocks are undated. All three wells penetrated Cenozoic sedimentary rocks that unconformably overlie the plutonic rocks.

Much of the Alaska Peninsula is underlain by plutonic rocks and by thick volcanic rocks of Oligocene age (Detterman et al., 1983). K-Ar dates from these plutonic and volcanic rocks cluster between late Eocene and early Miocene (40–20 Ma; Wilson, Gaum, and Herzon, 1981; Wilson and

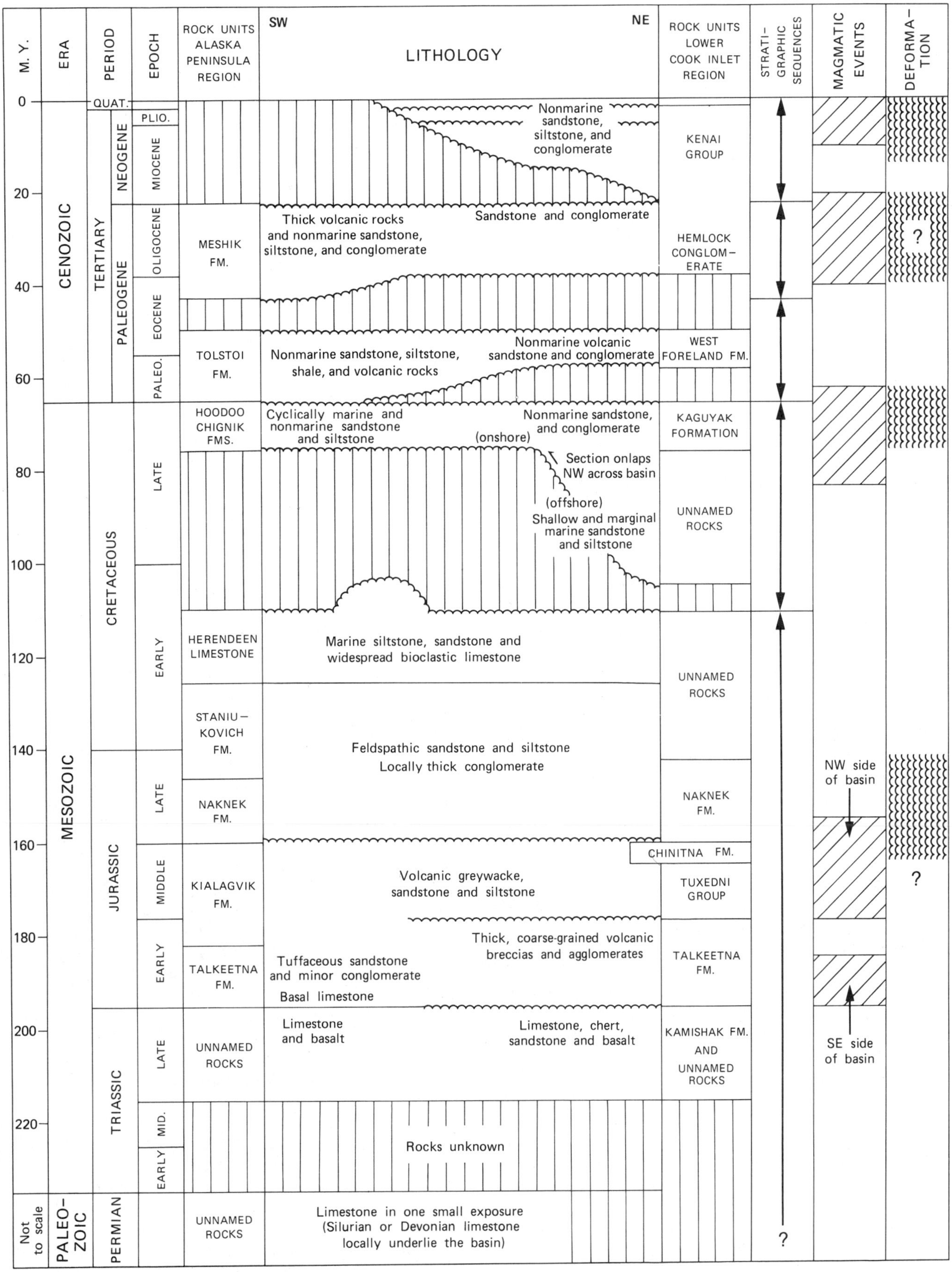

Figure 3. Geologic column illustrating the generalized lithology and major stratigraphic sequences of the Cook-Shelikof Basin, as well as magmatic and deformational events that affected the basin. Time scale is from van Eysinga (1975).

Shew, 1982). Southwest of the Cook-Shelikof Basin, these rocks cause strong magnetic anomalies, and many plutons are associated with molybdenum and copper sulfide deposits (Case et al., 1981). The youngest igneous rocks are late Miocene and younger (10–0 Ma). Large Quaternary volcanoes surmount the Alaska Peninsula, forming the southwestern extension of the magmatic arc that begins near Anchorage and extends along the northwest flank of the Cook-Shelikof Basin.

GEOLOGIC STRUCTURE

The structures that bound the Cook-Shelikof Basin on the northwest and southeast are major reverse or thrust faults. The extensive Alaska-Aleutian Range batholith lies along the northwest side of the Bruin Bay fault; together these features form the northwest basin boundary. Along the Alaska Peninsula, Cenozoic sedimentary rocks unconformably overlie plutonic rocks near the bottom of three exploratory wells, indicating that southwest of Becharof Lake the batholith continues to form the northwest limit of Mesozoic sedimentary rocks in the Cook-Shelikof Basin. The Bruin Bay fault extends in discontinuous outcrop southwestward from its intersection with the Castle Mountain fault, possibly as far southwest as Port Heiden on the Alaska Peninsula, a distance of 700 km (Figure 2; Burk, 1965; Detterman and Hartsock, 1966; Magoon, Adkison, and Egbert, 1976; Detterman et al., 1976; Detterman and Reed, 1980). Where the fault is exposed along the west coast of Cook Inlet, it cuts through Upper Triassic and Lower Jurassic rocks, and locally the fault offsets batholithic rocks. For most of its length, the Bruin Bay fault is a high-angle reverse fault that is downdropped on the southeast and dips between vertical and 45°NW; most dips, however, are between 60° and 70°NW. In places, this fault is composed of four or five subparallel strands that form a zone 6–8 km wide. Along the northwest coast of lower Cook Inlet, the fault throw amounts to as much as 3000 m; elsewhere, offset rocks contain few distinct marker beds from which to determine the throw. Southwest of Ugashik Lake, on the Alaska Peninsula, the throw may be as much as 3000 m, and the sense of throw may be opposite to that observed northeast of this lake in that Miocene rocks on the northwest side of this fault are juxtaposed against Middle Jurassic rocks on its southeast side, implying relative uplift of the Cook-Shelikof Basin side of this fault.

The Bruin Bay fault was active as early as the late Middle and Late Jurassic. The Alaska-Aleutian Range batholith was uplifted along the fault, shedding voluminous and locally very-coarse-grained granitic and metamorphic rock detritus now encased in upper Middle and mainly Upper Jurassic rocks of the Cook-Shelikof Basin. South of Iliamna Lake, the fault zone is intruded by a granodiorite stock that yielded an Oligocene (26 Ma) age; this stock shows no evidence of offset, suggesting that little or no post-Oligocene movement has occurred along the fault near this stock (Detterman and Reed, 1980).

The southeast boundary of the Cook-Shelikof Basin lies along Early Jurassic crystalline rocks exposed on the northwest side of the Border Ranges fault. This fault crops out discontinuously for 2400 km, from Chichagof Island in southeastern Alaska to near Sanak Island (MacKevett and Plafker, 1974; Plafker and Campbell, 1979; Fisher, 1980; Pavlis, 1982; Fisher and von Huene, 1984). The fault dips steeply north or northwest but its dip is poorly known along the Kodiak islands because of complex geology. This fault forms a structural separation of the fill in the Cook-Shelikof Basin, composed of middle(?) and upper Paleozoic and younger rocks, from the melange and deformed turbidite sequences, no older than Cretaceous, that lie southeast of the fault.

The Border Ranges fault is marked by conspicuous geophysical anomalies, such as the large magnetic anomaly that parallels the fault westward and southwestward from the Copper River Basin to the central Kenai lowland (Grantz, Zietz, and Andreasen, 1963; Fisher, 1980; Fisher and von Huene, 1984). Near lower Cook Inlet, other large magnetic anomalies terminate obliquely at this fault (Fisher, 1980; Fisher and von Huene, 1984). Between the south end of the Kenai Peninsula and Shuyak Island, marine gravity data and satellite geoid-height data show that the Border Ranges fault causes pronounced gradients with reliefs of 40 mGal and 3 m, respectively; thus, this fault forms a major disruption of rock-density structure along the southeast flank of the Cook-Shelikof Basin. Strong gravity and magnetic anomalies mark the fault along and southwest of Kodiak Island. These anomalies extend with sharp relief about 50 km southwest of this island, where both types of anomalies lose strength abruptly (at point C, Figure 2). Although reduced anomalies show that the fault continues at least as far southwest as Sutwik Island, modeling of this abrupt decrease in anomalies indicates that fault uplift diminishes southwestward from Kodiak Island (Fisher and von Huene, 1984).

The Border Ranges fault was apparently active mainly during the Cretaceous. MacKevett and Plafker (1974) proposed that the Border Ranges fault was active during the late Mesozoic and early Tertiary. Pavlis (1982) used regional geologic relationships and the Early Cretaceous age of plutons that cut the fault near Anchorage to propose that the fault was active primarily during the Early Cretaceous. On the basis of the regional stratigraphy and structure of the Cook-Shelikof Basin, Fisher and von Huene (1984) proposed that during the Late Jurassic or Early Cretaceous, mainly strike-slip motion occurred along this fault and that in the Late Cretaceous, renewed convergence at a subduction zone southeast of the fault changed the predominant fault motion from strike slip to reverse. Paleogene rocks southwest of lower Cook Inlet do not dip northwest near this fault, indicating that little or no vertical movement has occurred along the fault since the end of the Cretaceous or the early Paleocene. Seismic-reflection data reveal no offset beds in rocks of probable Pliocene and younger age, suggesting that little or no fault movement has occurred since the Miocene.

The Cook-Shelikof Basin itself can be divided into two structural levels that are separated by the unconformity between Mesozoic and Cenozoic rocks. Mesozoic rocks beneath lower Cook Inlet are deformed into a syncline that spans the entire width of the basin, whereas Cenozoic rocks in the upper structural level dip generally to the southeast without a major reversal in dip (A–A′, Figures 4, 5). The unconformity separating the structural levels cuts down section through Mesozoic rocks that form the southeast limb of the deep-basin syncline. Thus, belts of progressively older Mesozoic rocks are truncated by the base of Cenozoic rocks southeastward across the basin. This progression ends at the Border Ranges fault, where Upper Triassic rocks crop out.

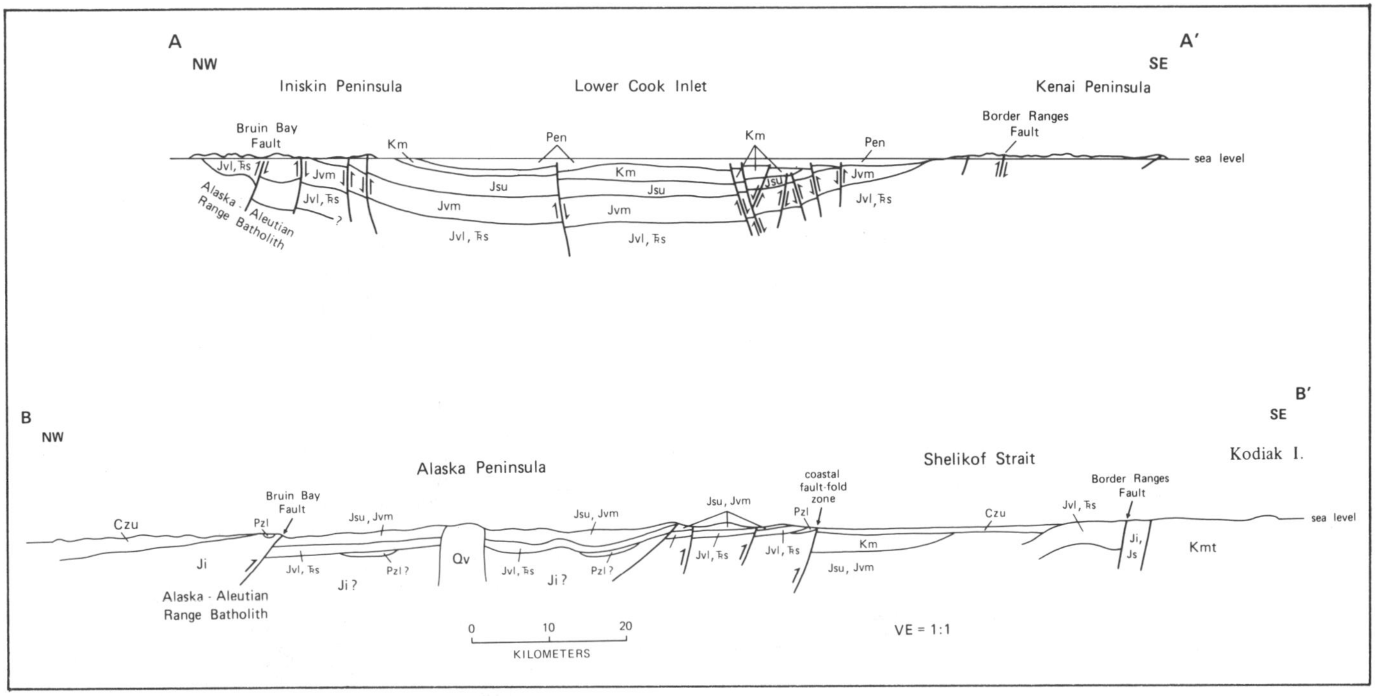

Figure 4. Cross-sections through the Cook-Shelikof Basin. Symbols are the same as those used in Figure 2. A–A′ is modified from Fisher and Magoon (1978), and B–B′ is modified from Detterman (in press). Locations shown in Figure 2.

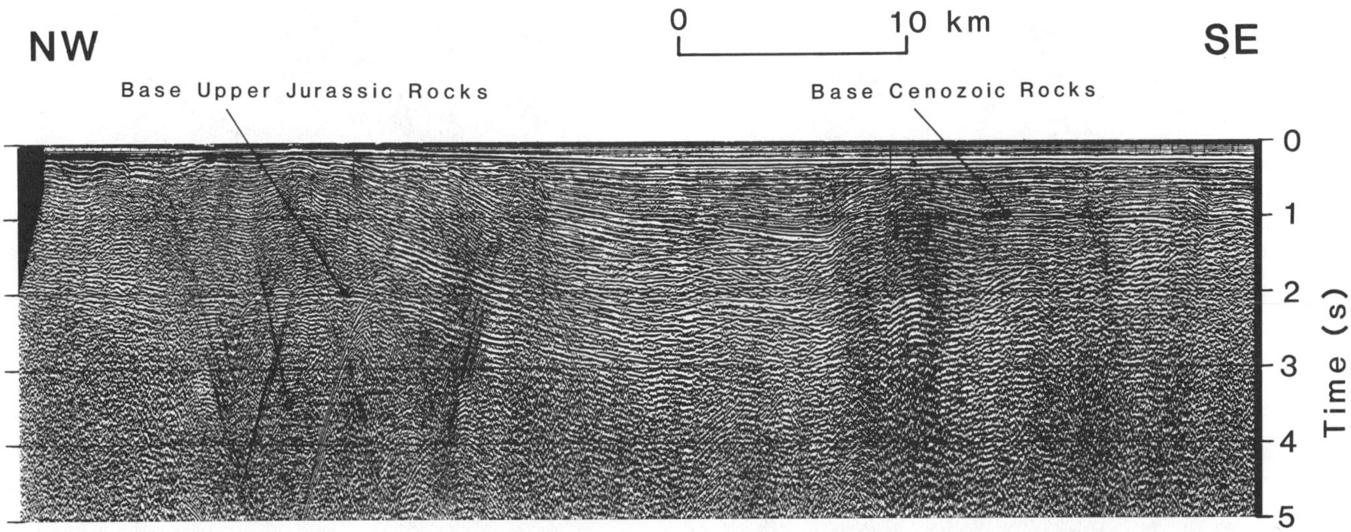

Figure 5. Seismic line 755, collected and processed for the U.S. Geological Survey. The unconformity at the base of Cenozoic rocks truncates the limbs of the basin syncline formed during the Late Cretaceous. Location of line shown in inset map, Figure 6.

Most of the Alaska Peninsula is underlain by rocks of the deeper structural level. The syncline formed by Mesozoic rocks beneath lower Cook Inlet and Shelikof Strait and the northeast Alaska Peninsula bifurcates southwest of Puale Bay, where a zone of en-echelon reverse faults and large anticlines along the Pacific shore of the Alaska Peninsula forms a coastal fault-fold zone that divides the basin longitudinally (Figure 2 and B–B′ of Figure 4). One coastal anticline deforming rocks at Wide Bay is 150 km long and exposes rocks as old as Middle Jurassic at its core. These coastal anticlines die out northeastward from Puale Bay.

Gravity values decrease by 40 mGal across the coastal fault-fold zone (M. A. Fisher, unpublished data, 1980), and modeling of these values indicates that faults within this coastal zone offset the basement by about 5 km. Thus, in northwest-southeast cross-section (B–B′, Figure 4), the Cook-Shelikof Basin is made up of two longitudinal segments tilted northwest and bounded by two major faults and a fault-fold zone. Between each pair of these structures, basement rocks rise abruptly toward the upthrown side and then dip gradually northwestward from each structure.

The three principal structures involved in longitudinal segmentation of the Cook-Selikof Basin formed at different times. The Border Ranges fault may have been active for a time spanning the late Mesozoic and earliest Cenozoic. Similarly, near Iliamna Lake, the Bruin Bay fault seems to have

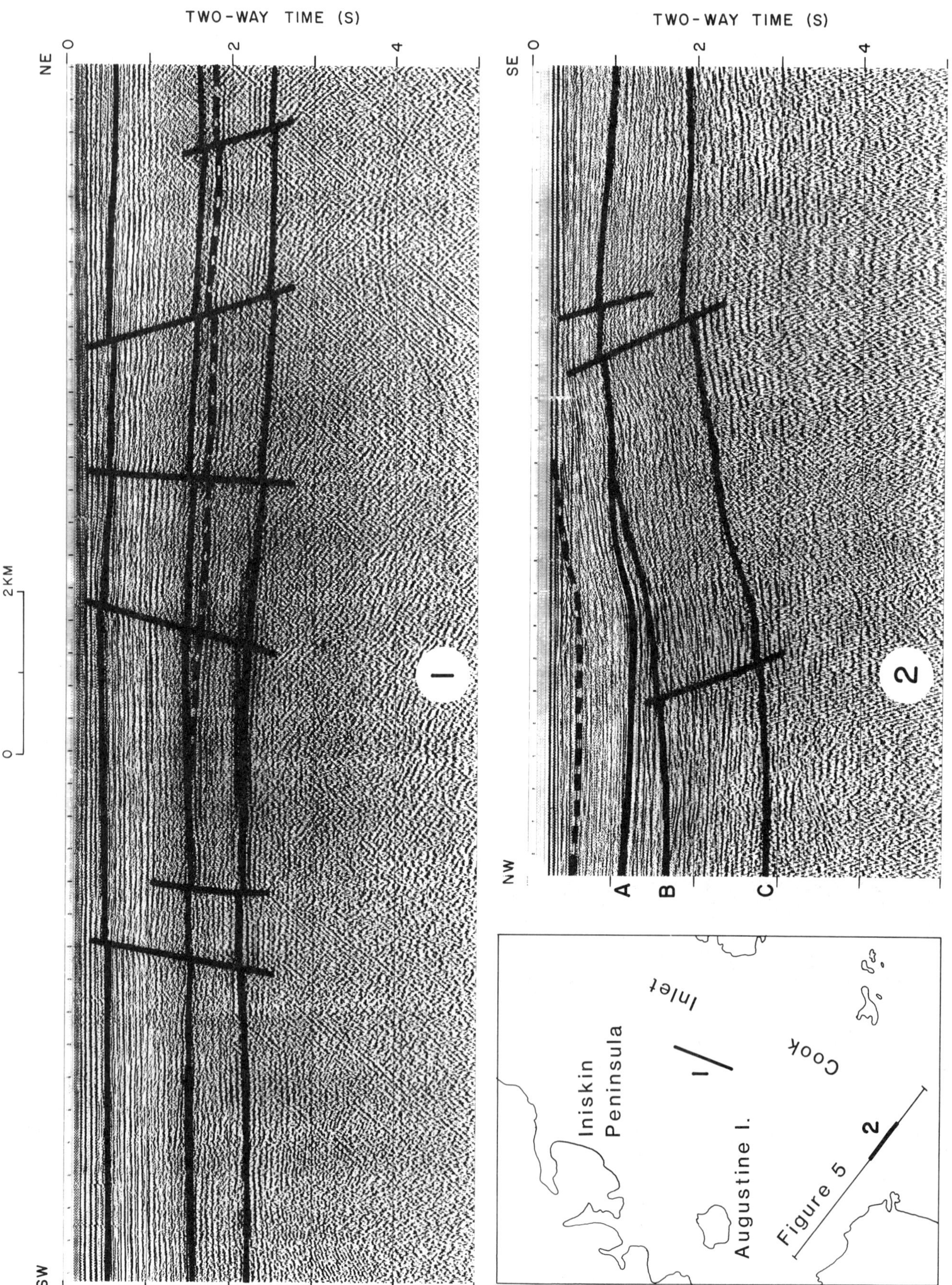
TWO-WAY TIME (S)
NE
SW
A
B
C
0
2
4
0
2 KM
1
SE
NW
2
Iniskin Peninsula
Cook
Inlet
Augustine I.
Figure 5

Figure 6. Seismic sections from lower Cook Inlet. Horizon C may be the base of Upper Jurassic rocks; Horizon B may be the base of the Cretaceous section; Horizon A is the base of the Cenozoic section. These sections show the broad, gentle Augustine-Seldovia arch (section 1) and one of the late Cenozoic (probably Miocene and younger), northeast-trending anticlines that deforms all rocks in the basin (section 2).

been active from the late Middle and Late Jurassic until the Oligocene. Farther southwest along the Alaska Peninsula, this fault may have been active during the Miocene. In contrast, the coastal fault-fold zone formed mainly during the Miocene (Burk, 1965). Along the coastal fault-fold zone near the southwest end of the Alaska Peninsula, Oligocene rocks within a narrow zone are deformed strongly, but a 10-m.y.-old pluton remains undeformed; these ages bracket the probable time of coastal deformation in this area (Wilson, Case, Detterman, 1985).

Local, late Cenozoic anticlines and reverse faults are superimposed over rocks in the two structural levels of the Cook-Shelikof Basin. These anticlines bow all strata within the basin except for a thin veneer of Quaternary(?) rocks (section 2, Figure 6). These features may have formed at the same time as the coastal anticlines along the Alaska Peninsula. Beneath Shelikof Strait, low-relief anticlines trend northeast and lie mainly beneath the axis of the strait. Small normal faults occur throughout offshore areas of the basin.

Some structures cut transversely across the Cook-Shelikof Basin. The Augustine-Seldovia arch trends northwestward from Seldovia, perpendicular to the basin axis (Figures 2, 7, 8, and section 1, Figure 6). Despite the gentle dips of its flanks, this arch has exerted profound influence on the geology of the basin, mainly by controlling the thickness of Cenozoic rocks. On the north side of the arch lies the thick (7 km) accumulation of Cenozoic rocks beneath most of upper Cook Inlet; Cenozoic rocks south of the arch, however, are mainly <2 km thick (Figure 8). This arch also causes a broad warp in the Mesozoic Basin fill of the lower structural level, which it divides into northeast- and southwest-plunging synclines

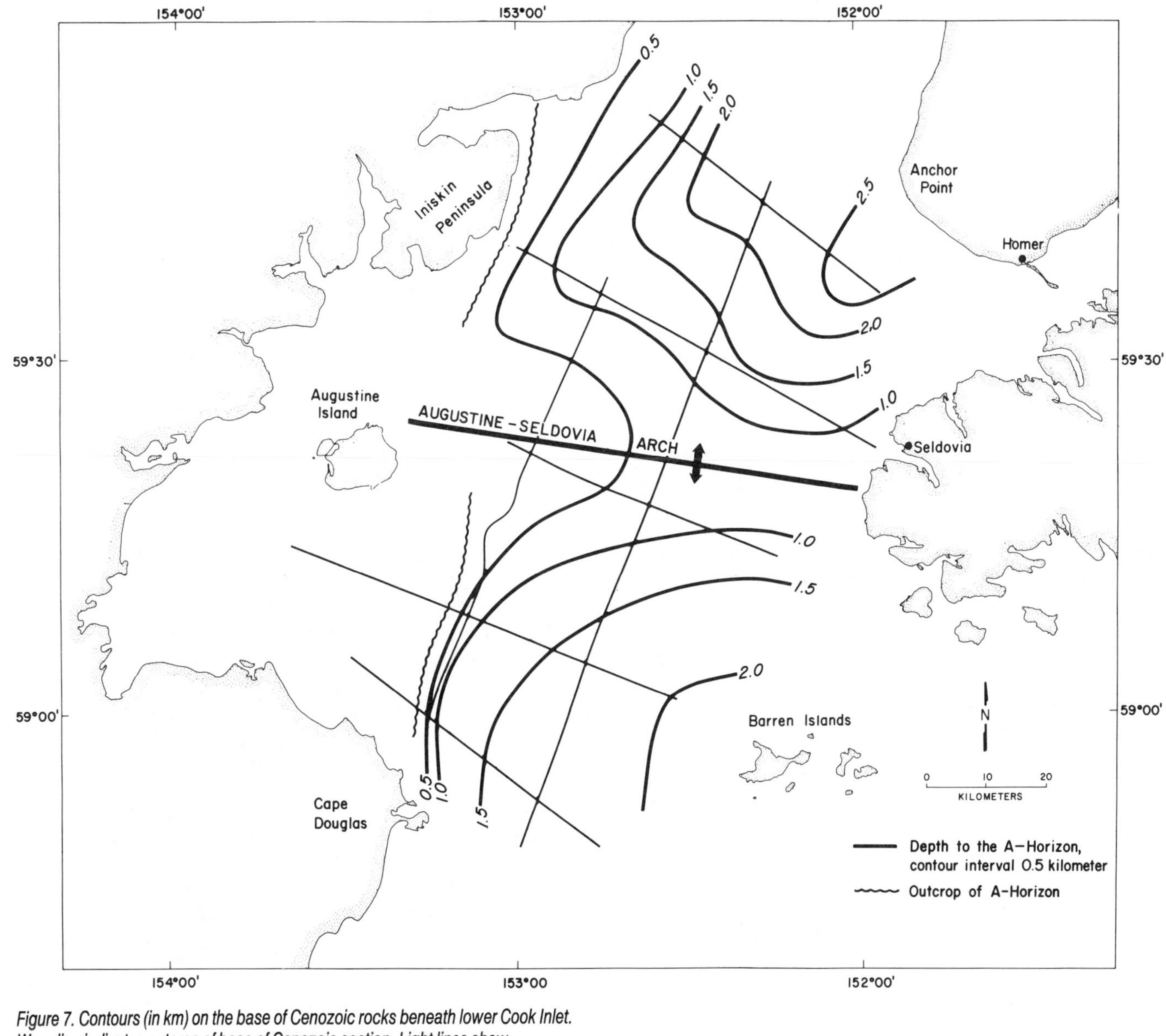

Figure 7. Contours (in km) on the base of Cenozoic rocks beneath lower Cook Inlet. Wavy line indicates outcrop of base of Cenozoic section. Light lines show tracklines of multichannel seismic data.

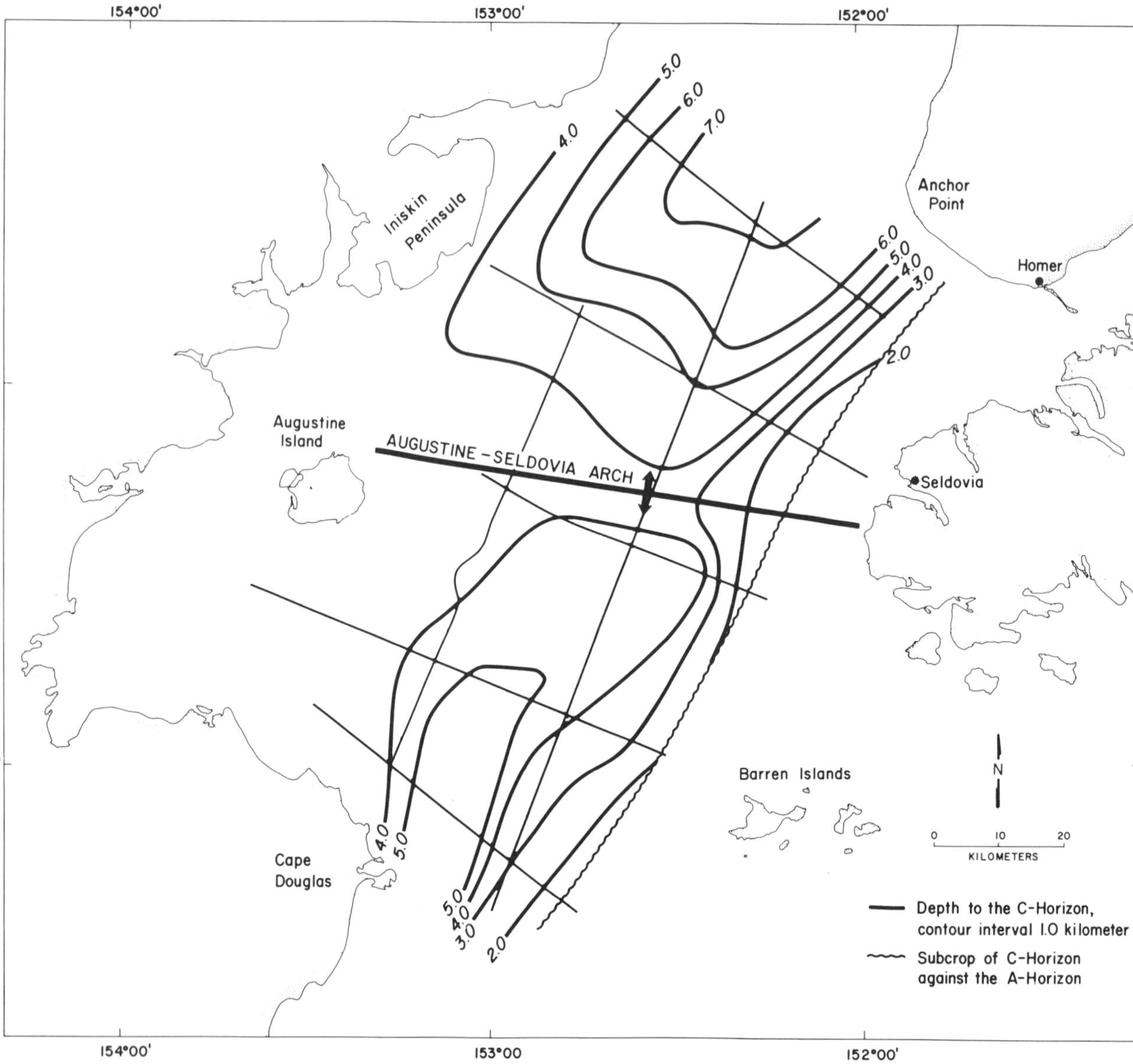

Figure 8. Contours (in km) on the base of probable Upper Jurassic rocks. Wavy line indicates truncation of this horizon by the unconformity at the base of the Cenozoic section. Light lines show tracklines of multichannel seismic data.

(Figure 9). Intersections between the northeast-trending late Cenozoic anticlines and the crossing Augustine-Seldovia arch may form domes that could be traps for hydrocarbons.

At the southwest end of Kodiak Island, another transverse structure cuts through the basin. We have insufficient seismic-reflection data to determine what, if any, deformation of basin rocks resulted from this transverse structure. However, it has left its structural imprint on the entire margin, from the lower slope adjacent to the Aleutian Trench to the magmatic arc on the Alaska Peninsula (Fisher, Bruns, and von Huene, 1981). Case et al. (1981) describe a persistent northwest-southeast grain of linear aeromagnetic anomalies spaced 5 to 25 km apart in data collected over the Alaska Peninsula southwest of the Cook-Shelikof Basin. They propose that these linear anomalies represent faults, transverse to the arc, that developed in response to subduction at the Aleutian Trench.

POTENTIAL–FIELD DATA

Rocks within the Cook-Shelikof Basin produce a large, broad-wavelength magnetic anomaly known as the Cook Inlet pattern (Grantz, Zietz, and Andreasen, 1963). On its northwest side, this anomaly is bounded by a steep gradient and by sharp, irregular anomalies caused by batholithic rocks. Another steep gradient bounds the Cook Inlet pattern on the southwest; however, a nearly featureless magnetic field was measured over the deformed Cretaceous turbidite sequences and melange southeast of this gradient. The Cook Inlet pattern does not extend southwest along the Alaska Peninsula but ends within northern Shelikof Strait, implying a considerable difference in the magnetic susceptibility of rocks beneath the inlet and peninsula (Fisher and von Huene, 1984). The source for the Cook Inlet pattern is still a matter of speculation. It may be due to either basic igneous rocks injected into lower levels of the basin during Paleocene crustal extension (Griscom and Case, 1982); or, Lower Jurassic volcanic rocks might be thickest and most strongly magnetic beneath the Cook Inlet (Grantz, Zietz, and

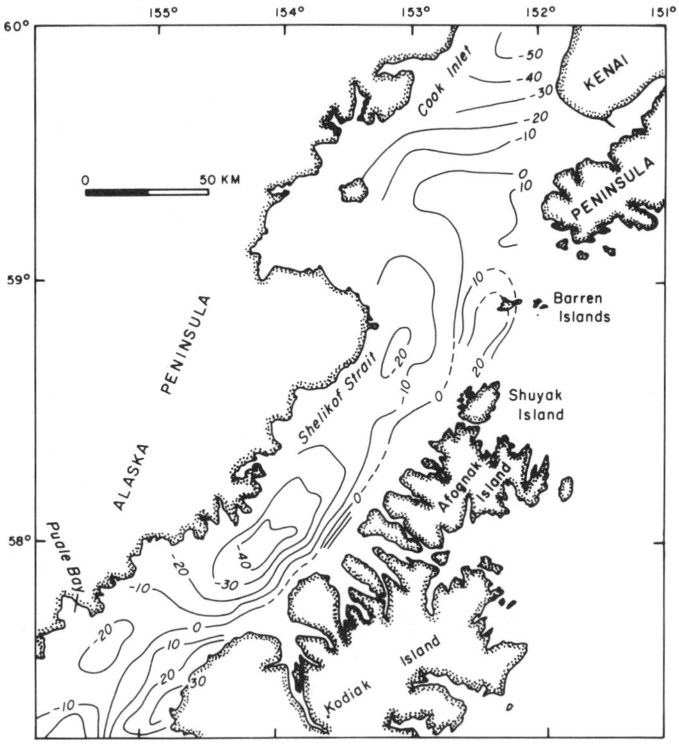

Figure 9. Marine free-air gravity map. Contours in mGals (adapted from Fisher, von Huene, and Steele, 1983).

Andreasen, 1963; Fisher and von Huene, 1984).

Free-air gravity data collected offshore within the Cook-Shelikof Basin probably respond most intensely to density changes within the Upper Cretaceous and Cenozoic section, because laumontite and other authigenic minerals cement rocks of pre-Late Cretaceous age and make these rocks nearly uniform in density. Gravity contours bulge to the west over the Augustine-Seldovia arch, and a sharply decreasing gradient into upper Cook Inlet most likely reflects the increasing thickness of Cenozoic rocks (Figure 9). Subdued gravity gradients over lower Cook Inlet and the northeastern Shelikof Strait agree qualitatively with seismic-reflection data, that Upper Cretaceous and Cenozoic rocks are uniformly thin. A −40 mGal gravity low measured over Shelikof Strait (at lat. 58°N and long. 154°W) may represent a local thickening of the Upper Cretaceous and Cenozoic rocks.

GEOLOGIC HISTORY

Paleomagnetic and stratigraphic data support the concept that many of the rocks that form Alaska south of the Denali fault were once in low magnetic latitudes and were subsequently transported northward by oceanic plates, ultimately to affix themselves to the growing mass of Alaska. In the terrane classification proposed by Jones et al. (1981), the rocks within the Cook-Shelikof Basin are part of the Peninsular terrane. This terrane has an obscure northwest boundary; in places, however, its southeastern boundary, the Border Ranges fault, is clearly exposed. Southeast of this fault lie the melange and deformed turbidites of the Chugach and Prince William terranes, which are combined into the Chugach-Prince William composite terrane in this chapter.

Early paleomagnetic data (Stone, Panuska, and Packer, 1982) suggested that the Peninsular terrane was not part of Alaska until the late Cenozoic. This date conflicted with the much older, middle to Late Cretaceous, date for the docking of this terrane proposed on the basis of the history of deformation of flysch belts and terranes of interior Alaska (Jones et al., 1982; Csejtey et al., 1982). Recent paleomagnetic data (Hillhouse and Gromme, 1982; Thrupp and Coe, 1984), however, suggest that the Peninsular terrane docked before the early Eocene. In their summary of paleomagnetic and geologic data, Hillhouse and Coe (in press) conclude that the Peninsular terrane docked against Alaska at some time during the Late Cretaceous and early Paleogene (100 to 55 Ma). According to their analysis, during the Paleocene, the Chugach-Prince William terrane was separated by 25° of latitude from the Peninsular terrane and apparently sutured against Alaska at some time after the early Eocene (55 Ma). This date is constrained mainly by paleomagnetic data from the Paleocene volcanic rocks exposed on Kodiak Island (Plumley, Coe, and Byrne, 1983; Moore et al., 1983).

Paleomagnetic data suggest that during the Paleozoic and early Mesozoic, the Peninsular terrane lay far to the south of its present latitude, where it was involved in convergent-margin tectonics, as suggested by the voluminous Early Jurassic plutonic and volcanic deposits preserved in and around the basin. One major outstanding geologic problem concerning the Cook-Shelikof Basin is the relationship between the two belts of Jurassic plutonic rocks that flank the basin on the northwest and southeast—were these rocks once continuous and then split apart as an ancestral Cook-Shelikof Basin formed, or did they arise in widely separated areas, having resulted from entirely different events, and come into proximity by dint of subsequent plate motion? Reed, Miesch, and Lanphere (1983) do not believe that batholithic rocks are continuous under lower Cook Inlet as suggested by Fisher and Magoon (1978).

Reed, Miesch, and Lanphere (1983) conclude, from alkali-oxide gradients measured in the Alaska-Aleutian Range batholith, that the oceanic plate that produced this batholith subducted toward the present south. If so, this subduction could have accommodated closure of the ocean basin that is hypothesized to have lain between Alaska and the Peninsular terrane. Pavlis (1983) suggested that the Lower Jurassic volcanic rocks are remnants of an island arc and that these rocks form the basement of the Peninsular terrane and the Cook-Shelikof Basin.

During the Middle Jurassic, convergence ceased, as suggested by the cessation of plutonism and the absence of primary volcanic debris within younger rocks (Egbert and Magoon, 1981). This magmatic quiescence lasted until the early Late Cretaceous. Structural activity near the Cook-Shelikof Basin, however, did not wane together with the magmatism; it intensified locally as uplift along the Bruin Bay fault exposed the Alaska-Aleutian Range batholith, the erosion of which provided abundant feldspar to Upper Jurassic rocks now in the Cook-Shelikof Basin. This tectonic activity may mean that plate motions continued near this basin, but relative motion may have been mostly transform. Fisher and von Huene (1984) propose that, during the Late Jurassic or Early Cretaceous, the Border Ranges fault was a strike-slip fault that truncated the southeast side of the Cook-Shelikof Basin, forming the abrupt structural contact between rocks as old as middle(?) or late Paleozoic in the basin and rocks no older than Cretaceous southeast of the fault.

During the Early Cretaceous, while the magmatic arc was still inactive, the Cook-Shelikof Basin was exposed and eroded as the hiatus between Lower and Upper Cretaceous rocks developed.

During the Late Cretaceous, plate convergence either became nearly normal to the margin or the rate of convergence increased, as suggested by the reactivated magmatic arc. Convergence during the Early Jurassic may have been toward the present south, whereas the renewed convergence had the opposite polarity, to the northwest, as inferred from the structural fabric of the rocks within the Chugach-Prince William terrane (Moore and Wheeler, 1978).

During the latest Cretaceous, rocks in the Cook-Shelikof Basin were bowed into the basin-wide syncline, and rocks forming the southeast limb of this syncline were uplifted and eroded. Uplift of areas around this basin is suggested by the coarsening of debris shed into the basin during the latest Cretaceous. Contemporaneously, Upper Cretaceous and lower Paleocene turbidite sequences in the Chugach-Prince William terrane were deformed and then intruded by magma that now forms an undeformed batholith and plutons yielding early Paleocene K-Ar dates (63–60 Ma; Kienle and Turner, 1976; Hudson, Plafker, and Peterman, 1979; J. M. Armentrout, 1979, written communication). These turbidite sequences underwent major uplift penecontemporaneously with deformation of rocks under lower Cook Inlet, as suggested by Late Cretaceous fission-track dates (68 ±8 Ma; Hawley, Bruhn, and Evans, 1984) obtained from apatite in Upper Cretaceous rocks of the Chugach-Prince William terrane. Fisher and von Huene (1984) suggested that renewed convergence during the Late Cretaceous changed the motion along the Border Ranges fault from strike slip to reverse, while the rocks of the Chugach-Prince William terrane were deformed against the truncated edge of the Cook-Shelikof Basin.

Paleomagnetic and geologic data conflict regarding the Paleocene geography of the Peninsular and Chugach-Prince William terranes. Paleomagnetic data suggest that during the Paleocene these two terranes were separated by an ocean basin about 3000 km wide. In contrast, 60-m.y-old dikes cut rocks of both the Peninsular and Chugach-Prince William terranes, suggesting that these terranes have been coupled since the middle Paleocene (Davis and Moore, 1984). Furthermore, we cannot reconcile this inferred separation with the synchronous deformation of the two terranes, which we interpret to mean that these terranes were amalgamated by the close of the Cretaceous.

Magmatic activity decreased during the late Paleocene and Eocene, when the Cook-Shelikof Basin was exposed and eroded. The magmatic arc resurged during the late Eocene and early Oligocene, leaving thick volcanic deposits on the Alaska Peninsula. Oligocene conglomerate exposed on the Alaska Peninsula and under offshore areas of Cook Inlet attests to at least local uplift around the basin; however, no structures in the Cook-Shelikof Basin are known to have formed at this time. Oligocene structures could be present but unrecognized because of later, overprinted structural fabrics.

The intensity of volcanic activity decreased during the early and middle Miocene and then increased again during the late Miocene to Holocene. During the Miocene, the basin beneath upper Cook Inlet began to subside rapidly. Contemporaneously, the Chugach Mountains and Alaska Range rose (Kirschner and Lyon, 1973), shedding coarse debris into the Cook-Shelikof Basin, mainly into that part beneath upper Cook Inlet. The Cook-Shelikof Basin southwest of the Augustine-Seldovia arch, however, subsided only slightly and was covered by thin Cenozoic deposits. At this time, the fault-fold zone along the Pacific shore of the Alaska Peninsula formed or was reactivated, and rocks beneath lower Cook Inlet were folded locally into the northeast-trending anticlines.

PETROLEUM GEOLOGY

Despite the successful exploration for hydrocarbons in the thick Cenozoic strata beneath upper Cook Inlet, drilling in Mesozoic rocks of the southwestern basin areas has yielded disappointing results. Near upper Cook Inlet, hydrocarbons are contained mainly in reservoirs of Cenozoic age, particularly in Oligocene conglomerate. Magoon and Claypool (1981) show that oils from upper Cook Inlet fields are similar chemically and isotopically. They found that the gas chromatograms of C_{15+} hydrocarbons extracted from marine Middle Jurassic rocks drilled beneath the Iniskin Peninsula (near lower Cook Inlet) are similar to chromatograms of the oils from upper Cook Inlet, and therefore they believe that these Middle Jurassic rocks are the likely source of the oil. These authors suggested that oil was not expelled from the probable Middle Jurassic source rocks until the rocks were buried by the very thick accumulation of Cenozoic strata beneath upper Cook Inlet. In some structures, potential source rocks in the Middle Jurassic section lie unconformably below Cenozoic reservoirs, enabling direct hydrocarbon migration from source to reservoir (Boss, Lennon, and Wilson, 1976; Magoon and Claypool, 1981).

Two lease sales were held covering the Lower Cook Inlet and Shelikof Strait (Figure 10). Ten wells drilled after these sales did not discover economic hydrocarbons, although oil shows were encountered in two wells. Specific data on the organic geochemistry and reservoir properties of basin rocks southwest of upper Cook Inlet can be found in McLean (1977), Lyle and Morehouse (1977), Claypool, Threlkeld, and Magoon (1980), Magoon and Claypool (1981), and Magoon (in press). Organic-geochemical data from Middle and Upper Jurassic rocks penetrated by the Koniag No. 1 well (located due north of Sutwik Island, near the shore of the Alaska Peninsula, Figure 2) show that these rocks are thermally mature to overmature for oil but generally contain <0.4 wt. % organic carbon (report on file with the State of Alaska, 1985).

Although oil seeps on the Alaska Peninsula and around lower Cook Inlet indicate that oil has formed in Mesozoic rocks of this region, geologic data suggest two possible reasons why exploration in lower Cook Inlet and on the Alaska Peninsula has been disappointing—not only are potential reservoirs in Cenozoic rocks southwest of upper Cook Inlet thin and at shallow depth or absent entirely, but the potential source rocks in the Middle Jurassic section have not been buried deeply and, therefore, may not be thermally mature. The lack of success in petroleum exploration in the thin Cenozoic section means that future exploration must focus, in part, on finding reservoirs in uncemented, pre-Cenozoic rocks.

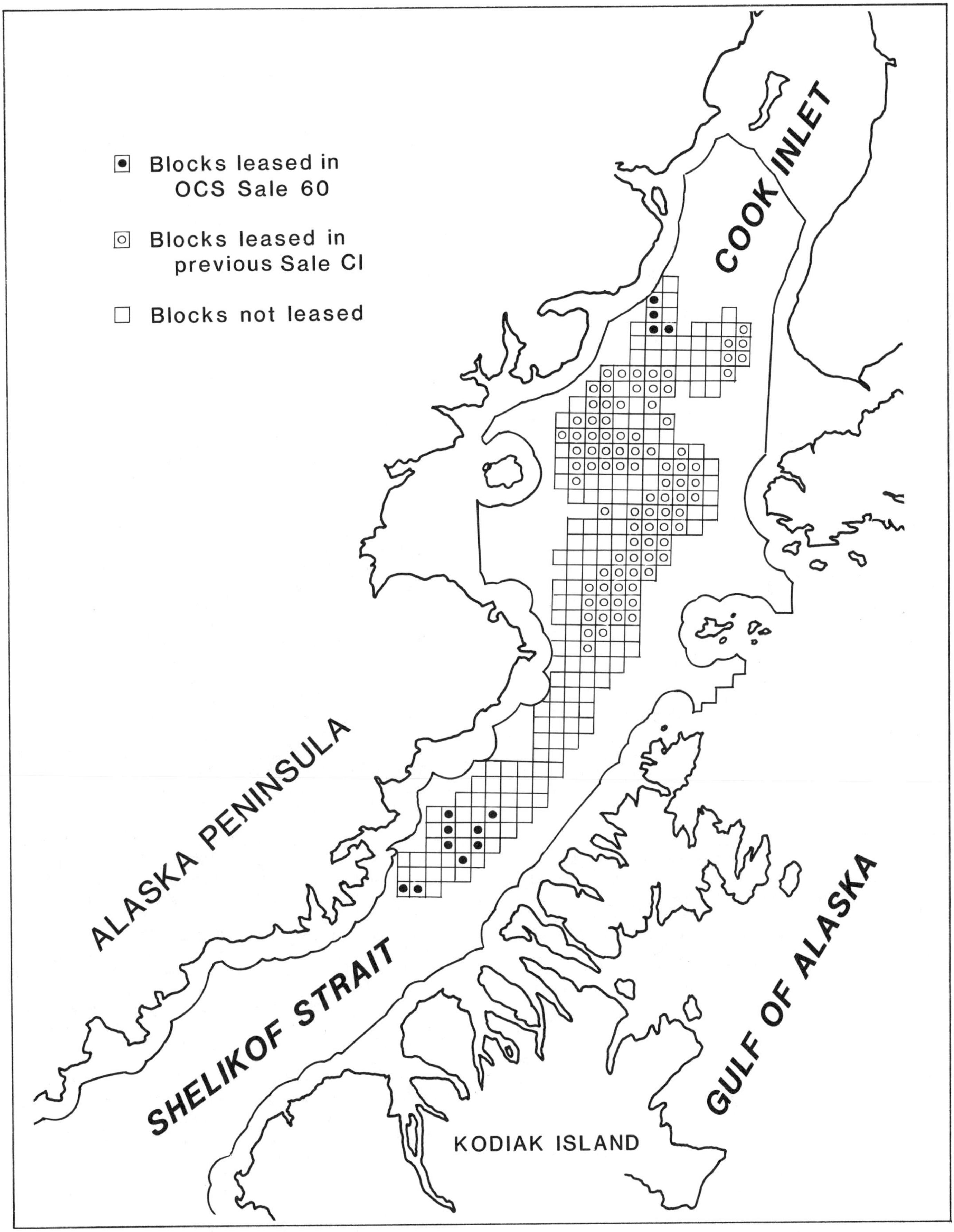

Figure 10. Offshore tracts leased during two sales covering the lower Cook Inlet and northern Shelikof Strait.

Although potential reservoirs of Late Cretaceous age were penetrated by the COST well in lower Cook Inlet, in wide areas, pre-Cretaceous rocks are cemented by laumontite and other diagenetic minerals to the extent that reservoir space in these rocks is occluded (Franks and Hite, 1980; Bolm, McCulloch, and Stewart, 1982). Bolm, McCulloch, and Stewart (1982) suggest that one strategy for exploring in pre-Cenozoic rocks is to search for reservoirs that may have accumulated hydrocarbons before laumontite formed, or for other reservoirs spared from laumontite cementation by the proximity of carbonate rocks. Environments favorable for the diagenetic formation of laumontite are characterized by high geothermal gradients, low salinity and high-pH pore water, and absence of carbon dioxide (McCulloch, Cashman, and Stewart, 1978; Bolm, McCulloch, and Stewart, 1982; Crossey, Frost, and Surdam, 1984). Future exploration strategies must include efforts to locate where environments antagonistic to laumontite cementation prevailed during the main period of hydrocarbon migration. One ray of hope is provided by Crossey, Frost, and Surdam (1984), who proposed that carbon dioxide and carboxylic acid produced by maturing kerogen would destroy existing laumontite cements; therefore, sandstone reservoirs with secondary porosity developed by leaching of laumontite could underlie cemented sandstones.

REFERENCES CITED

Bolm, J. G., T. H. McCulloch, and R. J. Stewart, 1982, Diagenesis of sandstones in lower Cook Inlet, Alaska and its implication for Mesozoic petroleum plays (abs.): Program and Abstracts, Alaska Geological Society Symposium, p. 8.

Boss, R. F., R. B. Lennon, and B. W. Wilson, 1976, Middle Ground Shoal oil field, Alaska: AAPG Memoir 24, p. 1–22.

Brockway, R., B. Alexander, P. Day, W. Lyle, R. Hiles, W. Decker, W. Polski, and B. Reed, 1975, Bristol Bay region, stratigraphic correlation section, southwest Alaska: Anchorage, Alaska Geological Society, 1 sheet.

Burk, C. A., 1965, Geology of the Alaska Peninsula—Island arc and continental margin (Parts 1, 2, and 3): GSA Memoir 99, 250 p.

Calderwood, K. W., and W. C. Fackler, 1972, Proposed stratigraphic nomenclature for Kenai Group, Cook Inlet Basin, Alaska: AAPG Bulletin, v. 56, p. 739–754.

Carden, J. R., and J. E. Decker, 1977, Significance of the Knik River schist terrane, south-central Alaska: Alaska Division of Geological and Geophysical Surveys Geologic Report 55, p. 7–9.

Carden, J. R., W. Connelly, R. B. Forbes, and D. L. Turner, 1977, Blueschists of the Kodiak Islands, Alaska: An extension of the Seldovia schist terrane: Geology, v. 5, p. 529–533.

Case, J. E., D. P. Cox, D. E. Detra, R. L. Detterman, and F. H. Wilson, 1981, Maps showing aeromagnetic survey and geologic interpretation of the Chignik and Sutwik Island quadrangles, Alaska: U.S. Geological Survey Miscellaneous Field Studies Map MF–1053B, 8 p., 2 sheets.

Claypool, G. E., C. N. Threlkeld, and L. B. Magoon, 1980, Biogenic and thermogenic origins of natural gas in Cook Inlet Basin, Alaska: AAPG Bulletin, v. 64, p. 1131–1139.

Connelly, W., 1978, Uyak Complex, Kodiak Islands, Alaska: A Cretaceous subduction complex: GSA Bulletin, v. 89, p. 755–769.

Connelly, W., and J. C. Moore, 1979, Geologic map of the northwest side of the Kodiak Islands, Alaska: U.S. Geological Survey Miscellaneous Field Studies Map MF–1057.

Cowan, D. S., and R. F. Boss, 1978, Tectonic framework of the southwestern Kenai Peninsula, Alaska: GSA Bulletin, v. 89, p. 155–158.

Crossey, L. J., B. R. Frost, and R. C. Surdam, 1984, Secondary porosity in laumontite-bearing sandstones, *in* D. A. McDonald and R. C. Surdam, eds., Clastic diagenesis: AAPG Memoir 37, p. 225–237.

Csejtey, B., Jr., D. P. Cox, R. C. Evarts, G. D. Stricker, and H. C. Foster, 1982, The Cenozoic Denali fault system and the Cretaceous accretionary development of southern Alaska: Journal of Geophysical Research, v. 87, p. 3741–3754.

Davis, D. L., and J. C. Moore, 1984, 60 m.y. intrusive rocks from the Kodiak Islands link the Peninsular, Chugach and Prince William terranes (abs.): GSA Abstracts with Programs, v. 16, p. 277.

Detterman, R. L., 1977, Interpretation of depositional environments in the Chignik Formation, Alaska Peninsula: U.S. Geological Survey Circular 772–B, p. B62–B63.

Detterman, R. L. (in press), Alaska Peninsula, *in* Continent-ocean transect A2: North Pacific Basin to Kuskokwim Mountains: GSA.

Detterman, R. L., and J. K. Hartsock, 1966, Geology of the Inisin-Tuxedni region, Alaska: U.S. Geological Survey Professional Paper 512, 78 p.

Detterman, R. L., and B. L. Reed, 1980, Stratigraphy, structure, and economic geology of the Iliamna quadrangle, Alaska: U.S. Geological Survey Bulletin 1368–B, p. B1–B86.

Detterman, R. L., J. E. Case, and F. E. Wilson, 1979, Paleozoic rocks on the Alaska Peninsula: U.S. Geological Survey Circular 804–B, p. B85–B86.

Detterman, R. L., T. P. Miller, M. E. Yount, and F. H. Wilson, 1981, Geologic map of the Chignik and Sutwik Island quadrangles, Alaska: U.S. Geological Survey Miscellaneous Investigations Map I-1250,000.

Detterman, R. L., J. E. Case, F. E. Wilson, M. E. Yount, and W. H. Allaway, Jr., 1983, Generalized geologic map of the Ugashik, Bristol Bay, and part of the Karluk quadrangles, Alaska: U.S. Geological Survey Miscellaneous Field Studies Map MF–1539–A, 1 sheet.

Detterman, R. L., T. Hudson, G. Plafker, R. G. Tysdal, and J. M. Hoare, 1976, Reconnaissance geologic map along Bruin Bay and Lake Clark faults in Kenai and Tyonek quadrangles, Alaska: U.S. Geological Survey Open-File Report 76–477, 1 sheet, 4 p.

Egbert, R. M., 1981, Petrography and provenance of Upper Cretaceous sandstone from Saddle Mountain in the Iniskin-Tuxedni area, lower Cook Inlet, Alaska: U.S. Geological Survey Circular 823–B, p. B88–B90.

Egbert, R. M., and L. B. Magoon, 1981, Petrology, provenance, and tectonic significance of Middle and Upper Jurassic sandstone from Tuxedni Bay, Cook Inlet, Alaska: U.S. Geological Survey Circular 823–B, p. B86–B88.

Fisher, M. A., 1980, Location of the Border Ranges fault southwest of Kodiak Island, Alaska: GSA Bulletin, v. 92, p. 19–30.

Fisher, M. A., and L. B. Magoon, 1978, Geologic framework of lower Cook Inlet, Alaska: AAPG Bulletin, v. 62, p. 373–402.

Fisher, M. A., and R. von Huene, 1984, Geophysical investigation of a suture zone: The Border Ranges fault of southern Alaska: Journal of Geophysical Research, v. 89, p. 11,333–11,351.

Fisher, M. A., T. R. Bruns, and R. von Huene, 1981, Transverse tectonic boundaries near Kodiak Island, Alaska: GSA Bulletin, Part 1, v. 92, p. 10–18.

Fisher, M. A., R. von Huene, and W. C. Steele, 1983, Marine free-air gravity map of the western Gulf of Alaska, lower Cook Inlet and Shelikof Strait, Alaska: U.S. Geological Survey Open-File Report 83–25, 2 sheets.

Forbes, R. B., and M. A. Lanphere, 1973, Tectonic significance of mineral ages of blueschists near Seldovia, Alaska: Journal of Geophysical Research, v. 78, p. 1383–1386.

Forbes, R. B., J. R. Carden, D. L. Turner, and W. Connelly, 1979, Regional tectonic implications of Alaskan blueschist terranes, *in* A. Sisson, ed., The relationship of plate tectonics to Alaskan geology and resources: Anchorage, Alaska Geological Society, p. L1–L28.

Franks, S. G., and D. M. Hite, 1980, Controls of zeolite cementation in Upper Jurassic sandstone, lower Cook Inlet, Alaska: AAPG Bulletin, v. 64, p. 708.

Grantz, A., I. Zietz, and G. E. Andreasen, 1963, An aeromagnetic reconnaissance of the Cook Inlet area, Alaska: U.S. Geological Survey Professional Paper 316–G, p. 117–134.

Griscom, A., and J. E. Case, 1982, Magnetic expression of geologic terranes in southern Alaska: Society of Exploration Geophysicists Technical Program, Abstracts and Biographies, 52nd Annual Meeting, p. 283–285.

Hansen, B. M., 1957, Middle Permian limestone on Pacific side of Alaska Peninsula: AAPG Bulletin, v. 41, p. 2376–2378.

Hartman, D. C., G. H. Pessel, and D. L. McGee, 1972, Preliminary report on stratigraphy of Kenai Group, upper Cook Inlet, Alaska: Alaska Department of Natural Resources Special Report 5, 4 p., 11 sheets.

Hastings, D. S., A. G. Robinson, and N. M. Robinson, 1983, Stratigraphy, depositional history, and reservoir potential of Cretaceous and early Tertiary rocks of lower Cook Inlet, Alaska: AAPG Bulletin, v. 63, p. 480.

Hawley, B. W., R. L. Bruhn, and E. H. Evans, Jr., 1984, Vertical tectonics in a fore arc region, southern Alaska, using fission track dating of apatitie grains and flexural beam modeling (abs.): GSA Abstracts with Programs, v. 16, p. 533.

Hillhouse, J. W., and C. S. Gromme, 1982, Limits to northward drift of the Paleocene Cantwell Formation, central Alaska: Geology, v. 10, p. 552–556.

Hudson, T., 1979, Mesozoic plutonic belts of southern Alaska: Geology, v. 7, p. 230–234.

Hudson, T., G. Plafker, and Z. E. Peterman, 1979, Paleogene anatexis along the Gulf of Alaska margin: Geology, v. 7, 573–577.

Imlay, R. W., and R. L. Detterman, 1977, Some Lower and Middle Jurassic beds in Puale Bay-Alinchak Bay area, Alaska Peninsula: AAPG Bulletin, v. 61, p. 607–611.

Jones, D. L., and R. L. Detterman, 1966, Cretaceous stratigraphy of the Kamishak Hills, Alaska Peninsula: U.S. Geological Survey Professional Paper 550–D, p. D53–D58.

Jones, D. L., and J. W. Miller, 1976, Preliminary geologic map of the Alaska Peninsula showing post-Callovian Mesozoic fossil localities: U.S. Geological Survey Open-File Report 76–76, 2 sheets, scale 1:500,000.

Jones, D. L., N. J. Silberling, H. C. Berg, and G. Plafker, 1981, Map showing tectonostratigraphic terranes of Alaska, columnar sections, and summary description of terranes: U.S. Geological Survey Open-File Report 81–792, 2 sheets.

Jones, D. L., N. J. Silberling, W. Gilbert, and P. Coney, 1982, Character, distribution, and tectonic significance of accretionary terranes in the central Alaska range: Journal of Geophysical Research, v. 87, p. 3705–3717.

Kienle, J., and D. L. Turner, 1976, The Shumagin-Kodiak batholith—A Paleocene magmatic arc?: Alaska Division of Geological and Geophysical Surveys, Short Notes on Alaskan Geology, Geologic Report 51, p. 9–15.

Kirschner, C. E., and C. A. Lyon, 1973, Stratigraphic and tectonic development of Cook Inlet petroleum province: AAPG Memoir 19, p. 396–407.

Lankford, S. M., and L. B. Magoon, 1977, Petrography of the Upper Jurassic through Oligocene sandstones in the Cape Douglas-Kamiskak Hills area, lower Cook Inlet: U.S. Geological Survey Circular 772–B, p. B60–B62.

Lyle, W. M., and J. A. Morehouse, 1977, Physical parameters of potential petroleum reservoir and source rocks in the Kamisha-Iniskin-Tuxedni region, lower Cook Inlet, Alaska: Alaska Division of Geological and Geophysical Surveys Open-File Report 104, 76 p.

MacKevett, E. M., Jr., and G. Plafker, 1974, The Border Ranges fault in south-central Alaska: Journal of Research of the U.S. Geological Survey, v. 2, p. 323–329.

Magoon, L. B., in press, Geological studies of the lower Cook Inlet COST no. 1 well, Alaska outer continental shelf: U.S. Geological Survey Bulletin 1596.

Magoon, L. B., and G. E. Claypool, 1981, Petroleum geology of Cook Inlet Basin—An exploration model: AAPG Bulletin, v. 65, p. 1043–1061.

Magoon, L. B., W. L. Adkison, and R. M. Egbert, 1976, Map showing geology, wildcat wells, Tertiary plant fossil localities, K-Ar age dates and petroleum operations, Cook Inlet area, Alaska: U.S. Geological Survey Miscellaneous Investigations Map I–1019, 3 sheets, scale 1:250,000.

Magoon, L. B., R. M. Egbert, and G. W. Petering, 1977, Upper Jurassic and Cretaceous rocks of the Kamishak Hills-Douglas River area, lower Cook Inlet: U.S. Geological Survey Circular 772–B, p. B57–B59.

Magoon, L. B., F. B. Griesbach, and R. M. Egbert, 1980, Nonmarine Upper Cretaceous rocks, Cook Inlet, Alaska: AAPG Bulletin, v. 64, p. 1259–1266.

Mancini, E. A., T. M. Deeter, and F. H. Wingate, 1978, Upper Cretaceous arc trench gap sedimentation on the Alaska Peninsula: Geology, v. 6, p. 437–439.

McCulloch, T. H., S. M. Cashman, and R. J. Stewart, 1978, Diagenetic baselines for interpretative reconstruction of maximum burial depths and paleotemperatures in clastic sedimentary rocks, *in* D. F. Olitz, ed., Low temperature metamorphism of kerogen and clay minerals: SEPM, Pacific Section, Los Angeles, p. 18–46.

McLean, H., 1977, Organic geochemistry, lithology, and paleontology of Tertiary and Mesozoic rocks from wells on the Alaska Peninsula: U.S. Geological Survey Open-File Report 77–813, 63 p.

McLean, H., 1979, Sandstone petrology: Upper Jurassic Naknek formation of the Alaska Peninsula and coeval rocks on the Bering Shelf: Journal of Sedimentary Petrology, v. 49, p. 1263–1268.

Miller, J. W., R. L. Detterman, and J. E. Case, 1980, Lower Cretaceous (Albian) rocks on the Alaska Peninsula: U.S. Geological Survey Circular 844, p. 83–86.

Molenaar, C. M., 1980, Cretaceous stratigraphy, Chignik area, Alaska Peninsula: AAPG Bulletin, v. 64, p. 752.

Moore, J. C., and R. L. Wheeler, 1978, Structural fabric of a melange, Kodiak Islands, Alaska: American Journal of Science, v. 278, p. 739–765.

Moore, J. C., T. Byrne, P. W. Plumley, M. Reid, H. Gibbons, and R. S. Core, 1983, Paleogene evolution of the Kodiak Islands, Alaska: Consequences of ridge trench interaction in a more southerly latitude: Tectonics, v. 2, p. 265–293.

Pavlis, T. L., 1982, Origin and age of the Border Ranges fault of southern Alaska and its bearing on the late Mesozoic tectonic evolution of Alaska: Tectonics, v. 1, p. 343–368.

Pavlis, T. L., 1983, Pre-Cretaceous crystalline rocks of the western Chugach Mountains, Alaska: Nature of the basement of the Jurassic Peninsular terrane: GSA Bulletin, v. 94, p. 1329–1344.

Petering, G. W., and T. N. Smith, 1981, Stratigraphic sections, Hallo Bay to Katmai Bay, Shelikof Strait, Alaska, 1979: U.S. Geological Survey Open-File Report 81–21, 1 sheet.

Plafker, G., and R. B. Campbell, 1979, The Border Ranges fault in the Saint Elias Mountains: U.S. Geological Survey Circular 804–B, p. B102–B104.

Plumley, P. W., R. S. Coe, and T. Byrne, 1983, Paleomagnetism of the Paleocene Ghost Rocks Formation, Prince William terrane, Alaska: Tectonics, v. 2, p. 295–314.

Reed, B. L., and M. A. Lanphere, 1972, Generalized geologic map of the Alaska-Aleutian range batholith showing potassium-argon ages of the plutonic rocks: U.S. Geological Survey Miscellaneous Field Studies Map MF–372, 2 sheets, scale 1:250,000.

Reed, B. L., and M. A. Lanphere, 1973, Alaskan-Aleutian Range batholith: Geochronology, chemistry and relation to circum-Pacific plutonism: GSA Bulletin, v. 84, p 2583–2610.

Reed, B. L., A. T. Miesch, and M. A. Lanphere, 1983, Plutonic rocks of Jurassic age in the Alaska-Aleutian Range batholith: Chemical variations and polarity: GSA Bulletin, v. 94, 1232–1240.

Stone, D. B., B. C. Panuska, and D. R. Packer, 1982, Paleolatitudes versus time for southern Alaska: Journal of Geophysical Research, v. 87, p. 3697–3707.

Thrupp, G. A., and R. S. Coe, 1984, Paleomagnetism of Tertiary volcanic rocks from the Alaska Peninsula: Is the Peninsular terrane a coherent tectonic entity? (abs.): GSA Abstracts with Programs, v. 16, p. 337.

van Eysinga, F. W. B., 1975, Geological time table: Amsterdam, Elsevier Science Publishing Company, 1 sheet.

Wilson, F. H., and N. Shew, 1982, Apparent episodicity of magmatic activity [in southern Alaska] deduced from radiometric age determinations: U.S. Geological Survey Circular 844, p. 13–15.

Wilson, F. H., J. E. Case, and R. L. Detterman, 1985, Preliminary description of a Miocene zone of structural complexity, Port Moller and Stepovak Bay quadrangles: U.S. Geological Survey Circular 945, p. 55–56.

Wilson, F. H., W. C. Gaum, and P. L. Herzon, 1981, Map and tables showing geochronology and whole-rocks geochemistry, Chignik and Sutwik Island quadrangles, Alaska: U.S. Geological Survey Miscellaneous Field Studies Map MF–103M, 3 sheets.

Wolfe, J. A., 1977, Paleogene floras from the Gulf of Alaska region: U.S. Geological Survey Professional Paper 997, 108 p.

Wolfe, J. A., and T. Tanai, 1980, The Miocene Seldovia Point flora from the Kenai Group, Alaska: U.S. Geological Survey Professional Paper 1105, 52 p.

REGIONAL GEOLOGY AND PETROLEUM POTENTIAL OF THE NORTHERN GULF OF ALASKA CONTINENTAL MARGIN

George Plafker

U.S. Geological Survey
Menlo Park, California

ABSTRACT

Cenozoic basins along the northern Gulf of Alaska continental margin occur within the tectonically complex and active transition zone from a transform boundary on the east to a convergent boundary on the west. Basinal sequences are developed within both the Prince William and Yakutat terranes, and their petroleum potential differs significantly because of marked differences in source terranes, depositional environment, underlying basement, and deformational history. The basinal strata of the Prince William terrane were deposited in their present position relative to the continental margin from the middle Eocene to the present as a slope terrace and in shelf basins overlying a highly deformed and slightly metamorphosed basement of the Orca Group. In contrast, the Paleogene sequence of the Yakutat terrane was deposited as a complex of coalescing deltas at least 600 km to the southeast along the continental margin of British Columbia and southeastern Alaska. Basement is in part Paleogene oceanic crust and in part a highly deformed and variably metamorphosed flysch and melange complex of Late Mesozoic age. The Paleogene sequence of the Yakutat terrane was transported northward to its present position by dextral strike-slip along the Queen Charlotte-Fairweather fault system during the past 25 m.y., and the thick overlying upper Cenozoic sequence was derived from the adjacent Chugach and Saint Elias Mountains. The fold-thrust belt developed in the western and northwestern part of the Yakutat terrane is a consequence of major late Cenozoic convergence with the Prince William terrane. The possibility of finding petroleum in commercial quan-

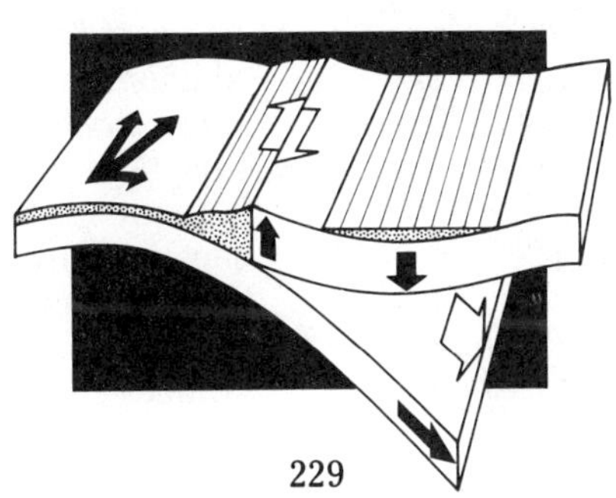

tities lies primarily within the Paleogene sequence of the western two-thirds of the Yakutat terrane, which includes suitable source rocks in close association with thick sandstone and large oil seeps onshore. The quality of the reservoir sandstone and the suitability of the structural traps at drillable depth have been, and will continue to be, the major obstacles to finding petroleum in commercial quantities. The possibility of petroleum in basinal strata in the eastern part of the Yakutat terrane, in the Prince William terrane, and on the deep-sea floor adjacent to these terranes is negligible, because the source rocks and/or maturation history are not suitable.

INTRODUCTION

This chapter outlines the geology and tectonic evolution of the northern Gulf of Alaska continental margin. Emphasis is placed on the onshore and offshore Cenozoic basinal sequences because they are of primary interest for petroleum potential. It amplifies a series of interim reports prepared before the selection of lease tracts for petroleum exploration on the Outer Continental Shelf (Plafker, Bruns, and Page, 1975; Plafker et al., 1978c, 1980; Bruns and Plafker, 1982). The present paper summarizes the earlier work and emphasizes relevant new data.

Physiography

The northern Gulf of Alaska continental margin, as used in this chapter, extends 600 km from Cross Sound on the east (long. 136°30′W) to Hinchinbrook Seavalley on the west (long. 147°00′W), and as much as 170 km from the base of the continental slope to the limit of Cenozoic outcrops along the southern flanks of the Fairweather Range and Chugach Mountains (Figure 1). The total area underlain by Cenozoic basinal strata is roughly 71,200 km^2, of which 18,750 km^2 is onshore and the remainder lies offshore to water depths of 1000 m.

The onshore part of the region consists of alluvium- and glacier-covered coastal lowlands, 0–40 km wide, backed by a belt as much as 40 km wide of rugged foothills that rise to elevations of about 2000 m. The Cenozoic strata are bordered to the north by the exceedingly steep and high Chugach and Saint Elias Mountains, which average >2000 m in elevation, and include numerous peaks >5000 m in elevation. All major drainages in the coastal mountains are occupied by glaciers except for the Alsek River, which drains across the Fairweather Range from Canada.

Details of the offshore topography are given by Atwood et al. (1981), and the generalized bathymetry is shown on Figure 1. The width of the continental shelf ranges from as little as 13 km at the eastern end of the area to 105 km at the west end. In general, the topography of the shelf and slope is gently undulated, except where it is broken by six major submarine valleys and a number of smaller valleys. East of Kayak Island, a relatively smooth and steep continental slope descends to a gentle continental rise at water depths of 2000 to 4000 m. West of Kayak Island the slope makes up the more irregular inner wall of the eastern Aleutian Trench, which has floor depths in excess of 4500 m. Yushin Ridge (part of a complex of three ridges) and Khitrov Ridge are prominent linear topographic highs along the base of the slope.

The northern gulf margin as used here corresponds closely to the Gulf of Alaska Tertiary Province, as described by Miller, Payne, and Gryc (1959), except that it extends west of Kayak Island to include Cenozoic basinal strata on the adjacent Middleton Shelf (Plafker, 1967). Basinal strata of the province are discussed under the following five major subdivisions, each of which differs significantly in structure, stratigraphy, and underlying basement, and is separated by major structural boundaries—(1) the eastern segment of the Yakutat terrane from Yakobi Valley to the Dangerous River zone, (2) the central segment of the Yakutat terrane from the Dangerous River zone to the Pamplona zone, (3) the western segment of the Yakutat terrane from the Pamplona zone to the Kayak Island zone (consisting of the Ragged Mountain-Wingham Island fault system and its inferred offshore continuation on the continental shelf and slope); (4) the eastern part of the Prince William terrane from the Kayak Island zone to Hinchinbrook Seavalley, and (5) the deep-sea floor of the Pacific plate south of the Transition fault system and Aleutian megathrust (Figure 1). The rationale for making these subdivisions is presented in the section on Tectonic Setting.

Data Sources

Sources of data for the onshore geology consist of published and unpublished maps by the U.S. Geological Survey and wells and core holes drilled for petroleum. Much of the earlier onshore data relevant to petroleum have been summarized by Miller, Payne, and Gryc (1959), Stoneley (1967), Gates, Grantz, and Patton (1968), and Plafker (1967, 1971).

As part of a U.S. Geological Survey program to evaluate the petroleum potential of the Outer Continental Shelf (OCS) before lease sales, marine geophysical and geologic studies were carried out along the northern Gulf of Alaska from 1974 to 1979 with the writer as project chief. Interpretation of the structure and seismic stratigraphy is primarily based on a regional network of multichannel data supplemented by selected single-channel and refraction seismic data. In addition, regional bathymetric, gravity, and magnetic maps of the continental margin were compiled from all available data. Age control for the structural horizons interpreted from the seismic data is from dredged rocks along the continental slope and from offshore wells. Data on source rocks, thermal maturity, and sandstone reservoir properties on the OCS are from analyses of offshore dredge and selected well samples.

TECTONIC SETTING

Mesozoic and lower Cenozoic rocks that make up the southern Alaska continental margin, upon which the basinal strata of the northern Gulf of Alaska margin were deposited, record an exceptionally complex history of crustal plate mobility and interactions that continues to the present. The entire continental margin south of the Denali fault is a collage of structurally bound tectonostratigraphic terranes that assembled into their present positions in the late Mesozoic and Cenozoic Eras (Figure 2; Jones et al., 1986). From south to north these terranes are (1) the Yakutat terrane; (2) the Prince William terrane; (3) the Ghost Rocks terrane; (4) the Chugach terrane; and (5) a collage of far-travelled microplates between the Border Ranges and Denali fault systems, the largest of which are the Wrangellia, Peninsular, and Alexander terranes. Structural relationships of these terranes to each other and to the subducting Pacific plate are depicted on Figure 2 and by schematic regional cross sections in Figure 3.

Yakutat Terrane

The Yakutat is a composite terrane 600 km long and 200 km wide along the northern Gulf of Alaska margin. It is bounded on the east and north by the Fairweather and Chugach-Saint Elias faults, on the west by the Kayak Island zone, and on the south by the Transition fault system (Plafker, 1983; Jones et al., 1986). Its basement east of the Dangerous River zone consists of upper Mesozoic flysch and melange of the Yakutat Group intruded by Eocene granitoid plutons. The melange contains a suite of exotic olistostromal blocks characteristic of the Wrangellia terrane in the Queen Charlotte Islands and Vancouver Island. The Yakutat Group is characteristically highly deformed and is locally metamorphosed to low greenschist facies. Northwest structural trends are obliquely truncated by the Dangerous River zone.

Lower Eocene and possible Paleocene oceanic crust underlies the western two-thirds of the block, which is subdivided into a relatively undeformed segment between the Dangerous River and Pamplona zones and a highly deformed segment from the Pamplona zone to the Kayak Island zone. Overlying the basement is a dominantly clastic, lower Eocene through Quaternary, sequence that shows marked differences in thickness, lithology, and age east and west of the Dangerous River zone. In general, sandstones of the Paleogene sequence reflect a predominantly igneous and high-grade metamorphic provenance not found in the adjacent mountains, whereas the upper Cenozoic sequence was derived locally.

Geologic and paleontologic studies suggest that the Yakutat terrane was transported to its present position during the late Cenozoic by dextral strike-slip along the Fairweather-Queen Charlotte transform fault system from a site along the continental margin at least 600 km to the southeast (see the section on Tectonic Evolution). Reliable paleomagnetic data on the displacement history of the terrane have not yet been obtained onshore (Hillhouse and Grommé, 1984). Preliminary data from the offshore Yakutat well suggest $13 \pm 3°$ post-early Eocene displacement relative to cratonal North America, with no more than 20° rotation (Van Alstine, Bazard, and Whitney, 1985). Paleontologic data on the terrane displacement relative to the continental margin are equivocal (compare Keller et al., 1984, with Rau, Plafker, and Winkler, 1983; Marincovich and McCoy, 1984; Wolfe and McCoy, 1984).

Prince William Terrane

The Prince William terrane forms an outcrop belt >100 km wide that extends from the Copper River and Prince William Sound areas south-westward beneath the continental shelf and slope off the Kodiak Islands (Jones et al., 1986). The eastern part of this terrane, of interest here, extends from the Kayak Island zone to Hinchinbrook Seavalley. The terrane basement consists of the Orca Group, an accreted late Paleocene and early through middle Eocene deep-sea-fan complex interbedded with oceanic volcanic rocks and minor pelagic sediment (Winkler, 1976; Winkler and Plafker, 1981; Plafker et al., 1985). Sandstone compositions reflect a dominantly plutonic and high-grade metamorphic provenance similar to that of the Paleogene section of the Yakutat terrane. The presence of conglomerate bodies that locally contain abundant siliceous-volcanic and orthoquartzite clasts suggests proximity to a cratonal sediment source for part of the unit.

The Orca Group is complexly folded and faulted; the entire sequence was "continentalized" during a major thermal event at about 50 Ma, which was accompanied by widespread plutonism and zeolite- to middle greenschist facies metamorphism. The overlying sequence, which occurs entirely offshore, consists of as much as 4 km of relatively undeformed siliciclastic sedimentary rocks, ranging in age from late Eocene or older to Quaternary, and was deposited in shelf and slope basins on a gradually subsiding continental margin. Paleontologic data support the conclusion that these strata were deposited at about their present paleolatitude relative to the North America continental margin (Rau, Plafker, and Winkler, 1983; Keller et al., 1984).

Paleomagnetic studies of pillow basalt in the Orca Group yield low-quality and contradictory results. Data from the Knight Island area in western Prince William Sound suggest little or no northward displacement (Hillhouse and Grommé, 1977), whereas preliminary analysis of data from the northeastern part of Prince William Sound suggest an origin possibly 40° to the south (Plumley and Plafker, 1985). Geologic data indicate that the Orca Group was accreted to the Chugach terrane by 51 ± 3 Ma and that there was no significant post-accretion horizontal displacement along this suture zone (Winkler and Plafker, 1981).

Ghost Rocks Terrane

The Ghost Rocks terrane as defined in this chapter includes an outcrop belt about 10 km wide of mafic volcanic rocks and melange along the southern margin of Kodiak Island and mafic pillow basalt and intrusive rocks on the Resurrection Peninsula on the mainland near Seward (Figure 2).

The Ghost Rocks Formation is a Cretaceous to Paleocene subduction complex (Byrne, 1982) that was previously correlated with the Orca Group (Plafker, 1971). Paleontologic and isotopic studies now show that the Orca Group is significantly younger than the Ghost Rocks Formation (Plafker et al., 1985) and was accreted to the continental margin some 11 m.y. after the Ghost Rocks terrane. Paleomagnetic data indicate northward translation of the Ghost Rocks Formation as much as 25° (Moore et al., 1983; Plumley, Coe, and Byrne, 1983).

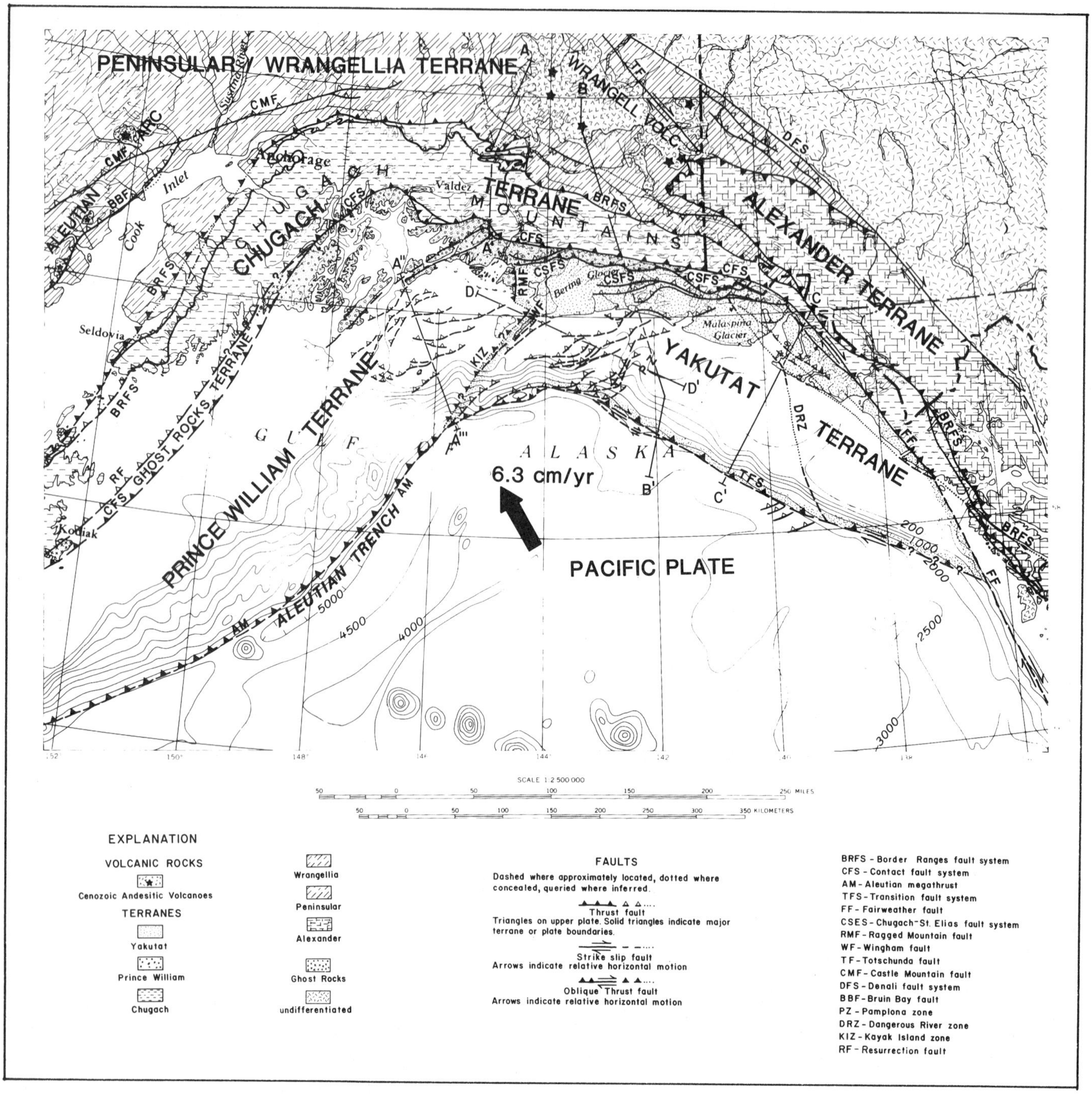

Figure 2. Tectonostratigraphic terranes of the northern Gulf of Alaska continental margin and surrounding areas. Modified from Plafker, Nokleberg, and Lull (1985), Jones et al. (1986), and Figure 1. Pacific plate relative motion from Engebretsen (1982). See Figure 3 for structure sections.

As shown on Figure 2, the Ghost Rocks terrane is tentatively inferred to include the undated pillow lava and related igneous rocks on the mainland that yield paleomagnetic poles closely comparable with those from the Ghost Rocks Formation (Grommé and Hillhouse, 1981; Plumley, Coe, and Byrne, 1983). If correct, this interpretation requires that the structural contact mapped by Nelson, Dumoulin, and Miller (1984) between the igneous complex and the adjacent flysch of the Valdez Group is the Ghost Rocks/Chugach terrane boundary.

Chugach Terrane

The Chugach terrane forms the "backstop" against which the Prince William and Yakutat terranes were accreted. The terrane extends for approximately 2000 km through the coastal mountains of southern Alaska, where it is from 60 to 100 km wide (Plafker, Jones, and Pessagno, 1977). In southcentral Alaska it is composed of at least three subterranes including (1) Late Cretaceous volcaniclastic flysch and oceanic basaltic rocks which comprise the great bulk of the terrane, (2) subordinate Late Jurassic to Early Cretaceous melange, and (3) minor Early(?) Jurassic or older blueschist-greenschist. The melange sequence in southeastern Alaska contains exotic blocks similar to those in the Yakutat Group melange that are suggestive of a Wrangellian provenance.

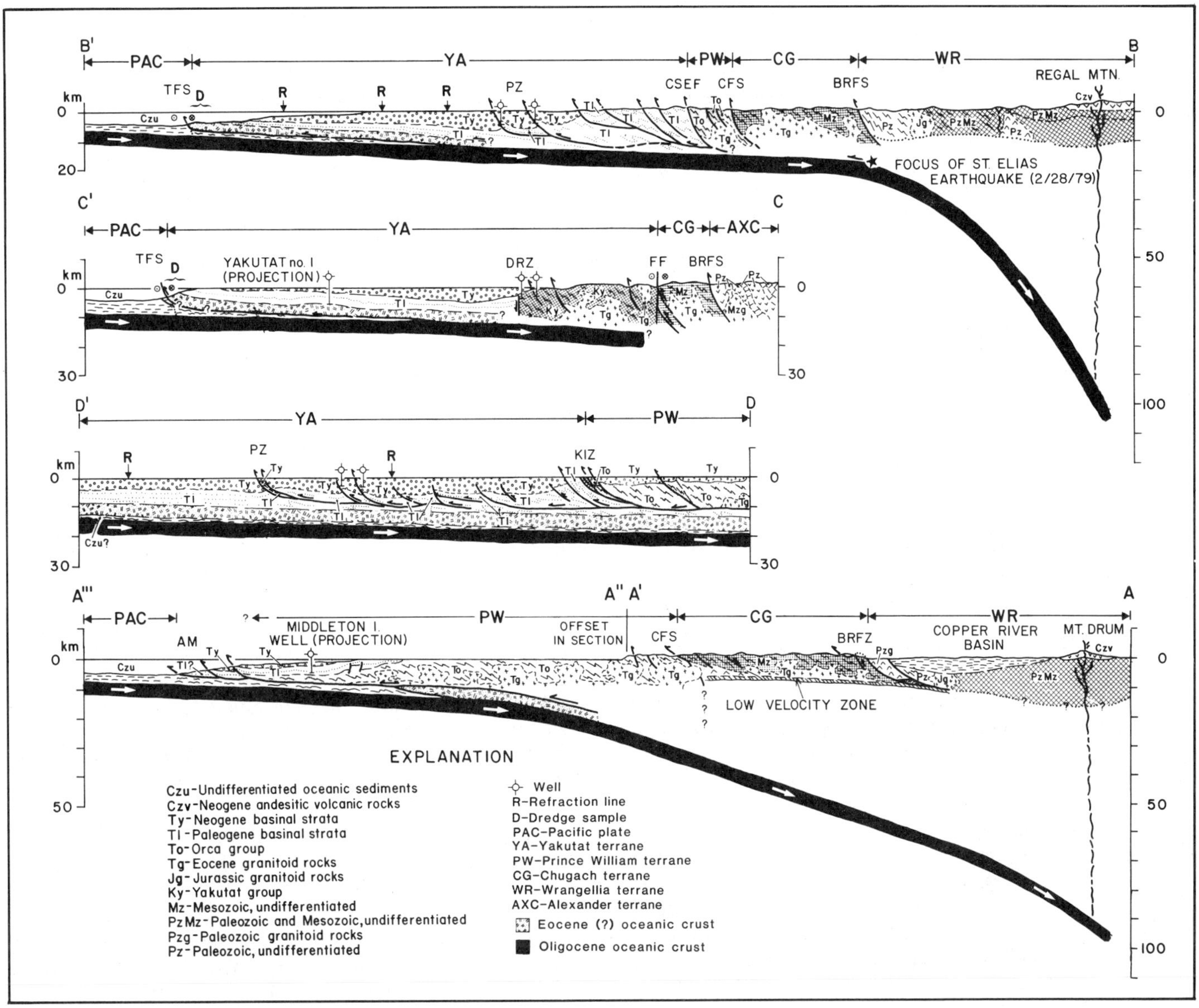

Figure 3. Diagrammatic regional structure sections across the northern Gulf of Alaska margin. Locations shown on Figure 2. Abbreviated fault names as in Figure 2. AXC, Alexander terrane; WR, Wrangellia; CG, Chugach terrane; PW, Prince William terrane; YA, Yakutat terrane; PAC, Pacific plate; and R, location of refraction line. No vertical exaggeration.

In the Chugach Mountains, the Chugach terrane was accreted and initially deformed in the latest Cretaceous and/or earliest Tertiary with formation of south-verging structures; these structures were subsequently overprinted by predominantly north-verging structures. The rocks are variably metamorphosed from zeolite to amphibolite facies and are intruded by pervasive 51 ± 3-m.y.-old plutons and dikes (Plafker, Nokleberg, and Lull, 1985).

If the Resurrection Peninsula is not part of the Chugach terrane (as suggested above), the only remaining paleomagnetic data for igneous rocks of the terrane are from the layered gabbro of the Crillon-La Perouse area (Grommé and Hillhouse, 1981). However, this body may have been affected by postemplacement northeastward tilting that could equal the apparent inclination anomaly of $19 \pm 8°$, as is suggested by the prominent decrease in metamorphic grade of country rocks from southwest to northeast across the gabbro body.

Terranes North of the Chugach Terrane

The Wrangellia and Peninsular terranes (Jones et al., 1986) consist primarily of Early Mississippian to Permian intraoceanic andesitic arc rocks, carbonate rocks, and minor amounts of siliceous metasedimentary rocks with associated mafic and ultramafic plutonic rocks. Paleozoic rocks are overlain by distinctive Triassic sedimentary and volcanic sequences, and superposed intrusive and extrusive magmatic rocks, of mainly Jurassic age. There are notable differences in the Mesozoic sequences that characterize the two terranes (Jones, Silberling, and Hillhouse, 1977). Some of the Paleozoic rocks are locally metamorphosed to amphibolite grade and are strongly mylonitized, particularly along the southern terrane boundaries. However, most rocks of these terranes are relatively unmetamorphosed and undeformed compared with the accreted sequences to the south. The continental margin of British Columbia is underlain mainly by rocks of the Wrangellia terrane (Sutherland Brown, 1968; Muller, 1977).

The Alexander terrane is characterized by variably metamorphosed Late Proterozoic to Triassic bedded strata and intrusive rocks (Jones et al., 1986). In the area shown on Figure 2, the rocks are predominantly Paleozoic metabasalt and marble intruded by a variety of plutonic rocks of late Paleozic and Mesozoic age. The Alexander terrane probably underlies the segment of the southeastern Alaska continental margin between Chatham Strait and Dixon Entrance, although offshore geologic data are unavailable in this region.

These older terranes are microplates made up mainly of oceanic plateaus and arcs. The Wrangellia and Peninsular terranes were about 30° south of their present latitude relative to cratonic North America during the Late Triassic (Hillhouse and Grommé, 1984). If an origin in the northern hemisphere is assumed, the Alexander terrane may have undergone about 15° of northward movement relative to North America since the Pennsylvanian (Van der Voo et al., 1980) but no apparent change in latitude since the Late Triassic (Hillhouse and Grommé, 1984).

Structural Relationships of Terranes

Inferred structural relationships between the various terranes are shown on the regional map and cross sections in Figures 2 and 3.

Combined geological, seismological, and geophysical studies indicate that the Yakutat terrane is moving northwestward relative to the Chugach terrane along the Fairweather fault. It is thrusting beneath the Chugach and Prince William terranes along the Chugach-Saint Elias fault system on the north and between the Pamplona zone and Kayak Island zone on the west.

The eastern, relatively undeformed part of the Yakutat terrane east of the Pamplona zone was originally referred to informally as the "Yakutat block" (Rogers, 1977; Plafker et al., 1978c). Subsequently, recognition that the Paleogene stratigraphy of the Yakutat block tracks continuously westward to the Kayak Island zone led to revision of the terrane boundaries in Jones et al. (1986). The term "Yakutat block" is retained in the original sense of Rogers (1977) and Plafker et al. (1978c). In this sense it refers only to the part of the Yakutat terrane east of the Pamplona zone that is now moving as a tectonic block with almost Pacific plate velocity relative to adjacent structural blocks (Lahr and Plafker, 1980).

As indicated in Figure 2, relative motion of the Pacific plate to the North America plate in the northern Gulf of Alaska is deduced as N16°W at about 6.3 cm/yr for roughly the past 5 m.y. and was about N35°W at 4.5 cm/yr in the previous 23 m.y. (Engebretsen, 1982). Coupling between the Pacific plate and Yakutat terrane has undoubtedly varied through time, depending upon how displacement was shared between the Fairweather and Transition fault systems. At present, the Yakutat block is moving with the Pacific plate, but with a slightly lower relative velocity (estimated at 5.9 cm/yr relative to the North America plate and 0.4 cm/yr relative to the Pacific plate). As a consequence, a large strike-slip relative motion occurs along the dextral Fairweather transform at the eastern boundary of the Yakutat block, and a large compressional component is taken up by folding and thrusting between the Pamplona zone and the Kayak Island zone along the western margin. Both boundaries are areas of high seismicity and active late Holocene deformation (Plafker, Bruns, and Page, 1975; Plafker et al., 1978b; Lahr and Plafker, 1980; Perez and Jacob, 1980). The Transition fault system between the Yakutat terrane and Pacific plate is relatively aseismic. However, young folds and faults reflecting oblique compression along the interface (Figure 1), and the 1973 Cross Sound earthquake (6.7 Ms)—a thrust event that appears to have ruptured about 40 km of the fault near its eastern end (Page, 1975)—attests to some ongoing deformation. Oblique relative motion along the Transition fault system implies underthrusting of the Pacific plate along a subhorizontal plane as shown diagrammatically in Figure 3 (B–B′, C–C′, and D–D′). If so, this low-angle detachment is relatively aseismic.

The Prince William terrane was thrust relatively northward beneath the Chugach terrane along the Contact fault system in the Cenozoic prior to about 50 Ma, and the Chugach terrane was thrust at least 40 km northward beneath the Wrangellia-Peninsular composite terrane along the Border Ranges fault system between the late Cretaceous and early Eocene (Fuis et al., 1985; Plafker, Nokleberg, and Lull, 1985; Page et al., 1986). The Contact and Border Ranges fault systems appear to merge into a subhorizontal, low-velocity zone, which may represent a decollement that underlies the Chugach and southern Wrangellia-Peninsular terranes at a depth of 5–9 km (Figure 3, A–A′). A few kilometers beneath the shallow low-velocity zone are eight northward-dipping layers about 30 km thick with alternating high and low velocity. These layers are tentatively interpreted as an imbricated stack of subducted and underplated oceanic crust and mantle rocks. Distribution of earthquake hypocenters suggests that active subduction involves at least the lowest two and possibly the lowest four layers.

REGIONAL STRATIGRAPHY OF CENOZOIC BASINAL SEQUENCES

Cenozoic bedded rocks are siliciclastic sediments, with minor amounts of volcanic rocks and coal. Each epoch from latest Paleocene through the Pleistocene is represented. The bedded sequence, which has a composite maximum thickness of 9500 m, includes both marine and nonmarine units. The following four major subdivisions of Cenozoic rocks are recognized on the basis of fossils and gross lithologic characteristics—(1) early to middle Eocene basalt and marine sedimentary section of the Samovar Hills and their offshore correlatives; (2) upper Paleocene and Eocene continental to marine clastic coal-bearing strata, (3) Oligocene and lower Miocene marine clastic, glauconitic, and basaltic strata, and (4) middle Miocene through Quaternary predominantly marine and glaciomarine clastic strata. Changes in depositional environment are characteristically gradational and appear to transgress time in different parts of the basin. Important regional unconformities occur at the base of (2) and (4).

Figures 4 through 6 show the generalized stratigraphy and correlations of stratigraphic units. Table 1 summarizes data on dredged outcrop samples from the continental slope and outer part of the continental shelf. Details of much of the onshore stratigraphy and biostratigraphy are presented elsewhere (Miller, Payne, and Gryc, 1959; Miller, 1971, 1975; Stoneley, 1967; Plafker, 1971, 1974; Tysdal, Hudson, and Plafker, 1976; Addicott, Winkler, and Plafker, 1978; Plafker, 1981; Winkler and Plafker, 1981; Rau, Plafker, and Winkler,

Table 1. Data for dredged outcrop samples. Locations shown on Figures 1 and 11. Fossil identifications and age assignments by D. Bukry (nannofossils), W. W. Rau and R. Z. Poore (foraminifers), Anderson-Warren Associates (palynomorphs), J. A. Barron and D. Bukry (diatoms and silicoflagellates).

Map No.	Dredge No.	Location (Depth and coordinates are given for start and finish of dredge casts): Water Depth (m)	Latitude (N.)	Longitude (W.)	Estimated Recovery (kg)	Sample Designation (Italic indicates thin section available)	Lithology (listed in order of relative abundance in dredge sample)	Fossils	Comments
1	S10-79-EG-16	3105- 2650	59°04.58′ 59°05.78′	142°33.37′ 142°34.22′	125	79-*16A, 16B, 16C, 16D, 16E*	*Sandstone*, lithofeldspathic, highly micaceous, hard, medium-light-gray to medium-dark-gray, fine- to medium-grained, moderately well sorted, in part calcareous and cut by calcite veinlets, local laminae of carbonized plant material, in part sheared and brecciated.	None	Slickensides in sandstone
						79-*16F, 16G, 16H, 16I, 16J*	*Claystone*, argillaceous, hard, brownish-black, with interbedded sandstone and partings of wispy black plant? material. Sandstone laminae 2 mm to 35 cm thick, medium-gray to yellow-gray, very-fine- to medium-grained, in part graded and cross-laminated.		
2	S5-78-EG-41	3110- 2700	59°02.94′ 59°05.02′	142°26.51′ 142°24.17′	20	78-41A, 41B, 41C	*Pebbly and sandy mudstone*, hard, light-gray, rusty-brown-weathering, attached to cobble-size gabbro clast.	Pliocene-Quaternary foraminifers Tertiary palynomorphs Late Miocene to Holocene diatoms	Mudstone-matrix conglomerate or diamictite in Yakatag (?) Formation
						78-*41D* through 41Z	*Cobbles and boulders*, loose, polymictic, subangular to rounded, largest is 18 × 15 × 12 cm. Includes graywacke, gabbro, gneiss, schist, altered basalt, quartz diorite, granodiorite, aplite, etc.		Diorite clast minimum age is 33.3 ±2.5 m.y. based on whole-rock K/Ar date
3	S5-78-EG-23	3010- 2890	58°55.96′ 58°54.66′	142°10.78′ 142°14.41′	2	78-*23A*	*Sandstone*, lithofeldspathic, hard, dense, light-gray, medium-grained, subangular to subrounded, poorly sorted, tightly compacted, slightly calcareous.		May be erratic
4	S5-78-EG-22	2470- 2430	59°00.44′ 58°57.63′	142°08.55′ 142°09.39′	300	78-*22C1, 22C2, 22C3*	*Sandstone*, lithofeldspathic, hard, light-gray to gray and greenish-gray, very-fine- to coarse-grained, in part laminated and graded, highly micaceous, veins of prehnite and carbonate.		Paleogene and (?) Mesozoic
						78-*22B2, 22B3*	*Siltstone*, hard, black, carbonaceous, in part with laminae of greenish-gray, graded and channeled, very-fine-grained sandstone.		
						78-22D1, 22D2, 22D3, 22D4	*Claystone*, medium-gray, brownish-gray and dark-brown, waxy, organic-rich, massive.	Pre-Pliocene foraminifers Small pelecypods	
						78-*22A1, 22A2, 22A3, 22A4, 22A5, 22A6, 22A7, 22A8*	*Basalt and diabase*, angular to subangular blocks (possibly talus), very hard, green, gray-green, and dark-green, in part veined, in part agglomeratic, slightly to highly altered.		Basalt minimum age 55.2 ±1 Ma based on whole-rock K/Ar data. Plafker et al., 1980
5	S10-79-EG-28	2760- 2225	58°56.16′ 58°55.20′	142°05.94′ 142°06.78′	25	79-*28A*	*Sandstone*, moderately indurated, friable, massive, pale-yellow-brown-weathering, medium- to coarse-grained with scattered very coarse grains, micaceous.		
6	S2-77-EG-20	3180- 3080	58°53.17′ 58°54.28′	142°03.21′ 142°02.90′	75	77-*20A*	*Sandstone-matrix conglomerate*, hard, fresh, polymictic, clast size from pebbles to blocks (35 × 25 × 20 cm), pebbles well-rounded and smooth, coarser clasts rounded to subrounded, matrix dirty, calcareous, medium-grained, lithofeldspathic.		Paleogene or Mesozoic
						77-20B	*Claystone*, hard, dark-gray, slightly calcareous, abundant finely divided pyrite.		
7	S2-77-EG-21	2610- 2600	58°55.83′ 58°52.31′	142°06.57′ 142°03.39′	140	78-*21D4, 21D5*	*Sandstone*, lithofeldspathic, moderately hard to hard, sheared, massive, light-gray to gray, medium- to coarse-grained, very poorly sorted, angular to subangular, in part pebbly, with abundant shale chips.		
						78-21A, 21B, 21C	*Claystone*, moderately hard, black, carbonaceous, in small fragments within sandstone.		

(continued).

8	S2-77-EG-22	2575- 2300	58°58.60′ 58°59.77′	141°59.68′ 141°58.46′	50	*77-22A*	*Sandstone*, moderately hard, friable, massive, slightly weathered, pale-yellowish-gray, fine-grained, micaceous, sparse glauconite, very dirty, with moderate secondary porosity.	Tertiary arenaceous foraminifers	
						77-22B, 22C, 22D, 22E	*Claystone*, soft to moderately hard, medium-light-gray, micaceous and carbonaceous.		
9	S10-79-EG-42	2600- 2160	58°47.10′ 58°48.44′	141°12.25′ 141°10.71′	25	79-*42A, 42B*	*Sandstone*, lithofeldspathic, slightly indurated and friable, in part calcareous and hard, medium-gray, medium-grained, moderately well sorted, angular to subrounded, micaceous, commonly massive but in part with dark-gray laminae of carbonaceous and micaceous siltstone.	Middle Eocene to early Oligocene nannofossils Eocene(?) foraminifers Eocene palynomorphs	
						79-42C, 42D, 42E	*Claystone*, moderately indurated to hard, brownish-gray to dusky-brown, massive, abundant kerogen, microfossils and carbonized plant remains.		
10	S10-79-EG-43	1980- 1518	58°48.55′ 58°50.05′	141°10.51′ 141°08.85′	75	79-43A, 43B *49D*, 43E	*Claystone*, soft, locally calcareous and moderately indurated, massive, grayish-brown to brownish-gray, abundant kerogen and microfossils, highly micaceous, local dusky-yellow-brown, very hard, calcareous "cannonball" concretions.	Early Eocene nannofossils Late early to early middle Eocene (Penutian to Ulatisian) foraminifers Late Eocene palynomorphs	Middle bathyal
						79-*43C*	*Sandstone*, lithofeldspathic, hard, massive, medium-gray, moderately well sorted, angular to subangular, abundant carbonate and minor clay cement.		
11	S10-79-EG-44	1450- 1312	58°51.11′ 58°52.48′	141°07.81′ 141°06.23′	10	79-44A, 44B, 44C	*Claystone*, slightly indurated, massive, dusky-brown, abundant kerogen, minor microfossils.	Probable late Eocene palynomorphs	
12	S10-79-EG-37	2870- 2560	58°44.91′ 58°46.80′	141°12.33′ 141°10.24′	375	79-37A, 37B, 37C	60% *Claystone*, slightly to moderately indurated, massive, grayish-brown to dusky-brown, in part carbonaceous, common microfossils, large calcareous concretions.	Middle Eocene to early Oligocene (possibly early or middle Eocene) nannofossils Middle to late Eocene (Narizian) foraminifers Eocene palynomorphs	Lower bathyal
						79-*37F*	35% *Sandstone*, lithofeldspathic, hard, massive, medium-light-gray, medium-grained, angular to subrounded, well-sorted, 30%–35% calcite cement, micaceous.		
						79-*37D*	*Palagonitic vitric crystal tuff*, massive, hard, grayish-blue, very-fine- grained, with angular to rounded green, gray, and violet clasts to 1 cm across of palagonite containing plagioclase, opaque crystals, and minor clear glass fragments. Vesicles and veinlets infilled with zeolites and minor hematite.		
						79-*37E*	*Siltstone*, slightly indurated, massive, medium-dark-gray, highly micaceous, abundant coarse carbonized plant material.		
13	S10-79-EG-38	2510- 1700	58°46.69′ 58°47.90′	141°09.99′ 141°08.44′	5	79-38A, 38B, 38C	*Siltstone*, argillaceous, slightly indurated to moderately indurated, massive, grayish-brown to dusky-brown.	Middle Eocene to early Oligocene nannofossils, Late Eocene foraminifers Eocene palynomorphs Fish scales, one small gastropod	Middle bathyal with transported upper shelf and slope material
14	S10-79-EG-39	2125- 1467	58°47.26′ 58°48.79′	141°09.70′ 141°07.62′	500	79-39D, 39E, 39F, 39G	*Claystone*, slightly indurated to hard, massive to laminated, medium-gray to medium-dark-gray and dusky-brown, in part calcareous, locally abundant carbonized plant material.	Middle Eocene to early Oligocene (possibly middle Eocene?) nannofossils Late Eocene (Narizian) foraminifers Eocene palynomorphs One small clam in claystone	
						79-*39A*	*Sandstone*, bioclastic and volcaniclastic, hard, bedded, medium-gray, coarse-grained. Clasts predominantly a calcareous reef hash with subordinate rounded basaltic rock fragments in a calcite matrix.	Warm-water reef assemblage in sandstone including *Amphistogina*, pelagic foraminifers, coral, coralline algae, bryozoans, echinoids, etc.	
						79-*39B*	*Interbedded claystone and sandstone*. Moderately indurated, laminated, brownish-gray, organic-rich claystone and medium-light-gray, medium-grained, feldspatholithic sandstone.		
						79-39C	*Siltstone*, moderately indurated, laminated, medium-dark-gray, highly micaceous and carbonaceous.		

(continued).

15	S5-78-EG-43	2640- 1915	58°45.60′ 58°47.50′	141°12.64′ 141°10.48′	300	78-43E, 43F	*Siltstone*, moderately hard, massive, light-greenish-gray, slightly calcareous.	Early Eocene (*D. lodoensis* zone) nannofossils Late early Eocene to early Eocene (Penutian-Ulatisian) foraminifers Late Eocene palyno-morphs	Some slickensides in siltstone Lower bathyal or deeper
						78-*43A*	*Palagonitic tuff*, hard, dark-green with white analcite amygdules and pore fillings.		
						78-43B	*Limestone*, very hard, massive, light-gray concretions or beds.		
16	S5-78-EG-44	1415- 1270	58°45.90′ 58°46.84′	141°08.49′ 141°07.45′	450	78-44D, 44E, 44F	*Claystone*, soft to moderately hard, finely laminated, dark-olive-gray to dusky-yellowish-brown, abundant microfossils and fish scales, minor glauconite?	Middle or late Eocene nannofossils Middle to late Eocene (Narizian) foraminifers Eocene palynomorphs Rare thin-shelled small pelecypods and abundant fish scales in claystone	Abundant slicken-sides Middle bathyal, low oxygen conditions
						78-*44A*	*Sandstone*, lithofeldspathic, moderately hard, friable, gray, fine- to medium-grained, angular to subangular, dirty.		
						78-*44B*	*Sandstone*, volcaniclastic, moderately hard, grayish-green, very-fine- to medium-grained, angular to rounded, clasts mostly altered basaltic glass, plagioclase, glauconite, and zeolites, abundant car-bonized plant material on bedding planes.		
						78-*44C*	*Vitric basaltic tuff*, hard, mottled brown and gray, vesicular, chloritic, with abundant disseminated pyrite.		
17	S5-78-EG-45	2955- 2625	58°42.40′ 58°43.83′	141°07.18′ 141°03.58′	5	78-*45A*, 45B	*Volcaniclastic pebbly sandstone*, hard, massive, dark-purplish-gray to reddish-black, subangular to rounded vesicular chloritized basaltic glass clasts with intersertal texture to 12 cm maximum diameter, in vitric tuff matrix.	Early Eocene nannofos-sils Early to middle Eocene (Penutian-Ulatisian) foraminifers Late Eocene palyno-morphs	Bathyal
						78-45C	*Siltstone*, soft, massive, waxy, gray.		
18	S10-79-EG-40	2780- 1980	58°42.61′ 58°44.08′	141°07.74′ 141°05.90′	10	79-*40A, 40C, 40D*	*Basalt*, hard, massive, commonly amygda-loidal, medium-bluish-gray to brownish-gray, possible pillow structure, slightly to extremely altered.		Basalt minimum age 29.7 ±3.6 Ma based on whole-rock K/Ar data (Plafker et al., 1980).
						79-*40B*	*Volcaniclastic pebbly sandstone*, hard, massive, olive-gray, subangular to sub-rounded basalt and zeolite clasts to 25 mm maximum diameter in tuffaceous matrix.		
						79-40E	*Claystone*, slightly indurated, weathered light-olive-brown.		
							Calcareous claystone, argillaceous, moder-ately hard, massive, light-yellowish-brown.	Early Eocene (*D. lodoensis* zone?) nannofossils Late Eocene palyno-morphs One small gastropod	
19	S19-79-EG-41	2170- 1850	58°43.36′ 58°44.43′	141°06.08′ 141°05.04′	25	79-41A, *41B*	*Claystone*, moderately indurated, massive, light-brownish-gray, calcareous, abun-dant kerogen and microfossils.	Early Eocene nanno-plankton Early to middle Eocene (early Ulatisian to early(?) Penutian) foraminifers Middle to late Eocene palynomorphs	Cold water, shelf depths.
						79-*41C*	*Claystone*, very hard, massive, medium-bluish-gray, siliceous, abundant kerogen and microfossils.		
						79-*41D*	*Basaltic lapilli tuff*, hard, mottled yellow, gray, and pink.		
						79-*33A*	*Palagonitic basaltic lapilli tuff*, hard, mas-sive, greenish-black, amygdaloidal, microlitic basaltic glass, extensively altered.		Basalt minimum age 26.2 ±3.0 Ma based on whole-rock K/Ar date (Plafker et al., 1980).
						79-*33G, 33H*	*Basalt*, hard, massive, dark-gray, amygda-loidal porphyritic texture, slightly to moderately altered.		

(continued).

20	S10-79-EG-33	3020- 1778	58°40.17′ 58°42.41′	140°00.67′ 140°56.63′	125	79-*33B, 33C, 33D, 33E, 33F*	*Banded calcite*, wavy bands of vuggy, coarsely crystalline, grayish-orange-pink to amber-brown with thin layers of light-greenish-gray, fine-grained calcite in massive chunks to 35 × 30 × 22 cm. Occurs interbedded with thin layers of palagonitized basaltic tuff from 3–15 mm thick, as matrix cementing basaltic lapilli, or with included basaltic rock fragments.		Deposited in fresh water (Barnes et al., 1981).
						79-*33I*	*Sandstone*, lithofeldspathic, hard, massive, medium-gray, medium-grained, angular to subrounded, well-sorted, with approximately 40% primary calcite matrix.		
						79-33J, 33K	*Glauconitic siltstone*, soft, medium-gray-weathering, 30% pelletal glauconite, abundant biotite, finely disseminated pyrite?, rare sand-size quartz grains, abundant opaline skeletal fragments.		Probably Eocene based on lithology
21	S10-79-EG-34	2405- 1679	58°41.13′ 58°42.53′	140°58.80′ 140°56.49′	75	79-*34B*, 34D	*Olivine basalt*, hard, fractured and veined, dusky-brown, cut by veins of green nontronite, zeolite, and yellow clay, moderately to highly altered.		
						79-*34C, 34E*	*Basalt*, hard, medium-bluish-gray, moderately to highly altered.		
						79-*34F*	*Volcaniclastic sandstone*, very hard, massive, medium-gray, medium- to very-coarse-grained, moderately well sorted. Mainly basaltic rock fragments and subordinate bioclastic detritus in a calcareous matrix.	Foraminifers and bioclastic detritus including large foraminifers *(Amphistogina?)*	
						79-34A	*Pebbly mudstone*, crumbly, massive, medium-dark-gray, waxy, with angular to subangular volcanic clasts to 6 cm in maximum size.	Middle Eocene (late Ulatisian) foraminifers including *Amphistogina* Eocene palynomorphs	
22	S10-79-EG-35	1300- 560	58°42.70′ 58°44.37′	140°56.50′ 140°54.92′	75	79-*35B*	*Amygdaloidal basalt*, hard, medium-bluish-gray, highly altered.		
						79-*35A*	*Basalt lapilli tuff*, hard, massive, grayish-blue palagonitized basaltic lapilli in altered fine-grained matrix.		
						79-35C	*Claystone*, moderately indurated, massive, brownish-gray, abundant kerogen.		
						79-35D, *35E*	*Siltstone*, moderately indurated, massive, brownish-gray, abundant microfossils and kerogen, in part calcareous.	Middle to late Eocene palynomorphs	
23	S10-79-EG-36	930- 390	58°44.14′ 58°46.11′	140°58.77′ 140°55.24′	475	79-36A, *36B*, 36C, *36D*	*Organic-rich claystone*, slightly to moderately indurated, laminated to massive, dusky-yellowish-brown, abundant microfossils and kerogen, mica and fish scales commonly on bedding planes, one large piece of carbonized wood 15 cm long, in part calcareous (concretions?), in part with beds to 1 cm thick of fine-grained, greenish-gray volcaniclastic, highly fossiliferous sandstone.	Middle to late Eocene nannofossils Late Eocene (upper Narizian) foraminifers including *Amphistogina* and *Discocyclina* Eocene (late Eocene(?) palynomorphs Siliceous sponge remains	Lower bathyal with transported shallow-water material
						79-*36F*	*Volcaniclastic sandstone*, hard, massive, mottled grayish-blue, medium- to very-coarse-grained (2 cm), mainly highly altered basaltic rock fragments and minerals, abundant megafossil fragments and large foraminifers, calcareous matrix.	*Amphistogina* sp. and *Discocyclina* sp.	
						79-*36E*	*Basaltic lapilli tuff*, hard, grayish-purple, angular lapilli to 4 cm in green altered tuff matrix.		
24	S10-79-EG-26	2300- 1500	58°40.70′ 58°41.06′	140°55.98′ 140°52.07′	170	79-*26C*, 26E, 26F, 26G, 26I, 26K, *26L*	*Basalt and olivine basalt*, hard, massive, gray-black, brown, medium-dark-gray, and olive-gray, slightly to moderately altered.		Basalt minimum age 38.9 ±2.8 Ma based on whole-rock K/Ar data (Plafker et al., 1980).
						79-*26A, 26B*	*Bioclastic sandstone*, very hard, white or weathered moderate red, angular to rounded calcareous clasts to 14 mm in medium-dark-gray or greenish-gray, fine- to medium-grained matrix of calcite-cemented bioclastic material and basaltic detritus. Bioclastic detritus consists mainly of calcareous algae, coral, bryozoans, large foraminifers, and shell fragments.	Foraminifers, including Eocene *Amphistogina* sp. and *Discocyclina* sp.	

(continued).

						79-*26J*	*Bioclastic volcaniclastic sandstone*, very hard, massive, fine- to medium-grained, angular to rounded reef detritus, foraminifers, and mafic volcanic rock and mineral fragments in calcite matrix. Rare silt-size quartz grains.		
25	S2-77-EG-8	2750- 2350	58°35.61′ 58°36.32′	140°38.40′ 140°38.77′	2	77-*8*	*Basalt*, hard, dark-green, calcite-veined, moderately altered.		
26	S2-77-EG-9	2125- 1700	58°37.61′ 58°38.87′	140°36.92′ 140°38.59′	3	77-*9*	*Basalt*, hard, dark-green, strongly vesiculated and possibly brecciated, thoroughly altered.		
27	S2-77-EG-11	890- 500	58°39.16′ 58°39.69′	140°36.89′ 140°36.03′	25	77-11A, 11B	*Siltstone*, moderately well indurated, slabby, light-olive-gray to dark-gray, abundant kerogen, moderately abundant microfossils, sparse pyrite, carbonaceous plant remains, and glauconite, in part sandy.	Eocene or Oligocene nannofossils Late Eocene (Refugian) foraminifers Eocene palynomorphs	
						77-*11C*	*Sandstone*, lithofeldspathic, moderately well indurated, slabby, gray, medium-grained, angular to subrounded, well-sorted, 23% carbonate matrix.		
28	S2-77-EG-10	1875- 1400	58°37.39′ 58°38.11′	140°35.71′ 140°35.29′	75	77-10A, 10B	*Siltstone*, soft to moderately indurated, slabby, gray, greenish-gray, or dark-gray, waxy, minor pyrite and glauconite?	Late Eocene to Oligocene foraminifers Eocene palynomorphs	
						77-*10C*	*Sandstone*, lithofeldspathic, friable, blocky, gray, fine- to medium-grained, angular to subrounded, pervasive siliceous cement and trace of calcite cement. Moderate secondary porosity.		
29	S10-79-EG-24	1340- 480	58°35.70′ 58°38.33′	140°21.66′ 140°20.47′	100	79-24A, 24B, 24C, *24D*	*Claystone*, slightly indurated, finely laminated, dusky-brown, brown, abundant microfossils, kerogen, and mica, in part calcareous and moderately indurated, rare fine-grained sandstone laminae.	Late to late middle(?) Eocene nannoplankton Late Eocene (Narizian(?) foraminifers Middle to late Eocene palynomorphs	Middle to lower bathyal
						79-*24E*	*Sandstone*, lithofeldspathic, very hard, massive, light-gray, medium-grained, clean, calcareous cement, abundant carbonaceous plant fragments and mica.		
30	S2-77-EG-17	1025- 925	58°32.62′ 58°33.02′	140°06.66′ 140°06.49′	75	77-17B, 17C, 17D, 17E	*Claystone*, soft to moderately hard, laminated to massive, grayish-orange, brownish-gray, and medium-gray, waxy, abundant microfossils, kerogen, mica, carbonaceous plant fragments, in part silty and interbedded with laminae of medium-gray siltstone or fine-grained highly micaceous sandstone.	Late Eocene or Oligocene (Refugian or Zemorrian) foraminifers Eocene palynomorphs Eocene(?) diatoms and silicoflagellates	
						77-*17A*	*Sandstone*, lithofeldspathic, moderately hard, friable, light-gray, medium-grained, subangular to subrounded, highly micaceous, siliceous cement, secondary porosity.		
31	S2-77-EG-18	510- 250	58°34.72′ 58°35.92′	140°05.46′ 140°04.67′	275	77-18A, 18B	*Claystone*, silty, soft to moderately indurated, blocky, laminated in part, tan to light-gray and yellowish-gray, abundant siliceous microfossils and mica, sparse pyritic micronodules and glauconite.	Eocene palynomorphs Late Oligocene diatoms and silicoflagellates	Abundant slickensides and iron-oxide-coated fractures
32	S10-79-EG-20	2040- 650	58°16.48′ 58°19.24′	139°32.42′ 139°28.37′	400	79-20B, 20D, 20E, *20I*	*Claystone*, silty, slightly to moderately indurated, massive, medium-dark-gray, locally abundant microfossils, mica, carbonaceous plant material, and fish scales, in part with very hard, light-gray calcareous concretions.	Early(?) Eocene nannoplankton Early to middle(?) Eocene (upper Penutian to lower? Ulatisian) foraminifers Eocene, possibly early Eocene palynomorphs	Abundant slickensides Middle to lower bathyal Eocene with transported upper slope and shelf material
						79-*20H*	*Sandstone*, volcaniclastic and bioclastic, very hard, medium-gray, fine- to coarse-grained, angular to subrounded; clasts are predominantly basaltic volcanic rock and mineral fragments with subordinate quartz, potash feldspar, chert, and other nonvolcanic detritus, and minor shelly material, calcareous cement.	Abundant *Amphistogina*	
						79-*20F*, *20G*	*Sandstone and siltstone*, hard, bedded, fissile, medium-dark-gray, fine- to medium-grained, feldspatholithic sandstone alternating with dark-gray to black, inch- to millimeter-scale laminae of carbonaceous and micaceous siltstone.		
						79-20A, 20C	*Claystone*, slightly indurated, massive, low density, yellowish-gray, abundant siliceous microfossils, very-fine-grained mica, carbonaceous plant fragments.	Late Oligocene to Recent diatoms	Yakataga Formation equivalent?

(continued).

33	S10-79-EG-21	735- 300	58°19.00′ 58°20.16′	139°28.80′ 139°27.10′	300	79-*21A*	*Sandstone*, lithofeldspathic, hard, friable, massive, medium-grained, angular to subrounded, abundant basaltic detritus.		Some blocks are slickensided
						79-*21B, 21C*, 21D, 21E, 21F	*Claystone*, slightly to moderately indurated, medium-dark-gray to dark-gray, sparse microfossils, disseminated carbonaceous plant material, some small trails or burrows; in part very hard and calcareous (concretions?).	Late to late middle(?) Eocene nannofossils Late Eocene (Narizian) foraminifers Middle to late Eocene palynomorphs	
34	S10-79-EG-17	2480- 710	58°09.62′ 58°13.10′	139°06.91′ 139°05.73′	1	79-*17B*	*Basalt*, hard, massive, greenish-black, moderately altered.		
35	S10-79-EG-18	1200- 930	58°11.27′ 58°11.94′	139°05.10′ 139°04.10′	170	79-*18C, 18D, 18E, 18F*	*Basalt*, hard, massive, medium-dark-gray to dark-gray, slightly to moderately altered.		Basalt minimum age 50.7 ±5.2 Ma based on whole-rock K/Ar data (Plafker et al., 1980)
						79-18A	*Claystone*, moderately indurated, homogeneous, medium-dark-gray.		
						79-18B	*Siltstone*, slightly indurated, deeply weathered to moderate-yellow-brown, medium-gray, abundant finely divided mica.	Tertiary palynomorphs	May be late Cenozoic based on lithology
36	S10-79-EG-19	780- 320	58°11.56′ 58°13.24′	139°04.09′ 139°02.40′	125	79-19A, 19B, 19C	*Siltstone*, crumbly to moderately indurated, massive, medium-dark-gray, highly micaceous, sparse fish scales and small megafossils.	Eocene palynomorphs	
37	S2-77-EG-3	1250- 750	58°00.62′ 58°00.92′	138°28.04′ 138°27.40′	250	77-*3A*	*Sandstone*, volcaniclastic, very hard, blocky, greenish-gray, in part pebbly.		Probable Yakutat Group equivalent
						77-*3B*	*Argillite*, hard, grayish-black.		
38	S2-77-EG-2	2090- 1500	57°58.99′ 57°59.59′	138°28.76′ 138°26.88′	250	77-2A	*Siltstone*, slightly indurated, dark-gray, sandy, with floating pebbles.	Quaternary(?) nannofossils; Pleistocene or Holocene foraminifers	Yakataga Formation equivalent Deep and cold water fossil assemblages
						77-2B	*Siltstone*, hard, light-gray, calcareous, minor glauconite.	Early Pliocene diatoms Tertiary palynomorphs	
39	S5-78-EG-14	510- 340	57°54.31′ 57°54.57′	138°02.45′ 138°01.23′	50	78-*14B, 14C*	*Sandstone*, lithofeldspathic, hard, slabby, medium-gray, medium-grained, angular to subangular, poorly sorted, micaceous, in part with calcite cement.		Sandstone intensely sheared
						78-14A	*Siltstone*, soft, weathered light-brown, waxy, abundant biotite, diatomaceous.	Late Pliocene to Holocene with reworked Miocene diatoms	Middle to outer shelf
40	S2-77-EG-6	157- 150	58°11.55′ 58°12.15′	138°23.44′ 138°23.12′	30	77-*6A*	*Metasandstone*, lithofeldspathic, hard, gray to greenish-gray, fine- to medium-grained, angular to subangular, poorly sorted, laumontite in patches and veinlets.	None	Rocks intensely sheared Probable Yakutat Group equivalent
41	S5-78-EG-40	3680- 2875	59°19.59′ 59°21.88′	143°43.81′ 143°43.48′	50	40A, 40B	*Siltstone*, soft to moderately well indurated, gray, abundant radiolarians.	Late Quaternary diatoms Plio-Pleistocene foraminifers	Bathyal, probably lower bathyal
42	S5-77-EG-36	3500- 2650	59°15.27′ 59°17.07′	143°52.76′ 143°53.47′	2	77-36A, B	*Siltstone*, olive-gray, siliceous, dense, with scattered "floating" sand grains and pebbles to 1 cm maximum diameter.	Early Pleistocene diatoms Reworked coccoliths Sponge spicules	
43	S5-78-EG-38	4046- 3150	59°13.04′ 59°14.64′	144°28.96′ 144°28.22′	20	78-38A	*Siltstone*, moderately hard, gray.	Middle to early Pliocene diatoms	Probable Yakataga Formation equivalent with reworked Paleogene
						78-38B	*Mudstone*, very soft, gray, sandy.	Late Eocene foraminifers Tertiary nonmarine palynomorphs	
44	S5-78-EG-39	2800- 2690	59°14.93′ 59°15.25′	144°30.59′ 144°31.95′	0.1	78-39	*Mudstone*, soft, brownish-gray, sandy.		Probable Yakataga Formation equivalent
45	S5-78-EG-37	4090- 2850	59°11.93′ 59°13.26′	144°28.96′ 144°28.22′	15	78-37	*Mudstone*, soft, brown-weathering, sandy, with smooth polished floating grains to granule size.	Pliocene to Holocene diatoms Eocene palynomorphs	Probable Yakataga Formation equivalent with reworked Paleogene
46	S5-78-EG-32	3430- 3450	59°08.29′ 59°08.29′	144°53.66′ 144°52.08′	80	78-32C	*Mudstone*, very soft, very-dark-gray, sandy.	Plio-Pleistocene foraminifers	Cold-water, shelf environment
						78-32D, 32E	*Siltstone*, soft, brownish-gray, waxy, sparse "floating" sand grains, highly micaceous and carbonaceous, fissile.	Eocene palynomorphs	Probable Yakataga Formation equivalent with reworked Paleogene
						78-*32B1, 32B, 32B2*	*Sandstone*, hard, gray, very-fine- to medium-grained, lithofeldspathic, abundant burrows, some thin graded layers with calcareous siltstone laminae.		
						78-*32A1*	*Limestone*, hard, dense, tan, concretions with disseminated very fine pyrite and glauconite.		
47	S5-78-EG-28	2630- 2060	59°14.52′ 59°15.77′	145°03.68′ 145°05.31′	80	78-28A, 28B, 28C	*Mudstone*, soft to moderately hard, dark-gray to greenish-gray, sandy and pebbly, with pebbles to 3 cm diameter.	Late Pliocene to Holocene diatoms Pleistocene foraminifers Eocene palynomorphs	Probable Yakataga Formation equivalent with reworked Paleogene

(continued).

48	S5-78-EG-31	3250- 2525	59°15.37′ 59°16.37′	145°00.12′ 145°02.32′	100	78-31F 78-31D 78-31C 78-*31A, 31B,* 31E	*Mudstone*, soft, very-dark-gray, very sandy. *Siltstone*, moderately hard, dark-gray. *Siltstone*, soft to hard, gray, calcareous, with minor disseminated pyrite and sand grains. *Limestone*, hard, brown, silty, with disseminated pyrite and "floating" sand grains and one pebble.	Rare probable Quaternary foraminifers in 78-31F Eocene palynomorphs in all samples	Probable Yakataga Formation equivalent with reworked Paleogene
49	S5-78-EG-27	990- 785	59°26.74′ 59°26.58′	145°14.43′ 145°23.59′	10	78-27A, 27B	*Mudstone*, soft, dark-gray, sandy, pebbly.	Late Pliocene to Holocene diatoms Quaternary nannoplankton Eocene to Oligocene palynomorphs	Probable Yakataga Formation equivalent with reworked Paleogene
50	S5-78-EG-36	3330- 2755	59°07.53′ 59°09.72′	145°43.23′ 145°43.11′	5	78-36A 78-*36B*	*Siltstone*, soft, gray, massive. *Siltstone*, hard, dark-gray, calcareous, with minor pyrite and carbonized plant material.	Quaternary and Eocene foraminifers Plio-Pleistocene and Tertiary marine palynomorphs	Probable Yakataga Formation equivalent with reworked Paleogene and (?) Paleogene (B)
51	S5-78-EG-33	4450- 4220	58°59.65′ 59°03.28′	145°55.76′ 145°59.52′	3	78-33B 78-33A 78-*33C*	*Siltstone*, soft, gray-brown, in part pebbly and sandy. *Siltstone*, moderately hard, gray. *Sandstone*, hard, medium-gray, micaceous, very-fine-grained, laminated.	Early Quaternary diatoms Quaternary nannoplankton Tertiary nonmarine palynomorphs	Probable Yakataga Formation equivalent with reworked Paleogene and (?) Paleogene (A, C)
52	S5-78-EG-34	3870- 3250	59°04.29′ 59°08.52′	145°58.47′ 145°59.52′	1/2	78-34A	*Siltstone*, soft, gray, abundant fine-grained mica.	Plio-Pleistocene and Tertiary nonmarine palynomorphs	Probable Yakataga Formation equivalent with reworked Paleogene
53	S5-78-EG-35	3350- 2650	59°05.75′ 59°07.64′	146°02.52′ 146°04.96′	4	78-35A, 35C 78-35B	*Mudstone*, black and brownish-gray, micaceous, in part sandy and pebbly. *Siltstone*, moderately hard, light-brown, abundant mica, rare glauconite.	Late Miocene or younger and late Eocene foraminifers Eocene marine palynomorphs	Probable Yakataga Formation equivalent with reworked Paleogene and (?) Paleogene (B)

1983; Keller et al., 1984). These data are summarized briefly below, emphasis being given to relevant unpublished data.

Yakutat Terrane

Late Paleocene to Middle Eocene

The most complete sedimentary sequences in the region are in the Yakutat terrane, especially in the segment west of the Dangerous River zone (Figures 5, 6). Paleocene coal-bearing nearshore marine strata occur at one isolated locality along the northern margin of the Tertiary basin 12 km northeast of the Samovar Hills. Although these strata have been correlated on the basis of lithology with the Kulthieth Formation, a sparse but distinctive megafauna suggests that these strata may be older than the predominantly Eocene Kulthieth Formation (Addicott and Plafker, 1971). Elsewhere, the oldest dated strata above pre-Tertiary basement rocks are early to middle Eocene marine mafic volcanic rocks and fine-grained clastic rocks that were deposited in a shallow to moderately deep marine environment. Strata of this age are known to be exposed onshore only in the Samovar Hills—a large nunatak north of the Malaspina Glacier (inset A, Figure 1; Figures 4, 5). The oldest of these units, informally designated here as the basalt of Hubbs Creek and siltstone of Oily Lake, are of special interest because they appear to correlate closely with strata sampled by dredging on the continental slope and because large seepages of oil and gas are associated with their outcrop distribution (inset A, Figure 1). The following descriptions of these two units are based on work done by the U.S. Geological Survey in the Samovar Hills during 1979 and 1982.

Basalt of Hubbs Creek—Basalt is exposed at Hubbs Creek, where it occurs as a fault-bounded block in contact with the Upper Cretaceous flysch facies of the Yakutat Group on the east and the younger Kulthieth Formation on the west, and in isolated outcrops along Marvitz Creek. Its distribution suggests that it may conformably overlie the Yakutat Group and is in turn overlain by the siltstone of Oily Lake (inset A, Figure 1). The basalt unit is several hundred meters thick and consists of approximately 15% basaltic flows, 75% agglomerate, and 10% tuff cut by numerous diabasic dikes and sills. Fragments of calcarenite as much as 2.5 cm in maximum dimension occur as patches within the tuff and sandstone. The shell fragments and algal and bryozoan remains are nondiagnostic as to age, but indicate a shallow subtropical marine environment. Whole-rock K-Ar ages on basalt fragments yield apparent minimum ages of 40.0 ± 4.8 and 50.0 ± 3.9 Ma (analyses by Krueger Enterprises, Inc., June 30, 1980). The older age probably is closest to the crystallization age of the rock, as it is similar to the paleontologically dated overlying marine unit.

The basaltic unit is correlated with unit C on the continental slope (Figure 6) because of similarities in lithology, age, and stratigraphic occurrence. Correlative basalt also occurs in the intervening area in the Yakutat #1 well, 6 km southeast of Yakutat at 5351–5424 m (American Stratigraphic Company well log). Undated, but lithologically similar, olivine basalt was penetrated in the Dangerous River well 13 km northeast of Yakutat at 442–527 m, immediately above the contact with the pre-Tertiary Yakutat Group; and in the Malaspina #1 well on the west side of Yakutat Bay at 3783–3785 m. These occurrences suggest that the basalt unit is widely distributed along, and west of, the Dangerous River zone.

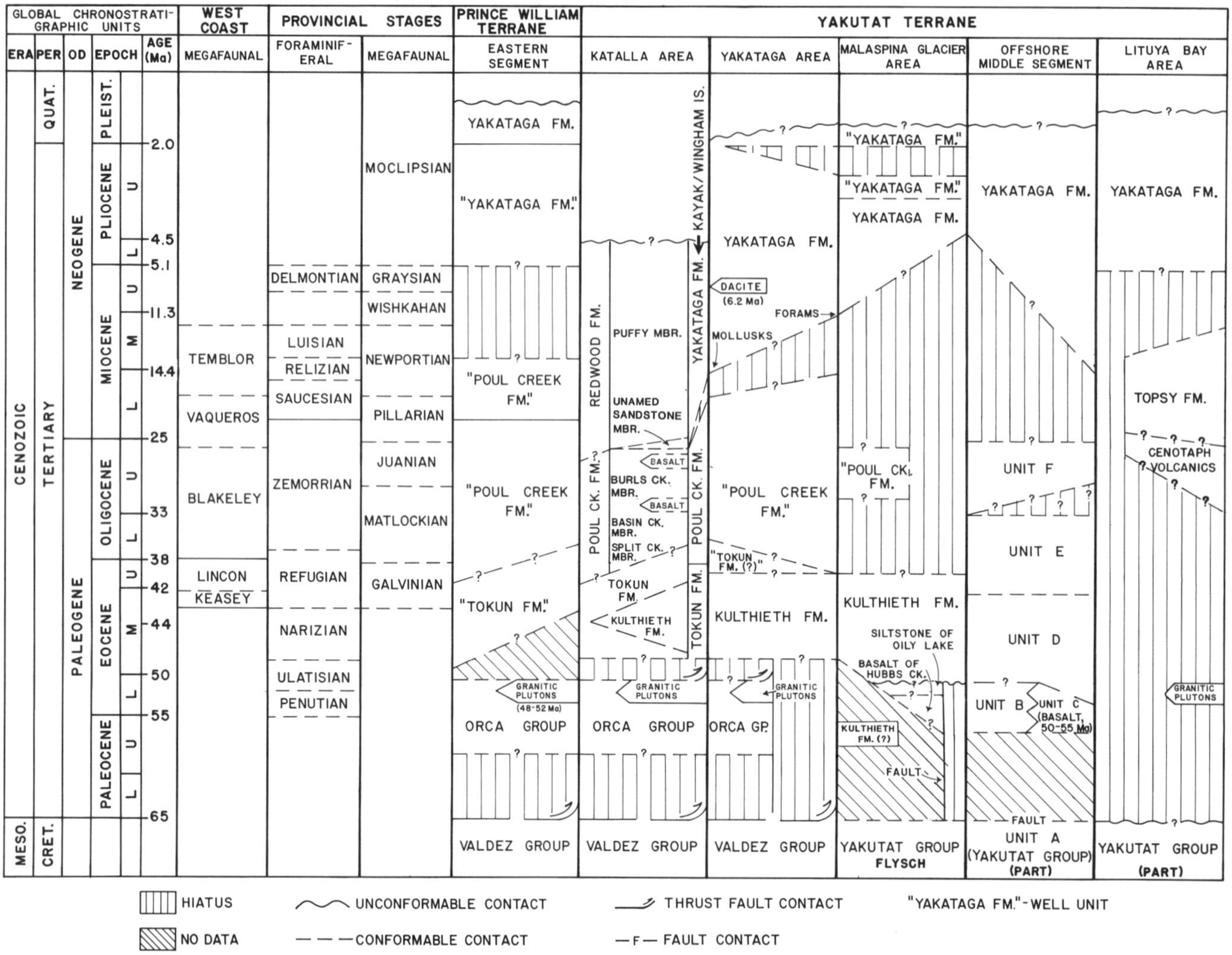

Figure 4. Stratigraphic correlation chart for the Gulf of Alaska margin. Modified from Plafker et al. (1978c, 1980), Rau, Plafker, and Winkler (1983), and G. Plafker, unpublished data. Time scale from Harland et al. (1982); provincial stages modified after Addicott, Winkler, and Plafker (1978), Rau, Plafker, and Winkler (1983), and unpublished worksheet of the Correlation of Stratigraphic Units of North America (Alaska) in preparation for AAPG.

Siltstone of Oily Lake—The siltstone consists of 100–200 m of dark-gray to greenish-gray, thick-bedded strata with a small percentage of basaltic lapilli tuff, vitric tuff, and very-fine-grained, calcareous, locally graded sandstone in beds 10–30 cm thick. The siltstone contains a sparse fauna of foraminifers that indicate a middle Eocene (Ulatisian) age and suggest outer-shelf to upper-slope conditions of deposition (Weldon W. Rau, 1983, written communication). The siltstone unit correlates closely in age and lithology with unit C on the continental slope (Figure 6).

Paleocene, Eocene, and Early Oligocene (?)

During the late Paleocene and Eocene, there was a general marine regression, seas were relatively warm, and climate was subtropical to paratropical. Strata deposited during this interval consist mainly of thick, interfingering, coal-bearing, alluvial-plain, delta-plain, barrier-beach, and shallow-marine deposits (Kulthieth Formation, including the now-abandoned Kushtaka Formation), the marine equivalents of these deposits deposited mainly in delta (Tokun Formation) and prodelta (Stillwater Formation) environments. These units crop out in a belt extending from the west side of Yakutat Bay to the western margin of the Yakutat terrane (Figure 1).

Distribution of the lithofacies indicates a generally east-west-trending paleoshoreline; the continental coal-bearing Kulthieth Formation on the mainland intertongues offshore and on Kayak and Wingham Islands with the marine Tokun Formation, and toward the west with the Stillwater Formation. Predominantly arkosic sandstone reflects derivation from a granitic and metamorphic source terrane that differs from the predominantly flysch and oceanic rocks that presently underlie the mountains to the north. In the Samovar Hills, a prominent thick basal chert-pebble conglomerate of variable thickness marks the contact of the Kulthieth Formation where it unconformably overlaps all the older sequences (inset A, Figure 1).

The sequence contains an abundant marine mollusk fauna and leaf flora of Eocene and early Oligocene(?) age (Figure 4). Megafossils in the Kulthieth Formation within several hundred meters of its base in the Samovar Hills are of late or latest early Eocene to early middle Eocene age (L. Marincovich, 1982, written communication).

Late Eocene, Oliocene and Early Miocene (?)

Transgression occurred during the latest Eocene, Oligocene, and possibly the early Miocene. Much of the deposition

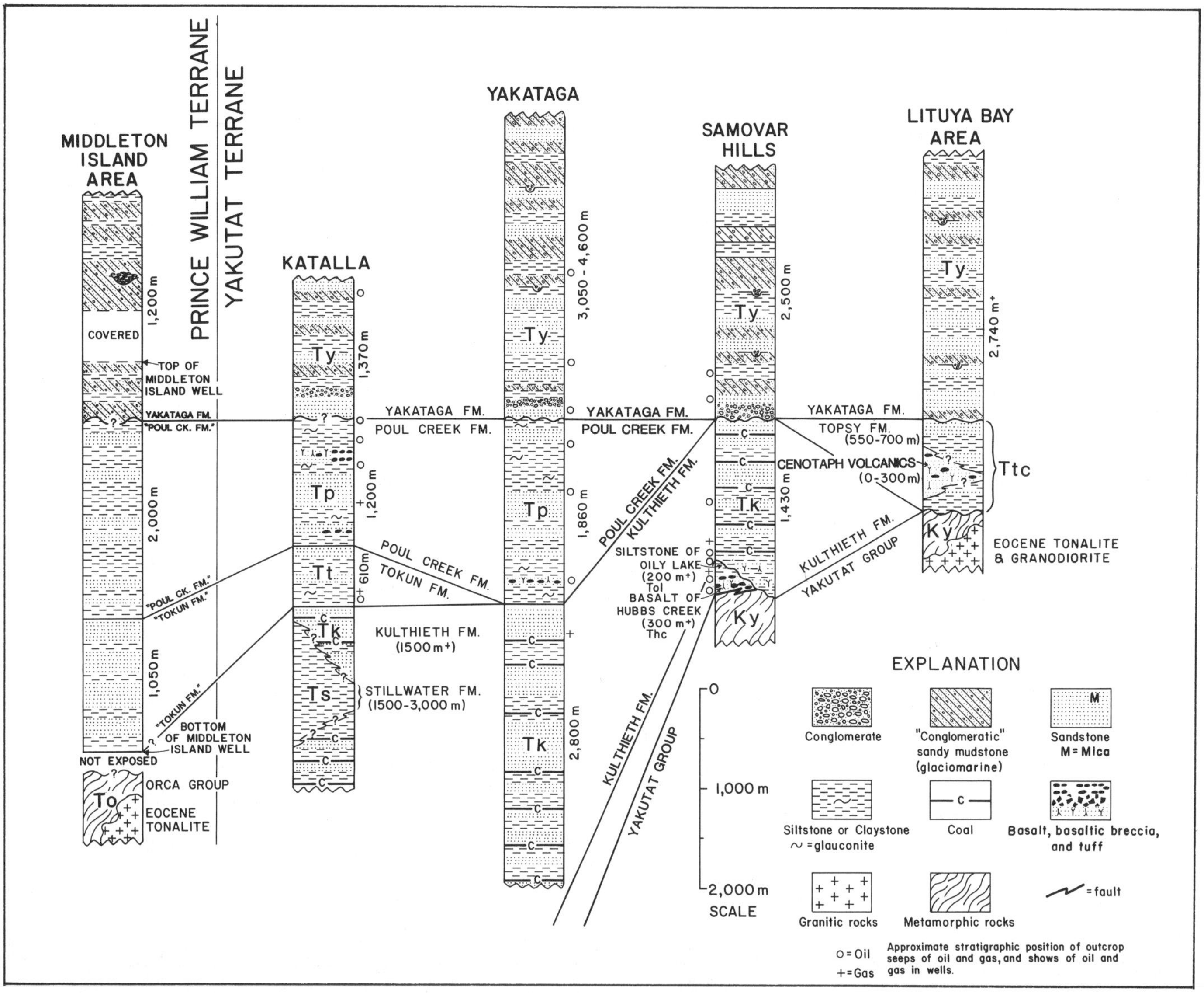

Figure 5. Generalized stratigraphic columns of onshore basinal strata along the northern Gulf of Alaska margin. After Miller, Payne, and Gryc (1959), Plafker (1981), and G. Plafker unpublished data.

Figure 6. Correlation of onshore and offshore stratigraphic units in the central part of the Yakutat terrane. Time scale used is that of Harland et al. (1982).

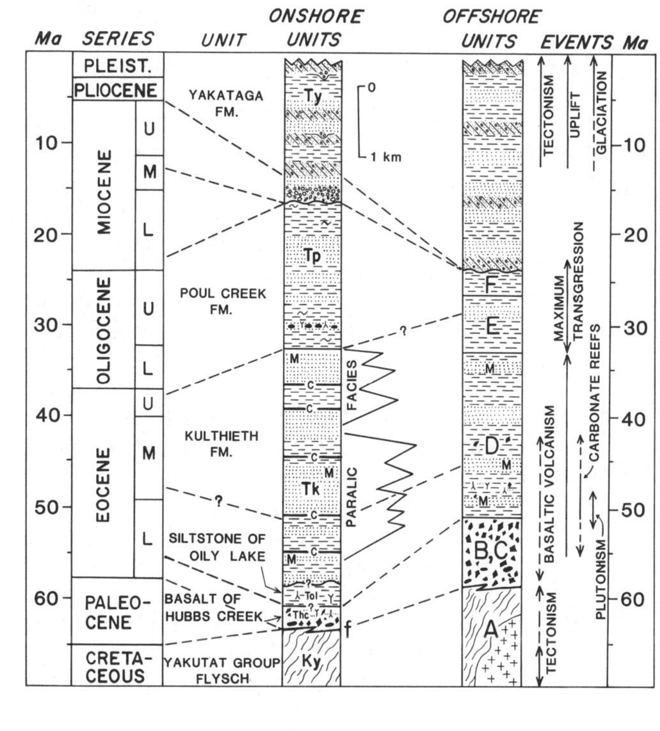

was in neritic to bathyal environments within the oxygen-minimum zone. Predominantly shaly sediment—in part glauconitic, organic-rich, and intercalated with intrabasinal water-laid alkalic basaltic tuff, breccia, and pillow lava—accumulated to form the Poul Creek Formation (which includes the now-abandoned Katalla Formation) and perhaps the lower sandy part of the Redwood Formation. Much of the glauconite in these strata formed by alteration of basaltic detritus (Plafker, 1974). These rocks are exposed in outcrop intermittently along the coastal belt from Icy Bay to Ragged Mountain and on Kayak Island. They were penetrated in wells drilled along the coast (Plafker, 1967) and in at least three wells drilled on the adjacent outer continental shelf (Table 2).

The volcaniclastic strata of the Cenotaph Volcanics, and the intercalated siltstone and sandstone of the Topsy Formation, which overlie pre-Tertiary basement in a relatively

Table 2. Data for petroleum exploration and COST wells drilled on the outer continental shelf.

Location No. on Map (Figure 1)	Company and Well Designation	Years	Total Depth (m)	Results	Units Penetrated
1	Middleton Island State 1	1969	3659	Plugged and abandoned	Yakataga Fm. Oligocene, Eocene and ?Paleocene
2	Exxon OCS Y-0080	1977	4117	Plugged and abandoned	Yakataga Fm.
3	Shell OCS Y-0011	1976–77	4135	Plugged and abandoned	Yakataga Fm.
4	Exxon COST 1	1975	1570	Stratigraphic hole; incomplete owing to drilling problems.	Yakataga Fm.
5	Exxon OCS Y-0050	1977	3961	Plugged and abandoned	Yakataga Fm.
6	Exxon OCS Y-0072	1978	2998	Plugged and abandoned	Yakataga Fm.
7	Gulf OCS Y-0059	1977	3709	Plugged and abandoned	Yakataga Fm.
8	Texaco OCS Y-0046	1977	4580	Plugged and abandoned	Yakataga Fm.
9	Shell OCS Y-0014-2	1977	4691	Plugged and abandoned	Yakataga Fm. Poul Creek Fm.
10	Arco OCS Y-0007	1976–77	5462	Plugged and abandoned	Yakataga Fm. Poul Creek Fm., Tokun Fm.
11	Texaco OCS Y-0032	1977	4768	Plugged and abandoned	Yakataga Fm. Poul Creek Fm.
12	Exxon OCS Y-0035	1978	3576	Plugged and abandoned	Yakataga Fm.
13	Arco Yakutat #1	1983	5430	Plugged and abandoned	Yakataga Fm. Paleogene undifferentiated

small basin between Lituya Bay and Cross Sound, may be partly equivalent in age to the Poul Creek Formation. Offshore distribution of the Cenotaph Volcanics and Topsy Formation is not known.

Late Oligocene (?), Miocene, Pliocene, and Quaternary

Uplift of the Chugach and Saint Elias Mountains along the northern margin of the Yakutat terrane resulted from collision of the Yakutat terrane with, and partial underthrusting beneath, the North American plate. Since the late Oligocene(?), an enormous thickness of clastic sediment of the Yakataga Formation and most or all of the Redwood Formation was shed onto the adjacent continental margin (Plafker, 1981). The Yakataga Formation is characterized by rapid deposition of siliciclastic sediments including an abundance of glacially-derived material from the Chugach and Saint Elias Mountains, whereas the Redwood Formation consists of nonglacial siliciclastic sediments with abundant conglomerate. The sediment was deposited in a predominantly shelf to upper bathyal environment, but locally in nearshore or even continental environments onshore in the Malaspina Glacier area and in deeper bathyal environments offshore. The Yakataga Formation is widely distributed onshore and underlies much of the continental shelf and slope, having thicknesses that locally exceed 5000 m; the Redwood Formation is 1130 m thick and crops out only in the Katalla area west of Bering Glacier.

A dacitic plug with a whole-rock K-Ar age of 6.2 ± 0.3 Ma (Nelson, Dumoulin, and Miller, 1985) was emplaced at Cape Saint Elias on Kayak Island, after folding of the Miocene and younger Yakataga Formation to vertical dips. This plug attests to continuing but minor magmatic activity during the late Neogene and provides a minimum age for the Yakataga Formation in the western part of the Yakutat terrane.

The basal part of the Yakataga Formation shows a general increase in age from east to west across the area (Figure 4). Precise dating of this interval is of special interest because it provides a control on the onset of the deformation that produced most of the structures in the basinal sequences along the Gulf of Alaska margin. Age assignments for the various faunal assemblages differ considerably; ages based on mollusks are generally older than ages based on foraminifers.

The abundant molluscan megafauna suggests that the base of the formation is probably late Oligocene (?) and early to middle Miocene at Kayak Island, and early middle Miocene at Yakataga reef (Addicott, Winkler, and Plafker, 1978; Ariey, 1978). Benthic foraminifers suggest a probable early to middle Miocene (Saucesian and Relizian) age at Kayak

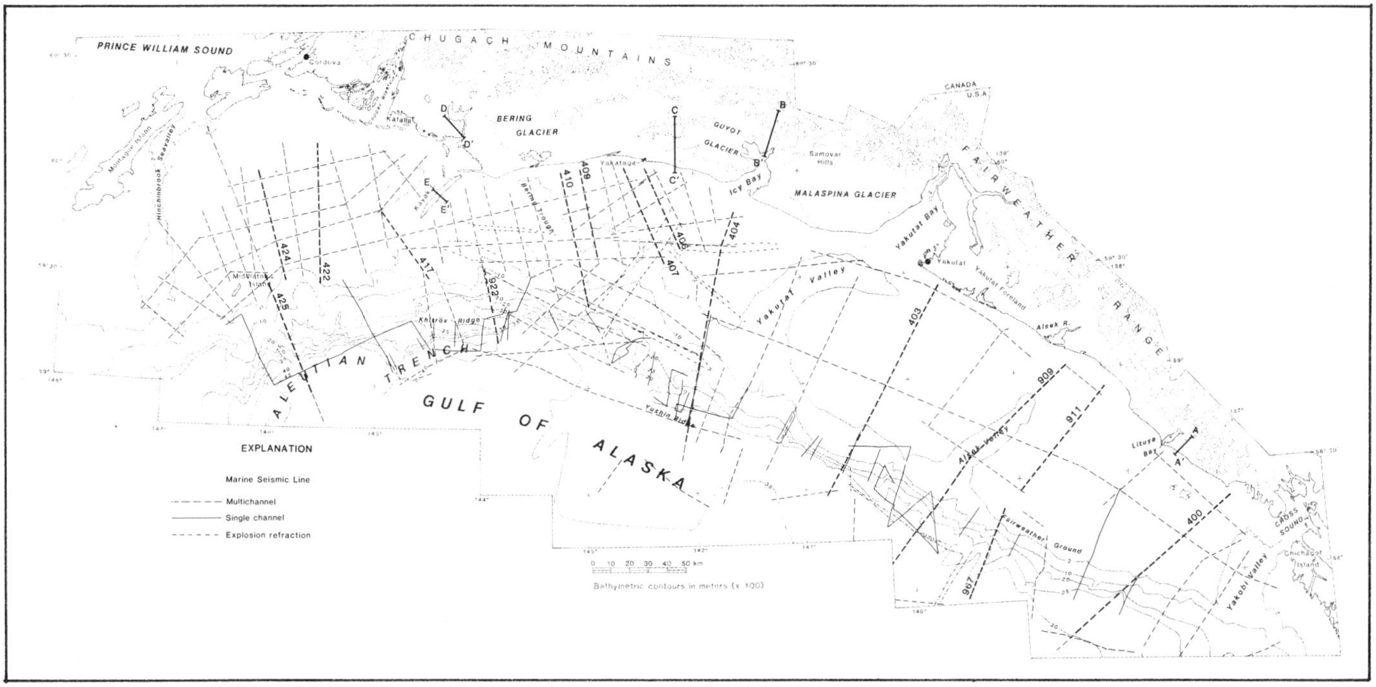

Figure 7. Trackline map for marine seismic-reflection and refraction surveys in the northern Gulf of Alaska. Heavy lines indicate structure sections shown on Figures 8 through 10.

Island, and late Miocene or younger at Yakataga reef (Lagoe, 1983; Rau, Plafker, and Winkler, 1983). Armentrout, Echols, and Nash (1978) concluded that the maximum age of the Yakataga Formation at Yakataga reef is middle Miocene (about 12 Ma) based on planktonic foraminifers; a K-Ar age of 5.1 ± 6.8 Ma obtained for two samples of glauconitic sandstone near the base is undoubtedly too young (owing to argon loss). A K-Ar age on glauconite that I collected from the top of the Poul Creek Formation at Yakataga reef (35 m below the samples collected by Armentrout, Echols, and Nash) yielded an apparent age of 14.2 Ma (unpublished analysis by M. Silberman, U.S. Geological Survey, 1972)—an age that is at least 10 m.y. younger than that indicated by the late Oligocene foraminifers in this part of the section. At Kayak Island the entire Yakataga section was deposited and deformed prior to intrusion of dacite at 6.2 Ma. Data from offshore wells suggest that Poul Creek sedimentation continued in structural lows of the deep offshore environment through the early Miocene. On structural highs, the lower part of the Yakataga Formation is locally missing, probably because of nondeposition; as a consequence, Pliocene strata locally disconformably overlie the Poul Creek Formation (Lattanzi, 1981).

Prince William Terrane

In the Middleton Shelf area of the Prince William terrane, the stratigraphy of post-Orca Group Paleogene basinal strata is known only from subsurface and limited offshore data; there are no known onshore outcrops. The most complete stratigraphic data are from the 3659-m-deep Middleton Island well and from the 1200 m of Yakataga Formation exposed on Middleton Island (Table 2; Figure 5).

A study of the abundant benthic foraminiferal fauna in the well was made by Rau (Rau, Plafker, and Winkler, 1983), as part of a comprehensive analysis of foraminifers throughout the region. The data indicate that the lower unit (3656–2600 m) is early(?), middle, and late Eocene (Penutian?, Ulatisian, and Narizian) in age and that the paleoenvironment was bathyal at the base to neritic in the upper part. If Penutian age fossils are indeed present, this would indicate that the well penetrated the Orca Group. The overlying middle unit (2600–620 m) is of late Eocene, Oligocene, and early to middle Miocene age (Refugian, Zemorrian, and undifferentiated Saucesian and Relizian) and was deposited in upper to middle bathyal water depths. The Yakataga Formation (620–0 m) is of undifferentiated Pliocene and Pleistocene age in the well and is of Pleistocene age at nearby Middleton Island. Sparse paleoecologic data suggest deposition in cool to cold water at shelf depths probably < 150 m.

Keller et al. (1984) interpret sparse coccolith and planktonic faunas near the bottom of the well as being late middle Eocene in age. This younger age assignment is compatible with geologic relationships, since it indicates that the basinal sequence postdates the Orca Group on which it is presumably deposited (Winkler and Plafker, 1981).

STRUCTURE

Structures in the Cenozoic basinal sequence along the northern Gulf of Alaska margin tend to parallel the northern margin of the basin, except near Ragged Mountain and on the shelf west of Kayak Island (Figure 1). In the following section the major structural features are described from east to west along the continental margin, followed by a brief section on the deep-sea floor and slope rise. The distribution of mapped folds and faults is shown in Figure 1, which also shows locations of the representative cross-sections on Figures 8, 9, and 10. Figures 1 and 7 show the locations of marine seismic lines on which the offshore structural interpretations are based.

Eastern Segment of Yakutat Terrane

The eastern segment of the Yakutat terrane is bounded on the east, west, and south by the Fairweather fault, Danger-

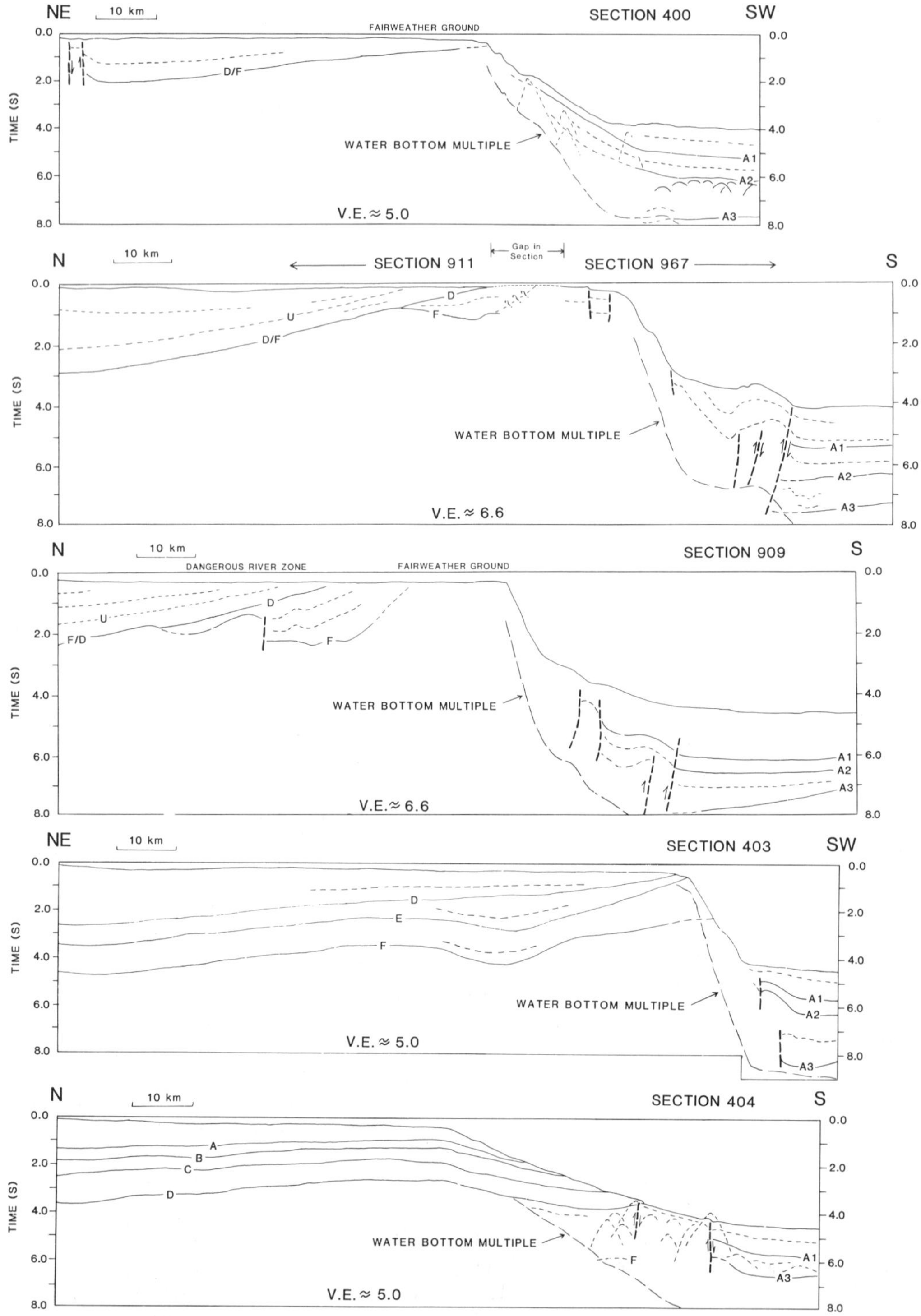

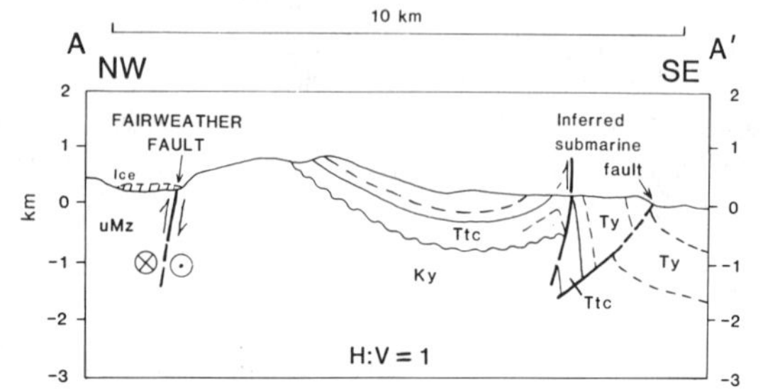

Figure 8. Structure sections in the eastern and central segments of the Yakutat terrane. Section locations on Figures 1 and 7. Inferred correlation of seismic horizons are: A, B, and C, within the Yakataga Formation; D, base of the Yakataga Formation; E, middle Oligocene time horizon; F, acoustic basement (Eocene basalt?), A1, earliest Pleistocene time horizon; A2, earliest Pliocene time horizon; A3, top of oceanic basalt. Unit symbols are: uMz, Upper Mesozoic; Ky, Yakutat Group; Ttc, Topsy Formation and Cenotaph Volcanics; Ty, Yakataga Formation. Offshore sections from Bruns (1983b); onshore section after Plafker (1971).

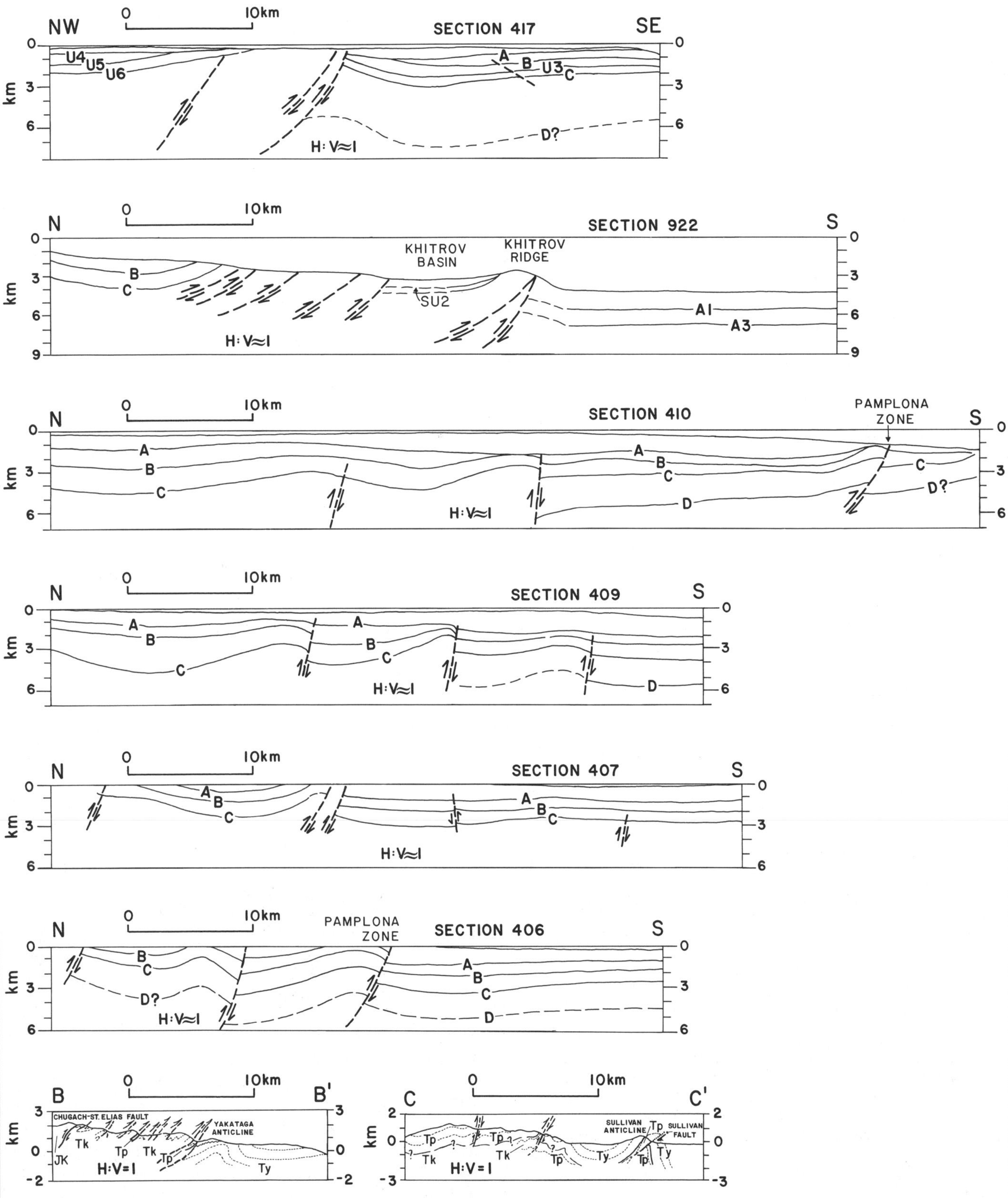

Figure 9. Structure sections in the western segment of the Yakutat terrane. Section locations on Figures 1 and 7. Seismic unconformity horizons U3–U6 and SU2 are within the upper part of the Yakataga Formation; all other reflector horizons as in Figure 8. Unit symbols are: JK, Jurassic and Cretaceous; Tk, Kulthieth Formation; Tp, Poul Creek Formation; and Ty, Yakataga Formation. Offshore sections after Bruns and Schwab (1983); onshore sections after Plafker (1971).

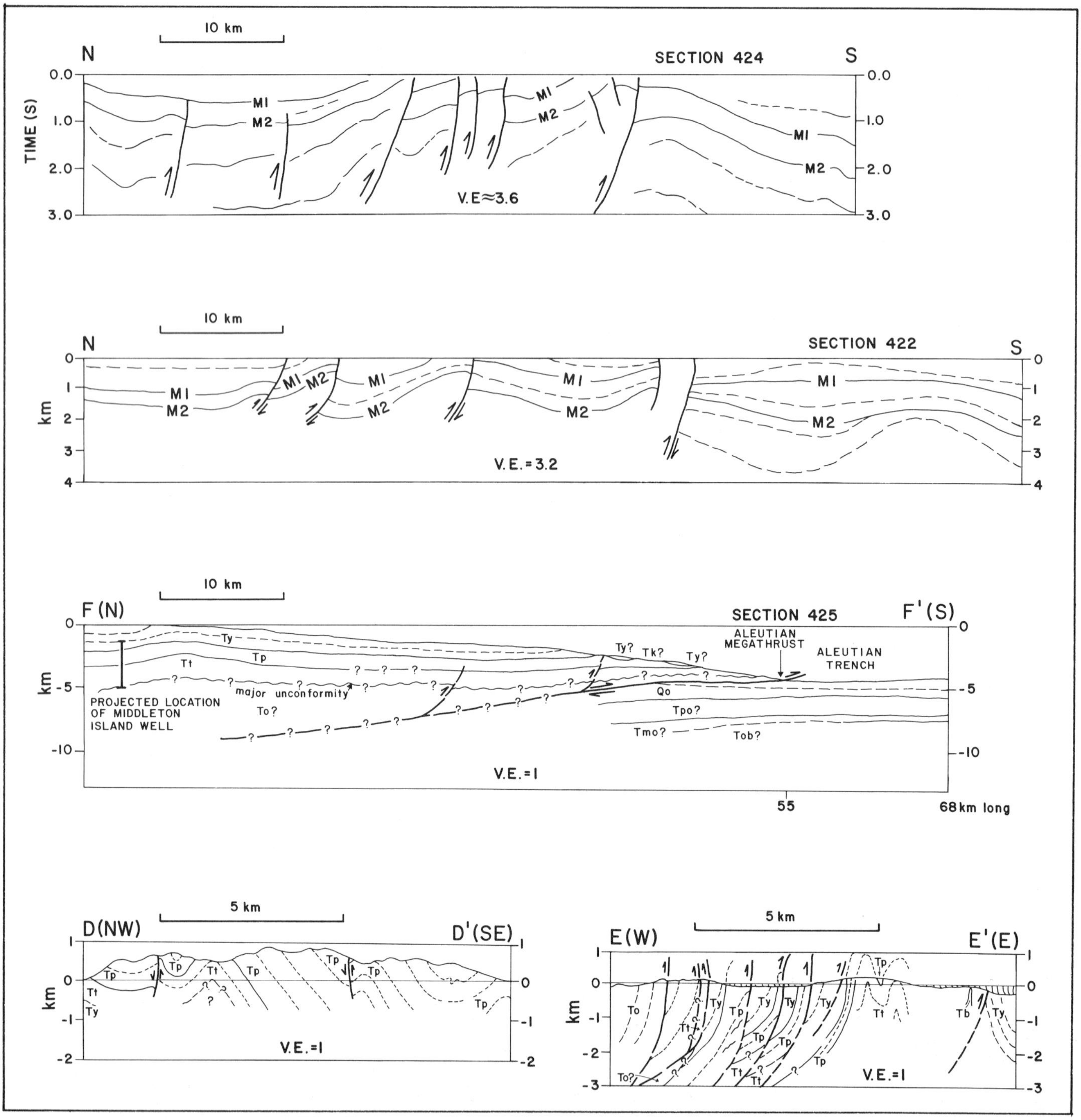

Figure 10. Structure sections in the western segment of the Yakutat terrane and the eastern part of the Prince William terrane. Section locations on Figures 1 and 7. Seismic horizons M1 and M2 are at an unknown stratigraphic level within the upper part of the Yakataga Formation. Unit symbols are: To, Orca Group; Tt, Tokun Formation; Tp, Poul Creek Formation; Ty, Yakataga Formation; Qo, Quaternary; Tpo(?), Pliocene(?); Tmo (?), Miocene(?); and Tob(?), Oligocene basalt(?). Seismic sections after Bruns (1985); D–D' after Plafker (1971); E–E' after Plafker (1974); F–F' after Plafker et al. (1982a).

ous River zone, and Transition fault system, respectively (Figure 1). Basement onshore is the pre-Tertiary Yakutat Group and Eocene granitoid plutons that intrude the Yakutat Group (Plafker, 1967). Structural trends, metamorphic isograds, and lithologic units in the Yakutat Group trend northwest roughly parallel to the trend of the Fairweather fault and are truncated obliquely on the west by the Dangerous River zone (Plafker, 1967; Hudson, Plafker, and Turner, 1977). The offshore Fairweather Ground is underlain by similar basement rocks, as indicated by samples dredged from the continental slope and shelf (nos. 39 and 40, Figures 1, 11; Table 1) and by aeromagnetic anomalies probably reflecting plutons within the sequence (Taylor and O'Neill, 1974). Offshore acoustic basement is correlated with the top of the Yakutat Group (Bruns, 1983b) and is mapped as the F horizon on the seismic sections (sections 400 and 911/967, and the northern part of section 909, Figure 8).

Basement in this segment is overlain unconformably by a relatively thin section of the Cenotaph Volcanics and Topsy Formation, which are in turn overlain by a thick section of

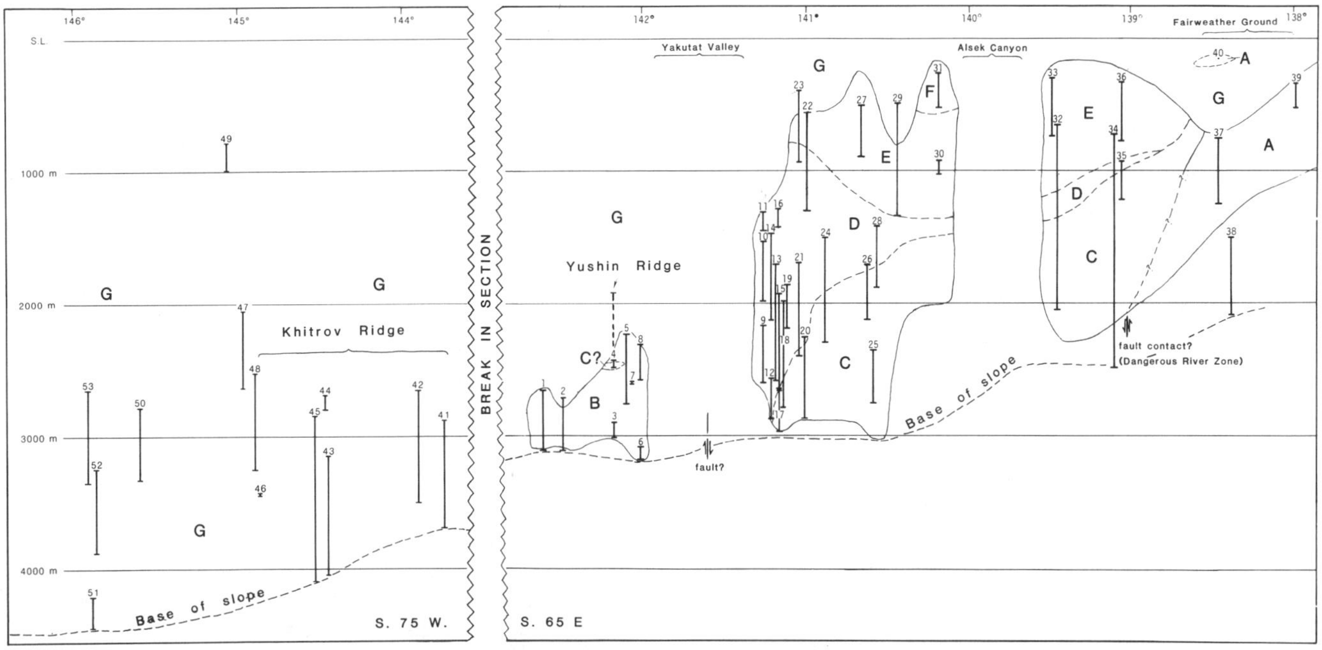

Figure 11.. Section parallel to the continental slope showing vertical interval of dredge samples (numbers) and the distribution of lithologic units (letters) deduced from the dredge and seismic data. Letters correspond with units shown in Figure 6. Dredge locations on Figure 1 and sample data in Table 1. From Plafker et al. (1978b, 1979, 1980).

the Yakataga Formation (Figures 4, 5). Contours in Figure 1 are isopachs on the approximate base of the Yakataga Formation, which is equivalent to the D horizon on the structure sections of Figures 8 and 9 (Bruns, 1983b).

The only known fold in this segment is the tightly appressed faulted anticline that trends roughly parallel to the coast south of Lituya Bay (section A-A′, Figure 8). The structure consists of a gently dipping north flank and a vertical to slightly overturned south limb. Along the south side of the crest of the structure the fold is cut by a thrust fault that has limited displacement. The south limb is inferred, from geometric considerations, to be offset by a major thrust fault just offshore from the coast (Plafker, 1967, 1971). This suggests a significant component of compression in a broad zone across the northeastern margin of the Yakutat terrane. Uplifted Holocene marine terraces on the south limb of this structure indicate continued active growth (Hudson, Plafker, and Rubin, 1976).

Nearshore deformation noted by Bruns (1983b) at the north end of seismic section 400 (Figure 8) is probably related to the postulated offshore fault. To the southeast, the fault appears to merge with the active offshore segment of the Fairweather fault as delineated by Carlson et al. (1979). The extent of this postulated offshore fault northwest of Lituya Bay is not known; it could conceivably extend northwest some 150 km to connect with active oblique thrust faults that occur on strike in the Yakutat area (Plafker and Thatcher, 1982).

A broad basin offshore and parallel to the coast contains as much as 4 km of bedded rocks above acoustic basement, much of which is probably the Yakataga Formation. To the south, the upper Cenozoic strata lap up onto the Fairweather Ground high and locally show evidence of uplift and truncation over the high (Bruns, 1983b; section 400, Figure 8).

Dangerous River Zone

The eastern and central segments of the Yakutat terrane are separated by the Dangerous River zone, across which the basin thickens markedly from east to west. The Dangerous River zone has been delineated (1) in the Samovar Hills, where Eocene volcanic rocks and marine sediment are faulted against the Yakutat Group (inset A, Figure 1); (2) between the Dangerous River and Yakutat wells near Yakutat, where the sedimentary section thickens 3 km in a horizontal distance of 8 km; (3) on the continental shelf, where the zone has been intersected along four seismic lines (Bruns, 1982), one of which is seismic section 909 (Figure 8); and (4) on the continental slope, where dredge samples demonstrate an abrupt change from the Yakutat Group in the east to Paleogene basalt in the west (Figure 11).

Data cited above, together with explosion refraction data (Bayer et al., 1978), suggest that the Dangerous River zone marks a transition from Mesozoic "continental" crust on the east to Paleogene oceanic crust on the west.

Central Segment of Yakutat Terrane

The roughly triangular central segment of the Yakutat terrane is bounded on the east, west, and south by the Dangerous River zone, the Pamplona zone, and the Transition fault system, respectively (Figure 1). The location of the northwestern boundary of this segment in the vicinity of the Samovar Hills is not entirely clear, because of the structural complexity and the large areas covered by ice. In this chapter, the western part of the Samovar Hills is included, because the older section here is closely comparable with Eocene sections on the continental slope between the Dangerous River zone and the Pamplona zone.

The central segment is underlain by a thick Cenozoic sequence onshore and offshore. The onshore sequence includes the informally designated basalt of Marvitz Creek and the siltstone of Oily Lake in the Samovar Hills, the Kulthieth Formation, and the Yakataga Formation (Figure 5). Five informally designated Tertiary stratigraphic units have been defined by samples dredged from the continental slope (Plafker et al., 1980; Figures 6, 11; Table 1). From oldest to youngest they are:

Unit B—900+ m of clastic sedimentary rocks and basalt of unknown age, but probably late Paleocene or early Eocene. Basalt whole-rock K-Ar ages are variable, the oldest being 55.2 ± 7.1 Ma. Probably in part laterally equivalent to Units C and D.

Unit C—1300 m of basaltic flows and pyroclastic rocks with associated clastic marine sedimentary rocks of late Paleocene(?) and early Eocene age. The basalt, which has geochemical similarities to both mid-ocean-ridge and continental-margin basalt (Davis and Plafker, 1984), has yielded a spread of whole-rock K-Ar minimum ages, the oldest of which is 50.7 ± 5.2 Ma.

Unit D—As much as 2100 m of lower and middle Eocene sandstone, conglomerate, and tuffaceous siltstone rich in organic material partly overlies, and partly intertongues with, the volcanic unit.

Unit E—800 m of lower Eocene to lower Oligocene siltstone rich in organic material, tuffaceous siltstone, siltstone, and sandstone that partly intertongues with, and partly overlies, units C and D.

Unit F—300+ m of siltstone, at least partly of late Oligocene age, that crops out locally along the upper slope, where it may disconformably overlie unit E.

Upper Cenozoic strata—As much as 2 km of upper Cenozoic siliciclastic sedimentary strata laterally equivalent to the Yakataga Formation overlies all of the five older units over most of the continental shelf and slope. This unit is not delineated on Figure 1, although thicknesses of part or all of the unit based on seismic-reflection data are indicated by structure contours.

Volcanic rocks of unit C are inferred to be the source of the slope magnetic anomaly (Plafker et al., 1978b, 1980)—a prominent linear magnetic anomaly 15–60 km wide that extends from the continental slope of the central segment of the Yakutat terrane west-northwest for about 500 km (Naugler and Wageman, 1973; Taylor and O'Neill, 1974; Schwab and Bruns, 1979). Model studies of the anomaly by A. Griscolm (in Plafker et al., 1980; and unpublished data, 1980) suggest that the most probable source body is the truncated southern edge of a subhorizontal, planar unit having the approximate thickness and magnetic properties of unit C. Along the Transition fault, which defines its southern margin, the magnetic source body is juxtaposed against oceanic sediment overlying the oceanic crust of the Pacific plate. If correct, this model implies underthrusting of Pacific plate oceanic crust beneath the Yakutat terrane along the Transition fault system, as shown diagrammatically in Figure 3.

The strata of Paleogene age extend beneath the adjacent continental shelf, and seismic data indicate that they are >4500 m thick near Icy Bay and Yakutat Bay (Bruns, 1983b; Figure 1). The configuration of acoustic basement is indicated by the F horizon in the seismic sections on Figure 8 and is interpreted by Bruns (1983b) as corresponding to the top of the basalt unit. Overlying strata, equivalent to the onshore Yakataga Formation, dip gently northward into an elongate basin, the axis of which is near the coast. The maximum thickness of these strata is 5000+ m between Yakutat and Icy Bays, as shown by isopach contours on the base of the formation (Figure 1).

Within this segment there are no mapped folds in the upper Cenozoic section. The only structure delineated is a large, gently dipping structural high in Paleogene strata that is unconformably overlain by undeformed strata of inferred late Oligocene and younger age (Figure 1; section 403, Figure 8). Most tracts leased in OCS lease sale no. 55 were on this structure, and it apparently was the target for the unsuccessful ARCO Yakutat #1 wildcat well (no. 13, Figure 1 and Table 2). The section penetrated in the well, as interpreted from lithologic and electric logs (American Stratigraphic Company), is shown below, together with possible onshore and offshore correlative units; paleontological data for the well were not publicly available at the time this chapter was written.

0–1495 m: Sandstone, conglomerate, claystone, mudstone, and conglomeratic sandy mudstone. Offshore equivalent of the Yakataga Formation.

1495–2613 m: Calcareous and glauconitic claystone and siltstone. Offshore equivalent of the Poul Creek Formation and laterally equivalent to unit F.

2613–4939 m: Micaceous and carbonaceous sandstone, organic-rich brown claystone, and micaceous and carbonaceous siltstone. Offshore equivalent of the Kulthieth Formation and laterally equivalent to unit E.

4939–5350 m: Calcareous claystone and minor siltstone, grading downward into tuffaceous siltstone. Possible offshore equivalent of the siltstone of Oily Lake and laterally equivalent to unit D.

5350–5430 m: Basalt, 74 m thick, with interbedded claystone at base. Offshore equivalent of the basalt of Hubbs Creek and laterally equivalent to unit C, and possibly unit D.

The basalt near the bottom of the well is about 1000 m above the F seismic horizon on acoustic basement (Bruns, 1983b). This relationship suggests that the interbedded basalt and sedimentary rocks are roughly 1000 m thick below the bottom of the well—a thickness that is reasonably close to the 1300 m deduced for the probably correlative unit C on the continental slope. Marine refraction data south of Icy Bay suggest that the Cenozoic sequence is 7–10 km thick above Paleogene basalt basement (Bayer et al., 1978). On the basis of the dredge data, the most likely age for the basalt in this area is early Eocene, but it could be as old as Paleocene.

West of Yakutat Seavalley, near the base of the slope, are three prominent ridges, including Yushin Ridge, with as much as 900 m of local relief (Figure 1; section 404, Figure 8; Figures 7 and 8, Plafker et al., 1980). These ridges are underlain by the basalt unit and well-indurated sedimentary rocks that are undated; the sedimentary rocks are at least as old as Eocene, based on K-Ar ages of associated basalt and the relative maturation of the argillaceous rocks. Because strata in these units appear as acoustic basement on seismic-reflection profiles, the internal structure cannot be determined. However, the isolated occurrence of these steep-sided ridges straddling the Transition fault system strongly suggests that they are tectonically controlled. They are tentatively interpreted as having been uplifted by oblique thrusting on faults along their southeast flanks, although they could represent slices offset from farther southeast by dextral slip. The location of Yushin Ridge 5 km seaward of the projected location of the Transition fault is anomalous. It suggests that the Transition fault in this area is probably a wide and complex system of faults, with Yushin Ridge as a fault-bounded sliver. In any case, similarities in geology between Yushin Ridge and Paleogene strata in the adjacent Yakutat terrane appear to preclude an origin for the ridge beyond the limits of the Yakutat terrane.

Pamplona Zone

The Pamplona zone marks the boundary between relatively undeformed strata of the seismically inactive central segment and the fold and thrust fault belt that constitutes the seismically active western part of the terrane (Plafker et al., 1978a, 1978c). The Pamplona zone, which is locally marked by prominent topographic highs on the sea floor, is the eastern deformational front of the fold and thrust belt (Rogers, 1977; Bruns and Schwab, 1983). The Pamplona zone extends onshore to the northeast along the mountain front to the Samovar Hills area.

Western Segment of Yakutat Terrane

The western segment of the Yakutat terrane is a fold and thrust fault belt that is bounded on the southeast, west, and north by the Pamplona zone, the Kayak Island zone, and the Chugach-Saint Elias fault, respectively (Figure 1). To the southwest, the fold-fault belt trends into the zone of deformation associated with the Aleutian megathrust (Figure 2). Large anticlinal structures of this fold and thrust belt were the major targets for the unsuccessful exploratory drilling onshore from 1954 to 1963 and offshore in OCS lease sale no. 39 from 1969 to 1983 (Figure 1 and Table 2). The offshore structure is shown in Figure 1 by structure contours in probable lower to middle Pliocene strata, and the structural style is illustrated by structure sections across this belt (Figures 9, 10). Information from the seismic data is limited to strata of late Cenozoic age, because acoustic penetration to the base of the Yakataga Formation is rare (Bruns and Schwab, 1983).

Extrapolation of the geology along the mainland coast and on Kayak and Wingham Islands (Plafker, 1974) suggests that a thick Paleogene sequence including the offshore equivalents of the Poul Creek, Tokun, and Kulthieth Formations underlies much of the continental shelf and slope. Of the 10 exploratory wells and one stratigraphic hole drilled offshore in this segment, one reached the probable offshore-equivalent Tokun Formation (no. 10, Figure 1 and Table 2), two others reached the offshore-equivalent Poul Creek or Poul Creek(?) Formation (nos. 9 and 11, Figure 1 and Table 2), and the remainder bottomed in the offshore-equivalent Yakataga Formation.

A maximum stratigraphic thickness of 10-11 km above probable Paleogene basalt basement is indicated by marine refraction data on the continental shelf (Bayer et al., 1978). Additional indirect evidence that the Paleogene basaltic sequence of the central segment extends beneath the western segment and the Prince William terrane comes from analysis of the slope magnetic anomaly—an anomaly which seems to be closely linked to outcrop distribution of the basalt unit in the central segment of the Yakutat terrane (Plafker et al., 1980). As shown on Figure 1, this magnetic anomaly extends northwestward approximately 220 km beneath the western segment of the Yakutat terrane and part of the Prince William terrane, providing direct evidence for subduction in this region (Case et al., 1979; Schwab and Bruns, 1979; Schwab, Bruns, and von Huene, 1980).

The structure of most of the western segment is dominated by broad synclines and tightly appressed anticlines that trend generally northeastward. Widths of individual structures range from about 4 to 10 km, and closure on the lower Yakataga horizon along strike is present for distances of 15 to 40 km. Anticlines are generally asymmetric and doubly plunging and are commonly bounded on the seaward side by high-angle thrust faults. Dips on the flanks of the anticlines commonly range from 5° to 45° on the landward side to vertical or overturned on the seaward side. Some folds have sharp, but unfaulted, flexures on the south limb suggestive of deformation above a structural ramp at depth (for example, section B–B′, Figure 9). The amount of deformation is greatest onshore and in the northern part of the shelf; it apparently decreases seaward (compare sections B–B′ and C–C′ to offshore sections 406, 407, 409 and 410, Figure 9). This difference is probably a function of the deeper structural level exposed in the onshore folds. Structural shortening perpendicular to the regional strike at surface level is estimated to be about 45% in the Paleogene sequence and <25% in the late Cenozoic units (Plafker, 1974). Exploration of these structures is complicated by repetition of the section due to thrusting, as was noted in several wells drilled in the Yakataga area (Rau, Plafker, and Winkler, 1983) and probably also in the Shell OCS Y-0011 offshore well (no. 3, Figure 1; and Table 2). Abnormally high fluid pressures encountered at depth in all offshore holes in this belt, together with bore-hole deformation, reflect northwest-southeast-directed lateral stresses (Hottman, Smith, and Purcell, 1979). High fluid pressures would probably facilitate large-scale bedding-plane detachment thrusts at depth, as depicted schematically in section D–D′, Figure 3.

It is not known how far southwestward the Paleogene sequence extends beyond the OCS wells on the continental shelf. Dredge samples along the south slope of Khitrov Ridge and along the slope between Khitrov Ridge and the Kayak Island zone are mainly slightly-indurated to well-indurated siliciclastic sediments and calcareous sediments that are probably equivalent to the Yakataga Formation on the basis of lithology (nos. 41–47, Figure 1, Table 1). Paleontologic data indicate early Pliocene through Pleistocene ages for those samples that can be conclusively dated. Dating of these samples has been a problem, because the hardest samples are unfossiliferous sandstone and limestone, and because some of the samples with microfossils contain abundant reworked Paleogene foraminifers and palynomorphs with pristine preservation. A few of the samples previously reported as Paleogene based on preliminary and incomplete paleontologic data (Plafker et al., 1979) are now known to be Pliocene or younger, and it is not certain that any of these samples are older than Pliocene. Strontium-isotopic analysis of the calcareous dredge samples in progress at the time this chapter was written may resolve the age of these rocks.

At least some of the Neogene samples on Khitrov Ridge appear to be accreted trench deposits that have been subjected to unusually rapid lithification, possibly due to tectonic deformation related to plate convergence in this area (Plafker et al., 1978b). Pelagic sediments have not been recovered, and even the relatively pure carbonate rocks contain reworked microfossils that indicate a terrigenous sediment source. Conceivably, all of Khitrov Ridge may be underlain by accreted deep-sea-floor deposits as suggested by Bruns (1985), and much or all of the continental slope between Khitrov Ridge and the Kayak Island zone (southwest of the projection of the Transition fault system) may also consist of accreted sediment.

Along the northern and western margins of the western segment of the Yakutat terrane, there is a significant increase in the intensity of folding and magnitude of fault displacements (sections D–D′ and E–E′, Figure 10). This complexity is attributed to deformation of the bedded sequence of the Yakutat terrane against the relatively resistant "backstop" of the Prince William terrane. On Kayak Island, the entire sequence was deformed virtually as a unit after the middle Miocene and prior to emplacement of a 6.2 Ma dacite plug into the sequence (Plafker, 1974; Nelson, Dumoulin, and Miller, 1985).

Rogers (1977) noted that seismic profiles show a sequential development of folds resulting from a "front" of deformation that migrated southeastward from the Kayak Island area to the Pamplona zone during the development of this late Cenozoic structural belt. Additional detail on the pattern and development of the offshore structures is given by Bruns and Schwab (1983). Ongoing deformation of structures within this segment of the terrane is indicated by (1) earthquake-related uplift of late Holocene marine terraces along the south flank of the Sullivan anticline west of Icy Bay, at an average rate of 1 cm/yr (Plafker et al., 1982b), (2) geodetically detectable horizontal deformation in the same region (Savage and Lisowski, 1985, unpublished data); (3) deformation of well bore holes (Hottman, Smith, and Purcell, 1979); and (4) seismic activity (Lahr and Plafker, 1980; Perez and Jacob, 1980).

Kayak Island Zone

The boundary between the Yakutat terrane and the Prince William terrane to the west is the Kayak Island zone, which consists of the onshore Ragged Mountain and Wingham Island thrust faults (Plafker, 1974; Winkler and Plafker, 1981) and their offshore extensions southwest of Kayak Island (Figure 1). Both the Ragged Mountain and Wingham Island faults juxtapose intensely deformed Paleogene (probably Eocene) marine sedimentary rocks on the east against intricately deformed and slightly metamorphosed pillow basalt and associated deep-sea sedimentary rocks of the Orca Group (section E–E′, Figure 10). The inferred location of the contact offshore along the north side of Kayak Island is approximately defined by the position of a large positive aeromagnetic anomaly associated with the Orca volcanic rocks. The southwestern extension of this contact across the continental shelf and slope is poorly delineated and is inferred to coincide with a broad zone of acoustic basement on the seismic-reflection records (section 417, Figure 9).

Eastern Part of Prince William Terrane

The eastern part of the Prince William terrane as defined here extends from the Kayak Island zone westward to Hinchinbrook Seavalley and southward to the Aleutian megathrust at the base of the continental slope. The onshore and continental shelf parts of this area, and perhaps part of the continental slope, are underlain by the Orca Group, which constitutes acoustic basement on the seismic-reflection profiles. In this segment, the basinal sequence is known primarily from the Middleton Island well (no. 1, Figure 1 and Table 2; Figure 5), which penetrated approximately 4000 m of section and bottomed in Eocene strata (Rau, Plafker, and Winkler, 1983; Keller et al., 1984), and from about 1100 m of Quaternary strata exposed on Middleton Island. In addition, dredging along the lower continental slope indicates that both indurated to moderately well indurated siltstone and sandstone of Paleogene(?) age (nos. 50, 51, 53, Figure 1; Table 1) and poorly consolidated late Cenozoic siltstone and mudstone are exposed at the surface (nos. 50–53, Figure 1; Table 1). The more indurated strata contain Paleogene palynomorphs and one assemblage of late Eocene foraminifers; however, the possibility cannot be discounted that these microfossils are reworked. The younger strata contain Pliocene and Quaternary foraminifers and siliceous microfossils and abundant reworked Paleogene palynomorphs. Gently dipping reflector horizons on seismic profile 425 suggest that the Paleogene(?) strata sampled on the slope may correlate with the sequence penetrated in the lower part of the Middleton Island well (section F–F′, Figure 10).

Seismic data indicate that the upper Cenozoic section beneath the shelf is relatively thin but variable in thickness (Bruns, 1979). The structure on a seismic horizon inferred to be in Pleistocene strata of the Yakataga Formation is shown on Figure 1, and sections illustrating the deformational style are shown in Figure 10. Anticlines are characteristically tight and extensively faulted and have irregular, but predominantly east-west to northeast-southwest, trends. Multiple minor unconformities and growth features within the upper Cenozoic section indicate that local deformation was contemporaneous with deposition.

Middleton Island is located on a shelf-edge structural high. The structure in this area is of particular interest, because it is the one part of the basin where a thick, gently dipping section can be observed on seismic-reflection profiles. Middleton Island has undergone Holocene northwestward tilting of as much as 28°, faulting, and earthquake-related tectonic uplift at a rate that averages close to 1 cm/yr and continues to the present time (Plafker, 1969; Plafker and Rubin, 1978). However, paleontologic and dipmeter data suggest that the section in the nearby well is relatively uncomplicated by steep dips or faulting, and the section can be correlated with seismic horizons that dip gently seaward on seismic profile 425. These relationships suggest that there may be a fault between Middleton Island and the well.

The anomalously modest deformation of the basinal sequence in this area of high tectonic activity may be attributable to the presence of a relatively rigid undeformable crust composed of the Orca Group beneath the basinal strata. Finally, the apparent absence of comparable deep reflectors in seismic profiles on the adjacent shelf suggests that the thick post-Orca basinal sequence is probably largely limited to the slope and that it may wedge out onto the Orca Group close to the shelf edge (section A–A′, Figure 3).

Transition Fault System and Eastern End of Aleutian Megathrust

The boundary of the Pacific plate with the eastern and central segments of the Yakutat terrane is the Transition fault system; and the boundary between the Pacific plate and both the western segment of the Yakutat terrane and the Prince William terrane is the Aleutian megathrust (Figures 1, 2).

The Transition fault system truncates the pre-Tertiary basement of the eastern segment of the Yakutat terrane, the Paleogene oceanic crust, and overlying clastic sequence of the central segment. This truncation, together with numerous slickensided rocks in the dredge samples (Table 1, nos. 1, 15, 16, 31–33, 39, and 40), suggests large-scale displacement along the Transition fault system—displacement that has affected strata of Eocene, Oligocene, and late Pliocene or younger age. As noted previously, truncation of the Paleogene basaltic basement of the central segment of the Yakutat terrane and underthrusting by the Pacific plate is believed to be the most probable source of the slope magnetic anomaly. As shown in Figure 1, apparent northwestward extension of the slope magnetic anomaly beneath the western segment of the Yakutat terrane and part of the Prince William terrane suggests a minimum of 220 km of subduction parallel to the Transition fault system (Case et al., 1979; Schwab and Bruns, 1979; Schwab, Bruns, and von Huene, 1980). This amount of subduction represents 3.6 m.y. of displacement parallel to the Transition fault system at present convergence rates and directions as given by Engebretsen (1982). The magnetic anomaly has been modelled as a basalt slab with its top at a depth of about 15 km south of Montague Island and a thickness of approximately 3 km at its southern margin as depicted in section A–A′, Figure 3 (Andrew Griscom, 1980, written communication). The present orientation of the Transition fault system relative to Pacific plate motion (Figure 2), to the orientation of the folds and faults adjacent to the fault system southwest of Fairweather Ground, and the focal mechanism solutions for the Cross Sound earthquakes along the fault trace near Yakobi Valley (Page, 1975) all suggest dextral oblique slip on the Transition fault system during the last 5 m.y. The folding of the late Quaternary strata, the seismicity, and the occurrence of sheared strata of probable late Pliocene to Holocene age along this margin are incompatible with the suggestions that the Transition fault has been inactive since the Miocene and that the segment of the Transition fault seaward of Fairweather Ground is probably a normal fault (Bruns, 1985, p. 28, 46).

As used herein, the Aleutian megathrust system of faults terminates at the east end of Khitrov Ridge near the eastern end of the Aleutian Trench as a topographic feature (Figures 1, 2). It extends northeastward across the continental slope and shelf as the Pamplona zone of thrust faults and related folds. Dextral offsets of 30–40 km of the structural trend of the Aleutian megathrust and Pamplona zone deformational fronts both east and west of Khitrov Ridge are inferred to reflect transcurrent faulting roughly parallel to the Transition fault system (Figure 1). Along seismic line 425 (Figure 10), the entire sedimentary sequence on the Pacific plate is underthrust beneath the Prince William terrane at least 14 km without any evidence of deformation at the surface (Plafker et al., 1982a; Plafker and Bruns, 1982). Insufficient acoustic penetration to the east precludes determination of whether comparable subduction is presently occurring either along the south flank of Khitrov Ridge or along the Transition fault system. Landward tilting of probable Quaternary strata in Khitrov Basin to the north of the ridge (Figure 9, section 922), suggests recent growth of that structure either by underplating of oceanic sediments beneath the structure or by imbrication along north-dipping thrust faults within the structure.

Deep-Sea Floor and Slope Rise

Data on the deep-sea floor and slope rise consist of widely spaced multichannel seismic lines, refraction data, and moderately abundant single-channel reflection data (Figure 7; Shor, 1965; von Huene, Shor, and Wageman, 1979; Plafker et al., 1978b, 1979, 1980; Bruns, 1985).

Sediment thickness near the base of the slope, indicated by horizon A3 on seismic sections, ranges from 6 km off the eastern segment of the Yakutat terrane (Figure 8) to 2–4 km westward to Khitrov Ridge (section 922, Figure 9). Along section 425 south of Middleton Island, depth to probable basaltic crust is 3 $^1/_2$ km (Figure 10). Shor (1965) described the part of the margin south of the Fairweather Ground, where the sedimentary section is >6 km thick, as a filled trench that would be comparable in depth with the Aleutian Trench if the sediment were removed. In my judgment, the pronounced local thickening of sediment in the vicinity of Yakobi Valley most probably is due to a flood of predominantly late Pleistocene sediment deposited by the trunk glacier system that flowed out of Cross Sound during glacial maxima. However, there are as yet no data on the age of the deposits to test this inference. The strata along the slope base thin seaward, and are <2 km thick over much of the abyssal plain farther than about 60 km from the base of the slope (von Huene, Shor, and Wageman, 1979).

The sediment wedge is probably composed of pelagic and hemipelagic deposits in the basal 1–2 km of the sequence, which are overlain near the base of the slope by thick turbidites and hemipelagic deposits containing abundant glacial detritus. The abyssal-plain sediments are younger than about middle to late Oligocene as indicated by the age of the closest ocean-floor magnetic anomalies (Naugler and Wageman, 1973). In the absence of direct sampling data, seismic-reflection horizons in Figures 8–10 have been assigned ages by correlation through the sparse seismic grid to DSDP hole 178, which is 150–500 km distant from the map area (von Huene, Shor, and Wageman, 1979; Bruns and Plafker, 1982; Bruns, 1985). The only structures detected in the deep-sea sediments are along the eastern segment of the Yakutat terrane, where a series of folds and probable thrust faults involving the youngest part of the sequence trend N60°–70°E (sections 967 and 909, Figure 8; Bruns, 1985). These structures reflect oblique compression across this segment of the Transition fault system. Atwood et al. (1981) noted that a subsea fan of unknown age from the mouth of Yakutat Valley has an apparent dextral offset of at least 50 km, an offset that could have resulted from displacement along the Transition fault system.

TECTONIC EVOLUTION OF THE NORTHERN GULF OF ALASKA CONTINENTAL MARGIN

The data presented in this chapter indicate a complex Mesozoic and Cenozoic tectonic evolution for the northern Gulf of Alaska—an evolution that continues to the present. No single model has yet been devised that satisfactorily explains all the geologic, paleontologic, geophysical, and paleomagnetic data, because some of these data are mutually contradictory and/or internally inconsistent. The following discussion of the tectonic evolution of the continental margin emphasizes the geologic data because they are the data with which I am most familiar and because they have the fewest uncertainties.

Major Constraints

Some of the more important constraints that must be considered in modelling the evolution of this margin are:

1. The late Mesozoic and Paleocene(?) continental margin had a relatively linear northwest trend, as indicated by the configuration and displacement history of major interior fault systems (Grantz, 1966), structural patterns in Mesozoic accreted rocks on both limbs of the bend (Moore and Connelly, 1977), and the distribution of accreted deep-sea sequences of Mesozoic and early Cenozoic age. Additionally, most, but not all, of the paleomagnetic data for Mesozoic and Paleocene rocks in western Alaska suggest counterclockwise rotations (Coe et al., 1985).
2. The Upper Mesozoic melange facies rocks of the eastern Yakutat terrane correlate with a closely comparable sequence in southeastern Alaska (Kelp Bay Group) as shown in Figure 12. Exotic blocks in the melange facies of the eastern Yakutat terrane require a Wrangellian source terrane comparable with that of the Queen Charlotte Islands (Sutherland Brown, 1968) and Vancouver Island (Muller, 1977). These blocks include Jurassic tonalite (Island? Intrusions), Sinemurian volcanogenic sandstone (Kunga? Group), and Upper Triassic carbonate (Quatsina? Limestone) associated with greenstone pillow basalt (Karmutsen? Formation). No suitable source terrane for this combination of exotic blocks is known south of Puget Sound. Furthermore, systematic regional compositional changes in the sandstone of the Upper Cretaceous flysch of the Chugach and Yakutat terranes (Zuffa, Nilsen, and Winkler, 1980) suggest a similar provenance, but with the Yakutat terrane to the southeast of the presumably coeval sequences in southeastern Alaska. These data, together with the conspicuous gap in the Mesozoic belt between Chatham Strait and Vancouver Island, require derivation of at least the eastern segment of the Yakutat terrane from somewhere along the continental margin of southeastern Alaska or British Columbia. The Kulthieth Formation overlap assemblage and correlative marine units tie together the entire Yakutat terrane by the middle Eocene.
3. The Ghost Rocks terrane was offscraped, accreted to the continental margin, and intruded by plutons by about 62 Ma (Moore et al., 1983).
4. The Prince William terrane was offscraped, accreted to the continental margin, and intruded by plutons by about 51 Ma (Plafker et al., 1985).
5. Counterclockwise oroclinal bending of southwestern Alaska is mainly post-Cretaceous and continued through at least the middle Eocene, based on deformation of accreted rocks in the Chugach (Moore and Connelly, 1977) and Prince William terranes (Winkler and Plafker, 1981; Plafker et al., 1986) and paleomagnetic evidence for counterclockwise rotation of the Ghost Rocks terrane (Grommé and Hillhouse, 1981; Plumley, Coe, and Byrne, 1983).
6. All terranes inboard of the Yakutat terrane (i.e., Prince William, Ghost Rocks, Chugach, Wrangellia, Peninsular) were essentially sutured together by about 51 Ma (Winkler and Plafker, 1981; Davies and Moore, 1984; Plafker, Nokleberg, and Lull, 1985).

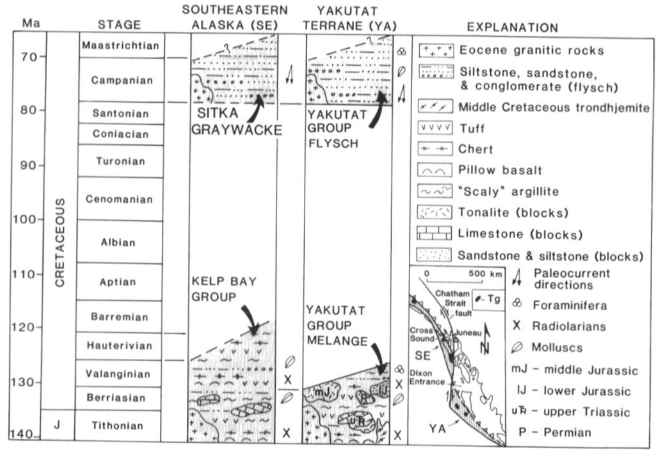

Figure 12. Columnar sections showing similarities between upper Mesozoic sequences in the eastern Yakutat terrane (Yakutat Group) and in southeastern Alaska (Kelp Bay Group and Sitka Graywacke). The inset map shows the preferred reconstructed position of the upper Mesozoic sequence of southeastern Alaska and the Yakutat terrane after offset along the Chatham Strait fault and before northward displacement of the Yakutat terrane. Time scale is from Harland et al. (1982).

7. Sutures between the Prince William, Chugach, and Wrangellia-Peninsular terranes have had no large post-emplacement dextral slip (Plafker, Nokleberg, and Lull, 1985; Plafker et al., 1986).
8. Wrangellia and all terranes attached to it were at their present latitude relative to the craton by about 52 Ma (Hillhouse, Grommé, and Csejtey, 1985).
9. Andesitic volcanism on the Alaska Peninsula from about 50 ± 2 Ma (Wilson, 1985; Scholl, Vallier, and Stevenson, this volume) to the present records the location and duration of underthrusting of the Pacific plate beneath the western Gulf of Alaska; comparable volcanism in the Wrangell Mountains from about 20 Ma (J. A. Smith, 1986, oral communication) to the present records underthrusting of the Yakutat terrane beneath the northern Gulf of Alaska continental margin.
10. The Transition fault was the ridge-trench transform prior to about 25 Ma, after which both the Transition and Queen Charlotte/Fairweather fault systems shared the Pacific-North America ridge-trench relative motions (Plafker, 1983).
11. Paleogene sandstones of the Yakutat terrane are most compatible with a provenance in the Coast Mountains of British Columbia or southeastern Alaska (Hollister, 1979; Plafker et al., 1980; Chisholm, 1985).
12. Displacement of the Yakutat terrane about 600 km northwestward with about 20° counterclockwise rotation relative to the continental margin is inferred from correlation of the Dangerous River zone with the Chatham Strait fault (Plafker, 1983, 1984).

Paleomagnetic data indicating as much as 40° northward displacement of the Ghost Rocks terrane relative to the craton since 62 Ma (Plumley, Coe, and Byrne, 1983; Grommé and Hillhouse, 1981) and foraminiferal data suggesting that the Yakutat terrane has been displaced some 30° northward relative to the continental margin since the early Eocene (Keller et al., 1984) conceivably can be accommodated in this model if the entire margin of southern Alaska south of the Denali fault was involved in the displacements. However, there are no compelling geologic data that require northward displacements of this magnitude and no known faults along which displacements of this magnitude have occurred.

In this model, the Ghost Rocks terrane is assumed to have been at about its present latitude by 52 Ma for reasons cited in item 8 above. The interpretation that foraminifers in middle Eocene strata of the Yakutat terrane indicate a paleolatitude of 30° south of the present position is not incorporated in the model, because it appears to be incompatible with the conclusions of other paleontologic studies of the foraminifers and megafossils suggesting limited or no northward displacement of Paleogene strata (Rau, Plafker, and Winkler, 1983; Marincovich and McCoy, 1984; Wolfe and McCoy, 1984), and it is contradicted by the paleomagnetic data (Van Alstine, Bazard, and Whitney, 1985; Engebretsen, 1982).

A Suggested Model

The suggested Cenozoic evolution of the Gulf of Alaska margin is shown diagrammatically in Figure 13 and is outlined below. Relative plate motions are after Engebretsen (1982), except that the demise of the Kula plate, which is poorly constrained, is assumed to approximately coincide with widespread plutonism along the eastern margin of the Gulf of Alaska at about 51 ± 3 Ma (Plafker et al., 1985) and the onset of arc volcanism in the Alaska Peninsula at 50 ± 2 Ma. The present coordinate grid is shown in Figure 13 for reference only; it does not imply knowledge of the actual position of southern Alaska relative to the North American continental margin during the Paleogene. The time scale used is that of Harland et al. (1982).

62 Ma (Figure 13A)

Accretion of the Chugach terrane occurred south of the Border Ranges fault system (BRF) during the middle and latest Cretaceous (Plafker, Jones, and Pessagno, 1977) with incorporation of olistostromal exotic blocks of Wrangellian affinity into Yakutat Group melange. This was followed by accretion of the Ghost Rocks terrane, emplacement of plutons, and metamorphism by 62 Ma (Moore et al., 1983). The Upper Mesozoic Chugach terrane is inferred to have been originally coextensive with the lithologically similar Pacific Rim and Leach River complexes of southern Vancouver Island (Fairchild and Cowan, 1982), and the gap between these complexes and Chatham Strait was in part filled by the eastern Yakutat terrane. According to this model, the Chugach terrane between Chatham Strait and Cross Sound was part of the same belt as the Leach River complex, but it is not the offset equivalent of the Leach River complex as postulated by Cowan (1982). Post-Cretaceous dextral displacements on the Peril Strait fault, which constitutes the landward margin of the Chugach terrane north of Chatham Strait, is about 11 km (Plafker et al., 1976), two orders of magnitude less than is required for these rocks to have been offset from the Leach River complex.

62 to 52 Ma (Figure 13B)

The main phase of uplift of the Coast Mountains of British Columbia and Alaska (Hollister, 1979) and counterclockwise oroclinal bending of western Alaska occurred during this interval. Deep-sea fans formed on oceanic crust and were displaced northwestward relative to the continental margin (Helwig and Emmett, 1981). Progressive accretion of the sedimentary and oceanic volcanic rocks that comprise the Orca Group of the Prince William terrane occurred south of the Contact fault.

52 to 48 Ma (Figure 13C)

Subduction of the Kula plate with a rapid change in relative motion of the oceanic plate from northeastward to northwestward occurred during this interval. Approximately 180 km of dextral offset is inferred on the Chatham Strait fault. Accretion of the Prince William terrane south of the Contact fault was completed. There was a major episode of plutonism and metamorphism throughout the Chugach and Prince William terranes east of the Kodiak Island area, possibly related to subduction of the Kula-Farallon Ridge (Marshak and Karig, 1977). Plutons sutured across the Contact and Border Ranges fault systems. Minor oroclinal bending continued that affected the Prince William terrane. Deposition of lower to middle Eocene marine clastic deposits from a crystalline-complex source with intermixed carbonate-reef detritus took place on predominantly oceanic crust of the Yakutat terrane.

48 to 25 Ma (Figure 13D)

Andesitic volcanism on the Alaska Peninsula associated with the Aleutian arc began at about 48 Ma. The Transition fault was formed as a ridge-trench transform system across the northeastern Gulf of Alaska, and the main deposition of the Paleogene alluvial fan-delta sequence took place on trapped oceanic crust northeast of the Transition fault system. Deposition of the locally derived Paleogene basinal sequence overlying the Orca Group occurred in the eastern part of the Prince William terrane. Possible middle to late Eocene dextral offset of up to 1000 km took place on intracontinental transcurrent faults (Gabrielse, 1985). Sediments of interior and shelf-margin basins, and possibly the Zodiac deep-sea fan, were derived mainly from the crystalline complex of the uplifted Coast Mountains.

25 to 20 Ma (Figure 13E)

The Queen Charlotte-Fairweather dextral transform fault was initiated to create the Yakutat terrane. Andesitic arc volcanism began in the Wrangell Mountains at 20 Ma following subduction of about 225 km of the Yakutat terrane beneath the Chugach-Saint Elias fault system and the Kayak Island zone. Slow clastic sedimentation took place along the continental margin.

20 to 0 Ma (Figure 13F)

About 1200 km of northwestward motion of the Pacific plate relative to the continental margin was shared approximately equally between the Queen Charlotte-Fairweather transform and the Transition fault system. Transpression along the northern margin of the terrane resulted in extreme uplift of the Chugach Mountains and Fairweather Range with concurrent deposition of thick, locally derived upper Cenozoic clastic sequences of the Yakataga and Redwood Formations in continental margin basins on the Yakutat and Prince William terranes and on the deep-sea floor. The progressively younger ages of coarse clastic deposits from west to east in the Yakutat terrane reflects the northwestward motion of the terrane relative to the uplifted

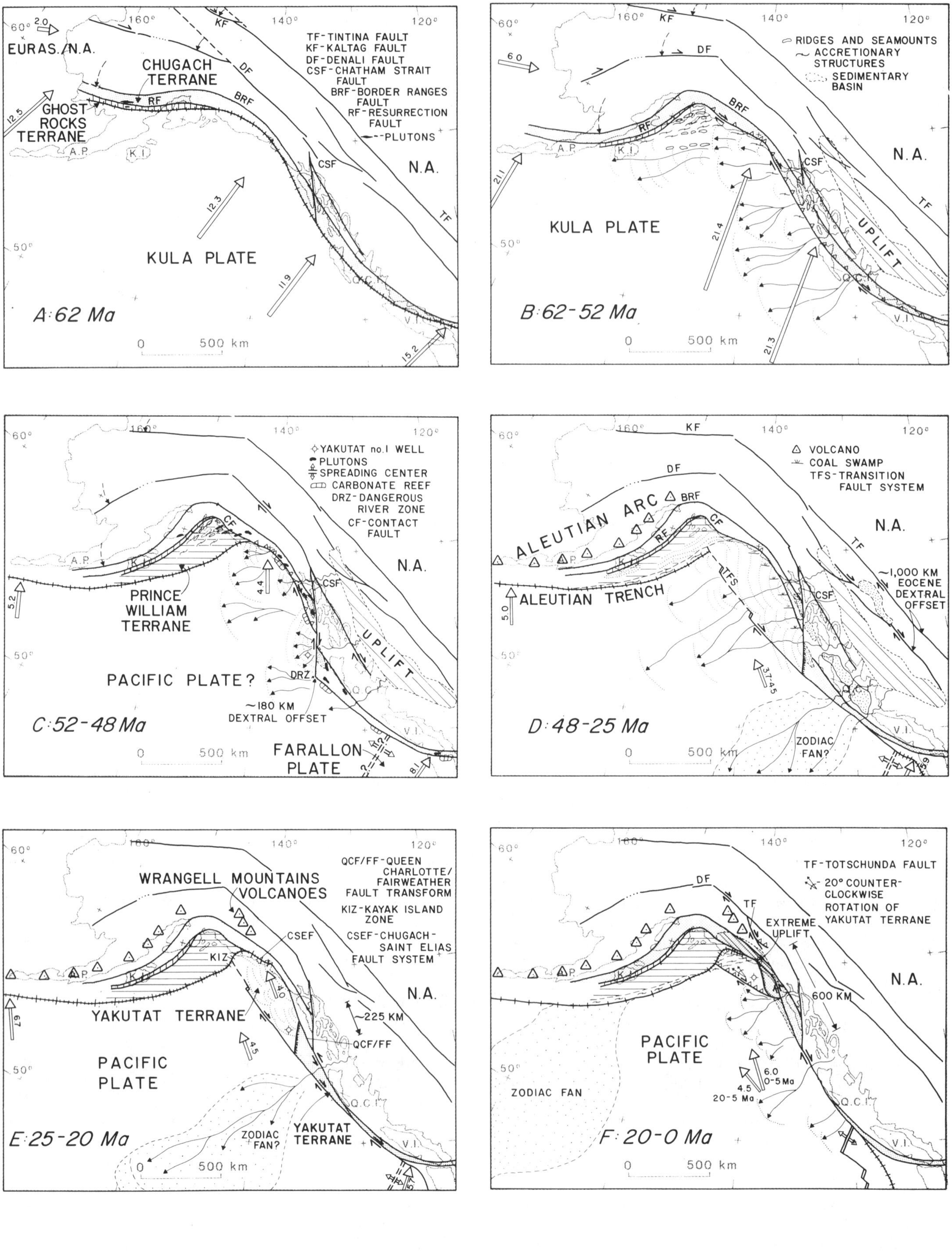

A: 62 Ma
EURAS./N.A.
CHUGACH TERRANE
GHOST ROCKS TERRANE
KULA PLATE
N.A.
TF-TINTINA FAULT
KF-KALTAG FAULT
DF-DENALI FAULT
CSF-CHATHAM STRAIT FAULT
BRF-BORDER RANGES FAULT
RF-RESURRECTION FAULT
PLUTONS
0 500 km
B: 62-52 Ma
RIDGES AND SEAMOUNTS
ACCRETIONARY STRUCTURES
SEDIMENTARY BASIN
KULA PLATE
UPLIFT
N.A.
0 500 km
C: 52-48 Ma
YAKUTAT no.1 WELL
PLUTONS
SPREADING CENTER
CARBONATE REEF
DRZ-DANGEROUS RIVER ZONE
CF-CONTACT FAULT
PRINCE WILLIAM TERRANE
PACIFIC PLATE?
UPLIFT
~180 KM DEXTRAL OFFSET
FARALLON PLATE
N.A.
0 500 km
D: 48-25 Ma
VOLCANO
COAL SWAMP
TFS-TRANSITION FAULT SYSTEM
ALEUTIAN ARC
ALEUTIAN TRENCH
~1,000 KM EOCENE DEXTRAL OFFSET
ZODIAC FAN?
N.A.
0 500 km
E: 25-20 Ma
WRANGELL MOUNTAINS VOLCANOES
QCF/FF-QUEEN CHARLOTTE/FAIRWEATHER FAULT TRANSFORM
KIZ-KAYAK ISLAND ZONE
CSEF-CHUGACH-SAINT ELIAS FAULT SYSTEM
YAKUTAT TERRANE
~225 KM
PACIFIC PLATE
ZODIAC FAN?
N.A.
0 500 km
F: 20-0 Ma
TF-TOTSCHUNDA FAULT
20° COUNTER-CLOCKWISE ROTATION OF YAKUTAT TERRANE
EXTREME UPLIFT
600 KM
PACIFIC PLATE
ZODIAC FAN
6.0 0-5 Ma
4.5 20-5 Ma
N.A.
0 500 km

Figure 13. Diagrams illustrating the inferred tectonic evolution of the northern Gulf of Alaska continental margin. Large open arrows and numerals indicate directions and rates of plate movement; solid arrows indicate relative fault displacement; dashed arrows show inferred direction and amount of oroclinal bending of western Alaska. Time scale used is that of Harland et al. (1982); relative plate motions from Engebretsen (1982). Geographic reference localities are: A. P., Alaska Peninsula; K. I., Kodiak Islands; Q. C. I., Queen Charlotte Islands; and V. I., Vancouver Island. The main points illustrated by the figures are described below; see text for details and data sources. A, Pre-bending configuration of southern Alaska margin showing Chugach terrane (shaded); counterclockwise oroclinal bending of SW Alaska; accretion of Ghost Rocks terrane and plutonism in Ghost Rocks and Chugach terranes of the Kodiak Islands area. B, Continued oroclinal bending of SW Alaska; uplift of Coast Mountains of British Columbia and Alaska with rapid sedimentation on continental margin and deep-sea floor; progressive northward displacement and accretion of Prince William terrane deep-sea deposits to northern and western Gulf of Alaska margin. C, Intrusion and metamorphism of accreted Prince William terrane and Chugach terrane along northern and eastern Gulf of Alaska margin; about 180 km offset of continental margin and Chugach terrane along Chatham Strait fault (CSF) to form the proto-Dangerous River zone (DRZ); probable completion of main oroclinal bending; local accumulation of sediments, including carbonate reef detritus, along continental margin. D, Initiation of northwestward relative motion of Pacific plate; beginning of arc volcanism on Alaska Peninsula and formation of the Transition fault system (TFS) as a ridge-trench transform; continued rapid sedimentation along continental margin and on ocean floor; possible large-scale dextral offset on Tintina fault (TF) and related faults of interior British Columbia; possible accumulation of Zodiac fan deposits on oceanic crust off southern British Columbia. E, Initiation of the Queen Charlotte/Fairweather fault system (QCF/FF) as a transform that shares relative Pacific-North America (N.A.) displacement with the TFS; formation of the Yakutat terrane situated between the QCF/FF and TFS; onset of Wrangell Mountains volcanism after about 225 km of northwestward subduction of the Yakutat terrane. F, Continued northward displacement, totalling 600 km for the Yakutat terrane, with about 20° counterclockwise rotation; approximately 600 km dextral oblique displacement along TFS; extreme uplift due to transpression along northern and northeastern margins of the Yakutat terrane; major pulse of clastic sedimentation on continental margins and deep-sea floor, culminating in the Pliocene and Quaternary.

coastal mountains. Folding and thrusting occurred along the northern and western margins of the Yakutat terrane, particularly between the Pamplona and Kayak Island zones, and progressed southward and eastward with time. Underthrusting and large-scale subduction of ocean crust and the overlying sediments took place along both the Aleutian megathrust and the Kayak Island zone, but there was relatively minor compressional deformation within the eastern Prince William terrane. Dextral displacement began on the intraplate Totschunda fault with concurrent reactivation of parts of the Denali fault system. The Yakutat terrane rotated counterclockwise approximately 20° relative to the continental margin as it moved into the bend of the northern Gulf of Alaska.

Discussion

The suggested model is compatible with most of the available geologic and paleontologic data, with the paleomagnetic data from the Yakutat no. 1 well, and with plate reconstructions for the northeast Pacific region. Some unresolved problems are:

1. The Orca Group locally contains channel-fill conglomerate that includes sparse orthoquartzite and abundant metarhyolite tuff for which the closest known source-terrane is the North American craton of southern California or Mexico. These channel-fill deposits suggest the possibility that the Orca Group had a mixed provenance, part of the sediments having been carried long distances northward on the Kula plate, and the bulk of it having been derived from the Coast Mountains of British Columbia and southeastern Alaska. However, a bimodal source implies significant differences in age for sediments from the two areas, a difference that is not apparent in the sparse paleontologic data for the Orca Group.
2. The Eocene and younger basinal strata penetrated in the Middleton Island well and interpreted on seismic-reflection profiles are surprisingly undeformed considering their location above an active subduction zone. Depending upon the position of the well relative to the intersection of the Transition fault transform with the eastern end of the Aleutian megathrust, reconstructions of plate motions require at least 1000 km, and possibly >2000 km, relative convergence since the middle Eocene (Figure 13). The only explanation I can offer for these relationships is that the relatively rigid and buoyant "continentalized" crust of the Orca Group beneath the basinal sequence has inhibited deformation of the upper plate. An argument against this interpretation is that Pleistocene strata are tilted as much as 28° at Middleton Island, and there is clear evidence for young deformation within the younger part of the Yakataga Formation on seismic profiles in this segment of the shelf.
3. The model predicts about 600 km of relative subduction along the Kayak Island zone. However, as noted by Rogers (1977), there is neither a greatly thickened sequence within the fold-thrust belt between the Kayak and Pamplona zones, as would be expected for large-scale offscraping and imbrication, nor is there a detectable gravity anomaly west of the Pamplona zone indicating a large volume of underthrust low-density sediment. Depending upon assumptions made regarding the thickness of the Paleogene section fed into the subduction zone and depth to a basement decollement, offscraping and structural shortening in the fold-fault belt may be as much as 150–200 km. The data, however, suggest deep subduction of the remaining 400–450 km of sediments that were carried into the subduction zone on the Yakutat terrane, as depicted diagrammatically in section D–D′ on Figure 3. This interpretation is required by the evidence for at least 220 km of subhorizontal subduction of the slope magnetic anomaly, together with refraction data that preclude significant thickening of subducted Paleogene strata in the western segment of the Yakutat terrane. Comparable subduction of oceanic crust and the overlying sedimentary sequence is now occurring at the eastern end of the Aleutian Trench (Plafker et al., 1982a). High fluid pressures are undoubtedly involved, but the mechanism for subhorizontal subduction of large volumes of sediments is not at all clear. It is noteworthy that this problem is more than five times as great for the much larger relative displacements deduced from study of planktonic foraminifers (Bruns, 1983a; Keller et al., 1984).
4. According to the model, the segment of the Yakutat terrane underlain by Mesozoic crust to the east of the Dangerous River zone, when restored to its inferred original position south of Chatham Strait, is about 350 km too short to fill the entire gap in the Mesozoic belt between Chatham Strait and Vancouver Island. The unaccounted-for Mesozoic rocks include those that were originally contiguous with the Pacific Rim and Leech River complexes of Vancouver Island (Cowan, 1982). Possible explana-

tions for the missing crust include any or all of the following: (1) subduction beneath the continental margin of southeastern Alaska and British Columbia prior to formation of the Yakutat terrane, (2) incorporation of at least some of this sequence in the continental shelf and slope seaward of the Queen Charlotte-Fairweather transform, and (3) subduction beneath the eastern segment of the Yakutat terrane along the Transform fault system.

5. Although the model requires about 600 km of late Cenozoic dextral displacement along the Queen Charlotte-Fairweather transform fault system, matching of drainage offsets across the Fairweather fault suggests limited (5–6 km) slip (Plafker et al., 1978a). The idea of limited displacement, which was based on the inference that major drainages were antecedent to uplift of the Fairweather Range during the Miocene, is undoubtedly incorrect; it is now considered likely that the offset drainages are much younger features, probably no older than the Sangamon (interglacial) stage of the Pleistocene.
6. Paleomagnetic data from lower and middle Eocene strata in the Yakutat no. 1 well indicate $13 \pm 3°$ northward displacement of the Yakutat terrane relative to the craton—a paleolatitude about opposite the Oregon-Washington border (Van Alstine, Bazard, and Whitney, 1985). However, if dextral displacement of >1000 km on intracontinental transcurrent faults (Gabrielse, 1985) was dominantly in the middle Eocene or later, the combined northward component of displacement on such faults (approximately 8°) and the northward component of displacement of the Yakutat terrane along the Queen Charlotte-Fairweather fault system (5°) would be virtually identical to the 13° displacement relative to the craton indicated by the paleomagnetic data. This apparent agreement between the fault offset data and the paleomagnetic data is undoubtedly coincidental. Nevertheless, both lines of evidence are entirely compatible with the independently derived provenance data for the Mesozoic and Paleogene sequences, which suggest that the Yakutat terrane originated along the continental margin between Chatham Strait and Puget Sound.

A consequence of approximately 600 km northwestward displacement and subduction is that less than half of the Yakutat terrane remains. Subduction of this remaining part of the terrane, which is caught in the junction between the Aleutian megathrust and the Queen Charlotte-Fairweather transform fault system, becomes increasingly difficult because it involves the relatively thick and indigestible eastern segment of the Yakutat terrane, which is underlain by Mesozoic "continental" crust. The complex and extremely active deformation that characterizes the late Cenozoic history of the northern Gulf of Alaska will continue until the remaining part of the Yakutat terrane is either completely subducted or is accreted to the continental margin.

PETROLEUM GEOLOGY

Rocks that have source and reservoir characteristics suitable for petroleum accumulation are known to occur only in the Paleogene basinal sequence of the central and western segments of the Yakutat terrane. The basinal sequences of the eastern Yakutat and Prince William terranes, and sediment on the deep-sea floor and rise, are considered to have little potential for petroleum in commercial quantities. This judgment is based primarily on the fact that the strata are largely thermally immature, and/or are too low in kerogen to be suitable as a petroleum source. The thickest known Paleogene basinal sequence in any of these less promising areas is the section of Oligocene to late Eocene and older(?) strata penetrated in the Middleton Island well (no. 1, Figure 1; Figure 5). Levels of maturity in this well ranged from immature to early stages of oil generation with a maximum vitrinite reflectance (R_o) of 0.6–0.7%. Rocks at this level of maturity are unlikely to have had significant petroleum expulsion (unpublished report for the Alaska Division of Geological and Geophysical Surveys, 11/29/83). Besides the source-rock considerations, there are additional serious problems in all of these areas with respect to the availability of suitable structural traps and reservoir rocks. For these reasons, the petroleum potential of areas other than the central and western segments of the Yakutat terrane will not be considered further.

Exploration History

The presence of many oil and gas seeps first called attention to the petroleum possibilities of the northern Gulf of Alaska. The geologic setting and locations of deep exploratory wells are shown on Figure 1, and the three main areas of oil and gas seepage are shown on inset maps A–C. One minor shallow-depth field at Katalla was developed in a seep area and produced between 1902 and 1933. Twenty-five other wells and core holes were drilled and abandoned onshore between 1954 and 1963, and one offshore well was drilled and abandoned in 1969 on state lands near Middleton Island. Information on these wells is summarized in Plafker (1971), Plafker, Bruns, and Page (1975), and Rau, Plafker, and Winkler (1983).

During 1975–1976, one stratigraphic test and ten offshore wildcat wells were drilled and abandoned in OCS lease sale no. 39—an area located in the fold-fault belt in the western segment of the Yakutat terrane. The locations of these wells are shown on Figure 1, and drilling information is summarized in Table 2. OCS lease sale no. 55 was held in 1981 for the central segment of the Yakutat terrane, and a reoffering sale was held in 1981. A test well was drilled during 1983 on a large deep structure without encountering oil or gas in commercial quantity (Figure 1; Table 2). There has been no subsequent exploration in the Gulf of Alaska region, although the Alaskan Crude Corporation began drilling in 1986 to reactivate the Katalla field.

Source Rocks, Thermal Maturity, and Organic Geochemistry

This section is based mainly on data from dredge samples on the continental slope (Plafker and Claypool, 1979; Plafker et al., 1980), five onshore wells and the Middleton Island well (Palmer, 1976), six offshore wells (Alaska Division of Oil and Gas, unpublished report), and analyses of outcrop samples and onshore oil seeps.

Source-Rock Characteristics

Analysis of selected dredge samples from the continental slope in the central segment of the Yakutat terrane indicates that rocks with favorable hydrocarbon source-rock characteristics are present in both Eocene and Oligocene sections.

Organic matter in these samples is predominantly herbaceous, with subordinate amounts of humic kerogen and amorphous kerogen of the type that can generate both liquid hydrocarbons and gas. Total organic-carbon content of the analyzed samples ranges from 0.42 to 1.87% and averages >1.0% (Table 3). This average is above the generally accepted minimum value of 0.5% for argillaceous hydrocarbon source rocks (Tissot and Welte, 1978). Comparable analyses have been obtained onshore for Paleogene strata penetrated in five wells drilled along the coast (Palmer, 1976) and from samples of organic-rich argillaceous rocks of Eocene and Oligocene age in the Samovar Hills, on Kayak and Wingham Islands, and in the Katalla area (Table 3). Oligocene samples from Kayak Island are exceptionally rich in organic matter with as much as 6.76% organic carbon, and they have a higher amorphous kerogen content than the Eocene rocks that were analyzed.

In contrast, samples from the Neogene and Quaternary

Table 3. Organic-carbon and visual-kerogen assessment of selected dredge samples from the continental slope and outcrop samples from Kayak Island. Analyses by Geochem Laboratories, Inc.

Map No.	Sample No.[1]	C[2] (%)	Visual Kerogen (%)				TAI[4]	Ro[5]	HI[6]
			A[3]	He[3]	Hu[3]	I[3]			
1	79-16F	0.8	0	38	31	31	2.9	1.4	
	16G	1.5	0	40	30	30	2.9	2.1	
	16H	1.1	0	27	36	36	3.1	2.2	
4	78-22D2	0.8	0	71	29	0	2.4	0.6	139
	22D3	1.1	22	44	33	0	2.6	.7	126
	22D4	n.d.	0	67	33	0	2.5	.9	73
6	77-20B	1.0	0	20	40	40	3.6	1.9	
8	22B	1.4	0	83	17	0	2.3	0.4	
9	79-42D	1.2	11	56	33	0	2.3	0.5	
	42E	1.1	0	56	33	11	2.3	.5	
10	43A	0.9	0	55	27	18	2.1	0.4	
	43E	1.2	20	60	10	10	2.1	.4	
11	44A	n.d.	9	45	36	9	1.6	0.3	
12	37A	1.5	37	63	0	0	2.0	0.4	
	37B	1.3	0	67	33	0	2.3	.5	
	37C	1.2	0	56	44	0	2.3	.5	
13	38A	1.5	31	54	15	0	2.0	0.5	
	38C	1.5	27	55	9	9	2.2	.4	
14	39D	1.6	11	67	22	0	2.3	0.6	
	39F	1.3	50	50	0	0	2.1	.5	
	39G	1.1	30	50	10	10	2.1	.5	
15	78-43F	n.d.	27	36	36	0	2.3	.4	
16	44D	1.6	0	100	0	0	1.9	.3	
	44E	1.5	0	83	17	0	1.9	.3	
	44F	1.4	22	67	11	0	1.9	.3	
23	79-36A	1.3	40	60	0	0	2.2	0.3	
	36C	1.4	30	60	10	0	2.1	.3	
	36D	1.9	27	55	9	9	2.1	.4	
27	77-11B	1.1	0	86	14	0	2.1	0.4	112
28	10B	n.d.	0	71	29	0	2.1	0.5	59
29	79-24A	1.6	0	75	25	0	1.8	0.3	
	24B	1.4	11	56	33	0	1.9	.3	
	24C	1.5	10	50	30	10	2.1	.3	
30	77-17B	1.2	0	83	17	0	2.2	0.4	50
	17E	0.6	11	56	33	0	1.2	.2	
31	18A	0.6	0	56	44	0	1.2	.2	
32	79-20B	0.7	0	50	40	10	2.3	0.6	50
33	21E	0.6	0	44	44	11	2.3	0.6	
35	18A	0.5	0	22	33	44	4.0	4.6	

[1]Complete sample numbers are: 1979, S10-79-EG; 1978, S5-78-EG; 1977, S2-77-EG.
[2]Weight % organic carbon.
[3]Weight % visual kerogen; A-amorphous; He-herbaceous; Hu-humic; I-inertinite.
[4]Thermal alteration index.
[5]Mean vitrinite reflectance.
[6]Hydrogen index.

Yakataga Formation are characteristically low in organic carbon with values rarely >0.55%, and most in the range of 0.2–0.3%.

Thermal Maturity

The only known mature or marginally mature outcrop samples, either onshore or offshore, are from the pre-Oligocene section. Analyzed coals from the Kulthieth Formation (Barnes, 1967, and unpublished data) range in composition from subbituminous (46.8% fixed carbon) to anthracite (95.7% fixed carbon). These coals correspond in rank to vitrinite reflectance (R_0) of about 0.4–0.6% (immature to slightly immature) between Icy Bay and Yakutat Bay, to about 0.9–2.0% from Icy Bay to Bering Glacier (peak wet-gas generation to wet-gas floor), and to 2.0–2.5% in the Controller Bay coal field area west of Bering Glacier (below wet-gas floor). Analyzed Eocene dredge samples along the continental slope are mostly immature to slightly mature, some are close to optimally mature, and two are clearly overmature (Figure 14). The overmature samples could have undergone local baking by hypabyssal volcanic rocks. Data are not available on thermal maturity in the Yakutat #1 well, but findings in the well are likely to be comparable with those from dredge samples on the nearby slope. Calculations using the method of Lopatin indicate that 25–70% of the Paleogene section in the central segment of the Yakutat terrane could be mature enough to generate hydrocarbons, depending on assumptions made regarding geothermal gradient (Bruns, 1983b). Data from offshore wells in the western segment of the Yakutat terrane indicate that all units penetrated that are late Eocene and younger in age are immature.

Maturation of kerogen in the Paleogene strata probably results from an exceptionally high geothermal gradient at the time these strata were deposited. High heat flow is suggested by the close stratigraphic and temporal relationship

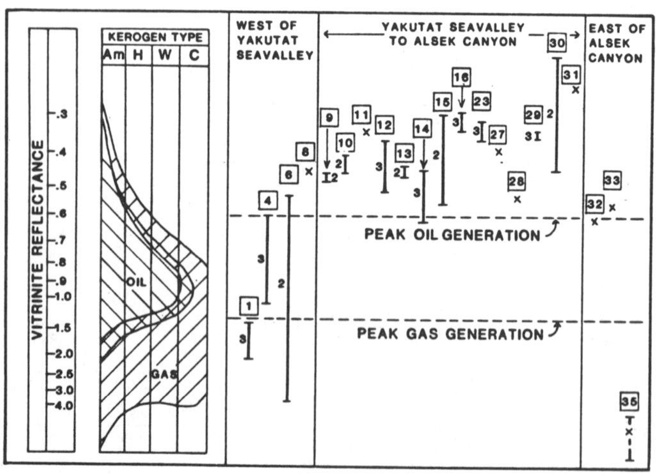

Figure 14. Vitrinite reflectance data for dredged samples from the continental slope. Sample numbers in squares correspond to those in Figure 11 and Table 3. Vertical bars show range of values for different samples in the same dredge haul; numeral beside bar indicates number of analyzed samples; an "X" indicates average values for one sample.

between thermally mature Paleogene strata and basaltic rocks. Regionally high heat flow during the early and middle Eocene is also reflected in the widespread occurrence of granitoid plutons of this age throughout the basement-rock sequences of the Chugach and Prince William terranes. The most likely heat source on a regional scale is subduction of the Kula Ridge. Hydrocarbon expulsion and updip migration is probably related to accumulation of the thick Yakataga Formation in a shelf basin during the late Cenozoic; there is no evidence of any earlier period of deformation or loading that might have induced significant migration of hydrocarbons.

Eocene and older(?) dredge samples from two localities on the continental slope (nos. 4 and 14, Figure 1, Table 4) are exceptionally rich in extractable organic matter and contain saturated hydrocarbons with a mature, waxy distribution similar to petroleum. The data indicate that petroleum hydrocarbons have been generated in sufficient quantities to consider that the samples are good source rocks and that they have undergone close to optimum maturation for liquid-hydrocarbon generation. The other three dredge samples analyzed are submature, but have above-average extractable organic matter contents with significant potential for liquid-hydrocarbon generation. Outcrop samples from the Oligocene section on Kayak Island (Table 4), although rich in extractable organic matter, are not mature enough to have converted a significant proportion of the kerogen to liquid petroleum.

Oil-Seep Analyses

Oil and gas seeps are abundant in the Samovar Hills, Yakataga, Katalla, and Wingham Island areas (insets A–C, Figure 1). Oils from all these areas have been analyzed for their group-type hydrocarbon composition, sulphur content, and carbon-isotope compositions. Two representative analyses of the freshest oil samples, one from Hubbs Creek in the Samovar Hills and the other from an oil-stained intertidal reef at the south end of Wingham Island 200 km to the west, are given in Table 5. These are light, waxy, low-sulphur, fairly mature oils that have strikingly similar hydrocarbon distributions and comparable carbon-isotope compositions. Analyses of extractable hydrocarbons from dredge and onshore outcrop samples are given in Table 4. In terms of gas chromatographic "fingerprints," dredge sample 78-22D3 (no. 4, Table 4) can be considered to be a good match as a source rock for the oil seeps, and sample 79-39F (no. 14, Table 4) a less likely, but still possible source rock (George Claypool, 1980 and 1981, written communication). According to Claypool, these oils are unlike other Alaska and west coast Tertiary oils, and most closely resemble oils from the Gulf of Mexico. The composition of Gulf of Alaska oils is compatible with generation from source rocks deposited in a river-delta complex, where there was a large amount of terrestrially derived organic matter.

The data suggest that petroleum source rocks of Eocene age with similar herbaceous-rich kerogen composition and similar depositional facies are present beneath much of the central and western segments of the Yakutat terrane. These rocks occur in a belt that extends at least 200 km east-west between Wingham Island and the Samovar Hills, and offshore as much as 150 km to the continental slope.

Table 4. Analyses of extractable hydrocarbons from five dredge samples from the continental slope and four outcrop samples from Kayak Island. Analyses by U.S. Geological Survey, Lakewood, Colorado.

Map No. (Fig. 1)	Sample No.[1]	Organic Carbon (wt. %)	Bitumen (ppm)	Hydc.[2] (ppm)	Sat./Arom[3]	Hydc./Bit. (%)	Hydc./C Org. (%)	CPI[4]	pr/ph[5]	^{13}C Sat. ‰	^{13}C Arom. ‰
CONTINENTAL SLOPE											
4	78-22D3	1.13	980	700	1.36	71	6.19	1.2	5.1	−27.7	−26.2
14	79-39F	0.92	1100	822	1.3	75	8.9	1.0	3.4	−28.2	−27.6
16	78-44D	1.64	224	101	0.55	45	0.61	.7	0.7	−30.6	−30.4
23	79-36CD	1.74	569	315	2.1	55	1.8	n.d.	n.d.	n.d.	n.d.
29	79-24B	1.78	312	152	1.8	49	0.85	n.d.	n.d.	n.d.	n.d.
KAYAK ISLAND											
Not shown on map	72APr 117A2	5.21	2653	831	0.61	31	0.16	n.d.	n.d.	n.d.	n.d.
	72APr 117A3	4.56	2426	750	0.59	31	0.16	n.d.	n.d.	n.d.	n.d.
	72APr 117B	6.76	2113	873	0.51	41	0.13	n.d.	n.d.	n.d.	n.d.
	72APr 119	1.78	463	218	1.06	47	0.12	n.d.	n.d.	n.d.	n.d.

[1]Complete samples numbers are: 1979, S10-79-EG; 1978, S5-78-EG; 1977, S2-77-EG. Applies to dredge samples only.
[2]Total hydrocarbons.
[3]Saturated/aromatic hydrocarbon ratio.
[4]Carbon preference index.
[5]Pristane/phytane ratio.

Table 5. Analyses of seep oil from Hubbs Creek in the Samovar Hills and from the south end of Wingham Island. (Analyses by Global Geochemistry, Inc., and U.S. Geological Survey, Lakewood, Colorado.)

Locality	Sample No.	API Gravity (%)	Sulphur (%)	[1]$C_{15}+$	$C_{15}+$ Normalized (%) Sat. Hydc.	Arom. Hydc.	Non-Hydc.	CPI[2]	pr/ph[3]	^{13}C Sat. ‰	^{13}C Arom. ‰
Hubbs Creek	80APr127	37.4	0.07	61.5	83.3	15.1	1.7	1.1	4.8	−26.9	−25.5
Wingham Island	81APr51C				67	23	10	1.2	5.7	−27.4	−26.3

[1]Nonvolatile at 40°C.
[2]Carbon preference index.
[3]Pristane/phytane ratio.

Reservoir Rocks

Sandstones are the only rocks in the area that are suitable as potential reservoirs for petroleum in commercial quantities. The reservoir characteristics of the Tertiary sandstones, as deduced from outcrop and well samples, have been presented by Winkler, McLean, and Plafker (1976), and data from dredge samples are given by Plafker et al. (1980).

The Tertiary sandstones in the basinal sequences characteristically are fine to medium grained and range in composition from feldspathic to lithofeldspathic, the means being clustered in the lithofeldspathic field (Figure 15). Lithic fragments average 21–26% in volume and are dominantly plutonic or volcanic. The sandstones are texturally immature and mineralogically unstable, but commonly are low in clayey matrix. Heavy minerals, in approximate order of abundance, include biotite, epidote, muscovite, amphibole, and garnet. Compositional and textural data indicate that the Paleogene sands were well washed when deposited and that deposition was rapid and from nearby source areas.

For major units onshore, the approximate percentage of sandstone, thickness range, porosity range, and permeability range are as follows (Winkler, McLean, and Plafker, 1976; Lyle and Palmer, 1976):

Kulthieth and Tokun Formations: 60%, 15–600 m, 1.8–22.7%, 0–43 md;
Poul Creek Formation: 1–30%, 15–350 m, 2.2–19.2%, 0–12 md;
Yakataga and Redwood Formations: 9–53%, 15–490 m, 1.2–32.2%, 0–597 md.

The percentage of sandstone in the section, as well as the average thickness of sandstone beds, decreases from northeast to southwest away from the sediment source. High porosity and permeability values, as reported by Lyle and Palmer (1976) in the Yakataga Formation, are restricted to the upper part of the section, at least 2000 m above the base. Porosity, permeability, and grain-density measurements of 15 dredge samples of Paleocene(?) and Eocene age are comparable with those from onshore (Plafker et al., 1980). Two samples might have high porosities (31.3% , 23.8%) and permeabilities (26.0 md, 35.9 md) due to secondary submarine leaching of the calcite matrix. Physical properties of the sandstones penetrated in wells in the area are roughly comparable with the data from outcrops and dredge hauls.

Diagenetic alteration of framework grains throughout the stratigraphic sequence has produced widespread zeolite cement or phyllosilicate grain coatings and pseudomatrix.

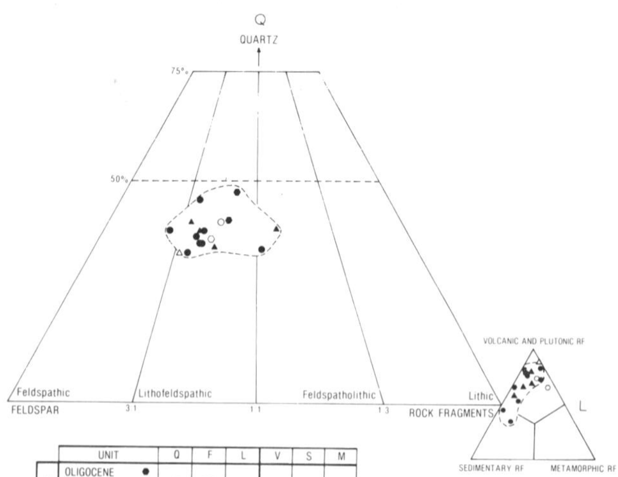

	UNIT	Q	F	L	V	S	M
OFFSHORE	OLIGOCENE ● n=2	45	32	23	73	25	2
OFFSHORE	EOCENE ● n=8	38	42	20	65	25	10
OFFSHORE	PRE-TERTIARY ▲ n=4	38	38	24	68	23	9
ONSHORE	Poul Creek ○ n=10 MEAN	37	41	22	67	6	27
ONSHORE	Tokun/Kulthieth ○ n=8 MEAN	41	34	25	74	11	15
ONSHORE	Yakutat △ n=35 MEAN	34	50	16	89	9	2

Q = total quartzose grains; F = total feldspar; L = total rock fragments (RF); V = volcanic and plutonic RF; S = sedimentary RF; M = metamorphic RF

Figure 15. Q-F-L and rock fragment ternary diagrams comparing sandstone compositions of dredged Paleogene and pre-Tertiary samples (solid symbols) with mean compositions of probably coeval onshore rocks (open symbols). From Winkler, McLean, and Plafker (1976) and Plafker et al. (1980).

Deep burial and strong deformation of the onshore Eocene and older sandstones have caused extensive alteration and grain interpenetration, resulting in postdepositional reduction of porosity and permeability. Less intense deformation and intermediate depth of burial of the middle Tertiary Poul Creek Formation have resulted in improved textural characteristics—better sorting, lower packing densities, and decreased matrix. Nonetheless, measured porosities and permeabilities are quite low, and grain densities and sonic velocities are high, perhaps because carbonate cement is present. Sandstones from the late Cenozoic Yakataga and Redwood Formations exhibit a greater range of textural properties. Although some sandstones are well sorted, poorly sorted and tightly packed sandstones with interpenetrating grains are common. In addition, unstable grains and primary matrix are more abundant than in the older units. Soft and mineralogically unstable framework grains have undergone considerable alteration, reducing pore space even in the youngest rocks. Thus, although there is a modest upsection improvement in sandstone-reservoir characteristics throughout the area, the occurrence of texturally mature sandstones with relatively little framework deterioration is very spotty. This condition is particularly true for the older part of the sequence—the part which has the best source-rock potential

Sandstone compositions suggest a predominantly metamorphic-plutonic source terrane for the Paleogene units in contrast with a volcanic-arc source terrane for the Cretaceous units (Zuffa, Nilsen, and Winkler, 1980; Lull and Plafker, 1985). Suitable source terranes for the pre-Tertiary and Paleogene sandstones are not known to occur along the northern Gulf of Alaska margin. The most probable provenance for them is the plutonic and high-grade metamorphic complex of the Coast Mountains of British Columbia and Alaska (Plafker et al., 1980; Zuffa, Nilsen, and Winkler, 1980). Petrologic and radiometric studies demonstrate that 10–20 km of overburden was stripped off these mountains during the early Tertiary (Hollister, 1979). As suggested by Hollister, much of this sediment, which was initially deposited along the continental margin as slope and deep-sea fan deposits, was subsequently displaced northwestward to the Gulf of Alaska by dextral transform faulting along the margin of the North American plate. Late Cenozoic deposits, in contrast, were derived by erosion of the Paleogene sequences and the pre-Tertiary rocks from the Chugach and Saint Elias Mountains along the northern Gulf of Alaska margin.

Traps

Exploration efforts to date along the northern Gulf of Alaska margin have been concentrated almost exclusively on finding and drilling large anticlinal structures. The 22 deep test wells onshore and the 12 exploratory wells offshore were located mainly on known or inferred structural highs. All of the major accessible structures have been tested by the exploration thus far carried out. In most of these wells drilling has been unsuccessful because of difficulty in reaching the middle Eocene and older section, which has the best source potential, because of the prevailing structural complexity near the anticlinal cores and/or the thick upper Cenozoic overburden. As a consequence, few wells have penetrated the older parts of the basinal sequence, and none of those that reached this part of the section were located on anticlines with definite structural closure. The Texaco Company has a lease on the Samovar Hills and proposes to drill a broad late Cenozoic anticline that has a core of complexly deformed Eocene strata in an area of numerous petroleum seeps (inset A, Figure 1). The surface structure is probably cut by a thrust at shallow depth, because oil seeps that issue from Cretaceous flysch must have their source in Paleogene strata that have been thrust relatively beneath the Samovar Hills. The Samovar Hills area is a nunatak in the Saint Elias National Park, and drilling has been suspended indefinitely pending resolution of problems related to production and transportation of any oil that might be discovered.

Future plays are likely to involve drilling subtle combination structural-stratigraphic traps in the central segment of the Yakutat terrane, anticlines in deep water on the continental slope of the western segment of the Yakutat terrane, or overthrusts such as those along the western margin of the Yakutat terrane. It will be a major technological challenge to define and drill such plays successfully.

In the central segment of the Yakutat terrane, fault traps, stratigraphic traps, and combination traps may be present along the Dangerous River zone where the Paleogene strata lap up onto, or are faulted against, pre-Tertiary basement rocks. Along this trend, opportunities could be enhanced for the occurrence of coarse clastic rocks, updip pinch-outs, possible Eocene carbonate patch reefs, and down-to-the-basin fault seals. Wells along the east and west sides of Yakutat Bay tested part of this zone unsuccessfully. However, the predominantly continental section penetrated in these wells suggests that potentially favorable areas along the Eocene paleoshoreline lay farther to the southwest.

Poor acoustic penetration precludes interpretations of traps where the fold-fault belt of the western Yakutat terrane extends southwestward across the continental slope; but the segment north of the projection of the Transition fault system is likely to be underlain by suitable Paleogene source rocks. A gigantic faulted anticline 5 km wide by 70 km long forms Khitrov Ridge at the base of the slope (Figure 1; section 922, Figure 9). Rocks dredged from the south flank of this feature indicate that it probably consists mainly or entirely of early Pliocene through Pleistocene strata laterally equivalent to the upper part of the Yakataga Formation. The structure is unlikely to include Paleogene rocks with petroleum potential, because it is south of the Transition fault system which truncates the older sequence of the Yakutat terrane along its southwestern margin.

At the western margin of the Yakutat terrane, petroliferous basinal strata have been underthrust northwestward beneath the Prince William terrane along the Ragged Mountain-Wingham Island faults and their offshore continuation into the Kayak Island zone. Petroleum seeps that issue from the upper-plate rocks along the west side of Ragged Mountain demonstrate that petroliferous strata have been thrust for at least 6 km beneath this plate (inset C, Figure 1; Tysdal, Hudson, and Plafker, 1976). Thus, the possibility exists for petroleum plays beneath the upper plate along the western margin of this overthrust belt.

Summary

Petroleum potential in the northern Gulf of Alaska is restricted to the central and western segments of the Yakutat terrane, where suitable source rocks are known to occur in the Paleocene(?) and Eocene strata and where maturity, migration, timing, and seals do not pose significant problems. However, reservoir quality and the occurrence of suitable structural traps in which the Eocene target horizons are at drillable depth have been, and will continue to be, the major obstacles to finding petroleum in commercial quantities.

ACKNOWLEDGMENTS

This paper has benefited from helpful discussions and technical reviews of all or parts of the manuscript by George Claypool, Andrew Griscom, Sherman C. Grommé, John W. Hillhouse, Charles E. Kirshner, Leslie B. Magoon III, and Parke D. Snavely, Jr., and from thorough editing by Catherine C. Campbell, J. R. LeCompte, and David W. Scholl.

Illustrations were drafted by Jeanette Smith and John S. Lull.

REFERENCES CITED

Addicott, W. O., and G. Plafker, 1971, Paleocene mollusks from the Gulf of Alaska Tertiary Province—A significant new occurrence on the north Pacific rim: U.S. Geological Survey Professional Paper 750-B, p. 848–852.

Addicott, W. O., G. R. Winkler, and G. Plafker, 1978, Preliminary megafossil biostratigraphy and correlation of selected stratigraphic sections of the Gulf of Alaska Tertiary Province: U.S. Geological Survey Open-File Report 78-491, 2 plates.

Ariey, C., 1978, Molluscan biostratigraphy of the upper Poul Creek and lower Yakataga Formations, Yakataga district, Gulf of Alaska (abs.), *in* Correlation of tropical through high latitude marine Neogene deposits of the Pacific Basin: Stanford University Publications, Geological Sciences, v. 14, p. 1–2.

Armentrout, J. M., R. J. Echols, and K. W. Nash, 1978, Late Neogene climatic cycles of the Yakataga Formation, Robinson Mountains, Gulf of Alaska area (abs.), *in* Correlation of tropical through high latitude marine Neogene deposits of the Pacific Basin: Stanford University Publications, Geological Sciences, v. 14, p. 3–4.

Atwood, T. J., T. R. Bruns, P. R. Carlson, B. F. Molnia, and G. Plafker, 1981, Bathymetric maps of the northern Gulf of Alaska: U.S. Geological Survey Miscellaneous Field Studies Map MF-859, 3 sheets, scale 1:250,000.

Barnes, F. F., 1967, Coal resources of Alaska: U.S. Geological Survey Bulletin 1242-B, 36 p.

Barnes, I., G. Plafker, D. L. White, and A. K. Armstrong, 1981, Potential natural gas in the Gulf of Alaska indicated by calcite depleted in carbon-13: U.S. Geological Survey Circular 844, p. 143–146.

Bayer, K. C., R. E. Mattick, G. Plafker, and T. R. Bruns, 1978, Refraction studies between Icy Bay and Kayak Island, eastern Gulf of Alaska: U.S. Geological Survey Journal of Research, v. 6, no. 5, p. 625–636.

Bruns, T. R., 1979, Late Cenozoic structure of the continental margin, northern Gulf of Alaska, *in* A. Sisson, ed., The relationship of plate tectonics to Alaskan geology and resources: Anchorage, Sixth Alaska Geological Society Symposium Proceedings, 1977, p. I1–I30.

Bruns, T. R., 1982, Structure and petroleum potential of the continental margin between Cross Sound and Icy Bay, northern Gulf of Alaska: U.S. Geological Survey Open-File Report 82-929, 63 p.

Bruns, T. R., 1983a, A model for the origin of the Yakutat block, an accreting terrane in the northern Gulf of Alaska: Geology, v. 11, p. 718–721.

Bruns, T. R., 1983b, Structure and petroleum potential of the Yakutat segment of the northern Gulf of Alaska continental margin: U.S. Geological Survey Miscellaneous Field Studies Map MF-1480, 3 sheets, scale 1:500,000, 22 p.

Bruns, T. R., 1985, Tectonics of the Yakutat block, an allochthonous terrane in the northern Gulf of Alaska: U.S. Geological Survey Open-File Report 85-13, 112 p.

Bruns, T. R., and G. Plafker, 1982, Geology, structure, and petroleum potential of the southeastern Alaska and northern Gulf of Alaska continental margins, *in* T. R. Bruns, ed., Hydrocarbon resource report for proposed OCS lease sale 88: Southeastern Alaska, northern Gulf of Alaska, Cook Inlet, and Shelikof Strait, Alaska: U.S. Geological Survey Open-File Report 82-928, p. 11–52.

Bruns, T. R., and W. C. Schwab, 1983, Structure maps and seismic stratigraphy of the Yakataga segment of the continental margin, northern Gulf of Alaska: U.S. Geological Survey Miscellaneous Field Studies Map, MF-1424, 2 sheets, scale 1:250,000, 25 p.

Byrne, T., 1982, Structural geology of coherent terranes in the Ghost Rocks Formation, Kodiak Islands, Alaska, in J. K. Legett, ed., Trench and forearc sedimentation and tectonics: Special Publications of the Geological Society of London, v. 10, p. 229–242.

Carlson, P. R., G. Plafker, T. R. Bruns, and W. P. Levy, 1979, Seaward extension of the Fairweather fault: U.S. Geological Survey Circular 804-B, p. B135–B139.

Case, J. A., R. J. Tysdal, J. W. Hillhouse, and C. S. Grommé, 1979, Geologic interpretation of aeromagnetic map of the Seward and Blying Sound quadrangles, Alaska: U.S. Geological Survey Miscellaneous Field Studies Map MF-880D, 2 plates, scale 1:250,000.

Chisholm, W. A., 1985, Comment on "Model for the origin of the Yakutat block, an accreting terrane in the northern Gulf of Alaska:" Geology, v. 13, no. 1, p. 87.

Coe, R. S., B. R. Globerman, P. W. Plumley, and G. A. Thrupp, 1985, Paleomagnetic results from Alaska and their tectonic implications: AAPG Circum-Pacific Earth Sciences Series 1, p. 85–108.

Cowan, D. S., 1982, Geological evidence for post-40 Ma large-scale northwestward displacement of part of southeastern Alaska: Geology, v. 10, no. 6, p. 309–313.

Davies, D. L., and J. C. Moore, 1984, 60 m.y. intrusive rocks from the Kodiak Islands link the Peninsular, Chugach, and Prince William terranes (abs.): GSA Abstracts with Programs, v. 16, p. 277.

Davis, A., and G. Plafker, 1984, Geochemistry and petrography of basalts recovered from the continental slope in the eastern Gulf of Alaska: U.S. Geological Survey Open-File Report 84-298, 24 p.

Engebretsen, D. C., 1982, Relative motions between oceanic and continental plates in the Pacific Basin: Stanford, California, Stanford University unpublished Ph.D. dissertation, 211 p.

Fairchild, L. H., and D. S. Cowan, 1982, Structure, petrology, and tectonic history of the Leech River complex northwest of Victoria, Vancouver Island: Canadian Journal of Earth Sciences, v. 19, no. 9, p. 1817–1835.

Fuis, G. S., E. L. Ambos, W. D. Mooney, R. A. Page, and D. L. Campbell, 1985, Preliminary results of TACT 1984 seismic-refraction survey of southern Alaska: U.S. Geological Survey Circular 967, p. 56–60.

Gabrielse, H., 1985, Major dextral transcurrent displacements along the Northern Rocky Mountain Trench and related lineaments in north-central British Columbia: GSA Bulletin, v. 96, no. 1, p. 1–14.

Gates, G. O., A. Grantz, and W. W. Patton, Jr., 1968, Geology and natural gas and oil resources of Alaska: AAPG Memoir 9, v. 1, p. 3–48.

Grantz, A., 1966, Strike-slip faults in Alaska: U.S. Geological Survey Open-File Report 267, 82 p.

Grommé, C. S., and J. W. Hillhouse, 1981, Paleomagnetic evidence for northward movement of the Chugach terrane, southern and southeastern Alaska: U.S. Geological Survey Circular 823-B, p. B70–B72.

Harland, W. B., A. Cox, P. G. Llewellyn, C. A. G. Pickton, A. G. Smith, and R. Walters, 1982, Geologic time scale: Cambridge, England, Cambridge University Press.

Helwig, J., and P. Emmett, 1981, Structure of the early Tertiary Orca Group in Prince William Sound and some implications for the plate tectonic history of southern Alaska: Journal of the Alaska Geological Society, v. 1, p. 12–35.

Hillhouse, J. W., and C. S. Grommé, 1977, Paleomagnetic poles from sheeted dikes and pillow basalt of the Valdez(?) and Orca Groups, southern Alaska (abs.): EOS (American Geophysical Union Transactions), v. 58, p. 1127.

Hillhouse, J. W., and C. S. Grommé, 1984, Northward displacement and accretion of Wrangellia: New paleomagnetic evidence from Alaska: Journal of Geophysical Research, v. 89, no. B6, p. 4461–4477.

Hillhouse, J. W., C. S. Grommé, and B. Csejtey, Jr., 1985, Tectonic implications of paleomagnetic poles from lower Tertiary volcanic rocks, south central Alaska: Journal of Geophysical Research, v. 90, no. B14, p. 12,523–12,535.

Hollister, L. S., 1979, Metamorphism and crustal displacements: New insights: Episodes, v. 1979, no. 3, p. 3–8.

Hottman, C. E., J. H. Smith, and W. R. Purcell, 1979, Relationship among earth stresses, pore pressure, and drilling problems offshore Gulf of Alaska: Journal of Petroleum Technology, v. 31, no. 11, p. 1477–1484.

Hudson, T., G. Plafker, and M. Rubin, 1976, Uplift rates of marine-terrace sequences in the Gulf of Alaska: U.S. Geological Survey Circular 733, p. 11–13.

Hudson, T., G. Plafker, and D. L. Turner, 1977, Metamorphic rocks of the Yakutat-Saint Elias area, south-central Alaska: U.S. Geological Survey Journal of Research, v. 5, no. 2, p. 173–184.

Jones, D. L., N. J. Silberling, and J. W. Hillhouse, 1977, Wrangellia—A displaced terrane in northwestern North America: Canadian Journal of Earth Sciences, v. 14, p. 2565–2577.

Jones, D. L., N. J. Silberling, P. J. Coney, and G. Plafker, 1986, Lithotectonic terrane map of Alaska (west of the 41st meridian): U.S. Geological Survey Miscellaneous Field Studies Map MF-1874, scale 1:2,500,000.

Keller, G., R. E. von Huene, T. R. Bruns, and K. McDougall, 1984, Paleoclimatic evidence for Cenozoic migration of Alaskan terranes: Tectonics, v. 3, p. 473–495.

Lagoe, M. B., 1983, Oligocene through Pliocene foraminifera from the Yakataga reef section, Gulf of Alaska Tertiary Province, Alaska: Micropaleontology, v. 29, no. 2, p. 202–222.

Lahr, J. C., and G. Plafker, 1980, Holocene Pacific-North American plate interaction in southern Alaska: Implications for the Yakataga seismic gap: Geology, v. 8, p. 483–486.

Lattanzi, R. D., 1981, Planktonic foraminiferal biostratigraphy of Exxon Company, U.S.A. wells drilled in the Gulf of Alaska during 1977 and 1978: Journal of the Alaska Geological Society, v. 1, p. 48–59.

Lull, J. S., and G. Plafker, 1985, Petrography of sandstone from the Yakutat Group, Malaspina district, southern Alaska: U.S. Geological Survey Circular 945, p. 73–77.

Lyle, W. M., and I. F. Palmer, Jr., 1976, A stratigraphic study of the Gulf of Alaska Tertiary Province, northern Gulf of Alaska area: Alaska Division of Geological and Geophysical Surveys Open-File Report 93, 24 p.

Marincovich, L., Jr., and S. McCoy, Jr., 1984, An overview of Paleogene molluscan biostratigraphy and paleoecology of the Gulf of Alaska region: Palaeogeography, Palaeoclimatology, Palaeoecology, v. 47, p. 91–102.

Marshak, R. S., and D. E. Karig, 1977, Triple junctions as a cause for anomalously near-trench igneous activity between the trench and volcanic arc: Geology, v. 5, no. 4, p. 233–236.

Miller, D. J., 1971, Geologic map of the Yakataga district, Gulf of Alaska Tertiary Province, Alaska: U.S. Geological Survey Miscellaneous Geologic Investigations Map I-610, scale 1:125,000.

Miller, D. J., 1975, Geologic map and sections of the central part of the Katalla District, Alaska: U.S. Geological Survey Miscellaneous Field Studies Map MF-722, scale 1:40,000.

Miller, D. J., T. G. Payne, and G. Gryc, 1959, Geology of possible petroleum provinces in Alaska, with annotated bibliography by E. H. Cobb: U.S. Geological Survey Bulletin 1094, 131 p.

Moore, J. C., and W. Connelly, 1977, Mesozoic tectonics of the southern Alaska margin: Journal of Geophysical Research, Maurice Ewing Series, v. 1, p. 71–82.

Moore, J. C., T. Byrne, P. W. Plumley, M. Reid, H. Gibbons, and R. Coe, 1983, The Paleocene evolution of the Kodiak Islands, Alaska: Consequences of ridge-trench interaction in a more southerly latitude: Tectonics, v. 2, no. 3, p. 265–293.

Muller, J. E., 1977, Geology of Vancouver Island: Geological Survey of Canada Open-File Map 463, scale 1:250,000.

Naugler, F. P., and J. M. Wageman, 1973, Gulf of Alaska: Magnetic anomalies, fracture zones, and plate interactions: GSA Bulletin, v. 84, no. 5, p. 1575–1584.

Nelson, S. W., J. A. Dumoulin, and M. L. Miller, 1985, Geologic map of the Chugach National Forest, south-central Alaska: U.S. Geological Survey Miscellaneous Field Studies Map MF-1645-B, scale 1:250,000.

Page, R. A., 1975, Evaluation of seismicity and earthquake shaking at offshore sites: Houston, Texas, 7th Offshore Technology Conference Proceedings, v. 3, p. 179–190.

Page, R. A., G. Plafker, G. A. Fuis, W. J. Nokleberg, E. L. Ambos, W. D. Mooney, and D. L. Campbell, 1986, Accretion and subduction tectonics in the Chugach Mountains and Cooper River Basin, Alaska: Initial results of TACT: Geology, v. 14, no. 6, p. 501–505.

Palmer, I. F., Jr., 1976, U.S.G.S. releases Alaskan well data: Oil and Gas Journal, v. 74, no. 12, p. 101–104.

Perez, O. J., and K. H. Jacob, 1980, Tectonic model and seismic potential of the eastern Gulf of Alaska and Yakataga seismic gap: Journal of Geophysical Research, v. 85, no. B12, p. 7132–7150.

Plafker, G., 1967, Geologic map of the Gulf of Alaska Tertiary Province, Alaska: U.S. Geological Survey Miscellaneous Geologic Investigations Map I-484, 1 sheet, scale 1:500,000.

Plafker, G., 1969, Tectonics of the March 27, 1964, Alaska earthquake: U.S. Geological Survey Professional Paper 543-I, 74 p.

Plafker, G., 1971, Pacific margin Tertiary basin, *in* I. H. Cram, ed.,

Petroleum provinces of North America: AAPG Memoir 15, p. 120–135.

Plafker, G., 1974, Preliminary geologic map of Kayak and Wingham Islands, Alaska: U.S. Geological Survey Open-File Map 74-82, 1 sheet, scale 1:31,860.

Plafker, G., 1981, Late Cenozoic glaciomarine deposits in the Yakataga Formation, Alaska, *in* M. J. Hambrey and W. B. Harland, eds., Earth's pre-Pleistocene glacial record: Cambridge, England, Cambridge University Press, p. 694–699.

Plafker, G., 1983, The Yakutat block—An actively accreting tectonostratigraphic terrane in southern Alaska (abs.): GSA Abstracts with Programs, v. 15, no. 5, p. 406.

Plafker, G., 1984, Comment on "A model for the origin of the Yakutat block, an accreting terrane in the northern Gulf of Alaska:" Geology, v. 12, no. 9, p. 563.

Plafker, G., and T. R. Bruns, 1982, Late Cenozoic subduction—rather than accretion—at the eastern end of the Aleutian Arc (abs.): GSA Abstracts with Programs, v. 14, no. 7, p. 569.

Plafker, G., and G. Claypool, 1979, Petroleum source rock potential of rocks dredged from the continental slope in the eastern Gulf of Alaska: U.S. Geological Survey Open-File Report 79-295, 24 p.

Plafker, G., and M. Rubin, 1978, Uplift history and earthquake recurrence as deduced from marine terraces on Middleton Island, Alaska, *in* Proceedings of Conference VI, Methodology for identifying seismic gaps and soon-to-break gaps: U.S. Geological Survey Open-File Report 78-943, p. 687-721.

Plafker, G., and W. Thatcher, 1982, Geological and geophysical evaluation of the mechanisms of the great 1899–1900 Yakutat Bay, Alaska, earthquakes (abs.): Program and Abstracts, AGU Conference on Fault Behavior and the Earthquake Generating Process, Snowbird, Utah.

Plafker, G., T. R. Bruns, and R. A. Page, 1975, Interim report on petroleum resource potential and geologic hazards in the Outer Continental Shelf of the Gulf of Alaska Tertiary Province: U.S. Geological Survey Open-File Report 75-592, 74 p.

Plafker, G., D. L. Jones, and E. A. Pessagno, Jr., 1977, A Cretaceous accretionary flysch and melange terrane along the Gulf of Alaska margin: U.S. Geological Survey Circular 751-B, p. B41–B43.

Plafker, G., W. J. Nokleberg, and J. S. Lull, 1985, Summary of the 1984 TACT geologic studies in the northern Chugach Mountains and southern Copper River Basin: U.S. Geological Survey Circular 967, p. 76–79.

Plafker, G., T. R. Bruns, G. R. Winkler, and R. G. Tysdal, 1982a, Cross section of the eastern Aleutian arc, from Mount Spurr to the Aleutian Trench near Middleton Island, Alaska: GSA Map and Chart Series MC-28P, 1 sheet, scale 1:250,000.

Plafker, G., T. Hudson, T. R. Bruns, and M. Rubin, 1978a, Late Quaternary offsets along the Fairweather fault and crustal plate interactions in southern Alaska: Canadian Journal of Earth Sciences, v. 15, no. 5, p. 805–816.

Plafker, G., T. Hudson, M. Rubin, and K. L. Dixon, 1982b, Holocene marine terraces and uplift history in the Yakataga seismic gap near Icy Cape, Alaska: U.S. Geological Survey Circular 844, p. 111–115.

Plafker, G., D. L. Jones, T. Hudson, and H. C. Berg, 1976, The Border Ranges fault system in the Saint Elias Mountains and Alexander Archipelago: U.S. Geological Survey Circular 733, p. 14–16.

Plafker, G., G. Keller, J. A. Barron, and J. R. Blueford, 1985, Paleontologic data on the age of the Orca Group, Alaska: U.S. Geological Survey Open-File Report 85-429, 24 p.

Plafker, G., W. J. Nokleberg, S. M. Roeske, and G. R. Winkler, 1986, Nature and timing of deformation along the Contact fault system in the Cordova, Bering Glacier, and Valdez quadrangles: U.S. Geological Survey Circular 978, p. 74–77.

Plafker, G., G. R. Winkler, W. L. Coonrad, and G. Claypool, 1980, Preliminary geology of the continental slope adjacent to OCS lease sale 55, eastern Gulf of Alaska; Petroleum Resource Implications: U.S. Geological Survey Open-File Report 80-1089, 72 p.

Plafker, G., P. R. Carlson, W. L. Coonrad, S. J. Hunt, and P. Quinterno, 1979, Geologic implications of 1978 outcrop sample data from the continental slope in the eastern Gulf of Alaska: U.S. Geological Survey Circular 804-B, p. B143–B146.

Plafker, G., G. R. Winkler, S. J. Hunt, S. Bartsch-Winkler, W. L. Coonrad, and P. Quinterno, 1978b, Outcrop samples from the continental slope in the eastern Gulf of Alaska: U.S. Geological Survey Circular 772-B, p. 97–101.

Plafker, G., T. R. Bruns, P. R. Carlson, B. F. Molnia, E. W. Scott, R. Kahler, and C. Wilson, 1978c, Petroleum potential, geologic hazards, and technology for exploration in the Outer Continental Shelf of the Gulf of Alaska Tertiary Province: U.S. Geological Survey Open-File Report 78-490, 33 p.

Plumley, P. W., and G. Plafker, 1985, Additional estimate of Paleolatitude for the Paleocene/Eocene? Prince William terrane–Orca volcanics (abs.): GSA Abstracts with Programs, v. 17, no. 6, p. 401.

Plumley, P. W., R. S. Coe, and T. Byrne, 1983, Paleomagnetism of the Paleocene Ghost Rocks Formation, Prince William terrane, Alaska: Tectonics, v. 2, no. 3, p. 295–314.

Rau, W. W., G. Plafker, and G. R. Winkler, 1983, Foraminiferal biostratigraphy and correlations in the Gulf of Alaska Tertiary Province: U.S. Geological Survey Oil and Gas Investigations Chart OC-120.

Rogers, J. F., 1977, Implications of plate tectonics for offshore Gulf of Alaska petroleum exploration: 9th Annual Offshore Technology Conference Proceedings, p. 11–16.

Scholl, D. W., T. L. Vallier, and A. J. Stevenson, 1987, Geologic evolution and petroleum geology of the Aleutian Ridge, this volume.

Schwab, W. C., and T. R. Bruns, 1979, Preliminary residual magnetic map of the northern Gulf of Alaska: U.S. Geological Survey Miscellaneous Field Studies Map MF-1054, scale 1:500,000.

Schwab, W. C., T. R. Bruns, and R. E. von Huene, 1980, Map showing structural interpretation of magnetic lineaments in the northern Gulf of Alaska: U.S. Geological Survey Miscellaneous Field Studies Map MF-1245, scale 1:500,000.

Shor, G. G., 1965, Continental margins and island arcs of western North America: Geological Survey of Canada Paper 66-15, p. 216–222.

Stonely, R., 1967, The structural development of the Gulf of Alaska sedimentary province in southern Alaska: Quarterly Journal of the Geological Society of London, v. 123, p. 25–57.

Sutherland Brown, A., 1968, Geology of the Queen Charlotte Islands, British Columbia: British Columbia Department of Mines and Petroleum Resources Bulletin 54, 225 p.

Taylor, P. T., and N. J. O'Neill, 1974, Results of an aeromagnetic survey of the Gulf of Alaska: Journal of Geophysical Research, v. 79, no. 5, p. 719–723.

Tissot, B. P., and D. H. Welte, 1978, Petroleum formation and occurrence: New York, Springer-Verlag, 538 p.

Tysdal, R. G., T. Hudson, and G. Plafker, 1976, Geologic map of the Cordova B-2 quadrangle and northern part of the Cordova A-2 quadrangle, south-central Alaska: U.S. Geological Survey Miscellaneous Field Studies Map MF-783, scale 1:63,360.

Van Alstine, D. R., D. R. Bazard, and J. W. Whitney, 1985, Paleomagnetism of cores from the Yakutat well, Gulf of Alaska (abs.): American Geophysical Union Transactions, v. 66, no. 46, p. 865.

Van der Voo, R. M., M. Jones, C. S. Grommé, G. D. Eberlein, and M. Churkin, Jr., 1980, Paleozoic paleomagnetism and the northward drift of the Alexander terrane, southeastern Alaska: Journal of Geophysical Research, v. 85, p. 5281–5296.

von Huene, R. E., G. G. Shor, Jr., and J. M. Wageman, 1979, Continental margins of the eastern Gulf of Alaska and boundaries of tectonic plates, *in* J. S. Watkins and L. Montadert, eds., Geological investigations of continental margins: AAPG Memoir 29, p. 273–290.

Wilson, F. H., 1985, The Meshik arc—An Eocene to earliest Miocene magmatic arc on the Alaska Peninsula: Alaska Division of Geological and Geophysical Surveys Professional Report 88, 14 p.

Winkler, G. R., 1976, Deep sea fan deposition of the Orca Group, eastern Prince William Sound, Alaska, *in* T. P. Miller, ed., Recent and ancient sedimentary environments in Alaska: Alaska Geological Society Symposium Proceedings, p. R1–R20.

Winkler, G. A. and G. Plafker, 1981, Preliminary geologic map of the Cordova and Middleton Island quadrangles, Alaska: U.S. Geological Survey Open-File Report 81-1164, scale 1:250,000.

Winkler, G. A., H. McLean, and G. Plafker, 1976, Textural and mineralogical study of sandstones from the onshore Gulf of Alaska Tertiary Province, southern Alaska: U.S. Geological Survey Open-File Report 76-198, 48 p.

Wolfe, J. A., and S. McCoy, Jr., 1984, Comment on "Model for the origin of the Yakutat block, an accreting terrane in the northern Gulf of Alaska": Geology, v. 12, no. 9, p. 564–565.

Zuffa, G. G., T. H. Nilsen, and G. R. Winkler, 1980, Rock-fragment petrography of Upper Cretaceous Chugach terrane, southern Alaska: U.S. Geological Survey Open-File Report 80-713, 28 p.

CHAPTER TWELVE

GEOLOGY AND PETROLEUM POTENTIAL OF THE SOUTHEAST ALASKA CONTINENTAL MARGIN

T. R. Bruns
P. R. Carlson

U.S. Geological Survey
Menlo Park, California

ABSTRACT

Single-channel and multichannel seismic-reflection data show the structure and sedimentary basins of the southeast Alaska continental margin and delineate the location of the Fairweather-Queen Charlotte fault zone between Dixon Entrance and the onshore Fairweather fault near Cross Sound. The Fairweather-Queen Charlotte fault truncates the Mesozoic and Paleozoic rocks of southeastern Alaska and marks the western edge of continental North America. Geological and geophysical data currently preclude significant Pliocene and Quaternary faulting landward or seaward of the fault zone. The fault system has therefore been the major late Cenozoic transform boundary between the Pacific and North American plates. The fault system has evolved to its present form during Miocene and younger time after passage along the margin of the Yakutat terrane, which is now colliding with and accreting to southern Alaska.

The continental shelf is underlain by several elongate basins containing as much as 3 km of bedded rocks, probably of Miocene and younger age. The slope and continental rise are underlain by strata as thick as 10 km, also of Miocene and younger age. South of Sitka, these strata are deformed in two structural zones along probable splays of the Fairweather-Queen Charlotte fault system. North of Sitka, the slope and rise strata are undeformed.

The petroleum potential of the southeast Alaska margin could be favorable, because thick sedimentary accumulations and possible structural traps are present. The source and reservoir rock potential is unknown.

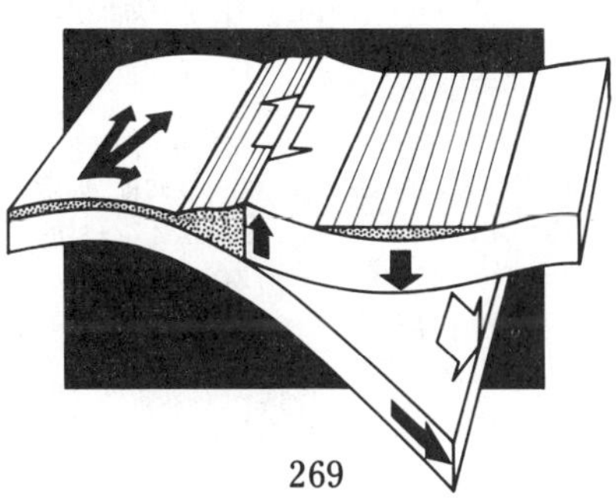

INTRODUCTION

In this chapter, we summarize the geology and petroleum potential of the southeast Alaska continental margin, which consists of the approximately 500-km-long continental shelf and slope between Dixon Entrance and Cross Sound in the eastern Gulf of Alaska (Figures 1, 2).

The major tectonic feature of the southeast Alaska continental margin is the mostly submerged Fairweather-Queen Charlotte fault system, which is the active transform boundary between the Pacific and North American plates. Movement along the transform system is estimated as about 6 cm/yr by Minster and Jordan (1978). The fault system runs northward along the Canadian margin as the Queen Charlotte fault from the intersection of the Juan de Fuca spreading ridge and the North American continental margin off Vancouver Island (see Yorath, this volume). The fault system continues along the Alaskan continental margin to Icy Point in the northern Gulf of Alaska. At Icy Point, the offshore fault system is interpreted to merge with the onshore Fairweather fault of southern Alaska (Figure 1). The fault system has dimensions similar to the better-exposed and better-studied San Andreas fault of California (Carlson, Plafker, and Bruns, 1985).

Most of the marine geophysical data collected over the southeast Alaska margin were acquired to study the offshore extension of the Fairweather fault and its presumed connection with the Queen Charlotte fault off British Columbia. Von Huene, Shor, and Wageman (1979) used widely spaced single-channel seismic-reflection profiles to identify faults and submarine scarps on the shelf south of Icy Point, which appeared to align with the onshore trace of the Fairweather fault. However, the seismic profiles were too widely spaced to allow detailed mapping of offshore faults. More closely spaced (2–10 km) single-channel seismic-reflection profiles collected in 1978 were used by Carlson et al. (1979), and by Carlson, Plafker, and Bruns (1985), to map the offshore trend of the Fairweather fault across the shelf and slope toward the Queen Charlotte fault.

These single-channel seismic-reflection data do not have sufficient areal coverage or acoustic penetration to delineate the structure or the thickness of sedimentary strata along the margin. However, about 1000 km of multichannel seismic-reflection data, acquired along widely spaced (about 25 to 50 km) tracklines in 1978 (Bruns, 1981, 1985; Bruns et al., 1984) provide regional control for the interpretation of structure and sediment thickness presented here. No significant geological samples have been collected from the sea floor along the southeast Alaska margin by dredging or bottom sampling.

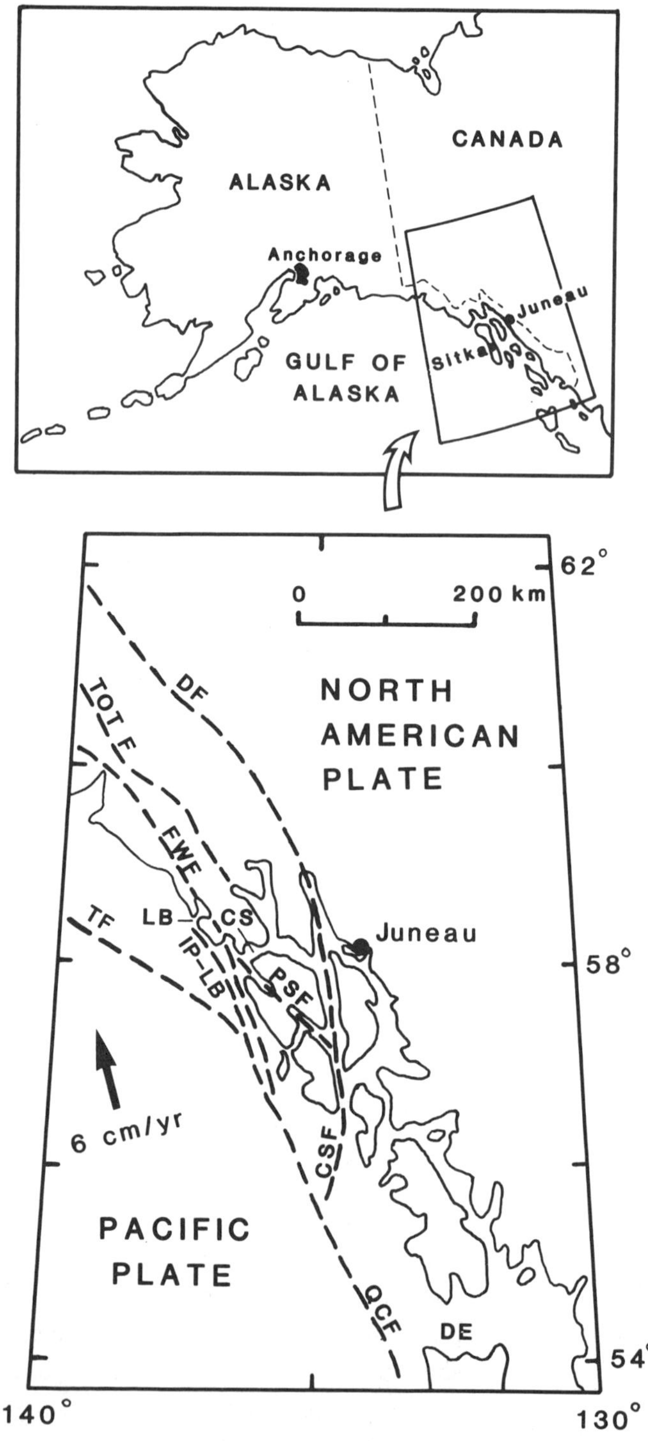

Figure 1. Location map and tectonic setting of southeastern Alaska continental margin. Black arrow indicates relative convergence vector for Pacific and North American plates (Minster and Jordan, 1978). Abbreviations: CS, Cross Sound; CSF, Chatham Strait fault; DE, Dixon Entrance; DF, Denali fault; FWF, Fairweather fault; IP-LB, Icy Point-Lituya Bay fault; LB, Lituya Bay; PSF, Peril Strait fault; QCF, Queen Charlotte fault; TF, Transition fault; TOTF, Totschunda fault.

REGIONAL SETTING

Terranes

Southern and southeastern Alaska and the adjacent Queen Charlotte Islands of British Columbia are underlain by a diverse assemblage of Paleozoic and Mesozoic rocks. These rocks comprise parts of five fault-bounded tectonostratigraphic terranes that were amalgamated by the early Tertiary (Figure 2; Berg, Jones, and Coney, 1978; Jones et al., 1981; Monger and Berg, 1984). The terranes include (1) the Chugach terrane, composed of Upper Jurassic and Cretaceous graywacke, argillite, and melange containing blocks of basic volcanic rocks, chert, limestone, and ultramafic rocks; (2) the Wrangellia terrane, composed of Upper Paleozoic to Jurassic volcanic rocks, limestone, pelite, and tholeiitic basalt overlain by carbonate rocks; (3) the Alexander terrane, composed of three distinctive subterranes, but in

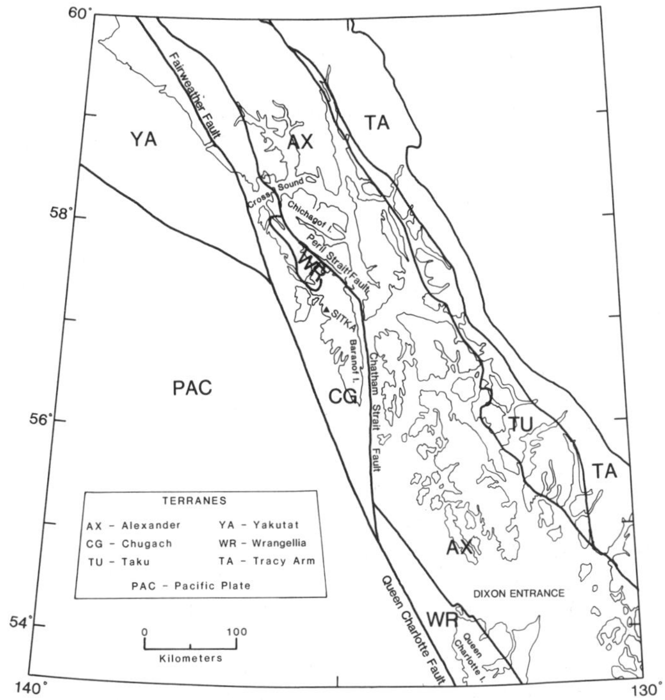

Figure 2. Terranes of southeast Alaska and northern British Columbia. From Monger and Berg (1984).

general formed of Precambrian to Triassic schist and gneiss, mafic to silicic volcanic rocks, and minor pelitic and shallow-water carbonate rocks; (4) the Taku terrane, composed of variably metamorphosed Upper Paleozoic and Triassic basalt, carbonate rock, and pelite, and (5) the Tracy Arm terrane, composed of varied metamorphic rocks of possible Late Proterozoic protolith age (Monger and Berg, 1984).

A sixth terrane, the Yakutat terrane, adjoins the northern part of the southeast Alaska margin, and underlies the continental margin between Cross Sound and Kayak Island. Basement rocks of this terrane include upper Mesozoic pelite, graywacke, and melange in the eastern part of the terrane, and Eocene basalt in the western part. The basement rocks are overlain by a thick (as much as 9 km) sedimentary section composed of Eocene and younger marine and nonmarine clastic rocks (Plafker et al., 1980; Bruns, 1979, 1983a, 1983b, 1985; Bruns and Schwab, 1983; Plafker, 1983, and this volume).

In southeast Alaska, Cenozoic tectonic activity after terrane amalgamation has included intrusion, thermal metamorphism, localized volcanism, and local deposition of nonmarine and deltaic volcaniclastic sedimentary rocks. Faulting has redistributed the Paleozoic and Mesozoic terranes along major fault zones such as the Chatham Strait, the Peril Strait, and the Fairweather-Queen Charlotte fault zones (Figures 1, 2; Berg, 1979).

In the Queen Charlotte Islands region, Cenozoic rocks include thick Paleogene(?) through lower Miocene subaerial alkalic basalt flows and rhyolite ash flows (Masset Formation; Sutherland Brown, 1968), overlain onshore by more than 2 km of Miocene and Pliocene marine and nonmarine sandstone and siltstone (Skonun Formation; Sutherland Brown, 1968). This section thickens offshore, where coeval rocks are as thick as 4.5 km in the Queen Charlotte Basin, which lies beneath Hecate Strait east of the Queen Charlotte Islands (Shouldice, 1973; Yorath, this volume). The basin evolved in a rift and flexural subsidence setting, and has been disrupted by faulting along several major faults during the Tertiary (Yorath and Chase, 1981; Yorath and Hyndman, 1983; Yorath, this volume).

The offshore geology of the southeast Alaska continental margin must be inferred by the projection offshore of known onshore geology and by inference from the limited offshore geophysical data described in this report. Basement rocks underlying the shelf are most likely composed of the continuation of the Paleozoic and Mesozoic rocks of the nearby islands (Figure 2). North of Chatham Strait and west of Baranof and Chichagof Islands, the basement probably consists of rocks of the Chugach terrane. South of Chatham Strait and the Chatham Strait fault, basement is likely composed of Alexander terrane rocks or of Wrangellia terrane rocks like those exposed on the Queen Charlotte Islands.

The various basement rocks can extend seaward no farther than the base of the continental slope, because oceanic crust of middle Miocene age (Chron 5a, 12 Ma) underlies the adjacent continental rise (Naugler and Wageman, 1973; Berggren et al., 1985; Wilson, 1985; Stevenson and Embley, this volume). Moreover, gravity modeling and seismic-refraction studies by von Huene, Shor, and Wageman (1979) suggest that the slope is underlain by low-density, low-velocity sedimentary rocks rather than by the basement rocks of the various terranes. Therefore, the rocks underlying the slope seaward of the Fairweather-Queen Charlotte fault are probably not the same as the terrane basement rocks exposed onshore. The Fairweather-Queen Charlotte fault forms a fundamental tectonic boundary along southeast Alaska and truncates the basement rocks of the North American continental margin.

Tectonic Evolution

The allochthonous terranes of southeastern and southern Alaska traveled northward on ocean plates along the North American continental margin, and accreted to the continent in the Mesozoic and Cenozoic (Jones et al., 1981). Several studies indicate that the last of the allochthonous terranes to pass the southeast Alaska margin is the Yakutat terrane, which is currently colliding with and accreting to southern Alaska (Figures 2, 3; Bruns, 1983a, 1985; Plafker, 1983; Keller et al., 1984; von Huene et al., 1985). The history of passage of the Yakutat terrane past the southeast Alaska continental margin therefore has implications on the geology, structure, and petroleum potential of the shelf basins.

Plafker et al. (1980) and Plafker (1983, 1984, and this volume) conclude that the source region for rocks of the Yakutat terrane was the high-grade plutonic and metamorphic rocks of coastal British Columbia, and that the Yakutat terrane lay west of Dixon Entrance during the Paleogene. In this model, the terrane moved 500 km northward into its present position by no later than 15 Ma. Since 15 Ma, the Transition fault, which bounds the south side of the terrane, would have been active as either a transform fault between the Pacific plate and the Yakutat terrane or as an oblique subduction zone, with the Pacific plate subducting beneath the Yakutat terrane.

Alternatively, von Huene, Shor, and Wageman (1979), Bruns (1983a, 1983b, 1984, and 1985), and Bruns et al. (1984), using geophysical evidence of undeformed strata

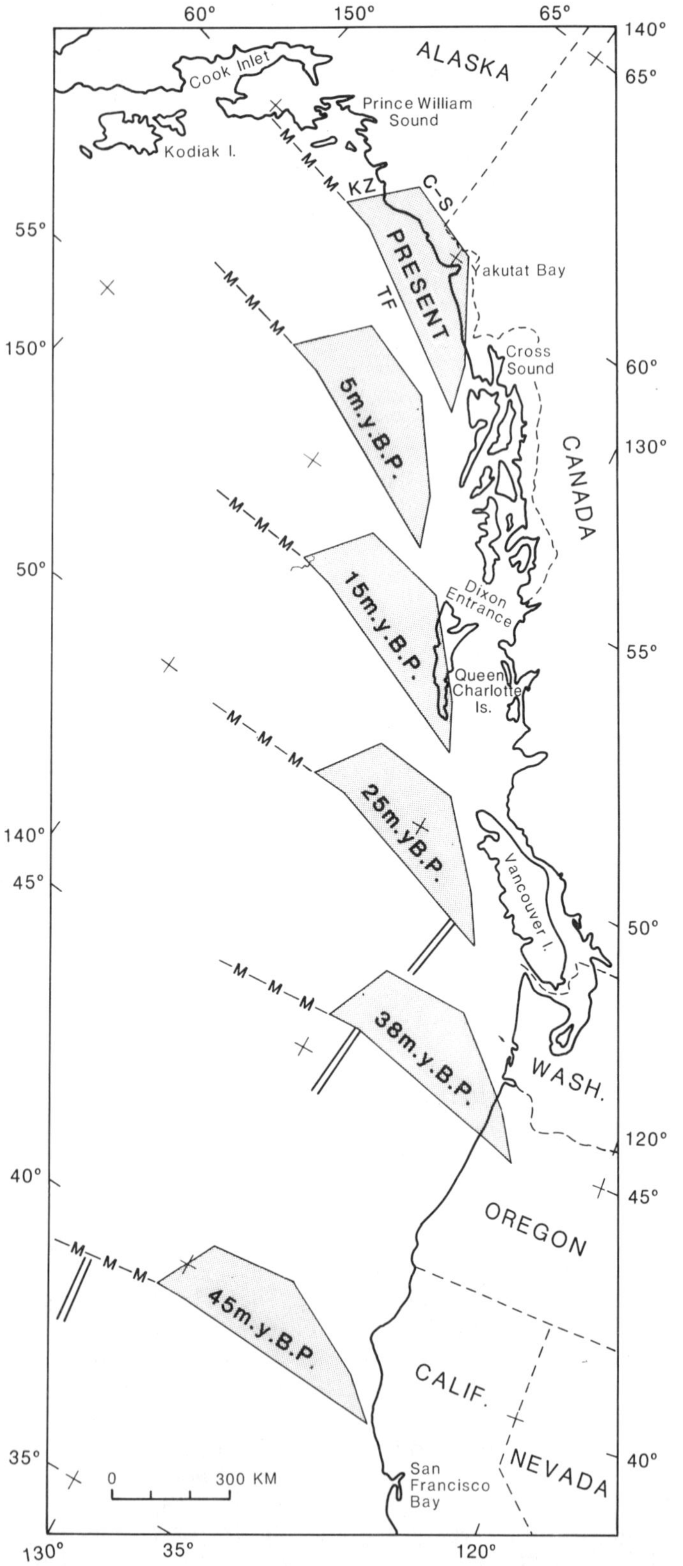

Figure 3. Travel path of Yakutat terrane with respect to North America based on model of Bruns (1983a). Model assumes that block originated during subduction of the Kula-Farallon spreading center and has since moved north with the Kula and Pacific plates. Double line indicates approximate position of spreading center. Dashed line with M's indicates approximate position of subducted part of Yakutat block. C-S, Chugach-Saint Elias fault system; KZ, Kayak zone; TF, Transition fault zone.

across and along the Transition fault, concluded that the Transition fault is presently inactive as either a transform fault or oblique subduction zone, has been inactive during at least the Pliocene and Quaternary, and has probably been inactive or at most weakly active since about Oligocene time. In this model, the Yakutat terrane has been moving with the Pacific plate past southeastern Alaska during the late Cenozoic (Figure 3).

Bruns (1983a, 1984, and 1985), Keller et al. (1984), and von Huene et al. (1985), using microfaunal evidence, further concluded that the Yakutat terrane has been moving northward with the Pacific plate during most of the Cenozoic and is an allochthonous terrane that originated off the western United States (Figure 3). The original location is unclear. Keller et al. (1984) found that Eocene microfauna from the Yakutat terrane correlate best with coeval West Coast fauna in southern California. They thus favor an origin for at least part of the terrane as far south as southern California. Bruns (1983a) shows an amalgamated terrane originating off northern California, and van Alstine, Bazard, and Whitney (1985) present paleomagnetic evidence for an origin off Washington or Oregon. In any of these cases, the Yakutat terrane would have moved past the southeast Alaska margin during the Miocene, and would have been colliding with and subducting beneath southern Alaska since the latest Miocene (Bruns, 1983a, 1985).

Whatever the movement history of the Yakutat terrane, the southeast Alaska continental margin evolved into its present form only after passage of the terrane northward along the Fairweather-Queen Charlotte fault system. The Fairweather-Queen Charlotte fault system is therefore a modern analog of the transform fault systems required for the Mesozoic and early Tertiary assembly of much of southern Alaska.

GEOPHYSICAL DATA

Single-channel seismic-reflection data presented in this chapter were acquired in 1978 from the U.S. Geological Survey research vessel SEA SOUNDER using a 40 cu. in. air gun (Carlson, Plafker, and Bruns, 1985). Multichannel seismic-reflection data were acquired in 1978 from the U.S. Geological Survey research vessel S.P. LEE using a tuned air-gun array with a capacity of 21.7 liters (1326 cu. in.), and a 2400-m, 24-group streamer. The data were processed by the U.S. Geological Survey in Menlo Park, California. Navigation on both ships was by means of an integrated satellite, Loran C, and doppler sonar navigation system. Gravity and magnetic data were simultaneously acquired; the gravity data are presented by Bruns, Atwood, and Childs (1981), and Bruns, Childs, and Atwood (1981).

Velocity information for converting two-way traveltime to depth on the seismic-reflection data was determined from stacking velocities obtained during processing of the data. In the continental shelf of the adjacent Yakutat terrane, stacking velocities and seismic-refraction velocities were used to obtain a time-to-depth conversion curve, an interval-velocity curve, and approximate error limits on these curves for sedimentary strata (Figure 4; Bruns, 1983b). Stacking velocities from the southeast Alaska margin are similar to those of the Yakutat Shelf, and we therefore adopt the conversion curve from the Yakutat Shelf as approximately valid

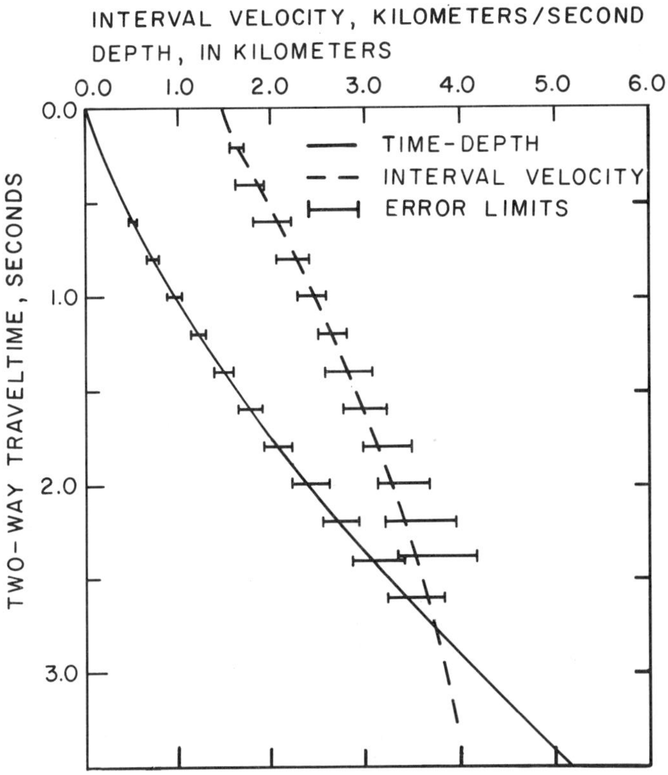

Figure 4. Reflection time versus depth conversion curve and interval-velocity curve for sedimentary rocks of the southeast Alaska margin, and approximate error limits on curves. The time-depth curve is used to convert two-way sub-sea-floor traveltime on seismic-reflection sections to depth; water layer thickness must be added for total depth. From Bruns (1983b).

for the southeast Alaska margin (Figure 4). The curve is valid only in basin areas of the shelf and in the sedimentary rocks of the slope and continental rise. The curve is invalid in areas of hard, well-lithified, or deformed rocks.

Bathymetric data along the southeast Alaska margin are sparse; for much of the segment, the bathymetry is from Chase, Menard, and Mammerickx (1970) and is reconnaissance only. More detailed bathymetric coverage is available in the vicinity of Dixon Entrance and southward (Seemann, 1982), and in the Cross Sound region and westward (Atwood et al., 1981).

STRUCTURE

Fairweather-Queen Charlotte Fault System

The Fairweather-Queen Charlotte fault system, as discussed herein, includes (1) the onshore Fairweather fault; (2) a parallel, inferred fault, the Icy Point-Lituya Bay fault, which lies just offshore of the Fairweather fault; and (3) the extension of these two features southeastward along a set of subparallel fault traces that extend along the outer shelf and upper slope of the southeast Alaska margin (Figure 5). This fault system continues into the Queen Charlotte fault off British Columbia (Yorath, this volume).

Offshore, the fault system has been mapped using bathymetric lineations, offsets of seismic reflectors, and sea-floor scarps. The fault is not a single trace, but instead seems to be composed of many traces. Closely spaced (2–10 km) single-channel seismic data (Carlson, Plafker, and Bruns, 1985) show evidence for two sets of subparallel fault traces that range from about 6 km apart off Icy Point in the north to about 12 km apart off southern Chichagof Island (Figures 5, 6). Between Chatham Strait and Cross Sound, the eastern set of up to five fault traces is located primarily beneath the middle shelf, then crosses the shelf beneath Yakobi Valley, and trends into the onshore Fairweather fault north of Cross Sound. The western set of two to three subparallel traces is located near the shelf break between Chatham Strait and Cross Sound, and north of Cross Sound trends across the shelf into the Icy Point-Lituya Bay fault (IP-LB fault, Figure 5A) offshore of the coastline between Icy Point and Lituya Bay (Plafker, 1967; von Huene, Shor, and Wageman, 1979; Bruns, 1983b, 1985; Bruns et al., 1984; Carlson, Plafker, and Bruns, 1985). An abrupt 15° to 25° change in strike occurs between the offshore fault zones and the onshore Fairweather fault and nearshore Icy Point-Lituya Bay fault (Figures 5, 6).

The western fault zone is the more clearly defined of the two offshore fault zones and shows the best evidence for late Pleistocene and Holocene displacement, as indicated by sea-floor scarps with 20 to 40 m of relief, and disruption of well-defined near-surface seismic reflections (Figure 7). In most of the crossings of the western set of fault traces, where sea-floor offsets are well displayed, the sense of movement is down on the northeast side (Figure 7), the same sense of vertical displacement exhibited onshore along the Fairweather fault during the 1958 Lituya Bay earthquake (Tocher, 1960). The eastern fault zone is discontinuous and sinuous, with few sea-floor scarps or bathymetric lineations visible on the seismic records.

We conclude from the seismic-reflection data that the western fault zone is relatively more active than the eastern zone, although it is the offshore eastern fault zone that trends into the presently active onshore Fairweather fault (Figure 5; Carlson, Plafker, and Bruns, 1985).

The onshore Fairweather fault lies in a trenchlike fault valley that may extend offshore. Seismic-reflection records show evidence of a buried trough beneath the shelf that extends at least 35 km southeastward of Icy Point. This 2- to 4-km-wide trough contains sedimentary strata at least 600 m thick and is aligned with the onshore trench of the Fairweather fault. These strata do not appear to be disrupted, but the seismic resolution within the trough strata is poor and obscured by near-surface gas (biogenic methane) anomalies. The Fairweather fault trace may extend offshore along the axis of the buried trough (Carlson, Plafker, and Bruns, 1985).

The two sets of subparallel fault traces extend across the shelf 225 km southeastward from Icy Point to just south of Sitka, where they merge (Figure 5). The fault system, as mapped on single-channel records (von Huene, Shor, and Wageman, 1979) and on widely spaced multichannel seismic-reflection lines (Bruns et al., 1984; Bruns, 1985), continues southeastward along the upper slope from near Chatham Strait to Dixon Entrance (Figures 5, 8), and along the west side of Queen Charlotte Island (Snavely et al., 1981; Yorath, this volume).

Evidence that the complex Fairweather-Queen Charlotte fault system is now accommodating the relative motion between the Pacific and North American plates includes recorded seismicity along the fault (Page, 1975), 5 to 6 cm/yr of offset during the Holocene on the onshore Fairweather fault (Plafker et al., 1978), and young structural and morpho-

Figure 5. Structure and sediment thickness of inferred Miocene and younger strata beneath the southeast Alaska continental margin: from Lituya Bay to Dixon Entrance. Sediment thickness is based on time-depth curve of Figure 4. Figure 5 also shows contours of Miocene and younger strata on and adjacent to the Yakutat terrane (from Bruns, 1983b, 1985). IP-LB, Icy Point-Lituya Bay fault.

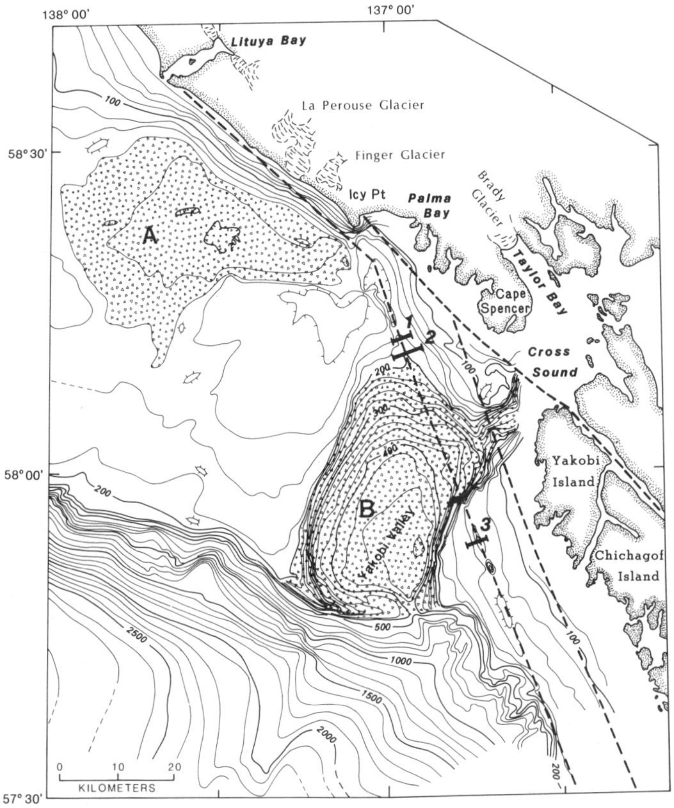

Figure 6. Detailed bathymetry of Lituya Bay-Cross Sound region (after Atwood et al., 1981). Note offsets of contours forming distorted sea valleys (patterned). A, ancestral Yakobi Valley; B, modern Yakobi Valley. Dashed lines indicate main traces of offshore faults (from Carlson, Plafker, and Bruns, 1985). Numbered thick lines show locations of seismic-reflection profiles illustrated on Figure 7.

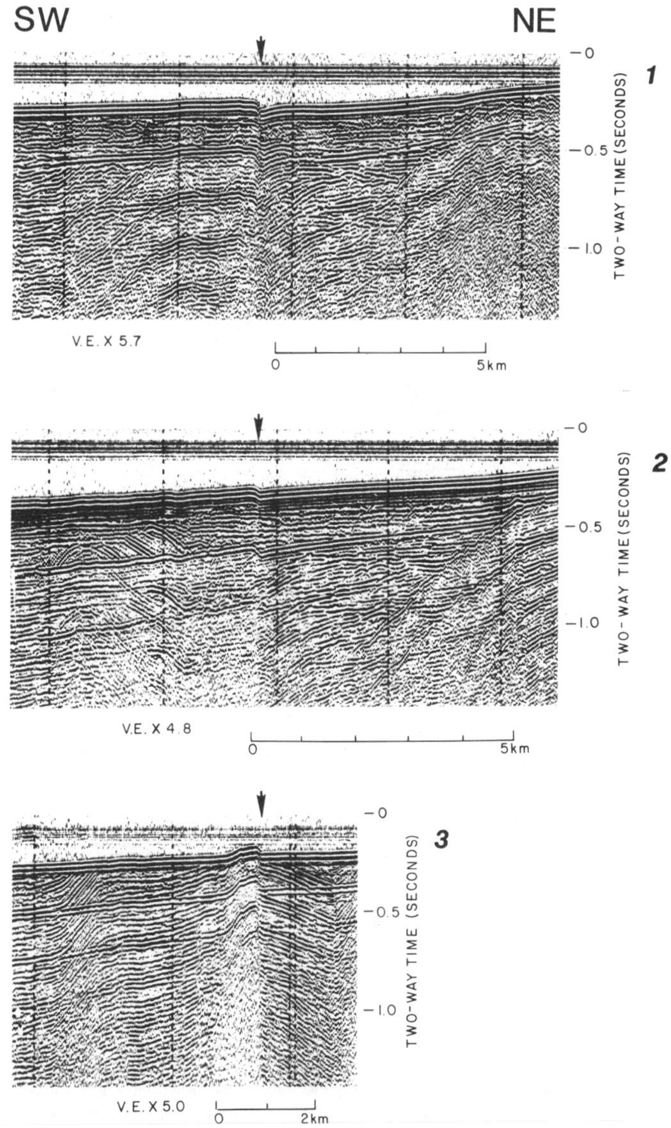

Figure 7. Single-channel seismic-reflection records across the shelf adjacent to Yakobi Valley, showing most active trace of Fairweather-Queen Charlotte fault system (arrow). Locations of profiles shown on Figure 6.

logic features observed both onshore and offshore along the fault (Plafker et al., 1978; von Huene, Shor, and Wageman, 1979; Carlson, Plafker, and Bruns, 1985).

Although the amount and rate of displacement along these offshore fault traces and the age of strata cut by them are unknown, evidence exists for late Pleistocene and Holocene motion. Yakobi Valley, located seaward of Cross Sound, is a glacially carved sea valley; the glacier flowed through Cross Sound and across the Fairweather-Queen Charlotte fault zone (Figure 6; Carlson et al., 1982). The 200–m isobath that outlines Yakobi Valley exhibits well-defined notches on both the northwest and southeast sides of the sea valley. Detailed bathymetry of Yakobi Valley (Figure 6) shows right-lateral displacement of 300 to 700 m of the southeast wall of the valley along the fault traces. This offset is about the amount that would accumulate as a result of Holocene displacement along the fault zone since retreat of the glacier that carved the valley (von Huene, Shor, and Wageman, 1979; Atwood et al., 1981; Carlson et al., 1982; Carlson, Plafker, and Bruns, 1985). Atwood et al. (1981) and Carlson et al. (1982) further suggest that the boxlike shape of Yakobi Valley results from a combination of glacial erosion and right-lateral displacement along the Fairweather-Queen Charlotte fault system, as the northwest wall of the glacial valley seaward of the fault system was systematically offset to the northwest.

Atwood et al. (1981) identified a depression on the continental shelf between Cross Sound and Lituya Bay that is similar in form to Yakobi Valley (Figure 6). This depression could be an ancestral Yakobi Valley—a partly buried, glacial valley that once lay in the present position of Yakobi Valley and has been offset to the northwest.

If the boxlike morphology and present position of Yakobi Valley and the ancestral depression are fault controlled, right-lateral offset could be about 20 km for Yakobi Valley, and about 70 km for the ancestral Yakobi Depression, largely during the Quaternary. However, known post-late-Miocene displacement along the connecting onshore segment of the Fairweather fault is only about 5.5 km (Plafker et al., 1978). Thus, either post-late-Miocene displacement onshore is greater than estimated by Plafker et al. (1978), or much Pliocene and Quaternary motion along the Fairweather-Queen Charlotte fault system may have occurred on the western (seaward) set of fault traces and their probable northern extension along the Icy Point-Lituya Bay fault (Figure 5; Bruns et al., 1984; Bruns, 1985; Carlson, Plafker, and Bruns, 1985).

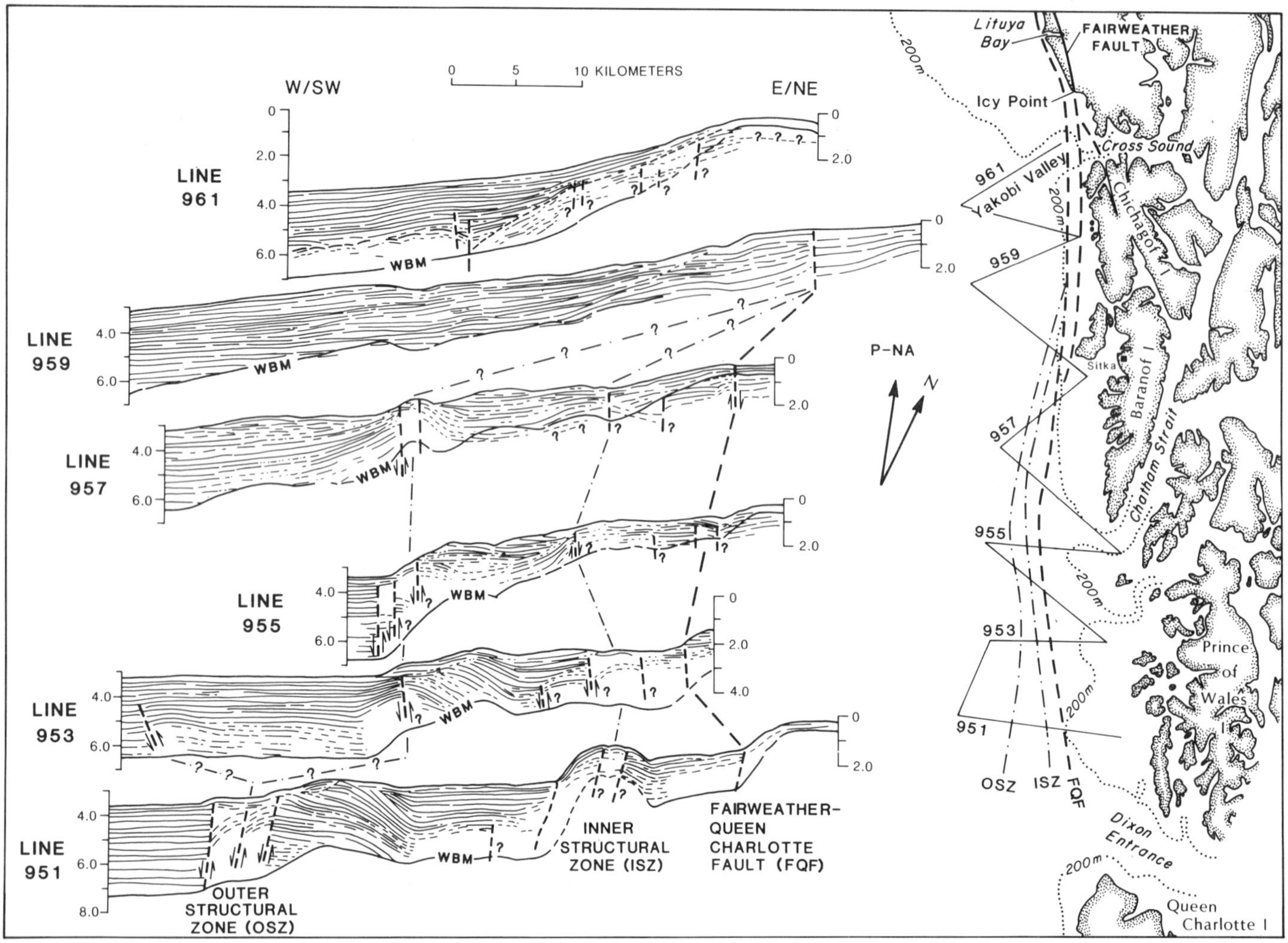

Figure 8. Simplified line drawings of multichannel seismic-reflection profiles across the southeast Alaska continental margin, showing locations of principal structural trends (from Bruns et al., 1984; Bruns, 1985). Trends shown are the Fairweather-Queen Charlotte fault traces (FQF), an outer structural zone (OSZ), and an inner structural zone (ISZ). Vertical scale on seismic-reflection lines is two-way time in seconds. WBM—water-bottom multiple. Pacific-North America (PNA) relative convergence direction (arrow) from Minster and Jordan (1978).

Southeast Alaska Shelf

The continental shelf landward of and adjacent to the Fairweather-Queen Charlotte fault zone is underlain by small, elongate basins containing at least 2.5 to 3.0 km of stratified rocks above poorly resolved acoustic basement (Figures 5, 9, 10). The basin configuration, the location of sediment depocenters, and the maximum sediment thicknesses in the basins are all poorly known because of the widely spaced seismic-reflection data coverage. Where seismic data exists, the sediment thickness is poorly constrained because acoustic basement reflections are poor and discontinuous, and cannot be easily identified on the seismic profiles.

Basement rocks beneath and surrounding the basins are probably highly indurated sedimentary, metamorphic, or igneous rocks. In shelf areas adjacent to the basins, refraction velocities measured on the 2400–m multichannel cable are commonly above 5 km/sec (Figure 5). These high-velocity rocks form the seismically nonreflective to weakly reflective rocks around and underlying the basins. These rocks correspond to the rocks exposed in the adjacent southeast Alaska islands.

The age of the bedded strata in the shelf basins is not known. Basement rocks beneath the basins are presumably Upper Cretaceous rocks of the Chugach terrane (Figure 2). Thus, the basin fill is unlikely to be older than early Tertiary. However, we feel that most of these strata are more likely Neogene and Quaternary in age, for the following reasons. (1) This is the age of the adjacent Queen Charlotte Basin fill. (2) Thick strata of this age are present beneath the slope and rise along southeast Alaska (next section, and Stevenson and Embley, this volume) and indicate a high rate of sedimentation during Miocene and younger time. Presumably, some of this young sediment would have been deposited in the shelf basins. (3) Stacking velocities from the multichannel seismic-reflection data indicate that the velocity of the basin fill rocks is relatively slow (Figure 4). (4) The seismic-refraction velocities of surface and near-surface sedimentary fill in the basins from the multichannel cable refraction breaks are also low, commonly ranging from about 1900 to 2300 m/sec. (5) Each of these basins is associated with a well-expressed low free-air gravity anomaly (Bruns, Childs, and Atwood, 1981). The velocity and gravity data suggest that the rocks in the basins are not highly indurated and that they have a low density relative to the surrounding and underlying basement rocks.

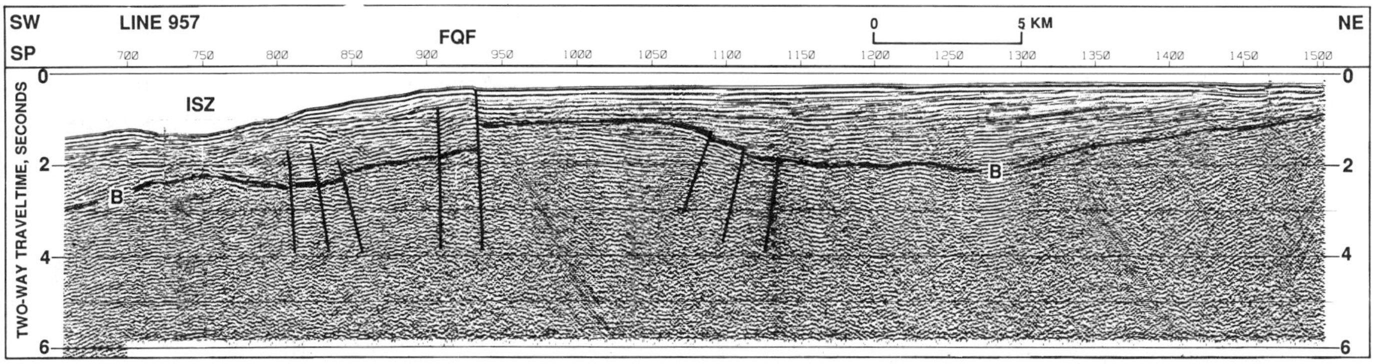

Figure 9. Portion of multichannel seismic line 957 across shelf basin, Fairweather-Queen Charlotte fault zone (FQF), and inner structural zone (ISZ). Horizon B is approximate top of high velocity (5–6 km/sec) acoustic basement rocks. Compare minor deformation observed on inner structural zone with deformation across the same zone on lines 951 and 955 (Figures 11, 12), and with lack of deformation on line 959 (Figure 10), and note how deformation in structural zones increases to the south. On this and other multichannel profiles, vertical exaggeration at the sea floor is 2:1; SP = shot point. Location of line shown on Figures 5, 8.

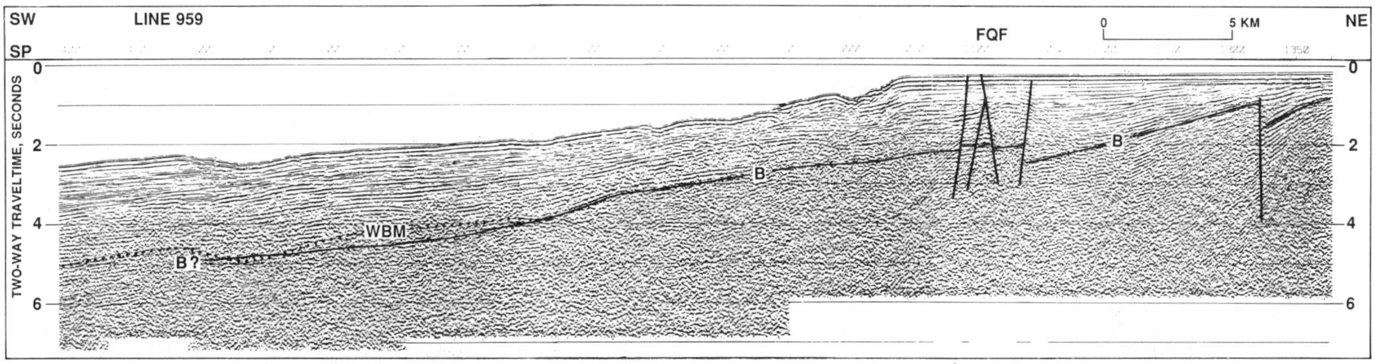

Figure 10. Portion of multichannel seismic line 959 showing shelf basin, Fairweather-Queen Charlotte fault zone (FQF), and thick, undeformed slope section seaward of the fault zone. Basement horizon B is inferred to continue seaward, but is obscured beneath the water-bottom multiple (WBM). Compare lack of deformation beneath the slope here with deformation seen on inner and outer structural zones to the south on lines 957 (Figure 9), 955 (Figure 12), and 951 (Figure 11). Also note that Transition fault trends into the line (Figure 5), but no evidence for faulting is seen. Location shown on Figure 5.

Hence, we suggest that the basins are filled with Neogene and Quaternary deposits rather than Paleogene and older rocks, which would presumably be more indurated.

At the southern end of the margin, the basin strata probably correlate with the Miocene and younger Skonun Formation of the Queen Charlotte Islands and Queen Charlotte Basin (Sutherland Brown, 1968; Shouldice, 1973; Yorath, this volume). At the northern end of the margin, the basin strata could be similar to the Miocene and younger glaciomarine rocks of the Yakutat terrane, which include the Yakataga Formation onshore (Plafker and Addicott, 1976) and the thick (as much as 6 km) correlative continental shelf and slope strata (Figures 2, 5; Bruns, 1983b, 1985).

Beneath the shelf adjacent to the basins, little subsurface structure can be interpreted on the seismic-reflection data. In two locations, we interpret faulting based partly on disrupted seismic reflections on the seismic data, and partly on velocity discontinuities of the near-surface rocks. The northern fault is located on the west side of the mouth of Chatham Strait (Figure 5B). The southern fault crosses the landward ends of lines 951, 953, and 954, south of the mouth of Chatham Strait (Figure 5B). Either fault could be the seaward extension of the Chatham Strait fault, although the northern fault is less likely to be because it cannot be followed southward onto line 954.

Slope and Continental Rise

Seismic-reflection data (Figures 5, 8-13) show that sediment thickness increases from at least 2 km near the shelf-slope break (200–m bathymetric contour) to at least 5 km at the base of the slope. The maximum thickness is unknown beneath the slope and rise because the acoustic basement is concealed beneath the water-bottom multiple on seismic profiles. Gravity modeling and refraction data indicate that low-density and low-velocity sedimentary rocks could be as much as 10 km thick beneath the slope near Dixon Entrance (von Huene, Shor, and Wageman, 1979).

The sedimentary section at the base of the slope is middle Miocene and younger, based on the middle Miocene age of oceanic magnetic anomalies on the underlying Pacific plate (Naugler and Wageman, 1973; Wilson, 1985; Berggren et al., 1985), and on correlations of reflections on single-channel and multichannel seismic-reflection data to Deep Sea Drilling Project (DSDP) hole 178 (von Huene, Shor, and Wageman, 1979; Bruns, 1983b, 1985). Near Cross Sound, Miocene strata may be as much as 3 km thick at the base of the slope and are overlain by an additional 3 km of probable Pliocene and younger strata (Figures 5, 13; Bruns, 1985). About 1.5 km of possible Miocene age strata is present at the base of the slope just south of line 951 near Dixon Entrance (Snavely et al., 1981).

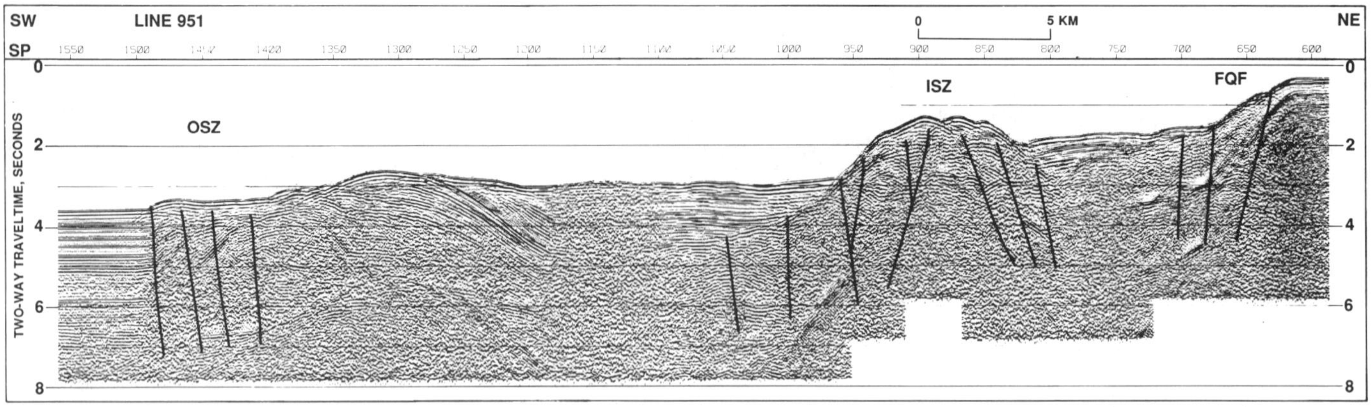

Figure 11. Portion of multichannel seismic line 951 just north of Dixon Entrance across Fairweather-Queen Charlotte fault zone (FQF), inner structural zone (ISZ), and outer structural zone (OSZ). Structural zones probably reflect wrench-fault deformation along strike-slip splays of the Fairweather-Queen Charlotte fault system. Location shown on Figures 5, 8.

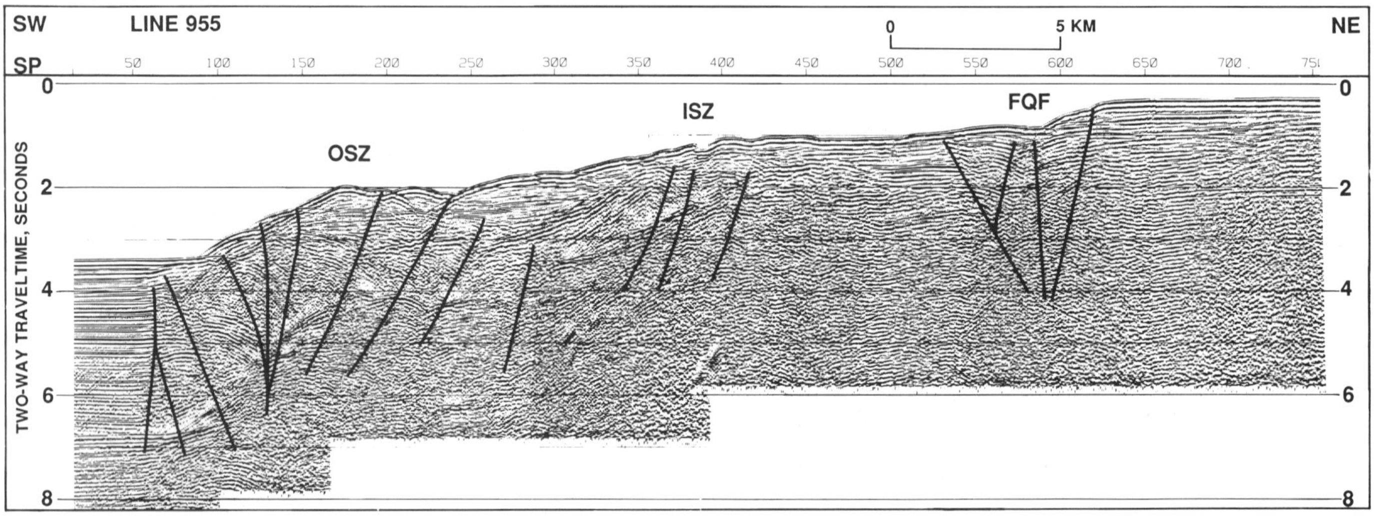

Figure 12. Portion of multichannel seismic line 955 west of south end of Baranof Island across Fairweather-Queen Charlotte fault zone, and inner and outer structural zones (ISZ and OSZ). As on line 951 (Figure 11), deformation probably reflects wrench-fault deformation along splays of Fairweather-Queen Charlotte fault (FQF), except that here the region affected by the structural zones narrows, and deformation becomes less intense on the inner structural zone. Location shown on Figures 5, 8.

These slope and rise strata are markedly deformed along the southern part of the margin, probably because of wrench-fault tectonics along the Fairweather-Queen Charlotte fault system (von Huene, Shor, and Wageman, 1979; Snavely et al., 1981; Bruns et al., 1984; Bruns, 1985). From Dixon Entrance to Chatham Straight, strata are deformed into broad, faulted folds along two structural zones, an outer structural zone (OSZ) and an inner structural zone (ISZ) (Figures 8, 9, 11, 12; see also Snavely et al., 1981). The western folds and faults in the outer structural zone deform the shallowest sedimentary strata and are likely Quaternary features. The folds of the inner structural zone are in places covered by as much as 0.5 km of undeformed strata (Figures 8, 9). These folds are therefore somewhat older than the western folds and perhaps developed in the Pliocene or early Pleistocene. Sparse bathymetric data (Chase, Menard, and Mammerickx, 1970; Seemann, 1982) suggest that individual structures within the zones form an en-echelon pattern, as is typical of deformation in strike-slip fault zones (Harding and Lowell, 1979). Here, this deformation reflects laterally distributed transform motion along the Fairweather-Queen Charlotte fault zone (Bruns, 1985; Snavely et al., 1981).

The degree of deformation in the two structural zones decreases from south to north (Figures 8-13), and may reflect the amount of oblique convergence between the Pacific and North American plates. The greatest amount of deformation in both structural zones occurs between Dixon Entrance and Chatham Strait, where the plate convergence direction is not quite parallel to the margin (lines 951–955, Figures 8, 11, 12). Between Chatham Strait and Sitka, only minor deformation has occurred (Figures 8, 9, line 957), and between Sitka and Cross Sound, where the plate convergence direction is parallel to the fault, strata seaward of the Fairweather-Queen Charlotte fault zone are undeformed (Figures 8, 10, line 959). Thus, north of Sitka, we do not find evidence for late Cenozoic faulting or for transform faults seaward of the shelf break. Faulting, if present, would have occurred prior to about Pliocene time and would have affected the lowermost part of the sedimentary section where the structure is

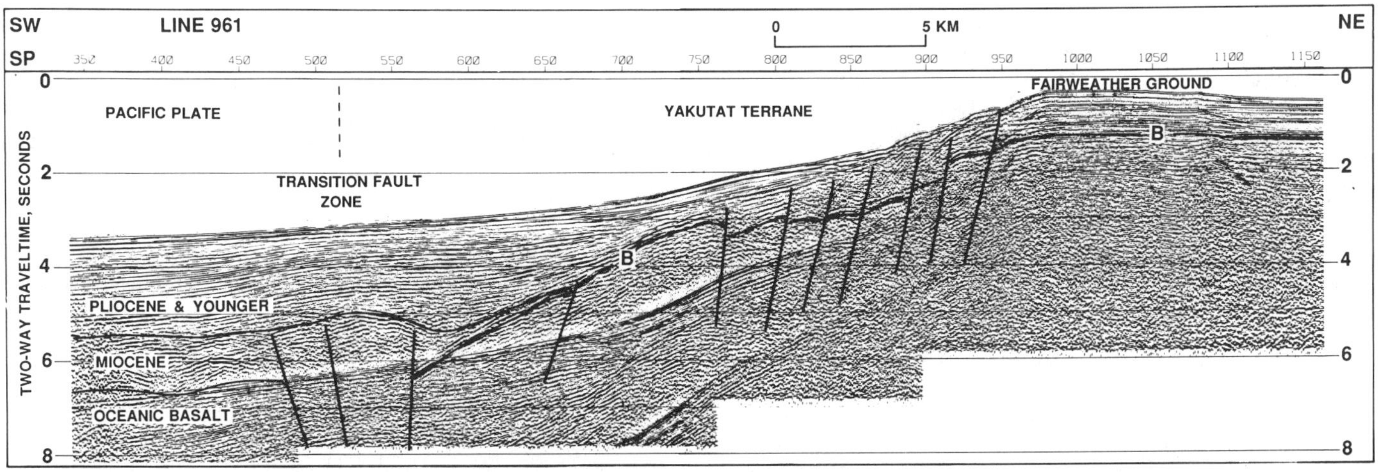

Figure 13. Portion of multichannel seismic line 961 across the southern margin of the Yakutat terrane, showing Fairweather Ground, the Transition fault, and the Pacific plate. Horizon B indicates approximate top of basement rocks, which beneath Fairweather Ground are of Cretaceous or older age (Plafker et al., 1980; Bruns, 1983b; 1985). Ages indicated for continental rise and oceanic rocks are after Bruns (1983b; 1985). Location of Transition fault is the region on the seismic line which separates continental margin basement rocks of the Yakutat terrane from oceanic basement rocks of the Pacific plate (Bruns, 1985). Minor faulting has affected the basement rocks of the Yakutat terrane, but strata beneath the rise are undeformed. This lack of deformation suggests that the Transition fault is not presently active as either a transform fault or a subduction zone fault and has not been throughout the Pliocene and Quaternary. There is thus no obvious active connection between the active Fairweather-Queen Charlotte fault system and the Transition fault. Location of line shown on Figures 5, 8.

obscured by the water-bottom multiple on the seismic-reflection data.

DISCUSSION

Faults along southeastern Alaska must have accommodated transform motion between the Pacific and North American plates during the late Cenozoic. Presently, this motion is being taken up almost entirely along the onshore Fairweather fault, but only about 5.5 km of offset has been recognized onshore (Plafker et al., 1978). Most of the late Cenozoic Pacific-North American plate motion may have been accommodated elsewhere than on the onshore Fairweather fault and its connection to the eastern set of offshore traces. Perhaps motion was largely taken up on the westernmost fault traces mapped offshore (trending into the Icy Point-Lituya Bay fault), or landward or seaward of the mapped traces of the Fairweather-Queen Charlotte fault system.

Faults located landward (east) of the Fairweather-Queen Charlotte fault system may have accommodated some late Cenozoic transform motion. At least two, the Chatham Strait and Peril Strait faults (Figures 1, 5) have moved since the Cretaceous. About 150 km of right-lateral offset has occurred on the Chatham Strait fault between post-middle-Cretaceous and pre-Holocene time. About 100 km of this offset may have occurred during post-Oligocene time, based on offset of an Oligocene volcanic sequence (Hudson, Plafker, and Dixon, 1982). About 11 km of right-lateral separation has occurred on the Peril Strait fault since the Late Cretaceous (Plafker et al., 1976). Thus, known offset on fault systems in southeast Alaska can accommodate only a small part of the Cenozoic Pacific-North American plate motion.

Seaward of the Fairweather-Queen Charlotte fault system, there has probably not been faulting during at least the Pliocene and Quaternary. Structures within the two structural zones south of Sitka are probably caused by displacements on seaward splays of the Fairweather-Queen Charlotte fault system. However, north of Sitka, there is no evidence for major faulting in strata beneath the slope or rise. These thick, undeformed strata suggest to us that transform faulting has not occurred seaward of the shelf-slope break during about the Pliocene and Quaternary.

Thus, the faults mapped as the Fairweather-Queen Charlotte fault system by Carlson, Plafker, and Bruns (1985) appear to have taken up much of the late Cenozoic motion between the Pacific and North American plates. As much as 70 km of possible late Pleistocene and Holocene offset can be inferred along the system based on offsets of glacially carved Pleistocene sea valleys (Carlson et al., 1982). In comparison, there is no evidence for throughgoing late Cenozoic faults seaward of these mapped traces, and limited Cenozoic offset is known landward on the Chatham Strait and Peril Strait faults. Thus, we conclude that the Fairweather-Queen Charlotte fault system, as mapped from seismic-reflection data, has been the major transform boundary between the Pacific and North American plates during about the Pliocene and Quaternary. We do not know how motion has been distributed along the various faults that make up the system, except that the western set faults has been the most active.

The movement history of the Yakutat terrane is in part constrained by the structure and stratigraphy southwest of Cross Sound. This area lies at the junction between the Transition fault, which separates the Yakutat terrane from the Pacific plate, and the Fairweather-Queen Charlotte fault system, which separates the southeast Alaska margin from the Pacific plate (Figures 1, 5A).

Seismic lines 959 and 961 (Figures 8, 10, 13) illustrate the thick (greater than 2 km) undeformed sedimentary section that suggests the Transition fault is inactive. On both lines, the undeformed section lies over the projected location of the Transition fault. Deformation similar to that observed in late Cenozoic strata along the active Fairweather-Queen Charlotte fault system south of Sitka is not seen along the Transition fault; such deformation might be expected if the Transition fault was active as a transform fault. Alternatively, if the Transition fault was an active oblique subduction zone, then, with a thick sedimentary section entering the subduction zone, an extensive accretionary wedge should exist at the base of the slope. The absence of an accretionary wedge and the lack of deformation along the base of the slope north of Sitka suggests that neither subduction nor transform mechanisms have been active along the Transition fault during the late Cenozoic, and that the Yakutat terrane has been moving with the Pacific plate past southeast Alaska for at least that time.

PETROLEUM GEOLOGY

Of the major criteria required for petroleum generation and accumulations—the presence of thick sedimentary rocks, good source and reservoir rocks, thermal maturity, and availability of traps and seals—we have direct information only about sediment thickness and traps. Thick accumulations of Miocene and younger strata, perhaps in excess of 10 km, are present beneath the slope and continental rise, and some potential anticlinal traps exist within these strata. No direct knowledge of thermal gradient, organic content of possible source rocks, or porosity and permeability of potential reservoir rocks exists. However, from the regional geology, we can speculate about these parameters. In adjacent areas, potentially petroliferous rocks are present beneath the Queen Charlotte Basin (Yorath and Hyndman, 1983; Yorath, this volume) and beneath the Yakutat terrane (Plafker and Claypool, 1979; Plafker et al., 1980; Bruns and Plafker, 1982; Bruns, 1983b, 1985).

In the Queen Charlotte Basin, thick (as much as 4.5 km) marine strata are present with favorable potential for both source and reservoir rocks. However, the strata of the basin become nonmarine and thin to the north, the organic matter becomes more terrigenous and gas prone, and the petroleum potential decreases (Yorath, this volume). Thus, if these strata continue into the basins off southeast Alaska, the petroleum potential may be poor.

The Yakutat terrane is underlain by as much as 4 km of Paleogene strata that have both source and reservoir potential (Plafker et al., 1980; Bruns, 1983b; Plafker, this volume). Exploratory drilling in these rocks has not discovered economically recoverable oil, but oil in onshore seeps probably originates from organic matter within these Paleogene strata (Bruns and Plafker, 1982). If the Yakutat terrane originated seaward of Dixon Entrance, as speculated by Plafker (1983, and this volume), Paleogene rocks equivalent to those of the Yakutat terrane may lie beneath the southeast Alaska margin. The presence of these rocks would significantly enhance the petroleum potential of the southeast Alaska margin because of a better possibility for finding source and reservoir rocks than we presently estimate. If, alternatively, the Yakutat terrane came from farther south, as discussed earlier (Bruns, 1983a, 1985; Keller et al., 1984), we cannot use the Yakutat terrane as an analog for the southeast Alaska margin. Thus, deciphering the movement history of the Yakutat terrane has immediate impact on assessing the petroleum potential of the southeast Alaska margin.

Source Rocks

The thick sedimentary section beneath the slope and rise is partly to largely composed of deep-sea turbidite fans (see Stevenson and Embley, this volume). The minimum of about 0.4% organic carbon content considered necessary for hydrocarbon genesis (Hunt, 1979) can be present in such fan sequences. For example, at Deep Sea Drilling Project sites 177 off Vancouver Island and 178 off Kodiak Island, organic carbon values were commonly around 0.5–0.6%, and maximum values of greater than 1.0% were found in a few places (Bode, 1973; Kvenvolden and von Huene, 1985). The organic carbon from site 178 was primarily from terrestrial sources, would generate mainly gas, and was immature with respect to gas generation (Kvenvolden and von Huene, 1985). Similar strata are likely to be present beneath the slope and rise off southeast Alaska.

The source-rock potential of the shelf basin strata may be poor. In adjacent areas, the glaciomarine Miocene and younger rocks of the Yakutat terrane have poor source-rock potential (Plafker and Claypool, 1979; Plafker et al., 1980; Plafker, this volume), and the Queen Charlotte Basin rocks are nonmarine and thin just south of Dixon Entrance (Yorath, this volume).

Thermal Maturity

The geothermal gradient of the southeast Alaska margin is unknown. In the nearby Yakutat terrane shelf, Bruns (1983b) estimated a geothermal gradient of around 25°C/km, with a range of 20° to 30°C/km easily possible. Yorath and Hyndman (1983) found that the geothermal gradient in Queen Charlotte Basin wells ranged from 22° to 31°C/km. If a 25°C/km gradient for the southeast Alaska margin is assumed, rocks below a depth of about 4 km could be thermally mature. Substantial areas of the slope and rise are probably underlain by more than 4 km of strata. Thus, large volumes of sediment are potentially mature for petroleum generation.

Reservoir Rocks

Thick turbidite packages are likely to be present in slope and rise rocks, and could possibly have reservoir potential. However, upper Miocene and younger strata in the northern Gulf of Alaska are dominantly glacially derived (Plafker and Addicott, 1976; Stevenson and Embley, this volume) and contain large amounts of glacial flour and unstable rock fragments that weather to clay during burial diagenesis (Winkler, McLean, and Plafker, 1976). For these reasons, drilling in the Yakutat terrane did not recover good reservoir rocks in the Miocene and younger sequences. Similar poor reservoir characteristics could be expected within the southeast Alaska shelf basin sedimentary sections. However, beneath the slope and rise, sands with good reservoir characteristics could be present within turbidite packages. Hydrocarbons could migrate along sand layers, and be trapped in sand layers sealed by fine-grained sediment.

Traps

The large structures beneath the continental slope south of Sitka could trap migrating hydrocarbons. Also, sand packages within turbidite sequences could form stratigraphic traps. No potential traps are known on the shelf. Too few data are available to determine the extent of or possible closure on the known structural traps, or to evaluate the possibility of structural or stratigraphic traps elsewhere.

Petroleum Potential Summary

Thick sedimentary strata (as much as 10 km) are present along the southeast Alaska margin, and even with fairly cool geothermal gradients, organic matter could be mature enough for petroleum generation within these sequences. Structures are present that could serve as traps. Little is known about other factors that may lead to petroleum generation or accumulation within the region. Adequate evaluation of the petroleum potential of the southeast Alaska margin requires further information about the stratigraphy, structure, organic geochemistry, thermal gradient, and reservoir rock potential of the margin.

ACKNOWLEDGMENTS

We thank Alan K. Cooper and Andrew J. Stevenson for critical reviews, Holly F. Ryan and Dennis M. Mann for aid in processing the multichannel seismic data, and Douglas M. Rearic for drafting aid.

REFERENCES CITED

Atwood, T. J., T. R. Bruns, P. R. Carlson, B. F. Molnia, and G. Plafker, 1981, Bathymetric maps of the northern Gulf of Alaska: U.S. Geological Survey Miscellaneous Field Studies Map MF-859, 3 sheets, scale 1:250,000.

Berg, H. C., 1979, Significance of geotectonics in the metallogenesis and resource appraisal of southeastern Alaska, a progress report: U.S. Geological Survey Circular 804–B, p. 116–118.

Berg, H. C., D. L. Jones, and P. J. Coney, 1978, Map showing pre-Cenozoic tectonostratigraphic terranes of southeastern Alaska and adjacent areas: U.S. Geological Survey Open-File Report 78–1085, 2 sheets, scale 1:1,000,000.

Berggren, W. A., D. V. Kent, J. J. Flynn, and J. A. van Couvering, 1985, Cenozoic geochronology: GSA Bulletin, v. 96, p. 1407–1418.

Bode, G. W., 1973, Appendix VI. Carbon and carbonate analysis, Leg 18, *in* L. D. Kulm, R. von Huene, et al., Initial reports of the Deep Sea Drilling Project (Vol. 18): Washington, D.C., U.S. Government Printing Office, p. 1069–1076.

Bruns, T. R., 1979, Late Cenozoic structure of the continental margin, northern Gulf of Alaska, *in* A. Sisson, ed., The relationship of plate tectonics to Alaskan geology and resources: Anchorage, Alaska Geological Society Symposium Proceedings 1977, p. I1–I30.

Bruns, T. R., 1981, Structure of the continental slope along the Fairweather fault system, southeastern Alaska: American Geophysical Union Transactions, v. 62, no. 6, p. 60.

Bruns, T. R., 1983a, A model for the origin of the Yakutat block, an accreting terrane in the northern Gulf of Alaska: Geology, v. 11, p. 718–721.

Bruns, T. R., 1983b, Structure and petroleum potential of the Yakutat segment of the northern Gulf of Alaska continental margin: U.S. Geological Survey Miscellaneous Field Studies Map MF–1480, 22 p., 3 sheets, scale 1:500,000.

Bruns, T. R., 1984, Reply concerning "Model for the origin of the Yakutat block, an accreting terrane in the northern Gulf of Alaska": Geology, v. 12, no. 9, p. 563–564.

Bruns, T. R., 1985, Tectonics of the Yakutat block, an allochthonous terrane in the northern Gulf of Alaska: U.S. Geological Survey Open-File Report 85–13, 112 p.

Bruns, T. R., and G. Plafker, 1982, Geology, structure, and petroleum potential of the southeastern Alaska and northern Gulf of Alaska continental margins, *in* T. R. Bruns, ed., Hydrocarbon resource report for proposed OCS lease sale 88: southeastern Alaska, northern Gulf of Alaska, Cook Inlet, and Shelikof Strait, Alaska: U.S. Geological Survey Open-File Report 82–928, 133 p.

Bruns, T. R., and W. C. Schwab, 1983, Structure and seismic stratigraphy of the Yakataga segment of the continental margin, northern Gulf of Alaska: U.S. Geological Survey Miscellaneous Field Studies Map MF–1424, 20 p., 4 sheets, scale 1:250,000.

Bruns, T. R., T. J. Atwood, and J. R. Childs, 1981, Free-air gravity anomaly map, Cross Sound to Icy Bay, northern Gulf of Alaska: U.S. Geological Survey Miscellaneous Field Studies Map MF–1306, 1 sheet, scale 1:500,000.

Bruns, T. R., J. R. Childs, and T. J. Atwood, 1981, Preliminary free-air gravity anomaly map, Dixon Entrance to Cross Sound, northeastern Gulf of Alaska: U.S. Geological Survey Miscellaneous Field Studies Map MF–1307, 1 sheet, scale 1:500,000.

Bruns, T. R., R. von Huene, P. R. Carlson, and G. Keller, 1984, The eastern Gulf of Alaska transcurrent plate boundary and the migration of terranes: Lille, France, Juin 1984, Annales de la Societe Geologique du Nord, v. CIII, p. 325–331.

Carlson, P. R., G. Plafker, and T. R. Bruns, 1985, Map and selected seismic profiles of the seaward extension of the Fairweather fault, eastern Gulf of Alaska: U.S. Geological Survey Miscellaneous Field Studies Map MF–1722, 2 sheets.

Carlson, P. R., T. R. Bruns, B. F. Molnia, and W. C. Schwab, 1982, Submarine valleys in the northeastern Gulf of Alaska: Characteristics and probable origin: Marine Geology, v. 47, p. 217–242.

Carlson, P. R., G. Plafker, T. R. Bruns, and W. P. Levy, 1979, Seaward extension of the Fairweather fault: U.S. Geological Survey Circular 804B, p. 135–139.

Chase, T. E., H. W. Menard, and J. Mammerickx, 1970, Bathymetry of the north Pacific, sheet 4: San Diego, University of California, Scripps Institute of Marine Resources, Technical Reports Series TR-9.

Harding, T. P., and J. D. Lowell, 1979, Structural styles, their plate tectonic habitats, and hydrocarbon traps in petroleum provinces: AAPG Bulletin, v. 63, p. 1016–1058.

Hudson, T., G. Plafker, and K. Dixon, 1982, Horizontal offset history of the Chatham Strait fault: U.S. Geological Survey Circular 844, p. 128–132.

Hunt, J. M., 1979, Petroleum geochemistry and geology: San Francisco, W. H. Freeman and Co., 617 p.

Jones, D. L., N. J. Silberling, H. C. Berg, and G. Plafker, 1981, Map showing tectonostratigraphic terranes of Alaska, columnar sections, and summary description of terranes: U.S. Geological Survey Open-File Report 81–792, scale, 1:2,500,000.

Keller, G., R. von Huene, K. McDougall, and T. R. Bruns, 1984, Paleoclimatic evidence for Cenozoic migration of Alaskan terranes: Tectonics, v. 3, no. 4, p. 473–495.

Kvenvolden, K. K., and R. von Huene, 1985, Natural gas generation in sediments of the convergent margin of the eastern Aleutian Trench area, *in* D. G. Howell, ed., Tectonostratigraphic terranes of the circum-Pacific region: Houston, Texas, Circum-Pacific Council for Energy and Mineral Resources Earth Science Series 1, p. 31–49.

Minster, J. B., and T. H. Jordan, 1978, Present-day plate motions: Journal of Geophysical Research, v. 83, no. 811, p. 5331–5354.

Monger, J. W. H., and H. C. Berg, 1984, Lithotectonic terrane map of western Canada and southeastern Alaska, *in* N. J. Silberling and D. L. Jones, eds., Lithotectonic terrane maps of the North American Cordillera: U.S. Geological Survey Open-File Report 84–523, scale, 1:2,500,000.

Naugler, F. P., and J. M. Wageman, 1973, Gulf of Alaska: Magnetic anomalies, fracture zones, and plate interactions: GSA Bulletin, v. 84, p. 1575–1584.

Page, R. A., 1975, Evaluation of seismicity and earthquake shaking at offshore sites: Houston, Texas, 7th Offshore Technology Conference Proceedings, v. 3, p. 179–190.

Plafker, G., 1967, Geologic map of the Gulf of Alaska Tertiary province, Alaska: U.S. Geological Survey Miscellaneous Geologic Investigations Map I–484, scale 1:500,000.

Plafker, G., 1983, The Yakutat block: An actively accreting tec-

tonostratigraphic terrane in southern Alaska (abs.): GSA Abstracts with Programs, v. 15, no. 5, p. 406.
Plafker, G., 1984, Comment on "Model for the origin of the Yakutat block, an accreting terrane in the northern Gulf of Alaska": Geology, v. 12, no. 9, p. 563.
Plafker, G., 1987, Regional geology and petroleum potential of the northern Gulf of Alaska continental margin, this volume.
Plafker, G., and W. O. Addicott, 1976, Glaciomarine deposits of Miocene through Holocene age in the Yakataga Formation along the Gulf of Alaska margin, *in* T. P. Miller, ed., Recent and ancient sedimentary environments in Alaska: Anchorage, Alaska Geological Society Symposium Proceedings 1975, p. Q1–Q23.
Plafker, G., and G. Claypool, 1979, Petroleum source potential of rocks dredged from the continental slope in the eastern Gulf of Alaska: U.S. Geological Survey Open-File Report 79–295, 24 p.
Plafker, G., T. Hudson, T. R. Bruns, and M. Rubin, 1978, Late Quaternary offset along the Fairweather fault and crustal plate interactions in southern Alaska: Canadian Journal of Earth Sciences, v. 15, no. 5, p. 805–816.
Plafker, G., D. L. Jones, T. Hudson, and H. C. Berg, 1976, The Border Ranges fault system in the Saint Elias Mountains and Alexander Archipelago: U.S. Geological Survey Circular 733, p. 14–16.
Plafker, G., G. R. Winkler, W. L. Coonrad, and G. Claypool, 1980, Preliminary geology of the continental slope adjacent to OCS lease sale 55, eastern Gulf of Alaska; petroleum resource implications: U.S. Geological Survey Open-File Report 80–1089, 72 p.
Seemann, D. A., 1982, Bathymetry off the coast of British Columbia: Sidney, B. C., Canada, Pacific Geoscience Centre, Energy, Mines and Resources, Earth Physics Branch Open-File 82–25.
Shouldice, D. N., 1973, Western Canadian continental shelf, *in* R. G. McGrossan, ed., Future petroleum provinces of Canada: Canadian Society of Petroleum Geologists Memoir 1, p. 7–35.
Snavely, P. D. Jr., H. C. Wagner, D. H. Tompkins, and D. L. Tiffin, 1981, Preliminary geologic interpretation of a seismic reflection profile across the Queen Charlotte Island fault system off Dixon Entrance, Canada-United States: U.S. Geological Survey Open-File Report 81–299, 12 p.
Stevenson, A. J., and R. Embley, 1987, Deep-sea-fan bodies, terrigenous turbidite sedimentation, and petroleum geology, Gulf of Alaska, this volume.
Sutherland Brown, A., 1968, Geology of the Queen Charlotte Islands, British Columbia: British Columbia Department of Mines and Petroleum Resources Bulletin 54, 226 p.
Tocher, D., 1960, The Alaska earthquake of July 10, 1958—Movement on the Fairweather fault and field investigation of southern epicentral region: Seismological Society of America Bulletin, v. 50, p. 267–292.
van Alstine, D. R., D. R. Bazard, and J. W. Whitney, 1985, Paleomagnetism of cores from the Yakutat well, Gulf of Alaska (abs.): American Geophysical Union Transactions, v. 66, no. 46, p. 865.
von Huene, R., G. G. Shor, Jr., and J. Wageman, 1979, Continental margins of the eastern Gulf of Alaska and boundaries of tectonic plates, *in* J. S. Watkins and L. Montadert, eds., Geological and geophysical investigations of continental margins: AAPG Memoir 29, p. 273–290.
von Huene, R., G. Keller, T. R. Bruns, and K. McDougall, 1985, Cenozoic migration of Alaska terranes indicated by paleontologic study, *in* D. G. Howell, ed., Tectonostratigraphic terranes of the circum-Pacific: Houston, Texas, Circum-Pacific Council for Energy and Mineral Resources Earth Science Series 1, p. 121–136.
Wilson, D. S., 1985, Tectonic history of the Juan de Fuca Ridge: Stanford, California, Stanford University unpublished Ph.D. dissertation, 105 p.
Winkler, G., H. McLean, and G. Plafker, 1976, Textural and mineralogical study of sandstones from the onshore Gulf of Alaska Tertiary province, southern Alaska: U.S. Geological Survey Open-File Report 76–198, 48 p.
Yorath, C. J., 1987, Petroleum geology of the Canadian Pacific continental margin, this volume.
Yorath, C. J., and R. L. Chase, 1981, Tectonic history of the Queen Charlotte Islands and adjacent areas—A model: Canadian Journal of Earth Sciences, v. 18, p. 1717–1739.
Yorath, C. J., and R. D. Hyndman, 1983, Subsidence and thermal history of Queen Charlotte Basin: Canadian Journal of Earth Sciences, v. 20, p. 135–159.

CHAPTER THIRTEEN

PETROLEUM GEOLOGY OF THE CANADIAN PACIFIC CONTINENTAL MARGIN

C. J. Yorath

Geological Survey of Canada
Pacific Geoscience Centre
Sidney, B.C.

ABSTRACT

The geological architecture of the Canadian Pacific continental margin is expressed by a northwesterly aligned structural and stratigraphic fabric impressed upon igneous, metamorphic, and sedimentary rocks. The fundamental crustal components are represented by Paleozoic rocks of the Alexander terrane and by Paleozoic and Mesozoic rocks of Wrangellia. Beneath Vancouver Island, Wrangellia is underlain by an underplated oceanic assemblage of probable early Tertiary age and by the subducting Juan de Fuca plate.

The contemporary structure comprises the North American, Pacific, and Juan de Fuca plates, which, adjacent to Queen Charlotte Sound, meet at a triple junction. North of the junction the North American and Pacific plates are separated by the Queen Charlotte transform fault, along which the Pacific plate moves northward at about 55 mm/yr. To the south, the Juan de Fuca plate converges upon the North American plate at about 4 mm/yr and descends beneath the continent along a subduction zone that extends from the triple junction southward to the Mendocino fracture zone adjacent to northern California.

Of the several sedimentary basins along the margin, the Tofino, Queen Charlotte, and Georgia basins are considered to be the best prospects for petroleum. Initial hydrocarbon estimates for the offshore basins are: natural gas = $265 \times 10^9 m^3$; oil = $38.5 \times 10^6 m^3$ (each representing an "average expectation" at a 34% level of probability).

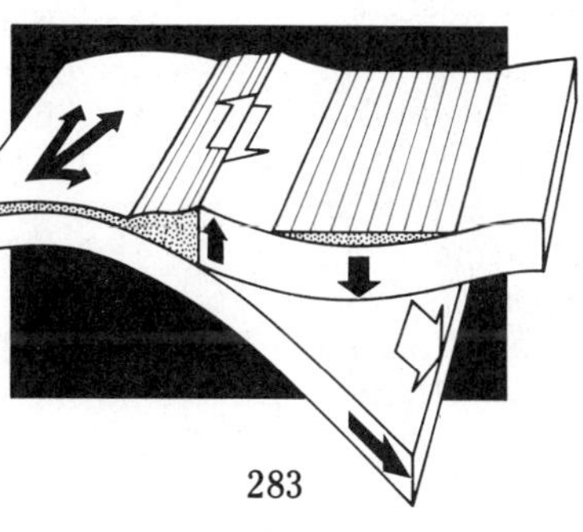

INTRODUCTION

The Canadian Pacific continental margin comprises the continental shelf and slope together with Vancouver Island and the Queen Charlotte Islands of the Insular Belt of the Canadian Cordillera. The region extends for about 1000 km from the Strait of Juan de Fuca northward to Dixon Entrance between the Queen Charlotte Islands and the Alexander Archipelago of southeastern Alaska; the enclosed area is about 450,000 km^2 (Figure 1).

The fundamental crustal architecture of the region is expressed by Paleozoic to mid-Mesozoic rocks of Wrangellia, which are thought to have amalgamated with the mainly Paleozoic rocks of the Alexander terrane in the latest Jurassic to earliest Cretaceous. The resultant Insular superterrane is then believed to have accreted to the Intermontane superterrane (mainly Stikinia, Quesnellia, and the Slide Mountain and Cache Creek terranes) in the mid-Cretaceous (Price, Monger, and Roddick, 1985). Other terranes such as the Pacific Rim and Olympic terranes are believed to have been juxtaposed adjacent to and beneath western and southern Wrangellia during the early Tertiary. Extensive discussion of the pre-mid-Cretaceous history of this region is beyond the scope of this chapter. Attention is devoted to the development of Cretaceous and Tertiary sedimentary basins that have potential for petroleum generation, entrapment, and recovery.

REGIONAL FRAMEWORK

The geological architecture of the continental margin is revealed by a northwesterly to northerly aligned stratigraphic, plutonic, and structural fabric. The contemporary crustal structure of the region is dominated by the Juan de Fuca, Pacific, and North American lithospheric plates which, adjacent to Queen Charlotte Sound, meet at a ridge-fault-trench triple junction. Beneath the continental shelf and slope adjacent to Vancouver Island, the Juan de Fuca plate descends beneath a subduction complex within the Tofino Basin. North of the triple junction the Queen Charlotte transform fault separates the North American plate from the northward-moving Pacific plate. Between the Queen Charlotte Islands and the mainland of British Columbia, the Queen Charlotte Basin contains Tertiary sedimentary and volcanic rocks that developed in a rift and flexural subsidence setting. Between Vancouver Island and the mainland, the Cretaceous to Tertiary Georgia Basin encloses a thick sedimentary section that has both fore-arc and foreland-basin characteristics.

QUEEN CHARLOTTE ISLANDS MARGIN

The Queen Charlotte Islands are traversed by three major fault systems (Figure 2). The Rennell Sound fault zone consists of a broadly curvilinear, northwesterly trending set of vertical to steep northeasterly dipping faults that display both dextral and east-side-down normal displacements (Sutherland Brown, 1968). Physiographically, the zone is most strongly expressed in the comparatively incompetent sediment of the middle to Upper Cretaceous Queen Charlotte Group, and by abrupt changes in stratigraphy across individual strands. Eastward, in the subsurface beneath Hecate Strait, some components of the system appear in single-channel and multichannel seismic profiles, where they dislocate Neogene and Pleistocene(?) sediment; the zone has been identified as coincident with the locus of suture between the Alexander terrane and Wrangellia (Figure 2; Yorath and Chase, 1981; Yorath et al., 1985a).

The Louscoone Inlet fault zone extends northwesterly from the southernmost Queen Charlotte Islands, through the eastern island passages to a point where it is truncated by the Rennel Sound fault (Figure 2). Described by Sutherland Brown (1968) as a component of the "Rennell Sound-Louscoone Inlet fault zone" the major strands of the system are vertical faults with apparent right-lateral displacements. On the basis of geophysical and stratigraphic data, Yorath and Chase (1981) proposed that the two components of the system are temporally and dynamically distinct, and that during the late Tertiary movement on the Louscoone Inlet system predated that on the Rennell Sound zone.

The Sandspit fault system extends northwesterly from northeastern Moresby Island to central Graham Island, beyond which the zone is masked by Quaternary drift and probably by the younger flows of the Neogene Masset Formation. Geophysical data support its continuation beyond northernmost Graham Island (Yorath and Chase, 1981). The youngest movements along the system are expressed by offset creeks, small scarplets in Pleistocene sediment, sag ponds, and inactive sea cliffs (Sutherland Brown, 1968). Older activity is indicated by the alignment of small, hypabyssal Tertiary plutons parallel with the system and by the dramatic increase in thickness of upper Neogene sediment of the Skonun Formation across the system into the subsurface beneath Hecate Strait. Yorath and Chase (1981) proposed that the Sandspit fault system was once continuous and colinear with the Louscoone Inlet fault zone, the two components having been dislocated by late Tertiary movement along the Rennell Sound fault zone; and, further, that it is colinear with the suture between Wrangellia and the Alexander terrane.

A few northeasterly trending vertical faults, principal among which is the Beresford Bay fault, cross northwestern Graham Island (Figure 2). These have a vertical component of separation and a significant horizontal sinistral displacement shown by the dislocation of Upper Cretaceous rocks of the Queen Charlotte Group.

In west-central Graham Island, H. W. Tipper (*in* Sutherland Brown, Yorath, and Tipper, 1983) suggested the presence of northeasterly dipping, fault-bounded panels that may reflect thrust separations.

The Queen Charlotte fault defines the edge of the North American craton in the Queen Charlotte Islands area (Figure 2). Although the fault primarily is the locus of strike-slip motion at an average rate of about 55 mm/yr (Riddihough, 1977; Keen and Hyndman, 1979), there is a discrepancy of between 10° and 20° between the strike of the fault off the southern islands and the direction of relative plate motion as predicted by global models (Minster and Jordan, 1978; Davis and Riddihough, 1982) that suggests a component of convergence along the boundary. Convergence and oblique underthrusting are supported by a wide range of data, including the presence of the Queen Charlotte terrace, an apparent accretionary sedimentary prism with arcuate, high-amplitude folds resting upon oceanic crust and bounded on its landward side by the Queen Charlotte fault (Figure 2).

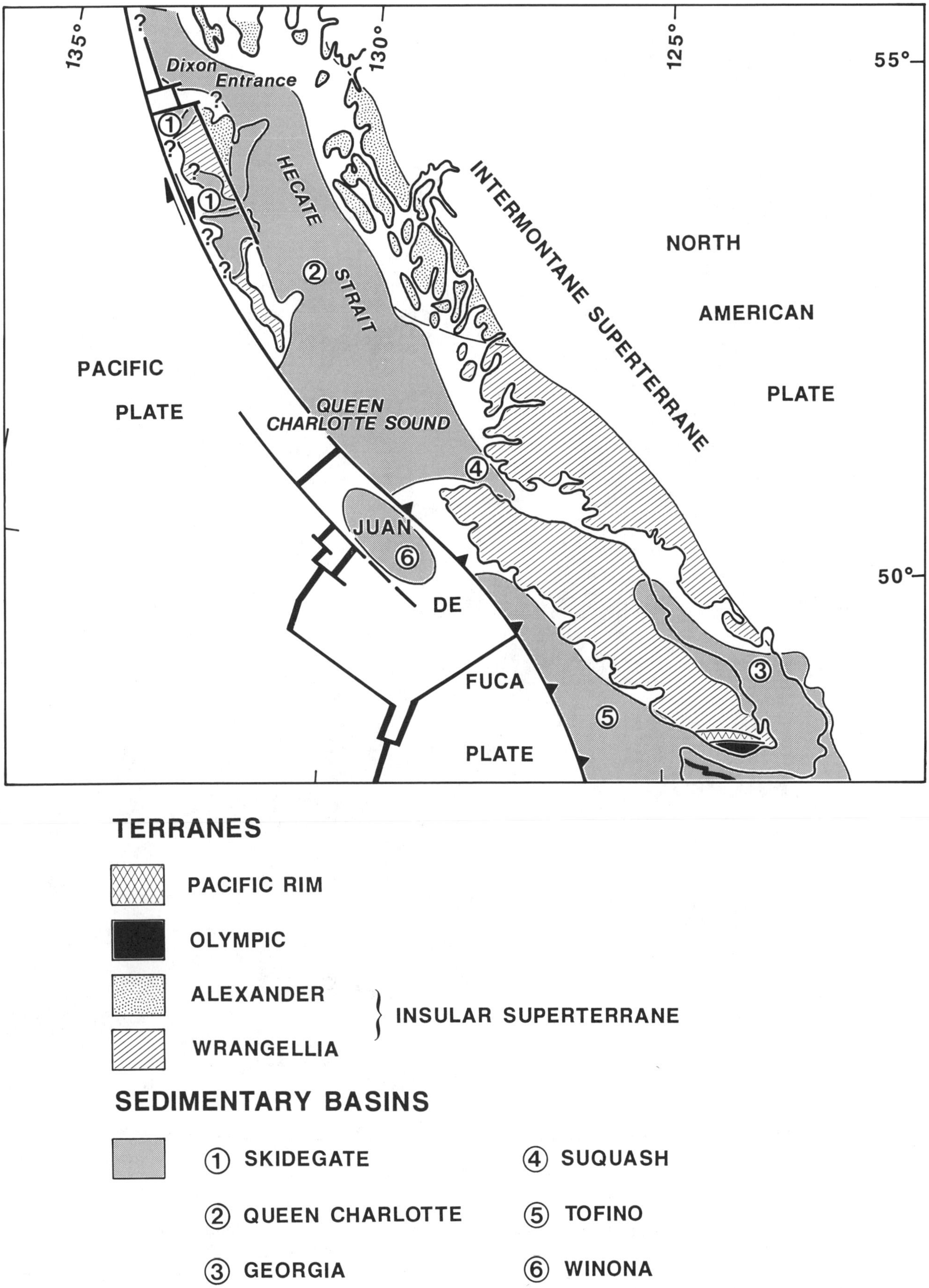

Figure 1. Location map showing the distribution of the principal terranes and sedimentary basins of the Canadian Pacific continental margin.

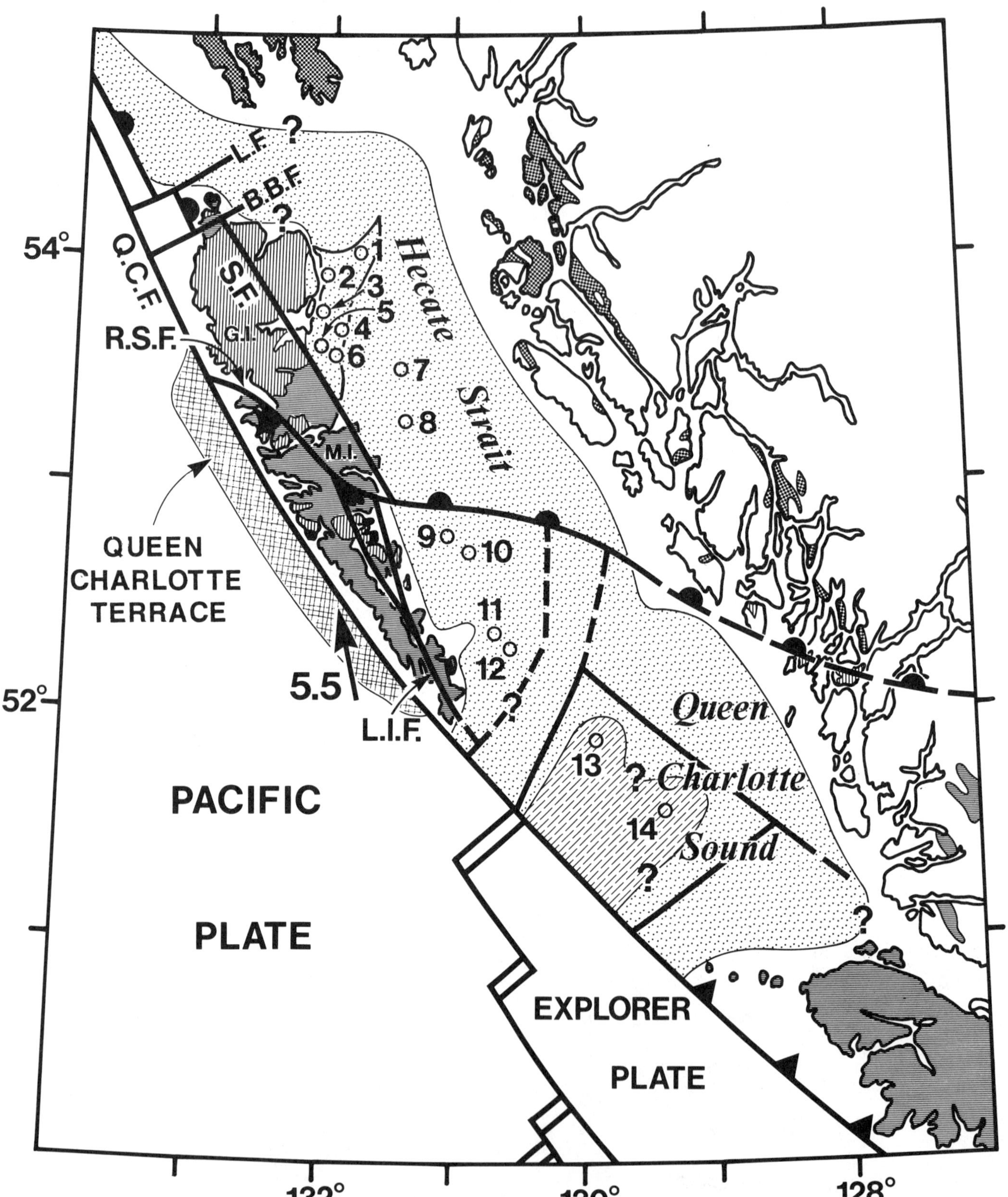

Figure 2. Simplified regional geology of the Queen Charlotte Islands continental margin (from Yorath and Hyndman, 1983).

On the Queen Charlotte Islands folds are subordinate to faults and, regionally, are not significant structures. Locally, however, and particularly within the Queen Charlotte Group adjacent to the Rennell Sound fault zone, steep upright and overturned panels attest to intense folding. Within the fault zone, Neocomian turbidite deposits of the Lower Cretaceous Longarm Formation are intensely sheared and bedding is transposed. Folds in older rocks are more westerly directed and are commonly truncated by the northwesterly trending

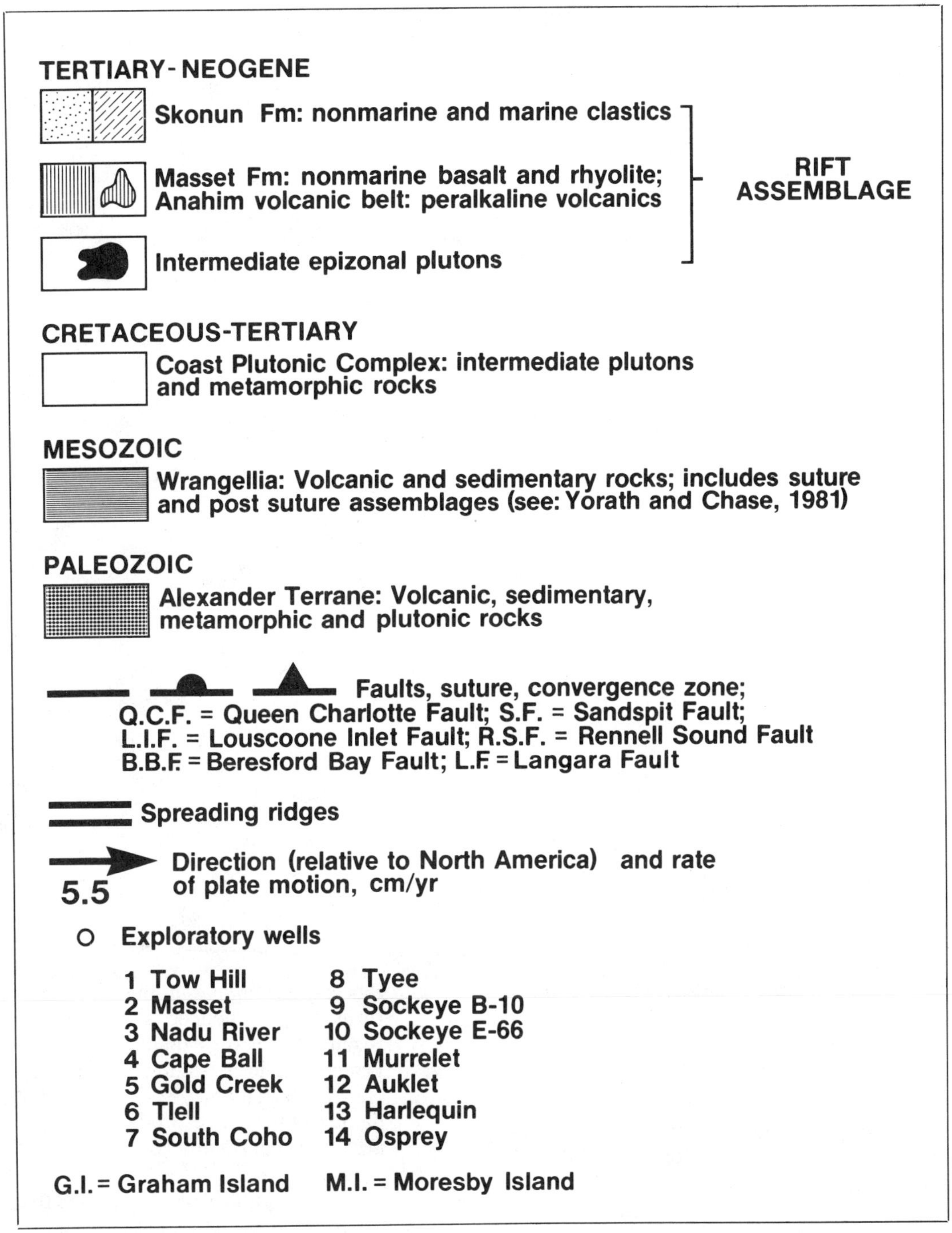

faults. Thick pillow lavas of the Upper Triassic Karmutsen Formation have been warped into broad, low-amplitude anticlines, whereas the immediately overlying thinly interbedded argillite and limestone of the Kunga Formation are commonly intensely deformed into tight chevron folds.

To express the tectonic history of the Queen Charlotte Islands, Yorath and Chase (1981) divided the stratigraphic succession into four tectonic assemblages:

(1) The allochthonous assemblages comprising Paleozoic rocks of the Alexander terrane and Upper Triassic and Jurassic rocks of Wrangellia. The distribution of these terranes beneath Hecate Strait and Queen Charlotte Sound (Figure 3) is inferred from differences in the geophysical potential-field signature north and south of the offshore continuation of the Rennell Sound fault zone; and from the surface and subsurface distribution of the two assemblages as revealed in outcrop and in exploratory bore holes.

(2) The suture assemblage represented by coarse conglomerate and turbidite deposits of Neocomian age (Longarm Formation) and possibly by Middle to Upper Jurassic meozonal plutons. The Longarm Formation represents the overlap assemblage between the Alexander terrane and Wrangellia (Figure 4).

(3) The postsuture assemblage expressed by conglomerate, sandstone, and shale of the mid- to Upper Cretaceous Queen Charlotte Group deposited in the

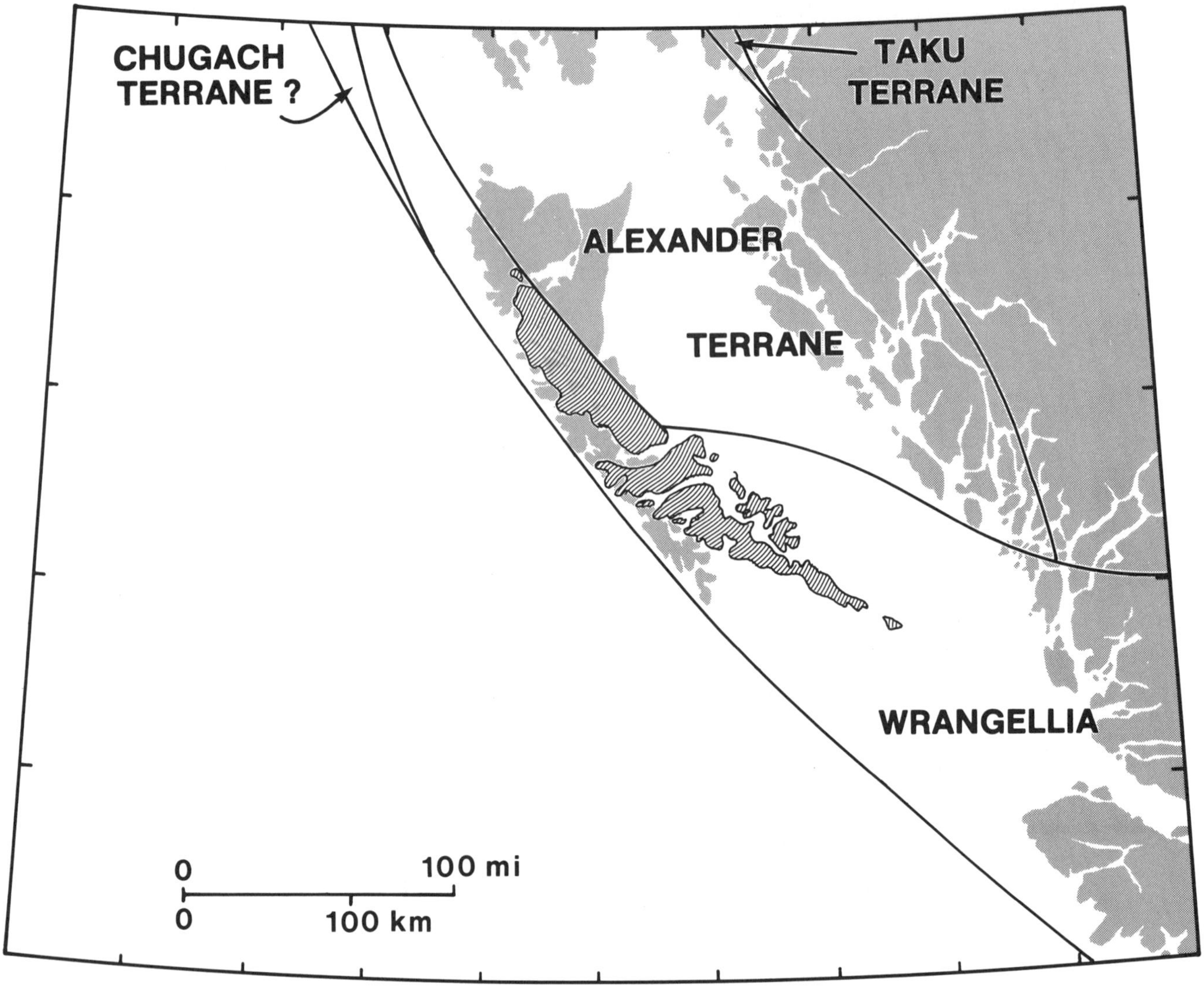

Figure 3. Locus of suture between the Alexander terrane and Wrangellia. The proto-Queen Charlotte Islands (wavy pattern) are shown in their prerift geometry and position (from Yorath and Chase, 1981).

Skidegate Basin (Figure 5). Yorath and Chase (1981) suggested that the conglomerate of the Honna Formation, probably of Coniacian age (J. Haggart and B. E. B. Cameron, oral communication, 1985), reflects the initial stages of accretion of the Insular to the Intermontane superterrane.

(4) The rift assemblages represented by mid- to upper Tertiary volcanic rocks, hypabyssal plutons, and marine and nonmarine clastic deposits (Figure 2). Rifting and crustal extension are believed to have occurred in Queen Charlotte Sound up to about 17 Ma, resulting in the proto-Queen Charlotte Islands being displaced about 70 km northward toward their present position by transcurrent motion along the Louscoone Inlet-Sandspit fault system (Figure 6). Subsequent reactivation of the ancient suture between Wrangellia and the Alexander terrane, manifest as the Rennell Sound fault, disrupted the continuity of the Louscoone Inlet-Sandspit fault system (Figure 7). The rifting generated a significant thermal anomaly and a restricted deep basin as a consequence of crustal thinning and subsequent thermal cooling. Beginning about 6 Ma, oblique underthrusting commenced along the continental margin, resulting in flexural uplift along the western part of the Queen Charlotte Islands and companion subsidence in Hecate Strait (Figure 8). The combination of rifting and flexural subsidence produced Queen Charlotte Basin (Figure 9; Yorath and Hyndman, 1983).

Sedimentary Basins

Two potentially important hydrocarbon-bearing sedimentary basins occur on and adjacent to the Queen Charlotte Islands—the mid-Cretaceous Skidegate Basin and the Neogene Queen Charlotte Basin.

Skidegate Basin

The Skidegate Basin (herein used for descriptive convenience) contains the Queen Charlotte Group of mid- to Late Cretaceous age. The group is exposed in three subbasinal areas in the northwestern, central, and south-central Queen Charlotte Islands (Figure 5); the original dimensions and geometry of the basin are poorly known, owing to erosion and subsequent development of the superimposed Queen Charlotte Basin beneath which equivalent strata presumably occur in Hecate Strait. Partly correlative strata occur on northwestern Vancouver Island in the Coal Harbour Group

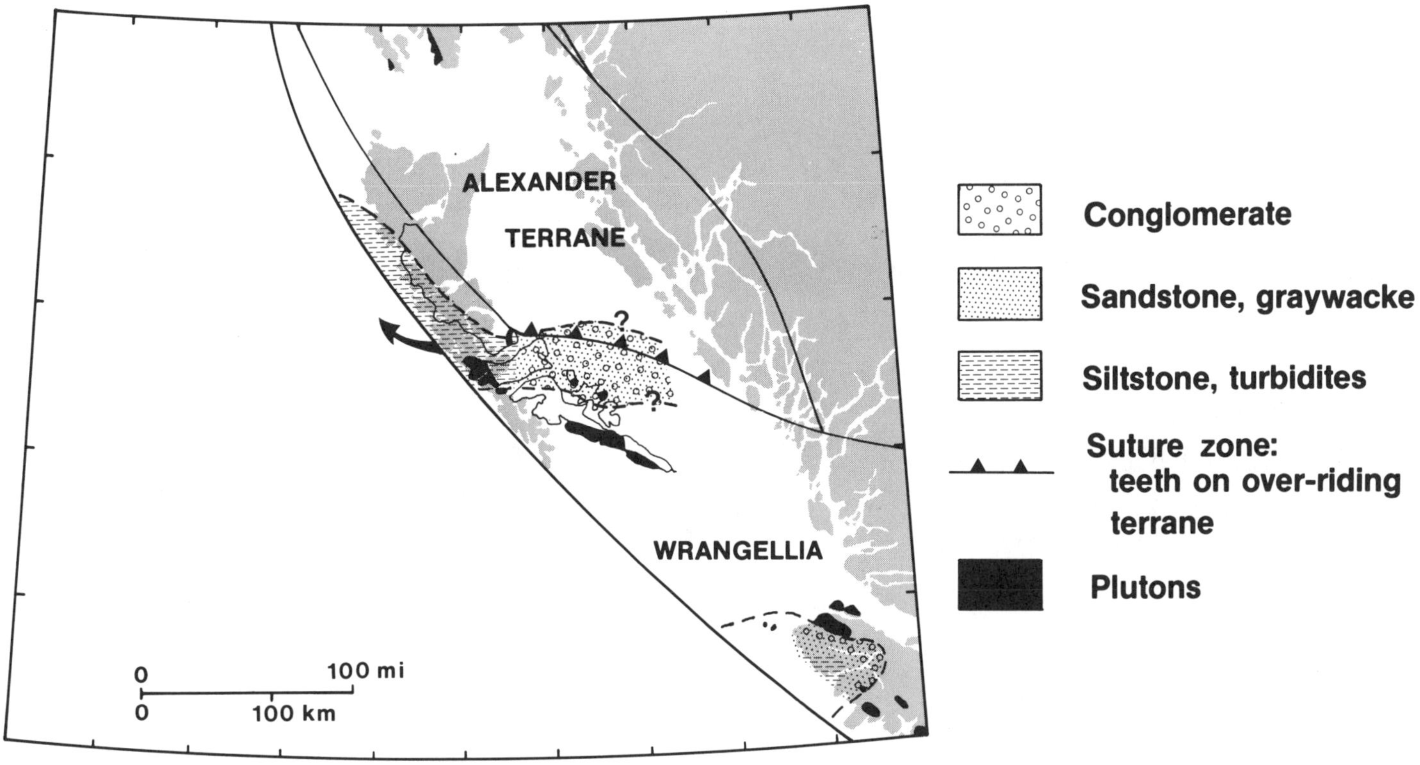

Figure 4. Paleogeography of the suture assemblage showing the distribution of Upper Jurassic plutons and lowermost Cretaceous sediment. The assemblage is shown as an overlap sequence across the suture between the Alexander terrane and Wrangellia. The proto-Queen Charlotte Islands are shown in their prerift geometry and position (from Yorath and Chase, 1981).

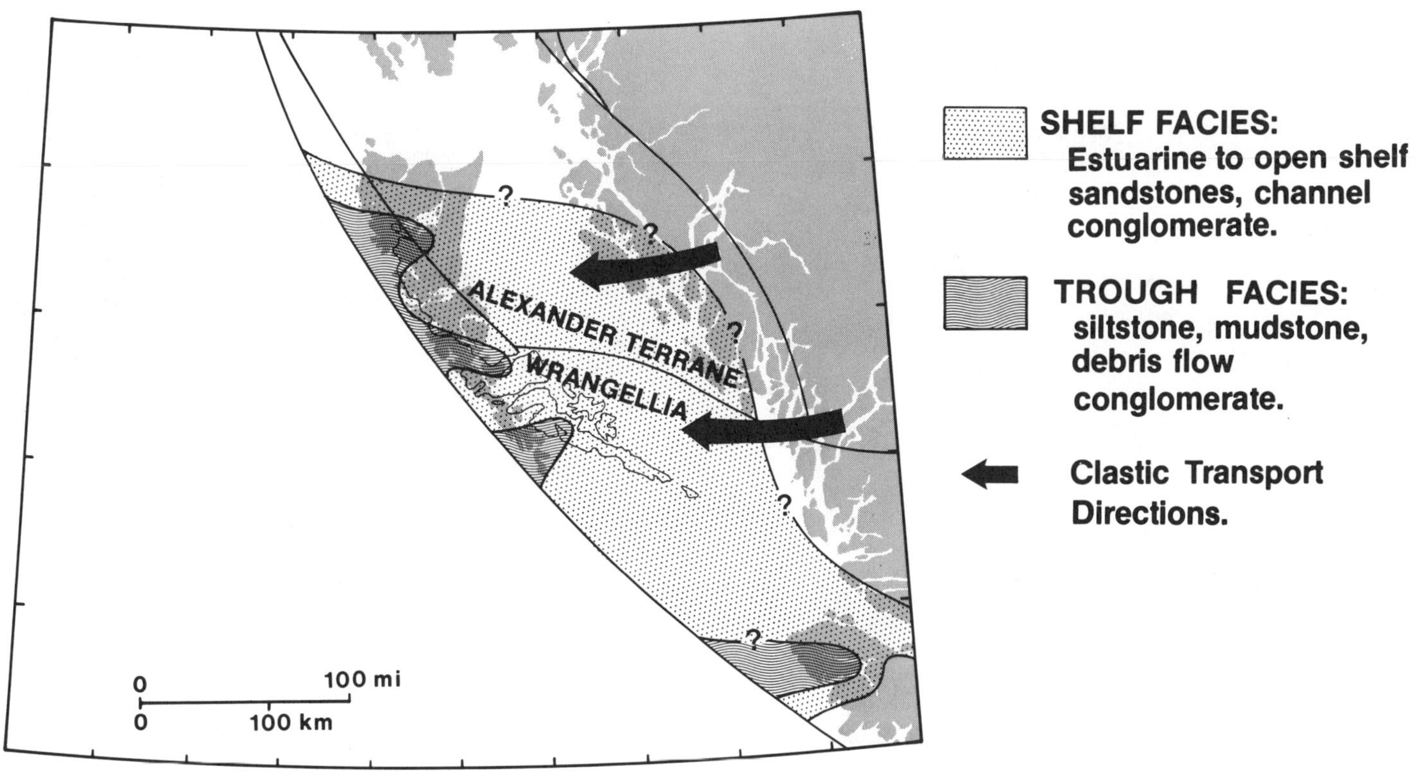

Figure 5. Paleogeography of the post suture assemblage (Queen Charlotte Group). The proto-Queen Charlotte Islands are shown in their prerift geometry and position (from Yorath and Chase, 1981).

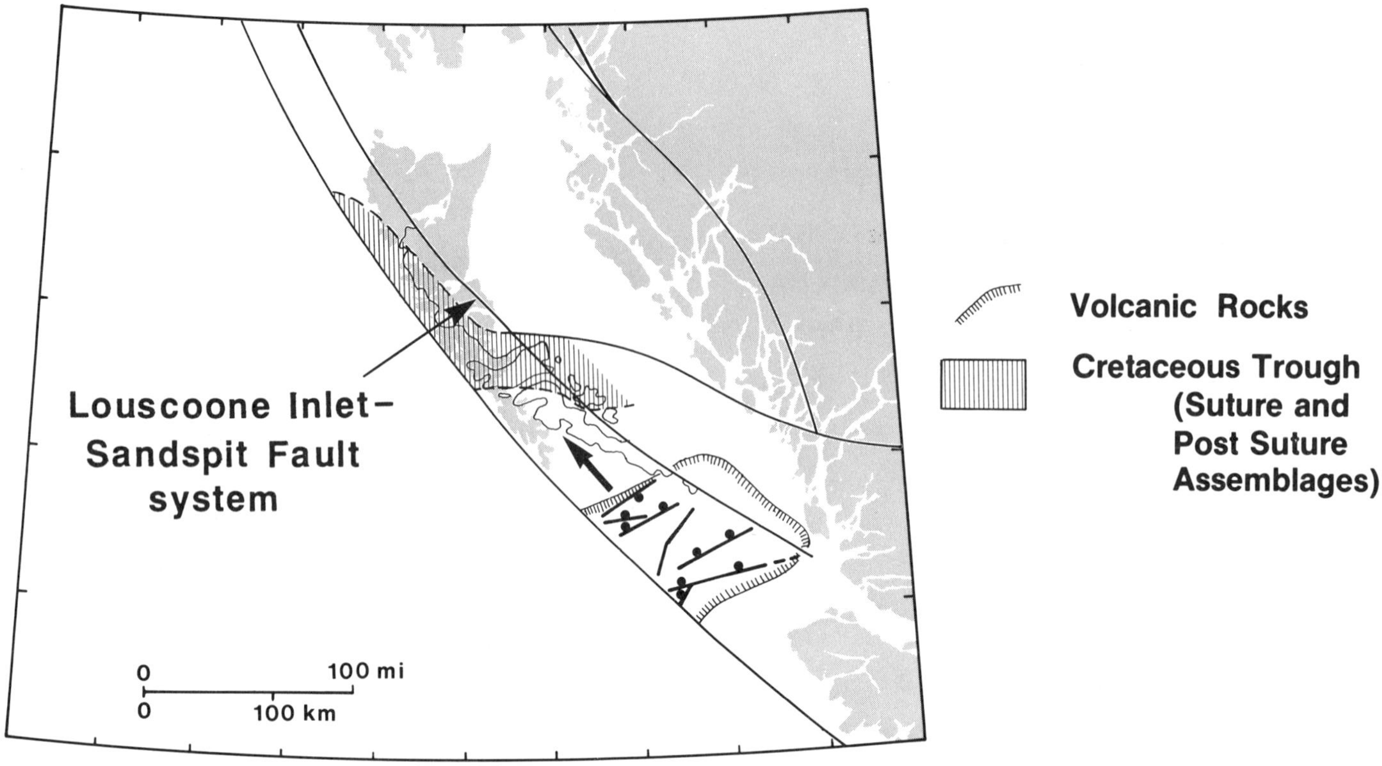

Figure 6. The rift assemblages—late Oligocene to early Miocene. The proto-Queen Charlotte Islands begin to move northward toward their present position by movement along the Louscoone Inlet-Sandspit fault system (from Yorath and Chase, 1981).

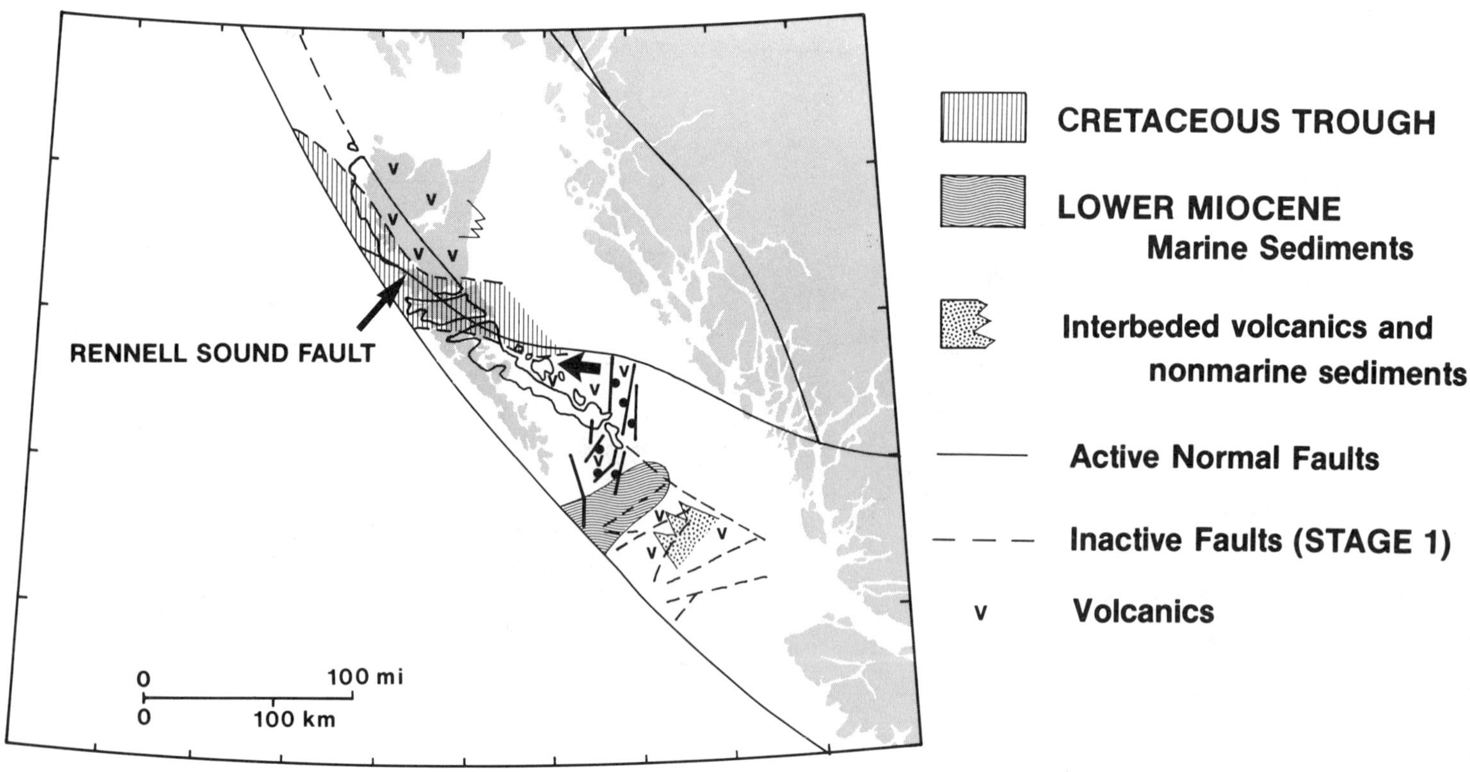

Figure 7. The rift assemblage—early Miocene. The proto-Queen Charlotte Islands have moved northward toward their present position. The Louscoone Inlet-Sandspit fault system is dislocated by movement along the Rennell Sound fault (from Yorath and Chase, 1981).

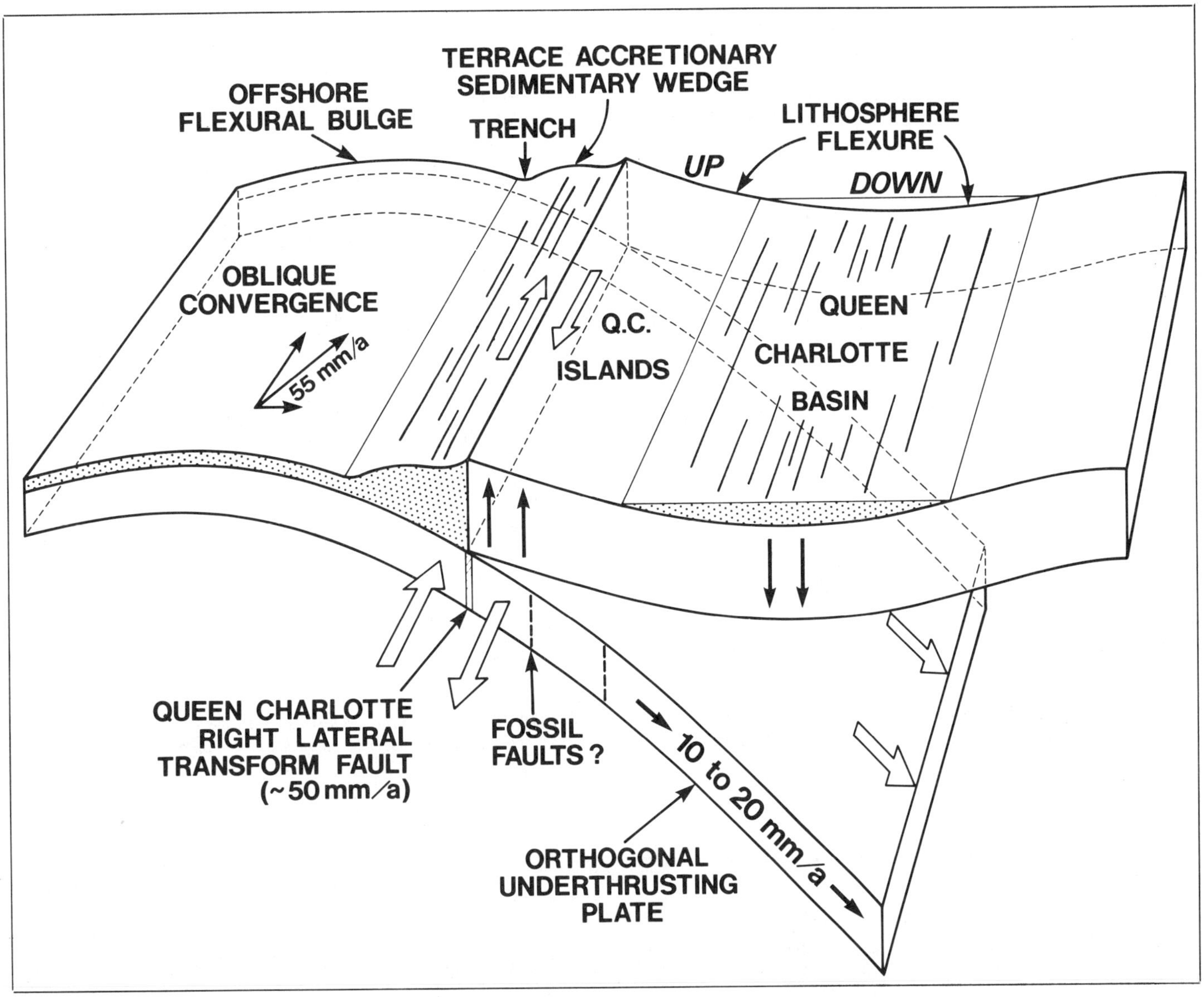

Figure 8. Oblique convergence and underthrusting beginning about 6 Ma. The western Queen Charlotte Islands are uplifted and Hecate Strait is depressed by lithospheric flexure, forming the Queen Charlotte Basin (from Yorath and Hyndman, 1983).

(Jeletzky, 1976). In the central Queen Charlotte Islands the Queen Charlotte Group consists of 2000 to 3000 m of conglomerate, sandstone, siltstone, mudstone, and shale that thicken westward toward the Rennell Sound fault zone. Sandstone varieties include black and white, coarse, pebbly, mature sandstone; grayish-green and green, fine- to medium-grained, immature glauconitic arkose; and grayish-brown, argillaceous and silty quartzose sandstone. Porosities are low but improve in the more mature types. Ironstone concretions are common as are small logs, coal pebbles derived from the Jurassic Yakoun Group, carbonized woody debris, pelecypod coquinas, and ammonites.

Sedimentary structures include low-angle cross-bedding, hummocky cross-stratification, convolute laminations, and grazing trails. The interpreted depositional environment of the sandstone is that of easterly derived clastic components deposited in a low-wave-energy, high-tidal-energy estuarine embayment surrounded by subdued topography underlain by Jurassic volcanic edifices (Sutherland Brown, 1968; Sutherland Brown, Yorath, and Tipper, 1983; J. Harper and B. D. Bornhold, oral communication, 1982). Intergradational, locally calcareous and argillaceous flyschoid siltstone and mudstone display grazing trails, burrows, channel scours, and rare ripple marks. The components appear to be deposited in an open-shelf setting. The conglomerates are polymictic. Clast compositions reflect the total Wrangellian assemblages; primary sedimentary structures, including imbrication and cross-bedding, as well as clast-size distribution and sorting, indicate a dominant eastern source for the Queen Charlotte Group. The conglomerate is commonly displayed in poorly graded to well-graded packages and includes resedimented components characteristic of submarine debris flows.

Queen Charlotte Basin

The Queen Charlotte Basin (Figure 1) includes mainly nonmarine and marine sedimentary rocks that underlie northeastern Graham Island, Hecate Strait, and Queen Charlotte Sound (Shouldice, 1971). The basin is approximately 450 km long and 130 km wide at its widest point in Queen Charlotte

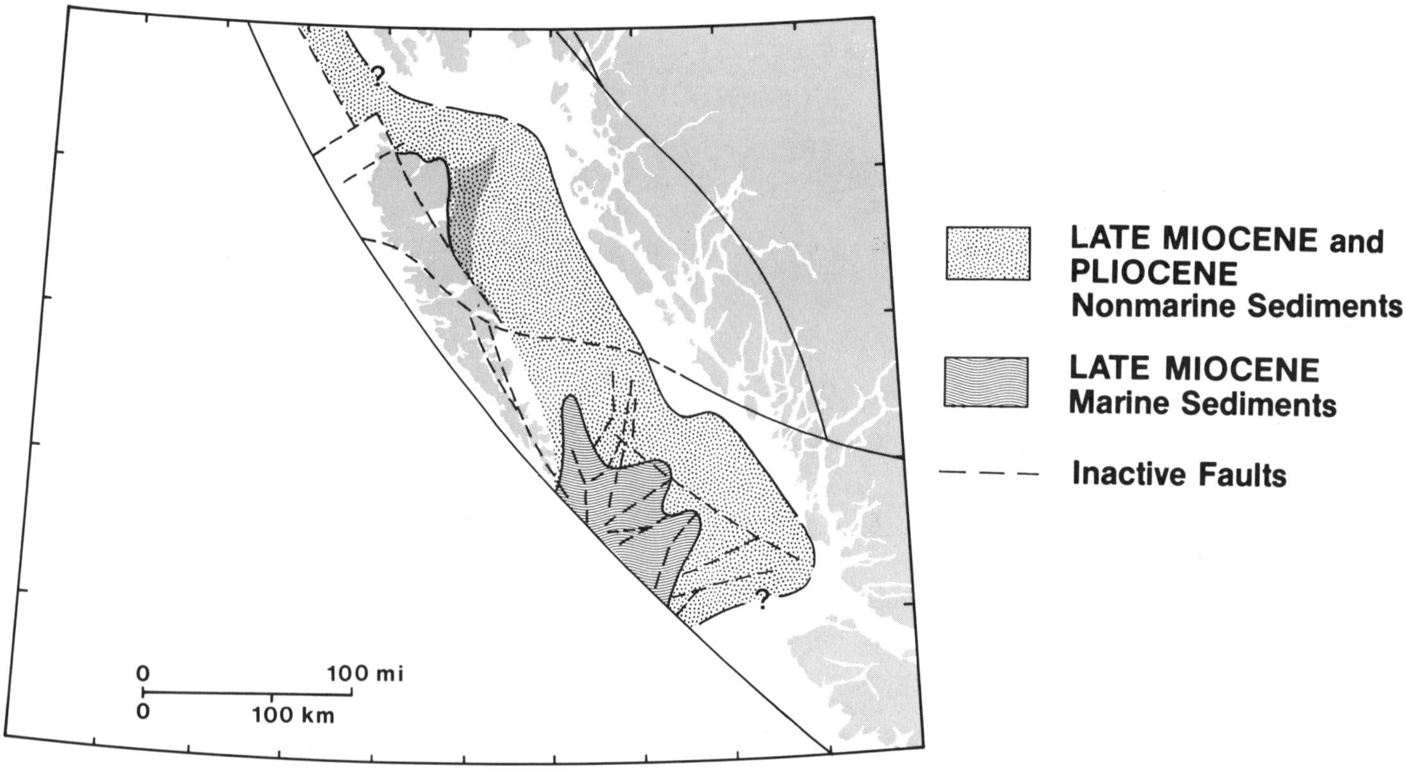

Figure 9. The rift assemblage—late Miocene and Pliocene. The Queen Charlotte Islands have rotated into their final position. The rift zone in Queen Charlotte Sound continues to fill with marine sediment while the flexural basin in Hecate Strait receives nonmarine sediment (from Yorath and Chase, 1981).

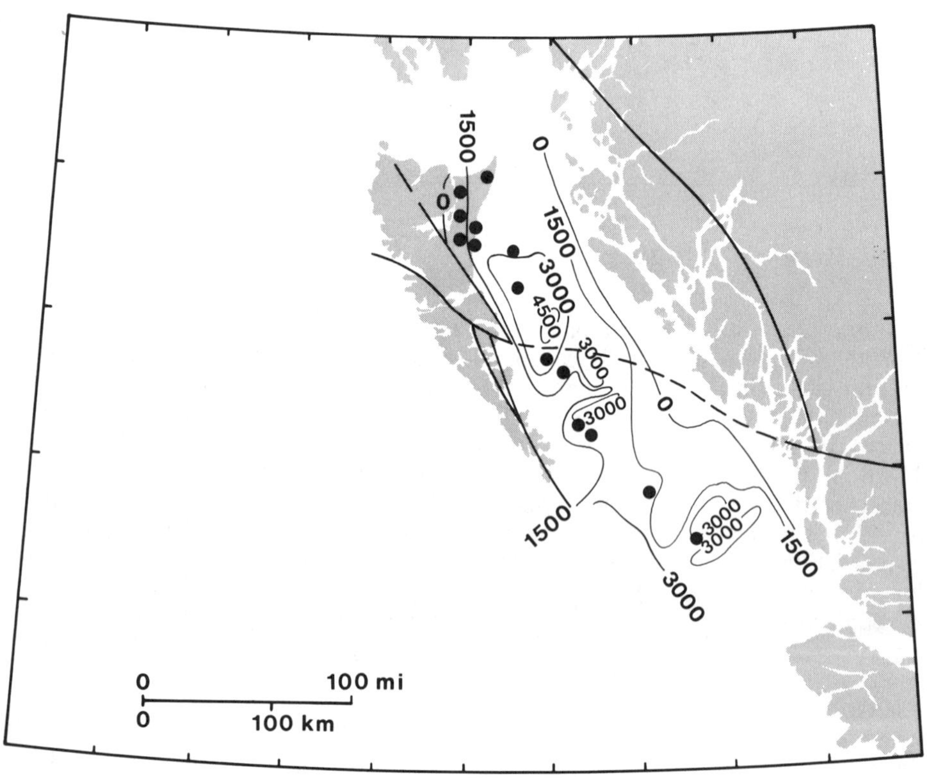

Figure 10. Isopachs (in meters) of Neogene sediment in Queen Charlotte Basin (from Shouldice, 1973).

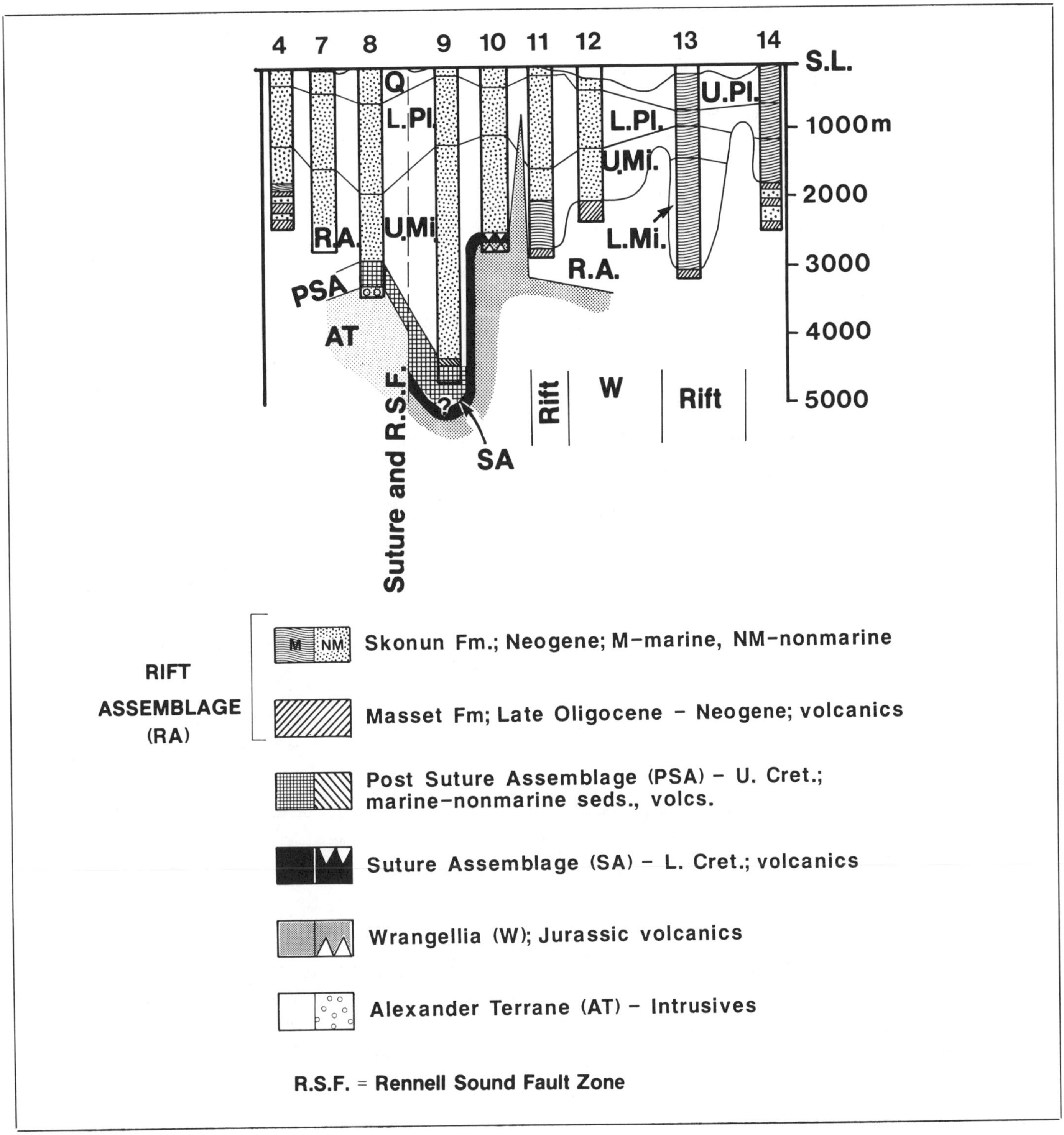

Figure 11. Stratigraphic cross-section through the offshore wells in Queen Charlotte Basin. Well locations shown on Figure 2 (modified from Shouldice, 1971).

Sound, where its western boundary is the Queen Charlotte fault (Figure 2). Within the basin, sediment reaches a maximum thickness of slightly more than 4500 m; however, for the most part thicknesses do not exceed 3000 m. Isopachs (Figure 10) show a linear trough extending through Hecate Strait, which is interrupted in the south by a northeasterly trending pre-Tertiary basement ridge over which Tertiary sediment is thin. North of the Rennell Sound fault zone in central Hecate Strait, and east of the Sandspit fault through central Graham Island, the basement of Queen Charlotte Basin is believed to comprise Paleozoic rocks of the Alexander terrane overlain by mid- to Upper Cretaceous postsuture assemblage clastic strata and by lower Neogene volcanic units. In southern Hecate Strait the Tertiary sediment is underlain by lower Neogene lava, which, beyond the rift zone in Queen Charlotte Sound, overlies postsuture assemblage strata and basement rocks of Wrangellia (Yorath and Chase, 1981; Yorath and Hyndman, 1983).

Tertiary sedimentary deposits form an eastward-thickening wedge on northeastern Graham Island, where

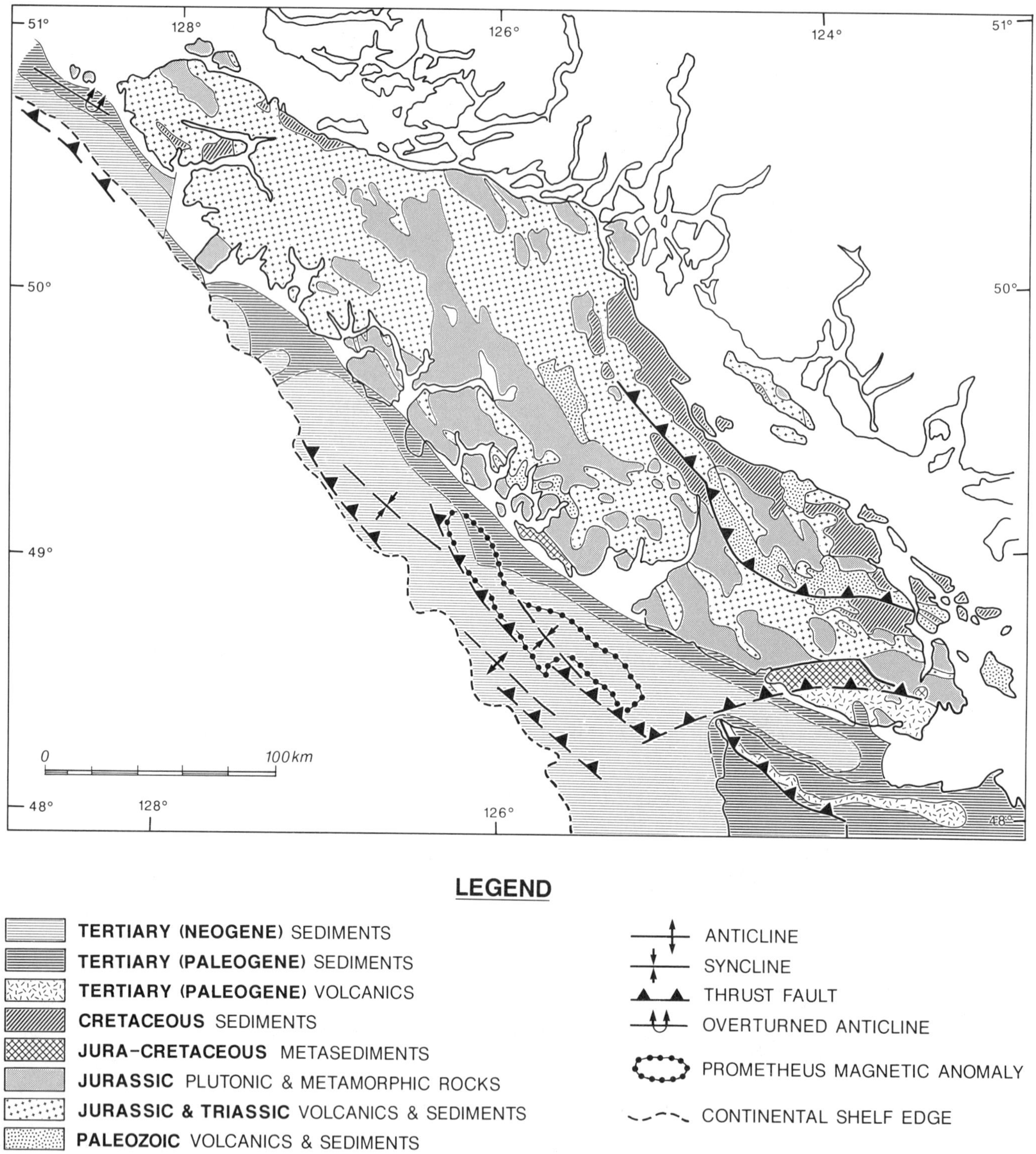

Figure 12. Simplified geology of the Vancouver Island continental margin.

they have been named the Skonun Formation by Sutherland Brown (1968). They comprise a succession of dominantly nonmarine, commonly cross-bedded, immature feldspathic sandstone, minor mudstone, lignite, and pebble conglomerate. Within subsurface sections, both on Graham Island and beneath Hecate Strait, the formation retains these general characteristics (Figure 11). Sandstone porosities range between 20% and 30%; permeabilities are low. Although northern sections of the formation are dominantly nonmarine, thin marine units occur locally near the top and bottom. In southern Hecate Strait and in Queen Charlotte Sound the formation becomes partly to entirely marine. The age of the Skonun Formation varies throughout Queen Charlotte Basin. On Graham Island outcrop sections range in age from late early to middle Miocene, whereas subsurface sections represent late Miocene to Pliocene time. In Queen Charlotte Sound, at the site of the Harlequin well, the formation ranges from early Miocene (Burdingalian) to Pliocene (B. E. B. Cameron, written communication, 1982).

VANCOUVER ISLAND MARGIN

The structural style of the Vancouver Island margin is expressed by northwesterly trending structural culminations, by northwesterly and northerly trending plutons, and by northwesterly and westerly directed primary faults (Figure 12); major westerly striking faults occur on southern Vancouver Island (Muller, 1977). As in the Queen Charlotte Islands, the Upper Triassic Karmutsen Formation appears to be the most competent unit which, between important faults, is warped into broad culminations and depressions. Less competent underlying and overlying units are more intensely and, locally, penetratively deformed. In the Gulf Islands, adjacent to the southeast coast, the Upper Cretaceous Nanaimo Group is deformed into northwesterly trending cylindrical and isoclinal folds and, perhaps, by easterly dipping thrust faults. The Cowichan and Nanoose uplifts are northwesterly trending structures underlain by Paleozoic Sicker Group sedimentary and volcanic strata. The Cowichan uplift and its colinear expression in the Beaufort Range to the northwest are bounded on the southwest by a major thrust system that involves Upper Cretaceous strata of the Nanaimo Group (Figures 12, 13; Sutherland Brown and Yorath, 1985). In the Nanoose uplift, on the east coast of the island, Sicker Group strata are isoclinally folded; northwesterly trending faults, some of which may be thrusts, disrupt the Paleozoic and Cretaceous sequence.

The geological structure of Vancouver Island is best illustrated by one of four recently obtained multichannel seismic-reflection profiles (Yorath et al., 1985b), the location of which is shown on Figure 14. In Figure 13 the lowermost reflector zone (A) is considered to represent the top of the Juan de Fuca plate, comprising a sequence (approximately 3 km thick) of interdigitated materials of probable sedimentary and volcanic origin resting upon presumed oceanic crust. Alternatively, this layer might represent an interval of decoupling between the descending plate and the overlying continental crust. The overall dip of the zone is northeasterly at between 10° and 12°; beyond this point, other geophysical data suggest a change in dip to as much as 30° (Rogers, 1983). Within this zone reflector discontinuities clearly indicate the presence of easterly dipping faults. Whether these faults penetrate the underlying crust or become listric within it is unknown.

Above the Juan de Fuca plate is a thick interval with fewer, less coherent, discontinuous and discordant reflectors (Figure 13). This "underplated zone" (B) may represent accreted oceanic material, either slices of oceanic crust or imbricated oceanic sediment such as that found in the core of the Olympic Mountains of northern Washington. Several of the faults that dislocate the lowest layer (A) extend upward into the underplated zone. A post-Eocene age for underplating is supported by the probable early Eocene age of the wedge of basalt identified as the Metchosin Formation at the western end of the profile (Brandon, 1984); Shouldice, 1971; Yorath, 1980; Yorath et al., 1985b). A corollary to this interpretation is the necessary removal of several tens of kilometers of lithosphere as a consequence of, or prior to, the emplacement of the underplated zone. Moreover, the upper and lower bounding detachment surfaces may define the layer as an interval of duplex structures; some evidence for internal thrust faults is provided by reflector discontinuities within the zone.

Above the underplated zone and extending eastward beneath most of the island is another layered interval that is interpreted as a decollement zone (C) at or beneath the base of Wrangellia (Figure 13). Similar to the lower layered interval, this zone shows numerous reflectors, the lateral continuities of which are interrupted by easterly dipping surfaces; some of these surfaces (faults) are listric within this zone, whereas others appear to penetrate the underlying underplated zone. Most of these faults which arise from the decollement zone, although not clearly evident in the shallower portion of the record section, can be projected to the surface, where they coincide with the positions of mapped faults or plutonic contacts (Sutherland Brown and Yorath, 1985).

Similar to the lower layered sequence, the decollement zone may represent interdigitated materials of sedimentary and volcanic origin that overlay and were emplaced with the underplated zone. At the western end of the profile, the interval underlies the wedge of Eocene Metchosin volcanic rocks from which it is probably separated by a fundamental detachment surface that represents the roof fault of the duplex structures discussed above. The upper surface of the basalt wedge is the Tofino fault (E; Brandon, 1984), which merges with the decollement zone and which defines the base of Wrangellia beneath western Vancouver Island.

In the upper part of the section, Wrangellia is represented by irregular areas of coherent and discoherent reflectors (J) as well as by integrated surface geological data (Sutherland Brown and Yorath, 1985). Surrounding areas are acoustically transparent (H,I) and are thought to represent the Lower to Middle Jurassic granodioritie masses of the Island Intrusions. The southwest flank of the Cowichan uplift is dislocated by easterly dipping thrust faults that have displaced strata of the Sicker Group and Karmutsen Formation, as well as the enclosed granodioritic rocks. The Beaufort Range-Cowichan Lake fault zone is probably broadly transpressive and shows a thrust displacement of about 1 km. Beneath and west of these comparatively high level faults (K), other faults that partition Wrangellia (F,G) and the intrusive rocks show separations of several kilometers and appear to control the structural level of the plutons.

Sedimentary Basins

The sedimentary basins of Vancouver Island and adjacent offshore regions that have significant petroleum potential are the Upper Cretaceous Georgia Basin and the Tertiary Tofino Basin. Other sedimentary accumulations of Cretaceous age in Suquash Basin and on northern Vancouver Island (Coal Harbour Group) are considered nonprospective (Figure 1). Likewise, Winona Basin, adjacent to the continental slope off northern Vancouver Island, is excluded from this discussion, even though some hydrocarbon potential is attributed to its uppermost Tertiary and Quaternary succession by Haimila and Procter (1982).

Georgia Basin

The Nanaimo Group of late Santonian to Maastrichtian, and possibly early Tertiary age, underlies the coastal plain of southeastern Vancouver Island, the adjacent Gulf Islands, and the Cowichan and Alberni Valleys of south-central Vancouver Island and the Strait of Georgia (Muller and Jeletzky, 1970). The group is enclosed mainly within two subdivisions of the Georgia Basin, the Comox Basin in the north and the

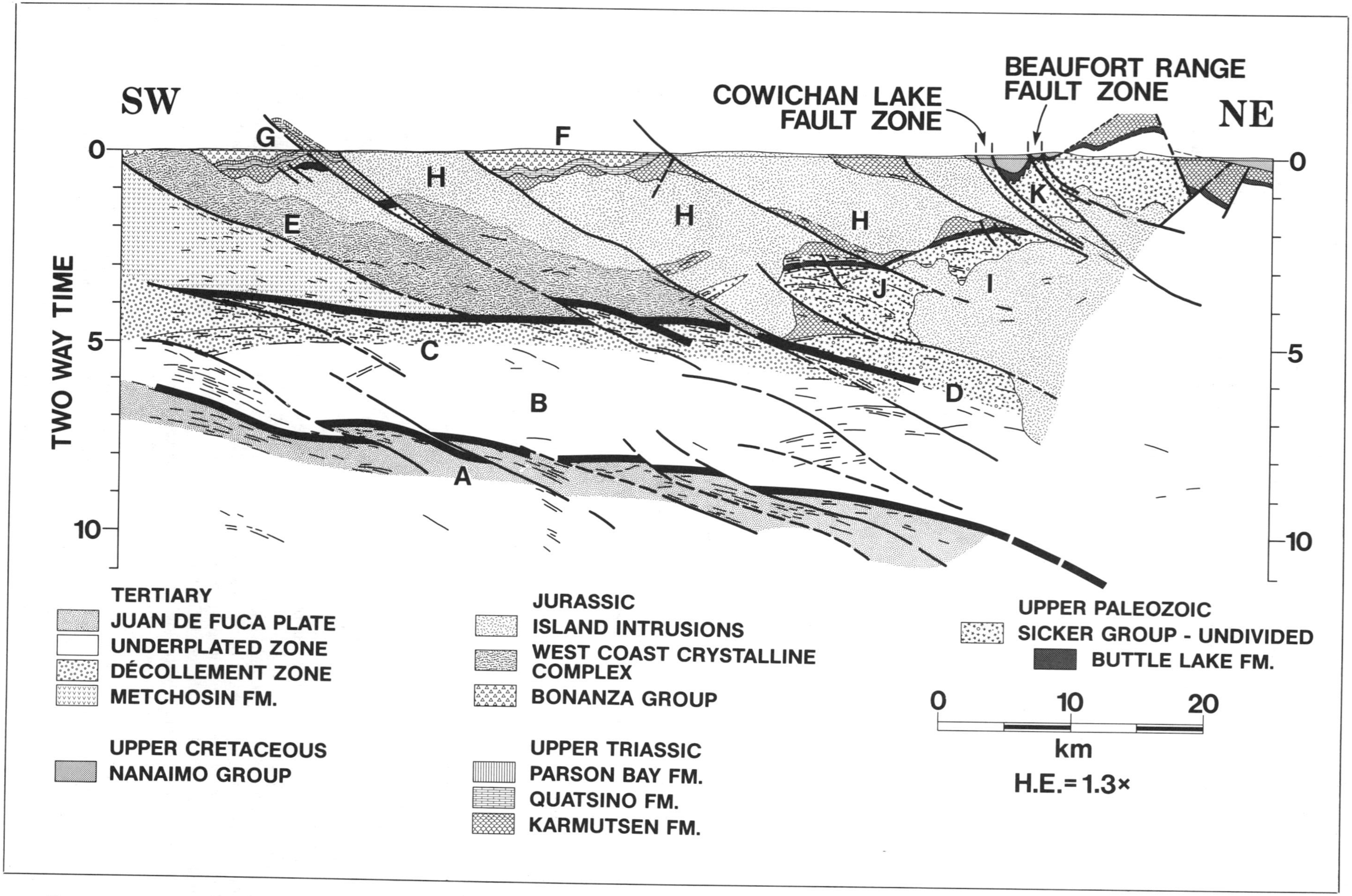

Figure 13. Interpretation of LITHOPROBE Line 1 (unmigrated) across southern Vancouver Island. Short line segments represent reflections. See text for descriptions of features A–K. Location of line shown on Figure 14 (from Yorath et al., 1985b).

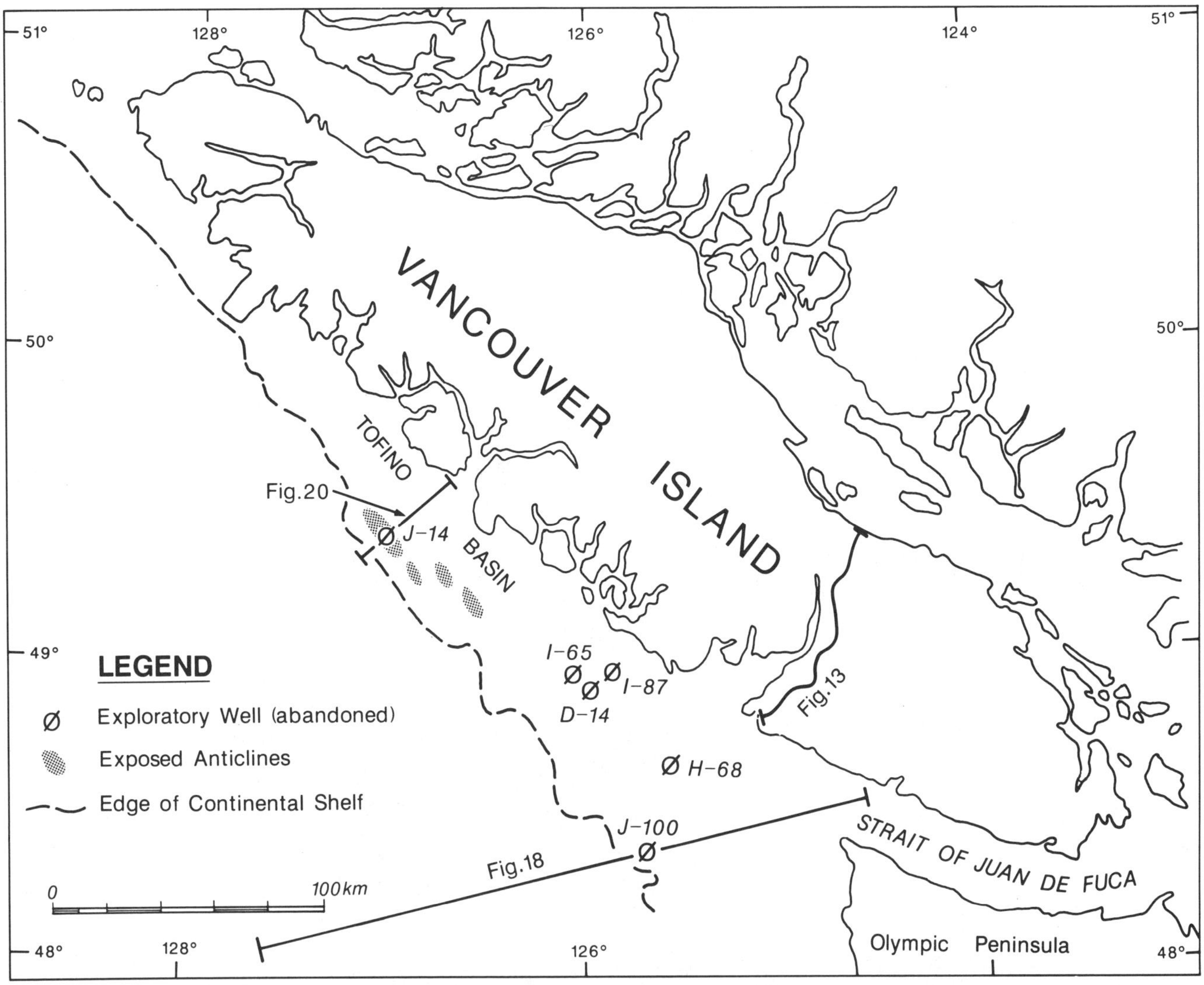

Figure 14. Locations of exploratory wells and seismic profiles in Tofino Basin and on Vancouver Island.

Nanaimo Basin in the south. Its nine component formations (in upward stratigraphic order: Comox, 350 m; Haslam, 200 m; Extension-Protection, 300 m; Cedar District, 350 m; De Courcy, 350 m; Northumberland, 250 m; Geoffrey, 150 m; Spray, 200 m; and Gabriola, 350 m) comprise variously alternating sequences of coarse- and fine-grained terrigenous clastic deposits (Figure 15). In the Comox Basin these are dominantly of neritic to outer-shelf affinity, whereas in the Nanaimo Basin, turbidite sequences with high-energy sedimentary structures suggest the presence of deeper water slope and fan environments adjacent to a narrow shelf zone (Sliter, 1973; Ward and Stanley, 1982; Pacht, 1980, 1984). During the Santonian and earliest Campanian the dominant source regions for Nanaimo Basin clastic beds are thought to have been local areas to the west and southeast, whereas during the remainder of the Late Cretaceous, eastern Coast Plutonic Belt sources are important (Figure 16; Pacht, 1984). In the Comox Basin the sediment is more or less undeformed; however, thrust faults probably dislocate the lower part of the Nanaimo Group in the southern part of the basin and in Alberni Valley (Sutherland Brown and Yorath, 1985). In the Nanaimo Basin all components of the group are deformed into linear, northwesterly trending cylindrical folds which are possibly associated with thrust faults. Along the northeast coast of the island the small Suquash Basin contains Campanian sandstone and shales that are assigned to the Nanaimo Group.

Near the center of the Georgia Basin the Richfield Pure Point Roberts well penetrated approximately 1400 m of middle and upper Eocene nonmarine sandstone, mudstone, and minor coal overlying the Nanaimo Group which, in turn, is unconformably overlain by 1600 m of nonmarine Miocene sediment (Figure 15). Hopkins (1966) assigned the Eocene strata to the Burrard and Kitsilano Formations of the Vancouver area, which are equivalent to the Chuckanut Formation of northwestern Washington (Johnson, 1984).

Tofino Basin

On the southwest coast of Vancouver Island, the middle Eocene to Oligocene Carmanah Group is exposed along the rim of Tofino Basin and on the north shore of the Strait of Juan de Fuca (Figures 12, 14); the seaward limit of the basin is the base of the continental slope. Along the southwest coast, the group is in both depositional and fault contact with Lower Jurassic volcanic, plutonic and metamorphic rocks (Muller, Cameron, and Northcote, 1981), and on the north shore of the strait of Juan de Fuca it is in stratigraphic contact

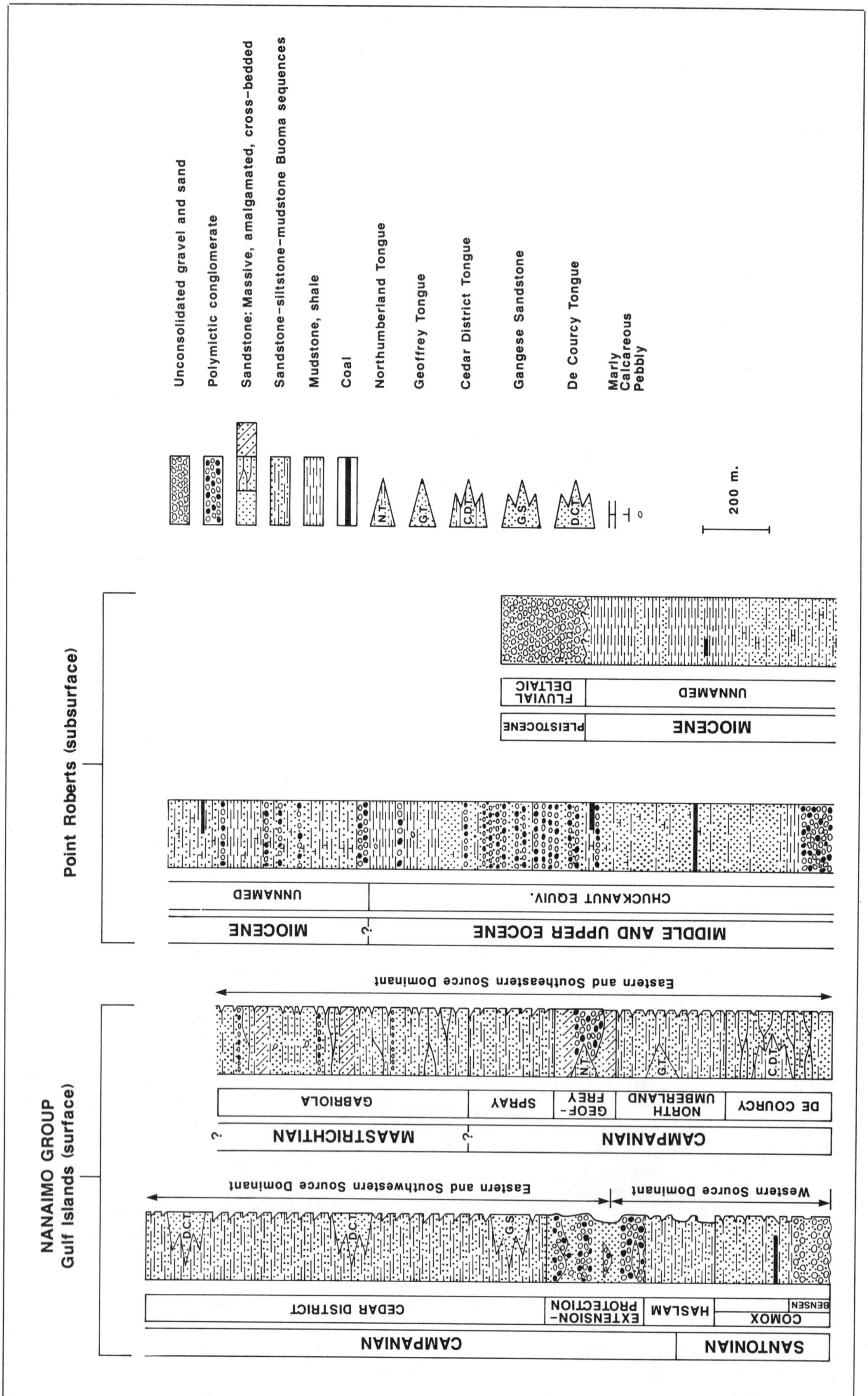

Figure 15. Generalized stratigraphy of the Upper Cretaceous Nanaimo Group and Tertiary sediment in southern Georgia Basin.

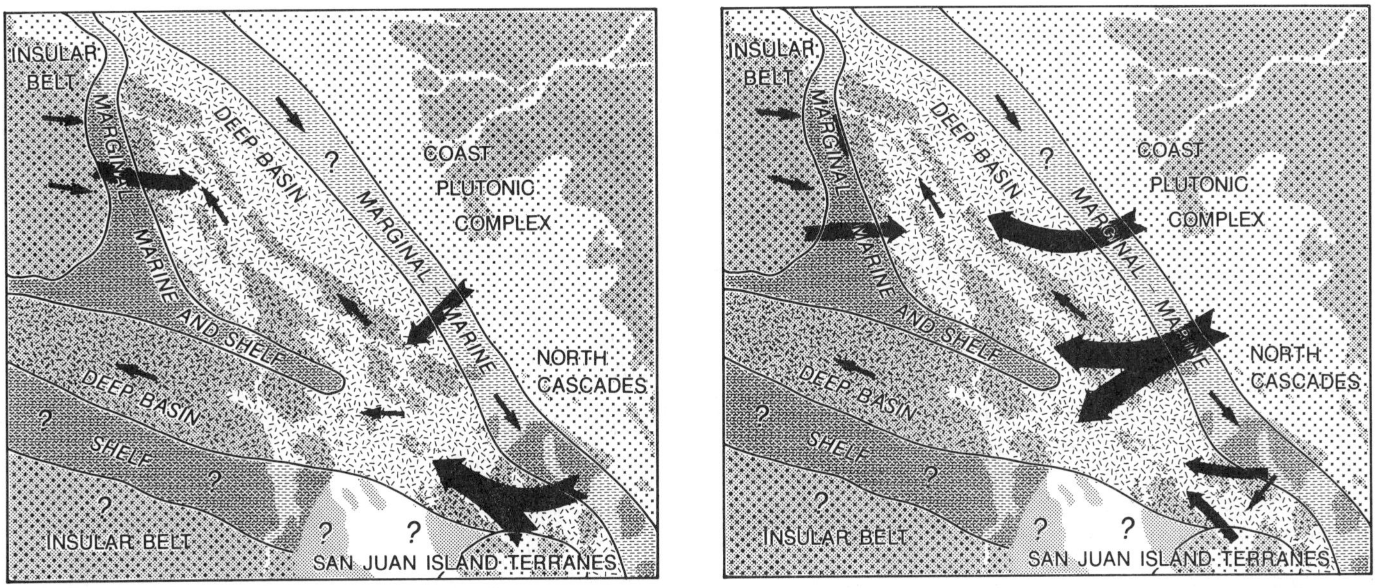

Figure 16. Paleogeography and sediment transport directions in southern Georgia Basin during (A) Santonian to earliest Campanian and (B) mid- to late Campanian and Maastrichtian (after Pacht, 1984).

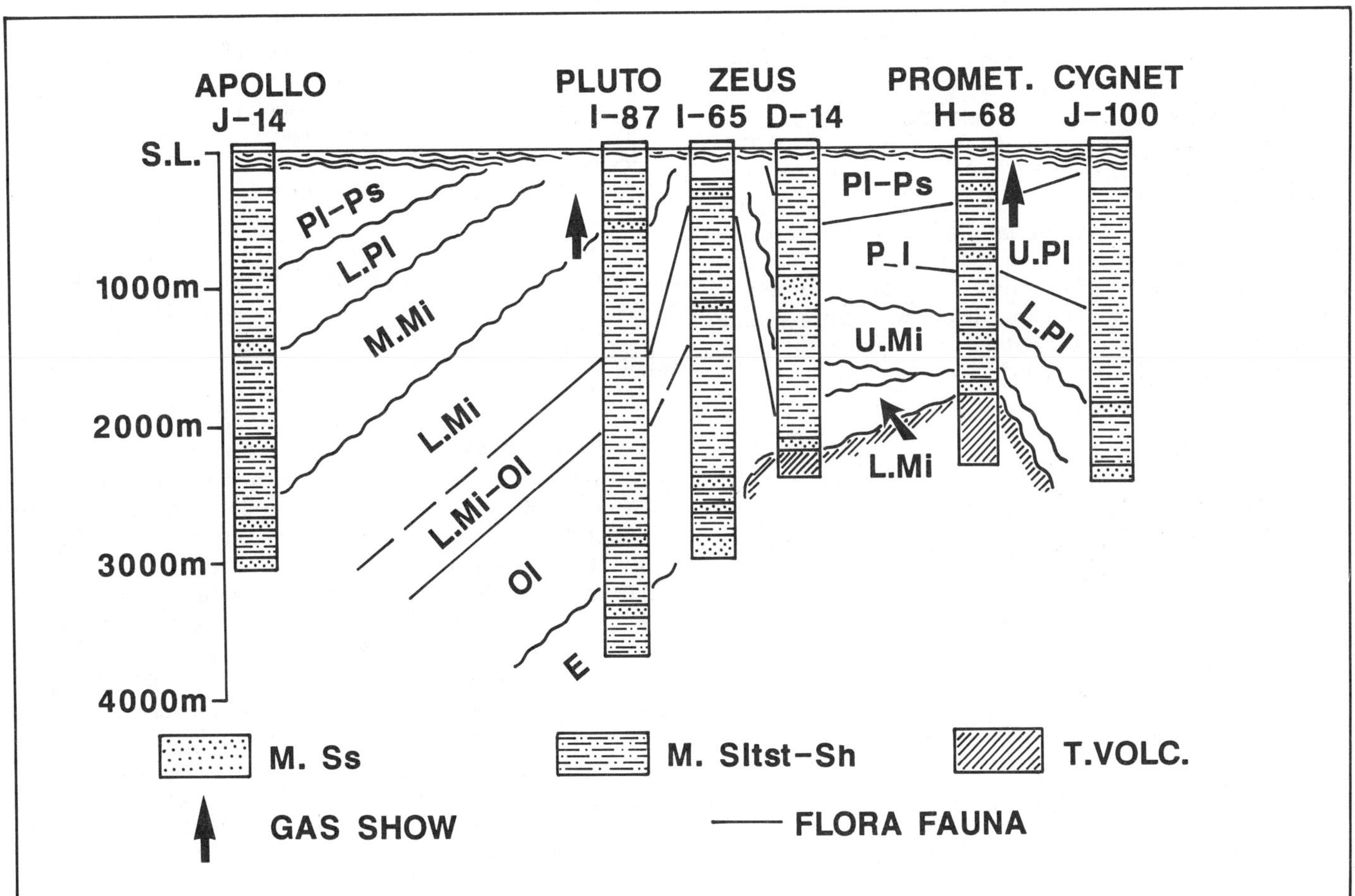

Figure 17. Stratigraphic cross-section through the offshore wells in Tofino Basin. Well locations shown on Figure 14 (from Shouldice, 1971).

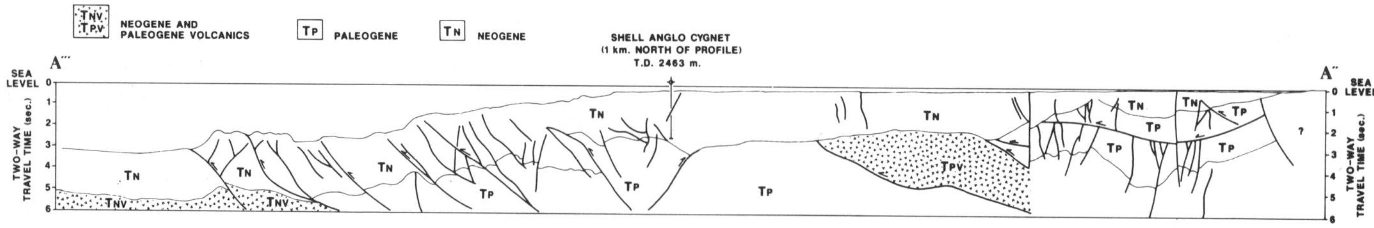

Figure 18. Interpretation of unmigrated seismic-reflection profile through Tofino Basin. Line location shown on Figure 14 (modified from Snavely and Wagner, 1981). Seaward to left.

with Jura-Cretaceous metasediment of the Leech River Formation and the lower Eocene Metchosin volcanics (Muller, 1982). The Carmanah Group comprises the Escalante, Hesquiat, and Sooke Formations, in upward stratigraphic order. The Escalante Formation (140 m) consists of outer-shelf to bathyal calcareous sandstone and minor polymictic conglomerate that grades upward into irregularly interbedded deep-water shale and sandstone of the Hesquiat Formation (1100 m; Cameron, 1980). The Sooke Formation (<200 m) consists of coarse littoral sandstone and conglomerate.

On the continental shelf six exploratory wells penetrated Neogene sediment, and four of these encountered sediment and volcanic rocks of Paleogene age (Figure 17; Shouldice, 1971). The volcanic units, as discussed previously, and shown on Figure 18 (Tpv), are equivalent to the Metchosin and Crescent Formations of southern Vancouver Island and Washington, respectively, and represent a detached slab of Eocene oceanic crust that is enclosed within Paleogene sediment and unconformably overlain by Neogene strata (Figure 18; Yorath, 1980). The Paleogene sediments encountered in the Pluto 1–87 and Zeuss 1–65 wells (Figure 17) are considered, at least in part, to be equivalent to the Carmanah Group. The overlying Neogene strata consist of a monotonous sequence of pale-gray, immature marine siltstone, shale, and minor sandstone within which are a number of local unconformities. The total known thickness of the Tertiary section in Tofino Basin is in excess of 3000 m (Figure 19).

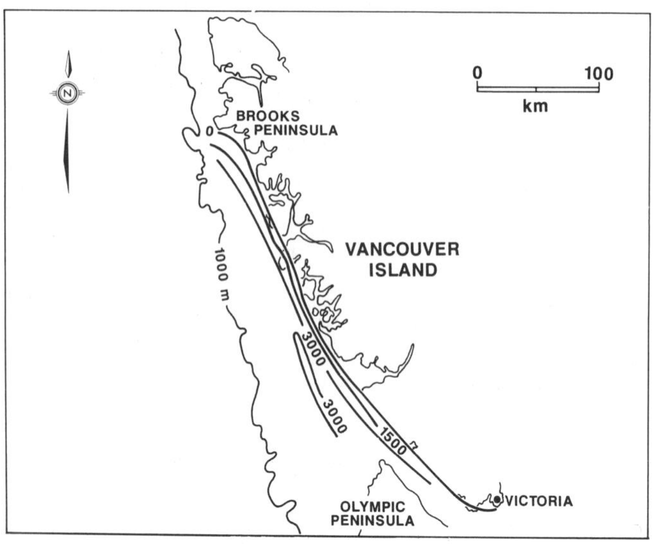

Figure 19. Isopachs (in meters) of Tertiary sediment in Tofino Basin (from Shouldice, 1973).

Beneath the continental slope the Paleogene and lower Neogene rocks of Tofino Basin comprise a thrust-imbricated subduction complex above the modern descending Juan de Fuca plate (Figure 18; for line location, see Figure 14; Snavely and Wagner, 1981). The subduction complex comprises seaward-verging thrust faults that extend progressively higher into the section toward and beyond the toe of the continental slope.

Above the subduction complex and beneath the continental shelf, Neogene sediments are essentially undeformed in the southern Tofino Basin, except near the edge of the shelf, where they are deformed into broad anticlines (Shouldice, 1971). In the central Tofino Basin, landward of the shelf-edge anticlines, a series of curvilinear folds interrupts the otherwise flat, gently sloping sea floor (Figure 14). These have been interpreted as anticlines induced by gravitational sliding on shallow detachment surfaces within Neogene and Quaternary sediment (Figure 20; Yorath, 1980).

HYDROCARBON POTENTIAL

The principal prospective petroleum basins of the continental margin are the Georgia, Tofino, and Queen Charlotte basins. The Winona Basin, located on the Explorer plate adjacent to the northern Vancouver Island continental margin, and the Suquash Basin, on and adjacent to northeastern Vancouver Island, might conceivably contain hydrocarbons but are not considered prospective because of difficulty of access to the former and lack of sufficient sediment volume in the latter (Figure 1).

Eight wells penetrated the thick, nonmarine deltaic Tertiary section in the Fraser lowlands in the southeastern Georgia Basin, and one was drilled into the Upper Cretaceous Nanaimo Group in the Gulf Islands. At the time of writing, exploration of the Nanaimo Group, including slim-hole drilling, is proceeding in the Nanaimo and Comox subbasins. In the Tertiary section abundant reservoirs are available, but neither adequate source rocks nor traps have been found. The nonmarine character of the succession suggests that gas would be the principal hydrocarbon. The largely marine Nanaimo Group offers several opportunities for the occurrence of hydrocarbons in both structural and stratigraphic traps, but porosities in the surface rocks are generally 5% or less. Moreover, vitrinite reflectance data and stratigraphic analyses suggest that much of the basin sustained deep burial within a high-thermal regime. Thus, dry gas would be the indicated resource.

Six wells tested the Tertiary succession in Tofino Basin (Shouldice, 1973). Gas shows were encountered in locally highly pressured Miocene and Pliocene marine strata. Potential reservoir rocks are commonly tight because of clay-mineral plugging of initial porosity in the largely immature clastic deposits. Although abundant opportunities exist for large structural traps, adequate source rocks have yet to be encountered. Moreover, because of the location of the

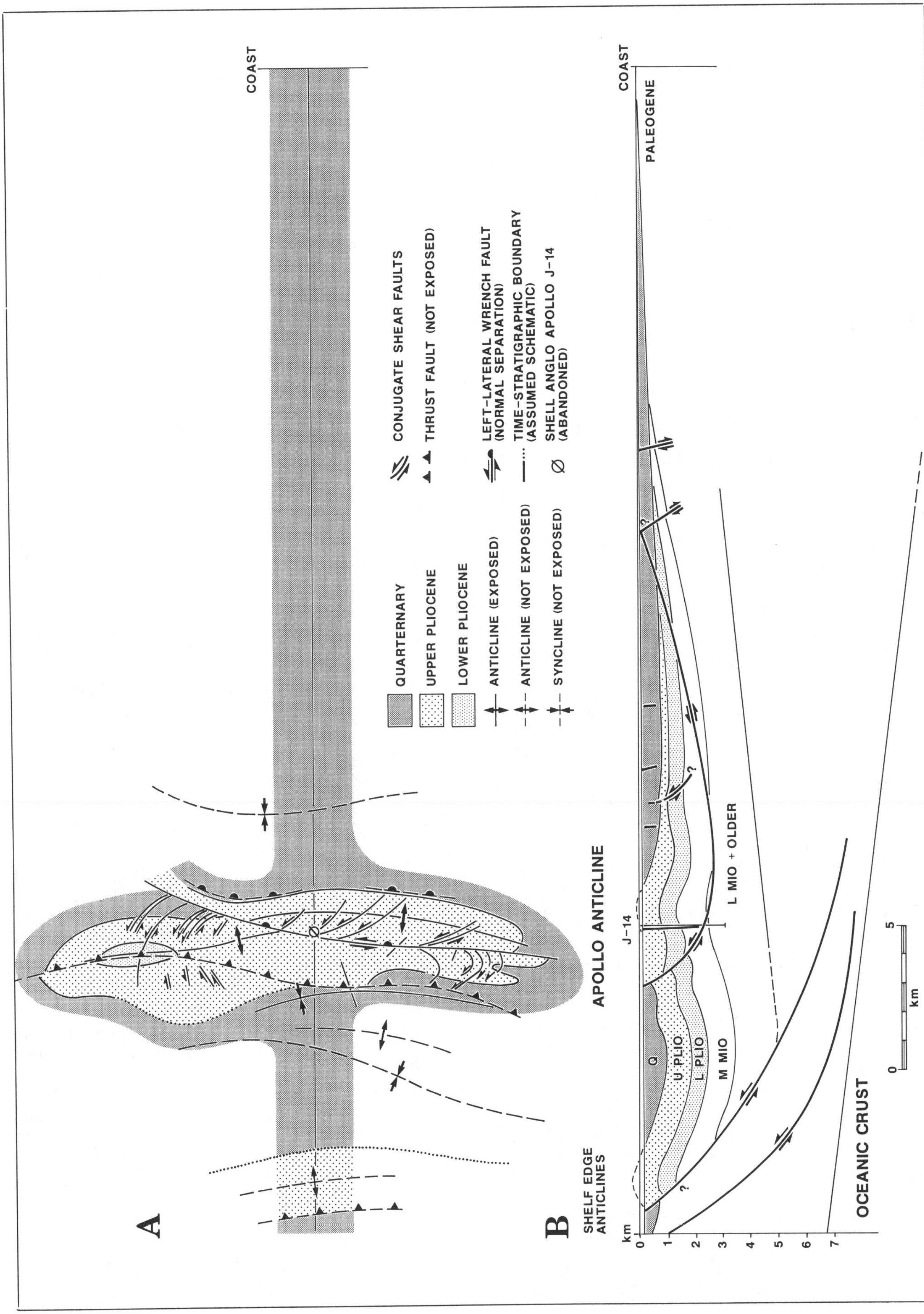

Figure 20. Structure of the continental margin across the Apollo anticline in the central Tofino Basin (from Yorath, 1980).

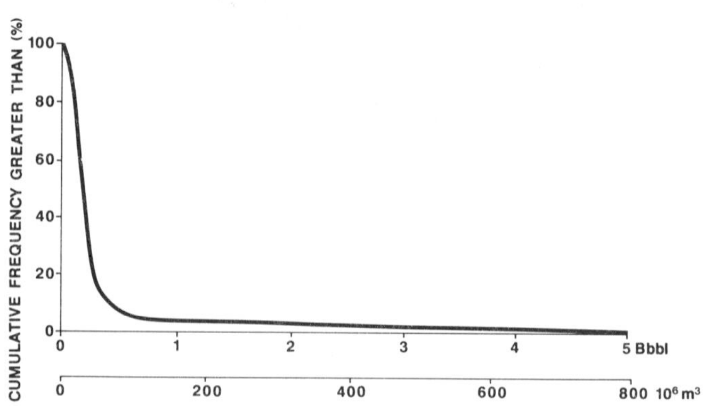

Figure 21. Oil potential of British Columbia offshore areas, excluding the Georgia Basin (from Haimila and Procter, 1982).

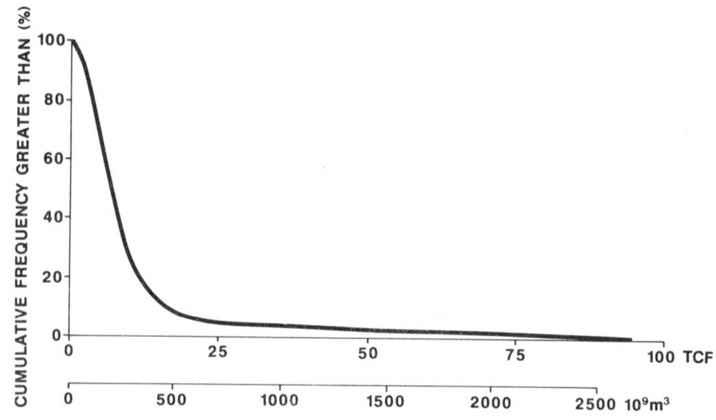

Figure 22. Natural gas potential of British Columbia offshore areas, excluding the Georgia Basin (from Haimila and Procter, 1982).

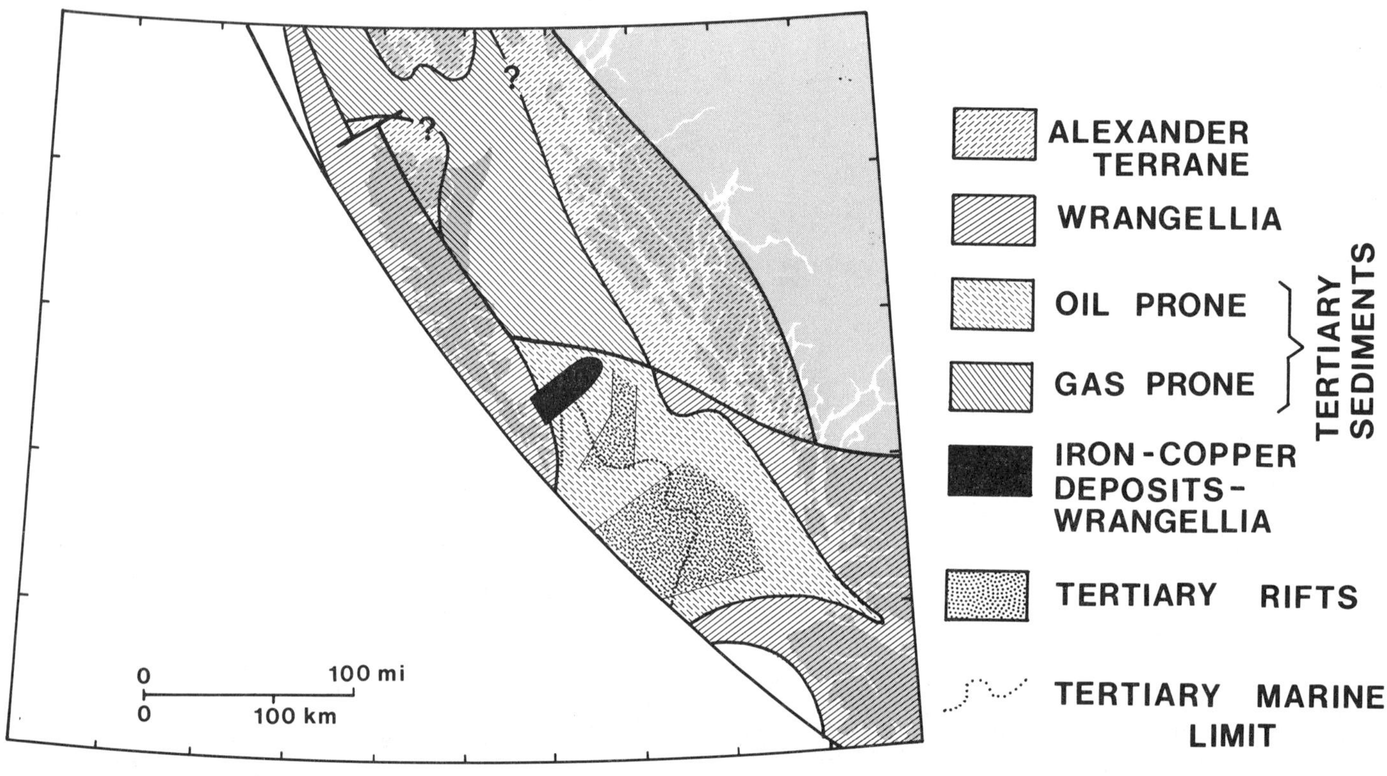

Figure 23. Possible hydrocarbon-type distribution in Queen Charlotte Basin.

basin above the descending Juan de Fuca plate, the depressed thermal regime may have been insufficient for maturation beyond the level of the immature gas.

Queen Charlotte Basin probably has the best potential for substantial petroleum resources. Fourteen wells penetrated the Tertiary section, both on northeastern Graham Island and offshore in Hecate Strait and Queen Charlotte Sound, where one oil show was encountered in Pliocene nonmarine strata (Shouldice, 1973). Since exploration ceased in the late 1960s, considerable new information has been acquired in the region that has contributed to renewed interest by petroleum companies.

Queen Charlotte Basin is presently believed to have formed through the combined processes of rifting and lithospheric flexure, and to have developed as a successor basin above and across the suture between Wrangellia and the Alexander terrane (Figure 8; Yorath and Chase, 1981). The rifting model suggests a favorable thermal regime and structural style for the maturation and entrapment of Tertiary hydrocarbons (Yorath and Hyndman, 1983). Source rocks for the basin may be found in the Queen Charlotte terrace adjacent to the west coast of the Queen Charlotte Islands. The terrace consists of a folded sedimentary prism overlying the Pacific plate which, adjacent to the Queen Charlotte transform fault, is moving northward at about 55 km/m.y. During the time of basin formation, it may have been situated off the mouth of Queen Charlotte Sound. Additional sources of hydrocarbons occur within Triassic and Jurassic rocks in the Queen Charlotte Islands (Macauley, 1983; Cameron and Tipper, 1985), which are believed to underlie the Tertiary sedimentary and volcanic succession of the southern part of the basin. On the Queen Charlotte Islands petroleum from these sources occurs in Tertiary vesicular basalt, and thus is likely to occur also in Tertiary sedimentary rocks offshore (Sutherland Brown, Yorath, and Tipper, 1983).

The Queen Charlotte Group of the Skidegate Basin has some potential, if adequate seals and source rocks can be found.

Figures 21 and 22 show the estimate of hydrocarbon potential for the continental margin excluding the Georgia Basin, Queen Charlotte Islands, and Vancouver Island. The estimate, described in Haimila and Procter (1982), considers stratigraphic and structural exploration plays associated with Tertiary rocks of the Tofino, Queen Charlotte, and Winona Basins and pre-Tertiary plays in the offshore part of the Skidegate Basin.

The following quotation, from Haimila and Procter (1982), includes an interpretation of Figures 21 and 22: "The shapes of the (cumulative frequency) curves for both natural gas and oil are characteristic of regions in which there is little exploration, sparse control, and high levels of uncertainty regarding the presence of pooled hydrocarbons. Curves of this type are highly skewed with very little potential associated with the higher levels of probability, but larger potentials are suggested at lower levels of probability. In the case of the natural gas potential, for example, slightly more than $3.18 \times 10^6 m^3$ are thought to exist at the 100% probability. The values of the potential increase gradually to an average expectation of $265 \times 10^9 m^3$ at the 34% level and increase rapidly in the lower probability ranges."

"Similarly, the distribution of estimates for the recoverable oil indicates that at the 100% probability only a nominal quantity of oil may exist. There is an average expectation that more than $39.5 \times 10^6 m^3$ exist."

The above estimates predate modern reconstructions of the tectonic history of the Queen Charlotte Islands margin region and, moreover, predate the calculation of thermal and subsidence models (Yorath and Chase, 1981; Yorath and Hyndman, 1983). When more quantitative data are available for detailed models, it may become apparent that the hydrocarbon potential of the Queen Charlotte Basin is greater than presently assumed. Figure 23 shows the possible hydrocarbon-type distribution within the Tertiary section. North of the suture between the Alexander terrane and Wrangellia, the Tertiary section is thin, entirely nonmarine, and overlies Paleozoic rocks with no known hydrocarbon potential for "leaked" resources into the Tertiary section. If hydrocarbons are present, they may be immature gas. South of the suture the increased thermal regime during the following rifting, the presence of marine sediment, and the known occurrence of petroleum within fractured Wrangellia basement rocks suggests that the southern Queen Charlotte Basin may contain significant oil accumulations in addition to natural gas.

REFERENCES CITED

Brandon, M. T., 1984, Deformational processes affecting unlithified sediments at margins: A field study and structural model: Seattle, University of Washington unpublished Ph.D. thesis, 160 p.

Cameron, B. E. B., 1980, Biostratigraphy and depositional environment of the Escalante and Hesquiat formations (early Tertiary) of the Nootka Sound area, Vancouver Island, British Columbia: Geological Survey of Canada Paper 78–9, 28 p.

Cameron, B. E. B., and H. W. Tipper, 1985, Jurassic stratigraphy of the Queen Charlotte Islands: Geological Survey of Canada Bulletin 365, 49 p.

Davis, E. E., and R. P. Riddihough, 1982, The Winona Basin: Structure and tectonics: Canadian Journal of Earth Sciences, v. 19, p. 767–788.

Haimila, N. E., and R. M. Procter, 1982, Hydrocarbon potential of offshore British Columbia: Geological Survey of Canada Open-File 824, 28 p.

Hopkins, W. S. Jr., 1966, Palynology of Tertiary rocks of the Whatcom Basin, southwestern British Columbia and northwestern Washington: Vancouver, University of British Columbia unpublished Ph.D. thesis, 184 p.

Jeletzky, J. A., 1976, Mesozoic and ? Tertiary rocks of Quatsino Sound, Vancouver Island, British Columbia: Geological Survey of Canada Bulletin 242, 243 p.

Johnson, S. Y., 1984, Stratigraphy, age, and petrography of the Eocene Chukanut Formation, northwest Washington: Canadian Journal of Earth Sciences, v. 21, p. 92–106.

Keen, C. E., and R. D. Hyndman, 1979, Geophysical review of the continental margins of eastern and western Canada: Canadian Journal of Earth Sciences, v. 16, p. 712–747.

Macauley, G., 1983, Source rock-oil shale potential of the Jurassic Kunge Formation, Queen Charlotte Islands: Geological Survey of Canada Open-File 921, 52 p.

Minster, J. B., and T. H. Jordan, 1978, Present day plate motions: Journal of Geophysical Research, v. 83, p. 5331–5354.

Muller, J. E., 1977, Evolution of the Pacific margin, Vancouver Island and adjacent regions: Canadian Journal of Earth Sciences, v. 14, p. 2062–2085.

Muller, J. E., 1982, Geology, Victoria, British Columbia: Geological Survey of Canada Map 1553A, Scale, 1:100,000.

Muller, J. E., and J. E. Jeletzky, 1970, Geology of the Upper Cretaceous Nanaimo Group, Vancouver Island and Gulf Islands, British Columbia: Geological Survey of Canada Paper 69–25, 77 p.

Muller, J. E., B. E. B. Cameron, and K. E. Northcote, 1981, Geology and mineral deposits of Nootka Sound map-area, Vancouver Island, British Columbia: Geological Survey of Canada Paper 80–16, 53 p.

Pacht, J. A., 1980, Sedimentology and petrology of the Late Cretaceous Nanaimo Group deposited in the Nanaimo Basin, western Washington and British Columbia: Implications for Cretaceous tectonics: Columbus, Ohio, The Ohio State University unpublished Ph.D. thesis, 368 p.

Pacht, J. A., 1984, Petrologic evolution and paleogeography of the Late Cretaceous Nanaimo Basin, Washington and British Columbia: Implications for Cretaceous tectonics: GSA Bulletin, v. 95, p. 766–778.

Price, R. A., J. W. H. Monger, and J. A. Roddick, 1985, Cordillera cross-section; Calgary to Vancouver, *in* Field guides to geology and mineral deposits in the southern Canadian Cordillera: GSA Cordilleran Section Meeting, Vancouver, British Columbia, p. 3-1-3-85.

Riddihough, R. P., 1977, A model for recent plate interactions off Canada's west coast: Canadian Journal of Earth Sciences, v. 14, p. 384-396.

Rogers, G. C., 1983, Some comments on the seismicity of the northern Puget Sound-Southern Vancouver Island region, *in* J. C. Yount and R. S. Crossan, eds., Earthquake hazards of the Puget Sound region, Washington State: U.S. Geological Survey Open-File Report 1983–19, p. 19–39.

Shouldice, D. N., 1971, Geology of the western Canadian continental shelf: Canadian Society of Petroleum Geologists Bulletin, v. 19, p. 405-436.

Shouldice, D. N., 1973, Western Canadian continental shelf, *in* R. G. McGrossan, ed., Future petroleum provinces of Canada: Canadian Society of Petroleum Geologists Memoir 1, p. 7–35.

Sliter, W. V., 1973, Upper Cretaceous foraminifers from the Vancouver Island area, British Columbia, Canada: Journal of Foraminiferal Research, v. 2, p. 167–183.

Snavely, P. D., and H. C. Wagner, 1981, Geologic cross-section across the continental margin off Cape Flattery, Washington and Vancouver Island, British Columbia: U.S. Geological Survey Open-File Report 81–978, p. 5.

Sutherland Brown, A., 1968, Geology of the Queen Charlotte Islands, British Columbia: British Columbia Department of Mines and Petroleum Resources Bulletin 43, 226 p.

Sutherland Brown, A., and C. J. Yorath, 1985, LITHOPROBE profile across southern Vancouver Island: Geology and tectonics, *in* Field guides to geology and mineral deposits in the southern Canadian Cordillera: GSA Cordilleran Section Meeting, Vancouver, British Columbia, p. 8-1 to 8-23.

Sutherland Brown, A., C. J. Yorath, and H. W. Tipper, 1983, Geology and tectonic history of the Queen Charlotte Islands: Field Trip Guide Book #8, Geological Association of Canada, Mineralogical Association of Canada, Canadian Geophysical Union Annual Meeting, Victoria, British Columbia, 21 p.

Ward, P., and K. O. Stanley, 1982, The Haslam Formation: A late Santonian-early Campanian forearc basin deposit in the Insular Belt of southwestern British Columbia and adjacent Washington: Journal of Sedimentary Petrology, v. 52, p. 975–990.

Yorath, C. J., 1980, The Apollo structure of Tofino Basin, Canadian Pacific continental shelf: Canadian Journal of Earth Sciences, v. 17, p. 758–775.

Yorath, C. J., and R. L. Chase, 1981, Tectonic history of the Queen Charlotte Islands and adjacent areas—A model: Canadian Journal of Earth Sciences, v. 18, p. 1717–1739.

Yorath, C. J., and R. D. Hyndman, 1983, Subsidence and thermal history of Queen Charlotte Basin: Canadian Journal of Earth Sciences, v. 20, p. 135–159.

Yorath, C. J., G. J. Woodsworth, R. P. Riddihough, R. G. Currie, R. D. Hyndman, G. C. Rogers, D. A. Seemann, and A. D. Collins, 1985a, B-1 Intermontane Belt (Skeena Mountains) to Insular Belt (Queen Charlotte Islands): GSA Centennial Continent/Ocean Transect #8, Decade of North American Geology.

Yorath, C. J., A. G. Green, R. M. Clowes, A. Sutherland Brown, M. T. Brandon, E. R. Kanasewich, R. D. Hyndman, and C. Spencer, 1985b, LITHOPROBE, southern Vancouver Island: Seismic reflection sees through Wrangellia to the Juan de Fuca plate: Geology, v. 13, p. 759–762.

CHAPTER FOURTEEN

TERTIARY GEOLOGIC FRAMEWORK, NEOTECTONICS, AND PETROLEUM POTENTIAL OF THE OREGON–WASHINGTON CONTINENTAL MARGIN

P. D. Snavely, Jr.

U.S. Geological Survey
Menlo Park, California

ABSTRACT

The Tertiary geologic evolution of the Oregon and Washington continental margin was molded by episodic periods of convergence between the Pacific oceanic plates and the North American plate. This margin is the site of a deep basin that is floored by Paleocene to lower Eocene oceanic crust and contains more than 7000 m of sedimentary and volcanic rocks. A middle Eocene turbidite sequence overlaps this oceanic basalt and pre-Tertiary rocks of the Klamath Mountains, thus indicating that suturing of the Coast Range-Olympic terrane to North America occurred about 50 Ma.

Oblique convergence between the Farallon and North American plates occurred during most of the middle Eocene to middle Miocene; and sedimentation, punctuated by episodes of volcanism, was essentially continuous in a fore-arc basin whose axis lay along the present inner continental shelf. Two periods of more head-on convergence occurred during the late Eocene and middle Miocene, and thick accretionary wedges of melange and broken formation were formed during these strongly compressive episodes. Geochemical analyses indicate that the melange, which crops out along the west side of the Olympic Peninsula and beneath the adjacent shelf, has the highest potential for oil and gas generation. These tectonically complex strata are undoubtedly the source rocks for gas seeps

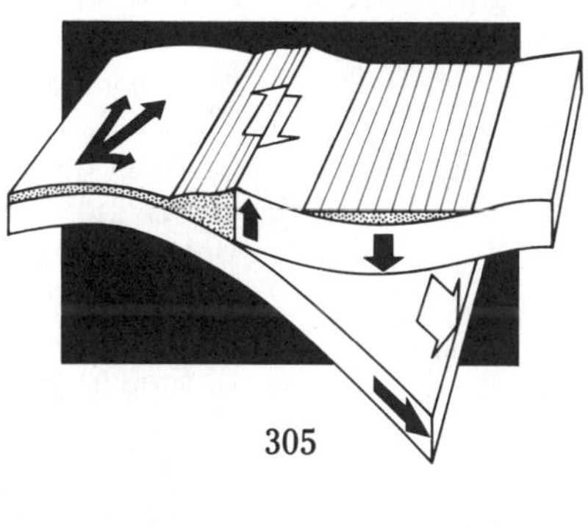

and oil and gas shows in exploratory wells in this area. Potential exploration targets exist where Eocene and/or Oligo–Miocene melanges are underplated beneath the Eocene oceanic basalt. Hydrocarbons generated in the melanges could migrate upward into structures in strata that overlie the basalt in the upper plate.

INTRODUCTION

The sedimentary and volcanic rocks in the Coast Ranges of Oregon and Washington, in the Olympic Mountains, and on the adjacent continental margin (Figure 1) provide a nearly complete record of the Tertiary history of this part of the Pacific Northwest. In this chapter, I attempt to show that the tectonic, depositional, and basin-forming events are related to the geometric framework of relative plate motion in the northeast Pacific as developed by Engebretson (1982) and outlined in Wells et al. (1984). Also, I appraise the petroleum potential of the Oregon and Washington continental margin within the framework of the Tertiary geologic history.

The Cenozoic history of the Oregon and Washington continental margin is interpreted as having been molded by episodic periods of underthrusting, transcurrent faulting, and extension as a result of interaction between oceanic plates and North America (Snavely and MacLeod, 1977; Snavely, Wagner, and Lander, 1980; Snavely, 1984; Wells et al., 1984). Interactions between the Kula or Farallon plates and North America from the Paleocene to the Miocene apparently were dominated by oblique convergence with a strong component of dextral faulting and attendant extension occurring along the continental margin. The oceanic basalt basement in western Oregon and Washington is interpreted to represent in situ eruption of basalt during rifting and extension of the continental margin in the Paleocene and early Eocene. The long period of transform faulting was interrupted by episodes of more head-on convergence in the late early Eocene, middle late Eocene, late middle Miocene, and Quaternary. Post-early-Eocene rifting and extension of the fore-arc basin is recorded by several major periods of igneous activity in the Oregon and Washington Coast Ranges. However, periods of compressional tectonics occurred in the Olympic Peninsula and in the Strait of Juan de Fuca concurrent with igneous events in the Coast Ranges. This marked difference in tectonic style is thought to have resulted from the collision of a northward-moving Coast Range-Olympic terrane against the buttress of thick continental crust on Vancouver Island.

The geologic evolution of the continental margin of Oregon and Washington has been revealed by the detailed geologic mapping of many workers (notably E. M. Baldwin, R. D. Brown, W. M. Cady, H. D. Gower, Linn Hoover, N. S. MacLeod, A. R. Niem, W. W. Rau, R. W. Tabor, R. E. Wells, and the writer). They established the basic onshore Tertiary stratigraphic and tectonic framework. Paleontological studies (in particular by W. O. Addicott, David Bukry, E. J. Moore, and W. W. Rau)[1] and petrochemical and geophysical studies also have contributed significantly to the interpretation of the geology based upon classic field mapping techniques. Offshore single-channel seismic-reflection data acquired by L. D. Kulm, E. A. Silver, S. H. Clarke, and Bobb Carson, broadly spaced multichannel seismic-reflection data acquired by the U.S. Geological Survey, and subsurface data from deep offshore exploration wells (Figure 1, Table 1) have provided a means for correlating onshore and offshore geology. The cross-feed between the onshore and offshore data sets has measurably improved the interpretation of the geologic framework and the Tertiary evolution of this active continental margin.

The figures and diagrams of the Oregon and Washington continental margin as presented in this report do not show the large-scale clockwise rotation of the terrane as documented by the paleomagnetic studies of Simpson and Cox (1977); Beck and Plumley (1980); Magill, Cox, and Duncan (1981); Magill et al. (1982); Wells and Coe (1985); and other workers. Some workers interpret the earliest rotation, which involved the lower to middle Eocene oceanic crust and seamount terrane and the overlying Tyee Formation, to have occurred during collision and accretion (Model One of Simpson and Cox, 1977; Magill, Cox, and Duncan, 1981; Duncan, 1982). Recent work suggests, however, that the rotation of Tertiary rocks in the Coast Ranges may be due entirely to postaccretion events, part of which resulted from rotation of blocks within a broad zone of right-lateral shear, and part from back-arc extensional deformation (Model Two of Simpson and Cox, 1977; Heller and Ryberg, 1983; Wells, 1984; Wells and Coe, 1985). If so, the continental margin would have had a more northwest trend than at present.

GEOLOGIC FRAMEWORK

Paleocene to Early Middle Eocene

The basement rocks in western Oregon and Washington, and on the inner continental shelf of Oregon, consist of pillow lava and breccia with minor interbeds of basaltic sedimentary rocks of Paleocene to early middle Eocene age (Figure 2). From north to south these include the Metchosin Volcanics (Clapp, 1917; Muller, 1977) on southern Vancouver Island; the Crescent Formation (Arnold, 1906; Brown, Gower, and Snavely, 1960; Cady, 1975) in western Washington; the Siletz River Volcanics (Snavely and Baldwin, 1948; Snavely, MacLeod, and Wagner, 1968) in the northern part of the Oregon Coast Range; and the Siletz River Volcanics

[1]Benthic Foraminifera stages used in this report are in part modified from Kleinpell, 1938; Mallory, 1959, and Rau, 1981.

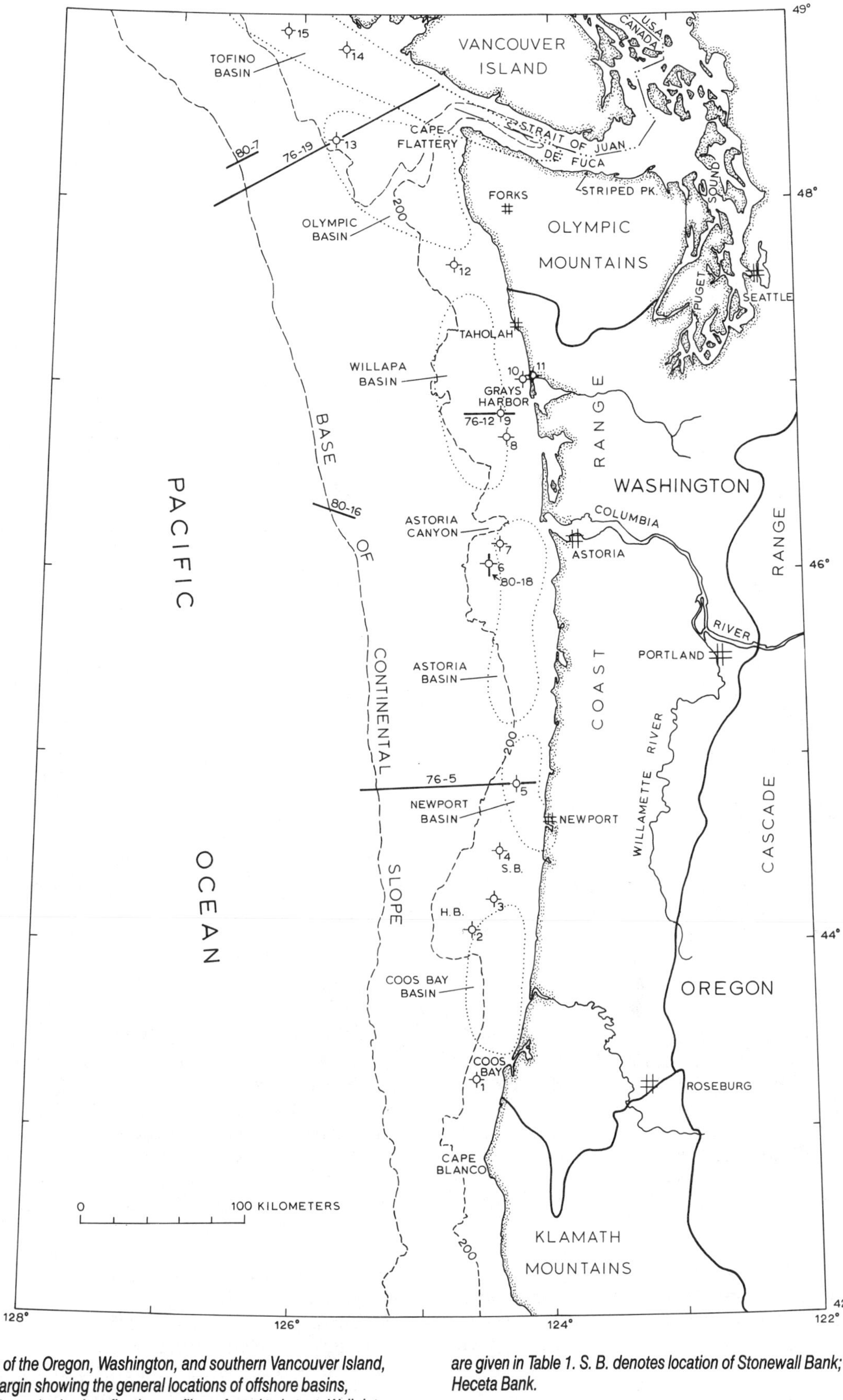

Figure 1. Index map of the Oregon, Washington, and southern Vancouver Island, B. C., continental margin showing the general locations of offshore basins, exploratory test wells, and seismic-reflection profiles referred to in text. Well data are given in Table 1. S. B. denotes location of Stonewall Bank; H. B. denotes Heceta Bank.

Table 1. Location, depths, and oldest rock penetrated in exploratory wells drilled on the Oregon and Washington continental shelves and selected wells drilled on the shelf off southern Vancouver Island, British Columbia.

Well Number (Figs. 1, 2)	Company, Name, and Number of Well	Year Drilled	Location: Latitude (North)	Location: Longitude (West)	Total Depth (m)	Oldest Rocks Penetrated
1	Pan American Oil Co. No. 1, P-0112	1967	43°14.8′	124°35.6′	1873	Lower Eocene arkosic wacke (Snavely et al., 1982)
2	Union Oil Co. Fulmar P-0130	1966	44°03.6′	124°38.8′	3725	Lower Eocene arkosic wacke (Snavely et al., 1982)
3	Shell Oil Co. P-087 1ET–2ET	1965	44°13.3′	124°28.2′	2546	Lower(?) Eocene basalt (Snavely et al., 1982)
4	Union Oil Co. Grebe P-093	1966	44°29.8′	124°24.9′	3051	Lower(?) Eocene basalt (Snavely et al., 1982)
5	Standard Oil Co. Nautilus No. 1 P-0103	1965	44°51.5′	124°16.7′	3849	Middle to upper Eocene basalt(?) (Snavely et al., 1982)
6	Shell Oil Co. P-072 1ET	1966	46°02.8′	124°29.9′	2505	Upper Oligocene to middle Miocene melange(?)
7	Shell Oil Co. P-075 1ET	1966	46°09.1′	124°24.5′	3097	Lower(?) Eocene basalt
8	Shell and Pan American Oil Co. P-0150	1966	46°43.5′	124°21.3′	4017	Upper Oligocene to lower Miocene melange and broken formation
9	Shell Oil Co. P-0155 1ET	1967	46°51.2′	124°24.5′	3402	Upper Oligocene melange (Snavely and Wagner, 1982)
10	Union Oil Co. Tidelands State No. 2	1962	47°03.22′	124°12.81′	1546	Lower to middle Miocene melange and broken formation
11	Sunshine Mining Co. Medina No. 1	1962	47°03.03′	124°10.02′	1262	Lower Miocene melange (onshore well produced 12,000 bbls of oil)
12	Pan American Oil Co. P-0141	1967	47°39.7′	124°47.5′	3160	Upper Oligocene(?) to middle Miocene melange
13	Shell Canada Ltd. Anglo Cygnet J-100	1969	48°19.67′	125°43.96′	2460	Lower to middle Miocene (Snavely and Wagner, 1981; Shouldice, 1971)
14	Shell Canada Ltd. Anglo Prometheus H-68	1967	48°48.9′	125°39.6′	2335	Lower(?) Eocene Crescent Formation (Shouldice, 1971)
15	Shell Canada Ltd. Anglo Zeus I-65	1968	48°54.60′	126°09.15′	3042	Lower(?) Eocene arkosic wacke (Shouldice, 1971)

(Molenaar, 1985) or the Roseburg Formation of Baldwin (1974) in the southern Oregon Coast Range (Figure 3). This basement volcanic sequence can be broadly divided into two petrochemical types that show differentiation trends similar to those of the Hawaiian tholeiitic and alkalic suites (Snavely, MacLeod, and Wagner, 1968; Snavely et al., 1976a). The older Paleocene and early Eocene pillow lavas are low-K_2O tholeiite that has a composition similar to oceanic-ridge basalts. On the Olympic Peninsula, the older unit contains thin interbeds of red coccolith-bearing pelagic limestone (Garrison, 1973) which were deposited in a deep oceanic environment (Bukry, 1974, written communication). The younger unit of early and middle Eocene age consists of alkalic basalts, tholeiitic basalt enriched in alkalies, and porphyritic augite basalt, picrite basalt, and feldsparphyric basalt. In places these porphyritic basalts form subaerial flows. The basalts of the younger unit are considered to represent a differentiated alkalic suite that formed oceanic islands and seamounts on the older tholeiitic oceanic crust (Snavely, MacLeod, and Wagner, 1968; Snavely et al., 1976a; Glassley, 1974; Muller, 1977). The younger volcanic sequences intertongue complexly with shallow-water basaltic sandstone which contains foraminifers, mollusks, corals, and algae of late early to early middle Eocene age (Penutian and Ulatisian Stages). In places arkosic sandstone, diorite conglomerate, and felsic tuff-breccia derived from a terrestrial source are interbedded with basalt flows (Brown, Gower, and Snavely, 1960; Tabor and Cady, 1978a; Wells, 1979; Snavely, Wagner, and Lander, 1980). Thus, these basalts are believed to represent oceanic islands and seamounts erupted on oceanic crust adjacent to the continental margin (Snavely, MacLeod, and Wagner, 1968; Snavely and MacLeod, 1974, 1977; Wells et al., 1984; Snavely, 1984).

Geophysical studies support the interpretation that oceanic basalts form the basement rocks in western Oregon and Washington. A north–south crustal refraction profile (Berg et al., 1966) near the axis of the coast ranges of northern Oregon and southern Washington indicates that rock with oceanic crust velocity is present to a depth of about 10 km and with mantle velocity at a depth of about 15.8 km.

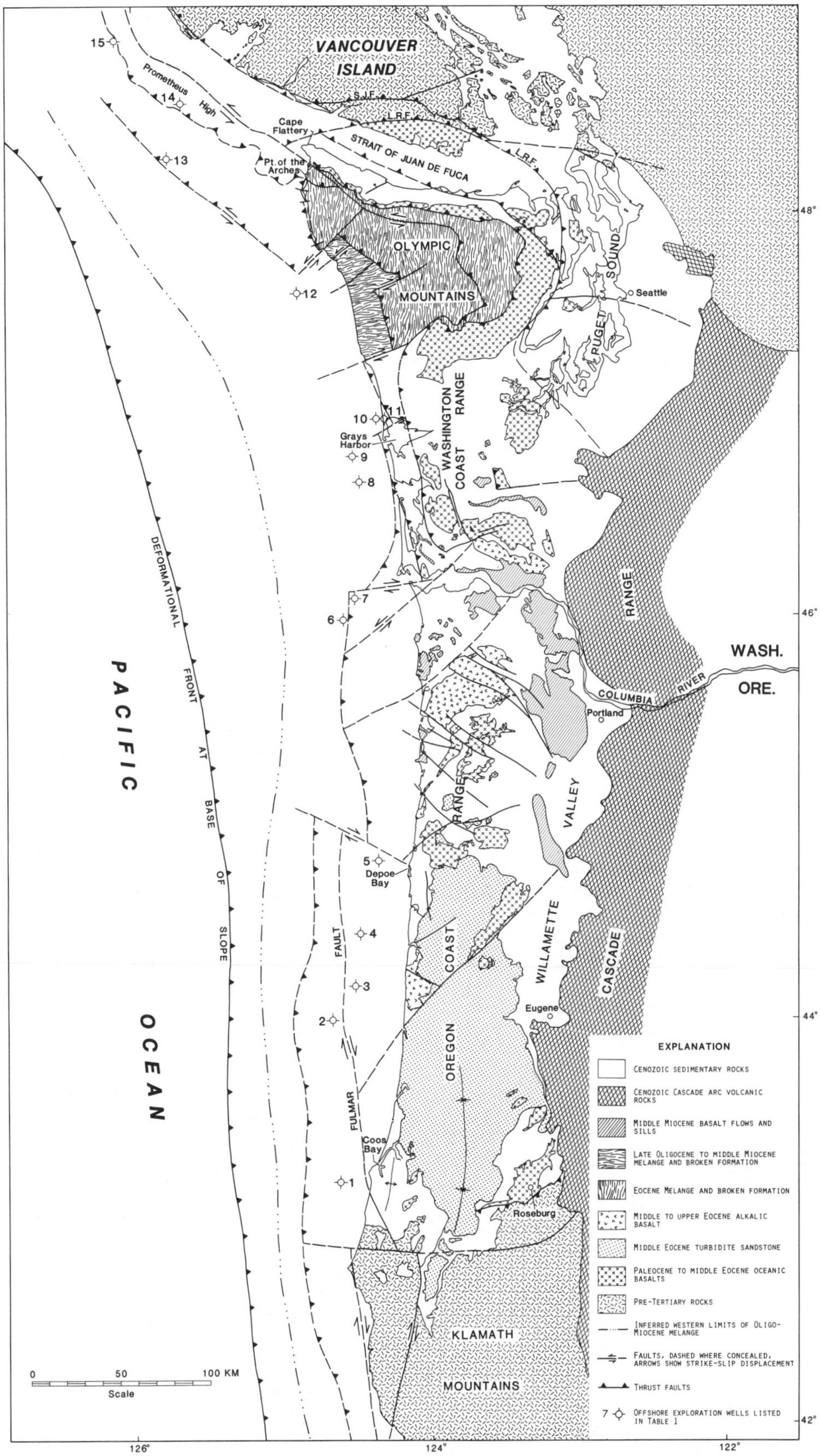

Figure 2. Generalized onshore geologic map of western Oregon, western Washington, and southern Vancouver Island, B. C., with inferred locations of major faults on the adjacent continental shelves. Offshore faults concealed by younger strata are indicated by short dashes. Modified from Snavely and Wells, 1984. LRF, Leech River fault; SJF, San Juan fault. Outcrops of middle Miocene basalt not shown in Willamette Valley; upper middle Eocene volcanics not shown in eastern part of Puget Sound.

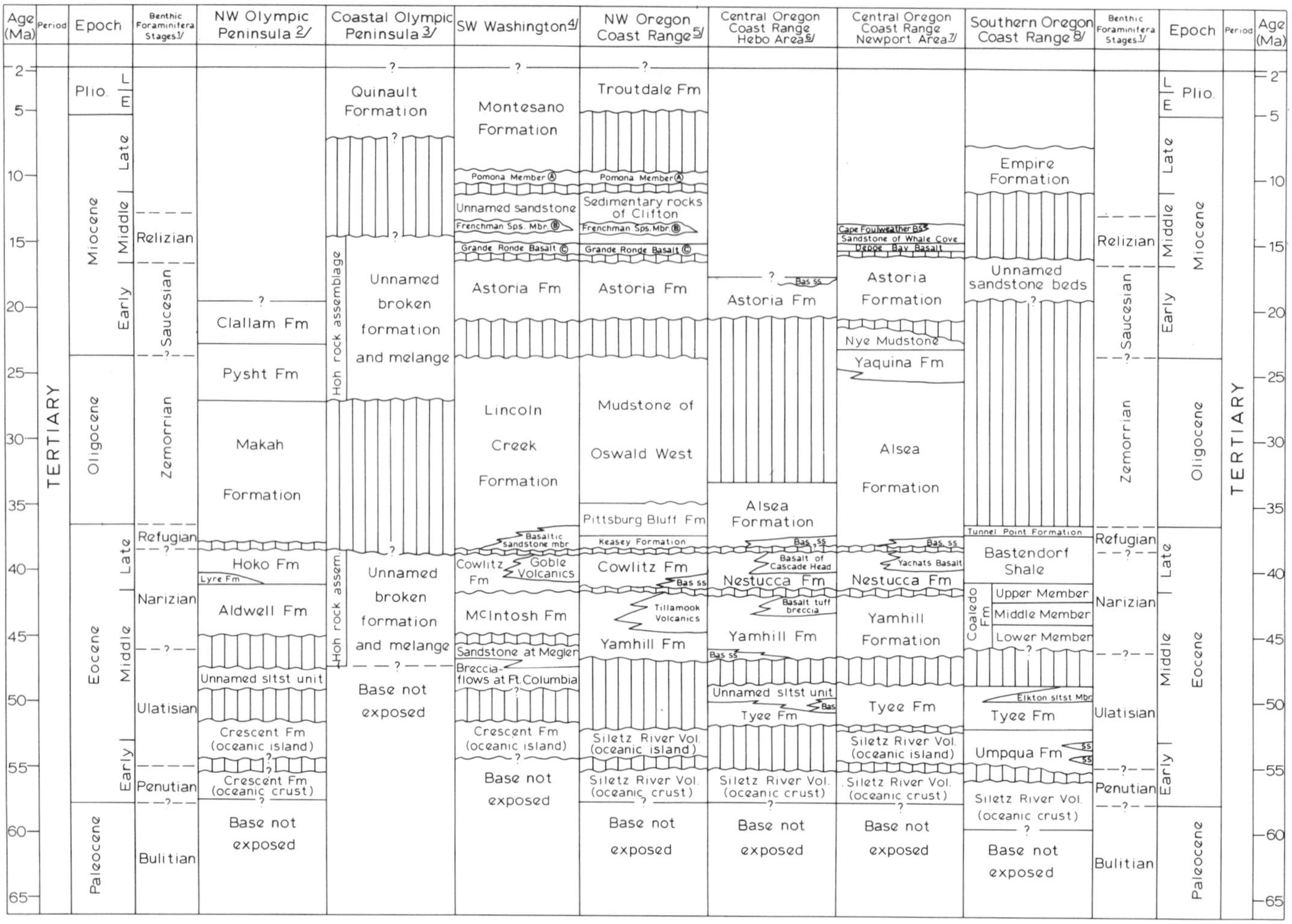

Figure 3. Correlation chart of principal Tertiary formations in western Oregon and Washington. Data in part modified from (1) Kleinpell, 1938; Mallory, 1959; Rau, 1981; (2) Snavely, Niem, and Pearl, 1978; Snavely et al., 1980, 1986; (3) Rau, 1973, 1975, 1979; Snavely, unpublished map; (4) Snavely et al., 1958; Gower and Pease, 1965; Pease and Hoover, 1957; Rau, 1967; Wagner, 1967a, 1967b; Wells, 1979, 1981, 1985; (5) Niem and Van Atta, 1973; Armentrout, 1980; Wells et al., 1983; (6) Snavely and Vokes, 1949; Wells et al., 1983; Snavely and MacLeod, unpublished map; (7) Snavely et al., 1976a, 1976b, 1976c; (8) Allen and Baldwin, 1944; Baldwin, 1974; Armentrout, 1980; Molenaar, 1985. In the southern Oregon Coast Range the Umpqua Formation is equivalent to the Flournoy and Lookingglass Formations of Baldwin, 1974; the Siletz River Volcanics is equivalent to the Roseburg Formation of Baldwin, 1974. In southwestern Washington and northwestern Oregon the Columbia River Basalt Group includes the Pomona Member of the Saddle Mountains Basalt (A), the Frenchman Springs Member of the Wanapum Basalt (B), and the Grande Ronde Basalt (C).

Several models have been proposed for the origin of the Oregon–Washington oceanic basalt basement. These include (1) accretion of a seamount-bearing oceanic plate during subduction (Snavely and MacLeod, 1974, 1977; Dickinson, 1976; Snavely, Wagner, and Lander, 1980), (2) accretion of hot-spot-generated aseismic ridges or seamount chains (Simpson and Cox, 1977; Duncan, 1982; Wells et al., 1984), (3) accretion of seamounts erupted along leaky fractures or transforms during Kula-Farallon ridge reorganization, or (4) in situ eruption of basalts during a period of continental-margin rifting and extension (Wells et al., 1984; Snavely, 1984). A model of rifting along the leading edge of the continental margin of western North America during a period of rapid, strongly oblique, northeast motion of the Kula and Farallon plates (Figure 4) appeals to the writer because:

(a) The continental crust would be attenuated and fragmented in this pull-apart regime, and diapirs of magma from the upper mantle might rise along this elongate zone of crustal thinning and provide the source for the great volumes of Paleocene to lower middle Eocene basalt extruded in western Oregon and Washington.

(b) The growth of seamounts and oceanic islands would be adjacent to the continental margin and thus could account for the terrestrially derived rocks found interbedded in places with the basalts.

(c) This model provides freedom for the northward migration of superterranes west of the rift basin, but only a small (350 km) northward translation of the Eocene oceanic basalts as deduced from paleomagnetic studies (Beck, 1984).

(d) The location of the oceanic basalts within a broad zone of dextral shear also may provide the kinematic setting for part of a clockwise rotation of the oceanic basalt and younger Tertiary rocks as described notably by Simpson and Cox (1977), Beck (1980), Magill and Cox (1980), Magill, Cox, and Duncan (1981), Globerman, Beck, and Duncan (1982), and Wells and Coe (1985).

(e) Igneous activity in the Oregon and Washington Coast

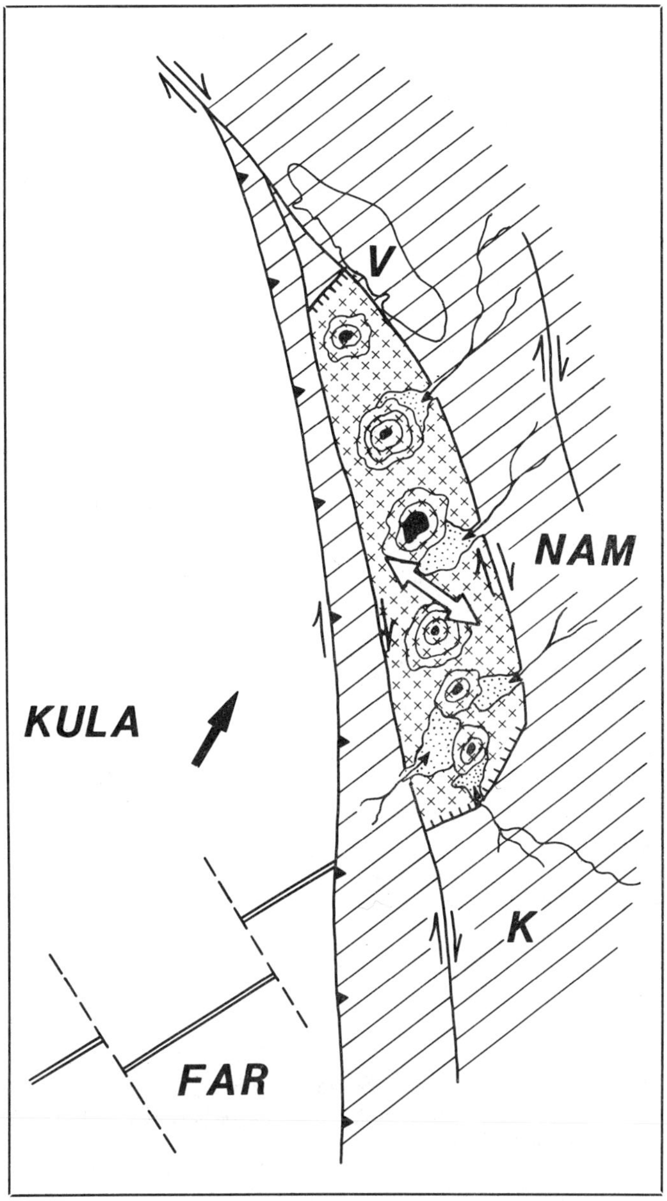

Figure 4. Diagram showing inferred continental-margin-rifting model for the origin of the Paleocene to lower middle Eocene oceanic basalt (x pattern) and oceanic islands (circular areas) in western Oregon and Washington. K, Klamath Mountains; V, Vancouver Island, B. C.; FAR, Farallon plate; KULA, Kula plate; NAM, North American plate. Arrow denotes direction of movement of the Kula plate relative to the North American plate (from Engebretson, 1982).

> Ranges has occurred throughout much of the Paleogene (Figure 2); consequently, long-lived "hot" mantle diapirs also may have been the magma sources for these episodic igneous events.

This model of an oblique rifting geometry perhaps is analogous to the Gulf of California-Salton Trough (Crowell, 1974) and the Andaman Sea north of Sumatra (Curray et al., 1979), where large components of transcurrent motion are evident in the rifting process.

In basinal areas adjacent to the seamount-island chain, thick (3000 m) sequences of siltstone, turbidite sandstone, and conglomerate were deposited on the Paleocene and lower Eocene oceanic basalt. Convergence between the Kula-Farallon and North American plates in the late early Eocene (~52 Ma) deformed the island chain and overlying marine sedimentary rocks against the North American buttress. In the southern part of the Oregon Coast Range, this compressional deformation produced a set of presently southeast-dipping imbricate thrust sheets and isoclinal folds which deformed the Paleocene and lower Eocene oceanic basalts and overlying lower Eocene sedimentary rocks (Baldwin, 1974; Molenaar, 1985). In the central part of the Coast Range, the Siletz River Volcanics were faulted and folded prior to rapid downwarping and the deposition of the overlying mildly deformed Tyee Formation of early middle Eocene age (Snavely et al., 1976a).

In places, erosion of uplifted or constructional highs of the older tholeiitic basalt occurred during the late early Eocene. In the northwest part of the Olympic Peninsula, conglomerate with clasts of red calcareous siltstone of early Eocene age (Bukry, 1985, written communication) derived from the lower part of the Crescent Formation are interbedded with pillow lavas and basaltic siltstone in the upper part of the Crescent Formation of middle Eocene age (W. W. Rau, 1977, written communication). In the southern part of the Oregon Coast Range, shallow-water limy crystal tuff of lower and earliest middle Eocene age unconformably overlies the lower Eocene basalts and breccias of the Roseburg Formation of Baldwin, 1974 (Hoover, 1963; Snavely, Wagner, and Rau, 1982).

The period of convergence also deformed and uplifted continental terranes adjacent to the marginal (rift?) basin and accelerated erosion of pre-Tertiary rocks in the Klamath terrane to the south and on Vancouver Island to the north. Thick-bedded lithic quartzose sandstone and mudflow conglomerates of middle Eocene age in the northwesternmost Olympic core rocks (Figure 5; Tabor and Cady, 1978a; Snavely et al., 1986) represent material eroded off the uplifted terrane of Vancouver Island. In the Puget Sound area, streams transported large quantities of arkosic and feldspathic material into the northern part of the basin across broad, low-lying flood plains. The coal-bearing continental fluvial deposits, represented by the lower part of the Puget Group (Wolfe, Gower, and Vine, 1961), the Chuckanut Formation (McLellan, 1927; Johnson, 1984), and the Swauk Formation (Russell, 1900; Tabor et al., 1984), formed in lowland areas. Some of this coarse clastic material probably was transported westward through submarine channels into the deeper parts of the basin to form turbidite sandstone deposits. Some of the highly deformed middle and upper Eocene thick-bedded carbonaceous, lithic, arkosic sandstone units within the Olympic core rocks (Tabor, Yeats, and Sorensen, 1972; Tabor and Cady, 1978b) and in the Cape Flattery coastal area (Snavely et al, 1986) may be marine correlatives of these continental deposits.

In the southern Oregon part of the marginal basin, 2000 m of arkosic and lithic turbidite sand and silt of the early middle Eocene Tyee Formation, derived in part from an uplifted Klamath terrane and from the Idaho batholith and adjacent terranes (Heller et al., 1985), buried pre-existing volcanic highs and deformed lower Eocene sedimentary rocks (Snavely, Wagner, and MacLeod, 1964; Lovell, 1969; Baldwin, 1974; Chan and Dott, 1983). Inasmuch as this sedimentary sequence laps across the tectonic boundary between the Klamath pre-Tertiary rocks and the Eocene

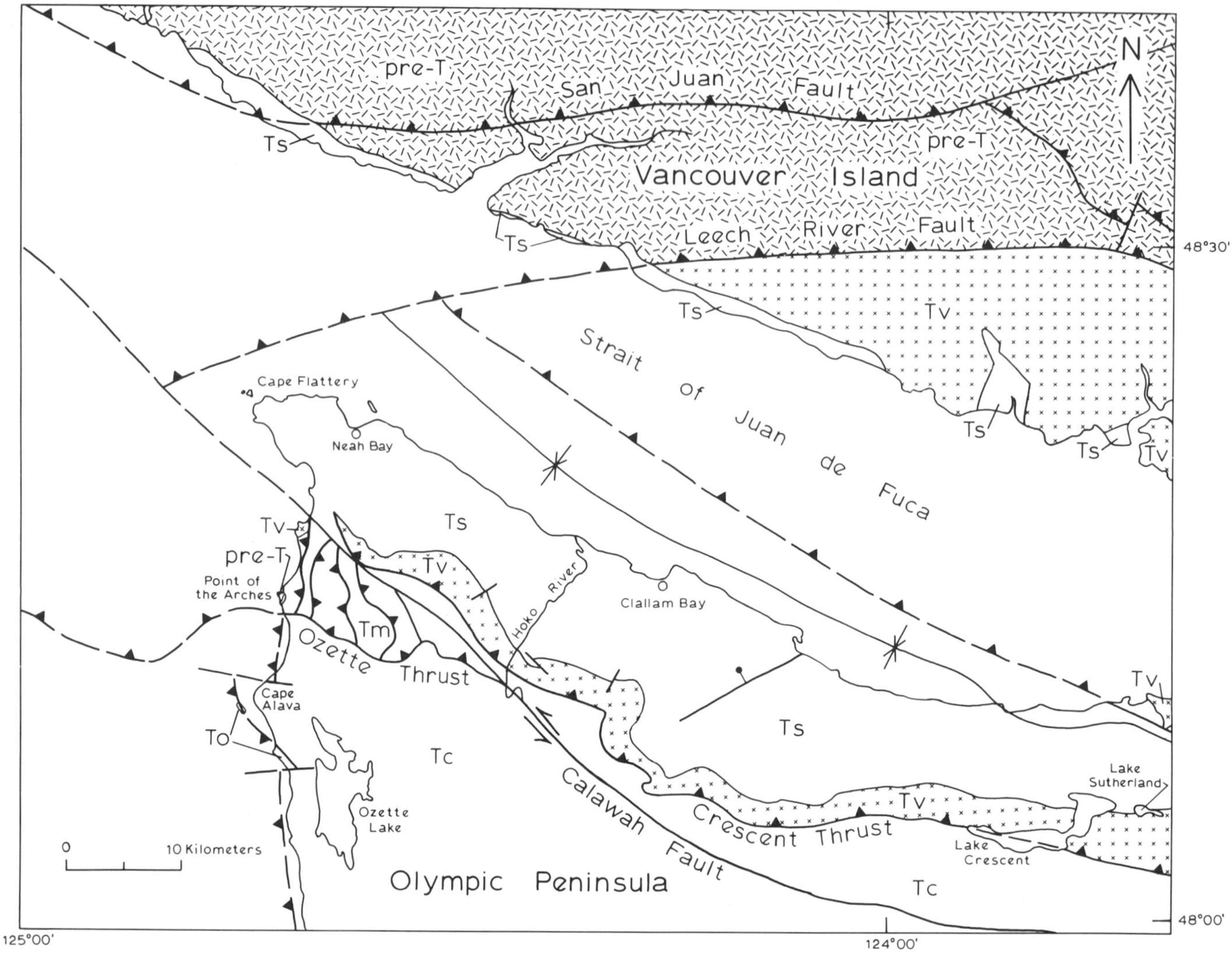

Figure 5. Generalized geologic map of the region adjacent to Cape Flattery, showing locations of major onshore faults and their offshore extensions (modified from MacLeod et al., 1977, and unpublished maps by Snavely). Ts, undivided Tertiary sedimentary rocks; To, upper Oligocene sandstone and conglomerate; Tm, overthrust pre-Tertiary to late Eocene broken formation above Ozette thrust; Tc, highly deformed Eocene sandstone and siltstone and other rocks of the Olympic core; Tv, Eocene pillow basalt, breccia, associated dikes and sills of the Metchosin Volcanics (Clapp, 1917) on Vancouver Island and the Crescent Formation in Washington; pre-T, undivided pre-Tertiary igneous and metamorphic rocks on Vancouver Island and at the Point of the Arches.

oceanic basalts to the north, the Coast Range-Olympic terrane was accreted to the North American plate in the early middle Eocene (about 50 Ma). On the Olympic Peninsula, suturing predates 41-Ma hornblende quartz diorite dikes that cut lower to middle Eocene sedimentary and volcanic rocks in the core of the Olympic Mountains (Tabor, Yeats, and Sorensen, 1972). Dikes and small plugs of dacite and andesite, dated about 44 Ma, cut the lower Eocene basalts of the Crescent Formation at Striped Peak on the south coast of the Strait of Juan de Fuca and the Crescent Formation west of Bremerton, Washington (Snavely, unpublished data). These postaccretion felsic rocks are interpreted as being related to the development of the Cascade arc, intruding the much older Eocene basalt and sedimentary rock.

The convergence velocity between the Farallon and North American plates was greatly reduced at about 43 Ma (Engebretson, Gordon, and Cox, 1980; Engebretson, Cox, and Thompson, 1984). This reduction generally coincides with renewed dextral faulting on the Oregon continental shelf, which juxtaposed a lower Eocene (Penutian Stage) graywacke sequence on the west against the lower Eocene oceanic crust of the Coast Range (Fulmar fault, Figure 2; Snavely, Wagner, and Rau, 1982). An aeromagnetic map (U.S. Geological Survey, 1970) shows a linear magnetic anomaly with a steep gradient on the inner continental shelf—a gradient which is believed to mark the contact between the wacke and the volcanic rocks. This gradient intersects the coast just north of Fivemile Point, about 15 km south of Coos Bay. The nearly vertical turbidite sandstone and siltstone strata at Fivemile Point are early Eocene in age. The sandstone is petrographically similar to the early Eocene wacke penetrated in Pan American test well P-0112 drilled west of the Fulmar fault on the continental shelf off Coos Bay (Snavely, Wagner, and Rau, 1982).

Farther south, the Fulmar dextral fault may correlate with the Coquille River fault that forms the contact between pre-Tertiary rock to the west and lower and middle Eocene Umpqua Formation to the east, at least as far south as lat.

42°30′N (Figure 2). Offshore seismic-reflection data and subsurface stratigraphic information from deep test wells (Snavely, Wagner, and Rau, 1982) indicate that the lower Eocene wacke and lower Eocene volcanic rocks have a minimum separation of at least 200 km (Snavely, Wagner, and Lander, 1980). However, the rate of dextral movement on the Fulmar fault during the late middle and late Eocene may have been high enough to transport the lower Eocene wackes northward to the latitude of the Olympic Mountains. The highly deformed and imbricate thrust-faulted Eocene strata in the eastern core rocks of the Olympic Peninsula (for example, the Needles-Gray Wolf Lithic Assemblage of Tabor and Cady, 1978a) may be correlative with the lower Eocene wackes on the Oregon shelf.

The lower Eocene oceanic basalt and sedimentary rocks of late early Eocene age (siltstone member of the Umpqua Formation) and early middle Eocene age (Tyee Formation) apparently do not extend west of the Fulmar transcurrent fault (Snavely, Wagner, and Rau, 1982). However, they are present on the inner continental shelf of Oregon, inasmuch as strata correlative to the Tyee Formation were penetrated in the Shell Oil Company P-087 well (Snavely, Wagner, and Lander, 1985), where they onlap the Eocene oceanic basalt.

The major movement along the Fulmar fault must have occurred from the late middle Eocene to the late Eocene, because on the Oregon shelf this fault is overlapped by strata of late Eocene age (Refugian Stage). However, minor movement apparently occurred along this dextral fault in post-late-Eocene time, because strata as young as Pleistocene, which overlie the fault, are cut by faults with small vertical offsets but an undetermined amount of strike-slip separation. The Fulmar fault, and other major dextral faults that probably existed farther to the east, formed a transtensional zone of right-lateral shear along the continental margin of Oregon and southwestern Washington. Magma generated in the mantle rose along deep shears within this zone to produce eruptive and intrusive events near the present coast. These igneous episodes include (1) Eocene (44–38 Ma) alkalic basalt flows and sills, (2) early Oligocene (33 Ma) nepheline syenite sills and stocks, and flows and intrusives of camptonite, (3) late Oligocene (30 Ma) ferrogabbro sills, and (4) middle Miocene (16–14 Ma) tholeiitic basalt flows and intrusives (MacLeod and Snavely, 1973; Snavely, MacLeod, and Wagner, 1973; Snavely, Wagner, and Lander, 1980; Wells et al., 1983).

Late Middle Eocene to Oligocene

Postaccretionary tectonic and magmatic events in western Oregon and Washington during the period ~44–38 Ma were contemporaneous with the start of Cascade volcanism (Wells et al., 1984). The northern and southern parts of the fore-arc basin were subjected to different styles of deformation, both interpreted to be related to dextral faulting along the continental margin initiated by movement along the Fulmar fault. In southwest Washington and Oregon, extension is inferred along a broad zone of right-lateral shear, and basaltic magma erupted at several major volcanic centers. In the northern part of the fore-arc basin, however, compression characterized the structural style, initiating major thrust faulting in the late middle Eocene along the Crescent and other major faults (Figures 2, 5; MacLeod et al., 1977; Snavely, 1983).

The alkalic basalts that erupted into the fore-arc basin are chiefly subaerial; however, pillow lava and breccia in the lower part of these units intertongue laterally with deep-water siltstone of the Yamhill and Nestucca Formations (Figure 3). These basalt flows and breccia include the Tillamook Volcanics in northwest Oregon (Warren, Grivetti, and Norbisrath, 1945; Snavely, MacLeod, and Rau, 1970; Magill, Cox, and Duncan, 1981; Wells et al., 1983). They were followed closely by the Goble Volcanics of southwest Washington (Livingston, 1966; Wells, 1981), the Yachats Basalt (Snavely and MacLeod, 1974), the basalts of Cascade Head (Snavely and Vokes, 1949) in central coastal Oregon. All of these volcanic rocks are porphyritic and have a wide range in composition ranging through alkalic basalt, basaltic andesite, and dacite. This range suggests that the basaltic magma that produced these volcanic rocks was differentiated in the upper crust before extrusion.

Although differential uplift appears to have been in the areas of subaerial volcanism, such as the Heceta Head-Cape Perpetua area on the central Oregon coast (Snavely and MacLeod, 1974), regional downwarping occurred in most places in the fore-arc basin. Bathyal thin-bedded siltstone and sandstone were unconformably deposited on lower and middle Eocene oceanic basalt and sedimentary rocks.

In the northern part of the Olympic Peninsula, the middle and upper Eocene Aldwell Formation (Brown, Gower, and Snavely, 1960) overlapped older sedimentary rock to rest upon basalts of the Crescent Formation. The basal contact is marked in places by beds 10 to 30 m thick of cobble and boulder conglomerate derived from the underlying Crescent Formation. In the north-central part of the Oregon Coast Range, bathyal siltstone of the middle and upper Eocene Yamhill Formation (Baldwin et al., 1955) unconformably overlaps folded strata of the Tyee Formation to rest upon basalt of the Siletz River Volcanics (Wells et al., 1983; Snavely, unpublished maps). In the northern part of the Oregon Coast Range, the Yamhill Formation intertongues with pillow lavas and breccia of the middle and late Eocene Tillamook Volcanics (Wells et al., 1983). In southwest Washington, the McIntosh Formation as defined by Snavely et al. (1958) is correlative with the Yamhill and Aldwell Formations (Figure 3). Some confusion exists, however, outside the McIntosh type area. For example, Pease and Hoover (1957) mapped an upper and lower member of the McIntosh Formation and stated that the lower member intertongues downward with basalt flows and breccia of the Crescent(?) Formation. However, a review of the structural data on their map indicates that an unconformity most likely is present between their upper and lower members, and this may be the same unconformity at the base of the Yamhill and Aldwell Formations. Thus, only their upper member is correlative with the type McIntosh. In southwest Washington, siltstone that Wells (1979; in press) correlates with the McIntosh Formation also unconformably overlies a siltstone and sandstone unit that interfingers with basalts that have been included in the upper part of the Crescent by other workers.

During the middle and late Eocene, streams transported large quantities of arkosic material to the broad, low-lying coastal plains that bordered the eastern and southern parts of the fore-arc basin. Coal-bearing continental beds > 3500 m thick are represented by the Puget Group (Wolfe, Gower, and Vine, 1961) and the Skookumchuck Formation (Snavely et al., 1958) in Washington, and the Coaledo Formation

(Allen and Baldwin, 1944; Dott, 1966) in southeast Oregon. Some of this coarse clastic material undoubtedly was transported along channels into shelf basins and formed submarine fans and turbidite deposits.

In the latest Eocene (~37 Ma), the fore-arc basin was deformed over a broad front. This regional orogenic event is believed to represent a period of head-on convergence, the locus of underthrusting lying along the present mid-shelf of Oregon and Washington. Regional uplift that accompanied this convergence segmented and shoaled the fore-arc basin and produced a number of shelf basins, some of which deepened westward into deep basins on the continental shelf. Erosion of uplifted middle and late Eocene subaerial volcanic piles, such as the Yachats Basalt and the Tillamook Volcanics, produced thick deposits of marine basaltic sandstone and conglomerate that fringed the basaltic highlands. Erosion of uplifted pre-Tertiary terranes on Vancouver Island supplied coarse clastic debris to the northern part of the foreland basin to form the conglomerate-rich Lyre Formation (Brown, Snavely, and Gower, 1956; Snavely et al., 1986) along the northern Olympic Peninsula. Melange and broken formation of middle and late Eocene age along the west side of the Olympic Peninsula probably records this period of plate convergence. Despite the fact that this period of apparently more normal plate convergence is not observed in plate models, the geologic record documents a period of regional tectonism in the late Eocene. Perhaps this period of more normal convergence was rapid and short lived, and thus not recognized in models of relative plate motion.

Interpretation of seismic-reflection profiles and subsurface data from deep test wells on the Oregon continental shelf (Figure 6; Snavely, Wagner, and Lander, 1980; Snavely, Wagner, and Rau, 1982) leads us to believe there is a regional unconformity at the base of siltstone strata of late Eocene (Refugian Stage) age. These strata unconformably overlie folded and faulted rocks as old as early Eocene and overlap the Fulmar fault (Snavely, Wagner, and Lander, 1980; Snavely, 1984). Also, in several areas in the northwesternmost part of the Olympic Peninsula, strata of Refugian age onlap melange and broken formation of middle and late Eocene age (Snavely et al., 1986).

Rapid subsidence occurred in the fore-arc basin during the latest Eocene to early Miocene, and >2500 m of bathyal tuffaceous siltstone and arkosic sandstone was deposited in axial parts of the basin. On the Oregon continental margin, these strata are represented by tuffaceous siltstone of the Alsea Formation of late Eocene and Oligocene age (Snavely et al., 1975) and the Nye Mudstone of early Miocene (Saucesian Stage) age (Vokes and Snavely, 1948; Snavely, Rau, and Wagner, 1964). In the late Oligocene, high-energy deltaic deposits of pumiceous coarse-grained sandstone, such as the Yaquina Formation (in part) (Snavely and Wagner, 1963; Goodwin, 1973), were deposited in places along the eastern margin of the basin. In a deep marginal Tofino-Fuca basin along the north side of the Olympic Peninsula, >2500 m of turbidite sandstone and siltstone of the Hoko, Makah, and Pysht Formations (Figure 3) were deposited from the late Eocene to the early Miocene (Snavely, Niem, and Pearl, 1978; Snavely et al., 1980, 1986). This basin shoaled in the early Miocene, and nearshore coal-bearing sandstone and conglomerate beds of the Clallam Formation were deposited (Gower, 1960; Addicott, 1976).

Volcanic activity in the Cascade arc contributed large quantities of ash and pumiceous tuff-breccia to the fore-arc basin. Oligocene Cascade-derived mudflow conglomerate and thick (30 m) pumice beds occur as far west as the present coastline (Snavely et al., 1975). Near the late Eocene to middle Oligocene strandline, which lay near the present foothills of the Cascade Range, andesitic and dacitic tuff-breccia and volcaniclastic strata are complexly intercalated throughout much of the marine strata.

Early and Middle Miocene

Regional uplift in the fore-arc basin in the early Miocene (about 20 Ma) restricted marine deposition to the west flank of the Oregon Coast Range and the adjacent continental shelf, and to the Coos Bay, Newport, Astoria, Grays Harbor, and Tofino-Fuca structural embayments (Figure 1). Nearshore deltaic and strandline sandstone and siltstone deposits of the early and middle Miocene Astoria Formation (Cooper, 1981) grade westward into predominantly bathyal siltstone beds in the deep marginal basin off Oregon (Snavely, Wagner, and Rau, 1982). Along the central and northern Oregon Coast, the Astoria Formation—a nearshore marine sandstone—rests on strata ranging in age from late Eocene to early Miocene. In the deep marginal basin on the Oregon continental shelf, seismic-reflection profiles and drill-hole data indicate that sedimentation was virtually continuous, and siltstone strata correlative with the Nye Mudstone and Astoria Formation form a rock-stratigraphic unit (Figure 6; Snavely, Wagner, and Lander, 1980).

On the Olympic Coast, lower Miocene thick-bedded sandstone and thin-bedded siltstone and sandstone of both shelf and bathyal depositional environments onlap the Eocene melange and broken formation (Rau, 1975, 1979). These deposits were intensely folded and faulted during plate convergence in the late middle Miocene.

In coastal central Oregon, two tholeiitic basalt units were erupted from local vents and are interbedded with lower and middle Miocene sandstone and siltstone (Snavely, MacLeod, and Wagner, 1973). The oldest—the Depoe Bay Basalt—is petrochemically identical to the Grande Ronde Basalt, and the younger Cape Foulweather Basalt is petrochemically identical to the Frenchman Springs Member of the Wanapum Basalt of the Columbia River Basalt Group. The Depoe Bay Basalt extends >16 km seaward and was penetrated in several of the test wells at depths as much as 2.5 km (Figure 6; Snavely, 1984). The Cape Foulweather Basalt, however, is restricted to the inner shelf. Sills of Depoe Bay Basalt are widespread in the northern Oregon Coast Range and on the continental shelf (Snavely and Wells, 1984). The stratigraphic and petrologic similarity between the coastal basalts and correlative units on the Columbia Plateau has led some workers (Beeson, Perttu, and Perttu, 1979) to suggest that the coastal basalts are invasive tongues of the Columbia River Basalt Group that erupted on the plateau. The widespread occurrence in coastal central Oregon of dikes and sills of Depoe Bay and Cape Foulweather Basalts, some of which intrude sedimentary and volcanic rocks as old as early Eocene, supports a local deep-seated origin for these coastal basalts rather than an invasive origin.

Late Miocene and Pliocene

In late middle Miocene (12 Ma) intense deformation occurred in the Coast Ranges and on the continental shelves of Oregon and Washington. Lower Eocene to upper Eocene rocks were thrust over the Eocene melange and broken formation near the coast along the northwest side of the Olympic Peninsula (Figure 5; Snavely et al., 1986). However, farther to the south in the western core of the Olympic Mountains (Tabor and Cady, 1978a), Oligocene to upper middle Miocene strata appear to underthrust older Eocene strata. On the southern Washington shelf and outer Oregon continental shelf, an upper Eocene to upper middle Miocene sequence also is interpreted to have been thrust beneath the Eocene oceanic crust (Figure 6; Snavely, Wagner, and Lander, 1980; Snavely and Wagner, 1982). Likewise, correlative strata probably were thrust beneath the basalt of the Crescent(?) Formation that formed the Prometheus high (Shouldice, 1971; MacLeod et al., 1977; Snavely et al., 1981) offshore from southern Vancouver Island (Figure 2). This period of deformation may represent more rapid or more normal convergence that is not observed in the plate models; or it may represent stronger coupling with the young downgoing slab. A close approach of the ridge at this time also may have caused uplift and tectonism along the continental margin (Wells et al., 1984).

Regional uplift of the Coast Range-Olympic Mountain terrane formed highland areas that were rapidly eroded and supplied large amounts of clastic debris to elongate basins on the continental shelf. On the Oregon inner shelf, strata as young as middle Miocene were folded and uplifted, truncated by erosion, and subsequently downwarped and overlapped unconformably by upper Miocene strata (Figure 6; Snavely, Wagner, and Lander, 1980). Uplift was greatest on the Olympic Peninsula, perhaps because of the isostatic uplift of the thick prism of melange and broken formation that was underthrust during the late middle Miocene convergence.

In the late Miocene and Pliocene, episodic downwarping of a deep marginal basin off Oregon was virtually continuous, and >2000 m of sand and silt were deposited. Upper Miocene and Pliocene deposits thinned against the eastern border of the marginal basin and against older mid-shelf structural highs (Snavely, Wagner, and Lander, 1980; Clarke, Field, and Hirozawa, 1981). Slope basins formed landward of folded and thrust-faulted upper Oligocene to upper middle Miocene melange welts. On the central and southern Washington shelf, as much as 2000 m of upper Miocene and Pliocene sediments accumulated on a thick accretionary wedge of melange and broken formation of late Oligocene to late middle Miocene age (Rau, 1975, 1979; Snavely and Wagner, 1982). The Miocene and Pliocene strata thin against, or are intruded by, diapirs whose core consists of upper Oligocene to middle Miocene melange and broken formation. Near these diapiric structures numerous unconformities, growth faults, and gravity slides occur within younger strata, all of which reflect episodic uplift (Figure 7).

Off northwest Washington and southern Vancouver Island, upper Miocene strata rest unconformably on older Tertiary rocks (Figure 8; Shouldice, 1971; Snavely and Wagner, 1980). Although most strata of this age are restricted to the Pacific shelf, upper Miocene shallow-water sandstone and siltstone occur in a small isolated basin in the lower Bogachiel River Valley near Forks, Washington (Rau, 1979), and in fault-bounded blocks along the coast north of Taholah, Washington (Rau, 1970, 1975). Upper Miocene strata also crop out on the southern Oregon coast near Coos Bay (Addicott, 1976; Armentrout, 1980).

On the abyssal plain, about 3500 m of strata occur above a late Miocene oceanic crust (Kulm and Fowler, 1974; Snavely, Wagner, and Lander, 1980; Carlson and Nelson, this volume) in a filled trench along the base of the slope (Figure 6). From velocity analysis, it is assumed that as much as half of this fill is of late Miocene and Pliocene age.

Pleistocene and Holocene

Pleistocene and Holocene sediments extend across most of the continental shelf of Oregon and Washington, and Neogene strata are only exposed on the sea floor on large banks such as Stonewall and Heceta Banks (Figure 1; Kulm and Fowler, 1974; Snavely and Wagner, 1980; Snavely, Wagner, and Lander, 1980; Clarke, Field, and Hirozawa, 1981; Carlson and Nelson, this volume). These Quaternary deposits are thickest (500 m) on the inner shelf and in basins between compressional folds on the outer shelf and lower slope. Several unconformities occur within this sequence, indicating episodic downwarping of the basins as well as uplift of diapiric structures during deposition (Figure 7). On the abyssal plain >1800 m of strata are probably of Pleistocene and Holocene age (Kulm and Fowler, 1974; Snavely, Wagner, and Lander, 1980; Carlson and Nelson, this volume).

REGIONAL STRUCTURE

The deep (pre-late Miocene) structural framework of the continental margin of Oregon and Washington is poorly known because of the broad reconnaissance nature of our multichannel seismic-reflection coverage. The shallow (post-middle-Miocene) structure is better understood because numerous single-channel surveys have been conducted by staff members and graduate students at Oregon State University (Kulm and Fowler, 1974), the University of Washington (Carson, 1977), and Scripps Institute of Oceanography (Silver, 1972). The U.S. Geological Survey also has conducted moderately detailed single-channel seismic-reflection surveys on the southern Oregon shelf (Clarke, Field, and Hirozawa, 1981) and off northwestern Washington, in part in cooperation with the Geological Survey of Canada (MacLeod et al., 1977; Snavely and Wagner, unpublished data).

The deep structure off Oregon was first interpreted by Braislin, Hastings, and Snavely (1971). Seely, Vail, and Walton (1974) made a geologic interpretation of a multichannel profile located near lat. 44°30′N, and for the first time established an underthrust tectonic model characterized by an imbricate set of landward-dipping thrust faults. Geologic cross-sections off the Oregon Coast, which generally confirm Seely's tectonic model, have been constructed by Snavely, Wagner, and Lander (1980, 1985; Figure 6). Other cross-sections (Figure 1) that illustrate different structural styles have been constructed off southwestern Washington (Figure 9; Snavely and Wagner, 1982) and off northwestern Washington (Figure 8; Snavely and Wagner, 1981). I believe that the Oregon cross-section is generally representative of the broad structural framework of the Oregon continental margin. The geologic cross-section constructed across the southwestern Washington inner shelf and the adjacent Coast

Range probably is representative of the structure of the Washington shelf from the Columbia River northward to about lat. 47°45′N. The cross-section off northwestern Washington probably characterizes in broad detail the tectonically complex offshore area from north of lat. 47°45′N to southern Vancouver Island.

In describing the structural framework of the continental margin, it is convenient to separate the region into four major tectonic terranes: (1) the Oregon continental margin, (2) the southern and central Washington continental margin, (3) the northwestern Washington continental margin, and (4) the Oregon and Washington continental slope and abyssal plain. Almost certainly the generalizations made concerning these structural frameworks will be substantially revised as new multichannel seismic data are obtained, and as large gaps in our reconnaissance surveys are filled in.

Oregon Continental Margin

The Cenozoic sequence on the Oregon continental margin is interpreted as being part of the upper plate above a low-angle, east-dipping megathrust along which the Juan de Fuca plate underthrusts the North American plate (Snavely, Wagner, and Lander, 1980). West of the Coast Range, four major offshore geologic terranes are present: (1) a deep marginal basin on the inner shelf, (2) a mid-shelf tectonic high, (3) a mid- to outer-shelf accretionary melange wedge, and (4) broad folds and an imbricate set of landward-dipping thrusts on the outer shelf and continental slope (Figure 6). The fold axis and traces of the thrust faults trend generally northward and are approximately parallel to the base of the continental slope. Major zones of vertical crustal movements in the upper plate in the late Cenozoic across this convergent margin appear to indicate (1) uplift along the continental slope and outer shelf along fold-and-thrust belts; (2) episodic subsidence that produced a deep marginal basin on the inner shelf; and (3) uplift in the Coast Range.

The principal structure on the southern part of the Oregon continental shelf is the buried north-trending dextral Fulmar fault (Figure 2). Major movement on this fault probably took place during the late middle to early late Eocene, juxtaposing the lower Eocene oceanic crust on the east with lower Eocene arkosic wacke on the west (Snavely, Wagner, and Lander, 1980). Although the Fulmar fault is overlapped by latest Eocene and younger strata, these strata are offset by small faults above the Fulmar fault, suggesting minor renewed movement along the fault (Snavely, Wagner, and Lander, 1985). The Fulmar fault is inferred to have had at least 200 km of dextral movement, but its total displacement may be much greater if lithologically similar Eocene micaceous wacke in the eastern core of the Olympic Peninsula (Tabor and Cady, 1978a) and the Eocene sandstone penetrated in the Shell-Canada Zeus 1-65 well on the Prometheus high off southern Vancouver Island are correlative with the Oregon shelf wackes.

North-trending seismic-reflection profiles indicate that broad major west-trending downwarps cross the generally north-trending structural grain of the Oregon inner shelf and mid shelf. Onshore structural embayments such as those at Newport, Tillamook, and Astoria may reflect this east–west warping. This east–west downwarping also is reflected in the regional warping of Pleistocene terraces along the coast.

On the southern Oregon shelf a broad, generally west-trending high formed at depth by the upper Eocene Yachats Basalt separates the Coos Bay basin on the south from the Newport basin on the north (Figure 1). This volcanic high may in large part be a constructional feature and may represent the seaward continuation of upper Eocene centers of volcanism.

On the Oregon shelf, south of lat. 45°N, folds in the pre-late-Miocene strata have been uplifted and truncated prior to downwarping and deposition of upper Miocene and younger strata (Figure 6). However, north of lat. 45°N, several anticlines, which have considerable structural relief in the Eocene and Oligocene strata, only broadly fold the Neogene strata above the late Miocene unconformity (see SP LEE line 77-11, Mann and Snavely, 1984). The trends of the fold axes in the older strata in most places do not parallel those in post-middle-Miocene strata. Because of the changes in the direction of plate convergence with respect to the continental margin during the Tertiary, changes in fold and fault trends that reflect these changes might be expected.

The large banks on the Oregon shelf, such as Heceta and Stonewall, probably are fault controlled (Kulm and Fowler, 1974; Clarke, Field, and Hirozawa, 1981; Mann and Snavely, 1984), perhaps uplifted along the east-dipping thrust faults (Snavely, Wagner, and Lander, 1980). These banks are composed of upper Miocene and Pliocene sedimentary rocks that rise above the surrounding Pleistocene sediments (Kulm and Fowler, 1974).

Fold axes in late Miocene and younger strata trend approximately north–south, but in places they may range 30°–40° from north–south. These folds are either broadly symmetrical or asymmetrical with east-dipping axial planes—a trend consistent with convergence from the west (Snavely, Wagner, and Lander 1980, 1985; Snavely et al., 1981; Clarke, Field, and Hirozawa, 1981). Some folds have sea-floor relief, and others on the shelf have been truncated during low stands of sea level in the Pleistocene. Other folds have been truncated by erosion before the deposition of overlying strata. In places, such as just south of the Columbia River, broad domal growth folds are flanked by marginal synclines, growth faults, and onlapping unconformity-bounded Pleistocene sedimentary units (Figure 7). All these features suggest that these folds are of diapiric origin.

An imbricate set of thrust faults and associated folds appears to dominate the slope and outer shelf of Oregon (Figure 6; Snavely, Wagner, and Lander, 1980). On the mid-shelf, the anticlinal structures are cut by a plexus of faults with chiefly normal displacements. Logs of test wells drilled on these structural highs, such as the Shell Oil Company P-087 well (Snavely, Wagner, and Lander, 1985), support the conclusion that faulting is common and probably is due to extensions. A series of north-trending normal faults occurs along the eastern margin of the deep marginal basin in the vicinity of Newport. These faults, some of which offset the sea floor as well as the coastal Pleistocene marine terraces, are downthrown to the east. These faults may have a component of strike-slip movement and perhaps are responsible for the small earthquakes in this area (Berg and Baker, 1963).

Although speculative, and based principally on an interpretation of aeromagnetic data, the Eocene basalt basement appears to be segmented by generally northwest- and northeast-trending faults that seem to offset these basalts landward (Figure 2). For example, the steep magnetic gradient along the Fulmar fault that forms the contact between

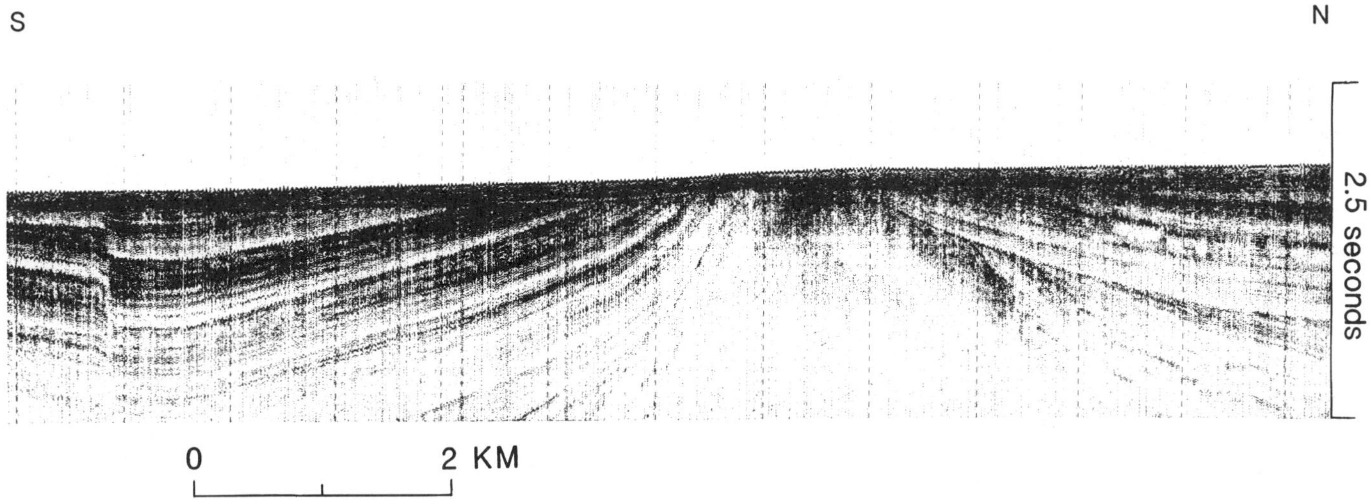

Figure 7. High-resolution (Uniboom) single-channel seismic-reflection profile 80–18 across an asymmetric diapiric structure off the mouth of the Columbia River (Figure 1). Strata of Pleistocene age onlap and thin against this growing structure. Marked unconformity on north side of diapiric structure reflects episodic uplift of the core, which appears to be bounded by a growth fault on its south side. Note fault near south side of profile that offsets uppermost Pleistocene strata but not sea-floor sediments. Vertical exaggeration approximately 12:1.

the Eocene oceanic crust on the east and the lower Eocene wacke on the west (Snavely, Wagner, and Lander, 1980) terminates abruptly near lat. 45°N. One interpretation is that the fault-bounded lower Eocene rocks are offset eastward by a northwest-trending right-lateral fault. It is difficult to trace the continuation of this fault on land with certainty. Where this fault projects on land in the Cape Foulweather and Euchre Mountain quadrangles, however, an en echelon set of northwest-trending normal faults, which offset middle Eocene to middle Miocene rocks, has been mapped by Snavely et al. (1976a).

North of lat. 45°N, northeast-trending right- and left-lateral tear faults may segment the thrust boundary between the deep marginal basin on the east and the upper Oligocene and upper middle Miocene melange wedge to the west. The subsurface data from two offshore test wells—Shell Oil Co. P-075, drilled south of the Astoria Canyon, and Shell Oil Co. P-0150, drilled north of the canyon (Figure 1)—appear to support right-lateral offset of the Eocene basalt basement. Shell P-075 penetrated Eocene basalt at 2970 m, whereas Shell P-0150 drilled 60 km to the north, bottomed in upper Oligocene to lower Miocene melange and broken formation at 4017 m (Table 1; Snavely, Pearl, and Lander, 1977).

Southern and Central Washington Continental Margin

Our multichannel seismic-reflection profiles and the single-channel records of Silver (1972), together with subsurface information from Shell Oil Company wells P-0155 and P-0150, and Pan American well P-0141 (Figure 1, and Table 1), indicate that the inner and mid-continental shelf of Washington is underlain by a thick accretionary wedge of melange and broken formation. This melange sequence is overlain by as much as 2000 m of late Miocene and younger strata, which in places are pierced by diapirs (Grim and Bennett, 1969; Silver, 1972; Snavely and Wagner, 1982). Rau and Grocock (1974) concluded that melange exposed along the Olympic coast immediately north of Point Grenville represents a piercement structure similar to those found on the continental shelf. The cores of these diapirs consist chiefly of melange of late Oligocene to late middle Miocene age (Snavely and Wagner, 1982); but near the coast, fault slices or olistostromal blocks of late Eocene sedimentary rocks may be incorporated in the melange (Rau and McFarland, 1982). On the east–west geologic cross-sections of the area just south of Grays Harbor, Snavely and Wagner (1982) inferred that the melange is underplated beneath oceanic basalt of the lower and middle Eocene Crescent Formation. Offshore magnetic anomalies characteristic of Eocene basalt indicate that the toe of this zone of underthrusting may lie only a few kilometers west of the southwestern Washington coast (Figure 2). North of Grays Harbor, however, surface and subsurface data suggest that this zone is offset landward by several northeast-trending tear faults. The upper Oligocene to upper middle Miocene melange (upper part of the Hoh rock assemblage of Rau, 1975, 1979) has been transported eastward onto the west flank of the Olympic Mountains by a miniplate, the inferred southern boundary of which generally lies along the Quinault River, and the northern boundary along the Hoh River (Figure 2). Although the field evidence is not conclusive, it appears that this melange and broken formation was subducted beneath the Eocene sedimentary sequence that constitutes the western core rocks of the Olympic Mountains (Tabor and Cady, 1978a, 1978b).

Many of the diapiric structures on the Washington shelf apparently are circular. Elongate diapiric ridges have not been recognized, perhaps because of the limited seismic-reflection coverage. Some of the diapirs breach the sea floor and have growth faults along their flanks. Others dome the overlying upper Miocene and Pliocene strata and commonly have marginal synclines (moats) bordering the uplift. High-resolution seismic profiles across several diapirs show them to have bald, ragged tops, suggesting that uplift is presently occurring and perhaps the melange is being extruded onto the sea floor. The thinning and onlap of post-middle-Miocene strata over or adjacent to these diapirs (Figure 7) indicate they are growing structures that were probably initiated during the late Miocene. In places where large blocks have been uplifted along thrust faults, or uplifted by diapirism, gravity sliding appears to have played an important role in the redistribution of strata on the shelf. Snavely and Wagner (1982) interpreted the structure near the west end of their cross-section off southwestern Washington as being a large fault block of upper Oligocene(?) strata that initially was uplifted and later slid eastward off this structural high to override middle Miocene(?) strata. This may explain some of the inverse stratigraphic relationships encountered in several test wells drilled near the coast north of Grays Harbor (Rau and McFarland, 1982). However, thrust faults may well account for most of the stratigraphic anomalies in these test wells.

The youngest set of faults mapped along the Washington coast are northeast-trending tear faults that offset strata as young as Pliocene (Rau, 1975). High-resolution seismic profiles on the shelf indicate that some faults reach the sea floor and in places offset it as much as 30 m (Snavely, Pearl, and Lander, 1977).

Northwestern Washington Continental Margin

The most tectonically disrupted segment of the Oregon and Washington continental margin—both onshore and offshore—is northwestern Washington (Figure 5). The Tertiary rocks in this region (MacLeod et al., 1977; Snavely et al., 1980; 1986) have been subjected to three principal compressional-stress regimes: (1) north–south compression that is indicated by north-dipping imbricate thrust sheets (Crescent thrust, Figure 5; MacLeod et al., 1977; Snavely et al., 1980; Snavely, 1983), (2) northeastward convergence and underthrusting in the mid-late-Eocene that produced a thick middle to upper Eocene accretionary wedge of broken formation and melange (Tc, Figure 5; Tabor and Cady, 1978b; Snavely et al., 1986), and (3) generally northeastward convergence in the late middle Miocene that in the northwesternmost part of the Olympic Peninsula obducted pre-Tertiary and lower to upper Eocene rocks (Tm, Figure 5) over the Eocene sequence subducted earlier (Tc, Figure 5; Snavely et al., 1986), and to the south in the west-central part of the Olympic Peninsula subducted lower and middle Miocene strata beneath early to middle Eocene core rocks (Figure 2; Snavely, unpublished maps). Post-middle Miocene northeastward convergence has produced left-lateral strike-slip movement such as the left-lateral Calawah fault (Figure 5; Gower, 1960; MacLeod et al., 1977; Snavely et al., 1980, 1986; Snavely, 1983). This faulting apparently was along a zone of weakness formed by an earlier thrust fault. Onshore, the Calawah fault offsets and shears glacial drift of late(?) Pleistocene age. On the shelf, the Calawah fault offsets sea-floor sediments and has bathymetric expression in the form of horst and graben structures (MacLeod et al., 1977; Snavely, 1979).

The major structures on the continental margin off Washington and southern Vancouver Island include (1) the Prometheus high (Figure 2; Shouldice, 1971; MacLeod et al., 1977), a northwest-trending pre-late-Miocene linear structural high in Canadian waters that is underlain by lower Eocene basalt equivalent to the lower part of the Crescent Formation (Snavely and Wagner, 1981); (2) the east-trending and northward-dipping Leech River thrust that extends along southern Vancouver Island, and forms the contact between the Eocene Metchosin Volcanics on the south (Crescent equivalent) and the pre-Tertiary Leech River Schist (MacLeod et al., 1977) to the north; (3) the northwest-trending left-lateral Calawah fault, which, northwest of the junction with the Leech River fault, is believed to form the contact between the Eocene oceanic crust on the south and pre-Tertiary rocks to the north (Figure 2; MacLeod et al., 1977; Snavely and Wagner, 1981). The Ozette thrust (Figure 5) mapped by Snavely et al., 1986 in the Ozette Lake and Lake Pleasant quadrangles extends offshore south of the Point of the Arches and forms the southern thrust contact between Eocene oceanic basalts and late Eocene strata to the south—strata that are underplated beneath the basalt.

Seismic profiles on the continental shelf off Hoh Head, on the northern part of the Olympic coast, indicate that a major tectonic boundary separates the northwest-trending predominantly Eocene and Oligocene folded sedimentary rocks in the Olympic basin to the north from the upper Oligocene to upper middle Miocene melange and broken formation to the south. This northwest-trending tectonic interface on the seismic profiles forms a broad zone of steeply northward-dipping thrust faults and shear zones. The upper Oligocene to middle Miocene melange, which was penetrated in Pan American well P-0141 about 24 km to the south of this zone, appears to underthrust strata as young as Pliocene(?). The southeastward (landward) extension of this fault zone would intersect the inner shelf near Hoh Head. Since no evidence of this broad shear zone was found along the coast north of Hoh Head, it may have been offset to a position south of Hoh Head; from this position, it may extend inland generally along the Hoh River (Rau, 1979) and form the contact between the Eocene core rocks on the north and the upper Oligocene to middle Miocene melange on the south (Figure 2).

In the northern part of the Olympic Peninsula, north–south compression of Olympic-Coast Range Paleogene terrane against the buttress of pre-Tertiary rocks on Vancouver Island produced a set of landward-dipping (northward) thrust faults from the early late Eocene to the early middle Miocene (Snavely, 1983). In contrast, thrust faulting along the west side of the Olympic Peninsula, in response to the late middle Miocene period of oblique convergence, resulted in overthrusting as well as underthrusting of pre-late-Miocene strata (Figure 2; Snavely et al., 1986 and unpublished maps).

Inasmuch as upper Oligocene to upper middle Miocene melange and broken formation (Hoh rock assemblage of Rau, 1979; Rau and McFarland, 1982) crop out only in the western part of the Olympic Peninsula but are widespread on the continental shelf, these strata may have been displaced eastward on a miniplate (Figure 2). The rate of northeast movement of this miniplate in the late middle Miocene may have been greater than that of adjacent areas, thus transporting the Hoh melange farther eastward.

Oregon and Washington Continental Slope and Abyssal Plain

The continental slope of Oregon and Washington is a deformational front characterized by a folded and thrust-faulted sequence of uplifted abyssal Pliocene and Pleistocene strata of the Cascadia basin (Silver, 1972; Kulm and Fowler, 1974; Seely, Vail, and Walton, 1974; Snavely, Wagner, and Lander, 1980; Carlson and Nelson, this volume). These uplifted and tectonically accreted strata overlie the subducting upper Miocene oceanic crust of the lower plate. South of lat. 45°30′N, the Oregon lower slope terrace consists of a set of imbricate, predominantly landward-dipping low-angle thrust faults that cut and uplift a thick prism of Pliocene and Pleistocene abyssal strata (Figure 6). In contrast, north of lat. 45°30′N on the northern Oregon and Washington continental slope, strata of the upper plate are cut by a set of seaward-dipping thrust faults (Figure 10). This change in structural style between the Oregon and Washington continental slopes was first recognized by Silver (1972), but he was unable to explain why thrust faults with different vergence should result from a similar pattern of subduction.

Seely (1977) applied laboratory models and theory, as well as geologic and geophysical data collected on the lower slope terrace of northwest Washington, to explain the landward vergence of thrust faults in this area. He suggested that zones of low shear strength (probably overpressured zones) in trench and slope sediments play a major role in the potential development of landward vergence (Seely, 1977, p. 197). Seely concluded that these overpressured zones are due to rapid deposition of the trench sediments.

Two of our multichannel seismic-refraction profiles across the abyssal plain and slope off Vancouver Island are located 55 and 90 km northwest of Seely's (1977) profile, and show seaward vergence of thrust faults (Figures 8, 11). Since the thickness of the section above the oceanic crust at the base of the slope in Seely's and our profiles is virtually the same, the rate of deposition in these two nearby areas of the Cascadia basin is probably the same. Therefore, factors other than rapid deposition and attendant overpressure may not play an important role in controlling the direction of vergence of thrust faults along the Oregon and Washington continental slope.

Geologic mapping in the northwestern part of the Olympic Peninsula documents both overthrust and underthrust terranes (Figure 5; Snavely et al., 1980; Snavely, 1983; Snavely et al., 1986). Although conclusive geologic evidence controlling vergence is lacking, the following generalizations can be made:

(1) underthrusting is associated with northwest-trending, steep northward-dipping, thrust faults, such as the Crescent thrust (Figure 5; MacLeod et al., 1977; Snavely et al., 1980, 1986; Snavely, 1983),
(2) overthrusting is associated with low-dipping northeastward-directed thrust faults (Ozette thrust, Figure 5; Snavely et al., 1986),
(3) the lithology of these Tertiary units apparently does not affect the vergence of the onland thrust faults.

I speculate that the direction of vergence of thrust faults along the continental slope of Oregon and Washington may have been controlled by the slope of the backstops against which the melange wedge accreted. The steep backstop off Oregon, formed by the fault-bounded block of lower Eocene wacke (Snavely et al., 1980), resulted in seaward vergence; whereas the gently dipping backstop of upper Oligocene to upper middle Miocene melange wedge off Washington resulted in landward vergence. An onland analogy for "seaward" vergence might be the north-dipping Crescent and coeval thrust faults (Figure 2) that formed in response to the deformation of the northward-moving Olympic terrane against the steep buttress of pre-Tertiary rocks on Vancouver Island.

NEOTECTONICS

Past Seismic-Related Events

Although current seismic activity is low on the continental margin of Oregon and Washington, the Cenozoic rocks in this region record a history of intense and episodic tectonism (Silver, 1972; Carson, 1977; Snavely, Wagner, and Lander, 1980; Snavely, 1984). Tectonic events that reflect convergence of the Pacific oceanic plates and the North American plate are recorded by major regional angular unconformities of early Eocene, late middle Eocene, late Eocene, middle Miocene, and late Miocene age (Figure 3). The stratigraphic record also documents episodic periods of uplift and erosion, followed closely by downwarping and renewed deposition. Extensive deposits of thick-bedded turbidite sandstone such as the Tyee Formation (Snavely, Wagner, and MacLeod, 1964), and the turbidite packets in the Makah Formation (Snavely et al., 1980), may have been produced by the slumping of delta fronts triggered by earthquakes. Ancient major landslide deposits, such as the Jansen Creek Member of the Makah Formation (Snavely et al., 1980), consist of upper Eocene nearshore sandstone that was uplifted along faults on southern Vancouver Island and transported by gravity slides into an Oligocene deep-marginal basin. In many places on the Olympic Peninsula and in the Coast Ranges, mudflows and conglomerate channel deposits interbedded with deep-water marine strata indicate rapid uplift and erosion of areas marginal to or within the marine basins.

Pleistocene and Holocene faults have been mapped along the coastal zone and on the continental shelf. In the western part of the Olympic Peninsula, late Pleistocene glacial drift is sheared and tectonically interleaved with siltstone beds of Eocene age (Snavely, 1983). North of Ozette Lake, a Holocene soil zone on late Pleistocene outwash gravels is offset as much as 2 m. On the continental shelf off Grays Harbor, Washington, sea-floor sediments are offset about 7 m by a "trap-door" type fault (Snavely, Pearl, and Lander, 1977). On the inner shelf and coastal zone of central Oregon, a 75-km-long north-trending zone of steeply dipping normal faults offsets Holocene(?) sediments (Snavely, Wagner, and Lander, 1980); onshore these faults offset late Pleistocene marine-terrace deposits (Snavely et al., 1976a, 1976b). These faults are downthrown to the east toward the uplifted Coast Range rather than toward the basin, as one would expect for seaward-verging thrust faults or gravity faults. Several earthquakes with intensities of III to IV have occurred in the vicinity of Newport (Berg and Baker, 1963) and may have been generated by movement (perhaps strike-slip) along faults within this zone.

Historic Earthquakes

Although most of the major earthquakes recorded recently are concentrated in the Puget Sound area (Crosson, 1980), there are a few historic accounts of strong earthquakes along the Washington coast. James G. Swan (1870), a historian and student of Indian culture, recounts a Makah Indian legend of a natural upheaval of land at Cape Flattery (Figure 5) that occurred a long time before his arrival in the early 1860s. Swan wrote, "A long time ago, said my informant, but not at a very remote period, the water of the Pacific flowed through what is now the swamp and prairie between Waatch village and Neeah Bay, making an island of Cape

Flattery. The water suddenly receded, leaving Neeah Bay perfectly dry. It was four days reaching its lowest ebb, and then rose again without any waves or breakers, till it had submerged the Cape, and in fact the whole country, excepting the tops of the mountains at Clyoquot[1] The water on its rise became very warm, and as it came up to the houses, those who had canoes put their effects into them, and floated off with the current, which set very strongly to the north. Some drifted one way, some another; and when the waters assumed their accustomed level, a portion of the tribe found themselves beyond Nootka[2] * * * Many canoes came down in the trees and were destroyed, and numerous lives were lost. The water was four days regaining its accustomed level." This Indian legend best describes an uplift of the coast and a major tsunami, including some exaggeration or misunderstanding regarding the duration of the event.

Major landslides such as the one that divided ancient Lake Crescent into the present two lakes—Lake Crescent on the west and Lake Sutherland on the east (Figure 5; Brown, Gower, and Snavely, 1960; Tabor, 1975)—may have been initiated by a seismic event during the early(?) Holocene. A Quileute Indian legend purports that during a great battle between the Clallam and Quileute Indians, Mount Storm King (on the south side of Lake Crescent) became angry and took a great piece of rock from his crest and hurled it down the valley, killing all who were fighting (Tabor, 1975). Although the debris that forms the older and major landslide came largely from the north, a large slide also originated from the ridge just east of Mount Storm King (Brown, Gower, and Snavely, 1960). The youthful nature of the older landslide, such as the lack of a thick soil zone on the landslide debris and hummocky topography with closed depressions that contain little recent fill, suggests that the older slide may have occurred relatively soon after the melting of the ice lobe that occupied the Lake Crescent-Lake Sutherland valley. A submerged tree on the south side of the older landslide, apparently in an upright (growth) position in Lake Crescent, has a ^{14}C age of 350 years (Snavely, unpublished data). However, this tree was probably transported into the lake by the younger slide, perhaps the one witnessed by the Indians.

Archaeological excavations at the Ozette site at Cape Alava on the northwest Olympic Coast (Figure 5) have yielded prehistoric and historic cultural materials associated with mudflow deposits (Mauger and Daugherty, 1979; Samuels, 1983). The oldest mudflow overlies or contains cultural organic material with ^{14}C dates of 790 $\pm$80 and 810 $\pm$ 70. The youngest and largest mudflow crushed and buried five longhouses, one of which contained an interior floor hearth with carbon that yielded a ^{14}C date of 440 $\pm$90 BP. It is believed that the house was occupied for about 100 years before the mudflow (Samuels, 1983); thus the mudflow is about 350 years old. Because the melange and outwash deposits in the sea cliff adjacent to the Ozette site are unstable, the mudflow deposits may simply be the normal product of mass wasting. However, I can speculate that the mudflows were triggered by earthquakes. Apparently they occurred suddenly, because the youngest mudflow rapidly buried a longhouse occupied by several Indians and their dogs. If, in fact, the two major mudflows at the Ozette site were triggered by earthquakes, a recurrence interval of about 450 years is suggested between the two seismic events. However, the writer is aware that it is highly speculative to presume that this interval is real, or representative.

[1]Mountains north of Ucluetet on southern Vancouver Island, west of Barkely Sound, B.C.

[2]Indian village on central west coast of Vancouver Island, B.C.

Earthquake Potential

The present seismic activity is low along the continental margin of Oregon and Washington; however, this region lies along a major subduction zone where the Juan de Fuca and North American plates are converging at the rate of 3.5 cm/yr (Riddehough, 1977). In most subduction zones, episodic strain release has generated large earthquakes (Plafker, 1972; Kanamori, 1977; Heaton and Kanamori, 1984).

The absence of a well-defined Benioff zone along the Washington and Oregon convergent margin has perplexed geologists and seismologists. They have proposed the following models to account for the seismically quiescent plate boundary:

(1) Convergence has ceased between the Juan de Fuca and North American plates.
(2) Convergence is occurring along the plate boundaries, but is aseismic because a young, warm, plastic oceanic crust is being subducted with ductile rather than brittle deformation of the plate boundary.
(3) The Juan de Fuca and North American plates presently are coupled, forming a seismic gap along this segment of the Northeast Pacific.

Geodetic measurements by Savage, Lisowki, and Prescott (1981) indicate that crustal strain is accumulating in the Puget Sound area and has an average principal direction of contraction of N71°E $\pm$6°. This direction of contraction is in general agreement with the N50°E direction of relative motion between the Juan de Fuca and North American plates. It also agrees closely with the direction of tectonic transport during the late middle Miocene period of plate convergence, as deduced from drag folds in the upper plate of the Ozette thrust fault in the northwesternmost part of the Olympic Peninsula (Tc, Figure 5; Snavely et al., 1986).

Landward tilting in the Oregon and Washington Coast Range and the Olympic Peninsula has been documented by using tide gauges and by geodetic leveling of uplifted marine terraces (Adams, 1984). This landward tilt, and crustal shortening shown by thrust faults and folds on the outer shelf and slope of Oregon (Kulm and Fowler, 1974; Seely, 1977; Snavely, Wagner, and Lander, 1980) and on the slope, continental shelf, and coastal zone of Washington (Silver, 1972; Carson, 1977; Snavely and Wagner, 1982; Rau, 1975, 1979; Snavely et al., 1986), is characteristic of many other subduction zones where major thrust earthquakes have occurred (Adams, 1984).

The geologic evidence outlined earlier strongly supports episodic tectonic events during the Cenozoic. Seismic-reflection profiles across the deformational front along the continental slope of Oregon and Washington support episodic underthrusting of the Juan de Fuca plate beneath the North American plate over a long time scale (Silver, 1972; Carson, 1977; Snavely, Wagner, and Lander, 1980). This convergence has produced a series of elongate en echelon anticlinal ridges bounded by thrust faults. These ridges have

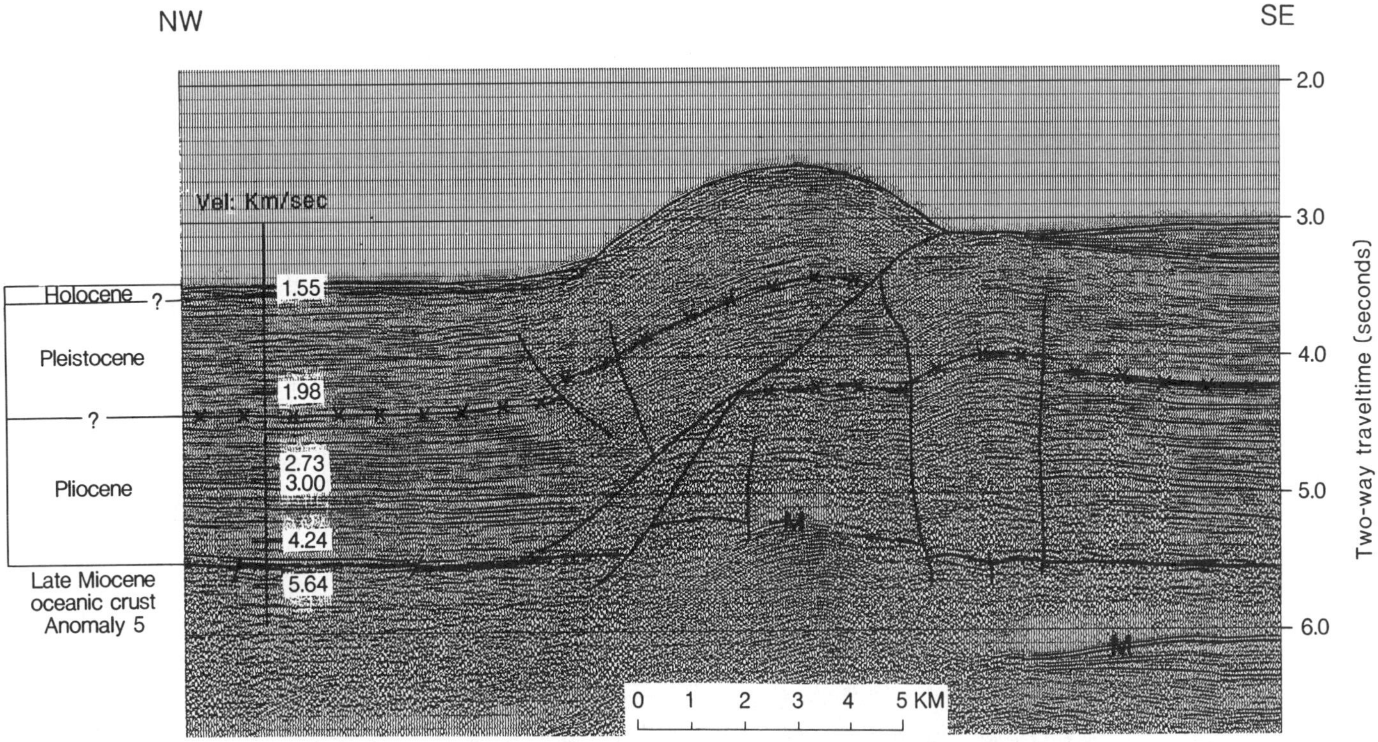

Figure 10. Twenty-four channel seismic-reflection profile, line 80-16 (Figure 1) across the abyssal plain and lower slope off southern Washington, showing folded and uplifted abyssal Pleistocene and Pliocene strata above seaward-dipping thrust fault and its branches. Velocities in km/sec are based on sonobuoy refraction data. M, indicates sea-floor multiple. Vertical exaggeration approximately 3:1, based on sea-water velocity.

bathymetric expression and uplift Pleistocene abyssal sediments as much as 1100 m (Byrne, Fowler, and Maloney, 1966; Carson et al., 1974; Kulm and Fowler, 1974; Snavely, Wagner, and Lander, 1980; Snavely and Wagner, 1981).

The profile off central Oregon (Figure 6) indicates that two lower slope basins formed landward of anticlinal folds in Pleistocene strata. The larger basin, which is near the base of the upper slope, contains as much as 800 m of Quaternary sediments and an unconformity within these basin deposits, thus indicating two distinct periods of uplift of the thrust-anticline that bounds the basin on the west.

Multichannel seismic-reflection profiles off southern Vancouver Island are similar to those off Oregon, in that sediments of the Cascadia basin have been uplifted as much as 600 m (Figure 8). The seaward segment of a profile (Figure 11) that crosses the base of the slope 35 km north of the profile in Figure 8 clearly shows a major northeast-dipping thrust fault (fault A, Figure 11) that uplifts virtually the entire section of strata above late Miocene oceanic crust. Fault A is overlain by a small basin which contains about 120 m of Holocene and late Pleistocene(?) sediments. The 120 m of sediment that overlies thrust fault A indicates that this fault has been inactive since the late(?) Pleistocene. Thrust fault B, which lies seaward (southwest) of the principal thrust fault A, also is overlain by a thick sequence of late Pleistocene and Holocene sediments. As the imbricate thrust faults along the outer shelf and slope of Oregon and Washington have migrated seaward with time, the next major displacement may be along fault B. This relationship of late Pleistocene sediments deposited over the trace of thrust fault A in Figure 11, and folded or tilted unconformities in lower slope basins (Figure 6), suggests episodic rather than continuous underthrusting at the base of the continental slope. This relationship suggests that at the present time the Juan de Fuca and North American plates are strongly coupled and that compressive stresses are accumulating across the continental margin (Heaton and Kanamori, 1984). On the basis of their interpretation of earthquake focal mechanism data, Weaver and Smith (1983) also concluded that the subduction zone is locked. Since there is no historic record of large shallow earthquakes along the subduction zone, a 700-km-long seismic gap seems to be present along the subduction zone off Oregon and Washington—the most remarkable gap to be found in the Circum-Pacific seismic belt (Heaton and Kanamori, 1984). Despite the fact that present-day seismic activity is low, the potential for a major subduction-type earthquake cannot be discounted (Weaver and Smith, 1983; Snavely, 1984; Heaton and Kanamori, 1984).

Tsunamis

If we accept the interpretation that faulting and folding along the continental slope are the result of episodic events that uplifted abyssal sediments 400–1100 m above their original depositional positions, the sudden uplift of long anticlinal ridges along the continental slope may be capable of generating destructive tsunamis. Tsunamis have not been reported following recorded earthquakes whose epicenters lie on the Oregon and Washington continental shelf and slope. However, this is to be expected because all instrumentally reported earthquakes in the region are far below the 7.0-Ms threshold required to generate a tsunami. A detailed study along coastal Washington and in the valleys of the large rivers and estuaries could conceivably uncover concentrations of driftwood or ancient forest trim-lines that would document ancient tsunamis.

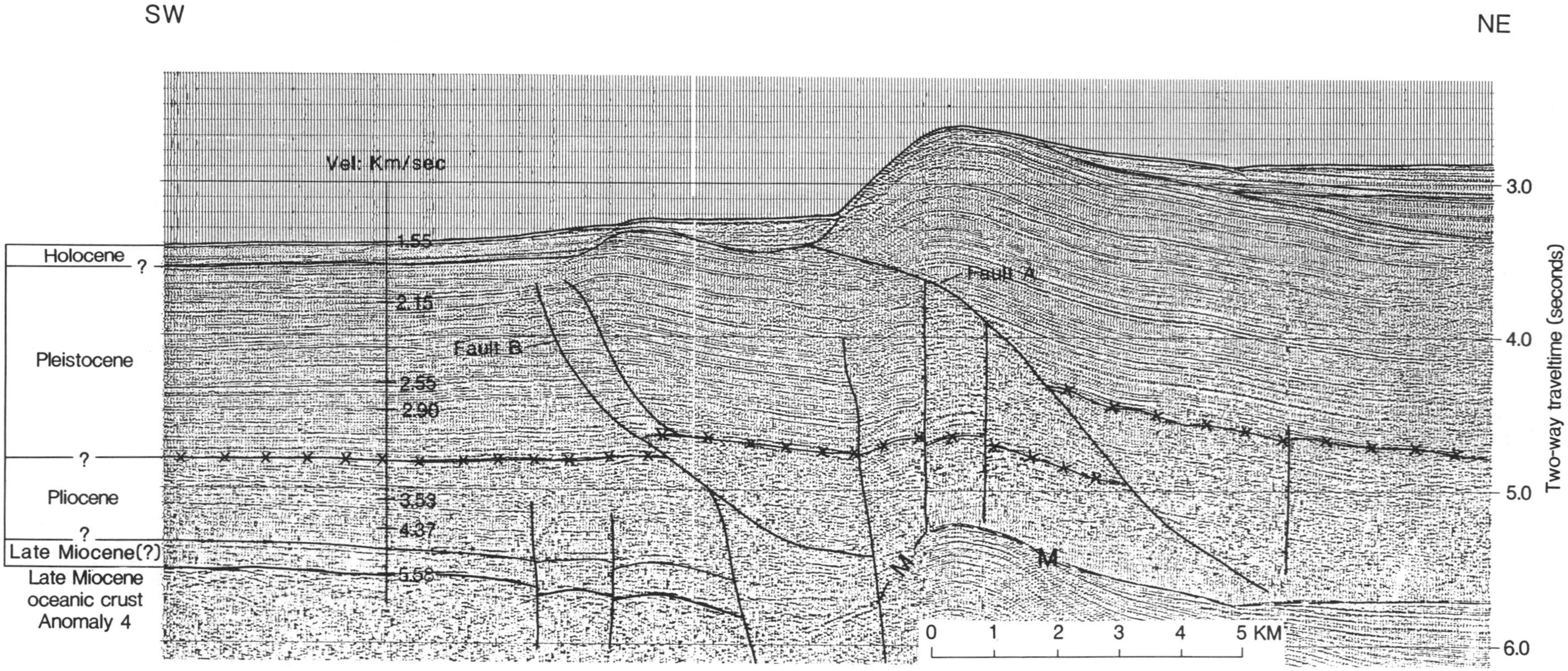

Figure 11. Twenty-four channel seismic-reflection profile, line 80-7 (Figure 1) across the abyssal plain and lower slope off southwestern Vancouver Island, B. C. Landward-dipping thrust fault A has folded and uplifted Pleistocene and Pliocene abyssal sediments approximately 700 m thick. This fault apparently has been inactive during the Holocene, because deposits of Holocene age in the small basin south of the thrust fold are not offset by fault A. Fault B, southwest of fault A, indicates that the deformation front at the base of the slope is migrating seaward. Velocities in km/sec are based on sonobuoy refraction data. M, indicates sea-floor multiple. Vertical exaggeration is approximately 3:1, based on sea-water velocity.

PETROLEUM POTENTIAL

Although the Oregon and Washington continental margin contains a thick sequence of strata that are potential source beds for petroleum, numerous structural and stratigraphic traps, and siltstone and mudstone beds that could provide effective seals, the region generally is deficient in sandstone beds that might be suitable reservoir rocks. Also, much of the section has not been subjected to temperatures high enough to generate hydrocarbons. However, these deficiencies may not apply to the entire margin, and probably some geologic settings may contain all the key elements (source beds, reservoir rocks, seals, thermal history, and structure) required for petroleum production.

Although the petroleum potential of the Oregon and Washington continental margin has not been adequately tested by drilling,[1] and only reconnaissance sampling for hydrocarbon analysis has been done, the potential for the generation of oil appears to be low, whereas that for gas is considered to be moderate. The presence of commercial quantities of gas was established in May 1979 when the Mist Gas Field was discovered in the northwestern part of the Oregon Coast Range (Newton, 1979). As of December 31, 1983, this field has produced 16,427,402 mcf of gas from upper Eocene sandstone reservoirs. Subcommercial quantities of oil were produced from the Sunshine Mining Company well, Medina No. 1, drilled on the Washington coast north of Grays Harbor (Figure 1). This well produced 12,000 bbl of 38.9° API gravity paraffin-base oil from fractured siltstone of middle Miocene age (Braislin, Hastings, and Snavely, 1971). Significant shows of gas were encountered in several of the deep test wells drilled on the continental shelf, notably Pan American well P-0112 in the Coos Bay basin and Shell well P-0150 in the Willapa basin (Figure 1; Ziegler and Cassell, 1978).

Gas, and less commonly traces of oil, have been reported as seeps and in wells drilled along the west coast of the Olympic Peninsula (Arnold, 1906; Lupton, 1915; Glover, 1936; McFarland, 1981). There are several active seeps of thermogenic gas near the coast, notably the Garfield gas mound north of Taholah, and the Pysht River gas seep south of Clallam Bay (Figure 2; Snavely, Pearl, and Lander, 1977).

Overpressured Melange and Broken Formation

The Tertiary sequences beneath the late Miocene unconformity in most places on the continental shelves of Oregon, Washington, and southern Vancouver Island are in an overpressured regime. High subsurface pressures are indicated by the presence of numerous shale diapirs and associated growing structures. Heavy muds (100–130 lb/cu ft) were required in drilling OCS wells in Oregon and Washington because of these abnormally high subsurface pressures in strata of pre-late-Miocene age. Of the 10 deep test wells drilled on the Oregon and Washington OCS, all but one (the Standard-Union Nautilus well P-0103 in the Newport basin) recorded problems with sloughing shales, and mud weights were raised continuously until stable hole conditions were attained (Ziegler and Cassell, 1978). Pressure gradients in the Tofino basin off southern Vancouver Island (Shouldice, 1971), computed from sonic logs and the mud weights required to drill the Pluto and Prometheus wells, show a marked buildup of pressure near the base of the late Miocene unconformity from hydrostatic pressure to "hard geopressure."

Abnormally high pressures are particularly common in areas underlain by upper Oligocene to upper middle Miocene melange on the Washington continental margin. The presence of numerous diapiric structures on the Washington shelf (Snavely and Wagner, 1982) and the adjacent coastal zone (Rau and Grocock, 1974) leads to the conclusion that these structures were formed by the flow of overpressured sediments. The reduction in overall density of the deeply buried upper Oligocene to upper middle Miocene melange, which caused it to rise diapirically, was most likely the result of two temperature-dependent events: (1) the expulsion of water molecules in the conversion of smectite to illite[1] and (2) the generation of gas from organic matter within the melange. These water- and gas-charged low-density siltstone "slurries" would then rise diapirically and intrude or deform the overlying blanket of upper Miocene and younger strata, much as a low-density salt diapir.

Geologic mapping along the west side of the Olympic Peninsula (Rau, 1975, 1979; Snavely unpublished data) indicates that diapirism probably was an important process in the tectonic mixing of both the Eocene and Miocene melanges and broken formations. The studies of the extensive melange zones of northern Irian Jaya by Williams, Pigram, and Dow (1984) also pointed out the importance of diapirism in the development of the internal structure of accretionary melange terranes. Seismic-reflection profiles on the continental shelves of northern Oregon and Washington show that major landslides occur along the flanks of some diapirs (Snavely and Wagner, 1982). Diapirism, therefore, may be an important element in the redistribution of post-middle-Miocene sediments on the shelf. Episodic diapiric uplift, indicated by the presence of numerous unconformities in upper Miocene to Pleistocene strata adjacent to these structures (Figure 7), suggests a possible relationship between accelerated diapiric uplift and episodes of plate convergence.

Source-Rock Potential

Thick sequences of siltstone and mudstone that range in age from early Eocene to Pliocene are present in the Tertiary sequence of western Oregon and Washington and on the adjacent continental shelf. Predominantly fine-grained clastic rocks that crop out in this region include the Umpqua, Yamhill, McIntosh, and Aldwell Formations that range in age from early to early late Eocene; the Nestucca, Alsea, Hoko, Makah, and Pysht (in part) Formations that range in age from late Eocene to Oligocene; the Nye Mudstone of

[1]Only 11 deep test wells have been drilled in this area of >46,000 km^2. The density of drilling is about one test well for an area the size of about seven 15-minute quadrangles.

[1]X-ray defraction analyses of six samples of lower and middle Miocene melange indicate that illite, and illite with a few mixed layers of smectite, are major to medium components of these argillaceous rocks. Chlorite and kaolinite, however, are the major clay minerals in the samples analyzed (James Hein, 1985, written communication).

Table 2. Maturity classification of selected samples of Olympic melange based upon vitrinite reflectance (R_o).[1]

Oil Generation					Gas Generation					
<0.5R_o Immature	0.5–0.6 R_o Marg. Mature	0.6–1.35 R_o Mature		>1.35 R_o Overmature	<0.6 R_o Immature	0.6–1.0 R_o Marg. Mature		1.00–2.00 R_o Mature		>2.00 R_o Overmature
	S84-20	S84-6	S84-33	S84-19	S84-20	S84-6	S84-33	S84-21	S84-25	S84-19
	S84-28	S84-9	S84-34	S84-21	S84-28	S84-9	S84-34	S84-22	S84-37	
		S84-18	S84-35	S84-24		S84-18	S84-35	S84-24		
		S84-22	S84-37	S84-25		S84-31	S84-38			
		S84-31	S84-38							

[1]Analyses performed by Clark Geological Services and funded under Department of Energy-U.S. Geological Survey Interagency Agreement No. DE-A121-83-MC20422. Also see Table 3 for additional data.

early Miocene age; and the Quinault Formation of late Miocene and Pliocene age (Figure 3).

Reconnaissance surface sampling for hydrocarbon analyses of many Tertiary stratigraphic units on the north flank of the Olympic Mountains (Table 3 in Snavely, Pearl, and Lander, 1977; Kvenvolden and Snavely, 1985) and in the Oregon Coast Range (Law et al., 1984; Snavely, unpublished data) indicates that most of these rocks are immature to marginally mature. The total organic carbon content in most samples is <1.0 weight-percent and is predominantly type III kerogen; therefore, these rocks have a potential for generating gas but only minor quantities of oil.

Despite the fact that a few outcrops of mudstones, such as the lower Miocene Nye Mudstone in the Newport basin (Figure 1), have a strong petroliferous odor when freshly broken, hydrocarbon analysis indicates that these rocks are thermally immature ($R_o = 0.44$). However, these mudstones and other siltstone units may be more mature downdip in the axial part of the offshore deep marginal basin of Oregon.

Heat released into the upper crust during extension and eruption of middle and late Eocene alkalic basalts of the Yachats Basalt and Tillamook Volcanics in coastal Oregon would appear to have played a role in the maturation of hydrocarbons. However, shows of hydrocarbons are lacking in the marine strata that underlie, or intertongue with, these volcanic rocks. Thermal maturation of siltstone occurred near, or adjacent to, thick middle Miocene basalt and Eocene gabbro sills, but the thermal effect of these sills is local (Snavely, unpublished data).

On the basis of limited geochemical analyses, I believe that rocks with the highest potential for oil and gas generation are the thick accretionary melange wedges, Eocene to Miocene in age, that crop out along the west side of the Olympic Peninsula (Figure 2). These tectonic melanges were formed during two separate periods of strong convergence between the Farallon and North American plates, one in the middle late Eocene and the other in the late middle Miocene. The upper Oligocene to middle Miocene melange is widespread on the Washington continental shelf and was penetrated in the Pan American well P-0141 and Shell Oil Company wells P-0155 and P-0150 (Figure 1, Table 1). Along the Olympic coast, some sheared siltstones in melanges of both middle Eocene and late Oligocene to late middle Miocene ages have strong petroliferous odors and are locally referred to as "smell muds." These intensely deformed Eocene to Miocene strata are probably the source rocks for gas and oil found in seeps encountered in test wells along the southwest Washington coast (Lupton, 1915; McFarland, 1981; Kvenvolden and Snavely, 1985).

In many parts of the world, where the shelf and slope are underlain by oceanic crust, exploration drilling has been unsuccessful (for example, in Indonesia; J. Kingston, 1980, oral communication). This failure may be ascribed to the relatively low heat flow as compared with the more radioactive continental crust. These heat-flow constraints, however, may not apply to the continental margin of Oregon and Washington, because young upper Miocene oceanic crust with high heat-flow values of 2 to 3 HFU (10^{-6} cal/cm^2-sec) underlies most of the continental margin (Dehlinger et al., 1970). The thermal gradient of about 1.75°F/100 ft (0.0274°C/m) in Shell Oil Company well P-0150, drilled to a depth of 4017 m off Grays Harbor (Figure 1), indicates that the heat at this depth is sufficient to mature the sediments thermally and generate hydrocarbons.

Despite the fact that melange is considered part of the economic basement in many regions of the world, the Olympic melange is thermally mature and of sufficient organic richness (total organic carbon ranges from 0.26 to 1.60 weight-percent) to be considered a potential source rock for hydrocarbon. Visual kerogen, vitrinite reflectance, and rock-evaluation pyrolysis of samples of sheared siltstone, collected from widely spaced localities along the Olympic coast between Portage Head and the Indian village of Taholah (Figure 1), suggest that most of these samples are mature with respect to the oil-generative thermal window, are marginally mature to mature with respect to the wet-gas window, and for the most part are gas prone (Tables 2, 3).

In general, samples of middle Eocene age have a higher thermal maturity than most younger samples; however, some of the late Oligocene to middle Miocene samples are as mature as those of middle Eocene age. Since the Oligocene and Miocene strata have suffered only a single period of major plate convergence, the tectonic and thermal overprint on these strata might be expected to be less than that for the middle Eocene strata that were involved in both the middle late Eocene and the late middle Miocene episodes of plate convergence and subduction. Although the tectonically complex Olympic melange terrane may have experienced unlike thermal histories that reflect different depths of subduction burial, the higher maturity values for some of the Oligocene and Miocene strata may be ascribed to localized heating, since these samples were collected near or within fault zones. However, the uniformity of kerogen type and the general lack of scatter in the thermal parameters of samples from the coastal Olympic melanges seem to contradict the geologic interpretation that the melanges that comprise this accretionary terrane were formed during two different episodes of subduction. If the analyses for thermal maturity

Table 3. Rock-evaluation pyrolysis of siltstone samples from Tertiary melange along the west coast of the Olympic Peninsula, Washington.[1]

Sample Age and Field Number	TOC	Mg/Gm Rock S_1	Mg/Gm Rock S_2	Mg/Gm Rock S_3	Hydrogen Index	Oxygen Index	Oil or Gas Shows	Oil or Gas Potential	Trans Ratio	T_{MAX} (°C)	R_o	TAI
1. Early mid Miocene S84-9	1.60	0.21	2.11	0.26	131.9	16.3	0.21	2.32	0.09	443	0.64	2.5
2. Early mid Miocene S84-21	0.89	0.05	0.02	0.13	2.2	14.6	0.05	0.07	0.71	459	1.50	3.5
3. Early mid Miocene S84-22	1.01	0.06	0.11	0.11	10.9	10.9	0.06	0.17	0.35	463	1.23	3.0
4. Early mid Miocene S84-24	0.74	0.05	—	0.12	—	16.2	0.05	0.05	—	—	1.98	3.0–3.5
5. Early mid Miocene S84-25	0.69	0.04	0.05	0.12	7.2	17.4	0.04	0.09	0.44	508	1.92	3.0–3.5
6. Early mid Miocene S84-28	0.50	0.02	0.20	0.15	40.0	30.0	0.02	0.22	0.09	438	0.50	2.5
7. Early mid Miocene S84-31	0.76	0.12	0.83	0.10	109.2	13.2	0.12	0.95	0.13	441	0.62	2.5
8. Early mid Miocene S84-33	0.98	0.34	0.97	0.11	99.0	11.2	0.34	1.31	0.26	459	0.87	2.7–3.3
9. Early mid Miocene S84-35	0.56	0.06	0.24	0.07	42.9	12.5	0.06	0.30	0.20	457	0.87	2.6
10. Early mid Miocene S84-7	0.61	0.01	0.24	0.62	39.0	101.0	0.01	0.25	0.04	478	0.83	2.6–2.7
11. Oligocene S84-37	0.49	0.09	0.02	0.06	4.1	12.2	0.09	0.11	0.82	467	1.18	2.7–3.3
12. Late Eocene S84-38	0.68	0.03	0.16	0.09	23.5	13.2	0.03	0.19	0.16	466	0.97	3.0
13. Late Eocene S84-34	0.26	0.05	0.02	0.13	7.7	50.0	0.05	0.07	0.71	442	0.61	2.6
14. Late Eocene S84-3	0.49	0.08	0.53	0.05	108.0	10.0	0.08	0.61	0.13	440	0.63	2.4–2.6
15. Middle Eocene S84-6	1.20	0.20	1.17	0.20	97.5	16.7	0.20	1.37	0.15	444	0.64	2.5
16. Middle Eocene S84-18	0.43	0.17	0.16	0.27	37.2	62.8	0.17	0.33	0.52	424	0.73	2.6
17. Middle Eocene S84-20	0.44	0.04	0.04	0.12	9.1	27.3	0.04	0.08	0.50	435	0.51	2.5
18. Cretaceous S84-19	0.70	0.05	—	0.09	—	12.9	0.05	0.05	—	—	2.29	3.3–3.7

TOC = Total Organic Carbon; Hydrogen Index = $(S_2/TOC) \times 100$; Oil or Gas Potential = $S_1 + S_2$; TAI = Thermal Alteration Index

Oil or Gas Shows = S_1; Oxygen Index = $(S_3/TOC) \times 100$; Transformation Ratio = $S_1/(S_1 + S_2)$; R_o = Vitrinite Reflectance

[1]Rock-evaluation pyrolysis performed by Clark Geological Services, supported by Department of Energy-U.S. Geological Survey Interagency Agreement No. DE-AI21-83-MC20422.

Table 4. Molecular composition of hydrocarbons from northwestern Olympic Peninsula gas seeps.

	Coastal Seep[1]	Pysht River Seep[2]	Wilson Ranch Well[3]
Methane	97.55%	94.33	95.8
Ethane	1.48	3.50	2.7
Propane	0.37	1.54	0.90
Butanes	0.22	0.43	0.30
Pentanes	0.13	0.14	0.27
Total C_6's	0.08	0.05	
$\delta^{13}C(CH_4)$‰		−31.5	−34.5

[1]Tidal pool seep along coast about 3.5 km northwest of Taholah, NE1/4 Sec. 27, T22S, R13W, Taholah quadrangle, Washington (Rau and Grocock, 1974).

[2]Seep on west bank of Pysht River, NE1/4 NE1/4 Sec. 27, T13N, R12W, Pysht quadrangle, Washington (Snavely, Pearl, and Lander, 1977). Analyst, A. J. Miller, Mobil Oil Company.

[3]Gas from Sol Duc Oil Co., Bogachiel No. 1 well, drilled in 1924 on Wilson Ranch, SE1/4 NW1/4 Sec. 22, T28N, R14W, Forks quadrangle, Washington. Analyst, K. A. Kvenvolden, U.S. Geological Survey.

and kerogen content of the analyzed Olympic melange samples are typical of the entire terrane, then it must be assumed that the depositional and burial histories of the middle and upper Eocene and of the upper Oligocene to middle Miocene melanges were similar. For example, the uniformity of kerogen type may be due to the fact that the strata that comprise both of the melange sequences were deposited in a deep fore-arc basin, and that the source areas for both were similar Eocene rocks that now crop out to the east along the foothills of the Cascade Range. As the middle Eocene melange was probably more deeply buried than the late Oligocene to middle Miocene melange, it might be expected that the middle Eocene melange would have a higher thermal history. Since this apparently is not the case, I speculate that a warm late Miocene oceanic crust may have been subducted beneath the entire Olympic melange, thus providing a thermal overprint that affected both the middle and upper Eocene and the late Oligocene to late middle Miocene accretionary melange wedges. Alternatively, convective heating of the thick (15–20 km?) accretionary wedge by hot fluids circulating through the melange and broken formation may have uniformly maturated these Eocene and Miocene rocks.

Several small gas seeps occur in the melange along the west side of the Olympic Peninsula. Many of the wells drilled in this area have shows of gas and, less commonly, oil (Lupton, 1915; Glover, 1936; McFarland, 1981). The largest of the coastal gas seeps forms the Garfield gas mounds about 0.8 km northwest of the Quinault Indian village of Taholah (Figure 1). The largest seep, located in the SE1/4 NW1/4 Sec. 35, T22N, R13W, forms an elliptical mud mound about 15 m in height and 75 m in length on a marine terrace 75 m above sea level. An analysis of gas collected from a coastal seep near Taholah is shown in Table 4, column 1. In the northwest part of the Olympic Peninsula south of Clallam Bay, a seep occurs along the Pysht River in Sec. 27, T13N, R12W. This area, which is in the upper plate of the Crescent thrust (Figure 5), is underlain by sandstone and siltstone beds of the upper Eocene Hoko Formation with near-vertical dips. As most of the strata in the upper plate of the Crescent thrust are immature (Snavely, Pearl, and Lander, 1977), the source of this gas is probably from Eocene melange in the lower plate of the Crescent thrust, and the gas has migrated into strata in the upper plate. An analysis of the Pysht River gas seep is shown in Table 4.

The substantial shows of gas reported in Pan American Co. well P-0112 on the southern Oregon continental shelf (Figure 1; Ziegler and Cassell, 1978) may also represent gas generated in Oligocene and Miocene melange in the lower plate of a major eastward-dipping thrust fault (Snavely, Wagner, and Lander, 1980); the gas later migrated into fractured lower Eocene wacke in the upper plate. Other areas in coastal western Washington may have a similar tectonic setting, where Eocene and upper Oligocene to upper middle Miocene melange and broken formations have been thrust beneath older rocks. Snavely and Wagner (1982) speculated that the subducting late Miocene Juan de Fuca plate may have underplated Oligocene and Miocene melange, and perhaps Eocene melange, beneath the middle to lower Eocene basalt of the Crescent Formation in the coast range of southwestern Washington. This implies that the structure in the upper plate exhibits thin-skin tectonics, which is characterized by a group of east-dipping imbricate thrust faults that deform the Tertiary strata in the upper plate to produce narrow, asymmetric anticlines separated by broad synclines.

Reservoir Rocks

Most of the strata encountered in test wells on the Oregon and Washington continental shelf are fine-grained silty sandstone and siltstone; coarser sandstone probably constitutes <20% of the pre-Pliocene strata (Snavely, Pearl, and Lander, 1977; Snavely, Wagner, and Rau, 1982). The sandstone units present in the shelf are probably turbidite sands with relatively low permeability and porosity, because clay minerals and calcareous cement commonly fill the pores between grains (Table 5). In areas where the sandstone units have been highly deformed tectonically, zeolite minerals and, in places, silica cement the sandstones and greatly reduce their porosity and thus their reservoir potential (Pearl, 1977; Snavely, Wagner, and Rau, 1982).

The nearshore Tertiary sandstone beds do not extend for any great distance beyond the present strandline (Snavely, Wagner, and Rau, 1982). However, some coarse clastic debris probably was transported into the deep marginal basins by density currents and mass flowage, or by the slumping of the delta fronts of the Coaledo, Yaquina, and Astoria Formations (Figure 3). Channel deposits of sandstone probably are present in places on the shelf, as they occur in middle and late Eocene and middle Miocene strata along the Oregon coast (Snavely et al., 1976b; Cooper, 1981).

Despite the fact that many turbidite sands have a high clay content (as much as 20%), which reduces effective porosity and permeability, some turbidite sands appear to have been "cleaned-up" as they were transported into the basins. As shown in Table 5, measurements on four samples of Oligocene turbidite sandstone from the upper part of the Makah Formation (Snavely, Wagner, and Lander, 1980) in the northwestern part of the Olympic Peninsula indicate that these sandstones have sufficient porosities and permeabilities to be considered potential reservoir rocks. Table 5 also lists porosities and permeabilities for selected Tertiary sandstones that crop out in western Oregon and Washington and were deposited either in a shelf environment or in a deep marginal basin. It is evident from the data in Table 5 in this region that sandstones deposited in a shelf environment are potentially better reservoir rocks than sandstones deposited in a deep marginal basin. Fracture porosity may exist in broken formations within the melange wedges of Eocene to Miocene ages, as well as in Eocene rocks in the upper plates of thrust faults on the continental shelf. For example, the lower Eocene well-cemented, but fractured, arkosic wacke penetrated in Pan American well P-0112 and Union Oil well P-0130 (Figure 1; Snavely, Wagner, and Rau, 1982) on the Oregon shelf are potential reservoir rocks for gas.

To define the offshore areas that have the greatest potential for sand deposition, the paleogeography of the bordering landmass must be considered in order to delineate areas such as ancient deltas, delta-front fans, or fan channels where coarse-grained clastic debris was available for transportation beyond the shelf edge and into the deeper water. Surface mapping and stratigraphic studies in coastal Oregon and Washington have indicated that environments of deposition were such that sand could be transported across the shelf into deeper offshore basins (Figure 1). These onshore areas include (1) Eocene deltaic deposits in the Coos Bay basin (Dott, 1966; Ryberg, 1978), (2) upper Oligocene deltaic deposits in the Newport basin (Snavely et al., 1976b; Goodwin, 1973), (3) middle Miocene delta-front fans and fan-channel deposits in the Astoria basin (Cooper, 1981), (4) Pliocene deltaic and channel deposits in the Willapa basin (Rau, 1975), and (5) Oligocene turbidite sands in the Olympic basin (Snavely et al., 1980, 1986; Rau, 1975).

Traps

Many structural and stratigraphic traps are present on the Oregon and Washington continental margin. The 10 deep test wells drilled on the OCS in the middle and late 1960s (Figure 1; Table 1) were located on anticlinal highs delineated by seismic-reflection surveys. These wells and many of those drilled on anticlinal highs onshore were unsuccessful, apparently because many of the structures were either complex at depth or their cores were formed of older lower and middle Eocene sandstone or basalt that was intensely deformed and had low porosity and permeability. Other wells drilled on anticlinal highs, such as the Grebe P-093 and Nautilus P-0103 wells, penetrated a predominantly fine-grained clastic sequence with few sands. Several test wells drilled on diapiric structures on the Washington shelf, such as Shell's P-0150 and P-0155 wells, encountered melange and broken formations that form the cores of these diapirs.

One observation that can be made in hindsight is that most of the exploration test wells drilled in the study area might have been too high on the structures. More rewarding plays might lie along the flanks of these structural highs, because the late Eocene regional unconformity (at the base of the Refugian Stage) results in the stratigraphic onlap of the upper Eocene and younger strata onto the "bald" high of older Eocene sedimentary and volcanic rocks. The flanks of the diapiric structures on the Washington shelf probably are better exploration targets than are the axial parts of the structures whose cores are melange and broken formation. Potential reservoir sands adjacent to diapiric structures are strata of late Miocene and Pliocene age. However, these strata provide little opportunity for the development of cap rock to prevent the loss and destruction of petroleum that may have migrated updip along the flanks of these structures. Growth faults and buttress unconformities along the margins of some of the diapirs may form traps for upward-

Table 5. Porosity and permeability for selected Tertiary sandstones—western Oregon and Washington

	Shelf Facies		Deep Marginal Basin	
Age	Porosity[1]	Permeability[2]	Porosity[1]	Permeability[2]
	22.1*	21.4	14.4	.05
	29.2*	1682.0	10.2	.16
	26.6*	31.2	18.2	234.0
	22.0*	20.1		
	25.6*	29.2		
	20.4*	3.0		
	20.7	34.9	24.6[3]	675.0
	23.4	15.6	20.2[3]	7.5
	25.3	319.0	20.7[3]	2.0
	24.5	18.6	14.7	.13
	27.1	69.7	24.3	14.0
			18.2[3]	17.0
	35.2*	3306.0	8.5	1.1
	19.4*	.05	9.4	0.7
	13.3*	8.5	7.4	0.3
	29.3*	558.0	9.7	1.4
	33.1*	530.0	8.7	0.9
	21.1*	1.8	6.6	2.3
	20.7*	3.4	9.2	0.4
	21.5*	82.1	5.1	0.2
	27.9*	478.0	16.9	14.0
	30.5*	1992.0		
			15.1	0.30
			19.4	47.0
			16.6	1.8
			9.3	4.1
			4.8	0.2
			16.1	3.3
			18.5	3.3
			16.4	2.5
			20.9	4.4
			10.4	0.4

*Drill hole cores—remainder, surface samples.
[1]Effective porosity in percent.
[2]Permeability in millidarcys.
[3]Turbidite sandstones in the upper part of the Makah Formation (Snavely et al., 1980), northwesternmost part of the Olympic Peninsula.

migrating petroleum. Along the eastern margins of the Newport and probably Astoria basins (Figure 1), north- and northwest-trending faults intersect the updip sequence of upper Eocene to middle Miocene strata that probably contain potential reservoir rocks. Here, combinations of stratigraphic and fault traps may exist.

Seismic-reflection profiles and geologic maps on the Washington continental margin indicate that strata of Eocene to late middle Miocene age are cut by thrust faults along which sedimentary sequences were overthrust or underthrust (Seely, 1977; Snavely and Wagner, 1981, 1982; Rau, 1975, 1979; Snavely et al., 1986). These thrust faults may seal the upward migration of petroleum generated in the lower plate. In some areas, however, as along the Crescent thrust (Figure 5) and its offshore extension in the Olympic basin (Snavely and Wagner, 1981), petroleum generated in the melange and broken formation in the lower plate may either be trapped at the base of the basalt or may migrate across the thrust into reservoir rocks of the upper plate.

Combinations of faulting and eastward-onlapping

sequences undoubtedly exist on the Washington shelf, particularly in areas where upper Miocene and Pliocene sandstones onlap the Eocene and Miocene melanges and are cut by thrust and normal faults (Rau, 1979; Snavely and Wagner, 1981; Snavely, unpublished data).

Major unconformities of middle and late Eocene and late Miocene age also could form effective traps where they truncate folded older strata. For example, the Standard-Union well P-0103 (Figure 1) was drilled on an anticline of middle Miocene and older strata that was beveled by the unconformity at the base of the late Miocene (Figure 6). The numerous gas shows in the vicinity of Forks, Washington probably occur along the unconformity between upper Miocene strata and the underlying Eocene melange and broken formation.

Extensive sills, as much as 300 m thick, of early Miocene basalt (Snavely, MacLeod, and Wagner, 1968) cut older strata in the northern Oregon Coast Range and extend seaward as much as 40 km (Snavely and Wells, 1984). As these sills cut across older folded and faulted strata and are surrounded by a zone of thermally metamorphosed rock, they possibly could act as traps for upward-migrating oil or gas.

ACKNOWLEDGMENTS

Norman S. MacLeod and Ray E. Wells reviewed the manuscript and offered valuable suggestions that improved its quality. The writer is indebted to Weldon W. Rau and David Bukry, who provided the age assignments for most of the Tertiary strata. John Kingston reviewed and improved the section on petroleum potential. I also wish to thank Norman MacLeod, Ray Wells, Rowland Tabor, George Plafker, Alan Niem, and Weldon Rau for stimulating discussions over the years. The cooperation of the oil companies, who provided invaluable subsurface data from their wells drilled on the continental shelf, is gratefully acknowledged. The writer also expresses his appreciation to Diane L. Minasian, who drafted the illustrations, and to Juanita Mascardo, who patiently typed many rough drafts as well as the final draft of the manuscript.

REFERENCES CITED

Adams, J., 1984, Active deformation of the Pacific northwest continental margin: Tectonics, v. 3, no. 4, p. 449–472.

Addicott, W. O., 1976, Neogene molluscan stages of Oregon and Washington: SEPM, Pacific Section, Neogene Symposium, p. 95–115.

Allen, J. E., and E. M. Baldwin, 1944, Geology and coal resources of the Coos Bay quadrangle: Oregon Department of Geology and Mineral Industries Bulletin 27, 153 p.

Armentrout, J. M., 1980, Cenozoic stratigraphy of Coos Bay and Cape Blanco, southwestern Oregon, *in* K. F. Oles, J. G. Johnson, A. R. Niem, and W. A. Niem, eds., Geologic field trips in western Oregon and southwestern Washington: Oregon Department of Geology and Mineral Industries Bulletin 101, p. 175–216.

Arnold, R., 1906, Geological reconnaissance of the coast of the Olympic Peninsula, Washington: GSA Bulletin, v. 17, p. 451–468.

Baldwin, E. M., 1974, Eocene stratigraphy of southwestern Oregon: Oregon Department of Geology and Mineral Industries Bulletin 83, 40 p.

Baldwin, E. M., R. D. Brown, Jr., J. E. Gair, and M. H. Pease, Jr., 1955, Geology of the Sheridan and McMinnville quadrangles, Oregon: U.S. Geological Survey Oil and Gas Investigations Map OM-155, scale 1:62,500.

Beck, M. E., Jr., 1980, Paleomagnetic record of plate-margin tectonic processes along the western edge of North America: Journal of Geophysical Research, v. 84, p. 7115–7131.

Beck, M. E., Jr., 1984, Has the Washington-Oregon Coast Range moved northward?: Geology, v. 12, p. 737–740.

Beck, M. E., Jr., and P. W. Plumley, 1980, Paleomagnetism of intrusive rocks in the coast range of Oregon—Microplate rotations in middle Tertiary time: Geology, v. 8, p. 573–577.

Beeson, M. H., R. Perttu, and J. Perttu, 1979, The origin of the Miocene basalts of coastal Oregon and Washington—An alternative hypothesis: Oregon Geology, v. 41, p. 159–166.

Berg, J. W., Jr., and C. D. Baker, 1963, Oregon earthquakes, 1841 through 1958: Seismological Society of America Bulletin, v. 53, p. 95–108.

Berg, J. W., Jr., L. Trembly, D. A. Emilia, J. R. Hutt, J. M. King, L. T. Long, W. R. McKnight, C. K. Sarmah, R. Souders, J. V. Thiruvathural, and D. A. Vossler, 1966, Crustal refraction profile, Oregon Coast Range: Seismological Society of America Bulletin, v. 56, p. 1357–1362.

Braislin, D. B., D. D. Hastings, and P. D. Snavely, Jr., 1971, Petroleum potential of western Oregon and Washington, *in* I. A. Cram, ed., Possible future petroleum provinces of North America: AAPG Memoir 15, p. 229–238.

Brown, R. D., Jr., H. D. Gower, and P. D. Snavely, Jr., 1960, Geology of the Lake Crescent-Port Angeles area, Washington: U.S. Geological Survey Oil and Gas Investigations Map OM-203, scale 1:62,500.

Brown, R. D., Jr., P. D. Snavely, Jr., and H. D. Gower, 1956, Lyre Formation (redefinition), northern part of Olympic Peninsula, Washington: AAPG Bulletin, v. 40, p. 94–107.

Byrne, J. V., G. A. Fowler, and N. J. Maloney, 1966, Uplift of the continental margin and possible continental accretion off Oregon: Science, v. 154, no. 3757, p. 1654–1656.

Cady, W. M., 1975, Tectonic setting of the Tertiary volcanic rocks of the Olympic Peninsula, Washington: U.S. Geological Survey Journal of Research, v. 3, p. 573–582.

Carlson, P. R., and C. H. Nelson, 1987, Marine geology and resource potential of the Cascadia basin, this volume.

Carson, B., 1977, Tectonically induced deformation of deep-sea sediments off Washington and northern Oregon—Mechanical consolidation: Marine Geology, v. 24, p. 289–307.

Carson, B., J. Yuan, P. B. Myers, Jr., and W. D. Barnard, 1974, Initial deep-sea sediment deformation at the base of the Washington continental slope—A response to subduction: Geology, v. 2, no. 11, p. 561–564.

Chan, M. A., and R. H. Dott, Jr., 1983, Shelf and deep-sea sedimentation in Eocene forearc basin, western Oregon, fan or non-fan?: AAPG Bulletin, v. 67, p. 2100–2116.

Clapp, C. H., 1917, Sooke and Duncan map area, Vancouver Island, British Columbia: Geological Survey of Canada Memoir 96, 445 p.

Clarke, S. H., M. F. Field, and C. A. Hirozawa, 1981, Reconnaissance geology and geologic hazards of offshore Coos Bay basin, central Oregon continental margin: U.S. Geological Survey Open-File Report 81-898, 84 p.

Cooper, M. D., 1981, Sedimentation, stratigraphy, and facies variations of the lower to middle Miocene Astoria Formation in Oregon: Corvallis, Oregon, Oregon State University unpublished Ph.D. dissertation, 524 p.

Crosson, R. S., 1980, Review of seismicity in the Puget Sound region from 1970 through 1978, *in* Proceedings of Workshop XIV, Earthquake hazards of the Puget Sound region, Washington: U.S. Geological Survey Open-File Report 83-19, p. 6–18.

Crowell, J. C., 1974, Origin of late Cenozoic basins in southern California: SEPM Special Publication 22, p. 190–204.

Curray, J. R., D. G. Moore, L. A. Lawver, F. J. Emmel, R. W. Raitt, M. Henry, and R. Kickkhefer, 1979, Tectonics of the Andaman Sea and Burma, *in* J. S. Watkins, L. Montadert, and P. W. Dickerson, eds., Geological and geophysical investigations of continental margins: AAPG Memoirs, v. 29, p. 189–198.

Dehlinger, P., R. W. Couch, D. A. McManus, and M. Gemperle, 1970, Oceanic structures—Northern California to British Columbia, *in* A. E. Maxwell, ed., The sea: New York, Wiley-Interscience, v. 4, p. 133–190.

Dickinson, W. R., 1976, Sedimentary basins developed during evolution of Mesozoic-Cenozoic arc-trench system in western North America: Canadian Journal of Earth Science, v. 13, p. 1268–1287.

Dott, R. H., Jr., 1966, Eocene deltaic sedimentation at Coos Bay, Oregon: Journal of Geology, v. 74, p. 373–420.

Duncan, R. A., 1982, A captured island chain in the Coast Range of Oregon and Washington: Journal of Geophysical Research, v. 87, p. 10,827–10,837.

Engebretson, D. C., 1982, Relative motions between oceanic and continental plates in the Pacific Basin: Stanford, California, Stanford University unpublished Ph.D. dissertation, 211 p.

Engebretson, D. C., A. Cox, and G. A. Thompson, 1984, Correlation of plate motions with continental tectonics—Laramide to Basin-Range: Tectonics, v. 3, no. 2, p. 115–119.

Engebretson, D. C., R. G. Gordon, and A. Cox, 1980, Relative motions between the North America, Pacific, Kula, and Farallon plates from 130 m.y. to the present (abs.): Eos (American Geophysical Union Transactions), v. 61, p. 947.

Garrison, R. E., 1973, Space-time relations of pelagic limestones and volcanic rocks, Olympic Peninsula, Washington: GSA Bulletin, v. 84, p. 583–594.

Glassley, W., 1974, Geochemistry and tectonics of the Crescent volcanic rocks, Olympic Peninsula, Washington: GSA Bulletin, v. 85, p. 785–794.

Globerman, B. R., M. E. Beck, Jr., and R. A. Duncan, 1982, Paleomagnetism and tectonic significance of Eocene basalts from the Black Hills, Washington Coast Range: GSA Bulletin, v. 93, p. 1151–1159.

Glover, S. L., 1936, Preliminary report on petroleum and natural gas in Washington: Washington Division of Geology Report of Investigations 4, 24 p.

Goodwin, C. J., 1973, Stratigraphy and sedimentation of the Yaquina Formation, Lincoln County, Oregon: Corvallis, Oregon, Oregon State University unpublished M.S. thesis, 121 p.

Gower, H. D., 1960, Geology of the Pysht quadrangle, Washington: U.S. Geological Survey Geologic Quadrangle Map GQ-129, scale 1:62,500.

Gower, H. D., and M. H. Pease, 1965, Geology of the Montesano quadrangle, Washington: U.S. Geological Survey Geologic Quadrangle Map GQ-374, scale 1:62,500.

Grim, M. S., and L. C. Bennett, 1969, Shallow seismic profiling of the continental shelf off Grays Harbor, Washington: University of Washington Department of Oceanography Special Report 41, p. 72–92.

Heaton, T. H., and H. Kanamori, 1984, Seismic potential associated with subduction in the northwestern United States: Seismological Society of America Bulletin, v. 74, no. 3, p. 933–941.

Heller, P. L., and P. T. Ryberg, 1983, Sedimentary record of subduction to forearc transition in the rotated Eocene basin of western Oregon: Geology, v. 11, p. 380–383.

Heller, P. L., Z. E. Peterman, J. R. O'Neil, and M. Shafiqullah, 1985, Isotopic provenance of sandstones from the Eocene Tyee Formation, Oregon Coast Range: GSA Bulletin, v. 96, p. 770–780.

Hoover, L., 1963, Geology of the Anlauf and Drain quadrangles, Douglas and Lane Counties, Oregon: U.S. Geological Survey Bulletin 1122-D, 62 p.

Johnson, S. Y., 1984, Stratigraphy, age, and paleogeography of the Eocene Chuckanut Formation, northwest Washington: Canadian Journal of Earth Sciences, v. 21, p. 92–106.

Kanamori, H., 1977, Seismic and aseismic slip along subduction zones and their tectonic implications, *in* M. Talwani and W. C. Pittman, III, eds., Island arcs, deep sea trenches and back-arc basins: American Geophysical Union, Maurice Ewing Series 1, p. 173–174.

Kleinpell, R. M., 1938, Miocene stratigraphy of California: London, T. Murby and Company, 450 p.

Kulm, L. D., and G. A. Fowler, 1974, Oregon continental margin model, *in* C. A. Burk and C. L. Drake, eds., The geology of continental margins: New York, Springer-Verlag, p. 261–283.

Kvenvolden, K. A., and P. D. Snavely, Jr., 1985, Hydrocarbon gas potential of accretionary melange terranes—An example from the Olympic Peninsula, Washington (abs.): GSA Abstracts with Programs, v. 17, no. 7, p. 636.

Law, B. E., D. E. Anders, T. D. Fouch, M. J. Pawlewicz, M. R. Lickus, and C. M. Molenaar, 1984, Petroleum source rock evaluations of outcrop samples from Oregon and northern California: Oregon Geology, v. 46, no. 7, p. 77–81.

Livingston, V. E., Jr., 1966, Geology and mineral resources of the Kelso-Cathlamet area, Cowlitz and Wahkiakum Counties, Washington: Washington Division of Mines and Geology Bulletin 54, 110 p.

Lovell, J. P. B., 1969, Tyee Formation—Undeformed turbidites and their lateral equivalents, mineralogy, and paleogeography: GSA Bulletin, v. 80, p. 9–22.

Lupton, C. T., 1915, Oil and gas in the western part of the Olympic Peninsula, Washington: U.S. Geological Survey Bulletin 581, p. 23–81.

MacLeod, N. S., and P. D. Snavely, Jr., 1973, Volcanic and intrusive rocks of the central part of the Oregon Coast Range: Oregon Department of Geology and Mineral Industries Bulletin 77, p. 47–74.

MacLeod, N. S., D. L. Tiffin, P. D. Snavely, Jr., and R. G. Currie, 1977, Geologic interpretation of magnetic and gravity anomalies in the Strait of Juan de Fuca, U.S.-Canada: Canadian Journal of Earth Sciences, v. 14, no. 2, p. 223–238.

Magill, J. R., and A. Cox, 1980, Tectonic rotation of the Oregon western Cascades: Oregon Department of Geology and Mineral Industries Special Paper 10, 67 p.

Magill, J. R., A. Cox, and R. Duncan, 1981, Tillamook Volcanic Series: Further evidence for tectonic rotation of the Oregon Coast Range: Journal of Geophysical Research, v. 86, p. 2953–2970.

Magill, J. R., R. E. Wells, R. W. Simpson, and A. V. Cox, 1982, Post 12 m.y. rotation of southwest Washington: Journal of Geophysical Research, v. 87, no. B5, p. 3761–3777.

Mallory, V. S., 1959, Lower Tertiary biostratigraphy of the California Coast Ranges: Tulsa, Oklahoma, AAPG, 416 p.

Mann, D., and P. D. Snavely, Jr., 1984, Multichannel seismic-reflection profiles collected in 1977 in the eastern Pacific Ocean off of the Washington-Oregon coast: U.S. Geological Survey Open-File Report 84-5, 1 p.

Mauger, J. E., and R. D. Daugherty, 1979, Ozette archaeological project, interim final report, Phase XI, Project Report Number 68: Pullman, Washington, Washington State University, 56 p.

McFarland, C. R., 1981, Oil and gas exploration in Washington 1900–1981: Washington Division of Geology and Earth Resources Information Circular 67R, 119 p.

McLellan, R. D., 1927, The geology of the San Juan Islands: Washington University Publications in Geology, v. 2, 185 p.

Molenaar, C. M., 1985, Depositional relationships of the Umpqua and Tyee Formations (Eocene), southwestern Oregon: AAPG Bulletin, v. 69, p. 1217–1229.

Muller, J. E., 1977, Evolution of the Pacific margin, Vancouver Island, and adjacent regions: Canadian Journal of Earth Sciences, v. 14, no. 9, p. 2062–2085.

Newton, V. C., 1979, Oregon's first gas wells completed: Oregon Geology, v. 41, no. 6, p. 87–90.

Niem, A. R., and R. O. Van Atta, 1973, Cenozoic stratigraphy of northwestern Oregon and adjacent southwestern Washington, *in* J. D. Beaulieu, ed., Geologic field trips in northern Oregon and southern Washington: Oregon Department of Geology and Mineral Industries Bulletin 77, p. 75–132.

Pearl, J. E., 1977, Petrology of Tertiary sedimentary rocks in the northernmost part of the Olympic Peninsula, Washington: San Jose, California, San Jose State University unpublished M.S. thesis, 91 p.

Pease, M. H., and L. Hoover, 1957, Geology of the Doty-Minot Peak area, Washington: U.S. Geological Survey Oil and Gas Investigations Map OM-188, scale 1:62,500.

Plafker, G., 1972, Alaskan earthquake of 1964 and Chilean earthquake of 1960—Implications for arc tectonics: Journal of Geophysical Research, v. 77, p. 901–925.

Rau, W. W., 1967, Geology of the Wynoochee Valley quadrangle, Grays Harbor County, Washington: Washington Division of Mines and Geology Bulletin 56, 51 p.

Rau, W. W., 1970, Foraminifera, stratigraphy and paleoecology of the Quinault Formation, Point Grenville-Raft River coastal area, Washington: Washington Division of Mines and Geology Bulletin 62, 40 p.

Rau, W. W., 1973, Geology of the Washington coast between Point Grenville and the Hoh River: Washington Division of Geology and Earth Resources Bulletin 66, p. 1–58.

Rau, W. W., 1975, Geological map of the Destruction Island and Taholah quadrangles, Washington: Washington Division of Geology and Earth Resources Geologic Map GM-13, scale 1:62,500.

Rau, W. W., 1979, Geologic map in the vicinity of the lower Bogachiel and Hoh River Valleys, and the Washington Coast: Washington Division of Geology and Earth Resources Geologic Map GM-24, scale 1:62,500.

Rau, W. W., 1981, Pacific Northwest Tertiary benthic foraminiferal biostratigraphic framework—An overview, *in* J. M. Armentrout, ed., Pacific Northwest Cenozoic biostratigraphy: GSA Special Paper 184, p. 67–84.

Rau, W. W., and G. R. Grocock, 1974, Piercement structure outcrops along the Washington Coast: Washington Division of Geology and Earth Resources Information Circular 51, 7 p.

Rau, W. W., and C. R. McFarland, 1982, Coastal wells of Washington: Washington Division of Geology and Earth Resources Report of Investigations 26, 4 sheets.

Riddehough, R. P., 1977, A model for recent plate interactions off Canada's west coast: Canadian Journal of Earth Sciences, v. 14, p. 384–396.

Russell, I. C., 1900, Geology of the Cascade Mountains in northern Washington: U.S. Geological Survey Annual Report 20, pt. 2, p. 83–210.

Ryberg, P. T., 1978, Lithofacies and depositional environments of the Coaledo Formation, Coos County, Oregon: Eugene, Oregon, University of Oregon unpublished M.S. thesis, 159 p.

Samuels, S. R., 1983, Spatial patterns and cultural processes in three northwest coast longhouse floor middens from Ozette: Pullman, Washington, Washington State University unpublished Ph.D. dissertation, 187 leaves.

Savage, J. C., M. Lisowki, and W. H. Prescott, 1981, Geodetic strain measurements in Washington: Journal of Geophysical Research, v. 86, p. 4929–4940.

Seely, D. R., 1977, The significance of landward vergence and oblique structural trends on trench inner slopes, *in* M. Talwani and W. C. Pitman, III, eds., Island arcs, deep sea trenches and back-arc basins: American Geophysical Union, Maurice Ewing Series 1, p. 187–198.

Seely, D. R., P. R. Vail, and G. G. Walton, 1974, Trench slope model, *in* C. A. Burk and C. L. Drake, eds., The geology of continental margins: New York, Springer-Verlag, p. 249–260.

Shor, G. C., Jr., P. Dehlinger, H. K. Kirk, and W. S. French, 1968, Seismic refraction studies off Oregon and Northern California: Journal of Geophysical Research, v. 73, p. 2175–2194.

Shouldice, D. H., 1971, Geology of the western Canadian continental shelf: Canadian Petroleum Geology Bulletin, v. 19, no. 2, p. 405–436.

Silver, E. A., 1972, Pleistocene tectonic accretion of the continental slope off Washington: Marine Geology, v. 13, p. 239–249.

Simpson, R. W., and A. Cox, 1977, Paleomagnetic evidence of tectonic rotation of the Oregon Coast Range: Geology, v. 5, p. 585–589.

Snavely, P. D., Jr., 1979, Stratigraphic and tectonic framework, Oregon-Washington continental margin: U.S. Geological Survey Professional Paper 1150, p. 149.

Snavely, P. D., Jr., 1983, Peripheral rocks—Tertiary geology of the northwestern part of the Olympic Peninsula, Washington, *in* J. E. Muller, P. D. Snavely, Jr., and R. W. Tabor, The Tertiary Olympic terrane, southwest Vancouver Island and northwest Washington: Geological Association of Canada, Mineralogical Association of Canada, Canadian Geophysical Union, Field Trip Guidebook, Trip 12, 59 p.

Snavely, P. D., Jr., 1984, Sixty million years of growth along the Oregon continental margin, *in* S. H. Clarke, ed., U.S. Geological Survey, Highlights in marine research: U.S. Geological Survey Circular 938, p. 9–18.

Snavely, P. D., Jr., and E. M. Baldwin, 1948, Siletz River volcanic series, northwestern Oregon: AAPG Bulletin, v. 32, p. 806–812.

Snavely, P. D., Jr., and N. S. MacLeod, 1974, Yachats Basalt—An upper Eocene differentiated volcanic sequence in the Oregon Coast Range: U.S. Geological Survey Journal of Research, v. 2, no. 4, p. 395–403.

Snavely, P. D., Jr., and N. S. MacLeod, 1977, Evolution of Eocene continental margins of western Oregon and Washington (abs.): GSA Bulletin, v. 9, no. 7, p. 1183.

Snavely, P. D., Jr., and H. E. Vokes, 1949, Geology of the coastal area between Cape Kiwanda and Cape Foulweather, Oregon: U.S. Geological Survey Oil and Gas Investigations Preliminary Map 97, scale 1:62,500.

Snavely, P. D., Jr., and H. C. Wagner, 1963, Tertiary geologic history of western Oregon and Washington: Washington Division of Mines and Geology Report of Investigations 22, 25 p.

Snavely, P. D., Jr., and H. C. Wagner, 1980, Generalized isopach map of Tertiary sedimentary rocks, western Oregon and Washington and adjacent continental margin: U.S. Geological Survey Open-File Report 80-889, scale 1:1,000,000.

Snavely, P. D., Jr., and H. C. Wagner, 1981, Geologic cross section across the continental margin off Cape Flattery, Washington, and Vancouver Island, British Columbia: U.S. Geological Survey Open-File Report 81-0978, 6 p.

Snavely, P. D., Jr., and H. C. Wagner, 1982, Geologic cross section across the continental margin of southwestern Washington: U.S. Geological Survey Open-File Report 82-459, 10 p.

Snavely, P. D., Jr., and R. E. Wells, 1984, Tertiary volcanic and intrusive rocks on the Oregon and Washington continental shelf: U.S. Geological Survey Open-File Report 84-282, 17 p.

Snavely, P. D., Jr., N. S. MacLeod, and W. W. Rau, 1970, Summary of the Tillamook area, northern Oregon Coast Range: U.S. Geological Survey Professional Paper 650-A, p. A47.

Snavely, P. D., Jr., N. S. MacLeod, and H. C. Wagner, 1968, Tholeiitic and alkalic basalts of the Eocene Siletz River Volcanics, Oregon Coast Range: American Journal of Science, v. 266, p. 454–481.

Snavely, P. D., Jr., N. S. MacLeod, and H. C. Wagner, 1973, Miocene tholeiitic basalts of coastal Oregon and Washington and their relations to coeval basalts of the Columbia Plateau: GSA Bulletin, v. 84, no. 2, p. 387–424.

Snavely, P. D., Jr., A. R. Niem, and J. E. Pearl, 1978, Twin River Group (upper Eocene to lower Miocene)—Defined to include the Hoko River, Makah, and Pysht Formations, Clallam County, Washington: U.S. Geological Survey Bulletin 1457-A, p. A111–A120.

Snavely, P. D., Jr., J. E. Pearl, and D. L. Lander, 1977, Interim report on petroleum resources potential and geologic hazards in the outer continental shelf—Oregon and Washington Tertiary province with a section on resource appraisal estimate, by E. W. Scott: U.S. Geological Survey Open-File Report 77-282, 64 p.

Snavely, P. D., Jr., W. W. Rau, and H. C. Wagner, 1964, Miocene stratigraphy of the Yaquina Bay area, Newport, Oregon: Ore Bin,

v. 26, no. 8, p. 133–151.
Snavely, P. D., Jr., H. C. Wagner, and D. L. Lander, 1980, Geological cross section of the central Oregon continental margin: GSA Map and Chart Series MC-28J, scale 1:250,000.
Snavely, P. D., Jr., H. C. Wagner, and D. L. Lander, 1985, Land-sea geologic cross section of the southern Oregon continental margin: U.S. Geological Survey Miscellaneous Investigations Series Map-1463, scale 1:125,000.
Snavely, P. D., Jr., H. C. Wagner, and N. S. MacLeod, 1964, Rhythmic-bedded eugeosynclinal deposits of the Tyee Formation, Oregon Coast Range, *in* D. F. Merriam, ed., The symposium on cyclic sedimentation: Kansas Geological Survey Bulletin 169, v. 11, p. 461–480.
Snavely, P. D., Jr., H. C. Wagner, and W. W. Rau, 1982, Biostratigraphy and correlation of Tertiary rocks in wells drilled on the southern Oregon continental margin: U.S. Geological Survey Miscellaneous Field Studies Map MF-1482.
Snavely, P. D., Jr., R. D. Brown, Jr., A. E. Roberts, and W. W. Rau, 1958, Geology and coal resources of the Centralia-Chehalis district, Washington: U.S. Geological Survey Bulletin 1053, 159 p.
Snavely, P. D., Jr., N. S. MacLeod, A. R. Niem, and D. L. Minasian, 1986, Geologic map of Cape Flattery area, Northwestern Olympic Peninsula, Washington: U.S. Geological Survey Open-File Report 86–344B, scale 1:48,000.
Snavely, P. D., Jr., N. S. MacLeod, and H. C. Wagner, and W. W. Rau, 1976a, Geologic map of the Cape Foulweather and Euchre Mountain quadrangles, Lincoln County, Oregon: U.S. Geological Survey Miscellaneous Investigations Map I-868, scale 1:62,500.
Snavely, P. D., Jr., N. S. MacLeod, and H. C. Wagner, and W. W. Rau, 1976b, Geologic map of the Yaquina and Toledo quadrangles, Lincoln County, Oregon: U.S. Geological Survey Miscellaneous Investigations Map I-867, scale 1:62,500.
Snavely, P. D., Jr., N. S. MacLeod, H. C. Wagner, and W. W. Rau, 1976c, Geologic map of the Waldport and Tidewater quadrangles, Lincoln, Lane, and Benton Counties, Oregon: U.S. Geological Survey Miscellaneous Investigation Series Map I-866, scale 1:62,500.
Snavely, P. D., Jr., H. C. Wagner, W. W. Rau, and D. Bukry, 1981, Geologic cross section of the southern Oregon Coast Range and adjacent continental shelf: U.S. Geological Survey Open-File Report 81-957.
Snavely, P. D., Jr., N. S. MacLeod, and W. W. Rau, W. O. Addicott, and J. E. Pearl, 1975, Alsea Formation—An Oligocene marine sedimentary sequence in the Oregon Coast Range: U.S. Geological Survey Bulletin 1395-F, 21 p.
Snavely, P. D., Jr., A. R. Niem, N. S. MacLeod, J. E. Pearl, and W. W. Rau, 1980, Makah Formation—A deep marginal basin sedimentary sequence of upper Eocene and Oligocene age in the northwestern Olympic Peninsula, Washington: U.S. Geological Survey Professional Paper 1162-B, 28 p.
Swan, J. G., 1870, The Indians of Cape Flattery, at the entrance to the Strait of Fuca, Washington Territory: Smithsonian Contributions to Knowledge, v. XVI, City of Washington, Smithsonian Institution MDCCCLXX, 108 p.
Tabor, R. W., 1975, Guide to the geology of Olympic National Park: Seattle, Washington, University of Washington Press, 144 p.
Tabor, R. W., and W. M. Cady, 1978a, The structure of the Olympic Mountains, Washington—Analysis of a subduction zone: U.S. Geological Survey Professional Paper 1033, 38 p.
Tabor, R. W., and W. M. Cady, 1978b, Geologic map of the Olympic Peninsula, Washington: U.S. Geological Survey Miscellaneous Investigations Series Map I-994, scale 1:125,000.
Tabor, R. W., R. S. Yeats, and M. L. Sorensen, 1972, Geologic map of the Mount Angeles quadrangle, Clallam and Jefferson Counties, Washington: U.S. Geological Survey Geologic Quadrangle Map GQ-958, scale 1:62,500.
Tabor, R. W., V. A. Frizzell, Jr., J. A. Vance, and C. W. Naeser, 1984, Ages and stratigraphy of lower and middle Tertiary sedimentary and volcanic rocks of the central Cascades, Washington: Application to the tectonic history of the Straight Creek fault: GSA Bulletin, v. 95, p. 26–44.
U.S. Geological Survey, 1970, Aeromagnetic survey composite map, Oregon Coast: U.S. Geological Survey Open-File Map, scale 1:500,000.
Vokes, H. E., and P. D. Snavely, Jr., 1948, The age and relationships of the Eugene and Fisher Formations: Geological Society Oregon County News Letters, v. 14, no. 5, p. 38–41.
Wagner, H. C., 1967a, Preliminary geologic map of the Raymond quadrangle, Pacific County, Washington: U.S. Geological Survey Open-File Report, scale 1:62,500.
Wagner, H. C., 1967b, Preliminary geologic map of the South Bend quadrangle, Pacific County, Washington: U.S. Geological Survey Open-File Report, scale 1:62,500.
Warren, W. C., R. M. Grivetti, and H. Norbisrath, 1945, Geology of northwestern Oregon, west of Willamette River and north of latitude 45°15′: U.S. Geological Survey Oil and Gas Investigations Preliminary Map OM-42.
Weaver, C. S., and S. W. Smith, 1983, Regional tectonic and earthquake hazard implications of a crustal fault zone in southwestern Washington: Journal of Geophysical Research, v. 88, no. B12, p. 10,371–10,383.
Wells, R. E., 1979, Geologic map of the Cape Disappointment-Naselle River area, Pacific County, Washington: U.S. Geological Survey Open-File Report 79-389, scale 1:48,000.
Wells, R. E., 1981, Geologic map of the eastern Willapa hills, Cowlitz, Lewis, Pacific, and Wahkiakum Counties, Washington: U.S. Geological Survey Open-File Map 81-674, scale 1:62,500.
Wells, R. E., 1984, Paleomagnetic constraints on the interpretation of early Cenozoic Pacific Northwest paleogeography, *in* T. H. Nilsen, ed., Geology of the Upper Cretaceous Hornbrook Formation, Oregon and California: SEPM, Pacific Section, v. 42, p. 231–237.
Wells, R. E., in press, Geologic map of the Cape Disappointment area, Pacific and Wahkiakum Counties, Washington: U.S. Geological Survey Miscellaneous Investigations Map 1832, scale 1:62,500.
Wells, R. E., and R. S. Coe, 1985, Paleomagnetism and geology of Eocene volcanic rocks in southwest Washington. Implications for mechanism of tectonic rotation: Journal of Geophysical Research, v. 90, p. 1925–1947.
Wells, R. E., D. C. Engebretson, P. D. Snavely, Jr., and R. S. Coe, 1984, Cenozoic plate motions and the volcano-tectonic evolution of western Oregon and Washington: Tectonics, v. 3, no. 2, p. 275–294.
Wells, R. E., A. R. Niem, N. S. MacLeod, P. D. Snavely, Jr., and W. A. Niem, 1983, Preliminary geologic map of the west half of the Vancouver (Wash-Ore) 1° × 2° sheet quadrangle, Oregon: U.S. Geological Survey Open-File Report 83-591, scale 1:250,000.
Williams, P. R., C. J. Pigram, and D. B. Dow, 1984, Melange production and the importance of shale diapirism in accretionary terrains: Nature, v. 309, p. 145–146.
Wolfe, J. A., H. D. Gower, and J. D. Vine, 1961, Age and correlation of the Puget Group, King County, Washington: U.S. Geological Survey Professional Paper 424-C, p. C230–232.
Ziegler, D. L., and J. K. Cassell, 1978, A synthesis of OCS wells information, offshore central and northern California, Oregon, and Washington, *in* F. L. Hill, coord., Energy Exploration and Politics: California Division of Oil and Gas, 27 p.

CHAPTER FIFTEEN

GEOLOGY OF THE CALIFORNIA CONTINENTAL MARGIN NORTH OF CAPE MENDOCINO

Samuel H. Clarke, Jr.

U.S. Geological Survey
Menlo Park, California

ABSTRACT

The Eel River Basin, situated on the California continental margin north of Cape Mendocino. is the site of a late Cenozoic fore-arc basin in which more than 3000 m of principally marine sediment was deposited on a basement of late Mesozoic and early Cenozoic allochthonous terranes. Major episodes of deformation occurred between the Eocene and the middle Miocene, during the middle to late Miocene, and during the middle and late Pleistocene. The Pleistocene deformation is characterized by generally northeast–southwest compression focused in the southern part of the basin, and probably reflects north–northwestward propagation of the San Andreas fault and encroachment by the Pacific plate.

The northeast margin of the offshore basin is cut by north–northwest-trending, high-angle faults that, in places, vertically displace basement rocks as much as 1300 m (west side down), offset Quaternary sediment, and show evidence of youthful right slip. However, faults that would extend San Andreas motion northwestward across the offshore basin to the margin of the North American plate, forming the northeast edge of the "Humboldt plate," are not apparent in acoustic reflection data. The western margin of the offshore basin is characterized by a zone of youthful, right-stepping en echelon folds that suggest right shear between the axial part of the basin and the uplifted accretionary prism underlying the continental slope to the west.

Commercial quantities of gas have been produced from the onshore Eel River Basin, and large areas in the onshore basin and most of the offshore basin are unexplored. Structural and stratigraphic traps probably are associated with youthful reverse and thrust faults and growth folds offshore, but the extent and quality of potential source and reservoir rocks are unknown.

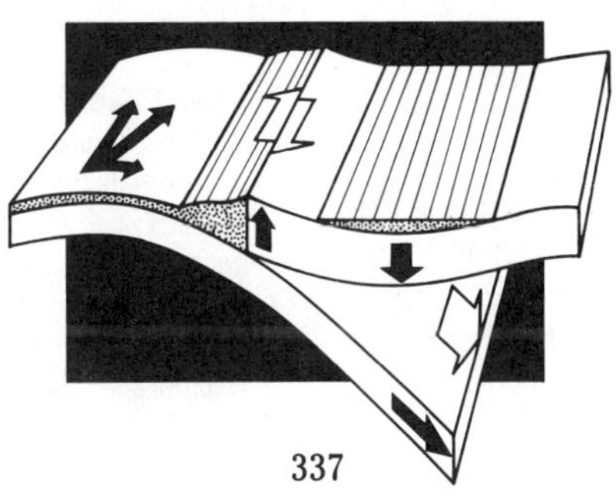

INTRODUCTION

The Eel River Basin is situated in the structurally complex region north of the Mendocino triple junction (Figure 1), which is formed by the intersection of three major crustal plates—the North American plate to the east, the Gorda plate to the north, and the Pacific plate to the south. The basin extends from Cape Mendocino (lat. 40°30′N) on the northern California coast northward for 200 km to Cape Sebastian (lat. 42°20′N) in southern Oregon, and from the coastline seaward to the continental slope, an average distance of about 70 km. The south end of the basin extends inland in continuous exposures for about 50 km in the Humboldt Bay–lower Eel River area; isolated exposures of strata coeval with the Neogene fill of the Eel River Basin are present 40 km farther to the south, indicating either that the basin was originally much more extensive or that several smaller coeval structural basins were present to the southeast.

The late Tertiary history of the Eel River Basin region was strongly influenced by subduction of the Gorda (Farallon) plate beneath the North American plate and development of a fore-arc basin in Neogene time (Dickinson, Ingersoll, and Graham, 1979), which formed the principal site for late Cenozoic deposition. The southern margin of the fore-arc basin underwent progressive deformation and destruction in the late Neogene as a result of north–northwestward migration of the Mendocino triple junction and propagation of the San Andreas fault. This interaction has overprinted west- to northwest-trending compressional faults and folds on predominantly north- to northwest-trending structures formed by convergence of the Gorda and North American plates.

The geology of the onland portion of the basin has been described by Ogle (1953), and that of the surrounding region by Irwin (1960) and Bailey, Irwin, and Jones (1964). Recent studies of regional structure and basinal and prebasinal sedimentation and tectonics have been published by Bachman (1982), Bachman, Underwood, and Menack (1984), McLaughlin et al. (1982), Underwood (1983, 1984, 1985), Menack (1984), Menack and Bachman (1983), and Menack, Roth, and Barron (1985), and the subsurface geology of the onshore basin was studied by Hopps and Horan (1984). Jachens and Griscom (1983) described the regional gravity field in relation to crustal plate geometry and the corresponding implications for the structure and tectonic history of northwestern California, and Smith and Knapp (1980) and Simila (1980) discussed deformation of the southern Gorda plate as inferred from seismicity data. The geology of the offshore basin was described generally by Hoskins and Griffiths (1971) and Field, Clarke, and White (1980), and the regional structure and tectonic history was discussed by Silver (1971a, 1971b).

GEOLOGIC SETTING

The onshore Eel River Basin (Figure 2) is a synclinorium that trends nearly west and is bounded on the north and east by northwest-trending, northeast-dipping high-angle reverse and thrust faults. Its upturned southern flank is abutted on the south by the Russ fault and "False Cape shear zone," which is a nearly west-trending melange in the coastal belt of the Franciscan Complex comprising a pervasively disrupted, heterogeneous assemblage of Tertiary and relatively minor Cretaceous rocks. (Note: The False Cape shear zone is not differentiated from the coastal belt on Figure 2, but lies between the Russ fault and a line extending generally eastward from Cape Mendocino.) The basin contains a thick sequence of Miocene and younger strata that unconformably overlie and locally are in fault contact with basement rocks of the central and coastal belts of the Franciscan Complex (Figure 3). This basement complex comprises late Mesozoic and early Tertiary rock assemblages that are broadly related to subduction and accretion settings and processes. Recent studies indicate that the Franciscan

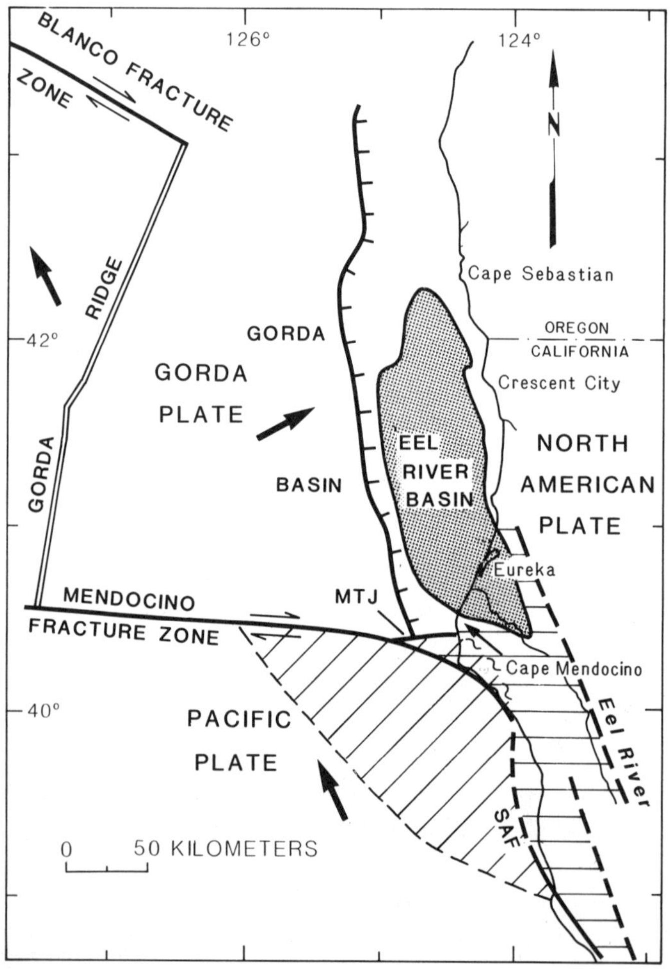

Figure 1. Sketch map showing plate tectonic setting of the Eel River Basin. Arrows indicate oceanic plate motions relative to North American plate. MTJ indicates approximate position of Mendocino triple junction. SAF indicates position of San Andreas fault. Hatched line shows position of base of continental slope—the approximate boundary between the Gorda and North American plates. Horizontal pattern indicates location of onshore part of the "Humboldt plate" of Herd (1978). Diagonal pattern depicts approximate extent of Vizcaino block (McCulloch, this volume) and adjacent onshore King Range terrane (McLaughlin et al., 1982).

Complex consists of terranes having complex tectonic histories involving large-scale lateral translation as well as simple convergence.

Along the eastern margin of the basin, basement rocks are assigned to the central belt, a melange comprising tectonic blocks of mid-Jurassic, Cretaceous and Paleogene age including graywacke, metagraywacke, blueschist, greenstone, chert, serpentinite, and minor limestone in a pervasively sheared matrix of graywacke and argillite (Blake and Jones, 1981). Paleomagnetic evidence (Alvarez et al., 1980; Tarduno et al., 1986) indicates that the limestone inclusions were formed at 14° ±5° S paleolatitude, were transported northward, and accreted to the North American margin between the latest Cretaceous and middle Tertiary (70 Ma–50 Ma). Elsewhere, the coastal and Yager (Yager Formation of Ogle, 1953) terranes of the coastal belt (Blake, Jayko, and McLaughlin, 1985) underlie the onshore Eel River Basin and probably form basement for most of the offshore basin as well. The coastal terrane is Late Cretaceous to Eocene in age (Evitt and Pierce, 1975) and consists principally of interbedded fine- to coarse-grained clastic rocks that show various degrees of stratal disruption, are slightly metamorphosed (zeolite facies), and locally contain exotic blocks of basaltic rock, pelagic limestone, and chert. The coastal terrane is interpreted by some to represent an accretionary complex comprising a variety of trench deposits, and perhaps lower slope deposits, formed on the seaward flank of a fore-arc setting (Bachman, 1982; Bachman, Underwood and Menack, 1984). Others (Harbert, McLaughlin, and Sliter, 1984; Sliter, 1984; McLaughlin et al., 1985b; Sliter et al., in press) have shown that oceanic elements of the coastal terrane formed on the Farallon plate at equatorial paleolatitudes and were translated northward in the Late Cretaceous and early Tertiary. Coastal terrane rocks are in contact with the Yager terrane along a narrow zone of structural transition interpreted by Underwood (1983, 1984) to represent a tectonized unconformity. The Yager terrane, Paleogene in age (Evitt and Pierce, 1975), consists principally of well-indurated marine mudstone and thin-bedded siltstone, with lesser graywacke and conglomerate. Yager strata are distinguished from lithologically similar coastal terrane rocks by their greater stratal coherence, multiple complex folding, the generally less pronounced penetrative shear fabric, and absence of exotic blocks, a characteristic prevalent in the coastal terrane (Underwood, 1983, 1984). The Yager is at least 765 m and perhaps as much as about 3000 m thick (Ogle, 1953), and is inferred to have been deposited in various slope (e.g., open-slope, channel, and basin) environments above broadly coeval, deformed accreted basement rocks of the coastal terrane (Underwood, 1983, 1984).

The Yager terrane lies in contact with rocks of the central belt of the Franciscan Complex along a folded, east-dipping zone of faulting referred to locally as the Garberville thrust (Underwood, 1982, 1983), or Freshwater fault (Ogle, 1953). Regionally, this complex system of faults was termed the coastal belt thrust by Jones et al. (1978). The age of suturing of the Yager and coastal terranes with the central belt of the Franciscan Complex, which occurred along the coastal belt thrust zone, is constrained to post-late Eocene to middle Miocene (probably Oligocene) time, as the Freshwater fault is overlapped by strata of the Wildcat Group east of Eureka (Ogle, 1953, McLaughlin et al., 1982, 1985b).

LATE CENOZOIC HISTORY

Regional deformation occurred intermittently between the Eocene and late Miocene, and Miocene strata overlie Paleogene and older rocks in the Eel River Basin with angular unconformity (Figure 3; Ogle, 1953; Hopps and Horan, 1984). Lower to middle Miocene (Relizian to Luisian) siliceous shale, siltstone, and lesser sandstone, informally termed the Bear River beds, occur within the "False Cape shear zone" about 2–5 km north of Cape Mendocino. These strata contain foraminiferal faunas generally indicative of bathyal depths and have a maximum thickness of about 900 m (Ogle, 1953; Haller, 1980). Lithologically similar coeval rocks unconformably overlie Yager strata in the subsurface in the axis of the Eel River Basin, where they are >600 m thick (Hopps and Horan, 1984), and in structurally isolated exposures about 30 km to the southeast (Kleinpell, 1938; Menack, 1984; Menack, Roth, and Barron, 1985). The principal fill of the onshore Eel River Basin comprises units of the Wildcat Group, middle or late Miocene (Mohnian) to Pleistocene in age, which overlie with angular unconformity strata of the Bear River beds in the subsurface of the central Eel River Basin and Paleogene and older units elsewhere (Ogle, 1953; Ingle, 1976; Hopps and Horan, 1984). The Wildcat Group comprises an essentially conformable sequence of finely to coarsely clastic, principally marine strata as much as about 3600 m thick. Predominant lithologies are weakly consolidated mudstone, siltstone, and claystone, with subordinate sandstone and conglomerate, and minor lignite and tuff. Studies of the age and correlation of Wildcat strata using the techniques of biostratigraphy (Faustman, 1964; Ingle, 1976; Haller, 1980; Burkle, Dodd, and Stanton, 1980; Keller and Ingle, 1981), sedimentary facies analysis (Piper, Normark, and Ingle, 1976; Bourgeois and Leithold, 1983), magnetostratigraphy (Dodd, Mead, and Stanton, 1977; Kodama, 1979), and tephrochronology (Sarna-Wojcicki et al., 1982) have added resolution to the stratigraphic framework originally established by Ogle (1953). Units of the Wildcat Group record, in general, an eastward transgression during late Miocene and early Pliocene; deepening of the basin to bathyal-abyssal depths with basin-plain, submarine-fan, and basin-slope deposition during the Pliocene and early Pleistocene; followed by westward regression with shoaling to shelf depths and ultimately by emergence and marginal-marine and nonmarine deposition in the early to middle Pleistocene. Lithologic units of the Wildcat Group are time-transgressive (Wagner, 1980; Sarna-Wojcicki et al., 1982); in general, the section coarsens upward, reflecting Pleistocene regression, and eastward, reflecting the presence of a landmass. Marine to nonmarine nearshore deposits of late Pliocene and Pleistocene age in isolated localities north of the Eel River Basin (e.g., Manning and Ogle, 1950; Back, 1957; Faustman, 1964, p. 112; Kelsey, 1982; Carver, 1985) suggest that the eastern margin of the basin extended generally north–northwestward from the vicinity of outliers of Wildcat-age strata on central belt rocks 10–15 km east of the Eel River Basin to the vicinity of Crescent City. The southeastward extent of the basin is uncertain; however, if coeval strata that unconformably overlie central belt rocks near Garberville (Menack, 1984; Menack, Roth, and Barron, 1985) were once continuous with the Neogene fill of the Eel River Basin, the basin may have extended at least 40 km south–southeastward from the present margin.

Deposition of Wildcat sediment ended in the mid-Pleistocene with basin-margin warping and uplift, which culminated in basin-wide deformation that produced broad, west- to northwest-plunging folds and northwest-trending faults. Pleistocene gravel, sand, silt, and clay having an aggregate thickness of as much as 1000 m and representing various shallow-marine and coastal-plain depositional environments unconformably overlie Wildcat strata in onshore parts of the basin. This deformation, which is ongoing, marked the close of a lengthy period of deposition characterized by progressive infilling and shallowing of the basin. The change from depths of about 2000 m to shallow-marine and nonmarine nearshore conditions occurred over a time span extending from early Pliocene (possibly late Miocene), to mid-Pleistocene (Ingle, 1976; Carver 1985). Deformation was most pronounced in the southern part of the basin and included rapid uplift of the basin margins, initiation of major folding (about 0.7 Ma), and initiation (or reactivation along pre-existing structural zones of weakness) of high-angle reverse and thrust faulting (about 1 ± 0.3 Ma) (Wagner, 1980; Carver, 1985; Carver, Burke, and Kelsey, 1986). This event reflects north–south to northeast–southwest compression of the southern part of the Eel River Basin, perhaps marking the arrival of the allochthonous Vizcaino block of McCulloch (this volume) and the adjacent King Range terrane (McLaughlin et al., 1982) in their present position south of Cape Mendocino (Figure 1).

CONTINENTAL MARGIN STRUCTURE AND STRATIGRAPHY

Figure 2 depicts the structure of the Eel River Basin, and Figure 4 summarizes major structural relationships within the offshore part of the basin. The axis of the Eel River Basin trends N80°W onshore (Ogle, 1953), but swings to a near-northerly orientation offshore (Figure 2). Major faults and fold axes are parallel or subparallel to the basin margins. The Russ fault, the major fault bounding the Eel River Basin on the south (RF, Figure 2), is a steeply south-dipping reverse fault. Although the basal member of the Wildcat Group was mapped by Ogle (1953) as resting unfaulted on the trace of the Russ fault, Wildcat strata in near-coastal exposures are intensely fractured and faulted along its trend (G. A. Carver, 1986, oral communication) and elements of this fault may cut Quaternary strata offshore. Other principal faults in the central part of the onshore basin include the Little Salmon and Freshwater faults (LSF and FWF, Figure 2). Major movement on the Freshwater fault—in this area a northeast-dipping, high-angle reverse fault—preceded deposition of the Wildcat Group (Ogle, 1953); recent mapping by Wagner (1980) suggests that it was reactivated following Wildcat deposition. The Little Salmon fault is a northwest-trending, northeast-dipping thrust fault with a total vertical displacement of about 2000 m and a history of late Quaternary activity (Wagner, 1980). Elements of this fault and late Quaternary thrust and high-angle reverse faults (Buhne Point fault, Bay Entrance fault, and faults underlying the Table Bluff anticline; not identified on Figure 2) in the Humboldt Bay area that are probably related to the Little Salmon fault have been traced northwestward into the offshore Eel River Basin, where Quaternary faults having similar trends and displacements have been identified (Figure 2).

The southern part of the offshore basin is subdivided by elements of major thrust and reverse faults and anticlinal folds mapped on land in the vicinity of Arcata and Humboldt Bays into three major synclines containing thick sequences of Neogene sediment (Figure 2). The southernmost syncline is the offshore extension of the onshore Eel River syncline; it is bounded on the north by faults and folds that extend seaward south of Humboldt Bay and that may be continuous as a zone with the onshore Table Bluff anticline and associated structures. Farther north, a syncline lies between these structures and offshore elements of the Little Salmon fault and associated folds and faults that extend seaward near the mouth of Humboldt Bay. Still farther north, a syncline lies between the latter structures and elements of the Mad River fault zone. These folds can be traced, more or less continuously, for a maximum distance of 50–65 km. Along much of their extent, major anticlines are expressed in the sea floor and Neogene strata thin and are upwarped along their flanks, attesting to their syndepositional growth and recent movement. In the southern part of the basin (A, Figure 4), anticlines trend west–northwest to northwest and commonly are thrust and reverse faulted; farther north (C, Figure 4), they trend northwest to north–northwest, oblique to the basin margin. Here, faulting is less prominent, and folds are commonly en echelon, right-stepping, suggesting that north–south compression characteristic of the southern part of the offshore basin in the Humboldt-Arcata Bay area is accompanied northward by a progressively greater component of right-lateral shear.

Figure 2. Map showing geology and structure of the Eel River Basin. Onshore geology is generalized from Strand (1962), Carver (1985), and Wagner and Saucedo (in press). Rock Units: JKf = Mesozoic basement rocks of the central belt, Franciscan Complex and related units; KTc = Cretaceous and Paleogene rocks of the coastal terrane of the coastal belt, Franciscan Complex; Ty = Paleogene rocks of the Yager terrane of the coastal belt, Franciscan Complex; TQw = upper Tertiary and Quaternary, principally marine sedimentary rocks of the Wildcat Group and related units; other, unpatterned = mostly nonmarine, Quaternary, undifferentiated. Structural features: SFMF = South Fork Mountain fault; LMF = Lost Man fault; GF = Grogan fault; BMF = Bald Mountain fault; TF = Trinidad fault; MRFZ = Mad River fault zone; FWF = Freshwater fault; LSF = Little Salmon fault; RF = Russ fault. Locations: CC = Crescent City; KR = Klamath River; TH = Trinidad Head; A = Arcata; E = Eureka; AB = Arcata Bay; HB = Humboldt Bay; ER = Eel River; CM = Cape Mendocino; numbers 7, 12, 14, and 19 = exploratory wells. Fold axes are indicated by arrows, solid where well defined, dashed where inferred, queried where continuity is questionable. Probable offshore extensions of onshore faults and fault zones are indicated by abbreviations. Bar and ball on downthrown side of offshore faults. Hatched areas offshore indicate fault zones. Apparent offsets between major onshore faults and their offshore extensions north of Trinidad Head are probably navigational in origin, resulting either from a systematic navigational error in this area, or from small differences in positions determined from USGS and non-USGS data sets.

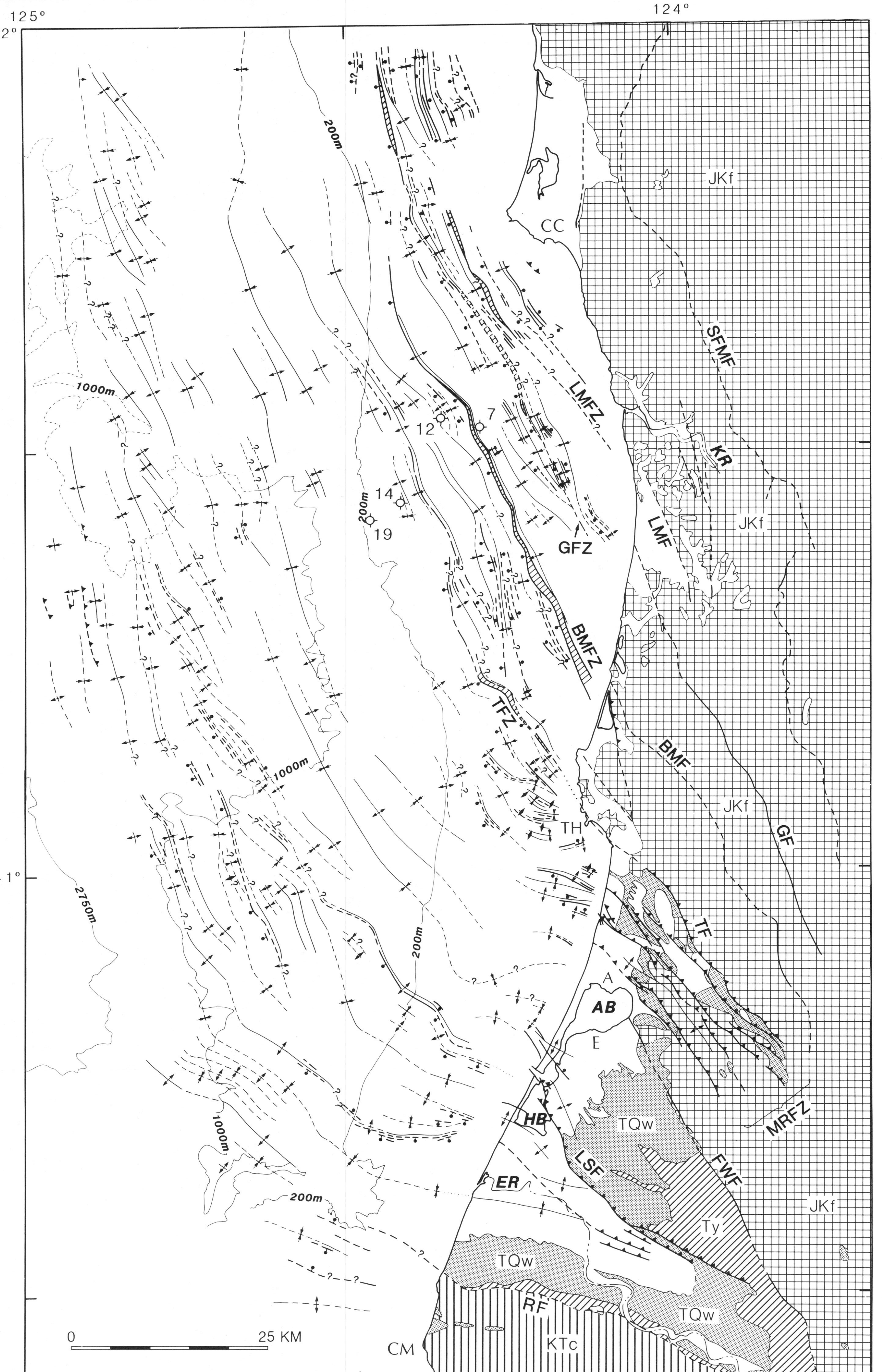
125°
124°
42°
41°
JKf
CC
SFMF
KR
LMF
LMFZ
GFZ
BMFZ
BMF
TFZ
GF
TH
TF
A
AB
E
HB
MRFZ
TQw
LSF
FWF
ER
Ty
RF
KTc
CM
7
12
14
19
200m
1000m
2750m
0
25 KM

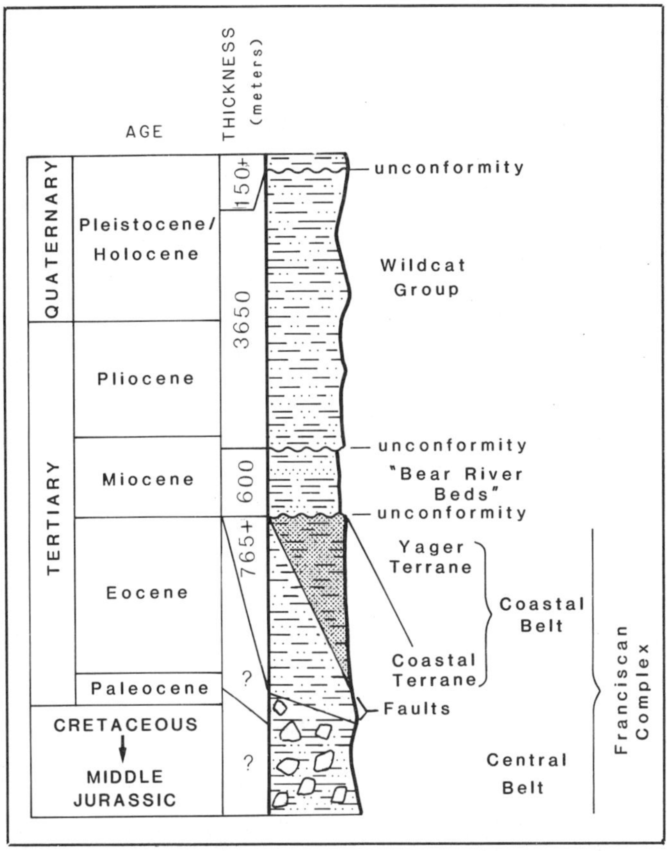

Figure 3. Composite stratigraphic column, onshore Eel River Basin.

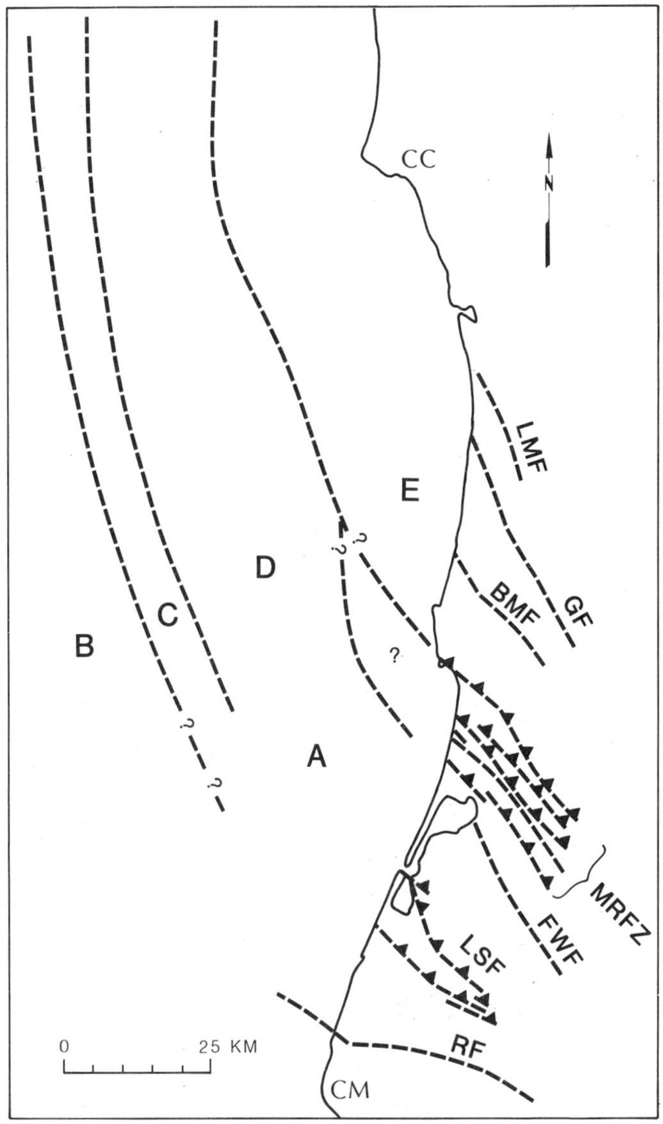

Figure 4. Principal structural domains in the offshore Eel River Basin, characterized by: A—compression across folds and faults that trend east-west to southeast-northwest; B—slope-parallel folds and faults that have north-south to north-northwest trends related to subduction/accretion; C—en echelon folds that have axial trends oblique to those in areas B and D, reflecting compression accompanied by right shear; D—broad, open folds having north-northwest trends (central basin); E—prominent basement faults that cut Quaternary strata, with suggestion of right slip. Queried area (offshore extension of Mad River fault zone) is complex, probably reflecting structural transition between areas A and E. Structures in all areas are expressed in youthful strata and/or the sea floor. Abbreviations explained in Figure 2.

Farther offshore (Figure 2; B, Figure 4), north- to northwest-trending folds involving Pliocene and younger strata are expressed in the sea floor along the western margin of the basin (marginal plateaus of Silver, 1971a) and the adjacent upper continental slope, and are cut by reverse and thrust faults having dip separations that are predominantly west side down. Folds are characteristically broadly symmetrical or asymmetrical with east-dipping axial planes, reflecting east–west compression. The uplifted western margin of the basin has been interpreted as a series of stacked, imbricated wedges of slope and offscraped Gorda Basin sediment related to oblique underthrusting of the Gorda plate during the late Cenozoic (Silver, 1971a). Numerous unconformities in the sedimentary fill of restricted basins along the west flank of the offshore Eel River Basin reflect concurrent episodic deformation of the basin margin (Figure 5). Some high-amplitude folds of the outer margin of the basin appear to be associated with shale flowage and diapirism, and some of the marginal ridges are piercement structures that show as much as 250 m of sea-floor relief (Figure 5). Samples from the crests of these structures are typically sheared, dewatered clay and silt containing lower to upper Pliocene foraminifers. Similar piercement folds have been mapped northward along the Oregon–Washington margin by Snavely, Pearl, and Lander (1977), Snavely and Wagner (1982), and Clarke, Field, and Hirozawa (1985). A possibly analogous structure was described from coastal exposures in northwestern Washington by Rau and Grocock (1974), where the core consisted of a melange of siltstone, sandstone, mudstone, and volcanic blocks in a matrix of mudstone containing lower to middle Miocene (Saucesian) foraminifera; this melange was intruded along faults into the Pliocene Quinault Formation and was associated with gas seeps and abnormally high subsurface pressures (Snavely, this volume).

The deep structural depression in the central part of the offshore basin (Figure 2; D, Figure 4) is filled by as much as about 3300 m of upper Tertiary and Quaternary hemipelagic sediment and turbidite deposits that have been deformed into gentle open folds (Figure 5). The maximum thickness of section occurs beneath the shelf about 25 km west of Trinidad Head. Little direct evidence exists concerning the age of the sedimentary rocks in the offshore Eel River Basin. Their

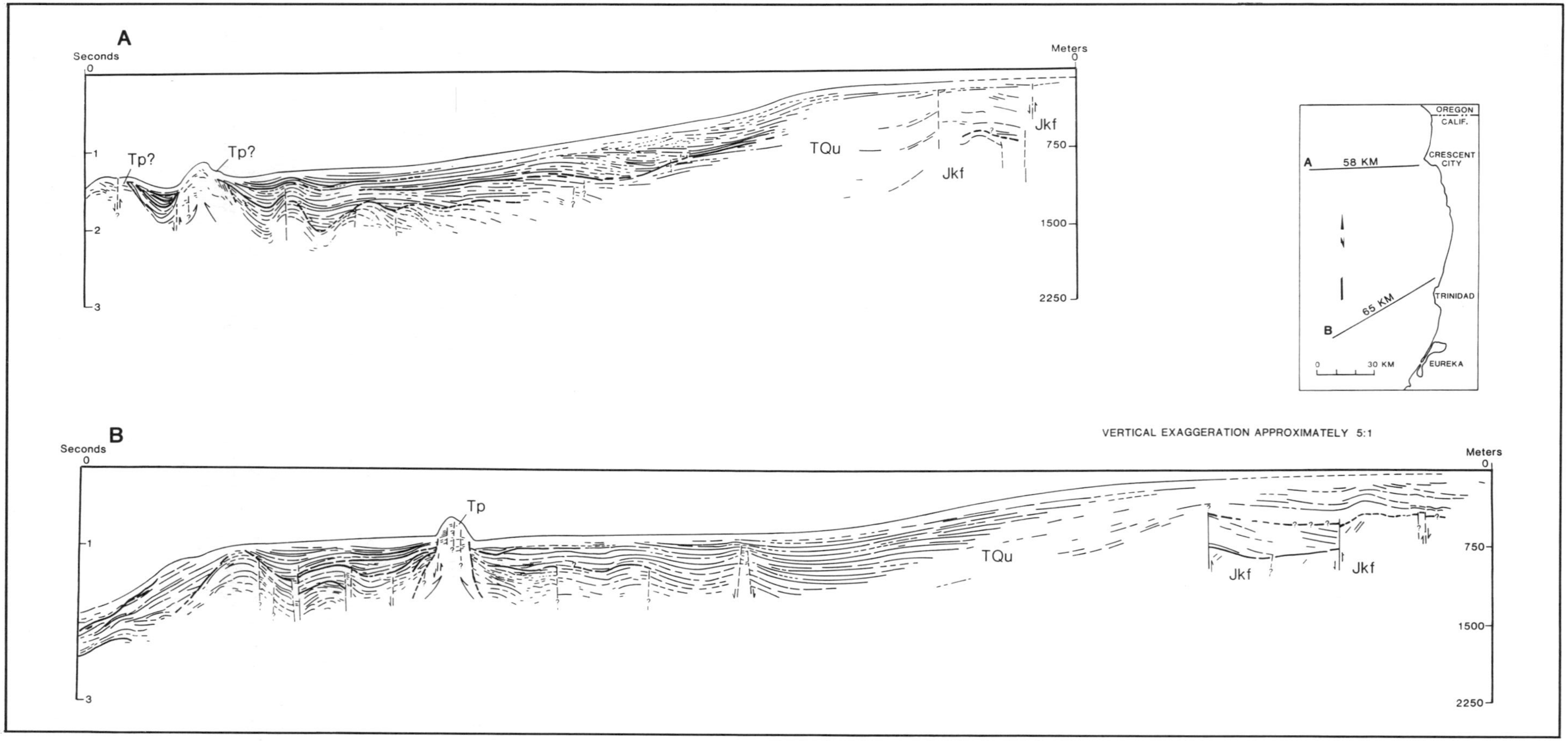

Figure 5. Line drawings of single-channel seismic-reflection profiles (USGS cruise S-9-77NC) across the Eel River Basin. JKf = Franciscan Complex, central belt; TQu = undifferentiated Tertiary and Quaternary strata, probably correlative with the Wildcat Group onshore; Tp = Pliocene, undifferentiated, ages based on sampling.

stratigraphic position unconformably above older, mildly to intensely deformed basement, lithologic information from wells drilled on this margin during the 1960s (Figure 6 and summary by Zieglar and Cassell, 1978), and a limited amount of acoustic-velocity data (summarized by Wagner, 1980) indicate that strata in the axial part of the offshore basin are broadly equivalent to the onshore late Cenozoic sedimentary section that overlies the coastal and Yager terranes of the coastal belt of the Franciscan Complex. As the lower part of the Wildcat Group records a generally eastward transgression onshore, equivalent rocks offshore are probably somewhat older than the Mohnian ages recorded for the lowermost strata of the Wildcat Group on the south flank of the onshore Eel River syncline. Undated strata offshore that unconformably overlie acoustic basement, and locally are unconformably overlain by younger Tertiary sedimentary rocks, may be equivalent to the Bear River beds (Figure 3). These strata are present on land in the subsurface along the axis of the Eel River syncline, where they are unconformably overlain by the Wildcat Group, and in isolated exposures along the coast south of the basin. Bear River deposits are accordant with overlying units in the central part of the offshore basin, but were folded and truncated along the basin margins prior to deposition of younger Tertiary rocks. The Bear River beds onshore also are transgressive (Haller, 1980,) so that the basal strata offshore may be somewhat older than correlative rocks on land.

Major structures associated with the northeast margin of the onshore Eel River Basin include northwest-trending faults of the Mad River fault zone (Carver, Stephens, and Young, 1982), the Bald Mountain fault, and the Grogan-Lost Man fault zone (Kelsey and Cashman, 1983) (Figure 2; E, Figure 4). The Mad River fault zone is 8–10 km wide north of Arcata Bay and comprises northeast-dipping high-angle reverse and thrust faults characterized by late Quaternary activity and predominantly dip-slip displacements that total >4000 m (Carver, Stephens, and Young, 1982; Coppersmith et al., 1982). The Bald Mountain fault juxtaposes Franciscan Complex central belt rocks to the southwest with accreted rocks, largely Upper Jurassic metagraywacke and metabasalt, of the Pickett Peak[1] terrane (Redwood Creek schist) to the northeast (Kelsey and Hagans, 1982; Blake, Howell, and Jones, 1982). Although the Bald Mountain fault is commonly depicted as a thrust, recent mapping (Cashman, Cashman, and Longshore, 1986) has shown that it is steeply dipping over much of its length.

Farther northeast, the Grogan fault apparently juxtaposes dissimilar late Mesozoic rocks of the Pickett Peak and Yolla Bolly terranes (Blake, Howell, and Jones, 1982). Late Tertiary right slip of at least 75 km has been proposed by Kelsey and Hagans (1982) and Cashman, Cashman, and Longshore (1986), based partly on the present position of the Pickett Peak assemblage with respect to potentially correlative rocks to the southeast. Kelsey and Cashman (1983) indicate that the Grogan and related faults have undergone vertical displacement of 300 m and possible dextral displacement during the Quaternary, and suggest that right-lateral shear associated with the transform boundary between the Pacific and North American plates may be transferred to this system of faults. This follows the concept of Herd (1978) in which a narrow, roughly triangular intracontinental plate—the "Humboldt plate"—is formed by continuing realignment of plate boundaries in the region north and east of Cape Mendocino (Figure 1). The "Humboldt plate," possibly more appropriately termed the "Humboldt block," encompasses the Eel river basin and is bounded on the west by the San Andreas fault south of Cape Mendocino and the Gorda plate north of it; on the east and north it is bounded by a system of youthful, right-stepping, en echelon right-slip faults—the Lake Mountain, Maacama, Rodgers Creek-Healdsburg and Hayward fault zones (not shown on Figure 2)—that join the San Andreas fault southeast of the San Francisco Bay. (More recently, McLaughlin et al., 1985a, have proposed that the northeast boundary of this structural block lies somewhat farther to the east and coincides with a major zone of youthful right slip comprising the Lake Mountain, Bartlett Springs, Green Valley, Concord, and Calavaras fault zones.) Right-lateral motion presumably could be transferred from the northernmost fault zone of this system—the Lake Mountain fault zone—into the offshore along any of several youthful major faults reaching the coast between Arcata Bay and the Klamath River. Probable large-scale late Tertiary dextral movement has been documented for only one of these, the Grogan fault, although the offset histories of other faults in this region are complex and incompletely known.

Faults apparently associated with the Mad River fault zone, Bald Mountain fault, and Grogan–Lost Man fault zone can be traced into the offshore, forming slivers bounded by lengthy and prominent basement offsets (Figure 2). They commonly appear on acoustic-reflection records as steeply east-dipping to vertical faults, downthrown to the west, that extend to or near the sea floor and offset Quaternary sediment or, locally, die out upward in folds. Basement offsets, as indicated by acoustic-reflection data along the offshore extensions of these faults, range from about 100–1300 m (based on an estimated acoustic velocity of 2700 m/sec in Neogene strata). Vertical offset of about 700 m across the offshore extension of the Bald Mountain fault is also indicated by depth to basement in two wells drilled about 20 km west of the mouth of the Klamath River (Figure 6, 007 and 012), both of which apparently bottomed in Franciscan central belt rocks. Components of right slip along these faults are suggested by their relatively straight traces, variations in apparent vertical offset along individual fault traces, and en echelon superimposed folds in Quaternary strata that locally are expressed in the sea floor. Right slip in this region is further suggested by a M.5.6 shallow-focus earthquake on the continental shelf about 15 km northwest of Cresent City for which Bolt, Lomnitz, and McEvilly (1968) calculated a right-lateral first-motion solution (assuming a northwest trend). These faults appear to merge along trend, but together they can be mapped northwestward to about lat. 42°N. They share the orientation and sense of motion of the San Andreas fault system and faults comprising the eastern boundary of Herd's (1978) "Humboldt plate." However, instead of extending northwestward to join the Blanco transform, these faults appear to assume a northward trend on the continental shelf near the Oregon border. Other youthful faults that would extend a zone of right shear northwestward across the Eel River Basin to the margin of the North American plate have not been identified from offshore acoustic-reflection data.

[1]Prior to their identification as separate allochthonous terranes, the Pickett Peak and Yolla Bolly assemblages were included with the central belt of the Franciscan Complex; for simplicity, they are so shown on Figure 2.

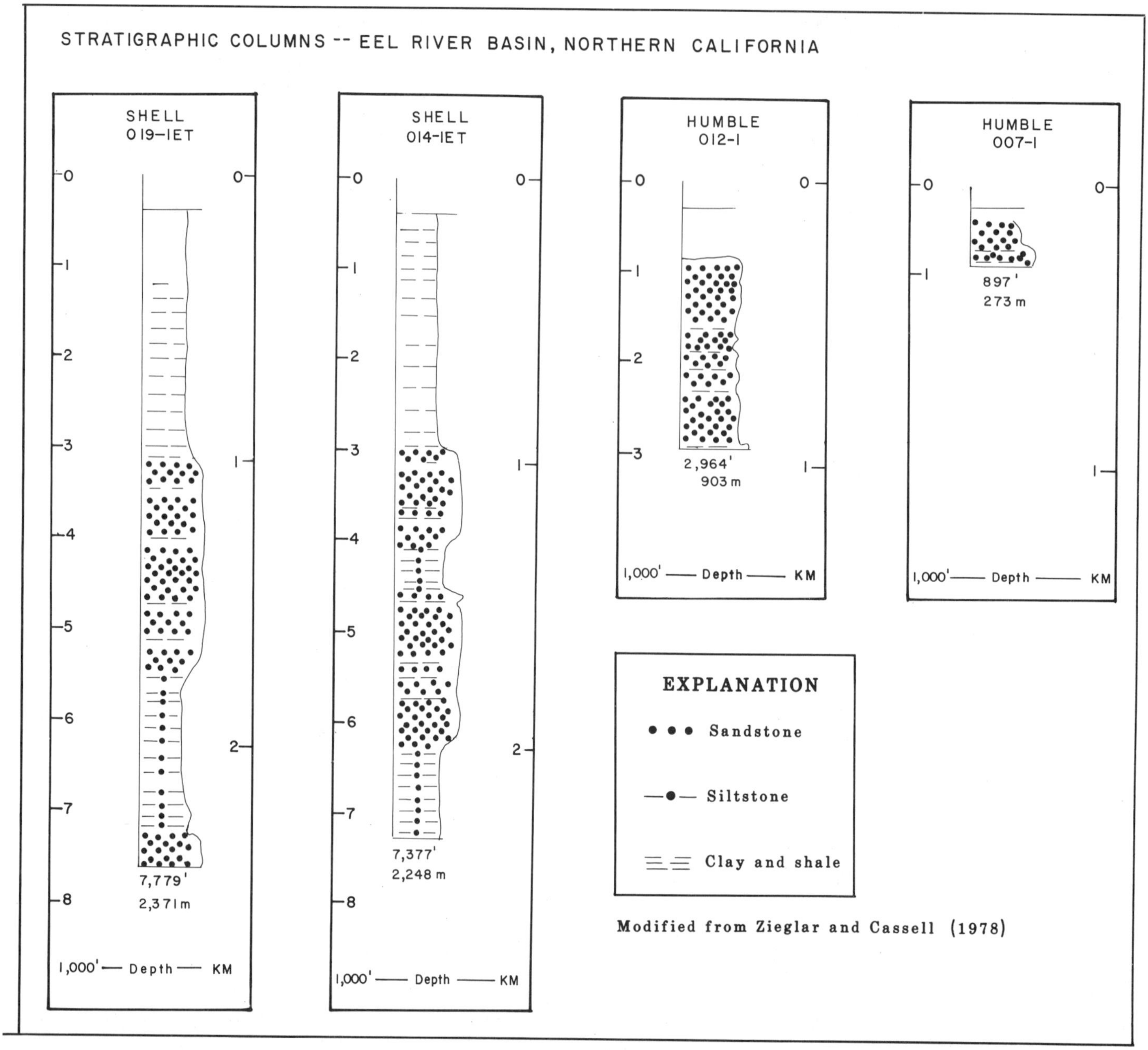

Figure 6. Generalized stratigraphic columns from exploratory test wells drilled during 1964–1965 in the offshore Eel River Basin. Well locations are shown on Figure 2.

GORDA PLATE

Studies of sea-floor structure and seismicity suggest that the Gorda plate (Figure 1) is being deformed, although the nature and timing of deformation is conjectural. Silver (1971b) concluded that the greater length of magnetic anomalies west of the Gorda Ridge with respect to correlative anomalies to the east suggests approximately 45 (±30) km of underthrusting by the Gorda plate along the east part of the Mendocino fracture zone during the past 5 m.y. Plots of hypocenters from microearthquakes in the vicinity of the Mendocino triple junction (Hutchings et al., 1981; Turcotte et al., 1980) provide evidence of 15 km of underthrusting of the Gorda plate beneath the Pacific plate east of about long. 125°W. Compression and overthrusting by the Pacific plate is further suggested by unusually high uplift rates (perhaps as much as 4 m/1000 yr) of Quaternary marine-terrace deposits in the region immediately south of Cape Mendocino (Lajoie, Sarna-Wojcicki, and Ota, 1982), elevation (as much as 1200 m) of the sea floor in the area south of the Mendocino triple junction with respect to its corresponding position in the southeastern Gorda Basin, and focal mechanisms indicating north–south compression and reverse faulting (Simila, Peppin, and McEvilly, 1975; Hutchings et al., 1981). Other workers (for example, Jachens and Griscom, 1983; Wilson, 1985) conclude that motion along the Mendocino fault is entirely transcurrent, with little or no overthrusting of the Pacific plate; this conclusion is based on (1) the absence along the Mendocino fault of features (e.g., flexure of oceanic lithosphere in the descending plate, presence of a sediment-filled trench, etc.) normally associated with subduction boundaries, (2) studies of the regional gravity field, and (3) models through time of the relative motion between the Gorda and Pacific plates. The balance of evidence, however, indicates a probability that the Pacific plate has overridden the Gorda plate by at least 15 km along the eastern end of the Mendocino fault, perhaps east of about long. 125°W, whereas farther to the west the motion between these two plates is problematic but may be mostly or wholly transcurrent. It is noteworthy that the Vizcaino block of McCulloch (this volume), probably an allochthonous block of continental affinity, extends westward along the Mendocino fault to approximately long. 126°W. Overthrusting and uplift in the Pacific plate along the eastern part of the Mendocino fault may result from compression associated with encroachment of this block as it travels relatively northward with the Pacific plate and interacts with the margin of the North American plate in the vicinity of the Mendocino triple junction (Figure 1).

The nature and timing of internal deformation of the Gorda plate is also controversial. Bending of magnetic anomalies in the southern Gorda Basin (Figure 1) may reflect an additional 50 km of north–south shortening of the Gorda plate during the past 2.5 m.y. (Silver, 1971b). Bolt, Lomnitz, and McEvilly (1968), noting the northwest trends of structures on land and focal mechanisms for earthquakes in the southern Gorda Basin that are consistent with right slip on northwest-trending faults, concluded that the San Andreas fault system was breaking northwestward across the Gorda plate. This would require the cessation of subduction of the Gorda plate. However, Silver (1971b) noted that youthful faults mapped from seismic-reflection data from the southern Gorda Basin have northeast trends, generally parallel to the magnetic anomaly pattern; these are consistent with left-lateral first-motion solutions and argue against a northwestward extension of the San Andreas fault system across the Gorda Basin.

This interpretation is supported by main and aftershock sequences and focal mechanisms for three recent earthquakes that indicate left-lateral slip along northeast-trending faults in the Gorda plate (Eaton, 1981; Smith, McPherson, and Severy, 1981). Two of these, the M_L 5.2 Ferndale earthquake (June 7, 1975) and M_S 6.8 Eureka earthquake (November 8, 1980) occurred in the Gorda plate beneath the edge of the North American continent; deformation related to these earthquakes is not reflected in the shallow structure of the overlying North American plate. The third earthquake (M_S 6.8, November 26, 1976) occurred farther offshore, beyond the seaward edge of the North American plate. It appears likely that this fracturing is occurring along ridge-parallel faults originally formed as normal faults at the Gorda Ridge and subsequently reactivated as left-lateral faults accompanying clockwise rotation and deformation of the Gorda plate. Moreover, a magnetic anomaly thought to reflect a buried mafic igneous intrusive mass that extends at least 10 km inland across the projected trace of the San Andreas fault in the vicinity of Cape Mendocino would seem to preclude a northwestward extension of the San Andreas fault, per se, into the Gorda Basin (Griscom, 1980).

Detailed analysis of seismicity in the region north of the Mendocino triple junction by Smith and Knapp (1980), Cockerham (1984), and Walter (in press) reveals (1) the existence of a well-defined band of seismicity dipping beneath the continent, at about 10° to 11° for a distance of 120 km, and at about 25° from 120 to 250 km, which is inferred to be within the upper part of the subducted Gorda plate, and (2) deformation of this dipping layer by northeast-trending strike-slip movement along near-vertical faults. Smith and Knapp (1980) concluded that the San Andreas fault terminates at the Mendocino triple junction in a fault-fault-trench triple junction, and that the Gorda plate is breaking along a complex series of northeast-trending, left-lateral faults while being emplaced beneath the North American plate, which is characterized by predominantly right-lateral faults that trend northwest. This interpretation is consistent with evidence from seismicity and from offshore and onshore structure, as mapped from acoustic-reflection and field studies.

RESOURCE POTENTIAL

The onshore part of the Eel River Basin, which constitutes approximately 10% of the total basin area, has produced an insignificant amount of oil from pre-Neogene rocks in the False Cape shear zone, and commercial quantities (85 Bcf; Hopps and Horan, 1984) of gas from predominantly structural traps in folded and faulted Pliocene rocks within about 10 km southeast of Humboldt Bay. Source rocks for these hydrocarbons have not been identified, but recent studies of organic carbon content and thermal history suggest that potential source beds occur in Cretaceous and Paleogene strata of the coastal belt of the Franciscan Complex (Underwood, 1985). Additionally, Miocene siliceous shale and siltstone lithologically similar to probable source rocks in basins farther to the south are present in the Bear River beds in largely unexplored parts of the onshore Eel River Basin. Coastal belt rocks probably have limited reservoir potential because of their diagenetic and deformational history. Reservoir-quality sandstone in the Wildcat Group within explored parts of the onshore basin tends to be fine grained, lenticular, and sporadic in distribution. Offshore, four wells (Figures 2, 6) drilled in the east-central part of the basin during the mid 1960s were dry; however, only two of these penetrated a significant thickness of Neogene strata.

Although the Eel River Basin would appear to have limited hydrocarbon potential, large areas of the onshore basin and virtually the entire offshore basin remain untested. Hopps and Horan (1984) note that, in the onshore basin, the lower part of the Neogene sedimentary section as well as major structural trends in the central and northern parts of the basin and most of the eastern part of the basin are relatively unexplored. Offshore, coastal belt rocks are widespread and strata possibly equivalent to the onshore Bear River beds are present locally; these strata have unknown potential as petroleum source rocks. If the onshore is representative of the basin as a whole, offshore gas prospects may have the highest potential, and the prospective section is probably within equivalents of the Neogene Wildcat Group.

Potential reservoir rocks may exist within coarser facies of the Wildcat Group, particularly in the thick Neogene section offshore from the Arcata-Humboldt Bay area in the southern part of the offshore basin. Thrust- and reverse-faulted anticlines were formed syndepositionally in this area, and it seems likely that potential structural and stratigraphic traps occur beneath fault planes and along the flanks of these structures. Conditions favorable for hydrocarbon entrapment also may be present where rocks equivalent to the Wildcat Group unconformably overlie and truncate folds in undated strata thought to be correlative with the Bear River beds. Additionally, there is evidence of shale flowage and diapirism, with the possibility of related structural traps, in deeper water along the western margin of the basin.

A sample taken from Quaternary sediment ponded atop a shale diapir located approximately 42 km west–southwest of Trinidad Head (water depth 500 m), and subjected to chromatographic analysis at sea, yielded traces of gasoline-range hydrocarbons and high-molecular-weight hydrocarbon gases (Kvenvolden, Field, and Clarke, 1980; McLean and Wiley, this volume). These hydrocarbons are thought to have formed within older Tertiary strata and to have migrated upward along fractures into the overlying unconsolidated sediment. In summary, much of this region is unexplored, and while it is likely that potential traps are present, the quality and distribution of source and reservoir rocks in the offshore are open questions pending the results of drilling and further stratigraphic and geochemical analysis.

SUMMARY

Two late Cenozoic marine sedimentary sequences separated by an angular unconformity overlie accreted basement terranes of the Franciscan Complex north of Cape Mendocino. The lower to middle Miocene Bear River beds consist largely of finely clastic rocks as much as 900 m thick, and are present in the subsurface of the Eel River Basin and in isolated exposures farther south. The Bear River beds are overlain by finely to coarsely clastic rocks of the Wildcat Group, as much as 3600 m thick, that comprise the principal fill of the onshore Eel River Basin. The Wildcat Group records a progressive shallowing and infilling of the basin during the late Miocene to mid-Pleistocene, succeeded by westward regression of the shoreline to its present position at the coast. Correlative units having a maximum thickness of about 3300 m are thought to be present offshore.

Structures of the onshore Eel River Basin and the southern part of the offshore basin are characterized by geologically youthful northwest-trending folds and northeast-dipping high-angle reverse and thrust faults. Farther north, the axial part of the offshore basin (characterized by north-trending open folds) is separated from the western basin margin and adjacent upper continental slope (characterized by slope-parallel structures related to subduction/accretion) by a narrow zone of north–northwest-trending, right-stepping en echelon folds that suggest youthful right-lateral shear. Along the northeast margin of the basin, north–northwest-trending faults vertically offset basement rocks as much as 1300 m (west side down), cut Quaternary sediment, and show evidence of youthful right slip. These faults appear to coincide with older terrane boundaries onshore, and to have been reactivated during the Quaternary. Offshore, they bend generally northward, parallel to the coast. Faults that would extend the eastern boundary of the "Humboldt plate" (Herd, 1978) northwestward to the margin of the North American plate were not evident in acoustic-reflection data.

Three major episodes of late Cenozoic deformation occurred in the Eel River Basin region—the first was after deposition and accretion of coastal belt rocks of the Franciscan Complex (McLaughlin et al., 1982, 1985b) and before deposition of the Bear River beds (between Eocene and middle Miocene time); the second was subsequent to deposition of the Bear River beds but prior to deposition of the Wildcat Group (during middle to late Miocene time); and the third was after deposition of the Wildcat Group (approximately in the mid-Pleistocene). Evidence of deformation during the middle to late Miocene and the mid-Pleistocene is widespread on the continental margin of Oregon and Washington (Snavely, Pearl, and Lander, 1977; Snavely, Wagner, and Lander, 1980; Clarke, Field, and Hirozawa, 1985; Kulm, Prince, and Snavely, 1973; Kulm and Fowler, 1974), and has been related to episodic changes in the rate of underthrusting of the Gorda plate (Silver, 1971a; Ingle, 1976). Mid-Pleistocene deformation in the Cape Mendocino-southern Eel River Basin area is characterized by rapid uplift of the basin margins and initiation of major folding and high-angle reverse and thrust faulting along west-to-northwest trends. This northeast–southwest compression probably is related to north–northwestward propagation of the San Andreas fault and encroachment of the Pacific plate, carrying with it the Vizcaino block (McCulloch, this volume) and the adjacent King Range terrane.

ACKNOWLEDGMENTS

I thank R. J. McLaughlin and P. D. Snavely, Jr., of the U.S. Geological Survey, and G. A. Carver of Humboldt State University for their helpful and critical reviews of the manuscript. In addition, I am grateful to McLaughlin, Carver, and D. L. McCulloch of the Geological Survey for generously sharing their ideas regarding the geologic evolution of the Eel River Basin and the adjacent region and for their ongoing constructive comments.

REFERENCES CITED

Alvarez, W., D. V. Kent, I. Premoli-Silva, R. A. Schweikert, and R. A. Larson, 1980, Franciscan Complex limestone deposited at 17° south paleolatitude: GSA Bulletin, v. 91, p. 476–484.

Bachman, S. B., 1982, The coastal belt of the Franciscan: Youngest phase of northern California subduction, *in* J. K. Leggett, ed., Trench-forearc geology: Sedimentation and tectonics on modern and ancient active plate margins: Geological Society of London Special Publication 10, p. 401–417.

Bachman, S. B., M. B. Underwood, and J. S. Menack, 1984, Cenozoic evolution of coastal northern California *in* J. K. Crouch and S. B. Bachman, eds., Tectonics and sedimentation along the California margin: SEPM, Pacific Section, v. 38, p. 55–56.

Back, W., 1957, Geology and ground-water features of the Smith River plain, Del Norte County, California: U.S. Geological Survey Water-Supply Paper 1254, 76 p.

Bailey, E. H., W. P. Irwin, and D. L. Jones, 1964, Franciscan and related rocks, and their significance in the geology of western California: California Division of Mines and Geology Bulletin 183, 177 p.

Blake, M. C., and D. L. Jones, 1981, The Franciscan assemblage in northern California: A reinterpretation, *in* W. G. Ernst, ed., The geotectonic evolution of California, Rubey Volume 1: Englewood Cliffs, N. J., Prentice-Hall, Inc., p. 306–328.

Blake, M. C., D. G. Howell, and D. L. Jones, 1982, Preliminary tectonostratigraphic terrane map of California: U.S. Geological Survey Open-File Report 82-593, 8 p., 3 map sheets.

Blake, M. C., A. S. Jayko, and R. J. McLaughlin, 1985, Tectonostratigraphic terranes of the northern Coast Ranges, California, *in* D. G. Howell, ed., Tectonostratigraphic terranes of the Circum-Pacific region: Circum-Pacific Council for Energy and Mineral Resources, Earth Science Series 1, p. 159–171.

Bolt, B. A., C. Lomnitz, and T. V. McEvilly, 1968, Seismological evidence on the tectonics of central and northern California and the Mendocino Escarpment: Seismological Society of America Bulletin, v. 58, no. 6, p. 1725–1767.

Bourgeois, J., and E. L. Leithold, 1983, Sedimentation, tectonics and sea-level change as reflected in four wave-dominated shelf sequences in Oregon and California, *in* D. K. Larue and R. J. Steel, eds., Cenozoic marine sedimentation, Pacific margin, U.S.A.: SEPM, Pacific Section, p. 1–16.

Burkle, L. H., J. R. Dodd, and R. J. Stanton, Jr., 1980, Diatom biostratigraphy and its relationship to paleomagnetic stratigraphy and molluscan distribution in the Neogene Centerville Beach section, California: Journal of Paleontology, v. 54, p. 664–674.

Carver, G. A., 1985, Quaternary tectonics north of the Mendocino triple junction—The Mad River fault zone, *in* H. M. Kelsey, T. E. Lisle, and M. E. Savina, conveners, Redwood Country: Guidebook, American Geomorphological Field Group 1985 Field Trip, p. 155–167.

Carver, G. A., R. M. Burke, and H. M. Kelsey, 1986, Deformation of late Pleistocene marine terraces along the California coast north of Cape Mendocino (abs.): GSA Abstracts with Programs, v. 18, no. 2, p. 93.

Carver, G. A., T. A. Stephens, and J. C. Young, 1982, Quaternary reverse and thrust faults, Mad River fault zone, *in* D. R. Harden, D. C. Marron, and A. MacDonald, conveners, Late Cenozoic history and forest geomorphology of Humboldt County, California: Guidebook, Friends of the Pleistocene 1982 Pacific Cell Field Trip, p. 93–98.

Cashman, P. H., S. M. Cashman, and J. D. Longshore, 1986, Deformational history and regional tectonic significance of the Redwood Creek schist, northwestern California: GSA Bulletin, v. 97, no. 1, p. 35–47.

Clarke, S. H., M. E. Field, and C. A. Hirozawa, 1985, Reconnaissance geology and geologic hazards of the offshore Coos Bay Basin, Oregon: U.S. Geological Survey Bulletin 1645, 41 p.

Cockerham, R. S., 1984, Evidence for a 180-km-long subducted slab beneath northern California: Seismological Society of America Bulletin, v. 74, no. 2, p. 569–576.

Coppersmith, K. J., T. A. Stephens, F. H. Swan, III, N. E. Denning, and K. A. Malek, 1982, Near-surface behavior of thrust faults in the Humboldt Bay area, California, *in* D. R. Harden, D. C. Marron, and A. MacDonald, conveners, Late Cenozoic history and forest geomorphology of Humboldt County, California: Guidebook, Friends of the Pleistocene 1982 Pacific Cell Field Trip, p. 63–92.

Dickinson, W. R., R. V. Ingersoll, and S. A. Graham, 1979, Paleogene sediment dispersal and paleotectonics in northern California: GSA Bulletin, Part II, v. 90, no. 10, p. 1458–1528.

Dodd, J. R., J. Mead, and R. J. Stanton, Jr., 1977, Paleomagnetic stratigraphy of Pliocene Centerville Beach section, northern California: Earth and Planetary Science Letters, v. 34, p. 381–386.

Eaton, J. P., 1981, Distribution of aftershocks of the November 8, 1980, Eureka earthquakes (abs.): Seismological Society of America Earthquake Notes, v. 52, no. 1, p. 44–45.

Evitt, W. R., and S. T. Pierce, 1975, Tertiary ages from the coastal belt of the Franciscan Complex, northern California: Geology, v. 3, no. 8, p. 433–436.

Faustman, W. F., 1964, Paleontology of the Wildcat Group at Scotia and Centerville Beach, California: Berkeley, University of California Publications in Geological Sciences, v. 41, no. 2, p. 97–160.

Field, M. E., S. H. Clarke, Jr., and M. E. White, 1980, Geology and geological hazards of offshore Eel River Basin, northern California continental margin: U.S. Geological Survey Open-File Report 80-1080, 80 p.

Griscom, A., 1980, Aeromagnetic interpretation of the Mendocino triple junction, *in* R. Streitz and R. Sherburne, eds., Studies of the San Andreas fault zone in northern California: California Division of Mines and Geology Special Report 140, p. 121–127.

Haller, C. R., 1980, Pliocene biostratigraphy of California: AAPG Studies in Geology 11, p. 183–341.

Harbert, W. P., R. J. McLaughlin, and W. V. Sliter, 1984, Paleomagnetic and tectonic interpretation of the Parkhurst Ridge limestone, coastal belt Franciscan, northern California, *in* M. C. Blake, Jr., ed., Franciscan geology of northern California: SEPM, Pacific Section, v. 43, p. 175–183.

Herd, D. G., 1978, Intracontinental plate boundary east of Cape Mendocino, California: Geology, v. 6, no. 12, p. 721–725.

Hopps, T., and E. Horan, 1984, Subsurface stratigraphy: Eel River Basin, *in* R. L. Hester and D. E. Hallinger, eds., Selected papers presented to the Pacific Section: Sacramento, California, AAPG 1983 Annual Meeting, v. 1, p. 1–29.

Hoskins, E. G., and J. R. Griffiths, 1971, Hydrocarbon potential of northern and central California offshore: AAPG Memoir 15, v. 1, p. 212–228.

Hutchings, L., T. Turcotte, M. Schnapp, and R. McPherson, 1981, Seismicity of the Mendocino triple junction (abs.): Seismological Society of America Earthquake Notes, v. 52, no. 1, p. 42–43.

Ingle, J. C., Jr., 1976, Late Neogene paleobathymetry and paleoenvironment of Humboldt Basin, northern California, *in* A. E. Fritsche, H. Ter Best, Jr., and W. W. Wornardt, eds., The Neogene Symposium: SEPM, Pacific Section, Annual Meeting, p. 53–61.

Irwin, W. P., 1960, Geologic reconnaissance of the northern Coast Ranges and Klamath Mountains, California, with a summary of mineral resources: California Division of Mines Bulletin 179, 80 p.

Jachens, R. C., and A. Griscom, 1983, Three-dimensional geometry of the Gorda plate beneath northern California: Journal of Geophysical Research, v. 88, no. B11, p. 9375–9392.

Jones, D. L., M. C. Blake, Jr., E. H. Bailey, and R. J. McLaughlin, 1978, Distribution and character of upper Mesozoic subduction complexes along the west coast of North America: Tectonophysics, v. 47, p. 207–222.

Keller, G., and J. C. Ingle, Jr., 1981, Planktonic foraminiferal biostratigraphy, paleoceanographic implications, and deep-sea correlation of the Plio-Pleistocene Centerville Beach section, northern California, *in* J. M. Armentrout, ed., Pacific northwest Cenozoic biostratigraphy: GSA Special Paper 184, p. 127–135.

Kelsey, H. M., 1982, Structure and tectonics of Plio-Pleistocene sediments exposed within and adjacent to Prairie Creek State Park, Humboldt County, north coastal California, *in* D. R. Harden, D. C. Marron, and A. MacDonald, conveners, Late Cenozoic history and forest geomorphology of Humboldt County, California: Guidebook, Friends of the Pleistocene 1982 Pacific Cell Field Trip, p. 130–150.

Kelsey, H. M., and S. M. Cashman, 1983, Wrench faulting in northern California and its tectonic implications: Tectonics, v. 2, no. 6, p. 565–576.

Kelsey, H. M., and D. K. Hagans, 1982, Major right-lateral faulting in the Franciscan assemblage of northern California in late Tertiary time: Geology, v. 10, no. 7, p. 387–391.

Kleinpell, R. M., 1938, Miocene stratigraphy of California: Tulsa, Oklahoma, AAPG, 450 p.

Kodama, K. P., 1979, New paleomagnetic results from the Rio Dell Formation, California: American Geophysical Union, Geophysical Research Letters, v. 6, no. 4, p. 253–256.

Kulm, L. D., and G. A. Fowler, 1974, Oregon continental margin structure and stratigraphy: A test of the imbricate thrust model, *in* C. A. Burk and C. L. Drake, eds., The geology of continental margins: New York, Springer-Verlag, p. 261–283.

Kulm, L. D., R. A. Prince, and P. D. Snavely, Jr., 1973, Site survey of the northern Oregon continental margin and Astoria fan, *in* L. D. Kulm and R. von Huene, and others, Initial reports of the Deep Sea Drilling Project, Vol. 18: Washington D.C., U.S. Government Printing Office, p. 979–987.

Kvenvolden, K. A., M. E. Field, and S. H. Clarke, 1980, Thermogenic hydrocarbon gases in unconsolidated seafloor deposits, northern California continental margin (abs.): AAPG Bulletin, v. 64, p. 736.

Lajoie, K. R., A. M. Sarna-Wojcicki, and Y. Ota, 1982, Emergent Holocene marine terraces at Ventura and Cape Mendocino, California—Indicators of high tectonic uplift rates (abs.): GSA Abstracts with Programs, v. 14, p. 178.

Manning G. A., and B. A. Ogle, 1950, Geology of the Blue Lake quadrangle, California: California Division of Mines Bulletin 148, 36 p.

McCulloch, D. S., 1987, Regional geology and hydrocarbon potential of offshore central California, this volume.

McLaughlin, R. J., S. A. Kling, R. Z. Poore, K. McDougall, and E. C. Beutner, 1982, Post-middle Miocene accretion of Franciscan rocks, northwestern California: GSA Bulletin, v. 93, no. 7, p. 595–605.

McLaughlin, R. J., N. H. Ohlin, D. J. Thormahlen, D. L. Jones, J. W. Miller, and C. D. Blome, 1985a, Geologic map and structure sections of the Little Indian Valley-Wilbur Springs geothermal area, northern Coast Ranges, California: U.S. Geological Survey Open-File Report 85-285, 24 p., 1 map sheet, scale 1:24,000.

McLaughlin, R. J., S. D. Ellen, W. V. Sliter, J. A. Tarduno, W. P. Harbert, A. S. Jayko, and M. C. Blake, Jr., 1985b, Interactions between tectonostratigraphic terranes of northwestern California adjacent to Mendocino triple junction (abs.): Cordilleran Section, GSA Abstracts with Programs, v. 17, no. 6, p. 369.

McLean, H., and T. J. Wiley, 1987, Geologic potential for hydrocarbons in unexplored offshore basins of western North America, this volume.

Menack, J. S., 1984, Non-marine to continental shelf sedimentation in the Garberville-Briceland forearc basins, northern Coast Ranges, California (abs.): Abstracts, SEPM 1st Annual Midyear Meeting, San Jose, California, August 10–13, 1984, p. 54–55.

Menack, J. S., and S. B. Bachman, 1983, Neogene fore-arc basin development in northern California: Eel River Basin (abs.): AAPG Bulletin, v. 67, p. 512–513.

Menack, J. S., B. Roth, and J. Barron, 1985, Tertiary terrane tectonics of northwestern California constrained by timing of Neogene shelf deposits (abs.): EOS (American Geophysical Union Transactions), v. 66, no. 46, p. 1090.

Ogle, B. A., 1953, Geology of Eel River Valley area, Humboldt County, California: California Division of Mines Bulletin 164, 128 p.

Piper, D. J. W., W. R. Normark, and J. C. Ingle, Jr., 1976, The Rio Dell Formation: A Plio-Pleistocene basin slope deposit in northern California: Sedimentology, v. 23, p. 309–328.

Rau, W. W., and G. R. Grocock, 1974, Piercement structure outcrops along the Washington coast: Washington Division of Geology and Earth Resources Information Circular 51, 7 p.

Sarna-Wojcicki, A. M., J. R. Wagner, M. E. Perkins, K. R. Lajoie, and S. D. Morrison, 1982, Introduction to the tectonic and stratigraphic setting of the Humboldt Basin, Humboldt County, northwestern California, *in* D. R. Harden, D. C. Marron, and A. McDonald, conveners, Late Cenozoic history and forest geomorphology of Humboldt County, California: Guidebook, Friends of the Pleistocene 1982 Pacific Cell Field Trip, p. 46–62.

Silver, E. A., 1971a, Transitional tectonics and late Cenozoic structure of the continental margin off northernmost California: GSA Bulletin, v. 82, no. 1, p. 1–22.

Silver, E. A., 1971b, Tectonics of the Mendocino triple junction: GSA Bulletin, v. 82, no. 11, p. 2965–2978.

Simila, G. W., 1980, Seismological evidence on the tectonics of the northern section of the San Andreas fault region, *in* R. Streitz and R. Sherburne, eds., Studies of the San Andreas fault in northern California: California Division of Mines and Geology Special Report 140, p. 131–137.

Simila, G. W., W. A. Peppin, and T. V. McEvilly, 1975, Seismotectonics of the Cape Mendocino, California area: GSA Bulletin, v. 86, no. 10, p. 1399–1406.

Sliter, W. V., 1984, Foraminifers from Cretaceous limestones of the Franciscan Complex, northern California, *in* M. C. Blake, Jr., ed., Franciscan geology of northern California: SEPM, Pacific Section, v. 43, p. 149–162.

Sliter, W. V., R. J. McLaughlin, G. Keller, and W. R. Evitt, in press, Paleogene accretion of Late Cretaceous oceanic limestone in northern California: Geology.

Smith, S. W., and J. S. Knapp, 1980, The northern termination of the San Andreas fault, *in* R. Streitz and R. Sherburne, eds., Studies of the San Andreas fault zone in northern California: California Division of Mines and Geology Special Report 140, p. 153–164.

Smith, S. W., R. C. McPherson, and N. I. Severy, 1981, The Eureka earthquake of 1980, breakup of the Gorda plate (abs.): Seismological Society of America Earthquake Notes, v. 52, no. 1, p. 44.

Snavely, P. D., Jr., 1987, Tertiary geologic framework, neotectonics, and petroleum potential of the Oregon-Washington continental margin, this volume.

Snavely, P. D., Jr., and H. C. Wagner, 1982, Geologic cross section across the continental margin of southwestern Washington: U.S. Geological Survey Open-File Report 82-459, 10 p.

Snavely, P. D., Jr., J. E. Pearl, and D. L. Lander, 1977, Interim report on petroleum resources potential and geologic hazards in the outer continental shelf—Oregon and Washington Tertiary province: U.S. Geological Survey Open-File Report 77-282, 64 p.

Snavely, P. D., Jr., H. C. Wagner, and D. L. Lander, 1980, Geological cross section of the central Oregon continental margin: GSA Map and Chart Series, MC-28J.

Strand, R. G., 1962, Geologic map of California, Redding sheet: California Division of Mines and Geology, scale 1:250,000.

Tarduno, J. A., M. McWilliams, W. V. Sliter, H. E. Cook, M. C. Blake, Jr., and I. Premoli-Silva, 1986, Southern hemisphere origin of the Cretaceous Laytonville Limestone of California: Science, v. 231, p. 1425–1428.

Turcotte, T., L. Hutchings, R. Simon, and P. Somerville, 1980, Summary of seismicity investigations, Appendix D, *in* Evaluation of the potential for resolving the geologic and seismic issues at the Humboldt Bay Power Plant Unit No. 3, prepared for the Pacific Gas and Electric Company: San Francisco, Woodward-Clyde Consultants, p. D1–D145.

Underwood, M. B., 1982, The Garberville thrust—A contact of probable Miocene age within the Franciscan Complex, northern California (abs.): GSA Abstracts with Programs, v. 14, no. 7, p. 635–636.

Underwood, M. B., 1983, Depositional setting of the Paleogene Yager Formation, northern Coast Ranges of California, *in* D. K. Larue and R. J. Steel, eds., Cenozoic marine sedimentation, Pacific margin, U.S.A.: SEPM, Pacific Section, p. 81–101.

Underwood, M. B., 1984, Franciscan and related rocks of southern Humboldt County, northern California Coast Ranges: Analysis of structure, tectonics, sedimentary petrology, paleogeography, depositional history, and thermal maturity: Ithaca, N.Y., Cornell University unpublished Ph.D. thesis, 354 p.

Underwood, M. B., 1985, Sedimentology and hydrocarbon potential of Yager structural complex—Possible Paleogene source rocks in Eel River Basin, northern California: AAPG Bulletin, v. 69, p. 1088–1100.

Wagner, D. L., and G. J. Saucedo, in press, Geologic map of California, Weed sheet: California Division of Mines and Geology, scale 1:250,000.

Wagner, J. R., 1980, Summary of regional stratigraphy and geologic structure, Appendix A, *in* Evaluation of the potential for resolving the geologic and seismic issues at the Humboldt Bay Power Plant Unit No. 3, prepared for the Pacific Gas and Electric Company: San Francisco, Woodward-Clyde Consultants, p. A1–A73.

Walter, S. A., in press, Intermediate-focus earthquakes associated with Gorda plate subduction in northern California: Seismological Society of America Bulletin.

Wilson, D. S., 1985, Tectonic history of the Juan de Fuca Ridge: Stanford, California, Stanford University unpublished Ph.D. thesis, 105 p.

Zieglar, D. L., and J. K. Cassell, 1978, A synthesis of OCS well information, offshore central and northern California, Oregon and Washington: Preprint of paper presented at 53rd Annual Meeting, AAPG Pacific Section, Sacramento, California, April 28–29, 1978, 27 p.

CHAPTER SIXTEEN

REGIONAL GEOLOGY AND HYDROCARBON POTENTIAL OF OFFSHORE CENTRAL CALIFORNIA

David S. McCulloch

U.S. Geological Survey (USGS)
Menlo Park, California

ABSTRACT

Geology of the offshore central California margin records Mesozoic and Cenozoic events attributable to continental-oceanic plate interactions. Basement blocks carried several thousands of kilometers northward on the Kula and/or Farallon oceanic plates were accreted to North America by late Eocene. Following Pacific-North American plate contact (~30 Ma), subduction changed to transform motion between the northward-migrating Mendocino and southward-migrating Rivera triple junctions. Oblique Pacific-North America plate convergence continued across the area of transform motion and produced a now inactive accretionary wedge on the Pacific plate beneath undeformed continental slope strata. The coastal margin was subaerial during Oligocene global low sea level, and erosion stripped Upper Cretaceous and Eocene strata from the elevated basement blocks. Collapse of the margin followed erosion. Basement blocks were rotated downward along large-displacement near-coastal faults, and shallow-water marine late late Oligocene–early Miocene volcanics were covered by hemipelagic sedimentation. Collapse coincided with tectonism and significant changes in water depth around the Pacific margin, and may have resulted from a change in Pacific plate motion that produced transtension along the California margin. Transtension may have initiated rupture and volcanism along the

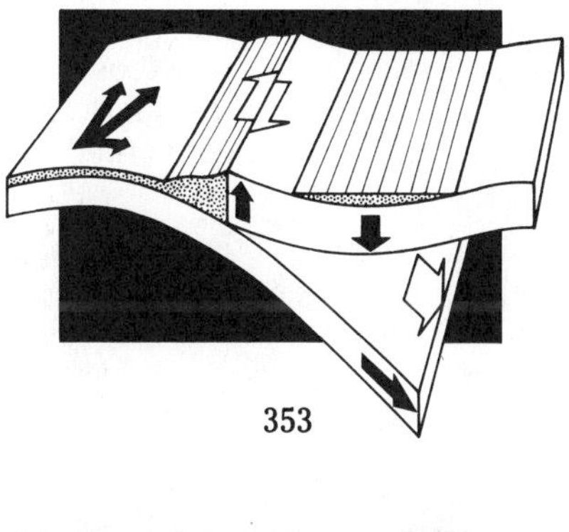

San Andreas fault system (~23 Ma) and wrench-tectonic deformation on the shelf. In the late Miocene, growing structural highs on the seaward shelf edge formed elongate shelf basins. In the late Miocene and early Pliocene, transtension was replaced by transpression with strong shoreward compression. Compression probably resulted from clockwise rotation of the Pacific plate (~5–3 Ma), and was accompanied by a marked increase in terrigenous debris deposited on the shelf and slope, indicating elevation of the adjacent coastline.

INTRODUCTION

The first descriptions of the structural evolution of the offshore central California active transform margin in the context of plate tectonics were made by Curray (1966), McKenzie and Morgan (1969), Hamilton (1969), Silver (1969, 1971), Silver, Curray, and Cooper (1971), and Atwater (1970). Industry exploration of this area preceded a 1963 Federal Outer Continental Shelf Leasing Program and the subsequent drilling of 14 offshore exploratory wells, after which Hoskins and Griffiths (1971) delineated the boundaries, stratigraphy, sediment thickness and major structures within several shallow late Tertiary shelf basins (Figure 1). More recent descriptions of the geologic framework and associated geologic hazards were prepared for subsequent OCS lease sales in these areas (e.g. McCulloch et al., 1977, 1980; Howell, McCulloch, and Vedder, 1978; Richmond et al., 1981). Four segments of the central California margin have been depicted on continent–ocean transects (Page et al., 1979a, b; Ross and McCulloch, 1979; Irwin and Dennis, 1979; Saleeby et al., 1986).

Recent drilling in the northwestern end of Santa Barbara Channel preceding a 1981 Outer Continental Lease sale indicated the presence of attractive petroleum targets in the offshore Santa Maria Basin. In the subsequent sale, industry bid a lease bonus total of 2.28 billion dollars for tracts in offshore Santa Maria Basin, the second highest such bid in federal lease sale history. Post-sale exploratory drilling has discovered several fields, the giant Point Arguello field being the largest (Figure 2) (Epstein and Nary, 1982; Ogle, 1984; Crain, Mero, and Patterson, 1985). Estimates of total recoverable hydrocarbons of one billion barrels of oil and a trillion cubic feet of gas (Ogle, 1984) in the southern part of offshore Santa Maria Basin were made with data available in 1984. More recent exploratory drilling in offshore Santa Maria Basin has made additional discoveries. Although industry exploration of the central California margin has continued intermittently since the late 1950s, the data remain proprietary, and the recent cancellation of proposed lease sales has generally halted public discussion of the geology of the central California shelf basins.

Geologic maps accompanying this chapter are primarily based on U.S. Geological Survey high-resolution and intermediate penetration single-channel seismic reflection surveys. These data are augmented by Scripps Institute of Oceanography and the U.S. Minerals Management Service single-channel surveys (Figure 3). McCulloch et al. (1980) give detailed track charts and system descriptions. A few U.S. Geological Survey multichannel reflection profiles cross the continental shelf and slope and are most useful in the deepwater areas, whereas the shelf and upper slope geology is largely interpreted from single-channel data.

Ages assigned to offshore strata are either based on a very limited number of dated sea-floor rock samples (McCulloch, Utter, and Menack, 1985) or are inferred by tying the geometry of reflector packets and unconformities on seismic reflection records to the geologic cross sections of Hoskins and Griffiths (1971) derived from seismic profiles and exploratory drilling. Logs of the exploratory wells are publicly available but contain no paleontologic data, making it difficult to evaluate ages assigned on the cross sections.

REGIONAL GEOLOGIC FRAMEWORK

Early views of California geology envisioned major structural blocks separated by faults that were thought to have undergone largely vertical displacement, and differences in the structure and stratigraphy of these blocks were explained accordingly (e.g. Clark, 1930). Recognition of continental drift and associated plate motion, the nearly worldwide demonstration of the present juxtaposition of mutually exotic tectonostratigraphic terranes, geographically displaced faunas, and rocks with paleomagnetic inclinations that show large-scale latitudinal translation, has led to a fundamental reinterpretation of California geology (Coney, Jones, and Monger, 1980; Blake, Howell, and Jones, 1982; Vedder, Howell, and McLean, 1983; Howell, Jones, and Schermer, 1985). Properties of these basement terranes, such as their thermal histories, the possible presence of northward-transported pre-Neogene strata, and the relation between the response of basement blocks to deformation and the style of deformation of prospective Neogene strata, may play significant roles in determining their hydrocarbon potential.

Basement rocks of onshore coastal central California (Figure 4) are now recognized as comprising tectonostratigraphic terranes that may have originated thousands of kilometers to the south. Most northward transport of the basement terranes—suturing to adjacent blocks and accretion to what was to become coastal California—were largely pre-Neogene events that probably resulted from northwest transport on the Kula and/or Farallon plates. These plates converged obliquely with the North American plate through

Figure 1. Map showing the generalized boundaries of late Tertiary shelf and slope basins and locations of offshore exploratory wells on the central California continental margin.

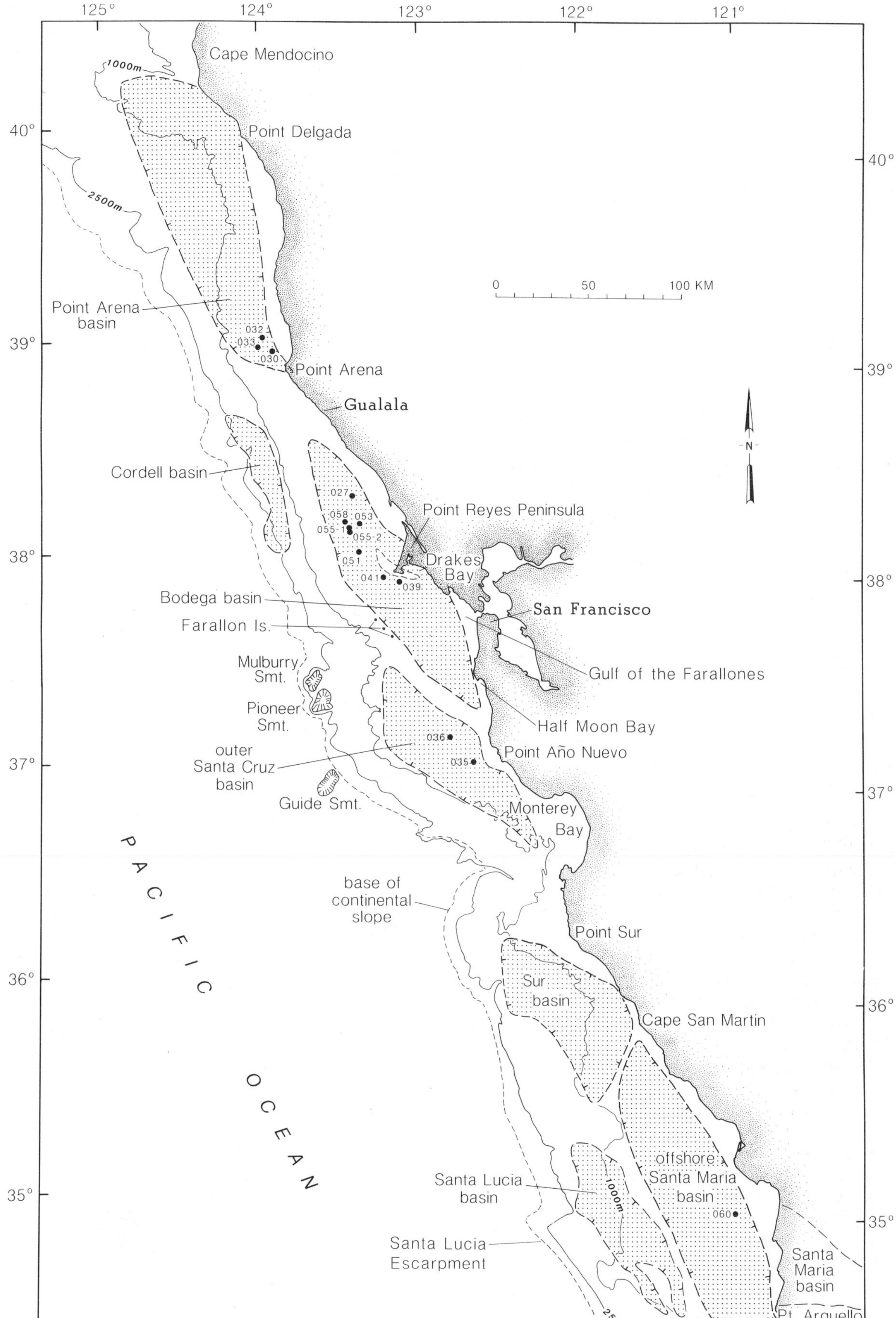
125°
124°
123°
122°
121°
40°
39°
38°
37°
36°
35°
Cape Mendocino
1000m
Point Delgada
2500m
0 50 100 KM
Point Arena basin
032
033
030
Point Arena
Gualala
N
Cordell basin
027
058
053
055-1
055-2
051
041
039
Point Reyes Peninsula
Drakes Bay
San Francisco
Bodega basin
Farallon Is.
Mulburry Smt.
Pioneer Smt.
Gulf of the Farallones
Half Moon Bay
036
035
Point Año Nuevo
outer Santa Cruz basin
Guide Smt.
Monterey Bay
PACIFIC
base of continental slope
Point Sur
Sur basin
Cape San Martin
OCEAN
offshore Santa Maria basin
060
Santa Lucia basin
1000m
Santa Lucia Escarpment
Santa Maria basin
2500m
78-164-1
Pt. Arguello
124°
123°
122°
121°

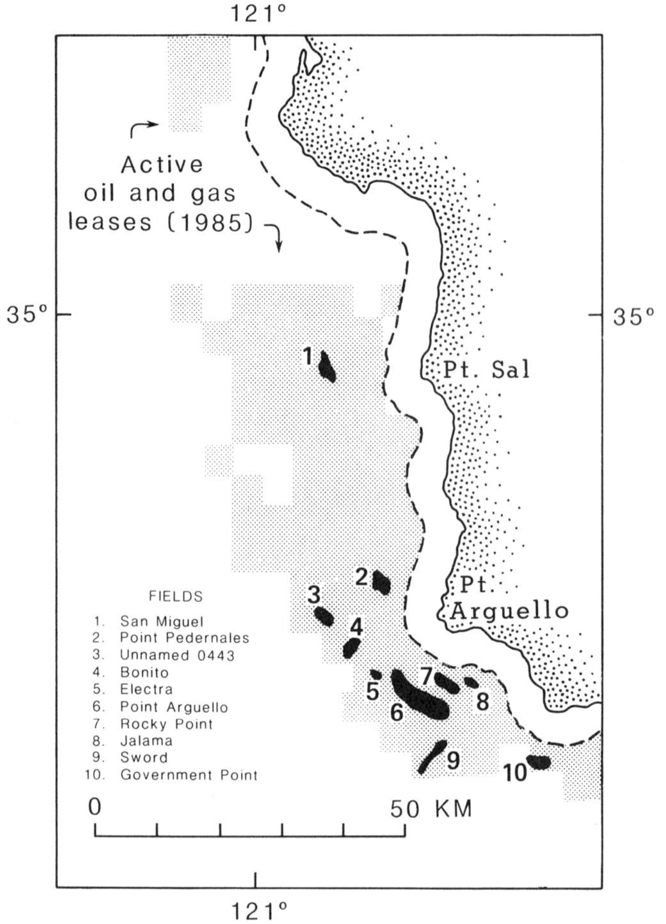

Figure 2. Oil and gas fields in western Santa Barbara Channel and southern offshore Santa Maria Basin, and areas of active offshore leases. Modified from Webster et al. (1985).

Late Cretaceous and into early Eocene time (Vedder, Howell, and McLean, 1983; Page and Engebretson, 1984). Basement terranes recognized onshore west of the San Andreas fault in coastal central California are the Salinia and Sur–Obispo composite terranes, which together constitute the larger Santa Lucia–Orocopia allochthon (Figure 4) (Vedder, Howell, and McLean, 1983). Stratigraphic, paleontologic, paleomagnetic and isotopic geochemical data (Vedder, Howell, and McLean, 1983; Kistler and Peterman, 1973; Champion, Howell, and Grommé, 1984; McWilliams and Howell, 1982) indicate that the Salinia and Sur Obispo composite terranes, both of which contain pre-lower Eocene rocks, were probably sutured to form the Santa Lucia-Orocopia allochthon in Late Cretaceous time. The allochthon was carried northward at least 1500 km between Late Cretaceous and early Eocene times, and was accreted to the North American plate by latest Eocene (Figure 4a) (Howell, Champion, and Vedder, 1987; Champion and Howell, 1986). Subsequent contact approximately 30 to 28 Ma (Atwater, 1970; Engebretson, 1982) between the Pacific and North American plates formed a triple junction, first along the Pioneer Fracture

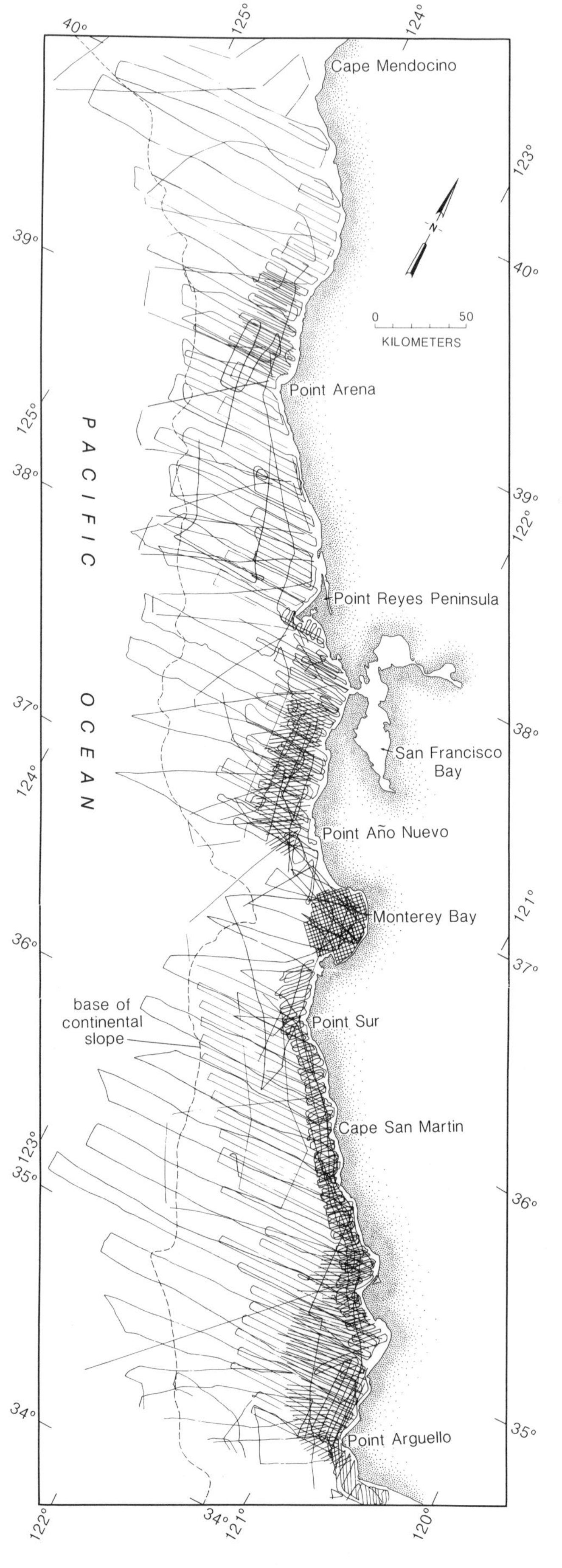

Figure 3. Map of seismic reflection track lines. In three areas of dense coverage (southern offshore Santa Maria, southern Bodega and southern Point Arena basins) only half the track lines are shown.

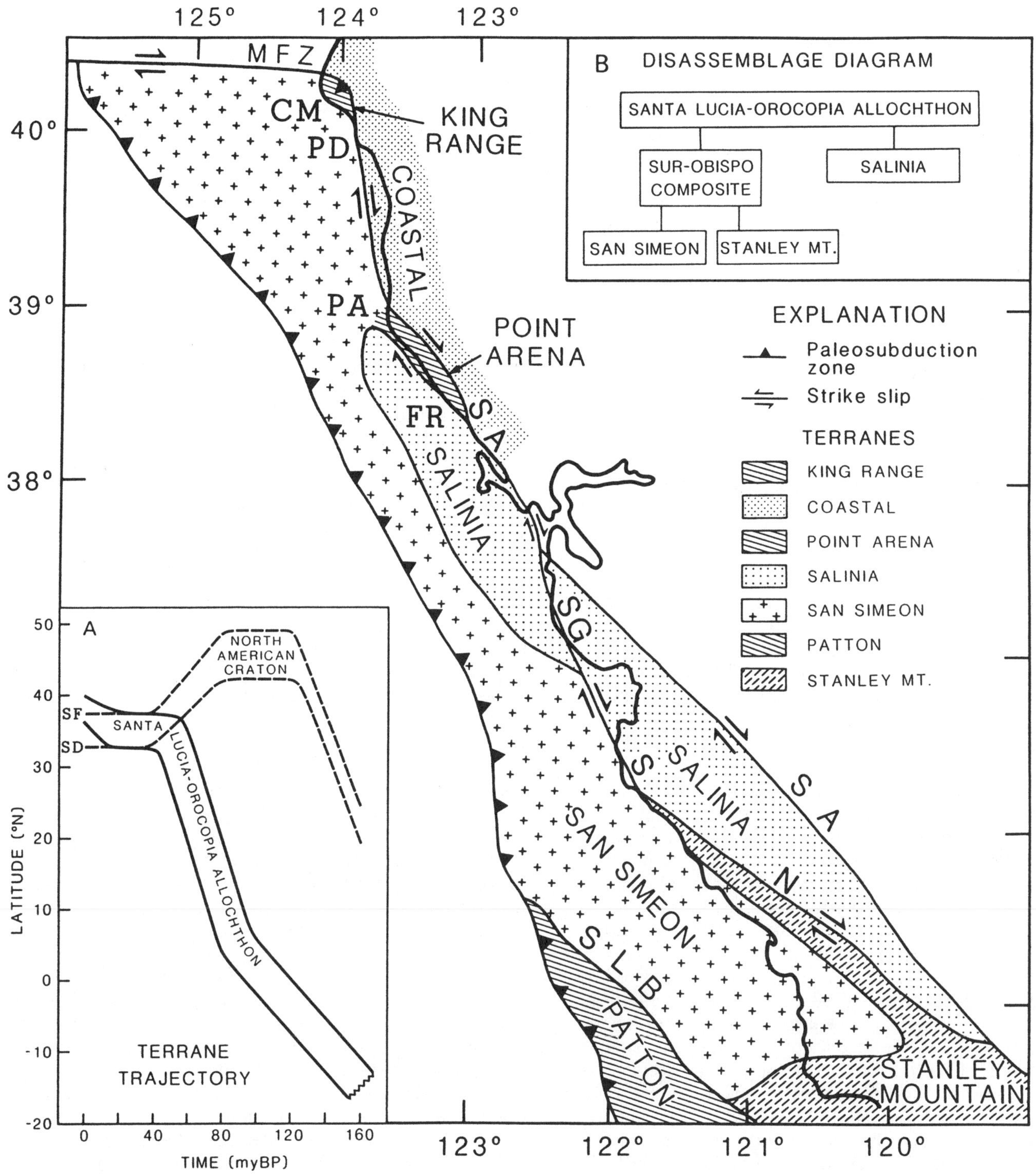

Figure 4. Generalized terrane boundaries of west central California (modified after Howell et al., 1986, and Silberling et al., 1984) and major related faults. A. Latitude trajectory history for the Santa Lucia-Orocopia allochthon and North American craton (modified after Howell, Champion and Vedder, 1987). Dashed lines indicate cratonal positions of San Francisco and San Diego through time. B. Diagram of terrane derivation as proposed by Vedder, Howell, and McLean (1983). Faults—MFZ, Mendocino Fracture Zone; SA, San Andreas; SG, San Gregorio; S, Sur; N, Nacimiento; SLB, Santa Lucia Bank. Place names—CM, Cape Mendocino; PD, Point Delgada; PA, Point Arena; FR, Fort Ross.

Zone (Wilson, 1985) and shortly thereafter along the Mendocino Fracture Zone. Subduction was displaced from south to north along the central California margin by right-slip transform faulting. Transform faulting on the San Andreas and related faults then segmented the allochthon (Johnson and Normark, 1974; Ross, 1973; Ross and Brabb, 1973; Kistler and Peterman, 1973) and produced the present distribution of recognized terranes (Figure 4). The Sur–Obispo composite terrane is composed of the exotic San Simeon and Stanley Mountain terranes (Blake, Howell, and Jones, 1982), which underlie the southern and southwestern central California coast and are separated from the Salinia terrane on the northeast by the Sur–Nacimiento fault zone. The San Simeon terrane, a subduction complex lithologically correlative with the Franciscan Complex, is typically a melange containing graywacke, serpentine, greenstone, chert, knockers of blue schist, and large coherent slabs of folded Upper Cretaceous graywacke in a sheared argillite matrix (Howell et al., 1977). The Stanley Mountain terrane lies structurally above the San Simeon terrane and, in outcrop, flanks it to the south and northeast. The Stanley Mountain terrane, a lithologic correlative of the Great Valley sequence, is composed of Middle(?) and Upper Jurassic ophiolite, Upper Jurassic tuffaceous chert, and a thick sequence of Upper Jurassic and Cretaceous forearc basin turbidites.

Northeast of the Sur–Nacimiento fault zone and extending into the offshore in the San Francisco Bay area is the magmatic Salinia terrane, which includes the Salinian block of Reed (1933) and Page (1970). Salinia terrane basement is composed of Mesozoic and older Sur Series metamorphic rocks intruded by felsic plutons radiometrically dated as Cretaceous (120–75 Ma) in age (Ross, 1984; Mattinson, 1978, 1983). Mattinson and James (1985) suggest that the Salinia terrane formed in middle Late Cretaceous time as a west-facing continental arc underlain by Precambrian rocks, and that northward migration of this terrane was initiated after the emplacement of the youngest (75 Ma) plutonic rocks. As noted, by latest Eocene time the terrane had been displaced generally northward perhaps 1500 km. If latitudinal displacements indicated by paleomagnetic inclinations are accepted, ensuing Neogene transform faulting may have brought the total northward displacement to 2600 km (Kanter and Debiche, 1985).

The Point Arena terrane lies between the northwest-trending offshore extension of Salinia terrane and the San Andreas fault north of Fort Ross. This terrane is an elongate sliver composed of Late Cretaceous to Eocene turbidites that are overlain by Neogene clastic strata and basalts, and are structurally underlain by spilitized basalt thought to represent oceanic crust (Weaver, 1943; Wentworth, 1968; Blake, Howell, and Jayko, 1984). Graham and Dickinson (1978), and Ross (1984) suggest that the spilitized basalt may belong to the Franciscan Complex. Paleomagnetic data suggest a total northward post-Late Cretaceous transport of 4000 km for the Point Arena terrane (Kanter and Debiche, 1985).

The Coastal terrane that underlies the coastline northward from Point Arena to the vicinity of Point Delgada is composed of highly distorted strata consisting of Late Cretaceous to late Eocene age arkosic sandstone, mudstone and conglomerate. The western part of the terrane is a chaotic melange that contains pillow basalt and foraminiferal limestone derived from mid-ocean seamounts and rare blocks of blue schist (Blake, Jayko, and McLaughlin, 1984; Sliter, 1984). Like the neighboring Point Arena terrane, the Coastal terrane has been transported about 4000 km to the north, as indicated by a faunal assemblage in the limestone that is thought to have originated near lat. 15 °N (Sliter, 1984).

The King Range terrane, the northernmost terrane beneath coastal California south of Cape Mendocino, is composed of oceanic basement overlain by Paleogene and Neogene sedimentary and igneous rocks of both continental and oceanic derivation (McLaughlin et al., 1982). The King Range terrane is bounded on the landward side by a west-dipping shear zone (Griscom, 1980), along which this terrane is thought to have been accreted by obduction.

Although terrane boundaries can be drawn with some assurance onshore, the lack of information on basement composition makes extension of these boundaries into the offshore speculative. In the following discussion of the shelf basins, evidence is presented for the recognition of offshore basement blocks that have independent structural histories, differing magnetic signatures, and boundaries that are reflected by discontinuities in Neogene structures. These boundaries differ from some of the inferred boundaries shown on Figure 4.

OFFSHORE SANTA MARIA, SUR AND SANTA LUCIA BASINS

Three structural basins containing upper Tertiary strata—offshore Santa Maria, Sur, and Santa Lucia Basins—lie offshore of the southern part of the central California margin (Figures 1 and 5). The largest of the three, offshore Santa Maria Basin, is approximately 5100 km^2 in area. It is elongate parallel to the coast, and is bounded on the northeast by the Hosgri fault (Wagner, 1974). Along this fault, down-to-basin vertical separation has juxtaposed a thick upper Tertiary basin section to the southwest against elevated basement rocks of the Franciscan Complex to the northeast. The Santa Maria Basin is bounded on the southwest by the Santa Lucia Bank fault, which extends along the elongate narrow Santa Lucia Bank structural high. The southeastern limit of the basin is placed along the axis of a southwest-trending Cretaceous through late Miocene structural high ('Amberjack high' of Crain, Mero, and Patterson, 1985). Upper Jurassic and Lower Cretaceous strata in the bottom of a COST well, OCS-CAL78-164 No.1 (Cook, 1979), drilled just northwest of the high (Figure 1), are similar to strata (Espada? Formation) found in the onshore Point Arguello area (western Santa Ynez Range). The Espada Formation is unreported beneath Santa Barbara Channel. The offshore Santa Maria Basin and the Santa Barbara Channel areas had independent Tertiary histories, as indicated by a major unconformity between Lower Cretaceous and lower Miocene strata in the COST well and throughout offshore Santa Maria Basin. The unconformity is not present in the unbroken sequence of lower Tertiary marine strata beneath the Santa Barbara Channel (McCulloch et al., 1979; Vedder, this volume).

Offshore Santa Maria Basin shallows to the northwest toward a structural discontinuity here named the San Martin structural discontinuity (shown as SMSD on Figure 5). The discontinuity separates the irregular basement beneath the offshore Santa Maria Basin from the relatively undeformed shoreward-dipping basement surface beneath the Sur Basin. Sur Basin (Blake et al., 1978; also informally called Partington basin) has an approximate area of 1800 km^2 and contains a thick (~ 3 km) section of upper Tertiary strata. It is bounded on the northeast by a nearshore fault along which down-to-basin vertical separation of at least 4 km has lowered the Tertiary basin strata against elevated basement of the Franciscan Complex. This fault, or an associated splay, bends to the northwest, and separates the basin from the adjacent structurally high Sur platform to the northeast. Basement beneath Sur Basin shallows westward and reaches the sea floor along the western edge of the basin.

An elongate structural basin, here informally named the Santa Lucia Basin (Figure 1), lies between the Santa Lucia

Bank high and an unnamed structural high to the southwest. The basin has an area of approximately 2000 km^2 and a sedimentary section commonly as thick as 0.75 sec (~ 1.5 km). For most of its length the basin abuts the Santa Lucia Bank high along faults that have eastside-up, high-angle reverse displacement. Comparable faults occur along the west side of the basin, but for most of its length the basin fill thins westward and is uplifted and truncated by erosion along the basin edge. The basin shallows to the south, where it is interrupted by several minor faultbounded structural highs that trend parallel to the basin axis.

Basement Rocks

Several lines of evidence suggest that the Hosgri and Santa Lucia Bank faults (Figure 5) bound discrete basement blocks with differing structural histories that have been juxtaposed by lateral displacement along the present traces of these faults. The Santa Lucia Bank fault has a long (140 km) linear trace typical of lateral-slip faults; on seismic-reflection profiles the dip is too steep to be resolved. The Tertiary sedimentary section changes thickness abruptly across the fault, consistent with lateral displacement (Figure 6). Tertiary strata within the offshore Santa Maria Basin generally thicken toward the Santa Lucia Bank fault, whereas on the structural high immediately across the fault not only is the total Tertiary section thinner, but probable correlatives of major packets of Miocene and Pliocene strata are also considerably thinner. Differences in the residual magnetic field across the Santa Lucia Bank fault also suggest lateral slip. Although the magnetic field is nearly flat southwest of the fault, anomalies of about 200 nT occur east of the fault (Figure 7). Differences in the magnetic field led Page et al. (1979a, b) to suggest that the Franciscan melange east of the fault contains large ultramafic bodies not present in a possibly younger melange facies beneath Santa Lucia Bank. A linear magnetic anomaly along the northeast side of Santa Lucia Bank trends obliquely toward the fault. The anomaly, which may reflect an ophiolitic body at depth, does not cross the Santa Lucia Bank fault, again suggesting lateral displacement on the fault.

Hoskins and Griffiths (1971) suggest that basement west of the Hosgri fault is granitic based on the presence of granite-derived coarse clastic rocks said to be of Late Cretaceous or Eocene age on Santa Lucia Bank and on the style of the deformation characteristic of the bank and adjacent offshore Santa Maria Basin. This would necessitate considerable revision of existing tectonic models that limit granitic basement to the Salinia terrane east of the Nacimiento fault (Figure 4). Others (McCulloch et al., 1977; Page et al., 1979a, b; Ross and McCulloch, 1979), suggest that the area west of the Hosgri fault is underlain by the Franciscan Complex, based on metasediments and altered basic igneous rocks dredged from Santa Lucia Bank, on the proximity of Franciscan rocks along the eastern shore of the basin, and on the presence of poorly defined acoustic reflections that suggest bedding within the basement.

The physiographic expression and the structure of Santa Lucia Bank are similar to Patton Ridge, a California borderland structure that lies to the south (Vedder, this volume). Howell, Jones, and Schermer (1985) suggest that Santa Lucia Bank is underlain by the Patton terrane (Figure 4), which on Patton Ridge is composed largely of greywacke and argillite (analogous to the coastal belt Franciscan).

Howell, Jones, and Schermer, (1985) extend the western boundary of the San Simeon terrane, a subduction complex lithologically correlative with the Franciscan Complex, westward across the Hosgri fault and beneath the offshore Santa Maria basin to the Santa Lucia Bank fault (Figure 4). Gray (1980) describes Mesozoic rocks beneath onshore Santa Maria Basin from well cores as an Upper Cretaceous sedimentary sequence (sandstone, siltstone, shale) overlying and incorporated into melanged ophiolitic rocks (ophicalcite, basalt, greenstone, intrusive rocks, gabbro, and serpentinized peridotite) of undetermined age. Gray concluded that the sedimentary strata are unlike nearby outcrops of relatively undisturbed rocks of the Great Valley sequence of the Stanley Mountain terrane, and are more like disrupted rocks of the Cambria, Port San Luis and Pfeiffer Beach slabs. It is speculated that these Late Cretaceous disrupted rocks accumulated in slope basins on subducting accretionary Franciscan melange, and that they were tectonically kneeded down into the melange (Howell et al., 1977; Underwood, 1977).

Structure contours on basement beneath the onshore Santa Maria Basin show irregular but generally northwest-trending basement structural panels bounded by lateral displacement faults to which Gray (1980) assigns right slip. Hall (1982), Kamerling and Luyendyk (1979), Hornafius (1985), and Kieniewicz and Luyendyk (1986) also indicate northwest trends in the basement structure, but in order to explain the observed clockwise rotation of the magnetic field in Tertiary rocks in this area Kamerling, Luyendyk, and Hornafius suggest that the western margin of the basin was dragged by right slip along the Hosgri fault, producing left slip between the basement panels.

Trends of the structural axes change abruptly at the Hosgri fault; linear magnetic anomalies approximately parallel to the northwest-trending onshore basement structure extend into the offshore and terminate at the fault (Figure 7). West of the Hosgri fault, structures in the acoustic basement are en echelon, doubly plunging porpoise-shaped sinuous highs and intervening lows that trend to the north; these structures lie oblique to, and are bounded by, the Hosgri and Santa Lucia Bank faults (Figure 8). The basement structures, previously noted in the central and southern parts of the basin (Page et al., 1979a, b; Crain, Mero, and Patterson, 1985), continue to the north, but with decreasing amplitude, to about the latitude of Cape San Martin. These structures generally do not cross the entire basin, but plunge toward a mid-basin fault that approximately parallels the Hosgri and Santa Lucia Bank faults. The mid-basin fault does not display the large vertical separation characteristic of the Hosgri and Santa Lucia Bank faults; however, it was apparently as important as the major bounding faults in determining the local stress field during development of the basement structures.

The lows between the en echelon basement structural highs acted as depocenters for the accumulation of marine strata that thin toward and onlap structural highs, suggesting that basement relief was developing throughout the Miocene. Small-displacement normal faults along the edges of the structural highs displace basement and rocks as young as late Miocene. Displacement decreases upward, suggesting that structural growth continued through middle Miocene time. Minor folding along these faults in the central and northern parts of the basin suggests some contemporaneous lateral slip.

The oblique trend of the basement structures between the

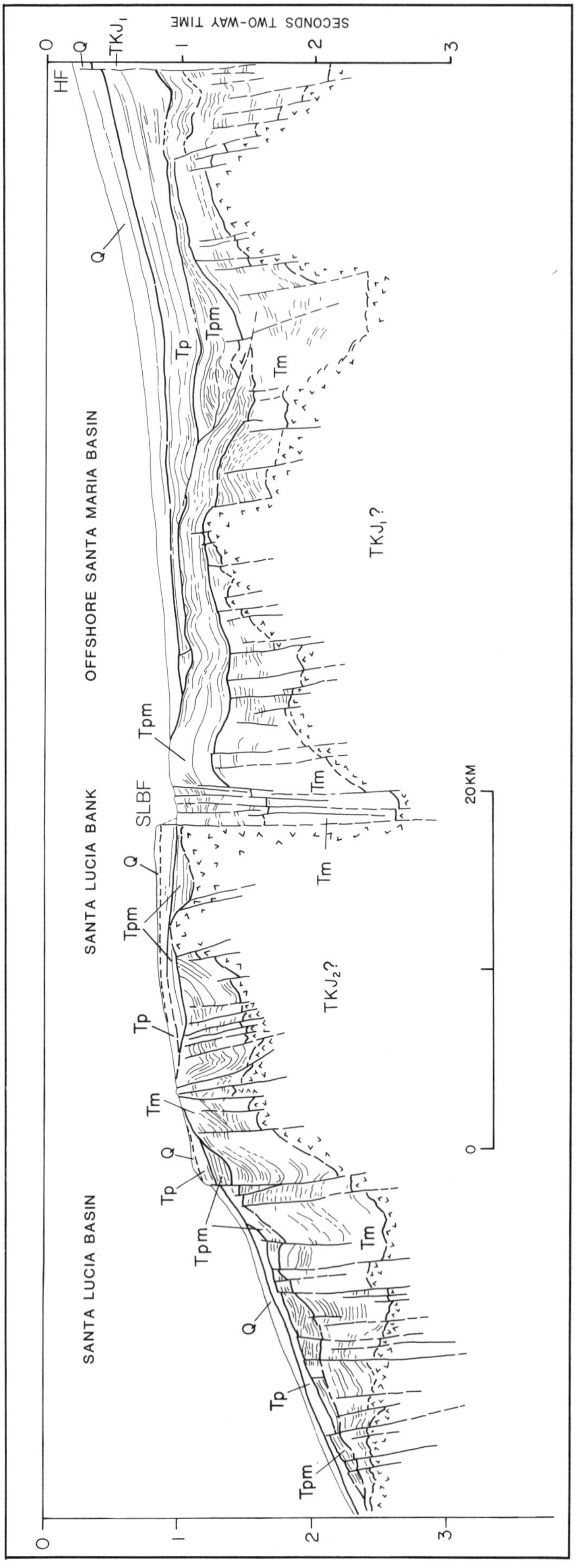

Figure 6. Interpretation of a single-channel seismic reflection profile across offshore Santa Maria Basin. Vertical exaggeration approximately 8:1 at the sea floor. Symbols—Q, Quaternary; Tp, Pliocene; Tpm, Pliocene and Miocene; Tm, Miocene; TKJ_1, Tertiary to Jurassic; TKJ_2, Tertiary to Jurassic. Faults—SLBF, Santa Lucia Bank; HF, Hosgri.

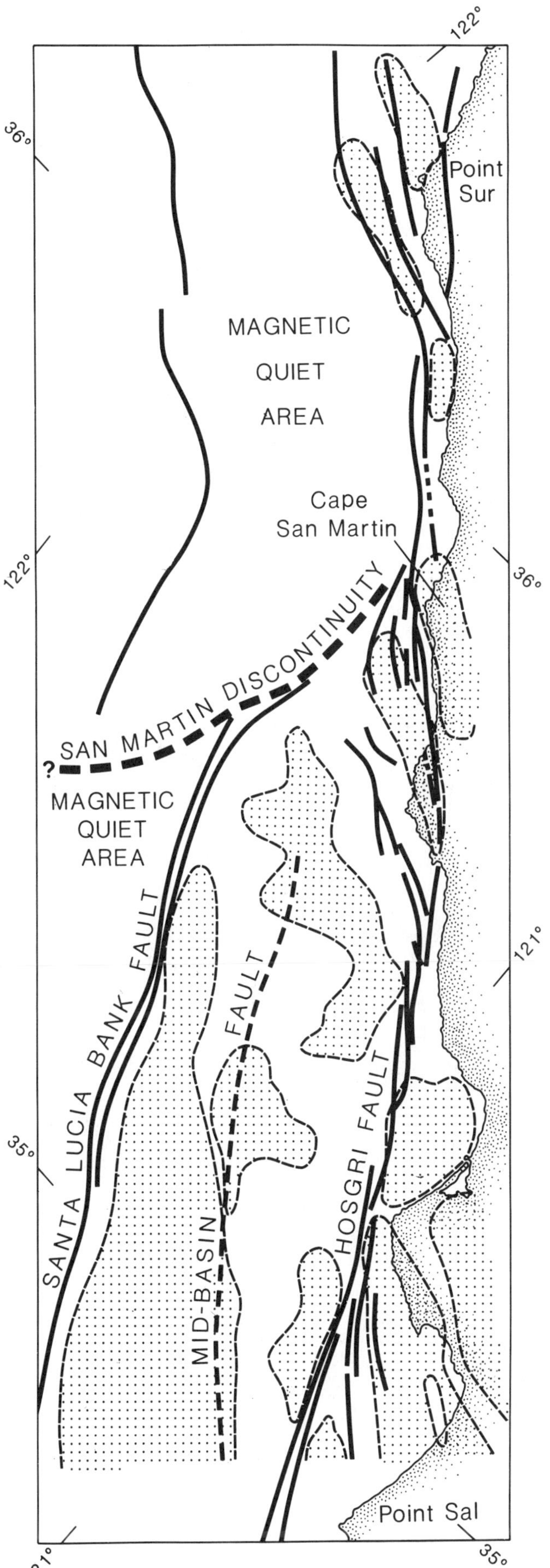

Figure 7.. Positive residual magnetic anomalies (patterned areas) and faults mapped from seismic-reflection profiles in the offshore Santa Maria Basin area. Magnetic data from McCulloch and Chapman (1977) and unpublished magnetic data collected by USNS Bartlett.

Hosgri and Santa Lucia Bank faults, and infilling during development of the structural lows accompanied by growth of normal faults and minor compression, suggest extension and sagging of the basement during transtensional wrenching. The age of deposits affected by this deformation suggests that the Hosgri, Santa Lucia Bank, and mid-basin faults may have experienced lateral slip throughout the Miocene.

At the north end of the offshore Santa Maria Basin is the San Martin structural discontinuity, along which there is a discontinuity in the Neogene structures. The discontinuity trends southwestward across the continental shelf from Cape San Martin and appears to join the Santa Lucia Bank fault, and together they trend northward toward Cape San Martin. The fault and the discontinuity probably lie along a change in the basement rocks. Both features coincide with a change in magnetic anomaly pattern and separate the basement with en echelon structures in the offshore Santa Maria Basin from the relatively smooth shoreward-dipping basement surface beneath Sur Basin to the north (Figures 1, 5, and 7). This suggests that they form a boundary between basement blocks having different structural histories and that the discontinuity in the Neogene structure may reflect independent movement between the basement blocks along an old suture.

Stratigraphy and Tectonics

Strata above basement in offshore Santa Maria Basin were penetrated by several exploratory wells (Figures 1, 9, and 10). Stratigraphic columns of Zieglar and Cassell (1978) are based on logs of the offshore wells, whereas those of Hoskins and Griffiths (1971) prepared for this basin and basins to the north are generalized basin-wide columns based in part on possible onshore-equivalent sections. An erosional unconformity separates Upper Cretaceous from lower Miocene rocks at the south end of the basin, and a probably comparable unconformity occurs throughout the offshore Santa Maria, Sur and Santa Lucia Basins. Little is known of the approximately 40-million-year history represented by the erosional unconformity. Hoskins and Griffiths (1971) report highly compact, coarse clastic Eocene rocks beneath an unconformity on Santa Lucia Bank. Erosional remnants of correlative rocks may have been protected from erosion in physiographic lows on the basin floors, but such rocks have not as yet been reported, and the early Tertiary stratigraphic record of these basins has been lost to erosion.

The earliest Tertiary depositional episode in the offshore Santa Maria Basin was volcanogenic. In the Standard-Humble P-060-1 well off Point Sal (060, Figure 1), an undated acoustically well-bedded unit at least 390 m thick and composed of tuffs, flow rocks, and volcaniclastic strata, rests unconformably on basement (Epstein and Nary, 1982). The driller's log reports that the lower part of the unit is dominated by bentonite, the middle by andesite, and the upper by rhyolite. The volcanic rocks were emplaced prior to the deformation that formed the en echelon basement structures, for Epstein and Nary (1982) show the volcanics resting on basement and having been displaced by the normal faults associated with the growth of the basement structures. In a well southwest of Point Arguello (Chevron Phillips P-0450-1), a 213-m section of rhyolite tuff, unwelded pumice and volcaniclastics rests unconformably on Cretaceous strata (Crain, Mero, and Patterson, 1985). The early Miocene age was assigned to the volcanic rocks, based upon interbedded marine strata that contain a Saucesian microfauna. If a tenuous correlation can be drawn between the rhyolitic facies in these two volcanic sequences, the volcanics off Point Sal may also be of early Miocene age. If so, the volcanism accompanying the tectonic event that initiated transtensional basement rifting and the change from nondeposition to marine sedimentation may also have occurred in early Miocene time.

Marine volcanism was followed in the offshore Santa Maria Basin by continued subsidence of the continental margin, growth of the basin wrench structures, and deep-water marine sedimentation in the structural lows and adjacent to the Hosgri and Santa Lucia Bank faults. Further north in Sur Basin, downward rotation of the basement along the major fault bounding the Sur Platform and down-to-the-east normal faults that cut the basin floor (Figure 5) was probably initiated at this time. Basement continued its downward rotation in the Sur Basin during the Miocene, and strata deposited on the eroded basement surface thicken and dip eastward (Figure 11).

Near the Point Arguello field at the south end of offshore Santa Maria Basin, lower Miocene fine-grained calcareous strata (Relizian Stage limey mudstone and dolostone) with an aggregate thickness of as much as 760 m, were deposited on the volcanics (Crain, Mero, and Patterson, 1985). Subsequent middle and early late Miocene sedimentation proceeded with deposition of well-bedded deep-water siliceous shale and chert, commonly called the Monterey Formation. These highly siliceous strata have a maximum basin thickness of 890 m. Basement wrench grabens continued to subside and the siliceous rocks were cut by associated growth normal faults. Deep, quiet-water sedimentation continued above the top of the Monterey(?) Formation with deposition of upper Mohnian Stage diatomaceous mudstone, siliceous shale and porcelanite.

The Santa Lucia high may have been relatively highstanding during deposition of the siliceous strata. Its structural and physiographic counterpart, Patton Ridge, the westernmost ridge in the California borderland to the south (Vedder, this volume), was sufficiently elevated to shed lithic clasts and a shallow-water fossil assemblage (barnacles, mollusks, and bryozoans) off the ridge to middle Miocene strata in DSDP Hole 469, which lies 20 km seaward of the toe of Patton Escarpment (Vedder, Crouch, and Lee-Wong, 1981).

A change in regional stress produced compression across offshore Santa Maria and Santa Lucia Basins in late Miocene or early Pliocene time (end of Sisquoc Formation deposition; Crouch, Bachman, and Shay, 1984). Miocene and lower Pliocene strata were folded and faulted, with the most severe deformation along the eastern margins of the Santa Lucia and offshore Santa Maria Basins where the basin fills were buttressed against elevated basement rocks. In the offshore Santa Maria Basin this deformation produced an eastern fold

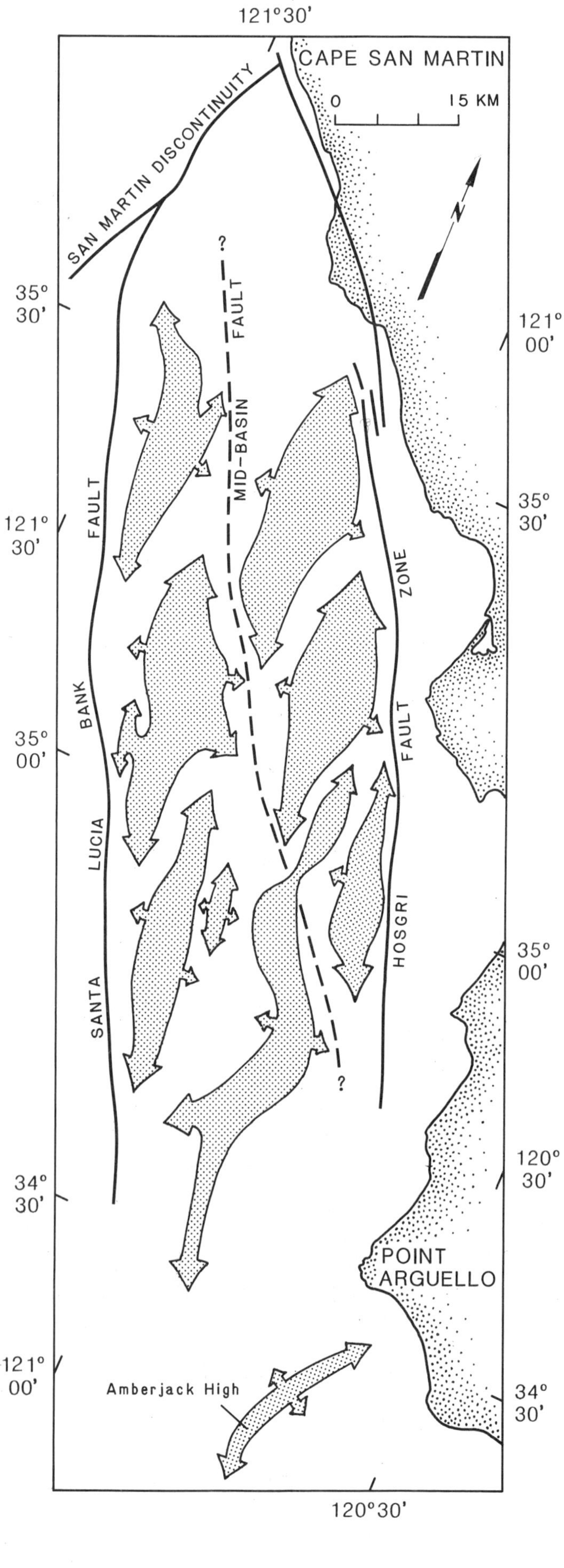

Figure 8. Structures on the acoustic basement beneath offshore Santa Maria Basin. Modified after Crain, Mero, and Patterson (1985).

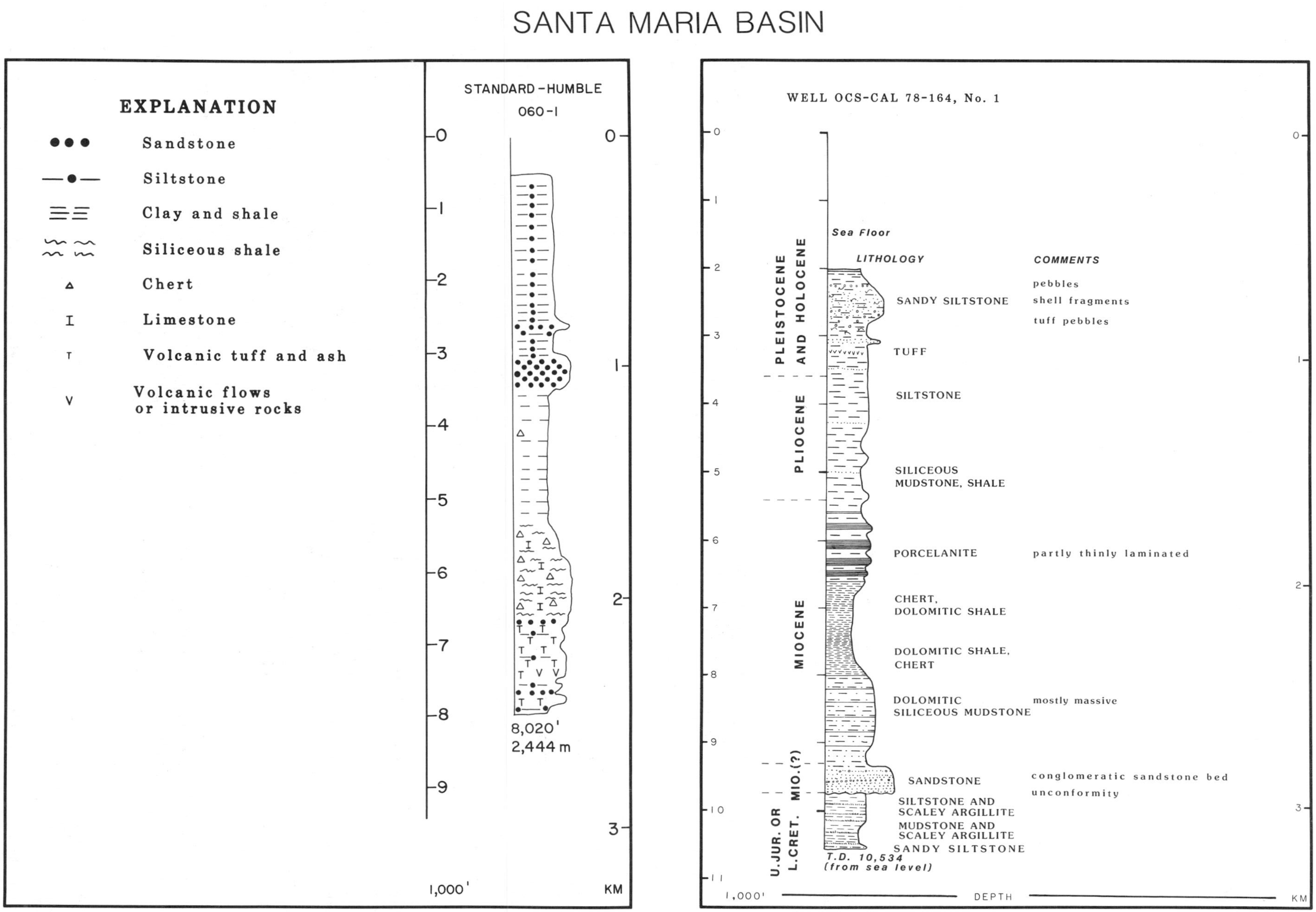

Figure 9.. Stratigraphic columns for exploratory well Standard Humble 060-1 (Zieglar and Cassell, 1978) and COST Well OCS-CAL 78-164, No. 1 (modified after Cook et al. (1979), and Isaacs et al. (1983)). Wells are located on Figure 1. The explanation for the Standard Humble Well also applies to stratigraphic columns of exploratory wells in succeeding figures. No correlations are implied for these stratigraphic columns.

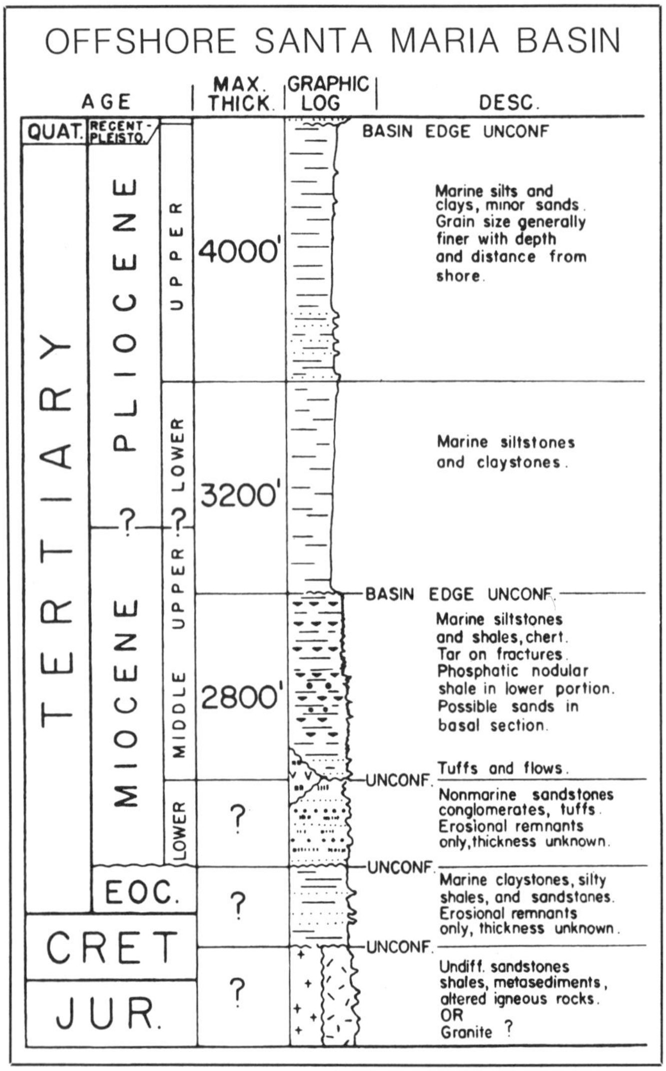

Figure 10. Generalized stratigraphic column for offshore Santa Maria Basin (from Hoskins and Griffiths, 1971). In this and following generalized stratigraphic columns from Hoskins and Griffiths, some units have been extrapolated from onshore sections thought to be equivalent to sections in the offshore.

belt, (the Hosgri or Eastern trend of Ogle, 1984), notable for associated west-directed thrusting (Crouch, Bachman, and Shay, 1984), that lies within and adjacent to the Hosgri fault zone. Farther west in the offshore Santa Maria Basin, a fold belt with a more westerly trend crosses the old basement structures (Figure 12) and probably indicates right shear across the basin. Folding in this belt also involved west-directed thrusting, especially where folds developed adjacent to basement highs (Figure 13). By contrast, the folds in the adjacent Sur Basin are of very low amplitude (Figure 11) and are best seen on single-channel seismic-reflection records with high vertical exaggeration.

An angular discordance that can be traced throughout the offshore Santa Maria, Santa Lucia, and Sur Basins also reflects this episode of late Miocene and/or early Pliocene deformation. Compression may have been superimposed and structurally overprinted on earlier normal faults and given them the present appearance of high-angle reverse displacement. Structures formed during this episode of compression are primary petroleum targets in the offshore Santa Maria Basin. In the Sur Basin, the axes of the the young folds approximately parallel the shoreline, suggesting shoreward-directed compression. Folds and faults along the San Martin discontinuity in the Sur and offshore Santa Maria basins appear to have been dragged by right-lateral displacement between the basement blocks.

The offshore episode of compression and folding accompanied the growth of physiographic relief on shore. Terrigenous Pliocene sediment derived from rising coastal mountains flooded the offshore basins, and generally overtopped the folds in the Miocene rocks. The Pliocene strata are sandy at the south end of offshore Santa Maria Basin, and grade laterally to siltstone and claystone in midbasin areas. Compression is still occurring in offshore Santa Maria Basin as indicated by deformation of strata nearly at the sea floor adjacent to a fault-propagated fold (Figure 13). As noted later, active compression is consistent with earthquake first motions and geodetic measurements.

OUTER SANTA CRUZ BASIN

Outer Santa Cruz Basin is a relatively shallow elongate syncline that trends northwest from Monterey Bay across the continental shelf to the toe of the continental slope (Figures 1 and 14). It is bounded on the northeast by the Pigeon Point high, on the southwest by the Santa Cruz high (Hoskins and Griffiths, 1971), and along the lower edge of the continental slope by an unnamed structural high of probable volcanic rocks along which the Mulburry, Guide, and Pioneer Seamounts form prominent topographic highs. Alkali basalt, basic tuff, and rhyolitic tuff have been dredged from the seamounts (McCulloch, Utter, and Menack, 1985), but the volcanic rocks have not been radiometrically dated.

Basement Rocks

Prior to the recognition of strike-slip faulting on the California margin Hoskins and Griffiths (1971) suggested that the basement throughout the Outer Santa Cruz Basin is granitic, based upon the proximity of onshore granite, and unstated geophysical evidence. However, Silver, Curray, and Cooper (1971) concluded on the basis of seismic profiles and mafic volcanic rocks dredged from Santa Cruz high, and from the heads of Pioneer and Ascension canyons just west of the Pigeon Point high, that the basin was largely underlain by Franciscan rocks. They also proposed that the Sur–Nacimiento fault, which onshore separates Franciscan rocks of the Stanley Mountain terrane on the west from Salinian granitic rocks on the east (Figure 4; Page, 1970), extends offshore along the western edge of the Pigeon Point and Farallon structural highs (Figure 14). More recent dredging near the twin heads of Ascension Canyon (Mullins and Nagle, 1981) has recovered serpentinites and metavolcanics that also indicate Franciscan Complex basement just west of the offshore Sur-Nacimiento fault.

The offshore extension of the seismically active San Gregorio fault has been mapped across the mouth of Monterey Bay (Hoskins and Griffiths, 1971; Greene et al., 1973; Coppersmith, and Griggs, 1978; Griggs, 1979). Late Neogene right-slip on the San Gregorio fault has offset the offshore extension of the Sur-Nacimiento fault to the northwest (Graham, 1976; Graham and Dickinson, 1978). Determination of the southern limit of granitic basement west of the San Gregorio fault should define the position of the displaced Sur-Nacimiento fault and the amount of its displacement. Although the Farallon high has yielded diorites and tonalites (Chesterman, 1952; Hanna, 1952; Ross, 1984) that can be followed in the subsurface southeastward toward Half Moon Bay (Figure 12), the limits of granitic basement have not been established.

South of the Farallon high, discontinuities in the Neogene structures suggest the presence of two major structural blocks—the Pigeon Point and Año Nuevo blocks (Figures 14 and 15). The Pigeon Point block includes the Pigeon Point high of Hoskins and Griffiths (1971) and is the offshore extension of the onshore Pigeon Point block of Clark (1930). The boundary between the Pigeon Point and Año Nuevo blocks, as defined by a structural discontinuity in Neogene rocks, lies along a magnetic anomaly (Figure 15) with an amplitude of about 150 nT, which suggests a possible difference in basement composition across the discontinuity. Basement rocks beneath the Pigeon Point block are unsampled. Offshore, sandstone containing late Oligocene(?) to early Miocene diatoms and cocoliths has been recovered in dart cores from the crest of the Pigeon Point high on the block (McCulloch, Utter, and Menack, 1985). Onshore, the oldest strata in the Pigeon Point block are highly deformed Upper Cretaceous submarine fan deposits that include mudstone, sandstone, and conglomerate (the Pigeon Point Formation). These Upper Cretaceous strata are unconformably overlain by shallow-marine strata containing a "Vaqueros" Stage marine molluscan fauna of late late Oligocene to early Miocene age (Clark and Brabb, 1978). The onshore Pigeon Point block was originally mapped as extending east to the San Gregorio fault (Clark, 1930), however, recent work (Lajoie, Weber, and Tinsley, 1972; Weber and Lajoie, 1979, 1980; Gibson, 1983) has shown that the block is segmented into two structurally and stratigraphically discrete blocks by major faulting within the wide San Gregorio fault zone (Figure 15). The Pomponio block is bounded on the east by the Coastways strand and on the west by the Frijoles strand, which separates it from the Pigeon Point block. Clark and Brabb (1978), extrapolating from westward dips in the Upper Cretaceous strata on the Pigeon Point block, conjecture that felsic volcanics that crop out east of the Frijoles strand might extend westward as basement for the Pigeon Point block. However, faulting along the Frijoles strand appears to confine the Upper Cretaceous strata to the Pigeon Point block. Mapping by Weber and Lajoie (1979, 1980) and Howell and Joyce (1981) shows that these felsic volcanics are separated from Upper Cretaceous strata of the Pigeon Point block by a fault, and lie either on the Pomponio block or on an independent fault sliver.

The Upper Cretaceous submarine fan deposits on the Pigeon Point block contain granitic clasts, indicating that they were derived in part from a granitic terrain. In addition, a granitic boulder from a basal marine boulder conglomerate (Vaqueros? Formation) that rests directly on upturned truncated Upper Cretaceous beds at Pescadero Beach, mineralogically and texturally resembles granodiorite exposed on southeast Farallon Island, approximately 70 km to the northwest (D.C. Ross, written communication, 1984). Granitic boulders in this basal conglomerate are relatively unweathered, suggesting that they were not recycled but were derived from granitic outcrops during the erosional episode that truncated the Upper Cretaceous strata.

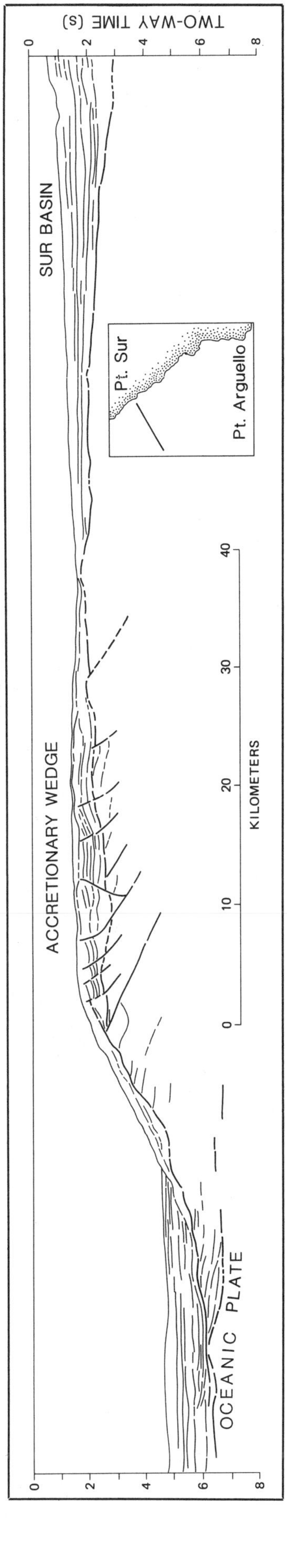

Figure 11. Interpreted time-section of a migrated multichannel seismic reflection profile across the Sur Basin and adjacent continental shelf and slope.

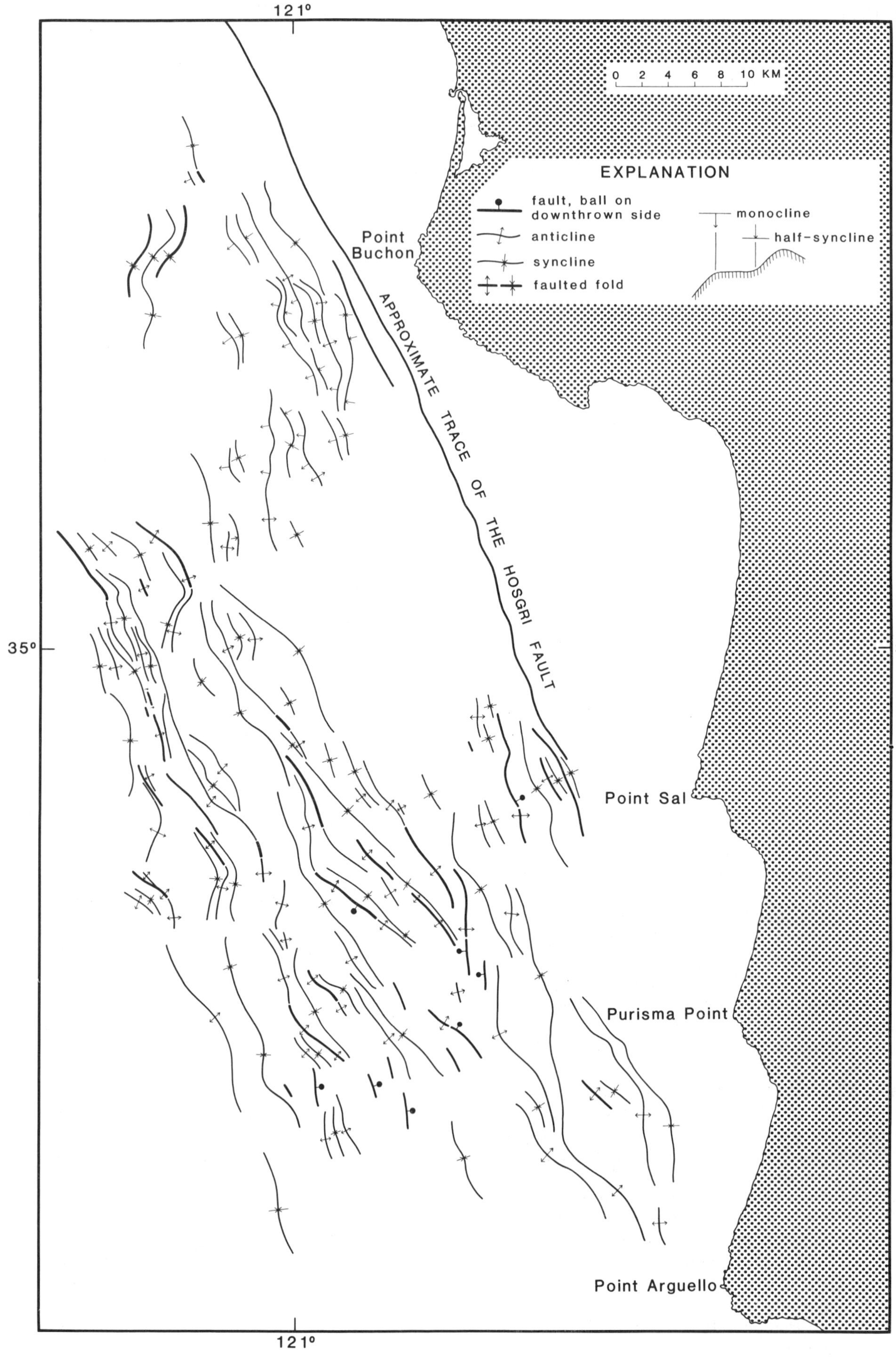
121°
0 2 4 6 8 10 KM
EXPLANATION
fault, ball on downthrown side
anticline
syncline
faulted fold
monocline
half-syncline
Point Buchon
APPROXIMATE TRACE OF THE HOSGRI FAULT
35°
35°
Point Sal
Purisma Point
Point Arguello
121°

Although it remains moot as to whether or not the Pigeon Point block is underlain by granitic basement, the block received granite-derived debris during Late Cretaceous fan deposition and was probably located adjacent to a granitic mass during the late Oligocene-early Miocene erosion.

The composition of basement rocks beneath the Año Nuevo block is also unknown. Offshore, this block was penetrated by two exploratory wells (Figure 1; 035, 036) that bottomed at depths of 2357 and 2283 m, respectively, in Upper Cretaceous marine sandstone without reaching basement. There is no strong evidence that the Año Nuevo block extends onland along the west side of the San Gregorio fault zone. Because the offshore boundaries of the Año Nuevo block are defined only generally by structural discontinuities, and seismic-reflection data are scant for the shallow near-shore, an onshore-offshore tie remains open to question. The oldest rocks exposed on land are submarine-fan deposits that crop out along the north shore of Point Año Nuevo. These strata have been tentatively correlated with similar Upper Cretaceous submarine fan deposits (Pigeon Point Formation) in the Pigeon Point block (Eric Taylor, oral communication, 1985).

Stratigraphy and Tectonics

Outer Santa Cruz Basin is an asymmetrical syncline (Figure 14) produced by deposition on a deforming erosional surface. The erosional surface is usually on the acoustic basement. Sporadic poorly developed deeper reflectors indicate that the erosional surface truncates strata deformed in an earlier tectonic episode. Based on mafic volcanics dredged from Santa Cruz high, where this erosional surface is exposed at the sea floor, the erosional event truncated Franciscan rocks. Although the erosional surface probably had an initial westward dip, it now dips landward from the Santa Cruz high to the axis of the syncline as a relatively unfaulted plane. The northeastern limb of the basement syncline dips more steeply and is commonly broken by high-angle eastdipping faults with apparent high-angle reverse displacement along the southwest sides of the Pigeon Point and Farallon highs.

The stratigraphy of the outer Santa Cruz Basin is known from two exploratory oil wells, 035 and 036 (Figures 1 and 16). Volcanic rocks (bentonites, tuffs, and flow rocks) are reported near the base of both wells. Driller's logs show interbedded volcanic rocks and shale over vertical intervals of 460 m and 335 m in 035 and 036, respectively. In well 035, volcanic rocks assigned an early Miocene age (Hoskins and Griffiths, 1971), presumably on the basis of benthic foraminifera, overlie sandy organic siltstone of probable Cretaceous age. The lateral extent of the volcanic rocks is unknown. Volcanics at about the same stratigraphic position are widespread in the offshore Santa Maria Basin, and may be related to extension of the basin floor. Volcanic rocks in outer Santa Cruz Basin also may be related to early extension, but possibly are localized along the steeply faulted northeastern limb of the syncline where faulting accompanied the downward rotation of basement. Volcanism was followed by the deposition of middle and lower upper Miocene chert and siliceous shale (Monterey? Formation) that indicate quiet-water conditions and mid- to lower-bathyal depths. Miocene strata onlap both the Santa Cruz and Pigeon Point structural highs and overtop the Pigeon Point high. Dips decrease progressively upward as these strata onlap the Santa Cruz high, and down-to-basin growth faults terminate near the top of the Miocene section, indicating that the basin continued to grow through early late Miocene time. Hoskins and Griffiths (1971) report a basin-wide unconformity of early late Miocene age. This unconformity is overlain by upper Miocene and lower Pliocene claystone and siltstone, and these units were also deposited across the Pigeon Point and Santa Cruz highs. Deformation continued through this depositional interval, but it was not accompanied by faulting within the basin. Late Pliocene and Quaternary sedimentation filled the basin and overtopped the Santa Cruz high with marine clay and silt and minor amounts of very fine sand that form the present depositional continental shelf and slope.

Figure 12. Map showing young Neogene structures that parallel the Hosgri fault and that trend across older structures on the acoustic basement of southern offshore Santa Maria Basin (compare with Figure 8).

BODEGA AND CORDELL BASINS

Bodega Basin, an elongate (approximately 180 km by 25 km) late Tertiary shelf basin extends northwest from the vicinity of Half Moon Bay to the latitude of Gualala, approximately 25 km south of Point Arena (Figure 1). Bodega Basin, as defined here, includes the Santa Cruz Basin of Hoskins and Griffiths (1971)—a basin that lies beneath the central and southern Gulf of the Farallones. Bodega Basin is bounded on the west by the Farallon and Pigeon Point highs and a subsurface ridge that extends to the vicinity of Point Arena. The basin is bounded on the east by the offshore extension of the San Gregorio fault, the San Andreas fault, and an eroded margin where older Tertiary rocks crop out at, or near, the sea floor.

Cordell Basin, named here from the adjacent Cordell Bank, like Santa Lucia Basin to the south, is an elongate, late Tertiary slope basin. It is approximately 80 km by 15 km, and it lies on the continental slope seaward of the structural high that forms the western boundary of Bodega Basin (Figure 1).

Basement Rocks

Granitic plutons and Sur Series metamorphic rocks of the Salinia terrane are thought to form the floor of Bodega Basin. This basement block, whose boundaries differ somewhat from the overlying basin, is bounded on the west by the offshore extension of the Sur-Nacimiento fault, which was displaced northwestward along the offshore San Gregorio fault. The block is bounded on the northeast by the offshore San Gregorio and San Andreas faults (Figure 14 and 15). In the offshore, the southernmost known granitic rocks (granodiorite; Ross, 1984) occur on the southeast Farallon Island, and the northwesternmost (tonalite) have been collected from Cordell Bank by scuba divers (R. Schmider, oral communication, 1984). Granitic rocks are also reported at depths of 1707 and 1423 m from the bottom of two exploratory wells (Figure 1; 039, 041) drilled south of the Point Reyes fault along the northern edge of the Gulf of the Farallones (Figures 1 and 14). The northernmost granitic rocks on land west of the San Andreas fault are tonalites at Bodega Head

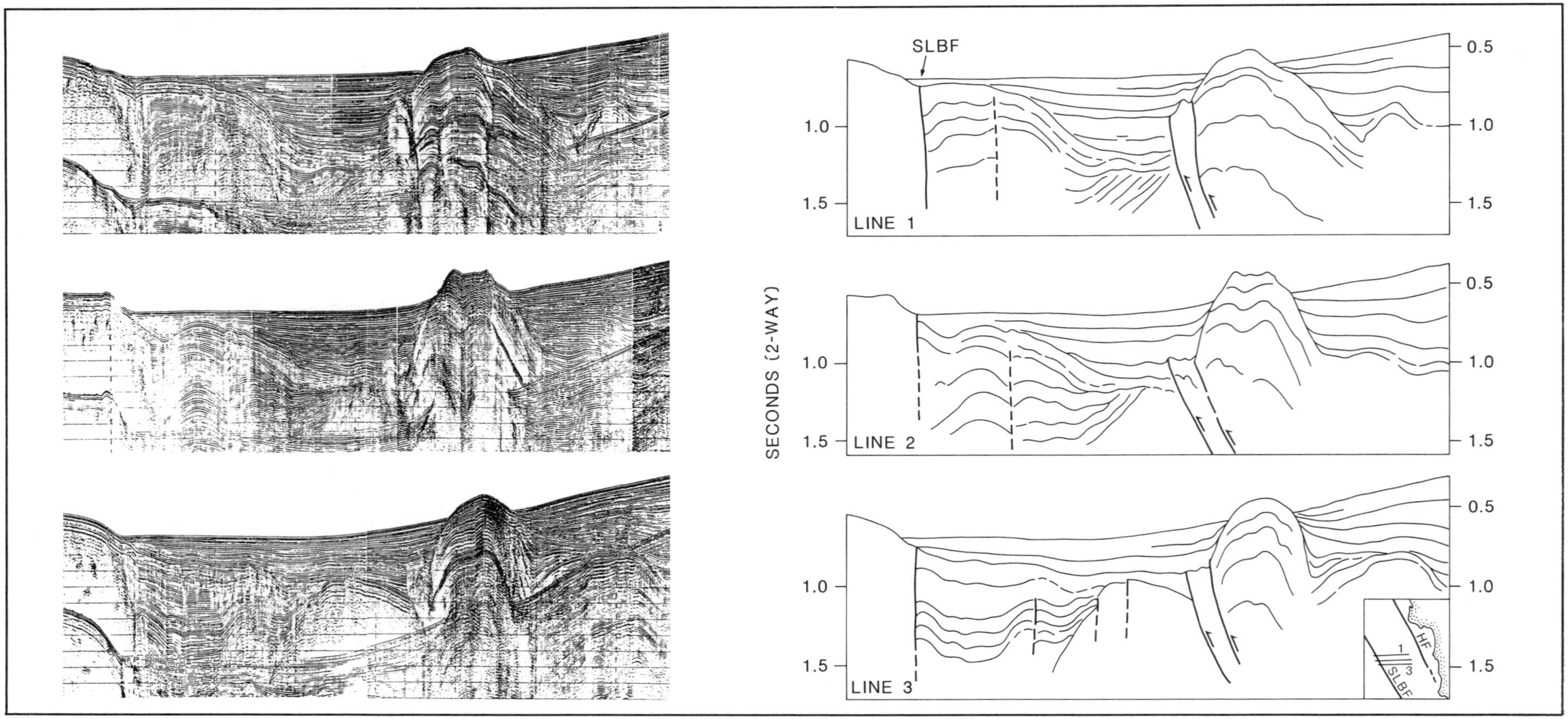

Figure 13. Single-channel seismic reflection profiles of a possible west-directed fault-propagation fold developed against a basement high in offshore Santa Maria Basin. Lower Pliocene rocks are exposed at the seafloor in the crest of the fold, and younger strata are deformed against the rising fold. SLBF, Santa Lucia Bank fault; HF, Hosgri fault.

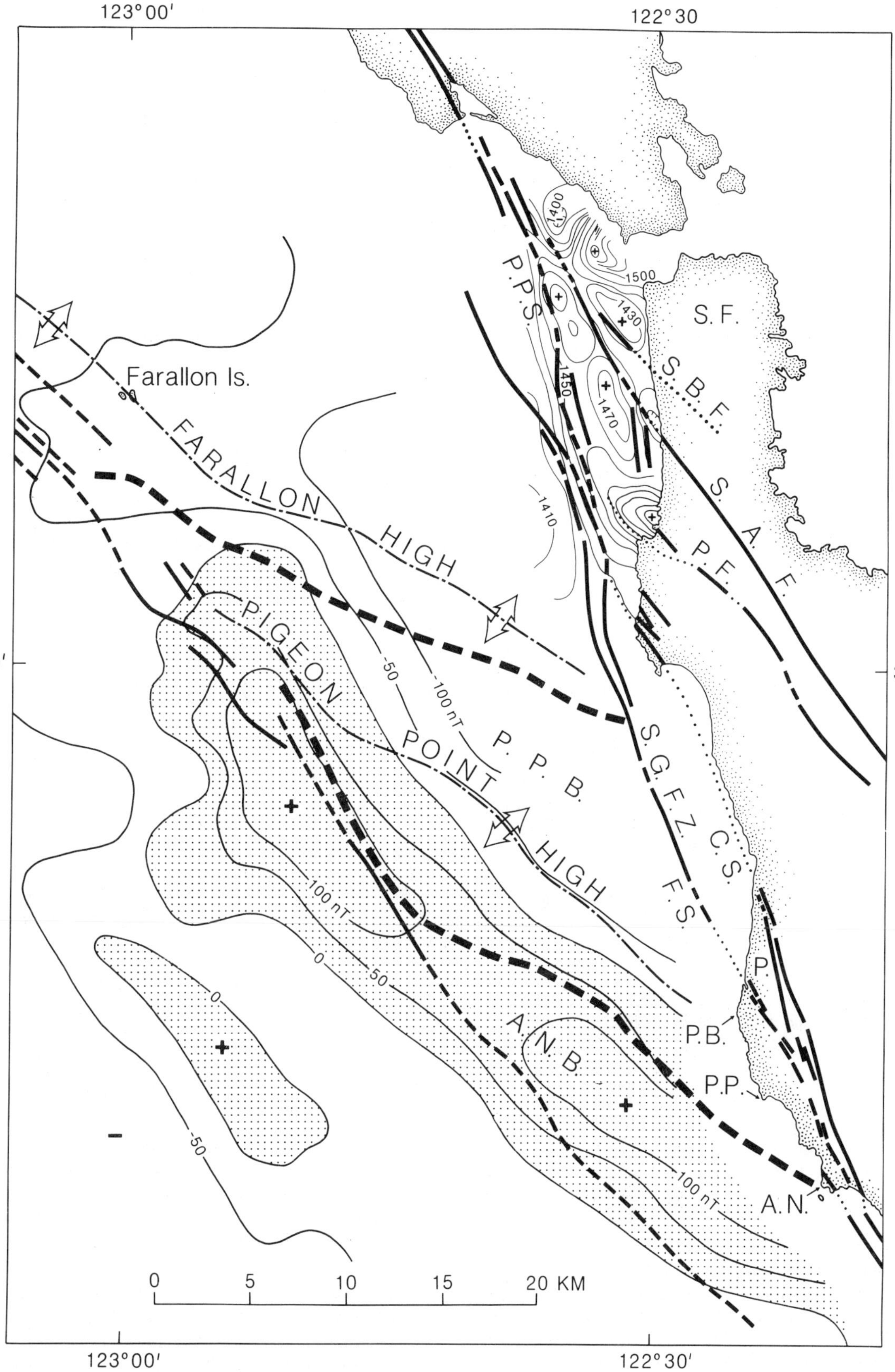

Figure 15. Major structural elements and residual magnetic anomalies in the area of the Pigeon Point and Farallon structural highs, and adjacent to the San Francisco peninsula. Heavy dashed lines are discontinuities in Neogene structures (see Figure 4). Onshore mapping of the San Gregorio fault zone (SGFZ) from Weber and Lajoie (1980). Magnetic data west of the San Gregorio fault zone from NOAA (1974) and west of the San Francisco peninsula from Brabb and Hanna (1981). Faults—SBF, San Bruno (Mustart and McCulloch, 1984); SAF, San Andreas; PPS, Potato Patch strand; PF, Pilarcitos; SGFZ, San Gregorio; CS, Coastways strand; FS, Frijoles strand. Structural blocks—PPB, Pigeon Point; ANB, Año Nuevo; P, Pomponio. Place names—SF, San Francisco; PB, Pescadero Beach; PP, Pigeon Point; AN, Point Año Nuevo.

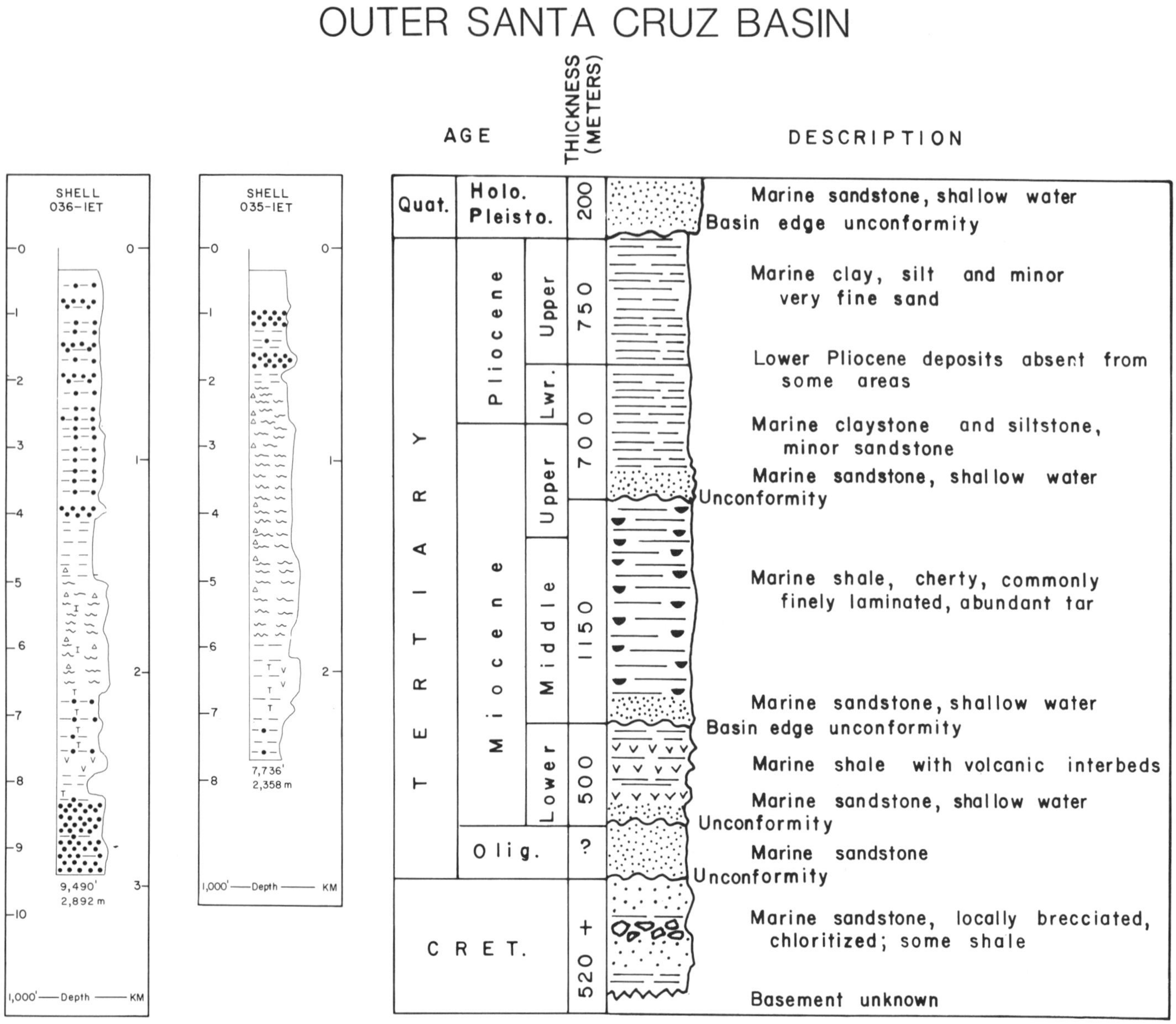

Figure 16. Stratigraphic columns for exploratory wells (Zieglar and Cassell, 1978) and a generalized stratigraphic column for Outer Santa Cruz Basin (from Hoskins and Griffiths, 1971). See Figure 1 for well locations and Figure 9 for explanation.

(Ross, 1984). Curray (1966) and Silver (1969) conjectured that the structural ridge that forms the oceanward edge of the Bodega Basin north of Cordell Bank is a continuation of the granitic Farallon high. This basin-bounding ridge continues northwestward to the vicinity of Point Arena (Figure 14) where it terminates against a discontinuity in Neogene structures. This structural discontinuity, named here the Navarro discontinuity, trends northwest from near Point Arena, approximately along the axis of Navarro Canyon. Southeast of the discontinuity, the San Andreas fault has a clearly defined offshore strand that forms the western boundary of a terrane mapped on land as the Point Arena terrane (Wentworth, 1968; Blake, Howell, and Jones, 1982). The western strand of the San Andreas fault is either absent or obscure northwest of the Navarro discontinuity, and structures to the north trend more northwesterly. The Navarro discontinuity also coincides with the northern boundary of a magnetic anomaly (Figure 17), suggesting a change in the character of the basement rocks across the discontinuity. The discontinuity may bound the northern limit of both the granitic Salinia terrane and the adjacent Point Arena terrane.

Cordell Basin lies west of the possible extension of the Sur-Nacimiento fault, which to the south separates granitic Salinia terrane on the east beneath Bodega Basin from Franciscan Complex on the west. Basement rocks beneath Cordell Basin are therefore inferred to be of the Franciscan Complex; however, no basement rocks have been recovered in the vicinity of Cordell Basin.

Stratigraphy and Tectonics

Bodega Basin (Figure 18) was the most drilled of the central California offshore basins following the Federal leasing program in the early 1960s. Eight wells were drilled off Point

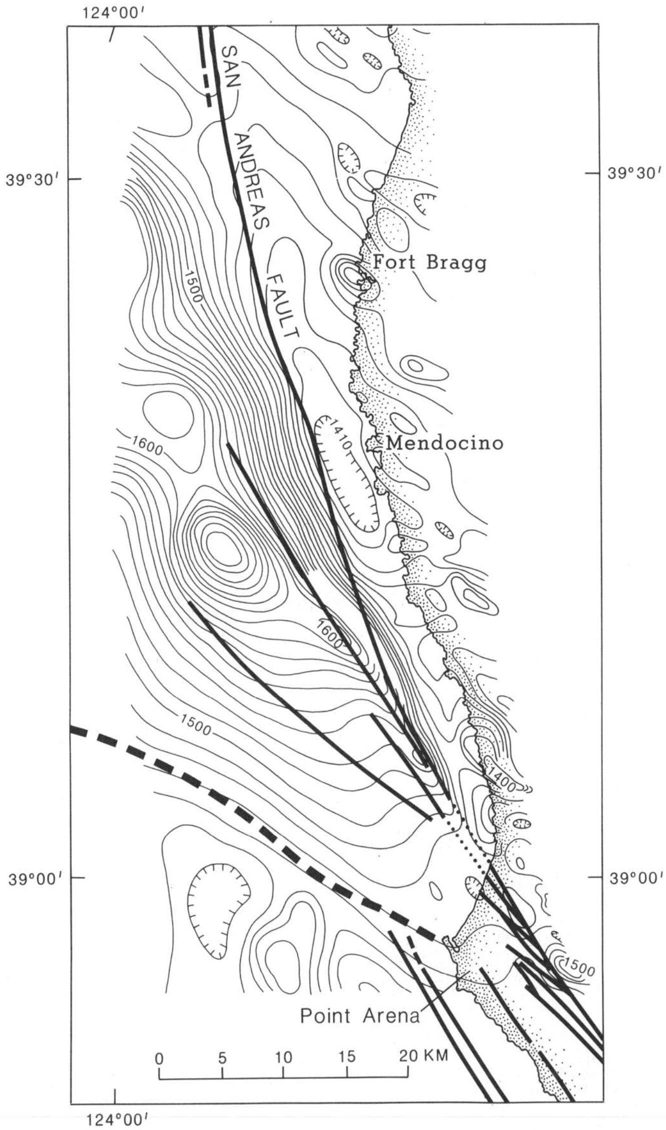

Figure 17. Residual magnetic anomalies (unpublished data, Gulf Oil Company, 1954) and faults mapped from seismic-reflection profiles from Point Arena to Fort Bragg. Navarro structural discontinuity shown by dashed line. Contour interval 10 nT.

Reyes (Figures 19 and 20). Unlike most of the central California offshore basins, the eastern margin of Bodega Basin is exposed onshore, on the Point Reyes peninsula (Figure 14), where the geology has been described by Weaver (1943), Galloway (1977), Ross (1977, 1984) and Clark et al. (1984). Granitic rocks (tonalite, granodiorite, porphyritic granodiorite) are exposed over the northeastern half of the peninsula and at Point Reyes and, as noted above, were reported at the bottom of two offshore wells (039, 041). At Point Reyes, an erosional surface on the granitic rocks is overlain by approximately 200 m of transgressive marine beds. These strata consist of interbedded arkosic sandstone conglomerates that contain a bathyal Paleocene fauna and are interpreted as having been deposited in an upper mid-fan channel complex. Limited paleocurrent data from conglomerate channels suggest local west- to northwest-trending flow, and the clast lithology indicates a source area containing porphyritic granodiorite and first or second erosion-cycle silicic volcanic rocks. Correlative rocks have not been reported in the adjacent offshore wells; however, six of the wells probably bottomed in younger rocks of early Eocene age. These younger strata are restricted to the offshore basin and are described by Hoskins and Griffiths (1971) as dense, deep-water marine sandstones of early and middle Eocene age. They suggest that these beds were deposited in a basin with a depocenter in the vicinity of Point Arena and that the Eocene basin shallowed southward and did not exist at the south end of the present Bodega Basin. Following the Eocene transgression and basin sedimentation, an interval of complex folding and faulting produced a structural grain that differs from that imposed by younger post-Miocene deformation. As Hoskins and Griffiths (1971) note, these old structures are difficult to define by available seismic-reflection data. Following deformation that produced the old structures the area was raised relative to sea level, and erosion exposed the once deeply buried dense Eocene sandstone. The absence of Oligocene strata, as reported from sea-floor samples, from offshore wells, or from onshore at Point Reyes, indicates that erosion persisted through late Oligocene time. Erosion also truncated granitic basement rocks that were later to rise as the Farallon Ridge.

This episode of Paleogene erosion ended with an early Miocene transgression that deposited a basal sandstone followed by fine bathyal sand and deep-sea mud on eroded Eocene rocks beneath the eastern part of the offshore basin. The same early Miocene transgressive event may be recorded at Pescadero Beach on the Pigeon Point block, where pillow lava of early Miocene age (22.0 ± 0.7 Ma) overlies a dolomite that in turn rests on a shallow-marine sandstone with a coarse basal conglomerate (Eric Taylor, oral communication, 1985). As noted in the preceding description of the Pigeon Point block, the basal conglomerate lies on an erosional surface cut across deformed Upper Cretaceous submarine-fan deposits (Pigeon Point Formation). Taylor has collected a late late Oligocene to early Miocene ("Vaqueros" Stage) marine molluscan fauna from the basal sandstone and dolomite, and from sandy stringers between the lava pillows.

In nearby outer Santa Cruz Basin and in offshore Santa Maria Basin, the cessation of early and middle Tertiary erosion and the beginning of downward displacement of the eroded basement floors to middle and lower bathyal depths was heralded by the accumulation of thick, locally derived, volcanic sequences possibly related to extension of the continental margin. The early Miocene volcanics at Pescadero Beach may reflect the same regional event. Although the volcanics on the Pigeon Point block are not within Bodega Basin, they lie along the San Gregorio fault that forms the eastern boundary of the basin, suggesting that both areas may have undergone middle Tertiary extension or transtension.

Marine volcanic rocks were reported in Bodega Basin proper by Hoskins and Griffiths (1971) (Figure 20), and are described as being overlain by middle and upper Miocene marine shale and chert (the Monterey? Formation), and underlain by middle Miocene basal marine sandstone that lies unconformably on lower Miocene shale. The occurrence of volcanic rocks in the basin is uncertain. With the exception of two cobble-sized fragments of basalt or andesite recovered from a depth of 2345 m in well 053, that could be

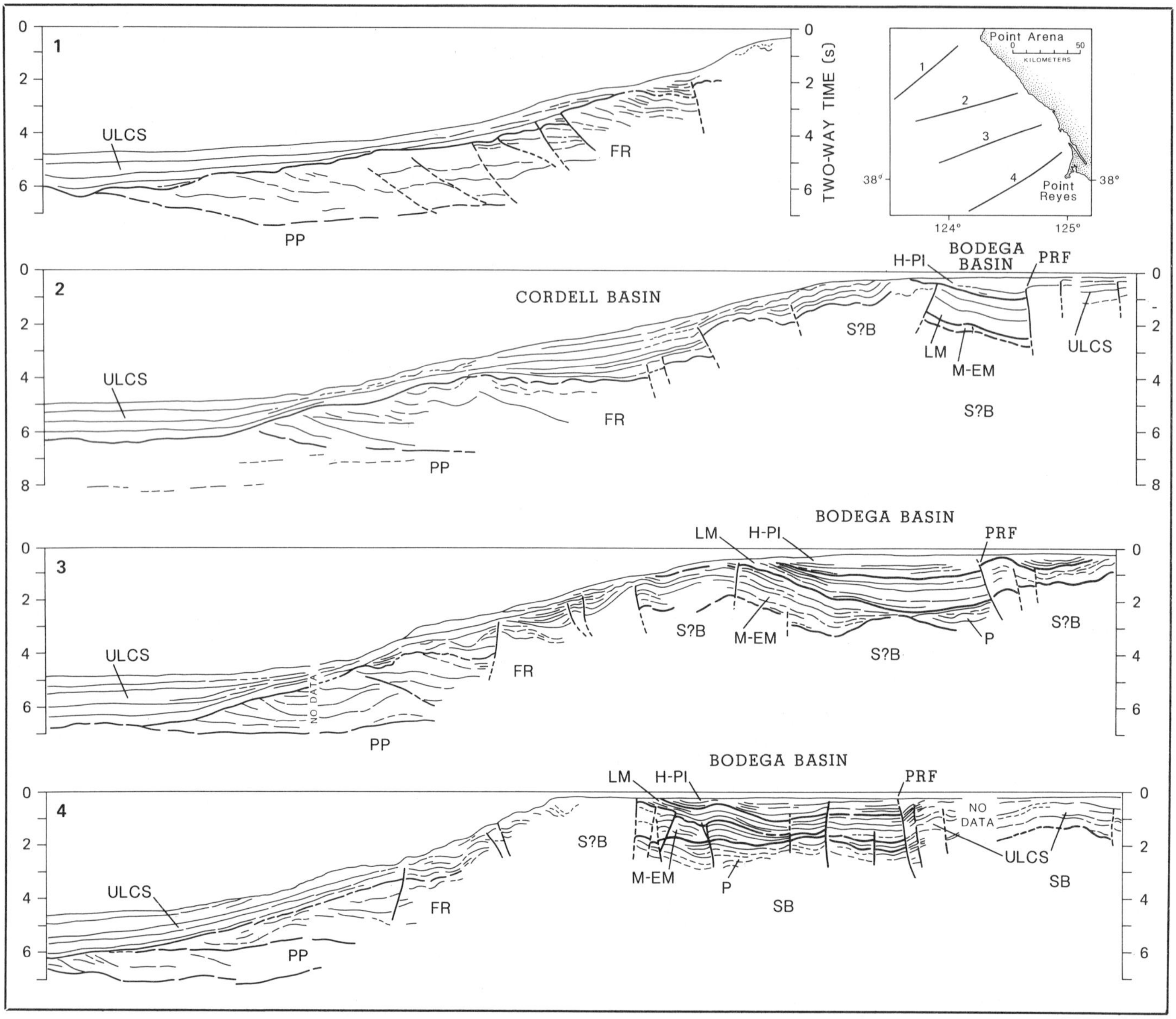

Figure 18. Interpreted time sections of migrated multichannel seismic-reflection profiles between Point Reyes and Point Arena. Symbols—H-Pl, Holocene to Pliocene; LM, late Miocene; M-EM, middle to early Miocene; P, Paleogene; ULCS, undifferentiated late Cenozoic strata; FR, Franciscan rocks of unknown age; SB, Salinian basement; PP, Pacific plate; PRFZ, Point Reyes fault zone.

clasts in sandstone of early(?) Eocene age, driller's logs make no mention of volcanic rocks from either cuttings or sidewall cores. Middle Miocene volcanics are absent from stratigraphic columns of Zieglar and Cassell (1978), constructed from the driller's logs (Figure 19). Clark et al. (1984) describe possibly correlative volcanics, or volcanics resulting from a continuing episode of extension along the San Andreas fault. They note that Standard-Tevis No. well near Double Point on the Point Reyes peninsula, approximately 4 km west of the San Andreas fault, penetrated basalt (at a depth interval of 1502 to 1506 m) between middle Miocene (Monterey? Formation) and lower Eocene or Paleocene strata. Clark et al. (1984) suggest that the volcanics occupy a stratigraphic position similar to volcanics they believe to be of Oligocene age; however, stratigraphic control in the well does not rule out a possible Oligocene to middle Miocene age.

The late Oligocene and early Miocene transgression that marked the close of Oligocene erosion is not recorded by deposits on Point Reyes peninsula. Erosion on the peninsula appears to have continued until a widespread middle Miocene transgression occurred across the entire offshore basin and the peninsula. In the offshore, the transgression deposited a basal sandstone overlain by deep-water shale and chert (Monterey? Formation) during the middle and early late Miocene. On the peninsula, the transgression deposited a similar sequence with a basal sandstone (Laird Sandstone) directly overlying eroded crystalline Salinian basement rocks in inland exposures. Sequentially, siliceous mudstone, porcelainite, chert and shale of the Monterey Formation conformably overlie the sandstone. Clark et al. (1984) assigned a middle to late Miocene age (Luisian to Mohnian Stages) to the basal sandstone and the overlying Monterey Formation, based on benthic foraminifera from the basal part of the Monterey Formation. The foraminiferal fauna

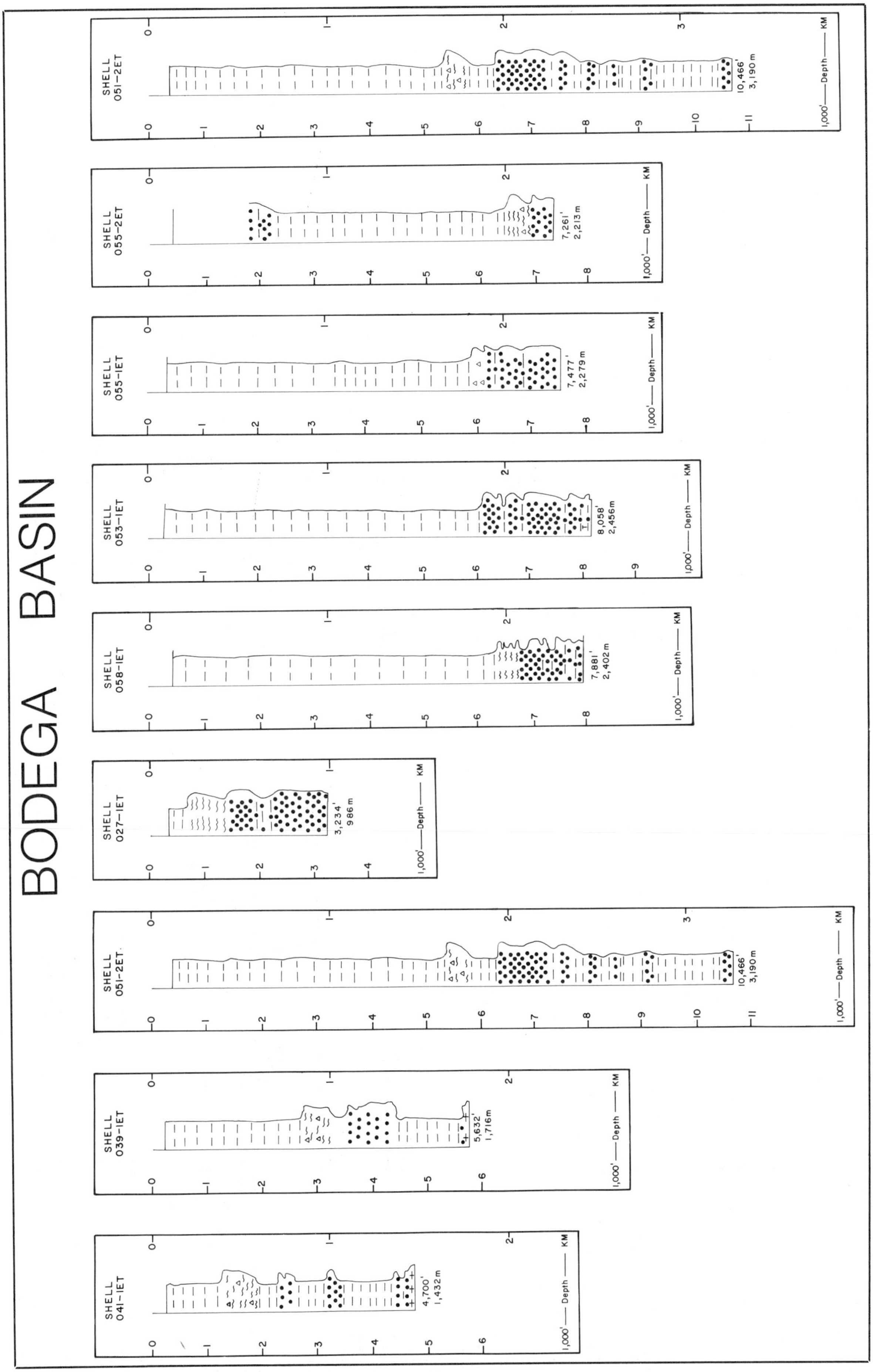

Figure 19. Stratigraphic columns for exploratory wells in Bodega Basin (Zieglar and Cassell, 1978). See Figure 1 for well locations and Figure 9 for explanation.

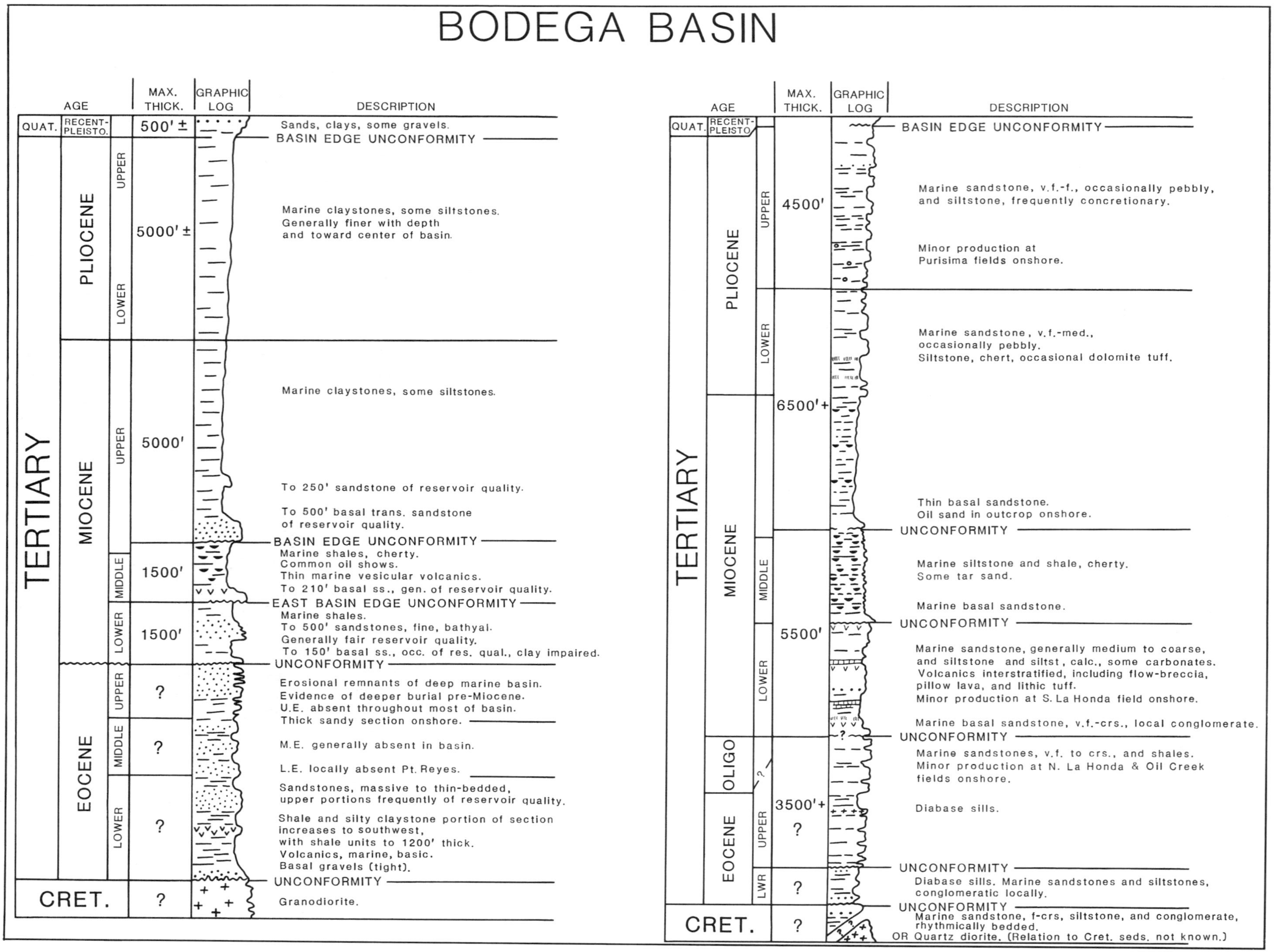
BODEGA BASIN
AGE
MAX. THICK.
GRAPHIC LOG
DESCRIPTION
QUAT.
RECENT-PLEISTO.
500' ±
Sands, clays, some gravels.
BASIN EDGE UNCONFORMITY
TERTIARY
PLIOCENE
UPPER
LOWER
5000' ±
Marine claystones, some siltstones.
Generally finer with depth
and toward center of basin.
MIOCENE
UPPER
5000'
Marine claystones, some siltstones.
To 250' sandstone of reservoir quality.
To 500' basal trans. sandstone
of reservoir quality.
BASIN EDGE UNCONFORMITY
MIDDLE
1500'
Marine shales, cherty.
Common oil shows.
Thin marine vesicular volcanics.
To 210' basal ss., gen. of reservoir quality.
EAST BASIN EDGE UNCONFORMITY
LOWER
1500'
Marine shales.
To 500' sandstones, fine, bathyal.
Generally fair reservoir quality.
To 150' basal ss., occ. of res. qual., clay impaired.
UNCONFORMITY
EOCENE
UPPER
?
Erosional remnants of deep marine basin.
Evidence of deeper burial pre-Miocene.
U.E. absent throughout most of basin.
Thick sandy section onshore.
MIDDLE
?
M.E. generally absent in basin.
L.E. locally absent Pt. Reyes.
LOWER
?
Sandstones, massive to thin-bedded,
upper portions frequently of reservoir quality.
Shale and silty claystone portion of section
increases to southwest,
with shale units to 1200' thick.
Volcanics, marine, basic.
Basal gravels (tight).
UNCONFORMITY
CRET.
?
Granodiorite.
AGE
MAX. THICK.
GRAPHIC LOG
DESCRIPTION
QUAT.
RECENT-PLEISTO
BASIN EDGE UNCONFORMITY
TERTIARY
PLIOCENE
UPPER
4500'
Marine sandstone, v.f.-f., occasionally pebbly,
and siltstone, frequently concretionary.
Minor production at
Purisima fields onshore.
LOWER
Marine sandstone, v.f.-med.,
occasionally pebbly.
Siltstone, chert, occasional dolomite tuff.
MIOCENE
6500'+
Thin basal sandstone.
Oil sand in outcrop onshore.
UNCONFORMITY
MIDDLE
Marine siltstone and shale, cherty.
Some tar sand.
Marine basal sandstone.
UNCONFORMITY
LOWER
5500'
Marine sandstone, generally medium to coarse,
and siltstone and siltst, calc., some carbonates.
Volcanics interstratified, including flow-breccia,
pillow lava, and lithic tuff.
Minor production at S. La Honda field onshore.
Marine basal sandstone, v.f.-crs., local conglomerate.
UNCONFORMITY
OLIGO
?
Marine sandstones, v.f. to crs., and shales.
Minor production at N. La Honda & Oil Creek
fields onshore.
EOCENE
UPPER
3500'+
?
Diabase sills.
UNCONFORMITY
LWR
?
Diabase sills. Marine sandstones and siltstones,
conglomeratic locally.
UNCONFORMITY
CRET.
?
Marine sandstone, f-crs, siltstone, and conglomerate,
rhythmically bedded.
OR Quartz diorite. (Relation to Cret. seds. not known.)

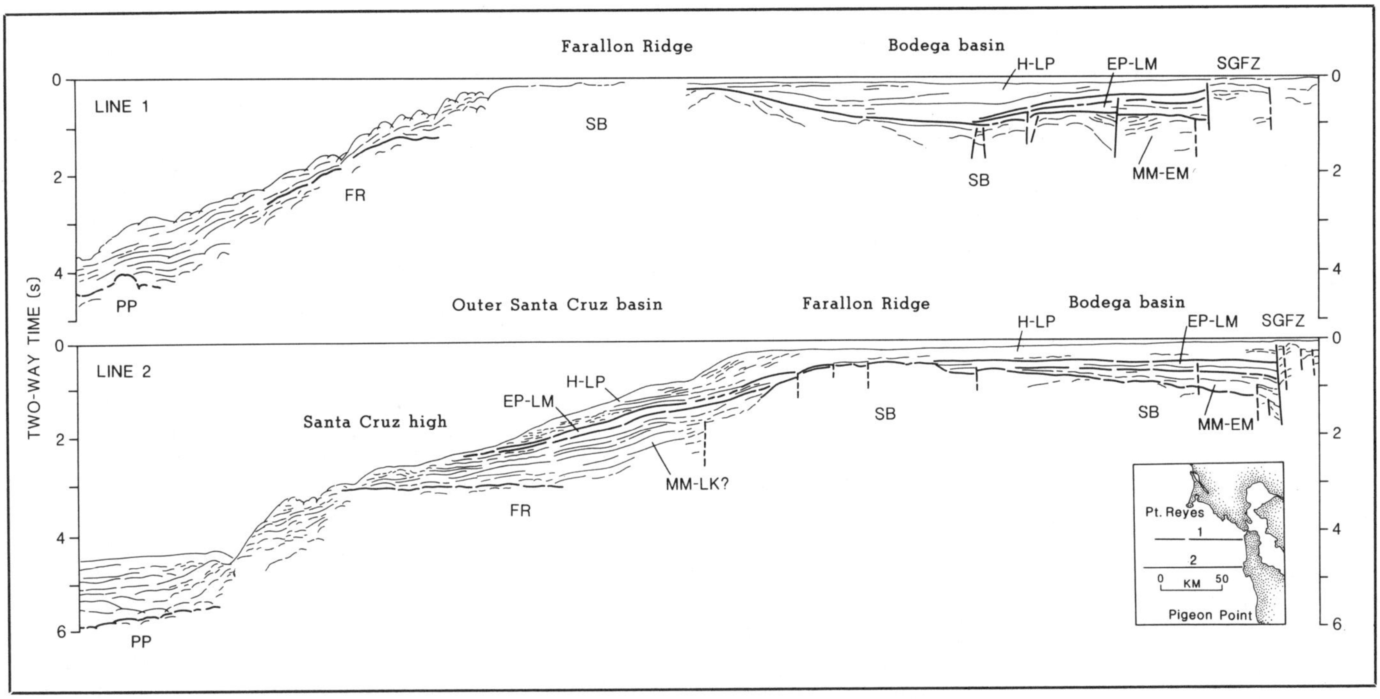

Figure 21. Interpretations of single-channel seismic reflection profiles across southern Bodega Basin and Outer Santa Cruz Basin. Symbols—H-LP, Holocene to late Pliocene; EP-LM, early Pliocene to late Miocene; MM-EM, middle to early Miocene; MM-LK, middle Miocene to late Cretaceous; FR, Franciscan rocks of unknown age; SB, Salinian basement; PP, Pacific plate; SGFZ, San Gregorio fault zone.

from the Monterey Formation described by Clark et al. (1984) indicate deposition in bathyal to mid-bathyal depths (K. McDougall, oral communication, 1986). Siliceous shale and chert presumed to be correlative in the offshore basin were probably deposited in comparable water depths. Deposition of early to middle Miocene strata in the offshore was not confined to the present basin, but was widespread. Westward across the basin, these rocks thicken slightly and were deposited across the granitic basement that was later uplifted to form the Farallon Ridge.

Deep-water sedimentation was followed by an interval of deformation and a subsequent marine transgression in the early late Miocene. The resulting erosional unconformity truncated folded strata of the Monterey Formation on the Point Reyes peninsula and can be traced into the offshore basin. The unconformity has remained remarkably undeformed in the central and southern Gulf of the Farallones, and led Cooper (1971) to call this area the Farallon Platform. The unconformity slopes very gently eastward across the gulf toward the offshore extension of the San Gregorio fault. Along the eastern margin of the basin it has been locally deformed by folds associated with compression that accompanied down-to-basin separation on the San Gregorio fault (Figure 21). Before early late Miocene erosion, the Farallon structural high was elevated relative to the Gulf of the Farallones, and the lower and middle Miocene rocks were tilted upward along the east side of the rising granitic ridge. Deformation was not confined to the ridge. Beneath the gulf, discontinuous acoustic penetration reveals broad elongate northwest-trending folds and faults that trend northwest, subparallel to the Farallon structural high (Figure 22). The folds lie oblique to, and are truncated by, the offshore San Gregorio fault to the east. The folds and faults beneath the gulf can be traced to the north, where they appear to have been reactivated by younger deformation that involved the

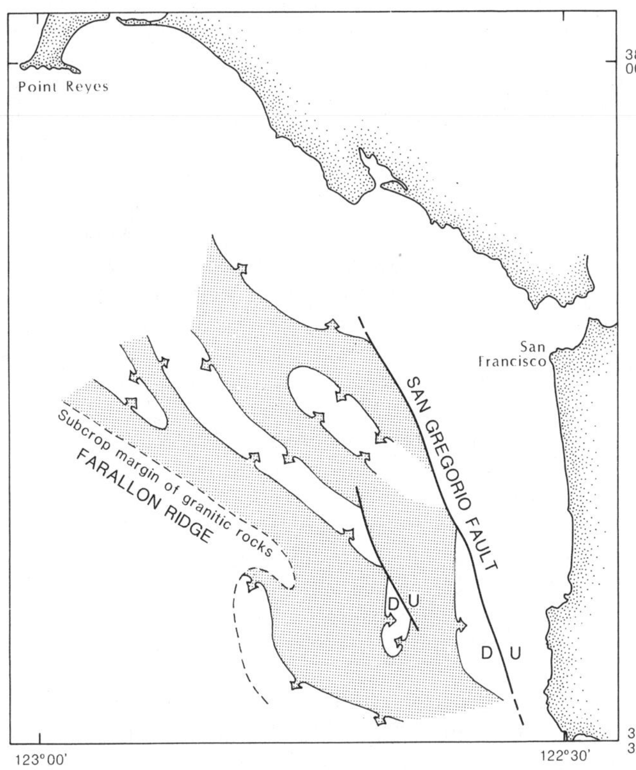

Figure 22. Pre-Miocene structures beneath the Gulf of the Farallones. Offshore patterned areas are structural highs.

Figure 20. Generalized stratigraphic columns for northern (left) and southern (right) Bodega Basin (from Hoskins and Griffiths, 1971).

overlying early late Miocene unconformity and younger strata. These structures converge with the eastern side of the Farallon structural high in the vicinity of Cordell Bank (Figure 14), where there is clear evidence for fault offset (down-to-the-east separation) between the basin and the Farallon structural high.

Development of the early late Miocene unconformity was followed by deposition of as much as 3 km of late Miocene and Pliocene marine claystone and siltstone. The geometry of these strata indicates a predominantly landward sediment source; the strata thin westward across the basin and onlap the upturned eroded edge of early and middle Miocene rocks and the Farallon structural high. On Point Reyes peninsula, deposits of the early late Miocene transgression and a subsequent basin-filling sequence are as much as 2.5 km thick. The basal transgressive sandstone (Santa Margarita Sandstone) of late Miocene age (Clark et al., 1984) has a several-meter-thick basal conglomerate that contains locally derived clasts of granite, Monterey Formation chert, and clasts of recycled porphyritic siliceous volcanics. The thick section above the basal sandstone is composed of a lower siliceous mudstone unit (Santa Cruz Mudstone, of late late Miocene age) and an upper siltstone, mudstone, and sandstone unit (Purisima Formation, late late Miocene and early Pliocene age; Clark et al., 1984). Diatom faunas from these two units indicate open-ocean conditions with water depths in excess of 500 m in late Miocene and early Pliocene time, with subsequent shallowing toward the top of the Purisima Formation (John Barron, oral communication, 1986).

The last major episode of deformation to affect Bodega Basin refolded folds and faults within the basin, and appears to have resulted from transpression that accompanied lateral slip along the offshore San Andreas fault (Figure 14). This folding is well developed in the northern part of the basin. Similar folding in the northern Gulf of the Farallones gives the impression that the Point Reyes peninsula acted as a buttress. During this deformation, the Point Reyes peninsula was elevated above the adjacent basin along the Point Reyes fault (Figure 14). The Point Reyes fault follows a sinuous westward trend across the mouth of Drakes Bay, and then turns northwestward. An en echelon continuation of the Drakes Bay fault that can be mapped to approximately the latitude of Gualala, appears to have been folded with the adjacent basin structures. The fault displays large vertical separation. Granitic basement is displaced vertically at least 1.4 km by the fault at Point Reyes, where granitic rocks are exposed on the peninsula and also occur in the bottom of exploratory offshore wells (039, 041) in the Gulf of the Farallones. Down-to-the-west vertical separation of similar magnitude occurs along most of the fault. The fault plane dips steeply everywhere, and has the appearance of high-angle reverse displacement (Figure 18). The age of fault offset is not firmly known. The youngest dated rocks (Purisima Formation) displaced by the elevation of the Point Reyes peninsula are late late Miocene in age (diatom Zone X); however, Clark et al. (1984) conjecture that the uppermost undated part of the formation may be as young as early Pliocene. Thus the displacement may have been early Pliocene or younger. Hoskins and Griffiths (1971) indicate that the fault offsets late Pliocene strata in the offshore, and is of Pleistocene age. The apparent high-angle reverse displacement on the Point Reyes fault, the variation in strike of the fault through nearly 45° of azimuth, and the uplift of the Point Reyes peninsula above the basin of which it was once a part, suggest that the fault may be a thrust at depth. Possibly comparable thrusts have been mapped elsewhere south of San Francisco along the east side of the San Andreas fault (McLaughlin, 1974; Sorg and McLaughlin, 1980). Alternatively, the Point Reyes fault may have originated as a normal or high-angle reverse fault during transtensional wrenching, and its sinuosity, which approximates the sinuosity of the folds and faults in Bodega Basin, may be due to the episode of late Neogene northeast–southwest compression.

Cordell Basin, which lies on the accretionary wedge west of the Farallon Ridge (Figures 1 and 18), contains a well-bedded sedimentary section at least 1-sec thick. The lowermost strata onlap the structurally high accretionary wedge beneath the western edge of the basin. Gentle shoreward dips in these basal strata suggest slight elevation of the western basin during the beginning of basin filling. High-angle reverse faults along the eastern side of the basin postdate most of the basin fill. Rocks in the basin fill are unsampled and their ages are unknown; however, if sedimentation started at the end of subduction, the section might contain early(?) Miocene through Pliocene age rocks.

POINT ARENA BASIN

Point Arena basin is fan shaped in plan (Figure 1). From its narrow southern end near Point Arena toward its northern boundary at the Mendocino escarpment, the basin widens between the San Andreas fault on the east and a northwest-trending partly buried wide structural high named the Oconostota ridge (Curray, 1966) on the west. Oconostota ridge does not appear to have been as elevated structurally as the highs that form the western boundaries of the shelf basins to the south, and Point Arena Basin probably was never as deep structurally as the southern basins.

Basement Rocks

Hoskins and Griffiths (1971) attribute the difference in the style of structural deformation between Bodega and Point Arena basins to differing basement rocks. An exploratory well (033) drilled just north of Point Arena bottomed in pre-Cretaceous (Jurassic?) metasedimentary rocks (Figure 1), whereas Bodega Basin may be underlain by granitic basement. As noted above, the boundary between differing basement rocks beneath these basins is probably expressed as the northwest-trending Navarro discontinuity along which there is a change in both the Neogene structures and the residual magnetic-anomaly pattern.

Basement rocks north of the discontinuity may constitute more than a single terrane, but are here referred to collectively as the Vizcaino block after the submarine canyon and offshore knoll of the same name. The northern edge of the Vizcaino block lies along the Gorda Escarpment, where the surface of the block stands as much as 1500 m higher than the sea floor on the adjacent Gorda plate. The Vizcaino block is separated from the Gorda plate by the active right-lateral Mendocino transform fault, along which the Gorda plate is moving eastward to be subducted beneath the North American plate (Figure 23). The block is bounded on the east by the offshore San Andreas fault.

The western margin of the block forms the continental slope that north from a projection of the Navarro discontinuity turns slightly westward and intersects the Mendocino Escarpment at approximately long. 126°W (Figure 23b). The residual magnetic anomaly pattern changes between the dominant northwest trend on the Vizcaino block and the north trend on the Pacific plate. The location of the pattern change cannot be accurately determined without additional detailed magnetic data; however, it is tentatively extended from the Navarro discontinuity (Figure 17), which is clearly apparent in detailed magnetic data, generally northward along the continental slope (Figure 23a). On the oceanic plate, the Pioneer Fracture Zone lies along a change in the trend of the anomalies. The fracture zone has surface expression on a side-scan sonar (GLORIA) survey to long. 126°W (EEZ-SCAN 84 Scientific Staff, 1986) and continues eastward as a sediment-covered bedrock ridge to lat. 36°N, long. 125° 30′W (Normark and Gutmacher, 1985). An east-trending magnetic high that lies south of the junction of the fracture zone and anomaly 7A (?) continues eastward on the oceanic plate and extends to within 15 km of the coast, suggesting that the fracture zone may lie at depth beneath the continental shelf.

A crustal density model based on Bouguer gravity and densities derived from seismic-refraction velocities (Griscom, 1973a) is satisfied by a landward-dipping moho beneath the Vizcaino block (Figure 23c). Subduction beneath the block is also suggested by the interpretation of migrated multichannel reflection profiles (Figure 23d). These records show a shoreward-thickening wedge of acoustically poorly defined rocks beneath the continental slope. This wedge rises to form a buried structural high at the outer edge of the shelf. Within the wedge, poorly defined landward-dipping reflections may indicate westward-directed low-angle thrust faults. Although the poor acoustic expression of rocks within the wedge obscures its internal structure, the geometry of the wedge and the possible internal low-angle thrusts suggest that it may be accretionary. Presumably, the toe of rocks in the wedge lies at the base of the continental slope. However, the poor acoustic signal that precludes differentiation of rocks within the wedge from the Pacific plate, as well as gently dipping reflecting horizons within the plate, make it difficult to identify the top of the subducted oceanic plate beneath the wedge.

Rocks collected from four localities on the Vizcaino block (Figure 23) include (1) pre-Cretaceous (Jurassic?) metasedimentary rocks noted by Hoskins and Griffiths (1971) from the bottom of exploratory well 033 northwest of Point Arena, (2) graywacke with nondiagnostic middle Eocene to Oligocene nannofossils dredged from the southeastern wall of Noyo Canyon (McCulloch, Utter, and Menack, 1985), (3) undated andesite overlain by upper Oligocene(?) marine nannofossil ooze recovered from the bottom of DSDP Hole 173 (Kulm and von Huene, 1973; Ingle, 1973), and (4) cobbles and boulders of pigeonite-bearing tholeiitic basalt, basalt(?) conglomerate, and single boulders of calcareous ooze and serpentine dredged from the northwest corner of the block (Krause, Menard, and Smith, 1964).

The andesite in DSDP Hole 173 probably originated in an island-arc setting. MacLeod and Pratt (1973) conclude that the andesite is similar to Tertiary and Quaternary basaltic andesites from the Cascade Range, and that it differs in composition from typical abyssal basalts and from most rocks of similar SiO_2 content known from seamounts and oceanic islands. High fragmentation of the volcanics and their texture (glassy to aphanitic matrix with vesicles and small feldspar phenocrysts) led MacLeod and Pratt to suggest that the andesite originated as a subaqueous breccia or a pillow-breccia flow. The western edge of the Vizcaino block appears to be a subduction melange, as seismic profiles suggest (Figure 23d); however, the reflection data are insufficient to define the basement structure of the remainder of the block. Although metasedimentary rocks, graywacke, andesite, and tholeiitic basalt could coexist in an island-arc terrane, the northwest trend of the Neogene structures and residual magnetic anomalies in the block generally parallel lateral-slip elements along the margin and suturing within the adjacent King Range terrane (McLaughlin et al., 1982), which together suggest the possibility of juxtaposed tectonic slivers.

Stratigraphy and Tectonics

Stratigraphic data for Point Arena Basin come from three offshore exploratory wells at the southern end of the basin (Figure 24), from DSDP Hole 173 (Figure 23b), and from the possible onshore extension of the basin at Point Arena. The Jurassic(?) metasedimentary basement rocks are overlain unconformably by Upper Cretaceous marine silty shale and fine sandstone within the offshore basin. Hoskins and Griffiths (1971) suggest that the dense Upper Cretaceous rocks are the erosionally thinned remnants of a formerly more extensive deposit that covered most of the shelf from Point Arena to Point Conception. However, recent tectonic models suggest that the tectonostratigraphic elements that constitute the present shelf were not assembled by this time. Lower Eocene marine shale and sandstone were deposited on the eroded Upper Cretaceous rocks in the Point Arena Basin. As in the offshore basins to the south, these Eocene rocks are reported to be dense erosional remnants of a once thicker section. If deposition continued throughout the Eocene, these rocks were removed by erosion probably related to Oligocene emergence because the only reliably dated Oligocene strata in the basin are of late Oligocene age (DSDP 173). The subsequent early Miocene transgression deposited a basal sand on the eroded lower Eocene rocks. Deposition in early through middle Miocene time suggests increasing water depth accompanied by a decrease in terrigenous contribution. This transition is reflected by a sedimentary section that changed in early Miocene from sandstone to shale, and in middle Miocene time to thin-bedded sandstone, shale and chert (Monterey Formation equivalent?). While the original extent of the Upper Cretaceous and lower Eocene rocks is unknown, Miocene strata were deposited across the basin, the Oconostota structural ridge along the western edge of the Vizcaino block (where they occur in DSDP 173), and seaward over the continental slope.

Folded strata of presumed early and middle Miocene age are truncated by an unconformity in the southern part of the Point Arena Basin (Figure 25). This unconformity has been dated as early late Miocene by Hoskins and Griffiths (1971). Deformation appears largely to have ceased during the final cutting of the unconformity. The overlying thick sections of presumed upper Miocene and lower Pliocene marine mudstone and claystone, which indicate increased contribution of terrigenous debris, were deposited conformably or by downlap onto the unconformity. Minor syndepositional deformation produced thickening of the upper Miocene and

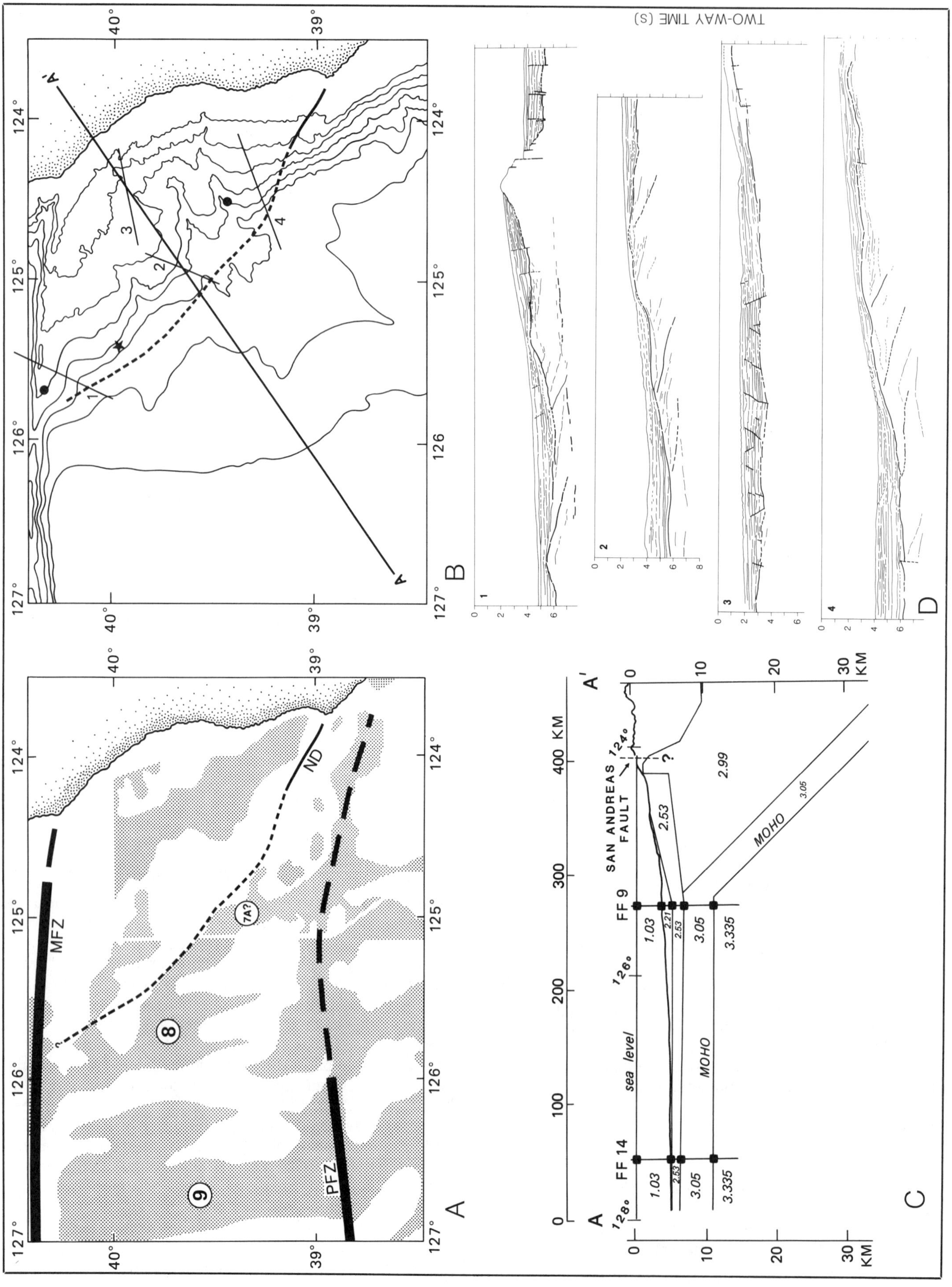
A
B
C
D
MFZ
PFZ
ND
7A?
8
9
SAN ANDREAS FAULT
FF 9
FF 14
sea level
MOHO
A
A'
KM
TWO-WAY TIME (S)
1.03
2.21
2.53
3.05
3.335
2.99

lower Pliocene strata in the downwarped areas and may reflect the initial phase of the subsequent northeast–southwest compression. This compression developed the northwest-trending folds that characterize the late Neogene structures in the basin, and it produced faults that generally are parallel to, and concentrated along, the anticlines (Figure 25). The northeast–southwest compression was probably accompanied by elevation of the adjacent land area, as indicated by coarsening of the sediment to include substantial amounts of sand, and by the change from blanket deposition to seaward progradation and thinning of foreset strata whose upper surface forms the present continental shelf.

A relatively complete late Neogene section was recovered from DSDP Hole 173, drilled at a water depth of 2927 m on Oconostota Ridge (Figures 14 and 23) (Kulm and von Huene, 1973; Ingle, 1973). The drill site is remote from the basin, and unlike the basin that underwent appreciable late Neogene deformation, paleobathymetric evidence from microfossils indicates that the drill site remained at mid- to lower- bathyal depths and above the CCD (carbonate compensation depth) throughout the late Neogene (Ingle, 1973). Nevertheless, tectonic events and global sea level changes that affected both the basin and the onshore Point Arena area appear to be recorded in the drill cores by depositional hiatuses or compressed biological zones and by changes in the kind and rates of sedimentation (Ingle, 1978; Barron, 1986a, b, and in press). Near the bottom of the 333.5-m-deep hole, upper Oligocene(?) pelagic nannofossil ooze resting directly on undated andesite at a core depth of 320 m may represent initial deposition following the period of Oligocene erosion that truncated Eocene strata in the basin and at Point Arena.

Pelagic sedimentation prevailed at DSDP Site 173 with the deposition of diatomite from early into early late Miocene time, to a core depth of 193 m. Deposition then changed to dominantly hemipelagic silt and clay and remained hemipelagic throughout the late Miocene, Pliocene, and Pleistocene.

Barron (in press) has recognized depositional hiatuses or biologically compressed zones above the basal Oligocene unconformity in Hole 173 across the early-middle Miocene boundary (17.5–15.5 Ma), at the middle-late Miocene boundary (10 Ma), at the Miocene-Pliocene boundary (5 Ma), and during early to late Pliocene time (4.3–3.2 Ma). On the basis of dated diatoms, radiolaria, nannofossils, and numerous volcanic tephra layers from Hole 173, depositional hiatuses occur from 20–17 Ma (A. Sarna-Wojcicki, written communication, 1986) and from 4–2.8 Ma (Sarna-Wojcicki et al., 1987). Of these depositional hiatuses, the 10 Ma event may be related to a regional tectonic event that produced the early late Miocene angular unconformity in Point Arena Basin and the other offshore basins. The change from pelagic- to hemipelagic-dominated sedimentation occurred at approximately 6.5 Ma in Hole 173 (Kulm and von Huene, 1973). Ingle (1978) attributed the change to the global sea level lowering (Vail and Mitchum, 1979) that produced the Messinian salinity crisis in the isolated Mediterranean (Hsu et al., 1977) and that accompanied the growth of the Antarctic ice sheet (Hodell, Elmstrom, and Kennett, 1986). Continued relatively slow deposition at Hole 173 across this sedimentologic boundary (Sarna-Wojcicki, written communication, 1986) and the lack of evidence for accompanying tectonism suggest lowering sea level accompanied by a seaward shift in the shoreline and distal edge of a land-derived terrestrial facies rather than coastal uplift.

Figure 23. Aspects of the Vizcaino block. A. Dashed light line joining the Navarro discontinuity (ND) indicates the approximate western boundary of the block as suggested by the magnetic anomaly pattern and the physiography. Positive magnetic anomalies are shaded. Heavy solid lines are the Pioneer Fracture Zone (PFZ) traced by surface and subsurface expression, and the Mendocino Fracture Zone (MFZ). Long dashes are possible eastward extensions. Magnetic data west of 25W from Mason and Raff (1961), and east of 125°W from Emilia, Berg and Bales (1968). B. Bathymetry with 500-m-contour interval (Wilde, Normark, and Chase, 1976; Wilde et al., 1978) and location of gravity and seismic profiles. Dredge sites indicated by dots, DSDP Site 173 indicated by star. C. Crustal density model based on Bouguer gravity and seismic refraction velocities across the Pacific plate and Vizcaino block (Griscom, 1973a). D. Interpreted time sections of migrated multichannel seismic-reflection profiles.

Sedimentation proceeded slowly, at approximately 16 m/m.y., through the change to hemipelagic sedimentation. It remained slow until the 4 to 2.8 Ma depositional hiatus, after which sedimentation increased to approximately 46 m/m.y., and remained high until 0.5 Ma (Sarna-Wojcicki et al., 1987). Within the nearby basin the period of slow sedimentation occurred during blanket deposition of lower Pliocene siltstone and claystone. Deposition was interrupted in the basin with deformation and minor erosion, presumably during the 4–2.8 Ma hiatus at Site 173. Deformation was followed by the building of local coastal relief, accompanied by the introduction of upper Pliocene sandstone and the seaward progradation of the shoreline. This uplift is reflected in the core at Site 173 by the period of rapid sedimentation that overlies the 4–2.8 Ma hiatus.

In the onshore Point Arena area west of the San Andreas fault (Weaver, 1943; Boyle, 1967; Addicot, 1967; Wentworth, 1968, 1972), Upper Cretaceous and Paleocene(?) to middle Eocene marine mudstone and sandstone turbidites (Gualala and German Rancho formations) underwent deformation that produced large west-trending folds intersecting the San Andreas fault. The folded section was eroded prior to extrusion of shallow water marine basalt flows and pillow lavas (Iverson Basalt; average age 23.8 m.y.; Turner, 1970; Dalrymple, 1979) and before the deposition of interbedded marine sandstone at about the Oligocene–Miocene boundary (23.7 Ma; Berggren et al., 1985). Thus, the Oligocene, and possibly the late Eocene, were also periods of erosion in the onshore Point Arena area. Boyle (1967) attributed the pre-Oligocene folding to drag associated with lateral slip on the San Andreas fault. However, this deformation was too early to be attributed to movement on the present San Andreas fault (Clarke and Nilsen, 1973). Mapping these early structures in the offshore might clarify their origin, but they are not defined by available seismic-reflection profiles. Onshore, the shallow-marine volcanism at about the Oligocene–Miocene boundary was followed by deep-water marine deposition of late late Oligocene–early Miocene ("Vaqueros" Stage of Addicott, 1972) glauconitic mudstone and sandstone (the Skooner Gulch Formation). An overlying early Miocene (upper Zemorrian or Saucesian) foraminiferal mudstone and sandstone (the Gallaway Formation) in turn was overlain by deep-water marine deposits (Point Arena Formation, possible Monterey Formation correlative). Jahns and Hamilton (1971) divide these deep-water deposits into a lower unit of lower Miocene diatomaceous porcelaneous shale and mudstone, and an upper unit of middle Miocene (Luisian) fine sandstone with asphaltic interbeds.

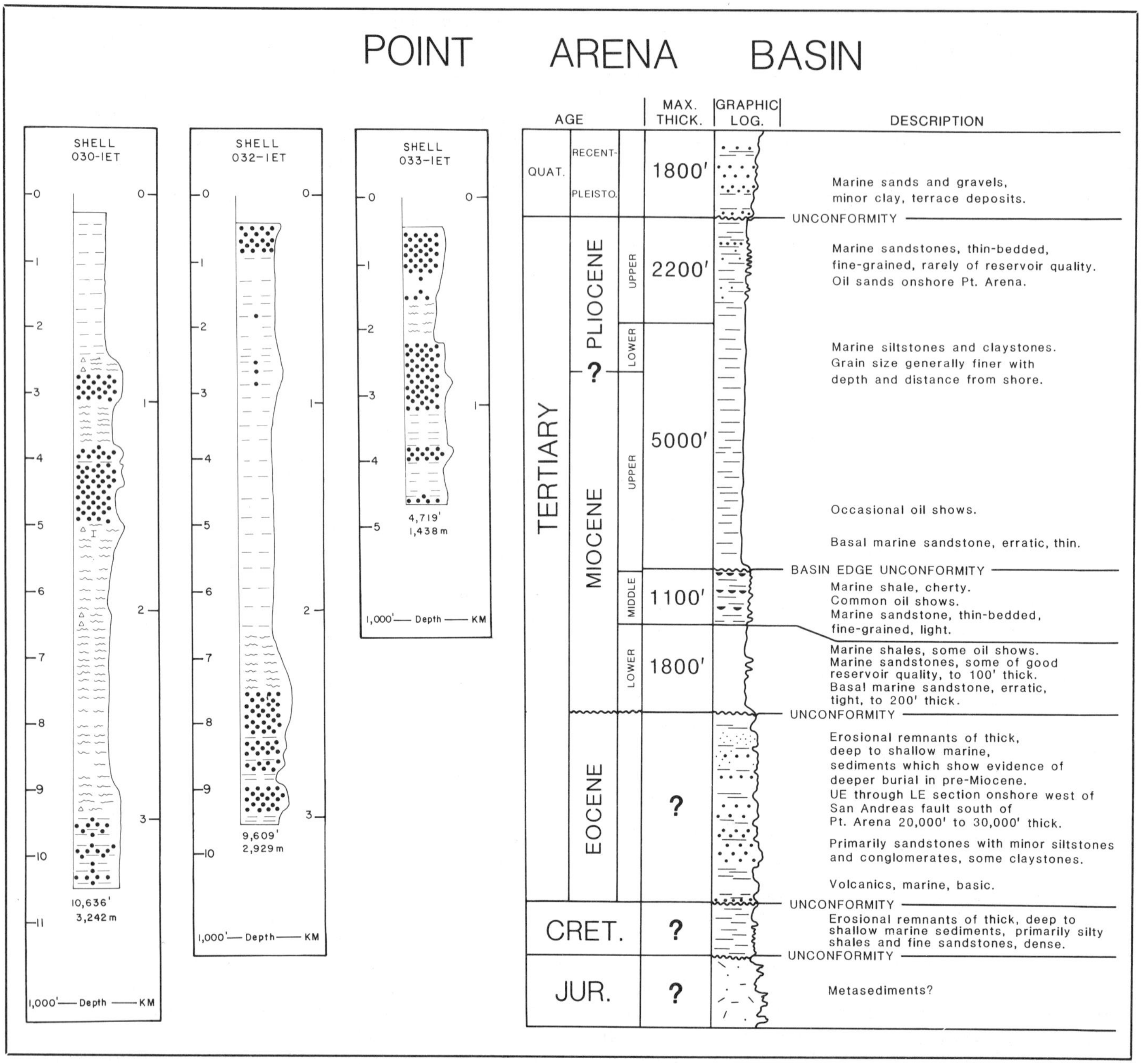

Figure 24. Stratigraphic columns of exploratory wells (from Ziegler and Cassell, 1978) and a generalized stratigraphic column for Point Arena Basin (from Hoskins and Griffiths, 1971). See Figure 9 for explanation.

After middle Miocene deposition, the entire section west of the San Andreas fault zone was deformed by northeast–southwest-directed compression—probably the same middle Pliocene tectonic event that deformed the adjacent offshore Point Arena Basin. Because there are no onshore Pliocene strata west of the San Andreas fault zone in the Point Arena area, this deformation is not precisely dated. Jahns and Hamilton (1971) suggested that Miocene strata onshore were emergent in Pliocene time; however, the presence of upper Pliocene marine strata for many miles to the south along the east side of the San Andreas fault (Higgins, 1960) and in terrace deposits in the adjacent San Andreas fault zone suggest that Pliocene strata were at one time more extensive, and have been removed by erosion. Boyle (1967) suggested that deformation of upper Pliocene marine terrace deposits within the San Andreas fault zone northeast of Point Arena indicated that northeast–southwest compression continued into post-late Pliocene time, but Jahns and Hamilton (1971) attribute terrace deformation to movement on the San Andreas fault.

Figure 25. Map of late Neogene structures and interpretations of single-channel seismic reflection profiles from Point Arena Basin. Structural highs are shaded. Symbols—H-UP, Holocene to upper Pliocene; LP-UM, lower Pliocene upper Miocene; MM-LM, middle and lower Miocene. S indicates surface or near-surface faulting.

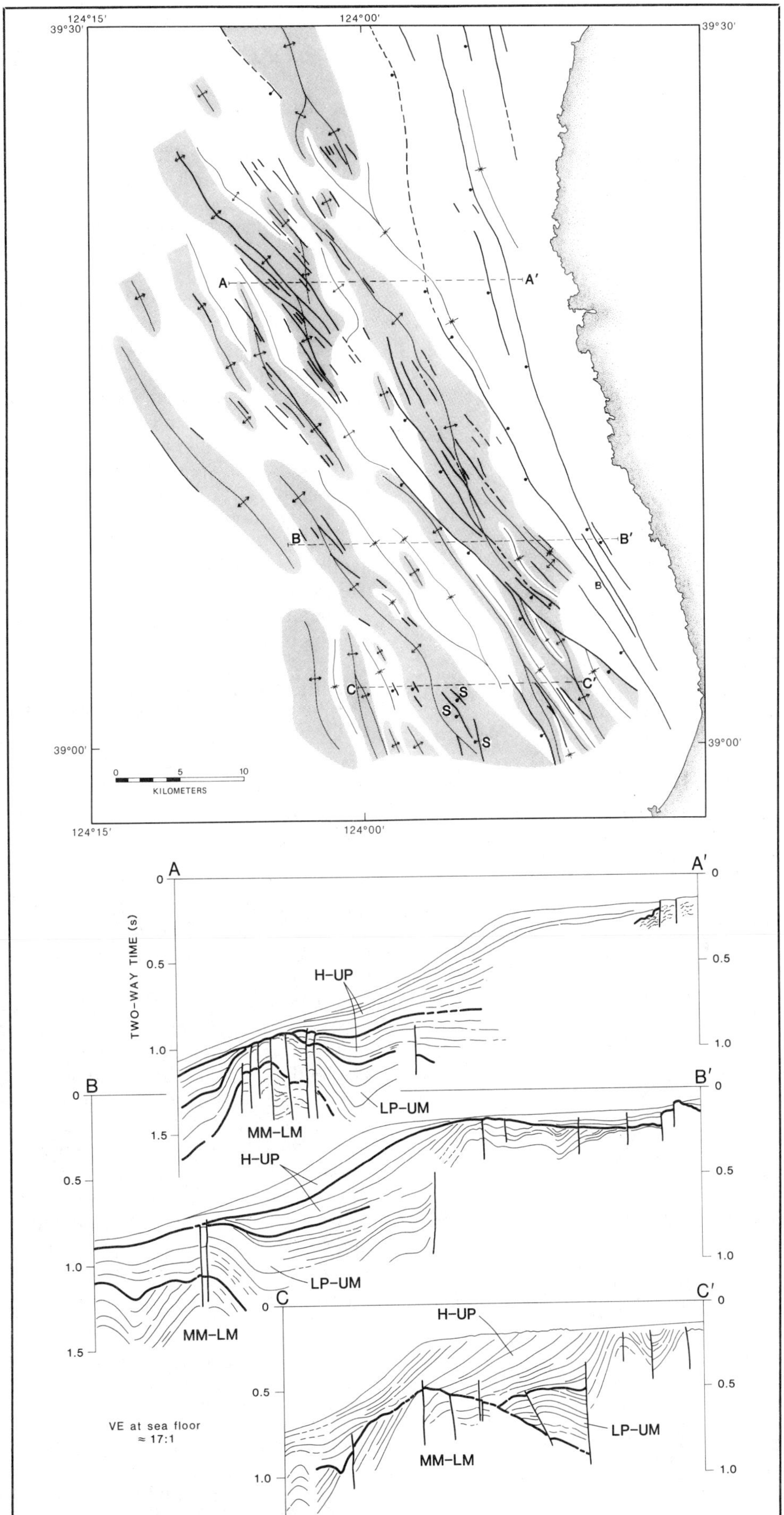
124°15'
124°00'
39°30'
39°00'
A
A'
B
B'
C
C'
S
0
5
10
KILOMETERS
TWO-WAY TIME (s)
0
0.5
1.0
1.5
H-UP
LP-UM
MM-LM
VE at sea floor
≈ 17:1

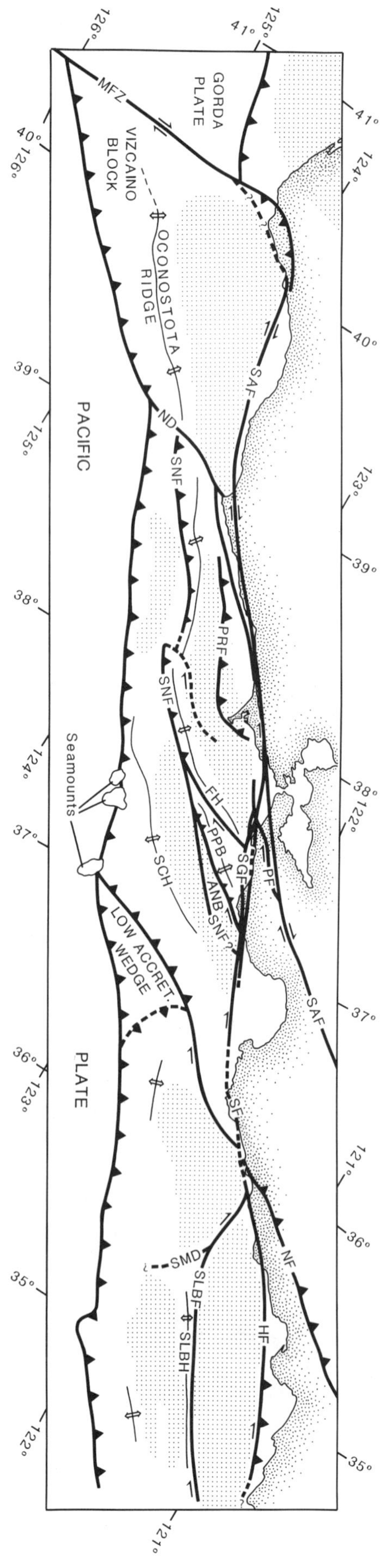

Figure 26. Major structural features of offshore central California. Teeth on upthrown side of high-angle reverse faults or upper plate of thrust faults. See text for discussion. Shaded offshore areas are late Tertiary basins. Faults—SAF, San Andreas; SF, Sur; NF, Nacimento; SNF, Sur-Nacimento; SGF, San Gregorio; HF, Hosgri; SLBF, Santa Lucia Bank; PRF, Point Reyes; PF, Pilarcitos; MFZ, Mendocino Fracture Zone. Structural discontinuities—SMD, San Martin; ND, Navarro. Structural highs—SLBH, Santa Lucia Bank; SCH, Santa Cruz; FH, Farallon. Blocks—PPB, Pigeon Point; ANB, Año Nuevo.

INFERRED STRUCTURAL RELATIONS OF OFFSHORE BASEMENT BLOCKS

We can suggest boundaries of discrete basement blocks beneath the offshore continental margin and outline their general displacement histories, based on the foregoing basin descriptions. Block boundaries differ in detail from terrane boundaries proposed earlier (Figures 4 and 26). Questions remain concerning the composition of some basement blocks.

A paleosubduction zone is preserved along the entire outer edge of the central California continental margin (Figure 26). On multichannel seismic-reflection profiles, with penetration limited by a 5-sec recording interval, the top of the Pacific plate can be followed as much as 30 km shoreward from the toe of the continental slope (Figures 11, 18, and 23). With longer recording, the plate could probably be followed farther landward. Rocks of the continental slope that overlie the subducted plate have been deformed and contain shoreward-dipping reflecting horizons that probably developed along thrust faults. This deformed wedge has been poorly sampled. Undated graywacke and granitic sandstone have been collected from sites on, and just above, the Santa Lucia Escarpment west of Santa Lucia Basin (Figure 1), and mafic volcanic rocks have been recovered from a single site on the Santa Cruz high just south of Pioneer Canyon (McCulloch, Utter, and Menack, 1985). Rocks collected from the deformed accretionary wedge at the outer margin of the Vizcaino block north of the Navarro discontinuity include graywacke (middle Eocene? to Oligocene?), undated island arc andesite (DSDP Hole 173), and undated tholeiitic basalt and serpentine from above the Gorda Escarpment (Figure 23). The internal deformation and possible thrusting within the accretionary wedge, its structural position on the subducted oceanic plate, and its fault juxtaposition with the Salinia terrane (as in onshore exposures) together suggest that it is the tectonic equivalent and probable correlative of the Franciscan subduction melange, as suggested by Curray (1966) and Silver (1969), and part of the San Simeon terrane (Figure 4).

The presence of the paleosubduction zone along the entire central California margin indicates that the transform motion between the North American and Pacific plates that was initiated at the time of their contact (~30 to 28 Ma; Atwater, 1970; Engebretson, 1982) was accompanied by oblique convergence that produced subduction of the Pacific plate. The end of the subduction episode is not closely dated. Undeformed strata overlie the subduction zone and postdate subduction (Figure 18). The lowermost of these undeformed strata, although undated, probably predate the oldest nearby rocks recovered from above the accretionary wedge, which are early Mohnian, middle and upper Miocene shales located ~13 km east of Guide Seamount (McCulloch, Utter, and Menack, 1985).

The paleosubduction zone appears to have been sediment starved north from the latitude of Monterey Bay, and rocks of the accretionary wedge lie directly on the oceanic plate with no intervening sedimentary layer (Figure 18). This may be explained if the northward-migrating fracture zone that was progressively consumed along the North American plate had sufficient topographic relief to act as a seafloor barrier to sediment derived from the continent while the accretionary wedge was forming on the subducting Farallon plate.

Within the offshore area shown in Figure 4 as San Simeon terrane, there are several discrete basement blocks. Offshore Santa Maria Basin at the south end of the central California shelf (Figure 1) lies on a block bounded by the Hosgri and Santa Lucia Bank faults and by the San Martin structural discontinuity (Figures 5 and 7). Lateral displacement along the Hosgri and Santa Lucia Bank faults is suggested by changes in the trends of the basement structures and changes in the residual magnetic-anomaly patterns at the block boundaries. The San Martin discontinuity may extend westward across the north end of Santa Lucia Bank, but the western end of the discontinuity is not defined by the available data.

North of the offshore Santa Maria block, and separated from it by the San Martin structural discontinuity, is a basement block that underlies the Sur Basin (Figure 1). This block is bounded on the northeast by a near-coastal fault with large (4 km) apparent down-to-basin vertical separation (Figures 5 and 26). The fault turns northwest along the southern edge of the physiographically high Sur platform offshore of Point Sur, and is inferred to join a thrust fault seaward of Monterey Bay. The western margin of the block is the paleosubduction zone, and the western part of the block is deformed by seaward-directed thrusting (Figure 11). The northern margin of the block may lie along a southwest-trending inferred fault west of Monterey Bay that marks the end of the well-developed Santa Lucia Escarpment, and a physiographic boundary with a lower accretionary wedge to the north (Figure 26). The fault is indicated as a thrust to account for northwest displacement along the San Martin discontinuity, as bending of the Santa Lucia Bank fault and adjacent Neogene structures suggests (Figure 5).

The low accretionary wedge north of the inferred thrust fault lies within a reentrant and at a bend in the paleosubduction zone. The wedge is structurally overlain along a thrust fault by a higher accretionary wedge that forms the basement beneath outer Santa Cruz Basin (Saleeby et al., 1986). The higher accretionary wedge, or for convenience the outer Santa Cruz block, becomes the frontal wedge of the paleosubduction zone north from near Guide Seamount (Figure 6). Serpentinites and metavolcanics dredged from Ascension Canyon and mafic volcanics from the Santa Cruz high suggest that this block is underlain by Franciscan rocks.

Onshore at Point Sur, the southeastern edge of the outer Santa Cruz block is defined by the Sur fault (at SF on Figure 26) (Trask, 1926), a zone of high-angle reverse faults and low-angle thrusts that juxtapose structurally elevated crystalline Salinian basement rocks to the northeast against Franciscan rocks to the southwest (Gilbert, 1973). Graham and Dickinson (1978) and Silver (1978) have suggested that although the Sur fault has a complex displacement history, including major Mesozoic underthrusting (Page, 1970), it has more recently become part of the right-slip San Gregorio-Hosgri fault zone. To the south, the San Gregorio-Hosgri fault zone includes the Hosgri fault (Hoskins and Griffiths, 1971; Wagner, 1974; Hall, 1975; Buchanan-Banks et al., 1978) and its probable northern extension through Cape San Martin (Leslie, 1981). The southern end of the Hosgri fault shown on Figure 15 is as mapped by data available to this author; however, Stiretz (1986) continues the fault to the south of Point Arguello, and Crain, Mero and Patterson (1985) suggest a similar southern continuation. To the north, the San Gregorio-Hosgri fault zone includes the San Gregorio fault zone and its offshore extensions (Hoskins and Griffiths, 1971; Greene et al., 1973) that cross Monterey Bay and extend north into the Gulf of the Farallones, where they join the San Andreas fault zone along the Potato Patch fault strand (Figures 14, 15, and 26).

The displacement history of the San Gregorio-Hosgri fault system is controversial. Some maintain that the fault has undergone only minor lateral displacement (Hamilton and Willingham, 1977); others present evidence for major offset of geologic units along the fault. Hall (1975, 1978) suggests 80–95 km of right slip in the last 5 m.y., and Graham and Dickinson (1978) suggest approximately 100 km in the last 8 m.y. At present, the fault is seismically active throughout its entire length (Greene et al., 1973; Gawthrop, 1978; Coppersmith and Griggs, 1978; Lindh et al., 1981; Eaton, 1984). Because it is a through-going fault, it may well account for some of the approximately ~2 cm/yr displacement between the Pacific and North American plates not presently accounted for by displacement along the San Andreas and associated onshore faults (Gawthrop, 1978; Weldon and Humphries, 1986; Minster and Jordan, 1987).

Some of the displacement between the Pacific and North American plates has occurred, and may now be occurring, along the Santa Lucia Bank fault (Figure 26). This fault approaches the San Gregorio-Hosgri fault zone near Cape San Martin (Figure 5). The fault experienced late Tertiary lateral displacement, and is now active seismically. Like most young faults, it has physiographic expression (fault strands on the bank have up to 40 km of continuous sea-floor offset), and a strand of one of these faults on the bank that is visible on GLORIA side-scan records extends south from the bank, and crosses Arguello Canyon at a right-lateral offset in the canyon axis. The surface displacement of the fault shows that it crosses the west end of Santa Barbara Channel and extends to Rodriguez Seamount (McCulloch et al., 1984).

In the offshore, the San Gregorio-Hosgri fault forms the northeastern edge of the outer Santa Cruz block to its intersection with the Nacimiento fault, which has been displaced to the northwest along the San Gregorio-Hosgri fault (Figure 26). The Año Nuevo and Pigeon Point basement blocks appear to have been shingled against the Farallon Ridge along dextral slip faults. The composition of basement beneath these two blocks is uncertain, which leads to uncertainty about the location of the Nacimiento fault offshore. The Nacimiento fault may lie along the southwest edge of the Año Nuevo block, as seismic reflection profiles suggest, along a structural discontinuity that coincides with a magnetic anomaly at the boundary between the Año Nuevo and Pigeon Point blocks (Figure 15), or along the structural discontinuity between the Pigeon Point block and the Farallon Ridge. Farther to the northwest, the offshore Nacimiento fault takes the form of en echelon faults along the west side of the Farallon Ridge (Figures 14 and 26). Although not well imaged by seismic reflection profiles, these en echelon faults have the general appearance of steeply landward-dipping high-angle reverse displacement. Continuation of the Nacimiento fault northwest from Cordell Bank as a boundary between the San Simeon and Salinia terranes (Figure 4) is likely, but conjectural. Although apparent east-dipping high-angle reverse faults mark the boundary between the melange wedge to the west and the structural high that forms the southwestern margin of Bodega Basin, the composition of basement rocks beneath Bodega Basin is unknown northwest of Cordell Bank. These reverse faults apparently do not cross the Navarro discontinuity at the north end of the block.

To the east, the structural block of known and conjectured Salinia terrane beneath Bodega Basin lies between the Navarro discontinuity on the north and the Pigeon Point block on the south. Its northeastern edge is defined by the San Gregorio and San Andreas faults in the Gulf of the Farallones, and to the northwest by an unnamed fault along the western side of the Point Arena terrane (Figure 14). Before the latter fault was mapped, Wentworth (1968) proposed the existence of such an offshore fault to account for locally derived granitic detritus in Cretaceous and lower Tertiary K-feldspar-bearing arkosic sandstone and conglomerate in the Point Arena terrane, which is otherwise of oceanic affinity. However, the appreciable Cretaceous and early Tertiary lateral transport of these blocks (Kanter and Debiche, 1985) makes it unlikely that they were juxtaposed along the present fault. Graham and Dickinson (1978) also suggest the existence of a fault along the west side of the Point Arena terrane. They propose that, if Point Arena terrane was once part of the sliver of Franciscan rocks between the San Andreas and Pilarcitos faults on the San Francisco peninsula (Figures 4 and 14), it was displaced northward by right slip along the San Gregorio-Hosgri fault. The offshore extension of the Pilarcitos fault should, then, be the boundary between Franciscan rocks and the Salinia terrane. The straight trace of the offshore fault suggests youthful displacement, especially in contrast to the adjacent sinuous late Neogene structure in Bodega Basin (Figure 14), and young growth folding and faulting along the fault is evident on high resolution seismic profiles (McCulloch et al., 1980). The offshore fault does not appear to cross the Navarro discontinuity to the north, but possibly unrelated faults just north of the discontinuity have sea-floor offset that indicates some distributed young displacement north of the discontinuity (Figures 17 and 25).

The northernmost block on the central California Shelf, the Vizcaino block, lies between the Pioneer and Mendocino Fracture Zones (Figures 23 and 26). Its southern boundary is the Navarro discontinuity, and it is bounded on the east and north by the San Andreas fault and the Mendocino Fracture Zone. Basement composition of the block is poorly known, and petrologic heterogeneity (Jurassic? metasedimentary rocks, undated island-arc andesite, pigeonite-bearing tholeiitic basalt). Although these rocks are compatible with an island-arc setting, this large block ($\sim 13{,}500$ km^2) is too poorly sampled to infer its composition, or to be sure that it is not composed of smaller sutured basement blocks.

The San Andreas fault makes a clockwise bend of 20° along the eastern side of the Vizcaino block (the bend at the Garlock-San Andreas fault junction is about 30°), and trends northward toward Point Delgada as a zone several kilometers wide, with two or more well-developed strands (Figure 14) (Curray and Nason, 1967). The juxtaposition of relatively highly deformed strata west of the fault against a 2-km-thick section of little-deformed strata east of the fault that dip gently toward the fault suggests repeated long-term lateral slip. The change in the grain of the residual magnetic-anomaly pattern across the fault (Figure 17) (Griscom, 1973b) also suggests substantial lateral displacement. The age of the gently dipping strata east of the San Andreas fault is not known. Hoskins and Griffiths (1971) suggest they may be Cretaceous in age, and Nason (1968) suggests that they may be of Pliocene age. The character of the acoustic reflectivity suggests that part of the section may be of Miocene age, and may be akin to middle Miocene strata described by Ogle (1953) from elongate fault-controlled basins on the Franciscan Complex, approximately 20 km to the east at Briceland.

The San Andreas fault is seismically active offshore along the edge of the Vizcaino block, and a sea-floor scarp along the fault indicates recent displacement. The scarp faces seaward with a maximum height of approximately 24 m south of Noyo Canyon (Figure 14), and faces shoreward with a maximum height of approximately 9 m north of the canyon. Several lines of evidence suggest a continuing dip-slip component of displacement. The fault cuts the gently sloping shelf obliquely, and pure right slip would produce apparent vertical separation opposite to that observed. Tolo Bank, a wave-truncated bedrock bank immediately west of the fault near Point Delgada (Figure 14), appears to have been elevated (Curray and Nason, 1967). The uplift probably extends onshore at Point Delgada where an exposed wave-cut terrace, radiocarbon dated at 44,800 ± 1300 ybp (probably cut during a −37-m middle Wisconsinan high sea-level stand) has been elevated 44 m (McLaughlin et al., 1983). Recent displacement is suggested by offset of a wave-cut surface that probably largely postdates the post-glacial rise in sea level and by the preserved scarp within the base of actively eroding present-day storm waves that sweep the entire surface of the continental shelf (Cacchione et al., 1987). It has also been suggested (McCulloch, 1985) that a short interval of lowland flooding at the mouth of Navarro River at the time of the 1906 San Andreas earthquake may have resulted from a tsunami generated by vertical displacement along the fault.

Although the location of the San Andreas fault is clearly defined south of Point Delgada, it is not well established to the north. Curray and Nason (1967) suggest that the major strand of the fault turns westward south of Point Delgada and skirts the coastline in a long, sweeping westward curve to join the Mendocino Fracture Zone. Lawson et al. (1908) and Brown and Wolf (1972) mapped a fault onshore that showed displacement in the 1906 earthquake. This fault aligns with the eastern strand of the San Andreas fault zone as mapped offshore to the south, and onshore it extends northward into the offshore approximately 6.5 km northwest of Point Delgada (Figure 14). Based on a near-shore residual magnetic anomaly and the edge of a steep Bouguer gravity anomaly, Griscom (1973a, 1980) suggested that the fault passes offshore north of Point Delgada, continues north-northwestward close to the shoreline, and turns northwest to join the Mendocino Fracture Zone. The offshore strand north of Point Delgada remains undemonstrated by seismic profiling.

Dickinson and Snyder (1979a) and Hsu and Wu (1979) suggested that north from Point Arena, where the San Andreas fault bends 20° clockwise and diverges from the slip vector of transform motion, there should be a region of extension that accounts for the vertical displacement along the fault and for a downwarp in the crust beneath the generally undeformed strata that dip toward the fault from the east. Although Hsu and Wu predicted compression along the fault as it turns westward across the slip vector to join the Mendocino Fracture Zone, they saw none. There is, however, evidence for onshore and offshore compression. The Vizcaino block has been tilted up to the northeast, which accounts for the high sea floor south of the Mendocino Fracture Zone near Cape Mendocino, and arching of the block has produced tension-failure normal faults in the overlying well-bedded marine strata (Profile 3, Figure 23d). Arching suggests that compression accompanied tilting. Northeast-directed compression may account for obduction of the King Range terrane suggested by McLaughlin, Sliter, and Frederiksen (1986) as having occurred in early Pleistocene time. Tilting and compression have produced high coastal uplift rates. Radiocarbon dates of 0.2-8.9 k.y. on fossil mollusks from a flight of beach ridges as much as 30 m in elevation located 40 km southwest of Cape Mendocino yield an uplift rate of ~5.0 m/ka (Lajoie, Sarna-Wojcicki, and Ota, 1982). Present-day northeast–southwest compression delivered across the Mendocino Fracture Zone is suggested by left slip on recent northeast-trending faults across the Gorda plate, by thrusting in earthquake focal mechanisms just west of Cape Mendocino, and by young northwest-trending thrusts and folds onshore and offshore north of the Mendocino Fracture Zone (Clarke, this volume).

SUMMARY OF MARGIN HISTORY

Several major events common to the central California shelf basins reflect events of regional scale, although the histories of individual basins differ in detail. After docking in Late Cretaceous or Eocene time, terranes accreted to the continental margin were uplifted and extensively eroded during the Oligocene. An episode of pre-Oligocene deformation possibly related to accretion is recorded by structures in remnants of Upper Cretaceous and lower to middle Eocene marine strata that escaped erosion. The orientation of these structures differs from superimposed late Neogene structures, but the geometry of the earlier structures is too poorly known to define the associated deformational regime. The Oligocene erosion was profound, and it exposed the underlying Franciscan rocks or Salinia basement rocks over much of the continental margin. Initially, the erosional surface probably sloped westward toward the oceanic plate; but beneath several of the basins, the surface now slopes shoreward toward near-coastal faults that have as much as several kilometers of down-to-the-west vertical separation.

Faulting and tilting of the Oligocene erosional surface was accompanied by marine volcanism and an early Miocene marine transgression. Some early volcanic flows on the basin floors are interbedded with shallow-water marine strata of late-late Oligocene ("Vaqueros" Stage) and early Miocene (Saucesian Stage) age. This episode of volcanism was followed in the offshore Santa Maria Basin by the development of transtensional structures. Water depth then increased over the margin, and by middle Miocene time the margin was at middle to lower bathyal depth and was receiving fine calcareous and siliceous hemipelagic sediment. The Miocene strata generally thicken eastward toward the near-coastal faults, which suggests continued downward rotation of the basement blocks. Although thickest in the shelf basins, and in the slope basins that must have been forming at this time, Miocene sediments were not confined to the basins, but were blanket deposits covering the structural highs that now form the western basin margins.

Local unconformities occur within some of the basins; however, the most widespread unconformity is of early late Miocene age. Its development was associated with the growth of structural highs that form the western margins of the shelf basins. Development of the unconformity was accompanied by a marked increase in terrigenous detritus. A final late Pliocene episode of deformation was accompanied by northeast–southwest compression and deposition of coarser clastic strata in prograded deposits that now form parts of the continental shelf.

The foregoing events can be placed in the more regional context of lithospheric plate interaction. Following contact between the Pacific and North American plates approximately 30 to 28 Ma (Atwater, 1970; Engebretson, 1982), and the formation of the southward-migrating Rivera triple junction and the northward-migrating triple junction located first along the Pioneer Fracture Zone and later along the Mendocino Fracture Zone, subduction changed to transform motion progressively from south to north along the central California margin. Onshore, northward migration of the triple junction was tracked by active volcanism that becomes younger to the north and is thought to have formed along the trailing edge of the northward-retreating margin of the subducting plate (Dickinson and Snyder, 1979a; Fox et al., 1985). Initial volcanism (the Morro Rock-Islay Hill complex of Ernst and Hall, 1974) identified with northward triple junction migration occurred at about 27.6 ± 0.8 Ma. Offshore basement was still elevated and eroding during this early stage of transform motion, and in the offshore this may have been a time of oblique subduction that accounts in part for the paleosubduction zone preserved beneath the continental margin. Subduction ceased several million years after initial contact at approximately the Oligocene–Miocene boundary, as suggested by the youngest oceanic magnetic anomaly (6c, 23.2 to 24.2 Ma; Berggren et al., 1985) at the paleosubduction zone (Atwater, 1970).

In the latest Oligocene or early Miocene, approximately 5 or 7 m.y. after initial plate contact, there was a collapse of the offshore margin accompanied by volcanism and downward rotation of basement blocks along near-coastal faults of large displacement. The once-emergent basement blocks were soon covered by deep-water marine strata, possibly due to the combination of collapse and rising late Oligocene global sea level. Collapse appears to have accompanied a change in relative Pacific–North American plate motion. Ingle (1973, 1980) noted that major changes in water depth and deformation of the continental margins around the Pacific occurred at about this time (Oligocene–Miocene boundary). He suggested that although generally attributed to local causes, these essentially synchronous changes were more likely caused by adjustment of the plate margins in response to a change in the vector and rate of subduction of the Pacific plate. This Pacific-wide event caused back-arc spreading in the western Pacific, and the initiation of translational tectonism that created the California borderland basins attributed by Vedder (this volume) and Howell et al. (1980) to wrench tectonics. Atwater and Molnar (1973) proposed an early Miocene change in relative Pacific–North American plate motion based on the geometry of dated oceanic magnetic anomalies. Jackson, Silver, and Dalrymple (1972) and Clague and Jarrard (1973) have presented evidence for regional effects of a change in plate motion at this time.

A change in plate motion at about this time (22–24 Ma) is suggested by a bend in the Hawaiian hot-spot trend (Figure 27). An age–distance plot of volcanics along the island chain suggests that plate rotation of $\sim 10°$ was accompanied by a change in the rate of Pacific plate motion. The indicated counterclockwise change in plate motion should have produced extension along the California margin. By matching paleobathymetric data from offset segments of the onshore La Honda Basin, Stanley (1986, 1987, and in press) has inferred that slip initiated along the central California segment of the San Andreas fault in the early Miocene. His interpretation is consistent with data from earlier studies (e.g. Clarke and Nilsen, 1973; Huffman, Turner, and Jack, 1973), but inconsistent with interpretations of these data that suggest slip initiated approximately 6 m.y. later (e.g. Graham and Dickinsen, 1978). The earlier slip is coincident with the suggested change in plate motion. Stanley has also shown that the initial episode of displacement was relatively rapid (not tightly constrained, but possibly 3 cm/yr) and that slip was accompanied by volcanism along the length of the central California segment of the San Andreas and related faults. He interprets the combination of volcanism and rapid slip as evidence for a period of transtension. Transtension is consistent with extension along the continental margin that would have resulted from clockwise plate rotation. Dating of the volcanics indicates that rapid slip lasted for approximately 2 m.y., between 22 and 24 Ma (across the Zemorrian-Saucesian boundary). Within the accuracy of available dating of volcanics in the offshore basins (late late Oligocene and early Miocene) the rapid slip may have occurred at the time the margin began to collapse. As dated rocks from within the undisturbed strata that blanket the paleosubduction zone suggest, the end of subduction was pre-lower Mohnian (middle-late Miocene). Dating the rocks from the base of the undisturbed strata immediately above the accretionary wedge (Figure 18) may show that subduction ceased earlier, possibly at the start of collapse.

The long period of time (13 to 11 m.y.) that was thought to have elapsed between Pacific-North American plate contact at 30 to 28 Ma (Atwater, 1970; Engebretson, 1982) and initial displacement on the San Andreas fault at ~ 17 Ma (Graham and Dickinson, 1978), has suggested to some that initial transform motion was probably accommodated by displacement on faults offshore of the San Andreas fault (Garfunkle, 1973; Beck, 1986). Dating of an episode of San Andreas fault displacement at 24–23 Ma (Stanley, 1986, 1987, and in press) shortens this interval to 7–4 m.y., and the interval may have been still shorter if this slip episode was not the initial fault displacement. The seismic reflection data do not show evidence of strike-slip faulting of the subducted Pacific plate, and from strength considerations (Vink, Morgan, and Zhao, 1984) initial transform faulting is more likely to have occurred on the relatively weaker North American plate.

There is no throughgoing fault on the offshore margin that could have acted as a single avenue for transform displacement (Figure 26), The interpreted pattern of displacement on the offshore faults indicates dextral shear; however, the lack of well-dated displacement histories and offset piercing points makes it difficult to test the proposition that initial Pacific–North America transform motion occurred on offshore faults. The start of displacement, possibly in late late Oligocene–early Miocene time on the large down-to-the-west near-coastal faults and on the Hosgri fault, suggests that, within the limits of available dating, this faulting may have coincided with early San Andreas fault displacement. In any case, the recognition of early displacement on the San Andreas fault decreases the possible importance of pre-San Andreas slip on offshore faults.

Hoskins and Griffiths (1971) indicate that the present shelf basins formed nearly synchronously in late Miocene as the result of uplift of the structural highs along their western margins. Blake et al. (1978) suggest that the synchroneity indicates a change in Pacific–North American plate motion, producing the basins by extension of the tectonically pliant margin, rather than by the progressive basin formation that might otherwise be related to the northward migration of the triple junction. Blake et al. date the change in plate motion and basin formation as middle to late Miocene. Howell et al. (1980) suggested that the structures within the late Tertiary basins in both the borderland to the south and on the central California margin indicate a wrench tectonic regime, rather than pure extension.

Miocene and Pliocene basin histories differ in detail, but all have an early late Miocene unconformity. As noted in the description of Point Arena Basin, the unconformity marks a change from sedimentation dominated by calcareous and siliceous pelagic sediment to an increase in terrigenous detritus, as shown at DSDP Site 173 on the western edge of the Vizcaino block. At Site 173, the increase in terrigenous sediment was not accompanied by an increase in sedimentation rate and the change probably reflects a westward shift in the shoreline and its terrigenous facies, rather than coastal uplift. This change is dated at Site 173 as 6.5 Ma, and Ingle (1978) suggests that the shelf-wide event reflects a pronounced global lowering of sea level.

The structural grain of the offshore basins is dominated by late Neogene folds and compressional faults that strike generally parallel to the coast, and indicate significant northeast compression. Some of the folds and faults appear to be developing at present. In addition, most of the major basin-bounding faults that developed large vertical separation during earlier collapse of the margin now have the appearance of high-angle reverse displacement, possibly because of structural overprinting during this late Neogene compression. In the onshore, evidence for late Neogene compression is widespread. Page (1981) and Aydin and Page (1984) note Pliocene-Pleistocene folds and compressional faults that parallel the San Andreas fault, which they attribute to compression across the fault. Geodetic measurements across the San Andreas fault (Burford, 1966) kinematic models based on fault slip rates (Weldon and Humphreys, 1986, 1987) combined with neotectonic and paleoseismic estimates of deformation rates, and rates of change of long baseline interferometry (Minister and Jordan, 1987) indicate active compression across the San Andreas fault. Wentworth and Zoback (1986) attribute recent earthquakes just east of the San Andreas fault (1983, Coalinga; 1986, Kettleman) to

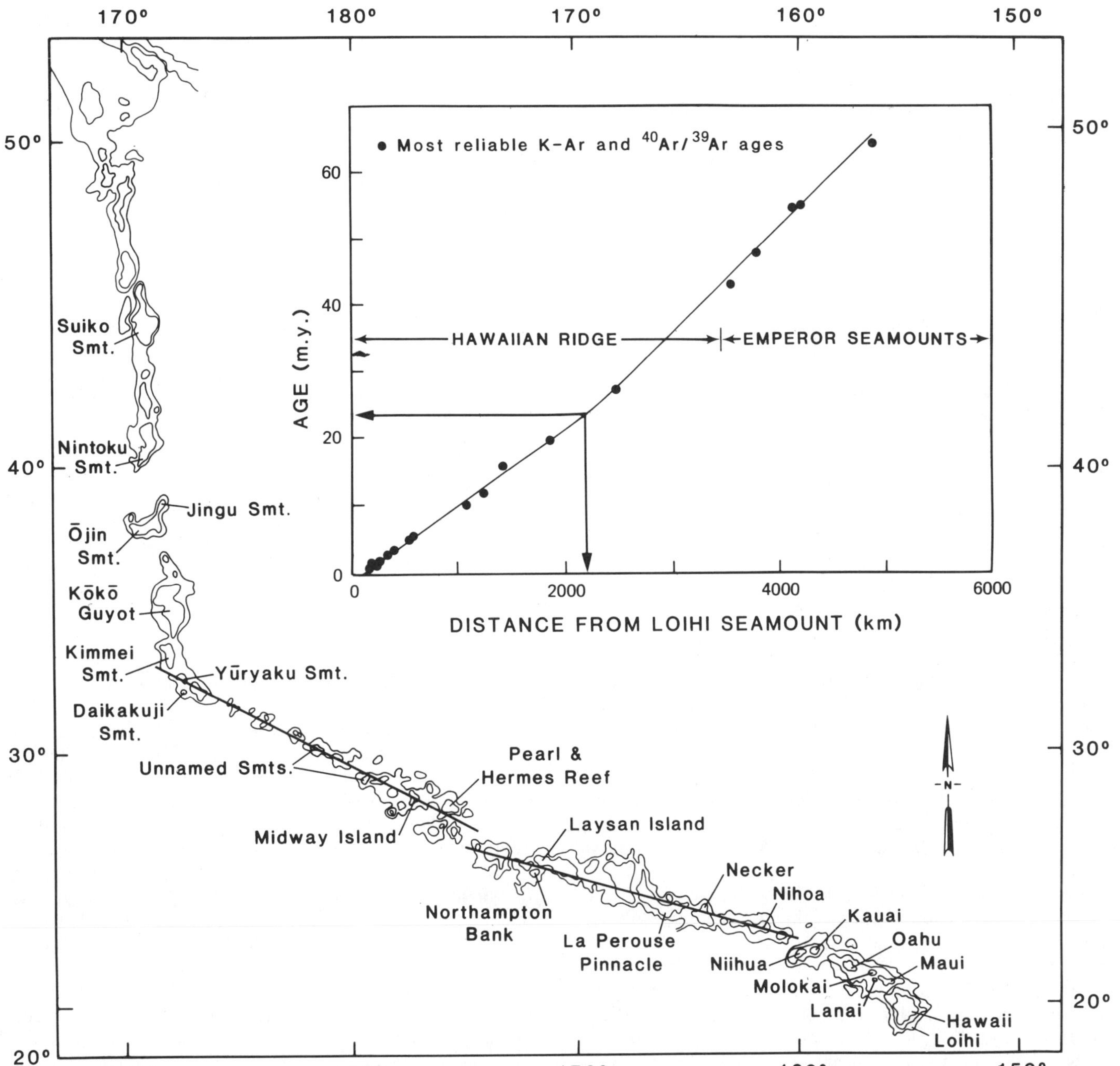

Figure 27.. Bathymetry of the Hawaiian-Emperor volcanic chain (Chase, Menard, and Mammerickx, 7970). Straight lines drawn through the volcanic edifices suggest a change in the direction of plate motion after the formation of Pearl and Hermes Reef. On the inset, the location of this change is plotted as the distance along the chain from Loihi. This distance is projected to linear regressions of the K-Ar and $^{40}Ar/^{39}Ar$ ages judged by Duncan and Clague (1985) to be the most reliable dates along the chain. The location of the bend approximately coincides with a change in the rate of plate motion at approximately 23 Ma. Clague and Dalrymple (in press) also indicate that there was a change in the rate of plate motion after the Hawaiian-Emperor bend (ca 43 Ma) and before or near the time the Laysan volcano formed (ca 20 Ma).

northeastward thrusting resulting from northeast compression. Zoback et al. (in preparation) have summarized stress-induced borehole elongations, earthquake focal mechanisms, hydraulic fracturing, in situ stress measurements and young (<2 Ma) volcanic alignments along the entire onshore San Andreas fault. They suggest that because the fault may have low shear strength, the motion between the Pacific and North American plates, which has a small compressional component, results in maximum compressive stress oriented normal to the fault. Study of Pacific–Antarctic relative plate motion, of the bends in hotspot trends, and of the absolute motion of the Pacific plate relative to the hotspot frame of reference, suggest that the Pacific plate underwent clockwise rotation at about the start of late Neogene compression, between 3 and 5 Ma (Cox and Engebretson, 1985; Pollitz, 1986; Harbert and Cox, 1986). Late Pliocene to Holocene tectonism along the California margin (Christensen, 1966; Aydin and Page, 1984) has been attributed to the transpression that should have resulted from this change in plate motion. In the shelf basins the amount of terrigenous sediment deposited increased markedly during this interval, indicating uplift of the adjacent coast. In DSDP Hole 173 adjacent to the Point Arena Basin, the increase in sedimentation rate is dated as having started between 4 and 2.8 Ma. This transpressive regime is the probable cause of the late Neogene structures that dominate the structural grain of the shelf basins.

ADDITIONAL QUESTIONS

Many questions remain about the structural development of the central California margin. Paleontologic data for age determinations for most of the offshore strata are proprietary, and the underlying basement rocks are inadequately sampled. Basement composition that might indicate the origin, trajectory, and displacement history of large areas of the offshore is unknown or inferred, and paleomagnetic data are not available. More fundamental, however, is our ignorance of the deeper structural framework. Are basement blocks or entire terranes structural flakes that lie on detachment surfaces or thrusts at depth? Trehu and Wheeler (in press) have identified a low-velocity layer (5 km/sec) that they suggest may be subducted sedimentary strata. The wedge lies onshore east of the Hosgri fault at a depth of 14 to 22 km beneath the surface at the shoreline. It thins eastward for approximately 50 km, and its thinning edge passes under the western edge of the Salinia terrane. Crouch, Bachman, and Shay (1984) propose that offshore Santa Maria Basin is underlain by westward-directed thrusts that sole out at depth along the top of an aseismic zone identified beneath the Coast Ranges, and they suggest that thrust displacement dominates the motion along the Hosgri fault. Most of the evidence for thrusting is south of the San Martin discontinuity (Figures 5 and 13).

Northwest of Cape San Martin, the western edge of the seismically active continental margin lies relatively close to shore. Earthquakes occur on offshore and onshore strands of the San Gregorio and San Andreas fault zones, in the Gulf of the Farallones, along the southern part of the Point Reyes fault, and along the western edge of the Farallon Ridge (Figure 14). First-motion determinations of earthquakes along the San Gregorio fault zone (Gawthrop, 1978; Coppersmith and Griggs, 1978) indicate that right slip parallel to the fault zone is the dominant motion, with only a minor oblique thrust component, and north-northeast compression. South of Cape San Martin, the seismically active margin widens to the western side of Santa Lucia Bank and present deformation is partitioned between right-slip and northeast–southwest compression. The Hosgri fault at San Simeon exhibits both right-slip and a smaller component of east-side-up vertical slip (Weber, 1983; Hall, et al., 1987), and compressive thrusting associated with the Hosgri fault and in the offshore Santa Maria Basin (Crouch, Bachman and Shay, 1984) appears to be ongoing (Figure 13). Earthquake first motions in this area include both right slip, and for larger events, nearly pure thrust along the Hosgri fault and on faults on Santa Lucia Bank (Figure 5) (Gawthrop, 1978; Coppersmith and Griggs, 1978; Lindh et al., 1981; Eaton, 1984). With the exception of a few small thrust earthquakes that lie on east-trending strands of the Hosgri fault just south of Cape San Martin, the indicated strike for the remaining earthquakes is parallel to the northwest-trending faults, and is accompanied by northeast-directed compression. To the south, there is considerable evidence for present-day compression across the northern edge of Santa Barbara Channel (Yerkes, 1985). Yeats (1981) and Howell et al. (1986) suggest that there is crustal shortening by underthrusting beneath the east–west-trending Transverse Ranges that lie across the right slip between the Pacific and North American plates. First-motion solutions in the western part of the Santa Barbara Channel and southwest of Point Conception show a change in the direction of compression. Compression is directed northward in the Santa Barbara Channel, and becomes northeastward in the vicinity of Point Conception (Yerkes and Lee, 1979). This suggests the possibility that part of the crust beneath the western end of the channel and offshore Santa Maria Basin, and possibly farther west, is accommodating the motion between the Pacific and North America plates by underthrusting and clockwise rotation. Alternatively, if one presumes that the Santa Lucia and Patton escarpments were once a continuous linear subduction margin, the Santa Lucia Escarpment has been rotated slightly counterclockwise and moved seaward. This suggests that westward motion of the continent, possibly in response to extension in the Great Basin, may contribute to compression.

HYDROCARBON POTENTIAL

Before discovery of the giant Point Arguello oil field and nearby smaller fields in offshore Santa Maria Basin (Figure 2), Zieglar and Cassell (1978) suggested that in light of the generally unexplored nature of the offshore central California basins and the encouraging hydrocarbon shows in the offshore wells drilled after the 1963 federal lease sale, a pessimistic view of the petroleum potential of these basins was unwarranted. Several factors contributed to early pessimism. Primary was the lack of a discovery thought to be producible with the available technology, geologic knowledge and economic factors of the 1960s. Eight dry holes in Bodega Basin were particularly disappointing. The basin, by virtue of its hydrocarbon shows, thick shale sections, reservoir-quality sandstone, and well-developed structures, appeared to have the prospect of becoming a major hydrocarbon-producing basin. Also disappointing was the relatively thin prospective section in the shelf basins and the lack of correlative strata with productive strata in Santa Barbara Channel fields (Vedder, this volume), where production comes from a nearly complete Tertiary (Eocene through Pliocene) section. By contrast, the offshore central California basins have comparatively thin sequences of Cretaceous sandstone, and Salinian basement or Franciscan rocks often represent the pre-Tertiary section. Furthermore, the overlying Tertiary section contains little or no Paleocene, and only remnants of formerly more extensive Eocene deposits, these being remnants that are usually described as sandstone too dense to act as reservoir rocks (Hoskins and Griffiths, 1971), and it usually lacks the Oligocene section.

Pessimism also had an historic basis. In the mid-1970s, a well in the Hondo offshore field in Santa Barbara Channel showed that high flow rates could be achieved from Miocene rocks, a discovery that fortuitously coincided with a rise in the price of oil. Prior to this discovery, however, there was a belief that the heavy hydrocarbons found in offshore Miocene rocks would not be economic to produce (Isaacs, 1984a). This view was based on experience with middle Miocene and Pliocene fractured-shale reservoirs in onshore Santa Maria Basin, where wells had early rapid production declines and yielded viscous oil that required dense well spacing, extensive well treatment, and long-term field operation to remain economic. Ogle (1984) and Crouch and Bachman (1985) suggest that this experience led to inadequate testing of offshore Miocene rocks. Even with the new view for potential Miocene production, the initial exploration that ultimately led to the discovery of the giant Point Arguello field may have been directed, at least in part, toward a buried erosional unconformity developed on Upper Cretaceous strata. However, with drilling, middle and late Miocene organic-rich carbonaceous and siliceous rocks similar to the Monterey Formation were targeted for the final drilling that led to the field discovery.

The Monterey Formation has received considerable study in the adjacent onshore regions (e.g. Bramlette, 1946). Interest in these rocks has recently been renewed because of their unique lithology, the paleoenvironmental information contained in their rich fossil record, their diagenetic alteration, and their hydrocarbon potential (Isaacs, 1981, Garrison and Douglas, 1981; Keller and Barron, 1983; Barron, 1986a, b). The principal sedimentary components of these rocks are biogenous silica and calcite (diatom frustules, coccoliths and foraminifers), fine terrigenous debris (clay, quartz and feldspar), and organic matter. Deposition occurred as a hemipelagic rain in mid-to-lower bathyal depths on slopes and basin floors. A depositional model used to explain the low terrigenous contribution envisions a series of silled basins in which the terrigenous detritus is trapped in the near shore, as in the modern southern California continental borderland (Gorsline and Emery, 1959) and Vedder, Crouch, and Lee-Wong (1981) have shown that some of the basement ridges in the borderland were probably within wave base during middle Miocene time. The high abundance of silica in these rocks has been attributed to increased upwelling that increased biological productivity (Ingle, 1973). On the other hand, it has been shown (Isaacs, 1985) that, although there are beds within the Monterey Formation that represent intervals of rapid silica accumulation, rates of silica accumulation in the Monterey Formation as a whole are actually lower than in overlying and underlying strata where the silica accumulation is masked by the greater volume of terrigenous detritus.

Many of the characteristics of oil from the Monterey, including its low API gravity (high viscosity), are attributed to early generation rather than to subsequent biodegradation (Isaacs and Petersen, in press). Thus, understanding the heat-flow history may be an important key to exploration. Although the sedimentary sections in the offshore central California basins are generally less than 3050 m (10,000 ft) thick—a rule-of-thumb burial depth generally thought necessary for hydrocarbon generation—unique properties of the Monterey-like rocks permit hydrocarbon generation at a shallower depth. Isaacs (1984a) has shown that, as a consequence of the combination of low thermal conductivity and large compaction of the diatomaceous rocks with burial, these strata follow a thermal history that produces not only a high thermal gradient across the strata, but ultimately a higher temperature. Her model indicates a temperature of about 100 °C by about 3.5 Ma, which lies within the normal oil-generating window (Hunt, 1979), and which exceeds the 85 °C temperature shown by Petersen and Hickey (in press) to be the threshold temperature for hydrocarbon generation in the COST well (OCS-CAL78-164 NO. 1) (Figure 1) at the southern end of offshore Santa Maria Basin.

Because of its importance to hydrocarbon generation and maturity, knowledge of heat flow and the geothermal gradients beneath the shelves would be extremely useful for exploration and for resource appraisal, but such data are only fragmentary. Elevated temperatures are encountered in wells near Point Arguello, as noted by Crain, Mero, and Patterson (1985), by McCulloh and Beyer (1979), and by McCulloh (1979). McCulloh further notes that the onshore Santa Maria Basin, westernmost Santa Ynez Mountains and western Santa Barbara Channel are characterized by persistent high geothermal gradients, and possibly by high heat flow, and that these temperatures exceed most estimates for petroleum generation in very young rocks. Few heat-flow data are available for the offshore basins to the north; however, Lachenbruch and Sass (1980) indicate that the Coast Ranges that border the eastern side of the offshore basins are characterized by a higher-than-normal heat flow of 1.5–2.5 HFU. They attributed the anomalous heat flow to passage of the Mendocino triple junction, and noted that the heat flow increased by a factor of 2 within 4 m.y. after the passage of the triple junction.

In order to evaluate the thermal evolution and the related hydrocarbon maturation of coastal California, Heasler and Surdam (1985) modeled temporal and spatial thermal effects of a northward-migrating triple junction (as suggested by Lachenbruch and Sass, 1980) accompanied by asthenospheric upwelling into the slabless window of Dickinson and Snyder (1979b). Their model is particularly relevant for areas east of the transform faults that formed following contact between the Pacific and North American plates, where northward migration of the triple junction is tracked by both northward-progressing volcanism (Dickinson and Snyder, 1979a; Fox et al., 1985) and by evolution of an onshore basin (Graham et al., 1984). East of the transform faults, basins closer to the initial point of plate contact will have received both a larger and longer thermal pulse, whereas the difference in the thermal histories in the offshore basins that formed nearly synchronously west of the transform faults and that were displaced with the Pacific plate should probably show less effect of the migrating triple junction. West of the San Andreas fault, at the north end of the area, the King Range terrane that trailed the migrating triple junction on the Pacific plate experienced a middle Miocene hydrothermal event that produced a geothermal gradient of perhaps 30 °C/km and local peak temperatures of 300 °C (Underwood, 1986). High heat flow is thought to have persisted here from middle Miocene through Pliocene time.

Fracture porosity and permeability are the major factors that make the Monterey-like siliceous shales a reservoir rock. The propensity for fracturing of these siliceous rocks is related to silica diagenesis and lithologic composition. Fracturing is promoted by the diagenetic alteration of silica in these rocks, but fracturing intensity is inversely related to clay content (Belfield et al., 1983). Diatomaceous Miocene rocks (Monterey Formation equivalents?) recovered from structural highs along the edges of outer California border-

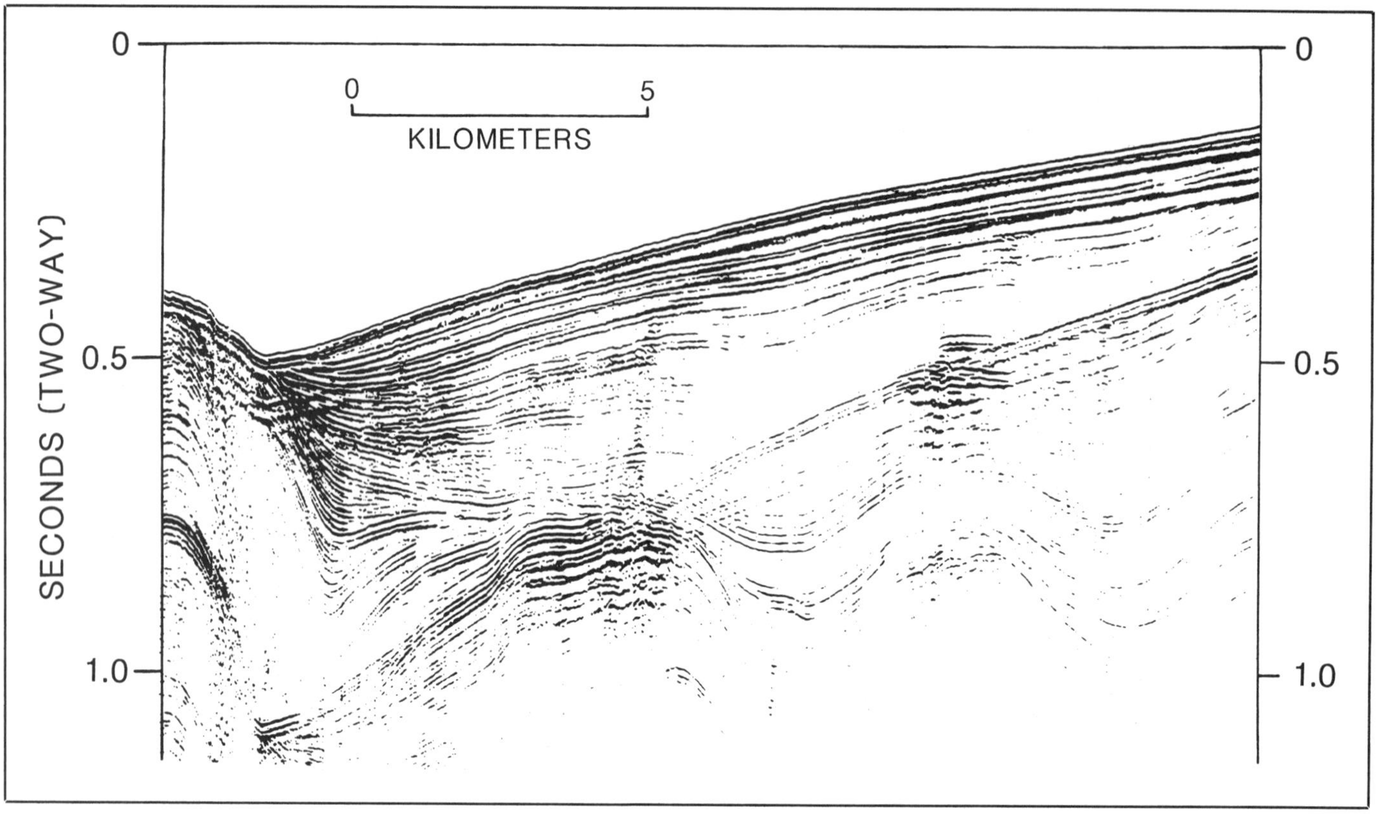

Figure 28. Single-channel seismic reflection record from offshore Santa Maria Basin, showing bright spots in the tops of anticlines, and chimneying of bright spots upward from the top of an anticline.

land basins have not been diagenetically altered; the opaline diatom frustules are still preserved and the rocks are nearly unlithified and moldable (Vedder, this volume). With progressive diagenetic alteration from opal to opal-CT to diagenetic quartz (Murata, Friedman, and Gleason, 1977; Isaacs, 1981; Isaacs, Pisciotto, and Garrison, 1983) and their diagenetic equivalents, chert, porcelanite and siliceous shale, there is an associated loss of volume and contraction that produces fracturing. As the rocks become more indurated, they become more susceptible to fracturing by deformation, and structural development may control the orientation and degree of fracturing in any area. In the Point Arguello field, fractures generally parallel the structural grain and crestal wells have better test rates and calculated permeability than flank wells in the same structural zone (Crain, Mero, and Patterson, 1985). Deformation fracturing may extend to appreciable depth below these structures, for as Ogle (1984) noted, some of the closed structures have an oil column greater than the apparent structural closure.

Strata assigned to the Monterey Formation (or Miocene Monterey-like rocks) are reported from the onshore extensions of the offshore basins at Point Año Nuevo (Clark, 1981), Point Reyes peninsula (Clark et al., 1984), and Point Arena (Boyle, 1967), and from all the offshore basins where they are described as chert, cherty shale, porcelaneous shale and siliceous shale. These lithologies suggest that the possibility for fracture reservoirs resulting from silica diagenesis is widespread. As shown by the 1960s drilling, these rocks are fractured and commonly contain high-gravity oil and tarry shows. Some shows are continuous for greater thicknesses (as much as 914 m) than geologically comparable, producing horizons in the onshore Santa Maria Basin and the Santa Barbara Channel (Zieglar and Cassell, 1978).

While success at the Point Arguello field and recognition of the importance of fracture reservoirs in the Monterey Formation may make Monterey-like rocks the primary exploration target in the offshore basins (Crouch and Bachman, 1985), other geologic strata may also be prospective. Zieglar and Cassell (1978) cite Cretaceous and lower Miocene sandstone of reservoir quality with continuous hydrocarbon shows through intervals 42- to 92-m thick in outer Santa Cruz and northern Bodega Basins. Crain, Mero and Patterson (1985) indicate potential gas and oil production from sandstone in both the upper Miocene Santa Margarita Formation and Pliocene strata in the Point Arguello field area.

The offshore basins also may have significant gas potential. As noted earlier, Ogle (1984) estimated from data then available that there were a trillion cubic feet of recoverable gas in the Point Arguello and associated fields in the southern part of offshore Santa Maria Basin. Shallow gas, as revealed by high-amplitude reflections on 2-sec air gun and higher resolution records, occurs in large areas (a total of $> 1500\ km^2$) in offshore Santa Maria and Sur basins, and to a lesser extent in offshore basins to the north (distribution mapped in McCulloch et al., 1980). In offshore Santa Maria Basin, the gas appears to have accumulated in subsurface structural crests (Figure 26), suggesting that it is thermogenic. Isotopic analysis of methane adsorbed to surface sediment in offshore Santa Maria Basin (Stahl and Faber, 1983; Faber, Rehder and Stahl, 1983) shows similarities to the methane from producing source rocks or petroleum accumulations. Some of the gas is migrating upward to the sea floor as chimneys from anticlines, and updip in the overlying less-deformed upper Pliocene and younger strata (Figure 28). Some faults in the area appear to be gas charged. Where gas reaches the sea floor, it has produced surface pits, some of which are actively venting (Richmond et al., 1981). Gas probably contributes to a decrease in strength of unconsolidated sea-floor sediments, which may account for several large retrogressive surface slumps (total area ~ 180 km^2) that have occurred on very gentle slopes ($<2°$). These slumps are restricted to areas where high-resolution acoustic profiling indicates laterally continuous gas within the unconsolidated surface sediments.

Gas in the offshore basins to the north occurs in three geologic settings: (1) along faults, especially along the offshore San Andreas fault on the east side of Bodega Basin north of Point Reyes peninsula; (2) in three small areas (each ~ 100 km^2) on the floor of southern and central Bodega Basin; and (3) as long, narrow continuous zones of shelf-edge gas located at the topographic break at the seaward edge of the continental shelf in outer Santa Cruz and Point Arena basins. No analyses of the gas in these northern basins are available, but the geologic setting suggests that the fault-related and basin-floor gas may be thermogenic. On the other hand, the shelf-edge gas that appears to be migrating up the continental slope from fine-grained organic-rich sediment that dominates modern slope deposition is probably biogenic.

A small amount of gas can produce an acoustic anomaly ($\sim 5\%$ by volume; Domenico, 1974), hence the presence of anomalies is not an indication of the total amount of gas present. However, in addition to the demonstrated gas potential in offshore Santa Maria Basin, gas has been reported from wells in Point Arena Basin (Zieglar and Cassell, 1978).

Primary exploration targets in the offshore basins probably will largely be late Neogene folds and faulted folds that involve Miocene and Pliocene strata. Such structures are common in all but outer Santa Cruz Basin. When new stratigraphic data from additional drilling become available, depositional stratigraphic traps and diagenetic traps developed in the silica-rich Miocene strata may become prospective (Ogle, 1984). Exploration for oil in the Monterey-like rocks in the offshore basins may be complicated. Normal well logs and production anomalies are inadequate to indicate potential production (Isaacs and Petersen, in press), because as Crain, Mero, and Patterson (1985) noted, no discovery was evident from such data after drilling the Point Arguello field. In addition, fracturing is facies sensitive (Isaacs, 1984b), the target facies being quartz, cherts, and dolostones, and subsurface continuity or patchiness of these rocks may be difficult to establish.

Ultimately, the presence or absence of gas and oil accumulations in these basins will have to be established by drilling, and producibility will be determined by the economics of the times and by technological development, including new extractive methods and innovative drill-rig design that are now evolving (Jennings, Alonzo, and Mitchell, 1985). For the interested reader, a summary of onshore coastal hydrocarbon production through 1980 can be found in McCulloch et al. (1980), and estimates of the total recoverable hydrocarbon potential for the offshore basins have been made by Dolton et al. (1981) and Cooke (1985).

ACKNOWLEDGMENTS

The author thanks S. H. Clarke, Jr., C. M. Isaacs, R. G. Stanley, E. A. Silver, D. S. Wilson, and J. G. Vedder for discussions of the geology of the California margin, and C. Campbell, S. H. Clarke, Jr., C. M. Isaacs, and J. G. Vedder for their very helpful review of this chapter. The author also appreciates the help of P. H. McClellan and H. F. Ryan for advanced processing of multichannel seismic reflection data, and S. C. Vath for preparation of most of the illustrations.

REFERENCES CITED

Addicott, W. O., 1967, Age of the Schooner Gulch Formation, Mendocino County, California: U.S. Geological Survey Bulletin 1254-C, 11 p.

——— 1972, Provincial middle and late Tertiary molluscan stages, Temblor Range, California, *in* E. H. Stinemeyer and C. C. Church, eds., Symposium on Miocene biostratigraphy of California: Society of Economic Paleontologists and Mineralogists, Pacific Section, Bakersfield, p. 1-26.

Atwater, T., 1970, Implications of plate tectonics for the Cenozoic tectonic evolution of western North America: Geological Society of America Bulletin, v. 81, p. 3513-3536.

——— and P. Molnar, 1973, Relative motions of the Pacific and North American plates deduced from seafloor spreading in the Atlantic, Indian, and South Pacific Oceans, *in* R. L. Kovach and A. Nur, eds., Proceedings of the conference on tectonic problems of the San Andreas Fault system: Stanford University, Publications in the Geological Sciences, v. 13, p. 136-148.

Aydin, A., and B. M. Page, 1984, Diverse Pliocene-Quaternary tectonics in a transform environment, San Francisco Bay region, California: Geological Society of America Bulletin, v. 95, p. 1303-1317.

Barron, J. A., 1986a, Paleoceanographic and tectonic controls on deposition of the Monterey Formation and related siliceous rocks in California: Palaeogeography, Palaeoclimatology, Palaeoecology, v. 53, p. 27-45.

——— 1986b, Updated diatom biostratigraphy of the Monterey Formation of California, *in* R. E. Casey, and J. A. Barron, eds., Siliceous microfossil and microplankton studies of the Monterey Formation and modern analogs: Society of Economic Paleontologists and Mineralogists, Pacific Section, Book 45, p. 105-119.

——— in press, The late Cenozoic stratigraphic record and hiatuses of the northeast Pacific: Results from the Deep Sea Drilling Project, *in* E. L. Winterer, D. M. Hussong, and R. W. Decker, eds., The eastern Pacific Ocean and Hawaii: Decade of North American Geology, v. N.

Beck, M. E., Jr., 1986, Model for late Mesozoic-early Tertiary tectonics of coastal California and western Mexico and speculations on the origin of the San Andreas fault: Tectonics, v. 5, n. 1 p. 49-64.

Belfield, W. C., J. Helwig, P. LaPointe, and W. K. Dahleen, 1983, South Elwood oil field, Santa Barbara Channel, California, a Monterey Formation fractured reservoir: Petroleum generation and occurrence in the Miocene Monterey Formation, California: Society of Economic Paleontologists and Mineralogists, Pacific Section, Bakersfield, California, p. 213-221.

Berggren, W. A., D. V. Kent, J. J. Flynn, J. A. Van Couvering, 1985, Cenozoic geochronology: Geological Society of America Bulletin, v. 96, p. 1407-1418.

Blake, M. C., Jr., R. H. Campbell, Jr., T. W. Dibblee, Jr., D. G. Howell, T. H. Nilsen, W. R. Normark, J. C. Vedder, and E. A. Silver, 1978, Neogene basin formation and relation to plate-tectonic evolution of San Andreas fault system, California: American Association of Petroleum Geologists Bulletin, v. 62, n. 3, p. 344-372.

——— D. G. Howell, and A. S. Jayko, 1984, Tectonostratigraphic terranes of the San Francisco Bay region, *in* M. C. Blake, Jr., ed., Franciscan geology of northern California: Society of Economic Paleontologists and Mineralogists, Pacific Section, Book 43, p. 5-22.

——— ——— and D. L. Jones, 1982, Preliminary tectonostratigraphic terrane map of California: U.S. Geological Survey Open-File Report 82-593, scale 1:500,000.

——— A. S. Jayko, and P. J. McLaughlin, 1984, Tectonostratigraphic terranes of the Northern California Coast Ranges, *in* D. G. Howell, ed., Tectonostratigraphic terranes of the Circum-Pacific region: Circum-Pacific Council for Energy and Mineral Resources, Earth Science Series, n. 1 p. 159-172.

Boyle, M. W., 1967, The stratigraphy, sedimentation, and structure of an area near Point Arena, California: University of California at Berkeley, M.S. Thesis, 71 p.

Brabb, E. E., and W. F. Hanna, 1981, Maps showing aeromagnetic anomalies, faults, earthquake epicenters, and igneous rocks in the southern San Francisco Bay region, California: U.S. Geological Survey Geophysical Investigations Map GP-932, 6 p., 3 sheets, scale 1:125,000.

Bramlette, M. N., 1946, The Monterey Formation of California and the origin of its siliceous rocks: U.S. Geological Survey Professional Paper 212, 57 p.

Brown, R. D., and E. W. Wolf, 1972, Map showing recently active breaks along the San Andreas fault between Point Delgada and Bolinas Bay, California: U.S. Geological Survey Miscellaneous Geologic Investigations Map 1-692, 2 sheets, scale 1:24,000.

Buchanan-Banks, J., E. H. Pampeyan, H. C. Wagner, and D. S. McCulloch, 1978, Preliminary map showing recency of faulting in coastal south central California: U.S. Geological Survey Miscellaneous Field Studies Map MF-910, 3 sheets, scale 1:250,000.

Burford, R. O., 1966, Strain analysis across the San Andreas Fault and Coast Ranges of California: Proceedings of the 2nd International Symposium on recent crustal movements, Aulanko, Finland, Aug. 3-7, 1965, Suomalaisen Tiedeakatemian Toimituksia (Acadamiae Scientiarum Fennicae, Annales), series A, III, Geologica-Geographica 90, p. 99-110, Helsinki, Finland.

Cacchione, D. A., W. D. Grant, D. E. Drake, and S. M. Glenn, 1987, Storm-dominated bottom boundary layer dynamics on the northern California continental shelf: Measurements and predictions: Journal of Geophysical Research, v. 92, n.C2, p. 1817-1827.

Champion, D. E., and D. G. Howell, 1986, Paleomagnetism of Cretaceous and Eocene strata, San Miguel Island, California Borderland and the northward translation of Baja, California: Journal of Geophysical Research, v. 91, n. B11, p. 11557-11570.

——— ——— and C. S. Grommé, 1984, Paleomagnetic and geologic data indicating 2,500 km of northward displacement for the Salinian and related terranes, California: Journal of Geophysical Research, v. 89, n. B9, p. 7736-7752.

Chase, T. E., H. W. Menard, and J. Mammerickx, 1970, Bathymetry of the north Pacific: La Jolla, Scripps Institution of Oceanography, Charts 2, 7, 8.

Chesterman, C. W., 1952, Descriptive petrology of rocks dredged off the coast of central California: California Academy of Sciences Proceedings, Fourth Series, v. 27, n. 10, p. 359-374.

Christensen, M. N., 1966, Late Cenozoic crustal movements in the Sierra Nevada of California: Geological Society of America Bulletin, v. 77, p. 163-182.

Clague, D. A., and G. B. Dalrymple, 1987, The geology of the Hawaiian–Emperor volcanic chain, *in* R. W. Decker, ed., Volcanism in Hawaii: U.S. Geological Survey Professional Paper 350, p. 5-54.

——— and R. D. Jarrard, 1973, Tertiary Pacific plate motion deduced from the Hawaiian–Emperor Chain: Geological Society of America Bulletin, v. 84, p. 1135-1154.

Clark, B. L., 1930, Tectonics of the Coast Ranges of middle California: Geological Society of America Bulletin, v. 41, p. 747-828.

Clark, J. C., 1981, Stratigraphy, paleontology, and geology of the central Santa Cruz Mountains, California Coast Ranges: U.S. Geological Survey Professional Paper 1168, 51 p.

——— and E. E. Brabb, 1978, Stratigraphic contrasts across the San Gregorio fault, Santa Cruz Mountains, west central California, *in* E. A. Silver, and W. R. Normark, eds., San Gregorio-Hosgri fault zone, California: California Division of Mines and Geology Special Report 137, p. 3-12.

——— ——— H. G. Greene, and D. C. Ross, 1984, Geology of Point Reyes Peninsula and implications for San Gregorio fault history, *in* J. K. Crouch and S. B. Bachman, eds., Tectonics and sedimentation along the California margin: Society of Economic Paleontologists and Mineralogists, Pacific Section, v. 38, p. 67-86.

Clarke, S. H., Jr., and T. H. Nilsen, 1973, Displacement of Eocene strata and implications for the history of offset along the San Andreas fault, central and northern California, *in* R. L. Kovach and A. Nur, eds., Proceedings of the conference on tectonic problems of the San Andreas fault system: Stanford University, Publications in the Geological Sciences, v. 13, p. 358-367.

Coney, P. J., D. L. Jones, and J. W. H. Monger, 1980, Cordilleran suspect terranes: Nature, v. 288, p. 329-333.

Cook, H. E., ed., 1979, Geologic studies of the Point Conception deep stratigraphic test well OCS-CAL 78-164 NO. 1, outer continental shelf, southern California, United States: U.S. Geological Survey Open-File Report 79-1218, 148 p.

Cooke, L. W., 1985, Estimates of undiscovered oil and gas resources for the outer continental shelf: U.S. Minerals Management Service, OCS Report, MMS 85-0012, 45 p.

Cooper, A. K., 1971, Structure of the continental shelf west of San Francisco, California: U.S. Geological Survey Open-File Report, 65 p.

Coppersmith, K. J., and G. B. Griggs, 1978, Morphology, recent activity and seismicity of the San Gregorio fault zone, *in* E. A. Silver and W. R. Normark, eds., San Gregorio-Hosgri fault zone, California: California Division of Mines and Geology Special Report 137, p. 33-43.

Cox, A., and D. Engebretson, 1985, Change in motion of Pacific plate at 5 Myr BP: Nature, v. 313, 7 February, p. 472-474.

Crain, W. E., W. E. Mero, and D. Patterson, 1985, Geology of the Point Arguello discovery: American Association of Petroleum Geologists Bulletin, v. 69, n. 4, p. 537-545.

Crouch, J. K., S. B. Bachman, and J. T. Shay, 1984, Post-Miocene compressional tectonics along the central California margin, *in* J. K. Crouch and S. B. Bachman, eds., Tectonics and sedimentation along the California margin: Society of Economic Paleontologists and Mineralogists, Pacific Section, v. 38, p. 37-54.

——— and ——— 1985, Monterey Formation may extend northward: California basins gain attention: American Association of Petroleum Geologists Explorer, v. 6, n. 4, p. 1, 8-9.

Curray, J. R., 1966, Geologic structure on the continental margin from subbottom profiles, northern and central California, *in* E. J. Bailey, ed., Geology of northern California: California Division of Mines and Geology Bulletin, v. 190, p. 337-342.

——— and R. D. Nason, 1967, San Andreas fault north of Point Arena, California: Geological Society of America Bulletin, v. 78, p. 413-418.

Dalrymple, 1979, Critical tables for conversion of K-Ar ages from old to new constants: Geology, v. 7, p. 558-560.

Dickinson, W. R., and W. S. Snyder, 1979a, Geometry of triple junctions related to San Andreas transform: Journal of Geophysical Research, v. 84, p. 561-572.

——— and ——— 1979b, Geometry of subducted slabs related to San Andreas transform: Journal of Geophysical Research, v. 87, p. 609-627.

Dolton, G. L., K. H. Carlson, R. R. Charpentier, A. B. Coury, R. A. Crovelli, S. E. Frezon, A. S. Khan, J. H. Lister, R. H. McMullin, R. S. Pike, R. B. Powers, E. W. Scott, and K. L. Varnes, 1981, Estimates of undiscovered recoverable conventional resources of oil and gas in the United States: U.S. Geological Survey Circular 860, 87 p.

Domenico, S. N., 1974, Effect of water saturation on seismic reflectivity of sand reservoirs encased in shale: Geophysics, v. 39, n. 6, p. 759-769.

Duncan, R. A., and D. A. Clague, 1985, Pacific plate motion recorded by linear volcanic chains, *in* A. E. M. Nairn, F. G. Stehli, and S. Uyeda, eds., The ocean basins and margins: New York and London, Plenum Publishing Corporation, v. 7A, p. 89-121.

Eaton, J. P., 1984, Focal mechanisms of near-shore earthquakes between Santa Barbara and Monterey: U.S. Geological Survey Open-File Report 84-477, 10 p.

EEZ-SCAN 84 Scientific Staff, 1986, Atlas of the Exclusive Economic Zone, Western Conterminous United States: U.S. Geological Survey Miscellaneous Investigations Series 1-1792, 52 p., scale 1:500,000.

Emilia, D. A., J. W. Berg, Jr., and W. E. Bales, 1968, Magnetic anomalies off the northwest coast of the United States: Geological Society of America Bulletin, v. 79, p. 1053-1062.

Engebretson, D. C., 1982, Relative motion between oceanic and continental plates in the Pacific basin: Stanford University, Ph.D. dissertation, 211 p.

Epstein, S. A., and Q. L. Nary, 1982, A look at the geology of the hot Santa Maria Basin off California: Oil and Gas Journal, November 5, p. 162-168.

Ernst, W. G., and C. A. Hall, Jr., 1974, Geology and petrology of the Cambrian felsite, a new Oligocene formation, west central California Coast Ranges: Geological Society of America Bulletin, v. 85, n. 4, p. 523-532.

Faber, E., S. Rehder, and W. Stahl, 1983, Surface sediment gas data from petroliferous regions of the North Sea and offshore California: Erdol und Kohle-Erdgas-Petrochemie vereinigt mit Brennstoff-Chemie, Bd. 36, Heft 8, p. 357-361.

Fox, K. F., Jr., R. J. Fleck, G. H. Curtis, and C. E. Meyer, 1985, Implications of the northwestwardly younger age of the volcanic rocks of west-central California: Geological Society of America Bulletin, v. 96, p. 647-654.

Galloway, A. J., 1977, Geology of the Point Reyes peninsula, Marin County, California: California Division of Mines and Geology, Bulletin 202, 72 p.

Garfunkel, Z., 1973, History of the San Andreas fault as a plate boundary: Geological Society of America Bulletin, v. 84, p. 2034-2042.

Garrison, R. E., and R. G. Douglas, eds., 1981, The Monterey Formation and related siliceous rocks of California: Society of Economic Mineralogists and Paleontologists Publication 15, Pacific Section, 327 p.

Gawthrop, W. H., 1978, Seismicity and tectonics of the central California coastal region, *in* E. A. Silver and W. R. Normark, eds., San Gregorio-Hosgri fault zone, California: California Division of Mines and Geology Special Report 137, p. 45-56.

Gibson, J. D., 1983, Tectonic setting and deformation associated with the San Gregorio fault system, San Francisco Peninsula, California, *in* D. W. Andersen and M. J. Rymer, eds., Tectonics and sedimentation along faults of the San Andreas system: Los Angeles, California, Society of Economic Paleontologists and Mineralogists, Pacific Section, p. 67-78.

Gilbert, W. G., 1973, Franciscan rocks near Sur fault zone, northern Santa Lucia Range, California: Geological Society of America Bulletin, v. 84, no. 10, p. 3317-3328.

Gorsline, D. S., and K. O. Emery, 1959, Turbidity current deposits in San Pedro and Santa Monica basins off southern California: Geological Society of America Bulletin, v. 70, p. 279-290.

Graham, S. A., 1976, Tertiary sedimentary tectonics of the central Salinian block of California: Ph.D. dissertation, Stanford University, Stanford, California, 216 p.

——— and W. R. Dickinson, 1978, Apparent offsets of onland geologic features across the San Gregorio-Hosgri fault trend, *in* E. A. Silver and W. R. Normark, eds., San Gregorio-Hosgri fault zone, California: California Division of Mines and Geology, Special Report 137, p. 13-23.

——— C. McCloy, H. Hitzman, R. Ward, and R. Turner, 1984, Basin evolution during change from convergent to transform continental margin in central California: American Association of Petroleum Geologists Bulletin, v. 68, n. 3, p. 233-249.

Gray, L. D., 1980, Geology of Mesozoic basement rocks in the Santa Maria Basin, Santa Barbara and San Luis Obispo counties, California: San Diego State University, M.S. Thesis, 78 p.

Greene, H. G., W. H. K. Lee, D. S. McCulloch, and E. E. Brabb, 1973, Faults and earthquakes in the Monterey Bay region, California: U.S. Geological Survey Miscellaneous Field Studies Map MF-518, 4 sheets, scale 1:200,000.

Griggs, G. B., 1979, Survey of the Alpine Geophysical Associates seismic profiling survey—Año Nuevo Point to Santa Cruz, *in* G. E. Weber, K. R. Lajoie, and G. B. Griggs, eds., Field Trip Guide Coastal Tectonics and Coastal Geologic Hazards in Santa Cruz and San Mateo Counties, California: Geological Society of America Cordilleran section, 75th Annual Meeting, p. 35-43.

Griscom, A., 1973a, Bouguer gravity map of California, Redding sheet: California Division of Mines and Geology, p. 1-13.

——— 1973b, Tectonics at the junction of the San Andreas fault and Mendocino fracture zone from gravity and magnetic data, *in* R. L. Kovach and A. Nur, eds., Proceedings of the conference on tectonic problems of the San Andreas fault system: Stanford University, Publications in the Geological Sciences, v. 13, p. 383-390.

——— 1980, Aeromagnetic map and interpretation of magnetic anomalies, King Range and Chemise Mountain instant study areas, Humboldt and Mendocino Counties, California: U.S. Geological Survey Miscellaneous Field Map, MF-1196B.

Hall, C. A., Jr., 1975, San Simeon-Hosgri fault system, coastal California: economic and environmental implications: Science, v. 190, n. 4221, p. 1291-1294.

——— 1978, Origin and development of the Lompoc-Santa Maria pull-apart basin and its relation to the San Simeon-Hosgri strike-slip fault, western California, *in* E. A. Silver and W. R. Normark, eds., San Gregorio-Hosgri Fault zone, California: California Division of Mines and Geology Special Report 137, p. 25-31.

——— 1982, Pre-Monterey subcrop and structure contour maps, south-central California: U.S. Geological Miscellaneous Investigation Series, MF 1384, 6 maps scale 1:6,250 and accompanying pamphlet, 28 p.

Hall, T. N., D. T. Hunt, P. A. Vaughan, and F. R. Bicker, 1987, Trenching and mapping investigations of the late Quaternary behavior of the San Simeon fault, San Luis Obispo County, California (abs.): Geological Society of America Abstracts with Programs, v. 19, p. 385.

Hamilton, D. H., and C. R. Willingham, 1977, Hosgri fault zone, amount of displacement and relationship to structures of the Western Transverse Ranges (abs.): Geological Survey of America Abstracts with Programs, v. 9, p. 429.

Hamilton, W., 1969, Mesozoic California and the underflow of Pacific mantle: Geological Society of America Bulletin, v. 80, p. 2409-2430.

Hanna, G. D., 1952, Geology of the continental slope off central California: California Academy of Sciences Proceedings, Fourth Series, v. 27, n. 9, p. 325-328.

Harbert, W., and A. Cox, 1986, Late Neogene motion of the Pacific plate (abs.): EOS, v. 67, n. 44, p. 1225.

Heasler, H. P., and R. C. Surdam, 1985, Thermal evolution of coastal California with application to hydrocarbon maturation: American Association of Petroleum Geologists Bulletin, v. 69, n. 9, p. 1386-1400.

Higgins, C. G., 1960, Ohlson Ranch Formation, Pliocene, northern Sanoma County, California: University of California Publications in Geological Sciences, v. 36, n. 3, p. 199-231.

Hodell, D. A., K. M. Elmstrom, and J. P. Kennet, 1986, Latest Miocene benthic $\delta^{18}O$ changes, global ice volume, sea level and the Messinian salinity crisis: Nature, v. 320, p. 411.

Hornafius, J. S., 1985, Neogene tectonic rotation of the Santa Ynez Range, western Transverse Ranges, California, suggested by paleomagnetic investigation of the Monterey Formation: Journal

of Geophysical Research, v. 90, n. B14, p. 12,503-12,522.
Hoskins, E. G., and J. R. Griffiths, 1971, Hydrocarbon potential of northern and central California offshore: American Association of Petroleum Geologists Memoir 15, p. 212-228.
Howell, D. G., D. E. Champion, and J. E. Vedder, 1987, Terrane accretion, crustal kinematics and basin evolution, southern California, *in* R. V. Ingersoll and W. G. Ernst, eds., Cenozoic basin development of coastal California: Englewood Cliffs, N.J., Prentice-Hall, Inc., Rubey, v. 6, p. 242-258.
——— J. K. Crouch, H. G. Greene, D. S. McCulloch, and J. G. Vedder, 1980, Basin development along the late Mesozoic and Cenozoic California margin: a plate tectonic margin of subduction, oblique subduction and transform tectonics, *in* P. F. Ballance and H. G. Reading, eds., Sedimentation in oblique-slip mobile zones: International Association of Sedimentologists Special Publication Number 4, Blackwell Scientific Publications, p. 43-62.
——— J. D. Gibson, G. S. Fuis, J. H. Knapp, G. B. Haxtal, B. R. Keller, L. T. Silver, and J. G. Vedder, 1986, Continent-Ocean Transect C-3, southern California to Socorro, New Mexico: Geological Society of America, 3 sheets, scale 1:500,000.
——— D. L. Jones, and E. R. Schermer, 1985, Tectonostratigraphic terranes of the Circum-Pacific region, *in* D. G. Howell, ed., Tectonostratigraphic terranes of the Circum-Pacific region: Circum-Pacific Council for Energy and Mineral Resources, Earth Science Series, n. 1, p. 3-30.
——— and J. M. Joyce, 1981, Field guide to the Upper Cretaceous Pigeon Point Formation, *in* V. Frizell, ed., Upper Cretaceous and Paleocene turbidites, central California coast: Pacific Coast Field Guide 6, Society of Economic Paleontologists and Mineralogists, Pacific Section, v. 4, p. 61-70.
——— D. S. McCulloch, and J. G. Vedder, 1978, General geology, petroleum appraisal, and nature of environmental hazards, eastern Pacific shelf, Latitude 28° to 38° north: U.S. Geological Survey Circular 78B, 29 p.
——— J. G. Vedder, H. McLean, J. M. Joyce, S. H. Clarke, Jr., and G. Smith, 1977, Review of Cretaceous geology, Salinian and Nacimiento blocks, Coast Ranges of California, *in* D. G. Howell, J. G. Vedder, and K. McDougall, eds., Cretaceous geology of the California Coast Ranges west of the San Andreas: Pacific Coast Field Guide 2, Society of Economic Paleontologists and Mineralogists, Pacific Section, p. 1-46.
Hsu, K. J., L. Montadert, D. Bernoulli, M. B. Cita, A. Erickson, R. E. Garrison, R. B. Kidd, F. Melieres, C. Muller, and R. Wright, 1977, History of the Mediterranean salinity crisis: Nature, v. 267, p. 399-403.
Hsu, I-C, and F. T. Wu, 1979, Normal faulting along San Andreas between Point Arena and Cape Mendocino (abs.): EOS Transactions, American Geophysical Union, v. 60, n. 18, p. 322-323.
Huffman, O. F., D. A. Turner, and R. N. Jack, 1973, Offset of late Oligocene-early Miocene volcanic rocks along the San Andreas fault in central California, *in* R. L. Kovach and A. Nur, eds., Proceedings of the conference on tectonic problems of the San Andreas fault system: Stanford University, Publications in the Geological Sciences, v. 13, p. 368-373.
Hunt, J. M., 1979, Petroleum Geochemistry and Geology: San Francisco, W. H. Freeman and Company, 617 p.
Ingle, J. C., Jr., 1973, Summary comments on Neogene biostratigraphy, physical stratigraphy, and paleo-oceanography in the marginal northeastern Pacific Ocean, *in* L. D. Kulm and R. von Huene et al., Initial Reports of the Deep Sea Drilling Project, v. 8, Washington, D.C., U.S. Government Printing Office, p. 949-960.
——— 1978, Evidence of latest Miocene refrigeration and associated eustatic events in California and the North Pacific (abs.): Geological Society of America, Abstracts with Programs, v. 10, n. 7, p. 427.
——— 1980, Cenozoic paleobathymetry and depositional history of selected sequences within the southern California continental borderland, *in* W. V. Sliter, ed., Studies in marine micropaleontology and paleoecology, a memorial volume to Orville L. Bandey: Cushman Foundation for Foraminiferal Research, Special Publication No. 19, p. 163-195.
Irwin, W. P., and M. D. Dennis, 1979, Geologic structure section across southern Klamath Mountains, Coast Ranges and seaward from Point Delgada, California: Geological Society of America Map and Chart Series MC-28D, scale 1:250,000.
Isaacs, C. M., 1981, Guide to the Monterey Formation in the California coastal area, Ventura to San Luis Obispo: American Association of Petroleum Geologists, Pacific Section, v. 92, 91 p.
——— 1984a, The Monterey—key to offshore California boom: Oil and Gas Report, January 9, p. 75-81.
——— 1984b, Geology and physical properties of the Monterey Formation, California: Society of Petroleum Engineers of American Institute of Mining Engineers, Regional Meeting in California, p. 83-92.
——— 1985, Abundance versus rates of accumulation in fine-grained strata of the Miocene Santa Barbara basin, California: Geo-Marine Letters, v. 5, p. 25-30.
——— M. A. Keller, V. A. Gennai, K. C. Stewart, and J. E. Taggert, Jr., 1983, Preliminary evaluation of Miocene lithostratigraphy in the Point Conception COST well OCS-CAL 78-164 No. 1, southern California, *in* C. M. Isaacs and R. E. Garrison, eds., Petroleum generation and occurrence in the Miocene Monterey Formation, California: Society of Economic Paleontologists and Mineralogists, Pacific Section, Special Publication, p. 99-110.
——— and N. F. Petersen, in press, Petroleum in the Miocene Monterey Formation, California, U.S.A., *in* J. R. Hein, ed., Siliceous sedimentary rock-hosted ores and petroleum: Von Nostrand-Reinhold Publishers, p. 83-116.
——— K. A. Pisciotto, and R. E. Garrison, 1983, Facies and diagenesis of the Miocene Monterey Formation, California: A summary, *in* A. Iijima, J. R. Kein, and R. Silver, eds., Siliceous deposits in the Pacific region: Elsevier Scientific Publishing Company, Amsterdam, p. 247-282.
Jackson, E. D., E. A. Silver, and G. B. Delrymple, 1972, Hawaiian-Emperor chain and its relation to Cenozoic circum-Pacific tectonics: Geological Society of America Bulletin, v. 83, p. 601-618.
Jahns, R. H., and D. H. Hamilton, 1971, Geology of the Mendocino power plant, *in* Mendocino Power Plant Preliminary Safety Analysis Report: submitted to the U.S. Atomic Energy Commission by Pacific Gas and Electric Company, v. 1, appendix 2.5.A.
Jennings, E. E., C. J. Alonzo, and M. E. Mitchell, 1985, Offshore California heavy crude: potential and production: Pacific Oil World, v. 77, n. 12, p. 36-40.
Johnson, J. D., and W. R. Normark, 1974, Neogene tectonic evolution of the Salinian Block, west-central California: Geology, v. 12, n. 1, p. 11-14.
Kamerling, M. J., and B. P. Luyendyk, 1979, Tectonic rotations of the Santa Monica Mountains region, western Transverse Ranges, California, suggested by paleomagnetic vectors: Geological Society of America Bulletin, v. 90, p. 331-337.
Kanter, L. R., and M. Debiche, 1985, Modeling of motion histories of the Point Arena and central Salinia terranes, *in* D. G. Howell, ed., Tectonostratigraphic terranes of the circum-Pacific region: Circum-Pacific Council for Energy and Mineral Resources, Earth Science Series, n. 1, p. 227-238.
Keller, G., and J. A. Barron, 1983, Paleoceanographic implication of Miocene deep-sea hiatuses: Geological Society of America Bulletin, v. 94, p. 590-613.
Kieniewicz, P. M., and B. P. Luyendyk, 1986, A gravity model of the basement structure in the Santa Maria Basin, California: Geophysics, v. 51, p. 1127-1140.
Kistler, R. W., and Z. E. Peterman, 1973, Variations in Sr, Rb, K, Na, and Initial Sr^{87}/Sr^{86} in Mesozoic granitic rocks and intruded wall rocks in central California: Geological Society of America Bulletin, v. 84, n. 11, p. 3489-3512.
Krause, D. C., H. W. Menard, and S. M. Smith, 1964, Topography and lithology of the Mendocino Ridge: Journal of Marine Research, v. 22, n. 3, p. 236-249.

Kulm, L. D., and R. von Huene, et al., 1973, Initial Reports of the Deep Sea Drilling Project, v. 18, Washington, D.C., U.S. Government Printing Office, p. 31-43.

Lachenbruch, A. H., and J. H. Sass, 1980, Heat flow and energetics of the San Andreas fault zone: Journal of Geophysical Research, v. 85, n. B11, p. 6185-6222.

Lajoie, K. R., G. E. Weber, and J. C. Tinsley, 1972, Marine terrace deformation: San Mateo and Santa Cruz counties, *in* U.S. Geological Survey Staff and J. C. Cummings, eds., Progress Report on the U.S.G.S. Quaternary studies in the San Francisco Bay Region: Guidebook for Friends of the Pleistocene, October 6-8, p. 100-111.

——— A. M. Sarna-Wojcicki, and Y. Ota, 1982, Emergent Holocene marine terraces at Ventura and Cape Mendocino, California—Indicators of high tectonic uplift rates (abs.): Geological Society of America, Cordilleran Section, Abstracts with Programs, v. 14, n. 4, p. 178.

Lawson, A. C., et al., 1908, The California earthquake of April 18, 1906-Report to the State Earthquake Commission: Carnegie Institute, Washington, Publication 87, v. 1, p. 56-57.

Leslie, R. B., 1981, Continuity and tectonic implications of the San Simeon-Hosgri fault zone, central California: U.S. Geological Survey Open-File Report 81-430, 59 p.

Lindh, A., C. Matooka, S. Ball, and R. Dollar, 1981, Current seismicity of the central California coastal region from Point Buchon to Point Piedras Blancas—A preliminary report: U.S. Geological Survey Open-File Report 81-44, 41 p.

MacLeod, N. S., and R. M. Pratt, 1973, Petrology of volcanic rocks recovered on Leg 18, *in* L. D. Kulm and R. von Huene et al., Initial Reports of the Deep Sea Drilling Project, v. 18, Washington, D.C., U.S. Government Printing Office, p. 935-945.

Mason, R. G., and A. D. Raff, 1961, A magnetic survey off the west coast of North America, 32°N to 42°N: Geological Society of America Bulletin, v. 72, p. 1259-1265.

Mattinson, J. M., 1978, Age, origin, and thermal histories of some plutonic rocks from the Salinian block of California: Contributions to Mineralogy and Petrology, v. 67, p. 233-245.

——— 1983, Basement rocks of the southeastern Salinian block, California: U-Pb isotopic relationships (abs.): Geological Society of America Abstracts with Programs, v. 15, n. 5, p. 414.

——— and E. W. James, 1985, Salinian block U/Pb age and isotopic variations: Implications for origin and emplacement of the Salinian terrain, *in* D. G. Howell, ed., Tectonostratigraphic terranes of the circum-Pacific region: Circum-Pacific Council for Energy and Mineral Resources, Earth Science Series, n. 1, p. 215-226.

McCulloch, D. S., 1985, Evaluating tsunami potential, *in* J. I. Ziony, ed., Evaluating earthquake hazards in the Los Angeles region: U.S. Geological Survey Professional Raper 1360, p. 375-413, 496-500.

——— and R.H. Chapman, 1977, Residual magnetic intensity maps along the California coast from 37°30′N lat to 34°30′N lat: In cooperation with the California Division of Mines and Geology: U.S. Geological Survey Open-File Report 77-79, 14 aeromagnetic maps, scale 1:250,000.

——— S. H. Clarke, Jr., M. E. Field, E. W. Scott, and P. Utter, 1977, A summary report on the regional geology, petroleum potential, and environmental geology of the southern proposed lease sale 53, central and northern California outer continental shelf: U.S. Geological Survey Open-File Report 77-593, 57 p.

——— J. V. Gardner, S. L. Eittreim, and D. G. Masson, 1984, GLORIA survey of the morphology of the mid-California continental margin (abs.): EOS Transactions, American Geophysical Union, v. 65, n. 45, p. 1083.

——— H. G. Greene, K. S. Heston, and D. M. Rubin, 1980, A summary report of the geology and geologic hazards in proposed lease sale 53, central California outer continental shelf: U.S. Geological Survey Open-File Report 80-1095, 76 p., 12 maps, scale 1:25,000.

——— J. G. Vedder, H. C. Wagner, and R. H. Brune, 1979, Geologic setting, *in* H. E. Cook, ed., Geologic studies of the Point Conception deep stratigraphic test well OCS-CAL 78-164 No. 1, outer continental shelf, southern California, United States: U.S. Geological Survey Open-File Report 79-1218, p. 10-25.

——— P. A. Utter, and J. S. Menack, 1985, Maps showing locations of selected pre-Quaternary rock samples from 34°30′N lat to 42°N lat, California continental margin: U.S. Geological Survey Miscellaneous Investigations Series, MF 1719, 38 p., 4 maps, scale 1:250,000.

McCulloh, T. H., 1979, Implications for petroleum appraisal, *in* H. E. Cook, ed., Geological studies of the Point Conception deep stratigraphic test well OCS-CAL 78-164 No. 1, outer continental shelf, southern California, United States: U.S. Geological Survey Open-File Report 79-1218, p. 26-42.

——— and L. A. Beyer, 1979, Geothermal gradients, *in* H. E. Cook, ed., Geologic studies of the Point Conception deep stratigraphic test well OCS-CAL 78-164 No. 1, outer continental shelf, southern California, United States: U.S. Geological Survey Open-File Report 79-1218, p. 43-48.

McKenzie, D. P., and W. J. Morgan, 1969, Evolution of triple junctions: Nature, v. 224, p. 125-133.

McLaughlin, R. J., 1974, The Sargent-Berrocal fault zone and its relation to the San Andreas system in the southern San Francisco Bay region and Santa Clara Valley, California: U.S. Geological Survey Journal of Research, v. 2, n. 5, p. 593-598.

——— W. V. Sliter, and N. O. Frederiksen, 1986, Plate motions recorded by tectonostratigraphic terranes adjacent to the Mendocino triple junction (abs.): EOS, v. 67, n. 44, p. 1219.

——— S. A. Kling, R. Z. Poore, K. McDougall, and E. C. Beutner, 1982, Post-middle Miocene accretion of Franciscan rocks, northwestern California: Geological Society of America Bulletin, v. 93, p. 595-605.

——— K. R. Lajoie, D. H. Sorg, S. D. Morrison, and J. A. Wolfe, 1983, Tectonic uplift of a middle Wisconsin marine platform near the Mendocino triple junction, California: Geology, v. 11, p. 35-39.

McWilliams, M. O., and D. G. Howell, 1982, An exotic origin for terranes of western California: Nature, v. 297, p. 215-217.

Minster, J. B., and T. H. Jordan, 1987, Vector constraints on western U.S. deformation from space geodesy, neotectonics, and plate motions: Journal of Geophysical Research, v. 92, n. 86, p. 4798-4804.

Mullins, H. T., and D. K. Nagle, 1981, Franciscan-type rocks off Monterey Bay, California: Implications for western boundary of Salinian block: Geo-Marine Letters, v. 1, n. 2, p. 135-139.

Murata, K. J., I. Friedman, and J. D. Gleason, 1977, Oxygen isotope relations between diagenetic silica minerals in Monterey Shale, Temblor Range, California: American Journal of Science, v. 277, p. 259-272.

Mustart, D. A., and D. S. McCulloch, 1984, The San Bruno fault-a possible eastern strand of the San Andreas fault in the San Francisco region (abs.): Geological Society of America, Abstracts with Programs, v. 16, n. 6, p. 605-606.

Nason, R. D., 1968, San Andreas fault at Cape Mendocino, *in* W. R. Dickinson and A. Grantz, eds., Proceedings of the conference on geologic problems of the San Andreas fault system: Stanford University, Publications in the Geological Sciences, v. 11, p. 231-241.

National Oceanic and Atmospheric Administration, 1974, Magnetic map, vicinity of Point Sur to Point Reyes: National Ocean Survey 1307N-11M, scale 1:250,000.

Normark, W. R., and C. E. Gutmacher, 1985, Delgada Fan, Pacific Ocean, *in* A. H. Bouma, W. R. Normark, and N. E. Barnes, eds., Submarine fans and related turbidite systems, New York, N.Y., Springer-Verlag, p. 59-64.

Ogle, B. A., 1953, Geology of Eel River Valley area Humboldt County, California: California Division of Mines Bulletin 164, 128 p.

——— 1984, Significant new developments in the Santa Maria/Santa Barbara area, offshore California—the geology and economic potential: Proceedings of the Southwestern Legal Foundation, Exploration and Economics of the Petroleum Industry, New York, N.Y., Matthew Bender and Company Inc., v. 22, p. 23-55.

Page, B. M., 1970, Sur-Nacimiento fault zone of California continental margin tectonics: Geological Society of American Bulletin, v. 81, n. 3, p. 667-690.

——— 1981, The southern Coast Ranges, *in* W. G. Ernst, ed., The tectonic development of California: Englewood Cliffs, N.J., Prentice-Hall Inc., Rubey, v. 1, p. 329-417.

——— 1982, Migration of the Salinian composite block, California, and disappearance of fragments: American Journal of Science, v. 282, p. 1694-1734.

——— and D. C. Engebretson, 1984, Correlation between the geologic record and computed plate motions for central California: Tectonics, v. 3, p. 133-155.

——— H. C. Wagner, D. S. McCulloch, E. A. Silver, and J. H. Spotts, 1979a, Geologic cross section of the continental margin off San Luis Obispo, southern Coast Ranges, and the San Joaquin Valley, California: Geological Society of America Map and Chart Series, MC-28G, 12 p., 1 map, scale 1:250,000.

——— ——— ——— ——— and ——— 1979b, Tectonic interpretation of a geologic section of the continental margin off San Luis Obispo, the Southern Coast Ranges, and the San Joaquin Valley, California: Cross-Section summary: Geological Society of America Bulletin, Part 1, v. 90, p. 808-812.

Petersen, N. F., and P. J. Hickey, in press, California Plio-Miocene oils: Evidence of early generation, *in* R. F. Meyer, ed., Exploration for heavy crude oil and bitumen: American Association of Petroleum Geologists, Studies in Geology 23.

Pollitz, F. F., 1986, Pliocene change in Pacific-plate motion: Nature, v. 320, p. 738-741.

Reed, R. D., 1933, Geology of California: American Association of Petroleum Geologists, Tulsa, Oklahoma, 355 p.

Richmond, W. C., L. J. Cummings, S. Hamlin, and M. E. Nagaty, 1981, Geological hazards and constraints in the area of proposed OCS oil and gas lease sale 48, southern California: U.S. Geological Survey Open-File Report 81-307, 37 p.

Ross, D. C., 1972, Petrographic and chemical reconnaissance study of some granite and gneissic rocks near the San Andreas Fault from Bodega Head to Cajon Pass, California: U.S. Geological Survey Professional Paper 698, 92 p.

——— 1977, Pre-intrusive metasedimentary rocks of the Salinian block, California—a tectonic dilemma, *in* J. H. Stewart, C. H. Stevens and A. E. Fritsche, eds., Paleozoic paleogeography of the western United States: Society of Economic Paleontologists and Mineralogists, Pacific Section, Pacific Coast Paleogeography Symposium 1, p. 371-380.

——— 1984, Possible correlations of basement rocks across the San Andreas, San Gregorio-Hosgri, and Rinconada-Reliz-King City faults, California: U.S. Geological Survey Professional Paper 1317, 37 p.

——— and E. E. Brabb, 1973, Petrography and structural relations of granitic basement rocks in the Monterey Bay area, California: U.S. Geological Survey Journal of Research, v. 1, n. 3, p. 273-283.

——— and D. S. McCulloch, 1979, Cross section of the southern Coast Ranges and San Joaquin valley from offshore Point Sur to Madera, California: Geological Society of America Map and Chart Series, MC-2BH, 4 p., 1 map, scale 1:250,000.

Saleeby, J. B., R. C. Speed, M. C. Blake, R. W. Allmendinger, P. B. Gans, R. W. Kistler, D. C. Ross, D. A. Stauber, M. L. Zoback, A. Griscom, D. S. McCulloch, A. H. Lachenbruch, R. B. Smith, and D. P. Hill, 1986, Continent-ocean transect, Corridor C2, central California offshore to Colorado Plateau: Geological Society of America, 2 sheets, scale 1:50,000.

Sarna-Wojcicki, A. M., S. D. Morrison, C. E. Meyer, and J. W. Hillhouse, 1987, Correlation of upper Cenozoic tephra layers between sediments of the western United States and eastern Pacific Ocean and comparison with biostratigraphic and magnetostratigraphic age data: Geological Society of America Bulletin, v. 98, p. 207-223.

Silberling, N. J., D. L. Jones, M. C. Blake, Jr., and D. G. Howell, 1984, Lithotectonic terrane maps of the North American Cordillera: U.S. Geological Survey Open File Report 84-523, 100 p., 4 maps, scale 1:2,500,000.

Silver, E. A., 1969, Structure of the continental margin off northern California, north of the Gorda Escarpment: Scripps Institution of Oceanography, University of California San Diego, Ph.D. dissertation, 123 p.

——— 1971, Small plate tectonics in the northeastern Pacific: Geological Society of America Bulletin, v. 82, p. 3491-3496.

——— 1977, Are the San Gregorio and Hosgri fault zones a single fault system (abs.): Geological Society of America Abstracts with Programs, v. 9, n. 4, p. 500.

——— 1978, The San Gregorio-Hosgri fault zone: An overview, *in* E. A. Silver and W. R. Normark, eds., San Gregorio-Hosgri fault zone: California Division of Mines and Geology, Special Report 137, p. 1-2.

——— J. R. Curray, and A. K. Cooper, 1971, Tectonic development of the continental margin off central California, *in* J. H. Lipps and E. M. Moores, eds., Geologic guide to the Northern Coast Ranges, Point Reyes Region, California: Geological Society of Sacramento Annual Field Trip Guidebook, p. 1-10.

Sliter, W. V., 1984, Foraminifers from Cretaceous limestone of the Franciscan Complex, northern California, *in* M. C. Blake, Jr., ed., Franciscan geology of northern California: Society of Economic Paleontologists and Mineralogists, Pacific Section, v. 43, p. 149-162.

Sorg, D. H., and R. J. McLaughlin, 1980, Geologic and environmental implications of the Monto Bello Ridge landslide complex near Stevens Creek Reservoir, Santa Clara County, California, *in* R. Streitz, and R. Sherburne, eds., Studies of the San Andreas fault zone in northern California: California Division of Mines and Geology Special Report 140, p. 51-58.

Stahl, W., and E. Faber, 1983, Carbon isotopes as a petroleum exploration tool: London, 11th World Petroleum Congress, London, Geology Exploration Resources, Proceedings v. 2, p. 147-159.

Stanley, R. G., 1986, Evidence for late Oligocene-early Miocene episode of transtension along San Andreas fault system in central California (abs.): American Association of Petroleum Geologists Bulletin, v. 70, n. 4, p. 480.

——— 1987, New estimates of displacement along the San Andreas fault in central California based on paleobathymetry and paleogeography: Geology, v. 15, p. 171-174.

——— in press, Implications of the northwestwardly younger age of the volcanic rocks of west-central California: Alternative interpretation: Geological Society of America Bulletin.

Stiretz, J., 1986, The southern termination of the Hosgri fault zone, offshore south-central California: M.S. Thesis, University of California at Santa Barbara, 78 p.

Trask, P. D., 1926, Geology of the Point Sur quadrangle, California: University of California, Department of Geological Science Bulletin, v. 13, n. 5, p. 133-174.

Trehu, A. M., and W. H. Wheeler IV, 1987, Possible evidence for subducted sedimentary materials beneath central California: Geology, v. 15, p. 254-258.

Turner, D. L., 1970, Potassium-argon dating of Pacific coast Miocene foraminiferal stages, *in* O. L. Bandy, ed., Radiometric dating and paleontologic zonation: Geological Society of America Special Paper 124, p. 91-129.

Underwood, M. B., 1977, The Pfeiffer Beach slab deposits, Monterey County, California: Possible trench-slope basin, *in* D. G. Howell, J. G. Vedder, and K. McDougall, eds., Cretaceous geology of the California Coast Ranges west of the San Andreas: Society of Economic Paleontologists and Mineralogists, Pacific Section, Pacific Coast Field Guide 2, p. 57-70.

——— 1986, Geothermal history of the Cape Mendocino region of northern California (abs.): EOS, v. 67, n. 44, p. 1219.

Vail, P. R., and R. M. Mitchum, Jr., 1979, Global cycles of relative changes of sea level from seismic stratigraphy, *in* J. S. Watkins, L. Montadert, and P. W. Dickenson, eds., Geological and geophysical investigations of continental margins: American Association of Petroleum Geologists Memoir 29, p. 469-472.

Vedder, J. G., D. G. Howell, and H. McLean, 1983, Stratigraphy, sedimentation, and tectonic accretion of exotic terranes, southern Coast Ranges, California, *in* J. S. Watkins and C. L. Drake, eds., Studies in continental margin geology: American Association of Petroleum Geologists Memoir 34, p. 471-496.

——— J. K. Crouch, and F. Lee-Wong, 1981, Comparative study of rocks from Deep Sea Drilling Project Holes 467, 468, and 469, and the California Borderland, *in* R. S. Yeats, B. V. Haq et al., eds., Initial Reports of the Deep Sea Drilling Project, v. 63, Washington, D.C., U.S. Government Printing Office, p. 907-918.

Vink, G. E., W. J. Morgan, and W-L. Zhao, 1984, Preferential rifting of continents: A source of displaced terranes: Journal of Geophysical Research, v. 89, n. B12, p. 10072-10076.

Wagner, H. C., 1974, Marine geology between Cape San Martin and Pt. Sal, south-central California offshore: U.S. Geological Survey Open-File Report 74-252, 17 p.

Weaver, C. E., 1943, Point Arena-Fort Ross region (California): California Division of Mines and Geology Bulletin 118, p. 628-632.

Weber, G. E., 1983, Geological investigation of the marine terraces of the San Simeon region and Pleistocene activity on the San Simeon fault zone, San Luis Obispo County, California: Technical report to U.S. Geological Survey under contract 14-08-0001-18230, 66 p.

——— and K. R. Lajoie, 1980, Map of Quaternary faulting along the San Gregorio fault zone, San Mateo and Santa Cruz Counties, California: U.S. Geological Survey Open-File Report 80-907, 3 maps, scale 1:24,000.

——— and——— 1979, Evidence for Holocene movement on the Frijoles fault near Point Año Nuevo, San Mateo County, California, *in* G. E. Weber, K. R. Lajoie, and G. B. Griggs, eds., Coastal tectonics and coastal geologic hazards in Santa Cruz and San Mateo Counties, California: Cordilleran Section, Geological Society of America Field trip guide for 75th Annual Meeting, p. 92-100.

Webster, F. L., D. J. Burdick, D. G. Griggs, and K. A. Yenne, 1985, Geologic report, proposed southern California Planning Area, OCS Lease Sale No. 95: Minerals Management Service Pacific OCS Region, OCS Report MMS 85-0042, 64 p.

Weldon, R., and E. Humphreys, 1986, A kinematic model of southern California: Tectonics, v. 5, n. 1 p. 33-48.

Wentworth, C. M., Jr., 1968, Upper Cretaceous and lower Tertiary strata near Gualala, California and inferred large right slip on the San Andreas fault, *in* W. R. Dickinson and A. Grantz, eds., Proceedings of conference on geologic problems of San Andreas fault system: Stanford University Publications in the Geological Sciences, v. 11, p. 30-142.

——— 1972, Geology, San Andreas offset, and seismic environment of the Gualala block, *in* E. M. Moores and R. A. Matthew, eds., Geologic guide to the northern coast ranges, Lake Mendocino and Sanoma Counties, California: Geological Society of Sacramento, Annual Field Trip Guidebook, p. 95-110.

——— and M. D. Zoback, 1986, An integrated faulting model for the Coalinga earthquake: A guide to thrust deformation along the Coast Range–Great Valley boundary in central California (abs.): EOS, v. 67, n. 44, p. 1222.

Wilde, P., T. E. Chase, M. L. Holmes, W. R. Normark, J. A. Thomas, D. S. McCulloch, and L. D. Kulm, 1978, Oceanographic data off northern California-southern Oregon 40° to 43° North including the Gorda deep sea fan: University of California, Berkeley, Lawrence Berkeley Laboratory Publication 251, 1 map, scale 1:815,482.

——— W. P. Normark, and T. E. Chase, 1976, Oceanographic data off central California 37° to 40° North including Delgada deep sea fan: University of California, Berkeley, Lawrence Berkeley Laboratory Publication 92, 1 map, scale 1:864,518.

Wilson, D. S., 1985, Tectonic history of the Juan de Fuca ridge: Stanford University, Ph.D. Dissertation, 105 p.

Yeats, R. S., 1981, Quaternary flake tectonics of the California Transverse Ranges: Geology, v. 9, no. 1, p. 16-20.

Yerkes, R. F., 1985, Geologic and seismologic setting, *in* J. I. Ziony, ed., Evaluating earthquake hazards in the Los Angeles Region—an earth-science perspective: U.S. Geological Survey Professional Paper 1360, p. 25-42.

——— and W. H. Y. Lee, 1979, Fault activity, epicenters, focal depths, and focal mechanisms, 1970-1975 earthquakes, western Transverse Ranges, California: U.S. Geological Survey Miscellaneous Field Studies Map, MF-1032, 2 sheets, scale 1:250,000.

Zieglar, D. L., and J. K. Cassell, 1978, A synthesis of OCS well information, offshore central and northern California, Oregon, and Washington: Sacramento, California, American Association of Petroleum Geologists, Pacific Section, preprint paper presented at the 53rd Annual Meeting, April 28-29, 27 p.

Zoback, M. D., M. L. Zoback, V. S. Mount, J. P. Eaton, J. H. Healy, D. Oppenheimer, P. Reasenberg, L. Jones, C. B. Raleigh, I. Wong, O. Scotti, and C. Wentworth, in preparation, State of stress of the San Andreas fault system.

CHAPTER SEVENTEEN

REGIONAL GEOLOGY AND PETROLEUM POTENTIAL OF THE SOUTHERN CALIFORNIA BORDERLAND

J. G. Vedder

U.S. Geological Survey
Menlo Park, California

ABSTRACT

The southern California borderland includes or adjoins the petroleum-rich Santa Barbara Channel, and the Ventura and Los Angeles basins. Borderland basement rocks are composed of four main types: (1) a Cretaceous(?) subduction complex, (2) Jurassic arc-volcanogenic rocks, (3) pre-Late Cretaceous high-pressure metavolcanic and metasedimentary rocks, and (4) a Triassic(?) and Jurassic graywacke-argillite broken formation overlain by Late Jurassic and Early Cretaceous arc-volcanogenic rocks and intruded by Cretaceous granitic plutons. These basement types constitute distinct allochthonous terranes, each of which is overlain by a contrasting stratigraphic sequence. The terranes were emplaced after long-range northward translation during the Cretaceous and Tertiary. After subduction of the Farallon plate, the Pacific plate collided with North America about 29 Ma, and a transform margin began to evolve. The resulting broad zone of right shear, which persisted through the late Tertiary, was conducive to the development of depositional basins. Eocene, Oligocene, and Miocene sea-level fluctuations influenced the nature of depositional systems and the coarseness of sedimentation in these basins. Crustal shortening and uplift during the late Pliocene and Pleistocene created compressive structures and high relief.

Because some basins farther offshore have had histories similar to those of the oil-producing basins, they may have high petroleum potential. Miocene sedimentary sequences include excellent source rocks, and Late Cretaceous and Paleogene

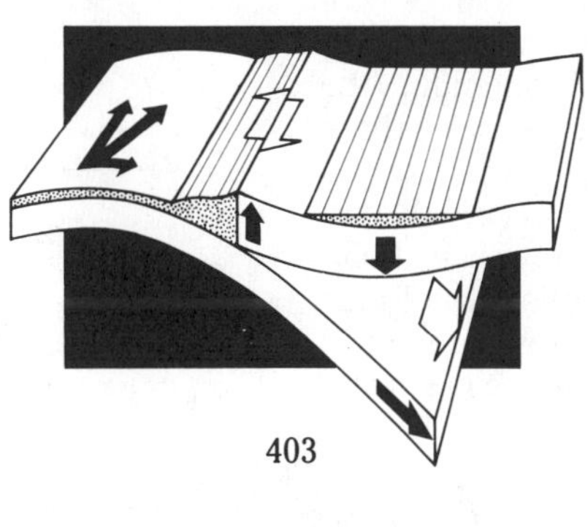

sandstone zones as well as fractured Miocene shale may provide suitable reservoirs. Favorable structural and stratigraphic traps occur along some basin flanks. However, Cretaceous and Paleogene strata are absent over large areas, and Neogene sequences are relatively thin in the outer basins and truncated or absent on most ridges. Paleogene beds generally have a low organic content, and most of the Neogene strata on the outer borderland have had insufficient burial and thermal histories for significant hydrocarbon generation.

INTRODUCTION

Purpose and Scope

Most of the work by the U.S. Geological Survey on the California Continental Borderland was done between 1973 and 1980. The results were released principally in circulars and open-file reports or in volumes of collected papers that were not widely distributed. The purpose of this chapter is to summarize these somewhat obscure earlier reports by freely excerpting from them and by updating where newer information is available.

A two-part format is used to deal with the main aspects of this chapter. The first part consists of a generalized description of the age, composition, distribution, and tectonic history of borderland rocks. The second part contains an area-by-area review of petroleum-related stratigraphy, structure, and hydrocarbon potential. Stratigraphic names used in both parts are those that are most commonly applied by petroleum geologists who work in the region.

Although more than half of the California Continental Borderland, hereinafter called "the borderland", lies south of the international boundary (Figure 1), only the region north of the boundary is described in detail. A line connecting Point Conception and Rodriguez Seamount is used as an arbitrary northwestern limit of the area discussed.

Geomorphic and Bathymetric Features

The borderland is a predominantly subsea geomorphic province that covers an area of approximately 150,000 km^2. Aligned topographic entities that have two distinct directional trends characterize the geomorphic provinces of mainland southern California (Jahns, 1954; Figure 1). West-trending features of the long, narrow Transverse Ranges province transect the dominant northwest grain of the Coast Ranges and Peninsular Ranges provinces and the borderland. If the conventional onshore provinces are extended offshore, the borderland seems to include submergent parts of both the Transverse Ranges and Peninsular Ranges with the result that all three provinces overlap. Although this apparent inconsistency has been a source of contention, the complex history and intersecting tectonic styles of these provinces require such an overlap.

As defined by Shepard and Emery (1941) and Moore (1969), the borderland is bounded on the northwest by Point Arguello and Arguello Canyon; on the southeast by Bahia Sebastian Vizcaino and Cedros Island. Its western edge is marked by the base of the continental slope at the north and by Cedros Deep at the south (Moore, 1969, Plate 1). The Baja California Seamount province of the Pacific Ocean Basin lies to the west.

North of lat. 31°N, the borderland is typified by elongate northwest-trending sea-floor basins and ridges, some of which protrude above sea level as islands. South of lat. 30°N, the borderland narrows, and the numerous basins and ridges coalesce at its southern end. Although most of the ridges are linear, a few are gently curved, and others tend to bifurcate southeastward. Offshore from southern California, the borderland differs from an ordinary continental shelf in that it encompasses large depressions as deep as 2100 m below sea level and island peaks as high as 750 m above sea level. In the southern part of the borderland, some basin floors are >2800 m below sea level. Topographic relief

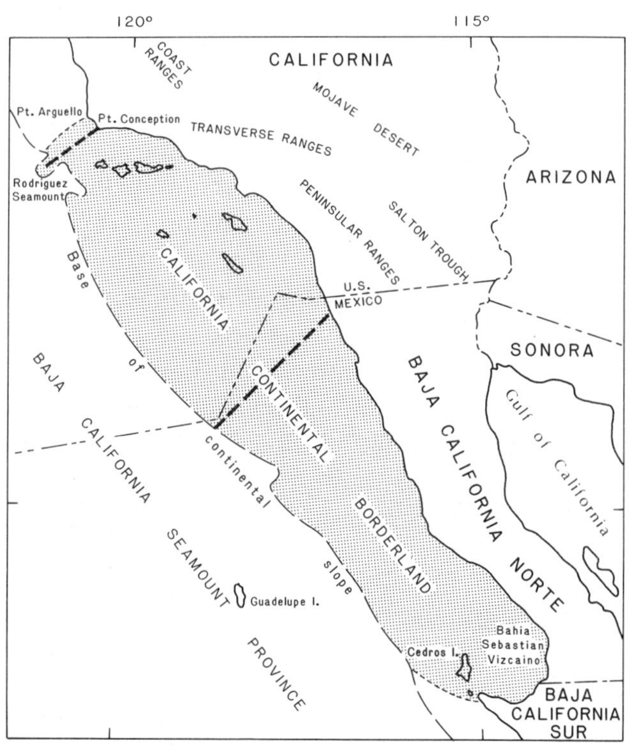

Figure 1. Outline map of the California Continental Borderland and adjoining geographic, geomorphic, and structural provinces. Dots indicate the northern and southern limits of the borderland, inferred from the definitions of Shepard and Emery (1941) and Moore (1969). Heavy dashed lines indicate northwestern and southeastern limits of the area that is discussed in detail.

within a single ridge-basin pair is as much as 2700 m (summit of Santa Cruz Island to the floor of Santa Cruz Basin) (Figure 2). Locally, the relief along the Patton Escarpment is >2750 m; yet the term escarpment is a misnomer because it generally has a slope gradient of <15°. North of the Mexico-U.S. boundary, there are eight large islands (San Miguel, Santa Rosa, Santa Cruz, Anacapa, Santa Barbara, San Nicolas, Santa Catalina, and San Clemente) and several small, isolated pinnacles (Richardson Rock, Wilson Rock, Begg Rock, and nearly emergent Bishop Rock).

REGIONAL GEOLOGIC FRAMEWORK

Introduction

Published descriptions of rocks and structures on the borderland (exclusive of the islands) are scant. Large amounts of

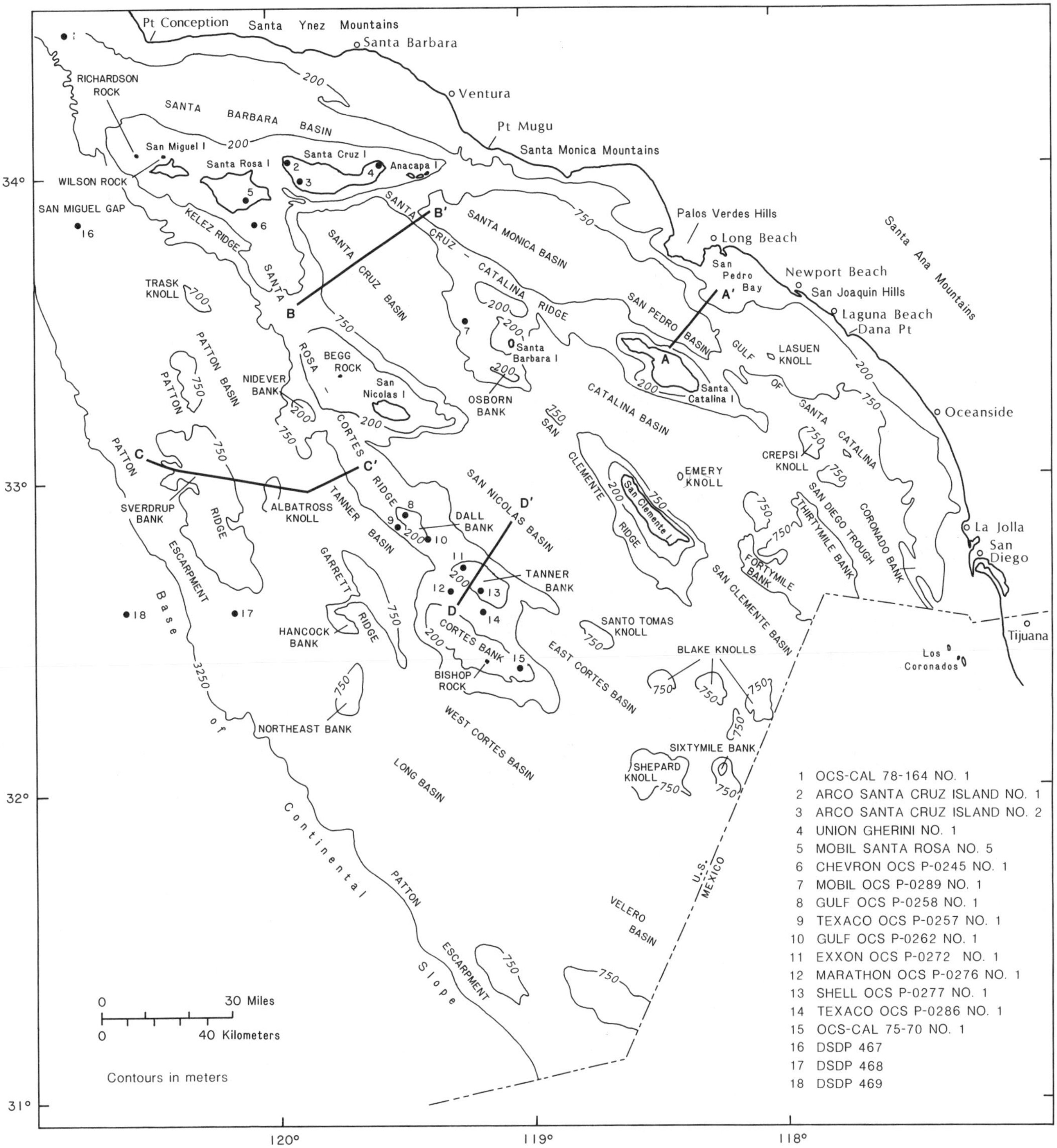

Figure 2. Map of the northern part of the borderland showing place names, subsea features, selected drill-hole sites, and locations of acoustic-reflection profiles. Dashed lines are generalized 250–m, 750–m, and 3250–m isobaths. Lines A–A′, B–B′, C–C′, and D–D′ are tracklines of acoustic-reflection profiles shown in Figures 12 and 20.

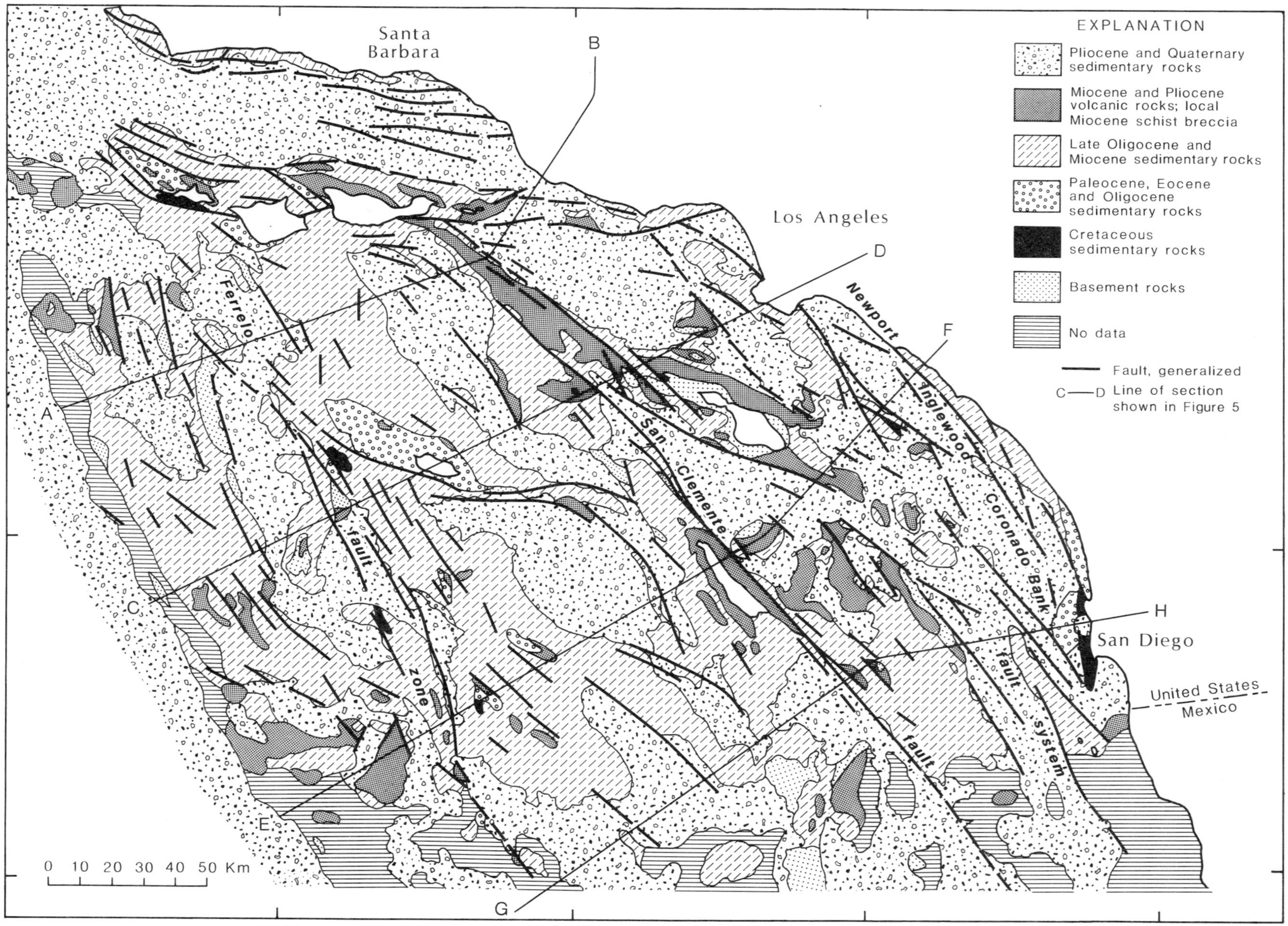

Figure 3. Schematic geologic map of the borderland north of lat. 32°N. The distribution of units is highly generalized, and only the dominant rock type is indicated within each patterned area. Modified from Vedder et al. (1974) and Crouch (1981).

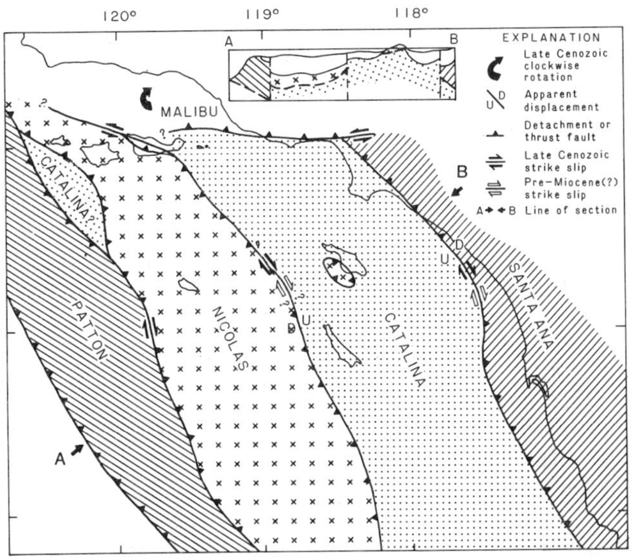

Figure 4. Diagram of inferred tectonostratigraphic terranes of the northern part of the borderland. Modified from Howell and Vedder (1981) and Howell, Champion, and Vedder (in press).

subsea data have been collected and analyzed by oil companies since the 1950s, but most of this information is proprietary. There is little doubt that the undersea geology is similar to that of the mainland coast, but the relatively simple structural and stratigraphic patterns that are depicted on published maps contrast strikingly with the complex relationships mapped on shore, where outcrop and subsurface studies have been in progress for more than 75 years. Only recently have detailed reports been released on small segments of the borderland, such as individual oil fields or nearshore basins.

The known geologic record of the borderland spans the Late Jurassic to Holocene. However, the age of some of the basement rocks is uncertain and could be older than Late Jurassic. The presence of Early Cretaceous rocks is by inference only. In contrast, Late Cretaceous and Paleogene strata are fairly well documented, although areally restricted, and Neogene rocks are widely distributed (Figure 3).

Interpretations of the tectonic history of the borderland are controversial, particularly with respect to pre-Miocene circumstances. By analogy with mainland California, tectonism on the borderland probably included episodes of late Mesozoic and early Cenozoic subduction, early to late Cenozoic translational accretion, and late Cenozoic transcurrent faulting and compression. It is generally agreed that since the Oligocene, the borderland evolved through an orderly succession of tectonic events that were governed by transform processes.

Basement Rocks

Mesozoic rocks exposed on Cedros Island probably are representative of most of the basement rocks that underlie the borderland farther north. Tectonically superimposed sequences on the island consist of three primary lithologic units (Jones, Blake, and Rangin, 1976; Rangin, 1978; Boles and Landis, 1984): (1) Early Jurassic blueschist-serpentine-graywacke melange and broken formation at the lowest structural level, (2) dismembered Middle Jurassic ophiolite depositionally overlain by epiclastic strata of Late Jurassic and Early Cretaceous age, and (3) an Early Jurassic andesitic arc complex that is intruded by a 142- to 148-Ma tonalite pluton. Late Cretaceous epiclastic rocks disconformably overlie the arc complex.

On the borderland north of lat. 32°N, the basement rocks include pre-Late Cretaceous high-pressure schistose rocks, undated zeolite-facies graywacke and argillite, Late Jurassic ophiolite, and possibly related schistose arc-volcanogenic rocks. The easternmost part of the borderland probably is underlain by Triassic(?) and Jurassic graywacke and Late Jurassic and Early Cretaceous andesitic arc-volcanogenic rocks that are intruded by Cretaceous granite plutons of the northern Peninsular Ranges.

On the basis of the inferred distribution of contrasting basement rocks and the different stratigraphic successions that overlie them, the southern California segment of the borderland is subdivided into five tectonostratigraphic terranes, three of which extend into the the mainland (Figure 4). Although the boundaries are indistinct, each terrane can be approximately delineated by integrating topographic, acoustic-reflection, gravity, magnetic, rock-sample, and exploratory-well data. The boundaries represent zones of regional structural dislocation; their nature, however, is not well known, and their depiction undoubtedly is oversimplified. Because the means of identification and the northward travel paths of these accreted terranes are described in detail elsewhere (Howell and Vedder, 1981; Blake, Howell, and Jones, 1982; Gibson et al., 1985; Howell, Champion, and Vedder, in press), they are noted only briefly here. From west to east, the north-trending terranes are named Patton, Nicolas, Catalina, and Santa Ana (Figure 4). A small fragment of the Catalina terrane apparently lies between the Patton and Nicolas terranes south of San Miguel Island. An east-trending terrane, the Malibu, borders the northwestern end of the borderland and may be related to the Santa Ana terrane. The northern edge of the Malibu terrane probably underlies the Santa Barbara Channel.

Subduction Complex and Related Rocks

Blueschist-facies metagraywacke and metavolcanic rocks exposed on Santa Catalina Island and in the Palos Verdes Hills (Woodford, 1924) generally are correlated with the Franciscan assemblage. On the island, these rocks structurally underlie greenschist- and amphibolite-facies rocks (Bailey, 1941; Platt, 1976a, 1976b; Sorensen, 1984). Metamorphism apparently took place about 110 Ma (Forman, 1970; Suppe and Armstrong, 1972). Collectively, these rocks are called the Catalina Schist, which forms the underpinnings of the Catalina terrane (Figure 4). Subsea exposures of blueschist and greenschist occur on Sixtymile Bank and the central ridge of Blake Knolls as well as on Thirtymile Bank. Similar rocks probably are present at Emery Knoll and Fortymile Bank (Vedder et al., 1979, 1981). The westernmost body of glaucophane schist is believed to be on Kelez Ridge, a narrow subsea ridge 10 km south of Santa Rosa Island (Vedder et al., 1974; Junger, 1979). Other rocks dredged from this ridge include quartz-crossite schist, actinolite schist, saussurite gabbro(?), and amphibolite. Although this suite of rocks may represent Miocene schist breccia, acoustic-reflection profiles, a gravity high, and reconstructions of Miocene paleogeography suggest that it is basement rather than coarse clastic detritus.

On northeastern Santa Cruz Island, the Union Gherini No. 1 well bottomed in melange that includes pillow basalt containing lawsonite and glaucophane and zeolite-facies metamorphic rocks. A whole-rock K/Ar age of 152 ±8 Ma was determined on greenstone from the bottom of the hole (Howell, McLean, and Vedder, 1976). Beneath the central Gulf of Santa Catalina, a buried, southeast-trending schist ridge is inferred from acoustic-reflection data (Junger, 1974), as well as from sedimentologic studies of Miocene schist breccia on shore (Stuart, 1974). Wells drilled for oil in San Pedro Bay as far southeast as Newport Beach and as far south as the northern edge of Lasuen Knoll bottomed in Catalina Schist. Thus, it is apparent that blueschist and related high-pressure metamorphic basement rocks are widespread on the inner part of the borderland between Santa Cruz Island and Sixtymile Bank, and it seems likely that they are locally present south of Santa Rosa Island (Figure 3). However, the suggestion by Yeats (1970) that Catalina Schist underlies most of the Santa Barbara Basin is not tenable if the reconstruction of the borderland by Howell et al. (1974) is correct.

Low-grade metasedimentary and altered volcanic rocks occur at places on Patton Ridge, Albatross Knoll, the unnamed ridge 12 km south of Nidever Bank, and the deeper parts of Patton Escarpment. These rocks include laumontitic graywacke, argillite, phyllite, chert, incipient greenstone, and metabasalt. Serpentinite and ultramafic rocks are present locally on Patton Ridge and Patton Escarpment (Emery and Shepard, 1945; Winterer et al., 1969; Vedder et al., 1974, 1977; Vedder et al., 1976a, 1976b; J. C. Taylor and F. Lee-Wong, unpub. data). Uchupi and Emery (1963) reported glaucophane schist and granite from the southern part of Patton Ridge, but their samples are believed to be kelp- and tree-root rafted erratics. The Patton Ridge-Albatross Knoll basement rocks, which are assigned to the Patton terrane and possibly are related to part of the Franciscan assemblage, may be analogous to the Coastal belt Franciscan of northern California.

Ophiolite

On Santa Cruz Island, the Willows Diorite of Weaver (1969) consists predominantly of hornblende diorite and hornblende gabbro, with subordinate leucotonalite and ultramafic rocks (Hill, 1976; Sorensen, 1985). Zircons from this body yield nearly concordant $^{205}Pb/^{238}U$ ages of 162 ±3 Ma (Mattinson and Hill, 1976). Both the age and composition of the dioritic rocks indicate a correlation with the Coast Range ophiolite of Bailey, Blake, and Jones (1970); and the chemistry suggests island-arc affinities (Shervais and Kimbrough, 1985; Sorensen, 1985). These ophiolitic rocks form the basement of the Nicolas terrane. Elsewhere in the borderland, scattered occurrences of saussurite gabbro and related rocks form a body on the south side of Santa Catalina Island (Platt, 1976b), subsea exposures on northern San Clemente Ridge 12–20 km southeast of Osborn Bank (Vedder et al., 1974; Sorensen, 1985), and possibly part of Kelez Ridge 20 km south of Santa Rosa Island. Samples from the crest of Trask Knoll include serpentinite, fractured pyroxenite, and actinolite schist(?) (Vedder et al., 1974, 1976b) that possibly represent ophiolite. Suspected ophiolite in the Long Beach oil field (Schoellhamer and Woodford, 1951; Sorensen, 1985), suggests that this type of basement rock as a broad areal extent.

Evidence for the structural superposition of ophiolite in relation to other basement rocks is contained in coarse detritus derived from nearby sources. On the southwest side of Santa Cruz Island, a sequential change of clast types in Miocene breccia indicates successive downward erosion of adjoining basement from saussurite gabbro and diorite into blueschist and greenschist (McLean, Howell, and Vedder, 1976). Clasts of these basement types abound in Miocene beds throughout the Santa Cruz-Catalina Ridge area and southward to Fortymile Bank, but apparently are limited to the northwestern part of the Santa Rosa-Cortes Ridge and to the Blake Knolls and Shepard Knoll in the region to the west (Vedder and Howell, 1976a). Similar Miocene breccia beds are widely distributed along the mainland coast (Woodford, 1925; Stuart, 1974, 1976). In a belt extending about 30 km northwest from Oceanside, fragments of metagabbro constitute as much as 50% of the clast assemblage in the basal part of the breccia sequence; but at higher levels, blueschist and greenschist clasts predominate. Farther northwest near Laguna Beach, this systematic change is not evident, although some breccia zones within the section contain as much as 40% metagabbro clasts. The ordered sequences of changing clast composition suggest that the basement source areas consisted of Catalina Schist structurally overlain by ophiolite in a relationship similar to that of the Franciscan assemblage and Coast Range ophiolite in central and northern California (Howell and Vedder, 1981).

Island-Arc Volcanogenic Rocks

The Santa Cruz Island Schist of Weaver (1969) contains low-grade greenschist-facies minerals that indicate protoliths that were principally spilitic to quartz-keratophyric volcanic rocks together with smaller amounts of epiclastic rocks (Hill, 1976; Sorensen, 1985). Intruding the Santa Cruz Island Schist is the Alamos Tonalite of Weaver (1969), an irregularly shaped and unsheared body of leucotonalite dated at 141 ±3 Ma (Mattinson and Hill, 1976). Although the North Valley Anchorage fault of Hill (1976) separates the Santa Cruz Island Schist and the Willows Diorite, the volcanic protolith of the schist may be consanguineous with the dioritic rocks (Sorensen, 1985).

On the mainland, basement rocks that probably are related to the Late Jurassic island arc are the Santa Monica Slate in the Malibu terrane of the south-central Transverse Ranges and the Santiago Peak Volcanics in the Santa Ana terrane of the northern Peninsular Ranges. Presumably both of these formations extend a short distance seaward beneath the inner borderland shelf. The Santa Monica Slate consists chiefly of meta-argillite and subordinate epiclastic and crystal-lithic tuffs (Hoots, 1931; Jones, Blake, and Rangin, 1976) that contain fossils of late Oxfordian to early Kimmeridgian age (Imlay, 1963). The Santiago Peak Volcanics unconformably overlies Middle Jurassic and older(?) graywacke, shale, and limestone of the Bedford Canyon Formation in the Santa Ana Mountains (Schoellhamer et al., 1981). Island-arc andesitic to dacitic flows constitute most of the Santiago Peak Volcanics, which includes argillite of Tithonian age near San Diego (Fife, Minch, and Crampton, 1967; Balch, Bartling, and Abbott, 1984). Probably correlative volcanic rocks were penetrated in an offshore exploratory well about 5 km west of Point Loma, suggesting that some of the Peninsular Ranges basement rocks extend as far west as the Newport Inglewood-Coronado Bank fault system (Figure 3). Although Early Cretaceous granitic plutons intrude both the

Santa Monica Slate and the Santiago Peak Volcanics, granitic rocks of this age have not been reported on the northern part of the borderland.

Cretaceous Strata

Early Cretaceous sedimentary rocks are sparsely distributed in coastal southern California. In contrast, Late Cretaceous fore-arc-basin strata are relatively widespread in the western Transverse Ranges and along the western flank of the Peninsular Ranges. On the borderland, however, Late Cretaceous beds are exposed subaerially only on San Miguel Island, where a 3000-m-thick marine sequence forms the western seacliffs, and on North Island of the Los Coronados island group, where about 200 m of nonmarine beds crop out (Lamb, 1978). Benthic foraminifers in the San Miguel Island section are assigned to the upper Turonian through lower Maestrichtian Stages (Weaver, 1969; Bartling and Abbott, 1983).

Exploratory wells that penetrated Cretaceous strata include one on the southern part of Santa Rosa Island, two on the western part of Santa Cruz Island, one on the shelf south of Santa Rosa Island, and seven in the Dall-Tanner-Cortes Bank area (Figure 2). Fossil assemblages as old as Cenomanian have been identified from Deep Stratigraphic Test Well OCS-CAL 75-70 No. 1 on southeastern Cortes Bank, where basement is inferred to be 300 m or more below the lowest beds drilled at a depth of 3330 m (Paul et al., 1976). The Cretaceous section higher in the hole ranges in age from Turonian to Maestrichtian. On northern Garrett Ridge, claystone samples yielded coccolithophorid assemblages that have age ranges of late Albian to early Turonian and Coniacian to Santonian (Vedder et al., 1976b, 1981). Unfossiliferous siltstone and sandstone of probable Cretaceous age are present at Nidever Bank, and Coniacian to Santonian siltstone and conglomerate are exposed near the northwest end of Cortes Bank (Vedder et al., 1976a, 1976b, 1979). Along the base of the slope 15 km southeast of San Nicolas Island, dredged samples of argillite and altered sandstone resemble rock types from the lower part of the Great Valley sequence and possibly represent sea-floor exposures of rocks that are older than Late Cretaceous.

Beneath the northwest end of Santa Cruz Island, north of the Santa Cruz Island fault, the Richfield (Arco) Santa Cruz Island No. 1 well penetrated Late Cretaceous beds directly below Miocene volcanic rocks. Near the east end of the same island, in the Union Gherini No. 1 well, correlative volcanic rocks overlie Miocene and older Tertiary clastic strata (Howell, McLean, and Vedder, 1976), but Cretaceous beds apparently are absent. South of the Santa Cruz Island fault, Late Cretaceous strata drilled in the Richfield (Arco) Santa Cruz Island No. 2 well lie below an exposed Paleogene clastic sequence. These diverse subsurface sections imply different geologic histories and suggest probable allochthonous relationships for the parts of the island.

Interpretations of acoustic-reflection data indicate that Cretaceous strata are widely distributed beneath Santa Rosa-Cortes Ridge and the adjoining Santa Cruz and San Nicolas Basins (Figure 5). Along the western edge of Santa Rosa-Cortes Ridge, these strata are truncated by a broad zone of high-angle faults; on the eastern side of these basins, the same sequence of beds locally pinches out eastward, apparently by onlap onto basement rocks. Possibly equivalent beds extend southward beneath East and West Cortes Basins and Velero Basin. The maximum age and thickness of the Late Cretaceous and older(?) fore-arc strata of the outer borderland are not known. Drilled sequences of Turonian(?) and younger Cretaceous strata beneath Santa Rosa-Cortes Ridge are as much as 1600 m thick, but they do not represent complete sections, which may be nearly twice as thick. Gravity data suggest that Cretaceous and Late Jurassic(?) strata may be 3000 m or more thick at places in the Nicolas terrane (L. A. Beyer, 1978, oral communication).

East of the Newport Inglewood-Coronado Bank fault system, Late Cretaceous beds are discontinuously exposed from the Santa Ana Mountains to northwestern Baja California and extend seaward at least as far as the edge of the shelf between Laguna Beach and San Diego. These strata are late Turonian and younger and were deposited in nonmarine to westward-prograding subsea-fan environments (Bottjer and Link, 1984).

At the southeast end of Santa Catalina Island, altered sedimentary rocks enveloped by Miocene igneous rocks have lithologic affinities to Late Cretaceous and Paleogene strata on the mainland (Vedder, Howell, and Forman, 1979). These isolated, pervasively intruded beds are the only known pre-Miocene sedimentary rocks in the inner part of the borderland west of the Newport Inglewood-Coronado Bank fault system. The stratigraphic and structural relationships between the Catalina Schist and these beds are obliterated by the Miocene intrusive rocks.

Late Cretaceous fore-arc-basin strata exposed on San Miguel Island consist of turbidite sequences that include granitic- and volcanic-cobble conglomerate. The largely buried Late Cretaceous strata of Santa Rosa-Cortes Ridge and adjoining Santa Cruz and San Nicolas Basins are interpreted to be parts of a large fore-arc basin that formed west of the Jurassic and Early Cretaceous arc complex and Cretaceous plutons of the Peninsular Ranges (Howell and Vedder, 1981). The nonmarine section on North Island of the Los Coronados island group presumably correlates with Late Cretaceous fluvial deposits near San Diego (Lamb, 1978) and may represent paralic environments shoreward from the fore-arc basin. The near absence of Cretaceous rocks in the inner part of the borderland may be an artifact of postdepositional erosion (Woodford and Gander, 1977) or the result of post-Cretaceous juxtaposing of allochthonous terranes. The rarity of Cretaceous(?) sandstone clasts in the eastward-transported Miocene schist breccia along the mainland coast (Vedder and Howell, 1976a) suggests that Cretaceous sediments, if deposited in the area of the Gulf of Santa Catalina, were thin or confined to small basins, only fragments of which remain.

In summary, a complex set of circumstances is required to explain regional stratigraphic and structural relationships of the Cretaceous sedimentary rocks. Along the western flank of the Peninsular Ranges, Late Cretaceous paralic strata lap onto the arc-volcanogenic and plutonic basement rocks of the Santa Ana terrane, whereas in the central borderland, Late Cretaceous and older(?) subsea-fan deposits are inferred to overlie ophiolite of the Nicolas terrane. Thus, these remnants of a fore-arc-basin sequence overlap juxtaposed basement rocks of both the Nicolas and Santa Ana terranes. The wide expanse of intervening Catalina Schist of the Catalina terrane implies tectonic dismemberment, probably involving uplift and/or large amounts of strike-slip along its edges.

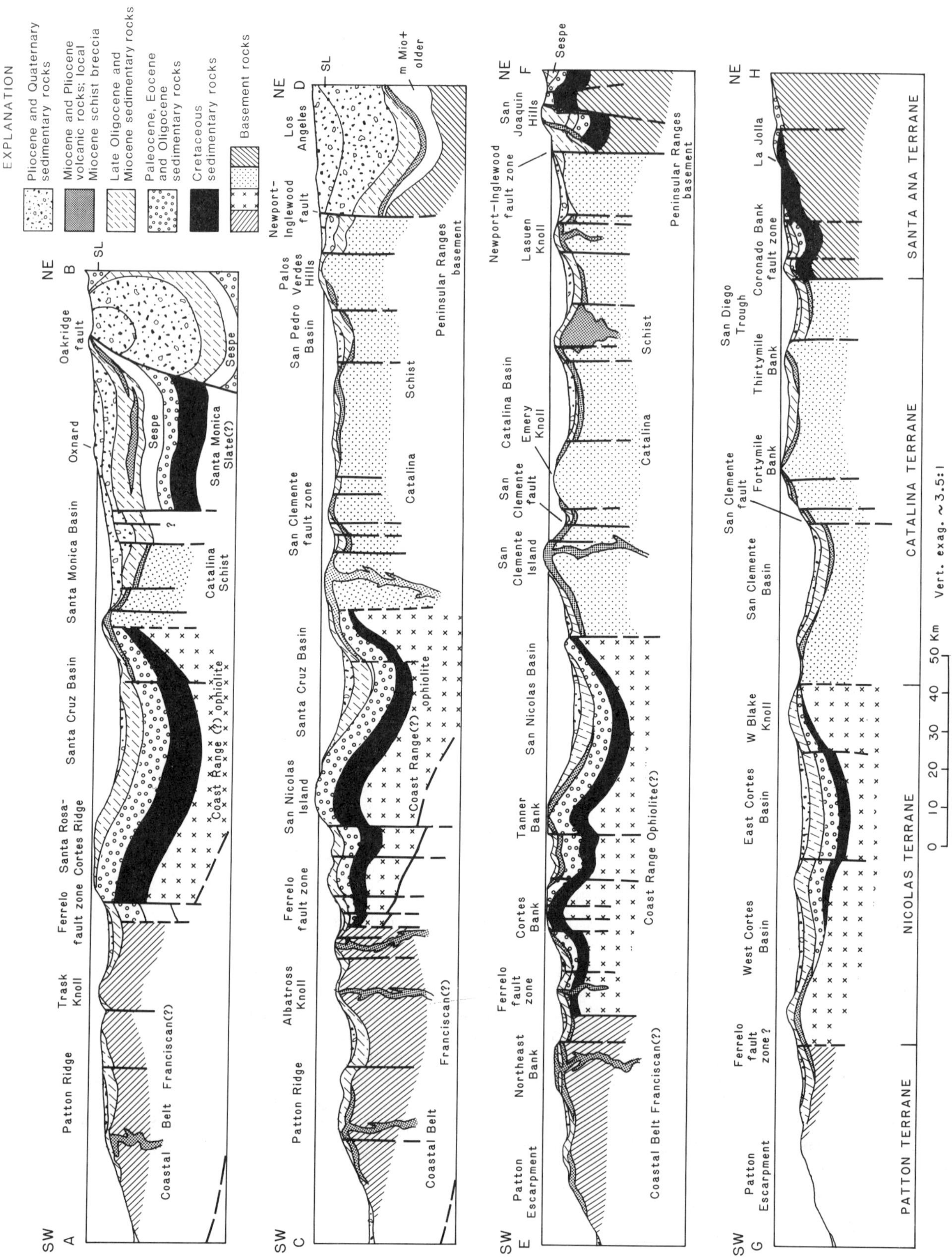

Figure 5. Schematic structure sections across the northern part of the borderland and adjoining mainland. Thicknesses of rock units and configuration of structures are based upon interpretations of acoustic-reflection profiles, dart-core samples, and sparse drill-hole data. Vertical exaggeration ~3.5:1. Sections A–B, C–D, and E–F modified from Vedder et al. (1974).

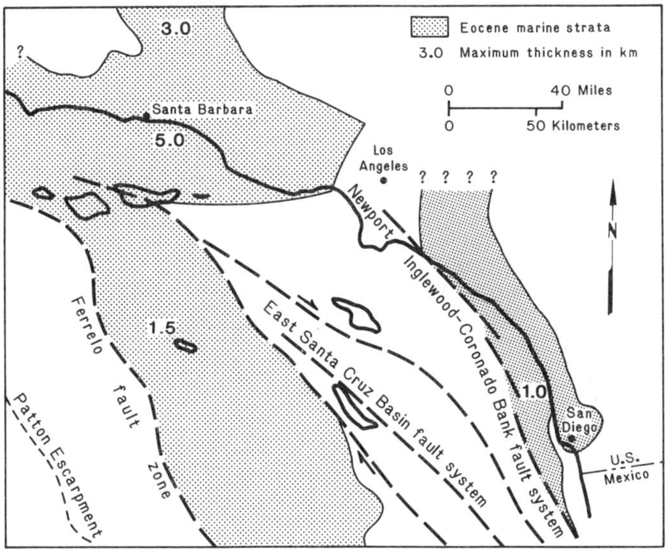

Figure 6. Map showing inferred distribution and thickness of Eocene sedimentary rocks on the northern part of the borderland and adjoining mainland. Modified from Blake et al. (1978).

Tertiary Strata

Paleocene

Nonmarine and shallow-marine Paleocene strata occur in the Santa Monica Mountains and flank the northernmost Santa Ana Mountains and parts of the west margin of Baja California. Within the borderland, the known and inferred distribution of Paleocene rocks is restricted to Santa Rosa-Cortes Ridge and adjoining basins on the east and south. A possible exception is in the small patch of sedimentary rocks that is enveloped in hypabyssal intrusive rocks on the southeastern end of Santa Catalina Island (Vedder, Howell, and Forman, 1979).

Paleocene beds are exposed on Santa Cruz and San Miguel Islands. On Santa Cruz Island, a 90-m-thick section consisting of predominantly medium- to fine-grained sandstone and siltstone is inferred to be shallow marine (Doerner, 1969). On San Miguel Island, Paleocene sandstone and siltstone turbidites occur in a faulted section that may be as thick as 460 m (Weaver and Doerner, 1969a); this section extends westward onto the island platform for an unknown distance. Other submergent outcrops are present on the northern part of Santa Rosa-Cortes Ridge near the shelf break about 30 km south of Santa Rosa Island. From well-log and acoustic-reflection data, Paleocene rocks are inferred in the subsurface section beneath most of Santa Rosa-Cortes Ridge; possibly they extend eastward under Santa Cruz and San Nicolas Basins.

These data are too few to accurately reconstruct Paleocene depositional patterns in the central part of the borderland. No unconformity has been recognized between Late Cretaceous and inferred Paleocene strata in the exploratory wells on Santa Rosa-Cortes Ridge. On Santa Cruz Island, however, sedimentary features, molluscan paleoecology, and the probable unconformity at the top of the Cretaceous section suggest Paleocene shoaling of parts of the preexisting fore-arc basin. Inferred major Paleocene events and their effects are summarized in Table 1.

Eocene

Eocene strata, chiefly marine sandstone and mudstone, are discontinuously distributed along the mainland coast from the western Transverse Ranges to the Peninsular Ranges near San Diego. These rocks are present on and beneath the shelf between Oceanside and the international boundary and are exposed on San Miguel, Santa Rosa, Santa Cruz, and San Nicolas Islands. Farther offshore, Eocene sequences that may be as much as 1400 m thick underlie large parts of Santa Rosa-Cortes Ridge and Santa Cruz and San Nicolas Basins. Sea-floor exposures of Eocene mudstone and siltstone are present on the San Miguel, Santa Rosa, and San Nicolas Island platforms and at Nidever, Dall, Tanner, and Cortes Banks (Vedder et al., 1974, 1976b, 1979). Exploratory wells on the northern and southern parts of Santa Rosa-Cortes Ridge penetrated similar Eocene strata. Beds of probable Eocene age were drilled in the Union Gherini No. 1 well at the east end of Santa Cruz Island and in the Mobil OCS P-0289 No. 1 well about 17 km northwest of Santa Barbara Island.

The Eocene section apparently is truncated by faults along the western edge of Santa Rosa-Cortes Ridge (Figure 5). Faults and depositional wedge-outs, both of which are inferred from acoustic-reflection data, mark the limits of Eocene rocks along the eastern margins of Santa Cruz and San Nicolas Basins (Vedder et al., 1974). Interpretations from acoustic-reflection data suggest that Eocene strata extend southward into East and West Cortes Basins and Velero Basin (Vedder, 1981). The inferred distribution of Eocene rocks on the northern part of the borderland is shown in Figure 6.

Conglomerate is a small, yet significant, constituent in Eocene rocks of the borderland, and the characteristic silicic metavolcanic cobbles closely resemble those in the Poway Group of the San Diego area (Merriam, 1972; Howell, 1975; Minch, Gibson, and Peterson, 1976; Abbott and Smith, 1978; Abbott et al., 1983). Most of the enclosing beds typically are medium- to coarse-grained arkosic sandstone and mudstone. Fluvial-channel, alluvial-fan, coastal-plain, fan-delta, shelf, and subsea-channel environments of deposition are represented in the Poway Group (Howell and Link, 1979).

The petrographic uniformity, the flyschlike character, and the systematic flow patterns all suggest that the Eocene strata of the borderland are part of a single large subsea-fan system that was influenced by eustatic sea-level change (Ferrelo fan of Howell and Vedder, 1984) (Table 1). Paleocurrent directions indicate a west-facing depositional system that fanned out from the Santa Cruz Island area toward the south, west, and northwest (Cole, 1975; Merschat, 1971; Howell, 1975; Erickson, 1975). Thick sandstone and conglomerate lenses in the northern Channel Islands indicate that the apex of this fan probably lay near the northeastern embayment of Santa Cruz Basin (Howell, 1975). An episode of middle Miocene right-slip on the East Santa Cruz Basin fault system (Figure 6) apparently moved this fan system northwest away from correlative shallower marine and nonmarine facies at San Diego (Howell et al., 1974).

Paleomagnetic data suggest that the northern Channel Islands have been tectonically rotated 69° to 81° clockwise since the early(?) Miocene (Kamerling and Luyendyk, 1985). A palinspastic correction for this amount of rotation reorients the Eocene Channel system at Santa Cruz Island to a southward flow direction. However, such rotation probably

Table 1. Chronology of major events and geologic evidence of Cenozoic tectonism on the northern part of the California Continental Borderland. Estimated ages of the epoch boundaries from Berggren et al. (1985). Table modified from Vedder and Howell (1980).

Period	Epoch Boundary ages in Ma	Major Events	Effects and Evidence
	Holocene	End of rapid sea-level rise ca. 6 ka	Drowned and backfilled drainage channels along mainland coast and at Santa Cruz and Santa Catalina Islands.
	0.011		
Quaternary	Pleistocene	Beginning of post-glacial sea-level rise ca. 17 ka	Eolianites on outer islands. Submergent surf-cut platforms around islands and along mainland coast.
		Filling of onshore basins; fluctuating eustatic sea levels	Emergent surf-cut platforms around islands and along mainland coast. Nonmarine sediments throughout large parts of onshore basins.
		Increase in tectonism with widespread intensive uplift and subsidence	Thick accumulations of marine and nonmarine sediments in onshore and nearshore basins. Deformed, deeply eroded basin margins and high relief on mainland. Faults throughout region.
	1.6		
	Pliocene	Eustatic lowering of sea level with onset of polar glaciation	Shallow-marine deposits on Santa Rosa, Santa Cruz, and San Clemente Islands and along basin margins at Santa Monica Mountains, Puente Hills, San Joaquin Hills, and San Diego.
		Crustal shortening, uplift, and accelerating deposition in nearshore basins accompanied by local subsidence and basin collapse	Deep-marine deposits in central Ventura and Los Angeles Basins. Increasing relief in the Transverse and Peninsular Ranges and basin-edge unconformities throughout the borderland.
		Cessation of volcanism	Basaltic flows and aquagene tuffs at Northeast Bank.
	5.3		
	Miocene	Regional basin subsidence and diminution of northward translation	Encroaching seas and deepening nearshore basins. Widespread deep-marine deposits in Ventura and Los Angeles Basins.
		Diminishing volcanism ca. 10–7 Ma	Local igneous intrusions, flows and ash falls at Santa Catalina Island, Santa Monica Mountains, Palos Verdes Hills, Puente Hills, and San Joaquin Hills.
		Vigorous wrench tectonism and peak volcanic activity ca. 16–12 Ma. Incipient rotation of western Transverse Ranges	Variable sea-floor relief. Shallow-marine deposits and tephra at Santa Cruz, Santa Catalina Islands. Marine-nonmarine breccias along southern mainland coast. Igneous intrusions, flows, and tephra in Santa Monica Mountains, margins of Los Angeles Basin, and Santa Barbara, Santa Catalina, and San Clemente Islands. Volcanic rocks throughout inner borderland ridges.
		Commencing basin development and initial "leaky-transform" volcanism ca. 24–20 Ma	Basement uplift and basin-margin breccia at Santa Cruz and Santa Catalina(?) Islands. Deep-water mudstone in Santa Barbara region. Igneous flows at San Miguel Island and Tanner-Cortes Bank area.
Tertiary		Emplacement of Catalina terrane between Nicolas and Santa Ana terranes(?)	No direct evidence.
	23.7		
	Oligocene	Exceptionally low-standing sea level	Nonmarine deposits along margins of Ventura and Los Angeles Basins and Santa Rosa and Santa Catalina Islands. Unconformities on San Miguel and Santa Cruz Islands. Marine deposits restricted to western Santa Ynez Mountains, southern Santa Rosa-Cortes Ridge, and Patton Ridge.
		Plate-tectonic shift from convergent to transform regime ca. 29 Ma	Disrupted magnetic-anomaly patterns. 26-23 Ma transgressive shallow-marine deposits at northern Channel Islands, Santa Monica Mountains, and northeast margins of Ventura and Los Angeles basins.

(continued).

Period	Epoch Boundary ages in Ma	Major Events	Effects and Evidence
	36.6		
	Eocene	Renewed northward crustal transport of borderland terranes and arrival at 20°N ca. 40 Ma	Paleomagnetic inclinations from Nicolas and Santa Ana terranes.
		Subdued tectonism	Continued peneplanation and fluvial transport of distinctive clast assemblages from areas now east of the Peninsular Ranges.
		Repeated eustatic sea-level changes	Progradational cycles of deposition in the San Diego area. Pulses of sandstone turbidite deposition, Santa Rosa-Cortes Ridge area and adjoining basins to the east.
	57.8		
	Paleocene	Regional fore-arc-basin deposystem inherited from Cretaceous precursor	Continued unroofing of Peninsular Ranges plutons and peneplanation. Nonmarine and subsequent marine sedimentation in the Santa Monica and Santa Ana Mountains. Submarine fan environments in the northern Channel Islands and Santa Rosa-Cortes Ridge area.
		Apparent vertical tectonism	Basal unconformities in Santa Monica and Santa Ana Mountains.
		Southward crustal transport of borderland terranes until ca.50 Ma	Paleomagnetic inclinations from Nicolas and Santa Ana terranes.
	66.4		

is restricted to the northwesternmost part of the borderland, because a palinspastic correction for all of the central borderland would result in an unlikely juxtaposition of east-flowing distal-fan facies at San Nicolas Island against nonmarine and west-flowing proximal-fan facies of the mainland.

Oligocene

Throughout much of onshore southern California, strata of Oligocene age are nonmarine. In the western Santa Ynez Mountains and Santa Monica Mountains, however, red beds intertongue westward with shallow-marine beds of Oligocene and/or early Miocene age.

On San Miguel and Santa Cruz Islands, early Miocene rocks lie unconformably on late Eocene strata, and Oligocene strata apparently are absent. Variegated mudstone locally caps the Eocene section on Santa Cruz Island, suggesting that nonmarine deposition continued into the Oligocene (Doerner, 1969). On Santa Rosa Island, a nonmarine Oligocene(?) section at least 150 m thick unconformably overlies late Eocene strata and is gradationally overlain by shallow-marine Miocene strata (Weaver and Doerner, 1969a). Abundant siliceous volcanic pebbles in the nonmarine section may indicate redeposition of clasts from emergent Eocene and possibly older strata. These nonmarine strata probably extend no farther south than Santa Barbara Island and northern Santa Rosa-Cortes Ridge.

Sea-floor exposures of Oligocene marine mudstone and sandstone are present at Dall, Tanner, and Cortes Banks and on the shelf north of San Nicolas Island. Although these beds may have been deposited in shelf and slope environments from the vicinity of Tanner Bank northward, coeval lower bathyal turbidites occur in the Deep Stratigraphic Test OCS-CAL 75-70 No. 1 well at the southeastern end of Cortes Bank, where the sequence is about 400 m thick (Paul et al., 1976). Presumably, correlative strata extend southward as far as Velero Basin (Vedder, 1981).

Farther east, at the southeastern end of Santa Catalina Island, a sequence of variegated beds consisting of mudstone, sandstone, and conglomerate lies stratigraphically between Miocene breccia and probable Paleogene or Late Cretaceous strata (Vedder, Howell, and Forman, 1979). This thin sequence is the only known occurrence of nonmarine Oligocene(?) on the borderland south of lat. 35°15′N. Pebbles and cobbles of gneissic, quartzitic, plutonic, and siliceous volcanic rocks in a conglomeratic zone near the top of the 30-m-thick section imply a source terrane in the Peninsular Ranges. The same conglomerate, however, contains clasts of ultramafic rocks, quartz keratophyre, and quartz schist that could have been derived locally.

At places along Patton Ridge, acoustic-reflection profiles indicate that bedded strata lie below sequences of late Oligocene and early Miocene marine mudstone and siltstone (Crouch, 1981). Cores of unfossiliferous quartzofeldspathic sandstone and micaceous siltstone from a few sites on the northern part of Patton Ridge (Vedder et al., 1976b) may correlate with the bedded sequence observed on the profiles. These undated strata possibly are early Oligocene in age.

From the available data, it seems likely that the Oligocene marine strata represent aprons of shelf and slope deposits. Presumably the distribution of these marine deposits was limited to the central and outer borderland as a result of the regional regression. These Oligocene marine strata may be the last vestiges of the pre-existing broad Cretaceous and early Tertiary fore-arc basin of the central borderland (Table 1). The northern Channel Islands region, and possibly the area of Trask Knoll and Kelez Ridge south of Santa Rosa Island, may have been emergent during the Oligocene.

Miocene

Rocks of Miocene age occur throughout the coastal margins of the western Transverse Ranges and Peninsular Ranges, as well as in the subsurface sections of the Ventura and Los Angeles basins. The stratigraphic sequences consist of diverse kinds of sedimentary rocks, chiefly marine, and igneous rocks, both intrusive and extrusive. The igneous rocks are described in a following section of this chapter. The sedimentary rocks range from massive boulder conglomerate to laminated shale and were deposited in varied environments.

Thickness and facies changes are characteristic, particularly in the coarse clastic sequences. Structural and depositional patterns of the present-day borderland probably began to form in the early Miocene (Crouch, 1979; Howell et al., 1980; Howell and Vedder, 1981), but little is known of the initial stages of basin development because of the fragmentary geologic record. Consequently, it is not yet possible to trace all phases of growth of the offshore basins, even though the general distribution and composition of middle and late Miocene rocks are fairly well known. Apparent post-early Miocene tectonic rotation along the northern margin is an additional complication to palinspastic reconstruction of the borderland, as is late-stage northward translation. By the end of the Miocene, well-defined ridges and basins characterized most of the region (Vedder and Howell, 1980) (Table 1).

Early Miocene strata are exposed on San Miguel, Santa Rosa, and Santa Cruz Islands and along the mainland coast in the Santa Ynez Mountains, Santa Monica Mountains, and San Joaquin Hills. These basal Miocene beds, together with most correlative strata elsewhere in southern California, mark the beginning of a regional marine transgression. Sea-floor samples and exploratory-well data indicate that early Miocene strata extend along the entire Santa Rosa-Cortes Ridge. Acoustic-reflection profiles suggest that equivalent beds underlie Santa Cruz and San Nicolas Basins, as well as East and West Cortes Basins and Velero Basin. Although two early Miocene sample sites are shown near Santa Barbara Island by Jennings (1962), rocks of this age have not been verified from any of the inner borderland ridges and basins that are underlain by Catalina Schist. However, strata of this age probably are present at places on the shelf east of the offshore segment of the Newport Inglewood-Coronado Bank fault system.

On Santa Rosa Island, recycled siliceous metavolcanic pebbles in early Miocene conglomeratic sandstone beds possibly signify uplift and erosion of underlying Paleocene and Eocene strata. On Santa Cruz Island, large clasts in early Miocene breccia beds indicate denudation of Paleogene sequences as well as the Willows Diorite. Ensuing early and middle Miocene erosion of a nearby blueschist basement source (no longer exposed) changed the clast composition of the breccia from diorite to schist (Yeats, 1968a, 1968b; Howell et al., 1974). Similar schist breccia was widely dispersed during the middle Miocene in the inner part of the borderland and along the mainland coast from Point Mugu to Oceanside and locally near Tijuana (Woodford, 1925; Stuart, 1974, 1976; Vedder and Howell, 1976a). The principal source for the schist breccia that extends southward from the San Joaquin Hills to Oceanside is believed to have been a north-trending Miocene ridge, now covered by younger rocks, along the inner edge of the Gulf of Santa Catalina (Junger, 1974). Other known sources of schist detritus were Kelez Ridge, Santa Cruz-Catalina Ridge, Thirtymile Bank, eastern Blake Knolls, Sixtymile Bank, and, possibly, Emery Knoll and Fortymile Bank. The widespread breccia beds record an episode of rapid differential uplift and erosion of a distinctive basement terrane in the area that extends from Santa Cruz Island southeastward at least as far as Sixtymile Bank.

At most places where early Miocene transgressive marine beds occur, they are superseded by early and middle Miocene sandstone and mudstone beds that contain successively deeper sublittoral and bathyal fossils in upward stratigraphic order. This regional subsidence seems to have affected not only the borderland but also the southern Coast Ranges. From Santa Cruz Island westward through Santa Rosa Island, middle Miocene strata are largely volcaniclastic and grade westward and southward from lenticular conglomerate and sandstone into thin-bedded sandstone and siliceous mudstone (Howell and McLean, 1976). Along Santa Cruz-Catalina Ridge and at San Clemente Island, similar volcaniclastic beds and schist breccia are interlayered with, and overlain by, thin-bedded sequences of middle and late Miocene claystone and siltstone (Vedder and Howell, 1976b; Vedder, Howell, and Forman, 1979).

Shale and laminated silty claystone of middle and late Miocene age cover large tracts of Santa Rosa-Cortes Ridge northwest of Begg Rock and drape the slopes of the ridge southward from San Nicolas Island to Cortes Bank and Santo Tomas Knoll. Similar strata extend intermittently from the platform northwest of Santa Barbara Island to the southeast end of San Clemente Ridge, and are inferred to pass beneath younger sedimentary rocks in the deep basins to the west. Much of the crest of Patton Ridge is underlain by correlative shaly strata.

The regional transgression and ocean deepening that occurred in the early Miocene ($\sim$24 to 20 Ma) apparently was interrupted by local growth of high-standing bathymetric features in the central part of Santa Rosa-Cortes Ridge during the middle and late Miocene (Arnal, 1976, Figure 8). The regional deepening that took place about 22 to 20 Ma (Howell, Champion, and Vedder, in press) may have resulted from a slight directional change in relative plate motions (Page and Engebretson, 1984). By the end of the Miocene, large parts of the Ventura and Los Angeles regions were deep marine basins that were connected to the borderland; and sea-floor relief of at least 2000 m developed on adjoining parts of the borderland (Vedder and Howell, 1980). A few islands survived from a subsided middle Miocene volcanic archipelago, remnants of which are preserved at Santa Cruz, San Clemente, and Santa Catalina Islands (Nolf and Nolf, 1969; Vedder and Howell, 1976b; Vedder, Howell, and Forman, 1979). A shallow submarine ridge at the present site of the southern Santa Rosa-Cortes Ridge seems to have subsided 300 to 500 m near the end of the Miocene. Deepening to abyssal depths ($>$2500 m), not recorded in middle Miocene rocks, occurred to the south and west of this area (Arnal and Vedder, 1976). Even though mid-bathyal depths apparently prevailed over most of the region between the subsiding ridge and the present shoreline, the water was shallower near the northern end of Coronado Bank, where depths probably were $<$500 m. In the Los Angeles and Ventura areas, the basin troughs subsided to lower bathyal and abyssal depths (1000 to $>$2500 m) by the end of the Miocene

(Natland, 1957; Ingle, 1973). Much of the clockwise rotation of crustal blocks in the western Transverse Ranges and northern Channel Islands presumably was completed before the beginning of the Pliocene (Hornafius, 1985; Kamerling and Luyendyk, 1985).

At most places on the Channel Islands, late Miocene strata either were removed completely by subsequent erosion or were never deposited, leaving an incomplete record of latest Miocene events. The assignment of volcaniclastic strata on Santa Rosa Island to the late Miocene (Jennings, 1959) is misleading, because microfossils from interlayered shale beds are no younger than middle Miocene (Avila and Weaver, 1969; J. A. Barron, 1976, written communication). The only known island exposures of late Miocene sedimentary rocks occur on Santa Catalina and San Clemente Islands as thin, discontinuous volcaniclastic sandstone beds. Fossil assemblages in these late Miocene beds indicate a range of depths from middle bathyal to littoral (Vedder and Howell, 1976b; Vedder, Howell, and Forman, 1979; Vedder and Moore, 1976). On Santa Catalina Island, the presence of depth-controlled assemblages of both mollusks and foraminifers in sediments deposited on irregular volcanic flow surfaces imply a steep bottom slope from <50 m to as deep as 1000 m. At San Clemente Island, only shallow-water mollusks are present in correlative beds, which accumulated as depression-filling veneers on the older lava surfaces. These limited island outcrops are insufficient to interpret the original distribution and thickness of the sedimentary bodies they represent. Nevertheless, the incomplete record of late Miocene deposition implies substantial sea-floor relief at the time.

In summary, the undulatory relief that seems to characterize the Miocene paleobathymetry of the borderland is attributable to transtension and transpression that developed in response to right shear in the transform system between the Pacific and North American plates.

Pliocene

On the mainland, coastal exposures of Pliocene rocks extend from the vicinity of Santa Barbara to Ventura and discontinuously from the Santa Monica Mountains, Palos Verdes Hills, and San Joaquin Hills to the San Diego area. Successions of Pliocene strata >4000 m thick occur in the subsurface sections of the Ventura and Los Angeles basins. Most of the outcropping strata consist of repetitiously interbedded marine mudstone and sandstone with subordinate amounts of conglomerate. Basinward facies in these onshore basins are composed largely of sandstone and mudstone turbidites that locally contain thin ash beds.

Remnants of eolian and shallow-marine calcarenite beds of probable late Pliocene age are present beneath Pleistocene terrace deposits on northeastern Santa Cruz Island (Rand, 1933; Weaver and Meyer, 1969). Small pockets of Pliocene shallow-marine sandstone overlie Miocene strata and volcanic rocks on San Clemente Island (Vedder and Howell, 1976b; Vedder and Moore, 1976; Stadum and Susuki, 1978), and lower bathyal tuffaceous sandstone of early Pliocene(?) age occurs on Santa Catalina Island (Vedder, Howell, and Forman, 1979). On Santa Rosa Island, calcareous sandstone beds, in part eolian, contain late(?) Pliocene foraminifers (Vedder and Howell, 1980). Pliocene mollusks embedded in volcanic breccia at Northeast Bank are shallow-water species that represent depths of <50 m. (Hawkins, Allison, and MacDougall, 1971). Outer sublittoral mollusks of possible Pliocene age are common in fine-grained sandstone that caps southeastern Coronado Bank (Vedder et al., 1976b). Each of these contrasting paleoenvironmental occurrences is stratigraphically incomplete and areally limited. None of the depositional remnants can be directly related to either the scattered outcrops on the mainland shelf or the nearly completely buried fine-grained turbidites (?) in the Santa Barbara, Santa Monica, and San Pedro Basins, where >1000 m of Pliocene sediments accumulated (Junger and Wagner, 1977).

In contrast to the exceptionally thick sequences in the Ventura and Los Angeles basins and the inner borderland basins, relatively thin accumulations of Pliocene sediment are present in the central and outer borderland basins, where the strata are composed mainly of mudstone. This seaward decrease in thickness suggests that submarine ridges and islands impeded dispersal of terrigenous detritus and that hemipelagic and locally transported material from the barrier ridges were the primary sources of sediment. The configuration of the Pliocene sea floor probably closely resembled that of the modern borderland (Blake et al., 1978; Vedder and Howell, 1980). Parts of Patton Ridge may have been above sea level, as latest Miocene and/or early Pliocene rocks seem to be limited to the northern and southern ends of the ridge and the adjoining basins, and truncated middle Miocene and older rocks predominate elsewhere on the ridge.

The nearshore basins received large amounts of sediment from the north and east, yet the sea floor in the subsiding axial parts remained in water as deep as 1500 m until the late Pliocene (Vedder and Howell, 1980). Unconformities on the flanks of these basins demonstrate that adjoining areas were being structurally deformed (Fischer, 1976; Greene, 1976; Junger and Wagner, 1977). Late Pliocene marine strata now at a sea-floor depth of 1250 m near the northwestern end of San Clemente Ridge contain foraminiferal assemblages that imply water depths nearly twice that deep at the time of deposition. Early Pliocene beds on Santa Catalina Island probably were deposited at depths >2000 m (R. E. Arnal, 1977, 1978, written communications). These depth changes indicate local tectonic uplifts in the central and inner borderland that may have been as much as 2000 m since the early part of the epoch and 1000 m since the late part. Other areas, such as Santa Rosa and Santa Cruz Islands, seem to have remained relatively stable. During the Pliocene, maximum rates of uplift on the borderland are estimated to have been between 0.5 and 0.7 m per 1000 years in the San Clemente Ridge and Santa Catalina Island areas (Vedder and Howell, 1980). In Pliocene basins throughout coastal southern California, subsidence rates (Table 1) began to accelerate about 4 ± 1 Ma (Yeats, 1978).

Quaternary Strata

Pleistocene

Thick sequences of Pleistocene marine and nonmarine beds are present in the Ventura and Los Angeles basins; but elsewhere along the mainland coast, correlative strata are thin or absent. On the islands, Pleistocene beds are limited chiefly to terrace deposits and dune remnants. Mainland outcrops generally consist of poorly consolidated sediments that range from boulder conglomerate to clayey mudstone.

In the onshore basins, depositional modes included fluvial, flood-plain, eolian, estuarine, and shelf environments. Correlative sequences in the offshore basins consist primarily of sand and mud that tend to be finer grained and to contain increasing amounts of pelagic material farther from the coast.

Sediments accumulated rapidly in the coastal basins during the Pleistocene, whereas rates of deposition in basins farther offshore were relatively slow. Unconformities are recognizable on acoustic-reflection profiles in the Santa Monica and San Pedro Basins (Greene, 1976; Fischer, 1976; Junger and Wagner, 1977; Nardin and Henyey, 1978; Osborne et al., 1980). Complex, overlapping submarine fan systems and associated channels beneath the Gulf of Santa Catalina apparently were structurally controlled (Graham and Bachman, 1983). In basins seaward from the inner islands, interruptions in sedimentation are less distinct (Teng, 1985), suggesting that tectonism diminished seaward and that the contribution of pelagic material tended to obscure erosional gaps. Seaward transport of terrigenous sediment into the outer basins was partly obstructed by ridges and basin sills.

Eustatic sea-level changes were partly responsible for the cutting of flights of terraces that locally are well displayed along the island and mainland coasts. Correlation of terraces throughout the borderland, based upon matching altitudes, is not reliable because of the effects of disharmonic deformation that may have surpassed all earlier episodes of tectonism in amount and rate (Vedder and Howell, 1980).

As many as five well-defined submergent terraces between the depths of 10 and 130 m were recorded by Emery (1960), who believed that they formed in Wisconsin time and were regionally tilted seaward. Some of these subsea platforms are more likely to be uncorrelatable local features caused by submarine slumping. The submergent terraces are both erosional and constructional (Ridlon, 1972; Fischer, 1976; Junger, 1979). Depositional terraces flanking the island platforms may have been built during regressive phases, effectively enlarging the shelves; but most of these terraces probably were destroyed during transgressive phases when erosional terraces were cut (M. E. Field, 1978, written communication). Terraces deeper than 120 m along the Santa Rosa-Cortes and Santa Cruz-Catalina Ridges (Uchupi, 1961; Junger and Wagner, 1977) and elsewhere (Emery, 1960) probably are older than late Pleistocene, and some may be as old as late Pliocene.

According to Curray (1965) and Milliman and Emery (1968), the sea began to return from approximately −120 m to its present level about 17,000 to 18,000 years ago. During this low stand and subsequent transgression, the unusual eolianites and caliche deposits on the upper surfaces of the outer islands may have begun to form (Olmsted, 1958; Vedder and Norris, 1963; Johnson, 1967).

That the borderland consisted of independent structural blocks during the Pleistocene is obvious from the lack of correlation between emergent surf-cut terraces on the islands and those along the mainland coast. Quaternary tectonism is further substantiated by sea-floor faults and folds that deform Pleistocene and Holocene sediments (Vedder et al., 1974; Greene et al., 1975; Fischer, 1976; Junger and Wagner, 1977; Nardin and Henyey, 1978). One of the most striking examples of deformation and erosion is evident in the vicinity of the Dos Cuadras and Carpinteria offshore oil fields southeast of Santa Barbara, where several thousand meters of Pliocene and early Pleistocene strata were removed from the crest of a growing anticline (Rincon trend) during the late Pleistocene and Holocene.

Preliminary amino-acid age determinations from fossil marine mollusks in terrace deposits at San Nicolas Island, Palos Verdes Hills, and San Joaquin Hills imply different amounts of cumulative uplift for each of these structural blocks as well as changing rates of uplift within each block during the last 500,000 years (Lajoie and Wehmiller, 1978). The highest estimated rate among these sites is slightly less than 1.0 m per 1000 years at Palos Verdes Hills. Intense deformation along the north side of the Santa Barbara Channel is indicated by uplifted and tilted marine terraces and north-dipping thrust faults (Yerkes and Lee, 1979; Yerkes et al., 1980; Yeats, 1981). Exceptionally high average uplift rates of 4 to 10 m per 1000 years occurred along the Ventura Coast during the last 100,000 years (Lajoie et al., 1979).

Holocene

Description of Holocene deposits and structural events is beyond the scope of this report. Recent papers that deal with these subjects include Clarke et al., (1983), Field and Edwards (1980), Gorsline (1980), Osborne et al. (1980), Graham and Bachman (1983), Nardin (1983), and Teng (1985).

Cenozoic Igneous Rocks

Volcanism on the borderland spanned most of the late Cenozoic, although Quaternary activity seems to have been limited to the Bahia San Quintin and Punta Santo Domingo areas south of lat. 31°N. This episodic activity began in the early Miocene, culminated in the middle Miocene, and diminished sharply during the late Miocene. The volcanic rocks consist chiefly of massive and fragmental flows that are intercalated with subordinate pyroclastic material. Shallow intrusive rocks are common on the islands and probably are widespread on the ridges. Petrographic examination shows that the effusive volcanic rocks range in composition from rhyolitic to basaltic, and that andesitic types are the most abundant (Vedder, Crouch, and Lee-Wong, 1981). The extent and volume of the volcanic rocks are not definitely known, but it is likely that they constitute large parts of the borderland south of lat. 31°N, where basalt apparently is the primary rock type (Krause, 1965; Crouch, 1979).

Little is known about the distribution and age of volcanic rocks in the outer part of the borderland other than that they are concentrated in the southern part and probably are mostly Miocene in age. Pliocene and late Miocene (?) submarine extrusive rocks form most of Northeast Bank (Hawkins, Allison, and MacDougall, 1971), and early(?) and middle(?) Miocene basaltic and andesitic rocks are present on Hancock Bank and at the southern end of Patton Ridge. Hawkins (1970) reported tholeiitic basalt from near the northern end of the Patton Escarpment. Pyroclastic rocks of Miocene and early Pliocene(?) age, and basaltic rocks of unknown age occur along the northwestern salient of Patton Ridge (Wright, 1967; Vedder, Crouch, and Lee-Wong, 1981). Middle and late Miocene lapilli tuff was drilled at DSDP Site 467, and middle Miocene volcaniclastic beds were penetrated at DSDP Site 468 (Yeats, Haq, et al., 1981).

Late Oligocene and/or early Miocene igneous rocks intrude older strata in the vicinity of San Miguel Island, where one of several pluglike bodies of dacite yielded a K/

Ar age of 24.4 ± 0.3 Ma (Weaver and Doerner, 1969b; Crowe et al., 1976). Similar dacite forms Richardson Rock west of San Miguel Island (Palmer, 1965) and is present on Kelez Ridge south of Santa Rosa Island, but these occurrences are undated. Volcanic rocks are sparse in the northern part of Santa Rosa-Cortes Ridge, although thin layers of vitric tuff commonly are interbedded in strata of early(?) and middle Miocene age. Dikes of presumed Miocene age, such as those exposed on San Nicolas Island and Begg Rock, probably are sporadically distributed elsewhere along the same ridge. Basalt of early Miocene age (23–20 Ma) is present in the subsurface section at Dall, Tanner, and Cortes Banks and possibly on Santa Rosa Island. Acoustic-reflection profiles indicate that the sedimentary sequences beneath Santa Cruz and San Nicolas Basins do not contain significant amounts of interlayered volcanic rocks. Magnetic data further suggest that Tertiary volcanic rocks with high contrasts in magnetic susceptibility do not underlie large parts of these basins (L. A. Beyer, 1978, oral communication).

In the inner part of the borderland, early Miocene magmatic activity included emplacement of a 19-Ma granodiorite-quartz diorite stock on Santa Catalina Island (Forman, 1970; Vedder, Howell, and Forman, 1979). This stock contains the only granitic rocks known to intrude the Catalina Schist. Extrusion occurred later over large expanses of Santa Cruz, Anacapa, Santa Barbara, Santa Catalina, and San Clemente Islands, where dacitic to basaltic flows yield K/Ar ages that range from 16 to 12 Ma. Subsea tracts of probably correlative volcanic rocks, chiefly andesitic, are spread along Santa Cruz-Catalina Ridge, the Santa Barbara Island-San Clemente Ridge area, and the Thirtymile–Fortymile Bank area, as well as Crespi Knoll in the Gulf of Santa Catalina. Acoustic-reflection profiles along the flanks of the Santa Monica and San Pedro Basins show that volcanic rocks extend downdip beneath the basins (Junger and Wagner, 1977). Andesitic, basaltic, and dacitic dikes of middle Miocene age extend onto the sea floor from Laguna Beach (Vedder, Yerkes, and Schoellhamer, 1957); and a Miocene(?) basaltic dike cuts Eocene strata at La Jolla. Middle Miocene basalt flows that apparently had an offshore source straddle the shoreline south of Tijuana (Minch, 1967; Scheidemann and Kuper, 1979). Andesite dikes and basalt and andesite breccia are common on the islands of Los Coronados (Lamb, 1978).

Chemical data indicate that the preponderance of volcanic rocks on Santa Cruz and Santa Catalina Islands are tholeiitic to calc-alkaline (Crowe et al., 1976; Wood, 1982; Hurst, Wood, and Hume, 1982). On these two islands, the rocks were largely subaerially emplaced, but on Santa Rosa, Santa Barbara, and San Clemente Islands, submarine volcanism predominated. Variation in rock types and complex contact relationships characterize igneous bodies on Santa Catalina Island, where emplacement modes include a stock, hypabyssal dike swarms, a near-surface dome, subaerial and submarine volcanic flows and breccias, as well as air-fall and reworked tephra deposits.

Thick aprons of coarse volcaniclastic debris were deposited along the flanks of the middle Miocene volcanic centers. These coarse clastic rocks commonly are intercalated with lava flows, and, at places on Santa Cruz and Santa Rosa Islands, they grade distally into finer grained subsea-fan facies and ultimately into laminated tuffaceous shale (Howell and McLean, 1976; McLean and Howell, 1984). Siliceous hemipelagic debris also accumulated on the flanks of volcanoes during periods of quiescence; for example, mid-bathyal foraminiferal shale is intercalated with basalt flows on Santa Barbara Island, neritic foraminiferal shale is interbedded with andesitic and dacitic rocks on Santa Catalina Island, and neritic to bathyal foraminiferal and diatomaceous shale occur with dacite on San Clemente Island (Vedder and Howell, 1976b).

The origin of the Miocene volcanic rocks on the borderland is debatable. Yeats (1968a) and Hawkins and Divis (1975) suggested that the magmatism resulted from the mantle-plus-sediment melting that accompanied passage of the ancestral East Pacific Rise. On the other hand, Crowe et al. (1976) and Weigand (1982) attributed the island arclike composition of the rocks to subduction, possibly of the Farallon plate. Crowell (1976) and Hall (1981) ascribed the onshore Miocene igneous rocks to an invasion of magmas into structures that formed in a transform fault system. The 16 to 12 Ma age of most of the borderland volcanic rocks seems too young to be subduction related. However, paleomagnetic data indicate that the northern part of the borderland could have been as much as 1500 km south of its present position in the middle Miocene (Luyendyk et al., 1985). Although such a southern position would place the extrusive centers in a region where young rocks might be expected, it does not resolve their origin.

It is noteworthy that the greatest encroachment of seas and some of the deepest subsidence of late Cenozoic basins in southern California occurred near the end of the Miocene (Natland, 1957; Ingle, 1973). Perhaps these events occurred partly in response to the diminished igneous activity and subsidence that accompanied thermal cooling within the crust. The Miocene clockwise rotation of crustal blocks in the northwestern part of the borderland probably slowed at the end of the epoch (Luyendyk et al., 1985).

PETROLEUM RESOURCE GEOLOGY

Introduction

Between 1974 and 1981, resource-oriented geologic reports were prepared by the U.S. Geological Survey in advance of the Call for Nominations for Outer Continental Shelf (OCS) Lease Sales 35, 48, 68, and 80 on the borderland. Regional geology, petroleum geology, environmental hazards, and operational measures were outlined in these reports, which were intended to update the petroleum resource papers of Curran, Hall, and Herron (1971) and Parker (1971).

Customarily, the southern California OCS region was divided into three approximately rectangular areas of unequal size (Figure 7). These three arbitrarily selected areas of the OCS conform neither to geologic provinces nor to formally defined petroleum provinces. To be consistent with the earlier reports, however, the same three OCS subdivisions are used here; they are designated Santa Barbara Channel, inner basins and banks, and outer basins and banks.

In this section, stratigraphy and structure pertaining to petroleum occurrences are described for each of the OCS subdivisions. These descriptions are followed by discussions of source rocks, reservoirs, and traps. Although some duplication between this section and the framework geology sec-

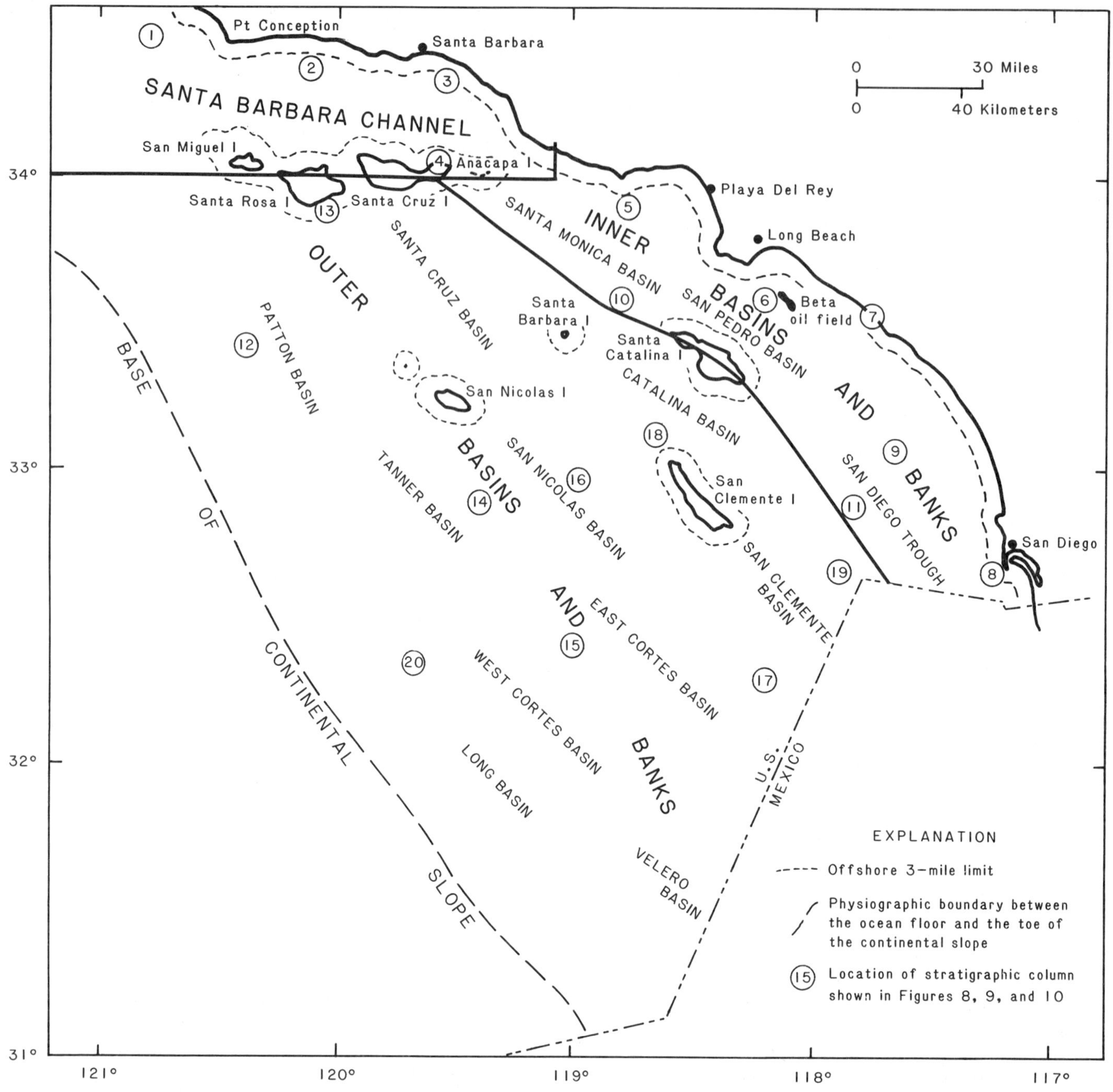

Figure 7. Map showing customary subdivisions of the Outer Continental Shelf (OCS) used in Summary Reports prepared in advance of Lease Sales 35, 48, 68, and 80.

tion is unavoidable, emphasis here is placed on thicknesses and rock types in basin sequences rather than on regional distribution, tectonic setting, and origin of all rock units. Descriptions of Quaternary strata are minimal except in cases that apply to structural deformation. Selected columns shown in Figures 8, 9, and 10 illustrate the variety of stratigraphy in each OCS area. A combination of onshore geology, exploratory-well data, sea-floor samples, and acoustic-reflection profiles were used to compile the stratigraphic columns. Sources of information for age and rock type of sea-floor outcrops are given in the framework geology section and are not routinely cited in this section. Much of the information used in the paragraphs that deal with petroleum generation and entrapment was excerpted from Taylor (1976).

Because the tectonic regimes in which the borderland basins evolved are described in the preceding section of this chapter and elsewhere (Crowell, 1976; Blake et al., 1978; Crouch, 1979; Gibson et al., 1985; Howell, Moore, and Wiley, this volume), details of the timing and nature of regional tectonism are not reiterated. Table 1 summarizes the major events and their effects. In brief, Neogene basin evolution on the borderland is attributed to right-lateral shear, which began after subduction of the Farallon plate about 30 Ma

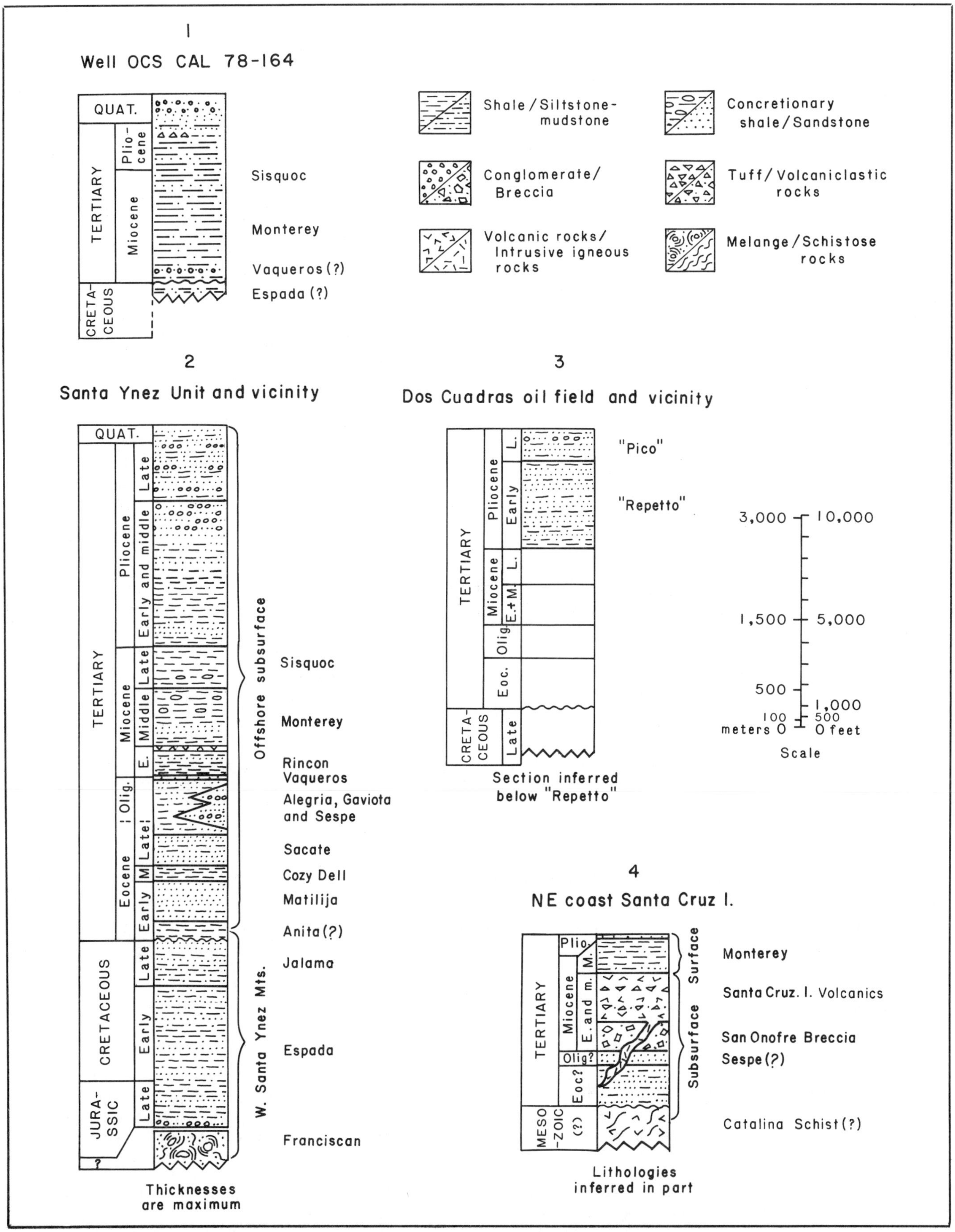

Figure 8. Stratigraphic columns, Santa Barbara Channel area. Sites of the columns are shown on Figure 7. The rock types, where known, are generalized. Stratigraphic names and age assignments are those in common usage in this area. Modified from Vedder et al. (1976c).

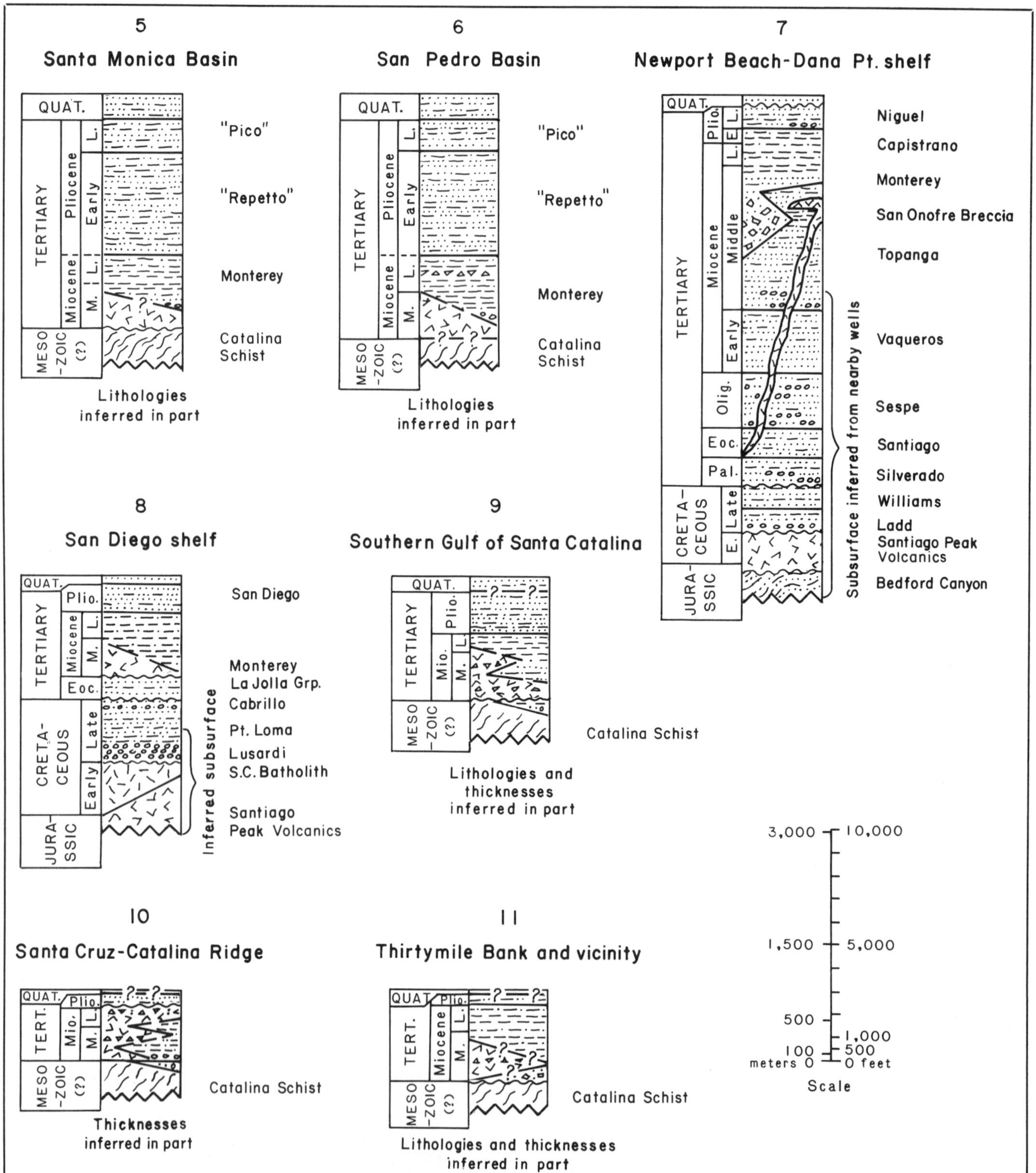

Figure 9. Stratigraphic columns, inner basins and banks area. Sites of the columns are shown on Figure 7. The rock types are generalized. Stratigraphic names and age assignments are those in common usage in this area. Lithologic symbols are the same as those explained on Figure 8. Modified from Vedder et al. (1976c).

(Atwater, 1970). Following subduction, a network of transtensional and transpressional structures began to develop in response to transform motion between the Pacific and North American plates. Uplift and erosion of structures during the Miocene resulted in thick accumulations of sediment in the rapidly subsiding basins. Volcanism began on the basin rims about 24 Ma and lasted more than 10 m.y. In the northwestern part of the borderland, crustal blocks were rotated clockwise during the Miocene and Pliocene. Regional deformation in response to continued right shear prevailed

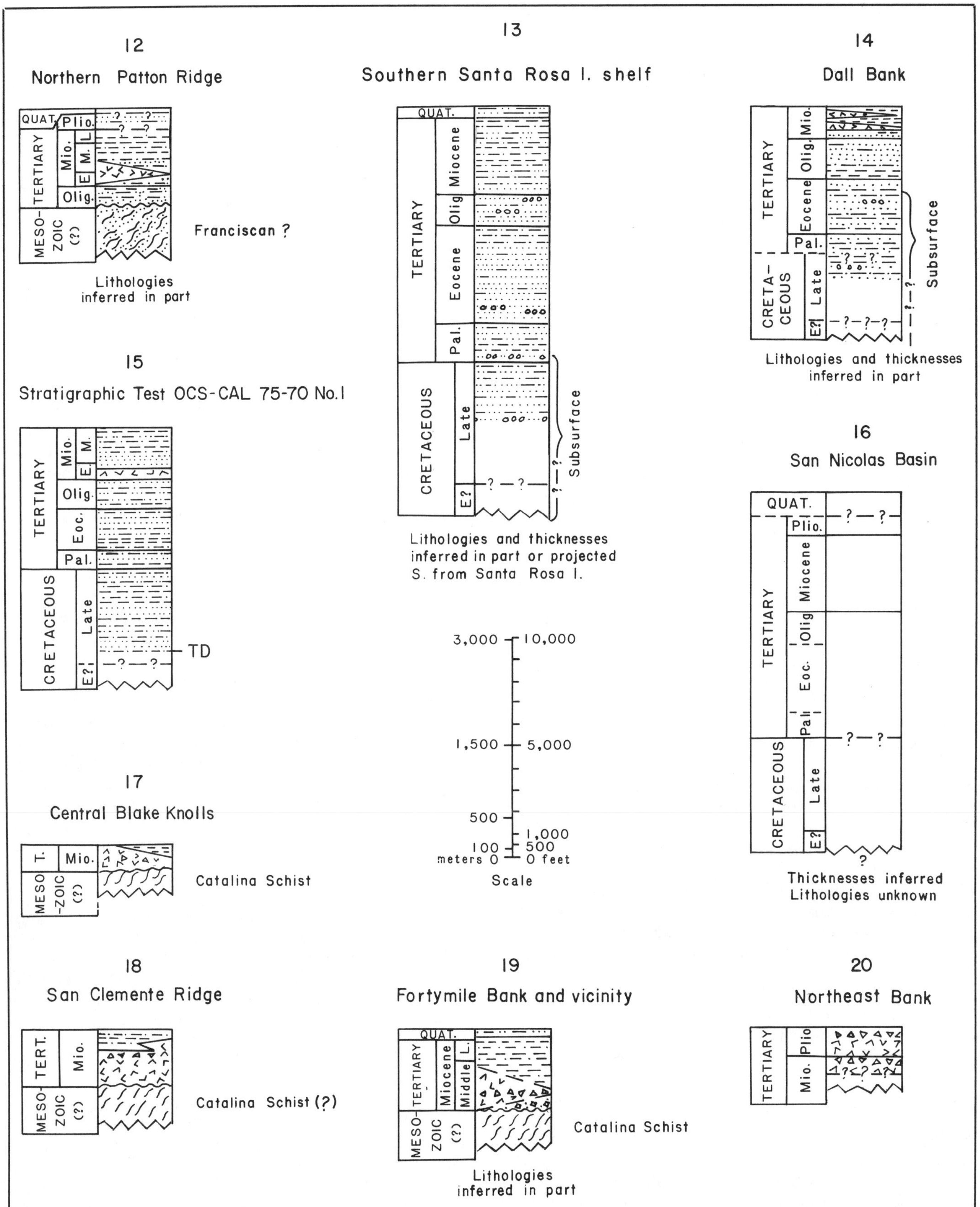

Figure 10. Stratigraphic columns, outer basins and banks area. Sites of columns are shown on Figure 7. Most rock units are unnamed; and rock types, where known, are generalized. Stratigraphic names and age assignments are those in common usage in this area. Lithologic symbols are the same as those explained on Figure 8. Except for Deep Stratigraphic Test Well OCS-CAL 75-70 No. 1, thicknesses are estimated from acoustic-reflection profiles. Modified from Vedder et al. (1976c).

throughout most of the Miocene and Pliocene and created zones of echelon folds and faults. A late Pliocene and Quaternary episode of compression, possibly related to a change in plate motion (Crouch, Bachman, and Shay, 1984; Page and Engebretson, 1984), caused parallelism in structures and high relief. It is these late Cenozoic sedimentary sequences and tectonic events that provided source rocks and traps for petroleum in the Neogene basins.

Petroleum-Resource-Related Stratigraphy and Structure

Santa Barbara Channel

The composition and distribution of rocks that form the basement floor of the Santa Barbara Channel are poorly known. About 122 m of metamorphosed mafic volcanic rocks that possibly are related to the Franciscan assemblage or the Catalina Schist were penetrated below a drilled depth of about 2130 m in the Union Gherini No. 1 well (Figure 2) on the east end of Santa Cruz Island (Howell, McLean, and Vedder, 1976). Unpublished reports of serpentine, gabbro, and "metamorphic rocks" from two wells in the central part of the channel north of Santa Cruz Island suggest the presence of ophiolite beneath Cretaceous strata, but these reports have not been substantiated. Just west of the arbitrary western boundary of the Santa Barbara Channel, Deep Stratigraphic Test Well OCS-CAL 78-164 No. 1 (Figure 2) drilled into Late Jurassic or Early Cretaceous mudstone and sandy siltstone at a depth of about 3040 m (McCulloch et al., 1979; McLean, 1979). Ophiolite presumably underlies this well section.

Late Cretaceous Rocks—Late Cretaceous sedimentary rocks are broadly distributed beneath the Santa Barbara Channel (Vedder, Wagner, and Schoellhamer, 1969; Weaver, 1969). A probable average thickness for Cretaceous sequences in the area was estimated to be in the range of 1370 to 1520 m by Curran, Hall, and Herron (1971). The Richfield (Arco) Santa Cruz Island No. 1 well, at the west end of the island and north of the Santa Cruz Island fault (Figure 2), drilled through 610 m of Late Cretaceous conglomerate, sandstone, and shale before reaching total depth (Weaver, 1969; Howell, McLean, and Vedder, 1976). Highly interpretive structure sections by Campbell et al. (1975) show questionable occurrences of Cretaceous marine strata beneath the channel. According to Ogle (1984), the Cretaceous section (commonly called Jalama Formation) under parts of the channel includes beds of indurated marine shale, mudstone, siltstone, sandstone, and conglomerate and may be exceptionally thick. Late Cretaceous sandstone under the east-central part of the channel has produced gas in the Santa Rosa unit (Figure 11; Williams, 1983, Ogle, 1984).

Paleogene Rocks—Paleogene rocks beneath the Santa Barbara Channel probably include Paleocene, Eocene, and Oligocene strata, although Paleocene beds have not been reported in the subsurface section. A complete Paleogene sequence directly offshore from the mainland may be as much as 3050 m thick. Farther inland in the Santa Ynez Mountains, the Eocene section alone is nearly 4000 m thick (Dibblee, 1966; Curran, Hall, and Herron, 1971; Campbell et al., 1975). Much of the insular platform west of San Miguel Island, at least as far as Richardson Rock, is underlain by Paleogene and older strata (Junger, 1979). The entire Paleogene section is absent in Deep Stratigraphic Test Well OCS-CAL 78-164 No. 1 west of the channel (McCulloch et al., 1979).

Along the mainland coast, marine sandstone and claystone are the chief constituents in Paleocene and Eocene formations. Conglomerate beds are interlayered in some of the sandstone sequences. On the northern Channel Islands, Paleocene and Eocene strata are exposed only south of the Santa Cruz Island fault and its westward offshore projection. Targets for petroleum exploration in the channel are Eocene sandstones (Camino Cielo Sandstone Member of the Juncal Formation, Matilija Sandstone, Sacate Formation, Gaviota Formation), which are as much as 600 m thick in the Santa Ynez unit (Figure 11; Ogle, 1984). Eocene strata in the north-central part of the channel are nearly 1500 m thick but apparently are completely truncated in the westernmost part of the area.

The late Eocene and early Oligocene section along the coastal side of the Santa Ynez Mountains grades westward from nonmarine to marine and is composed primarily of sandstone and siltstone and subordinate conglomerate. Nonmarine beds (Sespe Formation) that extend eastward from the vicinity of Santa Barbara form a sequence as much as 3270 m thick which is overlain by a 100-m-thick sequence of marine sandstone beds (Vaqueros Formation) of late Oligocene and early Miocene age (Dibblee, 1966). In the western Santa Ynez Mountains, the nonmarine section thins to < 100 m at Gaviota and tongues out westward, and the underlying late Eocene and Oligocene marine sandstone (Alegria Formation) is as much as 450 m thick. In the same area, the overlying marine sandstone (Vaqueros Formation) is < 100 m thick (Dibblee, 1950). Oligocene marine and nonmarine beds probably underlie most of the Santa Barbara Channel except in the extreme western part, where they are completely truncated by younger strata. On the northern Channel Islands, correlative nonmarine beds are present only on Santa Rosa Island (Weaver and Doerner, 1969a). Several offshore oil fields along the northern and eastern edges of the channel produce, or have the capacity to produce, from the nonmarine section (Curran, Hall, and Herron 1971). These fields include Elwood, Offshore Summerland, Montalvo, Hueneme, and Sockeye (Ogle, 1984).

Neogene Rocks—The composition and bedding of Neogene strata in the Santa Barbara Channel area (Figure 8) are relatively uniform in contrast with the diverse kinds of sedimentary rocks in adjoining regions to the south. In general, the early Miocene section (Rincon Formation) is composed of mudstone and minor sandstone; the middle Miocene (Monterey Formation), calcareous and siliceous shale and subordinate sandstone; and the late Miocene (Sisquoc Formation), siliceous claystone, siltstone, and minor sandstone. Pliocene formations (Repetto and Pico Formations of local usage) are typified by lateral lithofacies changes in sequences that tend to thicken and coarsen eastward. Igneous intrusives, flows, and tephra are interspersed in the early and middle Miocene successions, particularly along the southern margin of the area, where volcanic rocks form a large part of the middle Miocene section (Tranquillon, Conejo, San Miguel, and Santa Cruz Island Volcanics) (Dibblee, 1950; Weaver and Doerner, 1969a; Nolf and Nolf, 1969; Crowe et al., 1976).

Early Miocene mudstone and claystone beds (Rincon Formation) overlie late Oligocene and early Miocene(?) trans-

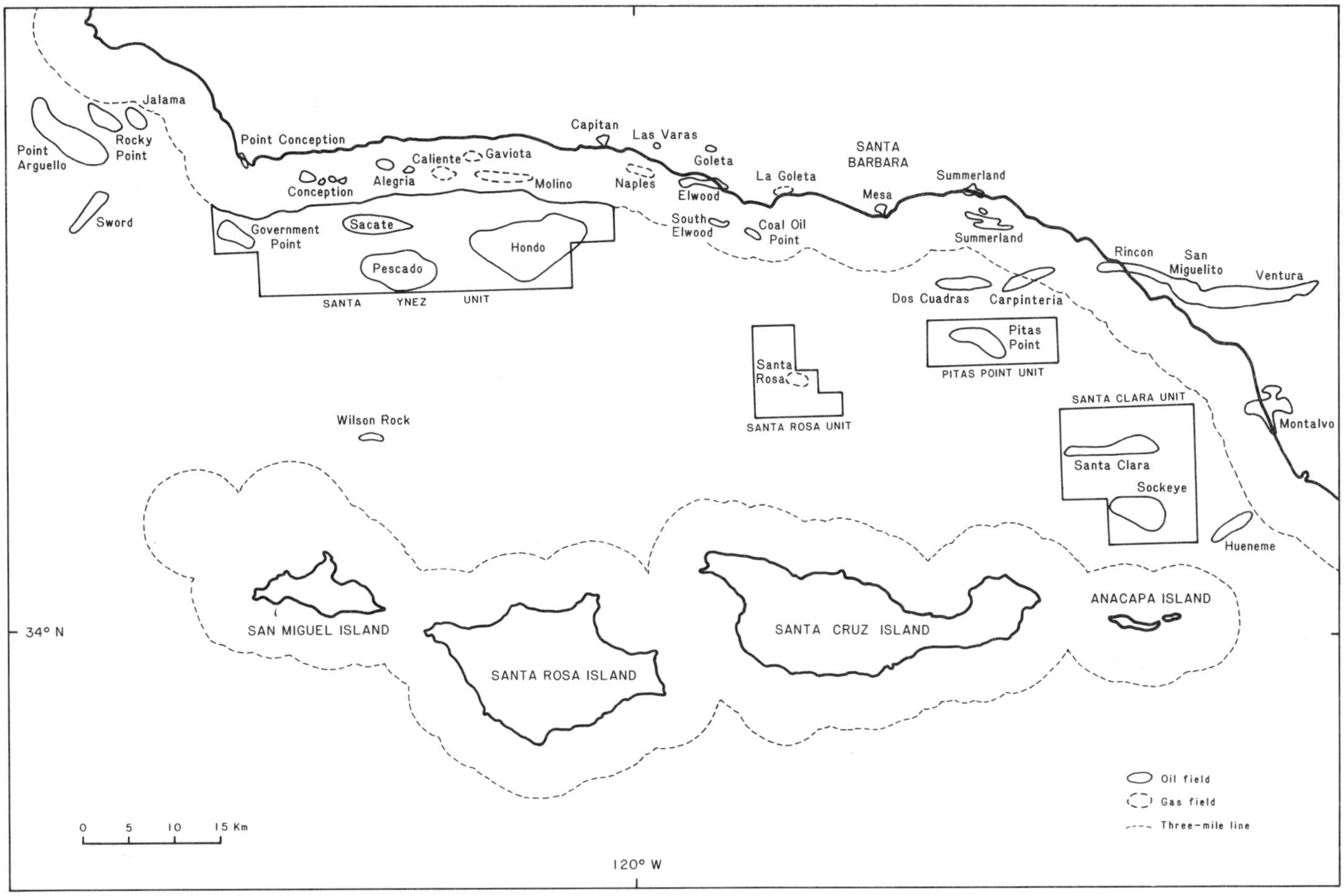

Figure 11. Map showing oil and gas fields and unit boundaries in the Santa Barbara Channel area.

gressive marine sandstone (Vaqueros Formation) throughout most of the channel area. The fine-grained, clay-rich strata locally contain ash beds near the base and top, and the succession thins westward from Ventura, where it is about 750 m thick, to Point Conception, where it is 260 m thick. In contrast, correlative beds on the northern Channel Islands generally are coarser grained and grade laterally into breccia and conglomerate. The early Miocene section is >800 m thick on the north side of Santa Rosa Island (Weaver, 1969, Correlation Chart). In the South Elwood oil field just west of Santa Barbara, sandstone zones in the mudstone sequence are productive (Ogle, 1984).

Early middle Miocene sandstone beds that correlate with a thick sequence of sandstone, mudstone, and intercalated volcanic rocks in the Santa Monica Mountains (Topanga Group of Yerkes and Campbell, 1979) produce oil in the Hueneme and Sockeye fields in the eastern part of the channel (Ogle, 1984). About 100 m of possibly equivalent sandstone was penetrated in the Union Gherini No. 1 well at the east end of Santa Cruz Island.

Middle and late Miocene calcareous and siliceous claystone and shale (Monterey Formation) compose a relatively uniform sequence of thin-bedded to laminated highly organic strata around much of the perimeter of the Santa Barbara Channel (Isaacs, 1983). The same shaly sequence persists beneath the channel. The thickness, however, is variable. Near Santa Barbara, it is about 700 m thick, but at Point Conception, it is only about half as thick (Dibblee, 1950, 1966). On eastern Santa Cruz Island, an equivalent sequence is about 490 m thick (Weaver, 1969, Correlation Chart; Weaver and Meyer, 1969). Similar thin-bedded rocks of the same age are present on the flanks of the San Miguel Island platform and extend as sea-floor outcrops at least 22 km northwest of the island (Vedder et al., 1981). Farther northwest beyond the sea-floor outcrops, probably correlative strata occur as continuous well-defined reflectors in acoustic-reflection profiles (Junger, 1979). Beneath the northern part of the channel, the shaly sequence is as much as 760 m thick (Ogle, 1984). Fractured siliceous and dolomitic shale zones form reservoirs in offshore oil fields along the entire northern side of the channel. Similar reservoirs recently have been reported from the Santa Rosa unit.

Late Miocene strata consist primarily of claystone and siltstone (Sisquoc Formation) together with subordinate amounts of sandstone (Santa Margarita Formation). Although late Miocene beds are extensive along the mainland coast, where they may be as thick as 600 m, correlative strata are not known on the northern Channel Islands. Sea-floor outcrops, chiefly silty claystone, are present on the northwestern part of the San Miguel Island platform. In the eastern part of the channel, hydrocarbons have been detected in subsurface sandstone zones that ultimately may prove to be productive. Farther west in the channel, the late Miocene sequence generally is finer grained and may include reservoirs in fractured siliceous claystone and permeable sandstone zones (Ogle, 1984).

Pliocene strata in the eastern part of the Santa Barbara Channel and adjoining onshore areas consist largely of turbidite sequences of mudstone, sandstone, and conglomerate. These strata are nearly 4000 m thick at places onshore. A conglomerate-free section 13 km south of Summerland is 3800 m thick (Wayland et al., 1978) and presumably is composed mainly of sandstone and mudstone turbidites. In mid-channel south of Santa Barbara, the Pliocene section is chiefly shale and is estimated to be about 1800 m thick (Curran, Hall, and Herron, 1971). Farther offshore, the same section thins westward and wedges out southward (Curran, Hall, and Herron, 1971; Campbell et al., 1975).

The early Pliocene section (commonly called Repetto Formation) in the eastern part of the channel ranges in thickness from about 600 to 1200 m. Sandstone zones within this formation in the Dos Cuadras and Offshore Carpinteria oil fields and in part of the Santa Clara unit yield large amounts of oil and gas. Similar sandstone zones in the Pitas Point unit probably are potential producers. Late Pliocene sandstone, siltstone, and mudstone beds (commonly called Pico Formation) have a composite thickness that ranges between 2750 and 3800 m in the eastern part of the channel, where they produce gas in the Pitas Point unit (Ogle, 1984). Equivalent strata farther west generally are <2000 m thick.

Quaternary sediments, which onshore grade westward from nonmarine to marine, are >1220 m thick directly offshore from Ventura (Greene, Wolf, and Blom, 1978). Thicknesses, however, are variable and generally <400 m in mid-channel. These generally coarse-grained strata (commonly called Santa Barbara, Saugus, and San Pedro Formations and unnamed Quaternary deposits) are composed chiefly of semiconsolidated sandstone and conglomerate together with subordinate amounts of siltstone and claystone, all of which show greater compaction downward in the section. Farther west and south beneath the channel, the section is increasingly finer grained and grades laterally into beds that consist mainly of mudstone. Quaternary sand zones that overlap some structural highs are charged with gas from underlying sources.

Structure—Structurally, the Santa Barbara Channel is a tectonic depression bounded on the north by a warped and faulted homocline along the south flank of the Santa Ynez Mountains, and bounded on the south by the faulted and folded insular platform of the northern Channel Islands. Structural relief across the region is estimated to exceed 18,300 m. Nearly continuous tectonism throughout the late Cenozoic has left a complex imprint on older structures. Most faults and folds in the channel are oriented east–west in a pattern similar to that of onshore parts of the western Transverse Ranges (Dibblee, 1950; Vedder, Wagner, and Schoellhamer, 1969; Campbell et al., 1975).

Folds occur in well-defined trends that are parallel or en echelon (Campbell et al., 1975). High-angle faults, with apparent normal and reverse separations, are interspersed with those that have strike-slip components of movement (Ellsworth et al., 1973; U.S. Geological Survey, 1975; Greene, Wolf, and Blom, 1978; Yerkes and Lee, 1979). North of mid-channel, north-dipping reverse faults with a left-lateral component of slip are dominant (Yerkes et al., 1980; Yeats, 1981). Older structures may have controlled sediment dispersal as early as the Eocene, and at some places faults cut strata no younger than late Miocene and early Pliocene, particularly along the southern margin of the channel (Junger, 1979). On the other hand, domed late Pleistocene alluvium, uplifted and tilted marine terrace platforms, and faults that cut Holocene sea-floor sediment (Vedder, Wagner, and Schoellhamer, 1969) attest to recent tectonic activity, particularly along the mainland edge of the channel. The uplift rate on one deformed terrace has been as much as 10 m per 1000 years for the past 45,000 years (Lajoie et al., 1979).

Faulted anticlines provide the trapping structures in the oil and gas fields along the northern and eastern margins of the channel. With two noteworthy exceptions, all of the these anticlines form well-defined, persistent trends that conform to the regional east–west structural grain. The exceptions are in the Santa Rosa unit north of Santa Cruz Island, where two north-trending anticlines reflect a pre-middle Miocene structural high (Ogle, 1984).

Inner Basins and Banks

The offshore area that extends southeastward from Anacapa Island and eastward from the crests of Santa Cruz-Catalina Ridge and Thirtymile Bank contains three large bathymetric basins. From northwest to southeast, these are the Santa Monica Basin, San Pedro Basin, and San Diego Trough. Outer Continental Shelf (OCS) lease sales since 1975 have resulted in only one oil-field development within this area, the Beta field, which lies near the shelf break on the inner edge of the San Pedro Basin (Figure 7).

On shore, Catalina Schist forms the basement west of the Newport-Inglewood fault zone, and oil-producing reservoirs occur in fractured schist between Playa del Rey and Long Beach. The eastern shelf of the Gulf of Santa Catalina probably is underlain by basement rocks that are related to metavolcanic and plutonic rocks of the Peninsular Ranges (Howell and Vedder, 1981). Miocene and younger sedimentary rocks in the Santa Monica and San Pedro Basins are underlain by middle Miocene volcanic rocks and/or Catalina Schist. Miocene and older(?) strata beneath the volcanic rocks are thin or absent (Junger and Wagner, 1977). In contrast, Miocene rocks along the easternmost margin of the borderland are underlain by relatively thick sequences of Late Cretaceous and Paleogene strata similar to those on the east side of the Los Angeles Basin (Yerkes et al., 1965). Columns 6 and 7 of Figure 9 illustrate the striking differences in stratigraphy on opposite sides of this basin.

Late Cretaceous and Tertiary Rocks—In the deep northwestern part of Santa Monica Basin, as much as 2500 m of latest Miocene, Pliocene, and younger sedimentary rocks (Malaga Mudstone, Fernando Formation, Santa Barbara Formation, and San Pedro Sand) may have accumulated above middle Miocene volcanic and sedimentary rocks (Zuma Volcanics and Monterey Formation). Along the northeast slope of Santa Monica Basin, Miocene strata composed chiefly of shale (Nardin and Henyey, 1978) (Monterey Formation) are as thick as 790 m; and along the southeast flank of the basin, they are 400 to 550 m thick (Junger and Wagner, 1977). On the northeast edge of San Pedro Basin, the Miocene sedimentary section is estimated to be 900 m thick and probably consists largely of claystone and siliceous shale (Monterey and Puente Formations). This well-bedded section thins southwestward and may be absent at places beneath the central part of the basin (Figure 12, profile A–A′). As much

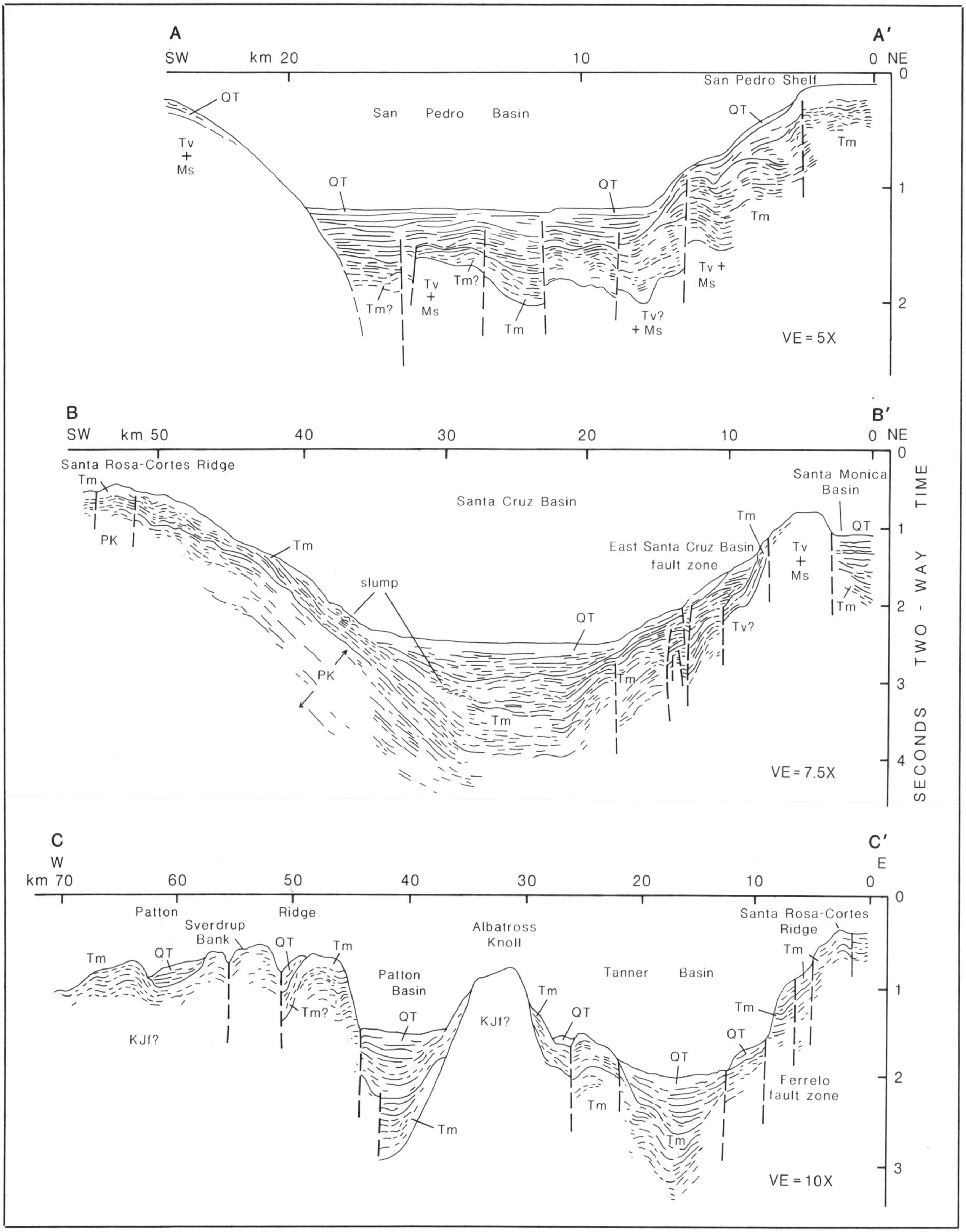

Figure 12. Line-drawing interpretations of single-channel acoustic-reflection profiles across major borderland features. Locations of the tracklines are shown on Figure 2. Letter symbols represent the following: Ms, Catalina Schist; KJf, Franciscan-like rocks; PK, Cretaceous and Paleogene strata; Tv, Miocene volcanic rocks; Tm, Miocene strata; QT, latest Miocene(?), Pliocene, and Quaternary strata. Modified from Howell et al. (1980, Figure 11).

as 1600 m of Pliocene strata, chiefly mudstone and sandy siltstone (Fernando Formation), are believed to overlie volcanic rocks and/or schist in the central part of San Pedro Basin (Junger and Wagner, 1977). Sandstone zones in the late Miocene and Pliocene sequences in the Wilmington oil field and other nearby fields are exceptionally productive (Ford, Union Pacific, Lower Terminal, Upper Terminal, Ranger, and Tar zones). These producing zones occur through a section nearly 1220 m thick (Mayuga, 1970). In the Beta oil field, seven late Miocene sandstone zones between depths of about 1000 m and 1500 m are productive (Williams, 1983). The isopach map of inferred post-Miocene strata (Figure 13) indicates that other nearshore basins contain substantial thicknesses of Pliocene strata beneath Quaternary sequences that generally are <400 m thick.

In the nearshore shelf area east of the Newport Inglewood-Coronado Bank fault system, Late Cretaceous and Paleogene strata (Ladd, Williams, Silverado, Santiago, and Sespe Formations) unconformably overlie Jurassic and Cretaceous basement rocks (Santiago Peak Volcanics and possibly Bedford Canyon Formation and Peninsular Ranges batholith). These Late Cretaceous and Paleogene strata consist chiefly of sandstone, conglomerate, and mudstone. Onshore near Laguna Beach, this pre-Miocene stratigraphic sequence is >2100 m thick in the subsurface section (Figure 9, col. 7). In the vicinity of San Diego, partly equivalent strata are estimated to have a maximum thickness of about 1500 m. Noncommercial amounts of oil formerly were produced from Late Cretaceous sandstone in the now-abandoned San Clemente oil field.

Nearly 3400 m of early and middle Miocene sandstone, shale, conglomerate, and breccia beds (Vaqueros, Topanga, and Monterey Formations and San Onofre Breccia) underlie the coastal area between Newport Beach and Laguna Beach (Vedder, Yerkes, and Schoellhamer, 1957; Vedder, 1975). Correlative strata on the San Diego Shelf are thin or absent as a result of nondeposition and/or erosion. Late Miocene mudstone and subordinate sandstone beds (Capistrano Formation, lower part) form a sequence about 500 m thick in the vicinity of San Juan Capistrano and San Clemente. Pliocene siltstone and sandstone beds together with thin conglomerate lenses (Capistrano Formation, upper part, and Niguel Formation) have a composite thickness of as much as 300 m east of Dana Point, and similar beds are about 380 m thick at San Diego (San Diego Formation).

In the Gulf of Santa Catalina, a combination of sea-floor samples and acoustic-reflection profiles show that relatively thin sequences of interlayered Miocene shale and volcanic flows are overlain by Pliocene siltstone and sandstone. The Pliocene sequence may be as much as 600 m thick beneath the central part of the Gulf of Santa Catalina and 300 m thick beneath the San Diego Trough. On Santa Catalina Island, however, only thin veneers and pockets of Miocene and Pliocene strata are preserved (Vedder, Howell, and Forman, 1979). Gravity and magnetic anomalies (Figures 14, 15) along the Santa Cruz-Catalina Ridge suggest that Miocene volcanic rocks rest directly on basement (Beyer, Sasnett, and Pisciotto, 1975; Beyer, 1980). A Bouguer gravity anomaly along the eastern part of the San Diego Trough and western flank of Coronado Bank (Figure 14) implies an eastward-thickening sequence of pre-Pliocene sedimentary rocks and/or low-density basement rocks (Beyer, 1980). Magnetic anomalies directly west of the San Diego area (Figure 15) may be caused by near-surface Miocene volcanic rocks or the Santiago Peak Volcanics.

Structure—The structure of the inner basins and banks is complex. Folds near the mainland coast along the offshore extension of the Newport-Inglewood fault zone are comparatively small and steep flanked, and faults occur both in patterns of echelon breaks or as single traces that bifurcate (Junger and Wagner, 1977; Nardin and Henyey, 1978; Legg and Kennedy, 1979; Clarke et al., 1983). Most folds and faults are oriented northwest. Some large fault-bounded antiforms, such as Redondo Knoll, which separates Santa Monica and San Pedro Basins, Lasuen Knoll, and Coronado Bank (Figure 2), are broad and nearly symmetrical; but internal structures in most cases consist of small faults and folds. The large folds apparently deform strata as young as Pliocene, and some small folds along the Palos Verdes and Newport-Inglewood faults show post-Pliocene warping. Faults range in age from middle Miocene to Quaternary and presumably include those with vertical and horizontal slip (Ziony et al., 1974; Junger and Wagner, 1977; Clarke et al., 1983). The age of most small folds probably is restricted to post-late Miocene time.

Outer Basins and Banks

The outer borderland is believed to consist of northwest-trending belts (terranes) of different basement rocks that are blanketed by contrasting stratigraphic successions (Figures 4, 5, 10). In west to east sequence, these belts of basement rocks are correlated with the Franciscan assemblage, the Coast Range ophiolite, and the Catalina Schist (Vedder and Howell, 1976a; Crouch, 1979, 1981; Howell and Vedder, 1981; Howell, Champion, and Vedder, in press). Franciscan-like argillite and graywacke that form the acoustic basement under the continental slope and Patton Ridge have compressional velocities of 5.1 to 6.2 km/sec and appear as discordant, discontinuous reflectors on acoustic-reflection profiles (Crouch, 1981). The ophiolite belt and overlying Great Valley-like sequence underlie Santa Rosa-Cortes Ridge and the large basins adjoining it on the east and south. In contrast with the Franciscan-like rocks, the Great Valley-like rocks are characterized by compressional velocities of 4.5 to 4.6 km/sec and concordant, continuous seismic reflectors. The Catalina Schist has a compressional velocity of 4.5 km/sec at Santa Catalina Island and generally displays nonpersistent, chaotic seismic reflectors. Bouguer gravity highs (Figure 14) at Kelez Ridge, Nidever Bank, Albatross Knoll, and Emery Knoll (Figure 2) imply shallow basement; lower Bouguer values over Trask Knoll, Hancock Bank, and Garrett Ridge suggest underlying rocks that are less dense (Beyer, 1980). A pronounced Bouguer gravity high extends southeastward from Osborn Bank to an area west of San Clemente Island and is believed to represent shallow basement rocks similar to those exposed on northern San Clemente Ridge. Late Jurassic and Early Cretaceous rocks have not been conclusively identified on the outer part of the borderland, although strata of this age possibly are present in the subsurface section beneath Santa Rosa-Cortes Ridge.

Late Cretaceous Rocks—Late Cretaceous strata have been penetrated in exploratory wells on Santa Cruz and Santa Rosa Islands and along Santa Rosa-Cortes Ridge as far south as Cortes Bank. The Richfield (Arco) Santa Cruz Island No. 2

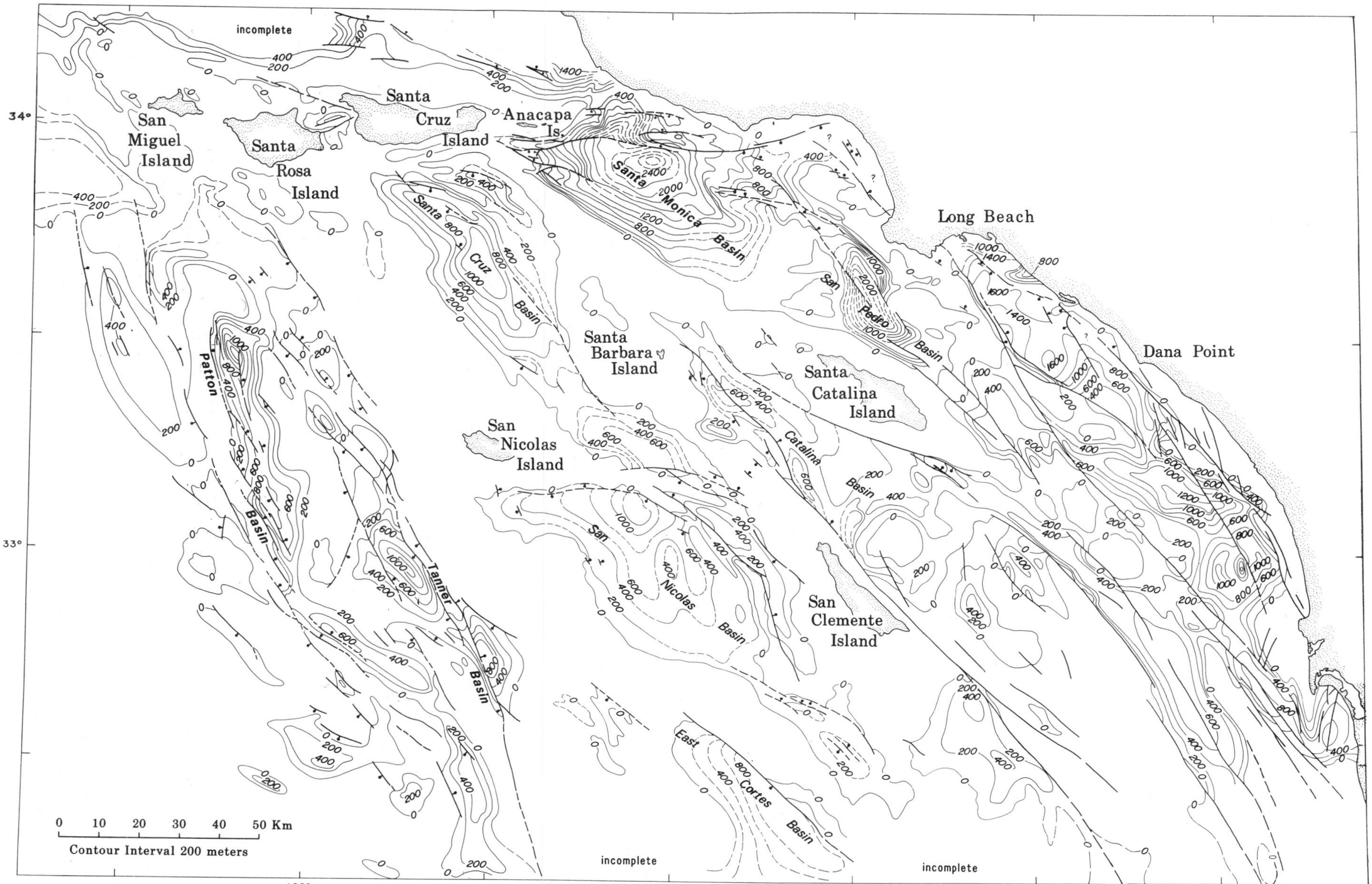

Figure 13. Isopach map of inferred post-Miocene strata based upon interpretations of acoustic-reflection profiles. Faults dashed where inferred; bar and ball on apparent downthrown side. Modified and revised from maps by Vedder et al. (1974), Junger and Wagner (1977), Junger (1979), and Crouch (1981). Late Miocene strata possibly are included at places where biostratigraphic data are uncertain.

well, located on the Christi anticline just south of the Santa Cruz Island fault, penetrated 623 m of Late Cretaceous mudstone, sandstone, and conglomerate beds above the bottom of the hole at a drilled depth of 689 m. The Chevron OCS P-0245 No. 1 well about 9 km south of Santa Rosa Island bottomed in Late(?) Cretaceous strata. The Mobil Santa Rosa No. 5 well south of the median fault on Santa Rosa Island spudded in Eocene strata and drilled through Late Cretaceous mudstone and sandstone between the depth of 1920 m (?) and the bottom of the hole at 3349 m. Nearly 1220 m of Late Cretaceous sandstone and mudstone turbidite beds (Figure 10) were penetrated in the Deep Stratigraphic Test Well OCS-CAL 75-70 No. 1 near the southeast end of Cortes Bank. Other exploratory wells in the Dall-Tanner-Cortes Bank area bottomed in similar Late Cretaceous beds that include conglomerate in some holes. Cretaceous sedimentary rocks are present as sea-floor outcrops in the vicinity of Nidever Bank, on Garrett Ridge, and at the northwest end of Cortes Bank. Interpretations of acoustic-reflection profiles suggest that correlative strata extend eastward beneath Santa Cruz and San Nicolas Basins (Figures 5, 10; Figure 12, profile B–B′) and possibly southward into East and West Cortes Basins and Velero Basin.

Paleogene Rocks—Although early Tertiary sedimentary rocks are sparsely distributed as sea-floor outcrops, they probably underlie younger rocks throughout at least one-third of the outer borderland (Figure 6). Paleocene strata are exposed on the northwestern part of Santa Rosa-Cortes Ridge and may be present on the shelf west of San Miguel Island. Nearly 1100 m of early and middle Eocene sandstone, siltstone, and subordinate conglomerate are exposed on San Nicolas Island (Vedder and Norris, 1963). Part of the platform west of San Miguel Island and the broad shelf around San Nicolas Island are composed of interbedded sandstone and siltstone of early and middle Eocene age, and similar Eocene strata probably constitute a thick subsurface section southeast of Santa Rosa Island. All of the Eocene sequences consist chiefly of turbidites. Late Eocene and early Oligocene(?) sandstone and mudstone beds are exposed on the sea floor at Cortes, Tanner, and Dall Banks. At a few places along Patton Ridge and the northern part of Santa Rosa-Cortes Ridge, Oligocene mudstone occurs as sea-floor outcrops.

Deep Stratigraphic Test Well OCS-CAL 75-70 No. 1 drilled through a nearly complete section of marine Oligocene, Eocene, and Paleocene strata approximately 1220 m thick (Paul et al., 1976). Other exploratory wells on Cortes, Tanner, and Dall Banks (Figure 2) penetrated correlative early Tertiary sequences, which at Tanner Bank are nearly 2500 m thick. Similar Paleogene strata probably extend southeastward beneath younger rocks as far as Velero Basin, where they may be 1500 m or more thick (Vedder, 1981).

The distribution of Paleocene and Eocene rocks in the subsurface section of the outer basins and banks is not known with certainty, but strata of this age are known to underlie younger rocks beneath most of Santa Rosa-Cortes Ridge, where they may range in thickness from <500 m to >1500 m. Poorly constrained biostratigraphic data make it difficult to differentiate Paleocene beds from lithologically similar Late Cretaceous beds in the exploratory well sections. Interpretations of acoustic-reflection profiles indicate that Paleocene (?) and Eocene turbidite sequences as much as 2000 m thick may underlie both Santa Cruz and San Nicolas Basins (Figures 5, 10). The same sequences thin eastward and wedge out or are locally fault truncated along the eastern basin flanks (Howell and Vedder, 1984). Oligocene rocks are poorly known except in the exploratory wells on the southeastern part of Santa Rosa-Cortes Ridge, where sandstone and subordinate mudstone beds form sequences that range in thickness from about 400 to 800 m. Sandstone constitutes $>60\%$ of the Oligocene section in Deep Stratigraphic Test Well OCS-CAL 75-70 No. 1 (Paul et al., 1976).

Neogene Rocks—Early Miocene strata occur at places along the crest of Santa Rosa-Cortes Ridge directly south of Santa Rosa Island and on the San Nicolas Island platform north of Begg Rock. These beds are composed chiefly of silty claystone. Strata of the same age and similar composition are present on the shelf west of San Miguel Island. Correlative sedimentary rocks probably are widely distributed in the subsurface section of Santa Rosa-Cortes Ridge southeast of San Nicolas Island and presumably underlie parts of Patton Ridge and the intervening basins. Sandstone of possible early(?) Miocene age occurs at Sverdrup Bank on Patton Ridge and in the Dall-Cortes-Tanner banks area. In their descriptions of Deep Stratigraphic Test Well OCS-CAL 75-70 No. 1, Paul et al. (1976) assigned about 220 m of porous, fine- to medium-grained sandstone to the early Miocene, although confirmatory fossil evidence is lacking.

Middle Miocene beds, composed largely of shale and claystone, blanket broad expanses of Santa Rosa-Cortes Ridge. Shaly beds of the same age occur on the flanks of the shelf west and northwest of San Miguel Island, on Shepard Knoll, on Garrett and Patton Ridges, and on ridges east of the northwestern part of Long Basin. Mudstone and diatomaceous shale of middle and late Miocene age are present locally on San Clemente Ridge, around Santa Barbara Island and Osborn Bank, and in the vicinity of Fortymile Bank. Sandstone, conglomerate, and breccia of early, middle, and late Miocene age are concentrated near the northern Channel Islands and along the crest of Santa Cruz-Catalina Ridge, where they consist chiefly of volcanic and schist detritus (Vedder and Howell, 1976a; McLean and Howell, 1984). Tuffaceous mudstone laminae and thin ash beds are common in both the middle and late Miocene shaly sequences. Late Miocene strata, which are composed chiefly of diatomaceous mudstone and silty claystone, generally are more restricted than similar middle Miocene strata along the tops of the outer ridges. However, the late Miocene sequences are inferred to drape most of the slopes and pass beneath younger sediments throughout the basins.

Data from core holes, exploratory wells, and acoustic-reflection profiles indicate that the predominantly fine-grained Miocene sedimentary sequences may be as thick as 1100 to 1500 m in Santa Cruz, San Nicolas, and Patton Basins (Figures 5, 10; Figure 12, profile C–C′). However, much of the Miocene section has been eroded from the bank tops. These shaly sequences possibly contain sandstone turbidites. The Miocene section is relatively thin in Catalina Basin, where the thickest succession probably is <600 m. In the region of Thirtymile and Fortymile Banks, where volcanic and basement rocks are exposed on the sea floor or are buried at shallow depth, the Miocene section probably is not more than 200 m thick.

Because they may have influenced regional thermal gradients at the time they were emplaced, Miocene igneous rocks

are important constituents of the stratigraphic successions in the outer basins and banks area. These igneous rocks represent diverse conditions of emplacement and include hyaloclastite extrusions, thick extensive flows, narrow near-vertical intrusions, sill-like bodies, and large stocklike masses. Their volume and areal distribution are difficult to estimate, and the depth to the magma chambers is unknown.

The composition of the volcanic rocks ranges from rhyolite to basalt (Vedder, Crouch, and Lee-Wong, 1981). They are widespread along Santa Cruz-Catalina Ridge and San Clemente Ridge and around Santa Barbara Island and Fortymile Bank. At least 300 m of basaltic flows are exposed in the seacliffs at Santa Barbara Island, and >400 m of volcanogenic rocks are present at Santa Catalina Island. The composite flow sequence on San Clemente Island is as much as 725 m thick (Olmsted, 1958), yet neither the base nor the top is exposed. Volcanic rocks occur sporadically on Santa Rosa-Cortes Ridge and form substantial parts of Dall, Cortes, and Tanner Banks and the Santa Barbara Island platform. Presumably, most of the volcanic rocks are middle Miocene in age. However, two flow sequences are interlayered with sedimentary rocks in the southern part of the Santa Rosa-Cortes Ridge; one sequence is early Miocene, the other, middle Miocene (for age relationships, refer to section on geologic framework). Along Patton Ridge and Patton Escarpment, volcanic rocks seem to increase in abundance southward; their thickness, however, is unknown, and their age range is uncertain. Pliocene and probable Miocene volcanic rocks are present at Northeast Bank. Although thick sequences of volcanic rocks apparently are not coextensive with the floors of the large outer basins (L. A. Beyer, 1978, oral communication), they may underlie much of southeastern Catalina Basin and parts of the eastern flanks of Santa Cruz and San Nicolas Basins.

Pliocene sedimentary rocks are sparse on the shelves and upper slopes of the outer borderland. Seaward from the islands, Pliocene strata are reported at only a few places in water <450 m deep, and sequences >100 m thick are restricted primarily to the basins (Figure 13). The dominant rock types among Pliocene samples are semiconsolidated mudstone and silty claystone, unconsolidated mud, and minor amounts of sand. Redeposited sediment in the form of slumped material or turbidites derived from bordering ridges, banks, and islands probably is present in many of the basins. From interpretations of acoustic-reflection profiles, thickness estimates range from 600 m in the central parts of Santa Cruz and San Nicolas Basins to <150 m on the flanks. Crouch (1981) reported a maximum of 500 m of Pliocene strata in the deeper part of Patton Basin. The Pliocene section in Catalina Basin is <300 m at most places. DSDP Hole 467 in San Miguel Gap penetrated about 355 m of Pliocene silty and calcareous claystone beds that include thin sandstone and dolomite zones (Yeats, Haq, et al., 1981). Regional thickness variations are evident from the isopachs on Figure 13, if the relatively thin (generally <300 m) Quaternary slope-and-basin sections are subtracted. These thicknesses contrast strikingly with those in the axial parts of the Los Angeles and Ventura basins, where as much as 4000 m of Pliocene section is present.

Structure—The dominant trend of faults in the outer basins and banks is northwest, but there are two conspicuous east-west zones—one in the vicinity of the northern Channel Islands, and the other south and east of San Nicolas Island. These faults show different kinds of apparent slip. Strike-slip is indicated on some, such as the San Clemente fault, Ferrelo fault zone, and the Miocene East Santa Cruz Basin fault system (Figures 3, 6). Normal displacements, basin side down, are indicated on many faults along the flanks of the main ridges (Junger 1976, 1979; Crouch, 1981). Some of these normal faults have apparent dip-slip of as much as 700 m. Northeast-dipping reverse faults along the southern part of Santa Rosa-Cortes Ridge and opposing reverse faults along the north flank of San Nicolas Basin were interpreted from acoustic-reflection profiles by Junger (1976). The ages of movement suggested by displaced reflectors range from late(?) Miocene to Holocene, and recurrent movement probably has taken place on many of the major fault zones. Thrust or detachment faults of unknown age cut the basement rocks on Santa Catalina Island (Platt, 1976b) and may disrupt or stack pre-Miocene rocks elsewhere in the outer basins and banks area. In contrast with the mainland coast of the Santa Barbara Channel and the Santa Monica Mountains where they are common, post-Miocene low-angle thrusts have not been detected.

In general, large folds trend west-northwest at angles oblique to the major fault zones and show an echelon arrangement at places. Many anticlinal structures are very large (>20 km long), symmetrical, and have low dips on their flanks. High-standing features along all the main ridge systems commonly reflect underlying anticlines (Figure 5); examples are those at Tanner Bank and San Nicolas Island. Smaller anticlines are present at Trask Knoll, Nidever Bank, and Sverdrup Bank (Crouch, 1981). Along major upwarps such as Santa Rosa-Cortes Ridge and Santa Cruz-Catalina Ridge, numerous small folds are superimposed on the larger features but seem to die out basinward. Broad, downwarped structural lows form both Santa Cruz and San Nicolas Basins. Similar but smaller downwarps are present in Patton and Tanner Basins. Some folds deform sediments as young as Pleistocene, as in the central San Nicolas Basin. Other folds probably are as old as Miocene, such as the main anticlinal structure on northwestern Santa Rosa-Cortes Ridge, where early or middle Miocene strata truncate Paleogene and older(?) strata on both limbs (Junger, 1979). A pronounced unconformity within the Miocene sequences on the flanks of central Santa Rosa-Cortes Ridge and on the crest of Patton Ridge suggests a widespread episode of deformation in the late middle or late Miocene. Other, less well-defined unconformities, which may or may not be coeval, separate early middle Miocene strata from early Miocene and/or Oligocene strata and early(?) Pliocene strata from late(?) Pliocene strata (Junger, 1976, 1979; Crouch, 1981).

Most of the late Cenozoic structures in the outer part of the borderland are transpressional and are attributed to wrench faulting across a broad zone of right shear between the North American and Pacific plates (Junger, 1976; Blake et al., 1978; Crouch, 1981; Howell and Vedder, 1981).

Potential Source Rocks, Reservoirs, and Traps

Santa Barbara Channel

As a result of exploratory drilling following OCS lease sales in 1968, 1979, and 1982, the Santa Barbara Channel is relatively well known to the petroleum industry. However, because much of the exploratory information is confidential, some potentially productive objectives will remain obscure to the general public until the data are released. The primary source rocks are believed to be Miocene shale sequences that have been buried deeply enough to be thermally mature. Known reservoirs beneath the eastern part of the channel include sandstone zones in Late Cretaceous, Eocene, and early Miocene formations and fractured shales in the Monterey Formation; other reservoirs are inferred to be present in some post-late Miocene sandstone zones (Ogle, 1984). Structural traps, chiefly faulted anticlines similar to those in leased areas, may occur in the few remaining unleased areas.

In the north-central part of the channel and westward to the Point Conception area, potential reservoirs presumably are similar to the Eocene through Miocene reservoirs in the Santa Ynez unit. Some potential reservoirs may be present in Pliocene rocks west of Point Conception. Thick sandstone zones interlayered in mudstone of the early Miocene Rincon Formation form the main reservoir in the South Elwood field (Dames & Moore, 1974), and these zones may extend both eastward and westward from there. Onshore, the nonmarine Sespe Formation grades westward into shallow-marine facies that may contain potentially high-quality sandstone reservoirs directly offshore. In the same area, the late Oligocene and early Miocene(?) Vaqueros Formation, which is the main sandstone reservoir in the coastal area west of Santa Barbara, is generally thin but locally may be thick. Many Eocene sandstone zones probably are mass-flow deposits that have low reservoir potential; correlative sandstone zones onshore have a poor productive history. The main Pliocene turbidite reservoirs are believed to be restricted to the northeastern edge of the channel in the Dos Cuadras field and areas to the east of it. Because these sandstone zones thin abruptly southward and westward, they are considered poor prospective reservoirs in the central part of the channel.

Inner and Outer Basins and Banks

The Beta field (Figure 7) is the only oil field that has been developed on the OCS part of the inner basins and banks. Nevertheless, most of Santa Monica Bay and much of the nearshore part of the Gulf of Santa Catalina have been fairly well explored. Furthermore, these nearshore areas are contiguous to productive areas that have been thoroughly investigated. Therefore, emphasis here is placed on little-known areas of sedimentary rocks farther offshore.

Deep Stratigraphic Test Well OCS-CAL 75-70 No. 1 drilled at Cortes Bank in 1975 provided data that seemed to enhance the petroleum potential of that part of the borderland. However, the results of later exploratory drilling at Dall, Tanner, and Cortes Banks, on the shelf south of Santa Rosa Island, and on a knoll northwest of Santa Barbara Island (Figure 2) were discouraging. Although potential source rocks occur in strata of late Eocene and Miocene age in the deep stratigraphic test well, organic matter in all analyzed Tertiary rocks is immature. Nevertheless, the same rocks might have generated petroleum in adjacent basins if sufficiently buried. Late Cretaceous strata are >1525 m thick in the well; but reservoir-quality rocks are present only in the upper part of the Late Cretaceous section, and the source-rock potential is low. The lower 914 m of section in the well, below an unconformity or fault within Late Cretaceous strata, contains small amounts of mature organic matter. Other exploratory wells on Santa Rosa-Cortes Ridge drilled into Late Cretaceous and Paleogene strata, but information on the reservoir- and source-rock potential of these beds has not been published. The Mobil OCS P-0289 No. 1 well, which was drilled 15 km northwest of Santa Barbara Island, proved unproductive and bottomed in Eocene(?) conglomerate, after penetrating middle Miocene volcanic rocks and an early Miocene (?) to early (?) Eocene sedimentary section. The distribution of potential reservoir sandstone zones on Patton Ridge and the region south of Cortes Bank is unknown.

Acoustic-reflection profiles and sonobuoy refraction data (Crouch et al., 1978) indicate that thicknesses of Miocene strata may be as much as 1500 m in some of the larger outer basins such as Patton Basin. Geochemical analyses of Miocene samples from DSDP hole 467 in San Miguel Gap (Simoneit, Meyers, and Summerhayes, 1981) and scattered bottom samples (Taylor, 1976) indicate that these strata may have a high source-rock potential. Moreover, late middle Miocene through Pliocene strata within these large outer basins may include potential reservoir rocks in the form of sandstone turbidites. For example, DSDP hole 467 penetrated late middle Miocene and early Pliocene sandstone beds (Yeats, Haq, et al., 1981). Pliocene and younger hemipelagic strata that overlie potential reservoir rocks of Miocene age are as much as 1100 m thick at places in the outer basins.

Source Rocks and Organic Maturity, Outer Borderland—Selected, paleontologically dated dart cores and island samples were analyzed to determine total organic carbon content, extractable hydrocarbons, and the characteristics of those hydrocarbons. The results are listed in Tables 2 and 3, and a summary is shown in Table 4. Oligocene and older shale samples in cores from the deep stratigraphic test well at Cortes Bank also were analyzed (Paul et al., 1976); these results are shown in Table 5.

Siltstone and mudstone samples from Eocene outcrops on San Nicolas Island (Table 4) give results that are consistent with those from Eocene cores from the sea floor (Figure 16). The mean value of the organic carbon from the 54 samples analyzed is 0.35 wt.%, indicating that few of the samples contain sufficient organic matter to be considered potential source rocks. Because most of the samples are limited to the vicinity of San Nicolas Island, it is possible that coeval strata elsewhere on the outer borderland may have been deposited under conditions more favorable for organic accumulation and/or preservation of such organic material.

Where exposed on the northern Channel Islands and San Nicolas Island, Eocene formations have high sand/shale ratios and include substantial amounts of conglomerate. Shale is more abundant in the basinward (southward) extensions of these submarine-fan deposits (basin-plain facies of Howell and Vedder, 1984), and therefore the sequence probably is organically richer. Potentially adequate to excellent source rocks are present in Oligocene to middle Eocene beds in the Deep Stratigraphic Test Well OCS-CAL 75-70 No. 1 (Table 5) (Paul et al., 1976).

Within the Neogene section, middle and late Miocene shale units are considered one of the most likely sources of the prolific oil in the onshore coastal basins (Barbat, 1958; Philippi, 1965). Sea-floor samples of correlative Miocene strata (Figure 17) indicate that good to excellent potential source rocks are present throughout the borderland (Tables 2, 3, 4). Although there is a progressive increase in organic content from a mean value of 3.2 wt. % in early Miocene samples to a mean value of 6.8 wt. % in late Miocene samples, preferential increase by area is not apparent throughout the outer basin region.

The paleogeography and environmental conditions that controlled the organic content in the Neogene sedimentary sequences are not very well known. However, the organic-rich Miocene strata have an organic content similar to that of sediments now being deposited on the modern basin floors (Figure 18). Nearshore basin sediments have lower values than those in the offshore basins and even lower values on the slope and deep sea floor. Emery (1960) attributed the high concentration of organic matter at intermediate distances from the mainland to relatively slow, detritus-poor deposition in euxenic environments. Samples of Holocene sediments from Cortes and Tanner Banks give low values of about 1% organic carbon.

Seventy Eocene, Oligocene, and Miocene samples from the outer borderland yielded a low hydrocarbon content, and their HC/C ratios are less than the 1% required for good- to excellent-quality source rocks (Figure 19). The Paleogene rocks differ in some ways from the Neogene ones; for example, saturated/aromatic hydrocarbon ratios in Paleogene rocks are >1, rather than <1, and the HC/C ratios are slightly higher. Although these ratios might suggest a higher temperature history and hence a greater maturity, the chromatograms show a marked odd-carbon predominance, a condition that characterizes thermal immaturity. The Paleogene samples probably contain a different type of organic matter and may not have had a history of significantly deeper burial than the Miocene samples. Most Paleogene outcrop samples are judged immature and are interpreted as having had a temperature history too low to generate petroleum (Taylor, 1976).

Along the outer banks and ridges in water depths <1000 m, Miocene and older strata are tilted and beveled. Acoustic-reflection profiles across the ridges indicate that the Miocene beds were covered by <300 m of younger sedimentary rocks before they were beveled. Supporting evidence for shallow burial is provided by the nature of dart-core samples—when extruded from the core barrels, Paleogene cores are consolidated and stiff to semibrittle, whereas the Neogene cores are only slightly consolidated, soft, and moldable. Because the pyrolysis data indicate immaturity and because x-rays of Miocene diatoms show no conversion from opal to cristobalite, shallow burial seems confirmed (Bramlette, 1946; Murata and Nakata, 1974).

Since the close of the Miocene, some of the large deep basins continued to subside relative to the uplifted banks and ridges, and the top of the Miocene section is inferred to be between 2500 m and 3000 m below sea level in parts of the Santa Cruz, San Nicolas, Tanner, and Santa Monica Basins (Vedder et al., 1974, sheet 5). Subsidence and burial by as much as 2100 m of post-Miocene sediments brought these above-average to excellent potential source rocks into a higher pressure and temperature regime than similar rocks on the banks and ridges. If the depth of burial, paleogeothermal gradient, conductivity of the sediments, duration of time, and total overburden pressures all were suitably balanced, a temperature regime sufficient to generate hydrocarbons from these potential Miocene source beds may have been present in some of the large borderland basins.

Table 2. Organic geochemistry of Eocene outcrop samples, San Nicolas Island (Analyses by Continental Oil Co.).

Unit No.[1]	USGS sample No.	Organic carbon (weight %)	Extractable hydrocarbons Total (ppm)	HC (ppm)	HC C
20	E-7	<0.1	210	84	>8.4
20	E-4	<.1	160	53	>5.3
20	E-1a	<.1	210	0	
18	55-211	.31	140	29	.94
18	55-216	.73	340	38	.52
18	55-204	.12	150	73	6.1
18	55-196	.22	350	102	4.6
18	55-190	.13	200	73	5.6
18	55-189	.12	80	<80	<6.7
16	55-163	.30	140	28	.93
16	55-157	.26	80	20	.77
16	55-152	.38	170	44	1.2
16	55-150	.48	510	90	1.9
16	55-144	.11	100	24	2.2
14	55-141	.12	30	<30	<2.5
12	55-135	.12	60	<60	<5.0
12	55-132	.40	260	32	.80
12	55-131	.28	300	39	1.4
10	55-130	.24	80	<80	<3.3
8	55-120	.12	50	<50	<4.2
6	55-116	.11	100	30	2.7
6	55-77	.12	180	60	5.0
6	55-69	.13	60	<60	4.6
6	55-64	.16	160	42	2.6
6	55-99	.12	90	28	2.3
6	55-55	.29	110	18	.62
6	AX-5	.38	180	22	.58
6	55-45	.49	230	33	.67
6	55-36	.37	160	39	1.1
6	AX-2	.32	90	13	.41
6	M-27	.45	180	41	.91
6	AX-1	.29	70	9	.31
6	55-25	.36	80	13	.36
6	55-21	.35	110	11	.31
6	55-18	.36	110	21	.58
6	55-12	.32	90	22	.69
6	M-16	.63	300	75	1.2
6	M-13	.78	600	180	2.3
6	M-11	.56	350	93	1.7
6	K-2	.14	520	222	15.9
6	M-7	.43	180	34	.79
6	M-5	.61	240	40	.66
6	M-3	.50	310	89	1.8
6	M-1	.21	140	32	1.5

Total samples———44
Mean value, organic content wt. %———0.30
Standard deviation———0.18

[1]Unit number shown in Vedder and Norris, 1963, pl. 5).

Less is known about the Paleogene rocks of the outer borderland. At the few places sampled, organic material in Eocene and Oligocene rocks is immature. However, the primary sampling area on the San Nicolas Island platform may never have been deeply buried. Organic matter is immature in the Oligocene, Eocene, and Paleocene beds in Deep Stratigraphic Test Well OCS-CAL 75-70 No. 1 at Cortes Bank (Paul et al., 1976). Nevertheless, coeval strata buried beneath Miocene and younger rocks in Santa Cruz and San Nicolas

Table 3. Organic geochemistry of dart samples from the southern California borderland (Analyses by U.S. Geological Survey Organic Geochemistry Laboratory, Lakewood, Colo.).

Age and Sample No.	Organic Carbon (weight %)	Carbonate Carbon (weight %)[2]	Extractable Hydrocarbons				Kerogen Isolation[7] H/C[8]	Location[9]	
			Bitumen (ppm)[3]	Total HC (ppm)[4]	S/A[5]	HC/C %[6]		x	y
Late Cretaceous									
KZ 73-1-39	0.2	5.0	—	—	—	—	—	904,687	413,034
Early Eocene									
LCB 111-10	.2	2.4	62	8	2.67	0.40		906,279	467,598
LCB 110-8	.5	.5	—	—	—	—	—	931,689	453,069
Early-Middle Eocene									
LCB 110-10	2.5	2.8	526	65	1.14	.26	—	929,349	459,303
KZ 73-1-22	.2	.6	—	—	—	—	—	951,083	430,043
KSB 697-13	<.1	1.0	—	—	—	—	—	728,407	729,556
Middle Eocene									
KZ 73-1-20	.2	2.6	—	—	—	—	—	956,411	435,628
KZ 73-1-36	.2	1.5	—	—	—	—	—	914,630	413,061
LCB 110-5	.4	.6	—	—	—	—	—	934,810	443,787
LCB 110-1	.5	1.0	—	—	—	—	—	937,777	431,667
Late Eocene or Oligocene									
LCB 130-8B	.7	.5	111	6	1.0	.09	—	1,060,612	160,792
Early Miocene									
LCB 106-4	1.5	8.1	184	50	.69	.33	1.20	900,668	597,430
LCB 111-1	4.1	1.6	441	101	.35	.25	—	932,889	478,393
LCB 133-6	4.1	.6	685	142	.01	.35	—	1,244,063	128,871
LCB 106-19	2.7	1.0	—	—	—	—	—	851,891	599,655
LCB 106-6	1.5	2.1	—	—	—	—	—	894,616	597,557
LCB 110-20	5.1	1.9	—	—	—	—	—	919,704	490,505
Early or Middle Miocene									
LCB 127-7	11.1	3.3	1,765	576	.13	.52	—	1,078,927	215,823
LCB 143-6	4.4	3.6	1,033	340	.27	.77	—	1,528,677	214,818
LCB 120-2	7.3	2.9	—	—	—	—	—	873,738	227,093
LCB 121-2	3.2	1.8	—	—	—	—	—	838,100	210,265
KSB 795-6	1.8	1.6	—	—	—	—	—	871,740	632,951
KSB 905-9	6.8	1.7	—	—	—	—	—	1,092,409	565,116
KSB 908-2	2.7	1.1	—	—	—	—	—	1,150,210	454,436
KSB 908-3	5.9	1.9	—	—	—	—	—	1,150,101	454,195
KSB 709A-6	3.7	.4	—	—	—	—	—	709,232	754,500
KSB 709A-10	4.2	3.5	—	—	—	—	—	699,012	742,959
LCB 105-12	2.0	.1	—	—	—	—	—	848,223	625,475
Middle Miocene									
LCB 117-5B	3.2	1.8	352	52	.04	.16	—	801,793	353,614
LCB 128-1	6.9	6.5	1,080	450	.36	.65	—	1,085,266	183,406
LCB 129-5B	6.1	2.8	551	116	.29	.19	—	1,093,549	165,988
LCB 133-7	6.5	1.3	1,419	228	.04	.35	—	1,241,643	127,062
SCSF 74AA-117	5.9	3.0	1,452	200	.27	.34	1.30	1,065,502	236,378
SCSF 74AA-121	4.7	4.0	1,450	289	.08	.61	—	1,057,617	227,512
SCSF 74AA-131	9.3	6.4	1,617	433	.02	.47	1.18	1,034,021	209,102
LCB 108-11	6.4	1.1	880	131	.01	.20	—	1,103,577	509,148
LCB 109-4B	8.5	2.8	953	190	.14	.22	1.37	1,095,178	496,678
LCB 135-3	6.7	2.7	1,015	288	.02	.43	—	1,337,926	20,150
LCB 134-6	1.9	7.9	207	63	.26	.33	—	1,309,649	21,952
LCB 135-11	3.9	7.3	364	127	.05	.33	—	1,315,337	34,418
LCB 134-14	3.3	.8	351	54	.06	.16	—	1,287,710	36,331
LCB 3-3	4.6	4.5	—	—	—	—	—	774,588	543,553
KSB 697-3	1.4	2.2	—	—	—	—	—	707,549	706,061
KSB 783-6	1.7	<.01	—	—	—	—	—	889,957	577,686
KZ 73-3-4	4.7	3.4	—	—	—	—	—	1,017,287	276,379
KZ 73-4-1	3.2	2.2	—	—	—	—	—	1,068,556	268,414
KZ 73-4-4	3.1	.6	—	—	—	—	—	1,062,962	262,757
KZ 73-7-2	5.6	1.2	—	—	—	—	—	1,183,961	186,532
KSB 907-1A	6.1	.05	—	—	—	—	—	1,182,699	538,333
KSB 906-7	7.5	2.7	—	—	—	—	—	1,121,873	533,709
KZ 73-9-10	5.3	2.3	—	—	—	—	—	1,323,239	201,483
KZ 73-10-27	6.3	.7	—	—	—	—	—	1,476,422	186,368

(continued).

Age and Sample No.	Organic Carbon (weight %)	Carbonate Carbon (weight %)[2]	Extractable Hydrocarbons: Bitumen (ppm)[3]	Total HC (ppm)[4]	S/A[5]	HC/C %[6]	Kerogen Isolation[7] H/C[8]	Location[9] x	y
LCB 123-4	0.5	2.0	—	—	—	—	—	978,459	200,393
LCB 103-5	3.5	5.1	—	—	—	—	—	797,531	652,121
Middle or Late Miocene									
LCB 112-1	11.2	3.2	1,374	571	.15	.51	—	902,125	432,699
SCSF 74AA-118	8.2	5.5	1,385	303	.08	.37	1.30	1,063,136	234,231
KSB 783-8	2.2	1.4	—	—	—	—	—	884,535	577,876
KZ 73-1-8	3.2	1.8	—	—	—	—	—	992,121	446,920
KZ 73-7-1	—	—	—	—	—	—	—	1,184,163	186,704
KZ 73-6-10	6.5	7.3	—	—	—	—	—	1,173,951	507,567
LCB 107A-5B	2.1	1.7	—	—	—	—	—	935,115	656,916
KZ 73-9-3	5.4	6.1	—	—	—	—	—	1,348,578	206,722
KZ 73-9-16	6.3	5.6	—	—	—	—	—	1,307,693	201,515
LCB 116-3B	5.7	3.2	—	—	—	—	—	789,797	377,707
LCB 116-5B	7.9	6.8	—	—	—	—	—	784,854	375,089
LCB 106-18	1.7	2.4	—	—	—	—	—	853,316	599,503
LCB 105-17	6.5	2.0	—	—	—	—	—	840,702	609,808
Late Miocene									
LCB 123-1	5.5	3.2	992	57	.48	.10	—	989,053	199,635
KSB 709A-8	7.0	5.4	2,123	56	.60	.08	—	704,101	748,697
SCAL 75KD 286[10]	7.4	3.3	883	43	.27	.06	—	1,118,588	227,887
SCAL 75KD 474[11]	12.0	5.3	1,154	54	.30	.04	—	1,017,028	263,057
LCB 106-1	6.1	3.5	721	185	.08	.30	1.18	910,466	596,386
LCB 125-4	6.3	7.2	695	207	.25	.33	—	979,596	337,137
LCB 126-12	6.3	7.7	772	347	.06	.55	—	1,022,557	281,661
LCB 130-13B	9.3	6.4	6,244	0	—	—	—	1,044,373	159,743
LCB 131-3	5.2	8.2	627	164	—	.32	—	1,052,880	137,830
LCB 109-1	2.3	3.5	260	119	.20	.52	1.32	1,100,093	488,784
SCSF 74AA-128	5.9	5.8	724	173	.09	.29	1.30	1,042,031	213,417
KSB 697-1	5.5	3.5	—	—	—	—	—	707,128	705,717
KZ 73-1-1	10.0	5.4	—	—	—	—	—	996,675	459,800
KZ 73-4-3	6.3	7.0	—	—	—	—	—	1,063,320	262,807
KZ 73-7-5	5.6	7.6	—	—	—	—	—	1,173,951	180,596
KZ 73-14-3	7.5	4.6	—	—	—	—	—	1,285,010	335,814
KZ 73-14-14	9.4	4.4	—	—	—	—	—	1,254,088	324,027
LCB 141-5	6.7	5.7	—	—	—	—	—	1,484,555	161,989
KSB 709A-22	5.1	.8	—	—	—	—	—	677,432	713,505
LCB 105-21	6.2	4.6	—	—	—	—	—	845,384	600,414
Late Miocene or Early Pliocene									
KZ 73-2-40	6.6	7.5	—	—	—	—	—	973,801	350,548
KZ 73-3-44	4.6	7.4	—	—	—	—	—	963,601	344,629
KZ 73-9-28	6.5	4.7	—	—	—	—	—	1,271,705	190,728
Holocene									
SCAL 75KD 416[10]	1.0	4.3	229	58	2.00	.58	—	1,122,207	92,169

[1]Acid-insoluble carbon by combustion.
[2]Total carbon minus acid-insoluble carbon.
[3]Total chloroform-extractable organic matter.
[4]Sum of organic extract eluted from silca-gel with (a) heptane and (b) benzene.
[5]Ratio of saturated hydrocarbons (heptane eluate) to aromatic hydrocarbons (benzene eluate).
[6]Ratio of total hydrocarbon to organic carbon, in percent.
[7]Nonextractable organic matter isolated by HCl-HF digestion, heavy-liquid flotation, visual examination, elemental analysis for C, H, N.
[8]Atomic ratio of hydrogen to carbon in kerogen.
[9]Lambert coordinates, California zone 6.
[10]Samples frozen upon collection and kept frozen until analyzed.

Basins could have been subjected to temperatures high enough to generate hydrocarbons.

Parts of the borderland may have had a higher than normal geothermal gradient during the Miocene. An early episode of volcanism apparently occurred between about 24 Ma and 20 Ma along the Santa Rosa-Cortes Ridge. Widespread volcanism began about 16 Ma and dwindled about 13 Ma. These magmatic episodes probably were preceded by a general rise in the geothermal gradient. Heat-flow along the present East Pacific Rise is more than eight times normal and is more than sufficient to cause elevated temperatures in the overlying sedimentary column. Furthermore, above-average heat flow and southward increase in thermal regime now characterize the borderland (Henyey, 1976).

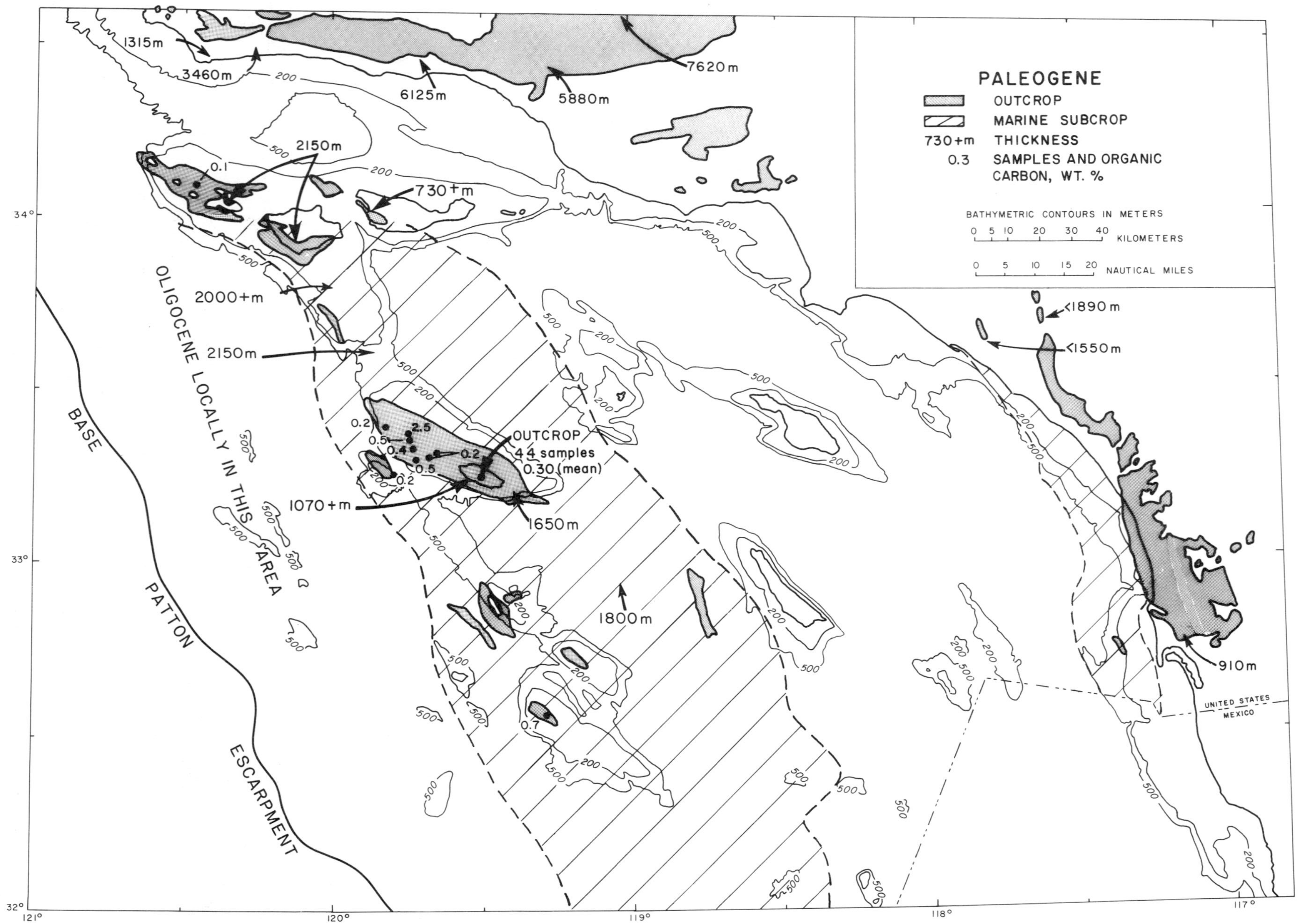

Figure 16. Sites of selected Eocene and Oligocene dart-core and outcrop samples and their organic carbon content (weight %). Paleocene strata in the offshore outcrop areas generally are thin or absent. Paleogene outcrop areas (including subsea) are shaded, and subcrop areas beneath the borderland are indicated by diagonal-line pattern. Modified from Taylor (1976).

Table 4. Summary of organic chemistry on dart samples[1] by stratigraphic age, southern California borderland. (Samples are listed in Tables 2 and 3).

	Organic Carbon			Bitumen			Total HC			HC/C %		
Sample Age	Mean (weight %)	Number of Samples	Standard Deviation	Mean (ppm)	Number of Samples	Standard Deviation	Mean (ppm)	Number of Samples	Standard Deviation	Mean (ppm)	Number of Samples	Standard Deviation
Late Cretaceous	0.2	1	—	—	—	—	—	—	—	—	—	—
Eocene to Oligocene[1]	.35	54	0.35	233	3	255	26	3	34	0.25	3	0.16
Early Miocene	3.2	6	1.5	437	3	251	98	3	46	.31	3	.10
Early or Middle Miocene	4.8	11	2.8	1399	2	518	458	—	—	.65	2	—
Middle Miocene	4.9	26	2.2	829	11	486	182	11	124	.34	11	.17
Middle or Late Miocene	5.6	12	2.9	—	—	—	—	—	—	—	—	—
Late Miocene	6.8	20	2.1	1447	10	1755	122	10	104	.25	9	.2
Late Miocene or Early Pliocene	5.9	3	1.1	—	—	—	—	—	—	—	—	—
Holocene	1.0	1	—	229	1	—	58	1	—	.58	1	—

[1]Eocene includes 44 outcrop samples from San Nicolas Island.

Table 5. Organic geochemistry of cores, Deep Stratigraphic Test Well OCS-CAL 75-70 No. 1. Modified from Paul et al. (1976, Table VII-I).

Core No.	Depth, meters (feet)	Age and Rock Type	Organic Carbon (weight %)	$CHCl_3$ Bitumen (ppm)	Total HC (ppm)	HC in Bitumen (%)	S/A[1]	HC/^{C}org (%)	Pyrolitic Oil Yield (weight %)	Oil Content (weight ppm)	T,°C Maximum Pyrolysis Yield
2	1008.0–1010.9 (3307–3316.5)	Oligocene shale	3.3	1392	289	21	0.28	0.88	0.64	732	475
2	1011.6–1012.7 (3319–3322.5)	Oligocene shale	4.3	1497	324	22	0.26	0.75	0.97	851	490
4	1301.2–1303.0 (4269–4275)	Eocene shale	2.7	559	185	33	0.41	0.69	0.54	188	490
5	1456.3–1458.2 (4778–4784)	Eocene shale	2.5	392	164	42	0.46	0.66	0.82	135	500
7	1639.2–1642.3 (5378–5388)	Eocene shale	0.4	43	17	40	1.4	0.43	0.01	10	504
7	1642.3–1644.4 (5388–5395)	Eocene shale	0.4	73	49	67	3.5	1.2	0.01	30	492
8	1831.2–1833.1 (6008–6014)	Eocene shale	0.7	75	37	49	1.4	0.53	0.01	16	496
11	2531.1–2533.8 (8304–8313)	Cretaceous shale	0.4	104	55	53	1.1	1.4	0.02	16	508
15	3260.1–3262.0 (10,696–10,702)	Cretaceous shale	0.6	143	101	71	0.98	1.7	0.04	20	515
16	3326.0–3328.4 (10,912–10,920)	Cretaceous shale	0.6	126	90	71	0.67	1.5	0.03	13	520

[1]Saturated hydrocarbons/aromatic hydrocarbons.

Temperatures sufficient for generating hydrocarbons in the buried Eocene rocks, therefore, seem plausible.

Reservoir Rocks of the Outer Borderland—Late Cretaceous sandstone beds occur in all of the exploratory wells along the Santa Rosa-Cortes Ridge. Details of the reservoir characteristics of these sandstone sequences are reported only for Deep Stratigraphic Test Well OCS-CAL 75-70 No. 1, where approximately 41 m of potential reservoir-quality sandstone occurs above other nonprospective sandstone beds below an unconformity (Paul et al., 1976).

Paleocene and Eocene sequences contain abundant sandstone beds throughout the borderland areas shown on Figure 6. Deep-marine sandstone deposited by mass-gravity flow (Cole, 1975) forms about half of the Eocene sedimentary sequence on San Nicolas Island, and strata deposited in environments ranging from bathyal to neritic are present on the northern Channel Islands (Doerner, 1969; Weaver and Doerner, 1969a). Throughout the Paleogene section, the quantity of coarse clastic material is sufficient to form potential reservoirs. All of the examined Paleogene sandstone samples were derived from predominantly continental terranes of siliceous plutonic rocks, quartz-rich metamorphic rocks containing muscovite and epidote, and resistant metavolcanic rocks with an equigranular matrix of quartz and feldspar. Ordinarily, stable-constituent sandstones such as

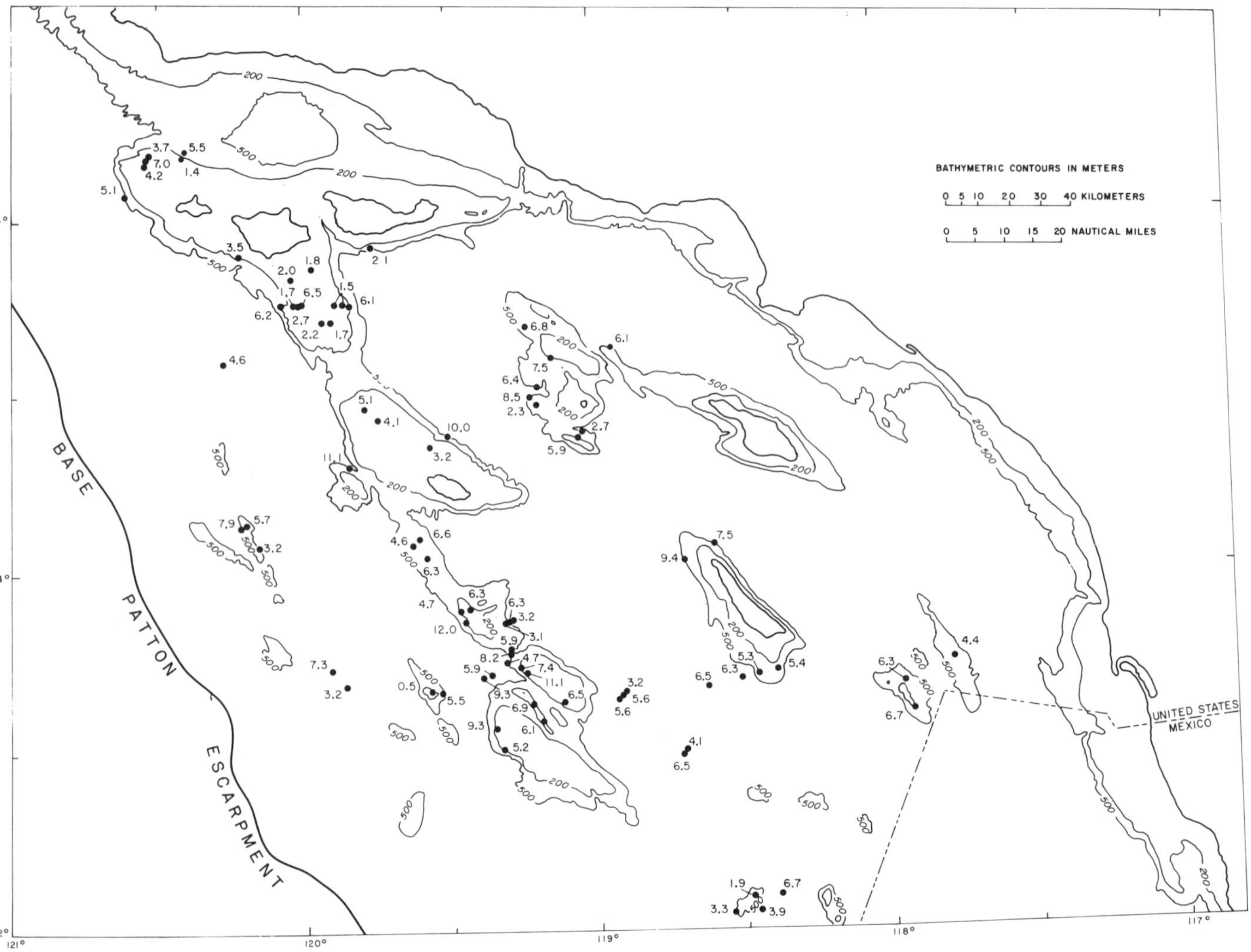

Figure 17. Sites of selected Miocene dart-core samples and their organic carbon content (weight %). Tables 3 and 4 show additional analytical data. Modified from Taylor (1976).

these form excellent reservoirs if predepositional sorting had removed silt- and clay-size components, and if deep burial, cementation, or porosity-destroying diagenesis did not occur. Samples from both onshore and sea-floor outcrops generally have moderate porosity (Taylor, 1976). Fifteen samples of sandstone from the Eocene sequence on San Nicolas Island have porosities that range from 10.9 to 27.8% (Vedder and Norris, 1963, Table 2).

About 215 m of sandstone occurs in the Eocene and Paleocene section in Deep Stratigraphic Test Well OCS-CAL 75-70 No. 1, but the beds are thin and separated by relatively thick zones of mudstone. Although the sandstone is well sorted and has moderate average porosity (17.3 % for Eocene sandstone), the mean permeability is low (< 10 millidarcies for Late Cretaceous, Paleocene, and Eocene sandstone) (Paul et al., 1976). Approximately 260 m of friable, well-sorted sandstone beds averaging 30% porosity were penetrated in the Oligocene interval, which is about 410 m thick (Paul et al., 1976). Similar sandstone sequences probably are present at Dall and Tanner Banks.

Sandstone beds of late Oligocene and early Miocene age occur on San Miguel and Santa Rosa Islands; they are interbedded with coarse, near-source breccia lenses on Santa Cruz Island. On Santa Rosa-Cortes and Patton Ridges, undated sandstone cores composed of well-sorted continental detritus commonly include broken fragments of neritic megafauna and may be of late Oligocene or early Miocene age. Some of these samples contain clasts of intermediate to basic volcanic rocks, which are more characteristic of Miocene sandstone than of Paleogene sandstone. Erosion and redeposition of Eocene sandstone and conglomerate during an early Neogene transgression could have resulted in widespread deposits of shallow-water, high-quality reservoir sands, but until additional subsurface data are available, such inferences will remain unsubstantiated.

Because middle and late Miocene fractured siliceous shales form highly productive reservoirs in the Santa Barbara Channel, it seems reasonable that similar beds should constitute potential reservoirs at other places on the borderland. Chert lenses and siliceous zones, some of which are fractured, occur in the diatomite sections on Santa Catalina and San Clemente Islands and the northern Channel Islands, suggesting that fracture porosity in Miocene shales extends well beyond the Santa Barbara Channel.

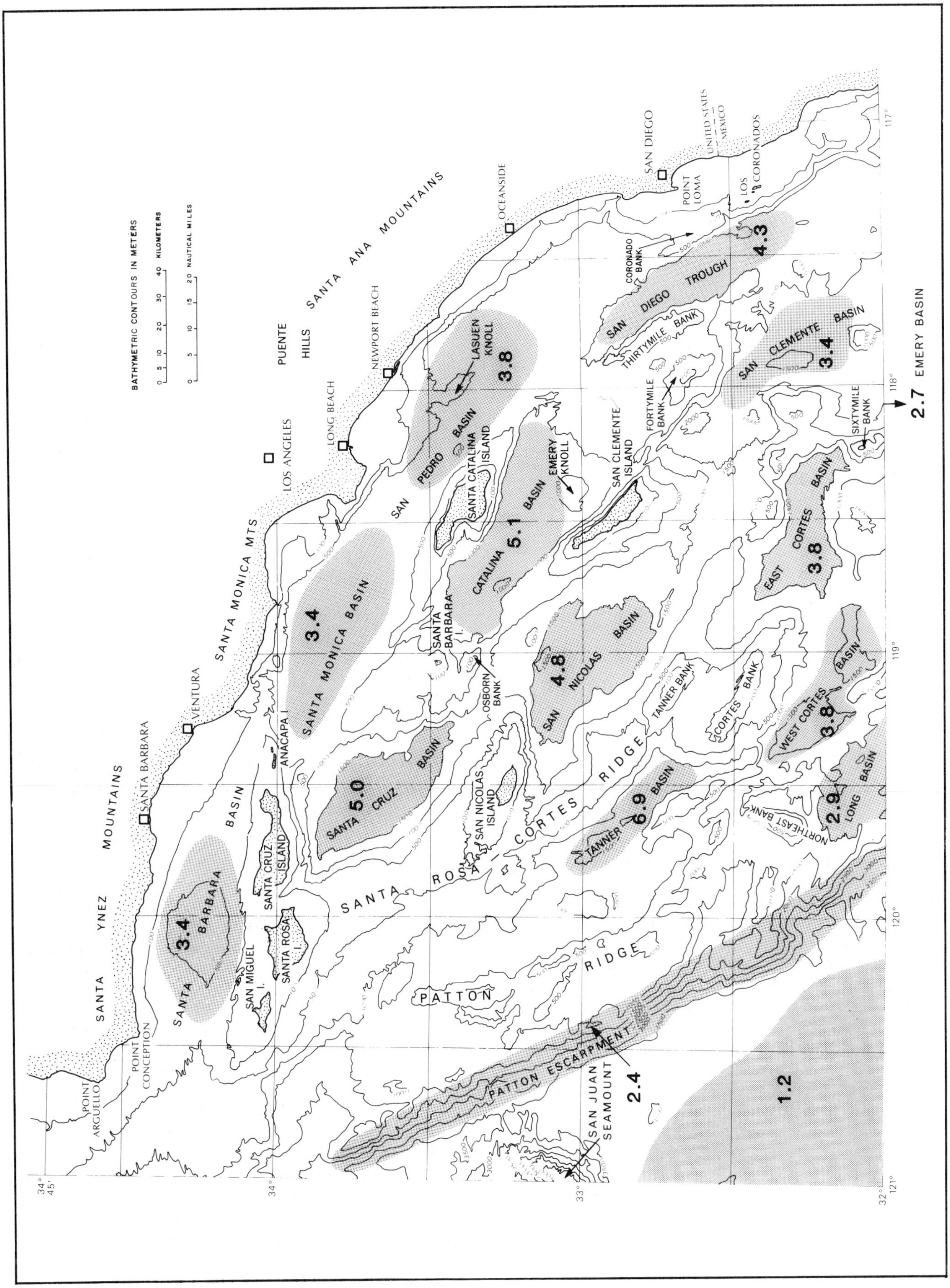

Figure 18. Organic carbon content (weight %) in Holocene sediments from basins, outer slope, and ocean floor. Converted from Emery's (1960) data on organic content by dividing by 1.7. Modified from Taylor (1976).

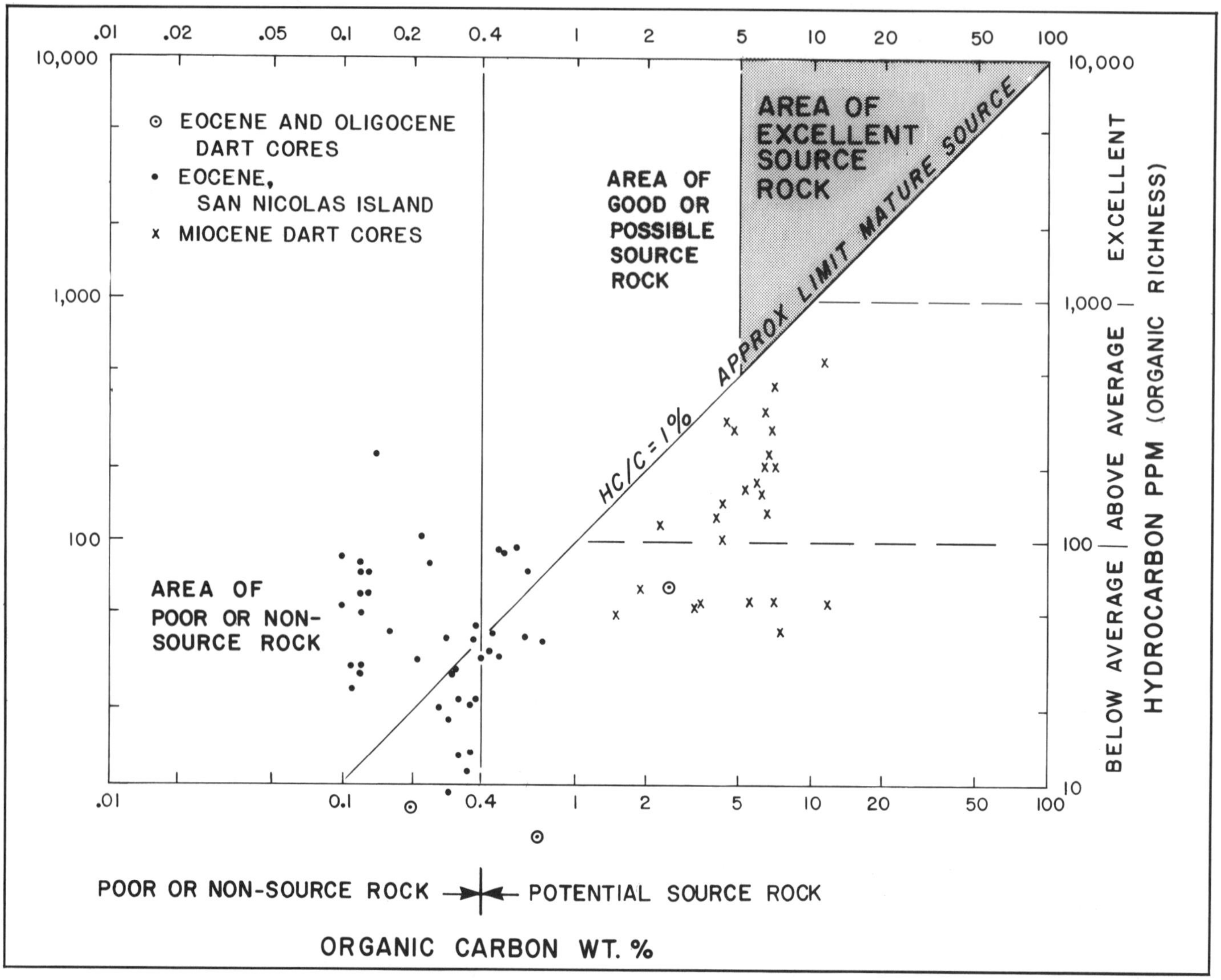

Figure 19. Source-rock analysis diagram showing organic carbon content (weight %) plotted against hydrocarbon (ppm) for Eocene, Oligocene, and Miocene samples. Sample sites are shown on Figures 15 and 16. Organic carbon content (weight %) of additional samples is listed in Tables 3 and 4. From Taylor (1976).

Over 80% of the total production from onshore coastal basins has come from thick, extensive, deep-water late Miocene and Pliocene sandstone zones in the Los Angeles and Ventura basins. These two subsided basins derived the bulk of their coarse clastic material from nearby uplifted areas. Borderland topography, however, was well defined by the early Pliocene, and it is unlikely that similar deep-water sands were transported very far seaward beyond obstructing ridges and nearshore basins (Santa Barbara Channel, Santa Monica and San Pedro Basins) in quantities large enough to create potential reservoirs. Although late Miocene and Pliocene sandstone sequences have not been reported from the outer part of the borderland, they may be present locally as turbidites derived from uplifted flanking ridges.

Traps in the Outer Borderland—Large and small anticlines are present throughout the outer borderland. However, several of the major anticlinal structures along the outer banks and ridges have been truncated by erosion so that some potential traps have been destroyed. Although many of the large productive structures on the mainland and near shore formed as a result of Pleistocene deformation, similar young structures apparently are uncommon in the outer basins where most of the folding seems to have taken place during the late Miocene and/or Pliocene.

Prospective homoclinal traps formed by basinward-thickening stratigraphic wedges occur in the Miocene and younger rocks on the flanks of the major ridges. For example, on the northeast flank of Tanner Bank, the character of seismic reflectors changes within the Miocene section, suggesting updip wedge-outs of nonbedded or coarse-grained strata within a well-bedded sequence (Figure 20). Similar basinward-thickening Miocene sequences are present along the margins of Patton, Tanner, West Cortes, and Velero Basins. Although potential traps within the Paleogene sequences are seismically obscure, the structural complexity undoubtedly is greater than in the Neogene rocks. Inferred Paleogene pinch-outs beneath the eastern flanks of Santa Cruz, San Nicolas, and East Cortes Basins are believed to have resulted from both eastward transgressive onlap and truncation. Similar Paleogene pinch-outs probably are present in the West Cortes Basin and the Velero Basin (Vedder, 1981).

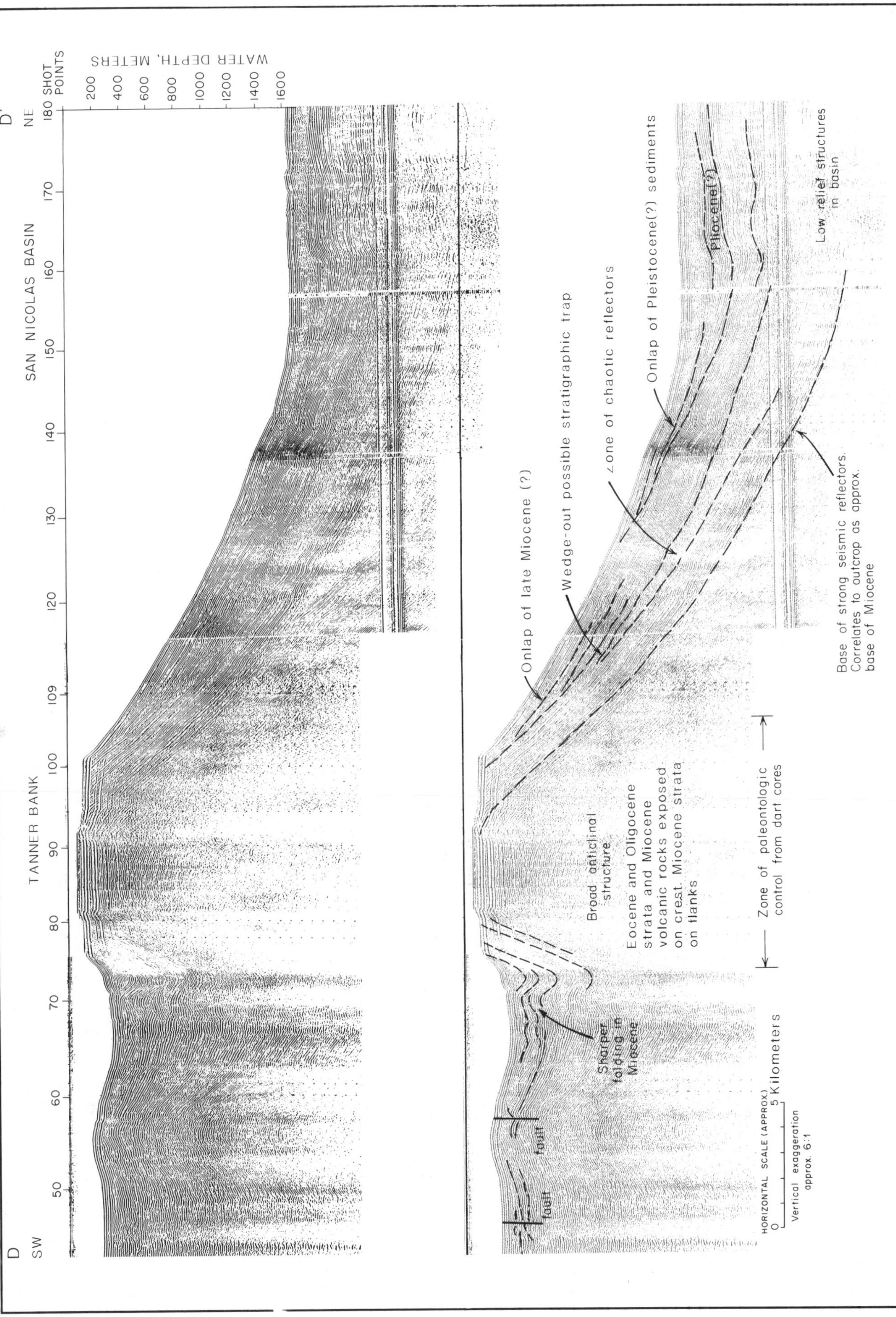

Figure 20. Single-channel acoustic-reflection profile across Tanner Bank and San Nicolas Basin, showing original record and interpreted structural and stratigraphic relationships. Modified from Taylor (1976). Location of profile is shown on Figure 2.

SUMMARY

Jurassic and younger rocks on the borderland and adjoining mainland provide a record of intermittent magmatism and nearly continuous tectonism. Juxtaposed Mesozoic basement rocks occur in four northwest-trending terranes, each of which is overlain by a different stratigraphic sequence. From west to east, these terranes are composed of (1) a Cretaceous(?) subduction complex, (2) a Jurassic arc-volcanogenic system, (3) a pre-Late Cretaceous assemblage of high-pressure metavolcanic and metasedimentary rocks, and (4) a Triassic(?) and Jurassic graywacke-argillite sequence that is overlain by Late Jurassic and Early Cretaceous arc-volcanogenic rocks and is intruded by Cretaceous granitic plutons. Because the fourth terrane lies east of the Newport Inglewood-Coronado Bank fault system, most of it is on shore. A fifth terrane, which contains rocks similar to those in the fourth, underlies the northernmost edge of the borderland and extends onto the mainland. Each terrane was emplaced after probable long-range northward translation, which began in the Cretaceous and continued into the late Tertiary.

Relatively thick sequences of Late Cretaceous and Paleogene fore-arc-basin turbidites accumulated on the Jurassic arc-volcanogenic basement of the central borderland, and correlative nonmarine and nearshore deposits were laid down along the western flank of the Cretaceous plutons to form the proximal rim of the same fore-arc basin. These sedimentary rocks either were never deposited on the third terrane or were completely eroded before the middle Miocene. Sea-level fluctuations accompanied tectonism and had a marked influence on Paleogene depositional systems.

Subduction of the Farallon plate brought the Pacific plate into contact with the North American plate near the present site of the Patton Escarpment about 29 Ma, and a transform margin began to develop. The resulting broad zone of right shear along the continental margin persisted through most of the late Tertiary. This long-term episode of wrench faulting created transtensional and transpressional structures that disrupted the Cretaceous and Paleogene fore-arc basin and formed new depositional basins. Paleomagnetic data indicate that post-middle Miocene clockwise rotation affected the northernmost edge of the borderland in the vicinity of the northern Channel Islands, the Santa Monica Mountains, and the Santa Ynez Range. Crustal shortening and uplift during the late Pliocene and Pleistocene were responsible for the compressive structures and high topographic relief that typify most of southern California.

The northern part of the borderland includes or adjoins the richly petroliferous Ventura and Los Angeles basins and the Santa Barbara Channel. Because some of the basins farther offshore have had tectonic and depositional histories similar to those of the oil-producing basins, they may have a better than average petroleum potential, particularly in places where the Neogene sedimentary sequences are thick. Positive factors with respect to potential are: (1) Miocene sedimentary sequences include excellent source rocks; (2) Late Cretaceous and Paleogene sandstone zones and fractured Miocene shale may provide suitable reservoirs at places, and (3) favorable structural and stratigraphic traps occur nearly everywhere along the flanks of the large borderland basins. Partly offsetting negative factors are: (1) the Neogene section is relatively thin in the outer basins and is truncated or absent along the crests of some of the ridges, (2) Cretaceous and Paleogene strata are absent over large areas, and structures that deform these sequences are obscure, (3) organic content of Paleogene beds generally is low, and (4) most of the Neogene strata on the outer borderland have had insufficient burial and thermal histories for hydrocarbon generation.

ACKNOWLEDGMENTS

Without the accomplishments of H. C. Wagner, A. Junger, J. C. Taylor, and L. A. Beyer during the early phases of the program (1973–1976), this summary of borderland geology would not have been possible. To these individuals, the writer owes an immeasurable debt of gratitude. Equally significant contributions were made during later phases (1975–1980) by project co-workers D. G. Howell, H. McLean, and J. K. Crouch. Special thanks are due to R. E. Arnal, J. A. Barron, and D. Bukry, who provided paleontologic data during the course of the work, and to F. Lee-Wong for petrographic identifications of rocks. Comments by D. G. Howell and H. McLean on a preliminary version of the report proved helpful in an attempt to improve the content and quality of the text. Illustrations were drafted by B. V. Fulop, and the manuscript was typed by D. L. Moore and J. S. Mascardo.

REFERENCES CITED

Abbott, P. L., and T. E. Smith, 1978, Trace-element comparison of clasts in Eocene conglomerates, southwestern California and northwestern Mexico: Journal of Geology, v. 86, p. 753–762.

Abbott, P. L., R. P. Kies, W. R. Bachman, and C. J. Natenstedt, 1983, A tectonic slice of Eocene strata, northern part of California Continental Borderland, *in* D. K. Larue and R. J. Steel, eds., Cenozoic marine sedimentation, Pacific margin, U.S.A.: SEPM, Pacific Section, p. 151–168.

Arnal, R. E., 1976, Miocene paleobathymetric changes of the Santa Rosa-Cortes Ridge area, California Continental Borderland, *in* D. G. Howell, ed., Aspects of the geologic history of the California Continental Borderland: AAPG, Pacific Section, Miscellaneous Publication 24, p. 60–79.

Arnal, R. E., and J. G. Vedder, 1976, Late Miocene paleobathymetry of the California Contenental Borderland north of 32°, *in* A. E. Fritsche, H. TerBest, Jr., and W. W. Wornardt, eds., The Neogene symposium: SEPM, Pacific Section, p. 1–8.

Atwater, T. M., 1970, Implications of plate tectonics for the Cenozoic tectonic evolution of western North America: GSA Bulletin, v. 81, p. 3513–3536.

Avila, F. A., and D. W. Weaver, 1969, Mid-Tertiary stratigraphy, Santa Rosa Island [California], *in* D. W. Weaver, assisted by D. P. Doerner and B. Nolf, eds., Geology of the northern Channel Islands: AAPG and SEPM, Pacific Sections, Special Publication, p. 48–67.

Bailey, E. H., 1941, Mineralogy, petrology, and geology of Santa Catalina Island, California: Stanford, Stanford University unpublished Ph.D. dissertation, 193 p.

Bailey, E. H., M. C. Blake, Jr., and D. L. Jones, 1970, On-land Mesozoic oceanic crust in the California Coast Ranges: U.S. Geological Survey Professional Paper 700–C, p. C70–C81.

Balch, D. C., S. H. Bartling, and P. L. Abbott, 1984, Volcaniclastic strata of the Upper Jurassic Santiago Peak Volcanics, San Diego, California, *in* J. K. Crouch and S. B. Bachman, eds., Tectonics and sedimentation along the California margin: SEPM, Pacific Section, p. 157–170.

Barbat, W. F., 1958, The Los Angeles basin area, California, *in* J. W.

Higgins, ed., A guide to the geology and oil fields of the Los Angeles and Ventura regions: AAPG, Pacific Section, Field Trip Guidebook for Annual Meeting, p. 37–49.

Bartling, W. A., and P. L. Abbott, 1983, Upper Cretaceous sedimentation and tectonics with reference to the Eocene, San Miguel Island, and San Diego area, California, *in* D. K. Larue and R. J. Steel, eds., Cenozoic marine sedimentation, Pacific margin, U.S.A.: SEPM, Pacific Section, p. 133–150.

Berggren, W. A., D. V. Kent, J. J. Flynn, and J. A. Van Couvering, 1985, Cenozoic geochronology: GSA Bulletin, v. 96, p. 1407–1418.

Beyer, L. A., 1980, Offshore southern California, *in* H. W. Oliver, ed., Interpretation of the gravity map of California and its continental margin: California Division of Mines and Geology Bulletin 205, p. 8–15.

Beyer, L. A., N. B. Sasnett, and K. A. Pisciotto, 1975, Interpretation of gravity and magnetic surveys of the California Continental Borderland (abs.): AAPG, SEG, SEPM, Pacific Sections, Abstracts with Programs, p. 8–9.

Blake M. C., Jr., D. G. Howell, and D. L. Jones, 1982, Preliminary tectonstratigraphic terrane map of California: U.S. Geological Survey Open-File Report 82-593, 10 p., 3 pls.

Blake, M. C., Jr., R. H. Campbell, T. W. Dibblee, Jr., D. G. Howell, T. H. Nilsen, W. R. Normark, J. G. Vedder, and E. A. Silver, 1978, Neogene basin formation in relation to plate-tectonic evolution of San Andreas fault system, California: AAPG Bulletin, v. 62, p. 344–372.

Boles, J. R., and C. A. Landis, 1984, Jurassic sedimentary melange and associated facies, Baja California, Mexico: GSA Bulletin, v. 95, p. 513–521.

Bottjer, D. J., and M. H. Link, 1984, A synthesis of Late Cretaceous southern California and northern Baja California paleogeography, *in* J .K. Crouch and S. B. Bachman, eds., Tectonics and sedimentation along the California margin: SEPM, Pacific Section, v. 38, p. 171–188.

Bramlette, M. N., 1946, The Monterey Formation of California and the origin of its siliceous rocks: U.S. Geological Survey Professional Paper 212, 57 p.

Campbell, R. H., S. C. Wolf, R. E. Hunter, H. C. Wagner, A. Junger, and J. G. Vedder, 1975, Geologic map and sections, Santa Barbara Channel region, California: U.S. Geological Survey Open-File Map 75-123, scale 1:250,000.

Clarke, S. H., H. G. Greene, M. E. Field, and W. H. K. Lee, 1983, Reconnaissance geology and geologic hazards of selected areas of the southern California continental borderland: U.S. Geological Survey Open-File Report 83-62, 78 p.

Cole, M. R., 1975, Eocene sedimentation and paleocurrents, San Nicolas Island, California: Guidebook, GSA Cordilleran Section Meeting, 32 p.

Crouch, J. K., 1979, Neogene tectonic evolution of the California Continental Borderland and western Transverse Ranges: GSA Bulletin, v. 90, no. 4, p. 1338–1345.

Crouch, J. K., 1981, Northwest margin of California Continental Borderland; Marine geology and tectonic evolution: AAPG Bulletin, v. 65, p. 191–218.

Crouch, J. K., S. B. Bachman, and J. T. Shay, 1984, Post-Miocene compressional tectonics along the central California margin, *in* J. K. Crouch and S. B. Bachman, eds., Tectonics and sedimentation along the California margin: SEPM, Pacific Section, p. 37–54.

Crouch, J. K., M. L. Holmes, T. H. McCulloch, A. T. Long, and R. H. Brune, 1978, Multichannel seismic reflection and sonobuoy data in the outer southern California borderland: U.S. Geological Survey Open-File Report 78-706, 25 p.

Crowe, B. M., H. McLean, D. G. Howell, and R. E. Higgins, 1976, Petrography and major element chemistry of the Santa Cruz Island volcanics, *in* D. G. Howell, ed., Aspects of the geologic history of the California Continental Borderland: AAPG, Pacific Section, Miscellaneous Publication 24, p. 196–215.

Crowell, J. C., 1976, Implications of crustal stretching and shortening of coastal Ventura Basin, California, *in* D. G. Howell, ed., Aspects of the geologic history of the California Continental Borderland: AAPG, Pacific Section, Miscellaneous Publication 24, p. 365–382.

Curran, J. F., K. B. Hall, and R. F. Herron, 1971, Geology, oil fields, and future petroleum potential of Santa Barbara Channel area, California, *in* I. H. Cram, ed., v. 1, region 2, Future petroleum provinces of the United States—Their geology and potential: AAPG Memoir 15, p. 192–211.

Curray, J. R., 1965, Late Quaternary history, continental shelves of the United States, *in* H. E. Wright, Jr., and D. C. Frey, eds., The Quaternary of the United States: Princeton, New Jersey, Princeton University Press, p. 723–735.

Dames & Moore, 1974, Resumption of drilling operations in the South Ellwood offshore oil field from platform Holly; final environmental impact report: California State Lands Commission, 3 vols., 1561 p.

Dibblee, T. W, Jr., 1950, Geology of southwestern Santa Barbara County, California; Point Arguello, Lompoc, Point Conception, Los Olivos and Gaviota quadrangles: California Division of Mines Bulletin 150, 95 p.

Dibblee, T. W, Jr., 1966, Geology of southwestern Santa Barbara County, California: California Division of Mines and Geology Bulletin 186, 99 p.

Doerner, D. P., 1969, Lower Tertiary biostratigraphy of southwestern Santa Cruz Island [California], *in* D. W. Weaver, assisted by D. P. Doerner and B. Nolf, eds., Geology of the northern Channel Islands: AAPG and SEPM, Pacific Sections, Special Publication, p. 17–29.

Ellsworth, W. L., R. H. Campbell, D. P. Hill, R. A. Page, R. W. Alewine, III, T. C. Hanks, T. H. Heaton, J. A. Hileman, H. Kanamori, B. Minster, and J. H. Whitcomb, 1973, Point Mugu, California, earthquake of 21 February 1973 and its aftershocks: Science, v. 182, no. 4, p. 1127–1129.

Emery, K. O., 1960, The sea off southern California, a modern habitat of petroleum: New York, John Wiley & Sons, 366 p.

Emery, K. O., and F. P. Shepard, 1945, Lithology of the sea floor off southern California: GSA Bulletin, v. 56, p. 431–478.

Erickson, J. W., 1975, Sedimentology of the South Point Formation (Eocene), Santa Rosa Island, California, *in* D. W. Weaver, G. R. Hornaday, and A. Tipton, eds., Paleogene symposium: SEPM, Pacific Section, p. 169–190.

Field, M. E., and B. D. Edwards, 1980, Slopes of the southern California continental borderland: A regime of mass transport, *in* M. E. Field, A. H. Bouma, I. P. Colburn, R. G. Douglas, and J. C. Ingle, eds., Quaternary depositional environments of the Pacific Coast: SEPM, Pacific Section, Pacific Coast Paleogeography Symposium 4, p. 169–184.

Fife, D. L., J. A. Minch, and P. J. Crampton, 1967, Late Jurassic age of the Santiago Peak Volcanics, California: GSA Bulletin 78, p. 299–304.

Fischer, P. J., 1976, Late Neogene–Quaternary tectonics and depositional environments of the Santa Barbara Basin, California, *in* A. E. Fritsche, H. TerBest, Jr., and W. W. Wornardt, eds., The Neogene symposium: SEPM, Pacific Section, p. 33–52.

Forman, J. A., 1970, Age of the Catalina Island pluton, California, *in* O. L. Bandy, ed., Radiometric dating and paleontologic zonation: GSA Special Paper 124, p. 37–45.

Gibson, J. D., D. G. Howell, G. S. Fuis, and J. G. Vedder, 1985, North American continent-ocean transect C-3: Pacific Abyssal Plain to the Rio Grande rift, *in* D. G. Howell, J. D. Gibson, G. S. Fuis, J. H. Knapp, G. B. Haxel, B. R. Keller, L. T. Silver, and J. G. Vedder: GSA Centennial Continent/Ocean Transect #5, 23 p.

Gorsline, D. S., 1980, Depositional patterns of hemipelagic Holocene sediments in borderland basins of an active margin, *in* M. E. Field, A. H. Bouma, I. P. Colburn, R. G. Douglas, and J. C. Ingle, eds., Quaternary depositional environments of the Pacific Coast: SEPM, Pacific Section, Pacific Coast Paleogeography Symposium 4, p. 185–200.

Graham, S. A., and S. B. Bachman, 1983, Structural controls on submarine-fan geometry and internal architecture: Upper La Jolla fan system, offshore southern California: AAPG Bulletin, v. 67, p. 83–96.

Greene, H. G., 1976, Late Cenozoic geology of the Ventura Basin, California, *in* D. G. Howell, ed., Aspects of the geologic history of the California Continental Borderland: AAPG, Pacific Section, Miscellaneous Publication 24, p. 499–529.

Greene, H. G., S. C. Wolf, and K. G. Blom, 1978, The marine geology of the eastern Santa Barbara Channel with particular emphasis on the ground water basins offshore from the Oxnard Plain, southern California: U.S. Geological Survey Open-File Report 78-305, 104 p.

Greene, H. G., S. H. Clarke, Jr., M. E. Field, F. I. Linker, and H. C. Wagner, 1975, Preliminary report on the environmental geology of selected areas of southern California borderland: U.S. Geological Survey Open-File Report 75-596, 70 p., 16 pls.

Hall, C. A., Jr. 1981, San Luis Obispo transform fault and middle Miocene rotation of the western Transverse Ranges, California: Journal of Geophysical Research, v. 86, no. B2, p. 1015–1031.

Hawkins, J. W., 1970, Petrology and possible tectonic significance of late Cenozoic volcanic rocks, southern California and Baja California: GSA Bulletin, v. 81, no. 11, p. 3323–3338.

Hawkins, J. W., and A. F. Divis, 1975, Petrology and geochemistry of mid-Miocene volcanism on San Clemente and Santa Catalina Islands and adjacent areas of the southern California borderland (abs.): GSA Abstracts with Programs, v. 7, no. 3, p. 323–324.

Hawkins, J. W., E. C. Allison, and D. MacDougall, 1971, Volcanic petrology and geologic history of Northeast Bank, southern California borderland: GSA Bulletin, v. 82, no. 1, p. 219–228.

Henyey, T. L., 1976, Heat flow and tectonic patterns on the southern California borderland, *in* D. G. Howell, ed., Aspects of the geologic history of the California Continental Borderland: AAPG, Pacific Section, Miscellaneous Publication 24, p. 427–448.

Hill, D. J., 1976, Geology of the Jurassic basement rocks, Santa Cruz Island, California, and correlation with other Mesozoic basement terranes in California, *in* D. G. Howell, ed., Aspects of the geologic history of the southern California Continental Borderland: AAPG, Pacific Section, Miscellaneous Publication 24, p. 16–46.

Hoots, H. W., 1931, Geology of the eastern part of the Santa Monica Mountains, Los Angeles County, California: U.S. Geological Survey Professional Paper 165–C, p. 83–134.

Hornafius, J. S., 1985, Neogene tectonic rotation of the Santa Ynez Range, western Transverse Ranges, California, suggested by paleomagnetic investigations of the Monterey Formation: Journal of Geophysical Research, v. 90, no. B14, p. 12,503–12,522.

Howell, D. G., 1975, Middle Eocene paleogeography of southern California, *in* D. W. Weaver, G. R. Hornaday, and A. Tipton, eds., Conference on future energy horizons of the Pacific Coast: AAPG, Pacific Section, Paleogene Symposium and Selected Technical Papers, p. 272–293.

Howell, D. G., and M. H. Link, 1979, Eocene conglomerate sedimentology and basin analysis, San Diego and the southern California borderland: Journal of Sedimentary Petrology, v. 49, no. 2, p. 517–540.

Howell, D. G., and H. McLean, 1976, Middle Miocene paleogeography, Santa Cruz and Santa Rosa Islands, *in* D. G. Howell, ed., Aspects of the geologic history of the California Continental Borderland: AAPG, Pacific Section, Miscellaneous Publication 24, p. 266–293.

Howell, D. G., and J. G. Vedder, 1981, Structural implications of stratigraphic discontinuities across the southern California borderland, *in* W. G. Ernst, ed., The geotectonic development of California, Rubey Volume 1: Englewood Cliffs, New Jersey, Prentice-Hall, Inc., p. 535–558.

Howell, D. G., and J. G. Vedder, 1984, Ferrelo fan, California, depositional system influenced by eustatic sea level changes: Geo-Marine Letters, v. 3, p. 187–192.

Howell, D. G., D. E. Champion, and J. G. Vedder, in press, Terrane accretion, crustal kinematics, and basin evolution, southern California, *in* R. V. Ingersoll and W. G. Ernst, eds., Cenozoic basin development of coastal California, Rubey, Volume 5: Englewood Cliffs, New Jersey, Prentice-Hall, Inc.

Howell, D. G., H. McLean, and J. G. Vedder, 1976, Cenozoic tectonism on Santa Cruz Island, *in* D. G. Howell, ed., Aspects of the geology history of the California Continental Borderland: AAPG, Pacific Section, Miscellaneous Publication 24, p. 392–417.

Howell, D. G., G. W. Moore, and T. J. Wiley, 1987, Tectonics and basin evolution of western North America—An overview, this volume.

Howell, D. G., C. J. Stuart, J. P. Platt, and D. J. Hill, 1974, Possible strike-slip faulting in the southern California borderland: Geology, v. 2, no. 2, p. 93–98.

Howell, D. G., J. K. Crouch, H. G. Greene, D. S. McCulloch, and J. G. Vedder, 1980, Basin development along the late Mesozoic and Cainozoic California margin: A plate tectonic margin of subduction, oblique subduction, and transform tectonics, *in* P. F. Ballance and H. G. Reading, eds., Sedimentation in oblique-slip mobile zones: International Association of Sedimentologists Special Publication 4, p. 43–62.

Hurst, R. W., W. R. Wood, and M. Hume, 1982, The petrologic and tectonic evolution of volcanic rocks in the southern California borderland, *in* E. G. Frost and D. M. Martin, eds., Mesozoic–Cenozoic tectonic evolution of the Colorado River region, California, Arizona, and Nevada: San Diego, Cordilleran Publishers, p. 287–297.

Imlay, R. W., 1963, Jurassic fossils from southern California: Journal of Paleontology, v. 37, no. 1, p. 97–107.

Ingle, J. C., Jr., 1973, Biostratigraphy and paleoecology of early Miocene through early Pleistocene benthonic and planktonic foraminifers, San Joaquin Hills-Newport Bay-Dana Point area, Orange County, California, *in* Miocene sedimentary environments and biofacies, southeastern Los Angeles Basin: Guidebook, SEPM Trip 1, 1973 Annual Meeting AAPG, SEPM, SEG, Anaheim, California, p. 18–38.

Isaacs, C. M., 1983, Compositional variation and sequence in the Miocene Monterey Formation, Santa Barbara coastal area, California, *in* D. K. Larue and R. J. Steel, eds., Cenozoic marine sedimentation, Pacific margin, U.S.A.: SEPM, Pacific Section, p. 117–132.

Jahns, R. H., 1954, Investigations and problems of southern California geology, *in* R. H. Jahns, ed., Geology of southern California: California Division of Mines Bulletin 170, chapter 1, p. 5–29.

Jennings, C. W., 1959, Geologic map of California, Olaf P. Jenkins edition, Santa Maria sheet: California Division of Mines, scale 1:250,000.

Jennings, C. W., 1962, Geologic map of California, Olaf P. Jenkins edition, Long Beach sheet: California Division of Mines and Geology, scale 1:250,000.

Johnson, D. L., 1967, Caliche on the Channel Islands: California Division of Mines and Geology, Mineral Information Service, v. 20, no. 12, p. 151–158.

Jones, D. L., M. C. Blake, Jr., and C. Rangin, 1976, The four Jurassic belts of northern California and their significance to the geology of the southern California borderland, *in* D. G. Howell, ed., Aspects of the geologic history of the California Continental Borderland: AAPG, Pacific Section, Miscellaneous Publication 24, p. 343–362.

Junger, A., 1974, Source of the San Onofre Breccia (abs.): GSA Abstracts with Programs, v. 6, no. 3, p. 199–200.

Junger, A., 1976, Tectonics of the southern California borderland, *in* D. G. Howell, ed., Aspects of the geologic history of the California Continental Borderland: AAPG, Pacific Section, Miscellaneous Publication 24, p. 486–529.

Junger, A., 1979, Maps and seismic profiles showing geologic structure of the northern Channel Islands platform, California Continental Borderland: U.S. Geological Survey Miscellaneous Field

Studies Map MF–991, 16 p., 5 sheets, scale 1:250,000.

Junger, A., and H. C. Wagner, 1977, Geology of the Santa Monica and San Pedro Basins, California Continental Borderland: U.S. Geological Survey Miscellaneous Field Studies Map MF–820, 10 p., 5 sheets, scale 1:250,000.

Kamerling, M. J., and B. P. Luyendyk, 1985, Paleomagnetism and Neogene tectonics of the northern Channel Islands, California: Journal of Geophysical Research, v. 90, no. B14, p. 12,485–12,502.

Krause, D. C., 1965, Tectonics, bathymetry, and geomagnetism of the southern continental borderland west of Baja California, Mexico: GSA Bulletin, v. 76, no. 6, p. 617–650.

Lajoie, K. R., and J. F. Wehmiller, 1978, Quaternary uplift rates, southern California borderland (abs.): Santa Barbara, California, Santa Barbara Museum of Natural History, Multidisciplinary Symposium on the California Islands, p. 3.

Lajoie, K. R., J. P. Kern, J. F. Wehmiller, G. L. Kennedy, S. A. Mathieson, A. M. Sarna-Wojcicki, R. F. Yerkes, and P. F. McCrory, 1979, Quaternary marine shorelines and crustal deformation, San Diego to Santa Barbara, California, *in* P. L. Abbott, ed., Geological excursions in the southern California area: San Diego State University, Department of Geological Sciences, p. 3–16.

Lamb, T. N., 1978, Geology, *in* H. T. Kuper, ed., Natural history of the Coronado Islands, Baja California, Mexico: San Diego Association of Geologists, Field Trip Guidebook, p. 12–44.

Legg, M., and M. P. Kennedy, 1979, Faulting offshore San Diego and northern Baja California, *in* P. L. Abbott and W. J. Elliott, eds., Earthquakes and other perils, San Diego region: GSA Annual Meeting Guidebook, p. 29–46.

Luyendyk, B. P., M. J. Kamerling, R. Terres, and J. S. Hornafius, 1985, Simple shear of southern California during Neogene time suggested by paleomagnetic declinations: Journal of Geophysical Research, v. 90, no. B14, p. 12,454–12,466.

Mattinson, J. M., and D. J. Hill, 1976, Age of plutonic basement rocks, Santa Cruz Island, California, *in* D. G. Howell, ed., Aspects of the geologic history of the California Continental Borderland: AAPG, Pacific Section, Miscellaneous Publication 24, p. 53–59.

Mayuga, M. N., 1970, Geology and development of California's giant—Wilmington oil field, *in* Geology of giant petroleum fields—Symposium: AAPG Memoir 14, p. 158–184.

McCulloch, D. S., J. G. Vedder, H. C. Wagner, and R. H. Brune, 1979, Geologic setting, *in* H. E. Cook, ed., Geologic studies of the Point Conception Deep Stratigraphic Test Well OCS-CAL 78-164 No. 1, Outer Continental Shelf, southern California: U.S. Geological Survey Open-File Report 79-1218, p. 10–25.

McLean, H., 1979, Lithologic descriptions, *in* H. E. Cook, ed., Geologic studies of the Point Conception Deep Stratigraphic Test Well OCS-CAL 78-164, No. 1, Outer Continental Shelf, southern California: U.S. Geological Survey Open-File Report 79-1218, p. 49–59.

McLean, H., and D. G. Howell, 1984, Miocene Blanca fan, northern Channel Islands, California: Small fans reflecting tectonism and volcanism: Geo-Marine Letters, v. 3, nos. 2–4, p. 161–166.

McLean, H., D. G. Howell, and J. G. Vedder, 1976, Miocene strata on Santa Cruz and Santa Rosa Islands—A reflection of tectonic events in the southern California borderland, *in* D. G. Howell, ed., Aspects of the geologic history of the California Continental Borderland: AAPG, Pacific Section, Miscellaneous Publication 24, p. 241–253.

Merriam, R., 1972, Reconnaissance geologic map of the Sonoyta quadrangle, northwest Sonora, Mexico: GSA Bulletin, v. 83, no. 11, p. 3533–3536.

Merschat, W. R., 1971, Lower Tertiary paleocurrent trends, Santa Cruz Island, California: Athens, Ohio University unpublished Master's thesis, 77 p.

Milliman, J. D., and K. O. Emery, 1968, Sea levels during the past 35,000 years: Science, v. 162, no. 3858, p. 1121–1123.

Minch, J. A., 1967, Stratigraphy and structure of the Tijuana-Rosarito Beach area, northwestern Baja California, Mexico: GSA Bulletin, v. 78, no. 9, p. 1155–1178.

Minch, J. A., K. N. Gibson, and G. L. Peterson, 1976, Clast populations in Sespe and Poway conglomerates and their possible bearing on the tectonics of the southern California borderland, *in* D. G. Howell, ed., Aspects of the geologic history of the California Continental Borderland: AAPG, Pacific Section, Miscellaneous Publication 24, p. 256–265.

Moore, D. G., 1969, Reflection profiling studies of the California Continental Borderland—Structure and Quaternary turbidite basins: GSA Special Paper 107, 142 p.

Murata, K. J., and J. K. Nakata, 1974, Cristobalite stage in the diagenesis of diatomaceous shale: Science, v. 184, p. 567–568.

Nardin, T. R., 1983, Late Quaternary depositional systems and sea level change, Santa Monica and San Pedro Basins, California Continental Borderland: AAPG Bulletin, v. 67, p. 1104–1124.

Nardin, T. R., and T. L. Henyey, 1978, Plio-Pleistocene diastrophism in the area of the Santa Monica and San Pedro shelves, California Continental Borderland: AAPG Bulletin, v. 62, p. 247–272.

Natland, M.L., 1957, Paleoecology of west coast Tertiary sediments, *in* H. S. Ladd, ed., Treatise on marine ecology and paleoecology: GSA Memoir 67, p. 543–571.

Nolf, B., and P. Nolf, 1969, Santa Cruz Island volcanics, *in* D. W. Weaver, assisted by D. P. Doerner and P. Nolf, eds., Geology of the northern Channel Islands: AAPG and SEPM, Pacific Sections, Special Publication, p. 91–94.

Ogle, B. A., 1984, Significant new developments in the Santa Maria/Santa Barbara area, offshore California—The geology and economic potential, *in* Exploration and economics of the petroleum industry: New York, Matthew Bender and Company, Inc., Proceedings of the Southwestern Legal Foundation, v. 22, p. 23–81.

Olmsted, F. H., 1958, Geologic reconnaissance of San Clemente Island, California: U.S. Geological Survey Bulletin 1071–B, p. B56–B68.

Osborne, R. H., R. C. Scheidemann, Jr., T. R. Nardin, and A. S. Harper, 1980, Quaternary stratigraphy and depositional environments, Santa Monica Bay, southern California, *in* M. E. Field, A. H. Bouma, I. P. Colburn, R. G. Douglas, and J. C. Ingle, eds., Quaternary depositional environments of the Pacific Coast: SEPM, Pacific Section, Pacific Coast Paleogeography Symposium 4, p. 143–156.

Page, B. M., and D. C. Engebretson, 1984, Correlation between the geologic record and computed plate motions for central California: Tectonics, v. 3, no. 2, p. 113–155.

Palmer, H. D., 1965, Geology of Richardson Rock, northern Channel Islands, Santa Barbara County, California: GSA Bulletin, v. 76, no. 10, p. 1197–1202.

Parker, F. S., 1971, Petroleum potential of southern California offshore, *in* I. H. Cram, ed., Future petroleum provinces of the United States—Their geology and potential: AAPG Memoir 15, v. 1 (Region 2), p. 178–191.

Paul, R. C., R. E. Arnal, J. P. Baysinger, G. E. Claypool, J. L. Holte, C. M. Lubeck, J. M. Patterson, R. Z. Poore, R. L. Slettene, W. V. Sliter, J. C. Taylor, R. B. Tudor, and F. L. Webster, 1976, Geological and operational summary, southern California Deep Stratigraphic Test OCS-CAL 75-70 No. 1, Cortes Bank area, offshore California: U.S. Geological Survey Open-File Report 76-232, 65 p.

Philippi, G. T., 1965, On depth, time, and mechanism of petroleum generation: Geochimica et Cosmochimica Acta, v. 29, no. 9, p. 1021–1049.

Pisciotto, K. A., N. B. Sasnett, and L. A. Beyer, 1974, Residual magnetic map of the California Continental Borderland, *in* J. G. Vedder, L. A. Beyer, A. Junger, G. W. Moore, A. E. Roberts, J. C. Taylor, and H. C. Wagner, Preliminary report on the geology of the continental borderland of southern California: U.S. Geological Survey Miscellaneous Field Studies Map MF–624, map sheet 8, scale 1:500,000.

Platt, J. P., 1976a, The significance of the Catalina Schist in the history of southern California borderland, *in* D. G. Howell, ed., Aspects of the geological history of the California Continental Borderland: AAPG, Pacific Section, Miscellaneous Publication 24, p. 47–52.

Platt, J. P., 1976b, The petrology, structure, and geologic history of the Catalina Schist terrain, southern California: University of California Publications in Geological Sciences, v. 112, 111 p.

Rand, W. W., 1933, The geology of Santa Cruz Island, California: Berkeley, University of California unpublished Ph.D. dissertation, 184 p.

Rangin, C., 1978, Speculative model of Mesozoic geodynamics, central Baja California to northeastern Sonora , *in* D. G. Howell and K. A. McDougall, eds., Mesozoic paleogeography of western United States: SEPM, Pacific Section, Pacific Coast Paleogeography Symposium 2, p. 85-106.

Ridlon, J. B., 1972, Pleistocene–Holocene deformation of the San Clemente Island crustal block, California: GSA Bulletin, v. 83, no. 6, p. 1831–1844.

Scheidemann, R. C., and H. T. Kuper, 1979, Stratigraphy and lithofacies of the Sweetwater and Rosarito Beach Formations, southwestern San Diego County, California and northwestern Baja California, Mexico, *in* C. J. Stuart, ed., A guidebook to Miocene lithofacies and depositional environments, coastal southern California and northwestern Baja California: SEPM, Pacific Section, p. 107–118.

Schoellhamer, J. E., and A. O. Woodford, 1951, The floor of the Los Angeles Basin, Los Angeles, Orange, and San Bernardino Counties, California: U.S. Geological Survey Oil and Gas Investigations Map OM-117, scale 1:62,500.

Schoellhamer, J. E., J. G. Vedder, R. F. Yerkes, and D. M. Kinney, 1981, Geology of the northern Santa Ana Mountains, California: U.S. Geological Survey Professional Paper 420-D, 109 p.

Shervais, J. W., and D. L. Kimbrough, 1985, Geochemical evidence for the tectonic setting of the Coast Range ophiolite: A composite island arc-oceanic crust terrane in western California: Geology, v. 13, no. 1, p. 35–38.

Shepard, F. P., and K. O. Emery, 1941, Submarine topography off the California Coast—Canyons and tectonic interpretation: GSA Special Paper 31, 171 p.

Simoneit, B. R. T., P. A. Meyers, and C. P. Summerhayes, 1981, Sources, preservation, maturation, and migration of organic matter in Neogene sediments from the continental margin off California and Baja California: A synthesis of organic geochemical studies from Deep Sea Drilling Project Leg 63, *in* R. S. Yeats and B. U. Haq, et al., Initial Reports of the Deep Sea Drilling Project Leg 63: Washington, D.C., U.S. Government Printing Office, p. 943–948.

Sorensen, S. S., 1984, Petrology of basement rocks of the California Continental Borderland and the Los Angeles Basin: Los Angeles, University of California unpublished Ph.D. dissertation, 423 p.

Sorensen, S. S., 1985, Petrologic evidence for Jurassic, island arc-like basement rocks in the southwestern Transverse Ranges and California Continental Borderland: GSA Bulletin, v. 96, no. 8, p. 997–1006.

Stadum, C. J., and T. Susuki, 1978, A Neogene section, northeastern San Clemente Island, California: Natural History Museum of Los Angeles County Contributions in Science, no. 299, 24 p.

Stuart, C. J., 1974, The San Onofre Breccia in northwestern Baja California and its regional tectonic implications, *in* G. Gastil and J. Lillegraven, eds., Geology of peninsular California: AAPG, SEPM, and SEG, Pacific Sections, Field Trip Guidebook, p. 80–90.

Stuart, C. J., 1976, Source terrane of the San Onofre Breccia—Preliminary notes, *in* D. G. Howell, ed., Aspects of the geologic history of the California Continental Borderland: AAPG, Pacific Section, Miscellaneous Publication 24, p. 309–325.

Suppe, J., and R. L. Armstrong, 1972, Potassium-argon dating of Franciscan metamorphic rocks: American Journal of Science, v. 272, p. 217–233.

Taylor, J. C., 1976, Geologic appraisal of the petroleum potential of offshore southern California: The borderland compared to onshore coastal basins: U.S. Geological Survey Circular 730, 43 p.

Teng, L. S-Y., 1985, Seismic stratigraphic study of the California Continental Borderland basins: Los Angeles, University of Southern California unpublished Ph.D. dissertation, 197 p.

Uchupi, E., 1961, Submarine geology of the Santa Rosa-Cortes Ridge: Journal of Sedimentary Petrology, v. 31, p. 534–545.

Uchupi, E., and K. O. Emery, 1963, The continental slope between San Francisco, California, and Cedros Island, Mexico: Deep-Sea Research, v. 10, no. 4, p. 397–448.

U.S. Geological Survey, 1975, Oil and gas development in the Santa Barbara Channel, Outer Continental Shelf, California: U.S. Geological Survey, Department of the Interior Final Environmental Statement, 3 vols., 934 p.

Vedder, J. G., 1975, Revised geologic map, structure sections, and well table, San Joaquin Hills-San Juan Capistrano area, California: U.S. Geological Survey Open-File Report 75-552, 5 sheets, scale 1:24,000.

Vedder, J. G., 1981, Geologic framework, *in* R. B. Powers, ed., Geologic framework, petroleum potential, petroleum estimates, environmental hazards and deep-water drilling technology of the maritime boundary region, offshore southern California: U.S. Geological Survey Open-File Report 81-264, p. 7–26.

Vedder, J. G., and D. G. Howell, 1976a, Review of the distribution and tectonic implications of Miocene debris from the Catalina Schist, California Continental Borderland and adjacent coastal areas, *in* D. G. Howell, ed., Aspects of the geologic history of the California Continental Borderland: AAPG, Pacific Section, Miscellaneous Publication 24, p. 326–342.

Vedder, J. G., and D. G. Howell, 1976b, Neogene strata of the southern group of Channel Islands, California, *in* D. G. Howell, ed., Aspects of the geologic history of the California Continental Borderland: AAPG, Pacific Section, Miscellaneous Publication 24, p. 80–106.

Vedder, J. G., and D. G. Howell, 1980, Topographic evolution of the southern California borderland during late Cenozoic time, *in* D. M. Power, ed., The California Island: Proceedings of a multidisciplinary symposium: Santa Barbara Museum of Natural History, p. 7–31.

Vedder, J. G., and E. J. Moore, 1976, Paleoenvironmental implications of fossiliferous Miocene and Pliocene strata on San Clemente Island, California, *in* D. G. Howell, ed., Aspects of the geologic history of the California Continental Borderland: AAPG, Pacific Section, Miscellaneous Publication 24, p. 107–135.

Vedder, J. G., and R. M. Norris, 1963, Geology of San Nicolas Island, California: U.S. Geological Survey Professional Paper 369, 65 p.

Vedder J. G., J. K. Crouch, and F. Lee-Wong, 1981, Comparative study of rocks from Deep Sea Drilling Project holes 467, 468, and 469 and the southern California borderland, *in* R. S. Yeats, B. U. Haq, et al., Initial reports of the Deep Sea Drilling Project: Washington, D.C. National Science Foundation, v. 63, p. 907–918.

Vedder, J. G., D. G. Howell, and J. A. Forman, 1979, Miocene rocks and their relation to older strata and basement, Santa Catalina Island, *in* J. M. Armentrout, M. R. Cole, and H. TerBest, Jr., eds., Cenozoic paleogeography of the western United States: SEPM, Pacific Section, Pacific Coast Paleogeography Symposium 3, p. 239–256.

Vedder, J. G., H. C. Wagner, and J. E. Schoellhamer, 1969, Geologic framework of the Santa Barbara Channel Region, *in* Geology, petroleum development, and seismicity of the Santa Barbara Channel region, California: U.S. Geological Survey Professional Paper 675-A, p. 1–11.

Vedder, J. G., R. F. Yerkes, and J. E. Schoellhamer, 1957, Geologic map of the San Joaquin Hills-San Juan Capistrano area, Orange County, California: U.S. Geological Survey Oil and Gas Investigations Map OM-193, scale 1:24,000.

Vedder, J. G., R. E. Arnal, D. Bukry, and J. A. Barron, 1976a, Preliminary descriptions of pre-Quaternary samples, R/V *Lee*,

March 1976, offshore southern California: U.S. Geological Survey Open-file Report 76-629, 15 p.

Vedder, J. G., J. C. Taylor, R. E. Arnal, and D. Bukry, 1976b, Map showing location of selected pre-Quaternary rock samples from California Continental Borderland: U.S. Geological Survey Miscellaneous Field Studies Map MF-737, 3 sheets, scale 1:250,000.

Vedder, J. G., H. G. Greene, E. W. Scott, J. C. Taylor, et al., 1976c, A summary report of the regional geology, petroleum potential, environmental geology, and technology for exploration and development in the area of proposed lease sale 48, California Continental Borderland: U.S. Geological Survey Open-File Report 76-787, 43 p.

Vedder, J. G., R. E. Arnal, D. Bukry, J. A. Barron, and F. Lee-Wong, 1979, Descriptions of dart core samples, R/V *SAMUEL P. LEE* cruise L2-78-SC, May 1978, California Continental Borderland: U.S. Geological Survey Open-File Report 79-936, 46 p.

Vedder, J. G., R. E. Arnal, J. A. Barron, D. Bukry, J. K. Crouch, and F. Lee-Wong, 1981, Composition and correlation of bedrock and sediment cores, R/V *SEA SOUNDER* cruise S3-79-SC, May 1979, California Continental Borderland: U.S. Geological Survey Open-File Report 81-0744, 75 p.

Vedder J. G., J. K. Crouch, R. E. Arnal, D. Bukry, J. A. Barron, and F. Lee-Wong, 1977, Descriptions of pre-Quaternary samples, R/V *ELLEN B. SCRIPPS*, September 1976, Patton Ridge to Blake Knolls, California Continental Borderland: U.S. Geological Survey Open-File Report 77-474, 19 p.

Vedder, J. G., L. A. Beyer, A. Junger, G. W. Moore, A. E. Roberts, J. C. Taylor, and H. C. Wagner, 1974, Preliminary report on the geology of the continental borderland of southern California: U.S. Geological Survey Miscellaneous Field Studies Map MF-624, 34 p., 9 sheets, scale 1:500,000.

Wayland, R. G., A. D. Acuff, T. H. McCulloh, C. B. Raleigh, J. G. Vedder, and K. A. Yenne, 1978, Facts relating to Well No. 5, Lease OCS-P 0234, Pitas Point unit area, and the earthquake of August 13, 1978, Santa Barbara Channel, California: U.S. Geological Survey Open-File Report 78-906, 59 p.

Weaver, D. W., 1969, Paleogeographic implications and geologic history, *in* D. W. Weaver, D. P. Doerner, and B. Nolf, eds., Geology of the northern Channel Islands: AAPG and SEPM, Pacific Sections, Special Publication, p. 115–131.

Weaver, D. W., and D. P. Doerner, 1969a, Lower Tertiary stratigraphy, San Miguel and Santa Rosa Islands, *in* D. W. Weaver, D. P. Doerner, and B. Nolf, eds., Geology of the northern Channel Islands: AAPG and SEPM, Pacific Sections, Special Publication, p. 30–47.

Weaver, D. W., and D. P. Doerner, 1969b, Mid-Tertiary stratigraphy, San Miguel Island, *in* D. W. Weaver, D. P. Doerner, and B. Nolf, eds., Geology of the northern Channel Islands: AAPG and SEPM, Pacific Sections, Special Publication, p. 80–83.

Weaver, D. W., and G. L. Meyer, 1969, Stratigraphy of northeastern Santa Cruz Island, *in* D. W. Weaver, D. P. Doerner, and B. Nolf, eds., Geology of the northern Channel Islands: AAPG and SEPM, Pacific Sections, Special Publication, p. 95–104.

Weigand, P. W., 1982, Middle Cenozoic volcanism of the western Transverse Ranges, *in* D. W. Fife and J. A. Minch, eds., Geology and mineral wealth of the California Transverse Ranges: South Coast Geological Society, Annual Symposium and Guidebook 10, p. 170–188.

Williams, B., 1983, Hot play in Miocene Monterey fans out to Santa Barbara Channel: Oil and Gas Journal, v. 81, no. 43, p. 49–52.

Winterer, E. L., J. W. Hawkins, J. L. Matthews, and E. C. Allison, 1969, Inverted metamorphic zonation of Franciscan rocks of the California borderland—Possible relation to Mesozoic continental underthrusting (abs.): GSA Abstracts with Programs, pt. 7, p. 240–241.

Wood, W. R., 1982, Geology, petrography, and geochemistry of the Santa Catalina Island volcanic rocks, Black Jack Peak to Whitleys Peak area: Los Angeles, California State University unpublished M.S. thesis, 146 p.

Woodford, A. O., 1924, The Catalina metamorphic facies of the Franciscan series: University of California Publications, Bulletin of the Department of Geological Sciences, v. 15, no. 3, p. 49–68.

Woodford, A. O., 1925, The San Onofre Breccia—Its nature and origin: University of California Publications, Bulletin of the Department of Geological Sciences, v. 15, no. 7, p. 159–280.

Woodford, A. O., and C. Gander, 1977, Los Angeles erosion surface of middle Cretaceous age: AAPG Bulletin, v. 61, p. 1979–1990.

Wright, F. F., 1967, The marine geology of San Miguel Gap off Point Conception, California: Los Angeles, University of Southern California unpublished Ph.D. dissertation, 271 p.

Yeats, R. S., 1968a, Southern California structure, seafloor spreading and history of the Pacific Basin: GSA Bulletin, v. 79, no. 12, p. 1693–1702.

Yeats, R. S., 1968b, Rifting and rafting in the southern California borderland, *in* W. R. Dickinson and A. Grantz, eds., Proceedings of conference on geologic problems of San Andreas fault system: Stanford University Publications, Geological Sciences, v. 11, p. 307–322.

Yeats, R. S., 1970, Stratigraphic evidence for Catalina Schist basement north of Channel Islands, California: GSA Bulletin, v. 81, no. 7, p. 2147–2152.

Yeats, R. S., 1978, Neogene acceleration of subsidence rates in southern California: Geology, v. 6, no. 8, p. 456–460.

Yeats, R. S., 1981, Quaternary flake tectonics of the California Transverse Ranges: Geology, v. 9, no. 1, p. 16–20.

Yeats, R. S., B. V. Haq, et al., 1981, Initial reports of the Deep Sea Drilling Project, Vol. 63: Washington, D.C., U.S. Government Printing Office, 967 p.

Yerkes, R. F., and R. H. Campbell, 1979, Stratigraphic nomenclature of the central Santa Monica Mountains, Los Angeles County, California; Contributions to stratigraphy, with a section on age of the Conejo Volcanics by D. L. Turner and R. H. Campbell: U.S. Geological Survey Bulletin 1457-E, 31 p.

Yerkes, R. F., and W. H. K. Lee, 1979, Late Quaternary deformation in the western Transverse Ranges, California: U.S. Geological Survey Circular 799-B, p. 27–31.

Yerkes, R. F., H. G. Greene, J. C. Tinsley, and K. R. Lajoie, 1980, Seismotectonic setting of the Santa Barbara Channel area, southern California: U.S. Geological Survey Miscellaneous Field Studies Map MF-1169, 1 plate, 10 p.

Yerkes, R. F., T. H. McCulloh, J. E. Schoellhamer, and J. G. Vedder, 1965, Geology of the Los Angeles Basin, California—An introduction: U.S. Geological Survey Professional Paper 420-A, 57 p.

Ziony, J. I., C. M. Wentworth, J. M. Buchanan-Banks, and H. C. Wagner, 1974, Preliminary map showing recency of faulting in coastal southern California: U.S. Geological Survey Miscellaneous Field Studies Map–585, 3 sheets, scale 1:250,000.

CHAPTER EIGHTEEN

GEOLOGY AND NEOGENE HISTORY OF THE PACIFIC CONTINENTAL MARGIN OF BAJA CALIFORNIA SUR, MEXICO

William R. Normark

U.S. Geological Survey
Menlo Park, California

Jon E. Spencer

Arizona Bureau of Geology and Mineral Technology
Tucson, Arizona

James C. Ingle, Jr.

Department of Geology, Stanford University
Stanford, California

ABSTRACT

Seismic-reflection profiling data, together with lithologic and biostratigraphic control from dredge samples and correlation with onshore geology, permit partial reconstruction of the Neogene history of the Pacific margin of the Baja California Peninsula south of lat. 28°N. Shallow sedimentary basins on the slope and shelf are characterized by folded and faulted sediments, 500 m to 2 km thick, as young as latest Miocene. Basin history appears to have involved initial subsidence in the early Miocene, basin filling during the middle and late Miocene, and widespread uplift and folding in the earliest Pliocene. A major lineament defined by a series of

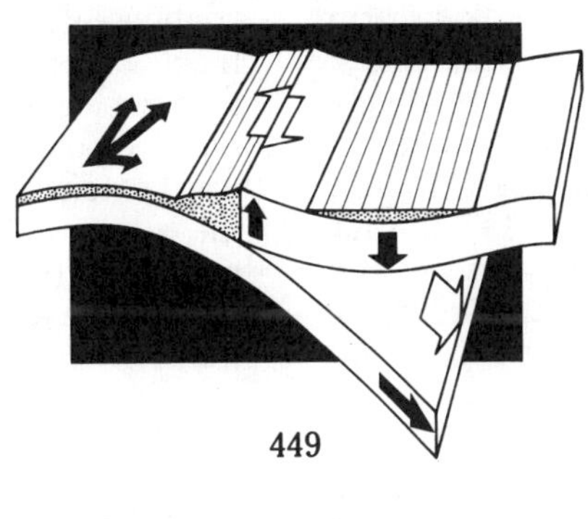

northwest-trending ridges, troughs, and faults extends for at least 400 km along the margin. Sea-floor morphology and deformation and/or truncation of subbottom reflectors along the lineament show that it is a continuous fault zone—the Tosco-Abreojos fault. We believe that this zone was a major part of the Pacific-North American transform plate boundary between 14 to 12 and 4 Ma. Disruption of the youngest sediments observed on some reflection profiles indicates that recent movement is probable in some areas, consistent with warping of Pleistocene terraces documented in adjacent onshore areas. A thick accumulation of late Mesozoic and Tertiary sediment in the Sebastian Vizcaino Bay area is the most likely setting for hydrocarbon resources where several unconformities in the uppermost 1 km of the section provide possible stratigraphic traps. Along the southern part of the margin, the thinner sedimentary section is more disrupted by faulting and folding with numerous, but small, possible structural traps.

INTRODUCTION

The continental margin along the Pacific side of the southern Baja California Peninsula is a poorly understood complex zone separating a Tertiary volcanic arc from a region of the northeast Pacific Basin that has undergone several major plate reorganizations during the last 25 m.y. (Klitgord and Mammerickx, 1982; Mammerickx and Klitgord, 1982). The geology of the Pacific continental margin of the Baja California Peninsula, especially Baja California Sur (the state south of lat. 28°N), has not been studied nearly as extensively as the rest of the Baja Peninsular margin, which includes the Gulf of California and the southern California Continental Borderland (Figure 1). Recent reinterpretations of the plate tectonic history of the northeast Pacific and results from numerous new on-land geologic studies provide the basis for a reinterpretation of seismic-reflection profiles obtained more than 10 years ago during several cruises to the Pacific margin of the Baja California Peninsula. The current plate tectonic interpretation and the work on the margin and adjacent peninsula are briefly reviewed below as background for our study.

Plate Tectonic Setting

Interpretation of the magnetic anomaly pattern of the northeast Pacific provided a plate tectonic framework for the evolution of the Gulf of California and the Baja California Peninsula (Atwater, 1970). Atwater's model showed that at about 29 Ma, the Pacific-Farallon spreading center encountered the Farallon-North American subduction zone south of the Mendocino fracture zone and created a transform boundary between the Pacific and North American plates. The two triple junctions thus created migrated relatively away from each other, lengthening the transform boundary, as subduction of the Farallon plate continued (Atwater, 1970; Atwater and Molnar, 1973; Blake et al., 1978). After subduction of the Farallon plate off Mexico ceased, and before the opening of the Gulf of California as a complex transform boundary, the Pacific margin of the Baja Peninsula was the probable site of transform displacement between the Pacific and North America plates (Spencer and Normark, 1979).

Recent reinterpretation of the magnetic and bathymetric data west and south of Baja California (Klitgord and Mammerickx, 1982) suggests both a revised timing for events along the margin and several complexities in plate boundaries not envisioned by Atwater (1970). Specifically, Mammerickx and Klitgord (1982) now suggest that the entire segment of the Pacific-Guadalupe spreading center (the new name for this part of the Pacific-Farallon Ridge of Atwater, 1970) died out at 12.5 Ma along the western edge of the Baja Peninsula between lat. 29°30′N and 23°30′N (Figure 1). This change occurred because a counterclockwise rotation of the spreading center between 25 and 12.5 Ma brought the spreading center parallel to the base of the Baja margin. Small offsets in the general trend of the spreading center probably resulted in stranding a few small slivers of the Guadalupe plate next to the Baja margin, if the spreading center was not entirely consumed by subduction.

North of 27°N, a westward jump of the spreading center at 12.5 Ma resulted in stranding a large fragment of the Guadalupe plate (Menard, 1978; Batiza and Chase, 1981). Between 12.5 and 11 Ma, the spreading center between 27°N and 29°30′N was abandoned, and this change probably resulted in transform-fault displacement along the Baja margin west of Sebastian Vizcaino Bay (Mammerickx and Klitgord, 1982). Ness et al. (1981) suggest that a similar ridge abandonment south of 23°30′N could explain a 14.5 Ma age for the crust at DSDP site 471 that is significantly older than predicted by continuous spreading. Whether or not this ridge abandonment occurred, the magnetic anomaly pattern indicates that counterclockwise rotation of the spreading center and continual subduction of the Guadalupe plate resulted in southward extension of the proposed transform boundary along the western Baja California margin (Klitgord and Mammerickx, 1982).

After the eastward jump of the Pacific-North American transform boundary to the Gulf of California about 4 Ma, the

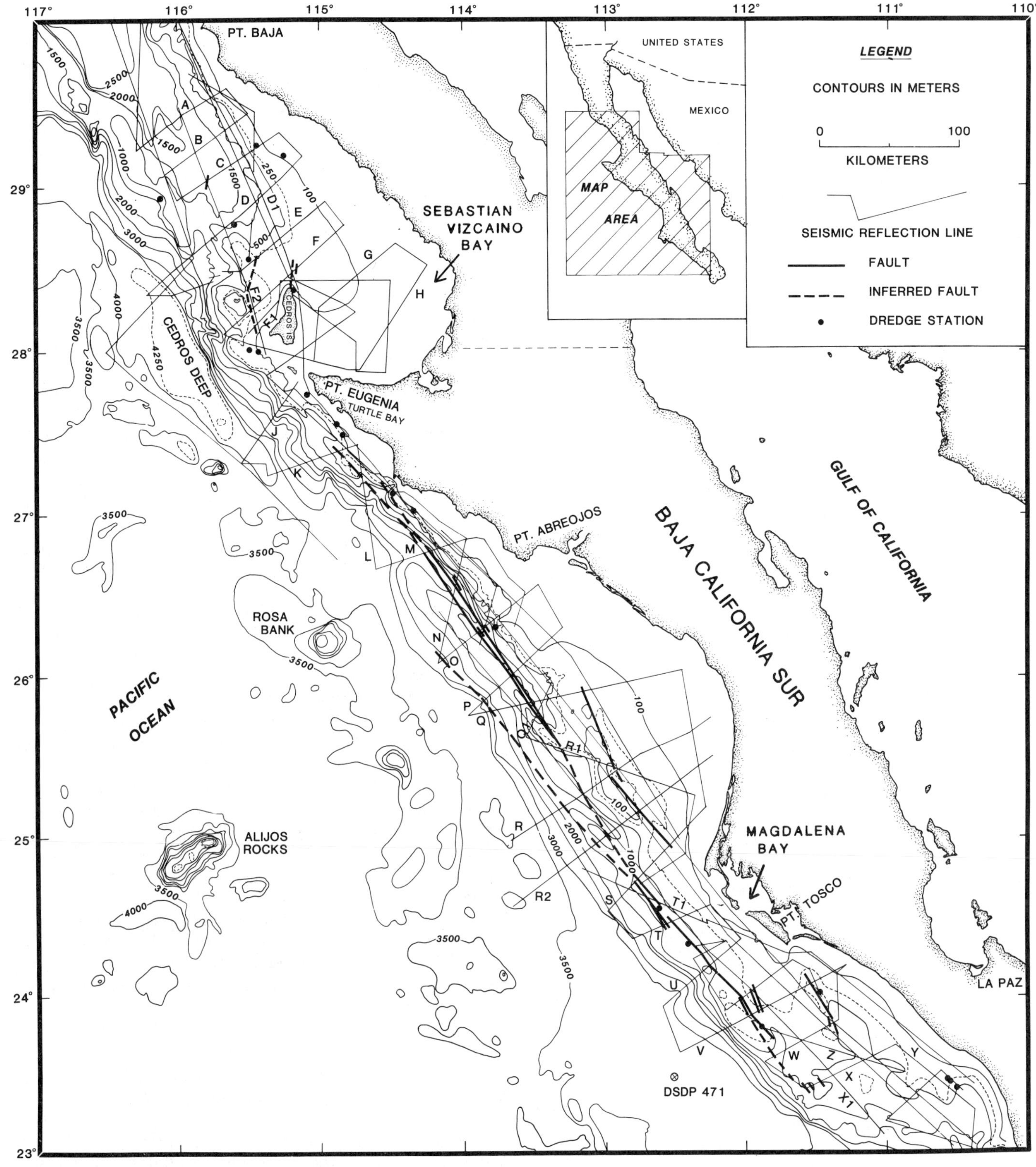

Figure 1. Bathymetry of the Pacific margin of central and southern Baja California showing location of seismic-reflection profiles and dredge samples in our data base, and major offshore faults, geographic features, and DSDP hole 471. Bathymetry not shown in Gulf of California. Depth contours in meters.

Pacific margin of Baja California Sur is presumed to have become an intraplate setting.

Prior Work

When the major data-collecting expeditions that provide the basis for this report were undertaken (pre-1970) and the preliminary report was being prepared (Normark, 1977), relatively little geologic background information was available for the Baja California Sur Peninsula. The works of Beal (1948) and Mina (1957) provided a general description of the volcanic arc and the major structure of the peninsula. A basement complex, including Franciscan-Complex-

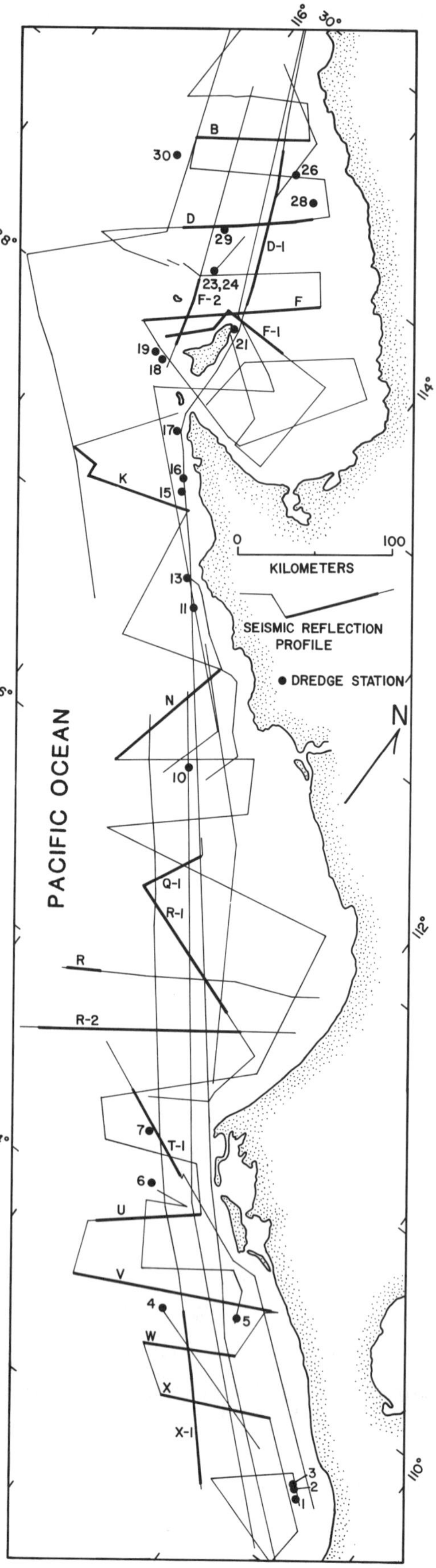

Figure 2. Locations of reflection profiles and dredge stations along central and southern Baja California. Profiles discussed in this chapter are indicated by heavy dark lines.

equivalent(?) metasedimentary rocks, blueschist-facies rocks, dismembered ophiolitic rocks, and some intrusive igneous rocks, had been recognized on the Cedros and the San Benitos Islands (32 km west of Cedros Island; Figure 1) (Cohen et al., 1963; Kilmer, 1969). By the mid 1970s, however, several groups had begun to study selected areas, especially in the Sebastian Vizcaino Bay region, but these studies were primarily focused on pre-Tertiary events (Minch et al., 1976; Jones, Blake, and Rangin, 1976).

Before 1970, the offshore area of the Pacific margin of the southern Baja California Peninsula south of the California Continental Borderland was little known (Krause, 1965; Moore, 1969; Doyle and Bandy, 1972). Fisher and Hess (1963) briefly discussed the topography and crustal structure of the Cedros Deep region. Emery (1948) discussed the geology of Ranger Bank, 35 km northwest of Cedros Island (Figure 1). Miocene phosphatic marine sediments off the southern part of the peninsula were described by D'Anglejan (1967). Several unpublished seismic-refraction profiles from near Cedros Island and south of Magdalena Bay provided the only other source of offshore geophysical or geologic information for the margin.

During the last 10 years, Baja California Sur has been the focus of numerous local studies encompassing structure, volcanic and sedimentary sequences, and biostratigraphy, especially in the area from Sebastian Vizcaino Bay south to lat. 25°N. Many of these studies are presented in the recent volume edited by Frizzell (1984) and in an earlier field guide edited by Abbott and Gastil (1979). These studies place new constraints on the pre-Neogene history of the peninsular area and provide new information on the age and extent of subduction-related volcanism and the development of the proto-Gulf of California (Hausback, 1984; Sawlan and Smith, 1984). A study of the Miocene Tortugas Formation on the Vizcaino Peninsula (Helenes, 1980) is especially pertinent to interpretation of Neogene basin history. These new results will be discussed in relation to our present study as necessary.

During the same 10-year period, little detailed geological work has been added in the offshore area. A comprehensive, regional-scale mapping of Mexico's 200-nautical-mile exclusive economic zone in a cooperative study with Oregon State University has provided new bathymetric, gravity, and magnetic maps but has added only a limited amount of seismic-reflection data for the continental shelf and slope areas.

Purpose

In this report, we reexamine the evidence for the transform regime first described by Spencer and Normark (1979). In addition, this report presents for the first time the biostratigraphic and sedimentologic information from 22 successful dredge hauls from within the study area (Figure 2). Despite the improved picture of the onshore geology for Baja California Sur, the timing for major tectonic events is still best constrained by the interpretation of the magnetic anomalies of the northeast Pacific Basin (Klitgord and Mammerickx, 1982). All models reconstructing plate motions for this area of the Pacific invoke transform displacement within the Pacific margin during the middle Neogene.

METHODS OF STUDY

The study area extends along the entire Pacific margin of Baja California Sur and includes Sebastian Vizcaino Bay north to Punta Baja (Figure 1). Single-channel seismic-reflection profiles were obtained during four expeditions to the area between 1967 and 1976 (Figures 1, 2). In addition, two sparker profiles across the margin were available from the 1964 cruise to the tip of the Baja California Peninsula by F. P. Shepard and J. R. Curray. We also reviewed several reflection profiles obtained by R. W. Couch and G. E. Ness of Oregon State University. Twenty-three days of reflection profiles obtained at survey speeds of 10 to 18 km/hr provide >7000 trackline kilometers of data. Sound sources included both 10-cu. in. airguns as well as 20- and 120-kjoule sparkers. Magnetometer data are available for most of the survey lines. An additional 17 days of rock dredging, sediment coring, high-resolution (3.5 kHz) profiles, and deep-tow surveying were completed during the four main expeditions and two brief surveys in parts of Sebastian Vizcaino Bay.

Scripps Institution of Oceanography vessels and equipment were used for all but the 1976 cruise. Position control for most of the surveys is based on celestial and radar fixes. Trackline positions have been adjusted to match satellite-navigated sounding lines obtained on post-1969 cruises and to provide the best match of water depths, subbottom structures, and magnetic field values at track crossovers. The length of the study area (>900 km) and the relatively wide spacing of the survey lines tend to minimize problems in geologic interpretation caused by uncertainties in the position control and to reconfirm the reconnaissance nature of this study.

SEDIMENTARY BASINS AND BIOSTRATIGRAPHY

Before detailing the structural evidence for major faulting within the margin, we will review briefly the major offshore sedimentary basins using the general compilation by Normark (1977). The dredge samples indicate that the exposed sections of these deformed basins are middle and late Neogene in age (Tables 1–3).

Sedimentary Basins

The basin shapes, continental-slope morphology, and structure along the Baja California Peninsula margin all show far more complexity than would be expected from the late Tertiary sea-floor spreading history immediately west of the 900-km-long study area.

The sedimentary basins are grouped into three fairly distinct physiographic provinces: (1) Sebastian Vizcaino Bay and the slope adjacent to Cedros Deep; (2) the slope plateau province between Punta Abreojos and Magdalena Bay; and (3) the Magdalena Borderland south of Magdalena Bay (Figure 3).

Sebastian Vizcaino Bay

The basin-and-ridge topography of the borderland offshore of the northern third of the Baja California Peninsula merges southeastward into a broad, shallow, sediment-filled trough beneath the Sebastian Vizcaino Bay. This northwest-trending trough is flanked to the southwest by the basement-cored San Benito ridge (Krause, 1965), which forms a largely submarine continuation of the Vizcaino Peninsula and includes Cedros Island, the San Benito Islands, and Ranger Bank. Seismic-reflection profiles across the continental margin in the area northwest of Sebastian Vizcaino Bay (Figures 2, 4, 5, lines B, D, D-1) reveal a broad synclinorium plunging gently northwest from the bay into the southernmost basin of the California Continental Borderland (Figure 6A) (Moore, 1969). A secondary basin with >1 km of fill lies west of Cedros Island, as shown on Figure 4, lines F and F-1 (also see Figure 7). The continental slope west of Sebastian Vizcaino Bay is relatively steep and is bordered by the partly sediment-filled trough known as Cedros Deep. The steep slope continues to the south adjacent to Punta Abreojos (Figure 1).

The uppermost sedimentary unit in the bay consists of late Pliocene and younger semiconsolidated clayey silts and is relatively undeformed except by slumping (Normark, 1974). A major slide scar is apparent on line D between km 58 and km 68 (Figure 4). The general instability of this unit is compounded by the irregular unconformable surface on which it rests (Figures 4 and 5, lines B, D, and D-1). The unconformity is cut into mudstone, siltstone, and fine sandstone of late Miocene or younger age (Table 2), and this section is underlain by older (Miocene?—not sampled by dredging) sediment that is slightly more deformed. Those sections below the uppermost unconformity are cut by numerous faults (Figure 6A), commonly with only 200 m or less of throw.

Seismic-refraction profiles in the southern part of Sebastian Vizcaino Bay suggest that the Cretaceous and younger sedimentary fill may be as much as 5 km thick. The single-channel profiles used in this report illustrate only the uppermost sedimentary units. The contours of Figure 6B show the depth of acoustic penetration that represents depth to acoustic basement only locally. Sedimentation within the Sebastian Vizcaino embayment has probably continued intermittently, possibly since the Late Jurassic or at least the Late Cretaceous (Minch et al., 1976; Helenes, 1984).

Slope Plateau Province

South of Punta Abreojos (Figures 1, 3), the continental margin is characterized by a broader, more gentle slope and a wide (up to 100 km) continental shelf. The slope between lat. 24°30′N and 26°20′N is constructed of ridges that act as sediment barriers, in some cases producing gently sloping plateaus (lines N and R-2, Figure 8). The ridges generally show discontinuous, shallow subbottom reflectors, and many may be constructed of deformed sedimentary rocks. No dredge samples are available from ridges in this province, but similar shallower features to the south are composed of sedimentary rocks. Sediments are ponded at the base of the slope, filling a broad trough that parallels the entire slope-plateau province (Figure 9). The sediment cover observed in these profiles rarely exceeds 500 m (Figure 10B), but gentle folding, faulting, and erosional truncation of slope and outer-shelf sediments are common.

Many of the ridge segments appear as acoustic basement on our low-power single-channel profiles. Sediments within the easternmost topographic basin (lines R-1 and R-2, Figure 8) are slightly deformed and appear to extend eastward

Table 1. Data for 22 dredge stations, Sebastian Vizcaino Bay and Baja California Sur.

Station	Location	Depth (m)	Sample No.	Sample Description
69–1	23°24.9′N 110°30.4′W	420	69-1	Very-fine-grained sandstone with yellow phosphatic mudstone matrix. Quartz and feldspar grains are coated with a light-yellow mineral.
69-2	23°27.3′N 110°33.3′W	195	69-2	Very-fine-grained sandstone with yellow phosphatic mudstone matrix. Sand grains are composed of quartz, feldspar, opaque minerals, glauconite, diatom tests, and pellets.
69-3	23°27.8′N 110°34.2′W	160	69-3	Foram tests, fish bones, pellets, oolites(?), and shell fragments, with fine-grained, slightly phosphatic calcite matrix.
69-4	23°45.0′N 111°55.2′W	420	69-4-A, B	Foram tests, fish bones, pellets, oolites(?), and shell fragments, with yellow phosphatic mudstone matrix. Brecciated(?) and bioturbated.
69-5	24°01.0′N 111°28.6′W	200	69-5-A	Foram tests, pellets, and miscellaneous bone and shell fragments with yellow phosphatic mudstone matrix.
			69-5-B, C, D, E, F	Very-fine-grained crystalline limestone.
69-6	24°26.5′N 112°22.0′W	730	69-6-A, B	Pellets and miscellaneous bone and shell fragments with yellow phosphatic matrix. Pellets, fragments, and void walls are coated with light-yelow mineral.
			69-6-C	Medium-grained sandstone with very-fine-grained crystalline-calcite matrix. Subangular sand grains are composed primarily of quartz and feldspar with minor polycrystalline quartz and lithic sedimentary fragments.
69-7	24°36.1′N 112°32.0′W	600	69-7-A	Brecciated phosphatic mudstone with pellets, bone and shell fragments, and phosphatized(?) foram tests.
			69-7-B	Calcareous, slightly phosphatic mudstone with sparse foram tests.
			69-7-C, E	Medium- to fine-grained sandstone with finely crystalline calcite matrix. Angular to subangular sand grains are composed of feldspar and monocrystalline and polycrystalline quartz.
			69-7-D	Fine-grained sandstone with slightly calcareous, phosphatic mudstone matrix.
69-10	26°03.6′N 113°27.2′W	320	69-10-A, G	Medium- to fine-grained sandstone with phosphatic mudstone matrix. Subangular to angular quartz and feldspar grains. Sand grains are coated with light-yellow mineral in sample 10-G.
			69-10-B, E	Coarse-grained calcareous sandstone. Subangular to subrounded grains include microcline, plagioclase, quartz, microcrystalline quartzite, and volcanic-lithic fragments. One volcanic-lithic fragment (in sample 10-B) contains a euhedral hornblende phenocryst, suggesting hornblende andesite composition.
			69-10-D	Medium- to fine-grained sandstone with very-fine-grained crystalline calcite matrix.
			69-10-F, H	Calcareous mudstone with sparse foram tests, diatom tests, and sponge spicules(?). Sample 10-H is dolomitic.
69-11	27°01.2′N 114°20.8′W	200	69-11	Fine-grained sandstone with calcareous mudstone matrix. Subangular to angular sand grains include quartz, feldspar, polycrystalline quartz, and sparse mafic minerals and volcanic-lithic fragments.
69-13	27°12.0′N 114°31.2′W	145	69-13	Calcareous mudstone with very sparse foram tests and fine grains of quartz and feldspar.
69-15	27°28.3′N 114°51.0′W	220	69-15	Calcareous mudstone with sparse, recrystallized foram tests.
69-16	27°33.0′N 114°53.2′W	220	69-16-A	Calcareous siltstone with sparse recrystallized foram tests and diatoms.
			69-16-B, C	Fine-grained calcareous sandstone with angular to subangular grains of feldspar and monocrystalline and polycrystalline quartz.

(continued).

			69-16-D	Tuff breccia. Plagioclase phenocrysts and microlites in a dark-gray nonvesicular groundmass.
69-17	27°43.0′N 115°07.0′W	45	69-17	Calcareous mudstone with sparse recrystallized foram tests and fine grains of quartz and feldspar.
69-18	28°00.0′N 115°25.5′W	90	69-18-A	Siltstone.
			69-18-B, C	Medium-grained sandstone, porous, with angular to subangular feldspar, monocrystalline quartz, polycrystalline quartz, volcanic-lithic fragments, and rounded sedimentary-lithic fragments.
69-19	28°00.6′N 115°29.5′W	110	69-19-A	Medium-grained sandstone with subangular grains of quartz, feldspar, and sparse polycrystalline quartz and volcanic-lithic fragments.
69-21	28°22.0′N 115°10.7′W	70	69-21	Medium-grained sandstone with coarsely crystalline calcite matrix. Sand grains include sparse volcanic-lithic fragments.
69-23	28°33.4′N 115°29.6′W	145	69-23-A	Bioturbated phosphatic mudstone with sparse, fine, subangular to subrounded grains of quartz and feldspar. Also contains pellets, sedimentary-lithic fragments, and miscellaneous bone(?) and shell fragments.
69-24	28°33.4′N 115°29.6′W	165	69-24-A	Bioturbated phosphatic mudstone with sparse, fine, subangular to subrounded grains of quartz and feldspar. Also contains pellets, sedimentary-lithic fragments, and miscellaneous bone(?) and shell fragments.
			69-24-B	Medium-grained sandstone with phosphatic mudstone matrix. Sand grains are coated with a light-yellow mineral. Sparse phosphatic oolites(?) have concentric black, orange, and yellow layers around cores of quartz, feldspar, and miscellaneous bone(?) and shell fragments.
			69-24-C	Bioturbated phosphatic mudstone with sparse miscellaneous bone(?) and shell fragments and fine grains of quartz and feldspar.
			69-24-D	Tuff breccia with plagioclase phenocrysts in altered groundmass.
69-26	29°10.5′N 115°27.7′W	110	69-26-A, B	Fine-grained, bioturbated sandstone with phosphatic mudstone matrix. Contains subangular quartz and sparse feldspar. Also contains sparse pellets, miscellaneous bone(?) and shell fragments, and oolites(?) with concentric phosphatic layers around cores of plagioclase and quartz. Light-yellow mineral coats sand grains in sample 69-26-A.
			69-26-C	Very-fine-grained quartzite (metachert?) with sparse opaques and medium- to fine-grained quartz veins.
69-28	29°11.1′N 115°14.3′W	90	69-28-A	Calcareous mudstone with recrystallized foram tests and sparse fine grains of quartz and feldspar.
CE-29	28°47.7′N 115°35.8′W	1100	CE-29	Friable, semiconsolidated green clayey silt, commonly with abundant open worm burrows.
BA-30	28°56.0′N 116°07.0′W	1460	BA-30-A, F	Serpentine (chrysotile).
			BA-30-B	Highly altered metavolcanic(?) rock with rounded and broken plagioclase phenocrysts, quartz, opaques, and plagioclase microlites.
			BA-30-C	Limestone composed of finely crystalline calcite with sparse, fine-grained, subangular to subrounded quartz, feldspar, and polycrystalline quartz.
			BA-30-D	Altered tuff breccia with plagioclase phenocrysts in a highly altered groundmass.
			BA-30-E	Serpentinized periodotite(?) with augite and opaque minerals.

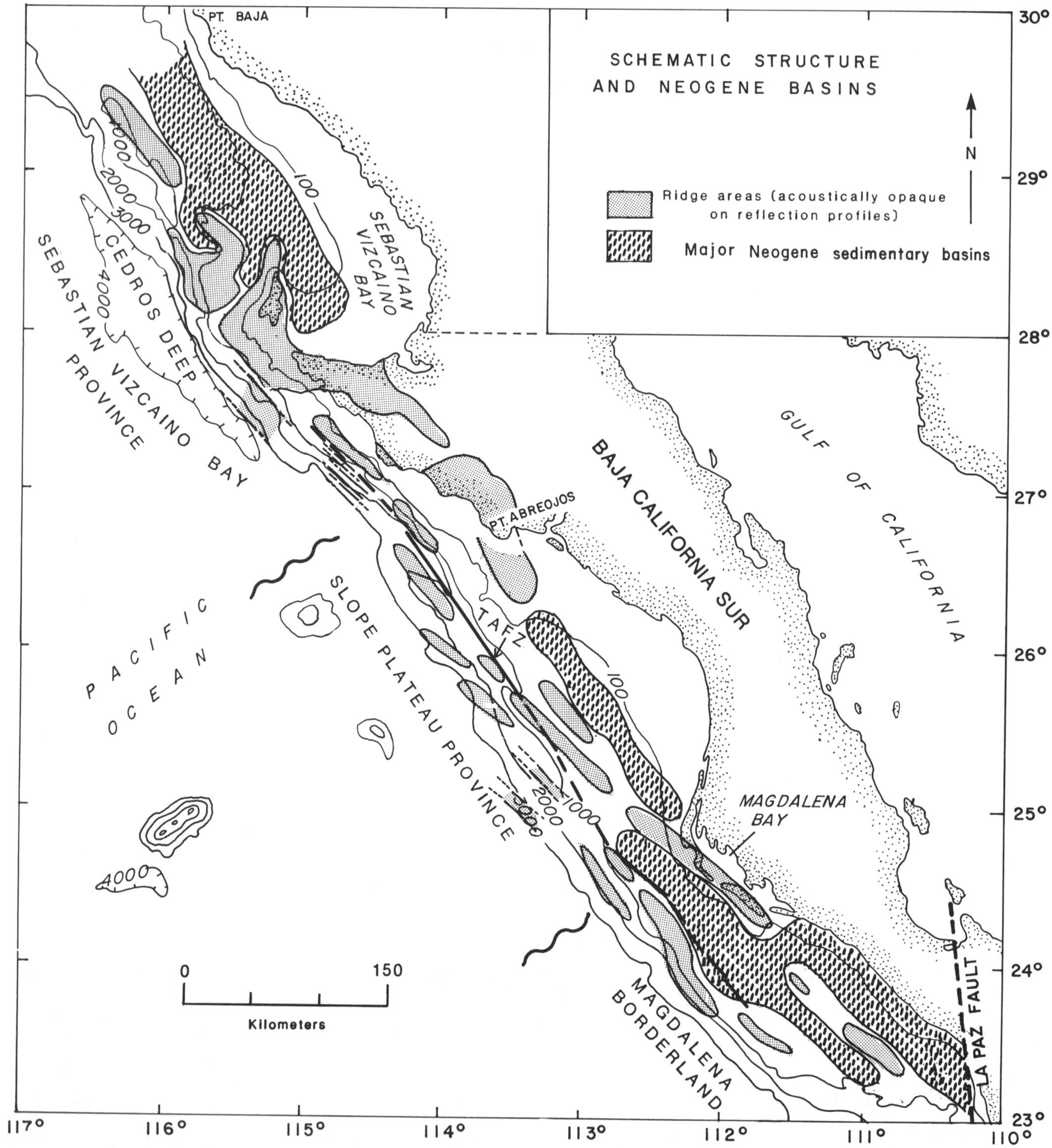

Figure 3. Schematic representation of areas underlain by rocks giving limited or no acoustic subbottom reflectors (ridges) and major Neogene sediment bodies (basins). In some places, the pattern of northwest-trending ridges is continuous with structures in pre-Neogene metasedimentary, metavolcanic, and intrusive rocks underlying onshore coastal uplands and islands. Tosco-Abreojos fault zone (TAFZ) serves as reference point for bathymetry and trackline charts (Figures 1, 2) and reflection profiles (Figures 4, 5, 7, 8, 9). Depth contours in meters. Modified from Normark (1977).

under the shelf. Acoustic reverberation in the shallow shelf area, however, precluded any effective mapping there.

Magdalena Borderland

South of lat. 24°30′N, the slope ridges and plateaulike character give way to the steepest continental slope along the Baja Peninsula (line V, Figure 5, and line U, Figure 8). Between the steep, high-relief continental slope on the west and the shelf break, a broad (> 100 km wide) area of numerous irregularly shaped banks and intervening basins resembles a shallow (< 1 km of water depth over most of the area) version of the California Continental Borderland. The reflection profiles (lines V and X-1, Figure 5, and lines X and W, Figure 8) show that gently folded and faulted strata underlie

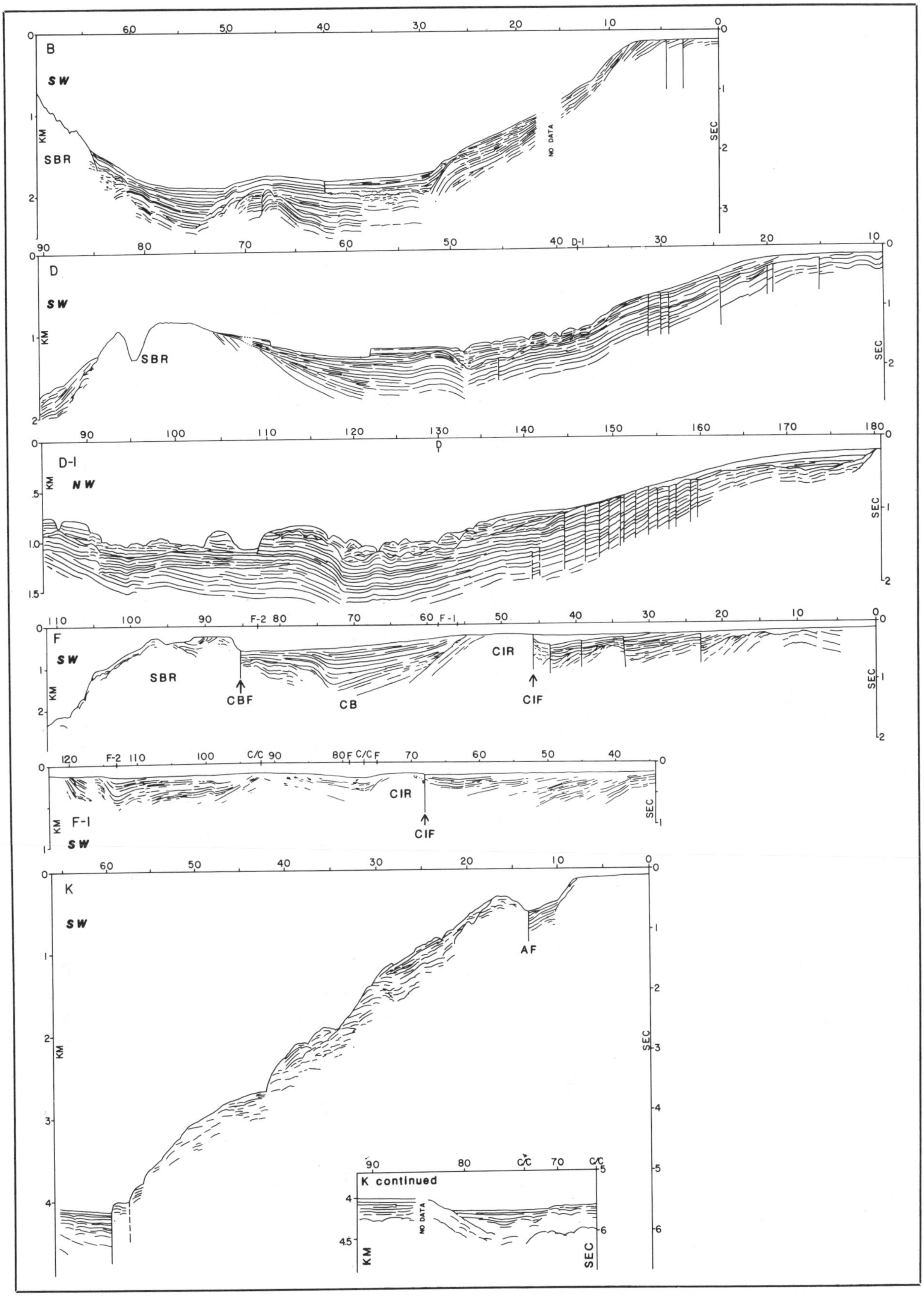

Figure 4. Line drawings of reflection profiles showing inferred faults in northern half of study area. Profile crossings are indicated at the top edge of several profiles. C/C indicates course change. Depth in kilometers based on water-sound velocity of 1467 m/sec. Horizontal scale in kilometers from eastern or northern end of each profile. Vertical exaggeration generally 8–12×. Locations of profiles in Figures 1 and 2. Major features include San Benito Ridge (SBR), Cedros Island Ridge (CIR), Cedros Island fault (CIF), Cedros Basin fault (CBF), Cedros Basin (CB), San Benito fault (SBF), and Abreojos fault (AF).

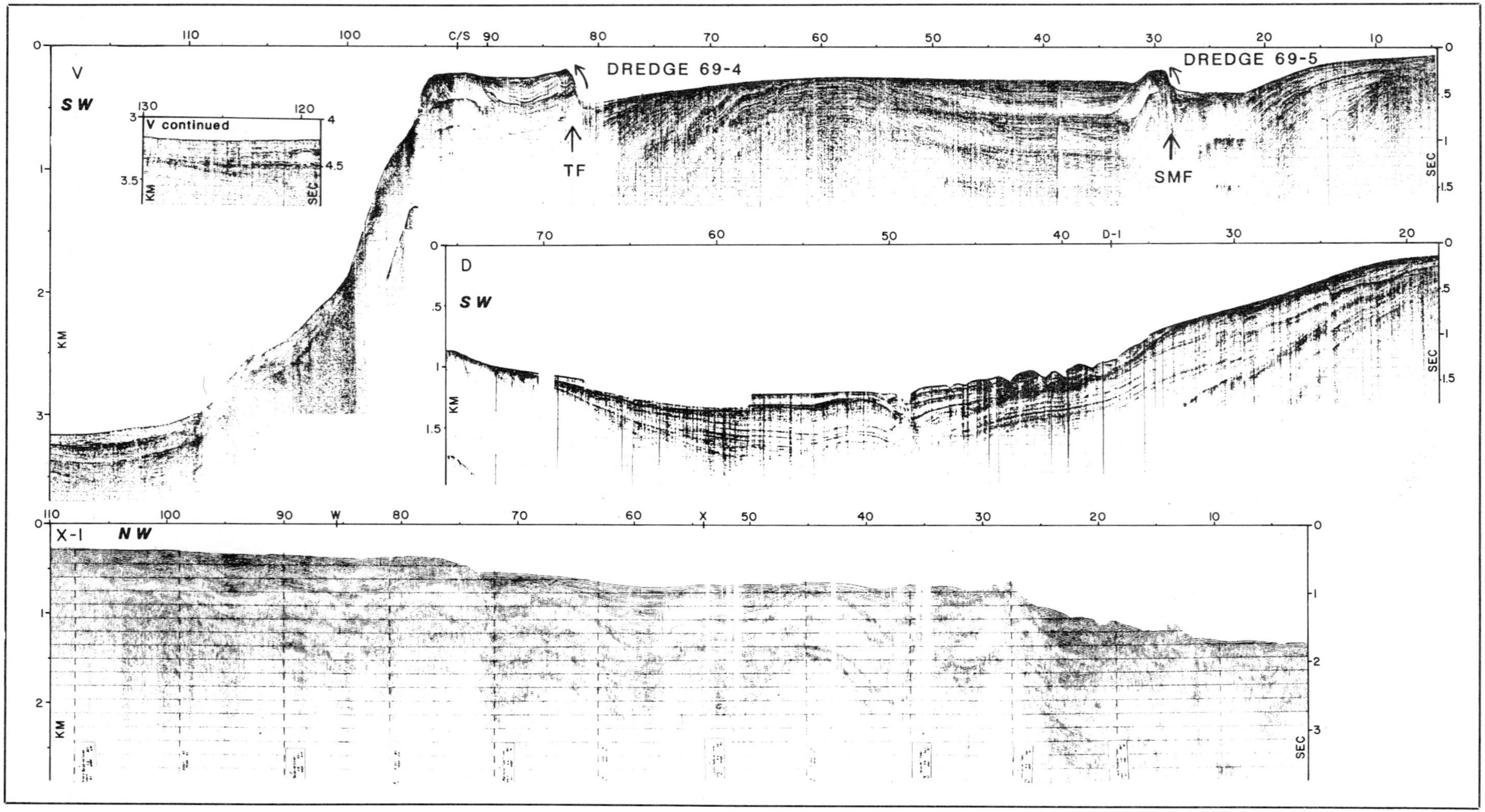

Figure 5. Seismic-reflection profiles from the Baja California margin. C/S indicates change in speed. Depth based on water-sound velocity of 1467 m/sec. Horizontal scale in kilometers. Profile crossings indicated along upper edge of profiles D and X-1. Vertical exaggeration is approximately 8:1. Location of profiles in Figures 1 and 2. Major features include the Tosco fault (TF) and Santa Margarita fault (SMF).

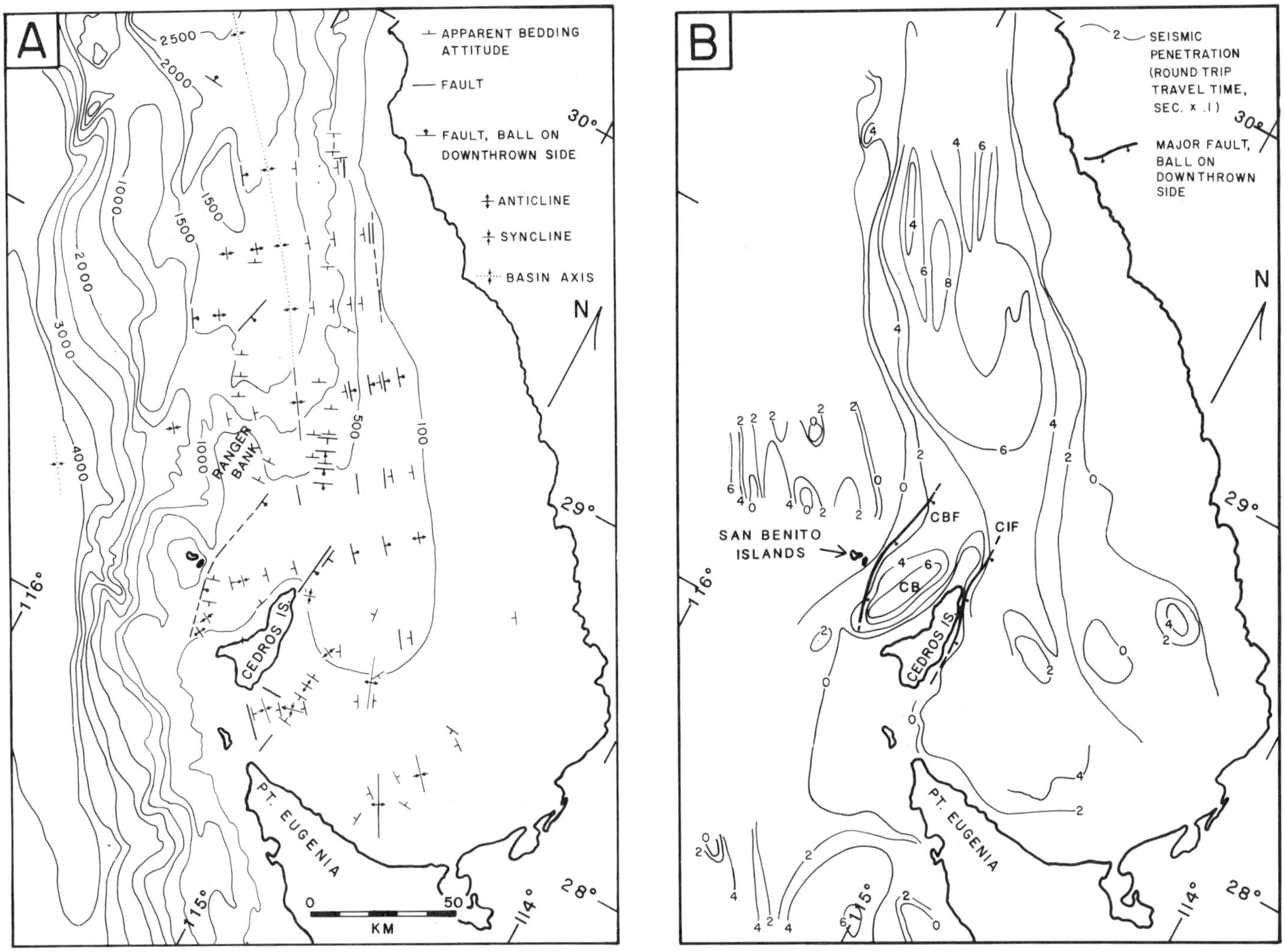

Figure 6. A. Bathymetric map of Sebastian Vizcaino Bay area showing structures observed on seismic-reflection profiles. Strike of structures is drawn perpendicular to trackline unless multiple tracklines allow more accurate determination of orientation. Depth contours in meters. B. Contour map of seismic penetration based on reflection profiles, showing location of inferred major faults. The Cedros Island tilted block, which includes Cedros Basin (CB) and Cedros Island, is bounded to the east by the Cedros Island fault (CIF) and to the west by the Cedros Basin fault (CBF).

much of the borderland topography. Angular unconformities are common between units, and on some profiles undisturbed sedimentary sections are evident only along the outer shelf. Although our profiles show < 1 km of sediment over this basin (Figure 10B), unpublished seismic-refraction data along the shelf south of Magdalena Bay suggest that the total sediment thickness locally exceeds 2 km.

The predominant rock types sampled in the Magdalena borderland are dark-brown phosphatic foraminiferal mudstone and light-gray to buff siltstone and fine sandstone (Table 1). Even many "acoustic basement" topographic highs (e.g., km 28–32, line V, Figure 5) consist of sedimentary rocks similar to those recovered along fault scarps and on the outer continental shelf. The sedimentary rocks are Miocene (Table 2), and the age of the folding and fault disruption of this broad basin may have begun by the latest Miocene.

Foraminiferal Biostratigraphy and Paleobathymetry of Dredge Samples

Samples of sedimentary rocks were obtained from 22 successful dredge stations, primarily from the Sebastian Vizcaino Bay and Magdalena borderland provinces (Figures 2, 3). Table 1 gives the location, water depth, and description of the rock types recovered. Many dredges recovered samples of more than one sedimentary unit, but not all sedimentary units are fossiliferous. Tables 2 and 3 summarize the biostratigraphic results and estimated paleobathymetry of the samples.

Faunal Analysis

Forty-three samples of dredged sedimentary rock were analyzed for foraminifera and other microfossils (Table 3) for the purpose of establishing the age of the units, correlating dredge samples with onshore marine sequences, and providing evidence of depositional history. The analyzed samples include hard siliceous and phosphatic siltstones, shales, and silty sandstones (Table 1), precluding normal extraction of microfossils. One exception, an upper Pliocene siltstone from the Ranger Bank area (Sample CE-29), yielded a well-preserved foraminiferal assemblage by flotation with carbon tetrachloride. Thin sections of the resistant rock types were analyzed for foraminifera and other microfossils using transmitted light at 50× to 100× magnification. Thus,

Table 2. Estimated ages and paleobathymetry of selected sedimentary rocks in Baja-69 and Ceride dredge samples, Baja California, Mexico[1].

Sample	Age[2]	Estimated Original Depth of Deposition (meters)	In Situ Depth of Dredge Sample (meters)
69-1-A	Middle Miocene; Relizian-Luisian	500–1500	420
69-1-B	Middle Miocene; Luisian-Mohnian	500–1500	420
69-2	Miocene	500+	420
69-4-A	Middle Miocene; Luisian(?)	500–1500	420
69-4-B	Middle Miocene; Luisian	500–1500	420
69-5-A	Middle-late Miocene; Luisian-Mohnian	500–1500	200
69-5-B	Early-middle Miocene; Relizian-early Mohnian	500–1500	200
69-5-C	Early-middle Miocene; Saucesian-Luisian	500–1500	200
69-5-D	Early-middle Miocene; Saucesian-Relizian	200–500	200
69-5-E	Early-middle Miocene; Saucesian-Luisian	500–1500	200
69-5-F	Early-middle Miocene; Saucesian-Luisian	500–1500	200
69-6-A	Middle Miocene; Relizian-Luisian	500–1500	730
69-6-B	Middle Miocene; Relizian-Luisian	500–1500	730
69-6-C	Middle Miocene; Luisian-Mohnian	500–1500	730
69-7-A	Middle-late Miocene; Luisian-Mohnian	500–1500	600
69-7-B	Middle-late Miocene; Luisian-Mohnian	500–1500	600
69-7-C	Middle Miocene(?)	500+	600
69-10-A	Middle Miocene; Luisian	500–1500	320
69-10-B	Middle Miocene(?)	500+	320
69-10-D	Middle Miocene(?)	500+	320
69-10-E	Middle Miocene(?)	500–1500	320
69-10-F	Middle-late Miocene; Luisian-Mohnian(?)	1000–2000	320
69-10-G	Middle Miocene(?)	200+	320
69-10-H	Middle-late Miocene(?)	500–1500	320
69-11	Middle-late Miocene; Luisian-Mohnian	500–1500	145
69-13	Middle-late Miocene	500–1500	220
69-15-A	Middle Miocene; Luisian-early Mohnian	500–1500	220
69-16-A	Late Miocene; Mohnian	500–1500	220
69-16-B	Miocene	500–1500	220
69-16-C	Middle Miocene; Luisian-early Mohnian	500–1500	220
69-17	Early Miocene; Saucesian-Relizian	500–1500	45
69-23	Middle Miocene; Luisian	500+	145
69-24-A	Miocene	500+	165
69-24-B	Miocene	500+	165
69-24-C	Miocene	500+	165
69-26-A	Miocene	500+	110
69-26-B	Miocene	500+	110
69-28-A	Middle Miocene; Luisian-early Mohnian	500–1500	90
CE-29	Late Pliocene	2000	1100

[1]Estimated ages and paleobathmetry based upon thin-section analysis of foraminifera and other microfossils, lithologic characteristics, and correlation with onshore Neogene units; see Table 3 and Figure 11 for listing of microfossils and stratigraphy of onshore section.
[2]Provincial Miocene ages of Kleinpell (1938, 1980).

despite the abundance of both benthonic and planktonic foraminifera in many of the samples, identification at the species level was difficult or impossible and generic identifications were the rule (Table 3).

Thin-section analysis limited the precision of age assignment and correlation of foraminiferal faunas. Nevertheless, distinctive associations of genera together with the identification of rare species provided sufficient data to generally assign faunas to the provincial Miocene stages of California (Kleinpell, 1938, 1980) or a range of these stages (Table 2). For example, large, robust species of *Valvulineria* and *Baggina* were assigned to the middle Miocene Luisian Stage; the presence of *Siphogenerina* was assumed to preclude a post-Luisian age; and large species of *Globobulimina* were assumed to be typical of the Luisian and Mohnian age faunas, following generic patterns common within Saucesian through Mohnian faunas in California.

In some samples, planktonic foraminifera also aided general age assignments through correlation with the distinctive distributional patterns of dominant planktonic species within Miocene and Pliocene sequences in the California province (Ingle, 1973; Keller, 1981). For example, abundance of *Globigerina angustiumbilicata* and *G. concinna* characterizes early and early middle Miocene Saucesian through Relizian planktonic faunas. In contrast, *Globigerina bulloides* becomes the dominant species with late-middle and late Miocene Luisian and Mohnian faunas.

The presence and abundance of other microfossil groups and phosphorite, together with lithology, provided additional means of correlating the samples with onshore Neogene marine sequences. The well-described Bahia Tortugas sequence of the northern Vizcaino Peninsula served as a biostratigraphic and lithostratigraphic standard (Helenes, 1980) with which to compare and correlate the isolated lithologies

Table 3. Distribution of foraminifera, other microfossils, and diagnostic detrital debris identified in selected dredge samples from the Baja-69 expedition, western slope of Baja California, Mexico[1].

SAMPLE NUMBERS	69-1-A	69-1-B	69-1-D	69-2	69-4-A	69-4-B	69-5-A	69-5-B	69-5-C	69-5-D	69-5-E	69-5-F	69-6-A	69-6-B	69-6-C	69-7	69-7-A	69-7-B	69-7-C	69-7-E	69-10-A	69-10-B	69-10-D	69-10-E	69-10-G	69-10-H	69-11	69-13	69-15-A	69-16-A	69-16-B	69-16-C	69-16-D	69-17	69-18-B	69-23	69-24-A	69-24-B	69-24-C	69-24-D	69-26-A	69-26-B	69-28-A
MICROFOSSILS																																											
Planktonic foraminifera																																											
Globigerina angustiumbilicata												C		P														R		R				A									
Globigerina bulloides							R	C	P								P				P						R		R	R							R						
Globigerina concinna					C						P	P									P									P				C									
Globigerina praebulloides	P				A	R										P																											
Globigerina spp.(small)	C					P	R			P	C							P			C											R				P							P
Globigerinoides spp.								R	R																																		
Globoquadrina dehiscens(?)					R					R	R	R																															
Orbulina spp.																					P								R	R													
Praeorbulina spp.	R																																										
Sphaeroidinella/Sphaeroidinellopsis spp.					P																																						
Turborotalia spp.									R	R																																	
Benthonic foraminifera			0																	0													0		0								
Baggina spp.																																											
Bolivina spp.	R																	R											R	P				P									
Bolivina spp.(small)					R			P				R																		R													
Bulimina spp.	P				R					P				R				P												P				R									
Cassidulina spp.										R																				R				R									
Chilostomella spp.																													R														
Florilus spp.								P		R								R											R	R													
Globobulimina spp.																		R			R								P														
Gyroidina spp.(large)	R																																										
Gyroidina spp.(small)								R																																			
"Hemicristellaria" beali																																											
Lenticulina spp.							R																																				
Nodosaria spp.																																											
Plectofrondicularia spp.									R																																		
Siphogenerina spp.	R								P																																		
Uvigerina spp.(costate)								C	R												R									R													P
Uvigerina spp.(hispid)						C	P	P									C																										
Uvigerinella spp.										R																			P														C
Valvulineria californica						P																																					
Valvulineria spp.																					R																						
unidentifiable arenaceous foraminifera																											R					P											
unidentifiable rotalid foraminifera	R															P			P																								
unidentifiable foraminiferal fragments				P									P						R					R							R												C
Miscellaneous																																											
diatoms								A										C				P				A			A	C													C
fish debris(bones and scales)													C	C		C			P													P					C		C		P		
radiolarians								P	P			C					P		R				R	P		P	A		C	A	A	A		C			R					C	
sponge spicules															P	P			P		P		C				P			C	C	A										P	C
Authigenic minerals																																											
glauconite pellets	P																C					P		P												P							
phosphatic pellets	C	A		A	C	C	A						A	A	P							C	P	C	C											A	A	A	A	R	P	P	

[1]Relative abundances as determined from thin section analysis: R = rare, P = present, C = common, A = abundant, 0 = no microfossils.

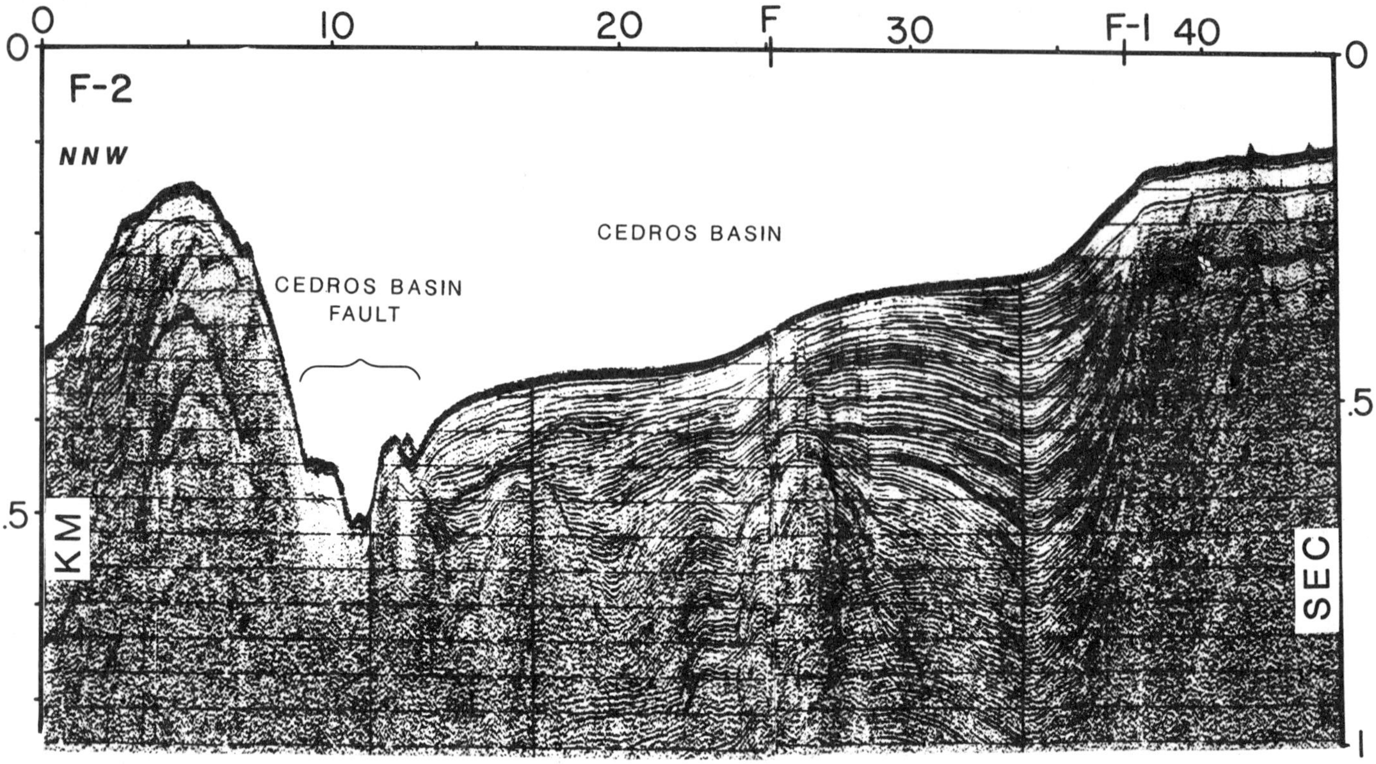

Figure 7. Reflection profile from area between Cedros and San Benito Islands. Crossings with other profiles shown along top edge of figure. Vertical exaggeration is approximately 30:1. Horizontal scale in kilometers. Water depth assumes sound velocity of 1467 m/sec.

and faunas encountered in the dredge samples (Figure 11).

By combining evidence provided by foraminiferal analysis, lithology, and correlation with the Bahia Tortugas sequence (Figure 11), a best estimate of age was made for each dredge sample analyzed (Table 2). Confidence of age assignments ranges from high, where specific identification (e.g., *Valvulineria californica*) allowed some precision, to low where only nonfaunal evidence was available (e.g., the presence of pelletal phosphorite). Correlation of the California Neogene stages with oceanic planktonic biostratigraphic zones and the radiometric time scale follows Ingle (1980), Barron (1981), and Keller and Barron (1981).

Paleobathymetry and Depositional Environments

Estimates of water depth at time of deposition (the paleobathymetry of dredged samples) are based on analyses of the generic composition of benthonic foraminiferal assemblages, general patterns of species morphology and ornamentation, and correlation of faunal compositions with recurring distributional patterns of living foraminifera along the Pacific Coast of North America following summaries by Ingle (1980), Ingle and Keller (1980), and Douglas (1981). Modern faunal associations in the borderland basins off southern California, in the Gulf of California, and along the continental margin off Central America serve as models for general paleobathymetric interpretation of the closely related Neogene assemblages present in both the dredged samples and onshore sequences. For example, common hispid and costate species of *Uvigerina* and *Uvigerinella* are assumed to represent mid-bathyal water depths (500–1500 m), as are large species of *Gyroidina* and *Globobulimina*. Furthermore, the co-occurrence of shallower water taxa (e.g., *Valvulineria* and *Florilus*) with bathyal species provides evidence of downslope transport and faunal mixing within Neogene basins. Paleobathymetric analysis of assemblages followed the concepts and procedures outlined by Ingle (1980)—assuming that the deepest-dwelling faunal elements in any assemblage represent a minimum depth of deposition.

Other paleobathymetric criteria that were applied in the absence of, or in addition to, foraminiferal evidence include abundance of radiolarians, laminated structure characteristic of anoxic depositional settings, and correlation of a given sample with the stratigraphic and paleobathymetric history of the Bahia Tortugas sequence (Figure 11).

Paleobathymetric analysis indicates that the majority of the Neogene assemblages present in the dredged samples were originally deposited at midbathyal water depths between 200 and 1500 m within basin slope, fan, and basin-floor environments similar to the environments represented in the Miocene Tortugas Formation (Figure 11). The laminated structure, disseminated phosphorite, and common organic matter (including fish debris in some samples) suggest deposition in suboxic to near anoxic subsill settings common to modern and Miocene borderland basins described from California (Garrison and Douglas, 1981; Ingle, 1981). Maximum water depth within the Neogene basins off Baja California Sur during the Miocene appears to have been 1500 m. One clearly lower bathyal assemblage containing *Melonis pompilioides*, however, was encountered in an upper Pliocene sample (CE-29) dredged from a depth of 1100 m, suggesting continued basin subsidence in the Pliocene followed by later uplift.

Comparison of in situ depths from which samples were dredged with estimated original depths of deposition demonstrates that the amount of vertical uplift associated with

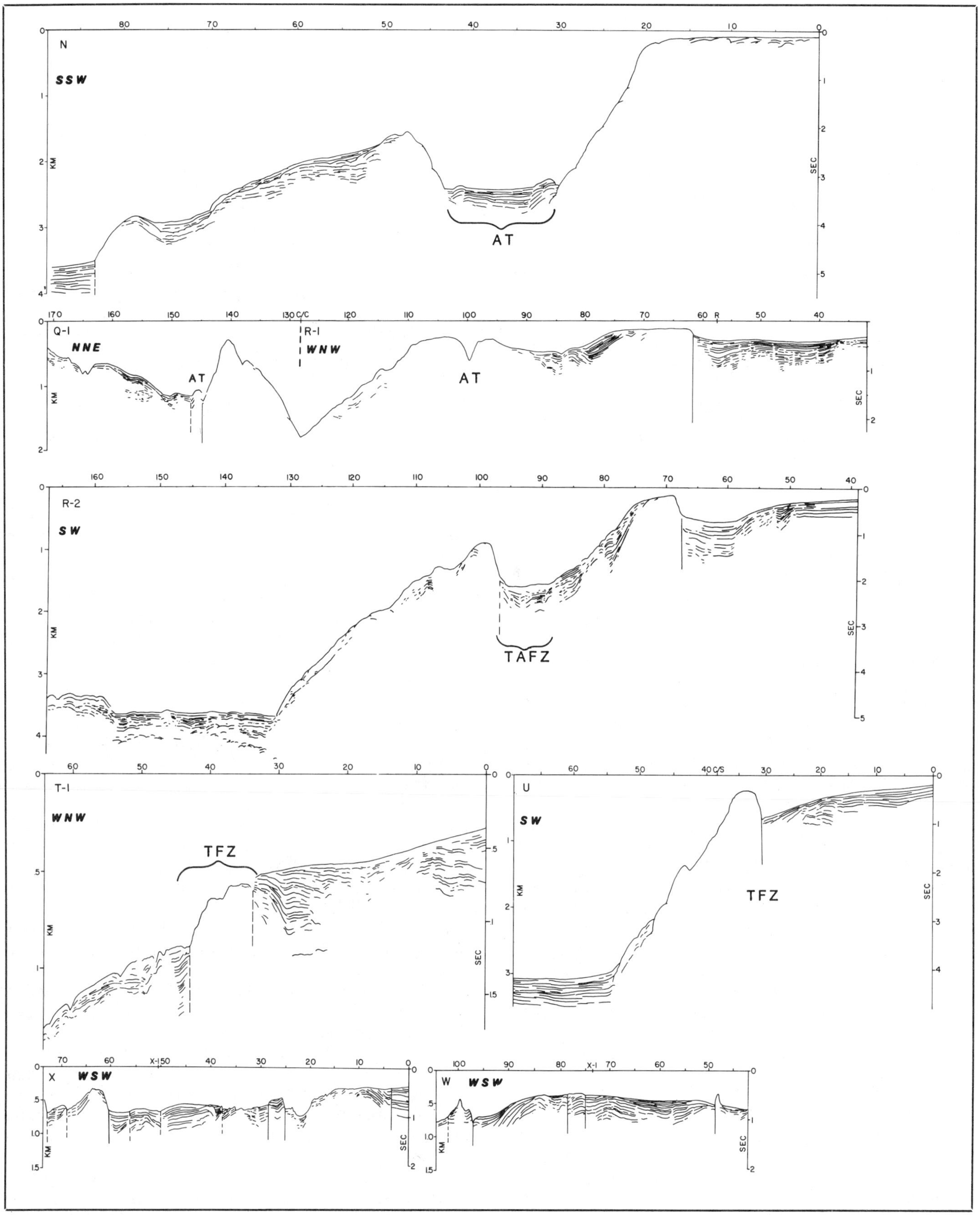

Figure 8. Reflection profiles showing inferred faults in southern half of study area. Horizontal scale in kilometers. Profile crossings are indicated along top edge of several profiles. C/C indicates course change, C/S indicates change in speed. Depth based on water-sound velocity of 1467 m/sec. Vertical exaggeration generally 8–12×, except for line T-1 where V. E. = 28:1. Locations of profiles in Figures 1 and 2. Major features include Abreojos Trough (AT), Tosco-Abreojos fault zone (TAFZ), and Tosco fault zone (TFZ) segment of TAFZ.

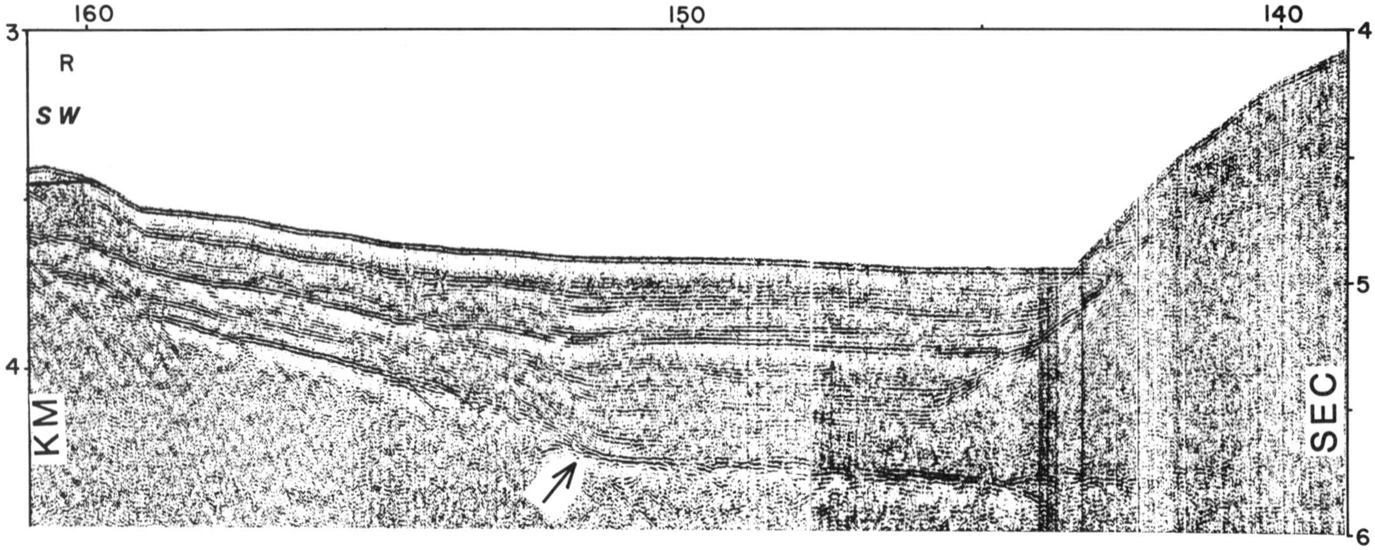

Figure 9. Seismic-reflection profile of the buried trench at the toe of the continental margin of southern Baja California. Note acoustic basement (bold arrow) beneath sediments, which projects to the east underneath the toe of the continental margin. Location on Figures 1 and 2. Horizontal scale in kilometers. Vertical exaggeration is approximately 5:1.

tectonic flexing of these units during the Pliocene and Pleistocene ranged from 100 m to > 1500 m (Table 2).

Correlation with Onshore Neogene Sequences

Neogene marine strata are exposed at scattered localities along the entire Pacific Coast of Baja California Sur, and early workers recognized that the Miocene and Pliocene faunas in this region are similar to those described from California (Allison, 1964). In addition, siliceous and diatomaceous Neogene shales present in this region were generally correlated with the Miocene Monterey Formation of California. More recent work has demonstrated that some of these siliceous units (San Gregorio Formation) are as old as upper Oligocene and hence are not correlative with the Monterey Formation (Hausback, 1984; Smith, 1984) but rather underlie Miocene volcanic units.

The best known Neogene marine sequence exposed along the Baja California Sur margin is in the Bahia Tortugas area of the Vizcaino Peninsula (Figure 11). Initial descriptions of this phosphatic and diatomaceous Miocene sequence were completed by Allison (1964) and Robinson (1975). Helenes (1980) described the stratigraphy, age, foraminiferal faunas, and depositional history of the Miocene Tortugas Formation in detail (Figure 11). This unit includes lower, middle, and upper Miocene shelf, slope, and basin-plain deposits that record the initial subsidence and subsequent filling of a portion of the so-called Tortugas Basin (Blake et al., 1978). Significantly, the largest portion of the Tortugas Basin lies offshore, with the onshore Bahia Tortugas sequence representing a rare opportunity to examine in detail the history of an offshore basin.

The close similarity between rock types, microfaunal distribution, foraminiferal biofacies, and the interpreted paleobathymetry of dredged samples of offshore Neogene units (Table 2) with the Tortugas Formation allows this sequence to serve as a general model of the Neogene basin evolution in this region.

Studies of onshore geology along the Pacific Coast of Baja California Sur suggest that pre-Neogene basement rocks in the adjacent offshore region include Mesozoic blueschists (Franciscan Complex), Triassic and Jurassic ophiolitic rocks, Jurassic and Cretaceous turbidite sandstone (Eugenia and Valle Formations), Paleocene and Eocene shelf and slope sediment (Bataque and Tepetate Formations), and Oligocene and Miocene volcanic and volcaniclastic units commonly assigned to the "Comondu Formation" of Beal (1948). Fragments of all of these "basement" rocks, with the exception of lower Tertiary units, appear to be present among samples dredged from the study region (Table 1).

The Neogene marine sequence in the Bahia Tortugas area rests with a sharp angular unconformity upon the Cretaceous sandstones of the Valle Formation (Figure 11). The overlying Neogene units include the Miocene Tortugas Formation, Pliocene Almejas Formation, and Pleistocene marine-terrace deposits totaling some 600–700 m in thickness. An angular unconformity between the bathyal deposits of the Tortugas Formation and littoral and shelf sediments of the Almejas Formation, however, suggests that significant erosion of the upper Tortugas occurred in conjunction with late Miocene uplift of this sequence. Hence, the original thickness of the Tortugas sequence may have been between 1 and 2 km, and thus similar to the thicknesses of Neogene basin fill interpreted from seismic records in the adjacent offshore region.

Helenes (1980) recognized five lithologic members within the Tortugas Formation that indicate variations in depositional environments and variations in rates of sediment accumulation and composition during basin evolution. A basal lower-to-middle Miocene siliceous silty mudstone unit deposited at bathyal depths (Figure 11) is overlain by middle Miocene turbidite sands, the upper portion of the formation being composed of laminated and massive diatomites, diatomaceous siltstones, and porcellanites of middle and late Miocene age. A well-cemented middle Miocene sandstone and sandy siltstone containing common to abundant pelletal phosphorite constitutes the basal member of the formation in marginal areas of the basin. Displaced pelletal phosphorite is present in middle Miocene turbidite sandstones. Sandy phosphatic rocks of middle Miocene age are present among the dredge samples and probably represent this same facies

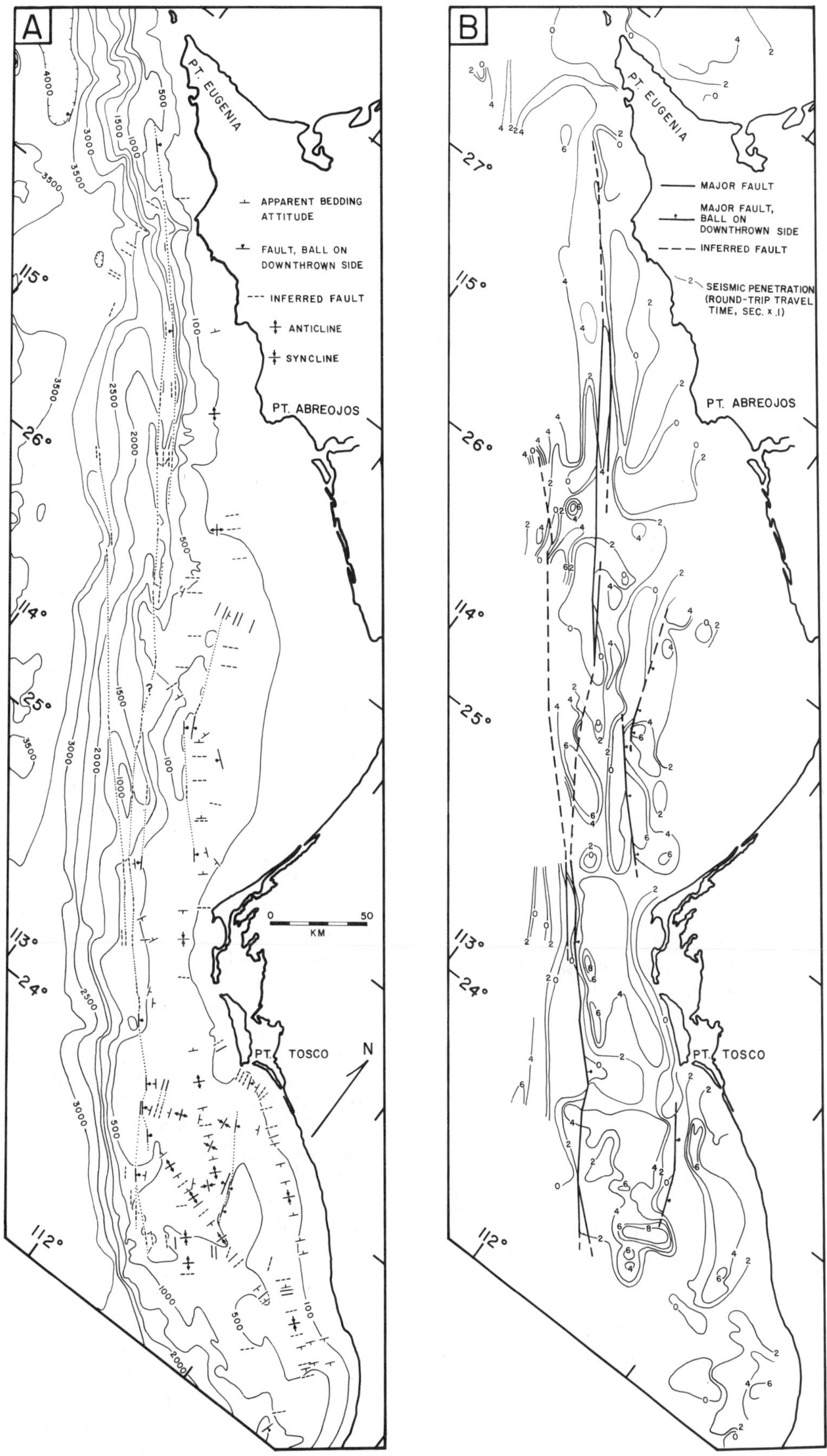

Figure 10. A. Bathymetric map of margin of southern Baja California showing structures observed on seismic-reflection profiles. Strike of structures is drawn perpendicular to tracklines unless multiple tracklines allow more accurate determination of orientation. Dotted lines connect possible major faults, queried where uncertain. B. Contour map of seismic penetration based on reflection profiles, showing location of inferred throughgoing faults.

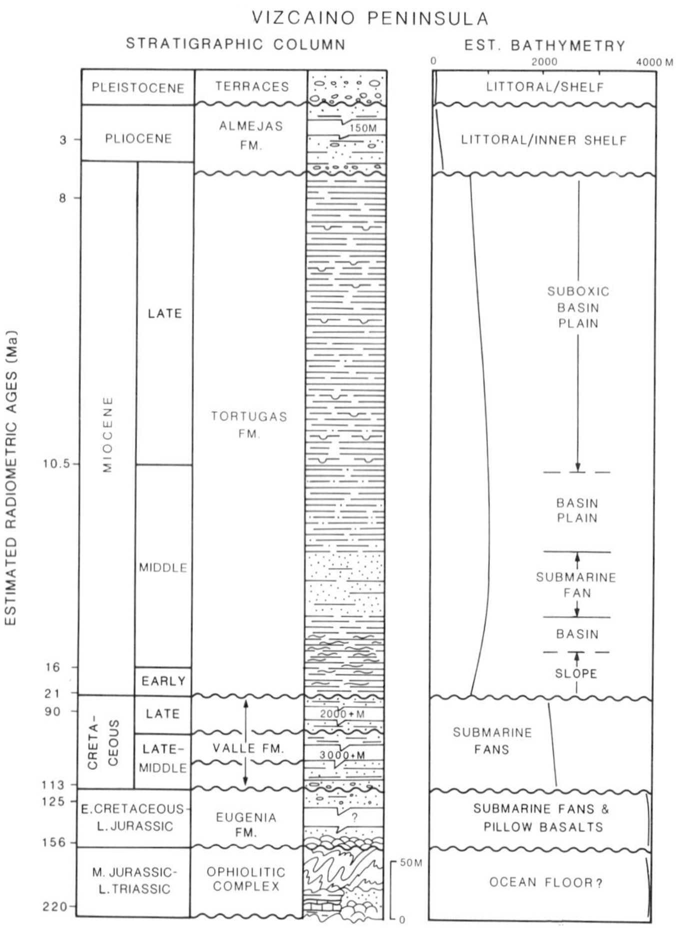

Figure 11. Composite stratigraphic column, depositional environments, and estimated paleobathymetric history of the northern Vizcaino Peninsula. Estimated radiometric ages taken from Berggren, Kent, and Couvering (in press). Paleobathymetric history of the Miocene Tortugas Formation modified from Helenes (1980). Thick Pliocene and Pleistocene deposits present offshore (Normark, 1974) are missing in this onshore sequence. Thickness of missing intervals given in meters. The Paleocene and Eocene Bateque Formation is not included in this column but is reported from the offshore and onshore area immediately north of the Vizcaino Peninsula (Helenes, 1984). The Triassic-Jurassic ophiolitic complex includes the San Hipolito Formation and the Sierra de San Andres Ophiolite (see Barnes, 1984). Pre-Neogene ages assigned in part from Harland et al., 1982.

in other Neogene basins offshore (Table 3).

Analysis of dredged samples indicates that the lithologic and faunal sequence present within the Tortugas Formation is duplicated in many of the offshore Neogene basins, as evidenced by (1) a similar early (Saucesian) through late (Mohnian) Miocene age range, (2) dominantly mid-bathyal biofacies, (3) common siliceous lithologies rich in radiolarian and diatom remains, and (4) common pelletal phosphorite, disseminated phosphorite, and fish debris indicative of deposition in suboxic basin settings.

STRUCTURE

The structural features of the southern Baja Peninsula are discussed in two parts—Sebastian Vizcaino Bay and the area south of Punta Abreojos.

Sebastian Vizcaino Bay and San Benito Ridge

The northwest-trending San Benito Ridge separates the Sebastian Vizcaino Bay region from Cedros Deep and is composed of rocks that are acoustically opaque on our profiles (Figures 1, 6). The ridge is overlain on its flanks by outward-dipping sedimentary rocks (Figure 4, lines B, D, F, F-1). Rocks thought to be correlative with the acoustic basement are exposed on the San Benito Islands (Cohen et al., 1963), Cedros Island (Kilmer, 1969; Rangin, 1978; Boles and Landis, 1984) and the Vizcaino Peninsula (Rangin, 1978; see also papers in Frizzell, 1984). Exposed basement in these areas is composed of a variety of rock types, including Franciscan-equivalent metasedimentary rocks, flysch, volcanic rocks, blueschist-facies rocks, dismembered ophiolitic rocks, and granitic intrusives. The San Benito Ridge is broken by two major north-trending faults in the vicinity of Cedros Island that are herein named the Cedros Basin and Cedros Island faults (Figure 4, line F, km 85 and km 46, respectively, and line F-1, km 68; Figure 6). Cedros Island is on the uplifted side of a tilted north-trending fault block that is bounded by these two faults. The north-trending Cedros Basin west of Cedros Island is on the down-tilted side of the fault block (Figure 6B).

Northwest-trending faults are present in subaerial exposures along the southwestern edge of the Vizcaino Peninsula and Cedros Island, and on the San Benito Islands. The northwest-trending San Benito shear zone was first recognized on the west island of the San Benito Islands, where a fault juxtaposes a graywacke sequence to the southwest with mixed rocks of Franciscan affinity to the northeast (Cohen et al., 1963). A northwest-trending fault at the southwest tip of Cedros Island juxtaposes dissimilar rock types (Kilmer, 1969; Rangin, 1978; Boles and Landis, 1984; Kimbrough, 1984) and is on strike with the San Benito shear zone. The northwest-trending Bahia Tortugas and La Bamba faults on the southwest side of the tip of the Vizcaino Peninsula are approximately on strike with the San Benito shear zone (Minch et al., 1976), as is a northwest-trending fault 20 km to the south near Morro Hermoso (Rangin, 1978). Right-lateral displacement is indicated by small-scale structures exposed along the fault zone on the San Benito Islands (Cohen et al., 1963) and on the Vizcaino Peninsula (Moore, 1984).

Correlation of exposed faults and linear geophysical features along the Pacific margin of the central part of the Baja California Peninsula defines a semicontinuous linear zone of significant lateral extent. Faults on the San Benito Islands, Cedros Island, and the Vizcaino Peninsula define a northwest-trending fault zone that extends for approximately 120 km. Krause (1965) and Calderon-Riveroll and Couch (1977) recognize a strong, linear, magnetic anomaly associated with the San Benito Ridge and propose that the anomaly is the result of a major fault extending southeastward for approximately 250 km from the San Quintin Basin to the Punta Eugenia area at the tip of the Vizcaino Peninsula. This inferred submarine fault is continuous with the northwest-trending faults exposed on land in the San Benito Islands-Cedros Island-Punta Eugenia area (Krause, 1965). The San Benito fault as defined by Krause (1965) was correlated with the Tosco-Abreojos fault zone off southern Baja California (Spencer and Normark, 1979). A 200 to 300 km right-lateral displacement during the middle to late Miocene was inferred for the combined San Benito-Tosco-Abreojos faults (Spencer and Normark, 1979) based on estimates of relative motion of the Pacific and North American plates (Atwater and Molnar, 1973).

The southern part of the northwest-trending San Benito fault, as located by Krause (1965), extends from the San Benito Islands to the western tip of Cedros Island. Three of our seismic-reflection profiles cross (but do not show) the inferred trace of the fault in this area (Figure 4, line F, km 85–95, and line F-1, km 105–120; Figure 7, line F-2, km 20–30). Folded but structurally continuous sedimentary strata occupy this interval on line F-2 (Figure 7). In addition, there is no evidence of faulting on line F-1 (Figure 4) in the vicinity of the inferred San Benito fault, although subbottom reflectors fade out at km 117. If the folded sedimentary rocks recognized on lines F-1 and F-2 (Figures 4, 7) in the area of the proposed San Benito fault are middle to late Miocene, as suggested by numerous dredge samples of rocks of this interval from other areas along the Baja California margin (Table 2, Figure 2), then profile F-2 prohibits a throughgoing Neogene fault in the area, as suggested by Krause (1965). Any throughgoing fault of major Neogene strike-slip displacement (Spencer and Normark, 1979) must therefore lie west of line F-2 and probably southwest of all on-land exposures, although on-land faults may be subsidiary splays from, or en echelon breaks within, a major throughgoing fault zone.

Tosco-Abreojos Fault Zone

The Tosco-Abreojos fault zone is the dominant structural and geomorphic feature of the central area and the Magdalena borderland, and is represented by two linear bathymetric and structural features that are inferred to be connected. The northern Abreojos Trough, as seen in the reflection profiles, is a well-defined bathymetric feature that we interpret as a fault zone; the southern Tosco fault appears to be an active fault, as indicated by both bathymetric relief and truncation of subbottom reflectors.

The Abreojos fault is represented by a gently northeast-plunging trough or sharp break in topography that is recognized on seven of the eight seismic-reflection profiles that cross it over a fault length of about 180 km. This trough reaches a depth of about 2.5 km at its northern end and opens to the southwest (Figure 2, line M, in Spencer and Normark, 1979). The northward projection of the Abreojos fault is traced onto the continental shelf and may be represented on line K (Figure 4, km 13) by a narrow, apparently faulted, valley <25 km from the Vizcaino Peninsula. To the southeast, the Abreojos Trough trends gradually up the continental slope, where it appears as a flat-bottomed valley as much as 10 to 15 km wide flanked to the southwest by a sea-floor ridge that is generally acoustically opaque on our seismic profiles (Figure 8, line N, km 30–43). The valley narrows to the southeast (Figure 8, line Q-1, km 145–150) and is last recognized as a V-shaped notch in the sea floor at a water depth of 300 to 600 m (Figure 8, line R-1, km 100).

Our seismic-reflection profiles suggest possible faulting of subbottom reflectors along the Abreojos Trough. Bathymetric evidence supports the interpretation that the trough is a relict strike-slip fault, because the valley defining the fault is as much as 1 km deep and defines a virtually straight line over a distance of at least 180 km (Figure 1).

The southern part of the Tosco-Abreojos fault zone is represented by the Tosco fault, which is recognized over a distance of about 130 km and is crossed by 7 seismic-reflection profiles. Unlike the Abreojos fault, the Tosco fault appears to be active, with sharp breaks in sea-floor morphology that separate rocks of contrasting acoustic character on reflection profiles (Figure 5, line V, km 81; Figure 8, line T-1, km 33–43; line U, km 31). The southern part of the Tosco fault separates a broad continental shelf from an uplifted bank or ridge at the edge of the shelf (Figure 5, line V; Figure 8, line U). To the north, the outer-shelf ridge narrows and disappears as the Tosco fault gradually descends over the edge of the shelf and down the slope, reaching a depth of about 1 km on line S (Figure 1). In northern areas, the Tosco fault appears to border narrow, acoustically opaque blocks that are internally disrupted and is marked by sharp breaks in slope (Figure 8, line T-1, km 33–43).

Northward continuation of the Tosco fault and possible connection with the Abreojos fault are suggested by features recognized on the two reflection profiles that cross the intervening area. On line R-2, a deep trough within the continental slope could be bounded on the west by a fault (Figure 8, line R-2, km 97). On line R (Figure 2 in Spencer and Normark, 1979), the sharpest morphologic feature within the slope is interpreted as a fault-bounded ridge. The fault that is inferred to cross lines R and R-2 connects the Tosco and Abreojos faults and defines a right-stepping en echelon offset of the combined Tosco-Abreojos fault zone. The data also permit a northward continuation of the Tosco fault that gradually descends to the foot of the slope (Figure 1; Figure 8, line N, km 83). Faults near the southwest end of lines X and W (Figure 8) may be the southward continuation of the Tosco fault.

Spencer and Normark (1979) proposed that the Tosco fault is an active remnant of the Neogene transform plate boundary that accommodated 200 to 300 km of right-lateral displacement. A possible problem with this interpretation concerns line V (Figure 5; also Figure 2 in Spencer and Normark, 1979). Two ridges, each bounded on the east by a fault, are visible on the profile. Dredge samples from the east flank of each ridge yielded middle or middle-to-late Miocene fossil ages—Luisian on the west ridge, Luisian to Mohnian on the east ridge (Figure 2; Table 2, samples from stations 69-4 and 69-5). Sediments on opposite sides of the Tosco fault, as seen on line V, have similar dip and seismic-penetration characteristics (Figure 5, line V, km 70–86). Correlation of the sediments on opposite sides of the fault would probably prohibit major displacement on the Tosco fault, since major displacement would most likely juxtapose rocks with contrasting structural and acoustic properties. However, the sediments appear to lap eastward onto the eastern basement ridge (Figure 5, line V, km 34–32) that yielded the Luisian-to-Mohnian-aged dredge sample; this suggests that the sediments between km 32 and km 81 are younger than, and not correlative with, those west of the Tosco fault. Dissimilar ages of sediment across the fault allows for (but is not evidence of) major lateral displacement on the Tosco fault.

The several-hundred-meter-thick blanket of sediments between km 32 and km 81 on line V (Figure 5) has been gently folded, truncated by erosion, and downdropped by normal movement on the Tosco fault. It seems unlikely that sediment deposition in this largely sediment-starved, outer-shelf environment, and subsequent folding, erosion, and normal faulting, all occurred since the end of major movement on the Neogene transform plate boundary 4 to 5 Ma.

Thus, although the sediments overlying the anticlinal ridge at km 60 on line V may be younger than Luisian or Mohnian (< 15 Ma), they are probably older than 4 to 5 Ma. Therefore, in the vicinity of line V, the Neogene transform plate boundary is probably located either at the Tosco fault, is farther west within the steep continental slope, or is east of km 28 (line V). The eastern location seems unlikely because the only recognized significant structure in this area is the Santa Margarita fault, which is not known to be laterally extensive and projects toward the land in the Magdalena Bay region, where there is no available evidence of strike-slip faulting (Rangin, 1978).

DISCUSSION

The seismic-reflection profile data and dredged samples from the Pacific continental margin of Baja California Sur provide new constraints for interpretation of the sedimentation history and tectonic events in this area during the Neogene. The reconnaissance nature of the study, however, precludes a firm reconciliation of the timing of basin development, filling, and later deformation with proposed plate tectonic evolution of the Baja California Peninsula and Gulf of California. The data can be used to make a few general observations about the resource potential of the offshore margin of Baja California Sur, but the lack of definition within sedimentary basins > 1 km thick makes this evaluation quite tenuous.

We will first review basin history based on correlation of offshore and onshore sedimentary sequences; then summarize the evidence for history of transform motion within the margin; and finally consider resource potential.

Basin History

Integration of the three-phase evolution of the Tortugas Formation (proposed by Helenes, 1980) with lithologic and faunal information obtained from dredge samples and seismic-stratigraphic patterns along the Baja California Sur margin allows a general history of Neogene basin evolution to be constructed. Such a history involves (1) initial rapid subsidence of the Pacific margin of Baja California between 22 and 20 Ma, (2) development of a number of discrete borderland basins between 20 and 15 Ma characterized by relatively low rates of sediment accumulation, anoxic to suboxic water, and precipitation of pelletal phosphorite on adjacent starved shelves and bank tops, (3) maximum Miocene subsidence to water depths between 1000 and 1500 m by 15 Ma, (4) development and encroachment of deep-sea fans across basin floors by the middle Miocene (ca. 15 to 11 Ma), carrying shelf-derived faunas and pelletal phosphorite onto basin slopes and floors, (5) rates of sediment accumulation outpacing rates of basin subsidence in later Miocene time (10 to 5 Ma), with increasing productivity of diatoms and deposition of diatomaceous sediment as rates of upwelling increased because of major climatic cooling during this period (Ingle, 1981; Keller and Barron, 1981), (6) widespread flexing, uplift, and erosion of Miocene basinal deposits during the latest Miocene and early Pliocene (ca. 5 to 4 Ma), (7) severely reduced delivery of terrigenous clastics to the off-shelf parts of the Baja California Sur margin during the late Pliocene and early Pleistocene, except for local subsiding depocenters such as the Ranger Bank area, (8) a widespread tectonic event in the middle Pleistocene represented by an angular unconformity between the upper Pleistocene terrace deposits and underlying Neogene units on the Vizcaino Peninsula (Figure 11) and on Cedros Island (Kilmer, 1979), and (9) continuing margin deformation, as documented by seismic evidence of Holocene faulting and warping of Pleistocene terraces onshore (Ortlieb, 1979). Table 4 summarizes the history of basin evolution.

Neogene Plate Boundary

The orientation and distribution of active and inactive faults within and adjacent to the Baja California Peninsula (Figure 12) represent a structural record of Neogene plate-tectonic interactions between the North American plate and the Pacific and Farallon-Guadalupe plates.

Plate-tectonic reconstructions indicate that about 200 to 300 km of displacement occurred between the Pacific and North American plates during the Neogene (Atwater, 1970; Atwater and Molnar, 1973). Relative motion was parallel to the continental margin of central and southern Baja California. There is little evidence that this displacement was accommodated by faulting within the Baja California Peninsula, Gulf of California, mainland Mexico, or within oceanic crust outboard of the continental margin. Therefore, the Neogene plate boundary is almost certainly within the Pacific margin of the Baja California Peninsula. This conclusion seems highly probable, since alternative locations are lacking, and linear bathymetric and structural features do exist within the margin that could represent the relict plate boundary (Spencer and Normark, 1979).

The exact location of the Neogene plate boundary within the continental margin is constrained by our data (Figure 12). If the relict plate boundary is within the continental slope west of the Tosco fault, it cannot project northward to the base of the slope in the vicinity of line R (Figure 1). The toe of the slope is thrust westward over oceanic crust, and both are buried by a thick sequence of sediments (Figure 9, km 144–147). The flat-lying sediments that overlap the toe of the margin are possibly correlative with sediments that form the Magdalena fan, which was drilled at DSDP hole 471 and is of middle Miocene age (Ness et al., 1981; Yeats and Haq, 1981). Such a correlation is supported by the similarity in seismic-reflection characteristics of sediments at the base of the slope on line R and on the DSDP drill-hole profile. A strong reflector slightly more than halfway above acoustic basement is present on both profiles. This reflector corresponds to a sharp drop in sedimentation rate at 13.0 Ma and marks a major lithologic boundary (Yeats and Haq, 1981). Therefore, it appears that the toe of the margin was tectonically attached to the Pacific-Magdalena plate, at least in the vicinity of line R, when these sediments were first deposited at about 14.5 Ma.

Thus, the Neogene transform plate boundary is either represented by the Tosco-Abreojos fault zone, which appears as a morphologic or structural feature on at least 15 seismic-reflection profiles, or it occurs within the continental slope and is largely invisible on our records. In either case, the overall orientation of the relict plate boundary is about N35°W ±3°.

Resource Potential

The existing seismic-reflection profiling data suggest that two offshore areas of Baja California Sur may have hydro-

Table 4. Summary of time constraints on Neogene tectonic activity and sedimentation.

From Plate Motion Reconstructions		From Biostratigraphic Synthesis for Baja California Sur	
29 Ma	Initiation of transform motion between Pacific and North American plates		
		22–20 Ma	Initial rapid subsidence
		20–15 Ma	Development of discrete basins and maximum subsidence (approximately 1500 m)
14 Ma	Age of oceanic crust at Site 471		
12.5 Ma	Cessation of ridge spreading between lat. 23°30′N and 29°30′N	15–11 Ma	Turbidites deposited in basins Implied slowing of rate of subsidence
12.5–11 Ma	Westward jump of spreading center between lat. 27°N and 29°30′N		
14/12–4/3.5 Ma	Proto-Gulf extension and Tosco-Abreojos activity		
		10–5 Ma	Sedimentation rate exceeds subsidence rate
5 Ma	Change in relative motion of Pacific and North American plates	5–4 Ma	Uplift and widespread erosion of Miocene basin fill
4/3.5 Ma	Initiation of activity in Gulf of California transform system	3–2 Ma	Local subsidence—not widespread
		≤1 Ma	Tectonic disruption causing unconformities between Neogene and upper Pleistocene terrace deposits

carbon resources—Sebastian Vizcaino Bay and the Magdalena borderland. Both the seismic-refraction data and the on-land mapping discussed earlier suggest an appreciable accumulation of Cretaceous and younger sediment in the synclinorium that underlies Sebastian Vizcaino Bay (Helenes, 1984). Drilling by PEMEX within the Sebastian Vizcaino Bay area provided the material for Helenes' 1980 study, but other well data apparently have not been released. On most profiles (Figures 4, 6A) our reflection data from the Miocene and younger units in the basin show numerous faults that cut possible stratigraphic traps in the uppermost sedimentary units. Thus, the potential for accumulation of hydrocarbons on the Neogene section appears low, although some faults might form traps rather than act as conduits.

Although our reflection profiles show several possible structural traps within the Magdalena borderland, its shallow depth and our lack of knowledge of basin thickness and pre-Miocene sediment type make it impossible to comment meaningfully on the hydrocarbon potential. Here, also, fault disruption of the folded sequences is common.

The common occurrence of phosphatic sediment in Miocene strata, especially from the southern half of the investigation area (D'Anglejan, 1967; Table 1), suggests that further investigation of this potential resource may be warranted.

CONCLUSIONS

Key time constraints on the Neogene history of the Baja California Sur Pacific margin are summarized in Table 4. We have already presented references for the ages given in Table 4, as well as an interpretation of the major structural features. Synthesis of the structural, biostratigraphic, and plate tectonic reconstructions suggests the following geologic history of the margin (see Figure 12 for the general setting).

Neogene basin development began while subduction of the Farallon plate was still active off that part of the Mexican margin that would eventually become Baja California Sur. The rapid subsidence around 20 to 22 Ma might reflect a change to subduction of older, deeper crust as major transform offsets of the Pacific-Guadalupe spreading center (such as the Molokai fracture zone-Shirly Trough feature) moved past the margin, reflecting Pacific-North American relative motion (e.g., see Figure 2C in Blake et al., 1978; Figure 3 in Menard, 1978). Alternatively, subsidence at this time might be related to reestablishment of the magmatic arc along the eastern margin of the Baja California Peninsula by about 22 Ma (Hausback, 1984). Southwestward migration of magmatism occurred in northwestern Mexico and southwestern United States in the middle Tertiary, and magmatism during this migration was associated with a strong pulse of extensional tectonism (Coney, 1978). The sweep of magmatism is attributed to steepening of the subducted slab (Coney and Reynolds, 1977). Subsidence and silled-basin formation within the continental margin might be the result of slab steepening and associated minor extension within the continental margin. Basin deepening appeared to slow after spreading-ridge abandonment and after subduction along the margin ceased about 12 Ma.

During formation of the proto-Gulf of California and transform motion on the Tosco-Abreojos fault zone (post-12 or post-14 Ma), the sedimentation rate generally exceeded subsidence rates in the basins. Southwestward tilting of the incipient peninsula, caused by the thermal effects of extension in the proto-Gulf, could have resulted in increased rates of clastic sediment delivery to the continental margin and

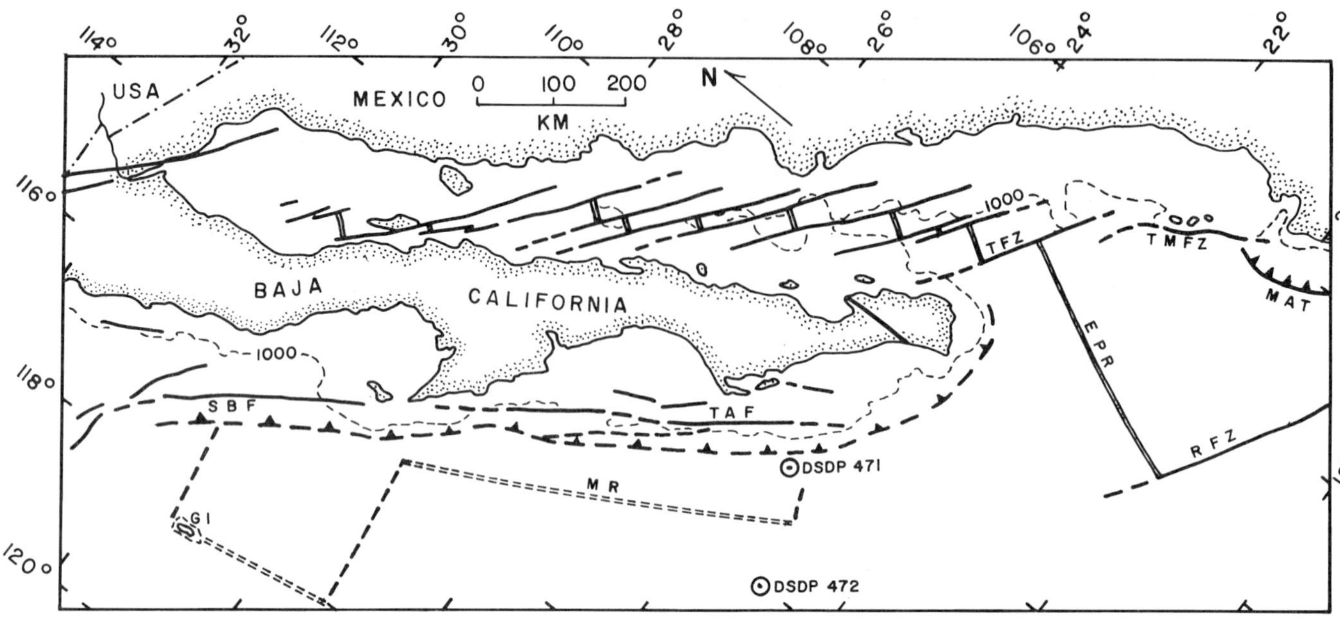

Figure 12. Tectonic map of Baja California and surrounding area. Active features include spreading centers and transform faults in the Gulf of California, the Tamayo fracture zone (TFZ), East Pacific Rise (EPR), Rivera fracture zone (RFZ), Tres Marias fracture zone (TMFZ), and Middle American Trench (MAT). Inactive features include the San Benito fault (SBF), Tosco-Abreojos fault zone (TAF) and the hypothetical Magdalena Ridge (MR). Also shown is an inactive subduction zone (dashed thrust fault symbol) at the base of the continental margin of Baja California and Guadalupe Island (GI), west of Sebastian Vizcaino Bay. Depth contour (dashed line) in meters.

corresponding increased rates of submarine-fan sediment accumulation.

The orientations of active and inactive faults on both sides of the Baja California Peninsula (Figure 12) indicate that Pacific-North America relative plate motion was more westward after about 4–5 Ma. Such movement contradicts the conclusion of Cox and Engebretson (1985), who recognize a change to more northward relative plate motion at about 5 Ma. Their conclusion is based on changes in orientation of the Hawaiian Seamount chain. Regardless of the exact nature of changes in relative plate motions at this time, the plate boundary jumped inland to form the Gulf of California, and this jump coincided with widespread uplift and erosion of Miocene basin-fill sediment along the Pacific margin of Baja California Sur.

After the present Gulf of California began to develop, the Pacific margin remained tectonically active, especially during the Pleistocene, as indicated by uplift and warping of marine terrace deposits and significant uplift of late Pliocene sediment in the Sebastian Vizcaino Bay area. The cause of this continued deformation is not obvious but could reflect some continued transform displacement between the Pacific plate and the Baja California Peninsula.

ACKNOWLEDGMENTS

Most of the data for this study reflect the support of many individuals from Scripps Institution of Oceanography during the senior writer's student and postgraduate studies at that institution. Special thanks are due J. R. Curray, who provided an introduction to the Baja California margin, and F. N. Spiess, on whose deep-tow cruises much of the initial data were collected. The late E. C. Allison of San Diego State University, who participated on the only rock-dredging cruise, passed away before having a chance to complete his biostratigraphic studies on the samples; his guidance during the early phase of this study is gratefully acknowledged. We also thank R. W. Couch and G. E. Ness for stimulating discussions and for access to unpublished data. G. G. Shor provided access to unpublished seismic-refraction data. The manuscript was improved after critical reviews from H. McLean and K. D. Klitgord.

REFERENCES CITED

Abbott, P., and G. Gastil (eds.), 1979, Baja California geology: San Diego State University, Department of Geology, 228 p. (prepared for 1979 Annual Meeting, Geological Society of America, San Diego).

Allison, E. C., 1964, Geology of areas bordering the Gulf of California, *in* Tj. van Andel and G. Shor, eds., The marine geology of the Gulf of California: AAPG Memoir 3, p. 3–29.

Atwater, T., 1970, Implications of plate tectonics for the Cenozoic tectonic evolution of western North America: GSA Bulletin, v. 81, no. 12, p. 3513–3536.

Atwater, T., and P. Molnar, 1973, Relative motion of the Pacific and North American plates deduced from sea-floor spreading in the Atlantic, Indian, and South Pacific Oceans, *in* R. L. Kovach and A. Nur, eds., Proceedings of the conference on tectonic problems of the San Andreas fault system, California: Stanford University Publications in the Geological Sciences, v. 13, p. 136–148.

Barnes, D. A., 1984, Volcanic arc derived Mesozoic sedimentary rocks, Vizcaino Peninsula, Baja California Sur, Mexico, *in* V. Frizzell, ed., Geology of the Baja California Peninsula: SEPM, Pacific Section, v. 39, p. 119–130.

Barron, J. A., 1981, Late Cenozoic diatom biostratigraphy of the middle latitude eastern North Pacific, Deep Sea Drilling Project Leg 63, *in* R. S. Yeats, B. U. Haq, et al., Initial Reports of Deep Sea Drilling Project (Vol. 63): Washington, D.C., U.S. Government Printing Office, p. 509–538.

Batiza, R., and C. G. Chase, 1981, Miocene spreading centre south of Isla Guadalupe: Nature, v. 289, p. 787–789.

Beal, C. H., 1948, Reconnaissance of the geology and oil possibilities of Baja California, Mexico: GSA Memoir, v. 31, 138 p.
Berggren, W. A., D. V. Kent, and J. A. Couvering, in press, Neogene geochronology and chronostratigraphy, *in* N. J. Snelling, ed., Geochronology and the geologic time scale: Geological Society of London Special Paper.
Blake, M. C., Jr., R. H. Campbell, T. W. Dibblee, Jr., D. G. Howell, T. H. Nilsen, W. R. Normark, J. G. Vedder, and E. A. Silver, 1978, Neogene basin formation in relation to plate tectonic evolution of San Andreas fault system, California: AAPG Bulletin, v. 62, p. 344–372.
Boles, J. R., and C. A. Landis, 1984, Jurassic sedimentary melange and associated facies, Baja California, Mexico: GSA Bulletin, v. 95, no. 5, p. 513–521.
Calderon-Riveroll, G., and R. W. Couch, 1977, A marine geophysical study of the Vizcaino Bay area of Baja California (abs.): EOS (American Geophysical Union Transactions), v. 58, p. 1151.
Cohen, L. H., K. C. Condie, L. J. Kuest, Jr., G. S. Mackenzie, F. H. Meister, P. Pushkar, and A. M. Stueber, 1963, Geology of the San Benitos Islands, Baja California, Mexico: GSA Bulletin, v. 74, p. 1355–1370.
Coney, P. J., 1978, The plate-tectonic setting of southeastern Arizona, *in* J. F. Callender, J. C. Wilt, and R. E. Clemons, eds., Land of Cochise: New Mexico Geological Society, Twenty-Ninth Field Conference Guidebook, p. 285–290.
Coney, P. J., and S. J. Reynolds, 1977, Cordilleran Benioff zones: Nature, v. 270, p. 403–406.
Cox, A., and D. Engebretson, 1985, Change in motion of Pacific plate at 5 m.y.B.P.: Nature, v. 313, p. 472–474.
D'Anglejan, B., 1967, Origin of marine phosphorites off Baja California, Mexico: Marine Geology, v. 5, p. 15–44.
Douglas, R. G., 1981, Paleoecology of continental margin basins; a modern case history from the borderland of southern California, *in* R. G. Douglas, I. T. Colburn, and D. S. Gorsline, eds., Depositional systems of active continental margin basins: SEPM, Pacific Section, Short Course Notes, v. 16, p. 121–156.
Doyle, L. J., and O. L. Bandy, 1972, Southern Continental Borderland, Baja California—Its tectonic and environmental development: GSA Bulletin, v. 83, no. 12, p. 3785–3794.
Emery, K. O., 1948, Submarine geology of Ranger Bank, Mexico: AAPG Bulletin, v. 32, p. 790–805.
Fisher, R. L., and H. H. Hess, 1963, Trenches, *in* M. N. Hill, ed., The Sea, The Earth Beneath the Sea (Vol. 3): New York, John Wiley and Sons, Inc., p. 411–436.
Frizzell, V. A., Jr., 1984, The geology of the Baja California Peninsula: An introduction, *in* V. A. Frizzell, Jr., ed., Geology of the Baja California Peninsula: SEPM, Pacific Section, v. 39, p. 1–7.
Garrison, R. E., and R. G. Douglas (eds.), 1981, The Monterey Formation and related siliceous rocks of California: SEPM, Pacific Section, v. 15, 327 p.
Harland, W. B., A. V. Cox, P. G. Llewellyn, G. A. Pickton, A. G. Smith, and R. Walters, 1982, A geologic time scale: London, Cambridge University Press, Cambridge Earth Sciences Series, 131 p.
Hausback, B. P., 1984, Cenozoic volcanic and tectonic evolution of Baja California Sur, Mexico, *in* V. A. Frizzell, Jr., ed., Geology of the Baja California Peninsula: SEPM, Pacific Section, v. 39, p. 219–236.
Helenes, J., 1980, Stratigraphy, depositional environments and foraminifera of the Miocene Tortugas Formation, Baja California Sur, Mexico: Mexicana Geologos Boletin, v. 41, nos. 1 and 2, p. 47–67.
Helenes, J., 1984, Dinoflagellates from Cretaceous to early Tertiary rocks of the Sebastian Vizcaino Basin, Baja California, Mexico, *in* V. A. Frizzell, Jr., ed., Geology of the Baja California Peninsula: SEPM, Pacific Section, v. 39, p. 89–106.
Ingle, J. C., Jr., 1973, Neogene foraminifera from the northeastern Pacific Ocean, Leg 18, Deep Sea Drilling Project, *in* L. D. Kulm, R. von Huene, et al., Initial Reports of the Deep Sea Drilling Project (Vol. 18): Washington D.C., U.S. Government Printing Office, p. 517–567.
Ingle, J. C., Jr., 1980, Cenozoic paleobathymetry and depositional history of selected sequences within the southern California Continental Borderland: Cushman Foundation of Foraminiferal Research Special Publication 19, p. 163–195.
Ingle, J. C., Jr., 1981, Origin of Neogene diatomites around the North Pacific rim, *in* R. E. Garrison and R. G. Douglas, eds., The Monterey Formation and related siliceous rocks of California: SEPM, Pacific Section, v. 15, p. 159–179.
Ingle, J. C., Jr., and G. Keller, 1980, Benthic foraminiferal biofacies of the eastern Pacific margin between 40°S and 32°N, *in* M. Field, A. H. Bouma, I. T. Colburn, R. G. Douglas, and J. C. Ingle, Jr., eds., Quaternary deposition environments of the Pacific Coast: SEPM, Pacific Section, v. 10, p. 341–355.
Jones, D. L., M. C. Blake, Jr., and C. Rangin, 1976, The four Jurassic belts of northern California and their significance to the geology of the southern California Borderland, *in* D. G. Howell, ed., Aspects of the geologic history of the California Continental Borderland: AAPG Miscellaneous Publication 24, p. 343–362.
Keller, G., 1981, Miocene biochronology and paleoceanography of the North Pacific: Marine Micropaleontology, v. 6, p. 535–551.
Keller, G., and J. A. Barron, 1981, Integrated planktic foraminiferal and diatom biochronology for the northeast Pacific and the Monterey Formation *in* R. E. Garrison and R. G. Douglas, eds., The Monterey Formation and related siliceous rocks of California: SEPM, Pacific Section, v. 15, p. 43–54.
Kilmer, F. H., 1969, Preliminary report on the geology of Cedros Island, Baja California, Mexico (abs.): GSA Special Paper 21, p. 521.
Kilmer, F. H., 1979, A geological sketch of Cedros Island, Baja California, Mexico, *in* P. Abbott and G. Gastil, eds., Baja California geology: San Diego State University, Department of Geology, p. 11–28.
Kimbrough, D. L., 1984, Paleogeographic significance of the Middle Jurassic Gran Canyon Formation, Cedros Island, Baja California Sur, *in* V. A. Frizzell, ed., Geology of the Baja California Peninsula: SEPM, Pacific Section, v. 39, p. 107–118.
Kleinpell, R. M., 1938, Miocene stratigraphy of California: Tulsa, AAPG, 350 p.
Kleinpell, R. M., 1980, The Miocene stratigraphy of California revisited: Tulsa, AAPG, 349 p.
Klitgord, K. D., and J. Mammerickx, 1982, Northern East Pacific Rise: Magnetic anomaly and bathymetric framework: Journal of Geophysical Research, v. 87, p. 6725–6750.
Krause, D. C., 1965, Tectonics, bathymetry, and geomagnetism of the southern continental borderland west of Baja California, Mexico: GSA Bulletin, v. 76, p. 617–650.
Mammerickx, J., and K. D. Klitgord, 1982, Northern East Pacific Rise—Evolution from 25 m.y.B.P. to the present: Journal of Geophysical Research, v. 87, p. 6751–6759.
Menard, H. W., 1978, Fragmentation of the Farallon plate by pivoting subduction: Journal of Geology, v. 86, p. 99–110.
Mina, F., 1957, Bosquejo geologico del Territorio Sur de la Baja California: Asociacion Mexicana de Geologos Petroleros Boletin, v. 9, p. 139–270.
Minch, J. A., G. Gastil, W. Fink, J. Robinson, and A. H. James, 1976, Geology of the Vizcaino Peninsula, *in* D. G. Howell, ed., Aspects of the geologic history of the California Continental Borderland: AAPG Miscellaneous Publication 24, p. 136–195.
Moore, D. G., 1969, Reflection profiling studies of the California Continental Borderland: Structure and Quaternary turbidite basins: GSA Special Paper 107, 142 p.
Moore, T. E., 1984, Sedimentary facies and composition of Jurassic volcaniclastic turbidites at Cerro el Calvario, Vizcaino Peninsula Baja California Sur, Mexico, *in* V. A. Frizzell, ed., Geology of the Baja California Peninsula: SEPM, Pacific Section, v. 39, p. 131–147.
Ness, G. E., Z. O. Sanchez, R. W. Couch, and R. S. Yeats, 1981, Bathymetry and oceanic crustal ages in the vicinity of the mouth

of the Gulf of California, illustrated using Deep Sea Drilling Project Leg 63 underway geophysical profiles, *in* R. S. Yeats, B. U. Haq, et al., Initial Reports of the Deep Sea Drilling Project (Vol. 63): Washington, D.C., U.S. Government Printing Office, p. 919–923.

Normark, W. R., 1974, Ranger submarine slide, northern Sebastian Vizcaino Bay, Baja California, Mexico: GSA Bulletin, v. 85, no. 5, p. 781–784.

Normark, W. R., 1977, Neogene basins and transform motion within the Pacific continental margin of Baja California: Houston, Texas, Offshore Technology Conference Proceedings, v. 3, p. 93–100.

Ortlieb, L., 1979, Quaternary marine terraces in southwestern Vizcaino Peninsula, Baja California, Mexico, *in* P. Abbott and G. Gastil, eds., Baja California geology: San Diego State University, Department of Geology, p. 89–93.

Rangin, C., 1978, Speculative model of Mesozoic geodynamics, central Baja California to northwestern Sonora (Mexico), *in* D. G. Howell and K. A. McDougall, eds., Mesozoic paleogeography of the western United States: SEPM, Pacific Section, Pacific Coast Paleogeography Symposium 2, p. 85–106.

Robinson, J. W., 1975, Reconnaissance geology of the northern Vizcaino Peninsula, Baja California Sur, Mexico: San Diego, San Diego State University unpublished Master's thesis, 111 p.

Sawlan, M. G., and J. G. Smith, 1984, Petrologic characteristics, age, and tectonic setting of Neogene volcanic rocks in northern Baja California Sur, Mexico, *in* V. A. Frizzell, Jr., ed., Geology of the Baja California Peninsula: SEPM, Pacific Section, v. 39, p. 237–251.

Smith, J. T., 1984, Miocene and Pliocene marine mollusks and preliminary correlations, Vizcaino Peninsula to Arroyo La Purisima, northwestern Baja California Sur, Mexico, *in* V. A. Frizzell, ed., Geology of the Baja California Peninsula: SEPM, Pacific Section, v. 39, p. 197–218.

Spencer, J. E., and W. R. Normark, 1979, Tosco-Abreojos fault zone—A Neogene transform plate boundary within the Pacific margin of southern Baja California, Mexico: Geology, v. 7, p. 554–557.

Yeats, R. S., and B. U. Haq, 1981, Deep-sea drilling off the Californias: Implications of Leg 63, *in* R. S. Yeats, B. U. Haq, et al., Initial Reports of the Deep Sea Drilling Project (Vol. 63): Washington, D.C., U.S. Government Printing Office, p. 949–961.

CHAPTER NINETEEN

STRUCTURAL FRAMEWORK, SEDIMENTARY SEQUENCES, AND HYDROCARBON POTENTIAL OF THE ALEUTIAN AND BOWERS BASINS, BERING SEA

A. K. Cooper
D. W. Scholl
M. S. Marlow

U.S. Geological Survey
Menlo Park, California

ABSTRACT

The abyssal (2600–3800 m) Aleutian and Bowers Basins are underlain by 2 to 10 km of Cenozoic sedimentary rocks mostly deposited after the Eocene development of the Aleutian Ridge. These rocks rest on Mesozoic (?) oceanic crust.

Large structural depressions (ancient trenches?) are buried in the Aleutian Basin at the base of Bowers Ridge and beneath the Beringian and Koryak continental slopes. Sedimentary prisms filling these depressions are deformed on their landward side by faulting (thrust, normal, wrench), folding (diapiric, compressional), and basement relief (intrusions, fault blocks, horsts). Similar sedimentary prisms and structures have not been found in Bowers Basin.

The central Aleutian Basin is characterized by 2 to 4 km of mostly flat-lying sedimentary rocks, as much as 1 to 2 km of basement relief, elevated heat flow, thinner crust, and numerous velocity-amplitude anomalies (VAMPs). Basement ridges were formed prior to sediment deposition, and there is no evidence for a Cenozoic sea-floor-spreading ridge beneath the Aleutian Basin.

The Aleutian Basin may contain economic hydrocarbon deposits because of the

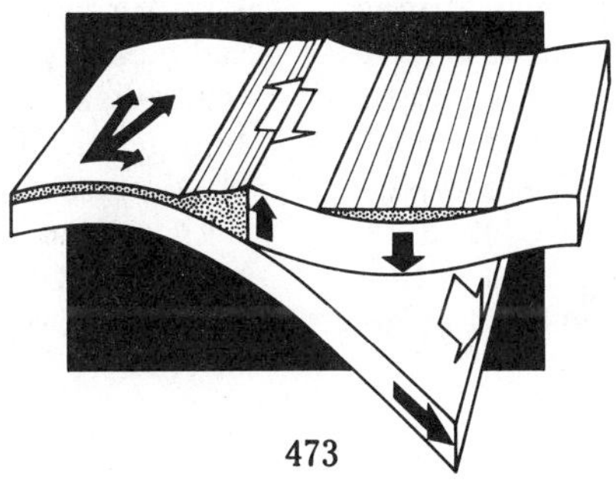

presence of adequate structures, traps, and temperatures, as well as potential reservoir beds and source beds, particularly along the Beringian and Koryak continental slope and rise. Bowers Basin has a lesser potential for economic hydrocarbons, because sedimentary rocks are thinner and basement relief is smaller than that in the Aleutian Basin.

INTRODUCTION

The Bering Sea Basin comprises three deep-water basins—the Aleutian, Bowers, and Komandorsky Basins (Figure 1). Two of these basins (Aleutian and Bowers) lie principally within United States waters, as defined by the 1867 treaty line. The Aleutian Basin, the largest of the three, lies adjacent to the broad Beringian continental shelf (Bering Shelf) and covers an area nearly the size of the state of Texas. Bowers Basin, about one-sixth the size of the Aleutian Basin, is flanked by the Bowers and Aleutian Ridges. Geophysical investigations, which have been conducted within these basins since the mid-1960s, show that they are underlain by thick and laterally extensive sedimentary deposits. In recent years these deposits have been investigated for their petroleum potential.

We summarize here previous geologic and geophysical studies in the Aleutian and Bowers Basins and discuss the implications of these findings on hydrocarbon prospects.

OVERVIEW

Syntheses of the Bering Sea abyssal basins and adjacent areas are given by Ewing, Ludwig, and Ewing (1965), Gnibidenko (1973, 1980), Ludwig (1974), Scholl, Buffington, and Marlow (1975), Cooper, Marlow, and Scholl (1976, 1977, and this volume), Cooper et al. (1979), Ben-Avraham and Cooper (1981), Bogdanov and Neprochnov (1984), and Stone (in press). A brief summary of these studies is given here.

The Aleutian and Bowers Basins are back-arc marginal basins that, unlike most other back-arc basins of the Pacific, contain thick sections of mostly Cenozoic sediment. The structure of the underlying igneous crust is geophysically similar to that of oceanic areas. Sea-floor spreading anomalies beneath the Aleutian Basin (M1 to M13, Figure 1) and regional geologic data provide evidence that the Aleutian Basin may have been part of the North Pacific region before the early Tertiary (55–50 Ma). The origin of Bowers Basin is uncertain, and it may be a younger feature.

The initial development of the Aleutian Ridge in the early Tertiary severed the oceanic (Kula?) plate and isolated a segment of oceanic crust on the north side of the newly formed ridge. Bowers and Shirshov Ridges (Figure 1) may also have begun as island arcs at this time; alternatively, the two ridges may have been existing features on the oceanic crust, which were isolated in the Bering Sea. Since the early Tertiary, terrigenous and pelagic sediments have draped the continental slopes and filled the Aleutian and Bowers Basins with a nearly flat-lying sedimentary section that is thickest at the margins of the basins.

STRUCTURAL FRAMEWORK

The Aleutian Basin is bordered on its northern and eastern sides by the Koryak and Beringian segments of the continental margin, and on the south and west by large bathymetric ridges—the Aleutian, Bowers, and Shirshov Ridges.

Central Basin Regions

Geologic sampling of the sedimentary section in the Aleutian Basin (Lisitsyn, 1966; Gershanovich, 1968; Creager, Scholl, et al., 1973) is limited to surface cores and DSDP site 190, which reached maximum depth of 627 m below the sea floor. The samples consist of diatomaceous mud with various amounts of interbedded terrigenous material. At DSDP site 190 (Figure 2A), three sedimentary units were penetrated—an upper unit of diatomaceous silty clay with interbedded turbidite beds, 375 m thick (Holocene through middle Pliocene in age); a middle unit of semi-indurated diatomaceous silty clay, 240 m thick (middle Pliocene through late Miocene in age); and a lower unit of indurated mudstone of unknown thickness (late middle Miocene in age). A similar, but more diatom-rich, section was found at DSDP site 188 on the lower slope of Bowers Ridge where it descends into Bowers Basin (Figure 2B).

Acoustic basement (igneous?) was not reached at sites in the Aleutian Basin or on the flank of the Bowers Basin. Consequently, the age and lithology of the lower and thicker part of the sedimentary sections are not known. Dredge samples along the Beringian margin of the Aleutian Basin range in age from late Jurassic to late Cenozoic (Jones et al., 1981; Marlow et al., 1983). These rocks consist of shallow-water or continental-slope deposits and are not necessarily similar to rocks in the adjacent basin. The age of the oldest sedimentary rocks overlying the acoustic basement of the Aleutian Basin is believed to be Cretaceous, on the basis of regional geologic and geophysical data, although the bulk of the section began to accumulate in the early Tertiary (Scholl, Buffington, and Marlow, 1975; Cooper and Ben-Avraham, 1981).

Seismic-reflection profiles recorded in the central Aleutian and Bowers Basins (Ewing, Ludwig, and Ewing, 1965; Ludwig, Houtz, and Ewing, 1971; Scholl, Buffington, and Marlow, 1976; Rabinowitz and Cooper, 1977; Cooper and Ben-Avraham, 1981; Figure 3A, B) reveal a 2–5-km-thick sedimentary section above an acoustic basement with irregular relief. The upper 0.5 to 1.5 km of the sedimentary section generally consists of a well-stratified and horizontally continuous sequence of reflectors that overlies a deeper unit with few such reflectors. A prominent refraction horizon lies near the acoustic boundary between the upper and lower parts of the sedimentary section and marks a sharp increase in velocity from 1.8–2.0 km/sec to 2.3–2.8 km/sec (Figure

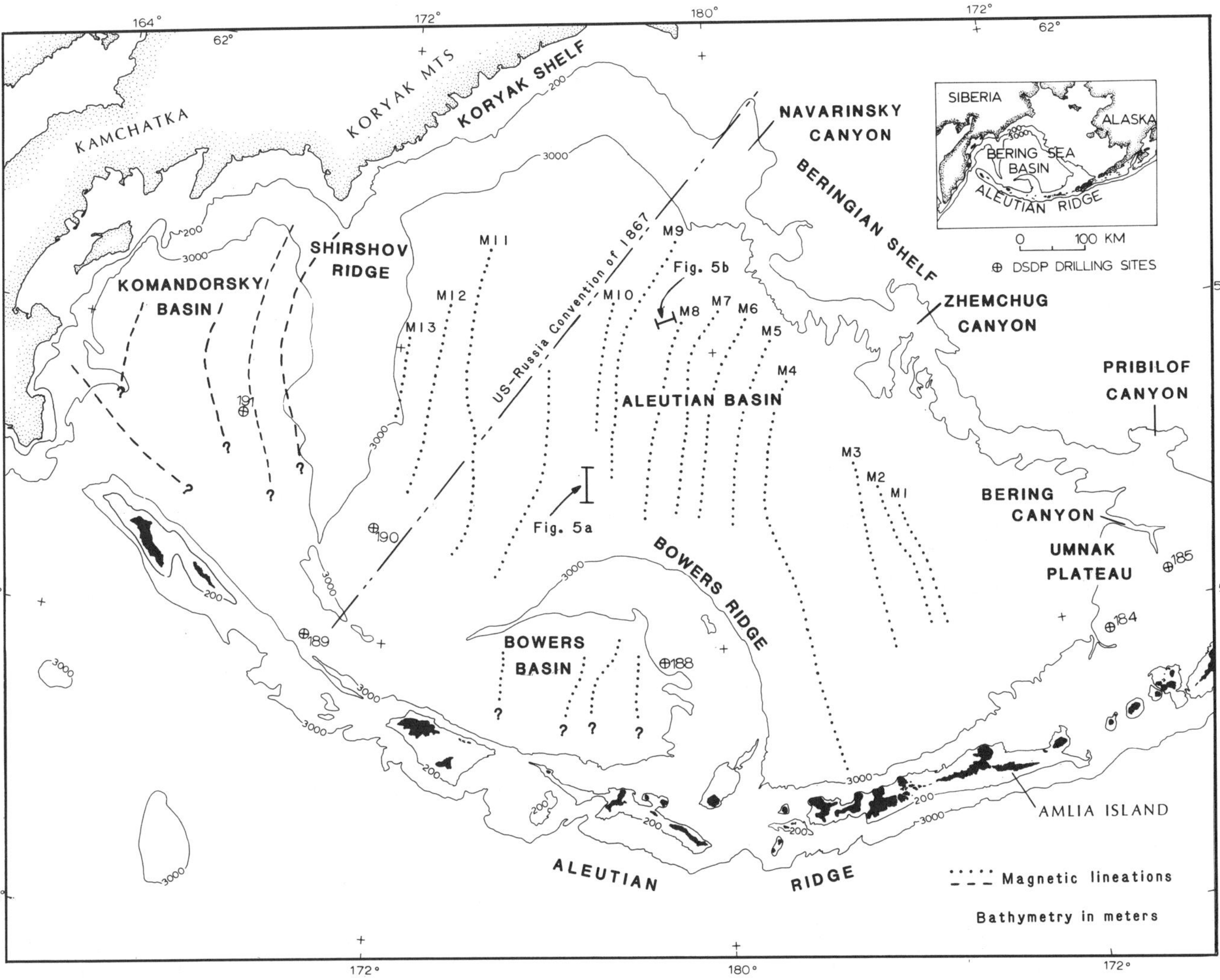

Figure 1.. Index map of the Bering Sea Basin showing magnetic lineations (from Cooper, Scholl, and Marlow, 1976), DSDP drilling sites, and US-Russia 1867 Treaty line. Albers equal area projection.

3C). The refraction horizon can be observed basin wide (Shor, 1964; Ludwig, Houtz, and Ewing, 1971; Rabinowitz and Cooper, 1977; Cooper et al., 1979). This horizon is believed to be a diagenetic transition from diatom-rich sediment containing much opal-A to underlying better-indurated rocks cemented by opal-CT (Hein et al., 1978; Cooper et al., 1979). A bottom-simulating reflector (BSR) commonly coincides with this refraction horizon.

The thickness of the sedimentary rocks shows a wide range throughout the Aleutian Basin (Figure 4), but the section is generally thinner in the central part of the basin and thicker toward its edges. At depth, the older sedimentary deposits are thickest at the base of the north side of Bowers Ridge and at the base of the Beringian margin (Cooper and Ben-Avraham, 1981). The younger sedimentary rocks are thickest adjacent to the Koryak margin (Cooper et al., 1979). The sedimentary section is generally undeformed except locally near some basement structures and along the basin margins, where tectonism has occurred (Scholl, Buffington, and Marlow, 1975; Cooper and Ben-Avraham, 1981).

The sedimentary sections within the Aleutian and Bowers Basins are underlain by oceanic igneous crust, which is 5 to 10 km thick (Figure 3C). This crust consists of a rock sequence 1 to 2 km thick that has a velocity of 5.0 to 6.0 km/sec (oceanic layer 2), and an underlying layer that is 2 to 6 km thick and has a velocity of 6.6 to 7.4 km/sec (oceanic layer 3). Mantle is characterized by velocities of 7.5 to 8.2 km/sec. Refractions from these igneous sections are easily measured in sonobuoy data, yet reflections from deep intracrustal horizons are rarely observed in multichannel seismic-reflection profiles (Cooper and Ben-Avraham, 1981). Consequently, little is known concerning the detailed geometry of layers and structures within the deeper igneous crust of the basin.

A few large basement ridges (with >1 km relief) have been identified beneath the thick sedimentary section of the Aleutian Basin (Figure 5). These buried ridges, which are generally recorded only in multichannel seismic-reflection (MCS) profiles (Cooper, Marlow, and Scholl, 1977; Cooper and Ben-Avraham, 1981), are old structures that have been

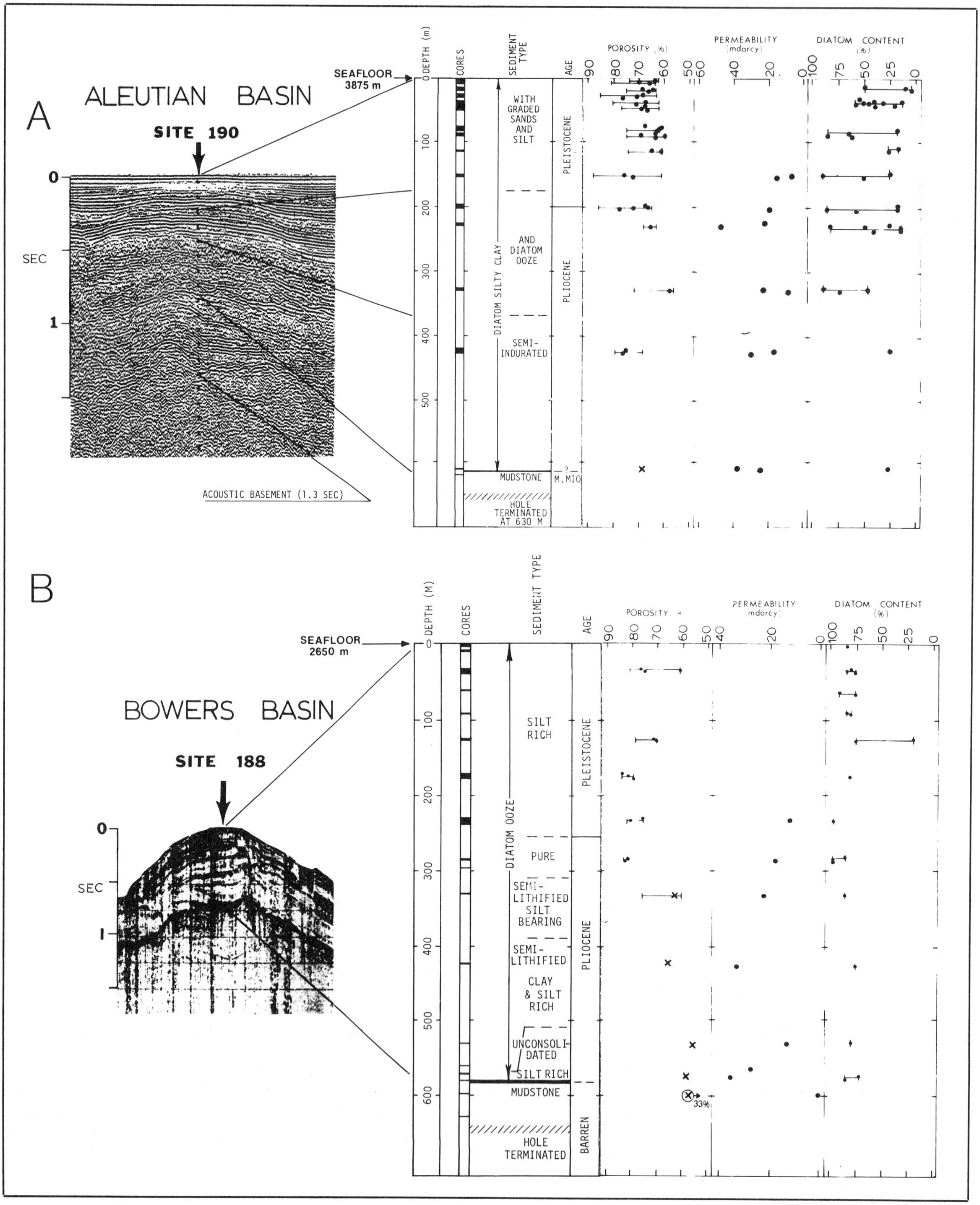

Figure 2. DSDP drilling results for sites 188 and 190 in the Bering Sea (Creager, Scholl, et al., 1973; Cooper et al., 1979). See Figure 1 for locations.

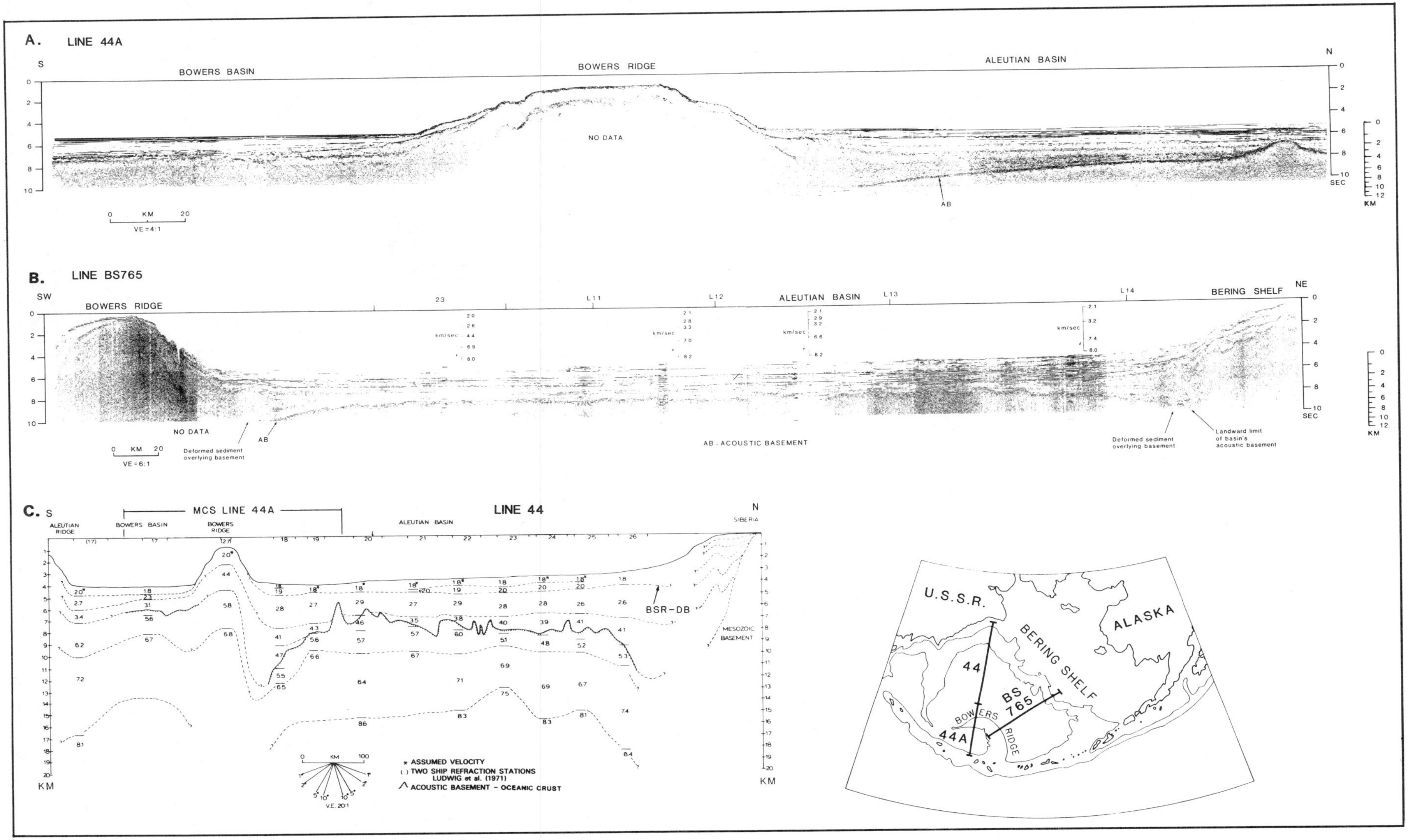

Figure 3. Multichannel seismic-reflection and refraction lines across the Aleutian and Bowers Basins. A and B are lines 44A and BS765 from Cooper and Ben-Avraham (1981). Thick, flat-lying sedimentary sections fill the central Aleutian Basin and the sediment wedges at the base of Bowers Ridge and the Bering Shelf. C, Refraction line 44 (from Cooper et al., 1979). There is a great difference in sediment thickness between Bowers and the Aleutian Basins, especially the buried trench on the north (but not on the south) side of Bowers Ridge.

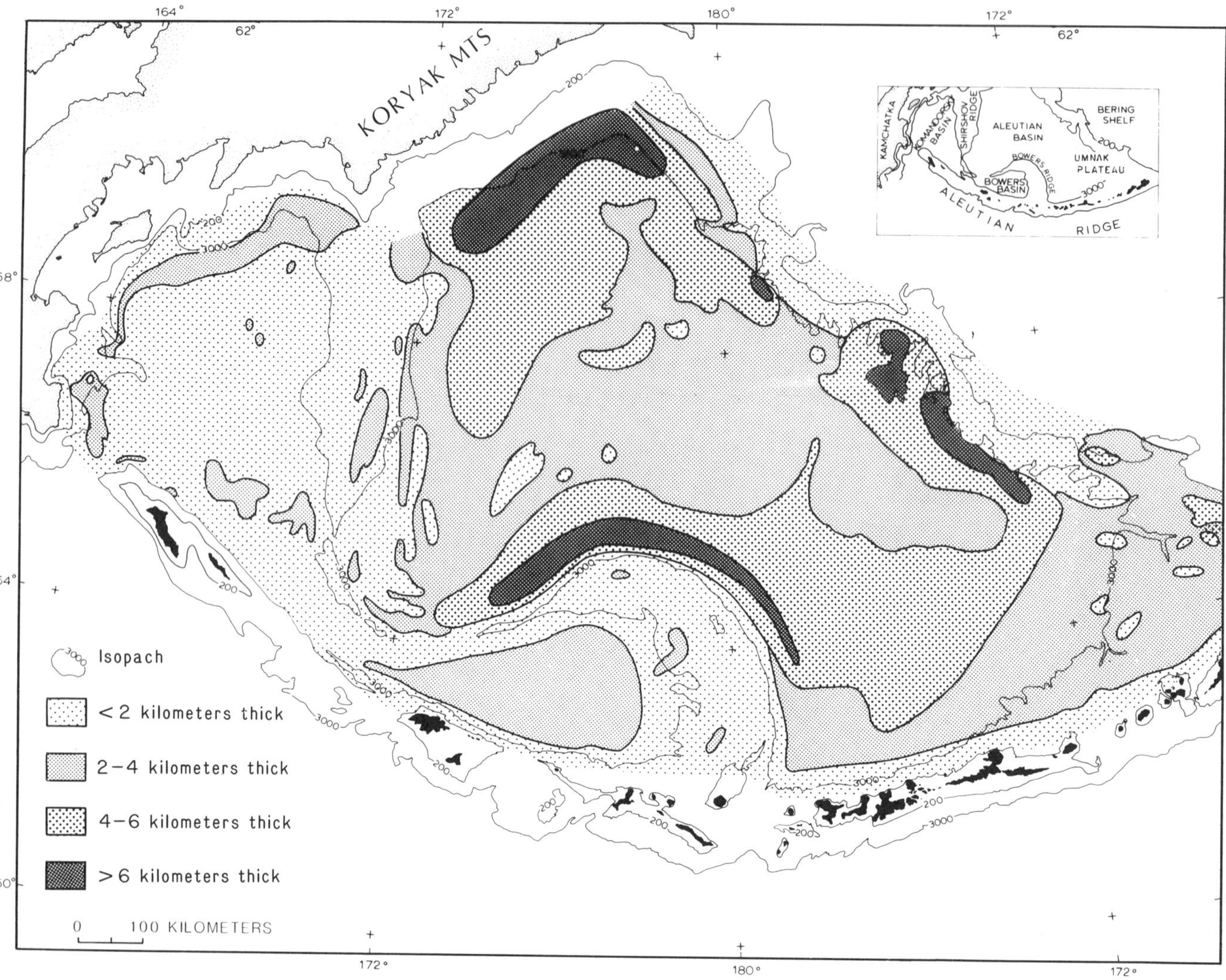

Figure 4. Isopach map showing thickness of sedimentary rocks overlying acoustic basement (basalt?) in the abyssal basins (from Marlow et al., 1979a). In some areas (Umnak Plateau, Shirshov Ridge, and possibly Bowers Basin) acoustic basement may be deformed sediment overlying igneous basement, and true sedimentary rock thickness will be greater than shown. Bathymetric contours in meters.

covered by more than 1 km of flat-lying or slightly draped sedimentary rocks. Two ridges (Bartlett and Sounder Ridges, Figure 5; Cooper, Marlow, and Scholl, 1976, 1977), however, exhibit minor deformation of shallow reflection horizons, possibly indicating differential compaction and sedimentation or slight uplift. Locally, positive gravity anomalies are associated with the ridges.

Data on the magnetic properties and suspected age of the igneous crust have been derived from magnetic surveys in the Bering Sea Basin (Soloveva, 1968; Vorobev, 1970; Kienle, 1971; Regan, Cain, and David, 1975; Cooper, Marlow, and Scholl, 1976, 1977; Cooper et al., 1976). Two distinct types of anomalies are found in the Aleutian Basin. The first is a set of narrow linear anomalies (25 to 100 km wide) that are believed to be the M1 to M13 (Lower Cretaceous) sequence of sea-floor-spreading anomalies (Figure 1; Cooper, Marlow, and Scholl, 1976). The second type (not shown) is a long-wavelength negative anomaly that is centered over the Aleutian Basin (Regan, Cain, and David, 1975; Cooper, Marlow, and Scholl, 1977). This anomaly is thought to be related to deep-seated upper-mantle thermal activity beneath the basin (Cooper, Marlow, and Scholl, 1977).

Studies of the deep structures of the lithosphere and asthenosphere beneath the Aleutian Basin, based on traveltimes of Sn waves (Bhattacharya, 1979) and models of long-wavelength, magnetic and heat-flow anomalies (Cooper, Marlow, and Scholl, 1977), indicate that the basin is not underlain by a lithospheric low-velocity zone. Incipient heating of the base of the lithosphere may be occurring, however, and causing the regionally higher heat-flow (Figure 6) and lower magnetic-field values that are observed in the central part of the basin. Slightly positive free-air gravity values, a thinner sedimentary section, and greater basement relief in the central basin region also suggest that this region is being affected by deep-seated processes that cause crustal thinning or uplift (Cooper, Marlow, and Scholl, 1977).

Basin Margins

The Aleutian and Bowers Basins are surrounded by large geologic structures of different ages and origins. Foldbelts of Mesozoic rocks extend beneath the outer edge of the Beringian continental margin from the Alaska Peninsula to the northern end of Shirshov Ridge. Massive submarine volcanic

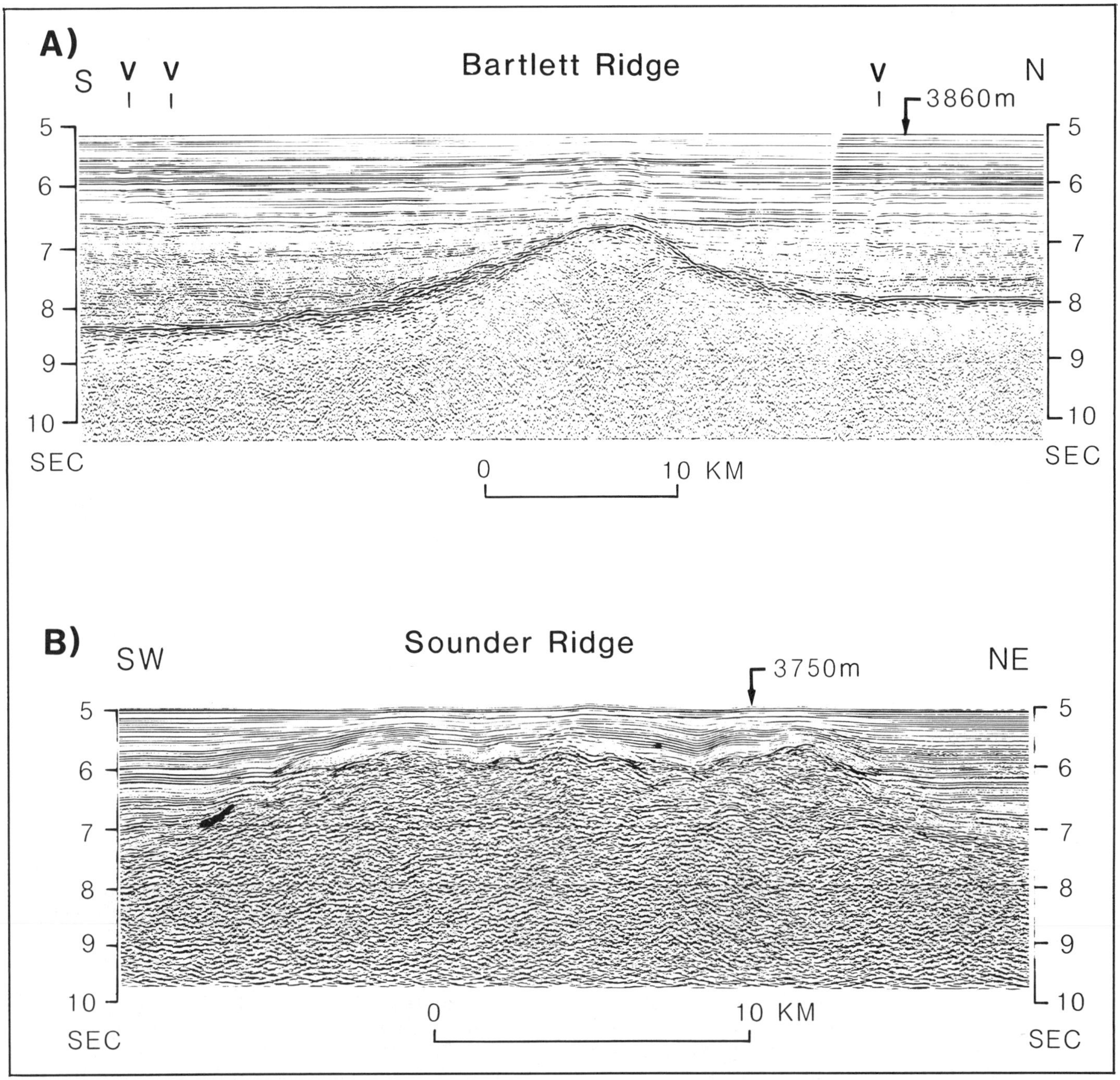

Figure 5. Multichannel seismic-reflection lines over buried basement ridges in the Aleutian Basin. Sedimentary layers over the ridges are deformed, which suggests differential sediment compaction and subsidence: A, Bartlett Ridge; B, Sounder Ridge. V indicates locations of velocity-amplitude anomalies (VAMPs). See Figure 1 for locations.

ridges of Mesozoic (?) and Tertiary age frame the western and southern boundaries of the basins.

Shirshov Ridge

Shirshov Ridge forms the western margin of the Aleutian Basin. This linear volcanic and sedimentary ridge has been called an ancient spreading center (Kienle, 1971), a remnant arc (Karig, 1972), and an ancient volcanic arc (Scholl, Buffington, and Marlow, 1975; Bogdanov and Neprochnov, 1984). Tertiary volcaniclastic basement rocks have been recovered from the ridge (Scholl, Buffington, and Marlow, 1975), but Bogdanov and Neprochnov (1984) report siliceous units of Cretaceous age. Thus the origin of the rocks of Shirshov Ridge is unclear. Seismic-refraction and gravity data (Rabinowitz and Cooper, 1977; Cooper, Marlow, and Scholl, this volume) show no evidence of a thick sedimentary section, such as a buried trench, beneath the eastern flank of the ridge. At the northern end of the ridge and along the Koryak segment of the connecting Beringian continental margin, however, a large sedimentary wedge does exist (Figure 4), and the oceanic crust beneath the Aleutian Basin appears to dip beneath both the ridge and the contiguous margin (Rabinowitz and Cooper, 1977).

The eastern flank of Shirshov Ridge is draped with 2 to 3

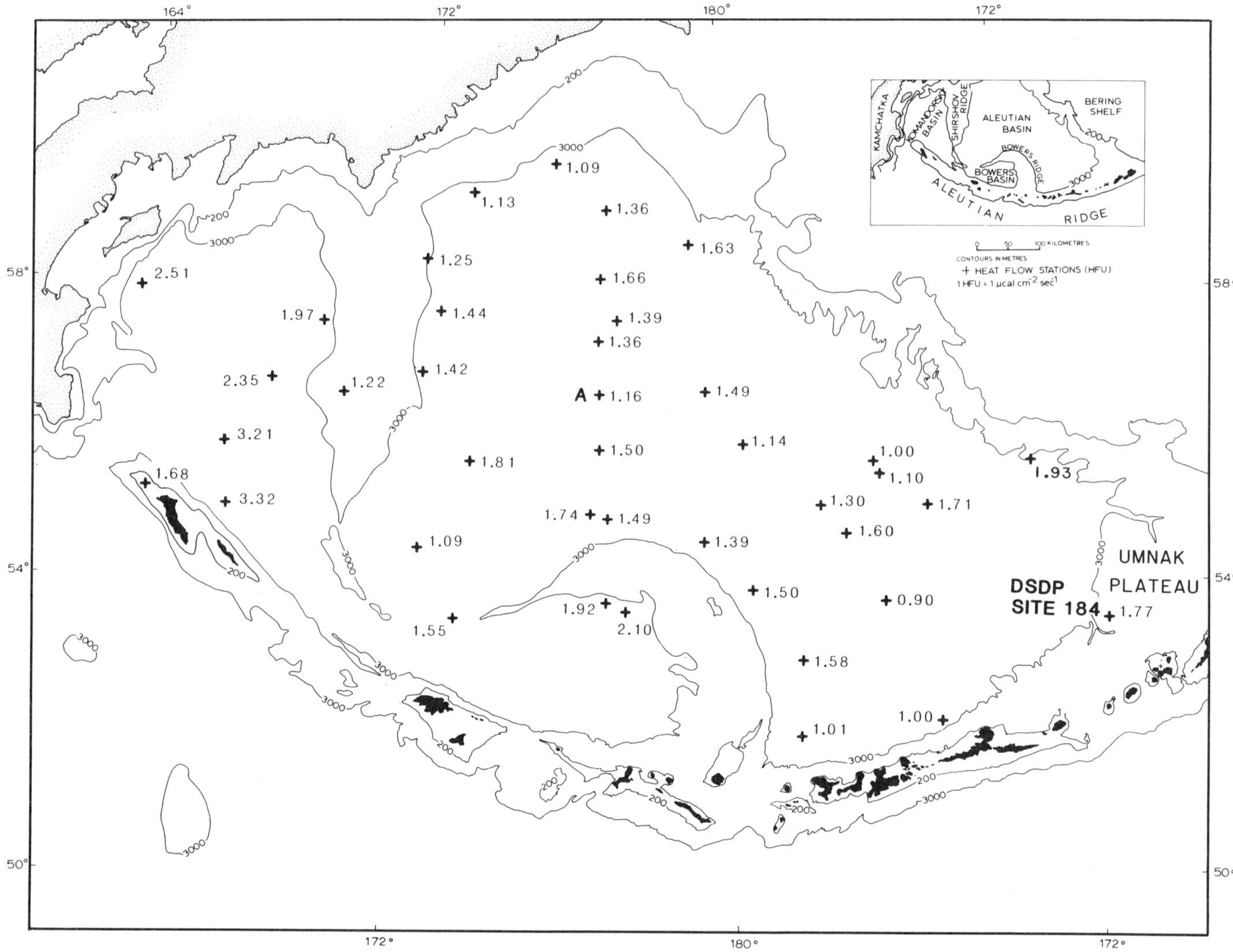

Figure 6. Shallow (3 to 10 m below sea floor) heat-flow measurements in the Bering Sea Basin (from Watanabe, Langseth, and Anderson, 1977; Langseth, Hobart, and Horai, 1980; and Cooper, 1978). Measurement at site 184 (Umnak Plateau) is from Erickson (1973). Value at site A was recorded directly above a VAMP (Cooper, 1978). Heat flow is in microcalories per square centimeter per second.

km of sediment that fills axial subbasins along the crest of the ridge and continues eastward into the Aleutian Basin (profiles A–K, Figure 7). Sedimentary rocks sampled on the ridge (Bogdanov and Neprochnov, 1984) and nearby at DSDP site 190 (Figure 2, Creager, Scholl, et al., 1973) indicate a probable Neogene and younger age for most of the upper 1 to 2 km of the sedimentary section. Multichannel seismic-reflection profiles have not been recorded across the ridge; consequently, little is known about deeply buried sediment bodies and basement structures adjacent to the ridge.

Bowers Ridge

Bowers Ridge is an arcuate crustal belt of arc-type volcanic rocks that is bordered on its north side by a thick sedimentary prism that extends from the western end of the ridge to a point near its structural connection with the Aleutian Ridge (Figure 4). Bowers Ridge is believed to be an ancient island arc with an associated buried trench (Kienle, 1971; Ludwig et al., 1971; Scholl, Buffington, and Marlow, 1975; Cooper and Ben-Avraham, 1981).

The buried trench beneath the north side of Bowers Ridge is filled with as much as 10 km of sedimentary rocks; these strata are deformed over a 20-km-wide zone adjacent to the ridge (profiles L and M, Figure 7). These folded rocks are buried beneath a 1-2-km-thick blanket of nearly flat-lying strata. Cooper and Ben-Avraham (1981) suggest that this sediment deformation occurred mostly during Paleogene and possibly earlier subduction of the Aleutian Basin crust beneath Bowers Ridge, although some deformation may have occurred during the Neogene.

Aleutian Ridge

The geology of the Aleutian Ridge, the southern structural boundary for the back-arc basins of the Bering Sea, is discussed by Scholl, Vallier, and Stevenson (1983; this volume). They describe three sedimentary sequences that, collectively, range in age from middle Eocene to Quaternary. North of Amlia Island (Figure 1), these sequences can be traced in seismic-reflection profiles down the north flank of the Aleutian Ridge to the Aleutian Basin. At the base of the ridge the thickness of the sedimentary section reaches 3 to 5 km, and rocks in the oldest sequence (Eocene) are believed

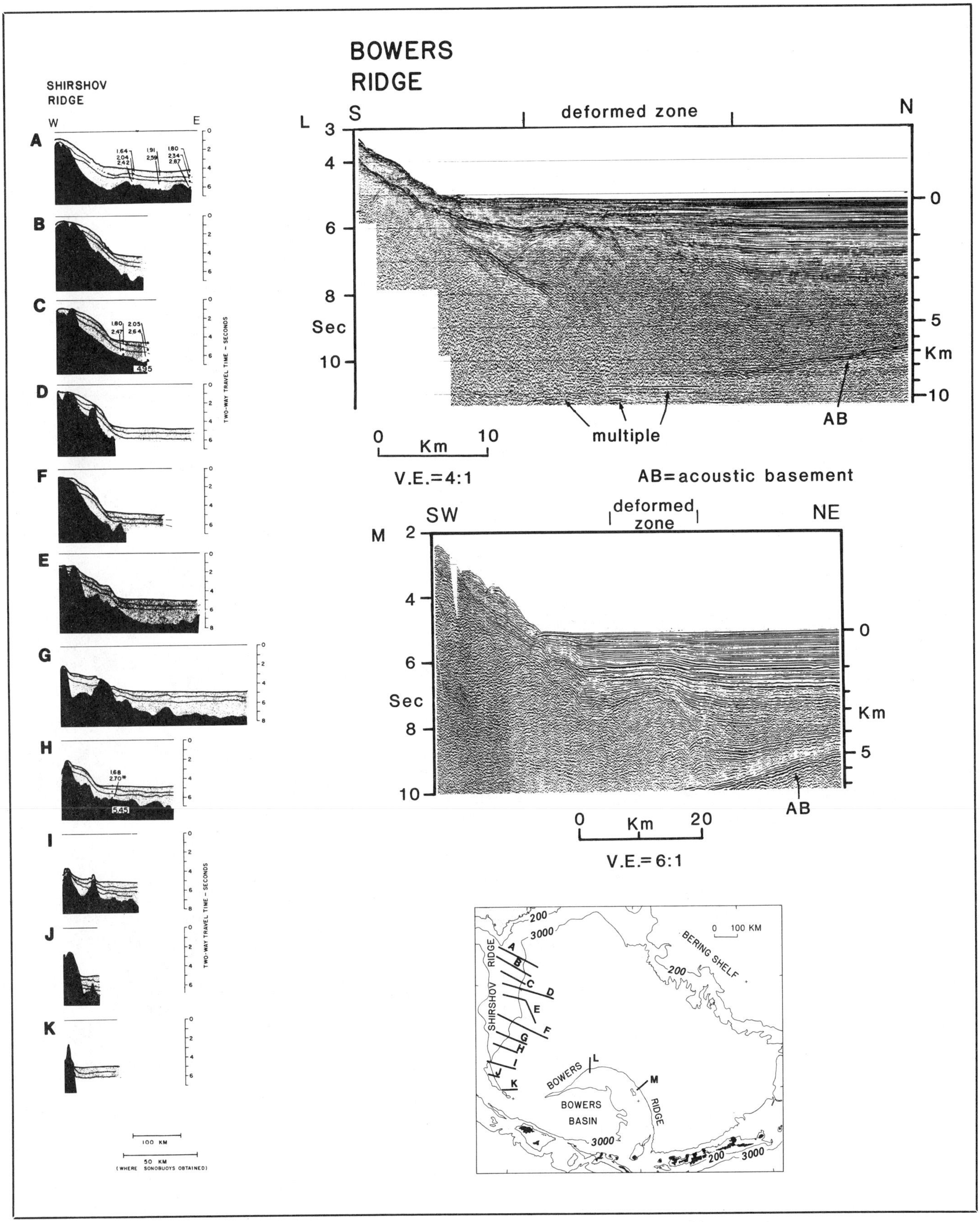

Figure 7. Seismic-reflection profiles across Shirshov Ridge (profiles A-K; Rabinowitz and Cooper, 1977) and Bowers Ridge (profiles L and M; Cooper and Ben-Avraham, 1981). Along Shirshov Ridge, acoustic basement is shown in black; numbers are sonobuoy interval and refraction (basement) velocities in kilometers per second. Thick sediment wedges are found along the base of Bowers Ridge but not Shirshov Ridge.

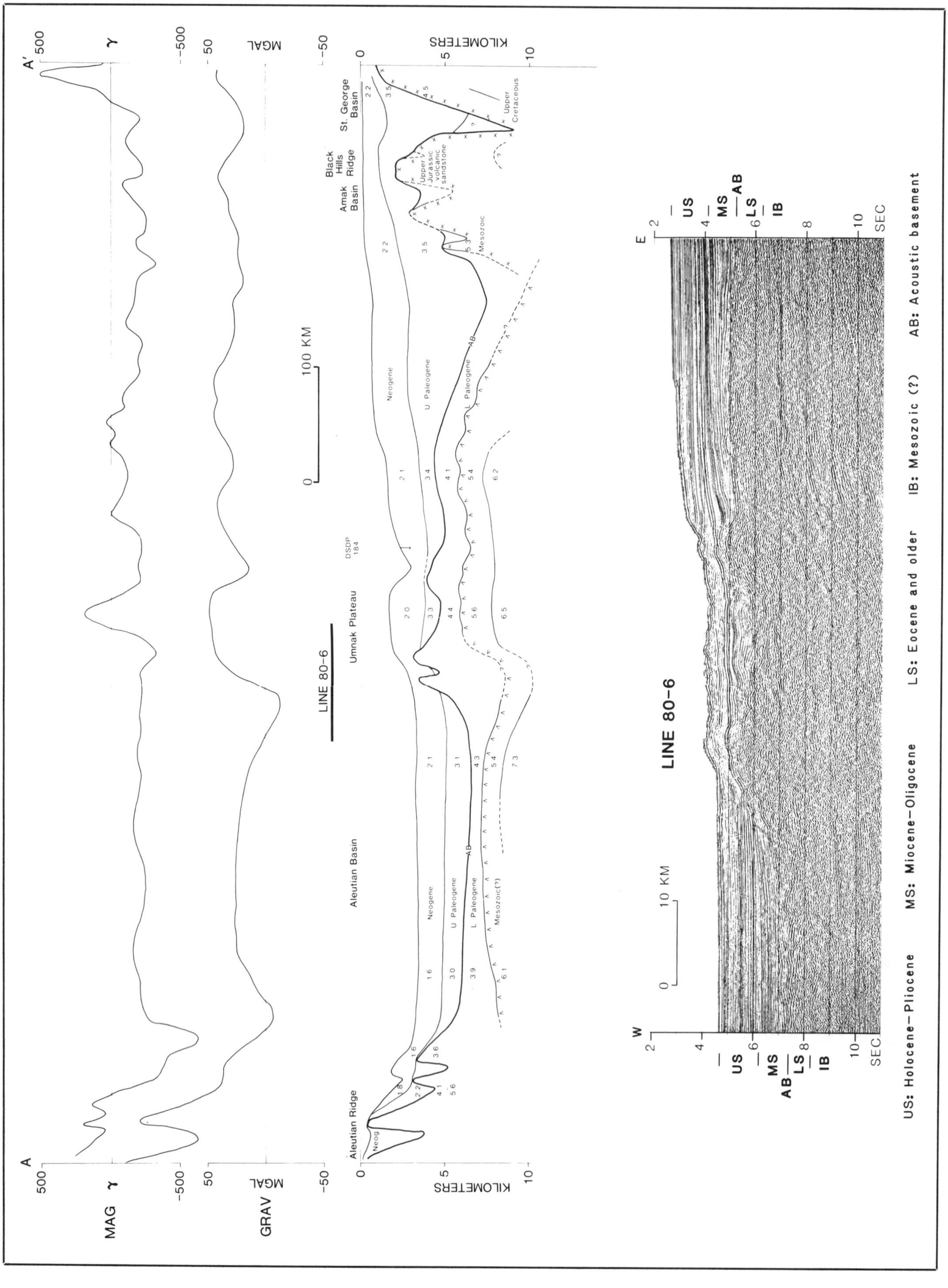
A
A'
MAG
GRAV
MGAL
KILOMETERS
Aleutian Ridge
Aleutian Basin
Umnak Plateau
DSDP 184
Amak Basin
Black Hills Ridge
St. George Basin
LINE 80-6
100 KM
Neogene
U Paleogene
L Paleogene
Mesozoic(?)
Mesozoic
AB
Upper Jurassic volcanic sandstone
Upper Cretaceous
LINE 80-6
10 KM
W
E
SEC
US
MS
LS
AB
IB
US: Holocene–Pliocene
MS: Miocene–Oligocene
LS: Eocene and older
IB: Mesozoic (?)
AB: Acoustic basement

to overlie igneous oceanic crust (Figure 8).

The oceanic crust of the Aleutian Basin may descend gently beneath the northern flank of the Aleutian Ridge, but it does not plunge into a buried trench similar to that bordering the north and east flanks of Bowers Ridge. Scholl, Vallier, and Stevenson (this volume) speculate that the constructional phase of the Aleutian Ridge may have thickened its igneous crust above older crust of the Aleutian Basin.

Umnak Plateau

Umnak Plateau lies at the southeastern corner of the Aleutian Basin in the triangular region between the converging trends of the Aleutian Ridge and the Beringian continental shelf (Figure 1). The plateau is a flat-topped feature at about 1600 m of water depth that is underlain by sedimentary deposits as much as 6 km thick (Figure 4). Because of the structural position of the plateau between the folded Mesozoic basement rocks of the Beringian Shelf, the Cenozoic island-arc massif of the Aleutian Ridge, and the Mesozoic oceanic crust of the Aleutian Basin, many authors have speculated about its origins. These speculations include (1) subsidence of the northern flank of the Aleutian Ridge (Scholl, Vallier, and Stevenson, 1983) or the outer Beringian margin (Scholl, Buffington, and Hopkins, 1968) in the early Tertiary, (2) collision of an allochthonous oceanic plateau with the Beringian margin before the formation of the Aleutian Ridge (Ben-Avraham and Cooper, 1981), or (3) uplift of the oceanic crust of the Aleutian Basin during initial construction of the Aleutian Ridge (Cooper et al., 1980).

Seismic-reflection profiles (Figures 8, 9) show a nearly continuous well-layered sedimentary section beneath the plateau and the adjacent slope and basin areas. The age of this well-layered sequence is believed to range from upper Paleogene to Holocene, based on Aleutian Ridge dredge samples (Scholl, Vallier, and Stevenson, 1983; this volume), DSDP drilling on the plateau (sites 184 and 185, Figure 10; Creager, Scholl, et al., 1973), and COST well drilling on the outer or southern edge of the Beringian Shelf (Turner et al., 1984). An acoustically incoherent sedimentary(?) unit lies beneath the well-layered section and forms an effective acoustic basement on seismic-reflection profiles. This lower unit, which is thought to be of lower Paleogene age and possibly older (Figure 8), lies directly above presumed igneous basement.

Large thicknesses (as much as 8 km) of Paleogene and younger sedimentary rocks fill deep structural troughs flanking the perimeters of the plateau adjacent to the Aleutian Ridge (north of Akutan Island, Figure 9) and the Beringian continental slope (Figure 4). These structural depressions are asymmetric in cross-sectional profile; the thickest sedimentary sections are beneath the fault-bounded landward edges. On the basis of magnetic models, the sediment-filled depression beneath the northeast side of the plateau is thought to be underlain by oceanic crust partly subducted or thrust beneath the margin (Cooper et al., 1980). This structural depression filled with sediment during the Paleogene and early Neogene (Cooper, Marlow, and Vallier, 1984) at the same time that the large basins of the outer Bering Shelf were acquiring their older sedimentary fill (Marlow, Cooper, and Fisher, this volume).

Figure 8. Generalized transect across Umnak Plateau from the Aleutian Ridge to the outer Bering Shelf (Cooper, Marlow, and Vallier, 1984). See Figure 9 for location. Postulated ages are based on Aleutian Ridge rock sequences (US, MS, LS; Scholl, Vallier, and Stevenson, 1983) and on Bering Shelf drilling (Marlow, Cooper, and Fisher, this volume). Seismic line 80–6 shows diapir-like structures underlying edge of Umnak Plateau. IB is igneous basement. Numbers are refraction velocities in kilometers per second.

Beringian and Koryak Continental Slope and Rise

The Beringian and Koryak segments of the continental slope and rise extend northwest nearly 1600 km, from near Umnak Plateau to the northern end of Shirshov Ridge (Figure 1). Large submarine canyons—such as Bering, Pribilof, Zhemchug, and Navarinsky—incise the Beringian margin and are the pathways along which large volumes of sediment have reached the Aleutian Basin. These deep, fault-controlled canyons are thought to have been mostly cut in the late Cenozoic (Scholl et al., 1970; Carlson and Karl, 1984). However, the sedimentary continental-rise prisms that lie at the base of the Beringian and Koryak margins, and are thickest near the mouths of these canyons, imply that these canyons or their predecessors have existed since at least the Paleogene (Cooper, Marlow, and Scholl, 1979).

The shape of these prisms, or sediment wedges, differs along the Beringian and Koryak margins. The continental rise of the Beringian margin is relatively narrow and is underlain by wedges in which the zone of greatest thickening lies within the deeper Paleogene (?) sedimentary sections. The rise prism for the Koryak margin is nearly twice as broad, and the thickest sedimentary units may lie within the shallower Neogene and Quaternary sections (Figure 11). We believe that a change in sediment distribution patterns resulting from late Cenozoic uplift and glaciation, especially in the Koryak Ranges (Nelson, Hopkins, and Scholl, 1974; Scholl, Buffington, and Marlow, 1975), may partly explain the thicker and wider late Cenozoic sections along the base of the Koryak margin.

The sedimentary wedges are mostly undeformed, except at the base of the Beringian continental slope and beneath the Koryak continental slope. Along the southern Beringian margin, from the Alaska Peninsula northwestward to Zhemchug Canyon (Figure 1), undeformed rise deposits thicken toward the margin (lines L1 and BS765; Figure 12, 13, 14) and they abut and overlap deeply buried and deformed sedimentary rocks at the base of the continental slope. These deeply buried Paleogene(?) rocks are compressionally or diapirically folded and faulted within a narrow, 10–20-km-wide zone that lies seaward of the extensionally faulted Mesozoic bedrock underlying the middle and upper reaches of the slope. At the base of the slope, the igneous basement and some sedimentary horizons appear to extend beneath (underthrust?) the folds and to the bedrock framework of the margin (line BS765, Figure 14).

In the vicinity of Zhemchug Canyon (line 77-4, Figure 15), the sedimentary section reaches thickness of 10 to 12 km above an ancient buried trench(?) near the base of the continental slope (Marlow and Cooper, 1985). Landward of this ancient trench, the entire sedimentary section is gently uplifted, folded, and faulted over a basement high. A large slump block of Neogene sedimentary rock appears to have slid from the seaward flank of the uplifted sedimentary section into the Aleutian Basin. The deformation and slumping probably occurred in the Neogene, based on the seismic-reflection data (Figure 15).

Figure 9.. Generalized transect and multichannel seismic-reflection line segment across the sedimentary wedge and basement faults along the northeast side of Umnak Plateau (Cooper, Marlow, and Vallier, 1984). IB is igneous basement and horizons 1 and 3 are acoustic basement. Velocities are in kilometer per second. N is Neogene; UP is upper Paleogene; LP is lower Paleogene.

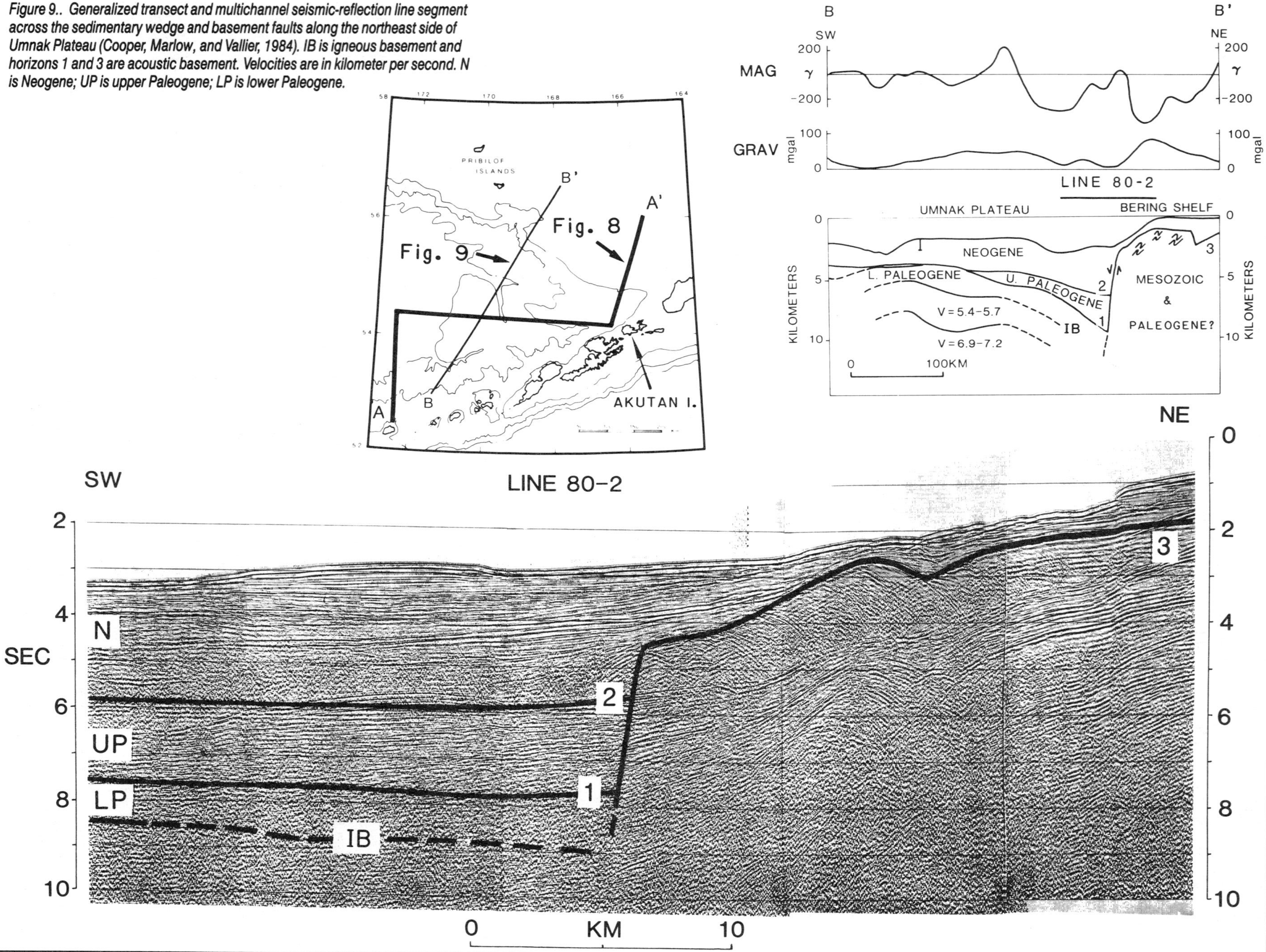

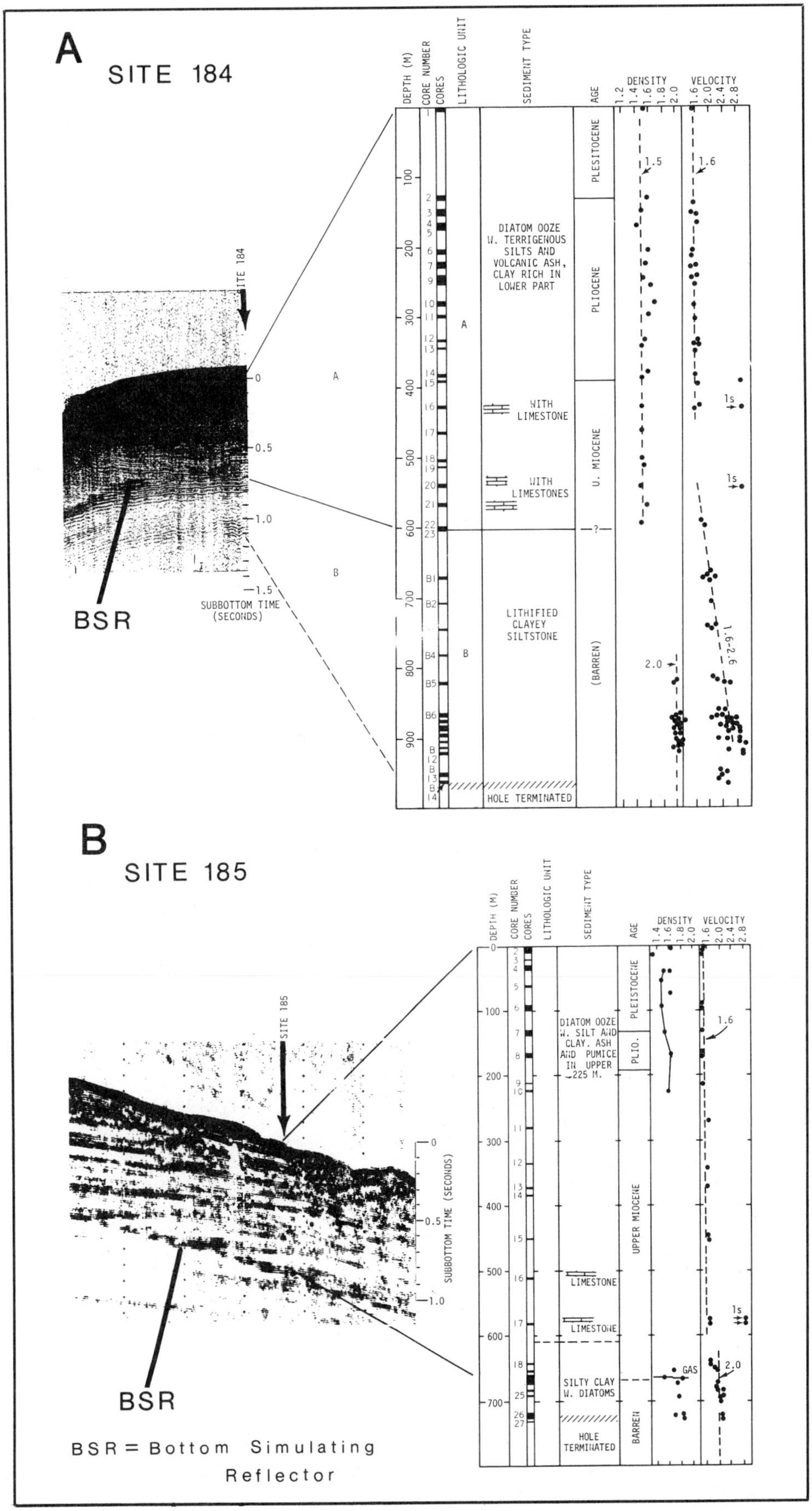

Figure 10. Summary of DSDP drilling results at sites 184 and 185 on Umnak Plateau (Creager, Scholl, et al., 1973). See Figure 1 for locations.

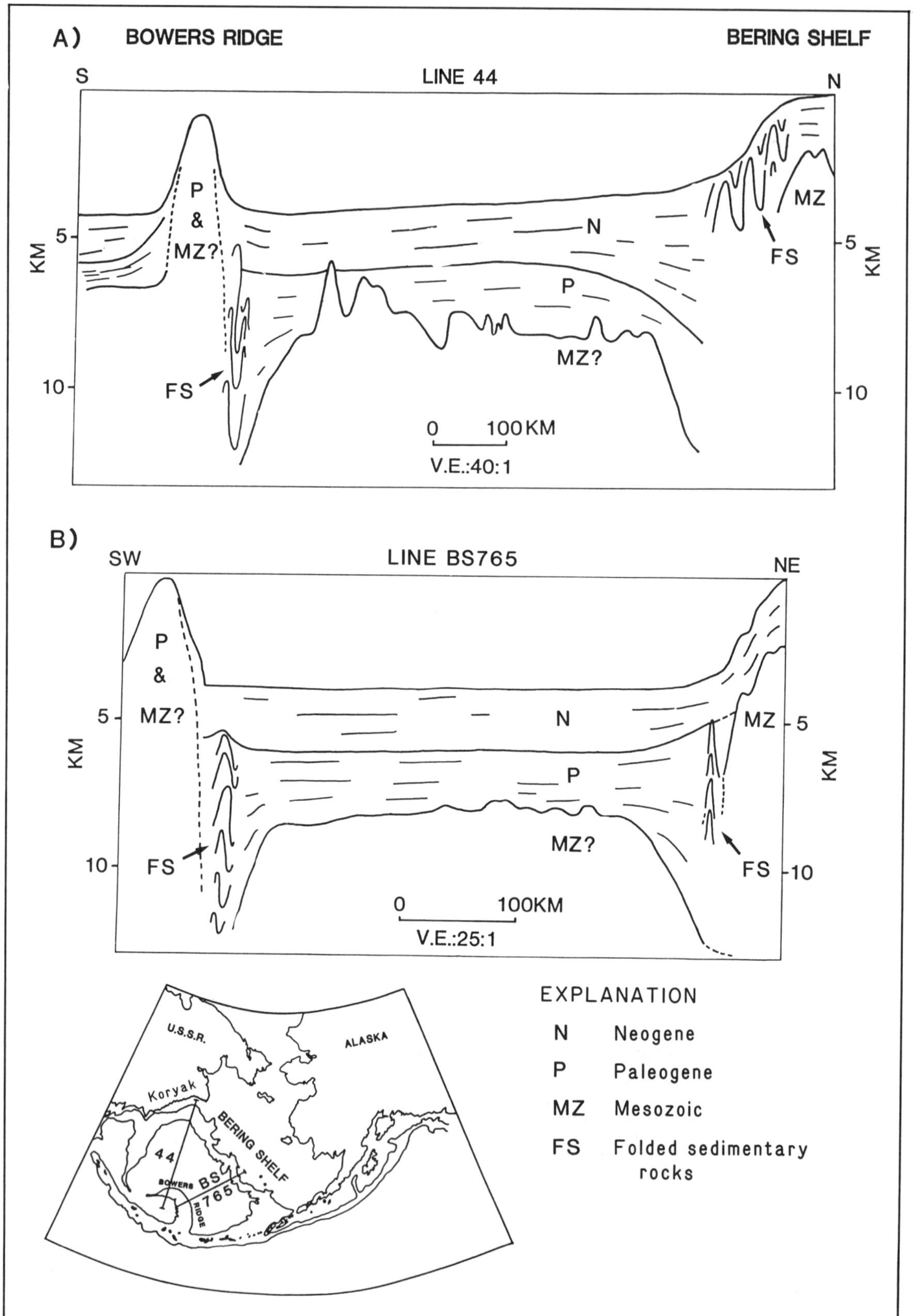

Figure 11. Generalized interpretation (in depth) of multichannel seismic-reflection lines 44 and BS765 across the Aleutian Basin. The thickening and folding of the rise prisms adjacent to Bowers Ridge (A and B) and the central Bering Shelf (B) is in Paleogene(?) sedimentary units. Thickening of the Koryak prism (A) is mostly in the Neogene section, and folding involves nearly the entire sedimentary section beneath the slope.

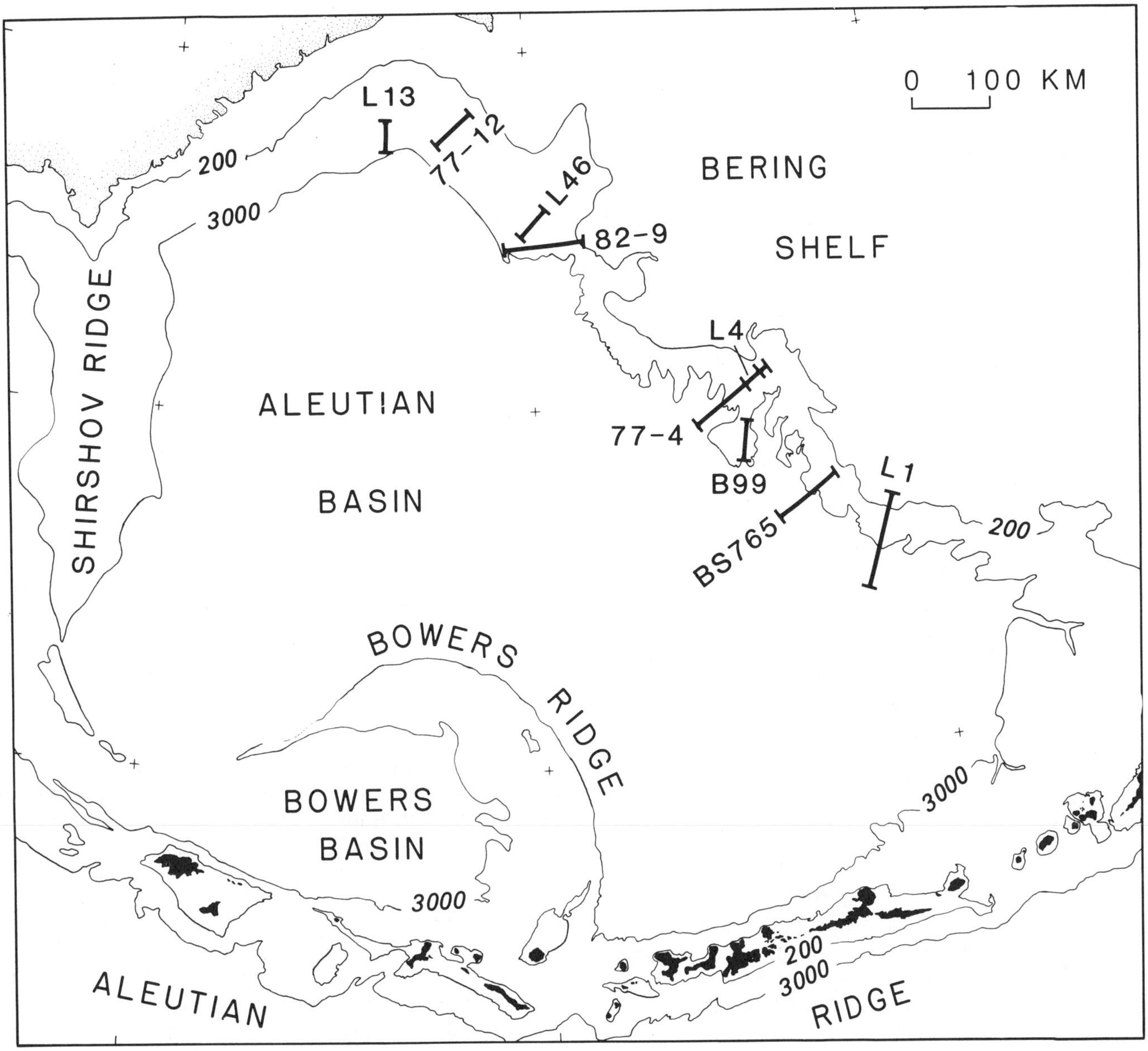

Figure 12. Index map showing location of seismic lines across the Beringian and Koryak margins illustrated in Figures 13–18, 20, 21.

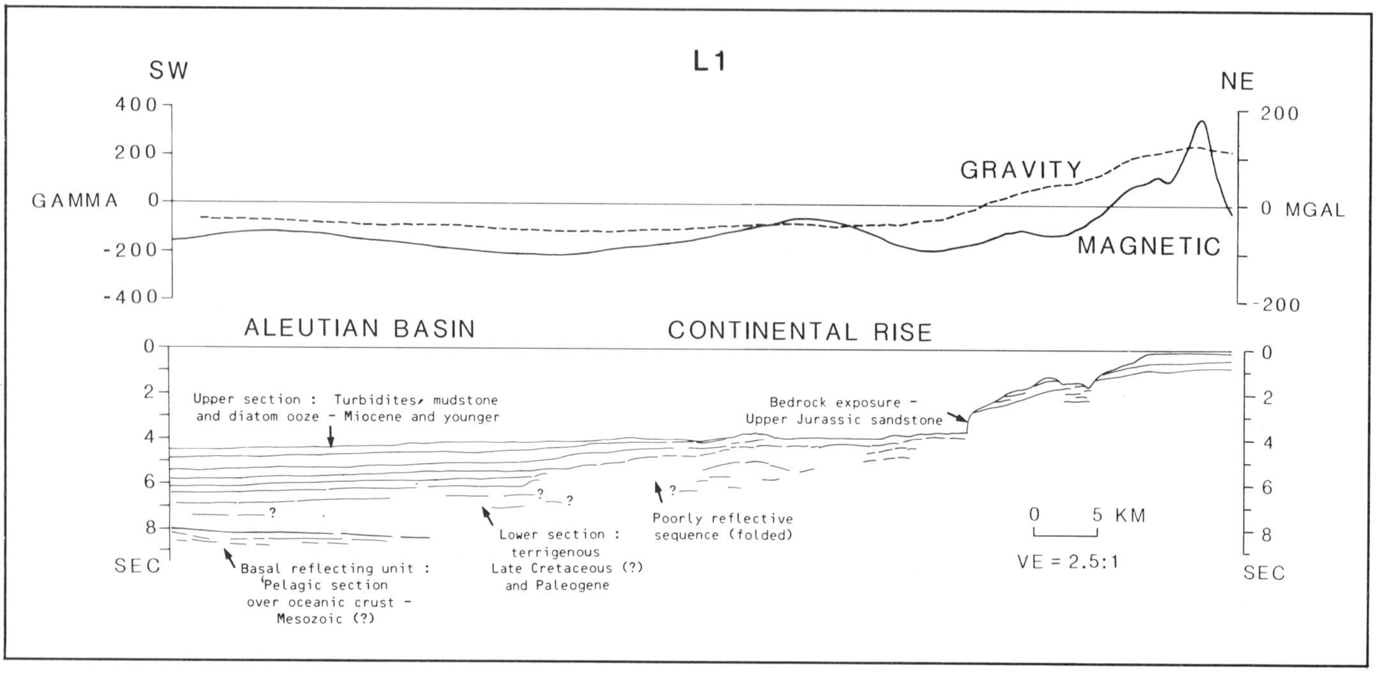

Figure 13. Interpretive line drawing for multichannel seismic-reflection line L1 across the southeastern part of the Beringian margin (from Marlow et al., 1982). The 20–30-km-wide zone of poorly reflective and folded sedimentary rocks lying beneath the continental rise is similar to other parts of the margin. See Figure 12 for location.

North of Zhemchug Canyon, the sedimentary wedges are locally thinner and are differentially warped or folded over acoustically disrupted basement or folded sedimentary units at the base of the slope (line 82-9, Figure 16). The acoustically disrupted units may have been deformed either by extensional faulting and collapse of outer-shelf basement rocks or by underthrusting and folding of Paleogene (?) rise-prism sedimentary units. Local faulting of younger Neogene rocks is also apparent in some places.

North of Navarinsky Canyon, the entire flat-lying rise prism is faulted and folded at the base of the slope (line 77-12, Figure 17). The abruptness of the deformation throughout the 7–9-km-thick section suggests that wrench faulting, probably striking parallel to the margin, is the cause.

Beneath the Koryak margin, the sedimentary section is too thick to resolve basement in available seismic-reflection profiles, but extensive folding of the upper 9 km of the Paleogene(?) and Neogene section is apparent beneath most of the continental slope (line 13, Figure 18). As along the northernmost Beringian margin, the onset of deformation occurs over a relatively short (5 to 10 km) distance, and fold structures extend nearly to the sea floor. The folded zone, however, is significantly wider beneath the Koryak margin than beneath the Beringian margin.

The deformation of the deeply buried Paleogene(?) sedimentary sections within the continental-rise prisms along the Beringian and Koryak margins, as well as along the north side of Bowers Ridge, probably occurred in response to convergence and underthrusting of Aleutian Basin crust as well as wrench faulting beneath these margins (Cooper, Marlow, and Scholl, this volume). Most deformation probably took place before the Eocene formation of the Aleutian Ridge; however, more recent but limited deformation of the rise prisms has continued into the late Cenozoic.

HYDROCARBON POTENTIAL

During the past decade, the Bering Sea Basin, like other ocean-margin basins of the Pacific (Schlanger and Combs, 1975), has been recognized as a frontier area for hydrocarbon resources, primarily because of the great thickness of Cenozoic sediment and the higher-than-normal heat flow. Geophysical and geological studies indicate that structural and stratigraphic traps, thermal gradients, source rocks, and reservoir beds all exist and favor the generation and entrapment of hydrocarbon deposits.

Structural and Acoustic Features

Within the generally flat-lying sedimentary section of the Aleutian Basin, numerous VAMPs (velocity-amplitude features), or deep-water seismic "bright spots" (Scholl and Cooper, 1978; Cooper et al., 1979), occur and provide the most convincing evidence that structural and stratigraphic traps are present at shallow depths below the sea floor (Figure 19). VAMPs are characterized by isolated, phase-inverted, high-amplitude reflections (bright spots) underlain by columns of concave-upward reflections that exhibit increased offset and curvature with depth (velocity pull-down). The strata above VAMPs are commonly structurally upbowed. The VAMPs are believed to result from a low-velocity zone of gas-charged deposits and possibly other hydrocarbon products within the upper (and possibly lower) part of the sedimentary section (Figure 19A; Scholl and Cooper, 1978; Cooper, Marlow, and Scholl, 1979).

Detailed surveys indicate that some VAMPs are associated with basement ridges, but others are not. Furthermore, not all basement ridges are overlain by VAMPs. In general, the central part of the Aleutian Basin has the greatest concentra-

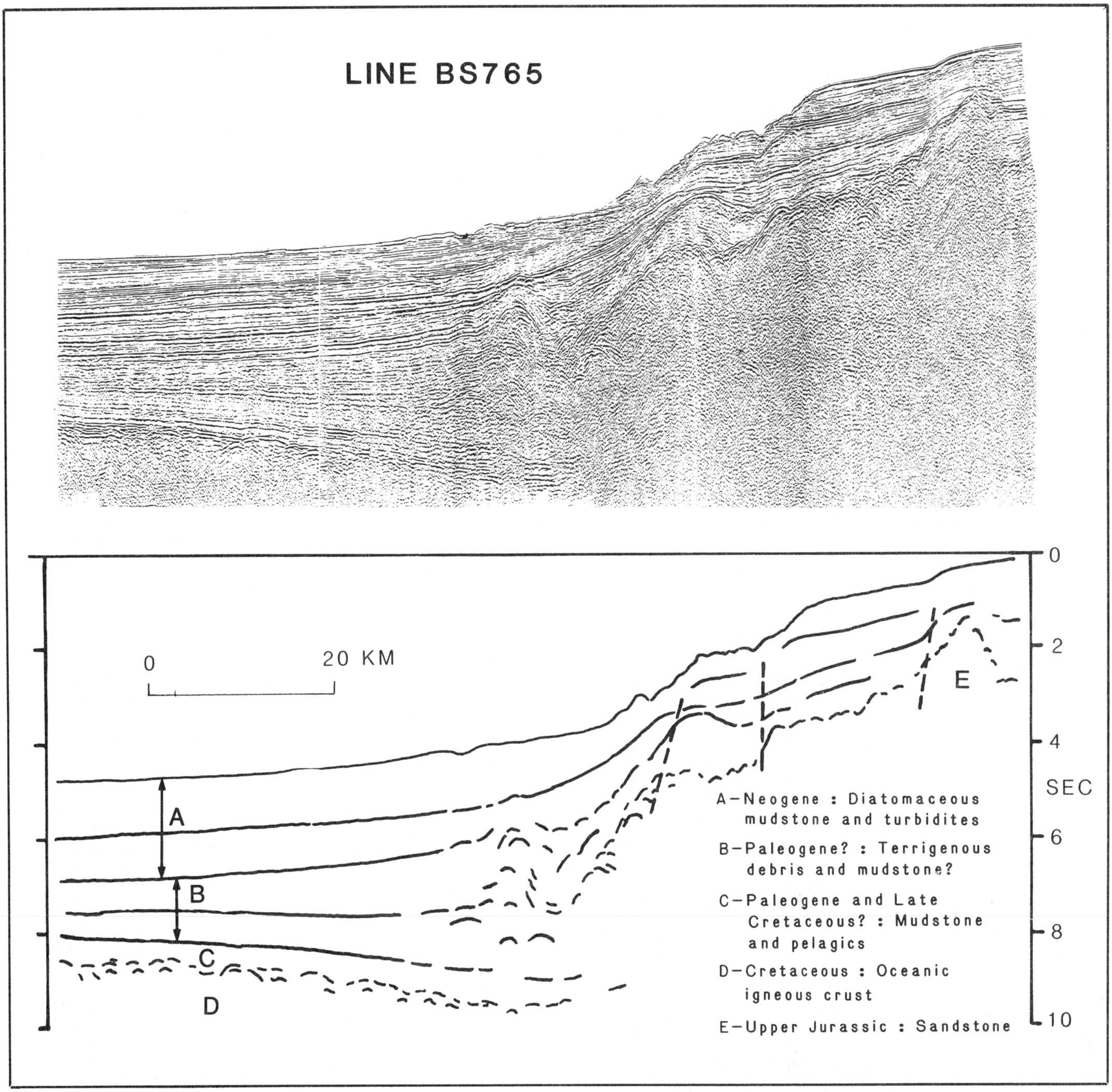

Figure 14. Multichannel seismic-reflection line BS765 across the central Beringian margin (modified from Cooper and Ben-Avraham, 1981; see also Figure 3). The thickening of the continental rise prism occurs in the lower (Paleogene?) part of the prism. Deeply buried sediment are deformed at the base of the slope. See Figure 12 for location.

tion of VAMPs (Figure 19B) and also basement ridges over which the sedimentary section thins (Figure 4).

The thickness of the sedimentary section beneath the Aleutian Basin ranges from 2 to 10 km, and sedimentary strata are commonly faulted and warped over deeply buried basement structures such as ridges, knolls, and fault-displaced basement rocks. Sounder Ridge (Figure 5B) is covered by sedimentary rocks that are folded and possibly faulted along its flanks and are differentially compacted or uplifted over its crest.

Sedimentation and Thermal History

Little is known of the history of sedimentation in the abyssal areas of the Aleutian and Bowers Basins, because subsurface drill information is limited to only the upper part of the sedimentary section where middle (?) to upper Miocene and younger rocks have been found. These diatom-rich pelagic and turbidite rocks were deposited rapidly at sedimentation rates of 70 to 140 m/m.y. (Creager, Scholl, et al., 1973). Large areal differences in sedimentation rates are

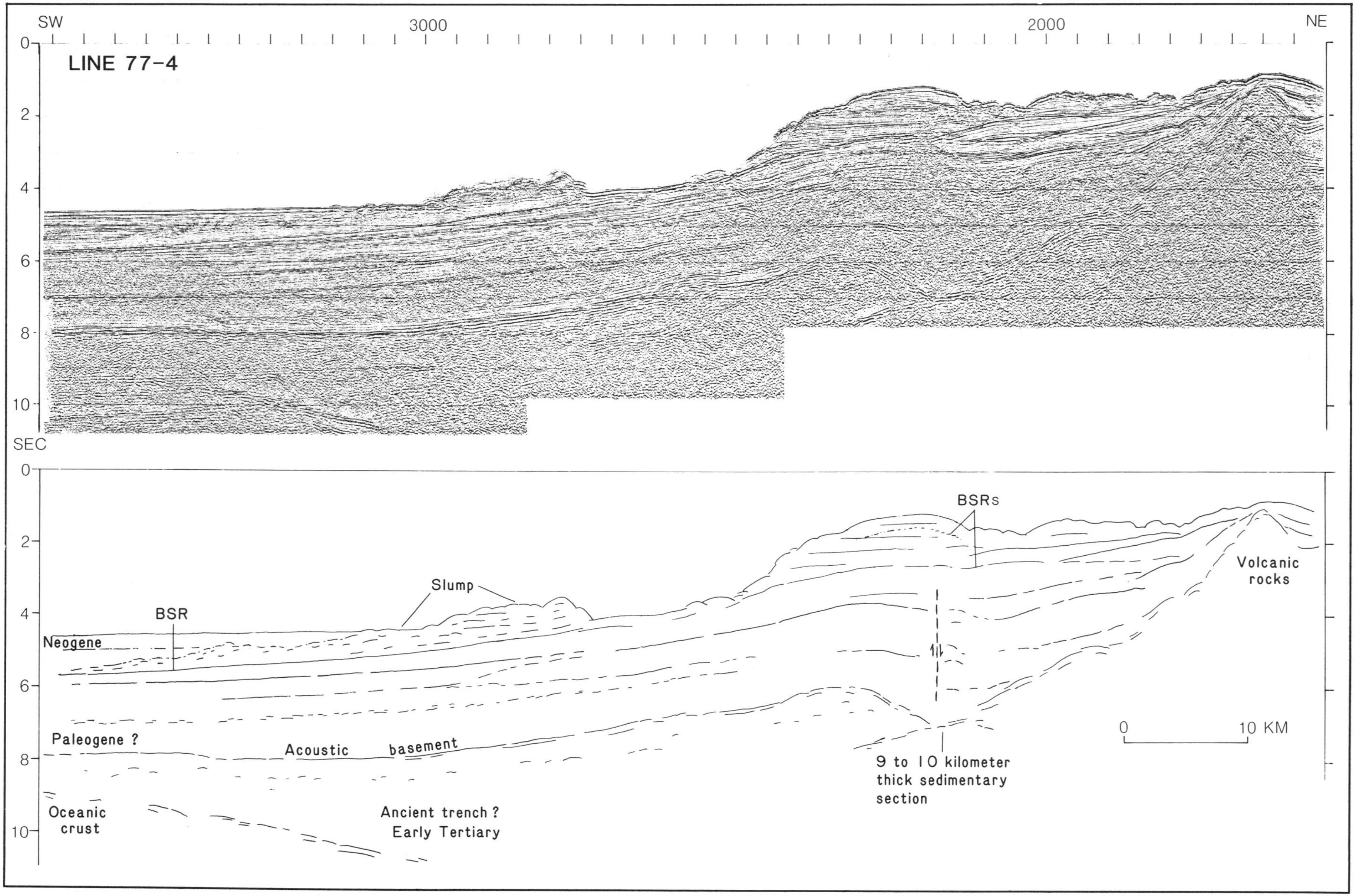

Figure 15. Multichannel seismic-reflection line 77-4 across the continental margin near the northern end of Zhemchug Canyon. An ancient trench? (Marlow and Cooper, 1985) lies seaward of uplifted and faulted continental-rise units 9 to 10 km thick. The slip surface for the major slump block at the sea floor may have been the bottom-simulating reflector (BSR-DB) at the base of the slump. Numbers at top of profile are shot points. See Figure 12 for location.

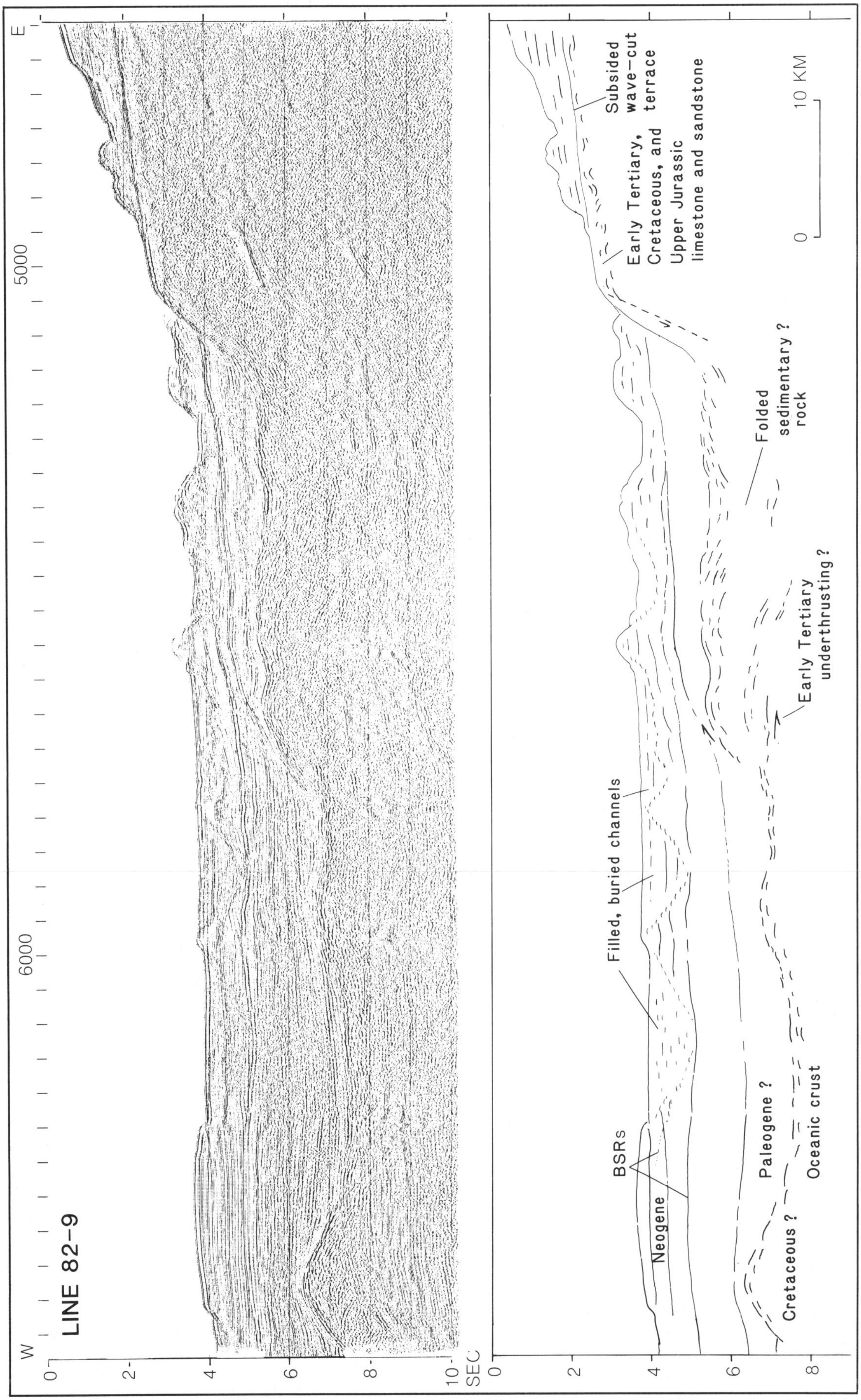

Figure 16. Multichannel seismic-reflection line 82-9 across the Beringian margin near Navarinsky Canyon. The deeper sedimentary section of the Aleutian Basin abuts (fault contact?) and covers a folded zone that may comprise deformed basinal units or subsided shelf bedrock units. Sediment-filled channels and BSRs commonly occur in this region (see text for explanation). Numbers at top of profile are shot points. See Figure 12 for location.

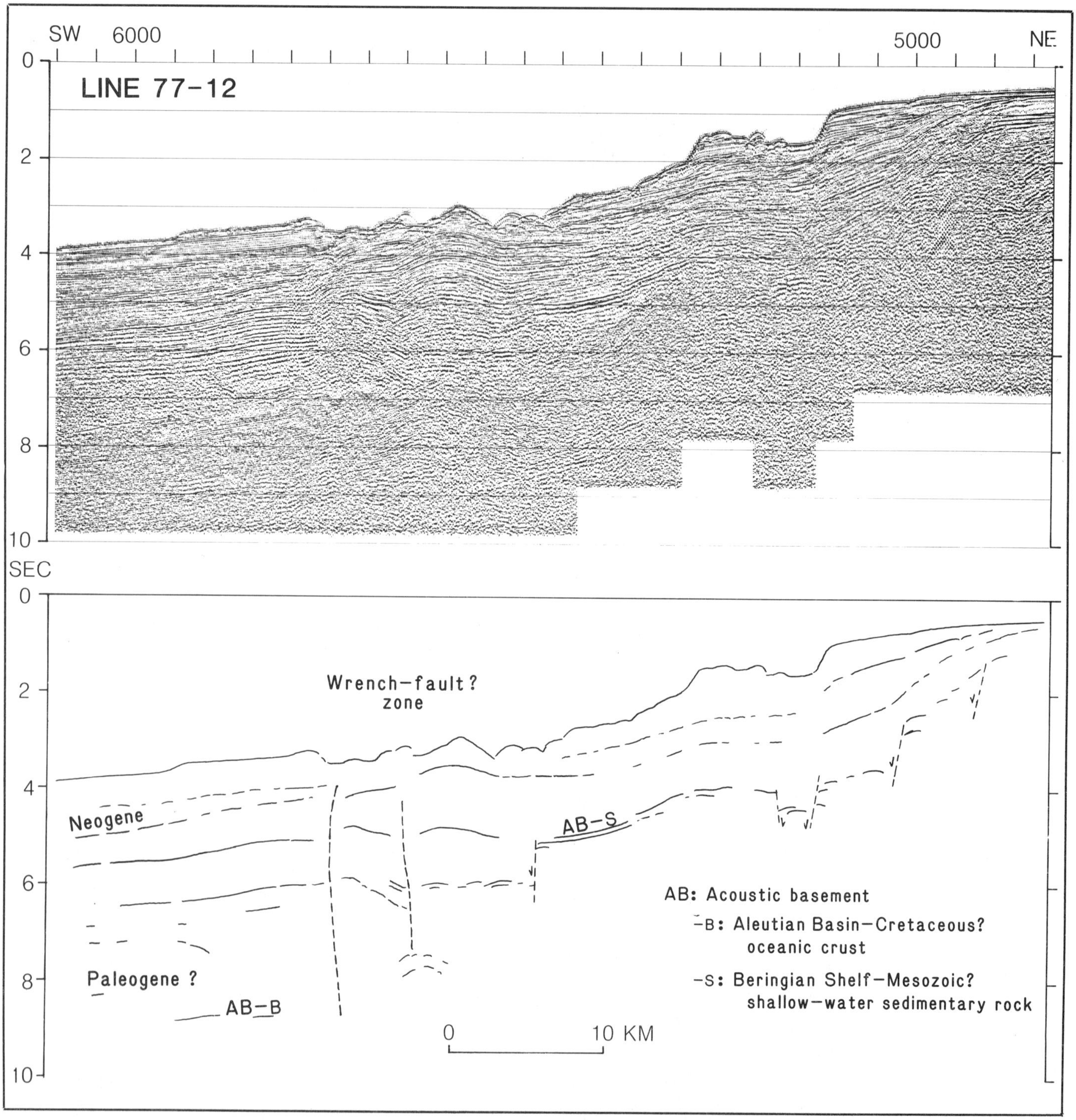

Figure 17. Multichannel seismic-reflection line 77-12 across the northwestern end of the Beringian margin. The thick, flat-lying rise prism above acoustic basement (AB-B; oceanic igneous crust?) is abruptly deformed (by wrench faults?) only 12 km from the suspected edge of subsided shelf bedrock (AB-S) of Mesozoic shallow-water deposits. Numbers at top of profile are shot points. See Figure 12 for location.

likely to be found within the Aleutian Basin because of variations in biogenic productivity, the amount of suspended sediment carried in from the Pacific Ocean, and the amount of terrigenous material contributed by the Siberian, Alaskan, and Aleutian drainages that by-passed the Bering shelves (Fullam et al., 1973).

Sea-floor heat-flow values in the Aleutian Basin average about 1.44 $\mu cal/cm^2/sec$ (Figure 6; Watanabe, Langseth, and Anderson, 1977; Langseth, Hobart, and Horai, 1980); they are generally highest in the central region of the basin. Based on this average heat flow and the sediment thermal conductivity measured at DSDP sites (2.5 mcal/cm sec °C; Erickson, 1973), an average temperature gradient of 58°C/km can be calculated for the upper part of the basin. Temperatures needed to cause the onset of hydrocarbon generation (50 to 100°C) could therefore be reached at relatively shallow subbottom depths (0.9 to 1.8 km). After corrections are made for the high sedimentation rates in the basin, the

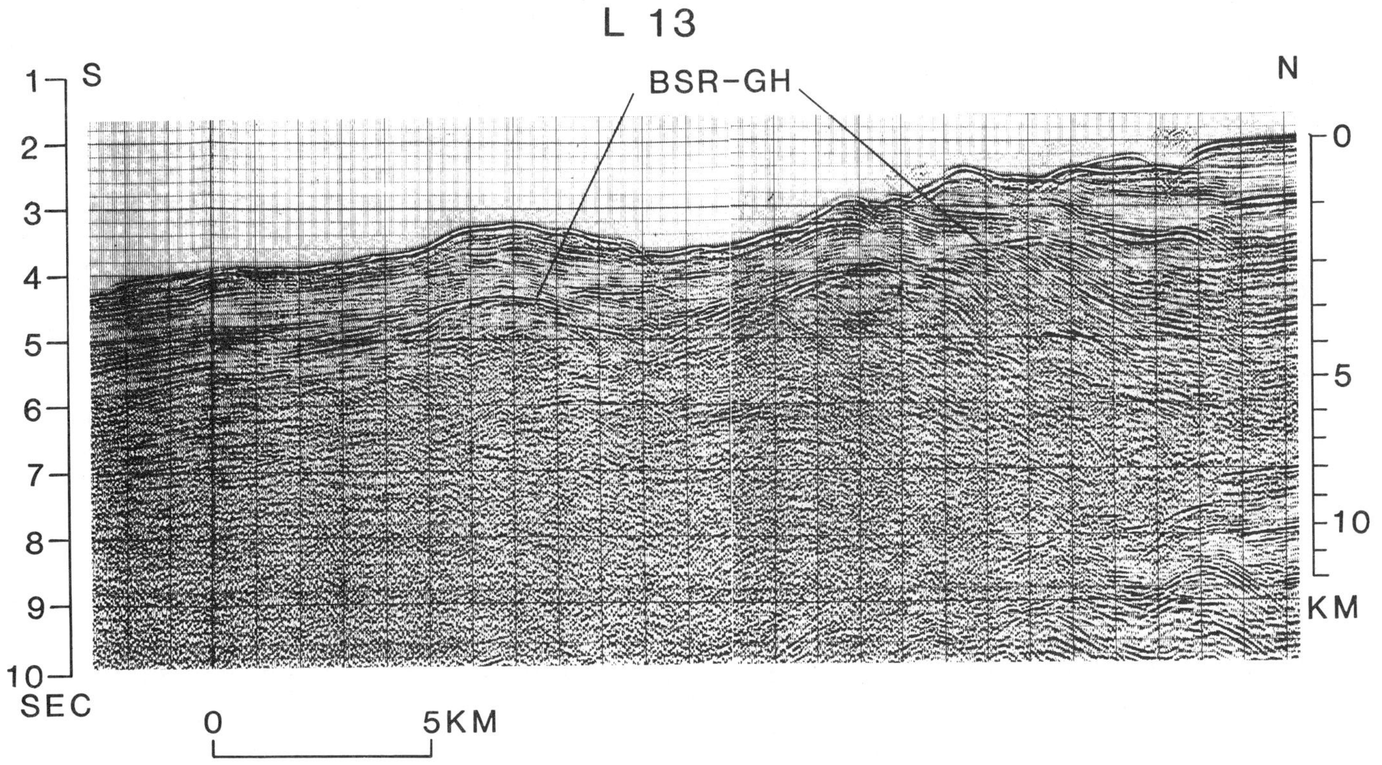

Figure 18. Multichannel seismic-reflection line L13 across the central Koryak margin. Folded sedimentary rocks are as much as 9 km thick beneath the mid-slope area. BSR-GH is thought to be the base of a gas-hydrate zone. See Figure 12 for location.

actual heat flow and temperature values may be 10 to 25% larger than observed. Locally higher heat flow or higher sedimentation rates can cause maturation temperatures to occur at shallower subbottom depths. However, higher thermal conductivity values (20 to 30% higher), such as those expected in the lithified lower part of the sedimentary section, would cause the maturation temperatures to occur at greater depths.

If the criteria of Schlanger and Combs (1975), concerning the temperature versus age and depth for known oil and gas producing basins, are applied to the Aleutian Basin, then the generation of hydrocarbons could occur in rocks as young as early Miocene (18 to 23 Ma). Rocks older than early Miocene are likely to be present within the 2–10-km-thick sedimentary section and within the ideal temperature window for oil and gas generation. The entire basin, and especially the sediment wedges of the rise prisms (Figure 20), may have a suitable thermal history for the generation of hydrocarbons.

Organic Carbon and Hydrocarbons

Subsurface organic-carbon values have been measured from core samples in the three abyssal- or basin-flanking DSDP sites (188, 189, and 190, Figure 1). The values for the diatomaceous silty clay and mudstone (Holocene to middle Miocene age) are nearly uniform with depth and range from 0.2 to 0.8% (Bode, 1973). Late Cenozoic sedimentary rocks cored (Gershanovich, 1968) and dredged (Underwood et al., 1979; Jones et al., 1981; Fischer, 1981; Cooper, Marlow, and Vallier, 1984) from the northern and eastern continental margin have a similar organic-carbon content.

Hydrocarbon gases in subsurface sediments are primarily methane in shallow sea-floor cores (Kvenvolden and Redden, 1980; Carlson et al., 1985) and deeper DSDP cores (sites 189 and 191; Claypool, Presley, and Kaplan, 1973; McIver, 1973). Significant ethane is reported only from DSDP site 189 near the base of the Aleutian Ridge. In 3-m sea-floor cores methane concentrations are small; they increase with depth and are slightly higher in areas where VAMPs are abundant (Cooper et al., 1979). The shallow hydrocarbon gases probably originate from microbiological processes operating near the sea floor (Kvenvolden and Redden, 1980); however, at depth the gases may be either of biogenic or thermogenic origin (Claypool, Presley, and Kaplan, 1973). The acoustic character of the VAMPs suggests that gas concentrations at depth are greater than those in the surface cores (Cooper, Marlow, and Scholl, 1979).

Potential Reservoir Beds and Seals

The characteristics of potential reservoir beds in the Aleutian Basin are likely to be controlled by geographic factors related to sediment distribution, diagenetic alteration, and local tectonism. Subsurface samples from which reservoir characteristics are known come only from DSDP sites 188 and 190 and from a few dredge samples along the Beringian continental margin. Hence, much of the information is indirect and derived from seismic-reflection data.

Three types of reservoirs may exist in the basin. They include the coarser grained deposits suspected to be interbedded within the thick sediment wedges at the base of the continental margin. These wedges are presumed to include sandy or coarser turbidite deposits contributed to the basin via the many submarine canyons that have incised the continental slope (Scholl et al., 1970; Carlson and Karl, 1984). Highly reflective horizons, typical of turbidite deposits and

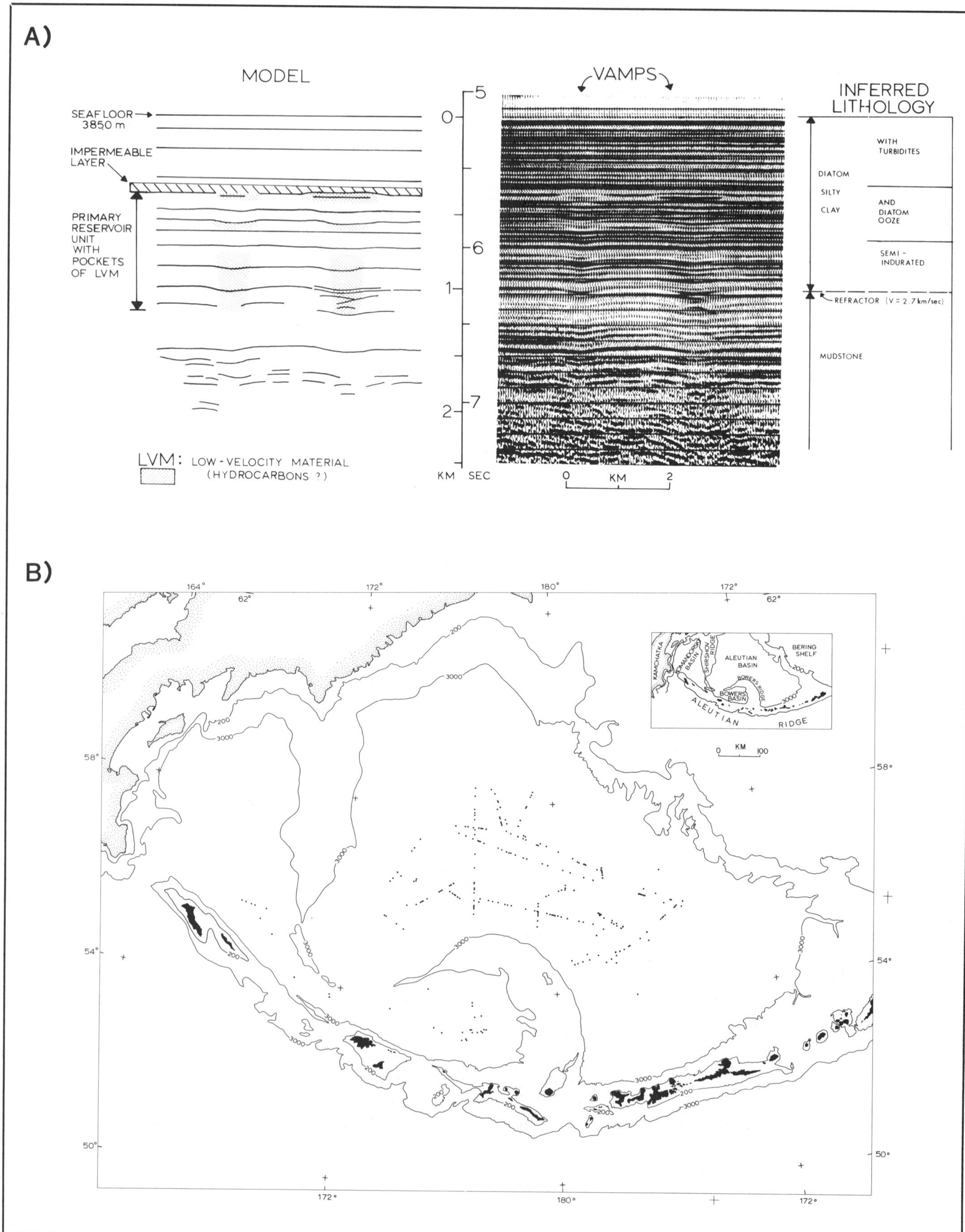

Figure 19. Seismic characterization and distribution of velocity-amplitude features (VAMPs) found in the Bering Sea Basin (Cooper et al., 1979). A, Multichannel seismic-reflection data, inferred lithology, and generalized model for VAMPs. B, Map showing locations of VAMPs discovered along seismic-reflection tracklines.

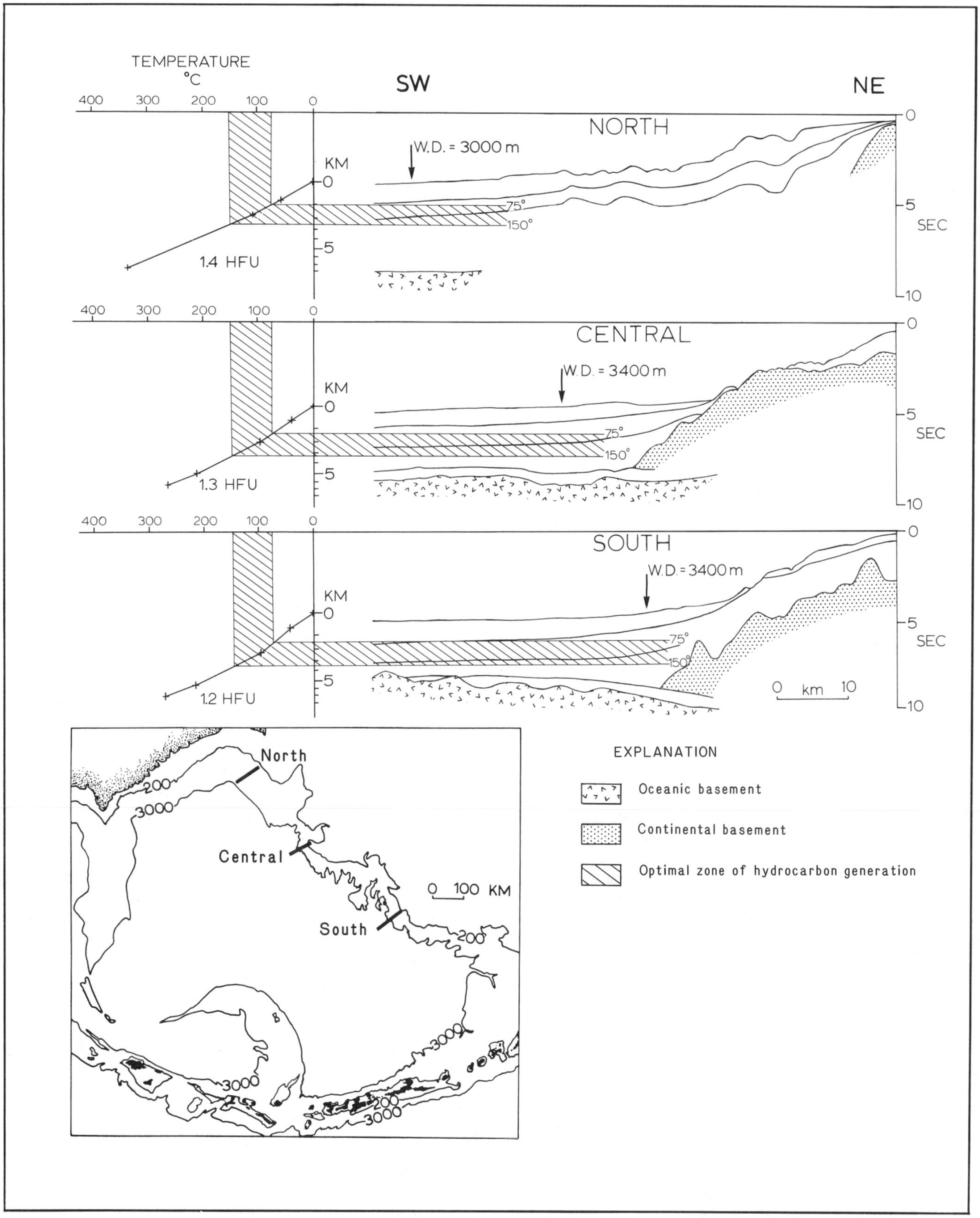

Figure 20. Simplified line drawings of multichannel seismic-reflection profiles showing estimated temperatures below the sea floor within sediment bodies beneath the continental rise. Thermal gradients are based on shallow heat-flow measurements (Figure 6) and on estimates for increase in thermal conductivity with depth (Erickson, 1973). Temperatures below the sea floor are not corrected for effects of rapid sedimentation; temperatures at depth could be higher than shown. The optimal temperatures for hydrocarbon generation probably occur within Paleogene and Neogene rocks buried beneath the continental rise adjacent to the entire Beringian margin.

other coarser grained deposits, are recorded throughout the sediment wedges (Figures 14-18). Marlow et al. (1981) report high porosities (45 to 68%) but low permeabilities (1 to 19 md) for four late Oligocene to middle Miocene samples dredged from the upper continental slope.

Two other types of reservoirs are suspected—those associated with diatom-rich rocks within the upper 1 to 1.5 km of the sedimentary section, and those related to fractures within the highly indurated siliceous rocks and mudstone at depths > 1.5 to 2 km. These potential reservoirs are associated with the silica-rich sedimentary rocks that have been deposited in the Bering Sea since the Oligocene, but principally in the Neogene.

Subsurface porosity and permeability values for unconsolidated and semi-indurated diatomaceous units in the upper 650 m of the sedimentary sections of the Aleutian and Bowers Basins are 58 to 85% and 8 to 49 md, respectively (Figure 2B). In these units, porosity increases with increasing amount of diatomaceous debris, and the porosity in the diatom-rich beds does not decrease significantly with increasing burial depth. Permeability also increases with diatom content, and permeability values actually increase with depth at DSDP sites 188 and 190. The upper diatomaceous sequence has adequate porosity and permeability to be a potential reservoir.

The base of the unconsolidated and semi-indurated units, is marked by a basin-wide diagenetic boundary separating overlying sedimentary rocks rich in biogenic silica (opal-A) from underlying indurated mudstone composed of crystalline silica (opal-CT; Hein et al., 1978). Below the boundary, measurements on a single DSDP core sample show that porosity drops from 58 to 33% and permeability decreases from 35 to < 0.01 md (Figure 2B; Lee, 1973). The highly indurated character of samples beneath the diagenetic front at all Bering Sea DSDP sites implies that the deeply buried rocks in the Aleutian Basin with high silica content are not likely to make good reservoirs unless they are highly fractured.

Fracture porosity and permeability, however, may indeed convert the deeply buried siliceous sedimentary section into a possible reservoir. Evidence for fractures at depth is suggested in seismic-reflection profiles. Laterally discontinuous reflection events, which are vertically offset or associated with weak hyperbolic reflections (below 7 sec, Figure 5A), are common at depth and suggest faulting and fracturing of the indurated sedimentary section. In addition, velocity pull-down effects at VAMPs are observed to increase with depth. This increase indicates that additional low-velocity material (gas?) may exist in the suspected siliceous deposits, possibly in fractures, at greater subsurface depth beneath the VAMPs (Figure 19A). Fractured siliceous rocks of the Monterey formation of California, which have been more highly deformed than the surrounding rocks, are reservoirs for economic hydrocarbon deposits (Isaacs, 1984). Similar siliceous rocks in the Aleutian Basin may also be potential reservoirs, especially near basement structures.

Measured permeability values and seismic-reflection data indicate that reservoir seals may exist basin-wide in the upper part of the sedimentary section. Cooper et al., (1979) suggest that low-permeability fine-grained turbidite beds, as well as other impermeable layers, may be the seal for underlying diatom-rich deposits (Figure 19B). The impermeable layer may also be associated with the base of a gas-hydrate zone (Scholl and Cooper, 1978), which is observed elsewhere along the margins of the Aleutian Basin. The suspected sealing layer overlies the numerous, phase-inverted "bright spots" (VAMPs) that occur over the central region of the Aleutian Basin. The seal-reservoir stratigraphy in the Aleutian Basin (Figure 19B) may be similar to that observed in the sea of Japan, where DSDP drilling (sites 299, 301; Ingle et al., 1975) encountered methane along with ethane at a subbottom depth of 500 m near the base of a turbidite section (seal) overlying diatomaceous sediment (reservoir).

Significance of Bottom–Simulating Reflectors (BSR)

Anomalous reflection horizons that cut across layered reflectors are commonly observed in seismic-reflection profiles recorded in the Bering Sea. Because they mimic the bathymetric relief of the sea floor, they were named bottom-simulating reflectors (BSRs) by Scholl and Creager (1973). The significance of the BSRs found on the Beringian Shelf has been discussed by Hammond and Gaither (1983) and Turner et al. (1984). During our studies of the Bering continental slope, rise, and abyssal basin, we suspected that at least two types of BSRs commonly occurred in the same area, as first documented by Cooper (in Marlow et al., 1981) and later by Carlson et al. (1985). We believe the shallower of the two BSRs (BSR-GH) follows the base of a gas hydrate zone, and the deeper BSR (BSR-DB) follows the silica-diagenetic boundary between overlying and incompletely consolidated opal-A-rich sediment and underlying more indurated mudstone rich in opal-CT. The characteristics of the two types of BSRs are given in Table 1.

Scholl and Creager (1973) first described the lithologic nature of the BSR, now named BSR-DB, at DSDP drilling sites 184 and 185 on Umnak Plateau, at site 188 on the flanks of Bowers Ridge, and at site 190 in the Aleutian Basin (reflector "P" of Ewing, Ludwig, and Ewing, 1965). The BSR-DB is a large-amplitude reflection horizon that corresponds to a basin-wide velocity discontinuity between overlying lower velocity and underlying higher velocity rocks (Figure 3C). This reflection horizon is commonly difficult to identify in sections of flat-lying sedimentary rocks, where the sea floor is also effectively a horizontal surface conformable with bedding. However, the BSR-DB can readily be identified cutting across bedding horizons along eroded parts of the sloping continental margin (Figures 21, 22; Carlson et al., 1985) and over buried structures beneath the continental shelf (Hammond and Gaither, 1983). At all locations where the BSR-DB has been sampled at depth (Scholl and Creager, 1973; Turner et al., 1984), and at outcrops (Marlow et al., 1979b, 1983; Carlson and Karl, 1984; Carlson et al., 1985), highly diatomaceous rocks have been found.

A second and shallower BSR (BSR-GH) has been identified recently in the Bering Sea. This reflection horizon has geophysical characteristics different from those of the BSR-DB but similar to those described for gas hydrates (Table 1; Shipley et al., 1979; MacLeod, 1982; Kvenvolden, this volume). The BSR-GH is most readily observed beneath eroded knolls on the continental slope and rise (Figures 16, 21, 22), where the sea-floor relief (and conformable BSR-GH) cuts across flat or gently dipping strata. The BSR-GH can also be seen beneath parts of the slope that are underlain by folded or deformed strata (Figure 18). Although not confirmed by sampling, the uniform subsurface depth to the tops of the VAMPs

(Scholl and Cooper, 1978) causes us to suspect that a BSR-GH exists beneath large areas of the abyssal floor of the Aleutian Basin.

Seismic-reflection profiles that cross some areas of the continental slope and rise show the stacked effect of both BSRs (Figures 21, 22). The BSR-GH is commonly a weak reflector that lies at reflection times of 0.3 to 0.6 sec below the sea floor. It can only be identified where the configuration of the sea floor is different from that of the bedding surfaces of the underlying sedimentary section. In contrast, the BSR-DB is a strong reflector that can be identified at reflection times of 0.9 to 1.2 sec below the sea floor. This horizon is recorded in nearly all areas of the Bering Sea Basin that have sufficient sediment thickness (700 to 1100 m) and thermal gradient for silica diagenesis to occur.

The widespread distribution of the two types of BSRs may be an important factor in future hydrocarbon exploration in the abyssal regions and along the margins of the Aleutian Basin, especially if hydrocarbons are found in association with the BSRs. The gas hydrate zone (whose base is marked by BSR-GH) may contain significant quantities of natural gas, and this zone may act as an impermeable barrier to the upward migration of other liquid or gaseous hydrocarbons. Similarly, the diagenetic boundary (BSR-DB), which is lithologically as well as thermally controlled, may be an important factor in determining migration paths, trapping, and sealing of moving hydrocarbons.

SUMMARY

Geologic and geophysical studies in the Aleutian and Bowers Basins have shown extensive accumulations of thick, flat-lying Paleogene to Holocene sedimentary rocks. These rocks are deformed locally over basement structures in the central Aleutian Basin, and deformed over a greater area beneath the sedimentary wedges along the north side of

Table 1. Characteristics of bottom-simulating reflectors (BSRs) found in the Bering Sea.

BSR-GH (Gas Hydrate)	BSR-DB (Diagenetic Boundary)
General Characteristics	
Description	
Reflector that marks the base of a gas hydrate zone—the transition from gas hydrate to free gas.	Reflector that marks a silica diagenetic boundary between overlying opal-rich (opal-A) sediment and underlying indurated cristobalite-bearing (opal-CT) mudstone.
Areal Distribution	
Presumed regional in extent, in water depths >500 m. Identifiable most easily in areas with bathymetric relief and undeformed sediment.	Regional extent where adequate siliceous diatomaceous sediments are found or are presumed to be present.
Controlling Factors	
Sufficient gas concentrations and suitable pressure-temperature conditions.	Appropriate conditions (pressure, temperature, geochemistry) for diagenesis of biogenic to crystalline silica.
Geophysical Signature	
Geometry	
Reflector that mimics sea-floor bathymetry and commonly cuts across other layered reflectors below the sea floor.	Same description as BSR-GH but found at greater depths below sea floor.
Depth Below Sea Floor	
Depth to BSR-GH INCREASES with increasing water depth—commonly found at 200–500 m (0.3–0.6 sec) below sea floor.	Depth to BSR-DB DECREASES or is the same with increasing water depth—commonly found 700–1000 m (0.9–1.2 sec) below sea floor.
Velocity Below Sea Floor	
Higher than typical sediment velocity ABOVE the BSR-GH.	Higher than typical sediment velocity BELOW the BSR-DB.
Phase of BSR Reflector	
INVERTED (negative) phase	NORMAL (positive) phase
Amplitude of BSR Reflector	
Highly variable, commonly faint in multichannel seismic-reflection data. Hydrate zone above BSR-GH commonly associated with diffuse and faint reflectors.	Commonly large amplitude and easily identifiable, except where BSR-DB is conformable with other reflectors.
Significance for Hydrocarbon Distribution	
General	
Gas hydrate zone may (1) contain significant included gas and (2) act as a seal that prevents upward migration of hydrocarbons.	Diagenetic boundary (1) has a major decrease in porosity and permeability beneath the boundary and (2) may influence migration path and hydrocarbon trapping.
Traps	
Hydrocarbons (liquid? and gas) may be trapped and accumulate beneath impermeable gas hydrate zone.	In areas of bathymetric relief, laterally migrating hydrocarbons may be trapped against diagenetic boundary in porous interbedded strata.
Migration of Hydrocarbons	
Possible lateral migration of hydrocarbons along base of gas hydrate zone.	Possible migration beneath BSR-DB via fractures in highly indurated rocks or in unaltered strata originally poor in biogenic silica.
Operational Hazards	
Pockets of overpressured gas may be trapped beneath gas hydrate zone and allow sediment failure.	BSR-DB may form a slip surface above which sediment bodies may slide (see Figure 15).

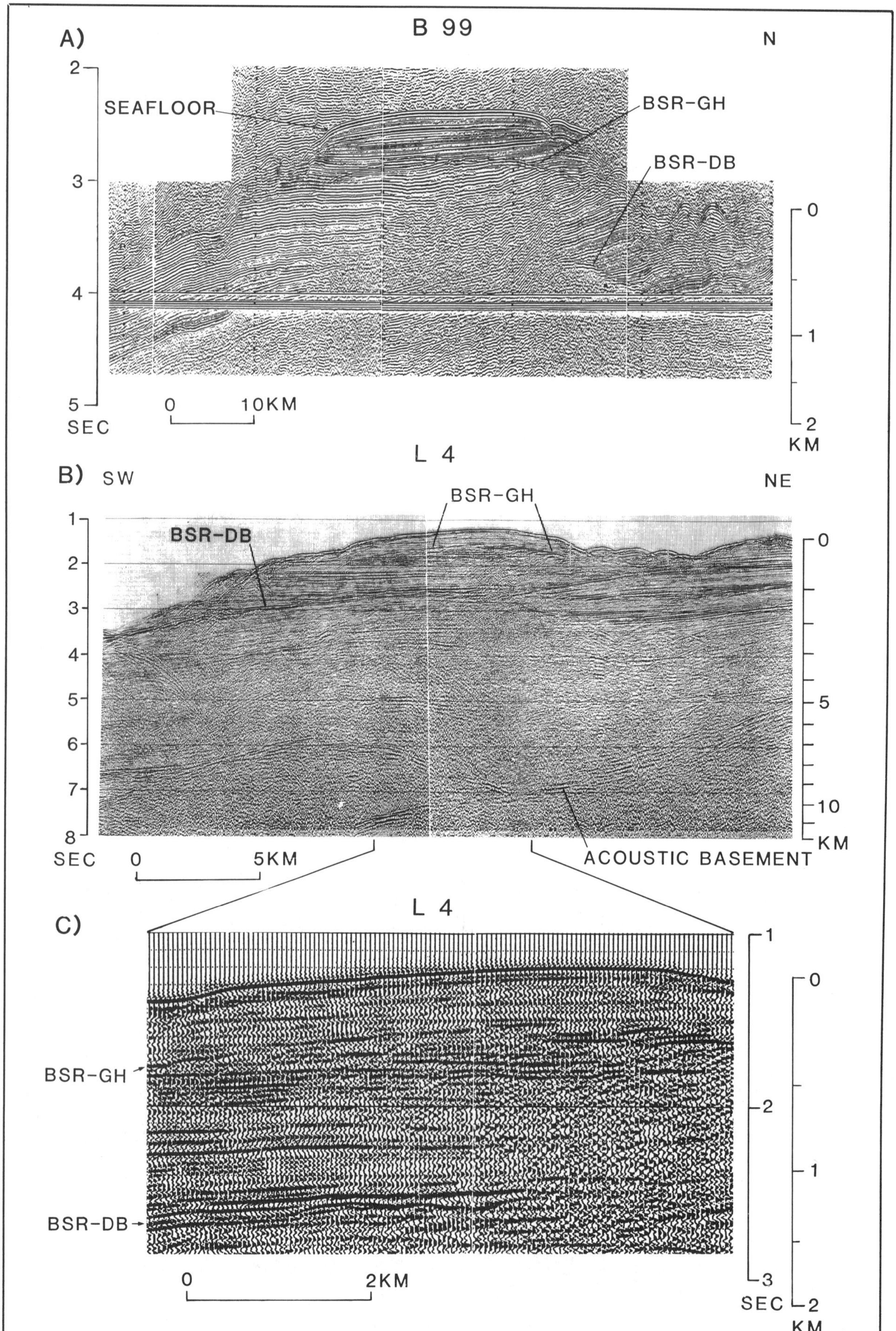
A)
B 99
N
SEAFLOOR
BSR-GH
BSR-DB
2
3
4
5
SEC
0 10KM
0
1
2
KM
B)
L 4
SW
NE
BSR-GH
BSR-DB
1
2
3
4
5
6
7
8
SEC
0 5KM
0
5
10
KM
ACOUSTIC BASEMENT
C)
L 4
BSR-GH
BSR-DB
1
2
3
SEC
0
1
2
KM
0 2KM

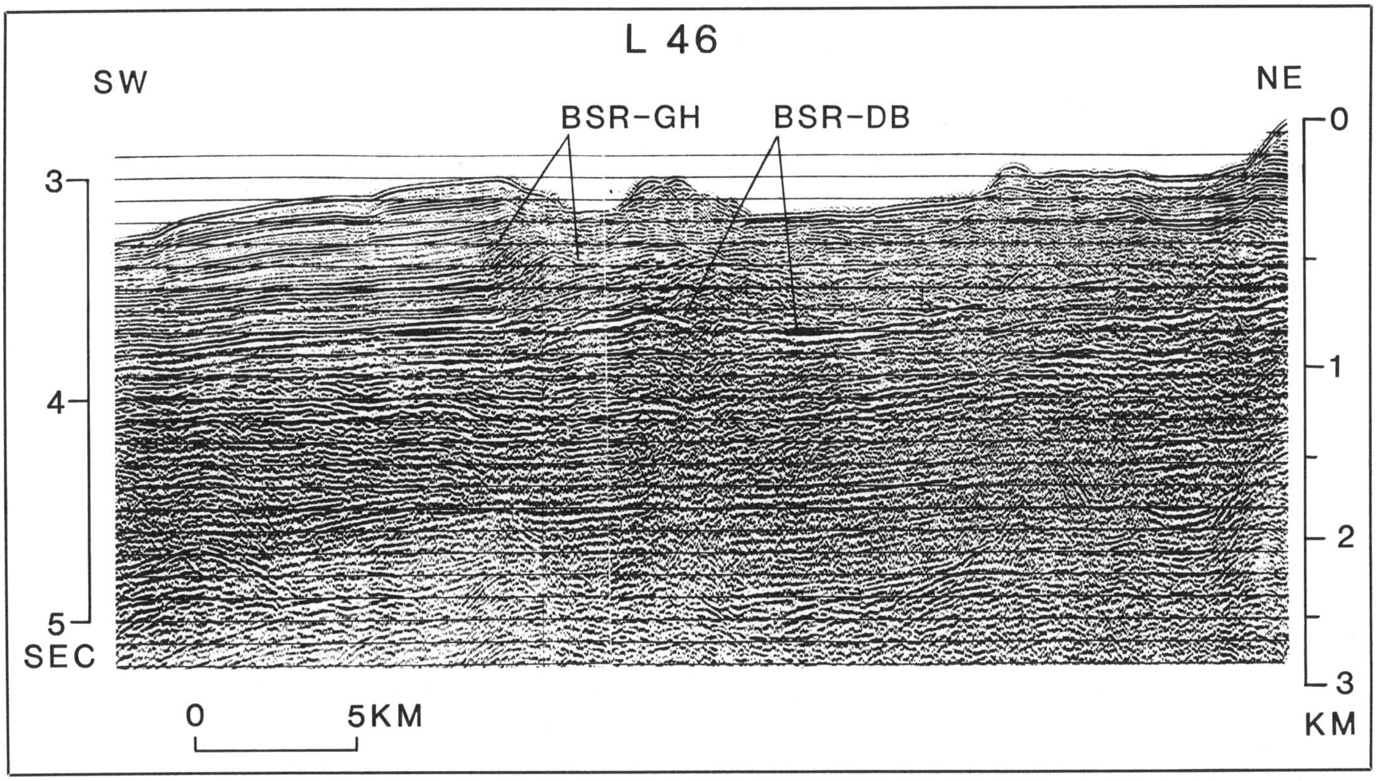

Figure 22. Multichannel seismic-reflection line L46 across the continental rise of the northern Beringian margin showing tiered occurrence of two BSRs. See Figure 21 for description of BSRs and Figure 12 for location of line. Other tiered BSRs are shown in Figures 16 and 21.

Bowers Ridge and beneath the Beringian and Koryak continental slopes and rises.

Both the style and the timing of sedimentation and deformation within the continental-rise sediment wedges are variable. Deposition and deformation of the wedges along the north side of Bowers Ridge and the southern half of the Beringian margin occurred mainly before the Neogene. Along the Koryak and northernmost Beringian margins, however, most of the sedimentary sections were deposited during the Neogene, and deformation has continued from the Paleogene into the Neogene. This northward shift in Neogene sediment depocenters along the continental slope was probably caused by changes in sediment distribution patterns reflecting the Cenozoic uplift and glaciation of the Koryak Ranges.

We suspect that the sedimentary section that is deeply buried beneath the Beringian and Koryak continental slope and rise was deformed in response to pre-Eocene convergence and underthrusting of the Aleutian Basin crust beneath these areas. More recent deformation, following the formation of the Aleutian Ridge in the Eocene, and continuing into the late Cenozoic, is recorded in these areas. Normal faulting and wrench faulting may be the principal mechanisms for the younger deformation.

Figure 21. Seismic-reflection data showing tiered occurrence of two BSRs beneath a knoll west of Zhemchug Canyon: A, Single-channel seismic line B99, B, Multichannel seismic-reflection line L4, C, Expanded view of multichannel seismic-reflection line L4. BSR-GH may lie at the base of a gas-hydrate zone and BSR-DB probably corresponds to a silica diagenetic boundary (see text and Table 1). BSRs are usually more easily identified in high-exaggeration, gain-ranged, single-channel seismic data such as B99. See Figure 12 for location.

In the central Aleutian Basin, basement structures with as much as 2 km of relief are common. However, there is no evidence, such as basement symmetry or deformed sedimentary rocks, for a relict sea-floor-spreading center, either here or beneath the Bowers Basin. Sedimentary rocks in the basins are of marine origin, and more than half of the sedimentary section may be Oligocene and younger rocks that have large concentrations of siliceous diatomaceous material. At subsurface depths less than about 1 km these rocks are highly porous, but at greater depths they could include fractured siliceous mudstone.

Two localities are thought to have future economic potential for hydrocarbons (Figure 23):

1. Along the northern base of Bowers Ridge and the base of the Beringian-Koryak continental margins. In these areas, thick sedimentary prisms, large buried structures, and adequate reservoir and marine organic source material are likely to be present.
2. The central part of the Aleutian Basin where thinner sedimentary sections, greater basement relief, higher heat flow, and numerous deep-water "bright spots" or VAMPs have been recorded.

Resource investigations in the abyssal regions of the Bering Sea have provided tantalizing evidence that hydrocarbons are present, yet many questions regarding the crustal structure and evolution of these regions remain unanswered. Future programs directed toward deep drilling and acquisition of deep-penetration multichannel seismic-reflection and wide-aperture seismic-refraction data are needed in order to acquire new information on the age, type,

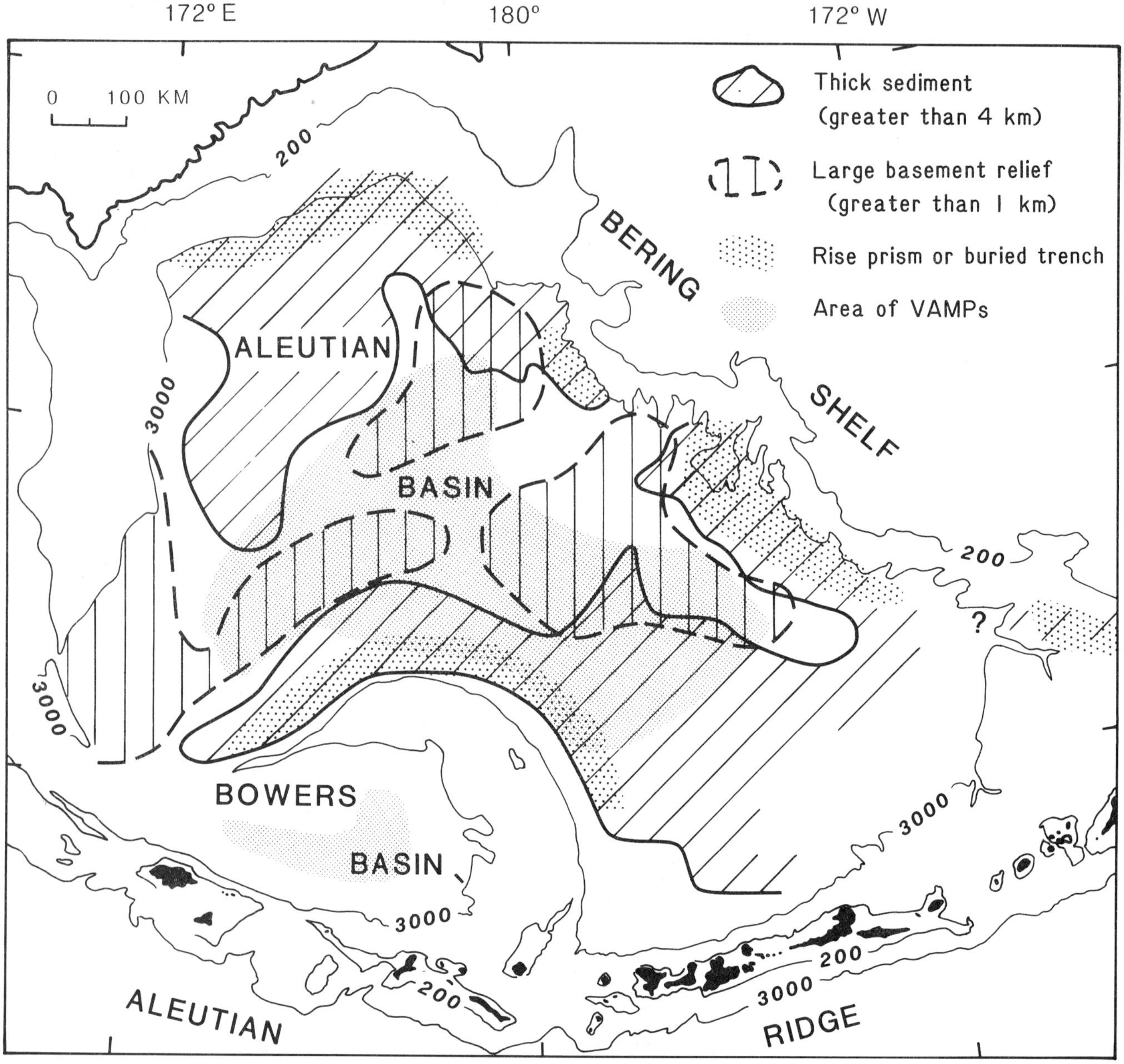

Figure 23. Summary map showing the areas of VAMPs, large basement relief, thick sediment, and sediment wedges (rise prisms). The central area of the Aleutian Basin, as well as peripheral areas along the Beringian margin and Bowers Ridge, probably have the greatest potential for hydrocarbon accumulations.

and origin of the deep sedimentary and igneous crustal rocks beneath the basins.

ACKNOWLEDGMENTS

We thank Paul Carlson and Keith Kvenvolden for their thoughtful reviews. We appreciate the assistance of Phyllis Swenson and Lee Clark in drafting and technical preparation of the manuscript.

REFERENCES CITED

Ben-Avraham, Z., and A. K. Cooper, 1981, Early evolution of the Bering Sea by collision of oceanic rises and North Pacific subduction zones: GSA Bulletin, v. 92, p. 485–495.

Bhattacharya, B., 1979, Propagation of Sn waves through the Bering Sea and the northwest Pacific: Tectonophysics, v. 60, p. T49–T53.

Bode, G. W., 1973, Carbon-carbonate, *in* Deep Sea Drilling Project Initial Reports (Vol. 19): Washington D.C., U.S. Government Printing Office, p. 663–666.

Bogdanov, N. A., and Yu. P. Neprochnov, 1984, Geology of the Bering Sea deep basins, *in* Origin and history of marginal and inland seas: Proceedings of the 27th International Geological Congress, VNU Science Press, v. 23, p. 1–17.

Carlson, P. R., and H. A. Karl, 1984, Discovery of two new large submarine canyons in the Bering Sea: Marine Geology, v. 56, p. 159–179.

Carlson, P. R., M. Golan-Bac, H. A. Karl, and K. A. Kvenvolden, 1985, Seismic and geochemical evidence for shallow gas in sediment on Navarin continental margin, Bering Sea: AAPG Bulletin, v. 69, p. 422–436.

Claypool, G. E., B. J. Presley, and I. R. Kaplan, 1973, Gas analyses in sediment samples from Legs 10, 11, 13, 14, 15, 18, and 19, *in* Deep Sea Drilling Project Initial Reports (Vol. 9): Washington D.C., U.S. Government Printing Office, p. 879–884.

Cooper, A. K., 1978, Heat flow and organic gas measurements from the Aleutian Basin, Bering Sea: U.S. Geological Survey Circular 772–B, p. 92–93.

Cooper, A. K., and Z. Ben-Avraham, 1981, Multichannel seismic evidence bearing on the origin of Bowers Ridge, Bering Sea: GSA Bulletin, v. 92, p. 474–484.

Cooper, A. K., M. S. Marlow, and D. W. Scholl, 1976, Mesozoic magnetic lineations in the Bering Sea marginal basin: Journal of Geophysical Research, v. 81, p. 1916–1934.

Cooper, A. K., M. S. Marlow, and D. W. Scholl, 1977, Bering Sea: A multifarious marginal basin, *in* Island arcs, deep sea trenches and back-arc basins: American Geophysical Union, v. 1, p. 437–450.

Cooper, A. K., M. S. Marlow, and D. W. Scholl, 1979, Thick sediment accumulations beneath continental margin of outer Bering Sea: AAPG Bulletin, v. 63, p. 436.

Cooper, A. K., M. S. Marlow, and D. W. Scholl, 1986, Geologic framework of the Bering Sea crust, this volume.

Cooper, A. K., M. S. Marlow, and T. L. Vallier, 1984, Summary Geologic Report for the St. George Outer Continental Shelf Planning Area, Bering Sea, Alaska: U.S. Geological Survey Open-File Report 84–418, 81 p.

Cooper, A. K., D. W. Scholl, and M. S. Marlow, 1976, Plate tectonic model for the evolution of the eastern Bering Sea Basin: GSA Bulletin, v. 87, p. 1119–1126.

Cooper, A. K., D. W. Scholl, T. L. Vallier, and E. W. Scott, 1980, Resource report for the deep-water areas of proposed Outer Continental Shelf lease sale #70, St. George basin, Alaska: U.S. Geological Survey Open-File Report 80–246, 90 p.

Cooper, A. K., K. A. Bailey, J. I. Howell, M. S. Marlow, and D. W. Scholl, 1976, Preliminary residual magnetic map of the Bering Sea Basin and Kamchatka Peninsula: U.S. Geological Survey Miscellaneous Field Studies Map MF–715, scale 1:2,500,000.

Cooper, A. K., D. W. Scholl, M. S. Marlow, J. R. Childs, G. D. Redden, K. A. Kvenvolden, and A. J. Stevenson, 1979, Hydrocarbon potential of Aleutian Basin, Bering Sea: AAPG Bulletin, v. 63, p. 2070–2087.

Creager, J. S., D. W. Scholl, et al., 1973, Initial reports of the Deep Sea Drilling Project (Vol. 19): Washington, D.C., U.S. Government Printing Office, 913 p.

Erickson, A., 1973, Initial report on downhole temperature and shipboard thermal conductivity measurements, Leg 19, Deep Sea Drilling Project, *in* Deep Sea Drilling Project Initial Reports (Vol. 19): Washington D.C., U.S. Government Printing Office, p. 643–656.

Ewing, M., W. J. Ludwig, and J. Ewing, 1965, Oceanic structural history of the Bering Sea: Journal of Geophysical Research, v. 70, p. 4593–4600.

Fischer, J. M., 1981, Carbon contents of Navarin Basin sediments, *in* P. R. Carlson and H. A. Karl, eds., Seafloor geologic hazards, sedimentology, and bathymetry: Navarin Basin province, northwestern Bering Sea: U.S. Geological Survey Open-File Report 81–1217, p. 44–57.

Fullam, T. J., P. R. Supko, R. E. Boyce, and R. J. Stewart, 1973, Some aspects of late Cenozoic sedimentation in the Bering Sea and north Pacific Ocean, *in* Deep Sea Drilling Project Initial Report (Vol. 19): Washington D.C., U.S. Government Printing Office, p. 887–896.

Gershanovich, D. E., 1968, New data on geomorphology and recent sediments of the Bering Sea and the Gulf of Alaska: Marine Geology, v. 6, p. 281–296.

Gnibidenko, H. S., 1973, Tectonics of the floor of the Bering Sea: Geotectonics, v. 4, p. 237–243.

Gnibidenko, H. S., 1980, Tectonics of the Komandorsky Basin, *in* V. B. Kurnosov and I. O. Murdmaa, eds., Geology of the Komandorsky Basin, Vladivostok, Akad. Nauk SSSR, Dal'nevostochnyi nauchnyi Tsentr (in Russian), 73 p.

Hammond, R. D., and J. R. Gaither, 1983, Anomalous seismic character of the Bering Sea Shelf: Journal of the Alaska Geological Society, v. 3, p. 47–54.

Hein, J. R., D. W. Scholl, J. A. Barron, M. G. Jones, and J. Miller, 1978, Diagenesis of late Cenozoic diatomaceous deposits and formation of the bottom simulating reflector in the southern Bering Sea: Sedimentology, v. 25, p. 155–181.

Ingle, J. C., Jr., et al., 1975, Initial reports of the Deep Sea Drilling Project (Vol. 31): Washington D.C., U.S. Government Printing Office, 927 p.

Issacs, C. M., 1984, The Monterey - key to offshore California boom: Oil and Gas Journal, v. 82, p. 75–81.

Jones, D. M., M. J. Kingston, M. S. Marlow, A. K. Cooper, J. A. Barron, F. H. Wingate, and R. E. Arnal, 1981, Age, mineralogy, physical properties, and geochemistry of dredge samples from the Bering Sea continental margin: U.S. Geological Survey Open-File Report 81–1297, 68 p.

Karig, D. E., 1972, Remnant arcs: GSA Bulletin, v. 83, p. 1057–1968.

Kienle, J., 1971, Gravity and magnetic measurements over Bowers Ridge and Shirshov Ridge, Bering Sea: Journal of Geophysical Research, v. 76, p. 7138–7153.

Kvenvolden, K. A., 1986, Gas hydrate offshore Alaska and western continental United States, this volume.

Kvenvolden, K. A., and G. D. Redden, 1980, Hydrocarbon gas in sediment from the shelf, slope and basin of the Bering Sea: Geochimica et Cosmochimica Acta, v. 44, p. 1145–1150.

Langseth, M. G., M. A. Hobart, and K. Horai, 1980, Heat flow in the Bering Sea: Journal of Geophysical Research, v. 85, p. 3740–3750.

Lee, H. J., 1973, Measurements and estimates of engineering and other physical properties, Leg 19, *in* Deep Sea Drilling Project Initial Reports (Vol. 19): Washington D.C., U.S. Government Printing Office, p. 701–719.

Lisitsyn, A. P., 1966, Recent sedimentation in the Bering Sea (in Russian): Institute of Okeanologi Akademii Nauk USSR (translated by Israel Program for Scientific Translation, available from Springfield, Va. U.S. Department of Commerce, Clearinghouse for Federal Scientific and Technical Information, 1969, 614 p.).

Ludwig, W. J., 1974, Structure of the Bering Sea Basins, *in* C. A. Burk and C. L. Drake, eds. The geology of continental margins: New York, Springer-Verlag, p. 661–668.

Ludwig, W. J., R. E. Houtz, and M. Ewing, 1971, Sediment distribution in the Bering Sea—Bowers Ridge, Shirshov Ridge, and enclosed basins: Journal of Geophysical Research, v. 76, p. 6367–6375.

Ludwig, W. J., S. Murauchi, N. Den, M. Ewing, H. Hotta, R. W. Houtz, T. Yoshii, T. Asanuma, K. Hagiwara, T. Saito, and S. Ando, 1971, Structure of Bowers Ridge, Bering Sea: Journal of Geophysical Research, v. 76, p. 6350–6366.

MacLeod, M. K., 1982, Gas hydrates in ocean bottom sediments: AAPG Bulletin, v. 66, p. 2649–2662.

Marlow, M. S., and A. K. Cooper, 1985, Regional geology of the Beringian continental margin, *in* N. Nasu et al., eds., Formation of active ocean margins: Tokyo, Terra Scientific Publishing Company, p. 497–515.

Marlow, M. S., A. K. Cooper, and M. A. Fisher, 1986, Petroleum geology of the Beringian continental shelf, this volume.

Marlow, M. S., A. K. Cooper, A. W. Parker, and J. R. Childs, 1979a, Isopach map of strata above acoustic basement in the Bering Sea: U.S. Geological Survey Miscellaneous Field Studies Map 1164, scale 1:2,500,000.

Marlow, M. S., A. K. Cooper, D. W. Scholl, and H. McLean, 1982, Ancient plate boundaries in the Bering Sea region, *in* J. K. Leggett, ed., Trench-forearc geology: Sedimentation and tectonics on modern and ancient active plate margins: Geological Society of London Special Publication 10, p. 201–211.

Marlow, M. S., A. K. Cooper, D. W. Scholl, T. L. Vallier, and H. McLean, 1979b, Description of dredge samples from the Bering

Sea continental margin: U.S. Geological Survey Open-File Report 79–1139, 5 p.

Marlow, M. S., T. L. Vallier, A. K. Cooper, J. A. Barron, and F. H. Wingate, 1983, A description of dredge samples collected in 1982 from the Bering Sea continental margin west of Navarin Basin: U.S. Geological Survey Open-File Report 83-325, 4 p.

Marlow, M. S., P. Carlson, A. K. Cooper, H. Karl, H. McLean, R. McMullin, and M. B. Lynch, 1981, Hydrocarbon resource report for proposed OCS sale no. 83, Navarin Basin, Alaska: U.S. Geological Survey Open-File Report 81–252.

McIver, R. D., 1973, Hydrocarbons in canned muds from sites 185, 186, 189, and 191, *in* Deep Sea Drilling Project Initial Reports (Vol. 19): Washington D.C., U.S. Government Printing Office, p. 875–878.

Nelson, C. H., D. M. Hopkins, and D. W. Scholl, 1974, Tectonic setting and Cenozoic sedimentary history of the Bering Sea, *in* Y. Herman, ed., Marine geology and oceanography of the Arctic Seas: New York, Springer-Verlag, p. 119–140.

Rabinowitz, P. D., and A. K. Cooper, 1977, Structure and sediment distribution in the western Bering Sea Basin: Marine Geology, v. 24, p. 309–320.

Regan, R. D., F. E. Cain, and W. M. David, 1975, A global magnetic anomaly map: Journal of Geophysical Research, v. 80, p. 794–802.

Schlanger, S. O., and J. Combs, 1975, Hydrocarbon potential of marginal basins bounded by an island arc: Geology, v. 3, p. 397–400.

Scholl, D. W., and A. K. Cooper, 1978, VAMPs: Possible hydrocarbon-bearing structures in the Bering Sea Basin: AAPG Bulletin, v. 62, p. 2481–2488.

Scholl, D. W., and J. S. Creager, 1973, Geologic synthesis of Leg 19 (DSDP) results: Far North Pacific, Aleutian Ridge and Bering Sea, *in* Deep Sea Drilling Project Initial Reports (Vol. 19): Washington D.C., U.S. Government Printing Office, v. 19, p.897–913.

Scholl, D. W., E. C. Buffington, and T. M. Hopkins, 1968, Geologic history of the continental margin of North America in Bering Sea: Marine Geology, v. 6, p. 297–330.

Scholl, D. W., E. C. Buffington, and M. S. Marlow, 1975, Plate tectonics and the structural evolution of the Aleutian-Bering Sea region, *in* R. B. Forbes, ed., Contributions to the geology of the Bering Sea Basin and adjacent regions: GSA Special Paper 151, p. 1–32.

Scholl, D. W., E. C. Buffington, and M. S. Marlow, 1976, Aleutian-Bering Sea region seismic reflection profiles: U.S. Geological Survey Open-File Report 76–748.

Scholl, D. W., T. L. Vallier, and A. J. Stevenson, 1983, Geologic evolution of the Aleutian Ridge—Implications for petroleum resources: Journal of the Alaska Geological Society, v. 3, p. 33–46.

Scholl, D. W., T. L. Vallier, and Stevenson, 1986, Geologic evolution and petroleum geology of the Aleutian Ridge, this volume.

Scholl, D. W., E. C. Buffington, D. M. Hopkins, and T. R. Alpha, 1970, The structure and origin of the large submarine canyons of the Bering Sea: Marine Geology, v. 8, p. 187–210.

Shipley, T. P., M. B. Houston, R. T. Buffler, F. J. Shaub, K. J. McMillen, J. W. Ladd, and J. L. Worzel, 1979, Seismic reflection evidence for the widespread occurrence of possible gas-hydrate horizons on continental slopes and rises: AAPG Bulletin, v. 63, p. 2204–2213.

Shor, G. G., Jr., 1964, Structure of the Bering Sea and the Aleutian Ridge: Marine Geology, v. 1, p. 213–219.

Soloveva, N. M., 1968, Karta Anomalnovo Magnitovo Polya Teritori SSR: Ministerstvo Geologii SSR.

Stone, D. B., in press, Bering Sea, Aleutian Arc, Alaska, *in* A. E. M. Nairn, F. G. Stehli, and S. Uyuda, eds., Ocean basin and margins, v. 7B–Pacific: New York, Plenum Press,.

Turner, R. F., C. M. McCarthy, D. A. Steffy, M. B. Lynch, G. C. Martin, K. W. Sherwood, T. O. Flett, and A. J. Adams, 1984, Geological and operational summary, Navarin Basin COST No. 1 well, Bering Sea, Alaska: Minerals Management Service OCS Report MMS 84–0031, 248 p.

Underwood, M. B, T. L. Vallier, J. V. Gardner, and J. A. Barron, 1979, Age, grain size, mineralogy, and carbon/carbonate content of Miocene and Pliocene samples from dredge hauls, DSDP holes 184b and 185, and the Sandy River well, southern Bering Sea continental margin and Alaska Peninsula: U.S. Geological Survey Open-File Report 79–450.

Vorobev, N. M., 1970, Results of marine magnetic surveys adjacent to the western parts of the Aleutian Island arc: Academy of Sciences USSR, Siberian Division of Translation, no. 24 (in Russian).

Watanabe, T., M. G. Langseth, and R. N. Anderson, 1977, Heat flow in back-arc basins of the western Pacific, *in* Island arcs, deep sea trenches, and back-arc basins: American Geophysical Union, v. 1, p. 137–162.

CHAPTER TWENTY

DEEP-SEA FAN BODIES, TERRIGENOUS TURBIDITE SEDIMENTATION, AND PETROLEUM GEOLOGY, GULF OF ALASKA

Andrew J. Stevenson

United States Geological Survey
Menlo Park, California

Robert Embley

National Oceanic and Atmospheric Administration
Newport, Oregon

ABSTRACT

The Pacific plate underlying the Gulf of Alaska has been the site of voluminous terrigenous turbidite sedimentation throughout the Cenozoic. The sediment can be grouped into three large fan bodies—the Zodiac, Surveyor, and Baranof fans. The depositional ages of sediment comprising these fans show overlapping ranges, with Zodiac fan as oldest (42–24? Ma), succeeded by the Surveyor fan (20?–0 Ma), and the Baranof fan (10?–0 Ma). Thus a depositional continuum has existed in the gulf since at least the late Eocene. Owing to the presence of a convergent and transform plate boundary between the sites of deposition and the sediment sources on the North American plate, these fans have moved close to their sources through time, the oldest having moved the most. Restoration of these bodies to

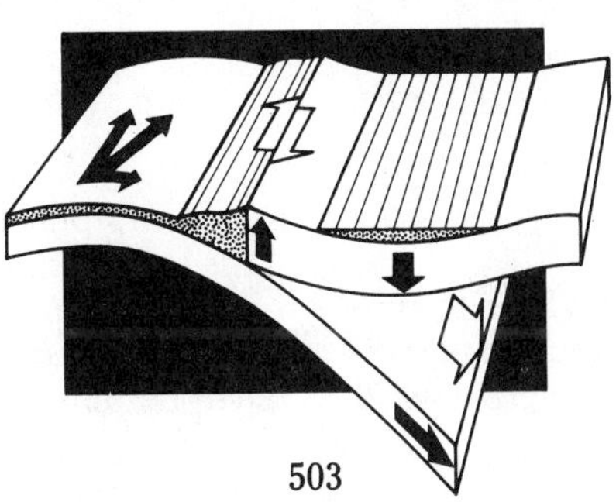

their depositional positions with respect to the continent demonstrates a complex pattern of long-distance sediment transport, shifting sediment sources, shifting depocenters, and tectonically controlled changes in sediment supply. Some of these changes can be related to specific tectonic events occurring along the margin, and can be used to constrain their timing and magnitude. The sediment of the fans contains a significant amount of organic carbon that might become available for hydrocarbon generation. The ultimate fate of these deposits is to encounter a subduction zone and to enter an environment (subduction zone/accretionary wedge) within which all the conditions necessary for the generation, migration, entrapment and preservation of hydrocarbons could be present.

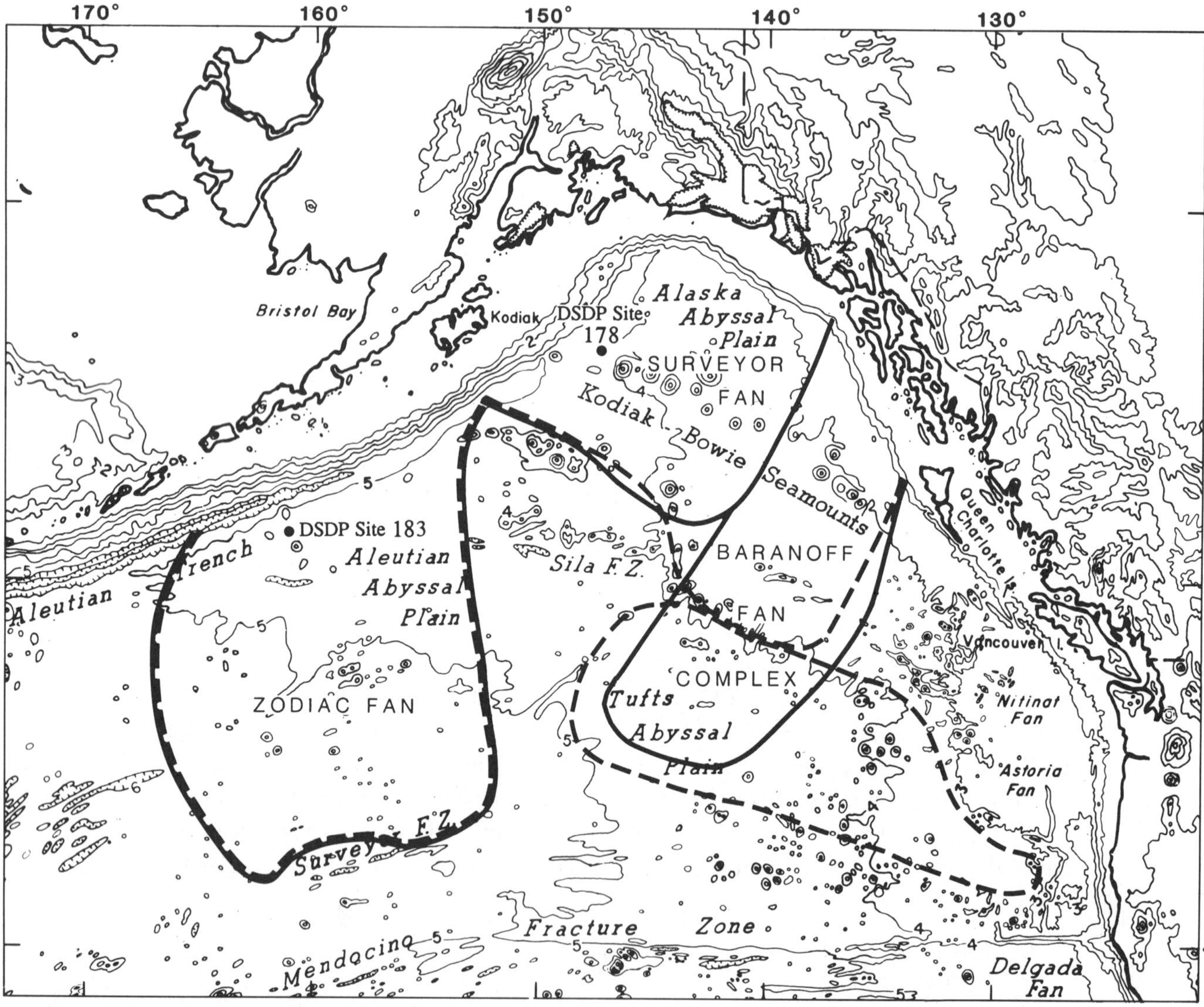

Figure 1. Index map of the Gulf of Alaska (from Chase, Menrad, and Mamerickx, et. al., 1970). Fan bodies are outlined in heavy lines, abyssal plains are outlined in dashed lines.

INTRODUCTION

The Gulf of Alaska, in the northeast Pacific, has been the site of voluminous deep-sea Cenozoic terrigenous turbidite sedimentation and is the only area in the Pacific where abyssal plain morphology is widespread. These deposits have buried the basement relief over much of the gulf and established an abyssal plain province that extends from long. 135 °W to long. 170 °W and from lat. 50 °N to lat. 60 °N (Figure 1). Several linear basement highs (formed predominantly by seamount

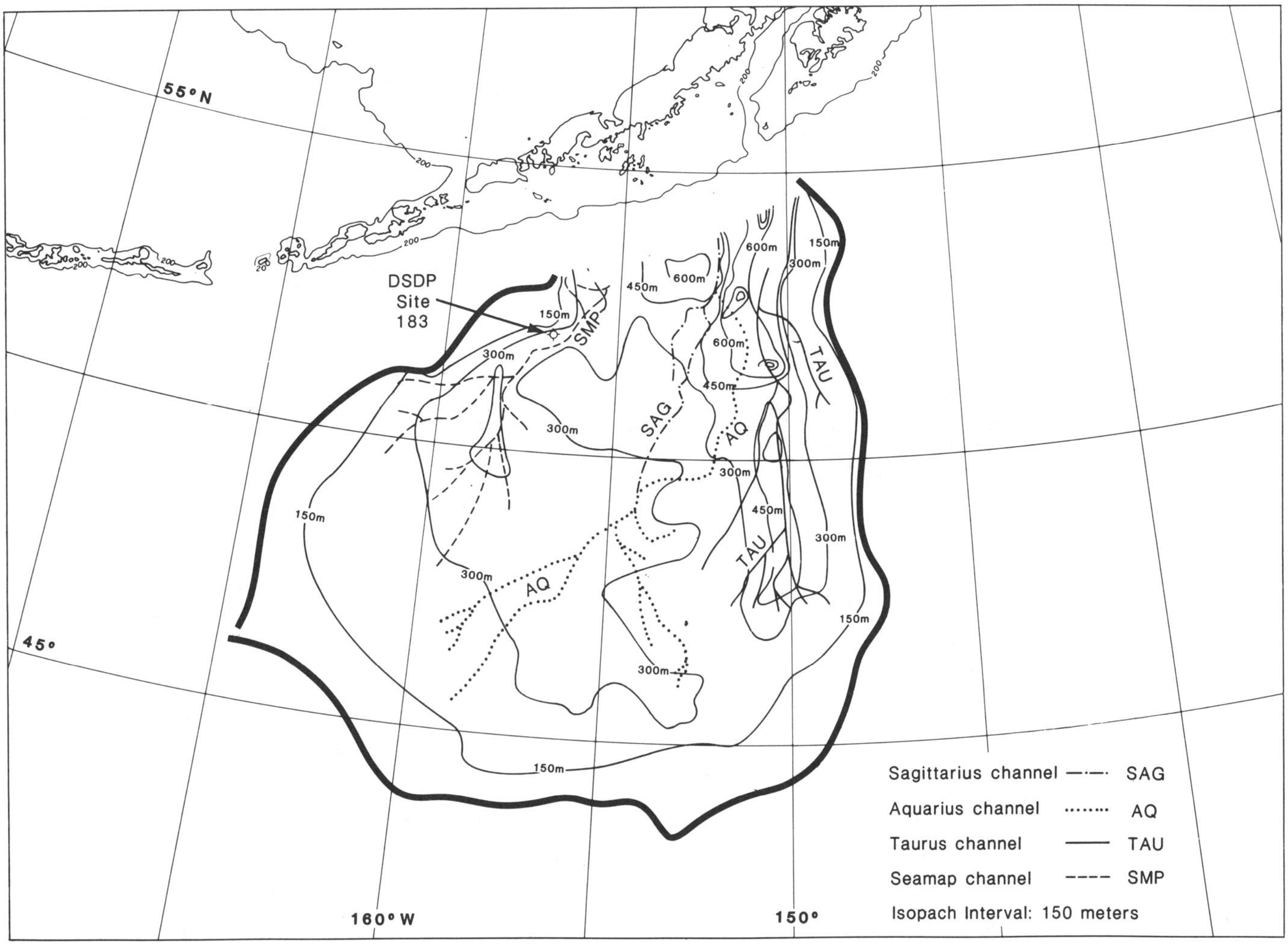

Figure 2. Zodiac fan, showing thickness isopachs in meters and channel courses in line patterns (from Stevenson, Scholl, and Vallier, 1983).

chains) divide the province into three abyssal plains (Hurley, 1960). The westernmost plain—the Aleutian Abyssal Plain (AAP)—lies south of the Aleutian Trench, east of the Patton and Gilbert Seamount chains, and north of the Surveyor Fracture Zone. Northeast of the AAP and east of the Patton and Gilbert chains, lies the Alaskan Abyssal Plain (ALAP). This plain is adjacent to the margin of North America. Southwest of the ALAP and east of the AAP is the Tufts Abyssal Plain (TAP), which is farthest from the continental margin and is relatively unexplored.

The geomorphic boundaries of these plains are poorly defined, and where no intervening basement highs exist, the plains grade into one another without obvious demarcation. The sedimentary sections underlying them differ in origins and ages, and deposits derived from a particular distributary system are not confined to a single abyssal plain. Consequently, the division of these deposits into three abyssal plains has geomorphic, but no genetic, significance.

Regional Sedimentary History

In a genetic sense, the deposits underlying these plains can be divided into two major sequences based on age. The Zodiac fan (Figures 1, 2), which lies beneath most of the AAP, is known to have received sediment from 40–32 Ma, and may have been active into the Miocene (Stevenson, Scholl, and Vallier, 1983). Sediment that floors the ALAP and TAP can be grouped into numerous fan complexes of roughly common age that relate to the onset of major alpine glaciation and the resultant increase in sediment supply around the gulf some time in the late Neogene (von Huene, Shor, and Wageman, 1979; von Huene and Kulm, 1973)

Deposits beneath the TAP and ALAP are subdivided based on the distributary system from which they are derived. The ALAP is underlain by two fan systems, the Surveyor system to the north and the Baranof system to the south (Figure 1). Deposits of the two systems interfinger along a front that trends southwestward across the central ALAP from Cross Sound to just north of Pathfinder Seamount (Figures 1, 3, 4, 5). The TAP is underlain by sediment derived from three fan systems—the distal Baranof complex to the north, the Scott-Moresby system to the east, and the Cascadia system to the south (Figure 4).

The Surveyor fan, which floors the northern ALAP, is divisible into two major depositional sequences (Figures 3, 6). The upper sequence is composed of sediment derived from the Surveyor Channel, which has controlled deposition in the northern Gulf during the Quaternary. This channel originates near the base of the slope offshore of Yakutat Bay and Dry Bay and transports sediment west and north to the Aleutian

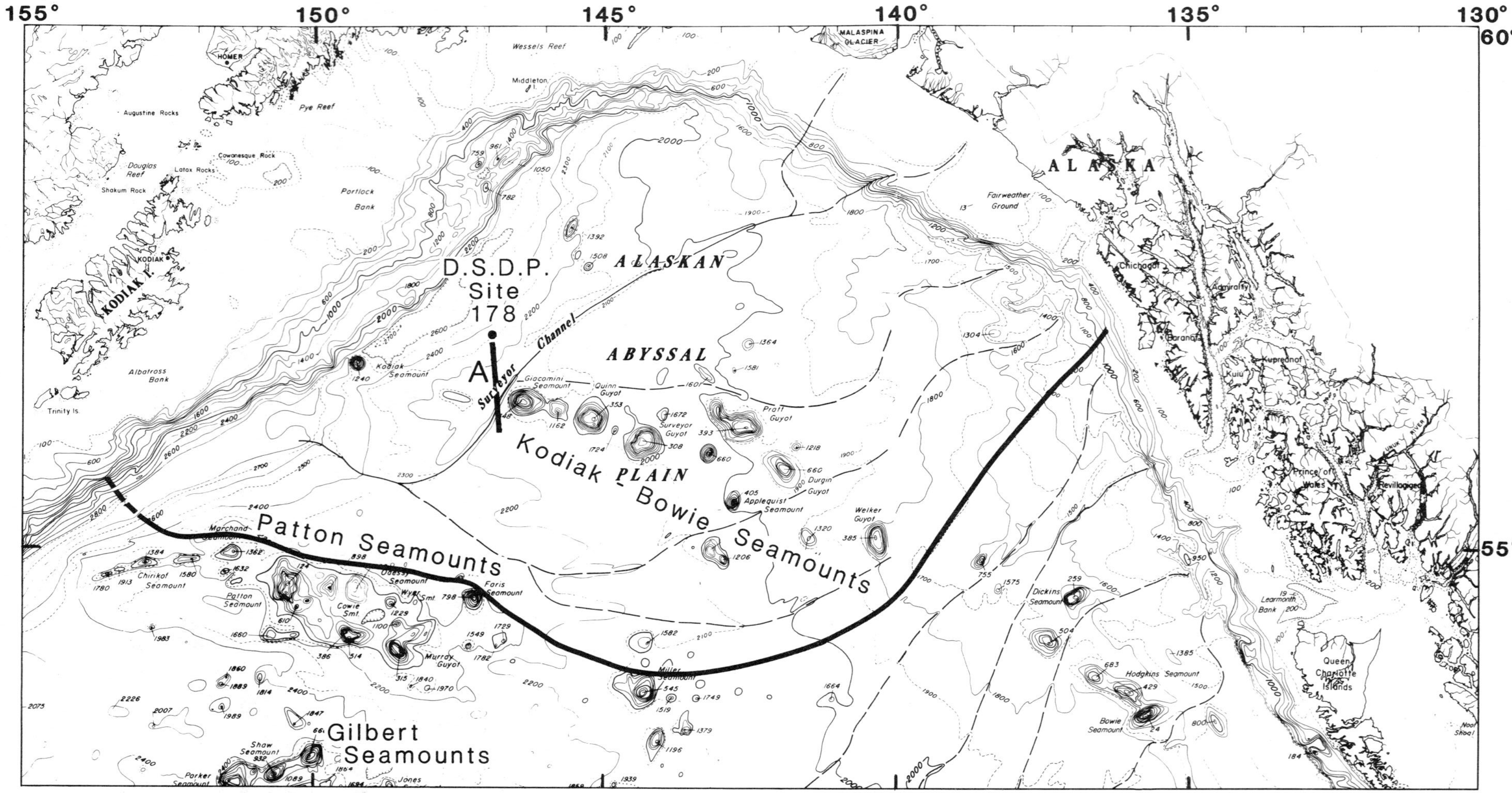

Figure 3. Surveyor fan limits and channel courses. Bathymetry and channel tracks from Chase, Menrad, and Mammerickx, 1970. Channel courses and plain limits are questionable where dashed (inferred or known only from bathymetry). Location of profile shown in Figure 12B indicated by line A.

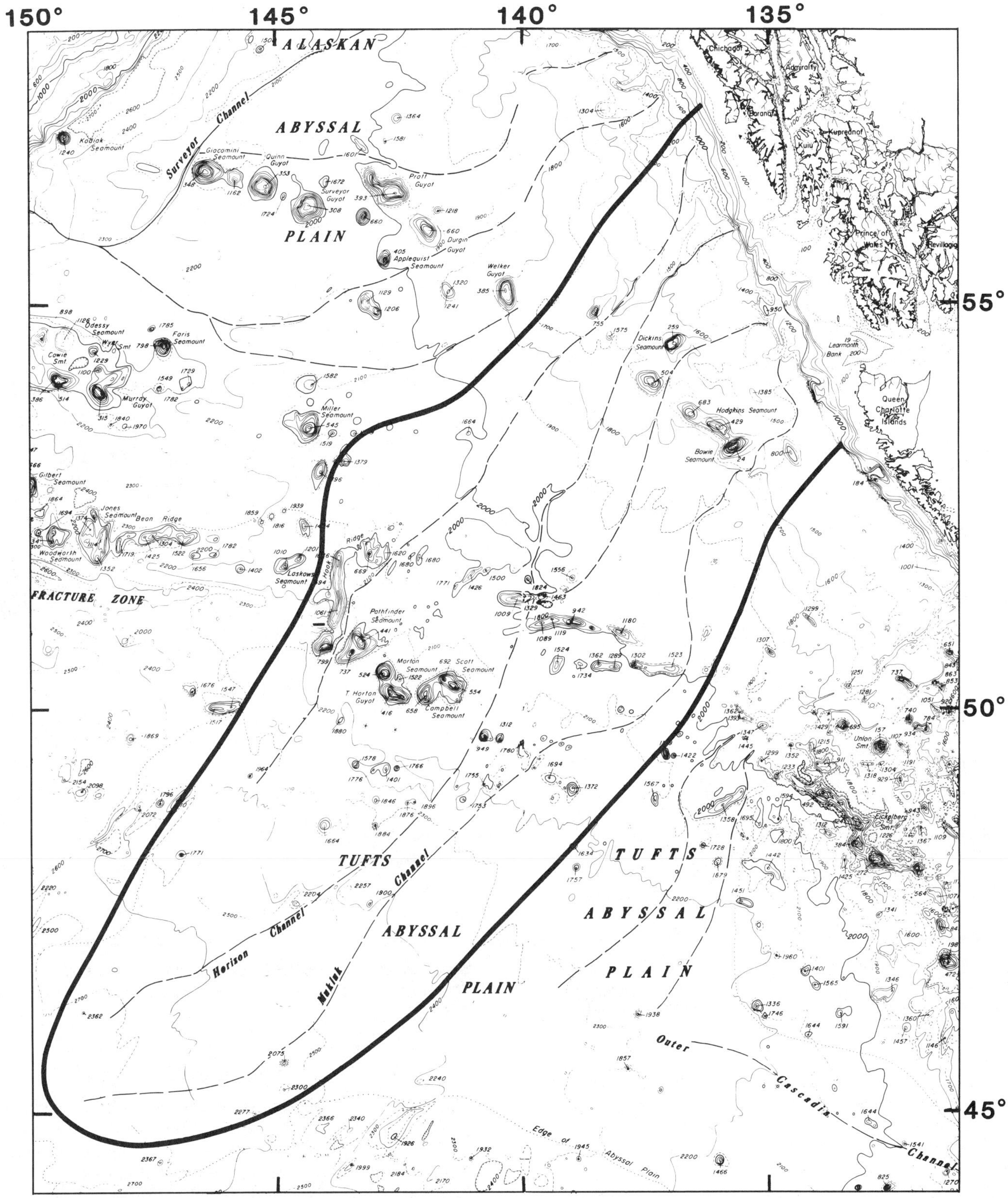

Figure 4. Baranof fan limits and channel courses as previously known. Bathymetry and channel tracks from Chase, Menard, and Mammerickx, 1970. Channel courses are questionable where dashed (inferred or known only from bathymetry).

Trench south of Kodiak Island. The lower sequence is of unknown but presumed northern origin. The fan is bounded on the north by the Aleutian Trench and on the west by the Gilbert and Patton Seamount chains; to the south, it merges with the Baranof fan as described above.

The southern ALAP is floored by the Baranof fan complex (Figure 4). The proximal fan consists of a number of coalescing cone-shaped sedimentary bodies of turbidite and debris flow origin which originate along the lower margin slope. The bathymetry of the cones suggests numerous points of sediment supply and rapid changes in transport systems. Seaward

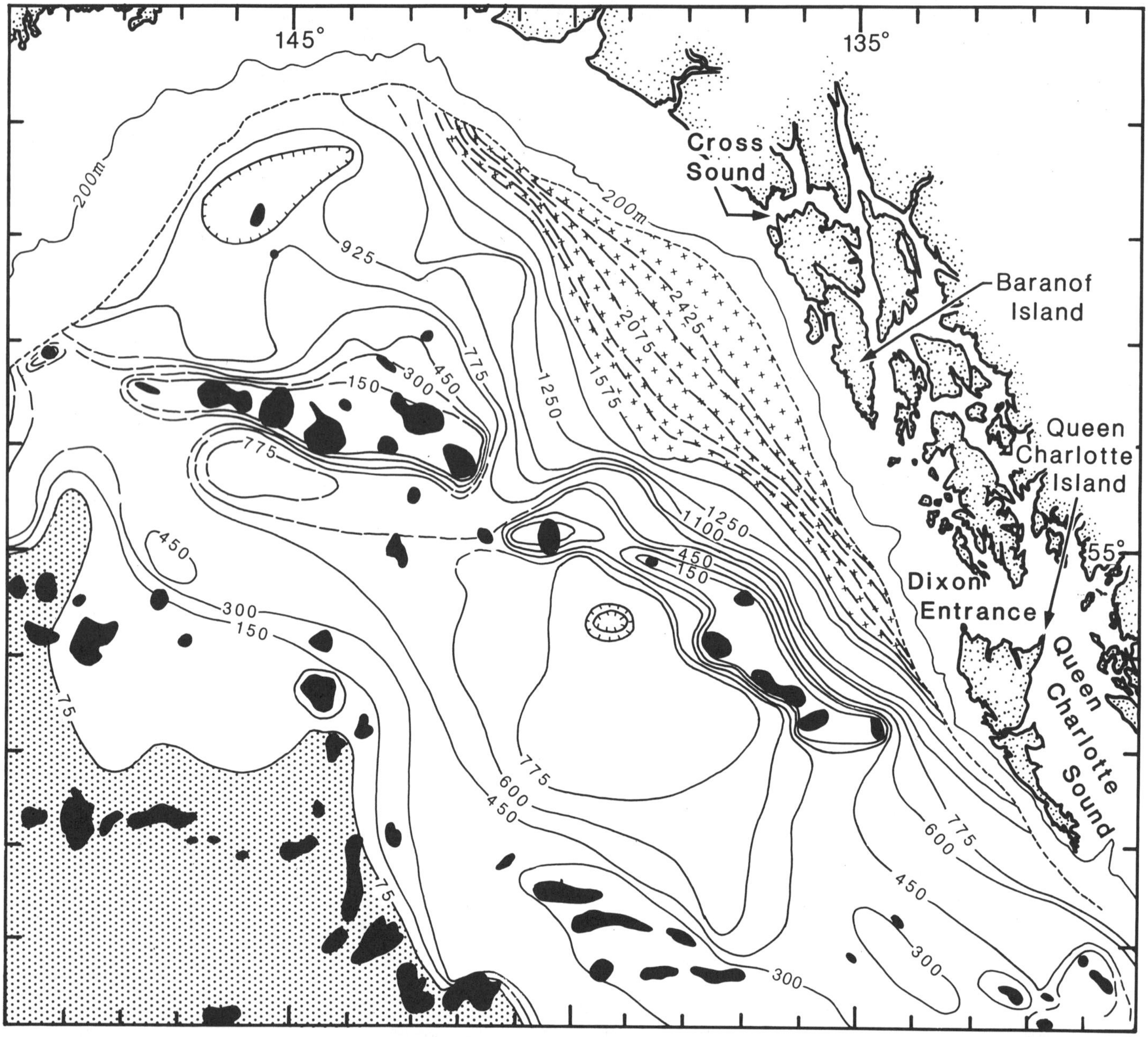

Figure 5. Total sediment thickness isopach map. Thicknesses were read every 2.5 km and averaged over 25 km segments. Data were contoured in time and converted to thickness using sonobuoy velocity data. Basement could not be resolved in the areas patterned with crosses owing to the thickness of the sedimentary section, so isopachs in this area are minimum values. Contours in meters.

of the cones the fan grades into a broad bajada-like sediment apron that extends west of the western margin of the ALAP into the northwestern area of the TAP. Deposition on the bajada is dominated by a few large submarine channels, notably the Horizon and the Mukluk.

The age of deposition of these fans is poorly constrained (see the following Sections on individual fans), but sufficient information exists to suggest a continuum of turbidite sedimentation on this portion of the Pacific plate from the late Eocene to the present. Certainly the bulk of the Surveyor and Baranof fan deposits are derived from the intense Plio-Pleistocene alpine glaciation and uplift of the coastal mountain belt surrounding the gulf, but Zodiac fan sediments clearly predate this event and record the effects of an earlier tectonic and climatic setting.

Dominant Sedimentary Style

The turbidite depositional system responsible for the delivery of sediment to the fans of the Gulf of Alaska is one common to the fans of abyssal deep-sea areas; however, this type of deposition is rarely found in the more often-described and better-known marginal fans, which are characterized by short, rapidly migrating distributional channels that terminate in a mid-fan depocenter or a suprafan lobe. Depositional processes on deep-sea fans are dominated by levee overbanking from large, long-lived constructional channels. These channels run the full length of the fan, and channelized turbidite flow predominates to the distal reaches of the sediment body. This depositional style gives rise to a typical fan morphology. The form of the fan is that of a shield of sediment, and the depositional channel is at the bathymetrically highest point on

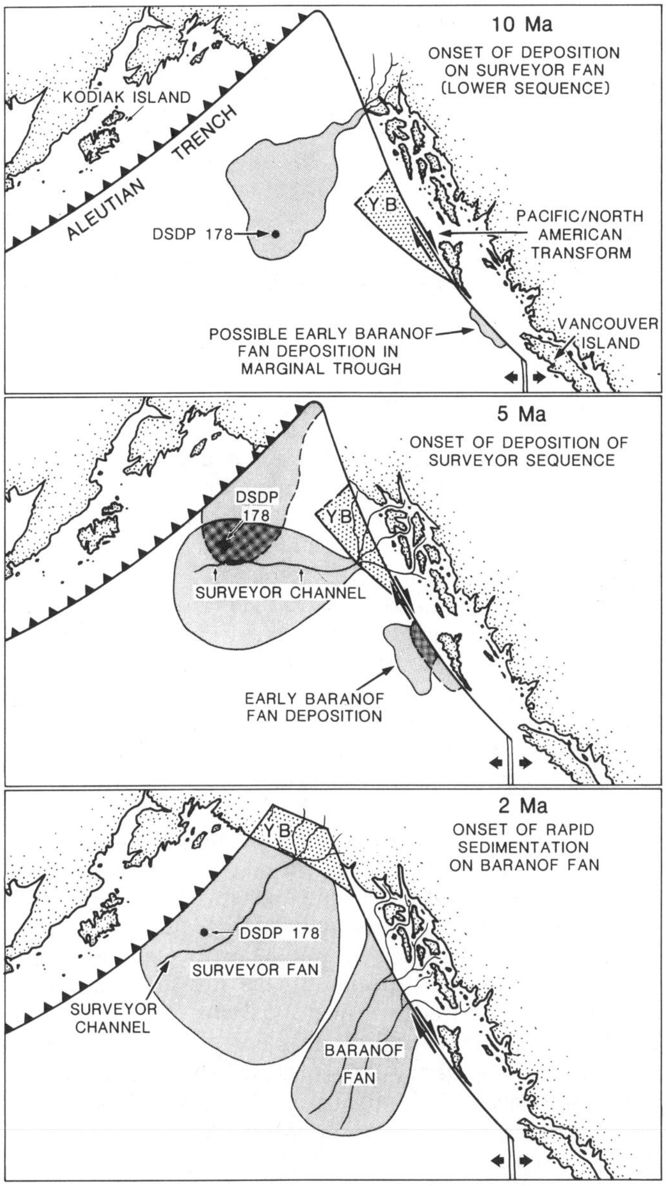

Figure 6. Speculative reconstruction of the effect of the Yakutat terrane (YB) on the development and sources of sediment for the Surveyor and Baranof fans. The northern boundary of the Yakutat terrane is poorly constrained owing to subduction of the northern margin of the block; it may extend as much as 200 km farther to the north, making the block much larger than shown. The outline used in this construction is a minimum size.

the shield (Figure 7). The seismic stratigraphy of these fans shows discrete stacked sequences of individual channels and their overbank deposits, rather than interfingering levees. Such a configuration indicates that it is rare for more than one channel to be active on the fan at the same time, and that channels tend to be long lived, changing their courses less often than on marginal fans. The surface of the fan slopes gently away from the channel axes, to the edge of the fan, or it may merge with the levee slope of another channel. As the fan grows and the channel changes course, a series of elevated fan channels separated by interchannel swales evolves. Viewed in seismic sections, the swales are underlain by stacked levee sequences from adjoining channels. Relative channel ages can be derived from the nature of these overlaps, and can be used to determine the growth history of the fan.

Because these fan bodies are underlain by oceanic crust of the Pacific plate, their original sites of deposition with respect to the location of the North American margin must be reconstructed based on Pacific-North American plate motion (Figures 6 and 8). This has a major effect on the depositional setting of the Zodiac fan, which reconstructs to an oceanic region 1500 km from the nearest margin. The effect of relative plate motion on the younger Surveyor and Baranof fans is not as dramatic but is still important. The Surveyor and Baranof fans are, and have been, separated from their sediment supply on the continent by a zone of transform motion. Total offset along this zone since the onset of deposition has been on the order of 300–400 km (Bruns et al., 1984), and must have caused occasional disruption of the sediment supply system.

FAN DESCRIPTIONS

Zodiac Fan

The Zodiac fan, which underlies all but possibly the northeastern corner of the Aleutian Abyssal Plain, is composed of roughly 280,000 km^3 of terrigenous sedimentary debris deposited by settling from turbidite flows. The fan covers an area $>1,000,000$ km^2 of Pacific crust (Figures 1, 2). Deposition is known to have begun by the late Eocene and to have continued at least to the end of the early Oligocene, and conceivably into the early Miocene (Stevenson, Scholl, and Vallier, 1983). The provenance of the sediment recovered from DSDP site 183, which represents the only physical samples of the fan, is constrained by detrital mineralogy to be a granitic-metamorphic terrane (Stewart, 1976). The fan lies buried

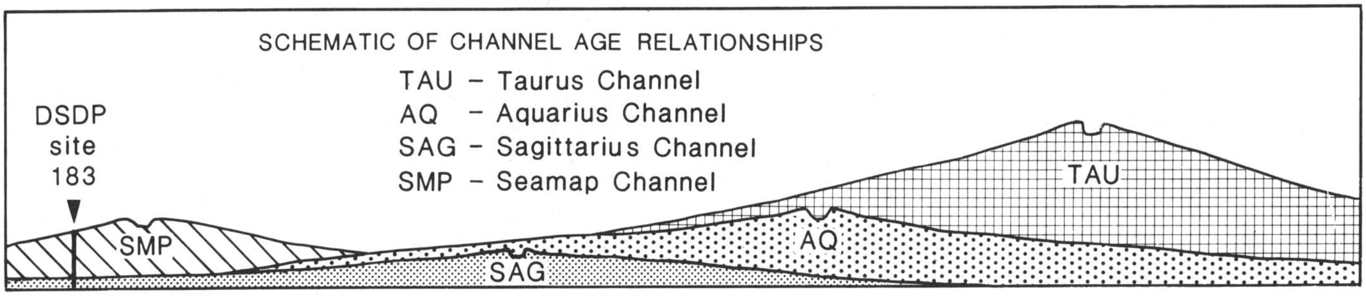

Figure 7. Schematic cross section of channel age relationships and growth patterns for the Zodiac fan. Composite from many seismic lines; not to scale.

beneath a thick blanket of late Tertiary and Quaternary pelagic debris (35–250 m thick) that thickens northward, and thus cannot be sampled by any means other than drilling. The proximal part of the fan lies on the lip of the Aleutian Trench, and an unknown amount has already passed landward of the trench and has been subducted beneath, or accreted to, the margin (Scholl, 1974).

Fan Morphology

The Zodiac fan is typical of a class of deep-sea fans dominated by channelized turbidite flow and levee overbank depositional processes. Four large, elevated channel systems cross the fan and persist to its distal edges (Figures 2, 7). These channel systems controlled deposition across the fan, which occurred primarily via levee overbanking, rather than by formation of a suprafan-fan lobe morphology. This channel-dominated type of growth pattern has been attributed to a primarily fine-grained sediment supply (Normark 1974; 1979, oral communication). The sediment supplied to the Zodiac fan is believed to have been fine grained based on its source and transport distance (see below). The common habitat for fans of this type is in the deep sea, seaward of large integrated drainages that contain significant delta or shelf-basin sediment traps to remove the coarse fraction from the sediment supply. Most of the larger modern fans show this channel-dominated morphology (i.e., the Bengal fan, the Indus cone, and the Amazon cone), and many base-of-slope fans develop a similar pattern near their distal margins.

The Zodiac fan consists of two major sediment lobes, each with associated leveed channels separated by an area of thinner turbidite section (Jones, Ewing, and Truchan, 1971). Both lobes are elongated southward parallel to the channel orientation and, to a lesser extent, to the topographic grain of the underlying oceanic basement (Figure 2). Areas of maximum sediment thickness coincide with channel edges, particularly where they are superimposed over basement depressions.

Bathymetric and seismic-reflection profiles that cross the fan reveal a series of coalescing and overlapping leveed channels. These channels have been grouped into three channel systems that show overlapping levee relationship and can be placed into a relative age sequence (Figure 7). Deposition on the fan began with the formation of Sagittarius Channel on the eastern lobe. This channel lies just above basement and is narrower than Aquarius Channel, which diverges from Sagittarius near the head of the fan and parallels it to the east. These two channels form the Sagittarius-Aquarius (SA) system. The younger Taurus Channel system approximately parallels the SA system to the east. Each channel built upon the eastern levee of its predecessor and, in many areas, the older channels are completely obscured by overbank deposits of the younger channels. The western lobe is dominated by Seamap Channel, which overlies distal parts of the western levee of Sagittarius Channel.

Overlap sequences, as shown in Figure 7, establish the following age progression among the channels—Sagittarius > Seamap > Aquarius > Taurus (the symbol > denotes older than).

Site of Deposition

The most intriguing circumstance of the deposition of the Zodiac fan becomes apparent when the fan is restored to its depositional position with respect to the North American margin at the time the fan was actively forming (Figure 8). The fan is not only separated from the continent by a minimum of 1500 km, but it is also isolated from turbidite flows on the east and north by linear regional basement highs. To the north lay the Kula-Pacific spreading center, which ceased activity between 56 and 43 Ma (Byrne, 1979; Engebretson, Cox, and Gordon, 1984). The spreading center, although inactive, would remain as a bathymetric high owing to the presence of older crust to the south and north and residual thermal effects. To the east lay the active Farallon-Pacific spreading center. Increasing crustal age to the north and east of these two spreading axes implies that turbidite flows trying to reach the site of the fan from the North American margin would have had to defeat a gentle eastward- and northward-dipping regional slope over a minimum distance of hundreds of kilometers, and possibly for as much as 2000 km. Despite this, the presence of the fan attests to the fact that turbidites did surmount these barriers, and in great quantity.

The depositional channels of Zodiac fan converge toward the northeastern margin or apex of the fan, indicating that a pass through the inferred bathymetric barriers existed somewhere in the vicinity of the former Kula-Pacific-Farallon triple junction (Figure 2). Elevating the turbidity currents to the level of this pass was accomplished in some unknown fashion. The most likely method of elevation was achieved by keeping the distance between the source and the topographic barriers as small as possible, thus creating a restricted basin that allowed turbidite deposits to fill this depression to its sill depth. A modern analog of this geometry is the Cascadia fan, which has filled the basin between the base of the margin and the Juan de Fuca Ridge, and has spilled over onto the Pacific plate via the Blanco fracture zone. If such a geometry existed during Zodiac time, a large landmass, moving in conjunction with the Pacific plate and lying seaward of the continental margin, is required. Recent investigations into the motion history of the Yakutat block (von Huene, Fisher, and Bruns, this volume) suggest that it may have fulfilled this role. A second possibility is that a part of southern Alaska occupied the proper position (Stone and Packer, 1977; Panuska, 1985; von Huene, Fisher, and Bruns, this volume), but this interpretation is clouded by conflicting paleomagnetic data and the need for an unrecognized subduction zone landward of this terrane to accommodate Pacific-North American plate motion.

Sediment Source

An analysis of sediment yield from modern drainages in temperate settings indicates that a drainage area of 500,000 km^2 would be necessary to supply the sediment deposited on Zodiac fan. This figure is based on the assumption that all sediment that reaches the river terminus is transported to the fan; this is an unreasonable assumption in light of the depositional setting. The only drainage approaching this size that is known to have existed in the area of the gulf is the ancestral Yukon, which entered the gulf from Cook Inlet (Kirschner and Lyon, 1973). This drainage, which passed through the ancestral Alaska Range and reached well into the Alaskan interior, would have derived the bulk of its sedimentary debris from the Yukon-Tanana terrane, which is compatible with the mineralogy of the deposits cored at site 183. The location of the mouth of the paleo-Yukon fits well with a reconstruction using a southern position for much of southern Alaska to account for Zodiac deposition, but it could not have been a source for the Zodiac fan if sediment transport to the fan via the more southerly Yakutat block is postulated. Other potential source

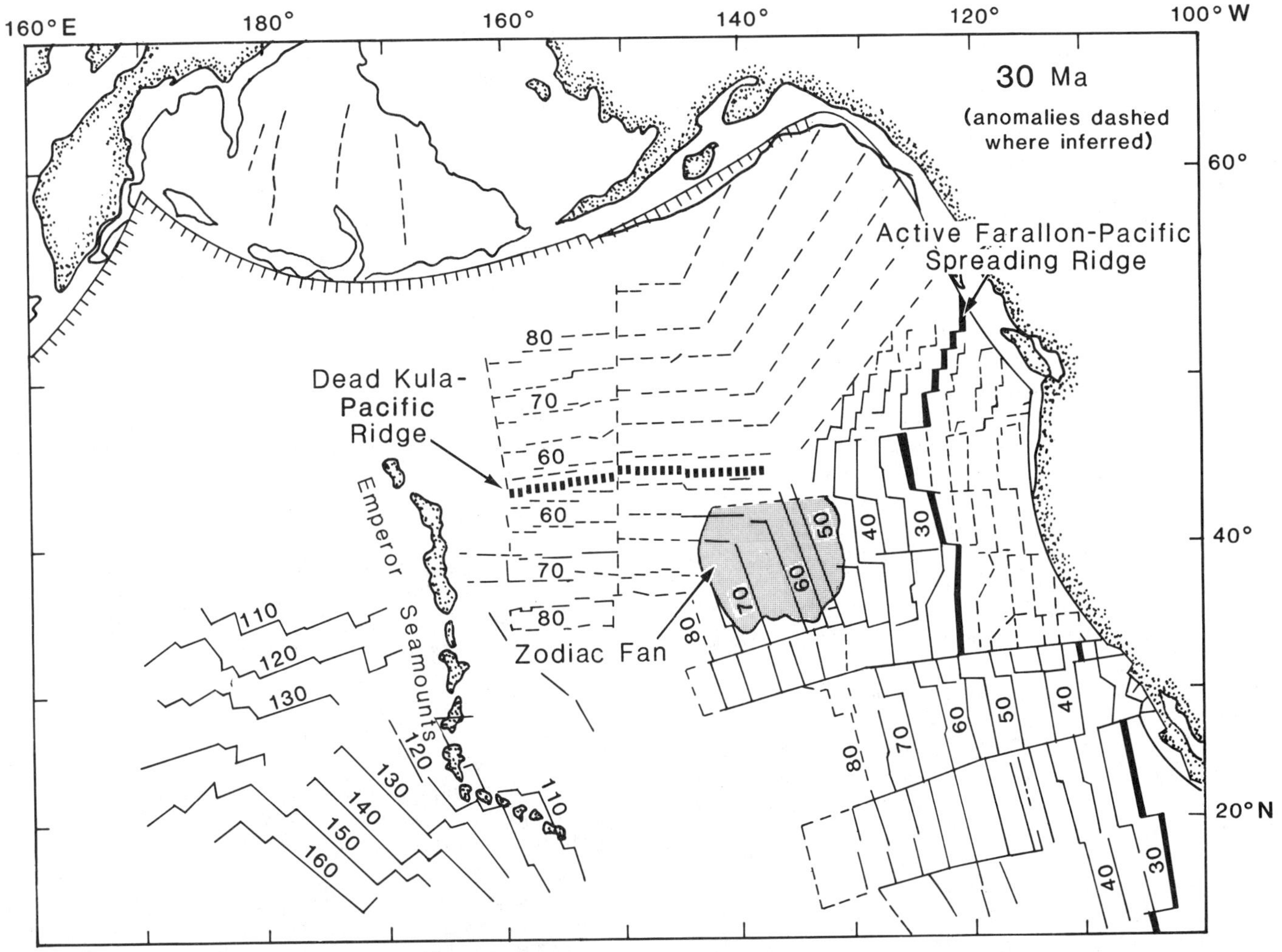

Figure 8. Thirty Ma reconstruction of the site of deposition of the Zodiac fan (last phase of fan growth). Model assumes a 56 Ma age for cessation of spreading on the Kula-Farallon Ridge. A change in this age to 42 Ma, would not affect the location of the fan but would affect the bathymetry north of the site. Chron ages shown in m.y.

drainages undoubtedly existed along the northwest margin of the North American continent, but information on age, size, and mineralogy is insufficient to evaluate their suitability as a source of detritus. Pollen and spores recovered from the fan interval at site 183 favor a northern source for the drainage (Wolfe, 1973; Evitt, 1973).

A characteristic of the sediment cored at site 183 is the uniformity of the mineralogy through the fan section. This implies a stable source and transport system over a period of possibly 10 m.y. (40–30 Ma). This stability is difficult to envision in light of the tectonic setting, which predicts that significant motion between source and site of deposition would have occurred during the growth of the fan. The maintenance of a supply system in the face of such motion reinforces the implication of a large integrated drainage capable of overcoming tectonic disruptions.

Surveyor Fan

The Surveyor fan, which underlies the northern part of the ALAP, can be divided into two major depositional sequences on the basis of seismic stratigraphy. The sequences are separated by a strong seismic reflector which coincides in depth with the base of the Surveyor Channel (Ness and Kulm, 1973; Figure 12, this paper). The total volume of sediment on the fan, including pelagic contributions, is in excess of 175,000 km^3 (Figure 5). Turbidite deposition on the fan began in the late Miocene and has continued into the Holocene (Hogan et al., 1978). The upper sequence (Surveyor sequence) contains Surveyor Channel, which dominates the morphology of the fan and is responsible for the bulk of post-Miocene deposition. The Surveyor Channel is >700 km long, originating at the base of the slope off Icy Bay and continuing westward to a location west of Giacomini Seamount, where it turns north and plunges into the Aleutian Trench (Figure 3). The channel has been active in Holocene time as a conduit for turbidity currents, at least along its upper reaches. Unlike the Zodiac fan, Surveyor fan is not covered with a blanket of pelagic sediment (Ness and Kulm, 1973). Owing to high sedimentation rates caused by late Cenozoic glaciation in gulf drainages, conventional surface cores are unable to sample more than a tiny fraction of the depositional record of the fan. Information on the growth and history of the fan is derived from seismic-reflection profiles over the ALAP and from subsurface sampling at DSDP site 178. The site, located on the north levee of Surveyor Channel about 50 km north of its axis, is roughly midway along the course of the channel (Kulm and von Huene, 1973).

The northwestern margin of the fan body impinges upon the eastern Aleutian Trench, and deposits of the fan have been traced beneath the landward trench slope (von Huene,

1979). An unknown amount of both lower and upper turbidite sequences have therefore been subducted beneath or accreted to the inner trench slope. The amount of fan lost down the subduction zone cannot be directly measured, but probably as much as 500 km of ocean crust has been subducted since the onset of turbidite sedimentation at site 178. If the subducted part of the Pacific plate hosted fan deposits similar to those at the drill site, a large amount of sediment has been subducted.

Fan Morphology

Significant terrigenous turbidite sedimentation commenced in the area of site 178 in the late Miocene, disrupting a pelagic sedimentary environment that had existed since crustal formation in the Eocene (Kulm and von Huene, 1973). Deposition of the lower turbidite sequence continued into early Pliocene time. Sediment derived from the Surveyor Channel began to arrive at the site very soon after cessation of deposition of the lower sequence, as evidenced by the lack of a recognizable hiatus in the cores recovered at site 178. The lower sequence can be described as a southward-thinning wedge of sediment. In seismic profiles it seems to thicken slightly between site 178 and the Aleutian Trench, and to thin rapidly south of the site. On seismic sections no major channel structures typical of channel-dominated fans have been linked to the lower fan section; but smaller distributaries have been noted (Ness and Kulm, 1973). Two possible explanations for this observation are: (1) the lower turbidite sequence was deposited by a type of depositional system different from that operating today in the gulf, which did not involve long distributary channels (sheetflow?), or (2) the lower sequence distributary system lay well to the north of site 178 and has since been carried into the Aleutian subduction zone, leaving only a portion of its southern levee exposed. Support for the latter explanation can be found in (1) the facies and depositional style of the lower sequence, which are similar to, although slightly coarser than, deposits of the overlying Surveyor Channel, and (2) in the stratigraphic similarities of Surveyor fan to Zodiac fan at DSDP site 183, which was deposited by a channel-dominated fan system (Piper, 1973). Further support is found in the asymmetric, wedge-shaped, southward-thinning geometry of the lower sequence as a whole, which can be interpreted as a single large levee deposit.

The overlying Surveyor sequence is clearly derived from the Surveyor Channel, and ranges in age from early Pliocene to Holocene. The two sequences are divided by a prominent seismic reflector coincident with the base of Surveyor Channel. At site 178, the reflector corresponds to a significant change in the lithology from a finer grained sequence (Surveyor) to a coarser (lower) sequence.

The Surveyor Channel is presently the most bathymetrically prominent channel in the gulf; deposits of its southern levee extend at least 500 km southward from the channel axis, where they merge with distal deposits of the Baranof fan complex (Ness and Kulm, 1973). The Surveyor Channel exhibits different channel morphologies along its course that have been related by Ness and Kulm (1973) to changes in depositional processes. The upper reaches of the channel, 0–150 km from its head, are characterized by modest levee heights, an aggraded channel floor, and gently rounded levee crests attributed to high sedimentation rates within and adjacent to the channel. The mid reaches, 150–400 km downchannel, are characterized by greater levee height, little channel aggradation, and more sharply rounded levee crests, attributed to little deposition within the channel and moderate overbank spillage. The lower channel course, 400–700 km, displays a steep-walled, narrow, and deeply incised channel that has downcut its floor into the underlying turbidites. This channel shape is attributed to erosion and nondeposition accompanying the increase in gradient as the channel turns west and flows parallel to the slope of the outer trench swell into the seaward trench axis.

This succession of depositional environments downchannel evolved only during the later phases of Surveyor's history. For example, a substantial thickness of sediment had been deposited in the area of the lower channel, presumably by the channel itself, before it entered an erosional phase. This is consistent with the northward motion of the Pacific plate, which carried the Surveyor Channel to the outer trench slope in the late Cenozoic.

Site of Deposition

Oceanic crust of the Pacific plate underlying the Surveyor fan ranges in age from chron 20 to chron 6 (44–24 Ma; Berggren et al., 1985). The crust becomes younger toward the east and is depressed into a deep, sediment-filled trough beneath the proximal fan and lower continental slope of the Yakutat block (von Huene, Shor, and Wageman, 1979; Figure 5, this chapter). Seaward of this trough the crust slopes gently westward to the Kodiak-Bowie Seamount chain, a northwest trending basement high which has been partly submerged beneath the turbidite blanket. The high divides the fan into east and west depocenters. The western edge of the fan is limited by the Patton Seamount group, which separates Surveyor fan from Zodiac fan.

Since its formation, the Surveyor fan has been moving steadily northwestward with respect to North America (Figure 6). The fan has therefore been adjacent to a transform plate boundary for much of its history. This circumstance has brought the site of deposition steadily closer to the northern rim of the gulf, and its source areas, during the past 43 m.y. Onset of deposition on the fan was controlled by two complementary circumstances—the approach of the site to the source area, and the increase in terrigenous debris accompanying the onset of glaciation around the gulf. The relative importance of these events cannot be evaluated, because the precise time of the beginning of mountain-valley glaciation around the Gulf of Alaska is not well known (for example, see Lagoe, 1983).

The head of the Surveyor Channel lies offshore of a major allochthonous terrane, the Yakutat block (Bruns et al., 1984; von Huene et al., 1985; von Huene, Fisher, and Bruns, this volume), which has been moving in concert with the Pacific plate during deposition of the fan. Consequently the Surveyor Channel and the Yakutat block have remained in the same place relative to each other (Figure 6). This circumstance has allowed the block to act as a collector and buffer for sediment derived from terranes along the North American margin and destined for the Surveyor sequence. The master channel that controlled deposition of the lower sequence must have headed along the transform margin northeast of the Yakutat block, as evidenced by the northwestward thickening of the sequence.

Sediment Source

The Surveyor Channel heads along the southeastern Alaska margin in the vicinity of Yakutat Bay and Dry Bay. An unusual feature of the channel head area is that it does not lie offshore of a drainage of appropriate size to have supplied the fan sediment. The channel does lie immediately offshore from the terminus of the Malaspina Glacier, and glacial outwash has been postulated as a source of the sediment for the Surveyor sequence (von Huene and Kulm, 1973). A large, bowl-shaped area of the continental slope landward of the head of the channel may have acted as a collector and funnel for glacial debris, at least during the later history of the fan.

During deposition of the Surveyor sequence, the location of the channel head and the Yakutat block have moved northward with respect to interior North American crust by about 125–150 km (Figure 6). This movement places the proximal fan opposite Baranof Island during the initial stages of Surveyor Channel deposition. No major drainages are found between this position and the current position of the fan head, so evidently most of the Surveyor sequence was derived from glaciers formed on or crossing the Yakutat block. During glacially depressed sea levels, glaciers probably transported sediment to the shelf edge and dumped it directly onto the slope above the fan head. Despite the short transport from tidewater to fan head, significant trapping of conglomeratic and coarse-sand fractions must have occurred before sediment entered the fan channel. This is evidenced by the internally reflective, closely spaced, laterally coherent acoustic character of the seismic reflections seen at site 178. These reflections remain constant to the proximal edge of the fan, which suggests little change in fan stratigraphy and depositional style toward the fan head. No seismic units with acoustic character indicative of coarse detritus (high-amplitude discontinuous, chaotic) have been observed on the proximal fan. The lack of a major integrated feeder drainage also implies that most of the sediment on the fan was derived locally from the coastal mountains of southeast Alaska, although the contributing area was constantly changing. The volume of sediment on the fan, approximately 175,000 km^3 (Figure 5), implies extremely high erosion rates.

The nature and location of the source of sediment for the lower sequence is unknown. The age and morphology of the sequence suggest that the channel responsible for its deposition lay to the north of the Yakutat block. The presence of the active Aleutian Trench along the northwestern fan perimeter throughout the depositional interval suggests that the source lay to the northeast, opposite the transform margin south of the trench (Figure 6). The Aleutian Trench may have been filled and overtopped, but a large supply of sediment is necessary to accomplish this feat. Cessation of deposition of the lower sequence is presumed to be a consequence of the collision of the Yakutat block with the northern margin of the gulf in the Pliocene, which occluded the northeast source of the lower sequence and formed the mountainous terrain that provided some of the glacial debris for the succeeding Surveyor sequence.

Baranof Fan Complex

The Baranof fan complex, which underlies the southern ALAP and the northwestern area of the TAP, is a large mass of terrigenous sediment deposited by turbidity currents. The amount of sediment in this complex is imprecisely quantified owing to inadequate seismic-reflection data over the TAP, and consists of >200,000 km^3, including pelagic contributions (Figures 4, 5). This volume represents most of the fan, because its distal TAP portion is thinner and of limited (but unknown) extent. Away from the continental margin the bulk of fan deposition is controlled by the two major fan channels, Horizon and Mukluk, which have been mapped >1200 km seaward to the westernmost limits of the TAP. These two channels are the longest in the Gulf of Alaska.

The initial age of deposition on the Baranof fan is unknown from direct sampling, but from regional considerations we infer that deposition began no earlier than late Miocene. Turbidite sedimentation has continued into the Holocene, according to evidence from conventional cores and the lack of channel infilling by pelagic sediment recorded on seismic profiles. Owing to the Pliocene and younger age of the sediment, and the proximity of the fan to the intensively glaciated coastal mountain belt of western Canada and southeast Alaska, the sediment source for the fan complex is presumed to be these glaciated drainages.

Fan Morphology

Knowledge of the depositional system and limits of the Baranof fan are uncertain owing to the complexity of the distribution system, the scarcity of good-quality low-exaggeration seismic profiles, and the complex relationship with its neighboring fans to the north (Surveyor) and south (Scott-Moresby).

The head of the Baranof fan lies along the southeast Alaskan margin, extending from south of Cross Sound to just south of Dixon Entrance. The proximal fan deposits fill a broad depression in the oceanic crust that extends seaward to the Kodiak-Bowie Seamount chain (Figure 5). Adjacent to the margin, a landward-dipping unconformity within the fan section (Embley, unpublished) and fanning of depositional surfaces indicate that some subsidence occurred during deposition of the fan. Gravity modelling further suggests that igneous crust descends to a depth of 8 km below sea level offshore of Dixon Entrance (von Huene, Shor, and Wageman, 1979). Overlying this filled depression are 6 km of sediment, which may all be part of the Baranof fan. At a minimum, the upper 4 km of section (above the unconformity) can be seen on seismic profiles to be part of the fan complex.

The juncture between the slope and the fan head off Baranof Island is marked by a depression partly filled with slumps and debris flows from the slope. Undisturbed banks in this depression appear to be levees; this feature is probably the head of one of the major fan channels (Figure 9). The presence of slumps and debris flows in the channel suggest that the turbidity currents responsible for the formation of the fan are, at least in part, formed directly by slope failure, possibly induced by earthquake activity. Seaward of the margin the Kodiak-Bowie Seamounts form a basement high that was a significant barrier to fan deposition until the trough lying between the seamounts and the margin was filled to its sill depth (Figure 5).

Sediment flooding across the Kodiak-Bowie basement high moved into the mid-fan area and continued to aggrade the sea floor until it encountered the next seamount barrier westward, which was in turn overtopped, providing access to the TAP. Thus the growth history of the Baranof fan consists of three phases, each coinciding with the progressive seaward filling of a crustal depression. Sediment thickness increases landward toward the margin, and local thicks overlie the axes of the depressions (Figure 5). The age of the basal terrigenous deposits is inferred to increase toward the margin.

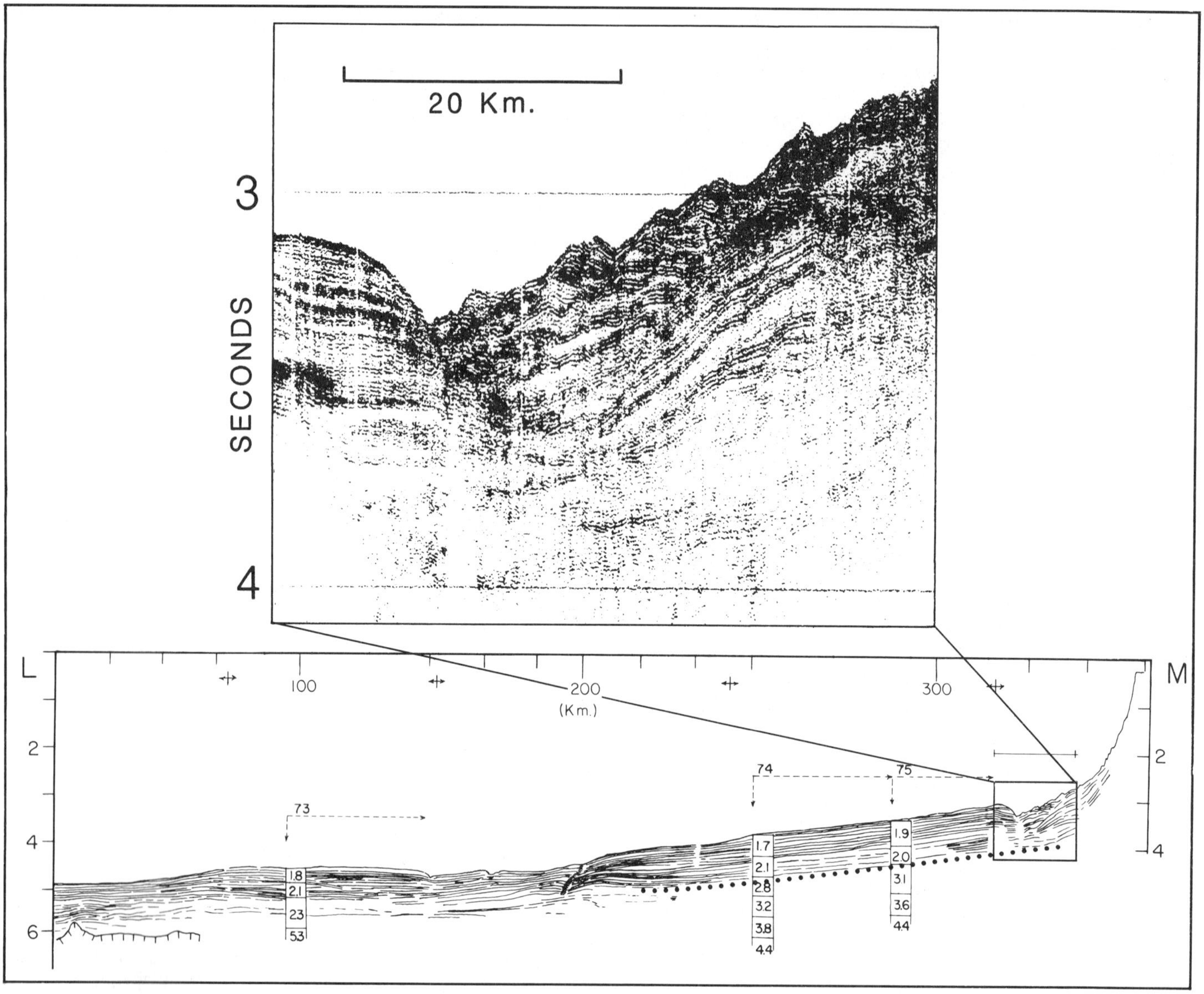

Figure 9. Seismic profile seaward of Baranof Island. Inset shows detail of channel head partly filled with debris flow. Depth shown in seconds (two-way time). Interval velocities from sonobuoys plotted on section. Vertical exaggeration 30:1.

Channel identification is difficult based on existing information, and even major channels identified on the mid-fan and lower fan cannot be confidently traced to channel systems known on the proximal fan. Recently collected seismic lines (Stevenson, unpublished) and new bathymetric compilations (Seemann, 1982) have been used to update the channel courses and identifications published by Chase, Menard, and Mammerickx, (1970). The course of the Mukluk Channel, which is a prominent bathymetric feature, and that of the Horizon Channel have been clarified (Figure 10). The new course shown for Horizon Channel in Figure 10 is still considered to be tentative between lat. 51 ° and 53 °N, owing to the poor bathymetric expression of the channel in this area.

The major channel shown on Figure 10 north of Pathfinder Seamount is an important element in the depositional history of the fan, because the channel has controlled deposition on much of the northern margin of the fan. In light of the reinterpretation of the course of the Horizon Channel, the unnamed channel cannot be connected to those on the proximal fan. One possible explanation of this channel pattern is that the northern channel is an early, buried course of the Horizon Channel; after the deposition of the northern lobe, this channel diverged southward to its present course just seaward of the Kodiak-Bowie basement high. Thus the two channels may share the same upper course, but may exist as separate channels seaward of the seamounts. This interpretation is consistent with the depositional relationship of the channels.

A partial history of fan growth can be derived from the overlap relationship of levees belonging to the various channels. Accordingly, the northern channel and Mukluk Channel are older than Horizon Channel. No direct contact between Mukluk and the northern channel levees has been recorded, but the height of their channel floors above basement (a crude estimate of channel age) indicates that Mukluk is older. Both the Mukluk and the northern channel are built upon turbidite deposits of unknown origin, but they may have been derived from the buried channel recorded beneath Horizon Channel (see Figures 11, 12). The origin and course of this channel are

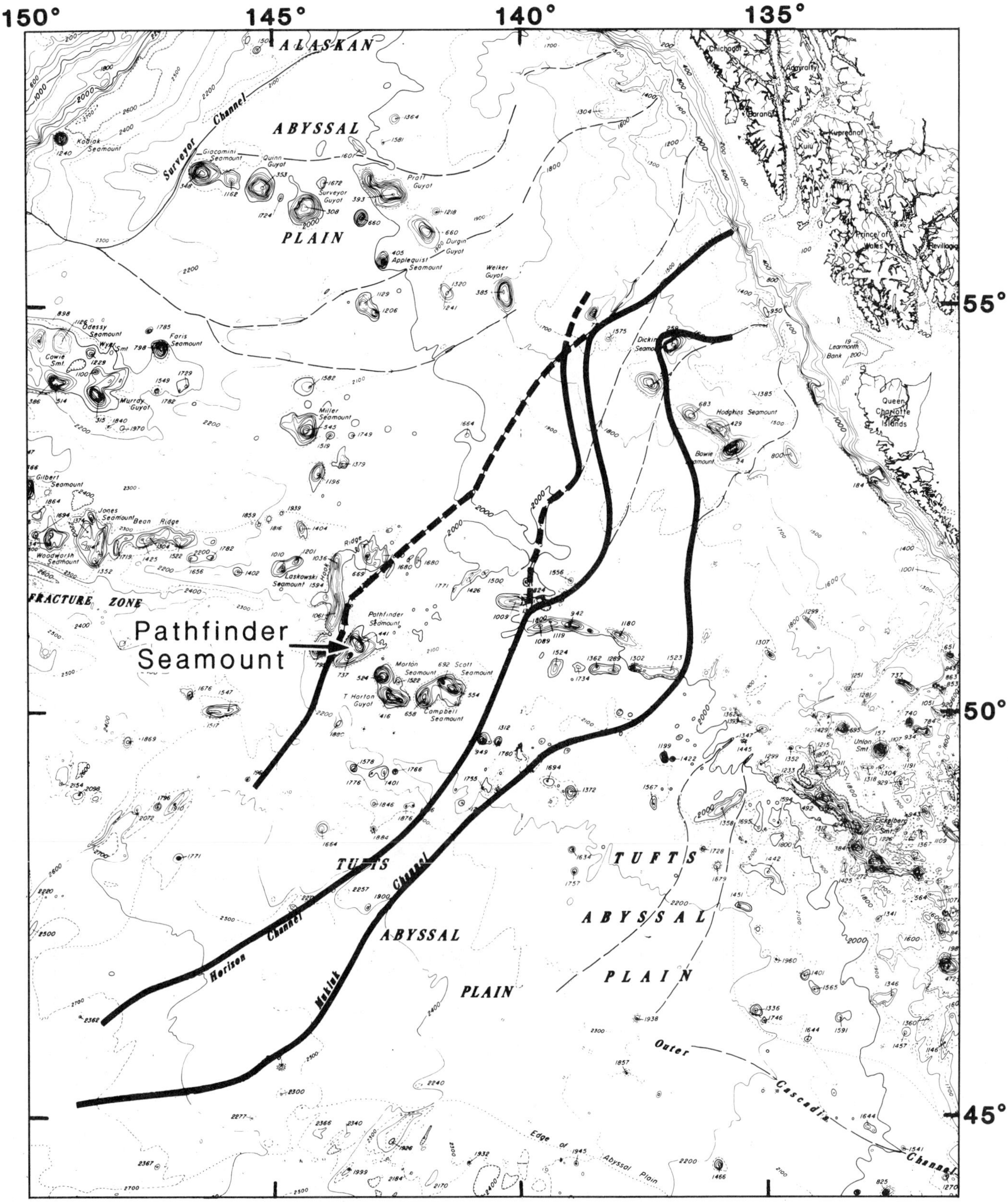

Figure 10. Same as Figure 4 but modified to indicate proposed new channel courses and associations.

unknown. The Horizon, the Mukluk, and the northern channel are therefore involved with the later phases of fan growth, and may not have any relationship to an older fan morphology.

Fan deposition modifies the original bathymetry of the depositional site and exerts a strong influence on later channel positions and fan growth. This is exemplified by Horizon Channel, which was trapped in the low between the levees of the northern channel and Mukluk Channel from seaward of the Kodiak-Bowie chain to the TAP and beyond. Deposits

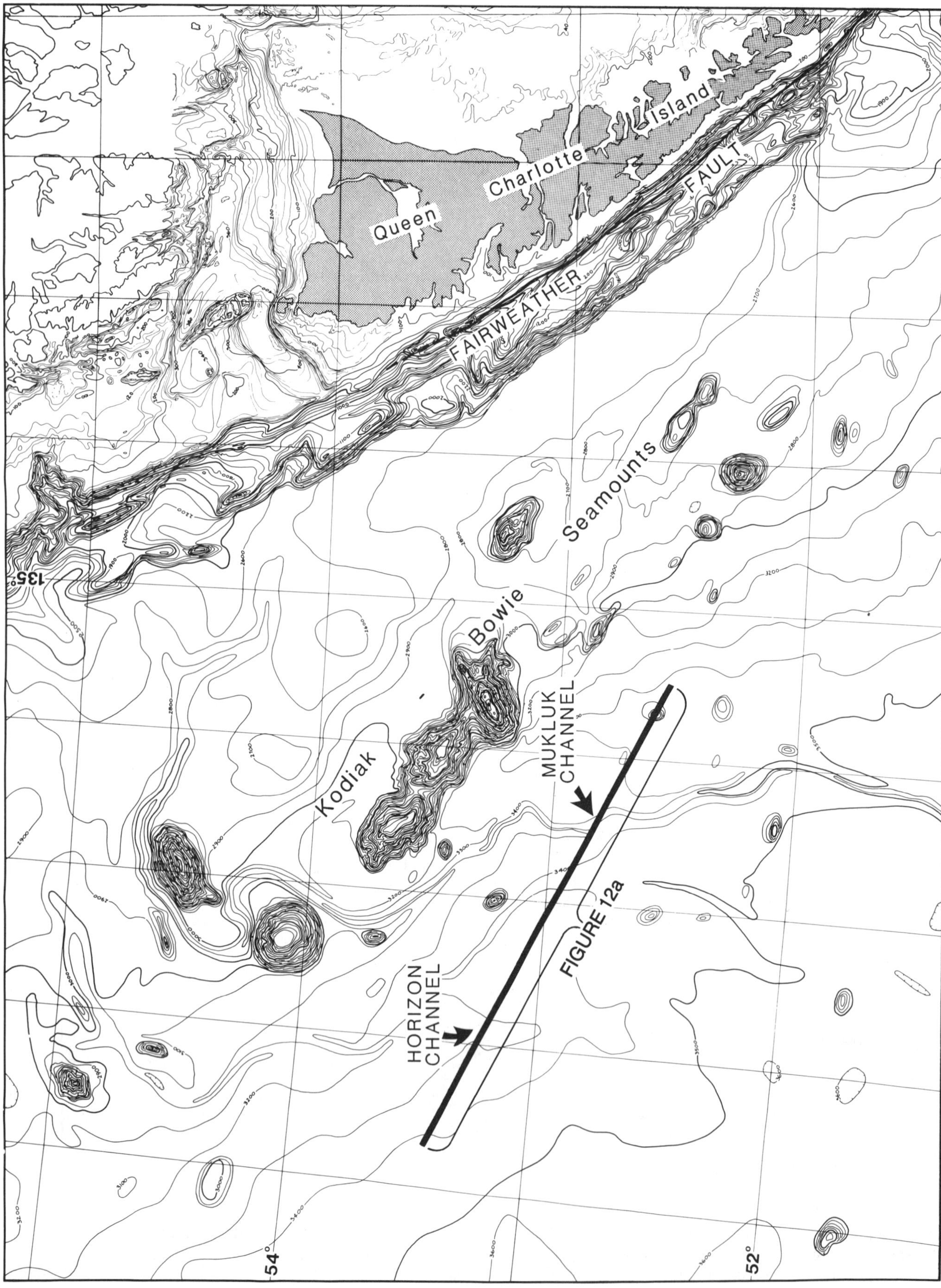

Figure 11. Index map showing the location of the single-channel profile in Figure 12A. Bathymetry from Seeman, 1982.

from the Horizon Channel eventually filled this depression and built the sea floor up above the level of the older channels, partly (and in some areas completely) obscuring the flanking channels beneath its levees (Figure 12).

Site of Deposition

The crust underlying the Baranof fan becomes progressively older downfan, ranging in age from chron 5a, 12 Ma, (Wilson, 1985; Berggren et al., 1985) just south of Dixon Entrance, to chron 21, 50, Ma at the western margin of the TAP. Reconstructions of plate interactions along the Oregon-Washington-British Columbia margin (Wilson, 1985) indicate that the youngest crust covered by the Baranof fan was generated along the western side of the Pacific-Farallon spreading ridge segment, which intersected the British Columbia margin at the latitude of Vancouver Island. Consequently, the subfan piece of crust has always been part of the Pacific plate and in transform contact with the North American margin. The fan and underlying crust have moved parallel to this margin and northward about 400 km to its present position (Figure 6).

The proximal region of the fan has been flanked to the northwest throughout its migration by the Yakutat block, an allochthonous terrane also moving in concert with the Pacific plate (Bruns et al., 1984; von Huene et al., 1985; von Huene, Fisher, and Bruns, this volume). The position of this block has affected the source terrains available to supply debris to the fan.

Sediment Source and Depositional Age

The onset of deposition on the Baranof fan cannot be stated with certainty. The youngest crust covered by fan deposits is 12 m.y. old, and these deposits belong to the Mukluk Channel, one of the oldest channels. In this area, because no acoustically transparent interval typical of pelagic beds can be identified underlying the fan deposits on seismic sections, no major hiatus is implied between crustal formation and onset of turbidite deposition. The maximum age for the fan is therefore roughly earliest late Miocene. Regional seismic correlations tied to DSDP site 178 by von Huene, Shor, and Wageman (1979) and Bruns et al. (1984) suggest that most of the proximal fan deposits lie above the Plio-Pleistocene boundary, and that almost all of the mid-fan and distal-fan deposits are post-Pliocene in age.

The presence of the Yakutat block immediately to the northwest of the Baranof fan may have played a major role in the initiation of fan growth by diverting the flow of sediment from the northeast, much of which must have been shunted to the Surveyor fan. Only after the block passed north of western Canadian sediment sources could they have become available to the Baranof fan. Support for this view can be found in the distinct change in the age and appearance of the base of slope apron across the southern margin of the block north of Cross Sound to the Baranof margin south of the Sound. The Yakutat block prism is much thicker and contains a larger and older pre-Pliocene wedge (von Huene, Shor, and Wageman, 1979). Thus the Baranof fan obtained the bulk of its sediment from sources abandoned by the Surveyor fan, owing to the northward migration of the Yakutat block.

Early in the history of the Baranof fan, it would have had access to sediment transported off the shelf through Queen Charlotte Sound north of Vancouver Island (Yorath, this volume); but once the fan head migrated north of the south end of Queen Charlotte Island, it would have received little or no sediment until it came within range of clastics being transported seaward via Dixon Entrance (Figure 6). Reconstructions (Wilson, 1985) indicate the proximal fan would have been shadowed by the island during the period 12–6 Ma. Subsequently, the fan has been the recipient of an ever-increasing supply of sediment as glacial erosion increased and a larger area of the coastal drainages became available to supply the fan as the Yakutat block moved north.

POTENTIAL PETROLEUM RESOURCES ASSOCIATED WITH DEEP SEA FAN DEPOSITS

Introduction

The abyssal turbidite deposits on the Pacific plate in the Gulf of Alaska represent a large reservoir of organic carbon derived from terrestrial and pelagic sources. Except near the margin, where sediment thickness approaches 10 km, these deposits are not in a geologic setting favorable for the generation of petroleum resources. The ultimate fate of these deposits, however, is to encounter the Aleutian subduction zone where a portion of them will be subducted beneath the Alaskan forearc and a portion will be incorporated into the accretionary wedge seaward of it. In these settings the potential for hydrocarbon generation is likely much enhanced (von Huene, Fisher, and Bruns, this volume). The northern parts of both the Zodiac fan and the Surveyor fan have already entered this zone, and represent the latest in a long series of terrigenous sediment bodies to encounter the trench. An examination of the interaction between the trench and the impinging fan deposits provides the basis for speculation on the petroleum potential of this zone.

Proximal Fan Deposits Abutting a Transform Margin

The eastern proximal margins of the Surveyor and Baranof fans contain a thick accumulation of sediment deposited in troughs on the oceanic crust immediately adjacent to the Yakutat block and Baranof margin, respectively. No heat-flow data are available over this wedge, but sediment thicknesses in excess of 10 km in places make it probable that at least the lower threshold of the hydrocarbon generation window occurs in the deeper sections of the wedge. It is important to recognize that the amount of sediment lying within the window is likely to be small, owing to the decrease in wedge volume downsection.

The origin, mineralogy, and organic content of the sediment forming the pre-late Miocene section of the fan wedge is unknown. If we consider that the wedge shares a similar origin and depositional history with the overlying section, assumptions can be made about its potential as a source and reservoir rock. A detailed source rock evaluation of sediment recovered at DSDP site 178 by Kvenvolden and von Huene (1985) indicates that the section has poor hydrocarbon potential. The average total organic carbon content is 0.36%, barely below the minimum considered necessary for hydrocarbon genesis (0.4% Hunt, 1979). The percentage of "live" (available for hydrocarbon generation) carbon is also low, averaging 18%. The organic matter is woody and of terrestrial origin and thus would most likely generate gas. Despite the poor average hydrocarbon potential of the section at site

178, Bode (1973) reports the presence of some organic-rich sediment intervals in the lower fan sequence. Maximum total organic carbon values exceed 1.0% in two samples from the early Miocene.

The fine grained nature of the sediment on the fans requires the removal of the coarse fraction from the sediment before it entered the fan channels. The most likely traps for the coarse debris are the marginal depressions underlying the sediment of the proximal fan wedge, which could contain thick coarse conglomeratic and coarse sand sequences. Certainly the post-late Pliocene section of the wedge received and trapped copious quantities of coarse debris during glacial maxima, when tidewater glaciers emptied onto (and perhaps beyond) the shelf adjacent to the proximal fan wedge. This likelihood implies that appropriate facies for reservoirs could exist in the wedge.

Suitable structures for the entrapment and storage of migrating hydrocarbons are revealed in seismic-reflection sections. Stratigraphic traps exist deep in the wedge where inferred coarse facies pinch out against the seaward lip of the trough and also against the foot of the continental slope. Higher in the section, early fan channels have deposited thick packages of channel sands that have been sealed by later finer grained sediment, creating potential high-capacity reservoirs. At least one discontinuity underlies the Plio-Pleistocene fan sequence, which is also a potential trap.

Owing to the great depth of water overlying these wedges (2250–3000 m), the marginal source rock potential, and the limited volume of sediment in the deeper part of the wedge, this area could not be considered as prospective today. But as economic circumstances change, the region is potentially prospective because the necessary conditions for the formation and entrapment of hydrocarbons (mainly gas) are believed to exist.

Accretionary Wedge and Forearc Basin

Sediment accumulating in active subduction zones has only recently been recognized as a potential site for hydrocarbon generation and entrapment (Thompson, 1976; Gwilliam, 1982; McCarthy et al., 1984) Kvenvolden and von Huene, 1985; von Huene, Fisher, and Bruns, this volume) One commercial field (in Barbados) and many shows of oil and gas (Makran, Andaman Islands) are known to be hosted, and presumably generated from, accretionary complexes fronting active subduction zones.

The large fans of the Gulf of Alaska are impinging on the Aleutian Trench, and the introduction of a large sedimentary sequence overlying oceanic crust into the subduction zone has been a major factor in creating a large accretionary complex fronting the Alaskan and Aleutian forearc (Scholl, Vallier, and Stevenson, this volume; von Huene, Fisher, and Bruns, this volume). Substantial portions of this overlying sediment load become incorporated into the accretionary complex; the remainder passes beneath the complex to an unknown fate, possibly involving underplating or consumption. It is therefore instructive to examine how these terrigenous fan deposits may contribute to the petroleum potential of the forearc region.

The structure of the accretionary complex and forearc basin of the central Aleutian Trench has been investigated using carefully processed multichannel seismic profiles by McCarthy and Scholl (1985) and McCarthy et al. (1984). No fan sequences are currently impinging on the trench in this area, but it is believed that fan bodies have been subducted in the recent past and others will enter this zone in the future (Zodiac fan). This makes the central Aleutians a good area to use as a model of the subduction environment into which fan sequences will be transported. The structure of the eastern Aleutian forearc and trench, where the Surveyor fan is being subducted, has been described by von Huene, Fisher, and Bruns (this volume); the geology relevant to hydrocarbon generation is summarized in Kvenvolden and von Huene (1985). Their interpretations show a number of structures suitable for the entrapment of hydrocarbons. Within the accretionary prism, fan sequences and trench fill have been incorporated into the margin as structurally discrete packets of sediment tens of kilometers long bounded by arcward-dipping thrust faults. Many packets are gently deformed by folding and tilting but are not greatly disrupted internally. The packets bear a similarity to the structural style of a continental thrust belt. Thrusts bounding the packets sole into a master decollement that underlies the complex. Some of these fault-bounded packets could contain thick channel sand sequences with excellent reservoir properties. Folds with a kilometer or more of relief and fault-bounded monoclines are also present within the complex. Landward of the accretionary complex, and partly overlying it, is a sequence of sedimentary basins developed on the fore arc which contains a thick fill of coarse and fine volcaniclastic sediment derived from the erosion of the arc. These basins have been gently deformed into broad folds and fault-bounded basement uplifts which also represent potential sites for hydrocarbon entrapment (Figure 13).

The source potential of the fan sequences lying seaward of the Aleutian Trench has been investigated by Kvenvolden and von Huene (1985). In addition to the organic carbon values derived for Surveyor fan from site 178 cores (see above), sediment from Zodiac fan cored at site 183 was investigated. Total organic carbon values in this fan section were slightly higher than those at site 178, but still were very near the minimum required for hydrocarbon generation (0.46%). The percentage of live carbon was also low (20%), and the organic material was of a gas-prone type.

The dynamic nature of the subduction zone insures that a continuous fresh supply of organic-rich sediment is being transported into and beneath the accretionary complex through time. The consequences of this constant rejuvenation of the organic component passing through the hydrocarbon generation window is unknown, but the potential for the generation of large quantities of hydrocarbons is apparent. The total amount of organic carbon passing through this system is very large. McCarthy et al. (1984) estimate that a 1-km-long segment of the central Aleutian Trench has consumed 10 billion tons of organic carbon in the last 5.5 m.y., and this number must be much greater where thick fan sequences are being subducted. Kvenvolden and von Huene (1985) have made a careful estimate of the potential for gas generation within the eastern Aleutian forearc and conclude that as much as 5×10^8 m^3 of methane can be produced along each 1 km of forearc each million years.

The location of the hydrocarbon generation window beneath the accretionary wedge and the forearc basin has been estimated by McCarthy et al. (1984) and Kvenvolden and von Huene (1985), who used geothermal gradients estimated from gas hydrates within the forearc (Figure 13). The petroleum window occurs between a depth of 3 and 6.5 km in the central Aleutian accretionary complex, well above the maxi-

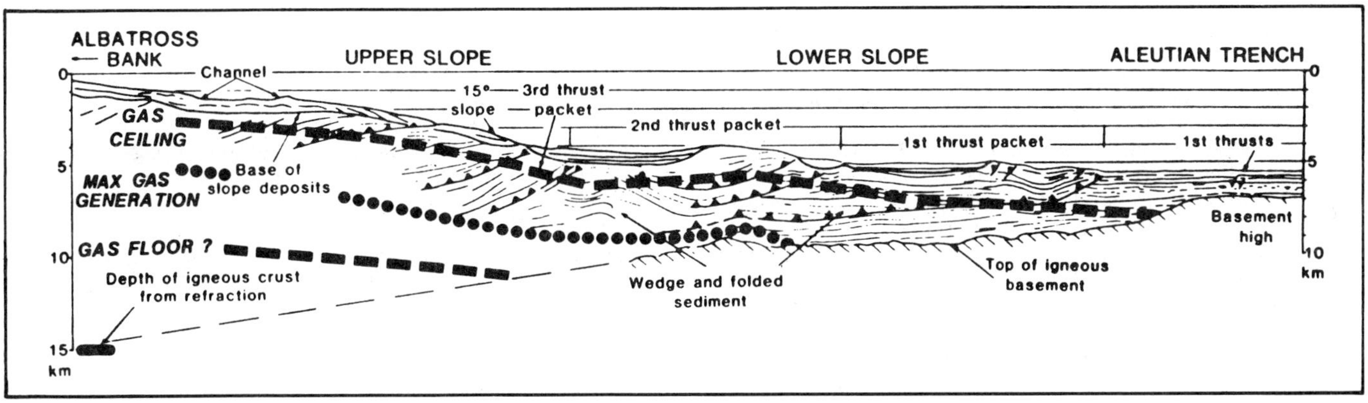

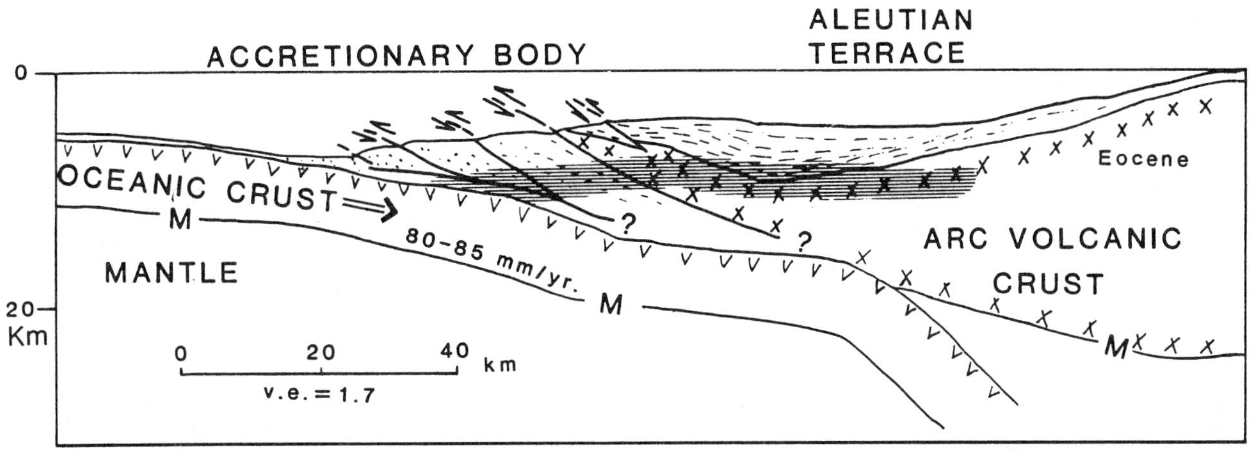

Figure 13. Schematics of the location of the hydrocarbon generation window in accretionary settings. A, Inferred generation window in the eastern Aleutian where fan sediments are currently being subjected (from Kvenvolden and von Huene, 1985). B, Location of inferred generation zone (ruled pattern) in the central Aleutians (from McCarthy et al., 1984). In this area, fan sediments are not presently impinging on the trench but will in the near future.

mum thickness (8 km) that has been acoustically imaged. Deeper parts of the complex could still lie within the gas generation zone. In the eastern Aleutian forearc, the zone of maximum gas generation is inferred to lie farther landward, under the forearc basins of the mid and upper slope. Hydrocarbons (primarily gas) generated in these zones would have ample opportunity to migrate into structures in the accretionary wedge and forearc basin along the many faults that disrupt the wedge and forearc.

The two main uncertainties surrounding the creation of hydrocarbon accumulations within the forearc-accretionary wedge setting are the longevity of the trapping structures and the quality of the reservoir seal. The presence of a strong water drive caused by the dewatering of subducted sediment, coupled with the numerous faults that reach deep into the accretionary complex, provides both an excellent environment for the migration of hydrocarbons into higher structures and at the same time provides numerous avenues by which these fluids could escape into the sea and be lost. The tectonic setting of the complex also mitigates against structures surviving for an extended period, except perhaps for those farthest from the trench. The fore-arc basins provide much better potential reservoirs, owing to the less intense tectonic deformation and greater longevity of the structures present, but communication with the underlying subduction zone source area is unproven. Despite these uncertainties, all the conditions necessary for the generation and entrapment of hydrocarbons exist in this setting (von Huene, Fisher, and Bruns, this volume).

For the same reasons the marginal wedges described above are considered not economically prospective, it is clear that hydrocarbon deposits of deeply submerged accretionary masses are not presently producible. Nonetheless, accreted masses of trench floor and abyssal fan deposits remain potentially prospective.

REFERENCES CITED

Berggren, W. A., D. V. Kent, J. J. Flynn, and J. A. Van Couvering, 1985, Cenozoic geochronology: GSA Bulletin, v. 96, p. 1407–1418.

Bode, G. W., 1973, Carbon and carbonate analyses-Leg 18, *in* L. D. Kulm, R. von Huene, et al., 1973, Initial Reports of the Deep Sea Drilling Project, Vol. 18: Washington, D.C., U.S. Government Printing Office, p. 1069–1077.

Bruns, T. R., R. Von Huene, P. R. Carlson, and G. Keller, 1984, The eastern Gulf of Alaska transcurrent plate boundary and the migration of terranes: Annales de la Societe Geologique du Nord, CIII, p. 325–331.

Byrne, T., 1979, Late Paleocene demise of the Kula-Pacific spreading center: Geology, v. 7, p. 341–344.

Chase, T. E., B. W. Menard, and J. Mammerickx, 1970, Bathymetry of the North Pacific: Institute of Marine Resources Technical

Report Series TR-8, 9 (Scripps Institution of Oceanography), 2 maps.

Engebretson, D. C., A. Cox, and R. G. Gordon, 1984, Relative motion between plates of the Pacific Basin: Journal of Geophysical Research, v. 89, p. 10,291–10,310

Evitt, W. R., 1973, Dinoflagellates from Leg 19, sites 183 and 192, Deep Sea Drilling Project, *in* J. S. Creager, D. W. Scholl, et al, eds., Initial Reports of the Deep Sea Drilling Project, Volume 19: Washington, D.C., U.S. Government Printing Office, p. 737–738.

Gwilliam, W. J., 1982, Subducted organic origin gas hypothesis for deep source methane, *in* W. J. Gwilliam, ed., Deep source gas workshop technical proceedings: Morgantown, West Virginia, Morgantown Energy Technology Center, U.S. Department of Energy, DOE/METC/82-50, UC-92a, p. 140–161.

Hogan, L. G., K. F. Scheidegger, L. D. Kulm, J. Dymond, N. Mikkelsen, 1978, Biostratigraphic and tectonic implications of ^{40}AR-^{39}Ar dates of ash layers from the northeast Gulf of Alaska: GSA Bulletin, v. 89, p. 1259–1264.

Hunt, J. M., 1979, Petroleum geochemistry and geology: San Francisco, W. H. Freeman Co., 617 p.

Hurley, R. J., 1960, The geomorphology of abyssal plains in the northeast Pacific Ocean: Scripps Institution of Oceanography Reference 60-7, unpublished manuscript, 105 p.

Jones, E. J. W., J. Ewing, and M. Truchan, 1971, Aleutian plain sediments and lithospheric plate motions: Journal of Geophysical Research, v. 76, p. 8121–8127.

Kirschner, C. E., and C. A. Lyon, 1973, Stratigraphic and tectonic development of Cook Inlet petroleum province, *in* Arctic geology: AAPG Memoir 19, p. 396–407.

Kulm, L. D., R. von Huene, et al., 1973, Initial reports of the Deep Sea Drilling Project, Vol. 18: Washington, D.C., U.S. Government Printing Office, 1077 p.

Kvenvolden, K. A., and R. von Huene, 1985, Natural gas generation in sediments of the convergent margin of the eastern Aleutian Trench area, *in* D. G. Howell, ed., Tectonostratigraphic terranes of the Circum-Pacific region: Houston, Texas Circum-Pacific Council for Energy and Mineral Resources Earth Science Series 1, p. 31–49.

Lagoe, M. B., 1983, Oligocene through Pliocene foraminifera from the Yakataga reef section, Gulf of Alaska Tertiary province, Alaska: Micropaleontology, v. 29, p. 202–222.

McCarthy, J., A. J. Stevenson, D. W. Scholl, and T. L. Vallier, 1984, Speculations on the petroleum geology of the accretionary body: An example from the central Aleutians: Marine and Petroleum Geology, v. 1, p. 151–167.

McCarthy, J., and D. W. Scholl, 1985, Mechanisms of subduction accretion along the central Aleutian Trench: GSA Bulletin, v. 96, p. 691–701.

Ness, G. E., and I. D. Kulm, 1973, Origin and development of Surveyor deep-sea channel: GSA Bulletin, v. 84, p. 3339–3354.

Normark, W. R., 1974, Submarine canyons and fan valleys: Factors affecting growth patterns of deep-sea fans, *in* R. H. Dott, Jr., and R. H. Shaver, eds., Modern and ancient geosynclinal sedimentation: SEPM Special Publication 19, p. 56–68.

Panuska, B. C., 1985, Paleomagnetic evidence for a post-Cretaceous accretion of Wrangellia: Geology, v. 13, p. 880–883.

Piper, D. J. W., 1973, The sedimentology of silt turbidites from the Gulf of Alaska, *in* L. D. Kulm, R. von Huene, et al., Initial Reports of the Deep Sea Drilling Project, Vol. 18: Washington, D.C., U.S. Government Printing Office, p. 847–867.

Scholl, D. W., 1974, Sedimentary sequences in the north Pacific trenches, *in* C. A. Burke and C. L. Drake, eds., The geology of the continental margins: New York, Springer-Verlag, p. 493–504.

Scholl, D. W., T. L. Vallier, and A. J. Stevenson, 1987, Geologic evolution and petroleum geology of the Aleutian Ridge, this volume.

Seemann, D. A., 1982, Bathymetry off the coast of British Columbia: Sidney, British Columbia, Pacific Geoscience Centre, Energy, Mines and Resources, Earth Physics Branch Open-File 82-25, map.

Stevenson, A. J., D. W. Scholl, and T. L. Vallier, 1983, Tectonic and geologic implications of the Zodiac fan, Aleutian Abyssal Plain, northeast Pacific: GSA Bulletin, v. 94, p. 259–273.

Stewart, R. J., 1976, Turbidites of the Aleutian Abyssal Plain: Mineralogy, provenance, and constraints for Cenozoic motion of the Pacific plate: GSA Bulletin, v. 87, p. 793–808.

Stone, D. B., and D. R. Packer, 1977, Tectonic implications of Alaska Peninsula paleomagnetic data: Tectonophysics, v. 37, p. 183–201.

Thompson, T. L., 1976, Plate tectonics in oil and gas exploration of continental margins: AAPG Bulletin, v. 60, p. 1463–1501.

von Huene, R., and L. D. Kulm, 1973, Tectonic summary of Leg 18, *in* L. O. Kulm and R. von Huene, et al., Initial Reports of the Deep Sea Drilling Project, Vol. 18: Washington, D.C., U.S. Government Printing Office, p. 961–976.

von Huene, R., 1979, Structure of the outer convergent margin off Kodiak Island, Alaska, from multichannel seismic records, *in* J. S. Watkins, L. Montsdert, and P. W. Dickerson, eds., Geological and geophysical investigations of continental margins: AAPG Memoir 29, p. 261–272.

von Huene, R., M. A. Fisher, and T. R. Bruns, 1987, Geology and evolution of the Kodiak margin, Gulf of Alaska, this volume.

von Huene, R., G. G. Shor, and J. Wageman, 1979, Continental margins of the eastern Gulf of Alaska and boundaries of tectonic plates, *in* J. S. Watkins, L. Montadert, and P. W. Dickerson, eds., Geological and geophysical investigations of continental margins: AAPG Memoir 29, Tulsa, Oklahoma, p. 273–290.

von Huene, R., G. Keller, T. R. Bruns, and K. McDougall, 1985, Cenozoic migration of Alaskan terranes indicated by paleontologic study, *in* D. G. Howell, ed., Tectonostratigraphic terranes of the circum-Pacific region: Houston, Texas, Circum-Pacific Council for Energy and Mineral Resources Earth Science Series 1, p. 121–136.

Wilson, D. S., 1985, Tectonic history of the Juan de Fuca Ridge: Stanford, CA, Stanford University unpublished Ph.D thesis, 105 p.

Wolfe, J. A., 1973, Reconnaissance spore and pollen examination, early Tertiary turbidite beds, Aleutian Abyssal Plain, site 183, *in* J. S. Creager and D. W. Scholl, et al., Initial Reports of the Deep Sea Drilling Project, Vol. 19: Washington, D.C., U.S. Government Printing Office, p. 739.

Yorath, C. J., 1987, Petroleum geology of the Canadian Pacific continental margin, this volume.

CHAPTER TWENTY-ONE

MARINE GEOLOGY AND RESOURCE POTENTIAL OF CASCADIA BASIN

Paul R. Carlson and C. Hans Nelson

U.S. Geological Survey
Menlo Park, California

ABSTRACT

Cascadia Basin, off the coast of the U.S. Pacific northwest, covers 163,000 km^2 of deep-sea floor and is bordered by the Oregon-Washington-Vancouver Island continental margin, the Blanco fracture zone, and the Juan de Fuca Ridge. The Juan de Fuca lithospheric plate, which is generated at the ridge and subducted beneath the North American plate, forms the basement floor of the basin. The wedge of Miocene to Quaternary sediment, which has accumulated to a maximum thickness of 2.5 km on the two large deep-sea fans (Astoria and Nitinat) adjacent to the base of the continental slope, may be partly subducted with the basaltic basement at the convergent margin, but more is apparently being accreted to the continental slope. The basin-filling sediment consists of interleaved pelagic and turbidite deposits. The turbidite beds, although present throughout, dominate the upper part of the section (Pleistocene); these graded sand to silt beds apparently accumulated at a rate of about 0.5 m per thousand years, accelerating during glacial stages when sea level was near the shelf-slope break. Even though the fan turbidite strata contain potential reservoirs, the petroleum prospects are diminished by the lack of large amounts of organic carbon in source beds, thick sedimentary sequences, and thermal maturity of the basin sediment.

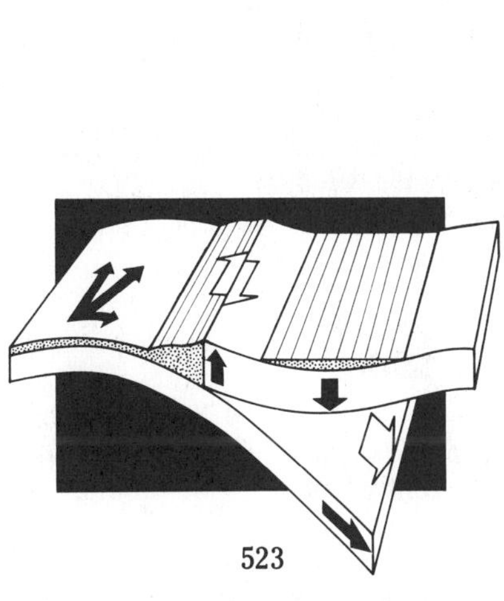

INTRODUCTION

Cascadia Basin is a nearly triangular shaped deep-sea basin located seaward of the continental margin of the U.S. Pacific northwest (Figure 1). This morphologic and tectonic basin was named after a hypothetical ancient landmass (McManus, 1964); other names previously suggested for the basin were the Great Trough (Menard, 1955) and Cascadia Abyssal Plain (Hurley, 1960).

The eastern and northern boundary of Cascadia Basin is the ~800-km-long section of continental margin stretching from Cape Blanco, Oregon, to nearly the northern end of Vancouver Island, British Columbia. The basin is bounded on the west by the ~600-km-long Juan de Fuca Ridge and on the south by the ~360-km-long Blanco Fracture Zone.

Although aspects of Cascadia Basin have been studied in considerable detail, especially the sedimentology of the fans and related deep-sea channels, the basin has not been studied as an entity. This chapter reviews the geological and geophysical data that have been collected in and around Cascadia Basin and comments on the resource potential.

TECTONIC SETTING

Symmetrical magnetic-anomaly patterns in the northeast Pacific (Raff and Mason, 1961) led to the discovery of the Juan de Fuca Ridge (Wilson, 1965), which is part of the mid-ocean-ridge system where ocean crust is formed. The dating of the magnetic stripes within the crust underlying Cascadia Basin led to the conclusion that the Juan de Fuca plate was being subducted beneath the North American plate (Vine, 1968; Atwater, 1970). Recent work by Nishimura, Wilson, and Hey (1984), shows that present subduction rates vary from 2 cm/yr at the southern part of the Juan de Fuca plate to 4 cm/yr at the northern end. Thus, the continental margin from Cape Blanco, Oregon, to northern Vancouver Island is being continually modified by convergence between the North American and Juan de Fuca lithospheric plates. According to Hey and Wilson (1982), oceanic lithosphere has been underthrusting this margin in an eastward direction for at least 45 m.y. They suggest that about 4.9 m.y. ago the relative motion of the Pacific-Juan de Fuca plates suddenly shifted from east to east–southeast, parallel to the present Blanco fracture zone. Because of the ongoing subduction, the oldest lithosphere underlying the present Cascadia Basin is about 9 m.y. old; however, because of spreading center jumps and realignments caused by propagating rifts, the Juan de Fuca plate has enlarged at the expense of the Pacific plate (Hey and Wilson, 1982).

Where the Juan de Fuca plate and its accumulation of sedimentary material of Cascadia Basin converges with the continental margin, at least some of the sediment is scraped off and accreted onto the lower continental slope (Figures 3, 4) (Byrne, Fowler, and Maloney, 1966; Kulm and Fowler, 1974; Carson, 1977; Snavely, Pearl, and Lander, 1977; Snavely and Wagner, 1981; Snavely, this volume).

MORPHOLOGY OF CASCADIA BASIN

Casadia Basin includes 163,000 km^2 of deep-sea floor at a depth of 2400 to 3000 m. The basin floor slopes gently from north to south at an average gradient of ~1:1000 (0.057°). Differences in the overall gradient are caused by the presence along the east side of the basin of two large deep-sea fans, Nitinat and Astoria (Figure 1), which have gradients of about 1:175 (0.33°) and 1:167 (0.34°), respectively. In addition to the fans, the nearly flat basin floor is also disrupted by three long deep-sea channels—Vancouver, Juan de Fuca, and Cascadia—and by isolated hills and seamounts that are most common in the southwestern part of the basin. Cascadia Basin has three openings to the adjacent Pacific sea floor—Vancouver gap, at the northern end of the basin; Blanco saddle, at the southern end adjacent to the continental slope; and Cascadia gap, a narrow passage through the Blanco fracture zone (McManus, 1964).

Deep-Sea Fans

The Nitinat and Astoria fans are both elongate fans (each ~20,000 km^2 in area) that formed at the base of the Washington-Oregon continental slope (Nelson et al., 1970; Carson, 1971; Nelson, 1983/84). The Nitinat fan, which dominates the northern part of Cascadia Basin, has its apex near the mouth of Juan de Fuca, Nitinat, and Barkley submarine canyons. The fan extends about 100 km west to Vancouver Valley and about 200 km south, where it abuts the Astoria fan that has comparable dimensions. Astoria fan, which has its apex at the mouth of Astoria Canyon, extends westward 100 km to Cascadia Channel and southward 200 km, where its basin plain becomes the southeastern end of Cascadia Basin (Figure 1).

Carson (1971) and Stokke (1976) divide the Nitinat fan morphologically into upper, middle, and lower parts. The upper part is concave upward and is dominated by leveed valleys with relief of 70–75 m and gradients of 1:25 to 1:100 (2.3 to 0.6°). The middle-fan depositional lobe creates a convex profile, which is characterized by hummocky topography caused by small distributary and remnant valleys. Gradients range from 1:75 to 1:200 (0.8 to 0.3°). The much flatter profile (slightly concave) of the lower fan has broad, shallow valleys flanked by poorly defined levees and separated by nearly featureless interchannel areas. The lower fan gradients range from 1:350 to 1:750 (0.2 to 0.08°). Carson (1971) also pointed out that the southern part of the Nitinat fan has a greater concentration of fan valleys, suggesting a southward migration of the distributary system. Cascadia Channel, which connects with Nitinat and Juan de Fuca Canyons (Carson, 1973) is the major deep-sea channel that trends along the eastern edge of the Nitinat fan (Hampton, Karl, and Kenyon, 1984)) this channel continues south across the basin and forms the western boundary of the Astoria fan (Griggs and Kulm, 1970).

The Astoria fan, like the Nitinat fan, can be divided into three morphologic parts (Nelson et al., 1970). The upper part is distinguished by a concave profile, deep, narrow leveed valleys with relief of 75–180 m, and surface gradients that range from 1:15 to 1:100 (3.8 to 0.6°). The middle fan surface has a convex profile, broader valleys that split into distributaries, and surface gradients that range from 1:100 to 1:400 (0.6 to 0.1°). The slightly concave lower fan contains many shallow distributary valleys and featureless interchannel areas with gradients of 1:400 to 1:1000 (0.1 to 0.06°). Below 3000 m, a basin plain with gradients of <1:1000 (<0.06°) extends southward of the Astoria fan. From the perspective of looking down-gradient, the fan-valley systems on the

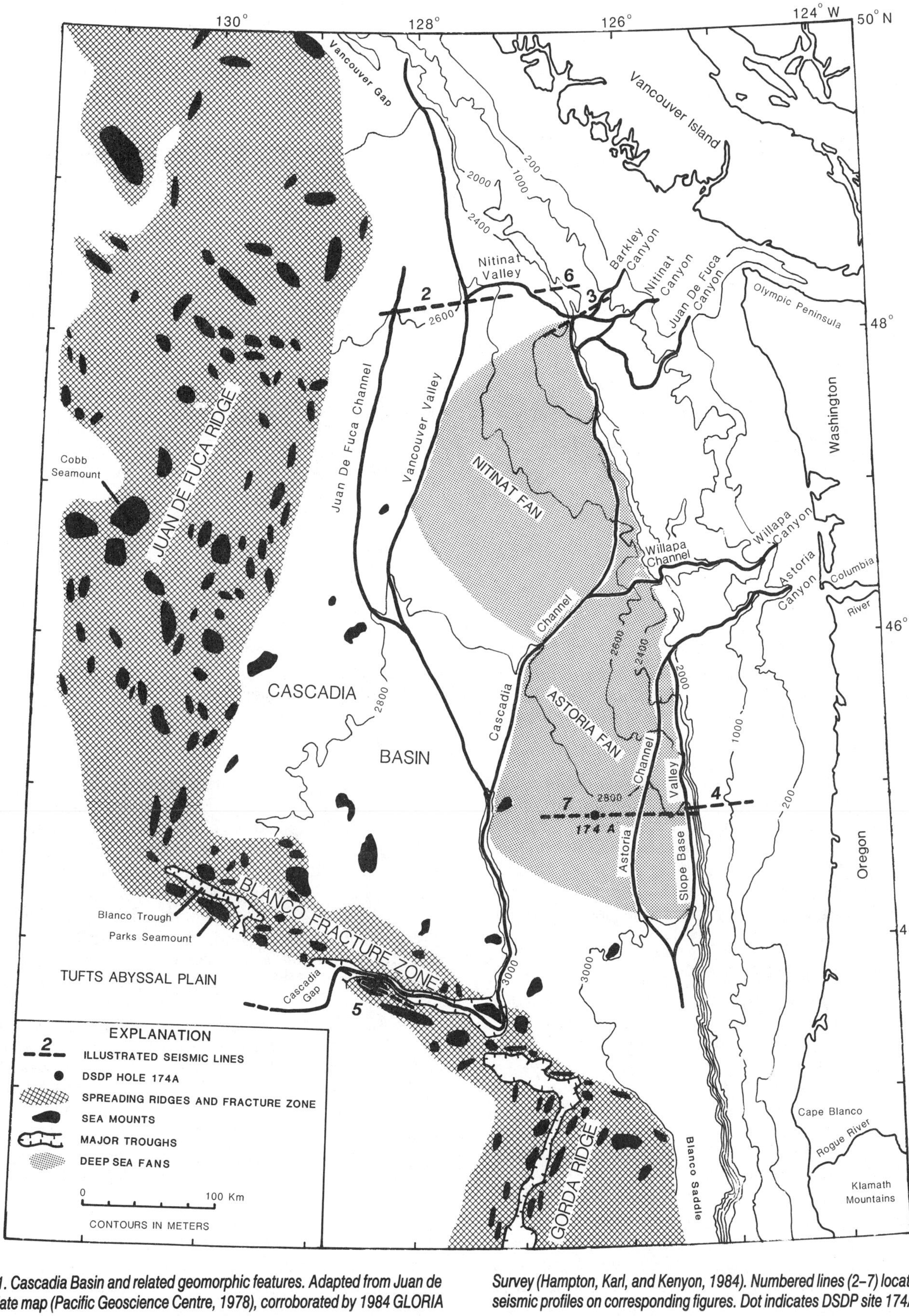

Figure 1. Cascadia Basin and related geomorphic features. Adapted from Juan de Fuca plate map (Pacific Geoscience Centre, 1978), corroborated by 1984 GLORIA Survey (Hampton, Karl, and Kenyon, 1984). Numbered lines (2–7) locate illustrated seismic profiles on corresponding figures. Dot indicates DSDP site 174A.

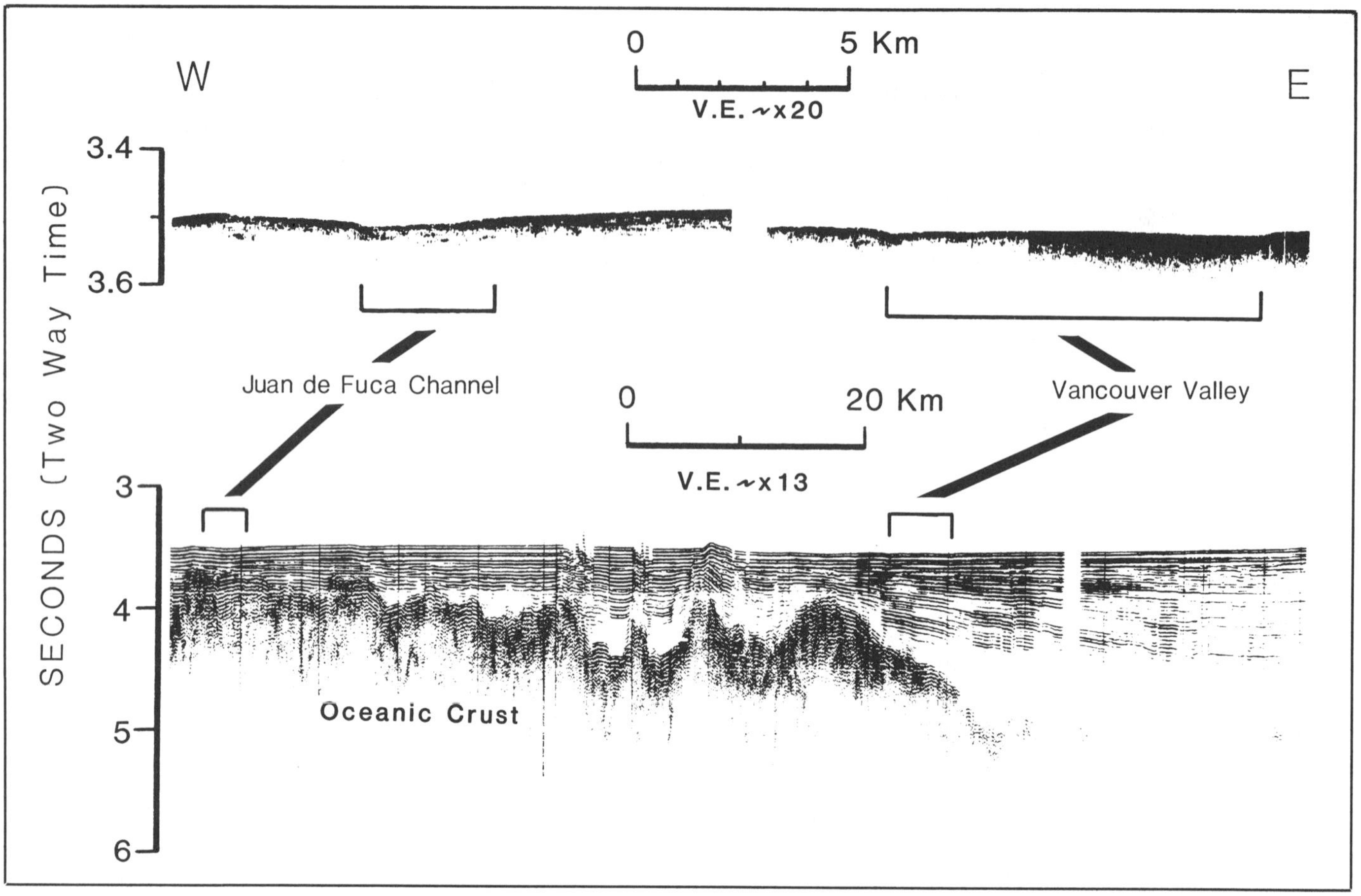

Figure 2. Single-channel air-gun (40 cu in.) profile (bottom), showing northern Cascadia Basin turbidite beds thinning over rugged acoustic basement of Juan de Fuca Ridge basalts. Accompanying 3.5 kHz profiles (top) show the broad, shallow deep-sea channels, Vancouver Valley and Juan de Fuca Channel. Collected on R/V KONA KEOKI cruise of 10/83.

Astoria fan have migrated to the left from north to south across the fan, a process attributed to the coriolis force which causes construction of higher right levees and more channel shifts over the lower left levees during fan development (Menard, 1955; Nelson et al., 1970). Astoria Channel, which connects with Astoria Canyon (Nelson et al., 1970), is the primary channel that crosses the Astoria fan and extends south to Blanco saddle (Duncan, 1968).

Deep-Sea Channels

Cascadia Basin is crossed by several large channels (Figure 1) but only one, Cascadia Channel, exits from the basin. All of the other channels, except the most recently active Astoria fan valleys, appear to be tributaries of Cascadia Channel.

Cascadia Channel, according to Carson (1973), forms the seaward extension of the Juan de Fuca-Nitinat Canyon system and extends south across the eastern edge of the Nitinat fan for 180 km, where it is joined by Willapa Channel. Upper Cascadia Channel with its well-developed levees is depositional, whereas the lower part of the channel is erosional. This deep-sea channel extends another 360 km south-southwest to the Blanco fracture zone, where it winds through the narrow Cascadia gap and out across the Tufts Abyssal Plain. Cascadia Channel has an average gradient of 1:1000 (0.06°), an average width of 1–2 km, and wall relief that ranges from 37 m to 300 m within Cascadia Basin (Griggs and Kulm, 1970).

Vancouver Valley and Juan de Fuca Channel occupy the west side of Cascadia Basin (Figure 1). Juan de Fuca Channel (~300 km long) appears to begin near the north end of Juan de Fuca Ridge, and Vancouver Valley (~600 km long) probably connects to one or more of the submarine canyons that dissect the continental margin off northern Vancouver Island. Carson (1973) showed Juan de Fuca Channel merging with Vancouver Valley at lat. ~46°N, and Griggs and Kulm (1970) reported that Vancouver Valley in turn joins Cascadia Channel at lat. 45N° as a hanging valley ~60 m above the main valley floor. This relationship has been verified by recent GLORIA side-scan sonographs (M. Hampton and H. Karl, 1984, oral communication). Both Juan de Fuca Channel and the Vancouver Valley have gradients of about 1:1100 (0.05°). Vancouver Valley is very broad (~33 km) and shallow in the northern part of the basin but narrows to < 10 km (Figure 2) as it approaches Cascadia Channel (McManus, 1964). Juan de Fuca Channel, although not completely delineated, is also broad and of low relief (Carson, 1973; and Figure 2).

Astoria Channel, and a fan valley located at the base of the slope, both begin at the mouth of Astoria Canyon as leveed, narrow, U-shaped channels (Nelson et al., 1970). These main channels widen, lose relief, and develop a series of abandoned distributary channels as they cross the southeastern part of the Astoria fan. Astoria Channel is ~250 km long, ~4 km wide near its head, and has an average thalweg gra-

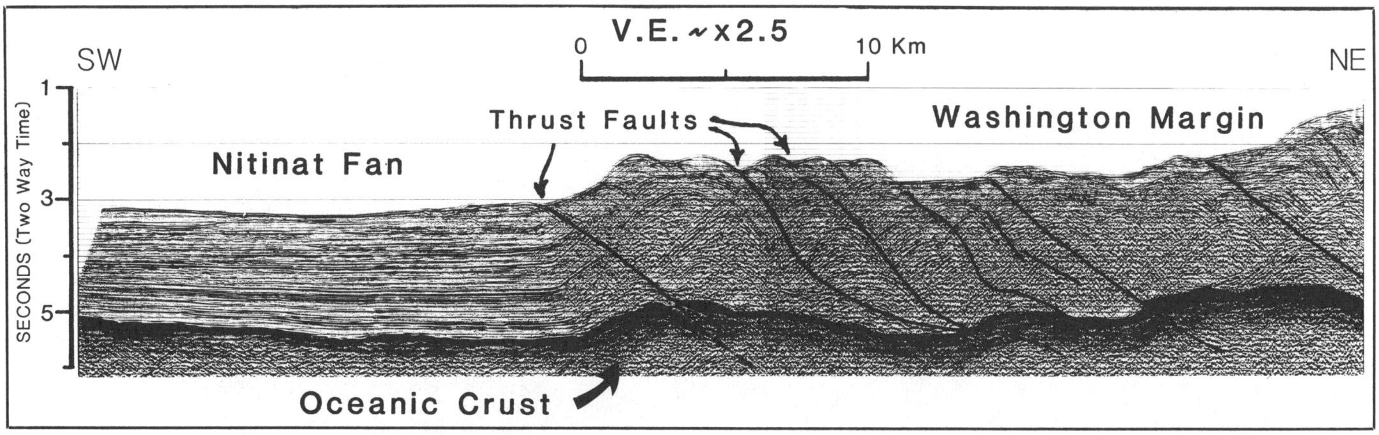

Figure 3. Multichannel seismic profile of lower continental slope off Washington, showing Nitinat fan sediment being thrust against the base of the margin by the underthrusting Juan de Fuca plate. From Snavely and Wagner (1981).

dient of 1:225 (0.3°); the slope-base fan valley is ~230 km long, ~2 km wide near its head, and has an average thalweg gradient of 1:180 (0.3°) (Nelson et al., 1970). Both the Astoria and the slope-base channels trend south and at lat. 44°N merge into the broad Blanco saddle, which also has been called Blanco Valley (Duncan, 1968) because of the continuous but low-gradient (1:5700 or 0.01°) southward slope that provides an exit for sediment from Cascadia Basin.

Other deep-sea channels in Cascadia Basin (Figure 1) include Nitinat Valley, which connects Vancouver Valley with Nitinat and other canyons at the apex of the Nitinat fan (Carson, 1973), and Willapa Channel, which joins Willapa Canyon to the Cascadia Channel system (McManus, 1964; Griggs and Kulm, 1970).

MORPHOLOGY AND STRUCTURE OF CASCADIA BORDERS

Adjacent Continental Margin

The Oregon-Washington-Vancouver Island continental margin, which forms the east side of Cascadia Basin, consists of a relatively narrow (20–70 km) shelf and fairly wide (30–90 km) slope that is disrupted in the lower part (>1500 m depth) by a series of north–northwest-oriented ridges and troughs. At least 15 submarine canyons dissect the Washington-Vancouver Island margin (McManus, 1964); however, no major canyons cut into the Oregon margin between Cape Blanco at the southern extremity of Cascadia Basin and Astoria Canyon, which is located seaward of the Columbia River (Byrne, 1962).

The continental slope off Oregon, Washington, and Vancouver Island contains numerous large-scale deformation features formed by the underthrusting of the Juan de Fuca plate beneath the North American plate (Figures 3, 4). Some variations in these structures occur on different sections of the slope. Much of the Oregon slope consists of structural benches, hills, ridges, and steep escarpments (Kulm and Fowler, 1974; Snavely, Wagner, and Lander, 1980). However, the character of the northern Oregon continental slope changes to one of broad anticlinal folds with intervening synclinal basins that are commonly bounded by faults (Kulm, Prince, and Snavely, 1973; Snavely, Pearl, and Lander, 1977). Several investigators (Silver, 1971; Carson, 1977; Seeley, 1977; Barnard 1978) described the Washington continental slope as consisting of en echelon anticlinal ridges, many of which are imbricate thrust folds with west-dipping bounding faults and less deformed intervening basins.

A multichannel seismic-reflection profile (Figure 3) collected off northern Washington shows the late Tertiary and Quaternary strata of the continental slope to be deformed into northeastward-dipping imbricate thrusts and asymmetric folds (Snavely and Wagner, 1981; Snavely, this volume). The slope off Vancouver Island consists of fault-bounded discontinuous ridges and blocks that have ponded sediment in large enough quantities to moderate the relief (Tiffin, Cameron, and Murray, 1972; Barr, 1974). These lower-slope basins formed between accretionary ridges along the Oregon slope (Figure 4) are generally small—tens of kilometers in length and a few kilometers in width (Kulm and Fowler, 1974). Some of these small basins are filled to overflowing with hemipelagic and turbidity-current deposits (Barnard, 1978; Kulm and Scheidegger, 1979). The composition of the folded strata forming the basin can range from pelagic to hemipelagic mudstone to turbidite beds, suggesting incorporation of abyssal-plain deposits into the continental slope (von Huene and Kulm, 1973). This hypothesis is supported by faunal data from DSDP hole 175 (Ingle, 1973) and dredge data from the slope (Byrne, Fowler, and Maloney, 1966; Barr, 1974; Barnard, 1978).

Adjacent Spreading Ridge and Fracture Zone

The western boundary of Cascadia Basin is marked by a group of seamounts and ridges that Wilson (1965) named the Juan de Fuca Ridge (Figure 1)—a feature he delineated largely from magnetic anomaly patterns (Morton et al., this volume). McManus (1967), on the other hand, outlined this seamount, ridge, and trough province on the basis of bathymetry and referred to it as Cobb Rise. The discontinuous crest is about 600 km long and has an average depth of about 2500 m, but it shoals to as little as 33 m at the summit of Cobb Seamount (McManus, 1967). Although a median valley and bordering ridges are discernible on some crossings of Juan de Fuca Ridge (Delaney, Johnson, and Karsten, 1981),

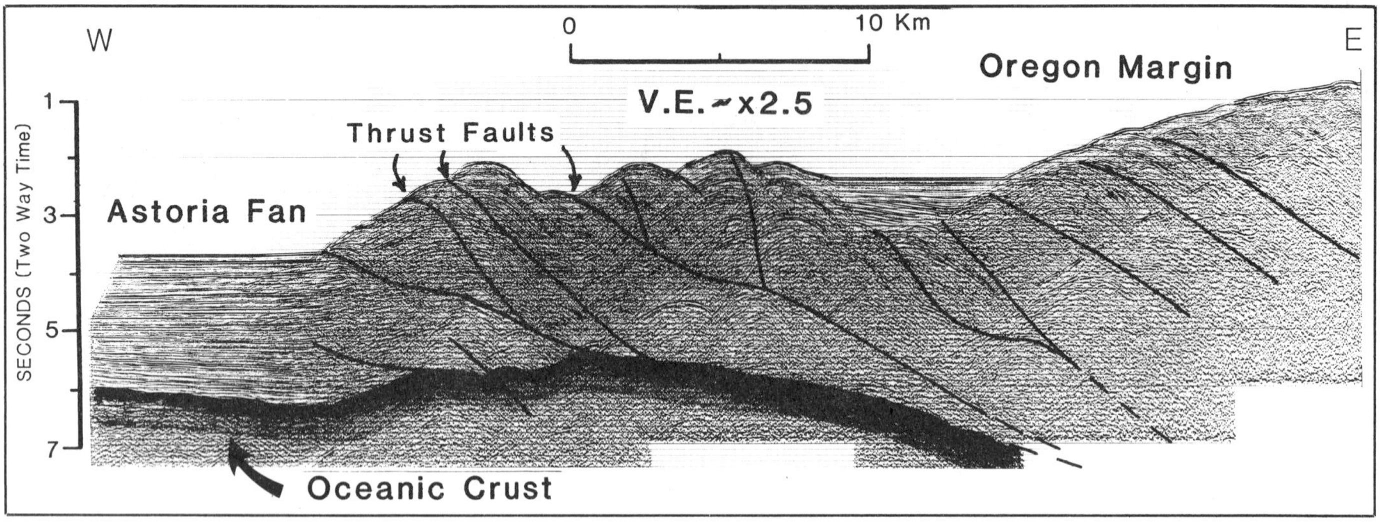

Figure 4. Multichannel seismic profile showing wedge of the Astoria fan turbidite deposits entering the subduction zone of the complexly thrust-faulted continental slope off central Oregon. From Snavely, Wagner, and Lander (1980).

they are not as prominent or continuous as on Gorda Ridge (Edwards et al., 1984).

The morphotectonic features of Juan de Fuca Ridge are quite different along its length (Hampton, Karl, and Kenyon, 1984; Holmes et al., 1984). This is because of the propagation history of the ridge segments (Hey and Wilson, 1982; Wilson, Hey, and Nishimura, 1984), as indicated by the oblique offsets of the symmetrical linear magnetic anomalies that delineate the ridge system, and because of the differences in sea-floor-spreading rates and magma supply from the Cobb hotspots.

The southern part of this medium rate (5–9 cm/yr) spreading ridge is typified by a narrow (~1 km) flat-floored axial valley, flooded by young ($<10^3$ yr B.P.) basalt flows (Lichtman and Eissen, 1983). The scarps that form the 100-m-high valley walls apparently are the product of steep normal faults (Normark, Morton, and Delaney, 1982; Morton et al., this volume).

The central third of the ridge lacks the bathymetric and magnetic symmetry of both the southern and northern thirds, does not appear to contain the fresh volcanic glass dredged both north and south, and includes a westerly offset active volcanic zone (Delaney, Johnson, and Karsten, 1981). GLORIA and SEAMARC II side-scan sonographs, however, show extensive flows and vents superimposed on the ridges (M. Hampton and H. Karl 1984, oral communication; Holmes, et al., 1984).

The northern part of the Juan de Fuca Ridge is dominated by three axial valleys and corresponding ridges (McManus et al., 1972). Turbidite deposits occupy most of the lows, and at least one of the axial valleys is completely filled (~3 km of sediment). The sediment of this valley fill has been offset as the underlying normal fault blocks rotated away from the ridge crests. Davis and Lister (1977) report that only the westernmost valley currently appears to be spreading at the rate assumed for the Juan de Fuca Ridge system.

The Blanco fracture zone, which forms the southern boundary of Cascadia Basin, is characterized by extremely rugged morphology (Figure 5; Embley et al., this volume). This 20- to 75-km-wide transform fault zone includes an elongate deep hole (Blanco Trough) that reaches a depth of >4500 m (McManus, 1965) and, at the other extreme, the adjacent Parks Seamount that shoals to ~1500 m and produces relief of 2700 m in a distance of about 10 km (Figure 1). The average height of the elongate ridges that mark this fracture zone is about 2200 m. Midway along the Blanco fracture zone, Cascadia Channel (Figure 5) winds through the ridges for about 150 km and then onto Tufts Abyssal Plain (Griggs and Kulm, 1970; Embley et al., this volume).

SEISMIC STRATIGRAPHY AND SEDIMENTARY FACIES

The Cascadia Basin floor has been receiving terrigenous and hemipelagic sediment since about the late Miocene. This date is based on an interpretation of magnetic anomalies indicating that oceanic basement adjacent to the continental margin is about 9 m.y. old (Silver, 1971; Snavely, Wagner, and Lander, 1980; Hey and Wilson, 1982). Corroboration of the date comes from Deep Sea Drilling Project (DSDP) hole 174A on the Astoria fan about 80 km west of the base of the slope (Kulm, Prince, and Snavely, 1973), which was completed to within 32 m of 8-Ma-old oceanic basement). This 879-m-deep hole bottomed in Pliocene thin-bedded silt turbidite and mud beds that contain planktonic foraminifers which accumulated during a warm event in the early Pliocene (Ingle, 1973).

Thickness of Basin Sediment

The Neogene and Quaternary sediment of Cascadia Basin has buried and effectively smoothed out the rugged oceanic crustal topography that originated at the Juan de Fuca spreading center. The sediment forms a wedge-shaped deposit that is thickest along the continental margin (Figure 6; Connard et al., 1984) and thins westward as it laps onto the modern ridge (Figure 2). Sediment that covers the floor of the basin is also beginning to "flood" the lows of the Blanco fracture zone (Embley et al., this volume). Seismic-refraction

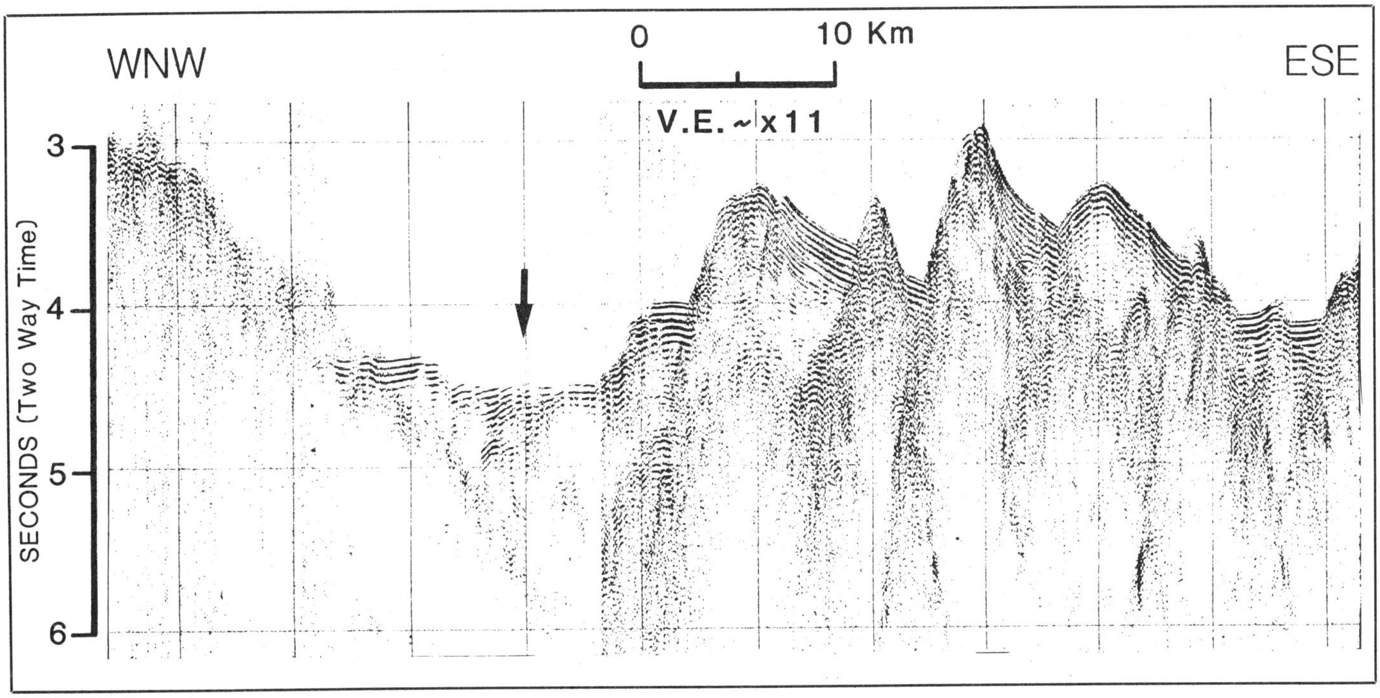

Figure 5. Single-channel air-gun (40 cu in.) profile along Blanco fracture zone. Arrow indicates Cascadia Channel. Collected on R/V KONA KEOKI cruise of 10/83.

data across the basin show that the sediment is thickest (2.5 km) near the apex of the the Astoria fan, and that it thins to 1.3 km west of Cascadia Channel and to < 0.4 km at the western side of the basin (Shor et al., 1968).

Seismic Stratigraphy of Basin Sediment

Published seismic-reflection records (Hamilton, 1967; McManus et al., 1972; Carson, 1973; Kulm, Prince, and Snavely, 1973; Kulm and Fowler, 1984; Seeley, 1977; Wilde et al., 1977, 1979; Snavely and Wagner, 1981; Snavely, Wagner, and Lander, 1980; McClain, Peper, and Holmes, 1984) from various parts of Cascadia Basin show numerous parallel, relatively flat-lying reflectors similar to those shown on Figure 2. All the authors mentioned above attribute this type of seismic signature to turbidity-current deposits. This interpretation has been verified at DSDP site 174A drilled in Cascadia Basin (Figure 7; Kulm et al., 1973). The 879-m sedimentary section, drilled in the distal part of the Astoria fan, has been divided into two major stratigraphic units. The upper unit (0–284 m) is dominated by upper Pleistocene, thick- to thin-bedded, medium- to fine-grained sand with a sand/shale ratio of 1.6:1 (Kulm et al., 1973). The lower unit (284–879 m) consists of thin beds of silty clay with thin basal silt or, rarely, fine sand of Pliocene and Pleistocene age. Von Huene and Kulm (1973) suggest that the section from site 174A indicates an initial environment of pelagic abyssal deposition interrupted by increasing numbers of deep-ocean turbidite deposits succeeded by the coarser deep-sea fan turbidite sediment that signaled the beginning construction of the Astoria fan system. They further suggest that a seismic discontinuity, which coincides closely with the stratigraphic unit boundary (Figure 7), marks the initiation of Astoria fan deposition at about 1 m.y. ago.

Piston cores from the Astoria fan show fewer turbidity current layers in the thin (~ 1 m thick interchannel areas) Holocene unit (sand/shale ratio = < 1:9) than in the uppermost Pleistocene section where the sand/shale ratios = 1:1 to 3:1 (Nelson and Kulm, 1973). This sequence indicates that turbidity-current activity on the fan has been much diminished since the beginning of the Holocene rise in sea level (Carlson and Nelson, 1969; Nelson, 1976). Carson (1971) has found similar changes in cores from the Nitinat fan, as have Griggs and Kulm (1970) in Cascadia Channel.

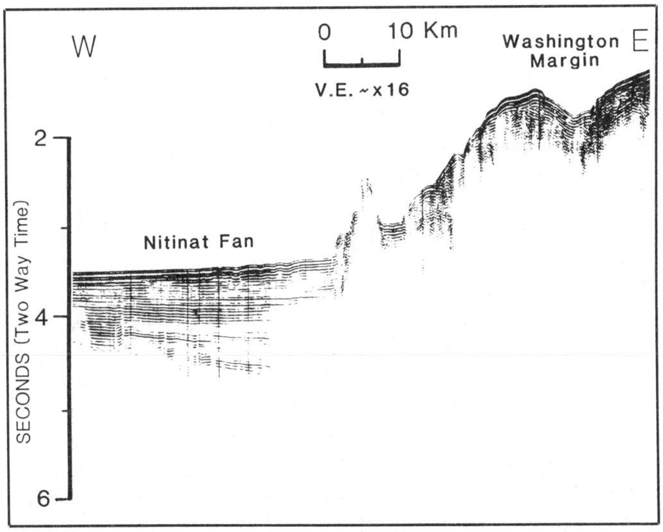

Figure 6. Single-channel air-un (40 cu in.) profile across the base of the Washington margin, showing the initial deformation of a thick wedge of turbidite beds of northern Nitinat fan. Note narrow interslope basin of ponded sediment. Collected on R/V KONA KEOKI cruise of 10/83.

The sedimentary material that fills the northern part of Cascadia Basin has been divided into four seismic-stratigraphic units (McManus et al., 1972; Carson, 1973). The

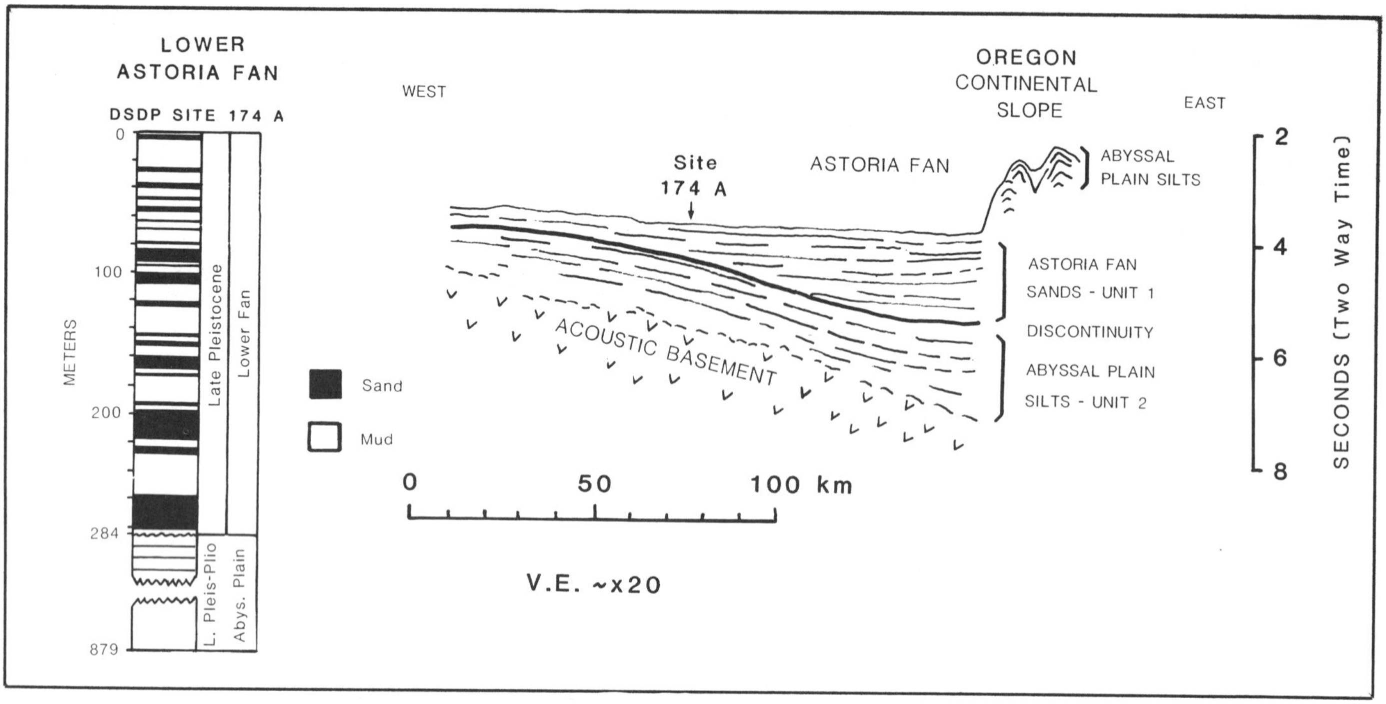

Figure 7. Diagrammatic sketch of seismic stratigraphy across the Astoria fan through DSDP site 174A. After von Huene and Kulm (1973); Kulm, Prince, and Snavely (1973).

lowermost unit (unit A), overlying the rugged relief of the Juan de Fuca Ridge basalt, is a thick acoustically transparent sequence (Figure 2) probably composed of thin turbidite beds interleaved with pelagic sediment. This acoustically transparent unit is more restricted than the overlying unit (unit B), which consists of many closely spaced strong reflectors. These two units (A and B) appear to correlate with the Pliocene and Pleistocene abyssal unit identified by Kulm et al. (1973) that underlies the the Astoria fan. The upper two units (C and D) correlate with the Pleistocene and Holocene section of the Astoria fan. McManus et al. (1972) and Carson (1973) combine the two upper units (C and D) into an upper turbidite unit consisting of a series of parallel reflectors that range from weak to strong. This upper turbidite unit covers much of northern Cascadia Basin and attains a thickness of as much as 1 km beneath the Nitinat fan. A seismic-reflection record recently collected across the northern part of the Nitinat fan shows Cascadia Basin sediment >2 sec (~2 km) thick terminating abruptly against the base of the slope (Figure 6). This basin section contains an acoustically transparent unit (~0.5 sec) overlying the irregular acoustic basement, which is in turn overlain by a thick sequence (1–1.5 sec) of parallel reflectors (turbidite beds).

Holocene and Pleistocene Sedimentary Facies

Although thin, abyssal-plain turbidite deposits are present in the Pliocene section of Cascadia Basin (Kulm, Prince, and Snavely, 1973; McManus et al., 1972; Kulm and Fowler, 1974), turbidity-current activity increased greatly during the Pleistocene low-stands of sea level. For example, more than one episode of turbidity-current activity occurred every 6 years on the Astoria fan during the late Pleistocene (~12,500–25,000 yr B.P.; Nelson, 1976). Increases in turbidity-current activity from the Holocene to the Pleistocene also have been reported for the Nitinat fan (Carson, 1971), Cascadia Channel (Griggs and Kulm, 1970), and the southern and western parts of Cascadia Basin (Duncan, 1968).

The modern sedimentation regime within Cascadia Basin is dominated by hemipelagic silty clay deposited at rates that range from ~10 cm/10^3 yr for the interchannel areas of the Astoria fan (Nelson, 1976) to 1–3 cm/10^3 yr in the western Cascadia Basin (Duncan, 1968). Sedimentation rates for the channel-floor areas of the Astoria fan, on the other hand, include more turbidite deposition in the Holocene and, as a result, the rates increase to about 25 cm/10^3 yr (Nelson, 1976). Carson (1971) and Stokke (1976) have calculated similar Holocene rates of sediment accumulation for the Nitinat fan. The highest reported rates of Holocene sediment accumulation in Cascadia Basin are from Cascadia Channel, where they range from 71 cm/10^3 yr in the upper channel to an exceptional 125 cm/10^3 yr in the lower channel, where coarse sediment is being ponded in a depression within Blanco fracture zone (Griggs and Kulm, 1970). These rates, however, include turbidity-current deposits, whereas rates of associated Holocene hemipelagic deposition range from 3–14 cm/10^3 yr.

Although piston cores collected from the Astoria fan only penetrate the upper 5–6 m of the sediment column, the Holocene sequence is thin enough in many places so that the underlying upper Pleistocene beds can be sampled. Nelson (1976) has estimated minimum sedimentation rates of 36–40 cm/10^3 yr for the upper Pleistocene sequence sampled on the Astoria fan. Kulm, Prince, and Snavely (1973) report a rate of 30–40 cm/10^3 yr for 284 m of turbidite-dominated sediment drilled at DSDP site 174A on the the distal part of the part of the Astoria fan. In addition to similar late Pleistocene sedimentation rates, similar sand/shale ratios (1:1 to 3:1) have been reported for the distal part of the Astoria fan from near-surface piston cores (Nelson, 1983/1984) and from DSDP site 174A subsamples (1.6:1) (Kulm, Prince, and Snavely, 1973).

Duncan (1968) calculated rates of sediment accumulation for eight cores collected from a wide variety of environments in Cascadia Basin and concluded that the average late Pleistocene sedimentation rate (44 cm/10^3 yr) was about six times greater than the Holocene rates (7 cm/10^3 yr).

Sources of Cascadia Basin Sediment

Sediment that has been accumulating since the late Miocene in Cascadia Basin consists principally of two general types, hemipelagic mud and sand- to silt-rich turbidity-current deposits. Except on channel floors, hemipelagic sediment has been the chief type of sediment accumulating in the basin during the Holocene (Carlson and Nelson, 1969; Duncan, Kulm, and Griggs, 1970; Stokke, 1976; Carson, 1977). Beneath the channel floors, lower Holocene hemipelagic sediment is interbedded with numerous graded sand and/or silt layers and laminae deposited by turbidity currents. In Cascadia Channel thick silt sequences deposited by turbidity currents have accumulated throughout the Holocene (Griggs and Kulm, 1970). The hemipelagic sediment consists of terrigenous poorly sorted silty clay or clay with a limited (< 10%) sand fraction consisting predominantly of radiolarians. On the Nitinat fan, flows of low-density turbid water appear to be the principal transport agent for the hemipelagic sediment (Stokke, Carson, and Baker, 1977). Particle-by-particle deposition dominates during the Holocene on the western two-thirds of the Astoria fan, whereas turbid-layer flow is dominant in Astoria Channel and on the eastern third of the fan during the Holocene and over the entire fan during the Pleistocene (Nelson, 1976).

The Columbia River is the source of the clay that makes up a large part of both Pleistocene and Holocene fine-grained sediment on the Astoria fan and in the southern Cascadia Basin; however, in the Holocene deposits of the southern part of the basin, clay from the Klamath Mountains introduced by the Rogue River dilutes the Columbia River contribution (Duncan, Kulm, and Griggs, 1970). Clay-rich sediment in northern Cascadia Basin has as its source the adjacent margins of Vancouver Island and northern Washington (Carson, 1971).

Sand- to silt-rich turbidite beds dominate the Pleistocene deposits of Cascadia Basin (Carlson and Nelson, 1969; Duncan and Kulm, 1970; Kulm, Prince, and Snavely, 1973) and have been important contributors to the entire basin's sedimentary sequence since the middle Miocene (Carson, 1973; Kulm, Prince, and Snavely, 1973). Scheidegger, Kulm, and Piper (1973) found two quite different heavy-mineral suites from DSDP hole 174A. Most of the Pliocene silt turbidite deposits that constitute the lower part of the cored section (370–879 m) contain a heavy-mineral assemblage dominated by amphiboles from either a northern (Vancouver Island, British Columbia) or a southern (Klamath Mountains) source. In the upper 370 m of section and in numerous piston cores, the nearly equal presence of pyroxene and amphibole shows a Columbia River source for the sand-rich turbidity currents that were funneled through Astoria Canyon and onto the Astoria fan (Carlson, 1968; Nelson, 1968, 1976; Duncan, 1968). Piston cores from Vancouver Valley and western Cascadia Basin contained heavy-mineral assemblages from the Vancouver Island-British Columbia area (Duncan and Kulm, 1970). Some heavy-mineral assemblages in the turbidite sands from the northern part of Cascadia Basin originated in the Vancouver Island-British Columbia provenance; however, much Nitinat fan sediment moved north from the Columbia River and was funneled down the numerous canyons of the Washington continental margin (Carson; 1971; Barnard, 1978).

An independent tracer of Columbia River sediment, transported into Cascadia Basin during the Holocene, is Mazama volcanic glass, a product of a cataclysmic eruption of Mt. Mazama 6845 ± 50 yr B.P. (Bacon, 1983). The volcanic glass, found associated with graded sand and silt layers in the deep-sea cores from Cascadia Basin, was carried by the Columbia River to the shelf and then by turbidity currents through Astoria and Willapa canyons to the Astoria fan and Cascadia Channel (Royse, 1967; Nelson et al., 1968; Griggs and Kulm, 1970).

RESOURCE POTENTIAL OF CASCADIA BASIN

Mineral Resources

Although economic recovery of polymetallic sulfide deposits is remotely possible from the Juan de Fuca Ridge (Koski, Clague, and Oudin, 1984; Koski et al., this volume; Morton et al., this volume), and placer deposits from the shelf are possibilities (Clifton and Luepke, this volume), there appears to be little mineral resource potential in Cascadia Basin. Basalts created at the Juan de Fuca spreading center move across and beneath Cascadia Basin. Consequently, sulfides similar to those discovered at the ridge crest possibly underlie the basin. Unless geothermal processes mineralize the sea floor (Lister, this volume; Clague and Holmes, this volume), the deep water (2400–2800 m) and thick sediment blanket (2500 m) make discovery and recovery remote possibilities. Some minerals present in the beach and shelf placer deposits—such as ilmenite, magnetite, and even gold (Clifton, 1968; Clifton and Luepke, this volume)—must have been carried to the shelf edge during low stands of sea level, slumped into canyons, and carried onto the deep-sea fans by turbidity currents. Even though present on the fans, the shelf placer minerals would be diluted by other mineral grains while being transported rather than concentrated by this process.

Petroleum Resources

On the basis of available refraction (Shor et al., 1968) and reflection data (McManus et al., 1972; Carson, 1973; Kulm and Fowler, 1974; Connard et al., 1984), we estimate that the average thickness of sediment in Cascadia Basin is about 1 km, although a thickness near 2.5 km occurs at the apexes of the fans. The volume of sediment in Cascadia Basin is approximately 163,000 km^3. Although this is not a particularly large volume or thick sequence of sedimentary material, Cascadia Basin, especially the large deep-sea fans, does contain most of the required elements for the collection of hydrocarbons—sufficient geothermal gradient (Lister, this volume), source rocks, reservoir rocks, and traps.

Source Rocks

The hemipelagic sediment that is interlayered with the thin silt turbidite layers and constitutes the lower half to two-thirds of the basin section (Pliocene and Pleistocene) appears to be the best source for petroleum hydrocarbon generation. At DSDP site 174A, the poorly sorted, compacted clay and silty clay of the lower section has accumulated in beds 20–60 cm thick commonly separated by thin basal silt and less commonly by sand laminae. The organic carbon content of the 21 analyzed intervals averages 0.4% and ranges from 0.2–0.8% (Bode, 1973). The upper unit (Pleistocene), which in hole 174A represents the Astoria fan (0–284 m), consists of about 30% hemipelagic sediment in beds that average 10–20 cm thick (Kulm, Prince, and Snavely, 1973). Organic carbon measurements from the upper stratigraphic unit range from 0.1 to 1.0%; values of 0.1–0.2% were obtained from sandy layers, and values that average 0.4% from silty clay units. Thus, hemipelagic layers interbedded with thicker turbidite beds of the deep-sea fans (Astoria and Nitinat) are marginal source beds. Higher values of organic carbon have been obtained from Holocene hemipelagic sediment from Cascadia Channel, which contains as much as 2.9% organic carbon (Griggs, Cary, and Kulm, 1969). From other parts of the basin, Holocene sediment yields organic carbon values that range from 2.0 to 2.5% near the mouth of Astoria Canyon (Carlson, 1968) and from 0.2 to 1.5% on Cascadia plain (Duncan, 1968).

Reservoir Rocks

The most likely reservoir rocks in the Cascadia Basin section are the turbidite sands of the large deep-sea fans (Astoria and Nitinat) that cover nearly 25% of the basin (Figure 1). Turbidite beds cored at DSDP site 174A range in thickness from 5 to 733 cm and grade upward from basal sand to silt to silty clay (Kulm, Prince, and Snavely, 1973). The sands range in modal size from medium to very fine and in sorting from well to poor, but most of the sands are fine grained and moderately well sorted. The areal distribution of these characteristics can be estimated from piston cores that sampled the top few meters of the Astoria fan. Carlson and Nelson (1969) found the sorting to change from very poor near the apex to moderate in the distal part of the fan; the grading of the beds improves, and their thickness appears to increase toward the distal part of the fan. Nelson (1976) reports that the coarsest material (poorly sorted muddy gravel) is found in the proximal parts of the channels, and the channels throughout the fan contain thicker and coarser sands than the adjacent interchannel areas.

Porosity of the sands from DSDP hole 174A was difficult to measure, because the best-sorted sands were commonly disturbed in the coring process; however, von Huene, Piper, and Duncan (1973) obtained porosity values that ranged from 37 to 52% in fine to medium sand. The thickest, coarsest, and most plentiful turbidite sands are present in the upper unit (0–284 m) on the Astoria fan and thus are less compacted than sediment deeper in the section.

Snavely, Pearl, and Lander (1977) report that Tertiary turbidite sandstone units present in test wells drilled on the continental shelf off Washington and Oregon have relatively low porosity and permeability because the pore spaces are commonly filled with calcareous cement and clay minerals, or, where tectonically deformed, with silica cement and zeolites. They do, however, mention four Oligocene samples of turbidite sandstones from the Olympic Peninsula that apparently became well sorted in transport and thus have sufficient porosity and permeability to be classified as potential reservoir rocks.

Although turbidite sandstone in the Pacific northwest has not yielded commercial quantities of hydrocarbons, turbidite deposits elsewhere have proved to be good reservoir rocks. For example, nearly all of the several billion barrels of oil recovered from the Los Angeles Basin have been pumped from turbidite sands (Sullwold, 1961), and much of the oil produced in the southern San Joaquin Valley comes from late Miocene turbidite sandstones that are believed to be deep-sea-fan deposits (MacPherson, 1978). Likewise, much of the natural gas produced in the Sacramento Valley comes from Paleogene turbidite beds deposited in canyon-fan settings (Almgren and Hacker, 1984). Turbidite deposits also are important reservoir rocks in the North Sea, as exemplified by the large Frigg gas field (Hertier, Lossel, and Wathne, 1979) and the Forties oil field (Thomas, Walmsley, and Jenkins, 1974; Parker, 1975). For comparison with the sandstone units of the Astoria fan, the turbidite sands of the Forties field are moderately to poorly sorted and have porosities from 20 to 30%; grading present in thinner beds is often lacking in massive beds, which attain thicknesses of as much as 60 m (Parker, 1975).

Traps

Stratigraphic traps are likely to occur in a deep-sea-fan turbidite sequence such as those of Cascadia Basin. For example, good stratigraphic traps are formed where coarse channel deposits on the upper fan are sealed off as the channel is bypassed, or where distal or mid-fan sand lobes are sealed by surrounding hemipelagic muds. Bloomer (1977) reports on two producing fields in west Texas that appear to have these characteristics—the Northeast Bloodworth field, in which the Lower Permian Cook Sandstone is a channel-fill deposit; and the Jameson field, where the Cook Sandstone consists of thin graded sandstone lenses deposited to form a submarine fan.

Two other possibilities for traps in the Cascadia Basin setting are structural or gas hydrate in origin. Structural traps would most likely form at the base of the slope where the underthrusting Juan de Fuca plate causes folding and faulting of offscraped basinal deposits (Seeley, 1977; Barnard, 1978; Snavely, Wagner, and Lander, 1980; Snavely, this volume). Also, at the base of the slope, given the proper pressure and temperature conditions, gas hydrates could form (Kvenvolden, this volume) and effect a stratigraphic seal for migrating hydrocarbons.

Problems

Although the fans in Cascadia Basin contain possible source and reservoir rocks, and the setting is favorable for adequate stratigraphic traps, several factors deter the likely occurrence of large accumulation of hydrocarbons. These interrelated factors include time, temperature, thickness of basinal sediment, and tectonic setting. The subduction margin that forms the eastern boundary of Cascadia Basin severely restricts the age and amount of sediment that can accumulate on a specific part of the floor of the basin before that part of the Juan de Fuca plate underthrusts the North American plate. As a result, the thickest part of the sediment wedge building in the basin is 2.5 km, overlying the 8-m.y.-

old conveyor-belt-like segment of the crust that abuts the continental margin. The average thickness of sediment in the basin is much thinner, about 1 km. Because of the relatively young age of the sediment (Pliocene and Pleistocene) and the limited depth of burial, the hemipelagic mud that is the most likely candidate for source rock is probably immature with respect to the temperature history that is required for hydrocarbon generation. However, the young oceanic crust probably supports an adequate thermal gradient (Lister, this volume). This less-than-positive scenario, together with the oceanic depths of the basin floor and the typically poor probability of hydrocarbon generation in this setting (Nelson and Nilsen, 1984), indicate that Cascadia Basin is not a very attractive target for future petroleum exploration. The adjacent continental margin is a more likely habitat for significant accumulations of hydrocarbons (Snavely, this volume).

ACKNOWLEDGMENTS

We thank Monty Hampton, Richard Hey, Mark Holmes, Herman Karl, William Normark, and Parke Snavely for reading the manuscript and offering helpful suggestions at various stages of manuscript preparation.

REFERENCES CITED

Almgren, A. A., and P. D. Hacker (eds.), 1984, Paleogene submarine canyons of the Sacramento Valley, California: AAPG, Pacific Section Symposium, v. 1, 187 p.

Atwater, T., 1970, Implications of plate tectonics for the Cenozoic tectonic evolution of western North America: GSA Bulletin, v. 81, p. 3513–3536.

Bacon, C. R., 1983, Eruptive history of Mount Mazama and Crater Lake caldera, Cascade Range, U.S.A.: Journal of Volcanology and Geothermal Research, v. 18, p. 57–115.

Barnard, W. D., 1978, The Washington continental slope—Quaternary tectonics and sedimentation: Marine Geology, v. 27, p. 79–114.

Barr, S. M., 1974, Structure and tectonics of the continental slope west of southern Vancouver Island: Canadian Journal of Earth Sciences, v. 11, p. 1187–1199.

Bloomer, R. R., 1977, Depositional environments of a reservoir sandstone in west-central Texas: AAPG Bulletin, v. 61, p. 344–359.

Bode, G. W., 1973, Carbon and carbonate analyses—Leg 18, *in* L. D. Kulm, R. von Huene, et al., Initial Reports of the Deep Sea Drilling Project (Vol. 18): Washington, D.C., U.S. Government Printing Office, p. 1069–1076.

Byrne, J. V., 1962, Geomorphology of the continental terrace off the central coast of Oregon: Ore Bin, v. 24, p. 65–74.

Byrne, J. V., G. A. Fowler, and N. J. Maloney, 1966, Uplift of the continental margin and possible continental accretion off Oregon: Science, v. 154, p. 1654–1656.

Carlson, P. R., 1968, Marine geology of Astoria submarine canyon: Corvallis, Oregon, Oregon State University unpublished Ph.D. dissertation, 259 p.

Carlson, P. R., and C. H. Nelson, 1969, Sediments and sedimentary structures of the Astoria submarine canyon-fan system, northeast Pacific: Journal of Sedimentary Petrology, v. 39, p. 1269–1282.

Carson, B., 1971, Stratigraphy and depositional history of Quaternary sediments in northern Cascadia Basin and Juan de Fuca Abyssal Plain, northeast Pacific Ocean: Seattle, Washington, University of Washington unpublished Ph.D. dissertation, 249 p.

Carson, B., 1973, Acoustic stratigraphy, structure, and history of Quaternary deposition in Cascadia Basin: Deep-Sea Research, v. 20, p. 387–396.

Carson, B, 1977, Tectonically induced deformation of deep-sea sediments off Washington and northern Oregon: Mechanical consolidation: Marine Geology, v. 24, p. 289–308.

Clague, D. A., and M. L. Holmes, 1987, Geology, petrology, and mineral potential of the Gorda Ridge, this volume.

Clifton, H. E., 1968, Gold distribution in surface sediments on the continental shelf off southern Oregon—A preliminary report: U.S. Geological Survey Circular 587, 6 p.

Clifton, H. E., and G. Luepke, 1987, Heavy-mineral placer deposits of the continental margin of Alaska and the Pacific Coast states, this volume.

Connard, G., R. Couch, K. Keeling, J. Roy, and S. Troseth, 1984, Abyssal plain and continental net-objective sedimentary thicknesses, *in* L. D. Kulm et al., eds., Western North American continental margin and adjacent ocean floor off Oregon and Washington: Marine Science International, Ocean Margin Drilling Program, Regional Atlas Series, Atlas 1, sheet 7.

Davis, E. E., and C. R. B. Lister, 1977, Tectonic structure on the Juan de Fuca Ridge: GSA Bulletin, v. 88, p. 346–363.

Delaney, J. R., H. P. Johnson, and J. L. Karsten, 1981, The Juan de Fuca Ridge-hot spot-propagating rift system: New tectonic, geochemical, and magnetic data: Journal of Geophysical Research, v. 86, p. 11,747–11,750.

Duncan, J. R., Jr., 1968, Late Pleistocene and postglacial sedimentation and stratigraphy of deep-sea environments off Oregon: Corvallis, Oregon, Oregon State University unpublished Ph.D. dissertation, 222 p.

Duncan, J. R., Jr., and L. D. Kulm, 1970, Mineralogy, provenance, and dispersal history of late Quaternary deep-sea sands in Cascadia Basin and Blanco fracture zone off Oregon: Journal of Sedimentary Petrology, v. 40, p. 874–887.

Duncan, J. R., Jr., L. D. Kulm, and G. B. Griggs, 1970, Clay mineral composition of late Pleistocene and Holocene sediments of Cascadia Basin, northeastern Pacific Ocean: Journal of Geology, v. 78, p. 213–221.

Edwards, B. D., P. R. Carlson, J. V. Gardner, M. Holmes, and H. A. Karl, 1984, Surface morphology of the Gorda Ridge as seen by Sea MARC II sea-floor mapping system (abs.): SEPM Abstracts, Annual Mid-Year Meeting, p. 29–30.

Embley, R. W., L. D. Kulm, G. Massoth, D. Abbott, and M. Holmes, 1987, Morphology, structure, and resource potential of the Blanco transform fault zone, this volume.

Griggs, G. B., and L. D. Kulm, 1970, Sedimentation in Cascadia deep-sea channel: GSA Bulletin, v. 81, p. 1361–1384.

Griggs, G. B., A. G. Carey, Jr., and L. D. Kulm, 1969, Deep-sea sedimentation and sediment-fauna interaction in Cascadia Channel and on Cascadia Abyssal Plain: Deep-Sea Research, v. 16, p. 157–170.

Hamilton, E. L., 1967, Marine geology of abyssal plains in the Gulf of Alaska: Journal of Geophysical Research, v. 72, p. 4189–4213.

Hampton, M. A., H. A. Karl, and N. H. Kenyon, 1984, Sonograph mosaic of the Oregon-Washington EEZ: EOS (American Geophysical Union Transactions), v. 65, p. 1083.

Hertier, F. E., P. Lossel, and E. Wathne, 1979, Frigg field—Large submarine-fan trap in lower Eocene rocks of North Sea Viking graben: AAPG Bulletin, v. 63, p. 1999–2020.

Hey, R. M., and D. S. Wilson, 1982, Propagating rift explanation for the tectonic evolution of the northeast Pacific—The pseudomovie: Earth and Planetary Science Letters, v. 58, p. 167–188.

Holmes, M., J. Gardner, H. Johnson, P. Carlson, B. Edwards, and H. Karl, 1984, SEAMARC II studies of the Juan de Fuca and Gorda Ridges: EOS (American Geophysical Union Transactions), v. 65, p. 279.

Hurley, R. J., 1960, The geomorphology of abyssal plains in the northeast Pacific Ocean: San Diego, California, Marine Physical Laboratory, Scripps Institute of Oceanography, 105 p.

Ingle, J. C., Jr., 1973, Neogene foraminifera from the northeastern Pacific Ocean, Leg 18, Deep-Sea Drilling Project, *in* L. D. Kulm, R. von Huene, et al., Initial Reports of Deep Sea Drilling Project (Vol. 18): Washington, D.C., U.S. Government Printing Office, p. 517–568.

Koski, R. A., D. A. Clague, and E. Oudin, 1984, Mineralogy and chemistry of massive sulfide deposits from Juan de Fuca Ridge: GSA Bulletin, v. 95, p. 930–945.

Koski, R. A., S. D. Scott, M. D. Hannington, J. R. Delaney, and M. K. Tivey, 1987, Hydrothermal processes and massive sulfide deposits on the Juan de Fuca Ridge and other northeast Pacific spreading axes, this volume.

Kulm, L. D., and G. A. Fowler, 1974, Cenozoic sedimentary framework of the Gorda-Juan de Fuca plate and adjacent continental margin—A review, *in* R. H. Dott, Jr., and R. H. Shaver, eds., Modern and ancient geosynclinal sedimentation: SEPM Special Publication 19, p. 212–229.

Kulm, L. D., and K. F. Scheidegger, 1979, Quaternary sedimentation of the tectonically active Oregon continental slope, *in* L. J. Doyle and O. H. Pilkey, Jr., eds., Geology of continental slopes: SEPM Special Publication 27, p. 247–264.

Kulm, L. D., R. A. Prince, and P. D. Snavely, Jr., 1973, Site survey of the northern Oregon continental margin and the Astoria fan, *in* L. D. Kulm, R. von Huene, et al., Initial Reports of Deep Sea Drilling Project (Vol. 18): Washington, D.C., U.S. Government Printing Office, p. 979–986.

Kulm, L. D., R. von Huene, J. R. Duncan, J. C. Ingle, Jr., S. A. Kling, L. F. Musich, D. J. W. Piper, R. M. Pratt, H. J. Schrader, O. E. Weser, and S. W. Wise, Jr., 1973, Site 174: Initial reports of Deep-Sea Drilling Project (Vol. 18): Washington, D.C., U.S. Government Printing Office, p. 97–169.

Kvenvolden, K. A., 1987, Gas hydrate offshore Alaska and western continental United States, this volume.

Lichtman, G. S., and J. P. Eissen, 1983, Time and space constraints on the evolution of medium-rate spreading centers: Geology, v. 11, p. 592–595.

Lister, C. R. B., 1987, Geothermal resources, this volume.

MacPherson, B. A., 1978, Sedimentation and trapping mechanism in Upper Miocene Stevens and older turbidite fans of southeastern San Joaquin Valley, California: AAPG Bulletin, v. 62, p. 2243–2274.

McClain, K. J., J. S. Peper, and M. L. Holmes, 1984, Single channel seismic reflection records of western Washington margin and Cascadia Basin, *in* L. D. Kulm et al., eds., Western North American continental margin and adjacent ocean floor off Oregon and Washington: Marine Science International, Ocean Margin Drilling Program, Regional Atlas Series, Atlas 1, sheet 26.

McManus, D. A., 1964, Major bathymetric features near the coast of Oregon, Washington, and Vancouver Island: Northwest Science, v. 38, p. 65–82.

McManus, D. A., 1965, Blanco fracture zone, northeast Pacific Ocean: Marine Geology, v. 3, 429–455.

McManus, D. A., 1967, Physiography of Cobb and Gorda Rises, northeast Pacific Ocean: GSA Bulletin, v. 78, p. 527–546.

McManus, D. A., M. L. Holmes, B. Carson, and S. M. Barr, 1972, Late Quaternary tectonics, northern end of Juan de Fuca Ridge: Marine Geology, v. 23, p. 141–164.

Menard, H. W., 1955, Deep-sea channels, topography and sedimentation: AAPG Bulletin, v. 39, p. 236–255.

Morton, J. L., W. R. Normark, D. M. Mann, and D. H. Tompkins, 1987, Geology, evolution, and mineral potential of the Juan de Fuca Ridge, this volume.

Nelson, C. H., 1968, Marine geology of Astoria deep-sea fan: Corvallis, Oregon, Oregon State University unpublished Ph.D. dissertation, 287 p.

Nelson, C. H., 1976, Late Pleistocene and Holocene depositional trends, processes, and history of Astoria deep-sea fan, northeast Pacific: Marine Geology, v. 20, p. 129–173.

Nelson, C. H., 1983/1984, The Astoria fan—An elongate type fan: Geo-Marine Letters, v. 3, p. 65–70.

Nelson, C. H., and L. D. Kulm, 1973, Submarine fans and channels, *in* Turbidites and deep water sedimentation: SEPM, Pacfic Section Short Course, p. 39–78.

Nelson, C. H., and T. H. Nilsen, 1984, Modern and ancient deep-sea sedimentation: SEPM Short Course 14, 404 p.

Nelson, C. H., P. R. Carlson, J. V. Byrne, and T. R. Alpha, 1970, Development of the Astoria Canyon-fan physiography and comparison with similar systems: Marine Geology, v. 8, p. 259–291.

Nelson, C. H., L. D. Kulm, P. R. Carlson, and J. R. Duncan, 1968, Mazama ash in the northeastern Pacific: Science, v. 161, p. 47–49.

Nishimura, C., D. S. Wilson, and R. N. Hey, 1984, Pole of rotation analysis of present-day Juan de Fuca plate motion: Journal of Geophysical Research, v. 89, p. 10,283–10,290.

Normark, W. R., J. L. Morton, and J. R. Delaney, 1982, Geologic setting of massive sulfide deposits and hydrothermal vents along the southern Juan de Fuca Ridge: U.S. Geological Survey Open-File Report 82–200A, 21 p.

Parker, J. R., 1975, Lower Tertiary sand development in the central North Sea, *in* A. W. Woodland, ed., Petroleum and the continental shelf of northwest Europe: New York, John Wiley and Sons, v. 1 (Geology), p. 447–453.

Pacific Geoscience Centre, 1978, Juan de Fuca plate map: Relief: Sidney, British Columbia, scale 1:2,000,000.

Raff, A. D., and R. G. Mason, 1961, Magnetic survey off west coast of North America, 40°N latitude to 42°N latitude: GSA Bulletin, v. 72, p. 1267–1270.

Royse, C. F., 1967, Mazama ash from the continental slope off Washington: Northwest Science, v. 41, p. 103–109.

Scheidegger, K. F., L. D. Kulm, and D. J. W. Piper, 1973, Heavy mineralogy of unconsolidated sands in northeastern Pacific sediments—Leg 18, Deep Sea Drilling Project, *in* L. D. Kulm, R. von Huene, et al., Initial Reports of Deep Sea Drilling Project (Vol. 18): Washington, D.C., U.S. Government Printing Office, p. 877–888.

Seeley, D. R., 1977, The significance of landward vergence and oblique structural trends on trench inner slopes, *in* M. Talwani and W. C. Pitman, III, eds., Island arcs and deep-sea trenches and back-arc basins: American Geophysical Union Maurice Ewing Series 1, p. 187–198.

Shor, G. G., Jr., P. Dehlinger, H. K. Kirk, and W. S. French, 1968, Seismic refraction studies off Oregon and northern California: Journal of Geophysical Research, v. 73, p. 2175–2194.

Silver, E. A., 1971, Small plate tectonics in the northeastern Pacific: GSA Bulletin, v. 82, p. 3491–3496.

Snavely, P. D., Jr., 1987, Tertiary geologic framework, neotectonics, and petroleum potential of the Oregon-Washington continental margin, this volume.

Snavely, P. D., Jr., and H. C. Wagner, 1981, Geologic cross section across the continental margin off Cape Flattery, Washington, and Vancouver Island, British Columbia: U.S. Geological Survey Open-File Report 81–978, 5 p.

Snavely, P. D., Jr., J. E. Pearl, and D. L. Lander, 1977, Interim report on petroleum resource potential and geologic hazards in outer continental shelf—Oregon and Washington Tertiary province: U.S. Geological Survey Open-File Report 77–282, 64 p.

Snavely, P. D., Jr., H. C. Wagner, and D. L. Lander, 1980, Geologic cross section of the central Oregon continental margin: GSA Map and Chart Series MC–28JH, 8 p., 1 map sheet, scale 1:250,000.

Stokke, P. R., 1976, Holocene and present day sedimentation on Nitinat deep-sea fan, northeast Pacific Ocean: Bethlehem, Pennsylvania, Lehigh University unpublished Ph.D. dissertation, 229 p.

Stokke, P. R., B. Carson, and E. T. Baker, 1977, Comparison of bottom nepheloid layer and late Holocene deposition on the Nitinat fan—Implications for lutite dispersal and deposition: GSA Bulletin, v. 88, p. 1586–1592.

Sullwold, H. H., Jr., 1961, Turbidites in oil exploration, *in* J. A.

Peterson and J. C. Osmond, eds., Geometry of sandstone bodies: Tulsa, Oklahoma, American Association of Petroleum Geologists, p. 63–81.

Thomas, A. N., P. J. Walmsley, and D. A. L. Jenkins, 1974, Forties field, North Sea: AAPG Bulletin, v. 58, p. 396–406.

Tiffin, D. L., B. E. B. Cameron, and J. W. Murray, 1972, Tectonics and depositional history of the continental margin off Vancouver Island, British Columbia: Canadian Journal of Earth Science, v. 9, p. 280–296.

Vine, F. J., 1968, Magnetic anomalies associated with mid-ocean ridges, *in* R. A. Phinney, ed., The history of the earth's crust: Princeton, New Jersey, Princeton University Press, p. 73–89.

von Huene, R., and L. D. Kulm, 1973, Tectonic summary of Leg 18, *in* L. D. Kulm, R. von Huene, et al., Initial reports of Deep Sea Drilling Project (Vol. 18): Washington, D.C., U.S. Government Printing Office, p. 961–976.

von Huene, R., D. J. W. Piper, and J. R. Duncan, 1973, Measurement of porosity in sediments of the lower continental margin, deep-sea fan, the Aleutian Trench, and the Alaskan Abyssal Plain, *in* L. D. Kulm, R. von Huene, et al., Initial reports of Deep Sea Drilling Project (Vol. 18): Washington, D.C., U.S. Government Printing Office, p. 889–898.

Wilde, P., T. E. Chase, M. L. Holmes, W. R. Normark, and J. A. Thomas, 1977, Oceanographic data off Washington 46° to 40° North including the Nitinat deep-sea fan: Berkeley, California, University of California, Lawrence Berkeley Laboratory, LBL Publication 233.

Wilde, P., T. E. Chase, M. L. Holmes, W. R. Normark, J. A. Thomas, D. S. McCulloch, P. R. Carlson, L. D. Kulm, and J. D. Young, 1979, Oceanographic data off Oregon 43° to 46° North including the Astoria deep-sea fan: Berkeley, California, University of California, Lawrence Berkeley Laboratory, LBL Publication 253.

Wilson, J. T., 1965, Transform faults, oceanic ridges, and magnetic anomalies southwest of Vancouver Island: Science, v. 150, p. 482–485.

Wilson, D. S., R. N. Hey, and C. Nishimura, 1984, Propagation as a mechanism of reorientation of the Juan de Fuca Ridge: Journal of Geophysical Research, v. 89, p. 9215–9225.

CHAPTER TWENTY-TWO

GEOLOGY, EVOLUTION, AND MINERAL POTENTIAL OF THE JUAN DE FUCA RIDGE

Janet L. Morton
William R. Normark
Dennis M. Mann

U.S. Geological Survey
Menlo Park, California

Donald H. Tompkins

Nekton, Inc.
San Diego, California

ABSTRACT

The Juan de Fuca Ridge is a medium-rate spreading center (total opening rate about 60 mm a^{-1}) located in the northeast Pacific Ocean. The ridge is bounded by the Sovanco fracture zone to the north and the Blanco fracture zone to the south. Bottom-water temperature anomalies indicate that hydrothermal vents occur at bathymetric rises along each of four ridge segments, and active vents have been mapped and sampled at three locations—southern Juan de Fuca Ridge, Axial Seamount, and Endeavour Ridge segment.

Factors affecting the location of hydrothermal vents seem to be the presence of fissures and faults as channelways for migrating fluids and the presence of a shallow heat source. At the southern Juan de Fuca Ridge, hydrothermal vents occur along a cleft in the axial valley floor—the source of the most recent lava flows at the axis. Multichannel seismic-reflection profiles across the axis indicate the presence of a crustal magma chamber 1 to 2 km wide and 2.3 to 2.5 km beneath the sea floor.

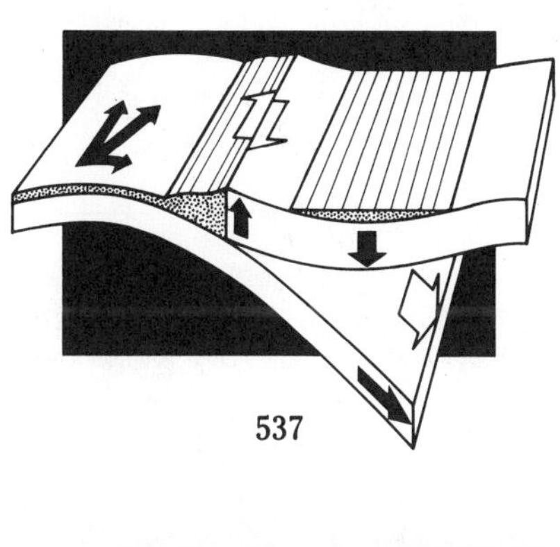

INTRODUCTION

The Juan de Fuca Ridge extends approximately 500 km, from the Blanco fracture zone to the Sovanco fracture zone in the northeast Pacific Ocean west of Oregon and Washington (Figure 1). The ridge forms the boundary between the Juan de Fuca plate and the Pacific plate and has a total opening rate of about 60 mm a^{-1} (Atwater, 1970; Riddihough, 1977), making it a medium-rate spreading axis. The Juan de Fuca plate is a remnant of the Farallon plate, most of which has been subducted beneath North America (Atwater, 1970). Other spreading centers in the northeast Pacific Ocean include the Gorda Ridge, offset from the Juan de Fuca Ridge by the Blanco fracture zone (see Embley et al., this volume; Clague and Holmes, this volume) and the Explorer Ridge, offset from the Juan de Fuca Ridge by the Sovanco fracture

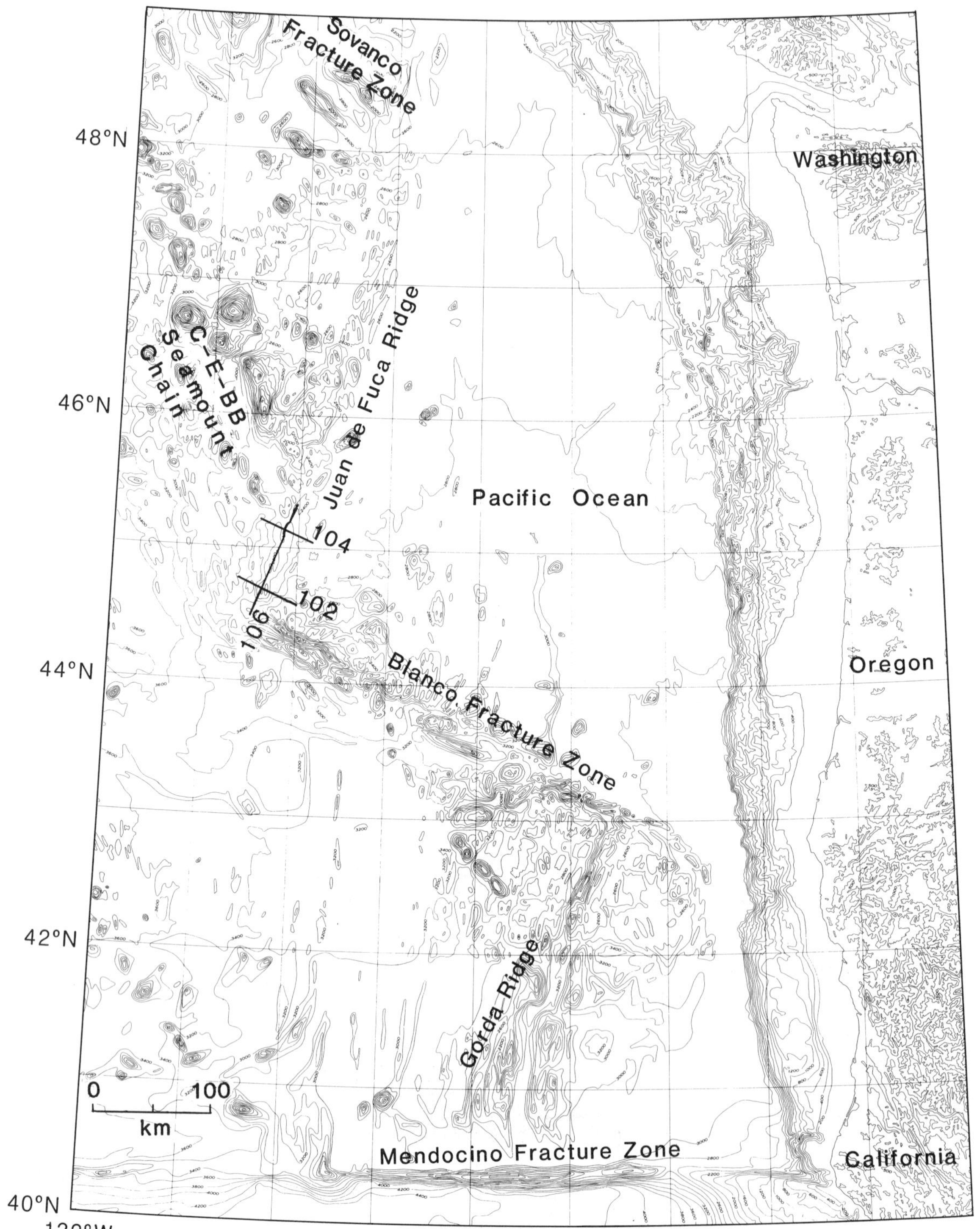

Figure 1. Bathymetry of the Juan de Fuca Ridge and surrounding regions (Pacific Geoscience Centre, 1978). Locations of the multichannel seismic-reflection lines discussed in this chapter are shown. C-E-BB denotes the Cobb-Eikelberg-Brown Bear seamount chain. Bathymetric contour interval 200 m.

zone (Riddihough, 1977).

The magnetic anomalies associated with the Juan de Fuca Ridge have been closely surveyed (Raff and Mason, 1961; National Oceanic and Atmospheric Administration, 1974) and have formed the basis of many models of plate interaction in the region (Vine, 1966; Elvers et al., 1973; Riddihough, 1977; Carlson, 1981). The magnetic anomalies indicate that the orientation of the spreading ridge has changed within the past 9 m.y. from northward to the present orientation of about N20°E (Menard and Atwater, 1968; Riddihough, 1977; Carlson, 1981). Wilson, Hey, and Nishimura (1984) have postulated that the change in spreading direction occurred by rift propagation (Hey, 1977; Hey and Wilson, 1982). One of these propagating rifts, the Cobb offset, displaces the ridge axis 30 km near lat. 47°40′N (Figure 2A; Johnson et al., 1983).

Recent detailed investigations of the axis of the Juan de Fuca Ridge using Seabeam swath bathymetry and conventional echosounding, magnetic profiling, dredging, and SeaMARC I and II side-scan imaging indicate that the axis can be divided into several segments (Delaney, Johnson, and Karsten, 1981; Embley et al., 1983; Crane et al., 1985; Kappel and Ryan, in press) (Figure 2A). The southern segment extends northward from the Blanco fracture zone to approximately lat. 45°40′N and displays a marked cross-axis symmetry. Near lat. 45°03′N the axis of the southern segment is offset about 3 km left laterally. The ridge segment south of the lat. 45°03′N offset has been named Rift A by Embley et al. (1983) or Cleft Ridge segment by Kappel and Ryan (in press), and the segment north of the offset has been named Rift B. The central segment of the ridge is influenced by the presence of a large seamount (Axial Seamount) and seems to be offset 10 to 20 km to the west between lat. 45°40′N and lat. 46°30′N (Delaney, Johnson, and Karsten, 1981; Embley et al., 1983). The northern symmetrical or split-volcano ridge segment extends north of Axial Seamount to the Cobb offset at lat. 47°50′N (Malahoff, Embley, and Hammond, 1983; Kappel and Ryan, in press). From the Cobb offset to the intersection with the Sovanco fracture zone, the spreading axis is referred to as the Endeavour Ridge segment (Hammond et al., 1983).

Axial depths along the Juan de Fuca Ridge generally range from 2200 to 2700 m, except at Axial Seamount, where the ridge rises to within 1400 m of the surface (Figure 2B). As seen in a simplified axial bathymetric profile (Figure 2B), each ridge segment, except Rift B, contains a prominent dome. An axial valley lies along most of the ridge axis (Figure 3).

Indications of hydrothermal activity have been observed along each of the four ridge segments. High-temperature venting and associated massive sulfide deposits have been described and sampled on the southern Juan de Fuca Ridge (Rift A; Normark et al., 1983), at Axial Seamount (Canadian American Seamount Expedition, 1985), and on the Endeavour Ridge segment (Merge Group, 1984; Hammond et al., 1984). In this chapter, we discuss differences in axial structure and volcanic processes along the ridge axis and their relationship to hydrothermal processes, using examples from the southern Juan de Fuca Ridge—an area that has been the focus of a 6-year study by the U.S. Geological Survey. Koski et al. (this volume) review the sulfide-mineral deposits along the Juan de Fuca Ridge.

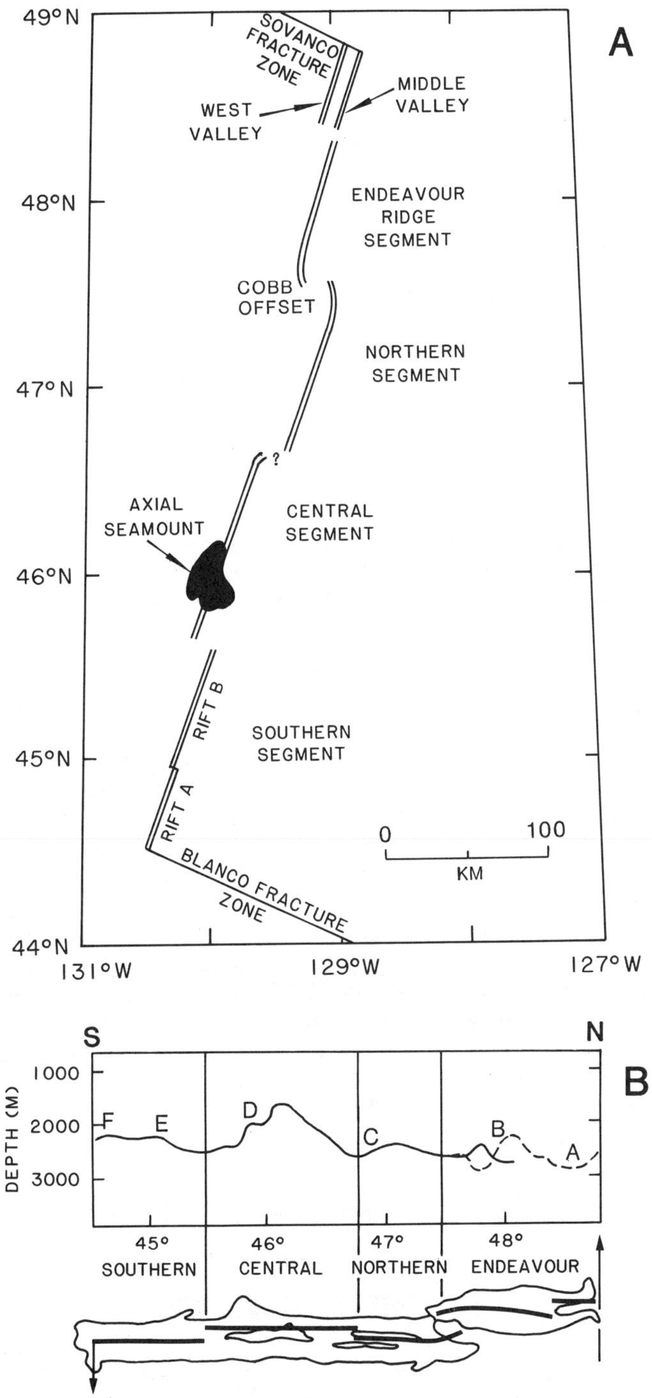

Figure 2. A, Stick diagram of the Juan de Fuca Ridge showing location of the various ridge segments discussed in this chapter. B, Longitudinal profile and schematic map view of segments of the Juan de Fuca Ridge. Vertical exaggeration is 46:1. Adapted from Figure 3 in Eaby, Clague, and Delaney (1984). Letters A–F along profile indicate location of bathymetric cross-sections shown in Figure 3.

RIDGE CRUSTAL STRUCTURE (SOUTHERN SEGMENT)

Two factors affecting the occurrence of hydrothermal systems at spreading axes are the presence or absence of a shallow heat source (such as an axial magma chamber) and the existence of faults or fissures that provide pathways for migrating fluids. Multichannel seismic-reflection (MCS) data over spreading centers have shown subbottom reflections from the top of axial magma chambers beneath the East Pacific Rise at lat. 9°N (Herron et al., 1978; Hale, Morton, and Sleep, 1982) and beneath the back-arc spreading ridge in the Lau Basin (Morton and Sleep, 1985). In September 1981, MCS data were collected across the Juan de Fuca Ridge by the U.S. Geological Survey research vessel *S.P. Lee* to determine whether a reflection could be detected from a magma chamber beneath the ridge axis. Two profiles perpendicular to the ridge axis at lat. 44°40′N (line 102; Rift A) and lat. 45°05′N (line 104; Rift B) and one profile along the ridge axis (106) were recorded (Figure 1). The southern profile (102) crosses the ridge at a site of known recent volcanic activity and active hydrothermal venting (Normark et al., 1983).

MCS data for lines 102 and 104 are shown in Figure 4. Rift A (line 102) shows a high degree of symmetry about the ridge axis, out to distances of as much as 30 km from the axis. The ridge has a small central valley about 80 to 100 m deep and about 1 km wide at the base of the inner valley walls. The inner valley walls, including a low terrace about 30 m above the valley floor, are formed by steep normal faults. Rift B (line 104) is less symmetrical, is deeper, and has a wider axial valley (about 5 km) than Rift A. A low volcanic ridge in the center of the axial valley becomes more pronounced to the north along Rift B.

A low-amplitude reflector appears 0.9 to 1.1 sec beneath the ridge axis (arrow B on Figures 4A, 4B). This reflector has been enhanced in processed MCS sections (Figure 5) by using the procedures of Hale, Morton, and Sleep (1982) and Morton and Sleep (1985). Additional velocity information for the reprocessing was obtained from sonobuoy seismic-refraction data at the southern Juan de Fuca Ridge crest and from ocean-bottom seismometer refraction data from the northern Juan de Fuca Ridge axis (Davis, Lister, and Lewis, 1976).

Reflection B is seen more clearly on the reprocessed lines, centered beneath the ridge axis (Figure 5). Away from the ridge axis, no clear reflections are seen below the sea floor. We used ray-tracing techniques to model the steeply dipping diffraction hyperbolas at the ends of reflection B. This modeling suggests that event B represents a structure 2.3 to 2.5 km below the sea floor with an across-axis full width of 1 to 2 km rather than sidescatter from sea-floor topography (see Appendix). Because reflection B appears only beneath the ridge axis and at a depth similar to that of seismically detected magma chambers at other ridge axes, we interpret it as representing the roof of an axial magma chamber.

During 1981, a seismic-refraction experiment was conducted at the ridge axis between lines 102 and 104 at lat. 44°48′N (Jung and Lewis, 1983). Both ocean-bottom seismometers and a deep-towed hydrophone array were used as receivers. No abrupt time delays across the ridge axis, indicative of a large (approximately 10 km wide) crustal low-velocity zone, were detected (Lewis, Johnson, and Jung, 1983). However, Jung and Lewis (1983) did observe attenuation of energy across the ridge axis in the data recorded by the towed array. This attenuation could have resulted from a small low-velocity zone, too narrow to produce a noticeable time delay in the refraction records.

The possibility that event B resulted from a high-velocity zone beneath the ridge, rather than a low-velocity zone, cannot be ruled out on the basis of the reflection data alone.

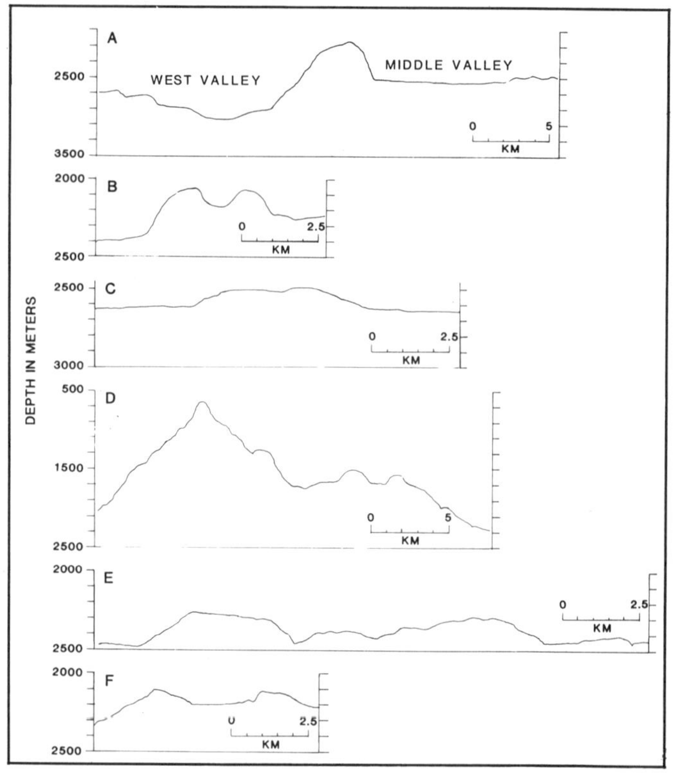

Figure 3. Bathymetric cross-sections perpendicular to each segment of the Juan de Fuca Ridge (see Figure 2B for location). Vertical exaggeration is 5:1. A, West Valley and Middle Valley on northern Endeavour Ridge segment. B, Southern Endeavour Ridge segment. C, Northern segment. D, Axial Seamount. E, Southern segment, Rift B. F, Southern segment, Rift A.

Lister (1983) has suggested that axial magma chambers may be short lived as a result of rapid cooling by hydrothermal circulation. Possibly event B represents the top of a high-velocity solidified magma chamber. Once a magma chamber has solidified, the roof becomes the boundary between oceanic layer 2 and layer 3 and should appear seismically the same as that boundary away from the ridge. Because event B appears only beneath the ridge axis and not beneath the ridge flanks on either line 102 or 104, we prefer the interpretation that reflection B represents the top of a partially molten magma chamber rather than a solidified chamber. The low amplitude of the reflection suggests that the velocity contrast across the interface is slight, that the interface is gradational, or both.

VOLCANIC AND STRUCTURAL DEVELOPMENT OF AXIAL VALLEY (SOUTHERN SEGMENT)

The volcanic and structural processes acting along the southern Juan de Fuca Ridge axis have been investigated using SeaMARC II, bottom photography, sampling, and observations from the *Alvin* submersible.

The linearity and symmetry of bathymetric features in the southernmost segment of the Juan de Fuca axial valley are clearly depicted (Figure 6A) by SeaMARC II imagery. The

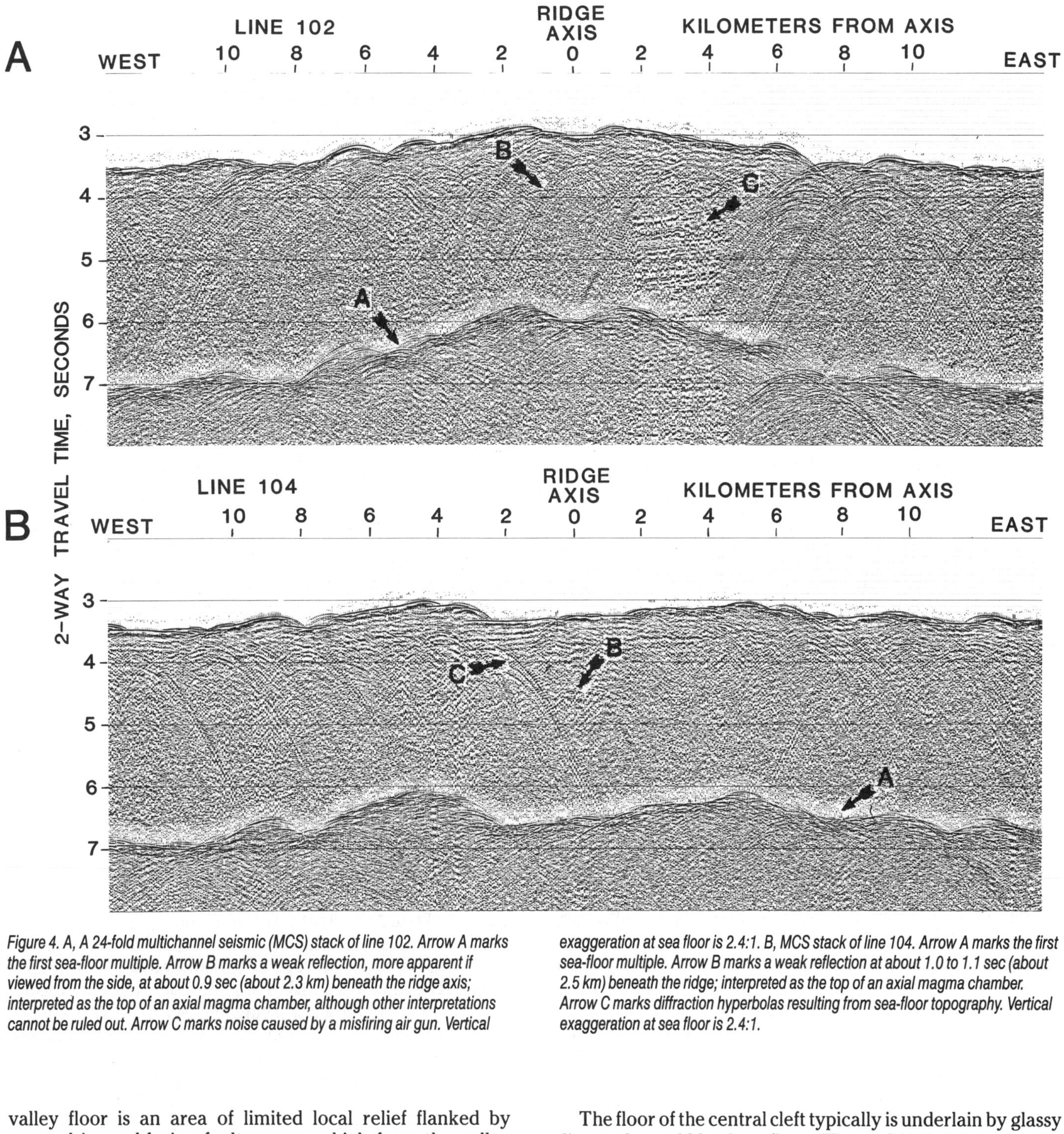

Figure 4. A, A 24-fold multichannel seismic (MCS) stack of line 102. Arrow A marks the first sea-floor multiple. Arrow B marks a weak reflection, more apparent if viewed from the side, at about 0.9 sec (about 2.3 km) beneath the ridge axis; interpreted as the top of an axial magma chamber, although other interpretations cannot be ruled out. Arrow C marks noise caused by a misfiring air gun. Vertical exaggeration at sea floor is 2.4:1. B, MCS stack of line 104. Arrow A marks the first sea-floor multiple. Arrow B marks a weak reflection at about 1.0 to 1.1 sec (about 2.5 km) beneath the ridge; interpreted as the top of an axial magma chamber. Arrow C marks diffraction hyperbolas resulting from sea-floor topography. Vertical exaggeration at sea floor is 2.4:1.

valley floor is an area of limited local relief flanked by stepped inward-facing fault scarps, which form the valley walls (Figure 6B). The valley floor narrows slightly to the north and is split lengthwise by a central depression or cleft (Figure 7). The cleft is not a single symmetrical notch but widens locally and comprises overlapping strands with small (< 30 m) offsets (Figures 6A, 6B). Between the central cleft and the inner valley walls, the axial valley floor is not broken by fissures or faults and is characterized by strong backscatter, suggesting that the floor is formed by fresh, unsedimented lava flows. A few less reflective areas of the axial valley floor, marked as backscatter contrast on Figure 6B, probably represent remnants of older, partially sedimented flows, or possibly a change in surface flow morphology.

The floor of the central cleft typically is underlain by glassy lineated or rubbly sheet flows (Figure 8A) that have locally subsided near the cleft walls. Observations from the submersible *Alvin* indicate that the walls of the cleft are formed in places by a single nearly-vertical scarp and elsewhere by a series of small stepped scarps and steep slopes. At a few localities, truncated flow units are exposed in the walls of the cleft. Elsewhere, lava-drainback features on the walls suggest that the cleft is the source of the recent lava flows that flood the axial valley (Figure 8B) (Normark et al., 1984; USGS Juan de Fuca Study Group, 1985).

The central cleft is flanked by a region of subsided fresh lava flows. Basaltic pillars are common along the boundary between the subsided flows and the cleft, and these pillars

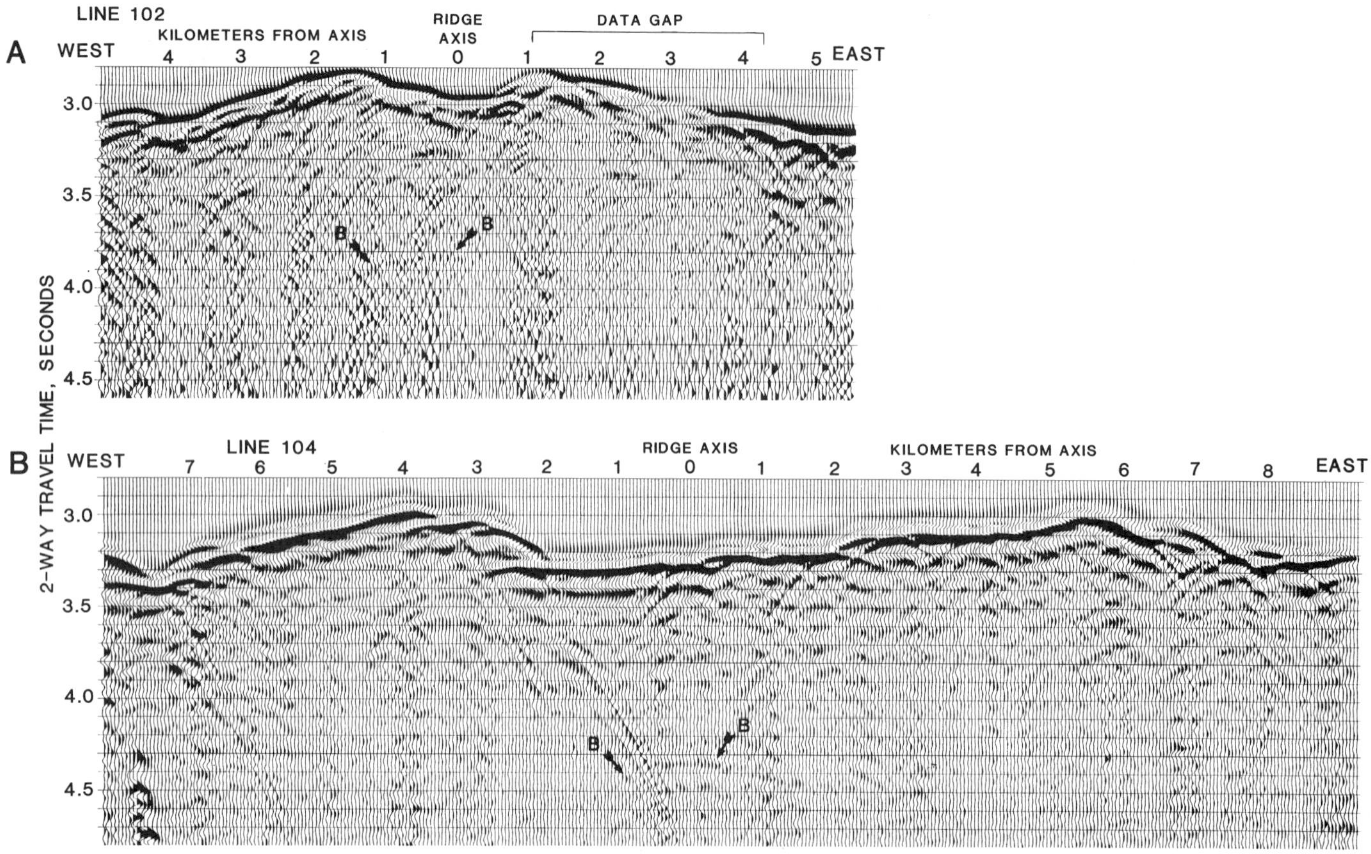

Figure 5. A, A 24-fold reprocessed MCS stack of the axial part of line 102. Arrow B marks the reflection at 0.9 sec beneath the ridge axis, which we interpret as reflecting from the top of an axial magma chamber. The amplitude of the reflection is low, suggesting that the velocity and density contrast across the interface is slight, that the interface is gradational, or both. The noisy shots in Figure 4A (arrow C) have been removed, leaving a 3-km segment east of the ridge axis that contains very few shots (marked as a data gap). Vertical exaggeration at sea floor is 3.3:1. B, Stacked time section of the axial part of line 104. Arrow B marks the reflection at about 1 sec beneath the axis. Vertical exaggeration at sea floor is 3.3:1.

also occur locally within the cleft. The dominant lava-flow morphology of the axial valley floor is lobate sheet flows (Lichtman et al., 1983). Toward the center of the axial valley, the lobate flows contain numerous collapse pits (Figures 7, 8C). Smooth, unpitted lobate sheet flows flank the region of pitted sheet flows (Figures 7, 8D). All flows within the axial valley seem fresh, but small variations in sediment cover indicate that the youngest flows are confined to the center of the valley (Figure 7; Lichtman et al., 1983). Pillow flows are rare within the axial valley, only occurring locally within the central cleft, but they are commonly exposed on the inner valley walls (Figure 7).

RIDGE PROCESSES AND MINERALIZATION

The morphologic character of the ridge axis differs from segment to segment (Figure 3; Kappel and Ryan, in press). The northern and southern segments and the southern part of the Endeavour Ridge segment have cross-axis symmetry and an axial valley. The width of the axial valley ranges from < 1 km along the northern segment (Malahoff, Embley, and Hammond, 1983; Kappel and Ryan, in press) to 6 to 14 km along Rift B of the southern segment (Embley et al., 1983; Kappel and Ryan, in press). In places the axial valley floor is dominated by extrusive volcanism, such as along Rift A of the southern segment and along the northern segment (Malahoff, Embley, and Hammond, 1983). Extensional fissures and faults dominate the valley floor, and evidence for recent volcanism is lacking in other places, such as the southern Endeavour Ridge segment (Davis et al., 1984; Merge Group, 1984). The ridges flanking the axial valley are formed by predominantly inward-facing fault scarps. Some workers have postulated that the axial valley along these ridge segments represents the collapsed summit of a volcanic constructional ridge (Kappel and Ryan, in press). Alternatively, the ridge crests could have formed by uplift along axial-facing normal faults.

Two sections of the Juan de Fuca Ridge display a morphology different from the symmetrical rifted axis described above—the central segment and the northern part of the Endeavour Ridge segment. The Cobb-Eikelberg-Brown Bear Seamount chain intersects the Juan de Fuca Ridge at the central offset segment (Delaney, Johnson, and Karsten, 1981). The exact location of the spreading axis along this segment is not well defined. Voluminous volcanism from Axial Seamount, the most recent member of the Cobb-Eikelberg-Brown Bear chain, seems to dominate the morphology of the ridge axis along at least part of the central segment, and an axial valley may be absent (Delaney, Johnson, and Karsten, 1981).

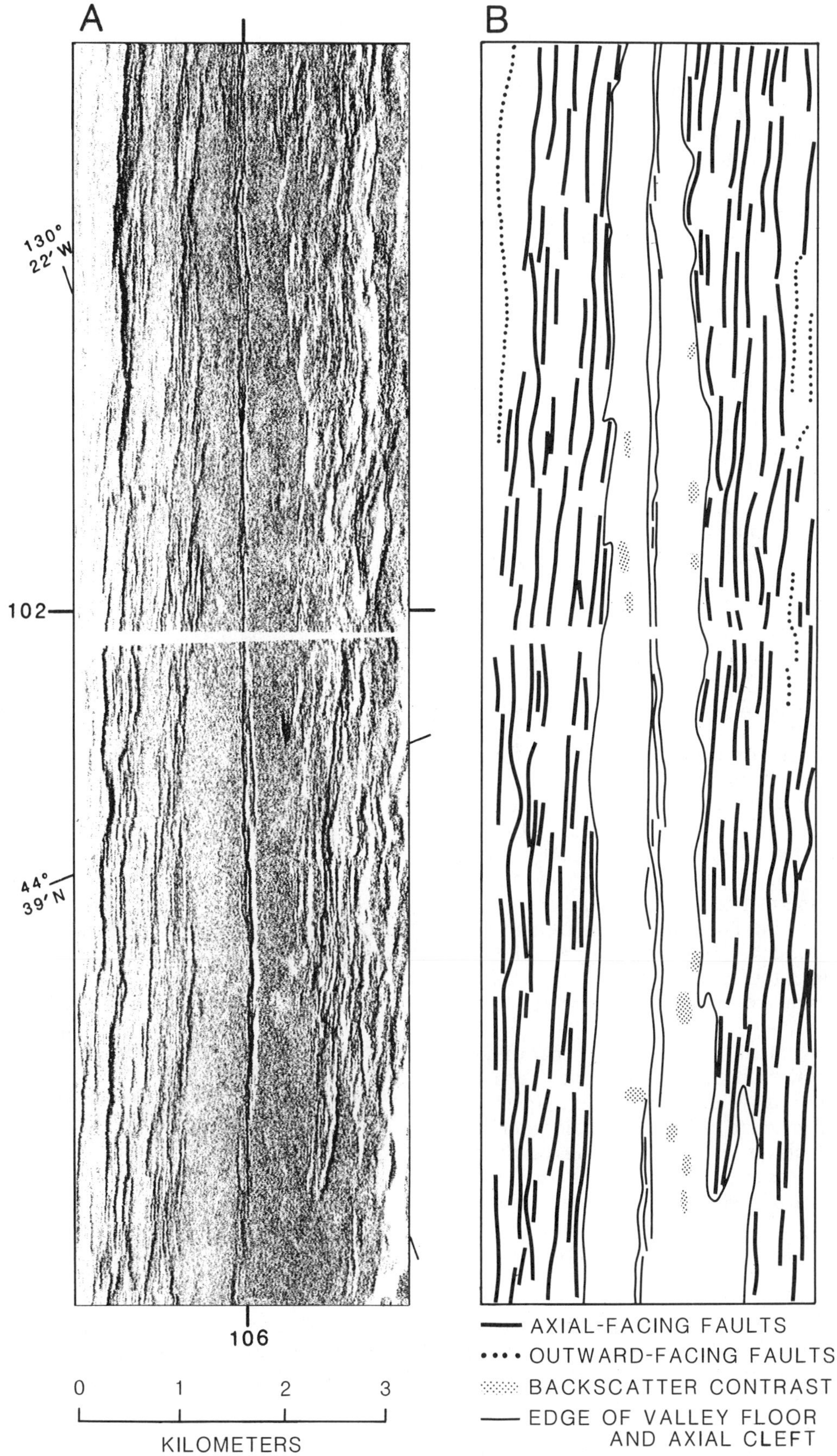

Figure 6. A, SeaMARC II (Hussong and Fryer, 1983) image of a 12-km segment of the southern Juan de Fuca Ridge axis. Image frame is aligned with ridge axis. Bold lines outside frame give approximate location of MCS profiles 102 and 106. See intersection of MCS profiles in Figure 1 for image location. B, Pictographic representation of SeaMARC II image. Note that a clear image of the axial cleft is visible on the SeaMARC II record, but resolution of features within the cleft is limited.

The axis at the northern end of the Endeavour Ridge segment comprises two parallel rift valleys, Middle Valley and West Valley. Active spreading seems to have recently shifted from Middle Valley to West Valley (Davis and Lister, 1977). West Valley is offset left laterally from the spreading axis of the southern part of the Endeavour Ridge segment (Hammond et al., 1984). Middle Valley is filled with as much as 1.5 km of turbidite sediment and is characterized by regions of high heat flow, indicative of igneous intrusions or circulation of hydrothermal fluids in the sediment (Villinger and Davis, 1984).

Near-bottom water-temperature measurements obtained with a deep-towed thermistor chain indicate that five regions of geothermal activity exist along the Juan de Fuca Ridge between the Blanco fracture zone and lat. 48°N (Crane et al., 1985). Each of these possible geothermal regions seems to coincide with a bathymetric rise along the ridge axis (Figure 2). A sixth region of hydrothermal activity is indicated in Middle Valley by elevated bottom-water temperatures (Villinger and Davis, 1984). Active high-temperature vents and associated massive sulfide deposits have been described and sampled at three of these localities—southern Juan de Fuca Ridge (Normark et al., 1983; USGS Juan de Fuca Study Group, 1985), Axial Seamount (Canadian American Seamount Expedition, 1985), and southern Endeavour Ridge segment (Merge Group, 1984; Hammond et al., 1984). At a fourth site along the northern segment, bottom photographs of vent-related biologic communities confirm the hydrothermal discharge (Malahoff, Embley, and Hammond, 1983). In this section, we discuss the relationship of the hydrothermal vents to the volcanic and structural processes.

At the southern Juan de Fuca Ridge, all seven known hydrothermal vent sites occur within the central cleft except for one small site photographed near the base of the eastern valley wall. Extensive photographic coverage revealed no indications of hydrothermal venting or mineral deposits on the unfissured axial-valley floor outside the central cleft. Features forming conduits for the hydrothermal fluids include fractures along the base of the walls of the central cleft, basaltic pillars, and hollow lobate flows (Figures 8B, 8E, 8F).

The hydrothermal vents at Axial Seamount occur within a flat-floored caldera at the summit of the seamount (Canadian American Seamount Expedition, 1985). The vents are aligned along a major fissure on the caldera floor and are associated with fresh glassy basalt. On the Endeavour Ridge segment, sulfide deposits around active and inactive vents were found only at the western edge of the axial valley and on the western valley wall. These vents are not associated with fresh volcanic rocks (Merge Group, 1984; Hammond et al., 1984).

The primary factors controlling the location of hydrothermal vents and associated sulfide deposits along the Juan de Fuca Ridge seem to be the presence of a shallow heat source and faults and fissures to provide circulation channels for the mineral-rich fluids. Each of the known hydrothermal-vent regions occurs on a bathymetric rise along the ridge axis. A similar relationship between water depth over the ridge axis and the occurrence of hydrothermal vents has been noted along the East Pacific Rise (Francheteau and Ballard, 1983). These shallow points could conceivably represent thermally elevated regions, resulting from a larger or shallower magma chamber, or a young not-yet-subsided edifice. At this time, MCS data are insufficient to establish whether axial magma chambers underlie all areas where thermal bulges are suspected. Sites of massive volcanism, such as at Axial Seamount, could conceivably sustain longer-lived hydrothermal systems and thereby result in larger mineral deposits.

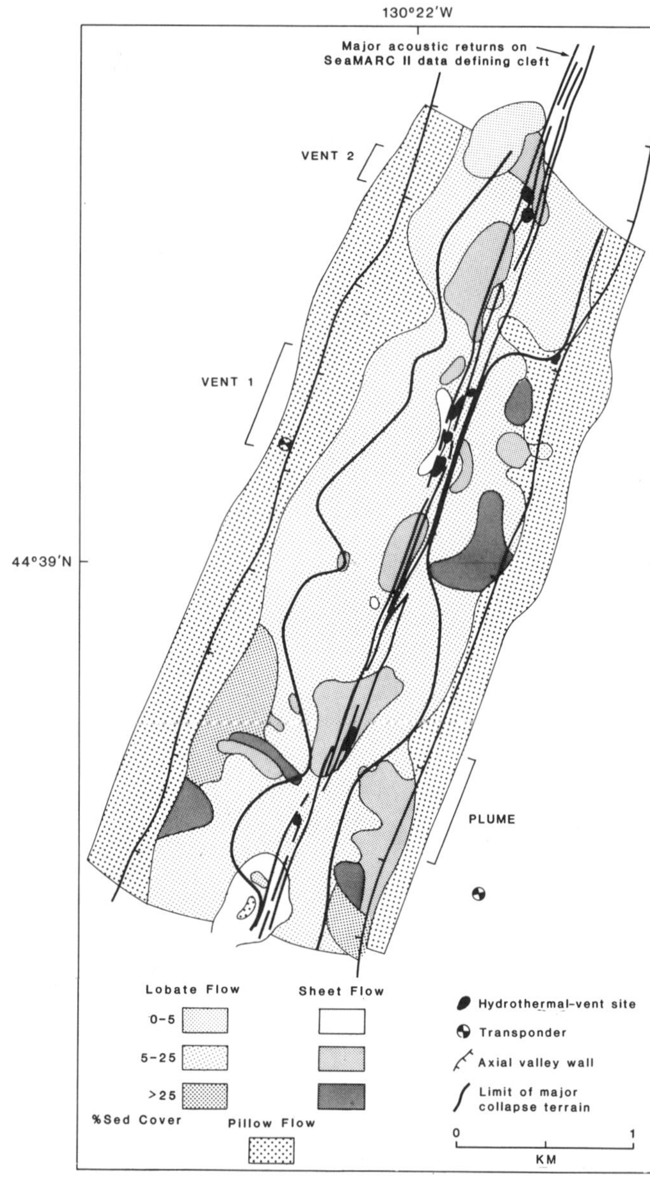

Figure 7. Geologic map of the axial valley of the southern Juan de Fuca Ridge based on sea-floor photographs obtained in 1981 (modified from Lichtman et al., 1983). Details of the central cleft were added from SeaMARC II imagery. Intensive sea-floor photography conducted in 1983 generally confirmed the pattern of lava-flow morphology depicted here and provided additional information on the nature of the central cleft. Latitude and longitude coordinates same as in Figure 6A.

ACKNOWLEDGMENTS

We thank the captain and crew of the *S.P. Lee* for their diligent support in collection of the seismic-reflection data. Jon Claerbout provided the computation facilities for processing the seismic data and Norm Sleep provided the ray-tracing

Figure 8. Selected photographs of lava-flow morphologies and hydrothermal deposits on the southern Juan de Fuca Ridge taken from Alvin (A, B, D, E, F) and the USGS towed camera (C). Frame width of each photograph is approximately 3 m. A, Glassy, lineated sheet flows on the floor of the central cleft. Photograph by J. Slack. B, Scalloped rim of the central cleft showing lava drainback selvages on vertical wall and sulfide chimneys constructed on a hollow lobate flow. C, Collapse pit in lobate sheet flow. D, Smooth, unpitted lobate sheet flows of the axial valley floor. Photograph by W. Normark. E, Massive sulfide deposits on wall of central cleft. Photograph by R. Koski. F, Sulfide chimney atop basaltic pillar within central cleft.

routines. Critical reviews by Alan Cooper and George Moore improved the manuscript.

REFERENCES CITED

Atwater, T., 1970, Implications of plate tectonics for the Cenozoic tectonic evolution of western North America: GSA Bulletin, v. 81, p. 3513–3536.

Canadian American Seamount Expedition, 1985, Hydrothermal vents on an axis seamount of the Juan de Fuca Ridge: Nature, v. 313, p. 212–214.

Carlson, R. L., 1981, Late Cenozoic rotation of the Juan de Fuca Ridge and the Gorda Rise—A case study: Tectonophysics, v. 77, p. 171–188.

Clague, D. A., and M. L. Holmes, 1987, Geology, petrology, and mineral potential of the Gorda Ridge, this volume.
Crane, K., F. Aikman, III, R. Embley, S. Hammond, A. Malahoff, and J. Lupton, 1985, The distribution of geothermal fields on the Juan de Fuca Ridge: Journal of Geophysical Research, v. 90, p. 727–744.
Davis, E. E., and C. R. B. Lister, 1977, Tectonic structures on the Juan de Fuca Ridge: GSA Bulletin, v. 88, p. 346–363.
Davis, E. E., C. R. B. Lister, and B. T. R. Lewis, 1976, Seismic structure of the Juan de Fuca Ridge—Ocean bottom seismometer results from the median valley: Journal of Geophysical Research, v. 81, p. 3541–3555.
Davis, E. E., R. G. Currie, R. P. Riddihough, W. B. F. Ryan, K. Kastens, D. M. Hussong, S. R. Hammond, and A. Malahoff, 1984, Geologic mapping of the northern Juan de Fuca and Explorer Ridges using SeaMARC I, SeaMARC II, and Seabeam acoustic imaging: EOS (American Geophysical Union Transactions), v. 65, p. 1110.
Delaney, J. R., H. P. Johnson, and J. L. Karsten, 1981, The Juan de Fuca Ridge- hot spot-propagating rift system—New tectonic, geochemical, and magnetic data: Journal of Geophysical Research, v. 86, p. 11747–11750.
Eaby, J., D. A. Clague, and J. R. Delaney, 1984, Sr isotope variations along the Juan de Fuca Ridge: Journal of Geophysical Research, v. 89, p. 7883–7890.
Elvers, D., S. P. Srivastasva, K. Potter, J. Morley, and D. Seidel, 1973, Asymmetric spreading across the Juan de Fuca and Gorda Rises as obtained from a detailed magnetic survey: Earth and Planetary Science Letters, v. 20, p. 211–219.
Embley, R. W., L. D. Kulm, G. Massoth, D. Abbott, and M. Holmes, 1987, Morphology, structure, and resource potential of the Blanco transform fault zone, this volume.
Embley, R. W., S. Hammond, A. Malahoff, W. B. F. Ryan, K. Crane, and E. Kappel, 1983, Rifts of the southern Juan de Fuca Ridge: EOS (American Geophysical Union Transactions), v. 64, p. 853.
Francheteau, J., and R. D. Ballard, 1983, The East Pacific Rise near 21°N, 13°N, and 20°S—Inferences for along-strike variability of axial processes of the mid-ocean ridge: Earth and Planetary Science Letters, v. 64, p. 93–116.
Hale, L. D., C. J. Morton, and N. H. Sleep, 1982, Reinterpretation of seismic reflection data over the East Pacific Rise: Journal of Geophysical Research, v. 87, p. 7707–7717.
Hammond, S. R., J. Karsten, E. Davis, R. Currie, and A. Malahoff, 1983, Bathymetric and hydrothermal characterization of the northernmost Juan de Fuca Ridge: EOS (American Geophysical Union Transactions), v. 64, p. 853.
Hammond, S. R., J. S. Lee, A. Malahoff, R. Feeley, R. W. Embley, and J. Franklin, 1984, Discovery of high-temperature hydrothermal venting on the Endeavour segment of the Juan de Fuca Ridge: EOS (American Geophysical Union Transactions), v. 65, p. 1111.
Herron, T. J., W. J. Ludwig, P. L. Stoffa, T. K. Kan, and P. Buhl, 1978, Structure of the East Pacific Rise from multichannel seismic reflection data: Journal of Geophysical Research, v. 85, p. 798–804.
Hey, R. N., 1977, A new class of "pseudofaults" and their bearing on plate tectonics: Earth and Planetary Science Letters, v. 37, p. 321–325.
Hey, R. N., and D. S. Wilson, 1982, Propagating rift explanation for the tectonic evolution of the northeast Pacific—The pseudomovie: Earth and Planetary Science Letters, v. 58, p. 167–188.
Hussong, D. M., and P. Fryer, 1983, Back-arc seamounts and the SeaMARC II sea floor mapping system: EOS (American Geophysical Union Transactions), v. 64, p. 627–632.
Johnson, H. P., J. L. Karsten, J. R. Delaney, E. E. Davis, R. G. Currie, and R. L. Chase, 1983, A detailed study of the Cobb offset of the Juan de Fuca Ridge—Evolution of a propagating rift: Journal of Geophysical Research, v. 88, p. 2297–2315.
Jung, H., and B. T. R. Lewis, 1983, Seismic refraction results from the southern Juan de Fuca Ridge: EOS (American Geophysical Union Transactions), v. 63, p. 1153.
Kappel, E. S., and W. B. F. Ryan, in press, Evidence for volcanic episodicity and a non-steady state rift valley: Journal of Geophysical Research.
Koski, R. A., S. D. Scott, M. D. Hannington, J. R. Delaney, and M. K. Tivey, 1987, Hydrothermal processes and massive sulfide deposits on the Juan de Fuca Ridge and other northeast Pacific spreading axes, this volume.
Lewis, B. T. R., D. Johnson, and H. Jung, 1983, Structure under the southern Juan de Fuca Ridge from geophysics: EOS (American Geophysical Union Transactions), v. 63, p. 1153.
Lichtman, G. S., W. R. Normark, J. R. Delaney, J. L. Morton, H. P. Johnson, and J. L. Karsten, 1983, Photogeology and evolution of the Juan de Fuca Ridge: U.S. Geological Survey Open-File Report 83–464, 16 p.
Lister, C. R. B., 1983, On the intermittency and crystallization mechanism of sub-seafloor magma chambers: Geophysical Journal of the Royal Astronomical Society, v. 73, p. 351–365.
Malahoff, A., R. Embley, and S. Hammond, 1983, Submarine volcanism and hydrothermal activity along the northern end of the Juan de Fuca Ridge axis: EOS (American Geophysical Union Transactions), v. 64, p. 889.
Menard, H. W., and T. M. Atwater, 1968, Changes in direction of sea floor spreading: Nature, v. 219, p. 463–467.
Merge Group, 1984, Regional setting and local character of a hydrothermal field/sulfide deposit on the Endeavour segment of the Juan de Fuca Ridge: EOS (American Geophysical Union Transactions), v. 65, p. 1111.
Morton, J. L., and N. H. Sleep, 1985, Seismic reflections from a Lau Basin magma chamber, *in* D. W. Scholl and T. L. Vallier, eds., Geology and offshore resources of Pacific island arcs—Tonga region: Houston, Circum-Pacific Council for Energy and Mineral Resources, Earth Science Series, v. 2, p. 441–453.
National Oceanic and Atmospheric Administration, 1974, National Ocean Survey Seamap Series, maps NOS-12042-12M and NOS-13242-12M, scale 1:1,000,000.
Normark, W. R., J. L. Morton, R. T. Holcomb, R. A. Koski, and S. L. Ross, 1984, A continuous linear depression in the axial valley of the southern Juan de Fuca Ridge—Tectonic or volcanic origin? (abs.): GSA Abstracts with Programs, v. 16, p. 611.
Normark, W. R., J. L. Morton, R. A. Koski, D. A. Clague, and J. R. Delaney, 1983, Active hydrothermal vents and sulfide deposits on the southern Juan de Fuca Ridge: Geology, v. 11, p. 158–163.
Pacific Geoscience Centre, 1978, Juan de Fuca plate map—Relief: Sidney, British Columbia, Pacific Geoscience Centre, Earth Physics Branch, scale 1:2,000,000.
Raff, A. D., and R. G. Mason, 1961, Magnetic survey off the west coast of North America, 40°N latitude to 50°N latitude: GSA Bulletin, v. 72, p. 1267–1270.
Riddihough, R. P., 1977, A model for recent plate interactions off Canada's west coast: Canadian Journal of Earth Science, v. 14, p. 384–396.
USGS Juan de Fuca Study Group, 1985, Submarine fissure eruption and associated hydrothermal vents along the southern Juan de Fuca Ridge: EOS (American Geophysical Union Transactions), v. 66, p. 116.
Villinger, H., and E. E. Davis, 1984, Heat flow and bottom water temperature measurements from the rift valley of the northern Juan de Fuca Ridge: EOS (American Geophysical Union Transactions), v. 65, p. 1111.
Vine, F. J., 1966, Spreading of the ocean floor—New evidence: Science, v. 154, p. 1405–1415.
Wilson, D. S., R. N. Hey, and C. Nishimura, 1984, Propagation as a mechanism of reorientation of the Juan de Fuca Ridge: Journal of Geophysical Research, v. 89, p. 9215–9225.

APPENDIX

On seismic-reflection records in an area of rugged topography, care must be exercised to distinguish between reflections resulting from structures below the sea floor and those resulting from out-of-plane sidescatter from sea-floor features. The steeply dipping ends of reflection B are diffraction hyperbolas that yield information about the velocity of the overlying material. A low apparent velocity (i.e., near a water velocity of 1490 m/sec) would suggest that event B represented sidescatter. We used a ray-tracing program to model the shape of the diffraction hyperbola on the west side of the ridge axis on line 102 using the western end of the reflector as a point source. We tried several two-layer models, varying the velocity of the material between the sea floor and reflector B, and a one-layer model at water velocity (1490 m/sec). The water-velocity model produces a steeply dipping hyperbola that does not fit the data. The best fit resulted from a model with a 4800-m/sec layer overlying reflector B, similar to the layer 2 velocity determined from the sonobuoy refraction records (4900–5400 m/sec). This result suggests that the event represents structure below the sea floor rather than sidescatter. If a velocity of 5000 m/sec is used for the overlying rock, the depth to reflector B is 2.3 to 2.5 km below the sea floor. The across-axis full width of the reflector is <2 km.

CHAPTER TWENTY-THREE

MORPHOLOGY, STRUCTURE, AND RESOURCE POTENTIAL OF THE BLANCO TRANSFORM FAULT ZONE

R. W. Embley

NOAA Marine Resources Research Division
Hatfield Marine Science Center
Newport, Oregon

L. D. Kulm

College of Oceanography
Oregon State University
Corvallis, Oregon

G. Massoth

NOAA, Pacific Marine Environmental Laboratory
Seattle, Washington

D. Abbott

College of Oceanography
Oregon State University
Corvallis, Oregon

M. Holmes

Marine Geology Division
U.S. Geological Survey
Menlo Park, California

ABSTRACT

A series of rhombohedral basins along the 350-km-long Blanco transform fault zone (BTFZ) are oceanic analogs to pull-apart basins along major divergent wrench fault systems on the continent. In plan view, the BTFZ is similar to the Gulf of California; a series of long strike-slip faults (as much as 100 km long) are separated by short extensional basins (< 20 km) that have internal structures oriented at a large angle to the master faults. In the Blanco region, continual reorientation of the transform fault in response to changes in plate motion during the past few million years provides a mechanism for the formation of these basins. A series of back-tilted, inward-facing fault blocks covered with uplifted turbidite beds brackets the largest of the depressions (Cascadia), which may be a sea-floor-spreading center. Deformed Holocene and Pleistocene turbidite sequences are present within the depression, and there is substantive morphologic and sedimentologic evidence of rapid Holocene subsidence of the floor of the Cascadia Depression. A large water-column methane signal of unknown origin is present, but to date there is no direct evidence of a sea-floor hydrothermal system. However, the probable presence of intrusive bodies within the sedimentary sequence of the Cascadia Depression could give rise to sediment-hosted sulfide bodies such as those of the Guaymas Basin in the Gulf of California. The depressions at either end of the BTFZ (the Gorda and Blanco Depressions) are largely unexplored, but sulfide-veined greenstone breccias and hydrothermal sediments have been recovered from them. Oceanic transform fault zones such as the Blanco may also have hydrothermal systems associated with deep fracturing and high topographic relief.

INTRODUCTION

The Blanco fracture zone (Figure 1) includes the transform fault zone that links the northern Gorda Ridge to the southern Juan de Fuca Ridge and a series of topographic highs that lie on a southeastern continuation of the trend. The bathymetry of the area was first described by McManus (1967), who identified and named some of the major structural components of the fracture zone. This chapter summarizes the data collected over the transform section of the fracture zone since 1967, including extensive NOAA and Navy multibeam coverage (Figure 2). We constructed a first-order structural map of the transform fault zone (Figure 2, inset) by relying on assumptions that (1) the basins are extensional features which offset the strike-slip faults, and (2) the plate boundary is simple in geometry, consisting of alternate strike-slip and extensional structures. The new map delineates four right-stepping segments which are bounded by scarps or ridges and separated by deep basins. This morphotectonic style is best formed along the eastern three-quarters of the transform; the western end near Parks Seamount is somewhat more complicated. The pattern of long transform segments separated by short spreading centers or pull-apart basins is similar to that in the Gulf of California (Elders et al., 1972) and to several East Pacific Rise transforms (Searle, 1983; Madsen, Fox, and MacDonald, in press), although the Blanco transform fault zone (hereafter abbreviated BTFZ) is intermediate in total offset (350 km) between these systems. In the parlance of continental geology, the long strike-slip segments are termed "master faults" (Mann et al., 1983). Extensional basins along major continental strike-slip faults have been generally referred to as pull-apart basins (Crowell, 1974; Aydin and Nur, 1982; Mann et al., 1983). Short extensional segments have been described along other oceanic transform fault zones. Macdonald et al. (1979) described a diffuse "extensional relay zone" along the Tamayo transform fault. Searle characterized the short extensional segments within the Quebrada, Discovery, and Gofar fracture zones as sea-floor-spreading centers. The terms "extensional relay zones" and "pull-apart basins" imply diffuse extension and an ephemeral nature; the term "sea-floor spreading" implies a more steady state process (and probably a crustal magma chamber) and a discrete neovolcanic

Figure 1. Bathymetry map of northeast Pacific showing location of Blanco fracture zone. Contour interval 600 m. Map is simplified from Anonymous (1978).

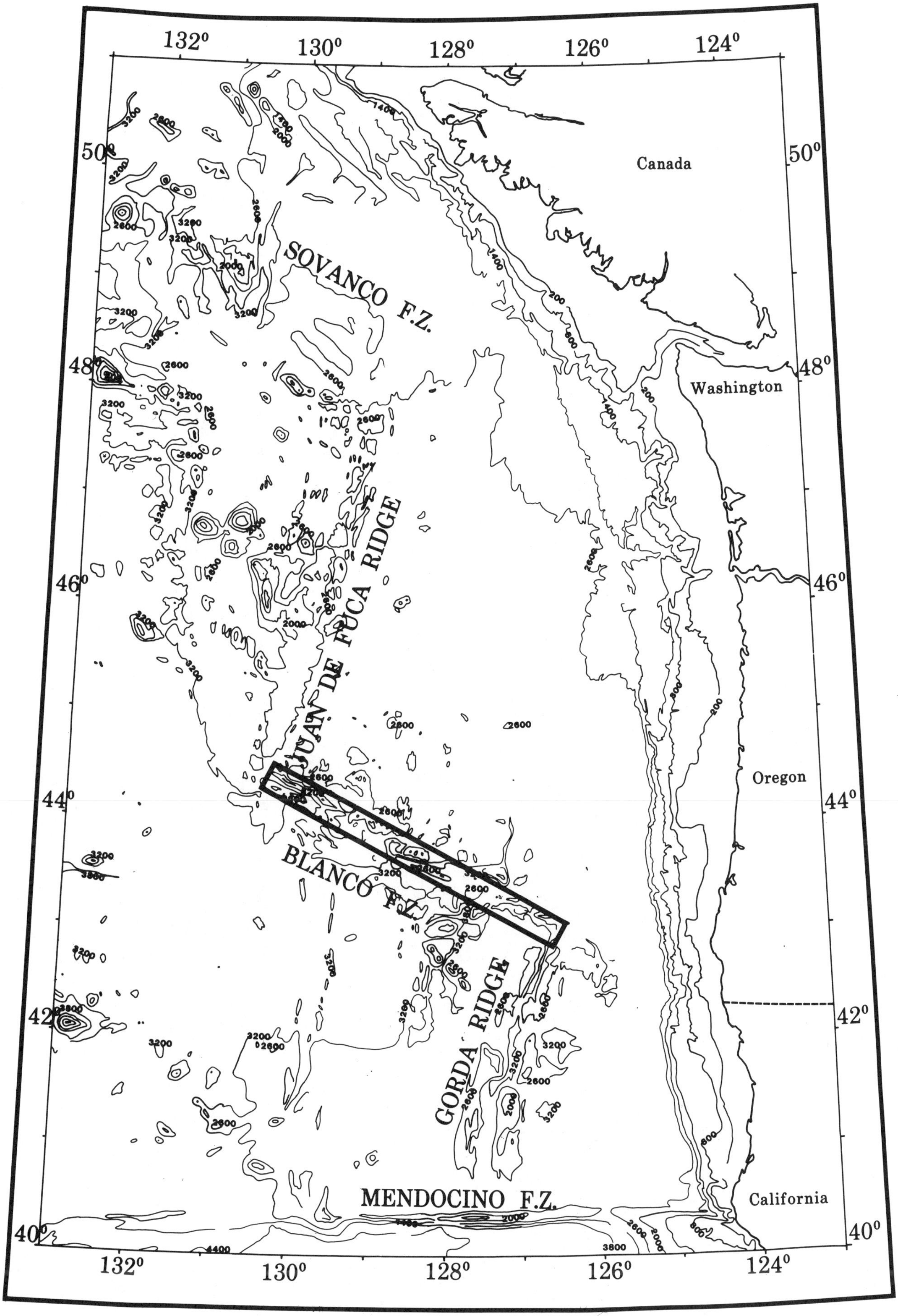
132°
130°
128°
126°
124°
50°
48°
46°
44°
42°
40°
Canada
Washington
Oregon
California
SOVANCO F.Z.
JUAN DE FUCA RIDGE
BLANCO F.Z.
GORDA RIDGE
MENDOCINO F.Z.

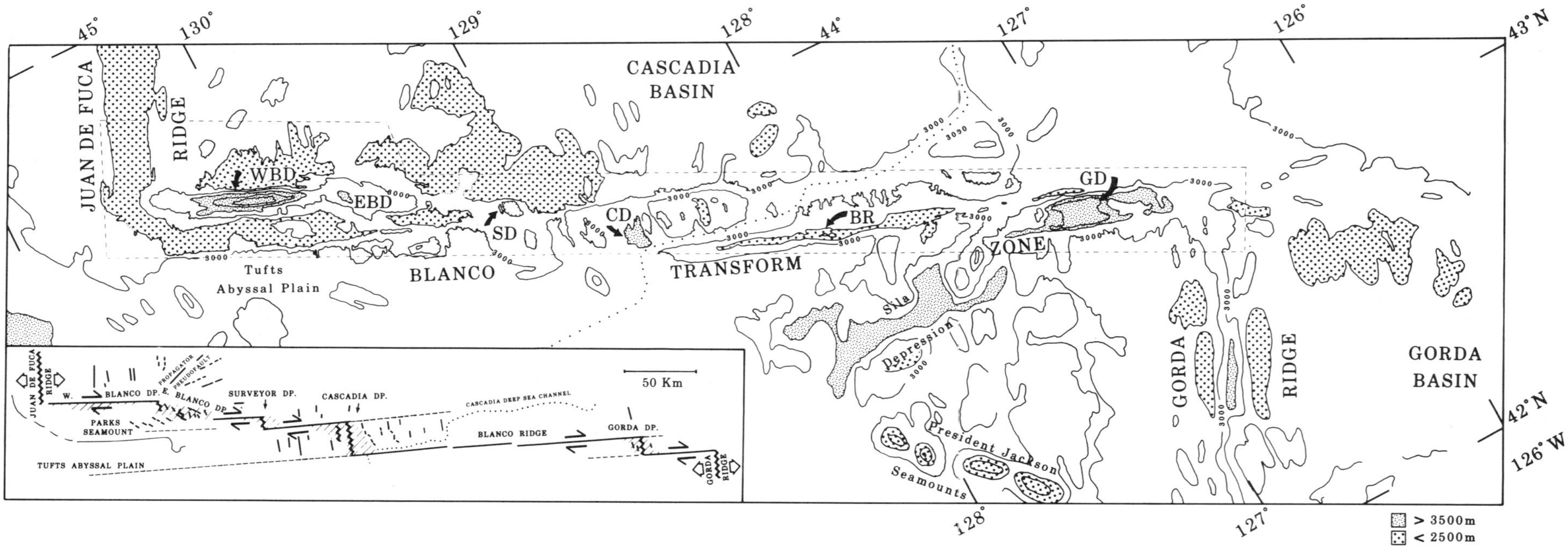

Figure 2. Bathymetry of Blanco transform fault zone. Bathymetry within lightly dashed area is based on essentially full coverage Sea Beam data and Navy multibeam data on easternmost end (Fox, 1984). Remainder of map after Anonymous (1974). Contour interval 500 m. Inset shows structural interpretation based on available Sea Beam data and information discussed in text. Wavy thick lines in inset are centers of extensional basins. Dashed lines show extensions of faults beyond basins. Long-dashed line extending northwest from Blanco Ridge is lineation from GLORIA side-scan coverage (M. Hampton, personal communication). WBD, West Blanco Depression. EBD, East Blanco Depression. SD, Surveyor Depression. CD, Cascadia Depression. BR, Blanco Ridge. GD, Gorda Depression. In inset, DP. is abbreviation for depression.

zone that gives rise to symmetrically distributed crustal generation process. In order to avoid confusion, and because the origin of these features is still speculative, we will refer to the Blanco basins simply as extensional basins.

The Blanco-Juan de Fuca intersection has been cited as an example of a nonorthogonal transform/ridge junction (Atwater and Macdonald, 1977). It is clear from the multibeam data (Figures 2, 3) that the trend of the individual strike-slip faults comprising the BTFZ (112°) is very close to orthogonal to the ridges (120°).

The Blanco fracture zone has long been identified as a high-seismicity plate boundary (Tobin and Sykes, 1968; Bolt, Lomnitz, and McEvilly, 1968; Chandra, 1974). However, owing to poor azimuthal distribution of seismograph stations and poorly constrained crustal velocity models, epicenters determined from P-phase data have routinely plotted 20 to 30 km from the topographic expression of the fracture zone (Northrup, 1970). The use of T-phase data greatly lessened the offset (Dunnebier and Johnson, 1967; Northrup, 1970), but these data were only available from 1965 to 1967. In any case, the seismicity data do not aid in interpretation of the Blanco transform fault structures.

This chapter summarizes what is presently known about the BTFZ, with emphasis on the largest of the extensional basins, the Cascadia Depression. The kinematic history of the BTFZ is discussed in some detail in other publications (Embley et al., in preparation; Hey and Wilson, 1982; Riddihough, 1984; Wilson, Hey, and Nishimura, 1984).

Speculation on the resource potential of the transform is discussed in the final section of this chapter.

MORPHOLOGY OF THE BLANCO TRANSFORM FAULT ZONE

Eastern Segment

The Gorda Ridge terminates in a triangular-shaped rift valley (Malahoff, 1981) whose northern end is characterized by a bending of the structural elements to the northwest against the NW–SE-trending fracture zone topography. The northern Gorda rift valley is morphologically similar to the Mid-Atlantic Rift; an average cross section (Malahoff, 1981) consists of high-relief (about 500 m) boundary faults, marginal lows, and a central volcanic high.

About 20 km west of the Gorda-Blanco intersection a deep rhombohedral basin (Gorda Depression) separates a southern, northward-facing scarp from the eastern Blanco Ridge (Figure 2). This depression is >4430 m deep at its center, 11 km long in the NE–SW dimension, and 18 km wide in the NW–SE dimension (Figure 3). Distinct troughs extending from its four corners are probably fault traces. Several earthquakes have been recorded on the fault scarp connecting the northern Gorda Ridge to the Gorda Depression using an ocean-bottom seismometer array on the northern Gorda Ridge (Bibee, 1985, personal communication). Distinct geomorphic similarities in shape are apparent between the Gorda Depression and the mature pull-apart basin or rhomb graben described along continental divergent wrench faults (Mann et al., 1983; Aydin and Nur, 1982). Clague and Holmes (this volume) report that SeaMarc II side-scan data over the Gorda Depression suggest the presence of lava flows emanating from fissures on the floor of the depression. However, dredging stations across the depression did not recover fresh basalts (D. A. Clague, 1986, personal communication).

Cascadia Segment

The steep ridge that forms the northern boundary of the Gorda Depression (Blanco Ridge) continues as a distinct feature to the limit of the Sea-Beam-mapped area at long. 128°40′W (Figure 2, inset). Recent GLORIA coverage of the area to the northwest (but south of the mapped area) shows a continuation of a fault trace to at least long. 130°30′W (Figure 2). The overall trend of the Blanco Ridge is about 108°E. However, a careful examination of the topography along the ridge reveals a series of closed depressions and small lineaments that appear to be aligned in en echelon fault traces striking at an azimuth of about 112°E. This may represent the most recent faulting.

North of the Blanco Ridge, between long. 127°00′W and 128°00′W, the Cascadia Channel cuts into flat-lying turbidite beds (Figure 2). Embley (1985) has described a secondary channel system draining into the Cascadia Channel in this area (apparent as the southward-pointing V-shaped 3000-m contour north of Blanco Ridge between long. 127°30′W and 128°00′W in Figure 2). These channels have probably formed as the result of a combination of (1) earthquake-induced slumping and headward erosion on the south wall of the Cascadia Channel and, (2) turbidity currents generated by slumps originating from a sediment bench on the north side of Blanco Ridge. Ibach (1981) has interpreted this bench as an uplifted turbidite sequence. This bench is outlined approximately by the 2500- to 2600-m contours (Embley, 1985), implying an uplift of approximately 500 m above the present basinal level of turbidity-current deposition.

The exact location of the active strike-slip fault at the westernmost end of the Blanco Ridge is not clear, but it presumably intersects the southern boundary of the Cascadia Depression, the largest of the Blanco depressions (Figures 2, 3, 4). This depression is 20 km long in its NE–SW dimension and, in its deepest area, about 8 km wide in the NW–SE dimension. The depression is bounded on the northwest and southeast by a series of inward-facing back-tilted blocks that are covered with a thick sequence of turbidite deposits (Duncan, 1968; Griggs and Kulm, 1973; Figures 5, 6). The normal faults are oriented within 5° to 10° of orthogonal to the strike of the south- (Blanco Ridge) and north-bounding master faults. These blocks extend outward as far as about 50 km on either side of the Cascadia Depression, and the blocks on the southeastern side appear to consistently have more relief than those to the northwest. A local seismicity study (Johnson and Jones, 1978) recorded a small earthquake swarm along the eastern boundary fault of the Cascadia Depression (along long. 128°40′W, Figure 4).

At least 500 m of silt turbidite beds are present in the Cascadia Depression (Griggs and Kulm, 1973), and the map clearly shows that the Cascadia Channel has been a major pathway for these sediments (Figures 2, 3, 4). However, the bathymetry also confirms Griggs and Kulms' (1973) conclusion that the Cascadia Depression currently forms a depositional endpoint for Cascadia Channel sediments, because the thalweg of the channel has a 130 m negative inflection coin-

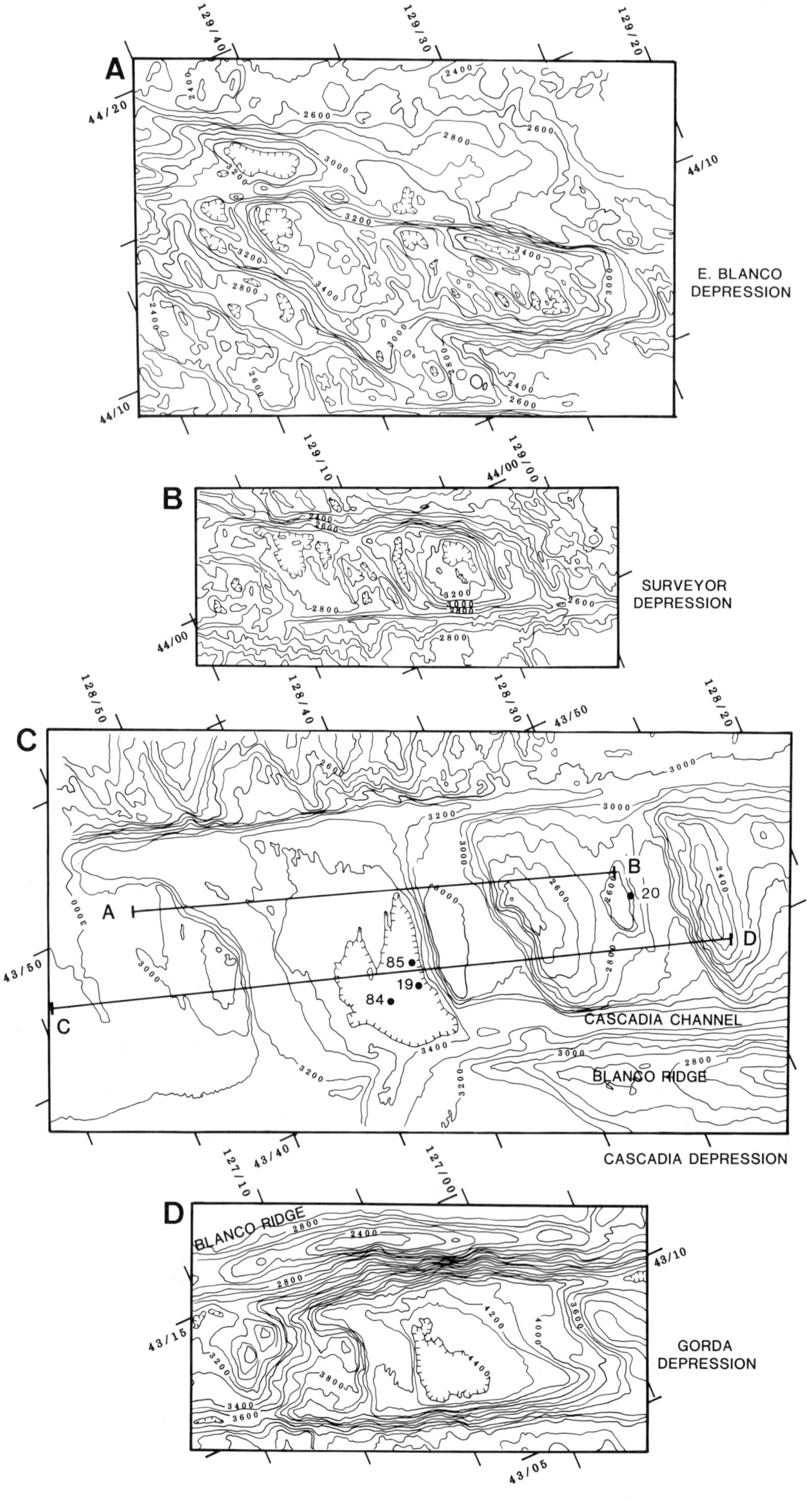
A
129/40
129/30
129/20
44/20
44/10
44/10
E. BLANCO
DEPRESSION
B
129/10
129/00
44/00
44/00
SURVEYOR
DEPRESSION
C
128/50
128/40
128/30
128/20
43/50
43/50
43/40
A
B
C
D
85
19
84
20
CASCADIA CHANNEL
BLANCO RIDGE
CASCADIA DEPRESSION
D
127/10
127/00
43/10
43/15
43/05
BLANCO RIDGE
GORDA
DEPRESSION

ciding with the depression (Figure 4). Pronounced Holocene subsidence of the Cascadia Depression is documented by core 6609-19 (Figure 3; Griggs and Kulm, 1970, 1973). Mt. Mazama ash was deposited 6600 years ago as a turbidite ash flow both upchannel and downchannel from the depression. Subsequent sediment ponding has occurred in the depression relative to the upchannel and downchannel areas, indicating rapid subsidence of the depression (Griggs and Kulm, 1970). The ash turbidity current was the last flow downchannel, but there have been two or more subsequent clastic turbidite flows in the depression and its upchannel areas. The calculated subsidence rate for the depression since about 6600 Y. B. P. is 1.8 cm/yr.

The turbidite deposits within the Cascadia Depression have undergone recent deformation. The detailed Sea Beam bathymetry (Figure 4) and the 3.5-kHz seismic-reflection records across the depression (Figure 5, top; Figure 7) reveal that the turbidite beds have been uplifted and backtilted. The deformation has created a series of small blocks 10 to 20 m in relief (appearing as southward-pointing fingerlike projections of the 3300- and 3400-m contours on Figure 4). The tops of these blocks give rise to corner reflections that appear on the 3.5-kHz seismic-reflection profiles (1–1′, 2–2′ and 3–3′) on Figure 7 as hyperbolic echos truncated on the steep side. The blocks face inward toward the middle of the basin. A 60- to 70-m-high northeast-trending block (Figure 7, profile 4–4′) is present at the center of the depression. The flat area with < 10 m of local relief south of the mound is underlain by slightly deformed turbidites (Figure 7, profiles 5–5′ and 6–6′). This sediment pond has received the most recent turbidity currents entering the depression via the Cascadia Channel, so the apparent lesser deformation here compared with the northern Cascadia Depression may be related to the higher sedimentation rate.

The doming in the center of the Cascadia Depression may be caused by the intrusion of sills into the sediments, which we believe are the sharply bounded, high-reflectivity zones noted on the seismic-reflection records (Figures 5, 6). Alternatively, this doming could be the result of serpentinite diapirism. However, the seismic-reflection records do not reveal direct evidence for diapirism (e.g., pull-up in deep reflectors).

At least one major structure is oblique to the primary north–south faults within the basin. The fault(?) appears to strike approximately SE–NW across the NW quadrant of the depression and is crossed on profile A-B (Figure 3C; Figure 5, top). Its nature is unknown, but it could be an oblique shear.

Figure 3. Sea Beam bathymetry at 100-m contour interval of four major extensional basins along Blanco transform. General locations shown on Figure 2. Seismic-reflection profiles A-B and C-D are located on Cascadia Depression map (C). Also shown on map C are locations of water-column profiles 84 and 85 (Figure 8) and cores 6609-19 and 6609-20, abbreviated as 19 and 20, respectively (Duncan, 1968).

Hydrocasts taken within the Cascadia Depression in 1984 and 1985 reveal large methane anomalies between water depths of 2600 and 3500 m (Figure 8). To our knowledge, these values (4–18 nmol/l) are in the high range when compared with those reported from other deep-water areas, which range from 0.3 to 12.0 nmol/l (Lilley, Baross, and Gordon, 1983; Campbell and Gieskes, 1984; Kim, 1983). For example, the bottom water methane values in Guaymas Basin are 1 to 2 nmol/kg (Campbell and Gieskes, 1984). However, the Cascadia methane is not correlated to high excess dissolved or particulate manganese that would be expected for an active hydrothermal system (Kim, 1983). The higher values measured for the 1985 hydrocast may be related to its proximity to the eastern boundary fault. The detection of high methane concentrations deserves further investigation, particularly in light of the strong evidence for active faulting and the likelihood of Holocene intrusive activity within the depression.

The Cascadia Depression is bounded on the north by a 500- to 600-m-high southward-facing scarp (Figures 3, 4). A channellike feature, which may have been either the former path of the Cascadia Channel into the depression or the present path of a tributary to it (which enters the area at about lat. 43°46′N, long. 128°18′W), separates the backtilted blocks of the eastern side of the Cascadia Depression from the scarp. This feature parallels the Cascadia Channel and may also be structurally controlled. The main depression turns, narrows, and shoals to the northwest.

The next depression separating the long faults occurs at about lat. 44°00′N, long. 129°10′W. The Surveyor Depression (Figures 2, 3B) is actually a double basin with a small ridge between. The western basin has a maximum depth of about 2900 m and the eastern basin about 3300 m, making the Surveyor Depression the shoalest of the Blanco basins. It is also the smallest; only about 5 km separates the master faults. The structural lineaments within the basin are about 20° from orthogonal (measured counterclockwise) to the north and south master faults.

Western Segment

The western segment of the BTFZ consists of several tectonic elements and appears to be more complex than the extensional basin and strike-slip fault pattern that characterize the Cascadia and eastern segments. The dominant tectonic element of the western segment is the Blanco Depression (informally called the Blanco Trough in Melson, 1969, and Selk, 1978), which we have divided into a shallower, 3600-m-deep eastern part, the East Blanco Depression (Figures 2, 3A), and the West Blanco Depression, which has a maximum depth of > 4800 m (Figure 2). The tectonic connection between the East Blanco Depression and the Surveyor Depression appears to be along a zone linking the northwestern corner of the Surveyor Depression with the southeastern corner of the East Blanco Depression. The morphologic trends within the East Blanco Depression are complex. The depression appears to consist of a series of small deeps (maximum depth 3600–3700 m) separated by intervening ridges (maximum depth 3300–3400 m). The East Blanco Depression is separated from the West Blanco Depression by a sill that shoals to about 2900 m. The West Blanco Depression consists of a shoaler, southern bench (2900 to 3600 m) and a deep trough to the north (4800 m). Its

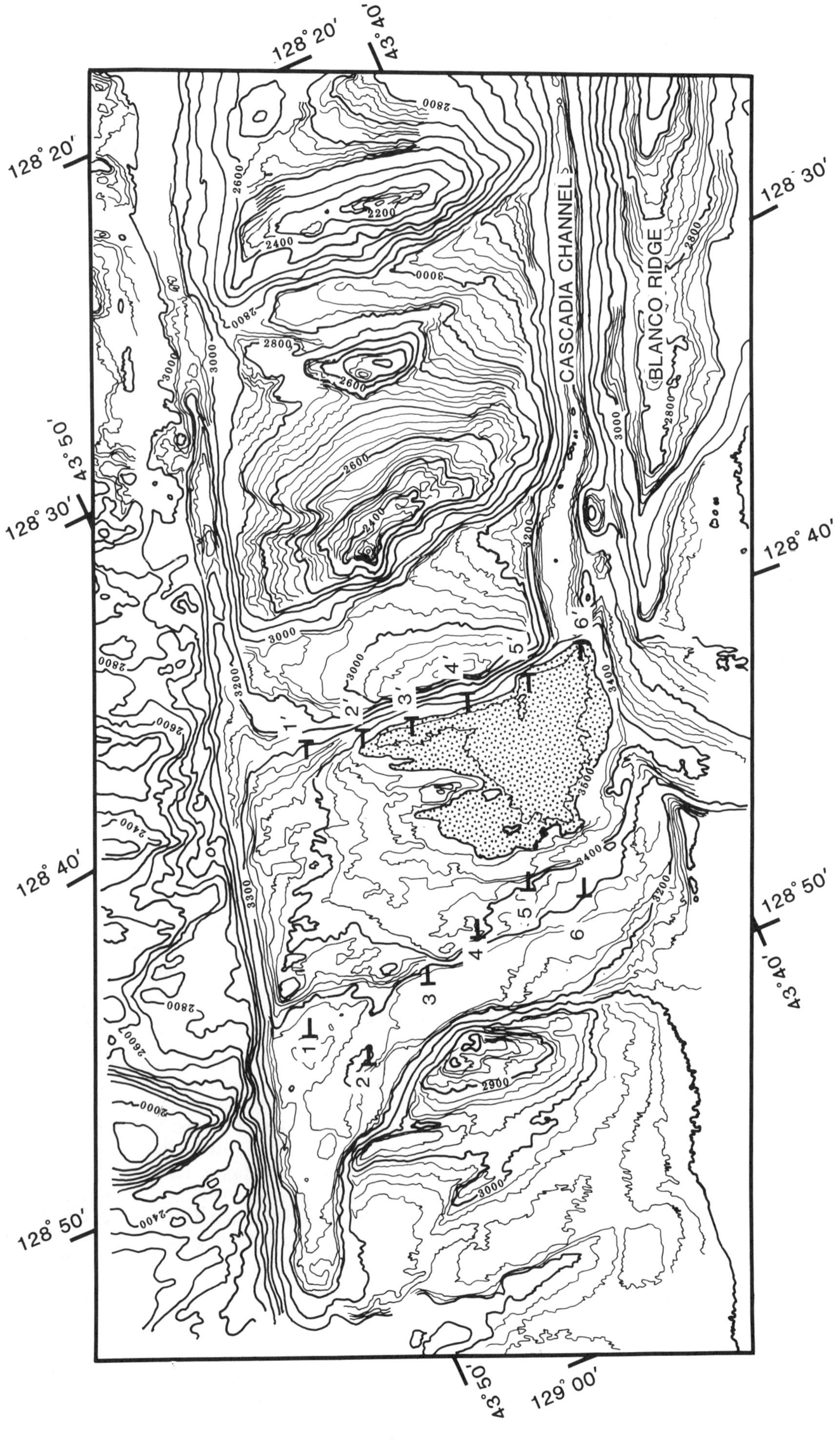

Figure 4. Detailed Sea Beam bathymetry of the Cascadia Depression (thick lines 100 m, thin lines 20 m). Stippled area encloses 3500-m contour of Cascadia Depression. Locations of seismic-reflection profiles A-B and C-D (Figures 5, 6) are shown.

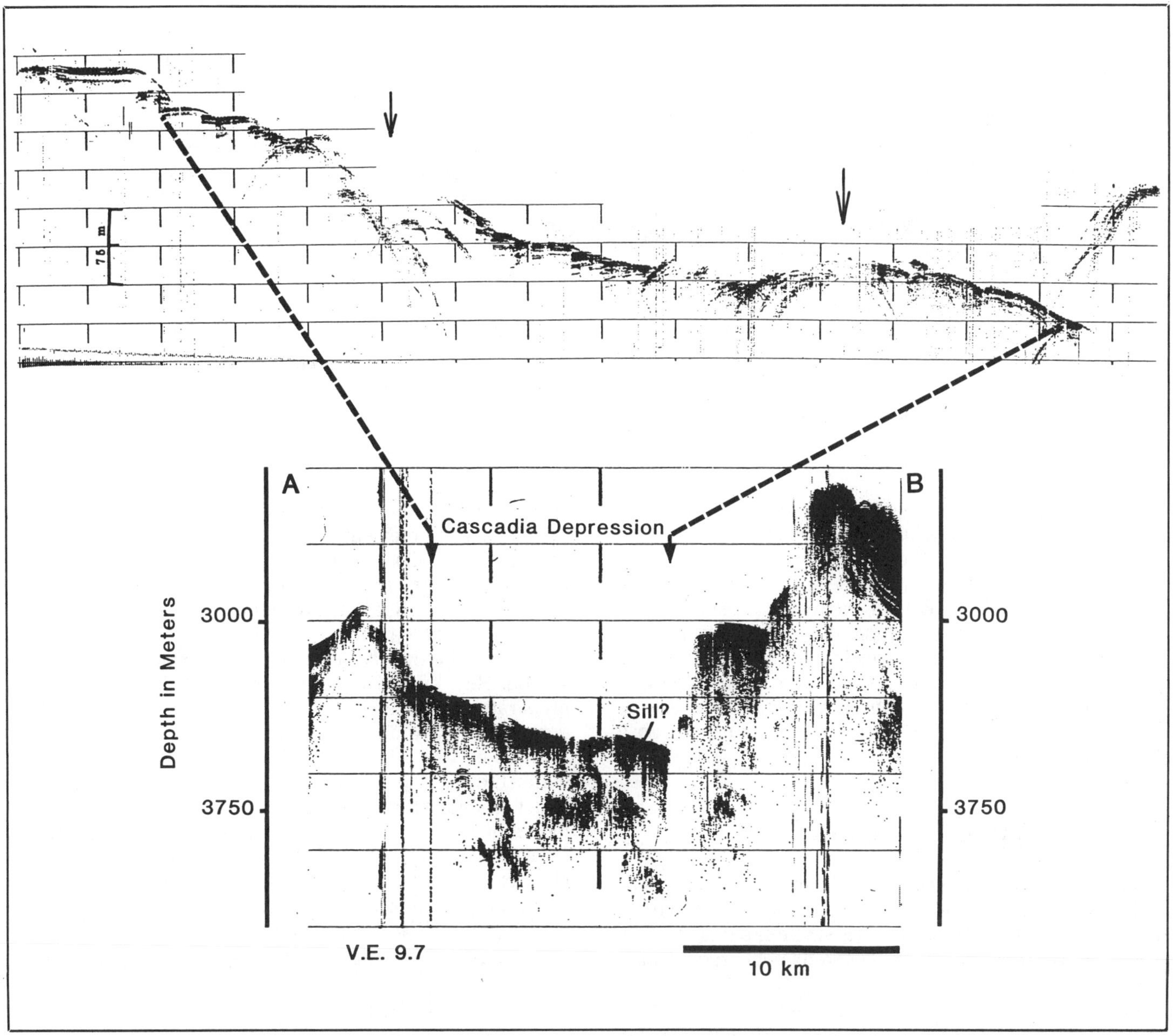

Figure 5. Top—3.5-kHz seismic-reflection record across oblique Cascadia Depression. Dashed lines locate position of profile with respect to water-gun record taken simultaneously. Arrow at right shows location of doming of turbidites above probable sills. Arrow at left marks location of crossing of northwest-trending fault in NW quadrant of Cascadia Depression. Bottom—Water-gun seismic-reflection profile A-B across Cascadia Depression (location shown on Figure 3C).

structure is different from the other Blanco depressions in that it has no consistent cross-cutting lineations. South of the East and West Blanco Depressions is a high area referred to on older maps as the Parks Seamount (Figure 2). Sea Beam bathymetry reveals that this area is not a seamount or series of seamounts but is rather a series of amorphous highs and lows at about the same average depth as the sea floor north of the Blanco depressions. However, one area south of the eastern side of the West Blanco Depression shoals to about 1600 m, the southward-facing scarp forming the southern boundary of the Parks area represents a major regional depth change of about 700 m and a basement depth change of about 1000 m. Dredged rocks from the north and south walls of the Blanco depressions (Melson, 1969) include dolerite, pillow basalt, basalt breccia, and hydrothermally altered basalt. Melson (1969) interpreted the West Blanco Depression as a grabenlike structure.

North of the Blanco depressions, the basement lineations strike parallel to the trend of the Juan de Fuca Ridge. However, at about lat. 44°18′N and long. 129°35′W, a northeast-striking lineation, which is actually a 100-m-high scarp on the detailed Sea Beam bathymetry, cuts obliquely (approximately 060°E) across the ridge-parallel trend. This feature is probably the pseudofault of a propagating rift which impinged on the Blanco fracture zone at about 1.5 Ma (Hey and Wilson, 1982). Basement lineations west of the pseudofault strike at more easterly trends. These lineations are within what Hey, Dunnebier, and Morgan (1980) characterize as the shear zone of a propagating rift, which lies between the pseudofault and the failed rift.

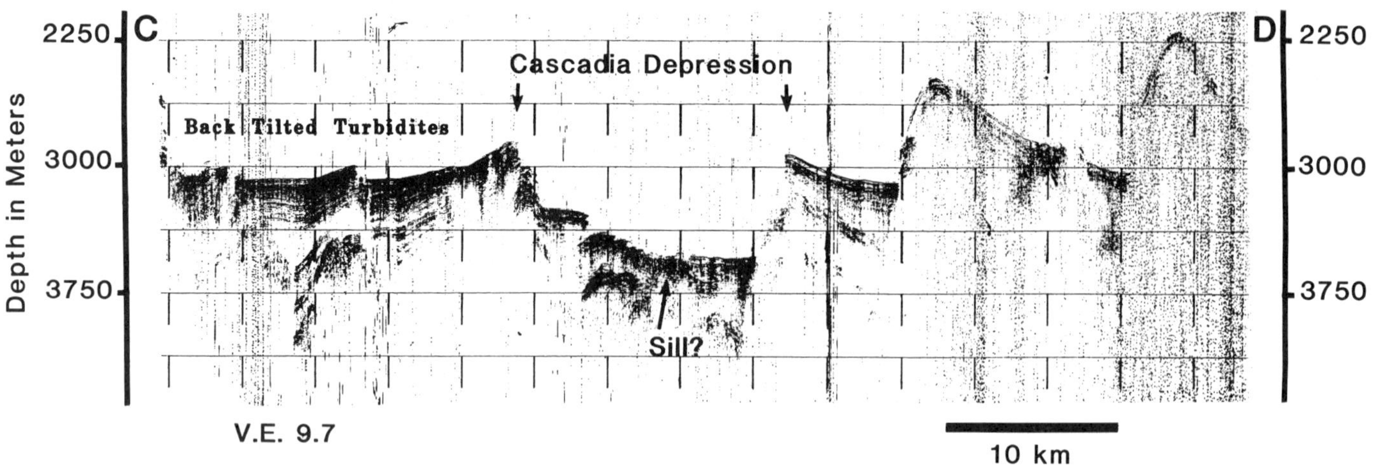

Figure 6. Water-gun seismic-reflection profile C-D across Cascadia Depression (location shown on Figure 3C).

The neovolcanic zone of the southern Juan de Fuca Ridge apparently ends at about lat. 44°30′N (unpublished Sea Beam data). The intersection of the ridge and transform occurs as the rift valley ends and the sea-floor fabric bends around toward the West Blanco Depression. A small basin about 3300 m deep separates the southern end of the rift valley from the West Blanco Depression across a sill depth of about 3500 m. The Juan de Fuca-Blanco Ridge-transform intersection is somewhat structurally analogous to several of the ridge-transform intersections along the northern East Pacific Rise (Gallo et al., 1984).

STRUCTURAL MODELS

The detailed bathymetry described in the previous sections reveals that the BTFZ consists of a series of short extensional basins connected by long strike-slip faults. The structures within the basins are generally not perpendicular to the master faults.

Searle (1983) has presented a model for the relationship between the angle of tectonic spreading fabric (measured relative to the strike of transform offsets) and the length of the spreading center. In effect, he suggests that the bending of the tectonic elements is due simply to the gradual rotation of the minimum compressive stress from the pure tensional to the pure strike-slip regimes. For long ridge segments, the bending only occurs near the transform/ridge intersection. For very short ridge segments (< 10 km), the fabric between master faults is always oblique to them. This model is generally consistent with the structures observed in the Cascadia, Gorda, and Surveyor Depressions. However, the model assumes a classic sea-floor-spreading origin for the tensional segments, whereas the literature on continental wrench faults (e.g., Mann et al., 1983) shows that the structural character observed between short offsets of continental strike-slip faults is complicated and depends on several factors. These include (1) the sense of lateral motion across the fault, (2) the sense of master fault offset along the zone (right stepping or left stepping), (3) the distance between the master faults, and (4) the amount of overlap between the active master faults. Without detailed mapping and ocean-bottom seismometer studies of the faults, it will remain uncertain which of the models fit the oceanic case. Some of the basins along the BTFZ may be pull-apart basins, whereas others may be undergoing sea-floor spreading (e.g., Cascadia). The West Blanco Depression may be something else entirely.

The development of the present morphology of the BTFZ is the result of the plate kinematics of the northeast Pacific over the past 5 m.y. as described in Hey and Wilson (1982), Wilson, Hey, and Nishimura (1984), and Embley et al. (in preparation). Embley et al. (in preparation) propose that the depressions have developed as the result of a major plate rotation at 4 to 5 Ma and at least two minor adjustments within the past 1 m.y. The complex nature of the small-scale structures along the transform also generally agrees with the model of Fox and Gallo (1985), which predicts that transforms connecting two medium-spreading-rate ridges will be characterized by extensional relay zones (pull-apart basins) developed in response to changes in plate motion.

RESOURCE POTENTIAL OF THE BLANCO TRANSFORM FAULT ZONE

The BTFZ appears to be an oceanic analog to continental "divergent wrench faults" (e.g., Wilcox, Harding, and Seely, 1973), which are defined as strike-slip systems dominated by extensional characteristics. These continental systems are recognized as proven and potential settings for geothermal and hydrocarbon resources (Harding and Lowell, 1979). In the deep ocean, three types of deposits have been seriously considered as potential resources—(1) manganese nodules, (2) manganese/cobalt crusts, and (3) polymetallic sulfides (Hein et al., this volume; Koski et al., this volume). Potentially commercial manganese nodules and manganese/cobalt crusts form over tens of millions of years on sea floor characterized by low sedimentation rates (< 1 cm/1000 yrs). Consequently, these deposits are not expected to be found in quantity along the BTFZ. The primary type of hydrothermal system found to date in the oceans is associated with the thinly sedimented neovolcanic zone of mid-ocean ridges. The hydrothermal systems and associated polymetallic sulfides of the Juan de Fuca and Gorda Ridges are discussed in other chapters in this volume. The sulfides formed in conjunction with these hydrothermal systems are rich in iron,

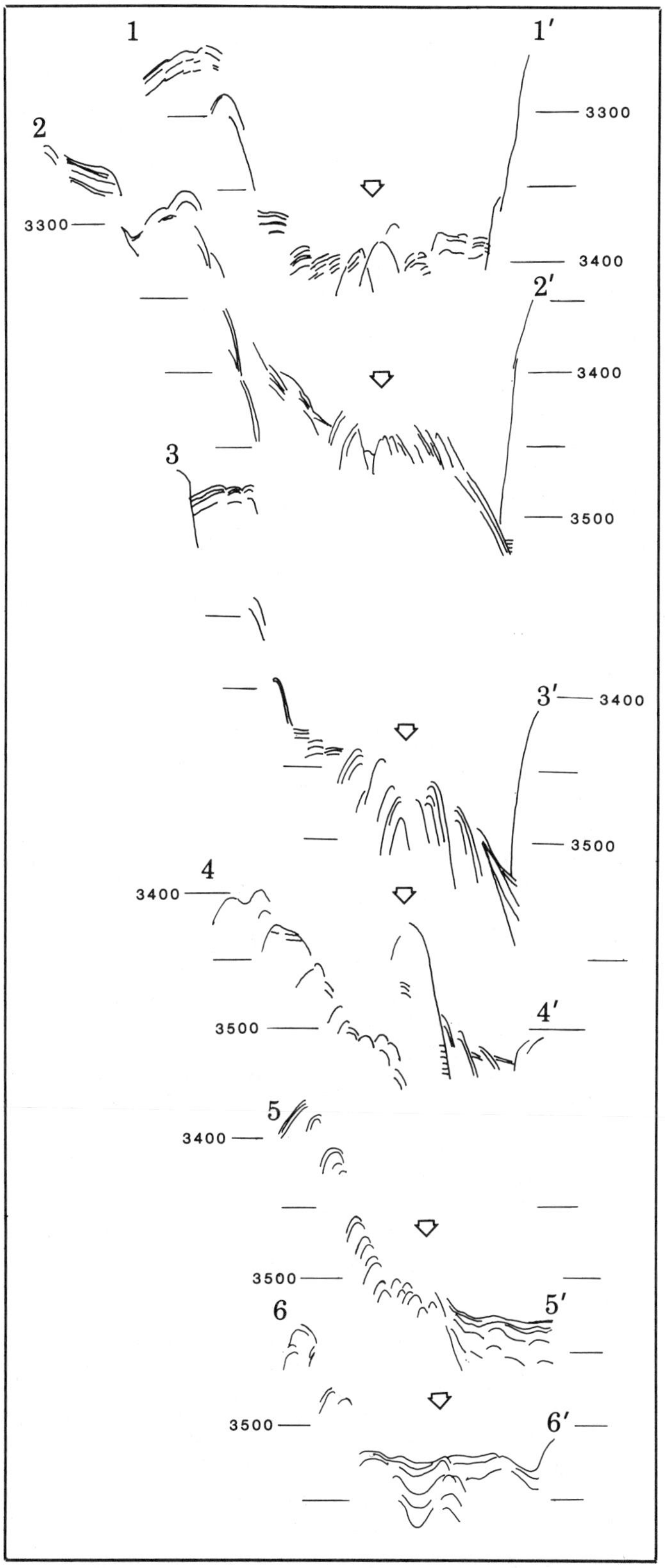

Figure 7. High-vertical-exaggeration 3.5-kHz records across Cascadia Depression. Arrows show location of top of central high. Locations shown on Figure 4.

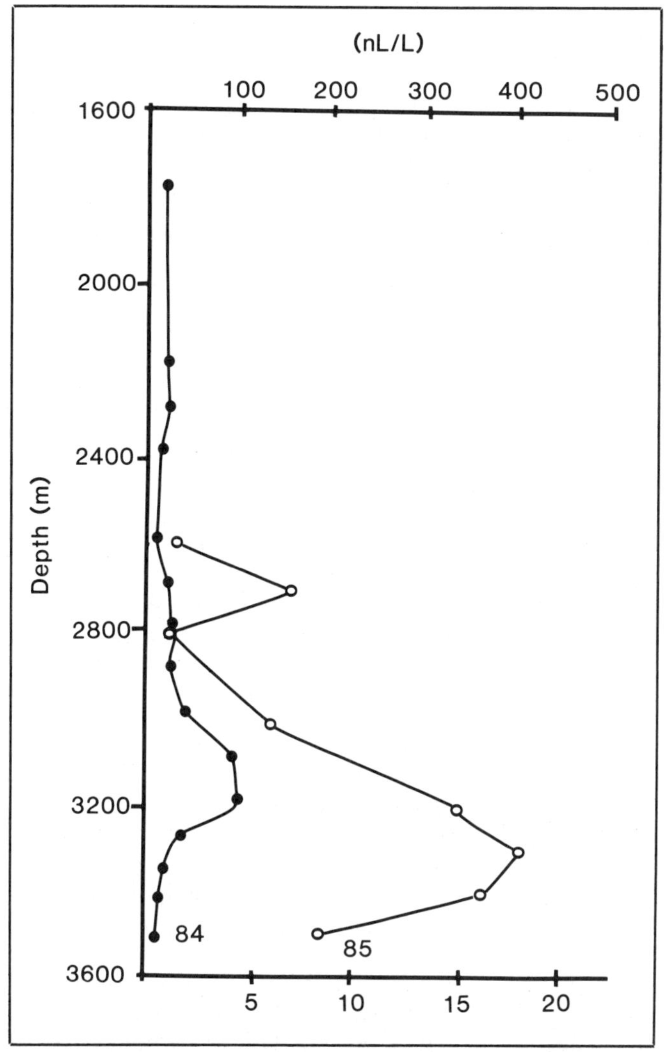

Figure 8. Water-column methane profiles from Cascadia Depression (locations shown on Figure 3C).

zinc, and copper (Koski et al., this volume).

The only direct evidence for hydrothermal deposits of this type in the BTFZ is the presence of iron- and manganese-rich deposits within the West Blanco Depression. Selk (1978) interpreted these as derivatives of a deep hydrothermal system associated with the shearing along the transform. An alternate interpretation for these deposits may be fallout from the southern Juan de Fuca Ridge hydrothermal plume (Normark et al., 1983).

The extensional origin of the basins along the BTFZ suggests other possible sites of hydrothermal activity. Hart, Pyle, and Robbins (1986) have reported the presence of sulfide veins in rocks dredged from the northern scarp of the Gorda Depression, but the origin of these is still speculative. To date, there is no direct evidence of volcanic activity within the extensional basins of the BTFZ, although very little bottom sampling has been done in the area.

We have noted that sills may be present within the sediments of the Cascadia Depression, which suggests the possibility of the presence of a sediment-hosted hydrothermal system. This type of system is the least understood and probably most difficult to examine because of the sediment cover. The only reasonably well-studied modern example of such a system is in the Guaymas Basin of the Gulf of California. The Guaymas Basin is one of a series of shallow (<2000-m depth) extensional basins that are generally <50 km long.

Drilling and submersible studies of the Guaymas Basin revealed two types of hydrothermal systems—a short-lived system associated with the intrusion of sills and consequent expulsion of pore water (no recharge), and a longer-lived system associated with a magma chamber or large intrusive body (Gieskes et al., 1982). The intrusive bodies drilled during Leg 64 range from about 5 m to > 100 m in thickness and have baked and dewatered zones above and below them ranging from 10 to 50 m in thickness.

In addition to this contact metamorphism, sill intrusions may also act as a cap for the longer-term recharged system associated with the primary magmatic body. Site 477 in the Guaymas Basin penetrates through a 160-m zone of greenschist sediment facies below a 50-m-thick sill. This zone probably formed as the result of a regional hydrothermal system (Gieskes et al., 1982). Although no massive sulfide bodies were sampled, analyses of sulfides taken from the hot spring areas in the southern rift zone of Guaymas by ALVIN indicate that they are metal-poor compared with their open-ocean equivalents (Koski et al., 1985). Von Damm et al. (1985) have suggested that sulfide bodies were formed at depth within the sediment body.

The Cascadia Depression exhibits some of the same morphologic and structural features as the Guaymas Basin, including probable sill intrusions. However, at present there is no direct evidence of hydrothermal venting from the sea floor within the basin. Thc high methane values measured over the Cascadia Depression are probably either the result of outgassing along faults of methane generated within the clastic sediments or of magmatic hydrothermal exhalations. There is no indication of the presence of hydrothermal petroleum, such as is found in the Guaymas Basin (Simoneit and Lonsdale, 1982), although the organic content of the sediment within the Cascadia Depression is as high as 2.4% (Duncan, 1968).

Magmatic intrusion into pull-apart basins such as the Cascadia Depression is one source of a hydrothermal deposit along the BTFZ. In addition to this type of system, three general characteristics of fracture zones could potentially give rise to hydrothermal systems different in character than those formed along extensional plate boundaries. These are (1) high total topographic relief, (2) deep fracturing and (3) thin oceanic crust (White et al., 1984). High topographic relief could provide the large pressure differential needed to drive a deep circulation system (Stefansson, 1983); deep fracturing could provide high permeability conduits; and thin crust could provide access of hydrothermal fluids to upper mantle rocks. These rocks may be enriched in nickel, cobalt, and chromium relative to crustal rocks. Because the thermal regime beneath a complex transform system is unknown, the existence of such systems is still highly speculative. However, greenschist-facies rocks, which have been proposed as originating within hydrothermal systems (e.g., Humphris and Thompson, 1978), have been dredged from Atlantic fracture zones (Bonatti and Honnorez, 1976), and Hart, Pyle, and Robbins (in press) have dredged hydrothermally altered rocks from the northern scarp of the Gorda Depression.

ACKNOWLEDGMENTS

The Sea Beam bathymetry and seismic-reflection data presented in this chapter were collected through the efforts of the crew and officers of the NOAA Ship SURVEYOR. We gratefully acknowledge A. Malahoff, chief scientist on one of these cruises, and R. Ganse and B. Edwards for valuable assistance at sea. We also acknowledge the Naval Oceanographic Office for the multibeam data collected on the eastern end of the BTFZ and northern Gorda Ridge, which were incorporated into our map. Discussions with R. Hart and D. Gallo were also very helpful.

REFERENCES CITED

Anonymous, 1974, NOS SEAMAP Series, Map 12042-128.

Anonymous, 1978, Relief map of the Juan de Fuca plate: Pacific Geoscience Centre and the Department of Energy, Mines and Resources, Canada.

Atwater, T., and K. C. MacDonald, 1977, Are spreading centers perpendicular to their transform faults?: Nature, v. 270, p. 715–719.

Aydin, A., and A. Nur, 1982, Evolution of pull-apart basins and their scale independence: Tectonics, v.1, p. 91–105.

Bolt, B. A., C. Lomnitz, and T. V. McEvilly, 1968, Seismological evidence on the tectonics of central and northern California and the Mendocino Escarpment: Seismological Society of America Bulletin, v. 58, p. 1725–1767.

Bonatti, E., and J. Honnorez, 1976, Section of the earth's crust in the equatorial Atlantic: Journal of Geophysical Research, v. 81, p. 4104–4116.

Campbell, A. C., and J. M. Gieskes, 1984, Water column anomalies associated with hydrothermal anomalies in the Guaymas Basin, Gulf of California: Earth and Planetary Science Letters, v. 68, p. 57–72.

Chandra, U., 1974, Seismicity, earthquake mechanisms and tectonics along the western coast of North America from 42°N to 61°N: Seismological Society of America Bulletin, v. 64, p. 1529–1549.

Clague, D. A., and M. L. Holmes, 1987, Geology, petrology, and mineral potential of the Gorda Ridge, this volume.

Crowell, J. C., 1974, Origin of Late Cenozoic basins in southern California, *in* W. R. Dickinson, ed., Tectonics and sedimentation: SEPM Special Publication 22, p. 190–204.

Duncan, J. P., 1968, Late Pleistocene and postglacial sedimentation and stratigraphy of deep sea environments off Oregon: Corvallis, Oregon, Oregon State University unpublished Ph.D. dissertation, 222 p.

Dunnebier, F. K., and R. H. Johnson, 1967, T-phase sources and earthquake epicenters in the Pacific Basin: Honolulu, Hawaii Institute of Geophysics Report HIG-67-24.

Elders, W. A., R. W. Rex, T. Maidav, P. T. Robinson, and S. Biehler, 1972, Crustal spreading in southern California: Science, v. 178, p. 15–24.

Embley, R. W., 1985, A locally formed deep-ocean canyon system along the Blanco transform: Geo-Marine Letters, v. 5, p. 99–104.

Embley, R. W., D. Wilson, A. Malahoff, and R. Ganse, in preparation, Tectonics of the Blanco transform.

Fox, C. G., 1984, A quantitative method for analyzing the roughness of the seafloor: New York, Columbia University unpublished Ph.D. dissertation, 217 p.

Fox, P. J., and D. Gallo, 1985, A tectonic model for ridge-transform-ridge plated boundaries: Implications for the structure of oceanic lithosphere, v. 104, p. 205–242.

Gallo, D. G., P. J. Fox, J. A. Madsen, and K. C. Macdonald, 1984, The morphotectonic signature of fast-slipping ridge-transform-ridge systems: A synthesis of Sea Beam bathymetry: EOS (American Geophysical Union Transactions), v. 65, p. 1103.

Gieskes, J. M., M. Kastner, G. Einsele, K. Kelts, and J. Niemitz, 1982, Hydrothermal systems in the Guaymas Basin: A synthesis, *in* J. R. Curray and D. G. Moore et al., Initial reports of the Deep Sea Drilling Project (Vol. 64): Washington D.C., U.S. Government Printing Office, p. 1159–1167.

Griggs, G. B., and L. D. Kulm, 1970, Sedimentation in Cascadia

deep-sea channel: GSA Bulletin, v. 81, p. 1361–1384.
Griggs, G. B., and L. D. Kulm, 1973, Origin and development of Cascadia deep-sea channel: Journal of Geophysical Research, v. 78, p. 6325–6339.
Harding, T. P., and J. D. Lowell, 1979, Structural styles, their plate tectonic habitats and hydrocarbon traps in petroleum provinces: AAPG Bulletin, v. 63, p. 1016–1058.
Hart, R., D. Pyle, and J. Robbins, in press, Hydrothermal sulphides, breccias and greenstones from the Gorda Depression: Department of Geology and Mineral Resources Open-File Report.
Hein, J. R., L. A. Morgenson, D. A. Clague, and R. A. Koski, 1987, Cobalt-rich ferromanganese crusts from the exclusive economic zone of the United States and nodules from the oceanic Pacific, this volume.
Hey, R. N., and D. S. Wilson, 1982, Propagating rift explanation for the tectonic evolution of the northeast Pacific—The pseudomovie: Earth and Planetary Science Letters, v. 58, p. 167–188.
Hey, R. N., F. K. Dunnebier, and W. J. Morgan, 1980, Propagating rifts on mid-ocean ridges: Journal of Geophysical Research, v. 85, p. 3647–3658.
Humphris, S. E., and G. Thompson, 1978, Hydrothermal alteration of basalts by seawater: Geochimica et Cosmochimica Acta, v. 42, p. 107–125.
Ibach, D. H., 1981, The structure and tectonics of the Blanco fracture zone: Corvallis, Oregon, Oregon State University unpublished M. S. thesis, 60 p.
Johnson, S. H., P. R., and Jones, 1978, Microearthquakes located on the Blanco fracture zone with sonobuoy arrays: Journal of Geophysical Research, v. 83, p. 255–261.
Kim, K., 1983, Methane and radioactive isotopes: University of California at San Diego unpublished Ph.D. dissertation, 206 p.
Koski, R. A., P. F. Lonsdale, W. C. Shanks, M. E. Berndt, and S. H. Have, 1985, Mineralogy and geochemistry of a sediment-hosted hydrothermal sulfide deposit from the southern trough of the Guaymas Basin, Gulf of California: Journal of Geophysical Research, v. 90, p. 6695–6707.
Koski, R. A., S. D. Scott, M. D. Hannington, J. R. Delaney, and M. K. Tivey, 1987, Hydrothermal processes and massive sulfide deposits on the Juan de Fuca Ridge and other northeast Pacific spreading axes, this volume.
Lilley, M. D., J. A. Baross, and L. I. Gordon, 1983, Reduced gases and bacteria in hydrothermal fluids: The Galapagos spreading center and 21°N East Pacific Rise, *in* P. A. Rona, K. Bostrum, L. Laubier, and K. L. Smith, eds., Hydrothermal processes at seafloor spreading centers, New York, Plenum Press, p. 411–449.
Macdonald, K. C., K. Kastens, F. N. Spiess, and S. P. Miller, 1979, Deep-tow studies of the Tamayo transform fault: Marine Geophysical Researches, v. 4, p. 37–70.
Madsen, J. A., P. J. Fox, and K. C. Macdonald, in press, The morphotectonic fabric of the Orozco transform fault: Results from a Sea Beam investigation: Journal of Geophysical Research.
Malahoff, A., 1981, Comparison between Galapagos and Gorda spreading centers: 13th Annual Offshore Technology Conference, no. 4129, p. 115–121.
Mann, P., M. R. Hempton, D. C. Bradley, and K. Burke, 1983, Development of pull-apart basins: Journal of Geology, v. 91, p. 529–554.
McManus, D. A., 1967, Blanco fracture zone, northeast Pacific Ocean: Marine Geology, v. 3, p. 429–455.
Melson, W. G., 1969, Preliminary results of a geophysical study of portions of the Juan de Fuca Ridges and Blanco fracture zone: Environmental Science Services Administration (ESSA) Technical Memorandum 6, U.S. Department of Commerce, 33 p.
Normark, W. R., J. L. Morton, R. A. Koski, and D. A. Clague, 1983, Active hydrothermal vents sulfide deposits on the southern Juan de Fuca Ridge: Geology, v. 11, p. 158–163.
Northrup, J., 1970, Accuracy of earthquake epicenters on the Gorda Ridge: Seismological Society of America Bulletin, v. 60, p. 265–267.
Riddihough, R. P., 1984, Recent movements of the Juan de Fuca plate system: Journal of Geophysical Research, v. 89, p. 6980–6994.
Searle, R. C., 1983, Multiple, closely spaced transform faults in fast-slipping fracture zones: Geology, v. 11, p. 607–610.
Selk, B. W., 1978, The manganese-enriched sediments of the Blanco Trough: Evidence for hydrothermal activity in a fracture zone: Corvallis, Oregon, Oregon State University unpublished M.S. thesis, p. 607–610.
Simoneit, B. R. T., and P. F. Lonsdale, 1982, Hydrothermal petroleum in mineralized mounds at the seabed of Guaymas Basin: Nature, v. 295, p. 198–202.
Stefansson, V., 1983, Physical environment of hydrothermal systems in Iceland and on submerged oceanic ridges, *in* P. A. Rona, K. Bostrom, L. Laubier, and K. Smith, eds., Hydrothermal processes at seafloor spreading centers: New York, Plenum Press, p. 321–360.
Tobin, D. G., and Sykes, L. R., 1968, Seismicity and tectonics of the northeast Pacific Ocean: Journal of Geophysical Research, v. 3, p. 3821–3845.
Von Damm, K., J. M. Edmond, C. I. Measures, and B. Grant, 1985, Chemistry of submarine hydrothermal solutions at Guaymas Basin, Gulf of California: Geochimica et Cosmochimica Acta, v. 49, p. 2221–2237.
White, R. S., R. S. Detrick, M. C. Sinha, and M. H. Cormier, 1984, Anomalous seismic crustal structure of oceanic fracture zones: Geophysical Journal of the Royal Astronomical Society, v. 79, p. 779–798.
Wilcox, R. E., T. P. Harding, and D. R. Seely, 1973, Basic wrench fault tectonics: AAPG Bulletin, v. 57, p. 74–96.
Wilson, D. S., R. N. Hey, and C. Nishimura, 1984, Propagation as a mechanism of reorientation of the Juan de Fuca Ridge: Journal of Geophysical Research, v. 89, p. 9215–9225.

CHAPTER TWENTY-FOUR

GEOLOGY, PETROLOGY, AND MINERAL POTENTIAL OF THE GORDA RIDGE

David A. Clague
Mark L. Holmes

U.S. Geological Survey
Menlo Park, California

ABSTRACT

The Gorda Ridge is an oceanic spreading center located within the U.S. Exclusive Economic Zone off the coast of northern California and southern Oregon. Analysis of underway geophysical data, including SeaMARC II side-scan sonar mosaics, indicates that the ridge consists of three short segments offset by transform faults or overlapping spreading centers. Samples recovered from along the ridge indicate that hydrothermal activity is widespread, although high-temperature discharge capable of producing metallic sulfide deposits is documented at only two areas along the ridge axis. The southernmost of these deposits lies within the Escanaba Trough and is sediment-hosted. Many large ophiolite-hosted sulfide deposits probably formed along back-arc spreading centers; some are hosted in volcanogenic sediment similar to that filling the Escanaba Trough. Further work is needed to determine the size and composition of the massive sulfide deposit discovered in the Escanaba Trough in 1985.

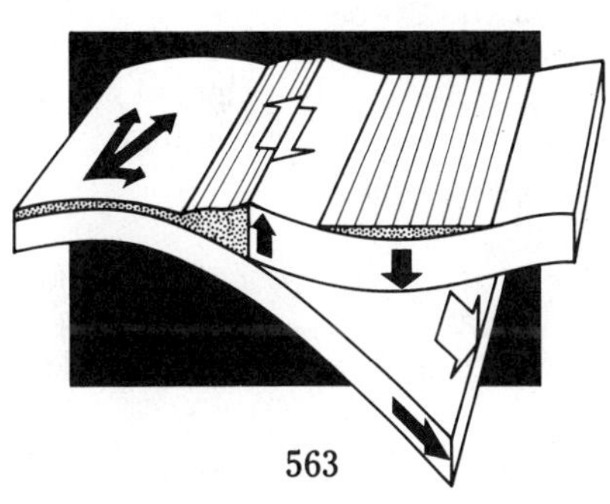

INTRODUCTION

The Gorda Ridge is an oceanic spreading center located within 370 km (200 n. miles) of the coast of northern California and southern Oregon (Figure 1). The ridge consists of three segments offset by short transform fault zones or overlapping spreading centers. It is bounded to the north by the Blanco fracture zone and to the south by the Mendocino fracture zone. A short chain of seamounts, named the President Jackson Seamounts, trends northwesterly from a point on the ridge at about lat. 42°N. The entire Gorda Ridge is about 300 km long, and the southern 90 km of axial valley is partly filled by turbidite deposits. This section is called the Escanaba Trough.

The Gorda Ridge is the only spreading center segment within the boundaries of the U.S. Exclusive Economic Zone. The earlier discoveries of polymetallic sulfide deposits on the Juan de Fuca Ridge (Figure 1) by U.S. Geological Survey and University of Washington scientists focused attention on the Gorda Ridge as a possible site for active hydrothermal systems and their associated mineral deposits. After declaration of the Exclusive Economic Zone, the U.S. Geological Survey began a program to assess the geologic framework and mineral resource potential of the Gorda Ridge. The U.S. Geological Survey conducted geological and geophysical surveys over the Gorda Ridge on four cruises during the period 1980 through 1983.

As a followup to the 1983 work, the U.S. Geological Survey conducted three short cruises to the Gorda Ridge in 1985. The 1985 cruises were coordinated through the Federal-State Gorda Ridge Technical Task Force. Four main work areas were examined including the Blanco fracture zone, the President Jackson Seamounts, an area on and adjacent to the ridge axis at about lat. 42°45′N, and the axis of the Escanaba Trough. Preliminary results from the 1985 cruises are included in the appropriate sections of this chapter.

This chapter summarizes the morphology of the Gorda Ridge based on underway acoustic geophysical data, describes the samples recovered, and discusses the implications of these data for hydrothermal activity and mineral potential along the Gorda Ridge.

GEOLOGICAL SETTING AND PREVIOUS WORK

The axis of the Gorda Ridge is 6 to 18 km wide at the top of the inward-facing fault scarps. The ridge axis is unusually deep compared with other spreading centers, ranging from about 3200 m to >3800 m. The axial graben has 800 to 2000 m of vertical relief. Numerous small uplifted fault blocks flank the ridge axis between lat. 41°N and 42°N. To the east of the ridge axis, all but the larger fault blocks are buried beneath turbidite sediment of the Astoria and Gorda deep-sea fans. The general morphology of the ridge was briefly described by Malahoff et al. (1982).

Previous studies of the magnetic lineations in the region show that the ridge offset near lat. 42°N is a triple junction (ridge–ridge-transform), and that the spreading rates north and south of this triple junction are different. To the north the full spreading rate is about 5.5 cm/yr, whereas to the south it averages 2.3 cm/yr (Atwater and Mudie, 1973; Riddihough, 1980). Earthquakes as deep as 14 km have occurred beneath the axis of the Gorda Ridge near lat. 43°N (Johnson and Lewis, 1984). The more southerly ridge offset, located near lat. 41°30′N, is characterized by swarms of small earthquakes (Northrup, Menard, and Duennebier, 1968) and may also be a ridge–ridge-transform triple junction with as much as 0.2 cm/yr left-lateral offset (Riddihough, 1980). The general tectonic features are shown in Figure 2.

Atwater and Mudie (1973) gave a detailed description of the topography, sediment distribution, and magnetization of the northern section of the Escanaba Trough. They concluded that the central rift valley resulted from large-scale normal faulting that broke the surface into long, narrow, tilted steps parallel to the spreading center. They described flat-lying sediment in the Escanaba Trough and noted that sediment also occurred on top of some fault blocks bounding the valley, thus implying that the fault blocks were uplifted to form the valley walls as they moved outward from the spreading axis.

The older bathymetric data, earthquake epicenters, locations of sediment samples and heat-flow stations, magnetics, gravity, and seismic-reflection and seismic-refraction studies in the Gorda Ridge area have been compiled in map form by Wilde et al. (1978). Recent studies (Wilson, 1984, 1985; Stoddard and Carlson, 1984) indicate that the Gorda "plate" is not separate from the Juan de Fuca plate but is instead a zone of continuous deformation between the Pacific and North American plates.

NEW DATA FROM U.S. GEOLOGICAL SURVEY CRUISES

A systematic survey of Gorda Ridge north of Escanaba Trough (Figure 2) was carried out in 1983 aboard R/V KANA KEOKI (cruise KK1-83-NP) using the SeaMARC II side-scan sonar system (Figure 3). Approximately 2200 km of acoustic imagery was obtained along tracklines spaced 8.5 km apart. After collection of the underway geophysical data, a sampling cruise (KK2-83-NP) recovered rocks from 15 of 17 dredge sites (Figure 3) along the Gorda Ridge axis from Escanaba Trough northward to the Blanco fracture zone (Figure 2). Underway geophysical data, including 24-fold multichannel seismic records, were collected from R/V S.P. LEE (cruise L11-81-NC) along widely spaced tracks in 1981 (Figure 3). The cruises conducted in 1985 emphasized bottom photography and dredging of northern Gorda Ridge in an area identified by scientists from the National Oceanographic and Atmospheric Administration and Oregon State University as the source of a hydrothermal plume. Other sampling and camera work was conducted along the Blanco fracture zone, the President Jackson Seamounts, and within the Escanaba Trough. A detailed geophysical survey of the Escanaba Trough was completed. Analysis of the data collected has just begun, so the results presented here are preliminary.

Seismic-Reflection Records

The 2200 km of single-channel seismic-reflection data from cruise KK1-83-NP were obtained using a 120 cu in. air gun towed at a speed of 7.5 knots. The data were processed into two frequency bands (25–75 and 75–250 Hz) and displayed on standard graphic recorders. The low-frequency records across Gorda Ridge are shown in Figure 4. The seismic records were collected along NW–SE tracks and have been "rotated" in Figure 4 approximately 45° about their crossing of the Gorda Ridge axis in order to simplify presentation. This process produces some spatial distortion in the apparent relationship of the Gorda Ridge and Blanco fracture zone but facilitates examination of morphological and structural changes along the strike of the major features.

Three 24-channel seismic-reflection records were obtained along and across southern Gorda Ridge (Escanaba Trough) during cruise L11-81-NC (Figure 3). These data were collected at 5.5 knots with a shot spacing of 50 m using a 2400–m-long streamer and an array of five air guns with a total capacity of 1326 cu in. Two tracks (lines 114 and 118) crossed the Escanaba Trough at lat. 41°N and lat. 40°30′N, respectively (Figure 3). A third line (111) was shot along the Escanaba Trough axis from lat. 41°30′N southward across the Mendocino fracture zone (Figures 2, 3). The 24-fold records from the two E–W transects (lines 114 and 118) and that portion of line 111 over Escanaba Trough are shown in Figure 5.

SeaMARC II Data

The SeaMARC II (*Sea Mapping And Remote Characterization*) is an intermediate range side-scan sonar system capable of obtaining high-quality acoustic imagery over a 10-km-wide swath at ship speeds up to 8 knots (Blackington, Hussong, and Kosalos, 1983; Kosalos and Chayes, 1983). This system was used on cruise KK1-83-NP to map >20,000 km^2 of Gorda Ridge from the intersection with the Blanco fracture zone to the northern end of Escanaba Trough (Figures 2, 3). The closely spaced tracks provided for overlap of the swaths from adjacent tracks and resulted in 100% coverage of the survey area. A mosaic of the acoustic imagery from the northern two-thirds of the Gorda Ridge is shown in Figure 6. Because both the field and shore-processed images are reversed in polarity, the highly reflective sea floor is displayed as dark areas and the low-reflectivity or shadow zones are displayed as light areas. SeaMARC II is unique among side-scan sonar systems in that it uses two rows of transducers on each side of the tow vehicle to measure phase differences in the back-scattered signals returning from the sea floor. Processing these phase data permits three-dimensional positioning of the pixels with respect to the tow-fish, and this information can be used to generate a bathymetric map (Figure 3) of the surveyed area. Because the acoustic imagery is more responsive to reflectivity changes in the sea floor than to specular reflections from bathymetric features, large regional slopes without significant reflectivity changes are difficult to distinguish from the imagery alone. The synoptically obtained phase (bathymetry) data can be used to greatly increase the utility of the side-scan imagery in the interpretation of structural, geological, and morphological features.

MORPHOLOGY AND STRUCTURE OF GORDA RIDGE

The Gorda Ridge is divided into three morphologically distinct segments by two prominent right-lateral offsets in the axial valley at lat. 42°26′N and 41°37′N (Figures 3, 6). These offsets represent either early stages in the formation of transform faults or the joining of discrete nonaligned spreading segments through ridge propagation to the north and south. The northern and central ridge segments were surveyed in detail during cruise KK1-83-NP, whereas the southern segment (Escanaba Trough) has been studied only in reconnaissance with a few multichannel reflection lines (Figure 5). The following discussion is based on single and multichannel seismic-reflection records (Figures 4, 5), the SeaMARC II acoustic imagery mosaic (Figure 6), and bathymetry (Figure 3).

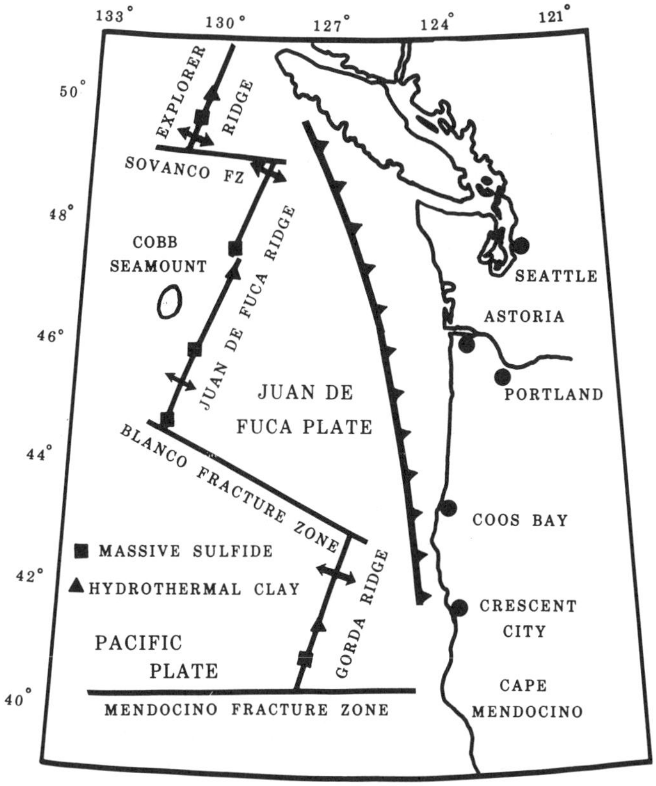

Figure 1. Regional location map showing the Gorda, Juan de Fuca, and Explorer Ridge spreading centers off the coast of California, Oregon, Washington, and British Columbia. The Gorda Ridge is bounded on the north by the Blanco fracture zone and on the south by the Mendocino fracture zone. Known locations of massive hydrothermal sulfide deposits and of hydrothermal nontronite are identified using different symbols.

Northern Gorda Ridge and Blanco Fracture Zone

The 70-km-long northern segment is characterized by a V-shaped cross-section from its intersection with the Blanco fracture zone to the northern axial offset, as shown on seismic-reflection lines 101 to 107 and 69 to 73 (Figure 4). The narrow valley floor ranges in depth from 3200 m along line 105 to >3800 m, the deepest part along the entire Gorda Ridge axis, at the line 101 crossing (Figure 4). A bathymetric profile constructed along the Gorda Ridge axis from the northern end of Escanaba Trough to the Blanco fracture zone is shown in Figure 7. The axial valley floor in the northern half of the north Gorda Ridge segment appears as a long

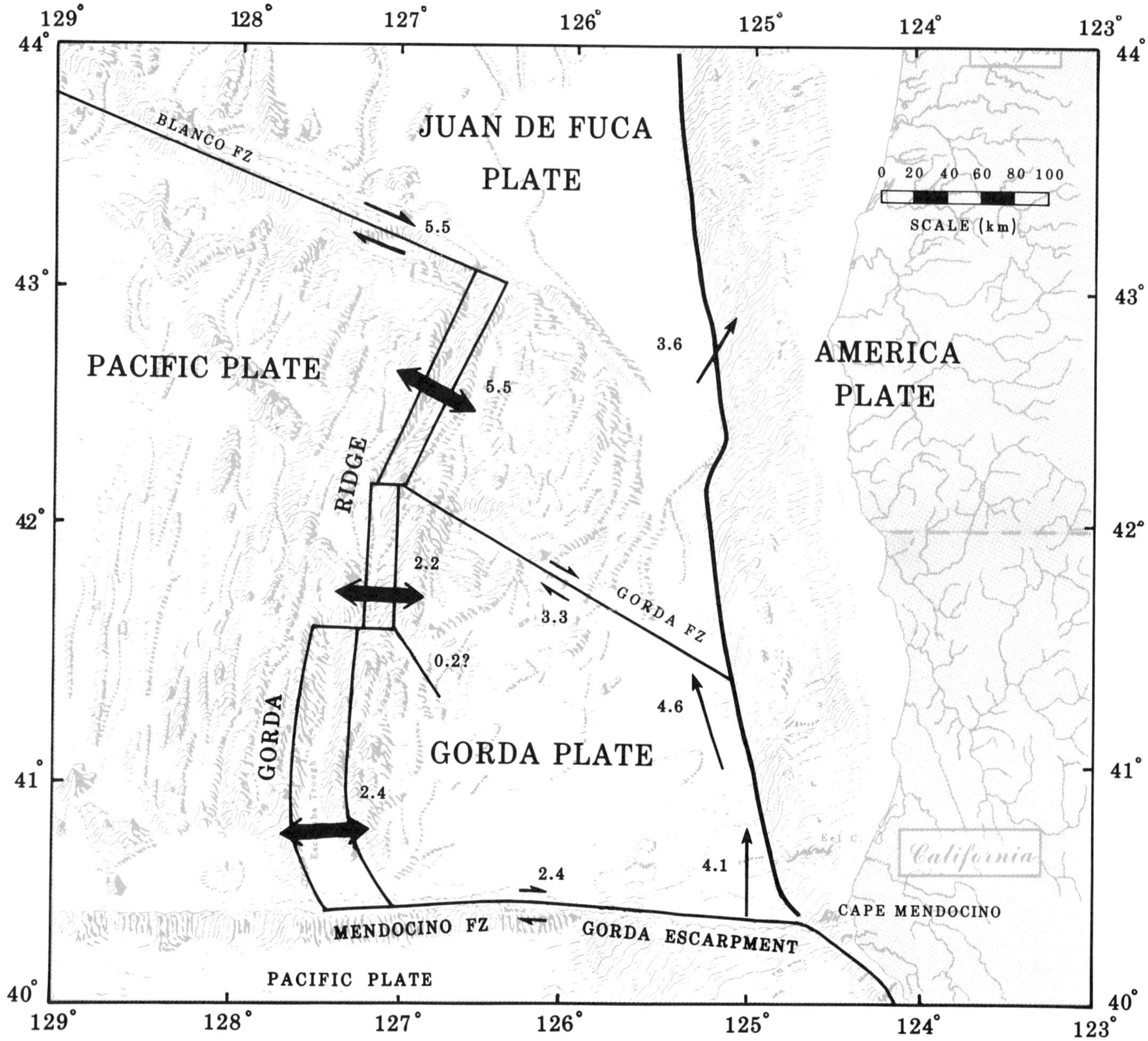

Figure 2. Tectonic setting of Gorda Ridge, showing spreading rates and plate interactions over the last 1 m.y. (after Riddihough, 1980). The spreading rate in the southern Gorda Ridge segment (Escanaba Trough) is less than half of the spreading rate in the northern part. Light arrows indicate motion vectors with rates indicated in cm/yr.

domal uplift in profile view. The greatest depths are found about 10 km north of the offset at lat. 42°26′N.

The SeaMARC II imagery shows a highly reflective valley floor with long linear fault scarps in the axis and along the base of the valley walls (Figure 6). The strong backscattering of acoustic energy is caused by the rough surfaces of relatively fresh pillow basalt and sheet flow extrusions. A small axial volcanic high was crossed by line 71 (Figure 4), and similar features can be seen at the sides of the axial valley on lines 101 and 103 (Figure 4).

The maximum relief in the northern segment is 1400 m from the axial graben floor to the top of the flanking ridges. The inward-facing slopes of the valley walls are relatively smooth, and narrow uplifted benches can be noted on the west valley wall (lines 101, 103, 105 on Figure 4). The reflectivity on the upper parts of these slopes is much lower than on the valley floor (Figure 6), indicating that sea-floor weathering of basalt and slowly accumulating pelagic deposits have reduced the microtopography of the volcanic rocks on the sea floor.

The ridges flanking the axial graben are continuous along the length of the northern Gorda Ridge segment (Figure 3); water depths range from 2200 to 2500 m. These ridges are discontinuous with those to the south of the ridge offset, and this break marks the southern end of this segment. At this ridge offset the valley axis bends westward for a distance of 6 to 7 km, and joins the axial valley of the central Gorda segment. The seismic record across the ridge axis offset (line 99, Figure 4) shows a very smooth V-shaped section with flanking slopes dipping inward at about 23°.

The outer flanks of the northern Gorda Ridge show a low-reflectivity surface broken by long, linear fractures that can be traced for distances of as much as 50 km (Figure 6). These scarps can also be seen on the seismic-reflection records (Figure 4), and some of the larger ridges on the flanks (lines 71, 73, 107) that have thin sediment cover appear as high-reflectivity areas on the SeaMARC imagery (Figure 6).

At the northern end of Gorda Ridge, the axial valley and the linear fractures on the ridge flanks bend sharply to the west, where they are truncated by the northwestward-trending Blanco fracture zone (Figures 3, 6). At this intersection the Blanco fracture zone is a 10-km-wide trough with a flat floor about 3000 m deep that dips gently to the northwest along the strike of the fracture zone. The low reflectivity in the trough (Figure 6) is the result of partial filling of this part of the Blanco fracture zone by turbidites from the Cascadia Abyssal Plain (line 69, Figure 4). The bathymetry (Figure 3) shows that the Gorda Ridge axis is about 500 m deeper than the Blanco fracture zone trough at their intersection. This transition is shown on the acoustic imagery by a sudden change from the highly reflective (dark) basalt in the axial basin at the northern end of Gorda Ridge to the low reflectivity of the floor of the Blanco fracture zone. Embley, Kulm, and Abbott (1984) describe the structure of the Blanco fracture zone based on seabeam bathymetry.

To the west of the Gorda-Blanco intersection, the floor of the Blanco fracture zone deepens and becomes progressively tilted toward the north (lines 71, 73 on Figure 4). Line 73 crosses the Gorda Depression, a deep pull-apart basin whose floor has subsided to a depth of 4500 m. This basin (Figure 3) clearly shows the effects of left-lateral shear on the floor of the Blanco fracture zone resulting from westward movement of the crust north of the Gorda Depression relative to the crust south of the Blanco fracture zone. The steep (30°) walls of the Gorda Depression are highly reflective on the SeaMARC imagery (Figure 6), indicating that they are comprised of relatively sediment-free rocks. The strong reflectivity of the basin floor appears to be caused by recent volcanic flows emanating from fractures on the basin floor and from the base of the southern wall. On the top of the northern wall, two extensive areas of slumping can be seen on both the acoustic imagery (Figure 6) and the bathymetry (Figure 3).

Central Gorda Ridge

Between the two major right-lateral offsets, the morphology of Gorda Ridge is more complex than the northern or southern (Escanaba Trough) sections. The axis is broadly sinuous in plan view. The bathymetric profile along the axis (Figure 7) is cyclical and similar to that of the northern ridge segment. The valley axis reaches its shallowest depth in the northern part of both segments, about 25 to 30 km south of their northern boundary. The axis then deepens to the south and shoals again at or near the offsets which mark the southern end of each segment. Another axial shoal at the northern end of Escanaba Trough appears to begin the cycle again.

The width and character of the central graben floor changes along its length. The northern third of the central Gorda segment is a narrow linear valley that widens from 10 to 18 km between lines 97 and 93 (Figure 4) and is similar in some characteristics to the central part of the northern Gorda Ridge. High sea-floor reflectivity (Figure 6) indicates relatively fresh volcanic rocks, and the valley axis has many narrow fault scarps parallel to the axis. The depth at line 95 is shallower than at any other place in the central segment (Figures 3, 7). Where line 91 (Figure 4) crosses the ridge, the axis widens and becomes more complex in cross-section; many large ridges and domes are seen on the seismic-reflection records. The SeaMARC imagery (Figure 6) shows faulted and fractured inward-facing scarps forming the boundary between the valley walls with low reflectivity and the valley floor with high reflectivity. There are no linear fractures on the valley floor as in the northern part of the segment. Instead, the axial zone is characterized by large cones, pits, flat-topped edifices, and areas of high reflectivity interpreted as fresh flows.

East of the right-lateral offset at the northern end of the central segment, the ridges and flank fractures are convex eastward (Figure 6), reflecting the separation at the offset and the more rapid spreading rate of the northern Gorda Ridge. The topographic grain west of the offset maintains a fairly straight trend parallel to the two ridge segments. This parallel trend indicates that the northern ridge segment is translating eastward at nearly the same rate as spreading is taking place.

The outer ridge flanks are more rugged in the central section than on the northern Gorda Ridge, as indicated on both the seismic-reflection records (Figure 4) and the acoustic imagery (Figure 6). Areas of exposed rock, indicated by dark highly reflective zones, are common along both ridge flanks. The northeast-trending ridges 20 km east of the axis near lat. 42°N (Figure 6) are more reflective than other bare rock areas, suggesting active off-axis volcanism along fractures that are not aligned with the Gorda Ridge axis. These areas correspond to the ridges shown on the eastern ends of reflection lines 93 and 95 (Figure 4).

At the southern end of the central segment, the Gorda Ridge axis is offset right-laterally almost 15 km by a large S-shaped bend that joins the central segment with the northernmost part of Escanaba Trough. The ridge flanks on opposite sides of this offset consist of large massifs (Figure 3) rising >1500 m above the axial valley floor. The tops of these blocks are rugged, and the SeaMARC II imagery (Figure 6) shows dense patterns of fractures trending parallel to the Gorda Ridge axis. The tectonic movement of these blocks is probably related to the formation and growth of the large right-lateral offset in the ridge axis. The axial valley becomes narrow at the offset (line 81, Figure 4) and widens into a depression at the northern end of Escanaba Trough. The highly reflective layered units at the base of the western valley wall on line 79 (Figure 4) may represent sheet flows erupted from the fault bounding the axial graben.

Escanaba Trough

Except for the northernmost segment of Escanaba Trough surveyed during cruise KK1-83-NP, details of the geological structure of this southern part of Gorda Ridge are poorly known. The Escanaba Trough differs markedly from the rest of the ridge because of the thick accumulation of hemipelagic and turbidite sediment on the floor of the axial graben. More than 390 m of Pleistocene sediment was drilled at site 35 (Figure 3) on leg 5 of the Deep Sea Drilling Project (McManus et al., 1970), and the hole did not reach basement.

The bathymetry of Escanaba Trough (Wilde et al., 1978) shows a uniform depth of about 3200 m along the 190-km-length of the axial valley floor. The axial valley widens from only a few km in the north at lat. 41°18′N to almost 18 km at lat. 40°37′N, where the flanking ridges begin to curve subtly to the east and are covered by a thick section of sediment north of the Mendocino fracture zone.

Along multichannel line 114, the spreading axis is marked by a 10-km-wide valley flanked by ridges which rise 1200 to 1300 m above the valley floor (Figure 5A). The sediment in the axial valley reaches a maximum thickness of 0.75 sec, or about 650 m. A large uplifted bench of turbidite sediment 3 km wide lies at the base of the eastern valley wall, and many small faults, shown by sea-floor offsets and diffraction patterns, cut the upper part of the sediment section in the central and western parts of the valley fill.

The strong reflection from the basement surface beneath the axial valley is quite irregular, with as much as 0.35 sec (300 m) of relief. The dome-shaped basement irregularities coincide with the more obvious sea-floor offsets, suggesting that the tectonic deformation is caused by emplacement of igneous intrusions into the sediment fill along the axis of Escanaba Trough. Disruption of the reflector sequences in the sediment section is especially prevalent near the center of the axial valley, and may be caused by sills or dikes intruding the sediment.

The outward-facing slopes of the flanking ridges, particularly those east of Escanaba Trough, are also characterized by the accumulation of sediment. These uplifted and tilted sections could represent parts of a former axial valley floor, now displaced through the process of crustal growth and spreading. Indeed, the bench of turbidite sediments along the eastern side of the axial valley and the uplifted basement surface beneath it may represent the initial stages in the formation of a new flanking ridge.

There is no large-scale surface expression of ridge structures along line 118, which crosses the southern end of Escanaba Trough only 6 km north of the Mendocino fracture zone (Figure 5B). A shallow graben about 50 m deep and 5 km wide and several other extensional faults mark the center of the spreading axis, which is displaced more than 10 km to the east of the topographic axis of Escanaba Trough. The sediment cover is considerably thicker along line 118 than it is to the north. Although it is difficult to trace the basement reflector, the sediment cover appears to be thickest (about 1 sec or 900 m) at the axis and thins both to the east and west. The faulted section and small sea-floor scarps demonstrate that this southernmost part of Gorda Ridge is undergoing active extensional tectonism.

A multichannel seismic-reflection line down the axis of Escanaba Trough (Figure 5C) shows some significant features not observed on either of the east–west transects. At the northern end of the line, near its crossing with line 114, the basement surface is irregular and lies at a depth of about 0.7 sec (about 600 m) beneath the valley floor. Doming of the sea floor over the uneven basement surface suggests a causal relationship, whereby recent intrusion of volcanic material along the spreading axis has faulted and uplifted the overlying sediment section.

To the south, the basement surface can be traced along an offset series of strong, nearly flat, reflector sequences. These could represent either sheet flow surfaces formed before deposition of the 500- to 600-m-thick sediment fill or sills that have been intruded into the sediment section.

A broad dome structure at least 5 km wide about 25 km from the northern end of the reflection record (Figure 5C) is underlain at a depth of 0.2 sec (180 m) by a strong reflector that closely mimics the sea floor. A deeper basement reflector trends beneath the edges of this strongly reflective horizon. However, it is uncertain whether the shallow reflectors represent the surface of a basement horst block or are caused instead by a sill-like or laccolithic structure within the sediment section.

The deep reflector sequences shown in the southern part of line 111 are offset and folded below a depth of about 0.6 sec (550 m), and the basement surface is difficult to discern. The thick sediment fill in this region has probably been subjected to intense tectonism related to the left-lateral relative movement between the southern Gorda Ridge and the Mendocino fracture zone.

In summary, the seismic-reflection data from the southern Gorda Ridge show that igneous activity and tectonic deformation have resulted in uplift of the sediment fill at several locations in Escanaba Trough and that numerous extensional faults occur at depth and cut the surficial sediment to produce sea-floor scarps. These faults could provide both the recharge and discharge paths for hydrothermal systems associated with active volcanism along the spreading axis.

Figure 3. SeaMARC II bathymetry of the Gorda Ridge at 250-m contour interval. Bathymetry south of the diagonal gap near lat. 41°N is from Wilde et al. (1978). Tracklines shown are from cruise KK1-83-NP. SeaMARC II side-scan sonar images and single-channel seismic-reflection data were obtained along 2200 km of trackline. The line numbers correspond to those shown in Figure 4. Tracklines for multichannel lines (MCL) 111, 114, and 118 (Figure 5) from cruise L11-81-NC are shown. Dredge locations from cruise KK2-83-NP (from Table 1) are shown and numbered. Dredges from the 1985 cruises are identified by letters. Locations A and B were the off- and on-axis site of a detailed survey consisting of 16 dredges, 2 gravity cores, and 9 camera stations. Two additional camera stations were completed at location E. Eight additional dredge locations in the President Jackson Seamounts (area D) and three in the Cascadia Depression are outside the area of the figure. The deep pull-apart basin (Gorda Depression) in the Blanco fracture zone shows the effects of left-lateral shearing and subsidence and is the site of three dredge stations (area C). The northern Gorda Ridge segment is narrow and gently curved, with unbroken flanking ridges. Maximum depth along the axis (>3750 m) is north of the ridge offset at lat. 42°24′N. The central Gorda Ridge is more structurally complex than the northern segment, with high broken ridges flanking the central valley. The largest right-lateral axial offset on Gorda Ridge, about 15 km, is at the southern end of the central Gorda segment.

Table 2. Normalized glass analyses of Gorda Ridge basalts recovered during cruise KK2–83–NP (in percent).

Flow Unit	Sample Number	SiO_2	Al_2O_3	FeO	MnO	MgO	CaO	Na_2O	K_2O	P_2O_5	TiO_2	SO_3
1	D1–21	51.1	15.4	9.61	0.19	7.57	11.2	2.65	0.15	0.20	1.62	0.29
1	D1–31	50.8	15.5	9.84	0.21	7.45	11.2	2.75	0.15	0.22	1.59	0.28
1	D1–12	50.3	15.5	9.66	0.20	7.44	11.7	2.89	0.15	0.22	1.62	0.35
1	D1–8	51.0	15.0	9.89	0.20	7.41	11.6	2.61	0.14	0.21	1.64	0.32
1	D2–3	50.6	15.1	9.93	0.21	7.60	11.9	2.62	0.09	0.17	1.60	0.25
2	D2–6	51.0	14.4	9.51	0.20	7.52	12.3	2.72	0.18	0.18	1.43	0.27
3	D2–5	51.1	15.1	9.80	0.19	7.27	11.4	2.84	0.13	0.20	1.75	0.32
1	D3–4c	51.2	15.0	9.36	0.21	7.46	12.0	2.68	0.17	0.19	1.49	0.26
1	D3–1	51.1	15.2	9.19	0.21	7.44	12.0	2.74	0.16	0.17	1.46	0.27
1	D3–4a	50.8	15.5	9.16	0.20	7.35	12.0	2.89	0.18	0.16	1.48	0.32
1	D3–6	50.9	14.9	9.24	0.18	7.34	12.4	2.86	0.18	0.19	1.49	0.33
1	D3–4b	50.7	14.9	9.28	0.19	7.33	12.5	2.89	0.19	0.17	1.49	0.36
2	D3–4	50.5	14.1	11.1	0.22	6.42	11.7	3.07	0.21	0.27	2.14	0.37
1	D4–15*	50.3	16.7	7.99	0.16	8.83	12.2	2.54	0.05	0.09	0.97	0.23
1	D4–8a	50.3	16.4	8.01	0.17	8.90	12.3	2.57	0.04	0.07	1.00	0.23
2	D4–5*	50.1	16.1	8.47	0.18	8.39	12.4	2.73	0.09	0.12	1.17	0.25
3	D6–18	51.0	14.2	10.7	0.22	7.41	11.0	2.87	0.16	0.21	1.94	0.30
1	D6–12	50.1	14.7	10.7	0.21	7.40	11.3	2.92	0.16	0.21	2.05	0.35
2	D6–4	50.8	15.1	10.1	0.20	7.26	11.4	2.69	0.13	0.18	1.81	0.32
4	D6–9*	50.8	15.1	10.1	0.22	7.16	11.5	2.75	0.14	0.21	1.76	0.30
3	D7–12*	49.9	15.6	9.78	0.21	7.89	12.0	2.65	0.09	0.16	1.43	0.29
2	D7–4	50.1	15.6	9.70	0.20	7.57	11.7	2.76	0.22	0.21	1.73	0.32
1	D7–8	51.1	15.3	9.37	0.19	7.40	11.9	2.71	0.17	0.18	1.43	0.27
1	D9–1	50.3	15.7	9.43	0.19	8.23	12.1	2.33	0.07	0.12	1.29	0.30
2	D9–2b	50.2	15.5	9.92	0.20	7.84	11.9	2.46	0.09	0.15	1.45	0.31
2	D9–3	50.0	15.1	10.1	0.21	7.66	12.3	2.64	0.09	0.11	1.49	0.33
1	D10–1	50.6	14.5	10.6	0.22	7.12	12.2	2.57	0.11	0.16	1.64	0.36
1	D11–9	50.3	15.4	9.28	0.20	8.24	12.11	2.63	0.08	0.11	1.39	0.29
2	D11–8	50.6	15.3	9.98	0.19	7.77	11.3	2.65	0.11	0.17	1.66	0.33
1	D12–2	51.0	14.5	10.3	0.20	7.94	11.1	2.74	0.10	0.15	1.68	0.33
2	D12–1	50.0	14.9	10.5	0.21	7.25	11.8	2.84	0.12	0.19	1.76	0.40
1	D13–11	50.6	15.5	9.44	0.20	8.18	12.0	2.43	0.09	0.12	1.25	0.25
1	D13–12*	50.4	15.4	9.30	0.19	8.06	12.3	2.55	0.10	0.12	1.30	0.32
1	D14–8	50.1	15.7	9.66	0.18	8.24	11.6	2.45	0.10	0.15	1.42	0.30
1	D14–13	50.3	15.5	9.77	0.21	8.16	11.6	2.53	0.11	0.14	1.42	0.24
1	D14–2	50.0	15.4	9.67	0.19	8.05	11.9	2.71	0.11	0.15	1.45	0.36
1	D14–18	49.5	15.9	9.56	0.20	7.96	12.1	2.72	0.11	0.14	1.44	0.34
1	D15–1	50.9	15.1	10.3	0.21	7.48	11.1	2.72	0.11	0.18	1.69	0.32
1	D16–3	50.1	15.8	9.71	0.20	7.36	11.5	3.00	0.14	0.16	1.67	0.34
2	D16–1	50.7	14.6	10.9	0.22	6.88	11.4	2.66	0.26	0.19	1.90	0.35
2	D16–2	50.5	14.0	11.0	0.21	6.78	11.9	2.63	0.27	0.25	1.97	0.37
1	D17–10	50.7	15.7	9.62	0.21	7.42	11.5	2.76	0.19	0.16	1.50	0.24

*Figures represent average of two separate analyses.
Analysts: D. A. Clague and Alice Davis.

Table 3. Chemical analyses of hydrothermal nontronite/Mn-oxide deposits recovered from dredge station 16, cruise KK2–83–NP.

Sample	Mn (%)	Fe (%)	Mo (ppm)	Zn (ppm)	Cu (ppm)	Ni (ppm)
1	0.14	28.75	<100	190	75	30
2	0.14	25.91	<100	45	20	<30
3	0.16	18.39	<100	120	45	70
4	0.14	17.94	<100	75	35	55
5	0.15	15.27	<100	110	50	65
6	5.80	14.31	200	130	60	70
7	21.95	7.62	330	115	45	65
8	37.09	2.13	385	170	85	175
9	43.19	1.14	1400	185	65	170
10	44.76	0.94	1700	110	250	975

Arranged in order of decreasing Fe. Pb and Co not detected at 50 ppm levels in any sample.
Analyst: R. Bouse; all analyses by atomic absorption spectrophotometry.

and Na but, as in the case of boehmite, the source of the Al is unclear. The Al is probably derived from the basaltic glass, as Staudigel and Hart (1983) show that Al is lost from sideromelane during formation of palagonite.

Quartz is generally present with chlorite in fractures. Disordered kaolinite with minor K-feldspar(?) occurs as a white coating on glass from dredge 14.

In addition to the hydrothermal nontronite-Mn-oxide samples described below, dredge 16 also contains altered sulfide. Many of the large pillow basalt fragments are coated on fractures with a yellowish material that smells like sulfur. X-ray diffraction analyses reveal that pyrite is present in some of this material. We interpret these deposits to be altered sulfide similar to that described below.

Massive Low-Temperature Hydrothermal Deposits

Dredge 16 recovered about 100 kg of hydrothermal nontronite and associated Fe and Mn oxides. Descriptions of 10 subsamples are given in Clague et al. (1984), and partial chemical analyses of samples of this deposit are presented in Table 3. Similar low-temperature hydrothermal nontronite deposits have been described from other locations including the Galapagos spreading center (Corliss et al., 1978), the Gulf of Aden (Cann, Winter, and Pritchard, 1977), the Explorer Ridge (Grill et al., 1981), and the Juan de Fuca Ridge (Murnane and Clague, 1983). The Gorda Ridge nontronite deposit has an upper Mn-oxide layer, an intermediate zone of Fe oxide, and an inner zone of bedded nontronite and hydrothermally altered detrital sediment. Unlike the other described deposits, the Gorda deposit is riddled by worm tubes indicating the presence of abundant fauna during or after formation of the deposit. No worms were recovered with the hydrothermal deposit, indicating that this location is not a site of present-day hydrothermal activity.

The chemical data in Table 3 demonstrate the low amounts of Zn, Cu, Ni, Pb, and Co present. Ni and Mo show a strong positive correlation with Mn concentration. However, all the analyzed samples are notably low in Co, Ni, and Cu. The Mn-oxide phases in the outer crust, identified by x-ray diffraction, include birnessite, natrobirnessite, todorokite, manjiroite (?), and manganite (?). The bulk of the deposit consists of interbedded nontronite and detrital sediment composed of illite, smectite, chlorite, quartz, and plagioclase. The high-Fe samples do not contain any well-crystallized Fe-oxide phase; amorphous $Fe(OH)_3$ probably constitutes the bulk of these samples.

Detailed studies of the nontronite and Mn oxides are in progress and should clarify the temperatures at which these hydrothermal deposits formed. Similar deposits elsewhere formed at 19° to 57°C (Murnane and Clague, 1983). These temperatures are well below the temperatures at which most sulfide deposition occurs (250° to 400°C) at submarine hydrothermal vents (Oudin, 1982). The nontronite/Mn-oxide deposits represent a low-temperature hydrothermal deposit that could stratigraphically overlie, or occur adjacent to, higher temperature sulfide deposits. In any case, hydrothermal activity persisted at this location for an extended time. The longevity of the hydrothermal discharge is evidenced by the interlayering of hydrothermal and detrital sediment.

Disseminated Sulfide Mineralization

Sulfide mineralization is present in a handful of samples from dredge 7. None of the hydrothermal sulfide found in the 1983 dredges is massive. Two sulfide assemblages are present. In the first type, pyrite and pyrrhotite occur with a Mn-Fe-Ca-carbonate (manganosiderite?) lining fractures in pillow basalt fragments. Pyrrhotite occurs as 30-micron flat hexagonal plates (Figure 8A) set in a clay matrix (Figure 8B). Figure 8C is a closeup of the manganosiderite(?) shown in the upper right hand corner of Figure 8A. Pyrite in these samples is present as < 10-micron to about 30-micron cubes that are partially corroded (Figures 8D, 8E).

The second sulfide-bearing assemblage in samples from dredge 7 consists of pyrite, sphalerite, and barite. This sulfide-sulfate assemblage lines open channels inside contorted basaltic fragments. Figure 8F shows euhedral barite crystals projecting into the open channel. Figure 8G shows cubic pyrite as 30-micron crystals and as smaller (about 2-micron) crystals growing on clay minerals. Figure 8H shows the entire assemblage of barite, sphalerite, and euhedral strongly corroded pyrite; Figure 8I is a closeup of the barite in Figure 8H.

The 1985 cruises recovered mineralized greenstone from fault scarps in the Blanco fracture zone (R. Hart and D. Clague, unpublished data). These samples contain rare to fairly common Fe and Cu sulfides in a matrix of mainly quartz,

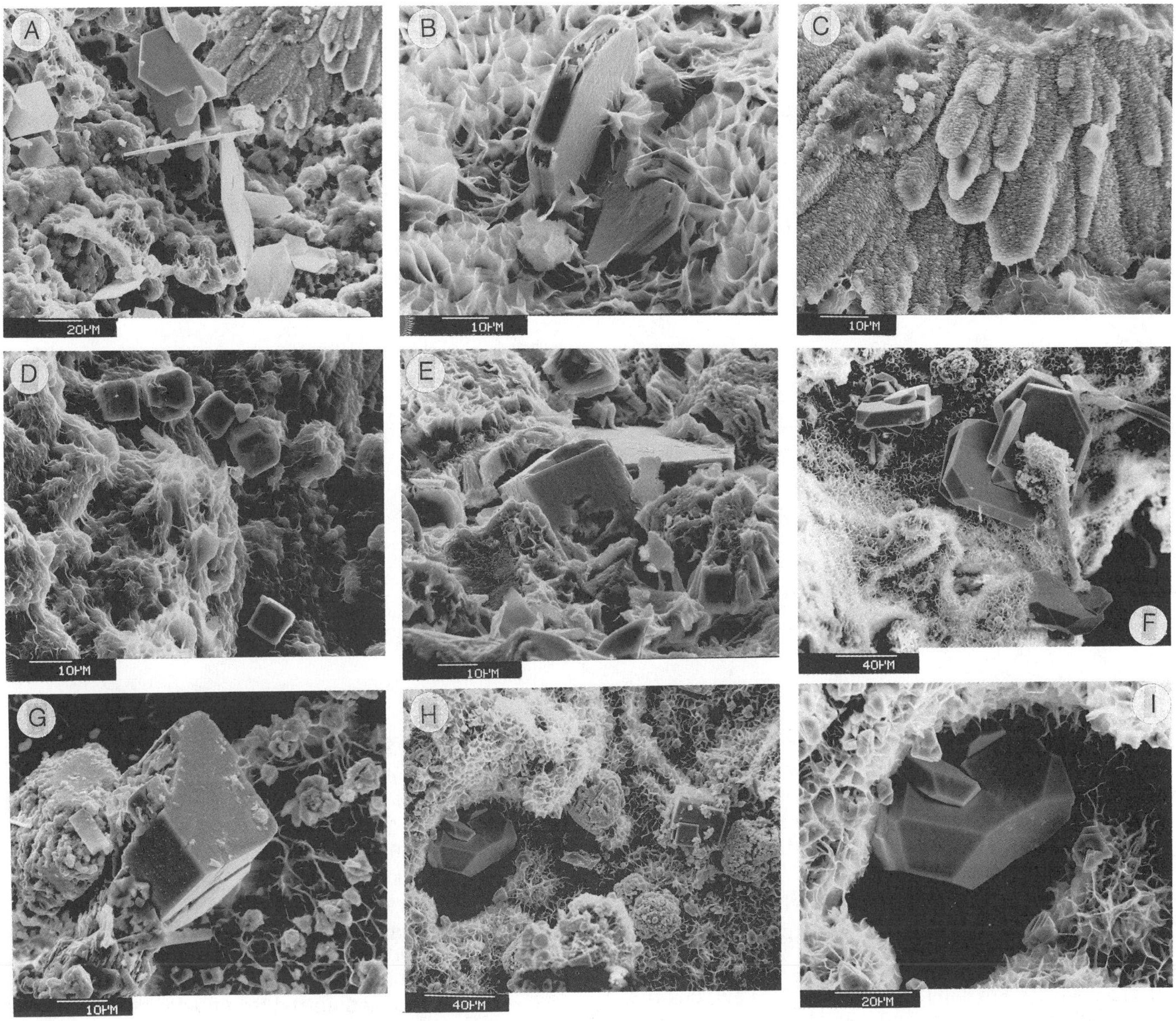

Figure 8. Scanning electron micrographs of hydrothermal phases in lavas from dredge 7. A, Unaltered hexagonal plates of pyrrhotite on smectite. Material in upper right-hand corner is manganosiderite (an Fe-Mn-Ca carbonate). B, Corroded hexagonal plates of pyrrhotite on smectite. C, Enlargement of manganosiderite shown in A. D, Rounded pyrite cubes on smectite. E, Corroded pyrite in smectite. F, Euhedral barite crystals projecting into open cavity. G, Euhedral pyrite crystal projecting into open cavity. H, Euhedral barite (left), corroded sphalerite (center), and euhedral and corroded pyrite (right center, right edge). I, Enlargement of euhedral barite in H. Note small pyrite crystals on the smectite.

chlorite, and albite. Many of the samples are breccia crosscut by small veins. The lavas, diabase, and microgabbro recovered have been modified by hydrothermal solutions to form the greenstone minerals assemblage.

Mn-Oxide Crusts

Dredge 17 sampled an uplifted off-axis fault block and recovered relatively fresh pillow basalt encrusted by layers of brown hydrogenous Mn oxide as much as 1 cm thick. X-ray diffraction shows the crusts are composed of δ-MnO, todorokite, birnessite(?), and abundant smectite. The bulk Mn-oxide crust was analyzed for Mn, Fe, Co, Ni, and Cu by atomic absorption spectrophotometry. The concentrations determined are: Mn = 14.59 wt%, Fe = 15.07 wt%, Co = 0.26 wt%, Ni = 0.21 wt%, and Cu = 230 ppm. The crust has a similar Co content, but lower Ni and Cu, than abyssal Fe-Mn nodules.

The samples dredged in 1985 from the President Jackson Seamounts are commonly encrusted by layers of pale-brown hydrogenous Mn oxide as much as about 1 cm thick. No analyses have been completed on these samples, which are similar in appearance to those described from dredge 17.

Sediment Samples

Sediment samples were collected during the 1983 cruise in a small (4″ × 15″) pipe dredge dragged behind the large chain bag dredge. The pipe dredge was lined with a canvas sample bag; we refer to these as "sock" samples. In addition, large chunks of sticky gray sediment were recovered in several rock dredges. Descriptions of the sediment samples and

Table 4. Composition of sediments recovered during cruise KK2-83-NP.

	Mn (%)	Fe (%)	Zn (ppm)	Ni (ppm)	Cu (ppm)	Co (ppm)	Pb (ppm)	Mo (ppm)
Gray Muds								
D7 MUD	.21	5.67	210	210	120	<50	270	<100
D3 MUD	.08	5.44	220	185	95	<50	<50	<100
D11 SOCK	.12	5.28	225	180	100	<50	<50	<100
D17 MUD	.07	6.40	190	170	60	<50	<50	<100
D2 SOCK	.15	5.52	215	175	90	<50	<50	<100
D15 TOP OF SOCK	.29	5.39	200	155	105	<50	<50	<100
Light-Brown Muds								
D6 SOCK	.25	5.81	185	170	90	<50	<50	<100
D3 SOCK	.27	5.57	215	180	90	<50	<50	<100
D10 SOCK	.43	5.47	200	190	105	<50	<50	<100
Dark-Brown Mud								
D7 SOCK	.36	5.64	200	190	105	<50	<50	<100
Chocolate-Brown Muds								
D1 SOCK	.53	5.70	275	200	100	<50	<50	<100
D15 BOTTOM OF SOCK	.41	5.24	150	145	80	<50	<50	<100
D12 SOCK	1.16	5.56	210	200	105	<50	<50	<100
D16 SOCK	1.43	5.43	170	155	95	<50	<50	<100

Analyst: Robin Bouse. All analyses by atomic absorption spectrophotometry. Samples arranged in increasing color intensity from top to bottom.

mineralogy are given in Clague et al. (1984); bulk chemical analyses determined by atomic absorption spectrophotometry are given in Table 4.

Most of the samples reacted very little when soaked in Morgan's solution and peroxide, indicating very low calcium carbonate and organic carbon contents. Samples D11(MUD), D17(MUD), and D17(SOCK) contain the most calcium carbonate and organic carbon.

The sediment samples range in color from a light olive gray to dusky yellow brown. Regardless of color, all the samples are composed of similar proportions of illite, chlorite, and smectite, and different amounts of quartz and plagioclase. The variation in color is apparently due solely to the addition of small (< 1.5 wt%) amounts of Mn oxide of probable hydrothermal origin. Fe concentration is constant through all the samples, as are Zn, Ni, and Cu. Co, Pb, and Mo were below detection levels of 50, 50, and 100 ppm, respectively, in all but the light-colored sediment from dredge 7 which contained 270 ppm Pb.

The brown sock samples generally contain abundant glass fragments and agglutinated benthic foraminifera. The gray samples generally have few glass fragments and agglutinated benthic foraminifera but contain abundant radiolaria and diatoms, commonly clumped together in fecal pellets. We interpret the brown sediment to be thin deposits on the basalt surface that are brown owing to the addition of hydrothermal Mn oxides. The abundant benthic foraminifers (described in Clague et al., 1984) probably reflect proximity to hydrothermal vents. Fe in hydrothermal vent waters precipitates nearer the vents than Mn, which is more mobile owing to its greater solubility in seawater and slow nucleation kinetics. The brown muds recovered from the Gorda Ridge are enriched in Mn but not in Fe, which suggests that they were deposited relatively far from hydrothermal vents.

Massive Sulfide Mineralization

Preliminary results from the 1985 U.S. Geological Survey cruises indicate that massive sulfide deposits occur in at least two areas of the Gorda Ridge. Dredging in the Escanaba Trough near lat. 40°45′N recovered small lumps of massive sulfide consisting almost entirely of pyrrhotite. The samples were recovered from the flank of one of four volcanic centers aligned along the axis of the Escanaba Trough (M. Holmes and J. Morton, unpublished data). High heat flow in the sediment surrounding these volcanic centers (D. Abbott, 1985, oral communication) and the unoxidized character of the recovered samples suggest that the deposits are the result of active hydrothermal discharge. These samples apparently are sediment hosted, in contrast with the chimney structures of the sulfides discovered at vent fields on the Juan de Fuca Ridge.

In addition to the dredged sulfide samples from the Escanaba Trough, the 1985 cruises returned with bottom photographs of two massive sulfide deposits located on lava flows in the center of the axial valley near lat. 42°45′N. The axial valley in this area is characterized by numerous young unaltered lava flows. Fissures are abundant, indicating tectonic activity since the last eruptive events. This site is near the off-axis source of a hydrothermal plume identified by both National Oceanographic and Atmospheric Administration (NOAA) and Oregon State University scientists (E. Baker, G. Massoth, and R. Collier, 1985, oral communication).

MINERAL POTENTIAL OF THE GORDA RIDGE

Hydrothermal activity appears to be widespread along the axis of the unsedimented northern two-thirds of the Gorda

Ridge. With the exception of the disseminated sulfides, sulfates, and talc deposits recovered from dredge 7 near lat. 42°28.4′N and the deposit photographed at lat. 42°45′N, the evidence indicates low-temperature hydrothermal activity or the interaction of seawater and hot rock on the sea floor. At these two sites, high-temperature hydrothermal fluids capable of precipitating sulfides apparently debouched on the seabed. We have no data that would suggest how long-lived these hydrothermal systems may be, if they are presently active, or whether massive sulfide deposits occur that are similar to those described from the East Pacific Rise at lat. 21°N (RISE Project Group, 1980; CYAMEX Scientific Team, 1979, 1981; Haymon and Kastner, 1981; Styrt et al., 1981), the Galapagos spreading center (Law et al., 1981), the Gulf of California (Lonsdale et al., 1980; Koski et al., 1985), and the Juan de Fuca Ridge (Normark et al., 1982, 1983; Koski, Clague, and Oudin, 1984; Koski et al., this volume). Biological density, particularly of benthic foraminifers, may indicate widespread hydrothermal activity. Chemical analyses of hydrothermal nontronite and associated Mn oxide, of Mn-oxide crusts (recovered in dredge 17 from an off-axis tilted fault block), and of brown muds all show low concentrations of Cu, Zn, Ni, Co, and Pb; Mn and Fe are the only metals present in significant concentrations.

The geophysical data demonstrate that the axial valley of the northern two-thirds of the Gorda Ridge is deep, narrow, and rough. Detailed camera surveys, used so successfully to locate massive sulfide deposits on the East Pacific Rise and Juan de Fuca Ridge, are difficult to perform in such rough terrain. Nevertheless, camera surveys along the axis near lat. 42°45′N show massive sulfide deposits at two sites. Clague et al. (1984) concluded that the Escanaba Trough, where little geological/geophysical surveying had been done, would prove to be the most promising part of the Gorda Ridge for sulfide exploration. This conclusion was based on the relative ease of exploration in the flat sediment-filled trough (using deep-tow side-scan and photographic techniques) and that any deposits might be higher grade because the sediment could serve as a source of metals in addition to the underlying basaltic crust. In addition, the multichannel seismic-reflection records obtained in 1981 indicated the existence of intrusive features beneath the sediments in Escanaba Trough.

Many of the large ophiolite-hosted massive sulfide deposits are now thought to have formed in back-arc basins, and some are hosted by volcanogenic sediment. Although the Gorda Ridge is not situated in a back-arc basin, the Escanaba Trough is filled with volcanogenic sediment that is geochemically similar to that shed into back-arc basins from the active volcanic arcs. The tectonic setting of the Gorda Ridge is not identical to that inferred for large ophiolite-hosted sulfide deposits, but the physical similarities between the Escanaba Trough and back-arc basins suggests that the Escanaba Trough may contain large sediment-hosted sulfide deposits. The massive sulfide samples recovered in 1985 from the Escanaba Trough, although of low grade, indicate that hydrothermal circulation and deposition of metals is occurring in the sediment filling the Escanaba Trough.

ACKNOWLEDGMENTS

The data from the 1985 cruises are the unpublished data of the writers and co-chief scientists R. A. Koski, J. L. Morton, and A. Stevenson; and Oregon State University guest scientists D. Abbott, D. Bibbee, M. Fisk, R. Hart, K. Howard, R. Karlin, and M. Lyle, who kindly allowed us to briefly discuss the preliminary results of these cruises. The success of the 1985 and earlier cruises was due in large part to the professionalism of the officers, crew, and scientific and technical staff of the R/V Kana Keoki and the R/V S.P. Lee. R. Embley kindly provided us with Seabeam bathymetry of the President Jackson Seamounts, which served as a base map for our dredge operations. R. A. Koski and A. Davis reviewed the manuscript and improved both the organization and content.

REFERENCES CITED

Atwater, T., and J. D. Mudie, 1973, Detailed near-bottom geophysical study of the Gorda Rise: Journal of Geophysical Research, v. 78, p. 8665–8686.

Blackington, J. G., D. M. Hussong, and J. G. Kosalos, 1983, First results from a combination side-scan sonar and sea-floor mapping system (SeaMARC II): 15th Annual Offshore Technology Conference Proceedings, v. 1, p. 307–311.

Cann, J. R., C. K. Winter, and R. G. Pritchard, 1977, A hydrothermal deposit from the floor of the Gulf of Aden: Mineral Magazine, v. 41, p. 193–199.

Clague, D. A., W. Friesen, P. Quinterno, M. Holmes, J. Morton, R. Bouse, L. Morgenson, and A. Davis, 1984, Preliminary geological, geophysical, and biological data from the Gorda Ridge: U.S. Geological Survey Open-File Report 84-364, 47 p.

Corliss, J. B., M. Lyle, J. Dymond, and K. Crane, 1978, The chemistry of hydrothermal mounds near the Galapagos Rift: Earth and Planetary Science Letters, v. 40, p. 12–24.

CYAMEX Scientific Team, 1979, Massive deep-sea sulfide ore deposits discovered on the East Pacific Rise: Nature, v. 277, p. 523–528.

CYAMEX Scientific Team, 1981, First manned submersible dives on the East Pacific Rise at 21N (Project RITA): General results: Marine Geophysical Research, v. 4, p. 345–379.

Davis, A., and D. A. Clague, in press, Geochemistry, mineralogy, and petrogenesis of lavas from the Gorda Ridge between latitude 43°N and 41.4°N: Journal of Geophysical Research.

Eaby, J. S., and D. A. Clague, 1982, Preliminary descriptions of basalt from the southern Juan de Fuca Ridge: U.S. Geological Survey Open-File Report 82-200C, 12 p.

Embley, R. W., L. Kulm, and D. Abbott, 1984, Kinematic and structural development of the Blanco fracture zone: EOS (American Geophysical Union Transactions), v. 65, p. 1110.

Grill, E. V., R. L. Chase, R. D. MacDonald, and J. W. Murray, 1981, A hydrothermal deposit from Explorer Ridge in the northeast Pacific Ocean: Earth and Planetary Science Letters, v. 52, p. 142–150.

Haymon, R. M., and M. Kastner, 1981, Hot-spring deposits on the East Pacific Rise at 21N: Preliminary description of mineralogy and genesis: Earth and Planetary Science Letters, v. 53, p. 363–381.

Johnson, D. A. L., and B. T. R. Lewis, 1984, Microearthquakes in the axial valley of the northern Gorda Ridge at 43N: EOS (American Geophysical Union Transactions), v. 65, p. 1111.

Kay, R., N. J. Hubbard, and P. W. Gast, 1970, Chemical characteristics and origin of oceanic ridge volcanic rocks: Journal of Geophysical Research, v. 75, p. 1585–1613.

Kosalos, J. G., and D. Chayes, 1983, A portable system for ocean bottom imaging and charting: Proceedings, Marine Technology Society, Oceans 83, p. 649–656.

Koski, R. A., D. A. Clague, and E. Oudin, 1984, Mineralogy and chemistry of massive sulfide deposits from the Juan de Fuca Ridge: GSA Bulletin, v. 84, p. 930–945.

Koski, R. A., P. F. Lonsdale, W. C. Shanks, M. E. Berndt, and S. S.

Howe, 1985, Mineralogy and geochemistry of a sediment-hosted hydrothermal sulfide deposit from the southern trough of Guaymas Basin: Gulf of California: Journal of Geophysical Research, v. 90, p. 6695–6707.

Koski, R. A., S. D. Scott, M. D. Hannington, J. R. Delaney, and M. K. Tivey, 1987, Hydrothermal processes and massive sulfide deposits on the Juan de Fuca Ridge and other northeast Pacific spreading axes, this volume.

Law, S., A. Malahoff, R. Embley, and D. Fornari, 1981, Massive polymetallic sulfides of the Galapagos rift: EOS (American Geophysical Union Transactions), v. 62, p. 1027.

Lonsdale, P. F., J. L. Bischoff, V. M. Burns, M. Kastner, and R. E. Sweeney, 1980, A high temperature hydrothermal deposit on the seabed at a Gulf of California spreading center: Earth and Planetary Science Letters, v. 49, p. 8–20.

Malahoff, A., R. Embley, S. Hammond, W. Ryan, and K. Crane, 1982, Juan de Fuca and Gorda Ridge axial ridge morphology and tectonics from combined Seabeam and SeaMARC data: EOS (American Geophysical Union Transactions), v. 63, p. 1147.

McManus, D. A., et al., 1970, Initial reports of Deep Sea Drilling Project (Vol. 5): Washington, D.C., U.S. Government Printing Office, 827 p.

Melson, W. G., G. R. Byerly, J. S. Nelen, T. O'Hearn, T. L. Wright, and T. Vallier, 1977, A catalog of the major element chemistry of abyssal volcanic glasses: Smithsonian Contributions to Earth Science, v. 19, p. 31–60.

Murnane, R., and D. A. Clague, 1983, Nontronite from a low-temperature hydrothermal system on the Juan de Fuca Ridge: Earth and Planetary Science Letters, v. 65, p. 343–352.

Normark, W. R., J. L. Morton, R. A. Koski, D. A. Clague, and J. R. Delaney, 1983, Active hydrothermal vents and sulfide deposits on the southern Juan de Fuca Ridge: Geology, v. 11, p. 158–163.

Normark, W. R., J. E. Lupton, J. W. Murray, J. R. Delaney, R. A. Koski, D. A. Clague, J. L. Morton, and H. P. Johnson, 1982, Polymetallic sulfide deposits and water column tracers of active hydrothermal vents on the southern Juan de Fuca Ridge: Marine Technology Society Journal, v. 16, p. 46–53.

Northrup, J., H. W. Menard, and F. K. Duennebier, 1968, Seismic and bathymetric evidence of a fracture zone on the Gorda Ridge: Science, v. 161, p. 688–690.

Oudin, E., 1982, Etudes mineralogique et geochemique des depots sulfures sousmarins actuels de la ride est-pacifique (21°N): Documents du Bureau de Recherches Geologiques et Minieres, France, 25, 241 p.

Riddihough, R. P., 1980, Gorda plate motions from magnetic anomaly analysis: Earth and Planetary Science Letters, v. 51, p. 163–170.

RISE Project Group, 1980, East Pacific Rise: Hot springs and geophysical experiments: Science, v. 207, p. 1421–1433.

Staudigel, H., and S. R. Hart, 1983, Alteration of basaltic glass: Mechanisms and significance for the oceanic crust-seawater budget: Geochimica et Cosmochimica Acta, v. 47, p. 337–350.

Stoddard, P. R., and R. L. Carlson, 1984, Deformation of the Gorda plate: EOS (American Geophysical Union Transactions), v. 65, p. 1100.

Styrt, M. M., A. J. Brackmann, H. D. Holland, B. C. Clark, V. Pusutha-Arnond, C. S. Eldridge, and H. Ohmoto, 1981, The mineralogy and isotopic composition of sulfur in hydrothermal sulfide/sulfate deposits on the East Pacific Rise at 21°N latitude: Earth and Planetary Science Letters, v. 53, p. 382–390.

Wakeham, S. E., 1978, Petrochemical patterns in young pillow basalts dredged from the Juan de Fuca and Gorda Ridges: Corvallis, Oregon, Oregon State University unpublished M. S. thesis, 95 p.

Wilde, P., T. E. Chase, M. L. Holmes, W. R. Normark, J. A. Thomas, D. S. McCulloch, and L. D. Kulm, 1978, Oceanographic data off northern California-southern Oregon, 40–43° North including the Gorda deep-sea fan: Lawrence-Berkeley Laboratory Publication 251, 1 mapsheet.

Wilson, D. S., 1984, The Gorda plate: When is a plate not a plate?: EOS (American Geophysical Union Transactions), v. 65, p. 1100.

Wilson, D. S., 1985, Tectonic history of the Juan de Fuca Ridge: Stanford, California, Stanford University unpublished Ph.D. dissertation, 105 p.

CHAPTER TWENTY-FIVE

GAS HYDRATES OFFSHORE ALASKA AND WESTERN CONTINENTAL UNITED STATES

K. A. Kvenvolden

U.S. Geological Survey
Menlo Park, California

ABSTRACT

Gas hydrates are inferred from marine seismic profiles to be present in outer continental margin sediments under the Beaufort Sea, the Bering Sea, the north Pacific Ocean, the Gulf of Alaska, and the eastern Pacific Ocean offshore northern California. The bases of these gas hydrates occur at subbottom depths ranging from 135 to 1000 m at water depths ranging from 400 to 4500 m. The areal extent of gas hydrates offshore northern Alaska is at least 7500 km^2 and offshore northern California at least 3000 km^2. Where the depth to the base of the gas hydrate is well defined, geothermal gradients ranging from 20° to 62°C/km have been estimated.

The geological consequences of gas-hydrate deposits in oceanic sediment are varied. Slumps and slides can possibly be triggered by decomposition of these deposits and the resulting release of gas; however, gas hydrates have not yet been implicated as causing sedimentary structures offshore Alaska and California. Gas hydrates can be a potential hazard during drilling and thus will need to be considered when exploration and production of petroleum extends to offshore areas where gas hydrates occur, such as in the Beaufort Sea adjacent to the North Slope of Alaska. Finally, gas hydrates may be an unconventional source of natural gas for the future. Estimates of the gas hydrate resources offshore Alaska and California are enormous (1800 to 4,500,000 trillion cu ft or 5.1×10^{13} to 1.3×10^{17} m^3), exceeding the estimated possible gas supply of the same region of 13 trillion cu ft (3.7×10^{11} m^3) by many orders of magnitude. However, the reality of the gas-hydrate resource estimates is very uncertain.

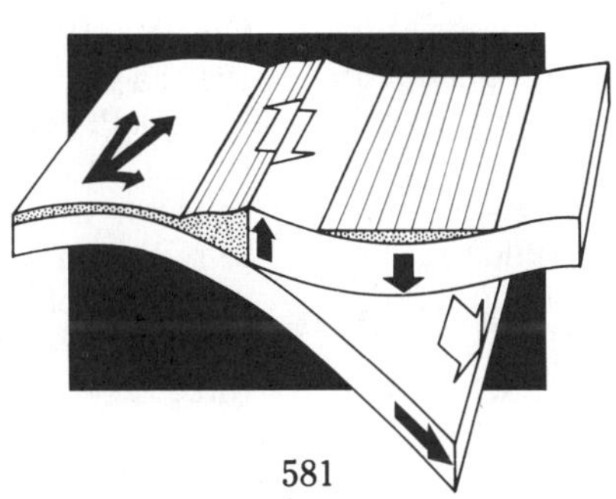

INTRODUCTION

Outer continental margins may contain vast quantities of natural gas locked in or trapped beneath gas hydrates. These solid, icelike substances, composed mainly of methane and water, are inferred to occur worldwide in oceanic sediments at water depths in excess of 300 to 500 m (Kvenvolden and Barnard, 1983a). At pressures greater than about 30 atm and temperatures <30°C, wherever large amounts of methane are present, natural gas hydrates should be the common form in which methane occurs in sediments. The actual pressure-temperature conditions under which gas hydrates are believed to be stable is shown in Figure 1. The geothermal gradient, however, limits the depths within sediment at which gas hydrates can occur. The gas hydrate base coincides with a pressure-temperature horizon that represents the maximum depth at which the gas hydrate is stable.

In gas hydrates, methane is condensed within a solid framework of water molecules so that the molar ratio of methane to water can be nearly 1:6. Thus, one volume of solid gas hydrate may have about 170 volumes of methane gas. Gas hydrates, therefore, may contain much more methane per unit volume than could be accommodated as free gas in the same space. In addition to serving as a kind of reservoir for methane, gas hydrates are also impermeable to free gas so that a layer of gas hydrate could trap free gas beneath it. These entrapping properties make gas hydrates important candidates for potential energy resources in outer continental margins (Hedberg, 1980). This chapter focuses on the occurrences and implications of gas hydrates inferred to be present offshore Alaska and the western conterminous United States.

The identification of gas hydrates in oceanic sediment has been based mainly on the appearance of an anomalous acoustic reflector on marine seismic records (Shipley et al., 1979). This reflector correlates with the base of the gas hydrate zone and approximately parallels the sea floor. If the geothermal gradient is known, the depths at which this reflector appears can be predicted on the basis of the pressure-temperature stability field for gas hydrates (Figure 1). This bottom-simulating reflector (BSR) probably results from the velocity contrast between sediment containing gas hydrates (higher velocities) and the underlying sediment (lower velocity). The lower velocities are believed to result from the absence of gas hydrate and the possible presence of free gas. BSRs attributed to gas hydrates have been observed on marine seismic records from the Blake Outer Ridge offshore Florida (Stoll, Ewing, and Bryan, 1971; Ewing and Hollister, 1972; Shipley et al., 1979; Dillon, Grow, and Paull, 1980; Paull and Dillon, 1981; Markl and Bryan, 1983), offshore North Carolina (Popenoe, Coward, and Cashman, 1982), and offshore New Jersey (Tucholke, Bryan, and Ewing, 1977). Shipley et al. (1979) also described BSRs attributed to gas hydrates in sediments in the western Gulf of Mexico, off the northern coasts of Colombia and Panama, and along the Pacific Coast of Central America from Panama to Acapulco, Mexico. That gas hydrates are present in sediments at some of these areas has been confirmed by drilling and sampling in the Blake Outer Ridge (Kvenvolden and Barnard, 1983b), offshore Guatemala (Harrison and Curiale, 1982; Kvenvolden and McDonald, 1985), and offshore Mexico (Shipley and Didyk, 1982).

Not all BSRs can be directly related to the presence of gas hydrates. In some areas, these acoustic features may also result from temperature-controlled diagenetic effects. During Deep Sea Drilling Project (DSDP) Leg 19, Scholl and Creager (1973) noted seismic reflectors that tend to parallel the sea floor in some Bering Sea sediments draped on the Umnak Plateau, and they first used the acronym BSR for the bottom-simulating reflector. At two sites, 184 and 185, the reflector was penetrated. Although methane was observed in sediment at site 185, no evidence for gas hydrates was noted. Scholl and Creager (1973) attribute this reflector to a lithologic transition from hemipelagic diatom ooze to indurated claystone. Hein et al. (1978) confirmed that opal-A is

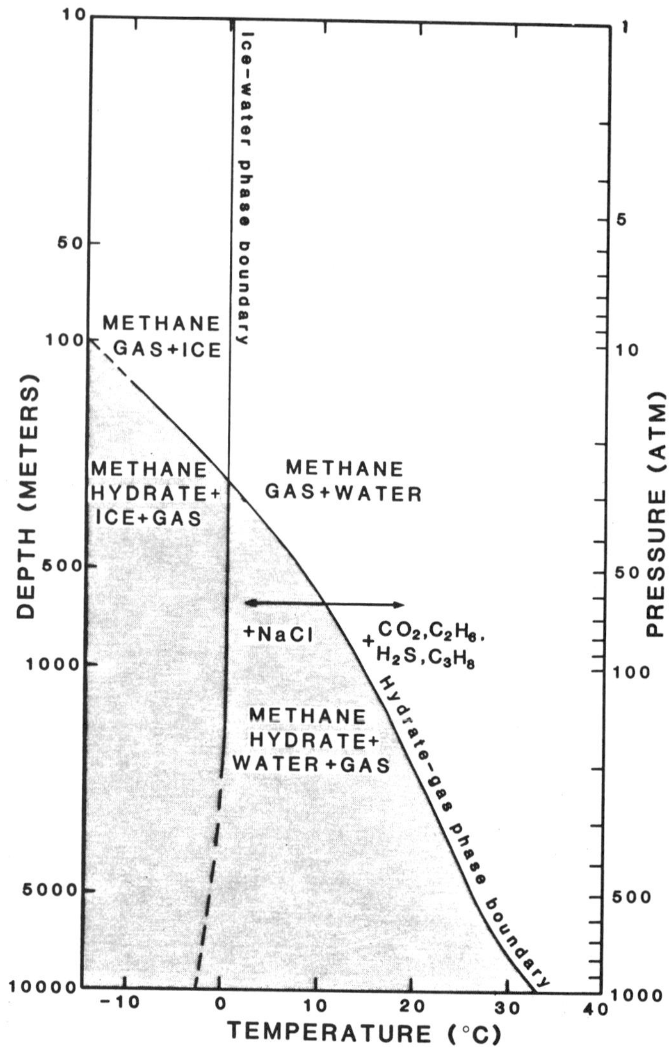

Figure 1. Phase diagram showing boundary between free methane gas (no pattern) and methane hydrate (pattern) for a pure-water and pure-methane system. Addition of NaCl to water shifts the hydrate-gas phase boundary to the left. Adding CO_2, H_2O, ethane, and propane to methane shifts that boundary to the right and increases the area of the gas-hydrate stability field. For natural gas in marine sediments, the shifts are of similar magnitude but in opposite directions. The effects approximately cancel each other. Thus, the boundary for a pure-water and pure-methane system provides a reasonable estimate of the pressure-temperature conditions under which natural gas hydrates, composed mainly of methane, will be stable on continental margins (Claypool and Kaplan, 1974). Depth scale assumes lithostatic and hydrostatic pressure gradients of 0.1 atm/m. Redrawn after Katz et al. (1959).

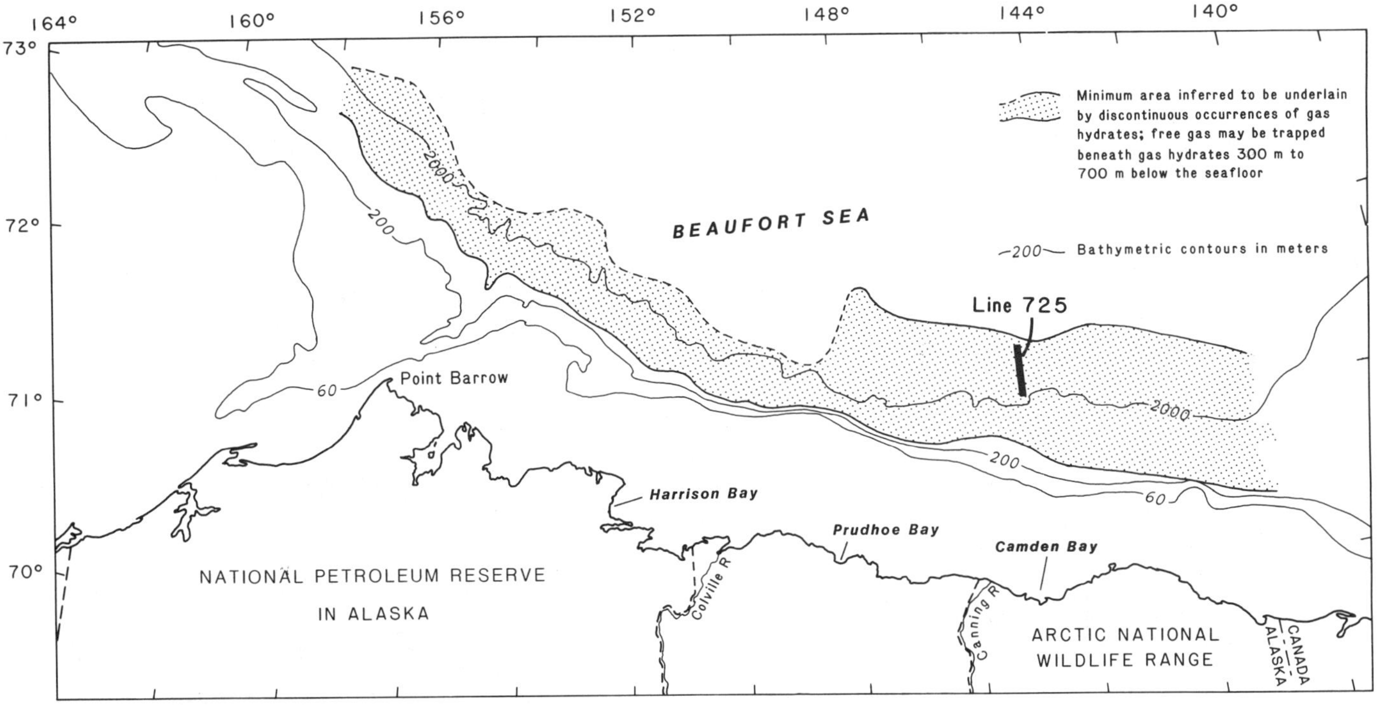

Figure 2. Map showing the minimal areal extent of gas hydrate offshore northern Alaska in the Beaufort Sea. Location of seismic line 725 (Figure 3) is given. Redrawn after Grantz et al. (1982).

transformed to opal-CT at the depth of the BSR in the Bering Sea sediment. Thus, there are two types of BSRs, one indicating the base of the gas hydrate zone and the other signaling a diagenetic boundary (Marlow et al., 1981). For the remainder of this chapter, unless otherwise indicated, BSR will refer to the anomalous acoustic reflector associated with gas hydrate occurrence. This reflector appears to parallel the sea floor at a depth that can be predicted from the physical-chemical properties of gas hydrates. The reflector is generally characterized by reflection polarity reversals and large reflection coefficients (Shipley et al., 1979).

GAS HYDRATES OFFSHORE ALASKA

Gas hydrates have been inferred to be present in sediment of the Beaufort Sea, the Bering Sea, the north Pacific Ocean, and the Gulf of Alaska because of the occurrence of BSRs on seismic records from marine geophysical surveys.

Beaufort Sea

The most extensive occurrence of gas hydrates thus far mapped offshore Alaska is found in sediment of the Beaufort Sea (Grantz, Boucher, and Whitney, 1976; Grantz and Dinter, 1980; Grantz et al., 1982; Grantz, May, and Dinter, this volume). Seismic surveys on the outermost continental shelf and continental slope north of Alaska show BSRs that lie between 300 and 700 m beneath the sea floor, where water depths range from about 400 to >2800 m. The area of continental slope over which the inferred gas hydrate occurs exceeds 7500 km^2 (Figure 2). The total possible extent of the gas hydrate may reach at least 30,000 km^2 and encompass the entire Arctic Basin where water is deeper than 400 to 600 m. The BSRs from sediment of the Beaufort Sea are most strongly developed beneath bathymetric highs (Figure 3; Grantz, May, and Dinter, this volume). These strong reflectors may result from the presence of free gas beneath an impermeable cap of gas hydrate (Grantz, Boucher, and Whitney, 1976). Drilling has not yet directly tested for gas hydrates in this area.

In considering gas hydrates in ocean-bottom sediments, MacLeod (1982) developed gas hydrate stability curves (Figure 4) from which the depth to the base of the gas hydrate or the geothermal gradient can be predicted. To illustrate the application of these curves, MacLeod used the large volume of data available from Grantz for the Beaufort Sea. A geothermal gradient in sediment beneath the Beaufort Sea was indicated to be 50 ±9°C/km.

Bering Sea

BSRs have been observed in marine seismic-reflection data recorded across the Bering Sea continental slope and rise near the Navarin Basin province in water depths ranging from 500 to 3000 m (Marlow et al., 1981). These reflectors have two possible causes. The shallower BSRs are believed to represent the base of a gas hydrate zone, whereas the deeper BSRs follow a diagenetic boundary (Marlow et al., 1981; Cooper, Scholl, and Marlow, this volume). Hammond and Gaither (1983) show a number of examples of the diagenetic BSR. Figure 5 gives the locations of seismically inferred gas hydrates offshore of Alaska, including those along the Beringian margin in the Bering Sea. Four seismic profiles along this margin, indicated on Figure 5, are discussed in detail by Cooper, Scholl, and Marlow (this volume), and one of these, profile L4, is shown in Figure 6 as an example of the simultaneous occurrence of

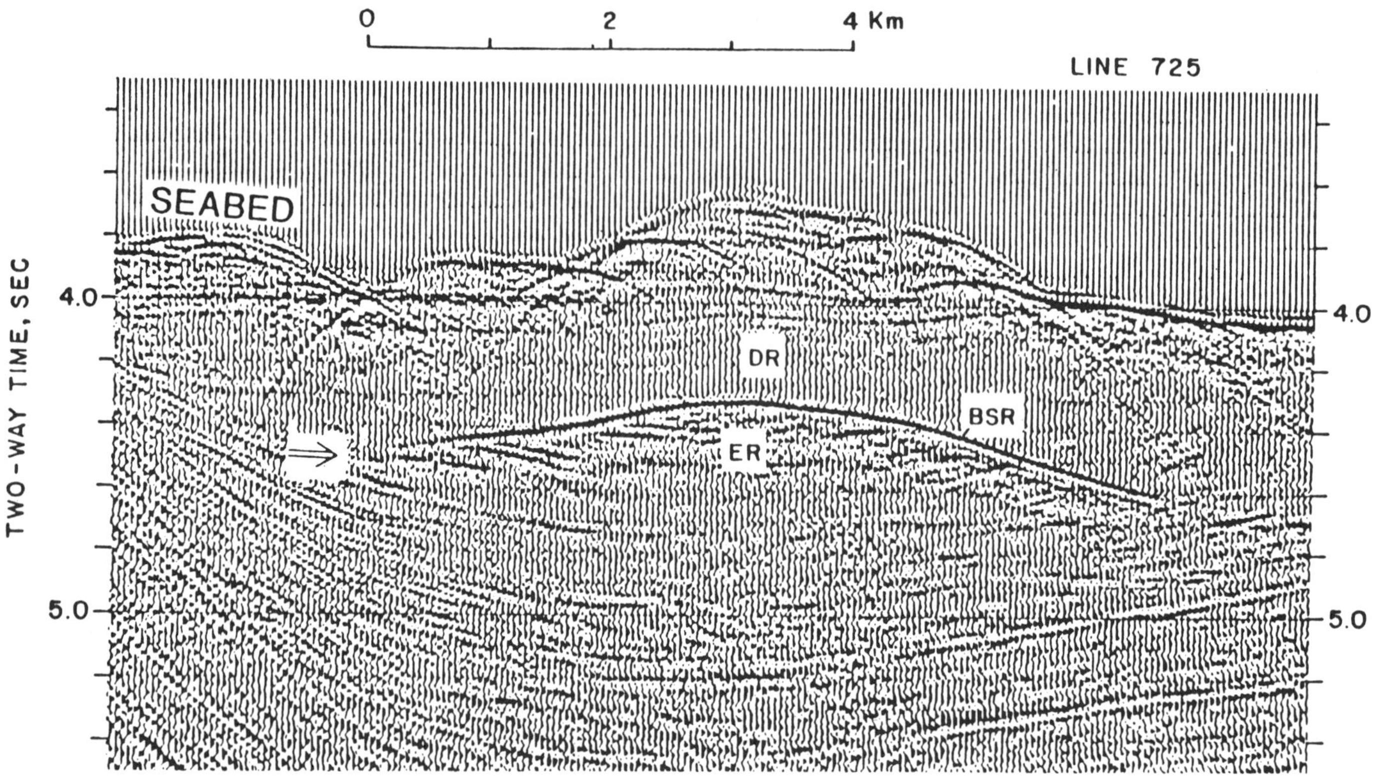

Figure 3. Segment of a multichannel seismic-reflection profile (line 725) on the northern Alaska continental slope at 2670 to 2935 m water depth off Camden Bay, Alaska (approximate location shown on Figure 2). The strong BSR (arrow) represents the interface between the zone of gas hydrate (DR) and the underlying free-gas-charged sediment (ER) (Grantz et al., 1982).

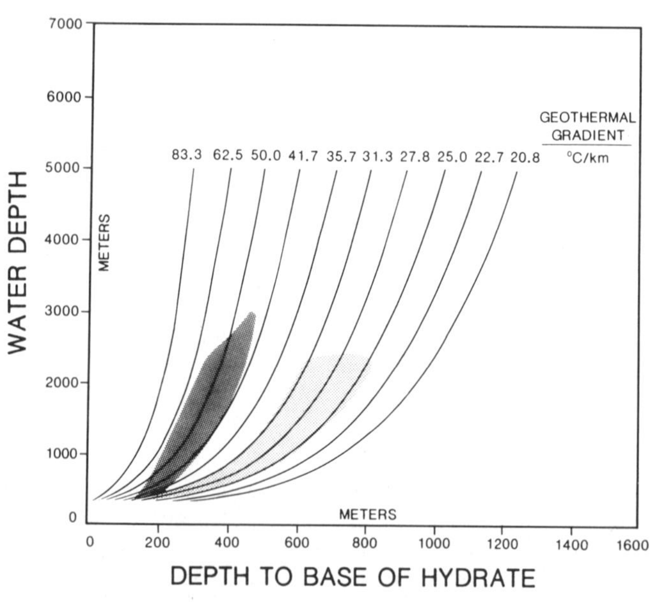

Figure 4. Graph for estimating geothermal gradients based on water depths and depths to the base of gas hydrate. The dark field shows the area where measurements plot for data from the Beaufort Sea. The light field shows the area where measurements plot for data from the Gulf of Alaska. Redrawn from MacLeod (1982).

both kinds of BSRs. The gas hydrate BSR is commonly a weak reflector, and it is difficult to identify except in areas with bathymetric relief and an undeformed sediment cover. The diagenetic BSR typically is a stronger reflector, which occurs at a depth different from that predicted from the conditions required by the physical chemistry of gas hydrates.

The existence of gas hydrates in the deep-water parts of the Bering Sea has been postulated by Scholl and Cooper (1978) during a study of bright spots, called velocity amplitude anomaly features (VAMPs), in the abyssal Aleutian Basin (Figure 7). VAMPs are acoustic anomalies showing narrow (1–2 km) subsurface columns of concave reflection horizons (velocity pulldowns) associated with flat-lying, high-amplitude horizons higher in the section. These anomalies typically occur in flat-lying beds at subbottom depths >500 m at water depths around 4000 m (Figure 8). Gas hydrates may be responsible for VAMPs by contributing to the restriction of upward-migrating gas and fluids.

Analyses of surface sediment from the Aleutian Basin (Cooper et al., 1979) show low concentrations of methane in the upper 2 m beneath the sea floor. These low concentrations imply that gas hydrates, if present in the sediment, are not immediately beneath the sea floor. Geochemical samplings of sediment from the Navarin Basin province in areas of seismically inferred gas hydrates also show low concentrations of methane, indicating that the gas hydrate is not present near the sea floor (Carlson et al., 1985).

Aleutian Ridge

Evidence for gas hydrates has been noted in sediments of the Aleutian Terrace (Scholl, Vallier, and Stevenson, 1983), which borders the Pacific side of the Aleutian Ridge (Figure 5). Seismic-reflection profiles, for example L9-7 (Figure 9), commonly reveal BSRs at a subbottom depth near 1000 m (Scholl, Vallier, and Stevenson, this volume). This BSR is believed to represent the base of the zone of gas hydrate.

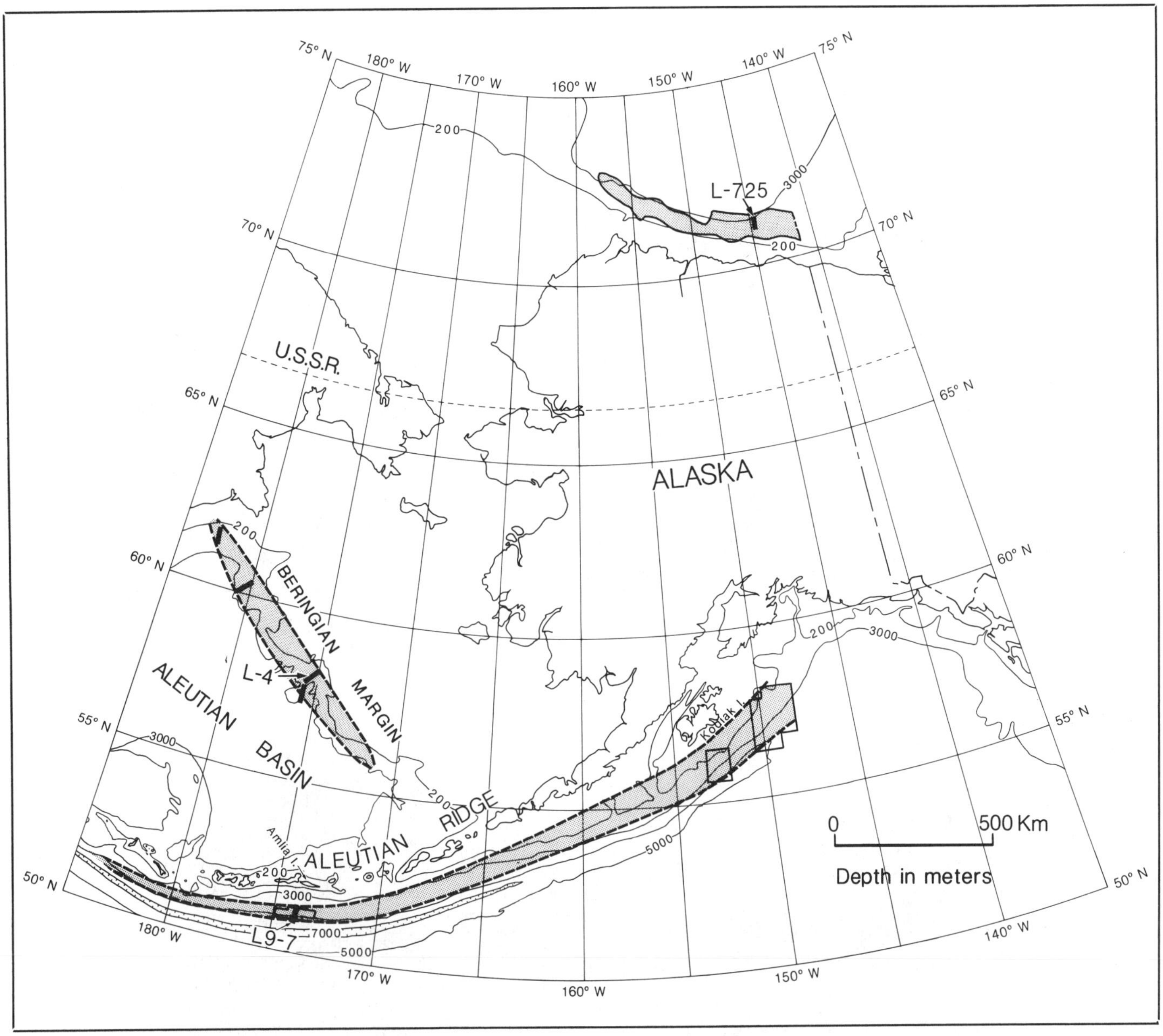

Figure 5. Map showing the locations of gas hydrates offshore of Alaska, as inferred from marine seismic-reflection profiles and the occurrence of BSRs. The extent of gas hydrate occurrence along the Beringian margin has been assumed from observations of BSRs on four seismic profiles and from the general physiography of this margin. Likewise, the occurrence of gas hydrates south of the Aleutian Ridge has been interpolated from seismic information obtained near Kodiak Island on the east and near Amlia Island on the west.

The depth of water overlying this basin is about 4500 m; the geothermal gradient, interpreted from the depth of the base of gas hydrate, is on the order of 20–25°C/km (McCarthy et al., 1984). Gassy cores from DSDP site 186 in the Atka Basin provide additional circumstantial evidence that gas hydrates are present.

Gulf of Alaska

In the area of the Gulf of Alaska, BSRs have been observed on seismic profiles of upper continental slope sediment bordering the Aleutian Trench southeast of Kodiak Island (Figure 5) at water depths from 1600 to 2300 m and at subbottom depths of 570 to 670 m (Kvenvolden and von Huene, 1985). These BSRs are believed to be caused by gas hydrates. The estimated geothermal gradient of about 30°C/km is based on the depths of these BSRs and on calculations made by using the method of Yamano et al. (1982). MacLeod (1982) also shows depths to BSRs in sediments of the Gulf of Alaska (Figure 4), but the locations are not given because his data are proprietary. MacLeod estimates a geothermal gradient of 27.8 ±3.5°C/km.

GAS HYDRATES OFFSHORE WESTERN CONTINENTAL UNITED STATES

The inner continental margin of northern California is underlain by a well-defined BSR (Field, Clarke, and White, 1980; Biddle and Seely, 1983; Field and Kvenvolden, 1985).

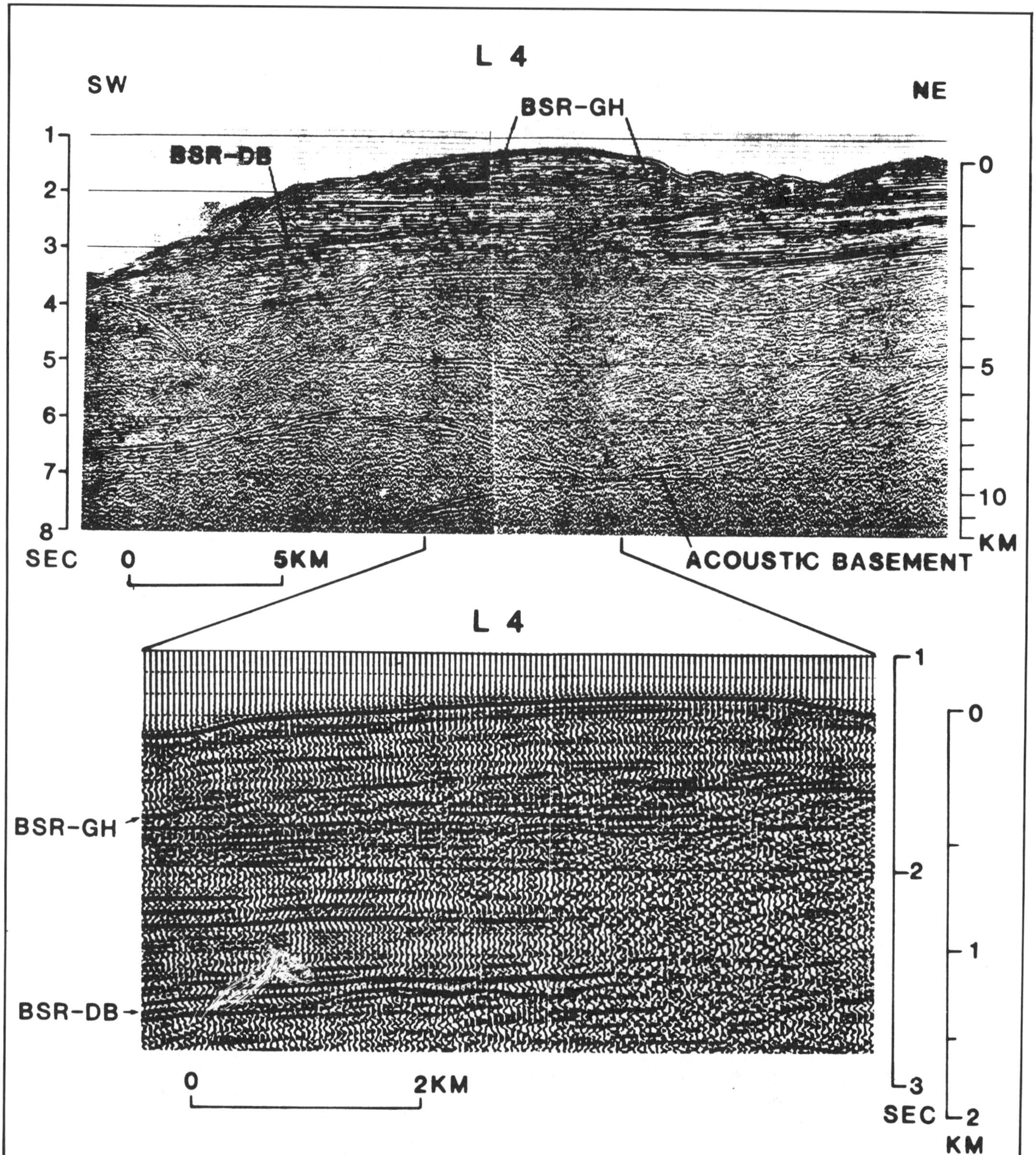

Figure 6. Seismic-reflection profile L4 along the slope of the Navarin Basin province. The profile shows two BSRs, one at the base of a gas hydrate (BSR-GH) and one at a diagenetic boundary (BSR–DB). From Marlow et al. (1981).

This anomalous acoustic reflector (Figure 10) lies at subsurface depths between 180 and 310 m beneath the Klamath Plateau off northernmost California and probably southernmost Oregon, in water depths of 800 m to >1200 m (Figure 11). The single seismic profile of Biddle and Seely (1983) shows that the BSR is present in sediment at water depths as great as about 2800 m. The BSR covers an area of at least 3000 km^2. The water and sediment depths and the pervasive nature of the BSR all indicate that it represents the base of the zone of gas hydrate. A geothermal gradient of 55°C/km

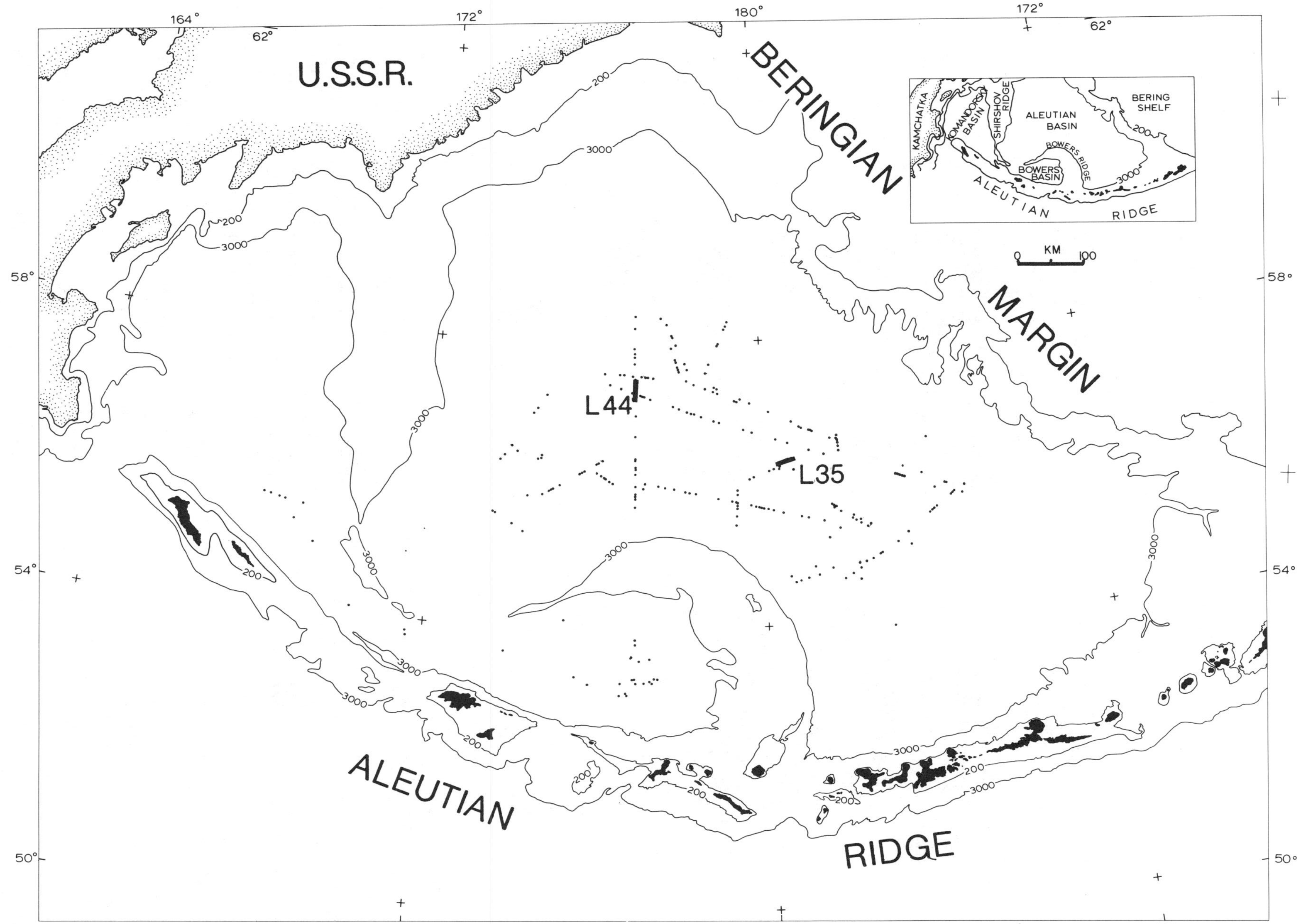

Figure 7. Map of the Bering Sea Basin. Closed circles show locations of VAMPs recognized along single-channel and multichannel seismic-reflection lines. L44 and L35 are locations of segments of seismic-reflection profiles illustrated in Figure 8. Redrawn from Scholl and Cooper (1978).

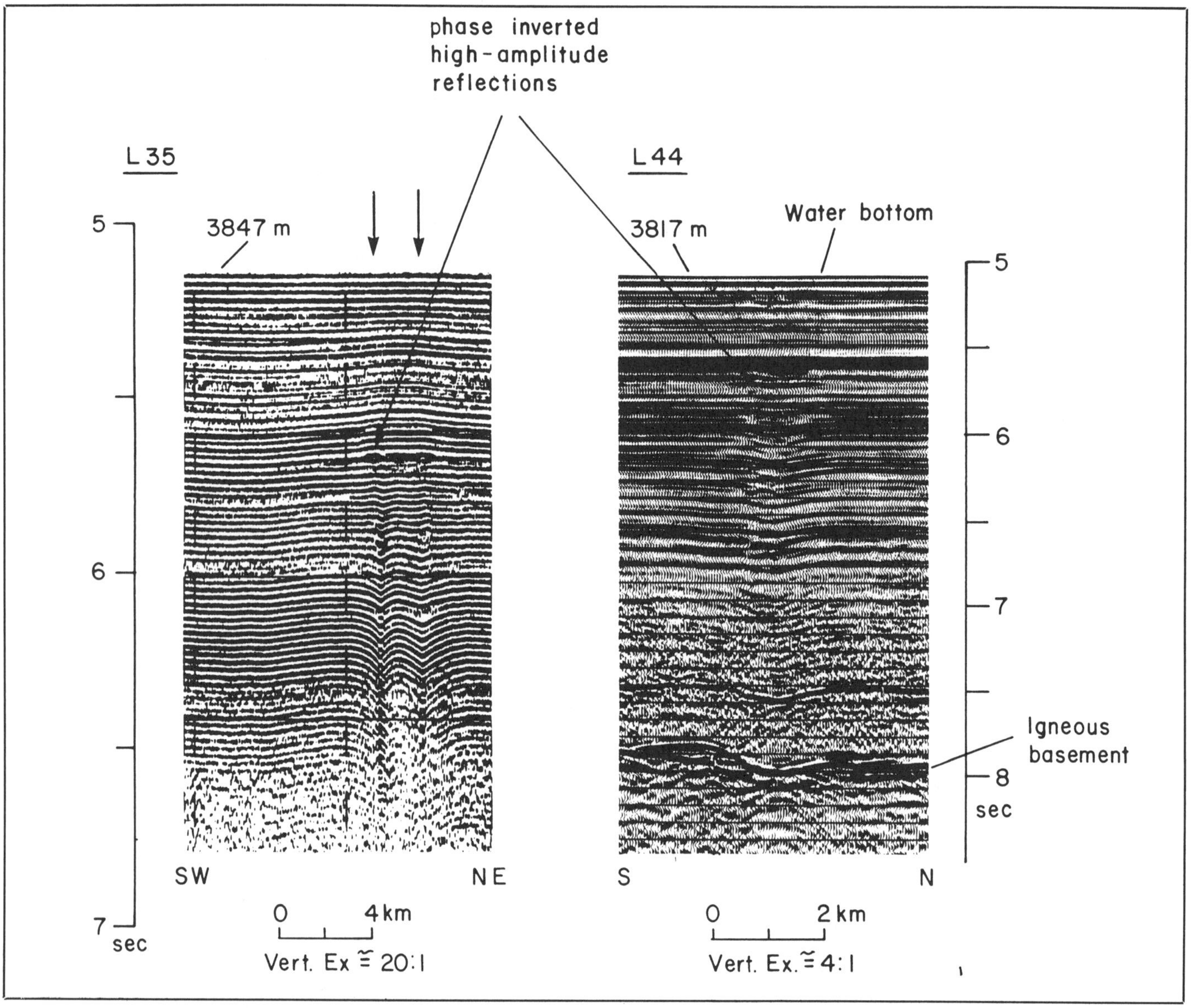

Figure 8. Profile L35 is a single-channel record of two VAMPs (vertical arrows). Profile L44 is a segment of a multichannel seismic-reflection profile that crosses a typical VAMP (Scholl and Cooper, 1978). Locations of profiles shown on Figure 7.

was estimated, based on methods used by Shipley et al. (1979), Yamano et al. (1982), and MacLeod (1982). This acoustically inferred gas hydrate is the first to be reported from the western conterminous U.S. outer continental margin.

SUMMARY OF OCCURRENCES

The occurrences of gas hydrates offshore Alaska and northern California have been inferred from anomalous acoustic reflectors on marine seismic profiles, yet confirmation of gas hydrates from drilling has not been obtained in these regions. DSDP hole 186 encountered gassy sediments with high concentrations of methane, but solid gas hydrates were not observed (Scholl and Creager, 1973), perhaps because they decomposed during the rotary coring process.

Table 1 summarizes the inferred occurrences of gas hydrates offshore Alaska and the western conterminous U.S., and Figures 5 and 7 show the distribution of possible gas hydrate occurrences offshore Alaska. The most extensively mapped gas hydrates are found in sediments under the Beaufort Sea north of Alaska, where at least 7500 km^2 of the continental margin is underlain by gas hydrates (Grantz, Boucher, and Whitney, 1976). Further seismic surveys along the continental margin under the Bering Sea and south of the Aleutian Ridge may show that the gas hydrates are more or less continuous between the areas where BSRs have already been observed.

CONSEQUENCES OF GAS HYDRATES

Some of the consequences of gas hydrate deposits in oceanic sediment are: (1) Where the depth to the base of the gas hydrate is known from well-developed BSRs, geothermal

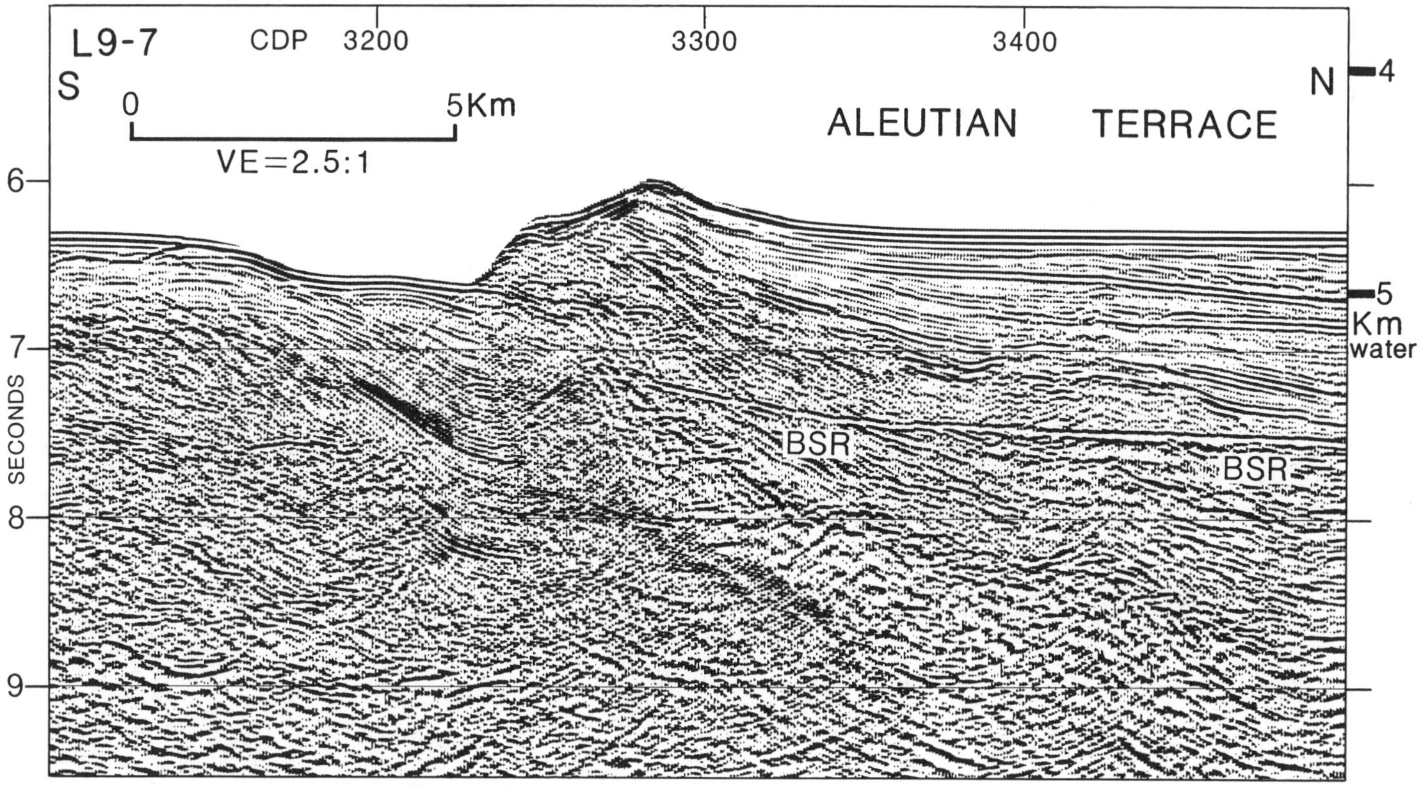

Figure 9. Segment of multichannel reflection profile (L9-7) on the slope of the Aleutian Terrace south of the Aleutian Ridge. The reflector inferred to be the base of the gas hydrate zone (BSR) is indicated. Location of profile shown on Figure 5.

Table 1. Inferred gas hydrates offshore Alaska and western conterminous United States.

Location	Water Depth (m)	BSR Subbottom Depth (m)	Calculated Geothermal Gradient (°C/km)	References
Continental slope north of Alaska, Beaufort Sea, Arctic Ocean	400-2800	~300-700	50±9	Grantz, Boucher, and Whitney, 1976 Grantz and Dinter, 1980 Grantz et al., 1982 MacLeod, 1982
Continental slope and rise, Navarin Basin province, Bering Sea	500-3000	300-1000 (est)	not calculated	Marlow et al., 1981
Aleutian Abyssal Basin, Bering Sea, "VAMPS"	~4000	>500	not calculated	Scholl and Cooper, 1978 Cooper et al., 1979
Aleutian Terrace, south of Aleutian Ridge, north of Aleutian Trench, at Atka Basin, North Pacific Ocean	~4500	~1000	20–25	Scholl, Vallier, and Stevenson, 1983
Continental slope of Aleutian Trench, south and east of Kodiak Island, Gulf of Alaska	1600-2300	570-670	~30	Kvenvolden and von Huene, 1985
			27.8±3.5	MacLeod, 1982
Continental margin, Klamath Plateau, east of northernmost California and southernmost Oregon, eastern Pacific Ocean	725-2800	135-315	55	Field, Clarke, and White, 1980 Biddle and Seely, 1983 Field and Kvenvolden, 1985

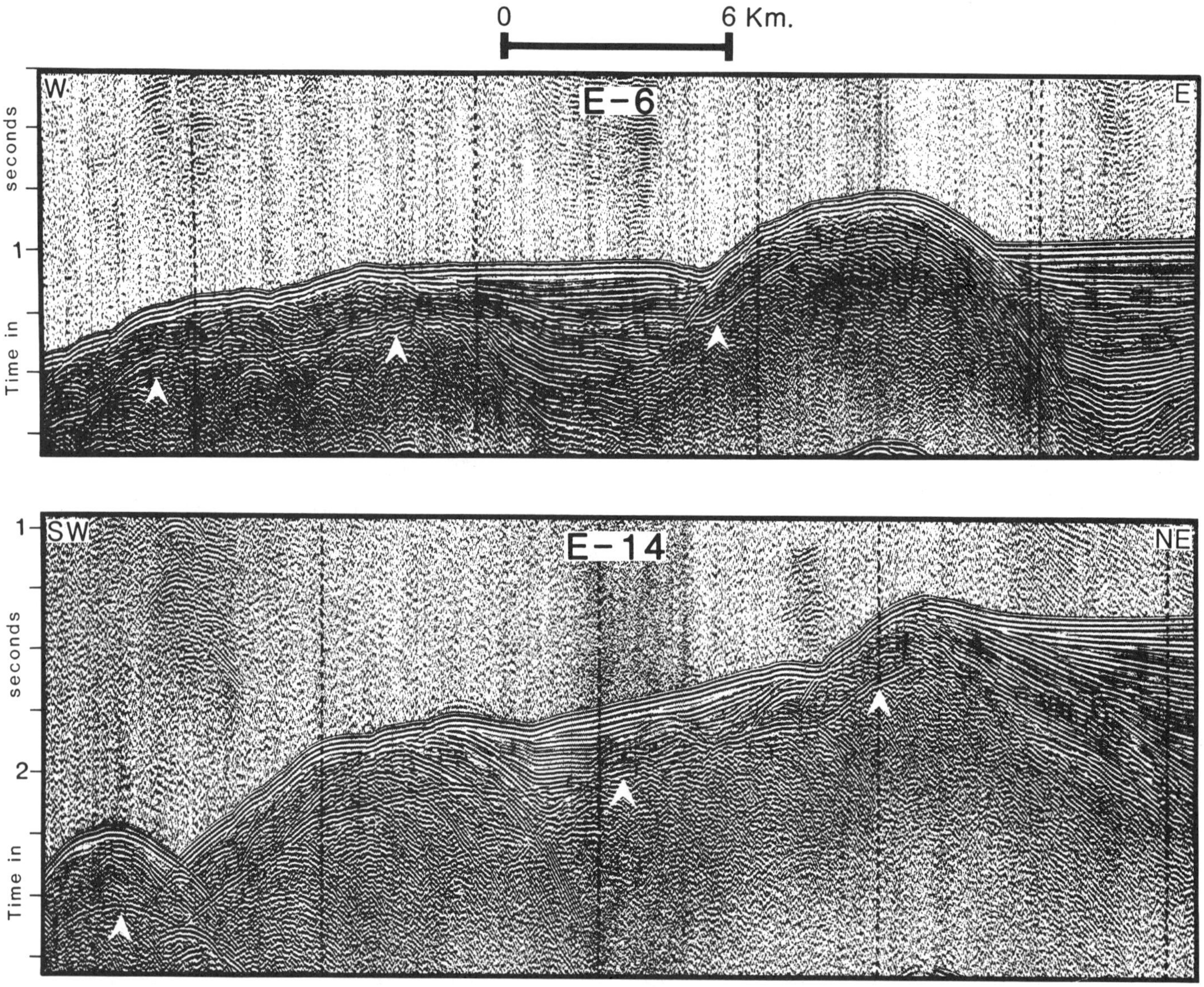

Figure 10. Single-channel, high-resolution seismic-reflection profiles illustrating the BSR (arrows) beneath the Klamath Plateau offshore northern California. Profile locations are shown on Figure 11 (Field and Kvenvolden, 1985).

gradient and heat flow can be calculated. (2) The presence of gas hydrates near the surface can cause sediment deformation wherever the gas hydrate decomposes because of changes in climate, sea level, or topography. (3) Penetration of gas hydrates during drilling could be hazardous if high-pressure gas is trapped beneath the gas hydrate, or if gas hydrates decompose behind emplaced casings. (4) Gas hydrates may constitute a potential resource for natural gas.

Geothermal Gradients

The physical-chemical properties determining gas hydrate stability are such that, given certain assumptions, both the geothermal gradient and the heat flow can be calculated (Shipley et al., 1979; Yamano et al., 1982; MacLeod, 1982). Variables affecting the calculations include bottom-water temperature, gas hydrate and sediment velocities, gas composition, and interstitial water salinity. Where thermal conductivity is known or assumed, heat flow can be determined. Geothermal gradients, ranging from 20° to 55°C/km, have been calculated from BSRs offshore Alaska and California (Table 1). These gradients are consistent with estimates of 22° to 77°C/km obtained elsewhere in the world (Table 2). The accuracy of the estimates of geothermal gradients is probably not high, but the values are consistent with geothermal gradients measured by conventional methods using temperature sensors.

Sediment Deformation

The movement of sediment as slumps or slides can possibly be caused by the decomposition of gas hydrates and the resulting expansion or release of gas. Summerhayes, Bornhold, and Embley (1979) suggest, as one possibility, that decomposition of gas hydrates may have been responsible for the surficial slides and slumps on the continental slope and rise of South West Africa; Carpenter (1981) likewise indicated that slumping on the continental slope of the east coast of the United States may have resulted from the release of gas from gas hydrates. Slope failure at or near the boundary

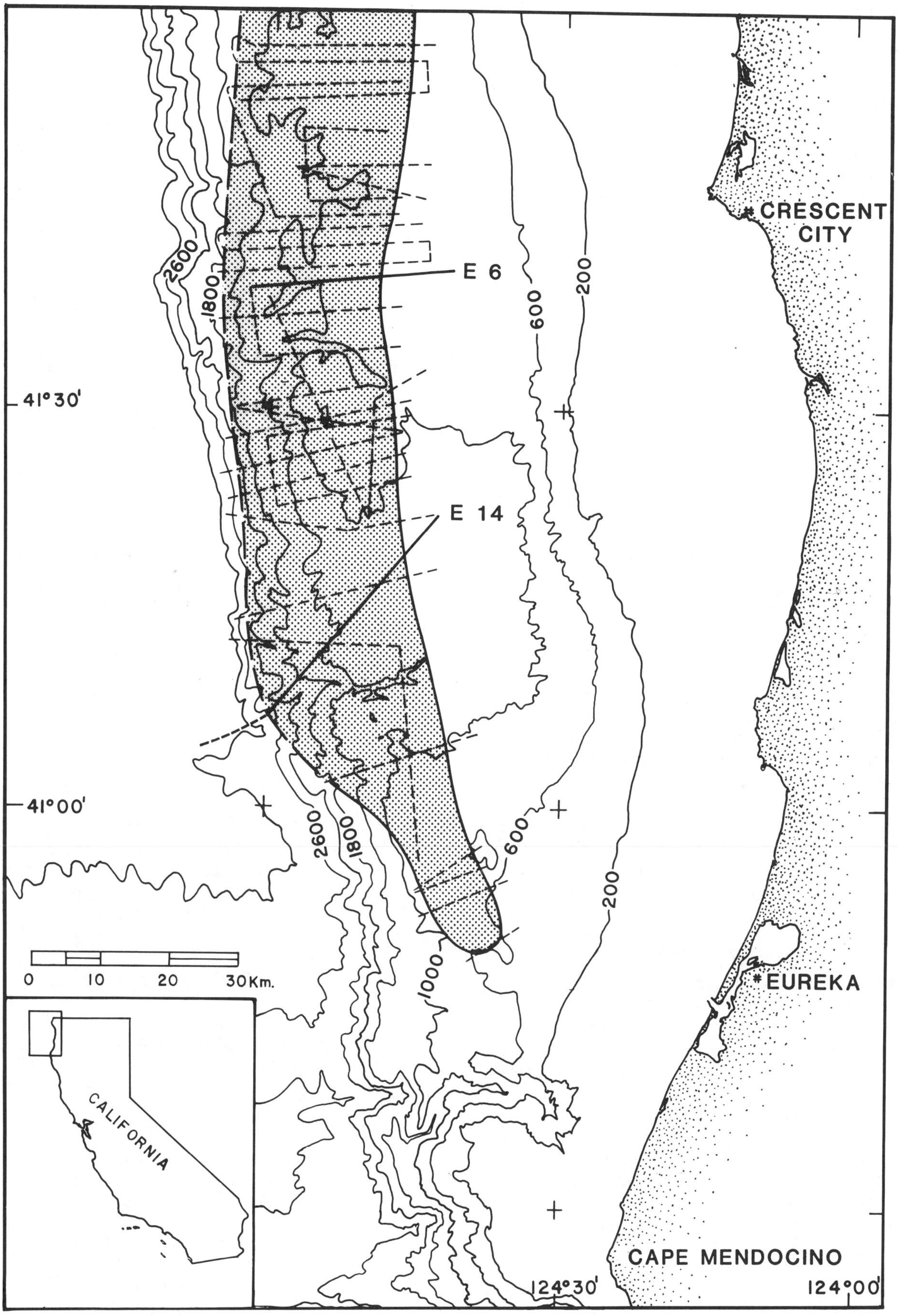

Figure 11. Map of the northern California continental margin showing tracklines of seismic-reflection profiles used to delineate the area of gas hydrates based on the occurrence of BSRs. The area boundary is dashed on the seaward side to indicate poor control. The area probably extends north into waters adjacent to Oregon. Contour interval 400 m, starting with the 200 m (shelf edge) isobath. Redrawn from Field and Kvenvolden (1985).

Table 2. Geothermal gradients estimated from depths to BSRs (Yamano et al., 1982)

Locality	Geothermal Gradient (°C/km)
Blake Outer Ridge	38.4–41.7
Eastern Pacific, Mexico	22.2–28.2
Western Gulf of Mexico	30.2–31.7
Eastern Pacific, Panama	66.0–77.3
Eastern Pacific, Costa Rica	30.1–35.1
Eastern Pacific, Nicaragua	30.2–32.5
Nankai Trough, Japan	41.5–65.8

between the gas hydrate and the underlying free gas may occur because of changes of pressure and temperature resulting from fluctuations in sea level or bottom-water temperatures, such as took place during Pleistocene time. These kinds of changes, although of long duration, may cause solid sediment to become gas-cut mud and result in mud diapirs, mud volcanoes, mud slides, and turbidite flows, depending on sediment configuration and bottom topography (McIver, 1982). Gas hydrate decomposition has not yet been implicated in sediment deformation offshore Alaska and northern California.

Drilling Hazards

The possible trapping of free gas beneath an impermeable gas hydrate cap has been suggested by Hedberg (1980) and Dillon, Grow, and Paull (1980). The safe drilling and production of this gas may prove to be difficult (Judge, 1982). Chilling of the drilling mud may be required to reduce the uncontrolled decomposition of the gas hydrate. Mud weight will also have to be adjusted to compensate for the rapid influx of gas during the drilling of gas hydrate intervals. Cement bonds between well casings and the formation containing the gas hydrate cap may be impossible to establish in order to prevent the escape of the gas in the reservoir. Because gas hydrates are a potential drilling hazard, mandatory precautions will be needed (Grantz and Dinter, 1980), for example, in the Beaufort Sea offshore of the North Slope of Alaska, where gas hydrates have been identified (Grantz, May, and Dinter, this volume).

Potential Resource

Gas hydrates may provide an unconventional source of natural gas for the future. Gas hydrates can trap methane within the hydrate structure as well as beneath the base of the gas hydrate zone. At depths of water plus sediment totaling >300 m but less than about 1700 m, gas hydrates can include greater volumes of methane within their lattice structure than can be trapped free in an equal volume of space. At total depths >1700 m the opposite is true. Thus, the volumetric advantage of gas-hydrate-trapped gas over free gas is limited to a specific depth range (Hunt, 1979). Even though large volumes of gas can be included within gas hydrates, current technology cannot produce the hydrated gas economically. The main economic asset of marine gas hydrates may be their impermeability, which provides a barrier or seal under which free gas can accumulate (Hedberg, 1980).

At least three types of structural or stratigraphic traps that involve impermeable gas hydrates are possible (Dillon, Grow, and Paull, 1980): (1) dome-shaped structures caused by the base of gas hydrate mimicking the sea-floor bathymetry under topographic highs; (2) dome-shaped structures at the base of the gas hydrate caused by local increases in heat flow, which in turn are caused by local changes in thermal conductivity such as the occurrence of salt domes; and (3) stratigraphic traps where gas-containing beds dip up toward the sea floor and are sealed by gas hydrates. To be effective traps for economically significant amounts of methane, gas hydrates must seal porous and permeable reservoir sediments in which free gas can accumulate and from which it can be produced.

The reservoir properties required for sediments in which gas hydrates accumulate offshore Alaska and California are not known, thus making resource assessment very inaccurate. Published estimates of gas hydrate resources in oceanic sediment along the continental margins of the United States range from 2700 to 6,700,000 trillion cu ft (7.6×10^{13} to 1.9×10^{17} m^3) (Potential Gas Committee, 1981). If this resource is distributed as a linear function along the continental margin, and because about two-thirds of the United States coastline is along the western conterminous United States and Alaska, then the estimates of gas hydrate resources for this offshore region range from 1800 to 4,500,000 trillion cu ft (Tcf) or 5.1×10^{13} to 1.3×10^{17} m^3. These estimates are orders of magnitude larger than the possible conventional gas supply for the same region of 13 Tcf (3.7×10^{11} m^3) or the speculative gas supply of 89 Tcf (2.5×10^{12} m^3) (Potential Gas Committee, 1981). Although the estimates of gas hydrate resources offshore Alaska and off the western continental United States are enormous, the reality of these estimates is very uncertain. Nevertheless, gas hydrates offshore are interesting geological substances that affect sedimentological processes and will influence future exploration strategy on outer continental margins.

ACKNOWLEDGMENTS

I thank the following colleagues who contributed information for the paper: P. R. Carlson, A. K. Cooper, M. E. Field, Arthur Grantz, D. W. Scholl, and Roland von Huene. This work was partially funded by the Morgantown Energy Technology Center under U.S. Geological Survey-Department of Energy Agreement No. DE-AI21-83MC20422.

REFERENCES CITED

Biddle, K. T., and D. R. Seely, 1983, Structure of a subduction complex, *in* A. W. Bally, ed., Seismic expression of structural styles: AAPG Studies in Geology 15, v. 3, p. 3.4.2-129 to 3.4.2-134.

Carlson, P. R., M. Golan-Bac, H. A. Karl, and K. A. Kvenvolden, 1985, Seismic and geochemical evidence for shallow gas in sediment of Navarin continental margin, Bering Sea: AAPG Bulletin, v. 69, p. 422–436.

Carpenter, G., 1981, Coincident sediment slump/clathrate complexes on the U.S. Atlantic continental slope: Geo-Marine Letters, v. 1, p. 29–32.

Claypool, G. E., and I. R. Kaplan, 1974, The origin and distribution of methane in marine sediments, *in* I. R. Kaplan, ed., Natural gases in marine sediments: New York, Plenum Press, p. 94-129.

Cooper, A. K., D. W. Scholl, and M. S. Marlow, 1987, Structural framework, sedimentary sequences, and hydrocarbon potential of the Aleutian and Bowers Basins, Bering Sea, this volume.

Cooper, A. K., D. W. Scholl, M. S. Marlow, J. R. Childs, G. D. Redden, K. A. Kvenvolden, and A. J. Stevenson, 1979, Hydrocarbon potential of Aleutian Basin, Bering Sea: AAPG Bulletin, v. 63, p.

2070–2087.
Dillon, W. P., J. A. Grow, and C. K. Paull, 1980, Unconventional gas hydrate seals may trap gas off southeast U.S.: Oil and Gas Journal, v. 78, no. 1, p. 124–130.
Ewing, J. I., and C. H. Hollister, 1972, Regional aspects of deep sea drilling in the western North Atlantic, *in* C. H. Hollister, J. I. Ewing, et al., Initial reports of the Deep Sea Drilling Project (Vol. 11): Washington, D.C., U.S. Government Printing Office, p. 951–973.
Field, M. E., and K. A. Kvenvolden, 1985, Gas hydrates on the northern California continental margin: Geology, v. 13, p. 517–520.
Field, M. E., S. H. Clarke, and M. E. White, 1980, Geology and geologic hazards of offshore Eel River Basin, northern California continental margin: U.S. Geological Survey Open-File Report 80–1080, 80 p.
Grantz, A., and D. A. Dinter, 1980, Constraints of geologic processes on western Beaufort Sea oil development: Oil and Gas Journal, v. 78, no. 18, p. 304–319.
Grantz, A., G. Boucher, and O. T. Whitney, 1976, Possible solid gas hydrate and natural gas deposits beneath the continental slope of the Beaufort Sea: U.S. Geological Survey Circular 733, p. 17.
Grantz, A., S. D. May, and D. A. Dinter, 1987, Regional geology and petroleum potential of the United States Beaufort and northeasternmost Chukchi Seas, this volume.
Grantz, A., D. A. Dinter, E. R. Hill, S. D. May, R. H. McMullin, R. L. Phillips, and E. Reimnitz, 1982, Geologic framework, hydrocarbon potential and environmental conditions for exploration and development of proposed oil and gas lease sale 87 in the Beaufort and northeast Chukchi Seas: U.S. Geological Survey Open-File Report 82–482, 71 p.
Hammond, R. D., and J. R. Gaither, 1983, Anomalous seismic character—Bering Sea Shelf: Geophysics, v. 48, p. 590–605.
Harrison, W. E., and J. A. Curiale, 1982, Gas hydrates in sediments of holes 497 and 498A, Deep Sea Drilling Project Leg 67, *in* J. Aubouin, R. von Huene, et al, Initial reports of the Deep Sea Drilling Project (Vol. 67): Washington, D.C., U.S. Government Printing Office, p. 591–594.
Hedberg, H. E., 1980, Methane generation and petroleum migration, *in* W. H. Roberts, III, and R. J. Cordell, eds., Problems of petroleum migration: AAPG Studies in Geology 10, p. 179–206.
Hein, J. R., D. W. Scholl, J. A. Barron, M. G. Jones, and J. Miller, 1978, Diagenesis of late Cenozoic diatomaceous deposits and formation of the bottom simulating reflector in the southern Bering Sea: Sedimentology, v. 25, p. 155–181.
Hunt, J. M., 1979, Petroleum geochemistry and geology: San Francisco, W. H. Freeman, 617 p.
Judge, A., 1982, Natural gas hydrates in Canada, *in* 4th Canadian Permafrost Conference Proceedings, Calgary, Alberta, Canada, p. 320–328.
Katz, D. L., D. Cornell, R. Kobayashi, F. H. Poettmann, J. A. Varg, J. R. Elenblass, and C. F. Weinaug, 1959, Handbook of natural gas engineering: New York, McGraw-Hill, 802 p.
Kvenvolden, K. A., and L. A. Barnard, 1983a, Hydrates of natural gas in continental margins, *in* J. S. Watkins and C. L. Drake, eds., Studies of continental margin geology: AAPG Memoir 34, p. 631–640.
Kvenvolden, K. A., and L. A. Barnard, 1983b, Gas hydrates of the Blake Outer Ridge Site 533, DSDP/IPOD Leg 76, *in* R. E. Sheridan, F. Gradstein, et al., Initial reports of the Deep Sea Drilling Project (Vol. 76): Washington, D.C., U.S. Government Printing Office, p. 353–365.
Kvenvolden, K. A., and T. J. McDonald, 1985, Gas hydrates of the Middle America Trench—DSDP Leg 84, in R. von Huene, J. Aubouin, et al., Initial reports of the Deep Sea Drilling Project (Vol. 84): Washington D.C., U.S Government Printing Office, p. 667–682.
Kvenvolden, K. A., and R. von Huene, 1985, Natural gas generation in sediments of the convergent margin of the eastern Aleutian Trench area, *in* D. G. Howell, ed., Tectonostratigraphic terranes of the Circum-Pacific Region: Circum-Pacific Council for Energy and Minerals Earth Sciences Series 1, p. 31–49.
Markl, R. G., and G. M. Bryan, 1983, Stratigraphic evolution of Blake Outer Ridge: AAPG Bulletin, v. 67, p. 666–683.
MacLeod, M. K., 1982, Gas hydrates in ocean bottom sediments: AAPG Bulletin, v. 66, p. 2649–2662.
Marlow, M. S., P. Carlson, A. K. Cooper, H. Karl, H. McLean, R. McMullin, and M. B. Lynch, 1981, Hydrocarbon resource report for proposed OCS Sale No. 83 Navarin Basin, Alaska: U.S. Geological Survey Open-File Report 81–252, 83 p.
McCarthy, J., A. J. Stevenson, D. W. Scholl, and T. L. Vallier, 1984, Speculations on the petroleum geology of the accretionary body: An example from the central Aleutians: Marine and Petroleum Geology, v. 1, p. 151–167.
McIver, R. D., 1982, Role of naturally occurring gas hydrates in sediment transport: AAPG Bulletin, v. 66, p. 789–792.
Paull, C. K., and W. P. Dillon, 1981, Appearance and distribution of the gas hydrate reflection in the Blake Ridge region, offshore southeastern United States: U.S. Geological Survey Miscellaneous Field Studies Map MF–1252.
Popenoe, P., E. L. Coward, and K. V. Cashman, 1982, A regional assessment of potential environmental hazards to and limitations on petroleum development off southeastern United States Atlantic continental shelf, slope, and rise offshore North Carolina: U.S. Geological Survey Open-File Report 82–136, 66 p.
Potential Gas Committee, 1981, Potential supply of natural gas in the United States (as of December 31, 1980): Golden, Colorado, Potential Gas Agency, Colorado School of Mines, 119 p.
Scholl, D. W., and A. K. Cooper, 1978, VAMPs—Possible hydrocarbon-bearing structures in Bering Sea Basin: AAPG Bulletin, v. 62, p. 2481–2488.
Scholl, D. W., and J. S. Creager, 1973, Geologic synthesis of Leg 19 (DSDP) results: Far North Pacific, and Aleutian Ridge, and Bering Sea, *in* J. S. Creager, D. W. Scholl, et al., Initial reports of the Deep Sea Drilling Project (Vol. 19): Washington, D.C., U.S. Government Printing Office, p. 897–913.
Scholl, D. W., T. L. Vallier, and A. J. Stevenson, 1983, Geologic evolution of the Aleutian Ridge—Implications for petroleum resources: Journal of the Alaska Geological Society, v. 3, p. 33–46.
Scholl, D. W., T. L. Vallier, and A. J. Stevenson, 1987, Geologic evolution and petroleum geology of the Aleutian Ridge, this volume.
Shipley, T. H., and B. M. Didyk, 1982, Occurrence of methane hydrate offshore southern Mexico, *in* J. S. Watkins, J. C. Moore, et al., Initial reports of the Deep Sea Drilling Project (Vol. 66): Washington, D.C., U.S. Government Printing Office, p. 547–555.
Shipley, T. H, M. H. Houston, R. T. Buffler, F. J. Shaub, K. J. McMillen, J. W. Ladd, and J. L. Worzel, 1979, Seismic reflection evidence for the widespread occurrence of possible gas-hydrate horizons on continental slopes and rises: AAPG Bulletin, v. 63, p. 2204–2213.
Stoll, R. D., J. I. Ewing, and G. M. Bryan, 1971, Anomalous wave velocities in sediments containing gas hydrates: Journal of Geophysical Research, v. 76, p. 2090–2094.
Summerhayes, C. P., B. D. Bornhold, and R. W. Embley, 1979, Surficial slides and slumps on the continental slope and rise of South West Africa: A reconnaissance study: Marine Geology, v. 31, p. 265–277.
Tucholke, B. F., G. M. Bryan, and J. I. Ewing, 1977, Gas-hydrate horizons detected in seismic-profiler data from the western North Atlantic: AAPG Bulletin, v. 61, p. 698–707.
Yamano, M., S. Uyeda, Y. Aoki, and T. H. Shipley, 1982, Estimates of heat flow derived from gas hydrates: Geology, v. 10, p. 339–343.

CHAPTER TWENTY-SIX

GEOLOGIC POTENTIAL FOR HYDROCARBONS IN UNEXPLORED OFFSHORE BASINS OF WESTERN NORTH AMERICA

Hugh McLean
Thomas J. Wiley

U.S. Geological Survey
Menlo Park, California

ABSTRACT

The enormous volumes of unexplored sedimentary rock that lie beneath and adjacent to the continental shelves of Alaska and the Pacific coast area of the lower 48 States provide an attractive target for exploration. The probability of producible hydrocarbons in the offshore basins depends largely on their tectonic setting. Basins that lie along passive margins such as the north coast of Alaska contain potentially productive strata that range in age from Late Paleozoic to early Tertiary. In areas dominated by Tertiary transform and convergent tectonic regimes such as western and southern Alaska and the Pacific Coast of the lower 48 States, however, potentially productive strata are mainly Eocene and younger, and in many areas along the San Andreas fault system are Miocene and younger. Timing between source-rock maturation and diagenetic reduction of reservoir quality is especially critical in Neogene basins, because rates of subsidence and deposition are extremely high.

Despite immense volumes of sedimentary rock in water that is <200 m deep, fewer than 150 exploratory wells have been drilled in Federal waters (excluding Cook Inlet, Alaska, and Santa Barbara Channel, California). In contrast, 550 wells have been drilled in the State of Oregon alone. Realistic assessments of petroleum potential can only be made when drilling provides data on source-rock volume, quality, and maturity, and on the spatial and temporal distribution of porous and permeable reservoir rocks.

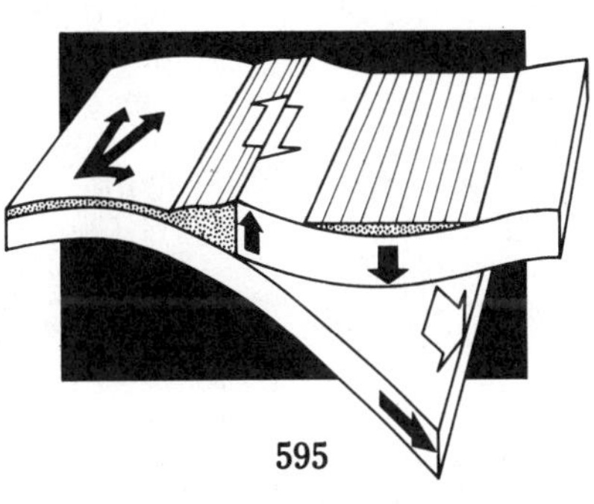

INTRODUCTION

Most of the offshore sedimentary basins along the western continental margin of North America have yet to be explored for hydrocarbons or are in the earliest stages of exploration. This chapter focuses on undrilled or poorly explored basins that probably contain sedimentary sections thick enough to generate hydrocarbons. Emphasis is placed on geologic parameters, rather than on environmental or economic problems. Evaluations of the potential for producing hydrocarbons are based on qualitative factors such as total thickness, types of structures, and probable lithologies, rather than on quantitative measurements of source-rock richness and maturity, and the properties of reservoir rocks. The geologic data used to evaluate the basins are mainly from published sources, and our interpretations rely heavily on interviews with geologists familiar with the local framework. Geological details of many of the basins can be found in other reports of this volume and in their reference citations.

Regions are discussed in geographic succession from north to south, beginning with the eastern Beaufort shelf and ending with the shelf adjacent to Baja California Sur, Mexico (Figure 1).

BEAUFORT SHELF

The Beaufort shelf and its adjacent slope and rise is the northernmost area in the United States with hydrocarbon potential. The shelf extends from near Point Barrow eastward to the Alaska-Canada border and averages about 100 km in width. Three sedimentary basins lie beneath the Beaufort shelf, each having significant geologic potential for petroleum resources. The Beaufort shelf is interpreted to be a rifted margin that opened in the Jurassic. Prospective sedimentary rocks range in age from Mississippian through Tertiary.

Each of the basins is discussed in detail by Grantz, May, and Dinter (this volume). From east to west, the basins are Kaktovik, Dinkum graben, and Nuwuk, with Nuwuk being the largest and Dinkum graben the smallest (Figure 2). The three basins rank highly in hydrocarbon potential because of their proximity to Prudhoe Bay and adjacent oil-producing areas onshore, and because each contains thick sedimentary sequences with seismically defined structures.

Kaktovik Basin

Kaktovik Basin is the easternmost basin on the Beaufort shelf and has an approximate area of 14,000 km^2. Basin fill is Jurassic and younger and has a maximum thickness of 12.5 km. Middle to upper Tertiary beds define Demarcation subbasin, which is 7.2 km thick, and Barter Island subbasin. Strata are deformed mainly by diapiric shale ridges, which are 40 to 70 km in length and 5 to 10 km in width, and by related normal faults. As much as 6.5 km of section is deformed. Diapirs have Jurassic, Cretaceous, and Paleocene clastic cores. Other major structures include northeast-trending compressional folds as much as 90 km long and 20 km wide, and north-dipping growth faults near the west side of Barter Island. Reservoir and source rocks, traps, and seals are considered to be highly favorable for petroleum generation and preservation.

Dinkum Graben

Dinkum graben is an elongate west–northwest-trending structure that underlies the western end of Kaktovik Basin. The structure is 170 km long, 10 to 40 km wide, and approximately 4250 km^2 in area. Approximately 4.7 km of Jurassic and Lower Cretaceous strata are overlain by as much as 5.9 km of younger Kaktovik Basin fill. In cross-section, the graben is strongly asymmetrical, with the thickest section on the northern side. Basement is composed of Franklinian (lower Paleozoic) sedimentary rocks that form a north-dipping slope beneath the basin fill. Normal and antithetic faults that parallel the long axis of the basin are the main structural elements. Faulting is thought to have started in the Jurassic and to have ended by Turonian time. At present depths, Cretaceous strata may lie within the oil-generation window. The Jurassic section is probably overmature for oil, although oil may have been generated in the past. Smaller and fewer structures, and an older section than in Nuwuk or Kaktovik basins, suggest that the petroleum potential of Dinkum graben is probably less than that of Nuwuk and Kaktovik basins.

Nuwuk Basin

Nuwuk Basin contains mainly Lower Cretaceous through Tertiary strata that have a maximum thickness of >12.6 km. The basin covers about 15,000 km^2. Structural elements include listric growth faults (down to the northeast) and large anticlines that are associated with backward rotation along the growth faults. The folds are best developed in Albian and Upper Cretaceous beds on the outer shelf and slope.

Reservoir rocks probably consist of Brookian sequence (Early Cretaceous and younger) clastic strata derived from uplift of the Brooks Range to the south. Source rocks at depths between 3 and 6 km probably lie within the zone of oil generation. The basin is underlain by the Mississippian to Early Cretaceous Ellesmerian sequence, which contain enormous reserves of oil and gas at Prudhoe Bay. The parameters of reservoir, source, trap, and seal within Nuwuk Basin favor petroleum accumulations.

CHUKCHI SHELF

The Chukchi shelf extends westward from the Alaskan coast between Point Barrow and Cape Lisburne to the U.S.-Russia convention line of 1867 (Figure 3). To the north, the shelf joins the Canada Basin of the Arctic Ocean, and to the south the shelf is separated from Hope Basin by a structural high known as Herald arch. Reconnaissance seismic data indicate at least seven structural zones beneath the Chukchi shelf. From north to south the structures are: Nuwuk Basin (described in the Beaufort shelf section), Chukchi platform, North Chukchi Basin, Hanna Trough, Barrow arch-Arctic platform, foreland foldbelt, and Herald arch (Figure 3). Each area is discussed in detail by Grantz, May, and Dinter (this volume). Their comments on petroleum potential are based mainly on extrapolation of trends known from exploratory work in the National Petroleum Reserve in Alaska (NPRA) onshore.

The Chukchi shelf is a region that was dominated by extensional tectonics related to continental breakup that culminated in the Jurassic with the formation of the Arctic Ocean. Economic basement for petroleum consists of the Franklinian sequence, a pre-Mississippian "argillite" overlain by the Ellesmerian and Brookian sequences, which produce oil and gas at Prudhoe Bay. The stratigraphy of the Chukchi shelf includes most of the formations that have been drilled along the north coast of Alaska and in the NPRA.

The prerifting strata of the Ellesmerian sequence onshore beneath the Alaskan North Slope consist of the Lisburne Group (Carboniferous and Permian), the Sadlerochit Group (Permian and Triassic), the Shublik Formation and Sag River Sandstone (Triassic), and the Kingak Shale (Jurassic and Lower Cretaceous). The provenance for these units is thought to lie north of the present Arctic shelf, and depositional rates are thought to have been low. Postrift strata, derived from uplift of the Brooks Range to the south and deposited in rapidly subsiding basins, are known as the Brookian sequence. This sequence includes the Torok Formation, the lower part of the Nanushuk Group (Lower Cretaceous), the Colville Group (Upper Cretaceous), and the Sagavanirktok Formation (Tertiary). Except for the Colville Group and the prerifting Lisburne Group, all of the units beneath the Arctic coastal plain of northern Alaska have shows of oil or gas (Grantz, May, and Dinter, this volume). Structural traps for petroleum on the Chukchi shelf are generally related to normal faults. Stratigraphic traps may lie within the sedimentary section.

The northern part of the Chukchi shelf is similar to that of the adjacent Beaufort shelf in terms of ages and thicknesses of the major sequences and the types of structural and stratigraphic traps. The southern part of the Chukchi shelf, however, contains the foreland foldbelt and Herald arch, two structural provinces dominated by low-angle thrusting and detachment structures. The structural complexity and estimates of low thermal gradients in these two areas reduces their hydrocarbon potential relative to the northern part of the Chukchi shelf.

Hope Basin

Hope Basin is an elongate, east–west-oriented trough that lies between Herald arch on the north and the Bering Strait on the south (Figure 3; Tolson, this volume). Basin fill consists of as much as 5.6 km of Upper Cretaceous(?) and Tertiary strata. The section thickens to the east and overlies steeply dipping basement reflectors along the northern side of the basin. Structural elements within the basin include east-trending elongate troughs and ridges, and asymmetrical ridges that resemble rotated fault blocks. An east–west-oriented middle Tertiary tectonic warp known as the Kotzebue arch is also a major structural feature, as well as small, deeply buried, east–west-trending anticlines that are either older than or coeval with the formation of the Kotzebue arch. Thrust faults, high-angle reverse faults, and high-angle normal faults are smaller scale features.

The basin probably formed on continental crust; the acoustic basement that forms some of the subsurface ridges may be strongly deformed Paleozoic(?) and Lower Cretaceous rocks of the Brooks Range orogen. Tectonic structures in the basin were dominated by north–south extension, and as a result most of the fill was derived from the east with some input from the west (Siberia).

Basin fill consists of Paleogene, Neogene, and Quaternary strata. The Paleogene depocenter was located west of Point Hope, and the Neogene depocenter was north of Kotzebue arch. Seismic reflections indicate that the Paleogene–Neogene boundary may be an unconformity. Reflectors in the Neogene sequence show laterally continuous bed forms with no channeling. The basin probably contains a predominantly marine sequence, although nonmarine rocks crop out along the basin margins and have been drilled near Kotzebue.

BERINGIAN SHELF

Geophysical surveys have defined several sediment-filled basins on the Beringian shelf, which extends south of the Bering Strait between Alaska and eastern Siberia, that presently are in the early stages of petroleum exploration. From north to south, the basins are Norton, Navarin, St. George, Amak, and Bristol Bay. Several Continental Offshore Stratigraphic Test (COST) wells have been drilled to examine the rocks that fill the basins, and lease sales have led to the first phases of exploratory drilling. We will cover only the general structural and stratigraphic framework of the basins. Details are discussed in Marlow, Cooper, and Fisher (this volume), and information is included in COST well reports that are published by the Minerals Management Service (see Steffy et al., 1985, and Turner, 1984).

Norton Basin

Norton Basin lies south of the Seward Peninsula and has an approximate area of 34,000 km^2. Maximum sediment thickness is approximately 6.5 km, and the basin is elongated east–west (Figure 4; Marlow, Cooper, and Fisher, this volume). Three structural areas are defined on the basis of sediment thickness and structural grain. The first area, which is north of St. Lawrence Island, contains as much as 2 km of section in a northwest-trending trough. A second area, located south and west of the Yukon horst, contains as much as 5 km of fill and also trends northwest. The third area or subbasin lies northeast of the Yukon horst and contains as much as 6 km of fill (Fisher, 1982; Fisher, Patton, and Holmes, 1982).

Much of the section in Norton Basin overlies Precambrian crystalline rocks and Paleozoic low-grade metasedimentary rocks (including carbonates). Cretaceous graywacke of the Yukon-Koyokuk province may underlie the southeastern part of the basin nearest the Yukon Delta. Sediment probably began filling the basin in the Eocene, a period of active faulting. Basement ridges create sediment-draped structures that have been compacted. Structural closure over the draped structures decreases upsection. Folds generated by basement ridges form broad anticlines with 2° to 9° flank dips. As a rule, onshore structures do not extend offshore, and little of the basin fill is exposed onshore.

Sediment that began filling the basin during early phases of extensional rifting in the Paleocene is probably derived from crystalline sources on the Seward Peninsula or from basement highs within the basin. The coarse-grained alluvium may be interbedded with coal. COST wells indicate that higher in the section marine strata interfinger with late Oligocene coal and sandstone. The coals contain kerogen that is predominantly type III humic with optimum vitrinite reflectance values (0.6–1.4%) reported from depths between 2.9 and 4.3 km; COST well No. 2 contains much amorphous kerogen (Type II) in the same depth range (Marlow, Cooper, and Fisher, this volume).

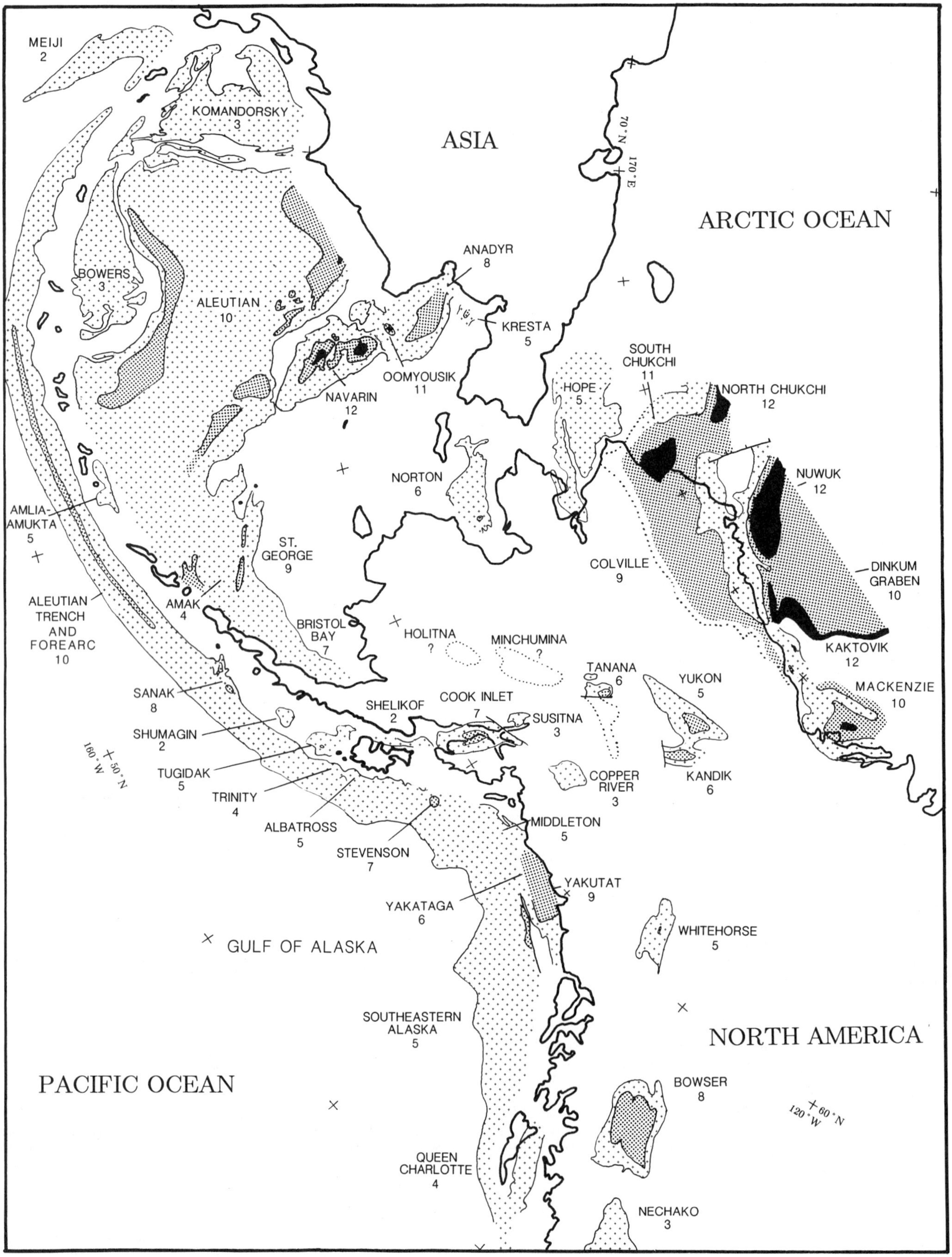

Figure 1. Index map showing prospective regions and sedimentary basins of western North America. The number below the basin name indicates thickness of basin fill in kilometers. Stipple patterns show approximate depth of basin-fill; broad (1–5 km), medium (5–10 km), dark (> 10 km).

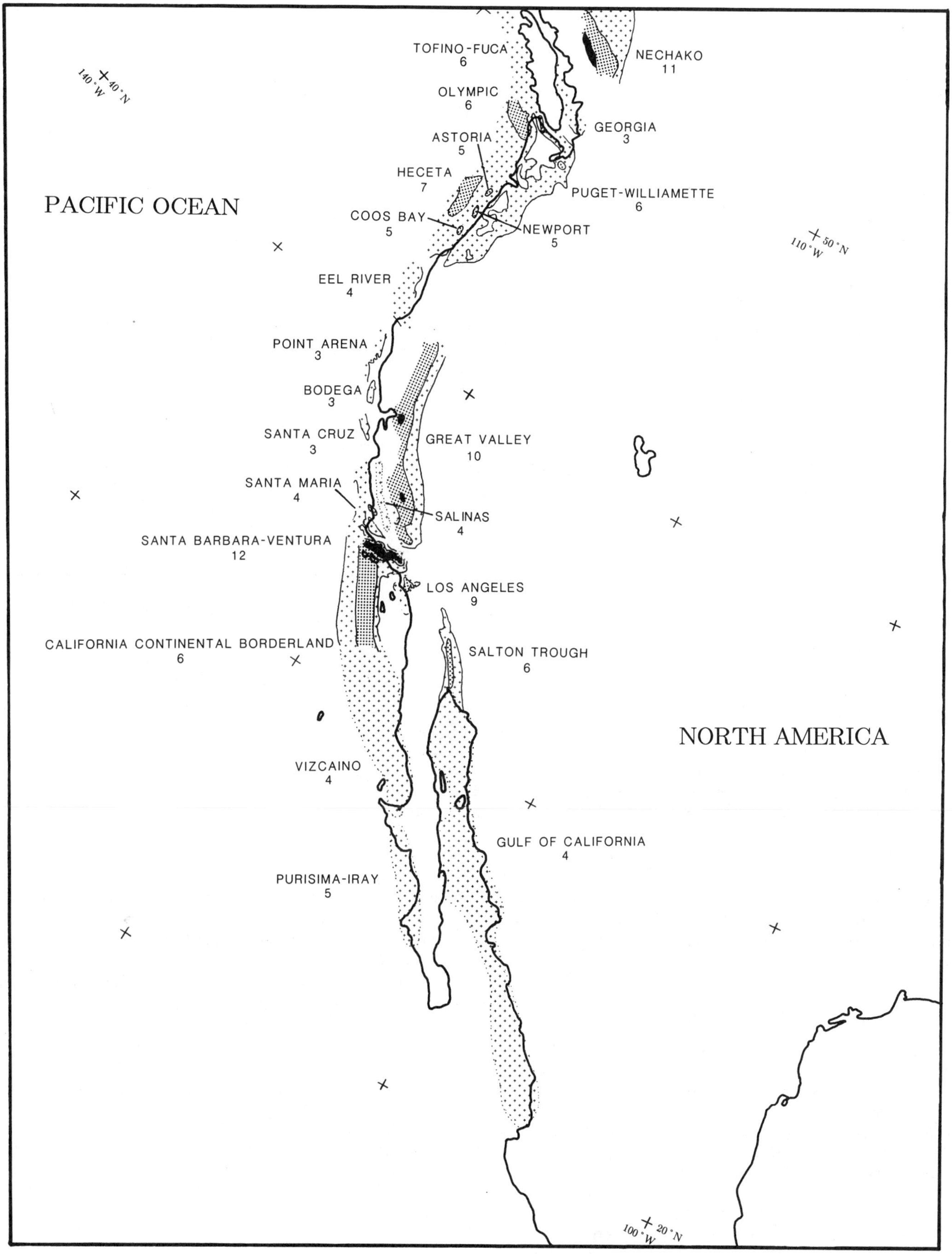

Figure 1.. Continued.

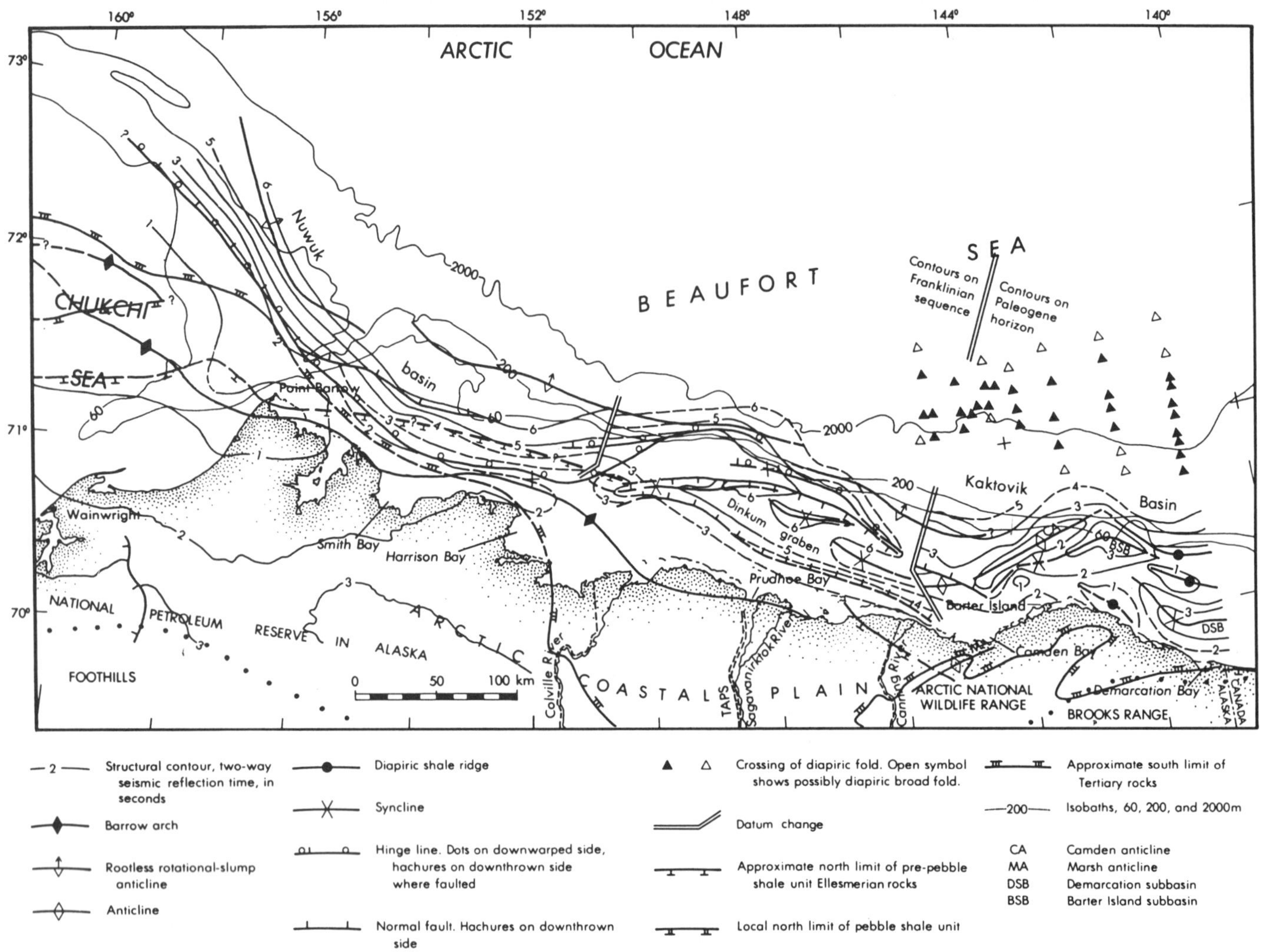

Figure 2. Location of Kaktovik Basin, Dinkum graben, and Nuwuk Basin, and thickness of post-Franklinian sequence fill in Nuwuk Basin and Dinkum graben. Modified from Grantz et al. (1982a) and Grantz, May, and Dinter (this volume).

The outlook for abundant oil deposits in Norton Basin is reduced by apparently widespread continental clastic strata interbedded with coal. The potential for gas may be greater than for oil.

Navarin Basin

Navarin Basin is the largest and probably the most prospective of the Bering Shelf basins, with an area of approximately 80,000 km^2 and a maximum sediment thickness of 15 km. The basin consists of northern, central, and southern subbasins (Figure 5). The oldest rocks penetrated in Navarin COST well No. 1 are Campanian or older and are overlain by Maastrichtian nonmarine strata. Tertiary sediments in the COST wells range from Eocene to Pliocene and are overlain by Pleistocene strata (Steffy et al., 1985; Turner, 1984).

Tertiary strata in the northern subbasin are folded into broad anticlines 10 to 15 km across. The folds are associated with gravity highs. In the southern part of the basin, elongate grabens create a series of northwest-trending ridges and troughs that parallel the basin margin. Growth faults also parallel the basin margin in the central and southern subbasins.

Dredge haul samples from the Bering margin adjacent to Navarin Basin have yielded Paleocene and Oligocene(?) sedimentary rocks (Marlow et al., 1981). Fossiliferous sandstone of Late Jurassic (Tithonian) age has also been dredged from the Bering slope, indicating that Mesozoic sedimentary rocks are present in the acoustic basement, at least in the southern subbasin. Sediment may have been supplied from the Yukon, Kuskokwim, and Anadyr (USSR) river systems.

The section drilled by the No. 1 COST well indicates that the zone of peak oil generation lies between 3125 m and approximately 4687 m (Steffy et al., 1985). The best source rocks are Eocene and early Oligocene in age and lie between drill depths of 3656 and 3994 m. Rocks below approximately

Figure 3. Basins and ridges of the Chukchi shelf. Isochrons are on base of the Ellesmerian sequence. Modified from Grantz et al. (1982b) and Grantz and May (this volume).

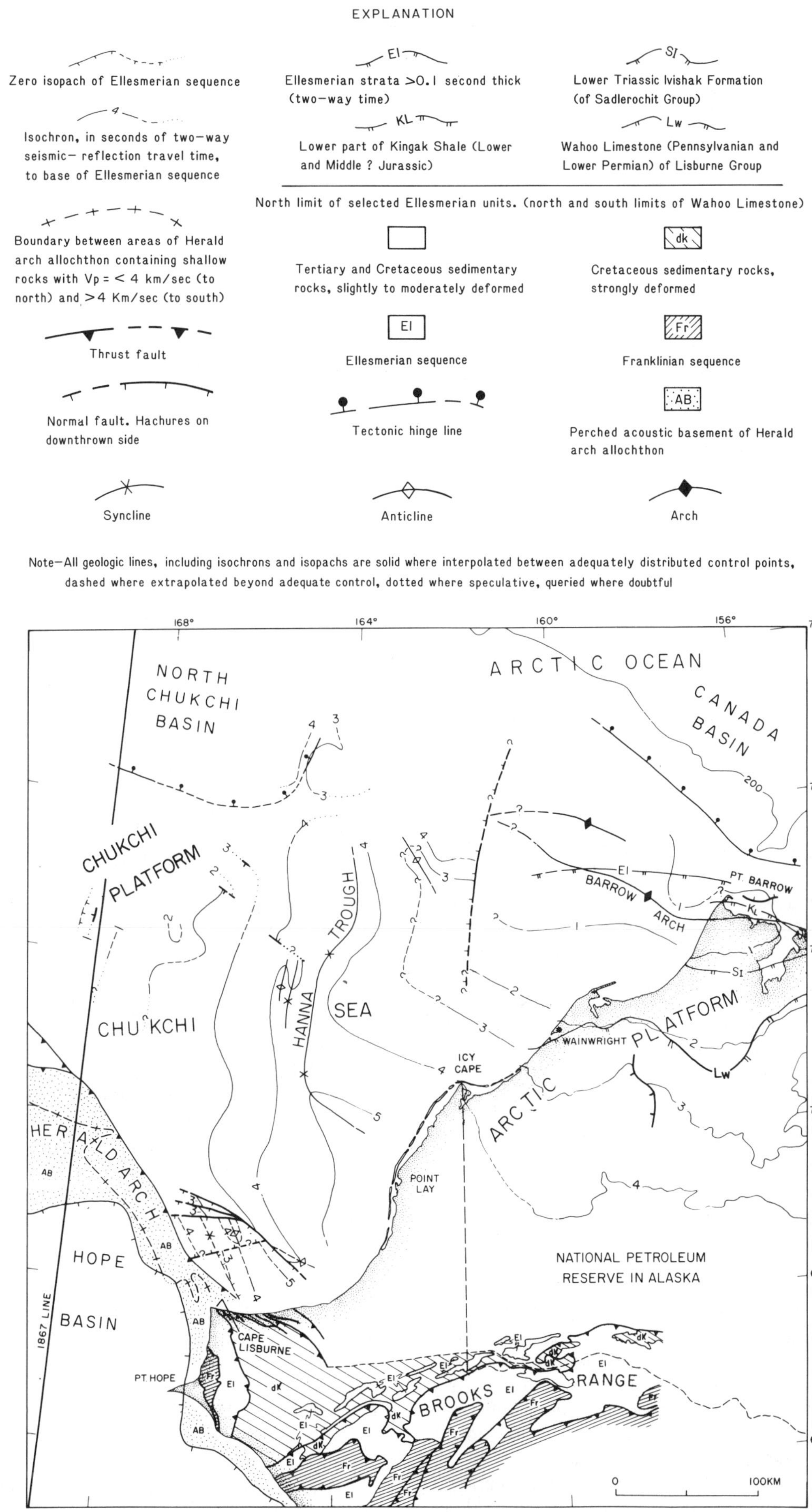
EXPLANATION
Zero isopach of Ellesmerian sequence
Ellesmerian strata >0.1 second thick (two-way time)
Lower Triassic Ivishak Formation (of Sadlerochit Group)
Isochron, in seconds of two-way seismic- reflection travel time, to base of Ellesmerian sequence
Lower part of Kingak Shale (Lower and Middle ? Jurassic)
Wahoo Limestone (Pennsylvanian and Lower Permian) of Lisburne Group
North limit of selected Ellesmerian units. (north and south limits of Wahoo Limestone)
Boundary between areas of Herald arch allochthon containing shallow rocks with Vp = < 4 km/sec (to north) and >4 Km/sec (to south)
Tertiary and Cretaceous sedimentary rocks, slightly to moderately deformed
Cretaceous sedimentary rocks, strongly deformed
Thrust fault
Ellesmerian sequence
Franklinian sequence
Normal fault. Hachures on downthrown side
Tectonic hinge line
Perched acoustic basement of Herald arch allochthon
Syncline
Anticline
Arch
Note—All geologic lines, including isochrons and isopachs are solid where interpolated between adequately distributed control points, dashed where extrapolated beyond adequate control, dotted where speculative, queried where doubtful
168°
164°
160°
156°
73°
72°
71°
70°
69°
68°
NORTH CHUKCHI BASIN
ARCTIC OCEAN
CANADA BASIN
CHUKCHI PLATFORM
HANNA TROUGH
CHUKCHI SEA
BARROW ARCH
PT. BARROW
WAINWRIGHT
ARCTIC PLATFORM
ICY CAPE
POINT LAY
HERALD ARCH
HOPE BASIN
1867 LINE
NATIONAL PETROLEUM RESERVE IN ALASKA
CAPE LISBURNE
PT. HOPE
BROOKS RANGE
0
100KM

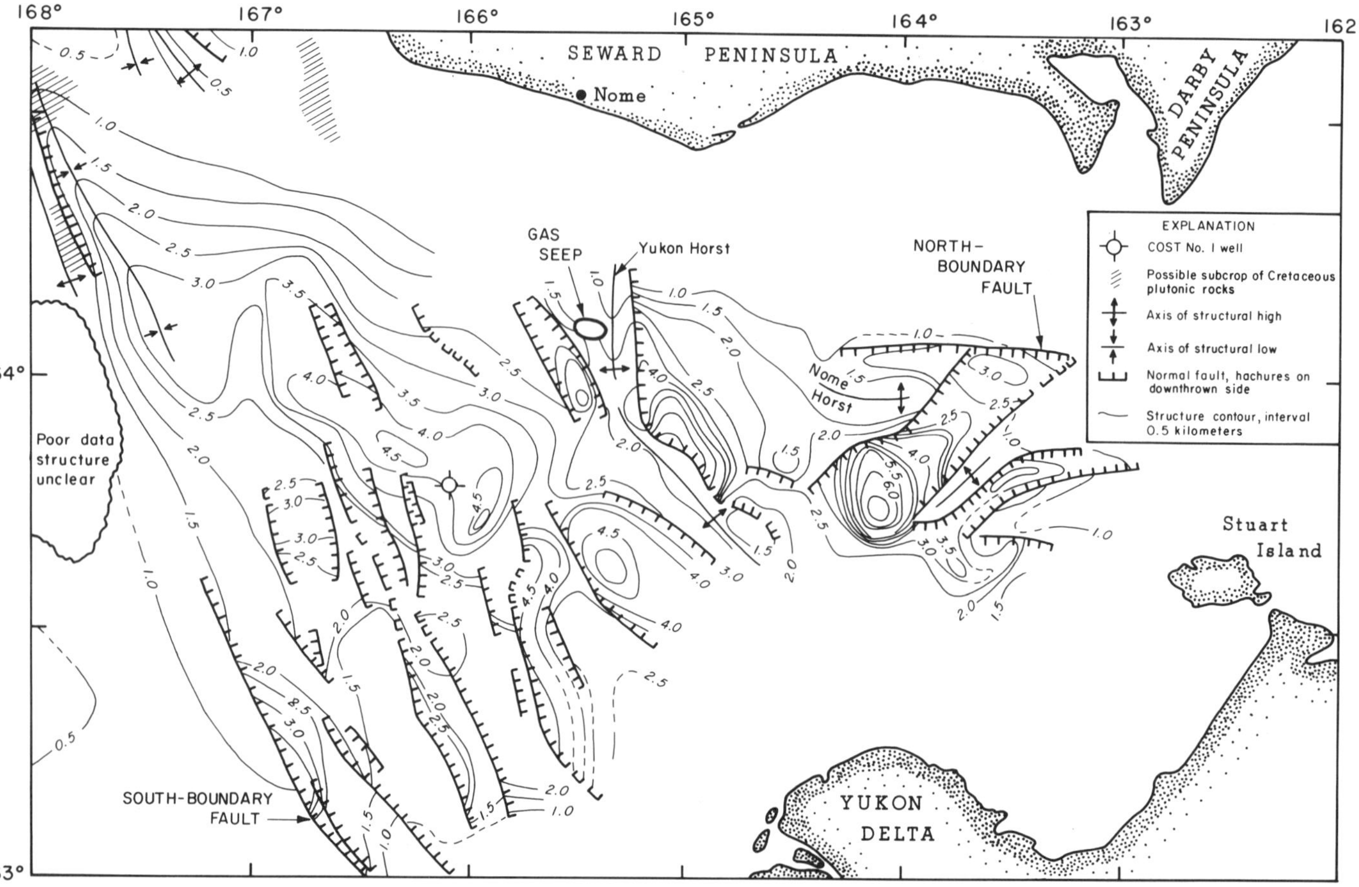

Figure 4. Structure map of Norton Basin with contours representing seismic horizon A—an unconformity between Precambrian(?) rocks below and uppermost Cretaceous or lower Paleogene rocks above. Modified from Fisher, Patton, and Holmes (1982); see also Marlow, Cooper, and Fisher (this volume).

4687 m (drill depth) are within the wet gas zone. Thermal and geochemical data from the No. 1 COST well indicate that the Eocene and early Oligocene marine source rocks are capable of generating petroleum, but reservoir rocks are not present within the source-rock sequence (Turner, 1984).

Large structures with seismically defined closure and seismic bright spots and wipeout zones make Navarin Basin the best prospect on the Bering shelf for overall petroleum potential. In addition, some parts of Navarin Basin might be analogous with Cook Inlet Basin, where Mesozoic strata are apparently the source of oil now being produced from upper Tertiary reservoir rocks.

St. George Basin

St. George Basin is a long, narrow, northwest-trending trough that contains > 10 km of sedimentary fill (Figure 5). A horst divides the basin into northwest and southeast subbasins. The basin, which began to subside about 50 to 70 Ma, is roughly symmetrical in cross-section and is bound by northwest-trending normal faults and growth faults. Locally, broad anticlines in the center of the basin are flanked by tighter folds associated with normal and growth faults along the edge of the basin. Drape folds that overlie fault blocks in the acoustic basement are the main type of structure in the lower part of the basin fill. Basement rocks are probably folded Jurassic and Cretaceous sedimentary rocks like those dredged from Zhemchug and Pribilof canyons.

The basin may have subsided in response to extension or transtension created by Mesozoic oblique-transform displacement along the Beringian margin (Marlow, Scholl, and Cooper, 1977; Marlow, Cooper, and Scholl, this volume). Source rocks in the basin are presumably marine shale and mudstone interbedded with feldspathic and lithic sandstone that was derived from major river systems flowing out of central Alaska. Average organic carbon in Tertiary mudstone dredged from the Beringian slope is 0.25 wt.%, in Cretaceous mudstone 1.0 wt.%, and in one Late Jurassic sample 0.27 wt.% (Marlow et al., 1979).

The petroleum potential of St. George Basin—in terms of sediment thickness, age of section, and quality of source and reservoir rocks—is probably similar to that of Navarin Basin, reflecting similar tectonic settings. The basins differ mainly in the size and amount of closure associated with individual folds. Both Navarin and St. George Basins have been drilled by COST wells, and several exploratory wells have been drilled and abandoned in both basins without commercial success.

Amak and Bristol Bay Basins

Amak and Bristol Bay basins are filled with sedimentary rocks that are probably similar in age and composition to the rocks that fill St. George Basin. The three basins are probably equivalent in their potential for hydrocarbon deposits. Crescent-shaped Amak Basin, which lies south of St. George Basin, has an area of about 2400 km^2 and a maximum of 4 km of sedimentary fill (Figure 5). Bristol Bay Basin, with an area of 18,900 km^2, is much larger than Amak Basin and has a maximum of 7 km of Tertiary fill. Sedimentary sections in

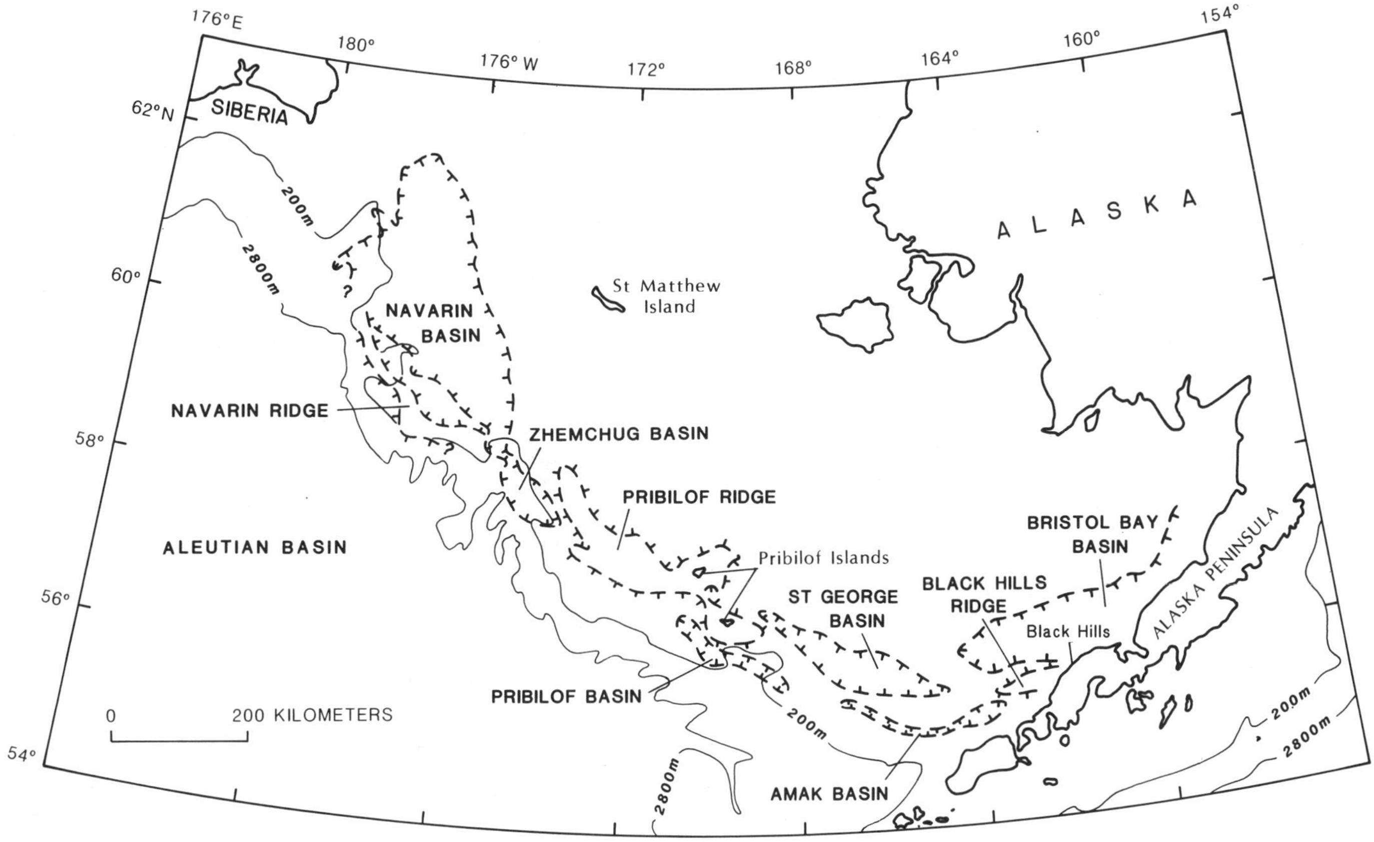

Figure 5. Location of Navarin Basin province, and Amak, St. George, and Bristol Bay (North Aleutian) basins. Modified from Marlow and Cooper (in press).

both basins are almost undeformed except for drape folds developed on acoustic basement that probably consists mainly of Mesozoic sedimentary rock. Basement rocks crop out in the Black Hills of the Alaska Peninsula, where an exploratory well (Amoco Cathedral River No. 1) drilled about 4600 m of Mesozoic sedimentary rock. The Black Hills Mesozoic strata extend northwestward into the offshore and separate Amak and Bristol Bay basins (Figure 5). Jurassic and Triassic(?) strata consist mainly of highly indurated sandstone, siltstone, mudstone, and some conglomerate. The Cathedral River well reached total depth in Triassic(?) limestone and chert. The Gulf Sandy River Federal No. 1 well, located on the north shore of the Alaska Peninsula, drilled through Pliocene and Miocene sedimentary rocks that may correlate with the sediment fill in Bristol Bay Basin. Details of the Sandy River drilling results are reported by McLean (1977).

Extrapolation northward from the Sandy River well into Bristol Bay Basin suggests that the best hydrocarbon potential appears to lie between depths of 1570 m and 3350 m. This interval is most likely within the window of thermal generation of in situ hydrocarbons, and also should have sufficient porosity and permeability for petroleum production. Source-rock thickness and quality remains problematic.

Aleutian Basin

The Aleutian Basin lies north of the Aleutian Islands and south and west of the Beringian continental margin. Several geological attributes make the basin a possible source of substantial petroleum accumulations, including a thick section, high heat flow, and velocity amplitude anomalies. Much of the basin has between 4 and 6 km of Eocene and younger sedimentary fill, and two subbasins have >9 km of seismically defined strata. Seismic reflectors in the upper part of the section are flat lying, with low-amplitude folds developed over basement ridges and protrusions in the lower part of the section. Seismic data also show more than 300 acoustic features in the central part of the basin that are known as VAMPs (velocity amplitude anomalies). These features, which may cover as much as 10% of the central part of the basin (Figure 6), may be due to gases entrapped in the section (Cooper et al., 1978; Cooper, Scholl, and Marlow, this volume). Deep Sea Drilling Project (DSDP) holes 188, 189, and 190 penetrate rocks ranging in age from Holocene through middle Miocene, and reach a maximum depth of 970 m. The holes barely penetrate beyond the opal A-opal CT diagenetic boundary, which commonly forms a prominent bottom-simulating reflector (BSR) (Cooper, Scholl, and Marlow, this volume).

Organic carbon measurements from DSDP cores are restricted to late Cenozoic deposits and are too few to provide any meaningful information for basin-wide generalizations about source-rock potential. However, the high percentage of pelagic organisms in the DSDP cores, and the presence of VAMPs, may be positive signs of thermogenic hydrocarbon generation. Heat flow that averages 1.5 heat-flow units (HFU) is also an encouraging factor, because temperatures high enough to generate thermogenic

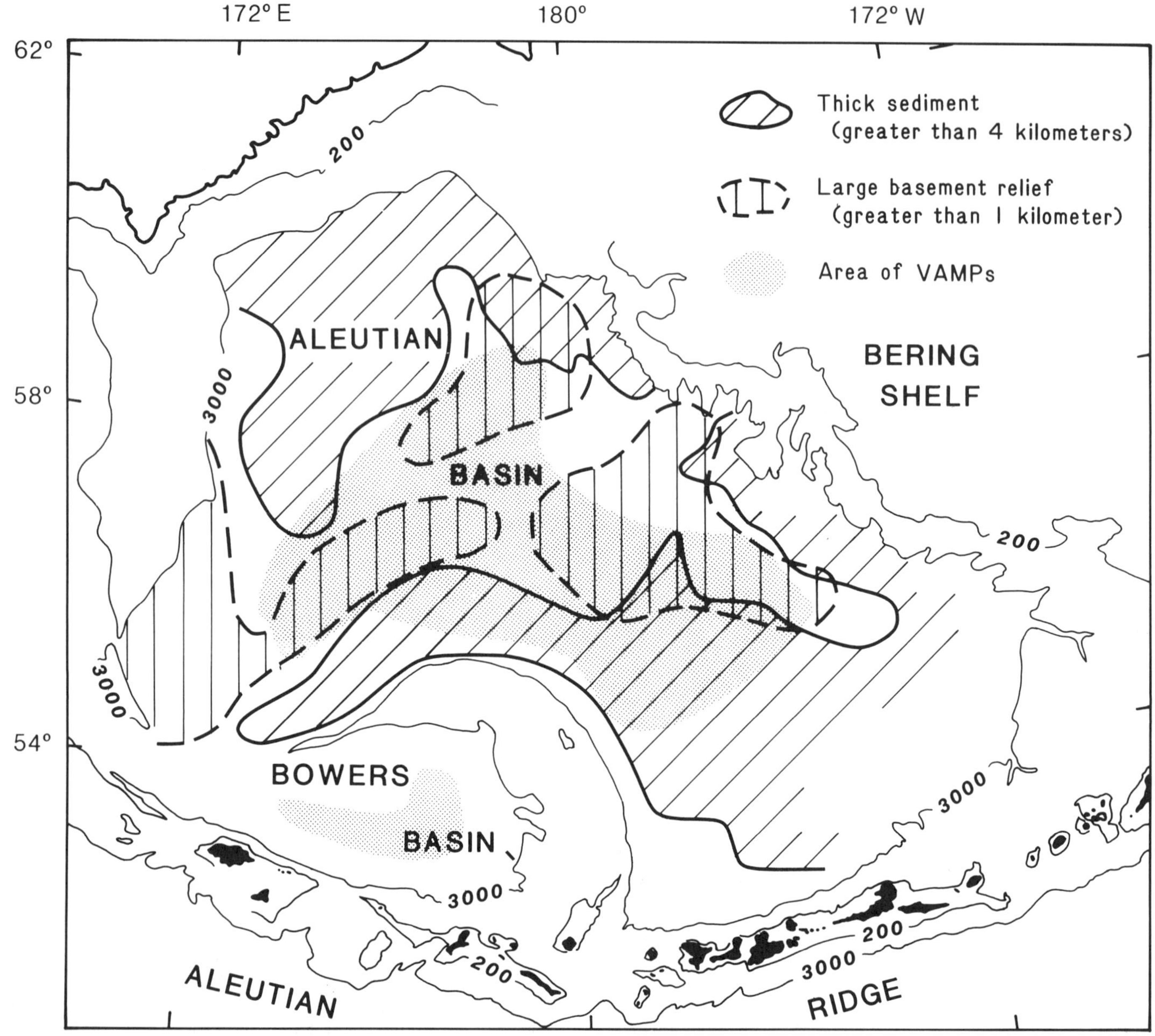

Figure 6. Aleutian Basin showing location and distribution of sediment >4 km thick, distribution of basement relief where >1 km, and area of VAMPs. Modified from Cooper et al. (1978); see also Cooper, Scholl, and Marlow (this volume).

hydrocarbons lie within depths that range from 2 to 6 km. Widespread high heat flow in the basin indicates that thermogenic degradation of organic matter (50–100°C) might occur throughout the basin.

ALEUTIAN RIDGE SUMMIT BASINS

Amlia and Amukta Basins

Several sediment-filled basins that underlie sea-floor depressions are located along the beveled ridge crest of the Aleutian Ridge (Figure 7). Two of the basins contain sections of mainly Pliocene and Quaternary deposits that are probably thick enough to generate hydrocarbons. Amlia Basin (long. 173° W) lies northeast of Amlia Island and is the largest of the so-called summit basins, with a surface area of 2700 km^2. Neighboring Amukta Basin (long. 171° W) lies between Sequam and Amukta Islands and has an area of 3000 km^2. Amlia Basin contains >5 km of fill and Amukta contains between 3 and 4 km of flat-lying to gently folded beds. The basins are bounded by faults and are considered to be grabens or half-grabens (Scholl, Marlow, and Buffington, 1975; Scholl, Vallier, and Stevenson, this volume).

Seismic acoustic patterns in Amlia and Amukta basins are similar to those at DSDP sites 184 and 185. Dredging and acoustic studies indicate that the sediment fill in the two basins consists of diatomaceous volcaniclastic sandstone, siltstone, and mudstone, with interbedded volcanic ash. Lithified sedimentary rocks of late Eocene to early Oligocene age form the basement of these basins and crop out on nearby Amlia and Atka Islands.

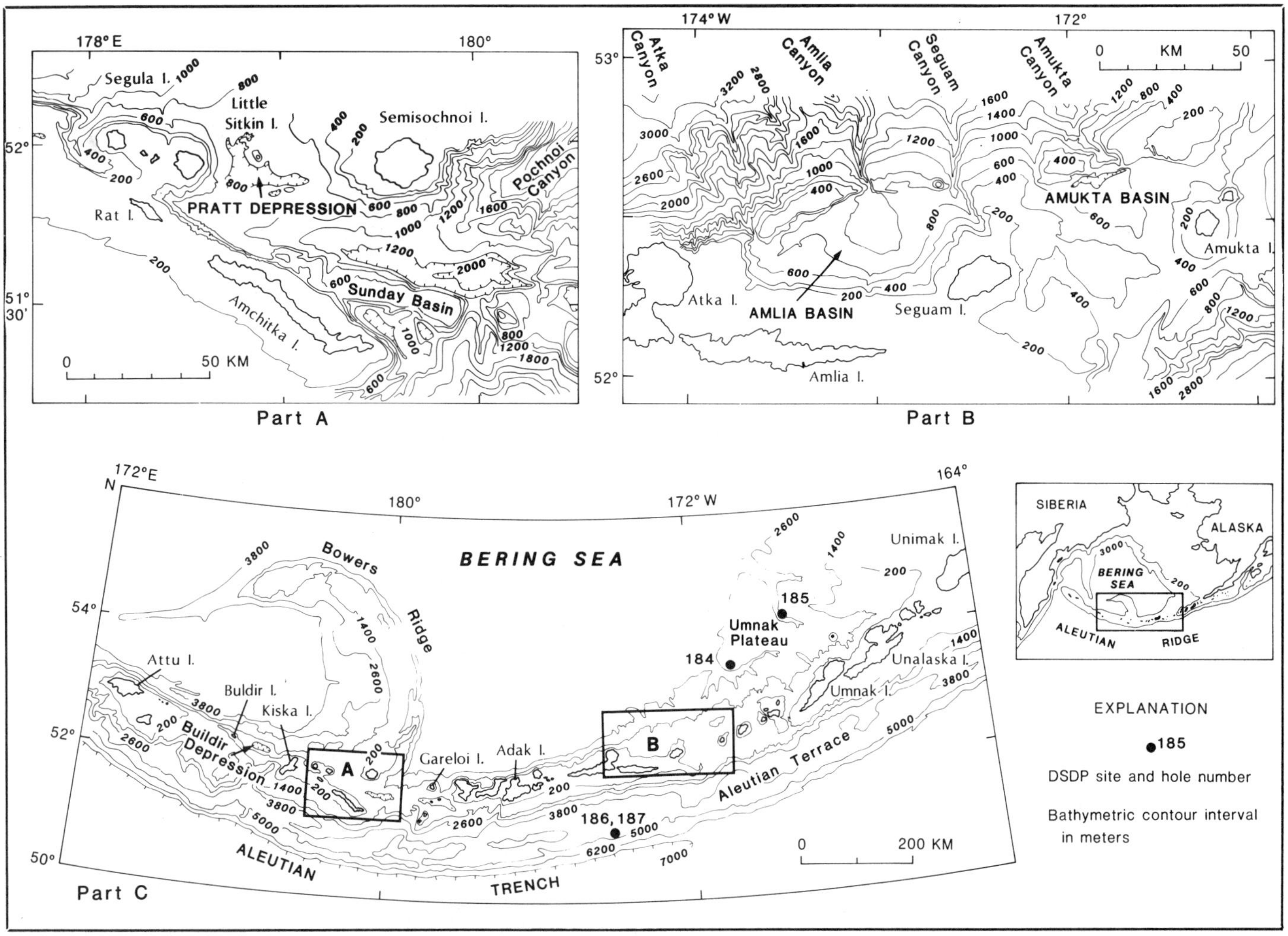

Figure 7. Location of Amlia and Amukta basins and Pratt Depression on the crest of the Aleutian Ridge. Aleutian accretionary body located between trench and southern edge of Aleutian Terrace. Modified from Scholl, Marlow, and Buffington (1975); see also Scholl, Vallier, and Stevenson (this volume).

The geology of other basins of the Aleutian Ridge, such as Pennant Basin and Pratt Depression, are described by Scholl, Vallier, and Stevenson (this volume).

Central Aleutian Accretionary Body

Seismic-reflection profiles indicate that a body of sedimentary rock 6 to 8 km thick makes up the central Aleutian accretionary body or subduction complex that lies between the trench floor and the southern edge of the arc massif (Figure 7). The accretionary body might contain the requisite properties for substantial accumulations of hydrocarbons, and a case for hydrocarbon potential has been presented by McCarthy et al. (1984). Seismic data show large structural antiforms, thrust and normal faults, and stratigraphic pinch-outs. The structures may include trench floor source rocks that possibly contain between 0.3 and 0.6 wt.% organic carbon, as well as older pelagic sediment also carried into the trench by underthrusting of oceanic crust. Accreted beds probably consist of diatom-rich sediment, fine-grained turbidite siltstone, and hemipelagic silty and clayey beds.

Potential reservoirs might be in fractured diatomaceous shales or in sandy turbidites. Seals might be formed by diagenetic alteration of ash-rich beds to smectitic clays. Heat-flow measurements on the adjacent Aleutian Terrace indicate that an oil-generation window could lie between 3 and 6.5 km.

An unresolved problem is how much of any generated hydrocarbons would survive the dewatering process that accompanies accretion of pelagic, hemipelagic, and terrigenous sediment. The many low-angle faults that rise to the surface of the accreting body might also be major pathways for the escape of both pore fluids and hydrocarbons. As with most hydrocarbon accumulations, the size and ultimate preservation of deposits in the accretionary wedge depends on the delicate balance between the processes of hydrocarbon generation and the structural disruption associated with the growth of most accretionary terranes.

PACIFIC SHELF OF ALASKA

Sanak and Shumagin Basins

Sanak and Shumagin basins are located on the Pacific side of the Alaska Peninsula (Figure 8). Each basin contains little-deformed mainly upper Cenozoic sedimentary fill that unconformably overlies acoustic basement composed of tightly folded late Mesozoic (probably Late Cretaceous) slope

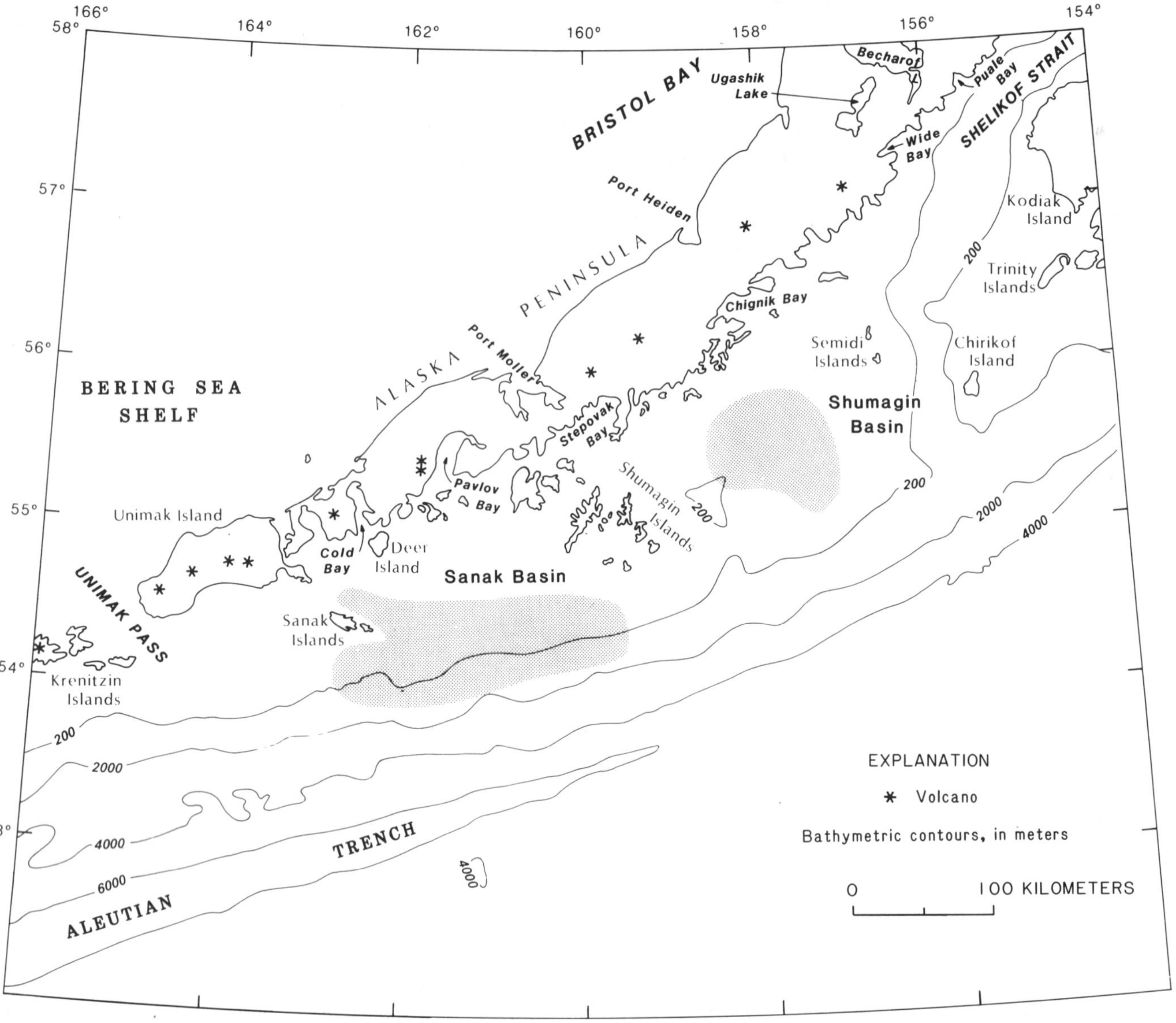

Figure 8. Shumagin Shelf and Alaska Peninsula showing location of Sanak and Shumagin basins. Modified from Bruns and von Huene (1977); see also Bruns et al. (this volume).

and trench sandstone and mudstone (Bruns et al., this volume).

Sanak Basin lies just east of Sanak Island, and has an area of about 7000 km^2. The basin contains >8 km of gently deformed strata that are believed to be mostly Neogene and Quaternary. Basement rocks exposed on nearby Sanak Island include isoclinally folded sandstone and siltstone flysch, which is in structural contact with broken formation containing beds of green and gray chert, pillow basalt, argillite, and sandstone. The only reported fossils from the island are radiolarians recovered from chert beds on Long Island that are Early Cretaceous in age, probably Valanginian (D. L. Jones, 1980, written communication). The isoclinally folded beds are intruded by a quartz diorite pluton with a K/Ar age of 59 Ma (Moore, 1974). The rocks exposed on Sanak Island most likely constitute the acoustic and economic basement of Sanak Basin.

Shumagin Basin has an area of 3800 km^2 and lies midway between the Shumagin and Semidi Islands. The basin contains two discrete depocenters, each with slightly >2.5 km of Tertiary sedimentary fill. Both Sanak and Shumagin Basins contain polymict sandstone, conglomerate, and pelitic strata probably derived from erosion of older rocks, as well as products of active volcanism on the nearby Alaska Peninsula.

The lower part of the Sanak Basin section may lie within the window for thermogenic gas, assuming an adequate thermal gradient has been maintained. The potential for hydrocarbons in Shumagin Basin is considered to be slight, because the Tertiary section is thin and basement rocks are nonprospective. Eocene and Oligocene strata on the Alaska Peninsula locally contain >1.0 wt.% organic carbon and might have the potential to generate gas if subjected to high enough temperatures, Bruns et al. (this volume). Although the geothermal gradient in Sanak Basin is not known, a gradient of 30° C/km has been estimated for the upper slope of

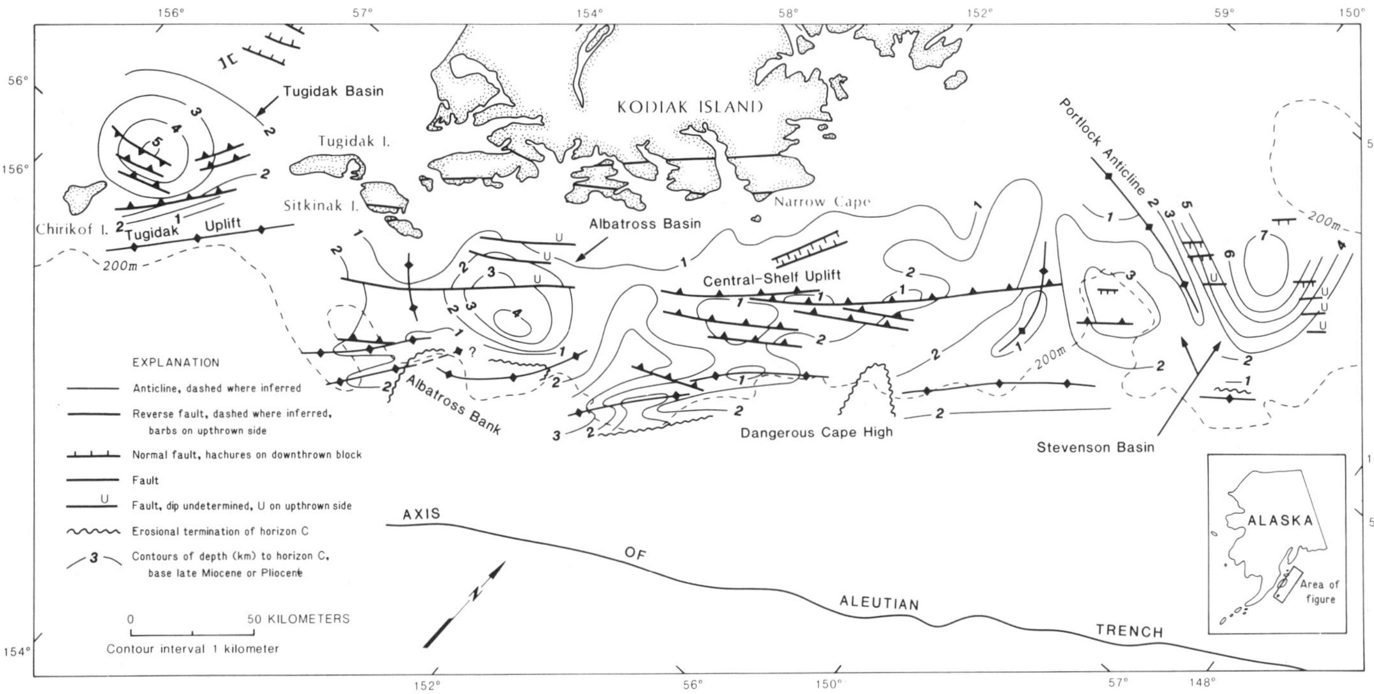

Figure 9. Location and structural elements of Tugidak, Albatross, and Stevenson basins. Contours represent base of late Miocene or Pliocene strata. Modified from Fisher (1980); see also von Huene, Fisher, and Bruns (this volume).

the Aleutian Trench by Kvenvolden and von Huene (1985), which indicates that petroleum might be generated at depths between 4 and 6 km. Folds within the basin fill and faults around the margins could act as trapping mechanisms in both basins.

Tugidak Basin

Tugidak Basin is located on the Shumagin shelf and lies approximately 90 km southwest of Kodiak Island (Figure 9; von Huene, Fisher, and Bruns, this volume). The circular-shaped basin contains as much as 5 km of late Miocene, Pliocene, and Pleistocene sedimentary fill that is gently folded and cut by seismically defined faults. The base of the basin sequence is probably marked by an erosional unconformity that formed during the late Miocene or early Pliocene. Subsequently, the uplifted area was buried and folded into a shelf-break anticline. Unconformities within the Neogene and Quaternary fill suggest that the depocenter shifted landward (northwest) as the basin was uplifted from slope to shelf depths (Fisher, 1979).

Acoustic basement is inferred to be isoclinally folded Upper Cretaceous and Paleogene turbidite beds that are exposed on the Trinity Islands and on Kodiak Island. Overlying Neogene strata probably consist of terrigenous marine strata derived in part from erosion of the Alaska Peninsula, as well as uplifted parts of the surrounding continental shelf. The proportions of sandstone, conglomerate, and pelitic strata are unknown.

Organic geochemical data from samples of Cenozoic strata exposed on Kodiak Island contain <0.5 wt.% organic carbon, and indicate that the kerogen is predominantly woody and coaly. Geochemical data from strata in nearby shelf basins indicate that source rocks are thermally immature, and source rocks in Tugidak Basin are inferred to be lean and thermally immature (Fisher, 1980). Hydrocarbons generated in Tugidak Basin would probably be gas and gas condensate that might be trapped in anticlinal structures and closures formed by high-angle normal and reverse faults. Reservoir properties of Miocene and younger sandstone may be adequate for hydrocarbon production, as indicated in detailed source and reservoir rock data by Fisher (1980, p. 1146).

KODIAK SHELF

Albatross and Stevenson Basins

Kodiak shelf lies between Kodiak Island and the Aleutian Trench and contains two sedimentary basins that have as much as 5 km of Neogene fill (von Huene, Fisher, and Bruns, this volume). Albatross Basin, which has a circular outline, lies directly south of Kodiak Island. It is asymmetrical from north to south, and the southern flank is steep (Figure 9).

Stevenson Basin lies northeast of Kodiak Island and is composed of two subbasins. The western subbasin contains 4 to 5 km of sedimentary fill, and the eastern subbasin contains >7 km; Portlock anticline separates the two subbasins. Albatross and Stevenson basins both contain gently deformed late Miocene or Pliocene and younger terrigenous strata that are gently folded except at the shelf edge and in the central-shelf uplift. Neogene strata overlie acoustic basement, which consists of moderately to strongly deformed slope and accreted trench sediment that is probably Paleocene in age. The seaward limit in both basins is formed by the present shelf break, which is undergoing rapid tectonic uplift. Rates of uplift near the shelf break in Albatross Basin average 300 m/Ma (Fisher and von Huene, 1980).

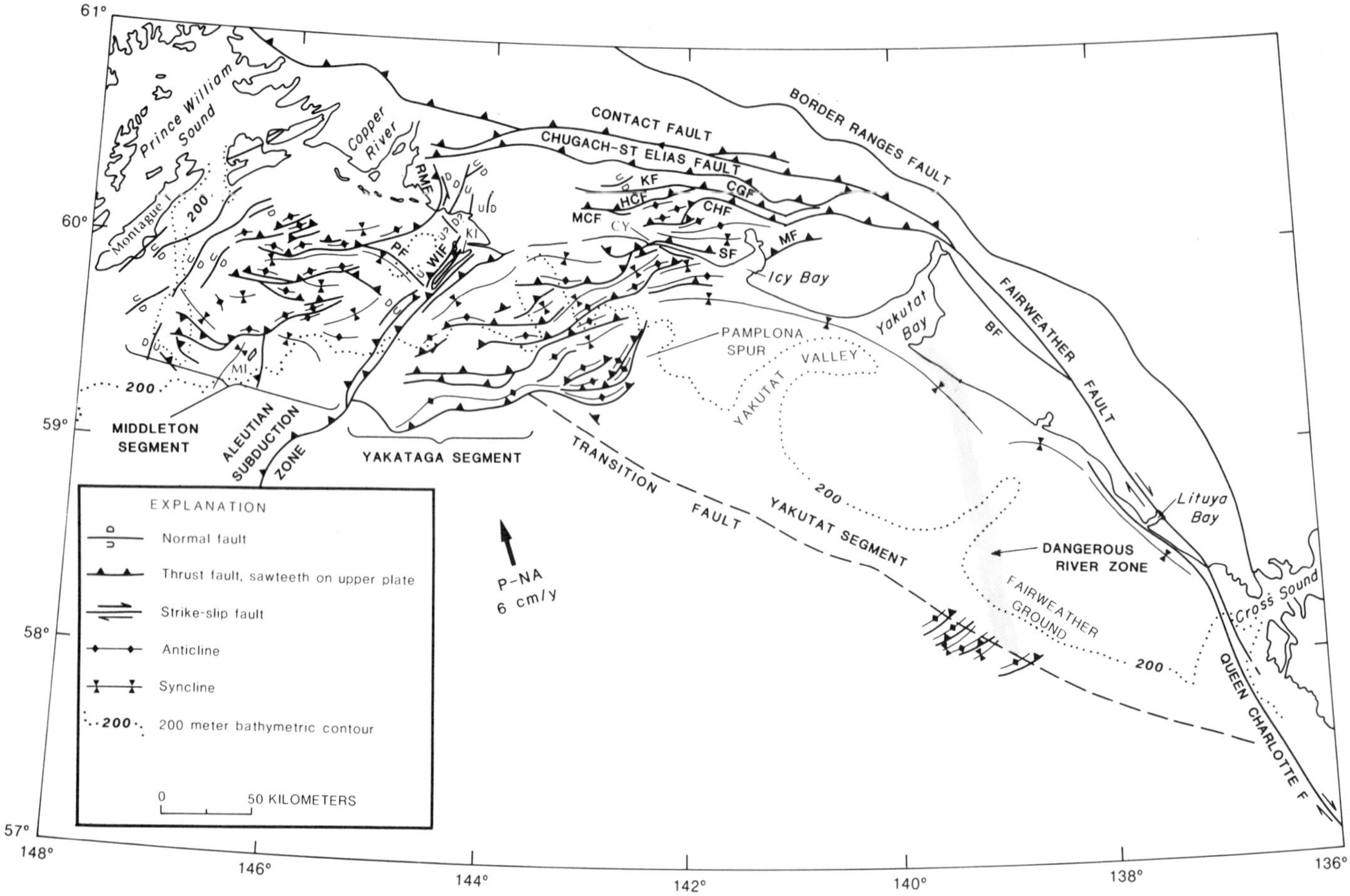

Figure 10. Principal geologic and tectonic elements of the northern Gulf of Alaska. CY–Cape Yakataga; KI–Kayak Island; MI–Middleton Island; BF–Boundary fault; CGF–Coal Glacier fault; CHF–Chaix Hills fault; HCF–Hope Creek fault; IPF–Icy Point–Lituya Bay fault; KF–Kosakuts fault; MCF–Miller Creek fault; MF–Malaspina fault; MI–Middleton Island; PF–Pinnacle fault; RMF–Ragged Mountain fault; SF–Sullivan fault; WIF–Wingham Island fault. After Bruns (1985); see also Bruns and Plafker (this volume).

Reservoir- and source-rock characteristics in both basins are believed to be similar to those in Tugidak Basin (see above). The Neogene sections are probably characterized by lean source rocks containing mainly Type III herbaceous kerogen. Six COST wells have been drilled along the Kodiak shelf, some of which have tested the sections in Albatross and Stevenson basins.

Gulf of Alaska

The Gulf of Alaska continental shelf extends from Dixon Entrance in southeast Alaska to Montague Island in the west. To simplify discussion of this geologically complex area, the shelf is subdivided into five tectonically defined segments. From east to west the segments are (1) the southeast Alaska segment, which extends from Dixon Entrance to Cross Sound, (2) the Yakutat segment, which extends from Cross Sound to Icy Bay, (3) the Yakataga segment, which extends from Icy Bay to Kayak Island, (4) the Middleton segment, which extends from Kayak Island to Montague Island, and (5) the Pacific Ocean Basin segment, which lies at the base of the continental shelf adjacent to the other segments (Figure 10). The geology of each segment is described by Bruns and Plafker (1982) and Bruns (1985). Most petroleum exploration to date has been in the Yakataga and western Yakutat segments.

Data from wells and outcrops both onshore and offshore indicate that the best source rocks are organic-rich mudstone and shale that range in age from early to late Eocene. Dredge samples of Eocene slope outcrops have thermal alteration indices (TAI) and vitrinite reflectance (R_o) values that indicate immature to marginally mature organic material (Bruns, 1982a, 1982b). However, in the central part of the basin, where many closed structures occur, the Eocene strata are buried to depths that are beyond the practical limits of drilling. A major problem for petroleum geology in the Gulf of Alaska is that the best source rocks are in the Paleogene sequence, which does not contain adequate reservoir rocks, and the fact that the Paleogene sequence is buried by an unusually thick section of thermally immature Neogene and younger glacial-marine strata. The thickest part of the Paleogene sequence lies about 75 km south of the mouth of Yakutat Bay, where the maximum thickness is about 5 km. The Neogene and Quaternary sequence has a maximum thickness of 5.5 km. The structural axis of the basin lies adjacent and parallel to the coastline between Icy Bay and Yakutat Bay. Many large, closed anticlinal folds are found between Kayak Island and Icy Bay, but they are geologically very young and primarily involve Neogene and Quaternary strata that lack favorable source or reservoir rocks.

Despite the onshore and offshore drilling activity in the Gulf of Alaska, only a small part of the offshore area has been thoroughly explored. Three of the five tectonic segments remain untested by drilling. The southeastern Alaska, Yakutat, and Pacific Ocean Basin segments may contain prospects for future exploration. The Pacific Ocean segment contains as much as 6 km of sedimentary fill which includes turbidite sands that may have flowed into the basin via submarine canyons, as well as hemipelagic and pelagic sediments (Bruns and Plafker, 1982).

The southeastern Alaska segment lies seaward of the Queen Charlotte-Fairweather fault system and has a sedimentary section ranging from 3 to 5 km thick. Transform folding has created a system of en echelon folds along the eastern edge of the basin. Seismic data suggest that the section is probably composed of Miocene and younger strata, but little is known about the lithology or sedimentary environments (Bruns and Plafker, 1982).

BASINS OF THE CANADIAN SHELF

Queen Charlotte Basin

Queen Charlotte Tertiary basin trends northwest and lies offshore of Queen Charlotte Island. The basin is approximately 360 km long and 60 km wide, and contains as much as 5 km of Tertiary sedimentary rocks. Fourteen offshore wildcat wells were drilled in the basin between 1967 and 1969. Although oil shows were encountered in Pliocene strata in one well, no commercial discoveries were made (Shouldice, 1973). Prospective strata consist of Miocene and Pliocene sandstone and shale that include marine and nonmarine facies. The sediments overlie a sequence of Tertiary volcanic rocks that in turn overlie a basement complex composed of a variety of Mesozoic and Paleozoic(?) sedimentary and volcanic rocks. Sandstone encountered during exploratory drilling had high porosity, but the permeability was low because of the filling of pore space by secondary clays derived from the alteration of feldspars (Shouldice, 1973).

Tofino Basin

Tofino Basin lies offshore from Vancouver Island. Basin dimensions are approximately 180 km by 50 km, and the Tertiary fill is as much as 4 km thick. Six offshore wildcat wells reveal a Tertiary section of marine siltstone, mudstone, and shale that ranges in age from Eocene through Pliocene. Sandstone was rarely encountered and porosities were low. The only significant hydrocarbon shows were shallow gas sands. Shales are light colored and organic contents are low (Shouldice, 1973).

Tofino and Queen Charlotte basins apparently have different lithologies, depositional histories, and styles of structural deformation. Significant differences in formation pressures also were noted during exploratory drilling. Normal pressures were encountered in Queen Charlotte Basin and strongly geopressured sections in Tofino Basin. The results of exploration to date suggest that the two basins of the Canadian shelf probably do not contain large hydrocarbon accumulations.

BASINS OF THE WASHINGTON–OREGON SHELF

The continental margin of Washington and Oregon is adjacent to an active subduction zone where oceanic crust of the Juan de Fuca plate is being thrust beneath the North American plate. This convergent plate boundary has been active since at least the early Eocene and has generated the Cascade Range of stratovolcanoes and a trench that lies along the base of the continental slope. The trench is now completely filled with sediment and has no bathymetric expression (Kulm and Fowler, 1974; Carlson and Nelson, this volume; Snavely, this volume). The long period of subduction along the Washington–Oregon coast produced a series of north–south-oriented trench-slope basins (Figure 11) that have been gradually uplifted from abyssal depths to shelf depths (uplift of 1 to 2 km is reported by Kulm and Fowler, 1974). Subduction-related underthrusting has created complex patterns of faulting and folding and has generated considerable structural relief in beds as young as Pleistocene (Snavely, this volume).

A major mid-late-Eocene unconformity beneath the Oregon shelf may reflect a period of intense underthrusting and uplift. Another intense period of underthrusting may have occurred in the mid-late Miocene, which uplifted the outer part of the continental margin and produced a major angular unconformity. Pliocene claystone and siltstone are widespread on the Oregon shelf; these deposits have been uplifted from upper shelf depths and now form submarine banks. Locally, diapirs bring Pliocene and Pleistocene deposits to shallow depths along the shelf.

Sediment thickness in most of the Washington–Oregon offshore basins exceeds 4 km and locally exceeds 7 km. Typically, the age of the basin fill is Eocene and younger, but in some basins the section is Miocene and younger. The basin fill is mostly sedimentary but locally includes basaltic volcanic rocks (Snavely, Pearl, and Lander 1977, Snavely et al., 1980; Snavely and Wagner, 1981, and Snavely et al., 1981). Basement rocks consist mainly of melange generated by Miocene underthrusting—the dominant deformational process associated with accretion. In some cases the basin fill is tectonically kneaded into the melange basement (Figure 12); the kneading process may help generate the high pore pressures that were encountered in several of the exploratory wells drilled during the 1960s.

A well drilled in the tidelands near Grays Harbor, Washington, produced 12,000 bbls of oil from sheared mudstone, which indicates that the Washington–Oregon shelf basins are capable of generating hydrocarbons (Snavely, in press; Snavely, this volume). The drilling results indicate temperatures and organic carbon levels are high enough for source-rock maturation. Turbidite and shelf sandstone beds throughout the sections provide potential reservoirs, but because much of the sand is volcaniclastic, pore space is susceptible to early precipitation of pore-filling cement.

The difficulties of finding commercial hydrocarbons in Washington–Oregon offshore basins may be analogous to the adjacent onshore basins, which after a protracted period of exploration (more than 550 wells have been drilled in Oregon) have produced gas from only one field, the Mist field in northwest Oregon. Complex structure resulting in small fault blocks with limited potential for migration and entrapment is probably the rule rather than the exception in

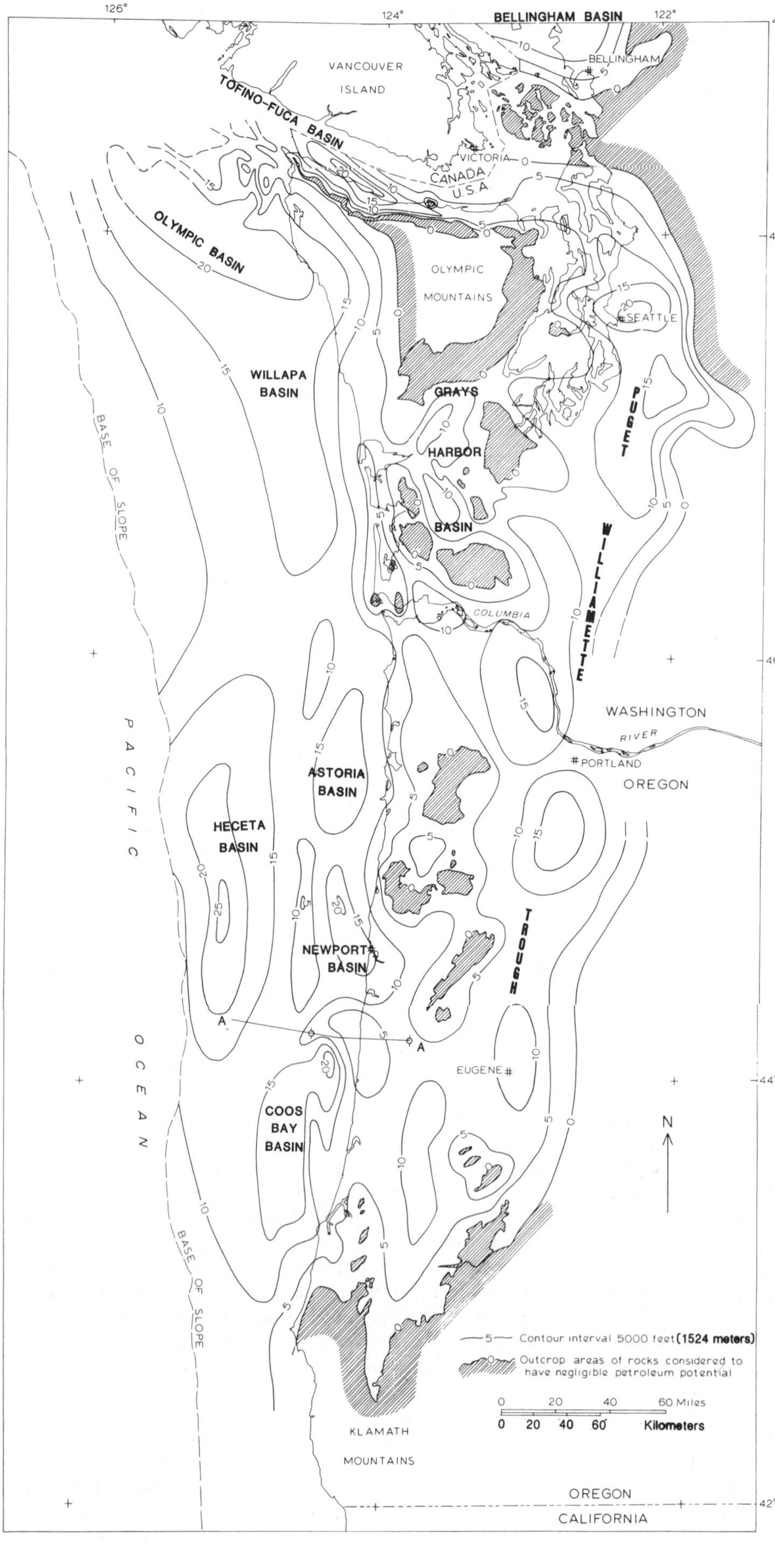

Figure 11. Distribution of Tertiary basins of western Washington-Oregon. Isopach thicknesses in thousands of feet. Modified from Snavely and Wagner (1980); see also Snavely (this volume). A–A′ is line of section shown in Figure 12.

Washington–Oregon offshore basins and unfortunately diminishes the prospects of finding giant fields.

NORTHERN AND CENTRAL CALIFORNIA SHELF

The offshore basins of the northern and central California coast have been the subject of periodic exploration for more than 20 years. From September 1963 to September 1967, 20 exploratory wells were drilled by the oil industry in the Eel River, Point Arena, Bodega, Outer Santa Cruz, and Santa Maria basins (Figure 13). All of the basins yielded shows of oil (Zieglar and Cassell, 1978) except the Eel River Basin, which had shows of gilsonite veining in the deepest well. Of particular interest to current exploration is the presence in most of the basins of oil-bearing siliceous shales and chert of middle Miocene age that are probably correlative with the Monterey Formation—the source and reservoir rock in the giant Arguello oil field recently discovered in the offshore Santa Maria Basin (Crain, Mero, and Patterson, 1985). Geology and petroleum resources for parts of the northern and central California shelf are reported by McCulloch et al. (1982).

Eel River Basin

The offshore Eel River Basin parallels the northern California coast from the mouth of the Eel River northward to the Oregon–California border (Figure 13). The basin occupies an area of 14,000 km and lies in water that ranges in depth from 0 to about 1440 m. Maximum thickness of the basin fill is about 4.5 km. Structures within the basin are parallel to the trend of basin elongation (Hoskins and Griffiths, 1971).

Basement rocks consist of metagraywacke, shales, and mafic igneous rocks of the coastal belt of the Franciscan assemblage. These rocks are structurally overlain by about 800 m of mostly Paleocene sandstone, siltstone, and shale. Onshore, the latter sequence is named the Yager Formation (Ogle, 1953) and consists of highly indurated and locally isoclinally folded submarine fan turbidite sequences and interbedded mudstone. Organic carbon in Yager shales ranges from 0.30 wt. % to 1.59 wt. %, and the mean vitrinite reflectance (R_o) ranges from 0.49 % to 1.08 % (Underwood, 1985). The thickest part of the section consists of Pliocene strata that overlie a thin section of Miocene claystone and siltstone. Both are part of the Wildcat Group, and unconformably overlie Cretaceous and Eocene strata. The Miocene section is overlain by lower Pliocene clayey siltstone, which is in turn overlain by clayey and silty marine strata of late Pliocene age.

Intrabasin structures are dominated by right-stepping en echelon folds, which are consistent with right-lateral transform motion. Other structures include high-angle reverse faults (most with the west side down) that cut Pliocene and younger rocks, and shale diapirs that intrude the Neogene section. Four wells were drilled on offshore structures in the 1960s, but no commercial hydrocarbons were discovered. The two deepest wells were drilled to depths of 2377 m and 2248 m; the wells bottomed in hard, dense Upper Cretaceous sandstone and middle Eocene siltstone, respectively. Drilling in the offshore part of the Eel River Basin has been stimulated by the discovery onshore of the now-abandoned Table Bluff field and the still-productive Tompkins Hill gas field.

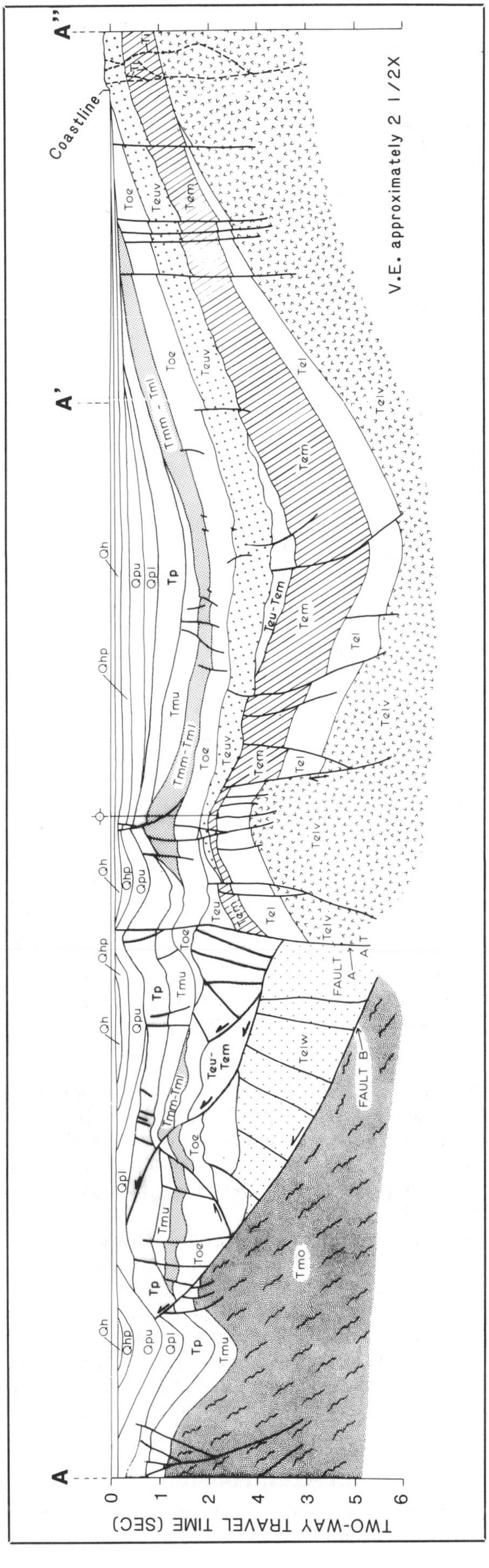

Figure 12. Geologic cross-section interpreted from seismic-reflection profiles. The westward increase in structural complexity is associated with proximity to the accretionary body. Symbols for rock units explained in Snavely et al. (1981). Line of section shown on Figure 11.

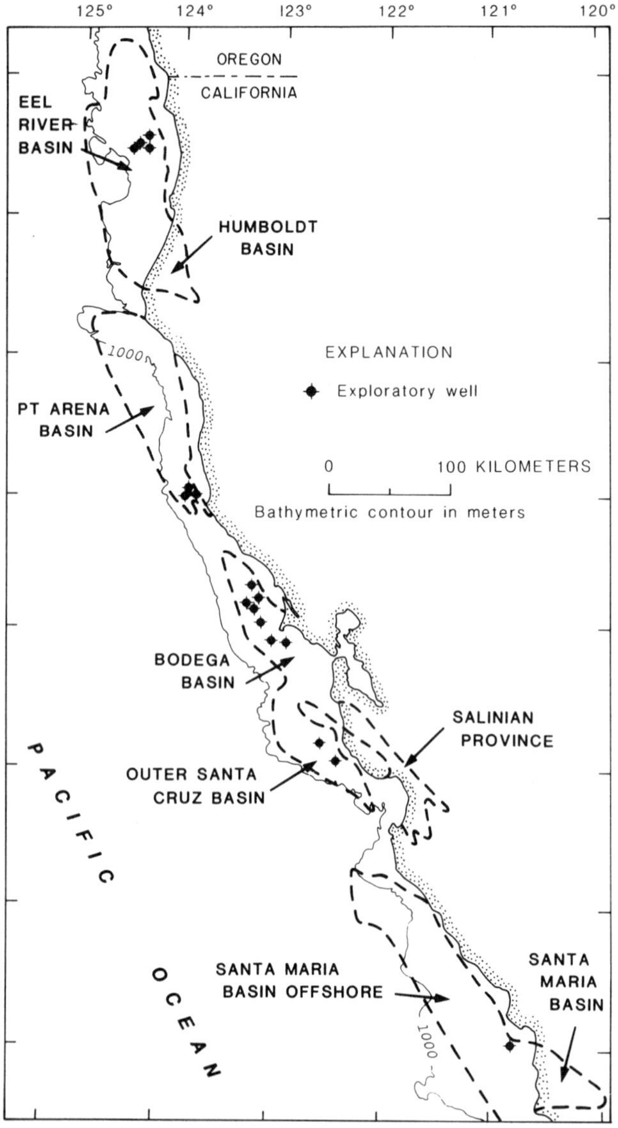

Figure 13. Distribution of Neogene basins of northern and central California, and locations of exploratory wells drilled in the 1960s. Modified from McCulloch et al. (1982); see also McCulloch and Clarke (this volume).

Cumulative production to 1983 for the two fields was 80.7 Bcf of natural gas. Most of the production is from the upper Pliocene Anderson sands in the lower part of the Rio Dell Formation (Hopps and Horan, 1984). The offshore part is probably also gas prone and has about the same potential as the onshore part.

A bottom-simulating reflector (BSR) has been mapped seismically beneath the Klamath Plateau at a subsurface depth of 250 m by Field and Kvenvolden (1985). They estimate a geothermal gradient within the gas hydrate associated with the BSR of 55°C/km, but the gradient within the basin fill is probably much lower. Kvenvolden and Field (1981) report indications of natural gas in samples of mudstone that may be from diapirs that crop out on the sea floor in the Eel River Basin.

Point Arena Basin

Point Arena Basin is a roughly triangular trough with its apex just offshore from Point Arena (Figure 13). The basin broadens and deepens toward the northwest. The eastern side of the basin is formed by the north-trending San Andreas fault system and its splays; the western side is formed by the edge of the continental shelf. Basement rocks are similar to those in the Eel River Basin; that is, Franciscan-assemblage rocks that are structurally overlain by erosional remnants of Upper Cretaceous and Eocene arkosic sandstone and siltstone. Three exploratory wells drilled in the 1960s were located on anticlines in the southern end of the basin (Hoskins and Griffiths, 1971). The wells, whose total depths ranged from 1438 m to 3242 m, indicate that the Neogene section consists of lower, middle, and upper Miocene, and lower and upper Pliocene marine deposits. Petroliferous cherts and siliceous shales of middle Miocene age (Monterey Formation equivalent?) probably have the greatest hydrocarbon potential of any rocks in the basin.

Santa Cruz and Bodega Basins

Santa Cruz and Bodega basins form a contiguous northwest-trending structure that extends from lat. 37°30′N to lat. 38°30′N (Figure 13). The axis lies 20 to 30 km west of, and parallel to, the San Andreas fault. Bodega Basin has an approximate area of 4500 km^2 and a maximum thickness of about 3.5 km; Santa Cruz Basin covers approximately 780 km^2 and has a thickness of 4690 m. Most of the fill in both basins is Neogene and younger. Granitic rocks as well as erosional remnants of Upper Cretaceous and Eocene strata underlie the Neogene fill.

Eight exploratory wells drilled in Bodega Basin during the 1960s ranged in total depth from 986 m to 3190 m. Data from the wells indicate that the Neogene sequence consists of lower Miocene marine shale and sandstone that has fair reservoir properties. Middle Miocene cherty shale is interbedded with vesicular volcanics, which overlie a basal sandstone. Upper Miocene strata consist of marine claystone and sandstone. The Miocene section is conformably overlain by clayey and silty marine Pliocene strata.

Santa Cruz Basin remains untested. Three of the exploratory wells in Bodega Basin had tarry oil shows in middle Miocene cherty shale (Hoskins and Griffiths, 1971); the shows indicate that source rocks have generated petroleum. Fractured chert sequences will be examined closely for their reservoir properties in any future wells that may be drilled.

Outer Santa Cruz Basin

Outer Santa Cruz Basin extends northwestward from near the tip of the Monterey peninsula to the edge of the continental shelf and covers an area of approximately 2500 km^2 (Figure 13). Water depths range from 0 to 2000 m; sedimentary fill is as much as 3 km thick. Like most of the other basins along the central and northern California coast, the part of the basin fill with the best prospects is Neogene in age. Two wells were drilled in the 1960s, to total depths of 2358 m and 2892 m. Middle Miocene cherty shale beds that lie within a 1200-m-thick sequence bounded by unconformities are attractive exploration targets. The middle Miocene sequence overlies shallow-water lower Miocene sandstone and volcanic strata, and is overlain by upper Miocene and lower Pliocene silty and clayey shallow-marine beds. The youngest part of the basin fill consists of upper Pliocene marine clayey and silty strata.

Santa Maria Basin Offshore

Santa Maria Basin is a north–south-oriented sediment-filled structural depression that extends from lat. 34°30′N, adjacent to Point Conception, to lat. 36°20′N (Figure 13). The basin is bounded on the west by the Santa Lucia Bank fault zone, and on the east by the Hosgri fault zone. Neogene basin fill locally exceeds 4.5 km and is folded into north-trending anticlines and synclines.

In 1981, Chevron U.S.A., Inc., and partners discovered the giant Point Arguello oil field at the southern edge of Santa Maria Basin; estimated recoverable reserves are in excess of 300 million bbls (Crain, Mero, and Patterson, 1985). Oil production is from fractured chert, porcellanite, siliceous mudstone, and dolostone of the middle and upper Miocene part of the Monterey Formation. Productive intervals are from subsea depths between 2031 m and 2421 m. Further potential oil production is reported in the lower Miocene Point Sal Formation and in upper Miocene marine strata (Sisquoc Formation?). Potential gas production is noted in the upper Pliocene part of the Pico Formation (Crain, Mero, and Patterson, 1985).

Oil in the Monterey Formation of the Point Arguello field was derived from maturation of biogenic laminae in siliceous and dolomitic beds. Maturation of organic matter was enhanced by above-normal temperature gradients that in the Point Arguello field range from 36.6°C to 45.6°C/km, (Crain, Mero, and Patterson, 1985). Gravity of pooled oil ranges from about 11° API to as high as 34° API.

The Point Arguello and nearby Point Pedernales fields, and the San Miguel field (discovered in 1983) 30 miles north of the Point Arguello field, offer encouragement for future offshore exploration in other central and northern California basins, particularly in view of the "dry hole" that was drilled and abandoned in the 1960s within a mile or two of the Arguello discovery well.

BASINS OF THE CALIFORNIA CONTINENTAL BORDERLAND

The part of the California Continental Borderland that we discuss below extends from lat. 34°N southward to the United States–Mexico boundary, and from the coastline westward to the Patton Escarpment, encompassing an area of roughly 50,000 km^2 (Figure 14). The Borderland is flanked on the north and northeast sides by highly productive oil and gas fields, but only the Beta field located just west of Huntington Beach produces hydrocarbons within the Federal offshore part of the area. No production or discoveries exist south of lat. 34°N or west of the Palos Verdes Hills fault, which is located on the east side of the Palos Verdes Peninsula.

The Borderland is dominated structurally and physiographically by a series of northwest-trending ridges and basins bounded by faults (Figure 14). Basement rocks include ophiolite (exposed on Santa Cruz Island), Catalina Schist (includes greenschist, blueschist, and amphibolite-facies rocks), and Franciscan graywacke. A sequence like the one in the Great Valley (consisting of Upper Cretaceous, Paleocene, and Eocene marine sandstone, siltstone, mudstone, and conglomerate) overlies the basement assemblage. This sequence may represent two or more tectonostratigraphic terranes (Howell and Vedder, 1981). Ridges within the Borderland are underlain by uplifted basement rocks and/or by rocks similar to those in the Great Valley; the basins typically contain sequences of Neogene and Quaternary strata that thin against the flanks of the ridges (Vedder, this volume).

Exploratory drilling to date has focused on sections of Upper Cretaceous, Paleocene, and Eocene rocks, with no commercial success; the wells have been drilled on and adjacent to Santa Cruz and Santa Rosa Islands and along the crest of Dall, Tanner, and Cortes Banks. Neogene sections that are highly productive in adjacent areas both onshore and offshore have yet to be tested, because most of the section lies in water deeper than 1000 m. Some promising nearshore areas have not been leased because of environmental concerns. Exploration within and adjacent to the Borderland indicates that the Upper Cretaceous to Eocene sequence has little chance of producing commercial quantities of oil or gas, mainly because of source-rock problems relating to type, quantity, and maturity of organic carbon. Rocks with good reservoir properties are not present, although most sections contain thick deposits of arkosic turbidite sandstone with a variety of submarine fan facies. Diagenesis has reduced intergranular porosity by means of various chemically precipitated cements; pore space has also been reduced by mechanical compaction.

Neogene strata hold the best promise for petroleum exploration within the Borderland. Nearly all of the adjacent production is from Miocene and Pliocene rocks. Strata within the Borderland that are correlative with the middle part of the Miocene Monterey Formation are targets for future exploration, as are correlatives of the early Miocene and Oligocene parts of the Sespe and Vaqueros Formations. Miocene rocks that are buried deep enough for maturation of organic carbon will probably attract attention as economic conditions become favorable for deep-water drilling. Vedder (this volume) fully explores the petroleum geology of the California Continental Borderland.

OFFSHORE BASINS OF NORTHERN BAJA CALIFORNIA, MEXICO

Fifteen relatively small sedimentary basins lie along the continental shelf between the United States-Mexico boundary and lat. 30°N. The California Continental Borderland extends southward into Mexican waters, but is narrower and deeper than the Borderland to the north. Figure 15 shows the distribution of the basins that lie in Mexican waters, which have been defined mainly by reconnaissance seismic surveys (Moore, 1969). None of the basins have been drilled for hydrocarbons, although samples have been dredged from sea-floor ridges located between some of the basins (Normark, Spencer, and Ingle, this volume). Basin areas range from as small as 200 km^2 to as large as 3800 km^2; thickness of sediment fill ranges from < 1 km to > 5 km. Extrapolation of basement-rock trends and onshore geology suggests that the western part of the Mexican borderland is underlain by structurally complex accretionary terranes that include Franciscan-like assemblages and Catalina Schist, and the eastern part is underlain by granitic rocks of the Peninsular batholith and prebatholithic metamorphic rocks. Basaltic volcanic rocks of Miocene age have been

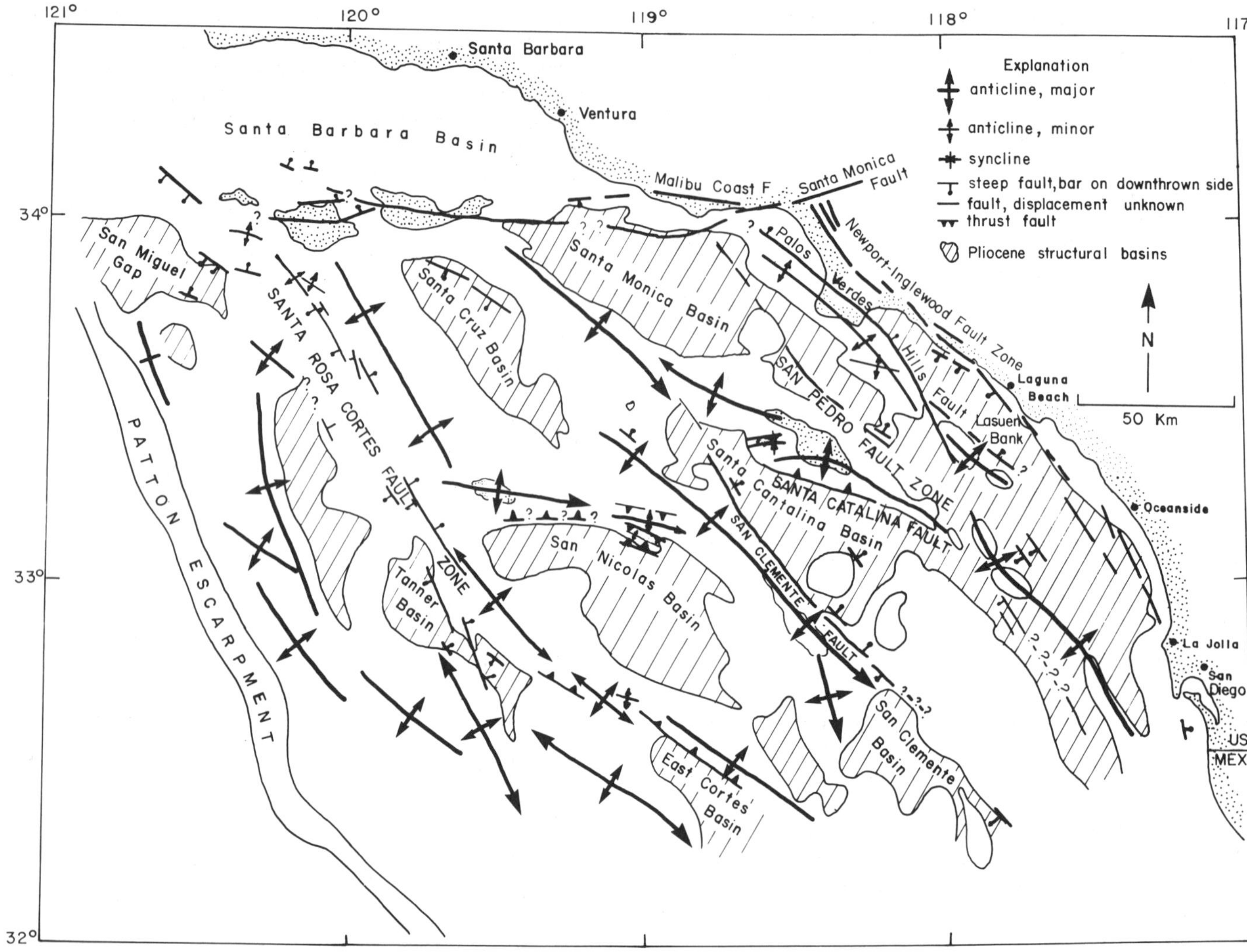

Figure 14. Distribution of bathymetric basins and selected tectonic elements of the California Continental Borderland. After Howell et al. (1980); see also Vedder (this volume).

dredged from some of the sea-floor ridges. The basement rocks have no known petroleum potential.

Accretionary basement is overlain in some areas by a succession of strata which resemble those of the Great Valley and which range in age from Late Cretaceous through Eocene. These strata, which overlap the Mexico-U.S. boundary, have been drilled by exploratory wells on Dall, Tanner, and Cortes Banks in U.S. waters, and in Mexican waters in Sebastian Vizcaino Bay (Figure 16); however, commercial shows of oil or gas have not been reported from either area.

Most of the sedimentary fill in the basins of the Mexican borderland is probably Miocene and younger, because most of the basins formed during the onset of wrench tectonics that began at about 18 to 20 Ma. Many of the basins are bounded by faults, some of which are major northwest-oriented structures of regional extent. Most of the folding in the basins is of local extent and of low amplitude.

Little is known or can be inferred about the composition or organic richness of the sedimentary fill. Onshore outcrops of Neogene strata consist of shallow-marine and nonmarine sandstone and conglomerate, and locally derived breccia (Normark, Spencer, and Ingle, this volume). Volcanic flows (mostly basalt) that intertongue with Neogene sedimentary rocks are well exposed in roadcuts and sea cliffs between Tijuana and Ensenada in Mexico. The strata in some of the basins may also consist of biogenous sediment that could be coeval with diatomaceous formations of Miocene age in California (Monterey Formation). Upper Oligocene to lowest Miocene diatomite and phosphatic strata (San Gregorio Formation) are known in the west-central part of Baja California Sur, Mexico (McLean and Hausback, 1984; and McLean, Hausback, and Knapp, 1985).

The only basins that are likely to contain strata that have reached burial depths sufficient for petroleum generation are those with sections thicker than 3 km, namely Punta Banda-Descanso shelf (4 km), "No Name" basin (5.3 ? km), and Velero Basin (5 km).

Cedros Deep

Cedros Deep Basin is a narrow, elongate structural trough about 35 km wide and 700 km long, which bathymetrically lies at the base of the continental slope due west of Cedros Island and Punta Eugenia on the Vizcaino Peninsula (Figure 16). The structural basin with extensions to the north and

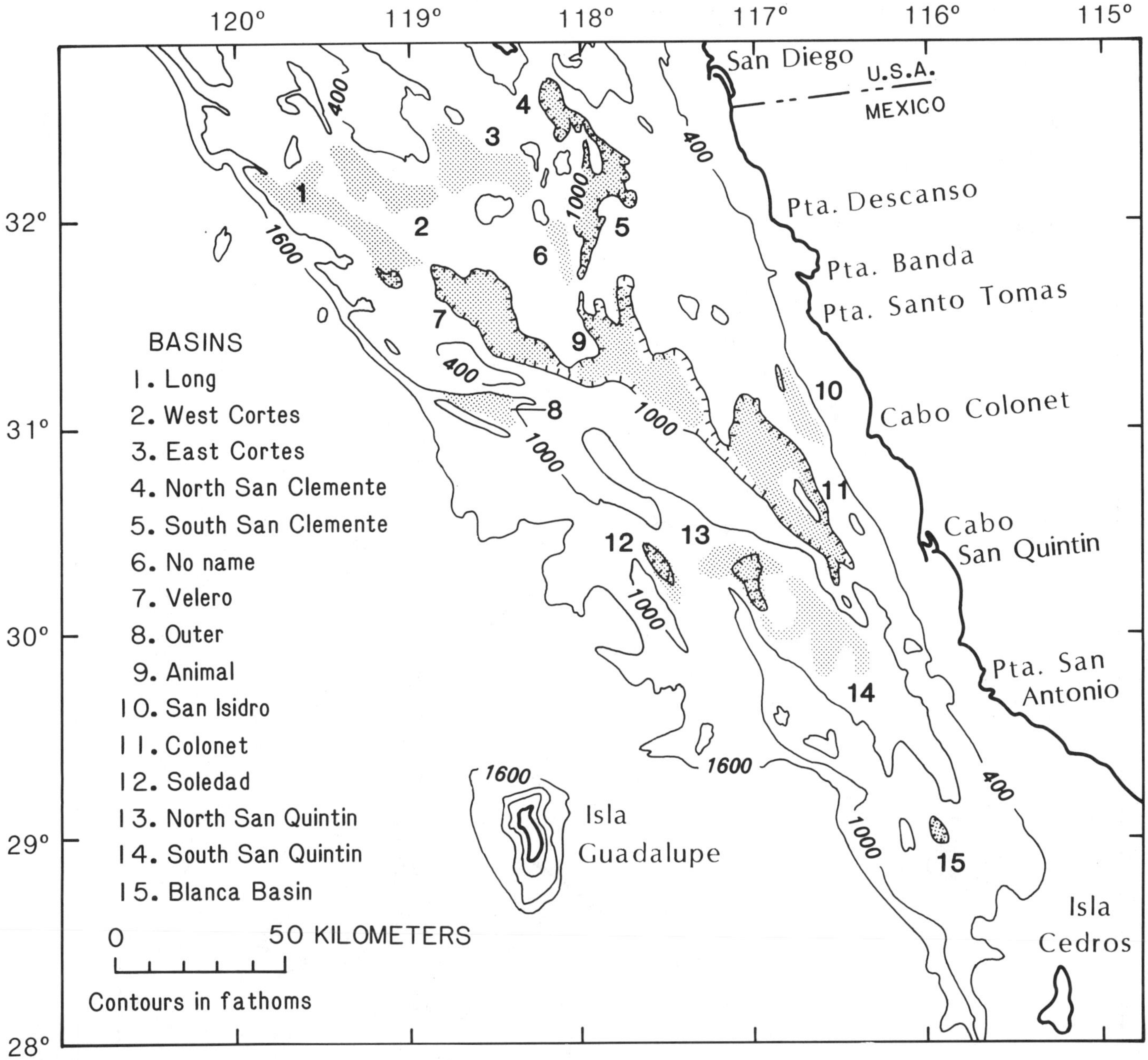

Figure 15. Locations of basins of the borderland adjacent to northern Baja California, Mexico. Modified from Moore (1969).

south covers an area of 29,000 km^2 and lies in waters that range in depth from about 2000 m to about 3500 m. Approximately 750 m of sediment overlies acoustic basement; the basement may be formed by trench fill related to pre-Miocene subduction along the Baja California margin.

The prominent topographic feature that forms the deep extends from about 29°00′N to 24°30′N latitude. Structurally, the deep appears to be a graben that is downfaulted on the east side against the continental slope, and against the sea floor on the west. A gravity low associated with the deep extends beyond the limits of the topographic depression. Seismic reflectors in the basin fill are flat lying, as shown by Yeats and Haq (1981, Figure 6).

Two DSDP holes have been drilled in the vicinity of Cedros Deep. Site 470 is 90 km west, and site 471 is about 500 km southeast, of the south end of the elongate basin. The sections cored in the two DSDP holes provide clues to the age and composition of the sedimentary section in Cedros Deep, but reveal little meaningful information about its petroleum potential.

Vizcaino Basin

Vizcaino Basin is a large northwest-oriented elongate structure that lies both onshore and offshore about half-way down the Baja California Peninsula (Figure 16). The onshore part underlies much of the vast arid lowland flanked on the west side by the mountains of the Vizcaino Peninsula, and on the east side by the crystalline backbone of the Baja California Peninsula. Length and width of the onshore basin are 100 km and 150 km, respectively.

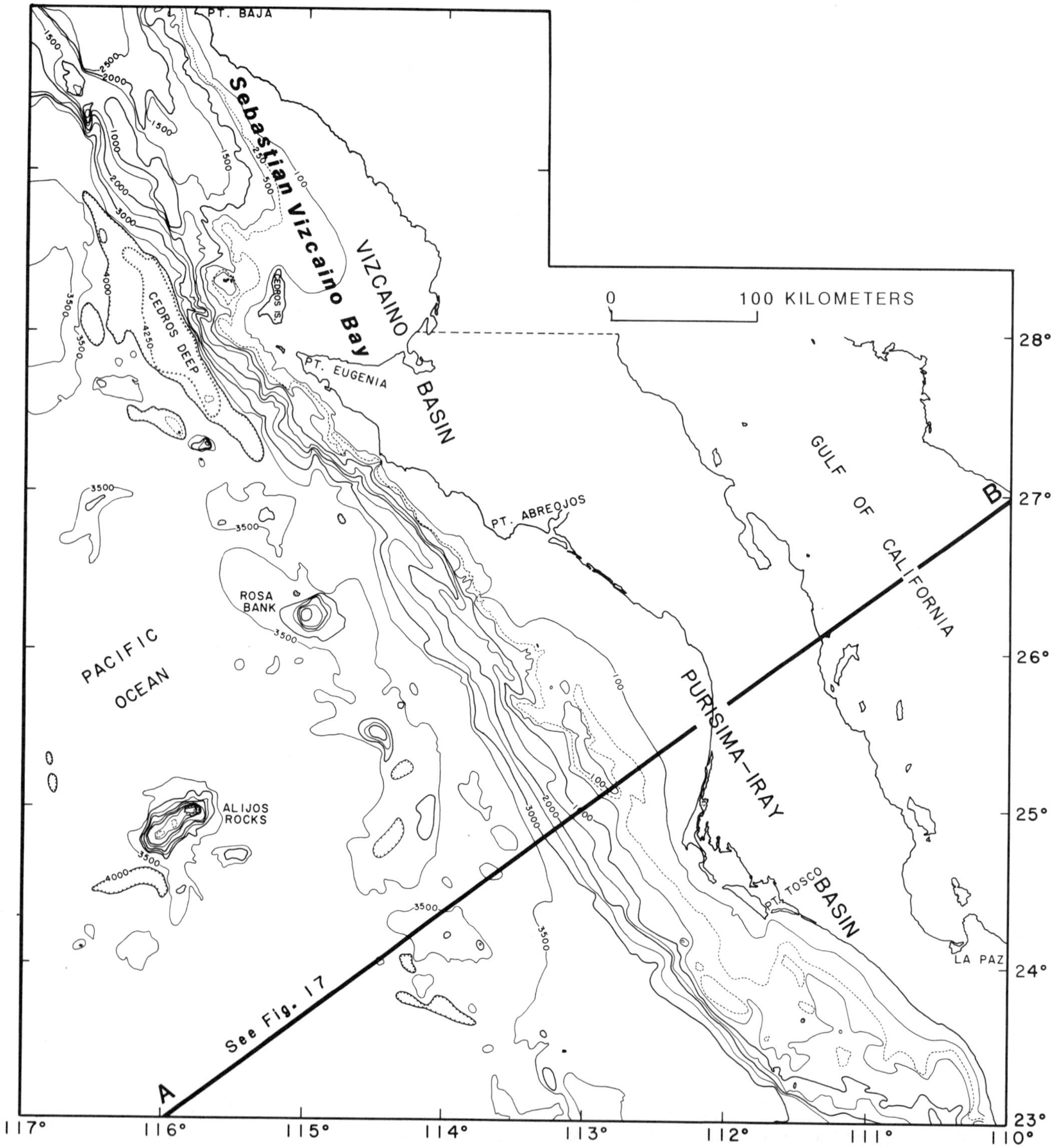

Figure 16. Bathymetric map of the continental shelf and slope of west-central Baja California, Mexico, showing location of Cedros Deep, Vizcaino Basin, and Purisima-Iray Basin. Section A–B is shown on Figure 17. Modified from Normark (1977); see also Normark, Spencer, and Ingle (this volume). Contour interval is 500 m; dashed line at 250 m.

Offshore, the basin underlies Sebastian Vizcaino Bay, between Cedros Island and the Pacific Coast of the Baja Peninsula. The basin is underlain by two principal basement rock assemblages similar to the basement rocks that underlie the San Joaquin Valley (Great Valley sequence) of California. The western basement consists of Franciscan-like rocks that include structurally complex Mesozoic ophiolite, graywacke, melange that contains blueschist, metagraywacke, and pillow basalt (Moore, 1984). Triassic pillow basalt, graywacke, and chert are also exposed along the coast. Basement rocks are well exposed on Cedros Island and along the west coast of the Vizcaino Peninsula. The eastern part of Vizcaino Basin is underlain by a mid-Cretaceous island-arc assemblage of volcanic, volcaniclastic, and carbonate rocks (Alisitos Formation) that is intruded by Late Cretaceous plutons of the Peninsular Ranges batholith.

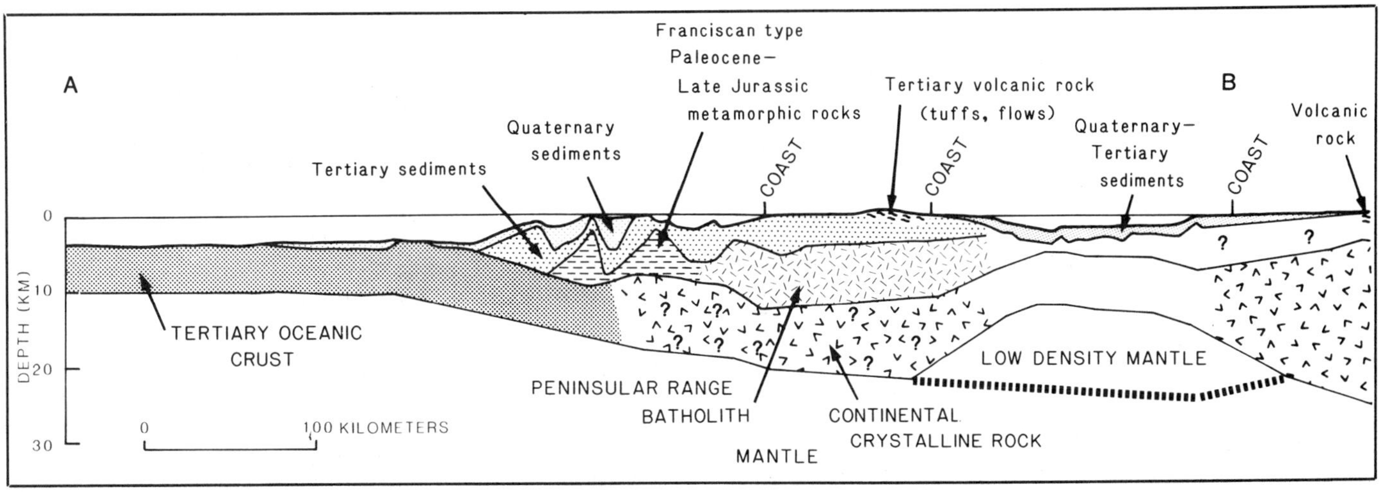

Figure 17. Geologic cross-section showing crust and subcrust across Baja California, Mexico, as interpreted from reconnaissance gravity and magnetic data. Line of section is shown on Figure 16. Modified from Coperude (1977).

Exploratory drilling of both onshore and offshore parts of Vizcaino Basin by Petroleos Mexicanos (the government-owned petroleum agency known as "Pemex") indicates that the sedimentary fill consists of Upper Cretaceous, Paleocene, and Eocene marine sandstone, conglomerate, and mudstone analogous to the Great Valley sequence of California in age, microfauna, and lithology. Total thickness of the section is about 3.5 km. A thin veneer of middle Miocene marine sandstone deposited in neritic depths overlies the Late Cretaceous and Paleogene section. The latter consists of arkosic detritus derived from erosion of plutonic rocks; onshore outcrops are loosely to moderately consolidated and appear to have good reservoir properties (Sorensen, 1982). Petroleum exploration has been unsuccessful in the basin, mainly because of scarce and immature organic source beds. Drilling to date has only produced noncommercial shows of gas.

Neogene structural stability has hampered the evolution of the Vizcaino Basin. Although flanked by areas of active Neogene tectonism—namely, the California Continental Borderland to the north and the Gulf of California to the east—the west-central part of Baja California Sur has experienced little vertical tectonic movement since the Paleocene (McLean, Hausback, and Knapp, 1985). Vertical stability is well illustrated along the southern edge of the Vizcaino Basin onshore, where virtually flat-lying Eocene turbidite sandstone is paraconformably overlain by upper Miocene (8 to 10 Ma) basalt. The flows cap an extensive area of plateaus that are about 100 m above present sea level. Locally, these flows have basal pillow sections that record a paleo sea level only 100 m above present sea level (McLean, Hausback, and Knapp, in press). Nearly flat-lying Paleocene and Eocene strata are locally exposed southward from the Vizcaino Basin for at least 400 km along the Pacific coastal plain.

Purisima-Iray Basin

The Purisima-Iray Basin extends southward from the Vizcaino Basin for a distance of approximately 375 km (Figure 16). The geology of the basement and sedimentary fill of Purisima-Iray Basin is similar to that of the Vizcaino Basin. Exploratory drilling indicates that most of the basin fill is Paleocene and Eocene sandstone and siltstone; the deepest well reported by Mina (1957) bottomed in Paleocene sandstone at a depth of 2905 m. Geophysical modeling by Coperude (1977) indicates that the Upper Cretaceous and Paleocene section overlying granitic basement locally may be at least 6000 m thick (Figure 17). At lat. 25°00′N, granitic basement appears to extend at least 50 km west of the shoreline.

The northern offshore extension of the Purisima-Iray Basin, informally referred to here as the Purisima shelf, extends from Punta Abreojos on the north to Punta Tosco on the south end of Santa Margarita Island (Figure 16). The area is roughly 250 km long and 90 km wide. Little has been published about the area, so estimates of petroleum potential cannot be made. Reconnaissance geophysical data do suggest, however, that the basin fill onlaps granitic basement to the east and a Franciscan-like assemblage of Mesozoic age to the west (Blake et al., 1984). Recent geologic mapping by Blake et al. (1984) indicates the complex nature of the structural relationships within Mesozoic basement rocks exposed on Magdalena and Santa Margarita Islands. The western edge of the Purisima shelf may contain a narrow elongate basin filled with Neogene and younger sediment; the basin may have formed during a period of Neogene wrench-type tectonism as suggested by Spencer and Normark (1979) and by Normark, Spencer, and Ingle (this volume).

The southern offshore extension of the Purisima-Iray Basin, informally referred to here as the Magdalena shelf, extends from Punta Tosco to the tip of the Baja Peninsula—an area about 150 km long and 120 km wide. Little is known about the geology of the Magdalena shelf in terms of basin size and age of sedimentary fill. Relations between Upper Cretaceous and Paleogene strata that resemble Great Valley rocks and basement rocks are probably similar to those described above for the Purisima shelf (Figure 17). To date, onshore exploratory drilling in the Purisima-Iray Basin has produced no commercial quantities of oil or gas, probably for the same reasons as outlined above for the Vizcaino Basin.

REFERENCES CITED

Blake, M. C., Jr., A. S. Jayco, T. E. Moore, V. Chavez, J. B. Saleeby, and K. Seel, 1984, Tectonostratigraphic terranes of Magdalena Island, Baja California Sur, *in* V. A. Frizzell, Jr., Geology of the Baja California Peninsula: SEPM, Pacific Section, v. 39, p. 183–192.

Bruns, T. R., 1982a, Hydrocarbon resource report for proposed OCS lease sale 88—Southeastern Alaska, northern Gulf of Alaska, Cook Inlet, and Shelikof Strait, Alaska: U.S. Geological Survey Open-File Report 82-928, 133 p.

Bruns, T. R., 1982b, Structure and petroleum potential of the continental margin between Cross Sound and Icy Bay, northern Gulf of Alaska: U.S. Geological Survey Open-File Report 82-929, 65 p.

Bruns, T. R., 1985, Tectonics of the Yakutat block, an allochthonous terrane in the northern Gulf of Alaska: U.S. Geological Survey Open-File Report 85-13, 112 p.

Bruns, T. R., and G. Plafker, 1982, Geology, structure, and petroleum potential of the southeastern Alaska and northern Gulf of Alaska continental margins, *in* T. R. Bruns, ed., Hydrocarbon resource report for proposed OCS lease sale 88—Southeastern Alaska, northern Gulf of Alaska, Cook Inlet, and Shelikof Strait, Alaska: U.S. Geological Survey Open-File Report 82-928, 133 p.

Bruns, T. R., and R. von Huene, 1977, Sedimentary basins on the Shumagin shelf, western Gulf of Alaska: Ninth Offshore Technology Conference, v. 1, p. 41–49.

Bruns, T. R., R. von Huene, R. C. Cullota, S. D. Lewis, and J. W. Ladd, 1987, Geology and petroleum potential of the Shumagin margin, Alaska, this volume.

Carlson, P. R., and C. H. Nelson, 1987, Marine geology and resource potential of the Cascadia Basin, this volume.

Cooper, A. K., D. W. Scholl, and M. S. Marlow, 1987, Structural framework, sedimentary sequences, and hydrocarbon potential of the Aleutian and Bowers basins, Bering Sea, this volume.

Cooper, A. K., D. W. Scholl, M. S. Marlow, J. R. Childs, G. D. Redden, and K. A. Kvenvolden, 1978, The Aleutian Basin, Bering Sea—A frontier area for hydrocarbon exploration: Houston, Texas, 10th Annual Offshore Technology Conference, p. 353–359.

Cooper, A. K., D. W. Scholl, M.S. Marlow, J. R. Childs, G. D. Redden, K. A. Kvenvolden, and A. J. Stevenson, 1979, Hydrocarbon potential of Aleutian Basin, Bering Sea: AAPG Bulletin, v. 63, p. 2070–2087.

Coperude, S. P., 1977, Geologic structure of the western continental margin of south central Baja California based on seismic and potential field data: Corvallis, Oregon, Oregon State University unpublished M. A. thesis, 66 p.

Crain, W. E., W. E. Mero, and D. Patterson, 1985, Geology of the Point Arguello discovery: AAPG Bulletin, v. 69, p. 537–545.

Field, M. E., and K. A. Kvenvolden, 1985, Gas hydrates on the northern California continental margin: Geology, v. 13, p. 517–520.

Fisher, M. A., 1979, Structure and tectonic setting of the continental shelf southwest of Kodiak Island, Alaska: AAPG Bulletin, v. 63, p. 301–310.

Fisher, M. A., 1980, Petroleum geology of Kodiak shelf, Alaska: AAPG Bulletin, v. 64, p. 1140–1157.

Fisher, M. A., 1982, Petroleum geology of Norton Basin, Alaska: AAPG Bulletin, v. 66, p. 286–301.

Fisher, M. A., and R. von Huene, 1980, Structure of Upper Cenozoic strata beneath Kodiak shelf, Alaska: AAPG Bulletin, v. 64, p. 1014–1033.

Fisher, M. A., W. W. Patton, Jr., and M. L. Holmes, 1982, Geology of Norton Basin and continental shelf beneath northwestern Bering Sea, Alaska: AAPG Bulletin, v. 66, p. 255–285.

Grantz, A., and S. D. May, 1987, Regional geology and petroleum potential of the United States Chukchi shelf north of Point Hope, this volume.

Grantz, A., S. D. May, and D. A. Dinter, 1987, Regional geology and petroleum potential of the United States Beaufort and northeasternmost Chukchi Seas, this volume.

Grantz, A., D. A. Dinter, E. A. Hill, S. D. May, R. H. McMullin, R. L. Phillips, and E. Reimnitz, 1982a, Geologic framework, hydrocarbon potential, and environmental conditions for exploration and development of proposed oil and gas lease sale 87 in the Beaufort and northeast Chukchi Seas: U.S. Geological Survey Open-File Report 82-482, 71 p.

Grantz, A., D. A. Dinter, E. A. Hill, R. E. Hunter, S. D. May, R. H. McMullin, and R. L. Phillips, 1982b, Geologic framework, hydrocarbon potential, and environmental conditions for exploration and development of proposed oil and gas lease sale 85 in the central and northern Chukchi Sea: U.S. Geological Survey Open-File Report 82-1053, 84 p.

Hopps, T., and E. Horan, 1984, Subsurface stratigraphy—Eel River Basin, *in* R. L. Hester and D. E. Hallinger, eds., Selected Papers Presented to Pacific Section AAPG 1983 Annual Meeting, Sacramento, California, v. 1, p. 1–29.

Hoskins, E. G., and J. R. Griffiths, 1971, Hydrocarbon potential of northern and central California offshore, *in* Future petroleum provinces of the United States—Their geology and potential: AAPG Memoir 15, p. 212–228.

Howell, D. G., and J. G. Vedder, 1981, Structural implications of stratigraphic discontinuities across the southern California borderland, *in* W. G. Ernst, ed., The geotectonic development of California: Rubey Volume 1, Prentice Hall, New York, Rubey Colloquium, v. 1, p. 535–558.

Howell, D. G., J. K. Crouch, H. G. Greene, D. S. McCulloch, and J. G. Vedder, 1980, Basin development along the late Mesozoic and Cainozoic California margin: A plate tectonic margin of subduction, oblique subduction and transform tectonics, *in* H. G. Reading and P. F. Balance, eds., Sedimentation along oblique slip mobile belts: International Association of Sedimentologists, v. 4, p. 43–62.

Kulm, L. D., and G. A. Fowler, 1974, Oregon continental margin structure and stratigraphy—A test of the imbricate thrust model, *in* C. A. Burk and C. L. Drake, eds., The geology of continental margins: New York, Springer-Verlag, p. 261–283.

Kvenvolden, K. A., and M. E. Field, 1981, Thermogenic hydrocarbons in unconsolidated sediment in the Eel River Basin, offshore northern California: AAPG Bulletin, v. 65, p. 1642–1646.

Kvenvolden, K. A., and R. von Huene, 1985, Natural gas generation in sediments of the convergent margin of the eastern Aleutian Trench area, *in* D. G. Howell, ed., Tectonostratigraphic terranes of the Circum-Pacific region: Circum-Pacific Council for Energy and Mineral Resources Earth Science Series 1, p. 31–49.

Marlow, M. S., and A. K. Cooper, 1985, Regional geology of the Beringian margin, *in* N. Nasu et al., eds., Formation of active ocean margins: Tokyo, Terra Scientific Publishing Co., p. 497–515.

Marlow, M. S., A. K. Cooper, and J. R. Childs, 1983, Tectonic evolution of Gulf of Anadyr and formation of Anadyr and Navarin basins: AAPG Bulletin, v. 67, p. 646–665.

Marlow, M. S., A. K. Cooper, and M. A. Fisher, 1987, Petroleum geology of the Beringian Continental Shelf, this volume.

Marlow, M. S., D. W. Scholl, and A. K. Cooper, 1977, St. George Basin, Bering Sea shelf—A collapsed Mesozoic margin, *in* Island arcs, deep sea trenches and back-arc basins: American Geophysical Union Maurice Ewing Series, v. 1, p. 211–219.

Marlow, M. S., J. V. Gardner, T. L. Vallier, H. McLean, E. W. Scott, and M. B. Lynch, 1979, Resource report for proposed OCS lease sale no. 70, St. George Basin, Alaska: U.S. Geological Survey Open-File Report 79-1650, 76 p.

Marlow, M. S., H. McLean, A. K. Cooper, T.L. Vallier, J. V. Gardner, R. McMullin, and M. B. Lynch, 1980, A preliminary summary of regional geology, petroleum potential, environmental geology, and technology for exploration and development for proposed OCS lease sale no. 75, northern Aleutian shelf, Bering Sea, Alaska: U.S. Geological Survey Open-File Report 80-653, 54 p.

Marlow, M. S., P. R. Carlson, A. K. Cooper, H. Karl, H. McLean, R.

McMullin, and M. B. Lynch, 1981, Hydrocarbon resource report for proposed OCS sale no. 83, Navarin Basin, Alaska: U.S. Geological Survey Open-File Report 81-252, 83 p.

McCarthy, J., A. J. Stevenson, D. W. Scholl, and T. L. Vallier, 1984, Speculations on the petroleum geology of the accretionary body: An example from the central Aleutians: Marine and Petroleum Geology, v. 1, p. 151–167.

McCulloch, D. S., S. H. Clarke, Jr., G. L. Dolton, M. E. Field, E. W. Scott, and P. A. Utter, 1982, Geology, environmental hazards, and petroleum resources for OCS lease sale 73, offshore central and northern California: U.S. Geological Survey Open-File Report 82-1000, 77 p.

McLean, H., 1977, Organic geochemistry, lithology, and paleontology of Tertiary and Mesozoic rocks from wells on the Alaska Peninsula: U.S. Geological Survey Open-File Report 77–813, 63 p.

McLean, H., 1979, Tertiary stratigraphy and petroleum potential of Cold Bay-False Pass area, Alaska Peninsula: AAPG Bulletin, v. 63, p. 1522–1526.

McLean, H., and B. P. Hausback, 1984, Reconnaissance geologic map of the La Purisima-Paso Hondo area, Baja California Sur, Mexico: U.S. Geological Survey Open File Map 84–93, scale 1:83,000.

McLean, H., B. P. Hausback, and J. A. Knapp, 1985, Reconnaissance geologic map of part of the San Isidro quadrangle, Baja California, Mexico: U.S. Geological Survey Miscellaneous Field Studies MF-1799, 1 sheet, scale 1:125,000.

McLean, H., B. P. Hausback, and J. A. Knapp, in press, The geology of west-central Baja California Sur, Mexico: U.S. Geological Survey Bulletin.

Mina, F. U., 1957, Bosquejo geologico del Territorio Sur de la Baja California: Boletin de la Asociacion Mexicana de Geologicos Petroleros, v. 9, p. 139–269.

Moore, D. G., 1969, Reflection profiling studies of the California Continental Borderland—Structure and Quaternary turbidite basins: GSA Special Paper 106, 135 p.

Moore, J. C., 1974, Geologic and structural map of the Sanak Islands, southwestern Alaska: U.S. Geological Survey Miscellaneous Investigations Series Map I-817, scale 1:63,360.

Moore, T. E., 1984, Sedimentary facies and composition of Jurassic volcaniclastic turbidites at Cerro El Calvario, Vizcaino Peninsula, Baja California Sur, Mexico, *in* V. A. Frizzell, Jr., The geology of the Baja California Peninsula: SEPM Publication, v. 39, p.131–147.

Normark W. R., 1977, Neogene basins and transform motion within the Pacific continental margin of Baja California: 9th Offshore Technology Conference, v. 1, p. 93–97.

Normark, W. R., J. E. Spencer, and J. C. Ingle, Jr., 1987, Geology and Neogene history of the Pacific continental margin of Baja California Sur, Mexico, this volume.

Ogle, B. A. 1953, Geology of the Eel River area, Humboldt County, California: California Division of Mines and Geology Bulletin 164, 128 p.

Scholl, D. W., M. S. Marlow, and E. C. Buffington, 1975, Summit basins of Aleutian Ridge, North Pacific: AAPG Bulletin, v. 59, p. 799–816.

Scholl, D. W., T. L. Vallier, and A. J. Stevenson, 1987, Geologic evolution and petroleum geology of the Aleutian Ridge, this volume.

Shouldice, D. H., 1973, Western Canadian continental shelf, *in* R. G. McCrossan, ed., The future petroleum provinces of Canada—Their geology and potential: Canadian Society of Petroleum Geologists Memoir 1, p. 7–35.

Snavely, P. D., Jr., 1987, Tertiary geological framework, neotectonics, and petroleum potential of the Oregon-Washington continental margins, this volume.

Snavely, P. D., Jr., in press, Tertiary evolution and petroleum potential of the Oregon-Washington continental margin (abs.): Fourth Circum-Pacific Energy and Mineral Conference, Singapore, August 16–17, 1986.

Snavely, P. D., Jr., and H. C. Wagner, 1980, Generalized isopach map of Tertiary sedimentary rocks, western Oregon and Washington and adjacent continental margin: U.S. Geological Survey Open-File Report 80-889, 2 p.

Snavely, P. D., Jr., and H. C. Wagner 1981, Geologic cross section across the continental margin off Cape Flattery, Washington, and Vancouver Island, British Columbia: U.S. Geological Survey Open-File Report 81-978, 5 p.

Snavely, P. D., Jr., J. E. Pearl, and D. L. Lander, 1977, Interim report on petroleum resources potential and geologic hazards in the outer continental shelf—Oregon and Washington Tertiary province: U.S. Geological Survey Open-File Report 77-282, 64 p.

Snavely, P. D., Jr., H. C. Wagner, W. W. Rau, and D. Bukry, 1981, Geologic cross section of the southern Oregon Coast Range and the adjacent continental shelf: U.S. Geological Survey Open-File Report 81-957, 11 p.

Snavely, P. D ., Jr., A. R. Niem, N. S. MacLeod, J. E. Pearl, and W. W. Rau, 1980, Makah Formation—A deep-marginal-basin sequence of late Eocene and Oligocene age in the northwestern Olympic Peninsula, Washington: U.S. Geological Survey Professional Paper 1162-B, 28 p.

Sorensen, P. A., 1982, Sedimentology and sedimentary petrology of a Paleogene basin near Laguna San Ignacio, Estado de Baja California Sur, Mexico: University of California, Santa Barbara, unpublished M. A. thesis, 114 p.

Spencer, J. E., and W. R. Normark, 1979, Tosco-Abreojos fault zone—A Neogene transform plate boundary within the Pacific margin of southern Baja California, Mexico: Geology, v. 7, p. 554–557.

Steffy, D. A., R. F. Turner, G. C. Martin, and T. O. Flett, 1985, Evolution and petroleum geology of the Navarin Basin, Bering Sea, Alaska: Oil and Gas Journal, v. 83, no. 31, p. 116–124.

Tolson, R. B., 1987, Structure and stratigraphy of the Hope Basin, southern Chukchi Sea, Alaska, this volume.

Turner, R. F., 1984, Summary and conclusions, in R. F. Turner, ed., Geological and operational summary, Navarin Basin COST No. 1 well, Bering Sea, Alaska: U.S. Department of the Interior Minerals Management Service, OCS Report MMS 84-0031, p. 230–232.

Underwood, M. B., 1985, Sedimentology and hydrocarbon potential of Yager structural complex—Possible Paleogene source rocks in Eel River Basin, northern California: AAPG Bulletin, v. 69, p. 1088–1100.

Vedder, J. G., 1987, Regional geology and petroleum potential of the southern California Borderland, this volume.

von Huene, R., M. A. Fisher, and T. R. Bruns, 1987, Geology and evolution of the Kodiak margin, Gulf of Alaska, this volume.

Yeats, R. S., and B. U. Haq, 1981, Deep-sea drilling off the Californias—Implications of Leg 63, *in* S. Orlofsky, ed., Initial reports of the Deep Sea Drilling Project (Vol. 63): Washington, D.C., U.S. Government Printing Office, p. 949–961.

Ziegler, D. L., and J. K. Cassell, 1978, Synthesis of OCS well information, offshore central and northern California, Oregon, and Washington: AAPG Bulletin, v. 62, p. 2364.

CHAPTER TWENTY-SEVEN

HYDROTHERMAL PROCESSES AND MASSIVE SULFIDE DEPOSITS ON THE JUAN DE FUCA RIDGE AND OTHER NORTHEAST PACIFIC SPREADING AXES

Randolph A. Koski

U.S. Geological Survey
Menlo Park, California

Steven D. Scott and Mark D. Hannington

University of Toronto
Ontario, Canada

John R. Delaney and Margaret K. Tivey

University of Washington
Seattle, Washington

ABSTRACT

Active hot springs and sulfide deposits are present in three contrasting settings on the Juan de Fuca Ridge (JFR)—the geomorphically simple axial valley near the Blanco fracture zone (the southern site), the caldera of Axial Seamount, and the fissured axial valley of the Endeavour segment north of the Cobb offset. Other sites

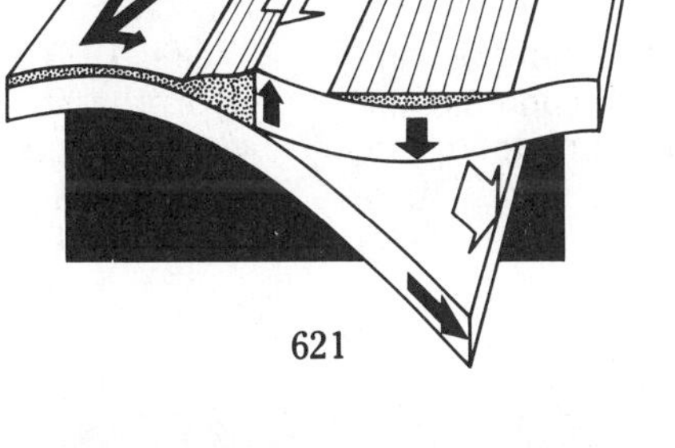

of hydrothermal discharge and sulfide deposition on northeast Pacific spreading axes include the axial grabens of the Explorer Ridge, sediment-floored troughs of the Gorda Ridge (Escanaba Trough) and Gulf of California (Guaymas Basin), and the upraised axial block at lat. 21°N on the East Pacific Rise.

The deposits on the JFR are Ag-rich and, in some cases, Au-rich accumulations of Zn and Fe sulfides with various amounts of barite and silica; Cu-Fe sulfides and anhydrite are concentrated in high-temperature chimneys. Hydrothermal fluids range in temperature from <40°C (Axial Seamount) to >400°C (Endeavour segment) and have compositions that indicate reaction with the underlying basalt and various degrees of subsurface mixing. Vent sites are also characterized by ferruginous sediment and an abundant biota including bacterial mats, vestimentiferan tube worms, limpets, gastropods, and crabs.

Although sulfide deposits may be expected in a variety of settings along the JFR, the intensity of vent activity and the size and composition of sulfide deposits may depend on whether extension-dominated or volcanism-dominated tectonics prevail. Preliminary estimates indicate the presence of 390,000 metric tons of massive sulfide at the southern site and 100,000 to 200,000 tons at the Endeavour site.

INTRODUCTION

The belts of crustal accretion encircling the globe as oceanic ridges are sites of sea-floor extension, volcanism, and geothermal activity. Hydrothermal convection and discharge along axes of spreading have a major effect on heat transfer through and alteration of the oceanic crust, the chemical budget of seawater, and the formation of metalliferous sediments and encrustations on the sea floor. Along the ridge axes, metal sulfide deposits form where hot fluids, saturated with metals and reduced sulfur, undergo rapid temperature and chemical changes upon discharge at the interface between the sea floor and the seawater. This chapter describes in detail the hydrothermal vent fields and sulfide deposits at three separate and distinct morphological and geologic/tectonic settings on the Juan de Fuca Ridge (JFR)—the southern site, Axial Seamount, and the Endeavour segment. A brief summary of the geology of sulfide deposits on the Explorer Ridge, the Gorda Ridge (Escanaba Trough), Gulf of California (Guaymas Basin), and the East Pacific Rise at lat. 21°N is also presented. A spatial and genetic model for sulfide deposits is presented and tested, and some resource implications are discussed. We conclude that hydrothermal systems occur in a variety of geologic and tectonic settings on spreading axes in the northeast Pacific, but the stage of morphological, volcanic, and tectonic evolution in any one ridge segment may determine the zones in which sulfide deposits are most probable.

HISTORICAL PERSPECTIVE

The association of metalliferous exhalations and volcanism on the sea floor has been described by economic and marine geologists for generations, but refined metallogenic concepts and major discoveries at oceanic ridges have occurred only in the last 20 years. Skornyakova (1965) and Bostrom and Peterson (1966, 1969) postulated that the anomalous metal content (Fe, Mn, Cu, Cr, Ni, Pb) of deep-sea sediments on the flanks of oceanic ridges is derived from metal-rich hydrothermal fluid debouching through newly formed oceanic crust. Variations in the major and trace-element content of glassy and holocrystalline basalt from the Mid-Atlantic Ridge led Corliss (1971) to conclude that penetration and alteration of cooling lavas by seawater could produce metal-rich solutions that later exit onto the sea floor. That metals and other ions can be leached from basalt during exchange reactions with heated seawater has been verified experimentally in several laboratories (Bischoff and Dickson, 1975; Mottl and Holland, 1978, 1979; Seyfried and Bischoff, 1977, 1981). Studies of Cyprus-type sulfide deposits and greenschist-facies alteration patterns in ophiolite complexes provided additional evidence that fluids circulating through oceanic crust can form deposits of valuable metals along zones of sea-floor spreading (Sillitoe, 1972; Constantinou and Govett, 1973; Spooner and Fyfe, 1973). Finally, the discovery of hot metalliferous brine pools in deep basins of the Red Sea resulted in direct sampling of active geothermal systems in a youthful oceanic rift environment (Bischoff, 1969).

The initial observation of hot springs in a mid-ocean-ridge setting occurred on the Galapagos spreading center near long. 86°W. Hydrothermal circulation within the crust and discharge of a warm-water plume above the sea floor were indicated by measurement of anomalous heat flow and elevated temperature, helium-3, and total dissolvable Mn values in bottom water sampled at the central axis (Sclater and Klitgord, 1973; Williams et al., 1974; Weiss et al., 1977; Lupton, Weiss, and Craig, 1977; Klinkhammer, Bender, and Weiss, 1977). In early 1977, low-temperature (3° to 13°C) hydrothermal vents and deposits of Fe silicate and Mn oxide were located with the submersible Alvin along a small volcanic ridge within the 3- to 4-km-wide, 250-m-deep, east–west-trending axial valley of the Galapagos spreading center

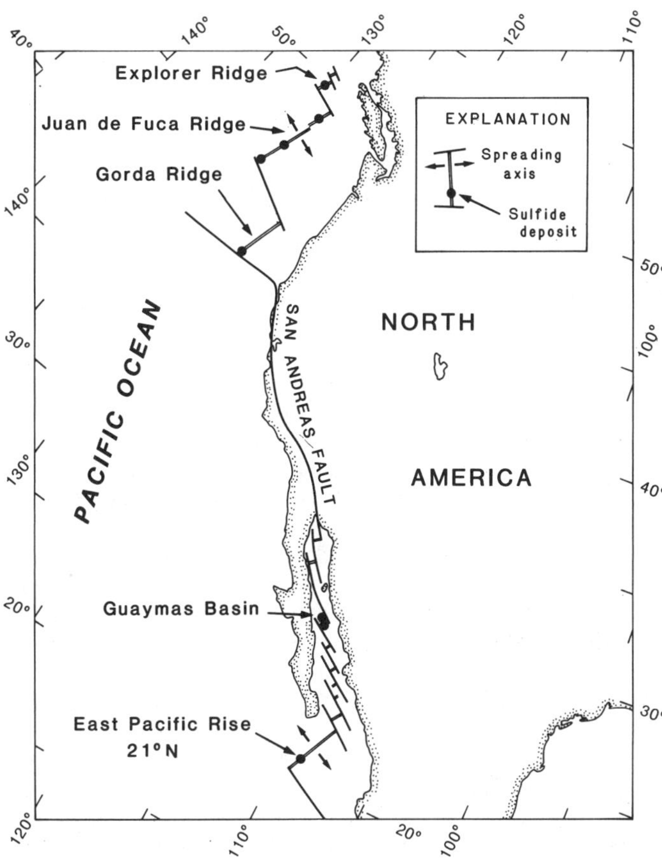

Figure 1. Location of massive sulfide deposits along spreading axes in the northeastern Pacific.

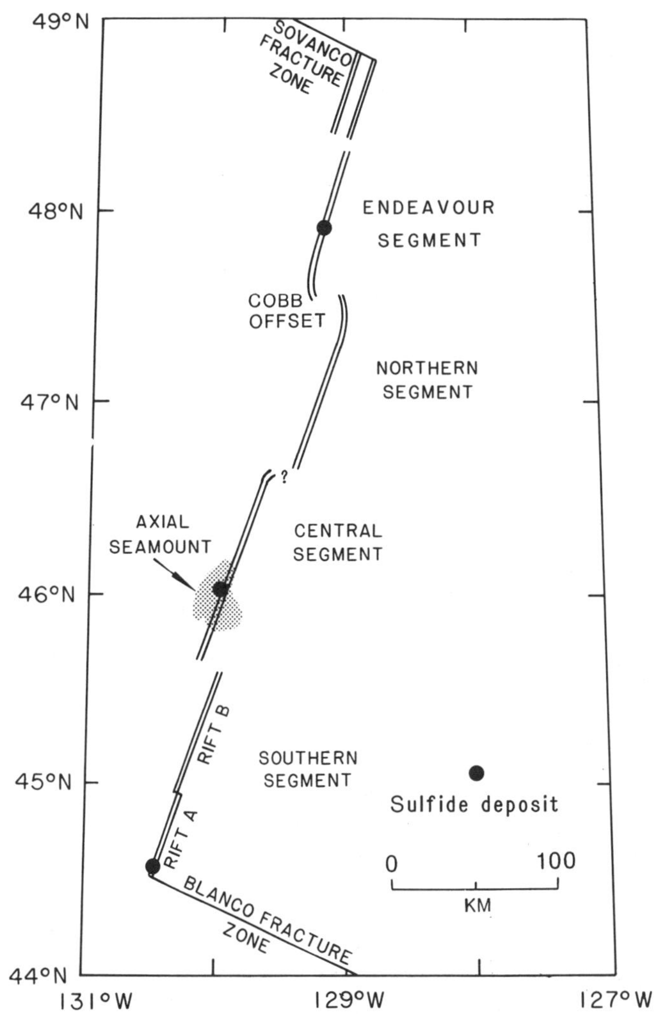

Figure 2. Generalized map showing areas of hydrothermal activity and sulfide deposits on the JFR.

(Corliss et al., 1979; Edmond et al., 1979a, 1979b).

In 1978 and 1979, further exploration with the submersibles Cyana and Alvin along the spreading axis of the East Pacific Rise near lat. 21°N resulted in the discovery of columnar and moundlike edifices composed of massive Fe-, Zn-, and Cu-Fe-sulfides and black smoker chimneys emitting hot solutions with temperatures of 350°C (Francheteau et al., 1979; Spiess et al., 1980, Hekinian et al., 1980). Since these remarkable discoveries, active thermal springs and sulfide deposits have been located at several other sites along the system of medium-rate spreading axes of the northeast Pacific (Figure 1).

GEOLOGIC SETTING OF THE JUAN DE FUCA RIDGE

The Gorda, Juan de Fuca, and Explorer Ridge system is the northern extension of the East Pacific Rise; the tectonic linkage is provided by the Mendocino Fracture Zone and the San Andreas transform fault (Figure 1). The 500-km-long JFR, located <500 km from the Oregon, Washington, and British Columbia coastline, is the accretionary plate boundary between the Pacific and Juan de Fuca plates; the half-spreading rate is 29 mm/yr. (Elvers et al., 1974). The morphology and tectonics of the JFR (Figure 2) are complicated by (1) the presence of seamounts on the Pacific plate that have formed above a hot spot that is now subjacent to the ridge at approximately lat. 46°N, (2) the westward displacement of the central segment of the ridge along transform faults (Delaney, Johnson, and Karsten, 1981), and (3) the formation of overlapping ridge segments (such as the Cobb offset) by complex rift propagation mechanisms (Hey and Wilson, 1982; Wilson, Hey, and Nishimura, 1984; Delaney, Johnson, and Karsten, 1981; Johnson et al., 1983). Morton et al. (this volume) discuss the tectonic evolution of the JFR in more detail.

HYDROTHERMAL ACTIVITY AND SULFIDE DEPOSITS ON THE JUAN DE FUCA RIDGE

The focal point of this chapter is to describe the geologic setting and composition of hydrothermal vent sites and sulfide deposits recently discovered in three contrasting geologic settings on the JFR (Figure 2)—the geomorphically simple axial valley in a normal segment of ridge at approximately lat. 44°40′N near the Blanco Fracture Zone (the southern site), the caldera of Axial Seamount straddling the ridge near lat. 46°N, and the fractured and fissured axial valley of the Endeavour segment at lat. 47°57′N near the Cobb offset. Thermal anomalies in the near-bottom water column

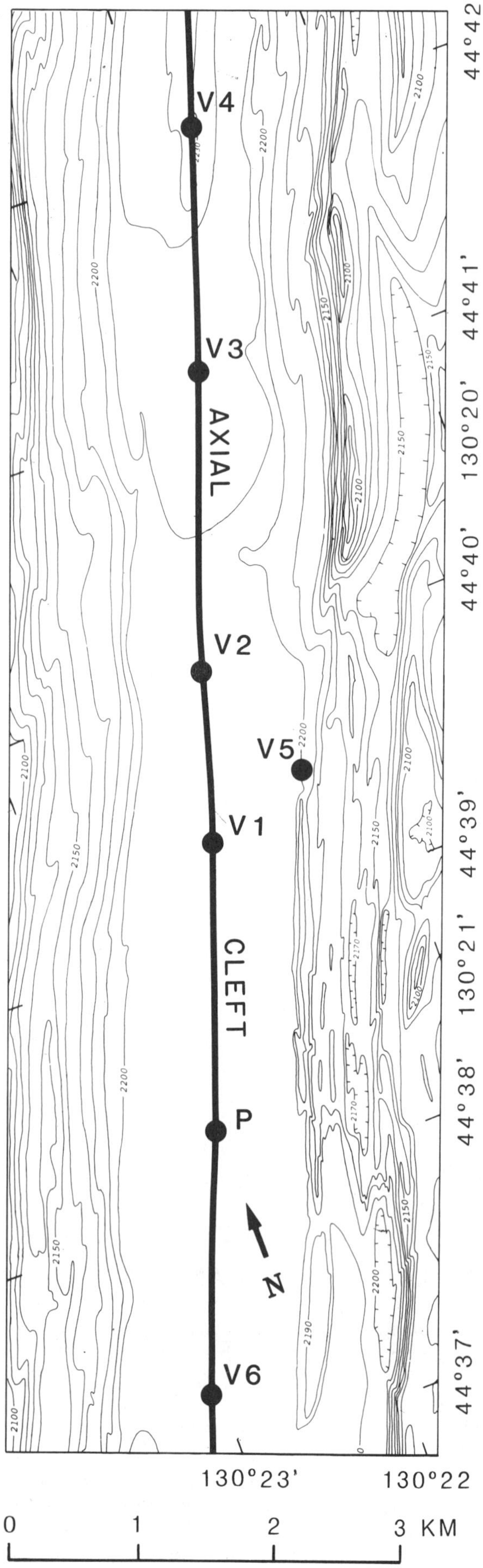

Figure 3. Bathymetric map of the southern site showing straight, flat-floored axial valley flanked by symmetrical, 100-m-high ridges. Location of the axial cleft and vent sites, V1 to V6 and plume (P), in the axial valley are also shown.

indicate that additional areas of hydrothermal discharge on the JFR occur 100 km north of Axial Seamount and on the southern flank of a volcano just south of the Endeavour segment (Crane et al., 1985).

Southern Site

The southern JFR between the Blanco transform fault and the Axial Seamount consists of three west-stepping en echelon rift segments (Embley et al., 1983). The southernmost segment, extending northward from the Blanco Fracture Zone, is characterized by a straight flat-floored axial valley at a depth of 2200 m flanked by symmetrical 100-m-high ridges (Figure 3). The relatively sediment-free valley floor is underlain by young glassy lobate and sheet flows of ferrobasaltic composition; pillow lavas are more abundant in the valley walls (Normark, Morton, and Delaney, 1982; Normark et al., 1983). The prolific sheet-flow lavas flooding the axial graben are a manifestation of the high-effusion-rate volcanic cycle that has continued for a maximum of 16,000 years (Lichtman and Eissen, 1983).

This segment is unfissured except for a narrow and nearly continuous cleft, 20 km long, 25 to 50 m wide, and 20 to 30 m deep, that bisects the axial valley floor (Normark et al., 1983; Kappel and Ryan, 1986; USGS JFR Study Group, 1986). Locally, the cleft is split into overlapping parallel strands by short offsets measuring a few tens of meters. The valley floor is marked by small collapse structures, generally in the form of craters 2 to 3 m in diameter, which increase in number toward the axial cleft. Larger collapse pits near the cleft have lava pillars with "bathtub-ring" markings. A large part of the floor of the axial cleft consists of lineated sheet flows, but sheet-flow rubble and talus are also abundant. The axial cleft and nearby collapse pits appear to represent subsidence of a very recent lava pool (Normark et al., 1983); the pool was part of a fissure eruption that fed the sheet and lobate flows in the axial valley. The subsidence resulted from lava drainback as the eruption cycle ceased; rubble in the collapse pits and axial trough are pieces of the chilled surface of the lava pond that have fallen into voids created as the lava drained away. Although the cleft is the locus of eruption (Normark et al., 1984), the linearity of the fissure and its location in the center of the axial valley suggest that this feature is, in part, a manifestation of extension at the ridge axis. Thus, an incipient graben, with dimensions an order of magnitude smaller, appears to be emerging within the main graben that forms the axial valley.

A series of surface-ship transponder-navigated photographic and sampling surveys conducted by the USGS between 1980 and 1983 (Koski, Normark, and Morton, 1985) located six hydrothermal vent sites within a 9-km-long zone of the axial cleft (Figure 3). A seventh site, seen in bottom photographs as craggy encrustations on talus at the base of the east wall of the axial valley, has not been investigated in detail and is not discussed further. Detailed observation and sampling of three hydrothermal vents (vent 1, vent 3, and plume site) and volcanic features in and near the axial cleft were conducted in 1984 (USGS JFR Study Group, 1986). In the axial cleft, vent sites are discontinuous zones of hydrothermal discharge

Figure 4. Characteristics of JFR vent sites. A, Clusters of small chimneys at southern site. Mound is approximately 2 m in width. B, Top of large (9 m high by 2 m wide) black smoker chimney at southern site. C, Ferruginous sediment apron on flanks of massive sulfide deposit, southern site. The field of view is approximately 2 m across. D, Moundlike colonies of vestimentiferan tube worms at low-temperature vent on Axial Seamount. Large tube worms are approximately 0.5 m high. E, Top of active 2-m-wide Lamphere chimney, Axial Seamount. F, Sulfide structure with black smoker side spout, Endeavour segment. Vertical field of view is approximately 3 m.

focused along the intersection of the cleft wall and floor. Although a linear feature in gross aspect, the axial cleft swells in width locally around hydrothermal vent sites and merges with the numerous large collapse pits.

The hydrothermal discharge zones are characterized by clusters of small solitary and coalesced sulfide spires (no central conduit) and chimneys (with open conduit) a few centimeters to 2 m high (Figure 4A), uncommon large chimneys as much as 9 m high (Figure 4B), blankets of unconsolidated yellow-orange ferruginous sediment and bacterial matter (Figure 4C), and concentrations of vent-specific fauna including tube worms, palm worms, limpets, gastropods, and crabs. Exposed sulfide surfaces are characteristically blue-black when observed in artificial light. Forests of chimneys

Table 1. Chemical composition of massive sulfide samples from the JFR and the Gorda Ridge.

	Southern Site, JFR			Axial Seamount	Endeavour Segment	Gorda Ridge
	1	2	3	4	5	6
Fe (%)	2.19	15.6	2.68	5.82	NA	47
Zn (%)	61.0	46.9	64.6	22.7	6.25	0.98
Cu (%)	0.08	0.35	0.03	0.43	0.51	0.87
Pb (ppm)	2,000	3,000	830	3,500	960	3,200
Mn (ppm)	90	720	20	597	NA	1,020
Co (ppm)	6.4	24	40	NA	25	<1
Cd (ppm)	1,060	490	1,000	493	260	48
As (ppm)	235	411	NA	579	NA	4,000
Ag (ppm)	230	290	240	186	29.5	43
Au (ppm)	<0.1	0.13	NA	4.9	0.08	<0.2

NA = not analyzed

Sample 1: porous, light-gray massive sulfide (Type B of Koski, Clague, and Oudin, 1984). Sample 2: dark-gray massive sulfide (Type A of Koski, Clague, and Oudin, 1984). Sample 3: inactive Zn-sulfide-rich chimney. Sample 4: average of 14 samples from Lamphere sulfide spires. Sample 5: homogenized subsample from dredge haul. Sample 6: average of three samples of pyrrhotite-rich massive sulfide from dredge haul.

Data for samples 1 and 2 from Bischoff et al. (1983). Data for sample 4 from Hannington and Scott (1984). Data for sample 5 from Tivey and Delaney (1985).

are nested within embayments of the cleft wall. Locally, chimneys have coalesced to form near-vertical sulfide sheets against the steep inward-facing slopes of the cleft.

The sulfide deposits sampled at four sites (vents 1, 2, 3, and plume) in the study area are Ag-bearing, Zn-sulfide-rich, and Cu-Fe-sulfide-poor (Table 1; Koski, Clague, and Oudin, 1984; Bischoff et al., 1983). This composition is similar to that of sulfide samples from lat. 21°N on the East Pacific Rise, but contrasts with the Cu-bearing, Fe-sulfide-rich, and Zn-sulfide-poor nature of massive sulfide deposits in ophiolites that appear to be ancient sea-floor analogs. Koski, Clague, and Oudin (1984) subdivided the sulfide samples dredged in 1981 into two types on the basis of texture and mineralogy. Type A samples (Figure 5A) have a layered appearance and are largely composed of Zn sulfide (sphalerite and wurtzite) and Fe sulfide (pyrite and marcasite). The complex layering appears to record an accretionary growth mechanism that involves continuous deposition, replacement, and remobilization under the influence of steep temperature gradients. Type B samples are spongy porous aggregates of dendritic and colloform Fe-poor sphalerite. This type of deposit lacks internal structure, and appears to form by simple outward growth mechanisms as low-temperature fluids emanate outward toward open space.

This basic classification is applicable to most of the chimney samples collected by Alvin in 1984 (USGS JFR Study Group, 1986). Most of the chimney samples are Zn-rich and are texturally similar to Types A and B; some have intermediate characteristics. In the Zn-rich chimney samples, anhydrite is present as channel infillings (Figure 5B) and as incomplete concentric rings within the chimney wall. Two chimney pieces have atypical compositions. The tip of a small spire has the assemblage marcasite + pyrite + barite, and is the only Fe-sulfide-rich sample collected from the southern site. A small tubular chimney (outside diameter 6 cm) collected from the side of a large (2 m wide × 9 m high) chimney at vent site 1 has a zoned construction consisting of cubic cubanite + chalcopyrite + pyrrhotite along the inner wall and sphalerite + anhydrite in the outer wall (Figure 5C). The tube was venting hot water and a black cloud of particulates at high velocity. The mineral assemblage of this black smoker suggests that the interior of the large chimney may contain a high-temperature Cu-rich sulfide assemblage.

Fluid emanations at the vent sites include clear diffuse flow from fissures between chimneys and more intense turbid discharge from chimney tops and the stumps of chimneys when the tops are removed. Chimney fluids sampled in 1984 by Alvin have a maximum temperature of 284 °C, high concentrations of chloride, silica, Zn, and Fe, and low pH relative to fluids sampled from other ridge settings (USGS JFR Study Group, 1986).

The blankets of ferruginous sediment that fill topographic depressions in the axial trough beyond and between sulfide deposits are particularly numerous near vent site 3. This sediment also forms apronlike accumulations around mounds and chimneys and resembles the amorphous Fe oxide flocs and bacterial mats that cling to chimney exteriors. The principal components of the sediment are amorphous Fe oxide and silica, along with bacteria, indicating that this material is derived in part from hydrothermal and biogenic sources. The sediment may also be derived from in-situ weathering of the sulfide deposits.

Evidence for hydrothermal activity extends beyond the vent sites. Many sheet-flow fragments recovered from the axial cleft have limonitic stains on fracture surfaces. Several sheet-flow slabs dredged in 1981 near vent site 2 have sooty coatings of magnetite, hematite, cristobalite, and tridymite on bottom surfaces (Eaby and Clague, 1982; Koski, Goodfellow, and Bouse, 1982). Furthermore, basalt samples dredged from the margins of the axial valley and the valley walls are covered with earthy, 1-cm-thick deposits of manganese oxide that appear to represent the rapid accumulation of hydrothermal Mn migrating away from the axial cleft.

Axial Seamount

Axial Seamount, centered at lat. 45°57′N, long. 130°02′W, is a large 1750-km^3 shield-type volcano on the

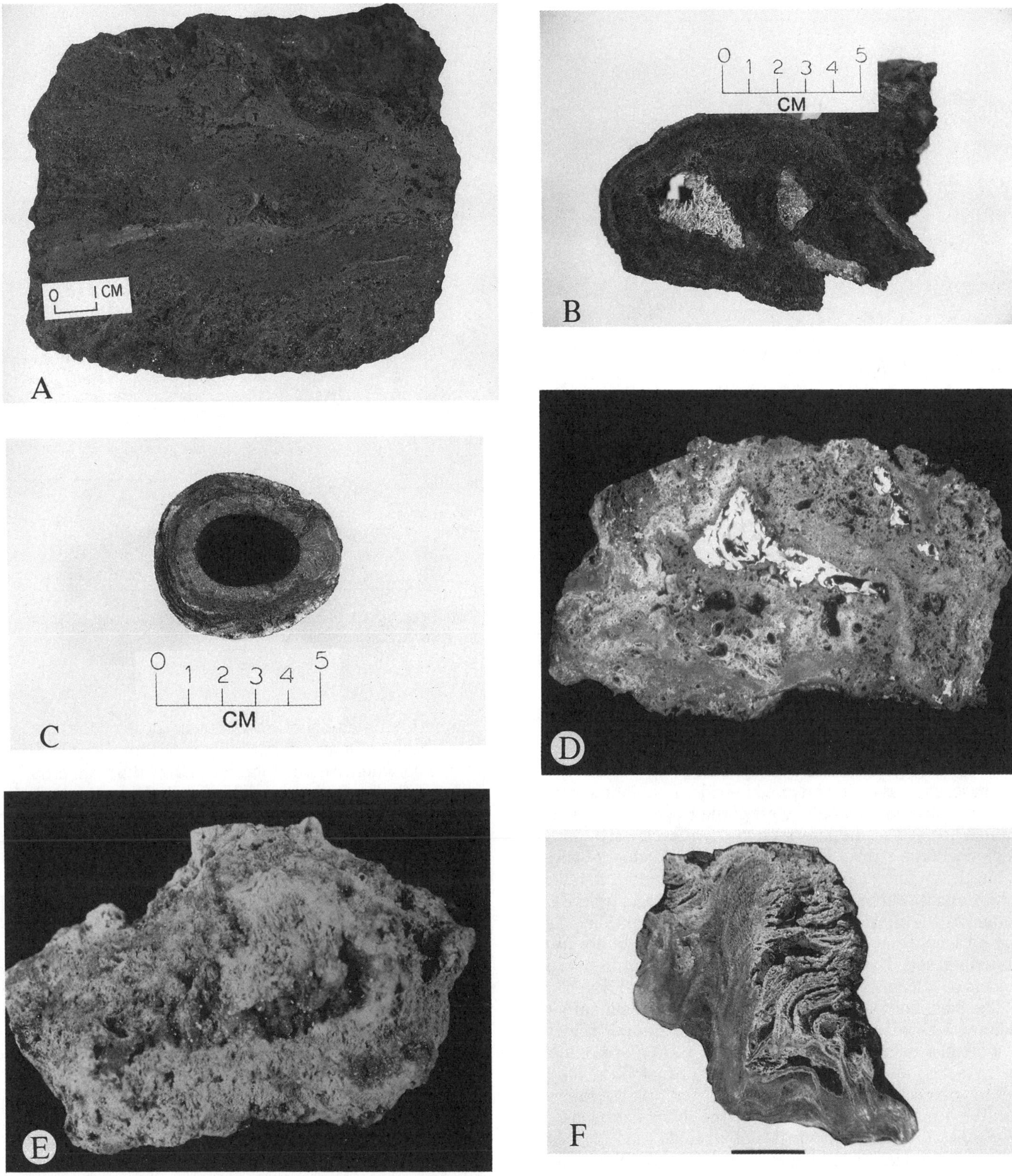

Figure 5. Sulfide samples from the JFR. A, Layered aggregate of Zn sulfide (sphalerite + wurtzite) and Fe sulfide (pyrite + marcasite in light-colored bands) from southern site. B, Anhydrite infilling in Zn-sulfide-rich chimney, southern site. C, Zoned section of tubular side chimney with Cu-rich (chalcopyrite + cubic cubanite + pyrrhotite) inner wall and anhydrite + sphalerite outer wall, southern site. D, Porous, Zn-sulfide-rich Lamphere chimney fragment from Axial Seamount, with worm tubes infilled by amorphous silica. E, Amorphous silica and barite inside Lamphere spire, Axial Seamount. F, Slab from massive sulfide structure showing complex mineral zonation and cross-cutting relationships, Endeavour segment. Scale bar is 10 cm.

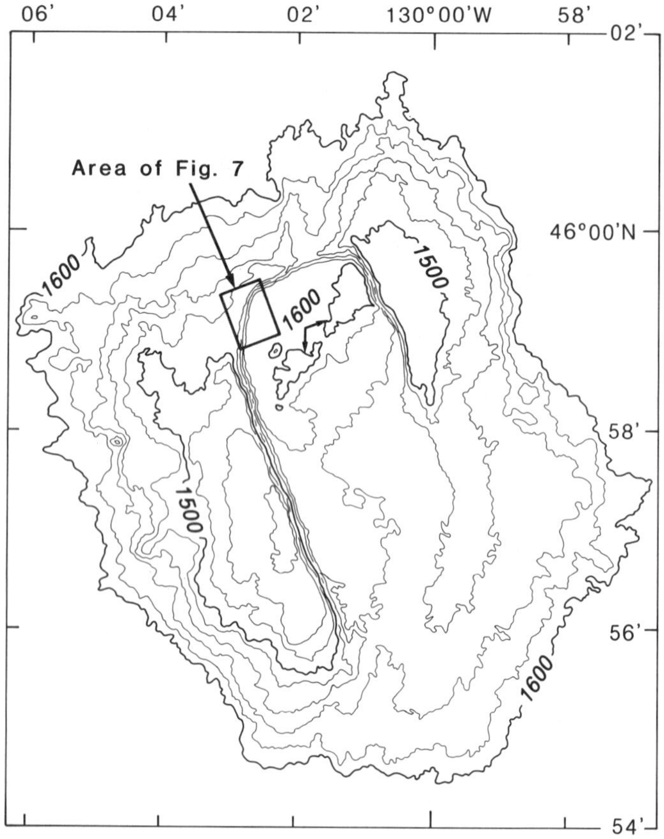

Figure 6. Simplified Seabeam bathymetry of Axial Seamount (CASM II, 1985), showing the steep-walled caldera breached at its southern end. Contour interval is 20 m. The mapped area is enlarged in Figure 7.

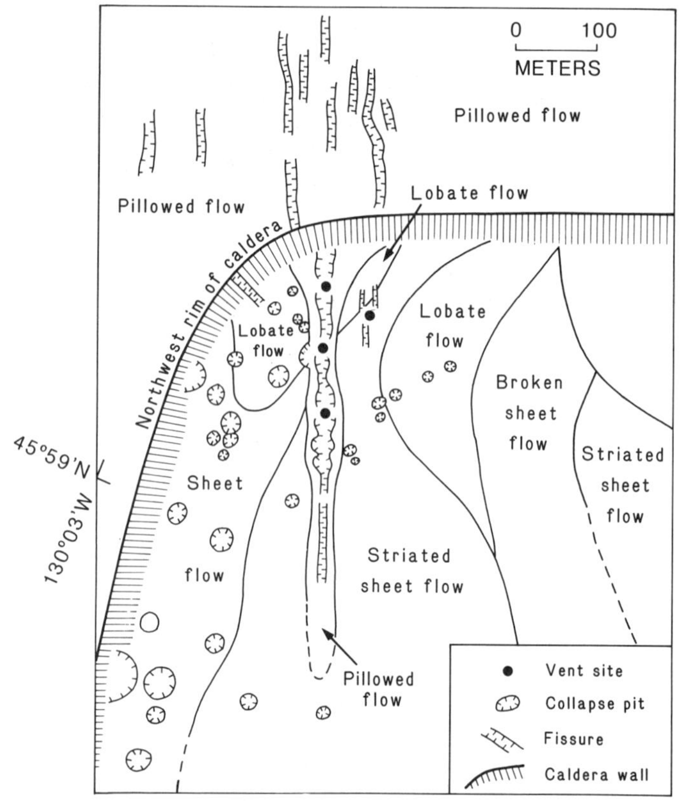

Figure 7. Geology of Axial Seamount caldera, northwest quadrant, observed during submersible observations.

central JFR (Figure 6). This volcanic edifice is 2100 m deep at the base, and has a rectilinear 3-km-wide, 7-km-long summit caldera oriented about 20° west of the ridge axis. The volcano is located at the intersection of the westward-shifted segment of the central JFR with the Cobb-Eickelberg-Brown Bear Seamount chain (Delaney, Johnson, and Karsten, 1981), and is split by the actively spreading ridge. Axial Seamount is the most recent member of the chain, and has grown where the ridge and an apparent hot spot are now superimposed. The seamount, therefore, represents a large excess of volcanism at an active segment of the JFR.

The main body of the seamount is built predominantly of pillow lavas which cover the flanks. The summit caldera region has been surveyed by Seabeam (high-resolution bathymetry: Malahoff et al., 1984) and SeaMARC I (high-resolution acoustic imagery: Malahoff et al., 1984; Crane et al., 1985) and has been explored by the submersibles Pisces IV (CASM II, 1985) and Alvin (Malahoff et al., 1984). The caldera is a vertical-walled structure as much as 100 m deep; the water depth of the caldera floor is approximately 1560 m. The walls are normal faults which truncate pillow, lobate, and sheet flows. Steep talus slopes bury the bottom third of the walls. In places, the youngest lava flows cover the lower talus slopes. The active spreading zone of the JFR intersects the northwest caldera wall and expresses itself as an eruptive fissure in the caldera floor, near its west wall, and as numerous north–south-trending fissures (<10 m wide) on top of the northwest wall (Figure 7).

Smooth striated sheet flows cover most of the caldera floor, resulting in an overall relief of <5 m. Lava collapse pits are common. Near the intersection of the main eruptive fissure with the northwest caldera wall, the lava morphologies include smooth, striated sheet flows and broken sheet flows; lobate and pillow lavas occur around and in the rift zone (Figure 7). At this location the fissure consists of a series of seven coalesced collapse pits, each 20 to 50 m in diameter and approximately 25 m deep, that extend over a distance of 300 m. These collapse features are cut by a north–south-trending cleft left open after the last pillow eruption. The distribution of lava morphologies, the orientation of sheet flow striae, and the occurrence of drainback features along the main rift zone indicate that this cleft is the local source for lavas on the caldera floor.

Axial Seamount has been recognized since 1981 as a geothermally active area and a potential hydrothermal vent site. The discovery of anomalous hydrothermal manganese in local bottom waters (Jones, Johnson, and Delaney, 1981), helium-3-rich plumes adjacent to the seamount (Lupton, 1983), and distinct thermal anomalies in bottom waters of the caldera (Crane et al., 1985) preceded exploration for hydrothermal vents in the caldera. The thermal anomalies in the caldera were found to correspond to sites of active hydrothermal venting near the northwest caldera wall (CASM II, 1985) and adjacent to the west wall in the southern end of the caldera (Malahoff et al., 1984).

Hydrothermal activity in the northwest quadrant of the caldera (Figure 7) occurs in the eruptive fissure within 200 m of the caldera wall, where three vent fields (Taylor, Shepherd, and Chambers vents) were discovered in the coalesced collapse pits. Each site consists of a cluster of warm-water vents with temperatures as high as 40°C that are heavily populated by vestimentiferan tube worms, which form moundlike colonies as much as 2 m in diameter (Figure 4D). No chimney structures were observed in the rift, but deposits of pale-green nontronite (Fe smectite) occur on the floor and walls of the rift adjacent to the vents.

Forty meters east of the main eruptive fissure, a series of small subparallel fissures in fresh broken sheet flows form the 30-m-diameter Lamphere vent field. Three large chimneylike structures aligned parallel to the main eruptive fissure occur at the north end of the field. Openings in broken basalt crusts are venting water only a few degrees above ambient and are fringed by white filamentous bacteria and small vestimentiferan worms. Broken sheet flows abut against the Lamphere chimneys and appear to bury their bases. Only one of the Lamphere chimneys, about 4 m high and 2 m in diameter at its base, is actively venting warm water (19°C) and is populated by vent-specific biota. The remaining spires, about 6 m high and 2 m in diameter at their bases, lack abundant biological inhabitants and seem to be inactive. However, small cracks in the chimney walls are lined with white filamentous bacteria sustained by vent fluid seeping from the porous interiors of the sealed chimneys. The occurrence of partly buried hydrothermal deposits in the Lamphere vent field suggests that a larger vent-producing fissure once existed east of the existing eruptive fissure. This older fissure has since been flooded by fresh sheet lavas, burying the hydrothermal mound and leaving only the tops of the tallest chimneys exposed.

Typical hydrothermal material from the Lamphere chimneys is very porous because of the rapid growth of skeletal crystals and the encrustation of sulfates and sulfides on worm tubes (Figure 5D). The chimneys have been sealed by precipitation of amorphous silica as hydrothermal fluid circulates through their porous interiors (Figure 5E). Each fossil chimney consists of a porous core of sphalerite, wurtzite, and amorphous silica inside less porous walls of marcasite and barite. Barite is the only sulfate mineral in the chimneys. Anhydrite may have been a primary precipitate, but is absent in the fossil chimneys because of its retrograde solubility. Trace amounts of pyrite, chalcopyrite, galena, and tetrahedrite occur as inclusions in sphalerite. Skeletal galena and argentiferous Pb-As-Sb sulfosalts form overgrowths on late-forming colloform sphalerite; all of these are covered by a thick accumulation of amorphous silica. The outer surfaces of the chimneys are coated with Fe and Mn oxides.

Major elements in the bulk chemical composition of the Lamphere chimneys reflect the mineral assemblage (Table 1). In addition to Zn and Fe, the chimneys are enriched in SiO_2 and Ba; the very high Au values, averaging 4.9 ppm, are associated with enrichments in Pb, Ag, As, and Sb and the presence of sulfosalt phases (Hannington, Peter, and Scott, 1986).

Vent fluid from one of the northernmost vents was sampled and analyzed by CASM II (1985). Dilution by entrained seawater is calculated to be 95%. Silica geothermometry, based on the total dissolved silica content of the fluid at 29°C, gives a temperature of 310°C for the end-member solution. The vent fluid has a low pH, possibly buffered by its high CO_2 and inorganic carbon content, which keeps the metals and H_2S in solution. High CO_2 and total He may be due to a magmatic component or a low water/rock ratio.

The vent fauna in the northwestern part of the caldera have been described in detail by CASM II (1985). The small active vents and the single active Lamphere chimney are populated by large vestimentiferan worms as much as 2 m in length and 2 cm in diameter. Other biota include alvinellid and ampharetid polychaete species, several gastropods, and a few tiny bivalves. Large quantities of sulfur-metabolizing filamentous bacteria form a mucus binding the worm tubes together. Analyses of the mucus material show that it scavenges significant amounts of Fe (>4000 ppm) and S (>60 ppm) from the hydrothermal fluid (Juniper, Thompson, and Calvert, 1986). Vent-associated fish and galatheid and spider crabs were abundant in the immediate vicinity of the colonized vents.

Adjacent to the west wall in the southern end of the caldera is a field of 1-m-high chimneys (McMurtry et al., 1984; Malahoff et al., 1984). The field measures 100 to 200 m on each side, and individual clusters of chimneys are as much as 30 m in diameter. Chimneys protrude from small subparallel fissures in the fresh sheet flows and have no visible basal mounds. Malahoff et al. (1984) have reported vent fluid temperatures as high as 293°C. Amorphous Fe-oxides, silica, and nontronite form friable precipitates and spires that are associated with inactive vents or vents with sluggish dispersed hydrothermal flow (McMurtry et al., 1984). The distribution of the active vents along recent fractures and the absence of large mature hydrothermal mounds suggests that the southern vent field on Axial Seamount may be either very young or an older reactivated vent field recently buried by fresh lavas. The highly oxidized and altered nature of the precipitates described by McMurtry et al. (1984) suggests that they are old.

Endeavour Segment

The Endeavour segment of the JFR is an elongate S-shaped geomorphic feature approximately 90 km in length that is bounded to the north and south by overlapping spreading centers (Figure 8). The northern third of the segment has relatively low relief both along and across the strike with only slight deviation (±100 m) from a depth of 2600 m. The southern two-thirds of the Endeavour segment is flanked on the east and west by ridges and basins which are elongate parallel to the overall 020° trend of the central ridge. The flanking off-axis ridges are asymmetrical, the steeper slopes being toward the zone of active spreading. This asymmetry also characterizes the southernmost part of the Endeavour segment, in that it has a steeper slope facing eastward. The central, shoalest part of the ridge is very nearly symmetrical and exhibits a well-developed tectonic axial valley with flanking bathymetric highs (Figure 9).

The southern part of the Endeavour segment is the northern limb of the Cobb offset, an overlapping spreading center located between lat. 47°40′ and 47°50′N. The offset has been identified by Hey (1977) as the current locus of a propagating rift. Modeling of magnetic anomaly patterns by Wilson, Hey, and Nishimura (1984) indicates that the offset has migrated northward over the past 4.2 m.y., although Johnson et al. (1983) presented models showing that the Cobb propagator has had an alternating north and south migration pattern in approximately its present position for as long as a million years. Present geologic information indicates

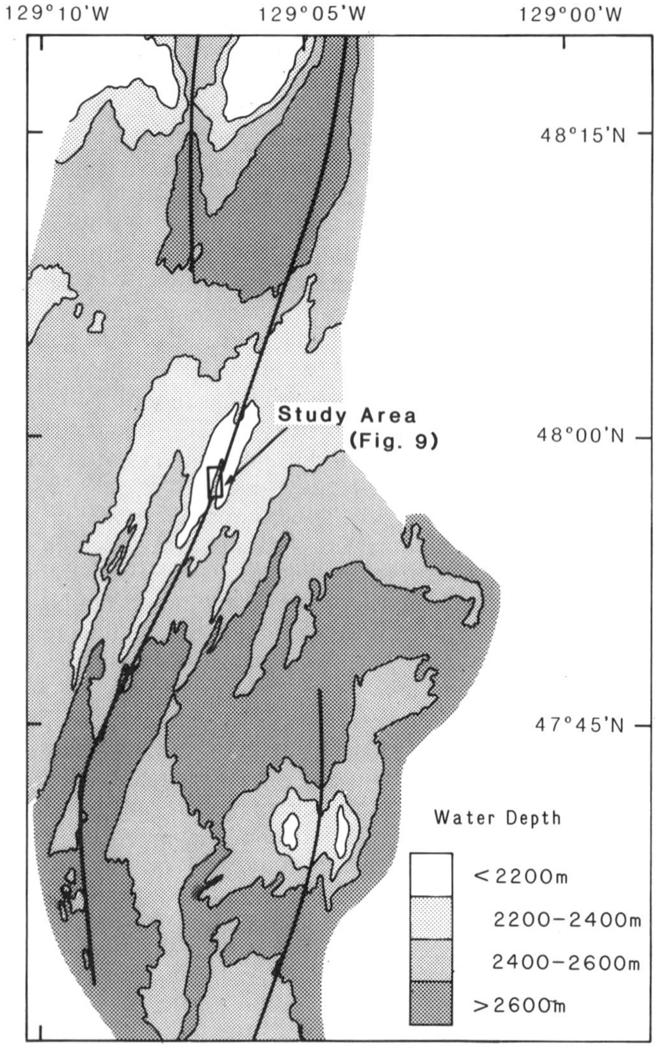

Figure 8. Seabeam bathymetry of southern part of Endeavour segment (after Karsten et al., 1986). Bold lines represent overlapping ridge axes.

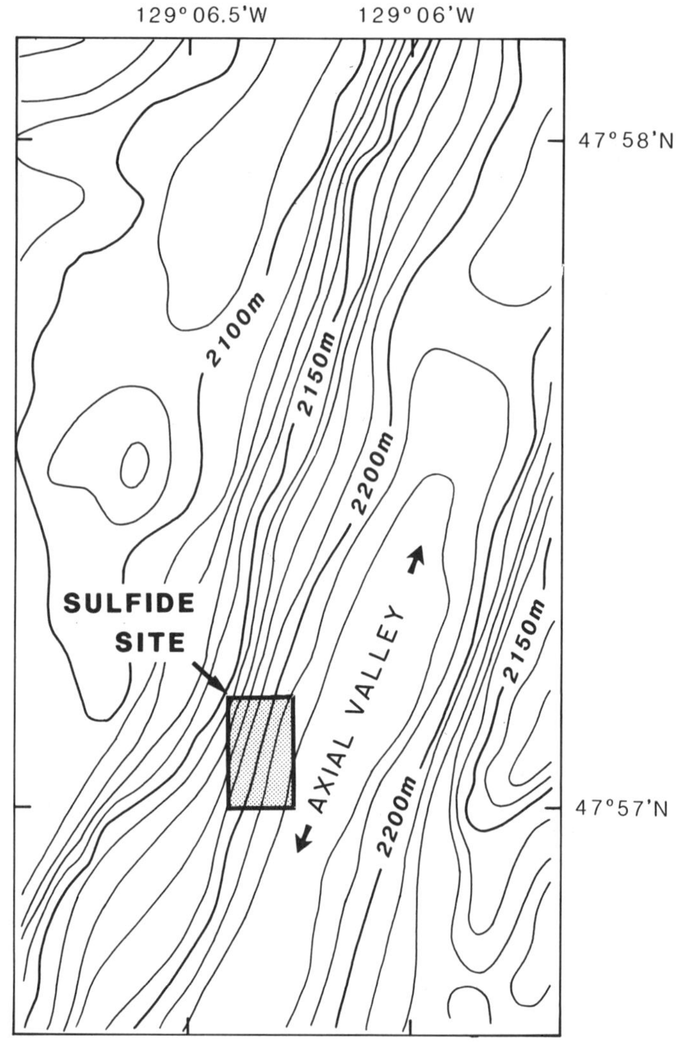

Figure 9. Seabeam map with 10-m contours showing location of vent sites and massive sulfide structures on the west side of the axial valley, Endeavour segment. After Tivey and Delaney (1986).

that both ridge segments bounding the overlapping zone between lat. 47°40′ and 47°50′N are actively spreading.

The symmetrical central third of the Endeavour segment lies just north of the zone of overlap between the two ridges bounding the Cobb offset. This segment is the site of the Endeavour sulfide zone (Figure 9). Regional and detailed mapping of this segment of the ridge crest using Seabeam (Karsten et al., 1986), SeaMARC II (Davis et al., 1984), deep-tow camera and dredging, and the Alvin submersible permit the following observations about the Endeavour site: (1) the steep, outward-facing flanks of the present ridge crest are unfractured, moderately sedimented pillow basalt; (2) the axial valley inner walls are intensely fractured, benched, and fissured parallel to the general 020° trend of the JFR; (3) the axial valley floor consists of moderately sedimented pillow and lineated sheet flows, with few through-going fractures; (4) evidence of very young volcanic activity was not observed; and (5) actively venting and inactive sulfide structures were observed only on the western margin of the valley floor near the break in slope.

Within the axial valley, the zone of current and recent hydrothermal activity extends for hundreds of meters along strike (MERGE Group, 1984). More than 40 large sulfide structures were observed and mapped. At their bases, these complexly coalesced Fe- and Zn-rich deposits emerge directly from deeply fissured lobate and pillow basalt substrate. The structures range in size from less than a meter in diameter and no more than a meter in height to as much as 30 m in diameter and 20 m in height. Above their bases, the sulfide structures generally taper upward into zones dominated by cantilevered constructional flanges of silicate- and sulfide-rich material. The upper third of the castlelike structures is composed of noncoalescing active and inactive chimneylike turrets as much as several meters in height.

More than 140 kg of massive sulfide was dredged from the Endeavour site in 1982; the texture and composition of representative samples have been described in detail by Tivey and Delaney (1985, 1986). These hydrothermal materials were apparently scraped from the large sulfide structures at the base of the western wall of the axial valley. Small live tube worms recovered with the sulfide indicate that the dredge sampled an active vent site; however, the samples contain sealed flow channels and were not associated with high-velocity fluid discharge at the time of sampling.

The predominant mineral phases in the Endeavour dredge samples are pyrite, marcasite, wurtzite, and amorphous silica. Pyrrhotite, sphalerite, Cu-Fe sulfides, galena, and barite are much less abundant. Although anhydrite is present in active chimney samples recovered with Alvin, it is absent in the dredge samples. The chemical analyses in Table 1 show the Zn- and Cu-poor nature of the Endeavour dredge samples. It should be noted, however, that actively venting conduits from this site are Cu rich, and have an inner lining of monomineralic chalcopyrite.

The Endeavour samples are characterized by their large size and textural complexity (Figure 5F; Tivey and Delaney, 1985, 1986). The zonation patterns, related to the proportion of Fe and Zn sulfide phases, grain size and crystallinity, the distribution of nonsulfide phases, and cross-cutting relationships, all suggest that the large sulfide structures contain a succession of superimposed and sealed channelways. Mineral deposition in each conduit involves the early deposition of sulfide and possibly anhydrite during rapid high-temperature fluid flow, closure of the channelway as flow is redirected, and finally sluggish diffuse flow through the pore space of closed channelways. During the early stages of deposition, cooling of the fluid occurs by mixing with seawater; this process yields to conductive cooling during sluggish flow, which promotes deposition of amorphous silica and further cementation of the deposit. Thus, growth patterns, based on flow redirection and reduced permeability, may account for the large size of Endeavour sulfide deposits.

Sulfide spouts extending from the tops and sides of the larger sulfide structures (Figure 4F) are jetting particulate-laden fluid at temperatures in excess of 400°C (Delaney, McDuff, and Lupton, 1984). These high-temperature solutions are characterized by alkalinity higher than seawater (McDuff et al., 1984) and the highest 222 radon values found thus far in submarine vents (Kadko and Lupton, 1984). The vigorous discharge from the Endeavour vent field has created a thermal plume rising approximately 200 m in the water column (Lupton et al., 1984). Diffuse lower temperature fluid flow through the sulfide structures and the surrounding sea floor supports dense colonies of tube worms, limpets, and gastropods.

OTHER SITES ON NORTHEAST PACIFIC SPREADING AXES

Explorer Ridge

The Explorer Ridge is a northeast-trending spreading axis displaced westward from the northern end of the JFR by the Sovanco Fracture Zone (Figure 1). Recent surveys by surface ship and submersible have located hydrothermal vent fields and sulfide deposits at depths of approximately 1800 m on the southern segment of Explorer Ridge between lat. 49°42′ and 49°46′N (Scott et al., 1984, 1985; Tunnicliffe et al., 1986). In this area, the ridge crest is formed by a horst and two grabens that together form a rugged ridge-and-valley topography. The east and west valleys are underlain by pillow and lobate lava flows; sheet flow units are more abundant on the intervening ridge. The somewhat thicker sediment cover in the west valley suggests that it is the older graben, although there are fresh sheet flows at its northern extremity.

Several active and extinct vent fields with more than 60 sulfide deposits have been located on the slopes flanking the east valley. As observed by towed camera systems and submersible, the sulfide deposits may have an aggregate mass of 3 to 5 million metric tons (Scott, 1987). Eight samples of massive sulfide from these deposits have an average composition of 8.5% Cu, 10.8% Fe, 9.0% Zn, 0.1% Pb, 112 ppm Ag, and 0.7 ppm Au. Ten of the deposits exceed 150 m in diameter and are 5 to 10 m in thickness. A large, active vent field referred to as "Magic Mountain" is about 250 m in diameter and is completely covered by sulfide sediment and chimney debris (Tunnicliffe et al., 1986). Numerous large conical chimneys, 10 to 20 m high, rise above the fragmental sulfide base. Fluids emanate at high velocity from small, cylindrical chimney structures growing from the large sulfide cones. These fluids have temperatures between 270° and 310°C. Other small chimneys and vents perched on the sulfide mound have fluids exiting at temperatures <100°C and are surrounded by clusters of vent-specific organisms. Another active vent field of unknown size, called AGOR 171, is located at the north end of the west valley.

A fault scarp through the large Parizeau deposit has exposed a stratigraphy that includes, from bottom to top, a stockwork mineralization in pillow lavas, a lens of massive sulfide which is Cu-rich at the base and Zn-rich at the top, Zn-sulfide-rich and barite-rich chimneys and spires, and a bright red capping of oxidized metalliferous sediment. This and other incised and displaced deposits predate recent tectonism and appear to be part of a mature hydrothermal system. Most active vents are associated with bacterial mats and dense colonies of vestimentiferan tube worms, polychaetes, limpets, gastropods, and pycnogonids, although most vents in which the fluid has very low H_2S are barren of organisms (Tunnicliffe et al., 1986).

Gorda Ridge

The 300-km-long Gorda Ridge is a medium- to slow-rate spreading axis displaced to the east from the south end of the JFR by the Blanco Fracture Zone (Figure 1; Clague and Holmes, this volume). The ridge is located approximately 250 km west of the coast of northern California and southern Oregon and is entirely within the Exclusive Economic Zone of the United States. The Gorda Ridge has greater relief (800 to 1400 m) and a deeper axial valley (3200 to 3800 m) than the JFR. Dredge samples of basalt with hydrothermal alteration and sulfide minerals along fractures and sediment containing nontronite and Fe-Mn oxides indicate that hydrothermal activity occurs along the northern two-thirds of the ridge (Clague et al., 1984).

The southernmost 90 km of the Gorda Ridge, referred to as the Escanaba Trough, is covered by >500 m of Pleistocene turbidite and hemipelagic sediment. Massive sulfide was recently dredged from the southern end of Escanaba Trough, near the intersection of Gorda Ridge with the Mendocino Fracture Zone. The recovery site at lat. 40°45′N is near one of seven volcanic centers, 3 to 6 km in diameter, that have intruded, uplifted, and pierced the sediment fill (Morton et al., 1987). Semilithified siltstone recovered with the sulfide fragments and interlayered sulfide and fine-grained sediment in one sample indicate that the dredged deposit occurs on or below the sediment surface.

The massive sulfide samples recovered in 1985 are aggregates of coarse-grained pyrrhotite with minor chalcopyrite, sphalerite, marcasite, and barite. The sample of interlayered sulfide and sediment described above is also impregnated with asphaltic petroleum (Kvenvolden et al., 1986). A partial chemical analysis of the pyrrhotite-rich massive sulfide is presented in Table 1. The samples have a strong mineralogical and textural similarity to massive sulfide from the sediment-covered floor of Guaymas Basin in the Gulf of California (Koski et al., 1985; also see following section). The geology and resource potential of the Gorda Ridge are discussed in greater detail by Clague and Holmes elsewhere in this volume.

Guaymas Basin, Gulf of California

The Gulf of California has formed by the oblique separation of Baja California from mainland Mexico where the East Pacific Rise intersects the continental margin of North America (Moore and Curray, 1982). Guaymas Basin is one of a series of tectonic depressions that result from spreading along short, northeast-striking axes that are separated by transform faults of much greater length (Figure 1). Guaymas Basin consists of two overlapping en echelon troughs oriented at 035°. Both troughs are floored by thick accumulations of semiconsolidated organic-rich turbidite deposits composed of diatomaceous ooze and terrigenous silt. Although no volcanic rocks are exposed, the sediment pile has been intruded and deformed by a series of basaltic sills and plugs (Lonsdale and Lawver, 1980; Einsele et al., 1980).

Hydrothermal vents, complex deposits of sulfide, sulfate, silicate, and carbonate minerals, colonies of vestimentiferan tube worms, and thick bacterial mats have been discovered at water depths near 2000 m in the troughs during a series of surface ship and submersible investigations. Samples recovered by submersible from ledges and terraces near the top of a fault scarp in the Northern Trough consist largely of talc and minor pyrrhotite and smectite (Lonsdale et al., 1980). In the Southern Trough, more than 100 hydrothermal mounds occur within an area of 12 km^2 on the smooth, flat sediment floor; fluids discharging from active vents have temperatures of 315°C (Scott, 1985; Lonsdale and Becker, 1985). Samples dredged from a 15-m-high mound are composed of pyrrhotite, chalcopyrite, sphalerite, galena, barite, calcite, talc, and opaline silica (Koski et al., 1985). Sulfide deposits in Guaymas Basin are also coated with petroliferous hydrocarbons, the product of hydrothermal reconstitution of organic matter in the sediment (Simoneit and Lonsdale, 1982; Simoneit, 1985). Hydrothermal fluid compositions are characterized by moderately high alkalinity and pH and very low transition metal contents, a result of the interaction between fluids circulating below the sea floor and the organic-rich sediment pile (Von Damm et al., 1985; Koski et al., 1985).

East Pacific Rise, at Lat. 21°N

The crest of the East Pacific Rise at lat. 21°N is a 5-km-wide axial block elevated about 80 m above the ridge flanks (Spiess et al., 1980). Water depth in the axial region is approximately 2600 m. Within the upraised block, a central zone of relatively undeformed, fresh glassy lava flows is flanked on both sides by a zone of extension characterized by narrow, axis-parallel horsts and grabens. Deeply weathered inactive sulfide edifices were located with the submersible Cyana on the flanks of steep-sided depressions in the horst-and-graben zone (Francheteau et al., 1979). A later expedition using the ANGUS camera sled (with temperature sensors) and the submersible Alvin located a field of 25 active vents with numerous sulfide mounds and chimneys in the central zone of recent lava flows (Spiess et al., 1980). Although pillow lavas and sheet flows are both abundant in the axial block, the active vents are spatially associated with a narrow ridge of pillow basalt.

The basal sulfide mounds are constructed on pillow basalt substrate from chimney debris and primary hydrothermal precipitates. This sulfide material is honeycombed by fossilized worm tubes. The basal deposits cover areas as large as 450 m^2 and have a maximum thickness of 2 m. Atop active mounds, black smoker chimneys discharge high-temperature (350°C) fluid containing dark particulates (mostly pyrrhotite) at high velocity (1 to 5 m/sec; Macdonald et al., 1980). Lower temperature chimneys emit clear or milky fluid and are commonly covered with tube worms and other organisms.

The mineralogy and chemical characteristics of sulfide deposits at lat. 21°N have been described by numerous authors (Hekinian et al., 1980; Haymon and Kastner, 1981; Styrt et al., 1981; Oudin, 1982; Goldfarb et al., 1983; Zierenberg, Shanks, and Bischoff, 1984), and the reader is referred to their publications for details on composition and genesis. In general, mound samples are aggregates of pyrite, sphalerite, and wurtzite with minor chalcopyrite and amorphous silica. Worm tube structure is ubiquitous. The active black smoker chimneys are concentrically zoned, with an exterior sulfide (sphalerite, pyrite) + sulfate (anhydrite) assemblage enclosing a Cu-Fe-sulfide-rich (chalcopyrite, cubic cubanite, bornite) interior. This zonation is the result of initial anhydrite deposition at the mixing interface between hydrothermal fluid and seawater and the subsequent replacement of anhydrite by sulfide phases as temperature and chemical gradients are established across the chimney wall (Haymon, 1983). Dead chimneys are internally sealed with the assemblage sphalerite + pyrite, and lack anhydrite. Older inactive chimney and mound material is an ocherous mixture of sulfide and the Fe oxyhydroxide products of sea-floor weathering.

METALLOGENETIC MODEL FOR THE JUAN DE FUCA RIDGE

A general model for hydrothermal circulation and formation of sulfide deposits at accretionary plate boundaries (shown diagrammatically in Figure 10) has evolved from many sources, including regional bathymetric, geochemical, and geophysical surveys of ocean ridges (Francheteau and Ballard, 1983; Edmond et al., 1979a, 1979b; Lupton et al., 1980; Lister, 1974; Wolery and Sleep, 1976; Davis and Lister, 1977; Fehn et al., 1983); studies of altered basalts and in-situ geothermal systems and deposits on the sea floor (Humphris and Thompson, 1978; Mottl, 1983; Macdonald et al., 1980; Spiess et al., 1980; Haymon and Kastner, 1981; Styrt et al., 1981); observations in ophiolite complexes on land (Adamides, 1980, 1984; Spooner and Fyfe, 1973); the experimental results of rock-water interaction (Bischoff and Dickson, 1975; Seyfried and Bischoff, 1977, 1981; Mottl and

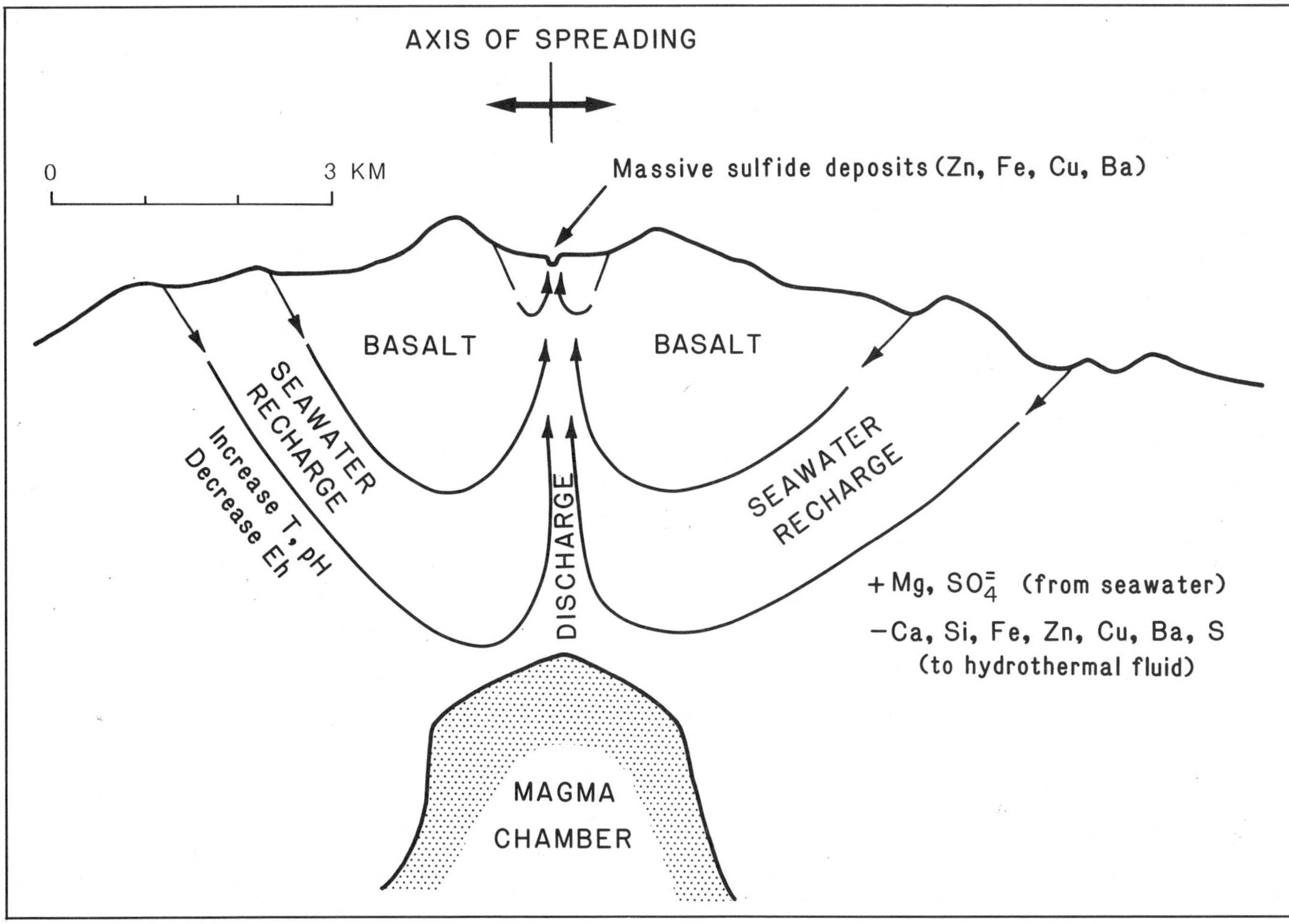

Figure 10. General model for fluid circulation, discharge, and hydrothermal deposits at accretionary plate boundary. The sea-floor profile is an east-west section across the southern site, JFR; vertical exaggeration 8. The depth to magma chamber is not to scale.

Holland, 1978, 1979); and calculations of theoretical chemical reactions from thermodynamic data (Janecky and Seyfried, 1984). In this model, high-level magma chambers emplaced at the ridge axis inflate the crust and elevate the sea floor, provide magma for sea-floor volcanism, and drive dynamic fluid convection cells. Cold oxidizing seawater in the downwelling limb penetrates the basaltic crust on the ridge flanks, becomes heated, and evolves into a highly reduced and acidic hydrothermal solvent during chemical exchange with the wallrock. Depending on the temperature and water/rock ratio, this fluid is capable of mobilizing Fe, Mn, and base metals from basalt; and dissolved seawater sulfate is reduced to sulfide. In the upwelling limb of circulation at the ridge axis, the buoyant hydrothermal fluid rises through permeable crust. Flow that is focused along extensional fractures at depth may become more diffuse while passing through porous lava near the sea floor. Metal sulfides are precipitated along channelways as the ascending fluid undergoes adiabatic expansion and then further cooling during mixing with ambient seawater below the sea floor. Vigorous discharge and instant mixing at the interface between the sea floor and the seawater results in the formation of high-temperature black smokers and the deposition of disequilibrium sulfide-sulfate assemblages in chimneys and mounds. At sediment-covered ridge axes such as Escanaba Trough and Guaymas Basin, venting fluids may have high alkalinity, moderately high pH, and low transition-metal contents after interaction of the circulating fluids with organic- and/or carbonate-rich sediment. Much of the metal content of the fluids may be precipitated at depth in the sediment column.

The hydrothermal systems on the JFR exhibit several variations on this general theme. For example, the three vent areas on the JFR occur in distinct tectono-volcanic settings. The southern site is characterized by a relatively simple ridge morphology—a symmetrical ridge-valley-ridge topography formed during an ongoing episode of sheet-flow-dominant volcanism that has nearly filled the axial graben. A narrow axial fissure is the only manifestation of active extension and is the locus for volcanism and hydrothermal activity. Recharge of seawater into the hydrothermal convection cell may occur along the inward-sloping faults bounding the axial valley, but the localized venting and small sulfide deposits suggest that circulation beneath the sea floor along still smaller scale structures may be responsible for the current hydrothermal discharge.

In contrast with the southern JFR site, the large volcanic edifice at Axial Seamount represents a voluminous outpouring of pillow lavas above the intersection of an apparent hot spot with the ridge axis. Despite the larger supply of magma at this junction, known hydrothermal activity is limited to narrow eruptive fissures in the caldera floor and a restricted zone at the intersection of the caldera wall and floor. However, other sulfide deposits may have been buried by recent volcanism. The most probable structures for recharge of the hydrothermal system are the perimeter faults bounding the caldera, and possibly the small fissures in the caldera floor. The association of hydrothermal discharge and small deposits of massive sulfide within collapse features related to lava drainback is also reported from the ridge axis at lat. 11°N on the East Pacific Rise (McConachy et al., 1986). The root zones of eruptive fissures appear to be capable of channeling upwelling hydrothermal fluids on a small scale.

The geologic relationships observed at the Endeavour segment indicate that the present ridge morphology developed relatively rapidly during a volcanic constructional episode that did not involve significant extension. Subsequently, a stage dominated by extension focused the zone of brittle failure on the crestal portion of the ridge and produced a well-defined axial valley and a well-fissured axial valley floor. The lack of fresh sheet flows on the axial valley floor is in sharp contrast with the southern and Axial Seamount sites.

Although the recent expression of magmatic activity is absent at the Endeavour site, the hydrothermal activity and sulfide deposits are substantial. Thus, hydrothermal activity at the Endeavour site is associated with a period of active extension and the formation of numerous axis-parallel fissures that apparently penetrate to considerable depth in the axial region above the magma chamber. Although the association of volcanism and hydrothermal activity at ridge axes is well established (Ballard et al., 1981), the more intense hydrothermal activity may be related to volcanically quiescent, rift-forming stages of ridge evolution at some medium- and fast-rate spreading axes.

The massive sulfide deposits from the three JFR sites exhibit differences in composition and internal structure that are related to the temperature, composition, and flow parameters of the vent fluids. Chimney samples from the southern site have very high Zn/Cu ratios, low total Cu, small quantities of nonsulfide phases, and consistent internal mineral zonations. Type A chimneys appear to grow during stages of an intensifying hydrothermal system (Koski, Clague, and Oudin, 1984) and may utilize anhydrite during repeated cycles of deposition, dissolution, and redeposition in their growth history, resulting in layered Fe- and Zn-sulfide assemblages. Conversely, Type B chimneys appear to form by a more passive growth mechanism involving slow outward growth at low temperature and formation of nearly monomineralic Zn-sulfide deposits. The spongy texture of the Type B sulfide is also related to habitation by organisms, especially worms, at low-temperature vents. The low pH, high salinity, and high silica, Zn, and Fe contents of the fluids (USGS JFR Study Group, 1986), and the low sulfate content of the chimney samples indicate relatively minor subsurface mixing between the endmember hydrothermal fluid and the seawater at some discharge points. The low Cu content of the chimney deposits is consistent with the low measured Cu in the fluid; Cu may be deposited in greater quantity in the underlying feeder system or in large sulfide chimneys.

The composition of the small sulfide deposits from the caldera floor of Axial Seamount (high Zn, Ag, Au, barite, and silica; low Fe and Cu) is largely influenced by the hydrothermal fluids and the chemistry of the underlying basalts. Sulfides in the interior of the Lamphere chimneys were deposited from a sustained hydrothermal flux at temperatures near 300°C. A late stage of hydrothermal activity deposited an Au- and Ag-enriched silica-barite assemblage at temperatures $<200°C$ (oxygen isotope geothermometry on barite yields a temperature of 185°C; Shanks, Koski, and Woodruff, 1984). The underlying mid-ocean-ridge basalt is enriched in large-ion lithophile elements (R. L. Chase, 1985, oral communication) and may account for the abundance of barite in the chimneys. The compositional variation within the chimneys implies that a relatively high-temperature geothermal cell has collapsed, and that increased mixing with seawater has produced lower temperature, more alkaline, sulfate-bearing fluids and deposits dominated by opaline silica and sulfate. The current low-temperature venting reflects seawater-dominated conditions.

An array of samples collected by dredge and submersible from the Endeavour segment vent field includes black-smoker chimneys with chalcopyrite- and anhydrite-rich interiors perched on basal Fe- and Zn-sulfide-rich edifices with abundant exterior amorphous silica. The growth of the sulfide structures occurs in phases (Tivey and Delaney, 1986). The early phase involves high-temperature, high-velocity throughflow and deposition of anhydrite and a mixed sulfide assemblage around the fluid jet in a manner similar to the construction of chimneys at lat. 21°N on the East Pacific Rise (Haymon, 1983). Clogging of this channelway and redirection of fluid flow results in cross-cutting mineral layers and a subsequent phase of diffuse sluggish flow in which anhydrite is dissolved and amorphous silica is deposited as the hydrothermal fluid is cooled by conduction. The amorphous silica is the cement that supports and stabilizes the large sulfide structures. The deposition of amounts of silica is consistent with the presence of abundant quartz at the base of some massive sulfide deposits in ophiolites on land (Constantinou and Govett, 1973).

CONCLUSIONS AND RESOURCE IMPLICATIONS

Within a span of 8 years (1978–1985), active hydrothermal vents and sulfide deposits have been found at more than 10 sites on spreading axes of the northeast Pacific, including three separate and distinct tectonic settings on the JFR. Although samples and geological data from these areas are accumulating in laboratories worldwide, it is apparent that our understanding of these sea-floor hydrothermal systems is fragmentary and our exploration of the total ocean-ridge system is in its infancy.

The preliminary studies of the JFR segments do permit the following general conclusions regarding metallogenesis at this, and perhaps other, sites of sea-floor spreading:

1. The gross ridge morphology and tectonic setting does not appear to be a significant factor in determining the presence or absence of hydrothermal activity, but it may affect the size of the hydrothermal system and deposits. Sulfide deposits may be expected in a variety of environments along the JFR axis.

2. The intensity of vent activity and the size of sulfide deposits seem to depend on whether extension-dominated

or volcanism-dominated ridge tectonics prevail. For example, the high-temperature vent field and large sulfide structures at the Endeavour site are associated with volcanic quiescence and development of fissures at the edge of the axial graben, whereas the small lower temperature deposits at the southern and Axial Seamount sites occupy narrow fissures in effusive sheetflow terranes.

3. Seamounts on or near the ridge axis are likely sites for prolonged hydrothermal activity because of the replenishment of upper-level magma chambers; however, at the shallower depths of seamount calderas (e.g., Axial Seamount at 1560 m), the boiling of hydrothermal fluids may preclude high-temperature transport of ore metals and may result in their concentration below the sea floor.

4. Individual sulfide deposits in volcanic settings on the JFR appear to be small in size with high Zn and Ag (at Axial Seamount high Au content); the Cu content is moderate to low but differs within the structure of individual chimneys. The high Zn/Cu ratios may be a function of temperature, which in turn is dependent on the degree of subsurface mixing. Larger deposits probably have Cu-rich interiors.

5. Most deposits are chimneylike or are composite structures, the chimneys being mounted on broader sulfide pedestals. Black smoker chimneys focus high-temperature and high-velocity fluid flow, but perhaps more fluid volume is expelled by diffuse emanations from the sides of the chimneys and from the basalt substrate.

6. Vigorous hydrothermal discharge in topographic depressions, such as the axial cleft at the southern JFR site, may produce colorful Fe-oxide-silicate halos that extend well beyond the limit of sulfide.

7. Biological and hydrothermal zonations form mappable units around active vent sites.

Although relationships between tectonic, volcanic, and hydrothermal processes along a complex mid-ocean-ridge system have been the principal concern of our research, a secondary objective is to develop a preliminary resource model. However, the data base, particularly with regard to deposit thickness and the difference in metal content within and between deposits, does not yet permit rigorous grade and tonnage estimates in the three JFR study areas. The extensive photographic surveys and recent submersible investigations are the basis for some general comments and calculations that may provide some perspective.

On the basis of evidence from early photographic surveys and assumptions of deposit thickness, linear continuity, and density, Koski, Normark, and Morton (1985) made a first-order estimate of 500,000 metric tons of sulfide with high-grade Zn and Ag in a 7-km-long zone within the axial cleft of the southern site between the plume site and vent site 3. For comparison, Cyprus-type massive sulfide deposits in ophiolites range in size from 50,000 to 20 million tons, and contain 0.5 to 4.5% Cu, 0.1 to 3% Zn, and 3 to 60 ppm Ag (Constantinou, 1980). Further observations from the submersible Alvin reveal that the deposits at each of the three vent sites (vent 1, vent 3, and plume) extend for at least 100 m and occur as clusters of as many as 30 small chimneys (≤ 2 m high) and a few large chimneys (≤ 9 m high) on a pad of sulfide as much as 10 m wide and 3 m high. In some areas, diffuse hydrothermal discharge and colonies of animals are observed, but no sulfide mound is present. Sulfides may be deposited in void spaces below the sea floor. If the surficial deposits are assumed to have a density of 3 g/cc, then each vent site contains approximately 9000 metric tons of sulfide. Furthermore, if such deposits are continuous (or nearly so) over half of the length of the study area between vent 4 and vent 6 (Figure 3), a total sulfide mass of approximately 390,000 metric tons is calculated. This value is less than the earlier estimate and is small compared with analogous deposits mined on land. To add additional perspective, the large cylindrical chimney (9 m in height, 2 m in diameter; Figure 4B) observed during the Alvin investigation has a mass of approximately 85 metric tons. Thus, it would require nearly 11,800 of these chimneys to produce a deposit of a million tons—the size of a small Cyprus-type deposit. Additional deposits probably occur in stepfaulted and fissured terrain north and south of the southern JFR site.

Deposits in the caldera of Axial Seamount also appear to have small tonnages, possibly because of rapid burial by recent lava flows (CASM II, 1985; Malahoff et al., 1984). At the Endeavour sulfide site, more than 40 massive sulfide structures have been observed along the west margin of the valley floor (Tivey and Delaney, 1986). The largest structure has dimensions of 30 m by 15 m by 20 m, but the average size is approximately 5 m across and 5 to 10 m high. These structures support black smoker spouts as much as 0.5 m high. An estimated 100,000 to 200,000 metric tons of sulfide may occur in this vent field.

The crude estimates stated here are preliminary; more data are necessary before rigorous grade-tonnage estimates can be made. It appears that massive sulfide mineralization on the JFR is characterized by clusters of small, high-grade (Zn and Ag, and in the case of Axial Seamount, Au) deposits; the deposits may occur within an array of tectonic settings, especially in elevated ridge segments. Additional photographic and hydrographic reconnaissance and site-specific sampling surveys using larger bulk sampling equipment and core drilling are required to refine these preliminary resource evaluations.

ACKNOWLEDGMENTS

Thoughtful reviews by Robert Zierenberg and Keith Bargar, and additional suggestions by William Normark, Janet Morton, and David Scholl improved the manuscript. Phyllis Swenson drafted the illustrations. The work of S. D. Scott has been supported by grants from the National Sciences and Engineering Research Council of Canada.

REFERENCES CITED

Adamides, N. G., 1980, The form and environment of formation of the Kalavasos ore deposits-Cyprus *in* A. Panayiotou, ed., Ophiolites: Cyprus Geological Survey Department, Proceedings of the International Ophiolite Symposium, Cyprus, 1979, p. 117–128.

Adamides, N. G., 1984, Cyprus volcanogenic sulphide deposits in relation to their environment of formation: Leicester University of Leicester, unpublished Ph.D. thesis, 383 p.

Ballard, R. D., J. Francheteau, T. Juteau, C. Rangin, and W. Normark, 1981, East Pacific Rise at 21°N: The volcanic, tectonic, and hydrothermal processes of the central axis: Earth and Planetary Science Letters, v. 55, p. 1–10.

Bischoff, J. L., 1969, Red Sea geothermal brine deposits: Their mineralogy, chemistry, and genesis, *in* E. T. Degens and D. A. Ross, eds., Hot brines and recent heavy metal deposits in the Red Sea: New York, Springer-Verlag, p. 368–401.

Bischoff, J. L., and F. W. Dickson, 1975, Seawater-basalt interaction at 200 °C and 500 bars: Implications for origin of seafloor heavy-metal deposits and regulation of seawater chemistry: Earth and Planetary Science Letters, v. 25, p. 385–397.

Bischoff, J. L., R. J. Rosenbauer, P. J. Aruscavage, P. A. Baedecker, and J. G. Crock, 1983, Sea-floor massive sulfide deposits from 21 °N, EPR; JdFR; and Galapagos Rift: Bulk chemical composition and economic implications: Economic Geology, v. 78, p. 1711–1720.

Bostrom, K., and M. N. A. Peterson, 1966, Precipitates from hydrothermal exhalations on the East Pacific Rise: Economic Geology, v. 61, p. 1258–1265.

Bostrom, K., and M. N. A. Peterson, 1969, The origin of aluminum-poor ferromanganoan sediments in the areas of high heat flow on the East Pacific Rise: Marine Geology, v. 7, p. 427–447.

CASM II (R. L. Chase, J. Delaney, H. P. Johnson, K. Juniper, J. Karsten, J. Lupton, S. D. Scott, V. Tunnicliffe, and R. E. McDuff), 1985, Hydrothermal vents on an axis seamount of the Juan de Fuca Ridge: Nature, v. 313, p. 212–214.

Clague, D. A., and M. L. Holmes, 1987, Geology, petrology, and mineral potential of the Gorda Ridge, this volume.

Clague, D., W. Friesen, P. Quinterno, L. Morgenson, M. Holmes, J. Morton, R. Bouse, L. Morgenson, and A. Davis, 1984, Preliminary geological, geophysical, and biological data from the Gorda Ridge: U.S. Geological Survey Open-File Report 84-364, 31 p.

Constantinou, G., 1980, Metallogenesis associated with the Troodos ophiolite, *in* A. Panayiotou, ed., Ophiolites—Cyprus Geological Survey Department, Proceedings of the International Ophiolite Symposium, Cyprus, 1979, p. 663–674.

Constantinou, G., and G. J. S. Govett, 1973, Geology, geochemistry, and genesis of Cyprus sulfide deposits: Economic Geology, v. 68, p. 843–858.

Corliss, J. B., 1971, The origin of metal-bearing submarine hydrothermal solutions: Journal of Geophysical Research, v. 76, p. 8128-8138.

Corliss, J. B., J. Dymond, L. I. Gordon, J. M. Edmond, R. P. von Herzen, R. D. Ballard, K. Green, D. Williams, A. Bainbridge, K. Crane, and T. H. van Andel, 1979, Submarine thermal springs on the Galapagos rift: Science, v. 203, p. 1073–1083.

Crane, K., F. Aikman, R. Embley, S. Hammond, A. Malahoff, and J. Lupton, 1985, The distribution of geothermal fields on the Juan de Fuca Ridge: Journal of Geophysical Research, v. 90, p. 727–744.

Davis, E. E., and C. R. B. Lister, 1977, Heat flow measured over the Juan de Fuca Ridge: Evidence for widespread hydrothermal circulation in a highly transportive crust: Journal of Geophysical Research, v. 82, p. 4845–4860.

Davis, E., R. G. Currie, B. S. Sawyer, and D. M. Hussong, 1984, Juan de Fuca Ridge atlas of SeaMarc II acoustic imagery: Pacific Geoscience Center, Earth Physics Branch, Open-File Report 84–17.

Delaney, J. R., H. P. Johnson, and J. L. Karsten, 1981, The Juan de Fuca Ridge—hot spot—propagating system: New tectonic, geochemical, and magnetic data: Journal of Geophysical Research, v. 86, p. 11,747–11,750.

Delaney, J. R., R. E. McDuff, and J. E. Lupton, 1984, Hydrothermal fluid temperatures of 400 °C on the Endeavour segment, northern Juan de Fuca: EOS (American Geophysical Union Transactions), v. 65, p. 973.

Eaby, J. S., and D. A. Clague, 1982, Preliminary description of basalt from the southern Juan de Fuca Ridge: U.S. Geological Survey Open-File Report 82-200C, 12 p.

Edmond, J. M., C. Measures, R. E. McDuff, L. H. Chan, R. Collier, and B. Grant, 1979a, Ridge crest hydrothermal activity and the balances of the major and minor elements in the ocean: The Galapagos data: Earth and Planetary Science Letters, v. 46, p. 1–18.

Edmond, J. M., C. Measures, B. Mangum, B. Grant, F. R. Sclater, R. Collier, A. Hudson, L. I. Gordon, and J. Corliss, 1979b, On the formation of metal-rich deposits at ridge crests: Earth and Planetary Science Letters, v. 46, p. 19–30.

Einsele, G., J. M. Gieskes, J. Curray, D. Moore, J. E. Aguayo, M. B. Aubry, D. J. Fornari, G. C. Guerrero, M. Kastner, K. Kelts, M. Lyle, Y. Matoba, A. Molina-Cruz, J. Niemitz, J. Rueda, A. Saunders, H. Schrader, B. Simoneit, and V. Vacquier, 1980, Intrusion of basaltic sills into highly porous sediments, and resulting hydrothermal activity: Nature, v. 283, p. 441–445.

Elvers, D., K. Potter, D. Seidel, and J. Morley, 1974, IDOE survey or seamap, N.O.S. seamap profiles plates BGM-1-71: Washington D.C., National Oceanographic and Atmospheric Administration Environmental Data Service.

Embley, R. W., S. Hammond, A. Malahoff, W. F. B. Ryan, K. Crane, and E. Kappel, 1983, Rifts of the southern Juan de Fuca: EOS (American Geophysical Union Transactions), v. 64, p. 853.

Fehn, U., K. E. Green, R. P. von Herzen, and L. M. Cathles, 1983, Numerical models for the hydrothermal field at the Galapagos spreading center: Journal of Geophysical Research, v. 88, p. 1033–1048.

Francheteau, J., and R. D. Ballard, 1983, The East Pacific Rise near 21 °N, 13 °N, and 20 °S: Inferences for along-strike variability of axial processes of the Mid-Ocean Ridge: Earth and Planetary Science Letters, v. 64, p. 93–116.

Francheteau, J., H. D. Needham, P. Choukroune, T. Juteau, M. Seguret, R. D. Ballard, P. J. Fox, W. Normark, A. Carranza, D. Cordoba, J. Guerrero, C. Rangin, H. Bougault, P. Cambon, and R. Hekinian, 1979, Massive deep-sea sulphide ore deposits discovered on the East Pacific Rise: Nature, v. 277, p. 523–528.

Goldfarb, M. S., D. R. Converse, H. D. Holland, and J. M. Edmond, 1983, The genesis of hot spring deposits on the East Pacific Rise, 21 °N, *in* H. Ohmoto and B. J. Skinner, eds., The Kuroko and related volcanogenic massive sulfide deposits: Economic Geology Monograph 5, p. 184–197.

Hannington, M. D., and S. D. Scott, 1984, A caldera-hosted silica-sulfate-sulfide deposit, Axial Seamount, central Juan de Fuca Ridge, N.E. Pacific Ocean (abs.): Geological Association of Canada-Mineralogical Association of Canada Program with Abstracts, v. 10, p. A48.

Hannington, M. D., J. M. Peter, and S. D. Scott, 1986, Gold in seafloor polymetallic sulfide deposits: Economic Geology, v. 81, p. 1867-1883.

Haymon, R. M., 1983, Growth history of hydrothermal black smoker chimneys: Nature, v. 301, p. 695–698.

Haymon, R. M., and M. Kastner, 1981, Hot spring deposits on the East Pacific Rise at 21 °N: Preliminary description of mineralogy and genesis: Earth and Planetary Science Letters, v. 53, p. 363–381.

Hekinian, R., M. Fevrier, J. L. Bischoff, P. Picot, and W. C. Shanks, 1980, Sulfide deposits from the East Pacific Rise near 21 °N: Science, v. 207, p. 1433–1444.

Hey, R. N., 1977, A new class of "pseudofaults" and their bearing on plate tectonics: A propagating rift model: Earth and Planetary Science Letters, v. 37, p. 321–325.

Hey, R. N., and D. S. Wilson, 1982, Propagating rift explanation for the tectonic evolution of the northeast Pacific—The pseudomovie: Earth and Planetary Science Letters, v. 58, p. 167–188.

Holmes, M. L., and J. L. Morton, 1986, Massive sulfide deposits from Escanaba Trough, Gorda Ridge (abs.): GSA Abstracts with Programs, v. 18, p. 118.

Humphris, S. E., and G. Thompson, 1978, Hydrothermal alteration of oceanic basalts by seawater: Geochimica et Cosmochimica Acta, v. 42, p. 107–125.

Janecky, D. R., and W. E. Seyfried, Jr., 1984, Formation of massive sulfide deposits on oceanic ridge crusts: Incremental reaction models for mixing between hydrothermal solutions and seawater: Geochimica et Cosmochimica Acta, v. 48, p. 2723–2738.

Johnson, H. P., J. L. Karsten, J. R. Delaney, E. E. Davis, R. G. Currie, and R. L. Chase, 1983, A detailed study of the Cobb offset of the Juan de Fuca Ridge: Evolution of a propagating rift: Journal of Geophysical Research, v. 88, p. 2297–2315.

Jones, C. J., H. P. Johnson, and J. R. Delaney, 1981, Distribution of hydrothermal manganese over the Juan de Fuca Ridge: Geophysical Research Letters, v. 8, p. 873–876.

Juniper, S. K., J. A. J. Thompson, and S. E. Calvert, 1986, Accumulation of minerals and trace elements in biogenic mucus at hydrothermal vents: Deep Sea Research, v. 33, p. 339–347.

Kadko, D., and J. E. Lupton, 1984, Radon-222 and temperature measurements on the Endeavor Ridge: EOS (American Geophysical Union Transactions), v. 65, p. 974.

Kappel, E. F., and W. B. F. Ryan, 1986, Volcanic episodicity and a non-steady state rift valley along northeast Pacific spreading centers: evidence from Sea Marc I: Journal of Geophysical Research, vol. 91, p. 13925–13940.

Karsten, J. L., S. R. Hammond, E. E. Davis, and R. G. Currie, 1986, Detailed geomorphology and neotectonics of the Endeavour segment, Juan de Fuca Ridge: New results from Seabeam swath mapping: GSA Bulletin, v. 97, p. 213–221.

Klinkhammer, G., M. Bender, and R. F. Weiss, 1977, Hydrothermal manganese in the Galapagos rift: Nature, v. 269, p. 319–320.

Koski, R. A., D. A. Clague, and E. Oudin, 1984, Mineralogy and chemistry of massive sulfide deposits from the Juan de Fuca Ridge: GSA Bulletin, v. 95, p. 930–945.

Koski, R. A., R. Goodfellow, and R. M. Bouse, 1982, Preliminary description of massive sulfide samples from the southern Juan de Fuca Ridge: U.S. Geological Survey Open-File Report 82-200B, 11 p.

Koski, R. A., W. R. Normark, and J. L. Morton, 1985, Massive sulfide deposits on the southern Juan de Fuca Ridge: Results of investigations in the USGS study area, 1980–83: Marine Mining, v. 5, p. 147–164.

Koski, R. A., P. F. Lonsdale, W. C. Shanks, M. E. Berndt, and S. S. Howe, 1985, Mineralogy and geochemistry of a sediment-hosted hydrothermal sulfide deposit from the Southern Trough of Guaymas Basin, Gulf of California: Journal of Geophysical Research, v. 90, p. 6695–6707.

Kvenvolden, K. A., J. B. Rapp, F. Hostettler, J. L. Morton, J. D. King, and G. E. Claypool, 1986, Petroleum associated with polymetallic sulfide in sediment from Gorda Ridge: Science, vol. 234, p. 1231–1234.

Lichtman, G. S., and J. P. Eissen, 1983, Time and space constraints on the evolution of medium-rate spreading centers: Geology, v. 11, p. 592–595.

Lister, C. R. B., 1974, On the penetration of water into hot rock: Geophysical Journal of the Royal Astronomical Society, v. 39, p. 465–509.

Lonsdale, P. F., and K. Becker, 1985, Hydrothermal plumes, hot springs, and conductive heat flow in the Southern Trough of Guaymas Basin: Earth and Planetary Science Letters, v. 73, p. 211–225.

Lonsdale, P. F., and L. A. Lawver, 1980, Immature plate boundary zones studied with a submersible in the Gulf of California: GSA Bulletin, v. 91, p. 555–569.

Lonsdale, P. F., J. L. Bischoff, V. M. Burns, M. Kastner, and R. E. Sweeney, 1980, A high-temperature hydrothermal deposit on the seabed at a Gulf of California spreading center: Earth and Planetary Science Letters, v. 49, p. 8–20.

Lupton, J. E., 1983, Helium isotopes and other volatiles as indicators of submarine hydrothermal activity (abs.): Geological Association of Canada–Mineralogical Association of Canada Program with Abstracts, no. 8, p. A43.

Lupton, J. E., R. F. Weiss, and H. Craig, 1977, Mantle helium in hydrothermal plumes in the Galapagos rift: Nature, v. 267, p. 603–604.

Lupton, J. E., J. Delaney, H. P. Johnson, and M. Tivey, 1984, Water-column temperature anomalies near an active deep-sea hydrothermal vent: EOS (American Geophysical Union Transactions), v. 65, p. 975.

Lupton, J. E., G. P. Klinkhammer, W. R. Normark, R. Haymon, K. C. Macdonald, R. F. Weiss, and H. Craig, 1980, Helium-3 and manganese at the 21 °N East Pacific Rise hydrothermal site: Earth and Planetary Science Letters, v. 50, p. 115–127.

Macdonald, K. C., K. Becker, F. N. Spiess, and R. D. Ballard, 1980, Hydrothermal heat flux of the "black smoker" vents on the East Pacific Rise: Earth and Planetary Science Letters, v. 48, p. 1–7.

Malahoff, A., G. McMurtry, S. Hammond, and R. Embley, 1984, High temperature hydrothermal fields—Juan de Fuca Ridge axial volcano: EOS (American Geophysical Union Transactions), v. 65, p. 1112.

McConachy, T. F., R. D. Ballard, M. J. Mottl, and R. P. von Herzen, 1986, Geological form and setting of a hydrothermal vent field at lat. 10 °56 'N, East Pacific Rise: a detailed study using *Angus* and *Alvin:* Geology, v. 14, p. 295–298.

McDuff, R. E., J. E. Lupton, D. Kadko, and M. D. Lilley, 1984, Chemistry of hydrothermal fluids, Endeavor Ridge, northeast Pacific: EOS (American Geophysical Union Transactions), v. 65, p. 975.

McMurtry, G. M., A. Malahoff, R. A. Feely, and G. J. Massoth, 1984, Geology and chemistry of hydrothermal nontronite deposits from the Juan de Fuca Ridge: EOS (American Geophysical Union Transactions), v. 65, p. 1112.

MERGE (Multidisciplinary Endeavour Ridge Geo-Expedition) GROUP: J. Karsten, J. Delaney, P. Johnson, M. Goldfarb, D. Kelley, R. McDuff, M. K. Tivey, M. Tivey, J. Dymond, D. Kadko, G. Taghon, M. Leinen, J. Lupton, M. Rhodes, D. Hey, V. Tunnicliffe), 1984, Regional setting and local character of a hydrothermal field/sulfide deposit on the Endeavour Segment of the Juan de Fuca Ridge: EOS (American Geophysical Union Transactions), v. 65, p. 1111.

Moore, D. G., and J. R. Curray, 1982, Geologic and tectonic history of the Gulf of California, in J. R. Curray, D. G. Moore, et al., Initial reports of the Deep Sea Drilling Project, Vol. 64 (pt. 2): Washington, D.C., U.S. Government Printing Office, p. 1279–1293.

Morton, J. L., M. L. Holmes, and R. A. Koski, 1987, Volcanism and massive sulfide formation at a sedimented spreading center, Escanaba Trough, Gorda Ridge, northeast Pacific Ocean: Geophysical Research Letters, vol. 14, p. 769–772.

Morton, J. L., W. R. Normark, D. M. Mann, and D. H. Tompkins, 1987, Geology, evolution, and mineral potential of the Juan de Fuca Ridge, this volume.

Mottl, M. J., 1983, Metabasalts, axial hot springs and the structure of hydrothermal systems at mid-ocean ridges: GSA Bulletin, v. 94, p. 161–180.

Mottl, M. J., and H. D. Holland, 1978, Chemical exchange during hydrothermal alteration of basalt by seawater I. Experimental results for major and minor components of seawater: Geochimica et Cosmochimica Acta, v. 42, p. 1103–1115.

Mottl, M. J., and H. D. Holland, 1979, Chemical exchange during hydrothermal alteration of basalt by seawater II. Experimental results for Fe, Mn and sulfur species: Geochimica et Cosmochimica Acta, v. 43, p. 869–884.

Normark, W. R., J. L. Morton, and J. R. Delaney, 1982, Geologic setting of massive sulfide deposits and hydrothermal vents along the southern Juan de Fuca Ridge: U.S. Geological Survey Open-File Report 82-200A, 10 p.

Normark, W. R., J. L. Morton, R. T. Holcomb, R. A. Koski, and S. L. Ross, 1984, A continuous linear depression in the axial valley of the southern Juan de Fuca Ridge: Tectonic or volcanic? (abs.): GSA Abstracts with Programs, v. 16, p. 611.

Normark, W. R., J. L. Morton, R. A. Koski, D. A. Clague, and J. R. Delaney, 1983, Active hydrothermal vents and sulfide deposits on the southern Juan de Fuca Ridge: Geology, v. 11, p. 158–163.

Oudin, E., 1982, Etudes mineralogique et geochimique des depots sulfures sous-marins actuels de la ride est-pacifique (21 °N): (France) Bureau de Recherches Geologiques et Minieres Documents 25, 241 p.

Sclater, J. G., and K. D . Klitgord, 1973, A detailed heat flow, topographic and magnetic survey across the Galapagos spreading center at 86 °W: Journal of Geophysical Research, v. 78, p. 6951–6975.

Scott, S. D., 1985, Seafoor polymetallic sulfide deposits: Modern and ancient: Marine Mining, v. 5, p. 191–212.

Scott, S. D., 1987, Seafloor polymetallic sulfide deposits: Scientific curiosities or mines of the future?, *in* P. G. Teleki, M. R. Dobson, J. R. Moore, and U. von Stackelberg, eds., Marine minerals: Advances in Research and Resource Assessment: Dordrecht, D. Reidel, p. 277–300.

Scott, S. D., T. J. Barrett, M. Hannington, R. L. Chase, Y. Fouquet, and K. Juniper, 1984, Tectonic framework and sulfide deposits of southern Explorer Ridge, northeastern Pacific Ocean: EOS (American Geophysical Union Transactions), v. 65, p. 1111.

Scott, S. D., R. L. Chase, P. J. Michael, G. T. Shea, T. J. Barrett, M. Gorton, M. Hannington, and J. Peter, 1985, Explorer Ridge and Tuzo Wilson Seamounts: Update: EOS (American Geophysical Union Transactions), v. 66, p. 925.

Seyfried, W. E., and J. L. Bischoff, 1977, Hydrothermal transport of heavy metals by seawater: The role of seawater/basalt ratio: Earth and Planetary Science Letters, v. 34, p. 71–77.

Seyfried, W. E., and J. L. Bischoff, 1981, Experimental seawater-basalt interaction at 300 °C, 500 bars, chemical exchange, secondary mineral formation and complications for the transport of heavy metals: Geochimica et Cosmochimica Acta, v. 45, p. 135–147.

Shanks, W. C., R. A. Koski, and L. C. Woodruff, 1984, Mineralogy and stable isotope systematics of sulfide deposits from the Juan de Fuca Ridge: EOS (American Geophysical Union Transactions), v. 65, p. 1113.

Sillitoe, R. H., 1972, Formation of certain massive sulphide deposits at sites of sea-floor spreading: Institute of Mining and Metallurgy Transactions, v. 81, sec. B, p. 141–148.

Simoneit, B. R. T., 1985, Hydrothermal petroleum: Genesis, migration, and deposition in Guaymas Basin, Gulf of California: Canadian Journal of Earth Sciences, v. 22, p. 1919–1929.

Simoneit, B. R. T., and P. F. Lonsdale, 1982, Hydrothermal petroleum in mineralized mounds at the seabed of Guaymas Basin: Nature, v. 295, p. 198–202.

Skornyakova, I. S., 1965, Dispersed iron and manganese in Pacific Ocean sediments: International Geology Review, v. 7, p. 2161–2174.

Spiess, F. N., K. C. Macdonald, T. Atwater, R. Ballard, D. Carranza, D. Cordoba, C. Cox, V. M. Diaz-Garcia, J. Francheteau, J. Guerrero, R. Hawkins, R. Haymon, R. Hessler, T. Juteau, M. Kastner, R. Larson, B. Luyendyk, J. D. Macdougall, S. Miller, W. Normark, J. Orcutt, and C. Rangin, 1980, East Pacific Rise: Hot springs and geophysical experiments: Science, v. 207, p. 1421–1433.

Spooner, E. T. C., and W. S. Fyfe, 1973, Sub-seafloor metamorphism, heat, and mass transfer: Contributions to Mineralogy and Petrology, v. 42, p. 287–304.

Styrt, M. M., A. J. Brackman, H. D. Holland, B. C. Clark, V. Pisutha-Arnond, C. S. Eldridge, and H. Ohmoto, 1981, The mineralogy and the isotopic composition of sulfur in hydrothermal sulfide/sulfate deposits on the East Pacific Rise, 21 °N latitude: Earth and Planetary Science Letters, v. 53, p. 382–390.

Tivey, M. K., and J. R. Delaney, 1985, Sulfide deposits from the Endeavour Segment of the Juan de Fuca Ridge: Marine Mining, v. 5, p. 165–179.

Tivey, M. K., and J. R. Delaney, 1986, Growth of large sulfide structures on the Endeavour Segment of the Juan de Fuca Ridge: Earth and Planetary Science Letters, v. 77, p. 303–317.

Tunnicliffe, V., M. Botros, M. De Burgh, A. Dinet, H. P. Johnson, S. K. Juniper, and R. E. McDuff, 1986, Hydrothermal vents of Explorer Ridge, northeast Pacific: Deep-Sea Research, v. 33, p. 401–412.

USGS Juan de Fuca Ridge Study Group (W. R. Normark, J. L. Morton, J. L. Bischoff, R. Brett, R. T. Holcomb, R. A. Koski, S. L. Ross, W. C. Shanks, J. F. Slack, K. L. Von Damm, R. A. Zierenberg, and E. S. Kappel), 1986, Submarine fissure eruptions and hydrothermal vents on the southern Juan de Fuca Ridge: Preliminary observations from the submersible ALVIN: Geology, vol. 14, p. 823–827.

Von Damm, K. L., J. M. Edmond, C. I. Measures, and B. Grant, 1985, Chemistry of submarine hydrothermal solutions at Guaymas Basin, Gulf of California: Geochimica et Cosmochimica Acta, v. 49, p. 2221–2237.

Weiss, R. F., P. Lonsdale, J. E. Lupton, A. E. Bainbridge, and H. Craig, 1977, Hydrothermal plumes in the Galapagos rift: Nature, v. 267, p. 600–603.

Williams, D. L., R. P. von Herzen, J. G. Sclater, and R. N. Anderson, 1974, The Galapagos spreading center: Lithospheric cooling and hydrothermal circulation: Geophysical Journal of the Royal Astronomy Society, v. 38, p. 587–608.

Wilson, D. S., R. N. Hey, and C. Nishimura, 1984, Propagation as a mechanism of reorientation of the Juan de Fuca Ridge: Journal of Geophysical Research, v. 89, p. 9215–9225.

Wolery, T. J., and N. H. Sleep, 1976, Hydrothermal circulation and geochemical flux at mid-ocean ridges: Journal of Geology, v. 84, p. 249–275.

Zierenberg, R. A., W. C. Shanks III, and J. L. Bischoff, 1984, Massive sulfide deposits at 21 °N, East Pacific Rise: Chemical composition, stable isotopes, and phase equilibria: GSA Bulletin, v. 95, p. 922–929.

Additional note: Since this paper was written, an area of high heat flow, active hydrothermal discharge and sulfide deposits has been located in the sediment-covered Middle Valley segment of the northern Juan de Fuca Ridge. The geologic setting and sulfide deposits at Middle Valley have been described by Davis et al. (1987).

Davis, E. E., W. D. Goodfellow, B. D. Bornhold, J. Adshead, B. Blaise, H. Villinger, and G. M. Le Cheminant, 1987, Massive sulfides in a sedimented rift valley, northern Juan de Fuca Ridge: Earth and Planetary Science Letters, vol. 82, p. 49–61.

CHAPTER TWENTY-EIGHT

OFFSHORE ALASKA COAL

Ronald H. Affolter
Gary D. Stricker

U.S. Geological Survey
Denver, Colorado

ABSTRACT

Evaluation of geologic data along the western continental margin of the United States indicates that coal may exist offshore of Alaska, Washington, and Oregon. The most probable areas for offshore coal development are along the continental shelf of Alaska. Thirteen areas on the shelf of Alaska have been identified as having the greatest potential for undiscovered coal resources, based on the interpretation of (1) onshore coal outcrops, (2) onshore-offshore oil and gas test wells, (3) COST (Continental Offshore Stratigraphic Test) wells, (4) offshore seismic lines, and (5) onshore to offshore stratigraphic trends. Because of limited data, undiscovered coal resources can be calculated only for the Chukchi Shelf, Beaufort Shelf, Norton Basin, Bristol Bay Basin, Navarin Basin, and Cook Inlet Basin. The quality of coal is an essential component for the U.S. Geological Survey's resource classification system. Since there are no data for offshore coal quality, it is assumed to be similar to onshore coal. Based on this assumption, undiscovered coal resources of more than 6 trillion metric tons are estimated for the continental shelf of Alaska. This estimate is three times greater than the undiscovered coal resource estimates for the onshore contiguous United States.

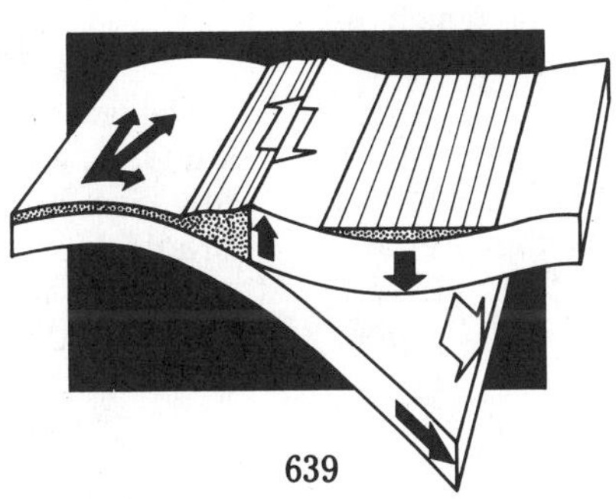

INTRODUCTION

Coal is an organic sedimentary rock which originated as an accumulation of terrestrial plant debris in fresh to brackish water environments. Almost 90% of the known economic deposits of coal have formed in paralic settings (Galloway and Hobday, 1983). During the formation of a major coal deposit, vast quantities of organic material accumulate in a strongly reducing peat swamp environment under restrictive conditions that include continued subsidence of the sedimentary basin, a rising water table to keep up with subsidence, and isolation from the influx of clastic material. When these conditions persist for long periods, the accumulating organic material is protected from oxidation and microbial decay, producing thick peat deposits. These organic accumulations are then transformed through time into coal by various biochemical and metamorphic processes.

Throughout time, the size and shape of continental land masses have changed because of sea-level fluctuations which resulted from continental movements and climatic changes. Because of these relative sea-level fluctuations, many continental shelves have been above sea level periodically and may contain significant amounts of nonmarine sediments including coal. Coal-bearing rocks along continental margins have been observed to extend offshore onto the present continental shelf. Offshore oil and gas test wells have also encountered nonmarine sediments, indicating the potential for the existence of offshore coal deposits. The evidence from drilling, seismic exploration, and onshore-offshore stratigraphic relationships indicates that offshore coal occurrences are worldwide (Wang and McKelvey, 1976). Offshore coal was mined as early as 1617 off the coast of Scotland (Austin, 1967). Since then, coal has been reported off the coastlines of Argentina, Australia, Brazil, Canada, Chile, Greece, Israel, Japan, Norway, Spain, Turkey, Arctic U.S.S.R., United Kingdom, and the United States (Wang and McKelvey, 1976). Japan, the United Kingdom, and Canada have mined some of their offshore coal deposits.

Exploration for coal fields in the United States is almost nonexistent today because of the abundance of onshore coal deposits that are in areas currently being mined or in fields that have not been developed. Evaluation of data along the western margin of the United States indicates that coal may exist offshore of Alaska, Washington, and Oregon (Wang and McKelvey, 1976). The Coos Bay field (Allen and Baldwin, 1944) in Coos County, Oregon and the Bellingham coal field (Beikman, Gower, and Dana, 1961) in Whatcom County, Washington probably have offshore coal, but data are insufficient to indicate offshore extensions of these fields. The most probable areas for offshore coal development are located along the continental shelf of Alaska. Wood and Bour (in press) have estimated offshore speculative coal resources for North America and adjacent islands to be approximately 118 billion metric tons, 40% of which is located offshore of Alaska near the North Slope and Cook Inlet areas.

Several Tertiary to Mesozoic submerged sedimentary basins off western Alaska have been observed to contain thick accumulations of nonmarine sediments (Steffy et al., 1985; Turner et al., 1983a, 1983b, 1984a, 1984b, 1984c; Marlow et al., 1980; McLean, 1977; Marlow and Cooper, 1984).

METHODS OF INVESTIGATION

This chapter will attempt to define potential offshore coal-bearing areas that may exist along the continental shelf of Alaska by summarizing the existing literature, reviewing current research projects, and interpreting (1) onshore coal outcrops, (2) onshore-offshore oil and gas test wells, (3) COST (Continental Offshore Stratigraphic Test) wells, (4) offshore seismic lines, and (5) onshore coal stratigraphic trends. Where sufficient data are available, undiscovered coal resources were calculated by:

(1) Estimating the areal extent of the coal-bearing unit (in km^2).
(2) Estimating the thickness and variability of the coal-bearing unit.
(3) Estimating the percentage of coal present in the coal-bearing unit.
(4) Calculating the coal volume based on areas and coal thickness within the areas.
(5) Multiplying areas of coal (in km^2) times the meters of coal under the areas times the weight of 1 m of coal per km^2. The weight of coal depends on coal rank, and this was estimated from onshore beds or from offshore data where available. The weight of 1 m of coal over an area of 1 km^2 of bituminous coal is 1.32 million metric tons, subbituminous coal is 1.30 million metric tons, and lignite is 1.29 million metric tons (Averitt, 1975).

Although this method of resource calculation does not conform to standards of the U. S. Geological Survey (Wood et al., 1983), the estimates in this chapter are the most complete and accurate for offshore Alaska coal to date. These undiscovered resources can be classified as either hypothetical or speculative depending on the amount of available data (Wood et al., 1983). Speculative offshore coal resources are those based upon coal that has been reported onshore near the continental margin, but because of the lack of detailed geological, stratigraphic, and structural data, offshore trends can only be estimated. Hypothetical offshore coal resources are based on strong evidence for projecting onshore coal-bearing rocks offshore, along with sufficient direct evidence in offshore drill holes that have penetrated coal. At present, offshore Alaska coal data are scarce and most of the coal resources there are speculative.

SUMMARY OF UNDISCOVERED OFFSHORE COAL AREAS

The following 13 areas that have been identified as having potential undiscovered offshore coal resources on the continental shelf of Alaska are shown in Figure 1. Onshore occurrences are from Barnes (1967) unless otherwise noted. Index map locations are keyed to area names in Figure 1 and in the following text.

Western Beaufort Shelf (1)

The western Beaufort Shelf is relatively narrow, 64 to 130 km wide, and is situated between the MacKenzie and Barrow Sea Valleys (Grantz, Eittreim, and Whitney, 1981; Grantz, May, and Dinter, this volume). Limited offshore data consists of single-channel seismic lines and proprietary drill-hole data that are not available. The possibility of offshore coal is indicated by the presence of coal chips from onshore

Figure 1. Index map showing location of potential undiscovered coal resource areas on the continental shelf of Alaska. Numbers are keyed to area names in text.

mud logs. Preliminary onshore investigations of the Tertiary Sagavanirktok Formation indicate a coal-bearing interval approximately 450 m thick, with coal beds ranging from 1.8 to 3.7 m in thickness onshore near Prudhoe Bay (S. B. Roberts, 1986, written communication). Our interpretation of the stratigraphic sections from S. B. Roberts and M. A. Carey (1985, written communication), Detterman et al. (1975), and Molenaar et al. (1984) indicates that the coal-bearing Sagavanirktok Formation is present beneath the Beaufort Sea, thins at approximately 5 m/km, and contains 8% coal. The shape of the coal-bearing basin under the Beaufort Sea was estimated from onshore structural contours (Roberts, 1986, written communication) based on Bird's (1982) studies of the Sagavanirktok Formation, Colville Sandstone, and Beikman and Lathram's (1976) geologic map of Alaska. The isopachs of the coal-bearing unit were then constructed on the basis of northward thinning of this unit (Figure 2). The rank of coal in the Sagavanirktok Formation has been reported to be lignite (Molenaar et al., 1984). The speculative coal resources for the Sagavanirktok Formation under the Beaufort Sea are estimated to be 300 billion metric tons of lignite.

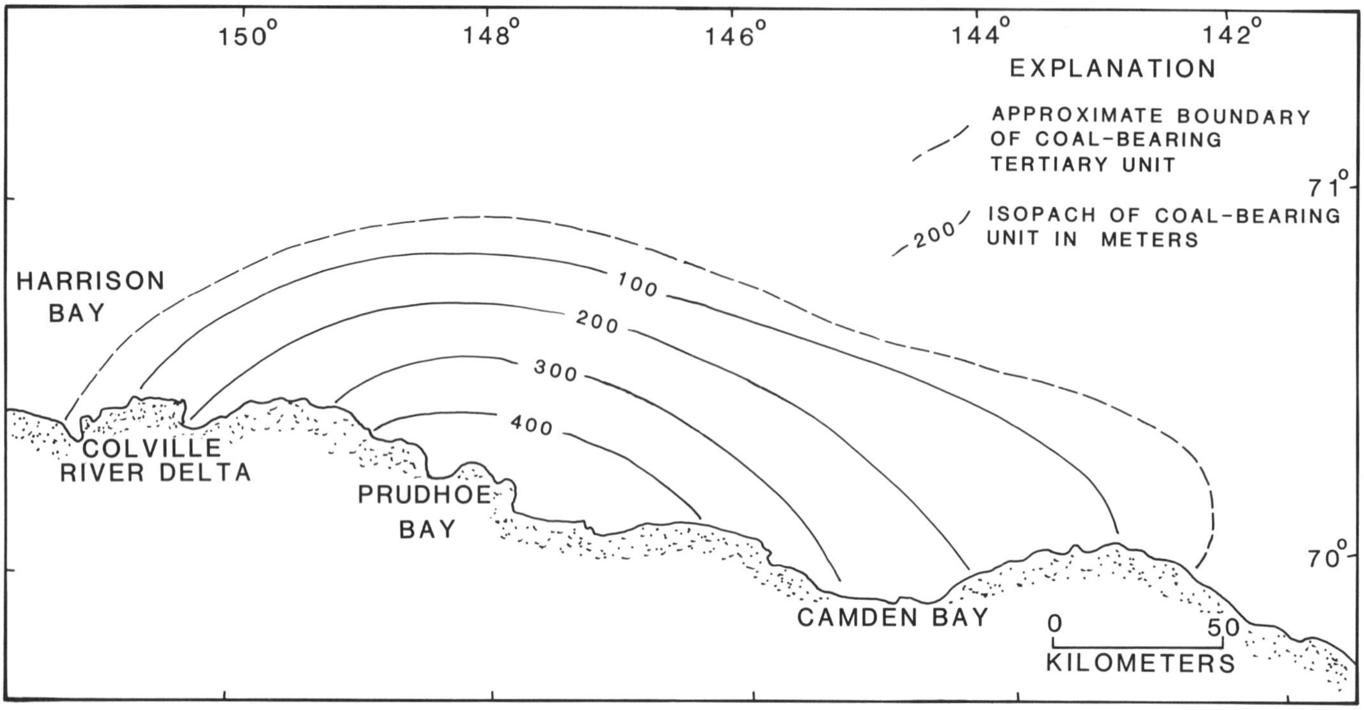

Figure 2. Isopachs of the coal-bearing Tertiary Sagavanirktok Formation, western Beaufort Shelf.

Chukchi Shelf (2)

The Chukchi Shelf, which is characterized by low bathymetric gradients, covers an area of approximately 518,000 km^2 (Grantz, Eittreim, and Whitney, 1981; Grantz and May, this volume). It lies north of Cape Lisburne, west of Barrow, and east of the longitude of Wrangel Island (180°W) (Figure 3). Because there are no offshore test wells in the Chukchi Sea, the onshore coal information must be extrapolated seaward. Tailleur and Brosge (1976) made preliminary estimates of offshore coal by extending onshore stratigraphic trends, based on outcrops and wells in the Meade-Kaolak region in the western North Slope, into the Chukchi Sea. This was one of the first published attempts to evaluate offshore coal resources west of northern Alaska. They estimated speculative resources between 45 and 900 billion metric tons. We have attempted to update this estimate by using recent stratigraphic, geologic, and depositional environment data (Callahan and Sloan, 1978; Martin and Callahan, 1978; Roehler and Stricker, 1979; Ahlbrandt et al., 1979; Huffman et al., 1985; and Sable and Stricker, 1985). Isopachs of the offshore equivalent of the nonmarine Nanushuk Group, Corwin Formation coal-bearing unit were constructed based on onshore thickness isopachs (Sable, Stricker, and Affolter, 1986), which were extrapolated offshore, and by interpreting seismic lines in the Chukchi Sea (A. Grantz, 1985, written communication) (Figure 3). Percent coal in the Corwin Formation equivalent (2–6%) was also extrapolated offshore and resources were estimated. Summaries of onshore North Slope coal (Affolter and Stricker, 1985) indicate that the rank ranges from lignite A to high-volatile A bituminous coal. Therefore, resource estimates were conservatively based on subbituminous coal. The amount of in situ speculative coal under the Chukchi Sea is estimated by these methods at 2.0 trillion metric tons.

Cape Lisburne Peninsula (3)

Outcrops of Mississippian coal, with an apparent rank of low-volatile bituminous, have been reported from several localities in the Cape Lisburne Peninsula. The coal is as thick as 3.3 m (Tailleur, 1965). The area is highly folded and faulted, and the coal-bearing rocks strike northward into the Chukchi Sea. This area has not been extensively explored, and little is known of the coal or of any offshore trends. However, coal of Mississippian age has been reported from deep oil and gas test wells near Barrow (Sable and Stricker, 1985), suggesting that offshore Mississippian coal may be present beneath the Chukchi Sea.

Hope Basin (4)

The 75,000-km^2 Hope Basin is located in the southern Chukchi Sea (Grantz, Holmes, and Kososki, 1976; Tolson, this volume). Multichannel seismic data indicate that 5 to 6 km of moderately consolidated Tertiary and possibly Upper Cretaceous sediments fill the Hope Basin (Tolson, this volume). Onshore data from the Nimiuk Point No. 1 and Cape Espenberg No. 1 wells from the northeast coast of the Seward Peninsula indicate the presence of Tertiary coal-bearing rocks (Meyerhoff, 1982). Hite and Nakayama (1980) have reported that part of the offshore sediments in the Hope Basin may be nonmarine. Data are insufficient at present to make any coal resource estimate, but coal probably is present in the Hope Basin.

Norton Basin (5)

The 45,000-km^2 Norton Basin (Marlow, Cooper, and Fisher, this volume) is located in Norton Sound east of St. Lawrence Island and south of Nome (Figure 4). The Norton Basin contains Cenozoic sediments as much as 5 to 6 km thick (Fisher, Patton, and Holmes, 1982; Marlow, Cooper,

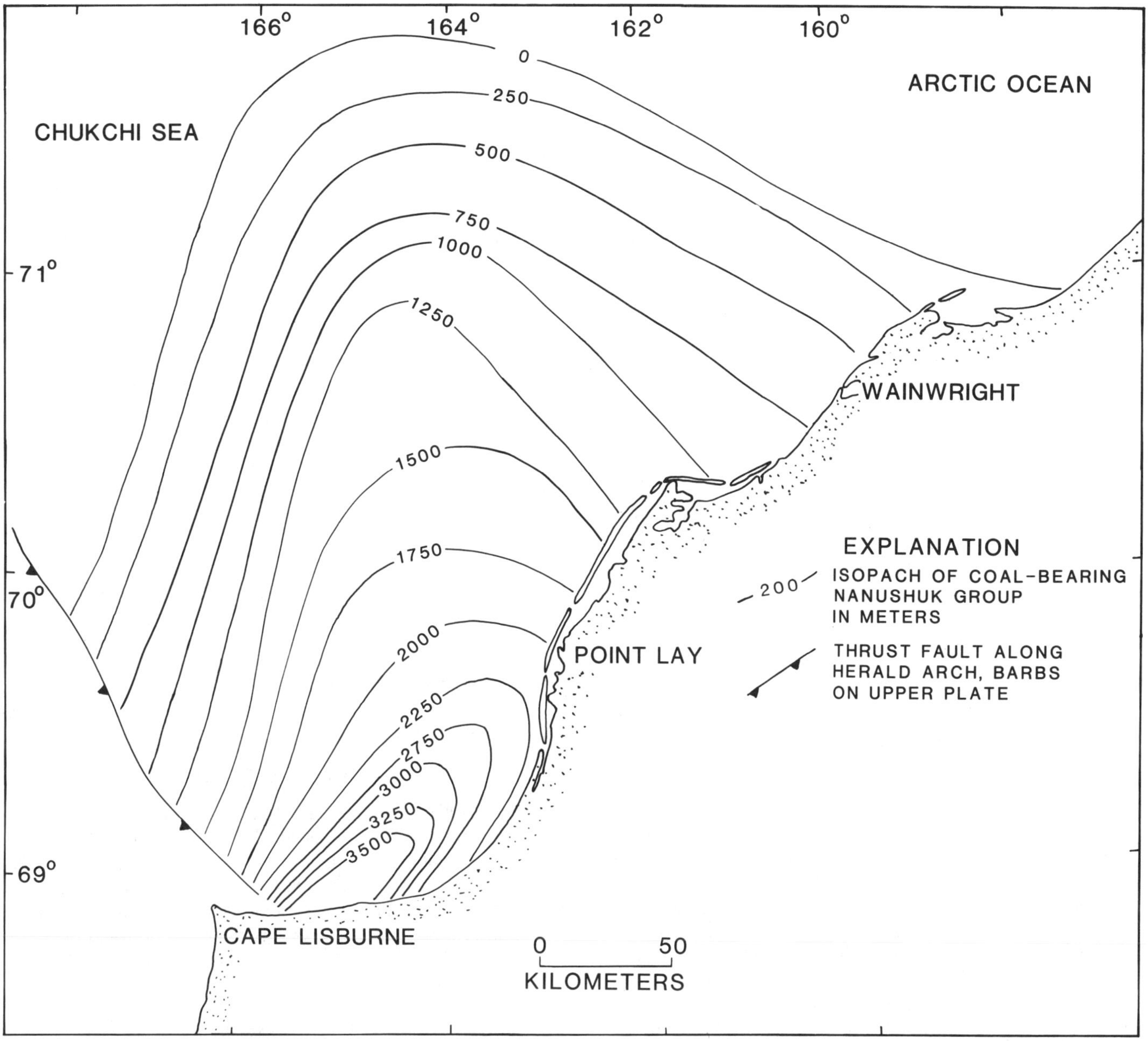

Figure 3. Isopachs of the coal-bearing Cretaceous equivalent Nanushuk Group, Chukchi Shelf.

and Fisher, this volume), which includes coal-bearing Paleogene sediments. Coal of Tertiary age has been found onshore east of the Norton Basin, approximately 16 km south of Unalakleet, and 64 km farther north above the mouth of the Unalakleet River. Coal has also been found at Elephant Point (Patton, 1973). West of the Norton Basin on St. Lawrence Island, Patton and Csejtey (1971, 1980) reported lignitic Tertiary coal on the western part of the island along the Aghnagluk Lagoon and on the Koozata River. Two COST wells were drilled in 1980 and 1982, and data was summarized by Turner et al. (1983a, 1983b). Both of the COST wells penetrated Tertiary coal. Coal thicknesses for each well were determined from the COST well E-logs and the AMSTRAT (American Stratigraphic) logs. The cumulative thickness of the nonmarine Tertiary section ranges from 0 to 1300 m (Figure 4). Coal resources were calculated based on an apparent rank of lignite. The amount of in situ speculative coal for the Norton Basin is 1.6 trillion metric tons.

Kuskokwim (6)

Coal has been reported on Nelson and Nunivak Islands. On Nelson Island the coal has been reported to be of Cretaceous age and bituminous, but is thin and has a high ash content. The coal on Nunivak Island is also thin and of Cretaceous age. The islands are approximately 32 km apart, and the Cretaceous coal may extend offshore beneath Etolin Strait between the islands.

Bristol Bay Basin (7)

The Bristol Bay Basin is an elongate structure (290 by 75 km) on the northern fringe of the Alaska Peninsula (Figure 5)

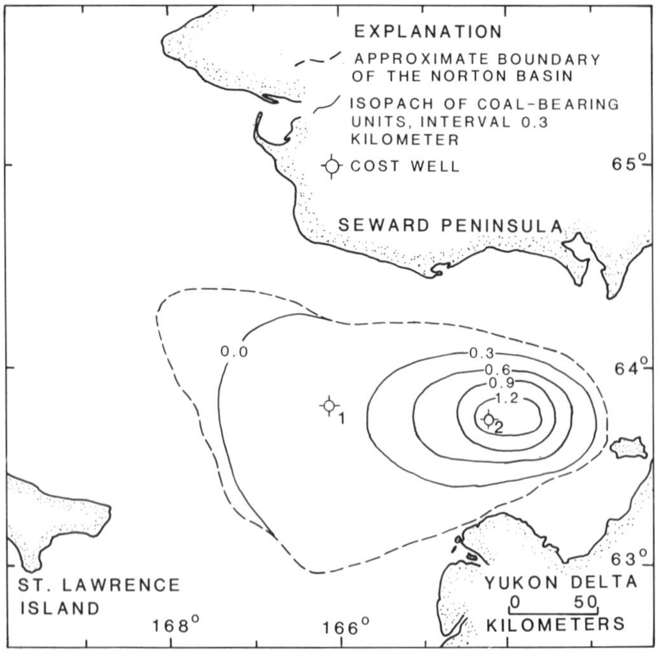

Figure 4. Isopachs of the coal-bearing Tertiary units, Norton Basin. (Basin boundary modified from Fisher, 1982.)

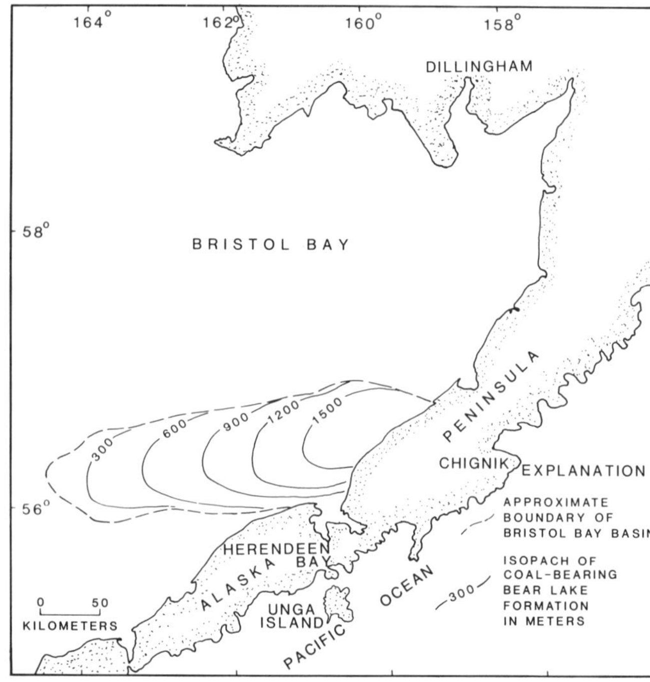

Figure 5. Isopachs of the coal-bearing Tertiary Bear Lake Formation, Bristol Bay Basin. (Basin boundary modified from Marlow and Cooper, 1984.)

(Marlow and Cooper, 1984; Marlow, Cooper, and Fisher, this volume). The 21,700-km^2 basin contains approximately 6100 m of Tertiary and Mesozoic sedimentary rocks, and the thickest sediments are in the southeastern part of the basin (Marlow and Cooper, 1984). One COST well has been drilled, but the information is proprietary and not available at this time. Nine oil and gas test wells have been drilled on the adjacent Alaska Peninsula. Some of the onshore drill holes have indicated the presence of coal of Tertiary age (McLean, 1977; Marlow et al., 1980; Marlow and Cooper, 1984). The Tertiary rocks on the northern part of the Alaska Peninsula contain approximately 1200–3300 m of marine to nonmarine rocks, some of which have indications of coal (Meyerhoff, 1982). In the southern part, coal occurs in the 1500-m-thick nonmarine and marine Bear Lake Formation (Marlow and Cooper, 1984). Tertiary and Cretaceous rocks exposed on the Alaska Peninsula possibly extend offshore (Marlow and Cooper, 1984). If so, and assuming that the percentage of coal and apparent rank in the Bear Lake Formation of the Bristol Bay Basin is similar to that calculated for the Norton Basin, we estimate speculative coal resources of 1.5 trillion metric tons for the Bristol Bay Basin.

Herendeen Bay, Unga Island, and Chignik Fields (8)

Bituminous coal beds in the Herendeen Bay field are from the Chignik Formation of Late Cretaceous age. The beds are moderately folded and are affected by several faults. Bed thicknesses range from a few centimeters to as much as 2 m. Tertiary lignite beds have been reported south to southeast of this area. Tertiary lignite beds a few centimeters to 1.2 m thick have also been found on the northwestern part of Unga Island, where they dip about 8°–10° to the west. Coal in the Chignik coal field is from the Chignik Formation. It is high-ash bituminous, ranging from 0.3 to 1.5 m thick, and is moderately folded and faulted. No data are available to indicate offshore extensions of these fields, although coal probably does occur offshore.

St. George Basin (9)

Two COST wells drilled in the St. George Basin have been summarized by Turner et al. (1984a, 1984b). A 2-mm-thick coal bed from an 18-m nonmarine core interval has been described at a depth of 3800 m in the COST No. 2 well and is reported to be of Early Cretaceous age. Although this basin contains coal-bearing rocks, it is currently not possible to calculate coal resources because of the lack of lateral control and thinness of beds.

Navarin Basin (10)

The Navarin Basin covers an area of approximately 80,000 km^2 and is located on the outer continental shelf of the Bering Sea, south of Cape Navarin (U.S.S.R.) and west of St. Matthew Island (Marlow, Cooper, and Fisher, this volume). Steffy et al. (1985) and Turner et al. (1984c) reported a single coal bed of Late Cretaceous age in the COST No. 1 well. The coal is approximately 0.6 m thick, is at a depth of 4200 m, and is overlain by diabase and basalt sills (Turner et al., 1984c). If we assume that this 0.6 m coal bed is bituminous and consistent throughout this basin, a speculative coal resource of 65 billion metric tons can be calculated for the Navarin Basin.

Cook Inlet Basin (11)

The offshore Cook Inlet Basin is a large 14,000-km^2, northeastward-trending estuary in south-central Alaska that extends from Anchorage to the Gulf of Alaska (Figure 6) (Fisher, Detterman, and Magoon, this volume). The Cook Inlet Basin contains significant amounts of onshore and offshore Tertiary coal. McGee and O'Connor (1975), after

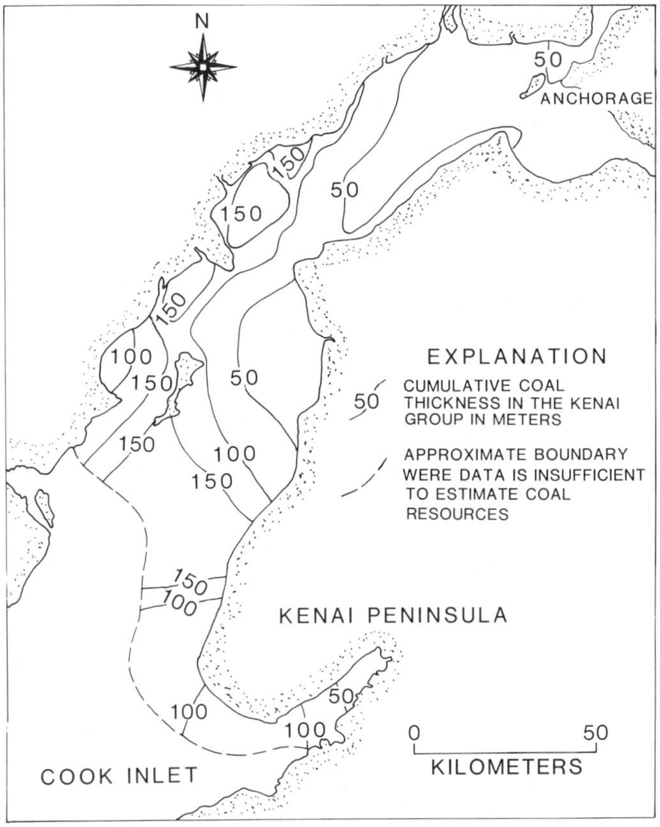

Figure 6. Cumulative coal thickness in the Tertiary Kenai Group, upper Cook Inlet. (Modified from McGee and O'Connor, 1975.)

reviewing 86 oil and gas drill holes, have estimated undiscovered coal resources for the onshore and offshore Cook Inlet Basin to be 1.2 trillion metric tons for the Tertiary Kenai Group. Their estimate is based on coal beds thicker than 0.6 m, from the surface to 3000 m, and an apparent rank of lignite. A COST well that was drilled in the lower Cook Inlet (Wills et al., 1978) penetrated Tertiary coals 0.3 to 1.5 m thick. Using McGee and O'Connor's (1975) isopachs of the thickness and distribution of coal for the onshore and offshore Cook Inlet Basin, we calculated the offshore coal resources. The hypothetical coal resources for the offshore portion of the Cook Inlet Basin are 0.8 trillion metric tons.

Gulf of Alaska Province (12)

Approximately 52,000 km^2 of the Gulf of Alaska province extends offshore and contains marine and nonmarine Eocene to Pliocene sediments as thick as 7300 m (McKelvey et al., 1969) (Figure 7). Coal occurs in the Bering River field from the Tertiary Kushtaka Formation. The geology of this area is complex, and coal beds which are overturned trend northeastward from the east shore of Bering Lake. The intensity of deformation increases to the northeast, as evidenced by an increase in rank from low-volatile bituminous coal in the southwestern part of the field to semianthracite and anthracite in the eastern part (Sanders, 1976). Analyses of some of the samples indicate that they may be of coking quality. Coal beds range from a few centimeters to 18 m thick. No coal was reported in a COST well drilled approximately 130 km southeast of the Bering River coal field (Bolm et al., 1976). Onshore information indicates that coal may be present in the Gulf of Alaska, but no resources could be estimated.

Kootznahoo Inlet (13)

Tertiary coal beds have been reported in a 52-km^2 area on the north and south sides of Kootznahoo Inlet, which lies on the west side of Admiralty Island approximately 96 km south of Juneau. The coal beds reportedly range from 0.6 to 0.9 m in thickness and the rank is bituminous. There are no data on the lateral extent of these coal beds offshore, and no resources could be estimated.

DISCUSSION

Because of the abundance of onshore coal deposits in the United States, very little exploration has been done to evaluate offshore coal potential. Basically, there are two major obstacles that inhibit our knowledge of offshore coal resources:

1. Lack of Data

Information on coal from wells is scarce to nonexistent in most offshore areas of the United States. Most data are derived from the relatively few oil and gas test wells and COST wells drilled along the continental shelf of Alaska. In only 6 of the 13 potential coal-bearing offshore areas in Alaska are data sufficient for resource calculations, and these estimates can only be considered to be in the undiscovered resource category as defined by Wood et al. (1983). Many references relating onshore coal to offshore coal trends are incomplete and would require further field investigations and sampling in order to adequately evaluate stratigraphic relationships. Although offshore seismic data are available for some areas, they are inadequate for determining individual coal bed thickness and areal extent. These data, therefore, do not meet coal-resource standards of the U.S. Geological Survey (Wood et al., 1983). Few attempts have been made to integrate the surface and subsurface data in a basin-wide approach in order to predict depositional settings. Very little to nothing is known about the depositional environment of the coal swamps, their lateral extent, compositional uniformity within the coal beds, or the nature of the overlying and underlying rocks. Although onshore stratigraphic trends can suggest the presence of offshore coal, such trends do not necessarily indicate that coal occurs there. Many coal beds are highly lenticular and discontinuous and therefore may not extend offshore. In order to reclassify offshore coal from unknown to identified resources, many additional offshore drill holes will be needed.

2. Coal Quality

Coal quality is an essential component of the U. S. Geological Survey's resource classification system (Wood et al., 1983). According to Hatch (1983), factors that control coal quality include geochemical conditions in the original swamp, elemental composition of roof rocks, rank, epigenic mineralization, ground-water composition, and weathering effects. However, little information has been obtained about these parameters for most U.S. coals and none for any offshore coal.

Hatch and Swanson (1977) list four basic reasons why coal

quality data are necessary for the evaluation and proper utilization of United States coal:

(1) Evaluation of the environmental impacts of mining and combustion of coal.

(2) Evaluation of the best and most effective technological use of coal (combustion, liquefaction, gasification, etc.).

(3) Determining the economical aspects of extracting elements such as germanium, selenium, uranium, vanadium, and zinc from coal.

(4) Development of geological and geochemical models to help interpret and predict coal quality as well as relating these factors to the stratigraphic and sedimentological framework.

With increasing emphasis on environmental concerns, the quality of coal (sulfur content, amount of trace elements, mineral matter, maceral content, etc.) is becoming almost as important as the resource itself. The best we can do in evaluating the quality of offshore coal is to assume that it is similar to that in onshore deposits. Onshore Alaska coals have some of the lowest-reported sulfur concentrations in the United States (Affolter and Stricker, 1984; Affolter, Simon, and Stricker, 1981). Lacking offshore data, we assume that offshore Alaska coal may also be low in sulfur, but we have little or no information to justify this assumption. Without actual chemical analyses of offshore coal samples, estimation of overall quality becomes difficult. Any evaluation of coal resource potential must take into consideration quality as well as quantity.

SUMMARY

The major constraint on the utilization of Alaska offshore coal deposits is economic. Offshore coal deposits have been or are presently being mined in other parts of the world. The United States has vast onshore coal deposits that are currently minable by today's technology and consist of developed deposits and fields that are known but undeveloped. Because of these onshore coal deposits, it is not economical to explore, assess, or utilize Alaska offshore coal. Therefore, these deposits should be classified as currently subeconomical.

The continental shelf of Alaska contains the greatest potential for future offshore coal exploration in the United States. Of the 13 possible locations for offshore coal, 6 have enough data from which to calculate hypothetical or speculative resources. The Chukchi Shelf, Beaufort Shelf, Norton Basin, Bristol Bay Basin, Navarin Basin, and Cook Inlet Basin have total undiscovered coal resources estimated to be greater than 6 trillion metric tons. A comparison of Averitt's 1975 estimates of 1.9 trillion metric tons of undiscovered coal for the onshore contiguous United States to the offshore estimates indicates that the offshore areas of Alaska have more than three times as much undiscovered coal as the lower 48 states.

REFERENCES CITED

Affolter, R. H., and G. D. Stricker, 1984, Geochemistry of selected coal beds from the Beluga coal field, Alaska (abs.): GSA Abstracts with Programs, v. 16, no. 5, p. 265.

Affolter, R. H., and G. D. Stricker, 1985, Geochemistry of coal from the Cretaceous Corwin and Chandler Formations, National Petroleum Reserve in Alaska (NPRA): AAPG Bulletin, v. 69, p. 655.

Affolter, R. H., F. O. Simon, and G. D. Stricker, 1981, Analysis of coal samples from the Healy, Kenai, Seldovia, and Utukok River quadrangles, Alaska, *in* Proceedings of the 1980 Focus on Alaskan Coal Symposium: University of Alaska Mineral Industry Research Laboratory Report 50, p. 236–266.

Ahlbrandt, T. S., A. C. Huffman, J. E. Fox, and I. Pasternack, 1979, Depositional framework and reservoir-quality studies of selected Nanushuk Group outcrops, North Slope, Alaska, *in* T. S. Ahlbrandt, ed., Preliminary geologic, petrologic, and paleontologic results of the study of Nanushuk Group rocks, North Slope, Alaska: U.S. Geological Survey Circular 794, p. 14–31.

Allen, J. E., and E. M. Baldwin, 1944, Geology and coal resources of the Coos Bay quadrangle, Oregon: Oregon Department of Geology and Mineral Industries Bulletin 27, 160 p.

Austin, C. F., 1967, In the rock—A logical approach for undersea mining of resources: Engineering and Mining Journal, v. 168, no. 8, p. 82–88.

Averitt, P., 1975, Coal resources of the United States, January 1, 1974: U.S. Geological Survey Bulletin 1412, 131 p.

Barnes, F. F., 1967, Coal resources of Alaska: U.S. Geological Survey Bulletin 1242–B, 36 p.

Beikman, H. M., and E. H. Lathram, 1976, Preliminary geologic map of northern Alaska: U.S. Geological Survey Miscellaneous Field Studies Map MF–789, 2 sheets, scale 1:1,000,000.

Beikman, H. M., H. D. Gower, and T. A. M. Dana, 1961, Coal reserves of Washington: Washington Division of Mines and Geology Bulletin 47, 115 p.

Bird, K. J., 1982, Rock-unit reports of 228 wells drilled on the North Slope, Alaska: U.S. Geological Survey Open-File Report 82-278, 106 p.

Bolm, J. G., F. B. Chmelik, G. H. Stewart, R. F. Turner, H. H. Waetjen, and J. C. Wills, 1976, Geological and operational summary, Atlantic Richfield Northern Gulf of Alaska COST Well No. 1: U.S. Geological Survey Open-File Report 76-635, 33 p.

Callahan, J. E., and E. G. Sloan, 1978, Preliminary report on analysis of Cretaceous coals from northwestern Alaska: U.S. Geological Survey Open-File Report 78-319, 29 p.

Detterman, R. L., H. N. Reiser, W. P. Brosge, and J. T. Dutro, Jr., 1975, Post-Carboniferous stratigraphy northeastern Alaska: U.S. Geological Survey Professional Paper 886, 46 p.

Fisher, M. A., 1982, Petroleum geology of Norton Basin, Alaska: AAPG Bulletin, v. 66, p. 286–301.

Fisher, M. A., R. L. Detterman, and L. B. Magoon, 1987, Tectonics and petroleum geology of the Cook-Shelikof Basin, southern Alaska, this volume.

Fisher, M. A., W. W. Patton, Jr., and M. L. Holmes, 1982, Geology of Norton Basin and continental shelf beneath northwestern Bering Sea, Alaska: AAPG Bulletin, v. 66, p. 255–285.

Galloway, W. E., and D. K. Hobday, 1983, Terrigenous clastic depositional systems: New York, Springer-Verlag, 423 p.

Grantz, A., and S. D. May, 1987, Regional geology and petroleum potential of the United States Chukchi Shelf north of Point Hope, this volume.

Grantz, A., S. Eittreim, and O. T. Whitney, 1981, Geology and physiography of the continental margin north of Alaska and implications for the origin of the Canada Basin, *in* A. E. M. Nairn, M. Churkin, Jr., and F. G. Stehli, eds., The ocean basins and margins: Geology of the Arctic Ocean Basin and its margins (Vol. 5): New York, Plenum Press, p. 439–492.

Grantz, A., M. L. Holmes, and B. A. Kososki, 1976, Geologic framework of the Alaskan continental terrace in the Chukchi and Beaufort Seas: Canadian Society of Petroleum Geologists Memoir 4, p. 669–699.

Grantz, A., S. D. May, and D. A. Dinter, 1987, Regional geology and petroleum potential of the United States Beaufort and northeasternmost Chukchi Seas, this volume.

Hatch, J. R., 1983, Geochemical processes that control minor and trace element composition of United States coals, *in* Cameron Volume on Unconventional Mineral Resources: AIME Special Publication, p. 89–98.

Hatch, J. R., and V. E. Swanson, 1977, Trace elements in Rocky Mountain coals, *in* D. K. Murray, ed., Geology of Rocky Mountain coal—A symposium: Colorado Geological Survey Resources Series 1, p. 143–165.

Hite, D. M., and E. M. Nakayama, 1980, Present and potential petroleum basins of Alaska, *in* M. L. Landwehr, ed., Exploration and economics of the petroleum industry (Vol. 18): Dallas, Southwestern Legal Foundation, p. 511–560.

Huffman, A. C., Jr., T. S. Ahlbrandt, I. Pasternack, G. D. Stricker, and J. E. Fox, 1985, Depositional and sedimentologic factors affecting the reservoir potential of the Cretaceous Nanushuk Group, central North Slope, Alaska, *in* A. C. Huffman, ed., Geology of the Nanushuk Group and related rocks, North Slope, Alaska: U.S. Geological Survey Bulletin 1614, p. 61–74.

Marlow, M. S., and A. K. Cooper, 1984, Summary geological report for the North Aleutian Shelf (OCS) Planning Area, Bering Sea, Alaska: U.S. Geological Survey Open-File Report 84-773, 29 p.

Marlow, M. S., A. K. Cooper, and M. A. Fisher, 1987, Petroleum geology of the Beringian continental shelf, this volume.

Marlow, M. S., H. McLean, A. K. Cooper, T. L. Vallier, J. V. Gardner, R. McMullin, and M. B. Lynch, 1980, A preliminary summary of regional geology, petroleum potential, environmental geology, and technology for exploration and development for proposed OCS lease sale #75, Northern Aleutian Shelf, Bering Sea, Alaska: U.S. Geological Survey Open-File Report 80-653, 54 p.

Martin, G. C., and J. E. Callahan, 1978, Preliminary report on the coal resources of the National Petroleum Reserve in Alaska: U.S. Geological Survey Open-File Report 78-1033, 29 p.

McGee, D. L., and K. M. O'Connor, 1975, Cook Inlet Basin subsurface coal reserve study: Alaska Division of Geological and Geophysical Surveys Open-File Report 74, 19 p.

McKelvey, V. E., F. F. H. Wang, S. P. Schweinfurth, and W. C. Overstreet, 1969, Potential mineral resources of the United States outer continental shelf, *in* Nossaman, Waters, Scott, Krueger and Riordan, Study of outer continental shelf lands of U.S.: Washington, D.C., U.S. Department of the Commerce, National Bureau of Standards, v. IV, app. 5-A, 117 p.

McLean, H., 1977, Organic geochemistry, lithology, and paleontology of Tertiary and Mesozoic rocks from wells on the Alaska Peninsula: U.S. Geological Survey Open-File Report 77-813, 63 p.

Meyerhoff, A. A., 1982, Hydrocarbon resources in arctic and subarctic regions: Canadian Society of Petroleum Geologists Memoir 8, p. 451–552.

Molenaar, C. M., A. R. Kirk, L. B. Magoon, and A. C. Huffman, 1984, Twenty-two measured sections of Cretaceous-Lower Tertiary rocks, eastern North Slope, Alaska: U.S. Geological Survey Open-File Report 84-695, 19 p.

Patton, W. W., Jr., 1973, Reconnaissance geology of the northern Yukon-Koyukuk province, Alaska: U.S. Geological Survey Professional Paper 774-A, 17 p.

Patton, W. W., Jr., and B. Csejtey, Jr., 1971, Preliminary geologic investigations of western St. Lawrence Island, Alaska: U.S. Geological Survey Professional Paper 684-C, p. C1–C15.

Patton, W. W., Jr., and B. Csejtey, Jr., 1980, Geologic map of St. Lawrence Island, Alaska: U.S. Geological Survey Miscellaneous Investigations Map I-1203, scale 1:250,000.

Roehler, H. W., and G. D. Stricker, 1979, Stratigraphy and sedimentation of the Torok, Kukpowruk, and Corwin Formations of Cretaceous age in the Kokolik-Utukok River region, National Petroleum Reserve in Alaska: U.S. Geological Survey Open-File Report 79-995, 80 p.

Sable, E. G., and G. D. Stricker, 1985, Coal in National Petroleum Reserve in Alaska (NPRA): Framework geology and resources: AAPG Bulletin, v. 69, p. 677.

Sable, E. G., G. D. Stricker, and R. H. Affolter, 1986, Coal investigations—North Slope of Alaska (abs.), *in* L. M. H. Carter, ed., U.S. Geological Survey research on energy resources—1986: Program and abstracts: U.S. Geological Survey Circular 974, p. 59–60.

Sanders, R. B., 1976, Geology and coal resources of the Bering River field: Focus on Alaska's coal '75: University of Alaska Mineral Industry Research Laboratory Report 37, p. 54–58.

Steffy, D. A., R. F. Turner, G. C. Martin, and T. O. Flett, 1985, Evolution and petroleum geology of the Navarin Basin, Bering Sea, Alaska: Oil and Gas Journal, v. 83, no. 31, p. 116–124.

Tailleur, I. L., 1965, Low-volatile bituminous coal of Mississippian age on the Lisburne Peninsula, northwestern Alaska, *in* Geological Survey research, 1965: U.S. Geological Survey Professional Paper 525-B, p. B34–B38.

Tailleur, I. L., and W. P. Brosge, 1976, Coal resources of northern Alaska may be nation's largest, *in* Focus on Alaska's Coal '75: University of Alaska Mineral Industry Research Laboratory Report 37, p. 219–226.

Tolson, R. B., 1987, Structure and stratigraphy of the Hope Basin, Southern Chukchi Sea, Alaska, this volume.

Turner, R. F., J. G. Bolm, C. M. McCarthy, D. A. Steffy, P. Lowry, and T. O. Flett, 1983a, Geological and operational summary, Norton Sound COST No. 1 well, Norton Sound, Alaska: U.S. Geological Survey Open-File Report 83-124, 164 p.

Turner, R. F., J. G. Bolm, C. M. McCarthy, D. A. Steffy, P. Lowry, T. O. Flett, and D. Blunt, 1983b, Geological and operational summary, Norton Sound COST No. 2 well, Norton Sound, Alaska: U.S. Geological Survey Open-File Report 83-557, p. 154.

Turner, R. F., C. M. McCarthy, C. D. Comer, J. A. Larson, J. G. Bolm, A. C. Banet, Jr., and A. J. Adams, 1984a, Geological and operational summary, St. George Basin COST No. 1 well, Bering Sea, Alaska: U.S. Minerals Management Service OCS Report 84-0016, 105 p.

Turner, R. F., C. M. McCarthy, C. D. Comer, J. A. Larson, J. G. Bolm, T. O. Flett, and A. J. Adams, 1984b, Geological and operational summary, St. George Basin COST No. 2 well, Bering Sea, Alaska: U.S. Minerals Management Service OCS Report 84-0018, 100 p.

Turner, R. F., C. M. McCarthy, D. A. Steffy, M. B. Lynch, G. C. Martin, K. W. Sherwood, T. O. Flett, and A. J. Adams, 1984c, Geological and operational summary, Navarin Basin COST No. 1 well, Bering Sea, Alaska: U.S. Minerals Management Service OCS Report 84-0031, 245 p.

Wang, F. F. H., and V. E. McKelvey, 1976, Marine mineral resources in world mineral supplies—Assessment and perspective: New York, Elsevier, p. 256–258.

Wills, J. C., J. G. Bolm, G. H. Stewart, R. F. Turner, M. B. Lynch, G. W. Petering, J. Parker, and B. Schoof, 1978, Geological and operational summary, Atlantic Richfield Lower Cook Inlet, Alaska, COST Well No. 1: U.S. Geological Survey Open-File Report 78-145, 46 p.

Wood, G. H., Jr., and W. V. Bour, III, in press, Coal map of North America: U.S. Geological Survey, scale 1:5,000,000.

Wood, G. H., Jr., T. M. Kehn, M. D. Carter, and W. C. Culbertson, 1983, Coal resource classification system of the U.S. Geological Survey: U.S. Geological Survey Circular 891, 63 p.

CHAPTER TWENTY-NINE

QUATERNARY DEPOSITIONAL HISTORY AND POTENTIAL SAND AND GRAVEL RESOURCES OF THE ALASKAN CONTINENTAL MARGINS

Peter H. Stauffer

U.S. Geological Survey
Menlo Park, California

ABSTRACT

Future mining of sand and gravel, the largest nonfuel mineral industry in the United States, will increasingly involve offshore deposits. Abundant potential resources exist on the Arctic and Pacific continental margins of Alaska, though exploitation faces problems of technology and environmental consequences. Many processes can produce or transport accumulations of coarse material, but water currents and waves are most active and widespread. Offshore sand and gravel occur both in active (presently subject to movement) and inactive (relict or fossil) deposits; most inactive deposits formed during Pleistocene periods of lower sea level, when much of the continental shelf was exposed and shallow-water processes affected areas now deep. Proper assessment of potential offshore sand and gravel resources of an area requires an understanding of the water and sediment dynamics and their changes through Quaternary time.

The Alaskan offshore regions are treated in four sections—Beaufort Sea, Chukchi Sea, Bering Sea and Aleutian Ridge, and Gulf of Alaska.

In the Beaufort Sea, ice is a dominating influence on sediment dynamics, especially in thick pressure ridges that gouge the sea bottom. Gravel occurs in barrier islands and beaches, sand in sand waves and delta margins. Gravel-island regimen

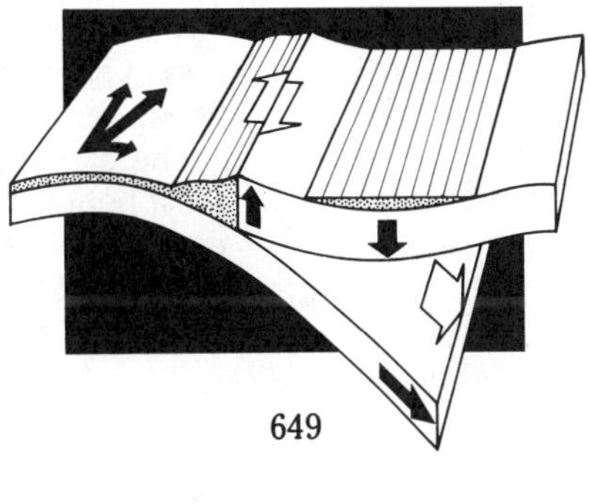

is not well understood, however, and rapid local erosion, partly from thawing of ice-cemented sediment, warns of possible consequences of mining.

On the broad Chukchi Shelf, sand and gravel are abundant mainly within about 100 km of the shore. Blossom Shoals off Icy Cape is a large accumulation that could probably be mined with minimal consequences. Farther offshore, other deposits occur as fillings of a poorly known system of paleovalleys.

The enormous Bering Sea contains a variety of large deposits of coarse material. Huge amounts of fine sand occur in the Yukon delta area and in sand-wave fields in the heads of submarine canyons along the Beringian margin. Coarse sand and gravel are widespread in the northern Bering Sea, but mainly as thin lag deposits or poorly sorted drowned glacial moraines.

The topographically complex Gulf of Alaska is subject to severe storm activity, strong tides, tectonism, and local direct glacial action. Sand and gravel accumulations are varied and some change rapidly. Fields of sand waves occur in lower Cook Inlet and on the Kodiak Shelf, and relict (mainly glacial) gravel is common on the shelf in the northern Gulf.

INTRODUCTION

This chapter attempts to survey sand and gravel deposits on the Arctic and Pacific Ocean continental margins of Alaska, to understand those deposits in the light of Quaternary history, and to assess their potential as resources. The area covered extends from the shoreline out to the deep sea beyond the continental slope. Most attention is paid to the shelf region, where more data are available and in some parts of which exploitation of sand and gravel is already technologically and perhaps economically feasible. Included in this coverage are the Pacific continental margins along the Gulf of Alaska and the Aleutian Ridge, the enormous shelf areas of the Bering and Chukchi Seas, and the northern Alaskan Arctic margin in the Beaufort Sea. These regions have a total area of perhaps 3×10^6 km^2, or about twice the land area of the State of Alaska. The remainder of the United States Pacific margin—off the coasts of Washington, Oregon, and California—is not dealt with in this chapter, which concentrates on the much more extensive frontier areas of the Alaskan offshore. Nonetheless, the proximity of densely populated markets and the more rapid depletion of convenient onshore resources make the conterminous United States margin likely to see significant offshore mining of sand and gravel sooner than the Alaskan margins. Readers interested in the sand and gravel resources offshore of Washington, Oregon, and California may consult Luken and Hess (1979), Martindale and Hess (1979), Moore and Luken (1979), and Osborne, Dango, and Scheidemann (1983).

Over large parts of these areas, few or no data are available on the kind of sediment exposed at the sea floor and, particularly, in the subsurface. Assessment of potential resources therefore involves a large amount of inference, extrapolation, and assumption. No attempt is made here to apply economic considerations or make economic appraisals. Instead, the focus is on trying to understand the sediment-transport processes operating in each region today and how these may have varied in the late Cenozoic, particularly during fluctuations of climate, glaciation, and sea level in the Quaternary. Gaps in understanding and some remaining unanswered questions are pointed out.

Definitions and Uses of Sand and Gravel

Although at times and in certain contexts the term "sand" has been taken to also imply a particular mineral composition, the terms "sand" and "gravel" are generally used to denote size classes of granular material, regardless of composition. Actual size boundaries used to define these terms, however, differ. In industry, the following definitions are used (Goldman and Raining, 1975): "Sand" is material retained on a No. 200 sieve (0.074 mm) and passing a No. 4 sieve (4.76 mm); "gravel" is material retained on a No. 4 sieve (4.76 mm), up to a maximum of 3½ in. (8.9 cm). The upper limit for gravel is sometimes also taken at 3 in. (7.6 cm) (Bates and Jackson, 1980). The definitions of sand and gravel used most commonly in geological studies (Bates and Jackson, 1980) follow the Udden-Wentworth scale of size grades; according to these, sand is material of size between 62.2 μm (0.062 mm) and 2 mm, whereas gravel is any material coarser than 2 mm (sometimes, however, the term gravel is limited to material smaller than 1 cm). In studies of soils, yet other, slightly different ranges are used to define sand and gravel.

The overlap and differences in size ranges for the industrial and scientific definitions of sand and gravel are shown in Figure 1. Even though this chapter deals with sand and gravel as potential resources for industry, the scientific (Udden-Wentworth) definitions will be used throughout. The main reason is that much of the data on offshore sediment have been reported in terms of those size ranges. Specification of actual sizes in mm where possible will allow evaluation in terms of the commercial definitions.

The sand and gravel industry, although considered by most people as an unglamorous type of mining, is an economic activity of great importance and monetary value. In the United States it is the largest nonfuel mineral industry, with an annual production (1970) of around 1 billion tons, valued at more than $1 billion (Goldman and Raining, 1975).

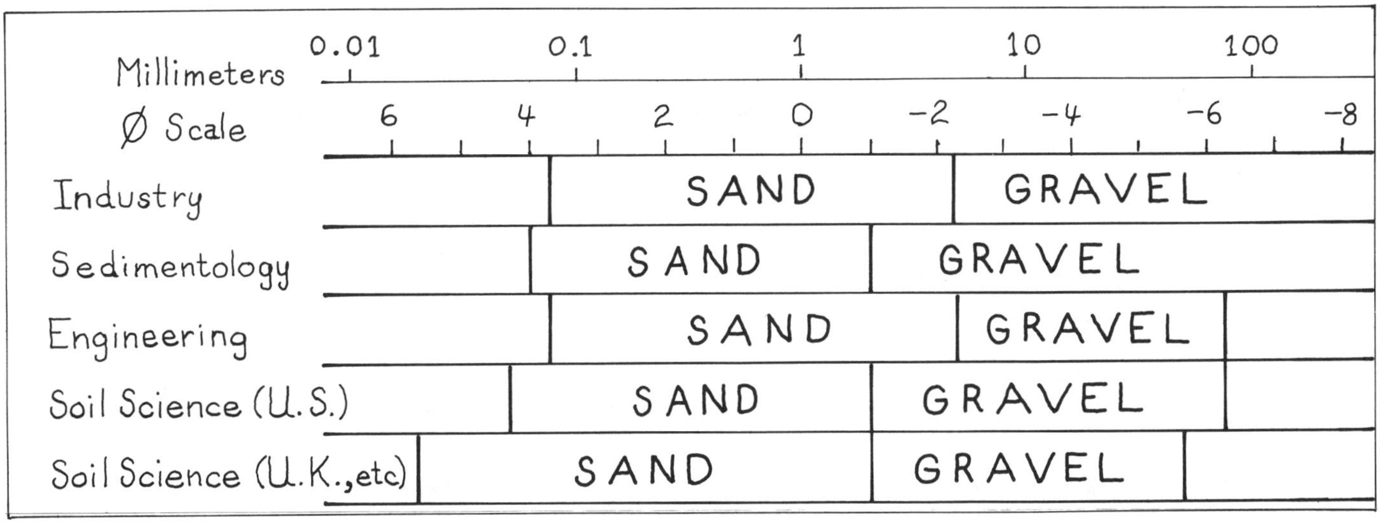

Figure 1. Industrial and scientific size definitions for sand and gravel, shown on a logarithmic scale in millimeters and on the ϕ scale commonly used in the scientific study of sediments. Industry definitions from Goldman and Raining (1975); others from Bates and Jackson (1980).

Nearly all of the sand and gravel produced is used in the construction industry, especially as concrete aggregate and in asphaltic mixtures for paving. A small amount is used for other purposes, such as making glass, for which particular compositions, purity, and other characteristics are required.

Besides being of appropriate grain size, material for concrete aggregate should possess certain other qualities: (1) it should not be too well sorted (<45% between two successive sieves); (2) it should, however, contain no more than 5% of material finer than the No. 200 sieve (0.074 mm); (3) it should not contain significant amounts of mechanically very weak or chemically reactive particles—micas, clays, salts, opal, and some volcanic glasses are among such undesirable constituents (Goldman and Raining, 1975). Most offshore sand and gravel discussed in this chapter probably meets these criteria, but some deposits, such as those rich in authigenic materials like glauconite, may be totally unsuitable.

Offshore exploitation of sand and gravel is still insignificant in United States waters, but some other countries having highly urbanized markets close to the sea and diminishing convenient resources on land are already turning to submarine mining on a large scale. One-eighth of the United Kingdom's production comes from offshore, and nearly one-fifth of Japan's (Bates and Jackson, 1982). Present operations are mostly in water depths of only a few tens of meters, mainly because of technological limitations, but with development of efficient machinery that operates on the sea bottom, extension to depths of 100 m or even more is likely in the near future. Therefore, although this chapter deals mainly with shallow areas of the continental shelf, the entire continental margin, down to even the deep sea floor, is considered.

The vast region of the Alaskan offshore areas will be examined in four segments: (1) the Beaufort Sea, (2) the Chukchi Sea, (3) the Bering Sea and Aleutian Ridge, and (4) the Gulf of Alaska.

Controls on the Distribution of Coarse Clastic Material

Intelligent evaluation of offshore sand and gravel deposits, extrapolations of their extent and shape, and predictions of related other occurrences depend on an understanding of the mode of origin of the deposits. Some accumulations of coarse material are residual and have formed in place by weathering and disintegration of outcropping rocks. Some others represent the coarse residuum of a deposit from which the finer material has been winnowed by currents. Both these types of accumulations tend to be thin (in some places, winnowed coarse material is only a pavement one clast thick), and residual accumulations are commonly very poorly sorted and angular. For these reasons, coarse materials of residual origin are in generally unlikely to represent potential sand and gravel resources (residual concentrations may, of course, have other economic value).

Most accumulations of loose coarse material on the sea floor represent clastic material that has been transported and deposited by a physical mechanism. A number of different mechanisms are able to transport sand and gravel (Table 1), and each tends to impart particular characteristics to the resulting deposit.

In terrestrial environments, the principal transport mechanisms for coarse material are moving water channelized in streams (easily the most important), various kinds of downslope gravitational movements (important in areas of steep slopes, but mainly serving to feed material to rivers), moving glacial ice (dominant in glaciated areas, but restricted to them), and wind action (important in areas lacking vegetative cover, but rarely moving material coarser than fine sand). Of these mechanisms, wind action imparts the highest degree of size sorting, producing deposits of very limited size ranges; gravitational mechanisms tend not to change the sorting of the material involved, merely transporting whatever assemblage of particles began to move downslope; and glacial ice imparts essentially no sorting to the material, characteristically depositing a mixture of all grain sizes from clay to large boulders. Water currents in rivers, by contrast, quickly impart at least a moderate degree of sorting to the material they transport—grains too large to move are left behind, whereas clay and silt tend to go into suspension and therefore continue moving when the bedload is deposited. The material transported by rivers to their mouths constitutes the main source of clastic sediment in the marine environment, where other mechanisms may continue

Table 1. Natural mechanisms for transporting and depositing sand and gravel.

Physical Agency	Maximum Particle Size	Sorting in Deposits	Depth Limits in Sea: Active	Depth Limits in Sea: Relict	Possible Examples of Deposits: Offshore	Possible Examples of Deposits: Shoreline/Nearshore
Grounded glacial ice	No limit	Very poor	9/10 ice thickness	~130 m + 9/10 ice thickness	Pleistocene moraines	Pleistocene and Holocene moraines
Floating glacial ice	No limit	Very poor for clasts larger than moved by currents	None	None	Quaternary glacial marine sediments	Pleistocene glacial marine sediments
River water flow	Commonly silt to gravel	Moderate	Channel depth	~130 m + channel depth	Pleistocene channel fill	Estuarine and river-mouth deposits
Marine bottom currents	Commonly silt to fine gravel	Good	Commonly 200 m	~130 m + 200 m	Sand waves, ripples, leeside shoals, winnowed lags	Intertidal sand bars
Surface wave action	Fine sand to cobbles	Good to excellent	'Wave base' (60 m?)	~130 m + 'wave base'	Symmetrical sand waves, ripples	Beaches, offshore bars
Internal waves	Commonly fine sand	Good to excellent	Unknown (none?)	None?	Sand-wave fields	None?
Subaqueous mass movements	No limit	Unchanged in cohesive slides to moderate in turbidity-current deposits	None	None	Turbidites, mass-flow deposits, slide masses	Sand flows
Wind action	Commonly silt to fine sand	Excellent	(None)	~130 m for dunes	Pleistocene dunes	Pleistocene dunes

transporting it to a final site of deposition. For most rivers at present the coarser (bedload) material is deposited in their estuaries and does not reach the marine environment proper.

For transporting and depositing coarse material in the marine environment, ice is important only as floating ice masses, from which sediment may rain onto the sea bottom as it is released by melting (although terrestrial glaciers can extend seaward until their base is a distance below sea level equal to about nine-tenths the ice thickness, at which point buoyancy equals the weight of the ice mass and it will float). Such ice-rafted sediment tends to be nearly as poorly sorted as direct glacial deposits, except that ambient marine currents will carry away the finest sizes up to a maximum that depends on the energy level; ice-rafted sediment is also typically diluted by other (finer grained) material transported to the site by other means.

Water in the sea transports and deposits coarse material by wave action, by bottom currents, and by involvement in sediment gravity flows. The action of surface waves, dominant along shorelines and in very shallow water, tends to impart good to excellent sorting to the sediment because it is continual and repetitive. Such wave action is limited to shallow depths down to wave base (60–100 m in most open marine areas, locally apparently deeper, but commonly shallower in sheltered sites), but internal waves at interfaces between water masses of differing densities can operate at any depth. Such internal waves are, however, generally able to move only very fine sand and silt (Karl, Cacchione, and Carlson, 1986). Bottom currents capable of moving sand or gravel include both oscillatory (tidal) currents and unidirectional (geostrophic) currents; these are also restricted to shallow water depths, rarely being significant deeper than a few tens of meters. They operate somewhat like river currents but are not channelized. Where the bottom topography provides an appropriate constriction, they can become concentrated and consequently accelerated to higher velocities. Like river currents, they tend to impart moderate to good sorting to the material transported.

Sediment gravity flows represent the only way in which coarse material can be transported into very deep water, other than by rafting. These mechanisms span a spectrum from dilute dispersions of particles in turbulent flows (turbidity currents), through several varieties of more concentrated and less turbulent flows, to flows moving largely as cohesive and rigid plugs (debris flows and slumps). Sediment gravity flows either do not change the sorting of the material involved (debris flows) or tend to impart improved sorting and separate material by size both laterally and vertically (turbidity currents).

Deposits found on the sea bottom today are not necessarily the products of marine mechanisms. During the Pleistocene, global sea level repeatedly fell to 100–200 m below its present position (an arbitrary average of 130 m is assumed in Table 1). At such times, much or all of the continental shelf was exposed and subject to terrestrial mechanisms of transport. These low sea levels were caused by glaciation, and glacial transport of material was most widespread at such times. Many fluvial, glacial, and eolian deposits that formed during low stands of sea level are now preserved at and below the sea bottom in shelf areas. Further complications have been introduced by later marine reworking of parts of such deposits, and by local tectonic uplift and subsidence, which have made the history of emergence and submergence different in different areas.

Active (that is, presently moving in transport) deposits of sand and gravel on the sea bottom are, by definition, the product (at least now) of marine forces, whereas inactive and relict deposits may have formed from marine or terrestrial activities. Each of these two types has certain advantages as potential sources of sand or gravel. Active deposits are loose and uncemented (in the Arctic this includes the possibility of ice-bonding); they are naturally replenishable, and mining them may therefore be less disruptive than mining fossil or relict deposits, because the changes produced by mining are reversible and similar to those already occurring naturally. Where a sediment transport system is active, however, the consequences of disturbing that system are not always predictable, and the processes of that system may complicate or even endanger the mining operations. Relict (inactive or fossil) deposits are, by contrast, partially cemented in some places; they are not naturally replenishable, and mining them is therefore an irreversible process. Although ongoing sedimentary processes may not be directly disrupted by mining of relict deposits, some such deposits may serve to protect underlying material, and hence mining them may initiate erosion of the sea bottom. Evaluating the probable environmental effects of exploiting any submarine sand or gravel deposit requires a good understanding of the water and sediment dynamics of the area.

BEAUFORT SEA

The continental margin along the Beaufort Sea, stretching from Point Barrow 600 km to the Canadian border and beyond, has a narrow (50–100 km) shelf, with an inner shelf break at about the 60–m depth, a further complex break in slope at 200–800 m, and a steep and fairly straight continental slope leading down to abyssal depths in the Canada Basin of the Arctic Ocean. Average gradients on the inner shelf and the adjoining coastal plain are generally 0.5 to 1 m/km, but they steepen to 3 to 10 m/km between 60 and 100 m and on the continental slope locally reach 100 m/km.

This continental margin is interpreted as an Atlantic-type margin produced by a rifting event in the Mesozoic (Grantz, Eittreim, and Dinter, 1979; Grantz and May, 1983; Grantz, May, and Dinter, this volume). Paleozoic and Mesozoic sedimentary rocks, partly derived from a northern (prerift) source, are overlain by a Mesozoic and Cenozoic sedimentary sequence (Brookian sequence) shed northward from the rising Brooks Range, mainly after the northern landmass rifted away to produce the present continental edge (Grantz, Eittreim, and Dinter, 1979; Grantz, May, and Dinter, this volume).

Several kilometers of northward-thickening and generally flat-lying Brookian sedimentary rocks underlie the Beaufort Shelf, giving way on the slope and rise to a complex and much disturbed section with large slumps and diapiric structures. This disturbed section in turn grades into a thick (minimum 4.5 km) flat-lying sedimentary section, the distal part of the Mackenzie sediment cone, under the deep Canada Basin (Eittreim and Grantz, 1979).

Quaternary deposits unconformably overlie the Brookian sequence. They are as thick as 100 m or more on the shelf and are characterized by regressive-transgressive cycles related to glacio-eustatic sea-level changes in the Pleistocene (Craig and Thrasher, 1982). There is no evidence that the Beaufort coastal plain or shelf was ever glaciated, and these Pleistocene sediments consist of alluvial, lacustrine, eolian, tundra, marine, and glacio-marine deposits, generally included within the late Pliocene and Pleistocene Gubik Formation onshore (Black, 1964; Hopkins and Hartz, 1978b). The Pleistocene Flaxman Member of the Gubik Formation crops out on some parts of the coastal plain and shelf (Dinter, 1985). This unit consists of glacio-marine sand and mud containing striated glacial megaclasts, as large as 2 m, of rocks exotic to Alaska and probably derived from the Amundsen Gulf in the Canadian Arctic (Rodeick, 1979; Hopkins and Hartz, 1978a).

Subaerial exposure of the shelf during low sea-level stands in the Pleistocene has led to overconsolidation of earlier Pleistocene mud, erosion and subsequent back-filling of alluvial channels, and formation of deep permafrost. Since the latest (Holocene) transgression, permafrost offshore has been thawed to depths of tens to hundreds of meters, and a variable but thin sequence of beach and marine sediment has been laid down. The Holocene sediment generally consists of unconsolidated mud (0 to 10 m) underlain by a transgressive basal layer of sandy gravel (beach?) typically only 1 to 2 m thick. In some areas, Holocene mud has also become overconsolidated and stiff, possibly from seasonal freezing of the bottom (Reminitz et al., 1980). Sets of active barrier islands and sandy shoals and bedforms have developed on the inner-shelf to mid-shelf area.

Present Sedimentary Processes on the Beaufort Shelf

The Beaufort Sea, particularly the shallow part of the shelf, is an environment of strange and seemingly paradoxical present-day sedimentary processes. Many sediment masses are relict from glacial times, when processes were quite different and, ironically, involved less direct ice action (Figures 2, 3). Understanding of these processes has come about through the work of Barnes, Reimnitz, and co-workers (e.g., Barnes and Reimnitz, 1974); Hopkins, Hartz, and co-workers (e.g., Smith, Hartz, and Hopkins, 1980); Rodeick (1979); and others.

Ice is a dominant influence in the Beaufort Sea (Barnes and Reimnitz, 1974; Reimnitz and Barnes, 1974; Barnes, Rearic, and Reimnitz, 1985). The sea is ice-covered for most of the year; open water is found on the shelf and sometimes beyond at most from mid-July to early October, and the deep Arctic Ocean is always covered by pack ice (Hopkins and Hartz, 1978b; Reimnitz, Toimil, and Barnes, 1978). Stationary winter ice on the inner shelf consists of bottom-fast ice out to a depth of about 2 m and floating fast ice out to about the 20–m isobath; the polar pack ice farther out moves steadily westward at 2 to 3 km/d as part of the clockwise Pacific

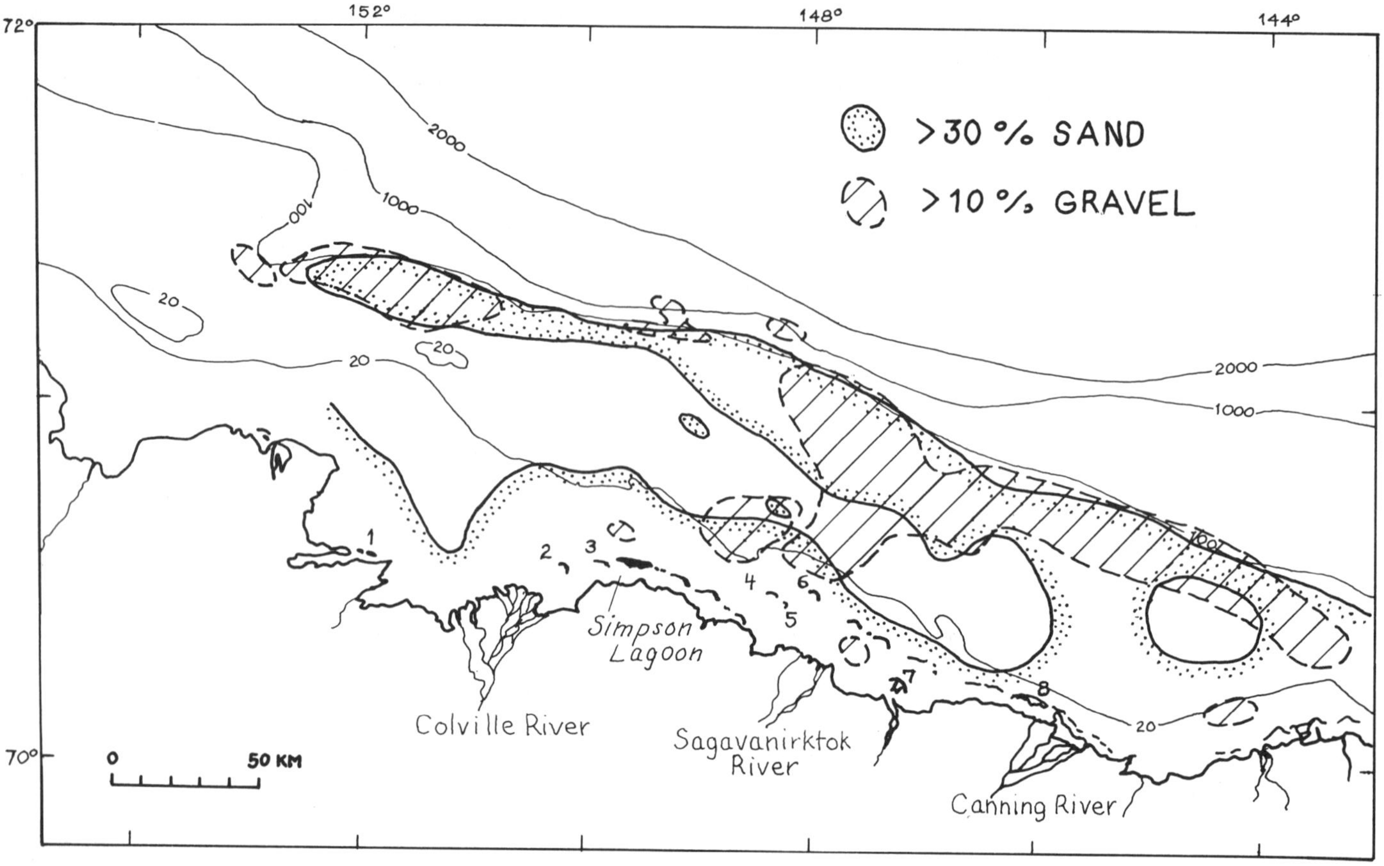

Figure 2. Beaufort Sea off the north coast of Alaska, showing areas of coarse bottom sediment. Areas where the sediment is >10% gravel after Barnes (1974) and Rodeick (1979); areas where it is >30% sand after unpublished work of P. W. Barnes, reproduced in Briggs (1983). Also shown are some localities mentioned in the text; islands are indicated by number, as follows: 1, Eskimo Islands; 2, Thetis Island; 3, Spy Island; 4, Reindeer Island; 5, Argo Island; 6, Cross Island; 7, Tigvariak Island; 8, Flaxman Island.

gyre in the Arctic Ocean. In the zone of slippage and shearing between polar pack and fast ice (the "stamukhi zone," as much as 50 km wide), pressure ridges and pileups of ice tend to become grounded. Though fast ice protects underlying sediment from wave action (not from currents) in winter, grounding of moving ice masses in the stamukhi zone and on the outer shelf, and farther inshore after ice breakup, results in intense gouging of the bottom and produces furrows and gouges as much as 5.5 m deep (though commonly only about 0.5 m). Gouges in shallow water (<20 m) can be smoothed and filled in a single season (Reimnitz and Barnes, 1974), and the recurrence interval for gouging is estimated at about 25 years (E. Reimnitz, 1983, oral communication). The upper few meters of sediment (in most places, this means all or most of the Holocene), from about the 20–m isobath out to the shelf edge at 50 to 70 m, consists entirely of ice-disturbed material and fillings of gouges (Reimnitz and Barnes, 1974). Ice pileups and beached ice masses also push up sea-bottom sediment and add it to the beaches and barrier islands, as well as reshaping the beach itself (Kovacs, 1983). During the annual breakup, ice also transports sediment incorporated in it, but very little coarse-grained material is transported in this way (Reimnitz and Ross, 1979). Ice-rafting of coarse material into the Beaufort Sea from distant areas occurs (Kovacs, 1983) but is insignificant at present (Barnes and Reimnitz, 1974) and made no real contribution to the Holocene transgressive sequence, in sharp contrast with the Flaxman Member, which formed during a Pleistocene transgression to about present sea level. Reasons for such differences are not clear.

Despite the controlling role of the ice regime, however, erosion and sediment transport by waves and currents are important. Although the fetch for wind-driven waves during the few months of open water is rarely as much as 100 km and the tidal range is only 10 to 20 cm (U.S. Department of Commerce, 1985), coastal erosion is rapid. Melting of ice in otherwise unconsolidated materials leads to collapse and lateral flow (thermokarst erosion) (Hopkins and Hartz, 1978b), and much erosion occurs during late summer storms that can generate a 2–m swell and storm surges that reach 3 m above normal sea level. Shoreline retreat rates are generally on the order of 1 m/yr; however, Lewellen (1977) measured rates of 3 to 4 m/yr for a 20–year period at Flaxman Island, and short-term rates on some points and capes reach 30 m/yr (Leffingwell, 1919; Hopkins and Hartz, 1978b). Despite this rapid coastal erosion, the beaches that derive their material mainly from it remain narrow and thin, because most of the eroded material is fine grained and goes off in suspension, and coarse material on the lower shoreface is presumably left behind as the beach migrates landward.

Bottom currents erode potholes ("ice-wallows") as much as to 2 to 3 m deep and 25 to 50 m across around grounded ice masses (Reimnitz and Kempema, 1982) and smooth and fill gouges formed by grounded ice in shallow depths (<20 m) during a single season (Barnes and Reimnitz, 1974). Ice

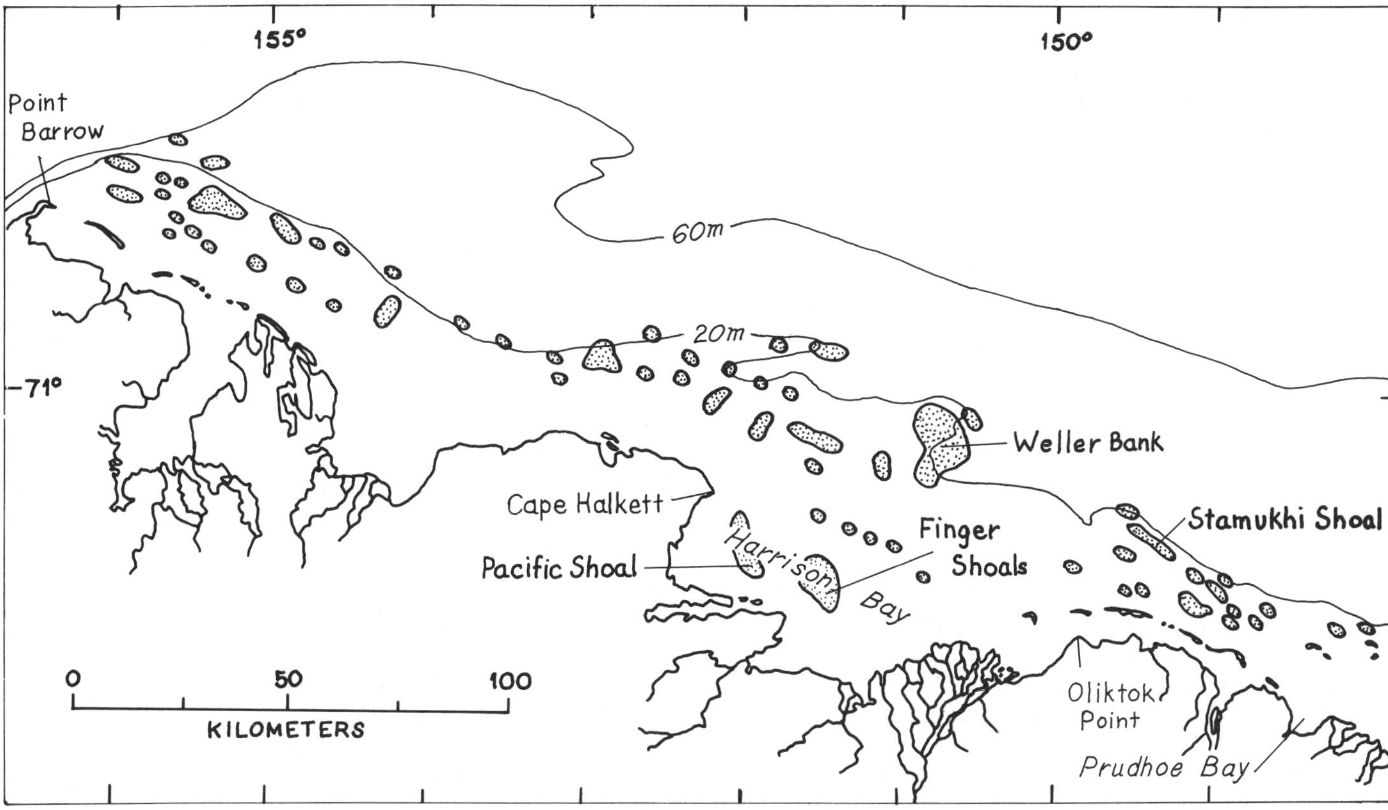

Figure 3. Beaufort Sea off the north coast of Alaska, showing prominent belt of shoals near the 20-m isobath. Shoals were mapped by tracing all isolated shoals covered by <11 m of water on NOS Chart No. 16004, except for Stamukhi Shoal, which was separately mapped. After Reimnitz and Kempema (1985).

gouging dominates in the stamukhi zone around the 20–m isobath and seaward, and nearly all box cores taken there bring up disturbed material. Beyond depths of about 50 m, bedded sediment in box cores shows reworking (Reimnitz and Barnes, 1974), and near the shelf edge sporadically strong water movement is indicated by apparently active bedforms and measured current pulses up to 70 cm/sec at 10 m off the bottom.

Ridges and dunes of clean sand on the inner shelf lack ice gouges on their crests and appear to represent active hydraulic bedforms (Barnes, Reminitz, and Naidu, 1981). These forms, which occur mostly in water depths of <5 m, include sand ridges 1 to 2 m high and 100 m wide in a belt near the west end of Simpson Lagoon (Figure 2) (Reimnitz et al., 1980), and the linear sand waves 1.5 m high and 200 m wide on Finger Shoals (Figure 3) in Harrison Bay (E. Reimnitz and P. Minkler, unpublished work).

Evidence of active sediment transport by bottom currents and waves is also provided by the disappearance of ice gouges from one year to the next, as observed in Harrison Bay, and by the filling of "strudel scours" produced by rapid violent draining of spring fluvial runoff through holes in the sea ice in delta-front areas as much as 10 km from shore. Strudel scours as deep as 6 m and as wide as 25 m fill by bedload transport in 2 to 3 years, indicating a net bedload transport rate of about 9 m^3/m/yr (Reimnitz and Kempema, 1983).

Transport of sediment is dominantly westward because of the prevailing easterly winds and consequent water currents. Brief reversals of direction during some storms may produce unusual amounts of erosion but are probably insignificant in the overall movement of material.

Rivers at present bring virtually no coarse sand or gravel onto the Beaufort Shelf. Though major streams have their headwaters in the high terrain of the Brooks Range and their present beds contain coarse sand and gravel well down onto the coastal plain, all coarse material is apparently deposited 1 to 2 km short of the coast (D. M. Hopkins, cited in Briggs, 1983). During spring runoff, as much as 75% of the annual discharge of rivers occurs in a few weeks (Carlson, 1977); flood waters spread out on top of the ice-covered shelf and eventually drain through scattered holes. The small sediment load consists entirely of fine sand and mud (E. Reimnitz, 1983, oral communication). This absence of supply of coarse sediment to the sea by rivers implies that sand and gravel shoals, barrier islands, and beaches, if eroded or quarried, can be renewed only from the sea bottom itself or from erosion of low coastal bluffs.

Any attempt to assess gravel and sand resources in the Beaufort Sea must take account of these peculiar conditions of sediment dynamics and the major uncertainties concerning them. Previous assessments have been made by Molnia (1979a), Hopkins (1981), and Briggs (1983).

Active Sand and Gravel Bodies on the Beaufort Shelf

Sand and gravel occur in numerous dynamically active accumulations from the shoreline out to the midshelf in the Beaufort Sea and in sparser and sometimes doubtfully active bodies in the outer-shelf and shelf-break regions.

Beaches

Along the mainland of the Beaufort coast, beaches are thin and narrow and are composed of coarse-grained gravelly material. Most of the coast is retreating at least 1 m/y, and in many sections the rate is several times as fast. At Esood (given as "Ksood" by Kovacs), west of Cape Halkett (Figure 3), air photos prove 410 m of retreat in 32 years, an average rate of 12.8 m/yr (Kovacs, 1983).

Longshore movement of beach sediment by waves is limited to the short open-water season and occurs mainly during storms. Estimated annual rates on the beach itself are on the order of 10^5 m^3 (Short, 1975; Nummedal, 1979). Ice push also affects beaches and coasts (Leffingwell, 1919), though commonly only in a small way. Perhaps once or twice in a century, major storms drive ice onshore, leaving scours and pushed-up ridges and mounds of material as far as 100 m inland (Kovacs, 1983). Such events cause disruption and reorganization of beaches and hasten coastal erosion; ice-push scours may even end up below sea level and so become ponds behind a reconstructed beach (Kovacs, 1983).

Barrier Islands

About one-third of the 600 km of coast from Point Barrow to the Canadian border is fronted by discontinuous chains of barrier islands, mostly 5 to 20 km offshore (Figures 2, 3). The main chains are (1) from Cape Simpson to Point Barrow (70 km); (2) from Prudhoe Bay along Simpson Lagoon possibly to Thetis Island (70 km); and (3) from Brownlow Point to Cross Island, with possible western outliers of Argo and Reindeer Islands (total 95 km). Smaller groups and isolated islands occur in other areas.

These islands have various origins. Flaxman and Tigvariak Islands, erosional remnants of the Pleistocene Flaxman Member, stand 3 to 10 m above sea level. Erosion of the low bluffs provides the coarse material that forms their beaches (Hopkins and Hartz, 1978b). This coarse-grained material is mainly of the exotic rock types in the Flaxman, also characteristic of gravel on the middle and outer Beaufort Shelf; Rodeick (1979) called this the "dolomite facies." Such gravel characterizes the whole chain from Flaxman to Reindeer Island, though coarse cobbles are known only at Flaxman Island and a few other islands (Hopkins and Hartz, 1978b), on the inshore sides of which are sea-bottom boulders interpreted as lag deposits from eroded Flaxman Member (Reimnitz and Ross, 1979). Constructional processes and longshore transport have probably helped form the other islands in this chain. Reindeer and Argo Islands at the west end of the chain are displaced about 7 km landward from the main line and appear to be rapidly migrating toward the shore after being freed by complete erosion of their original Flaxman cores. Location of outcrops of the Flaxman and derived islands may have been determined depositionally. Leffingwell (1919), noting that Flaxman Island stands higher than the adjacent mainland shore, suggested that the Flaxman represents the patchy, local deposits of a shore-parallel belt of grounded icebergs; this theory has yet to be improved upon.

The Eskimo Islands in Harrison Bay and part of the chain along Simpson Lagoon appear to be fragmented and partly drowned offshore continuations of mainland gravel ridges representing barrier beaches of the Pelukian (Sangamon interglaciation) transgression (Hopkins and Hartz, 1978b). The other offshore islands along the Beaufort coast are low (< 3 m) and generally are believed to be of constructional origin; Short's (1979) interpretation of the chain from Point Barrow to Cape Simpson as an early Holocene shoreline, now isolated by thaw erosion behind the beach ridges, assumes slow or nonexistent landward migration of the islands, an assumption not supported by the data. The clasts in these islands are dominantly of Brooks Range derivation, belonging to Rodeick's (1979) "chert facies," though they contain various amounts of clasts characteristic of the Flaxman.

The regimen of these islands is not well understood. Tides are too small for ebb-tide deltas to form at channels. All dynamic processes affecting the islands have a net landward component, and so the islands should be highly transgressive and migrate rapidly shoreward (Nummedal, 1979). Measurements indicate migration rates of 19 to 30 m/yr westward and 3 to 7 m/yr landward, or 4 to 7 m/yr to the southwest for isolated, arcuate islands (Hopkins and Hartz, 1978b). Rapid landward migration is supported by the presence beneath Reindeer Island of a subsurface "tail" of bonded permafrost that reaches depths greater than 30 m and distances more than 2 km seaward of the island (Smith, Hartz, and Hopkins, 1980).

If such rapid migration has been going on since sea level reached approximately its present level more than 5,000 years ago, the islands' original positions could have been at least 25–30 km seaward of their present locations. Because this migration is generally faster than that of the mainland coast, the islands would have to have formed not as coastal beaches but as offshore features, either relict or constructional.

Barrier island formation by upward growth of offshore submarine bars is unlikely in the Beaufort Sea, at least at present, because of the low level of wave energy. This low wave energy also poses the problem of the coarseness of the material in barrier islands of the Arctic coast. Whereas beaches and barriers outside of the Arctic are typically composed of unimodal sand (Bascom, 1959), the Arctic barrier islands of Alaska are made of a distinctively bimodal sediment—dominant gravel and subordinate sand. Why is it so coarse grained? Reimnitz and Maurer (1979) ruled out energetic wave action, supply of gravel by rivers, ice-rafting of coarse material, and Pleistocene glaciation, all of which are either absent or quantitatively unimportant on the Beaufort Shelf. Wind removal of sand, noticed by Leffingwell (1919) and emphasized by Reimnitz and Maurer (1979), undoubtedly accomplishes some coarsening of the material.

Supply of coarse-grained material to the islands may be mainly by the bulldozing action of grounded sea-ice masses, which is known to be an important process on the Beaufort Shelf (Reimnitz and Barnes, 1974; Barnes, Rearic, and Reimnitz, 1985) and is able to push sediment up onto the shore (Kovacs, 1983). However, vibracores taken seaward of Reindeer Island show gravel only close inshore, with clean sand and overconsolidated silty clay farther offshore. Reasons for the prominent bimodality of the Beaufort barrier beaches remain unknown.

Sea-Bottom Sediment on the Beaufort Shelf

Sampling programs on the Beaufort Shelf since the early 1970s have yielded a fairly detailed picture of the surface sediment. In general, the mean grain size decreases from nearshore areas, where it is mostly in the range of very fine sand to coarse silt (0.015 to 0.125 mm) to the continental slope, seaward of the 1000-m isobath, where it is near the silt/clay boundary (0.004 mm) or finer. Nearshore and slope sediments typically have moderate sorting, but the middle and outer shelf are characterized by mixed sediments with very poor sorting (Barnes, Reimnitz, and Ross, 1980; Naidu et al., 1981). Areas of lag gravel on the outer shelf are exceptions to both these generalizations (Figure 2). The highly mixed sediments of the middle and outer shelf commonly contain scattered pebbles in a structureless sandy mud (Barnes et al., 1979) and are a result of intense gouging and overturning by drifting ice masses (Barnes, Rearic, and Reimnitz, 1985).

Transport and winnowing of coarse sediment by currents and wave action are known, in general, only indirectly. While the sea is frozen (normally October to June), water currents under the ice of the inner shelf are extremely weak (Thomas, 1981). During the short open-water season, however, currents are quite active, as shown by movement of drifters (P. W. Barnes and E. Reimnitz, unpublished work) and the rapid infilling of strudel-scour depressions (Reimnitz and Kempema, 1983). Very strong water movement during storms is evidenced by erosion of ice-gouged material (Reimnitz et al., 1980) and by coarse-grained hydraulic bedforms (Barnes, Reimnitz, and Naidu, 1981).

E. Reimnitz and E. W. Kempema (unpublished work) estimated that maximum wave conditions (swell of 2 to 2.5 m) could cause grains 7 mm in size to move at water depths of 15 m. Bouldery lag deposits in some nearshore areas (Reimnitz and Ross, 1979), even in front of the Sagavanirktok River mouth (Figure 2), indicate active erosion and movement of sediment; this conflicts with Naidu and Weiss' (1981) ^{210}Pb-activity determinations of high sedimentation rates (about 1 cm/yr in Simpson Lagoon), unless those are short-term rates compensated by occasional major erosional events.

Delta platforms marked by a prominent 2-m bench off the Colville and several smaller rivers are covered by well-sorted sands, which are commonly cross-bedded and entirely lacking gravel (Barnes et al., 1979); about 1.5×10^6 m^3 of muddy sand occurs along the outer fringe of the 2-m bench in the Colville delta (Hopkins, 1981).

Sand Waves and Shoals

Hydraulic bedforms of sand form the characteristic minor relief features on the Beaufort Shelf landward of the 12-m isobath (Craig and Thrasher, 1982). A sand-wave field 5 km north of Spy Island (Figure 2), studied in detail since 1973 (Reimnitz et al., 1980), has north–south waves with crest lengths of 100 to 300 m, wavelengths of 10 to 25 m, and heights of about 1.5 m. The pattern of individual waves, each of which may contain from 500 m^3 to >5000 m^3 of sand, changes from year to year, but the field remains in the same place, and there is no known source of supply. Between sand bodies, the sea bottom is stiff silty clay, commonly having a fluted surface, and rounded mud balls. Similar sand waves occur seaward of Reindeer and Argo Islands (Reimnitz et al., 1980), whereas others having crest lengths 125 to 250 m, wave lengths 50 to 200 m, and heights 0.5 to 1 m are found in patches in the lee of Stamukhi Shoal (Figure 3) (Reimnitz and Kempema, 1985).

Finger Shoals, in about 5-m water depth in western Harrison Bay (Figure 3), is a field of linear sand waves (E. Reimnitz and P. Minkler, unpublished work) composed of fine-grained sand, oriented north–south, and having wavelengths of about 200 m and heights of about 1.5 m. They are symmetrical in cross-section and are locally branching, which implies that the field is related to wave action or alternating currents and is not actively migrating. The troughs between the waves expose stiff silty clay—despite the proximity to the mouth of the Colville River, there would appear to be no net deposition. Clean, cross-bedded sands, possibly from similar hydraulic bedforms, are commonly seen in vibracores seaward of barrier islands (Barnes et al., 1979).

Larger bodies of coarse material form shoals of generally uncertain origin with relief as much as 10 m on the Beaufort Shelf in water depths mainly <20 m. A series of linear shoals just landward of the 20-m isobath, typified by Stamukhi Shoal (Reimnitz and Kempema, 1985), localizes the formation of grounded ice ridges and thereby importantly influences the ice zonation of the shelf and the position of the stamukhi zone. The crest of Stamukhi Shoal, at water depths of 12 to 15 m, shows fields of slightly sinuous ripples (wavelength, 1 to 1.5 m; height, 8 to 10 cm) in fine gravel (pebbles 2 to 3 cm, rarely 6 cm), interspersed with areas of medium-grained sand in symmetrical ripples. These bedforms are clearly active, because they change from year to year and would be soon destroyed by ice gouging if relict. Scouring of the tops of these shoals by passing ice masses must be compensated by some constructional process. Most likely the shoals are maintained by ice-push bringing up mixed gravel-bearing sediments, which are then winnowed of fine-grained material by waves and currents. The original formation of these shoals remains obscure (Reimnitz and Kempema, 1985). The gravel clasts in them belong to Rodeick's (1979) "dolomite facies" of exotic rock types, and so their original deposition was probably by ice rafting.

Pacific Shoal, northwest of Finger Shoals (Figure 3), is a broad body composed mainly of sand (Hopkins, 1981), and several smaller shoals adjacent to or as much as 5 km seaward of some of the islands are covered by partly rippled sand and gravel. Sea-bottom ridges seaward of Cottle Island are capped by sand and gravel (Barnes et al., 1979). Other shoals, including a few in water >30 m deep, are known only from bathymetric data (Reimnitz et al., 1982).

The largest of the known shoals is Weller Bank (Figure 3), which straddles the 20-m isobath in outer Harrison Bay. It lies within the general belt of possibly ice-related shoals (Reimnitz and Kempema, 1985) but is much more extensive, having an area of >100 km^2 above the 18-m isobath. Its top, at a water depth of about 10 m, consists mainly of clean sand, with limited areas of rippled gravel on the east side (P. W. Barnes and T. Reiss, unpublished work). Although the gravel may have the same history as that in the smaller shoals, the great extent of Weller Bank has led to the suggestion (D. M. Hopkins, 1984, oral communication) that its core may be a field of Pleistocene eolian dunes similar to the large field on the mainland south of Teshekpuk Lake (Black, 1951; Carter and Robinson, 1978), which became stabilized during Holocene sea-level rise (Carter and Galloway, 1979). A single industry bore hole on Weller Bank, however, found mainly mud beneath only 2 m of sand, which suggests a different origin (P. W. Barnes, 1984, oral communication).

Surface Gravel on the Outer Shelf

Surface gravel not associated with conspicuous shoals has been reported from a number of sites on the Beaufort Shelf, especially eastward from Prudhoe Bay and along the shelf break near the 60-m isobath (Figure 2). On the outer and middle shelf, this gravel is always the exotic "dolomite facies" of Rodeick (1979), which consists of angular, commonly striated clasts having a dominant size of about 10 cm. Especially widespread near the shelf break at about 50 to 70 m, it has also been sampled at a depth of >100 m (Barnes and Reimnitz, 1974). Where penetrated by box cores, this gravel is thin (commonly no more than 10 cm) and overlies laminated to massive marine mud. Mowatt and Naidu (1974) interpreted this gravel as a lag deposit of ice-rafted clasts; however, if it is of Holocene age, as implied by a 15,000-yr age obtained for underlying marine mud at one site (Rodeick, 1979), it would require rates of ice-rafting transport higher than present rates (Barnes and Reimnitz, 1974).

That currents along the shelf edge can be strong enough to winnow finer material from this gravel and prevent its burial by mud is indicated by current pulses, measured by Aagard, of up to 70 cm/sec 10 m off the bottom in a water depth of 60 m north of Prudhoe Bay (Reimnitz, Barnes, and Phillips, 1985). Holocene sediment accumulations near the shelf edge must be very small.

Why this Holocene lag gravel is most conspicuous at depths of 50 to 70 m remains unclear. Perhaps bottom currents strong enough to produce a lag gravel are concentrated in that depth zone, or the deposit may represent an ice-rafted deposit similar to the Pleistocene Flaxman Member.

Fossil Sand and Gravel Bodies in the Beaufort Sea

Our knowledge of fossil sand and gravel accumulations beneath the sediment surface in the Beaufort Sea, both as to their origin and their present distribution, is still largely speculative and is based on seismic profiles, scattered bore holes, and inferences concerning the origin of the bodies and the geologic history of the area. Several distinct types of deposits, however, are clearly present.

Holocene Transgressive Gravel Sheet

The base of the Holocene marine sequence offshore is typically a thin (0 to 2 m) layer of sandy gravel and pebbly sand, commonly containing shells, which is interpreted to represent the discontinuous diachronous basal beach deposit of this transgressive sequence. In nearshore areas, it is difficult to distinguish from underlying fluvial outwash gravels, from which it was probably largely derived (Hopkins and Hartz, 1978a; Smith and Hopkins, 1979; P. A. Smith, unpublished work). Overburden on top of this deposit is thin (0 to 10 m) and generally consists of soft or locally overconsolidated mud (Reimnitz et al., 1980).

Glacial-Period Eolian Dune Fields

During Pleistocene low sea-level stands, fields of eolian dunes are known to have formed downwind from major river valleys on the present Beaufort coastal plain, and similar fields may have formed on the then-exposed continental shelf. The 100-km^2 Weller Bank (Figure 3) has been suggested to represent one such field, though the first drill-hole data do not support this opinion. A shoal of this size, if underlain by a few meters of clean sand, would represent a mass on the order of 10^8 m^3 of such material, with essentially no overburden.

Interglacial Beach/Barrier Ridges

Gravel ridges on the coastal plain west of Cape Halkett may represent (barrier?) beaches formed during the Pelukian transgression (Sangamon interglaciation) about 120,000 years ago (Hopkins and Hartz, 1978b). This line of ancient beach ridges is inferred to continue in the Eskimo Islands of western Harrison Bay (Figure 2) and again in some of the islands north of Simpson Lagoon (Hopkins and Hartz, 1978b), where it has been modified by modern barrier-island processes. Submerged parts of this string might be searched for in the gaps and extensions of this trend.

Pleistocene Fluvial (Outwash) Deposits

The Beaufort coastal plain is largely built of coalescing fluvioglacial outwash fans derived from the glaciated Brooks Range during the Pleistocene, and similar coarse deposits underlie at least some nearshore parts of the Beaufort Shelf (Hopkins and Hartz, 1978b). During glacial periods, streams evidently carried gravel north of the present shoreline, and such Brooks Range gravel ("chert facies" of Rodeick, 1979) is encountered in bore holes offshore as well as on land.

In general, however, such subsurface gravel gives way to finer grained deposits within a short distance of the present shoreline. Numerous shallow (25 m) industry bore holes in the Colville delta area show gravel giving way to sand and fine material about 5 km offshore. Several deeper (30 to 90 m) bore holes 1 to 5 km seaward of Flaxman Island and those as far as 20 km west (i.e., 10 to 20 km offshore of the Canning River fan) show 1 to 6 m of Holocene marine sediment underlain by a series of stacked similar marine interglacial/interstadial sequences—below the Flaxman(?) Member are deposits representing the Pelukian (Sangamon), Kotzebuan (Yarmouth? interglaciation), and Fishcreekian transgressions, with little or no intervening nonmarine deposits (P. A. Smith, unpublished work). The more distal parts of outwash fans appear to have been completely eroded or reworked during the interglacial transgressions. The marine sequences are mostly silt and clay, which commonly include thin basal transgressive beach gravel. Nonmarine gravel is first encountered in these bore holes at a subbottom depth of about 60 m.

Although the Colville and Canning River fans thus appear to have no significant offshore expression, abundant subsurface river gravel occurs off the Sagavanirktok River in the Prudhoe Bay area, and this has been interpreted as a gravel-filled Pleistocene paleovalley (Hopkins et al., 1979). Such valleys could have been eroded across the shelf during low sea-level stands and backfilled with sediment, perhaps including gravel and sand, during deglaciation episodes. The distance such subsurface Pleistocene coarse sediment might extend beyond the present shoreline is problematical, however, considering the low gradients.

At deeper levels gravel is probably more abundant. Beneath a prominent reflector that lies about 100 m below sea level and is taken to be the base of the Quaternary, the seismic character of the sediment suggests prograding alluvial fans or deltas, inferred to be the upper part of the Brookian sequence (Craig and Thrasher, 1982). A deep drill hole (Humble C-1) through Reindeer Island shows mostly gravel and sand from a depth of about 25 m to hole bottom at 155 m. The upper 40 m of these strata is interpreted as glacial outwash; the remainder may be pre-Quaternary alluvial deposits (Hopkins and Hartz, 1978a). The position and character of pre-Quaternary sediment under the middle and outer shelf are, however, essentially unknown.

Deep-Sea Turbidite Sands

The continental slope in the Beaufort Sea is mantled by silt and clay (Barnes, Reimnitz, and Ross, 1980; Naidu et al., 1981) and is characterized by large-scale slumps and faults (Eittreim and Grantz, 1979). It is extremely unlikely, therefore, that any significant masses of clean, coarse-grained material exist on the slope. The abyssal floor of the Canada Basin beyond the slope, however, is underlain by a flat-lying stack of sedimentary layers representing a distal part of the Mackenzie deep-sea fan (Eittreim and Grantz, 1979) and undoubtedly including numerous turbidites of probably clean sand. The great water depth (>3 km) and the unknown overburden make these inferred sand layers a potential resource of only problematical value at the present time.

CHUKCHI SEA

The continental shelf in the Chukchi Sea, unlike the Beaufort Shelf, is extremely broad. It extends north from the Bering Strait for more than 750 km, and from Alaska westward it continues without a break through the East Siberian and Laptev Seas for a total distance of about 2500 km. Westward at least to Wrangel Island, about 500 km from the Alaskan coast, it is almost flat. Gradients are generally < 1 m/km and often very much less, and water depths on most of the shelf are 30 to 60 m (Figure 4). Far offshore, Herald Shoal and Hanna Shoal rise to within 20 and 25 m of the surface, respectively (Grantz et al., 1982; Hill et al., 1984). Valley-like depressions thought to represent former drainage paths (Figure 4) include Hope Sea Valley and its tributaries (Scholl and Sainsbury, 1961, 1966; Creager, 1963; Creager and McManus, 1965) leading down to Herald Canyon, which is a sharper feature that cuts the shelf edge north and east of Herald Island, and Barrow Sea Valley, which forms a narrow re-entrant in the shelf edge (Phillips et al., 1984; Hill et al., 1984; Phillips and Reiss, 1985b; Grantz and May, this volume).

The Chukchi Shelf is in part a drowned subaerial erosion surface (peneplain) carved during the late Cenozoic (Hopkins, 1967b) on underlying bedrock that mainly continues the east–west structural and stratigraphic belts of the Alaskan mainland (Grantz, Holmes, and Kososki, 1976; Grantz, Eittreim, and Dinter, 1979; Grantz and May, 1983). The Wrangel-Herald arch forms a structural high connecting the Paleozoic and early Mesozoic rocks of Wrangel Island with similar rocks in the Lisburne Hills of Alaska (Grantz, Holmes, and Kososki, 1976; Grantz and May, this volume). South of this high, the filled Hope Basin contains a thick sequence of chiefly Tertiary sedimentary rocks (Tolson, this volume).

North of the Wrangel-Herald arch, the eastern Chukchi Shelf is a continuation of the Arctic platform, carved on gently folded Cretaceous rocks of the Colville foredeep. But northwest of the Barrow arch is the deep filled North Chukchi Basin, lying mostly north of lat. 72°N. This basin is thought to have formed by pull-apart during the Mesozoic rifting that created the Arctic continental margin of Alaska (Grantz et al., 1982; Grantz and May, this volume). To the northeast of the North Chukchi Basin lies the Chukchi borderland (Grantz, Eittreim, and Dinter, 1979), and beyond lies the deep Arctic Ocean. The deep Arctic, the Chukchi borderland, and most of the North Chukchi Basin are permanently ice covered (Ahlnäs and Wendler, 1979) and are still poorly known geologically.

Over most of the Chukchi Shelf, unconsolidated Quaternary sediment is very thin or absent. Linear ridges and other rocky patches on the sea bottom have been interpreted as exposures of resistant bedrock layers in the northeast Chukchi Sea (Phillips et al., 1984), in the Cape Thompson area (Scholl and Sainsbury, 1966), in the sides of Barrow Sea Valley (Hanna, 1954; Phillips and Reiss, 1985b), and on Hanna Shoal (Toimil and Grantz, 1976), and similar outcrops undoubtedly occur in other less studied areas.

Quaternary cover near shore is as thick as 8 to 15 m in the northeast (Phillips et al., 1984; Phillips and Reiss, 1985b) and southeast (Moore, 1964) Chukchi Sea, but thins offshore to 1 to 3 m. In the northeast Chukchi Sea, the Quaternary deposits are < 1 m thick in water depths > 24 m, except on benches on the side of Barrow Sea Valley, where they are as much as 10 m thick (Phillips and Reiss, 1985b). In the North Chukchi Basin, more than 100 km offshore, thicknesses of Quaternary sediment exceed 10 m and increase to the northwest (Grantz et al., 1982). Persistence of unfilled or only partly filled former stream-valley systems (McManus et al., 1983), including even low-relief valleys near shore (Scholl and Sainsbury, 1961), indicates that at least the Holocene deposits are very thin.

The Quaternary sediment and the topography of this region are the products of a complicated series of transgressions and regressions (Hopkins, 1967a, 1973; McManus et al., 1983) controlled mainly by glacioeustatic sea-level fluctuations, but influenced also by regional tectonic uplift and subsidence. Low sea levels during the Illinoian (−135 m, or lower) and Wisconsin (−120 m, or lower) glaciations exposed virtually the entire vast extent of the Chukchi (and Bering) Shelf, creating a large land bridge between North America and Asia (Hopkins, 1967b).

Quaternary sediment in coastal exposures is mainly transgressive beach and nearshore marine deposits (McCulloch, 1967; Williams, 1979), and high sea-level stands, particularly in the last (Sangamon) interglaciation, have left beach ridges and wave-cut cliffs on shore (Moore, 1966; Phillips et al., 1984). Similar or related offshore features will be discussed later. During the Illinoian glaciation, glaciers from the Brooks Range reached the present Chukchi Sea only in the Kotzebue Sound area, where the Baldwin Peninsula represents a terminal moraine (McCulloch, 1967), and glaciers from the mountains of Chukotka extended beyond the present shoreline in the southwestern Chukchi Sea (Petrov, 1967). Elsewhere there is no evidence that the Chukchi Shelf was ever glaciated.

Present Sedimentary Processes on the Chukchi Shelf

The Chukchi Shelf represents a vast and still poorly known environment, but two major points are clear: (1) the range and kinds of processes operating here are much the same as in the Beaufort Sea, including those uniquely high latitude processes related to ice action; (2) the general thinness of the Quaternary sediment shows that this region has been one of extremely small net deposition over the past several million years, partly because of paucity of sediment supply, but also because strong currents move sediment parallel to the coast rather than offshore.

Ice regimen and ice zonation in the Chukchi Sea are known in outline (Ahlnäs and Wendler, 1979; Grantz et al., 1982). The sea freezes over rapidly and completely between October and January and remains ice covered, except for intermittent leads in the drifting pack ice, until break-up begins in May or June. The edge of the pack ice then retreats northward to about lat. 73°N by September or October (only to 70° or 71° in bad years). During winter, fast ice extends 20 to 50 km from the Chukchi coast, where a zone of intense ice-ridging (stamukhi zone) marks the transition to the moving pack ice. Farther offshore, drift patterns, ridging, and opening of leads are irregular and unpredictable because of the interaction of varying wind directions and several major water currents.

Ice-gouging of the sea bottom has been documented near Barrow (Rex, 1955), on Hanna Shoal (Toimil and Grantz, 1976), in nearshore areas of the northeast Chukchi Sea (Phillips and Reiss, 1985b) and the eastern Chukchi Sea (Toimil, 1978), but its intensity and patterns are still poorly known. Gouging is generally intense in shallow nearshore areas, especially under the stamukhi zone at about 20-m depth, on major shoals such as Hanna Shoal, and on slopes facing the Arctic Ocean. Gouges are sparse at depths > 50 m (the deepest noted was at 58 m north of Hanna Shoal). Sparsity of gouges near Bering Strait and in some other areas may be because of rapid reworking of the bottom sediments by currents. Though gouging should be ineffective in areas where bedrock is at the surface, the irregular 1- to 3-m relief on Herald Shoal may be caused by ice-push (Grantz et al., 1982). Gouge relief is mostly < 1 m (3 to 4 m in the intensely gouged zone at 15- to 30-m depth); orientation of gouges tends to parallel isobaths or the shoreline in nearshore areas but is erratic farther out.

Ice action on Chukchi Sea beaches is more severe than in the Beaufort Sea because strong water currents (and longer wind fetch?) drive ice masses. Ice-pushed ridges are as high as 4 m (Grantz et al., 1982), and locally as much as 10% of the beach material has been added to the beach by ice-push (Hume and Schalk, 1964). Ride-up of floes has reached 35 m inland near Point Barrow, and on occasion ice has overtopped the 9-m bluff at Barrow (Kovacs and Sodhi, 1980).

Ice-rafting in the Chukchi Sea is probably insignificant, inasmuch as large icebergs and ice islands carrying glacial rock debris are rare (Grantz et al., 1982). Earlier reports of ice-rafted pebbles and cobbles, before recognition of the mixing effects of ice-gouging, must be considered suspect, but some rafting of adhered beach material (rounded pebbles) by sea ice probably does occur. Large erratics on the tundra along the Chukchi coast are probably pre-Holocene in age (P. A. Smith, 1983, oral communication).

Water currents (Figure 4) are a dominant influence on sediment movement in the Chukchi Sea. Tides are generally small (< 30 cm), and so tidal currents reach significant speeds only in inlets and passes (Grantz et al., 1982). Of the major oceanic currents in the Chukchi Sea, the most conspicuous is the Alaska Coastal Current, first noted in 1778 by Captain James Cook at the Bering Strait (Aagard and Coachman, 1964). Northward surface velocities as high as 2 m/sec have been measured there (Creager, 1963; Fleming and Heggarty, 1966). From Bering Strait, this current generally flows northward and northeastward along the coast, then bifurcates at Cape Lisburne, one branch flowing offshore to the northwest (Figure 4). In the northeast Chukchi Sea, the Alaska Coastal Current forms a definite zone tens of kilometers wide with surface velocities as high as 1 m/sec (Hufford, 1977). North of Barrow it interacts with the westward flow in the Arctic Ocean, producing large gyres (Grantz et al., 1982) seen from satellites as ice spirals (Solomon and Ahlnäs, 1980). Clockwise gyres or eddies within embayments of the Chukchi coast north of Cape Lisburne, Icy Cape, and Point Franklin produce southward nearshore currents that make those capes depocenters and sediment sinks (Aagard and Coachman, 1964; Hufford, 1977; Phillips et al., 1984; Phillips and Reiss, 1985a).

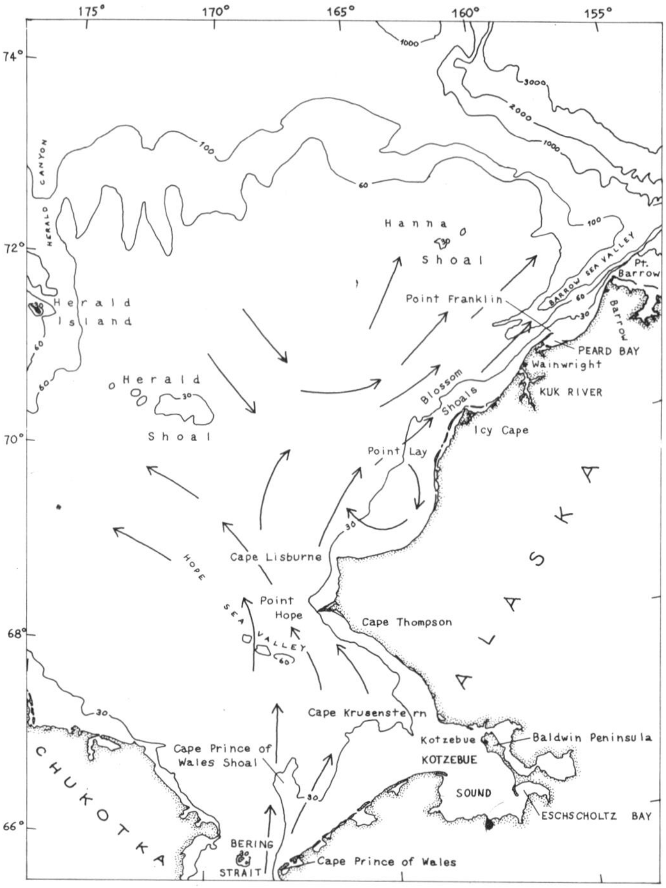

Figure 4. Map of the Chukchi Sea and its major water currents. Selected bathymetric contours, in meters, from Hill et al. (1984). Generalized upper-layer currents after Grantz et al. (1982). Chukchi borderland lies just north of the map edge in the central part.

Seaward of the Alaska Coastal Current in the northeast Chukchi Sea, equally strong southwestward currents have been observed (Hufford, 1977; Grantz et al., 1982). In the east-central Chukchi Sea, the northeast-directed Alaska Coastal Current is about 50 km wide, has water speeds 2 m from the bottom up to 20 cm/sec, and is flanked by comparable southwestward currents both inshore and offshore from it (Barnes, 1972). Relatively weak southeastward currents flow along the Chukotka coast (Aagard and Coachman, 1964).

Indirect evidence of bottom currents strong enough to erode sediment and transport coarse material is provided by erosional furrows, lag gravel, and sand waves and other bedform accumulations. Parallel furrows 5 to 10 m apart in the mud bottom of a channel near Kotzebue, and also in the strait into Eschscholtz Bay in southeast Kotzebue Sound (Figure 4), are interpreted as current-carved features (Hunter and Reiss, 1985).

Bottom gravel north and east of Cape Lisburne containing angular and fragile clasts of rock types similar to those of nearby sea-bottom outcrops was interpreted by Barnes (1972) as residual material, winnowed or at least protected from burial by currents. Similar gravel occurs on the inner shelf in the Cape Thompson area (Scholl and Sainsbury, 1966), in the Peard Bay area at depths of 15 to 25 m (Phillips et al., 1984), and in shallow water between Icy Cape and Wainwright (Phillips and Reiss, 1985a) in the northeast Chukchi Sea.

Shoals of sand on the north side of Bering Strait (McManus and Creager, 1963), in Kotzebue Sound (Hunter and Reiss, 1985), and at Blossom Shoals off Icy Cape (Phillips and Reiss, 1985a) are the products of active bottom currents. Their locations support Creager and McManus' (1966) contention that the two main sediment sources in the Chukchi Sea today are a northward influx through the Bering Strait and erosion along the coast, especially from sea cliffs on the Lisburne Peninsula. Other evidence of bottom currents includes fields of northward-migrating sand waves at depths of 15 to 30 m near Point Franklin (Phillips et al., 1984) and elsewhere along the northeastern Chukchi coast (Grantz et al., 1982), and sand ripples in 30-m depth on Hanna Shoal (Toimil and Grantz, 1976).

Winds on the Chukchi Sea normally blow from the northeast (bringing ice in from the Beaufort Sea), but summer storm winds blow from the southwest (Hufford, 1977). Most of the Chukchi Shelf is too deep (about 50 m) to be much affected by wave action, but symmetrical ripples in pebble gravel near Point Lay and in 16- to 18-m depths toward Wainwright (R. E. Hunter, P. W. Barnes, E. W. Kempema, and T. Reiss, unpublished work) are interpreted as the product of storm waves.

Along the coast, waves have produced chains of barrier islands (Short, 1979) and sets of offshore bars, commonly attached to shore at one end (Short, 1975). Longshore migration in the barriers is toward the major capes, away from the centers of large embayments (Grantz et al., 1982). Migration of material in mainland beaches is to the northeast near Barrow but to the southeast from Point Hope, because the dominant waves are from the west. Moore (1966) has shown the decline of less resistant clasts from the beaches near Point Hope's coastal cliffs to Cape Krusenstern to the southeast. Estimates of longshore sediment-movement rates are 5–25 $\times 10^3$ m^3/yr in general (Short, 1979), 22 $\times 10^3$ m^3/yr over the past 1000 years at Cape Krusenstern (Moore, 1966), and about 9 $\times 10^3$ m^3/yr at Barrow (Hume and Schalk, 1967).

The beaches are composed of coarse-grained material, with median and modal sizes commonly in the range of 1 to 5 cm (Moore, 1966; Hume and Schalk, 1967). The beach material is thought to be almost entirely derived by erosion of the coast because of lithological similarity and absence of other likely sources, but despite the longer open-water season and longer wind fetch, the rates of retreat along the Chukchi coast appear to be much lower than on the Beaufort Sea coast. The reason may be, in part, a lower ice content in the Chukchi coastal bluffs (P. A. Smith, 1983, oral communication). Although higher rates (4 to 6 m/yr) apparently occur near Point Barrow (Rex, 1964), a 27-year study of retreat between Barrow and Peard Bay showed rates mostly around 0.3 m/yr (Harper, 1978). The rate may, however, be strongly influenced by rare severe storms, which produce large (2 to 3 m) waves; storm surges have produced waves as high as 4 m in historical times (Hume and Schalk, 1967) and, from evidence in tidal river deposits, in prehistoric times also (Hopkins, Hartz, and Robinson, 1979).

Assessments of possible sand and gravel resources have been made by Molnia (1979a) for the Chukchi Shelf and by J. K. Brigham (unpublished work) for onshore areas adjacent to the north Chukchi coast.

Active Sand and Gravel Bodies on the Chukchi Shelf

Abundant sand and gravel are present in active sediment bodies along the Chukchi coast of Alaska and on the inner part of the (eastern) Chukchi Sea Shelf.

Beaches, Offshore Bars, and Barrier Islands

Beaches along the Chukchi coast of Alaska are coarse grained and contain large amounts of gravel and sand. For the 200-km stretch from Point Hope to Cape Krusenstern, Moore (1966) estimated a typical width of 100 m and a thickness at the shoreline of about 6 m, implying a total volume on the order of 10^8 m^3. Cape Krusenstern itself contains a volume of material on the same order, but its archeological importance removes it from consideration as a potential resource. Grantz et al. (1982) estimated barrier beaches between Barrow and Point Lay to be at least 10 m thick (the depth of migrating channels), but they are described as narrow (Short, 1975), and their total volume over that roughly 375-km distance may also be on the order of 10^8 m^3.

The beach material is derived by erosion from coastal bluffs of mainly unconsolidated Quaternary deposits in the northern Chukchi Sea and from cliffs of Paleozoic and Triassic bedrock in the area from Cape Lisburne to Point Hope. The northern beaches are coarser grained, perhaps because the bedrock on the Lisburne Peninsula tends to break into small fragments, whereas ready-made large clasts of resistant rock types are provided by the Quaternary exposures. Beaches at the southern end, near Bering Strait, are of fine-grained sand (Hunter and Reiss, 1985) brought northward through the strait.

Long, parallel (en echelon) offshore bars, trending at low angles to the coast and having one end attached to the beach, characterize the Chukchi coast from Icy Cape to Barrow (Short, 1975). The bars are as much as 8–10 km long, have ends as much as 800 m offshore, and usually overlap, so that 2 or 3 are present at any one place. Three such bars on the north side of Peard Bay rise from water depths of 7 to 8 m to within 1 to 2 m of the surface (Phillips et al., 1984). These numbers would imply that the amount of material in offshore bars is perhaps one-third that in the beaches.

Shoals and Sand-Wave Fields

The largest active accumulation of sand on the Chukchi Shelf is Cape Prince of Wales Shoal, formed of sediment transported through the Bering Strait by the strong and persistently northward currents and stretching 50 km north–northeast from the cape (Dietz et al., 1964). Seismic data indicate that the shoal deposit is about 30 m thick (Hunter

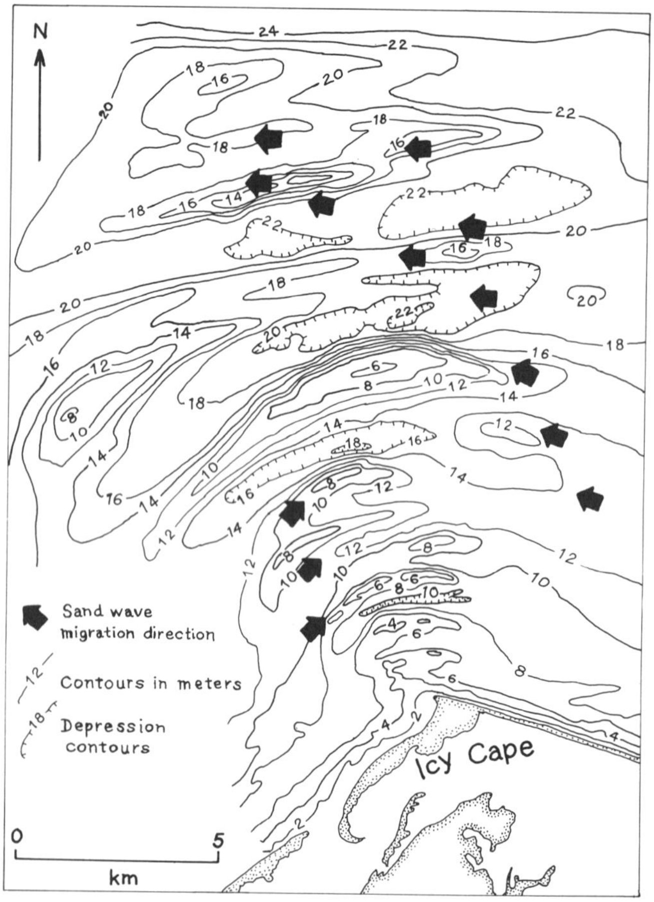

Figure 5. Blossom Shoals off Icy Cape, northeast Chukchi Sea. Most of the bathymetric relief is formed by constructional sand ridges. Note the convergent pattern of currents shown by the migration directions of sand waves. After Phillips and Reiss (1985a).

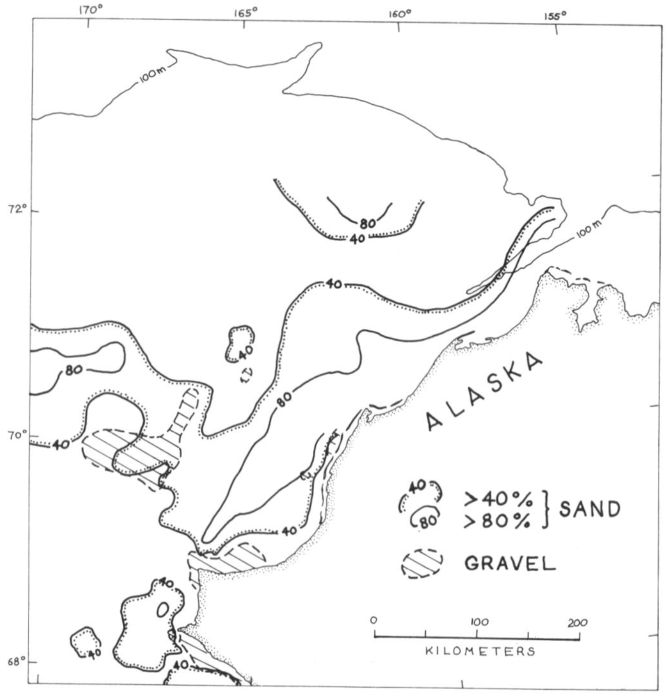

Figure 6. Areas of sand and gravel concentrations in sea-floor sediments of the northeast Chukchi Sea. Contoured sand values are actual percentages in the bottom sediments; gravel concentrations are based on factor analysis of sediment size distributions. Both sets of data are from Grantz et al. (1982).

and Reiss, 1985), and it has been estimated to contain about 2×10^{10} m^3 of coarse- to fine-grained sand (McManus and Creager, 1963).

Blossom Shoals off Icy Cape (Figure 5) comprise a series of nested arcuate ridges that rise from water depths of 10 to 22 m to within 4 to 6 m of the surface and are covered by abundant fields of sand waves. These sand waves show a converging movement pattern (Phillips and Reiss, 1985a)—directly off Icy Cape, the waves migrate northeastward, whereas farther offshore to the northeast, they migrate westward. Convergence of sediment movement at Icy Cape is also implied by the sense of overlapping in offshore bar systems on either side of the cape (Short, 1975) and by measurements of current directions (Phillips and Reiss, 1985a). Blossom Shoals thus seems to represent a sand sink and dead-end accumulation, the mining of which is unlikely to produce any significant disturbance of the present dynamic processes. Seaward-dipping internal surfaces on seismic sections through Blossom Shoals indicate the sand ridges migrate slowly northward (offshore), probably by erosion of the south side, progression of sand waves across the top, and deposition on the north side (Phillips and Reiss, 1985a).

The largest sand ridge in Blossom Shoals has a relief of about 12 m, a length of >10 km, and a maximum width of about 3 km (Figure 5). Its volume can be estimated at 10^8 m^3; the total volume of sand in Blossom Shoals may be two or three times that amount. If these shoals have been accumulating steadily over the last 5000 years, that volume would imply an average rate of accretion of about 2×10^4 m^3/yr (or half that from each of the two sides). These figures are consistent with estimates of present-day rates of longshore sediment movement, quoted above.

A similar sand sink may exist 8 to 15 km off Point Franklin at Peard Bay, where a large field of northward-migrating sand waves occurs. The full extent of this field remains undefined on the north and west sides (Phillips et al., 1984). Localization of this accumulation may be caused by a clockwise current gyre in the embayment east of Point Franklin; westward currents northeast of the point are shown by asymmetric patchy sand waves.

A large, very shallow shoal of fine-grained sand off Kotzebue in Kotzebue Sound has minimum water depth <2 m, is cut by a single large channel, and has been interpreted as an ebb-tide delta (Hunter and Reiss, 1985). Inner Kotzebue Sound may be one of the few areas of the Chukchi coast where tides, amplified by the physiographic setting, produce significant tidal currents; the channel bottom shows current marks (Hunter and Reiss, 1985).

Two large shoals are found far offshore in the Chukchi Sea—Herald Shoal, 275 km northwest of Cape Lisburne, and Hanna Shoal, 150 km northwest of Point Franklin (Figure 4); these probably do not contain significant amounts of unconsolidated coarse-grained material. Exposed bedrock forms the seabed on Herald Shoal (Grantz et al., 1982) and has been inferred at Hanna Shoal (Toimil and Grantz, 1976). However, the flanks of these shoals may contain sand and gravel accumulations resulting from winnowing of ice-gouged sediment by strong currents.

Sand waves have also been noted on the shallow nearshore part of the Chukchi Shelf. Near Bering Strait, southward-migrating sand waves occur just north of the westernmost bulge of the shoreline (local counter-current?), and northeastward-migrating waves occur between Cape Prince of Wales Shoal and the shore (Hunter and Reiss, 1985). Sand waves have also been identified in some shallow (< 10 m deep) localities northeast of Icy Cape (Phillips and Reiss, 1985a), in 10 to 20 m of water between Wainwright and Point Franklin (Phillips et al., 1984), and in some nearshore areas 2 to 15 m deep, where they parallel the shore and are probably produced by storm waves rather than currents (Grantz et al., 1982).

Other Sea-Bottom Occurrences of Coarse Sediment

Gravel mantles the sea bottom adjacent to gravel beaches and on parts of the shallow shelf in the Chukchi Sea (Figure 6), including areas near Cape Lisburne (Barnes, 1972), southeast of Point Hope (Scholl and Sainsbury, 1966; McManus, Kelley, and Creager, 1969), in Kotzebue Sound (Hunter and Reiss, 1985), near Icy Cape (Phillips and Reiss, 1985a), and near Peard Bay (Phillips et al., 1984). Though rippled locally, the gravel is commonly angular and associated with bedrock outcrops (Scholl and Sainsbury, 1966). These occurrences are interpreted as thin lag gravel, and they generally are found where the total Holocene sediment thickness is < 2 m (Grantz et al., 1982). Similar gravel on Herald Shoal (Grantz et al., 1982) and Hanna Shoal (Toimil and Grantz, 1976) is probably related to ice action as well as to current winnowing.

Sand covers the sea bottom in a nearshore belt as wide as about 80 km from near Cape Lisburne north to Barrow, on the outer flanks of Herald Shoal, and around Hanna Shoal (McManus, Kelley, and Creager, 1969; Barnes, 1972). Although the sand is commonly rippled, indicating active movement (Phillips et al., 1984; Phillips and Reiss, 1985b), its thickness outside areas of sand waves or shoals is slight, as indicated by nearby bedrock outcrops or patches of lag gravel.

Fossil Sand and Gravel Bodies in the Chukchi Sea

Several types of inactive, relict, or buried accumulations of coarse sediment occur or are suspected in various parts of the Chukchi Sea. Far from shore these may assume some importance as the only likely sources of sand and gravel.

Glacial Deposits

The Baldwin Peninsula in Kotzebue Sound (Figure 4), which is a terminal moraine of Illinoian age (McCulloch, 1967), marks the only place along the Chukchi coast of Alaska where Pleistocene glaciers are known to have reached or passed the present shoreline. This partly submerged moraine is unlikely to be mined for sand or gravel because cleaner deposits are present nearby, both in the nearshore zone and onshore.

Pleistocene Transgressive Deposits and Former Shorelines

Pleistocene sedimentary sequences exposed along the Chukchi coast include several transgressive gravel horizons as thick as 4–5 m (McCulloch, 1967). These must originally have extended offshore, but reworking during repeated sea-level fluctuations, implied by the thinness of the Pleistocene sediments on the shelf, may have made them unrecognizable. Their remnants are probably part of the widespread lag gravel on the shelf, and some thicker gravel layers may be concealed beneath Holocene mud on the middle and outer shelf.

Former shorelines are known onshore, where well-defined old sea cliffs with beach deposits of late Pleistocene (mainly Sangamon interglaciation) age occur inland from Peard Bay, intersecting the present coast on both north and south ends (Phillips et al., 1984), in the Point Hope area (Moore, 1966), and elsewhere. These old shorelines commonly have > 10 m of well-sorted coarse-grained material, ranging from sand to cobble gravel (J. K. Brigham, unpublished work). A gravel ridge of about 3-m relief, rising from 15-m water depth off Skull Cliff, may be a tectonically offset continuation of the Sangamon beach at Peard Bay (Phillips et al., 1984). Though most submerged old shoreline deposits are likely to have been disrupted, spread out, or destroyed during later transgressions, possible examples have been inferred in the southeastern Chukchi Sea (outside Kotzebue Sound) at several levels down to −38 m (Hopkins, 1967b). Hunter and Reiss (1985) report subbottom reflectors in southern Kotzebue Sound whose shapes suggest old wave-cut cliffs, now at 12 m and more below sea level.

Submerged Deltas in Bays and Paleovalleys

Outlines of the subaerial drainage system that evolved on the Chukchi Shelf at times of lower sea level are preserved in the form of sea-bottom valleys (Scholl and Sainsbury, 1961; Creager and McManus, 1967). Parts of these valleys have been filled by deltaic (and shoreline?) deposits in the course of the Holocene transgression. In southeastern Kotzebue Sound, a number of foreset sequences 5 to 15 m thick have prograded toward the centers of bays; most have merged with ones building from other directions and are capped by the flat present sea bottom (Hunter and Reiss, 1985). These are probably Holocene tidal deltas or early Holocene estuarine fills (Creager and McManus, 1967).

Farther out on the Chukchi Shelf, the Hope Sea Valley is well delineated at about 50-m depth, but segments of the valley center are > 60 m deep (Hill et al., 1984) (Figure 4). This segmentation has been attributed to estuarine and deltaic deposition into the main valley by tributaries heading in the Siberian uplands on the southwestern side of the Chukchi Sea during a sea-level stillstand near −55 m at the beginning of the Holocene (Creager and McManus, 1965). The characteristics of these deposits are not known, but they could include deltaic or shoreline sands.

Pleistocene Eolian Sands

Wind-blown sand dunes and sand blankets in onshore areas adjoining the Chukchi coast (Hopkins, 1967b) date from glacial intervals of low sea level, and equivalents probably formed on the now-submerged shelf. Such bodies might be preserved on the sea bottom, thinly mantled by later Quaternary sediment, though none has yet been identified.

Filled Paleovalleys on the Chukchi Shelf

Fillings of former river channels carved in the late Tertiary or during Pleistocene glacial intervals are inferred to form an extensive network that may contain large amounts of sand and/or gravel fill on the Chukchi Shelf. A well-defined paleochannel of the Kuk River, traced about 6 km offshore near Wainwright (Phillips and Reiss, 1985a), is > 600 m wide and contains as much as 23 m of sedimentary fill. A similar but smaller paleochannel, with as much as 11 m of fill, has been identified at Walapka Bay south of Barrow (Phillips and Reiss, 1985b). Other paleochannels outside Kotzebue Sound have as much as 10 m of fill (Moore, 1964).

Farther offshore on the Chukchi Shelf, filled channels and valleys with as much as 50 m of former relief have been recorded on seismic profiles north of Herald arch and Hanna Shoal (Grantz et al., 1982; Stephen Eittreim, unpublished data). These paleochannels are commonly associated with shallow free gas, possibly from organic material in the channel fill. The distribution of such shallow gas, as inferred from "turbid" zones on seismic sections (Grantz et al., 1982), indicates that filled paleochannels are widely distributed on the northern Chukchi Shelf.

BERING SEA AND ALEUTIAN RIDGE

The Bering Sea is one of the largest semienclosed seas in the world, having a total area (2.3×10^6 km^2) more than half again as large as the land area of the State of Alaska (Figure 7). It is about evenly divided between continental shelf (depth < 200 m) and deep sea (depth > 3000 m) by the relatively linear continental slope crossing its middle from northwest to southeast (the "Beringian margin"). Steeper continental slopes are present on the western (USSR) side and on the north flank of the Aleutian Ridge, and two major topographic features, the curving Bowers Ridge and the linear Shirshov Ridge, divide the abyssal area into three separate basins—Komandorski (or Kamchatka) Basin, Bowers Basin, and Aleutian Basin (by far the largest) (Cooper, Scholl, and Marlow, this volume).

The Bering or Beringian Shelf, 500 km wide and extending nearly 1500 km from the Alaska Peninsula to the northeastern USSR, is generally very flat, with a gradient typically < 1 m/km. North of St. Lawrence Island, the shelf is mostly < 50 m deep, and local relief is rarely as much as 10 m. At its northern end, the Bering Sea connects with the Chukchi Sea through the Bering Strait, which has a sill depth of 48 m (Creager and McManus, 1967). South of St. Lawrence Island the shelf slopes very gently to the southwest.

From the shelf break at about 150-m depth, the relatively steep (50–150 m/km) continental slope and gentler (15–25 m/km) continental rise extend down to abyssal plains deeper than 3500 m (Carlson and Karl, 1984b). Submarine canyons that incise the slope and shelf edge have been described along the Beringian margin (Scholl et al., 1970; Carlson and Karl, 1984b) and on the north side of the Aleutian Ridge (Cooper et al., 1980).

The geologic structure and evolution of this vast and complex region are still imperfectly known. The Aleutian Ridge, which bounds the Bering Sea on the southern side, is a volcanic arc begun in the earliest Tertiary and related to the convergent plate boundary marked by the Aleutian Trench on its southern side (Scholl, Vallier, and Stevenson, 1983a, 1983b, and this volume). The Aleutian and Bowers(?) Basins to the north are remnants of trapped Mesozoic ocean floor (Cooper, Marlow, and Scholl, 1977; Cooper, Scholl, and Marlow, this volume). Bowers Ridge is clearly a former volcanic arc (Cooper, Marlow, and Ben-Avraham, 1981), but the origin of Shirshov Ridge remains somewhat enigmatic (Cooper, Marlow, and Scholl, 1977; Ben-Avraham and Cooper, 1981). In the Alaska Peninsula, Cenozoic Aleutian arc volcanic activity is overprinted onto older rocks of the Peninsular terrane (Marlow and Cooper, 1983). The Bering Sea Shelf is underlain by a geologically complex basement. The Beringian margin is thought to have been an actively subducting plate boundary during the Mesozoic (Cooper, Scholl, and Marlow, 1976, and this volume; Cooper, Marlow, and Ben-Avraham, 1981), and Mesozoic arc rocks are locally exposed on Nunivak and St. Matthew Islands and the western part of St. Lawrence Island (Marlow et al., 1981).

Large elongate sedimentary basins that began to form beneath the shelf by the end of the Mesozoic (Scholl and Hopkins, 1969; Marlow et al., 1976, 1980; Marlow, Cooper, and Childs, 1983) are filled with as much as 10 to 12 km of Tertiary and Quaternary sedimentary deposits and are prospective petroleum areas. Somewhat similar basins on the Aleutian Ridge are only partly filled with sediment (Scholl, Marlow, and Buffington, 1975; Scholl, Vallier, and Stevenson, 1983a, and this volume).

The Bering Shelf became a submerged epicontinental sea in the early Tertiary. Since then, although basins continued to subside (Marlow and Cooper, 1980), tectonic activity has been minimal, and marine depositional environments near the shelf edge have changed little since at least the middle Miocene (Vallier et al., 1980).

The Quaternary history of the Bering Sea region was dominated by four major factors—(1) repeated glaciation of adjacent land areas and some parts of the present shelf, (2) glacio-eustatic sea-level fluctuations, alternately exposing and drowning vast areas of shelf, (3) persistent and often explosive volcanism in the Aleutian Arc and other areas, and (4) influx of fluvial sediment, especially from the Yukon River.

Pleistocene glaciers from the highlands of the northeastern USSR left terminal moraines in the northwestern Bering Sea as far east as the western end of St. Lawrence Island (Figure 8), 150 km from the present Chukotka Peninsula coast (Lisitsyn, 1966; Petrov, 1967; Nelson, Hopkins, and Scholl, 1974). Farther southwest, submerged moraines are reported to be widespread along the Koryak coast and the northern Kamchatka coast (Figure 7), mostly on the inner shelf and near coastal fjords, but to the shelf edge in Olyutorskii Bay, where the shelf is < 50 km wide (Lisitsyn, 1966).

On the Alaskan side, glaciers from small ice caps reached the present shore locally in the western Seward Peninsula (Sainbury, 1967), in the Bristol Bay area (Porter, 1967; Nelson, Hopkins, and Scholl, 1974), and undoubtedly also along the Aleutian Island chain. Submerged morainal deposits are found a few km offshore of the Seward Peninsula (Nelson and Hopkins, 1972) but have not been documented in other areas.

Sea level was lowered to at least −100 m in Wisconsin time and to about −135 m in Illinoian time (Hopkins, 1973). Submarine terraces on the Beringian margin suggest an earlier lowering as deep as −240 m (Pratt and Dill, 1974). Exposure of the Bering Shelf led to widespread formation of tundra peat, grassland soil, and eolian and fluvial deposits; beaches and related accumulations were laid down at now-submerged shorelines. During each interglacial stage, a transgressive marine sequence was laid down across the

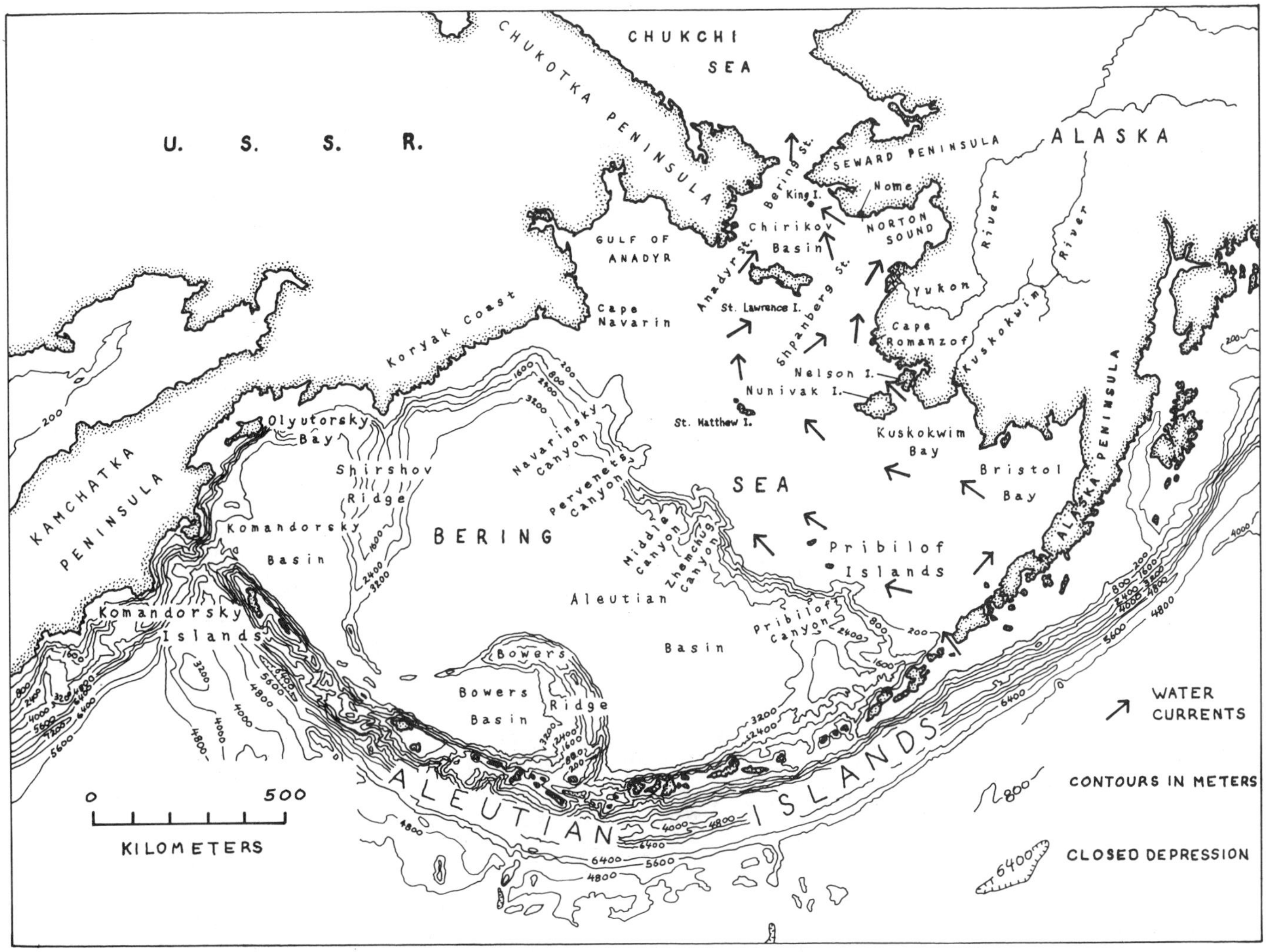

Figure 7. Map of the Bering Sea region, showing major coastal and bathymetric features and water currents. Submarine contours slightly generalized from Alpha and Swenson (1978). Major water currents in the northern and eastern Bering Sea shown diagrammatically, after Kinder and Schumacher (1981) and Nelson (1982).

shelf, with nearshore sand and gravel characteristic of its lower part. The latest Pleistocene and Holocene advance (Krusensternian transgression) is well documented (Hopkins, 1973), but at least four earlier Quaternary transgressions are known from fragmentary evidence (Hopkins, 1967a).

The latest transgression left deposits of various thicknesses in the Bering Sea. Between St. Lawrence Island and Bering Strait (Chirikov Basin), Holocene sediment generally is absent or consists of only a thin veneer of lag gravel or sand over Pleistocene or older deposits (Nelson and Creager, 1977; Nelson, 1982). South of Nome, Pleistocene peaty mud is exposed on the sea bottom between actively migrating sand bedforms (Nelson et al., 1982). Near the Yukon Delta shore, Holocene Yukon River silt and fine sand are as much as 14 m thick, but they thin rapidly to only 1–2 m over northern Norton Sound (Hess and Nelson, 1982) and are entirely absent in areas of strong tidal currents (Nelson, 1982).

Elsewhere in the Bering Sea, Holocene sediment has been reported to be mostly 3 to 4 m thick south of St. Lawrence Island (Knebel, 1972; Knebel, Creager, and Echols, 1974), about 3 m (range 0 to 10 m) in the southeast Bering Sea, >30 m near St. Matthew and Pribilof Islands (Moore, 1964), 2 to 5 m (maximum 10 m) southeast of St. Matthew Island (Askren, 1972), and 2 to 3 m at the base of continental slope (3000-m depth), but only 25 to 35 cm near the shelf edge (138–m depth) and on the abyssal plain of the Aleutian Basin (3800-m depth) (Baldauf, 1982).

The entire Quaternary section is commonly only a few meters thick on the Bering Shelf; it is totally missing in parts of the Chirikov Basin (Lisitsyn, 1966) and in some nearshore and insular areas (Nelson, Hopkins, and Scholl, 1974). Although gravity cores on the continental slope normally bring up Pleistocene and Holocene clayey silt (Carlson and Karl, 1984b), dredged pre-Quaternary rocks show that Quaternary sediment is locally absent. However, Quaternary sediment is as thick as 82 m in the middle of the Anadyr Basin in the Gulf of Anadyr (McLean, 1979) and probably comparable over other shelf basins, about 100 m thick near the Pribilof Islands (Moore, 1964), as much as 300 m thick locally on the Aleutian Ridge (Scholl and Creager, 1973), and probably several hundred meters thick in the fans at the base of the continental slope.

Volcanism has been important in the Quaternary history of the southern Bering Sea. There are more than 35 active volcanoes in the Aleutian Islands and adjacent parts of the Alaskan mainland, and the basaltic volcanoes of the Pribilof Islands may still be active (Marlow et al., 1979). Historical

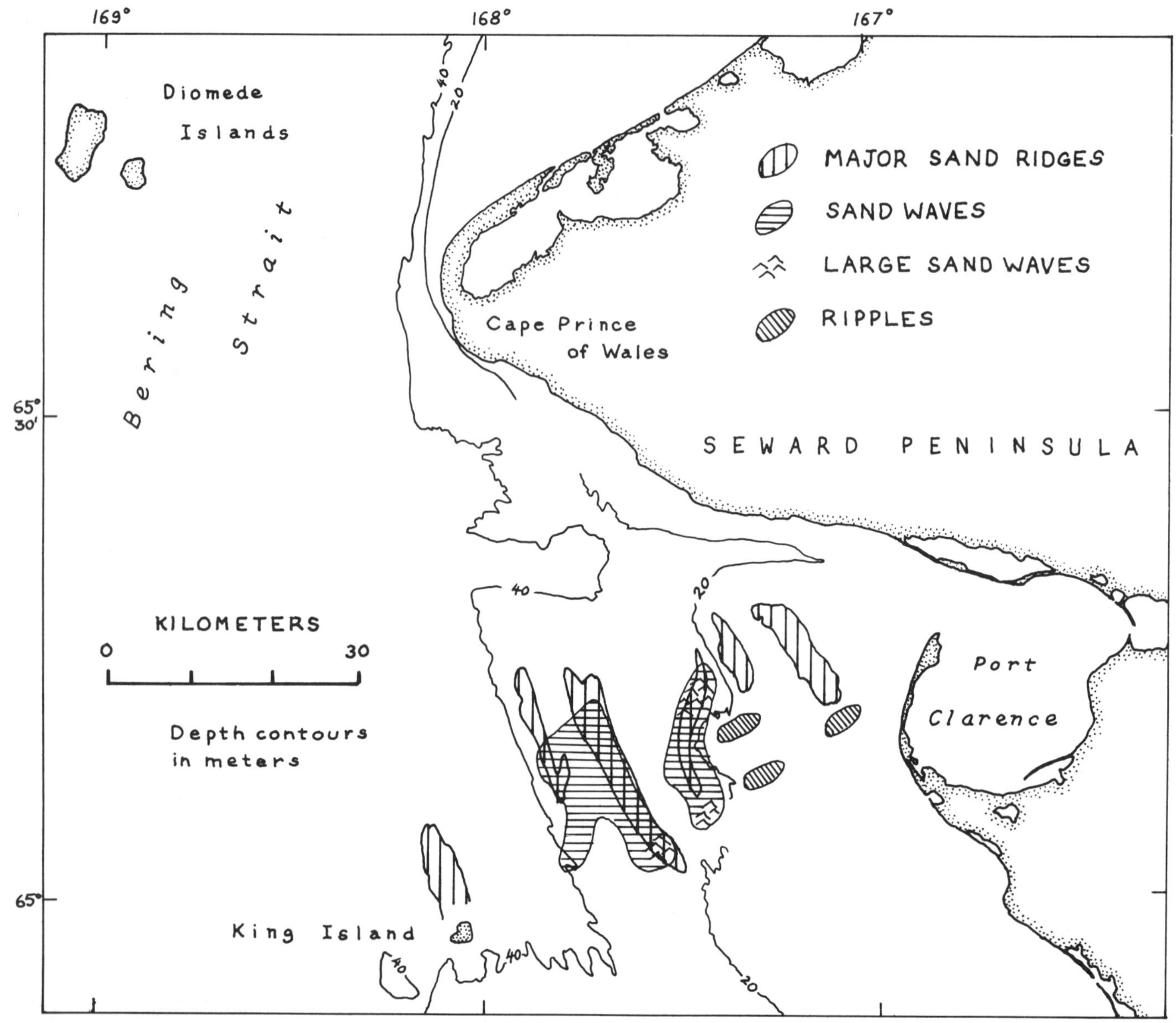

Figure 8. Areas of concentrations of gravel and of sand in bottom sediments of the northern Bering Sea. Selected contours from size analyses of bottom-sediment samples. After McManus et al. (1977).

eruptions have been frequent along the entire Aleutian chain and in Kamchatka (Lisitsyn, 1966). Major eruptions in the Aleutian arc spread ash about 500 km from the source (Marlow et al., 1979), and in particle size this ash is dominantly fine sand to coarse silt (Lisitsyn, 1966). Ash layers have been reported in cores from south of the Aleutian Ridge, in parts of the southern Bering Sea (Scholl and Creager, 1973), and also from a lake on the Yukon Delta (Dupré, 1977).

Another major source of sediment to the Bering Sea is the Yukon River. Major drainage diversions that occurred in the Miocene are thought to have doubled the drainage area of the Yukon (Nelson, Hopkins, and Scholl, 1974), and today its estimated total sediment load of almost 10^8 tons per year represents about 85% of the entire river-sediment flux into the Bering Sea from all adjoining land areas (Lisitsyn, 1966). Not surprisingly, the Yukon is the only river to have built a large delta into the Bering Sea. Position of the active delta has shifted widely and a number of times; the present mouth, on the south side of Norton Sound, may have been occupied for only the last 1000 to 2000 years (Dupré, 1977, 1982), shifting from an earlier position 80 to 160 km to the south (McManus et al., 1977). There is evidence that the Yukon River built a delta on the shelf south of St. Lawrence Island about 16,000 years ago and that during the lowest sea-level stand it flowed south of Nunivak Island (Figure 7) and entered the sea near the Pribilof Islands (Knebel and Creager, 1973). Most Yukon sediment seems to have bypassed directly into the deep sea via slope canyons, leaving little trace on the outer shelf or slope (Vallier et al., 1980).

Present Sedimentary Processes in the Bering Sea

Processes important in shallow parts of the Bering Sea include dominantly northward geostrophic currents, locally strong tidal currents, and periodic intense storm-wave activity. The Bering Sea also spans the southern limit of significant sea ice, and its northern half (essentially the Bering Shelf) experiences some gouging of the bottom and other ice processes. The continental slopes may be subject to slumping and possibly significant action by internal waves, whereas the deep-sea areas are generally quiet regions of little present activity.

Tidal amplitudes are generally small to moderate, amounting to only 1 to 2 m in the Aleutian Islands and 0.5 m or less over much of the northern Bering Sea (Lisitsyn, 1966; Hunter, Thor, and Swisher, 1982; U.S. Department of Commerce, 1985). Amplitudes rise to about 2 m in inner Norton Sound and in the Yukon Delta region, and they increase southward along the coasts, reaching 5 m in Kuskokwim Bay (Nelson et al., 1982) and >8 m at the head of Bristol Bay in the southeastern Bering Sea.

These tidal amplitudes produce tidal currents up to about 30 cm/sec moving parallel to shore off the Nome area (Hunter, Thor, and Swisher, 1982) and in outer Norton Sound (Field et al., 1981; Cacchione, Drake, and Wiberg, 1982), as high as 90 cm/sec following the shoreline near Nunivak Island in the east-central Bering Sea (Knebel, 1972), and 15 to 30 cm/sec generally on the southeastern Bering Shelf (Kinder and Schumacher, 1981). Much faster tidal currents flow in Bristol Bay, and in the passes between islands of the eastern Aleutians tidal currents reach speeds of 500 cm/sec (Cooper et al., 1980).

Because of the elevation difference between the Pacific and Arctic Oceans at the Bering Strait, ocean currents in the Bering Sea are dominantly northward, especially along the eastern margin (Alaska Coastal Current), and accelerate around promontories and in straits, particularly Bering Strait. In summer a strong southwestward flow develops along the western margin (East Kamchatka Current), forming a counterclockwise gyre in the southern Bering Sea (Lisitsyn, 1966; Brower et al., 1977; Kinder, 1981). Deep waters are deflected northwest by the Beringian continental slope to form the Bering slope current (Kinder, 1981).

Velocities of surface currents reach 100 to 200 cm/sec in and near Bering Strait, and bottom currents there have been measured or estimated at 30 cm/sec or more (Brower et al., 1977; McManus et al., 1977; Field et al., 1981). Northward surface flow through Shpanberg and Anadyr Straits has normal speeds of 50 to 75 cm/sec (Brower et al., 1977; Nelson et al., 1982), but currents out in the Chirikov Basin are only about 20 to 25 cm/sec (Nelson and Hopkins, 1972; McManus et al., 1977); the net northward flow in outer Norton Sound (after cancelling out the tidal oscillation) is <5 cm/sec (Cacchione, Drake, and Wiberg, 1982).

Currents in the wider southern Bering Sea are slow, ranging from about 30 cm/sec in the southward East Kamchatka Current and about 15 cm/sec just inside the Aleutian Islands (Askren, 1972), to net flow rates of <5 cm/sec near the Pribilof Islands (Gardner, Dean, and Vallier, 1980), in the Navarinsky Canyon area (Karl, Cacchione, and Carlson, 1986), and on the southeastern Bering Shelf (Kinder and Schumacher, 1981).

The strong northward water currents play an important role in sediment dynamics in the northern Bering Sea, producing irregular large (50–150 m diameter), shallow (60–80 cm deep) scour depressions in eastern Shpanberg Strait and northward toward Nome (Larsen, Nelson, and Thor, 1979), carrying away sediment disturbed by venting of biogenic gas to produce the numerous small (1–10 m diameter), shallow (<50 cm deep) circular craters in central and eastern Norton Sound (Nelson et al., 1979), and building fields of sand waves and other active bedforms in the northeastern Bering Sea (Field et al., 1981).

Currents in the Bering Sea also move suspended sediment, which is mainly silt and fine sand; little clay is contributed by the rivers on account of minimal chemical weathering in the Arctic climate (Lisitsyn, 1966). The most prominent turbid plume ls carried northward from the mouth of the Yukon River across Norton Sound and northwestward to Bering Strait (Cacchione and Drake, 1978; Feely, Massoth, and Paulson, 1981) and is present even in winter under the sea ice (Drake, Totman, and Wiberg, 1979). Tidal currents and storm waves (Field et al., 1981) assist currents to export possibly 50% or more of all modern Yukon sediment in this way through the Bering Strait to the Chukchi Sea (Nelson and Creager, 1977; Nelson, 1982).

Strong currents help account for low sedimentation rates and local absence of Quaternary sediment in the northern Bering Sea. Holocene sedimentation rates have been inferred to be <2 cm/1000 yr in parts of Norton Sound (Howard and Nelson, 1982) and as low as 8 cm/1000 yr in the east-central Bering Sea (Knebel, Creager, and Echols, 1974), but late Pleistocene and Holocene rates 10 to 50 times higher have been derived for other nearby areas (Knebel, 1972; Knebel, Creager and Echols, 1974). Low sedimentation rates, commonly <10 cm/1000 yr, on the outer shelf and slope of the Beringian margin are probably due to lack of sediment supply (Gardner, Dean, and Vallier, 1980; Carlson and Karl, 1984a).

The Bering Sea's long wind fetch and the storminess of its weather produce strong wave action along its shores and in shallow water. Bottom sediment is routinely affected by wave action to water depths of about 30 m (McManus et al., 1977), but storm waves probably do more work and certainly affect larger areas. Waves reach heights of 4 to 5 m in the north and 8 to 9 m in the southern Bering Sea in typical storms (Lisitsyn, 1966; Askren, 1972), whereas unusual storms produce 15-m waves every 10 years or so and 20-m waves once or twice a century (Brower et al., 1977). These largest storm waves have wave lengths of 150 to 250 m and should stir bottom sediment in 100 m (Lisitsyn, 1966) or even 200 m (Karl and Carlson, 1982) of water.

Larsen, Nelson, and Thor (1981) estimated that more than half the sediment transport in the northern Bering Sea was accomplished during storms representing only 10% of the time. Southwesterly winds during most major storms in the Bering Sea produce a strong surge in the northeast and intensify northward current flow (Nelson et al., 1982). Storm waves presumably produce and periodically activate the abundant symmetrical bedforms on the northeastern Bering Shelf that do not seem to be active under normal conditions (Hunter, Thor, and Swisher, 1982). Sharma, Naidu, and Hood (1972) and Sharma (1974, 1975) attributed the grading of the Bristol Bay Shelf sediment from clean sand to muddy sand deeper than 80 to 100 m to the effectiveness of waves in stirring the bottom and winnowing fines to that depth. Lisitsyn (1966) reported disturbance of bottom sediments by storm waves at 90-m depth in the Gulf of Anadyr. Storm surges have also been credited with liquefaction of fine sands off the Yukon Delta, leading to their redeposition as extensive graded sand beds (Nelson, 1982).

The important role of waves and storm surges in shoreline processes in the Bering Sea has been little studied. A major 1974 storm left debris lines as high as 4.75 m above sea level on the Seward Peninsula coast and caused total rearrangement of giant beach cusps and erosional retreat of coastal bluffs by as much as 45 m, but it also produced local shore progradation (Sallenger, Dingler, and Hunter, 1978). This 1974 storm produced waves of 3 to 4 m on top of a surge height of 3 to 5 m in Norton Sound (Larsen, Nelson, and Thor, 1981).

Much of the coast in the Yukon Delta region is retreating under wave attack at rates of several meters per year, though the rate differs greatly with orientation of coast (storm waves attack mainly from the southwest), proximity to river mouths, and presence or absence of barrier islands (Dupré. 1977). In the Chagvan Bay area, southeast of the Kuskokwim River mouth, Porter (1967) used the truncation of outwash-fan slopes to derive average rates of coastal retreat for the late Holocene (since completion of the sea-level rise) of about 0.4 m/yr or more.

The deep parts of the Bering Sea are not ice covered in winter, but sea ice does play a role in sediment dynamics on the Bering Shelf, especially in the northern part. Freezing normally begins in the north in October, and by January the ice edge lies near the Pribilof Islands and along the Beringian Shelf edge. Breakup and melting typically occur from April to June (Brower et al., 1977; Ahlnäs and Wendler, 1979). In a warm year, the southern ice limit may be nearer St. Matthew and Nunivak Islands (Niebauer, 1981).

Shore-fast ice forms at some bays and coastal stretches, usually to no more than 30 km from shore (McNutt, 1981), but rarely to 80 km from shore (Marlow et al., 1979). Sea ice reaches thicknesses of 1 to 2 m on the southeastern Bering Shelf (Marlow et al., 1979), but only 30 to 70 cm on the southwestern shelf (Marlow et al., 1981). The Bering Sea has significant tides, and fast ice commonly shows prominent tidal cracks (Stringer, 1981). Winter winds tend to push ice southwestward, leaving intermittent open water near suitably oriented stretches of coast (McNutt, 1981; Niebauer, 1981; Stringer, 1981).

The stamukhi zone of sea-ice pressure ridges is discontinuous and weakly developed in the Bering Sea, and significant pileups tend to occur only when storms reverse the usual winds and drive ice onshore (Stringer, 1981). However, thicker ice masses may also come into the Bering Sea from the north when, as occasionally happens, strong northerly winds reverse the usual flow through the Bering Strait (Ahlnäs and Wendler, 1979; Stringer, 1981).

Gouging of the bottom by thickened sea-ice masses is common in the area north of the Yukon Delta, where westward-drifting ice from Norton Sound meets southward-drifting ice and creates many pressure ridges (Larsen, Nelson, and Thor, 1981). Gouges here occur mainly at depths of 10 to 20 m and are not known at depths >30 m or in areas of shore-fast ice. Sparse ephemeral gouges have been reported off the Seward Peninsula (Hunter, Thor, and Swisher, 1982). In both areas, gouges are 5 to 20 m wide (multiples as wide as 100 m), and the maximum depth of incision (gouge relief) is about 1 m (Larsen, Nelson, and Thor, 1981; Hunter, Thor, and Swisher, 1982). Disruption of the bottom sediment by ice gouging is thought to be the main triggering mechanism that leads to the erosion of scour depressions on the sea bottom off the Yukon Delta (Larsen, Nelson, and Thor, 1979).

Winter ice plays a role in the redistribution of sediment from the Yukon River, which brings an annual load of nearly 10^8 tons of silt and fine sand to its delta in Norton Sound. From freezing (October/November) to breakup (May), sea ice covers the very shallow (1 to 3 m) delta platform of extremely low slopes (<1 m/km), but sub-ice distributary channels that extend to 25 km offshore remain active in carrying sediment, as evidenced by continuation of the suspended sediment plume trailing northward (Drake, Totman, and Wiberg, 1979; Ray and Dupré, 1981). In spring, river flood waters may transport much sediment on top of the sea ice (Dupré, 1982).

Sea ice can directly transport gravel that freezes to it in very shallow water and along shores. Most such ice-rafted pebbles are carried only short distances, and petrographic studies in the western Bering Sea show dispersal trends radiating southwestward near the shore (East Kamchatka current) and northward from the Komandorsky Islands (Lisitsyn, 1966). Rare rounded (shore-derived) pebbles found in cores in Norton Sound (Howard and Nelson, 1982) and south of St. Lawrence Island (Knebel, 1972) are considered to be ice rafted. There are no data on the amount of sand transported by sea ice in the Bering Sea—the sand would not be easily recognized as ice rafted and would more readily be reworked by waves and currents.

In the deeper parts of the Bering Sea, slow settling of suspended silt, clay, and biogenic particles ls the dominant process. Large slumps and slides observed on the Beringian continental slope deeper than 200 m (Marlow et al., 1981; Carlson, Karl, and Edwards, 1982) may date from before the last sea-level rise, inasmuch as present sedimentation rates on the slope and rise appear to be low (Carlson and Karl, 1984a). The large submarine canyons have clearly acted as conduits for mass transport of sediment during past low sea-level stands, and the flat-lying fills of the abyssal basins have largely been built by turbidity currents from adjacent shelf areas. Such processes may be only rarely active today. Fields of sand waves in the heads of some canyons, however, are thought to have been formed by the action of internal waves (Karl, Cacchione, and Carlson, 1986) and may still be active.

Understanding the sediment dynamics throughout the Quaternary is crucial to reasonable assessment of potential sand and gravel resources and of possible problems in their exploitation. Attempts at such assessments have been made by Molnia (1979a), who covered all of offshore Alaska, and by Hess and Nelson (1982) for the Norton Sound area.

Active Sand and Gravel Bodies of the Bering Shelf Region

Active bodies of sand and gravel related to fluvial, marine-current, and wave action are common in the Bering Sea, especially on the northern Bering Shelf. Some doubtfully active deposits on the shelf and upper continental slope are also included here for discussion.

Beaches, Barrier Islands, and Offshore Bars

In the Bering Sea, few data are available on these shoreline sediment masses, which are normally among the most accessible and readily utilized sources of coarse material. Coarse gravel beaches are common on the western shores, from Bering Strait toward Kamchatka (Lisitsyn, 1966), and steep gravel beaches occur locally on the eastern coasts, such as at the cliffed shores of Cape Romanzof and Nelson Island in the Yukon Delta region (Dupré, 1977). Gravel beaches undoubtedly also occur in the Aleutian and Komandorsky Islands.

Sand beaches are common around the Yukon Delta on the eastern shores of the Bering Sea (Dupré, 1977), and barrier islands of sand occur near Cape Romanzof and west of the Kuskokwim River mouth (Nelson et al., 1982), where their extensions exposed at low tide are covered by sand waves and current ripples. Dupré (1977) found migration rates of 15 to 20 m/yr for barrier islands in the Yukon-Kuskokwim region.

With increasing tidal range, barrier islands give way to offshore bars and shoals flanking extensive tidal flats, as for instance south of Hooper Bay (Dupré, 1977) and near the Kuskokwim River mouth (Nelson et al., 1982).

Data on dimensions, grain size, migration rates, and sediment dynamics of these shoreline deposits are very limited.

Yukon Delta

The dominant source of terrigenous sediment to the Bering Sea is the Yukon River, which has been building a delta into the south side of Norton Sound for at least 1200 years and possibly for several thousand years (Dupré, 1977). Because this site is underlain by nonsubsiding older rocks covered by only thin relict nonmarine deposits (Howard and Nelson, 1982), the river has built an extensive sheetlike deposit. A flat delta platform extends under 1 to 3 m of water from the shoreline out as far as 30 km, where a steep delta front drops down to about 10-m depth (Larsen, Nelson, and Thor, 1981). The deltaic sediment of the platform is about 14 m thick and consists of very fine sand and coarse silt (Hess and Nelson, 1982). Several channels, 2 to 4 km wide and 5 to 15 m deep, cross the west side of the delta platform. The channels carry mainly very fine sand (Hess and Nelson, 1982), and their lateral migration therefore produces tabular sand bodies about 10 m thick (Nelson et al., 1982). No channels cross the north side of the delta, which appears to be eroding (Hunter, Thor, and Swisher, 1982).

Beyond the delta front, a sheetlike deposit of Yukon River sand a few meters thick extends through central Norton Sound but pinches out northward and also westward into Chirikov Basin (Hess and Nelson, 1982). Roughly half the annual increment of Yukon sand and silt is, however, transported by currents through the Chirikov Basin and out the Bering Strait into the Chukchi Sea (Nelson and Creager, 1977; Nelson, 1982). The prodelta sediment is actually somewhat coarser farther from the delta, probably reflecting the winnowing effect of these currents (Dupré, 1982).

The Yukon Delta deposit represents a very large quantity of almost clay-free very fine sand to coarse silt—an estimated 50 km^3 in the delta platform and perhaps another 20 km^3 in the sediment sheet beyond the platform. The effects of possible mining on ecology and sediment regimen could be far reaching, however, and are not presently predictable (Hess and Nelson, 1982).

Sand Waves, Shoals, and Other Bedforms

The strong water movements on the shallow Bering Shelf produce a variety and abundance of active coarse-sediment bedforms ranging from symmetrical and asymmetrical ripples to large leeside shoals. Such deposits have been most studied in Norton Sound, the Yukon Delta region, and other areas in the northeastern Bering Sea (Field et al., 1981; Dupré, 1982; Hunter, Thor, and Swisher, 1982; Nelson et al., 1982).

Large ridges of medium and fine sand off Port Clarence on the southwestern Seward Peninsula (Figure 9) are 15 to 30 km long, 3 to 7 km wide, and have 10 to 15 m relief. Oriented roughly parallel to shore (NW–SE), they have been interpreted as relict shoreline or nearshore sand bodies formed during the Holocene transgression (Field et al., 1981; Hess and Nelson, 1982). The ridges are being reworked and display a hierarchy of bedforms—large sand waves (spacing 200 to 500 m, height 2 m), smaller sand waves (spacing 10 to 20 m, height about 1 m), and linguoid current ripples (spacing <50 cm, height about 5 cm). All these bedforms are asymmetrical, with their steep (downcurrent) sides facing north (Field et al., 1981). Actively moving ripples have been observed by camera at 18-m depth, beneath a current of 23 cm/sec measured 1 m above the bed (Field et al., 1981). Sand waves 2 m thick have also been observed to migrate over a year's time (Larsen, Nelson, and Thor, 1981), even in the absence of major storms (Field et al., 1981).

Symmetrical ripples of coarse sand to pebble gravel, with wavelengths of 0.5 to 2.0 m, have been noted near the entrance to Port Clarence and off the Nome-Solomon area farther east along the Seward Peninsula. They have not been observed to be active and probably respond only to strong waves during major storms (Hunter, Thor, and Swisher, 1982). In the Port Clarence area, the ripples trend NW–SE, implying a relationship to storm waves from the southwest, whereas in the Nome area three different sets of ripples of different sizes (wavelengths from 0.4 to 2.0 m) have orientations from E–W to NW–SE and represent different wave patterns (Hunter, Thor, and Swisher, 1982).

Transverse asymmetrical bedforms partly of pebble gravel in 12 to 15 m of water off Safety Sound, west of Solomon, are of uncertain origin (Hunter, Thor, and Swisher, 1982). As high as 2.5 m and spaced about 200 m apart, they trend at high angles to the shore and their steep sides face west—possibly they are formed and moved under rare extreme current conditions. Ridges of boulder-size material in similar depths off Nome, however, are almost certainly relict and will be discussed later.

As a result of the strong northward currents through the northern Bering Sea, large sediment bodies have formed downcurrent from major land projections. Such leeside shoals occur on both east and west flanks of the north side of Shpanberg Strait, and to the north of the Yukon Delta, King Island (Figure 9), and Cape Prince of Wales at Bering Strait (Hess and Nelson, 1982; Nelson et al., 1982). They are 25 to 100 km long, 5 to 25 km wide, generally have 10 to 20 m of relief, and consist of very fine sand (about 0.1 mm) and some silt. Though they have not been deeply cored, these leeside shoals are probably reasonably uniform through their thickness, and they appear to be active and continuing to form today. They therefore represent a very large and theoretically replenishable source of very fine sand (Hess and Nelson, 1982). The several shoals together may contain >10 km^3 of clean, well-sorted sand.

Dupré (1982) described a series of long (as much as 50 km) but low (2 to 3 m relief) shoals on the northwest side of the

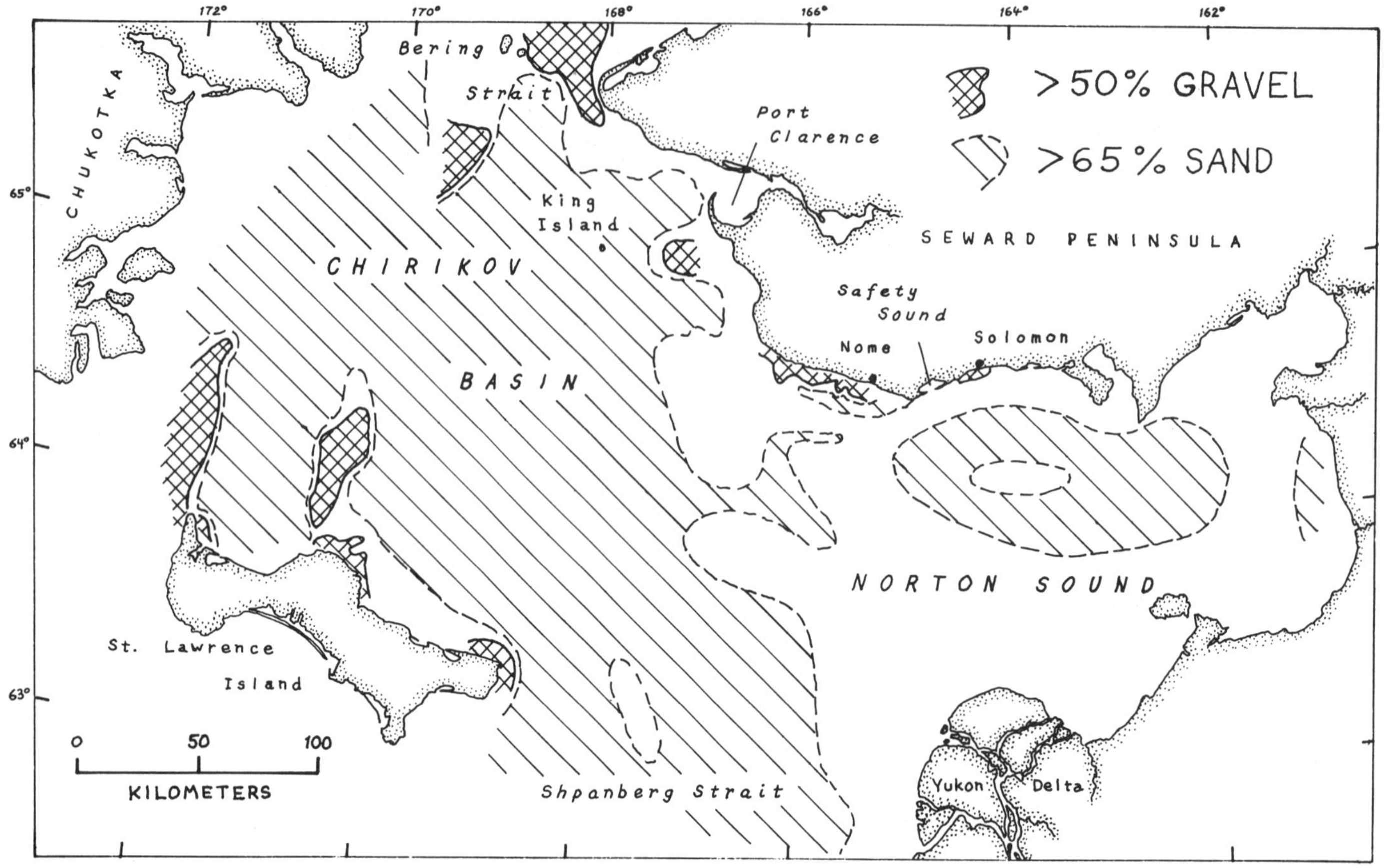

Figure 9. Areas of active bedforms of coarse material in the northeastern Chirikov Basin, Bering Sea. After Nelson et al. (1978).

Yukon Delta, which he interpreted as migrating northeastward into Norton Sound in response to storms or tidal currents. A variety of active sand waves and ripples occur along the delta front of the Yukon Delta and on the bottoms and flanks of the western distributary channels (Hunter, Thor, and Swisher, 1982).

In the funnel-shaped estuary of the Kuskokwim River, linear sand ridges attributed to the action of tidal currents are 4 to 50 km long, 0.5 to 4.0 km wide, and have relief of 4 to 10 m near the river mouth, increasing to 30 m offshore (Nelson et al., 1982). Some are asymmetrical and some slightly sigmoidal, and the general orientation is subparallel to the tidal currents in the estuary. As far as is known, the ridges consist of well-sorted fine sand, and sand waves and current ripples occur on their surfaces (Nelson et al., 1982).

Other areas of the Bering Sea are less well known. Active bedforms are widespread in the Chirikov Basin, especially north of western St. Lawrence Island (Nelson et al., 1978), but are apparently rare or absent in deeper shelf areas and in the deep Bering Sea basins. Tidal sand ridges and other active bedforms surely occur in shallow and nearshore areas along the Bristol Bay, Alaska Peninsula, Aleutian Island, and western Bering coasts, but data are lacking.

Other Sea-Bottom Occurrences of Coarse Material

Large areas of gravel on the northern Bering Sea floor are probably mainly relict Pleistocene deposits, but extensive areas of sand bottom may represent material in equilibrium with modern water dynamics. This sand is commonly rippled and in some areas forms irregular patches and ribbons alternating with (apparently underlying) exposed gravel (Hunter, Thor, and Swisher, 1982). In the northern Bering Sea, the sand is typically fine to very fine, but coarser sand occurs near the Seward Peninsula coast and around the eastern and western ends of St. Lawrence Island (Hess and Nelson, 1982). The sand probably is a very thin layer, except in the large-scale bedforms already described.

A zone of well-sorted fine sand at intermediate depths in Bristol Bay has been interpreted as the product of active winnowing mainly through storm waves (Sharma, Naidu, and Hood, 1972; Sharma, 1974, 1975), but this sand forms only a thin veneer and probably undergoes little real migration. The provenance of sand on the outer shelf in the southeast Bering region (Marlow et al., 1979) indicates that its distribution is not related to present current patterns (Gardner, Dean, and Vallier, 1980) and suggests that it is a relict deposit.

Along the northwestern part of the Beringian margin, the typical bottom sediment is mud, but sand and even gravel occur locally on the outer shelf, especially near the shelf edge (Marlow et al., 1981; Karl and Carlson 1982, 1984); this presumably reflects localized stronger bottom currents. The extent and thickness of these deposits are poorly known.

Extensive areas of sand are found along the western margin of the Bering Sea (Lisitsyn, 1966). Coarse- to medium-grained sands are found down to depths of 50 to 90 m, and fine sands occupy a belt as much as 170 km wide in the Gulf of Anadyr and extend to the shore in the west end of the

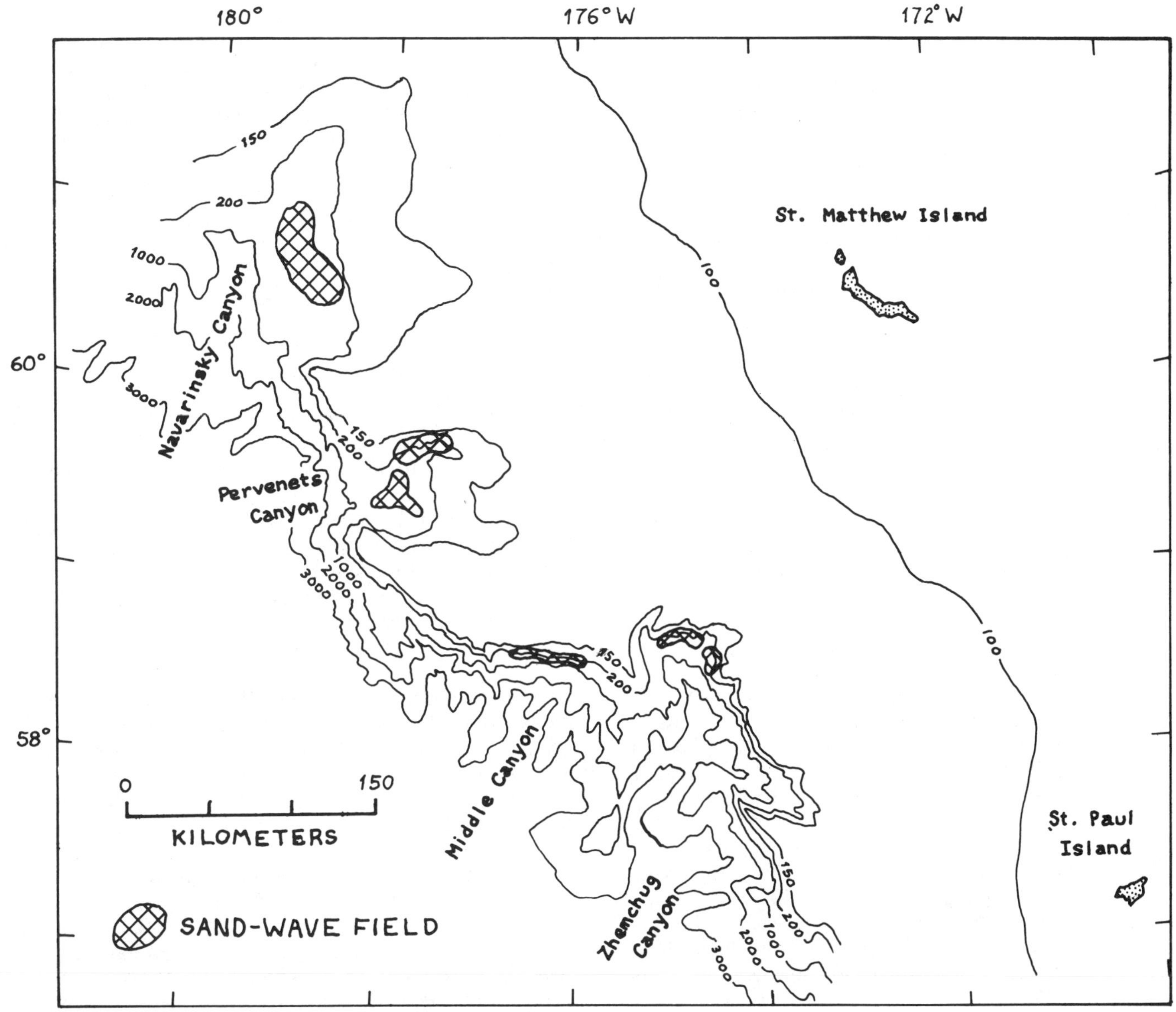

Figure 10. Sand-wave fields in submarine canyons of the Bering Sea continental slope. The deeper field in Pervenets Canyon consists almost entirely of buried bedforms. After Karl, Cacchione, and Carlson (1986).

Gulf. Along the stretch from Cape Navarin to Kamchatka, fine sand occurs down to depths of 80 to 100 m and locally much deeper. This fine sand is well sorted and probably frequently in motion, at least in depths of < 100 m (Lisitsyn, 1966).

Sand Bodies in Canyon Heads and Other Deep Sites

Fields of large bedforms of sand have been found in the heads of at least four of the large canyons that incise the Beringian margin (Karl and Carlson, 1982; Karl, Cacchione, and Carlson, 1986). The sand bodies are at depths of 175 to 490 m, but the bedforms are best developed in the range of 300 to 375 m (Figure 10). The largest and best studied of these remarkable sand bodies is the enormous field in the head of Navarinsky Canyon, which covers an area of about 1400 km^2. Its surface shows rhythmic waveforms, with typical wavelength about 650 m and typical height about 5 m; this set of waveforms is underlain by as many as six earlier similar wave sets, each set forming a cross-bedded packet of sediment as thick as 20 m, the lowest in turn underlain by flat reflectors. Surface samples and shallow cores show the sediment is fairly well sorted, fine to very fine sand. If this material is uniform throughout the body of sediment, the sand-wave field in Navarinsky Canyon alone must contain on the order of 75 km^3 of such material; the combined total in other canyon heads (Pervenets, Zhemchug, and Middle Canyons) may represent a comparable amount of material (Figure 10).

The origin of these sand waves is still uncertain, but their depth and morphology argue against both normal bottom currents and wave action. The most compelling suggestion is that they form in response to internal waves, amplified where they impinge on the continental slope (Karl, Cacchione, and Carlson, 1986). Tracing dated reflectors into the Navarinsky Canyon field suggests that most or all of the sand waves formed in the Pleistocene—the original source of the sand probably was rivers crossing the emerged shelf at times of low sea level. Even though no surface ripples have been observed, most of the large bedforms are not buried and may, therefore, still be active. Estimated wave migration rates of about 1 cm/yr over the last 30,000 years (Karl, Cacchione, and Carlson, 1986) suggest that the sand moves only occasionally. These unique sand bodies deserve further study, both for their scientific interest and as potential resources.

Possibly similar large sediment waveforms have been observed on seismic profiles in the Amukta Basin, on the north flank of the Aleutian Ridge, at about 600-m depth (Cooper et al., 1980).

Fossil Sand and Gravel Bodies in the Bering Sea

The climatic fluctuations of the Quaternary involved the movement of glaciers onto parts of the present Bering Shelf, the rise and fall of sea level, and the attendant passage of the shoreline repeatedly across most or all of that shelf. These processes have left behind a variety of abandoned or relict bodies of sand and gravel in the Bering Sea, including glacial, fluvial, shoreline, inner-shelf, and other types of deposits. Many of these have some overburden of more recent finer grained sediments.

Holocene Transgressive Sediment Sheet

The most widespread fossil deposit of coarse sediment in the Bering Sea is the transgressive sheet of sand and gravel laid down across the shelf during the last major rise of sea level in latest Pleistocene and early Holocene (hereafter referred to, for convenience, as the Holocene transgression). On the eastern Bering Shelf, cores that penetrate the Holocene sediment south and east of St. Matthew Island (Askren, 1972) and south and southwest of St. Lawrence Island (Knebel, 1972; Knebel, Creager, and Echols, 1974) commonly show a coarse basal transgressive layer generally 0.5–1.0 m thick, which is mostly sand but locally includes gravel. Norton Sound was rapidly transgressed in a few hundred years starting about 10,000 years ago, and only a very thin transgressive deposit of very fine sand and sandy silt was formed and has since been partly covered by sandy silt from the Yukon River (Hess and Nelson, 1982).

In the Chirikov Basin west of Norton Sound, however, the last transgression reworked coarse morainal and outwash deposits, producing a transgressive sediment sheet consisting in places of medium- and coarse-grained basal sand covered by a widespread transgressive deposit of fine sand and coarse silt generally <1 m thick (Hess and Nelson, 1982). Because strong water movements have persisted in the Chirikov Basin, in some areas this transgressive deposit has never been buried and may in fact have been extensively reworked since the transgression.

The Quaternary section is less well known in other areas, but a thin transgressive sheet of at least fine sand may be expected to be widespread on the outer Bering Shelf, to water depths of at least 100 m; coarser transgressive sediment probably occurs in rocky and glaciated areas near the coast and around islands. Sea-floor gravel near the Pribilof Islands may be a relict deposit of this type.

Submerged Pleistocene Shoreline Deposits

During periods of stable lower sea level, shoreline processes formed beaches and offshore bars in areas now submerged. Some of these features, particularly those formed during temporary stillstands of the last transgression, are still preserved on or under the sea bottom in the Bering Sea. The best documented are off the coast of the Seward Peninsula, where groups of parallel submarine ridges (Figure 9) have been mapped off both Port Clarence (Nelson et al., 1978) and Nome (Nelson and Hopkins, 1972; Tagg and Greene, 1973; Nelson et al., 1982). Such ridges occur at various water depths between 11 and 38 m, but some of them are small. The material in the ridges is mostly gravel in the Nome area (Nelson and Hopkins, 1972), but is well-sorted fine to medium sand off Port Clarence (Nelson et al., 1982), where the tops of the ridges are reworked into sand waves and ripples, described earlier.

A submerged gravel ridge extending 30 km northeastward from near the northwestern cape of St. Lawrence Island (Figure 8; McManus et al., 1977) has been interpreted as an ancient shoreline deposit formed when sea level was 30 m lower than at present (Hess and Nelson, 1982; Nelson et al., 1982). The fine gravel to coarse sand of this ridge aggregate to at least several million m^3 of reasonably well-sorted coarse material. Fine sand rich in heavy minerals on a ridge top at a water depth of 20 m in Shpanberg Strait may represent an ancient shoreline deposit (McManus et al., 1977).

On the southeastern Bering Shelf, a former shore zone formed by a sea level 74 m below present has been proposed from bathymetric data (Askren, 1972), but there is no evidence from sediment to confirm this interpretation. Near the shelf edge in the same region, longshore transport at a sea level perhaps 130 m below the present one has been inferred on the basis of sand provenance, but no specific shoreline deposits have been identified (Gardner et al., 1979; Gardner, Dean, and Vallier, 1980).

Topographic terraces on the Beringian continental slope as deep as 240 m have been suggested to represent Pleistocene shoreline features (Pratt and Dill, 1974). This suggestion also has no confirmation from sediment data, and any beach or nearshore coarse sediment on such terraces would probably now be buried under 1 to 3 m of Holocene mud.

Pleistocene Morainal and Outwash Deposits

Where Pleistocene glaciers reached areas below the present shoreline level of the Bering Sea, their moraines were left behind on what is now the sea bottom. Such submerged glacial ridges are widespread in the western Chirikov Basin and near the USSR coasts (Lisitsyn, 1966). In areas off the Koryak coast and off northern Kamchatka, morainal ridges are associated with fjords and generally limited to the inner shelf, but an extensive terminal-moraine system as much as 75 km offshore stretches from St. Lawrence Island to Bering Strait. The surface material in these deposits is angular coarse gravel, which locally shows bedforms and cross lamination (Nelson et al., 1982) and may be a lag gravel derived by winnowing of fines.

Boulder-gravel ridges trend at high angles to the shoreline in 12 to 15 m of water off the Nome area of the Seward Peninsula (Hunter, Thor, and Swisher, 1982). Composed of glacial till, a gravelly sandy silt of unknown thickness, they are covered by a thin (<30 cm) coarse lag gravel (Hess and Nelson, 1982), and near Nome they extend only a few kilometers offshore. A submerged outwash fan has been mapped from the shoreline out to about 5 km seaward east of Nome (Nelson and Hopkins, 1972; Nelson et al., 1982). The fan consists of gravel and is at least several meters thick, but its volume, size distribution, and stratigraphy are essentially unknown (Hess and Nelson, 1982).

Glaciers also reached the present shoreline in the Bristol Bay area (Porter, 1967; Nelson, Hopkins, and Scholl, 1974) and probably locally along the Alaska Peninsula and Aleutian Island coasts, but no submerged moraines or outwash deposits have been documented in those areas.

Buried Channel Deposits

During low sea-level stands, rivers draining adjacent land masses and meltwater streams from glaciers carved fluvial and deltaic channels across areas now submerged. Such channels, now filled and buried, have been documented on seismic sections, but their connections and former drainage patterns have in general not been mapped.

On the northern Bering Shelf, subsurface alluvial channels have been recorded in Norton Sound (Nelson, 1982), off Nome (Tagg and Greene, 1973), and in parts of the Chirikov Basin (Hess and Nelson, 1982). The channels off Nome are discontinuous, have as much as 10 m of fill, and commonly (curiously) run nearly parallel to shore (Tagg and Greene, 1973). Large and deeply incised (as much as 100 m) filled channels in the Chirikov Basin indicate drainage west from Port Clarence, south from King Island, and a coalescing system in the central part of the basin that drained northward toward Bering Strait (Gordon R. Hess, 1984, oral communication). Most of the fill of these central Chirikov Basin channels shows sharp and continuous seismic reflectors and is, therefore, probably fine-grained material (clay and silt). Some coarse sediments (channel-lag and point-bar deposits) presumably occur in the seismically structureless lowermost parts of the fill sections, but these appear to be too remote and uncertain to be regarded as resources. Former channels have been inferred in the east-central Bering Sea (Knebel, 1972; Knebel and Creager, 1973), but data for these are sparse.

During at least some Pleistocene low sea-level stands, the Yukon River flowed southward and entered the sea near the Pribilof Islands, where marine mass transport of its sediments helped to cut Pribilof and perhaps Zhemchug Canyons (Scholl et al., 1970; Knebel and Creager, 1973). Nonetheless, seismic traverses on the outer shelf in the Pribilof region have failed to reveal any major former alluvial channels either on the surface or buried (Gardner et al., 1979; Gardner, Dean, and Vallier, 1980). Similarly, there is little evidence of major buried alluvial channels on the outer Bering Shelf near the other large canyons farther to the northwest. Some small, scattered channels are seen near the actual canyon heads (Carlson and Karl, 1984b), and seismic traverses in the Gulf of Anadyr record crossings of buried channels that trend toward Navarinsky Canyon (Paul R. Carlson, 1983, oral communication). The apparent absence of deeply incised channels on the outer Bering Shelf may be a result of the morphology of the rivers during low sea-level stands. Because the gradient of the shelf is extremely low, rivers may have crossed it in very broad, shallow, even perhaps braided and bifurcating channel systems; thus no deep channels may ever have been cut.

Other Fossil or Relict Deposits

Gravity cores 3 to 5 m long from the floors of canyons incised into the Berinian margin in some places show thin layers of sand or silt or even intervals of pebbly sediment (Carlson and Karl, 1984a). These coarse materials are thought to be mainly Pleistocene in age and probably do not reflect present sediment dynamics. The distribution of these materials appears to be discontinuous and narrow and the quantity very limited.

Larger amounts of coarse material undoubtedly exist in the form of turbidite beds (and also ice-rafted debris) in the deep-sea fans of the continental rises and the flat fillings of the abyssal plains in the deep Bering Sea basins. Our knowledge of the shelf sediment and fluvial sources, however, suggests that very little of this is coarser than fine to very fine sand, except possibly that adjacent to the Kamchatka and Koryak coasts.

A large, gentle lobate shoal in the east-central Bering Sea, referred to as St. Lawrence Bank by Knebel (1972), has been interpreted by him to be a pre-Holocene sedimentary deposit related to a former Yukon River course entering the sea in the Cape Romanzof region. It might, therefore, contain a large quantity of probably rather fine-grained sand.

Stabilized Pleistocene eolian sand dunes occur onshore in the Yukon Delta (Dupré, 1977) and also on St. Paul Island in the Pribilofs (Knebel and Creager, 1973). During glacial low sea-level periods, similar eolian sand dunes probably formed on parts of the exposed Bering Shelf, but no preserved examples have been identified. They may have been mostly destroyed by reworking during the Holocene transgression.

GULF OF ALASKA

The Pacific Ocean margin of Alaska is a region of varied physiography and geology. It stretches more than 2000 km in a gentle arc around the Gulf of Alaska, from the tip of the Alaska Peninsula in the southwest to Dixon Entrance at the Canadian border in the southeast (Figure 11). The actual coastline is much longer, especially in southeastern Alaska, which has been estimated to have a coastline length of about 50,000 km (Sharma, 1979).

The width of the continental shelf in this region, as defined by the 200-m isobath, ranges from only 20 to 30 km in parts of southeastern Alaska to as much as 300 km in the northwest and off the Alaska Peninsula. The shelf is indented and crossed by broad sea valleys, and glacially carved depressions on the inner shelf and in embayments such as Prince William Sound reach depths of several hundred meters. The physiography of the shelf is varied and is well known only in a few areas.

The Kodiak Shelf off Kodiak Island (Figure 12) is as much as 100 km wide and shows large flat banks at depths of 50 to 100 m, separated by transverse troughs with water depths to nearly 200 m (Turner et al., 1979; Dunlavey, Childs, and von Huene, 1980). Between Kodiak Island and the Alaska Peninsula, the 50-km-wide Shelikof Strait slopes gently downward from 150-m depth at its northeastern end to >200 m in the southwestern part. Narrow marginal channels on the northwest and southeast sides reach maximum depths of 270 m and 240 m, respectively, and local depressions and rises have relief of 20 to 100 m (Whitney and Holden, 1980; Hampton et al., 1981). Depths in Cook Inlet, which extends from the northeast end of Shelikof Strait about 300 km between the Kenai Peninsula and the mainland, are mostly

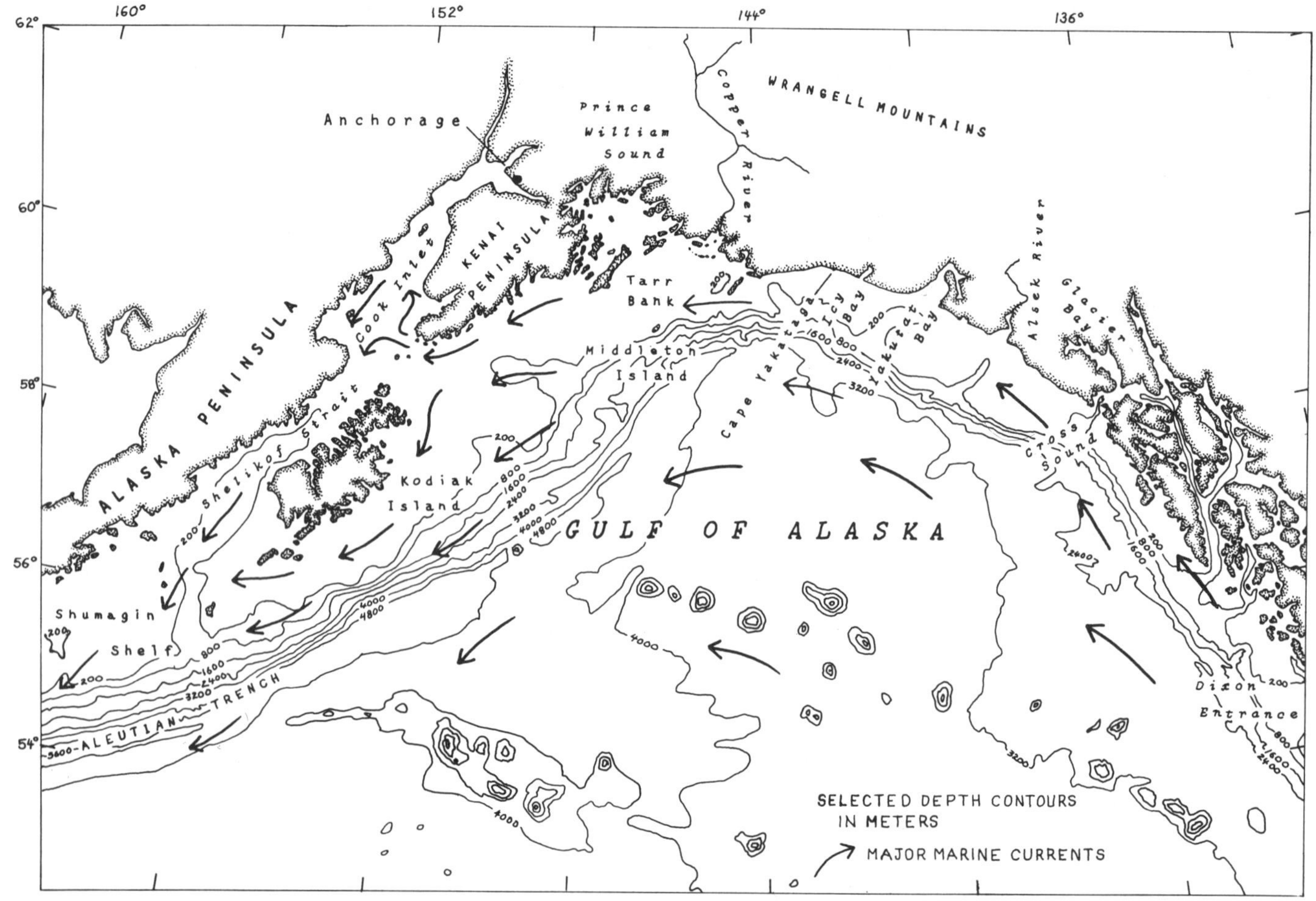

Figure 11. Geographic and bathymetric index and pattern of major oceanic currents for the Gulf of Alaska region. Base and bathymetry from Alpha and Swenson (1978). Marine current pattern generalized in part from Muench and Schumacher (1980) and Reimnitz (1966).

< 100 m but increase to 150 m southward toward Shelikof Strait. The physiography of the bottom is locally complex (Bouma, 1981; Rappeport, 1982).

Prince William Sound forms a nearly enclosed complex fjord east of the Kenai Peninsula (Figure 11). The sills connecting it with the open shelf are < 150 m deep, but troughs and depressions within the sound reach maximum depths as great as 800 m (Muench and Schmidt, 1975).

In the 650-km-long stretch between Prince William Sound and Cross Sound, the continental shelf is generally 50 to 100 km wide but is crossed by several broad submarine troughs (Carlson et al., 1982). The coastline also is indented by embayments, the largest being the 60-km-long Yakutat Bay, which has a maximum depth of 290 m (Wright, 1972b).

In southeastern Alaska, the continental shelf seaward of the main islands is only 20 to 40 km wide and has been little studied (Carlson and Schwab, 1982). Landward of the outer islands, numerous inlets and channels with maximum water depths of as much as 835 m form a complex fjord system, also as yet poorly known (Sharma, 1979).

The continental slope is steep in the western Gulf of Alaska; gradients on the upper slope along the Kodiak Shelf are as high as 20° (Hampton, 1982a), but are generally less steep in the northern and eastern Gulf. The slope leads down to abyssal plains and deep-sea floor at depths of 2500 to 3500 m, except in the west, where the Aleutian Trench reaches depths of 5000 to 6000 m.

The bedrock geology and tectonic history of the Gulf of Alaska region are as complex as its topography (Bruns et al., this volume; Fisher, Detterman, and Magoon, this volume; von Huene, Fisher, and Bruns, this volume; Plafker, this volume). Unlike the other continental margins around Alaska, the Gulf of Alaska continues to be the site of active tectonism, seismicity, and volcanism—oceanic crust of the Pacific plate is subducting and underthrusting Alaska in a northwestward direction, giving rise to the volcanic arc of the Alaska Peninsula and the Aleutian Islands. In southeastern Alaska, the Queen Charlotte-Fairweather fault system appears to represent a strike-slip boundary between the Pacific and North American plates (Carlson et al., 1979a).

Parts of Alaska adjoining the Pacific Ocean are thought to be formed of separate terranes, accreted to North America at various times and redistributed by strike-slip motion (Jones, Silberling, and Hillhouse, 1977; Jones et al., 1981; Bruns, 1983a). These consist dominantly of Paleozoic and Mesozoic metasedimentary and metavolcanic rocks, granitic intrusive rocks, late Mesozoic flysch, and Cenozoic sedimentary sequences (Burk, 1965; Moore, 1969; Reimnitz and Plafker, 1976).

Two belts of Cenozoic sedimentary basins are spread along the western and central Gulf of Alaska margin (Plafker and Rubin, 1967; Fisher, Bruns, and von Huene, 1981; Ehm, 1983). The inner belt includes the Cook Inlet area, where the thickness of basin fill reaches > 7 km (Magoon et al., 1979;

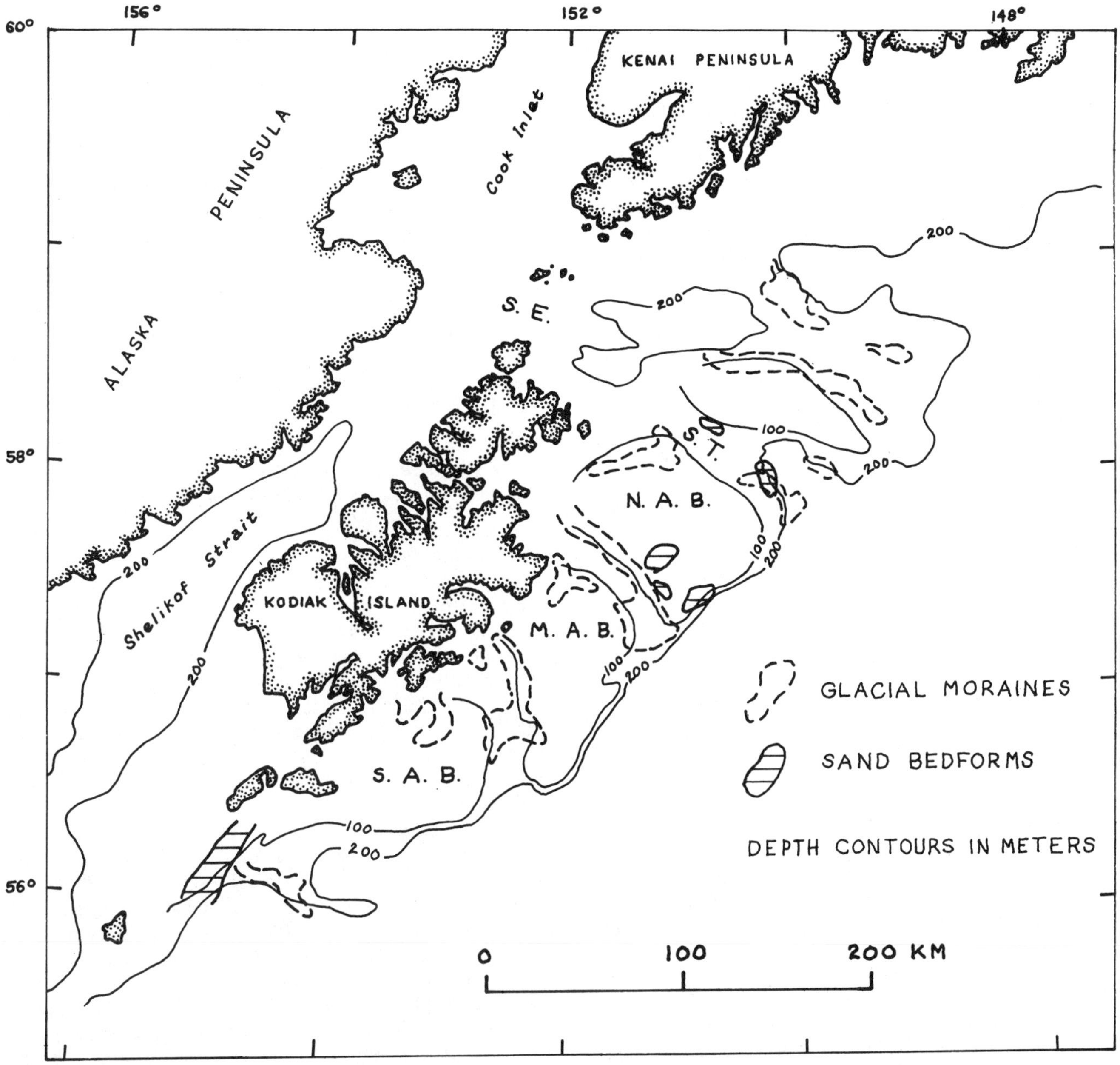

Figure 12. Areas of surficial sand and gravel accumulations on the Kodiak Shelf. Compiled from several maps in Thrasher (1979). The lateral and terminal glacial moraines are mainly gravelly sand (Thrasher's unit Qgm); the sand bedforms consist of clean, well-sorted sand of various grain sizes (Thrasher's unit Qb). S.E., Stevenson Entrance; S.T., Stevenson Trough; N.A.B., Northern Albatross Bank; M.A.B., Middle Albatross Bank; S.A.B., Southern Albatross Bank.

Fisher, Detterman, and Magoon, this volume), Shelikof Strait, and several shallow basins on the shelf to the southwest (Figure 11). The more continuous outer belt of basins includes the Shumagin and Kodiak Shelf basins with Neogene sedimentary deposits 1 to 7 km thick (Fisher and von Huene, 1980; Bruns et al., this volume) and basins in the Middleton Island to Yakutat sector with Cenozoic fill as thick as 9 km (Bruns, 1983a; Ehm, 1983; Plafker, this volume).

The Quaternary history of the Gulf of Alaska region has been dominated by active tectonics and fluctuating glaciation. The geography and climate of the region strongly favor development of glaciers that reach the sea—high mountains near the coast receive heavy precipitation (commonly >2.5 m per year, which is at least five times that in interior Alaska). This high precipitation, unusual at such latitudes, makes the snowline >100 m lower on the coastal side of the mountains than on the inland side (Karlstrom, 1964) and feeds large areas of ice, including a few glaciers that reach the sea even today.

Because of this favorable setting, Cenozoic glaciation began here well before the Quaternary. Evidence from the northwest Pacific Ocean of cooling since the mid-Tertiary (Addicott, 1969) is supported by the occurrence of ice-rafted debris in the Yakataga Formation of the central gulf coast (Plafker and Addicott, 1976): Tidewater glaciers were evidently present intermittently in the late Miocene and essentially continuously since the early Pliocene. During glacial maxima, when sea level was much lower, ice covered much or all of the present continental shelf in the gulf. Relict nonglacial depositional and erosional features are rare compared with features produced by glacial action. Thick, uplifted Pliocene and Pleistocene glacial-marine strata attest to Quaternary tectonics (Plafker and Addicott, 1976; Plafker, this volume).

Because of differential tectonics, some parts of the shelf have undergone erosion during the Quaternary and others have accumulated thick sedimentary sections. In nearshore areas, deep troughs were cleaned out to bedrock in places during glacial advances, but rapid deposition near the fronts of glaciers has produced local thick sections.

On the Kodiak Shelf, glacial moraines are present out to the shelf edge (Thrasher, 1979). The large banks in this area expose truncated, gently folded Pliocene and Pleistocene rocks (Fisher and von Huene, 1980), locally covered by glacial deposits (Thrasher, 1979), whereas soft Holocene mud blankets troughs between banks. The unconsolidated sediment (latest Pleistocene? and Holocene) ranges in thickness from zero over much of the banks to >40 m in the troughs, and it thickens locally to > 100 m both shoreward and onto the continental slope (Hampton, 1982a).

Pleistocene glaciers entered Shelikof Strait from both the Kodiak side (Maddren, 1919) and the Alaska Peninsula (Burk, 1965), and a Pleistocene glacial unit inferred from seismic profiles has a thickness of 200 to 800 m in the southwestern part of the strait (Hampton, 1985). This unit pinches out to the northeast under a pervasive blanket of inferred Holocene sediment, which shows flat and regular bedding and is generally > 100 m thick, except near the northwestern and southeastern coasts (Hampton, 1985).

Quaternary sediment in lower Cook Inlet thickens from 20 to 40 m in the northeast and west to 60 to 100 m in the central part and locally to >200 m in the south (Hampton, 1982b). Rappeport (1980) recognized four units—a basal morainal unit as much as 60 m thick; a thin (1 to 3 m) gravelly layer of possible outwash; local large-scale sand-wave complexes; and a discontinuous but locally thick (as much as 75 m) unit of sandy, water-laid glacial debris. The last may be chiefly lacustrine deposits, because other evidence indicates a large proglacial lake occupied central and upper Cook Inlet during the last glacial maximum (Karlstrom, 1964).

In Prince William Sound, glaciers sculpted troughs in the pre-Quaternary bedrock down to 400–800 m below present sea level; these were later modified by infilling of glacial drift and a coating of Holocene marine sediment, including some ice-rafted debris (von Huene, Shor, and Reimnitz, 1967). Fresh-water peat found beneath marine deposits 16 m below present sea level in a northern arm of the sound was dated at about 9.5 ka, showing that glacial retreat preceded the Holocene transgression there (Williams and Coulter, 1981).

Between Prince William Sound and Cross Sound, at least eight glacial valleys plastered with morainal material cross the shelf (Carlson et al., 1982). The thickness of Quaternary sediment is varied and difficult to determine, because glacial-marine deposits extend down well into the Tertiary section (Plafker and Addicott, 1976; Plafker, 1981; Armentrout, 1983), and local thickness is much affected by differential tectonics; Bruns and Schwab (1983) inferred thicknesses from a few hundred meters to 3000 m on the shelf near Yakataga. Pre-Quaternary bedrock is exposed in some areas on the outer shelf and the continental slope (Plafker et al., 1980). Holocene shelf sediment includes nearshore glacial-marine deposits as much as 100–300 m thick and local morainal deposits near the ends of glaciers (Molnia and Carlson, 1978, 1980). The Holocene Copper River deltaic deposit is as much as 180 m thick (Reimnitz, 1966). Benthic foraminiferal faunas indicate Holocene sediments on the outer shelf are typically < 1 m thick and may include a thin transgressive gravel (Quinterno, Carlson, and Molnia, 1981).

In southeastern Alaska, Holocene sediment is thin or absent over most of the continental shelf, and consolidated older materials are widely exposed (Carlson and Schwab, 1982), but data on Pleistocene sediments are still sketchy. In the fjords and inlets, Holocene mud occurs in the basins, but coarser sediment is locally present on the fjord flanks and near sediment sources (Sharma, 1979). Some Glacier Bay inlets deglaciated during the past 200 years have accumulated 100 to 200 m of post-glacial sediment (Carlson et al., 1979b, 1983).

Present Sedimentary Processes on the Gulf of Alaska Shelves

Ocean-water circulation in the Gulf of Alaska is dominated by a counterclockwise gyre manifested on the shelf and in coastal areas as generally westward currents, including the Alaska Stream, which impinges on the outer edge of the shelf with velocities of 30 to 50 cm/sec (Favorite, 1967; Muench and Schumacher, 1980), and westward currents nearer shore. This general westward movement is reflected in clay-mineral distribution on the shelf (Molnia and Fuller, 1977) and in westward transport of fine sediment from the Copper River (Reimnitz, 1966), which supplies clays to the shelves off Kodiak Island and Kenai Peninsula (Hein et al., 1979).

A westward flow, called the Kenai current, enters lower Cook Inlet south of the Kenai Peninsula. In lower Cook Inlet, this current typically has velocities of 10 to 30 cm/sec, though speeds as great as 100 cm/sec have been reported in the fall (Schumacher and Reed, 1980). At the west side of Cook Inlet, this current joins the net southward flow produced by river input to the upper inlet to form a strong southward current through the Shelikof Strait (Muench and Schumacher, 1980), at whose entrance an average velocity of about 50 cm/sec has been measured (Muench, Mofjeld, and Charnell, 1978).

Southwestward flow produced by the Alaska Stream dominates on the Kodiak Shelf (Muench and Schumacher, 1980), where thickness variations of the 1912 Katmai volcanic ash show erosion from the high areas of banks and redeposition in depressions and troughs (Hampton et al., 1979).

In embayments and coastal areas, tidal currents tend to dominate the short-term water movements. Average tidal ranges are moderate (2 to 4 m) on open coastal stretches, but rise in embayments and fjords to 5 to 6 m and at Anchorage, near the head of Cook Inlet, to 8 to 10 m (U.S. Department of Commerce, 1985). Tidal currents are strong in embayments and passages, such as the entrances to Prince William Sound, and inlets in the Copper River Delta area, where they reach 1 to 1.5 m/sec (Reimnitz, 1966). At the upper end of Cook Inlet, the maximum spring tide range reaches 11.4 m (Ovenshine et al., 1976), and strong tidal currents are probably the dominant factor in sediment movement. The flood is strongest in the eastern part of the lower inlet (Muench and Schumacher, 1980), where velocities commonly reach 60 to 70 cm/sec near the bottom (Rappeport, 1980; Mahmood, Ehlers, and Cilweck, 1981), and surface currents as fast as 7 m/sec have been reported (Hayes and Michel, 1982). In the narrow upper inlet, the flood becomes a tidal bore, with a velocity of 3 m/sec or more (Ovenshine et al., 1976). The ebb flow is weaker than the flood and is concentrated toward the west side of the lower inlet (Rappeport, 1982). Nonetheless, the sand waves in the central inlet show asymmetry, implying southward migration with the ebb direction (Hampton, 1982b).

Peak currents during the tidal cycle are able to move sand in most of Cook Inlet, as locally confirmed by bottom photographs (Bouma et al., 1979, 1980). Bottom sediments in Cook Inlet, mainly Pleistocene glacial deposits, are being eroded and reworked, leaving behind locally a coarse lag deposit. Most winnowed sand is redeposited in the inlet, but winnowed mud is carried into Shelikof Strait (Hampton, 1982b, 1985).

Elsewhere in the Gulf of Alaska, tidal currents are generally weaker. Their effect on the bottom is especially reduced where cold saline water has become stagnant in closed depressions on the shelf or in fjords (Sharma, 1979).

Winter cyclonic storms are frequent and severe in the Gulf of Alaska, carrying wind velocities of 135 km/hr or more (Rappeport, 1982) and producing waves as high as 15 m (Carlson et al., 1977). Although such waves occasionally disturb the bottom to depths of about 200 m (Carlson et al., 1977), the seaward limit, at about 66-m depth, of a nearshore belt of sand in the Yakataga-Yakutat area probably marks the limit of effective wave action in the movement of bottom sediment in that area (Armentrout, 1980); similar evidence suggests that effective wave base in front of the Copper River Delta is about 40 m (Reimnitz, 1966).

Nearshore easterly winds from cyclonic winter storms centered in the gulf and large storm waves arriving from the south and southeast cause strong erosion and westward longshore sediment transport on exposed sections of coast. The 500-km stretch between Cross Sound and the entrances to Prince William Sound (Figure 11) is characterized by coarse-grained beaches and rapid sediment movement and coastal changes; about half of this stretch of coast is retreating. Erosional scarps backing beaches of coarse sand and gravel show recent rates of retreat as high as 1.5 to 2 m/yr (Molnia and Wheeler, 1978). Some areas have retreated >2 km in 80 years (Molnia, 1980), and in the Icy Bay area local rates of retreat reach a phenomenal 90 m/yr (Molnia, 1977b). The fate of all the eroded material is not documented. Despite such short-term retreat, however, the central Gulf of Alaska is fundamentally an accreting coast because of periodic tectonic uplift (Molnia, 1977b).

At the Copper River Delta, wave action and a large fluvial sediment supply have built an 85-km-long string of barrier islands composed of clean sand and subordinate gravel. The material moves westward in longshore transport, and the barrier system progrades seaward (Reimnitz, 1966). The difference between this barrier system and the rapidly retreating coastline to the east reflects the large volume of sediment supplied by the Copper River, estimated to be about 6×10^7 m^3/yr (Reimnitz, 1966).

Wave action is comparatively weak in Cook Inlet, protected from oceanic swell and large storm waves by islands and narrow passages, and typical waves there are only about 0.5 m high (Rappeport, 1982). However, storm waves can reach 4-m height and may move fine sand on the bottom to depths of about 70 m (Rappeport, 1982). Large, essentially tidal sand waves in the central part of the inlet are probably activated by major storms (Mahmood, Ehlers, and Cilweck, 1981), and their occurrence to depths as great as 120 m (Whitney and Thurston, 1981) suggests either that some of them are relict or that unusually strong storms are important activating agents.

Longshore sediment movement in lower Cook Inlet tends to be dominantly south to north along the northeastern shore, but also into all major embayments from both sides, because of wave refraction (Hayes and Michel, 1982). About 45% of the coast in the lower inlet is erosional (Hayes and Michel, 1982), but no data are available on retreat rates or the fate of the eroded material.

Sandy beaches along the west coast of Kodiak Island contain gold placers derived from eroding coastal bluffs of glacial till. Inferred high rates of coastal retreat there (geomorphic features suggest 1 to 5 km since deglaciation) are supported by the low concentration of gold in the till, implying erosion of large amounts of it to produce the placers (Maddren, 1919).

Winds in the coastal areas of the Gulf of Alaska are frequently strong enough to transport fine sand, and strong northerly winds from the interior blow offshore in winter (Reimnitz, 1966). These winds have formed eolian dunes in the lower Copper River Valley (Reimnitz, 1966), and at times significant amounts of silt are blown at least several kilometers offshore (Swift, Molnia, and Jackson, 1978).

The role of ice in the marine dynamics of the Gulf of Alaska is limited. The open sea does not freeze; only parts of Cook Inlet (Hayes and Michel, 1982) and other embayments and some nearshore areas (Reimnitz, 1966) are known to freeze over in winter. Dominant easterly winds blow rafts of ice broken by tides and wave action westward and southwestward, causing sea ice to accumulate in the southwestern part of Cook Inlet (Hayes and Michel, 1982). Drifting ice is known to carve furrows in the tidal flats of the Copper River Delta (Reimnitz, 1966), but ice gouging is an extremely minor process in the gulf, because deep-draft pressure ridges of sea ice generally do not form.

Deep-draft icebergs are, however, produced by La Perouse Glacier, northwest of Glacier Bay, the only tidewater glacier in North America calving directly into open ocean, and by other glaciers in fjords and embayments (Molnia and Wheeler, 1978). In Prince William Sound, ice-rafting occurs now only in the northeastern part (von Huene, Shor, and Reimnitz, 1967), and ice rafting of sediment is probably a minor process on the open shelf, where icebergs are rare (Reimnitz, 1966). A small advance of the glaciers could increase ice-rafting manyfold, because numerous glaciers, including some large piedmont glaciers, have termini close to the coast now.

Pebbles, cobbles, and boulders attributed to ice-rafting have been observed scattered in lower Cook Inlet (Rappeport, 1980) and on the open shelf in the head of the gulf (Armentrout, 1980). The characteristic presence of ice-rafted coarse material throughout the Yakataga Formation indicates that that mode of transport has been important on the shelf from the Miocene to the Holocene (Plafker and Addicott, 1976).

Sediment supply to the Gulf of Alaska marine environment is from several sources. Large rivers are few because the drainage divide is generally not far inland, and major fluvial sediment influx is seen only at the Copper River Delta, the head of Cook Inlet, and at the mouth of the Alsek River. Copper River sand and gravel build an asymmetric prograding delta; carried mainly westward, they have started to fill one of the eastern entrances to Prince William Sound at Middle Ground Shoal (Reimnitz, 1966). Fine material is transported to the southwest, and a Copper River assemblage of clay minerals dominates modern sediment off the Kenai Peninsula, on the Kodiak Shelf, and in the eastern part of lower Cook Inlet (Hein et al., 1979).

Large amounts of coarse-grained material are dumped near the shoreline of the central and eastern Gulf of Alaska by tidewater glaciers and short outwash streams. Water-washed coarse material and unsorted glacial debris accumulate near where they are dumped, whereas glacial fines, mainly silt, go into suspension and later deposit by fallout from turbid plumes (Hoskin and Burrell, 1972). These suspended fines are moved westward parallel to the shore by the dominant currents, and hence the thickness of the Holocene accumulation of such deposits decreases rapidly offshore. Absence of gravel on the floor of a fjord in Glacier Bay suggests that neither ice-rafting nor slumping from marginal outwash deltas is an important process today in moving coarse material (Hoskin and Burrell, 1972).

Sedimentation rates differ greatly. Short-term rates of 1 to 5 m/yr have been determined in fjords and embayments still backed by active glaciers (Carlson et al., 1979b, 1983; Molnia, 1979b). Such high rates are local and temporary; glacial retreat or filling of the fjords can soon reduce them, while glacial readvance would destroy the sedimentary accumulations.

Long-term sedimentation rates on the shelf can be estimated by noting the total thickness of unconsolidated sediment deposited since the Holocene transgression (10,000 to 12,000 years). This method yields rates ranging from <0.2 cm/yr on most of the outer shelf to 1 to 4 cm/yr close to shore, and 5 to 175 cm/yr in bays and inlets (Molnia and Carlson, 1978; Molnia, Levy, and Carlson, 1980).

Pre-Holocene rocks and sediments crop out on the sea floor over much of Tarr Bank (Figure 11) and on some shelf areas in the central gulf (Carlson et al., 1977; Molnia and Carlson, 1978), and also on banks of the Kodiak Shelf (Hampton, 1982a), indicating net erosion or nondeposition during the Holocene. Sea-floor expression of folds in Cenozoic sediments on the Kodiak Shelf (Hampton, 1982a) and the northern Gulf of Alaska (Bruns, 1983b; Bruns and Schwab, 1983) and vertical fault displacement during earthquakes (Plafker and Rubin, 1967) indicate continuing tectonic control on net sedimentation rates.

An additional source of sediment in the western Gulf of Alaska is pyroclastic material from the Aleutian arc volcanoes. The 1912 Katmai ash, which includes pumice up to coarse sand size, is recognizable on the Kodiak Shelf as a thin layer, locally washed into accumulations of 1 m or more. Because little other sediment is supplied to the Kodiak Shelf, volcanic ash forms a major part of the modern sediment there (Hampton et al., 1979).

Cook Inlet is mainly an area of nondeposition at present, except for the uppermost parts, the sloping southern edge of the sand-wave field, and local sheltered nearshore areas of mud deposition (Hampton, 1982b). In Shelikof Strait, a thick wedge of Holocene sediment yields maximum sedimentation rates on the order of 1 cm/yr (Hampton et al., 1981).

Other processes are active in the Gulf of Alaska offshore. Slumps, often of large scale, occur on low slopes (Carlson and Molnia, 1977; Carlson, 1978; Hampton et al., 1978; Hoose and Whitney, 1980). Turbidity currents and other gravity flows carry sand from shallow water down into the Aleutlan Trench and sea valleys of the central gulf shelf (Armentrout, 1980; Carlson et al., 1982). Sea-floor "pockmarks" and craters form by escape of gas from underlying sediment (Molnia, 1979b, 1981); similar craters form also in gas-poor areas, possibly by different processes involving local liquefaction and compaction of volcanic ash (Hampton et al., 1981) or dewatering of underlying sediments (M. A. Hampton, 1984, oral communication).

Active Sand and Gravel Bodies on the Gulf of Alaska Shelves

Most sediment supplied to the Gulf of Alaska is glacial debris consisting largely of silt and coarse material up to very large size. Reworking by marine currents, tides, and wave action sorts the coarser material into accumulations nearly free of silt and clay. Coarse clastic material is abundant in dynamically active bodies on the continental shelves in the Gulf of Alaska, especially in the nearshore region, glacial areas, and areas of strong water movement such as Cook Inlet.

Beaches and Barrier Islands

Beaches are sporadically developed around the Gulf of Alaska coasts, where rocky fjords are common. Although some very large beaches are present, many are thin and rapidly changing.

On the southwest corner of Kodiak Island is a sandy shoreline about 48 km long and as much as 150 m wide (Maddren, 1919). Shoreline sediment in lower Cook Inlet has been examined by Hayes and Michel (1982). Although most of the coast there is retreating, beaches along the fronts of small fan deltas have an aggregate length of nearly 100 km; sand and gravel spits formed by longitudinal transport total about 65 km. Narrow beaches of gravel and sand also line many erosional parts of the shore in the inlet (Rappeport, 1980).

In the northern Gulf of Alaska, between Prince William Sound and Cross Sound, much of the coast is nearly linear and is openly exposed to the intense wave action resulting from Pacific Ocean swell and Gulf of Alaska storms. The beaches fronting this coast are generally composed of well-sorted medium-grained sand, and gravel is abundant only where coastal cliffs of morainal or other very coarse material are being eroded (Reimnitz and Plafker, 1976). The beaches are generally well over 100 m wide (Thomas and Berryhill, 1962), though sometimes less (Molnia and Wheeler, 1978), but many are suspected to be thin, especially in areas of rapidly retreating coast such as near Yakataga (Molnia, 1977b). In recently deglaciated Icy Bay, a spit >6 km long, built entirely in this century, must contain on the order of 10^7 m^3 of coarse material (Molnia, 1977a).

Wave action has built a string of barrier islands along 85 km of the Copper River Delta front (Reimnitz, 1966); they consist dominantly of clean sand (Molnia and Carlson, 1980) and are flanked by finer sediment on their landward side (Reimnitz, 1966). No estimate has been made of the amount of coarse material in these islands, but it may be on the order of 10^8 m^3 or more.

Such deposits as the spit in Icy Bay and the Copper River Delta barrier islands, which are actively prograding, probably could be mined for material with minimal environmental dangers, at least if the rate of extraction remained below the rate of replenishment (estimated roughly from progradation rates at 10^5 m^3/yr and 2.5×10^6 m^3/yr, respectively).

The beaches of southwestern and southeastern Alaska have not been much studied, and published data are limited (Thomas and Berryhill, 1962). In southeastern Alaska, coasts facing the open sea are mainly bare rock faces, though small beaches, in part of modern biogenic carbonate of sand and gravel size, do occur in sheltered sites on other sides (Hoskin and Nelson, 1969, 1971).

Sand Waves and Shoals

Active bedforms of sand at various scales occur on parts of the Gulf of Alaska shelves but are, in general, poorly known. Large sand waves near the 200-m depth in Stevenson Trough (Figure 12) on the Kodiak Shelf (Hampton et al., 1979) have heights up to 8 m, wavelengths to about 300 m, and are asymmetrical with the steep sides mostly facing southeast toward open sea (Hampton, 1982a). Scattered sand waves, ripples, and other bedforms are also found, at depths <100 m, south of Stevenson Trough on Northern Albatross Bank and Southern Albatross Bank (Figure 12) (Thrasher, 1979). Sand waves on Northern Albatross Bank are small, but those on Southern Albatross Bank reach heights of 15 m and wavelengths of 300 m. These form a distinct acoustic unit as much as 40 m thick that pinches out abruptly at the edges of the sand-wave field (Hampton, 1982a). The symmetry and restricted extent of the Southern Albatross Bank field suggest that it formed in response to either storm waves or internal waves. The bedforms on these banks form local accumulations on a surface of Pleistocene glacial-marine sediment (Thrasher, 1979). In Shelikof Strait, the bottom generally is smooth and lacks bedforms, and only one minor occurrence of small sand waves has been reported (Hampton, 1985).

The best-studied sand waves in the Gulf of Alaska are in the large field in lower Cook Inlet (Figure 13) (Bouma, Hampton, and Orlando, 1977; Bouma et al., 1977). They cover about 850 km^2 of the sea bottom in an area of about 3500 km^2 in the central part of lower Cook Inlet (Whitney et al., 1979). Bedforms there range from ripples, sand ribbons, and small sand waves <2 m high to large sand waves 5 to 14 m high and with wavelengths as much as 1000 m that dominate in the south-central part of the field (Figure 13) (Bouma et al., 1978, 1980; Hampton, 1982b). The material is well-sorted sand with a median grain diameter of 0.25 to 1.0 mm in the north, fining southward to about 0.2 mm in the south (Bouma et al., 1978; Mahmood, Ehlers, and Cilweck, 1981).

Asymmetry suggests that the large bedforms in this field migrate southward; presently documented is only ebb-dominated movement of ripples in the tidal currents (Rappeport, 1980). The sand waves themselves are probably mainly tidal in origin but appear to move very little at present (Hampton, 1982b). Perhaps they move only during extreme storms, or they may be in part or all inactive at present (Whitney et al., 1979; Mahmood, Ehlers, and Cilweck, 1981).

The thickness of sand in this sand-wave field ranges from <1 m to at least 20 m (Rappeport, 1980), and the total amount of sand in the unit may therefore be several times 10^9 m^3. The greatest thicknesses are near the southern edge, where the field has prograded onto Cook Ramp (Figure 13) and where buried older sand waves occur (Whitney et al., 1979). Whether or not migration of the bedforms is still active, this edge at least probably represents a dead-end sediment accumulation.

Fine- to medium-grained sand forms the dominant bottom sediment in the northermost part of Cook Inlet (Sharma and Burrell, 1970), and this sediment is probably in active equilibrium with tidal currents. Intertidal sand shoals are found in the upper ends of narrow branches of the inlet, where tidal ranges are high (Hayes and Michel, 1982), and probably in other similar inlets elsewhere on the Gulf of Alaska coastline. Lunate sand bars whose crests reach within a few meters of the sea surface have been noted seaward of the barrier islands and tidal flats in the Copper River Delta area (Reimnitz, 1966).

At the western end of the Copper River Delta, where Copper River material moved parallel to shore by waves and currents enters the quieter waters of Prince William Sound, a large shoal (tidal delta) known as Middle Ground Shoal has formed (Reimnitz, 1966). This shoal, whose top is covered by only a few meters of water, has a prograding front slope as steep as 4.5° and is locally >100 m thick. The volume of coarse material in this shoal must exceed 10^{10} m^3.

Other active bodies of sand and gravel on the Gulf of Alaska shelves have been little studied. The prograding ramp in Shelikof Strait could be regarded as a giant active body, but its material is no coarser than fine-grained muddy sand (Hampton et al., 1981). Similarly, the active accumulations of partly coarse material deposited from ice at the front of tidewater glaciers (Molnia and Wheeler, 1978) are poorly sorted, and the rapid changes and numerous hazards in the environment render their exploitation doubtful.

Fossil Sand and Gravel Bodies in the Gulf of Alaska

Exposed Neogene deposits around the Gulf of Alaska indicate that processes and environments on this continental margin have been basically similar to those of the present since well before the Quaternary. There should, therefore, be many buried, fossil, and relict (inactive) accumulations of coarse clastic material; however, persistent tectonic activity, repeated glacial fluctuations, and sea-level oscillations make the distribution of such fossil bodies complex and not easily predictable. Apart from Cook Inlet, there are few deep offshore drill holes in this region, and our knowledge of the subsurface is largely based on seismic sections and correlations with surface exposures.

Buried Sand Waves

The southwestern side of the large field of sand waves in lower Cook Inlet (Figure 13) is overlain by younger, possibly glacial-marine sediment (Rappeport, 1980); toward its southern edge, near Cook Ramp, the sand unit is thicker (>20 m) than the maximum sand-wave height in the area, indicating that the lower part represents buried material no longer actively moved. Some seismic sections show buried waves 10 to 20 m below the modern ones and with opposite asymmetry (Rappeport, 1982). Buried sand waves have been noted on seismic sections on Cook Ramp (Whitney et al., 1979), but their extent is not known.

Former Shorelines and Deltaic Deposits

Behind the barrier islands at the Copper River Delta stretches a belt of tidal flats, 6 to 15 km wide and about 85 km long, composed of sandy sediment at least 6 m deep. Remnants of former barrier beaches are found as much as 6 km behind the modern islands (Reimnitz, 1966). This extensive body of sand may contain several cubic kilometers (10^9 m^3) of sand.

On an even larger scale, beneath the modern delta surface, is a body of primarily sandy sediment 50 to 180 m thick that represents the progradation of the Copper River delta during the Holocene (Reimnitz, 1966). This body of generally well-sorted medium- to fine-grained sand should contain at least several tens of cubic kilometers of material.

Former shoreline deposits submerged and fossilized by changes in relative sea level have so far not been reported in the Gulf of Alaska. Old shorelines at shallow depths may have been destroyed by subsequent advances of ice masses, but some could have been preserved by relatively recent tectonic subsidence, such as has been observed historically (Plafker and Rubin, 1967).

Buried Channels

Areas of the continental shelves not occupied by ice during glacial low stands of sea level were subject to fluvial erosion, and channels formed in this way, subsequently filled with sediment, appear on seismic records as buried troughs. A filled channel of this type about 2 km wide and >12 km long was reported by Hoose and Whitney (1980) in the center of northern Shelikof Strait. Numerous partly or completely filled fluvial channels in lower Cook Inlet (Rappeport, 1980, 1982) have fill as thick as 50 m, and some have delta-like giant foreset structures. This fill is inferred to consist mainly of well-sorted sands.

Buried channels identified on seismic sections on the continental shelf between Yakutat Bay and Cross Sound (Carlson and Schwab, 1982) are seen mainly off present major streams and glacial valleys and generally within 30 km of the shore, though some extend at least 50 km offshore and into water >100 m deep. The paleodrainage systems of these ancient filled valleys have not been completely mapped; individual channels are generally 0.5 to 2.0 km wide and contain 25 to 100 m or more of fill.

A 10-km-wide buried trough containing 150 m of fill angles across the Kayak Trough, a modern sea valley east of the Copper River mouth, from which the buried trough appears to come (Carlson et al., 1982). The age of this buried trough and the nature of its fill are unknown.

Earlier Holocene Marine Sediment

Possible remnants of early Holocene nearshore transgressive sediment include small areas of shelly sand at 70- to 100-m depth on the Kodiak Shelf; radiocarbon ages on two samples, however, are <4 ka (Hampton, 1982a).

In lower Cook Inlet, a thin (1 to 3 m), poorly sorted layer of gravel and sand that underlies Holocene glacial-marine sediment and the active sand-wave field may be outwash or glacial-marine material deposited as the glaciers receded and sea level rose in the early Holocene (Rappeport, 1980). Similar layers are interbedded with morainal glacial debris in the underlying Pleistocene.

Some thin, graded sand layers, interpreted as turbidites derived from adjacent shallow shelf areas, have been cored

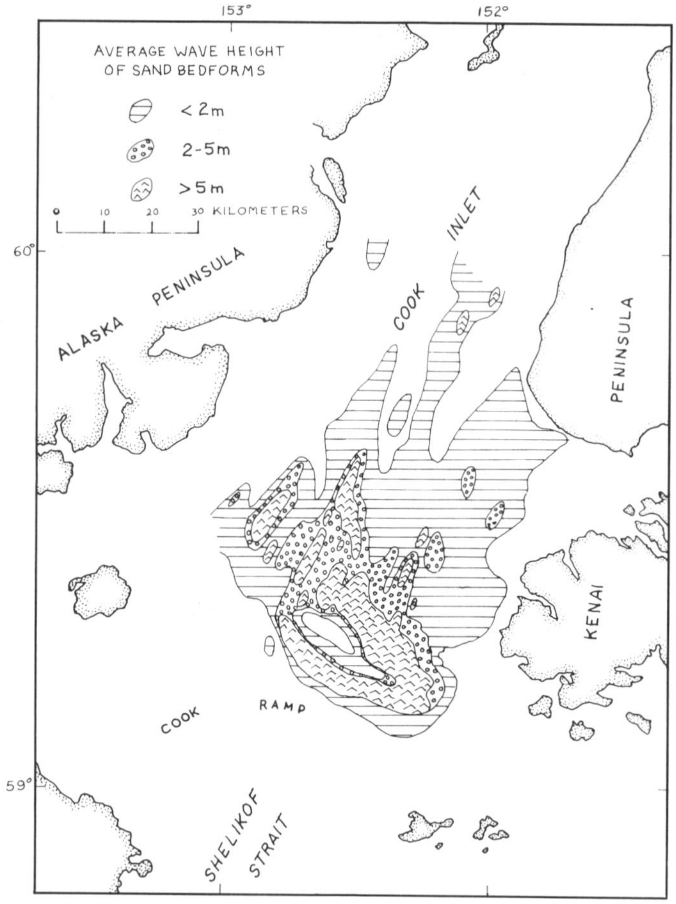

Figure 13. Large field of sand waves and other sand bedforms in lower Cook Inlet. After Hampton (1982b).

on the floors of sea valleys that cross the central Gulf of Alaska Shelf (Carlson et al., 1982), but these are probably insignificant in quantity and resource potential.

Relict Glacial Sediments

Relict Pleistocene morainal, outwash, and glacial-marine sediments, or gravelly lag deposits derived by winnowing of these, form the surface material in many offshore areas in the Gulf of Alaska and underlie thin Holocene mud in parts of the middle and outer shelves (Figure 14). Though poorly sorted (apart from the winnowed gravels), these Pleistocene deposits include large amounts of coarse and very coarse debris and therefore are a potential source of sand and gravel.

In the northern part of Cook Inlet, wave action and strong tidal currents have winnowed fines from Pleistocene sediment in many areas to leave a lag of coarse gravel and sandy gravel as the dominant bottom material (Sharma and Burrell, 1970; Rappeport, 1980). In the wider and deeper southern end of the inlet, gravel gives way to Holocene sand and silt as the usual bottom material (Hampton, 1982b).

On the open continental shelf in the central Gulf of Alaska, gravelly material (as much as 75% gravel) is found widely on the bottom on Tarr Bank (Figure 11), on Pamplona spur (Figure 14), and in patches near the shelf edge and the edges of sea valleys (Carlson et al., 1977; Wright, 1972a). Shelf sediments off southeastern Alaska include patches of coarse

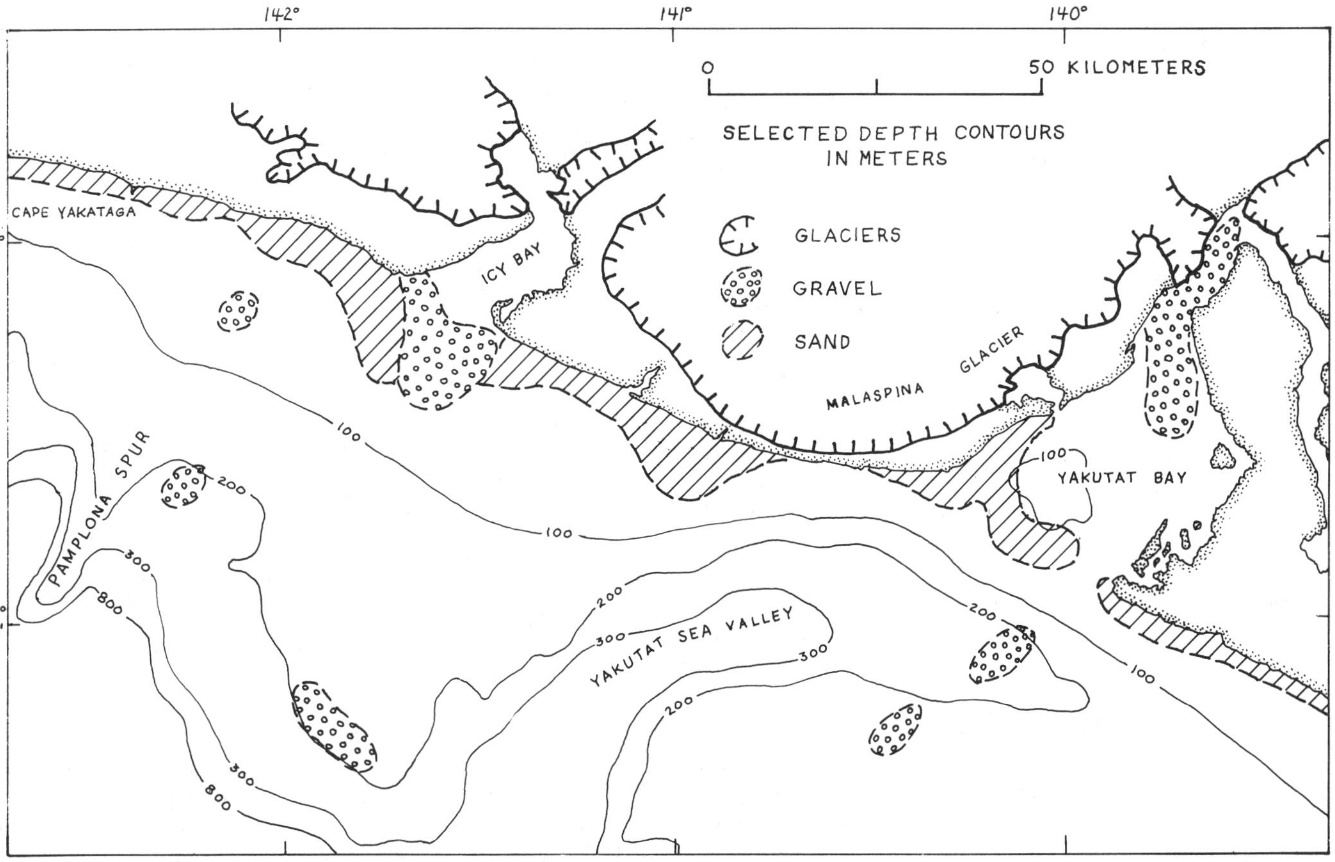

Figure 14. Areas of surficial sand and gravel on the continental shelf in the northeastern Gulf of Alaska. After Armentrout (1980). Continental slope indicated by 800-m contour; deeper contours omitted.

gravel and gravelly mixtures, presumably winnowed, and relict Pleistocene glacial debris (Sharma, 1979).

Some relict glacial deposits are of Holocene age. Submerged morainal accumulations of unsorted glacial debris, some of them representing glacial advances in the last few hundred years, are known in many bays and inlets along the coast, including Icy Bay and Yakutat Bay (Figure 14) (Carlson et al., 1982), the fjords and inlets of southeastern Alaska (Sharma, 1979), parts of Prince William Sound (von Huene, Shor, and Reimnitz, 1967), and other inlets (Reimnitz, von Huene, and Wright, 1970). Such deposits are irregular and extremely poorly sorted, however, and their usefulness is limited.

Sedimentary deposits bearing coarse clastic material extend to great depths in the Gulf of Alaska continental margin. The Yakataga Formation (Miocene to Quaternary glacial and glacial-marine sedimentary deposits) under much of the shelf in the central gulf is as much as several thousand meters thick and in parts contains 5 to 20% gravel (Plafker and Addicott, 1976; Armentrout, 1983). If this poorly sorted material could be used, either onshore or offshore, the amount of gravel available in it is measurable in cubic kilometers.

Beyond the shelf, coarse material is present on the sea floor as scattered ice-rafted clasts, and beneath the deep-sea floor in the form of turbidites. Neither of these can at present be regarded as resources, however, because recovery would be prohibitively expensive.

CONCLUSIONS

Abundant quantities of sand and gravel are present in a variety of deposits on the offshore continental margins of Alaska. Such deposits are most common in the shallow waters of the continental shelves, but they also occur on the slope and are widely present in rather inaccessible form in the deep-sea turbidite basins. Exploitation of these potential resources will have to face three types of problems—(1) many of the deposits are in areas where active marine processes may pose a threat to persons and equipment; (2) disturbance of the substrate by mining of sand or gravel on the sea bottom will have environmental consequences that are in general poorly understood but clearly may be severe in some areas; and (3) many of the deposits are in areas remote from present possible markets, for example, the major deposits in the Bering and Chukchi Seas. Nonetheless, future needs and uses cannot be safely predicted, and therefore the resource value of these deposits remains an open question.

Beaufort Sea

Exploitation of sand and gravel resources in the Beaufort Sea area faces potentially serious problems, including both hazards to the structures and people involved in such work and also dangers of damage to the environment and disruption of as yet poorly understood dynamic systems.

The regimen and mass budget of the barrier islands on the

inner shelf should be studied before any large-scale removal of material from those islands is attempted, because such removal could have far-reaching effects on the islands, the coast, and even perhaps on the ice zonation and general sediment movement pattern on the shelf.

Further testing of the subsurface structure of Weller Bank by drill holes seems advisable. If even part of this extensive shoal is underlain by coarse material, it could supply large amounts of such material with minimal effect on water circulation and sediment movement. Similar testing of other shoals may also be advisable, although environmental effects of mining any active sediment body will be a crucial consideration.

If gravel-filled or sand-filled paleovalleys can be located under the shelf, such subsurface deposits could be mined on a fairly large scale in the mid-shelf region of intense ice-gouging with minimal adverse environmental effects. Cavities produced would be quickly filled, and ice action has in any case a general character very similar to possible mining operations. Mining in this zone of intense ice-gouging would therefore introduce few new elements into the regimen, though obviously such operations would face problems from ice action.

Chukchi Sea

In the nearshore part of the eastern Chukchi Sea, sand and gravel are abundant. Some deposits appear to be end-point sediment accumulations, such as Blossom Shoals off Icy Cape, and could probably be mined without too great an effect on sediment dynamics in the surrounding areas. Mining of some other deposits, however, particularly shoreline deposits, is likely to cause serious coastal erosion in adjoining areas.

On the outer shelf, more than about 100 km from shore, gravel and sand are much more sparsely distributed. The largest and most minable bodies are probably in channels of filled paleovalleys, but these cannot be assessed until the paleovalley systems are mapped and bore holes determine the character of the channel fill.

The northern Chukchi Sea, including the shelf edge and slope, and the extensive Chukchi borderland, are very poorly known. On the sea bottom near the shelf edge, active sand waves or rippled lag gravel may form in response to localized strong bottom currents, and shoreline deposits could have formed down to 125 m or more below present sea level during Pleistocene glacial periods. However, the Arctic Ocean and northermost Chukchi Sea are perennially frozen now, and we know too little of the ocean's condition during those glacial stages to say what types of shoreline deposits would have formed then.

In the Chukchi borderland, even the highest of the plateaus lies too deep (>200 m) to have been exposed during glacial low sea levels. Because this borderland probably subsided to its present level soon after its formation in the Mesozoic, it is likely to have a thick cover of marine mud, with various amounts of ice-rafted debris and organic tests, and it is very unlikely that any usable sand or gravel accumulations are present near the surface. Even if such bodies are present, their exploitation, like that of the probable turbidite sands in the deep basins of the Arctic Ocean adjacent to the borderland, seems highly problematical under the moving Arctic pack ice and hundreds of meters of water.

Bering Sea and Aleutian Ridge

Exploitation of sand and gravel resources in the Bering Sea region faces a variety of problems and uncertainties. If very-fine-grained sand is desired, then availability of material is by itself not a limitation. In two areas, the Yukon Delta and the heads of submarine canyons along the Beringian margin, very large deposits of such material contain amounts measured in tens of cubic kilometers. Other than possible ecological effects, there would not appear to be any environmental danger in mining the sand-wave fields in the canyon heads. Mining of Yukon Delta sand, however, involves other possible dangers because of the intensity of wave and current action in that shallow water area. The outer parts of the delta deposit, where water depths would allow reasonable access by vessels, are quite thin, and exposure of underlying Pleistocene peaty clay might lead to serious erosion, triggered by ice-gouging and gas escape. In addition, large-scale removal of the delta sand could deprive the northward currents of their usual sediment load, causing erosion of the sea bottom and of bedforms in Norton Sound and the Chirikov Basin.

Deposits of coarser material (coarse sand and gravel) are more restricted, occurring mainly in shallow-water regions near coasts and in parts of the Chirikov Basin. These deposits are in areas of strong wave and current action and are generally thin, commonly only a veneer much less than 1 m thick. Hence, exploitation would involve disturbance of local water and sediment dynamics and should not be undertaken until those are well understood.

Shoreline deposits are poorly known in the Bering Sea, and further work is needed to document the distribution, character, and dynamics of beaches, barrier islands, and offshore bars. More detailed study, including drilling, of the large sand ridges off the Seward Peninsula would seem useful, as these accessible bodies of sediment are likely targets for exploitation. In areas far from shore, the best chance for supplies of sand (and possibly gravel) may be submerged former shoreline deposits, and such occurrences should be mapped in detail.

In the deeper parts of the Bering Sea, coarse sediment could be sought on the tops of the Bowers and Shirshov Ridges (perhaps under a cover of Holocene mud). Finally, the remarkable sand-wave fields in the heads of the Beringian submarine canyons deserve further study. They are the only known significant deposits of clean sand-size sediment that are close to some of the potentially petroliferous Tertiary basins on the outer Bering Shelf, and so they may assume importance one day as a resource.

Gulf of Alaska

Offshore mining of sea-bottom sand and gravel bodies in the Gulf of Alaska would face the problems of severe winter storms in the Gulf, ocean currents, locally very strong tidal currents and bores, tidewater glaciers and icebergs spawned by them, and frequent and sometimes severe earthquakes. Environmental effects of such mining are perhaps less likely to be serious in the Gulf than in Arctic Alaska, because the gravelly nature of the substrate in most areas would serve to limit possible erosion, and large amounts of suspended sediment are naturally introduced into the sea, at least to the nearshore parts, by glaciers and meltwater streams. However, sea-bottom excavation on the outer shelf, where the water normally contains little such glacial rock flour, could produce a local plume of turbid water that might have biological consequences.

Our knowledge of the Quaternary history of the Gulf of Alaska is still sketchy. Further search for Pleistocene shoreline, nearshore, and deltaic deposits in the outer-shelf and upper-slope regions could be useful. Questions remain concerning glaciation of the shelf—if the sea valleys in the central Gulf represent paths of ice flow, were the areas between the valleys dry land with outwash plains during low sea-level stands? If so, where are the subaerial deposits from these times? Was the entire shelf overridden and eroded by ice in a late phase of the last glacial period? Such questions deserve study.

ACKNOWLEDGMENTS

In compiling this review of a vast and somewhat nebulous topic, I have had the benefit of the advice and discussions of numerous geologists and geophysicists in the U.S. Geological Survey, especially Erk Reimnitz, Peter Barnes, Art Grantz, David Hopkins, David Dinter, George Moore, Michael Field, Hans Nelson, Parke Snavely, Monty Hampton, Peggy Smith, Thomas Chase, Stephen Wolf, and David Scholl, some of whom also freely shared with me unpublished information on the sediments and processes in the offshore regions. Portions of an earlier version of the manuscript were critically reviewed by P. A. Smith, Erk Reimnitz, R. L. Phillips, P. R. Carlson, M. S. Marlow, M. A. Hampton, Terry Bruns, and D. W. Scholl. Finally, I am grateful to David Scholl for first suggesting this study, and to Madeline for tolerating the midnight oil required to bring it to completion.

REFERENCES CITED

Aagard, K., and L. K. Coachman, 1964, Notes on the physical oceanography of the Chukchi Sea, *in* Oceanographic cruise USCGC Northwind, Bering and Chukchi Seas, July–September, 1962: U.S. Coast Guard Oceanographic Report 1, p. 13–16.

Addicott, W. O., 1969, Tertiary climatic change in the marginal northeastern Pacific Ocean: Science, v. 196, p. 583–586.

Ahlnäs, K., and G. Wendler, 1979, Sea-ice observations by satellite in the Bering, Chukchi, and Beaufort Seas: Fifth International Conference on Port and Ocean Engineering under Arctic Conditions (POAC) Proceedings, Norwegian Institute of Technology, August 13–18, 1979, 1, p. 313–329.

Alpha, T. R., and P. Swenson, 1978, Equal-area base map of Alaska and northeast Siberia: U.S. Geological Survey Open-File Map 78-82, 1 sheet, scale 1:5,000,000.

Armentrout, J. M., 1980, Surface sediments and associated faunas of upper slope, shelf and bay environments, Yakataga-Yakutat area, northern Gulf of Alaska, *in* M. E. Field, A. H. Bouma, I. P. Colburn, R. G. Douglas, and J. C. Ingle, eds., Quaternary depositional environments of the Pacific Coast: SEPM, Pacific Section, Pacific Coast Paleogeography Symposium 4, p. 241–255.

Armentrout, J. M., 1983, Glacial lithofacies of the Neogene Yakataga Formation, Robinson Mountains, southern Alaska Coast Range, Alaska, *in* B. F. Molnia, ed., Glacial-marine sedimentation: New York, Plenum Press, p. 629–665.

Askren, D. R., 1972, Holocene stratigraphic framework—Southern Bering Sea continental shelf: Seattle, University of Washington unpublished M.S. thesis, 104 p.

Baldauf, J. G., 1982, Identification of the Holocene-Pleistocene boundary in the Bering Sea by diatoms: Boreas, v. 11, p. 113–118.

Barnes, P. W., 1972, Preliminary results of geologic studies in the eastern central Chukchi Sea.: U.S. Coast Guard Oceanographic Report 50, p. 87–110.

Barnes, P. W., 1974, Preliminary results of marine geologic studies off the northern coast of Alaska: An ecological survey in the Beaufort Sea, WEBSEC 71–72: U.S. Coast Guard Oceanographic Report CG 373–64, p. 183–227.

Barnes, P. W., and E. Reimnitz, 1974, Sedimentary processes on Arctic shelves off the northern coast of Alaska, *in* J. C. Reed and J. E. Sater, eds., The coast and shelf of the Beaufort Sea: Arlington, Virginia, Arctic Institute of North America, p. 439–476.

Barnes, P. W., D. M. Rearic, and E. Reimnitz, 1985, Ice gouging characteristics and processes, *in* P. W. Barnes, E. Reimnitz, R. E. Hunter, R. L. Phillips, and S. Wolf, Geologic processes and hazards of the Beaufort and Chukchi Sea shelf and coastal regions: U.S. Department of Commerce, NOAA, Outer Continental Shelf Environmental Assessment Program Final Report 34, p. 183–221.

Barnes, P. W., E. Reimnitz, and A. S. Naidu, 1981, Sediments, *in* D. W. Norton and W. M. Sackinger, eds., Beaufort Sea (sale 71) synthesis report: U.S. Department of Commerce, p. 99–100.

Barnes, P. W., E. Reimnitz, and C. R. Ross, 1980, Nearshore surficial sediment textures—Beaufort Sea, Alaska: U.S. Geological Survey Open-File Report 80-196, 41 p.

Barnes, P. W., E. Reimnitz, L. J. Toimil, D. K. Maurer, and D. McDowell, 1979, Core descriptions and preliminary observations of vibracores from the Alaskan Beaufort Sea Shelf: U.S. Geological Survey Open-File Report 79-351, 17 p.

Bascom, W. N., 1959, The relationship between sand size and beach-face slope: American Geophysical Union Transactions, v. 32, p. 866–874.

Bates, R. L., and J. A. Jackson, 1980, Glossary of geology, 2nd ed.: Falls Church, Virginia, American Geological Institute, 751 p.

Bates, R. L., and J. A. Jackson, 1982, Our modern stone age: Los Altos, California, William Kaufman Inc., 136 p.

Ben-Avraham, Z., and A. K. Cooper, 1981, Early evolution of the Bering Sea by collision of oceanic rises and North Pacific subduction zones: GSA Bulletin, v. 92 (pt. I), p. 485–495.

Black, R. F., 1951, Eolian deposits of Alaska: Arctic, v. 4, p. 89–111.

Black, R. F., 1964, Gubik Formation of Quaternary age in northern Alaska: U.S. Geological Survey Professional Paper 302-C (Part 2), Regional Studies, p. 59–87.

Bouma, A. H., 1981, Submarine topography and physiography of lower Cook Inlet, Alaska: U.S. Geological Survey Open-File Report 81-1335, 28 p.

Bouma, A. H., M. A. Hampton, and R. C. Orlando, 1977, Sand waves and other bedforms in lower Cook Inlet, Alaska: Marine Geotechnology, v. 2, p. 291–308.

Bouma, A. H., M. A. Hampton, M. P. Wennekens, and J. A. Dygas, 1977, Large dunes and other bedforms in lower Cook Inlet, Alaska: Ninth Annual Offshore Technology Conference Proceedings, v. 1, p. 79–89.

Bouma, A. H., M. L. Rappeport, R. C. Orlando, and M. A. Hampton, 1980, Identification of bedforms in lower Cook Inlet, Alaska: Sedimentary Geology, v. 26, p. 157–177.

Bouma, A. H., M. L. Rappeport, R. C. Orlando, D. A. Cacchione, D. E. Drake, L. E. Garrison, and M. A. Hampton, 1979, Bedform characteristics and sand transport in a region of large sand waves, lower Cook Inlet, Alaska: Eleventh Annual Offshore Technology Conference Proceedings, p. 1083–1094.

Bouma, A. H., M. A. Hampton, M. L. Rappeport, J. W. Whitney, P. G. Teleki, R. C. Orlando, and M. E. Torresan, 1978, Movement of sand waves in lower Cook Inlet, Alaska: Tenth Annual Offshore Technology Conference Proceedings, v. 4, p. 2271–2284.

Briggs, S. R., 1983, Geology report for proposed Beaufort Sea OCS sand and gravel lease sale: U.S. Geological Survey Open-File Report 83-606, 25 p.

Brower, W. A., Jr., H. W. Searby, J. L. Wise, H. F. Diaz, and A. S. Prechtel, 1977, Climatic atlas of the outer continental shelf waters and coastal regions of Alaska, v. 2, Bering Sea: Anchorage, Arctic Environmental Information and Data Center, University of Alaska, 443 p.

Bruns, T. R., 1983a, Model for the origin of the Yakutat block, an accreting terrane in the northern Gulf of Alaska: Geology, v. 11, p. 718–721.

Bruns, T. R., 1983b, Structure and petroleum potential of the Yakutat segment of the northern Gulf of Alaska continental margin: U.S. Geological Survey Miscellaneous Field Studies MF-1480, 22 p., 2 map sheets (scale 1:500,000), 1 sheet of seismic sections.

Bruns, T. R., and W. C. Schwab, 1983, Structure maps and seismic stratigraphy of the Yakataga segment of the continental margin, northern Gulf of Alaska: U.S. Geological Survey Miscellaneous Field Studies MF-1424, 20 p., 3 map sheets (scale 1:250,000), 1 sheet of sections.

Bruns, T. R., R. von Huene, R. C. Cullota, S. D. Lewis, and J. W. Ladd, 1987, Geology and petroleum potential of the Shumagin margin, Alaska, this volume.

Burk, C. A., 1965, Geology of the Alaska Peninsula—Island arc and continental margin: GSA Memoir 99, 250 p.

Cacchione, D. A., and D. E. Drake, 1978, Sediment transport in Norton Sound—Northern Bering Sea, Alaska: U.S. Department of Commerce, NOAA, Environmental Assessment of the Alaskan Continental Shelf Annual Report, v. 12, p. 308–450.

Cacchione, D. A., D. E. Drake, and P. Wiberg, 1982, Velocity and bottom-stress measurements in the bottom boundary layer, outer Norton Sound, Alaska: Geologie en Mijnbouw, v. 61, p. 71–78.

Carlson, P. R., 1977, Effects of seasonability and variability of streamflow on nearshore coastal areas: U.S. Department of Commerce, NOAA, Environmental Assessment of the Alaskan Continental Shelf Annual Report, v. 14, p. 96–250.

Carlson, P. R., 1978, Holocene slump on continental shelf off Malaspina Glacier, Gulf of Alaska: AAPG Bulletin, v. 62, p. 2412–2426.

Carlson, P. R., and H. A. Karl, 1984a, Rates of sediment accumulation, *in* H. A. Karl and P. R. Carlson, eds., Surface and near-surface geology, Navarin province: Results of the 1980–1981 field seasons: U.S. Geological Survey Open-File Report 84-89, p. 21–27.

Carlson, P. R., and H. A. Karl, 1984b, Discovery of two new large submarine canyons in the Bering Sea: Marine Geology, v. 56, p. 159–179.

Carlson, P. R., and B. F. Molnia, 1977, Submarine faults and slides on the continental shelf, northern Gulf of Alaska: Marine Geotechnology, v. 2, p. 275–290.

Carlson, P. R., and W. C. Schwab, 1982, Northern Gulf of Alaska environmental geology, *in* T. R. Bruns, ed., Hydrocarbon resource report for proposed OCS lease sale 88: Southeastern Alaska, northern Gulf of Alaska, Cook Inlet, and Shelikof Strait, Alaska: U.S. Geological Survey Open-File Report 82-928, p. 73–86.

Carlson, P. R., H. A. Karl, and B. D. Edwards, 1982, Puzzling mass movement features in the Navarinsky Canyon head, Bering Sea: Geo-Marine Letters, v. 2, p. 275–290.

Carlson, P. R., T. R. Bruns, B. F. Molnia, and W. C. Schwab, 1982, Submarine valleys in the northeastern Gulf of Alaska: Characteristics and probable origin: Marine Geology, v. 47, p. 217–242.

Carlson, P. R., B. F. Molnia, S. C. Kittelson, and J. C. Hampson, Jr., 1977, Map of distribution of bottom sediments on the continental shelf, northern Gulf of Alaska: U.S. Geological Survey Miscellaneous Field Studies MF-876, 13 p., 2 map sheets (scale approx. 1:500,000 and 1:1,000,000).

Carlson, P. R., G. Plafker, T. R. Bruns, and W. P. Levy, 1979a, Seaward extension of the Fairweather fault, *in* K. M. Johnson and J. R. Williams, eds., U.S. Geological Survey in Alaska: Accomplishments during 1978: U.S. Geological Survey Circular 804-B, p. B136–139.

Carlson, P. R., M. C. Wheeler, B. F. Molnia, and T. J. Atwood, 1979b, Neoglacial sedimentation in Glacier Bay, Alaska, *in* K. M. Johnson and J. R. Williams, eds., U.S. Geological Survey, Alaska: Accomplishments during 1978: U.S. Geological Survey Circular 804–B, p. B114–116.

Carlson, P. R., M. C. Wheeler, B. F. Molnia, A. Post, and R. D. Powell, 1983, Maps showing post-Neoglacial sediment thickness and bathymetry in Tarr Inlet, Glacier Bay, Alaska: U.S. Geological Survey Miscellaneous Field Studies MF-1456, 1 sheet (scale 1:31,680) and seismic profiles.

Carter, L. D., and J. R. Galloway, 1979, Southward-progressing stabilization of Holocene eolian sand on the western Arctic coastal plain: U.S. Geological Survey Circular 804-B, p. 37–39.

Carter, L. D., and S. W. Robinson, 1978, Eolian sand and interbedded organic horizons at Kealok Creek on the Arctic coastal plain of Alaska: Possible regional implications: U.S. Geological Survey Open-File Report 78-320, 26 p.

Cooper, A. K., M. S. Marlow, and Z. Ben-Avraham, 1981, Multichannel seismic evidence bearing on the origin of Bowers Ridge, Bering Sea: GSA Bulletin, v. 92 (Part I), p. 474–484.

Cooper, A. K., M. S. Marlow, and D. W. Scholl, 1977, The Bering Sea—A multifarious marginal basin, *in* M. Talwani and W. C. Pitman, III, eds., Island arcs, deep sea trenches and back-arc basins: American Geophysical Union Maurice Ewing Series, v. 1, p. 437–450.

Cooper, A. K., D. W. Scholl, and M. S. Marlow, 1976, Plate tectonic model for the evolution of the eastern Bering Sea Basin: GSA Bulletin, v. 87, p. 1119–1126.

Cooper, A. K., D. W. Scholl, and M. S. Marlow, 1987, Structural framework, sedimentary sequences, and hydrocarbon potential of the Aleutian and Bowers Basins, Bering Sea, this volume.

Cooper, A. K., D. W. Scholl, T. L. Vallier, and E. W. Scott, 1980, Resource report for the deep-water areas of proposed OCS lease sale No. 70, St. George Basin, Alaska: U.S. Geological Survey Open-File Report 80-246, 91 p.

Craig, J. D., and G. P. Thrasher, 1982, Environmental geology of Harrison Bay, northern Alaska: U.S. Geological Survey Open-File Report 82-35, 25 p.

Creager, J. S., 1963, Sedimentation in a high energy, embayed, continental shelf environment: Journal of Sedimentary Petrology, v. 33, p. 815–830.

Creager, J. S., and D. A. McManus, 1965, Pleistocene drainage patterns on the floor of the Chukchi Sea: Marine Geology, v. 3, p. 279–290.

Creager, J. S., and D. A. McManus, 1966, Geology of the southeastern Chukchi Sea, *in* N. J. Wilimovsky and J. N. Wolfe, eds., Environment of the Cape Thompson region, Alaska: U.S. Atomic Energy Commission, p. 755–786.

Creager, J. S., and D. A. McManus, 1967, Geology of the floor of the Bering and Chukchi Seas—American studies, *in* D. M. Hopkins, ed., The Bering land bridge: Stanford, California, Stanford University Press, p. 7–31.

Dietz, R. S., A. J. Carsola, E. C. Buffington, and C. J. Shipek, 1964, Sediments and topography of the Alaskan shelves, *in* R. L. Miller, ed., Papers in marine geology—Shepard Commemorative Volume: New York, The Macmillan Company, p. 241–256.

Dinter, D. A., 1985, Quaternary sedimentation of the Alaska Beaufort Shelf: Influence of regional tectonics, fluctuating sea levels, and glacial sediment sources: Tectonophysics, v. 114, p. 133–161.

Drake, D. E., C. E. Totman, and P. L. Wiberg, 1979, Sediment transport during the winter on the Yukon prodelta, Norton Sound, Alaska: Journal of Sedimentary Petrology, v. 49, p. 1171–1180.

Dunlavey, J. M., J. R. Childs, and R. E. von Huene, 1980, Bathymetric map of the western Gulf of Alaska: U.S. Geological Survey Open-File Report 80-1093, 13 sheets (index at scale 1:500,000, 12 areas at 1:250,000).

Dupré, W. R., 1977, Yukon delta coastal processes study: U.S. Department of Commerce, NOAA, Environmental Assessment of

the Alaskan Continental Shelf Annual Report, v. 14, p. 508–553.
Dupré, W. R., 1982, Depositional environments of the Yukon delta, northeastern Bering Sea: Geologie en Mijnbouw, v. 61, p. 63–70.
Ehm, A., 1983, Oil and gas basins map of Alaska: Alaska Division of Geological and Geophysical Surveys Special Report 32, scale 1:2,500,000.
Eittreim, S., and A. Grantz, 1979, CDP Seismic Sections of the western Beaufort continental margin: Tectonophysics, v. 59, p. 251–262.
Favorite, F., 1967, The Alaska stream: Bulletin of the International North Pacific Fisheries Commission, v. 21, p. 1–20.
Feely, R. A., G. J. Massoth, and A. J. Paulson, 1981, The distribution and elemental composition of suspended particulate matter in Norton Sound and the northeastern Bering Sea Shelf: Implications for Mn and Zn recycling in coastal waters, *in* D. W. Hood and J. A. Calder, eds., The eastern Bering Sea Shelf: Oceanography and Resources: Washington, D.C., U.S. Department of Commerce, p. 321–337.
Field, M. E., C. H. Nelson, D. A. Cacchione, and D. E. Drake, 1981, Sand waves on an epicontinental shelf: Northern Bering Sea: Marine Geology, v. 42, p. 233–258.
Fisher, M. A., and R. von Huene, 1980, Structure of upper Cenozoic strata beneath Kodiak Shelf, Alaska: AAPG Bulletin, v. 64, p. 1014–1033.
Fisher, M. A., T. R. Bruns, and R. von Huene, 1981, Transverse tectonic boundaries near Kodiak Island, Alaska: GSA Bulletin, v. 92 (Part I), p. 10–18.
Fisher, M. A., R. L. Detterman, and L. B. Magoon, 1987, Tectonics and petroleum geology of the Cook-Shelikof Basin, southern Alaska, this volume.
Fleming, R. H., and D. Heggarty, 1966, Oceanography of the southeastern Chukchi Sea, *in* N. J. Wilimovsky and J. M. Wolfe, eds., Environment of the Cape Thompson region, Alaska: U.S. Atomic Energy Commission, p. 697–754.
Gardner, J. V., W. E. Dean, and T. L. Vallier, 1980, Sedimentology and geochemistry of surface sediments, outer continental shelf, southern Bering Sea: Marine Geology, v. 35, p. 299–329.
Gardner, J. V., T. L. Vallier, W. E. Dean, K. A. Kvenvolden, and G. D. Redden, 1979, Sedimentology and geochemistry of surface sediments and the distribution of faults and potentially unstable sediments, St. George Basin region of the outer continental shelf, southern Bering Sea: U.S. Geological Survey Open-File Report 79-1562, 89 p.
Goldman, H. B., and D. Raining, 1975, Sand and gravel, *in* S. J. Lefond, ed., Industrial minerals and rocks, 4th edition: New York, AIME, p. 1027–1042.
Grantz, A., and S. D. May, 1983, Rifting history and structural development of the continental margin north of Alaska, *in* J. S. Watkins and C. L. Drake, eds., Studies in continental margin geology: AAPG Memoir, v. 34, p. 77–100.
Grantz, A., and S. D. May, 1987, Regional geology and petroleum potential of the United States Chukchi Shelf north of Point Hope, this volume.
Grantz, A., S. Eittreim, and D. A. Dinter, 1979, Geology and tectonic development of the continental margin north of Alaska: Tectonophysics, v. 59, p. 263–291.
Grantz, A., M. L. Holmes, and B. A. Kososki, 1976, Geologic framework of the Alaskan continental terrace in the Chukchi and Beaufort Seas, *in* C. M. Yorath, E. R. Parker, and D. J. Glass, eds., Canada's continental margins and offshore petroleum exploration: Canadian Society of Petroleum Geologists Memoir, v. 4, p. 669–700.
Grantz, A., S. D. May, and D. A. Dinter, 1987, Regional geology and petroleum potential of the United States Beaufort and northeasternmost Chukchi Seas, this volume.
Grantz, A., D. A. Dinter, E. R. Hill, R. E. Hunter, S. D. May, R. H. McMullin, and R. L. Phillips, 1982, Geologic framework, hydrocarbon potential, and environmental conditions for exploration and development of proposed oil and gas lease sale 85 in the central and northern Chukchi Sea: U.S. Geological Survey Open-File Report 82-1053, 84 p.
Hampton, M. A., 1982a, Synthesis report: Environmental geology of Kodiak Shelf, Alaska: U.S. Geological Survey Open-File Report 82-59, 76 p.
Hampton, M. A., 1982b, Synthesis report: Environmental geology of lower Cook Inlet, Alaska: U.S. Geological Survey Open-File Report 82-197, 55 p.
Hampton, M. A., 1985, Quaternary sedimentation in Shelikof Strait, Alaska: Marine Geology, v. 62, p. 213–253.
Hampton, M. A., A. H. Bouma, T. P. Frost, and I. P. Colburn, 1979, Volcanic ash in surficial sediments of the Kodiak Shelf—An indicator of sediment dispersal patterns: Marine Geology, v. 29, p. 347–356.
Hampton, M. A., J. H. Johnson, M. E. Torresan, and W. J. Winters, 1981, Description of sea floor sediment and preliminary geoenvironmental report, Shelikof Strait, Alaska: U.S. Geological Survey Open-File Report 81-1133, 87 p.
Hampton, M. A., A. H. Bouma, D. A. Sangrey, P. R. Carlson, B. F. Molnia, and E. C. Clukey, 1978, Quantitative study of slope instability in the Gulf of Alaska: Tenth Annual Offshore Technology Conference Proceedings, v. 4, p. 2307–2318.
Hanna, G. D., 1954, Submarine geological investigations off Point Barrow, Alaska, 1954: San Francisco, California Academy of Sciences, 10 p.
Harper, J. R., 1978, Coastal erosion rates along the Chukchi Sea coast under Barrow, Alaska: Arctic, v. 31, p. 428–433.
Hayes, M. O., and J. Michel, 1982, Shoreline sedimentation within a forearc embayment, lower Cook Inlet, Alaska: Journal of Sedimentary Petrology, v. 52, p. 251–263.
Hein J. R., A. H. Bouma, M. A. Hampton, and C. R. Ross, 1979, Clay mineralogy, fine-grained sediment dispersal, and inferred current patterns, lower Cook Inlet and Kodiak Shelf, Alaska: Sedimentary Geology, v. 24, p. 291–306.
Hess, G. R., and C. H. Nelson, 1982, Geology report for proposed Norton Sound OCS sand and gravel lease sale: U.S. Geological Survey Open-File Report 82-997, 25 p.
Hill, E. R., A. Grantz, S. D. May, and M. Smith, 1984, Bathymetric map of the Chukchi Sea: U.S. Geological Survey Miscellaneous Investigations Map I-1182-D, scale 1:1,000,000.
Hoose, P. J., and J. W. Whitney, 1980, Map showing selected geologic features of outer continental shelf, Shelikof Strait, Alaska: U.S. Geological Survey Open-File Report 80-2035, 1 sheet, scale 1:250,000.
Hopkins, D. M., 1967a, Quaternary marine transgressions in Alaska, *in* D. M. Hopkins, ed., The Bering land bridge: Stanford, California, Stanford University Press, p. 47–90.
Hopkins, D. M., 1967b, The Cenozoic history of Beringia—A synthesis, *in* D. M. Hopkins, ed., The Bering land bridge: Stanford, California, Stanford University Press, p. 451–484.
Hopkins, D. M., 1973, Sea level history in Beringia during the past 250,000 years: Quaternary Research, v. 3, p. 520–540.
Hopkins, D. M., 1981, Gravel sources and management options, *in* D. W. Norton and W. M. Sackinger, eds., Beaufort Sea sale 71 synthesis report: U.S. Department of Commerce, p. 169–178.
Hopkins, D. M., and R. W. Hartz, 1978a, Offshore permafrost studies, Beaufort Sea: U.S. Department of Commerce, NOAA, Environmental Assessment of the Alaskan Continental Shelf Annual Report, v. 11, p. 75–146.
Hopkins, D. M., and R. W. Hartz, 1978b, Shoreline history of Chukchi and Beaufort Seas as an aid to predicting offshore permafrost conditions: U.S. Department of Commerce, NOAA, Environmental Assessment of the Alaskan Continental Shelf Annual Report, v. 12, p. 503–574.
Hopkins, D. M., R. W. Hartz, and S. W. Robinson, 1979, Record of a prehistoric storm surge in the Wainwright Inlet—Kuk River area: U.S. Geological Survey Circular 804-B, p. 29–31.
Hopkins, D. M., P. V. Sellman, E. Chamberlain, R. E. Lewellen, and W. W. Robinson, 1979, Buried valleys as a possible determinant

of the distribution of deeply buried permafrost on the continental shelf of the Beaufort Sea: U.S. Department of Commerce, NOAA, Environmental Assessment of the Alaskan Continental Shelf Annual Report, v. 9, p. 135–141.

Hoskin, C. W., and D. C. Burrell, 1972, Sediment transport and accumulation in a fjord basin, Glacier Bay, Alaska: Journal of Geology, v. 80, p. 539–551.

Hoskin, C. W., and R. V. Nelson, Jr., 1969, Modern marine carbonate sediment, Alexander Archipelago, Alaska: Journal of Sedimentary Petrology, v. 39, p. 581–590.

Hoskin, C. W., and R. V. Nelson, Jr., 1971, Size modes in biogenic carbonate sediment, southeastern Alaska: Journal of Sedimentary Petrology, v. 41, p. 1026–1037.

Howard, J. D., and C. H. Nelson, 1982, Sedimentary structures on a delta-influenced shallow shelf, Norton Sound, Alaska: Geologie en Mijnbouw, v. 61, p. 29–36.

Hufford, G. L., 1977, Northeast Chukchi Sea coastal currents: Geophysical Research Letters, v. 4, p. 457–460.

Hume, J. D., and M. Schalk, 1964, The effects of ice-push on Arctic beaches: American Journal of Science, v. 262, p. 267–273.

Hume, J. D., and M. Schalk, 1967, Shoreline processes near Barrow, Alaska: A comparison of the normal and the catastrophic: Arctic, v. 20, p. 86–103.

Hunter, R. E., D. R. Thor, and M. L. Swisher, 1982, Depositional and erosional features of the inner shelf, northeastern Bering Sea: Geologie en Mijnbouw, v. 61, p. 49–62.

Hunter, R. E., and T. E. Reiss, 1985, Inner-shelf geology of southeastern Chukchi Sea, *in* P. W. Barnes, E. Reimnitz, R. E. Hunter, R. L. Phillips, and S. C. Wolf, eds., Geologic processes and hazards of the Beaufort and Chukchi Sea Shelf and coastal regions: U.S. Department of Commerce, NOAA, Outer Continental Shelf Environmental Assessment Program, Final Report 34, p. 97–124.

Jones, D. L., N. J. Silberling, and J. Hillhouse, 1977, Wrangellia—A displaced terrane in northwestern North America: Canadian Journal of Earth Sciences, v. 14, p. 2565–2577.

Jones, D. L., N. J. Silberling, H. C. Berg, and G. Plafker, 1981, Map showing tectonostratigraphic terranes of Alaska, columnar sections, and summary description of terranes: U.S. Geological Survey Open-File Report 81-792, 12 p. and map sheet.

Karl, H. A., and P. R. Carlson, 1982, Large sand waves in Navarinsky Canyon head, Bering Sea: Geo-Marine Letters, v. 2, p. 157–162.

Karl, H. A., and P. R. Carlson, 1984, Textural variation of surficial bottom sediment, *in* H. A. Karl and P. R. Carlson, eds., Surface and near-surface geology, Navarin Basin province: Results of the 1980–81 field seasons: U.S. Geological Survey Open-File Report 84-49, p. 15–20.

Karl, H. A., D. A. Cacchione, and P. R. Carlson, 1986, Internal-wave currents as a mechanism to account for large sand waves in Navarinsky Canyon head, Bering Sea: Journal of Sedimentary Petrology, v. 56, p. 706–714.

Karlstrom, T. N. V., 1964, Quaternary geology of the Kenai lowland and glacial history of the Cook Inlet region, Alaska: U.S. Geological Survey Professional Paper 443, 69 p.

Kinder, T. H., 1981, A perspective of physical oceanography in the Bering Sea, 1979, *in* D. W. Hood and J. A. Calder, eds., The eastern Bering Sea Shelf: Oceanography and Resources, v. 1: Washington, D.C., U.S. Department of Commerce, p. 5–13.

Kinder, T. H., and J. D. Schumacher, 1981, Circulation over the continental shelf of the southeastern Bering Sea, *in* D. W. Hood and J. A. Calder, eds., The eastern Bering Sea Shelf: Oceanography and resources, v. 1: Washington, D.C., U.S. Department of Commerce, p. 53–75.

Knebel, J. H., 1972, Holocene sedimentary framework of the east-central Bering Sea continental shelf: Seattle, University of Washington unpublished Ph.D. thesis, 186 p.

Knebel, J. H., and J. S. Creager, 1973, Yukon River: Evidence for extensive migration during the Holocene transgression: Science, v. 179, p. 1230–1232.

Knebel, J. H., J. S. Creager, and R. J. Echols, 1974, Holocene sedimentary framework, east-central Bering Sea continental shelf, *in* Y. Herman, ed., Marine geology and oceanography of the Arctic seas: New York, Springer-Verlag, p. 157–172.

Kovacs, A., 1983, Shore ice ride-up and pile-up features. Part I: Alaska's Beaufort Sea coast: U.S. Army Corps of Engineers Cold Regions Research and Engineering Laboratory (CRREL) Report 83-9, 59 p.

Kovacs, A., and D. S. Sodhi, 1980, Shore ice pile-up and ride-up: field observations, models, theoretical analyses: Cold Regions Science and Technology, v. 2, p. 209–288.

Larsen, M. C., C. H. Nelson, and D. R. Thor, 1979, Geologic implications and potential hazards of scour depressions on Bering Shelf, Alaska: Environmental Geology, v. 3, p. 39–47.

Larsen, M. C., C. H. Nelson, and D. R. Thor, 1981, Sedimentary processes and potential hazards on the sea floor of the northern Bering Sea, *in* D. W. Hood and J. A. Calder, eds., The eastern Bering Sea Shelf: Oceanography and Resources, v. 1: Washington, D.C., U.S. Department of Commerce, p. 247–261.

Leffingwell, E. de K., 1919, The Canning River region, northern Alaska: U.S. Geological Survey Professional Paper 109, 251 p.

Lewellen, R., 1977, A study of Beaufort Sea coastal erosion, northern Alaska: U.S. Department of Commerce, NOAA, Environmental Assessment of the Alaskan Continental Shelf Annual Report, v. 15, p. 491–527.

Lisitsyn, A. P., 1966, Recent sedimentation in the Bering Sea: Instityut Okeanologi Akademia Nauk, U.S.S.R., English translation (Israel program for scientific translation), 1969, 614 p.

Luken, M. D., and H. D. Hess, 1979, Sand, gravel, and shell deposits of the southern California borderland: OCS Mining Policy Phase II Task Force Program Feasibility Document, OCS Hard Minerals Leasing, Appendix 10, 69 p.

Maddren, A. G., 1919, The beach placers of the west coast of Kodiak Island, Alaska: U.S. Geological Survey Bulletin 692, p. 299–319.

Magoon, L. B., A. H. Bouma, M. A. Fisher, M. A. Hampton, E. W. Scott, and C. L. Wilson, 1979, Resource report for proposed OCS sale no. 60, lower Cook Inlet - Shelikof Strait, Alaska: U.S. Geological Survey Open-File Report 79-500, 38 p.

Mahmood, A., C. J. Ehlers, and B. A. Cilweck, 1981, Sand waves in lower Cook Inlet, Alaska: Journal of the Geotechnical Engineering Division, Proceedings of the American Society of Civil Engineers, v. 107, no. GT10, p. 1293–1307.

Marlow, M. S., and A. K. Cooper, 1980, Mesozoic and Cenozoic structural trends under southern Bering Sea Shelf: AAPG Bulletin v. 64, p. 2139–2155.

Marlow, M. S., and A. K. Cooper, 1983, Wandering terranes in southern Alaska: The Aleutia microplate and implications for the Bering Sea: Journal of Geophysical Research, v. 88, p. 3439–3446.

Marlow, M. S., A. K. Cooper, and J. R. Childs, 1983, Tectonic evolution of Gulf of Anadyr and formation of Anadyr and Navarin Basins: AAPG Bulletin, v. 67, p. 646–665.

Marlow, M. S., D. W. Scholl, A. K. Cooper, and E. C. Buffington, 1976, Structure and evolution of Bering Sea Shelf south of St. Lawrence Island: AAPG Bulletin, v. 60, p. 161–183.

Marlow, M. S., J. V. Gardner, T. L. Vallier, Hugh McLean, E. W. Scott, and M. B. Lynch, 1979, Resource report for proposed OCS lease sale No. 70, St. George Basin, shelf area, Alaska: U.S. Geological Survey Open-File Report 79-1650, 79 p.

Marlow, M. S., P. R. Carlson, A. K. Cooper, H. A. Karl, H. McLean, R. McMullin, and M. B. Lynch, 1981, Hydrocarbon resource report for proposed OCS sale No. 83, Navarin Basin, Alaska: U.S. Geological Survey Open-File Report 81-252, 81 p.

Marlow, M. S., H. McLean, A. K. Cooper, T. L. Vallier, J. V. Gardner, R. McMullin, and M. B. Lynch, 1980, A preliminary summary of regional geology, petroleum potential, environmental geology, and technology for exploration and development for proposed OCS lease sale No. 75, northern Aleutian Shelf, Bering Sea,

Alaska: U.S. Geological Survey Open-File Report 80-653, 54 p.

Martindale, S. G., and H. D. Hess, 1979, Resource assessment: Sand, gravel, and shell deposits on the continental shelf of northern and central California: OCS Mining Policy Phase II Task Force Program Feasibility Document, OCS Hard Minerals Leasing, Appendix 9, 39 p.

McCulloch, D. S., 1967, Quaternary geology of the Alaskan shore of Chukchi Sea, *in* D. M. Hopkins, ed., The Bering land bridge: Stanford, California, Stanford University Press, p. 91–210.

McLean, H., 1979, Review of petroleum geology of Anadyr and Khatyrka basins, U.S.S.R.: AAPG Bulletin, v. 63, p. 1467–1477.

McManus, D. A., and J. S. Creager, 1963, Physical and sedimentary environments on a large spit-like shoal: Journal of Geology, v. 71, p. 498–512.

McManus, D. A., J. C. Kelley, and J. S. Creager, 1969, Continental shelf sedimentation in an Arctic environment: GSA Bulletin, v. 80, p. 1961–1983.

McManus, D. A., J. S. Creager, R. J. Echols, and M. L. Holmes, 1983, The Holocene transgression on the Arctic flank of Beringia: Chukchi Valley to Chukchi Estuary to Chukchi Sea, *in* P. M. Masters and N. C. Flemming, eds., Quaternary coastlines and marine archeology: London, Academic Press, p. 365–388.

McManus, D. A., V. Kolla, D. M. Hopkins, and C. H. Nelson, 1977, Distribution of bottom sediments on the continental shelf, northern Bering Sea: U.S. Geological Survey Professional Paper 759-C, 31 p.

McNutt, S. L., 1981, Remote sensing analysis of ice growth and distribution in the eastern Bering Sea, *in* D. W. Hood and J. A. Calder, eds., The eastern Bering Sea Shelf: Oceanography and Resources, v. 1: Washington, D.C., U.S. Department of Commerce, p. 141–165.

Molnia, B. F., 1977a, Rapid shoreline erosion and retreat at Icy Bay, Alaska—A staging area for offshore petroleum development: Ninth Annual Offshore Technology Conference Proceedings, v. 3, p. 115–126.

Molnia, B. F., 1977b, Coastal morphology, erosion, and accretion of the Gulf of Alaska coastline, Cape Suckling to Yakutat Bay (abs.): GSA Abstracts with Programs, v. 9, no. 7, p. 1098.

Molnia, B. F., 1979a, Sand and gravel resources of the continental shelf of Alaska: OCS Mining Policy Phase II Task Force Program Feasibility Document, OCS Hard Minerals Leasing, Appendix 4, 70 p.

Molnia, B. F., 1979b, Origin of gas pockmarks and craters (abs.): GSA Abstracts with Programs, v. 11, no. 7, p. 481–482.

Molnia, B. F., 1980, Twentieth-century history of the Gulf of Alaska coastline, Cape Suckling to Cape Spencer, *in* M. E. Field, A. H. Bouma, I. P. Colburn, R. G. Douglas, and J. C. Ingle, eds., Quaternary depositional environments of the Pacific Coast: SEPM, Pacific Section, Pacific Coast Paleogeography Symposium 4, p. 121–141.

Molnia, B. F., 1981, Distribution of gas-charged sediment and pockmarks in the northeastern Gulf of Alaska, Yakutat Bay to Cross Sound, *in* N. R. D. Albert and T. Hudson, eds., The United States Geological Survey in Alaska: Accomplishments during 1979: U.S. Geological Survey Circular 823-B, p. B125.

Molnia, B. F., and P. R. Carlson, 1978, Surface sedimentary units of northern Gulf of Alaska continental shelf: AAPG Bulletin, v. 62, p. 633–643.

Molnia, B. F., and P. R. Carlson, 1980, Quaternary sedimentary facies on the continental shelf of the northeast Gulf of Alaska, *in* M. E. Field, A. H. Bouma, I. P. Colburn, R. G. Douglas, and J. C. Ingle, eds., Quaternary depositional environments of the Pacific Coast: SEPM, Pacific Section, Pacific Coast Paleogeography Symposium 4, p. 157–168.

Molnia, B. F., and P. T. Fuller, 1977, Clay mineralogy of eastern Gulf of Alaska: AAPG Bulletin, v. 61, p. 815.

Molnia, B. F., and M. C. Wheeler, 1978, Report on the beach dynamics, geology and oil spill susceptibility of the Gulf of Alaska coastline in Glacier Bay National Monument—Sea Otter Creek to Icy Point: U.S. Geological Survey Open-File Report 78-282, 192 p.

Molnia, B. F., W. P. Levy, and P. R. Carlson, 1980, Map showing Holocene sedimentation rates in the northeastern Gulf of Alaska: U.S. Geological Survey Miscellaneous Field Studies Map MF-1170, scale 1:500,000.

Moore, D. G., 1964, Acoustic-reflection reconnaissance of continental shelves, eastern Bering and Chukchi Seas, *in* R. L. Miller, ed., Papers in marine geology—Shepard Commemorative Volume: New York, The Macmillan Company, p. 319–362.

Moore, G. W., 1966, Arctic beach sedimentation, *in* N. J. Wilimovsky and J. N. Wolfe, eds., Environment of the Cape Thompson region, Alaska: U.S. Atomic Energy Commission, p. 587–608.

Moore, G. W., 1969, New formations on Kodiak and adjacent islands, Alaska, *in* G. V. Cohee, R. G. Bates, and W. B. Wright, eds., Changes in stratigraphic nomenclature by the U.S. Geological Survey 1967: U.S. Geological Survey Bulletin 1274-A, p. A27–A35.

Moore, G. W., and M. D. Luken, 1979, Offshore sand and gravel resources of the Pacific Northwest: Oregon Geology, v. 41, p. 143–151.

Mowatt, T. C., and A. S. Naidu, 1974, Gravels from the Alaskan continental shelf, Beaufort Sea, Arctic Ocean: Petrographic character and implications for sediment source and transport: Alaskan Geology and Geophysical Survey Open-File Report AOF43, 40 p.

Muench, R. D., and C. M. Schmidt, 1975, Variations in the hydrographic structure of Prince William Sound: University of Alaska, Institute of Marine Science IMS Report 75-1, 135 p.

Muench, R. D., and J. D. Schumacher, 1980, Physical oceanographic and meteorological conditions in the northwest Gulf of Alaska: National Oceanic and Atmospheric Administration Technical Memorandum ERL PMEL-22, 147 p.

Muench, R. D., H. O. Mofjeld, and R. L. Charnell, 1978, Oceanographic conditions in lower Cook Inlet: Spring and summer 1973: Journal of Geophysical Research, v. 83, p. 5090–5098.

Naidu, A. S., and H. V. Weiss, 1981, Sedimentation rate in an arctic coastal region, *in* A. S. Naidu, L. H. Larsen, M. D. Sweeney, and H. V. Weiss, Sources, transport pathways, depositional sites and dynamics of sediments in the lagoon and adjacent shallow marine region, northern arctic Alaska: U.S. Department of Commerce, NOAA, Environmental Assessment of the Alaskan Continental Shelf Annual Report, v. 5, App. 4, p. 252–269.

Naidu, A. S., L. H. Larsen, M. D. Sweeney, and H. V. Weiss, 1981, Sources, transport pathways, depositional sites and dynamics of sediments in the lagoon and adjacent shallow marine region, northern Arctic Alaska: U.S. Department of Commerce, NOAA, Environmental Assessment of the Alaska Continental Shelf Annual Report, v. 5, p. 151–297.

Nelson, C. H., 1982, Late Pleistocene-Holocene transgressive sedimentation in deltaic and non-deltaic areas of the northeastern Bering epicontinental shelf: Geologie en Mijnbouw, v. 61, p. 5–18.

Nelson, C. H., and J. S. Creager, 1977, Displacement of Yukon-derived sediment from Bering Sea to Chukchi Sea during Holocene time: Geology, v. 5, p. 141–146.

Nelson, C. H., and D. M. Hopkins, 1972, Sedimentary processes and distribution of particulate gold in the northern Bering Sea: U.S. Geological Survey Professional Paper 689, 27 p.

Nelson, C. H., D. M. Hopkins, and D. W. Scholl, 1974, Tectonic setting and Cenozoic sedimentary history of the Bering Sea, *in* Y. Herman, ed., Marine geology and oceanography of the Arctic Seas: New York, Springer-Verlag, p. 119–140.

Nelson, C. H., W. R. Dupré, M. E. Field, and J. D. Howard, 1982, Variation in sand body types on the eastern Bering Sea epicontinental shelf: Geologe en Mijnbouw, v. 61, p. 37–48.

Nelson, C. H., M. E. Field, D. A. Cacchione, and D. E. Drake, 1978, Areas of active, large scale sand wave and ripple fields with scour potential on the Norton Basin sea floor: U.S. Department of Commerce, NOAA, Environmental Assessment of the Alaskan Continental Shelf Annual Report, v. 12, p. 291–307.

Nelson, C. H., D. R. Thor, M. W. Sandstrom, and K. A. Kvenvolden, 1979, Modern biogenic gas-generated craters (sea-floor "pockmarks") on the Bering Shelf, Alaska: GSA Bulletin, v. 90, p. 1144–1152.

Niebauer, H. J., 1981, Recent fluctuations in sea ice distribution in the eastern Bering Sea, *in* D. W. Hood and J. A. Calder, eds., The Eastern Bering Sea Shelf: Oceanography and Resources, v. 1: Washington, D.C., U. S. Department of Commerce, p. 133–140.

Nummedal, D., 1979, Coarse grained sediment dynamics—Beaufort Sea, Alaska: Fifth International Conference on Port and Ocean Engineering under Arctic Conditions (POAC) Proceedings, Norwegian Institute of Technology, v. 2, p. 845–858.

Osborne, R. H., N. J. Dango, and R. C. Scheidemann, Jr., 1983, Report of potential offshore sand and gravel resources of the inner continental shelf of southern California: Sacramento, State of California Department of Boating and Waterways, 302 p.

Ovenshine, A. T., S. Bartsch-Winkler, N. R. O'Brien, and D. E. Lawson, 1976, Sediment of the high tidal range environment of upper Turnagain Arm, Alaska, *in* T. P. Miller, ed., Recent and ancient sedimentary environments in Alaska: Anchorage, Alaska Geological Society Symposium Proceedings (April 1975), p. M1–M26.

Petrov, O. M., 1967, Paleogeography of Chukotka during late Neogene and Quaternary time, *in* D. M. Hopkins, ed., The Bering land bridge: Stanford, California, Stanford University Press, p. 144–171.

Phillips, R. L., and T. E. Reiss, 1985a, Nearshore marine geologic investigations, Icy Cape to Wainwright, northeast Chukchi Sea, *in* P. W. Barnes, E. Reimnitz, R. E. Hunter, R. L. Phillips, and S. C. Wolf, eds., Geologic processes and hazards of the Beaufort and Chukchi Sea Shelf and coastal regions: U.S. Department of Commerce, NOAA, Outer Continental Shelf Environmental Assessment Program Final Report 34, p. 125–155.

Phillips, R. L., and T. E. Reiss, 1985b, Nearshore marine geologic investigations, Point Barrow to Skull Cliff, northeast Chukchi Sea, *in* P. W. Barnes, E. Reimnitz, R. E. Hunter, R. L. Phillips, and S. C. Wolf, eds., Geologic processes and hazards of the Beaufort and Chukchi Sea Shelf and coastal regions: U.S. Department of Commerce, NOAA, Outer Continental Shelf Environmental Assessment Program Final Report 34, p. 157–181.

Phillips, R. L., T. E. Reiss, E. W. Kempema, and E. Reimnitz, 1984, Nearshore marine geologic investigations, Wainwright to Skull Cliff, northeast Chukchi Sea: U.S. Geological Survey Open-File Report 84-108, 33 p.

Plafker, G., 1981, Late Cenozoic glaciomarine deposits of the Yakataga Formation, Alaska, *in* M. J. Hambrey and W. B. Harland, eds., Earth's pre-Pleistocene glacial record: Cambridge, Cambridge University Press, p. 694–699.

Plafker, G., 1987, Regional geology and petroleum potential of the northern Gulf of Alaska continental margin, this volume.

Plafker, G., and W. O. Addicott, 1976, Marine glacier deposits of Miocene through Holocene age along the Gulf of Alaska margin, *in* T. P. Miller, ed., Recent and ancient sedimentary environments in Alaska: Anchorage, Alaska Geological Society Symposium Proceedings (April 1975), p. Q1–Q23.

Plafker, G., and M. Rubin, 1967, Vertical tectonic displacements in south-central Alaska during and prior to the great 1964 earthquake: Osaka City University, Journal of Geoscience, v. 10, p. 53–66.

Plafker, G., G. R. Winkler, W. L. Coonrad, and G. Claypool, 1980, Preliminary report on the geology of the continental slope adjacent to OCS lease sale 55, eastern Gulf of Alaska: Petroleum resource implications: U.S. Geological Survey Open-File Report 80-1089, 72 p.

Porter, S. C., 1967, Glaciation of Chagvan Bay area, southwestern Alaska: Arctic, v. 20, p. 227–246.

Pratt, R. M., and R. F. Dill, 1974, Deep eustatic terrace levels: Further speculations: Geology, v. 2, p. 155–159.

Quinterno, P., P. R. Carlson, and B. F. Molnia, 1981, Benthic foraminifers as indicators of the Pleistocene-Holocene boundary in the eastern Gulf of Alaska, *in* N. R. D. Albert and T. Hudson, eds., The United States Geological Survey in Alaska: Accomplishments during 1979: U.S. Geological Survey Circular 823-B, p. B128–B130.

Rappeport, M. L., 1980, Depositional environments and Quaternary sedimentary units within lower Cook Inlet, Alaska—A high-energy tidally dominated embayment along the Pacific margin of the United States, *in* M. E. Field, A. H. Bouma, I. P. Colburn, R. G. Douglas, and J. C. Ingle, eds., Quaternary depositional environments of the Pacific Coast: SEPM, Pacific Section, Pacific Coast Paleogeography Symposium 4, p. 73–88.

Rappeport, M. L., 1982, An analysis of oceanographic and meteorological conditions for central lower Cook Inlet, Alaska: U.S. Geological Survey Open-File Report 82-128, 46 p.

Ray, V. M., and W. R. Dupré, 1981, The ice-dominated regimen of Norton Sound and adjacent areas of the Bering Sea, *in* D. W. Hood and J. A. Calder, eds., The Eastern Bering Sea Shelf: Oceanography and resources: Washington, D.C., U.S. Department of Commerce, p. 263–278.

Reimnitz, E., 1966, Late Quaternary history and sedimentation of the Copper River delta and vicinity, Alaska: San Diego, University of California unpublished Ph.D. dissertation, 160 p.

Reimnitz, E., and P. W. Barnes, 1974, Sea ice as a geologic agent on the Beaufort Sea Shelf of Alaska, *in* J. C. Reed and J. E. Sater, eds., The coast and shelf of the Beaufort Sea: Arlington, Virginia, Arctic Institute of North America, p. 301–353.

Reimnitz, E., and E. W. Kempema, 1982, Dynamic ice-wallow relief of northern Alaska's nearshore: Journal of Sedimentary Petrology, v. 52, p. 451–461.

Reimnitz, E., and E. W. Kempema, 1983, High rates of bedload transport measured from infilling of large strudel-scour craters in the Beaufort Sea, Alaska: Continental Shelf Research, v. 1, p. 237–251.

Reimnitz, E., and E. W. Kempema, 1985, Pack ice interaction with Stamukhi Shoal, Beaufort Sea, Alaska, *in* P. W. Barnes, E. Reimnitz, R. E. Hunter, R. L., Phillips, and S. C. Wolf, eds., Geologic processes and hazards of the Beaufort and Chukchi Sea Shelf and coastal regions: U.S. Department of Commerce, NOAA, Outer Continental Shelf Environmental Assessment Program Final Report 34, p. 251–282.

Reimnitz, E., and D. K. Maurer, 1979, Eolian sand deflation—A cause for gravel barrier islands in Arctic Alaska?: Geology, v. 7, p. 507–510.

Reimnitz, E., and G. Plafker, 1976, Marine gold placers along the Gulf of Alaska margin: U.S. Geological Survey Bulletin 1414, 16 p.

Reimnitz, E., and C. R. Ross, 1979, Lag deposits of boulders in Stefansson Sound, Beaufort Sea, Alaska: U.S. Geological Survey Open-File Report 79-1205, 26 p.

Reimnitz, E., P. W. Barnes, and R. L. Phillips, 1985, Sixty-meter-deep pressure-ridge keels in the Arctic Ocean suggested from geological evidence, *in* P. W. Barnes, E. Reimnitz, R. E. Hunter, R. L. Phillips, and S. C. Wolf, eds., Geologic processes and hazards of the Beaufort and Chukchi Sea Shelf and coastal regions: U.S. Department of Commerce, NOAA, Outer Continental Shelf Environmental Assessment Program Final Report 34, p. 295–312.

Reimnitz, E., L. Toimil, and P. W. Barnes, 1978, Arctic continental shelf morphology related to sea-ice zonation, Beaufort Sea, Alaska: Marine Geology, v. 28, p. 179–210.

Reimnitz, E., R. von Huene, and F. F. Wright, 1970, Detrital gold and sediments in Nuka Bay, Alaska: U.S. Geological Survey Professional Paper 700-C, p. C35–C42.

Reimnitz, E., E. W. Kempema, C. R. Ross, and P. W. Minkler, 1980, Overconsolidated surficial deposits on the Beaufort Sea Shelf: U.S. Geological Survey Open-File Report 80-2010, 37 p.

Reimnitz, E., P. W. Barnes, D. M. Rearic, P. W. Minkler, E. W. Kempema, and T. Reiss, 1982, Marine geological investigations in the Beaufort Sea in 1981 and preliminary interpretations for region from the Canning River to the Canadian border: U.S. Geological Survey Open-File Report 82-974, 64 p.

Rex, R. W., 1955, Microrelief produced by sea ice grounding in the Chukchi Sea near Barrow, Alaska: Arctic, v. 8, p. 177–186.

Rex, R. W., 1964, Arctic beaches, Barrow, Alaska, *in* R. L. Miller, ed., Papers in marine geology—Shepard Commemorative Volume: New York, The Macmillan Company, p. 384–400.

Rodeick, C. A., 1979, The origin, distribution, and depositional history of gravel deposits on the Beaufort Sea continental shelf, Alaska: U.S. Geological Survey Open-File Report 79-234, 87 p.

Sainsbury, C. L., 1967, Quaternary geology of western Seward Peninsula, Alaska, *in* D. M. Hopkins, ed., The Bering land bridge: Stanford, California, Stanford University Press, p. 121–143.

Sallenger, A. H., J. R. Dingler, and R. H. Hunter, 1978, Coastal processes and morphology of the Bering Sea coast of Alaska: U.S. Department of Commerce, NOAA, Environmental Assessment of the Alaskan Continental Shelf Annual Report, v. 12, p. 451–502.

Scholl, D. W., and J. S. Creager, 1973, Geologic synthesis of Leg 19 (DSDP) results; far north Pacific and Aleutian Ridge and Bering Sea, *in* J. S. Creager, D. W. Scholl, et al., Initial reports of the Deep Sea Drilling Project (Vol. 19): Washington, D.C., U. S. Government Printing Office, p. 897–913.

Scholl, D. W., and D. M. Hopkins, 1969, Newly discovered Cenozoic basins, Bering Sea shelf, Alaska: AAPG Bulletin, v. 53, p. 2067–2078.

Scholl, D. W., and C. L. Sainsbury, 1961, Subaerially carved arctic seavalley under a modern epicontinental sea: GSA Bulletin, v. 72, p. 1433–1435.

Scholl, D. W., and C. L. Sainsbury, 1966, Marine geology of the Ogotoruk Creek area, *in* N. J. Wilimovsky and J. N. Wolfe, eds., Environment of the Cape Thompson region, ALaska: U.S. Atomic Energy Commission, p. 787–806.

Scholl, D. W., M. S. Marlow, and E. C. Buffington, 1975, Summit basins of Aleutian Ridge, North Pacific: AAPG Bulletin, v. 59, p. 799–816.

Scholl, D. W., T. L. Vallier, and A. J. Stevenson, 1983a, Geologic evolution of the Aleutian Ridge—Implications for petroleum resources: Journal of the Alaska Geological Society, v. 3, p. 33–46.

Scholl, D. W., T. L. Vallier, and A. J. Stevenson, 1983b, Arc, forearc, and trench sedimentation and tectonics; Amlia corridor of the Aleutian Ridge, *in* J. S. Watkins and C. L. Drake, eds., Studies in continental margin geology: AAPG Memoir 34, p. 413–439.

Scholl, D. W., T. L. Vallier, and A. J. Stevenson, 1987, Geologic evolution and petroleum geology of the Aleutian Ridge, this volume.

Scholl, D. W., E. C. Buffington, D. M. Hopkins, and T. R. Alpha, 1970, The structure and origin of the large submarine canyons of the Bering Sea: Marine Geology, v. 8, p. 187–210.

Schumacher, J. D., and R. K. Reed, 1980, Coastal flow in the northwest Gulf of Alaska: The Kenai current: Journal of Geophysical Research, v. 85, p. 6680–6688.

Sharma, G. D., 1974, Contemporary depositional environment of the eastern Bering Sea. Part 1. Contemporary sedimentary regimes of the eastern Bering Sea, *in* D. W. Hood and E. J. Kelley, eds., Oceanography of the Bering Sea, with emphasis on renewable resources: Fairbanks, University of Alaska Institute of Marine Science, p. 517–540.

Sharma, G. D., 1975, Contemporary epicontinental sedimentation and shelf grading in the southeast Bering Sea, *in* R. B. Forbes, ed., Contributions to the geology of the Bering Sea basin and adjacent regions: GSA Special Paper 151, p. 33–48.

Sharma, G. D., 1979, The Alaska Shelf: Hydrographic, sedimentary, and geochemical environment: New York, Springer-Verlag, 498 p.

Sharma, G. D., and D.C. Burrell, 1970, Sedimentary environment and sediments of Cook Inlet, Alaska: AAPG Bulletin, v. 54, p. 647–654.

Sharma, G. D., A. S. Naidu, and D. W. Hood, 1972, Bristol Bay: Model contemporary graded shelf: AAPG Bulletin, v. 56, p. 2000–2012.

Short, A. D., 1975, Offshore bars along the Alaska Arctic coast: Journal of Geology, v. 83, p. 209–221.

Short, A. D., 1979, Barrier island development along the Alaskan-Yukon coastal plains: GSA Bulletin, v. 90, p. 3–5.

Smith, P. A., and D. M. Hopkins, 1979, Offshore permafrost studies and shoreline history of Chukchi and Beaufort Seas as an aid to predicting offshore permafrost: U.S. Department of Commerce, NOAA, Environmental Assessment of the Alaskan Continental Shelf Annual Report, v. 9, p. 116–163.

Smith, P. A., R. W. Hartz, and D. M. Hopkins, 1980, Offshore permafrost studies and shoreline history as an aid to predicting offshore permafrost conditions: U.S. Department of Commerce, NOAA, Environmental Assessment of the Alaskan Continental Shelf Annual Report, v. 4, p. 159–255.

Solomon, H., and K. Ahlnäs, 1980, Ice spirals off Barrow as seen by satellite: Arctic, v. 33, p. 184–188.

Stringer, W. J., 1981, Nearshore ice characteristics in the eastern Bering Sea, *in* D. W. Hood and J. A. Calder, eds., The Eastern Bering Sea Shelf: Oceanography and resources, v. 1: Washington, D.C., U.S. Department of Commerce, p. 167–187.

Swift, D. J. P., B. F. Molnia, and R. G. Jackson, II, 1978, Intermittent structure in the atmospheric boundary layer made visible by entrained sediment—Example from the Copper River delta, Alaska: Journal of Sedimentary Petrology, v. 48, p. 897–900.

Tagg, A. R., and H. G. Greene, 1973, High-resolution seismic survey of an offshore area near Nome, Alaska: U.S. Geological Survey Professional Paper 759-A, 23 p.

Thomas, B. I., and R. V. Berryhill, 1962, Reconnaissance studies of Alaskan beach sands, eastern Gulf of Alaska: U.S. Bureau of Mines Report of Investigations 5986, 40 p.

Thomas, D. R., 1981, Behavior of oil spills under sea ice, Prudhoe Bay: U.S. Department of Commerce, NOAA, Environmental Assessment of the Alaskan Continental Shelf Annual Report, v. 6, p. 61–187.

Thrasher, G. P., 1979, Geologic map of the Kodiak outer continental shelf, western Gulf of Alaska: U.S. Geological Survey Open-File Report 79-1267, 2 map sheets, scale 1:250,000.

Toimil, L. J., 1978, Ice gouged microrelief on the floor of the eastern Chukchi Sea, Alaska: A reconnaissance survey: U.S. Geological Survey Open-File Report 78-693, 48 p.

Toimil, L. J., and A. Grantz, 1976, Origin of a bergfield in the northeastern Chukchi Sea and its influence on the sedimentary environment: AIDJEX (Arctic Ice Dynamics Joint Experiment) Bulletin 34, p. 1–42.

Tolson, R. B., 1987, Structure and stratigraphy of the Hope Basin, southern Chukchi Sea, Alaska, this volume.

Turner, B. W., G. P. Thrasher, G. B. Shearer, and K. D. Holden, 1979, Bathymetric maps of the Kodiak outer continental shelf, western Gulf of Alaska: U.S. Geological Survey Open-File Report 79-263, 13 sheets, scale 1:250,000.

U.S. Department of Commerce, 1985, Tide tables 1986: High and low water predictions, west coast of North and South America, including the Hawaiian Islands: Washington, D.C., National Ocean Service, 234 p.

Vallier, T. L., M. B. Underwood, J. V. Gardner, and J. A. Barron, 1980, Neogene sedimentation on the outer continental margin, southern Bering Sea: Marine Geology, v. 36, p. 269–287.

von Huene, R., M. A. Fisher, and T. R. Bruns, 1987, Geology and evolution of the Kodiak margin, Gulf of Alaska, this volume.

von Huene, R., G. G. Shor, Jr., and E. Reimnitz, 1967, Geological interpretation of seismic profiles in Prince William Sound, Alaska: GSA Bulletin, v. 78, p. 259–268.

Whitney, J. W., and K. D. Holden, 1980, Bathymetric map of the outer continental shelf of Shelikof Strait, Alaska: U.S. Geological Survey Open-File Report 80-2031, 1 sheet, scale 1:250,000.

Whitney, J. W., and D. K. Thurston, 1981, Geologic constraints for petroleum development of the lower Cook Inlet, Alaska, outer continental shelf lease area, *in* N. R. D. Albert and T. Hudson, eds., The United States Geological Survey in Alaska: Accomplishments during 1979: U.S. Geological Survey Circular 823-B, p.

B122–B125.
Whitney, J. W., W. G. Noonan, D. Thurston, A. H. Bouma, and M. A. Hampton, 1979, Lower Cook Inlet, Alaska: Do those large sand waves migrate?: Eleventh Annual Offshore Technology Conference Proceedings, p. 1071–1082.
Williams, J. R., 1979, Stratigraphy of the Gubik Formation at Skull Cliff, northern Alaska: U.S. Geological Survey Circular 804-B, p. 31–33.
Williams, J. R., and H. W. Coulter, 1981, Deglaciation and sea-level fluctuations in Port Valdez, Alaska, *in* N. R. D. Albert and T. Hudson, eds., The United States Geological Survey in Alaska: Accomplishments during 1979: U.S. Geological Survey Circular 823-B, p. B78–B80.
Wright, F. F., 1972a, Geology and geomorphology of the central Gulf of Alaska continental shelf, *in* D. H. Rosenberg, ed., A review of the oceanography and renewable resources of the northern Gulf of Alaska: Fairbanks, University of Alaska Institute of Marine Science (IMS) Report R72-23, p. 153–160.
Wright, F. F., 1972b, Marine geology of Yakutat Bay, Alaska: U.S. Geological Survey Professional Paper 800-B, p. B9–B15.

CHAPTER THIRTY

HEAVY-MINERAL PLACER DEPOSITS OF THE CONTINENTAL MARGIN OF ALASKA AND THE PACIFIC COAST STATES

H. E. Clifton
G. Luepke

U.S. Geological Survey
Menlo Park, California

ABSTRACT

The potential for economic heavy-mineral placers on the continental shelf and coastal areas of Alaska and the Pacific Coast states remains uncertain. With few exceptions, sampling for heavy-mineral content has been inadequate for economic evaluations of the total heavy-mineral assemblage (few three-dimensional studies), and few areas have been sampled sufficiently to assess the potential for a specific commodity. A further problem is that the heavy-mineral studies are commonly confined to the sand-size fraction of a sediment, and only a limited grain-size range is examined.

The tectonic and climatic settings of the western margin of North America produce a characteristically immature (and economically less interesting) assemblage of heavy minerals. The active tectonism and commonly arid climates are not conducive to development of heavy-mineral placers such as those presently mined from the Atlantic Coast of the United States or the coasts of Brazil, Australia, or Africa.

Mining of marine placers along the margin of Alaska and the Pacific Coast states has been directed largely toward the recovery of gold, with minor emphasis on platinum, chromite, and titanium-bearing minerals. This mining has historically been restricted to the most highly enriched natural concentrations. As a result, the mining operations have traditionally been of very small scale. The potential for many

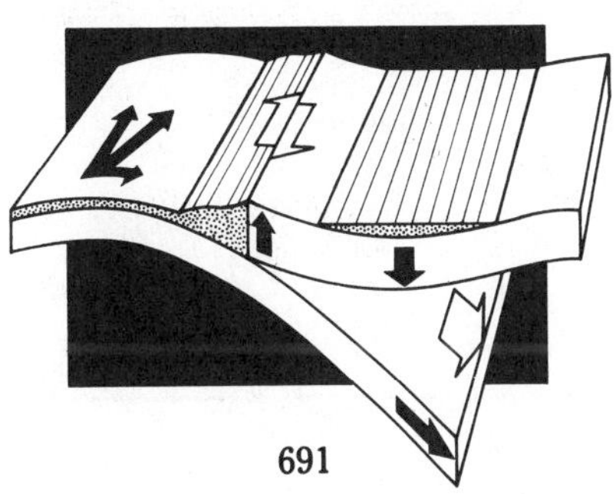

undiscovered deposits, particularly of gold, remains fairly high around the Seward Peninsula, along the Gulf of Alaska, and locally in the inland waterways of southeastern Alaska. Coastal primary sources and alluvial placers of cassiterite and platinum suggest a possibility of undiscovered resources of these minerals locally in sediment of the Bering Sea. A modest potential for gold and chromite exists along the coast of southern Oregon and northern California. In all areas, an adequate three-dimensional sampling program is a requisite for further economic evaluation.

INTRODUCTION

Placer deposits are mechanical concentrations of mineral grains that historically have been considered to be of high specific gravity (> 3.0), chemical stability, and physical durability (Bateman, 1955). These "heavy minerals" (Table 1) generally are disseminated in trace amounts in source rocks. Because heavy minerals are physically and/or chemically resistant and differ hydraulically in behavior from less dense minerals, the combined processes of weathering and erosion of source rocks and subsequent transport concentrate them in placer deposits.

Placer deposits in the sediment of a continental margin typically occur in coastal areas or on the adjacent shelf; the deposits may be modern (accumulating under the influence of contemporary processes) or relict (formed by such processes in the past). Relict deposits may lie today above sea level in emergent terraces, or they may be submerged on the continental shelf. Modern or relict, emergent or submerged placers may be of distinct economic interest. World annual production of placer minerals from beach and shelf areas was worth approximately $50 million in 1970 (United Nations, 1970).

The earliest reference to potentially economic placer minerals in Pacific Coast sands is in Shellrocke's *Voyages to California*, published in London in 1742. The author referred to the difficulty at that time of separating gold from the other heavy minerals; in fact, California prospectors more than a century later avoided the naturally occurring black sands (so-called for the concentrations of magnetite and ilmenite; see Clifton and Mason, 1969) for that very reason (Day and Richards, 1906a, p. 150). Not until 1905 did Congress direct the U.S. Geological Survey to include an investigation of the Pacific slope black sands in its yearly report of mineral resources and to suggest how the valuable minerals might be recovered.

In spite of the problems of separating gold from the other heavy minerals, beach placers were mined on the Pacific Coast below the Gold Bluffs north of Redwood Creek in Humboldt County, California, as early as 1851 (Hornor, 1918; see also Chase, 1874). A bizarre effort by an anonymous miner in primitive diving gear to find gold in a California river in 1849 may have been the first underwater attempt to extract gold (Sloane and Sloane, 1970).

Offshore placer mining on the Pacific Coast has not been significant. In the winter of 1905–1906, Captain Henry Finch prospected for gold near Nome, Alaska, at temperatures of −40°F through 6 feet of ice (Sloane and Sloane, 1970, p. 120). Offshore placer mining took place on a small scale before 1940 in Golovin Bay, Alaska, through holes cut in the ice, when that ice was thick enough to support a clamshell scoop and truck. The dredged material was then transported to shore for sluicing during the summer months (Conwell, 1976). A gold deposit in the submerged channel of Daniels Creek, near Bluff, Alaska, was similarly mined during the late 1930s and early 1940s by Joseph T. Sullivan (Mining World, 1941).

The value of placer deposits depends largely on the economics of their recovery. Numerous factors enter into such an economic assessment (Macdonald, 1971). Some relate to the deposits themselves—their volume, composition and grain size. Others relate to the existing technology of recovery and to market factors that change with world demand. Marine law may be a determining factor in many areas (Duane, 1976). This chapter deals only with the distribution of the deposits and their physical and chemical characteristics.

We first discuss those factors that determine the distribu-

Table 1. Heavy minerals of possible economic interest classified by stability. Specific gravity in parentheses. After Brenninkmeyer (1978).

STABLE	STABLE	STABLE	MODERATELY STABLE	UNSTABLE
Anatase (3.9)	Garnet (3.5–4.3)	Sillimanite (3.2)	Andalusite (3.2)	Apatite (3.2)
Barite (4.5)	Gold (15.0–19.3)	Spinel (3.5–4.0)	Ilmenite (4.5–5.0)	Pyrite (5.0–5.1)
Brookite (3.9–4.1)	Kyanite (3.5–3.7)	Staurolite (3.6–3.7)	Marcasite (4.8–4.9)	Pyrrhotite (4.6–4.7)
Cassiterite (6.8–7.1)	Leucoxene (3.5–4.5)	Topaz (3.4–3.6)	Pyrolusite (4.7–4.8)	
Chromite (4.3–4.6)	Magnetite (5.2–6.5)	Tourmaline (3.0–3.3)		
Columbite (5.3–7.3)	Monazite (4.9–5.3)	Wolframite (7.0–7.5)		
Corundum (4.0–4.1)	Platinum group (14–19)	Xenotime (4.5–4.6)		
Diamond (3.5)	Scheelite (5.9–6.1)	Zircon (4.2–4.9)		
Fluorite (3.2)	Rutile (4.2–4.3)			

tion and character of placers on the continental margin. We next consider the character of the western margin of North America as it relates to these factors. Then we apply the factors to specific heavy minerals of possible economic importance on this margin to analyze their potential for concentration, and to project the distribution and characteristics of such concentrations. These analyses and projections are applied to a regional examination of the conditions that influence the formation of economic placer deposits on the coast and shelf of Alaska and the conterminous United States. We conclude with a section evaluating the economic potential for heavy-mineral placers in these areas.

FACTORS THAT DETERMINE THE DISTRIBUTION AND CHARACTER OF CONTINENTAL MARGIN PLACER DEPOSITS

To understand the distribution and character of placer deposits on the shelves and coastal areas of a continental margin, one must consider the factors responsible for their formation. These factors include sources for the minerals, mechanisms whereby they are transported to the margin, processes of concentration, and preservation potential of the resulting deposit.

Sources

The sources of heavy minerals may be primary or secondary (Anderson, 1971). Primary sources include the rocks in which the minerals crystallized. The minerals may originally occur in minable concentrations, or they may be so disseminated as to be uneconomical. Commodity maps that locate prospects or mines of a particular mineral in the host rock may be a useful guide to its distribution as a placer mineral. Such maps are of less use, however, and may even be misleading for identifying primary sources for a mineral that is partly or totally disseminated in trace amounts throughout a host rock.

Secondary sources include sedimentary deposits that bear detrital grains of the mineral in question. The different concentrating mechanisms can cause secondary sources to be enriched relative to primary sources. Where sedimentary accumulations of different ages exist on a continental shelf, it has been argued that they become progressively enriched, whereby the youngest, which potentially can draw on the other deposits as sources, is likely to have the greatest economic promise (Bowman, 1972).

Transportation Mechanisms

Source rocks for heavy-mineral placers may be exposed underwater on the continental shelves, crop out at the shoreline, or lie well inland from the present shelf area. In the latter case particularly, some mechanism is required to transport the heavy minerals to the sites of placer concentration. Possible mechanisms include fluvial processes, glaciers, landslides, longshore or shelf currents, and wind. Because each of these mechanisms differs in competence and capacity for carrying heavy minerals, the physical characteristics of the transporting system can significantly influence the distribution and nature of placer deposits.

River systems are perhaps the most widespread mechanism for carrying heavy minerals from inland sources to the margin of a continent. The competence and capacity of the streams depend largely on their gradient and volume of discharge. A coastal plain (which would include portions of the present-day shelf exposed during eustatically lowered sea level) typically has a very low gradient; it is unlikely that streams will transport significant quantities of the heavier heavy minerals (specific gravity >6.0) far in such a setting. Emery and Noakes (1968) suggest that economic placer deposits of gold, platinum, and tin generally lie less <15 km from the source rock. The smallest particles of these minerals may be carried much farther, but their potential for accumulating in recoverable economic concentrations is questionable.

Glacial processes and landslides, in contrast, provide a mechanism for introducing coarse-grained placer minerals directly onto the continental shelf during glacio-eustatic lowstands of the sea. The placer minerals, typically scattered initially within glacial till or landslide debris, are available for subsequent reworking and concentration by marine processes following a subsequent rise in sea level. Ice rafting is another process whereby coarse placer minerals may be carried onto a high-latitude continental shelf (Nelson and Hopkins, 1972).

Marine processes such as littoral longshore transport or shelf currents may be effective in transporting heavy minerals emplaced on the continental shelf to a site of concentration. The lighter placer minerals (specific gravity <6.0; see Table 1) are more susceptible to this form of transport than are the heavier placer minerals (Emery and Noakes, 1968). Longshore transport is, however, considered to be primarily responsible for dispersing fine-grained gold flakes throughout the bottom sediment of the northern Bering Sea (Nelson and Hopkins, 1972).

Wind is an important process in some arid or semiarid coastal settings, such as the west coast of North Africa (Mero, 1965). Eolian transport may not, however, be a particularly effective means of transporting heavy minerals; their typically small particle size and high density render them relatively difficult to entrain (Land, 1964). Nonetheless, Bradley (1957) argued for selective transportation of heavy minerals on Mustang Island by the wind; he suggested that the higher degree of roundness of the heavy minerals (especially the opaque minerals) may have promoted eolian transport.

Mechanisms of Concentration

Heavy minerals must be concentrated into a placer deposit before it arouses economic interest. The degree of concentration required depends upon the combination of minerals present, their tenor, and also upon market factors such as price structure and cost of extraction. Several different mechanisms for concentrating heavy minerals on a continental margin can greatly influence the composition, distribution, and geometry of the resulting deposit.

Many of the placer minerals, especially zircon, rutile, monazite, and the precious metals, are chemically stable under conditions of weathering (Brenninkmeyer, 1978) and therefore can be partly concentrated within the source area. This preliminary concentration can be further enhanced by additional weathering and mechanical concentration of the placer minerals as they recycle from the initial source through secondary host rocks. Such a process is generally critical to the development of economic marine placer deposits (Emery and Noakes, 1968). Sutherland (1985) provides an analysis of the effects of different climatic regimes on the development of heavy-mineral placers. The chemical decay of a host rock can lead to residual placers on the bedrock surface or to eluvial placers on hill slopes, where transport by gravity may lead to further concentration (Hails, 1976). Table 1 presents a listing by stability of heavy minerals that may be of possible economic interest.

Streams provide an effective means of concentrating heavy minerals, particularly those of high specific gravity such as gold, platinum, and cassiterite (Emery and Noakes, 1968). Stream placers are by far the most important type of placer deposits for precious metals and gems on a worldwide basis, commonly forming high concentrations in gravel on the channel floor or atop natural riffles (Bateman, 1955). The physical mechanisms involved in producing alluvial placers are discussed by Tourtelot (1968) and Brady and Jobson (1973). Ancient river channels cut during lowstands of the sea on high-gradient shelves yield submerged economic concentrations of cassiterite in Malaysia (Anderson, 1982), and similar concentrations of gold and platinum may occur on the inner shelf where alluvial placer deposits of these minerals occur onshore.

Beaches have long been recognized as important sites of placer mineral accumulation (Bateman, 1955). The lighter heavy minerals (particularly magnetite, ilmenite, monazite, and rutile) are the most likely to form economic beach placer deposits (Emery and Noakes, 1968). Gold, however, was extensively mined in small-scale operations from the beaches of Alaska, Oregon, and California in the last century, and cassiterite has been mined from the beaches of Malaysia (Anderson, 1982). The richest concentrations of placer minerals on a beach typically occur at its landward edge, in berms, or in the basal layers of the beach sand where it overlies bedrock (Mero, 1965). Reid and Frostick (1985a) document high concentrations (locally as much as 90%) of heavy minerals on the crests of swash bars at the margin of Lake Turkana in northern Kenya.

The process of heavy-mineral concentration on beaches is not well understood. Noting that the quartz sand within the heavy-mineral-rich layers is finer than that elsewhere on the beach, Stapor (1973) suggested that heavy minerals were not concentrated on the beach face. Instead, he believed that the concentrations developed offshore and were delivered as a slug to the beach face. May (1973) argued that selective shoreward transportation of heavy minerals occurs under conditions of medium-sized waves—that these minerals are entrained preferentially by the higher water velocity developed shoreward under the crest of a shoaling wave. It is not clear to what degree such a process actually contributes to the accumulation of heavy minerals on a beach. Although beaches in many areas are enriched in heavy minerals relative to offshore areas (Bradley, 1957), Tanner, Mullins, and Bates (1961) note that the concentration of heavy minerals by weight, relative to quartz, is about four times greater in offshore sands than it is on adjacent beaches of the Florida panhandle.

The restricted distribution of heavy minerals on the beach itself and the observation that heavy-mineral concentrations are well developed on erosional beaches (Everts, 1972) suggest that concentration does occur on the beach face. Slingerland (1977) postulated that the combination of boundary flow turbulence and the ratio of grain size to bed roughness determines the entrainment and deposition of sand-size grains of different density in the surface layer. The larger, hydraulically equivalent quartz grains would be susceptible to being selectively rolled away from the relatively smooth bed formed by the finer heavy-mineral grains (Everts, 1973; Slingerland, 1977). The threshold stresses that generate selective entrainment of heavy-mineral grains by this process on Oregon beaches are analyzed by Komar and Wang (1984).

Another process, however, may also create beach placers. The dominant sorting mechanism during wave backwash is a shear sorting within a dense concentration of grains moving just above the bed in a fluid-driven grain flow (Clifton, 1969). The mechanism of dispersive pressure in such a flow appears to promote the transport of the larger and less dense grains (Sallenger, 1979), leading to a selective concentration of the heavy minerals in the upper reaches of a flow. Thus a model of concentration can be envisioned whereby the fine size of the heavy-mineral grains promotes their transport in suspension within the wave swash but inhibits their transport in the fluid-driven grain flow under the backwash. This combination of processes would lead to the observed distribution of placer concentrations on the upper beach. Reid and Frostick (1985b) suggest that dispersive stress contributes as an ancillary process to placer development.

Other factors undoubtedly contribute to the concentration of heavy minerals on beaches, including the selective removal of light minerals by winds (Twenhofel, 1946b). Selective longshore transport may lead to local concentrations; it has long been noted that the richest concentrations of heavy minerals in southern Oregon beaches lie on the southern side of headlands (Twenhofel, 1943). This distribution is thought to result from differential seasonal longshore transport—northward in the higher energy winter conditions, which move both heavy and light minerals, and southward during the summer when the light minerals are selectively winnowed away (Twenhofel, 1943; Peterson, Komar, and Scheidegger, 1986). Concentrations may also occur at inflection points in coastlines which are associated with the convergence of longshore flow (Peterson, Komar, and Scheidegger, 1986).

Eolian processes are known to concentrate heavy minerals, which have been observed to move more slowly under the influence of winds than do light-mineral grains (Land, 1964; Hails, 1972). Selective winnowing by winds may enrich a beach placer (Twenhofel, 1946b; Woolsey, Henry, and Hunt, 1975; Slingerland, 1977). Several studies have shown that the heavy-mineral content of a coastal dune field may substantially exceed that of an adjacent beach (Bradley, 1957; Cordes, 1966). Whether this enrichment results from selective eolian transport of the heavy minerals from the beach (Bradley, 1957) or a winnowing away of the light minerals by the wind (Woolsey, Henry, and Hunt, 1975) remains an open question.

Other types of currents, such as tidal or thermohaline flow, could concentrate heavy minerals into economic marine placer deposits. Little is known about the mechanisms involved. R. L. Phillips (1976, oral communication) has observed rich concentrations of opaque minerals on the top of subtidal sand shoals in Willapa Bay, Washington. The concentrations apparently develop under sheet-flow conditions when the shoals are nearly exposed during low tide and subjected to strong tidal currents. Connah (1962) indicates a similar origin for heavy-mineral deposits along the coast of Queensland, Australia, and notes that placers can form where tidal currents carry beach and dune deposits into estuaries. Griggs (1945) interpreted a large concentration of chromite-bearing black sand in elevated marine terrace deposits on the coast of southern Oregon as being the result of submarine concentration. Tidal or other persistent coastal currents may be locally important in creating placer deposits of the lighter heavy minerals.

Preservation

The mechanisms of concentration described above have operated at least locally on the continental margins throughout much of geologic time. All placer deposits older than those actively accumulating at present are susceptible to partial or total loss by processes of erosion. The possible preservation of heavy minerals depends on a number of factors, including tectonic regime, rate and direction of sea-level change, sedimentation rate, and the erosional forces within the environmental setting.

Tectonic forces can greatly influence the distribution of ancient placers. Tectonically passive margins maintain a coastal-plain continental-shelf setting over extended periods of time. The eastern margin of North America, for example, has existed more or less in its present form since the Cretaceous; placer deposits developed in this interval may exist in the shallow subsurface throughout its length (Grosz and Escowitz, 1983). In contrast, on tectonically active margins the paleogeography can change markedly over relatively short intervals of time; in these areas, coastal-plain continental-shelf environments have existed in their present location for no more than a few million years. The potential for ancient placers in tectonically active margins, either as economic deposits or as secondary sources for younger deposits, is reduced relative to the potential in passive margins.

A dominant factor in the preservation of placer deposits of Pleistocene age is glacio-eustatic fluctuation of sea level. The exchange of water between the ocean and the ice of continental glaciers may have caused sea level to fluctuate frequently during the past 2.5 m.y. by 50 to 100 m (Shackleton and Opdyke, 1977). The preservation and distribution of placer deposits formed before the last major oscillation depend strongly on the local tectonic setting. In areas of emergence, as for example the southern coast of Oregon (Griggs, 1945), a series of elevated marine terraces stands above sea level, where they are subject to subaerial erosion and dissection. Ancient placer deposits that remain intact within such terrace deposits, however, are generally accessible to recovery. In areas of tectonic subsidence, for example the central part of the Monterey embayment (Dupré, 1975), placer deposits are more likely to be preserved within the downwarped succession, but they are also likely to be buried too deeply for economic recovery.

Relict placer deposits that date from the most recent oscillation of sea level (within the past 100,000 years) are of particular economic interest in that, where preserved, they tend to lie at or near the present sediment surface. In the past 35,000 years, concomitant with the last stage of Wisconsin glaciation, sea level on the east coast of the United States has fluctuated markedly. It fell from a position near its present level to a lowstand about 130 m below, 15,000 to 20,000 years ago, then returned to a position slightly below its present level about 7000 years ago (Milliman and Emery, 1968). The youngest relict placers on the continental shelves formed during this excursion of the sea. Such placers are so young that, in most places, their distribution is not influenced by tectonics.

A variety of types of placer deposits may form during a single eustatic fall and rise of the sea. Residual or eluvial placers may form on an exposed shelf if the weathering during exposure is sufficiently intense. Streams crossing an exposed shelf can be the site of placer accumulations of the precious metals and tin. However, as noted, such deposits may be less valuable than those in adjacent onshore areas, if the shelf gradient is so low that the streams can carry only the finest, less economically attractive placer mineral grains. Beach, eolian, and tidal-inlet placers may form concentrations of the lighter heavy minerals at any particular stand of the sea during the oscillation.

The preservation of placer deposits formed during the late Wisconsin excursion of sea level (like those of earlier excursions) depends on the rates of sea-level change, slope gradient, sediment supply, and the erosive capability of the environment. Typically, the deposits at the shore of a rapidly rising sea have limited potential for being preserved. Owing largely to the shoreward translation of the concave-up nearshore equilibrium profile, the shoreface deposits are eroded and shifted progressively landward (Bruun, 1962; Everts, 1972). Nonetheless, examples of relict-beach placer deposits exist on continental shelves in different parts of the world. On the east coast of Africa, heavy minerals concentrated in beach placers at the lowstand were disseminated by the encroaching sea into widespread deposits of economic potential (Beiersdorf, Kudrass, and von Stackelberg, 1980). On the southern Oregon shelf, linear shore-parallel concentrations of fine gold in the surface sediment on the shelf are enclosed by broader halos of lighter heavy minerals; these likewise are interpreted to be relict-beach placers that have been partly dispersed by subsequent rise of sea level (Clifton, 1968; Kulm et al., 1968a).

A rising sea may create economic placer deposits if it encounters glacial till that contains the heavier heavy minerals (precious metals, tin, or tungsten minerals). Although these minerals may be so dispersed within the till as to be uneconomical, subsequent reworking by waves and currents may produce economic concentrations. Stream placers of these same minerals would be unaffected by the rising sea where the concentrations fill channels that lie below the level of wave planation.

The rate of sea-level rise can significantly influence the preservation of beach placers that form during a marine transgression. Everts (1972) suggests that a variable rate of rise may cause local aggradation of coastal sediment that can be isolated without total reworking. The extent of preservation in such a case depends on the erosive capability of the waves, the shape of the sediment wedge, the rate of sedimentation, and the shelf gradient.

Characteristics of the Western Margin of North America

The western margin of North America comprises a broad range of tectonic and environmental settings. The margin in the Arctic and northern Bering Sea areas is largely tectonically passive. Elsewhere the margin of the continent is tectonically active, either engaged in subduction as in the northern Pacific or in strike-slip motion as with much of California. The potential source rocks for heavy minerals reflect

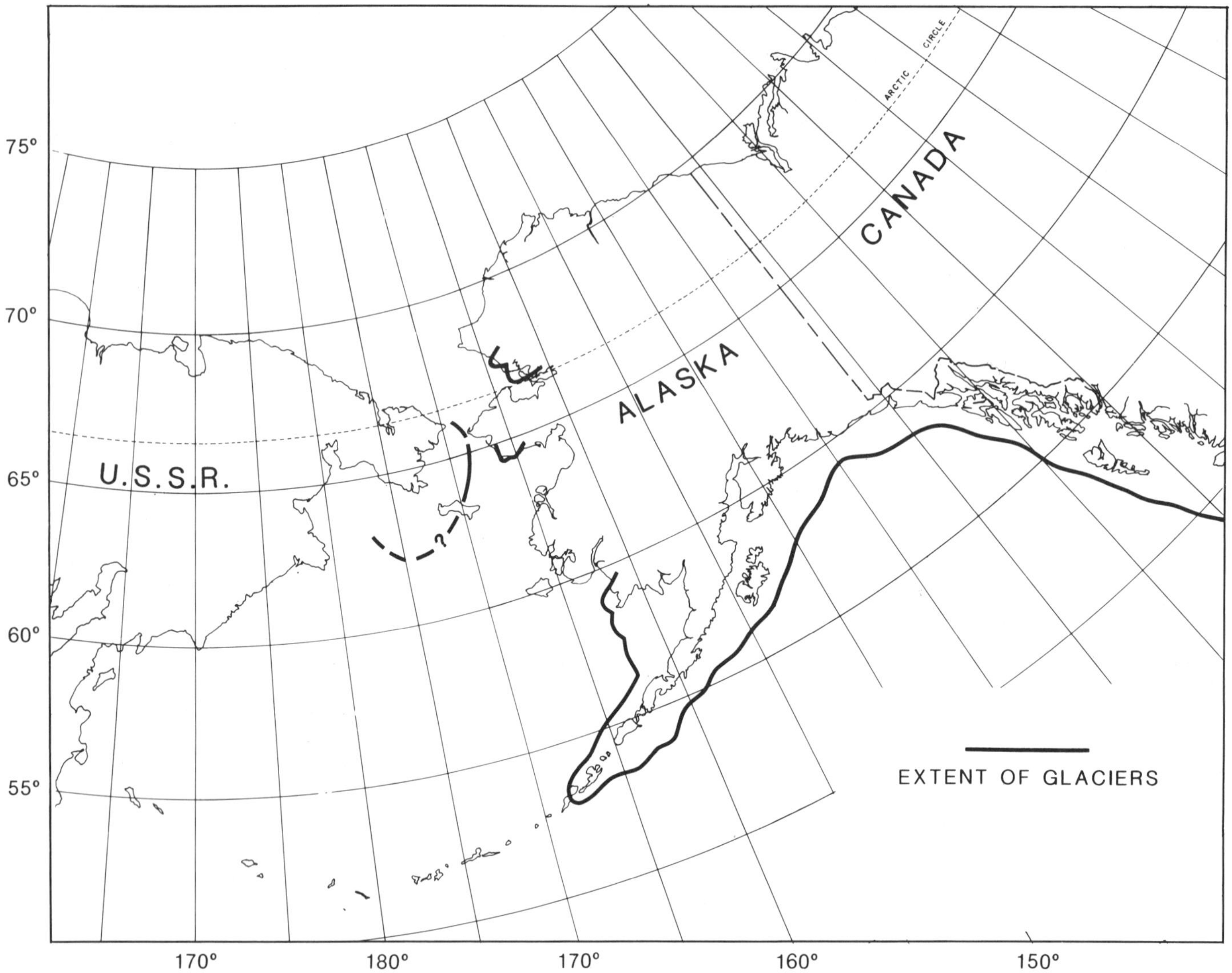

Figure 1. Extent of glaciation (heavy lines) on the Alaskan continental shelf (after Coulter et al., 1965, and Grim and McManus, 1970).

this tectonic regime. Volcanic rocks, particularly of andesitic and basaltic composition, are abundant in coastal areas, and plutonic granodioritic rocks are also prevalent there. Hydrothermal activity related to the emplacement of these igneous bodies is responsible for much of the metalliferous deposits onshore. Ultramafic rocks occur locally in the coast ranges of California and Oregon. Metamorphism is largely of low grade (greenstone, slate, or phyllite) or indicative of high-pressure, low-temperature conditions.

The climate here is not conducive to the deep weathering that pre-concentrates economic placer minerals elsewhere in the world. The warmest areas tend to be semiarid, and the areas with significant rainfall (the Pacific Northwest and Alaska) are cool to frigid. As a consequence of the combination of climate and tectonic activity, the heavy-mineral assemblages of the western North American margin are typified by large proportions of the less mature, less economically interesting minerals such as pyroxenes, amphiboles, and epidote.

A wide range of processes that can transport or concentrate heavy minerals have been (and are) active on this margin. Glacial processes have played a major role in sedimentation in the northern Bering Sea and in the Gulf of Alaska (Figure 1). Ice-related sedimentary processes are important on the northern Alaskan Shelf, where ice dominates most of the year, and in the northern Bering Sea. Waves and wave-related processes dominate most of the Pacific Coast. Tidal currents are potentially effective mechanisms for concentrating placer minerals in the southeastern Bering Sea, in Cook Inlet, and near the inlets or narrows of major coastal embayments (such as in southeastern Alaska, the Columbia River mouth, or San Francisco Bay). A consistent, northward thermohaline flow through the Bering Straits may concentrate the lighter heavy minerals in the northern Bering and southern Chukchi Seas. These potential concentrating mechanisms will be discussed in the regional evaluations in this chapter.

HEAVY-MINERAL PLACER COMMODITIES OF THE CONTINENTAL MARGIN OF WESTERN NORTH AMERICA

In this section we discuss the placer commodities of potential economic interest on the continental margin of Alaska

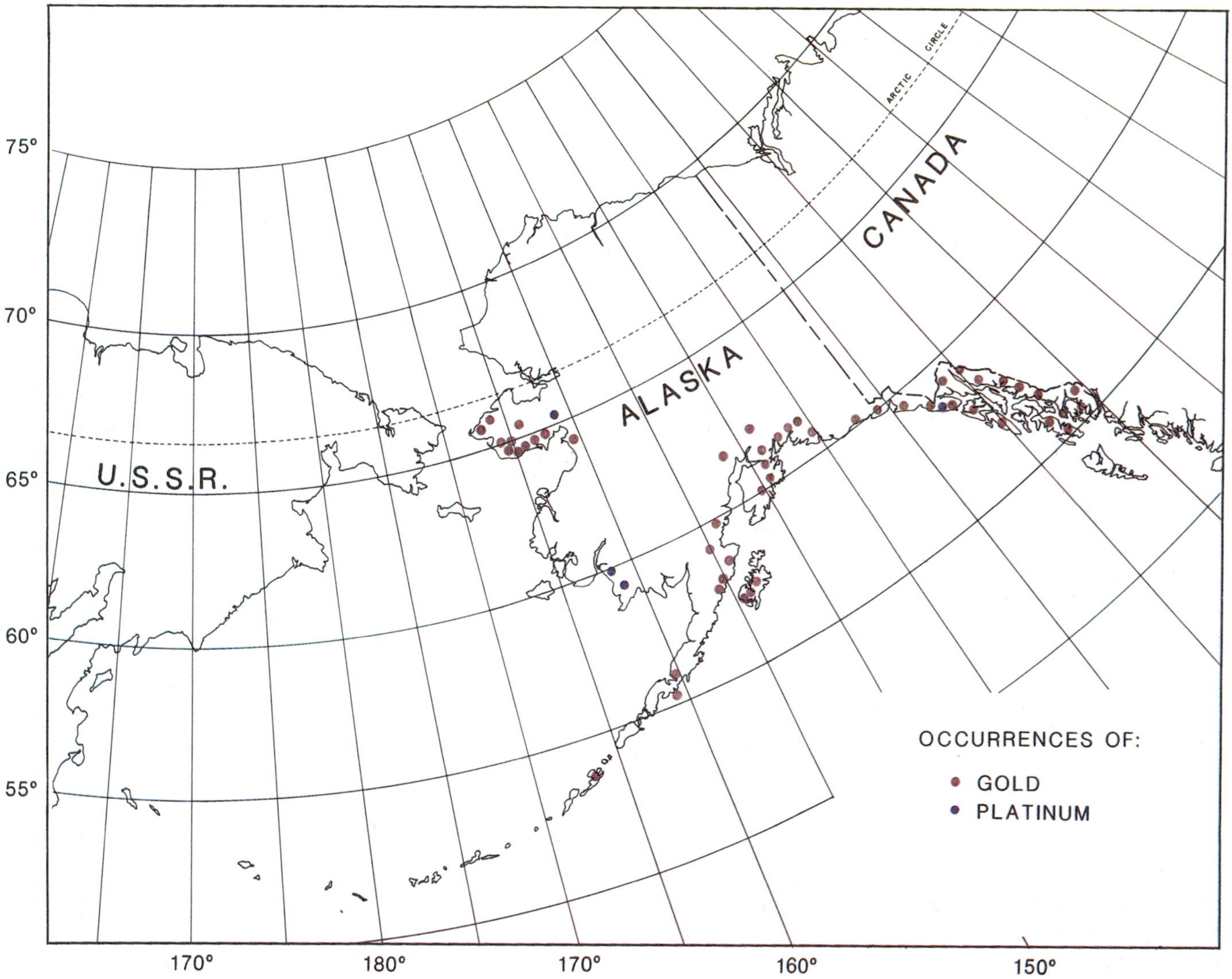

Figure 2. Coastal occurrences (lode and placer) of gold (red) and platinum (blue) in Alaska (Cobb, 1960, 1984a, 1984b).

and the Pacific Coast states that have been identified in earlier reports. The discussion of each commodity includes an analysis of potential sources and any special characteristics of the commodity minerals that bear on their accumulation in placer deposits. The discussion does not include commonly occurring heavy minerals of no present economic value (such as amphiboles or pyroxenes) or minerals that occur so rarely as to have no economic significance (such as diamonds).

Gold

Historically the most important placer mineral along the western margin of North America has been gold, the target of mining and prospecting activities since its discovery in California in 1848. About two-thirds of the gold production in Alaska and the Pacific states has been from placer deposits (Bateman, 1955). Most of this placer production has been from fluvial deposits; marine deposits are yet to yield significant production of placer gold (Simons and Prinz, 1973). Nonetheless, coastal deposits near Nome, Alaska, have been a source of gold since it was discovered in the present beach in 1899 (Cobb, 1973), and small-scale mining operations have produced gold in many beaches and elevated marine terraces around the Pacific and Bering Sea margins of North America (Cobb, 1973; Phillips, 1979; Tagg, 1979). Identified sources of lode and placer gold adjacent to the margin are shown in Figures 2 and 3.

The gold on the present continental shelf typically occurs as very small (<0.25 mm) flakes that are difficult to recover. Gold in many coastal drainage systems of high relief probably follows the pattern of the Sixes River, Oregon, in which coarse grains of gold are generally restricted to the upper reaches (Boggs and Baldwin, 1970).

Gold in marine deposits in the continental shelf offshore from the Sixes River tended to occur in flakes 0.04 to 0.25 mm in diameter (Clifton, 1968). Coarser grained gold flakes (as much as 0.5 mm in diameter) were found in adjacent modern beach deposits but were uncommon. Phillips (1979), in a detailed analysis of gold on an Oregon beach, found that 78% of the gold flakes were <0.177 mm in diameter, but these fine-sized flakes accounted for $<50\%$ of the total gold value. The gold in modern marine sediments in the northern

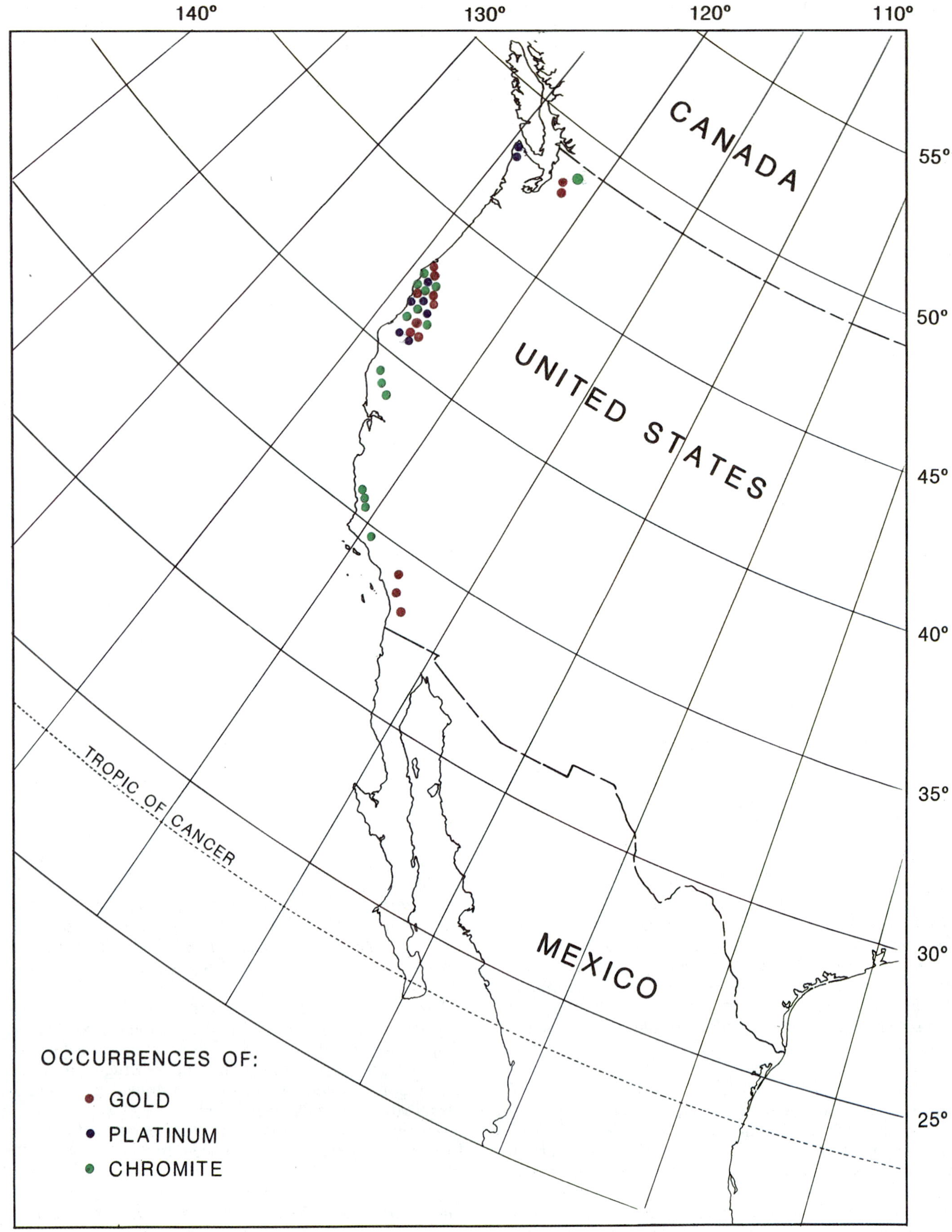

Figure 3. Coastal occurrences (lode and placer) of gold (red), platinum (blue), and chromite (green) in the Pacific Coast states (Koschmann and Bergendahl, 1962; Thayer and Miller, 1962; Brooks and Ramp, 1968; Clark, 1970; Blair, Page, and Johnson, 1977).

Bering Sea is similarly fine grained. Except along the shoreline, Holocene sand and muds of the open Bering Sea contain only flakes 0.25 mm or smaller (Nelson and Hopkins, 1972). Here, however, the fine-grained gold may contribute significantly to the overall grade. The beach sands between Nome and Bluff, Alaska, commonly contain gold flakes between 0.5 and 1.0 mm in diameter, but the richest concentrations lie in the very fine sand fraction (about 0.062 mm in diameter) (Luepke and Leong, 1981).

Where glacial drift is deposited on an emerged continental shelf, coarse-grained gold is available for reworking into potentially rich concentrates during the ensuing marine transgression. Gold particles in excess of 1 mm in diameter on the northern Bering Shelf are concentrated in the gravels that overlie glacial drift (Nelson and Hopkins, 1972). The extent of glaciation along the coasts of Alaska, and therefore the areas where coarse gold could occur in glacial drift, is shown in Figure 1.

The high specific gravity and great economic value of gold combine to create a particular problem in sampling placer deposits for analysis. The problem arises where the chance occurrence of one or more particles can produce misleading analytical results (the "particle-sparsity effect" of Clifton, Hubert, and Phillips, 1967). The size of a raw sample must be large enough to include a statistically significant number of particles for an analysis to be meaningful. If the gold flakes are 1 mm in diameter, an initial sample of about 25 kg is required (Nelson and Hopkins, 1972).

Platinum–Group Metals

The platinum-group minerals (platinum, palladium, iridium, osmium, rhodium, and ruthenium) also occur in placer deposits on the western margin of North America, although much more sporadically than does gold. Although minor amounts of the platinum group have been recovered as part of the mining of placer gold in the Pacific states and Alaska, significant amounts have been produced only from the Goodnews Bay district on the Bering Sea coast of Alaska (Page et al., 1973). Figures 2 and 3 show the coastal occurrences of platinum-group metals in Alaska and the Pacific Coast states.

Having specific gravities comparable with gold, the platinum-group metals are most likely to form economic placer concentrations in streams <15 km from the source rock (Emery and Noakes, 1968). Glacial transport can carry coarse particles of the group farther, and any subsequent reworking of glacial drift by marine processes could lead to coarse-grained placer deposits on a shelf. Page et al. (1973) speculate that 4 million troy ounces of undiscovered platinum-group metals may exist on the Alaskan continental shelf.

The problems of sampling for placer gold also attend the exploration for the platinum-group metals. The platinum group may be associated with gold and may constitute an economically significant by-product to gold placer mining (Page et al., 1973). Platinum is also commonly found interlocked with magnetite grains (Rosenblum et al., 1986).

Tin

Most of the world's production of tin comes from placer deposits (Sainsbury and Reed, 1973). Emery and Noakes (1968) suggest that the high specific gravity of cassiterite, the main commercial tin mineral, should cause it to accumulate in alluvial placers. Submerged residual, eluvial, and alluvial tin placer deposits are of major economic importance in southeast Asia, where they are presently mined by seagoing dredges (Hosking, 1971; Sainsbury and Reed, 1973). Beach placers have also yielded large amounts of cassiterite in Indonesia, and large reserves are thought to lie offshore on the Sunda Shelf (Sainsbury and Reed, 1973). Emery and Noakes (1968) suggest that the tendency of cassiterite to fracture during transport into fine, presently unrecoverable particles may limit the distance between economic placer deposits and the source rocks to only 8 km.

The only known source of tin on the western North American margin is on the Seward Peninsula of Alaska (Figure 4), where placer cassiterite has been commercially extracted (Economic Associates, Inc., 1968; Cobb, 1973). Sainsbury and Reed (1973), in estimating United States resources, included an assumption that tin deposits may be found on the Aleutian Islands where the Cenozoic geology roughly resembles that of the Japanese Islands (Scholl, Vallier, and Stevenson, this volume)—an area from which moderate amounts of lode tin have been produced.

Chromium

A strategic industrial metal for which the United States has very limited reserves, chromium is one of the few commodities, other than the precious metals and tin, that have been commercially mined from marine placer deposits along the western margin of North America. The mineral, chromite, which is the only commercial source of chromium (Thayer, 1973), is concentrated in the "black sands" (modern and ancient beach and nearshore marine concentrations of dark, opaque heavy minerals) of southern Oregon. The raised Pleistocene beaches in this area (Griggs, 1945) contain the second largest known chromium resource in the United States (Thayer, 1973).

The specific gravity of chromite is such that it is more likely to be concentrated in beach and nearshore marine placers than in alluvial placers. Although possible sources of chromite proximal to coastal areas occur in Alaska and throughout the Pacific Coast states (Figures 3, 4), potentially economic placer concentrations are known to exist only in southwestern Oregon. Chromium might, however, be an economically significant by-product in the development of other marine placer deposits.

Titanium Minerals

Titanium is a metal that is currently mined on a large scale from modern and ancient beach placer deposits in many parts of the world (e.g., India, Australia, and the eastern coast of the United States; Force, 1976). Commercially important placer deposits of titanium minerals are likely to be found only where rocks of high-grade metamorphism (particularly the granulite facies) are a major source rock (Force, 1976) and where a very mature topography exists (Economic Associates, Inc., 1968). Neither of these conditions is met particularly well on the western margin of North America. Nonetheless, titanium extraction from marine placer deposits is a possibility, if only as a by-product.

The heavy minerals ilmenite, rutile, and ilmenite's alteration product, leucoxene, are possible sources of titanium in placer deposits on the western margin of North America. Rutile, the preferred raw material for the commercial chemical extraction of titanium (Klemic, Marsh, and Cooper, 1973),

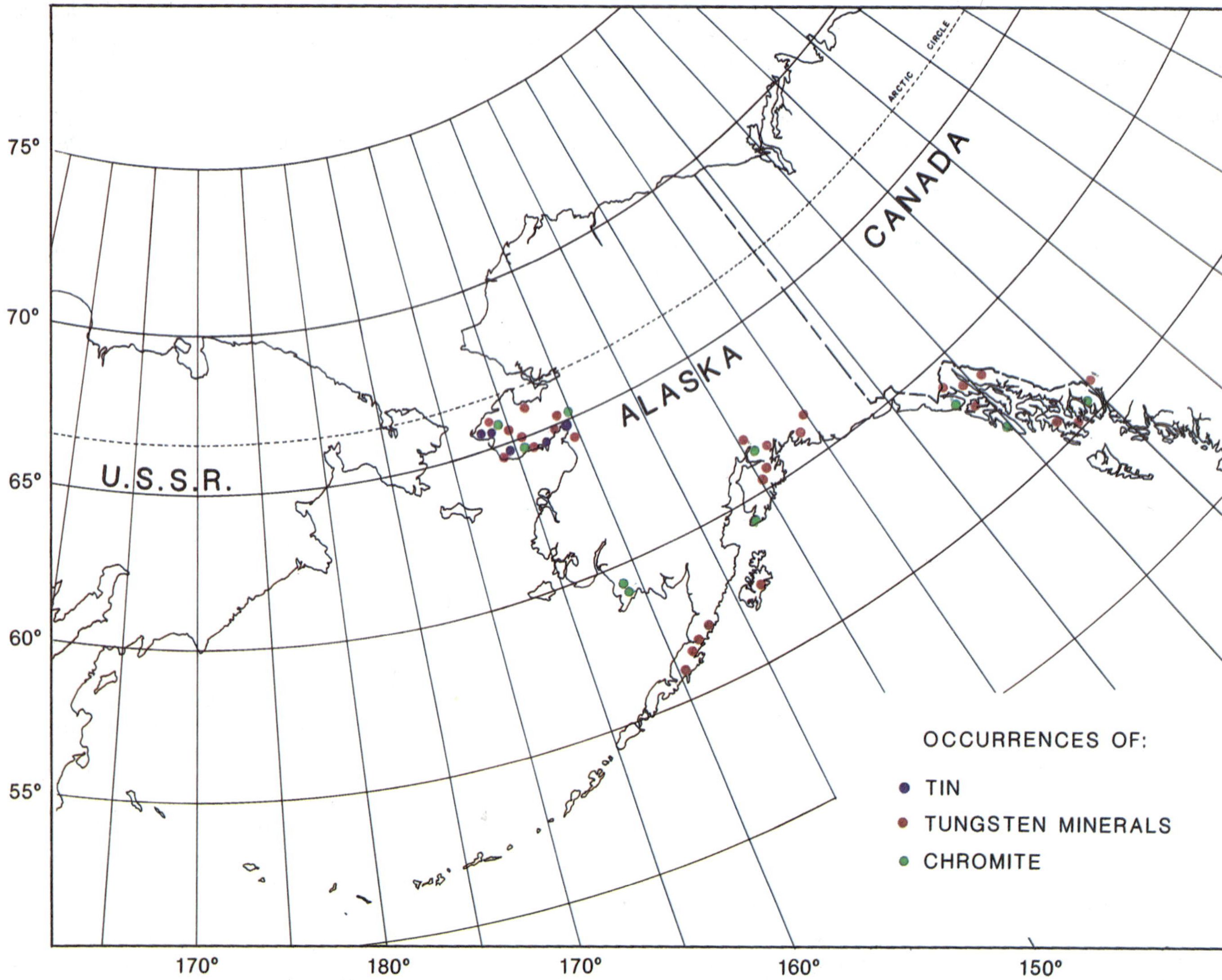

Figure 4. Coastal occurrences of tin (blue), tungsten minerals (red), and chromite (green) in Alaska (Cobb, 1975, 1984c; Cruz and Cobb, 1984).

tends to be reported as a minor component of heavy-mineral placer deposits in Alaska and the Pacific Coast states, having an economic potential, at best, of a by-product (Griggs, 1945; Thorsen, 1964; Sharma, 1978). Ilmenite, in contrast, is an abundant and widespread mineral in marine placer concentrations on the western margin, and titaniferous magnetite is locally abundant along the Alaskan coast (Sharma, 1978). Few data exist regarding leucoxene on this margin; the general lack of deep weathering profiles suggests that it may not contribute significantly to placer deposits.

Other Placer Minerals

A number of other heavy minerals of possible economic importance exist in marine placers in the western margin of North America. Many of these minerals are mined on other margins but here, because of low concentrations, limited size of deposit, or other economic factors, they are likely to be, at best, only by-products. Such minerals include zircon, monazite, garnet, magnetite, and the tungsten and vanadium minerals.

The principal commercial source of the world's supply of zirconium is zircon derived from placer deposits, particularly in modern and ancient beaches and sand dunes (Klemic, et al., 1973). Zircon is ubiquitous in the beach sands of the West Coast, but almost always occurs in concentrations of 2% or less (Cook, 1969; Sharma, 1978; Phillips, 1979).

Monazite, a source of rare-earth elements and thorium, is mined primarily from placer deposits. Beach placers on the coast of Brazil, the west coast of India, and Australia are among the largest producers of monazite (Adams and Staatz, 1973). The mineral occurs in trace amounts on many Pacific Coast beaches (Phillips, 1979), but generally it is not mentioned as a constituent of Alaskan beach placers (Sharma, 1978). The related yttrium phosphate mineral, xenotime, is found in the same environments as monazite but is even more rare (Adams and Staatz, 1973).

Because of the abundance of higher grade ores, placer magnetite is generally not considered as a source for iron. Nonetheless, magnetite is a target for exploration in the coastal waters of Asia, and has been mined off the coast of Japan for the manufacture of special steels (Hess, 1965; Padan, 1971). Magnetite-rich beach deposits are noted along many parts of the Pacific states (Hess, 1965) and in southeastern Alaska (Klemic, James, and Eberlein, 1973). Such deposits may be of interest because of other metals included in the magnetite, such as titanium or vanadium. About 75% of the world's reported and estimated vanadium production during 1970 was from titaniferous magnetite (Fischer, 1973). Placer deposits of this mineral are common along the Pacific and Bering Sea margins (Hess, 1965; Sharma, 1978).

Minerals of the garnet group are common components of many beaches in the western margin of North America and could serve as a source of commercial abrasive. Typically, however, the garnet is discarded from placer operations as economically insignificant (Thaden, 1973). Beach placers in southeastern Alaska have recently been assessed as sources of industrial quality almandine garnet (Eakins et al., 1983).

The tungsten minerals, wolframite and scheelite, have been commercially derived from stream placers from a few places in the United States (Hobbs and Elliott, 1973). For many years, wolframite placers were the world's principal source of tungsten (Stewart, 1957). As might be expected from the high specific gravity of these minerals (Table 1), placer accumulations are generally restricted to areas close to the source rocks. Moreover, the brittleness of scheelite generally restricts its placer occurrence (Stewart, 1957). Placers concentrated by eolian, beach, and other marine processes are unlikely. Submerged alluvial placers may be present on the continental shelf where sources of tungsten minerals occur in coastal areas. Such occurrences (Figure 4) exist on the Seward and Alaska Peninsulas and in southeastern Alaska (Cobb, 1984c).

Other heavy minerals of potential economic significance have been reported locally as traces in many placer deposits on the coasts of Alaska and the Pacific states. These will be noted in the following section, which describes the placer potential on a regional basis.

REGIONAL DESCRIPTIONS

In this section we examine the potential for heavy-mineral placer deposits on or at the margin of the continental shelf on a regional basis. For each region we summarize the existing data on the general provenance of the sediment; the nature of the coastal and shelf environments and their sediment, including its heavy-mineral content; potential sources of placer minerals; and the known marine and coastal alluvial placer deposits and their mineral associations.

The division here of the continental margin of Alaska and the Pacific states into regions does not follow a rigid format. For the Pacific Coast of the conterminous United States we use political (state) boundaries. We divide the Alaska margin into regions primarily on the basis of source terranes (Figures 5–10). This division differs from those based on oceanographic or geographic considerations (Sharma, 1979). For example, our "Arctic region" does not extend to the Bering Strait; instead it ends at a line extending west from Cape Krusenstern, thereby avoiding the splitting of the economically important Seward Peninsula into two different regions.

A number of general sources have proved invaluable in assembling the following material. Shepard and Wanless (1971) provided a useful summary of the coastal physiography over the entire margin. Sharma (1979) described the sediment and depositional processes over the whole of the Alaskan Shelf. The compilation of Alaskan geology by Beikman (1980) was useful for describing source terranes. Cobb (1973) summarizes in detail the known placer deposits of Alaska. Tagg (1979) and Phillips (1979) discuss the marine placer potential of Alaska and the Pacific Coast states, respectively. Kulm et al (1986) provide an inventory of geophysical surveys and sediment sampling on the continental shelf, beaches, and coastal terraces of southern Washington, Oregon, and northern California. The Quaternary deposits and depositional history are summarized in greater detail, particularly for the Alaskan shelves, by Stauffer (this volume).

The mineralogic data presented in the following sections were collected for a variety of purposes, including provenance studies and (less commonly) economic analyses. As a result, a considerable disparity exists in the sampling technique, analytical procedures, and reporting style of different workers. Most samples represent surface grabs; very few explore the mineralogy to depths within the substrate. Sample selection may be (1) random; (2) biased toward high concentrations (as where the visibly rich concentrations of heavy minerals were selectively sampled); or (3) biased toward low concentration (as where samples come from locations typically low in heavy minerals, such as the middle part of the beach foreshore). Some workers report heavy-mineral species as a percentage of the bulk sample, others as a percentage of some fraction thereof (such as a specific grain size component), as a percentage of the total heavy-mineral assemblage, or even as part of that assemblage (such as the nonopaque fraction). Analyses based on a specific size-fraction may be useful only as representing maximum observed values. Analyses based only on the heavy-mineral assemblage (or some portion thereof) may be useful primarily for establishing the expected composition of the assemblage. Where possible, we attempt to present the heavy-mineral percentages on a bulk-sample basis. Otherwise, we present the percentages as stated by the initial author. For each area that follows, we tabulate the sampling, analytical, and reporting approaches of those who provided mineralogic data (Tables 2–5).

Arctic Region

In this chapter, the Arctic continental shelf is defined as the region extending from the Alaska-Canadian border westward to the U.S.-U.S.S.R. 1867 Convention Line, and southward to Cape Krusenstern, Alaska (Figure 5). Two seas lie off the Arctic coast—the Beaufort, which extends from Point Barrow eastward to the islands of the Canadian Arctic Archipelago, and the Chukchi, which extends from Point Barrow southwestward to the Bering Strait (Grantz, May, and Dinter, this volume; Grantz and May, this volume). Only one-third of the area of the Chukchi Sea lies in U.S. territorial waters.

The region bordering the Arctic continental shelf from the Alaska-Canadian border to the Lisburne Peninsula is part of the Arctic Coastal Plain—a gently sloping surface of low relief. Gravel and sand beaches fringe the coastal plain; barrier islands and spits are well developed southwest of Point Barrow and at Point Hope. The foothills of the Brooks Range border the coast from Cape Krusenstern to Cape Thompson (113 km), the entire Lisburne Peninsula except for Point

Table 2. Comparison of available analyses of heavy minerals in Alaska.

Reference	Purpose of Analysis	Sample Type	Sample Size	Sample Representation	Fraction For Which HM Are Reported	Fraction From Which HM Are Separated	Type of Analysis	Specific Gravity of Separating Medium	Other Notes
Berryhill, 1963	r	s, d	not given	rm	sand only	2.0–0.062 mm	o, c	—	pan-concentrated in field
Cook, 1969	r	s, d	several kg	bh	sand only	2.0–0.062 mm	x	—	pan-concentrated in field
Gardner, Vallier, and Dean, 1978	p	s	not given	rm	sand only	0.088–0.063 mm	o	2.85	
Knebel and Creager, 1974	p	s	30–50 g	rm	sand only	0.14–0.062 mm	o	3.14–3.17	
Luepke, 1975	p	s	20–60 g	rm	sand only	0.125–0.062 mm	o	2.96	
McManus et al., 1974	p	s	not given	rm	sand only	0.15–0.062 mm 0.5–0.15 mm	o	3.14–3.17	
Thomas and Berryhill, 1962	r	s, d	not given	bh	sand only	2.0–0.062 mm	o, c		pan-concentrated in field
Wong, 1983	p	s	not given	rm	sand only	0.25–0.062 mm	o	2.89	
VenkataRathnam, 1971	p	s	not given	rm	sand only	0.5–0.062 mm	o	3.14–3.17	

KEY TO ABBREVIATIONS

Purpose of Analysis: r = resource potential studies; p = provenance/transport studies

Sample Type: s = surface grab samples; d = samples cored at depth

Sample Representation: rm = random sample; bh = biased sample, high heavy-mineral concentration; bl = biased sample, low heavy-mineral concentration

Type of Analysis: o = optical; c = chemical; x = x-ray; sp = spectrographic; em = electron microprobe

Table 3. Comparison of available analyses of heavy minerals in Washington.

Reference	Purpose of Analysis	Sample Type	Sample Size	Sample Representation	Fraction For Which HM Are Reported	Fraction From Which HM Are Separated	Type of Analysis	Specific Gravity of Separating Medium	Other Notes
Hands, 1968	p	s	2–3 g	rm	sand only	0.149–0.062 mm	o	3.17	
Kelly, 1947	r	d	not given	bh	sand only	2.0–0.062 mm	o, c	—	
Luepke and Clifton, 1983	p	s	20–60 g	rm	sand only	0.125–0.062 mm	o	2.96	
Martinez, Gomes, and Wong, 1981	r	s	40–50 lb	rm	sand and gravel	4.0–0.062 mm	o, c, sp, em	3.30	spiral-concentrated in field
Scheidegger and Phipps, 1976	p	s	not given	rm	sand only	0.5–0.062 mm	o	2.96	
Thorsen, 1964	r	d	35 lb	bh	sand only	0.354–0.074 mm	o, c	2.82 & 3.5–4.5	
VenkataRathnam, 1969	p	s	not given	rm	sand only	0.149–0.062 mm	o	3.17	
Whetten, Kelley, and Hanson, 1969	p	s	not given	rm	sand only	2.0–0.062 mm	o, sp	3.17	
White, 1967	p	s	50 g	rm	sand only	2.0–0.062 mm	o	3.14	
VenkataRathnam and McManus, 1973	p	s	not given	rm	sand only	0.149–0.062 mm	o	3.17	

KEY TO ABBREVIATIONS

Purpose of Analysis: r = resource potential studies
p = provenance/transport studies

Sample Type: s = surface grab samples
d = samples cored at depth

Sample Representation: rm = random sample
bh = biased sample, high heavy-mineral concentration
bl = biased sample, low heavy-mineral concentration

Type of Analysis: o = optical
c = chemical sp = spectrographic
x = x-ray em = electron microprobe

Table 4. Comparison of available analyses of heavy minerals in Oregon.

Reference	Purpose of Analysis	Sample Type	Sample Size	Sample Representation	Fraction For Which HM Are Reported	Fraction From Which HM Are Separated	Type of Analysis	Specific Gravity of Separating Medium	Other Notes
Chambers, 1969	p	s, d	not given	rm	sand only	2.0–0.062 mm	o	2.96	
Hundhausen, 1947	r	d	not given	rm	sand only	2.0–0.062 mm	o, c, sp	—	
Kulm et al., 1968b	p	s	not given	bl	sand only	2.0–0.062 mm	o	2.96	
Lissner, 1971	p	s	100 g	bl	sand only	0.25–0.35 mm	o	2.96	
Runge, 1966	p	s	not given	rm	sand only	2.0–0.062 mm	o	2.96	
Twenhofel, 1946a	p	s	20 g	bh	sand only	2.0–0.062 mm	o		
Twenhofel, 1946b	p	s	20 g	bh	sand only	2.0–0.062 mm	o		

KEY TO ABBREVIATIONS

Purpose of Analysis: r = resource potential studies
p = provenance/transport studies

Sample Type: s = surface grab samples
d = samples cored at depth

Sample Representation: rm = random sample
bh = biased sample, high heavy-mineral concentration
bl = biased sample, low heavy-mineral concentration

Type of Analysis: o = optical
c = chemical sp = spectrographic
x = x-ray em = electron microprobe

Table 5. Comparison of available analyses of heavy minerals in California.

Reference	Purpose of Analysis	Sample Type	Sample Size	Sample Representation	Fraction For Which HM Are Reported	Fraction From Which HM Are Separated	Type of Analysis	Specific Gravity of Separating Medium	Other Notes
Azmon, 1960	p	s	20–500 g	rm	sand only	2.0–0.062 mm	o	2.96	
Cherry, 1965	p	s	1/3–2/3 gal	bl	sand and coarse silt	0.25–0.053 mm	o	2.89	
Dingler et al., 1982	p	s	20–60 g	rm	sand only	0.125–0.062 mm	o	2.96	
Gorsline, Grant, and Booth, 1971	p	s	1 kg	rm	sand only	2.0–0.062 mm	o	2.90	
Gomes, Martinez, and Wong, 1980	r	s	40–50 lb	rm	sand and gravel	4.0–0.062 mm	o, c, em	3.30	spiral-concentrated in field
Handin, 1951	p	s	200 g	bl	sand only	0.125–0.062 mm	o	2.85	
Hutton, 1959	p	s	not given	bh, bl	sand only	0.50–0.062 mm	o	—	
Inman, 1953	p	s	15 g	bl	sand only	2.0–0.062 mm	o	2.82–2.84	
Judge, 1970	p	s, d	not given	bl	sand only	0.125–0.062 mm	o	2.87	
Klise, 1984	p	s	not given	rm	coarse silt	0.063–0.044 mm	o	2.90	
Sayles, 1966	p	s	1 quart	bl	not given	0.25–0.074 mm	o	2.89	
Trask, 1952	p	s	not given	rm	sand only	2.0–0.062 mm	o	2.85	
Yancey, 1968	p	s	not given	bl	sand and coarse silt	0.5–0.052 mm	o	2.88	

KEY TO ABBREVIATIONS

Purpose of Analysis: r = resource potential studies; p = provenance/transport studies

Sample Type: s = surface grab samples; d = samples cored at depth

Sample Representation: rm = random sample; bh = biased sample, high heavy-mineral concentration; bl = biased sample, low heavy-mineral concentration

Type of Analysis: o = optical; c = chemical; sp = spectrographic; x = x-ray; em = electron microprobe

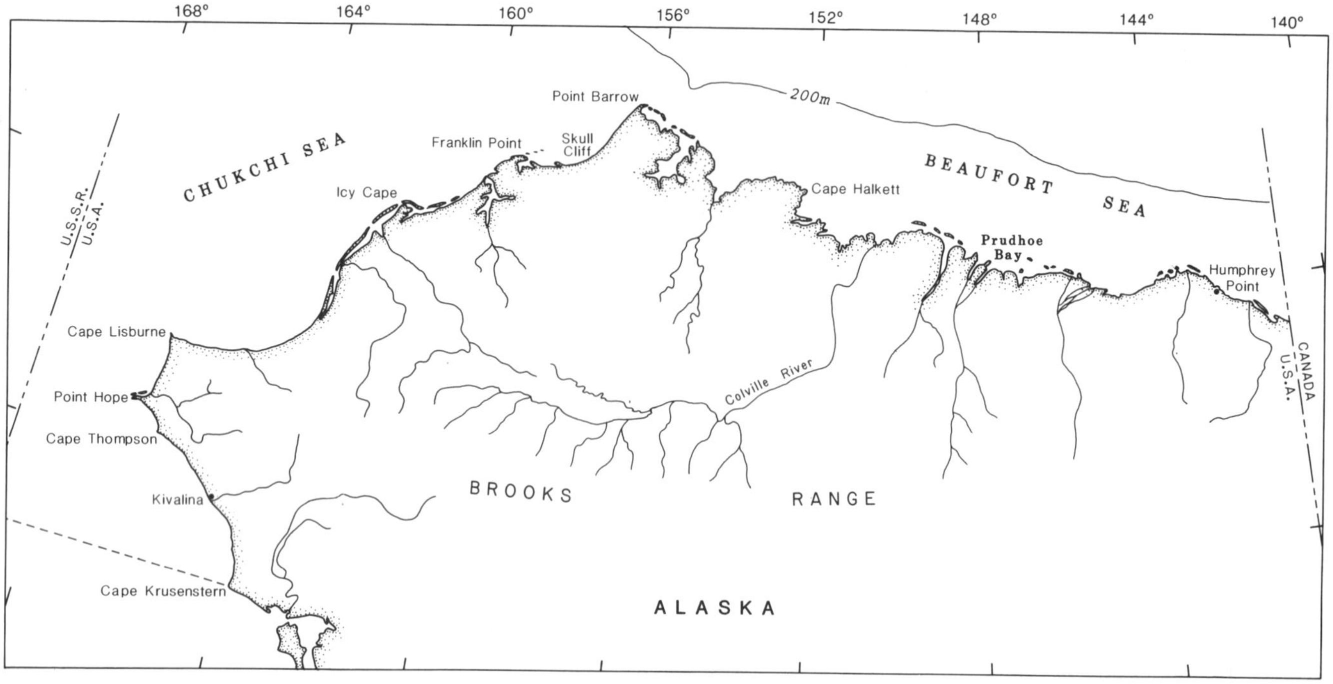

Figure 5. Arctic region. Dotted lines outline specific area covered.

Hope, and the coast north and east of Cape Lisburne for 137 km.

Sedimentary rocks crop out along almost all of the coast of the Arctic region. The dominantly marine Pleistocene Gubik Formation, consisting of loosely consolidated cross-bedded gravel, sand, silt, and clay, underlies the coastal plain from Skull Cliff (Figure 5) on the Chukchi coast to Humphrey Point on the Beaufort coast. Cretaceous sedimentary rocks are exposed along the remainder of the coast, and at Skull Cliff under the Gubik Formation, except for two areas. Paleozoic and Mesozoic sedimentary rocks crop out from Cape Lisburne to Cape Thompson. Paleozoic carbonate and clastic rocks and, less commonly, Lower Mesozoic basalt crop out from Cape Krusenstern to Kivalina (Beikman, 1980).

No glaciers extended to the Arctic Coastal Plain during the Pleistocene (Péwé, 1975). Only one major river, the 600-km-long Colville, exists in this region; it drains 78% of the Alaska North Slope and empties into the Beaufort Sea (Walker, 1974). This drainage area includes the Brooks Range, a series of parallel belts of Paleozoic sedimentary rocks locally metamorphosed and intruded by granitic bodies (Patton, 1964).

Barnes and Reimnitz (1974), Reimnitz and Barnes (1974), and Grantz et al. (1980, 1982) summarize oceanographic conditions in the Arctic region. Sediment transport on the shelf takes place mainly during major summer and fall storms, when waves and currents scour the sea floor. Direction of longshore transport is toward major capes and away from the center of intervening embayments in the Chukchi Sea (Grantz et al., 1982); in the Beaufort Sea, longshore transport generally is from east to west in response to the Pacific gyre (Barnes and Reimnitz, 1974). Pack ice covers the Arctic Shelf for 9 to 10 months; therefore, the effects of waves, wind-driven currents, and storm tides are minimal through most of the year. During the open-water period, waves commonly have periods of 2–3 sec and heights of 20–30 cm (Wiseman, Suhayda, and Hsu, 1974). Astronomical tides rarely exceed 0.5 m. These low-energy conditions probably allow only for the development of small placers, which would be further subject to continuous reworking by pack ice.

The Beaufort Shelf is relatively narrow, approximately 10 km off Point Barrow (Greenberg, Hart, and Grantz, 1981); elsewhere it ranges from 50 to 100 km in width (Tagg, 1979). The Chukchi Sea exceeds 100 m in depth only in the Barrow Sea Valley, which extends from Franklin Point to north of Point Barrow (Hill et al., 1984). Elsewhere, its sea floor, with depths ranging from 30 to 60 m, is nearly flat except at Blossom Shoals off Icy Cape, where large sand banks or ridges exist (Grantz and May, 1984). In the Chukchi nearshore in depths from 2 to 15 m, there appear to be numerous shore-parallel small-scale symmetrical sand waves (Grantz et al., 1982). Linear, northeast-trending parallel shoals commonly occur off capes, reflecting the northward sediment transport by the Alaskan Coastal Current (Grantz et al., 1982); these shoals lie at depths of 6 to 22 m.

The northern Chukchi Shelf is covered by a thin veneer (2–5 m) of unconsolidated sediment over well-indurated bedrock; in the southern Chukchi Sea, the thickness reaches 11 m (Grantz et al., 1982). This sediment cover is mostly silt and clay. A gravel accumulation is found off Cape Lisburne, its probable source being the bedrock cliffs of the cape. Sand is concentrated beneath the main course of the Alaskan Coastal Current from Point Hope north and northeast to the head of Barrow Sea Valley off Franklin Point.

The Beaufort Shelf is probably underlain by an offshore extension of the Gubik Formation; Pleistocene (?) sediments crop out in ice gouges, in current-scour depressions, and where Holocene sediments are absent (Grantz et al., 1980). Surficial Holocene deposits, mostly fine-grained marine sand, silt and clay (Barnes and Reimnitz, 1974), range in thickness from 5 m to more than 45 m near the shelf break (Dinter, 1982). These sediment types form an irregular patchwork. Sand is more prevalent on the shelf than at the shelf break; gravel concentrations occur at the shelf break and in the area just north of Prudhoe Bay (Rodeick, 1979). Surficial sediments are frequently reworked by keels of deep-draft sea-ice (Barnes and Reimnitz, 1974). Seasonal ice gouges nearshore may be smoothed over in a single storm (Grantz et al., 1980). The poorly sorted sediments plus the continuous mixing processes are probably not conducive to placer formation in the region.

Only surficial sediment samples have been analyzed for heavy-mineral content. Preliminary examination of heavy minerals from the Chukchi Sea sediments show concentrations as high as 10% of the total sample along the coast from Kotzebue Sound (Cape Krusenstern) to Cape Thompson (Creager and McManus, 1966). Weight percentages of heavy minerals in the sand-size fraction of Beaufort Sea sediments range from < 1% to approximately 3%; only two of 153 samples exceed 4% of heavy minerals. These values must be regarded as maximum percentages because of the the different amounts of sand and gravel in the sediment.

No published studies on heavy-mineral species in the Chukchi Sea exist; unpublished data indicate that the assemblage in the nonopaque fraction is dominated by clinopyroxene (Mark Holmes, U.S.G.S., 1985, oral communication). Clinopyroxene also dominates the heavy-mineral assemblage in the nonopaque sand-size fraction of the Beaufort samples (Luepke, 1975). The heavy-mineral fraction of two of 42 samples in the Beaufort Sea contains > 10% of chrome spinel. Zircon constitutes as much as 2% of the fraction, and rutile occurs as a trace. Garnet percentages range from < 1% to 11%, exceeding 5% only in samples west of Cape Halkett. Opaque minerals constitute between 6 and 30% of the total heavy-mineral fraction (Luepke, 1975).

No minerals capable of forming marine placers have been reported in the Arctic region in either lode or placer deposits (Cobb, 1973). Likewise, no economic concentrations of valuable heavy minerals are known on present beaches or the shelf.

Seward Peninsula–Northern Bering Sea

The Seward Peninsula–Northern Bering Sea region encompasses the continental margin under the domain of the United States between lines extending west from Cape Krusenstern on the north and Cape Vancouver on the south (Figure 6). The margin here consists of a broad, flat shelf that is > 500 km wide in places. The shelf is relatively shallow, and the 50-m bathymetric contour lies 250 km or more from the shoreline. The shelf platform is broken by a few banks and depressions. One large island (St. Lawrence Island) protrudes from the central part of the shelf, and a few small islands (Sledge Island, King Island, and the Diomede Islands) lie off the western side of the Seward Peninsula. Two broad, shallow open embayments, Kotzebue Sound and Norton Sound, lie north and south, respectively, of the Seward Peninsula.

The coastal topography within this region is variable (Hunter, Sallenger, and Dupré, 1979). Extensive attached or barrier beaches dominate much of the northern coast of the Seward Peninsula and parts of St. Lawrence Island. However, a number of broad headlands project without beaches into the sea on both the peninsula and the island. Along the southeastern coast of the region, the Yukon River has built a large delta, mostly fringed by prograding distributary-mouth bars and interdistributary tidal flats (Dupré, 1982).

Waves, geostrophic currents, and glaciation have combined to shape the character of the coastal and shelf sediment. The coastal waters are free of ice from May through October (Hunter, Thor, and Swisher, 1982). During the summer months, waves are generally no more than a few decimeters high, but late fall storms may cause sea level to rise several meters and can generate waves 3 to 4 m high (Sallenger, Dingler, and Hunter, 1978). A strong northward-flowing geostrophic current, which ultimately feeds through the Bering Strait, influences the northeastern Bering Shelf and transports much of the Yukon River sediment into the southern Chukchi Sea (Nelson and Creager, 1977). At least two stages of glaciation are recorded on the shelf (Sharma, 1979). Glacial sediment on the shelf north of St. Lawrence Island was derived from Siberia, and local valley glaciers extended a few kilometers seaward of the present shoreline on the southern side of the Seward Peninsula (Figure 1) (Nelson and Hopkins, 1972). During the lowstands of the sea that accompanied the glacial intervals, rivers cut channels into the exposed shelf. Buried alluvial channels between the present Yukon Delta and St. Lawrence Island and north of the delta suggest older positions of the Yukon River (Moore, 1964; Nelson, 1982).

The sediment on the northern Bering Shelf is strongly influenced by the Yukon River, which has deposited 1 to 14 m (average 4.5 m) of very fine sand and silt in a wedge atop peaty Pleistocene deposits on the southern side of Norton Sound (Steffy et al., 1981; Nelson, 1982). West of Norton Sound, shelf currents inhibit the deposition of the fine Yukon sediment, and a thin sheet of inner shelf sand overlies the limnic Pleistocene peat (Nelson, 1982).

Lag deposits of gravel overlie glacial till or bedrock near the Seward Peninsula, St. Lawrence Island, and in the southern part of the Bering Strait (Nelson, 1982). Storms significantly affect the sediment on the northern Bering Shelf (Cacchione and Drake, 1979; Cacchione, Drake, and Wiberg, 1982), and the nearshore sand and gravels are clearly responsive to the effects of waves and currents (Hunter, Thor, and Swisher, 1982). Gouging of the sea floor by drifting ice is prevalent, particularly on the southern side of Norton Sound (Thor and Nelson, 1981).

The eustatic fluctuations of sea level during the Pleistocene have produced a series of relict-beach ridges and/or deposits both above and below sea level. Well-developed, submerged shoreline ridges have been delineated at water depths of minus 10–12, 20–24, 30, and 38 m (Nelson et al., 1982) offshore from Port Clarence and Nome. McManus et al. (1977) interpret a ridge that crests at about 20 m below sea level southeast of St. Lawrence Island to be an ancient beach ridge, on the basis of enriched concentrations of heavy minerals and finer grained sand on its crest relative to its flanks. They suggest that this ridge formed at a sea-level stillstand at minus 20 m, which also produced shoreline deposits off Nome (Nelson and Hopkins, 1972) and Cape Spenser (Hopkins, 1967). A relict-beach deposit is also identified on the northern side of the western end of St. Lawrence Island at a water depth of 36–38 m (McManus et al., 1977). We describe more fully the economically important ancient beach deposits in the Nome area in a subsequent part of this section.

Streams flowing across the shelf during the Pleistocene lowstands of sea level cut numerous valleys that have been partly or totally filled, and in some cases buried, beneath a blanket of sediment (Tagg and Greene, 1973; Stauffer, this volume). In the Nome area, the channels appear to be discontinuous overall, perhaps due to the meandering nature of the incising streams (Tagg and Greene, 1973). Weak returns of the seismic signal (acoustical sinks) in the records suggest the presence of gravels in many of the channels (Tagg and Greene, 1973).

A number of source terranes contribute to the coastal and shelf sediment. The Noatak and Kobuk Rivers discharge sediment into Kotzebue Sound. This sediment is derived mostly from Precambrian and lower Paleozoic sedimentary rocks; Permian, Triassic, and Jurassic mafic volcanic rocks; and upper Cretaceous clastic deposits in the southern Brooks Range. The Seward Peninsula, underlain largely by Paleozoic metamorphic and sedimentary rocks intruded by felsic stocks and batholiths of Cretaceous age, sheds sediment to the coast to the north, south, and west. Paleozoic carbonate and clastic rocks, locally cut by Jurassic mafic intrusive bodies, underlie the western Brooks Range and provide a source of sediment to Kotzebue Sound. The Yukon River, the fifth largest stream in the United States, drains a vast area of central Alaska between the crests of the Brooks Range and the Alaskan Range. The Chukotsk Peninsula on the Asian mainland is a potentially significant source for sediment in the northwestern part of the region. The peninsula consists of Precambrian or Paleozoic sedimentary and metamorphic (greenschist to amphibolite facies) rocks intruded by Cretaceous granitic bodies and mantled by Upper Mesozoic volcanic deposits (Fisher, Patton, and Holmes, 1982). St. Lawrence Island, an upwarped platform of mostly Upper Paleozoic or Mesozoic sedimentary rocks intruded by Cretaceous quartz monzonite and locally overlain by Quaternary basalt (Patton and Csejtey, 1980), is another potential source of sediment.

McManus et al. (1974) examined the heavy-mineral assemblages of sand and coarse silt contributed by the Yukon River and other sources to the shelf of the northern Bering Sea. Sediment derived from the Yukon River has an assemblage dominated by pyroxenes, amphiboles, and epidote. Garnet occurs in the general range of 5–15% of the Yukon-derived heavy-mineral fraction, and zircon and kyanite constitute 1% or less. Heavy-mineral samples of sediment derived from the southwestern Seward Peninsula contain >50% garnet. Near the coast of St. Lawrence Island, the heavy-mineral assemblage contains as much as 90% olivine, probably reflecting the proximity of a large Cenozoic basalt outcrop (Beikman, 1980). North of St. Lawrence Island a broad area of the shelf is floored by sand and gravel derived from the Chukotsk Peninsula. Zircon and kyanite are present in this sediment in very low percentages (1% or less of the heavy-mineral fraction). The heavy-mineral assemblage in shelf sediment south of St. Lawrence Island (to St. Matthew and Nunivak Islands) is relatively homogeneous and is dominated by pyroxenes, amphiboles, and epidote (Knebel and Creager, 1974).

The Seward Peninsula–Northern Bering Sea region includes an area (the southern Seward Peninsula) long known for its placer mineral potential. As of 1961, the Seward Peninsula has been the source of nearly one-third of the placer gold produced in the state; placer tin, platinum, and scheelite have also been recovered (Cobb, 1973). Gold found on the beaches of Norton Sound in 1898 triggered a large gold rush. The Nome gold coast yielded $100 million in gold (Hess, 1965), and sporadic exploration for offshore gold placers continues today (Eakins et al., 1983).

Lode-gold occurrences are scattered over much of the Seward Peninsula but are concentrated on its southern side, particularly in the vicinity of Nome (Cobb, 1984a). Tin minerals also occur near Nome at the western tip of the peninsula and on its northern side (Cruz and Cobb, 1984), and tungsten minerals follow a similar pattern (Cobb, 1984c). Lode occurrences of beryllium, lead, zinc, fluorite, copper, and silver also lie on the Seward Peninsula (Tagg, 1979; Eakins et al., 1983), but of these only the beryllium minerals seem to exist in significant amounts (Cobb, 1973). Interest has developed recently in yttrium-bearing xenotime associated with tin occurrences on the Seward Peninsula (Eakins et al., 1983).

Many alluvial placers occur on the Seward Peninsula. Most of the gold placers are on the southern side of the continental divide, and a large number are concentrated in the Nome area (Cobb, 1984b). About 2000 tons of tin has been recovered from stream placers on the peninsula, and platinum has locally been a minor by-product from its eastern part (Cobb, 1973). Small amounts of scheelite have also been recovered, in part during the mining for gold. A detailed description of the alluvial placer deposits on the Seward Peninsula and the associations of minerals found there (including magnetite, hematite, and uranothorianite) is given by Cobb (1973).

Given the abundance of placer gold in the streams of the Seward Peninsula, it is not surprising that beach deposits have been a major source of gold in the Nome area (Location 1, Figure 6) since gold was discovered on the beach in 1899 (Cobb, 1973). Eustatic fluctuations of sea level have produced five distinct beach deposits at or near the present depositional surface—at about 24 m, 21 m, and 11 m below sea level; at the present shoreline; and onshore in a deposit (locally called Second Beach) at about 9 m above sea level (Nelson and Hopkins, 1972). The ages of the submerged beaches are uncertain, but they may have formed during the late Wisconsin and Holocene rise in sea level; Second Beach appears to have formed during the Sangamon Interglaciation (Hopkins, 1967). In addition, several still older beaches lie onshore buried beneath a thin mantle of glacial drift (Cobb, 1973).

Both the modern beach and Second Beach have been extensively mined for gold (Nelson and Hopkins, 1972), as have some of the beaches buried onshore beneath glacial drift (Cobb, 1973; U.S. Bureau of Mines, 1987). The mining of these beaches may account for as much as half of the gold recovered in the entire Seward Peninsula (Cobb, 1973). Large-scale operations have occurred only intermittently since 1962, owing to unfavorable economic factors rather than exhaustion of resources (Cobb, 1973).

Nelson and Hopkins (1972) summarize the distribution of gold in the northern Bering Sea. It occurs in richest concentrations and largest particles (>1 mm) where relict (transgressive) gravels lie atop lobes of glacial drift in the Nome area (Location 1, Figure 6) or above bedrock in the Sledge Island area to the west. The surface gravels of the submerged beaches contain low concentrations of gold, and the Holocene sands and muds contain only sparse flakes of fine-grained gold (<0.25 mm). Nelson and Hopkins (1972) suggest that gold coarser than 1 mm is not transported by normal marine processes, and that the fine-grained gold (<0.25 mm) is dispersed primarily by longshore currents adjacent to the modern and ancient shorelines. They con-

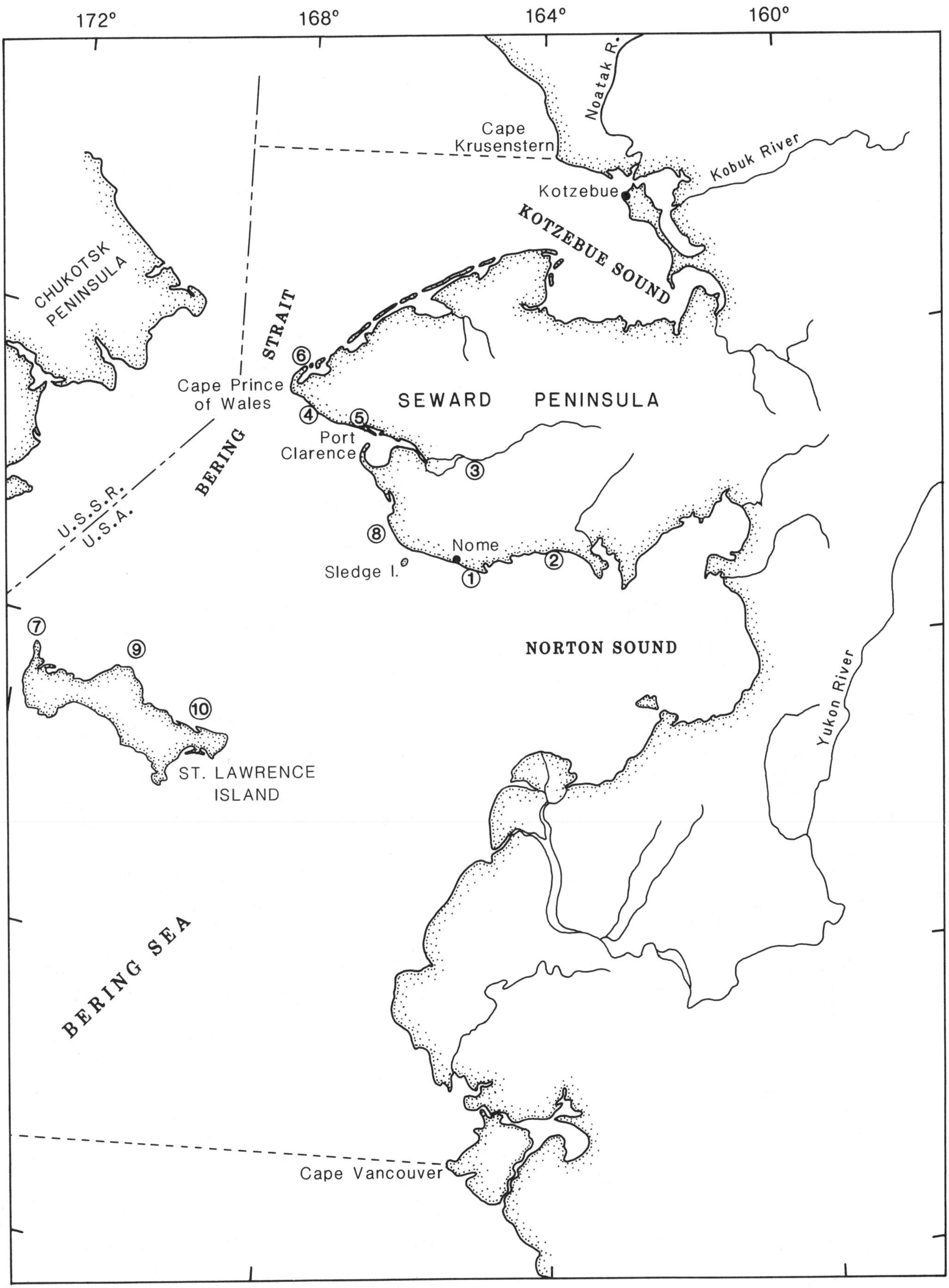

Figure 6. Seward Peninsula-northern Bering Sea region. Dotted lines outline specific area covered. Numbers refer to locations cited in text.

clude that economic placer deposits may exist in the gravel deposits above glacial drift (or bedrock in the Sledge Island area), at depths beneath the submerged beach deposits, and within buried outwash and alluvial channels that cut into auriferous glacial drift. Greene (1970) used a portable refraction seismograph to identify buried channels in the beach and tundra areas near Nome. The survey indicated channel trends onto the shelf that could be sites of alluvial placers.

Evaluation of the economic potential of the beaches and offshore areas near Nome has focused almost exclusively on gold. Cobb (1973) notes that the most common heavy minerals that accompany gold in placer concentrations in the Nome area are scheelite, magnetite, ilmenite, hematite, and garnet. Cassiterite is present in alluvial placers here but not in sufficient amounts to warrant recovery attempts. Cobb also mentions chalcopyrite and arsenopyrite in the older buried beach deposits near Nome. Cinnabar is mentioned as a possible by-product of gold mined offshore in this area (Economic Associates, Inc., 1968).

Placer gold has also been mined from two generations of beaches in the vicinity of Bluff (Location 2, Figure 6), about 50 miles east of Nome (Cobb, 1973; U.S. Bureau of Mines, 1987). Cinnabar and scheelite have been reported from placers in Daniels Creek—the apparent conduit for gold transport to the coast here. During the 1960s private companies prospected extensively for submarine gold placers offshore from Bluff, but their findings have remained proprietary (Cobb, 1973).

Other than near Nome and Bluff, there has been very little mining of beach placers on the Seward Peninsula. Cobb (1973) notes that about 20 tons of garnet sand was extracted in 1920 from Imuruk Basin at the eastern end of Port Clarence (Location 3, Figure 6). The possibility exists that economically significant beach and nearshore placer deposits may be found in several places around the peninsula (Nelson, 1971). Placer gold and cassiterite have been mined for a long time in the Anikovik River (Location 4, Figure 6), and placer deposits of cassiterite are found a short distance from the coast in many of the small streams that drain the western end of the peninsula as well as in the Lost River area (Location 5, Figure 6) to the east (Cobb, 1973; U.S. Bureau of Mines, 1987). Potentially economic concentrations of cassiterite just offshore from the mouth of Cape Creek may be the result of the reworking of old mine tailings; two anomalously high, but subeconomic, concentrations occur offshore from the western tip of the peninsula (Nelson, 1971; U.S. Bureau of Mines, 1987).

Streams that drain the north flank of Cape Mountain contain cassiterite, scheelite, monazite, xenotime, and zircon (Cobb, 1973). Some of these minerals may be concentrated in beaches and shoals along the extensive barrier island system that extends along the northwestern margin of the peninsula (Location 6, Figure 6). Strong northerly currents through the Bering Strait and along the coast could concentrate the placer minerals in the nearshore shoals. However, very little is known about the heavy minerals in this area (Tagg, 1979).

Elsewhere in the northern Bering Sea, the potential for economic placer deposits is uncertain. The Yukon Delta dominates the shoreline south of Norton Sound, but the only known placer deposits in the lower Yukon River system (gold and small amounts of platinum) lie more than 100 miles from the coast (Cobb, 1973). Within the outer northern Bering Sea, possible bedrock sources for placer minerals are mantled by a thick layer of fine-grained sediment (Nelson and Hopkins, 1972; Tagg, 1979). Gravels off the western coast of St. Lawrence Island and morainal gravels to the north may contain fairly coarse-grained gold particles (Location 7, Figure 6), and native copper of limited commercial significance may exist in the morainal gravels (Nelson and Hopkins, 1972). St. Lawrence Island contains minor amounts of metal and might serve as a source for marine placer deposits in the waters around its shoreline (Tagg, 1979), although none are known.

VenkataRathnam (1971) describes the distribution of different heavy minerals in the surface sediment of the northern Bering Sea. His data do not indicate percentages on a total sediment basis, but some generalizations can be drawn from his figures that show contours of (1) the percentage of sand in the surface sediment; (2) the percentage of heavy minerals in the sand-size fraction; and (3) the percentage of different minerals in the heavy-mineral fraction. On the basis of these figures, two areas can be delineated where heavy minerals constitute as much as 5 to 7% of the surface sediment. One area, about 20 mi in diameter, lies along the western coast of the Seward Peninsula (Location 8, Figure 6); another of similar dimensions lies along the northern coast off St. Lawrence Island (Location 9, Figure 6). The sediment off the Seward Peninsula contains 1% or less of undifferentiated opaque minerals, approximately 2% garnet, and <0.1% each of staurolite and zircon. The sediment on the north side of St. Lawrence Island contains a maximum of 2–3% olivine, 2% or less of undifferentiated opaque minerals, 0.5% or less of garnet, and traces of zircon. Zircon values that approach 0.1% of the total sediment were found on the northeast end of St. Lawrence Island (Location 10, Figure 6).

Heavy-mineral concentrations of 6% of the total sediment also occur just offshore from Nome. Garnet may total 2 to 3% of the sediment here; otherwise, the distribution of mineral species resembles that off the western end of the Seward Peninsula. No other significant heavy-mineral concentrations appear in VenkataRathnam's (1971) data.

Southern Bering Sea

The southern Bering Sea encompasses the continental shelf under the domain of the United States between a line extending west from Cape Vancouver and the western end of Unimak Island (Marlow, Cooper, and Fisher, this volume; Figure 7). It includes a very broad flat shelf from which Nunivak, St. Matthew, and the Pribilof Islands rise. The northern part of the coastline is dominated by the Yukon River delta and Kuskokwim Bay, and the southern part by Bristol Bay.

The character of the coastline is summarized by Hunter, Sallenger, and Dupré (1979) and by Sallenger and Hunter (1984). Beaches are common along the coastline that divides Bristol and Kuskokwim Bays and along the coast north of Kuskokwim Bay. Broad tidal flats flank the macrotidal Kuskokwim Delta and the upper reaches of Bristol Bay. Large tidal ranges as much as 5–6 m generate strong tidal currents in excess of 3 knots (National Ocean Survey, 1976) in these funnel-shaped bays. The heavy-mineral distribution on the shelf adjacent to the Alaska Peninsula suggests the presence of a net flow to the east along the peninsula, forming a clockwise gyre in Bristol Bay (Sharma, 1974). Evidence for at least four episodes of glaciation exists on the southern Bering Sea Shelf (Sharma, 1979) where, south of Cape Newenham, middle Pleistocene glaciation reached about halfway across the shelf (Coulter et al., 1965). The glaciers advanced southward from the Kuskokwim Mountains and to the north and west, respectively, from the Alaska Peninsula and Alaska Range.

The sediment that floors the shelf between St. George Island and Unimak Island is largely silt to silty sand (Gardner, Dean, and Vallier, 1980); sand is largely restricted to the vicinity of the islands, where gravel is locally abundant (Gardner, Vallier, and Dean, 1978; Marlow et al., 1979). This sediment is thought to have accumulated during lower stands of the sea during the Pleistocene and has been reworked by recent processes (Marlow et al., 1979). To the north, west of St. Matthew Island, silty sediment prevails, except at the shelf break where the sediment is sandy (Carlson and Karl, 1981; Karl and Carlson, 1984). The shelf sediment progressively coarsens shoreward from silt on the outer shelf to mostly sand at depths shallower than about 60 m, and continues to coarsen within Bristol Bay to coarse to very coarse sand at its head (Sharma, 1979). Sharma further notes that between depths of 60 and 75 m the textural distribution is complicated and does not follow any particular trend.

Much of the sediment on the central and outer shelf was probably deposited from streams that crossed the shelf during the Pleistocene lowstands of the sea. The Yukon River appears to have crossed the shelf at numerous positions northwest of the Pribilof Islands, and streams from the the Asian mainland may have contributed sediment to the northwesternmost part of the region (Carlson and Karl, 1984).

Source rocks for sediment on the coastal parts of the southern Bering Shelf include the Cretaceous sedimentary rocks and lower Paleozoic sedimentary and metamorphic rocks that are drained by the Kuskokwim River and, in the basins of streams that empty into Bristol Bay, Paleozoic and Mesozoic clastic and volcanic rocks; plutonic rocks of Jurassic and Tertiary ages; and volcanic rocks of Tertiary age. The Kuskokwim Mountains approach the coast in the series of headlands that separate Kuskokwim and Bristol Bays. The coastal Kuskokwim Mountains include Precambrian schist and gneiss, sedimentary rocks of Paleozoic and Mesozoic age, a few mafic intrusive rocks of early Mesozoic age, and ultramafic rocks of unknown age. The islands that protrude from the shelf are mostly composed of volcanic rock (Beikman, 1980).

Most sources for potential marine placer minerals are onshore from Kuskokwim Bay. The only significant U.S. production of platinum-group minerals has been from the alluvial placers in the Goodnews Bay mining district (Page et al., 1973). Production from these placers, which were discovered in 1926, exceeded half a million troy ounces in the period 1934–1966 (Cobb, 1973; U.S. Bureau of Mines, 1987). The source of the platinum is an ultramafic body that underlies Red Mountain (Location 1, Figure 7), a topographic feature that separates the placer area from Kuskokwim Bay, 3 to 5 km to the west. No workable lode deposits have been found in the dunite of the Red Mountain Ridge, suggesting that the platinum minerals are sparsely distributed in the source rock (Mertie, 1976).

Some placer gold has also been extracted from tributaries to the Goodnews and Arolik Rivers, northwest of Goodnews Bay. Some platinum was recovered with the gold in the Arolik River Basin. Otherwise, no rich placers have been found near the coast of the southern Bering Sea (Cobb, 1973). The potential for finding placer platinum-group minerals on the continental shelf off Goodnews Bay is unknown. A reconnaissance sampling using a hand auger located very little platinum (at most a few hundredths of an ounce per cubic yard of beach sand) in the beach sands bordering Kuskokwim Bay, including the beach on the west side of Red Mountain—the inferred source for the Goodnews Bay alluvial placer platinum (Berryhill, 1963). Pleistocene glaciation may have carried platinum to the west ,where it was concentrated on beaches formed at lower stands of the sea (Mertie, 1969). Such deposits, however, have neither been found nor indicated by offshore sampling. A thick overburden of glacial debris is considered to make the possibility of economic offshore deposits unlikely (Economic Associates, Inc., 1968).

Parts of the inner shelf (Kuskokwim Bay, Bristol Bay) are influenced by strong tidal currents that have produced large linear ridges composed of well-sorted fine sand (Hunter, Sallenger, and Dupré, 1979; Nelson et al., 1982). The winnowing on these ridges could produce concentrations of heavy minerals. A few samples collected from Bristol Bay show that heavy minerals constitute on the order of 20 to 70% of the bulk sediment on its southern and eastern sides (Sharma, 1979, p. 301). The heavy-mineral assemblage in these samples is dominated by pyroxenes and amphiboles; opaque minerals (undifferentiated magnetite and ilmenite) constitute about 4 to 7% of the total sediment. The titanium content of these samples ranges from 0.3 to 0.8% (Sharma, 1979, p. 312) Sillimanite is noted in one sample south of Cape Newenham, where it is present at the 0.2% level. This assemblage resembles that described from the central Bering Sea Shelf by Knebel and Creager (1974). Neither is promising in terms of economic placer minerals. (It should be noted, however, that these studies were not directed toward economic applications.)

A series of north-trending ridges lies off the coast near Goodnews Bay (National Oceanic and Atmospheric Administration, 1976). The largest of these ridges is about 40 km long and 2 km wide, and has a relief of >20 m. Possibly these ridges are part of the tidal sand-ridge system that is well developed to the north in Kuskokwim Bay. Alternatively, they may be moraines modified by waves or tides or drowned barriers. In any event, an understanding of their origin is likely to bear on the offshore potential for platinum placers in this area.

Reconnaissance hand-auger sampling of the beaches bordering Bristol Bay (Berryhill, 1963) indicated no major concentrations of valuable minerals. Ilmenite and titaniferous magnetite (and traces of fine gold) exist to depths of 9 m in the beach sands on the south side of Bristol Bay at Port Heiden, Port Moller, Nelson Lagoon, and Moffett Point (Berryhill, 1963). These heavy minerals are presumed to be concentrated from mainly glacial deposits on the adjacent coastal plain (Cobb, 1973). Berryhill (1963) concluded that the concentration of placer minerals in these sands was too low to warrant further consideration.

Sampling by Berryhill (1963) indicated the presence of small quantities of gold in a few Bristol Bay beaches. A single anomalous sample from a beach south of Egegik (Location 2, Figure 7) contained high concentrations of iron and traces of

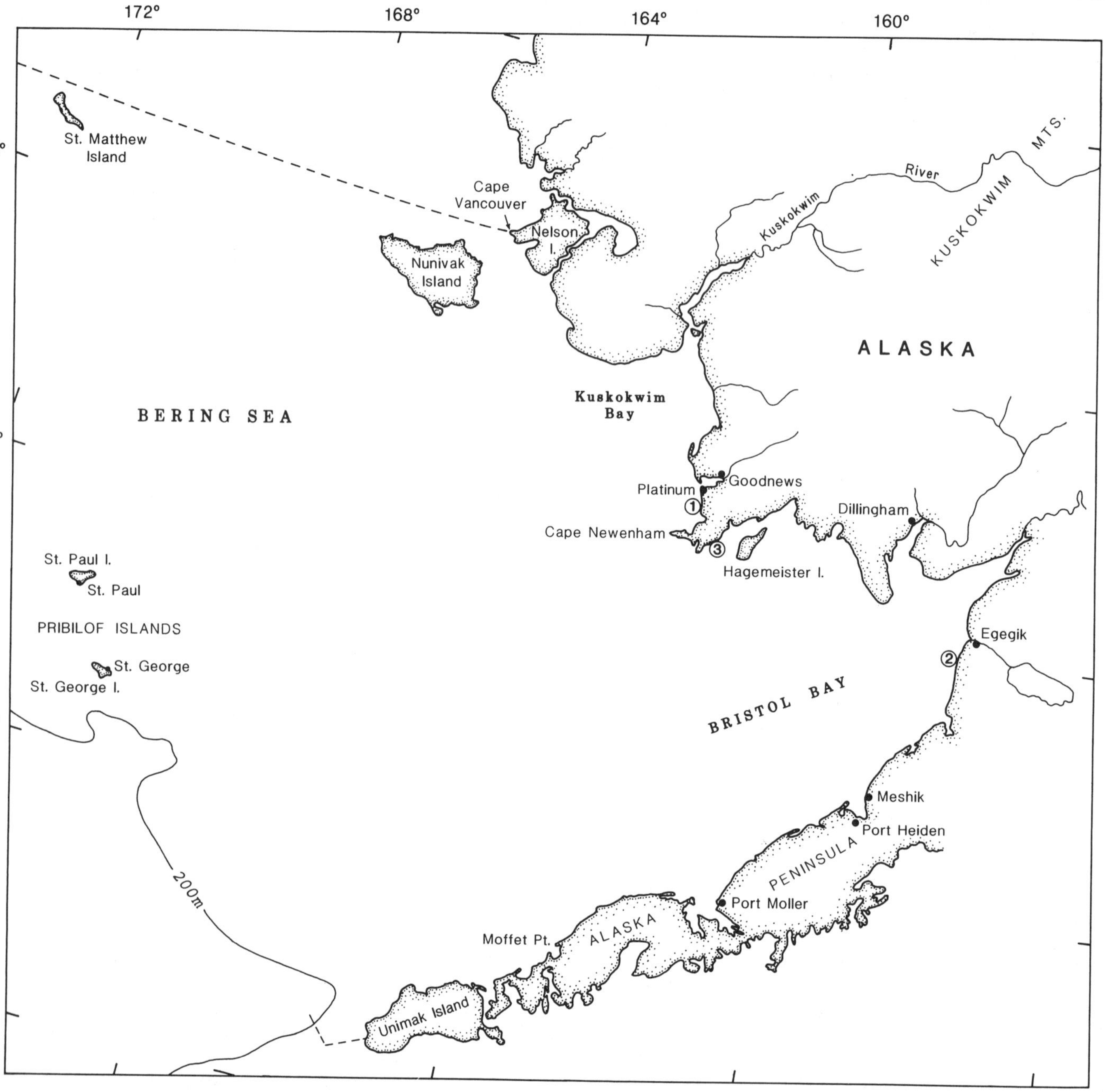

Figure 7. Southern Bering Sea region. Dotted lines outline specific area covered. Numbers refer to locations cited in text.

fine-grained gold. Low concentrations of fine-grained gold are also present in the beaches on the northwestern shore of Hagemeister Strait (Location 3, Figure 7), where a flurry of mining took place in 1937, but amount of production is unknown (U.S. Bureau of Mines, 1987). Cobb (1973) estimates that no more than a few hundred dollars worth of gold has been recovered from the beach deposits around Bristol Bay.

Alaska Peninsula–Aleutian Islands

The area consists of the shelf and coastal areas on the north and south sides of the Aleutian Island chain, and on the south side of the Alaska Peninsula (Figure 8). The area lies southwest of the western end of Unimak Island (on the north side), and west of a north–south line through the Semidi Islands (on the south side). The continental shelf narrows westward along the south side of the Alaska Peninsula from a maximum of about 200 km at the eastern boundary. Within the Aleutian chain the shelf width is irregular; at most it is generally <50 km across, and in many places it is virtually nonexistent. Volcanic and plutonic rock of Mesozoic and younger age form most of the peninsula, and Tertiary volcanic rocks form the islands (Scholl, Vallier, and Stevenson, this volume). Tertiary granitic plutons occur locally on the peninsula (Burk, 1965).

The Bristol Bay side of the Alaska Peninsula is fronted almost everywhere by beaches, many of which are backed by bluffs or cliffs cut into unconsolidated sediment (Sallenger

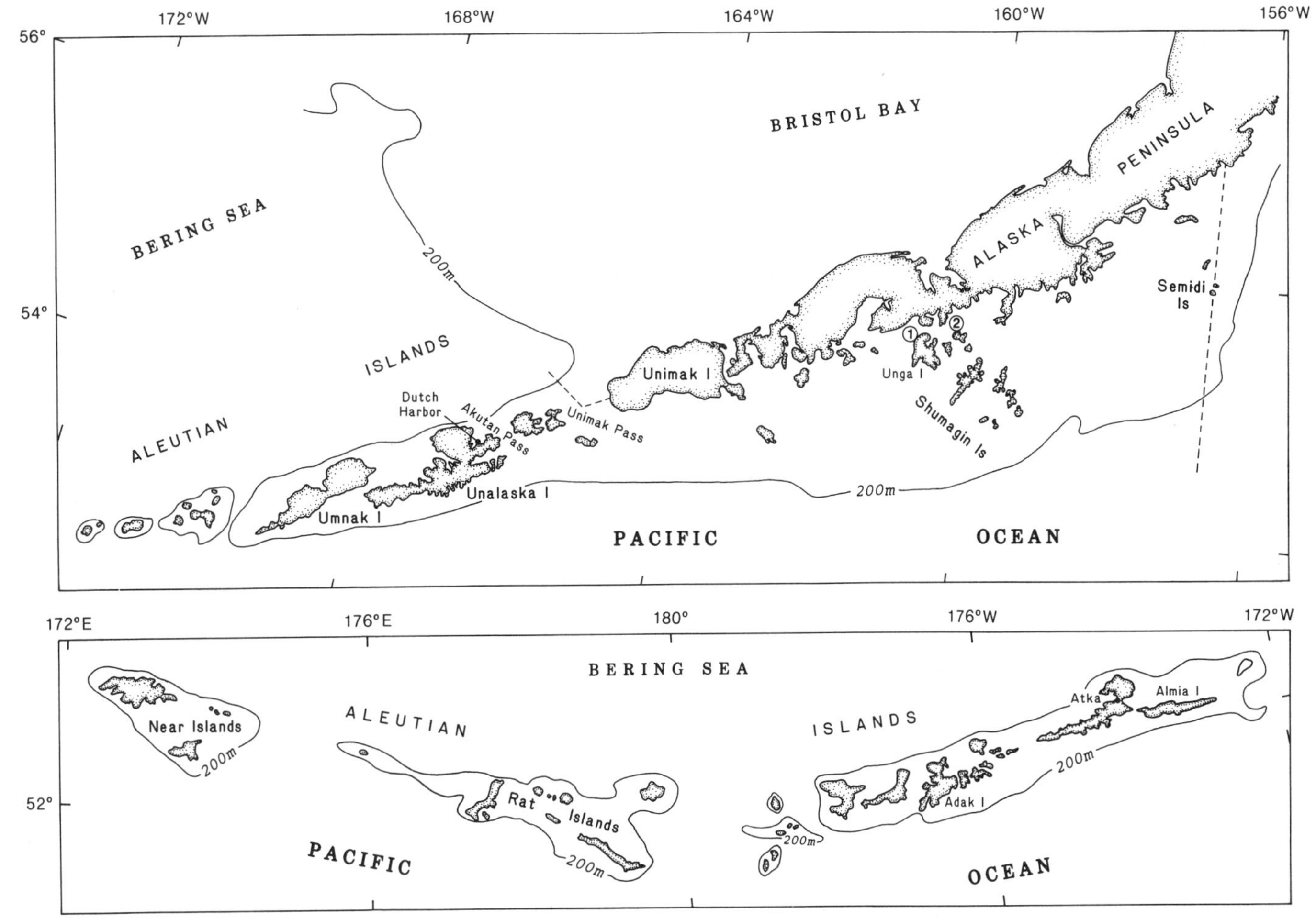

Figure 8. Alaska Peninsula-Aleutian Islands region. Dotted lines outline specific area covered. Numbers refer to locations cited in text.

and Hunter, 1984). In contrast, the Pacific Ocean side is steep, rugged, and deeply indented by fjords and embayments indicating repeated episodes of Pleistocene glaciation (Sharma, 1979). Both the Pacific Ocean and Bering Sea coasts of the Aleutians are rugged and generally devoid of beaches other than those in pockets. The passes between the Aleutian Islands are swept by exceptionally strong tidal flow; spring tidal currents in Akutan Pass can reach 9 knots and those in Unimak Pass 6.5 knots (National Ocean Survey, 1976).

Little is known about the sediment on the shelf in this area. On the south side of the Alaska Peninsula it would probably resemble that of the Kodiak Shelf slightly to the east. The mineralogy of the sediment adjacent to the Aleutians undoubtedly reflects that of the volcanic rocks that form them.

Deposits of lode gold occur sporadically on both the Alaska Peninsula and the Aleutian Island chain (Berg and Cobb, 1967; Cobb, 1984a). Only those on Unga Island (Location 1, Figure 8), where about $2 million worth of gold and silver ore was produced between 1891 and 1904, have been worked commercially (Cobb, 1973). Occurrences of tungsten minerals are reported from the Pacific side of the peninsula just north and east of Sutwik Island (Cobb, 1984c).

Placer gold was mined from the beach at Popof Island (Location 2, Figure 8) in 1904 and 1905 and reportedly from the beaches of Unga Island (Cobb, 1973). The only placer gold reported in the Aleutian Islands is a minor occurrence on Unalaska Island (Cobb, 1973).

Gulf of Alaska

The Gulf of Alaska, as we use the term, incorporates the Pacific Shelf and coastline between a north–south line through the Semidi Islands on the west and a line extending southwest from Icy Point on the east. The area includes Kodiak Island, Shelikof Strait, Cook Inlet and Prince William Sound (Figure 9).

The topography of the shelf of the Gulf of Alaska is varied west of Kayak Island. It is about 150 km wide (about 100 km wide southeast of Kodiak Island). In this area the shelf is characterized by plateaulike surfaces, sea valleys, and numerous islands (Sharma, 1979; Carlson et al., 1982). The sea valleys may be broad shore-parallel features (such as Shelikof Strait), formed by tectonic processes and modified by glaciers; or smaller shore-normal valleys, typically U-shaped, formed principally by glacial erosion (Sharma, 1979; Carlson et al., 1982). East of Kayak Island the shelf is narrower (mostly <80 km across) and is crossed by broad, U-shaped sea valleys that extend seaward to the shelf-slope break (Carlson et al., 1982).

Two large embayments, Cook Inlet and Prince William Sound, indent the Gulf of Alaska coastline. Cook Inlet is an

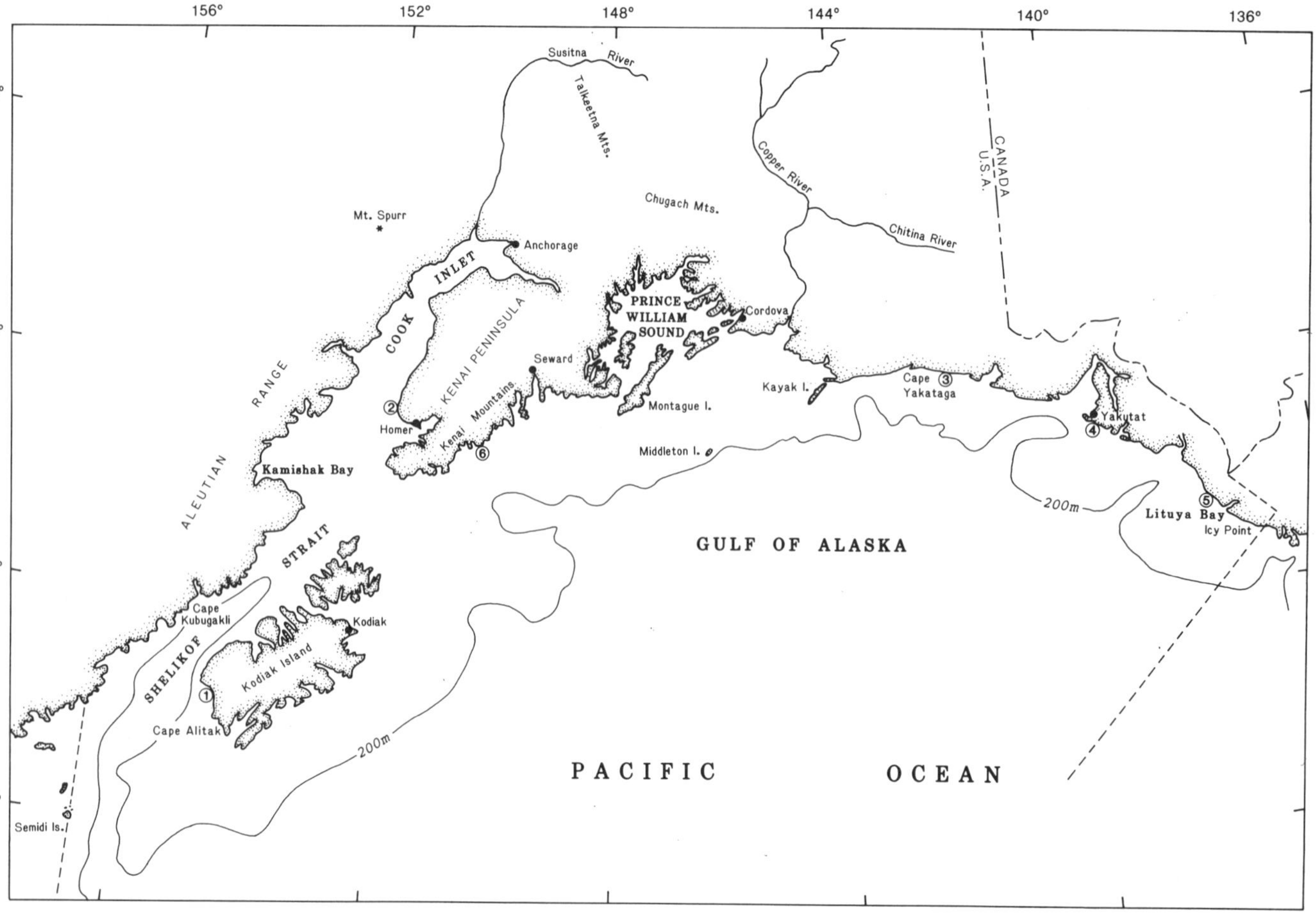

Figure 9. Gulf of Alaska region. Dotted lines outline specific area covered. Numbers refer to locations cited in text.

elongate structural basin filled with about 20,000 m of sediment and volcanic rocks (Kirschner and Lyon, 1973; Fisher and Magoon, 1978; Fisher, Detterman, and Magoon, this volume). The inlet gradually shallows from depths in excess of 100 m at its mouth to an average depth of <20 m in its upper reaches. Numerous shoals occur throughout the inlet and extensive mud flats border its upper part (Sharma, 1979). Prince William Sound is a locally deep (<800 m) extension of the Gulf of Alaska that lies semiprotected from the open gulf by several large islands. The sound consists of a series of smaller basins of probable glacial origin (von Huene, Shor, and Reimnitz, 1967).

The coastal topography and oceanographic climate within this area is varied. West of Cordova the coast is steep and rugged and cut by numerous embayments and fjords. East of Cordova, beaches and low-lying bays characterize the coastline (Tagg, 1979). Many of the beaches in this area are about half a mile wide (Thomas and Berryhill, 1962). The differences in physiography reflect both the intensity of wave activity in the eastern Gulf of Alaska and the large amount of sediment supplied to the coast by outwash from active glaciers. Four uplifted marine terraces lie onshore between Lituya Bay and Icy Point. These terraces may be quite young. Hudson, Plafker, and Rubin (1976) suggest that terraces as high as 85 m developed within the past 10,000 years. Tidal currents dominate the processes in Cook Inlet (Bouma et al., 1977) and attain velocities of several knots locally around Kodiak Island and in the straits of Prince William Sound (National Ocean Survey, 1976). The Gulf of Alaska area was extensively glaciated during the late Pleistocene, when glaciers extended over much of the existing shelf area (Karlstrom, 1964; Coulter et al., 1965; Carlson et al., 1982; Hampton, 1983; Hampton et al., in press). These glaciers removed most of the unconsolidated material and any placer deposits that may have formed in preglacial valleys (Cobb, 1973). In the eastern Gulf of Alaska, large glaciers extend to, or approach, the present shoreline. Sedimentation and erosion in this area locally are extraordinarily rapid (Molnia, 1977). The beaches consist mostly of well-sorted medium-grained sand (Reimnitz and Plafker, 1976). Only in the easternmost part of the area are beaches composed of coarse sand and gravel (Molnia and Wheeler, 1978). Storm effects here extend as high as 15 m above mean sea level.

The source areas for Gulf of Alaska sediment are complex and varied. The Aleutian Range on the northwest margin of Shelikof Strait (and Cook Inlet) is composed mostly of Mesozoic sedimentary rocks and Cenozoic volcanic rocks and volcaniclastic sediment (Fisher, Detterman, and Magoon, this volume). On the northwest margin of Cook Inlet this range is largely underlain by Jurassic and Tertiary intrusive rocks and some Mesozoic sedimentary rocks. The Talkeetna Mountains north of Cook Inlet contain Paleozoic sedimentary and volcanic rocks. The Kenai and Chugach Mountains are largely composed of Cretaceous and Jurassic clastic and volcanic rocks. Ultramafic bodies occur on the southwestern end of the Kenai Peninsula. Tertiary clastic and felsic intrusive rocks are found around the coasts of Prince William Sound and the area to the east. Kodiak Island consists mostly of Mesozoic sedimentary rocks; ultramafic intrusive bodies occur near the coast on its northwestern side, and felsic rocks of Tertiary age intrude the older sedimentary rock.

Sediment of the Gulf of Alaska Shelf has been examined in reconnaissance fashion by several workers. The shelf off Kodiak Island is floored by sediment that ranges from silt to gravel (Hampton, 1983). The textural distribution relates in part to the physiography of the shelf—elevated plateaus are covered by sand and gravel, whereas troughs and valleys are floored mostly by silt and clay (Bouma and Hampton, 1976). The heavy-mineral assemblage on the northwestern Gulf of Alaska Shelf consists primarily of opaque minerals (magnetite, ilmenite, and titanomagnetite), pyroxenes, amphiboles, epidote, and zoisite; garnet and zircon are minor constituents (Bortnikov, 1970; Sharma, 1979). Heavy minerals generally compose between 2 and 10% of the bottom surface sediment, although they locally can exceed 50% (Sharma, 1979). The amount of titanium in sediment from the northwestern and coastal Gulf of Alaska Shelf ranges between 0.3 and 0.8% (Sharma, 1979).

In Shelikof Strait, north of Kodiak Island, the surficial sediment is much finer grained overall than that on the open shelf; it grades from silty sand at the northeastern end to slightly sandy silt at the southwestern end (Hampton et al., 1981). The Holocene sediment in Shelikov Strait ranges from tens of meters to >200 m thick (Hoose, Holden, and Lybeck, 1980; Hampton, 1985).

Cook Inlet is floored by sediment that ranges from sandy silt to gravel (Sharma and Burrell, 1970; Orlando and Martin, 1981; Hampton, 1982a). Sharma (1979) delineates three sediment types within the inlet—well-sorted, fine-grained sand in the upper reaches, sandy gravel and gravel in the central stretch, and silty sand in the lower part of the inlet. Tidal currents of 2–4 knots shape the sand over much of the lower inlet into large sand waves (Bouma et al., 1977; Hampton, 1982a). The Quaternary section within the inlet consists of as much as 75 m of glacial till overlain by about 3 m of glacial outwash that is locally covered by as much as 20 m or more of recent sand, mostly in the form of sand waves (Hampton, 1982b). The heavy-mineral assemblage in Cook Inlet is dominated by amphiboles and pyroxenes; opaque minerals generally constitute 15 to 40%, and garnet and zircon mostly $<1\%$ of the total heavy minerals (Wong, 1983). The combination of heavy minerals and rock fragments in sand in lower Cook Inlet commonly accounts for half or more of the total sediment (Bouma and Hampton, 1976).

Sediment in Prince William Sound, the other inshore body of water, differs markedly from that in Cook Inlet. Most of the surficial sediment in the sound is fine grained, mostly clay (Sharma, 1979). Coarse-grained sediment occurs only in the shallowest portions or atop the sills of the fjords that fringe the sound (Sharma, 1979).

The Gulf of Alaska Shelf southeast of Prince William Sound is floored primarily by two types of material—Holocene marine sand and silt, and Pleistocene glacial marine gravelly mud (Carlson et al., 1977; Molnia and Carlson, 1977). The Holocene sediment occurs primarily on the inner shelf and the Pleistocene deposits on the outer shelf. Holocene sand is restricted to a zone generally no more than 10 km wide along the shoreline and grades seaward into silt (Carlson et al., 1977). The Holocene deposits form a wedge, thinning seaward, that locally attains thicknesses of >300 m (Carlson and Molnia, 1975). Such thicknesses reflect the rapid rate of sedimentation currently on the shelf. The average rate is about 4.5 mm per year; the maximum rate, near the Copper River Delta, exceeds 28 mm per year (Molnia, Levy, and Carlson, 1980). The heavy-mineral content of the shelf sand in the eastern Gulf of Alaska generally ranges from 4 to 10%; the assemblage is dominated by pyroxenes and amphiboles (Sharma, 1979).

The rocks onshore from the Gulf of Alaska are potential sources for a variety of placer minerals. Lode gold occurrences near the coast are found on Kodiak Island, in the Kamishak Bay area, on the Kenai Peninsula, in the upper reaches of Cook Inlet, and around the margins of Prince William Sound (Cobb, 1984a). Platinum minerals occur on the western end of Kodiak Island, and chromite has been produced from the western part of the Kenai Peninsula (Cobb, 1960). The glacial-marine late Cenozoic Yakataga Formation is probably a secondary source of gold on beaches at Middleton Island, near Lituya Bay, and at Cape Yakataga (Reimnitz and Plafker, 1976).

Onshore placer deposits occur sporadically near the coast (Cobb, 1973). The most important area for alluvial placers lies at the upper end of Cook Inlet (the Cook Inlet-Susitna River region of Cobb, 1973), which has been mined continuously since the early 1900s. In addition to gold, monazite and zircon have been reported in placers near Mount Spurr, northwest of upper Cook Inlet (Bates and Wedow, 1953). Gold has been recovered from a few small-scale alluvial placer mines on the Kenai Peninsula, the White River near Yakataga, and streams near Cape Kubugakli on the northern side of Shelikof Strait (Cobb, 1973). Despite the presence of major lode gold mines, no valuable placer deposits have been found in the Prince William Sound area, presumably because of the intense glaciation.

Beaches have been mined on a small scale for gold along many parts of the Gulf of Alaska shoreline. Cobb (1973) and the U.S. Bureau of Mines (1987) cite operations on the west end of Kodiak Island (Location 1, Figure 9), the eastern shore of Cook Inlet (Location 2, Figure 9), Middleton Island, and the beaches near Yakataga (Location 3, Figure 9), Yakutat (Location 4, Figure 9), and near Lituya Bay (Location 5, Figure 9). Mertie (1931) notes that the elevated beaches near Lituya Bay have been mined to a small extent. The concentrates from Kodiak Island consist primarily of magnetite but also include chromite and traces of platinum (Maddren, 1919). A small amount of gold was recovered from eolian sand at Cape Alitak on the western end of Kodiak Island (Cobb, 1973). The beaches at Yakataga include fine-grained gold that locally is sufficiently concentrated to be minable.

Small quantities of gold occur in most of the beach sands along the eastern Gulf of Alaska. Reimnitz and Plafker (1976) sampled about 400 km of coastline, taking the richest available concentrates of heavy minerals from 55 beaches. All of these contained at least traces of gold, and 36 held 0.01 ppm or more. One sample from a beach east of Cape Yakataga (Location 3, Figure 9) contained nearly 2 ppm of gold. The authors note, however, that where gold values were high, the volume of beach material containing gold was small.

Reimnitz, von Huene, and Wright (1970) found small quantities of gold in the sand and gravel of small deltas along the

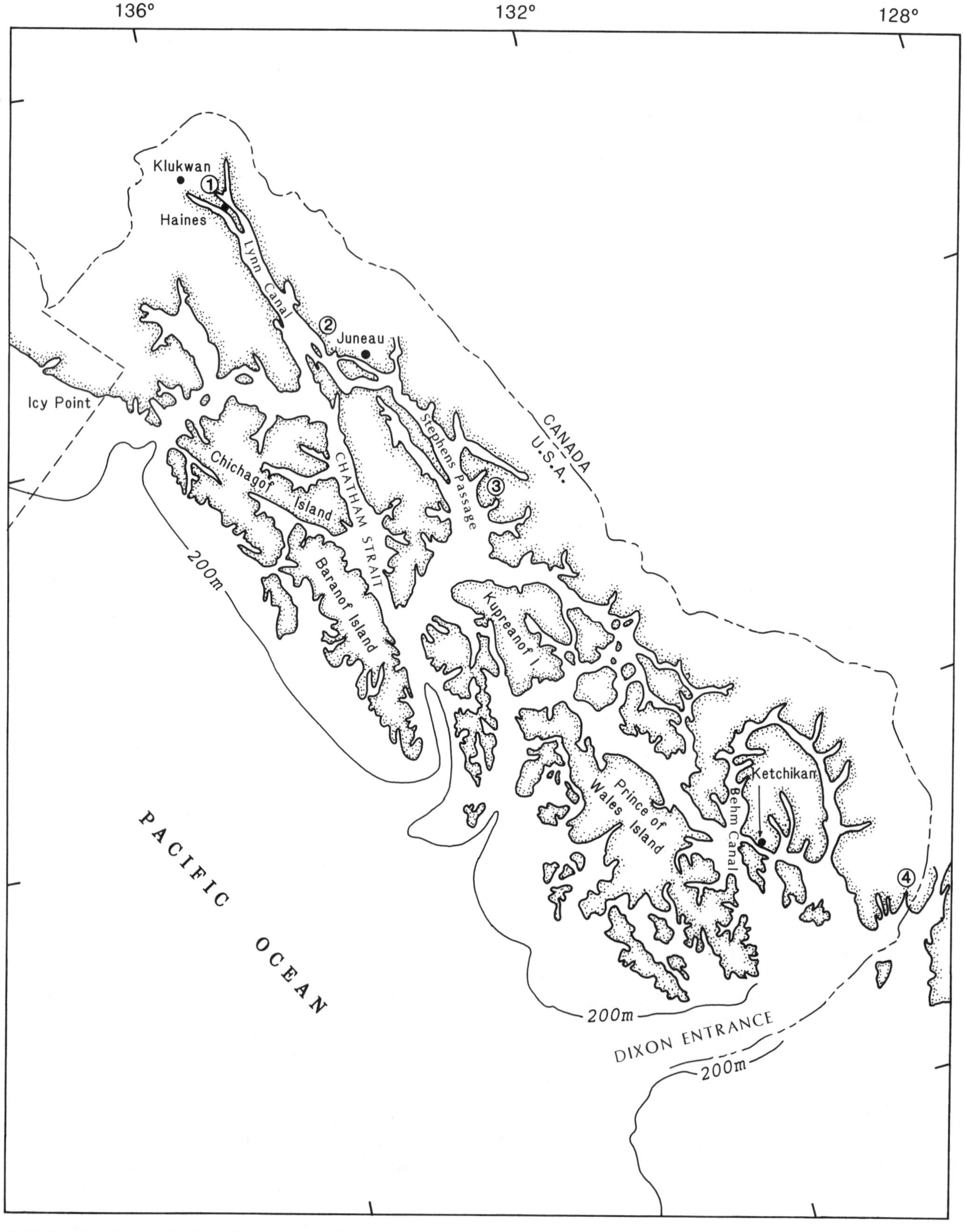

Figure 10. Southeastern Alaska region. Dotted lines outline specific area covered. Numbers refer to locations cited in text.

inner portion of Nuka Bay (Location 6, Figure 9), on the east side of the Kenai Peninsula, and in relict glacial gravel on the sea floor outside the bay, but none in the fine sediment that floors the bay. Wright (1969) detected particulate gold in a few samples taken from Yakutat Bay near the village of Yakutat.

A few studies have examined the assemblage of heavy minerals that occurs in the placer deposits. Thomas and Berryhill (1962) used a hand auger to sample the sandy beaches along 247 miles of the eastern Gulf of Alaska coastline; they report that the principal heavy minerals in Yakutat and Lituya Bay beaches are magnetite, garnet, and ilmenite, with lesser to trace amounts of rutite, zircon, chromite, and gold. Ilmenite is the dominant heavy mineral in the beach deposits of Lituya Bay. Rossman (1957) describes concentrations of 5 to 40% heavy minerals on modern beaches as much as 100 m wide and 3 km long. Cook (1969) found a similar assemblage in his analysis of heavy-mineral concentrates from beaches in the Yakataga and Lituya Bay areas. Platinum has also been reported in trace amounts from the Yakataga-Lituya Bay beaches (Thomas and Berryhill, 1962; Cobb, 1973). During the 1950s a private concern investigated the magnetite content of beach sand near Yakutat and Lituya Bay (Thomas and Berryhill, 1962). An analysis of heavy-mineral concentrates from Cook Inlet indicates that the sand on the north shore locally contains as much as 6.7% ilmenite and 0.6% zircon, and "minor amounts" of the mineral euxenite, a source of columbium, tantalum, and thorium (Cook, 1969). Sediment near the inlet of Yakutat Bay contains 5 to 10% heavy minerals composed of chlorite, amphibole, and pyroxene, with minor quantities of magnetite and ilmenite and traces of zircon (Wright, 1972).

Southeastern Alaska

The area of southeastern Alaska includes the coastline and shelf from a line extending southwest from Icy Point to the Canadian border in Dixon Entrance (Figure 10). The coastal region consists of rugged mountains and numerous narrow straits and fjords. The area owes much of its relief to intense alpine glaciation during the late Pleistocene (Coulter et al., 1965), and active glaciers are found in the northern and eastern parts. The open shelf, about 50 km wide, has been glaciated recently and is mostly covered with glacial sediment that was little reworked during the Holocene transgression (Sharma, 1979).

The open coast is exposed to the northern Pacific waves but, owing to the general absence of sediment, has few small pocket beaches. The inland waterways are narrow and deep. Many are silled as a consequence of their glacial origin (Sharma, 1979). Tidal currents within the inland waterways range from sluggish to several knots (National Ocean Survey, 1976).

The sediment being deposited at present in the inland waters of southeastern Alaska may differ substantially from that deposited over large areas of the open shelf. The modern sediment in the Inside Passage and other inland waterways is dominated by silt and clay; sand and gravelly sand accumulate only along the shoreline or atop sills (Sharma, 1979). In contrast, the sediment of the open coast and inner shelf is locally (and perhaps regionally) composed of a high proportion (40 to 90%) of biogenic carbonate grains (Hoskin and Nelson, 1969), perhaps because much of the terrigenous sediment supplied to the area is presently being trapped within the inland waterways. The few seismic records that exist over the continental shelf indicate that very little Holocene sediment has accumulated above the pre-Holocene erosional surface (Carlson and Schwab, 1982). The heavy-mineral assemblage is strikingly uniform over large areas of southeastern Alaska; it consists mostly of a combination of pyroxene, amphiboles, chlorite, and magnetite and contains traces of garnet, sillimanite, apatite, tourmaline, and staurolite (Sharma, 1979).

The source terrane for sediment in southeastern Alaska is a complex assemblage of sedimentary, volcanic, metamorphic, and plutonic rocks that range in age from Paleozoic to Tertiary (Plafker, this volume). A large, generally felsic composite intrusive body (the Coast Range batholith) extends along most of the eastern margin of the area.

Numerous potential sources for placer minerals are known. Occurrences of lode gold are scattered throughout the area, but are particularly concentrated west of Ketchikan (Cobb, 1984a). Lode platinum has been mined on Prince of Wales Island and is known to occur on Kupreanof Island (Cobb, 1960). Chromite is found on Baranof Island (Cobb, 1960), and tungsten minerals are known to exist sporadically in the area (Cobb, 1984c).

Placer mines are not particularly numerous, however, because all other deposits were removed by late Pleistocene glaciation (Cobb, 1973). Only gold and a small amount of platinum have been extracted from placer deposits in southeastern Alaska (Cobb, 1973). Placer gold has been mined from streams at the upper end of Lynn Canal (Location 1, Figure 10) and in the vicinity of Juneau (Location 2, Figure 10), and on a smaller scale from streams near Windham and on eluvial deposits on western Chichagof Island (Cobb, 1973). Placer gold has also been reported from a locality on Baranof Island and at Cape Fox (near Location 4, Figure 10) (U.S. Bureau of Mines, 1987). Placer occurrences of other minerals have been reported from Baranof Island (allanite and traces of monazite and scheelite) and of magnetite from the Takshanuk Mountains near Klukwan (Cobb, 1973). Conwell (1976) notes that there has been periodic interest in both offshore and onshore deposits of allegedly titaniferous magnetite near Haines.

Few beach or marine placer deposits are recorded in southeastern Alaska. Gold has been successfully mined on a small scale from the Stephens Passage area (Location 3, Figure 10) at Windham Bay (Hess, 1965). Sainsbury (1957) describes occurrences of detrital garnet in the beaches of Kanagunut Island (near Location 4, Figure 10) that have possible economic potential. A large sample taken from Bradfield Canal near Ketchikan indicated quantities of 68 pounds of magnetite, 6 pounds of ilmenite, and trace amounts of gold and zircon per cubic yard (Cook, 1969). Intertidal sands along the Behm Canal contain small quantities of gold and the heavy minerals garnet, zircon, and ilmenite (Conwell, 1976). Cobb (1973) noted commercial interest in submarine placer deposits in the Juneau area, but no results were available.

Washington

The continental margin of the State of Washington includes the coastal area of the Puget lowlands and the southern margin of the Strait of Juan de Fuca, as well as the open Pacific Coast from Cape Flattery to the mouth of the Columbia River (Figure 11). The Puget lowlands area is a

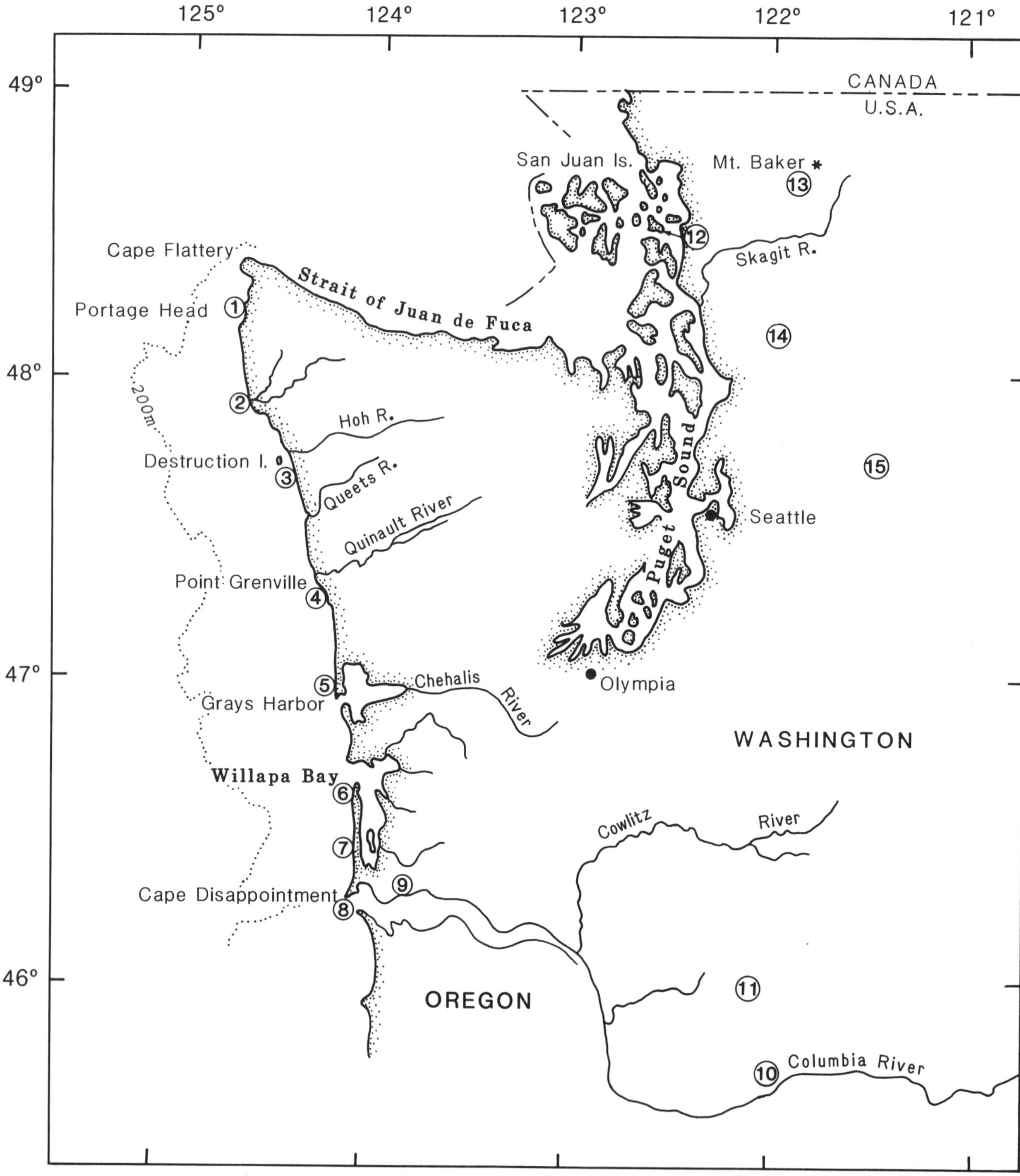

Figure 11. Coastal Washington region. Numbers refer to locations cited in text.

series of north–south-trending inlets and irregularly shaped islands. The open shelf ranges in width from 15 km off the Columbia River to > 52 km off Cape Flattery (Phillips, 1979).

The southwestern coast of Washington is a coastal plain dominated by two large tidal estuaries, Willapa Bay and Grays Harbor; Pleistocene terraces fringe these bays. From Cape Disappointment on the north side of the Columbia River mouth, broad beaches, dunes, and ridges stretch north almost continuously to the mouth of the Hoh River. Sea stacks composed of sandstones and shales begin to appear north of Point Grenville, some as far as 5 km from shore. The largest, Destruction Island, is 1.5 km long.

For 97 km from Grays Harbor north to Cape Flattery, the coast is almost straight; steep coastal cliffs extend from Point Grenville northward except at river mouths. The coast then turns sharply eastward at Cape Flattery and extends to Puget Sound, following the Strait of Juan de Fuca. This narrow glacial trough extends from Puget Sound across the entire shelf. The Olympic Mountains border the coast from Cape Flattery to the Puget Trough. The remaining coastline to the Canadian border turns northward once more, flanked by a flat to gently rolling glaciated plain.

The Washington coast has been influenced by both the Cordilleran continental glacier and by localized valley glaciers. The continental glacier covered the northernmost 48 km of open coast as it advanced along the Strait of Juan de Fuca (Shepard and Wanless, 1971). Valley glaciers from the Olympic Mountains flowed to the coast in the Quinault, Queets, and Hoh River valleys. Valley glaciers also extended from the northern Cascades to the Puget lowlands.

The open coast is dominated by waves. Surficial coastal currents flow northward during the winter months; bottom currents appear to move northward for most of the year. In March, shelf currents reverse and flow southward (Gross and Nelson, 1966). Storm-generated waves also produce bottom currents strong enough to concentrate heavy minerals (Phillips, 1979); Smith and Hopkins (1972) measured current velocities up to 70 cm/sec at 80-m depths. Wave and current activity are prevalent in depths to 40 m on the southern shelf; at depths >40 m, bioturbation is the dominant process (Nittrouer and Sternberg, 1981). Tidal currents are strong (2 to 8 knots) in the Strait of Juan de Fuca and in the many fjords of the Puget lowlands. A net eastward flow occurs at the bottom of the Strait of Juan de Fuca (Anderson, 1968).

Surficial sediment data on the Washington continental shelf is summarized by Roberts (1974b) and Phillips (1979). The shelf is generally smooth; the dominant surficial sediment on the inner shelf is sandy, and on the outer shelf is silt-clayey (Nittrouer and Sternberg, 1981). Prominent gravel patches lie on the inner shelf off Cape Flattery (Hands, 1968) and off Grays Harbor and the mouth of the Quinault River (VenkataRathnam, 1969). Stillstands of sea level are marked by the presence of discontinuous, surficial gravel and heavy-mineral concentrations ($>10\%$) at depths of 18–33 m, 44–51 m, and 73–82 m; a fourth concentration lies at the shelf break (165–180 m) (VenkataRathnam and McManus, 1973).

Surficial sediment data from inland waters (Strait of Juan de Fuca and Puget Sound) are summarized by Roberts (1974a). In Puget Sound and other inlets, sand with local gravel concentrations lies in the narrow band of shallow water near the coast and the islands. Silt and clay dominate the deeper parts of the fjords. Sand and/or gravel concentrations are also present near river mouths. The central area of the Strait of Juan de Fuca is floored with coarse-grained sand and common granitic pebbles; the sand becomes progressively finer toward the margins of the strait (Anderson, 1968).

Bedrock geology along the Washington coast is varied. The San Juan Islands, in the Puget Trough, lie in a complex terrane of Paleozoic and Mesozoic sedimentary rocks, pre-Devonian granitic rocks, mafic and ultramafic intrusives, and Quaternary glacial till and alluvium (McLellan, 1927; Wagner and Snavely, 1966; Snavely, this volume). The granitic pebbles in the Strait of Juan de Fuca are thought to have originated in the North Cascades of British Columbia (Anderson, 1968). Early Tertiary sedimentary rocks border the coast north of Grays Harbor and along the strait (Wagner and Snavely, 1966). Tertiary basalts crop out in a few places on the coast—at Cape Disappointment next to the Columbia River, at Cape Flattery, and at points at Agate, Crescent, and Freshwater Bays along the Strait of Juan de Fuca. No Tertiary volcanic or intrusive rocks are known to outcrop offshore (Snavely and Wells, 1984). As much as 3000 m of upper Miocene and Pliocene sediment underlies the shelf (Snavely, Pearl, and Lander, 1977). A major sediment source for the southern Washington coast and shelf is the Columbia River, which drains 670,810 km^2, making it the largest river in the northwestern United States (Rand McNally, 1980).

Heavy-mineral studies for sediment transport and provenance studies on the Washington continental shelf have been conducted by White (1967), Hands (1968), VenkataRathnam (1969), and VenkataRathnam and McManus (1973); in Grays Harbor by Scheidegger and Phipps (1976); and in Willapa Bay by Luepke and Clifton (1983). The Columbia River, whose sediment dominates the southern shelf, is characterized by high percentages of hypersthene and hornblende (Whetten, Kelley, and Hanson, 1969). The shelf sediment is lower overall in hypersthene, with clinopyroxene prevalent everywhere and hornblende more abundant on the inner shelf (VenkataRathnam, 1969). However, the heavy-mineral-rich zone at 44–51 m depth has relatively abundant hypersthene all along the shelf, particularly in the southern part (VenkataRathnam and McManus, 1973). Two factors may be responsible for this: (1) more orthopyroxene-rich sediment supplied in the past, and (2) stronger northward sediment drift, if headlands were largely absent.

Lesser amounts of epidote and garnet are present in the nearshore sands, commonly reflecting the mineralogy of adjacent beaches. Opaque minerals (magnetite and ilmenite) range from 15% of the sand fraction near Cape Flattery to 34% off the mouth of the Columbia River (Hands, 1968; VenkataRathnam, 1969). Garnet and hornblende dominate all samples from the Strait of Juan de Fuca (Anderson, 1968). Zircon generally does not exceed 1% in any of the offshore samples, although relatively large amounts of zircon and sphene are commonly present where garnet and hornblende are abundant (VenkataRathnam, 1969). Rutile and xenotime have been identified in trace amounts off Cape Flattery (Hands, 1968). No studies of heavy minerals from bottom sediment in the Puget Sound region are known.

Primary sources for valuable heavy minerals include lode deposits on the western slope of the Cascades. Gold deposits are found east of Everett (Location 15, Figure 11) and in a small area near Mount Baker (Koschmann and Bergendahl, 1962). Thayer and Miller (1962) report chromite in podiform deposits on Cypress Island (Location 12, Figure 11) and at the Twin Sisters Mountain deposits (Location 13, Figure 11). Two hundred tons of chromite ore was produced on Cypress Island in 1918; total chromite production from Washington was <300 tons from 1900 to 1905 (Huntting, 1956). Sources for monazite and other rare-earth minerals are found in eastern Washington in the uppermost reaches of the Columbia River (Olson and Adams, 1962), but these are presumably too distant to be significant.

The Olympic Mountains do not contain significant mineralization. Rutile and magnetite are found in the Pilchuck Creek deposit (Location 14, Figure 11)—a fossil placer in the basal beds of the Eocene Chuckanut Formation (Huntting,

1956). But for the most part, the Tertiary sedimentary rocks that border the major part of the Washington coast are unlikely secondary sources of valuable heavy minerals.

Alluvial placers of ilmenite and magnetite are found on the Skatgit River (Youngman, 1930) and near the Chehalis and Cowlitz Rivers (Huntting, 1956). Similar placers at Camas and Brush Prairie (Location 10, Figure 11) are also gold bearing (Rogers and Jaster, 1962). The only platinum-producing stream placer in Washington is on the South Fork of the Lewis River (Location 11, Figure 11) (Pardee, 1929; Huntting, 1956). Selected stream samples from the Puget Sound region show weight percentages of heavy minerals ranging from 0.2 to 2.7 (Martinez, Gomes, and Wong, 1981).

Beach placers of precious metals on the Washington coast are numerous (Blair, Page, and Johnson, 1977). The gold and platinum mined earlier in this century on the northern Washington coast is thought to have been derived from glacial sediment brought southward from British Columbia by the Cordilleran continental glacier; the ratio of gold to platinum was estimated to have been from 15:1 to 5:1 (Mertie, 1969). Gold was discovered on the coast south of Cape Flattery in 1894; approximately $15,000 in gold was extracted sporadically from that time until 1908 (Arnold, 1905). Placer gold mining in Washington has from its inception been intermittent and small scale. By 1900, most gold placer deposits had been worked out; between 1900 and 1955 gold production has been estimated to have been around 50,000 ounces ($1.25–$1.45 million; Huntting, 1955).

Gold and platinum have been produced at ShiShi and Ozette beaches (Location 1, Figure 11) and at Sunset Creek (Location 2, Figure 11) (Pardee, 1929; Huntting 1955). These metals have also been reported as traces from Ruby Beach, (Location 3, Figure 11), Moclips (Location 4, Figure 11), Point Brown/Damon's Point (Location 5, Figure 11), Leadbetter Point (Location 6, Figure 11), Ocean Park/Nahcotta (Location 7, Figure 11), and Fort Canby/Sand Island (Location 8, Figure 11). Gold has been reported from placers in the Puget lowlands near Anacortes and Day Creek (Huntting, 1955), and platinum from Cypress Island (Location 12, Figure 11) in amounts too small to be economic (Pardee, 1921, 1929). There is no current mining activity at any beach placer deposit in Washington.

Other minerals in beach placers on the Washington coast include ilmenite, magnetite, chromite, zircon, rutile, and sphene. Ilmenite is a major constituent in both modern and ancient beach sands in southwestern Washington; rutile and sphene occur as trace amounts in beach sediment (Thorsen, 1964). Chemical analyses of sand from Moclips Beach (Location 4, Figure 11) showed values of 21 pounds of magnetite and 50 pounds of ilmenite per ton (Martinez, Gomes, and Wong, 1981). Chemical analyses of raw samples from the McGowan deposit (Location 9, Figure 11) near the mouth of the Columbia River have yielded values of 21–43% iron and 14% titanium; a preconcentrated sample contained 53% magnetite and 18% titanium (Kelly, 1947). In general, the percentage of ilmenite increases and magnetite decreases progressively in beach sediment northward from the Columbia River (Pardee, 1929).

Chromite is generally a minor component of Washington beach deposits. Zircon, a trace ($<0.5\%$) constituent of most Washington black sands, is reported at 1% in a preconcentrated sample at the McGowan deposit (Location 9, Figure 11; Kelly, 1947) and 0.7% at both Damon's Point (Location 5, Figure 11; Thorsen, 1964) and Moclips Beach (Location 4, Figure 11; Martinez, Gomes, and Wong, 1981). Day and Richards (1906a) report 71.5 tons of monazite in a beach sand from Moclips near Grays Harbor. More recent chemical analyses have shown a weight percent of 0.18 ThO_2 from Moclips Beach and 0.17 ThO_2 from Copalis Beach (Martinez, Gomes, and Wong, 1981).

Heavy minerals in Washington Shelf sediment have been examined mostly in the sand-sized fraction. On the southern shelf, White (1967) shows that overall heavy-mineral percentages in the sand-sized fraction exceed 10% only in one small area south of Leadbetter Point (Location 6, Figure 11). On the northern inner shelf south of Cape Flattery, Hands (1968) found heavy-mineral concentrations between 1 and 5%, relatively higher in the nearshore off headlands, and in three samples far from shore. An anomalously high percentage of heavy minerals, 40%, was noted off Portage Head. Heavy minerals do not exceed 10% by weight in the Strait of Juan de Fuca (Anderson, 1968; Hands, 1968). VenkataRathnam's (1969) studies of the Washington Shelf from Grays Harbor to Cape Flattery indicated gravel deposits and concentrations of heavy minerals $>10\%$ off Grays Harbor and opposite the Hoh River.

Oregon

The continental margin of the state of Oregon is relatively uncomplicated—a straight, narrow (mostly <30 km wide) shelf backed by an upland coastal area of locally rugged relief (Figure 12). The shelf surface is generally flat; small islands protrude from the southern Oregon Shelf within 5 km of the shoreline, and gullies and a few canyons incise the shelf at its seaward break. Broad banks underlain by late Tertiary sedimentary rocks (Maloney and Byrne, 1964; Maloney, 1965) rise locally on the outer shelf.

The coast is dominated by long period (10- to 12-sec) waves that average about 2 m high in the summer and 3 m high in the winter (Komar, Neudeck, and Kulm, 1972). Storms play an important role in the distribution and character of the shelf sediment and in the development of coastal placer deposits (Twenhofel, 1943; Komar and Wang; 1984; Peterson, Komar, and Scheidegger, 1986). The southern part of the coast is generally rugged, and beaches tend to lie between fairly closely spaced headlands. North of Cape Blanco, to Heceta Head, fewer headlands separate more extensive beaches. North of Heceta Head, beaches are again somewhat localized. Tidal effects are significant only near the inlets to the many small bays that dot the coast and near the mouth of the Columbia River.

Uplifted coastal marine terraces are well developed along the tectonically active central and southern coast (Kulm et al., 1975). Terraces south of Coos Bay range to elevations of 500 m (Diller, 1902; Twenhofel, 1943). Three terraces, at about 100 to 125 m, 50 to 70 m, and 15 to 30 m above sea level, are well defined in the area between Coos Bay and Port Orford (Griggs, 1945). Tilting of the terraces about local east–west axes along the coast (Baldwin, 1976) makes the correlation of terrace surfaces elsewhere along the coast nearly impossible.

Kulm et al. (1975) summarize the character of modern sediment and sedimentary processes on the Oregon continental shelf. The Holocene transgression of the sea left a basal sand facies over the shelf. As sea level approached its present position, silt and clay began to accumulate over the mid- and

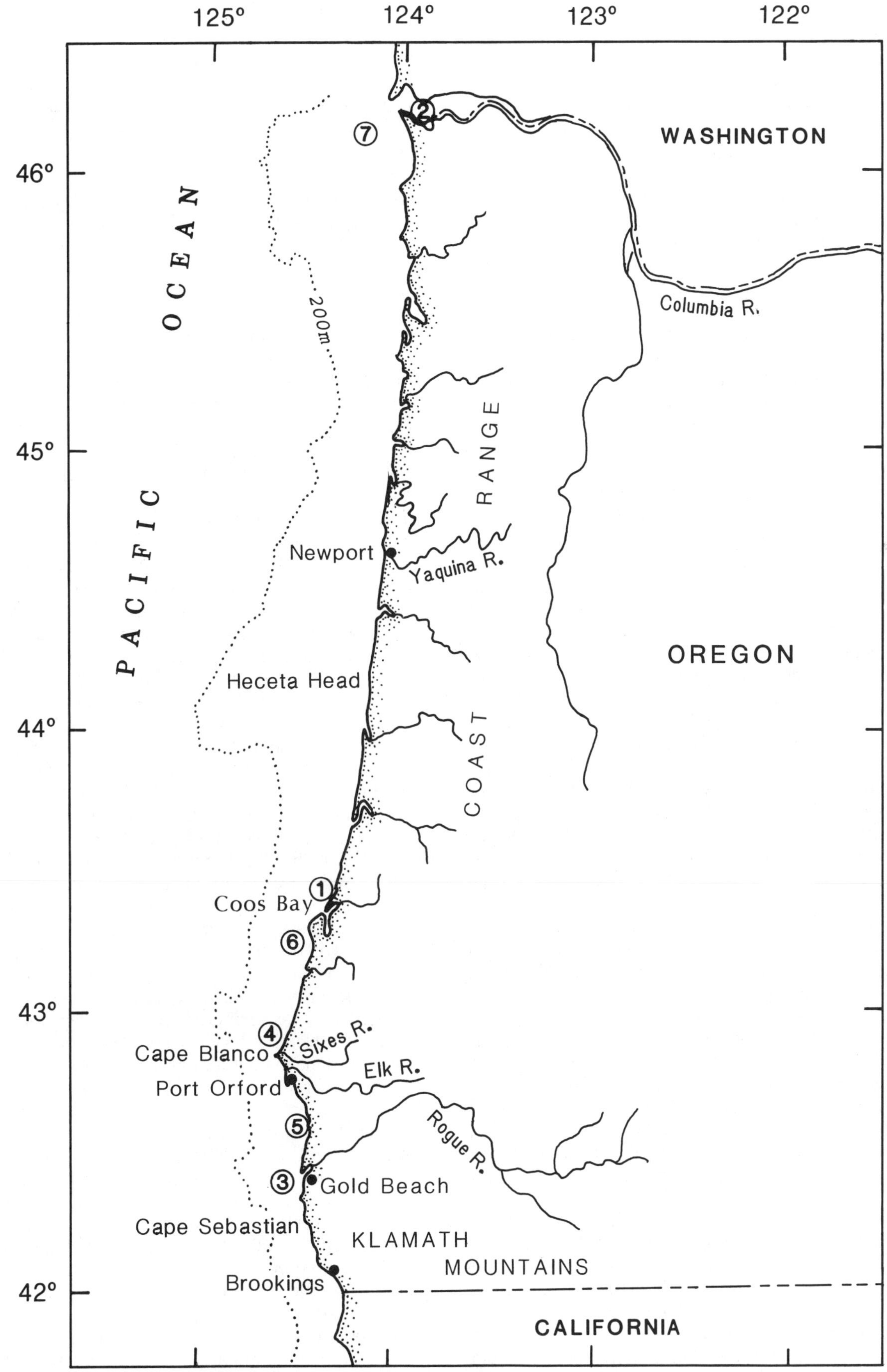

Figure 12. Coastal Oregon region. Numbers refer to locations cited in text.

outer-shelf areas. This muddy material forms a relatively thin patchy layer over the underlying sand (or sand and mud) facies. Sand is contributed at the present time to the shelf by coastal erosion and streams. Only very fine sand is delivered to the shelf by streams, and that occurs only at flood stage when the sand can bypass the coastal estuaries. The sand facies (which includes sand and gravel) on the inner shelf is either relict or a combination of relict and modern sediment. The depth at which the mud accumulates is determined partly by interaction of summer and winter waves (which stir the sediment on the sea floor to water depths of about 90 and 200 m, respectively) and partly by the volume of suspended sediment supplied by the rivers.

A number of rivers feed sediment onto the shelf. The Columbia River drains an area nearly four times as large as that drained by all the other coastal streams combined (Kulm et al., 1968b). A series of smaller streams transects a coast range underlain by Tertiary sedimentary and volcanic rocks; in the southwestern part of the state, the Rogue and smaller coastal rivers and creeks drain the Klamath Mountains—a terrane composed largely of Mesozoic sedimentary, volcanic, and plutonic rocks (Baldwin, 1976).

The heavy-mineral assemblage of the shelf sands is dominated by opaque minerals (magnetite, ilmenite, and chromite) and amphiboles and pyroxenes; garnet is locally abundant (Runge, 1966; Scheidegger, Kulm, and Runge, 1971). Heavy minerals in the sediment on the shelf and beaches can be related to onshore sources. The mineral assemblages of beach sands differ somewhat from those on the shelf. The Columbia River, for example, influences the composition of beach sands over the northernmost third of the state's coastline (Kulm et al., 1968b) but dominates the shelf sand mineralogy primarily to the north of the river mouth (Scheidegger, Kulm, and Runge, 1971). The Klamath River in California influences beach sands as far north as the Yaquina River (Kulm et al., 1968b), but it influences sand on the outer shelf nearly to the Columbia River area. On the central Oregon coast, in contrast, these studies show that garnet-rich sands occur both on the beach and on the inner shelf.

The mineralogic patterns of the shelf sand indicate a northerly transport of sand on the order of 100 km on the outer shelf. Longshore drift during the Wisconsin lowstand of the sea, when rivers presumably discharged their bed load directly onto an extensive uninterrupted beach, is a plausible mechanism for this transport (Scheidegger, Kulm, and Runge, 1971). A similar direction of longshore drift is indicated by the mineralogy of present-day beach sand (Kulm et al., 1968b), although existing headlands may present barriers to long-term transport (Luepke, 1980).

Primary sources for economically significant placer minerals on the Oregon coast are restricted to the Klamath Mountains and the drainage basin of the Columbia River. Tertiary sedimentary rocks of the Coast Range could, however, serve as a secondary source for placer minerals. The Klamath Mountains are a potential source for a variety of heavy minerals. Gold, mostly associated with Late Jurassic diorite intrusives, has been mined from a number of localities in the Rogue River drainage basin and nearer the coast in the Sixes and Elk Rivers basins (Brooks and Ramp, 1968). Although there are no lode deposits of platinum-group metals in the Klamath Mountains, the ultramafic rocks there are considered to be sources for placer occurrences (Ramp, 1969; Page et al., 1973). Chromite is more or less widespread in these ultramafic rocks (Ramp, 1961; Thayer and Miller, 1962; Thayer, 1969). Scheelite is found in the upper Rogue River drainage (Hobbs, 1969), presumably too far upstream for this high-specific-gravity heavy mineral to be considered as a source for tungsten in the coastal area. The rocky shoals and islands near the coast are composed of Late Jurassic sedimentary and volcanic rocks or Cretaceous sandstone and shale (Hunter, Clifton, and Phillips, 1970), and are unlikely sources for placer minerals. The offshore banks on the outer shelf to the north, which are underlain by fine-grained sedimentary rocks of Tertiary age (Maloney and Byrne, 1964), similarly have little potential as source rocks.

Alluvial placer mining in the Oregon coastal area is restricted to gold and platinum-group metals in the Klamath Mountains. More than three-fourths of the gold produced in the Klamath Mountains is from placer deposits (Brooks and Ramp, 1969). Accordingly, marine placers of gold and platinum-group metals are virtually restricted to the southwestern corner of the state (from Coos Bay to the state line). Gold was discovered in the present beaches in 1852 (Pardee, 1934) and was mined extensively for the next 10 years or so (Hornor, 1918). Early studies showed these deposits to be enriched in chromite, ilmenite, garnet, zircon, gold, and platinum (Day and Richards, 1906b). The amount of production in this early mining, which also included deposits in the elevated marine terraces, is unrecorded. After the turn of the century, the production was on a relatively small scale—about 2200 ounces of gold and 100 ounces of platinum (about \$2000 worth of platinum at about \$20 per ounce) in the period from 1903 to 1929 (Pardee, 1934). Small, sporadic attempts to mine beach gold continue to the present day.

When chromium became a critical commodity during World War II, the beaches and marine terrace deposits of southern Oregon (Location 1, Figure 12) were carefully examined and the terrace deposits were ultimately mined (Griggs, 1945). In excess of 50,000 tons of concentrate averaging 37 to 39% Cr_2O_3 (with a Cr:Fe ratio of 1.6:1) was produced from these mines (Thayer, 1969). After the war, the economics of mining and processing and the low grade of the ore rendered this chromite uncompetitive with foreign sources, and the mines were abandoned (Griggs, 1945).

No other significant mining of marine heavy-mineral placer deposits has taken place on the Oregon coast. A small amount of gold and platinum was reportedly recovered years ago from the beaches near the mouth of the Yaquina River, and the magnetite-rich sands of Clatsop Spit (Location 2, Figure 12) southwest of the mouth of the Columbia River have been tested by exploratory drilling. The Hammond deposit on Clatsop Spit consists of a body of black sand approximately 300 m long and 10 m wide, which averages 1 m in thickness. Chemical analyses indicate that the sand contains 20 to 30% Fe and 6 to 10% TiO_2 (Kelly, 1947).

The mineralogy of the southern Oregon beaches and marine terrace deposits has been studied in detail. Twenhofel (1946a) examined the mineralogy of sand on beaches in uplifted marine terraces and in selected streams from Coos Bay to just south of Cape Sebastian. He found that the opaque minerals (magnetite plus undifferentiated ilmenite and chromite) dominate the heavy-mineral suite. The content of either magnetite or undifferentiated ilmenite and chromite ranged from 0 to $>60\%$ of the total sample. Garnet was common in nearly all samples containing a significant proportion of heavy minerals, and accounted for nearly 30% of the total sample from some of the uplifted marine terraces. Zircon was ubiquitous in the black sands; it typically occurred as 1% or less of the total sample, and constituted as much as 4 to 6% of a few samples. Undifferentiated epidote and olivine composed >20% of one sample, but generally did not exceed 10%. Griggs (1945) notes that in the chromite mines the associated heavy minerals include, on a total-sample basis, about 10 to 30% garnet, 3 to 10% ilmentite, 1 to 3% zircon, and traces (<0.5%) of rutile. Chemical analyses of the ore indicate the presence of about 3 to 10% Cr_2O_3, 1 to 3% TiO_2, 0.24 to 1% ZrO_2, and traces of gold and platinum (Hundhausen, 1947). Lissner (1971) reports a surprising amount of topaz (as much as several percent of the total non-opaque heavy-mineral fraction) in beach, stream, and terrace sands in the vicinity of Gold Beach; even if his identification is correct, however, this mineral would appear to be economically insignificant in the bulk sediment. Fischer (1969) suggests that, although the element vanadium has not been reported from the black sands of southern Oregon, it is likely to be present but at grades too low to warrant economic consideration.

Gold on some of the beaches reaches its highest concentrations on the bedrock surface near the back of the beach (Phillips, 1979). Pardee (1934) states that the paystreak at the backshore of present and ancient beaches generally ranges from a few feet to 200 to 300 feet wide and is a maximum of 3 to 4 feet thick. The grain size is small—typical flakes are on the order of 0.01 mm thick and between 0.04 and 0.25 mm in diameter (Clifton, Hubert, and Phillips, 1967; Boggs and Baldwin, 1970; Phillips, 1979). The tenor of stream placers adjacent to the coast is related to the grain size of the gold, which decreases markedly in the downstream reaches (Boggs and Baldwin, 1970). Platinum-group metals are detectable on the modern beaches of southern Oregon, but at very low grades (Phillips, 1979).

The potential for submerged placer deposits on the continental shelf off southern Oregon has been examined by reconnaissance surface sampling (Clifton, 1968; Kulm et al., 1968a; Chambers, 1969; Bowman, 1972; U.S. Bureau of Mines, 1987). The sand-size fraction over a sizable area of the shelf off the Rogue River (Location 3, Figure 12) contains 10 to 20% heavy minerals, and locally the concentration ranges from 20 to 30% (Kulm et al., 1968a). A higher concentration (as much as 33%) occupies a smaller area in water depths of 18 to 55 m off Cape Blanco (Location 4, Figure 12). Kulm et al. (1968a) also report several anomalous concentrations of smaller extent, including one that contained 43% heavy minerals at a water depth of 150 m between Cape Blanco and the Rogue River. Unfortunately, the percentage of sand in the bulk sediment is not given, and the heavy-mineral percentages must be considered as maximum values only.

Associated with the heavy-mineral concentrations on the shelf are magnetic anomalies that may represent relict beach deposits (Kulm et al., 1968a). Linear shore-parallel concentrations (>5 ppb) of fine-grained gold in the surface sediment (Locations 5 and 6, Figure 12) in the area of the heavy-mineral and magnetic anomalies tend to support this conclusion (Clifton, 1968). The largest and richest of these anomalous concentrations of gold (5 to 150 ppb) lies at a water depth of about 40 m off Cape Blanco (Location 4, Figure 12). The stratigraphy and composition of the sediment beneath these anomalies is unknown.

Elsewhere in Oregon the potential for economic marine placer deposits appears to be limited. Twenhofel (1946b) describes the mineralogy of samples collected from modern beaches between Coos Bay and the mouth of the Columbia River. Black sands on these beaches commonly contained 10 to 50% of a combination of ilmenite and chromite, and traces to several percent of zircon. Other than variable quantities (typically 10 to 20%) of garnet (which was undifferentiated from olivine and epidote), no other economically significant minerals were noted. Surface sand from Sand Island in the mouth of the Columbia River contains high concentrations (locally >70%) of magnetite. Twenhofel (1946b) estimates the presence of ". . . thousands of tons of magnetic black sands on Sand Island." Ilmenite is common in these sands, but chromite and zircon are rare. Rogers and Jaster (1962) cite six occurrences of titanium minerals in beach or terrace sands outside southern Oregon; most are based on Twenhofel's (1946b) study. Monazite occurs at most as a trace mineral on Oregon beaches (Phillips, 1979).

The offshore areas of most of central and northern Oregon are no more promising than the beaches. Runge (1966) analyzed the heavy-mineral assemblage in surface-sediment samples on the continental shelf from north of Cape Blanco to north of the mouth of the Columbia River. Opaque minerals (magnetite, ilmenite, and chromite) typically constitute 5 to 25% of the heavy-mineral fraction; garnet a few percent; zircon, hematite, and rutile a trace to 1%; and monazite, sillimanite, and kyanite a trace at most. The concentrations of heavy minerals in the shelf-sand fraction typically are no more than a few percent in this area (Runge, 1966). The shelf area south of the Columbia River appears to be unusually rich in heavy minerals; a broad area (Location 7, Figure 12) in which the heavy-mineral concentration exceeds 25% extends from the shore to a depth of nearly 80 m (White, 1967). Phillips (1979) considers this deposit to be the largest concentration of heavy minerals along the Pacific Coast. Analyses of heavy minerals from samples taken near the mouth of the Columbia River (Kelly, 1947) suggests that the sand fraction of this concentration would consist of about 10 to 15% magnetite, 5% ilmenite, 1 to 3% garnet, and 0.2% zircon, as well as traces of rutile, olivine, hematite, and kyanite.

California

Off the state of California, the continental margin takes two disparate forms. North of Point Conception the coastline is fronted in most places by a shelf 10 to 40 km wide, the outer edge of which is cut by numerous submarine canyons and gullies (Figures 13, 14; McCulloch, this volume; Clarke, this volume). The shelf typically is flat and slopes gently seaward. Numerous small islands (sea stacks) dot the rocky segments of the coast, and the Farallon Islands protrude from the outer shelf off San Francisco. South of Point Conception, in contrast, the margin is broken into a borderland of basins separated by broad ridges; the Channel Islands emerge from the higher of the ridges (Vedder, this volume). The narrow (5- to 20-km-wide) shelf that fronts the coastline is transected by several submarine canyons.

The coast of central and northern California (i.e., north of Point Conception) is generally rugged and is broken only by a few broad river valleys or the entrances to coastal embayments such as San Francisco Bay and Tomales Bay. Extensive beaches are best developed along the lowland parts of the coast that front the river valleys. The coast is almost entirely dominated by waves. Tidal currents are important only near the inlets of the coastal embayments.

No clearly dominant general direction of longshore transport is discernible along the central California coastline; in the northern part longshore drift is to the north, as it is along the coast of Oregon. South of Point Conception (Figure 14), the coast takes the form of broad lowlands separated by widely spaced headlands. Littoral drift along the beaches that front the commonly cliffed lowland areas is to the south. The coastal zone and shelf here comprise five cells or compartments, each of which is bounded on the southern side by a constriction in the shelf caused by the coincidence of a promontory and one or more submarine canyons (Rice, Gorsline, and Osborne, 1976). Wave processes dominate the southern California coast; its configuration relative to the open Pacific Ocean, however, and the protection afforded by the Channel Islands combine to reduce the intensity of the waves compared with the rest of the Pacific Coast of the United States.

The California coast reflects the active tectonism of this margin. Elevated wave-cut terraces are found all along the California coast (Shepard and Wanless, 1971) and are particularly well developed in southern California north of San Diego. The terraces extend to elevations of >250 m. In the Santa Cruz area the terraces are tilted and change laterally from elevated wave-cut platforms in an area of uplift to sedimentary sequences that dip beneath the coastal plain in an adjacent area of subsidence (Bradley and Griggs, 1976; Dupré, Clifton, and Hunter, 1980).

A number of studies address the sedimentology of the modern shelf areas of California. Welday and Williams (1975) provide a general summary of the surficial deposits over the entire margin of the state. Most parts of the shelf follow a pattern whereby a wedge of modern sediment occupies the inner shelf and relict deposits are exposed on the outer shelf. The modern sediment generally changes from sand to mud seaward; the relict deposits typically are coarser (sand or gravel) than the modern mud inshore. The depth of transition from sand to mud differs. On the northern and central California Shelf, it occurs at 50 to 70 m (offshore Eel River Basin—Field, Clarke, and White, 1980; Russian River Shelf—Drake and Cacchione, 1985; Monterey Bay—Yancey, 1968; Chin, 1984). On the southern California Shelf, which is a somewhat more protected setting, the depth of transition occurs at about 20 m (Ventura-Port Hueneme Shelf—Howard and Reineck, 1981; San Pedro Shelf—Gorsline and Grant, 1972; Drake, Cacchione, and Karl, 1985).

The distribution of relict sediment on the outer shelf of California depends in part on the sedimentation rate. On the Eel River Shelf, the sedimentation rate is particularly high, and mud blankets all but a few isolated patches of coarser sediment (Field, Clarke, and White, 1980). On the southern California Shelf relict sands may be absent, as in the Ventura-Port Hueneme area (Howard and Reineck, 1981), or sporadic in occurrence, as on the San Pedro Shelf (Gorsline and Grant, 1972). Elsewhere (as on the Russian River Shelf and in Monterey Bay) the relict deposits form a band along the outer shelf at depths of 100 m or more. In areas of rapid sedimentation, as off the river mouths in northern California, the thickness of Holocene sediment may exceed 50 m (Field, Clarke, and White, 1980). Off San Diego County, Holocene sediment on the shelf forms a shore-parallel prism that reaches a maximum thickness of 10 to 15 m at a water depth of about 20 m and wedges out toward the shelf break (Henry, Berry, and Fischer, 1976).

Where examined in detail with high-resolution acoustical seismic profiling systems, the Quaternary depositional record can be complex. Depositional lobes occur off the mouths of streams such as the Klamath River (Moore and Silver, 1968a) and the Salinas River. The Salinas River lobe displays a complex internal stratigraphy of foresets and erosional surfaces and probably is partly pre-Holocene in origin (Chin, 1984). In Santa Monica Bay in southern California, a complex array of Pleistocene and Holocene depositional units is indicated on high-resolution profiles. Vibracores through some of the Pleistocene units that are exposed at the sea floor show magnetite and ilmenite laminae that suggest a beach origin (Osborne et al., 1980).

Fine- to coarse-grained sand floors the shallow (< 100 m deep) passages between the northern Channel Islands (Booth, 1973). This sediment apparently is derived largely from the erosion of sedimentary deposits on the islands. Wind-driven and tidal currents through these passages can exceed 50 cm/sec and provide a potential mechanism for concentrating heavy minerals into placers.

The continental margin along the state of California receives sediment from streams that drain a variety of terranes and also from wave erosion of coastal rocks. In the north, the Smith and Klamath Rivers drain the Klamath Mountains—an area underlain by a combination of Mesozoic ultramafic and granitic intrusives, and metavolcanic and metasedimentary rocks. To the south, rivers such as the Eel and Russian, flow through the northern Coast Ranges, which are mostly composed of Franciscan rocks; ultramafic intrusive rocks are a minor component of these drainage basins. The Golden Gate marks the mouth of the combined Sacramento and San Joaquin River systems, which drain nearly 40% of the state of California. Most of the sand carried by the rivers is deposited in their lower reaches, and at present only a small amount of fine sand enters San Francisco Bay (Krone, 1979). The Central Valley drainage has issued from the vicinity of San Francisco for the past 600,000 years or so (Sarna-Wojcicki et al., 1985) and, given the changes in sea level during this time, must strongly influence the heavy-mineral assemblage of the shelf and coastline of this area.

Streams to the south of San Francisco reflect more local coastal drainage basins. Streams such as the Salinas River that issue along the coast between the Golden Gate and Point Conception drain areas of the southern Coast Ranges underlain primarily by Tertiary sedimentary rocks and Franciscan lithologies that include a few small ultramafic and Cretaceous granitic intrusives. Streams draining the Transverse Ranges enter the ocean south of Point Conception. These streams drain areas largely underlain by Tertiary sedimentary (and locally volcanic) rocks and, in the San Gabriel Mountains area, Mesozoic granitic intrusives in Precambrian metamorphic rocks. South of Los Angeles in the Peninsular Ranges, the source terrane is dominated by granitic intrusive rock. Coastal erosion of the Tertiary sedimentary rocks that form sea cliffs contributes significantly to the beach and inner shelf sand in this area (Shepard and Wanless, 1971).

In the offshore borderland area of the margin, the rocks that compose the Channel Islands provide the only source of shallow marine sediment. The islands consist of a diversity

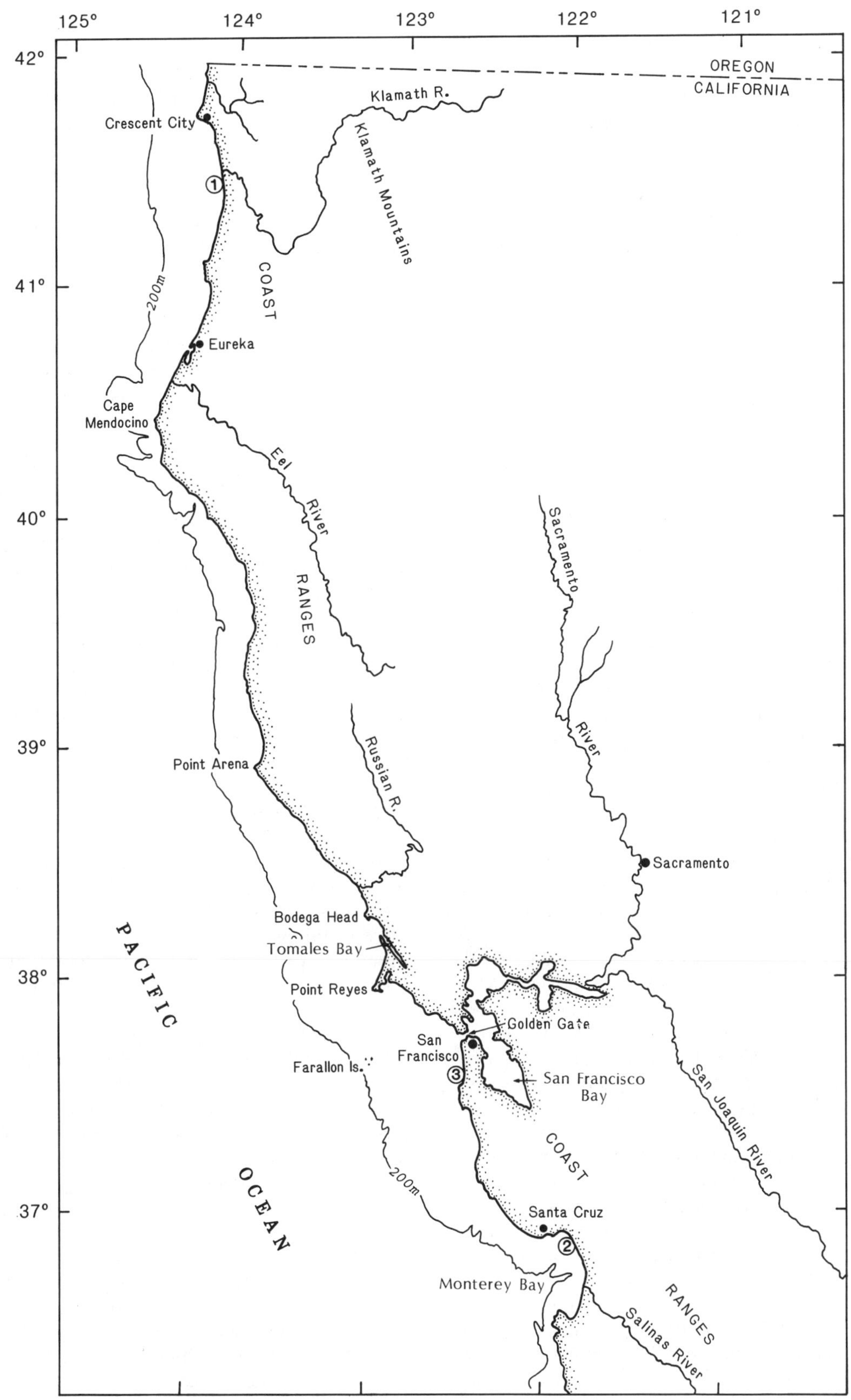

Figure 13. Coastal northern California region. Numbers refer to locations cited in text.

of rock types that reflect their origin in two different tectonostratigraphic terranes (Howell and Vedder, 1981; Vedder, this volume). San Miguel and Santa Rosa Islands are composed of Tertiary and (on San Miguel) Upper Cretaceous sedimentary rocks and Miocene andesitic to dacitic volcanic rocks. San Nicolas Island is formed primarily of sedimentary rocks of Eocene age. Santa Cruz Island is a complex assemblage of metamorphic rocks of the greenschist facies, dioritic intrusive rocks, Tertiary sedimentary rocks derived in part from a basement composed of ophiolite and rocks of greenschist and blueschist facies, and Miocene volcanic rock. Anacapa, Santa Barbara, and San Clemente Islands are formed almost entirely of Miocene volcanic rock. Santa Catalina Island is a complex assemblage of metamorphic rocks of the blueschist, greenschist, and amphibolite facies, quartz dioritic and dacitic intrusive rocks, serpentinite, Franciscan rocks, and Tertiary sedimentary and volcanic rocks (Howell and Vedder, 1981).

The heavy minerals in the modern marine sediment on the central and southern California margin have received much study, both on regional (Azmon, 1960; Judge, 1970; Gorsline, Grant, and Booth, 1971) and local scales. Fewer studies have been made in northern California. Kulm et al. (1968b) examined the heavy-mineral suite from samples of the Klamath and Smith Rivers. The nonopaque fraction from the Klamath River was dominated (76%) by amphiboles; garnet-group minerals and rutile constituted 1% each of the assemblage. The nonopaque fraction from the Smith River contained less amphibole (42%), more pyroxene (21%), and much more olivine (28%, as compared with a trace in the Klamath River suite). The Smith River suite also included 2% of the garnet-group minerals and a trace of zircon. Surprisingly few data seem to exist for the heavy-mineral assemblage contributed by the Eel River. The heavy-mineral fraction of samples collected offshore on the Eel River Shelf is dominated by amphiboles, epidote, opaque minerals, and pyroxenes; other minerals constitute fewer than 7% of the assemblage (Field, Clarke, and White, 1980).

In central and southern California, the heavy-mineral assemblages define mineralogical provinces (Yancey and Lee, 1972). In southern California, the mineralogy of the beach sand more or less parallels that of adjacent coastal streams (Gorsline, Grant, and Booth, 1971), whereas on parts of the central California coast the correlation between beach- and river-sediment mineralogy is less well defined (Judge, 1970).

A number of studies have established the mineralogy of beach and shelf sands in this area. On the shelf off the Russian River, the total sediment typically contains a few percent each of pyroxenes, amphiboles, and epidote, 1 to 2% each of zircon and opaque minerals, and a trace of chromite (Klise, 1984). Shelf sediment west of the Golden Gate resembles that within San Francisco Bay—the nonopaque fraction is dominated by amphiboles and pyroxenes, zircon and garnet occur as traces, and opaque minerals compose < 10% of the heavy-mineral fraction (Hall, 1965). Cherry (1965) examined the mineralogy of coastal sediment north of San Francisco and concluded that Point Reyes and Bodega Head serve as barriers to longshore transport. Hutton (1959) provides a detailed analysis of the heavy-mineral assemblage in beach sands from the coast south of San Francisco to Monterey Bay—an area marked by variable and locally high concentrations of zircon in the range of 10 to 15% of the total beach-sand sample at several locations (based on a combination of his Figures 2 and 3). Sayles (1966) and Yancey (1968) analyzed the heavy minerals in beach and shelf sands, respectively, in Monterey Bay; the nonopaque assemblages consist mostly of pyroxenes and amphiboles and small amounts (1 to 3%) of zircon. Although quantitative data are not available, the zircon probably constitutes only a trace in the bulk sample.

Judge (1970) provides a detailed analysis of the heavy-mineral assemblages in the fine (0.063–0.125 mm) fraction of beach and nearshore sand at 18 locations between Monterey Bay and Los Angeles. On beaches north of Lompoc, zircon and monazite each constitute several percent of the heavy-mineral assemblage calculated; on a total sample basis, however, the proportion of these minerals diminishes to <0.1%. Opaque minerals typically constitute 20 to 50% of the fine heavy-mineral fraction, but these fine opaque minerals everywhere compose < 1% of the total sample. No data are presented on the heavy-mineral content of the coarser sand fractions. Trask (1952) distinguished between ilmenite, magnetite, and leucoxene composites in his analysis of heavy-mineral assemblages in stream and beach sands between Monterey Bay and Santa Barbara. Several streams in this area (Pismo Creek, Arroyo Grande, and Jalama River) are notable in containing 40 to 50% of ilmenite, 5 to 20% leucoxene composites, and about 10% zircon in the fine (0.063–0.125 mm) heavy-mineral fraction. The heavy-mineral fraction in these samples ranged mainly between 0.2 and 3.0% of the total sample (Trask, 1952). Opaque minerals in offshore sand in Estero Bay typically constitute 20 to 40% of the heavy-mineral fraction (Dingler et al., 1982); analyses by the second author indicate that the opaque minerals consist mostly of chrome spinel.

Distinct trends are seen in the heavy-mineral assemblages on beaches between Point Conception and the United States-Mexican border (Gorsline, Grant, and Booth, 1971). Hornblende increases to the south, whereas epidote decreases. Zircon diminishes from approximately 10% of the heavy-mineral suite in the beaches near Point Conception to 1% or less in the beaches south of Point Fermin. Unfortunately from an economic standpoint, this pattern is superimposed on a trend whereby the total heavy-mineral percent increases from less than 3% in the beaches near Point Conception to values in excess of 15% in beaches south of Point Fermin (Gorsline, Grant, and Booth, 1971).

Several studies describe the heavy-mineral assemblages in beach sands at various locations on the borderland. Trask (1952) notes that ilmenite consistently occurs in the range of 10 to 30% of the total fine heavy minerals in beaches near the Santa Barbara harbor. His samples showed also that "leucoxene composites" could constitute 25% of the heavy minerals, and that zircon, although typically in the range of a few percent, could locally exceed 20%. Handin (1951) examined beach sands between the city of Ventura and Palos Verdes Hills. Samples from beaches in the south end of Santa Monica Bay contained as much as 90% of heavy minerals in the 0.063- to 0.125-mm size fraction. His data do not indicate what percentage of the total sample this represents. Magnetic grains accounted for 20 to 40% of the heavy-mineral fraction, and the nonmagnetic fraction consisted of 50 to 75% of opaque grains that may be ilmenite. Zircon in these samples occurs in the range of 1 to 10% of the nonmagnetic heavy-mineral fraction.

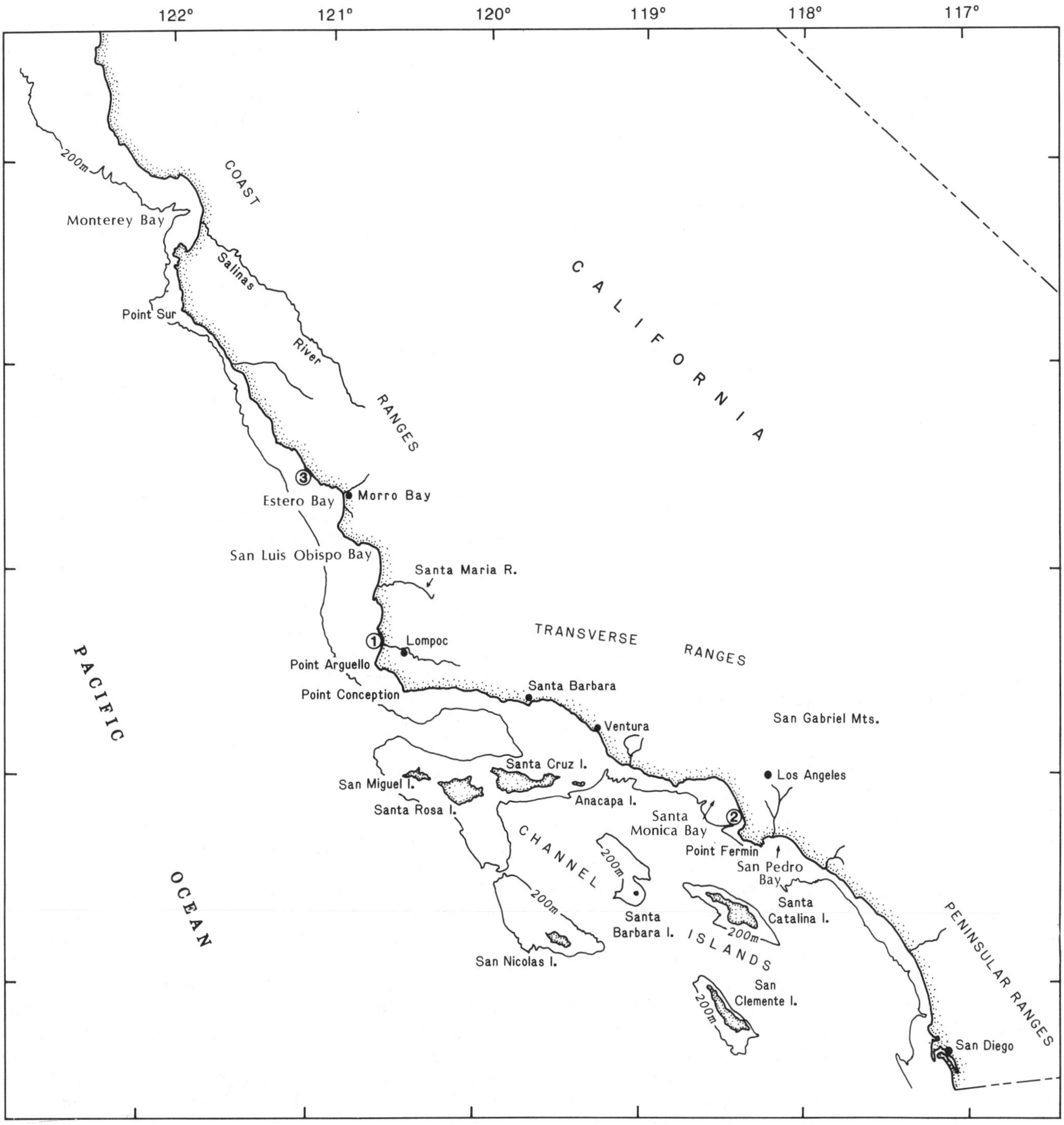

Figure 14. Coastal southern California region. Numbers refer to locations cited in text.

Elsewhere in southern California the beach and nearshore sands contain about 10% heavy minerals in bulk samples (Inman, 1953). Hornblende constitutes about 60% of the heavy-mineral fraction, and opaque minerals, epidote, sphene, and garnet are also abundant. Zircon, rutile, and tourmaline are found in minor amounts.

The heavy-mineral composition of beach sands on the Channel Islands has received relatively little study. Scholl (1960) noted that augite was the predominant heavy mineral in shelf sands off Anacapa Island. Zircon, garnet, ilmenite, and magnetite are found only in trace amounts in this area. Azmon (1960), in contrast, found that epidote and opaque minerals were the principal heavy minerals in beach and offshore samples collected in the vicinity of Santa Rosa and Santa Cruz Islands.

Sources for economically significant placer minerals on the continental margin of California are rather scattered (Smith et al., 1967, 1971, 1974; Albers and Fraticelli, 1984). The Klamath Mountains provide a source of gold, platinum, and chromite to the adjacent coastal areas. A small amount of lode gold has been found in the Santa Cruz Mountains (California Mining Bureau, 1888b). The northern Coast Ranges contain numerous sparsely scattered occurrences of chromite. The Sacramento and San Joaquin Rivers drain well-established gold-mining districts, but at such distances from the coast as to render them unlikely as sources for coastal placers. The southern Coast Ranges, south of San Francisco Bay, contain numerous occurrences of chromite (Walker and Griggs, 1953) near the ocean north of San Luis Obispo Bay and localized occurrences of gold. Pods of nephrite jade crop out in coastal cliffs between Morro Bay and Point Sur (Wilson and Mero, 1966), where they provide a source for this semiprecious mineral. The San Gabriel Mountains provide a distinct source of gold. If they are also the source of the titaniferous magnetite in beach placers at the southern end of Santa Monica Bay (Hess, 1927), the mountains are a potential source of the vanadium in titaniferous magnetite there (Fischer, 1962). The rocks associated with the southern California batholith in the Peninsular Ranges are potential sources of gold and tungsten minerals. A variety of gemstones are found in pegmatites in the Peninsular Range (Wright, 1957) but presumably are too far from the coast to serve as sources for any placer gem minerals.

A number of alluvial placers of gold occur in the California streams that drain the Klamath Mountains (Averill, 1941, 1946). Many also contain some platinum-group metals (Lowell, 1916). Gold was mined on a small scale from alluvial placers in the Santa Cruz Mountains area during the last century (California Mining Bureau, 1888b). Few alluvial placers have been noted in coastal areas elsewhere in the state. The Peninsular Ranges, despite having a number of mineable lode occurrences of gold (Clark, 1970), seem to lack alluvial placers. The only reported stream placer is in the Santa Ana Mountains about 20 km from the coast. The apparent absence of alluvial placers in this area may be due to the diminished rainfall in this part of the state.

Gold and, to a lesser extent, platinum have been mined from a number of beaches in California. The richest was in the area of the Gold Bluffs south of the mouth of the Klamath River (Location 1, Figure 13). Eyewitness accounts describe visible fine-grained gold on the surface of the beach here after a severe storm (Chase, 1874; Lowell, 1916). The beaches evidently derived their gold from the auriferous Pleistocene alluvial gravel that formed the bluffs behind the beach. The California Mining Bureau (1888a) estimated that a million dollars worth of gold may have been extracted from these beaches, but it was noted even at this date that production was diminishing. Attempts at large-scale operations with machinery proved unprofitable (Averill, 1941). An interesting attempt at offshore mining is described by Chase (1874), who noted a belief in the possibility of quantities of coarser grained gold in the surf zone and beyond and states, "I believe an expedition was fitted out to obtain this sand by means of a diving-bell or some such apparatus, which did not result favorably." The early mining efforts appear to have extracted much of the beach gold; little can be found here today.

Other attempts to recover gold and platinum from the beaches of Humboldt and Del Norte Counties were largely unprofitable (Logan, 1919). Lowell (1916) describes an ambitious attempt in 1913–1914 to dredge gold from black sands that lay landward from the present driftwood line 2 miles south of Crescent City. This effort, like many others, ended in failure (O'Brien, 1952). Averill (1941) notes that, although some platinum was extracted with the gold from the beach placers at the Gold Bluffs, most attempts at platinum recovery proved uneconomical because of the large amount of labor involved.

Elsewhere in the state, auriferous sands were mined from the modern beach near Aptos (Location 2, Figure 13), southeast of Santa Cruz, and from an ancient beach in a downwarped Pleistocene deposit (California Mining Bureau, 1888b). Hubbard (1943) notes that small amounts of gold and platinum were washed for many years from the beaches in this area, although probably with very little profit. Gold was also mined from 1938 to 1950 (by individuals) from the beach south of San Francisco (Location 3, Figure 13). The recorded production in the years 1938 to 1940 was valued at about $13,000 (Clark, 1970). Appreciable quantities of gold were mined in the last century from beaches (Location 1, Figure 14) between Point Arguello and the mouth of the Santa Maria River in Santa Barbara County. Clark (1970) notes that most of the county's production during its peak year ($41,000 in 1889) came from these beaches.

The other commodity mined from California beaches is a combination of titaniferous magnetite and ilmenite. Beaches in northern Monterey Bay and southern Santa Monica Bay served sporadically as a source of iron and titanium in the first half of this century (Youngman, 1930). The beaches in northern Monterey Bay (Location 2, Figure 13) consisted mostly of magnetite and were said to carry about 16% of TiO_2 in the form of ilmenite (Hess, 1924). In the Redondo Beach deposit (Location 2, Figure 14) in Santa Monica Bay, the sand averaged 7% ilmenite for a distance of 3000 ft (Tucker, 1927). In 1927, concentrated ilmenite shipped from this beach was equivalent to 3500 tons carrying 50% of TiO_2 (Hess, 1927). The primary purpose of this production was to provide a source for pigment. According to Oakeshott (1950), none of the operations using sand for its titanium content was active by 1948. In 1955, one plant was producing granules for roofing paper from the black sand at Redondo Beach (Stinson, 1957).

Other minerals having possible economic potential have been noted in California beaches. Chromite occurs in local placer deposits on the coast of San Luis Obispo County (Walker and Griggs, 1953) and was noted in the black sands at Aptos (Tucker, 1927) and in Del Norte and San Mateo Counties (Wilson and Mero, 1966). In a study of possible recovery of by-product heavy minerals from sand and gravel operations in central and southern California (Gomes, Martinez, and Wong, 1980), chromite was found in the amount of 3 pounds per ton of beach sand from Morro Bay. Nephrite jade occurs as pebbles on the beach at and near Cape San Martin (Location 3, Figure 14) in southern Monterey County (Wright, 1957), where it is collected by hobbyists. Other minor occurrences of economically interesting minerals are described by Wilson and Mero (1966) and Phillips (1979).

EVALUATIONS

In this section we evaluate the potential for economic deposits of placer minerals in each of the areas discussed in the previous section. We also assess the existing data base and indicate needs for additional information to establish an adequate understanding of the placer mineral potential.

Arctic Region

The extensive nearshore sands and relatively strong coastal currents during times of open water in the Chukchi Sea provide an area where the mechanisms for placer concentration are well developed. However, the lack of economic heavy-mineral sources in the neighboring source terrane renders this area unlikely to contain valuable heavy-mineral placers. In the Beaufort Sea, coastal currents are not as strong as in the Chukchi Sea, and sand concentrations tend to be patchy. Ice-gouging in both the Chukchi and Beaufort Seas is the principal factor in thoroughly mixing the bottom sediment, thereby destroying any placers that might have begun to concentrate.

Northern Bering Sea–Seward Peninsula

The most obvious potential for placer mining on the continental margin is in the northern Bering Sea and on the Seward Peninsula. The modern-beach and older, emergent-beach deposits near Nome have been significant sources of gold, and activity continues to the present time. Concentrations of gold have been identified in offshore gravel (Nelson and Hopkins, 1972), and these deposits very likely will be actively explored in the future. As of 1976, two offshore leases were held for gold exploration in the Nome area and a third was held in the Golovin Bay area to the east (Conwell, 1976).

Other areas with economic potential adjacent to the Seward Peninsula include the inner shelf southeast of Cape Prince of Wales, where submerged alluvial and current-winnowed sea-floor deposits may contain cassiterite, gold, and possibly tungsten minerals (Nelson, 1971). The shoal and barrier northeast of Cape Prince of Wales, as noted by Tagg (1979), may contain significant quantities of minerals derived from the Seward Peninsula. The Yukon-Kuskokwim Delta area has little potential for placer deposits because of the distance from source areas, the fine size of the sediment, and the general absence of placer-producing processes along the present shoreline.

Needs for further study of offshore placer deposits in the Nome area are listed by Nelson and Hopkins (1972). In particular, the auriferous gravels need to be sampled throughout their thickness and particularly at their bases. High-resolution acoustic seismic profiles may delineate the locations of submerged alluvial channels on the inner shelf southeast of Cape Prince of Wales, and the depth of burial by subsequent deposition by glaciers or marine processes. The barrier islands and shoals northeast of Cape Prince of Wales should be sampled to establish their heavy-mineral content.

Southern Bering Sea

Marine placer deposits of platinum may occur on the shelf west of the Goodnews Bay mining district. It seems generally accepted that Pleistocene glaciation has stripped any placer platinum from the western side of Red Mountain—the source of the alluvial placers worked a short distance inland—and carried it onto the shelf. Reworking of these glacial deposits during the ensuing rise in sea level may have produced offshore concentrations of platinum analogous to those of gold that lie on the northern Bering Sea Shelf near Nome. Submerged alluvial deposits, if not deeply buried by glacial drift, also may hold a potential for offshore platinum resources. As of 1976, a lease was held covering nearly 3000 acres of offshore area near Goodnews Bay (Conwell, 1976).

A high-resolution acoustic profiling survey could possibly resolve the Quaternary depositional record on the shelf and delineate targets for detailed sampling. Although small quantities of gold have been found in beaches facing Hagemeister Strait (Berryhill, 1963), the general absence of onshore alluvial placers or lode deposits suggests a low potential for economically significant marine gold placers in this area.

Reconnaissance sampling of the beaches along the northern side of the Alaskan Peninsula has indicated the presence of placer minerals but not in sufficient concentrations to stimulate economic interest. These samples extended to depths of about 9 m but did not reach bedrock (Berryhill, 1963); it is questionable whether greater concentrations will be found in future sampling programs.

Alaska Peninsula–Aleutian Island Area

The shelf area in the immediate vicinity of Unga Island on the south side of the Alaskan Peninsula is a potential site for submerged gold placers derived from the lode deposits that have been commercially worked extensively on the island. A combined sampling and acoustic profiling survey might further delineate the possibility for such placers.

The rest of the area, particularly the Aleutian Island chain, would seem to have a very low potential for economically significant concentrations of placer minerals because of the predominance of mafic volcanic rocks.

Gulf of Alaska

The process of glaciation largely controls the possibility of economic placer deposits in the Gulf of Alaska. In the western part of the area, glaciation has stripped most of the loose sediment from the coastal area. Any coastal mining of modern placers is on a small scale. The material removed by glaciers now lies offshore on the shelf, where subsequent reworking by marine processes may have formed economically interesting concentrations. The gravels and coarse sands reported to be present on elevated platforms on the sea floor southeast of Kodiak Island are likely to be auriferous. A comparison of such deposits with those in the Nome area illustrates the limited possibility for economically recovering this gold. The concentrations offshore of Nome are proximal to a particularly rich source terrane, are reasonably well defined, and occur in shallow water in a relatively benign oceanographic setting. In contrast, the western shelf of the Gulf of Alaska lies off a source area in which gold is much more diffuse, concentrations of gold on the shelf are but conjectural at present, and such concentrations would occur in several tens of meters of water in a particularly inhospitable marine environment. Inasmuch as the deposits offshore from Nome have yet to be mined, any future production from the Kodiak Shelf seems highly unlikely.

In the eastern Gulf of Alaska, glaciation presents a different problem. Modern glaciers extend into the present coastal area and contribute a large volume of fine sediment. Owing to the high rate of sedimentation here, any inner-shelf placers formed by the transgressive reworking of glacial drift are likely to be buried. The extensive beaches in the Yakutat and Lituya Bay areas will probably remain of moderate interest and warrant a careful regional examination.

Sampling of the coarser sediment in lower Cook Inlet might indicate interesting concentrations of placer minerals.

Such sampling would be particularly useful where it is clear that the glacial till and outwash had been eroded and reworked into the modern deposits.

Southeastern Alaska

Southeastern Alaska contains numerous sources for placer minerals, but the intense alpine glaciation has removed most of the unconsolidated material that could serve as secondary sources and has created a topography in which concentrating mechanisms are limited. Areas of potential interest include locations where strong tidal flow occurs above shoal areas, particularly those composed of glacial till. Reconnaissance seismic profiling of the inland waterways could probably delineate prospective areas for further sampling.

Washington

The Washington coastal area is rather limited in primary sources for economic placer minerals. In spite of known gold- and chromite-bearing deposits in the Cascade Range, no extensive economic coastal placers have been found in the Puget lowlands region. The deep fjords appear to have made this area unsuitable for developing placers. Similarly, no beaches along the Strait of Juan de Fuca have shown any development of placers.

All known high-grade black sand deposits on the Washington coast are volumetrically small. Huntting (Foreword in Thorsen, 1964) indicates that as of 1964 no production of black sands on the Washington coast has been recorded. Most of the outer coast of Washington is now in national parks and wildlife refuges, although not all potential beach-placer areas lie within these boundaries.

The Washington Shelf from the Columbia River to Cape Flattery contains several surficial heavy-mineral concentrations that may have formed as beach deposits during the Holocene sea-level rise (Phillips, 1979); most of these are seaward of present rivers and estuaries. However, it is unknown whether these concentrations persist to significant depths. A drilling or coring program in the areas of highest surficial heavy-mineral concentrations, particularly on the shelf north of the mouth of the Columbia River, is necessary to determine both the lateral and vertical extent of heavy-mineral concentrations.

Oregon

The greatest potential for marine placer minerals is in southwestern Oregon, adjacent to the Klamath Mountains. It is likely that, in a national emergency, the elevated terrace deposits might be mined again for chromite. However, these deposits are relatively small, and a significant amount of the richest concentrate has been removed by previous mining.

The modern beaches may retain some potential for concentrations of gold and other minerals. Although many of the beaches were extensively worked during the 19th century, much gold possibly remains. Drilling in 1966 with a truck-mounted auger indicated depths to bedrock in excess of 10 m on the beach at the mouth of Whisky Run, where much of the mining occurred. The primitive techniques used in early mining probably were not effective in extracting gold from the deeper beach sand. It should be noted that today these beaches are a popular tourist attraction, and competing land uses would undoubtedly complicate attempts at mining.

The potential for economic deposits on the continental shelf off southern Oregon is problematic. Anomalous concentrations of both heavy minerals and fine gold occur locally in the surficial sediment and may mark the site of ancient beach development during the pre-Holocene rise in sea level. No information is available on the composition of the sediment beneath these surficial concentrations and, accordingly, it is not clear whether they overlie potentially rich concentrations of black sand or are merely veneers on the sea floor.

Bowman (1972) argues that any concentrations produced during the most recent transgression of the sea are likely to be richer than the marine-terrace deposits which have served as secondary sources since their emergence. On the other hand, Komar and Wang (1984) suggest that the placers on modern beaches probably have accumulated over a long period of time because of a present balance between tectonic uplift and sea-level rise. If so, the modern beaches may be more enriched in placer minerals than those that formed during shorter intervals of time during the transgression.

The quantity of gold in the offshore deposits may not be sufficient to warrant attempts to recover it. An estimated 2200 oz of gold was recovered during the years 1903–1929 from modern and uplifted beaches in southern Oregon (Pardee, 1934); if this constitutes a fourth or more of the total amount of gold removed from these deposits since the initiation of mining, then the maximum amount of total gold recovered in southern Oregon is 10,000 oz. If a similar amount of gold remains in shelf deposits, its value at today's prices would be several million dollars. Because any offshore gold may be disseminated among a number of relict beach deposits, and offshore mining is a high-cost operation, it is questionable whether economic recovery of offshore gold is a realistic possibility. The reconnaissance surveys for offshore black-sand deposits in this area begun in the 1960s need to be completed and, most importantly, a coring or drilling program in prospective target areas must be undertaken before the economic potential of the shelf deposits can be adequately evaluated.

The other area in Oregon with possible economic placer deposits is on the shelf south of the Columbia River. Again, a coring or drilling program is necessary to determine whether the relatively high concentrations of heavy minerals in the surface sediment indicate a significant deposit in the sediment below, or are a veneer concentrated today by waves and currents as suggested by Phillips (1979).

California

The potential for economic placer deposits in northern California resembles that of southern Oregon. Beaches have been mined for gold, but the resources appear to be exhausted. As with southern Oregon, anomalous concentrations of heavy minerals and fine gold are found in surface sediment on the inner shelf, but at grades too low to be of economic significance. As noted by Moore and Silver (1968b), drilling or coring is necessary to establish the distribution of gold and other heavy minerals at depths within the deposits. The inner shelf north of the Klamath River, where large volumes of auriferous gravel have apparently been eroded by an encroaching sea, is particularly promising. Onshore black sands in the Crescent City area will probably continue to be of intermittent interest because of their accessibility.

Elsewhere in the state, beach deposits that have been mined in the past in southern California (southern Santa Monica Bay) are now located in residential or recreational areas and are unlikely to be further developed. Pleistocene deposits exposed on the inner shelf of southern Santa Monica Bay, probably the immediate source of heavy minerals in the modern deposits, may have some potential. Further vibracoring of the type conducted by Osborne et al. (1980) would be required to determine if the placer minerals are sufficiently concentrated in these deposits to warrant further interest.

The coastal area north of Point Conception deserves further examination. Studies of the heavy-mineral suites in beach and stream sediment between Point Conception and Point Sal indicate a high percentage of ilmenite relative to magnetite, and a relative abundance of zircon and monazite within the heavy-mineral fraction. Sources of chromite are found in the headwaters of the Santa Ynez River, and small quantities of chromite have been noted in the beach sands of Morro Bay. No economically interesting concentrations of heavy minerals are noted from the beaches in this area, but a further analysis, particularly of the richest concentrations on or within the beaches, may be indicated. If economically interesting concentrations of heavy minerals are found in present beaches in any of these areas, an investigation of the offshore possibilities would seem worthwhile.

SUMMARY

The potential for placer deposits on the continental margin of Alaska and the Pacific Coast states has been explored in only the most preliminary way. Initial indications are that the potential in different areas ranges from high (the Seward Peninsula) to low (the Arctic and Aleutian shelves) along Alaska and most of the margin of the Pacific Coast states. Moderate possibilities exist for platinum placers in southern Kuskokwim Bay and for heavy-mineral concentrates on beaches in the eastern Gulf of Alaska and in the inland passages of southeastern Alaska. Gold placers may yet be found in areas of reworked Pleistocene deposits on the shelf of southeastern Alaska and possibly on the southern Oregon and northern California shelves.

Very few areas have been systematically sampled in three dimensions, and few of the analytical data are relevant to resource analysis (see Tables 2–5). A telling commentary on the state of sampling lies in the fact that the analyses reported by Day and Richards (1906b) for black sands of Oregon and California continue to be cited as a standard reference on the economic mineralogy of these sands. A rigorous sampling program is critical to evaluating the potential for heavy-mineral placers in most of the coastal and shelf areas of Alaska and the Pacific States.

Placer deposits on the Pacific and Alaskan margins have historically been mined for a single commodity or, at most, a combination of the precious metals. As a consequence, the focus has been almost entirely on the most enriched concentrations, and the recovery operations have typically been of small scale. Recovery on a large scale awaits economic developments that lie beyond the scope of this report. Consideration of by-product minerals may determine whether any of the areas described here are mined for placer minerals in the future.

ACKNOWLEDGMENTS

This paper has benefited from useful comments and suggestions from many people. We thank A. E. Grosz and R. L. Phillips for their reviews of the entire manuscript, and P. R. Carlson, M. A. Hampton, R. E. Hunter, H. A. Karl, C. H. Nelson, and T. L. Vallier for their reviews of specific sections.

REFERENCES CITED

Adams, J. W., and M. H. Staatz, 1973, Rare-earth elements, *in* D. A. Brobst and W. P. Pratt, eds., United States mineral resources: U.S. Geological Survey Professional Paper 820, p. 547–556.

Albers, J. P., and L. A. Fraticelli, 1984, Preliminary mineral resource assessment map of California: U.S. Geological Survey Mineral Investigations Resource Map MR-88, scale 1:1,000,000.

Anderson, F. E., 1968, Seward terminus of the Vashon continental glacier in the Strait of Juan de Fuca: Marine Geology, v. 6, no. 6, p. 419–438.

Anderson, R. J., 1971, The role of paleogeography in the location of marine placers: Parkville, Victoria, Australia, 1971 Annual Conference Australasian Institute of Mining and Metallurgy, New Zealand, Paper 16, 18 p.

Anderson, R. J., 1982, Mineral deposits, *in* M. L. Schwartz, ed., The encyclopedia of beaches and coastal environments: Stroudsburg, Pennsylvania, Hutchinson Ross Publishing Company, p. 544–545.

Arnold, R., 1905, Gold placers on the coast of Washington: U.S. Geological Survey Bulletin 260, p. 154–157.

Averill, C. V., 1941, Mineral resources of Humboldt County: California Journal of Mines and Geology, v. 37, p. 499–528.

Averill, C. V., 1946, Placer mining for gold in California: California Division of Mines Bulletin 135, 377 p.

Azmon, E., 1960, Heavy minerals in sediments off southern California: Los Angeles, University of Southern California unpublished Ph.D. thesis, 148 p.

Baldwin, E. W., 1976, Geology of Oregon (revised edition): Dubuque, Iowa, Kendall/Hunt Publishing Company, 147 p.

Barnes, P. W., and E. Reimnitz, 1974, Sedimentary processes on Arctic shelves off the northern coast of Alaska, *in* J. C. Reed and J. E. Sater, eds., The coast and shelf of the Beaufort Sea: Arlington, Virginia, Arctic Institute of North America, p. 439–476 (with discussion).

Bateman, A. M., 1955, Economic mineral deposits: New York, John Wiley and Sons, 916 p.

Bates, R. G., and H. Wedow, Jr., 1953, Preliminary summary report of thorium-bearing mineral occurrences in Alaska: U.S. Geological Survey Circular 202, 13 p.

Beiersdorf, H., H. R. Kudrass, and U. von Stackelberg, 1980, Placer deposits of ilmenite and zircon on the Zambesi Shelf: Geologisches Jahrbuch, Reihe D, heft 36, 85 p.

Beikman, H. M. (compiler), 1980, Geologic map of Alaska: U. S. Geological Survey, 2 sheets, scale 1:2,500,000.

Berg, H. C., and E. H. Cobb, 1967, Metalliferous lode deposits of Alaska: U.S. Geological Survey Bulletin 1246, 254 p.

Berryhill, R. V., 1963, Reconnaissance of beach sands, Bristol Bay, Alaska: U.S. Bureau of Mines Report of Investigations 6214, 48 p.

Blair, W. N., N. J. Page, and M. G. Johnson, 1977, Map and list of reported occurrences of platinum-group metals in the conterminous United States: U.S. Geological Survey Miscellaneous Field Studies Map, MF-861, 2 sheets, scale 1:5,000,000.

Boggs, S., Jr., and E. M. Baldwin, 1970, Distribution of placer gold in the Sixes River, Oregon—A preliminary report: U.S. Geological

Survey Bulletin 1312-I, 27 p.

Booth, J. S., 1973, Textural changes as an indicator of sediment dispersion in the northern Channel Island passages, California: Journal of Sedimentary Petrology, v. 43, p. 238–250.

Bortnikov, V. S., 1970, Mineral composition of the coarse silt fraction of recent sediments in the northwestern part of the Gulf of Alaska, *in* P. A. Moiseev, ed., Soviet investigations in the northeast Pacific, Part V, p. 77-82.

Bouma, A. H., and M. A. Hampton, 1976, Preliminary report on the surface and shallow subsurface geology of lower Cook Inlet and Kodiak Shelf, Alaska: U.S. Geological Survey Open-File Report 76-695, 36 p.

Bouma, A. H., M. A. Hampton, M. P. Wennekens, and J. A. Dygas, 1977, Large dunes and other bedforms in lower Cook Inlet, Alaska: Houston, Texas, 9th Annual Offshore Technology Conference Proceedings, v. I, p. 79–90.

Bowman, K. C., Jr., 1972, Sedimentation, economic enrichment and evaluation of heavy mineral concentrations on the southern Oregon continental margin: Corvallis, Oregon, Oregon State University unpublished Ph.D. thesis, 136 p.

Bradley, J. S., 1957, Differentiation of marine and sub-aerial sedimentary environments by volume percentage of heavy minerals, Mustang Island, Texas: Journal of Sedimentary Petrology, v. 27, p. 116–125.

Bradley, W. C., and G. B. Griggs, 1976, Form, genesis and deformation of central California wave-cut platforms: GSA Bulletin, v. 84, p. 1685–1698.

Brady, L. L., and H. E. Jobson, 1973, An experimental study of heavy-mineral segregation under alluvial-flow conditions: U.S. Geological Survey Professional Paper 562-K, 38 p.

Brenninkmeyer, B. M., 1978, Heavy minerals, *in* R. W. Fairbridge and J. Bourgeois, eds., Encyclopedia of sedimentology: Stroudsburg, Pennsylvania, Dowden, Hutchinson and Ross, Inc., p. 400–402.

Brooks, H. C., and L. Ramp, 1968, Gold and silver in Oregon: Oregon Department of Geology Mineral Industries Bulletin 61, 337 p.

Brooks, H. C., and L. Ramp, 1969, Gold and silver, *in* Mineral and water resources of Oregon: U.S. 90th Congress, 2nd Session, Committee on Interior and Insular Affairs committee print, p. 135–143.

Bruun, P., 1962, Sea level rise as a cause of shore erosion: Journal of Waterways and Harbors (American Society of Civil Engineers Proceedings), v. 88, p. 117–130.

Burk, C. A., 1965, Geology of the Alaska Peninsula—Island arc and continental margin: GSA Memoir 99, 250 p.

Cacchione, D. A., and D. E. Drake, 1979, Sediment transport in Norton Sound, Alaska: U.S. Geological Survey Open-File Report 79-1555, 90 p.

Cacchione, D. A., D. E. Drake, and P. Wiberg, 1982, Velocity and bottom stress measurements in the bottom boundary layer, outer Norton Sound, Alaska: Geologie en Mijnbouw, v. 61, p. 71–78.

California Mining Bureau, 1888a, Humboldt County: California Journal of Mines and Geology, v. 8, p. 216–223.

California Mining Bureau, 1888b, Santa Cruz County: California Journal of Mines and Geology, v. 8, p. 550–556.

Carlson, P. R., and H. A. Karl, 1981, Seafloor geologic hazards, sedimentology and bathymetry: Navarin Basin province northwestern Bering Sea: U.S. Geological Survey Open-File Report 81-1217, 149 p.

Carlson, P. R., and H. A. Karl, 1984, Discovery of two new large submarine canyons in the Bering Sea: Marine Geology, v. 56, p. 159–179.

Carlson, P. R., and B. F. Molnia, 1975, Preliminary isopach map of Holocene sediments, northern Gulf of Alaska: U.S. Geological Survey Open-File Report 75-507, scale 1:500,000.

Carlson, P. R., and W. C. Schwab, 1982, Northern Gulf of Alaska environmental geology, *in* T. R. Bruns, ed., Hydrocarbon resource report for proposed OCS lease sale 88, southeastern Alaska, northern Gulf of Alaska, Cook Inlet, and Shelikov Strait, Alaska: U.S. Geological Survey Open-File Report 82-928, p. 73–86.

Carlson, P. R., T. R. Bruns, B. F. Molnia, and W. C. Schwab, 1982, Submarine valleys in the northeastern Gulf of Alaska: Characteristics and probable origin: Marine Geology, v. 47, p. 217–242.

Carlson, P. R., B. F. Molnia, S. C. Kittleson, and J. C. Hampson, Jr., 1977, Map of distribution of bottom sediments on the continental shelf, northern Gulf of Alaska: U.S. Geological Survey Miscellaneous Field Studies Map MF-876, 2 sheets.

Chambers, D. M., 1969, Holocene sedimentation and potential placer deposits on the continental shelf off the Rouge River, Oregon: Corvallis, Oregon, Oregon State University unpublished M.S. thesis, 83 p.

Chase, A. W., 1874, The auriferous gravel deposit of Gold Bluffs: American Journal of Science, Third Series, v. 7, no. 40, p. 379–384.

Cherry, J., 1965, Sand movement along a portion of the northern California coast: U.S. Army Coastal Engineering Research Center Technical Memorandum 14, 125 p.

Chin, J. L., 1984, Late Quaternary coastal sedimentation and depositional history, south-central Monterey Bay, California: San Jose, California, San Jose State University unpublished M.S. thesis, 130 p.

Clark, W. B., 1970, Gold districts of California: California Division of Mines and Geology Bulletin 193, 186 p.

Clarke, S. H., Jr., 1987, Geology of the California continental margin north of Cape Mendocino, this volume.

Clifton, H. E., 1968, Gold distribution in surface sediments on the continental shelf off southern Oregon: A preliminary report: U.S. Geological Survey Circular 587, 6 p.

Clifton, H. E., 1969, Beach lamination, nature and origin: Marine Geology, v. 7, p. 553–559.

Clifton, H. E., and R. S. Mason, 1969, Black sands, *in* Mineral and water resources of Oregon: U.S. 90th Congress, 2nd Session, Committee on Interior and Insular Affairs committee print, p. 102–107.

Clifton, H. E., A. Hubert, and R. L. Phillips, 1967, Marine sediment sample preparation for analysis for low concentrations of fine detrital gold: U.S. Geological Survey Circular 545, 11 p.

Cobb, E. H., 1960, Chromite, cobalt, nickel, and platinum occurrences in Alaska: U.S. Geological Survey Mineral Investigations Resource Map MR-8, scale 1:2,500,000.

Cobb, E. H., 1973, Placer deposits of Alaska: U.S. Geological Survey Bulletin 1374, 213 p.

Cobb, E. H., 1975, Chromite occurrences in Alaska: U.S. Geological Survey Mineral Investigations Resource Map MR-68, scale 1:2,500,000.

Cobb, E. H., 1984a, Map showing occurrences of lode gold and silver in Alaska: U.S. Geological Survey Mineral Investigations Resource Map MR-84, scale 1:2,500,000.

Cobb, E. H., 1984b, Map showing occurrences of placer gold in Alaska: U.S. Geological Survey Mineral Investigations Resource Map MR-83, scale 1:2,500,000.

Cobb, E. H., 1984c, Map showing occurrences of tungsten minerals in Alaska: U.S. Geological Survey Mineral Investigations Resource Map MR-85, scale 1:2,500,000.

Connah, T. H., 1962, Beach sand heavy mineral deposits of Queensland: Queensland Department of Mines Publication 302, 31 p.

Conwell, C. N., 1976, Progress and prospects in marine mining in Alaska: Alaska Department of Natural Resources Information Circular 22, 5 p.

Cook, D. J., 1969, Heavy minerals in Alaskan beach sand deposits: College, Alaska, University of Alaska, Mineral Industry Research Laboratory, Report 20, 114 p.

Cordes, E., 1966, Aufbau und Bildungsbedingungen der Schwermineralseifen bei Skagen (Danemark): Meyniana, v. 16, p. 1–35.

Coulter, H. W., D. M. Hopkins, T. N. V. Karlstrom, T. L. Péwé, C. Wahrhaftig, and J. R. Williams, 1965, Map showing extent of glaciations in Alaska: U.S. Geological Survey Miscellaneous Geo-

logic Investigations Map I-415, scale 1:2,500,000.
Creager, J. S., and D. A. McManus, 1966, Geology of the southeastern Chukchi Sea, *in* N. J. Wilimovsky and J. N. Wolfe, eds., Environment of the Cape Thompson region, Alaska: U.S. Atomic Energy Commission, p. 755–786.
Cruz, E. L., and E. H. Cobb, 1984, Map showing occurrences of tin minerals in Alaska: U.S. Geological Survey Mineral Investigations Resource Map MR-89, scale 1:2,500,000.
Day, D. T., and R. H. Richards, 1906a, Investigations of black sands from placer mines: U.S. Geological Survey Bulletin 285, p. 150–164.
Day, D. T., and R. H. Richards, 1906b, Useful minerals in the black sands of the Pacific slope: U.S. Geological Survey, Mineral Resources of the United States 1905, p. 1175–1246.
Diller, J. S., 1902, Topographic development of the Klamath Mountains: U.S. Geological Survey Bulletin 196, p. 26–28.
Dingler, J. R., R. J. Anima, D. E. Molzan, G. Luepke, and C. L. Peterson, 1982, A field study of littoral processes in Estero Bay, California: U.S. Geological Survey Open-File Report 82-537, 33 p.
Dinter, D. A., 1982, Holocene marine sediments on the middle and outer shelf of the Beaufort Sea north of Alaska: U.S. Geological Survey Miscellaneous Geologic Investigations Map I-1182-B, scale 1:500,000.
Drake, D. E., and D. A. Cacchione, 1985, Seasonal variation in sediment transport on the Russian River Shelf, California: Continental Shelf Research, v. 4, p. 495–514.
Drake, D. E., D. A. Cacchione, and H. A. Karl, 1985, Bottom currents and sediment transport on San Pedro Shelf, California: Journal of Sedimentary Petrology, v. 55, p. 15–28.
Duane, D. B., 1976, Sedimentation and ocean engineering: Placer mineral resources, *in* D. J. Stanley and D. J. P. Swift, eds., Marine sediment transport and environmental management: New York, John Wiley and Sons, p. 535–556.
Dupré, W. R., 1975, Quaternary history of the Watsonville lowlands, central Monterey Bay region, California: Stanford, California, Stanford University unpublished Ph.D. thesis, 232 p.
Dupré, W. R., 1982, Depositional environments of the Yukon delta, northeastern Bering Sea: Geologie en Mijnbouw, v. 61, p. 63–70.
Dupré, W. R., H. E. Clifton, and R. E. Hunter, 1980, Modern sedimentary facies of the open Pacific Coast and Pleistocene analogs from Monterey Bay, California, *in* M. E. Field, A. H. Bouma, I. P. Colburn, R. G. Douglas, and J. C. Ingle, eds., Quaternary depositional environments of the Pacific Coast: SEPM, Pacific Coast Paleogeography Symposium 4, p. 105–120.
Eakins, G. R., T. K. Bundtzen, M. S. Robinson, J. G. Clough, C. B. Green, K. H. Clautice, and M. A. Albanese, 1983, Alaska's mineral industry, 1982: Alaska Division of Geology and Geophysics Surveys Special Report 31, 63 p.
Economic Associates, Inc., 1968, The economic potential of the mineral and botanical resources of the U.S. continental shelf and slope: U.S. Department of Commerce, National Bureau of Standards, Institute for Applied Technology, PB 180 118, 520 p.
Emery, K. O., and L. C. Noakes, 1968, Economic placer deposits of the continental shelf: United Nations Economic Commission for Asia and the Far East Technical Bulletin, v. 1, p. 95–111.
Everts, C. H., 1972, Exploration for high energy marine placer sites: Part I—Field and flume tests, North Carolina coast: Madison, Wisconsin, The University of Wisconsin Sea Grant Program Report WIS-SG-72-210, 179 p.
Everts, C. H., 1973, Particle overpassing on flat granular boundaries: American Society of Civil Engineers Proceedings, v. 99, p. 425–438.
Field, M. E., S. H. Clarke, Jr., and M. E. White, 1980, Geology and geologic hazards of offshore Eel River Basin, northern California continental margin: U.S. Geological Survey Open-File Report 80-1080, 80 p.
Fischer, R. P., 1962, Vanadium in the United States: U.S. Geological Survey Mineral Investigations Resource Map MR-16, 8 p., 1 sheet, scale 1:3,168,000.
Fischer, R. P., 1969, Vanadium, *in* Mineral and water resources of Oregon: U.S. 90th Congress, 2nd Session, Committee on Interior and Insular Affairs committee print, p. 185–186.
Fischer, R. P., 1973, Vanadium, *in* D. A. Brobst and W. P. Pratt, eds., United States mineral resources: U.S. Geological Survey Professional Paper 820, p. 679–688.
Fisher, M. A., and L. B. Magoon, 1978, Geologic framework of lower Cook Inlet, Alaska: AAPG Bulletin, v. 62, p. 373–402.
Fisher, M. A., R. L. Detterman, and L. B. Magoon, 1987, Tectonics and petroleum geology of the Cook-Shelikof Basin, this volume.
Fisher, M. A., W. W. Patton, Jr., and M. L. Holmes, 1982, Geology of Norton Basin and continental shelf beneath northwestern Bering Sea, Alaska: AAPG Bulletin, v. 66, p. 255–285.
Force, E. R., 1976, Metamorphic source rocks of titanium placer deposits—A geochemical cycle: U.S. Geological Survey Professional Paper 959-B, 16 p.
Gardner, J. V., W. E. Dean, and T. L. Vallier, 1980, Sedimentology and geochemistry of the surface sediments, southern Bering Sea: Marine Geology, v. 35, p. 299–329.
Gardner, J. V., T. L. Vallier, and W. E. Dean, 1978, Grain-size total carbon, mineralogy and inorganic geochemical data from surface sediments of the southern Bering Sea outer continental shelf: U.S. Geological Survey Open-File Report 78-923, 5 p.
Gomes, J. M., G. M. Martinez, and M. M. Wong, 1980, Recovery of byproduct heavy minerals from sand and gravel operations in central and southern California: U.S. Bureau of Mines Report of Investigations 8471, 20 p.
Gorsline, D. S., and D. J. Grant, 1972, Sediment textural patterns on San Pedro Shelf, California (1951–1971): Reworking and transport by waves and currents, *in* D. J. P. Swift, D. B. Duane, and O. H. Pilkey, eds., Shelf sediment transport—Process and pattern: Stroudsburg, Pennsylvania, Dowden, Hutchinson and Ross, Inc., p. 575–600.
Gorsline, D. S., D. J. Grant, and J. S. Booth, 1971, Annual report on marine geologic research in the California Continental Borderland: Heavy mineral content of river, beach and nearshore sediments, southern California: Los Angeles, University of Southern California Technical Report USC-GEOL 71-5, 76 p.
Grantz, A., and S. D. May, 1984, Summary geologic report for Barrow arch outer continental shelf (OCS) planning area, Chukchi Sea, Alaska: U.S. Geological Survey Open-File Report 84-0395, 40 p.
Grantz, A., and S. D. May, 1987, Regional geology and petroleum potential of the United States Chukchi Shelf north of Point Hope, this volume.
Grantz, A., S. D. May, and D. A. Dinter, 1987, Regional geology and petroleum potential of the United States Beaufort and northeasternmost Chukchi Seas, this volume.
Grantz, A., P. W. Barnes, D. A. Dinter, M. B. Lynch, E. Reimnitz, and E. W. Scott, 1980, Geologic framework, hydrocarbon potential, environmental conditions and anticipated technology for exploration and development of the Beaufort Shelf north of Alaska: U.S. Geological Survey Open-File Report 80-94, 46 p.
Grantz, A., D. A. Dinter, E. R. Hill, R. E. Hunter, S. D. May, R. H. McMullen, and R. L. Phillips, 1982, Geologic framework, hydrocarbon potential and environmental conditions for exploration and development of proposed oil and gas lease sale 85 in the central and northern Chukchi Sea—A summary report: U.S. Geological Survey Open-File Report 82-1053, 89 p.
Greenberg, J., P. E. Hart, and A. Grantz, 1981, Bathymetric map of the continental shelf, slope, and rise of the Beaufort Sea, north of Alaska: U.S. Geological Survey Miscellaneous Geologic Investigations Map I-1182-A, scale 1:500,000.
Greene, H. G., 1970, A portable refraction seismographic survey of gold placer areas near Nome, Alaska: U.S. Geological Survey Bulletin 1312-B, 29 p.
Griggs, A. B., 1945, Chromite-bearing sands of the southern part of the coast of Oregon: U.S. Geological Survey Bulletin 945-E, p. 113–150.

Grim, M. S., and D. A. McManus, 1970, A shallow seismic-profiling survey of the northern Bering Sea: Marine Geology, v. 8, p. 293–320.

Gross, M. G., and J. L. Nelson, 1966, Sediment movement on the continental shelf near Washington and Oregon: Science, v. 151, p. 1–3.

Grosz, A. E., and E. C. Escowitz, 1983, Economic heavy minerals of the United States Atlantic continental shelf, *in* W. F. Tanner, ed., Nearshore sedimentology: Florida State University, 6th Symposium on Coastal Sedimentology Proceedings, p. 231–242.

Hails, J. R., 1972, The problem of recovering heavy minerals from the sea floor—An appraisal of depositional processes: Montreal, Canada, 24th International Geological Congress Proceedings, Section 8, p. 157–164.

Hails, J. R., 1976, Placer deposits, *in* K. H. Wolf, ed., Handbook of strata-bound and stratiform ore deposits, Vol. 3—Supergene and surficial ore deposits, textures and fabrics: New York, Elsevier Scientific Publishing Company, p. 213–244.

Hall, N. T., 1965, Petrology of the type Merced Group, San Francisco Peninsula, California: Berkeley, University of California unpublished M.A. thesis, 127 p.

Hampton, M. A. 1982a, Lower Cook Inlet environmental geology, *in* T. R. Bruns, ed., Hydrocarbon resource report for proposed OCS lease sale 88: Southeastern Alaska, northern Gulf of Alaska, Cook Inlet, and Shelikov Strait, Alaska: U.S. Geological Survey Open-File Report 82-928, p. 95–105.

Hampton, M. A., 1982b, Synthesis report—Environmental geology of lower Cook Inlet, Alaska: U.S. Geological Survey Open-File Report 82-197, 55 p.

Hampton, M. A., 1983, Geology of the Kodiak Shelf, Alaska—Environmental considerations for resource development: Continental Shelf Research, v. 1, p. 253–281.

Hampton, M. A., 1985, Quaternary sedimentation in Shelikof Strait, Alaska: Marine Geology, v. 62, p. 213–253.

Hampton, M. A., P. R. Carlson, H. J. Lee, and R. A. Feeley, 1986, North-central Gulf of Alaska continental margin geomorphology, sediment and sedimentary processes, *in* D. W. Hood, ed., Gulf of Alaska, Anchorage, Alaska, National Oceanic and Atmospheric Administration, Outer Continental Shelf Environmental Assessment Program, p. 93-143.

Hampton, M. A., K. H. Johnson, M. E. Torresan, and W. J. Winters, 1981, Description of seafloor sediment and preliminary geoenvironmental report, Shelikov Strait, Alaska: U.S. Geological Survey Open-File Report 81-1133, 87 p.

Handin, J. W., 1951, The source, transportation and deposition of beach sediment in southern California: U.S. Army Beach Erosion Board Technical Memorandum 22, 113 p.

Hands, E. B., 1968, Inner continental shelf sediments in the vicinity of Cape Flattery, Washington: Seattle, University of Washington Department of Oceanography Special Report 40, 39 p.

Henry, M. J., R. V. Berry, and P. J. Fischer, 1976, Holocene sediment deposits on the continental shelf, San Diego County (abs.): GSA Abstracts with Programs, v. 8, no. 3, p. 381.

Hess, F. L., 1924, Rare metals: U. S. Bureau of Mines, Mineral Resources of the United States 1924, Part 1, Metals, p. 451–476.

Hess, F. L., 1927, Rare metals: U.S. Bureau of Mines, Mineral Resources of the United States, 1927, Part 1, Metals, p. 393–453.

Hess, H. D., 1965, The ocean—Mining's newest frontier: Engineering and Mining Journal, v. 166, no. 8, p. 79–96.

Hill, E. R., A. Grantz, S. D. May, and M. Smith, 1984, Bathymetric map of the Chukchi Sea: U.S. Geological Survey Miscellaneous Geologic Investigations Map I-1182-D, scale 1:1,000,000.

Hobbs, S. W., 1969, Tungsten, *in* Mineral and water resources of Oregon: U.S. 90th Congress, 2nd Session, Committee on Interior and Insular Affairs committee print, p. 175–184.

Hobbs, S. W., and J. E. Elliott, 1973, Tungsten, *in* D. A. Brobst and W. P. Pratt, eds., United States mineral resources: U.S. Geological Survey Professional Paper 820, p. 667–678.

Hoose, P. J., K. D. Holden, and L. Lybeck, 1980, Isopach map of Holocene marine sediments, outer continental shelf, Shelikov Strait, Alaska: U.S. Geological Survey Open-File Report 80-2033, scale 1:250,000.

Hopkins, D. M., 1967, Quaternary marine transgressions in Alaska, *in* D. M. Hopkins, ed., The Bering land bridge: Stanford, California, Stanford University Press, p. 47–90.

Hornor, R. R., 1918, Notes on the black sand deposits of southern Oregon and northern California: U.S. Bureau of Mines Technical Paper 196, 42 p.

Hoskin, C. M., and R. V. Nelson, Jr., 1969, Modern marine carbonate sediment, Alexander Archipelago, Alaska: Journal of Sedimentary Petrology, v. 39, p. 581–590.

Hosking, K. F. G., 1971, The offshore tin deposits of southeast Asia: United Nations Economic Commission for Asia and the Far East Technical Bulletin, v. 5, p. 112–129.

Howard, J. D., and H. E. Reineck, 1981, Depositional facies of high-energy beach-to-offshore sequence—Comparison with low-energy sequence: AAPG Bulletin, v. 65, p. 807–830.

Howell, D. G., and J. G. Vedder, 1981, Structural implications of stratigraphic discontinuities across the southern California borderland, *in* W. G. Ernst, ed., The geotectonic development of California: Englewood Cliffs, New Jersey, Prentice-Hall, Inc., p. 535–558.

Hubbard, H. G., 1943, Mines and mineral resources of Santa Cruz County: California Journal of Mines and Geology, v. 39, p. 11–52.

Hudson, T., G. Plafker, and M. Rubin, 1976, Uplift rates of marine terrace sequences in the Gulf of Alaska, *in* E. H. Cobb, ed., The United States Geological Survey in Alaska—Accomplishments during 1975: U.S. Geological Survey Circular 733, p. 11–13.

Hundhausen, R. J., 1947, Chromiferous sand deposits in the Coos County area, Coos County, Oregon: U.S. Bureau of Mines Report of Investigations 4001, 13 p.

Hunter, R. E., H. E. Clifton, and R. L. Phillips, 1970, Geology of the stacks and reefs off the southern Oregon coast: Ore Bin, v. 32, p. 185–201.

Hunter, R. E., A. H. Sallenger, and W. R. Dupré, 1979, Maps showing directions of longshore sediment transport along the Alaskan Bering Sea coast: U.S. Geological Survey Miscellaneous Field Studies Map MF-1049, 5 sheets, scale 1:400,000.

Hunter R. E., D. R. Thor, and M. L. Swisher, 1982, Depositional and erosional features of the inner shelf, northeastern Bering Sea: Geologie en Mijnbouw, v, 61, p. 49–62.

Huntting, M. T., 1955, Gold in Washington: Washington Division of Mines and Geology Bulletin 42, 158 p.

Huntting, M. T., 1956, Inventory of Washington minerals: Washington Division of Mines and Geology Bulletin 37, pt. 2, Metallic minerals, v. I (text), 428 p; v. II (maps), 67 p.

Hutton, C. O., 1959, Mineralogy of beach sands between Half Moon and Monterey Bays, California: California Division of Mines Special Report 59, 32 p.

Inman, D. L., 1953, Areal and seasonal variations in beach and nearshore sediments at La Jolla, California: U.S. Army Beach Erosion Board Technical Memorandum 39, 82 p.

Judge, C. W., 1970, Heavy minerals in beach and stream sediments as indicators of shore processes between Monterey and Los Angeles, California: U.S. Army Coastal Engineering Research Center Technical Memorandum 33, 44 p.

Karl, H. A., and P. R. Carlson, 1984, Surface and near-surface geology, Navarin Basin province: Results of the 1980–81 field seasons: U.S. Geological Survey Open-File Report 84-89, 118 p.

Karlstrom, T. N. V., 1964, Quaternary geology of the Kenai lowland and glacial history of the Cook Inlet region, Alaska: U.S. Geological Survey Professional Paper 443, 69 p.

Kelly, J. V., 1947, Columbia River magnetite sands, Clatsop County, Oregon, and Pacific County, Washington, Hammond and McGowan deposits: U.S. Bureau of Mines Report of Investigations 4011, 7 p.

Kirschner, C. E., and C. A. Lyon, 1973, Stratigraphic and tectonic development of Cook Inlet petroleum province, *in* M. G. Pitcher,

ed., Arctic geology: AAPG Memoir 19, p. 396–407.
Klemic, H., H. L. James, and G. D. Eberlein, 1973, Iron, *in* D. A. Brobst and W. P. Pratt, eds., United States mineral resources: U.S. Geological Survey Professional Paper 820, p. 291–306.
Klemic, H, S. P. Marsh, and M. Cooper, 1973, Titanium, *in* D. A. Brobst and W. P. Pratt, eds., United States mineral resources: U.S. Geological Survey Professional Paper 820, p. 653–665.
Klemic, H., D. Gottfried, M. Cooper, and S. P. Marsh, 1973, Zirconium and hafnium, *in* D. A. Brobst and W. P. Pratt, eds., United States mineral resources: U.S. Geological Survey Professional Paper 820, p. 713–722.
Klise, D. H., 1984, Modern sedimentation on the California continental margin adjacent to the Russian River: San Jose, California, San Jose State University unpublished M.S. thesis, 120 p.
Knebel, H. J., and J. S. Creager, 1974, Heavy minerals on the east-central Bering Sea continental shelf: Journal of Sedimentary Petrology, v. 44, p. 553–561.
Komar, P. D., and C. Wang, 1984, Processes of selective grain transport and the formation of placers on beaches: Journal of Geology, v. 92, p. 637–655.
Komar, P. D., R. H. Neudeck, and L. D. Kulm, 1972, Observations and significance of deep-water oscillatory ripple marks on the Oregon continental shelf, *in* D. J. P. Swift, D. B. Duane, and O. H. Pilkey, eds., Shelf sediment transport—Process and pattern: Stroudsburg, Pennsylvania, Dowden, Hutchinson and Ross, Inc., p. 601–619.
Koschmann, A. H., and M. H. Bergendahl, 1962, Gold in the United States exclusive of Alaska and Hawaii: U.S. Geological Survey Mineral Investigations Resource Map MR-24, 22 p., 1 sheet, scale 1:3,168,000.
Krone, R. B., 1979, Sedimentation in the San Francisco Bay system, *in* T. J. Conomos, ed., San Francisco Bay—The urbanized estuary: San Francisco, Pacific Division, American Association for the Advancement of Science, p. 85–96.
Kulm, L. D., C. D. Peterson, and M. C. Stribling, 1986, Inventory of heavy minerals and metals, southern Washington, Oregon, and northern California continental shelf and coastal region: Oregon Department of Geology and Mineral Industries Open-File Report 0-86-10, 111 p.
Kulm, L. D., D. F. Heinrichs, R. M. Buerhrig, and D. M. Chambers, 1968a, Evidence for possible placer accumulations on the southern Oregon continental shelf: Ore Bin, v. 30, p. 81–104.
Kulm, L. D., K. F. Scheidegger, J. V. Byrne, and J. J. Spigai, 1968b, A preliminary investigation of the heavy mineral suites of the coastal rivers and beaches of Oregon and northern California: Ore Bin, v. 30, p. 165–184.
Kulm, L. D., R. C. Rouse, J. C. Harlett, R. H. Nuedeck, D. M. Chambers, and E. J. Runge, 1975, Oregon continental shelf sedimentation: Interrelationships of facies distribution and sedimentary processes: Journal of Geology, v. 83, p. 145–175.
Land, L. S., 1964, Eolian cross-bedding in the beach dune environment, Sapelo Island, Georgia: Journal of Sedimentary Petrology, v. 34, p. 389–394.
Lissner, F. G., 1971, Sources of littoral-zone sands in the vicinity of Gold Beach, Oregon: Eugene, Oregon, University of Oregon unpublished M.S. thesis, 86 p.
Logan, C. A., 1919, Platinum and allied metals in California: California State Mining Bureau Bulletin 85, 120 p.
Lowell, F. L., 1916, The counties of Del Norte, Humboldt, Mendocino: California Journal of Mines and Geology, v. 14, p. 371–425.
Luepke, G., 1975, Heavy-mineral trends in the Beaufort Sea: U.S. Geological Survey Open-File Report 75-667, 32 p.
Luepke, G., 1980, Opaque minerals as aids in distinguishing between source and sorting effects on beach-sand mineralogy in southwestern Oregon: Journal of Sedimentary Petrology, v, 50, p. 489–496.
Luepke, G., and H. E. Clifton, 1983, Heavy-mineral distribution in modern and ancient bay deposits, Willapa Bay, Washington, U.S.A.: Sedimentary Geology, v. 35, no. 4, p. 233–247.
Luepke, G., and K. W. Leong, 1981, Areal and textural distribution of particulate gold in sediments from Bluff Beach, Alaska: U.S. Geological Survey Open-File Report 81-1085, 39 p.
Macdonald, E. H., 1971, Detrital heavy mineral deposits in eastern Asia: United Nations Economic Commission for Asia and the Far East Technical Bulletin, v. 5, p. 13–31.
Maddren, A. G., 1919, The beach placers of the west coast of Kodiak Island, Alaska: U.S. Geological Survey Bulletin 692, p. 299–319.
Maloney, N. J., 1965, Geology of the continental terrace off the central coast of Oregon: Corvallis, Oregon, Oregon State University unpublished Ph.D. thesis, 233 p.
Maloney, N. J., and J. V. Byrne, 1964, Sedimentary rocks from the continental shelf and slope off the central coast of Oregon: Ore Bin, v. 26, p. 77–81.
Marlow, M. S., A. K. Cooper, and M. A. Fisher, 1987, Petroleum geology of the Beringian continental shelf, this volume.
Marlow, M. S., J. V. Gardner, T. L. Vallier, H. McLean, E. W. Scott, and C. L. Wilson, 1979, Resource report and environmental geology for proposed lease sale no. 70, St. George Basin, Alaska: U. S. Geological Survey Open-File Report 79–1650, 76 p.
Martinez, G. M., J. M. Gomes, and M. M. Wong, 1981, Recovery of byproduct heavy minerals from sand and gravel operations in Oregon and Washington: U.S. Bureau of Mines Report of Investigations 8563, 14 p.
May, J. P., 1973, Selective transport of heavy minerals by shoaling waves: Sedimentology, v. 20, p. 203–211.
McCulloch, D. S., 1987, Regional geology and hydrocarbon potential of offshore central California, this volume.
McLellan, R. D., 1927, The geology of the San Juan Islands: University of Washington Publications in Geology, v. 2, 185 p., map.
McManus, D. A., K. VenkataRathnam, D. M. Hopkins, and C. H. Nelson, 1974, Yukon River sediment on the northernmost Bering Sea Shelf: Journal of Sedimentary Petrology, v. 44, p. 1052–1060.
McManus, D. A., K. VenkataRathnam, D. M. Hopkins, and C. H. Nelson, 1977, Distribution of bottom sediments on the continental shelf, northern Bering Sea: U.S. Geological Survey Professional Paper 759-C, 31 p.
Mero, J. L., 1965, Mineral resources of the sea: New York, American Elsevier Publishing Company, 312 p.
Mertie, J. B., Jr., 1931, Notes on the geography and geology of Lituya Bay, Alaska: U.S. Geological Survey Bulletin 836, p. 117–135.
Mertie, J. B., Jr., 1969, Economic geology of the platinum metals; U.S. Geological Survey Professional Paper 630, 120 p.
Mertie, J. B., Jr., 1976, Platinum deposits of the Goodnews Bay district, Alaska: U.S. Geological Survey Professional Paper 938, 42 p.
Milliman J., and K. O. Emery, 1968, Sea levels during the past 35,000 years: Science, v. 162, p. 1121–1123.
Mining World, 1941, Mining through the ice: vol. 3, no. 3, p. 3–7.
Molnia, B. F., 1977, Rapid shoreline erosion and retreat at Icy Bay, Alaska—A staging area for offshore petroleum development: 9th Offshore Technology Conference Proceedings, v. III, p. 115–126.
Molnia, B. F., and P. R. Carlson, 1977, Surface sedimentary units of the northern Gulf of Alaska continental shelf: AAPG Bulletin, v. 62, p. 633–643.
Molnia, B. F., and M. C. Wheeler, 1978, Report on the beach dynamics, geology, oil spill susceptibility of the Gulf of Alaska coastline in Glacier Bay National Monument—Sea Otter Creek to Icy Point: U.S. Geological Survey Open-File Report 78-284, 191 p.
Molnia, B. F., W. P. Levy, and P. R. Carlson, 1980, Map showing sedimentation rates in the northern Gulf of Alaska: U.S. Geological Survey Miscellaneous Field Studies, Map MF-1170, scale 1:500,000.
Moore, D. G., 1964, Acoustic-reflection reconnaissance of continental shelves: Eastern Bering and Chukchi Seas, *in* R. L. Miller, ed., Papers in marine geology: New York, The McMillan Company, Shepard Commemorative Volume, p. 319–362.
Moore, G. W., and E. A. Silver, 1968a, Geology of the Klamath

River Delta: U.S. Geological Survey Professional Paper 600-C, p. C144–C148.

Moore, G. W., and E. A. Silver, 1968b, Gold distribution on the sea floor off the Klamath Mountains, California: U.S. Geological Survey Circular 605, 9 p.

National Ocean Survey, 1976, Tidal current tables 1977: Pacific Coast of North America and Asia: Rockland, Maryland, National Ocean Survey, 255 p.

National Oceanic and Atmospheric Administration, 1976, Kuskokwin Bay: National Ocean Survey Hydrographic Chart 16300, scale 1:200,000.

Nelson, C. H., 1971, Northern Bering Sea, a model for depositional history of Arctic Shelf placers; ecological impact of placer development: 1st International Conference on Port and Ocean Engineering under Arctic Conditions Proceedings, v.1, p. 246–254.

Nelson, C. H., 1982, Late Pleistocene-Holocene transgressive sedimentation in deltaic and non-deltaic areas of the northeastern Bering epicontinental shelf: Geologie en Mijnbouw, v. 61, p. 5–18.

Nelson, C. H., and J. S. Creager, 1977, Displacement of Yukon-derived sediment from Bering Sea to Chukchi Sea during the Holocene: Geology, v. 5, p. 141–146.

Nelson, C. H., and D. M. Hopkins, 1972, Sedimentary processes and distribution of particulate gold in the northern Bering Sea: U.S. Geological Survey Professional Paper 689, 27 p.

Nelson, C. H., W. Dupré, M. Field, and J. D. Howard, 1982, Variation in sand body types on the eastern Bering Sea epicontinental shelf: Geologie en Mijnbouw, v. 61, p. 37–48.

Nittrouer, C. A., and R. W. Sternberg, 1981, The formation of sedimentary strata in an allochthonous shelf environment—The Washington continental shelf: Marine Geology, v. 42, no. 1-4, p. 201–232.

Oakeshott, G. B., 1950, Black sands: California Division of Mines Bulletin 156, p. 133–136.

O'Brien, J. C., 1952, Mines and mineral resources of Del Norte County: California Journal of Mines and Geology, v. 48, p. 261–309.

Olson, J. C., and J. W. Adams, 1962, Thorium and rare earths in the United States: U.S. Geological Survey Mineral Investigations Resource Map MR-28, scale 1:3,168,000.

Orlando, R. C., and E. A. Martin, 1981, Grain size compilation and parameters of sediment samples—Lower Cook Inlet, Alaska: U.S. Geological Survey Open-File Report 81-827, 11 p.

Osborne, R. H., R. C. Scheidemann, Jr., T. R. Nardin, and A. S. Harper, 1980, Quaternary stratigraphy and depositional environments, Santa Monica Bay, southern California, *in* M. E. Field, A. H. Bouma, I. P. Colburn, R. G. Douglas, and I. C. Ingle, eds, Quaternary depositional environments of the Pacific Coast: SEPM, Pacific Coast Paleogeography Symposium 4, p. 143–156.

Padan, J. W., 1971, Marine mining and the environment, *in* D. Hood, ed., Impingement of man on the oceans: New York, John Wiley and Sons, p. 553–561.

Page, N. J., A. L. Clark, G. A. Desborough, and R. L. Parker, 1973, Platinum-group metals, *in* D. A. Brobst and W. P. Pratt, eds., United States mineral resources: U.S. Geological Survey Professional Paper 820, p. 537–545.

Pardee, J. T., 1921, Chromite ores in Washington: U.S. Geological Survey Bulletin 725–A, p. 61–66.

Pardee, J. T., 1929, Platinum and black sand in Washington: U.S. Geological Survey Bulletin 805, p. 1–15.

Pardee, J. T., 1934, Beach placers of the Oregon coast: U.S. Geological Survey Circular 8, 41 p.

Patton, W. W., Jr., 1964, Mineral and fuel resources—Northern and interior Alaska, *in* Mineral and water resources of Alaska: Report of U.S. Geological Survey to U.S. Senate Committee on Interior and Insular Affairs, 88th Congress, 2nd Session, p. 62–94.

Patton, W. W., Jr., and B. Csejtey, Jr., 1980, Geologic map of St. Lawrence Island, Alaska: U.S. Geological Survey Miscellaneous Investigations Map I-1203, scale 1:250,000.

Peterson, C. D., P. D. Komar, and K. F. Scheidegger, 1986, Distribution, geometry, and origin of heavy mineral placer deposits on Oregon beaches: Journal of Sedimentary Petrology, v. 56, no. 1, p. 67–77.

Péwé, T. L., 1975, Quaternary geology of Alaska: U.S. Geological Survey Professional Paper 835, 145 p.

Phillips, R. L., 1979, Heavy minerals and bedrock minerals on the continental shelf of Washington, Oregon, and California, *in* Outer Continental Shelf Mining Policy Phase II Task Force, Feasibility Document: Outer Continental Shelf Hard Minerals Leasing, U.S. Geological Survey, Appendix 8, 56 p.

Plafker, G., 1987, Regional geology and petroleum potential of the northern Gulf of Alaska continental margin, this volume.

Ramp, L., 1961, Chromite in southwestern Oregon: Oregon Department of Geology and Mineral Resources Bulletin 52, 169 p.

Ramp, L., 1969, Platinum-group metals, *in* Mineral and water resources of Oregon: U.S. 90th Congress, 2nd Session, Committee on Interior and Insular Affairs committee print, p. 167–171.

Rand McNally, 1980, Encyclopedia of world rivers: Chicago, Rand McNally and Company, 352 p.

Reid, I., and L. E. Frostick, 1985a, Beach orientation, bar morphology and the concentration of metalliferous placer deposits—A case study, Lake Turkana, N. Kenya: Journal of the Geological Society of London, v. 142, p. 837–848.

Reid, I., and L. E. Frostick, 1985b, Role of settling, entrainment and dispersive equivalence and of interstice trapping in placer formation: Journal of the Geological Society of London, v. 142, p. 739–746.

Reimnitz, E., and P. W. Barnes, 1974, Sea ice as a geologic agent on the Beaufort Sea Shelf of Alaska, *in* J. C. Reed and J. E. Sater, eds., The coast and shelf of the Beaufort Sea: Arlington, Virginia, Arctic Institute of North America, p. 301–353 (with discussion).

Reimnitz, E., and G. Plafker, 1976, Marine gold placers along the Gulf of Alaska margin: U.S. Geological Survey Bulletin 1415, 16 p., 1 sheet, scale 1:1,000,000.

Reimnitz, E., R. von Huene, and F. F. Wright, 1970, Detrital gold and sediments in Nuka Bay, Alaska: U.S. Geological Survey Professional Paper 700-C, p. C35–C42.

Rice, R. M., D. S. Gorsline, and R. H. Osborne, 1976, Relationships between sand input from rivers and the composition of sands from the beaches of southern California: Sedimentology, v. 23, p. 689–703.

Roberts, R. W., 1974a, Marine sedimentological data of the inland waters of Washington State (Strait of Juan de Fuca and Puget Sound): University of Washington Department of Oceanography Special Report 56, 120 p.

Roberts, R. W., 1974b, Marine sedimentological data of the Washington continental shelf: University of Washington Department of Oceanography Special Report 57, 129 p.

Rodeick, C. A., 1979, The origin, distribution and depositional history of gravel deposits on the Beaufort Sea continental shelf, Alaska: U.S. Geological Survey Open-File Report 79-234, 87 p.

Rogers, C. L., and M. C. Jaster, 1962, Titanium in the United States, exclusive of Alaska and Hawaii: U.S. Geological Survey Mineral Investigations Resource Map MR-29, 18 p., 1 sheet, scale 1:3,168,000.

Rosenblum, S., R. R. Carlson, J. M. Nishi, and W. C. Overstreet, 1986, Platinum-group elements in magnetic concentrates from the Goodnews Bay district, Alaska: U.S. Geological Survey Bulletin 1660, 38 p.

Rossman, D. L., 1957, Ilmenite-bearing beach sands near Lituya Bay, Alaska: U.S. Geological Survey Open-File Report 149, 10 p.

Runge, E. J., Jr., 1966, Continental shelf sediments, Columbia River to Cape Blanco, Oregon: Corvallis, Oregon, Oregon State University unpublished Ph.D. thesis, 139 p.

Sainsbury, C. L., 1957, Some pegmatite deposits in southeastern Alaska: U.S. Geological Survey Bulletin 1024-G, p. 141–161.

Sainsbury, C. L., and B. L. Reed, 1973, Tin, *in* D. A. Brobst and W. P. Pratt, eds., United States Mineral Resources, U.S. Geological Survey Professional Paper 820, p. 637–651.

Sallenger, A. H., Jr., 1979, Inverse grading and hydraulic equivalence in grain flow deposits: Journal of Sedimentary Petrology, v. 49, p. 553–562.

Sallenger, A. H., Jr., and R. E. Hunter, 1984, Coastal morphology maps of the Bristol Bay coast: U.S. Geological Survey Miscellaneous Field Studies Map MF-1672, scale 1:400,000.

Sallenger, A. H., Jr., J. R. Dingler, and R. E. Hunter, 1978, Coastal processes and morphology of the Bering Sea coast of Alaska: National Oceanic and Atmospheric Administration, Annual Reports of Principal Investigators for the Year Ending March 1978, v. 12, Hazards, p. 451–502.

Sarna-Wojcicki, A. M., C. E. Meyer, H. R. Bowman, N. T. Hall, P. C. Russell, M. J. Woodward, and J. L. Slate, 1985, Correlation of the Rockland ash bed, a 400,000-year-old stratigraphic marker in northern California and western Nevada, and implications for middle Pleistocene paleogeography of central California: Quaternary Geology, v. 23, p. 236–257.

Sayles, F. L., 1966, A reconnaissance heavy mineral study of Monterey Bay beach sediments: University of California Hydraulic Engineering Laboratory Technical Report HEL-2-16, 20 p.

Scheidegger, K. F., and J. B. Phipps, 1976, Dispersal patterns of sands in Grays Harbor estuary, Washington: Journal of Sedimentary Petrology, v. 46, p. 163–166.

Scheidegger, K. F., L. D. Kulm, and E. J. Runge, 1971, Sediment sources and dispersal patterns of Oregon continental shelf sands: Journal of Sedimentary Petrology, v. 71, p. 1112–1120.

Scholl, D. W., 1960, Relationship of the insular shelf sediments to the sedimentary environments and geology of Anacapa Island, California: Journal of Sedimentary Petrology, v. 30, p. 123–139.

Scholl, D. W., T. L. Vallier, and A. J. Stevenson, 1987, Geologic evolution and petroleum geology of the Aleutian Ridge, this volume.

Shackleton, N. J., and N. D. Opdyke, 1977, Oxygen isotope and paleomagnetic evidence for early Northern Hemisphere glaciation: Nature, v. 270, p. 216–219.

Sharma, G. D.. 1974, Contemporary depositional environment of the eastern Bering Sea—Part 1. Contemporary sedimentary regimes of the eastern Bering Sea, *in* D. W. Hood and E. J. Kelley, eds., Oceanography of the Bering Sea with emphasis on renewable resources: Fairbanks, Alaska, Institute of Marine Science, University of Alaska, p. 517–540.

Sharma, G. D., 1978, Marine placers of the Alaskan Shelf: St. Johns, Newfoundland, Memorial University of Newfoundland, Fourth International Conference on Port and Ocean Engineering under Arctic Conditions Proceedings, v. 2, p. 885–891.

Sharma, G. D., 1979, The Alaskan Shelf—Hydrographic, sedimentary and geochemical environment: New York, Springer-Verlag, 498 p.

Sharma, G. D., and D. C. Burrell, 1970, Sedimentary environment and sediments of Cook Inlet, Alaska: AAPG Bulletin, v. 54, p. 647–654.

Shepard, F. P., and H. R. Wanless, 1971, Our changing coastlines: New York, McGraw-Hill, Inc., 579 p.

Simons, F. S., and W. C. Prinz, 1973, Gold, *in* D. A. Brobst and W. P. Pratt, eds., United States mineral resources, U.S. Geological Survey Professional Paper 820, p. 263–275.

Slingerland, R. L., 1977, The effects of entrainment on the hydraulic equivalence relationships of light and heavy minerals in sands: Journal of Sedimentary Petrology, v. 47, p. 753–770.

Sloane, H. N., and L. L. Sloane, 1970, A pictorial history of American mining: New York, Crown Publishers, Inc., 342 p.

Smith, J. D., and T. S. Hopkins, 1972, Sediment transport on the continental shelf off of Washington and Oregon in light of recent current measurements, *in* D. J. P. Swift, D. B. Duane, and O. H. Pilkey, eds., Shelf sediment transport—Process and pattern: Stroudsburg, Pennsylvania, Dowden, Hutchinson, and Ross, Inc., chapter 7, p. 143–180.

Smith, M. B., V. L. Engler, D. L. Lee, K. J. Horn, and R. G. Wayland, 1967, Reported occurrences of selected minerals in the northern third of California: U.S. Geological Survey Mineral Investigations Resource Map MR-47, scale 1:500,000.

Smith, M. B., V. L. Engler, D. L. Lee, K. J. Horn, and R. G. Wayland, 1971, Reported occurrences of selected minerals in the central third of California: U.S. Geological Survey Mineral Investigations Resource Map MR-48, scale 1:500,000.

Smith, M. B., V. L. Engler, D. L. Lee, K. J. Horn, and R. G. Wayland, 1974, Reported occurrences of selected minerals in the southern third of California: U.S. Geological Survey Mineral Investigations Resource Map MR-49, scale 1:500,000.

Snavely, P. D., Jr., 1987, Tertiary geologic framework, neotectonics, and petroleum potential of the Oregon-Washington continental margin, this volume.

Snavely, P. D., Jr., and R. E. Wells, 1984, Tertiary volcanic and intrusive rocks on the Oregon and Washington continental shelf: U.S. Geological Survey Open-File Report 84-282, 25 p.

Snavely, P. D., Jr., J. E. Pearl, and D. L. Lander, 1977, Interim report on petroleum resources potential and geologic hazards in the outer continental shelf—Oregon and Washington Tertiary province: U.S. Geological Survey Open-File Report 77-282, 64 p.

Stapor, F. W., 1973, Heavy mineral concentrating processes and density/shape/size equilibria of marine and coastal dune sands from the Apalachicola, Florida, region: Journal of Sedimentary Petrology, v. 43, p. 396–407.

Stauffer, P. H., 1987, Quaternary depositional history and potential sand and gravel resources of the Alaskan continental margins, this volume.

Steffy, D. A., B. W. Turner, L. D. Lybeck, and J. T. Roe, 1981, Isopach map of Holocene sedimentary units, Norton Sound, Alaska: U.S. Geological Survey Open-File Report 81-720, 1 sheet, scale 1:250,000.

Stewart, R. M., 1957, Tungsten: California Division of Mines Bulletin 176, p. 655–667.

Stinson, M. C., 1957, Black sands, *in* L. A. Wright, ed., Mineral commodities of California: California Division of Mines and Geology Bulletin 176, p. 83–85.

Sutherland, D. G., 1985, Geomorphological controls on the distribution of placer deposits: Journal of the Geological Society of London, v. 142, p. 727–737.

Tagg, A. R., 1979, Offshore heavy metals resources of Alaska, *in* Outer Continental Shelf Mining Policy Phase II Task Force, Program Feasibility Document: Outer Continental Shelf Hard Minerals Leasing, U.S. Geological Survey, Appendix 6, 42 p.

Tagg, A. R., and H. G. Greene, 1973, High-resolution seismic survey of an offshore area near Nome, Alaska: U.S. Geological Survey Professional Paper 759-A, 23 p.

Tanner, W. F., A. Mullins, and J. D. Bates, 1961, Possible masked heavy mineral deposit, Florida panhandle: Economic Geology, v. 56, no. 6, p. 1079–1087.

Thaden, R. E., 1973, Abrasives, *in* D. A. Brobst and W. P. Pratt, eds., United States mineral resources: U.S. Geological Survey Professional Paper 820, p. 27–33.

Thayer, T. P., 1969, Chromite, *in* Mineral and water resources of Oregon: U.S. 90th Congress, 2nd Session, Committee on Interior and Insular Affairs committee print, p. 108–116.

Thayer, T. P., 1973, Chromium, *in* D. A. Brobst and W. P. Pratt, eds., United States mineral resources: U.S. Geological Survey Professional Paper 820, p. 111–121.

Thayer, T. P., and M. H. Miller, 1962, Chromite in the United States, exclusive of Alaska and Hawaii: U.S. Geological Survey Mineral Investigations Resource Map MR-26, 5 p. 1 sheet, scale 1:3,168,000.

Thomas, B. I., and R. V. Berryhill, 1962, Reconnaissance studies of Alaskan beach sands, eastern Gulf of Alaska: U.S. Bureau of Mines Report of Investigations 5986, 40 p.

Thor, D. R., and C. H. Nelson, 1981, Ice gouging on the subarctic Bering Shelf, *in* D. W. Hood and J. A. Calder, eds., The eastern Bering Sea Shelf, Oceanography and Resources, Vol. 1: Seattle, Washington, University of Washington Press, p. 279–291.

Thorsen, J. W., 1964, Mineralogy of black sands at Grays Harbor, Washington: Washington State Division of Mines and Geology Report of Investigations 23, 29 p.

Tourtelot, H. A., 1968, Hydraulic equivalence of grains of quartz and heavier minerals, and implications for the study of placers: U.S. Geological Survey Professional Paper 594-F, 13 p.

Trask, P. D., 1952, Source of beach sand at Santa Barbara, California, as indicated by mineral grain studies: U.S. Army Beach Erosion Board Technical Memorandum 28, 24 p.

Tucker, W. B., 1927, District reports of mining engineers, Los Angeles Field Division: California Journal of Mines and Geology, v. 23, p. 287–345.

Twenhofel, W. H., 1943, Origin of black sands on the coast of southwest Oregon: Oregon Department of Geology and Mineral Industries Bulletin 24, 25 p.

Twenhofel, W. H., 1946a, Beach and river sands of the coastal region of southwest Oregon with particular reference to the black sands—Part II, Minerals of the sands: American Journal of Science, v. 244, p. 200–214.

Twenhofel, W. H., 1946b, Mineralogical and physical composition of the sands of the Oregon coast from Coos Bay to the mouth of the Columbia River: Oregon Department of Geology and Mineral Industries Bulletin 30, 64 p.

United Nations, 1970, Mineral resources of the sea: United Nations Publication E.70.11.B.4-I-VI, p. 1–49.

U.S. Bureau of Mines, 1987, An economic reconnaissance of selected heavy mineral placer deposits in the U.S. Exclusive Economic Zone: U.S. Bureau of Mines Open-File Report 4-87, 112 p.

Vedder, J. G., 1987, Regional geology and petroleum potential of the southern California borderland, this volume.

VenkataRathnam, K., 1969, Size distribution and mineral composition of the sediments of the inner continental shelf off northern Washington: University of Washington Department of Oceanography Special Report 41, p. 93–113.

VenkataRathnam, K., 1971, Heavy minerals on the continental shelf of the northern Bering Sea: U.S. Geological Survey Open-File Report 1553, 93 p.

VenkataRathnam, K., and D. A. McManus, 1973, Origin and distribution of sands and gravels on the northern continental shelf off Washington: Journal of Sedimentary Petrology, v. 43, p. 799–811.

von Huene, R., G. G. Shor, Jr., and E. Reimnitz, 1967, Geological interpretation of seismic profiles in Prince William Sound, Alaska: GSA Bulletin, v. 78, p. 259–268.

Wagner, H. C., and P. D. Snavely, Jr., 1966, Geology—Western Washington, *in* Mineral and water resources of Washington: Report of the U.S. Geological Survey to U.S. Senate Committee on Interior and Insular Affairs, 88th Congress, 2nd Session, p. 37–46.

Walker, G. W., and A. B. Griggs, 1953, Chromite deposits of the southern Coast Ranges of California: California Division of Mines Bulletin 134, Part II, p. 39–88.

Walker, H. J., 1974, The Colville River and Beaufort Sea—Some interactions, *in* J. C. Reed and J. E. Sater, eds., The coast and shelf of the Beaufort Sea: Arlington, Virginia, Arctic Institute of North America, p. 513–540 (with discussion).

Welday, E. E., and J. W. Williams, 1975, Offshore surficial geology of California: California Division of Mines Geology Map Sheet 26, scale 1:500,000.

Whetten J. T., J. C. Kelley, and L. G. Hanson, 1969, Characteristics of Columbia River sediment and sediment transport: Journal of Sedimentary Petrology, v. 39, no. 3, p. 1149–1166.

White, S. M., 1967, The mineralogy and geochemistry of the sediments on the continental shelf off the Washington-Oregon coast: Seattle, University of Washington unpublished Ph.D. thesis, 213 p.

Wilson, T. A., and J. L. Mero, 1966, Economic deposits of the California offshore area, *in* E. H. Bailey, ed., Geology of northern California: California Division of Mines and Geology Bulletin 190, p. 343–353.

Wiseman, W. J., Jr., J. N. Suhayda, and S. A. Hsu, 1974, Characteristics of nearshore oceanographic environment of Arctic Alaska, *in* J. C. Reed and J. E. Sater, eds., The coast and shelf of the Beaufort Sea: Arlington, Virginia, Arctic Institute of North America, p. 49–64.

Wong, F. L., 1983, Heavy mineral data from lower Cook Inlet, Alaska: U.S. Geological Survey Open-File Report 83-782, 4 p.

Woolsey, J. R., V. J. Henry, and J. L. Hunt, 1975, Backshore heavy-mineral concentration on Sapelo Island, Georgia: Journal of Sedimentary Petrology, v. 45, p. 280–284.

Wright, F. F., 1969, Sedimentation and gold distribution, Yakutat Bay, Alaska: University of Alaska Institute of Marine Science Report R69-9, 12 p.

Wright, F. F., 1972, Marine geology of Yakutat Bay, Alaska: U.S. Geological Survey Professional Paper 800-B, p. B9–B15.

Wright, L. A., 1957, Gem stones: California Division of Mines Bulletin 176, p. 205–214.

Yancey, T. E., 1968, Recent sediments of Monterey Bay: University of California Hydraulic Engineering Laboratory Technical Report HEL-2-18, 145 p.

Yancey, T. E., and J. W. Lee, 1972, Major heavy mineral assemblages and heavy mineral provinces of the central California coast region: GSA Bulletin, v. 83, p. 2099–2104.

Youngman, E. P., 1930, Deposits of titanium-bearing ores (domestic and foreign): U.S. Bureau of Mines Information Circular 6386, 41 p.

CHAPTER THIRTY-ONE

GEOTHERMAL RESOURCES

C. R. B. Lister

School of Oceanography
University of Washington

ABSTRACT

The major geothermal resources that can be exploited for power generation commonly are associated with rifting, when large volumes of hot rock are emplaced at depth by the upwelling. Off the west coast of the United States, the prime sites are therefore the Juan de Fuca and Gorda Ridges, and areas of minor potential are the Buldir and Sunday Depressions in the Aleutians.

An ocean-floor resource with no counterpart on land occurs when thick sediment encroaches on young ocean floor. The oceanic crust is cracked, becomes highly permeable, and has a high heat flow. The blanket of poorly conducting sediment causes high temperatures to be developed in the permeable zone, and leads to a possibly producible reservoir of hot fluid. The prime areas for this type of resource are the sediment fans off Washington, Oregon, and northern California.

When sedimentary rocks are compressed and heated, hot fluids can be given off and may be the source of most of the small hot springs on land. An extreme example of this process occurs at depth in a subduction zone, where the fluid can become superheated during its rise to the surface. Such hot-fluid-driven geothermal areas may exist on the Aleutian Ridge and, to a much lesser degree, on the continental margin of northern California. Miscellaneous young volcanic rocks offshore in the Bering Sea and Gulf of Alaska could also be associated with some hydrothermal prospects.

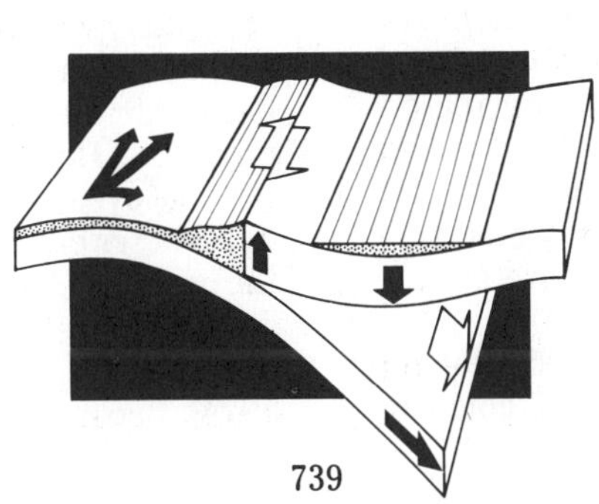

INTRODUCTION

A geothermal resource can be characterized by the product of available flow rate and the temperature difference between the fluids and a nearby sink for the heat. This product is the primary thermal power of the site: V (liters per second) × ΔT (° Celsius) × C_p (specific heat of fluid) = Power (Megawatts thermal). If the application is the generation of electrical power, the thermal power must be reduced by two factors: the maximum Carnot thermodynamic efficiency $(T_1-T_2)/T_1$, where T_1 is the effective source temperature and T_2 the sink temperature, both measured on the absolute or Kelvin scale, and the expected efficiency of the conversion machinery, typically about 50%. The effective source temperature may be that of the raw fluid if it can be run directly into a turbine; if the hydrothermal fluid is a corrosive brine, substantial heat-exchanger losses may be incurred before the heat can be used.

The available flow rate is the most difficult parameter to determine at any geothermal site. A drill hole into the hot region will define the available temperature quite well, but the actual producibility of fluid is not so easily measured. One concept that has been introduced into the literature is that of a geothermal "reservoir," which treats the hot region reached by drilling as if it were a static pool of desirable fluid like an oil field. The total amount of heat stored in a hot region near the surface can be calculated, but this is not necessarily all the heat that is available. By its very nature, a hydrothermal system is dynamic; even if the cap of surface rocks is relatively impermeable, heat would leak out by conduction over time periods significant for geologic processes. In practice, all the areas currently exploited for substantial electrical power generation had very large fluxes of heat through the surface in their natural states (Ellis and Wilson, 1955); in the case of Wairakei (New Zealand), the subsurface thermal distribution defined an upwelling plume of hot fluid (Elder, 1965). The extraction of heat through bore holes from 1951 to 1958 had not even reduced the natural output below the original level because of the reservoir effect of the heated near-surface volume (Fisher, 1964). This natural output, of about 400MW (thermal), was considered a minimum for further geothermal prospecting in New Zealand (W. A. J. Mahon, 1974, oral communication).

Whereas onshore geothermal areas are considered mainly as potential sources of electrical power and, to a lesser extent, sources of domestic or industrial process heat, the potential uses of offshore geothermal heat are less easily predictable. The highest capacity areas are in deep water, and the costs of constructing a deep-sea power plant are far above those of the most expensive forms of onshore power, even before the transmission costs are included. However, it is a well-established fact that power sources for remote areas can be intrinsically very expensive, and yet economically justifiable relative to the cost of a cable connection to a grid—the first uses of solar photovoltaic power were in satellites and remote signalling equipment. If, for any reason, long-term electrical power is ever required on the deep-sea floor, to power instruments or even to mine hydrothermal sulphides, a geothermal station using a hot vent would have a lower delivered cost than connection to a shore grid, or regularly replacing large battery packs.

The potential uses of lower-grade geothermal heat offshore are even less easily imagined. The heat could be applied to processing the fluids themselves for some valuable chemical species that is dissolved or suspended in them. Or warmth and chemical energy could conceivably be needed as nutrients for the production of exotic pharmaceuticals by the strange biota that are being discovered on the ocean floor—drugs that for some reason cannot be made by the ordinary bacteria available for genetic engineering. Such uses would be expected to be far in the future; therefore, even though areas of low-grade resources are included in this survey, they are not emphasized as much as the high-temperature areas.

NATURE OF THE SOURCES

Hot Rock and Magmas

As has been mentioned, heat tends to leak out of the earth by conduction at a rate that is fast relative to the widespread geological processes of sedimentation, erosion, and overthrusting/uplift. There is some outward heat flow almost everywhere, but the normal geothermal gradient is only about 20°C/km, so that little useful heat is available at economic drilling depths. Some more rapid process must emplace very hot material close to the surface if useful geothermal power is to exist. Two such processes are presently at least partially understood—rifting associated with plate motion and continental drift, and volcanism. Both processes involve the migration of magmas toward the earth's surface, but the distribution of the emplacements is quite different. In rifting, a whole new surface layer has to form, and the same volume of fresh material must be emplaced at all depths down to where the motion does not occur. In volcanism, magma must come to the surface to build the volcano, but, in the absence of rifting, the magma plumbing is small in diameter. The difficulty is that of making a small round hole in a slab of stone, relative to breaking it and moving the halves apart. In the absence of an abrasive mechanism (such as drilling) there is only one possibility: melting, or the blowtorch effect.

Many volcanic regions are also associated with some degree of rifting, and in these areas the volcanoes tend to lie on a line, as in North Island, New Zealand. However, a truly isolated volcano would generate a distribution of hot material, as shown in Figure 1A; the bulk of the material is on the surface, and only a small feeder plug is underground. Contrast this with Figure 1B, where a rift zone is filled with hot rock to an indefinite depth. The surface volcano is built by a series of eruptions, each normally adding a relatively thin layer of material to the surface, a layer that cools before the next eruption. There is very little stored heat in such a volcano. The contrast with the rift zone is obvious—a large volume of hot rock has been emplaced beneath the surface, essentially waiting for a hydrothermal system to extract it.

Thus far the analysis is very simple and would probably elicit a fair measure of agreement among geologists and geophysicists. However, there is no consensus as to how hydrothermal systems acquire their heat. The currently proposed hypotheses can be summarized by the sketches in Figure 2—the box model (Elder, 1965), the deep-crack model (Bodvarsson and Lowell, 1972), the magma chamber minibox model (Sleep, 1984), and the water-penetration theory (Lister, 1974). Because I originated the water-penetration theory, I

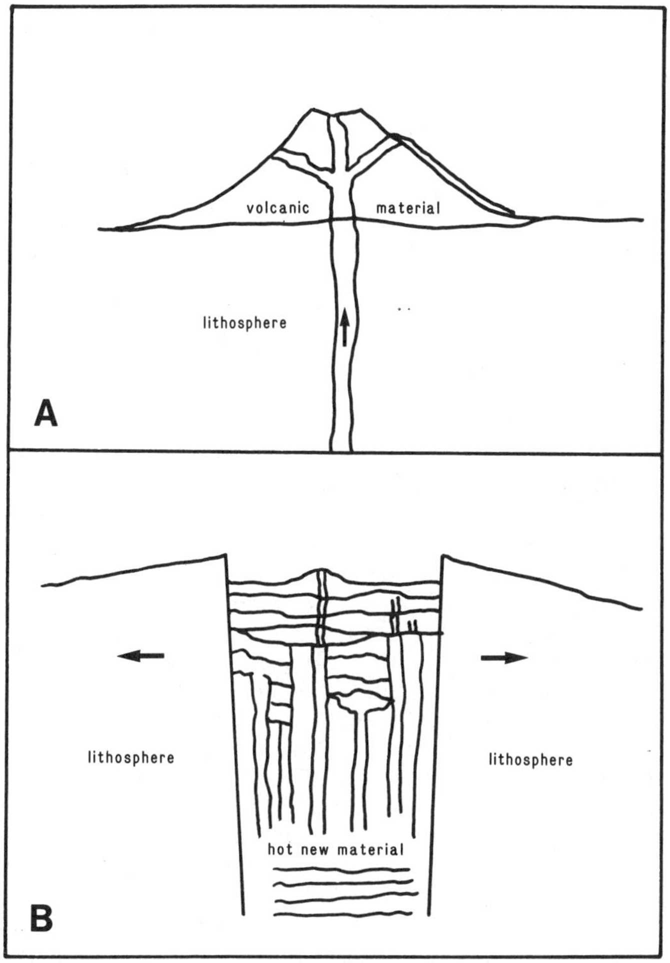

Figure 1. Diagram of the emplacement modes of hot new material in (A) an isolated volcano built up on the surface, and (B) a rift zone filling from below. Major volumes of buried hot rock are absent in (A) but present all along the rift zone in (B), leading to a much greater geothermal potential for the latter.

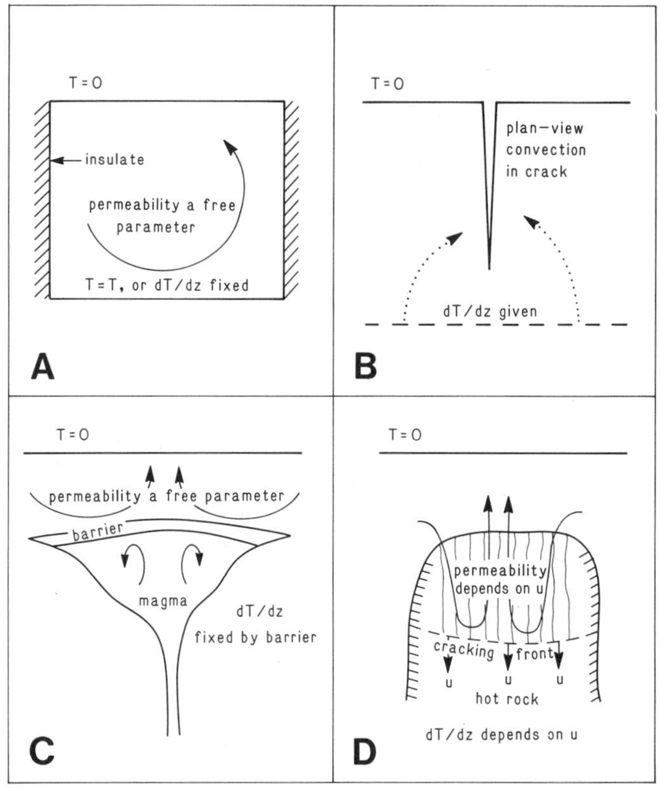

Figure 2. Essential features of the four major concepts of the source of geothermal heat. T = temperature, dT/dz = vertical temperature gradient. A, Basic box model of convection in a segment of porous slab; all parameters are artificially introduced. B, Macrocrack physical model, one that can neither generate the water temperatures nor the heat flux output of existing geothermal areas. C, Magma chamber heat source model, a more physical variation of the box model whose primary problem is the matching of observed plume fluxes. D, Water penetration theory, an attempt to develop a complete physics for the system, even though inexact and one-dimensional. The process is controlled by the velocity of the cracking front u, and stabilized by negative feedback between the heat output, which is proportional to u, and the permeability, which is approximately proportional to $1/u^2$. A separate feedback mechanism, the opening and closing of the cracks, stabilizes the hot-water temperature (see Lister 1974, 1984).

am naturally somewhat biased in believing that there are several reasons why it will probably prove to be the most nearly correct; this accounts for the use of the word "theory" rather than "model" in the description.

We have seen that the thermal delivery rate is the most important feature of a hydrothermal system, and that the best prospects have natural rates of at least 400 MW. Somewhere in the system, this heat must be conducted through a surface area of rock into the circulating fluids. For example, if the heat were to be collected from the ordinary heat flow of the earth—about 80 mW/m^2 (Sclater, Jaupart, and Galson, 1980)—an area of 5000 km^2 would be needed. If the fluid temperature is elevated, there must still remain a thermal gradient to the surface, and some conductive heat loss. Thus, if half the normal heat flow were to be extracted at 300°C, an area of 10,000 km^2 would have to be contacted by the fluid, and at a depth of at least 23 km. This is clearly absurd, both as to the area of coordinated flow and the depth to which surface-connected cracks must penetrate.

Higher local heat flows reduce the scale of the problem, which results in the popularity of the magma chamber source in current literature (Figure 2C). Suppose we take a very large magma chamber, like the Stillwater complex (McCallum, Raedeke, and Mathez, 1980), and model the top surface by a disk 10 km in diameter. The area is 78 km^2, requiring the heat flow to be 5 W/m^2 and the thermal gradient in basaltic rocks (conductivity 3 W/m°C) to be 1.7°C/m. From the 300°C water to the 1250°C magma the temperature drop is 950°C, implying that the boundary layer is 560 m thick. This is not incompatible with the exposed geology of such magma systems, or with the ability of a chamber of basaltic magma to convect heat to the top boundary (Lister, 1983).

The primary difficulty is to organize the percolating water into a single upwelling plume, such as that at Wairakei (Elder, 1965), and to obtain the observed water temperature of 300°–400°C, common to all such systems. The hydrothermal part of the system is still a "box model" because, although there is now a real source for the heat, the permeability of the box is still a free parameter, which is adjusted to match the water temperature. Convection tends to generate "square" cells, the horizontal extent of which are roughly equal to their vertical layer thickness (Hartline and Lister, 1981), so that the top of the magma chamber would have to

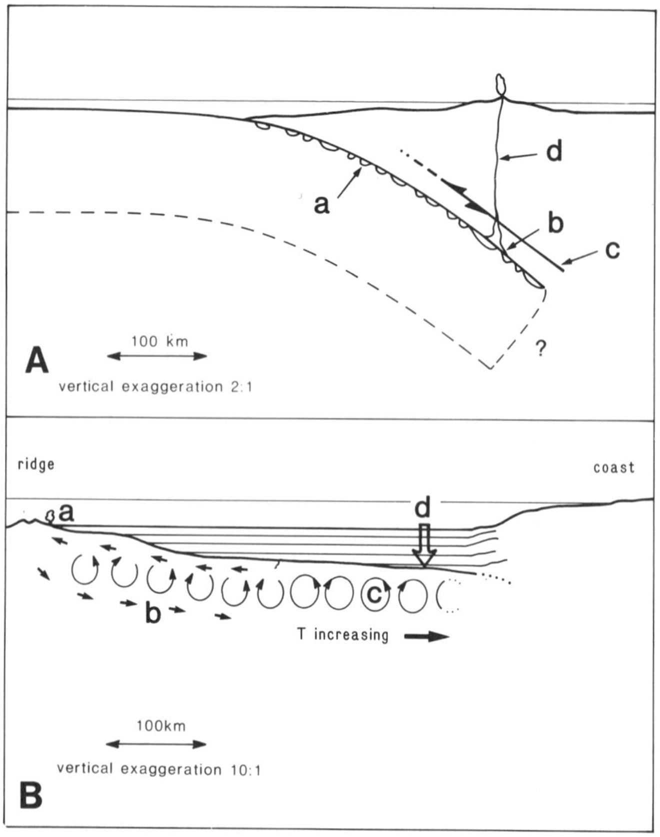

Figure 3. An attempt to illustrate two unusual sources of potential geothermal heat. A, A subduction zone, where marine sediment containing both silica and carbonate (a) are drawn down to substantial depth (b) and heated by heat conduction from dissipative shear zone (c). The upwelling CO_2-rich fluids (d) are heated further by near-isenthalpic decompression and can drive near-surface geothermal systems. B, Process by which permeable oceanic basement can become buried by detrital sediment and be reheated by the insulating effect of the sediment blanket. Direct venting of fluids to the sea floor at (a) can induce a large-scale cooling for some distance beneath the blanket (b). At (c), however, the trapped fluids are fully reheated, and could be somewhat overpressured by the weight of the sediment pile (d) if communication is lost to the vents near the ridge crest.

be at least 5 km deep to organize into a single hot plume. The restrictions needed to match the observations are all artificial—extra-large and extra-deep magma chambers, a carefully matched permeability. In addition, the latent heat of crystallization of the magma is only about a quarter of the total heat available from the intrusion.

How does a hard igneous rock become permeable? It cracks. Why do the cracks form? Because rock shrinks as it cools. The cooler the rock, the wider the cracks and the more permeable the rock becomes. This is an almost perfect feedback mechanism for controlling the hot water temperature, and it only needs to be coupled with an original crack spacing to solve the whole problem (in one dimension, at least). The mechanical properties of a pristine hot rock are not well enough known to solve the crack-spacing problem, although several mechanisms have been proposed (Lister, 1974, 1982, 1984). Major uncertainties remain in the heat transport capability of porous-medium convection, but all versions of the theory predict water temperatures in a narrow range around those observed and very high heat outputs per unit area. This is possible because the cracks are advancing into the hot rock, effectively "mining" the heat, and the conductive boundary layer is determined by the velocity of advance, not by some quasistatic mechanical system.

What this means for geothermal prospecting is that any area where hot rock is emplaced at moderate depth can be a major hydrothermal source. There is no need for a set of special conditions, such as extra-large magma chambers at great depths, or special permeabilities. All that is required is a body of hot rock, and initial access of water through tectonic action, itself an integral component of rift zones. In the resource-distribution maps that follow, major areas of rifting are the geothermal highlights.

Trench-Related Volcanism

Deep-sea trenches are the sites of subduction of oceanic lithosphere, where the surface plate sinks back into the earth's interior. In this process, some sediments and a considerable volume of weathered crustal rocks are dragged down to regions where the temperature becomes high enough to release the volatiles (Anderson, Delong, and Schwarz, 1978). The fluids produced at great depth should be a mixture of water and carbon dioxide, because there is bound water in clay minerals and a plentiful supply of both calcite and silica in marine sediments. At high enough temperatures the latter two react to liberate CO_2, probably at about 250°C in the presence of water, as estimated from the thermodynamic parameters (Robie, 1966).

It is generally accepted that arc volcanism is due to melting of hot rocks triggered by the depression of the melting point by the volatiles. The source of the heat is a matter of conjecture, but for volcanoes very near the trench, the heat is probably frictional and is produced by the mechanical action of dense lithosphere sinking into lighter mantle. As long as subduction continues, this source is inexhaustible, and can account for the impressive volume of volcanic constructs and eroded sediment observed in the west Pacific (Karig, 1971). Farther behind the trench than the volcanic arc are the so-called back-arc basins, which apparently have been formed by intermittent sea-floor spreading. Areas of back-arc rifting should be targets for careful geothermal prospecting, but in the offshore areas of the United States, the only position for such a region would be on the southern Bering Shelf. It is not clear whether the current episode of subduction at the Aleutian Trench is old enough to generate back-arc rifting on a large scale.

Another aspect of arc volcanism, and the reason for this separate section, is the rise and effusion of the volatiles themselves. The distinguishing characteristic of arc volcanoes, as opposed to those not associated with subduction, is that they are regularly explosive. Violent eruptions of the St. Helens (1980) type are impressive examples of geothermal output but are not very useful. The effusions are not always of this highly intermittent nature, and sometimes maintain a modest geothermal area, such as at Mt. Rainier (Frank, 1985), or produce a moderate event such as that at Mt. Baker (Malone, 1979; Friedman and Frank, 1980). The interesting aspect of these effusions is that their heat is not necessarily associated with that of magma. Once liberated at a depth of about 150 km by the heating of subducted material, a rising light fluid can heat itself substantially by release of gravitational energy (Figure 3A). The release of pressure, when the fluid is much less dense than the surrounding rocks, can be close to an isenthalpic expansion (Waldbaum, 1971; Ramberg, 1971). From the pressure equivalent to a depth of 150 km, the raw temperature rise could be as much as 700°C before low-pressure effects induce cooling. There is some evidence that the Mt. Rainier hydrothermal system is powered by a hot CO_2-rich fluid that mixes with glacial meltwater beneath the summit caldera, generating steam fumaroles at the summit and hot springs around the base (Frank, 1985).

What this means for geothermal prospecting is that there is an entirely different source of hydrothermal heat available above subduction zones. Moderate flows of very hot fluids may emanate in areas where there is no hot rock reservoir anywhere near the surface. The fluids themselves come from great depths and are not surface waters responding to simple thermal convection. Normally, the fluids will be associated with arc volcanoes, because a conduit must be available for the fluids to migrate to the surface. Because nothing is understood about such conduits at present, volatile-driven hydrothermal areas may exist separate from volcanoes, but none have yet been found. The appropriate zone in the Aleutians is marked on the regional map (Figure 4) as a potential high-temperature geothermal source.

Fluids Below Sediment Layers

Much of the ocean floor near the coast is covered with thick deposits of terrigenous sediment. These sediments are rich in clay minerals and have a very low permeability (Pearson and Lister, 1973) so that, if they form a complete cover for the basement, hydrothermal fluids can be trapped below. The sediments also have a low thermal conductivity, so that the temperature at their base can be substantially elevated. An estimate of basal temperature can be obtained by multiplying the heat flow by the thermal resistance of the sediment pile (Davis and Lister, 1977). The latter has to be based on assumptions about the degree of compaction and alteration of the sediment in places where the temperatures become elevated at depth, unless there is a drill hole to provide samples. To date, no such drill holes exist.

Data obtained near the northern end of the Juan de Fuca Ridge, where young ocean floor is covered by thick turbidite deposits, show that basal temperatures of 150°C are common, and a few temperatures of 200°C have been recorded. There tends to be an inverse relationship between heat flow and sediment thickness, implying that fluids circulate freely in the crust (Davis and Lister, 1977). The upper crustal rocks may therefore form a true "reservoir" of heat and hot fluid, because the poorly conducting sediment limits the total heat flux through the system. The permeability is in the form of cracks in the rock, whose distribution under these conditions is not known. Producibility of the fluids could be difficult, if drill holes happen to miss the available fractures. One such drill hole, reported by Anderson et al. (1985), gives the impression that the oceanic crust becomes impermeable at a depth of 1 km. This is not in agreement with the much more general results of heat flow measurement (Davis and Lister, 1977).

As sediment thickens toward the continental margin, the thermal resistance of the pile continues to increase. The estimated basal temperatures are high, even when the heat flows are moderate. Whether open-crack permeability exists in such deeply buried rocks can only be conjectured at present. If it does, there are more hot fluids available to be tapped by drilling. The pressure of those fluids depends on whether open permeability is retained laterally to where the oceanic basement outcrops offshore. If it does, the pressure will be hydrostatic, and will be due to the density of hot water, not cold water. A drill hole filled with cold water would therefore flow vigorously downward, as if the basement were underpressured (Anderson et al., 1985). If large-scale lateral permeability can be closed up without losing local permeability and porosity, the local zones could be substantially overpressured by the overburden and could also be hot (Figure 3B). Answers to these questions probably await actual drilling of typical sites.

REGIONAL ASSESSMENTS

For reasons of geography and tectonics, the regions covered by location maps are divided into the Aleutian chain (Figure 4), Gulf of Alaska and Bering-Chukchi Seas (Figure 5), Washington-Oregon offshore (Figure 6), and California offshore (Figure 7). Each region is unique in both tectonic style and the age of the most recent major activity, so that it is not possible to simply overlay symbols on the maps to indicate areas of geothermal interest. For a better understanding of the maps and the map symbols, the reader is referred to the text discussion that follows.

Aleutian Chain and Alaska Peninsula

The Aleutian Islands follow the line of the Aleutian Trench, a recently reactivated subduction zone whose Benioff zone, or dipping plane of earthquakes, is only 250 km wide (Engdahl and Scholz, 1977). Current interpretation equates the plane of earthquakes with the location of the cold, subducted lithospheric slab, which is therefore too narrow to induce the back-arc spreading observed in the west Pacific (Taylor and Karner, 1983). It is wide enough to produce the full panoply of volcanic effects, however, since these occur only 100–150 km beyond the trench axis.

Most of the major volcanoes have formed on the islands of the arc, but it is possible that any isolated topographic high point on the Aleutian Ridge is an underwater volcano. However, this is not necessarily the case, as sedimentary and intrusive rocks also crop out on the islands. Since large volumes of volcanic material are available to fill grabens, any closed depression near the volcanic line (about 100 km behind the trench; Scholl, Vallier, and Stevenson, this volume) has significant geothermal potential. The most recent examples are the Buldir Depression and the Sunday Basin, both of which have some unknown relationship to the Bowers Ridge, the sunken remains of a former north-facing arc (Figure 4). More sediment-filled, and therefore probably older, are the Amlia and Amukta Basins.

Hot springs are quite common on the Aleutian Islands, the main ones being marked by dots on Figure 4 (Motyka, Moorman, and Liss, 1981). Whether there are more hot springs on the submerged parts of the ridge depends on whether or not subduction-derived fluids upwell between volcanoes. In the case of the Cascade Range, which consists of mountains containing volcanoes associated with a former trench system, all the significant geothermal areas are associated with active volcanoes (Oregon Department of Geology and Mineral Industries, 1982; Washington Department of Natural Resources, Division of Geology and Earth Resources, 1981).

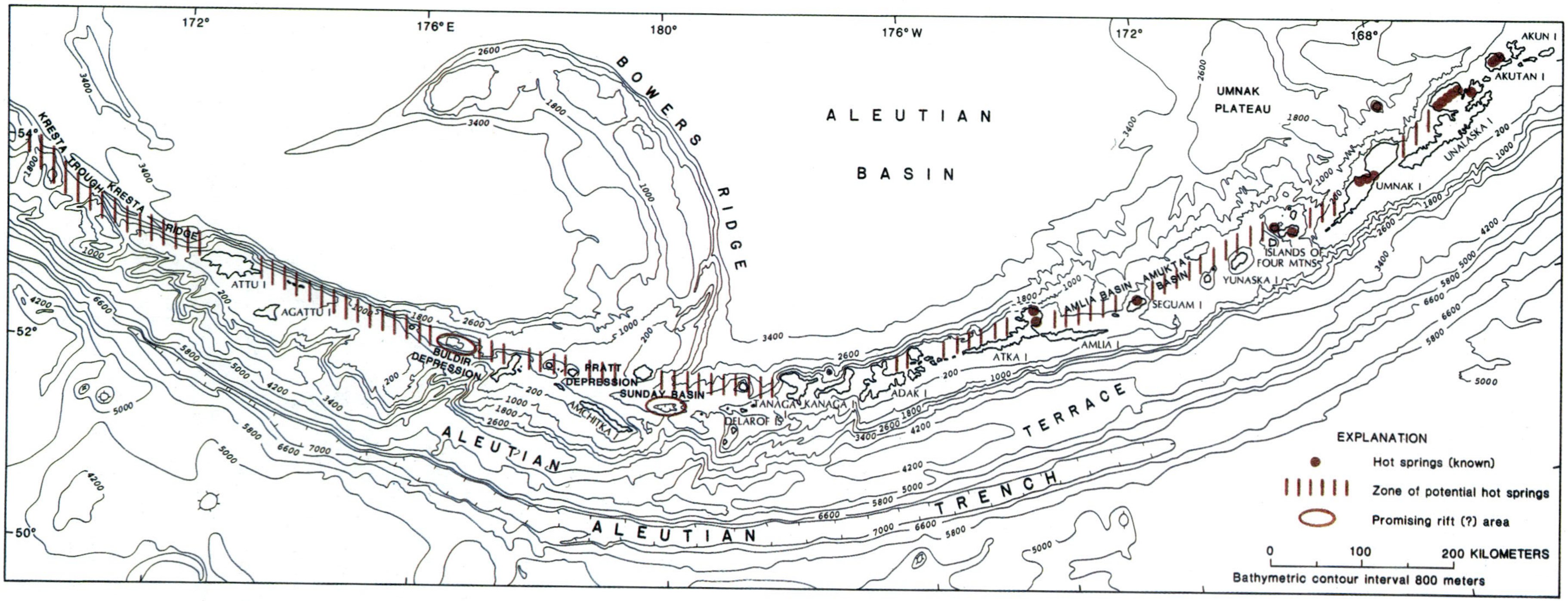

Figure 4. Map of the Aleutian Island Arc showing the sites of known hot springs (dots), areas of possible rifting (ovals), and zone of potential undersea hot springs similar to those found on land. Trench-arc hot springs seem to be powered by hot fluids rising from the subducted slab, even though there is usually substantial mixing with surface waters (see Figure 3A). They are normally found on or near arc volcanoes, and such volcanoes should break the sea surface on an arc as shallow as this. The rifted basins may or may not be recent enough, and magma-filled enough, to be prime geothermal prospects.

At the east end of the Aleutian chain lies the Alaska Peninsula (Bruns et al., this volume; Fisher, Detterman, and Magoon, this volume). The tectonic ridge is higher and wider here than farther west, partly because of the great thickness of sediment on the oceanic plate. All but one of the volcanoes are therefore above sea level, and the potential for significant offshore geothermal sources is very low. Cool springs, caused by the dewatering of the sediment, may be fairly common.

Gulf of Alaska and Bering-Chukchi Seas

The margins of continental Alaska on the west and north do not appear to have been tectonically active in the last 10 m.y. Conductive cooling would penetrate 17 km (to half the initial temperature) in this time in normal crystalline rocks. No massive geothermal resources should be expected, although there could be minor hot springs associated with localized recent volcanism, such as near some islands on the Bering Shelf (Cooper and Marlow, 1983; Cooper, Marlow, and Scholl, this volume). Several deep sedimentary basins are present (Marlow, Cooper, and Fisher, this volume), which could contain both oil and geopressured fluids.

Arctic Coast of Alaska

Alaska is bounded to the north and northwest by a broad continental shelf sloping gently into the Canada Basin of the Arctic Ocean. The basin is floored by a 4- to perhaps 12-km-thick sedimentary layer, and the oceanic crustal rocks below have the seismic signature of very old material (Mair and Forsyth, 1982). Aeromagnetic anomalies have been tentatively identified as the M-series, and are dated as 153–127 Ma (Taylor et al., 1981). Studies of the margin attribute the anomalies to the opening of the Canada Basin, mainly in the Early Cretaceous, and there are no recent volcanic rocks on the surface (Grantz, Eittreim, and Dinter, 1979; Grantz, May, and Dinter, this volume). There are thus no obvious geothermal prospects, or probabilities of sediment-dewatering springs over 20°C in temperature.

Bering Shelf

The constructional history of the oceanic part of the Bering Sea is not as agreed upon as that of the Canada Basin, but seismic-reflection lines indicate that the floor is covered by about 4 km of sediment (Cooper, Marlow, and Ben-Avraham, 1981; Cooper, Scholl, and Marlow, this volume). It is thus also very old, and the Bering Shelf margin has the thick sedimentary constructions one would expect from long-term quiescence (Marlow and Cooper, 1980; Cooper and Marlow, 1983). The difference, from the geothermal point of view, is that there are islands of Quaternary volcanic rocks on the shelf, and short-wavelength and high-amplitude magnetic anomalies on the shelf near some of these islands (Figure 5).

The geothermal prospects depend, in part, on the nature of the source for these volcanic piles—are they related to the present Aleutian Trench, a former Aleutian Trench, or something else altogether? All the islands are beyond the 250-km width of the present descending slab. The trend of the volcanic edifices between the Pribilof Islands and the Nunivak Island/Yukon Delta area follows the general trend of the trench and other Alaskan structures, whereas a somewhat older band of volcanic masses follows a similar general trend between St. Lawrence Island and the Seward Peninsula, still farther removed from the present trench (Beikman, 1980). The younger band lies about 500 km from the trench axis, which is about the current distance of back-arc spreading centers from their respective trenches (Taylor and Karner, 1983), but those spreading centers started within the island arcs and grew to their present positions (Karig, 1971). Whatever the source, one would be wise to assume that these magmas do not contain the high proportion of volatiles that can sustain geothermal areas of the Rainier type (Frank, 1985). Any geothermal activity would have to depend on magmatic heat, and the most likely subsurface storage would be in sills within the sediment pile on the shelf. On this basis, the Pribilof Islands would be the most promising area.

Gulf of Alaska

At its eastern end, the Aleutian Trench becomes buried by an enormous pile of continentally derived sediment, old examples of which are being upthrust into nonvolcanic islands along the trend of the arc (Beikman, 1980). The tectonic deformation seems to accelerate the dewatering of the sediment, and many springs are probably present on the continental slope from Kodiak to Cordova. Most of these will be cool, because the ocean floor beneath the Gulf is relatively old and therefore the thermal gradients are low—corresponding to a heat flow of about 80 mw/m^2 (Pitman and Hayes, 1968). This (theoretical) heat flow corresponds to a gradient of about 80°C/km in unconsolidated sediment, and about 40°C/km in consolidated rocks. Upon rapid tectonic compression, the sediment pile thickens without increasing the basal temperature, and so the thermal gradient is decreased in proportion to the thickening. Actual thermal gradients of about 30°C/km have been measured in the upper slope region, both in a drill hole near Middleton Island, and from the depths of anomalous seismic reflectors interpreted as gas hydrate phase boundaries (Kvenvolden and von Huene, 1985; von Huene, Fisher, and Bruns, this volume). Spring temperatures of 10°–15°C may therefore be expected.

South of Yakutat Bay, the continental margin is tectonically a transform fault, where the ocean floor is slipping northwestward relative to the land. Sometimes such simple strike-slip motion does not seem to have any thermal effect at all (Henyey and Wasserberg, 1971). However, if there is any misalignment between the fault and the motion, substantial compression or extension can occur over time, and the extension is a form of rifting which would be very promising geothermally. Most of the motion is inferred to occur along the Fairweather fault, and its offshore extension in southeastern Alaska (Beikman, 1980). At the northern end, any misalignment must be expected to be compressive, as there is an uplifted coastline and high mountains close to the shore. Farther south are fjords, which are characteristic of a sinking shoreline, or at least of no substantial uplift since the glaciations of the Pleistocene. An extraordinary feature is an isolated Quaternary volcano, Mount Edgecumbe, on Kruzof Island very close to the Fairweather fault, where the azimuth of the fault line changes slightly (Figure 5). In spite of the generally compressive nature of the continental margin, there is a possibility of other igneous effusions to the south, and the fault zone itself could be a zone of enhanced permeability, and therefore of some hot spring activity.

Figure 5.. Generalized map of Alaska showing Quaternary volcanoes (ovals), older volcanics (dashed ovals), hot springs on the Alaska Peninsula (dots), and the Fairweather fault (dashed line). The latter is included as a geothermal prospect because the active strike-slip motion can provide enhanced permeability, because of minor rifting at misalignments, and because of the possible relationship of the fault with the Quaternary volcano Mt. Edgecumbe. Other recent volcanic sites may be promising geothermally, depending upon the distribution of the volcanic intrusions (see Figure 1).

Washington-Oregon-Cape Mendocino Coastline

The west coast of the continental United States is split into two sections for discussion of geothermal resources. This is due to the tectonic break at Cape Mendocino, with an inland strike-slip fault to the south and actively spreading ridges offshore to the north (Figures 6, 7). The two ridges are the prime sites of the whole region for major offshore geothermal areas—rifting is rapid and hot rock is known to fill the gap up to the surface. Four active vent areas have been found on the Juan de Fuca Ridge so far (Kingston, Delaney, and Johnson, 1983; Normark et al., 1983; Chase et al., 1985; Morton et al., this volume). None have yet been located on

Figure 6. Canadian border to Cape Mendocino. There are two actively spreading ridges offshore. These fulfill all the requirements for highly active geothermal systems to develop, and a number of hot springs have been found (dots). In addition, young ocean floor is buried by thick sequences of continentally derived sediment. Much of this area is a prime geothermal prospect, because permeable ocean floor has been reheated by the insulating effect of the sediment pile on the high heat flow of young lithosphere (shadings). See also Figure 3B.

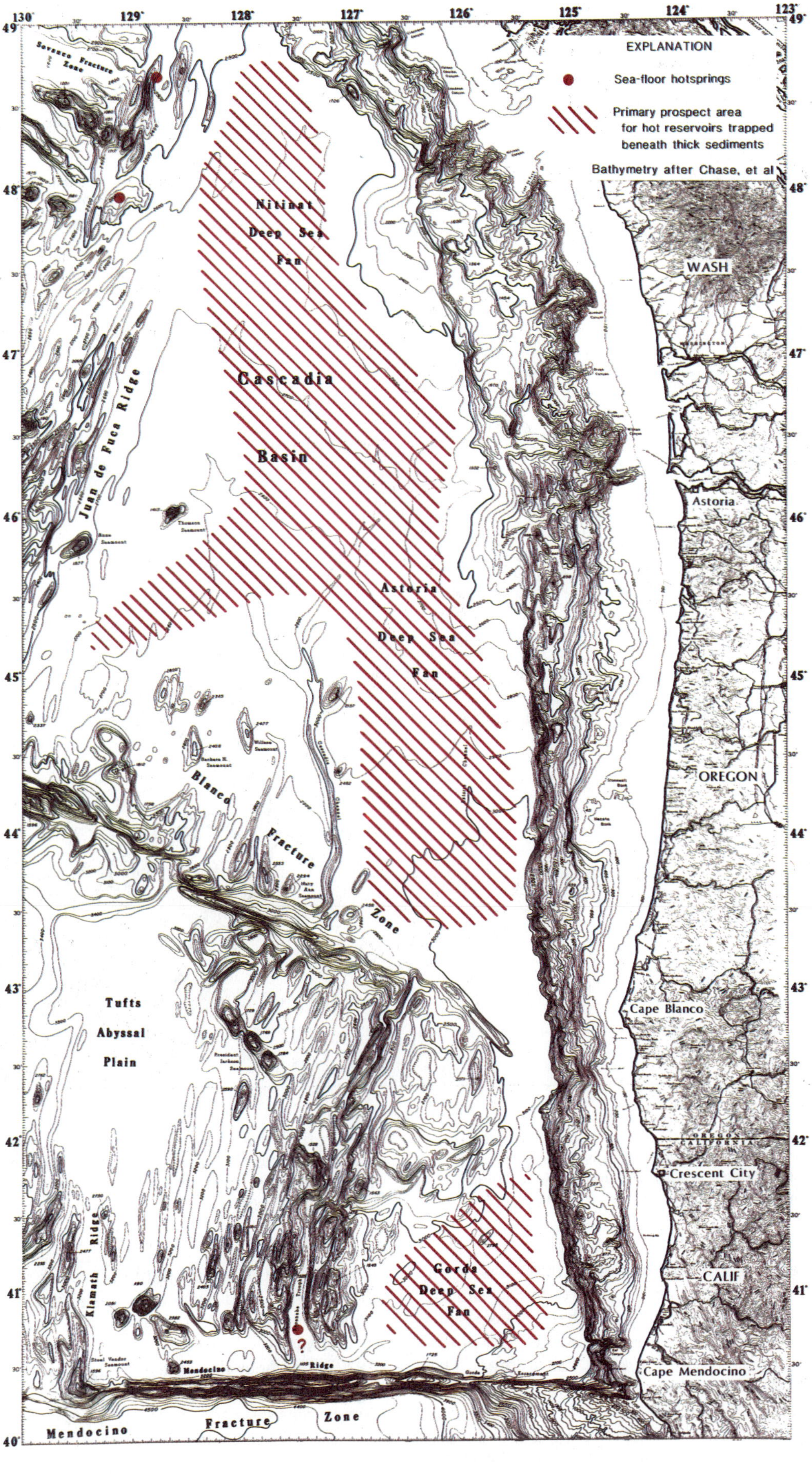
EXPLANATION
Sea-floor hotsprings
Primary prospect area for hot reservoirs trapped beneath thick sediments
Bathymetry after Chase, et al
Sovanco Fracture Zone
Nitinat Deep Sea Fan
Cascadia Basin
Juan de Fuca Ridge
Astoria Deep Sea Fan
Blanco Fracture Zone
Tufts Abyssal Plain
Klamath Ridge
Gorda Deep Sea Fan
Mendocino Ridge
Mendocino Fracture Zone
WASH
Astoria
OREGON
Cape Blanco
Crescent City
CALIF
Cape Mendocino
130°
129°
128°
127°
126°
125°
124°
123°
49°
48°
47°
46°
45°
44°
43°
42°
41°
40°

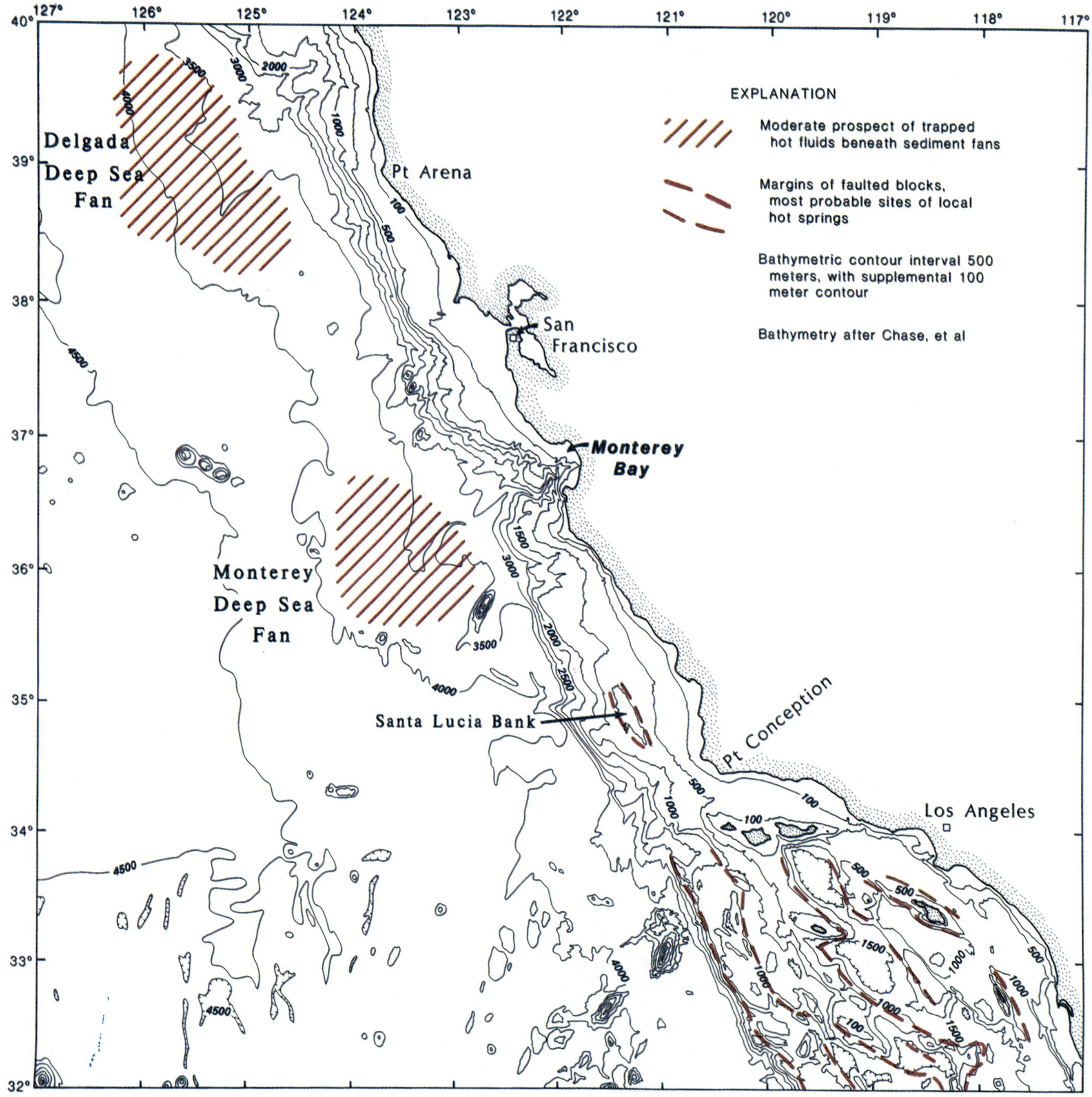

Figure 7. Cape Mendocino to Mexican border. The ocean floor is older but still has a relatively high heat flow. Areas buried by thick sediment are moderate geothermal prospects of the reservoir type (shading) (see also Figure 3B). Off southern California the offshore area is characterized by faulted blocks of foundered continental crust. Permeability should be enhanced along the faults, and some hot springs may occur along these (dashed lines). Because of the age of the stretching event, significant geothermal powers are unlikely, but chemically rich waters are probable.

the Gorda Ridge, but a hydrothermal plume has been reported near the northern rift zone (cruise communication, 1985) and, at a probable hot spring, unoxidized massive sulfides have been recovered from the sedimented floor of the ridge's southern axis (Clague and Holmes, this volume). The data gap of the Gorda Ridge is being filled in rapidly because of leasing prospects (Zippin, 1983).

East of the Juan de Fuca Ridge, Cascadia Basin is deeply filled with terrigenous sediment that is only now finding a way out via Cascadia Channel through the Blanco Fracture Zone (Figure 6; Carlson and Nelson, this volume). The ocean crust under the sediment is young (Vine, 1968), and so the basic heat flow from the oceanic lithosphere is high (Lister, 1977). The heat flow measured on the surface of the sediment is depressed in some parts of the central basin and some areas near the ridge (Korgen, Bodvarsson, and Mesecar, 1971; Moran and Lister, 1986), but has a peak approaching the theoretical about 100 km from the continental slope.

The inferred temperatures at the base of the sediment here are high, and if rock permeability is maintained, a large area of the "reservoir" type of geothermal resource may exist. Drillers should take account of the fact that a hole filled with cold water is an artificial downwelling convection limb because of the high density of the cold water. Considerable pumping may be required to reverse the flow and bring hot water up the hole, after which the natural flow may become upward through a hole that is now hotter than its surroundings.

The area of the Gorda deep-sea fan is also a good prospect for trapped high-temperature geothermal fluids. The ocean floor is even younger than that east of the Juan de Fuca Ridge (Vine, 1968), and there is plentiful sediment cover (Carlson and Nelson, this volume). Present heat-flow measurements (Von Herzen, 1964) indicate that the fan area has high heat flow; in conjunction with the thick sediment, basal temperatures should be high.

Cape Mendocino to the Mexican Border

The California margin south of Cape Mendocino is generally divided into the section that extends to Point Conception and the southern California borderland which extends well down into offshore Baja California. North of Point Conception the continental margin seems fairly normal, except perhaps for the narrowness of the shelf, whereas to the south there is a broad zone of blocks and basins extending offshore at depths midway between sea level and the deep ocean (Figure 7). A more careful contemplation of the map shows that the margin looks normal north of Monterey Bay, becomes transitional in character, and then blossoms into the full rifted complexity south of Point Conception. The intermediate zone appears to be a buried version of the openly stretched continental crust to the south, and could be merely the older result of the migration of a triple junction southward along the coast (Howell et al., 1980). However, this does not explain the dramatic increase in the amount of stretching, and also it does not take into account the coincidence between the boundary and a major current disturbance in the line of the San Andreas fault (Vedder and Howell, 1980).

The tectonic history is important to the geothermal interpretation, because prospects in the obviously rifted southern borderland depend on how recently the rifting took place. The oceanic crust presently off the borderland is at least as old as 15 Ma, and becomes older to the north (Klitgord and Mammericx, 1982). According to the model of Atwater (1970), the age of the residual oceanic crust corresponds closely to the conversion of a subducting margin to one of strike-slip motion. The Miocene volcanic rocks found on the ridges of the borderland confirm this date for the initial tectonic event (Vedder et al., 1974; Vedder, this volume).

If a continental crust is stretched, the prior thermal gradient is enhanced, and the decay of the elevated temperatures causes a subsidence that continues long after the loss of elevation due to the stretching itself (e.g., Sclater and Christie, 1980). The above scenario would predict a continued general subsidence, which is not in agreement with the existence of raised terraces on the offshore islands (Vedder and Howell, 1980). Are these raised terraces associated with the compression of the transverse ranges (Howell et al., 1980), or are they due to rejuvenated rifting associated with the opening of the Basin and Range province on land? The undisturbed nature of the Quaternary sediments on seismic profiles points to the former explanation.

Cape Mendocino to Monterey Bay

This quiescent margin borders ocean floor of 15–20 Ma age. Permeable oceanic crustal rocks beneath the Delgada and Monterey deep-sea fans may constitute a moderate geothermal resource, as the lithospheric heat flow is still above 120 mW/m^2 and thermal gradients through the sediment could approach 100°C/km (Figure 7).

There are many more hot springs on the California coast than on the coasts of Washington and Oregon, and the heat flow is much higher (California Department of Conservation, Division of Mines and Geology, 1980). The coastal terrain is also rising instead of being reasonably stable. These features can be due either to the thermal rebound after cessation of subduction, with the heat releasing fluids from subducted sediment in the Franciscan Formation, or to compressive tectonics related to the expansion of the Basin and Range province, with both the new heat and the fluids coming from diapirs of Franciscan material. The latter explanation seems more likely when one compares the California situation with that of the British Columbia inlets, where the coast switched from subduction to transform motion at about the same time (e.g., Hyndman, 1976). Similar moderate-sized hot springs should be expected on the continental shelf, where sediment cover is not too thick.

Monterey Bay to Point Conception

South of the Monterey deep-sea fan, the ocean floor is not sufficiently sediment-covered to constitute a probable geothermal resource, although moderate-temperature fluids undoubtedly circulate beneath the sediment. The coastal hot-spring activity is similar to that farther north, and thus some springs may be expected on the continental shelf. Deep sediment-filled basins are not good prospects, because this type of spring is associated with local permeability in a generally rather impermeable matrix. Random drilling would be unlikely to intersect such a conduit, so that fault zones in the seismic basement cropping out near the sea floor are the best prospects. The edges of Santa Lucia Bank would be one obvious place to look (Figure 7).

Southern California Borderland

Because this is a stretched area rather than a compressed area, diapiric activity is unlikely. A few hot springs have been found in the Los Angeles area, associated with an active fault zone (California Department of Conservation, Division of Mines and Geology, 1980). This type of activity might be expected in the offshore area if any of the faults associated with the stretching remain active. The heat flow is moderate throughout the borderland (Lee and Henyey, 1975; Henyey, 1976) and is commensurate with the date of the original tectonic event. High temperatures can be expected at the base of thickly filled sedimentary basins; however, because the rocks in this area are continental crust rather than young ocean floor, the permeability needed for extraction is unlikely to be present. Minor hot springs, which are therefore the only probable geothermal activity, can be expected to occur where major basin-forming faults crop out on the ridges. Barite has been found on steep basin slopes and is also present in veins in the Palos Verdes Hills (Emery, 1960). Although the origin and deposition mechanisms are not understood, upward-migrating fluids are the most probable source. The chemistry of barium suggests that the flows were most likely slow and diffuse, more from the dewatering of heated sediment than the direct effect of magmatic intrusion.

REFERENCES CITED

Anderson, R. N., S. E. Delong, and W. M. Schwarz, 1978, Thermal model for subduction with dehydration in the downgoing slab: Journal of Geology, v. 86, p. 731-739.

Anderson, R. N., M. D. Zoback, S. H. Hickman, and R. L. Newmark, 1985, Permeability versus depth in the upper oceanic crust: In situ measurements in Deep Sea Drilling Project hole 504B, eastern equatorial Pacific: Journal of Geophysical Research, v. 90, p. 3659-3669.

Atwater, T., 1970, Implications of plate tectonics for the Cenozoic tectonic evolution of western North America: GSA Bulletin, v. 81, p. 3513-3536.

Beikman, H. M., 1980, Geologic map of Alaska: State of Alaska Department of Natural Resources, Division of Geological and Geophysical Surveys, scale 1:2,500,000.

Bodvarsson, G., and R. P. Lowell, 1972, Ocean floor heat flow and the circulation of interstitial waters: Journal of Geophysical Research, v. 77, p. 4472-4475.

Bruns, T. R., R. von Huene, R. C. Culotta, S. D. Lewis, and J. W. Ladd, 1987, Geology and petroleum potential of the Shumagin margin, Alaska, 1987.

California Department of Conservation, Division of Mines and Geology, 1980, Geothermal resources of California: Sacramento, California, scale 1:750,000.

Carlson, P. R., and C. H. Nelson, 1987, Marine geology and resource potential of the Cascadia Basin, 1987.

Chase, R. L., J. R. Delaney, J. L. Karsten, H. P. Johnson, S. K. Juniper, J. E. Lupton, S. D. Scott, V. Tunnicliffe, S. R. Hammond, and R. E. McDuff, 1985, Hydrothermal vents on an axis seamount of the Juan de Fuca Ridge: Nature, v. 313, p. 212-214.

Clague, D. A., and M. L. Holmes, 1987, Geology, petrology, and mineral potential of the Gorda Ridge, 1987.

Cooper, A. K., and M. S. Marlow, 1983, Preliminary results of geophysical and geological studies of the Bering Sea Shelf during 1982: U.S. Geological Survey Open-File Report 83-222, 8 p.

Cooper, A. K., M. S. Marlow, and Z. Ben-Avraham, 1981, Multichannel seismic evidence bearing on the origin of Bowers Ridge, Bering Sea: GSA Bulletin, v. 92, p. 474-484.

Cooper, A. K., M. S. Marlow, and D. W. Scholl, 1987, Geologic framework of the Bering Sea crust, 1987.

Cooper, A. K., D. W. Scholl, and M. S. Marlow, 1987, Structural framework, sedimentary sequences, and hydrocarbon potential of the Aleutian and Bowers Basins, Bering Sea, 1987.

Davis, E. E., and C. R. B. Lister, 1977, Heat flow measured over the Juan de Fuca Ridge: Evidence for widespread hydrothermal circulation in a highly heat transportive crust: Journal of Geophysical Research, v. 82, p. 4845-4860.

Elder, J. W., 1965, Physical processes in geothermal areas: American Geophysical Union Monograph, v. 8, p. 211-239.

Ellis, A. J., and S. H. Wilson, 1955, The heat from the Wairakei-Taupo thermal district calculated from the chloride output: New Zealand Journal of Science and Technology, v. 36B, p. 622-631.

Emery, K. O., 1960, The sea off southern California: New York, John Wiley & Sons, 365 p.

Engdahl, E. R., and C. H. Scholz, 1977, Double Benioff zone beneath the central Aleutians: An unbending of the lithosphere: Geophysical Research Letters, v. 4, p. 473-476.

Fisher, M. A., R. L. Detterman, and L. B. Magoon, 1987, Tectonics and petroleum geology of the Cook-Shelikof Basin, southern Alaska, 1987.

Fisher, R. G., 1964, Geothermal heat flow at Wairakei during 1958: New Zealand Journal of Geology and Geophysics, v. 7, p. 172-184.

Frank, D. G., 1985, Hydrothermal processes at Mount Rainier, Washington: Seattle, University of Washington unpublished Ph.D. thesis, 195 p.

Friedman, J. D., and D. Frank, 1980, Infrared surveys, radiant flux and total heat discharge at Mount Baker volcano, Washington, between 1970 and 1975: U.S. Geological Survey Professional Paper 1022-D, 40 p.

Grantz, A., S. Eittreim, and D. A. Dinter, 1979, Geology and tectonic development of the continental margin north of Alaska: Tectonophysics, v. 59, p. 263-291.

Grantz, A., S. D. May, and D. A. Dinter, 1987, Regional geology and petroleum potential of the United States Beaufort and northeasternmost Chukchi Seas, 1987.

Hartline, B. K., and C. R. B. Lister, 1981, Topographic forcing of super-critical convection in a porous medium such as the oceanic crust: Earth and Planetary Science Letters, v. 55, p. 75-86.

Henyey, T. L., 1976, Heat flow and tectonic patterns on the southern California borderland, *in* D. G. Howell, ed., Aspects of the geologic history of the California continental borderland: AAPG, Pacific Section, Miscellaneous Publication 24, p. 427-448.

Henyey, T. L., and G. J. Wasserberg, 1971, Heat flow near major strike-slip faults in California: Journal of Geophysical Research, v. 76, p. 7924-7946.

Howell, D. G., J. K. Crouch, H. G. Greene, D. S. McCulloch, and J. G. Vedder, 1980, Basin development along the late Mesozoic and Cenozoic California margin: A plate tectonic margin of subduction, oblique subduction and transform tectonics: International Association of Sedimentologists Special Publication, v. 4, p. 43-62.

Hyndman, R. D., 1976, Heat flow measurements in the inlets of southwestern British Columbia: Journal of Geophysical Research, v. 81, p. 337-349.

Hyndman, R. D., 1983, Juan de Fuca plate map, JFP-10, Geothermal heat flux: Pacific Geoscience Centre, Earth Physics Branch, Department of Energy, Mines and Resources, Sidney, British Columbia, scale 1:2,000,000.

Karig, D. E., 1971, Structural history of the Mariana island arc system: GSA Bulletin, v. 82, p. 323-344.

Kingston, M. J., J. R. Delaney, and H. P. Johnson, 1983, Sulphide deposits from the Juan de Fuca Ridge at 47°57′N, 129°06′W: Marine Technology Society Proceedings of Oceans 83, v. 2, p. 811-815.

Klitgord, K. D., and J. Mammericx, 1982, Northern East Pacific Rise: Magnetic anomaly and bathymetric framework: Journal of Geophysical Research, v. 87, p. 6725-6750.

Korgen, B. J., G. Bodvarsson, and R. S. Mesecar, 1971, Heat flow through the floor of Cascadia Basin: Journal of Geophysical Research, v. 76, p. 4758-4774.

Kvenvolden, K. A., and R. von Heune, 1985, Natural gas generation in sediments of the convergent margin of the eastern Aleutian Trench area, *in* D. G. Howell, ed., Tectonostratigraphic terranes of the Circum-Pacific region: Circum-Pacific Council for Energy and Minerals, Earth Science Series 1, p. 31-49.

Lee, T., and T. L. Henyey, 1975, Heat flow through the southern California borderland: Journal of Geophysical Research, v. 80, p. 3733-3743.

Lister, C. R. B., 1974, On the penetration of water into hot rock: Geophysical Journal of the Royal Astronomical Society, v. 39, p. 465-509.

Lister, C. R. B., 1977, Estimators for heat flow and deep rock properties based on boundary-layer theory: Tectonophysics, v. 41, p. 157-171.

Lister, C. R. B., 1982, "Active" and "passive" hydrothermal systems in the oceanic crust: Predicted physical conditions, *in* K. A. Fanning and F. T. Manheim, eds., The dynamic environment of the ocean floor: Lexington, Massachusetts, D. C. Heath, p. 441-470.

Lister, C. R. B., 1983, On the intermittency and crystallization

mechanisms of sub-seafloor magma chambers: Geophysical Journal of the Royal Astronomical Society, v. 73, p. 351–366.
Lister, C. R. B., 1984, The basic physics of water penetration into hot rock, *in* P. A. Rona, K. Bostrom, L. Laubier, and K. L. Smith, Jr., eds., Hydrothermal processes at sea-floor spreading centers: New York, Plenum Press, North Atlantic Treaty Organization Conference Series IV, v. 12, p. 141–168.
Mair, J. A., and D. A. Forsyth, 1982, Crustal structures of the Canada Basin near Alaska, the Lomonosov Ridge and adjoining basins near the North Pole: Tectonophysics, v. 89, p. 239–253.
Malone, S. D., 1979, Gravity changes accompanying increased heat emission at Mount Baker, Washington: Journal of Volcanology and Geothermal Research, v. 6, p. 241–256.
Marlow, M. S., and A. K. Cooper, 1980, Mesozoic and Cenozoic structural trends under the southern Bering Sea Shelf: AAPG Bulletin, v. 64, p. 2139–2155.
Marlow, M. S., A. K. Cooper, and M. A. Fisher, 1987, Petroleum geology of the Beringian continental shelf, 1987.
McCallum, I. S., L. D. Raedeke, and E. A. Mathez, 1980, Investigations of the Stillwater complex: Part 1, Stratigraphy and structure of the banded zone: American Journal of Science, v. 280-A, p. 59–87.
Moran, J. E., and C. R. B. Lister, 1986, Heat flow across Cascadia Basin: in preparation.
Morton, J. L., W. R. Normark, D. M. Mann, and D. H. Tompkins, 1987, Geology, evolution, and mineral potential of the Juan de Fuca Ridge, 1987.
Motyka, R. J., M. A. Moorman, and S. A. Liss, 1981, Assessment of thermal springs sites Aleutian Arc, Atka Island to Becherof Lake—Preliminary results and evaluation: Alaska Department of Natural Resources, Division of Geological and Geophysical Surveys Open File Report 144, 173 p.
Normark, W. R., J. L. Morton, R. A. Koski, D. A. Clague, and J. R. Delaney, 1983, Active hydrothermal vents and sulphide deposits on the southern Juan de Fuca Ridge: Geology, v. 11, p. 158–163.
Oregon Department of Geology and Mineral Industries, 1982, Geothermal resources of Oregon: Salem, Oregon, scale 1:500,000.
Pearson, W. C., and C. R. B. Lister, 1973, Permeability measurements on a deep sea core: Journal of Geophysical Research, v. 78, p. 7786–7787.
Pitman, W. C., III, and D. E. Hayes, 1968, Sea floor spreading in the Gulf of Alaska: Journal of Geophysical Research, v. 73, p. 6571–6580.
Ramberg, H., 1971, Temperature changes associated with adiabatic decompression in geologic processes: Nature, v. 234, p. 539–540.
Robie, R. A., 1966, Thermodynamic properties of minerals, *in* S. P. Clark, Jr., ed., Handbook of physical constants: GSA Memoir 97, p. 437–458.
Scholl, D. W., T. L. Vallier, and A. J. Stevenson, 1987, Geologic evolution and petroleum geology of the Aleutian Ridge, 1987.
Sclater, J. G., and P. A. F. Christie, 1980, Continental stretching: An explanation of the post-mid-Cretaceous subsidence of the central North Sea Basin: Journal of Geophysical Research, v. 85, p. 3711–3739.
Sclater, J. G., C. Jaupart, and D. Galson, 1980, The heat flow through oceanic and continental crust and the heat loss of the earth: Reviews of Geophysics and Space Physics, v. 18, p. 269–311.
Sleep, N. H., 1984, Hydrothermal convection at ridge axes, *in* P. A. Rona, K. Bostrom, L. Laubier, and K. L. Smith, Jr., eds., Hydrothermal processes at seafloor spreading centers: New York, Plenum Press, North Atlantic Treaty Organization Conference Series IV, v. 12, p. 71–82.
Taylor, B., and G. D. Karner, 1983, On the evolution of marginal basins: Reviews of Geophysics and Space Physics, v. 21, p. 1727–1741.
Taylor, P. T., L. C. Kovacs, P. R. Vogt, and G. L. Johnson, 1981, Detailed aeromagnetic investigation of the Arctic Basin, 2: Journal of Geophysical Research, v. 86, p. 6323–6333.
Vedder, J. G., 1987, Regional geology and petroleum potential of the southern California borderland, 1987.
Vedder, J. G., and D. G. Howell, 1980, Topographic evolution of the southern California borderland during late Cenozoic time, *in* D. M. Power, ed., The California Islands: Proceedings of a Multidisciplinary Symposium, Santa Barbara, California, Santa Barbara Museum of Natural History, p. 7–31.
Vedder, J. G., L. A. Beyer, A. Junger, G. W. Moore, A. E. Roberts, J. C. Taylor, and H. C. Wagner, 1974, Preliminary report on the geology of the continental borderland of southern California, to accompany map MF-624: U.S. Geological Survey, 34 p.
Vine, F. J., 1968, Magnetic anomalies associated with mid-ocean ridges, *in* R. H. Phinney, ed., The history of the earth's crust, a symposium: Princeton, New Jersey, Princeton University Press, p. 73–89.
Von Herzen, R. P., 1964, Ocean-floor heat-flow measurements west of the United States and Baja California: Marine Geology, v. 1, p. 225–239.
von Huene, R., M. A. Fisher, and T. R. Bruns, 1987, Geology and evolution of the Kodiak margin, Gulf of Alaska, 1987.
Waldbaum, D. R., 1971, Temperature changes associated with adiabatic decompression in geologic processes: Nature, v. 234, p. 539–540.
Washington Department of Natural Resources, Division of Geology and Earth Resources, 1981, Geothermal resources of Washington: Olympia, Washington, scale 1:500,000.
Zippin, J. P. (coordinator), 1983, Draft environmental impact statement: Proposed polymetallic sulphide minerals lease offering, Gorda Ridge area, offshore Oregon and northern California: Minerals Management Service, 580 p.

CHAPTER THIRTY-TWO

COBALT-RICH FERROMANGANESE CRUSTS FROM THE EXCLUSIVE ECONOMIC ZONE OF THE UNITED STATES AND NODULES FROM THE OCEANIC PACIFIC

James R. Hein
Lisa A. Morgenson
David A. Clague
Randolph A. Koski

U.S. Geological Survey
Menlo Park, California

ABSTRACT

We compiled data from the published literature and from unpublished U.S. Geological Survey files on the chemistry and distribution of cobalt-rich ferromanganese crusts in the Pacific. Single-element and element-ratio maps are presented for Mn, Fe, Fe/Mn, Co, Ni, Cu, and Pb.

Ferromanganese crusts coat most hard substrates on the flanks of seamounts, ridges, and plateaus. Crusts commonly average 2 to 4 cm in thickness. Growth rates range from 2.5 to 5 mm/m.y. Crusts are generally younger than the substrates upon which they rest.

Cobalt contents as high as 2.5 weight % have been measured. In addition, crusts are rich in platinum (to 1.3 ppm), nickel (to 1.0%) and Pb, Ti, and Ce. Crusts are composed predominantly of $\delta - MnO_2$ and $FeOOH \cdot x H_2O$. Co, Ni, Pb, Pt, Ce, As, Mo, Cd, and Zn are positively correlated with the manganese phase, and Cu and Be with the iron phase. Fe, Cu, Be, K, Ce, Al, Ti, Si, and Ba increase in crusts with increasing depth of water, whereas, Mn, Co, Ni, Pb, Mo, Cd, As, and Sr show the inverse relationship.

The metals in the crusts are derived from seawater. Once the first layer of ferromanganese oxide is deposited, the addition of metals to the oxide surface becomes autocatalytic. The relative amounts of oxide phases in the crusts is determined by their ratios in colloidal flocs in the surrounding seawater. The crusts rich with cobalt occur in water depths of <2500 m and lie beneath the oxygen-minimum zone where the Fe/Mn ratio is small. Cobalt and other metals are fixed by lattice substitution for Mn^{+4} in the MnO_2 or by coprecipitation of cobalt oxide with manganese oxide.

Criteria to emphasize that may be important in locating areas of thick cobalt-rich crusts include (1) volcanoes and ridges shallower than 1500 to 1000 m, (2) substrates older than about 20 Ma, (3) areas of strong current activity, (4) volcanic structures not capped by large modern atolls or reefs, (5) a shallow and well-developed oxygen-minimum zone, and (6) slope stability and absence of local volcanism.

These criteria and high concentrations of cobalt, nickel, and platinum can be used to outline areas of high economic potential, which include the Kiribati, Marshall, and Micronesia island groups, and Johnston and Kingman-Palmyra Islands.

INTRODUCTION

Ferromanganese crusts coat most hard substrates within the Pacific Basin. Crusts from water depths shallower than about 2500 m are enriched in cobalt and differ in composition from abyssal nodules, hydrothermal crusts, and hydrogenetic crusts occurring in deeper water. This report emphasizes these shallow-water cobalt-rich hydrogenous crusts, hereafter referred to as crusts. Crusts occur on the flanks of seamounts, islands, plateaus, and other topographically positive areas in the Pacific. Sediment-covered flat-topped guyots and plateaus commonly contain cobalt-rich manganese nodules resting on the sediment surface in addition to crusts on nearby rock outcrops.

A first-order distribution of ferromanganese crusts in the Pacific can be related to three tectonic settings. Cobalt-rich ferromanganese crusts occur on mid-plate seamounts (linear volcanic chains), ridges, and plateaus, such as the Line Islands, Necker Ridge, the Hawaiian Ridge, and Manihiki Plateau. The edifices must be older than about 20 m.y. for crusts to be of significant thickness for consideration as a resource, as the rate of growth of these hydrogenous crusts is very slow. In the second tectonic setting, manganese-rich crusts occur along spreading axes such as the Galapagos Ridge and Juan de Fuca Ridge, where they grow rapidly because of the enhanced metal supply from hydrothermal fluids. Under this influence, thick manganese crusts can form on oceanic crust of essentially zero age. These crusts are impoverished in trace metals such as Co, Ni, Cu, and Pb. In the third tectonic setting, ferromanganese crusts occur on the flanks of island arcs such as the Tonga Islands and Mariana Islands, and are the least studied of the three types. Island-arc-hosted crusts appear to be mostly hydrogenous and have a small hydrothermal component. The crusts contain moderate concentrations of trace metals but can be enriched in Co, Ni, Cu, and especially Ti in some places.

Although crusts were noted on the expedition of H.M.S. *Challenger* in 1872–1876, a cruise dedicated to their study was not forthcoming until 1981. Scientists aboard the R/V *Sonne* of the Federal Republic of Germany studied crusts from the mid-Pacific mountains (MPM) and Line Islands in 1981, 1983, and 1984 (Halbach, Manheim, and Otten, 1982; Halbach et al., 1983, and work in progress). In 1983 and 1984, scientists aboard the U.S. Geological Survey's R/V *S.P. Lee* studied crusts from Necker Ridge in the Hawaiian Exclusive Economic Zone (EEZ), MPM, Line Islands, and Marshall Islands (Hein et al., 1985a, 1985b; Schwab et al., 1985). During this latter period the University of Hawaii conducted several cruises to study crusts within the Hawaiian EEZ. Crusts collected prior to these pioneering cruises were incidental to other studies. However, several publications have been devoted solely to analysis and interpretation of data pertaining to the crusts collected prior to 1981 (Aumento, Lawrence, and Plant, 1968; Craig, Andrews, and Meylan, 1982; Aplin and Cronan, 1985). This contrasts with the extensive literature dealing with abyssal ferromanganese nodules, literature which has been reviewed many times (see, for example, Glasby, 1977; Heath, 1981). Because of these previous reviews, we will only briefly summarize the data for nodules at the end of this chapter.

Crusts contain three strategic and economically important metals in abundance: Mn, Co, and Pt. In addition, Ni, Pb, Ti, and Ce are significantly enriched in crusts. Abyssal nodules are enriched in Cu, Ni, and Co in that order, but cobalt is 3 to 6 times more abundant in crusts than in prime abyssal nodules or in ores mined on land, primarily in Zaire and Zambia in Africa (M. Foose, U.S. Geological Survey, 1984, written communication). Crusts occur predominantly within the

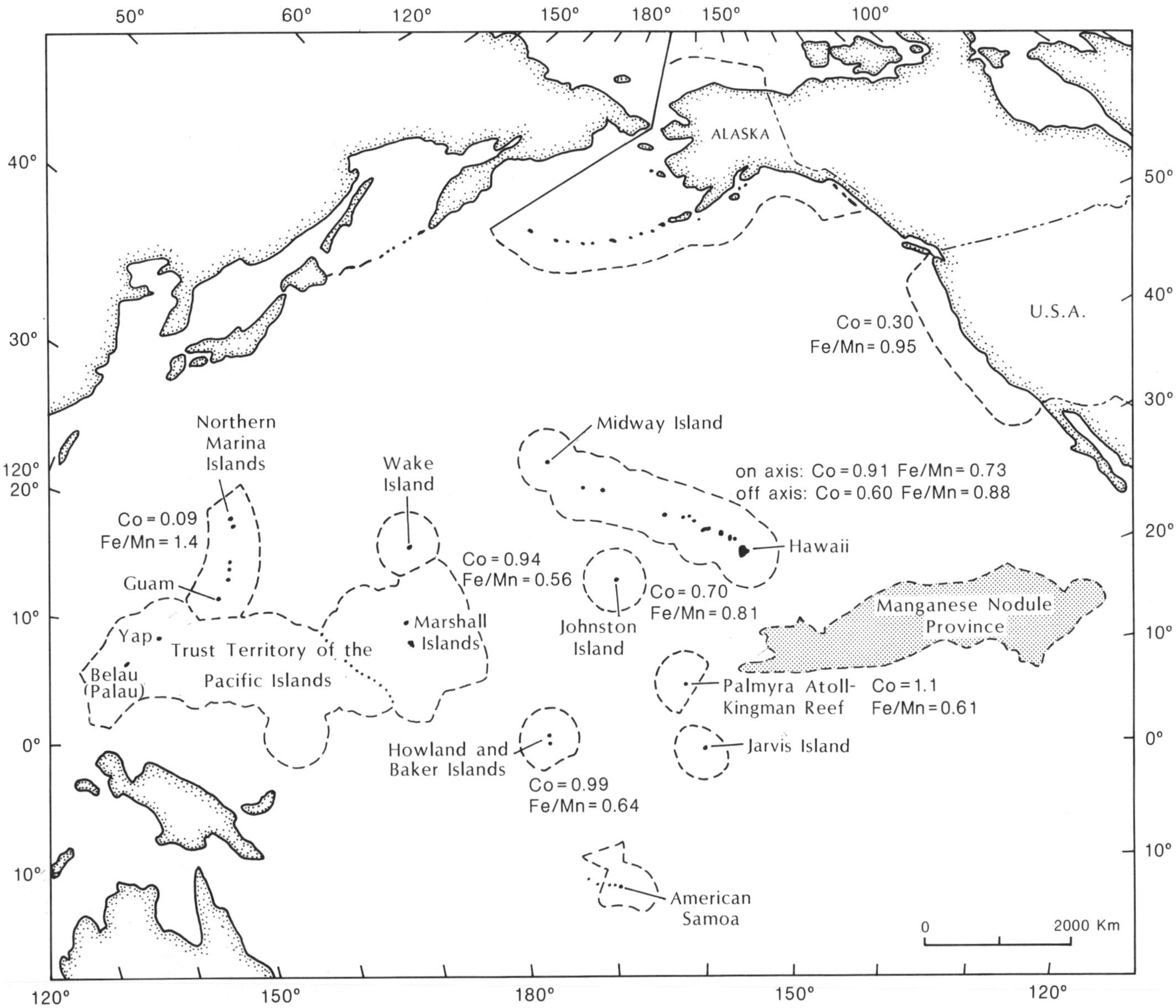

Figure 1. Exclusive Economic Zone (dashed lines) of the United States and its territories, showing average cobalt value (wt.%) and Fe/Mn ratio for each area (from Table 2). The Clarion-Clipperton prime manganese nodule region is stippled.

EEZ of the United States including its possessions and trust territories (Figure 1), and within the EEZ of other Pacific nations. In contrast, prime abyssal ferromanganese nodules are found within international waters. This is one reason why the climate for crust exploitation is better than for nodules. Another reason is the price of metals on the world market, where the value of cobalt is greater than nickel and much greater than copper. Countries actively involved in the exploration for crusts include the United States, Federal Republic of Germany, U.S.S.R., Japan, China, France, and India, whereas the Pacific nations and territories likely to benefit from exploitation of this resource by way of mining leases are the United States, Kiribati, Northern Mariana Islands, Marshall Islands, Federated States of Micronesia, Cook Islands, Tuvalu, and French Polynesia.

In the Pacific, the U.S. Geological Survey has collected crusts from the EEZ offshore of the western continental United States, Johnston Island, Palmyra Island, Marshall Islands, Tongan Islands, Fiji Islands, Mexico, and from the Gulf of Alaska Seamount province. This chapter will summarize and include new data from these U.S. Geological Survey cruises and will also include other pertinent published data.

ANALYTICAL METHODS

We have compiled all available published chemical data on crusts from water depths <2500 m and with manganese concentrations >10 weight % for the Pacific Basin (Figures 2–8). We chose this subset of data because cobalt enrichment occurs in crusts formed at <2500 m water depth and because variation in crust chemistry is primarily depth related (Aplin and Cronan, 1985); consequently, regional differences may become more apparent as the depth range

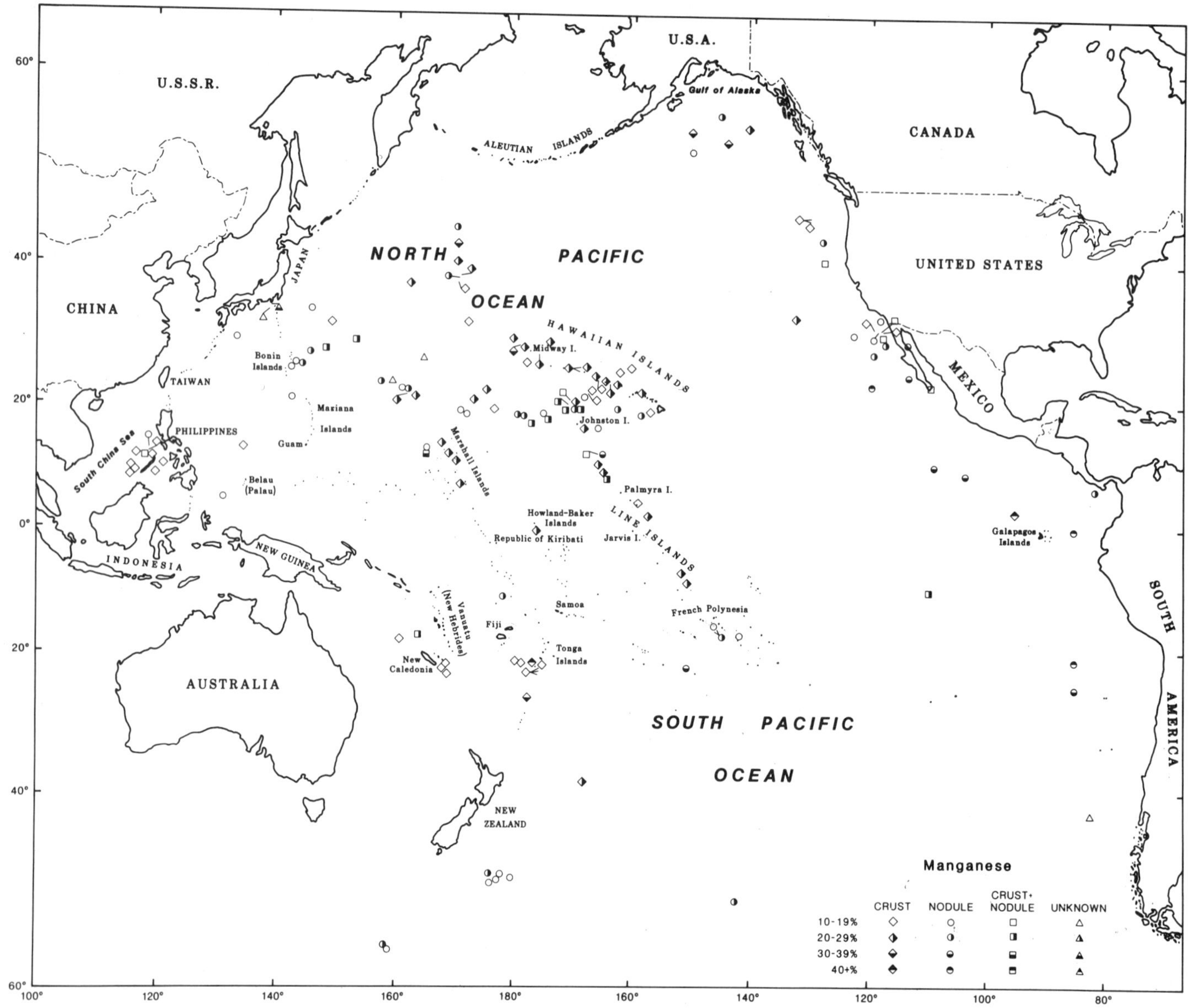

Figure 2. Variation of Mn in ferromanganese crusts and nodules in the Pacific. The diamond symbol represents crusts, the circle nodules, the square both crusts and nodules, and the triangle unknown sample type. Where sample locations are too closely spaced to plot all the symbols, leaders point from the appropriate symbol to the sample location.

Table 1. Mineralogy of cobalt-rich ferromanganese crusts.

Dominant: $\delta-MnO_2$ (vernadite), $FeOOH \cdot x H_2O$

Rarely Dominant: Todorokite, carbonate fluorapatite

Commonly Present: Quartz, plagioclase, carbonate fluorapatite, calcite, geothite, amorphous aluminosilicate

Less Commonly Present: Barite, clinoptilolite, smectite, phillipsite, natrojarosite, pyroxene, chlorite, manjiroite, K-feldspar

Presence Uncertain: Manganosite, Mn-calcite, maghemite, hematite

considered is narrowed. We arbitrarily chose samples with a minimum manganese concentration of 10 weight % in order to eliminate analyses from samples possibly contaminated with substrate material. We have purposefully gone back to the original literature for all the data used, because a number of errors were found in previously compiled data bases. Where depth, geographic location, or conditions of chemical analyses were not clear from the published reports, we contacted the author when possible for additional information. Thus, we hope to have a relatively tightly controlled data set. Our maps include analyses from 468 samples listed in 43 references. The number of samples available is limited considering the size of the Pacific, and far more analyses are being prepared by F. T. Manheim of the U.S. Geological Survey in Woods Hole (see Manheim, Ling, and Lane, 1983). Manheim's data base will include crusts from all ocean basins and depths of water.

Compiled chemical data were divided into water-corrected and uncorrected categories, although both types of data are plotted on Figures 2–8. Water-corrected data are from samples dried at 110°C. For each element, separate plots were made for bulk samples, inner layers, and outer

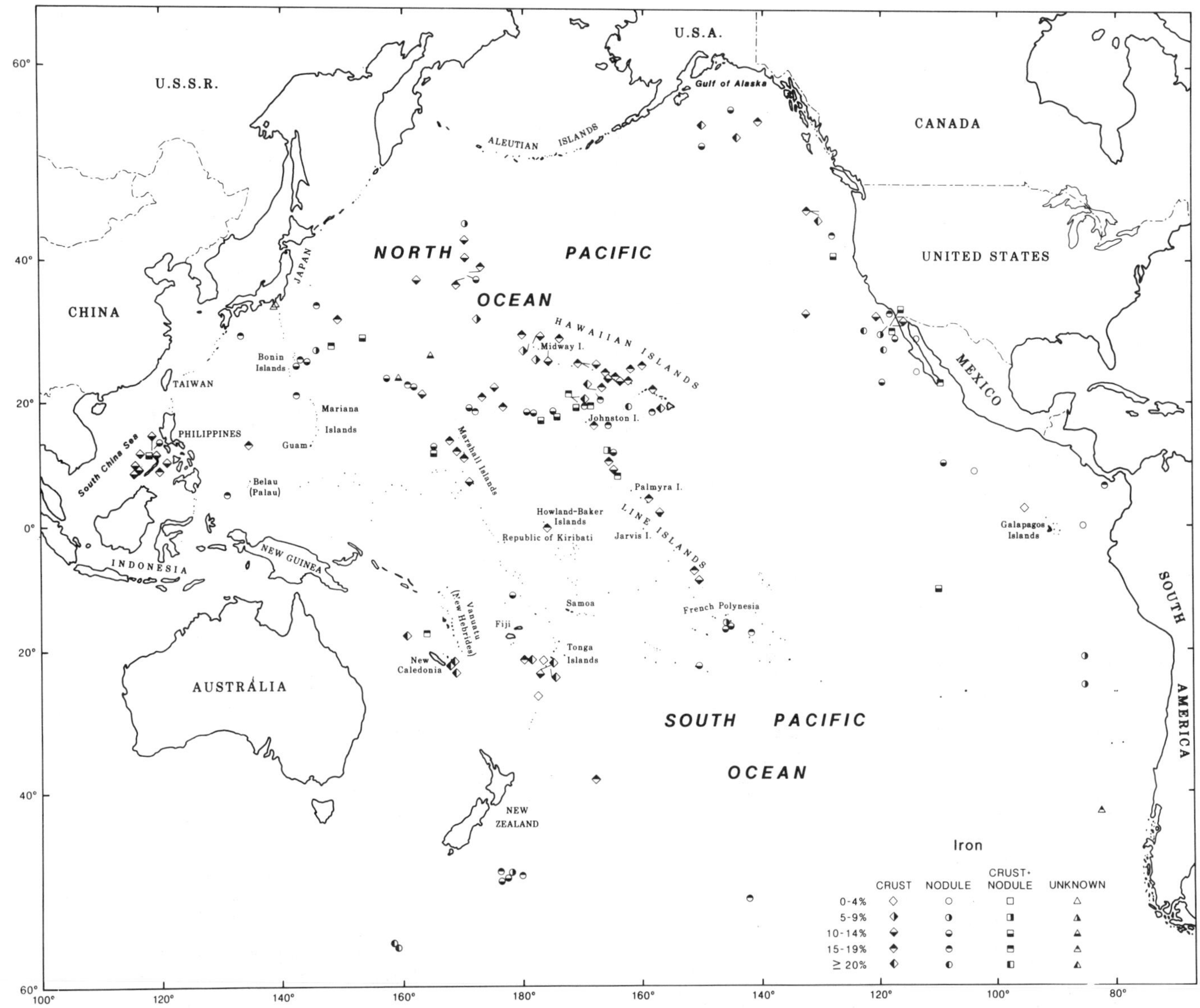

Figure 3. Variation of Fe in ferromanganese crusts and nodules in the Pacific. Symbols same as for Figure 2.

layers, and each sample was coded as either a crust or a nodule. Only data for bulk samples are plotted in Figures 2–8. Analyses from U.S. Geological Survey laboratories are by induction-coupled plasma emission spectroscopy (ICP) and atomic absorption (AA). Compiled data include these techniques as well as x-ray spectroscopy, polarography, neutron-activation analysis, microprobe, and wet-chemical techniques.

COBALT-RICH FERROMANGANESE CRUSTS

Mineralogy

The dominant mineral in the vast majority of crusts is $\delta-MnO_2$ (vernadite) with only two x-ray reflections at 2.40 and 1.42 A. Todorokite is rarely dominant, and we have found on the average one todorokite-rich sample per seamount studied. Carbonate fluorapatite can be abundant in the inner part of many crusts (Table 1). X-ray amorphous $FeOOH \cdot \times H_2O$ is the other dominant phase present along with $\delta-MnO_2$ in all crusts. $\delta-MnO_2$ forms epitaxial intergrowths with the $FeOOH \cdot \times H_2O$ (Burns, 1976). Other common minerals include plagioclase, quartz, and goethite, and minor minerals include barite, K-feldspar, calcite, manjiroite (an Na-analog of cryptomelane with the formula $[Na,K]_{1-2}Mn_8O_{16} \cdot \times H_2O$), zeolites, and clay minerals (Table 1).

$\delta-MnO_2$ is the most oxidized phase of the three common deep-sea manganese minerals, todorokite, birnessite, and $\delta-MnO_2$. The source of the manganese is the water column where manganese oxide occurs mixed with iron compounds in suspended colloidal flocs (Aplin and Cronan, 1985). The $\delta-MnO_2/FeOOH \cdot \times H_2O$ composition of the crusts reflects

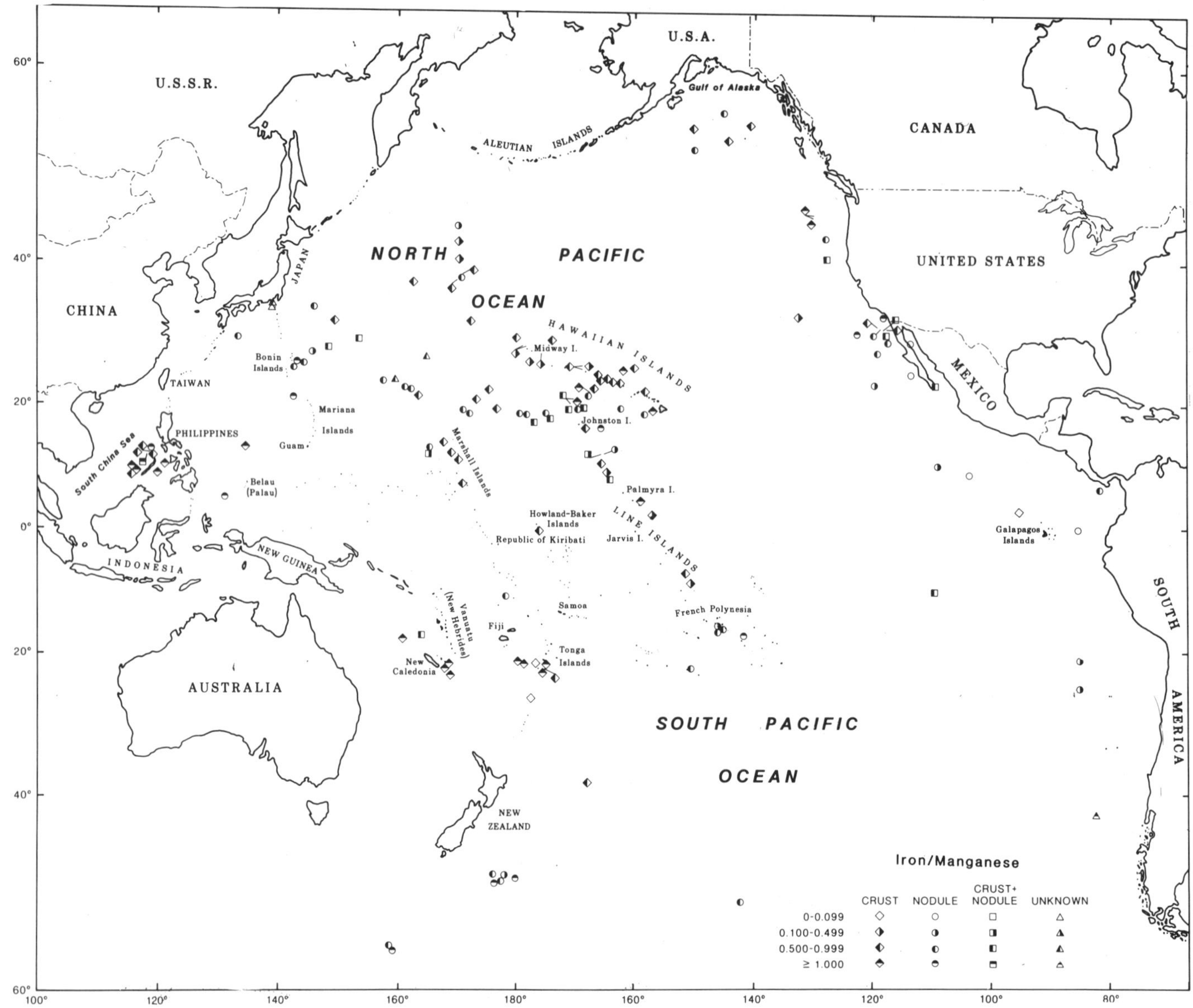

Figure 4. Variation of Fe/Mn ratio in ferromanganese crusts and nodules in the Pacific. Symbols same as for Figure 2.

their ratio in the water column. Goethite probably crystallized from the amorphous $FeOOH \cdot \times H_2O$ phase in some places.

The quartz and some of the plagioclase are eolian in origin and reflect the position of the crust-coated seamount relative to atmospheric wind belts (Hein et al., 1985a). The rest of the plagioclase as well as pyroxene, K-feldspar, zeolites, and others are derived from the substrate, probably by resuspension of weathered material. Calcite and phosphorite are derived ultimately from biological activity in the surface waters. The amount of phosphate present may reflect the intensity of upwelling around the seamount (Hein et al., 1985a).

Geochemistry

The chemical compositions of individual samples are significantly different, but the average abundances of metals are remarkably similar for large areas of the central Pacific (Table 2; Figures 2–8). For the central Pacific (Hawaiian, Pal myra, Johnston, Marshall, and Howland-Baker Islands' EEZ), values range mostly between 20 and 30% Mn, 15 and 20% Fe, 0.5 and 1.1% Co, 0.30 and 0.70% Ni, 0.10 and 0.25% Pb, 1.0 and 1.5% Ti, and 0.05 to 0.15% Cu (see Table 2 for averages). For borderland and marginal basin areas, these metals are diluted with a significant contribution of detrital minerals, which increases the Si and Al contents. This is especially evident in the South China Sea and off the western United States. The Fe/Mn ratio is greatest for island arcs and borderland areas, which may partly explain the lower amounts of Co, Ni, and Pb, which are associated with Mn. The Ti values are also low for these areas. The average Co and Ni contents are highest for the EEZ in the central Pacific (0.78%; 0.44%) and French Polynesia (1.15%; 0.60%). Overall, the central Pacific, the area with the most available analyses, and French Polynesia have the most promising grade of crusts as determined by contents of Mn, Co, Ni, and Pb. Plati-

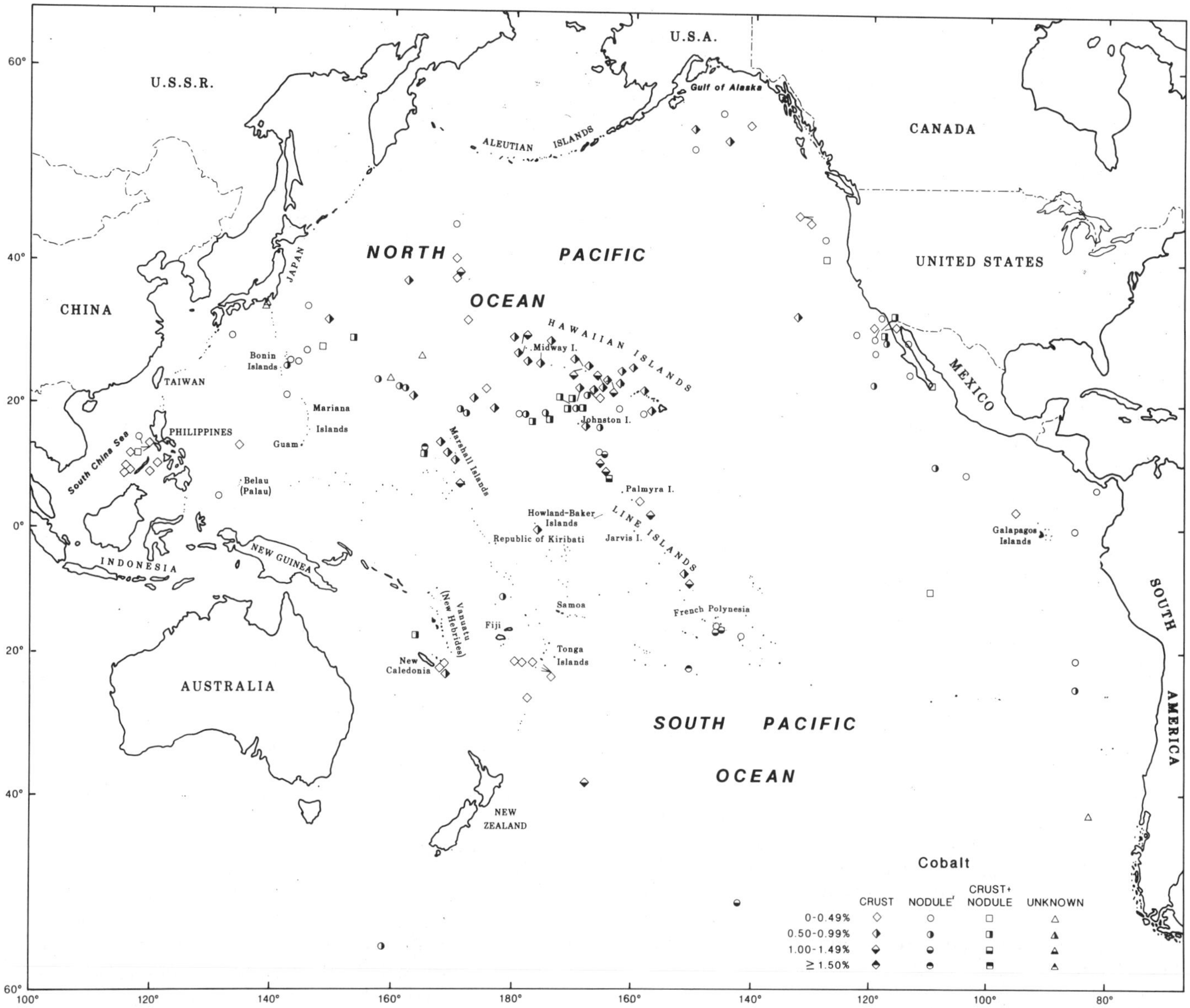

Figure 5. Variation of Co in ferromanganese crusts and nodules in the Pacific. Symbols same as for Figure 2.

num values are available for the Marshall Islands, northern Line Islands, and MPM (Halbach, Puteanus, and Manheim, 1984; Hein, unpublished data). Platinum values range up to 0.9 ppm, although recent analyses from Line Islands samples have produced values up to 1.3 ppm (P. Halbach, Technical University of Clausthal, 1985, oral communication).

No published chemical data within the subset of data we have chosen to use are available for the EEZ of American Samoa, Jarvis and Wake Islands, and the Federated States of Micronesia.

Cobalt is commonly enriched in the outer layers of the crusts, whereas platinum is enriched in the inner layers. Co, Pt, Pb, Ni, Ce, As, Mo, Cd, and Zn concentrations are positively correlated with Mn content, and Cu and Be are positively correlated with Fe (Glasby, 1977; Halbach, Manheim, and Otten, 1982; Halbach, Puteanus, and Manheim, 1984; Aplin, 1984; Aplin and Cronan, 1985; Hein et al., 1985b). Al, Si, Ti, and K compose the aluminosilicate fraction, with part of the Ti as well as V, Sr, and Pb partitioned between the iron- and manganese-oxide phases (Hein et al., 1985b; Aplin and Cronan, 1985).

In addition to these interelement correlations, correlations exist between various elements and depth of water as well as latitude. Of the elements that increase in abundance in crusts with increasing depth of water, copper and iron show the strongest correlations; other elements that show a positive correlation are Be, K, Ce, Al, Ti, Si, and Ba. Elements that correlate negatively with depth of water (decrease with depth) are Mn, Co, Ni, Pb, Mo, Cd, As, and Sr (Barnes, 1967; Hein et al., 1985b; Aplin and Cronan, 1985). Much of these differences with depth of water can be explained by the trace elements being associated with either of the dominant mineralogical phases, $\delta-MnO_2$ or $FeOOH \cdot \times H_2O$.

Frank et al. (1976) first suggested that crust chemistry changes regularly with latitude and longitude. They suggested that Mn, Ni, Cu, and Co increase and Fe decreases to

Table 2. Average chemical composition for various elements of crusts from <2500 m water depth from the EEZ of the United States and other Pacific nations. All data are in weight percent.

Areas	n	Mn	Fe	Co	Ni	Cu	Pb	Ti	SiO_2	P_2O_5	Fe/Mn
Hawaii and Midway (on axis)	2–38	24	16.	0.91	0.45	0.05	—	1.1	7.9	—	0.73
Hawaii and Midway (off axis)	4–15	21	18.	0.60	0.37	0.10	0.18	1.3	16.	1.0	0.88
Johnston Island	12–40	22	17.	0.70	0.43	0.11	0.17	1.3	12.	1.2	0.81
Palmyra Atoll-Kingman Reef	7–8	27	16.	1.1	0.51	0.06	0.17	1.1	5.5	1.9	0.61
Howland-Baker Islands*	3	29	18.	0.99	0.63	0.08	0.14	1.2	6.2	1.7	0.64
Marshall Islands	55–63	21	13.	0.74	0.45	0.08	0.14	0.9	4.4	1.5	0.61
Average central Pacific crusts	34–167	22	15.	0.78	0.44	0.08	0.16	1.1	6.3	1.5	0.71
Northern Mariana Is. (and Guam)	6–7	12	16.	0.09	0.13	0.05	0.07	—	—	—	1.41
Western U.S. borderland	2–5	19	16.	0.30	0.30	0.04	0.15	0.31	17.	—	0.95
Gulf of Alaska Seamounts	3–6	26	18.	0.47	0.44	0.15	0.17	0.57	—	0.98	0.72
Lau Basin (hydrothermal)	2	46	0.60	0.007	0.005	0.02	0.006	0.005	—	0.05	0.01
Tonga Ridge and Lau Basin (hydrogenous)	6–9	16	20.	0.33	0.22	0.05	0.16	1.0	—	1.0	1.26
South China Sea**	14	13	13.	0.13	0.34	0.04	0.08	—	14.	—	1.07
Bonin Island area (Japan)	1–10	21	13.	0.41	0.55	0.06	0.12	0.67	4.9	0.82	0.70
French Polynesia	2–9	23	12.	1.2	0.60	0.11	0.26	1.0	6.5	0.34	0.56
Average for Pacific hydrogenous crusts. All data from Figures 2–8	56–319	22	15.	0.63	0.44	0.08	0.16	0.98	11.	1.1	0.81

n = number of analyses for various elements.
— means no data.
*Data provided by P. Halbach, Technical University of Clausthal.
**Data provided by H. R. Kudrass, Bundesanstalt fur Geowissenschaften und Rohstoffe.
No data have been published for the EEZ of the Federated States of Micronesia, American Samoa, Jarvis Island, or Wake Island.

the northwest along the Hawaiian chain; that is, with increasing age of the substrate. This trend was again reported by Craig, Andrews, and Meylan (1982), although they suggested that only Ni, Co, and crust thickness increase and Ti and Fe decrease to the northwest. Co and Ni are enriched in crusts of the Line Islands relative to the more northerly MPM (Halbach, Manheim, and Otten, 1982; Halbach and Puteanus, 1984). A strong negative latitudinal correlation exists for Mn, Co, Mo, Ni, and Cd in the outer layers (younger generation) of crusts. These metals increase with proximity to the equator from the EEZ of Hawaii to that of Johnston Island and finally to that of Palmyra Island (Hein et al., 1985b). Apatite within the crusts also increases toward the equator. The reasons for the variations with geographic position require further study. Processes likely to control these regional trends include deep and intermediate-depth current systems, biological productivity, differences in the position of the carbonate compensation depth, and the degree of depletion of oxygen in the oxygen-minimum zone as well as the vertical extent of the zone.

Processes of Formation

Cobalt-rich ferromanganese crusts are hydrogenetic, and all of the metals (except minor contributions from detrital phases) are derived from seawater. Substrates can be ruled out as contributors to crust chemistry because of the lack of correspondence between substrate type and crust composition. Hydrothermal contributions can be ruled out because of the abundant minor metals (Co, Ni, etc.) present, the lack of hydrothermal manganese phases such as todorokite and birnessite, and the lack of a more complete fractionation between iron and manganese that is characteristic of hydrothermal deposits. However, Koski et al. (1985) reported that crusts found in island arcs with low to moderate cobalt contents may have some hydrothermal component.

What initiates precipitation of the first molecular layer from seawater onto volcanic or other substrate is unclear. Once the process of precipitation begins, however, it becomes autocatalytic. Suggestions have been made that iron was deposited first, either on organisms encrusting substrates or by some other mechanism; the iron in turn catalyzed the deposition of Mn (Burns and Brown, 1972). This process of early-formed iron oxide, however, has not been demonstrated for crust deposits. Hem (1978) stressed the possibility that Mn^{+3} oxide may form first at the crust surface, followed by conversion to MnO_2 through disproportionation.

The relative amount of the oxide phases in crusts is determined by their ratio in colloidal flocs suspended in the surrounding seawater (Aplin and Cronan, 1985). The major-element crust chemistry reflects both the particulate and dissolved seawater chemistry. In the water column, manganese and iron oxides probably occur together in the colloidal flocs. Within the oxygen-minimum zone, which ranges from 200 to 2000 m in depth, manganese is highly soluble and thus also occurs in the dissolved form. The Fe/Mn ratio is smaller in the oxygen-minimum zone than it is below that zone. Consequently, the manganophile elements Co, Ni, Pb, etc., should be more concentrated within crusts and flocs from the oxygen-minimum zone.

The ultimate controls on the concentration of minor metals in crusts are (1) concentration of each metal in seawater, (2) amount of scavenging by the iron and manganese oxide phases, (3) mineralogy, which in turn depends on the

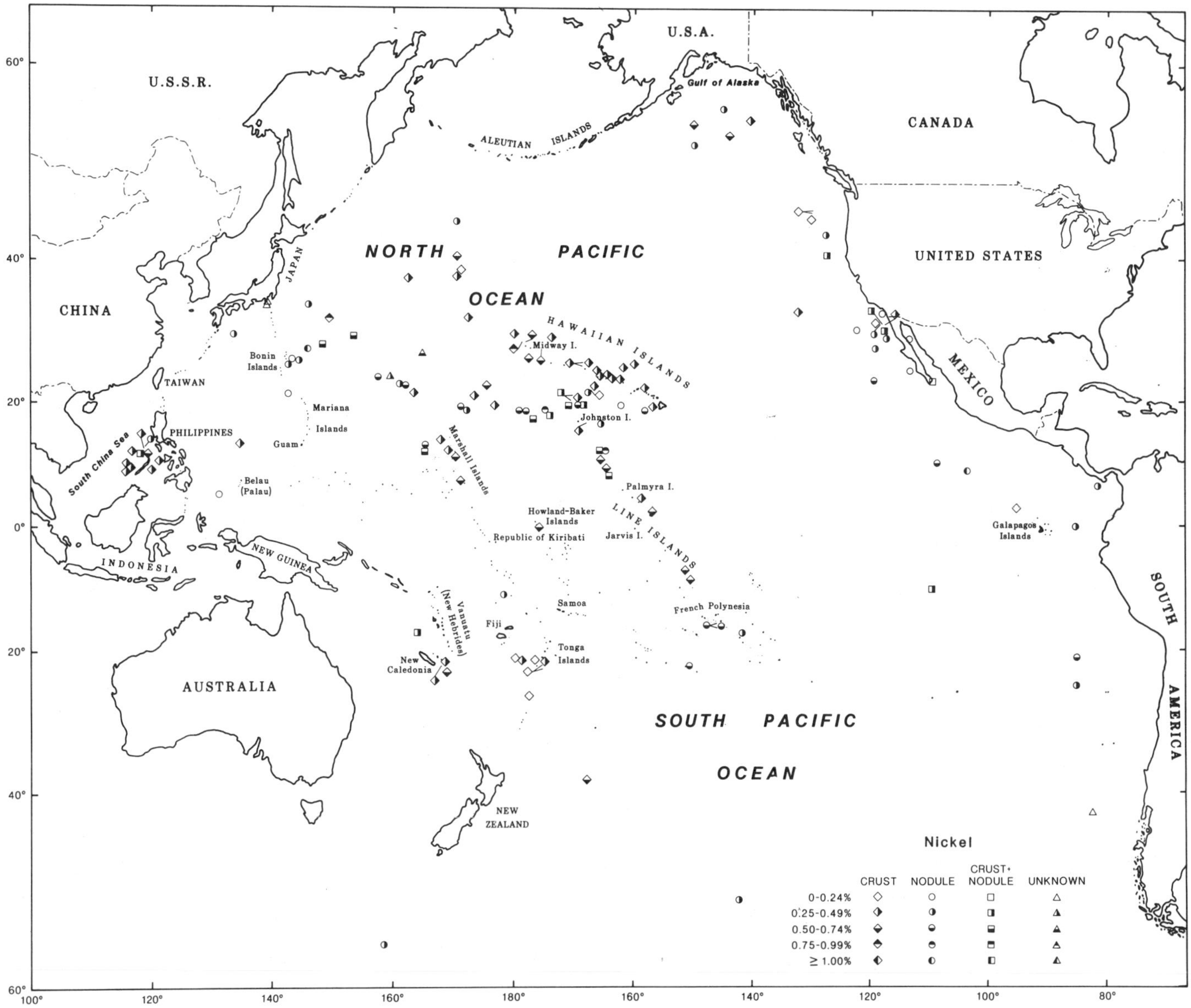

Figure 6. Variation of Ni in ferromanganese crusts and nodules in the Pacific. Symbols same as for Figure 2.

Eh of the environment, and (4) dilution by detrital phases (Goldberg, 1954; Bonatti, Kraemer, and Rydell, 1972). A slow-growth rate is conducive to a greater degree of scavenging by the major oxides (Goldberg, 1954).

Cobalt and other metals are fixed by lattice substitution for Mn^{+4} in the MnO_2 or by coprecipitation of cobalt-oxide with manganese oxide (Hem, 1978). The reason that cobalt is enriched in the $\delta - MnO_2$ to a greater degree than are Ni, Zn, Cd, and other metals is because Co^{+2} is oxidized to Co^{+3} on the surface of the MnO_2; Co^{+3} is less soluble and therefore more stable (Goldberg, 1961; Dillard, Crowther, and Murray, 1982; Halbach et al., 1983; Aplin and Cronan, 1985). The crust-seawater partition coefficients calculated by Aplin and Cronan (1985) indicate that lead may also be oxidized at the MnO_2 surface and is likewise relatively enriched over the other metals. The process is autocatalytic and trace metals and ferromanganese compounds precipitate onto the crust surface, releasing and incorporating ions to and from the water column. The slower the growth rate, the more trace metals that can be fixed; therefore, the crusts richest in trace metals are those that grew at the slowest rate. The flux of cobalt at all depths in the water column is apparently constant (Halbach et al., 1983), and the primary control on the degree of cobalt-enrichment is the rate of growth. Hydrogenetic crusts are among the slowest formed deposits in the geologic environment.

The higher cobalt contents in the outer part of the crusts imply that the more recent growth rates were slower than the earlier growth rates. The decrease in cobalt with depth of water can also be explained by faster rates of growth due to additional amounts of iron incorporated into the crusts. The increased supply of iron has been related to the dissolution of iron-bearing biogenic calcite below the calcite compensation depth (Halbach and Puteanus, 1984). Aplin and Cronan (1985), however, reject calcite dissolution as the source of iron below the 2000-m depth of water. Whatever

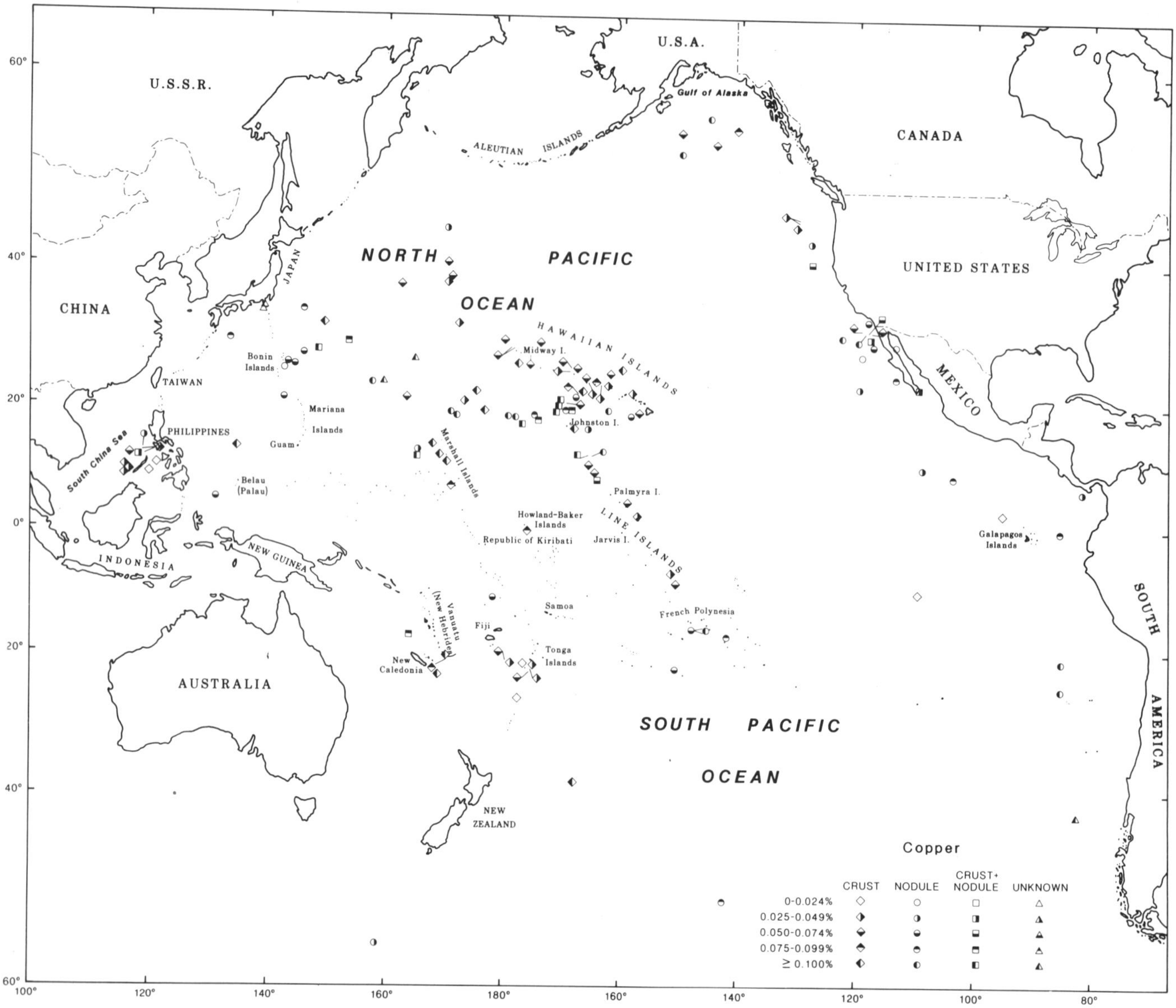

Figure 7. Variation of Cu in ferromanganese crusts and nodules in the Pacific. Symbols same as for Figure 2.

the source of the additional iron in the deeper water crusts, it does act as a diluent to the minor metals.

Crust Texture, Thickness, and Rates of Growth

Crusts are commonly laminated and show a textural change that delineates an inner older generation of growth and an outer younger generation of growth. The textural change is commonly marked by a paper-thin layer of phosphorite, which marks a depositional hiatus of ferromanganese compounds. The inner generation may be more dense compared with the outer layers. Also, the inner layers may be impregnated by phosphorite, contain laminae of detrital minerals, or contain larger clasts of substrate rock. The botryoidal surface texture is an expression of the undulatory and columnar growth patterns of the crust. Individual laminae may be truncated and are commonly distinguished from adjacent layers by more or less iron relative to manganese (Koski et al., 1985). In some places, botryoidal surface textures dominate, and at other places, crust surfaces are smooth. Botryoids can be various sizes and are pronounced or subdued. In a few examples, surfaces are polished and fluted from working by bottom currents or gravity currents carrying suspended sediment (for example, samples D6 and D15 in Schwab et al., 1985). Dark granular material occurs on the undersides, sides, or in the low areas of the botryoidal surfaces; this granular ferromanganese oxide is richer in cobalt than is the surrounding oxide.

The thickest hydrogenetic crust reported in the literature is 24 cm (von Stackelberg et al., 1984). This crust, dredged from 4830-m water depth, must have grown for most of the past 65 m.y., the age of the substrate upon which it rests. ^{10}Be dating revealed that the outer 16 mm grew at a rate of 2.7 mm/m.y., whereas the interval 16–38 mm grew at 4.8 mm/m.y. (Mangini et al., 1983); the age of the crust at 38 mm depth is 11 Ma.

Most crusts, however, have maximum thicknesses of 8 to

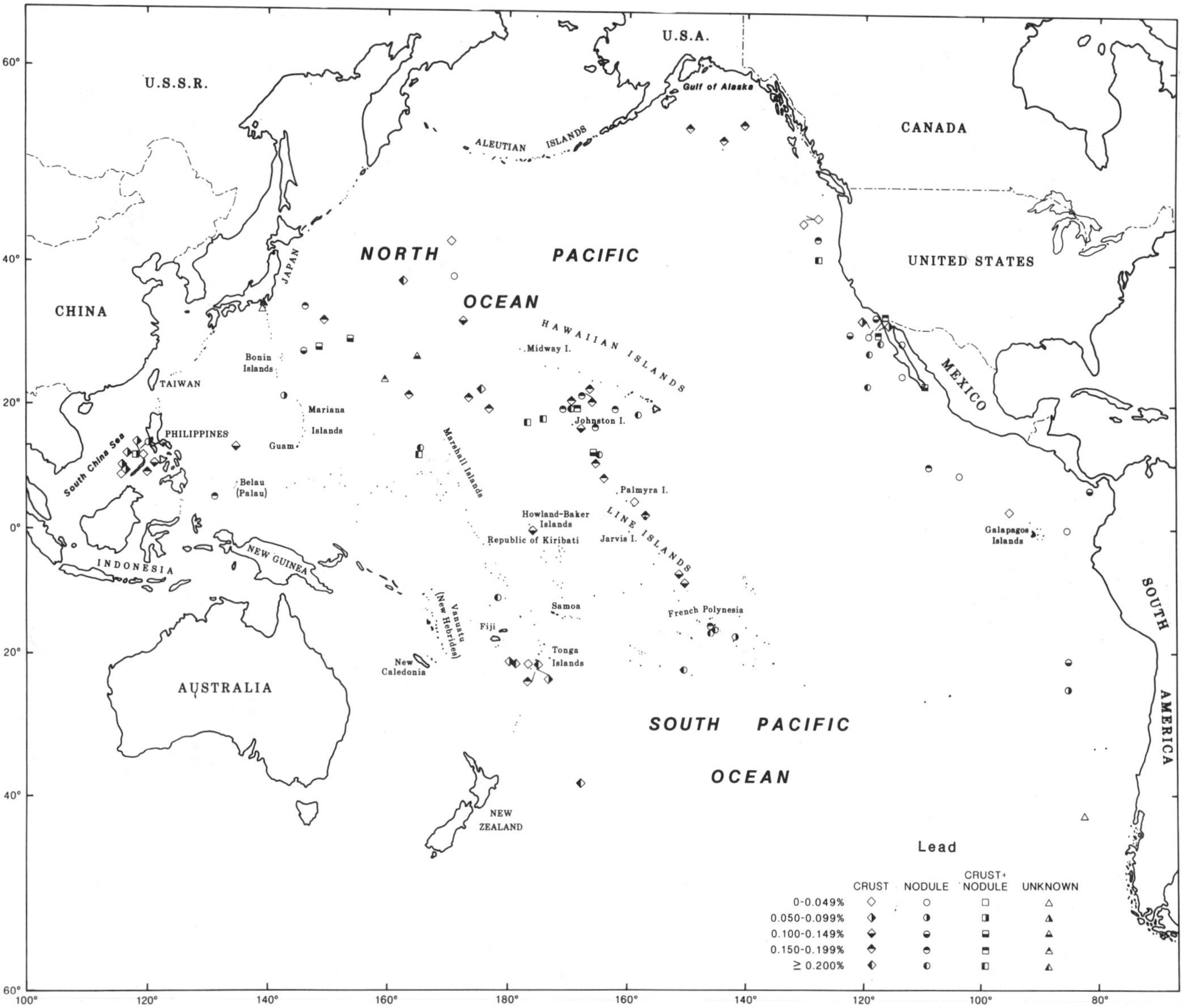

Figure 8. Variation of Pb in ferromanganese crusts and nodules in the Pacific. Symbols same as for Figure 2.

10 cm and average thicknesses of 2 to 4 cm (Figure 9). The thickness of crusts is controlled by the length of time of growth and the rate of growth, which in turn depends on the oceanographic and geochemical conditions at the site of formation. Although few age data are available, it is becoming evident that crusts are not as old as the substrates on which they grow. The 24-cm-thick crust just cited is the exception rather than the rule. Crusts should begin to grow without interruption once volcanism on the seamount ceases; growth continues unless the crust enters a reducing environment, either by burial by sediments or changes in the chemistry of the surrounding water. The oldest parts of most crusts recovered, however, may be 10 to 20 m.y. younger than the edifices upon which they rest. One reason for this age gap is that crusts grow in an unstable environment—the slopes of seamounts and ridges—where gravity flow and mass movement destroy crusts. Crusts and substrates may show two or three episodes of reworking, erosion, and regrowth (Hein et al., 1985a). Some gravity-flow deposits are clearly defined on seismic-reflection records, and disruption is recorded in multistage breccias.

More growth-rate data are available for abyssal nodules than for hydrogenetic crusts. The outer layers or the younger generation of crusts commonly formed at 1 to 3 mm/m.y. in contrast with nodules (1 to 10 mm/m.y.) and hydrothermal crusts (1000 to 2000 mm/m.y.) (Moore, 1973; Ku, 1977; Moore and Vogt, 1976; Ku et al., 1982; Halbach et al., 1983). The inner-crust layers or older generation may have grown at rates of up to 5 mm/m.y. (Segl et al., 1984). Growth rates for less than 20 crusts have been published. Many more crusts need to be analyzed, and more detailed sampling of crusts needs to be completed in order to delineate fine-scale changes in growth rates. In addition, other radiometric or isotopic dating techniques need to be applied to crusts, because U-series is applicable only for the outer millimeters and ^{10}Be for the outer centimeters. Parts of crusts older than 15 Ma cannot be dated directly by the techniques used thus far, and older ages must be extrapolated.

Figure 9. Photographs of slabbed rocks encrusted with ferromanganese oxides. a. Laminated crust, largely δ – MnO_2, on pillow lava substrate from lower western slope of Miller Seamount, Gulf of Alaska; lat. 53°32.6′N, long. 144°22.4′W; 2100 m water depth; thickness 4.0 cm. Crust is laminated and columnar to botryoidal, with aggregates of detrital grains between columns. Detrital grains are quartz, plagioclase, amphibole, magnetite, and altered volcanic glass. b. From S.P. Lee Guyot, central Pacific; lat. 8°20.8′N, long. 164°19.0′W; 1125–1150 m water depth. Average crust thickness 2.0 cm. Phosphorite clasts containing cavities filled with calcite are enclosed in a matrix of volcaniclastic sandstone with admixed fine-grained manganese oxide. This substrate is encrusted with Fe-Mn oxide that is laminated and columnar, is more porous near the outer surface, and contains phosphorite layers. Note botryoidal surface texture. c. Volcanic breccia cemented with phosphorite and Fe-Mn oxide and encrusted with Fe-Mn oxide from Horizon Guyot, central Pacific; lat. 19°30′N, long. 168°51.5′W, 1780–1799 m water depth. Average crust thickness is 2.5 cm, but crust is highly fragmented because of dehydration. Note phosphate veins in crust. d. Fe-Mn oxide encrusted basalt from Horizon Guyot. Same location and water depth as (C). The outer third of this laminated crust is more highly porous than the inner part. Veins in basalt are filled with carbonate and phosphorite. In places the veins extend into the crust. Note the 3-cm-long basalt inclusion in the crust.

SUBSTRATE CHARACTERISTICS

In approximate order of decreasing abundance, substrate rocks include volcanic breccia, hyaloclastite, basalt, phosphorite, limestone, volcaniclastic sandstone, mudstone, and chert (Craig, Andrews, and Meylan, 1982; Halbach, Manheim, and Otten, 1982; Aplin and Cronan, 1985; Hein et al., 1985a, 1985b; Schwab et al., 1985; Koski et al., 1985). The type of substrate depends on the part of the edifice that is sampled and the geographic location. For example, rocks dredged from the Hawaiian Ridge contain more basalt and mudstone than rocks recovered from the MPM and Line Islands. Phosphorite substrate is more abundant in the Line Islands and Necker Ridge than in the MPM. Volcaniclastic sandstone substrate is more common in island arcs. Most phosphorite is phosphatized limestone. In the northwestern Hawaiian Ridge, basalts are phosphatized and contain as much as 5% P_2O_5.

As expected, the mineralogy of substrates differs with rock type. Volcanic rocks, including breccia and volcaniclastic sandstone, contain plagioclase and pyroxene with minor olivine, magnetite, and the alteration products smectite, phillipsite, clinoptilolite, hematite, goethite, maghemite, and calcite. Smectite and/or phillipsite can be the dominant minerals in highly altered rocks. Phosphorites consist of carbonate fluorapatite and calcite with traces of quartz. Fine-grained sedimentary rocks are mostly smectite, zeolites, and plagioclase. Other minerals that occur less commonly are K-feldspar, illite, and chlorite. More complete mineralogy of substrates immediately underlying crusts can be found in Frank et al. (1976) and Hein et al. (1985b).

Basalt substrates are commonly enriched in MgO and CaO and to a lesser degree in Na_2O and Fe_2O_3, and are depleted in K_2O relative to volcanic breccia and hyaloclastite, which probably reflects the gain and loss of these elements during alteration (Hein et al., 1985b). Phosphatized rocks have moderate to high CaO and P_2O_5 contents depending on the degree of phosphatization. One rock dredged from S.P. Lee Guyot in the northern Line Islands contains >76% Fe_2O_3 (Hein et al., 1985b).

Unless cemented by phosphorite or calcite, the volcanic breccias, hyaloclastites, sandstones, and mudstones are friable, and commonly are highly altered. These structurally weak rocks are prone to slumping and to other gravity-flow processes (see next section). Many of these rocks are highly porous, having as much as 50% porosity, and are commonly saturated with water (Hein et al., 1985b).

VOLCANIC EDIFICES

Volcanic seamounts, ridges, and plateaus where crusts are thick enough to be considered as potential ores for mining must be older than about 20 m.y. Volcanic edifices in the central Pacific, including the Marshall Islands, Line Islands, MPM, and old seamounts and ridges surrounding the Hawaiian Ridge formed 85 to 105 Ma (perhaps as long ago as 128 Ma). Crusts have been most thoroughly studied in these areas (Figure 10). This area underwent epeirogenic uplift and post-edifice-building volcanism from 80 to 85 Ma, which allowed for the growth of reefs and for subaerial exposure (Jackson and Schlanger, 1976). A recurrent episode of volcanism may have occurred during the Eocene (Haggerty, Schlanger, and Silva, 1982). All volcanism ended by late Eocene, thus providing a maximum of 40 m.y. for the growth of ferromanganese crusts, uninterrupted by volcanism. In places, reefs developed concurrently with volcanism if volcanic edifices reached into the photic zone. The presence of modern reefs, especially large atolls, is unfavorable for the formation of crusts on the upper slopes of the volcanoes because large quantities of reefal debris are shed down the flanks. Consequently, within the upper 500 to 800 m of the reef-capped structures, ferromanganese occurs only as a stain or patina on the rocks.

The age of initiation of volcanism of Circum-Pacific island arcs varies, but most are at least 40 m.y. old. Volcanism has been episodic and continues to the present in most of the arcs including, for example, Tonga, Vanuatu, Solomon Islands, and the Mariana Islands. The active volcanism may explain both the relatively thin crusts in the arc environment and their apparent hydrothermal component.

The summit area of sediment-capped guyots supports ferromanganese-nodule fields and outcrops coated with crusts (Figure 10a–f); sediment thicknesses to 500 m have been measured (Winterer, Ewing, et al., 1973). The chemical composition of the nodules and adjacent crusts is the same. These shallow-water nodules are chemically and texturally different from abyssal nodules. Seamount nodules have a large nucleus and thin crust, whereas abyssal nodules have a small nucleus relative to crust thickness. The seamount nodule fields are lag deposits of crust-coated cobbles and pebbles that rest on the current-swept sediment surface (Hein et al., 1985a). Migrating sand waves and ripples periodically bury and exhume the nodule fields (Figures 10b, c, and e). The effect of burial on the oxides is not known—whether growth stops, dissolution occurs, or growth continues diagenetically.

Sediment also occurs on the flanks of the seamounts in ponds and on terraces. Triaxial compression tests performed on sediment from the summit of Horizon Guyot yielded an average friction angle (maximum stable slope) of 27°, which may explain why the sediment rests on the steep guyot flanks (Hein et al., 1985a). The slope angle of seamount flanks ranges from 10° to 20° and averages about 14° for the central Pacific. Locally, slopes are near vertical, and stepped or terraced flanks are seen on bottom-camera surveys and on seismic-reflection profiles. Little is known about the microtopography (decimeter- to meter-scale), which is information needed to develop mining strategies and ore-collecting equipment. Deep-towed side-scan surveys and more bottom-camera video surveys are needed to delineate the small-scale structures.

Determination of the percentage of sediment-covered versus crust-covered flanks of seamounts is needed in order to calculate the amount of crust present on the edifice and the length of time the feature could sustain a mining operation. Quantitative determinations are not available, but semi-quantitative estimates from photographs have been made for the summit of Horizon Guyot and S.P. Lee Guyot in the central Pacific, where the slopes are 50% and 80% sediment covered, respectively (Hein et al., 1985a, 1985b). The few available calculations of resource estimates of seamounts have used slope-sediment cover of 40 to 60% (see next section).

EXPLORATION AND RESOURCE POTENTIAL

Many seamounts and ridges can be eliminated as targets on exploratory cruises by following several basic guidelines. Inasmuch as cobalt increases with decreasing water depth, shallow features that rise to within 1500 to 1000 m or less of the sea surface should be emphasized. Emphasis should also be placed on sampling structures older than 20 Ma, because at least that much time is needed to provide sufficient crust thickness for exploitation. Areas of strong current activity and consequently little sediment cover provide for extensive crust-covered outcrops and constant introduction of undepleted midlevel and deep water. Edifices capped by modern large reefs and atolls have not yielded crusts as thick as those from volcanoes without the caps. A shallow, and well-developed oxygen-minimum zone may also be important; the O_2-minimum is well developed regionally in the north Pacific, and to a lesser extent in the south Pacific to lat. 15°S (see also Cronan, 1984). However, the regional distribution of the oxygen-minimum zone may not be crucial if upwelling is intrinsic to seamounts and ridges that rise to near the sea surface. Dynamic or obstructional upwelling may provide for at least a weakly developed O_2-minimum zone over most seamounts and ridges of interest. The richest cobalt crusts recovered thus far have come from within 20° latitude of the equator, although isolated high-cobalt crusts occur in other areas such as the Hawaiian Ridge, French Polynesia, and the southwest Pacific. Although few data are available, crusts from island arcs are thinner and lower grade than crusts from mid-plate seamounts (compare northern Mariana Islands, Bonin Islands, and Tonga Ridge-Lau Basin with other areas in Table 2). Continental borderland and marginal basin seamounts likewise do not contain crusts as thick or as high grade as mid-plate edifices (compare South China Sea,

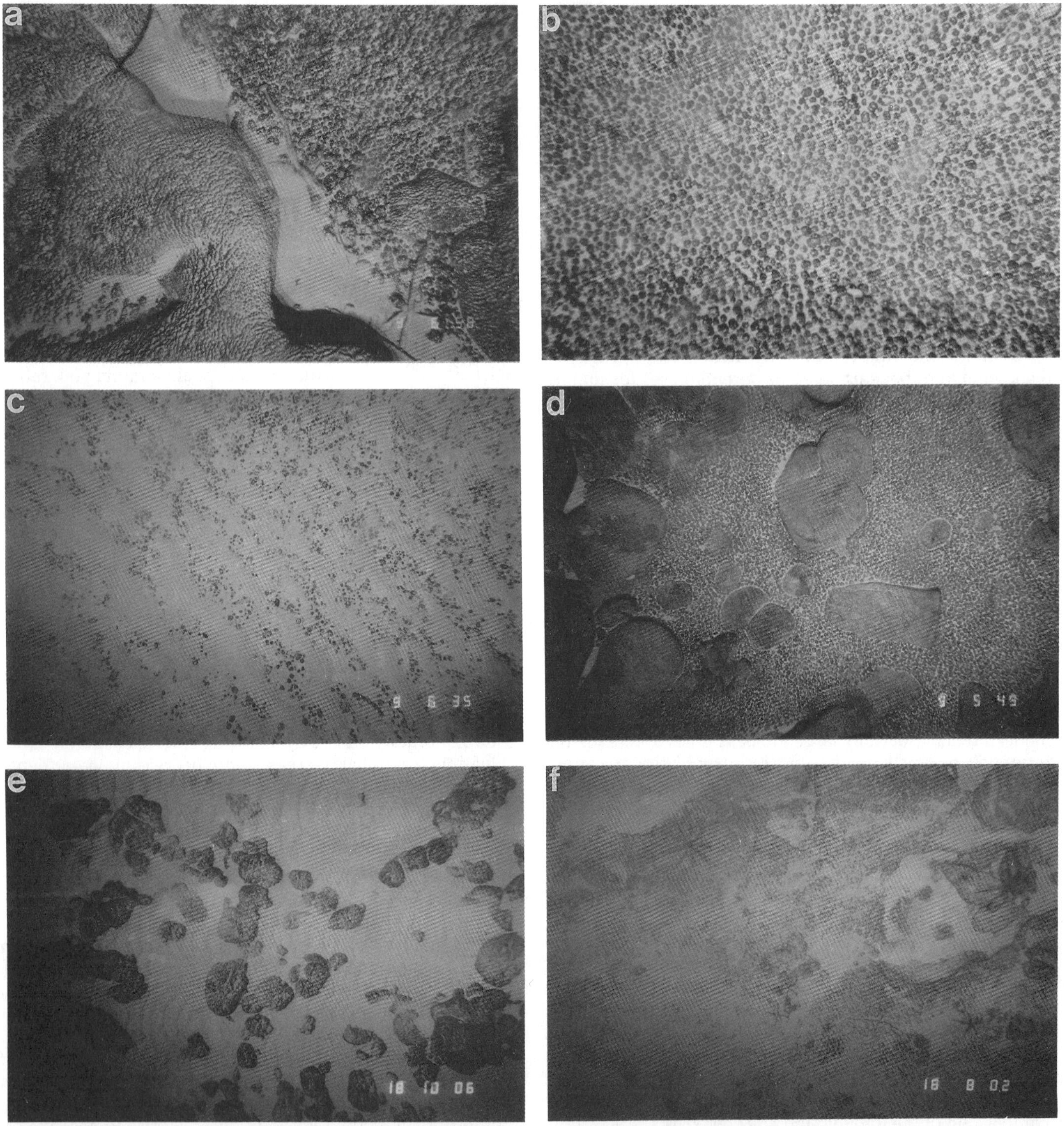

Figure 10. Bottom photographs of Horizon Guyot and S.P. Lee Guyot, central Pacific Ocean. Field of view ranges from 2 to 3 m on a side. All water depths given below are ± 10 m. a. Ferromanganese-encrusted outcrop surrounded by foraminiferal sand on Horizon Guyot. Note the botryoidal texture of the crusts, and siphons of infauna protruding from the sand. Water depth 1997 m. b. Nodule field on foraminiferal sand, Horizon Guyot. Water depth 1997 m. c. Ferromanganese nodules aligned along troughs of sand ripples, Horizon Guyot. Water depth 1975 m. d. Sea floor covered with ferromanganese-encrusted cobbles surrounded by nodules that rest on foraminiferal sand, Horizon Guyot. Water depth 1965 m. e. Ferromanganese-coated cobbles from S.P. Lee Guyot. Currents indicated by sand ripples. Water depth 1296 m. f. Encrusted outcrop, S.P. Lee Guyot, with a thin sediment cover and abundant organisms. Water depth 1753 m.

western U.S. borderland, and Gulf of Alaska with the others in Table 2).

Once a promising area is located, detailed studies are required to determine the grade, thickness, and areal extent of the prospective ore deposit. Work should include extensive sampling, seismic surveys, bottom photography, and deep-towed side-scan surveys. New equipment designs including a multifrequency device that could continuously measure crust thickness in situ would be a boon to the exploration for cobalt crusts.

Resource estimates that have been produced to date are based on very limited data, except perhaps for the Hawaiian

Table 3. Estimated total resource potential of crusts within the EEZ of Hawaii, U.S. Trust and affiliated territories, Kiribati, and French Polynesia.

Pacific Area (Johnson, Clark, and Otto)	Relative Ranking	Potential	Pacific Area (This Work)	Relative Ranking	Potential
Micronesia	1	High	Kiribati Is.	1	High
Marshall Is.	2	High	Marshall Is.	2	High
Northern Mariana Is.	3	High	Micronesia	3	High
Kingman-Palmyra	4	High	Johnston Is.	4	High
Johnston Is.	5	High	Kingman-Palmyra	5	High
Hawaii-Midway	6	Medium	French Polynesia	6	Medium
Wake	7	Medium	Hawaii-Midway	7	Medium
			Wake	8	Medium
			Howland-Baker	9	Medium
Jarvis	8	Low	Northern Mariana Is.	10	Low
Belau (Palau)	9	Low	Jarvis	11	Low
Guam	10	Low	Samoa	12	Low
Howland-Baker	11	Low	Belau (Palau)	13	Low
Samoa	12	Low	Guam	14	Low

Islands EEZ (Commeau et al., 1984; Halbach, 1984; Halbach and Manheim, 1984). In addition, most estimates have been based solely on a price per ton of metal calculation, and have not considered such necessary parameters as the amount of substrate recovered with the crust, the amount of sediment versus outcrop, and the percentage of the flanks of an edifice that could not be mined because of the roughness of small-scale topography.

Johnson, Clark, and Otto (1985) estimated the total resource potential within the EEZ of Hawaii and United States trust and affiliated territories (Table 3). They based their estimate primarily on the total area within the 800- to 2400-m depth range on seamounts with an estimated age of >40 Ma. We rearranged their ranking of areas, as shown in Table 3, on the basis of the geologic and oceanographic criteria we set forth in this section. In addition, we included two other Pacific nations, the Kiribati Islands and French Polynesia, in the table. Other factors such as nearness to port facilities and processing plants and the cost of transportation need to be considered for a more complete view of the economic potential of an area. The rankings in Table 3 are based on areal, geologic, and oceanographic considerations and may be significantly modified when these other economic considerations are evaluated.

The biological communities that inhabit seamounts are little known. Bottom photographs show an abundance of sea-floor life on both hard and soft substrates, including both sessile and mobile organisms (Figures 10a and f). Ferromanganese crusts are in turn encrusted with more than 30 taxa of organisms, mostly foraminifera (Lauren Mullineaux, Scripps Institution of Oceanography, 1985, written communication). The waters above seamounts commonly support planktonic and nektonic communities as the result of local upwelling. Although some data are available on the near-surface and bottom-dwelling organisms, the mid-level water column communities are virtually unknown for seamount areas.

The study of cobalt-rich ferromanganese crusts on seamounts and ridges has just begun. Five years of study has answered some broad questions, but has introduced many more specific questions. The kinds of information needed to make intelligent choices regarding potential mine sites, the development of mining equipment, and the exploitation of deposits is still far from being in hand.

DISTRIBUTION OF FERROMANGANESE NODULES

Many summary papers and books have been written concerning the distribution, chemistry, physical characteristics, and origin of ferromanganese nodules (for example, Mero, 1965; Horn, 1972; Glasby, 1977; Heath, 1981; Halbach, 1984). In addition, a recent map of the distribution of nodules in the Pacific has been published by the Circum-Pacific Map Project (Piper et al., 1985). Consequently, we will concentrate here on a brief description of nodule distribution, with emphasis on areas of economic importance.

Nodules occur at most depths of water from the continental shelf to the abyssal plain. However, nodule formation is limited to areas where rates of sedimentation are low, and consequently nodules are most abundant in abyssal areas of clay and siliceous clay deposition. These nodules from abyssal depths are the only ones of economic importance at this time.

The Pacific Basin is divided into four regions of distinctive nodule chemistry (Mero, 1965). Nodules are rich in iron from the northwest, eastern borderland, southwest, and south-central Pacific (region A, Table 4). Nodules are rich in manganese from the north equatorial Pacific, immediately offshore Japan, the Nazca plate off South America, and the Gulf of California (region B, Table 4). Nodules are rich in nickel and copper in the northeast, southwest equatorial, and most of the abyssal east Pacific (region C, Table 4); this area includes the prime site between the Clarion and Clipperton fracture zones considered for exploitation of nodules (Figure 1). Nodules from topographic highs are rich in cobalt (region D, Table 4) and were emphasized along with crusts in the previous sections of this chapter. Other workers have also attempted regional delineation of nodules based on chemical composition, and have made some refinements to Mero's scheme (for example, Cronan, 1972; Piper and Williamson, 1977). Table 5 lists chemical data from Cronan (1972) and McKelvey, Wright, and Rowland (1979) for abyssal nodules.

Grade (content of copper plus nickel) and the area of sea floor covered by nodules determines the potential of a region for economic exploitation. The cut-off grade has been determined to be 1.8 weight % copper plus nickel with a sea-floor coverage of 5 kg/m^2 of nodules (Kaufman, 1974; Frazer, 1977). A mine site should have an average grade of about 2.27% Cu + Ni with 10 kg/m^2 of nodules to be profitable. These criteria limit the significant potential economic regions of the Pacific to one, the Clarion-Clipperton region in the eastern equatorial Pacific (Figure 1; Frazer, 1977; Heath, 1981). Several other small areas may contain some mine sites (for example, southwest of Hawaii), but discovery of other regions of economic interest the size of the Clarion-Clipperton region is unlikely (Frazer, 1977).

The Clarion-Clipperton region (C–C region) lies on the northern edge of the equatorial zone of high biological productivity beneath 3200 to 5900 m of water. Bottom sediments are dominantly brown pelagic clay and siliceous clay. The high copper and nickel nodules in this region (Table 5) can be partly explained by their proximity to the equatorial zone of high productivity. In regions of lower productivity (pelagic clay regions to the north) the minor elements in the nodules are diluted by high-background iron, whereas in areas of higher productivity (calcareous and siliceous ooze regions to the south) the minor elements are diluted with diagenetically remobilized manganese (Heath, 1981). In addition, much of the iron in the C–C region is fixed by the formation of authigenic iron-rich smectite in the upper millimeters to centimeters of the sediment and is not available to significantly dilute the copper and nickel in the nodules (Hein et al., 1979).

The total area of the C–C region is 2.25 × 10^6 km^2. Of this area, about 1.25 to 1.37 × 10^6 km^2 contain nodules of high enough grade and coverage to be considered minable (Frazer, 1977; McKelvey, Wright, and Rowland, 1979). If 20% of the nodules are recoverable, this area would support 14 to 56 first-generation mine sites processing 3 million tons of dry nodules per year for 25 years (Frazer, 1977). A 20% recovery from 1.25 × 10^6 km^2 would produce 2.1 billion dry metric tons of nodules averaging about 1.3% Ni, 1.0% Cu, 25% Mn, 0.22% Co, and 0.05% Mo (McKelvey, Wright, and Rowland, 1979).

At this time, the exploitation of nodules from the C–C region is not attractive because of the uncertainties brought about by the United Nations Law of the Sea Treaty on mining in international waters, and because of the low price of copper on the world market. Even though the technologies for the recovery and processing of nodules exist, the potential and incentives for development of this deep-sea nodule resource are marginal at best.

Table 4. Chemical composition of manganese nodules from various regions in the Pacific (wt.%).

	A[1]	B	C	D	E
Mn	21.7	49.8	33.3	28.5	32.4
Fe	28.3	2.3	17.7	22.6	18.5
Co	0.35	0.06	0.39	1.20	0.47
Ni	0.46	0.26	1.52	0.66	1.14
Cu	0.32	0.14	1.13	0.21	0.80
Pb	0.21	0.05	0.18	0.30	0.19

[1]A, B, C, and D are from Mero's (1965) iron-rich, manganese-rich, copper- and nickel-rich, and cobalt-rich areas of the Pacific, respectively (see discussion in the text). E is the average for the entire Pacific. All analyses on a detrital mineral-free basis.

ACKNOWLEDGMENTS

We are grateful to Professor Peter Halbach, Technical University of Clausthal, and to Dr. H. R. Kudrass, Bundesanstalt fur Geowissenschaften und Rohstoffe, for allowing us to use, respectively, the Howland-Baker and South China Sea crust data before it was published. We thank Dr. Mary Fisk and Dr. Carla Moore for help in locating data and references. We thank Drs. Peter Halbach, Vesna Marchig, Doris Puteanus, Frank Manheim, William Schwab, and Alice Davis for helpful discussions and exchange of data. Charlaine Daniel-Fleishman provided much needed technical assistance. Drs. W. Schwab and J. D. Hem, U.S. Geological Survey, provided valuable reviews.

Table 5. Average chemical composition of abyssal nodules from various regions of the Pacific (Cronan, 1972; McKelvey, Wright, and Rowland, 1979).

	1	2	3	4	5	6	7
Mn	22.33	19.81	15.71	16.61	16.87	12.29	28.80
Fe	9.44	10.20	9.06	13.92	13.30	12.00	6.64
Ni	1.080	0.961	0.956	0.433	0.564	0.422	1.22
Co	0.192	0.164	0.213	0.595	0.395	0.144	0.23
Cu	0.627	0.311	0.711	0.185	0.393	0.294	0.99
Pb	0.028	0.030	0.049	0.073	0.034	0.015	0.048
Ba	0.381	0.145	0.155	0.230	0.152	0.196	0.32
Mo	0.047	0.037	0.041	0.035	0.037	0.018	0.048
V	0.041	0.031	0.036	0.050	0.044	0.037	0.03
Cr	0.0007	0.0005	0.0012	0.0007	0.0007	0.0044	0.02
Ti	0.425	0.467	0.561	1.007	0.810	0.634	0.61
L.O.I.	24.75	27.21	22.12	28.73	25.50	22.52	23.91
Depth (m)	4537	4324	5049	3539	5001	4990	—

1 = Northeast Pacific
2 = Southeast Pacific
3 = Central Pacific
4 = South Pacific
5 = West Pacific
6 = North Pacific
7 = Clarion-Clipperton prime area, based on from 43 samples (V) to 409 samples (Ni) (McKelvey, Wright, and Rowland, 1979).

REFERENCES CITED

Aplin, A. C., 1984, Rare earth element geochemistry of Central Pacific ferromanganese encrustations: Earth and Planetary Science Letters, v. 71, p. 13–22.

Aplin, A. C., and D. S. Cronan, 1985, Ferromanganese oxide deposits from the central Pacific Ocean, I. Encrustations from the Line Islands Archipelago: Geochimica et Cosmochimica Acta, v. 49, no. 2, p. 427–436.

Aumento F., D. E. Lawrence, and A. G. Plant, 1968, The ferromanganese pavement on San Pablo Seamount: Canada Geological Survey Paper 68-32, p. 1–30.

Barnes, S. S., 1967, Minor element composition of ferromanganese nodules: Science, v. 157, p. 63–65.

Bonatti, E., T. Kraemer, and H. Rydell, 1972, Classification and genesis of submarine iron-manganese deposits, *in* D. R. Horn, ed., Ferromanganese deposits on the ocean floor: Washington, D.C., National Science Foundation, p. 149–166.

Burns, R. G., 1976, The uptake of cobalt into ferromanganese nodules, soils and synthetic manganese (IV): Oxides: Geochimica et Cosmochimica Acta, v. 40, p. 95–102.

Burns, R. G., and B. A. Brown, 1972, Nucleation and mineralogical controls on the composition of manganese nodules, *in* D. R. Horn, ed., Ferromanganese deposits on the ocean floor: Washington, D.C., National Science Foundation, p. 51–61.

Commeau, R. F., A. Clark, C. Johnson, F. T. Manheim, P. J. Aruscavage, and C. M. Lane, 1984, Ferromanganese crust resources in the Pacific and Atlantic Oceans: Proceedings of Oceans, September 10–12, p. 421–430.

Craig, J. D., J. E. Andrews, and M. A. Meylan, 1982, Ferromanganese deposits in the Hawaiian Archipelago: Marine Geology, v. 45, p. 127–157.

Cronan, D. S., 1972, Regional geochemistry of ferromanganese nodules in the world ocean, *in* D. R. Horn, ed., Ferromanganese deposits on the ocean floor: Washington, D.C., National Science Foundation, p. 19–30.

Cronan, D. S., 1984, Criteria for the recognition of areas of potentially economic manganese nodules and encrustations in the CCOP/SOPAC region of the central and southwestern tropical Pacific: South Pacific Marine Geological Notes, v. 3, no. 1, 17 p.

Dillard, J. G., D. L. Crowther, and J. W. Murray, 1982, The oxidation state of cobalt and selected metals in Pacific ferromanganese nodules: Geochimica et Cosmochimica Acta, v. 46, p. 755–759.

Frank, D. J., M. A. Meylan, J. D. Craig, and G. P. Glasby, 1976, Ferromanganese deposits of the Hawaiian Archipelago: Hawaii Institute of Geophysics Report HIG-76-14, 71 p.

Frazer, J. Z., 1977, Manganese nodule reserves: an updated estimate: Marine Mining, v. 1, p. 103–123.

Glasby, G. P. (editor), 1977, Marine manganese deposits: Amsterdam, Elsevier, Elsevier Oceanography Series 15, 523 p.

Goldberg, E. D., 1954, Marine geochemistry 1: Chemical scavengers of the sea: Journal of Geology, v. 62, no. 3, p. 249–265.

Goldberg, E. D., 1961, Chemistry in the oceans, *in* M. Sears, ed., Oceanography: American Association for the Advancement of Science Publication, v. 67, p. 583–587.

Haggerty, J. A., S. O. Schlanger, and I. P. Silva, 1982, Late Cretaceous and Eocene volcanism in the southern Line Islands and implications for hotspot theory: Geology, v. 10, p. 433–437.

Halbach, P., 1984, Deep-sea metallic deposits: Ocean Management, v. 9, p. 35–60.

Halbach, P., and F. T. Manheim, 1984, Potential of cobalt and other metals in ferromanganese crusts on seamounts of the Central Pacific Basin: Marine Mining, v. 4, no. 4, p. 319–336.

Halbach, P., and D. Puteanus, 1984, The influence of the carbonate dissolution rate on the growth and composition of Co-rich ferromanganese crusts from Central Pacific Seamount areas: Earth and Planetary Science Letters, v. 68, p. 73–87.

Halbach, P., F. T. Manheim, and P. Otten, 1982, Co-rich ferromanganese deposits in the marginal seamount regions of the Central Pacific Basin—Results of the Midpac '81: Erzmetall, v. 35, no. 9, p. 447–453.

Halbach, P., D. Puteanus, and F. T. Manheim, 1984, Platinum concentrations in ferromanganese seamount crusts from the Central Pacific: Naturwissenschaften, v. 71, p. 577–579.

Halbach, P., M. Segl, D. Puteanus, and A. Mangini, 1983, Co-fluxes and growth rates in ferromanganese deposits from Central Pacific Seamount areas: Nature, v. 304, p. 716–719.

Heath, G. R., 1981, Ferromanganese nodules of the deep sea: Economic Geology, 75th anniversary volume, p. 736–765.

Hein, J. R., F. T. Manheim, W. C. Schwab, and A. S. Davis, 1985a, Ferromanganese crusts from Necker Ridge, Horizon Guyot and S. P. Lee Guyot: Geological considerations: Marine Geology, v. 69, p. 25–54.

Hein, J. R., C. R. Ross, E. Alexander, and H.-W. Yeh, 1979, Mineralogy and diagenesis of surface sediments from DOMES areas A, B, and C, *in* J. L. Bischoff and D. Z. Piper, eds., Marine geology and oceanography of the Pacific manganese nodule province: New York, Plenum Press, Inc., p. 365–396.

Hein, J. R., F. T. Manheim, W. C. Schwab, A. S. Davis, C. L. Daniel, R. M. Bouse, L. A. Morgenson, R. E. Sliney, D. Clague, G. B. Tate, and D. A. Cacchione, 1985b, Geological and geochemical data for seamounts and associated ferromanganese crusts in and near the Hawaiian, Johnston Island, and Palmyra Island Exclusive Economic Zones: U.S. Geological Survey Open-File Report 85-292, 129 p.

Hem, J. D., 1978, Redox processes at surfaces of manganese oxide and their effects on aqueous metal ions: Chemical Geology, v. 21, p. 199–218.

Horn, D. R. (editor), 1972, Ferromanganese deposits of the ocean floor: Washington, D.C., National Science Foundation, 293 p.

Jackson, E. D., and S. O. Schlanger, 1976, Regional synthesis, Line Islands chain, Tuamota Island chain, and Manihiki Plateau, central Pacific Ocean, *in* S. O. Schlanger, E. D. Jackson, et al., Initial reports of the Deep Sea Drilling Project (Vol. 33): Washington, D.C., U.S. Government Printing Office, p. 915–927.

Johnson, C. J., A. L. Clark, and J. M. Otto, 1985, Pacific Ocean minerals: The next twenty years: East-West Center, Resource Systems Institute, Honolulu, Hawaii (unpublished).

Kaufman, R., 1974, The selection and sizing of tracts comprising a manganese nodule ore body: Offshore Technology Conference Proceedings, v. 11, Paper OTC 2059, p. 283–298.

Koski, R. A., J. R. Hein, R. M. Bouse, and R. E. Sliney, 1985, The composition and origin of ferromanganese crusts from Tonga platform, southwest Pacific, *in* D. W. Scholl and T. L. Vallier, eds., Geology and offshore resources of Pacific island arcs—Tonga region: AAPG Earth Science Series, v. 2, p. 179–186.

Ku, T. L., 1977, Rates of accretion, *in* G. P. Glasby, ed., Marine manganese deposits: Amsterdam, Elsevier, p. 249–267.

Ku, T. L., M. Kusakabe, D. E. Nelson, J. R. Southon, R. G. Korteling, J. Vogel, and I. Nowikow, 1982, Constancy of oceanic deposition of ^{10}Be as recorded in manganese crusts: Nature, v. 299, p. 240–242.

Mangini, A., M. Segl, W. Wolfli, M. Suter, and G. Bonani, 1983, ^{10}Be-dating of the inner structure of Mn-crust VA 13-2 to 11 My. BP (abs.): EOS (American Geophysical Union Transactions), v. 64, p. 284.

Manheim F. T., T. H. Ling, and C. M. Lane, 1983, An extensive data base for cobalt-rich ferromanganese crusts from the world oceans: Proceedings of Oceans, August 29–September 1, p. 828–831.

McKelvey, V. E., N. A. Wright, and R. W. Rowland, 1979, Manganese nodule resources in the northeastern equatorial Pacific, *in* J. L. Bischoff and D. Z. Piper, eds., Marine geology and oceanography of the Pacific manganese nodule province: New York, Plenum Press, Inc., p. 747–762.

Mero, J. L., 1965, The mineral resources of the sea, Oceanography Series 1: Amsterdam, Elsevier, p. 182–219.

Moore, W. S., 1973, Accumulation rates of manganese crusts on rocks exposed on the sea floor, *in* Inter-university program of research on ferromanganese deposits of the ocean floor, phase report. Photocopy.

Moore, W. S., and P. R. Vogt, 1976, Hydrothermal manganese crusts from two sites near the Galapagos spreading axis: Earth and Planetary Science Letters, v. 29, p. 349–356.

Piper, D. Z., and M. E. Williamson, 1977, Composition of Pacific Ocean ferromanganese nodules: Marine Geology, v. 23, p. 285–303.

Piper, D. Z., T. R. Swint, F. W. McCoy, and L. G. Sullivan, 1985, Manganese nodule, seafloor sediment, and sedimentation rates of the Circum-Pacific Region: Tulsa, Oklahoma, AAPG, scale 1:17,000,000.

Schwab, W. C., A. Davis, J. Haggerty, T. Ling, and J. Commeau, 1985, Geologic reconnaissance and geochemical analysis of ferromanganese crusts of the Radak chain, Marshall Islands: U.S. Geological Survey Open-File Report 85-18, 10 p.

Segl, M., A. Mangini, G. Bonani, H. J. Hofmann, M. Nessi, M. Suter, W. Wolfli, G. Friedrich, W. L. Pluger, A. Wiechowski, and J. Beer, 1984, ^{10}Be-dating of a manganese crust from central North Pacific and implications for ocean palaeocirculation: Nature, v. 309, p. 540–543.

von Stackelberg, U., H. Kunzendorf, V. Marchig, and R. Gwozdz, 1984, Growth history of a large ferromanganese crust from the equatorial North Pacific nodule belt: Geologisches Jahrbuch, bd. 75, Reihe A., p. 213–235.

Winterer, E. L., J. I. Ewing, and Shipboard Scientific Party, 1973, Initial reports of the Deep Sea Drilling Project (Vol. 17): Washington, D.C, U.S. Government Printing Office, 930 p.

REFERENCES USED FOR MAP COMPILATIONS

Ahrens, L. H., J. P. Willis, and C. O. Oosthuizen, 1967, Further observations on the composition of manganese nodules, with particular reference to some of the rarer elements: Geochimica et Cosmochimica Acta, v. 31, p. 2169–2180.

Anderson, M. E., 1978, Accumulation rates of manganese nodules and sediments: An alpha track method: San Diego, University of California, Scripps Institution of Oceanography Dissertations, unpublished M.A. thesis, 113 p.

Andrews, J. E., M. Morgenstein, C. D. Fein, M. A. Meylan, S. V. Margolis, G. Andermann, and G. P. Glasby, 1972, Investigations of ferromanganese deposits from the central Pacific: Hawaii Institute of Geophysics Report HIG-72-23, 133 p.

Aplin, A. C., 1984, Rare earth element geochemistry of central Pacific ferromanganese encrustations: Earth and Planetary Science Letters, v. 71, p. 13–22.

Aplin, A. C., and D. S. Cronan, 1985, Ferromanganese oxide deposits from the central Pacific Ocean, I. Encrustations from the Line Islands Archipelago: Geochimica et Cosmochimica Acta, v. 49, no. 2, p. 427–436.

Bacon, J. I., 1967, A geochemical study of some manganese nodules: Tulsa, University of Tulsa unpublished M.A. thesis, 53 p.

Barnes, S. S., 1967, The formation of oceanic ferromanganese nodules: San Diego, University of California at San Diego, unpublished Ph.D. thesis, 59 p.

Bezrukov, P. L. (editor), 1976, Ferromanganese nodules of the Pacific Ocean: Transactions of the P. P. Shirshov Institute of Oceanology, v. 109, 301 p. (in Russian).

Bonatti, E., and O. Joensuu, 1966, Deep-sea iron deposit from the South Pacific: Science, v. 154, p. 643–645.

Craig, J. D., J. E. Andrews, and M. A. Meylan, 1982, Ferromanganese deposits in the Hawaiian Archipelago: Marine Geology, v. 45, p. 127–157.

Cronan, D. S., and J. S. Tooms, 1969, The geochemistry of manganese nodules and associated pelagic deposits from the Pacific and Indian Oceans: Deep-Sea Research, v. 16, p. 335–359.

Dietz, R. S., 1955, Manganese deposits on the northeast Pacific sea floor: California Journal of Mines and Geology, v. 51, no. 3, p. 209–220.

Fisk, M. B., J. Z. Frazer, J. S. Elliott, and L. L. Wilson, 1979, Availability of copper, nickel, cobalt and manganese from ocean ferromanganese nodules (II): Scripps Institute of Oceánography Reference 79-17, 46 p.

Frazer, J. Z., M. B. Fisk, J. Elliott, M. White, and L. Wilson, 1978, Availability of copper, nickel, cobalt, and manganese from ocean ferromanganese nodules: Scripps Institute of Oceanography Reference 78-25, 45 p.

Goldberg, E. D., 1954, Marine geochemistry 1: Chemical scavengers of the sea: Journal of Geology, v. 62, no. 3, p. 249–265.

Greenslate, J., M. Fisk, and D. Felix, 1979, Systematics in manganese concretion composition in the northeastern equatorial Pacific, *in* La Genese des nodules de manganese, no. 289: Paris, Colloques Internationaux du Centre National de la Recherche Scientifique, p. 39–54.

Halbach, P., unpublished chemistry of the Howland-Baker Islands.

Halbach, P., and D. Puteanus, 1984, The influence of the carbonate dissolution rate on the growth and composition of Co-rich ferromanganese crusts from Central Pacific Seamount areas: Earth and Planetary Science Letters, v. 68, p. 73–87.

Halbach, P., F. T. Manheim, and P. Otten, 1982, Co-rich ferromanganese deposits in the marginal seamount regions of the Central Pacific Basin—Results of the Midpac '81: Erzmetall, v. 35, no. 9, p. 447–453.

Halbach, P., D. Puteanus, and F. T. Manheim, 1984, Platinum concentrations in ferromanganese seamount crusts from the Central Pacific: Naturwissenschaften, v. 71, p. 577–579.

Halbach, P., M. Segl, D. Puteanus, and A. Mangini, 1983, Co-fluxes and growth rates in ferromanganese deposits from Central Pacific Seamount areas: Nature, v. 304, p. 716–719.

Han, K. N., M. P. Hoover, and D. W. Fuerstenau, 1979, The effect of temperature on the physio-chemical characteristics of deep-sea manganese nodules: Marine Mining, v. 2, no. 1–2, p. 131–149.

Hawaii Institute of Geophysics Resource assessment report 1984.

Hein, J. R., chemistry from the Lau Basin, 1986, manuscript in preparation.

Hein, J. R., F. T. Manheim, W. C. Schwab, A. S. Davis, C. L. Daniel, R. M. Bouse, L. A. Morgenson, R. E. Sliney, D. Clague, G. B. Tate, and D. A. Cacchione, 1985b, Geological and geochemical data for seamounts and associated ferromanganese crusts in and near the Hawaiian, Johnston Island, and Palmyra Island Exclusive Economic Zones: U.S. Geological Survey Open-File Report 85-292, 129 p.

Hewett, D. F., M. Fleischer, and N. Conklin, 1963, Deposits of the manganese oxides: Supplement: Economic Geology, v. 58, no. 1, p. 1–51.

Karpoff, A. M., I. Peterschmitt, and M. Hoffert, 1977, Mineralogy and geochemistry of sedimentary deposits on Emperor Seamounts, sites 430, 431, and 432: Authigenesis of silicates, phosphates, and ferromanganese oxides, *in* E. D. Jackson, I. Koizumi, et al., Initial reports of the Deep Sea Drilling Project (Vol. 55): Washington, D.C., U.S. Government Printing Office, p. 463–489.

Koski, R. A., in press, Ferromanganese deposits from the Gulf of Alaska Seamount Province: Mineralogy, chemistry, and origin: Canadian Journal of Earth Sciences.

Koski, R. A., J. R. Hein, R. M. Bouse, and R. E. Sliney, 1985, The composition and origin of ferromanganese crusts from Tonga platform, southwest Pacific, *in* D. W. Scholl and T. L. Vallier, eds., Geology and offshore resources of Pacific island arcs—Tonga region: AAPG Earth Science Series, v. 2, p. 179–186.

Kudrass, H. R., unpublished chemistry from the South China Sea.

Mangini, A., M. Segl, H. Kudrass, M. Wiedicke, G. Bonani, H. J. Hofmann, E. Morenzoni, M. Nessi, M. Suter, and W. Wolfli, 1986, Diffusion and supply rates of ^{10}Be and ^{230}Th radioisotopes in two manganese encrustations from the South China Sea: Geochimica et Cosmochimica Acta, v. 50, no. 1, p. 149–156.

Mero, J. L., 1965, The mineral resources of the sea, Oceanography Series 1: Amsterdam, Elsevier, p. 182–219.

Meylan, M. A., and H. G. Goodell, 1976, Chemical composition of manganese nodules from the Pacific-Antarctic Ocean, Drake Passage and Scotia Sea: Relation to ferromanganese oxide mineralogy and nucleus type, *in* G. P. Glasby and H. R. Katz, eds., UN ESCAP CCOP/SOPAC Technical Bulletin 2, p. 99–117.

Monzier, M., 1976, Manganese nodules and encrustations in the vicinity of New Caledonia and the Loyalty Islands, *in* G. P. Glasby and H. R. Katz, eds., UN ESCAP CCOP/SOPAC Technical Bulletin 2, p. 124–128.

Moorby, S. A., D. S. Cronan, and G. P. Glasby, 1984, Geochemistry of hydrothermal Mn-oxide deposits from the S.W. Pacific island arc: Geochimica et Cosmochimica Acta, v. 48, p. 433–441.

Morgan, C. L., and J. R. Moore, 1975, Role of the nucleus in formation of ferromanganese nodules: Processing guidelines for the marine miner: Offshore Technology Conference Proceedings, v. 1, Paper OTC 2243, p. 943–953.

Okada, A., and M. Shima, 1970, Study on the manganese nodule: Journal of the Oceanographical Society of Japan, v. 26, no. 3, p. 151–158. (in Japanese)

Piper, D. Z., and M. E. Williamson, 1977, Composition of Pacific Ocean ferromanganese nodules: Marine Geology, v. 23, p. 285–303.

Schwab, W. C., A. Davis, J. Haggerty, T. Ling, and J. Commeau, 1985, Geologic reconnaissance and geochemical analysis of ferromanganese crusts of the Radak Chain, Marshall Islands: U.S. Geological Survey Open-File Report 85-18, 10 p.

Smith, J. D., and J. D. Burton, 1972, The occurrence and distribution of tin with particular reference to marine environments: Geochimica et Cosmochimica Acta, v. 36, no. 5, p. 621–629.

Summerhayes, C. P., 1967, Manganese nodules from the southwestern Pacific: New Zealand Journal of Geology and Geophysics, v. 10, no. 6, p. 1372–1381.

Toth, J. R., 1980, Deposition of submarine crusts rich in manganese and iron: GSA Bulletin, pt. I, v. 91, p. 44–54.

Toyoda, K., Y. Nakamura, Y. Sano, and H. Wakita, 1985, Chemical composition of Mn nodules from Ogasawara and Mariana regions, *in* K. Kobayashi, ed., Preliminary report of the Hakuho Maru Cruise KH 84-1: Westpac, O.D.P., p. 240–243.

von Stackelberg, U., 1978, A polygenetic manganese nodule from the Scott Plateau off northwest Australia: BMR Journal of Australian Geology and Geophysics, v. 3, p. 349–351.

CHAPTER THIRTY-THREE

PHOSPHORITE DEPOSITS ALONG THE WESTERN NORTH AMERICAN CONTINENTAL MARGIN

B. J. Loebner

U.S. Geological Survey
Menlo Park, California

D. Z. Piper

U.S. Geological Survey
Menlo Park, California 94025

J. G. Vedder

U.S. Geological Survey
Menlo Park, California 94025

ABSTRACT

Nodular and sandy submarine phosphorite deposits occur along the western North American continental margin from San Francisco to the southern tip of Baja California, Mexico. These deposits have biostratigraphic ages of middle to late Miocene and Pliocene, and are associated with high-standing bathymetric features such as tops and flanks of shelves, banks, ridges, and seamounts. Phosphate deposits of the southern California borderland contain an estimated 54 to 117 million metric tons of phosphate nodules and sand. Nodules contain 19 to 37 weight

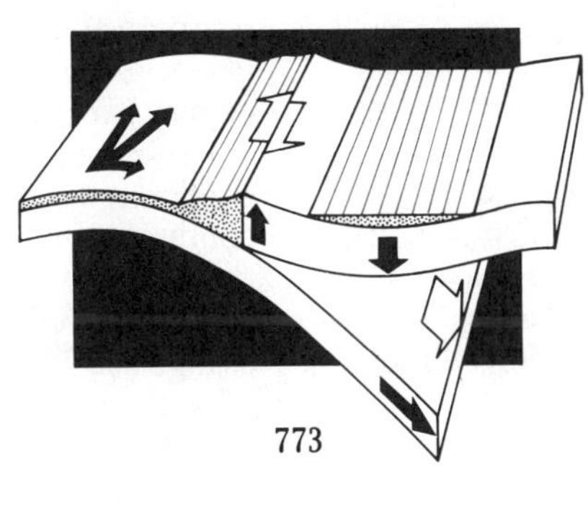

% P_2O_5, and average 27.6 weight % P_2O_5. Sand contains 0.3 to 9.7 weight % P_2O_5. Phosphate nodules occur within the Baja California borderland, but their extent is not known. Sandy phosphorite deposits occur within bays along the southern Baja California margin, and the largest of these is estimated to contain 3 billion tons of phosphate. Phosphate occurs along the central California margin but amounts are negligible. Phosphate deposits along the western North American continental margin are low in P_2O_5 content, but this factor may be offset by their favorable geographic position and the large market along the U.S. west coast and around the Pacific.

INTRODUCTION

Submarine phosphorite deposits along the western North American continental margin were first reported by Dietz, Emery, and Shepard (1942). These authors document the widespread occurrence of phosphate nodules on ridges, banks, and escarpments at shallow depths within the California Continental Borderland province, and describe the external and internal characteristics of nodules recovered in dredge hauls. Since this discovery, phosphorite deposits have been dredged, cored, or photographed on the sea floor along the continental margin from as far north as Cordell Bank near Point Reyes (Hanna, 1952; Chesterman, 1952) to as far south as the shelf off southern Baja California, Mexico (D'Anglejan, 1967)—a region which spans 2200 km of coastline (Figure 1). These deposits include phosphate nodules and phosphatic sands (Emery and Dietz, 1950; Emery et al., 1957; D'Anglejan, 1967; Inderbitzen, Carsola, and Everhart, 1970; Mullins and Rasch, 1985) and thin phosphate coatings on boulders of various lithologies (Greene, 1977).

The aim of this study is to evaluate the resource potential of phosphorite deposits along the western North American continental margin. This evaluation incorporates published and unpublished studies of phosphorite deposits and of the geology where they occur. Published chemical data on the phosphorite deposits are sparse and commonly incomplete, in spite of existing economic interest and numerous studies on these deposits. Nearly all the published chemical data are whole-rock analyses of phosphate nodules. These chemical analyses represent not only the phosphate present in nodules, but also nonphosphatic material such as glauconite, quartz, aluminosilicates, calcite, pyrite, biogenic silica, and lithic fragments of various chemical compositions that commonly are present in nodules. Commonly, P_2O_5 content alone or with CaO and F contents may be the only chemical data reported. New chemical data obtained using microprobe techniques are presented here for the phosphatic and nonphosphatic components in southern California borderland nodules.

For the purposes of the following presentation, we divide the continental margin between Point Reyes and the southern tip of Baja California into three regions—the central California continental margin, the southern California borderland, and the Baja California Pacific margin (Figure 1). The southern California borderland has been investigated to a much greater extent than the other two regions, and phosphorite deposits within its boundaries have received more attention because they could become an exploitable seafloor resource within the Exclusive Economic Zone of the United States. Preliminary investigations suggest that deposits offshore Baja California, Mexico may also be exploitable.

ANALYTICAL TECHNIQUES

Thin sections of phosphate nodules obtained from the U.S. Geological Survey were evaluated petrographically by Pasho (1972). These thin sections were polished and carbon-coated for chemical analysis. Chemical studies were performed by means of an ARL-SEMQ electron microprobe. An operating voltage of 15 kV and a specimen beam current of 0.02μA were used. To minimize loss of volatile elements, the beam was defocused to a diameter of approximately 20 microns. The following standards were used: fluorapatite for P, Ca, and F; andesite glass for Si, Al, Fe, Mg, K, and Na; barite for S; sodalite for Cl; and strontianite for Sr. Data reduction was performed using a modified version of Bence and Albee (1968). Analytical precision is better than 2% for all elements except F, where low counting statistics reduce the precision to about 8%.

SOUTHERN CALIFORNIA BORDERLAND

General Geology

The shoreline of the southern California borderland is bounded to the north by Point Conception and to the south by Point Descanso at approximately lat. 32°N (near the U.S.-Mexico international border; Figure 2). The borderland extends east–west from the mainland coast to the base of the Patton Escarpment, a distance of up to 250 km offshore. The southern boundary of the area described herein is arbitrary with respect to geologic provinces, and is placed at the southern edge of the region in which the majority of geologic investigations and economic appraisals of phosphorite deposits are concentrated.

The complex topography of the region is well known. The continental borderland differs from typical active or passive continental margins in that, beyond a narrow continental shelf, it consists of a network of basins as deep as 2100 m below sea level separated by shallow banks and island ridges as high as 750 m above sea level (Emery, 1960; Vedder et al.,

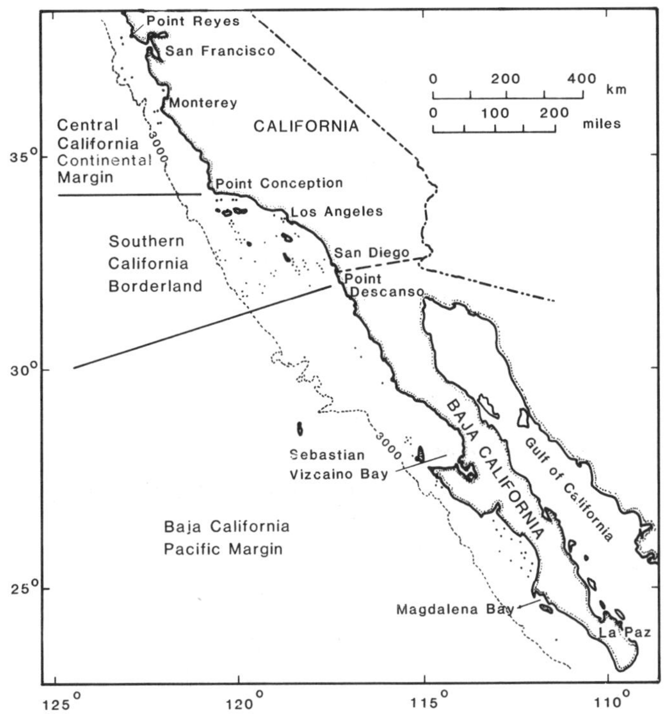

Figure 1. Distribution of submarine phosphorite (shown by dots) along the western North American continental margin. The 3000-m bathymetric contour defines the base of the continental slope.

1974). Basin areas at sill depth range from 655 to 2660 km^2 (Emery, 1960). Basins and ridges typically trend northwest–southeast, except for the northern Channel Islands, the Santa Barbara Basin, and the northern edge of the San Nicolas Basin, which trend east–west.

The regional geology of the southern California borderland is known from exposures on the Channel Islands, continuous seismic-reflection profiling, deep-penetration seismic-refraction studies, bottom sampling, and drilling. The regional geology of this area is described by Vedder (this volume) and will only be summarized here.

Deep basement beneath the borderland is interpreted to be oceanic crust (Shor and Raitt, 1958). Acoustic basement, however, may consist of severely deformed metasedimentary and blueschist facies rocks of Jurassic to early Tertiary (?) age, marine sedimentary rocks of Late Jurassic (?) to early Tertiary age, or various types of intrusive and extrusive volcanic rocks primarily of Miocene age (Emery, 1960; Vedder et al., 1974; Hill, 1976; Mattinson and Hill, 1976; Platt, 1976; Vedder, this volume). The severely deformed metasedimentary rocks are lithologically and structurally similar to the Franciscan assemblage of northern and central California (Crouch, 1979, 1981). This basement occurs in the Patton Ridge-Patton Escarpment area of the borderland (Emery, 1960; Vedder et al., 1974, 1976; Platt, 1975; Howell, McLean, and Vedder, 1976; Vedder, this volume). Marine sedimentary rocks that locally constitute acoustic basement range in age from Late Jurassic (?) to early Tertiary and consist of sandstone, shale, conglomerate, siltstone, and claystone (Vedder et al., 1974, 1976; Vedder, this volume). These rocks are slightly metamorphosed at places and are lithologically similar to the Great Valley sequence of northern California (Crouch, 1979, 1981). Volcanic rocks, predominantly of Miocene age, make up a third type of acoustic basement. They have been recovered from many localities throughout the borderland (Emery, 1960; Vedder et al., 1974, 1976; Vedder, this volume).

Sedimentary rocks overlying acoustic basement range from Cretaceous to Holocene in age. Outcrops of Late Cretaceous and early Tertiary sediments are common on the northern Channel Islands, but sea-floor outcrops are relatively rare and sparsely distributed (Vedder, this volume).

Miocene sedimentary and volcanic rocks are by far the most voluminous basinal deposits and form most of the high-standing submarine topography in the borderland (Emery, 1960; Moore, 1969; Vedder et al., 1974). Sea-floor exposures of these rocks are widely distributed along ridges and on knolls (Figure 3). Over much of the area west of the Santa Rosa-Cortes Ridge, Oligocene and/or younger strata are believed to lie directly on basement (Crouch, 1981; Vedder, this volume). Sedimentary rock types include mudstone, claystone, shale, phosphatic shale, chert, and limestone (Emery, 1960; Vedder et al., 1974). Volcanic rocks, most of which are believed to be of early and middle Miocene age (Vedder et al., 1974), have been recovered throughout the borderland (Figure 3). Andesitic types predominate in the inner borderland, forming nearly continuous exposures along the ridge that extends from the east end of Santa Cruz Island to Santa Catalina Island. Basalt and andesite occur as more isolated exposures on Fortymile and Thirtymile Banks. In the outer borderland, basalt has been recovered from Cortes, Tanner, and Northeast Banks and along the Patton Ridge (Vedder et al., 1974). Thicknesses of Miocene rocks range from < 300 m on the ridges to > 1500 m in some of the larger outer basins (Vedder et al., 1974; Vedder, this volume).

Exposures of Pliocene rocks, predominantly semiconsolidated mudstone and unconsolidated mud, are much less common than Miocene strata on the outer borderland shelves and slopes and are restricted primarily to the mainland shelf off Ventura and Los Angeles. Subbottom acoustic profiles indicate that Pliocene sediment has accumulated in all of the borderland basins, ranging in estimated thicknesses from < 300 m in Catalina Basin to as much as 2500 m in Santa Monica Basin (Vedder et al., 1974).

Unconsolidated sand and mud of Quaternary age make up relatively thick (300 to 1200 m) and extensive exposures on nearshore shelves and slopes and in nearshore basins when compared with accumulations of Quaternary hemipelagic mud in the outer borderland (Vedder et al., 1974).

Several models describing the geologic evolution of the southern California borderland within the context of plate-tectonic processes have been developed based on the distribution of basement complexes, stratigraphic and structural relationships, paleocurrent trends, and paleomagnetic data (Blake et al., 1978; Howell, McCulloch, and Vedder, 1978; Crouch, 1979, 1981; Howell and Vedder, 1981). A conclusion common to all of these models is that the present regional structure resulted from a major tectonic reorganization involving the Pacific, North American, and Farallon plates. Prior to 29 Ma the California continental margin was characterized by oblique subduction of the Farallon plate beneath the North American plate. Approximately 29 Ma the edge of the Pacific plate came into contact with the North American plate along southern California, forming a triple junction (ridge, trench, transform). This contact changed the nature

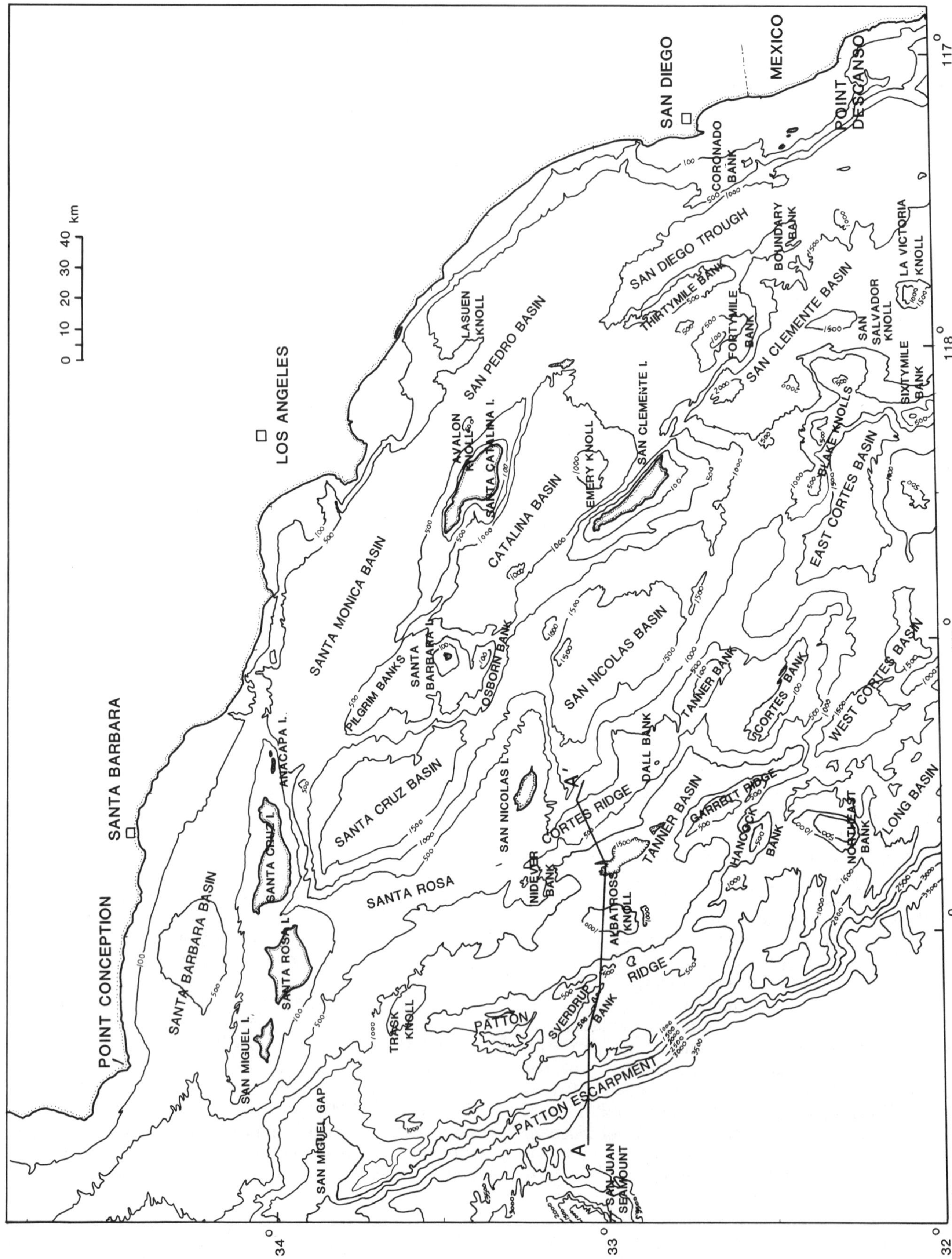

Figure 2. Bathymetry of the southern California borderland. A–A′ is a seismic-reflection line shown in profile on Figure 5. Contours are in meters. Modified from Vedder et al., 1974.

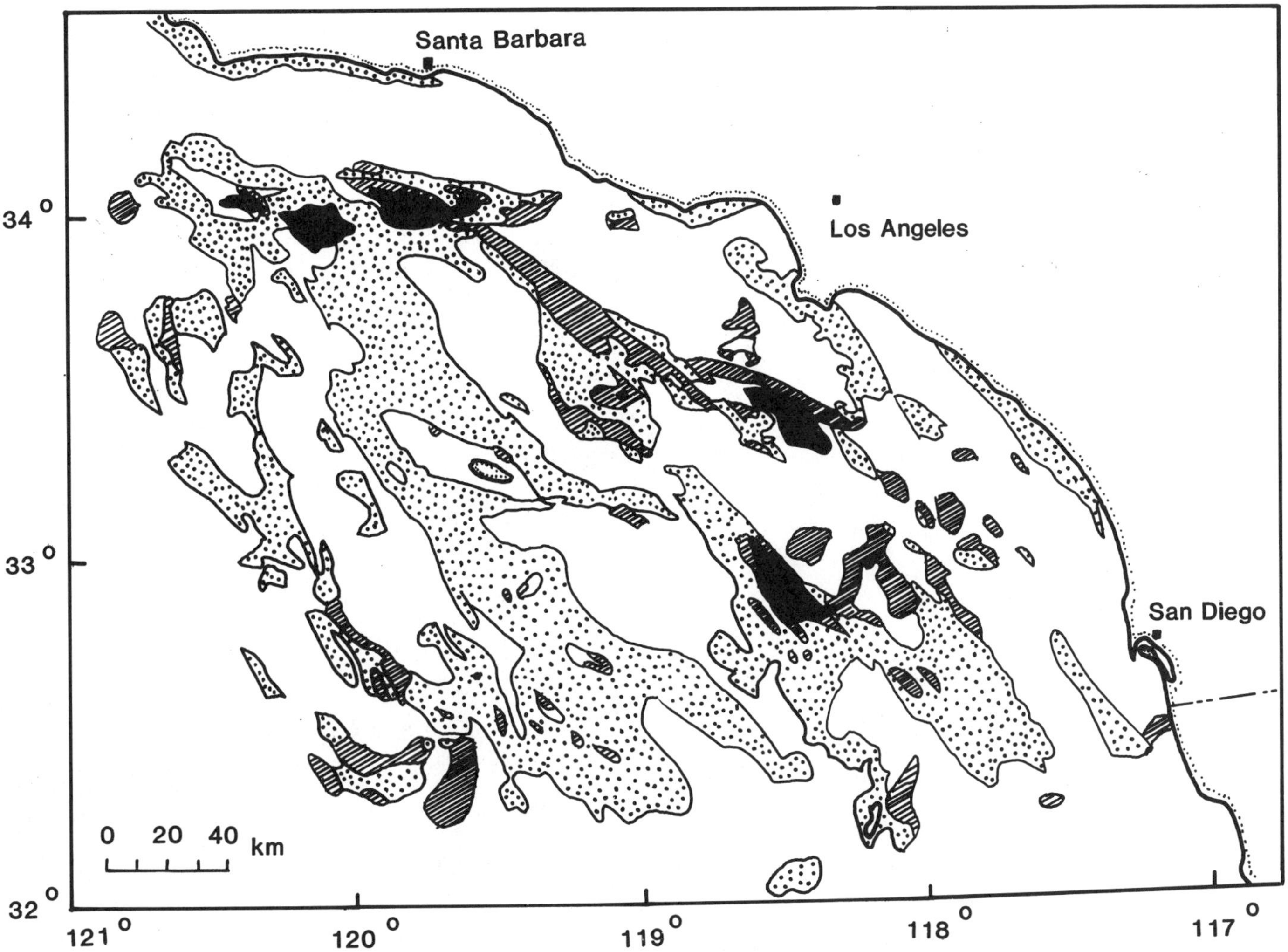

Figure 3. Generalized geologic map of the southern California borderland, showing the distribution of Miocene rocks. Stippled pattern = Miocene sedimentary rocks; lined pattern = volcanic and volcaniclastic rocks, chiefly Miocene; black areas = islands. Modified from Vedder, this volume.

of the western margin of North America from one characterized by oblique subduction to one characterized by right-lateral shear. The initial position of the boundary between the Pacific and North American plates was offshore southern California, but this boundary gradually shifted eastward to faults located inland, such as the San Andreas and related offsets. Blake et al. (1978) suggest that this gradual shifting caused local extension and compression within a broad zone of right-lateral shear and initiated formation of the basin-and-ridge borderland in the late Oligocene and early Miocene. A major phase of basin formation appears to coincide with a change in relative plate motion between the Pacific and North American plates about 10 Ma. The direction of relative motion became more westerly, imposing a component of extension (Blake et al., 1978).

Phosphorite Deposits

Phosphorites have been recovered in the southern California borderland from the tops and flanks of submerged banks, steep escarpments interpreted as fault scarps, submarine canyon walls, deep hills, basin slopes, and parts of continental shelves shallower than their surroundings (Figure 4). At the present time, these sites all represent essentially nondepositional environments (Dietz, Emery, and Shepard, 1942) generally characterized by surface exposures of Miocene strata (Emery, 1960; Vedder et al., 1974).

In spite of complex structural relationships and uncertainties in the interpretation of acoustic data within the southern California borderland, there appears to be a strong association between the abundance of phosphate nodules and the distribution of Miocene sedimentary rocks (Figures 3, 4). Most of the recovery sites for phosphate nodules that lie outside areas of Miocene strata occur downslope from sea-floor exposures of Miocene strata and may in fact represent transported material. Seismic-reflection profiles along bottom-sampling transects emphasize the association of phosphate nodules with high-standing areas that expose Miocene strata (Figure 5). Most phosphorite samples have been recovered from depths ranging between 30 and 330 m, but a few have been recovered from depths as great as 2800 m (Emery, 1960).

Phosphorite deposits typically consist of nodules, but several phosphatic sandy deposits are also present within the borderland. The following description of phosphorite nodules is summarized from Dietz, Emery, and Shepard, (1942), Emery and Dietz (1950), Emery (1960), and Pasho (1972). Nodules are hard, dense rocks of various colors, shapes and

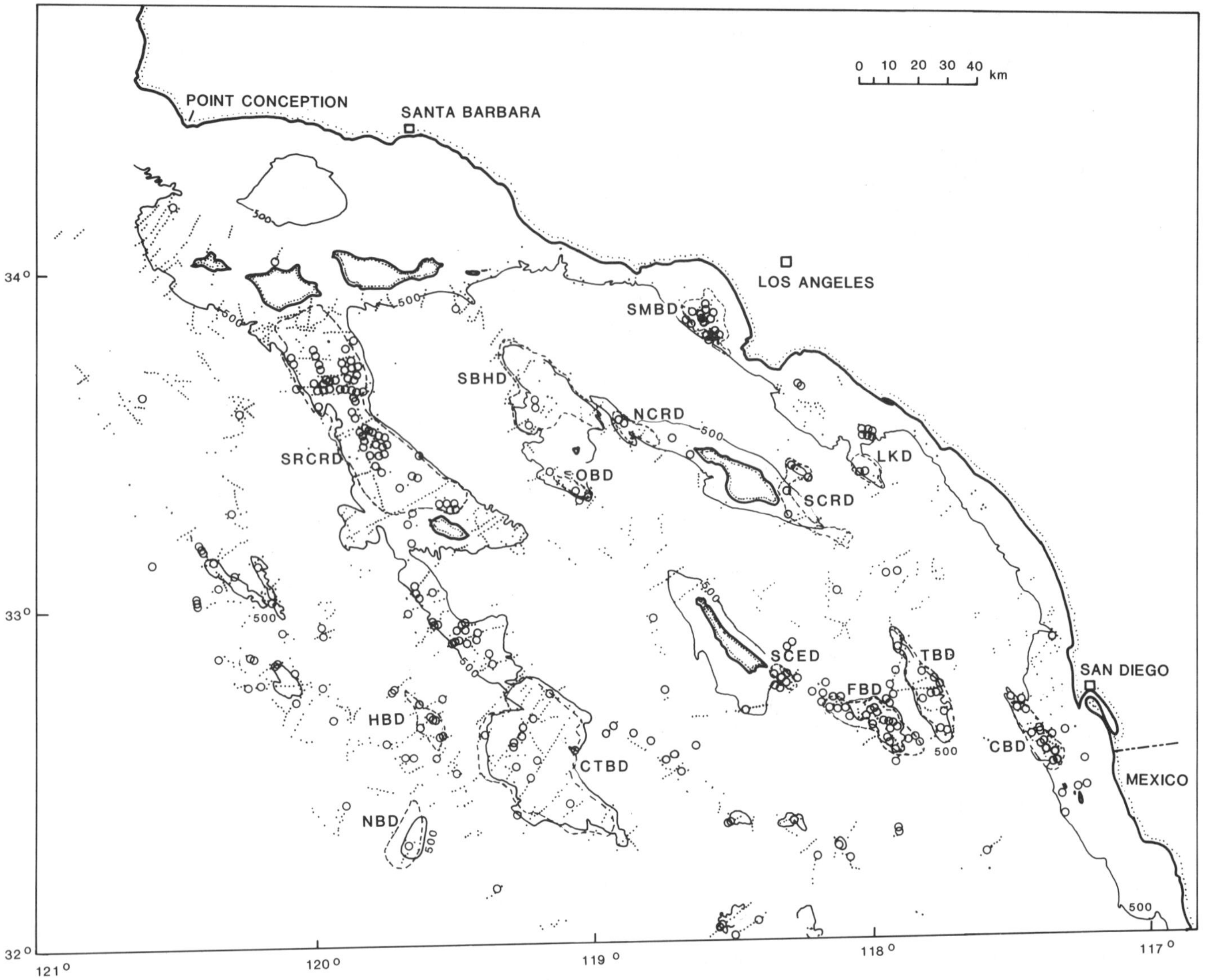

Figure 4. Distribution of phosphate in the southern California borderland. Dots = bottom samples without phosphorite; open circles = recovery of phosphate nodules or phosphatic sand in bottom samples. Approximately 2400 bottom samples from the following sources: U.S.G.S. cruises from 1968 to 1979; Dietz, Emery, and Shepard, 1942; Emery, 1960; Barnes, 1970; Inderbitzen, Carsola, and Everhart, 1970; Pasho, 1972. Dashed lines represent phosphate deposits as defined by Hess (1978). SMBD = Santa Monica Bay deposit; LKD = Lasuen Knoll deposit; SCRD = Southeast Catalina Ridge deposit; NCRD = Northwest Catalina Ridge deposit; OBD = Osborn Bank deposit; SBHD = Santa Barbara High deposit; SNRD = San Nicolas Ridge deposit; CBD = Coronado Bank deposit; TBD = Thirtymile Bank deposit; FBD = Fortymile Bank deposit; SCED = San Clemente Escarpment deposit; CTBD = Cortes-Tanner Banks deposit; HBD = Hancock Bank deposit; NBD = Northeast Bank deposit. 500–m contour is shown.

sizes, although nodules from a single small area tend to be similar in appearance. For example, nodules from Thirtymile Bank are large, dark brown, and contain numerous enclosed fragments of igneous rocks, whereas nodules from the Santa Monica Shelf are small, black, and contain little lithogenic material. Most nodules are tan to dark brown, but in some areas they are black. Recovered nodules range in size from pebbles a few cm in diameter to boulders >60 cm across, weighing up to 70 kg. Even larger nodules have been identified in sea-bottom photographs (Mero, 1961). Nodule surfaces may be mammillated, angular, or surmounted by knobby protuberances (Pasho, 1972). This surface is characterized by a glazed, polished appearance. Nodules from deeper waters are commonly coated on one side with manganese oxide.

Many marine organisms encrust nodule surfaces. Living and dead forms are found on opposite sides, indicating that nodules have occasionally been overturned.

Phosphate nodules are classified according to their internal structure as layered, conglomeratic, homogeneous, or hybrids of these (Dietz, Emery, and Shepard, 1942; Emery and Dietz, 1950). Layered nodules are the most common type recovered. Layers are irregular, nonconcentric, and range in thickness from a few millimeters to a few centimeters. Individual layers pinch out laterally, are rarely traceable for more than 5 or 6 cm (Pasho, 1972), and are truncated at nodule edges. Layers are commonly separated by a thin, dark-brown zone and small lenses of glauconitic foraminiferal sand. These thin, dark zones may consist of phosphate or, less commonly, manganese oxide. Several of these zones analyzed by microprobe during this study were all found to be composed of phosphate. Conglomeratic nodules are com-

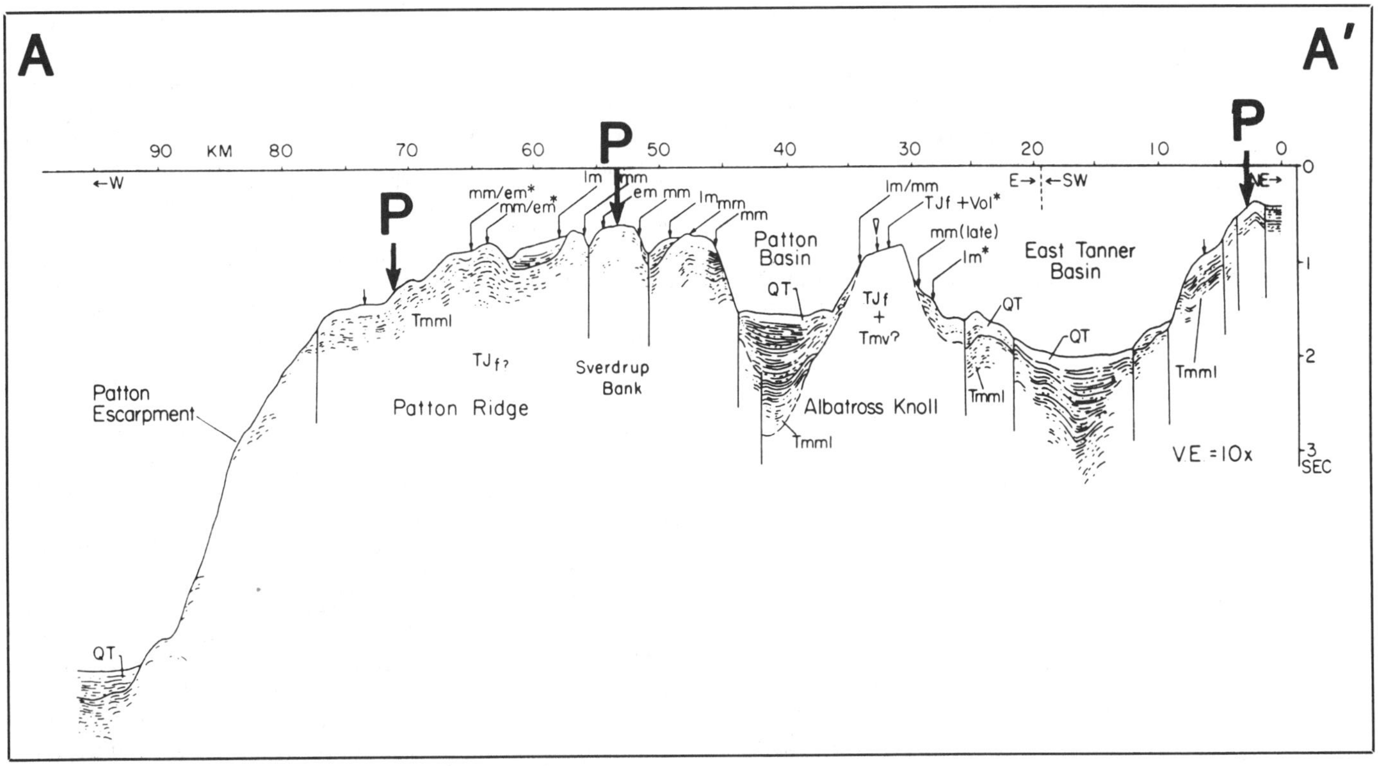

Figure 5. Line-drawing interpretation of seismic-reflection profile across Patton Ridge, Albatross Knoll, and part of Santa Rosa-Cortes Ridge (see Figure 2 for location). Age designations on dart cores are as follows: lm, mm, em = late, middle, and early Miocene; Tmml = middle and lower Miocene strata; Tmv = Miocene volcanic rocks; TJf = lower Miocene(?) to Jurassic rocks (Franciscan complex); QT = Quaternary to upper middle Miocene sediments. Vertical scale represents two-way traveltime, in seconds; 1 sec = 750 m water depth. Phosphate recoveries noted by 'P'. From Crouch, 1981.

mon, although less frequently recovered than layered nodules. Typically, light-colored lutitic clasts are enclosed in a darker colored matrix. In some cases the matrix is lighter than the clasts. Internally homogeneous nodules, with either lutitic or arenaceous texture, are less common than conglomeratic types. Hybrid nodules are present in several samples. A common type of hybrid has crude layering which is influenced to some degree by clasts similar to those found in conglomeratic nodules.

Surface protuberances reflect the internal character of each particular nodule and thus may be layered, homogeneous, or a resistant clast (Pasho, 1972). Internal vugs may be empty or filled with uncemented clastic grains, clay, or glauconitic sand. Vugs are commonly oriented parallel to the fabric of the nodule.

Nodule constituents include matrix, pellets, clastic mineral grains, bone fragments, glauconite, foraminifera, and lithic clasts. Relative percentages of southern California borderland nodule constituents determined by modal analyses of thin sections (D. Pasho, unpublished data) are presented in Table 1.

Matrix consists of cryptocrystalline francolite (carbonate fluorapatite) and accounts for $>50\%$ of the material in all types of nodules.

Phosphate pellets are common in nodules and range in size from 0.15 mm to 0.28 mm in diameter (Pasho, 1972). Two types of pellets are distinguishable, ovules and oolites. Ovules lack any primary internal structure, and are typically composed of material identical to the surrounding matrix. Both nucleated and nonnucleated ovules are found, commonly within the same nodule. Nuclei include foraminifera, angular clastic grains, and glauconite. Most nodules contain from 1 to 20% ovules (Pasho, 1972). Oolites, the second type of pellets, are generally well defined by alternating dark- and light-colored concentric bands, which may be due to the random and preferred orientation of submicroscopic francolite crystals. Both nonnucleated and nucleated varieties are found, and nuclei are similar to those found in ovules. Oolites are relatively rare in nodules, typically accounting for $<1\%$ and never $>5\%$ of nodule components.

Clastic grains, including angular quartz, feldspar, and minor hornblende and biotite, are present in most nodules, comprising 2 to 20% of the total constituents. Grain size typically ranges from .05 mm to .10 mm and is consistently smaller than pellets in the same nodule.

Glauconite occurs as dark-green subrounded to rounded grains in most nodules and averages from 0 to 5% of nodule constituents. Outer portions of glauconite grains are commonly replaced by apatite.

Fossils present in nodules include bone fragments (1% to rarely 4%), shark teeth, foraminifera tests (1–2% to rarely 14%), sponge spicules (2%), and echinoid spines and plates. Foraminifera tests are either well preserved or replaced by collophane and francolite.

Lithic clasts are angular to subangular, and range in size from 7 mm to 2 cm. They are composed of lutitic material with angular clastic grains, and may represent fragments of reworked phosphatic material.

Composition

Published chemical analyses of borderland nodules, representing only the most complete analyses, are presented in

Table 2. These include whole-rock wet-chemical analyses of nodules from Santa Monica Bay, Coronado Bank, Santa Barbara High, Fortymile Bank, Santa Rosa-Cortes Ridge, and Patton Ridge. In addition to these data, many less complete analyses or only P_2O_5 contents of nodules from southern California borderland localities have been published by Barnes (1970), Inderbitzen, Carsola, and Everhart (1970), Felix (1968), and Pasho (1972).

As part of an investigation into the petrogenesis of sedimentary phosphate deposits, phosphate nodules from the southern California borderland are being analyzed in our laboratory using microprobe techniques. It is possible in this way to isolate the composition of the phosphate material from other constituents within the nodule and evaluate the chemical composition of the main phosphate-bearing phase, apatite. The following discussion is preliminary because analysis of nodules has not been completed.

Within any given nodule, analyses of phosphatic components (matrix, ovules, halos around ovules, and different layers within oolites) yield apatite of similar composition. Apatite compositions of phosphatic components in nodules from several borderland deposits are presented in Table 3. CaO, P_2O_5, and F contents make up >80 weight % of these compositions. Analyses of nonphosphatic components such as aluminosilicates and glauconite are presented in Table 4.

Whole-rock chemical compositions of phosphate nodules from different localities within the southern California borderland are generally quite similar (Pasho, 1972). Microprobe data presented here support this observation. However, important differences exist between the microprobe results (Table 3) and the whole-rock compositions (Table 2). Contents of Si, Al, Fe, and K in microprobe analyses of phosphatic phases are lower than those reported for whole-rock nodule compositions. This is expected because these elements are principally contained within the nonphosphatic components in nodules such as quartz, feldspars, and glauconite, which the microprobe analysis technique excludes. For the same reason, the contents of Ca, P, and F, the main elements of apatite, are higher in the isolated phosphatic components relative to the whole-rock compositions. This is true for Ca and P, but the average F content of the phosphatic components (3.00 weight %) is lower than the average F content of the whole-rock compositions (3.22 weight %). The lower F content in microprobe analyses is most likely the result of problems with the analytical technique. Low counting statistics for F reduce the precision to about 8%. At this level of precision, the difference in F contents cannot be considered significant. Another factor that might contribute to the difference in F contents is the presence of another fluorine-bearing mineral in the nodules. Fluorite, a likely candidate, was not detected in borderland nodules studied by Dietz, Emery, and Shepard (1942) nor in those studied by Pasho (1972).

Elemental CaO/P_2O_5 ratios of phosphatic-component analyses range from 1.47 to 1.62 (mean 1.55), similar to the values of whole-rock analyses, which range from 1.44 to 1.70 (mean 1.58). Elemental F/P_2O_5 ratios of phosphatic-component analyses range from 0.079 to 0.109 (mean 0.096), whereas those of the whole-rock analyses range from 0.096 to 0.130 (mean 0.113). With the exception of sample 14135 from San Pedro Bay, the phosphatic-component F/P_2O_5 ratios are greater than the ideal ratio (0.089) of noncarbonate fluorapatite, and are similar to published F/P_2O_5 ratios of marine phosphorites off the coasts of California (Burnett, 1974), South West Africa (Baturin, Kochenov, and Petelin, 1970; Price and Calvert, 1978), and South America (Baturin and Petelin, 1972; Burnett, 1977).

Contents of S and Na are higher in phosphatic-component analyses than in bulk-rock compositions. Mg content is roughly the same in both types of analyses. This indicates that an important portion of these three elements is present in the apatite mineral. In the apatite structure, Na and Mg can substitute for Ca, and S can substitute for P (Gulbrandsen, 1966; McConnell, 1973).

All available data show that nodule P_2O_5 content ranges from 19 to 37 weight % and averages 27.6 weight %. The highest P_2O_5 analyses (31 to 37 weight %) are from the Santa Barbara High (Pasho, 1972).

Table 1. Relative percentages of constituents in southern California borderland phosphorite nodules. Unpublished data from D. Pasho. Range is shown in parentheses.

	SMB	SPB	SBH	SR–CR N of Saddle	SR–CR S of Saddle	SR–CR S of SNI	HB	TB	FB	CB
No. of stations	9	1	1	4	4	2	4	4	2	5
No. of nodules	13	1	2	9	5	2	5	8	4	5
Matrix	79 (60–99)	99	86 (80–89)	68 (56–82)	77 (70–87)	63 (58–68)	84 (74–93)	74 (64–97)	75 (64–87)	80 (64–81)
Ovules	13 (1–36)	—	1 (0–2)	17 (10–22)	17 (10–27)	20 (11–29)	12 (4–25)	4 (0–17)	17 (6–23)	8 (3–17)
Oolites	—	—	—	1 (0–5)	—	—	(0–2)	(0–1)	1 (0–2)	(0–1)
Mineral grains	7 (0–14)	—	12 (9–20)	10 (0–22)	1 (0–2)	12 (11–13)	2 (0–6)	20 (3–30)	2 (1–3)	9 (1–27)
Bone fragments	1 (0–2)	1	—	1 (0–4)	(0–1)	—	(0–1)	—	(0–1)	(0–1)
Glauconite	1 (0–6)	—	—	1 (0–3)	1 (0–4)	5 (0–10)	—	(0–5)	4 (0–12)	1 (0–3)
Foraminifera	—	—	1 (0–2)	2 (0–14)	3 (0–10)	—	1 (0–4)	—	(0–1)	(0–1)

SMB = Santa Monica Bay
SPB = San Pedro Bay
SBH = Santa Barbara High
SR–CR = Santa Rosa-Cortes Ridge
SNI = San Nicolas Island
HB = Hancock Bank
TB = Thirtymile Bank
FB = Fortymile Bank
CB = Coronado Bank

Table 2. Whole-rock chemical analyses (weight %) of selected southern California borderland phosphate nodules

	1	2	3	4	5	6	7	8	9	10	11	12	13	14	15	16	17	18	19	20
CaO	45.43	45.52	44.76	49.94	44.78	44.18	46.58	47.41	43.8	43.7	44.6	43.4	43.5	40.6	38.9	47.1	45.6	47.35	45.3	44.9
P_2O_5	29.19	28.96	26.3	31.1	30.61	30.65	29.09	29.66	28.1	28.1	28.6	27.6	26.2	24.9	25.3	30.0	28.8	29.56	27.7	28.4
F	3.12	3.07	3.06	3.56	2.95	3.98	3.15	3.36	3.21	3.14	3.34	3.13	3.11	2.89	2.78	3.49	3.38	3.31	—	3.22
Al_2O_3	0.30	2.03	0.67	0.11	1.06	1.71	—	—	0.74	0.85	0.79	0.50	0.94	0.61	1.06	0.36	0.23	—	1.66	0.85
Fe_2O_3	—	—	2.88	1.06	1.00	0.99	—	—	2.62	1.60	1.72	2.57	2.15	3.40	3.64	1.09	2.0	—	1.32	2.00
SO_3	—	—	2.11	1.67	0.34	0.36	—	—	1.62	1.85	2.05	1.88	1.71	1.82	1.94	2.18	2.04	—	1.60	1.90
CO_2	4.01	4.30	4.53	5.17	—	—	4.54	4.87	5.12	5.09	5.11	—	4.55	4.43	3.94	5.06	5.34	3.91	4.2	4.64
SiO_2	—	—	—	—	8.17	9.70	—	—	—	—	—	—	—	—	—	—	—	—	8.40	8.76
MgO	—	—	—	—	0.64	0.78	—	—	—	—	—	—	—	—	—	—	—	—	0.87	0.76
Na_2O	—	—	—	—	0.80	0.87	—	—	—	—	—	—	—	—	—	—	—	—	1.07	0.91
K_2O	—	—	—	—	0.64	0.56	—	—	—	—	—	—	—	—	—	—	—	—	0.30	0.50
Insol	3.57	4.45	8.49	1.10	—	—	3.57	2.12	7.15	7.93	5.37	8.17	11.1	13.7	12.6	1.12	2.54	2.59	—	5.97
L.O.I.	1.90	2.25	9.86	9.68	9.34	—	—	—	9.95	9.68	9.51	9.78	9.93	9.87	9.55	11.20	11.14	0.10	—	8.25
CaO/P_2O_5	1.56	1.57	1.70	1.61	1.46	1.44	1.60	1.60	1.56	1.56	1.56	1.57	1.66	1.63	1.54	1.57	1.58	1.60	1.64	1.58
F/P_2O_5	.107	.106	.116	.114	.096	.130	.108	.113	.114	.112	.117	.113	.119	.116	.110	.116	.117	.112	—	.113

Santa Monica Bay
1 = Station 106, Dietz, Emery, and Shepard (1942)
2 = Station 127, Dietz, Emery, and Shepard (1942)
3 = LP 14(R), Inderbitzen, Carsola, and Everhart (1970)
Santa Barbara High
4 = LP 5(R), Inderbitzen, Carsola, and Everhart (1970)
Santa Rosa-Cortes Ridge
5 = 14415, Burnett (1974)
6 = 14002, Burnett (1974)
Patton Ridge
7 = Station 158, Dietz, Emery, and Shepard (1942)
8 = Station 183, Dietz, Emery, and Shepard (1942)
Coronado Bank
9 = CB3-46, Butcher (1951)
10 = CB3-6, Butcher (1951)
Coronado Bank
11 = CB4-20, Butcher (1951)
12 = SL 240-63, Butcher (1951)
13 = SL 240-56, Butcher (1951)
14 = SL 266, Butcher (1951)
15 = SL 1220, Butcher (1951)
16 = Station 70, Butcher (1951)
17 = Station 159, Butcher (1951)
Fortymile Bank
18 = Station 69, Butcher (1951)
General Location
19 = California continental margin, McArthur (1978)
20 = Average of compositions 1-19

Table 3. Composition (weight %) of phosphatic components in southern California borderland nodules. The number of microprobe spot analyses in each composition is in parentheses. Total iron reported as FeO. Sample designations are from Pasho (1972).

	1 (57)	2 (36)	3 (44)	4 (21)	5 (63)	6 (30)	7 (67)	8 (45)	9 (50)	10 (70)	11 (90)	12 (573)
CaO	49.1	47.3	47.8	48.3	48.5	49.3	47.1	48.0	47.6	47.6	49.5	48.3
P_2O_5	33.4	30.2	30.5	30.9	31.1	31.7	29.9	30.2	29.6	29.7	30.6	31.2
F	2.64	3.01	2.99	3.04	3.07	3.31	3.01	3.11	3.19	3.11	3.33	3.00
SiO_2	0.01	1.75	1.64	1.31	0.62	0.27	1.18	0.69	0.66	0.06	0.09	0.56
Al_2O_3	0.07	0.76	0.65	0.73	0.41	0.39	0.59	0.59	0.35	0.21	0.23	0.34
FeO	0.18	0.50	0.46	0.47	0.27	0.21	0.43	0.78	0.49	0.14	0.28	0.35
MgO	0.89	0.85	0.85	0.89	0.89	0.75	0.95	0.86	0.76	0.78	0.85	0.87
Na_2O	0.89	1.16	1.19	1.11	1.13	1.12	1.27	1.19	1.36	1.35	1.40	1.17
K_2O	0.02	0.17	0.14	0.14	0.08	0.06	0.09	0.05	0.09	0.04	0.05	0.06
Cl	0.12	0.03	0.06	0.03	0.04	0.12	0.11	0.11	0.04	0.16	0.12	0.10
SrO	0.26	0.20	0.16	0.18	0.22	0.18	0.22	0.21	0.20	0.22	0.23	0.22
SO_3	0.95	2.77	2.67	2.35	2.82	2.30	3.03	2.42	3.83	2.94	3.25	2.43
CaO/P_2O_5	1.47	1.57	1.57	1.56	1.56	1.56	1.58	1.59	1.61	1.60	1.62	1.55
F/P_2O_5	.079	.100	.098	.098	.099	.104	.101	.103	.108	.105	.109	.096

San Pedro Bay
1 = 14135
Santa Monica Bay
2 = 14016A
3 = 14016C
4 = 14018A

Santa Monica Bay
5 = 14021A2
6 = 14021B
Santa Barbara High
7 = 11876A2
8 = 11876B1

Santa Rosa-Cortes Ridge
9 = 14001
10 = 14002
11 = 14005
General
12 = Average of all samples

Age

Micropaleontological, radiometric, and stratigraphic evidence suggest that most phosphorite deposits along the California continental margin formed in the Miocene but that some may have formed in the Pliocene. Foraminifera enclosed in many dark-brown nodules are middle Miocene to early late Miocene in age (Emery and Dietz, 1950; Emery, 1960). Pasho (1972) suggests that these phosphate nodules are direct erosional products (lag deposits) from outcropping Miocene phosphatic shales on submarine highs.

Other nodules have a different or more complex history. Foraminifera in light-brown nodules are late Pliocene to Pleistocene in age (Emery, 1960). Conglomeratic nodules from Coronado, Tanner, Fortymile, and Thirtymile Banks are composed of clasts containing Miocene foraminifera, and matrix containing late Pliocene to Pleistocene foraminifera (Emery, 1960; Pasho, 1972). Using Coronado Bank deposits as an example, Pasho (1972) suggests that the Miocene clasts weathered out of phosphatic shales and were incorporated in post-late Pliocene sediment, which was subsequently uplifted and phosphatized by leaching of Miocene phosphatic material. An alternative explanation for matrix and nodules of Plio-Pleistocene age is an episode of phosphate deposition during this period, as suggested by Barnes (1970). However, there is no evidence for this event in the Pliocene sections of California (Pasho, 1972).

Attempts to date recent formation of California borderland phosphate nodules using the $^{234}U/^{238}U$ disequilibrium method have led to the conclusion that these nodules are older than the dating limit of the $^{234}U/^{238}U$ method, 8×10^5 years (Kolodny, 1969).

Phosphatic Sands

Sandy phosphorite deposits have been reported from Santa Monica Bay and the Santa Barbara High (Inderbitzen, Carsola, and Everhart 1970). In Santa Monica Bay, this sand is composed of phosphate, glauconite, shell fragments, quartz, mica, and rock fragments. The phosphate is present as oolites and pellets of silt to pebble sizes (from <0.06 mm to 6 mm). Many pellets consist of a phosphate veneer which coats a mineral grain. P_2O_5 content of the bulk sand ranges from 0.35 to 9.68 weight %. Hand-picked pebbles from the coarse size fraction (>1.4 mm) have a P_2O_5 content of 27.7 % (Inderbitzen, Carsola, and Everhart, 1970), similar to nodule P_2O_5 contents.

Phosphate Resources

Estimates of commercially exploitable marine phosphate within the southern California borderland range from 58 to 117 million metric tons (Mero, 1961, 1967; Inderbitzen, Carsola, and Everhart, 1970; Hess, 1978; Chambers Consultants and Planners, 1979). A summary of individual deposit characteristics and tonnage estimates is presented in Table 5. These tonnage figures are rough estimates of phosphate resources because of the many uncertainties involved in calculation. These uncertainties include deposit area boundaries, the patchiness of phosphate distribution within these areas, and deposit thickness. Deposit boundaries are based on widely spaced bottom samples, except for the Coronado Bank deposit, which has been sampled on a uniform, close-grid basis (Barnes, 1970). Deposit area is computed using a compensating polar planimeter. The percentage of the sea floor covered by phosphate nodules and phosphatic sand is estimated from bottom photographs, dredge haul statistics, and general knowledge of the areas. Average sea-floor coverage by nodules in different deposits ranges from 1 to 10%, whereas average phosphatic sand sea-floor coverage ranges from 1 to 15%. However, the distribution of nodules and sand within specific deposits is patchy, which increases the uncertainty of these estimates (Hess, 1978). The greatest uncertainty in tonnage estimates involves deposit thickness. Sampling of marine phosphorite has been primarily limited to surface (dredge) and near-surface (dart core) collection of nodules and sediments. Estimates of deposit thickness are

Table 4. Composition (weight %) of glauconite and feldspars in southern California borderland nodules. The number of microprobe spot analyses are in parentheses. Total iron reported as FeO. Sample designations are from Pasho (1972).

	1 (30)	2 (3)	3 (3)	4 (2)	5 (2)	6 (3)	7 (4)	8 (3)	9 (7)	10 (3)	11 (1)	12 (2)
SiO_2	47.8	53.9	53.4	61.2	61.1	38.3	55.9	32.1	25.4	59.8	61.2	68.2
Al_2O_3	6.42	29.5	27.2	24.5	22.3	2.07	27.4	2.98	2.76	27.6	19.2	20.3
FeO	21.0	0.32	1.30	0.28	0.00	21.1	0.46	15.2	10.8	0.31	1.11	0.01
MgO	3.72	0.24	0.16	0.02	0.00	2.88	0.14	3.63	2.64	0.05	0.88	0.03
Na_2O	0.05	4.27	5.04	6.71	8.94	0.11	5.02	0.34	0.63	6.93	9.72	0.20
K_2O	6.82	0.21	0.31	0.85	0.08	5.52	0.28	5.55	3.80	0.40	0.68	16.6
CaO	1.31	12.8	11.5	6.40	4.61	1.10	12.0	18.6	27.9	7.27	6.49	0.11
P_2O_5	0.51	0.42	0.74	0.23	0.05	0.58	0.71	11.2	17.3	0.00	3.36	0.02
F	0.14	0.04	0.06	0.03	0.01	0.05	0.05	1.19	1.89	0.01	0.27	0.01
Cl	0.01	0.00	0.00	0.01	0.00	0.01	0.00	0.01	0.04	0.01	0.04	0.01
SrO	0.00	0.00	0.03	0.20	0.04	0.00	0.00	0.01	0.08	0.01	0.02	0.00
SO_3	0.51	0.04	0.12	0.03	—	0.20	0.05	0.66	1.14	0.01	1.13	0.00

Santa Monica Bay
1 = glauconite, 14016A and 14018A
2-5 = feldspars in phosphate pellets
2 = calcic plagioclase, 14016C
3 = calcic plagioclase, 14016C
4 = calcic plagioclase, 14018A
Santa Barbara High
6 = glauconite, 11876–B1
7 = calcic plagioclase, 11876–B1

Santa Rosa-Cortes Ridge
8 = glauconite, 14001-1
9 = glauconite, 14005-2
10-12 = feldspars in phosphate pellets
10 = plagioclase, pellet nucleus, 14002-1
11 = plagioclase, pellet nucleus region, 14005-2
12 = K-feldspar, near 11, 14005-2

Table 5. Southern California borderland phosphorite deposits.

Deposit	Water Depth (m)[1] Range	Average	Deposit Area (km^2) Inderbitzen	Hess	Estimated Bottom Coverage[2]	Wt. % P_2O_5 in Samples[3]	Estimated Phosphate in Millions of Metric Tons Inderbitzen	Hess	CCP[3]
Santa Monica Bay	40–100	75	106	200	1% Nodules	23–31	.6	1.1	.5
					15% Sand	0.3–10	7.7	43.3	7.5
Lasuen Knoll	110–500	300	67	67	10% Nodules	—	3.5	3.5	3.5
SE Catalina Ridge	220–500	300	31	150	1% Nodules	—	.2	.8	.2
NW Catalina Ridge	320–500	400	57	75	10% Nodules	—	3.0	4.0	—
Santa Barbara High	35–600	200	326	600	1% Nodules	31–37	1.7	3.2	1.7
					1% Sand	—	4.7	8.7	4.5
Osborn Bank	50–350	150	—	35	1% Nodules	—	—	.2	—
Santa Rosa-Cortes Ridge	20–500	130	1194	1650	1% Nodules	25–32	6.3	8.7	6.0
Coronado Bank	100–650	200	225	168	6% Nodules	22–30	7.2	5.3	7.1
Thirtymile Bank	240–600	400	264	264	10% Nodules	21–28	14.0	14.0	13.7
Fortymile Bank	80–700	340	166	270	10% Nodules	28	8.8	14.3	9.0
San Clemente Escarpment	250–600	400	—	30	10% Nodules	—	—	1.6	—
Cortes Bank and Tanner Bank	4–500	200	777	1400	1% Nodules	26–28	4.1	7.4	4.0
Hancock Bank	130–600	380	—	94	1% Nodules	—	—	.5	—
Northeast Bank	360–600	450	—	174	1% Nodules	30	—	.9	—
San Pedro Bay*									
Blake Knolls*									
Garrett Ridge*									
Albatross Knoll*									
Trask Knoll*									
Patton Ridge*									
								Total:	
						Nodules	46.4	65.5	45.7
						Sand	12.4	52.0	12.0

[1]Hess (1978)
[2]Inderbitzen, Carsola, and Everhart (1970)
[3]Fischer and Richmond, *in* Chambers Consultants and Planners (1979)
*Phosphate nodules recovered by U.S.G.S.; deposit size unknown.

therefore based on general observations and random samples. Inderbitzen, Carsola, and Everhart (1970) used an average thickness of .33 m for nodule deposits and 1 m for sandy deposits, recognizing that these may be optimistic estimates. More accurate deposit thickness data are not known, so these figures have been used in more recent tonnage estimate calculations. Once the volume of phosphate has been calculated by multiplying deposit area times percentage of sea-floor coverage times thickness, phosphate tonnage is computed using volume-to-weight factors of 1600 kg/m^3 for nodules and 1440 kg/m^3 for sand (Inderbitzen, Carsola, and Everhart, 1970).

Tonnage estimates of phosphate nodules are similar, ranging from 46 to 66 million metric tons. The higher estimates by Hess (1978) result from the expansion of several deposit areas and the addition of several new deposits. Tonnage estimates of phosphate sand range from 12 to 52 million metric tons. According to Hess (1978), the shallow Santa Monica Bay deposit may contain 43 million metric tons of phosphate sand. This single deposit thus accounts for most of the differences between the various phosphate sand tonnage estimates.

The phosphate nodule and pellet deposits of the southern California borderland form a low-grade source of fertilizer raw material when compared with other deposits being mined within the continental U.S. Most commercial-grade phosphate rock in the U.S. averages 32 to 35% P_2O_5. Borderland nodules average about 28% P_2O_5 in whole-rock analyses, and any upgrading to concentrate the apatite fraction (mean of roughly 32% from microprobe analyses) would probably be too costly to be economic. The P_2O_5 contents of bulk sand phosphorite deposits range from 0.35 to 9.68%. A mineralogical analysis by Inderbitzen, Carsola, and Everhart (1970) indicates that most of the phosphate in these sandy deposits is present in the coarser size fractions. Sand larger than 0.5 mm represents only 23% of the total weight but includes 64% of the phosphate present. Sand larger than 1.4 mm represents only 14% of the total weight but includes 45% of the phosphate present. Upgrading of sandy deposits by screening should therefore be possible.

Gangue minerals and impurities appear to be within acceptable limits for furnace feed. Elemental CaO/P_2O_5 ratios >1.55 indicate the presence of carbonate minerals, but only ratios >1.60 are considered excessive. Blending to achieve ratios of about 1.5 or less is possible with borderland nodules (J. Cathcart, written communication). MgO contents of borderland nodules are well within acceptable limits ($<1\%$), as are combined iron and alumina content ($<4.5\%$).

The economic advantage of these offshore deposits is their proximity to a large agricultural market and the ability to transport them on water. These deposits could be an important source of phosphate at some time in the future (Cathcart, 1980).

BAJA CALIFORNIA PACIFIC MARGIN

General Geology

The Baja California Pacific margin extends from approximately lat. 32°N to the tip of Baja California at approximately lat. 23°N. The margin offshore of the northern third of the Baja California Peninsula (lat. 32°N to lat. 29°N) displays the ridge-basin structure typical of the southern California borderland (Shepard and Emery, 1941; Krause, 1964, 1965; Moore, 1969; Doyle and Bandy, 1972). However, the Baja borderland is only about half as wide as the borderland to the north (Howell, McCulloch, and Vedder, 1978), depths in the south are about 450 m greater (Doyle and Gorsline, 1977), and the continental slope is much more abrupt and distinct off southern California than it is off Baja California (Krause, 1964, 1965).

The Baja borderland is dominated by three major troughs, each of which contains two or more basins separated by sills (Figure 6). Ridges and banks between these troughs have sharp summits as opposed to the flat-topped ridges to the north. Dredge hauls from ridge tops and sides have recovered silicic and mafic plutonic rocks, metasedimentary rocks, phosphorite, and late Cenozoic sedimentary rocks (Doyle and Bandy, 1972). Seismic-reflection profiles across basins show that sedimentary strata range in thickness from <100 m (Animal Basin) to about 1 km (North and South San Quintin Basins) (Doyle and Gorsline, 1977). Sedimentary rocks range in age from late Paleocene to Holocene and resemble rocks in stratigraphic sections of the southern California borderland (Doyle and Gorsline, 1977).

Doyle and Gorsline (1977) suggest that the east–west-trending Popcorn Ridge represents the boundary between the Baja borderland and Guadalupe Arrugado and Cedros Trench on the south. Closer to the coastline, borderland topography merges into a broad (80-km-wide) sediment-filled trough beneath Sebastian Vizcaino Bay. This is the first of three regions which make up the remainder of the Baja California Pacific margin south of the Baja Borderland (Figure 7).

Sebastian Vizcaino Bay is underlain by a northwest-plunging synclinorium filled with Upper Cretaceous (?) to Holocene strata as thick as 5 km (Fisher and Hess, 1963; Moore, 1969; Normark, Spencer, and Ingle, this volume). The upper two units in seismic-reflection profiles (Figure 8A) have been sampled by dredging. The upper unit consists of slightly consolidated late Pliocene and younger silty clay, whereas the second unit consists of indurated, phosphatized siltstone and sandstone of late Miocene or younger age (Blake et al., 1978; Normark, Spencer, and Ingle, this volume). A northwest-trending ridge that includes Cedros Island, the San Benito Islands, and Ranger Bank forms the continental slope and separates Sebastian Vizcaino Bay from the Cedros Trench. A complex of metasedimentary and blueschist-facies rocks, considered to be equivalent to rocks of the Franciscan assemblage, underlies this ridge and is exposed on the islands (Blake et al., 1978; Normark, Spencer, and Ingle, this volume).

South of Sebastian Vizcaino Bay, the continental margin in the Slope Plateau Province (lat. 26°20′N to 24°30′N) consists of a wide, shallow shelf, a faulted basin, and a series of en echelon northwest-trending ridges along the continental slope (Normark, Spencer, and Ingle, this volume). These ridges form shallow banks <50 m deep and appear to have been leveled off at approximately 100 m, suggesting subaerial and/or wave erosion (D'Anglejan, 1967). No dredge samples are available from ridges in this province, but similar features to the south are composed of sedimentary rocks (Normark, Spencer, and Ingle, this volume). Sediment ponded behind and in between these ridges has produced a series of terraces forming a continental slope and outer shelf,

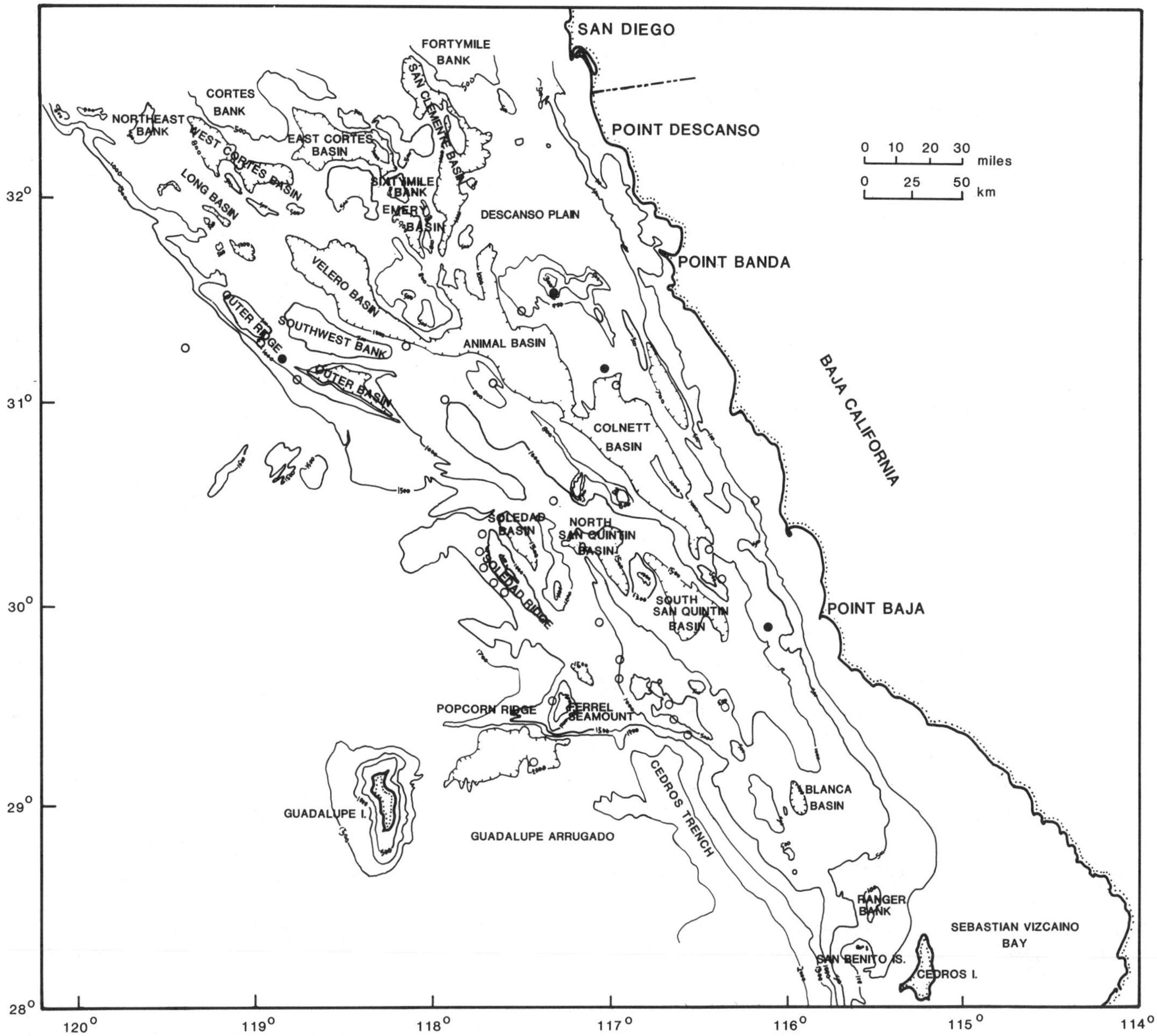

Figure 6. Distribution of phosphate nodules in the Baja borderland. Open circles = dredge hauls with no phosphate recovered; dots = dredge hauls with phosphate recovered. Bathymetry from Moore (1969). Contours in fathoms (1 fathom = 1.83 m).

which is wider than those to the north or south of this region (Blake et al., 1978; Normark, Spencer, and Ingle, this volume).

South of lat. 24°30′N, the Magdalena borderland is broad (100 km), relatively shallow (up to 1000 m), and characterized by numerous irregularly shaped banks and basins. Reflection profiles show that gently folded and faulted sedimentary rocks underlie most of this area (Figure 8B). Angular unconformities are common between stratigraphic units. Seismic-refraction data suggest that the sediments are about 2 km thick. Dredge samples from the cores of anticlinal folds along the outer shelf indicate that the upper 1.5 km of sedimentary material is of late Miocene or younger age (Blake et al., 1978; Normark, Spencer, and Ingle, this volume). The predominant rock types sampled in the Magdalena borderland are phosphatic foraminiferal mudstone, siltstone, and sandstone (Normark, Spencer, and Ingle, this volume).

Along the Pacific continental margin of Baja California, basin history appears to have involved initial subsidence in early Miocene, basin filling during middle and late Miocene, and widespread uplift and folding in earliest Pliocene (Normark, Spencer, and Ingle, this volume).

Phosphorite Deposits

Phosphorites have been recovered from many localities along the Baja California Pacific margin. Phosphate nodules are present on bank tops and escarpments within the Baja borderland, where they comprise a few percent of most dredge hauls but represent as much as 30% of some recoveries (Doyle and Gorsline, 1977). Partial chemical analyses reported by Pasho (1972) for 31 phosphate nodules from 10 banks within the Baja borderland are presented in Table 6.

Contents of CaO, P_2O_5, and F show more variation than that found in southern California borderland nodules. CaO contents range from 22.6 to 47.0 weight %, P_2O_5 contents range from 9.0 to 28.7 weight %, and F contents range from 1.1 to 3.4 weight %. High elemental CaO/P_2O_5 ratios (up to 3.91) in roughly one-third of the analyses indicate that there is abundant carbonate present in addition to apatite. Elemental F/P_2O_5 ratios are uniform, and their mean value (.113) is identical to that of southern California borderland nodule whole-rock analyses. Sampling density is too sparse to delimit nodule deposits.

Phosphatic Sands

Sandy phosphorite is widespread throughout San Cristobal Bay (Figure 9), although few samples contain more than a few weight % of phosphate pellets (Emery et al., 1957).

South of San Cristobal Bay, between lat. 24° and lat. 26°N, is the large sandy phosphorite deposit of Santo Domingo Bay that has been studied in some detail (D'Anglejan, 1967). The phosphate in this deposit occurs as well-sorted, sand-sized particles within a fine sand to silt matrix. Phosphate particles are made up of about 90 weight % carbonate-fluorapatite; the remainder consists of detrital inclusions, sulfide minerals, organic matter, opaline silica, and disseminated calcite. The mean P_2O_5 content of two analyses of the phosphate particles is 30.2 oxide weight % (D'Anglejan, 1967).

The area of highest phosphate concentration (15 to 40 weight % of the total sediments) coincides with an area of coarse sediment on the continental shelf (Figure 10) in water depths of 50 to 100 m. Seaward of the 100-m contour, phosphate content drops abruptly to near zero. Cores taken to a depth of 1.3 m show no significant difference in the percentage of phosphate pellets with sediment depth (D'Anglejan, 1967).

Phosphate Resources

D'Anglejan (1967) has estimated that the phosphorite deposit of Santo Domingo Bay contains 1.5 to 3 billion metric tons of P_2O_5, based on an average thickness of 6.5 m over an area of 1800 km^2 with an average apatite concentration of 5 weight % of the total sediment.

Sparse sampling and chemical data preclude estimating tonnage and grade of phosphate nodules within the Baja borderland.

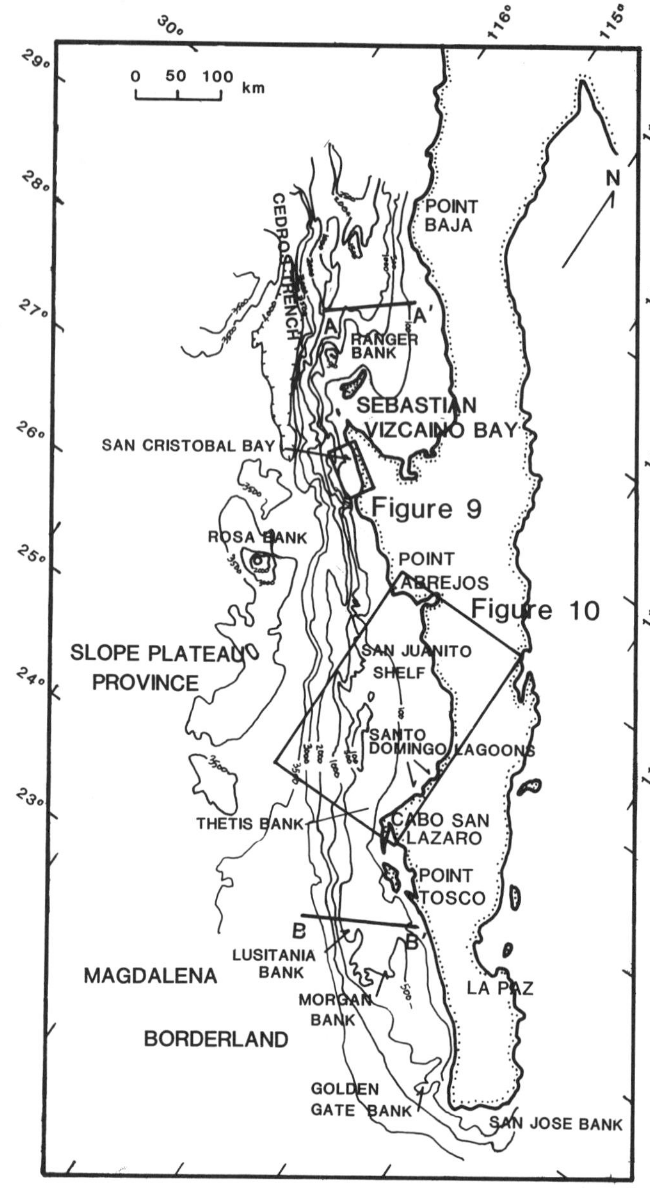

Figure 7. Index map and bathymetry of the southern portion of the Baja California Pacific margin. A–A′ and B–B′ are seismic-reflection lines shown in profile on Figure 8. Bathymetric contours in meters. Modified from Normark, Spencer, and Ingle, this volume.

CENTRAL CALIFORNIA MARGIN

General Geology

The central California margin extends about 500 km from Point Reyes in the north to Point Conception in the south. The shelf ranges in width from 30 km offshore of the Monterey Bay region to 80–100 km along the rest of the region. Underlying the shelf, three large Neogene basins are situated behind three major structural highs, all of which trend northwest and parallel the coastline (Figure 11) (Howell, McCulloch, and Vedder, 1978; Blake et al., 1978). Basin margins are fault controlled. Another major feature of this margin is the east–west-trending submarine Monterey Canyon, which cuts across Monterey Bay.

West of San Francisco, the subsurface Farallon Ridge can be traced offshore in seismic profiles as far north as Point Arena, and it intersects the coast at Point Año Nuevo in the south (Figure 11). This ridge is composed of quartz diorite in the Farallon Islands (Blake et al., 1978).

East of the Farallon Ridge, Bodega Basin contains nearly 3 km of Neogene strata (Blake et al., 1978). Seismic-reflection data and drilling show that uplift of the Farallon Ridge and subsidence of Bodega Basin began in the middle Miocene (Hoskins and Griffiths, 1971).

Offshore from Santa Cruz and southwest of the Farallon Ridge is the Santa Cruz structural high. Between these two subsurface ridges is the Outer Santa Cruz Basin, which contains at least 3 km thickness of Neogene strata (Blake et al., 1978). Dredging on Santa Cruz high has yielded mafic volcanic rocks.

South of Monterey, the Santa Lucia Bank forms the third structural high along the cental California margin. Dredge

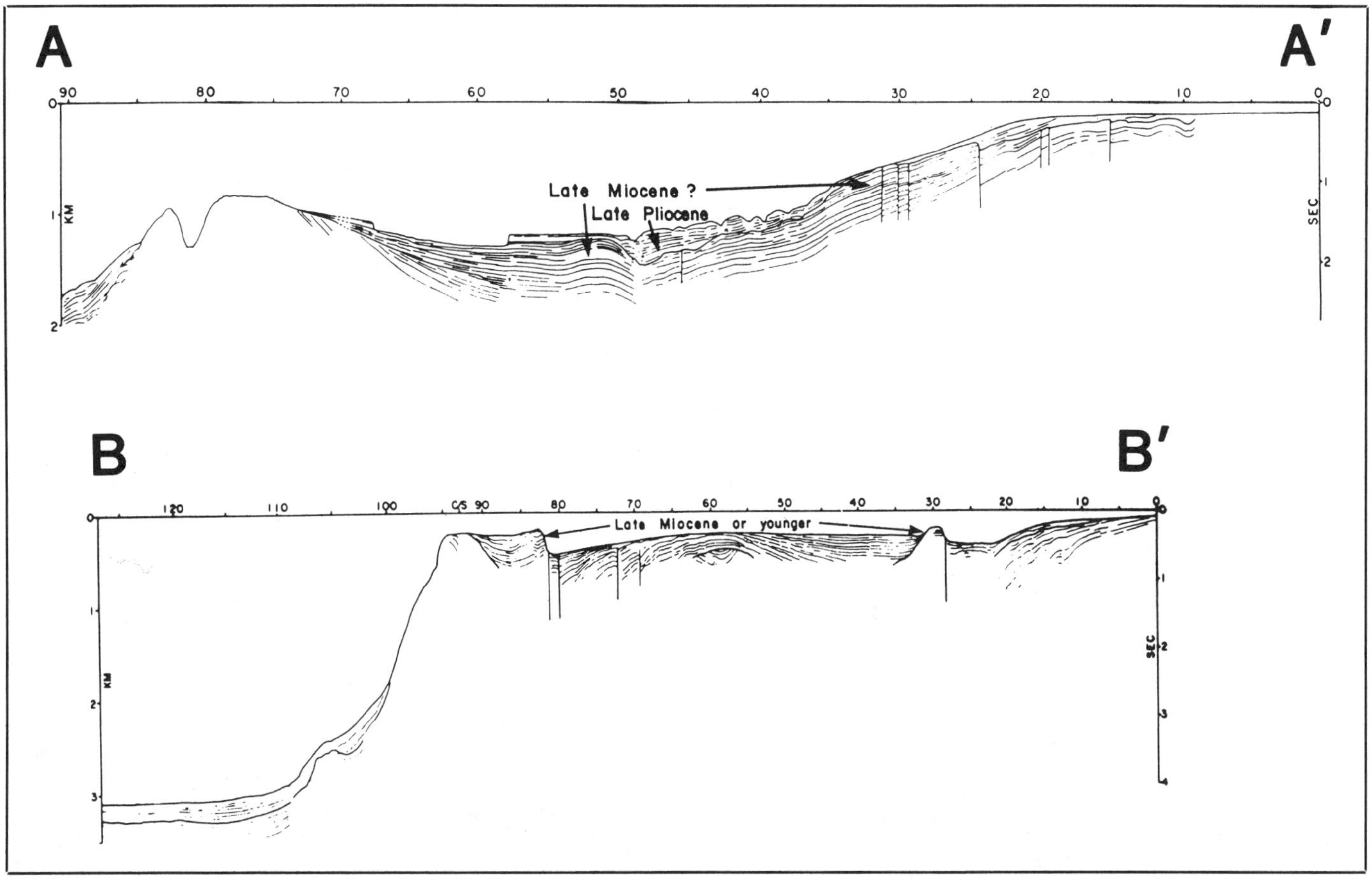

Figure 8. Line-drawing interpretations of seismic-reflection profiles across Sebastian Vizcaino Bay (A–A′) and Magdalena borderland (B–B′); see Figure 7 for locations. Horizontal scale in kilometers; vertical scale in seconds of two-way traveltime and kilometers of water depth. From Blake et al., 1978.

Table 6. Partial chemical analysis (weight %) of phosphate nodules off Baja California; from Pasho (1972).

	1	2	3	4	5	6	7	8	9	10	11	12	13	14	15
CaO	38.8	41.7	39.2	46.4	35.2	44.6	30.6	43.4	41.7	41.8	25.0	37.8	32.4	23.4	34.7
P_2O_5	23.0	25.8	24.7	28.7	20.1	28.2	18.9	26.2	25.0	26.2	10.1	21.7	11.9	9.00	14.0
F	2.48	—	—	—	—	—	—	—	—	3.10	—	—	—	—	—
FeO_3	2.03	—	—	—	—	—	—	—	—	0.70	—	—	—	—	—
Insol.	12.8	11.5	15.7	3.96	21.3	4.54	32.9	9.43	10.9	10.7	34.5	14.7	10.9	21.2	6.13

	16	17	18	19	20	21	22	23	24	25	26	27	28	29	30	31
CaO	40.6	42.4	35.8	39.7	43.9	42.6	26.0	47.0	46.8	37.6	30.2	35.0	42.7	35.2	41.4	22.6
P_2O_5	19.0	25.1	16.8	11.0	26.9	10.9	8.80	27.6	28.7	22.7	15.2	18.4	27.5	18.1	24.4	9.80
F	—	—	—	—	—	—	—	—	3.40	—	—	—	3.10	—	—	1.10
FeO_3	—	—	—	—	—	—	—	—	1.36	—	—	—	1.52	—	—	1.09
Insol.	6.60	6.92	11.2	5.25	7.51	6.70	28.6	3.10	3.69	12.4	27.9	17.1	7.85	18.1	8.99	54.4

San Jose Bank
1 = 14R
2 = 16R
Ranger Bank
3 = 18R
4 = 19R
5 = 20R
6 = 22R
7 = 23R
8 = 40R
9 = 41R
10 = 42R

Grand Bank
11 = 43R
12 = 79R
13 = 80R
14 = 83R
15 = 84R
Tasco Bank
16 = 93R
17 = 96R
18 = 98R

Morgan Bank
19 = 102
20 = 102A
Lusitania Bank
21 = 94R
22 = 95R
Golden Gate Bank
23 = 105R
24 = 106R
25 = 107

San Benito Island Shelf
26 = 139R
Cedros Island Shelf
27 = 36R
28 = 37R
29 = 38R
Thetis Bank
30 = 82R
San Juanito Shelf
31 = 57

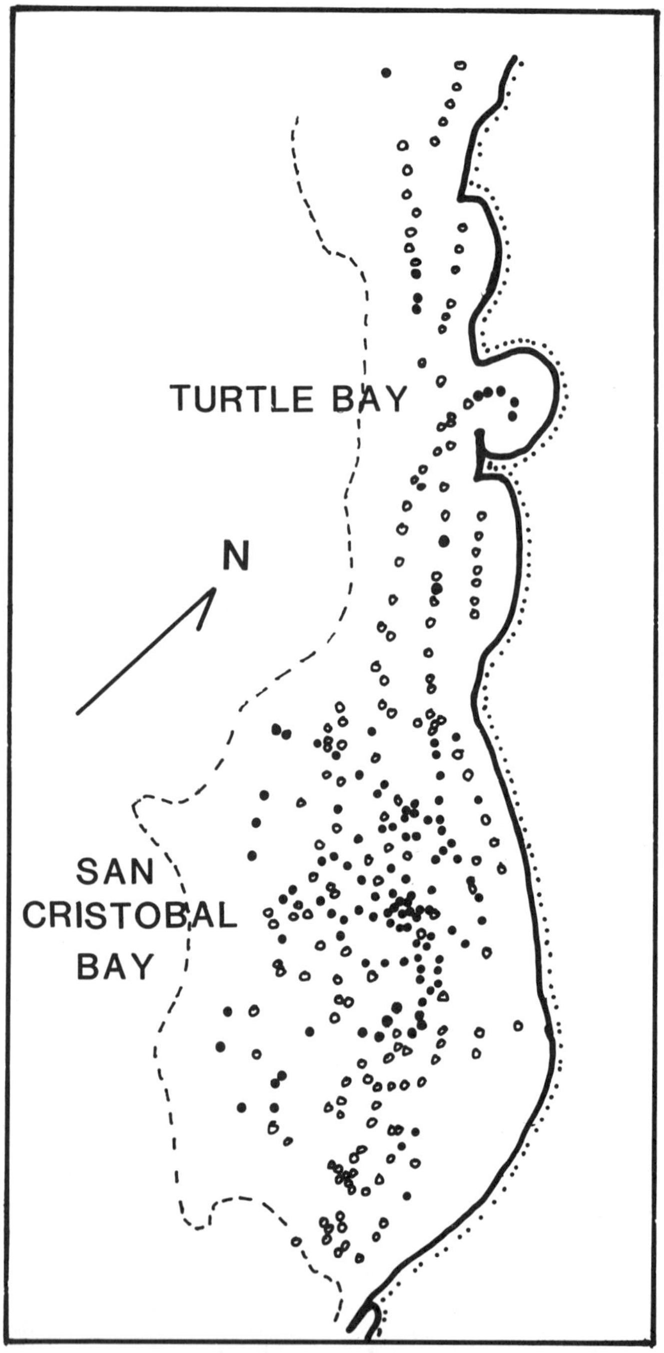

Figure 9. Distribution of phosphorite in San Cristobal Bay, Baja California, Mexico (see Figure 7 for location). Open circles = bottom samples with no phosphate recovered; dots = bottom samples with phosphate recovered. Dashed line = 200-m bathymetric contour. From Emery et al., 1957.

hauls from this bank contain a variety of rock types including metasediment, siliceous shale of middle Miocene age—perhaps equivalent to the Monterey Formation—altered mafic igneous rocks, and granitic rocks (Howell, McCulloch, and Vedder, 1978). East of Santa Lucia Bank, Santa Maria Basin contains approximately 2 km of Neogene sediment.

Bodega, Outer Santa Cruz, and Santa Maria Basins all began to accumulate thick piles of sediment in the late middle Miocene (Blake et al., 1978). Blake et al. (1978) suggest that a basin-forming mechanism acted almost synchronously on the continental margin of central California between 14 and 10 Ma, and that the change in relative motion between the Pacific and North American plates from northerly to more northwesterly by about 10 Ma may explain basin formation by oblique extension across north-trending faults.

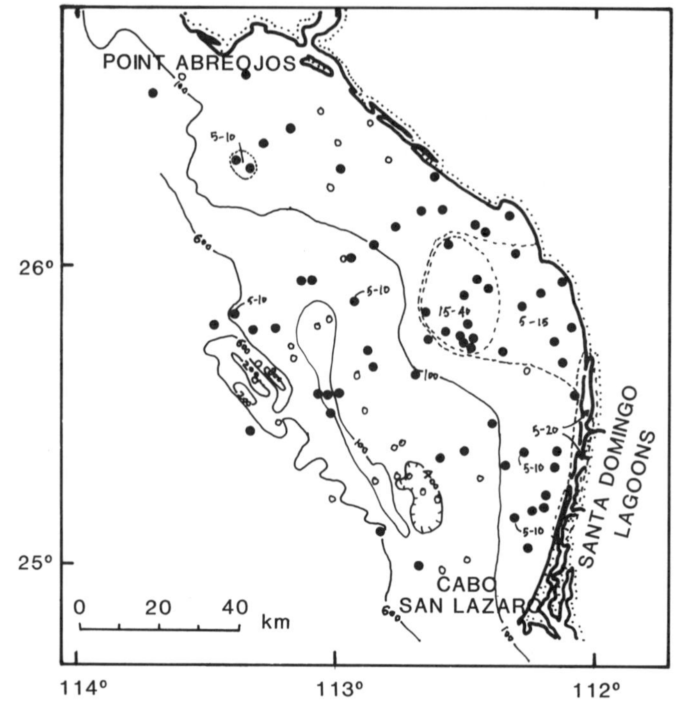

Figure 10. Distribution of phosphorite on the continental shelf of the Santo Domingo region (see Figure 7 for location). Open circles = bottom samples with no phosphate recovered; dots = bottom samples with phosphate recovered. Numbers represent weight % of fine-grained phosphorite in the surface sediments, contoured with dashed lines where appropriate. From D'Anglejan, 1967.

Phosphorite Deposits

Several phosphorite localities have been reported along the central California margin (Figure 12). Phosphate nodules have been dredged from the area around Cordell Bank, west of Point Reyes, and from seamount slopes west of San Francisco, in water depths between 130 and 1100 m (Hanna, 1952; Uchupi and Emery, 1963). Nodules and slabs of phosphate have been recovered from the lower continental slope off Pescadero Point, as well as from the slopes of Sur Knoll and Twin Knolls, in water depths of 1000 to 1400 m (Mullins and Rasch, 1985). Nodule petrography is similar to that described for southern California Borderland nodules.

Rock fragments coated with phosphate have been dredged from the walls of Monterey Canyon in Monterey Bay in water depths of 100 to 500 m (Greene, 1977). This occurrence appears to be younger than other phosphate deposits found along the margin, because some of the rock fragments are Pliocene in age (Greene, 1977).

In general, phosphorite along the central California margin is low in abundance and patchy in distribution. Mullins and Rasch (1985) found large differences between closely spaced dredge sites, and new bottom samples from phosphorite localities previously described in the literature by Hanna (1952), Uchupi and Emery (1963), and Greene (1977) contained no phosphate.

Whole-rock major element analyses of nodules from Sur Knoll, Twin Knolls, and the continental slope off Pescadero Point, reported by Mullins and Rasch (1985), show both similarities and differences in chemical composition between the samples (Table 7). Nodules from off Pescadero Point are uniform in composition and are similar to nodules from the southern California borderland. Nodules from Sur Knoll and Twin Knolls show considerable range in chemical composition. Nodules similar to those found off Pescadero Point are present at these localities. Other nodules have less CaO and P_2O_5, and greater SiO_2, Al_2O_3, Fe_2O_3, and K_2O, suggesting larger amounts of nonphosphatic constituents such as quartz, aluminosilicates, and glauconite.

Phosphate Resources

Phosphate deposits along the central California continental margin contain nodules which are similar in ore grade to southern California borderland nodules. Low sampling density precludes the establishment of deposit boundaries. Patchy distribution, low abundance, and occurrence in relatively deep water (1000 to 1400 m) limit their economic potential.

CONCLUSIONS

Phosphate deposits are present along the western continental margin of North America from San Francisco to the southern tip of Baja California, Mexico. The northernmost area—the central California continental margin—contains small phosphate deposits characterized by low abundance, variable chemical composition, and patchy distribution in relatively deep water, all of which limit their economic potential. Within the central area—the southern California borderland region—phosphate nodule and sand deposits are potential sources of fertilizer material for western markets. However, much additional information is required for an accurate appraisal of the tonnage, chemical composition, and distribution of the deposits. The southernmost area—the Baja California Pacific margin—contains the sandy phosphorite of Santo Domingo, which may be an extremely large deposit, but more detailed information is needed to accurately determine reserves.

The deposits of the southern California borderland region and the Baja California Pacific margin will be important sources of phosphate for fertilizer manufacture at some time in the future.

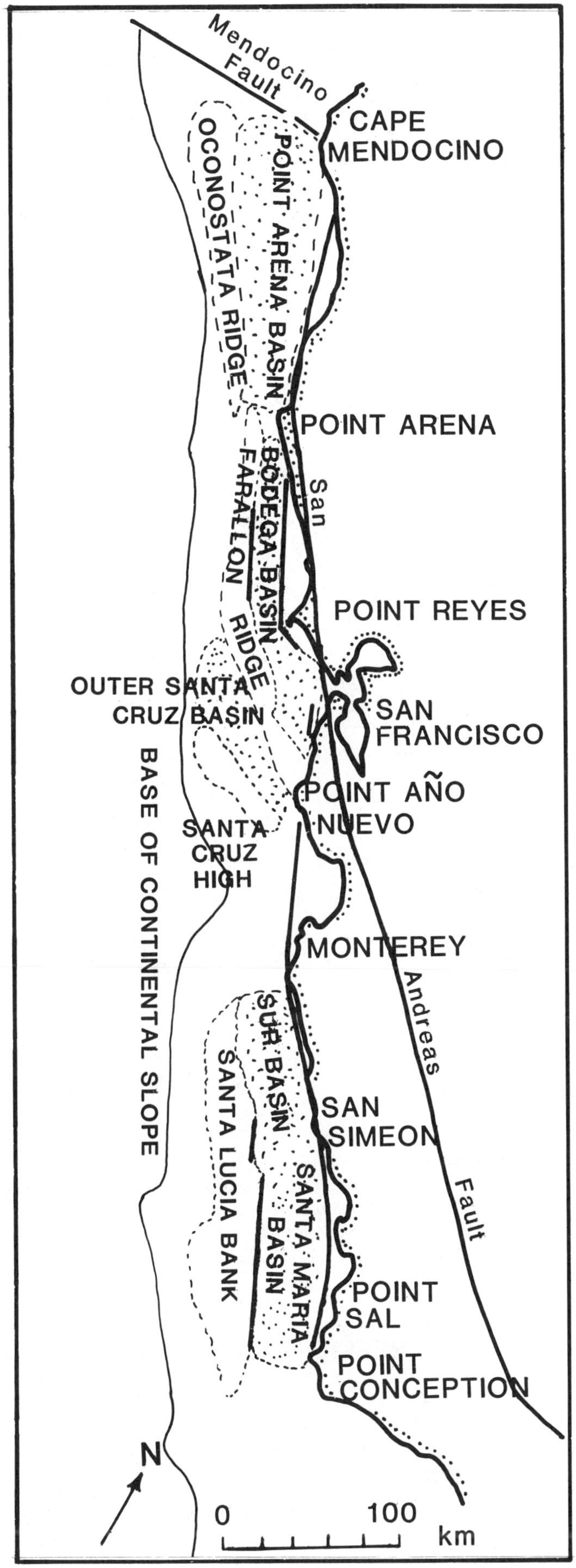

Figure 11. Location of major structural trends offshore northern and central California. Heavy lines represent faults. From Blake et al., 1978.

Table 7. Whole-rock chemical composition (weight %) of central California continental margin phosphorites. Data from Mullins and Rasch (1985).

	Pescadero continental slope			Sur Knoll			Twin Knolls		
Sample No.	CP-3-42C	CP-3-42FD	CP-3-42FL	OMB-11	CP-2-3	OMB-3	CP-2-11M	CP-2-13	CP-2-12H
CaO	46.4	42.8	43.4	41.9	36.1	33.0	44.1	40.2	32.7
P_2O_5	29.7	27.9	28.1	28.5	24.7	22.3	28.0	26.1	22.3
SiO_2	8.81	7.51	6.31	8.17	17.1	26.4	7.10	14.9	24.1
Al_2O_3		1.00			2.89	3.70		3.17	3.27
Fe_2O_3	2.30	1.57	2.07	7.62	5.02	4.98	3.96	3.07	4.93
MgO					2.77			1.11	
K_2O	0.58	0.54	0.40	0.48		2.20	0.73	1.30	1.30
CO_2	4.9	5.5	5.1			5.9		5.9	4.8
CaO/P_2O_5	1.56	1.53	1.54	1.47	1.46	1.48	1.58	1.54	1.47

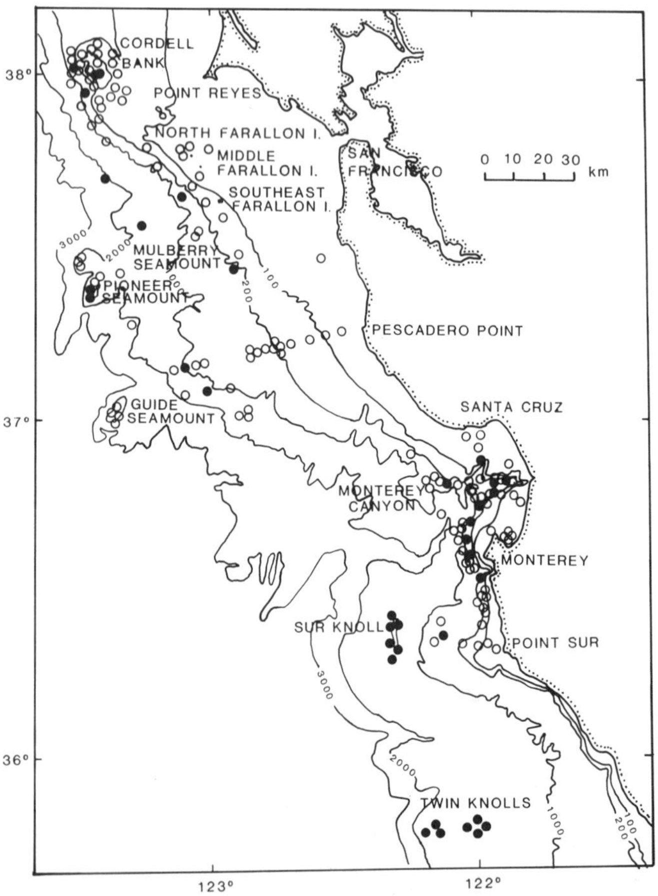

Figure 12. Distribution of phosphorite deposits along the central California continental margin. Open circles = bottom samples with no phosphate recovered; dots = bottom samples with phosphate recovered. Bathymetric contours in meters. Data from Hanna, 1952; Uchupi and Emery, 1963; Greene, 1977; and Mullins and Rasch, 1985.

REFERENCES CITED

Barnes, B., 1970, Marine phosphorite deposit delineation techniques tested on the Coronado Bank, Southern California: Houston, Texas, 1970 Offshore Technology Conference Preprints, v. 11, p. 315–350.

Baturin, G. N., and V. P. Petelin, 1972, Phosphorite concretions on the Chilean Shelf: Litologiia i Polezny Iskopaemye, no. 7, p. 261–267 (English translation).

Baturin, G. N., A. V. Kochenov, and V. P. Petelin, 1970, Phosphorite formation on the shelf of South-west Africa: Litologiia i Polezny Iskopaemye, no. 3, p. 266–276 (English translation).

Bence, A. E., and A. E. Albee, 1968, Empirical correction factors for the electron microanalysis of silicates and oxides: Journal of Geology, v. 76, p. 382–403.

Blake, M. C., Jr., R. H. Campbell, T. W. Dibblee, Jr., D. G. Howell, T. H. Nilsen, W. R. Normark, J. G. Vedder, and E. A. Silver, 1978, Neogene basin formation in relation to plate-tectonic evolution of San Andreas fault system, California: AAPG Bulletin, v. 62, p. 344–372.

Burnett, W. C., 1974, Phosphorite deposits from the sea floor off Peru and Chile: radiochemical and geochemical investigations concerning their origin: Honolulu, University of Hawaii, unpublished Ph.D. Dissertation, 164 p.

Burnett, W. C., 1977, Geochemistry and origin of phosphorite deposits from off Peru and Chile: GSA Bulletin, v. 88, p. 813–823.

Butcher, W. S., 1951, Lithology of the offshore San Diego area: University of California, Scripps Institution of Oceanography, Submarine Geology Report 20, 16 p.

Cathcart, J. B., 1980, The phosphate industry of the United States, *in* F. E. Khasawneh, E. C. Sample, and E. J. Kamprath, eds., The role of phosphorus in agriculture: Madison, Wisconsin, American Society of Agronomy, p. 19–42.

Chambers Consultants and Planners, 1979, Economic feasibility of outer continental shelf mining of phosphorites offshore southern California, *in* Outer Continental Shelf Mining Policy Phase II Task Force, Program Feasibility Document, Outer Continental Shelf Hard Minerals Leasing, v. 13, 109 p.

Chesterman, C. W., 1952, Descriptive petrography of rocks dredged off the coast of central California: Proceedings of the California Academy of Sciences, v. 27, p. 359–374.

Crouch, J. K., 1979, Neogene tectonic evolution of the California Continental Borderland and western Transverse Ranges: GSA Bulletin, v. 90, p. 338–345.

Crouch, J. K., 1981, Northwest margin of California Continental Borderland: Marine geology and tectonic evolution: AAPG Bulletin, v. 65, p. 191–218.

D'Anglejan, B. F., 1967, Origin of marine phosphorites off Baja California, Mexico: Marine Geology, v. 5, p. 15–44.

Dietz, R. S., K. O. Emery, and F. P. Shepard, 1942, Phosphorite deposits on the seafloor off southern California: GSA Bulletin, v. 53, p. 815–848.

Doyle, L. J., and O. L. Bandy, 1972, Southern continental borderland, Baja California: Its tectonic and environmental development: GSA Bulletin, v. 83, p. 3785–3794.

Doyle, L. J., and D. S. Gorsline, 1977, Marine geology of Baja California Continental Borderland, Mexico: AAPG Bulletin, v. 61, p. 903–917.

Emery, K. O., 1960, The sea off southern California: New York, John Wiley and Sons, Inc., 366 p.

Emery, K. O., and R. S. Dietz, 1950, Submarine phosphorite deposits off California and Mexico: California Division of Mines and Geology Bulletin, v. 46, p. 7–15.

Emery, K. O., D. S. Gorsline, E. Uchupi, and R. D. Terry, 1957, Sediments of three bays of Baja California: Sebastian Viscaino, San Cristobal and Todos Santos: Journal of Sedimentary Petrology, v. 27, p. 95–115.

Felix, D., 1968, Economic evaluation of Coronado Bank phosphorite deposit: Los Angeles, Department of Geological Sciences, University of Southern California, unpublished technical report, 23 p.

Fisher, R. L., and H. H. Hess, 1963, Trenches, *in* M. N. Hill, ed., The Sea, Vol. 3: New York, John Wiley & Sons, Inc., p. 411–436.

Greene, H. G., 1977, Geology of Monterey Bay region: U.S. Geological Survey Open-file Report 77-718, 347 p.

Gulbrandsen, R. A., 1966, Chemical composition of phosphorites of the Phosphoria Formation: Geochimica et Cosmochimica Acta, v. 30, p. 769–778.

Hanna, G. D., 1952, Geology of the continental slope off central California: Proceedings of the California Academy of Sciences, v. 27, p. 325–358.

Hess, H. D., 1978, Preliminary resource assessment—Phosphorites of the southern California borderland, *in* Outer Continental Shelf Mining Policy Phase II Task Force, Program Feasibility Document, Outer Continental Shelf Hard Minerals Leasing, v. 9, 51 p.

Hill, D. J., 1976, Geology of the Jurassic basement rocks, Santa Cruz Island, California, and correlation with other Mesozoic basement terranes in California, *in* D. G. Howell, ed., Aspects of the geologic history of the California Continental Borderland: AAPG, Pacific Section, Miscellaneous Publication 24, p. 16–46.

Hoskins, E. G., and J. R. Griffiths, 1971, Hydrocarbon potential of northern and central California offshore, *in* I. H. Cram, ed., Future petroleum provinces of the United States—Their geology and potential: AAPG Memoir 15, v. 1, p. 212–228.

Howell, D. G., and J. G. Vedder, 1981, Structural implication of stratigraphic discontinuities across the southern California borderland, *in* W. G. Ernst, ed., The geotectonic development of California (Rubey, Vol. 1): Englewood Cliffs, New Jersey, Prentice-Hall, Inc., p. 535–558.

Howell, D. G., D. S. McCulloch, and J. G. Vedder, 1978, General geology, petroleum appraisal, and nature of environmental hazards, eastern Pacific Shelf, latitude 28° to 38° North: U.S. Geological Survey Circular 786, 29 p.

Howell, D. G., H. McLean, and J. G. Vedder, 1976, Cenozoic tectonism on Santa Cruz Island, *in* D. G. Howell, ed., Aspects of the geologic history of the California Continental Borderland: AAPG, Pacific Section, Miscellaneous Publication 24, p. 392–417.

Inderbitzen, A. L., A. J. Carsola, and D. L. Everhart, 1970, The submarine phosphate deposits off southern California: Houston, Texas, 1970 Offshore Technology Conference Preprints, v. II, p. 287–304.

Kolodny, Y. 1969, Are marine phosphorites forming today?: Nature, v. 224, p. 1017–1019.

Krause, D. C., 1964, Lithology and sedimentation in the southern continental borderland, *in* R. L. Miller, ed., Papers in marine geology, Shepard commemorative volume: New York, Macmillan, p. 274–318.

Krause, D. C., 1965, Tectonics, bathymetry, and geomagnetism of the southern continental borderland west of Baja California, Mexico: GSA Bulletin, v. 76, p. 617–650.

Mattinson, J. M., and D. J. Hill, 1976, Age of plutonic basement rocks, Santa Cruz Island, California, *in* D. G. Howell, ed., Aspects of the geologic history of the California Continental Borderland: AAPG, Pacific Section, Miscellaneous Publication 24, p. 53–58.

McArthur, J. M., 1978, Systematic variations in the contents of Na, Sr, CO_3, and SO_4 in marine carbonate-fluorapatite and their relation to weathering: Chemical Geology, v. 21, p. 89–112.

McConnell, D., 1973, Apatite. Its crystal chemistry, mineralogy, utilization, and geologic and biologic occurrences: New York, Springer-Verlag, 111 p.

Mero, J. L., 1961, Sea floor phosphorites: California Division of Mines and Geology, v. 14, p. 1–12.

Mero, J. L., 1967, Ocean mining—A potential major new industry: New York, First World Dredging Conference Proceedings, p. 625–641.

Moore, D. G., 1969, Reflection profiling studies of the California Continental Borderland: Structure and Quaternary turbidite basins: GSA Special Paper 107, 142 p.

Mullins, H. T. and R. F. Rasch, 1985, Sea-floor phosphorites along the central California continental margin: Economic Geology, v. 80, p. 696–715.

Normark, W. R., J. E. Spencer, and J. C. Ingle, Jr., 1987, Geology and Neogene history of the Pacific Continental Margin of Baja California Sur, Mexico, this volume.

Pasho, D. W., 1972, Character and origin of marine phosphorites: Los Angeles, University of Southern California Report USC-Geol-72-5, 154 p.

Platt, J. P., 1975, Metamorphic and deformational processes in the Franciscan Complex, California: Some insights from the Catalina Schist terrane: GSA Bulletin, v. 86, p. 1337–1347.

Platt, J. P., 1976, The significance of the Catalina Schist in the history of the southern California borderland, *in* D. G. Howell, ed., Aspects of the geological history of the California Continental Borderland: AAPG, Pacific Section, Miscellaneous Publication 24, p. 47–52.

Price, N. B., and S. E. Calvert, 1978, The geochemistry of phosphorites from the Namibian Shelf: Chemical Geology, v. 23, p. 151–170.

Shepard, F. P., and K. O. Emery, 1941, Submarine topography off the California coast: Canyons and tectonic interpretations: GSA Special Paper 31, 171 p.

Shor, G. G., and R. W. Raitt, 1958, Seismic studies in the southern California borderland: Mexico, XX International Geological Congress Proceedings, p. 243–259.

Uchupi, E., and K. O. Emery, 1963, The continental slope between San Francisco, California and Cedros Island, Mexico: Deep-Sea Research, v. 10, p. 397–447.

Vedder, J. G., 1987, Regional geology and petroleum potential of the southern California borderland, this volume.

Vedder, J. G., L. A. Beyer, A. Junger, G. W. Moore, A. E. Roberts, J. C. Taylor, and H. C. Wagner, 1974, Preliminary report on the geology of the continental borderland of southern California: U.S. Geological Survey Miscellaneous Field Studies MF-624.

Vedder, J. G., H. G. Greene, E. W. Scott, J. C. Taylor, and others, 1976, A summary report of the regional geology, petroleum potential, environmental geology, and technology for exploration and development in the area of proposed lease sale 48, California Continental Borderland: U.S. Geological Survey Open-File Report 76-787, 43 p.

Index

A reference is indexed according to its important, or "key" words.

Four columns are to the left of a keyword entry. The first column, a letter entry, represents the Circum-Pacific Council for Energy and Mineral Resources book series from which the reference originated. In this case, CP stands for the Earth Science Series

The second column indicates the series number. In this case, 6 represents a reference from Earth Science Series 6. The third column lists the page number of this volume on which the reference can be found. The fourth column represents the type of entry: K = keyword, A = author, and T = title.

Index entries without page numbers represent the title or compilation editors of this volume.